Handbook of
CARCINOGENIC POTENCY and GENOTOXICITY DATABASES

Edited by
Lois Swirsky Gold
Errol Zeiger

CRC Press
Boca Raton   New York   London   Tokyo

Senior Editor:	Paul Petralia
Project Editor:	Carrie L. Unger
Editorial Assistant:	Cindy Carelli
Marketing Manager:	Susie Carlisle
Direct Marketing Manager:	Becky McEldowney
Page design:	Carrie L. Unger
Cover design:	Dawn Boyd
PrePress:	Greg Cuciak
Manufacturing:	Sheri Schwartz

Library of Congress Cataloging-in-Publication Data

Handbook of carcinogenic potency and genotoxicity databases / edited by Lois Swirsky Gold, Errol Zeiger
      p. cm.
    Includes bibliographical references and index.
    ISBN 0-8493-2684-2
    1. Carcinogenicity testing. 2. Mutagenicity testing. I. Gold, Lois Swirsky. II. Zeiger, Errol.
    [RC268.4.H355 1997]
    616.99′4071—dc20

96-34482
CIP

This book is dedicated to our families:
Jenny, Alissa, and Stuart Gold and
Marion, Asher, Brian, and Anita Zeiger,
with the hope that our work
contributes to a better future.

EDITORS

Lois Swirsky Gold, Ph.D., has developed and directed the Carcinogenic Potency Database at the University of California at Berkeley and the E.O. Lawrence Berkeley National Laboratory for 18 years. She has examined the scientific assumptions and data that underlie regulatory policy, emphasizing a broadened perspective on cancer risk assessment. In more than 70 papers, she has addressed such issues as setting priorities among possible cancer hazards, the background of human exposures to naturally occurring rodent carcinogens, validity problems associated with extrapolation from high doses in rodent bioassays to the low doses of most human exposures, and causes of human cancer with emphasis on mechanism. Dr. Gold has served on the panel of expert reviewers for the National Toxicology Program and on the boards of the Harvard Center for Risk Analysis and the Annapolis Center and was a member of the Harvard Risk Management Reform Group.

Errol Zeiger, Ph.D., J.D., is at the National Institute of Environmental Health Sciences (NIEHS), National Institutes of Health (NIH), in Research Triangle Park, North Carolina. He developed and implemented the *in vitro* genetic toxicity testing program for the National Toxicology Program and has been involved with the development and validation of genetic toxicity tests for more than 25 years. Much of his research has been on the improvement of *in vitro* tests for identifying chemical mutagens, the improvement of methods for evaluating genetic toxicity data, the study of mechanisms of the genetic activity of chemicals, and the development and evaluation of predictive systems for carcinogens and mutagens. He is the Editor-in-Chief of *Environmental and Molecular Mutagenesis* and serves on several other editorial boards.

PREFACE

In efforts to identify causes of human cancer, to understand the carcinogenic process, and to meet regulatory requirements, thousands of long-term carcinogenesis bioassays have been conducted in rodents, and thousands of *in vitro* tests have been conducted to assess genetic toxicity. This Handbook is designed as a general reference source for genetic toxicity and carcinogenicity test results, bringing together in a single volume two databases that present experimental results by chemical in easily accessible formats and provide citations to the original experiments. The two databases are the Carcinogenic Potency Database (CPDB), developed at the University of California at Berkeley and the E.O. Lawrence Berkeley National Laboratory, and the Genetic Toxicity database (GT), developed by the U.S. National Toxicology Program (NTP). Results on 2300 chemicals are reported, including cancer tests on 1298 and genetic toxicity tests on 1525. For 523 chemicals, results are given in both the carcinogenicity and genetic toxicity databases. Subsets of the CPDB and the GT have been published in several journal articles, but never before have all results on all chemicals in each database been presented together. For the CPDB, standardized results of all experiments on a chemical are reported together in plot format, whereas results published earlier may have appeared in six different publications; additionally, analyses of 300 cancer tests are presented for the first time. The GT test systems are described, and summary data are presented so as to allow qualitative comparisons among test systems.

The chapters are organized by database. Chapter 1 presents the inclusion rules and methods of the CPDB and a guide to facilitate use of the plot of cancer test results. The CPDB presents the analyses of 5152 cancer tests, including for each qualitative information on strain, sex, target organ, histopathology, and author's opinion about carcinogenicity, as well as quantitative information on carcinogenic potency (TD_{50}), statistical significance, tumor incidence, dose-response curve shape, length of experiment, dose-rate, and duration of dosing. Data are included for mice, rats, hamsters, nonhuman primates, and dogs. Chapter 2 is a compendium of CPDB results organized by target organ for 684 chemicals that are positive in at least one species. Chapter 3 gives an overview of CPDB results on each chemical, including the sex-species groups that have been tested, the strongest level of evidence for carcinogenicity based on the opinion of a published author, carcinogenic potency, target organs in each species, and mutagenicity in *Salmonella typhimurium*. Chapter 4 is an overview and update of analyses reported in earlier publications that include the data most recently added to the database. Examples of these analyses are reproducibility and validity, positivity rates, comparisons of carcinogenicity between rats and mice and between mutagens and nonmutagens, and ranking possible carcinogenic hazards to humans from rodent carcinogens.

Chapter 5 describes and presents the *in vitro* genetic toxicity database of the NTP, which contains qualitative results of experiments on the effects of chemicals on cellular DNA and chromosomes. Data included in this compilation are from the *Salmonella typhimurium* mutation test, tests for induction of chromosome aberrations and sister chromatid exchanges in cultured Chinese hamster ovary cells, a test for mutation in the mouse lymphoma L5178Y cell line, and from the sex-linked recessive lethal test in the fruit fly *Drosophila melanogaster*. Following a description of the test protocols, a table summarizes the results in each test and references the original publications. Chapter 6 is an index to all chemicals in the CPDB and GT by Chemical Abstract Services (CAS) registry number.

Both of the databases presented in this Handbook are dynamic. As results are added to each they will continue to be reported in future publications.

Lois Swirsky Gold
Berkeley, California

Errol Zeiger
Research Triangle Park, North Carolina

CONTRIBUTORS

Lois Swirsky Gold, Ph.D.
Division of Biochemistry and Molecular Biology
University of California at Berkeley and
Life Sciences Division
E.O. Lawrence Berkeley National Laboratory
Berkeley, California

Errol Zeiger, Ph.D., J.D.
Environmental Toxicology Program
National Institute of Environmental Health Sciences
Research Triangle Park, North Carolina

Bruce N. Ames, Ph.D.
Division of Biochemistry and Molecular Biology
University of California at Berkeley
Berkeley, California

Georganne Backman Garfinkel, A.B.
Division of Biochemistry and Molecular Biology
University of California at Berkeley
Berkeley, California

Neela B. Manley, M.D., Ph.D.
Division of Biochemistry and Molecular Biology
University of California at Berkeley
Berkeley, California

Lars Rohrbach, A.B.
Division of Biochemistry and Molecular Biology
University of California at Berkeley
Berkeley, California

Thomas H. Slone, M.S.
Division of Biochemistry and Molecular Biology
University of California at Berkeley and
Life Sciences Division
E.O. Lawrence Berkeley National Laboratory
Berkeley, California

TABLE OF CONTENTS

Chapter 6

ACKNOWLEDGMENTS

We thank the many researchers who have provided us with unpublished results and evaluations for the Carcinogenic Potency Database. We are grateful to J. Ward, R. Peto, M. Pike, L. Bernstein, and D. Freedman, who have provided us with advice on pathology and statistics for many years. We also thank the many people at the National Toxicology Program who have helped us by creating computer files of bioassays and advising us on details of the Technical Reports. A. Auletta assisted in providing us with unpublished genotoxicity evaluations. S. Sieber and U. Thorgeirsson have given us their time and expertise in our analyses of the experiments in nonhuman primates. For more than 18 years, many people at the National Institute of Environmental Health Sciences, the National Cancer Institute, and the Environmental Protection Agency have given us much appreciated advice. Collaborations with several investigators were involved in the publications that are updated in Chapter 4. We thank them all, especially L. Bernstein, D. Freedman, D. Gaylor, R. Peto, M. Pike, and J. Ward.

The Carcinogenic Potency Database is supported through the University of California, Berkeley by National Institute of Environmental Health Sciences Center Grant ESO1896 and through the E.O. Lawrence Berkeley National Laboratory by U.S. Department of Energy, contract DE-AC-03-76SFO0098. In earlier years, support was also provided by NIEHS/DOE Interagency Agreement 222-Y01-AS-10066 and EPA-NCI/DOE Interagency Agreement Y01-CP-15791 through the E.O. Lawrence Berkeley National Laboratory and by DOE Contract DE-AT03-80EV70156.

The genetic toxicology test results presented in Chapter 5 were developed for the U.S. National Toxicology Program by a number of excellent laboratories under contract to the National Institutes of Health. We are grateful to the following individuals in these laboratories for their efforts in this continually rewarding process: A. D. Bloom, D. Brusick, R. Combes, D. Dillon, P. Foureman, S. M. Galloway, D. K. Gulati, S. Haworth, J. L. Ivett, C. M. King, T. Lawlor, K. S. Loveday, E. McCoy, D. B. McGregor, A. D. Mitchell, K. Mortelmans, B. C. Myhr, T. M. Reid, H. S. Rosenkranz, C. J. Rudd, V. F. Simmon, W. Speck, R. Valencia, R. C. Woodruff, and S. Zimmering. Thanks are also due to our colleagues in the government for the design and management of the genetic toxicity testing contracts and the evaluation of the data: B. E. Anderson, W. J. Caspary, V. C. Dunkel, B. H. Margolin, J. M. Mason, M. A. Resnick, and M. D. Shelby.

Lois Swirsky Gold
Berkeley, California

Errol Zeiger
Research Triangle Park, North Carolina

CHAPTER 1

THE CARCINOGENIC POTENCY DATABASE

Lois Swirsky Gold, Thomas H. Slone, Neela B. Manley,
Georganne Backman Garfinkel, Lars Rohrbach, and Bruce N. Ames*

SECTION A: METHODS OF THE CARCINOGENIC POTENCY DATABASE

Epidemiological studies have identified several factors that are likely to have a major effect on lowering rates of human cancer: reduction of smoking, increased consumption of fruits and vegetables, and control of infections. Other factors include avoidance of intense sun exposure, reduction of alcohol consumption, reduction of high occupational exposures, and increased physical activity. In addition to epidemiological investigations, during the past 50 years, thousands of chronic long-term animal cancer tests have been conducted for the general purposes of understanding the carcinogenic process and identifying chemicals that might cause cancer in humans. These laboratory results are the main source of information that is used to set regulatory standards for potential human exposures. The published results of these experiments constitute a diverse literature that varies widely with respect to experimental and histological protocols as well as to how and which information is reported in published articles.

The Carcinogenic Potency Database (CPDB) presented in this handbook is a systematic and unifying analysis of animal cancer tests. It standardizes the published literature and creates an easily accessible research resource that can be and has been used to address a wide variety of research and regulatory issues in carcinogenesis. A measure of carcinogenic potency, TD_{50} (tumorigenic dose-rate for 50% of experimental animals), is estimated for the tumor incidence at each site for which results are reported in the database. The CPDB includes results reported in 1002 papers in the general literature through 1992 and 403 Technical Reports of the National Cancer Institute/National Toxicology Program (NCI/NTP) through 1994. Results are examined for 5152 experiments on 1298 chemical agents; these are displayed in a plot format organized by chemical name. Detailed information that is important in the interpretation of bioassays is reported on each experiment, (whether positive or negative for carcinogenicity) including qualitative information on strain, sex, target organ, histopathology, and author's opinion, as well as quantitative information on carcinogenic potency, statistical significance, tumor incidence, dose-response curve shape, length of experiment, dose-rate, and duration of dosing. Each set of experimental results references the original published paper. A word of caution is necessary about the limitations of the database. No attempt has been made to evaluate whether or not a compound induced tumors in any given experiment; rather, the opinion of the published authors is presented as well as the statistical significance of the TD_{50} calculated from their results. Moreover, the database contains only long-term tests which fit a set of criteria designed to measure potency and therefore does not cover all cancer tests.

This chapter describes the methods used to develop the CPDB (Section A), a guide to each of the variables in the plot of cancer test results (Section B), the plot itself (Section C), appendices of codes and definitions (Section D), and a bibliography of the original papers and NCI/NTP Technical Reports that serve as the sources of data (Section E).

Two additional tables systematically summarize the CPDB results (1) by target organ, listing all chemicals in each species that induce tumors at a specific site such as kidney[1] and (2) by chemical name summarizing for each chemical in each species the carcinogenic potency, positivity, and target organs from all experiments combined.[2]

---

* The following individuals have been authors on previous publications that included parts of the plot of the Carcinogenic Potency Database: Leslie Berstein, Mark Blumenthal, Ken Chow, Maria Da Costa, Margarita de Veciana, Susan Eisenberg, William R. Havender, N. Kim Hooper, Peggy Lopipero, Robert Levinson, Renae Magaw, Richard Peto, Malcolm C. Pike, Charles B. Sawyer, Mark Smith, Bonnie R. Stern, Jerrold M. Ward, and Michael Wong.

These summary tables also report mutagenicity in *Salmonella* for each chemical. Together, the plot, bibliography, and the two summary tables provide a guide to the literature of animal cancer tests as well as an accessible research resource.

In a series of papers published since 1984, our group has analyzed the CPDB to address many issues in carcinogenesis and interspecies extrapolation. The "Overview and Update of Analyses of the Carcinogenic Potency Database"[3] updates some of these analyses and tables using the data in this handbook, which includes results of 300 experiments never before included in the CPDB.[4-9] We refer the reader to our original papers for the detailed analyses.

CARCINOGENIC POTENCY

A numerical description of carcinogenic potency, the TD_{50}, is estimated for each set of tumor incidence data reported in the CPDB, thus providing a standardized quantitative measure for comparisons and analyses of many issues in carcinogenesis. In a simplified way, TD_{50} may be defined as follows: for a given target site(s), if there are no tumors in control animals, then TD_{50} is that chronic dose-rate in mg/kg body wt/day which would induce tumors in half the test animals at the end of a standard lifespan for the species. Since the tumor(s) of interest often does occur in control animals, TD_{50} is more precisely defined as that dose-rate in mg/kg body wt/day which, if administered chronically for the standard lifespan of the species, will halve the probability of remaining tumorless throughout that period. TD_{50} is analogous to LD_{50}, and a low value of TD_{50} indicates a potent carcinogen, whereas a high value indicates a weak one. TD_{50} can be computed for any particular type of neoplasm, for any particular tissue, or for any combination of these. Our numerical index of carcinogenic potency, the TD_{50}, and the statistical procedures adopted for estimating it from experimental data, have been described in detail elsewhere[10,11] and are briefly summarized in Appendix 8.

One goal of the CPDB has been to obtain data which would give the best estimates of carcinogenic potency, i.e., experiments for which detailed time-to-tumor data would permit adjustment for the gross effects of intercurrent mortality.[11,12] Full lifetable information was available for the calculations of potency from 403 bioassays of the NCI/NTP, from a set of bioassays on 33 aromatic amines,[13-15] and from studies on 17 chemicals tested in nonhuman primates by the NCI Laboratory of Chemical Pathology.[16] For estimates of potency from the general literature, we have calculated TD_{50} using the final proportions of animals with tumors, since only this summary information is consistently published.

SOURCES OF DATA

The inclusion criteria for the CPDB are designed to identify reasonably thorough, chronic long-term tests of single chemical agents, that permit estimation of carcinogenic potency. The two sources of data are the bioassays of the NCI/NTP and the general published literature. Thirty percent of the chemicals in the CPDB have been tested by NCI/NTP.

NCI/NTP BIOASSAYS

The CPDB provides a concise summary of results of the NCI/NTP bioassay program. All Technical Reports are included except a few, for which no comparable dose in mg/kg/day could be calculated because the test agent was a particulate or the route of administration was dermal. The NCI/NTP program was designed to use a similar experimental protocol for a large number of compounds and to report results in a consistent format including full histopathology. Since the 1970s, the NCI/NTP has provided us with computer files that contain information about the time of death and full histopathology for each animal in each experiment. By using these files in combination with the series of Technical Reports published by NCI/NTP, we have been able to estimate a TD_{50} for each site in each Technical Report that was evaluated as treatment-related, as well as for each site found to be statistically significant but not considered evidence of carcinogenicity.

The standard NCI/NTP bioassay protocol recommended that tests be conducted in two species of rodents (rats and mice) with both sexes tested individually at the maximally tolerated dose (MTD) and half that dose, using a control group and a vehicle control where appropriate.[17] The MTD is generally accepted to be defined as the maximum dose level which is not expected to shorten the normal longevity from non-neoplastic causes, and which is expected to result in no more than a 10% weight decrement in animals receiving this dose when compared to controls.[17]

The actual conduct of the bioassays published prior to July 1980 varied from one experiment to another, particularly with respect to the number of animals per group and the experiment length. Details about each experiment are reported in the plot of the CPDB in Section C. (An experiment is defined here as the dose groups and control group for one sex in one species from a single research report.) In Technical Reports published in 1993 and 1994, the NTP bioassays more often include three dose groups and a control, with testing at the MTD and below; the range of doses tested varies by experiment. Comparisons of results from the NCI/NTP experiments are particularly useful for the following reasons: each compound was usually tested in both sexes of rats and mice; the same mouse hybrid, B6C3F₁, was used throughout; the rat bioassays utilized mostly the Fischer F344 strain and less frequently either the Osborne-Mendel or Sprague-Dawley stocks; dosing continued for the majority of the animals' lives (18 to 24 months); terminal sacrifice was usually performed after 21 to 25 months on test; and our analyses utilized full lifetable data.

BIOASSAYS IN THE PUBLISHED LITERATURE

In the general literature, experimental designs as well as the authors' choice of information to report are quite diverse. For the CPDB we have developed a set of standard criteria, and experiments from the literature have been included only if they meet all of the following conditions:

1. Animals on test were mammals.
2. Administration was begun early in life (100 days of age or less for rats, mice, and hamsters).
3. Route of administration was diet, water, gavage, inhalation, or intravenous or intraperitoneal injection (i.e., where the whole body was more likely to have been exposed rather than only a specific site, as with subcutaneous injection or skin painting).
4. Test agent was administered alone, rather than in combination with other chemicals.
5. Exposure was chronic, with not more than 7 days between administrations.
6. Duration of exposure was at least one fourth the standard lifespan for that species.
7. Duration of experiment was at least half the standard lifespan for that species.
8. Research design included a concurrent control group.
9. Research design included at least five animals per group.
10. Surgical intervention was not performed.
11. Pathology data were reported for the number of animals with tumors rather than the total number of tumors.
12. Results reported were original data, rather than secondary analyses of experiments already reported by other authors.

A search of the published literature has been conducted through 1992 for all bioassays which met the standard criteria, whether or not the authors considered the test agent related to tumor induction. The literature search covered the *Survey of Compounds Which Have Been Tested for Carcinogenic Activity* (formerly PHS 149),[18] *Medline, Current Contents*, and the monographs on chemical carcinogens prepared by the International Agency for Research on Cancer. Additionally, a separate search has been done of the journals that most often report cancer test results, including the following: *British Journal of Cancer, Cancer Letters, Cancer Research, Carcinogenesis, Chemosphere,*

Environmental Health Perspectives, European Journal of Cancer and Clinical Oncology, Food and Chemical Toxicology, Fundamental and Applied Toxicology, International Journal of Cancer, Japanese Journal of Cancer Research, Journal of Cancer Research and Clinical Oncology, Journal of Environmental Pathology and Toxicology, Journal of Toxicology and Environmental Health, Journal of the National Cancer Institute, Regulatory Toxicology and Pharmacology, Toxicology, and *Toxicology and Applied Pharmacology.* From the literature search, 3573 experiments on 1027 chemicals met the inclusion criteria and are in the database.

Because we have adhered quite strictly to the standard inclusion criteria, bioassays of particulate or fibrous matters are not included, e.g., asbestos, cigarette smoke, and dusts. There are no single injection or skin painting tests and no co-carcinogenesis studies in which more than one chemical is administered to the same animal. However, with increasing attention in the published literature to mechanism of carcinogenesis, more of the recent experiments in the CPDB are tests of a single compound administered chronically as part of an experimental design to investigate the effects of co-administration of two or more chemicals.

Although we have excluded co-carcinogenesis experiments from the database, some mixtures to which humans are exposed are included, e.g., commercial preparations and technical grades including pesticides, industrial compounds, and drugs. For the relatively small literature on nonhuman primates, we have relaxed some of our standard rules to include tests of 17 chemicals in these long-lived animals by the NCI Chemical Pathology Laboratory (see Appendix 1).

Including experiments in both NCI/NTP Technical Reports and the general literature, results of 5152 experiments are summarized in the CPDB. About 3% of the experiments were conducted in hamsters, less than 1% in nonhuman primates or dogs, and the rest are about evenly divided between rats and mice. Experiments with 112 different mouse strains and 78 rat strains are included. For any single chemical, the number of experiments in the database may vary. Some chemicals have only one test in one sex of one species, while others have multiple tests including both sexes of a few strains of rats and mice, possibly using quite different protocols. For example, among the 997 chemicals tested in rats, 27% have only one test and 53% have two tests; however, 18 chemicals have more than ten tests. For the 847 chemicals tested in mice, the parallel numbers are 9% with one test, 61% with two tests, and 15 chemicals with more than ten tests. Additionally, the number of TD₅₀s reported for an experiment may vary, depending upon the results of the test and the extent of detail in the published paper.

A wide variety of experimental protocols meet the inclusion rules of the CPDB, e.g., with respect to the number of doses, the range of doses tested, and group size. The standard NCI/NTP protocol is two dose groups and a control (12% have three groups). In contrast, among experiments in rats or mice from the general literature, 63% have only one dose group, 17% have two dose groups and a control, and 20% have three or more. There is variation in the CPDB with respect to the range of doses tested within individual experiments: among NCI/NTP bioassays, 78% are tested within a 2-fold range of dose, 19% within a >2- to 10-fold range, and 3% more than 10-fold. The parallel percentages for literature experiments in rats or mice are: 63% have only one dose group, 10% are within a 2-fold range, 18% >2- to 10-fold, and 9% greater than 10-fold. The standard NCI/NTP protocol is 50 animals per group of dosed animals, but in the literature the group sizes vary widely: 42% of experiments have fewer than 20 animals per dose group, and 30% have more than 40 per group.

SELECTION OF TISSUE AND TUMOR TYPES FOR CALCULATION OF CARCINOGENIC POTENCY

In order to standardize the diverse bioassay literature, a great many decisions have had to be made in the construction of the database. These decisions were based on two major considerations: (1) to provide estimates of TD_{50} that are based on the most complete information available and (2) to provide a large resource with comparable data to facilitate the analysis of the results of chronic, long-term animal cancer tests.

Our choice of which sites and pathology to report from the published literature is limited by what individual authors have reported, and this varied considerably from paper to paper. We have been able to augment the published results through personal communication with authors of published papers, particularly for experiments conducted recently, e.g., to clarify opinions as to carcinogenicity at particular sites, to obtain additional tumor data or separate data for interim sacrifice experiments.

For reasons of both accuracy and consistency throughout the database, our general approach has been to calculate a TD_{50} for each category of neoplasm, benign or malignant, which an author evaluated as treatment-related, regardless of the statistical or biological basis for the evaluation. (Hyperplasia and nonneoplastic lesions are not included in the database.) In addition, a TD_{50} has been estimated for the category "all tumor-bearing animals" (TBA) wherever this was reported. We have not attempted to include results that indicate a reduction of tumor incidence in dosed animals.

In order to provide information which would permit comparisons of target sites across experiments, an additional category of histopathology was developed for the database which we call "mandatory sites." Whenever there is adequate documentation in the published paper or report, a TD_{50} is estimated for tumors of the liver in rats and mice and for tumors of the lung in mice. These tissues were selected as mandatory because they occurred most often in a frequency count of the positive sites in the NCI/NTP bioassays before 1980. From the general literature, whenever possible, we have taken the liver as a mandatory site in rats, mice, and hamsters; the lung is also a mandatory site in mice and hamsters.

In summary, for each experiment in the database, a TD_{50} is calculated whenever documentation is adequate for the following categories: (1) each target site evaluated by the author as treatment-related; (2) mandatory sites; and (3) all tumor-bearing animals.

Some special considerations follow about selection of pathology for the calculation of TD_{50} from each source of data.

NCI/NTP BIOASSAYS

From the NCI/NTP Technical Reports, TD_{50}s have been calculated for all categories of tissues and tumors listed in the statistical tables as positively related to dose. (See description of statistical tests in Technical Reports.) All sites are reported in the CPDB for which the NCI/NTP evaluation was "carcinogenic," "associated with carcinogenicity," "clear evidence," "some evidence," or "equivocal evidence" of carcinogenic activity. In the absence of such an evaluation, sites which are reported in the statistical tables but which are not considered treatment-related, are included only if the TD_{50} itself is significant at the $p < 0.05$ level (two-tailed p-value for the test that the slope of the dose-response is different from zero). We refer to these cases as "statistical sites" to indicate that the results were not evaluated in the Technical Report as evidence of carcinogenicity.

Because the computer files from the individual animal pathology tables of the NCI/NTP contain pathology results on each animal, it has been possible to calculate a TD_{50} for a composite category of all tumors which the NCI/NTP evaluated as treatment-related. It has generally not been possible to formulate such a composite category for other experiments in the database, with the exception of a group of studies on aromatic amines, bioassays in monkeys, and a few other experiments in the database.

In the NCI/NTP data, for each mandatory site, a TD_{50} is estimated for a specific combination of tumors. These were selected because they are the common histopathology reported for the specified tissue by the NCI/NTP. For the liver, neoplastic nodule, hepatocellular adenoma, and hepatocellular carcinoma are combined in the mandatory TD_{50} for Technical Reports prior to 1993; subsequently, the liver mandatory site for NTP includes hepatocellular adenoma, hepatoblastoma, and hepatocellular carcinoma. This change reflects diagnoses given in the Technical Reports. For the lung, alveolar/bronchiolar adenoma and alveolar/bronchiolar carcinoma are combined. In all cases, the incidence represents the proportion of animals with any of the tumors; animals with multiple tumors of the given tissue are counted only once. For the category TBA, we have excluded interstitial-cell tumors of the testis for Fischer 344 rats, because these tumors occur spontaneously in nearly all male animals by the end of their standard lifespan.

For several recent bioassays in which NTP suspected a potential for chemically induced renal tubule neoplasms, multiple additional sections were taken of the kidneys. We have estimated one TD_{50} value using data from the standard protocol and one that includes the step section data, and indicate the step section results on the plot by the words "with step."

BIOASSAYS IN THE PUBLISHED LITERATURE

Authors of papers in the general literature rarely indicate which animals were diagnosed as having more than one tumor. Therefore, it has been generally impossible to combine data on various tumors within a single tissue, or to specify particular tumor types which should be included in mandatory sites. To attempt to combine incidence data would risk multiple counting of animals.

For the mandatory sites from the published literature, a TD_{50} is estimated for individual tumor types reported in the mandatory tissue, as well as for any combination of tumors which is reported or could be created for that tissue without risking multiple counting.

In the interest of completeness, we have added to the database results for tissues and tumors in the literature which the authors did not consider treatment-related, but which we calculate as statistically significant (standard chi-square test, one-sided $p < 0.05$).

In the general literature, when we report results separately for benign and malignant tumors at a given site in a single experiment, we often do not know whether the author counted animals only for the most malignant tumor. For example, we may not know whether the proportion of animals with adenomas represents only those animals which did not have carcinomas. Therefore, in such cases no mix could be created of benign and malignant tumors at a given site.

ESTIMATION OF AVERAGE DAILY DOSE LEVEL

Because a variety of routes of administration, dosing schedules, species, strains, and sexes are used in carcinogenesis bioassays, some standardization of dose is required for a single index of potency. Our convention in estimating TD_{50} is to determine for each dose group in an experiment the daily dose-rate in mg/kg of body weight averaged over the duration of the experiment.

To convert ppm or percent in food, water, or air into mg/kg body weight during the dosing period, we assume 100% absorption and then use a set of standard values for each sex of each species, including body weight and average intake per day (Table 1). Using standard values, the daily dose-rate is calculated as follows:

$$\text{Dose rate} = \frac{\text{dose} \times \text{intake/day} \times \text{number of doses/week}}{\text{animal weight} \times 7 \text{ days/week}}$$

or equivalently,

$$\text{Dose-rate} = \text{dose} \times \text{intake/day as a proportion of body weight} \times \text{proportion of week dose is administered}$$

In an experiment where the animals were dosed the entire time on test this value would equal the average daily dose level. For example, in a bioassay of male mice fed 50 ppm of some test agent in the diet for the entire time on test, the calculation would be:

$$\text{Dose-rate} = 50 \text{ ppm} \times 0.12 \times 1$$
$$= 6 \text{ mg/kg body weight/day}$$

In this example, the exposure time is equal to the experiment time.

In using standard values we recognize that there is no single factor that precisely reflects the entire experimental literature. For example, strains within a species will vary in weight; younger animals have lower body weight and food intake than adults; animals within a group differ in food consumption and body weight; some test agents will reduce appetite due to taste; or illness may result in loss of appetite. However, the values used here for proportion of body weight consumed daily are within reasonable limits of those usually found in the published literature and are unlikely to produce substantial error. We note that bioassays in the general literature and NCI/NTP provide food *ad libitum*, which can itself affect tumor rates compared to calorie-restricted diets.[27,28]

TABLE 1
Standard Values for Dose Calculation: Animal Lifespans, Weights, and Intake by Diet, Water, and Inhalation[a]

Experimental animal	Sex	Standard lifespan (year)[b]	Weight (kg)[c]	Food (g)[c]	Food as % body weight/day	Water (mL/day)[d]	Inhalation volume (L/min)[e]
Rodents							
Mouse	Male	2	0.03	3.6	12.00	5	0.03
	Female	2	0.025	3.25	13.00	5	0.03
Rat	Male	2	0.5	20	4.00	25	0.10
	Female	2	0.35	17.5	5.00	20	0.10
Hamster	Male	2	0.125	11.5	9.20	15	0.06
	Female	2	0.110	11.5	10.45	15	0.06
Monkeys							
Cynomolgus (*Macaca fascicularis*)	Both	20					
Rhesus (*Macaca mulatta*)	Both	20					
Prosimians							
Bush babies (*Galago crassicaudatus*)	Both	10					
Tree shrews (*Tupaia glis*)	Both	4.5					
Dog	Both	11	16	400			

[a] Although values sometimes vary depending on the source, those given here are within reasonable limits of those usually found in the published literature. No value is given when this information was not necessary for our dose calculation.

[b] Rat and mouse: based on NCI trichloroethylene bioassay;[19] hamster: data of Williams;[20] nonhuman primates: data of S. M. Sieber (Laboratory of Chemical Pharmacology, NCI, National Institute of Health, Bethesda, MD), personal communication; bush babies: ages adapted from Dittmer;[21] tree shrews: data of D. J. Reddy (Northwestern University, Chicago, IL), personal communication; dog: data of M. S. Redfearn (Division of Animal Resources, University of California, Berkeley), personal communication.

[c] Rat and mouse: based on NCI trichloroethylene bioassay;[19] hamster and dog: data of D. Brooks (University of California, Davis), personal communication.

[d] Mouse, rat, and dog: data from NIOSH;[22] hamster: data from Hoeltge, Inc.[23]

[e] Mouse: data of Sanockij;[24] rat: data of Baker et al.;[25] hamster: data of Guyton.[26]

In many experiments the administration of the test compound is stopped before the terminal sacrifice or before the death of the last animal. In such cases, averaging the dose over the course of the experiment will lower the daily dose level. By convention we take the total dose administered and spread this over the entire experimental period; thus, for male mice dosed at 50 ppm in the diet for 70 weeks and then continued on test to 100 weeks, the dose is considered as equivalent to 35 ppm for the entire 100 weeks (i.e., 50 ppm × 70/100), resulting in a 4.2 mg/kg body wt/day average dose. In calculating the average daily dose-rate, we utilize the concept of exposure time (the period of active treatment) and experiment time (the actual time on test). When terminal sacrifice is performed, the experiment time is the length of time from the start of the experiment to sacrifice. When, however, animals are per-mitted to survive to natural death, experiment time is defined as the time of death of the last dosed animal.

EXTRAPOLATION OF TD$_{50}$ TO THE STANDARD LIFESPAN

TD$_{50}$ is defined in terms of the dose-rate that will halve the probability of remaining tumor free at the end of a standard lifespan. For each species the assumed value for the standard lifespan is given in Table 1; these values are within reasonable limits usually found in the published literature. The use of 2 years as standard for rats and mice reflects both the standard NCI/NTP protocol and values that appear frequently in the literature.

When an experiment is terminated before the standard lifespan, animals are not at risk of developing tumors later in life. Thus, the number of tumors found will be reduced, and the TD_{50} will be greater than the true TD_{50}, i.e., the compound will appear to be less potent than it actually is. Because tumor incidence increases markedly with age, our convention has been to adopt as a correction factor f^2, where f = experiment time/standard lifespan.[11]

Note that the correction factor f^2 is based on the time the animals are on test, rather than upon age. In an experiment which began when the animals were 6 weeks of age, and which terminated when the animals were 100 weeks of age, the experiment time is 94 weeks. Thus, TD_{50} is defined in terms of the dose-rate which would be administered throughout life, from birth to death, or the entire standard lifespan.

Taking the example above of male mice fed some test agent at 50 ppm for 70 weeks and then continued on test for 30 more weeks, the experiment time would be 100 weeks. The standard lifespan for mice is 104 weeks, so the extrapolation factor would be $(100/104)^2$ or 0.92.

By omitting from the database any experiments lasting less than half the standard lifespan for that species, the necessity for great extrapolation has been reduced. Bioassays are usually not continued for much longer than the period we have adopted for a standard lifespan, so the reverse correction is minimal. For nearly all NCI/NTP bioassays and 80% of literature tests, the value of the extrapolated TD_{50} is within a factor of 2 of the unextrapolated value, and for the remaining 20% of literature tests it is within a factor of 4.

RANGE OF CARCINOGENIC POTENCY

The TD_{50}s we have estimated according to the rules and conventions just described are presented in the "Plot of Experimental Results". The plot format provides a systematic means of distinguishing among the carcinogenic potencies of a variety of chemicals. The range of TD_{50}s is at least 10-million-fold for carcinogens in each sex of rat or mouse.

For female rats, the range of carcinogenic potency is shown in Figure 1, where we present the most potent TD_{50}s for a selected group of compounds which were evaluated as tumorigens in either an NCI/NTP Technical Report or the general literature. In each case, we have indicated the value for the most potent TD_{50} for a target site(s) which was considered positive, and for which the statistical significance of TD_{50} is less than 0.01. The range is more than 100-million-fold in female rats.

REFERENCES

1. Gold, L.S., Manley, N.B., and Slone, T.H. Summary of Carcinogenic Potency Database by target organ. In: Gold, L.S. and Zeiger, E., Eds. *Handbook of Carcinogenic Potency and Genotoxicity Databases*. Boca Raton, FL: CRC Press, 1997, pp. 1–605.
2. Gold, L.S., Slone, T.H., and Ames, B.N. Summary of Carcinogenic Potency Database by chemical. In: Gold, L.S. and Zeiger, E., Eds. *Handbook of Carcinogenic Potency and Genotoxicity Databases*. Boca Raton, FL: CRC Press, 1997, pp. 621–660.
3. Gold, L.S., Slone, T.H., and Ames, B.N. Overview and update of analyses of the Carcinogenic Potency Database. In: Gold, L.S. and Zeiger, E., Eds. *Handbook of Carcinogenic Potency and Genotoxicity Databases*. Boca Raton, FL: CRC Press, 1997, pp. 661–685.
4. Gold, L.S., Sawyer, C.B., Magaw, R., Backman, G.M., de Veciana, M., Levinson, R., Hooper, N.K., Havender, W.R., Bernstein, L., Peto, R., Pike, M.C., and Ames, B.N. A Carcinogenic Potency Database of the standardized results of animal bioassays. *Environ. Health Perspect.* 58: 9-319 (1984).

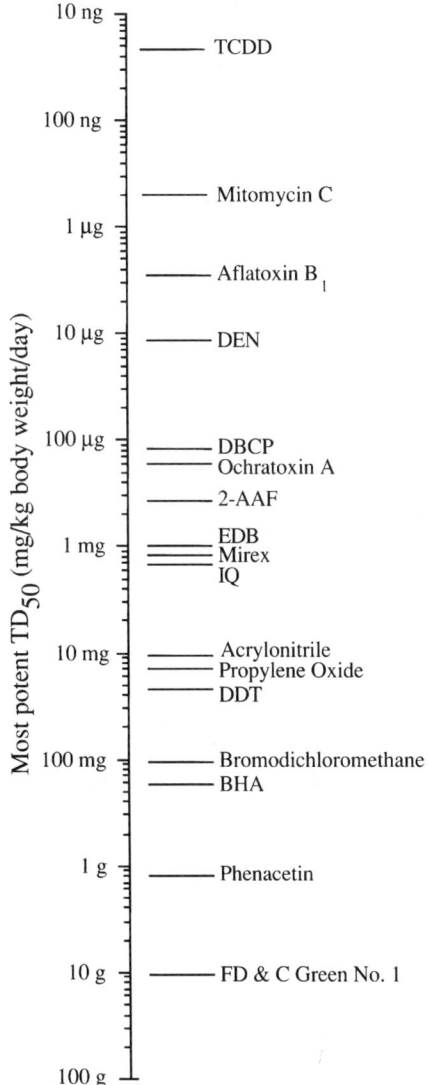

FIGURE 1 Range of carcinogenic potency in female rats.

5. Gold, L.S., de Veciana, M., Backman, G.M., Magaw, R., Lopipero, P., Smith, M., Blumenthal, M., Levinson, R., Bernstein, L., and Ames, B.N. Chronological supplement to the Carcinogenic Potency Database: Standardized results of animal bioassays published through December 1982. *Environ. Health Perspect.* 67: 161-200 (1986).

6. Gold, L.S., Slone, T.H., Backman, G.M., Magaw, R., Da Costa, M., Lopipero, P., Blumenthal, M., and Ames, B.N. Second chronological supplement to the Carcinogenic Potency Database: Standardized results of animal bioassays published through December 1984 and by the National Toxicology Program through May 1986. *Environ. Health Perspect.* 74: 237-329 (1987).

7. Gold, L.S., Slone, T.H., Backman, G.M., Eisenberg, S., Da Costa, M., Wong, M., Manley, N.B., Rohrbach, L., and Ames, B.N. Third chronological supplement to the Carcinogenic Potency Database: Standardized results of animal bioassays published through December 1986 and by the National Toxicology Program through June 1987. *Environ. Health Perspect.* 84: 215-286 (1990).

8. Gold, L.S., Manley, N.B., Slone, T.H., Garfinkel, G.B., Rohrbach, L., and Ames, B.N. The fifth plot of the Carcinogenic Potency Database: Results of animal bioassays published in the general literature through 1988 and by the National Toxicology Program through 1989. *Environ. Health Perspect.* 100: 65-135 (1993).

9. Gold, L.S., Manley, N.B., Slone, T.H., Garfinkel, G.B., Ames, B.N., Rohrbach, L., Stern, B.R., and Chow, K. Sixth plot of the Carcinogenic Potency Database: Results of animal bioassays published in the general literature 1989-1990 and by the National Toxicology Program 1990-1993. *Environ. Health Perspect.* 103 (Suppl. 8): 3-122 (1995).

10. Sawyer, C., Peto, R., Bernstein, L., and Pike, M.C. Calculation of carcinogenic potency from long-term animal carcinogenesis experiments. *Biometrics* 40: 27-40 (1984).

11. Peto, R., Pike, M.C., Bernstein, L., Gold, L.S., and Ames, B.N. The TD_{50}: A proposed general convention for the numerical description of the carcinogenic potency of chemicals in chronic-exposure animal experiments. *Environ. Health Perspect.* 58: 1-8 (1984).

12. Gold, L.S., Bernstein, L., Kaldor, J., Backman, G., and Hoel, D. An empirical comparison of methods used to estimate carcinogenic potency in long-term animal bioassays: Lifetable vs. summary incidence data. *Fundam. Appl. Toxicol.* 6: 263-269 (1986).

13. Russfield, A.B., Homburger, F., Boger, E., van Dongen, C.G., Weisburger, E.K., and Weisburger, J.H. *Carcinogenicity of Chemicals in Man's Environment.* Final Report, Contract No. NIH-NCI-E-68-1311. Cambridge, MA: Bio-Research Consultants, Inc., 1973.

14. Russfield, A.B., Homburger, F., Boger, E., van Dongen, C.G., Weisburger, E.K., and Weisburger, J.H. The carcinogenic effect of 4,4'-methylene-*bis*-(2-chloroaniline) in mice and rats. *Toxicol. Appl. Pharmacol.* 31: 47-54 (1975).

15. Weisburger, E.K., Russfield, A.B., Homburger, F., Weisburger, J.H., Boger, E., van Dongen, C.G., and Chu, K. Testing of twenty-one aromatic amines or derivatives for long-term toxicity or carcinogenicity. *J. Environ. Pathol. Toxicol.* 2: 325-356 (1978).

16. Adamson, R.H. and Sieber, S.M. Chemical carcinogenesis studies in nonhuman primates. In: Langenbach, R., Nesnow, S., and Rice, J.M., Eds. *Organ and Species Specificity in Chemical Carcinogenesis.* New York: Plenum Press, 1982, pp. 129-156.

17. Sontag, J.A., Page, N.P., and Saffiotti, U. *Guidelines for carcinogen bioassay in small rodents.* 1976. (Carcinogenesis Tech. Rep. Ser. No. 1, DHEW Pub. No. (NIH) 76-801.)

18. National Cancer Institute. *Survey of Compounds which have been Tested for Carcinogenic Activity, 1948-1992.* Bethesda, MD: NCI, 1994. (formerly PHS 149.)

19. National Cancer Institute (NCI). *Carcinogenesis Bioassay of Trichloroethylene.* Bethesda, MD: DHEW, Public Health Service, National Institutes of Health, 1976. (NCI Tech. Rep. Ser. No. 2.)

20. Williams, C.S.F. *Practical Guide to Laboratory Animals.* St. Louis, MO: C.V. Mosby, 1976.

21. Dittmer, D.S. and Altman, P., Eds. *Biology Data Book.* 2nd ed., Vol. I. Bethesda, MD: Federation of American Societies for Experimental Biology, 1973.

22. Sweet, D.V., Ed. *Registry of Toxic Effects of Chemical Substances (RTECS): Comprehensive Guide to the RTECS.* Cincinnati, OH: National Institute for Occupational Safety and Health, 1993. (Publication 93-130.)

23. Hoeltge, Inc. Animal Care Equipment Catalog. Cincinnati, OH, p. 15.

24. Methods for determining toxicity and hazards of chemicals. Sanockij, I.V., Ed. *Medicina.* Moscow (1970): 62-63. (in Russian, as cited in Principles and Methods for Evaluating the Toxicity of Chemicals, Part I. World Health Organization, Geneva, 1971).

25. Selected normative data. In: Baker, H.J., Lindsey, H.R., and Weisbroth, S.H., Eds. *The Laboratory Rat.* Vol. I. New York: Academic Press, 1979.

26. Guyton, A.C. Measurement of the respiratory volumes of laboratory animals. *Am. J. Physiol.* 150: 70-77 (1947).

27. National Toxicology Program (NTP). *Effect of Dietary Restriction on Toxicology and Carcinogenesis Studies in F344/N Rats and B6C3F₁ Mice. Board Draft.* Research Triangle Park, NC: DHHS, Public Health Service, National Institutes of Health, 1995. (NTP Tech. Rep. Ser. No. 460.)

28. Hart, R., Neumann, D., and Robertson, R. *Dietary Restriction: Implications for the Design and Interpretation of Toxicity and Carcinogenicity Studies.* Washington, D.C.: ILSI Press, 1995.

29. Haseman, J.K., Huff, J.E., Zeiger, E., and McConnell, E.E. Comparative results of 327 chemical carcinogenicity studies. *Environ. Health Perspect.* 74: 229-235 (1987).

SECTION B: GUIDE TO THE PLOT OF THE CARCINOGENIC POTENCY DATABASE

The results of our estimates of TD_{50} and the standardization of the bioassay literature are summarized and presented graphically in the plot of the Carcinogenic Potency Database (CPDB) (see Section C). The plot includes information on many aspects of the 5152 experiments it summarizes. The following description of the plot, in conjunction with the appendices (Section D), is intended as a guide for the reader to facilitate use of the data. The plot format is the same as that of earlier publications of the CPDB;[1-6] however, access to the material is facilitated here because all results from all earlier plots are organized together for each chemical.

The plot covers two facing pages and is organized alphabetically by chemical name. The left side includes a logarithmic scale of TD_{50}s on which is plotted the TD_{50} for the most potent site in each experiment. Experiments are listed under the name of the test agent, and each experiment can be identified by a unique number in the plot.

Immediately below is an example from the plot of one experiment on 1,2-dibromoethane from the NCI/NTP bioassays, which will be used to describe the variables included in the plot, the codes and conventions, and the appendices. The left side of the plot appears above the right side in the example. At the top of the example, as at the top of each page of the plot of the entire database, is a two-line header describing the type of information in each field. The header should be read across, alternating between the top and bottom line: first "Spe" then "Sex" then "Strain," etc. (see below).

Immediately beneath the header in this example, we have inserted a set of numbers (1)–(28) which will be used in the text to give details of the information in each field; this set of numbers does not appear in the large plot in Section C.

For any given chemical, to determine whether results are reported in the CPDB, the chemical name can be searched in the plot in Section C, in the section "Summary of Carcinogenic Potency Database by Chemical" in Chapter 3, or by synonym in Appendix 12. To search for a chemical by Chemical Abstracts Service (CAS) registry number, see the index in Chapter 6.

(1) The *chemical name* in upper case letters is indicated under (1) in the top line for a set of experiments. In the plot, common synonyms are also provided on the right side under (17), i.e., in the example, ethylene dibromide, EDB. Also under the number (1), immediately below the chemical name, is the unique plot number for each experiment, i.e., one sex of one species from one research report. In this example, we use the line number 1517 which corresponds to the number in the large plot. In the large plot, each new number indicates a separate experiment. A set of lower case letters, in the example "a" through "f" identifies each TD_{50} calculated for that same experiment; each TD_{50} uses a different set of sites and/or histopathology. Only the most potent site is plotted on the graph.

(2) The *species* used in this experiment is indicated under (2) and the column is headed by "Spe". The letter "M" refers to mice, "R" to rats, "H" to hamsters, "D" to dogs, "N" for prosimians, and "P" for monkeys.

LEFT SIDE

```
      Spe Strain  Site    Xpo+Xpt                                                                    TD50    2Tailpvl
          Sex  Route   Hist    Notes                                                                  DR      AuOp
  (1)  (2)(3) (4)(5)  (6)  (7) (8)(9)(10)                              (11)        (12)               (13)   (14)(15)16)

  1,2-DIBROMOETHANE              100...:.1ug....:.10......:.100....:.1mg....:.10.....:.100....:.1g......:.10
  1517 M f b6c gav MXB MXB 53w78 sv                                      :+  :                        3.74mg * P<.0005
   a   M f b6c gav sto sqc 53w78 sv                                                                   4.07mg * P<.0005c
   b   M f b6c gav lun MXA 53w78 sv                                                                   15.4mg * P<.04   c
   c   M f b6c gav lun a/a 53w78 sv                                                                   17.3mg * P<.03   c
   d   M f b6c gav TBA MXB 53w78 sv                                                                   3.52mg * P<.0005
   e   M f b6c gav liv MXB 53w78 sv                                                                   no dre   P=1.
   f   M f b6c gav lun MXB 53w78 sv                                                                   15.4mg * P<.04
```

RIGHT SIDE

```
      RefNum LoConf UpConf  Cntrl 1Dose  1Inc   2Dose   2Inc                       Citation or Pathology
                                                                                                Brkly Code
  (17)(18)(19) (20)    (21)    (22)  (23)   (24)   (25)    (26)                      (27)         (28)

  1,2-DIBROMOETHANE  (ethylene dibromide, EDB) 106-93-4
  1517 TR86 2.58mg 6.93mg  0/20  26.0mg 46/50  52.0mg  30/50               lun:a/a,a/c; sto:sqc. C
   a   TR86 2.81mg 8.09mg  0/20  26.0mg 46/50  52.0mg  28/50
   b   TR86 8.06mg n.s.s.  0/20  26.0mg 11/50  52.0mg   6/50                         lun:a/a,a/c.
   c   TR86 8.85mg n.s.s.  0/20  26.0mg 10/50  52.0mg   6/50
   d   TR86 2.42mg 6.31mg  0/20  26.0mg 47/50  52.0mg  31/50
   e   TR86 n.s.s. n.s.s.  0/20  26.0mg  1/50  52.0mg   0/50               liv:hpa,hpc,nnd.
   f   TR86 8.06mg n.s.s.  0/20  26.0mg 11/50  52.0mg   6/50                         lun:a/a,a/c.
```

(3) The *sex* is indicated by "f" for female, "m" for male under (3). Occasionally, an author will only report data for both sexes together, and in these cases the code used is "b" for both.

(4) The *strain* or *stock* of animal is reported as a three-letter-code under strain (4). Appendix 2 lists strain codes and definitions. Strains are coded just as they are referred to in the original publication. No attempt has been made to standardize the strain names; therefore, if different nomenclature is used by two authors who actually tested the same strain, then two different codes are used in the database. For monkeys and prosimians, this column is used for the species code, e.g., "rhe" for rhesus.

(5) The *route of administration* is indicated in the header line by "Route" and reported as a three-letter-code under (5). In the example, "gav" stands for gavage. Route codes are listed in Appendix 3 and use mnemonic codes like "eat" for administration in the diet.

(6 and 7) The *site* and *histopathology* used in the calculation of the TD_{50} are reported on this line under (6) and (7), and are marked in the header line by "Site" and "Hist". Each is indicated by a three-letter-code, and the respective codes and definitions are provided in Appendices 4 and 5. Three-letter-codes have been created so that they are similar to the words they represent; for example, line 1517a reports "sto sqc" which stands for stomach, squamous-cell carcinoma; and line 1517c reports "lun a/a" for lung, alveolar/bronchiolar adenoma.

For the NCI/NTP bioassays and the general literature, the nomenclature reflects the terminology used in the Technical Report or the published paper. The operational rule has been to retain what is published and not reinterpret or rename diagnostic categories. Thus, when various authors use different nomenclature for the same tissue or morphologic type of tumor, two different codes are used in the database. Occasionally it has been necessary to replace an adjective used for a tissue with a noun, e.g., the database uses kidney when renal is used in a paper. Some special considerations about the reporting of site and histopathology information from each source of data are as follows.

NCI/NTP BIOASSAYS

In the 1,2-dibromoethane example above, certain tissue and tumor codes are given in upper case letters; these denote particular mixes of sites or tumor types from the NCI/NTP bioassays. (Upper case letters are not used for other sources of data.) When these upper case letters appear, additional information about the specific pathology for these TD_{50}s is presented on the right side of the plot under (27) where the header line reads "Citation or Pathology". These special capitalized codes are used in the plot for TD_{50}s based on

special mixes of tissue and tumor types from the NCI/NTP bioassays:

The mandatory sites are denoted by "MXB" (for "Mix Berkeley") to indicate that the site was created especially for the CPDB and is not based upon the NCI/NTP evaluations in the Technical Reports. For every NCI/NTP experiment, the same sites are given per species: for mice, liv MXB (liver mandatory), lun MXB (lung mandatory), and TBA MXB (all tumor-bearing animals); for rats, liv MXB and TBA MXB.

For the NCI/NTP bioassays, these mandatory sites are always listed as the last TD_{50}s for the experiment, in the order TBA, liv, and lun. Thus in the example given here, lines 1517d, e, and f, report for mice the TBA MXB, liv MXB, and then lun MXB. The specific pathology is given for liv MXB and lun MXB under (27) on the right side.

"MXA" (for "Mix Author") is used to denote a combination of sites or tumor types which is taken directly from the Technical Report tables of primary tumors, and also denotes a mix of tissues or tumors created in those tables. In the example, the site and histopathology for line 1517b, "lun MXA" are listed under (27). Whenever MXA appears on the left side, as in line 1517b "lun MXA", the sites and/or histopathology which were combined are listed under (27) on the right side.

"MXB MXB" denotes that a combination of tissues and tumors has been created by our group, which consists of the aggregates of sites and histopathology evaluated in the Technical Report as "carcinogenic" or "clear" or "some" evidence of carcinogenic activity. In the example, under (27) for line 1517, MXB MXB is described as a combination of "lun: a/a, a/c; sto: sqc." (This stands for lung alveolar/bronchiolar adenoma, lung alveolar/bronchiolar carcinoma, and stomach squamous-cell carcinoma). See Appendix 10: Berkeley Codes.

BIOASSAYS IN THE PUBLISHED LITERATURE

The site and pathology information from the literature experiments is given in the plot for individual tissues and tumors just as it is for the NCI/NTP bioassays. As described earlier, it is usually not possible to combine sites from the published literature because, unlike the data available from the NCI/NTP bioassays, information is seldom reported about multiple tumor incidence in the same animal. When an author does give information about aggregated tissue or tumor types, the code "mix" is used in the plot to denote that specific sites and tumors are described in the paper. When the tumor types are not specified, the code "tum" is used. Mandatory sites from experiments in the literature are included in the database for the same tissues as the NCI/NTP bioassays. A TD_{50} is calculated for any mix of tumors reported in the mandatory site and for individual tumor types as well. All codes are in lower case letters. Whenever an author reported results for all animals with malignant tumors separately from all animals with benign tumors, we have included a TD_{50} for each category of "tba".

(8 and 9) The *exposure* and *experiment time* are indicated in the header line by "Xpo+Xpt" under **(8)** and **(9)**. Exposure time is the period over which the test agent is administered; if administration was once a week for 40 weeks, for example, the exposure time is 40 weeks. Experiment time is the total time on test; it is not the age of the animals. It is measured from the start of the experiment to the time of death of the last dosed animal. Within a single experiment, all TD_{50}s have the same exposure time and the same experiment time. Both times are always reported in the same units. When both are less than 100 weeks, exposure and experiment time are reported as "w" for weeks; when greater than this, "m" for months is used. For tests in long-lived experimental animals like dogs and nonhuman primates, "y" for years may also be used. When exposure time and experiment time are equal, then the duration of dosing was for the entire experiment. In the example, lines 1517a-f, the mice were dosed for 53 weeks, and the experiment lasted a total of 78 weeks.

(10) *Notes*, indicated in the header line by "Notes" under **(10)**, provide additional information about the experiment in single-letter-codes which are defined in Appendix 6: Notecodes. This supplementary information is helpful in evaluating the experimental data. In the example, the notecode "s" is used to denote that *survival* was poor due to toxicity or disease, and the notecode "v" denotes that dosing was *variable* or irregular, e.g., dose level changed during the course of the experiment. Other notecodes indicate such factors as: the experiment was a serial sacrifice in a longer study (notecode "k"), or that histopathological examination was restricted to only a few tissues (notecode "r"). In some recent NTP bioassays, results for the kidney were reported in the Technical Reports for the standard histopathology protocol and separately for results including additional sections of the kidney. In such cases the word "with step" appears under **(10)**, and a new line number is assigned in the plot for these results.

(11) The *logarithmic scale* under **(11)** is used for presenting the values of TD_{50} and its confidence limits (in units/kg body wt/day). The plot extends from 100 ng to 10 g. On the scale itself, the location of 100 ng, 1 μg, 10 μg, etc. is indicated by underscoring; the points for 5, 50, and 500 are denoted by a ":". For each experiment, only the TD_{50} for the "most potent site" is plotted; this TD_{50} is listed first. For other sites within an experiment, the TD_{50} is not plotted, but all other information about it is given in the plot.

The "most potent site" is determined by ordering the TD_{50}s in each experiment by statistical significance. If any TD_{50}s are significant at the $p < 0.01$ level, then these are listed first, in order of potency. Then follow all TD_{50}s with $p < 0.10$ sorted in order of potency. Last, all other TD_{50}s are listed in order of potency. For literature experiments, we have excluded the category "tba" from this sorting of the target sites, and have listed it last. For the NCI/NTP bioassays, the mandatory sites are also excluded from this sorted order and are listed at the end in the order TBA MXB, liv MXB, and lun MXB.

In the example of 1,2-dibromoethane, there are two TD_{50}s (excluding TBA) with statistical significance of $p < 0.01$. The TD_{50} for MXB MXB is the more potent and thus appears first. The plotted point for this most potent site lies between 1 and 10 mg on the scale, and the estimated TD_{50}, 3.74 mg, appears in the column to the right of the scale under the header title "TD50". These results indicate that 3.74 mg/kg body wt/day would halve the proportion of tumorless survivors at the end of a standard lifespan for mice (in the absence of all other causes of death).

(12) The TD_{50} value on the plot is indicated by one of three symbols: "+", "±", or ">", depending upon the p value associated with the TD_{50} for the most potent site. In an experiment where the statistical significance of a TD_{50} is $p < 0.01$, the symbol "+" is used, and the plotted point is the most potent estimate of TD_{50}, i.e., the smallest value of TD_{50} from among those with $p < 0.01$, as in the example of 1,2-dibromoethane. In an experiment where the statistical significance is between 0.01 and 0.10 a "±" symbol is used to plot the most potent site. If there is no evidence for any statistical association, i.e., for all TD_{50}s, $p > 0.10$, then the symbol ">" appears just to the right of the lower confidence limit. For these statistically nonsignificant experiments, the symbol ">" provides a lower bound for TD_{50}, i.e., it is extremely unlikely that TD_{50} could be less than (more potent than) the value plotted.

A special symbol "<+" is used for cases where 100% of all dosed animals had the tumor(s) of interest and the TD_{50} was calculated with summary data; the "<+" appears at the upper confidence limit. With summary data, only an upper bound, but no TD_{50}, can be estimated when all dosed animals have tumors.

Occasionally, the symbol for the most potent TD_{50} is plotted with parentheses around it to indicate that the test did not meet our standard criteria; usually the experiment time was too short. A few NCI/NTP bioassays do not meet the length criteria, e.g., "C.I. Direct Brown 95" for female rats, line 723 in the large plot. Also, these criteria have been relaxed for some experiments in nonhuman primates. (See Appendix 1: Species.)

Because there are both statistical and nonstatistical uncertainties in the estimation of TD_{50}, we have calculated *99% confidence intervals* for it. (See Appendix 8: Statistical Methods for Estimating TD_{50}.) Whenever the TD_{50} calculation is based on lifetable data, the symbol ":" denotes the confidence limits. When the TD_{50} calculation is based on summary data, the symbol "." denotes the confidence limits.

In an experiment where the statistical significance of TD_{50} is p <0.01, both the *lower* and *upper confidence limits* are plotted. The calculated values for these intervals are presented on the right side of the plot under **(20)** and **(21)** where the header line reads "LoConf" and "UpConf" for lower confidence and upper confidence limits. In the example, the plotted confidence interval is 2.58 to 6.93 mg indicating that the lower value would not reduce the proportion of tumorless survivors by half, while 6.93 mg/kg/body wt/day would more than halve it.

In an experiment where the statistical significance of TD_{50} is p >0.01, the confidence interval will be open at the upper end. In such a case, the lower limit dose-rate would be unlikely to halve the proportion of tumor-free survivors, but no statement can be made about the carcinogenic effect of higher doses. In such cases, there is no ":" or "." to the right of the TD_{50} on the plot.

When the plotted symbol is "<+" because 100% of the dosed animals had tumors and the TD_{50} calculation was based on summary data, the "<+" appears at the upper confidence limit. In such cases the upper limit dose-rate would induce tumors in all animals, but the rate which would reduce the proportion of tumorless survivors by half cannot be estimated with summary data.

Occasionally, values of TD_{50} or confidence limits that should be plotted are too large to be printed within the range up to 10 g; in such unusual cases no plotted symbol appears, but the value will be given in the field for TD_{50} **(13)** or confidence limits **(20)** or **(21)**.

(13) The *value of each TD_{50}* is presented just to the right of the plot range **(13)**, and includes the appropriate units (per kilogram) of body weight per day. The symbol "noTD$_{50}$" appears instead of a numerical value whenever 100% of the dosed animals had the tumor(s) of interest and the TD_{50} was calculated with summary data. The symbol "no dre", for *no dose-related effect,* indicates that TD_{50} is not estimable for some other reason. For statistically nonsignificant TD_{50}s the numerical value may be impossibly large.

(14) The *shape of the dose-response curve* for each TD_{50} appears in **(14)** under the header "DR". The codes and definitions for the curvatures are listed in Appendix 7: Dose-Response Curve Symbols. The shape of the dose-response has been determined by a test for departure from

linearity. (See Appendix 8.) If there was no significant departure from a linear dose-response, then the curve shape is listed as linear, and the symbol "*" appears. For experiments with three groups of animals including controls, a significant departure from linearity with upward curvature is denoted by the symbol "/". If there was a significant departure from linearity with downward curvature, then the TD_{50} is calculated without the data from the highest dose group, and the symbol "\" appears. We have adopted this convention to obtain the best estimate of TD_{50} by using only the linear portion of the dose-response curve in the calculation.

When there are more than three dose groups (including controls) in the TD_{50} calculation and there is a significant departure from linearity, the symbol "Z" is listed under **(14)**. When there is a blank space for the shape of the dose-response, there are two possible reasons. First, there may be only one dose group and a control group in the experiment, in which case there is not enough information to determine a curve shape. Second, there may be no dose-related effect, in which case the code "no dre" appears in **(13)** and "P=1." appears in **(15)**.

(15) The *two-tailed p* value appears in **(15)**, under the header "2Tailpvl". (See Appendix 8: Statistical Methods for Estimating TD_{50}.) This value indicates the statistical significance associated with testing whether the slope of the dose-response curve is different from zero. All values are given to one significant figure. When there is no dose-related effect or the slope is negative, then "P=1." appears in **(15)**. The lowest p-value reported is p <0.0005. Note that the significance level listed in this column determines which symbol will be plotted for the most potent site in each experiment.

(16) The *opinion of the original author*, as to the tumorigenicity of the test agent at the site for which the TD_{50} was calculated, is given in the last column on the left side of the plot **(16)** under the heading "AuOp" for author's opinion. Our rule for reporting opinions from all sources of data has been to record all clearly stated evaluations of tumorigenicity at the site(s) included in a TD_{50} calculation (See Appendix 9: Author's Opinion.)

Some special considerations about our codes for author's opinion are as follows.

NCI/NTP BIOASSAYS

Our conventions for reporting the author's opinions from the NCI/NTP Technical Reports are based upon the text of the report and the statistical analysis tables. An author's opinion is listed for all TD_{50}s except Berkeley Mixes (MXB) and the statistical sites, i.e., those included in the tables but not considered evidence for carcinogenicity in the text of the Technical Report. For these cases, the opinion column is blank, as in lines 1517, 1517d, e, and f of the 1,2-dibromoethane example.

A "c" in the author's opinion column indicates that the text of the report stated that at the site(s) on which TD_{50} is based, the compound was *carcinogenic* under the conditions of the bioassay or the abstract indicated that there was "clear evidence of carcinogenic activity." See, for example, the opinion column in the example of 1,2-dibromoethane for lines 1517a, b, and c. In earlier Technical Reports an "a" indicates an opinion that the incidence of tumors at that site(s) was evaluated as *associated* with administration of the compound under the conditions of the bioassay, or that the evidence for carcinogenicity was suggestive; these opinion codes in the CPDB are consistent with updated evaluations by NTP.[7] For all NTP Technical Reports that use the current evaluation methodology, we report all sites listed in the summary table in the abstract with the opinion for carcinogenic activity: "c" for clear, "p" for some, and "e" for equivocal. (See Appendix 9: Author's Opinion.)

The symbol "–" will appear in the opinion column for the most potent site in an NCI/NTP bioassay to denote the evaluation that the compound was not carcinogenic in that sex of that species under the conditions of the bioassay. In most cases, the "–" appears for the TD_{50} calculated for TBA, which is our convention whenever there is no evidence for carcinogenicity. For experiments evaluated as inadequate in the Technical Report, there is an "i" in the opinion column for the most potent site.

There are some cases when the "–" appears for a statistical site, i.e., one not evaluated as evidence for carcinogenicity in the report, but which was statistically significant according to the statistical tables in the report, and also had a TD_{50} significance level of $p < 0.05$. When this TD_{50} is the only evidence for a treatment-related effect, and thus the most potent site, we have indicated this by placing a "–" in the opinion column and flagging the TD_{50} with a # sign in the plot just to the left of the TD_{50} value. See, for example, "calcium cyanamide," for female mice, line 1281 in the plot. Note that the p-value associated with the TD_{50} determines the symbol plotted for the TD_{50}; this is independent of the author's opinion.

For bioassays in which some target sites were evaluated as treatment-related, the statistical sites are also reported, but the opinion column is left blank. In order to make it clear that NCI/NTP did not evaluate these statistical sites as evidence of carcinogenicity, we put an "S" for *statistical* in the last column on the right side of the plot (**28**), under the header "Brkly Code." (See Appendix 10: Berkeley Codes.)

Bioassays in the Published Literature

In the general literature whenever the author evaluated the proportion of animals with tumors at a particular site as treatment-related, a "+" will appear in the opinion column (16) for the TD_{50} calculated for that site. Such stated opinions as "positive," "carcinogenic," "induced," "treatment-related," and "tumorigenic," fit this category. The symbol "+" will only appear in the opinion column for a TD_{50} in one of the following two cases: (1) the author gave a positive opinion for the particular target sites included in the TD_{50} or (2) the occasional case where an author evaluated the compound as carcinogenic without specifying the target site, and we have indicated this with a "+" symbol in the opinion column for the category "all tumor bearing animals (tba)", e.g., line 3734a for "*N*-nitrosopiperazine".

Similarly, the opinion column will contain a "–" only when either (1) the author stated an opinion that there was no carcinogenic effect at the particular sites included in the TD_{50} or (2) the author concluded that there was no treatment-related effect in the experiment, in which case all TD_{50}s for the experiment have a "–" in the opinion column.

Sites which an author did not evaluate as positive are included in the database only when the statistical significance associated with an increased percentage of dosed animals with tumors is $p < 0.05$ (standard chi-square, one-sided p-value), or when the tissue is a mandatory site.

When no opinion about carcinogenicity is stated for sites which are, nevertheless, used for TD_{50} calculations, the author's opinion column is left blank. This may occur either for mandatory sites or for included sites which were not unequivocally evaluated by the author. If additional information about evaluations was obtained directly from the author, then "pers.comm." appears after the citation on the right side of the plot under (**27**).

In summary, the symbol for the author's opinion column in the general literature reflects what the author actually stated. Sites evaluated as positive are given a "+". Sites evaluated as negative are given a "–". The symbol "+" is used for tba when the compound was evaluated as positive, and no specific target site was evaluated as positive. For all other opinions the author's opinion column is blank.

Occasionally, when there is a "c", "p", "a", or "e" in the opinion column for an NCI/NTP bioassay, or a "+" for a test from the general literature, the positive evaluation was made because the incidence among dosed animals was high in comparison to historical control incidences; this occurs, for example, when there is a rare tumor among dosed animals. The actual numbers of animals bearing such tumors may be quite low, thus making the estimate of TD_{50} unreliable. In such cases, we have indicated that the author's opinion was based on historical control comparisons by putting "+hist" to the left of the TD_{50} value. See, for example, 5-nitro-*o*-toluidine for female mice, line 3444b in the large plot.

(17 and 18) The plot continues on the right side, by first repeating the *identification numbers* given for each line on the left side of the output, e.g., 1517, 1517a, b, etc. In **(17)** the *chemical name* is also repeated, followed by *common synonyms*, in some cases. In the example, "ethylene dibromide" and "EDB" are given as synonyms for 1,2-dibromoethane. The *Chemical Abstracts Service registry number* (CAS#) is reported under **(18)** in the example, 106-93-4. To find a chemical in the CPDB when the CAS number is known, see the index at the end of the book, Chapter 6. This index also indicates whether results for the chemical are reported in the Table of Genotoxicity Results in Chapter 5, Section B.

(19) Under the header "RefNum" in **(19)**, is the unique *reference number* assigned to each paper in the database. For NCI/NTP bioassays, this is the Technical Report number, TR86 in the example. For literature tests the number is a chronologically assigned paper identification number for the CPDB.

(20 and 21) The *lower* and *upper confidence limits* for each TD_{50} are presented in **(20)** "LoConf" and **(21)** "UpConf", respectively. When the abbreviation "n.s.s." appears for either the lower or upper confidence limit, it denotes *not statistically significant*. Whenever the statistical significance of TD_{50} is $p > 0.01$, then the upper 99% confidence limit will not be calculated. See, for example, line 1517f in the 1,2-dibromoethane example. When the lower confidence limit is "n.s.s." this usually indicates that there were no tumors or only one tumor of the specified type in the experiment, and the lower confidence limit was not estimable; most often this occurs for mandatory sites as in 1517e in the 1,2-dibromoethane example. Occasionally, the "n.s.s." occurs for the lower confidence limit because 100% of dosed animals had the tumor(s) of interest and hence no lower confidence limit could be estimated with summary data. This is also indicated by the plot symbol "<+".

(22–26) Beginning in **(22)** on the right side and extending through **(26)**, we report the *proportion of animals with tumors* and the *average dose-rate* in mg/kg body wt/day which we have calculated for each dose group in the experiment. The proportion of animals with tumors for TD_{50}s which have been calculated with lifetable data are presented here in summary form, i.e., the number of animals with tumors by the end of the experiment.

Reading across **(22)–(26)**, under "Cntrl" **(22)**, we list the proportion of *control* animals with the tumor types in the TD_{50} calculation; the average dose in the lowest dose group is given under "1Dose" **(23)**; the proportion of animals in that group with the tumor(s) is listed under "1Inc",

for *incidence*, **(24)**; the next highest average dose "2Dose" **(25)** and the proportion of those animals with the tumor(s) "2Inc" **(26)** are given next, etc. for as many dose groups as there are in the experiment.

Whenever the TD_{50} was calculated without the data from the highest dose group(s) (i.e., there was a significant downward departure from linearity), we have indicated this fact with parentheses around the data which were omitted from the final calculation. (See Appendix 8.) Thus, whenever the shape of the dose-response on the left side of the plot is "\" under **(14)**, the parentheses appear on the right side around the appropriate data. See, for example, "benzofuran" for female mice, line 554 in the large plot. Whenever there were more than three groups (including controls) and the dose-response was nonlinear, there is a "Z" under **(14)** on the left side. If the departure from linearity in such cases was downward, then this fact is indicated by parentheses around the group that was excluded from the TD_{50} calculation, e.g., "1-amyl-1-nitrosourea" in male rats, line 352. If there are no parentheses and a "Z" appears in the curve-shape column, then the dose-response had at least 3 doses and a control and curved upward, e.g., "butylated hydroxyanisole" in male rats, line 790.

NCI/NTP BIOASSAYS

For the proportion of animals with tumors, the number of animals reported in each group is the number at the *start* of the experiment; the TD_{50} was estimated with lifetable data. In the example, there were 20 in the control group and 50 in both the low and high dose groups. For some early NCI experiments, pooled controls were used in evaluating evidence for carcinogenicity in the Technical Report, and we have calculated TD_{50}s with those pools as well as the matched controls. (This convention has also been followed for the lifetable calculations of the experiments of aromatic amines.) Data using the pooled controls are indicated by the letter "p" following the control incidence in column **(22)**, by the word "pool" on the left side of the plot following the "Notes" in **(10)**, and by assigning a different line (experiment) number to the pooled data. See, for example, "aldrin" for male mice in the large plot, line 215.

BIOASSAYS IN THE PUBLISHED LITERATURE

The proportion of animals with tumors presented here is the number of animals used in the TD_{50} calculation. Many authors have reported only the starting number of animals. Whenever the published paper had additional information, i.e., the number of animals alive at the time of appearance of the first tumor or number examined histologically, then this number is used in the denominator of the proportion of animals with tumor. This is a more accurate description of the number of animals at risk of tumor. These data were used in the TD_{50} calculation and are reflected in columns (22), (24), and (26). In these cases, the notecode "e" for *effective number* appears on the left side of the plot under "Notes" in (10). Otherwise, the data reflect the number of animals started in each group. Because experimental designs vary in the literature, the incidence and dose-rate data may include a control and only one dose group or perhaps, a control and several dose groups.

In the general literature, use of pooled control data is rare, and is indicated on the plot in the same way as for NCI/NTP bioassays. In literature, pools are used to estimate TD_{50} only when no matched data were reported. Whenever vehicle control data were available, either in NCI/NTP or general literature, these were used in the CPDB.

(27) Reading across the right side of the plot, under "Citation or Pathology" (27) for the NCI/NTP bioassays, we present the three-letter codes for all *sites and histopathology* included in TD_{50}s which are MXA or MXB. This includes the MXB mandatory liver and lung sites, our combinations of sites which were individually evaluated as positive in the Report (MXB), the statistical sites which included more than one site in the Tables of Analyses (MXA), and any combination of sites evaluated as treatment-related in the Technical Report (MXA). The three-letter code for each tissue in the TD_{50} calculation is followed by a ":" and then by the three-letter codes for each category of neoplasm included in the calculation. A "." follows the last three-letter tumor code in each mix. The definitions for these codes are given in Appendices 4 and 5.

For the published literature, a citation is provided, listing the first author, code for the journal or book title, volume number, pages, and year of publication. The full titles of the four-letter codes for the names of references are listed in Appendix 11: Journals. The abbreviation *pers.comm.* indicates that additional data for the TD_{50} calculation or author's opinion were obtained through personal communication with the author(s). On the plot a new citation for a published paper is listed whenever experimental results are from a different paper, i.e., if several experiments are reported consecutively in the CPDB from a single paper, then the citation will not repeat. When the next experiments are from a different paper, the new citation will be reported. If additional experiments are later reported from an earlier citation, the citation will then be repeated. If pathology codes appear under (27), results are for an NCI/NTP bioassay.

A complete bibliography of all papers in the CPDB is given in Section E. Part 1 lists the NCI/NTP Technical Reports by chemical name and indicates Technical Report number and year of publication. Part 2 gives complete citations to articles, books, and reports in the general literature.

(28) The last column of the plot is used only for NCI/NTP bioassays. Under the header "Brkly Code", we indicate that a TD_{50} has been included in the database because of a decision by the Carcinogenic Potency Project (Berkeley) rather than because the sites were evaluated as treatment-related or combined in the NCI/NTP Technical Report. (See Appendix 10: Berkeley Codes.)

The upper case letters "C", "M", and "P" are used in the Berkeley Code column for MXB. The letter "C" denotes a TD_{50} calculated for the combination of all sites evaluated in the Technical Report as clear evidence for *carcinogenicity* in this experiment. See line 1517 in the 1,2-dibromoethane example. The letter "P" denotes a combination of tumors evaluated as *some* evidence. The letter "M", for *mix,* is uncommon and denotes a combination of all sites evaluated as "C" or "P" in a bioassay where "clear" and "some" evidence were evaluations for different target sites in the same experiment. The letter "S" is not an MXB, but rather indicates that the TD_{50} has been included in the plot because the sites were *statistically* significant in the tables of the Technical Report and the TD_{50} was significant at the $p < 0.05$ level; however, NCI/NTP did not evaluate the site as evidence of carcinogenicity. For all mandatory sites, the column for "Brkly Code" is blank.

The TD_{50}s in the plot are ordered systematically to facilitate use of the data. All bioassays of a particular chemical are organized under the chemical name, and these names are ordered alphabetically. (Colors are sorted by color, ignoring prefixes.) Within each compound, the experiments are ordered alphabetically by species code, so that dogs would appear first, then hamsters, mice, prosimians, monkeys, and finally rats. Within a species, the bioassays are ordered by the code for the strain or stock. If there is an NCI/NTP bioassay of the chemical, then all experiments using that strain are reported first, followed by the strain used in any other experiments providing lifetable data, and finally by any remaining strains ordered alphabetically. Within the strain, the bioassays of females are reported first. Thus, when there is an NCI/NTP bioassay, "b6c" mice will appear first, and all experiments using "b6c" female mice would be reported before any experiments using "b6c" males.

To facilitate easy reference back to this guide when using the large plot, we have described each variable here in the order in which it appears in the plot. The titles in the header of the large plot also refer back to the categories in this guide. Abbreviations and symbols are defined in detail in the appendices.

REFERENCES

1. Gold, L.S., Sawyer, C.B., Magaw, R., Backman, G.M., de Veciana, M., Levinson, R., Hooper, N.K., Havender, W.R., Bernstein, L., Peto, R., Pike, M.C., and Ames, B.N. A Carcinogenic Potency Database of the standardized results of animal bioassays. *Environ. Health Perspect.* 58: 9-319 (1984).

2. Gold, L.S., de Veciana, M., Backman, G.M., Magaw, R., Lopipero, P., Smith, M., Blumenthal, M., Levinson, R., Bernstein, L., and Ames, B.N. Chronological supplement to the Carcinogenic Potency Database: Standardized results of animal bioassays published through December 1982. *Environ. Health Perspect.* 67: 161-200 (1986).

3. Gold, L.S., Slone, T.H., Backman, G.M., Magaw, R., Da Costa, M., Lopipero, P., Blumenthal, M., and Ames, B.N. Second chronological supplement to the Carcinogenic Potency Database: Standardized results of animal bioassays published through December 1984 and by the National Toxicology Program through May 1986. *Environ. Health Perspect.* 74: 237-329 (1987).

4. Gold, L.S., Slone, T.H., Backman, G.M., Eisenberg, S., Da Costa, M., Wong, M., Manley, N.B., Rohrbach, L., and Ames, B.N. Third chronological supplement to the Carcinogenic Potency Database: Standardized results of animal bioassays published through December 1986 and by the National Toxicology Program through June 1987. *Environ. Health Perspect.* 84: 215-286 (1990).

5. Gold, L.S., Manley, N.B., Slone, T.H., Garfinkel, G.B., Rohrbach, L., and Ames, B.N. The fifth plot of the Carcinogenic Potency Database: Results of animal bioassays published in the general literature through 1988 and by the National Toxicology Program through 1989. *Environ. Health Perspect.* 100: 65-135 (1993).

6. Gold, L.S., Manley, N.B., Slone, T.H., Garfinkel, G.B., Ames, B.N., Rohrbach, L., Stern, B.R., and Chow, K. Sixth plot of the Carcinogenic Potency Database: Results of animal bioassays published in the general literature 1989-1990 and by the National Toxicology Program 1990-1993. *Environ. Health Perspect.* 103 (Suppl. 8): 3-122 (1995).

7. Haseman, J.K., Huff, J.E., Zeiger, E., and McConnell, E.E. Comparative results of 327 chemical carcinogenicity studies. *Environ. Health Perspect.* 74: 229-235 (1987).

SECTION C: PLOT OF CANCER TEST RESULTS

Quick Reference Guide to the Plot

```
        Spe Strain Site  Xpo+Xpt                                                                    TD50    2Tailpvl
            Sex   Route Hist   Notes                                                                       DR   AuOp
A-alpha-C                         100ng..:..1ug....:..10....:..100....:..1mg....:..10....:..100....:..1g.....:..10
1     M f cdf eat liv mix 97w97                                                     . + .              35.6mg    P<.0005+
a     M f cdf eat liv hpc 97w97                                                                        44.7mg    P<.0005
b     M f cdf eat blv hms 97w97                                                                        381.mg    P<.004  +
c     M f cdf eat lun ade 97w97                                                                        795.mg    P<.04
d     M f cdf eat lun mix 97w97                                                                        1.15gm    P<.4
2     M m cdf eat blv mix 97w97                                                 . + .                  82.6mg    P<.0005+
a     M m cdf eat blv hms 97w97                                                                        103.mg    P<.0005
b     M m cdf eat liv mix 97w97                                                                        122.mg    P<.0005
c     M m cdf eat liv hpc 97w97                                                                        225.mg    P<.0005
d     M m cdf eat liv hpa 97w97                                                                        352.mg    P<.004
e     M m cdf eat lun ade 97w97                                                                        no dre    P=1.
f     M m cdf eat lun mix 97w97                                                                        no dre    P=1.

ACETALDEHYDE                      100ng..:..1ug....:..10....:..100....:..1mg....:..10....:..100....:..1g.....:..10
3     H f syg inh lar mix 52w81 er                                                        .   ±        728.mg    P<.04  +
a     H f syg inh lar adq 52w81 er                                                                     1.14gm    P<.1   +
b     H f syg inh lar sqc 52w81 er                                                                     2.36gm    P<.3   +
4     H m syg inh res mix 52w81 er                                                        .   ±        461.mg    P<.02  +
a     H m syg inh lar mix 52w81 er                                                                     461.mg    P<.02  +
b     H m syg inh lar cic 52w81 er                                                                     641.mg    P<.04  +
c     H m syg inh nse apc 52w81 er                                                                     2.07gm    P<.3   +
d     H m syg inh lar sqc 52w81 er                                                                     2.07gm    P<.3   +
5     R f wis inh nse adc 26m28 erv                                                  .+ .               370.mg  * P<.0005+
a     R f wis inh nse sqc 26m28 erv                                                                    1.03gm  Z P<.0005+
b     R f wis inh nse cic 26m28 erv                                                                    3.00gm  * P<.007 +
6     R m wis inh nse adc 26m28 erv                                                .+ .                 185.mg  * P<.0005+
a     R m wis inh nse sqc 26m28 erv                                                                    627.mg  Z P<.0005+
7     R f wsr inh nac mix 12m24 erv                                               . + .                 148.mg  * P<.0005+
a     R f wsr inh nac adc 12m24 erv                                                                    201.mg  * P<.0005+
b     R f wsr inh nac sqc 12m24 erv                                                                    574.mg  * P<.01  +
8     R m wsr inh nac mix 12m24 erv                                             . + .                   88.5mg  * P<.0005+
a     R m wsr inh nac adc 12m24 erv                                                                    190.mg  * P<.002 +
b     R m wsr inh nac sqc 12m24 erv                                                                    200.mg  Z P<.0005+

ACETALDEHYDE  METHYLFORMYLHYDRAZONE    ..:..1ug....:..10....:..100....:..1mg....:..10....:..100....:..1g......:..10
9     M f swa gav lun mix 12m25 es                                        . + .                        5.66mg    P<.0005+
a     M f swa gav lun ade 12m25 es                                                                     7.37mg    P<.0005
b     M f swa gav lun adc 12m25 es                                                                     19.6mg    P<.002
c     M f swa gav for mix 12m25 es                                                                     28.0mg    P<.002 +
d     M f swa gav for sqp 12m25 es                                                                     32.4mg    P<.003
e     M f swa gav cli mix 12m25 es                                                                     38.2mg    P<.005 +
f     M f swa gav cli sqc 12m25 es                                                                     46.4mg    P<.01
g     M f swa gav liv hpt 12m25 es                                                                     28.0mg    P<.02  -
10    M m swa gav pre mix 52w79 es                                   . + .                             1.61mg    P<.0005+
a     M m swa gav pre sqc 52w79 es                                                                     2.17mg    P<.0005
b     M m swa gav pre fbs 52w79 es                                                                     21.3mg    P<.003
c     M m swa gav liv hpt 52w79 es                                                                     11.0mg    P<.03  -
d     M m swa gav lun ade 52w79 es                                                                     13.7mg    P<.03
e     M m swa gav lun mix 52w79 es                                                                     14.4mg    P<.07  +

ACETALDOXIME                      100ng..:..1ug....:..10....:..100....:..1mg....:..10....:..100....:..1g......:..10
11    R m f34 wat liv hnd 26m30 e                                                      .>              445.mg  * P<.4
a     R m f34 wat adr coa 26m30 e                                                                      no dre    P=1.

ACETAMIDE                         100ng..:..1ug....:..10....:..100....:..1mg....:..10....:..100....:..1g......:..10
12    M f cb6 eat lun a/a 52w69 e   pool                                                         .>    no dre    P=1.   -
a     M f cb6 eat liv hpc 52w69 e                                                                      no dre    P=1.
13    M m cb6 eat sto sqp 52w69 e   pool                                                    .   +   .  1.89gm  \ P<.0005
a     M m cb6 eat mln mno 52w69 e                                                                      2.66gm  \ P<.002
b     M m cb6 eat --- mix 52w69 e                                                                      3.01gm  * P<.0005+
c     M m cb6 eat --- mno 52w69 e                                                                      3.05gm  \ P<.002
d     M m cb6 eat --- mlh 52w69 e                                                                      6.24gm  * P<.002
e     M m cb6 eat mln mlh 52w69 e                                                                      9.52gm  * P<.02
f     M m cb6 eat liv hpc 52w69 e                                                                      57.5gm  * P<.7   -
g     M m cb6 eat lun a/a 52w69 e                                                                      no dre    P=1.   -
14    R f f34 eat liv hpc 52w69 e                                                     . + .            230.mg    P<.0005+
a     R f f34 eat liv nnd 52w69 e                                                                      4.15gm    P<.04  +
15    R m f34 eat liv hpc 52w69 es                                                  . + .              104.mg    P<.0005+
a     R m f34 eat liv nnd 52w69 es                                                                     9.96gm    P<.3   +
16    R m wis eat liv hpt 52w65 ek                                                      .   ±          372.mg    P<.02  +

ACETAMINOPHEN                     100ng..:..1ug....:..10....:..100....:..1mg....:..10....:..100....:..1g......:..10
17    M f b6c eat TBA MXB 24m24                                                              :>        no dre    P=1.   -
a     M f b6c eat liv MXB 24m24                                                                        no dre    P=1.
b     M f b6c eat lun MXB 24m24                                                                        11.0mg  * P<.6
18    M f b6c eat liv mix 31m31 e                                                          . ±8.32gm  * P<.05  -
a     M f b6c eat lun mix 31m31 e                                                                      no dre    P=1.
b     M f b6c eat tba tum 31m31 e                                                                      42.7gm  * P<1.
19    M m b6c eat TBA MXB 24m24                                                              :>        no dre    P=1.   -
a     M m b6c eat liv MXB 24m24                                                                        no dre    P=1.
```

RefNum	LoConf	UpConf	Cntrl	1Dose	1Inc	2Dose	2Inc		Citation or Pathology
									Brkly Code

A-alpha-C   (2-amino-9H-pyrido(2,3-b)indole)   26148-68-5

1	1616	21.8mg	60.7mg	0/40	104.mg	33/40				Ohgaki;carc,5,815-819;1984
a	1616	27.6mg	77.0mg	0/40	104.mg	30/40				
b	1616	155.mg	2.30gm	0/40	104.mg	6/40				
c	1616	241.mg	n.s.s.	0/40	104.mg	3/40				
d	1616	234.mg	n.s.s.	2/40	104.mg	4/40				
2	1616	48.1mg	158.mg	0/40	96.0mg	20/40				
a	1616	58.1mg	209.mg	0/40	96.0mg	17/40				
b	1616	66.3mg	259.mg	0/40	96.0mg	15/40				
c	1616	105.mg	657.mg	0/40	96.0mg	9/40				
d	1616	143.mg	2.12gm	0/40	96.0mg	6/40				
e	1616	288.mg	n.s.s.	3/40	96.0mg	3/40				
f	1616	390.mg	n.s.s.	11/40	96.0mg	4/40				

ACETALDEHYDE   75-07-0

3	1766	219.mg	n.s.s.	0/14	391.mg	3/15						Feron;ejca,18,13-31;1982/pers.comm.
a	1766	279.mg	n.s.s.	0/14	391.mg	2/15						
b	1766	383.mg	n.s.s.	0/14	391.mg	1/15						
4	1766	158.mg	n.s.s.	0/15	344.mg	4/15						
a	1766	158.mg	n.s.s.	0/15	344.mg	4/15						
b	1766	193.mg	n.s.s.	0/15	344.mg	3/15						
c	1766	337.mg	n.s.s.	0/15	344.mg	1/15						
d	1766	337.mg	n.s.s.	0/15	344.mg	1/15						
5	1757	263.mg	540.mg	0/50	101.mg	6/48	202.mg	26/53	209.mg	21/53		Woutersen;txcy,41,213-231;1986
a	1757	624.mg	1.92gm	0/50	101.mg	0/50	202.mg	5/53	209.mg	17/53		
b	1757	1.36gm	33.3gm	0/50	101.mg	0/48	202.mg	3/53	209.mg	5/53		
6	1757	137.mg	259.mg	0/49	70.8mg	16/52	142.mg	31/53	147.mg	21/49		
a	1757	380.mg	1.38gm	1/49	70.8mg	1/52	142.mg	10/53	147.mg	15/49		
7	1863	85.3mg	288.mg	0/18	50.1mg	1/20	100.mg	7/18	149.mg	11/17		Woutersen;txcy,47,295-305;1987/1986/pers.comm.
a	1863	110.mg	439.mg	0/18	50.1mg	0/20	100.mg	7/18	149.mg	8/17		
b	1863	234.mg	38.9mg	0/18	50.1mg	1/20	100.mg	0/18	149.mg	5/17		
8	1863	54.4mg	157.mg	0/19	35.1mg	2/20	70.1mg	8/20	104.mg	15/22		
a	1863	102.mg	710.mg	0/19	35.1mg	2/20	70.1mg	6/20	104.mg	6/22		
b	1863	105.mg	463.mg	0/19	35.1mg	0/20	70.1mg	2/20	104.mg	11/22		

ACETALDEHYDE METHYLFORMYLHYDRAZONE   16568-02-8

9	1267	3.21mg	14.0mg	13/50	6.78mg	35/50		Toth;jnci,67,881-887;1981
a	1267	4.01mg	22.5mg	12/50	6.78mg	31/50		
b	1267	9.49mg	93.0mg	2/50	6.78mg	13/50		
c	1267	12.7mg	101.mg	0/44	6.78mg	8/48		
d	1267	14.0mg	150.mg	0/44	6.78mg	7/48		
e	1267	15.6mg	292.mg	0/44	6.78mg	6/48		
f	1267	17.6mg	3.78gm	0/44	6.78mg	5/48		
g	1267	8.45mg	n.s.s.	0/32	6.78mg	3/18		
10	1267	1.01mg	2.62mg	0/37	9.40mg	45/50		
a	1267	1.40mg	3.49mg	0/37	9.40mg	41/50		
b	1267	9.65mg	98.0mg	0/37	9.40mg	8/50		
c	1267	2.68mg	n.s.s.	0/14	9.40mg	2/7		
d	1267	5.79mg	n.s.s.	8/47	9.40mg	18/49		
e	1267	5.57mg	n.s.s.	11/47	9.40mg	20/49		

ACETALDOXIME   107-29-9

11	1853	100.mg	n.s.s.	2/20	19.5mg	3/20	50.3mg	4/20		Lijinsky;txih,3,337-345;1987/pers.comm.
a	1853	323.mg	n.s.s.	1/20	19.5mg	1/20	50.3mg	0/20		

ACETAMIDE   60-35-5

12	1343	2.22gm	n.s.s.	2/89p	1.15gm	0/34	2.31gm	0/28		Fleischman;jept,3,149-170;1980
a	1343	6.96gm	n.s.s.	1/89p	1.15gm	2/41	2.31gm	0/46		
13	1343	717.mg	7.92gm	0/82p	1.07gm	5/32	(2.13gm	0/22)		
a	1343	1.01gm	13.2gm	0/74p	1.07gm	5/44	(2.13gm	0/39)		
b	1343	1.62gm	6.73gm	0/95p	1.07gm	7/50	2.13gm	7/46		
c	1343	1.16gm	14.1gm	0/95p	1.07gm	5/50	(2.13gm	1/46)		
d	1343	2.70gm	23.9gm	0/95p	1.07gm	2/50	2.13gm	5/46		
e	1343	3.29gm	n.s.s.	0/79p	1.07gm	1/44	2.13gm	3/39		
f	1343	6.31gm	n.s.s.	2/91p	1.07gm	0/50	2.13gm	2/46		
g	1343	1.55gm	n.s.s.	1/87p	1.07gm	0/24	2.13gm	0/24		
14	1343	147.mg	382.mg	0/49	888.mg	33/48				
a	1343	1.26gm	n.s.s.	0/49	888.mg	3/48				
15	1343	65.2mg	171.mg	0/50	710.mg	41/47				
a	1343	1.62gm	n.s.s.	0/50	710.mg	1/47				
16	2	158.mg	n.s.s.	0/7	800.mg	7/16				Weisburger;txap,14,163-175;1969

ACETAMINOPHEN   (tylenol, paracetamol) 103-90-2

17	TR394	955.mg	n.s.s.	35/50	76.8mg	28/50	384.mg	24/50	770.mg	34/50	
a	TR394	2.21gm	n.s.s.	3/50	76.8mg	4/50	384.mg	7/50	770.mg	3/50	liv:hpa,hpc,nnd.
b	TR394	1.83gm	n.s.s.	1/50	76.8mg	4/50	384.mg	4/50	770.mg	4/50	lun:a/a,a/c.
18	1955	3.29gm	n.s.s.	2/49	390.mg	2/46	780.mg	8/50			Amo;jjhg,40,567-574;1985
a	1955	12.0gm	n.s.s.	3/49	390.mg	1/46	780.mg	1/50			
b	1955	1.12gm	n.s.s.	32/49	390.mg	33/46	780.mg	33/50			
19	TR394	717.mg	n.s.s.	35/50	70.6mg	29/50	354.mg	31/50	710.mg	28/50	
a	TR394	2.27gm	n.s.s.	16/50	70.6mg	9/50	354.mg	10/50	710.mg	7/50	liv:hpa,hpc,nnd.

	Spe Strain Site Xpo+Xpt				TD50	2Tailpvl
	Sex Route Hist Notes				DR	AuOp
b	M m b6c eat lun MXB 24m24				3.56gm *	P<.3
20	M m b6c eat liv hpc 70w70 e			.>	no dre	P=1. -
a	M m b6c eat liv hpa 70w70 e				no dre	P=1. -
b	M m b6c eat lun ald 70w70 e				no dre	P=1. -
21	M m b6c eat liv mix 31m31 e			.>	no dre	P=1. -
a	M m b6c eat lun mix 31m31 e				no dre	P=1. -
b	M m b6c eat tum 31m31 e				no dre	P=1. -
22	M f ifm eat liv mix 78w78 e		. +		4.14gm /	P<.0005+
a	M f ifm eat lun tum 78w78 e				no dre	P=1.
23	M m ifm eat liv mix 78w78 e		. + .		1.01gm /	P<.0005+
a	M m ifm eat liv hpc 78w78 e				4.90gm /	P<.003
b	M m ifm eat lun tum 78w78 e				no dre	P=1.
24	M f swi eat --- leu 32w52 v		.>		no dre	P=1.
a	M f swi eat liv tum 32w52 v				no dre	P=1.
b	M f swi eat lun tum 32w52 v				no dre	P=1.
25	R f f34 eat --- mnl 24m24		: ±		481.mg *	P<.02 e
a	R f f34 eat TBA MXB 24m24				1.01gm *	P<.7
b	R f f34 eat liv MXB 24m24				no dre	P=1.
26	R m f34 eat zym car 24m24		: ±		#1.49gm *	P<.02 -
a	R m f34 eat TBA MXB 24m24				591.mg *	P<.6
b	R m f34 eat liv MXB 24m24				no dre	P=1.
27	R f f3d eat --- mnl 24m30 e			.>	6.59gm *	P<.5 -
a	R f f3d eat liv tum 24m30 e				no dre	P=1. -
b	R f f3d eat nas tum 24m30 e				no dre	P=1. -
c	R f f3d eat tba mix 24m30 e				no dre	P=1. -
28	R m f3d eat liv tum 24m30 e			.>	no dre	P=1.
a	R m f3d eat nas tum 24m30 e				no dre	P=1. -
b	R m f3d eat tba mix 24m30 e				no dre	P=1. -
29	R f lee eat liv nnd 78w78 e		. + .		1.34gm /	P<.0005+
a	R f lee eat ubl mix 78w78 e				1.73gm *	P<.08 +
b	R f lee eat ubl pam 78w78 e				2.33gm *	P<.2 +
c	R f lee eat ubl car 78w78 e				7.11gm *	P<.3 +
d	R f lee eat liv hpc 78w78 e				no dre	P=1.
e	R f lee eat tba mix 78w78 e				565.mg *	P<.0005+
30	R m lee eat liv nnd 78w78 e		. + .		1.05gm *	P<.002 +
a	R m lee eat ubl mix 78w78 e				1.19gm *	P<.007 +
b	R m lee eat ubl pam 78w78 e				1.34gm *	P<.02 +
c	R m lee eat ubl car 78w78 e				11.2gm *	P<.3 +
d	R m lee eat liv hpc 78w78 e				no dre	P=1.
e	R m lee eat tba mix 78w78 e				440.mg *	P<.0005+
31	R m sda eat liv tum 27m27			.>	no dre	P=1. -

ACETOHEXAMIDE   100ng..:..1ug....:..10.....:..100....:..1mg....:..10.....:..100....:..1g.....:..10

					TD50	2Tailpvl
32	M f b6c eat TBA MXB 24m25 sv			:>	no dre	P=1. -
a	M f b6c eat liv MXB 24m25 sv				no dre	P=1. -
b	M f b6c eat lun MXB 24m25 sv				no dre	P=1. -
33	M m b6c eat TBA MXB 24m25 v			:>	2.37gm \	P<.7 -
a	M m b6c eat liv MXB 24m25 v				no dre	P=1. -
b	M m b6c eat lun MXB 24m25 v				no dre	P=1. -
34	R f f34 eat TBA MXB 24m25			:>	no dre	P=1. -
a	R f f34 eat liv MXB 24m25				no dre	P=1. -
35	R m f34 eat --- leu 24m24		: +		:#527.mg \	P<.008 -
a	R m f34 eat TBA MXB 24m24				2.12gm *	P<.6
b	R m f34 eat liv MXB 24m24				16.6gm *	P<.4

ACETONE[4-(5-NITRO-2-FURYL)-2-THIAZOLYL]HYDRAZONE..:..1_0....:..100....:..1mg....:..10.....:..100....:..1g.....:..10

					TD50	2Tailpvl
36	R f hza eat for sqp 36w54 es		. ±		11.0mg	P<.05
a	R f hza eat liv tum 36w54 es				no dre	P=1.
37	R f hza eat for sqp 44w60 es		. + .		6.05mg	P<.0005+
a	R f hza eat liv tum 44w60 es				no dre	P=1.

ACETOXIME   100ng..:..1ug....:..10.....:..100....:..1mg....:..10.....:..100....:..1g.....:..10

					TD50	2Tailpvl
38	R f mrw wat liv hpa 18m30 e		. ±		127.mg	P<.03 -
a	R f mrw wat liv mix 18m30 e				127.mg	P<.03 -
b	R f mrw wat liv hem 18m30 e				408.mg	P<.2 -
c	R f mrw wat tba mix 18m30 e				91.5mg	P<.7
39	R m mrw wat liv hpa 18m26 e		. + .		12.1mg	P<.0005+
a	R m mrw wat liv mix 18m26 e				12.1mg	P<.0005-
b	R m mrw wat liv hem 18m26 e				136.mg	P<.05
c	R m mrw wat tba mix 18m26 e				12.8mg	P<.003

1'-ACETOXYSAFROLE   100ng..:..1ug....:..10.....:..100....:..1mg....:..10.....:..100....:..1g.....:..10

					TD50	2Tailpvl
40	M m cd1 eat liv car 56w69 s			.>	no dre	P=1.
a	M m cd1 eat lun tum 56w69 s				no dre	P=1.
b	M m cd1 eat tba mix 56w69 s				no dre	P=1.
41	M m cd1 eat liv car 30w69 s			.>	no dre	P=1.
a	M m cd1 eat lun tum 30w69 s				no dre	P=1.
b	M m cd1 eat tba mix 30w69 s				no dre	P=1.
42	R m cdr eat for pam 47w69 sv		. + .		30.7mg	P<.0005+
a	R m cdr eat liv car 47w69 sv				no dre	P=1. -
b	R m cdr eat tba mix 47w69 sv				32.4mg	P<.0005

	RefNum	LoConf	UpConf	Cntrl	1Dose	1Inc	2Dose	2Inc			Citation or Pathology / Brkly Code
b	TR394	1.05gm	n.s.s.	8/50	70.6mg	7/50	354.mg	5/50	710.mg	12/50	lun:a/a,a/c.
20	1779	867.mg	n.s.s.	0/30	600.mg	0/32	1.20gm	0/15			Hagiwara;faat,7,376-386;1986
a	1779	1.64gm	n.s.s.	3/30	600.mg	2/32	1.20gm	1/15			
b	1779	2.58gm	n.s.s.	2/30	600.mg	1/32	1.20gm	0/15			
21	1955	4.85gm	n.s.s.	13/43	360.mg	12/39	720.mg	6/45			Amo;jjhg,40,567-574;1985
a	1955	7.41gm	n.s.s.	5/43	360.mg	3/39	720.mg	2/45			
b	1955	1.99gm	n.s.s.	27/43	360.mg	21/39	720.mg	23/45			
22	1579	1.95gm	12.4gm	0/51	650.mg	0/58	1.30gm	9/50			Flaks;carc,4,363-368;1983
a	1579	2.77gm	n.s.s.	0/51	650.mg	0/58	1.30gm	0/50			
23	1579	591.mg	1.97gm	1/52	600.mg	1/57	1.20gm	20/27			
a	1579	1.86gm	28.7gm	0/52	600.mg	0/57	1.20gm	5/27			
b	1579	1.93gm	n.s.s.	0/52	600.mg	0/57	1.20gm	0/27			
24	1118	805.mg	n.s.s.	1/30	863.mg	1/30					Cohen;canr,38,1398-1405;1978
a	1118	1.33gm	n.s.s.	0/30	863.mg	0/30					
b	1118	1.33gm	n.s.s.	0/30	863.mg	0/30					
25	TR394	220.mg	n.s.s.	9/50	29.4mg	17/50	147.mg	15/50	296.mg	24/50	
a	TR394	154.mg	n.s.s.	47/50	29.4mg	46/50	147.mg	49/50	296.mg	46/50	
b	TR394	n.s.s.	n.s.s.	0/50	29.4mg	0/50	147.mg	0/50	296.mg	0/50	liv:hpa,hpc,nnd.
26	TR394	580.mg	n.s.s.	0/50	23.5mg	0/50	118.mg	4/50	237.mg	2/50	s
a	TR394	106.mg	n.s.s.	48/50	23.5mg	50/50	118.mg	48/50	237.mg	48/50	
b	TR394	1.04gm	n.s.s.	0/50	23.5mg	1/50	118.mg	1/50	237.mg	0/50	liv:hpa,hpc,nnd.
27	1712	1.56gm	n.s.s.	7/50	260.mg	12/50	520.mg	10/50			Hiraga;gann,76,79-85;1985
a	1712	2.79gm	n.s.s.	0/50	260.mg	0/50	520.mg	0/50			
b	1712	2.79gm	n.s.s.	0/50	260.mg	0/50	520.mg	0/50			
c	1712	885.mg	n.s.s.	40/50	260.mg	42/50	520.mg	35/50			
28	1712	1.55gm	n.s.s.	0/50	144.mg	0/50	288.mg	0/50			
a	1712	1.55gm	n.s.s.	0/50	144.mg	0/50	288.mg	0/50			
b	1712	90.8mg	n.s.s.	50/50	144.mg	49/50	288.mg	49/50			
29	1753	651.mg	3.83gm	0/40	250.mg	10/50	500.mg	10/50			Flaks;apms,93,367-377;1985/pers.comm.
a	1753	782.mg	n.s.s.	0/40	250.mg	5/49	500.mg	3/50			
b	1753	950.mg	n.s.s.	0/40	250.mg	4/49	500.mg	2/50			
c	1753	1.75gm	n.s.s.	0/40	250.mg	1/49	500.mg	1/50			
d	1753	953.mg	n.s.s.	0/40	250.mg	0/49	500.mg	0/50			
e	1753	341.mg	1.08gm	0/40	250.mg	6/49	500.mg	16/50			
30	1753	512.mg	3.35gm	0/40	200.mg	1/48	400.mg	9/49			
a	1753	558.mg	14.3gm	0/40	200.mg	3/48	400.mg	6/49			
b	1753	608.mg	n.s.s.	0/40	200.mg	3/48	400.mg	5/49			
c	1753	1.82gm	n.s.s.	0/40	200.mg	0/48	400.mg	1/49			
d	1753	747.mg	n.s.s.	0/40	200.mg	0/48	400.mg	0/49			
e	1753	265.mg	820.mg	0/40	200.mg	5/48	400.mg	17/49			
31	1459	1.65gm	n.s.s.	0/30	214.mg	0/30					Johansson;ijcn,27,521-529;1981

ACETOHEXAMIDE 968-81-0

	RefNum	LoConf	UpConf	Cntrl	1Dose	1Inc	2Dose	2Inc	Citation or Pathology / Brkly Code
32	TR50	939.mg	n.s.s.	2/15	568.mg	7/35	1.14gm	3/35	
a	TR50	1.78gm	n.s.s.	1/15	568.mg	1/35	1.14gm	1/35	liv:hpa,hpc,nnd.
b	TR50	n.s.s.	n.s.s.	0/15	568.mg	1/35	1.14gm	0/35	lun:a/a,a/c.
33	TR50	348.mg	n.s.s.	6/15	525.mg	14/35	(1.05gm	5/35)	
a	TR50	3.66gm	n.s.s.	3/15	525.mg	2/35	1.05gm	1/35	liv:hpa,hpc,nnd.
b	TR50	2.34gm	n.s.s.	2/15	525.mg	3/35	1.05gm	3/35	lun:a/a,a/c.
34	TR50	751.mg	n.s.s.	9/15	348.mg	23/35	694.mg	18/35	
a	TR50	n.s.s.	n.s.s.	0/15	348.mg	0/35	694.mg	0/35	liv:hpa,hpc,nnd.
35	TR50	256.mg	7.11gm	0/15	278.mg	10/35	(561.mg	4/35)	s
a	TR50	386.mg	n.s.s.	7/15	278.mg	19/35	561.mg	21/35	
b	TR50	2.70gm	n.s.s.	0/15	278.mg	0/35	561.mg	1/35	liv:hpa,hpc,nnd.

ACETONE[4-(5-NITRO-2-FURYL)-2-THIAZOLYL]HYDRAZONE 18523-69-8

	RefNum	LoConf	UpConf	Cntrl	1Dose	1Inc	Citation or Pathology / Brkly Code
36	1063m	3.23mg	n.s.s.	0/5	33.3mg	3/7	Morris;canr,29,2145-2156;1969
a	1063m	13.0mg	n.s.s.	0/5	33.3mg	0/7	
37	1063n	3.10mg	13.3mg	0/16	36.8mg	15/20	
a	1063n	50.5mg	n.s.s.	0/16	36.8mg	0/20	

ACETOXIME 127-06-0

	RefNum	LoConf	UpConf	Cntrl	1Dose	1Inc	Citation or Pathology / Brkly Code
38	1480	38.2mg	n.s.s.	0/20	24.6mg	3/16	Mirvish;jnci,69,961-962;1982
a	1480	38.2mg	n.s.s.	0/20	24.6mg	3/16	
b	1480	66.3mg	n.s.s.	0/20	24.6mg	1/16	
c	1480	11.3mg	n.s.s.	15/20	24.6mg	13/16	
39	1480	5.59mg	30.1mg	0/23	25.4mg	12/15	
a	1480	5.59mg	30.1mg	0/23	25.4mg	12/15	
b	1480	33.4mg	n.s.s.	0/23	25.4mg	2/15	
c	1480	4.79mg	106.mg	9/23	25.4mg	13/15	

1'-ACETOXYSAFROLE 34627-78-6

	RefNum	LoConf	UpConf	Cntrl	1Dose	1Inc	Citation or Pathology / Brkly Code
40	1042b	92.9mg	n.s.s.	3/35	29.3mg	0/35	Borchert;canr,33,590-600;1973
a	1042b	92.9mg	n.s.s.	0/35	29.3mg	0/35	
b	1042b	92.9mg	n.s.s.	3/35	29.3mg	0/35	
41	1042c	83.3mg	n.s.s.	3/35	26.3mg	0/35	
a	1042c	83.3mg	n.s.s.	0/35	26.3mg	0/35	
b	1042c	83.3mg	n.s.s.	3/35	26.3mg	0/35	
42	1042b	14.2mg	81.2mg	0/18	112.mg	10/15	
a	1042b	152.mg	n.s.s.	0/18	112.mg	0/15	
b	1042b	14.5mg	107.mg	1/18	112.mg	10/15	

```
      Spe Strain Site   Xpo+Xpt                                          TD50    2Tailpvl
        Sex  Route  Hist    Notes                                          DR      AuOp
```

			TD50	2Tailpvl
43	R m cdr eat for pam 36w52 s	. + .	21.1mg	P<.0005+
a	R m cdr eat liv hpc 36w52 s		348.mg	P<.3 -
b	R m cdr eat tba mix 36w52 s		21.1mg	P<.0005

```
N'-ACETYL-4-(HYDROXYMETHYL)PHENYLHYDRAZINE..1_ug....:..10....:..100....:..1mg....:..10....:..100....:..1g......:..10
```

			TD50	2Tailpvl
44	M f swa wat blv mix 27m27 e	. + .	287.mg	P<.0005+
a	M f swa wat lun mix 27m27 e		329.mg	P<.004 +
b	M f swa wat liv tum 27m27		no dre	P=1.
45	M m swa wat lun mix 27m27 e	. + .	208.mg	P<.003 +
a	M m swa wat blv mix 27m27 e		252.mg	P<.0005+
b	M m swa wat liv hpt 27m27 e		no dre	P=1. -

```
1-ACETYL-2-ISONICOTINOYLHYDRAZINE 100ng..:..1ug....:..10....:..100....:..1mg....:..10....:..100....:..1g......:..10
```

			TD50	2Tailpvl
46	M f swa wat lun mix 94w94 e	. + .	319.mg	P<.0005+
a	M f swa wat liv mix 94w94 e		55.7gm	P<.9
47	M m swa wat lun mix 92w92 e	. + .	342.mg	P<.0005+
a	M m swa wat liv hem 92w92 e		no dre	P=1. -

```
3-ACETYL-6-METHYL-2,4-PYRANDIONE 100ng..:..1ug....:..10....:..100....:..1mg....:..10....:..100....:..1g......:..10
```

			TD50	2Tailpvl
48	M f b6a orl liv hpt 76w76 evx	.>	no dre	P=1. -
a	M f b6a orl lun ade 76w76 evx		no dre	P=1. -
b	M f b6a orl tba mix 76w76 evx		no dre	P=1. -
49	M m b6a orl lun ade 76w76 evx	.>	no dre	P=1. -
a	M m b6a orl liv hpt 76w76 evx		no dre	P=1. -
b	M m b6a orl tba mix 76w76 evx		no dre	P=1. -
50	M f b6c orl liv hpt 76w76 evx	.>	no dre	P=1. -
a	M f b6c orl lun mix 76w76 evx		no dre	P=1. -
b	M f b6c orl tba tum 76w76 evx		no dre	P=1. -
51	M m b6c orl liv hpt 76w76 evx	.>	no dre	P=1. -
a	M m b6c orl lun mix 76w76 evx		no dre	P=1. -
b	M m b6c orl tba mix 76w76 evx		202.mg	P<.3 -

```
1-ACETYL-2-PHENYLHYDRAZINE    100ng..:..1ug....:..10....:..100....:..1mg....:..10....:..100....:..1g......:..10
```

			TD50	2Tailpvl
52	M f swa wat liv mix 26m26 e	. + .	44.8mg	P<.0005+
a	M f swa wat blv mix 26m26 e		49.0mg	P<.0005+
b	M f swa wat liv agm 26m26 e		105.mg	P<.0005
c	M f swa wat liv ang 26m26 e		105.mg	P<.0005
d	M f swa wat blv ang 26m26 e		117.mg	P<.005
e	M f swa wat blv agm 26m26 e		117.mg	P<.005
f	M f swa wat lun ade 26m26 e		413.mg	P<.5 -
g	M f swa wat lun mix 26m26 e		no dre	P=1.
53	M m swa wat blv mix 25m25 e	. + .	59.8mg	P<.0005+
a	M m swa wat liv mix 25m25 e		62.1mg	P<.0005+
b	M m swa wat blv agm 25m25 e		113.mg	P<.01
c	M m swa wat liv agm 25m25 e		128.mg	P<.01
d	M m swa wat liv ang 25m25 e		146.mg	P<.008
e	M m swa wat blv ang 25m25 e		161.mg	P<.03
f	M m swa wat lun adc 25m25 e		no dre	P=1. -
g	M m swa wat lun ade 25m25 e		no dre	P=1. -
h	M m swa wat lun mix 25m25 e		no dre	P=1.

```
4-ACETYLAMINOBIPHENYL         100ng..:..1ug....:..10....:..100....:..1mg....:..10....:..100....:..1g......:..10
```

			TD50	2Tailpvl
54	R f nss eat mgl adc 43w65	. + .	1.18mg	P<.0005+
a	R f nss eat liv tum 43w65		no dre	P=1.

```
1-ACETYLAMINOFLUORENE         100ng..:..1ug....:..10....:..100....:..1mg....:..10....:..100....:..1g......:..10
```

			TD50	2Tailpvl
55	R f buf eat pit ade 47w88 e	.>	43.5mg	P<.6
a	R f buf eat adr car 47w88 e		54.8mg	P<.3
b	R f buf eat --- ile 47w88 e		54.8mg	P<.3
c	R f buf eat ova gcc 47w88 e		54.8mg	P<.3
d	R f buf eat ute cas 47w88 e		54.8mg	P<.3

```
2-ACETYLAMINOFLUORENE         100ng..:..1ug....:..10....:..100....:..1mg....:..10....:..100....:..1g......:..10
```

			TD50	2Tailpvl
56	H m nss eat liv bdc 32w56	±	17.4mg	P<.04 +
57	H m nss ipj smi adc 34w69	.>	7.26mg	P<.3
a	H m nss ipj per sar 34w69		no dre	P=1.
b	H m nss ipj for pam 34w69		no dre	P=1.
58	H f syg eat liv cac 26w68 e	. ±	17.3mg	P<.03
59	H m syg eat liv cac 40w84 e	. + .	15.3mg	P<.003
60	H m syg eat liv cac 26w80 e	.>	31.5mg	P<.2
61	M f asw eat tba mix 84w84 e	.>	no dre	P=1.
a	M f asw eat tba mal 84w84 e		51.1mg	P<.3 +
62	M m asw eat tba mal 84w84 e	. + .	17.4mg	P<.006 +
a	M m asw eat tba mix 84w84 e		21.7mg	P<.2
63	M f ays eat liv mix 30m30 er	: + :	2.27mg Z	P<.0005+
a	M f ays eat liv hpc 30m30 er		3.31mg Z	P<.0005+
b	M f ays eat ubl mix 30m30 er		43.0mg Z	P<.0005+
c	M f ays eat ubl tcc 30m30 er		54.1mg Z	P<.0005+
64	M m ays eat ubl mix 30m30 er	: + :	2.45mg Z	P<.0005+
a	M m ays eat ubl tcc 30m30 er		4.32mg Z	P<.0005+
b	M m ays eat liv mix 30m30 er		4.61mg Z	P<.0005+

	RefNum	LoConf	UpConf	Cntrl	1Dose	1Inc	2Dose	2Inc			Citation or Pathology	Brkly Code
43	1042c	10.2mg	52.6mg	0/18	116.mg	11/18						
a	1042c	56.7mg	n.s.s.	0/18	116.mg	1/18						
b	1042c	10.2mg	52.6mg	0/18	116.mg	11/18						
N'-ACETYL-4-(HYDROXYMETHYL)PHENYLHYDRAZINE 65734-38-5												
44	410	143.mg	950.mg	8/96	125.mg	16/44					Toth;canr,38,177-180;1978	
a	410	148.mg	3.05gm	15/96	125.mg	17/44						
b	410	1.57gm	n.s.s.	0/99	125.mg	0/50						
45	410	98.6mg	1.38gm	22/92	104.mg	24/48						
a	410	127.mg	762.mg	5/88	104.mg	15/45						
b	410	997.mg	n.s.s.	2/62	104.mg	0/38						
1-ACETYL-2-ISONICOTINOYLHYDRAZINE 1078-38-2												
46	1055	196.mg	567.mg	14/104	800.mg	37/47					Toth;ejca,9,285-289;1973	
a	1055	2.66gm	n.s.s.	3/73	800.mg	2/41						
47	1055	202.mg	664.mg	11/91	667.mg	29/42						
a	1055	968.mg	n.s.s.	2/40	667.mg	0/9						
3-ACETYL-6-METHYL-2,4-PYRANDIONE (dehydroacetic acid) 520-45-6												
48	1294	67.2mg	n.s.s.	0/17	33.9mg	0/18					Innes;ntis,1968/1969	
a	1294	67.2mg	n.s.s.	1/17	33.9mg	0/18						
b	1294	32.0mg	n.s.s.	2/17	33.9mg	2/18						
49	1294	38.9mg	n.s.s.	2/18	31.6mg	1/17						
a	1294	59.1mg	n.s.s.	1/18	31.6mg	0/17						
b	1294	41.6mg	n.s.s.	3/18	31.6mg	1/17						
50	1294	67.2mg	n.s.s.	0/16	33.9mg	0/18						
a	1294	67.2mg	n.s.s.	0/16	33.9mg	0/18						
b	1294	67.2mg	n.s.s.	0/16	33.9mg	0/18						
51	1294	62.6mg	n.s.s.	0/16	31.6mg	0/18						
a	1294	62.6mg	n.s.s.	0/16	31.6mg	0/18						
b	1294	32.9mg	n.s.s.	0/16	31.6mg	1/18						
1-ACETYL-2-PHENYLHYDRAZINE 114-83-0												
52	1054	23.8mg	106.mg	4/94	30.0mg	16/37					Toth;bjca,39,584-587;1979	
a	1054	24.9mg	140.mg	8/94	30.0mg	16/37						
b	1054	45.1mg	463.mg	2/94	30.0mg	8/37						
c	1054	45.1mg	463.mg	2/94	30.0mg	8/37						
d	1054	46.9mg	1.19gm	4/94	30.0mg	8/37						
e	1054	46.9mg	1.19gm	4/94	30.0mg	8/37						
f	1054	65.8mg	n.s.s.	10/89	30.0mg	5/31						
g	1054	78.8mg	n.s.s.	15/89	30.0mg	5/31						
53	1054	27.8mg	245.mg	5/87	25.0mg	12/40						
a	1054	29.1mg	217.mg	3/87	25.0mg	11/40						
b	1054	42.9mg	9.84gm	3/87	25.0mg	7/40						
c	1054	47.1mg	9.32gm	2/87	25.0mg	6/40						
d	1054	52.1mg	3.67gm	1/87	25.0mg	5/40						
e	1054	53.7mg	n.s.s.	2/87	25.0mg	5/40						
f	1054	158.mg	n.s.s.	6/87	25.0mg	1/40						
g	1054	97.2mg	n.s.s.	17/87	25.0mg	5/40						
h	1054	112.mg	n.s.s.	22/87	25.0mg	5/40						
4-ACETYLAMINOBIPHENYL (4'-phenylacetanilide) 4075-79-0												
54	1424	.444mg	3.21mg	0/15	11.3mg	12/13					Miller;jnci,15,1571-1590;1955	
a	1424	11.9mg	n.s.s.	0/15	11.3mg	0/13						
1-ACETYLAMINOFLUORENE (N-1-fluorenylacetamide) 28314-03-6												
55	144	6.02mg	n.s.s.	2/18	6.77mg	3/17					Morris;jnci,24,149-180;1960	
a	144	8.92mg	n.s.s.	0/18	6.77mg	1/17						
b	144	8.92mg	n.s.s.	0/18	6.77mg	1/17						
c	144	8.92mg	n.s.s.	0/18	6.77mg	1/17						
d	144	8.92mg	n.s.s.	0/18	6.77mg	1/17						
2-ACETYLAMINOFLUORENE (N-2-fluorenylacetamide) 53-96-3												
56	308m	5.24mg	n.s.s.	0/17	15.9mg	3/18					Miller;canr,24,2018-2026;1964	
57	308n	1.18mg	n.s.s.	0/8	3.21mg	1/8						
a	308n	2.33mg	n.s.s.	0/8	3.21mg	0/8						
b	308n	1.34mg	n.s.s.	1/8	3.21mg	1/8						
58	347n	4.02mg	n.s.s.	1/59	16.0mg	2/8					Della Porta;jnci,22,463-471;1959	
59	347m	3.65mg	211.mg	0/39	17.5mg	2/5						
60	347n	5.10mg	n.s.s.	0/16	12.0mg	1/7						
61	213b	8.63mg	n.s.s.	14/16	39.0mg	15/18					Prier;txap,5,526-542;1963	
a	213b	13.2mg	n.s.s.	6/16	39.0mg	10/18						
62	213b	7.28mg	200.mg	1/10	36.0mg	9/14						
a	213b	6.89mg	n.s.s.	4/10	36.0mg	10/14						
63	2034n	.869mg	4.73mg	4/24	6.50mg	11/23	13.0mg	17/24	26.0mg	23/23	Wolff;jtxe,33,327-348;1991/pers.comm.	
a	2034n	1.02mg	8.56mg	3/24	6.50mg	7/23	13.0mg	10/24	26.0mg	19/23		
b	2034n	15.5mg	118.mg	0/24	6.50mg	0/23	13.0mg	0/24	26.0mg	9/23		
c	2034n	16.7mg	193.mg	0/24	6.50mg	0/23	13.0mg	0/24	26.0mg	7/23		
64	2034m	1.43mg	4.15mg	0/24	6.00mg	15/24	12.0mg	22/24	24.0mg	24/24		
a	2034m	2.47mg	7.13mg	0/24	6.00mg	9/24	12.0mg	18/24	24.0mg	21/24		
b	2034m	2.16mg	17.1mg	6/24	6.00mg	14/24	12.0mg	10/24	24.0mg	5/24		

	Spe	Sex	Strain	Route	Site	Hist	Xpo+Xpt	Notes	plot	TD50	DR	2Tailpvl / AuOp
65	M	f	b6c	eat	liv	mix	82w82	er	. +.	4.00mg	*	P<.0005+
a	M	f	b6c	eat	liv	mal	82w82	er		5.38mg	*	P<.0005
b	M	f	b6c	eat	ubl	mix	82w82	er		38.3mg	Z	P<.0005+
c	M	f	b6c	eat	ubl	mal	82w82	er		64.9mg	Z	P<.0005
d	M	f	b6c	eat	liv	hpb	82w82	er		526.mg	*	P<.06
66	M	f	b6c	eat	liv	mix	82w82	er	.+.	8.80mg	*	P<.0005+
a	M	f	b6c	eat	liv	mal	82w82	er		15.9mg	*	P<.0005
b	M	f	b6c	eat	ubl	mix	82w82	er		71.8mg	Z	P<.0005+
c	M	f	b6c	eat	ubl	mal	82w82	er		201.mg	Z	P<.0005
67	M	f	b6c	eat	liv	mix	30m30	er	: +:	3.99mg	*	P<.0005+
a	M	f	b6c	eat	liv	hpc	30m30	er		9.49mg	Z	P<.0005+
b	M	f	b6c	eat	ubl	mix	30m30	er		47.7mg	Z	P<.0005+
c	M	f	b6c	eat	ubl	tcc	30m30	er		89.2mg	Z	P<.0005+
68	M	m	b6c	eat	liv	mix	82w82	er	.+.	2.35mg	*	P<.0005+
a	M	m	b6c	eat	liv	mal	82w82	er		5.46mg	*	P<.0005
b	M	m	b6c	eat	ubl	mix	82w82	er		7.90mg	Z	P<.0005+
c	M	m	b6c	eat	ubl	mal	82w82	er		21.2mg	Z	P<.0005
d	M	m	b6c	eat	liv	hpb	82w82	er		57.3mg	*	P<.004
69	M	m	b6c	eat	ubl	mix	82w82	er	.+ .	7.29mg	Z	P<.0005+
a	M	m	b6c	eat	liv	mal	82w82	er		40.8mg	*	P<.01
b	M	m	b6c	eat	ubl	mal	82w82	er		107.mg	*	P<.005
c	M	m	b6c	eat	liv	mix	82w82	er		32.9mg	*	P<.04 +
70	M	m	b6c	eat	liv	mix	30m30	er	: + :	4.34mg	Z	P<.0005+
a	M	m	b6c	eat	ubl	mix	30m30	er		5.02mg	Z	P<.0005+
b	M	m	b6c	eat	ubl	tcc	30m30	er		7.52mg	Z	P<.0005+
c	M	m	b6c	eat	liv	hpc	30m30	er		14.4mg	Z	P<.0005+
71	M	f	bal	eat	liv	tum	52w52	ek	.>	no dre		P=1.
a	M	f	bal	eat	ubl	car	52w52	ek		no dre		P=1. -
72	M	m	bal	eat	ubl	car	52w52	ek	. + .	9.09mg	Z	P<.0005+
73	M	f	bcn	eat	ubl	tcc	65w78	ekr	. + .	66.5mg	Z	P<.0005+
a	M	f	bcn	eat	liv	hpc	65w78	ekr		120.mg	*	P<.0005+
74	M	f	bcn	eat	liv	hpc	15m24	ekr	. + .	30.6mg	*	P<.0005+
a	M	f	bcn	eat	ubl	tcc	15m24	ekr		95.5mg	*	P<.0005+
75	M	f	bcn	eat	ubl	tcc	52w52	ekr	. + .	150.mg	Z	P<.0005+
a	M	f	bcn	eat	liv	hpc	52w52	ekr		226.mg	*	P<.2 +
76	M	f	bcn	eat	ubl	tcc	60w60	ekr	. + .	68.8mg	Z	P<.0005+
a	M	f	bcn	eat	liv	hpc	60w60	ekr		747.mg	*	P<.2 +
77	M	f	bcn	eat	ubl	tcc	65w65	ekr	. + .	45.8mg	Z	P<.0005+
a	M	f	bcn	eat	liv	hpc	65w65	ekr		199.mg	*	P<.07 +
78	M	f	bcn	eat	ubl	tcc	69w69	ekr	.+ .	62.1mg	Z	P<.0005+
a	M	f	bcn	eat	liv	hpc	69w69	ekr		218.mg	*	P<.05 +
79	M	f	bcn	eat	ubl	tcc	73w73	ekr	. +.	64.1mg	Z	P<.0005+
a	M	f	bcn	eat	liv	hpc	73w73	ekr		147.mg	*	P<.05 +
80	M	f	bcn	eat	ubl	tcc	78w78	ekr	. +.	101.mg	Z	P<.0005+
a	M	f	bcn	eat	liv	hpc	78w78	ekr		154.mg	*	P<.0005+
81	M	f	bcn	eat	liv	hpc	24m24	ekr	.+.	33.6mg	Z	P<.0005+
a	M	f	bcn	eat	ubl	tcc	24m24	ekr		96.0mg	Z	P<.0005+
82	M	f	bcn	eat	liv	hpc	33m33	ekr	. + .	19.6mg	*	P<.008 +
a	M	f	bcn	eat	ubl	tcc	33m33	ekr		82.1mg	Z	P<.0005+
83	M	f	bcn	eat	liv	hpc	39w78	ekr	. + .	110.mg	*	P<.002 +
a	M	f	bcn	eat	ubl	tcc	39w78	ekr		142.mg	*	P<.004 +
84	M	f	bcn	eat	liv	hpc	9m24	ekr	. + .	19.1mg	*	P<.0005+
a	M	f	bcn	eat	ubl	tcc	9m24	ekr		97.2mg	Z	P<.0005+
85	M	f	bcn	eat	ubl	tcc	52w78	ekr	. + .	82.8mg	Z	P<.0005+
a	M	f	bcn	eat	liv	hpc	52w78	ekr		156.mg	Z	P<.009 +
86	M	f	bcn	eat	liv	hpc	12m24	ekr	. + .	24.3mg	*	P<.0005+
a	M	f	bcn	eat	ubl	tcc	12m24	ekr		101.mg	Z	P<.0005+
87	M	f	bcn	eat	liv	mix	23m24	aers	.+ .	11.6mg	Z	P<.0005+
a	M	f	bcn	eat	ubl	mix	23m24	aers		12.8mg	Z	P<.0005+
b	M	f	bcn	eat	ubl	car	23m24	aers		21.9mg	Z	P<.0005
c	M	f	bcn	eat	liv	mal	23m24	aers		30.0mg	Z	P<.0005
d	M	f	bcn	eat	liv	hpb	23m24	aers		3.20gm	*	P<.5
88	M	f	bcn	eat	ubl	mix	23m24	aer	.+.	10.4mg	Z	P<.0005+
a	M	f	bcn	eat	ubl	car	23m24	aer		21.7mg	Z	P<.0005
b	M	f	bcn	eat	liv	mix	23m24	aer		29.8mg	*	P<.0005+
c	M	f	bcn	eat	liv	mal	23m24	aer		58.3mg	*	P<.0005
89	M	m	bcn	eat	ubl	mix	22m24	aers	.+.	3.87mg	Z	P<.0005+
a	M	m	bcn	eat	ubl	car	22m24	aers		9.41mg	Z	P<.0005

	RefNum	LoConf	UpConf	Cntrl	1Dose	1Inc	2Dose	2Inc	3Dose	3Inc	4Dose	4Inc	5Dose	5Inc	6Dose	6Inc	7Dose	7Inc	Citation or Pathology / Brkly Code
65	2020m	2.97mg	5.45mg	43/117	19.5mg	67/70	26.0mg	89/92	32.5mg	91/95									Fullerton;faat,16,51-60; 1991/Nonoyama 1988/pers.comm.
a	2020m	4.31mg	6.80mg	14/117	19.5mg	60/70	26.0mg	81/92	32.5mg	87/95									
b	2020m	28.0mg	54.3mg	0/108	19.5mg	6/67	26.0mg	18/85	32.5mg	38/93									
c	2020m	44.0mg	101.mg	0/108	19.5mg	3/67	26.0mg	10/85	32.5mg	26/93									
d	2020m	215.mg	n.s.s.	0/96	19.5mg	2/96	26.0mg	1/96	32.5mg	3/96									
66	2020n	7.12mg	11.1mg	9/118	19.5mg	46/71	26.0mg	69/96	32.5mg	78/95									
a	2020n	12.6mg	20.5mg	3/118	19.5mg	21/71	26.0mg	48/96	32.5mg	65/95									
b	2020n	48.3mg	113.mg	0/117	19.5mg	0/69	26.0mg	5/95	32.5mg	32/93									
c	2020n	108.mg	464.mg	0/117	19.5mg	0/69	26.0mg	1/95	32.5mg	13/93									
67	2034r	2.89mg	5.76mg	4/48	6.50mg	31/47	13.0mg	45/47	26.0mg	44/48									Wolff;jtxe,33,327-348;1991/pers.comm.
a	2034r	6.56mg	14.2mg	1/48	6.50mg	12/47	13.0mg	28/47	26.0mg	33/48									
b	2034r	26.0mg	86.6mg	0/48	6.50mg	0/47	13.0mg	0/47	26.0mg	20/48									
c	2034r	45.5mg	198.mg	0/48	6.50mg	0/47	13.0mg	0/47	26.0mg	13/48									
68	2020m	1.85mg	3.11mg	21/95	4.80mg	53/94	7.20mg	79/94	9.60mg	83/94									Fullerton;faat,16,51-60; 1991/Nonoyama 1988/pers.comm.
a	2020m	4.11mg	7.92mg	12/95	4.80mg	28/94	7.20mg	49/94	9.60mg	59/94									
b	2020m	6.02mg	10.6mg	0/93	4.80mg	0/91	7.20mg	31/92	9.60mg	53/89									
c	2020m	14.2mg	33.6mg	0/93	4.80mg	0/91	7.20mg	10/92	9.60mg	26/89									
d	2020m	31.4mg	276.mg	0/96	4.80mg	4/96	7.20mg	5/96	9.60mg	6/96									
69	2020n	5.23mg	10.6mg	0/94	4.80mg	1/96	7.20mg	54/96	(9.60mg	17/94)									
a	2020n	21.0mg	2.69gm	4/96	4.80mg	7/96	7.20mg	8/96	9.60mg	15/94									
b	2020n	48.7mg	772.mg	0/94	4.80mg	0/96	7.20mg	2/96	9.60mg	6/94									
c	2020n	15.2mg	n.s.s.	13/96	4.80mg	14/96	7.20mg	15/96	9.60mg	26/94									
70	2034o	2.78mg	7.41mg	13/47	6.00mg	29/48	12.0mg	24/48	24.0mg	17/48									Wolff;jtxe,33,327-348;1991/pers.comm.
a	2034o	3.48mg	7.16mg	0/47	6.00mg	11/48	12.0mg	45/48	24.0mg	46/48									
b	2034o	5.00mg	11.0mg	0/47	6.00mg	6/48	12.0mg	33/48	24.0mg	40/48									
c	2034o	7.90mg	31.8mg	6/47	6.00mg	9/48	12.0mg	14/48	24.0mg	9/48									
71	1665	.969mg	n.s.s.	0/24	1.04mg	0/48	3.12mg	0/48	5.85mg	0/24	7.67mg	0/16	11.2mg	0/8					Haley;pseb,152, 156-159;1976
a	1665	.969mg	n.s.s.	0/24	1.04mg	0/48	3.12mg	0/48	5.85mg	0/24	7.67mg	0/16	11.2mg	0/8					
72	1665	4.11mg	26.7mg	0/23	.960mg	0/41	2.88mg	0/44	5.40mg	1/22	7.08mg	3/13	10.3mg	4/8					
73	1344a	40.3mg	123.mg	1/400	6.50mg	1/196	8.13mg	0/130	10.8mg	0/64	16.3mg	22/65							Littlefield;jept,3,17-34; 1980/pers.comm.
a	1344a	61.6mg	389.mg	1/401	6.50mg	4/196	8.13mg	5/130	10.8mg	1/64	16.3mg	4/65							
74	1344b	19.7mg	55.2mg	9/383	4.88mg	15/114	6.09mg	14/86	8.13mg	6/35	12.2mg	6/28							
a	1344b	48.6mg	234.mg	0/384	4.88mg	0/114	6.09mg	0/86	8.13mg	1/35	12.2mg	11/28							
75	1344m	70.7mg	457.mg	0/140	7.80mg	0/268	9.75mg	0/137	13.0mg	0/138	19.5mg	9/141							
a	1344m	92.4mg	n.s.s.	0/140	7.80mg	2/268	9.75mg	0/137	13.0mg	3/138	19.5mg	1/141							
76	1344n	41.1mg	128.mg	0/113	7.80mg	0/221	9.75mg	0/110	13.0mg	0/117	19.5mg	21/114							
a	1344n	184.mg	n.s.s.	0/113	7.80mg	0/224	9.75mg	0/110	13.0mg	1/117	19.5mg	1/114							
77	1344o	29.4mg	77.0mg	0/88	7.80mg	0/181	9.75mg	1/94	13.0mg	0/89	19.5mg	28/90							
a	1344o	86.2mg	n.s.s.	0/88	7.80mg	1/182	9.75mg	1/94	13.0mg	4/90	19.5mg	1/90							
78	1344r	42.0mg	97.3mg	0/183	5.85mg	1/271	7.80mg	0/265	9.75mg	0/175	13.0mg	1/90	19.5mg	36/85					
a	1344r	97.9mg	n.s.s.	0/183	5.85mg	2/272	7.80mg	4/265	9.75mg	5/174	13.0mg	1/90	19.5mg	3/86					
79	1344s	44.1mg	98.2mg	0/127	4.55mg	1/389	5.85mg	1/264	7.80mg	0/206	9.75mg	0/134	13.0mg	4/67	19.5mg	36/65			
a	1344s	86.0mg	n.s.s.	0/128	4.55mg	2/389	5.85mg	5/264	7.80mg	6/206	9.75mg	5/134	13.0mg	1/67	19.5mg	0/65			
80	1344t	76.4mg	139.mg	1/400	3.90mg	4/1573	4.55mg	1/796	5.85mg	1/383	7.80mg	3/269	9.75mg	1/267	13.0mg	5/131	19.5mg	62/121	
a	1344t	105.mg	310.mg	1/401	3.90mg	17/1573	4.55mg	7/792	5.85mg	7/383	7.80mg	7/268	9.75mg	6/267	13.0mg	6/131	19.5mg	7/121	
81	1344u	28.9mg	40.1mg	9/383	3.90mg	55/900	4.55mg	55/639	5.85mg	57/445	7.80mg	71/415	9.75mg	62/311	13.0mg	47/160	19.5mg	56/130	
a	1344u	77.4mg	121.mg	1/384	3.90mg	0/900	4.55mg	2/638	5.85mg	1/445	7.80mg	3/415	9.75mg	3/311	13.0mg	25/160	19.5mg	100/130	
82	1344v	10.1mg	409.mg	8/23	3.90mg	44/92	4.55mg	20/45	5.85mg	5/12	7.80mg	7/11	9.75mg	8/12	13.0mg	8/10			
a	1344v	44.2mg	179.mg	0/24	3.90mg	1/92	4.55mg	0/45	5.85mg	0/12	7.80mg	1/11	9.75mg	4/12	13.0mg	8/10			
83	1344w	49.1mg	585.mg	1/401	3.90mg	1/186	4.88mg	3/128	6.50mg	1/64	9.75mg	4/63							
a	1344w	57.7mg	1.16gm	1/400	3.90mg	0/184	4.88mg	2/128	6.50mg	1/64	9.75mg	4/63							
84	1344x	12.1mg	35.2mg	9/383	2.93mg	13/108	3.66mg	10/66	4.88mg	5/35	7.31mg	9/33							
a	1344x	40.8mg	388.mg	1/384	2.93mg	1/108	3.66mg	0/66	4.88mg	0/35	7.31mg	6/33							
85	1344y	44.8mg	185.mg	1/400	5.20mg	0/190	6.50mg	0/132	8.67mg	1/65	13.0mg	14/63							
a	1344y	67.6mg	4.45gm	1/401	5.20mg	1/190	6.50mg	6/132	8.67mg	0/65	13.0mg	2/63							
86	1344z	15.6mg	43.4mg	9/383	3.90mg	11/118	4.88mg	14/74	6.50mg	5/33	9.75mg	9/29							
a	1344z	46.6mg	306.mg	1/384	3.90mg	0/118	4.88mg	1/74	6.50mg	1/33	9.75mg	7/29							
87	2066m	7.97mg	18.3mg	5/92	13.0mg	51/91	(16.2mg	39/93	19.5mg	26/82)									Fullerton;faat,18,193-199; 1992/Nonoyama 1988/pers.comm.
a	2066m	10.3mg	16.1mg	1/87	13.0mg	29/90	16.2mg	58/90	19.5mg	63/82									
b	2066m	17.1mg	28.7mg	0/87	13.0mg	14/90	16.2mg	40/90	19.5mg	48/82									
c	2066m	21.2mg	46.3mg	1/92	13.0mg	31/91	16.2mg	23/93	(19.5mg	13/82)									
d	2066m	521.mg	n.s.s.	0/96	13.0mg	0/96	16.2mg	1/96	19.5mg	0/96									
88	2066n	8.47mg	12.8mg	1/92	13.0mg	29/91	16.2mg	74/95	19.5mg	76/91									
a	2066n	17.0mg	28.6mg	1/92	13.0mg	16/91	16.2mg	46/95	19.5mg	49/91									
b	2066n	22.7mg	41.8mg	1/93	13.0mg	28/94	16.2mg	35/96	19.5mg	27/92									
c	2066n	41.2mg	88.9mg	0/93	13.0mg	16/94	16.2mg	19/96	19.5mg	14/92									
89	2066m	3.09mg	4.93mg	0/88	2.40mg	4/86	4.80mg	50/86	7.20mg	82/90									
a	2066m	7.03mg	13.0mg	0/88	2.40mg	1/86	4.80mg	13/86	7.20mg	59/90									

	Spe	Sex	Strain	Route	Site	Hist	Xpo+Xpt	Notes	plot	TD50	DR	2Tailpvl / AuOp
b	M	m	bcn	eat	liv	mix	22m24	aers		3.26gm	*	P<1.
c	M	m	bcn	eat	liv	mal	22m24	aers		no dre		P=1.
d	M	m	bcn	eat	liv	hpb	22m24	aers		no dre		P=1.
90	M	m	bcn	eat	ubl	mix	22m24	aer	.+.	9.05mg	Z	P<.0005+
a	M	m	bcn	eat	ubl	car	22m24	aer		26.4mg	Z	P<.0005
b	M	m	bcn	eat	liv	mix	22m24	aer		69.0mg	Z	P<.2
c	M	m	bcn	eat	liv	mal	22m24	aer		no dre		P=1.
91	M	f	cd1	eat	liv	hpt	78w78	e	. + .	21.3mg		P<.0005
a	M	f	cd1	eat	liv	nod	78w78	e		76.2mg		P<.0005
b	M	f	cd1	eat	lun	ade	78w78	e		354.mg		P<.6
92	M	f	cd1	eat	liv	hpc	53w92	e	.+ .	7.65mg		P<.0005+
93	M	m	cd1	eat	liv	hpt	78w78	e	. + .	22.5mg		P<.0005
a	M	m	cd1	eat	lun	ade	78w78	e		no dre		P=1.
94	M	m	cd1	eat	liv	hpc	53w92	e	. + .	13.5mg		P<.0005+
a	M	m	cd1	eat	ubl	pam	53w92	e		34.5mg		P<.0005+
b	M	m	cd1	eat	ubl	tcc	53w92	e		34.5mg		P<.0005
c	M	m	cd1	eat	kid	mix	53w92	e		86.5mg		P<.003
d	M	m	cd1	eat	liv	hpa	53w92	e		639.mg		P<.8 +
95	M	m	cen	eat	liv	mix	52w52	kr	.>	5.61mg		P<.3
96	M	f	cf1	eat	mgl	car	52w52	e	. + .	14.6mg		P<.0005
a	M	f	cf1	eat	liv	lcc	52w52	e		16.5mg		P<.0005+
b	M	f	cf1	eat	lun	tum	52w52	e		no dre		P=1.
97	M	m	cf1	eat	liv	lcc	52w52	e	. + .	31.4mg		P<.006 +
a	M	m	cf1	eat	lun	ade	52w52	e		149.mg		P<.5
98	M	f	cva	eat	liv	mix	30m30	er	: + :	4.36mg	Z	P<.0005+
a	M	f	cva	eat	liv	hpc	30m30	er		10.3mg	*	P<.0005+
b	M	f	cva	eat	ubl	mix	30m30	er		54.8mg	Z	P<.0005+
c	M	f	cva	eat	ubl	tcc	30m30	er		63.5mg	Z	P<.0005+
99	M	m	cva	eat	ubl	mix	30m30	er	: + :	3.59mg	Z	P<.0005+
a	M	m	cva	eat	ubl	tcc	30m30	er		5.20mg	Z	P<.0005+
b	M	m	cva	eat	liv	mix	30m30	er		7.43mg	Z	P<.0005+
100	M	f	cvy	eat	liv	mix	30m30	er	: + :	2.02mg	Z	P<.0005+
a	M	f	cvy	eat	liv	hpc	30m30	er		7.94mg	Z	P<.0005+
b	M	f	cvy	eat	ubl	mix	30m30	er		69.0mg	Z	P<.0005+
c	M	f	cvy	eat	ubl	tcc	30m30	er		69.0mg	Z	P<.0005+
101	M	m	cvy	eat	ubl	mix	30m30	er	: + :	1.56mg	Z	P<.0005+
a	M	m	cvy	eat	liv	mix	30m30	er		4.23mg	*	P<.0005+
b	M	m	cvy	eat	ubl	tcc	30m30	er		4.30mg	Z	P<.0005+
102	M	f	ifc	eat	liv	mix	64w64		. + .	4.78mg		P<.0005+
a	M	f	ifc	eat	liv	hpt	64w64			5.95mg		P<.0005
b	M	f	ifc	eat	ubl	car	64w64			565.mg		P<.4
c	M	f	ifc	eat	ubl	pam	64w64			565.mg		P<.4 +
103	M	m	ifc	eat	ubl	car	64w64		. + .	36.1mg		P<.005 +
a	M	m	ifc	eat	liv	mix	64w64			36.1mg		P<.005 +
b	M	m	ifc	eat	liv	hpt	64w64			43.7mg		P<.01
c	M	m	ifc	eat	ubl	pam	64w64			69.8mg		P<.04 +
104	M	f	ifm	eat	liv	mix	60w60		. + .	19.6mg		P<.0005+
a	M	f	ifm	eat	liv	hpt	60w60			31.5mg		P<.0005
b	M	f	ifm	eat	ubl	mix	60w60			44.9mg		P<.0005+
105	M	f	ifm	gav	liv	ade	60w60		.>	91.2mg		P<.2
106	M	m	ifm	eat	liv	mix	60w60		. + .	42.9mg		P<.002 +
a	M	m	ifm	eat	liv	hpt	60w60			77.8mg		P<.004
107	M	m	ifm	gav	ubl	mix	60w60		. + .	12.9mg		P<.0005+
a	M	m	ifm	gav	liv	mix	60w60			57.2mg		P<.3
108	M	f	r3m	eat	ubl	mix	60w60		. ±	104.mg		P<.07 +
a	M	f	r3m	eat	liv	mix	60w60			no dre		P=1.
109	M	m	r3m	eat	ubl	mix	60w60		. ±	51.0mg		P<.04 +
a	M	m	r3m	eat	ubl	car	60w60			86.0mg		P<.1
b	M	m	r3m	eat	liv	mix	60w60			no dre		P=1. +
110	P	b	rhe	eat	npl	adc	5y26	w	: ±	33.0mg		P<.1
a	P	b	rhe	eat	tba	mal	5y26	w		no dre		P=1.
111	R	f	cdr	eat	mgl	mix	15m24	ae	. + .	4.75mg	*	P<.0005+
a	R	f	cdr	eat	liv	hpc	15m24	ae		26.9mg	*	P<.02 +
112	R	m	cdr	eat	liv	hpc	15m24	ae	. + .	3.78mg	*	P<.0005+
113	R	m	f3d	eat	liv	hpc	24m24	e	. +.	.640mg	*	P<.0005+
a	R	m	f3d	eat	liv	thc	24m24	e		.697mg	*	P<.0005
b	R	m	f3d	eat	liv	ghc	24m24	e		no dre		P=1.
c	R	m	f3d	eat	liv	clc	24m24	e		no dre		P=1.
114	R	f	hza	eat	mam	tum	36w54	es	. + .	.841mg		P<.0005+
a	R	f	hza	eat	eac	car	36w54	es		13.8mg		P<.4
b	R	f	hza	eat	liv	tum	36w54	es		no dre		P=1.
115	R	f	hza	eat	mam	tum	44w60	es	. + .	.981mg		P<.0005+
a	R	f	hza	eat	eac	car	44w60	es		8.75mg		P<.02
b	R	f	hza	eat	liv	tum	44w60	es		no dre		P=1.
116	R	f	lev	eat	liv	mix	56w56		. + .	1.53mg		P<.0005+
a	R	f	lev	eat	ski	mix	56w56			12.3mg		P<.0005+
b	R	f	lev	eat	mgl	mix	56w56			no dre		P=1.
117	R	m	lev	eat	liv	mix	56w56	s	.+ .	1.17mg		P<.0005+
a	R	m	lev	eat	ski	mix	56w56	s		20.1mg		P<.09 +
b	R	m	lev	eat	mgl	tum	56w56	s		54.4mg		P<.06 +

	RefNum	LoConf	UpConf	Cntrl	1Dose	1Inc	2Dose	2Inc			Citation or Pathology / Brkly Code
b	2066m	27.1mg	n.s.s.	19/92	2.40mg	18/90	4.80mg	15/88	7.20mg	19/87	
c	2066m	64.3mg	n.s.s.	9/92	2.40mg	5/90	4.80mg	7/88	7.20mg	5/87	
d	2066m	25.9mg	n.s.s.	1/96	2.40mg	0/96	4.80mg	0/96	7.20mg	0/96	
90	2066n	6.85mg	12.3mg	0/90	2.40mg	0/91	4.80mg	23/91	7.20mg	57/95	
a	2066n	17.3mg	43.2mg	0/90	2.40mg	0/91	4.80mg	9/91	7.20mg	23/95	
b	2066n	22.2mg	n.s.s.	5/93	2.40mg	4/94	4.80mg	10/94	(7.20mg	1/95)	
c	2066n	164.mg	n.s.s.	3/93	2.40mg	1/94	4.80mg	2/94	7.20mg	0/95	
91	66a	13.5mg	37.3mg	1/57	32.5mg	30/66					Epstein(review) {irdc};stev,6,103-154;1976
a	66a	37.1mg	218.mg	0/57	32.5mg	10/66					
b	66a	56.0mg	n.s.s.	6/57	32.5mg	9/66					
92	1635m	5.10mg	11.9mg	0/60	22.5mg	46/58					Weikel;jcph,19,591-604;1979/pers.comm.
93	66a	12.7mg	56.5mg	4/47	30.0mg	24/53					Epstein(review) {irdc};stev,6,103-154;1976
a	66a	155.mg	n.s.s.	8/47	30.0mg	1/53					
94	1635m	8.67mg	22.7mg	1/57	20.7mg	33/58					Weikel;jcph,19,591-604;1979/pers.comm.
a	1635m	19.2mg	71.1mg	0/57	20.7mg	16/58					
b	1635m	19.2mg	71.1mg	0/57	20.7mg	16/58					
c	1635m	37.3mg	375.mg	0/57	20.7mg	7/58					
d	1635m	53.1mg	n.s.s.	6/57	20.7mg	7/58					
95	1477	1.13mg	n.s.s.	5/8	36.0mg	7/8					Becker;canr,42,3918-3923;1982
96	469	7.28mg	36.3mg	0/29	39.0mg	11/30					Newberne;txap,41,535-546;1977
a	469	7.98mg	43.7mg	0/29	39.0mg	10/30					
b	469	60.3mg	n.s.s.	0/29	39.0mg	0/30					
97	469	11.9mg	290.mg	0/31	36.0mg	5/28					
a	469	22.3mg	n.s.s.	1/31	36.0mg	2/28					
98	2034u	2.73mg	7.59mg	4/24	6.50mg	14/24	13.0mg	22/24	26.0mg	23/24	Wolff;jtxe,33,327-348;1991/pers.comm.
a	2034u	5.71mg	23.4mg	2/24	6.50mg	7/24	13.0mg	14/24	26.0mg	9/24	
b	2034u	26.5mg	125.mg	0/24	6.50mg	0/24	13.0mg	0/24	26.0mg	12/24	
c	2034u	28.8mg	161.mg	0/24	6.50mg	0/24	13.0mg	0/24	26.0mg	10/24	
99	2034s	2.13mg	5.95mg	0/24	6.00mg	9/24	12.0mg	22/24	24.0mg	24/24	
a	2034s	2.95mg	8.96mg	0/24	6.00mg	5/24	12.0mg	17/24	24.0mg	21/24	
b	2034s	3.53mg	22.2mg	8/24	6.00mg	6/24	12.0mg	10/24	24.0mg	9/24	
100	2034w	1.12mg	4.04mg	16/24	6.50mg	23/24	13.0mg	23/24	26.0mg	24/24	
a	2034w	4.06mg	14.7mg	0/24	6.50mg	6/24	13.0mg	10/24	26.0mg	11/24	
b	2034w	31.1mg	188.mg	0/24	6.50mg	0/24	13.0mg	1/24	26.0mg	8/24	
c	2034w	31.1mg	188.mg	0/24	6.50mg	0/24	13.0mg	1/24	26.0mg	8/24	
101	2034v	.643mg	3.35mg	0/24	6.00mg	8/24	12.0mg	15/21	24.0mg	18/23	
a	2034v	1.76mg	15.9mg	14/24	6.00mg	11/24	12.0mg	11/21	24.0mg	7/23	
b	2034v	1.81mg	8.38mg	0/24	6.00mg	3/24	12.0mg	12/21	24.0mg	15/23	
102	1447	2.16mg	9.44mg	0/17	65.0mg	33/34					Wood;ejca,6,433-440;1970
a	1447	3.13mg	11.1mg	0/17	65.0mg	32/34					
b	1447	92.0mg	n.s.s.	0/17	65.0mg	1/34					
c	1447	92.0mg	n.s.s.	0/17	65.0mg	1/34					
103	1447	15.5mg	260.mg	0/13	60.0mg	7/20					
a	1447	15.5mg	260.mg	0/13	60.0mg	7/20					
b	1447	17.7mg	1.94gm	0/13	60.0mg	6/20					
c	1447	24.0mg	n.s.s.	0/13	60.0mg	4/20					
104	1069m	10.9mg	39.8mg	0/31	65.0mg	17/32					Wood;ejca,5,41-47;1969
a	1069m	16.1mg	74.4mg	0/31	65.0mg	12/32					
b	1069m	21.1mg	131.mg	0/31	65.0mg	9/32					
105	1069m	14.8mg	n.s.s.	0/31	22.9mg	1/18					
106	1069m	19.3mg	183.mg	1/42	60.0mg	9/31					
a	1069m	29.5mg	475.mg	0/42	60.0mg	5/31					
107	1069m	5.23mg	46.5mg	0/42	19.0mg	6/21					
a	1069m	11.2mg	n.s.s.	1/42	19.0mg	2/21					
108	1069m	35.7mg	n.s.s.	0/15	65.0mg	4/30					
a	1069m	134.mg	n.s.s.	0/15	65.0mg	0/30					
109	1069m	23.0mg	n.s.s.	0/10	60.0mg	8/34					
a	1069m	32.6mg	n.s.s.	0/10	60.0mg	5/34					
b	1069m	54.4mg	n.s.s.	1/10	60.0mg	3/34					
110	2001	5.37mg	n.s.s.	0/15	6.77mg	1/5					Adamson;ossc,129-156; 1982/Thorgeirsson 1994/Dalgard 1991/Thorgeirsson&Seiber pers.comm.
a	2001	9.47mg	n.s.s.	5/76	6.77mg	1/10					
111	1112	2.48mg	10.4mg	0/32	3.00mg	8/16	12.5mg	7/10			Weisburger;jnci,67,75-88;1981
a	1112	9.24mg	n.s.s.	0/32	3.00mg	2/16	12.5mg	2/10			
112	1112	1.78mg	9.52mg	1/32	2.40mg	3/16	10.0mg	10/10			
113	1756	.385mg	1.11mg	0/30	.320mg	3/29	1.60mg	26/28	8.00mg	23/23	Ogiso;txpy,13,257-265;1985
a	1756	.421mg	1.21mg	0/30	.320mg	3/29	1.60mg	25/28	8.00mg	23/23	
b	1756	28.2mg	n.s.s.	0/30	.320mg	0/29	1.60mg	1/28	8.00mg	0/23	
c	1756	1.52mg	n.s.s.	0/30	.320mg	0/29	1.60mg	0/28	8.00mg	0/23	
114	1063m	.286mg	2.98mg	0/5	10.0mg	8/9					Morris;canr,29,2145-2156;1969
a	1063m	2.24mg	n.s.s.	0/5	10.0mg	1/9					
b	1063m	5.00mg	n.s.s.	0/5	10.0mg	0/9					
115	1063n	.367mg	2.96mg	3/16	11.0mg	15/16					
a	1063n	3.00mg	n.s.s.	0/16	11.0mg	4/16					
b	1063n	12.1mg	n.s.s.	0/16	11.0mg	0/16					
116	1635m	1.05mg	2.28mg	0/60	15.0mg	60/70					Weikel;jcph,19,591-604;1979/pers.comm.
a	1635m	6.60mg	33.3mg	1/60	15.0mg	16/70					
b	1635m	16.4mg	n.s.s.	16/60	15.0mg	14/70					
117	1635m	.796mg	1.76mg	1/60	12.0mg	61/70					
a	1635m	7.46mg	n.s.s.	5/60	12.0mg	13/70					
b	1635m	16.5mg	n.s.s.	0/60	12.0mg	3/70					

```
     Spe Strain  Site   Xpo+Xpt                                                          TD50    2Tailpvl
       Sex  Route   Hist      Notes                                                        DR     AuOp
4-ACETYLAMINOFLUORENE            100ng..:..1ug....:..10....:..100....:..1mg....:..10....:..100....:..1g.....:..10
118  R f buf eat --- ile 62w97 e                                          .>               59.4mg   P<.2
 a   R f buf eat mgl adf 62w97 e                                                           59.4mg   P<.2
 b   R f buf eat ute sar 62w97 e                                                           59.4mg   P<.2

ACETYLATED DIAMYLOPECTIN PHOSPHATE    ..:..1ug....:..10....:..100....:..1mg....:..10....:..100....:..1g.....:..10
119  R f wis eat liv tum 24m24 e                                                           no dre   P=1.   -
 a   R f wis eat tba mix 24m24 e                                                           no dre   P=1.   -
120  R m wis eat liv tum 24m24 e                                                           no dre   P=1.   -
 a   R m wis eat tba mix 24m24 e                                                           no dre   P=1.   -

ACETYLATED DISTARCH ADIPATE      100ng..:..1ug....:..10....:..100....:..1mg....:..10....:..100....:..1g.....:..10
121  R f ofs eat liv tum 24m24 eg                                                          no dre   P=1.   -
 a   R f ofs eat tba mix 24m24 eg                                                          no dre   P=1.   -
122  R m ofs eat liv hem 24m24 e                                                           no dre   P=1.   -
 a   R m ofs eat tba mix 24m24 e                                                           14.0gm   P<.06  -

ACETYLATED DISTARCH GLYCEROL     100ng..:..1ug....:..10....:..100....:..1mg....:..10....:..100....:..1g.....:..10
123  R f ofs eat liv tum 24m24 eg                                                          no dre   P=1.   -
 a   R f ofs eat tba mix 24m24 eg                                                          no dre   P=1.   -
124  R m ofs eat liv hem 24m24 e                                                           no dre   P=1.   -
 a   R m ofs eat tba mix 24m24 e                                                           26.1gm   P<.3   -

ACETYLATED DISTARCH PHOSPHATE    100ng..:..1ug....:..10....:..100....:..1mg....:..10....:..100....:..1g.....:..10
125  R f wis eat liv tum 24m24 e                                                           no dre   P=1.   -
 a   R f wis eat tba mix 24m24 e                                                           20.4gm   P<.4   -
126  R m wis eat liv tum 24m24 e                                                           no dre   P=1.   -
 a   R m wis eat tba mix 24m24 e                                                           37.6gm   P<.8   -

ACIFLUORFEN                      100ng..:..1ug....:..10....:..100....:..1mg....:..10....:..100....:..1g.....:..10
127  M f b6c eat liv mix 78w78 e                                          . + .            328.mg * P<.0005+
 a   M f b6c eat liv hpa 78w78 e                                                           434.mg * P<.0005+
 b   M f b6c eat sto pam 78w78 e                                                           1.05gm * P<.002 +
 c   M f b6c eat liv hpc 78w78 e                                                           1.38gm * P<.007 +
128  M m b6c eat liv mix 78w78 es                                      . + .               149.mg * P<.0005+
 a   M m b6c eat liv hpa 78w78 es                                                          331.mg * P<.003 +
 b   M m b6c eat liv hpc 78w78 es                                                          395.mg * P<.0005+
 c   M m b6c eat sto pam 78w78 es                                                          2.05gm * P<.008 +
129  M f cd1 eat liv mix 24m24 ev                                     . + .                126.mg * P<.003 +
 a   M f cd1 eat liv hpa 24m24 ev                                                          160.mg * P<.005 +
 b   M f cd1 eat liv hpc 24m24 ev                                                          715.mg * P<.3
130  M m cd1 eat liv mix 24m24 ev                                        . ±               93.8mg * P<.06  +
 a   M m cd1 eat liv hpa 24m24 ev                                                          187.mg * P<.2
 b   M m cd1 eat liv hpc 24m24 ev                                                          307.mg * P<.4

ACROLEIN                         100ng..:..1ug....:..10....:..100....:..1mg....:..10....:..100....:..1g.....:..10
131  R f f34 wat adr coa 24m31 e                                          . ±              93.9mg   P<.07
 a   R f f34 wat liv mix 24m31 e                                                           188.mg   P<.4
 b   R f f34 wat liv hnd 24m31 e                                                           388.mg   P<.7
 c   R f f34 wat liv hpc 24m31 e                                                           433.mg   P<.3
132  R m f34 wat for tum 27m31 ae                                   . + .                  14.4mg Z P<.002
 a   R m f34 wat liv hpc 27m31 ae                                                          1.32gm * P<.9
 b   R m f34 wat adr coa 27m31 ae                                                          no dre   P=1.
 c   R m f34 wat liv hnd 27m31 ae                                                          no dre   P=1.
 d   R m f34 wat liv mix 27m31 ae                                                          no dre   P=1.

ACROLEIN DIETHYLACETAL           100ng..:..1ug....:..10....:..100....:..1mg....:..10....:..100....:..1g.....:..10
133  R f f34 wat liv mix 24m30 ae                                        . ±               123.mg * P<.06
 a   R f f34 wat liv hnd 24m30 ae                                                          131.mg * P<.06
 b   R f f34 wat adr coa 24m30 ae                                                          no dre   P=1.
 c   R f f34 wat liv hpc 24m30 ae                                                          no dre   P=1.
134  R m f34 wat liv hpc 24m30 ae                                           . ±            262.mg * P<.1
 a   R m f34 wat liv mix 24m30 ae                                                          124.mg * P<.2
 b   R m f34 wat liv hnd 24m30 ae                                                          266.mg * P<.4
 c   R m f34 wat adr coa 24m30 ae                                                          no dre   P=1.

ACROLEIN OXIME                   100ng..:..1ug....:..10....:..100....:..1mg....:..10....:..100....:..1g.....:..10
135  R f f34 wat liv hnd  7m29 e                                          .>               41.9mg   P<.3
 a   R f f34 wat adr coa  7m29 e                                                           no dre   P=1.
136  R f f34 wat liv mix 24m30 e                                       . ±                 27.1mg   P<.03
 a   R f f34 wat liv hpc 24m30 e                                                           104.mg   P<.1
 b   R f f34 wat liv hnd 24m30 e                                                           43.8mg   P<.2
 c   R f f34 wat adr coa 24m30 e                                                           no dre   P=1.
137  R m f34 wat liv mix  7m29 e                                      . ±                  16.4mg   P<.06
 a   R m f34 wat liv hpc  7m29 e                                                           50.7mg   P<.1
 b   R m f34 wat liv hnd  7m29 e                                                           29.3mg   P<.3
 c   R m f34 wat adr coa  7m29 e                                                           no dre   P=1.
138  R m f34 wat liv hpc 24m30 e                                       . ±                 72.6mg   P<.1
 a   R m f34 wat liv mix 24m30 e                                                           30.4mg   P<.2
 b   R m f34 wat liv hnd 24m30 e                                                           64.9mg   P<.4
 c   R m f34 wat adr coa 24m30 e                                                           no dre   P=1.
```

RefNum	LoConf	UpConf	Cntrl	1Dose	1Inc	2Dose	2Inc			Citation or Pathology
										Brkly Code

4-ACETYLAMINOFLUORENE   (N-4-fluorenylacetamide)   28322-02-3

										Morris;jnci,24,149-180;1960
118	144	9.67mg n.s.s.	0/18	7.98mg	1/13					
a	144	9.67mg n.s.s.	0/18	7.98mg	1/13					
b	144	9.67mg n.s.s.	0/18	7.98mg	1/13					

ACETYLATED DIAMYLOPECTIN PHOSPHATE   68130-14-3

										de Groot;fctx,12,651-663;1974
119	1407	86.6gm n.s.s.	0/29	15.0gm	0/28					
a	1407	12.6gm n.s.s.	21/29	15.0gm	18/28					
120	1407	69.2gm n.s.s.	0/30	12.0gm	0/28					
a	1407	4.62gm n.s.s.	26/30	12.0gm	24/28					

ACETYLATED DISTARCH ADIPATE   ---

										Truhaut;fctx,17,11-17;1979
121	1408	179.gm n.s.s.	0/24	31.0gm	0/28					
a	1408	20.6gm n.s.s.	24/24	31.0gm	23/28					
122	1408	107.gm n.s.s.	1/25	24.8gm	0/21					
a	1408	4.44gm n.s.s.	17/25	24.8gm	19/21					

ACETYLATED DISTARCH GLYCEROL   53123-84-5

										Truhaut;fctx,17,11-17;1979
123	1408	172.gm n.s.s.	0/24	31.0gm	0/27					
a	1408	20.9gm n.s.s.	24/24	31.0gm	22/27					
124	1408	123.gm n.s.s.	1/25	24.8gm	0/24					
a	1408	7.28gm n.s.s.	17/25	24.8gm	20/24					

ACETYLATED DISTARCH PHOSPHATE   68130-14-3

										de Groot;fctx,12,651-663;1974
125	1407	92.7gm n.s.s.	0/29	15.0gm	0/30					
a	1407	5.12gm n.s.s.	21/29	15.0gm	25/30					
126	1407	69.2gm n.s.s.	0/30	12.0gm	0/28					
a	1407	3.48gm n.s.s.	26/30	12.0gm	25/28					

ACIFLUORFEN   50594-66-6

										Quest;rtxp,10,149-159;1989
127	2129m	212.mg 599.mg	1/55	81.4mg	6/59	162.mg	5/57	325.mg	24/58	
a	2129m	265.mg 926.mg	1/55	81.4mg	5/59	162.mg	4/57	325.mg	19/58	
b	2129m	493.mg 3.74gm	0/45	81.4mg	0/48	162.mg	3/44	325.mg	6/45	
c	2129m	594.mg 17.3gm	0/45	81.4mg	1/47	162.mg	1/44	325.mg	5/46	
128	2129m	97.9mg 283.mg	9/58	75.1mg	21/60	150.mg	16/56	300.mg	40/59	
a	2129m	174.mg 1.91gm	8/58	75.1mg	18/60	150.mg	12/56	300.mg	25/59	
b	2129m	229.mg 917.mg	1/48	75.1mg	3/50	150.mg	4/45	300.mg	15/44	
c	2129m	709.mg 39.7gm	0/49	75.1mg	0/46	150.mg	0/43	300.mg	4/40	
129	2129n	57.4mg 892.mg	7/80	.975mg	5/69	5.85mg	4/80	29.6mg	15/66	
a	2129n	69.1mg 1.93gm	5/80	.975mg	2/69	5.85mg	4/80	29.6mg	11/66	
b	2129n	157.mg n.s.s.	2/80	.975mg	3/69	5.85mg	0/80	29.6mg	4/66	
130	2129n	35.4mg n.s.s.	19/79	.900mg	18/69	5.40mg	28/80	27.3mg	27/70	
a	2129n	61.3mg n.s.s.	9/79	.900mg	3/69	5.40mg	14/80	27.3mg	12/70	
b	2129n	66.6mg n.s.s.	10/79	.900mg	15/69	5.40mg	14/80	27.3mg	15/70	

ACROLEIN   107-02-8

										Lijinsky;txih,3,337-345;1987/pers.comm.
131	1853	30.8mg n.s.s.	1/20	20.1mg	5/20					
a	1853	39.2mg n.s.s.	2/20	20.1mg	4/20					
b	1853	48.5mg n.s.s.	2/20	20.1mg	3/20					
c	1853	70.4mg n.s.s.	0/20	20.1mg	1/20					
132	1853	7.36mg 55.4mg	0/20	2.64mg	5/20	6.81mg	7/20	(14.1mg	3/20)	
a	1853	57.7mg n.s.s.	0/20	2.64mg	2/20	6.81mg	0/20	14.1mg	1/20	
b	1853	128.mg n.s.s.	1/20	2.64mg	1/20	6.81mg	0/20	14.1mg	0/20	
c	1853	64.1mg n.s.s.	2/20	2.64mg	6/20	6.81mg	0/20	14.1mg	2/20	
d	1853	52.1mg n.s.s.	2/20	2.64mg	8/20	6.81mg	0/20	14.1mg	3/20	

ACROLEIN DIETHYLACETAL   3054-95-3

										Lijinsky;txih,3,337-345;1987/pers.comm.
133	1853	47.6mg n.s.s.	2/20	15.4mg	5/20	39.5mg	7/20			
a	1853	50.7mg n.s.s.	2/20	15.4mg	4/20	39.5mg	7/20			
b	1853	129.mg n.s.s.	1/20	15.4mg	3/20	39.5mg	1/20			
c	1853	197.mg n.s.s.	0/20	15.4mg	1/20	39.5mg	0/20			
134	1853	79.3mg n.s.s.	0/20	10.8mg	1/20	27.6mg	2/20			
a	1853	42.6mg n.s.s.	2/20	10.8mg	3/20	27.6mg	6/20			
b	1853	60.9mg n.s.s.	2/20	10.8mg	2/20	27.6mg	4/20			
c	1853	49.9mg n.s.s.	1/20	10.8mg	0/20	27.6mg	0/20			

ACROLEIN OXIME   5314-33-0

										Lijinsky;txih,3,337-345;1987/pers.comm.
135	1853m	11.3mg n.s.s.	2/20	7.96mg	5/20					
a	1853m	45.9mg n.s.s.	0/20	7.96mg	0/20					
136	1853n	10.4mg n.s.s.	2/20	10.4mg	8/20					
a	1853n	25.7mg n.s.s.	0/20	10.4mg	2/20					
b	1853n	13.8mg n.s.s.	2/20	10.4mg	6/20					
c	1853n	66.2mg n.s.s.	1/20	10.4mg	0/20					
137	1853m	5.80mg n.s.s.	2/20	5.57mg	7/20					
a	1853m	12.5mg n.s.s.	0/20	5.57mg	2/20					
b	1853m	7.88mg n.s.s.	2/20	5.57mg	5/20					
c	1853m	19.2mg n.s.s.	1/20	5.57mg	1/20					
138	1853n	17.8mg n.s.s.	0/20	7.37mg	2/20					
a	1853n	9.61mg n.s.s.	2/20	7.37mg	6/20					
b	1853n	13.5mg n.s.s.	2/20	7.37mg	4/20					
c	1853n	46.0mg n.s.s.	1/20	7.37mg	0/20					

```
     Spe Strain  Site   Xpo+Xpt                                                                         TD50    2Tailpvl
         Sex  Route   Hist    Notes                                                                      DR      AuOp
ACRONYCINE                    100ng..:...1ug....:...10.....:...100....:...1mg....:...10.....:...100....:...1g......:..10
139  M f b6c ipj TBA MXB 48w78 as                                              :>                        no dre  P=1.    i
  a  M f b6c ipj liv MXB 48w78 as                                                                        2.46mg * P<.3
  b  M f b6c ipj lun MXB 48w78 as                                                                        2.46mg * P<.3
140  M m b6c ipj TBA MXB 66w78 as                                              :>                        no dre  P=1.    i
  a  M m b6c ipj liv MXB 66w78 as                                                                        no dre  P=1.
  b  M m b6c ipj lun MXB 66w78 as                                                                        no dre  P=1.
141  R f sda ipj mgl MXA 52w80 e                               : + :                                     .395mg Z P<.003 c
  a  R f sda ipj mgl fba 52w80 e                                                                         .455mg Z P<.005 c
  b  R f sda ipj MXB MXB 52w80 e                                                                         .873mg Z P<.0005
  c  R f sda ipj per MXA 52w80 e                                                                         5.86mg * P<.0005c
  d  R f sda ipj per srn 52w80 e                                                                         11.4mg Z P<.003 c
  e  R f sda ipj TBA MXB 52w80 e                                                                         .568mg Z P<.0005
  f  R f sda ipj liv MXB 52w80 e                                                                         8.75mg * P<.06
142  R m sda ipj MXB MXB 51w80 ae                              : + :                                     .497mg Z P<.0005
  a  R m sda ipj --- ost 51w80 ae                                                                        .698mg Z P<.0005c
  b  R m sda ipj MXA ost 51w80 ae                                                                        1.03mg * P<.0005c
  c  R m sda ipj per MXA 51w80 ae                                                                        1.49mg * P<.0005c
  d  R m sda ipj adr coa 51w80 ae                                                                        1.85mg Z P<.0005
  e  R m sda ipj TBA MXB 51w80 ae                                                                        .227mg Z P<.0005
  f  R m sda ipj liv MXB 51w80 ae                                                                        2.11mg * P<.06

ACRYLAMIDE                    100ng..:...1ug....:...10.....:...100....:...1mg....:...10.....:...100....:...1g......:..10
143  R f f34 wat mgl ben 24m24 e                                           .  +   .                      4.21mg * P<.002 +
  a  R f f34 wat mgl fib 24m24 e                                                                         20.9mg * P<.0005+
  b  R f f34 wat orc sqp 24m24 e                                                                         16.2mg * P<.02  +
  c  R f f34 wat mgl adc 24m24 e                                                                         17.8mg * P<.02  +
  d  R f f34 wat tyf mix 24m24 e                                                                         24.4mg * P<.03
  e  R f f34 wat pit ade 24m24 e                                                                         13.2mg * P<.5   +
  f  R f f34 wat mgl fba 24m24 e                                                                         16.5mg * P<.3   +
144  R m f34 wat tyf ade 24m24 e                                                  .   +    .             13.2mg * P<.003
  a  R m f34 wat tnv msm 24m24 e                                                                         11.4mg * P<.03  +
  b  R m f34 wat spd ast 24m24 e                                                                         39.7mg * P<.05  +
  c  R m f34 wat adr phe 24m24 e                                                                         16.6mg * P<.2   +
  d  R m f34 wat spl leu 24m24 e                                                                         45.8mg * P<.8

ACRYLONITRILE                 100ng..:...1ug....:...10.....:...100....:...1mg....:...10.....:...100....:...1g......:..10
145  R m cdr wat zym sqc 24m24 es                                                   .  +   .             30.1mg * P<.0005+
  a  R m cdr wat for pam 24m24 s                                                                         97.4mg * P<.003
  b  R m cdr wat pit ade 24m24 es                                                                        no dre  P=1.
  c  R m cdr wat tba mal 24m24 es                                                                        31.3mg * P<.02
146  R f sda inh adr phe 12m24                                       .  ±                                1.49mg Z P<.02
  a  R f sda inh bra gli 12m24                                                                           41.2mg * P<.2   +
  b  R f sda inh liv hpt 12m24                                                                           no dre  P=1.
  c  R f sda inh tba mix 12m24                                                                           2.09mg Z P<.3
  d  R f sda inh tba mal 12m24                                                                           no dre  P=1.
147  R f sda gav liv hpt 12m24                                              .>                           no dre  P=1.    -
  a  R f sda gav tba mix 12m24                                                                           2.97mg  P<.3    -
  b  R f sda gav tba mal 12m24                                                                           no dre  P=1.    -
148  R f sda inh mam mix 24m24 g                                        .  +   .                         11.7mg  P<.003
  a  R f sda inh bra gli 24m24 g                                                                         132.mg  P<.04   +
  b  R f sda inh liv hpt 24m24 g                                                                         no dre  P=1.
  c  R f sda inh tba mal 24m24 g                                                                         25.2mg  P<.008 +
  d  R f sda inh tba mix 24m24 g                                                                         10.6mg  P<.02
149  R m sda inh bra gli 12m24                                              .   ±                        19.1mg * P<.04  +
  a  R m sda inh liv hpt 12m24                                                                           no dre  P=1.
  b  R m sda inh tba mix 12m24                                                                           1.30mg * P<.002
  c  R m sda inh tba mal 12m24                                                                           1.43mg Z P<.04
150  R m sda gav liv hpt 12m24                                              .>                           no dre  P=1.    -
  a  R m sda gav tba mix 12m24                                                                           5.59mg  P<.3    -
  b  R m sda gav tba mal 12m24                                                                           33.4mg  P<.8    -
151  R f sss inh cns ast 24m24 aes                                                .  +  .                49.6mg * P<.0005+
  a  R f sss inh ccx ast 24m24 aes                                                                       62.0mg * P<.0005
  b  R f sss inh zym mix 24m24 aes                                                                       81.5mg * P<.0005+
  c  R f sss inh zym ssc 24m24 aes                                                                       89.0mg * P<.0005
  d  R f sss inh brs ast 24m24 aes                                                                       123.mg * P<.002
  e  R f sss inh ntu rsc 24m24 aes                                                                       52.0mg * P<.06  +
  f  R f sss inh mgl adc 24m24 aes                                                                       69.1mg * P<.02  +
  g  R f sss inh liv clc 24m24 aes                                                                       362.mg * P<.9
  h  R f sss inh liv cho 24m24 aes                                                                       362.mg * P<.9
  i  R f sss inh mgl fba 24m24 aes                                                                       no dre  P=1.    +
  j  R f sss inh tba mix 24m24 aes                                                                       no dre  P=1.
152  R f sss wat cns ast 24m25 aesv                                          .  +  .                     5.31mg Z P<.0005+
  a  R f sss wat smi muc 24m25 aesv                                                                      8.19mg Z P<.0005+
  b  R f sss wat brs ast 24m25 aesv                                                                      11.4mg Z P<.0005
  c  R f sss wat stn mix 24m25 aesv                                                                      13.3mg * P<.0005+
  d  R f sss wat ccx ast 24m25 aesv                                                                      13.6mg * P<.0005
  e  R f sss wat stn sqp 24m25 aesv                                                                      13.8mg * P<.0005
  f  R f sss wat zym mix 24m25 aesv                                                                      22.6mg * P<.0005+
  g  R f sss wat zym ssc 24m25 aesv                                                                      31.4mg * P<.0005
  h  R f sss wat mgl mal 24m25 aesv                                                                      42.6mg * P<.004 +
```

	RefNum	LoConf	UpConf	Cntrl	1Dose	1Inc	2Dose	2Inc					Citation or Pathology	Brkly Code

ACRONYCINE 7008-42-6

	RefNum	LoConf	UpConf	Cntrl	1Dose	1Inc	2Dose	2Inc	3Dose	3Inc	Citation or Pathology	Brkly Code
139	TR49	.866mg	n.s.s.	20/30	.850mg	9/40	(2.57mg	0/35)				
a	TR49	.400mg	n.s.s.	0/30	.850mg	1/40	2.57mg	0/35			liv:hpa,hpc,nnd.	
b	TR49	.400mg	n.s.s.	0/30	.850mg	1/40	2.57mg	0/35			lun:a/a,a/c.	
140	TR49	1.13mg	n.s.s.	18/30	.850mg	14/40	(2.57mg	0/35)				
a	TR49	2.77mg	n.s.s.	2/30	.850mg	0/40	2.57mg	0/35			liv:hpa,hpc,nnd.	
b	TR49	n.s.s.	n.s.s.	0/30	.850mg	1/40	2.57mg	0/35			lun:a/a,a/c.	
141	TR49	.200mg	2.10mg	4/20	1.04mg	22/35	(2.09mg	16/35	4.20mg	5/35)	mgl:acn,ccn,cyn,fba,pac.	
a	TR49	.222mg	3.99mg	4/20	1.04mg	20/35	(2.09mg	13/35	4.20mg	3/35)		
b	TR49	.495mg	3.02mg	4/20	1.04mg	23/35	2.09mg	17/35	4.20mg	12/35	mgl:acn,ccn,cyn,fba,pac; per:fbs,msm,srn.	C
c	TR49	2.80mg	17.7mg	0/20	1.04mg	1/35	2.09mg	2/35	4.20mg	7/35	per:fbs,msm,srn.	
d	TR49	4.20mg	61.8mg	0/20	1.04mg	0/35	2.09mg	0/35	4.20mg	5/35		
e	TR49	.353mg	1.24mg	7/20	1.04mg	31/35	2.09mg	20/35	4.20mg	25/35		
f	TR49	2.98mg	n.s.s.	0/20	1.04mg	0/35	2.09mg	4/35	4.20mg	0/35	liv:hpa,hpc,nnd.	
142	TR49	.290mg	.843mg	0/20	1.04mg	9/35	2.09mg	13/35	6.40mg	13/35	---:ost; bon:ost; per:fbs,men,msm,srn; ver:ost.	C
a	TR49	.390mg	1.21mg	0/20	1.04mg	4/35	2.09mg	13/35	6.40mg	12/35		
b	TR49	.551mg	1.95mg	0/20	1.04mg	3/35	2.09mg	11/35	6.40mg	8/35	bon:ost; ver:ost.	
c	TR49	.573mg	6.91mg	0/20	1.04mg	5/35	2.09mg	1/35	6.40mg	2/35	per:fbs,men,msm,srn.	
d	TR49	.624mg	9.40mg	1/20	1.04mg	1/35	2.09mg	2/35	6.40mg	4/35		S
e	TR49	.135mg	.412mg	4/20	1.04mg	18/35	2.09mg	16/35	6.40mg	16/35		
f	TR49	.519mg	n.s.s.	0/20	1.04mg	2/35	2.09mg	0/35	6.40mg	0/35	liv:hpa,hpc,nnd.	

ACRYLAMIDE (2-propenamide) 79-06-1

	RefNum	LoConf	UpConf	Cntrl	1Dose	1Inc	2Dose	2Inc	3Dose	3Inc	4Dose	4Inc	Citation or Pathology
143	1787	2.08mg	22.0mg	10/60	10.0ug	11/60	.100mg	9/60	.500mg	19/58	2.00mg	23/61	Johnson;txap,85,154-168;1986
a	1787	7.94mg	87.1mg	0/60	10.0ug	0/60	.100mg	0/60	.500mg	0/58	2.00mg	5/61	
b	1787	6.10mg	n.s.s.	0/60	10.0ug	3/60	.100mg	2/60	.500mg	1/60	2.00mg	7/61	
c	1787	6.46mg	n.s.s.	2/60	10.0ug	1/60	.100mg	1/60	.500mg	2/60	2.00mg	6/61	
d	1787	7.94mg	n.s.s.	1/58	10.0ug	0/59	.100mg	1/59	.500mg	1/58	2.00mg	4/60	
e	1787	2.50mg	n.s.s.	25/59	10.0ug	30/60	.100mg	32/60	.500mg	27/60	2.00mg	32/60	
f	1787	4.19mg	n.s.s.	9/60	10.0ug	10/60	.100mg	8/60	.500mg	14/58	2.00mg	13/61	
144	1787	5.50mg	89.4mg	1/60	10.0ug	0/58	.100mg	2/59	.500mg	1/59	2.00mg	7/59	
a	1787	4.29mg	n.s.s.	2/60	10.0ug	0/60	.100mg	5/60	.500mg	8/60	2.00mg	7/60	
b	1787	11.0mg	n.s.s.	1/60	10.0ug	0/60	.100mg	0/60	.500mg	0/60	2.00mg	3/60	
c	1787	5.04mg	n.s.s.	3/60	10.0ug	7/59	.100mg	7/60	.500mg	5/60	2.00mg	10/60	
d	1787	4.68mg	n.s.s.	10/60	10.0ug	20/60	.100mg	14/60	.500mg	14/60	2.00mg	16/60	

ACRYLONITRILE 107-13-1

	RefNum	LoConf	UpConf	Cntrl	1Dose	1Inc	2Dose	2Inc	3Dose	3Inc	4Dose	4Inc	Citation or Pathology
145	1881	14.5mg	77.9mg	0/18	1.00mg	0/20	5.00mg	1/19	25.0mg	9/18			Gallagher;jact,7,603-615;1988
a	1881	33.6mg	591.mg	0/20	1.00mg	0/20	5.00mg	0/20	25.0mg	4/20			
b	1881	15.4mg	n.s.s.	5/18	1.00mg	3/20	5.00mg	1/19	(25.0mg	0/18)			
c	1881	12.6mg	n.s.s.	3/18	1.00mg	1/20	5.00mg	8/20	25.0mg	8/18			
146	BT201	.679mg	n.s.s.	1/30	.271mg	5/30	.542mg	7/30	(1.08mg	2/30	2.17mg	0/30)	Maltoni;anya,534,179-203;1988
a	BT201	10.1mg	n.s.s.	0/30	.271mg	0/30	.542mg	0/30	1.08mg	1/30	2.17mg	1/30	
b	BT201	.893mg	n.s.s.	0/30	.271mg	0/30	.542mg	0/30	1.08mg	0/30	2.17mg	0/30	
c	BT201	.623mg	n.s.s.	9/30	.271mg	23/30	.542mg	15/30	1.08mg	17/30	(2.17mg	10/30)	
d	BT201	4.00mg	n.s.s.	3/30	.271mg	12/30	.542mg	6/30	1.08mg	7/30	2.17mg	6/30	
147	BT203	8.83mg	n.s.s.	0/75	1.07mg	0/40							
a	BT203	.782mg	n.s.s.	39/75	1.07mg	25/40							
b	BT203	2.33mg	n.s.s.	17/75	1.07mg	9/40							
148	BT4003	5.82mg	75.7mg	24/60	11.1mg	37/54							
a	BT4003	40.1mg	n.s.s.	0/60	11.1mg	3/54							
b	BT4003	123.mg	n.s.s.	0/60	11.1mg	0/54							
c	BT4003	11.7mg	495.mg	9/60	11.1mg	20/54							
d	BT4003	4.69mg	n.s.s.	35/60	11.1mg	43/54							
149	BT201	5.77mg	n.s.s.	0/30	.190mg	0/30	.379mg	0/30	.759mg	1/30	1.52mg	2/30	
a	BT201	.625mg	n.s.s.	0/30	.190mg	0/30	.379mg	0/30	.759mg	0/30	1.52mg	0/30	
b	BT201	.674mg	6.77mg	8/30	.190mg	7/30	.379mg	19/30	.759mg	15/30	1.52mg	19/30	
c	BT201	.565mg	n.s.s.	3/30	.190mg	0/30	.379mg	10/30	(.759mg	9/30	1.52mg	10/30)	
150	BT203	8.83mg	n.s.s.	0/75	1.07mg	0/40							
a	BT203	1.55mg	n.s.s.	13/75	1.07mg	11/40							
b	BT203	3.14mg	n.s.s.	6/75	1.07mg	4/40							
151	1251n	29.6mg	94.2mg	0/99	3.24mg	4/100	13.0mg	17/99					Quast;dcrp;1980
a	1251n	35.2mg	130.mg	0/99	3.24mg	3/99	13.0mg	14/99					
b	1251n	42.0mg	193.mg	0/93	3.24mg	1/98	13.0mg	11/89					
c	1251n	44.7mg	217.mg	0/93	3.24mg	0/98	13.0mg	11/89					
d	1251n	55.6mg	444.mg	0/93	3.24mg	1/96	13.0mg	7/88					
e	1251n	12.8mg	n.s.s.	0/11	3.24mg	0/10	13.0mg	2/10					
f	1251n	31.7mg	13.1gm	9/99	3.24mg	8/100	13.0mg	20/99					
g	1251n	58.9mg	n.s.s.	0/72	3.24mg	1/68	13.0mg	0/23					
h	1251n	58.9mg	n.s.s.	0/72	3.24mg	1/68	13.0mg	0/23					
i	1251n	21.6mg	n.s.s.	79/99	3.24mg	96/100	13.0mg	75/99					
j	1251n	9.07mg	n.s.s.	97/99	3.24mg	99/100	13.0mg	93/99					
152	1268n	3.52mg	8.79mg	1/80	2.00mg	17/48	5.69mg	22/48	(15.4mg	24/47)			Quast;dcfr;1980
a	1268n	3.07mg	34.3mg	0/79	2.00mg	1/7	5.69mg	4/10	(15.4mg	4/34)			
b	1268n	6.74mg	22.5mg	0/78	2.00mg	9/48	5.69mg	11/47	(15.4mg	8/46)			
c	1268n	9.08mg	20.8mg	1/80	2.00mg	1/47	5.69mg	12/48	15.4mg	30/47			
d	1268n	9.11mg	24.1mg	1/80	2.00mg	10/48	5.69mg	17/47	15.4mg	21/47			
e	1268n	9.38mg	21.7mg	1/80	2.00mg	1/47	5.69mg	12/48	15.4mg	29/47			
f	1268n	14.1mg	43.5mg	1/80	2.00mg	5/48	5.69mg	8/48	15.4mg	18/47			
g	1268n	18.3mg	71.7mg	1/80	2.00mg	4/48	5.69mg	6/48	15.4mg	14/47			
h	1268n	20.2mg	353.mg	6/80	2.00mg	7/48	5.69mg	9/48	15.4mg	13/47			

	Spe	Strain	Site	Xpo+Xpt			TD50		2Tailpvl
	Sex	Route	Hist	Notes			DR		AuOp
i	R f	sss wat	stn sqc	24m25	aesv		46.1mg	Z	P<.0005
j	R f	sss wat	crl ast	24m25	aesv		65.9mg	*	P<.002
k	R f	sss wat	spd ast	24m25	aesv		90.8mg	*	P<.005
l	R f	sss wat	zym ade	24m25	aesv		93.7mg	*	P<.004
m	R f	sss wat	liv mix	24m25	aesv		169.mg	*	P<.2
n	R f	sss wat	mgl mix	24m25	aesv		no dre		P=1. +
o	R f	sss wat	tba mix	24m25	aesv		1.75mg	*	P<.1
153	R f	sss wat	ton mix	22m25	esv	. + .	34.5mg		P<.0005+
a	R f	sss wat	ton sqc	22m25	esv		42.6mg	*	P<.0005
154	R m	sss inh	cns ast	24m24	aes	. + .	32.4mg	*	P<.0005+
a	R m	sss inh	itl mix	24m24	aes		37.8mg	*	P<.0005
b	R m	sss inh	ccx ast	24m24	aes		38.8mg	*	P<.0005
c	R m	sss inh	zym ssc	24m24	aes		50.1mg	*	P<.0005
d	R m	sss inh	zym mix	24m24	aes		53.0mg	*	P<.003 +
e	R m	sss inh	stn mix	24m24	aes		99.0mg	*	P<.02 +
f	R m	sss inh	liv hpc	24m24	aes		403.mg	*	P<.2
g	R m	sss inh	tba mix	24m24	aes		no dre		P=1.
155	R m	sss inh	smi muc	24m24	es	. + .	37.7mg		P<.0005+
a	R m	sss inh	ton sqp	24m24	es		73.0mg		P<.007
b	R m	sss inh	ton mix	24m24	es		89.1mg		P<.02 +
156	R m	sss wat	stn mix	24m25	aesv	. +.	6.36mg	*	P<.0005+
a	R m	sss wat	cns ast	24m25	aesv		10.7mg	*	P<.0005+
b	R m	sss wat	ccx ast	24m25	aesv		12.0mg	*	P<.0005
c	R m	sss wat	stn sqp	24m25	aesv		12.2mg	*	P<.0005
d	R m	sss wat	stn sqc	24m25	aesv		14.1mg	*	P<.0005
e	R m	sss wat	brs ast	24m25	aesv		18.6mg	Z	P<.0005
f	R m	sss wat	zym ssc	24m25	aesv		34.1mg	*	P<.0005+
g	R m	sss wat	smi muc	24m25	aesv		128.mg	Z	P<.3
h	R m	sss wat	liv hpc	24m25	aesv		no dre		P=1.
i	R m	sss wat	tba mix	24m25	aesv		5.55mg	*	P<.005
157	R m	sss wat	ton mix	96w96	esv	. ±	66.6mg		P<.02 +
a	R m	sss wat	ton sqc	96w96	esv		66.6mg		P<.02

ACTINOMYCIN C   100ng..:..1ug....:..10......:..100.....:..1mg....:..10......:..100.....:..1g.....:..10

	Spe Sex	Strain Route	Site Hist	Xpo+Xpt Notes		TD50 DR	2Tailpvl AuOp
158	R m	b46 ivj	tba mix	12m24 es	. ±	1.68mg	P<.08
a	R m	b46 ivj	tba ben	12m24 es		2.23mg	P<.07
b	R m	b46 ivj	tba mal	12m24 es		10.8mg	P<.7

ACTINOMYCIN D   100ng..:..1ug....:..10......:..100.....:..1mg....:..10......:..100.....:..1g.....:..10

	Spe Sex	Strain Route	Site Hist	Xpo+Xpt Notes		TD50 DR	2Tailpvl AuOp
159	R f	cdr ipj	per sar	26w78 e	. + .	1.99ug *	P<.0005+
a	R f	cdr ipj	lun ade	26w78 e		50.1ug *	P<.03
b	R f	cdr ipj	liv tum	26w78 e		no dre	P=1.
c	R f	cdr ipj	tba mal	26w78 e		1.21ug *	P<.0005
d	R f	cdr ipj	tba mix	26w78 e		1.45ug *	P<.003
e	R f	cdr ipj	tba ben	26w78 e		no dre	P=1.
160	R m	cdr ipj	per sar	26w78 e	. + .	769.ng	P<.0005+
a	R m	cdr ipj	--- plc	26w78 e		20.1ug	P<.008
b	R m	cdr ipj	--- lys	26w78 e		20.1ug	P<.008
c	R m	cdr ipj	liv hem	26w78 e		40.7ug	P<.06
d	R m	cdr ipj	tba mix	26w78 e		501.ng	P<.0005
e	R m	cdr ipj	tba mal	26w78 e		708.ng	P<.0005
f	R m	cdr ipj	tba ben	26w78 e		no dre	P=1.

ADIPAMIDE   100ng..:..1ug....:..10......:..100.....:..1mg....:..10......:..100.....:..1g.....:..10

	Spe Sex	Strain Route	Site Hist	Xpo+Xpt Notes			TD50 DR	2Tailpvl AuOp
161	M f	cb6 eat	lun a/a	52w69 e	pool	.>	8.55gm \	P<.2 -
a	M f	cb6 eat	liv hpc	52w69 e			no dre	P=1.
162	M m	cb6 eat	liv hpc	52w69 e	pool	.>	no dre	P=1.
a	M m	cb6 eat	lun a/a	52w69 e			no dre	P=1. -
163	R f	f34 eat	liv tum	52w69 e		.>	no dre	P=1. -
164	R m	f34 eat	liv hpc	52w69 e		. +	6.42gm *	P<.007

AF-2   100ng..:..1ug....:..10......:..100.....:..1mg....:..10......:..100.....:..1g.....:..10

	Spe Sex	Strain Route	Site Hist	Xpo+Xpt Notes		TD50 DR	2Tailpvl AuOp
165	H f	syg eat	for mix	94w94 e	. + .	59.1mg *	P<.0005
a	H f	syg eat	eso pam	94w94 e		2.34gm *	P<.3
b	H f	syg eat	for sqc	94w94 e		2.34gm *	P<.3 +
c	H f	syg eat	liv tum	94w94 e		no dre	P=1.
d	H f	syg eat	lun tum	94w94 e		no dre	P=1.
166	H m	syg eat	for mix	94w94 e	. + .	30.5mg *	P<.0005
a	H m	syg eat	for sqc	94w94 e		164.mg *	P<.0005+
b	H m	syg eat	eso mix	94w94 e		328.mg *	P<.003
c	H m	syg eat	eso sqc	94w94 e		2.18gm *	P<.3 +
d	H m	syg eat	liv tum	94w94 e		no dre	P=1.
e	H m	syg eat	lun tum	94w94 e		no dre	P=1.
167	M f	cdf eat	for pam	18m24	. + .	714.mg *	P<.0005+
a	M f	cdf eat	for sqc	18m24		2.54gm *	P<.002 +
b	M f	cdf eat	lun tum	18m24		16.4gm *	P<.7
168	M m	cdf eat	for pam	18m24	. + .	753.mg *	P<.0005+
a	M m	cdf eat	for sqc	18m24		928.mg *	P<.0005+
b	M m	cdf eat	lun tum	18m24		9.27gm *	P<.5
169	M m	ddy eat	for mix	44w80	. + .	60.8mg	P<.005
a	M m	ddy eat	for sqc	44w80		72.9mg	P<.002 +

	RefNum	LoConf	UpConf	Cntrl	1Dose	1Inc	2Dose	2Inc			Citation or Pathology
											Brkly Code
i	1268n	23.7mg	108.mg	0/79	2.00mg	0/46	5.69mg	0/45	15.4mg	12/34	
j	1268n	33.1mg	285.mg	0/79	2.00mg	2/47	5.69mg	3/48	15.4mg	6/46	
k	1268n	41.1mg	280.mg	0/78	2.00mg	0/48	5.69mg	2/47	15.4mg	6/46	
l	1268n	42.4mg	798.mg	0/80	2.00mg	1/48	5.69mg	3/48	15.4mg	4/47	
m	1268n	34.0mg	n.s.s.	1/68	2.00mg	0/34	5.69mg	1/25	15.4mg	1/11	
n	1268n	11.0mg	n.s.s.	57/80	2.00mg	42/48	5.69mg	42/48	15.4mg	35/47	
o	1268n	.287mg	n.s.s.	78/80	2.00mg	47/48	5.69mg	48/48	15.4mg	48/48	
153	1268o	17.7mg	80.9mg	0/78	15.4mg	12/44					
a	1268o	20.7mg	110.mg	0/78	15.4mg	10/44					
154	1251m	18.9mg	63.8mg	0/96	2.27mg	4/93	9.10mg	15/82			Quast;dcrp;1980
a	1251m	20.0mg	117.mg	4/96	2.27mg	3/93	9.10mg	17/82			
b	1251m	21.7mg	80.1mg	0/96	2.27mg	2/93	9.10mg	14/82			
c	1251m	25.3mg	174.mg	1/96	2.27mg	3/93	9.10mg	11/82			
d	1251m	25.4mg	310.mg	2/96	2.27mg	4/93	9.10mg	11/82			
e	1251m	38.9mg	n.s.s.	1/95	2.27mg	2/93	9.10mg	6/81			
f	1251m	65.7mg	n.s.s.	0/74	2.27mg	0/64	9.10mg	1/48			
g	1251m	10.9mg	n.s.s.	86/100	2.27mg	75/100	9.10mg	80/100			
155	1251o	19.0mg	127.mg	2/96	9.10mg	14/81					
a	1251o	25.2mg	1.03gm	0/72	9.10mg	4/48					
b	1251o	35.1mg	n.s.s.	1/95	9.10mg	7/89					
156	1268m	4.58mg	9.11mg	0/79	1.75mg	3/46	4.98mg	23/47	14.9mg	39/47	Quast;dcfr;1980
a	1268m	7.33mg	17.9mg	1/73	1.75mg	8/45	4.98mg	19/47	14.9mg	23/44	
b	1268m	8.03mg	21.0mg	1/73	1.75mg	8/45	4.98mg	18/47	14.9mg	21/44	
c	1268m	8.40mg	18.6mg	0/79	1.75mg	2/46	4.98mg	17/47	14.9mg	25/47	
d	1268m	9.38mg	22.6mg	0/73	1.75mg	1/44	4.98mg	10/47	14.9mg	25/43	
e	1268m	9.34mg	49.5mg	0/73	1.75mg	3/44	4.98mg	8/47	(14.9mg	6/44)	
f	1268m	18.3mg	93.5mg	3/73	1.75mg	4/45	4.98mg	3/47	14.9mg	15/44	
g	1268m	29.9mg	n.s.s.	3/73	1.75mg	6/16	4.98mg	1/11	14.9mg	6/42	
h	1268m	82.9mg	n.s.s.	1/50	1.75mg	1/31	4.98mg	0/31	14.9mg	0/18	
i	1268m	2.31mg	70.8mg	67/80	1.75mg	37/47	4.98mg	47/48	14.9mg	46/48	
157	1268r	23.5mg	n.s.s.	1/68	14.9mg	5/37					
a	1268r	23.5mg	n.s.s.	1/68	14.9mg	5/37					

ACTINOMYCIN C (sanamycin) 8052-16-2

	RefNum	LoConf	UpConf	Cntrl	1Dose	1Inc	2Dose	2Inc			Citation or Pathology
158	1017	.504mg	n.s.s.	7/65	.500mg	6/22					Schmahl;arzn,20,1461-1467;1970
a	1017	.635mg	n.s.s.	3/65	.500mg	4/22					
b	1017	1.03mg	n.s.s.	4/65	.500mg	2/22					

ACTINOMYCIN D 50-76-0

	RefNum	LoConf	UpConf	Cntrl	1Dose	1Inc	2Dose	2Inc			Citation or Pathology
159	1336	1.14ug	3.87ug	0/182	3.15ug	18/30	7.93ug	1/5			Skipper;srfr;1976/Weisburger;canc;1977/Prejean pers.comm.
a	1336	8.16ug	n.s.s.	0/182	3.15ug	0/30	7.93ug	1/5			
b	1336	3.24ug	n.s.s.	0/182	3.15ug	0/30	7.93ug	0/5			
c	1336	658.ng	2.74ug	44/182	3.15ug	25/30	7.93ug	3/5			
d	1336	618.ng	10.0ug	103/182	3.15ug	26/30	7.93ug	4/5			
e	1336	11.8ug	n.s.s.	59/182	3.15ug	1/30	7.93ug	1/5			
160	1336	456.ng	1.38ug	0/177	3.15ug	27/34					
a	1336	4.93ug	.571mg	0/177	3.15ug	2/34					
b	1336	4.93ug	.571mg	0/177	3.15ug	2/34					
c	1336	6.63ug	n.s.s.	0/177	3.15ug	1/34					
d	1336	244.ng	1.10ug	59/177	3.15ug	32/34					
e	1336	394.ng	1.41ug	32/177	3.15ug	29/34					
f	1336	6.99ug	n.s.s.	27/177	3.15ug	3/34					

ADIPAMIDE 628-94-4

	RefNum	LoConf	UpConf	Cntrl	1Dose	1Inc	2Dose	2Inc			Citation or Pathology
161	1343	1.99gm	n.s.s.	2/89p	1.57gm	3/40	(2.35gm	0/39)			Fleischman;jept,3,149-170;1980
a	1343	8.83gm	n.s.s.	1/89p	1.57gm	1/40	2.35gm	0/39			
162	1343	3.05gm	n.s.s.	2/91p	1.44gm	0/38	2.17gm	0/40			
a	1343	8.13gm	n.s.s.	1/87p	1.44gm	1/38	2.17gm	0/40			
163	1343	2.15gm	n.s.s.	0/49	903.mg	0/34	2.18gm	0/48			
164	1343	2.44gm	71.7gm	0/50	722.mg	0/35	1.75gm	5/49			

AF-2 (furylfuramide) 3688-53-7

	RefNum	LoConf	UpConf	Cntrl	1Dose	1Inc	2Dose	2Inc			Citation or Pathology
165	1315	34.5mg	110.mg	0/16	83.6mg	10/17	167.mg	13/17			Kinebuchi;fctx,17,339-341;1979
a	1315	381.mg	n.s.s.	0/16	83.6mg	0/17	167.mg	1/17			
b	1315	381.mg	n.s.s.	0/16	83.6mg	0/17	167.mg	1/17			
c	1315	160.mg	n.s.s.	0/16	83.6mg	0/17	167.mg	0/17			
d	1315	160.mg	n.s.s.	0/16	83.6mg	0/17	167.mg	0/17			
166	1315	17.6mg	55.0mg	0/19	73.6mg	13/20	147.mg	17/17			
a	1315	81.7mg	452.mg	0/19	73.6mg	3/20	147.mg	8/17			
b	1315	133.mg	1.66gm	0/19	73.6mg	0/20	147.mg	6/17			
c	1315	355.mg	n.s.s.	0/19	73.6mg	0/20	147.mg	1/17			
d	1315	156.mg	n.s.s.	0/19	73.6mg	0/20	147.mg	0/17			
e	1315	156.mg	n.s.s.	0/19	73.6mg	0/20	147.mg	0/17			
167	1316	415.mg	1.38gm	0/50	78.0mg	2/50	390.mg	17/50			Takayama;clet,3,115-120;1977
a	1316	1.03gm	10.2gm	0/50	78.0mg	0/50	390.mg	6/50			
b	1316	1.73gm	n.s.s.	1/50	78.0mg	2/50	390.mg	2/50			
168	1316	426.mg	1.52gm	0/50	72.0mg	2/50	360.mg	15/50			
a	1316	499.mg	2.03gm	0/50	72.0mg	0/50	360.mg	14/50			
b	1316	1.53gm	n.s.s.	2/50	72.0mg	1/50	360.mg	3/50			
169	455	22.7mg	629.mg	1/10	165.mg	7/10					Sano;zkko,89,61-68;1977
a	455	28.2mg	309.mg	0/10	165.mg	6/10					

	Spe Strain Site Xpo+Xpt				TD50	2Tailpvl
	Sex Route Hist Notes				DR	AuOp
b	M m ddy eat lun ade 44w80				299.mg	P<.09
c	M m ddy eat liv tum 44w80				no dre	P=1.
d	M m ddy eat tba mix 44w80				60.8mg	P<.005
170	M f icr eat for mix 62w62		.+ .		95.0mg *	P<.0005+
a	M f icr eat for sqc 62w62				225.mg /	P<.0005+
b	M f icr eat --- leu 62w62				1.80gm *	P<.3
c	M f icr eat liv tum 62w62				no dre	P=1.
171	M f icr eat liv hct 24m24			.>	7.07mg *	P<.9 -
172	M m icr eat for mix 62w62		.+ .		90.3mg *	P<.0005+
a	M m icr eat for sqc 62w62				208.mg /	P<.0005+
b	M m icr eat liv tum 62w62				no dre	P=1.
173	M m icr eat liv hct 24m24			.>	no dre	P=1. -
174	R f sda eat mgl mix 46w66 e		. + .		11.4mg	P<.0005+
a	R f sda eat mgl adc 46w66 e		. + .		26.4mg	P<.0005+
175	R f wis eat mgl mix 18m24		. + .		74.7mg *	P<.0005+
a	R f wis eat for pam 18m24				424.mg *	P<.0005
b	R f wis eat liv hms 18m24				2.00gm *	P<.02
176	R m wis eat for pam 18m24		. + .		409.mg *	P<.0005
a	R m wis eat mgl mix 18m24				1.20gm *	P<.05 +
b	R m wis eat liv hms 18m24				1.60gm *	P<.02

AFLATOXICOL   100ng..:..1ug...:..10.....:..100....:..1mg....:..10.....:..100...:..1g.....:..10

177	R m f34 eat liv hpc 12m24	. + .			2.47ug *	P<.0005+
a	R m f34 eat kid adc 12m24				32.9ug *	P<.06

AFLATOXIN B1   100ng..:..1ug...:..10.....:..100....:..1mg....:..10.....:..100...:..1g.....:..10

178	M f c5v ipj lmr tum 52w92	. + .			4.02ug	P<.007 -
a	M f c5v ipj lun ben 52w92				75.3ug	P<.3 -
b	M f c5v ipj liv tum 52w92				no dre	P=1. -
c	M f c5v ipj tba tum 52w92				4.49ug	P<.03 -
179	M m c5v ipj lun ben 52w92	.>			.245mg	P<.9 -
a	M m c5v ipj liv mal 52w92				no dre	P=1. -
b	M m c5v ipj liv ben 52w92				no dre	P=1. -
c	M m c5v ipj tba tum 52w92				no dre	P=1. -
180	N b tst eat liv hpc 40m40 er	. + .			26.9ug	P<.0005+
181	P b cym mix MXB MXB 17y17 Wmw	: + :			9.25ug	P<.0005
a	P b cym mix bil mix 17y17 Wmw				20.1ug	P<.0005+
b	P b cym mix gab adc 17y17 Wmw				40.8ug	P<.0005+
c	P b cym mix --- ost 17y17 Wmw				62.7ug	P<.007
d	P b cym mix liv hes 17y17 Wmw				73.4ug	P<.002 +
e	P b cym mix k/p tcc 17y17 Wmw				44.6ug	P<.04
f	P b cym mix liv clc 17y17 Wmw				44.6ug	P<.04 +
g	P b cym mix liv hpc 17y17 Wmw				44.6ug	P<.04 +
h	P b cym mix pel fbs 17y17 Wmw				.104mg	P<.03
i	P b cym mix pnd adc 17y17 Wmw				.342mg	P<.06
j	P b cym mix bil scc 17y17 Wmw				.371mg	P<.06 +
k	P b cym mix tba mal 17y17 Wmw				8.06ug	P<.0005
182	P b rhe mix MXB MXB 15y15 Wmw	: + :			5.95ug	P<.0005
a	P b rhe mix liv clc 15y15 Wmw				8.20ug	P<.0005+
b	P b rhe mix bil mix 15y15 Wmw				11.5ug	P<.0005+
c	P b rhe mix gab adc 15y15 Wmw				14.8ug	P<.0005+
d	P b rhe mix liv hes 15y15 Wmw				16.3ug	P<.0005
e	P b rhe mix pnd adc 15y15 Wmw				23.3ug	P<.0005
f	P b rhe mix liv hpc 15y15 Wmw				.104mg	P<.005 +
g	P b rhe mix --- ost 15y15 Wmw				.159mg	P<.05
h	P b rhe mix tba mal 15y15 Wmw				6.78ug	P<.0005
183	R m buf eat liv hpc 35w65 ekr	<+			noTD50	P<.004 +
a	R m buf eat liv mix 35w65 ekr	<+			noTD50	P<.004 +
b	R m buf eat liv nnd 35w65 ekr	<+			noTD50	P<.004 +
184	R m buf eat liv hpc 26w52 ekr	. + .			5.73ug	P<.002 +
a	R m buf eat liv nnd 26w52 ekr	. + .			7.95ug	P<.004 +
185	R m cdr eat liv mix 24m24 r	. + .			4.19ug	P<.0005+
186	R f f34 eat col mix 75w75 e	. ±			.148mg	P<.05 +
187	R f f34 eat liv hpc 24m24 er	. ±			52.7ug *	P<.06 -
188	R m f34 eat col mix 75w75 e	. ±			55.8mg	P<.0005+
189	R m f34 eat liv hpc 12m24	. + .			1.34ug	P<.0005+
a	R m f34 eat kid adc 12m24				13.4ug	P<.3
190	R m f34 eat liv hpc 24m24 er	.>			49.9ug *	P<.3 -
191	R f fis eat liv hpt 24m24 e	.>			9.93ug *	P<.3 -
192	R m fis eat liv hpc 22m24 ae	. + .			932.ng *	P<.0005+
193	R m fis eat liv hpt 24m24 e	. + .			1.13ug	P<.003 +
194	R f por eat liv hpc 24m24 e	. + .			12.5ug *	P<.0005+
195	R m por eat liv hpc 24m24 e	. + .			3.52ug *	P<.0005+
196	R b sda eat liv sar 95w95 e	.>			12.0ug	P<.3
a	R b sda eat tba mal 95w95 e				+hist 3.96ug	P<.03 +
197	R m wag eat liv hpt 69w69 er	<+			noTD50	P<.0005+
198	R f wio eat liv hpt 24m24 e	. + .			7.49ug *	P<.0005+
199	R m wio eat liv hpt 24m24 e	. + .			6.76ug *	P<.0005+
200	R m wis eat kid car 55w55 er	. + .			36.6ug *	P<.006 +

	RefNum	LoConf	UpConf	Cntrl	1Dose	1Inc	2Dose	2Inc						Citation or Pathology
														Brkly Code
b	455	73.2mg	n.s.s.	0/10	165.mg	2/10								
c	455	201.mg	n.s.s.	0/10	165.mg	0/10								
d	455	22.7mg	629.mg	1/10	165.mg	7/10								
170	359	65.6mg	143.mg	0/65	104.mg	13/50	520.mg	36/50						Yokoro;gann,68,825-828;1977
a	359	140.mg	393.mg	0/65	104.mg	1/50	520.mg	25/50						
b	359	464.mg	n.s.s.	2/65	104.mg	5/50	520.mg	5/50						
c	359	317.mg	n.s.s.	0/65	104.mg	0/50	520.mg	0/50						
171	362	501.mg	n.s.s.	6/30	16.3mg	2/30	65.0mg	6/30	260.mg	5/30				Miyaji;tjem,103,331-369;1971
172	359	62.6mg	151.mg	0/25	96.0mg	15/50	480.mg	34/50						Yokoro;gann,68,825-828;1977
a	359	129.mg	363.mg	0/25	96.0mg	1/50	480.mg	25/50						
b	359	293.mg	n.s.s.	0/25	96.0mg	0/50	480.mg	0/50						
173	362	489.mg	n.s.s.	9/30	15.0mg	6/30	60.0mg	14/30	240.mg	6/30				Miyaji;tjem,103,331-369;1971
174	517	6.31mg	22.8mg	2/29	69.7mg	24/29								Cohen;gann,68,473-476;1977
a	517	14.2mg	56.5mg	0/29	69.7mg	15/29								
175	1316	49.9mg	125.mg	3/50	30.0mg	17/50	150.mg	37/50						Takayama;clet,3,115-120;1977
a	1316	223.mg	962.mg	0/50	30.0mg	1/50	150.mg	12/50						
b	1316	606.mg	n.s.s.	0/50	30.0mg	0/50	150.mg	3/50						
176	1316	205.mg	1.04gm	0/50	24.0mg	1/50	120.mg	10/50						
a	1316	415.mg	n.s.s.	0/50	24.0mg	1/50	120.mg	3/50						
b	1316	485.mg	n.s.s.	0/50	24.0mg	0/50	120.mg	3/50						

AFLATOXICOL　29611-03-8

	RefNum	LoConf	UpConf	Cntrl	1Dose	1Inc	2Dose	2Inc						Citation or Pathology
177	1083	1.38ug	4.97ug	0/20	1.00ug	4/20	4.00ug	14/20						Nixon;jnci,66,1159-1163;1981
a	1083	8.08ug	n.s.s.	0/20	1.00ug	0/20	4.00ug	2/20						

AFLATOXIN B1　1162-65-8

	RefNum	LoConf	UpConf	Cntrl	1Dose	1Inc	2Dose	2Inc						Citation or Pathology
178	1636	1.62ug	67.6ug	3/11	9.69ug	12/15								Griciute;iarc,813-822;1980
a	1636	12.3ug	n.s.s.	0/11	9.69ug	1/15								
b	1636	23.4ug	n.s.s.	0/11	9.69ug	0/15								
c	1636	1.68ug	n.s.s.	4/11	9.69ug	12/15								
179	1636	11.3ug	n.s.s.	1/20	8.07ug	1/15								
a	1636	19.5ug	n.s.s.	2/20	8.07ug	0/15								
b	1636	19.5ug	n.s.s.	2/20	8.07ug	0/15								
c	1636	8.52ug	n.s.s.	11/20	8.07ug	4/14								
180	1439	11.5ug	80.9ug	0/8	99.7ug	9/12								Reddy;canr,36,151-160;1976
181	2001	3.25ug	26.5ug	0/67	56.0ug	10/13								Adamson;ossc,129-156;
														1982/Thorgeirsson 1994/Dalgard 1991/Thorgeirsson&Seiber pers.comm.
a	2001	6.34ug	68.3ug	0/67	56.0ug	6/13								
b	2001	12.9ug	.203mg	0/61	56.0ug	4/12								
c	2001	12.0ug	1.98mg	0/69	56.0ug	2/14								
d	2001	21.4ug	.526mg	0/54	56.0ug	3/10								
e	2001	7.26ug	n.s.s.	0/15	56.0ug	1/2								
f	2001	7.26ug	n.s.s.	0/15	56.0ug	1/2								
g	2001	7.26ug	n.s.s.	0/15	56.0ug	1/2								
h	2001	16.9ug	n.s.s.	0/45	56.0ug	1/4								
i	2001	55.7ug	n.s.s.	0/61	56.0ug	1/12								
j	2001	60.5ug	n.s.s.	0/67	56.0ug	1/13								
k	2001	3.08ug	20.3ug	0/69	56.0ug	12/14								
182	2001	1.71ug	20.1ug	1/87	45.0ug	10/14								
a	2001	1.52ug	.185mg	0/46	45.0ug	2/4								
b	2001	2.11ug	71.9ug	1/76	45.0ug	5/11								
c	2001	2.24ug	.130mg	1/76	45.0ug	4/11								
d	2001	4.10ug	.139mg	0/86	45.0ug	3/13								
e	2001	6.39ug	.171mg	0/56	45.0ug	3/9								
f	2001	25.2ug	1.90mg	0/87	45.0ug	2/14								
g	2001	25.9ug	n.s.s.	0/56	45.0ug	1/9								
h	2001	2.06ug	22.3ug	2/87	45.0ug	10/14								
183	1071	n.s.s.	6.75ug	0/8	21.5ug	6/6								Angsubhakorn;bjca,43,881-883;1981
a	1071	n.s.s.	6.75ug	0/8	21.5ug	6/6								
b	1071	n.s.s.	6.75ug	0/8	21.5ug	6/6								
184	1664	2.66ug	18.7ug	0/14	20.0ug	9/20								Angsubhakorn;ijcn,28,621-626;1981
a	1664	3.40ug	47.3ug	0/14	20.0ug	7/20								
185	1454	2.55ug	7.52ug	0/50	4.00ug	24/50								Newberne;jnci,50,439-444;1973
186	17	44.5ug	n.s.s.	0/12	.100mg	3/14								Ward;jnci,55,107-110;1975
187	1824	8.58ug	n.s.s.	0/144	250.ng	0/24	750.ng	0/24	2.25ug	1/24				Elashoff;jnci,79,509-526;1987
188	17	13.3ug	n.s.s.	0/6	80.0ug	2/5								Ward;jnci,55,107-110;1975
189	1083	600.ng	4.24ug	0/20	1.00ug	8/20								Nixon;jnci,66,1159-1163;1981
a	1083	2.17ug	n.s.s.	0/20	1.00ug	1/20								
190	1824	6.83ug	n.s.s.	1/144	200.ng	0/23	600.ng	0/24	1.80ug	1/23				Elashoff;jnci,79,509-526;1987
191	1041	1.62ug	n.s.s.	0/15	1.00ug	1/15								Nixon;jnci,53,453-458;1974
192	18	614.ng	1.49ug	0/18	40.0ng	2/22	200.ng	1/22	600.ng	4/21	2.00ug	20/25	4.00ug	28/28　Wogan;fctx,12,
														681-685;1974
193	1041	422.ng	6.52ug	0/16	800.ng	5/13								Nixon;jnci,53,453-458;1974
194	13	7.83ug	21.2ug	0/34	5.00ug	5/30	25.0ug	26/33						Butler;fctx,6,135-141;1968
195	13	2.11ug	6.01ug	0/46	4.00ug	17/34	20.0ug	25/25						
196	1070	1.96ug	n.s.s.	0/90	279.ng	1/76								Fong;fctx,19,179-183;1981
a	1070	1.20ug	n.s.s.	0/90	279.ng	3/76								
197	1075	n.s.s.	2.13ug	0/10	10.0ug	12/12								Burtin;bjca,43,684-688;1981
198	1041	3.36ug	22.1ug	0/18	1.00ug	0/19	5.00ug	8/18						Nixon;jnci,53,453-458;1974
199	1041	2.89ug	21.9ug	0/17	800.ng	0/20	4.00ug	7/17						
200	15	18.7ug	.392mg	0/9	40.0ug	3/12	80.0ug	3/12	.120mg	6/12				Merkow;canr,33,1608-1614;1973

```
    Spe Strain Site   Xpo+Xpt                                                      TD50   2Tailpvl
    Sex Route  Hist   Notes                                                               DR  AuOp

AFLATOXIN, CRUDE       100ng..:..1ug....:..10......:..100....:..1mg...:..10....:..100....:..1g......:..10
201 M m swi gav --- lyk 52w52 e                        .  +    .                      .343mg   P<.002 +
  a M m swi gav liv hpt 52w52 e                                                      no dre   P=1.
202 R m cdr eat liv tum 52w52 e               .   +    .                             7.48ug   P<.002 +
203 R m chm eat liv car 68w68           .  +  .                                      1.87ug   P<.0005+
  a R m chm eat liv nod 68w68                                                        6.96ug   P<.002 +
204 R m lev eat ktu ade 58w78                             .  +   .                   3.44mg * P<.0005
  a R m lev eat liv nod 58w78                                                        6.18mg * P<.002
  b R m lev eat liv hpc 58w78                                                        14.5mg * P<.05

AGAR                   100ng..:..1ug....:..10......:..100....:..1mg...:..10....:..100....:..1g......:..10
205 M f b6c eat TBA MXB 24m24                                                   :> no dre   P=1.  -
  a M f b6c eat liv MXB 24m24                                                      no dre   P=1.
  b M f b6c eat lun MXB 24m24                                                      no dre   P=1.
206 M m b6c eat liv hpa 24m24                                                     #18.8gm * P<.002 -
  a M m b6c eat TBA MXB 24m24                                                      54.9gm * P<.8
  b M m b6c eat liv MXB 24m24                                                      33.1gm * P<.4
  c M m b6c eat lun MXB 24m24                                                      102.gm * P<.8
207 R f f34 eat adr coa 24m24                                                     #25.8gm * P<.03 -
  a R f f34 eat TBA MXB 24m24                                                      no dre   P=1.
  b R f f34 eat liv MXB 24m24                                                      no dre   P=1.
208 R m f34 eat TBA MXB 24m24                                                :> 39.3gm * P<1.  -
  a R m f34 eat liv MXB 24m24                                                      25.9gm * P<.3

ALDICARB               100ng..:..1ug....:..10......:..100....:..1mg...:..10....:..100....:..1g......:..10
209 M f b6c eat TBA MXB 24m24                                 :>                   no dre   P=1.  -
  a M f b6c eat liv MXB 24m24                                                      52.7mg / P<.9
  b M f b6c eat lun MXB 24m24                                                      no dre   P=1.
210 M m b6c eat TBA MXB 24m24                            :>                        1.74mg * P<.3
  a M m b6c eat liv MXB 24m24                                                      2.83mg * P<.4
  b M m b6c eat lun MXB 24m24                                                      14.8mg * P<.8
211 R f f34 eat TBA MXB 24m24                        :>                            no dre   P=1.
  a R f f34 eat liv MXB 24m24                                                      no dre   P=1.
212 R m f34 eat TBA MXB 24m24                        :>                            10.4mg / P<1.  -
  a R m f34 eat liv MXB 24m24                                                      1.33mg * P<.2

ALDRIN                 100ng..:..1ug....:..10......:..100....:..1mg...:..10....:..100....:..1g......:..10
213 M f b6c eat TBA MXB 80w90 s                          :>                        no dre   P=1.  -
  a M f b6c eat liv MXB 80w90 s                                                    no dre   P=1.
  b M f b6c eat lun MXB 80w90 s                                                    9.02mg * P<.6
214 M m b6c eat liv hpc 80w90                       :  +      :                    .741mg * P<.007 c
  a M m b6c eat lun MXA 80w90                                                      3.20mg * P<.05
  b M m b6c eat TBA MXB 80w90                                                      1.01mg * P<.08
  c M m b6c eat liv MXB 80w90                                                      .741mg * P<.007
  d M m b6c eat lun MXB 80w90                                                      3.20mg * P<.05
215 M m b6c eat liv hpc 80w89 pool                  :  +    :                      .861mg * P<.0005c
216 M b c3e eat liv hpa 24m24 e                           .  +   .                 4.41mg   P<.0005+
  a M b c3e eat lun car 24m24 e                                                    129.mg   P<.3
  b M b c3e eat lun ade 24m24 e                                                    no dre   P=1.
217 M b c3h eat liv hpt 24m24                          .  +   .                    3.45mg   P<.0005
  a M b c3h eat liv hpc 24m24                                                      no dre   P=1.
  b M b c3h eat tba ben 24m24                                                      4.25mg   P<.0005
  c M b c3h eat tba mal 24m24                                                      no dre   P=1.
218 R b osm eat lun lys 24m24 es                              .>                   42.3mg * P<.5

  a R b osm eat tba mix 24m24 es                                                   no dre   P=1.  +

219 R f osm eat TBA MXB 19m26                               :>                     no dre   P=1.  -
  a R f osm eat liv MXB 19m26                                                      466.mg * P<1.
220 R f osm eat adr coa 19m25 pool                   :  +    :                     #3.12mg \ P<.0005-
221 R f osm eat liv tum 25m25                         .>                           no dre   P=1.  -
  a R f osm eat tba mix 25m25                                                      no dre   P=1.  -
222 R f osm eat liv tum 26m27 ev                         .>                        no dre   P=1.  -
  a R f osm eat tba mix 26m27 ev                                                   no dre   P=1.  -
223 R f osm eat lmr tum 24m24 es                      .   ±                        4.35mg \ P<.03 -
  a R f osm eat liv tum 24m24 es                                                   no dre   P=1.  -
  b R f osm eat tba tum 24m24 es                                                   1.69mg \ P<.3
  c R f osm eat tba mal 24m24 es                                                   no dre   P=1.
224 R m osm eat TBA MXB 17m26                             :>                       16.4mg * P<.9  -
  a R m osm eat liv MXB 17m26                                                      no dre   P=1.
225 R m osm eat liv tum 25m25                          .>                          no dre   P=1.  -
  a R m osm eat tba mix 25m25                                                      4.21mg   P<.6  -
226 R m osm eat liv hem 29m31 ev                         .>                        no dre   P=1.  -
  a R m osm eat tba mix 29m31 ev                                                   no dre   P=1.  -
227 R f nss eat tba tum 24m24                               .>                     no dre   P=1.  -
228 R m nss eat tba tum 24m24                               .>                     no dre   P=1.  -
229 R f sda eat liv tum 24m24 es                         .>                        no dre   P=1.  -
  a R f sda eat tba tum 24m24 es                                                   no dre   P=1.
  b R f sda eat tba mal 24m24 es                                                   no dre   P=1.
```

	RefNum	LoConf	UpConf	Cntrl	1Dose	1Inc	2Dose	2Inc					Citation or Pathology	Brkly Code
AFLATOXIN, CRUDE		---												
201	309	.133mg	2.34mg	11/37	2.38mg	11/14							Louria;sabo,12,371-375;1974	
a	309	1.72mg	n.s.s.	0/37	2.38mg	0/14								
202	16	2.90ug	31.7ug	0/10	40.0ug	6/10							Newberne;livt,15,962-969;1966	
203	333	1.12ug	3.38ug	0/20	8.00ug	25/35							Newberne;aenh,19,489-498;1969	
a	333	3.38ug	25.8ug	0/20	8.00ug	10/35								
204	1514	1.92mg	6.89mg	0/10	.368mg	0/10	2.45mg	1/10	7.36mg	7/10	9.81mg	10/15	Lee;jnci,43,1037-1041;1969	
a	1514	3.05mg	23.5mg	1/10	.368mg	1/10	2.45mg	1/10	7.36mg	5/10	9.81mg	7/15		
b	1514	5.91mg	n.s.s.	0/10	.368mg	0/10	2.45mg	2/10	7.36mg	0/10	9.81mg	4/15		
AGAR	9002-18-0													
205	TR230	4.42gm	n.s.s.	26/50	3.19gm	24/50	(6.38gm	16/50)						
a	TR230	34.9gm	n.s.s.	4/50	3.19gm	5/50	6.38gm	1/50					liv:hpa,hpc,nnd.	
b	TR230	13.7gm	n.s.s.	7/50	3.19gm	3/50	(6.38gm	1/50)					lun:a/a,a/c.	
206	TR230	9.17gm	79.1gm	0/50	2.94gm	3/50	5.89gm	7/50						S
a	TR230	5.67gm	n.s.s.	24/50	2.94gm	24/50	5.89gm	25/50						
b	TR230	8.46gm	n.s.s.	9/50	2.94gm	8/50	5.89gm	13/50					liv:hpa,hpc,nnd.	
c	TR230	12.1gm	n.s.s.	6/50	2.94gm	6/50	5.89gm	7/50					lun:a/a,a/c.	
207	TR230	8.92gm	n.s.s.	0/50	1.21gm	0/50	2.45gm	4/50						S
a	TR230	2.79gm	n.s.s.	47/50	1.21gm	43/50	2.45gm	43/50						
b	TR230	n.s.s.	n.s.s.	0/50	1.21gm	0/50	2.45gm	0/50					liv:hpa,hpc,nnd.	
208	TR230	1.67gm	n.s.s.	32/50	981.mg	38/50	1.96gm	35/50						
a	TR230	7.81gm	n.s.s.	0/50	981.mg	2/50	1.96gm	1/50					liv:hpa,hpc,nnd.	
ALDICARB	(temik)	116-06-3												
209	TR136	1.27mg	n.s.s.	15/25	.260mg	21/50	.780mg	22/50						
a	TR136	2.83mg	n.s.s.	3/25	.260mg	0/50	.780mg	4/50					liv:hpa,hpc,nnd.	
b	TR136	3.30mg	n.s.s.	1/25	.260mg	4/50	.780mg	1/50					lun:a/a,a/c.	
210	TR136	.508mg	n.s.s.	8/25	.240mg	27/50	.720mg	30/50						
a	TR136	.776mg	n.s.s.	5/25	.240mg	14/50	.720mg	18/50					liv:hpa,hpc,nnd.	
b	TR136	1.53mg	n.s.s.	1/25	.240mg	6/50	.720mg	5/50					lun:a/a,a/c.	
211	TR136	.178mg	n.s.s.	21/25	.100mg	45/50	.300mg	46/50						
a	TR136	n.s.s.	n.s.s.	0/25	.100mg	0/50	.300mg	0/50					liv:hpa,hpc,nnd.	
212	TR136	.194mg	n.s.s.	16/25	80.0ug	23/50	.240mg	28/50						
a	TR136	.411mg	n.s.s.	1/25	80.0ug	1/50	.240mg	5/50					liv:hpa,hpc,nnd.	
ALDRIN	309-00-2													
213	TR21	.844mg	n.s.s.	1/10	.330mg	11/50	.640mg	5/50						
a	TR21	1.14mg	n.s.s.	0/10	.330mg	5/50	.640mg	2/50					liv:hpa,hpc,nnd.	
b	TR21	2.22mg	n.s.s.	0/10	.330mg	1/50	.640mg	1/50					lun:a/a,a/c.	
214	TR21	.411mg	10.0mg	3/20	.470mg	16/50	.840mg	25/50						
a	TR21	1.45mg	n.s.s.	0/20	.470mg	3/50	.840mg	5/50					lun:a/a,a/c.	S
b	TR21	.433mg	n.s.s.	7/20	.470mg	18/50	.840mg	28/50						
c	TR21	.411mg	10.0mg	3/20	.470mg	16/50	.840mg	25/50					liv:hpa,hpc,nnd.	
d	TR21	1.45mg	n.s.s.	0/20	.470mg	3/50	.840mg	5/50					lun:a/a,a/c.	
215	TR21	.467mg	3.01mg	17/95p	.470mg	16/50	.840mg	25/50						
216	20a	2.55mg	12.3mg	9/134	1.25mg	35/151							Davis;txap,4,187-189;1962	
a	20a	21.0mg	n.s.s.	0/134	1.25mg	1/151								
b	20a	28.4mg	n.s.s.	3/134	1.25mg	1/151								
217	22a	2.13mg	7.96mg	27/200	1.25mg	65/200							Epstein(review) {H J Davis};stev,4,1-52;1975	
a	22a	24.4mg	n.s.s.	4/200	1.25mg	3/200								
b	22a	2.43mg	13.8mg	30/200	1.25mg	61/200								
c	22a	24.7mg	n.s.s.	21/200	1.25mg	9/200								
218	23	7.89mg	n.s.s.	1/17	22.5ug	5/19	90.0ug	2/19	.450mg	2/22	2.25mg	2/18	4.50mg 4/11	
				6.75mg	1/9								Fitzhugh;fctx,2,551-562;1964	
a	23	10.0mg	n.s.s.	3/17	22.5ug	10/19	90.0ug	7/19	.450mg	8/22	2.25mg	5/18	4.50mg 5/11	
				6.75mg	1/9									
219	TR21	1.55mg	n.s.s.	5/10	1.07mg	35/50	2.14mg	32/50						
a	TR21	9.99mg	n.s.s.	1/10	1.07mg	1/50	2.14mg	3/50					liv:hpa,hpc,nnd.	
220	TR21	1.41mg	10.2mg	0/60p	1.07mg	8/50	(2.14mg	1/50)						S
221	21	1.67mg	n.s.s.	0/30	.250mg	0/30							Deichmann;txap,11,88-103;1967	
a	21	.298mg	n.s.s.	13/30	.250mg	13/30								
222	1004	5.05mg	n.s.s.	0/88	.956mg	0/47	1.44mg	0/44	2.38mg	0/31			Deichmann;imed,39,426-434;1970	
a	1004	5.18mg	n.s.s.	60/88	.956mg	20/47	1.44mg	24/44	(2.38mg	11/31)				
223	1040	1.76mg	n.s.s.	2/50	1.00mg	9/50	(2.50mg	4/50)					Deichmann;txoc,407-413;1979	
a	1040	7.36mg	n.s.s.	0/50	1.00mg	0/50	2.50mg	0/50						
b	1040	.487mg	n.s.s.	35/50	1.00mg	40/50	(2.50mg	18/50)						
c	1040	5.80mg	n.s.s.	10/50	1.00mg	14/50	2.50mg	10/50						
224	TR21	.875mg	n.s.s.	9/10	.800mg	30/50	1.60mg	30/50						
a	TR21	6.34mg	n.s.s.	1/10	.800mg	1/50	1.60mg	1/50					liv:hpa,hpc,nnd.	
225	21	1.33mg	n.s.s.	0/30	.200mg	0/30							Deichmann;txap,11,88-103;1967	
a	21	.579mg	n.s.s.	1/30	.200mg	2/30								
226	1004	5.75mg	n.s.s.	1/75	.765mg	0/45	1.15mg	0/46	1.93mg	0/45			Deichmann;imed,39,426-434;1970	
a	1004	22.9mg	n.s.s.	19/75	.765mg	5/45	1.15mg	7/46	1.93mg	4/45				
227	1002	8.27mg	n.s.s.	6/60	.125mg	1/40	.625mg	3/40	1.25mg	1/40			Cleveland;aenh,13,195-198;1966	
228	1002	8.67mg	n.s.s.	3/60	.100mg	1/40	.500mg	0/40	1.00mg	1/40				
229	1040	7.36mg	n.s.s.	0/50	1.00mg	0/50	2.50mg	0/50					Deichmann;txoc,407-413;1979	
a	1040	11.5mg	n.s.s.	9/50	1.00mg	8/50	2.50mg	5/50						
b	1040	17.8mg	n.s.s.	2/50	1.00mg	2/50	2.50mg	1/50						

```
      Spe Strain Site   Xpo+Xpt                                                   TD50   2Tailpvl
         Sex  Route  Hist   Notes                                                    DR    AuOp

ALKYLBENZENESULFONATE, LINEAR   100ng..:..1ug....:..10......:..100....:..1mg...:..10....:..100....:..1g......:..10
230  R f sls eat pre mix 33m34 er                                          .>       no dre  P=1.  -
231  R m sls eat pre mix 30m32 er                                          .>       no dre  P=1.  -

ALKYLDIMETHYLAMINE OXIDES, COMMERCIAL GRADE.1ug....:..10......:..100....:..1mg...:..10....:..100....:..1g......:..10
232  R f cdr eat pit ade 24m24 er                                      .>           230.mg * P<.2  -
233  R m cdr eat pit ade 24m24 er                                   .  ±            236.mg Z P<.03 -

ALLANTOIN   100ng..:..1ug....:..10......:..100....:..1mg...:..10....:..100....:..1g......:..10
234  R f f34 eat liv nnd 25m30 e                                         .>         no dre  P=1.  -
235  R m f34 eat liv nnd 25m30 e                                         .>         no dre  P=1.  -

ALLYL ALCOHOL   100ng..:..1ug....:..10......:..100....:..1mg...:..10....:..100....:..1g......:..10
236  R f f34 wat liv hpc 25m29 e                          .     ±                   64.2mg  P<.04
a    R f f34 wat liv mix 25m29 e                                                    41.5mg  P<.2
b    R f f34 wat liv hnd 25m29 e                                                    182.mg  P<.7
c    R f f34 wat adr coa 25m29 e                                                    no dre  P=1.
237  R m f34 wat liv mix 25m29 e                                .>                  128.mg  P<.7
a    R m f34 wat liv hpc 25m29 e                                                    142.mg  P<.3
b    R m f34 wat liv hnd 25m29 e                                                    no dre  P=1.
c    R m f34 wat adr coa 25m29 e                                                    no dre  P=1.

ALLYL CHLORIDE   100ng..:..1ug....:..10......:..100....:..1mg...:..10....:..100....:..1g......:..10
238  M f b6c gav sto MXA 78w91 dv                               :>         +hist 706.mg * P<.2  a
a    M f b6c gav TBA MXB 78w91 dv                                                   343.mg * P<.3
b    M f b6c gav liv MXB 78w91 dv                                                   2.18gm * P<.4
c    M f b6c gav lun MXB 78w91 dv                                                   1.25gm * P<.6
239  M m b6c gav sto MXA 67w91 adsv                                :>         +hist 727.mg * P<.4  a
a    M m b6c gav TBA MXB 67w91 adsv                                                 184.mg * P<.4
b    M m b6c gav liv MXB 67w91 adsv                                                 457.mg * P<.6
c    M m b6c gav lun MXB 67w91 adsv                                                 no dre  P=1.
240  R f osm gav TBA MXB 18m26 sv                         :>                        35.2mg * P<.2  i
a    R f osm gav liv MXB 18m26 sv                                                   no dre  P=1.
241  R m osm gav TBA MXB 18m26 sv                        :>                         28.1mg * P<.2  i
a    R m osm gav liv MXB 18m26 sv                                                   no dre  P=1.

ALLYL GLYCIDYL ETHER   100ng..:..1ug....:..10......:..100....:..1mg...:..10....:..100....:..1g......:..10
242  M f b6c inh hag ade 24m24                          :    ±                      121.mg * P<.02
a    M f b6c inh nre ade 24m24                                                      614.mg * P<.3  e
b    M f b6c inh TBA MXB 24m24                                                      no dre  P=1.
c    M f b6c inh liv MXB 24m24                                                      no dre  P=1.
d    M f b6c inh lun MXB 24m24                                  ±                   123.mg * P<.03
243  M m b6c inh nre ade 24m24                          :    ±                      182.mg * P<.06 p
a    M m b6c inh TBA MXB 24m24                                                      no dre  P=1.
b    M m b6c inh liv MXB 24m24                                                      no dre  P=1.
c    M m b6c inh lun MXB 24m24                                                      727.mg * P<1.
244  R f osm inh TBA MXB 24m25                   :>                                 25.9mg * P<.9  -
a    R f osm inh liv MXB 24m25                                                      no dre  P=1.
245  R m osm inh MXA MXA 24m25 s                      :    ±                        14.9mg * P<.04 e
a    R m osm inh TBA MXB 24m25 s                                                    1.01mg * P<.02
b    R m osm inh liv MXB 24m25 s                                                    no dre  P=1.

ALLYL ISOTHIOCYANATE   100ng..:..1ug....:..10......:..100....:..1mg...:..10....:..100....:..1g......:..10
246  M f b6c gav TBA MXB 24m24                             :>                       no dre  P=1.  -
a    M f b6c gav liv MXB 24m24                                                      no dre  P=1.
b    M f b6c gav lun MXB 24m24                                                      632.mg * P<.9
247  M m b6c gav lun a/c 24m24                          :    ±                      #118.mg * P<.04 -
a    M m b6c gav TBA MXB 24m24                                                      no dre  P=1.
b    M m b6c gav liv MXB 24m24                                                      no dre  P=1.
c    M m b6c gav lun MXB 24m24                                                      106.mg * P<.4
248  R f f34 gav sub fbs 24m24                               :    ±                 207.mg * P<.04 a
a    R f f34 gav TBA MXB 24m24                                                      157.mg * P<.9
b    R f f34 gav liv MXB 24m24                                                      568.mg * P<.3
249  R m f34 gav --- MXA 24m24 e                         :    ±                     54.3mg * P<.03
a    R m f34 gav --- ule 24m24 e                                                    57.2mg * P<.03
b    R m f34 gav ubl tpp 24m24 e                                                    96.0mg * P<.02 c
c    R m f34 gav TBA MXB 24m24 e                                                    38.0mg * P<.5
d    R m f34 gav liv MXB 24m24 e                                                    174.mg * P<.2

ALLYL ISOVALERATE   100ng..:..1ug....:..10......:..100....:..1mg...:..10....:..100....:..1g......:..10
250  M f b6c gav --- MXA 24m25                          :    ±                      62.8mg * P<.03 c
a    M f b6c gav --- mlm 24m25                                                      102.mg * P<.08 c
b    M f b6c gav --- mlh 24m25                                                      238.mg * P<.02 c
c    M f b6c gav TBA MXB 24m25                                                      65.5mg * P<.3
d    M f b6c gav liv MXB 24m25                                                      no dre  P=1.
e    M f b6c gav lun MXB 24m25                                                      4.45gm * P<1.
251  M m b6c gav gam sqp 24m25                            :    ±                    #295.mg * P<.04 -
a    M m b6c gav TBA MXB 24m25                                                      206.mg * P<.8
b    M m b6c gav liv MXB 24m25                                                      no dre  P=1.
c    M m b6c gav lun MXB 24m25                                                      no dre  P=1.
252  R f f34 gav TBA MXB 24m25                          :    ±                      43.5mg * P<.08 -
```

RefNum	LoConf	UpConf	Cntrl	1Dose	1Inc	2Dose	2Inc		Citation or Pathology
									Brkly Code

ALKYLBENZENESULFONATE, LINEAR   ---

| 230 | 570 | 1.09gm | n.s.s. | 1/13 | 20.0mg | 3/15 | 80.0mg | 3/13 | 300.mg | 1/15 | Hiraga;gann,68,369-370;1977 |
| 231 | 570 | 1.04gm | n.s.s. | 1/12 | 16.0mg | 2/12 | 64.0mg | 1/12 | 240.mg | 0/12 | |

ALKYLDIMETHYLAMINE OXIDES, COMMERCIAL GRADE   mixture

| 232 | 1697 | 73.8mg | n.s.s. | 32/49 | 5.00mg | 26/50 | 50.0mg | 28/50 | 100.mg | 36/50 | Cardin;faat,5,869-878;1985 |
| 233 | 1697 | 103.mg | n.s.s. | 16/50 | 4.00mg | 4/50 | 40.0mg | 11/50 | 80.0mg | 20/49 | |

ALLANTOIN   97-59-6

| 234 | 1654 | 217.mg | n.s.s. | 4/24 | 82.5mg | 3/20 | | | Lijinsky;fctx,22,715-720;1984 |
| 235 | 1654 | 241.mg | n.s.s. | 5/24 | 66.0mg | 2/20 | | | |

ALLYL ALCOHOL   107-18-6

236	1853	19.4mg	n.s.s.	0/20	10.4mg	3/20			Lijinsky;txih,3,337-345;1987/pers.comm.
a	1853	13.1mg	n.s.s.	2/20	10.4mg	6/20			
b	1853	22.8mg	n.s.s.	2/20	10.4mg	3/20			
c	1853	62.7mg	n.s.s.	1/20	10.4mg	0/20			
237	1853	16.0mg	n.s.s.	2/20	7.26mg	3/20			
a	1853	23.2mg	n.s.s.	0/20	7.26mg	1/20			
b	1853	20.8mg	n.s.s.	2/20	7.26mg	2/20			
c	1853	43.9mg	n.s.s.	1/20	7.26mg	0/20			

ALLYL CHLORIDE   (chloropropene) 107-05-1

238	TR73	288.mg	n.s.s.	0/20	78.0mg	3/50	156.mg	3/50	sto:sqc,sqp.
a	TR73	106.mg	n.s.s.	4/20	78.0mg	18/50	156.mg	15/50	
b	TR73	536.mg	n.s.s.	0/20	78.0mg	1/50	156.mg	1/50	liv:hpa,hpc,nnd.
c	TR73	258.mg	n.s.s.	1/20	78.0mg	5/50	156.mg	4/50	lun:a/a,a/c.
239	TR73	176.mg	n.s.s.	0/20	106.mg	2/50	198.mg	0/50	sto:sqc,sqp.
a	TR73	46.9mg	n.s.s.	5/20	106.mg	19/50	198.mg	3/50	
b	TR73	83.8mg	n.s.s.	2/20	106.mg	8/50	198.mg	1/50	liv:hpa,hpc,nnd.
c	TR73	144.mg	n.s.s.	3/20	106.mg	6/50	198.mg	0/50	lun:a/a,a/c.
240	TR73	13.4mg	n.s.s.	12/20	27.9mg	26/50	37.1mg	11/50	
a	TR73	175.mg	n.s.s.	2/20	27.9mg	0/50	37.1mg	0/50	liv:hpa,hpc,nnd.
241	TR73	8.98mg	n.s.s.	9/20	28.8mg	15/50·	39.8mg	4/50	
a	TR73	n.s.s.	n.s.s.	0/20	28.8mg	0/50	39.8mg	0/50	liv:hpa,hpc,nnd.

ALLYL GLYCIDYL ETHER   106-92-3

242	TR376	46.0mg	n.s.s.	0/50	7.13mg	0/50	14.3mg	5/50	S
a	TR376	100.mg	n.s.s.	0/50	7.13mg	0/50	14.3mg	1/50	
b	TR376	25.0mg	n.s.s.	35/50	7.13mg	21/50	14.3mg	31/50	
c	TR376	84.8mg	n.s.s.	6/50	7.13mg	5/50	14.3mg	2/50	liv:hpa,hpc,nnd.
d	TR376	46.7mg	n.s.s.	0/50	7.13mg	1/50	14.3mg	4/50	lun:a/a,a/c.
243	TR376	55.0mg	n.s.s.	0/50	5.94mg	0/50	11.9mg	3/50	
a	TR376	9.02mg	n.s.s.	32/50	5.94mg	26/50	(11.9mg	21/50)	
b	TR376	20.6mg	n.s.s.	23/50	5.94mg	11/50	(11.9mg	6/50)	liv:hpa,hpc,nnd.
c	TR376	27.8mg	n.s.s.	7/50	5.94mg	9/50	11.9mg	9/50	lun:a/a,a/c.
244	TR376	1.78mg	n.s.s.	43/50	1.70mg	47/50	3.40mg	44/50	
a	TR376	29.7mg	n.s.s.	1/50	1.70mg	0/50	3.40mg	0/50	liv:hpa,hpc,nnd.
245	TR376	3.48mg	n.s.s.	0/50	1.19mg	0/50	2.38mg	3/50	nof:adc; nre:ppa,sqc.
a	TR376	.457mg	n.s.s.	30/50	1.19mg	37/50	2.38mg	38/50	
b	TR376	n.s.s.	n.s.s.	0/50	1.19mg	1/50	2.38mg	0/50	liv:hpa,hpc,nnd.

ALLYL ISOTHIOCYANATE   57-06-7

246	TR234	15.1mg	n.s.s.	18/50	8.41mg	20/50	17.5mg	20/50	
a	TR234	56.1mg	n.s.s.	2/50	8.41mg	3/50	17.5mg	1/50	liv:hpa,hpc,nnd.
b	TR234	45.8mg	n.s.s.	2/50	8.41mg	2/50	17.5mg	3/50	lun:a/a,a/c.
247	TR234	40.7mg	n.s.s.	0/50	8.41mg	1/50	17.5mg	3/50	S
a	TR234	17.0mg	n.s.s.	33/50	8.41mg	22/50	17.5mg	26/50	
b	TR234	18.6mg	n.s.s.	21/50	8.41mg	14/50	17.5mg	19/50	liv:hpa,hpc,nnd.
c	TR234	26.9mg	n.s.s.	4/50	8.41mg	4/50	17.5mg	7/50	lun:a/a,a/c.
248	TR234	61.9mg	n.s.s.	0/50	8.41mg	0/50	17.5mg	3/50	
a	TR234	10.3mg	n.s.s.	42/50	8.41mg	43/50	17.5mg	42/50	
b	TR234	92.6mg	n.s.s.	0/50	8.41mg	0/50	17.5mg	1/50	liv:hpa,hpc,nnd.
249	TR234	24.1mg	n.s.s.	2/50	8.41mg	7/50	17.5mg	8/50	---:lhc,ule. S
a	TR234	25.0mg	n.s.s.	2/50	8.41mg	6/50	17.5mg	8/50	S
b	TR234	39.1mg	n.s.s.	0/50	8.41mg	2/50	17.5mg	4/50	
c	TR234	8.77mg	n.s.s.	38/50	8.41mg	45/50	17.5mg	39/50	
d	TR234	47.7mg	n.s.s.	2/50	8.41mg	0/50	17.5mg	5/50	liv:hpa,hpc,nnd.

ALLYL ISOVALERATE   2835-39-4

250	TR253	26.6mg	n.s.s.	11/50	21.5mg	11/50	43.2mg	18/50	---:mlh,mlm,mlp.
a	TR253	37.7mg	n.s.s.	6/50	21.5mg	5/50	43.2mg	10/50	
b	TR253	88.9mg	n.s.s.	0/50	21.5mg	1/50	43.2mg	1/50	
c	TR253	19.8mg	n.s.s.	31/50	21.5mg	20/50	43.2mg	30/50	
d	TR253	197.mg	n.s.s.	3/50	21.5mg	0/50	43.2mg	1/50	liv:hpa,hpc,nnd.
e	TR253	90.6mg	n.s.s.	4/50	21.5mg	4/50	43.2mg	3/50	lun:a/a,a/c.
251	TR253	100.mg	n.s.s.	0/50	21.5mg	1/50	43.2mg	1/50	S
a	TR253	25.0mg	n.s.s.	36/50	21.5mg	33/50	43.2mg	34/50	
b	TR253	67.8mg	n.s.s.	23/50	21.5mg	14/50	43.2mg	15/50	liv:hpa,hpc,nnd.
c	TR253	134.mg	n.s.s.	13/50	21.5mg	6/50	43.2mg	5/50	lun:a/a,a/c.
252	TR253	17.3mg	n.s.s.	38/50	21.5mg	43/50	43.2mg	43/50	

```
       Spe Strain  Site    Xpo+Xpt                                              TD50    2Tailpvl
           Sex   Route   Hist    Notes                                            DR     AuOp
  a     R f f34 gav liv MXB 24m25                                          no dre   P=1.
253     R m f34 gav --- MXA 24m25                        :   +        :    123.mg * P<.004 c
  a     R m f34 gav pre MXA 24m25                                          136.mg * P<.03
  b     R m f34 gav --- mle 24m25                                          149.mg * P<.02  c
  c     R m f34 gav TBA MXB 24m25                                          50.5mg * P<.2
  d     R m f34 gav liv MXB 24m25                                          264.mg * P<.2

1-ALLYL-1-NITROSOUREA               100ng..:..1ug....:..10.....:..100....:..1mg....:..10.....:..100....:..1g.....:..10
254     R f f34 gav for tum 7m24                      .    +    .                .476mg  P<.0005+
  a     R f f34 gav mgl adc 7m24                                                 .601mg  P<.002 +
  b     R f f34 gav ute mix 7m24                                                 1.87mg  P<.4  +
  c     R f f34 gav liv tum 7m24                                            no dre   P=1.
  d     R f f34 gav lun mix 7m24                                            no dre   P=1.
255     R m f34 gav for tum 7m24                  .    +    .                    .266mg  P<.0005+
  a     R m f34 gav lun mix 7m24                                                 .421mg  P<.002 +
  b     R m f34 gav col tum 7m24                                                 .541mg  P<.005 +
  c     R m f34 gav zym tum 7m24                                                 .541mg  P<.005

ALLYLHYDRAZINE.HCl                  100ng..:..1ug....:..10.....:..100.....:..1mg....:..10.....:..100....:..1g.....:..10
256     M f swa wat lun mix 24m24 e                        .    +    .           38.4mg  P<.0005+
  a     M f swa wat lun ade 24m24 e                                              46.1mg  P<.002
  b     M f swa wat lun adc 24m24 e                                              96.0mg  P<.003
  c     M f swa wat blv mix 24m24 e                                              119.mg  P<.02  +
  d     M f swa wat blv agm 24m24 e                                              163.mg  P<.02
  e     M f swa wat liv mix 24m24 e                                              234.mg  P<.09
257     M m swa wat lun ade 93w93 e                     .    +    .              28.9mg  P<.0005
  a     M m swa wat lun mix 93w93 e                                              30.9mg  P<.005 +
  b     M m swa wat liv hpt 93w93 e                                              274.mg  P<.04  -
  c     M m swa wat liv mix 93w93 e                                         no dre   P=1.

ALUMINUM POTASSIUM SULFATE          100ng..:..1ug....:..10.....:..100.....:..1mg....:..10.....:..100....:..1g.....:..10
258     M f cd1 wat --- 1km 31m31 e                  .    ±                      5.32mg  P<.02  -
  a     M f cd1 wat lun tum 31m31 e                                              11.4mg  P<.4   -
  b     M f cd1 wat tba mal 31m31 e                                              4.42mg  P<.02  -
  c     M f cd1 wat tba mix 31m31 e                                              4.23mg  P<.2   -
259     M m cd1 wat lun tum 35m35 e                          .>                  11.4mg  P<.4   -
  a     M m cd1 wat tba mal 35m35 e                                              6.29mg  P<.03  -
  b     M m cd1 wat tba mix 35m35 e                                              10.7mg  P<.5   -
260     R f leb wat tba mix 38m38 e                 .>                           4.68mg  P<.9   -
  a     R f leb wat tba mal 38m38 e                                         no dre   P=1.
261     R m leb wat tba mix 36m36 e              .    +    .                     .663mg  P<.006 -
  a     R m leb wat tba mal 36m36 e                                              1.93mg  P<.2   -

3-AMINO-4-ETHOXYACETANILIDE         100ng..:..1ug....:..10.....:..100.....:..1mg....:..10.....:..100....:..1g.....:..10
262     M f b6c eat pit MXA 78w93                                    :    +      #1.05gm \ P<.009 -
  a     M f b6c eat TBA MXB 78w93                                                569.mg \ P<.02
  b     M f b6c eat liv MXB 78w93                                           no dre   P=1.
  c     M f b6c eat lun MXB 78w93                                                8.63gm * P<.4
263     M m b6c eat thy MXA 78w93                                      :    +  : 2.07gm / P<.0005c
  a     M m b6c eat thy fcc 78w93                                                3.77gm / P<.0005c
  b     M m b6c eat thy MXA 78w93                                                4.02gm * P<.02  c
  c     M m b6c eat TBA MXB 78w93                                           no dre   P=1.
  d     M m b6c eat liv MXB 78w93                                                8.99gm / P<.9
  e     M m b6c eat lun MXB 78w93                                           no dre   P=1.
264     R f f34 eat adr MXA 18m25                                    :    +    : #415.mg \ P<.005 -
  a     R f f34 eat TBA MXB 18m25                                           no dre   P=1.
  b     R f f34 eat liv MXB 18m25                                                717.mg \ P<.002
265     R m f34 eat thy ccr 18m25                                    :    ±      #610.mg \ P<.05  -
  a     R m f34 eat TBA MXB 18m25                                           no dre   P=1.
  b     R m f34 eat liv MXB 18m25                                           no dre   P=1.

3-AMINO-9-ETHYLCARBAZOLE.HCl        100ng..:..1ug....:..10.....:..100.....:..1mg....:..10.....:..100....:..1g.....:..10
266     M f b6c eat liv hpc 78w95                              :  +  :              33.0mg  P<.0005c
  a     M f b6c eat TBA MXB 78w95                                                41.8mg  P<.0005
  b     M f b6c eat liv MXB 78w95                                                33.0mg  P<.0005
  c     M f b6c eat lun MXB 78w95                                                1.78gm  P<.6
267     M m b6c eat liv hpc 78w95                          :  +  :                46.4mg  P<.0005c
  a     M m b6c eat liv MXA 78w95                                                49.8mg  P<.0005c
  b     M m b6c eat TBA MXB 78w95                                                70.6mg  P<.005
  c     M m b6c eat liv MXB 78w95                                                49.8mg  P<.0005
  d     M m b6c eat lun MXB 78w95                                           no dre   P=1.
268     R f f34 eat MXB MXB 18m25                              :  +  :            55.1mg  P<.0005
  a     R f f34 eat ute acn 18m25                                                106.mg  P<.0005c
  b     R f f34 eat zym MXA 18m25                                                151.mg  P<.0005c
  c     R f f34 eat zym sqc 18m25                                                212.mg  P<.0005c
  d     R f f34 eat liv MXA 18m25                                                237.mg  P<.002 c
  e     R f f34 eat TBA MXB 18m25                                                44.7mg  P<.007
  f     R f f34 eat liv MXB 18m25                                                237.mg  P<.002
269     R m f34 eat MXB MXB 18m25                              :  +  :            28.1mg  P<.0005
  a     R m f34 eat liv MXA 18m25                                                39.2mg  P<.0005c
  b     R m f34 eat liv hpc 18m25                                                156.mg  P<.007 c
```

	RefNum	LoConf	UpConf	Cntrl	1Dose	1Inc	2Dose	2Inc	Citation or Pathology
									Brkly Code
a	TR253	276.mg	n.s.s.	1/50	21.5mg	1/50	43.2mg	0/50	liv:hpa,hpc,nnd.
253	TR253	59.7mg	934.mg	1/50	21.5mg	4/50	43.2mg	9/50	---:mle,mlh.
a	TR253	58.6mg	n.s.s.	0/50	21.5mg	5/50	43.2mg	2/50	pre:adn,can,sqc. S
b	TR253	66.7mg	n.s.s.	1/50	21.5mg	4/50	43.2mg	7/50	
c	TR253	17.3mg	n.s.s.	38/50	21.5mg	37/50	43.2mg	37/50	
d	TR253	79.8mg	n.s.s.	1/50	21.5mg	2/50	43.2mg	3/50	liv:hpa,hpc,nnd.

1-ALLYL-1-NITROSOUREA   760-56-5

	RefNum	LoConf	UpConf	Cntrl	1Dose	1Inc			Citation or Pathology
254	1915	.197mg	1.66mg	0/12	.608mg	7/12			Lijinsky;txih,5,925-935;1989
a	1915	.238mg	2.70mg	0/12	.608mg	6/12			
b	1915	.399mg	n.s.s.	2/12	.608mg	4/12			
c	1915	1.50mg	n.s.s.	0/12	.608mg	0/12			
d	1915	1.50mg	n.s.s.	1/12	.608mg	0/12			
255	1915	.113mg	.812mg	0/12	.426mg	8/12			
a	1915	.167mg	1.89mg	0/12	.426mg	6/12			
b	1915	.201mg	4.34mg	0/12	.426mg	5/12			
c	1915	.201mg	4.34mg	0/12	.426mg	5/12			

ALLYLHYDRAZINE.HCl   52207-83-7

	RefNum	LoConf	UpConf	Cntrl	1Dose	1Inc			Citation or Pathology
256	1051	19.3mg	156.mg	21/99	25.0mg	25/50			Toth;bjca,34,90-93;1976
a	1051	22.3mg	232.mg	18/99	25.0mg	22/50			
b	1051	41.4mg	688.mg	4/99	25.0mg	10/50			
c	1051	46.5mg	n.s.s.	5/99	25.0mg	9/50			
d	1051	59.1mg	n.s.s.	2/99	25.0mg	6/50			
e	1051	69.6mg	n.s.s.	3/99	25.0mg	5/50			
257	1051	14.4mg	112.mg	15/99	20.8mg	21/49			
a	1051	14.2mg	312.mg	23/99	20.8mg	23/49			
b	1051	67.4mg	n.s.s.	0/99	20.8mg	2/49			
c	1051	95.1mg	n.s.s.	6/99	20.8mg	2/49			

ALUMINUM POTASSIUM SULFATE   10043-67-1

	RefNum	LoConf	UpConf	Cntrl	1Dose	1Inc			Citation or Pathology
258	1395	2.19mg	n.s.s.	3/47	1.00mg	10/41			Schroeder;jnut,105,452-458;1975
a	1395	2.55mg	n.s.s.	9/47	1.00mg	11/41			
b	1395	1.90mg	n.s.s.	4/47	1.00mg	12/41			
c	1395	1.48mg	n.s.s.	14/47	1.00mg	19/41			
259	1395	2.95mg	n.s.s.	5/38	.833mg	9/41			
a	1395	2.52mg	n.s.s.	2/38	.833mg	9/41			
b	1395	2.21mg	n.s.s.	11/38	.833mg	15/41			
260	1456	.305mg	n.s.s.	17/24	.286mg	14/19			Schroeder;jnut,105,421-427;1975
a	1456	.795mg	n.s.s.	8/24	.286mg	6/19			
261	1456	.295mg	7.63mg	4/26	.250mg	13/25			
a	1456	.615mg	n.s.s.	2/26	.250mg	6/25			

3-AMINO-4-ETHOXYACETANILIDE   17026-81-2

	RefNum	LoConf	UpConf	Cntrl	1Dose	1Inc	2Dose	2Inc	Citation or Pathology
262	TR112	452.mg	39.2gm	7/100	432.mg	14/50	(845.mg	3/50)	pit:adn,can,cra. S
a	TR112	253.mg	n.s.s.	27/100	432.mg	32/50	(845.mg	21/50)	
b	TR112	2.94gm	n.s.s.	5/100	432.mg	4/50	845.mg	2/50	liv:hpa,hpc,nnd.
c	TR112	1.92gm	n.s.s.	4/100	432.mg	6/50	845.mg	3/50	lun:a/a,a/c.
263	TR112	1.04gm	6.70gm	2/100	398.mg	2/50	780.mg	12/48	thy:acn,adn,fca,fcc.
a	TR112	1.63gm	12.6gm	0/100	398.mg	0/50	780.mg	7/48	
b	TR112	1.60gm	n.s.s.	2/100	398.mg	1/50	780.mg	7/48	thy:acn,fcc.
c	TR112	879.mg	n.s.s.	41/100	398.mg	28/50	780.mg	19/48	
d	TR112	660.mg	n.s.s.	20/100	398.mg	13/50	(780.mg	2/48)	liv:hpa,hpc,nnd.
e	TR112	3.47gm	n.s.s.	17/100	398.mg	7/50	780.mg	3/48	lun:a/a,a/c.
264	TR112	175.mg	5.02gm	5/100	138.mg	11/48	(552.mg	0/50)	adr:coa,coc. S
a	TR112	763.mg	n.s.s.	80/100	138.mg	37/48	552.mg	30/50	
b	TR112	272.mg	3.45gm	0/100	138.mg	5/48	(552.mg	2/50)	liv:hpa,hpc,nnd.
265	TR112	213.mg	n.s.s.	3/99	110.mg	7/50	(442.mg	4/50)	S
a	TR112	142.mg	n.s.s.	62/99	110.mg	34/50	(442.mg	24/50)	
b	TR112	2.66gm	n.s.s.	6/99	110.mg	5/50	442.mg	1/50	liv:hpa,hpc,nnd.

3-AMINO-9-ETHYLCARBAZOLE.HCl   6109-97-3

	RefNum	LoConf	UpConf	Cntrl	1Dose	1Inc			Citation or Pathology
266	TR93	22.2mg	54.1mg	1/50	127.mg	43/50			
a	TR93	25.4mg	86.2mg	10/50	127.mg	46/50			
b	TR93	22.2mg	54.1mg	1/50	127.mg	43/50			liv:hpa,hpc,nnd.
c	TR93	256.mg	n.s.s.	3/50	127.mg	4/50			lun:a/a,a/c.
267	TR93	28.6mg	95.0mg	6/50	118.mg	41/50			
a	TR93	29.8mg	111.mg	8/50	118.mg	41/50			liv:hpa,hpc.
b	TR93	34.7mg	688.mg	22/50	118.mg	45/50			
c	TR93	29.8mg	111.mg	8/50	118.mg	41/50			liv:hpa,hpc,nnd.
d	TR93	277.mg	n.s.s.	10/50	118.mg	7/50			lun:a/a,a/c.
268	TR93	32.7mg	105.mg	1/50	73.0mg	25/50			liv:hpc,nnd; ute:acn; zym:can,sqc. C
a	TR93	50.9mg	339.mg	1/50	73.0mg	11/50			
b	TR93	75.5mg	355.mg	0/50	73.0mg	12/50			zym:can,sqc.
c	TR93	96.9mg	616.mg	0/50	73.0mg	9/50			
d	TR93	93.9mg	1.04gm	0/50	73.0mg	6/50			liv:hpc,nnd.
e	TR93	21.2mg	699.mg	38/50	73.0mg	45/50			
f	TR93	93.9mg	1.04gm	0/50	73.0mg	6/50			liv:hpa,hpc,nnd.
269	TR93	17.7mg	49.7mg	1/49	58.4mg	33/50			liv:hpc,nnd; ski:bcc,sqc,sqp; sub:sqc; zym:can,sqc. C
a	TR93	22.8mg	82.3mg	1/49	58.4mg	22/50			liv:hpc,nnd.
b	TR93	64.8mg	2.66gm	1/49	58.4mg	8/50			

```
        Spe Strain  Site    Xpo+Xpt                                                                    TD50    2Tailpvl
            Sex  Route  Hist     Notes                                                                    DR    AuOp
   c    R m f34 eat MXA MXA 18m25                                                                       203.mg   P<.003 c
   d    R m f34 eat zym MXA 18m25                                                                       240.mg   P<.003 c
   e    R m f34 eat MXA MXA 18m25                                                                       257.mg   P<.007 c
   f    R m f34 eat TBA MXB 18m25                                                                       42.7mg   P<.03
   g    R m f34 eat liv MXB 18m25                                                                       39.2mg   P<.0005

3-AMINO-9-ETHYLCARBAZOLE  MIXTURE    100ng..:..1ug....:..10......:..100....:..1mg....:..10......:..100....:..1g......:..10
 270  M f b6c eat liv hpc 78w95                                                     :+  :                30.5mg   P<.0005c
   a  M f b6c eat TBA MXB 78w95                                                                          59.7mg   P<.008
   b  M f b6c eat liv MXB 78w95                                                                          30.5mg   P<.0005
   c  M f b6c eat lun MXB 78w95                                                                          7.49gm   P<1.
 271  M m b6c eat liv hpc 78w94                                                         :  +  :          50.4mg   P<.0005c
   a  M m b6c eat TBA MXB 78w94                                                                          64.9mg   P<.007
   b  M m b6c eat liv MXB 78w94                                                                          50.4mg   P<.0005
   c  M m b6c eat lun MXB 78w94                                                                          4.13gm   P<1.
 272  R f f34 eat zym MXA 18m24                                                            :  +  :       58.0mg   P<.0005c
   a  R f f34 eat zym sqc 18m24                                                                          95.4mg   P<.006 c
   b  R f f34 eat zym can 18m24                                                                          148.mg   P<.01  c
   c  R f f34 eat TBA MXB 18m24                                                                          18.2mg   P<.04
   d  R f f34 eat liv MXB 18m24                                                                          no dre   P=1.
 273  R m f34 eat MXB MXB 18m24                                                  :  +  :                 11.8mg   P<.0005
   a  R m f34 eat liv MXA 18m24                                                                          17.1mg   P<.0005c
   b  R m f34 eat liv hpc 18m24                                                                          42.6mg   P<.002 c
   c  R m f34 eat ski MXA 18m24                                                                          45.0mg   P<.0005c
   d  R m f34 eat ski MXA 18m24                                                                          62.8mg   P<.003 c
   e  R m f34 eat zym MXA 18m24                                                                          104.mg   P<.009 c
   f  R m f34 eat TBA MXB 18m24                                                                          10.6mg   P<.0005
   g  R m f34 eat liv MXB 18m24                                                                          17.1mg   P<.0005

1-AMINO-2-METHYLANTHRAQUINONE    100ng..:..1ug....:..10......:..100....:..1mg....:..10......:..100....:..1g......:..10
 274  M f b6c eat liv MXA 70w97 aesv                                                       :  ±          174.mg * P<.07  c
   a  M f b6c eat TBA MXB 70w97 aesv                                                                     no dre   P=1.
   b  M f b6c eat liv MXB 70w97 aesv                                                                     174.mg * P<.07
   c  M f b6c eat lun MXB 70w97 aesv                                                                     no dre   P=1.
 275  M m b6c eat TBA MXB 70w97 aesv                                                           :>         no dre   P=1.   -
   a  M m b6c eat liv MXB 70w97 aesv                                                                     no dre   P=1.
   b  M m b6c eat lun MXB 70w97 aesv                                                  '                   no dre   P=1.
 276  R f f34 eat liv MXA 18m24 v                                                       :  +   :          115.mg * P<.003 c
   a  R f f34 eat liv hpc 18m24 v                                                                        182.mg * P<.002 c
   b  R f f34 eat TBA MXB 18m24 v                                                                        no dre   P=1.
   c  R f f34 eat liv MXB 18m24 v                                                                        115.mg * P<.003
 277  R m f34 eat MXB MXB 18m24 v                                                  :  +  :               34.1mg * P<.0005
   a  R m f34 eat liv MXA 18m24 v                                                                        39.9mg * P<.0005c
   b  R m f34 eat kid MXA 18m24 v                                                                        131.mg * P<.0005c
   c  R m f34 eat pit MXA 18m24 v                                                                        136.mg * P<.006
   d  R m f34 eat liv hpc 18m24 v                                                                        150.mg * P<.008 c
   e  R m f34 eat kid tla 18m24 v                                                                        196.mg * P<.003
   f  R m f34 eat thy MXA 18m24 v                                                                        283.mg * P<.007
   g  R m f34 eat kid MXA 18m24 v                                                                        441.mg * P<.02
   h  R m f34 eat TBA MXB 18m24 v                                                                        64.4mg * P<.07
   i  R m f34 eat liv MXB 18m24 v                                                                        39.9mg * P<.0005

2-AMINO-5-(5-NITRO-2-FURYL)-1,3,4-OXADIAZOLE      ...:..10....:..100....:..1mg....:..10....:..100....:..1g.....:..10
 278  R f sda eat mgl mix 46w66 ev                                          .  +  .                     3.67mg   P<.0005+
   a  R f sda eat mgl adc 46w66 ev                                                                       7.98mg   P<.0005
   b  R f sda eat for sqp 46w66 ev                                                                       37.0mg   P<.009 +
   c  R f sda eat k/p tcc 46w66 ev                                                               +hist 45.2mg   P<.02  +
   d  R f sda eat lun alc 46w66 ev                                                                       242.mg   P<.3   +
   e  R f sda eat liv tum 46w66 ev                                                                       no dre   P=1.
   f  R f sda eat tba mix 46w66 ev                                                                       2.74mg   P<.0005

2-AMINO-5-(5-NITRO-2-FURYL)-1,3,4-THIADIAZOLE     ...:..10....:..100....:..1mg....:..10....:..100....:..1g.....:..10
 279  R f sda eat mgl mix 45w66 ev                                     .  +  .                           .662mg   P<.0005+
   a  R f sda eat mgl adc 45w66 ev                                                                       1.20mg   P<.0005
   b  R f sda eat for sqp 45w66 ev                                                                       23.7mg   P<.07  +
   c  R f sda eat k/p tcc 45w66 ev                                                               +hist 73.4mg   P<.3   +
   d  R f sda eat lun alc 45w66 ev                                                                       73.4mg   P<.3   +
   e  R f sda eat liv tum 45w66 ev                                                                       no dre   P=1.
   f  R f sda eat tba mix 45w66 ev                                                                       .662mg   P<.0005

2-AMINO-4-(5-NITRO-2-FURYL)THIAZOLE     ..:..1ug....:..10....:..100....:..1mg....:..10....:..100....:..1g......:..10
 280  M f swi eat for mix 46w55 e                                           .  +  .                      7.87mg   P<.0005+
   a  M f swi eat for sqp 46w55 e                                                                        11.5mg   P<.0005
   b  M f swi eat liv tum 46w55 e                                                                        no dre   P=1.
   c  M f swi eat lun tum 46w55 e                                                                        no dre   P=1.
   d  M f swi eat tba mix 46w55 e                                                                        6.83mg   P<.0005
 281  R f fis eat for mix 52w68                                             .  +  .                      5.85mg   P<.0005+
   a  R f fis eat for sqp 52w68                                                                          8.94mg   P<.0005
   b  R f fis eat ubl mix 52w68                                                                          30.3mg   P<.003 +
```

	RefNum	LoConf	UpConf	Cntrl	1Dose	1Inc	2Dose	2Inc	Citation or Pathology	Brkly Code
c	TR93	81.4mg	1.13gm	0/49	58.4mg	6/50			ski:bcc,sqc,sqp; sub:sqc.	
d	TR93	102.mg	1.06gm	0/49	58.4mg	7/50			zym:can,sqc.	
e	TR93	95.5mg	3.17gm	0/49	58.4mg	5/50			ski:sqc,sqp; sub:sqc.	
f	TR93	18.7mg	n.s.s.	32/49	58.4mg	46/50				
g	TR93	22.8mg	82.3mg	1/49	58.4mg	22/50			liv:hpa,hpc,nnd.	

3-AMINO-9-ETHYLCARBAZOLE MIXTURE   (3-amino-9-ethylcarbazole and 3-amino-9-ethylcarbazole.HCl. CAS# 132-32-1 and 6109-97-3) mixture

	RefNum	LoConf	UpConf	Cntrl	1Dose	1Inc	2Dose	2Inc	Citation or Pathology	Brkly Code
270	TR93	20.0mg	52.1mg	1/50	85.8mg	37/50				
a	TR93	28.1mg	1.26gm	20/50	85.8mg	38/50				
b	TR93	20.0mg	52.1mg	1/50	85.8mg	37/50			liv:hpa,hpc,nnd.	
c	TR93	204.mg	n.s.s.	4/50	85.8mg	4/50			lun:a/a,a/c.	
271	TR93	28.7mg	130.mg	7/50	79.2mg	32/50				
a	TR93	30.9mg	914.mg	17/50	79.2mg	35/50				
b	TR93	28.7mg	130.mg	7/50	79.2mg	32/50			liv:hpa,hpc,nnd.	
c	TR93	165.mg	n.s.s.	6/50	79.2mg	6/50			lun:a/a,a/c.	
272	TR93	26.2mg	163.mg	0/50	30.0mg	10/50			zym:can,sqc.	
a	TR93	33.1mg	1.05gm	0/50	30.0mg	5/50				
b	TR93	55.4mg	7.12gm	0/50	30.0mg	5/50				
c	TR93	7.42mg	n.s.s.	32/50	30.0mg	40/50				
d	TR93	95.3mg	n.s.s.	2/50	30.0mg	1/50			liv:hpa,hpc,nnd.	
273	TR93	6.73mg	22.6mg	0/50	24.0mg	22/50			liv:hpc,nnd; ski:bcc,sqc,sqp; zym:can,sqc. C	
a	TR93	8.64mg	41.7mg	0/50	24.0mg	12/50			liv:hpc,nnd.	
b	TR93	16.8mg	189.mg	0/50	24.0mg	6/50				
c	TR93	19.2mg	150.mg	0/50	24.0mg	8/50			ski:bcc,sqc,sqp.	
d	TR93	24.0mg	324.mg	0/50	24.0mg	6/50			ski:sqc,sqp.	
e	TR93	37.4mg	3.31gm	0/50	24.0mg	5/50			zym:can,sqc.	
f	TR93	5.45mg	41.8mg	20/50	24.0mg	37/50				
g	TR93	8.64mg	41.7mg	0/50	24.0mg	12/50			liv:hpa,hpc,nnd.	

1-AMINO-2-METHYLANTHRAQUINONE   82-28-0

	RefNum	LoConf	UpConf	Cntrl	1Dose	1Inc	2Dose	2Inc	Citation or Pathology	Brkly Code
274	TR111	68.1mg	n.s.s.	4/50	61.1mg	12/49	58.5mg	2/50	liv:hpc,nnd.	
a	TR111	112.mg	n.s.s.	22/50	61.1mg	16/49	58.5mg	4/50		
b	TR111	68.1mg	n.s.s.	4/50	61.1mg	12/49	58.5mg	2/50	liv:hpa,hpc,nnd.	
c	TR111	325.mg	n.s.s.	1/50	61.1mg	0/49	58.5mg	1/50	lun:a/a,a/c.	
275	TR111	89.9mg	n.s.s.	21/50	56.4mg	18/50	54.0mg	1/50		
a	TR111	129.mg	n.s.s.	10/50	56.4mg	8/50	54.0mg	1/50	liv:hpa,hpc,nnd.	
b	TR111	189.mg	n.s.s.	11/50	56.4mg	5/50	(54.0mg	0/50)	lun:a/a,a/c.	
276	TR111	60.8mg	602.mg	2/50	37.0mg	11/45	74.0mg	11/48	liv:hpc,nnd.	
a	TR111	89.9mg	832.mg	1/50	37.0mg	3/45	74.0mg	10/48		
b	TR111	78.4mg	n.s.s.	45/50	37.0mg	31/45	74.0mg	28/48		
c	TR111	60.8mg	602.mg	2/50	37.0mg	11/45	74.0mg	11/48	liv:hpa,hpc,nnd.	
277	TR111	22.8mg	64.6mg	3/50	30.4mg	27/50	60.0mg	28/50	kid:acn,tla,uac; liv:hpc,nnd. C	
a	TR111	25.8mg	83.7mg	3/50	30.4mg	25/50	60.0mg	24/50	liv:hpc,nnd.	
b	TR111	73.2mg	270.mg	0/50	30.4mg	6/50	60.0mg	10/50	kid:acn,tla,uac.	
c	TR111	70.5mg	1.61gm	1/50	30.4mg	10/50	60.0mg	8/50	pit:adn,cra. S	
d	TR111	72.9mg	3.56gm	2/50	30.4mg	7/50	60.0mg	10/50		
e	TR111	98.4mg	929.mg	0/50	30.4mg	5/50	60.0mg	6/50		S
f	TR111	127.mg	3.06gm	0/50	30.4mg	3/50	60.0mg	5/50	thy:cca,ccr. S	
g	TR111	167.mg	n.s.s.	0/50	30.4mg	1/50	60.0mg	4/50	kid:acn,uac. S	
h	TR111	26.1mg	n.s.s.	26/50	30.4mg	42/50	60.0mg	35/50		
i	TR111	25.8mg	83.7mg	3/50	30.4mg	25/50	60.0mg	24/50	liv:hpa,hpc,nnd.	

2-AMINO-5-(5-NITRO-2-FURYL)-1,3,4-OXADIAZOLE   3775-55-1

	RefNum	LoConf	UpConf	Cntrl	1Dose	1Inc	2Dose	2Inc	Citation or Pathology	Brkly Code
278	1126	2.06mg	7.05mg	2/24	26.9mg	29/33			Cohen;jnci,54,841-850;1975	
a	1126	4.59mg	15.4mg	0/24	26.9mg	20/33				
b	1126	15.1mg	855.mg	0/24	26.9mg	6/33				
c	1126	17.1mg	n.s.s.	0/24	26.9mg	5/33				
d	1126	39.3mg	n.s.s.	0/24	26.9mg	1/33				
e	1126	73.8mg	n.s.s.	0/24	26.9mg	0/33				
f	1126	1.41mg	5.35mg	2/24	26.9mg	31/33				

2-AMINO-5-(5-NITRO-2-FURYL)-1,3,4-THIADIAZOLE   712-68-5

	RefNum	LoConf	UpConf	Cntrl	1Dose	1Inc	2Dose	2Inc	Citation or Pathology	Brkly Code
279	1126	.293mg	1.36mg	2/24	8.18mg	32/33			Cohen;jnci,54,841-850;1975	
a	1126	.695mg	2.16mg	0/24	8.18mg	28/33				
b	1126	7.17mg	n.s.s.	0/24	8.18mg	3/33				
c	1126	11.9mg	n.s.s.	0/24	8.18mg	1/33				
d	1126	11.9mg	n.s.s.	0/24	8.18mg	1/33				
e	1126	22.4mg	n.s.s.	0/24	8.18mg	0/33				
f	1126	.293mg	1.36mg	2/24	8.18mg	32/33				

2-AMINO-4-(5-NITRO-2-FURYL)THIAZOLE   38514-71-5

	RefNum	LoConf	UpConf	Cntrl	1Dose	1Inc	2Dose	2Inc	Citation or Pathology	Brkly Code
280	1076	4.20mg	15.3mg	0/29	90.2mg	24/27			Cohen;canr,33,1593-1597;1973	
a	1076	6.42mg	22.3mg	0/29	90.2mg	21/27				
b	1076	140.mg	n.s.s.	0/29	90.2mg	0/27				
c	1076	140.mg	n.s.s.	0/29	90.2mg	0/27				
d	1076	3.40mg	14.0mg	2/29	90.2mg	25/27				
281	1423	2.49mg	12.7mg	0/10	63.5mg	23/24			Wang;carc,3,275-277;1982	
a	1423	4.66mg	18.1mg	0/10	63.5mg	21/24				
b	1423	15.0mg	122.mg	0/10	63.5mg	11/24				

```
         Spe Strain Site   Xpo+Xpt                                                      TD50    2Tailpvl
             Sex  Route   Hist       Notes                                                DR    AuOp

trans-5-AMINO-3[2-(5-NITRO-2-FURYL)VINYL]-1,2,4-OXADIAZOLE...:..1 _00.....:..1mg...:..10....:..100....:..1g....:..10
282  M f cd1 eat for sqc 31w75 es                                           . + .        105.mg   P<.0005+
  a  M f cd1 eat thm lym 31w75 es                                                        262.mg   P<.005 +
  b  M f cd1 eat for sqp 31w75 es                                                        1.15gm   P<.1  +
  c  M f cd1 eat lun ade 31w75 es                                                        1.01gm   P<.5
  d  M f cd1 eat liv hem 31w75 es                                                        no dre   P=1.
  e  M f cd1 eat rel mix 31w75 es                                                        no dre   P=1.  +
  f  M f cd1 eat tba mix 31w75 es                                                        no dre   P=1.
283  M m cd1 eat thm lym 38w75 es                                           . + .        121.mg   P<.0005+
  a  M m cd1 eat for sqc 38w75 es                                                        130.mg   P<.0005+
  b  M m cd1 eat for sqp 38w75 es                                                        514.mg   P<.008 +
  c  M m cd1 eat rel mix 38w75 es                                                        492.mg   P<.08  +
  d  M m cd1 eat liv hem 38w75 es                                                        no dre   P=1.
  e  M m cd1 eat lun ade 38w75 es                                                        no dre   P=1.
  f  M m cd1 eat tba mix 38w75 es                                                        29.2mg   P<.0005

2-AMINO-4-NITROPHENOL            100ng..:..1ug....:..10....:..100....:..1mg...:..10....:..100....:..1g......:..10
284  M f b6c gav TBA MXB 24m24                                        :>                 no dre   P=1.  -
  a  M f b6c gav liv MXB 24m24                                                           2.16gm * P<.5
  b  M f b6c gav lun MXB 24m24                                                           2.18gm * P<.5
285  M m b6c gav MXA MXA 24m24                                      : +    :            #798.mg * P<.007 -
  a  M m b6c gav TBA MXB 24m24                                                           582.mg * P<.6
  b  M m b6c gav liv MXB 24m24                                                           no dre   P=1.
  c  M m b6c gav lun MXB 24m24                                                           no dre   P=1.
286  R f f34 gav TBA MXB 24m24                                        :>                 no dre   P=1.  -
  a  R f f34 gav liv MXB 24m24                                                           no dre   P=1.
287  R m f34 gav tes ict 24m24 s                                  : +    :               68.2mg * P<.003
  a  R m f34 gav kcx adn 24m24 s                                                         839.mg * P<.01  p
  b  R m f34 gav sub MXA 24m24 s                                                         458.mg * P<.05
  c  R m f34 gav sub fib 24m24 s                                                         584.mg * P<.05
  d  R m f34 gav liv MXA 24m24 s                                                         1.30gm * P<.02
  e  R m f34 gav TBA MXB 24m24 s                                                         84.1mg * P<.02
  f  R m f34 gav liv MXB 24m24 s                                                         1.30gm * P<.02

2-AMINO-5-NITROPHENOL            100ng..:..1ug....:..10....:..100....:..1mg...:..10....:..100....:..1g......:..10
288  M f b6c gav TBA MXB 24m24 ns                                     :>                 5.76gm * P<.9  -
  a  M f b6c gav liv MXB 24m24 ns                                                        no dre   P=1.
  b  M f b6c gav lun MXB 24m24 ns                                                        807.gm * P<1.
289  M m b6c gav TBA MXB 24m24 ns                                        :>              no dre   P=1.  -
  a  M m b6c gav liv MXB 24m24 ns                                                        no dre   P=1.
  b  M m b6c gav lun MXB 24m24 ns                                                        no dre   P=1.
290  R f f34 gav TBA MXB 24m24                                       :>                  no dre   P=1.  -
  a  R f f34 gav liv MXB 24m24                                                           no dre   P=1.
291  R m f34 gav tes ict 24m24 s                                 : +  :                  28.6mg * P<.0005
  a  R m f34 gav pan MXA 24m24 s                                                         107.mg * P<.0005
  b  R m f34 gav pan ana 24m24 s                                                         111.mg * P<.0005p
  c  R m f34 gav pre can 24m24 s                                                         562.mg / P<.004
  d  R m f34 gav amd MXA 24m24 s                                                         101.mg * P<.02
  e  R m f34 gav pre MXA 24m24 s                                                         392.mg * P<.04
  f  R m f34 gav pni isc 24m24 s                                                         393.mg * P<.04
  g  R m f34 gav TBA MXB 24m24 s                                                         33.6mg * P<.0005
  h  R m f34 gav liv MXB 24m24 s                                                         1.56gm * P<.05

4-AMINO-2-NITROPHENOL            100ng..:..1ug...:..10....:..100....:..1mg...:..10....:..100....:..1g....:..10
292  M f b6c eat TBA MXB 24m24                                        :>                 no dre   P=1.  -
  a  M f b6c eat liv MXB 24m24                                                           3.65gm * P<.3
  b  M f b6c eat lun MXB 24m24                                                           no dre   P=1.
293  M m b6c eat TBA MXB 24m24                                      :>                   no dre   P=1.  -
  a  M m b6c eat liv MXB 24m24                                                           2.25gm * P<.7
  b  M m b6c eat lun MXB 24m24                                                           no dre   P=1.
294  R f f34 eat ubl tcc 24m24                                      :>      +hist 1.88gm * P<.3  a
  a  R f f34 eat TBA MXB 24m24                                                           no dre   P=1.
  b  R f f34 eat liv MXB 24m24                                                           no dre   P=1.
295  R m f34 eat ubl tcc 24m24                                  : +    :                 309.mg / P<.002 c
  a  R m f34 eat TBA MXB 24m24                                                           270.mg / P<.4
  b  R m f34 eat liv MXB 24m24                                                           no dre   P=1.

2-AMINO-4-(p-NITROPHENYL)THIAZOLE 100ng..:..1ug....:..10....:..100....:..1mg...:..10....:..100....:..1g....:..10
296  M f swi eat --- lle 46w66 e                            . ±                          9.95mg   P<.03 +
  a  M f swi eat liv tum 46w66 e                                                         no dre   P=1.
  b  M f swi eat lun tum 46w66 e                                                         no dre   P=1.
  c  M f swi eat tba mix 46w66 e                                                         9.95mg   P<.03

2-AMINO-5-NITROTHIAZOLE          100ng..:..1ug....:..10....:..100....:..1mg...:..10....:..100....:..1g....:..10
297  M f b6c eat TBA MXB 24m24                                        :>                 no dre   P=1.  -
  a  M f b6c eat liv MXB 24m24                                                           109.mg * P<.4
  b  M f b6c eat lun MXB 24m24                                                           63.5mg * P<.09
298  M m b6c eat TBA MXB 24m24                                     :>                    no dre   P=1.  -
  a  M m b6c eat liv MXB 24m24                                                           no dre   P=1.
  b  M m b6c eat lun MXB 24m24                                                           no dre   P=1.
299  R f f34 eat ute esp 26m26                              : ±                         #57.2mg \ P<.03 -
```

RefNum	LoConf	UpConf	Cntrl	1Dose	1Inc	2Dose	2Inc	Citation or Pathology
								Brkly Code

trans-5-AMINO-3[2-(5-NITRO-2-FURYL)VINYL]-1,2,4-OXADIAZOLE   (SQ 18506) 28754-68-9

RefNum	LoConf	UpConf	Cntrl	1Dose	1Inc	2Dose	2Inc	Citation or Pathology / Brkly Code
282	1601	59.8mg	206.mg	0/50	134.mg	18/49			Dunsford;jnci,73,151-160;1984
a	1601	115.mg	2.33gm	1/50	134.mg	9/49			
b	1601	283.mg	n.s.s.	0/50	134.mg	2/49			
c	1601	188.mg	n.s.s.	4/50	134.mg	6/49			
d	1601	705.mg	n.s.s.	3/50	134.mg	0/49			
e	1601	234.mg	n.s.s.	9/50	134.mg	7/49			
f	1601	138.mg	n.s.s.	32/50	134.mg	24/49			
283	1601	69.5mg	240.mg	0/50	152.mg	18/50			
a	1601	73.6mg	263.mg	0/50	152.mg	17/50			
b	1601	195.mg	9.94gm	0/50	152.mg	5/50			
c	1601	171.mg	n.s.s.	2/50	152.mg	7/50			
d	1601	586.mg	n.s.s.	3/50	152.mg	1/50			
e	1601	652.mg	n.s.s.	6/50	152.mg	1/50			
f	1601	16.6mg	59.4mg	18/50	152.mg	45/50			

2-AMINO-4-NITROPHENOL   99-57-0

RefNum	LoConf	UpConf	Cntrl	1Dose	1Inc	2Dose	2Inc	Pathology / Brkly Code
284	TR339	181.mg	n.s.s.	32/50	87.6mg	27/50	175.mg	29/50	
a	TR339	449.mg	n.s.s.	2/50	87.6mg	2/50	175.mg	4/50	liv:hpa,hpc,nnd.
b	TR339	450.mg	n.s.s.	2/50	87.6mg	2/50	175.mg	4/50	lun:a/a,a/c.
285	TR339	323.mg	2.80gm	0/50	87.6mg	1/50	175.mg	5/50	abc:hem; liv:hes; mln:hes; pan:hem; spl:hes; sub:hem. S
a	TR339	103.mg	n.s.s.	34/50	87.6mg	39/50	175.mg	35/50	
b	TR339	292.mg	n.s.s.	15/50	87.6mg	18/50	175.mg	10/50	liv:hpa,hpc,nnd.
c	TR339	363.mg	n.s.s.	9/50	87.6mg	8/50	175.mg	6/50	lun:a/a,a/c.
286	TR339	146.mg	n.s.s.	42/50	87.6mg	39/50	175.mg	39/50	
a	TR339	n.s.s.	n.s.s.	0/50	87.6mg	1/50	175.mg	0/50	liv:hpa,hpc,nnd.
287	TR339	35.2mg	387.mg	39/50	87.6mg	39/50	175.mg	36/50	S
a	TR339	286.mg	62.3gm	0/50	87.6mg	1/50	175.mg	3/50	
b	TR339	171.mg	n.s.s.	2/50	87.6mg	6/50	175.mg	4/50	sub:fbs,fib,nfs,srn. S
c	TR339	213.mg	n.s.s.	1/50	87.6mg	5/50	175.mg	3/50	S
d	TR339	368.mg	n.s.s.	0/50	87.6mg	0/50	175.mg	3/50	liv:hpc,nnd. S
e	TR339	39.6mg	13.1gm	45/50	87.6mg	43/50	175.mg	37/50	
f	TR339	368.mg	n.s.s.	0/50	87.6mg	0/50	175.mg	3/50	liv:hpa,hpc,nnd.

2-AMINO-5-NITROPHENOL   121-88-0

RefNum	LoConf	UpConf	Cntrl	1Dose	1Inc	2Dose	2Inc	Pathology / Brkly Code
288	TR334	364.mg	n.s.s.	29/50	283.mg	30/50	566.mg	8/50	
a	TR334	1.48gm	n.s.s.	5/50	283.mg	3/50	566.mg	1/50	liv:hpa,hpc,nnd.
b	TR334	1.14gm	n.s.s.	4/50	283.mg	4/50	566.mg	1/50	lun:a/a,a/c.
289	TR334	550.mg	n.s.s.	31/50	283.mg	32/50	566.mg	8/50	
a	TR334	497.mg	n.s.s.	17/50	283.mg	16/50	(566.mg | 1/50)	liv:hpa,hpc,nnd.
b	TR334	1.02gm	n.s.s.	7/50	283.mg	8/50	566.mg	2/50	lun:a/a,a/c.
290	TR334	105.mg	n.s.s.	46/50	70.7mg	45/50	142.mg	38/50	
a	TR334	n.s.s.	n.s.s.	0/50	70.7mg	0/50	142.mg	0/50	liv:hpa,hpc,nnd.
291	TR334	16.8mg	65.2mg	42/50	70.7mg	40/50	142.mg	39/50	
a	TR334	52.7mg	328.mg	1/50	70.7mg	11/50	142.mg	3/50	pan:acc,ana. S
b	TR334	53.8mg	359.mg	1/50	70.7mg	10/50	142.mg	3/50	S
c	TR334	163.mg	5.02gm	0/50	70.7mg	0/50	142.mg	4/50	S
d	TR334	43.3mg	n.s.s.	20/50	70.7mg	16/50	142.mg	12/50	amd:phe,phm. S
e	TR334	126.mg	n.s.s.	3/50	70.7mg	2/50	142.mg	5/50	pre:adn,can. S
f	TR334	118.mg	n.s.s.	0/50	70.7mg	3/50	142.mg	0/50	S
g	TR334	19.0mg	88.7mg	42/50	70.7mg	44/50	142.mg	34/50	
h	TR334	332.mg	n.s.s.	0/50	70.7mg	0/50	142.mg	2/50	liv:hpa,hpc,nnd.

4-AMINO-2-NITROPHENOL   119-34-6

RefNum	LoConf	UpConf	Cntrl	1Dose	1Inc	2Dose	2Inc	Pathology / Brkly Code
292	TR94	558.mg	n.s.s.	8/20	159.mg	17/50	319.mg	16/50	
a	TR94	1.26gm	n.s.s.	0/20	159.mg	2/50	319.mg	2/50	liv:hpa,hpc,nnd.
b	TR94	1.73gm	n.s.s.	2/20	159.mg	3/50	319.mg	2/50	lun:a/a,a/c.
293	TR94	301.mg	n.s.s.	12/20	147.mg	32/50	294.mg	28/50	
a	TR94	336.mg	n.s.s.	6/20	147.mg	18/50	294.mg	19/50	liv:hpa,hpc,nnd.
b	TR94	863.mg	n.s.s.	5/20	147.mg	10/50	294.mg	7/50	lun:a/a,a/c.
294	TR94	563.mg	n.s.s.	0/20	61.3mg	1/49	123.mg	2/50	
a	TR94	140.mg	n.s.s.	13/20	61.3mg	34/49	123.mg	28/50	
b	TR94	n.s.s.	n.s.s.	0/20	61.3mg	0/49	123.mg	0/50	liv:hpa,hpc,nnd.
295	TR94	155.mg	1.14gm	0/20	49.0mg	0/50	98.1mg	11/50	
a	TR94	71.2mg	n.s.s.	12/20	49.0mg	22/50	98.1mg	35/50	
b	TR94	604.mg	n.s.s.	0/20	49.0mg	1/50	98.1mg	0/50	liv:hpa,hpc,nnd.

2-AMINO-4-(p-NITROPHENYL)THIAZOLE   2104-09-8

RefNum	LoConf	UpConf	Cntrl	1Dose	1Inc	2Dose	2Inc	Citation / Brkly Code
296	1076	3.42mg	n.s.s.	1/28	9.06mg	5/20			Cohen;canr,33,1593-1597;1973
a	1076	15.0mg	n.s.s.	0/28	9.06mg	0/20			
b	1076	15.0mg	n.s.s.	0/28	9.06mg	0/20			
c	1076	3.42mg	n.s.s.	1/28	9.06mg	5/20			

2-AMINO-5-NITROTHIAZOLE   121-66-4

RefNum	LoConf	UpConf	Cntrl	1Dose	1Inc	2Dose	2Inc	Pathology / Brkly Code
297	TR53	15.3mg	n.s.s.	26/50	6.50mg	31/50	13.0mg	28/50	
a	TR53	29.0mg	n.s.s.	2/50	6.50mg	6/50	13.0mg	5/50	liv:hpa,hpc,nnd.
b	TR53	24.6mg	n.s.s.	2/50	6.50mg	4/50	13.0mg	8/50	lun:a/a,a/c.
298	TR53	10.5mg	n.s.s.	39/50	6.00mg	32/50	12.0mg	34/50	
a	TR53	19.5mg	n.s.s.	20/50	6.00mg	16/50	12.0mg	15/50	liv:hpa,hpc,nnd.
b	TR53	20.8mg	n.s.s.	14/50	6.00mg	12/50	12.0mg	12/50	lun:a/a,a/c.
299	TR53	22.8mg	n.s.s.	2/50	14.9mg	9/50	(29.7mg | 3/50)	S

#	Spe	Sex	Strain	Route	Site	Hist	Xpo+Xpt	Notes	Plot	TD50	DR	2Tailpvl	AuOp
a	R	f	f34	eat	TBA	MXB	26m26			326.mg	*	P<.9	
b	R	f	f34	eat	liv	MXB	26m26			no dre		P=1.	
300	R	m	f34	eat	---	MXA	26m26		: + :	27.6mg	*	P<.003	a
a	R	m	f34	eat	mul	grl	26m26			90.2mg	*	P<.02	a
b	R	m	f34	eat	MXA	MXA	26m26			445.mg	*	P<.7	a
c	R	m	f34	eat	TBA	MXB	26m26			32.3mg	*	P<.2	
d	R	m	f34	eat	liv	MXB	26m26			318.mg	*	P<.2	
301	R	f	sda	eat	mgl	mix	46w66	ev	. ±	44.6mg		P<.03	+
a	R	f	sda	eat	k/p	tcc	46w66	v	+hist	157.mg		P<.09	+
b	R	f	sda	eat	lun	alc	46w66	v		157.mg		P<.09	+
c	R	f	sda	eat	liv	tum	46w66	v		no dre		P=1.	
d	R	f	sda	eat	tba	mix	46w66	v		32.5mg		P<.006	

2-AMINO-5-PHENYL-2-OXAZOLIN-4-ONE + Mg(OH)2.1ug....:...10.......:...100....:...1mg....:...10.......:...100....:...1g.....:...10

#	Spe	Sex	Strain	Route	Site	Hist	Xpo+Xpt	Notes	Plot	TD50	DR	2Tailpvl	AuOp
302	R	f	sda	eat	liv	tum	46w66	e	.>	no dre		P=1.	
a	R	f	sda	eat	tba	mix	46w66	e		3.01mg		P<.04	-
303	R	f	sda	eat	liv	tum	46w66	e	.>	no dre		P=1.	
a	R	f	sda	eat	tba	mix	46w66	e		no dre		P=1.	-

2-AMINOANTHRAQUINONE 100ng..:...1ug.....:...10.......:...100....:...1mg....:...10.......:...100....:...1g......:...10

#	Spe	Sex	Strain	Route	Site	Hist	Xpo+Xpt	Notes	Plot	TD50	DR	2Tailpvl	AuOp
304	M	f	b6c	eat	MXB	MXB	79w94		: + :	1.49gm	*	P<.0005	
a	M	f	b6c	eat	liv	hpc	79w94			2.82gm	*	P<.004	c
b	M	f	b6c	eat	---	lym	79w94			3.16gm	*	P<.02	c
c	M	f	b6c	eat	TBA	MXB	79w94			1.62gm	/	P<.02	
d	M	f	b6c	eat	liv	MXB	79w94			2.82gm	*	P<.004	
e	M	f	b6c	eat	lun	MXB	79w94			33.4gm	*	P<.7	
305	M	m	b6c	eat	liv	hpc	79w93		: + :	755.mg	*	P<.0005c	
a	M	m	b6c	eat	liv	MXA	79w93			798.mg	*	P<.0005c	
b	M	m	b6c	eat	TBA	MXB	79w93			952.mg	/	P<.008	
c	M	m	b6c	eat	liv	MXB	79w93			798.mg	*	P<.0005	
d	M	m	b6c	eat	lun	MXB	79w93			19.4gm	*	P<.8	
306	R	f	f34	eat	TBA	MXB	18m25	s	:>	124.mg		P<.6	-
a	R	f	f34	eat	liv	MXB	18m25	s		no dre		P=1.	
307	R	m	f34	eat	liv	MXA	18m25	v	: + :	101.mg	\	P<.0005c	
a	R	m	f34	eat	liv	hpc	18m25	v		268.mg	\	P<.002	c
b	R	m	f34	eat	TBA	MXB	18m25	v		759.mg	*	P<.6	
c	R	m	f34	eat	liv	MXB	18m25	v		101.mg	\	P<.0005	

o-AMINOAZOTOLUENE 100ng..:...1ug.....:...10.......:...100....:...1mg....:...10.......:...100....:...1g......:...10

#	Spe	Sex	Strain	Route	Site	Hist	Xpo+Xpt	Notes	Plot	TD50	DR	2Tailpvl	AuOp
308	M	m	cf1	eat	lun	ade	26w65	e	.>	80.2mg		P<.6	
a	M	m	cf1	eat	liv	mix	26w65	e		no dre		P=1.	
309	R	f	alb	eat	liv	mix	63w63	ems	. + .	3.70mg		P<.002	+
a	R	f	alb	eat	liv	hpn	63w63	ems		8.49mg		P<.02	+
b	R	f	alb	eat	liv	lca	63w63	ems		21.4mg		P<.2	+
310	R	m	alb	eat	liv	mix	67w67	ems	. + .	4.46mg		P<.005	+
a	R	m	alb	eat	liv	hpn	67w67	ems		11.5mg		P<.06	+
b	R	m	alb	eat	liv	lca	67w67	ems		15.2mg		P<.1	+

4-AMINODIPHENYL 100ng..:...1ug.....:...10.......:...100....:...1mg....:...10.......:...100....:...1g......:...10

#	Spe	Sex	Strain	Route	Site	Hist	Xpo+Xpt	Notes	Plot	TD50	DR	2Tailpvl	AuOp
311	M	f	abi	gav	ubl	car	39w94		. ±	5.94mg		P<.05	+
a	M	f	abi	gav	liv	hpt	39w94	e		no dre		P=1.	-
312	M	m	abi	gav	liv	hpt	39w94	e	.>	3.84mg		P<.5	-
a	M	m	abi	gav	ubl	car	39w94	e		11.6mg		P<.4	+
313	M	f	cif	gav	liv	mix	50w70	r	. + .	.993mg		P<.0005+	
a	M	f	cif	gav	liv	mhp	50w70	r		3.05mg		P<.0005	
314	M	m	cif	gav	liv	mix	50w70	r	. + .	3.91mg		P<.002	+
a	M	m	cif	gav	liv	mhp	50w70	r		7.50mg		P<.02	
b	M	m	cif	gav	ubl	car	50w70	r		32.5mg		P<.3	+

4-AMINODIPHENYL.HCl 100ng..:...1ug.....:...10.......:...100....:...1mg....:...10.......:...100....:...1g......:...10

#	Spe	Sex	Strain	Route	Site	Hist	Xpo+Xpt	Notes	Plot	TD50	DR	2Tailpvl	AuOp
315	R	f	sda	gav	mgl	car	31w52	ev	. + .	.980mg		P<.0005+	
a	R	f	sda	gav	mam	mix	31w52	ev		.984mg		P<.0005+	
b	R	f	sda	gav	liv	tum	31w52	ev		no dre		P=1.	-
c	R	f	sda	gav	tba	mix	31w52	ev		.897mg		P<.0005	

2-AMINODIPHENYLENE OXIDE 100ng..:...1ug.....:...10.......:...100....:...1mg....:...10.......:...100....:...1g......:...10

#	Spe	Sex	Strain	Route	Site	Hist	Xpo+Xpt	Notes	Plot	TD50	DR	2Tailpvl	AuOp
316	M	f	cif	eat	liv	mhp	52w58	r	<+	noTD50		P<.0005	
a	M	f	cif	eat	liv	mix	52w58	r		noTD50		P<.0005+	
317	M	m	cif	eat	liv	mix	52w67	r	. + .	5.73mg		P<.0005+	
a	M	m	cif	eat	liv	mhp	52w67	r		8.67mg		P<.0005	
b	M	m	cif	eat	ubl	car	52w67	r		22.3mg		P<.004	+

1-(AMINOMETHYL)CYCLOHEXANEACETIC ACID ..:...1ug.....:...10.......:...100....:...1mg....:...10.......:...100....:...1g......:...10

#	Spe	Sex	Strain	Route	Site	Hist	Xpo+Xpt	Notes	Plot	TD50	DR	2Tailpvl	AuOp
318	R	f	wis	eat	kid	nep	24m24	r		111.gm	*	P<.2	-
a	R	f	wis	eat	kid	tcc	24m24	r		111.gm	*	P<.2	-
b	R	f	wis	eat	kid	msm	24m24	r		no dre		P=1.	
c	R	f	wis	eat	kid	rcc	24m24	r		no dre		P=1.	
319	R	m	wis	eat	kid	lip	24m24	r		55.2gm	*	P<.2	-
a	R	m	wis	eat	kid	lps	24m24	r		349.gm	*	P<.8	-

	RefNum	LoConf	UpConf	Cntrl	1Dose	1Inc	2Dose	2Inc		Citation or Pathology
										Brkly Code
a	TR53	21.1mg	n.s.s.	40/50	14.9mg	44/50	29.7mg	44/50		
b	TR53	n.s.s.	n.s.s.	0/50	14.9mg	0/50	29.7mg	0/50		liv:hpa,hpc,nnd.
300	TR53	14.6mg	147.mg	13/50	11.9mg	19/50	23.8mg	28/50		---:leu,lym.
a	TR53	40.1mg	n.s.s.	2/50	11.9mg	4/50	23.8mg	9/50		
b	TR53	56.5mg	n.s.s.	6/50	11.9mg	4/50	23.8mg	6/50		---:lle,mly; mul:ule.
c	TR53	11.7mg	n.s.s.	36/50	11.9mg	37/50	23.8mg	39/50		
d	TR53	78.3mg	n.s.s.	0/50	11.9mg	1/50	23.8mg	1/50		liv:hpa,hpc,nnd.
301	1126	17.4mg	n.s.s.	2/39	33.4mg	8/35				Cohen;jnci,54,841-850;1975
a	1126	38.6mg	n.s.s.	0/39	33.4mg	2/35				
b	1126	38.6mg	n.s.s.	0/39	33.4mg	2/35				
c	1126	97.1mg	n.s.s.	0/39	33.4mg	0/35				
d	1126	14.2mg	371.mg	2/39	33.4mg	10/35				

2-AMINO-5-PHENYL-2-OXAZOLIN-4-ONE + Mg(OH)2 (magnesium pemoline) 18968-99-5

	RefNum	LoConf	UpConf	Cntrl	1Dose	1Inc				Citation or Pathology
302	200am	8.10mg	n.s.s.	0/35	2.79mg	0/35				Cohen;jnci,51,403-417;1973
a	200am	1.18mg	n.s.s.	4/35	2.79mg	11/35				
303	200an	162.mg	n.s.s.	0/35	55.8mg	0/35				
a	200an	122.mg	n.s.s.	4/35	55.8mg	1/35				

2-AMINOANTHRAQUINONE   117-79-3

	RefNum	LoConf	UpConf	Cntrl	1Dose	1Inc	2Dose	2Inc		Citation or Pathology
304	TR144	805.mg	4.96gm	11/100	539.mg	9/50	1.08gm	23/50		---:lym; liv:hpc. C
a	TR144	1.33gm	23.7gm	5/100	539.mg	5/50	1.08gm	12/50		
b	TR144	1.38gm	n.s.s.	7/100	539.mg	6/50	1.08gm	12/50		
c	TR144	742.mg	n.s.s.	27/100	539.mg	16/50	1.08gm	29/50		
d	TR144	1.33gm	23.7gm	5/100	539.mg	5/50	1.08gm	12/50		liv:hpa,hpc,nnd.
e	TR144	3.75gm	n.s.s.	4/100	539.mg	2/50	1.08gm	3/50		lun:a/a,a/c.
305	TR144	449.mg	1.83gm	18/100	498.mg	20/50	1.01gm	36/50		
a	TR144	463.mg	2.11gm	20/100	498.mg	20/50	1.01gm	36/50		liv:hpa,hpc.
b	TR144	458.mg	20.0gm	41/100	498.mg	28/50	1.01gm	42/50		
c	TR144	463.mg	2.11gm	20/100	498.mg	20/50	1.01gm	36/50		liv:hpa,hpc,nnd.
d	TR144	1.94gm	n.s.s.	17/100	498.mg	8/50	1.01gm	11/50		lun:a/a,a/c.
306	TR144	18.9mg	n.s.s.	19/25	71.0mg	17/50				
a	TR144	89.0mg	n.s.s.	2/25	71.0mg	1/50				liv:hpa,hpc,nnd.
307	TR144	57.8mg	204.mg	0/50	102.mg	18/50	(203.mg	18/50)		liv:hpc,nnd.
a	TR144	121.mg	1.13gm	0/50	102.mg	8/50	(203.mg	5/50)		
b	TR144	145.mg	n.s.s.	20/50	102.mg	26/50	203.mg	30/50		
c	TR144	57.8mg	204.mg	0/50	102.mg	18/50	(203.mg	18/50)		liv:hpa,hpc,nnd.

o-AMINOAZOTOLUENE   97-56-3

	RefNum	LoConf	UpConf	Cntrl	1Dose	1Inc				Citation or Pathology
308	103	11.5mg	n.s.s.	95/288	28.8mg	9/23				Walker;fctx,11,415-432;1973
a	103	22.7mg	n.s.s.	58/288	28.8mg	4/23				
309	30	1.16mg	16.9mg	0/5	28.6mg	6/7				Waters;yjbm,10,179-184;1937
a	30	2.77mg	n.s.s.	0/5	28.6mg	4/7				
b	30	5.18mg	n.s.s.	0/5	28.6mg	2/7				
310	30	1.91mg	27.7mg	0/4	21.7mg	9/12				
a	30	4.27mg	n.s.s.	0/4	21.7mg	5/12				
b	30	5.20mg	n.s.s.	0/4	21.7mg	4/12				

4-AMINODIPHENYL   92-67-1

	RefNum	LoConf	UpConf	Cntrl	1Dose	1Inc				Citation or Pathology
311	1165	.957mg	n.s.s.	0/28	2.37mg	1/5				Clayson;bjca,19,297-310;1965/Williams 1962
a	1165	1.99mg	n.s.s.	4/26	2.37mg	0/5				
312	1165	.761mg	n.s.s.	5/13	1.97mg	7/13				
a	1165	1.88mg	n.s.s.	0/5	1.97mg	1/11				
313	31	.544mg	1.94mg	1/31	6.12mg	24/28				Clayson;bjca,21,755-762;1967
a	31	1.58mg	6.92mg	0/31	6.12mg	13/28				
314	31	1.67mg	16.2mg	0/19	5.10mg	7/21				
a	31	2.58mg	n.s.s.	0/19	5.10mg	4/21				
b	31	5.28mg	n.s.s.	0/19	5.10mg	1/21				

4-AMINODIPHENYL.HCl   2113-61-3

	RefNum	LoConf	UpConf	Cntrl	1Dose	1Inc				Citation or Pathology
315	1721	.553mg	1.95mg	0/27	4.73mg	18/32				Tanaka;gann,76,570-576;1985/pers.comm.
a	1721	.535mg	2.41mg	2/27	4.73mg	19/32				
b	1721	7.80mg	n.s.s.	0/27	4.73mg	0/32				
c	1721	.494mg	2.08mg	2/27	4.73mg	20/32				

2-AMINODIPHENYLENE OXIDE   3693-22-9

	RefNum	LoConf	UpConf	Cntrl	1Dose	1Inc				Citation or Pathology
316	31	n.s.s.	3.31mg	0/31	35.0mg	30/30				Clayson;bjca,21,755-762;1967
a	31	n.s.s.	3.36mg	1/31	35.0mg	30/30				
317	31	2.94mg	12.6mg	0/19	27.9mg	15/20				
a	31	4.32mg	20.7mg	0/19	27.9mg	12/20				
b	31	9.02mg	126.mg	0/19	27.9mg	6/20				

1-(AMINOMETHYL)CYCLOHEXANEACETIC ACID gabapentin 60142-96-3

	RefNum	LoConf	UpConf	Cntrl	1Dose	1Inc	2Dose	2Inc	3Dose	3Inc	Citation or Pathology
318	2026	18.0gm	n.s.s.	0/50	250.mg	0/50	1.00gm	0/50	2.00gm	1/50	Dominick;txap,111,375-387;1991
a	2026	18.0gm	n.s.s.	0/50	250.mg	0/50	1.00gm	0/50	2.00gm	1/50	
b	2026	23.7gm	n.s.s.	0/50	250.mg	1/50	1.00gm	0/50	2.00gm	0/50	
c	2026	23.7gm	n.s.s.	0/50	250.mg	1/50	1.00gm	0/50	2.00gm	0/50	
319	2026	13.6gm	n.s.s.	0/50	250.mg	0/50	1.00gm	1/50	2.00gm	1/50	
a	2026	19.8gm	n.s.s.	1/50	250.mg	0/50	1.00gm	0/50	2.00gm	1/50	

	Spe	Sex	Strain	Route	Site	Hist	Xpo+Xpt	Notes	TD50	DR	2Tailpvl	AuOp

Dose scale: 100ng..:..1ug....:..10......:..100....:..1mg....:..10......:..100....:..1g......:..10

3-AMINOTRIAZOLE

	Spe	Sex	Strain	Route	Site	Hist	Xpo+Xpt	Notes	TD50	DR	2Tailpvl	AuOp
320	H	f	syg	eat	lun	tum	28m28	ae	no dre		P=1.	
a	H	f	syg	eat	liv	ben	28m28	ae	no dre		P=1.	
b	H	f	syg	eat	tba	mix	28m28	ae	no dre		P=1.	-
c	H	f	syg	eat	tba	mal	28m28	ae	570.mg	*	P<.6	-
321	H	m	syg	eat	liv	ben	31m32	ae	no dre		P=1.	-
a	H	m	syg	eat	lun	tum	31m32	ae	no dre		P=1.	-
b	H	m	syg	eat	tba	mix	31m32	ae	no dre		P=1.	-
c	H	m	syg	eat	tba	mal	31m32	ae	no dre		P=1.	-
322	M	f	b6a	orl	liv	hpt	59w59	evx	24.5mg		P<.0005+	
a	M	f	b6a	orl	lun	mix	59w59	evx	no dre		P=1.	
b	M	f	b6a	orl	tba	mix	59w59	evx	25.6mg		P<.0005	
323	M	m	b6a	orl	liv	hpt	52w52	evx	26.0mg		P<.0005+	
a	M	m	b6a	orl	lun	ade	52w52	evx	no dre		P=1.	
b	M	m	b6a	orl	tba	mix	52w52	evx	28.0mg		P<.0005	
324	M	f	b6c	orl	liv	hpt	57w57	evx	noTD50		P<.0005	
a	M	f	b6c	orl	lun	ade	57w57	evx	no dre		P=1.	
b	M	f	b6c	orl	tba	mix	57w57	evx	noTD50		P<.0005	
325	M	m	b6c	orl	liv	hpt	53w53	evx	25.4mg		P<.0005+	
a	M	m	b6c	orl	lun	ade	53w53	evx	no dre		P=1.	
b	M	m	b6c	orl	tba	mix	53w53	evx	30.8mg		P<.002	
326	M	f	nmr	eat	liv	tum	33m33	ae	no dre		P=1.	-
a	M	f	nmr	eat	lun	tum	33m33	ae	no dre		P=1.	-
b	M	f	nmr	eat	tba	mal	33m33	ae	6.92mg	Z	P<.2	-
c	M	f	nmr	eat	tba	mix	33m33	ae	no dre		P=1.	-
327	M	m	nmr	eat	liv	mal	34m34	ae	2.76gm	*	P<.8	-
a	M	m	nmr	eat	lun	tum	34m34	ae	no dre		P=1.	-
b	M	m	nmr	eat	tba	mal	34m34	ae	347.mg	*	P<.7	-
c	M	m	nmr	eat	tba	mix	34m34	ae	no dre		P=1.	-
328	R	b	nss	eat	thy	ade	24m24	er	3.74mg	*	P<.0005	
329	R	f	wis	wat	tyf	mal	16m24	r	102.mg		P<.002	
330	R	f	wis	eat	thy	ben	38m38	ae	11.5mg	*	P<.0005+	
a	R	f	wis	eat	thy	mal	38m38	ae	17.7mg	*	P<.0005+	
b	R	f	wis	eat	pit	ben	38m38	ae	20.8mg	*	P<.0005+	
c	R	f	wis	eat	liv	tum	38m38	ae	no dre		P=1.	
d	R	f	wis	eat	tba	mix	38m38	ae	7.55mg	*	P<.01	
e	R	f	wis	eat	tba	mal	38m38	ae	17.7mg	*	P<.0005	
331	R	m	wis	eat	thy	ben	38m38	ae	8.75mg	Z	P<.0005+	
a	R	m	wis	eat	thy	mal	38m38	ae	27.0mg	*	P<.0005+	
b	R	m	wis	eat	liv	hpc	38m38	ae	no dre		P=1.	
c	R	m	wis	eat	tba	mix	38m38	ae	13.8mg	*	P<.009	
d	R	m	wis	eat	tba	mal	38m38	ae	135.mg	*	P<.6	

11-AMINOUNDECANOIC ACID

Dose scale: 100ng..:..1ug....:..10......:..100....:..1mg....:..10......:..100....:..1g......:..10

	Spe	Sex	Strain	Route	Site	Hist	Xpo+Xpt	Notes	TD50	DR	2Tailpvl	AuOp
332	M	f	b6c	eat	TBA	MXB	24m25	es	15.6gm	*	P<.8	-
a	M	f	b6c	eat	liv	MXB	24m25	es	37.9gm	*	P<.8	
b	M	f	b6c	eat	lun	MXB	24m25	es	18.8gm	*	P<.4	
333	M	m	b6c	eat	---	mly	24m25	es	#4.97gm	*	P<.05	-
a	M	m	b6c	eat	TBA	MXB	24m25	es	6.50gm	*	P<.6	
b	M	m	b6c	eat	liv	MXB	24m25	es	6.25gm	*	P<.4	
c	M	m	b6c	eat	lun	MXB	24m25	es	no dre		P=1.	
334	R	f	f34	eat	TBA	MXB	24m25		1.25gm	\	P<.7	
a	R	f	f34	eat	liv	MXB	24m25		44.6gm	*	P<.9	
335	R	m	f34	eat	MXB	MXB	24m25	es	833.mg	*	P<.0005	
a	R	m	f34	eat	liv	MXA	24m25	es	1.10gm	*	P<.002 c	
b	R	m	f34	eat	liv	nnd	24m25	es	1.29gm	*	P<.004 c	
c	R	m	f34	eat	ubl	tcc	24m25	es	3.17gm	/	P<.002 c	
d	R	m	f34	eat	mgl	fba	24m25	es	2.91gm	*	P<.05	
e	R	m	f34	eat	TBA	MXB	24m25	es	802.mg	*	P<.1	
f	R	m	f34	eat	liv	MXB	24m25	es	1.10gm	*	P<.002	

AMMONIUM CHLORIDE

Dose scale: 100ng..:..1ug....:..10......:..100....:..1mg....:..10......:..100....:..1g......:..10

	Spe	Sex	Strain	Route	Site	Hist	Xpo+Xpt	Notes	TD50	DR	2Tailpvl	AuOp
336	M	f	ifc	wat	ubl	car	93w93	r	no dre		P=1.	
337	M	f	ifc	wat	ubl	tum	52w52	ekr	no dre		P=1.	

AMMONIUM CITRATE

Dose scale: 100ng..:..1ug....:..10......:..100....:..1mg....:..10......:..100....:..1g......:..10

	Spe	Sex	Strain	Route	Site	Hist	Xpo+Xpt	Notes	TD50	DR	2Tailpvl	AuOp
338	R	m	wis	eat	liv	tum	52w65	ek	no dre		P=1.	-

AMMONIUM HYDROXIDE

Dose scale: 100ng..:..1ug....:..10......:..100....:..1mg....:..10......:..100....:..1g......:..10

	Spe	Sex	Strain	Route	Site	Hist	Xpo+Xpt	Notes	TD50	DR	2Tailpvl	AuOp
339	M	f	c3h	wat	mgl	adc	24m24		no dre		P=1.	-
340	M	f	swi	wat	lun	mix	28m28	ae	no dre		P=1.	-
a	M	f	swi	wat	liv	mix	28m28	ae	no dre		P=1.	-
341	M	m	swi	wat	lun	adc	28m29	ae	7.33gm	*	P<.03	-
a	M	m	swi	wat	lun	mix	28m29	ae	15.3gm	*	P<.7	-
b	M	m	swi	wat	liv	mix	28m29	ae	26.7gm	*	P<.7	

AMOBARBITAL

Dose scale: 100ng..:..1ug....:..10......:..100....:..1mg....:..10......:..100....:..1g......:..10

	Spe	Sex	Strain	Route	Site	Hist	Xpo+Xpt	Notes	TD50	DR	2Tailpvl	AuOp
342	R	m	f34	wat	liv	hct	72w72	e	no dre		P=1.	

DL-AMPHETAMINE SULFATE

Dose scale: 100ng..:..1ug....:..10......:..100....:..1mg....:..10......:..100....:..1g......:..10

	Spe	Sex	Strain	Route	Site	Hist	Xpo+Xpt	Notes	TD50	DR	2Tailpvl	AuOp
343	M	f	b6c	eat	TBA	MXB	24m24		no dre		P=1.	-

	RefNum	LoConf	UpConf	Cntrl	1Dose	1Inc	2Dose	2Inc			Citation or Pathology	Brkly Code
3-AMINOTRIAZOLE (amitrol) 61-82-5												
320	1557o	2.00mg	n.s.s.	0/76	.105mg	0/76	1.05mg	0/76	10.5mg	0/76	Steinhoff;txap,69,161-169;1983	
a	1557o	242.mg	n.s.s.	1/76	.105mg	1/76	1.05mg	0/76	10.5mg	0/76		
b	1557o	59.5mg	n.s.s.	13/76	.105mg	10/76	1.05mg	12/76	10.5mg	11/76		
c	1557o	81.3mg	n.s.s.	1/76	.105mg	2/76	1.05mg	3/76	10.5mg	3/76		
321	1557o	2.29mg	n.s.s.	0/76	92.0ug	0/76	.920mg	0/76	9.20mg	0/75		
a	1557o	2.29mg	n.s.s.	0/76	92.0ug	0/76	.920mg	0/76	9.20mg	0/75		
b	1557o	85.6mg	n.s.s.	20/76	92.0ug	11/76	.920mg	13/76	9.20mg	11/75		
c	1557o	160.mg	n.s.s.	7/76	92.0ug	2/76	.920mg	5/76	9.20mg	2/75		
322	35a	9.88mg	58.2mg	0/17	321.mg	17/18					Innes;ntis,1968/1969	
a	35a	384.mg	n.s.s.	0/17	321.mg	0/18						
b	35a	10.0mg	66.8mg	2/17	321.mg	17/18						
323	35a	11.5mg	72.3mg	3/18	306.mg	16/18						
a	35a	283.mg	n.s.s.	2/18	306.mg	0/18						
b	35a	11.8mg	98.4mg	5/18	306.mg	16/18						
324	35a	n.s.s.	37.3mg	0/18	323.mg	18/18						
a	35a	359.mg	n.s.s.	1/18	323.mg	0/18						
b	35a	n.s.s.	42.2mg	3/18	323.mg	18/18						
325	35a	11.6mg	61.4mg	1/17	305.mg	16/18						
a	35a	294.mg	n.s.s.	2/17	305.mg	0/18						
b	35a	12.5mg	146.mg	6/17	305.mg	16/18						
326	1557n	3.29mg	n.s.s.	0/73	.130mg	0/73	1.30mg	0/74	13.0mg	0/74	Steinhoff;txap,69,161-169;1983	
a	1557n	3.29mg	n.s.s.	0/73	.130mg	0/73	1.30mg	0/74	13.0mg	0/74		
b	1557n	2.32mg	n.s.s.	46/73	.130mg	26/73	1.30mg	46/74	(13.0mg	38/74)		
c	1557n	26.3mg	n.s.s.	60/73	.130mg	48/73	1.30mg	59/74	13.0mg	55/74		
327	1557n	200.mg	n.s.s.	1/75	.120mg	0/73	1.20mg	1/73	12.0mg	1/72		
a	1557n	3.20mg	n.s.s.	0/75	.120mg	0/73	1.20mg	0/73	12.0mg	0/72		
b	1557n	41.1mg	n.s.s.	27/75	.120mg	38/73	1.20mg	26/73	12.0mg	32/72		
c	1557n	43.7mg	n.s.s.	52/75	.120mg	56/73	1.20mg	47/73	12.0mg	46/72		
328	1513	2.17mg	8.86mg	0/5	.450mg	1/10	2.25mg	2/15	4.50mg	17/26	Jukes;scie,132,296-297;1960	
329	35	58.9mg	299.mg	0/10	96.2mg	19/40					Tsuda;jnci,57,861-864;1976	
330	1557m	7.53mg	19.8mg	7/74	50.0ug	12/75	.500mg	8/75	5.00mg	44/75	Steinhoff;txap,69,161-169;1983	
a	1557m	11.4mg	30.1mg	0/74	50.0ug	1/75	.500mg	4/75	5.00mg	28/75		
b	1557m	11.3mg	61.3mg	14/74	50.0ug	20/75	.500mg	15/75	5.00mg	36/75		
c	1557m	1.71mg	n.s.s.	0/74	50.0ug	0/75	.500mg	0/75	5.00mg	0/75		
d	1557m	3.11mg	626.mg	59/74	50.0ug	67/75	.500mg	60/75	5.00mg	71/75		
e	1557m	9.13mg	73.9mg	20/74	50.0ug	34/75	.500mg	29/75	5.00mg	45/75		
331	1557m	5.85mg	14.2mg	5/75	40.0ug	9/74	.400mg	4/75	4.00mg	45/75		
a	1557m	15.2mg	60.5mg	3/75	40.0ug	0/74	.400mg	3/75	4.00mg	18/75		
b	1557m	115.mg	n.s.s.	0/75	40.0ug	0/74	.400mg	1/75	4.00mg	0/75		
c	1557m	6.31mg	441.mg	36/75	40.0ug	41/74	.400mg	44/75	4.00mg	53/75		
d	1557m	21.4mg	n.s.s.	19/75	40.0ug	20/74	.400mg	23/75	4.00mg	23/75		
11-AMINOUNDECANOIC ACID 2432-99-7												
332	TR216	1.76gm	n.s.s.	28/50	926.mg	27/50	1.85gm	19/50				
a	TR216	3.83gm	n.s.s.	7/50	926.mg	8/50	1.85gm	5/50			liv:hpa,hpc,nnd.	
b	TR216	4.63gm	n.s.s.	2/50	926.mg	3/50	1.85gm	3/50			lun:a/a,a/c.	
333	TR216	2.09gm	n.s.s.	2/50	854.mg	9/50	1.71gm	4/50				s
a	TR216	1.20gm	n.s.s.	30/50	854.mg	29/50	1.71gm	18/50				
b	TR216	1.56gm	n.s.s.	17/50	854.mg	18/50	1.71gm	12/50			liv:hpa,hpc,nnd.	
c	TR216	4.88gm	n.s.s.	10/50	854.mg	3/50	1.71gm	4/50			lun:a/a,a/c.	
334	TR216	202.mg	n.s.s.	46/50	358.mg	47/50	(716.mg	37/50)				
a	TR216	2.10gm	n.s.s.	5/50	358.mg	5/50	716.mg	6/50			liv:hpa,hpc,nnd.	
335	TR216	494.mg	1.96gm	1/50	286.mg	10/50	573.mg	15/50			liv:hpc,nnd; ubl:tcc.	c
a	TR216	606.mg	4.27gm	1/50	286.mg	10/50	573.mg	10/50			liv:hpc,nnd.	
b	TR216	666.mg	9.39gm	1/50	286.mg	9/50	573.mg	8/50				
c	TR216	1.36gm	12.8gm	0/50	286.mg	0/50	573.mg	7/50				
d	TR216	1.26gm	n.s.s.	0/50	286.mg	5/50	573.mg	2/50				s
e	TR216	304.mg	n.s.s.	31/50	286.mg	37/50	573.mg	37/50				
f	TR216	606.mg	4.27gm	1/50	286.mg	10/50	573.mg	10/50			liv:hpa,hpc,nnd.	
AMMONIUM CHLORIDE 12125-02-9												
336	142a	8.57gm	n.s.s.	0/26	2.00gm	0/26					Flaks;bjca,31,585-587;1975	
337	142b	2.68gm	n.s.s.	0/26	2.00gm	0/26					Flaks;jnci,51,2007-2008;1973	
AMMONIUM CITRATE 3012-65-5												
338	2	2.10gm	n.s.s.	0/7	1.54gm	0/17					Weisburger;txap,14,163-175;1969	
AMMONIUM HYDROXIDE 1336-21-6												
339	1117	264.mg	n.s.s.	23/30	200.mg	24/40					Toth;ijcn,9,109-118;1972/1969a	
340	1117	2.78gm	n.s.s.	14/109	200.mg	12/50	400.mg	8/48	600.mg	4/42		
a	1117	7.33gm	n.s.s.	3/89	200.mg	1/47	400.mg	2/41	600.mg	0/42		
341	1117	2.99gm	n.s.s.	0/86	167.mg	2/49	333.mg	2/48	500.mg	2/44		
a	1117	2.36gm	n.s.s.	10/86	167.mg	5/49	333.mg	5/48	500.mg	7/44		
b	1117	3.47gm	n.s.s.	2/67	167.mg	1/46	333.mg	3/28	500.mg	1/43		
AMOBARBITAL 57-43-2												
342	1690	29.6mg	n.s.s.	0/13	25.0mg	0/12					Diwan;jnci,74,509-516;1985	
DL-AMPHETAMINE SULFATE 60-13-9												
343	TR387	3.80mg	n.s.s.	40/50	2.56mg	27/50	(12.8mg	15/50)				

	Spe	Sex	Strain	Route	Site	Hist	Xpo+Xpt	Notes		TD50		2Tailpvl DR / AuOp
a	M	f	b6c	eat	liv	MXB	24m24			no dre		P=1.
b	M	f	b6c	eat	lun	MXB	24m24			no dre		P=1.
344	M	m	b6c	eat	thy	MXA	24m24		: ±	#116.mg	*	P<.05 -
a	M	m	b6c	eat	TBA	MXB	24m24			no dre		P=1.
b	M	m	b6c	eat	liv	MXB	24m24			no dre		P=1.
c	M	m	b6c	eat	lun	MXB	24m24			no dre		P=1.
345	R	f	f34	eat	TBA	MXB	24m24	:>		no dre		P=1.
a	R	f	f34	eat	liv	MXB	24m24			no dre		P=1.
346	R	m	f34	eat	TBA	MXB	24m24	:>		no dre		P=1. -
a	R	m	f34	eat	liv	MXB	24m24			107.mg	*	P<.2

AMPICILLIN TRIHYDRATE   100ng..:...1ug....:...10......:...100.....:...1mg...:...10......:...100....:...1g......:..10

	Spe	Sex	Strain	Route	Site	Hist	Xpo+Xpt	Notes		TD50		2Tailpvl
347	M	f	b6c	gav	TBA	MXB	24m24		:>	24.9gm	*	P<.9 -
a	M	f	b6c	gav	liv	MXB	24m24			no dre		P=1.
b	M	f	b6c	gav	lun	MXB	24m24			17.7gm	*	P<.3
348	M	m	b6c	gav	TBA	MXB	24m24		:>	10.2gm	*	P<.7 -
a	M	m	b6c	gav	liv	MXB	24m24			67.4gm	*	P<.9
b	M	m	b6c	gav	lun	MXB	24m24			no dre		P=1.
349	R	f	f34	gav	pta	can	24m24		: ±	#5.90gm	*	P<.05 -
a	R	f	f34	gav	TBA	MXB	24m24			11.0gm	*	P<.9
b	R	f	f34	gav	liv	MXB	24m24			no dre		P=1.
350	R	m	f34	gav	amd	MXA	24m24		: ±	1.31gm	*	P<.04 e
a	R	m	f34	gav	mul	MXA	24m24			1.77gm	*	P<.02
b	R	m	f34	gav	mul	mnl	24m24			1.91gm	*	P<.03 e
c	R	m	f34	gav	mul	MXA	24m24			1.95gm	*	P<.04
d	R	m	f34	gav	TBA	MXB	24m24			1.95gm	*	P<.5
e	R	m	f34	gav	liv	MXB	24m24			34.5gm	*	P<.3

1-AMYL-1-NITROSOUREA   100ng..:...1ug....:...10......:...100.....:...1mg...:...10......:...100....:...1g......:..10

	Spe	Sex	Strain	Route	Site	Hist	Xpo+Xpt	Notes		TD50		2Tailpvl
351	R	f	don	wat	eso	tum	65w65		. + .	.967mg	Z	P<.0005
a	R	f	don	wat	dgt	mix	65w65			1.15mg	*	P<.0005+
b	R	f	don	wat	for	tum	65w65			1.35mg	*	P<.0005
c	R	f	don	wat	liv	hpa	65w65			13.3mg	*	P<.0005
d	R	f	don	wat	---	mye	65w65			19.9mg	*	P<.005 +
e	R	f	don	wat	mgl	mix	65w65			11.9mg	*	P<.3 +
f	R	f	don	wat	tba	mix	65w65			1.93mg	*	P<.002
352	R	m	don	wat	for	tum	65w65		. + .	.532mg	Z	P<.0005
a	R	m	don	wat	dgt	mix	65w65			.532mg	Z	P<.0005+
b	R	m	don	wat	eso	tum	65w65			1.44mg	Z	P<.0005
c	R	m	don	wat	---	mye	65w65			9.17mg	*	P<.005 +
d	R	m	don	wat	liv	hpa	65w65			no dre		P=1.
e	R	m	don	wat	tba	mix	65w65			.541mg	Z	P<.0005
353	R	f	f34	gav	for	tum	12m24		. + .	.478mg		P<.0005+
a	R	f	f34	gav	lun	mix	12m24			1.09mg		P<.008 +
b	R	f	f34	gav	ute	mix	12m24			.934mg		P<.02 +
c	R	f	f34	gav	mgl	adc	12m24			2.98mg		P<.04 +
d	R	f	f34	gav	liv	tum	12m24			4.70mg		P<.09
354	R	m	f34	gav	for	tum	12m24		<+	noTD50		P<.0005+
a	R	m	f34	gav	lun	mix	12m24			.546mg		P<.0005+
b	R	m	f34	gav	---	mso	12m24			.865mg		P<.002
c	R	m	f34	gav	liv	tum	12m24			no dre		P=1.

AMYLOPECTIN SULFATE   100ng..:...1ug....:...10......:...100.....:...1mg...:...10......:...100....:...1g......:..10

	Spe	Sex	Strain	Route	Site	Hist	Xpo+Xpt	Notes		TD50		2Tailpvl
355	R	m	f34	eat	clr	adc	26w52	r	. + .	287.mg		P<.0005+
356	R	m	f34	eat	clr	mix	39w52	r	. + .	280.mg		P<.0005+

ANETHOLE   100ng..:...1ug....:...10......:...100.....:...1mg...:...10......:...100....:...1g......:..10

	Spe	Sex	Strain	Route	Site	Hist	Xpo+Xpt	Notes		TD50		2Tailpvl
357	M	f	cd1	eat	liv	tum	50w78	v	.>	no dre		P=1. -
a	M	f	cd1	eat	lun	ade	50w78	v		no dre		P=1. -

trans-ANETHOLE   100ng..:...1ug....:...10......:...100.....:...1mg...:...10......:...100....:...1g......:..10

	Spe	Sex	Strain	Route	Site	Hist	Xpo+Xpt	Notes		TD50		2Tailpvl
358	R	f	cdr	eat	liv	hpc	28m28	e	. +	7.71gm	Z	P<.005
a	R	f	cdr	eat	liv	hpa	28m28	e		7.16gm	*	P<.03
b	R	f	cdr	eat	liv	cho	28m28	e		no dre		P=1.
359	R	m	cdr	eat	lmr	rts	28m28	e	. ±	3.05gm	Z	P<.06
a	R	m	cdr	eat	tes	ict	28m28	e		4.28gm	*	P<.03
b	R	m	cdr	eat	liv	hpa	28m28	e		12.1gm	*	P<.4
c	R	m	cdr	eat	liv	cho	28m28	e		21.6gm	*	P<.7
d	R	m	cdr	eat	liv	hpc	28m28	e		26.9gm	*	P<.7
e	R	m	cdr	eat	liv	clc	28m28	e		no dre		P=1.

ANHYDROGLUCOCHLORAL   100ng..:...1ug....:...10......:...100.....:...1mg...:...10......:...100....:...1g......:..10

	Spe	Sex	Strain	Route	Site	Hist	Xpo+Xpt	Notes		TD50		2Tailpvl
360	M	f	b6a	orl	lun	ade	76w76	evx	.>	19.5mg		P<.6 -
a	M	f	b6a	orl	liv	hpt	76w76	evx		no dre		P=1. -
b	M	f	b6a	orl	tba	mix	76w76	evx		17.3mg		P<.6 -
361	M	m	b6a	orl	lun	ade	76w76	evx	.>	no dre		P=1. -
a	M	m	b6a	orl	liv	hpt	76w76	evx		no dre		P=1. -
b	M	m	b6a	orl	tba	mix	76w76	evx		no dre		P=1. -
362	M	f	b6c	orl	liv	hpt	76w76	evx	.>	no dre		P=1. -
a	M	f	b6c	orl	lun	mix	76w76	evx		no dre		P=1. -
b	M	f	b6c	orl	tba	tum	76w76	evx		no dre		P=1. -

	RefNum	LoConf	UpConf	Cntrl	1Dose	1Inc	2Dose	2Inc		Citation or Pathology
										Brkly Code
a	TR387	16.0mg	n.s.s.	5/50	2.56mg	1/50	(12.8mg	1/50)		liv:hpa,hpc,nnd.
b	TR387	6.80mg	n.s.s.	8/50	2.56mg	6/50	(12.8mg	1/50)		lun:a/a,a/c.
344	TR387	40.1mg	n.s.s.	0/50	2.37mg	1/50	11.9mg	3/50		thy:fca,fcc. S
a	TR387	3.75mg	n.s.s.	30/50	2.37mg	25/50	(11.9mg	18/50)		
b	TR387	5.75mg	n.s.s.	14/50	2.37mg	12/50	(11.9mg	2/50)		liv:hpa,hpc,nnd.
c	TR387	60.4mg	n.s.s.	8/50	2.37mg	3/50	11.9mg	4/50		lun:a/a,a/c.
345	TR387	1.14mg	n.s.s.	48/50	.985mg	42/50	(4.93mg	30/50)		
a	TR387	n.s.s.	n.s.s.	0/50	.985mg	0/50	4.93mg	0/50		liv:hpa,hpc,nnd.
346	TR387	.480mg	n.s.s.	46/50	.790mg	45/50	(3.95mg	37/50)		
a	TR387	17.5mg	n.s.s.	0/50	.790mg	0/50	3.95mg	1/50		liv:hpa,hpc,nnd.

AMPICILLIN TRIHYDRATE   7177-48-2

	RefNum	LoConf	UpConf	Cntrl	1Dose	1Inc	2Dose	2Inc		Citation or Pathology
347	TR318	1.65gm	n.s.s.	32/50	1.06gm	21/50	2.12gm	28/50		
a	TR318	n.s.s.	n.s.s.	0/50	1.06gm	0/50	2.12gm	0/50		liv:hpa,hpc,nnd.
b	TR318	4.60gm	n.s.s.	2/50	1.06gm	3/50	2.12gm	4/50		lun:a/a,a/c.
348	TR318	1.48gm	n.s.s.	23/50	1.06gm	21/50	2.12gm	18/50		
a	TR318	3.47gm	n.s.s.	9/50	1.06gm	4/50	2.12gm	7/50		liv:hpa,hpc,nnd.
b	TR318	3.93gm	n.s.s.	6/50	1.06gm	6/50	2.12gm	3/50		lun:a/a,a/c.
349	TR318	2.39gm	n.s.s.	0/50	531.mg	3/50	1.06gm	3/50		S
a	TR318	660.mg	n.s.s.	44/50	531.mg	41/50	1.06gm	45/50		
b	TR318	n.s.s.	n.s.s.	0/50	531.mg	2/50	1.06gm	0/50		liv:hpa,hpc,nnd.
350	TR318	560.mg	n.s.s.	13/50	531.mg	16/50	1.06gm	23/50		amd:phe,phm.
a	TR318	830.mg	n.s.s.	5/50	531.mg	14/50	1.06gm	14/50		mul:lle,mnl. S
b	TR318	859.mg	n.s.s.	5/50	531.mg	14/50	1.06gm	13/50		
c	TR318	851.mg	n.s.s.	6/50	531.mg	16/50	1.06gm	14/50		mul:lle,mlh,mlm,mnl. S
d	TR318	468.mg	n.s.s.	39/50	531.mg	42/50	1.06gm	42/50		
e	TR318	5.63gm	n.s.s.	0/50	531.mg	0/50	1.06gm	1/50		liv:hpa,hpc,nnd.

1-AMYL-1-NITROSOUREA   10589-74-9

	RefNum	LoConf	UpConf	Cntrl	1Dose	1Inc	2Dose	2Inc		Citation or Pathology
351	1328	.559mg	1.86mg	0/20	3.06mg	20/35	(6.12mg	15/33	12.2mg 18/33)	Fujii;gann,71,464-470;1980
a	1328	.835mg	1.77mg	0/20	3.06mg	24/35	6.12mg	27/33	12.2mg 27/33	
b	1328	.982mg	2.17mg	0/20	3.06mg	21/35	6.12mg	27/33	12.2mg 25/33	
c	1328	7.01mg	37.2mg	0/20	3.06mg	0/35	6.12mg	5/33	12.2mg 8/33	
d	1328	9.35mg	172.mg	0/20	3.06mg	1/35	6.12mg	2/33	12.2mg 6/33	
e	1328	3.80mg	n.s.s.	6/20	3.06mg	12/35	6.12mg	15/33	12.2mg 15/33	
f	1328	1.01mg	10.2mg	8/20	3.06mg	27/35	6.12mg	28/33	12.2mg 28/33	
352	1328	.367mg	.840mg	0/14	2.14mg	29/35	4.29mg	27/35	(8.57mg 20/35)	
a	1328	.367mg	.840mg	0/14	2.14mg	29/35	4.29mg	27/35	(8.57mg 22/35)	
b	1328	.926mg	3.19mg	0/14	2.14mg	13/35	4.29mg	18/35	(8.57mg 13/35)	
c	1328	4.93mg	69.6mg	0/14	2.14mg	1/35	4.29mg	6/35	8.57mg 7/35	
d	1328	21.7mg	n.s.s.	0/14	2.14mg	5/35	4.29mg	0/35	8.57mg 0/35	
e	1328	.359mg	1.03mg	1/14	2.14mg	29/35	4.29mg	28/35	(8.57mg 26/35)	
353	1915	.199mg	1.33mg	0/12	1.25mg	10/12				Lijinsky;txih,5,925-935;1989
a	1915	.420mg	21.2mg	1/12	1.25mg	7/12				
b	1915	.353mg	n.s.s.	2/12	1.25mg	8/12				
c	1915	.894mg	n.s.s.	0/12	1.25mg	3/12				
d	1915	1.15mg	n.s.s.	0/12	1.25mg	2/12				
354	1915	n.s.s.	.422mg	0/12	.875mg	12/12				
a	1915	.232mg	1.67mg	0/12	.875mg	8/12				
b	1915	.343mg	3.88mg	0/12	.875mg	6/12				
c	1915	1.27mg	n.s.s.	1/12	.875mg	1/12				

AMYLOPECTIN SULFATE   9047-13-6

	RefNum	LoConf	UpConf	Cntrl	1Dose	1Inc		Citation or Pathology
355	1716m	133.mg	830.mg	0/18	1.00gm	9/20		Ishioka;clet,26,277-282;1985
356	1716n	140.mg	669.mg	0/18	1.50gm	12/20		

ANETHOLE   (p-propenylanisole) 104-46-1

	RefNum	LoConf	UpConf	Cntrl	1Dose	1Inc		Citation or Pathology
357	1582	563.mg	n.s.s.	0/30	162.mg	0/30		Miller;canr,43,1124-1134;1983
a	1582	563.mg	n.s.s.	1/30	162.mg	0/30		

trans-ANETHOLE   (trans-p-propenylanisole) 4180-23-8

	RefNum	LoConf	UpConf	Cntrl	1Dose	1Inc	2Dose	2Inc		Citation or Pathology
358	1843	2.97gm	82.2gm	1/104	125.mg	0/78	250.mg	0/52	500.mg 6/52	Truhaut;fctx,27,11-19;1989/pers.comm.
a	1843	2.66gm	n.s.s.	2/104	125.mg	2/78	250.mg	0/52	500.mg 6/52	
b	1843	9.49gm	n.s.s.	4/104	125.mg	1/78	250.mg	1/52	500.mg 0/52	
359	1843	1.07gm	n.s.s.	2/104	100.mg	2/78	200.mg	5/52	(400.mg 0/52)	
a	1843	1.94gm	n.s.s.	0/104	100.mg	3/78	200.mg	3/52	400.mg 2/52	
b	1843	2.72gm	n.s.s.	3/104	100.mg	3/78	200.mg	0/52	400.mg 4/52	
c	1843	4.29gm	n.s.s.	0/104	100.mg	1/78	200.mg	1/52	400.mg 0/52	
d	1843	3.31gm	n.s.s.	2/104	100.mg	1/78	200.mg	3/52	400.mg 1/52	
e	1843	6.06gm	n.s.s.	0/104	100.mg	1/78	200.mg	0/52	400.mg 0/52	

ANHYDROGLUCOCHLORAL   (alpha-chloralose) 15879-93-3

	RefNum	LoConf	UpConf	Cntrl	1Dose	1Inc		Citation or Pathology
360	1297	2.86mg	n.s.s.	1/17	3.89mg	2/16		Innes;ntis,1968/1969
a	1297	6.85mg	n.s.s.	0/17	3.89mg	0/16		
b	1297	2.41mg	n.s.s.	2/17	3.89mg	3/16		
361	1297	4.75mg	n.s.s.	2/18	3.62mg	1/18		
a	1297	7.17mg	n.s.s.	1/18	3.62mg	0/18		
b	1297	3.75mg	n.s.s.	3/18	3.62mg	2/18		
362	1297	7.28mg	n.s.s.	0/16	3.89mg	0/17		
a	1297	7.28mg	n.s.s.	0/16	3.89mg	0/17		
b	1297	7.28mg	n.s.s.	0/16	3.89mg	0/17		

```
    Spe Strain  Site   Xpo+Xpt                                          TD50    2Tailpvl
        Sex  Route  Hist   Notes                                         DR      AuOp
363  M m b6c orl liv hpt 76w76 evx              .        ±            7.27mg  P<.05  -
a    M m b6c orl lun ade 76w76 evx                                    11.2mg  P<.2   -
b    M m b6c orl tba mix 76w76 evx                                    4.07mg  P<.009 -

ANILAZINE                      100ng..:..1ug....:..10.....:..100....:..1mg..:..10.....:..100....:..1g.....:..10
364  M f b6c eat TBA MXB 24m25                              :   ±       99.2mg  \ P<.02  -
a    M f b6c eat liv MXB 24m25                                         2.55gm  * P<.5
b    M f b6c eat lun MXB 24m25                                         1.69gm  * P<.2
365  M m b6c eat TBA MXB 24m25                         :>              no dre   P=1.   -
a    M m b6c eat liv MXB 24m25                                        no dre   P=1.
b    M m b6c eat lun MXB 24m25                                        no dre   P=1.
366  R f f34 eat TBA MXB 24m24                       :>              no dre   P=1.
a    R f f34 eat liv MXB 24m24                                        1.05gm  * P<.5
367  R m f34 eat TBA MXB 24m24                       :>              246.mg  * P<.7
a    R m f34 eat liv MXB 24m24                                       no dre   P=1.

ANILINE                        100ng..:..1ug....:..10.....:..100....:..1mg....:..10.....:..100....:..1g.....:..10
368  R m wis wat for pam 80w80                           .>           587.mg  * P<.2   -
a    R m wis wat ubl tum 80w80                                        no dre   P=1.   -

ANILINE.HCl                    100ng..:..1ug....:..10.....:..100....:..1mg....:..10.....:..100....:..1g.....:..10
369  M f b6c eat TBA MXB 24m25                              :>         9.15gm  * P<.6   -
a    M f b6c eat liv MXB 24m25                                        8.85gm  * P<.2
b    M f b6c eat lun MXB 24m25                                        no dre   P=1.
370  M m b6c eat TBA MXB 24m25                                :>       no dre   P=1.   -
a    M m b6c eat liv MXB 24m25                                        no dre   P=1.
b    M m b6c eat lun MXB 24m25                                        no dre   P=1.
371  R f f34 eat MXA MXA 24m25                            :    ±       1.74gm  * P<.02  c
a    R f f34 eat TBA MXB 24m25                                        no dre   P=1.
b    R f f34 eat liv MXB 24m25                                        13.0gm  * P<.4
372  R m f34 eat MXB MXB 24m25                      :+ :              88.0mg  * P<.0005
a    R m f34 eat MXA MXA 24m25                                        146.mg  * P<.0005c
b    R m f34 eat spl hes 24m25                                        160.mg  * P<.0005c
c    R m f34 eat MXA MXA 24m25                                        263.mg  * P<.0005c
d    R m f34 eat spl MXA 24m25                                        451.mg  * P<.006 c
e    R m f34 eat MXA MXA 24m25                                        572.mg  * P<.002 c
f    R m f34 eat adr MXA 24m25                                        556.mg  * P<.05
g    R m f34 eat TBA MXB 24m25                                        124.mg  * P<.005
h    R m f34 eat liv MXB 24m25                                        no dre   P=1.

o-ANISIDINE.HCl                100ng..:..1ug....:..10.....:..100....:..1mg....:..10.....:..100....:..1g.....:..10
373  M f b6c eat ubl MXA 24m24                             :  +  :      1.00gm  / P<.0005c
a    M f b6c eat ubl tcc 24m24                                        1.34gm  / P<.0005c
b    M f b6c eat TBA MXB 24m24                                        21.1gm  * P<1.
c    M f b6c eat liv MXB 24m24                                        no dre   P=1.
d    M f b6c eat lun MXB 24m24                                        no dre   P=1.
374  M m b6c eat ubl MXA 24m24                             :  +  :      935.mg  / P<.0005c
a    M m b6c eat ubl tcc 24m24                                        1.62gm  / P<.0005c
b    M m b6c eat TBA MXB 24m24                                        no dre   P=1.
c    M m b6c eat liv MXB 24m24                                        no dre   P=1.
d    M m b6c eat lun MXB 24m24                                        no dre   P=1.
375  R f f34 eat ubl MXA 22m24 as             :    +  :               27.8mg  / P<.0005c
a    R f f34 eat ubl tcc 22m24 as                                    29.1mg  / P<.0005c
b    R f f34 eat ubl tpp 22m24 as                                    693.mg  * P<.002
c    R f f34 eat TBA MXB 22m24 as                                    37.0mg  / P<.0005
d    R f f34 eat liv MXB 22m24 as                                    no dre   P=1.
376  R m f34 eat MXB MXB 22m24 as                  :  +:              31.9mg  / P<.0005
a    R m f34 eat ubl MXA 22m24 as                                    31.9mg  / P<.0005c
b    R m f34 eat ubl tcc 22m24 as                                    32.7mg  / P<.0005c
c    R m f34 eat thy MXA 22m24 as                                    235.mg  / P<.0005c
d    R m f34 eat MXA MXA 22m24 as                                    535.mg  * P<.0005c
e    R m f34 eat k/p tcc 22m24 as                                    1.04gm  * P<.002 c
f    R m f34 eat TBA MXB 22m24 as                                    30.9mg  / P<.0005
g    R m f34 eat liv MXB 22m24 as                                    483.mg  * P<.002

p-ANISIDINE.HCl                100ng..:..1ug....:..10.....:..100....:..1mg....:..10.....:..100....:..1g.....:..10
377  M f b6c eat TBA MXB 24m24                               :>        no dre   P=1.   -
a    M f b6c eat liv MXB 24m24                                        no dre   P=1.
b    M f b6c eat lun MXB 24m24                                        no dre   P=1.
378  M m b6c eat TBA MXB 24m24                                :>       no dre   P=1.   -
a    M m b6c eat liv MXB 24m24                                        no dre   P=1.
b    M m b6c eat lun MXB 24m24                                        12.6gm  * P<.6
379  R f f34 eat TBA MXB 24m24                            :>           no dre   P=1.
a    R f f34 eat liv MXB 24m24                                        6.83gm  * P<.5
380  R m f34 eat pre MXA 24m24                         :   ±          493.mg  \ P<.02  a
a    R m f34 eat lun a/a 24m24                                        3.59gm  * P<.06  a
b    R m f34 eat ski sqc 24m24                                        3.72gm  * P<.06  a
c    R m f34 eat TBA MXB 24m24                                        no dre   P=1.
d    R m f34 eat liv MXB 24m24                                        1.56gm  * P<.03
```

	RefNum	LoConf	UpConf	Cntrl	1Dose	1Inc	2Dose	2Inc			Citation or Pathology
											Brkly Code
363	1297	2.19mg	n.s.s.	0/16	3.62mg	3/18					
a	1297	2.76mg	n.s.s.	0/16	3.62mg	2/18					
b	1297	1.54mg	89.3mg	0/16	3.62mg	5/18					

ANILAZINE   101-05-3

	RefNum	LoConf	UpConf	Cntrl	1Dose	1Inc	2Dose	2Inc			Citation or Pathology
364	TR104	47.6mg	n.s.s.	4/25	62.0mg	22/50	(124.mg	18/50)			
a	TR104	628.mg	n.s.s.	0/25	62.0mg	1/50	124.mg	1/50			liv:hpa,hpc,nnd.
b	TR104	511.mg	n.s.s.	0/25	62.0mg	1/50	124.mg	2/50			lun:a/a,a/c.
365	TR104	224.mg	n.s.s.	15/25	57.7mg	20/50	114.mg	25/50			
a	TR104	365.mg	n.s.s.	9/25	57.7mg	6/50	114.mg	12/50			liv:hpa,hpc,nnd.
b	TR104	391.mg	n.s.s.	4/25	57.7mg	7/50	114.mg	6/50			lun:a/a,a/c.
366	TR104	32.9mg	n.s.s.	22/25	25.0mg	42/50	50.0mg	43/50			
a	TR104	257.mg	n.s.s.	0/25	25.0mg	1/50	50.0mg	1/50			liv:hpa,hpc,nnd.
367	TR104	34.2mg	n.s.s.	12/25	20.0mg	26/50	40.0mg	28/50			
a	TR104	n.s.s.	n.s.s.	0/25	20.0mg	0/50	40.0mg	0/50			liv:hpa,hpc,nnd.

ANILINE   62-53-3

	RefNum	LoConf	UpConf	Cntrl	1Dose	1Inc	2Dose	2Inc	3Dose	3Inc	Citation or Pathology
368	1460	144.mg	n.s.s.	0/28	15.0mg	0/28	30.0mg	1/28	60.0mg	1/28	Hagiwara;txlt,6,71-75;1980
a	1460	29.3mg	n.s.s.	0/28	15.0mg	0/28	30.0mg	0/28	60.0mg	0/28	

ANILINE.HCl   142-04-1

	RefNum	LoConf	UpConf	Cntrl	1Dose	1Inc	2Dose	2Inc			Citation or Pathology	Brkly Code
369	TR130	1.61gm	n.s.s.	21/50	741.mg	22/50	1.50gm	28/49				
a	TR130	3.27gm	n.s.s.	1/50	741.mg	5/50	1.50gm	5/49			liv:hpa,hpc,nnd.	
b	TR130	n.s.s.	n.s.s.	0/50	741.mg	0/50	1.50gm	0/49			lun:a/a,a/c.	
370	TR130	2.45gm	n.s.s.	24/50	693.mg	27/50	1.39gm	22/50				
a	TR130	5.78gm	n.s.s.	12/50	693.mg	9/50	1.39gm	7/50			liv:hpa,hpc,nnd.	
b	TR130	5.15gm	n.s.s.	4/50	693.mg	8/50	1.39gm	3/50			lun:a/a,a/c.	
371	TR130	786.mg	248.gm	0/25	144.mg	1/50	286.mg	7/50			mul:fbs,srn; spl:fbs,srn.	
a	TR130	265.mg	n.s.s.	17/25	144.mg	32/50	286.mg	35/50				
b	TR130	2.12gm	n.s.s.	0/25	144.mg	0/50	286.mg	1/50			liv:hpa,hpc,nnd.	
372	TR130	64.2mg	133.mg	0/25	115.mg	27/50	229.mg	38/50			bod:men; mul:fbs,hes,srn; spl:fbs,hes,srn.	C
a	TR130	99.9mg	277.mg	0/25	115.mg	19/50	229.mg	23/50			bod:men; mul:hes; spl:hes.	
b	TR130	108.mg	344.mg	0/25	115.mg	19/50	229.mg	20/50				
c	TR130	168.mg	565.mg	0/25	115.mg	11/50	229.mg	18/50			bod:men; mul:fbs,srn; spl:fbs,srn.	
d	TR130	251.mg	4.10gm	0/25	115.mg	7/50	229.mg	9/50			spl:fbs,srn.	
e	TR130	312.mg	2.01gm	0/25	115.mg	4/50	229.mg	11/50			bod:men; mul:fbs,srn.	
f	TR130	249.mg	n.s.s.	2/25	115.mg	6/50	229.mg	12/50			adr:phe,phm.	S
g	TR130	65.9mg	1.25gm	10/25	115.mg	42/50	229.mg	45/50				
h	TR130	2.38gm	n.s.s.	1/25	115.mg	0/50	229.mg	0/50			liv:hpa,hpc,nnd.	

o-ANISIDINE.HCl   134-29-2

	RefNum	LoConf	UpConf	Cntrl	1Dose	1Inc	2Dose	2Inc			Citation or Pathology	Brkly Code
373	TR89	611.mg	1.81gm	0/55	319.mg	1/55	638.mg	22/55			ubl:tcc,tpp.	
a	TR89	771.mg	2.64gm	0/55	319.mg	0/55	638.mg	18/55				
b	TR89	761.mg	n.s.s.	34/55	319.mg	20/55	638.mg	33/55				
c	TR89	4.37gm	n.s.s.	11/55	319.mg	1/55	638.mg	4/55			liv:hpa,hpc,nnd.	
d	TR89	4.67gm	n.s.s.	4/55	319.mg	2/55	638.mg	1/55			lun:a/a,a/c.	
374	TR89	575.mg	1.67gm	0/55	297.mg	2/55	589.mg	22/55			ubl:tcc,tpp.	
a	TR89	888.mg	3.46gm	0/55	297.mg	0/55	589.mg	15/55				
b	TR89	1.14gm	n.s.s.	43/55	297.mg	27/55	589.mg	30/55				
c	TR89	1.08gm	n.s.s.	28/55	297.mg	13/55	(589.mg	7/55)			liv:hpa,hpc,nnd.	
d	TR89	824.mg	n.s.s.	12/55	297.mg	9/55	(589.mg	2/55)			lun:a/a,a/c.	
375	TR89	11.0mg	48.6mg	0/55	248.mg	46/55	500.mg	50/55			ubl:tcc,tpp.	
a	TR89	11.2mg	52.9mg	0/55	248.mg	41/55	500.mg	50/55				
b	TR89	208.mg	3.85gm	0/55	248.mg	5/55	500.mg	0/55				S
c	TR89	21.0mg	62.0mg	52/55	248.mg	50/55	500.mg	51/55				
d	TR89	n.s.s.	n.s.s.	1/55	248.mg	0/55	500.mg	0/55			liv:hpa,hpc,nnd.	
376	TR89	20.6mg	48.8mg	0/55	198.mg	52/55	400.mg	52/55			k/p:tcc; thy:cyn,fca,fcc,pcn,pcy; tyf:pcn,pcy; ubl:tcc,tpp.	C
a	TR89	20.6mg	48.8mg	0/55	198.mg	52/55	400.mg	52/55			ubl:tcc,tpp.	
b	TR89	21.0mg	50.4mg	0/55	198.mg	50/55	400.mg	51/55				
c	TR89	98.9mg	595.mg	0/55	198.mg	7/55	400.mg	6/55			thy:cyn,fca,fcc,pcn,pcy.	
d	TR89	191.mg	1.90gm	0/55	198.mg	4/55	400.mg	4/55			thy:fca; tyf:pcn,pcy.	
e	TR89	304.mg	5.61gm	0/55	198.mg	2/55	400.mg	4/55				
f	TR89	20.3mg	49.2mg	39/55	198.mg	53/55	400.mg	53/55				
g	TR89	148.mg	3.15gm	0/55	198.mg	4/55	400.mg	0/55			liv:hpa,hpc,nnd.	

p-ANISIDINE.HCl   20265-97-8

	RefNum	LoConf	UpConf	Cntrl	1Dose	1Inc	2Dose	2Inc			Citation or Pathology
377	TR116	2.32gm	n.s.s.	34/55	638.mg	33/55	1.28gm	24/55			
a	TR116	4.69gm	n.s.s.	11/55	638.mg	10/55	1.28gm	6/55			liv:hpa,hpc,nnd.
b	TR116	5.15gm	n.s.s.	4/55	638.mg	5/55	1.28gm	3/55			lun:a/a,a/c.
378	TR116	2.19gm	n.s.s.	43/55	589.mg	29/55	1.18gm	36/55			
a	TR116	2.78gm	n.s.s.	28/55	589.mg	22/55	1.18gm	23/55			liv:hpa,hpc,nnd.
b	TR116	2.35gm	n.s.s.	12/55	589.mg	8/55	1.18gm	17/55			lun:a/a,a/c.
379	TR116	568.mg	n.s.s.	52/55	146.mg	38/55	291.mg	42/55			
a	TR116	1.41gm	n.s.s.	1/55	146.mg	1/55	291.mg	3/55			liv:hpa,hpc,nnd.
380	TR116	202.mg	n.s.s.	1/55	118.mg	8/55	(233.mg	3/55)			pre:adn,can.
a	TR116	1.09gm	n.s.s.	0/55	118.mg	0/55	233.mg	3/55			
b	TR116	1.12gm	n.s.s.	0/55	118.mg	0/55	233.mg	3/55			
c	TR116	424.mg	n.s.s.	39/55	118.mg	36/55	233.mg	31/55			
d	TR116	673.mg	n.s.s.	0/55	118.mg	3/55	233.mg	4/55			liv:hpa,hpc,nnd.

```
       Spe Strain Site  Xpo+Xpt                                                          TD50    2Tailpvl
           Sex  Route Hist  Notes                                                         DR      AuOp

ANTHRANILIC ACID        100ng..:..1ug....:..10......:..100.....:..1mg....:..10.....:..100...:..1g......:..10
381  M f b6c eat TBA MXB 18m24                                                     :>  68.4gm * P<.9    -
a    M f b6c eat liv MXB 18m24                                                         26.4gm * P<.4
b    M f b6c eat lun MXB 18m24                                                         80.7gm * P<.4
382  M m b6c eat TBA MXB 18m24                                             :>          no dre  P=1.     -
a    M m b6c eat liv MXB 18m24                                                         no dre  P=1.
b    M m b6c eat lun MXB 18m24                                                         no dre  P=1.
383  R f f34 eat TBA MXB 18m24                                         :>              no dre  P=1.     -
a    R f f34 eat liv MXB 18m24                                                         no dre  P=1.
384  R m f34 eat TBA MXB 18m24                                             :>          no dre  P=1.     -
a    R m f34 eat liv MXB 18m24                                                         no dre  P=1.

9,10-ANTHRAQUINONE      100ng..:..1ug....:..10......:..100.....:..1mg....:..10.....:..100...:..1g......:..10
385  M f b6a orl lun ade 76w76 evx                                            .>      847.mg  P<.6     -
a    M f b6a orl liv hpt 76w76 evx                                                    no dre  P=1.
b    M f b6a orl tba mix 76w76 evx                                                    749.mg  P<.6
386  M m b6a orl liv hpt 76w76 evx                                              .>    no dre  P=1.
a    M m b6a orl lun ade 76w76 evx                                                    no dre  P=1.
b    M m b6a orl tba mix 76w76 evx                                                    no dre  P=1.
387  M f b6c orl lun ade 76w76 evx                                        .     ±     494.mg  P<.1
a    M f b6c orl liv hpt 76w76 evx                                                    no dre  P=1.
b    M f b6c orl tba mix 76w76 evx                                                    494.mg  P<.1
388  M m b6c orl liv hpt 76w76 evx                                        .     ±     460.mg  P<.1
a    M m b6c orl lun mix 76w76 evx                                                    no dre  P=1.
b    M m b6c orl tba mix 76w76 evx                                                    460.mg  P<.1

ANTIMONY POTASSIUM TARTRATE  100ng..:..1ug....:..10......:..100.....:..1mg....:..10.....:..100...:..1g......:..10
389  M b cd1 wat lun tum 33m33 e                             .>                       no dre  P=1.
a    M b cd1 wat liv tum 33m33 e                                                      no dre  P=1.
b    M b cd1 wat tba tum 33m33 e                                                      no dre  P=1.
c    M b cd1 wat tba mal 33m33 e                                                      no dre  P=1.

ARAMITE                 100ng..:..1ug....:..10......:..100.....:..1mg....:..10.....:..100...:..1g......:..10
390  M f b6a orl lun ade 80w80 evx                                           .>       432.mg  P<.3
a    M f b6a orl liv hpt 80w80 evx                                                    no dre  P=1.
b    M f b6a orl tba mix 80w80 evx                                                    391.mg  P<.4
391  M m b6a orl liv hpt 77w77 evx                                              .>     15.9gm  P<1.
a    M m b6a orl lun ade 77w77 evx                                                    no dre  P=1.
b    M m b6a orl tba mix 77w77 evx                                                    no dre  P=1.
392  M f b6c orl liv hpt 77w77 evx                                             .>      973.mg  P<.3
a    M f b6c orl lun ade 77w77 evx                                                    973.mg  P<.3
b    M f b6c orl tba mix 77w77 evx                                                    92.7mg  P<.0005
393  M m b6c orl liv mix 80w80 evx                                        .   +   .    126.mg  P<.003
a    M m b6c orl liv hpt 80w80 evx                                                    158.mg  P<.006  +
b    M m b6c orl lun mix 80w80 evx                                                    no dre  P=1.
c    M m b6c orl tba mix 80w80 evx                                        .   +   .    103.mg  P<.002
394  R b cfn eat liv hnd 24m24                                      .   +   .          61.8mg * P<.0005+
a    R b cfn eat bil ade 24m24                                                        1.01gm * P<.05   +
395  R b fds eat liv mix 22m24 agrs                                          .   +     430.mg * P<.002  +
396  R f osm eat liv hem 27m27                                          .>            256.mg  P<.3     -
a    R f osm eat tba mix 27m27                                                        no dre  P=1.     -
397  R f osm eat liv tum 24m24 e                                      .>              no dre  P=1.     -
a    R f osm eat tba ben 24m24 e                                                      31.5mg  P<.6     -
b    R f osm eat tba mal 24m24 e                                                      no dre  P=1.     -
398  R m osm eat liv tum 27m27                                          .>            no dre  P=1.     -
a    R m osm eat tba mal 27m27                                                        101.mg  P<.1     -
b    R m osm eat tba mix 27m27                                                        198.mg  P<.6     -
399  R m osm eat liv tum 24m24 e                                      .>              no dre  P=1.     -
a    R m osm eat tba ben 24m24 e                                                      64.7mg  P<.3     -
b    R m osm eat tba mal 24m24 e                                                      no dre  P=1.     -
400  R b sda eat liv hnd 24m24                                              .>        760.mg * P<.3     -
401  R b wid eat liv hnd 24m24                                        .   +   .        79.9mg Z P<.0005+
a    R b wid eat bil ade 24m24                                                        397.mg * P<.002  +
b    R b wid eat liv hpc 24m24                                                 +hist 1.00gm * P<.04   +

ARECOLINE.HCl           100ng..:..1ug....:..10......:..100.....:..1mg....:..10.....:..100...:..1g......:..10
402  M f swi gav liv tum 25m25                                                  .>    no dre  P=1.     -
a    M f swi gav lun tum 25m25                                                        no dre  P=1.     -
b    M f swi gav tba tum 25m25                                                        no dre  P=1.     -
403  M f swi gav lun adc 25m25 b                                          .   ±        61.9mg  P<.07   +
a    M f swi gav liv hem 25m25 b                                                      116.mg  P<.06   +
b    M f swi gav tba mix 25m25 b                                                      33.6mg  P<.008  +
404  M m swi gav liv hem 25m25                                            .   +   .    67.8mg  P<.006  +
a    M m swi gav sto sqc 25m25                                                        196.mg  P<.1    +
b    M m swi gav lun adc 25m25                                                        251.mg  P<.5    +
c    M m swi gav tba mix 25m25 r                                                      34.6mg  P<.002  +
405  M m swi gav liv hem 25m25 b                                          .   ±        114.mg  P<.04   +
a    M m swi gav lun adc 25m25 b                                                      325.mg  P<.7    +
b    M m swi gav tba mix 25m25 b                                                      57.6mg  P<.06   +
```

	RefNum	LoConf	UpConf	Cntrl	1Dose	1Inc	2Dose	2Inc		Citation or Pathology
										Brkly Code
ANTHRANILIC ACID		**118-92-3**								
381	TR36	4.53gm	n.s.s.	1/15	1.73gm	5/35	3.45gm	5/35		
a	TR36	7.83gm	n.s.s.	0/15	1.73gm	1/35	3.45gm	2/35		liv:hpa,hpc,nnd.
b	TR36	13.1gm	n.s.s.	0/15	1.73gm	0/35	3.45gm	1/35		lun:a/a,a/c.
382	TR36	1.48gm	n.s.s.	0/15	1.59gm	13/35	3.18gm	6/35		
a	TR36	1.79gm	n.s.s.	0/15	1.59gm	6/35	3.18gm	5/35		liv:hpa,hpc,nnd.
b	TR36	n.s.s.	n.s.s.	0/15	1.59gm	5/35	(3.18gm	0/35)		lun:a/a,a/c.
383	TR36	361.mg	n.s.s.	10/15	402.mg	19/35	(804.mg	11/35)		
a	TR36	n.s.s.	n.s.s.	0/15	402.mg	0/35	804.mg	0/35		liv:hpa,hpc,nnd.
384	TR36	867.mg	n.s.s.	6/15	318.mg	14/35	643.mg	11/35		
a	TR36	n.s.s.	n.s.s.	0/15	318.mg	0/35	643.mg	0/35		liv:hpa,hpc,nnd.
9,10-ANTHRAQUINONE		**84-65-1**								
385	1229	124.mg	n.s.s.	1/17	169.mg	2/16				Innes;ntis,1968/1969
a	1229	297.mg	n.s.s.	0/17	169.mg	0/16				
b	1229	104.mg	n.s.s.	2/17	169.mg	3/16				
386	1229	312.mg	n.s.s.	1/18	157.mg	0/18				
a	1229	312.mg	n.s.s.	2/18	157.mg	0/18				
b	1229	312.mg	n.s.s.	3/18	157.mg	0/18				
387	1229	121.mg	n.s.s.	0/16	169.mg	2/17				
a	1229	316.mg	n.s.s.	0/16	169.mg	0/17				
b	1229	121.mg	n.s.s.	0/16	169.mg	2/17				
388	1229	113.mg	n.s.s.	0/16	157.mg	2/17				
a	1229	294.mg	n.s.s.	0/16	157.mg	0/17				
b	1229	113.mg	n.s.s.	0/16	157.mg	2/17				
ANTIMONY POTASSIUM TARTRATE			**28300-74-5**							
389	1036	9.12mg	n.s.s.	15/71	.877mg	10/76				Kanisawa;canr,29,892-895;1969
a	1036	19.7mg	n.s.s.	4/71	.877mg	1/76				
b	1036	6.83mg	n.s.s.	24/71	.877mg	18/76				
c	1036	9.91mg	n.s.s.	8/71	.877mg	6/76				
ARAMITE		**140-57-8**								
390	40a	99.0mg	n.s.s.	1/17	157.mg	3/16				Innes;ntis,1968/1969
a	40a	305.mg	n.s.s.	0/17	157.mg	0/16				
b	40a	85.9mg	n.s.s.	2/17	157.mg	4/16				
391	40a	166.mg	n.s.s.	1/18	146.mg	1/17				
a	40a	281.mg	n.s.s.	2/18	146.mg	0/17				
b	40a	145.mg	n.s.s.	3/18	146.mg	2/17				
392	40a	158.mg	n.s.s.	0/16	157.mg	1/17				
a	40a	158.mg	n.s.s.	0/16	157.mg	1/17				
b	40a	41.1mg	295.mg	0/16	157.mg	8/17				
393	40a	50.7mg	617.mg	0/16	146.mg	6/16				
a	40a	59.4mg	1.56gm	0/16	146.mg	5/16				
b	40a	285.mg	n.s.s.	0/16	146.mg	0/16				
c	40a	43.6mg	375.mg	0/16	146.mg	7/16				
394	42	38.4mg	122.mg	5/180	4.50mg	3/93	9.00mg	10/90	18.0mg 22/96	Popper;canc,13,1035-1046;1960
a	42	248.mg	n.s.s.	0/180	4.50mg	0/93	9.00mg	0/90	18.0mg 2/96	
395	1133	202.mg	1.99gm	0/22	22.5mg	1/20	71.1mg	2/21	225.mg 6/20	Oser;txap,2,441-457;1960
396	21	41.7mg	n.s.s.	0/30	10.0mg	1/30				Deichmann;txap,11,88-103;1967
a	21	16.1mg	n.s.s.	13/30	10.0mg	12/30				
397	84a	24.7mg	n.s.s.	1/30	4.00mg	0/30				Radomski;txap,7,652-656;1965
a	84a	5.40mg	n.s.s.	6/30	4.00mg	8/30				
b	84a	13.0mg	n.s.s.	6/30	4.00mg	3/30				
398	21	62.6mg	n.s.s.	0/30	8.00mg	0/30				Deichmann;txap,11,88-103;1967
a	21	24.7mg	n.s.s.	0/30	8.00mg	2/30				
b	21	27.2mg	n.s.s.	1/30	8.00mg	2/30				
399	84a	19.8mg	n.s.s.	0/30	3.20mg	0/30				Radomski;txap,7,652-656;1965
a	84a	10.5mg	n.s.s.	0/30	3.20mg	1/30				
b	84a	10.6mg	n.s.s.	3/30	3.20mg	2/30				
400	42	187.mg	n.s.s.	0/82	4.50mg	0/46	9.00mg	1/41	18.0mg 1/92	Popper;canc,13,1035-1046;1960
401	42	48.2mg	155.mg	2/193	4.50mg	2/93	9.00mg	3/100	18.0mg 20/90	
a	42	151.mg	1.86gm	0/193	4.50mg	0/93	9.00mg	0/100	18.0mg 5/90	
b	42	246.mg	n.s.s.	0/193	4.50mg	0/93	9.00mg	0/100	18.0mg 2/90	
ARECOLINE.HCl		**61-94-9**								
402	1659m	114.mg	n.s.s.	0/20	28.6mg	0/18				Bhide;zkko,107,169-171;1984/pers.comm.
a	1659m	114.mg	n.s.s.	0/20	28.6mg	0/18				
b	1659m	114.mg	n.s.s.	0/20	28.6mg	0/18				
403	1659n	18.6mg	n.s.s.	1/16	28.6mg	4/12				
a	1659n	28.4mg	n.s.s.	0/16	28.6mg	2/12				
b	1659n	12.5mg	690.mg	1/16	28.6mg	6/12				
404	1659m	30.6mg	524.mg	0/20	23.8mg	8/35				
a	1659m	59.4mg	n.s.s.	0/20	23.8mg	3/35				
b	1659m	55.5mg	n.s.s.	1/20	23.8mg	4/35				
c	1659m	17.8mg	143.mg	1/20	23.8mg	15/35				
405	1659n	34.5mg	n.s.s.	0/21	23.8mg	3/21				
a	1659n	40.6mg	n.s.s.	2/21	23.8mg	3/21				
b	1659n	20.3mg	n.s.s.	2/21	23.8mg	7/21				

```
      Spe Strain Site   Xpo+Xpt                                              TD50     2Tailpvl
          Sex  Route Hist    Notes                                           DR       AuOp

AROCLOR 1254            100ng..:..1ug...:..10....:..100....:..1mg...:..10....:..100....:..1g.....:..10
406  M m baj eat liv hpt 47w47 e                          .  (+)  .                   9.58mg   P<.0005+
407  M m baj eat liv hpt 26w47 e                             (.>)                     64.6mg   P<.2
408  R f f34 eat liv adn 24m24 a                               :      ±       +hist 34.8mg  *  P<.07   a
a    R f f34 eat sto acn 24m24 a                                             +hist no dre   P=1.    a
b    R f f34 eat TBA MXB 24m24 a                                                    no dre   P=1.
c    R f f34 eat liv MXB 24m24 a                                                    no dre   P=1.
409  R m f34 eat --- MXA 24m24                            :    +        :           5.94mg  *  P<.004
a    R m f34 eat --- leu 24m24                                                      6.75mg  *  P<.009
b    R m f34 eat liv MXA 24m24                                               +hist 15.1mg  *  P<.008  a
c    R m f34 eat MXA acn 24m24                                               +hist 34.6mg  *  P<.3    a
d    R m f34 eat TBA MXB 24m24                                                      3.23mg  *  P<.004
e    R m f34 eat liv MXB 24m24                                                      20.8mg  *  P<.03

AROCLOR 1260            100ng..:..1ug...:..10....:..100....:..1mg...:..10....:..100....:..1g.....:..10
410  R f shc eat liv mix 89w95 e                          .+ .                        1.04mg   P<.0005+
a    R f shc eat liv nnd 89w95 e                                                      1.76mg   P<.0005+
b    R f shc eat liv hpc 89w95 e                                 . +.                 18.3mg   P<.0005+
411  R m wis eat liv hpc 27m27 er                          . +.                       5.39mg   P<.0005+

ARSENATE, SODIUM        100ng..:..1ug...:..10....:..100....:..1mg...:..10....:..100....:..1g.....:..10
412  P b cym eat kcx ade 14y19 Ww              :      ±                               50.3ug   P<.08

a    P b cym eat tba ben 14y19 Ww                                                     87.5ug   P<.5
413  R b osm eat tba mal 24m24 s                                       .>             no dre   P=1.    -

a    R b osm eat tba mix 24m24 s                                                      no dre   P=1.    -

ARSENIOUS OXIDE         100ng..:..1ug...:..10....:..100....:..1mg...:..10....:..100....:..1g.....:..10
414  M b cbl wat liv tum 23m23 sv                                      .>             no dre   P=1.
a    M b cbl wat tba mix 23m23 sv                                                     no dre   P=1.
415  M m ssa wat sbg pam 24m24                                  .>                    54.8mg   P<.3
416  M f swi wat lun ade 65w65 e                                .>                    215.mg   P<.9    -
a    M f swi wat liv hpt 65w65 e                                                      no dre   P=1.    -
417  M m swi wat lun ade 65w65 e                             .    ±                   31.4mg   P<.06
a    M m swi wat liv hpt 65w65 e                                                      no dre   P=1.

ARSENITE, SODIUM        100ng..:..1ug...:..10....:..100....:..1mg...:..10....:..100....:..1g.....:..10
418  M b cd1 wat lun mix 24m24 e                                      .>              no dre   P=1.    -
a    M b cd1 wat liv mix 24m24 e                                                      no dre   P=1.    -
b    M b cd1 wat tba mal 24m24 e                                                      no dre   P=1.    -
c    M b cd1 wat tba mix 24m24 e                                                      no dre   P=1.    -
d    M b cd1 wat tba ben 24m24 e                                                      no dre   P=1.    -
419  R b leb wat liv tum 39m39 e                                .>                    10.6mg   P<.2
a    R b leb wat tba mal 39m39 e                                                      no dre   P=1.
b    R b leb wat tba tum 39m39 e                                                      no dre   P=1.
420  R b osm eat tba mal 24m24 s                                .>                    626.mg Z P<1.

a    R b osm eat tba mix 24m24 s                                                      no dre   P=1.

L-ASCORBATE, SODIUM     100ng..:..1ug...:..10....:..100....:..1mg...:..10....:..100....:..1g.....:..10
421  R m f3d eat eso tum 52w52 er                                     .>              no dre   P=1.    -
a    R m f3d eat for tum 52w52 er                                                     no dre   P=1.    -
b    R m f3d eat liv tum 52w52 er                                                     no dre   P=1.    -

L-ASCORBIC ACID         100ng..:..1ug...:..10....:..100....:..1mg...:..10....:..100....:..1g.....:..10
422  M f b6c eat TBA MXB 24m24                                                       :no dre   P=1.
a    M f b6c eat liv MXB 24m24                                                        no dre   P=1.
b    M f b6c eat lun MXB 24m24                                                        no dre   P=1.
423  M m b6c eat TBA MXB 24m24                                                        no dre   P=1.
a    M m b6c eat liv MXB 24m24                                                        no dre   P=1.
b    M m b6c eat lun MXB 24m24                                                        115.gm *  P<.7
424  R f f34 eat TBA MXB 24m24                                              :>        no dre   P=1.
a    R f f34 eat liv MXB 24m24                                                        no dre   P=1.    -
425  R m f34 eat TBA MXB 24m24                                              :>        no dre   P=1.    -
a    R m f34 eat liv MXB 24m24                                                        no dre   P=1.

ASPARTAME               100ng..:..1ug...:..10....:..100....:..1mg...:..10....:..100....:..1g.....:..10
426  R f sls eat bra tum 52w52 ekr                                    .>              no dre   P=1.    -
427  R f sls eat bra mix 24m24 er                                                     no dre   P=1.    -
428  R m sls eat bra tum 52w52 ekr                              .>                    no dre   P=1.    -
429  R m sls eat bra mix 24m24 er                                                     186.gm *  P<.5    -

ASPIRIN                 100ng..:..1ug...:..10....:..100....:..1mg...:..10....:..100....:..1g.....:..10
430  M f cb6 eat liv tum 77w77 e                                        .>            no dre   P=1.    -
431  M m cb6 eat liv tum 77w77 e                                        .>            no dre   P=1.    -
432  R b alb gav for tum 78w78 r                                  .>                  no dre   P=1.    -
a    R b alb gav stg tum 78w78 r                                                      no dre   P=1.    -
433  R m f34 eat ubl pam 68w68 e                                    .>                1.67gm   P<.3
a    R m f34 eat liv tum 68w68 e                                                      no dre   P=1.
```

RefNum	LoConf	UpConf	Cntrl	1Dose	1Inc	2Dose	2Inc					Citation or Pathology
												Brkly Code

AROCLOR 1254 (PCBs, polychlorinated biphenyls) 11097-69-1

RefNum	LoConf	UpConf	Cntrl	1Dose	1Inc	2Dose	2Inc	3Dose	3Inc	4Dose	4Inc	Citation or Pathology
406	1029	4.46mg	26.3mg	0/58	36.0mg	9/22							Kimbrough;jnci,53,547-549;1974
407	1029m	10.5mg	n.s.s.	0/58	19.6mg	1/24							
408	TR38	10.5mg	n.s.s.	0/24	1.30mg	0/24	2.50mg	1/24	5.00mg	2/24			
a	TR38	15.2mg	n.s.s.	0/24	1.30mg	1/24	2.50mg	1/24	5.00mg	0/24			
b	TR38	4.25mg	n.s.s.	11/24	1.30mg	18/24	2.50mg	13/24	5.00mg	12/24			
c	TR38	n.s.s.	n.s.s.	0/24	1.30mg	0/24	2.50mg	0/24	5.00mg	0/24			liv:hpa,hpc,nnd.
409	TR38	2.78mg	52.3mg	3/24	1.00mg	2/24	2.00mg	5/24	4.00mg	9/24			---:leu,lym. S
a	TR38	3.01mg	275.mg	3/24	1.00mg	2/24	2.00mg	5/24	4.00mg	8/24			S
b	TR38	5.21mg	338.mg	0/24	1.00mg	0/24	2.00mg	1/24	4.00mg	3/24			liv:adn,hpc.
c	TR38	8.40mg	n.s.s.	0/24	1.00mg	0/24	2.00mg	2/24	4.00mg	0/24			cec:acn; jej:acn; sto:acn.
d	TR38	1.60mg	23.4mg	7/24	1.00mg	7/24	2.00mg	12/24	4.00mg	13/24			
e	TR38	6.29mg	n.s.s.	0/24	1.00mg	0/24	2.00mg	1/24	4.00mg	2/24			liv:hpa,hpc,nnd.

AROCLOR 1260 (PCBs, clophen A 60, polychlorinated biphenyls) 11096-82-5

RefNum	LoConf	UpConf	Cntrl	1Dose	1Inc	Citation or Pathology
410	1320	.804mg	1.35mg	0/173	4.69mg	170/184	Kimbrough;jnci,55,1453-1456;1975
a	1320	1.39mg	2.24mg	0/173	4.69mg	144/184	
b	1320	11.3mg	34.3mg	1/173	4.69mg	26/184	
411	1605	3.90mg	7.75mg	1/131	4.00mg	61/129	Schaeffer;txap,75,278-288;1984

ARSENATE, SODIUM   7631-89-2

RefNum	LoConf	UpConf	Cntrl	1Dose	1Inc	2Dose	2Inc	3Dose	3Inc	4Dose	4Inc	5Dose	5Inc	Citation or Pathology
412	2001	12.4ug	n.s.s.	0/13	16.5ug	2/11									Adamson;ossc,129-156;
															1982/Thorgeirsson 1994/Dalgard 1991/Thorgeirsson&Seiber pers.comm.
a	2001	14.2ug	n.s.s.	1/19	16.5ug	2/12									
413	1507	116.mg	n.s.s.	8/50	1.41mg	14/50	2.81mg	8/50	5.63mg	10/50	11.3mg	6/50	18.0mg	5/50	Byron;txap,10,
															132-147;1967
a	1507	121.mg	n.s.s.	17/50	1.41mg	21/50	2.81mg	15/50	5.63mg	16/50	11.3mg	13/50	18.0mg	6/50	

ARSENIOUS OXIDE   1327-53-3

RefNum	LoConf	UpConf	Cntrl	1Dose	1Inc	Citation or Pathology
414	1505	39.4mg	n.s.s.	0/50	4.13mg	0/50	Hueper;aenh,5,445-462;1962
a	1505	39.4mg	n.s.s.	1/50	4.13mg	0/50	
415	1506	12.1mg	n.s.s.	1/50	3.33mg	3/50	Sanderson;becc,39,628-629;1961
416	1509	8.52mg	n.s.s.	34/137	20.0mg	4/15	Baroni;aenh,7,668-674;1963/Shubik 1962
a	1509	24.2mg	n.s.s.	4/137	20.0mg	0/15	
417	1509	11.4mg	n.s.s.	18/133	16.7mg	15/60	
a	1509	80.5mg	n.s.s.	2/133	16.7mg	0/60	

ARSENITE, SODIUM   7784-46-5

RefNum	LoConf	UpConf	Cntrl	1Dose	1Inc	2Dose	2Inc	3Dose	3Inc	4Dose	4Inc	Citation or Pathology
418	1512	14.2mg	n.s.s.	26/170	.877mg	3/103							Kanisawa;canr,27,1192-1195;1967
a	1512	14.6mg	n.s.s.	7/170	.877mg	1/103							
b	1512	7.97mg	n.s.s.	15/170	.877mg	6/103							
c	1512	10.9mg	n.s.s.	55/170	.877mg	11/103							
d	1512	14.5mg	n.s.s.	29/170	.877mg	3/103							
419	1036	3.35mg	n.s.s.	1/82	.265mg	5/91							Kanisawa;canr,29,892-895;1969
a	1036	8.08mg	n.s.s.	9/82	.265mg	3/91							
b	1036	2.90mg	n.s.s.	31/82	.265mg	25/91							
420	1507	13.8mg	n.s.s.	8/50	.703mg	8/50	1.41mg	9/50	2.81mg	15/50	5.63mg	7/50	(11.3mg 1/50) Byron;txap,10,
													132-147;1967
a	1507	60.2mg	n.s.s.	16/50	.703mg	14/50	1.41mg	15/50	2.81mg	18/50	5.63mg	15/50	11.3mg 6/50

L-ASCORBATE, SODIUM (vitamin C, sodium) 134-03-2

RefNum	LoConf	UpConf	Cntrl	1Dose	1Inc	Citation or Pathology
421	1900	206.mg	n.s.s.	0/10	400.mg	0/10	Hirose;carc,8,1731-1735;1987/pers.comm.
a	1900	206.mg	n.s.s.	0/10	400.mg	0/10	
b	1900	206.mg	n.s.s.	0/10	400.mg	0/10	

L-ASCORBIC ACID (vitamin C) 50-81-7

RefNum	LoConf	UpConf	Cntrl	1Dose	1Inc	2Dose	2Inc	Citation or Pathology
422	TR247	8.40gm	n.s.s.	30/50	3.19gm	28/50	6.38gm	27/50	
a	TR247	30.6gm	n.s.s.	3/50	3.19gm	1/50	6.38gm	3/50	liv:hpa,hpc,nnd.
b	TR247	24.1gm	n.s.s.	1/50	3.19gm	4/50	6.38gm	1/50	lun:a/a,a/c.
423	TR247	11.9gm	n.s.s.	29/50	2.94gm	31/50	5.89gm	24/50	
a	TR247	16.4gm	n.s.s.	16/50	2.94gm	16/50	5.89gm	13/50	liv:hpa,hpc,nnd.
b	TR247	16.5gm	n.s.s.	5/50	2.94gm	4/50	5.89gm	8/50	lun:a/a,a/c.
424	TR247	1.86gm	n.s.s.	41/50	1.23gm	45/50	2.45gm	40/50	
a	TR247	28.7gm	n.s.s.	2/50	1.23gm	0/50	2.45gm	0/50	liv:hpa,hpc,nnd.
425	TR247	2.05gm	n.s.s.	37/50	981.mg	41/50	1.96gm	39/50	
a	TR247	16.7gm	n.s.s.	2/50	981.mg	0/50	1.96gm	1/50	liv:hpa,hpc,nnd.

ASPARTAME   22839-47-0

RefNum	LoConf	UpConf	Cntrl	1Dose	1Inc	2Dose	2Inc	3Dose	3Inc	Citation or Pathology
426	1327m	471.mg	n.s.s.	0/16	1.00gm	0/16	2.00gm	0/16	4.00gm	0/16	Ishii;txlt,7,433-437;1981
427	1327n	45.8gm	n.s.s.	1/60	1.00gm	0/60	2.00gm	2/60	4.00gm	0/60	
428	1327n	471.mg	n.s.s.	0/16	1.00gm	0/16	2.00gm	0/16	4.00gm	0/16	
429	1327n	35.2gm	n.s.s.	0/59	1.00gm	1/59	2.00gm	0/60	4.00gm	1/60	

ASPIRIN   50-78-2

RefNum	LoConf	UpConf	Cntrl	1Dose	1Inc	Citation or Pathology
430	1028	1.77gm	n.s.s.	0/36	382.mg	0/41	Macklin;dact,3,135-163;1980
431	1028	1.94gm	n.s.s.	0/35	382.mg	0/45	
432	1571	125.mg	n.s.s.	0/23	26.9mg	0/40	Tsung-Hsien;jnci,70,1067-1069;1983
a	1571	125.mg	n.s.s.	0/23	26.9mg	0/40	
433	1786	272.mg	n.s.s.	0/36	200.mg	1/29	Sakata;canr,46,3903-3906;1986
a	1786	511.mg	n.s.s.	0/36	200.mg	0/29	

```
         Spe Strain  Site   Xpo+Xpt                                                                      TD50    2Tailpvl
         Sex  Route  Hist   Notes                                                                             DR  AuOp
ASPIRIN, PHENACETIN, AND CAFFEINE  100ng..:..1ug....:..10.....:..100....:..1mg....:..10.....:..100....:..1g.....:.10
434  M f b6c eat TBA MXB 78w94                                                                :>          no dre  P=1.   -
 a   M f b6c eat liv MXB 78w94                                                                            18.0gm * P<.4
 b   M f b6c eat lun MXB 78w94                                                                            no dre  P=1.
435  M m b6c eat TBA MXB 78w94                                                           :>              3.01gm * P<.3
 a   M m b6c eat liv MXB 78w94                                                                            44.5gm * P<.9
 b   M m b6c eat lun MXB 78w94                                                                            3.68gm * P<.07
436  M f cb6 eat liv hnd 77w77 e                                                               .>        8.76gm  P<.3   -
 a   M f cb6 eat liv hpc 77w77 e                                                                         no dre  P=1.   -
437  M m cb6 eat liv hnd 77w77 e                                                               .>        9.54gm  P<.3   -
 a   M m cb6 eat liv hpc 77w77 e                                                                         no dre  P=1.   -
438  R f f34 eat TBA MXB 18m25                                                         :>               no dre  P=1.   -
 a   R f f34 eat liv MXB 18m25                                                                           3.05gm * P<.07
439  R m f34 eat TBA MXB 18m25                                                             :>           no dre  P=1.   -
 a   R m f34 eat liv MXB 18m25                                                                           no dre  P=1.

ATRAZINE                           100ng..:..1ug....:..10.....:..100....:..1mg....:..10.....:..100....:..1g.....:.10
440  M f b6a orl lun ade 76w76 evx                                    .>                                 no dre  P=1.   -
 a   M f b6a orl liv hpt 76w76 evx                                                                       no dre  P=1.   -
 b   M f b6a orl tba mix 76w76 evx                                                                       58.8mg  P<.7   -
441  M m b6a orl lun ade 76w76 evx                               .>                                      28.2mg  P<.4   -
 a   M m b6a orl liv hpt 76w76 evx                                                                       no dre  P=1.   -
 b   M m b6a orl tba mix 76w76 evx                                                                       54.6mg  P<.7   -
442  M f b6c orl liv hpt 76w76 evx                                   .>                                  no dre  P=1.   -
 a   M f b6c orl lun mix 76w76 evx                                                                       no dre  P=1.   -
 b   M f b6c orl tba mix 76w76 evx                                                                       71.0mg  P<.3   -
443  M m b6c orl liv hpt 76w76 evx                               .  ±                                    15.0mg  P<.02  -
 a   M m b6c orl lun mix 76w76 evx                                                                       no dre  P=1.   -
 b   M m b6c orl tba mix 76w76 evx                                                                       9.29mg  P<.004 -
444  R f f31 eat ute mal 28m29 ae                                          .  ±                          197.mg * P<.09  +
 a   R f f31 eat --- mix 28m29 ae                                                                        158.mg * P<.2   +
 b   R f f31 eat --- leu 28m29 ae                                                                        267.mg * P<.2
 c   R f f31 eat --- lym 28m29 ae                                                                        698.mg * P<.5
 d   R f f31 eat tba mal 28m29 ae                                                                        30.0mg * P<.0005+
445  R m f31 eat --- mix 29m29 ae                                            .  ±                        68.2mg * P<.06
 a   R m f31 eat --- leu 29m29 ae                                                                        109.mg * P<.1
 b   R m f31 eat mgl mix 29m29 ae                                                                        262.mg * P<.02  +
 c   R m f31 eat mgl ben 29m29 ae                                                                        304.mg * P<.03  +
 d   R m f31 eat --- lym 29m29 ae                                                                        791.mg * P<.7
 e   R m f31 eat mgl mal 29m29 ae                                                                        2.17gm * P<.3
 f   R m f31 eat liv hpc 29m29 ae                                                                        2.19gm * P<.3
 g   R m f31 eat tba mal 29m29 ae                                                                        33.6mg * P<.0005+

ATROPINE                           100ng..:..1ug....:..10.....:..100....:..1mg....:..10.....:..100....:..1g.....:.10
446  R f sda ipj tba mal 24m24 e                                        .>                               90.7mg  P<1.   -
447  R m sda ipj liv hae 24m24 e                                      .>                                 no dre  P=1.
 a   R m sda ipj tba mal 24m24 e                                                                         no dre  P=1.   -

AURAMINE-O                         100ng..:..1ug....:..10.....:..100....:..1mg....:..10.....:..100....:..1g.....:.10
448  M f alb eat liv hpt 52w79 e                                          .  ±                           94.3mg  P<.08  +
449  M m alb eat liv hpt 52w84 e                                         .  +      .                      39.2mg  P<.008 +
450  M f cba eat liv hpt 52w99 e                                            .  +  .                       67.7mg  P<.0005+
451  M m cba eat liv hpt 52w79 e                                            .  +      .                   77.5mg  P<.004 +
452  R m wsw eat liv hpt 20m29 e                                       .  +  .                           11.0mg  P<.0005+

AURANOFIN                          100ng..:..1ug....:..10.....:..100....:..1mg....:..10.....:..100....:..1g.....:.10
453  M f cd1 gav liv mix 80w81 erv                                                      .>               no dre  P=1.   -
454  M m cd1 gav liv mix 80w81 erv                                                   .>                  42.6mg * P<.2   -

5-AZACYTIDINE                      100ng..:..1ug....:..10.....:..100....:..1mg....:..10.....:..100....:..1g.....:.10
455  M f b6c ipj MXA MXA 52w81                             :  +  :                                       .256mg * P<.0005a
 a   M f b6c ipj mul MXA 52w81                                                                           .520mg * P<.008 a
 b   M f b6c ipj --- lym 52w81                                                                           .615mg * P<.03  a
 c   M f b6c ipj TBA MXB 52w81                                                                           .219mg * P<.003
 d   M f b6c ipj liv MXB 52w81                                                                           no dre  P=1.
 e   M f b6c ipj lun MXB 52w81                                                                           no dre  P=1.
456  M m b6c ipj TBA MXB 52w81 s                              :>                                         1.05mg * P<.2   i
 a   M m b6c ipj liv MXB 52w81 s                                                                         15.7mg * P<.9
 b   M m b6c ipj lun MXB 52w81 s                                                                         61.4mg * P<1.
457  M f bal ipj lmr mix 50w72 e                         .  +  .                                         56.9ug  P<.0005+
 a   M f bal ipj mgl mix 50w72 e                                                                         .432mg  P<.002
 b   M f bal ipj mgl adb 50w72 e                                                                         .618mg  P<.008 +
 c   M f bal ipj ski mix 50w72 e                                                                         .499mg  P<.03  +
 d   M f bal ipj lun ade 50w72 e                                                                         no dre  P=1.
 e   M f bal ipj liv tum 50w72 e                                                                         no dre  P=1.
 f   M f bal ipj tba tum 50w72 e                                                                         36.4ug  P<.0005
458  M m bal ipj lun ade 50w67 e                         .  +    .                                       .121mg  P<.003 +
 a   M m bal ipj lmr mix 50w67 e                                                                         .285mg  P<.01  +
 b   M m bal ipj ski mix 50w67 e                                                                         .980mg  P<.04  +
 c   M m bal ipj liv hpa 50w67 e                                                                         1.49mg  P<.1
 d   M m bal ipj tba tum 50w67 e                                                                         53.9ug  P<.0005
```

RefNum	LoConf	UpConf	Cntrl	1Dose	1Inc	2Dose	2Inc	Citation or Pathology / Brkly Code

ASPIRIN, PHENACETIN, AND CAFFEINE   (APC) 8003-03-0

	RefNum	LoConf	UpConf	Cntrl	1Dose	1Inc	2Dose	2Inc	Citation / Brkly Code
434	TR67	2.92gm	n.s.s.	20/50	755.mg	17/50	1.51gm	14/50	
a	TR67	4.46gm	n.s.s.	1/50	755.mg	2/50	1.51gm	3/50	liv:hpa,hpc,nnd.
b	TR67	3.65gm	n.s.s.	4/50	755.mg	7/50	1.51gm	4/50	lun:a/a,a/c.
435	TR67	975.mg	n.s.s.	17/50	697.mg	24/50	1.39gm	21/50	
a	TR67	2.35gm	n.s.s.	7/50	697.mg	11/50	1.39gm	6/50	liv:hpa,hpc,nnd.
b	TR67	1.46gm	n.s.s.	6/50	697.mg	9/50	1.39gm	12/50	lun:a/a,a/c.
436	1028	1.43gm	n.s.s.	0/36	696.mg	1/34			Macklin;dact,3,135-163;1980
a	1028	2.67gm	n.s.s.	0/36	696.mg	0/34			
437	1028	1.55gm	n.s.s.	0/35	696.mg	1/37			
a	1028	2.91gm	n.s.s.	0/35	696.mg	0/37			
438	TR67	222.mg	n.s.s.	42/50	245.mg	41/50	(485.mg	32/49)	
a	TR67	1.32gm	n.s.s.	0/50	245.mg	4/50	485.mg	3/49	liv:hpa,hpc,nnd.
439	TR67	406.mg	n.s.s.	30/50	196.mg	37/50	388.mg	33/50	
a	TR67	2.75gm	n.s.s.	5/50	196.mg	4/50	388.mg	1/50	liv:hpa,hpc,nnd.

ATRAZINE   1912-24-9

	RefNum	LoConf	UpConf	Cntrl	1Dose	1Inc	2Dose	2Inc	Citation / Brkly Code
440	1244	12.4mg	n.s.s.	1/17	11.1mg	1/17			Innes;ntis,1968/1969
a	1244	20.7mg	n.s.s.	0/17	11.1mg	0/17			
b	1244	7.41mg	n.s.s.	2/17	11.1mg	3/17			
441	1244	5.90mg	n.s.s.	2/18	10.3mg	4/18			
a	1244	20.4mg	n.s.s.	1/18	10.3mg	0/18			
b	1244	6.40mg	n.s.s.	3/18	10.3mg	4/18			
442	1244	22.0mg	n.s.s.	0/16	11.1mg	0/18			
a	1244	22.0mg	n.s.s.	0/16	11.1mg	0/18			
b	1244	11.6mg	n.s.s.	0/16	11.1mg	1/18			
443	1244	5.16mg	n.s.s.	0/16	10.3mg	4/18			
a	1244	20.4mg	n.s.s.	0/16	10.3mg	0/18			
b	1244	3.75mg	56.8mg	0/16	10.3mg	6/18			
444	1957	77.3mg	n.s.s.	7/45	19.2mg	10/52	38.2mg	14/45	Pinter;nplm,35,533-544;1990/pers.comm.
a	1957	60.4mg	n.s.s.	12/44	19.2mg	16/52	38.2mg	22/51	
b	1957	87.9mg	n.s.s.	8/44	19.2mg	10/52	38.2mg	15/51	
c	1957	147.mg	n.s.s.	4/44	19.2mg	6/52	38.2mg	7/51	
d	1957	18.0mg	77.1mg	21/50	19.2mg	40/53	38.2mg	44/55	
445	1957	28.6mg	n.s.s.	22/47	15.3mg	26/47	30.6mg	32/48	
a	1957	42.4mg	n.s.s.	13/47	15.3mg	21/47	30.6mg	21/48	
b	1957	113.mg	n.s.s.	1/45	15.3mg	1/52	30.6mg	8/45	
c	1957	124.mg	n.s.s.	1/45	15.3mg	1/52	30.6mg	7/45	
d	1957	112.mg	n.s.s.	9/47	15.3mg	5/47	30.6mg	11/48	
e	1957	354.mg	n.s.s.	0/45	15.3mg	0/52	30.6mg	1/45	
f	1957	356.mg	n.s.s.	0/47	15.3mg	0/47	30.6mg	1/48	
g	1957	19.5mg	96.2mg	24/56	15.3mg	30/55	30.6mg	43/53	

ATROPINE   51-55-8

	RefNum	LoConf	UpConf	Cntrl	1Dose	1Inc	2Dose	2Inc	Citation / Brkly Code
446	1134	2.23mg	n.s.s.	3/33	.857mg	3/31			Schmahl;zkko,86,77-84;1976
447	1134	5.30mg	n.s.s.	1/36	.857mg	0/30			
a	1134	5.30mg	n.s.s.	1/36	.857mg	0/30			

AURAMINE-O   2465-27-2

	RefNum	LoConf	UpConf	Cntrl	1Dose	1Inc	2Dose	2Inc	Citation / Brkly Code
448	45	28.2mg	n.s.s.	0/6	85.0mg	3/10			Williams;bjca,16,87-91;1962
449	45	12.8mg	860.mg	0/7	74.3mg	4/7			
450	45	30.5mg	197.mg	3/41	136.mg	11/15			
451	45	30.3mg	666.mg	1/14	157.mg	7/12			
452	45	4.06mg	31.2mg	0/12	27.7mg	11/12			

AURANOFIN   ((2,3,4,6-tetra-O-acetyl-1-thio-1-beta-D-glucopyranosato-S)   (triethylphosphine) gold) 34031-32-8

	RefNum	LoConf	UpConf	Cntrl	1Dose	1Inc	2Dose	2Inc	Citation / Brkly Code
453	1870	130.mg	n.s.s.	4/220	1.00mg	3/110	3.00mg	0/110 7.45mg 0/110	Markiewicz;faat,11,277-284;1988
454	1870	13.8mg	n.s.s.	24/220	1.00mg	21/110	3.00mg	17/110 7.45mg 20/110	

5-AZACYTIDINE   320-67-2

	RefNum	LoConf	UpConf	Cntrl	1Dose	1Inc	2Dose	2Inc	Citation / Brkly Code
455	TR42	.138mg	.716mg	0/14	.610mg	17/35	1.60mg	0/35	---:lym; mul:grl,gsa.
a	TR42	.243mg	6.56mg	0/14	.610mg	10/35	1.60mg	0/35	mul:grl,gsa.
b	TR42	.249mg	n.s.s.	0/14	.610mg	7/35	1.60mg	0/35	
c	TR42	.112mg	1.35mg	2/14	.610mg	22/35	1.60mg	1/35	
d	TR42	n.s.s.		0/14	.610mg	0/35	1.60mg	0/35	liv:hpa,hpc,nnd.
e	TR42	1.65mg	n.s.s.	2/14	.610mg	0/35	1.60mg	0/35	lun:a/a,a/c.
456	TR42	.360mg	n.s.s.	2/16	.600mg	6/35	1.20mg	5/35	
a	TR42	.919mg	n.s.s.	1/16	.600mg	1/35	1.20mg	1/35	liv:hpa,hpc,nnd.
b	TR42	1.36mg	n.s.s.	1/16	.600mg	0/35	1.20mg	1/35	lun:a/a,a/c.
457	1819	35.2ug	.104mg	6/50	.198mg	36/50			Cavaliere;clet,37,51-58;1987
a	1819	.186mg	1.82mg	0/50	.198mg	7/50			
b	1819	.235mg	12.0mg	0/50	.198mg	5/50			
c	1819	.196mg	n.s.s.	1/50	.198mg	7/50			
d	1819	.329mg	n.s.s.	9/50	.198mg	7/50			
e	1819	.980mg	n.s.s.	0/50	.198mg	0/50			
f	1819	21.5ug	69.4ug	14/50	.198mg	44/50			
458	1819	60.4ug	.719mg	12/50	.213mg	27/50			
a	1819	.126mg	41.7mg	3/50	.213mg	12/50			
b	1819	.297mg	n.s.s.	0/50	.213mg	3/50			
c	1819	.365mg	n.s.s.	0/50	.213mg	2/50			
d	1819	31.5ug	.116mg	13/50	.213mg	38/50			

```
      Spe Strain Site  Xpo+Xpt                                               TD50     2Tailpvl
          Sex  Route  Hist   Notes                                                DR   AuOp
459   R f sda ipj TBA MXB 34w80 s                        :    ±                  .487mg /  P<.09  i
a     R f sda ipj liv MXB 34w80 s                                               no dre    P=1.
460   R m sda ipj TBA MXB 34w80 s                     :    +        :            .328mg *  P<.004 i
a     R m sda ipj liv MXB 34w80 s                                               no dre    P=1.
461   R m f34 ipj tes tum 52w52 e               .   +   .                        .222mg *  P<.0005+
a     R m f34 ipj liv tum 52w52 e                                               no dre    P=1.
b     R m f34 ipj tba tum 52w52 e                                                .170mg *  P<.0005+

6-AZACYTIDINE            100ng..:..1ug....:..10....:..100...:..1mg..:..10...:..100....:..1g...:..10
462   R m f34 ipj liv tum 52w52 e                             .>                no dre    P=1.
a     R m f34 ipj tba tum 52w52 e                                               no dre    P=1.   -

AZASERINE               100ng..:..1ug....:..10....:..100...:..1mg..:..10...:..100....:..1g...:..10
463   R b wis ipj pae car 26w78 e                  .   +    .                    .793mg    P<.0005+
a     R b wis ipj kid tum 26w78 e                                                .942mg    P<.0005
b     R b wis ipj liv tum 26w78 e                                               3.02mg     P<.03
464   R m wis ipj pan tum 24m24 r                       .>                      no dre    P=1.   -

AZATHIOPRINE            100ng..:..1ug....:..10....:..100...:..1mg..:..10...:..100....:..1g...:..10
465   M f b6c eat liv tum 86w86 k                    .>                         no dre    P=1.
a     M f b6c eat lun tum 86w86 k                                               no dre    P=1.
466   M f b6c eat lun tum 94w94 k                          .>                   4.89mg *  P<.2
a     M f b6c eat --- lsl 94w94 k                                               7.59mg *  P<.3
b     M f b6c eat liv tum 94w94 k                                               no dre    P=1.
467   M f b6c eat ute hae 23m23                   .   +   .              +hist   8.92mg *  P<.0005+
a     M f b6c eat --- lsl 23m23                                                 43.8mg *  P<.7   +
b     M f b6c eat liv hpa 23m23                                                 67.5mg *  P<.2
c     M f b6c eat lun tum 23m23                                                 no dre    P=1.
d     M f b6c eat liv hem 23m23                                                 no dre    P=1.
468   M m b6c eat liv tum 86w86 k                    .>                         no dre    P=1.
a     M m b6c eat lun tum 86w86 k                                               no dre    P=1.
469   M m b6c eat lun ade 94w94 k                          .>                   no dre    P=1.
a     M m b6c eat liv hpa 94w94 k                                               no dre    P=1.
b     M m b6c eat --- lsl 94w94 k                                               no dre    P=1.
470   M m b6c eat --- lsl 23m23                            .>                   9.66mg *  P<.2
a     M m b6c eat liv hpc 23m23                                                 55.9mg *  P<.2
b     M m b6c eat lun car 23m23                                                 55.9mg *  P<.2
c     M m b6c eat liv hpa 23m23                                                 no dre    P=1.
d     M m b6c eat lun ade 23m23                                                 no dre    P=1.
471   R f f34 eat edu sqc 52w52 ekr                       .>                    10.1mg    P<.2
a     R f f34 eat liv tum 52w52 ekr                                             no dre    P=1.   -

AZIDE, SODIUM           100ng..:..1ug....:..10....:..100...:..1mg..:..10...:..100....:..1g...:..10
472   R f f34 gav TBA MXB 24m24                         :>                      19.1mg *  P<.6   -
a     R f f34 gav liv MXB 24m24                                                 no dre    P=1.
473   R m f34 gav tes MXA 24m24                      :    ±                     #3.79mg / P<.03  -
a     R m f34 gav TBA MXB 24m24                                                 8.26mg *  P<.4
b     R m f34 gav liv MXB 24m24                                                 no dre    P=1.
474   R f cdr eat mgl tum 18m24 e                   .   +   .                   3.01mg \  P<.003
a     R f cdr eat pit cra 18m24 e                                               16.1mg *  P<.5
475   R m cdr eat pit cra 18m24 e                          .>                   no dre    P=1.   -

AZINPHOSMETHYL          100ng..:..1ug....:..10....:..100...:..1mg..:..10...:..100....:..1g...:..10
476   M f b6c eat TBA MXB 80w92                            :>                   2.58gm /  P<1.   -
a     M f b6c eat liv MXB 80w92                                                 no dre    P=1.
b     M f b6c eat lun MXB 80w92                                                 118.mg *  P<.3
477   M m b6c eat liv hpc 80w92                    :    +        :              #14.3mg * P<.007 -
a     M m b6c eat TBA MXB 80w92                                                 46.6mg *  P<.8
b     M m b6c eat liv MXB 80w92                                                 12.6mg *  P<.2
c     M m b6c eat lun MXB 80w92                                                 no dre    P=1.
478   R f osm eat TBA MXB 19m27                      :>                         no dre    P=1.
a     R f osm eat liv MXB 19m27                                                 164.mg /  P<.9
479   R f osm eat pit acn 19m25 pool               :    +    :                  #6.93mg \ P<.0005-
a     R f osm eat thy MXA 19m25                                                 15.9mg *  P<.004
b     R f osm eat adr coa 19m25                                                 16.4mg *  P<.02
480   R m osm eat pni MXA 19m27 v                      :    ±                   22.9mg *  P<.04  a
a     R m osm eat TBA MXB 19m27 v                                               5.42mg *  P<.3
b     R m osm eat liv MXB 19m27 v                                               52.7mg *  P<.6
481   R m osm eat pit cra 19m25 v pool             :    +        :              3.99mg \  P<.008
a     R m osm eat pit MXA 19m25 v                                               3.99mg \  P<.008
b     R m osm eat thy MXA 19m25 v                                               6.06mg *  P<.0005a
c     R m osm eat thy --- 19m25 v                                               7.87mg *  P<.003
d     R m osm eat thy MXA 19m25 v                                               10.5mg *  P<.006 a
e     R m osm eat adr MXA 19m25 v                                               15.3mg *  P<.006
f     R m osm eat pit MXA 19m25 v                                               7.43mg *  P<.02
g     R m osm eat liv hpa 19m25 v                                               23.3mg *  P<.02
h     R m osm eat pni MXA 19m25 v                                               30.6mg *  P<.03  a
i     R m osm eat adr acn 19m25 v                                               44.3mg *  P<.02

AZOBENZENE              100ng..:..1ug....:..10....:..100...:..1mg..:..10...:..100....:..1g...:..10
482   M f b6c eat TBA MXB 24m24 v                          :>                   no dre    P=1.   -
```

	RefNum	LoConf	UpConf	Cntrl	1Dose	1Inc	2Dose	2Inc			Citation or Pathology
											Brkly Code
459	TR42	.164mg	n.s.s.	7/15	.470mg	15/35	1.68mg	3/35			
a	TR42	n.s.s.	n.s.s.	0/15	.470mg	0/35	1.68mg	0/35			liv:hpa,hpc,nnd.
460	TR42	.156mg	2.32mg	1/15	.470mg	11/35	1.72mg	1/35			
a	TR42	n.s.s.	n.s.s.	0/15	.470mg	0/35	1.72mg	0/35			liv:hpa,hpc,nnd.
461	1906	.144mg	.404mg	10/49	10.7ug	1/10	.107mg	2/10	1.07mg	56/87	Carr;bjca,57,395-402;1988
a	1906	5.01ug	n.s.s.	0/49	10.7ug	0/10	.107mg	0/10	1.07mg	0/87	
b	1906	.114mg	.282mg	10/49	10.7ug	1/10	.107mg	2/10	1.07mg	63/87	

6-AZACYTIDINE   3131-60-0

	RefNum	LoConf	UpConf	Cntrl	1Dose	1Inc					Citation or Pathology
462	1906	.662mg	n.s.s.	0/49	1.07mg	0/12					Carr;bjca,57,395-402;1988
a	1906	.340mg	n.s.s.	10/49	1.07mg	2/12					

AZASERINE   115-02-6

	RefNum	LoConf	UpConf	Cntrl	1Dose	1Inc					Citation or Pathology
463	1172	.342mg	2.54mg	0/76	.475mg	7/34					Longnecker;canc,47,1562-1572;1981
a	1172	.384mg	3.38mg	0/76	.475mg	6/34					
b	1172	.742mg	n.s.s.	0/76	.475mg	2/34					
464	1746	.736mg	n.s.s.	0/9	.714mg	0/5					McGuinness;sjge,18,189-192;1983

AZATHIOPRINE   446-86-6

	RefNum	LoConf	UpConf	Cntrl	1Dose	1Inc	2Dose	2Inc			Citation or Pathology
465	1935m	.366mg	n.s.s.	0/5	.650mg	0/5	2.60mg	0/5			Ito;gann,80,419-423;1989/pers.comm.
a	1935m	.366mg	n.s.s.	0/5	.650mg	0/5	2.60mg	0/5			
466	1935n	1.68mg	n.s.s.	0/13	.650mg	2/15	2.60mg	2/11			
a	1935n	2.20mg	n.s.s.	0/13	.650mg	2/15	2.60mg	2/15			
b	1935n	1.22mg	n.s.s.	0/13	.650mg	0/15	2.60mg	0/11			
467	1935o	3.85mg	30.8mg	0/32	.650mg	0/30	2.60mg	7/34			
a	1935o	4.89mg	n.s.s.	2/32	.650mg	5/30	2.60mg	4/34			
b	1935o	11.0mg	n.s.s.	0/32	.650mg	0/30	2.60mg	1/34			
c	1935o	3.04mg	n.s.s.	0/32	.650mg	0/30	2.60mg	0/34			
d	1935o	3.04mg	n.s.s.	1/32	.650mg	0/30	2.60mg	0/34			
468	1935m	.338mg	n.s.s.	0/5	.600mg	0/5	2.40mg	0/5			
a	1935m	.338mg	n.s.s.	0/5	.600mg	0/5	2.40mg	0/5			
469	1935n	3.18mg	n.s.s.	1/15	.600mg	3/16	2.40mg	1/15			
a	1935n	5.97mg	n.s.s.	1/15	.600mg	1/16	2.40mg	0/15			
b	1935n	5.97mg	n.s.s.	1/15	.600mg	1/16	2.40mg	0/15			
470	1935o	3.65mg	n.s.s.	0/30	.600mg	3/29	2.40mg	3/30			
a	1935o	9.10mg	n.s.s.	0/30	.600mg	0/29	2.40mg	1/30			
b	1935o	9.10mg	n.s.s.	0/30	.600mg	0/29	2.40mg	1/30			
c	1935o	2.69mg	n.s.s.	7/30	.600mg	1/29	(2.40mg	1/30)			
d	1935o	2.67mg	n.s.s.	1/30	.600mg	0/29	2.40mg	0/30			
471	1695	3.04mg	n.s.s.	0/12	7.50mg	3/25					Frankel;txap,17,462-480;1970
a	1695	9.66mg	n.s.s.	0/12	7.50mg	0/25					

AZIDE, SODIUM   26628-22-8

	RefNum	LoConf	UpConf	Cntrl	1Dose	1Inc	2Dose	2Inc			Citation or Pathology
472	TR389	3.58mg	n.s.s.	53/60	3.53mg	54/60	7.06mg	39/60			
a	TR389	64.7mg	n.s.s.	1/60	3.53mg	0/60	7.06mg	0/60			liv:hpa,hpc,nnd.
473	TR389	1.68mg	n.s.s.	52/60	3.53mg	51/60	7.06mg	44/60			tes:iab,ica. S
a	TR389	2.23mg	n.s.s.	56/60	3.53mg	55/60	7.06mg	36/60			
b	TR389	20.4mg	n.s.s.	3/60	3.53mg	3/60	7.06mg	0/60			liv:hpa,hpc,nnd.
474	1112	1.45mg	19.1mg	3/16	3.75mg	17/26	(7.50mg	10/26)			Weisburger;jnci,67,75-88;1981
a	1112	3.92mg	n.s.s.	6/16	3.75mg	19/26	7.50mg	14/26			
475	1112	13.5mg	n.s.s.	3/16	3.00mg	6/26	6.00mg	3/26			

AZINPHOSMETHYL   (gusathion) 86-50-0

	RefNum	LoConf	UpConf	Cntrl	1Dose	1Inc	2Dose	2Inc			Citation or Pathology
476	TR69	18.2mg	n.s.s.	5/10	7.02mg	10/50	14.3mg	17/50			
a	TR69	96.3mg	n.s.s.	1/10	7.02mg	0/50	14.3mg	1/50			liv:hpa,hpc,nnd.
b	TR69	40.6mg	n.s.s.	0/10	7.02mg	1/50	14.3mg	3/50			lun:a/a,a/c.
477	TR69	7.82mg	169.mg	0/10	3.24mg	12/50	6.48mg				S
a	TR69	4.69mg	n.s.s.	4/10	3.24mg	23/50	6.48mg	23/50			
b	TR69	5.18mg	n.s.s.	2/10	3.24mg	11/50	6.48mg	19/50			liv:hpa,hpc,nnd.
c	TR69	16.8mg	n.s.s.	2/10	3.24mg	8/50	6.48mg	4/50			lun:a/a,a/c.
478	TR69	2.58mg	n.s.s.	7/10	2.17mg	37/49	4.34mg	26/50			
a	TR69	12.0mg	n.s.s.	2/10	2.17mg	2/49	4.34mg	5/50			liv:hpa,hpc,nnd.
479	TR69	3.14mg	20.4mg	0/105p	2.17mg	8/49	(4.34mg	1/50)			S
a	TR69	7.20mg	123.mg	1/105p	2.17mg	6/49	4.34mg	4/50			thy:adn,cyn. S
b	TR69	6.72mg	n.s.s.	5/105p	2.17mg	4/49	4.34mg	8/50			S
480	TR69	9.81mg	n.s.s.	0/10	2.18mg	1/50	4.35mg	6/50			pni:isa,isc.
a	TR69	1.68mg	n.s.s.	7/10	2.18mg	40/50	4.35mg	41/50			
b	TR69	10.5mg	n.s.s.	1/10	2.18mg	3/50	4.35mg	5/50			liv:hpa,hpc,nnd.
481	TR69	1.77mg	90.0mg	15/105p	2.18mg	21/50	(4.35mg	13/50)			S
a	TR69	1.77mg	90.0mg	15/105p	2.18mg	21/50	(4.35mg	15/50)			pit:cra,crc. S
b	TR69	3.69mg	11.0mg	0/105p	2.18mg	10/50	4.35mg	13/50			thy:acn,cyn,pcn.
c	TR69	3.98mg	44.0mg	8/105p	2.18mg	14/50	4.35mg	14/50			S
d	TR69	4.92mg	143.mg	7/105p	2.18mg	10/50	4.35mg	12/50			thy:adn,cyn,fca.
e	TR69	6.93mg	183.mg	4/105p	2.18mg	4/50	4.35mg	10/50			adr:acn,coa. S
f	TR69	3.32mg	n.s.s.	20/105p	2.18mg	21/50	4.35mg	20/50			pit:adn,cra,crc,cyn. S
g	TR69	9.63mg	n.s.s.	1/105p	2.18mg	3/50	4.35mg	5/50			S
h	TR69	11.3mg	n.s.s.	2/105p	2.18mg	1/50	4.35mg	6/50			pni:isa,isc.
i	TR69	15.3mg	n.s.s.	0/105p	2.18mg	1/50	4.35mg	3/50			S

AZOBENZENE   103-33-3

	RefNum	LoConf	UpConf	Cntrl	1Dose	1Inc	2Dose	2Inc			Citation or Pathology
482	TR154	78.4mg	n.s.s.	9/20	27.0mg	20/50	70.8mg	12/50			

	Spe	Sex	Strain	Route	Site	Hist	Xpo+Xpt	Notes	DR plot	TD50	DR	2Tailpvl AuOp
a	M	f	b6c	eat	liv	MXB	24m24	v		no dre		P=1.
b	M	f	b6c	eat	lun	MXB	24m24	v		no dre		P=1.
483	M	f	b6c	orl	lun	ade	76w76	evx	.>	47.3mg		P<.3
a	M	f	b6c	orl	liv	hpt	76w76	evx		no dre		P=1.
b	M	f	b6c	orl	tba	mix	76w76	evx		47.3mg		P<.3
484	M	m	b6c	eat	TBA	MXB	24m24		:>	no dre		P=1. -
a	M	m	b6c	eat	liv	MXB	24m24			no dre		P=1.
b	M	m	b6c	eat	lun	MXB	24m24			no dre		P=1.
485	M	m	b6c	orl	liv	mix	76w76	evx	. + .	3.85mg		P<.0005
a	M	m	b6c	orl	liv	hpt	76w76	evx		4.54mg		P<.0005
b	M	m	b6c	orl	lun	mix	76w76	evx		no dre		P=1.
c	M	m	b6c	orl	tba	mix	76w76	evx		3.85mg		P<.0005
486	M	f	b6a	orl	lun	ade	76w76	evx	.>	39.3mg		P<.6
a	M	f	b6a	orl	liv	hpt	76w76	evx		no dre		P=1.
b	M	f	b6a	orl	tba	mix	76w76	evx		343.mg		P<1.
487	M	m	b6a	orl	liv	hpt	76w76	evx	.>	44.1mg		P<.6
a	M	m	b6a	orl	lun	ade	76w76	evx		no dre		P=1.
b	M	m	b6a	orl	tba	mix	76w76	evx		no dre		P=1.
488	R	f	f34	eat	MXA	MXA	24m24		: + :	32.3mg	*	P<.0005c
a	R	f	f34	eat	MXB	MXB	24m24			32.3mg	*	P<.0005
b	R	f	f34	eat	spl	ost	24m24			90.5mg	*	P<.008 c
c	R	f	f34	eat	spl	fbs	24m24			102.mg	*	P<.02 c
d	R	f	f34	eat	TBA	MXB	24m24			51.5mg	*	P<.5
e	R	f	f34	eat	liv	MXB	24m24			276.mg	*	P<.3
489	R	m	f34	eat	MXA	MXA	24m24	a	: + :	19.2mg	/	P<.0005c
a	R	m	f34	eat	MXB	MXB	24m24	a		19.2mg	/	P<.0005
b	R	m	f34	eat	mul	fbs	24m24	a		43.9mg	/	P<.0005
c	R	m	f34	eat	spl	fbs	24m24	a		44.4mg	/	P<.004 c
d	R	m	f34	eat	TBA	MXB	24m24	a		11.4mg	/	P<.0005
e	R	m	f34	eat	liv	MXB	24m24	a		122.mg	*	P<.07

AZOXYMETHANE
`100ng..:..1ug....:..10....:..100....:..1mg....:..10...:..100...:..1g.....:..10`

	Spe	Sex	Strain	Route	Site	Hist	Xpo+Xpt	Notes	DR plot	TD50	DR	2Tailpvl AuOp
490	R	m	f34	wat	liv	mix	7m23	e	. + .	30.2ug	\	P<.0005+
a	R	m	f34	wat	liv	hpc	7m23	e		67.7ug	\	P<.0005+
b	R	m	f34	wat	kid	mix	7m23	e		.388mg	*	P<.0005+
c	R	m	f34	wat	col	mix	7m23	e		.585mg	*	P<.0005+
d	R	m	f34	wat	liv	hes	7m23	e		.982mg	*	P<.2
e	R	m	f34	wat	liv	nnd	7m23	e		28.6mg	*	P<1.
f	R	m	f34	wat	tba	mix	7m23	e		noTD50		P=1.
491	R	m	f34	gav	col	tum	30w65	e	. + .	.102mg		P<.0005+
a	R	m	f34	gav	col	mal	30w65	e		.171mg		P<.0005
b	R	m	f34	gav	zym	car	30w65	e		.204mg		P<.0005+
c	R	m	f34	gav	kid	mnp	30w65	e		.300mg		P<.002 +
d	R	m	f34	gav	liv	tum	30w65	e		no dre		P=1. -

1-AZOXYPROPANE
`100ng..:..1ug....:..10....:..100....:..1mg....:..10...:..100...:..1g.....:..10`

	Spe	Sex	Strain	Route	Site	Hist	Xpo+Xpt	Notes	DR plot	TD50	DR	2Tailpvl AuOp
492	R	m	sda	gav	ski	mix	26w77	ev	<+	noTD50		P<.0005+
a	R	m	sda	gav	ski	ker	26w77	ev		380.ng		P<.0005+
b	R	m	sda	gav	nas	mix	26w77	ev		596.ng		P<.0005+
c	R	m	sda	gav	nas	ene	26w77	ev		723.ng		P<.0005+
d	R	m	sda	gav	nas	pam	26w77	ev		7.36ug		P<.1 +
e	R	m	sda	gav	liv	hpc	26w77	ev		7.36ug		P<.1
f	R	m	sda	gav	liv	hpa	26w77	ev		no dre		P=1.

2-AZOXYPROPANE
`100ng..:..1ug....:..10....:..100....:..1mg....:..10...:..100...:..1g.....:..10`

	Spe	Sex	Strain	Route	Site	Hist	Xpo+Xpt	Notes	DR plot	TD50	DR	2Tailpvl AuOp
493	R	m	sda	gav	ski	ker	26w77	ev	. ±	2.68ug		P<.04 +
a	R	m	sda	gav	liv	hpc	26w77	ev		7.36ug		P<.1
b	R	m	sda	gav	liv	hpa	26w77	ev		no dre		P=1.

BARBITAL, SODIUM
`100ng..:..1ug....:..10....:..100....:..1mg....:..10...:..100...:..1g.....:..10`

	Spe	Sex	Strain	Route	Site	Hist	Xpo+Xpt	Notes	DR plot	TD50	DR	2Tailpvl AuOp
494	R	m	f34	eat	ubl	tum	56w56	e	.>	no dre		P=1. -
495	R	m	f34	eat	ktu	tum	52w52	kr	.>	no dre		P=1. -
496	R	m	f34	eat	k/p	tpp	72w72	er	. + .	105.mg		P<.0005+
a	R	m	f34	eat	ktu	tla	72w72	er		341.mg		P<.03 +
497	R	m	f34	eat	kid	mix	24m24		.>	719.mg		P<.2
a	R	m	f34	eat	kid	tla	24m24			1.11gm		P<.3
b	R	m	f34	eat	kid	uac	24m24			2.37gm		P<.3

BARBITURIC ACID
`100ng..:..1ug....:..10....:..100....:..1mg....:..10...:..100...:..1g.....:..10`

	Spe	Sex	Strain	Route	Site	Hist	Xpo+Xpt	Notes	DR plot	TD50	DR	2Tailpvl AuOp
498	R	m	f34	wat	liv	hct	72w72	e	.>	no dre		P=1.

BARIUM ACETATE
`100ng..:..1ug....:..10....:..100....:..1mg....:..10...:..100...:..1g.....:..10`

	Spe	Sex	Strain	Route	Site	Hist	Xpo+Xpt	Notes	DR plot	TD50	DR	2Tailpvl AuOp
499	M	f	cd1	wat	---	lkm	27m27	e	. ±	4.28mg		P<.05 -
a	M	f	cd1	wat	lun	tum	27m27	e		no dre		P=1.
b	M	f	cd1	wat	tba	mal	27m27	e		7.21mg		P<.3
c	M	f	cd1	wat	tba	mix	27m27	e		no dre		P=1.
500	M	m	cd1	wat	lun	tum	27m27	e	.>	no dre		P=1.
a	M	m	cd1	wat	tba	mix	27m27	e		14.6mg		P<.8
b	M	m	cd1	wat	tba	mal	27m27	e		5.98mg		P<.2
501	R	f	leb	wat	tba	mix	37m37	e	.>	no dre		P=1. -
a	R	f	leb	wat	tba	mal	37m37	e		no dre		P=1. -

	RefNum	LoConf	UpConf	Cntrl	1Dose	1Inc	2Dose	2Inc	Citation or Pathology
									Brkly Code
a	TR154	247.mg	n.s.s.	0/20	27.0mg	2/50	70.8mg	0/50	liv:hpa,hpc,nnd.
b	TR154	232.mg	n.s.s.	2/20	27.0mg	2/50	70.8mg	2/50	lun:a/a,a/c.
483	1104	7.70mg	n.s.s.	0/16	7.84mg	1/17			Innes;ntis,1968/1969
a	1104	14.7mg	n.s.s.	0/16	7.84mg	0/17			
b	1104	7.70mg	n.s.s.	0/16	7.84mg	1/17			
484	TR154	34.6mg	n.s.s.	13/20	24.0mg	25/50	(48.0mg	20/50)	
a	TR154	43.6mg	n.s.s.	8/20	24.0mg	16/50	(48.0mg	2/50)	liv:hpa,hpc,nnd.
b	TR154	287.mg	n.s.s.	2/20	24.0mg	4/50	48.0mg	1/50	lun:a/a,a/c.
485	1104	1.78mg	11.1mg	0/16	7.30mg	9/18			Innes;ntis,1968/1969
a	1104	2.02mg	14.9mg	0/16	7.30mg	8/18			
b	1104	14.5mg	n.s.s.	0/16	7.30mg	0/18			
c	1104	1.78mg	11.1mg	0/16	7.30mg	9/18			
486	1104	5.76mg	n.s.s.	1/17	7.84mg	2/16			
a	1104	13.8mg	n.s.s.	0/17	7.84mg	0/16			
b	1104	6.39mg	n.s.s.	2/17	7.84mg	2/16			
487	1104	6.12mg	n.s.s.	1/18	7.30mg	2/18			
a	1104	9.58mg	n.s.s.	2/18	7.30mg	1/18			
b	1104	5.74mg	n.s.s.	3/18	7.30mg	3/18			
488	TR154	19.8mg	69.9mg	0/20	10.0mg	5/50	20.0mg	19/50	jej:srn; mul:sar; spc:fbs; spl:hpm,ost,sar.
a	TR154	19.8mg	69.9mg	0/20	10.0mg	5/50	20.0mg	19/50	jej:srn; mul:sar; spc:fbs; spl:fbs,hpm,ost,sar. C
b	TR154	42.7mg	1.39gm	0/20	10.0mg	1/50	20.0mg	8/50	
c	TR154	46.2mg	n.s.s.	0/20	10.0mg	1/50	20.0mg	7/50	
d	TR154	13.0mg	n.s.s.	10/20	10.0mg	32/50	20.0mg	33/50	
e	TR154	83.5mg	n.s.s.	0/20	10.0mg	1/50	20.0mg	2/50	liv:hpa,hpc,nnd.
489	TR154	11.4mg	39.1mg	0/20	8.00mg	6/50	16.0mg	16/50	abc:hes; liv:hes; mey:srn; mul:msm; spl:npm,sar.
a	TR154	11.4mg	39.1mg	0/20	8.00mg	6/50	16.0mg	16/50	abc:hes; liv:hes; mey:srn; mul:msm; spl:fbs,npm,sar. C
b	TR154	22.9mg	110.mg	0/20	8.00mg	0/50	16.0mg	13/50	S
c	TR154	20.8mg	121.mg	0/20	8.00mg	2/50	16.0mg	7/50	
d	TR154	6.61mg	37.1mg	10/20	8.00mg	19/50	16.0mg	43/50	
e	TR154	41.4mg	n.s.s.	0/20	8.00mg	1/50	16.0mg	3/50	liv:hpa,hpc,nnd.

AZOXYMETHANE (Z-methyl-O,N,N-azoxymethane) 25843-45-2

	RefNum	LoConf	UpConf	Cntrl	1Dose	1Inc	2Dose	2Inc	Citation or Pathology
490	1641	14.5ug	73.4ug	2/20	85.3ug	17/20	(.361mg	16/20)	Lijinsky;clet,24,273-280;1984/1985a
a	1641	33.1ug	.168mg	0/20	85.3ug	11/20	(.361mg	13/20)	
b	1641	.192mg	.957mg	0/20	85.3ug	0/20	.361mg	11/20	
c	1641	.263mg	1.73mg	0/20	85.3ug	0/20	.361mg	8/20	
d	1641	.354mg	n.s.s.	0/20	85.3ug	3/20	.361mg	3/20	
e	1641	.465mg	n.s.s.	2/20	85.3ug	5/20	.361mg	3/20	
f	1641	n.s.s.	n.s.s.	20/20	85.3ug	20/20	.361mg	20/20	
491	1864	48.5ug	.249mg	0/19	.527mg	12/16			Lijinsky;canr,47,3968-3972;1987/pers.comm.
a	1864	77.9ug	.474mg	0/19	.527mg	9/16			
b	1864	90.0ug	.610mg	0/19	.527mg	8/16			
c	1864	.121mg	1.26mg	0/19	.527mg	6/16			
d	1864	.679mg	n.s.s.	0/19	.527mg	0/16			

1-AZOXYPROPANE 17697-55-1

	RefNum	LoConf	UpConf	Cntrl	1Dose	1Inc	2Dose	2Inc	Citation or Pathology
492	1837	n.s.s.	241.ng	1/29	1.40ug	29/29			Fiala;carc,8,1947-1949;1987/pers.comm.
a	1837	214.ng	751.ng	1/29	1.40ug	22/29			
b	1837	330.ng	1.22ug	0/29	1.40ug	17/29			
c	1837	390.ng	1.55ug	0/29	1.40ug	15/29			
d	1837	1.81ug	n.s.s.	0/29	1.40ug	2/29			
e	1837	1.81ug	n.s.s.	0/29	1.40ug	2/29			
f	1837	2.77ug	n.s.s.	1/29	1.40ug	1/29			

2-AZOXYPROPANE 17697-53-9

	RefNum	LoConf	UpConf	Cntrl	1Dose	1Inc	2Dose	2Inc	Citation or Pathology
493	1837	971.ng	n.s.s.	1/29	1.40ug	6/29			Fiala;carc,8,1947-1949;1987/pers.comm.
a	1837	1.81ug	n.s.s.	0/29	1.40ug	2/29			
b	1837	2.77ug	n.s.s.	1/29	1.40ug	1/29			

BARBITAL, SODIUM 144-02-5

	RefNum	LoConf	UpConf	Cntrl	1Dose	1Inc	2Dose	2Inc	Citation or Pathology
494	1942	35.9mg	n.s.s.	0/15	40.0mg	0/15			Diwan;txap,98,269-277;1989
495	1993m	41.2mg	n.s.s.	0/5	160.mg	0/5			Konishi;carc,11,2149-2156;1990/pers.comm.
496	1993n	52.4mg	266.mg	0/24	160.mg	11/28			
a	1993n	118.mg	n.s.s.	0/24	160.mg	4/28			
497	2085	203.mg	n.s.s.	1/23	160.mg	4/23			Kurata;fctx,30,251-259;1992/pers.comm.
a	2085	246.mg	n.s.s.	1/23	160.mg	3/23			
b	2085	386.mg	n.s.s.	0/23	160.mg	1/23			

BARBITURIC ACID 67-52-7

	RefNum	LoConf	UpConf	Cntrl	1Dose	1Inc	2Dose	2Inc	Citation or Pathology
498	1690	32.1mg	n.s.s.	0/13	25.0mg	0/13			Diwan;jnci,74,509-516;1985

BARIUM ACETATE 543-80-6

	RefNum	LoConf	UpConf	Cntrl	1Dose	1Inc	2Dose	2Inc	Citation or Pathology
499	1395	1.34mg	n.s.s.	3/47	1.00mg	5/21			Schroeder;jnut,105,452-458;1975
a	1395	2.63mg	n.s.s.	9/47	1.00mg	3/21			
b	1395	1.65mg	n.s.s.	4/47	1.00mg	4/21			
c	1395	2.10mg	n.s.s.	14/47	1.00mg	5/21			
500	1395	3.28mg	n.s.s.	5/38	.833mg	4/37			
a	1395	1.53mg	n.s.s.	11/38	.833mg	12/37			
b	1395	1.88mg	n.s.s.	2/38	.833mg	6/37			
501	1456	1.14mg	n.s.s.	17/24	.286mg	15/33			Schroeder;jnut,105,421-427;1975
a	1456	1.18mg	n.s.s.	8/24	.286mg	9/33			

```
     Spe Strain  Site   Xpo+Xpt                                                    TD50     2Tailpvl
         Sex   Route  Hist    Notes                                                      DR      AuOp
502  R m leb wat tba mix 37m37 e                              .>                    2.87mg   P<.4   -
  a  R m leb wat tba mal 37m37 e                                                    2.87mg   P<.2   -

BARIUM CHLORIDE DIHYDRATE        100ng..:..1ug...:..10....:..100...:..1mg...:..10....:..100...:..1g....:..10
503  M f b6c wat TBA MXB 24m24 s                                        :>          806.mg * P<.3   -
  a  M f b6c wat liv MXB 24m24 s                                                    4.41gm * P<.8
  b  M f b6c wat lun MXB 24m24 s                                                    13.4gm * P<.8
504  M m b6c wat --- hem 24m24 s                                        :   ±      #3.47gm * P<.02  -
  a  M m b6c wat TBA MXB 24m24 s                                                    no dre   P=1.
  b  M m b6c wat liv MXB 24m24 s                                                    no dre   P=1.
  c  M m b6c wat lun MXB 24m24 s                                                    no dre   P=1.
505  R f f34 wat TBA MXB 24m24                              :>                      no dre   P=1.   -
  a  R f f34 wat liv MXB 24m24                                                      no dre   P=1.
506  R m f34 wat TBA MXB 24m24                                :>                    no dre   P=1.   -
  a  R m f34 wat liv MXB 24m24                                                      no dre   P=1.

BEMITRADINE                      100ng..:..1ug...:..10....:..100...:..1mg...:..10....:..100...:..1g....:..10
507  R f cdr eat liv hpa 23m24 e                              . +   .               307.mg Z P<.0005+
  a  R f cdr eat liv hpc 23m24 e                                                    646.mg * P<.0005+
  b  R f cdr eat mgl car 23m24 e                                                    669.mg * P<.002  +
  c  R f cdr eat mgl ade 23m24 e                                                    3.01gm * P<.6    +
  d  R f cdr eat thy ade 23m24 e                                                    9.71gm * P<.5    +
508  R m cdr eat liv hpa 23m24 e                                   .   ±            2.53gm * P<.08   +
  a  R m cdr eat liv hpc 23m24 e                                                    4.16gm * P<.03   +
  b  R m cdr eat thy car 23m24 e                                                    12.4gm * P<.03   +
  c  R m cdr eat thy ade 23m24 e                                                    11.0gm * P<.6    +

BENZALDEHYDE                     100ng..:..1ug...:..10....:..100...:..1mg...:..10....:..100...:..1g....:..10
509  M f b6c gav for sqp 24m24                                        :  +          1.31gm * P<.008 p
  a  M f b6c gav TBA MXB 24m24                                                      no dre   P=1.
  b  M f b6c gav liv MXB 24m24                                                      3.42gm * P<.4
  c  M f b6c gav lun MXB 24m24                                                      no dre   P=1.
510  M m b6c gav for sqp 24m25                                        :   ±         1.73gm * P<.08   p
  a  M m b6c gav TBA MXB 24m25                                                      985.mg / P<.6
  b  M m b6c gav liv MXB 24m25                                                      3.77gm * P<.8
  c  M m b6c gav lun MXB 24m25                                                      no dre   P=1.
511  R f f34 gav TBA MXB 24m24                              :  +   :                101.mg \ P<.002 -
  a  R f f34 gav liv MXB 24m24                                                      91.7gm * P<1.
512  R m f34 gav tes MXA 24m24                            :  +  :                  #70.1mg \ P<.0005-
  a  R m f34 gav pit ade 24m24                                                      183.mg \ P<.002
  b  R m f34 gav amd MXA 24m24                                                      193.mg * P<.0005
  c  R m f34 gav --- mnl 24m24                                                      244.mg * P<.0005
  d  R m f34 gav thy MXA 24m24                                                      448.mg * P<.002
  e  R m f34 gav thy cca 24m24                                                      478.mg * P<.0005
  f  R m f34 gav pan ade 24m24                                                      755.mg * P<.0005
  g  R m f34 gav ski ker 24m24                                                      1.86gm * P<.009
  h  R m f34 gav amd phm 24m24                                                      888.mg * P<.02
  i  R m f34 gav pre MXA 24m24                                                      907.mg * P<.02
  j  R m f34 gav --- msm 24m24                                                      1.08gm * P<.03
  k  R m f34 gav ski pam 24m24                                                      2.18gm * P<.03
  l  R m f34 gav TBA MXB 24m24                                                      109.mg * P<.0005
  m  R m f34 gav liv MXB 24m24                                                      9.67gm * P<.7

BENZENE                          100ng..:..1ug...:..10....:..100...:..1mg...:..10....:..100...:..1g....:..10
513  M f b6c gav MXB MXB 24m24                                  :  +   :            23.7mg * P<.0005

  a  M f b6c gav MXB MXB 24m24                                                      24.6mg * P<.0005

  b  M f b6c gav liv hpa 24m24                                                      38.1mg Z P<.008
  c  M f b6c gav liv MXA 24m24                                                      53.6mg Z P<.009
  d  M f b6c gav ova mtb 24m24                                                      66.5mg Z P<.0005c
  e  M f b6c gav lun MXA 24m24                                                      83.6mg * P<.002 c
  f  M f b6c gav lun a/c 24m24                                                      117.mg * P<.0005c
  g  M f b6c gav ova MXB 24m24                                                      119.mg * P<.0005
  h  M f b6c gav mgl MXA 24m24                                                      132.mg * P<.0005c
  i  M f b6c gav ova MXA 24m24                                                      146.mg * P<.002
  j  M f b6c gav ova gct 24m24                                                      157.mg * P<.002 c
  k  M f b6c gav ova tua 24m24                                              +hist   295.mg * P<.006 p
  l  M f b6c gav mgl cas 24m24                                                      382.mg * P<.005 c
  m  M f b6c gav hag can 24m24                                                      509.mg * P<.006
  n  M f b6c gav MXA MXA 24m24                                                      79.1mg * P<.04

  o  M f b6c gav MXA MXA 24m24                                                      94.2mg * P<.07 c

  p  M f b6c gav hag MXA 24m24                                                      128.mg * P<.02
  q  M f b6c gav lun a/a 24m24                                                      169.mg * P<.02 c
  r  M f b6c gav MXA sqp 24m24                                                      173.mg * P<.02
  s  M f b6c gav ova lut 24m24                                              +hist   245.mg * P<.05  p
  t  M f b6c gav zym sqc 24m24                                                      421.mg * P<.02 c
  u  M f b6c gav ova pcy 24m24                                              +hist   574.mg * P<.08  p
  v  M f b6c gav TBA MXB 24m24                                                      30.4mg * P<.002
```

	RefNum	LoConf	UpConf	Cntrl	1Dose	1Inc	2Dose	2Inc	Citation or Pathology	Brkly Code
502	1456	.738mg	n.s.s.	4/26	.250mg	8/30				
a	1456	.844mg	n.s.s.	2/26	.250mg	6/30				
BARIUM CHLORIDE DIHYDRATE		10326-27-9								
503	TR432	227.mg	n.s.s.	34/50	98.9mg	32/53	247.mg	39/50	495.mg	17/54		
a	TR432	483.mg	n.s.s.	16/50	98.9mg	14/53	247.mg	19/50	495.mg	6/54	liv:hpa,hpb,hpc.	
b	TR432	1.07gm	n.s.s.	3/50	98.9mg	3/53	247.mg	5/50	495.mg	1/54	lun:a/a,a/c.	
504	TR432	1.31gm	524.gm	0/51	82.4mg	0/50	207.mg	2/49	414.mg	3/50		S
a	TR432	506.mg	n.s.s.	45/51	82.4mg	33/50	207.mg	32/49	414.mg	26/50		
b	TR432	1.06gm	n.s.s.	27/51	82.4mg	24/50	207.mg	18/49	414.mg	11/50	liv:hpa,hpb,hpc.	
c	TR432	1.03gm	n.s.s.	15/51	82.4mg	10/50	207.mg	8/49	414.mg	9/50	lun:a/a,a/c.	
505	TR432	110.mg	n.s.s.	48/50	28.2mg	44/50	70.7mg	44/50	142.mg	45/50		
a	TR432	1.11gm	n.s.s.	0/50	28.2mg	1/50	70.7mg	0/50	142.mg	0/50	liv:hpa,hpb,hpc.	
506	TR432	140.mg	n.s.s.	47/50	24.9mg	46/50	62.3mg	43/50	125.mg	43/50		
a	TR432	n.s.s.	n.s.s.	0/50	24.9mg	0/50	62.3mg	0/50	125.mg	0/50	liv:hpa,hpb,hpc.	
BEMITRADINE		(8-(2-ethoxyethyl)-phenyl-1,2,4-triazolo[4,3-c]pyrimidine-5 amine) 88133-11-3									
507	2103	171.mg	832.mg	9/120	46.2mg	9/60	139.mg	20/60	(416.mg	18/59)	Gad;japt,12,157-164;1992/pers.comm.	
a	2103	417.mg	1.10gm	2/120	46.2mg	1/60	139.mg	6/60	416.mg	26/59		
b	2103	340.mg	3.32gm	33/120	46.2mg	20/60	139.mg	24/60	416.mg	31/59		
c	2103	535.mg	n.s.s.	48/120	46.2mg	36/60	139.mg	29/60	416.mg	29/59		
d	2103	1.65gm	n.s.s.	4/120	46.2mg	5/60	139.mg	4/60	416.mg	4/59		
508	2103	894.mg	n.s.s.	16/120	46.2mg	7/60	139.mg	7/60	416.mg	14/60		
a	2103	1.47gm	n.s.s.	1/120	46.2mg	3/60	139.mg	1/60	416.mg	5/60		
b	2103	3.06gm	n.s.s.	0/120	46.2mg	0/60	139.mg	0/60	416.mg	2/60		
c	2103	1.58gm	n.s.s.	7/120	46.2mg	5/60	139.mg	6/60	416.mg	5/60		
BENZALDEHYDE		100-52-7								
509	TR378	653.mg	20.8gm	0/49	210.mg	5/49	420.mg	6/50		
a	TR378	528.mg	n.s.s.	28/49	210.mg	23/49	420.mg	27/50		
b	TR378	881.mg	n.s.s.	2/49	210.mg	4/49	420.mg	5/50	liv:hpa,hpc,nnd.	
c	TR378	2.33gm	n.s.s.	0/49	210.mg	1/49	420.mg	0/50	lun:a/a,a/c.	
510	TR378	617.mg	n.s.s.	1/50	140.mg	2/50	280.mg	5/50		
a	TR378	198.mg	n.s.s.	36/50	140.mg	26/50	280.mg	40/50		
b	TR378	380.mg	n.s.s.	19/50	140.mg	12/50	280.mg	20/50	liv:hpa,hpc,nnd.	
c	TR378	867.mg	n.s.s.	8/50	140.mg	4/50	280.mg	6/50	lun:a/a,a/c.	
511	TR378	51.9mg	491.mg	46/50	140.mg	48/50	(280.mg	43/50)		
a	TR378	1.81gm	n.s.s.	5/50	140.mg	0/50	280.mg	1/50	liv:hpa,hpc,nnd.	
512	TR378	39.6mg	181.mg	46/50	140.mg	47/50	(280.mg	31/50)	tes:iab,ica.	S
a	TR378	91.3mg	882.mg	15/50	140.mg	22/50	(280.mg	11/50)		S
b	TR378	117.mg	456.mg	19/50	140.mg	23/50	280.mg	19/50	amd:pbb,phm,pob.	S
c	TR378	141.mg	668.mg	10/50	140.mg	17/50	280.mg	16/50		S
d	TR378	240.mg	1.65gm	5/50	140.mg	10/50	280.mg	8/50	thy:cca,ccr.	S
e	TR378	260.mg	1.18gm	4/50	140.mg	8/50	280.mg	7/50		S
f	TR378	354.mg	2.36gm	3/50	140.mg	2/50	280.mg	7/50		S
g	TR378	632.mg	55.0gm	1/50	140.mg	0/50	280.mg	4/50		S
h	TR378	382.mg	n.s.s.	2/50	140.mg	5/50	280.mg	5/50		S
i	TR378	390.mg	n.s.s.	2/50	140.mg	6/50	280.mg	7/50	pre:ade,car.	S
j	TR378	458.mg	n.s.s.	0/50	140.mg	5/50	280.mg	2/50		S
k	TR378	735.mg	n.s.s.	0/50	140.mg	1/50	280.mg	3/50		S
l	TR378	67.0mg	241.mg	44/50	140.mg	45/50	280.mg	41/50		
m	TR378	969.mg	n.s.s.	2/50	140.mg	1/50	280.mg	1/50	liv:hpa,hpc,nnd.	
BENZENE		71-43-2								
513	TR289	14.9mg	50.7mg	17/50	17.7mg	28/50	35.4mg	41/50	70.7mg	42/50	lun:a/a,a/c; mgl:can,cas,sqc; mul:mlh,mlm,mlp, mlu,mno; ova:gct,lut,mtb,pcy,tua; spl:mlm,mlu,mno; ute:mlh; zym:sqc.	M
a	TR289	15.3mg	54.6mg	17/50	17.7mg	28/50	35.4mg	40/50	70.7mg	41/50	lun:a/a,a/c; mgl:can,cas,sqc; mul:mlh,mlm,mlp, mlu,mno; ova:gct,mtb; spl:mlm,mlu,mno; ute:mlh; zym:sqc.	C
b	TR289	15.9mg	880.mg	1/50	17.7mg	8/50	(35.4mg	5/50	70.7mg	4/50)		S
c	TR289	26.3mg	2.27gm	4/50	17.7mg	12/50	35.4mg	13/50	(70.7mg	7/50)	liv:hpa,hpc.	S
d	TR289	34.6mg	156.mg	0/50	17.7mg	1/50	35.4mg	12/50	(70.7mg	7/50)		
e	TR289	44.0mg	354.mg	4/50	17.7mg	5/50	35.4mg	10/50	70.7mg	13/50	lun:a/a,a/c.	
f	TR289	63.4mg	390.mg	0/50	17.7mg	3/50	35.4mg	6/50	70.7mg	6/50		
g	TR289	63.8mg	374.mg	0/50	17.7mg	2/50	35.4mg	7/50	70.7mg	5/50	ova:lut,pcy,tua.	P
h	TR289	73.1mg	294.mg	0/50	17.7mg	2/50	35.4mg	5/50	70.7mg	10/50	mgl:can,sqc.	
i	TR289	74.0mg	571.mg	1/50	17.7mg	1/50	35.4mg	6/50	70.7mg	8/50	ova:gcc,gct.	S
j	TR289	77.1mg	793.mg	1/50	17.7mg	1/50	35.4mg	6/50	70.7mg	7/50		
k	TR289	120.mg	3.25gm	0/50	17.7mg	0/50	35.4mg	3/50	70.7mg	3/50		
l	TR289	143.mg	3.12gm	0/50	17.7mg	0/50	35.4mg	1/50	70.7mg	4/50		
m	TR289	174.mg	5.94gm	0/50	17.7mg	0/50	35.4mg	0/50	70.7mg	4/50		S
n	TR289	34.3mg	n.s.s.	15/50	17.7mg	25/50	35.4mg	26/50	70.7mg	22/50	mul:lkn,mlh,mlm,mlp,mlu,mno,ule; spl:mlm,mlu,mno; ute:mlh.	S
o	TR289	37.2mg	n.s.s.	15/50	17.7mg	24/50	35.4mg	24/50	70.7mg	20/50	mul:mlh,mlm,mlp,mlu,mno; spl:mlm,mlu,mno; ute: mlh.	S
p	TR289	57.4mg	n.s.s.	5/50	17.7mg	6/50	35.4mg	10/50	70.7mg	10/50	hag:adn,can.	S
q	TR289	72.8mg	n.s.s.	4/50	17.7mg	2/50	35.4mg	5/50	70.7mg	9/50		
r	TR289	77.4mg	n.s.s.	1/50	17.7mg	3/50	35.4mg	6/50	70.7mg	5/50	cst:sqp; sto:sqp.	S
s	TR289	106.mg	n.s.s.	0/50	17.7mg	2/50	35.4mg	3/50	70.7mg	2/50		
t	TR289	146.mg	n.s.s.	0/50	17.7mg	0/50	35.4mg	1/50	70.7mg	3/50		
u	TR289	174.mg	n.s.s.	0/50	17.7mg	0/50	35.4mg	2/50	70.7mg	1/50		
v	TR289	16.1mg	158.mg	36/50	17.7mg	40/50	35.4mg	48/50	70.7mg	48/50		

	Spe	Sex	Strain	Route	Site	Hist	Xpo+Xpt	Notes			TD50	DR	2Tailpvl	AuOp
w	M	f	b6c	gav	liv	MXB	24m24				53.6mg	Z	P<.009	
x	M	f	b6c	gav	lun	MXB	24m24				83.6mg	*	P<.002	
514	M	m	b6c	gav	MXB	MXB	24m24		:+	:	15.1mg	*	P<.0005	
a	M	m	b6c	gav	pre	MXA	24m24				21.8mg	Z	P<.0005	
b	M	m	b6c	gav	pre	sqc	24m24				25.5mg	*	P<.0005c	
c	M	m	b6c	gav	hag	MXA	24m24				39.5mg	*	P<.0005	
d	M	m	b6c	gav	hag	adn	24m24				40.7mg	*	P<.0005c	
e	M	m	b6c	gav	lun	MXA	24m24				42.5mg	*	P<.0005c	
f	M	m	b6c	gav	MXA	MXA	24m24				50.9mg	*	P<.0005	
g	M	m	b6c	gav	MXA	MXA	24m24				51.8mg	*	P<.0005c	
h	M	m	b6c	gav	zym	sqc	24m24				56.3mg	Z	P<.0005c	
i	M	m	b6c	gav	lun	a/c	24m24				59.5mg	*	P<.0005c	
j	M	m	b6c	gav	lun	a/a	24m24				103.mg	*	P<.004	
k	M	m	b6c	gav	adr	phe	24m24				116.mg	Z	P<.01	
l	M	m	b6c	gav	ski	MXA	24m24				262.mg	*	P<.004	
m	M	m	b6c	gav	MXA	mlh	24m24				269.mg	*	P<.005 c	
n	M	m	b6c	gav	MXA	MXA	24m24				169.mg	*	P<.02	
o	M	m	b6c	gav	cst	sqp	24m24				222.mg	*	P<.03	
p	M	m	b6c	gav	TBA	MXB	24m24				17.1mg	*	P<.0005	
q	M	m	b6c	gav	liv	MXB	24m24				102.mg	*	P<.2	
r	M	m	b6c	gav	lun	MXB	24m24				42.5mg	*	P<.0005	
515	M	m	aks	inh	---	mly	72w72	e	.>		114.mg		P<.3	-
516	M	m	c56	inh	thm	lym	69w69		.	+	+hist 466.mg		P<.004	+
a	M	m	c56	inh	---	mix	69w69				+hist 441.mg		P<.04	+
b	M	m	c56	inh	---	lcl	69w69				+hist 681.mg		P<.2	+
517	M	f	swi	gav	mam	car	18m24		. +	.	279.mg		P<.0005+	
a	M	f	swi	gav	lun	mix	18m24				453.mg		P<.004	+
b	M	f	swi	gav	lun	ata	18m24				1.02gm		P<.004	
c	M	f	swi	gav	lun	ade	18m24				1.10gm		P<.2	
d	M	f	swi	gav	zym	car	18m24				6.52gm		P<.3	+
e	M	f	swi	gav	liv	hpt	18m24				no dre		P=1.	
f	M	f	swi	gav	tba	mix	18m24				150.mg		P<.0005	
g	M	f	swi	gav	tba	mal	18m24				187.mg		P<.0005+	
518	M	m	swi	gav	lun	mix	18m24		. +	.	382.mg		P<.0005+	
a	M	m	swi	gav	lun	ade	18m24				811.mg		P<.02	
b	M	m	swi	gav	lun	ata	18m24				1.20gm		P<.04	
c	M	m	swi	gav	zym	car	18m24				1.57gm		P<.02	+
d	M	m	swi	gav	liv	hpt	18m24				6.19gm		P<.7	
e	M	m	swi	gav	lun	adc	18m24				6.52gm		P<.3	
f	M	m	swi	gav	tba	mix	18m24				370.mg		P<.05	
g	M	m	swi	gav	tba	mal	18m24				939.mg		P<.3	+
519	R	f	f34	gav	MXB	MXB	24m24		: +	:	55.1mg	*	P<.0005	
a	R	f	f34	gav	zym	MXA	24m24				97.9mg	*	P<.0005	
b	R	f	f34	gav	zym	can	24m24				109.mg	*	P<.0005c	
c	R	f	f34	gav	MXA	MXA	24m24				113.mg	*	P<.0005c	
d	R	f	f34	gav	ute	esp	24m24				158.mg	*	P<.008	
e	R	f	f34	gav	MXA	sqp	24m24				186.mg	*	P<.01 c	
f	R	f	f34	gav	MXA	sqc	24m24				290.mg	*	P<.002 c	
g	R	f	f34	gav	ton	MXA	24m24				295.mg	*	P<.003 c	
h	R	f	f34	gav	ton	sqc	24m24				368.mg	*	P<.003 c	
i	R	f	f34	gav	pal	MXA	24m24				288.mg	*	P<.06 c	
j	R	f	f34	gav	MXA	MXA	24m24				364.mg	*	P<.04	
k	R	f	f34	gav	TBA	MXB	24m24				46.7mg	*	P<.006	
l	R	f	f34	gav	liv	MXB	24m24				no dre		P=1.	
520	R	m	f34	gav	MXB	MXB	24m24		: +	:	51.1mg	*	P<.0005	
a	R	m	f34	gav	MXA	MXA	24m24				91.5mg	*	P<.0005c	
b	R	m	f34	gav	MXA	sqp	24m24				140.mg	*	P<.0005c	
c	R	m	f34	gav	zym	MXA	24m24				155.mg	*	P<.0005	
d	R	m	f34	gav	zym	can	24m24				166.mg	*	P<.0005c	
e	R	m	f34	gav	ski	MXA	24m24				181.mg	*	P<.0005	
f	R	m	f34	gav	ski	MXA	24m24				191.mg	*	P<.0005	
g	R	m	f34	gav	lpp	MXA	24m24				220.mg	*	P<.0005c	
h	R	m	f34	gav	pal	MXA	24m24				243.mg	*	P<.0005c	
i	R	m	f34	gav	pal	sqp	24m24				250.mg	*	P<.0005c	
j	R	m	f34	gav	ski	sqc	24m24				260.mg	*	P<.0005c	
k	R	m	f34	gav	lpp	sqp	24m24				284.mg	*	P<.002 c	
l	R	m	f34	gav	MXA	sqc	24m24				295.mg	*	P<.0005c	
m	R	m	f34	gav	ski	sqp	24m24				427.mg	*	P<.003 c	
n	R	m	f34	gav	ton	MXA	24m24				382.mg	*	P<.02 c	
o	R	m	f34	gav	ton	sqc	24m24				430.mg	*	P<.02 c	
p	R	m	f34	gav	lpp	sqc	24m24				1.15gm	*	P<.02 c	
q	R	m	f34	gav	TBA	MXB	24m24				68.1mg	*	P<.006	
r	R	m	f34	gav	liv	MXB	24m24				1.91gm	*	P<.7	
521	R	f	sda	inh	mgl	mal	24m35	egv	.	±	907.mg		P<.1	
a	R	f	sda	inh	orc	car	24m35	egv			2.02gm		P<.09	
b	R	f	sda	inh	zym	car	24m35	egv			1.89gm		P<.3	
c	R	f	sda	inh	liv	hpt	24m35	egv			4.07gm		P<.3	
d	R	f	sda	inh	nas	car	24m35	egv			4.07gm		P<.3	

	RefNum	LoConf	UpConf	Cntrl	1Dose	1Inc	2Dose	2Inc		Citation or Pathology
										Brkly Code
w	TR289	26.3mg	2.27gm	4/50	17.7mg	12/50	35.4mg	13/50	(70.7mg 7/50)	liv:hpa,hpc,nnd.
x	TR289	44.0mg	354.mg	4/50	17.7mg	5/50	35.4mg	10/50	70.7mg 13/50	lun:a/a,a/c.
514	TR289	10.1mg	26.0mg	13/50	17.7mg	28/50	35.4mg	35/50	71.4mg 42/50	hag:adn; lun:a/a,a/c; mul:mlh,mlm,mlp,mlu,mno;
										pre:sqc; spl:mlh,mno; zym:sqc. C
a	TR289	15.1mg	32.6mg	0/50	17.7mg	5/50	35.4mg	19/50	71.4mg 31/50	pre:can,sqc. S
b	TR289	17.3mg	38.9mg	0/50	17.7mg	3/50	35.4mg	18/50	71.4mg 28/50	
c	TR289	24.5mg	79.0mg	1/50	17.7mg	10/50	35.4mg	13/50	71.4mg 14/50	hag:adn,can. S
d	TR289	25.7mg	73.0mg	0/50	17.7mg	9/50	35.4mg	13/50	71.4mg 11/50	
e	TR289	24.1mg	121.mg	10/50	17.7mg	16/50	35.4mg	19/50	71.4mg 21/50	lun:a/a,a/c.
f	TR289	28.4mg	141.mg	4/50	17.7mg	10/50	35.4mg	10/50	71.4mg 15/50	mul:lkn,mlh,mlm,mlp,mlu,mno; spl:mlh,mno. S
g	TR289	28.8mg	141.mg	4/50	17.7mg	9/50	35.4mg	9/50	71.4mg 15/50	mul:mlh,mlm,mlp,mlu,mno; spl:mlh,mno.
h	TR289	33.8mg	101.mg	0/50	17.7mg	1/50	35.4mg	4/50	71.4mg 21/50	
i	TR289	31.8mg	214.mg	5/50	17.7mg	11/50	35.4mg	12/50	71.4mg 14/50	
j	TR289	48.7mg	900.mg	6/50	17.7mg	6/50	35.4mg	8/50	71.4mg 12/50	S
k	TR289	47.9mg	14.1gm	1/50	17.7mg	1/50	35.4mg	7/50	(71.4mg 1/50)	S
l	TR289	94.4mg	1.96gm	0/50	17.7mg	0/50	35.4mg	2/50	71.4mg 3/50	ski:sqc,squ. S
m	TR289	98.0mg	2.24gm	0/50	17.7mg	0/50	35.4mg	3/50	71.4mg 3/50	mul:mlh; spl:mlh.
n	TR289	66.5mg	n.s.s.	2/50	17.7mg	2/50	35.4mg	3/50	71.4mg 5/50	cst:sqc,sqp; sto:sqc. S
o	TR289	80.3mg	n.s.s.	2/50	17.7mg	1/50	35.4mg	2/50	71.4mg 5/50	S
p	TR289	10.4mg	38.6mg	33/50	17.7mg	40/50	35.4mg	48/50	71.4mg 46/50	
q	TR289	35.1mg	n.s.s.	15/50	17.7mg	17/50	35.4mg	22/50	71.4mg 11/50	liv:hpa,hpc,nnd.
r	TR289	24.1mg	121.mg	10/50	17.7mg	16/50	35.4mg	19/50	71.4mg 21/50	lun:a/a,a/c.
515	1048	32.4mg	n.s.s.	24/50	83.7mg	29/49				Snyder;txap,54,323-331;1980
516	1048	190.mg	2.81gm	0/40	251.mg	6/40				
a	1048	167.mg	n.s.s.	2/40	251.mg	8/40				
b	1048	212.mg	n.s.s.	2/40	251.mg	6/40				
517	BT908	154.mg	655.mg	2/40	241.mg	19/40				Maltoni;anya,534,412-426;1988
a	BT908	214.mg	3.43gm	4/40	241.mg	15/40				
b	BT908	414.mg	6.13gm	0/40	241.mg	6/40				
c	BT908	362.mg	n.s.s.	4/40	241.mg	9/40				
d	BT908	1.06gm	n.s.s.	0/40	241.mg	1/40				
e	BT908	1.99gm	n.s.s.	0/40	241.mg	0/40				
f	BT908	77.1mg	523.mg	16/40	241.mg	32/40				
g	BT908	99.1mg	589.mg	11/40	241.mg	28/40				
518	BT908	193.mg	1.39gm	3/40	241.mg	16/40				
a	BT908	330.mg	n.s.s.	2/40	241.mg	9/40				
b	BT908	436.mg	n.s.s.	1/40	241.mg	6/40				
c	BT908	541.mg	n.s.s.	0/40	241.mg	4/40				
d	BT908	756.mg	n.s.s.	2/40	241.mg	3/40				
e	BT908	1.06gm	n.s.s.	0/40	241.mg	1/40				
f	BT908	149.mg	n.s.s.	15/40	241.mg	24/40				
g	BT908	279.mg	n.s.s.	9/40	241.mg	14/40				
519	TR289	37.2mg	96.2mg	1/50	17.7mg	10/50	35.4mg	16/50	70.7mg 21/50	cst:sqp; lpp:sqp; pal:sqc,sqp; ton:sqc,sqp; zym:
										can. C
a	TR289	60.8mg	179.mg	0/50	17.7mg	5/50	35.4mg	6/50	70.7mg 15/50	zym:adn,can. S
b	TR289	66.3mg	207.mg	0/50	17.7mg	5/50	35.4mg	5/50	70.7mg 14/50	
c	TR289	65.7mg	333.mg	1/50	17.7mg	5/50	35.4mg	12/50	70.7mg 9/50	cst:sqp; lpp:sqp; pal:sqc,sqp; ton:sqc,sqp.
d	TR289	73.3mg	4.18gm	7/50	17.7mg	7/50	35.4mg	7/50	70.7mg 14/50	S
e	TR289	90.9mg	13.1gm	1/50	17.7mg	4/50	35.4mg	8/50	70.7mg 5/50	cst:sqp; lpp:sqp; pal:sqp; ton:sqc.
f	TR289	141.mg	1.14gm	0/50	17.7mg	1/50	35.4mg	4/50	70.7mg 5/50	pal:sqc; ton:sqc.
g	TR289	143.mg	1.56gm	0/50	17.7mg	1/50	35.4mg	5/50	70.7mg 4/50	ton:sqc,sqp.
h	TR289	166.mg	1.76gm	0/50	17.7mg	0/50	35.4mg	4/50	70.7mg 4/50	
i	TR289	118.mg	n.s.s.	1/50	17.7mg	4/50	35.4mg	5/50	70.7mg 4/50	pal:sqc,sqp.
j	TR289	157.mg	n.s.s.	0/50	17.7mg	2/50	35.4mg	3/50	70.7mg 2/50	cvu:adq; utm:acn,can. S
k	TR289	23.3mg	567.mg	38/50	17.7mg	39/50	35.4mg	41/50	70.7mg 42/50	
l	TR289	238.mg	n.s.s.	0/50	17.7mg	3/50	35.4mg	1/50	70.7mg 0/50	liv:hpa,hpc,nnd.
520	TR289	35.9mg	83.8mg	3/50	35.4mg	21/50	70.7mg	27/50	142.mg 37/50	lpp:sqc,sqp; pal:sqc,sqp; ski:sqc,sqc, ton:sqc,
										sqp; zym:can. C
a	TR289	59.4mg	174.mg	1/50	35.4mg	9/50	70.7mg	16/50	142.mg 19/50	lpp:sqc,sqp; pal:sqc,sqp; ton:sqc,sqp.
b	TR289	84.2mg	334.mg	1/50	35.4mg	6/50	70.7mg	11/50	142.mg 13/50	lpp:sqp; pal:sqp; ton:sqp.
c	TR289	92.8mg	363.mg	2/50	35.4mg	7/50	70.7mg	10/50	142.mg 18/50	zym:adn,can. S
d	TR289	98.3mg	401.mg	2/50	35.4mg	6/50	70.7mg	10/50	142.mg 17/50	
e	TR289	102.mg	544.mg	1/50	35.4mg	7/50	70.7mg	5/50	142.mg 12/50	ski:adq,sqc,sqp,ulc. S
f	TR289	105.mg	656.mg	1/50	35.4mg	7/50	70.7mg	5/50	142.mg 11/50	ski:adq,sqc,sqp. S
g	TR289	120.mg	519.mg	0/50	35.4mg	2/50	70.7mg	5/50	142.mg 8/50	lpp:sqc,sqp.
h	TR289	135.mg	635.mg	0/50	35.4mg	4/50	70.7mg	5/50	142.mg 9/50	pal:sqc,sqp.
i	TR289	138.mg	668.mg	0/50	35.4mg	4/50	70.7mg	4/50	142.mg 9/50	
j	TR289	143.mg	859.mg	0/50	35.4mg	5/50	70.7mg	3/50	142.mg 8/50	
k	TR289	145.mg	986.mg	0/50	35.4mg	5/50	70.7mg	5/50	142.mg 5/50	
l	TR289	156.mg	977.mg	0/50	35.4mg	3/50	70.7mg	5/50	142.mg 7/50	lpp:sqc; pal:sqc; ton:sqc.
m	TR289	192.mg	2.56gm	0/50	35.4mg	2/50	70.7mg	1/50	142.mg 5/50	
n	TR289	173.mg	n.s.s.	1/50	35.4mg	3/50	70.7mg	6/50	142.mg 6/50	ton:sqc,sqp.
o	TR289	206.mg	n.s.s.	0/50	35.4mg	3/50	70.7mg	4/50	142.mg 4/50	
p	TR289	349.mg	n.s.s.	0/50	35.4mg	0/50	70.7mg	0/50	142.mg 3/50	
q	TR289	33.7mg	799.mg	39/50	35.4mg	44/50	70.7mg	45/50	142.mg 47/50	
r	TR289	261.mg	n.s.s.	2/50	35.4mg	2/50	70.7mg	5/50	142.mg 1/50	liv:hpa,hpc,nnd.
521	BT4004	291.mg	n.s.s.	2/60	53.4mg	6/54				Maltoni;ajim,7,415-446;1985
a	BT4004	496.mg	n.s.s.	0/60	53.4mg	2/54				
b	BT4004	433.mg	n.s.s.	1/60	53.4mg	3/54				
c	BT4004	663.mg	n.s.s.	0/60	53.4mg	1/54				
d	BT4004	663.mg	n.s.s.	0/60	53.4mg	1/54				

```
    Spe Strain Site   Xpo+Xpt                                    TD50    2Tailpvl
        Sex  Route  Hist    Notes                                    DR    AuOp
  e     R f sda inh tba car 24m35 egv                             947.mg   P<.07
  f     R f sda inh tba mal 24m35 egv                             553.mg   P<.2  +
522     R f sda gav zym car 12m33 e          .  +  .             222.mg * P<.0005+
  a     R f sda gav orc car 12m33 e                              945.mg * P<.04 +
  b     R f sda gav mgl mal 12m33 e                              517.mg * P<.3
  c     R f sda gav liv hpt 12m33 e                              no dre   P=1.
  d     R f sda gav ski car 12m33 e                              no dre   P=1.
  e     R f sda gav --- leu 12m33 e                    *         no dre   P=1.
  f     R f sda gav tba mal 12m33 e                              75.3mg * P<.0005
523     R f sda gav orc car 24m33 e           .  +  .            415.mg   P<.0005+
  a     R f sda gav zym car 24m33 e                              585.mg   P<.0005+
  b     R f sda gav for cic 24m33 e                              787.mg   P<.002 +
  c     R f sda gav liv hpt 24m33 e                              4.03gm   P<.3
  d     R f sda gav liv ang 24m33 e                              7.91gm   P<.2  +
  e     R f sda gav nas car 24m33 e                              8.51gm   P<.2  +
  f     R f sda gav ski car 24m33 e                              no dre   P=1.
  g     R f sda gav tba mal 24m33 e                              161.mg   P<.0005
  h     R f sda gav tba car 24m33                                248.mg   P<.0005
524     R m sda gav --- leu 12m33 e        .  +                  .506mg * P<.008
  a     R m sda gav liv hpt 12m33 e                              3.09gm * P<.2
  b     R m sda gav sub ang 12m33 e                              3.09gm * P<.2
  c     R m sda gav mgl mal 12m33 e                              no dre   P=1.
  d     R m sda gav ski car 12m33 e                              no dre   P=1.
  e     R m sda gav orc car 12m33 e                              no dre   P=1.
  f     R m sda gav zym car 12m33 e                              no dre   P=1.
  g     R m sda gav tba mal 12m33 e                              232.mg * P<.004
525     R m sda gav orc car 24m33 e          .  +  .             386.mg   P<.0005+
  a     R m sda gav zym car 24m33 e                              501.mg   P<.0005+
  b     R m sda gav ski car 24m33 e                              904.mg   P<.0005+
  c     R m sda gav nas car 24m33 e                              3.53gm   P<.05 +
  d     R m sda gav liv ang 24m33 e                              4.48gm   P<.09 +
  e     R m sda gav for ivc 24m33 e                              6.12gm   P<.2  +
  f     R m sda gav liv hpt 24m33 e                              no dre   P=1.
  g     R m sda gav tba mal 24m33 e                              131.mg   P<.0005
  h     R m sda gav tba car 24m33                                219.mg   P<.0005
526     R m sda inh liv tum 98w98                .>              no dre   P=1.  -
  a     R m sda inh tba tum 98w98                                no dre   P=1.  -
527     R f wis gav zym sqc 24m24               .  +            .1.36gm   P<.004
  a     R f wis gav orc sqc 24m24                                2.09gm   P<.02
  b     R f wis gav nas ulc 24m24                                8.70gm   P<.3
  c     R f wis gav tba mal 24m24                                482.mg   P<.02
  d     R f wis gav tba mix 24m24                                no dre   P=1.
528     R m wis gav zym sqc 24m24              .  +  .           1.14gm   P<.002
  a     R m wis gav nas ulc 24m24                                4.29gm   P<.1
  b     R m wis gav orc sqc 24m24                                8.48gm   P<.6
  c     R m wis gav tba mal 24m24                                523.mg   P<.01
  d     R m wis gav tba mix 24m24                                no dre   P=1.


BENZENEDIAZONIUM  TETRAFLUOROBORATE    ..:..1ug....:..10......:..100....:..1mg....:..10......:..100....:..1g......:..10
529     H f syg gav liv kcs 90w90 es                    .>       66.1mg * P<.2  -
  a     H f syg gav liv hem 90w90 es                             66.1mg * P<.2  -
  b     H f syg gav liv cho 90w90 es                             293.mg * P<.9  -
  c     H f syg gav lun tum 90w90 es                             no dre   P=1.
530     H m syg gav liv hem 90w90 es                    .>       293.mg * P<.9  -
  a     H m syg gav lun tum 90w90 es                             no dre   P=1.


BENZENESULPHONOHYDRAZIDE           100ng..:..1ug....:..10......:..100....:..1mg....:..10......:..100....:..1g......:..10
531     M f swi gav liv lct 50w60 ev                       ±     46.6mg   P<.1
  a     M f swi gav lun tum 50w60 ev                             46.6mg   P<.1


BENZIDINE                          100ng..:..1ug....:..10......:..100....:..1mg....:..10......:..100....:..1g......:..10
532     R b alb inh --- myl 20m28 e          .   ±              2.08mg   P<.02 +
  a     R b alb inh mgl car 20m28 e                              5.51mg   P<.2  +
  b     R b alb inh liv hpt 20m28 e                              11.2mg   P<.3  +
  c     R b alb inh tba mix 20m28 e                              1.73mg   P<.1  +


BENZIDINE.2HC1                     100ng..:..1ug....:..10......:..100....:..1mg....:..10......:..100....:..1g......:..10
533     M f cbn wat liv hpc 33m33 e                 .+.          18.9mg * P<.0005+

  a     M f cbn wat hag ade 33m33 e                              41.1mg Z P<.0005+

  b     M f cbn wat ute agm 33m33 e                              402.mg * P<.005 +

534     M f cbn wat liv hpc 60w60 ek              .  +  .        17.9mg * P<.0005+

  a     M f cbn wat liv hpa 60w60 ek                             30.2mg * P<.0005+
535     M f cbn wat liv hpc 80w80 e              .+.             9.60mg * P<.0005+
  a     M f cbn wat liv hpa 80w80 e                              78.7mg * P<.0005+
536     M m cbn wat liv hpc 33m33 e                .+.           60.5mg * P<.0005+
```

	RefNum	LoConf	UpConf	Cntrl	1Dose	1Inc	2Dose	2Inc	Citation or Pathology / Brkly Code
e	BT4004	314.mg	n.s.s.	1/60	53.4mg	5/54			
f	BT4004	181.mg	n.s.s.	9/60	53.4mg	14/54			
522	BT901	108.mg	716.mg	0/30	11.6mg	2/28	58.0mg	8/28	Maltoni;ajim,7,415-446;1985/1979
a	BT901	232.mg	n.s.s.	0/29	11.6mg	0/24	58.0mg	2/21	
b	BT901	141.mg	n.s.s.	4/30	11.6mg	4/28	58.0mg	7/28	
c	BT901	74.1mg	n.s.s.	0/13	11.6mg	0/30	58.0mg	0/7	
d	BT901	117.mg	n.s.s.	0/30	11.6mg	0/30	58.0mg	0/35	
e	BT901	382.mg	n.s.s.	1/30	11.6mg	2/30	58.0mg	1/29	
f	BT901	39.4mg	237.mg	7/30	11.6mg	10/28	58.0mg	21/29	
523	BT902	242.mg	794.mg	0/49	237.mg	20/39			Maltoni;ajim,7,415-446;1985
a	BT902	324.mg	1.21gm	0/49	237.mg	16/40			
b	BT902	318.mg	3.25gm	0/24	237.mg	6/19			
c	BT902	655.mg	n.s.s.	0/15	237.mg	1/14			
d	BT902	1.29gm	n.s.s.	0/35	237.mg	1/27			
e	BT902	1.39gm	n.s.s.	0/40	237.mg	1/29			
f	BT902	2.42gm	n.s.s.	1/35	237.mg	0/27			
g	BT902	92.3mg	312.mg	10/49	237.mg	35/40			
h	BT902	152.mg	432.mg	0/50	237.mg	28/40			
524	BT901	175.mg	8.25gm	0/22	11.6mg	0/23	58.0mg	4/24	Maltoni;ajim,7,415-446;1985/1979
a	BT901	503.mg	n.s.s.	0/30	11.6mg	0/30	58.0mg	1/35	
b	BT901	503.mg	n.s.s.	0/30	11.6mg	0/30	58.0mg	1/35	
c	BT901	84.1mg	n.s.s.	0/22	11.6mg	0/22	58.0mg	0/22	
d	BT901	117.mg	n.s.s.	0/30	11.6mg	0/30	58.0mg	0/35	
e	BT901	74.2mg	n.s.s.	0/19	11.6mg	0/20	58.0mg	0/17	
f	BT901	84.1mg	n.s.s.	0/22	11.6mg	0/22	58.0mg	0/22	
g	BT901	101.mg	1.63gm	1/22	11.6mg	1/23	58.0mg	8/24	
525	BT902	227.mg	727.mg	0/45	237.mg	21/39			Maltoni;ajim,7,415-446;1985
a	BT902	280.mg	1.07gm	1/45	237.mg	18/39			
b	BT902	424.mg	2.57gm	0/34	237.mg	9/32			
c	BT902	1.07gm	n.s.s.	0/35	237.mg	3/37			
d	BT902	1.10gm	n.s.s.	0/32	237.mg	2/31			
e	BT902	996.mg	n.s.s.	0/44	237.mg	1/21			
f	BT902	523.mg	n.s.s.	3/15	237.mg	3/15			
g	BT902	70.6mg	260.mg	11/45	237.mg	36/39			
h	BT902	134.mg	382.mg	1/50	237.mg	30/40			
526	518	415.mg	n.s.s.	0/27	50.4mg	0/45			Snyder;jtxe,4,605-618;1978
a	518	415.mg	n.s.s.	0/27	50.4mg	0/45			
527	BT907	552.mg	8.18gm	0/40	321.mg	6/40			Maltoni;anya,534,412-426;1988
a	BT907	722.mg	n.s.s.	0/40	321.mg	4/40			
b	BT907	1.42gm	n.s.s.	0/40	321.mg	1/40			
c	BT907	217.mg	n.s.s.	10/40	321.mg	21/40			
d	BT907	408.mg	n.s.s.	34/40	321.mg	27/40			
528	BT907	494.mg	4.73gm	0/40	321.mg	7/40			
a	BT907	1.06gm	n.s.s.	0/40	321.mg	2/40			
b	BT907	1.16gm	n.s.s.	1/40	321.mg	2/40			
c	BT907	238.mg	26.9gm	8/40	321.mg	19/40			
d	BT907	490.mg	n.s.s.	30/40	321.mg	23/40			

BENZENEDIAZONIUM TETRAFLUOROBORATE   369-57-3

	RefNum	LoConf	UpConf	Cntrl	1Dose	1Inc	2Dose	2Inc	3Dose	3Inc	Citation or Pathology
529	1329	10.8mg	n.s.s.	0/15	1.25mg	0/15	2.50mg	0/15	5.00mg	1/15	Gold;clet,15,289-300;1982
a	1329	10.8mg	n.s.s.	0/15	1.25mg	0/15	2.50mg	0/15	5.00mg	1/15	
b	1329	11.9mg	n.s.s.	1/15	1.25mg	0/15	2.50mg	0/15	5.00mg	1/15	
c	1329	1.65mg	n.s.s.	0/15	1.25mg	0/15	2.50mg	0/15	5.00mg	0/15	
530	1329	11.9mg	n.s.s.	1/15	1.25mg	0/15	2.50mg	0/15	5.00mg	1/15	
a	1329	1.65mg	n.s.s.	0/15	1.25mg	0/15	2.50mg	0/15	5.00mg	0/15	

BENZENESULPHONOHYDRAZIDE   (BSH) 5351-65-5

	RefNum	LoConf	UpConf	Cntrl	1Dose	1Inc	Citation or Pathology
531	1060	13.7mg	n.s.s.	1/29	30.1mg	4/24	Cremlyn;fctx,9,319-321;1971
a	1060	13.7mg	n.s.s.	1/29	30.1mg	4/24	

BENZIDINE   92-87-5

	RefNum	LoConf	UpConf	Cntrl	1Dose	1Inc	Citation or Pathology
532	598	.786mg	n.s.s.	0/21	.440mg	5/28	Zabezhinskii;bebm,69,72-74;1970
a	598	1.35mg	n.s.s.	0/21	.440mg	2/28	
b	598	1.83mg	n.s.s.	0/21	.440mg	1/28	
c	598	.610mg	n.s.s.	2/21	.440mg	8/28	

BENZIDINE.2HC1   531-85-1

	RefNum	LoConf	UpConf	Cntrl	1Dose	1Inc	2Dose	2Inc	3Dose	3Inc	4Dose	4Inc	Citation or Pathology
533	1563n	15.2mg	25.0mg	10/125	4.00mg	54/119	6.00mg	43/95	8.00mg	31/71	12.0mg	37/72 16.0mg 51/69	
					24.0mg	56/72							Littlefield;jtxe,12,671-685;1983/1984/Nelson pers.comm.
a	1563n	25.9mg	104.mg	6/121	4.00mg	28/117	6.00mg	16/94	8.00mg	17/71	12.0mg	10/70 16.0mg 9/67	
					24.0mg	8/68)							
b	1563n	201.mg	3.98gm	3/124	4.00mg	1/114	6.00mg	3/94	8.00mg	2/71	12.0mg	5/70 16.0mg 7/68	
					24.0mg	5/71							
534	1577o	12.3mg	27.5mg	0/45	6.00mg	1/69	12.0mg	2/48	24.0mg	10/41	40.0mg	17/35 80.0mg 13/14	Schieferstein;nctr;
													1982/Nelson 1982
a	1577o	17.6mg	80.9mg	1/45	6.00mg	6/69	12.0mg	8/48	24.0mg	6/41	40.0mg	12/35 80.0mg 5/14	
535	1577r	7.42mg	12.6mg	0/50	6.00mg	6/45	12.0mg	19/47	24.0mg	39/48	40.0mg	30/34 80.0mg 25/30	
a	1577r	40.5mg	338.mg	0/50	6.00mg	7/46	12.0mg	4/47	24.0mg	11/50	40.0mg	10/35 80.0mg 7/31	
536	1563n	43.3mg	97.6mg	17/123	5.00mg	20/118	6.67mg	20/95	10.0mg	23/72	13.3mg	24/71 20.0mg 37/71	
					26.7mg	32/71							Littlefield;jtxe,12,671-685;1983/1984/Nelson pers.comm.

```
       Spe Strain  Site   Xpo+Xpt                                               TD50    2Tailpvl
           Sex   Route   Hist      Notes                                           DR    AuOp
 a   M m cbn wat hag ade 33m33 e                                                208.mg * P<.02  +

537  M m cbn wat liv hpa 60w60 ek                            .   +      .        67.7mg * P<.002 +

 a   M m cbn wat liv hpc 60w60 ek                                               74.3mg * P<.009 +
538  M m cbn wat liv hpc 80w80 e                                     .+  .       39.0mg * P<.0005+
 a   M m cbn wat liv hpa 80w80 e                                  .+.            186.mg * P<.09  +
539  M f cff wat liv hpc 33m33 e                              .+.                10.9mg * P<.0005+

 a   M f cff wat hag ade 33m33 e                                                51.7mg Z P<.0005+

 b   M f cff wat ute agm 33m33 e                                                335.mg * P<.006 +

540  M f cff wat liv hpc 60w60 ek                          .  +  .               17.1mg Z P<.0005+

 a   M f cff wat liv hpa 60w60 ek                                               24.7mg * P<.0005+
541  M f cff wat liv hpc 79w80 ae                       .+.                      8.99mg * P<.0005+
 a   M f cff wat liv hpa 79w80 ae                                               41.9mg * P<.0005+
542  M m cff wat liv hpc 33m33 e                              .+ .               31.2mg * P<.0005+

 a   M m cff wat hag ade 33m33 e                                                129.mg * P<.0005+

543  M m cff wat liv hpa 60w60 ek                            .  +   .            60.8mg * P<.0005+

 a   M m cff wat liv hpc 60w60 ek                                               97.5mg * P<.0005+
544  M m cff wat liv hpc 80w80 e                              .  +  .            33.2mg * P<.0005+
 a   M m cff wat liv hpa 80w80 e                                                94.8mg * P<.0005+

BENZO(a)PYRENE                    100ng..:..1ug....:..10.....:..100....:..1mg....:..10.....:..100....:..1g...:..10
545  M f alb eat for pam 60w60 er                                  .   ±        1.86mg   P<.05
546  M f c5v gav lun ben 52w92            .>                                    12.1ug   P<.2   -
 a   M f c5v gav liv tum 52w92                                                  no dre   P=1.   -
 b   M f c5v gav tba mix 52w92                                                  2.68ug   P<.2   -
547  M m c5v gav lun ben 52w92            .>                                    15.7ug   P<.4   -
 a   M m c5v gav liv mal 52w92                                                  no dre   P=1.   -
 b   M m c5v gav liv ben 52w92                                                  no dre   P=1.   -
 c   M m c5v gav tba mix 52w92                                                  no dre   P=1.   -
548  M m cf1 wat eso pam 95w95                            .  +   .              11.0mg   P<.0005+
 a   M m cf1 wat for pam 95w95                                                  12.4mg   P<.03
 b   M m cf1 wat eso car 95w95                                                  119.mg   P<.3   +
549  R b sda eat mix mix 30m30 r                   .      ±                     .956mg   P<.04  +
 a   R b sda eat for pam 30m30 r                                                .972mg   P<.03  +

BENZOATE, SODIUM                  100ng..:..1ug....:..10.....:..100....:..1mg....:..10.....:..100....:..1g.....:..10
550  M f swa wat lun tum 31m31                                                 no dre   P=1.   -
 a   M f swa wat liv tum 31m31                                                 no dre   P=1.   -
551  M m swa wat liv hpt 26m26                                                 133.gm   P<.2   -
 a   M m swa wat lun tum 26m26                                                 no dre   P=1.   -
552  R f f34 eat tba mix 24m25 e                                      .>       24.9gm * P<.9   -
 a   R f f34 eat tba mal 24m25 e                                               104.gm * P<1.   -
553  R m f34 eat tba mix 24m25 e                                       .>      no dre   P=1.   -
 a   R m f34 eat tba mal 24m25 e                                               no dre   P=1.   -

BENZOFURAN                        100ng..:..1ug....:..10.....:..100....:..1mg....:..10.....:..100....:..1g.....:..10
554  M f b6c gav MXB MXB 24m24                              :  +  :             33.4mg \ P<.0005
 a   M f b6c gav liv hpa 24m24                                                  34.4mg \ P<.0005c
 b   M f b6c gav liv MXA 24m24                                                  36.4mg \ P<.0005
 c   M f b6c gav for MXA 24m24                                                  145.mg \ P<.003 c
 d   M f b6c gav for sqp 24m24                                                  172.mg \ P<.005
 e   M f b6c gav lun MXA 24m24                                                  174.mg * P<.0005c
 f   M f b6c gav lun a/a 24m24                                                  224.mg * P<.0005
 g   M f b6c gav MXA mlu 24m24                                                  322.mg \ P<.04
 h   M f b6c gav TBA MXB 24m24                                                  45.9mg \ P<.002
 i   M f b6c gav liv MXB 24m24                                                  36.4mg \ P<.0005
 j   M f b6c gav lun MXB 24m24                                                  174.mg \ P<.04
555  M m b6c gav liv MXA 24m24                             :  +  :              19.8mg \ P<.0005c
 a   M m b6c gav liv hpa 24m24                                                  20.5mg \ P<.0005c
 b   M m b6c gav liv MXA 24m24                                                  21.3mg \ P<.0005
 c   M m b6c gav MXB MXB 24m24                                                  23.9mg \ P<.002
 d   M m b6c gav liv hpb 24m24                                                  102.mg \ P<.0005c
 e   M m b6c gav for MXA 24m24                                                  108.mg \ P<.002 c
 f   M m b6c gav liv MXA 24m24                                                  114.mg \ P<.006
 g   M m b6c gav lun a/a 24m24                                                  136.mg * P<.004
 h   M m b6c gav for sqp 24m24                                                  154.mg * P<.007
 i   M m b6c gav lun MXA 24m24                                                  160.mg * P<.04  c
 j   M m b6c gav for sqc 24m24                                                  383.mg * P<.03
 k   M m b6c gav mul mlp 24m24                                                  919.mg * P<.04
 l   M m b6c gav TBA MXB 24m24                                                  54.3mg * P<.02
 m   M m b6c gav liv MXB 24m24                                                  21.3mg \ P<.0005
 n   M m b6c gav lun MXB 24m24                                                  160.mg * P<.04
556  R f f34 gav sub nlm 24m24                              :  +      :         98.8mg \ P<.005
```

	RefNum	LoConf	UpConf	Cntrl	1Dose	1Inc	2Dose	2Inc							Citation or Pathology
															Brkly Code
a	1563n	99.0mg	n.s.s.	10/122	5.00mg	19/113	6.67mg	17/92	10.0mg	13/72	13.3mg	13/68	20.0mg	16/69	
					26.7mg	13/66									
537	1577o	35.9mg	414.mg	0/46	5.00mg	2/63	10.0mg	1/44	20.0mg	5/47	33.3mg	2/37	66.7mg	4/21	Schieferstein;nctr;
															1982/Nelson 1982
a	1577o	38.4mg	3.64gm	0/46	5.00mg	1/63	10.0mg	3/44	20.0mg	3/47	33.3mg	4/37	66.7mg	2/21	
538	1577r	25.6mg	63.8mg	0/47	5.00mg	1/49	10.0mg	5/45	20.0mg	7/45	33.3mg	8/24	66.7mg	12/22	
a	1577r	65.4mg	n.s.s.	2/49	5.00mg	2/50	10.0mg	2/47	20.0mg	9/46	33.3mg	1/25	66.7mg	3/23	
539	1563m	9.29mg	13.0mg	3/124	4.00mg	51/120	6.00mg	52/95	8.00mg	45/72	12.0mg	55/71	16.0mg	60/69	
					24.0mg	64/72						Littlefield;jtxe,12,671-685;1983/1984/Nelson pers.comm.
a	1563m	33.3mg	130.mg	5/123	4.00mg	21/118	6.00mg	22/95	8.00mg	20/70	12.0mg	11/68	(16.0mg	12/68	
					24.0mg	8/71)									
b	1563m	163.mg	4.82gm	4/122	4.00mg	5/117	6.00mg	5/95	8.00mg	5/68	12.0mg	4/72	16.0mg	5/71	
					24.0mg	10/71									
540	1577m	11.8mg	26.0mg	1/48	6.00mg	2/69	12.0mg	2/48	24.0mg	6/45	40.0mg	29/45	80.0mg	9/10	Schieferstein;nctr;
															1982/Nelson 1982
a	1577m	16.4mg	39.3mg	0/48	6.00mg	1/69	12.0mg	1/48	24.0mg	8/45	40.0mg	21/45	80.0mg	5/10	
541	1577n	6.90mg	11.9mg	0/48	6.00mg	6/51	12.0mg	16/47	24.0mg	39/50	40.0mg	26/26	80.0mg	35/38	
a	1577n	29.4mg	78.7mg	0/47	6.00mg	3/50	12.0mg	9/47	24.0mg	12/50	40.0mg	8/26	80.0mg	16/37	
542	1563m	24.9mg	41.2mg	14/125	5.00mg	24/119	6.67mg	30/96	10.0mg	23/71	13.3mg	35/71	20.0mg	51/71	
					26.7mg	49/71						Littlefield;jtxe,12,671-685;1983/1984/Nelson pers.comm.
a	1563m	73.0mg	462.mg	6/124	5.00mg	21/117	6.67mg	19/93	10.0mg	18/70	13.3mg	16/70	20.0mg	13/67	
					26.7mg	17/68									
543	1577m	33.9mg	153.mg	0/47	5.00mg	0/70	10.0mg	3/46	20.0mg	5/44	33.3mg	3/42	66.7mg	5/22	Schieferstein;nctr;
															1982/Nelson 1982
a	1577m	46.4mg	353.mg	1/47	5.00mg	0/70	10.0mg	1/46	20.0mg	2/44	33.3mg	2/42	66.7mg	6/22	
544	1577n	22.7mg	53.2mg	0/49	5.00mg	5/48	10.0mg	2/50	20.0mg	12/52	33.3mg	7/28	66.7mg	16/24	
a	1577n	52.9mg	302.mg	0/47	5.00mg	0/47	10.0mg	4/50	20.0mg	3/49	33.3mg	7/26	66.7mg	2/21	

BENZO(a)PYRENE   50-32-8

	RefNum	LoConf	UpConf	Cntrl	1Dose	1Inc	2Dose	2Inc	Citation or Pathology
545	1274	.707mg	n.s.s.	0/40	.520mg	5/81			Chouroulinkov;bdca,54,67-78;1967
546	1636	2.97ug	n.s.s.	0/11	3.23ug	2/15			Griciute;iarc,813-822;1980
a	1636	7.81ug	n.s.s.	0/11	3.23ug	0/15			
b	1636	863.ng	n.s.s.	4/11	3.23ug	10/15			
547	1636	2.66ug	n.s.s.	1/20	2.69ug	2/15			
a	1636	6.51ug	n.s.s.	2/20	2.69ug	0/15			
b	1636	6.51ug	n.s.s.	2/20	2.69ug	0/15			
c	1636	3.15ug	n.s.s.	11/20	2.69ug	4/15			
548	1129	5.37mg	29.3mg	0/67	3.34mg	10/63			Horie;gann,56,429-441;1965
a	1129	5.10mg	n.s.s.	5/67	3.34mg	13/63			
b	1129	19.4mg	n.s.s.	0/67	3.34mg	1/63			
549	1326	.376mg	n.s.s.	3/64	.107mg	10/64			Brune;zkko,102,153-157;1981
a	1326	.393mg	n.s.s.	2/64	.107mg	9/64			

BENZOATE, SODIUM   532-32-1

	RefNum	LoConf	UpConf	Cntrl	1Dose	1Inc	2Dose	2Inc	Citation or Pathology
550	1996	18.3gm	n.s.s.	21/100	4.00gm	10/50			Toth;faat,4,494-496;1984
a	1996	68.4gm	n.s.s.	0/100	4.00gm	0/50			
551	1996	21.7gm	n.s.s.	0/100	3.33gm	1/50			
a	1996	17.5gm	n.s.s.	23/100	3.33gm	7/50			
552	1319	2.03gm	n.s.s.	8/43	480.mg	16/52	960.mg	11/52	Sodemoto;jept,4,87-95;1980
a	1319	4.00gm	n.s.s.	3/43	480.mg	5/52	960.mg	4/52	
553	1319	2.78gm	n.s.s.	9/25	384.mg	7/50	768.mg	11/50	
a	1319	7.17gm	n.s.s.	0/25	384.mg	1/50	768.mg	0/50	

BENZOFURAN   271-89-6

	RefNum	LoConf	UpConf	Cntrl	1Dose	1Inc	2Dose	2Inc	Citation or Pathology	Brkly Code
554	TR370	18.8mg	73.8mg	5/50	84.9mg	27/50	(170.mg	29/50)	for:sqc,sqp; liv:hpa; lun:a/a,a/c.	C
a	TR370	19.7mg	71.5mg	1/50	84.9mg	22/50	(170.mg	21/50)		
b	TR370	20.4mg	80.8mg	4/50	84.9mg	25/50	(170.mg	22/50)	liv:hpa,hpc.	S
c	TR370	61.8mg	890.mg	2/50	84.9mg	9/50	(170.mg	5/50)	for:sqc,sqp.	
d	TR370	69.8mg	1.88gm	2/50	84.9mg	8/50	(170.mg	5/50)		S
e	TR370	98.1mg	449.mg	2/50	84.9mg	9/50	170.mg	14/50	lun:a/a,a/c.	
f	TR370	122.mg	565.mg	1/50	84.9mg	5/50	170.mg	13/50		S
g	TR370	101.mg	n.s.s.	1/50	84.9mg	4/50	(170.mg	1/50)	mul:mlu; spl:mlu.	S
h	TR370	22.3mg	255.mg	27/50	84.9mg	35/50	(170.mg	35/50)		
i	TR370	20.4mg	80.8mg	4/50	84.9mg	25/50	(170.mg	22/50)	liv:hpa,hpc,nnd.	
j	TR370	98.1mg	449.mg	2/50	84.9mg	9/50	170.mg	14/50	lun:a/a,a/c.	
555	TR370	10.7mg	52.5mg	12/50	42.4mg	31/50	(84.9mg	40/50)	liv:hpa,hpb,hpc.	
a	TR370	11.5mg	46.9mg	4/50	42.4mg	24/50	(84.9mg	34/50)		
b	TR370	11.4mg	58.6mg	12/50	42.4mg	30/50	(84.9mg	37/50)	liv:hpa,hpc.	S
c	TR370	11.7mg	108.mg	20/50	42.4mg	32/50	(84.9mg	45/50)	for:sqc,sqp; liv:hpa,hpb,hpc; lun:a/a,a/c.	C
d	TR370	60.1mg	191.mg	0/50	42.4mg	3/50	84.9mg	18/50	for:sqc,sqp.	
e	TR370	58.7mg	438.mg	2/50	42.4mg	11/50	84.9mg	13/50	liv:hpb,hpc.	S
f	TR370	55.7mg	1.46gm	9/50	42.4mg	10/50	84.9mg	22/50		S
g	TR370	67.8mg	1.08gm	4/50	42.4mg	7/50	84.9mg	15/50		S
h	TR370	74.8mg	2.64gm	2/50	42.4mg	7/50	84.9mg	10/50		
i	TR370	67.6mg	n.s.s.	10/50	42.4mg	9/50	84.9mg	19/50	lun:a/a,a/c.	
j	TR370	163.mg	n.s.s.	0/50	42.4mg	4/50	84.9mg	3/50		S
k	TR370	272.mg	n.s.s.	0/50	42.4mg	0/50	84.9mg	3/50		S
l	TR370	25.8mg	n.s.s.	29/50	42.4mg	32/50	84.9mg	45/50		
m	TR370	11.4mg	58.6mg	12/50	42.4mg	30/50	(84.9mg	37/50)	liv:hpa,hpc,nnd.	
n	TR370	67.6mg	n.s.s.	10/50	42.4mg	9/50	84.9mg	19/50	lun:a/a,a/c.	
556	TR370	42.7mg	939.mg	1/50	42.4mg	9/50	(84.9mg	3/50)		S

```
        Spe Strain  Site    Xpo+Xpt                                              TD50    2Tailpvl
            Sex  Route  Hist     Notes                                             DR     AuOp

  a   R f f34 gav lun MXA 24m24                                               418.mg * P<.04
  b   R f f34 gav kid uac 24m24                                               424.mg * P<.02  p
  c   R f f34 gav lun a/a 24m24                                               530.mg * P<.04
  d   R f f34 gav ton sqp 24m24                                               574.mg * P<.04
  e   R f f34 gav TBA MXB 24m24                                               577.mg * P<.9
  f   R f f34 gav liv MXB 24m24                                               2.09gm * P<.3
557   R m f34 gav tes ict 24m24            :      +         :              #8.03mg \ P<.004 -
  a   R m f34 gav pit adn 24m24                                               45.8mg * P<.03
  b   R m f34 gav MXA mnl 24m24                                               55.3mg * P<.02
  c   R m f34 gav lun a/c 24m24                                               166.mg * P<.03
  d   R m f34 gav mgl fba 24m24                                               271.mg * P<.03
  e   R m f34 gav thy fcc 24m24                                               389.mg * P<.03
  f   R m f34 gav TBA MXB 24m24                                               22.5mg * P<.02
  g   R m f34 gav liv MXB 24m24                                               no dre   P=1.

BENZOGUANAMINE   100ng..:..1ug....:..10.....:...100.....:..1mg.....:..10.....:...100....:...1g.....:10
558   M f chi eat liv mix 77w94                                     :>        2.04gm * P<.3  -
  a   M f chi eat lun mix 77w94                                               9.34gm * P<.9  -
  b   M f chi eat tba mix 77w94                                               932.mg * P<.5  -
559   M m chi eat lun mix 73w94 a                           :    ±            304.mg * P<.03 -
  a   M m chi eat liv mix 73w94 a                                             3.20gm * P<.7  -
  b   M m chi eat tba mix 73w94 a                                             439.mg * P<.2  -
560   R m cdr eat liv mix 77w94                          :>                   no dre   P=1.  -
  a   R m cdr eat tba mix 77w94                                               16.3mg \ P<.2  -

BENZOIN   100ng..:..1ug....:..10.....:...100.....:..1mg.....:..10.....:...100....:...1g.....:10
561   M f b6c eat TBA MXB 24m24                                     :>        6.83gm * P<.9  -
  a   M f b6c eat liv MXB 24m24                                               8.42gm * P<.4  -
  b   M f b6c eat lun MXB 24m24                                               no dre   P=1.
562   M m b6c eat TBA MXB 24m24                                 :>            2.46gm * P<.6  -
  a   M m b6c eat liv MXB 24m24                                               3.56gm * P<.5  -
  b   M m b6c eat lun MXB 24m24                                               2.90gm * P<.3  -
563   R f f34 eat TBA MXB 24m24                      :>                       no dre   P=1.  -
  a   R f f34 eat liv MXB 24m24                                               no dre   P=1.
564   R m f34 eat liv MXA 24m24                    :    ±                    #88.2mg * P<.03 -
  a   R m f34 eat TBA MXB 24m24                                               no dre   P=1.
  b   R m f34 eat liv MXB 24m24                                               88.2mg * P<.03

1,4-BENZOQUINONE   100ng..:..1ug....:..10.....:...100.....:..1mg.....:..10.....:...100....:...1g.....:10
565   M f swa gav liv lle 56w56 es               .  +  .                      5.03mg   P<.0005+
  a   M f swa gav spl lle 56w56 es                                            5.03mg   P<.0005+
  b   M f swa gav tba mix 56w56 es                                            5.03mg   P<.0005+
566   M m swa gav liv lle 56w56 es            .  +  .                         5.12mg   P<.0005+
  a   M m swa gav spl lle 56w56 es                                            5.12mg   P<.0005+
  b   M m swa gav tba mix 56w56 es                                            5.12mg   P<.0005+

BENZOTHIAZYL DISULFIDE   100ng..:..1ug....:..10.....:...100.....:..1mg.....:..10.....:...100....:...1g.....:10
567   M f b6a orl lun ade 76w76 evx                              .>           no dre   P=1.  -
  a   M f b6a orl liv hpt 76w76 evx                                           no dre   P=1.  -
  b   M f b6a orl tba mix 76w76 evx                                           1.38gm   P<.7  -
568   M m b6a orl lun ade 76w76 evx                           .>              1.13gm   P<.7  -
  a   M m b6a orl liv hpt 76w76 evx                                           no dre   P=1.  -
  b   M m b6a orl tba mix 76w76 evx                                           511.mg   P<.5  -
569   M f b6c orl lun ade 76w76 evx                                .>         669.mg   P<.2  -
  a   M f b6c orl liv hpt 76w76 evx                                           no dre   P=1.  -
  b   M f b6c orl tba mix 76w76 evx                                           432.mg   P<.05  p
570   M m b6c orl liv hpt 76w76 evx                           .>              621.mg   P<.2  -
  a   M m b6c orl lun mix 76w76 evx                                           no dre   P=1.  -
  b   M m b6c orl tba mix 76w76 evx                                           401.mg   P<.05  -

1H-BENZOTRIAZOLE   100ng..:..1ug....:..10.....:...100.....:..1mg.....:..10.....:...100....:...1g.....:10
571   M f b6c eat lun MXA 24m25 v                             :  +  :         3.89gm \ P<.0005a
  a   M f b6c eat lun a/c 24m25 v                                             4.39gm \ P<.002 a
  b   M f b6c eat TBA MXB 24m25 v                                             18.9gm \ P<.8
  c   M f b6c eat liv MXB 24m25 v                                             no dre   P=1.
  d   M f b6c eat lun MXB 24m25 v                                             3.89gm \ P<.0005
572   M m b6c eat TBA MXB 24m25 v                                  :>         no dre   P=1.
  a   M m b6c eat liv MXB 24m25 v                                             no dre   P=1.
  b   M m b6c eat lun MXB 24m25 v                                             no dre   P=1.
573   R f f34 eat bra gln 18m24 v                                       +hist 20.3gm * P<.3  a
  a   R f f34 eat TBA MXB 18m24 v                                             no dre   P=1.
  b   R f f34 eat liv MXB 18m24 v                                             10.1gm * P<.2
574   R m f34 eat liv nnd 18m24 v                             :    ±          2.72gm * P<.02
  a   R m f34 eat bra MXA 18m24 v                                       +hist no dre   P=1.  a
  b   R m f34 eat TBA MXB 18m24 v                                             no dre   P=1.
  c   R m f34 eat liv MXB 18m24 v                                             2.72gm * P<.02

BENZOYL HYDRAZINE   100ng..:..1ug....:..10.....:...100.....:..1mg.....:..10.....:...100....:...1g.....:10
575   M f swa wat lun ade 26m26 e                .  +  .                      13.4mg   P<.0005
  a   M f swa wat lun mix 26m26 e                                             13.8mg   P<.0005+
  b   M f swa wat lun adc 26m26 e                                             39.0mg   P<.0005
```

	RefNum	LoConf	UpConf	Cntrl	1Dose	1Inc	2Dose	2Inc	Citation or Pathology
									Brkly Code
a	TR370	159.mg	n.s.s.	0/50	42.4mg	2/50	84.9mg	3/50	lun:a/a,a/c. S
b	TR370	161.mg	n.s.s.	0/50	42.4mg	1/50	84.9mg	4/50	
c	TR370	183.mg	n.s.s.	0/50	42.4mg	1/50	84.9mg	3/50	S
d	TR370	194.mg	n.s.s.	0/50	42.4mg	1/50	84.9mg	3/50	S
e	TR370	40.0mg	n.s.s.	46/50	42.4mg	48/50	84.9mg	42/50	
f	TR370	341.mg	n.s.s.	0/50	42.4mg	0/50	84.9mg	1/50	liv:hpa,hpc,nnd.
557	TR370	3.77mg	68.2mg	42/50	21.2mg	40/50	(42.4mg	41/50)	S
a	TR370	19.5mg	n.s.s.	18/50	21.2mg	16/50	42.4mg	22/50	S
b	TR370	25.4mg	n.s.s.	10/50	21.2mg	13/50	42.4mg	17/50	liv:mnl; mul:mnl. S
c	TR370	60.0mg	n.s.s.	0/50	21.2mg	3/50	42.4mg	2/50	S
d	TR370	77.8mg	n.s.s.	0/50	21.2mg	0/50	42.4mg	3/50	S
e	TR370	103.mg	n.s.s.	0/50	21.2mg	0/50	42.4mg	3/50	S
f	TR370	10.4mg	n.s.s.	45/50	21.2mg	42/50	42.4mg	46/50	
g	TR370	147.mg	n.s.s.	1/50	21.2mg	1/50	42.4mg	0/50	liv:hpa,hpc,nnd.

BENZOGUANAMINE   91-76-9

	RefNum	LoConf	UpConf	Cntrl	1Dose	1Inc	2Dose	2Inc	Citation or Pathology
558	381	503.mg	n.s.s.	0/20	223.mg	1/17	446.mg	1/18	Weisburger;jept,2,325-356;1978/pers.comm./Russfield 1973
a	381	433.mg	n.s.s.	6/20	223.mg	2/17	446.mg	3/18	
b	381	200.mg	n.s.s.	17/20	223.mg	8/17	446.mg	9/18	
559	381	119.mg	n.s.s.	5/18	240.mg	2/13	411.mg	7/14	
a	381	378.mg	n.s.s.	3/18	240.mg	0/13	411.mg	2/14	
b	381	131.mg	n.s.s.	12/18	240.mg	3/13	411.mg	9/14	
560	381	144.mg	n.s.s.	1/22	17.1mg	0/20	34.3mg	0/21	
a	381	5.02mg	n.s.s.	14/22	17.1mg	13/20	(34.3mg	9/21)	

BENZOIN   (2-hydroxy-1,2-diphenylethanone) 119-53-9

	RefNum	LoConf	UpConf	Cntrl	1Dose	1Inc	2Dose	2Inc	Citation or Pathology
561	TR204	572.mg	n.s.s.	27/50	322.mg	35/50	644.mg	28/50	
a	TR204	1.93gm	n.s.s.	2/50	322.mg	3/50	644.mg	4/50	liv:hpa,hpc,nnd.
b	TR204	2.85gm	n.s.s.	6/50	322.mg	5/50	644.mg	3/50	lun:a/a,a/c.
562	TR204	451.mg	n.s.s.	31/50	297.mg	27/50	594.mg	32/50	
a	TR204	716.mg	n.s.s.	16/50	297.mg	12/50	594.mg	18/50	liv:hpa,hpc,nnd.
b	TR204	846.mg	n.s.s.	5/50	297.mg	10/50	594.mg	8/50	lun:a/a,a/c.
563	TR204	22.3mg	n.s.s.	42/50	12.5mg	41/50	25.0mg	39/50	
a	TR204	n.s.s.	n.s.s.	0/50	12.5mg	0/50	25.0mg	0/50	liv:hpa,hpc,nnd.
564	TR204	30.5mg	n.s.s.	0/50	5.00mg	0/50	10.0mg	4/50	liv:hpc,nnd. S
a	TR204	10.3mg	n.s.s.	36/50	5.00mg	32/50	10.0mg	35/50	
b	TR204	30.5mg	n.s.s.	0/50	5.00mg	0/50	10.0mg	4/50	liv:hpa,hpc,nnd.

1,4-BENZOQUINONE   106-51-4

	RefNum	LoConf	UpConf	Cntrl	1Dose	1Inc	2Dose	2Inc	Citation or Pathology
565	1944	2.97mg	9.35mg	0/30	22.9mg	22/37			El-Mofty;nutc,17,97-104;1992/pers.comm.
a	1944	2.97mg	9.35mg	0/30	22.9mg	22/37			
b	1944	2.97mg	9.35mg	0/30	22.9mg	22/37			
566	1944	3.08mg	9.33mg	0/28	19.0mg	23/44			
a	1944	3.08mg	9.33mg	0/28	19.0mg	23/44			
b	1944	3.08mg	9.33mg	0/28	19.0mg	23/44			

BENZOTHIAZYL DISULFIDE (altax) 120-78-5

	RefNum	LoConf	UpConf	Cntrl	1Dose	1Inc	2Dose	2Inc	Citation or Pathology
567	1301	256.mg	n.s.s.	1/17	215.mg	1/18			Innes;ntis,1968/1969
a	1301	426.mg	n.s.s.	0/17	215.mg	0/18			
b	1301	155.mg	n.s.s.	2/17	215.mg	3/18			
568	1301	142.mg	n.s.s.	2/18	200.mg	3/18			
a	1301	396.mg	n.s.s.	1/18	200.mg	0/18			
b	1301	102.mg	n.s.s.	3/18	200.mg	5/18			
569	1301	164.mg	n.s.s.	0/16	215.mg	2/18			
a	1301	426.mg	n.s.s.	0/16	215.mg	0/18			
b	1301	130.mg	n.s.s.	0/16	215.mg	3/18			
570	1301	153.mg	n.s.s.	0/16	200.mg	2/18			
a	1301	396.mg	n.s.s.	0/16	200.mg	0/18			
b	1301	121.mg	n.s.s.	0/16	200.mg	3/18			

1H-BENZOTRIAZOLE   95-14-7

	RefNum	LoConf	UpConf	Cntrl	1Dose	1Inc	2Dose	2Inc	Citation or Pathology
571	TR88	2.21gm	7.77gm	0/50	1.50gm	10/50	(2.99gm	4/50)	lun:a/a,a/c.
a	TR88	2.43gm	9.15gm	0/50	1.50gm	9/50	(2.99gm	3/50)	
b	TR88	2.21gm	n.s.s.	21/50	1.50gm	25/50	(2.99gm	14/50)	
c	TR88	25.8mg	n.s.s.	1/50	1.50gm	2/50	2.99gm	1/50	liv:hpa,hpc,nnd.
d	TR88	1.89gm	11.6gm	0/50	1.50gm	10/50	(2.99gm	4/50)	lun:a/a,a/c.
572	TR88	1.68gm	n.s.s.	24/50	1.38gm	23/50	(2.76gm	18/50)	
a	TR88	9.88gm	n.s.s.	12/50	1.38gm	12/50	2.76gm	7/50	liv:hpa,hpc,nnd.
b	TR88	7.62gm	n.s.s.	4/50	1.38gm	7/50	2.76gm	7/50	lun:a/a,a/c.
573	TR88	3.31gm	n.s.s.	0/50	250.mg	0/50	450.mg	1/50	
a	TR88	535.mg	n.s.s.	27/50	250.mg	27/50	450.mg	30/50	
b	TR88	2.48gm	n.s.s.	0/50	250.mg	0/50	450.mg	2/50	liv:hpa,hpc,nnd.
574	TR88	1.03gm	n.s.s.	0/50	202.mg	0/50	360.mg	1/50	S
a	TR88	n.s.s.	n.s.s.	0/50	202.mg	3/50	360.mg	0/50	bra:gln,oli.
b	TR88	368.mg	n.s.s.	28/50	202.mg	20/50	(360.mg	18/50)	
c	TR88	1.03gm	n.s.s.	0/50	202.mg	0/50	360.mg	5/50	liv:hpa,hpc,nnd.

BENZOYL HYDRAZINE   613-94-5

	RefNum	LoConf	UpConf	Cntrl	1Dose	1Inc	2Dose	2Inc	Citation or Pathology
575	48	8.18mg	24.9mg	12/89	20.0mg	34/47			Toth;ejca,8,341-345;1972/1969a
a	48	8.29mg	26.1mg	14/89	20.0mg	34/47			
b	48	21.0mg	89.4mg	2/89	20.0mg	16/47			

	Spe	Strain	Site	Xpo+Xpt			TD50		2Tailpvl
	Sex	Route	Hist	Notes				DR	AuOp
c	M f	swa	wat	--- mly 26m26 e			52.1mg	P<.004	+
d	M f	swa	wat	liv mix 26m26 e			no dre	P=1.	
576	M m	swa	wat	lun mix 92w92 e	. + .		7.35mg	P<.0005	+
a	M m	swa	wat	lun ade 92w92 e			8.52mg	P<.0005	
b	M m	swa	wat	lun adc 92w92 e			24.1mg	P<.0005	
c	M m	swa	wat	--- mly 92w92 e			52.0mg	P<.07	+
d	M m	swa	wat	liv hem 92w92 e			no dre	P=1.	-
577	M f	swi	gav	lun tum 40w55	.>		no dre	P=1.	-

BENZYL ACETATE 100ng..:..1ug....:..10.....:..100....:..1mg..:..10....:..100....:..1g.....:..10

578	M f	b6c	gav	MXB MXB 24m24 s	: +	:	1.75gm	/ P<.003	
a	M f	b6c	gav	liv MXA 24m24 s			2.03gm	/ P<.02	
b	M f	b6c	gav	for sqp 24m24 s			2.94gm	* P<.04	p
c	M f	b6c	gav	liv hpa 24m24 s			3.21gm	* P<.02	p
d	M f	b6c	gav	TBA MXB 24m24 s			4.27gm	* P<.9	
e	M f	b6c	gav	liv MXB 24m24 s			2.03gm	/ P<.02	
f	M f	b6c	gav	lun MXB 24m24 s			no dre	P=1.	
579	M f	b6c	eat	TBA MXB 24m24	:>		107.mg	Z P<.5	-
a	M f	b6c	eat	liv MXB 24m24			no dre	P=1.	
b	M f	b6c	eat	lun MXB 24m24			26.4gm	* P<.9	
580	M m	b6c	gav	MXB MXB 24m24	: +	:	702.mg	* P<.0005	
a	M m	b6c	gav	liv MXA 24m24			758.mg	* P<.004	
b	M m	b6c	gav	liv hpa 24m24			955.mg	* P<.0005	p
c	M m	b6c	gav	for MXA 24m24			1.91gm	* P<.03	p
d	M m	b6c	gav	for sqp 24m24			2.14gm	* P<.03	p
e	M m	b6c	gav	TBA MXB 24m24			829.mg	* P<.2	
f	M m	b6c	gav	liv MXB 24m24			758.mg	* P<.004	
g	M m	b6c	gav	lun MXB 24m24			no dre	P=1.	
581	M m	b6c	eat	TBA MXB 24m24	:>		no dre	P=1.	-
a	M m	b6c	eat	liv MXB 24m24			no dre	P=1.	
b	M m	b6c	eat	lun MXB 24m24			418.mg	Z P<.7	
582	R f	f34	gav	TBA MXB 24m24	:	±	285.mg	* P<.06	-
a	R f	f34	gav	liv MXB 24m24			6.48gm	* P<.6	
583	R f	f34	eat	TBA MXB 24m24	:>		7.03gm	* P<.9	-
a	R f	f34	eat	liv MXB 24m24			6.55gm	* P<.2	
584	R m	f34	gav	pan ana 24m24	: +	:	282.mg	/ P<.003	e
a	R m	f34	gav	pre MXA 24m24			2.17gm	* P<.04	
b	R m	f34	gav	pre MXA 24m24			2.47gm	* P<.02	
c	R m	f34	gav	pre ccn 24m24			4.27gm	* P<.04	
d	R m	f34	gav	TBA MXB 24m24			376.mg	/ P<.2	
e	R m	f34	gav	liv MXB 24m24			6.47gm	* P<.5	
585	R m	f34	eat	TBA MXB 24m24	:>		no dre	P=1.	-
a	R m	f34	eat	liv MXB 24m24			no dre	P=1.	
586	R m	f34	eat	pan cic 24m24 er	.	±	2.67mg	P<.03	
a	R m	f34	eat	pan ade 24m24 er			27.0mg	P<1.	

BENZYL ALCOHOL 100ng..:..1ug....:..10.....:..100....:..1mg..:..10....:..100....:..1g.....:..10

587	M f	b6c	gav	TBA MXB 24m24	:>		57.5mg	* P<1.	-
a	M f	b6c	gav	liv MXB 24m24			1.04gm	* P<.4	
b	M f	b6c	gav	lun MXB 24m24			no dre	P=1.	
588	M m	b6c	gav	adr coa 24m24	:	±	#1.62gm	* P<.05	-
a	M m	b6c	gav	TBA MXB 24m24			no dre	P=1.	
b	M m	b6c	gav	liv MXB 24m24			588.mg	* P<.4	
c	M m	b6c	gav	lun MXB 24m24			546.mg	* P<.3	
589	R f	f34	gav	TBA MXB 24m24 s	:>		885.mg	* P<.7	-
a	R f	f34	gav	liv MXB 24m24 s			no dre	P=1.	
590	R m	f34	gav	mgl MXA 24m24	:	±	#1.14gm	* P<.03	-
a	R m	f34	gav	mgl fba 24m24			1.38gm	* P<.04	
b	R m	f34	gav	TBA MXB 24m24			no dre	P=1.	
c	R m	f34	gav	liv MXB 24m24			no dre	P=1.	

BENZYL CHLORIDE 100ng..:..1ug....:..10.....:..100....:..1mg..:..10....:..100....:..1g.....:..10

591	M f	b6c	gav	for mix 24m25 e	. + .		81.0mg	* P<.0005	+
a	M f	b6c	gav	lun mix 24m25 e			337.mg	* P<.04	
b	M f	b6c	gav	for car 24m25 e			445.mg	* P<.04	+
c	M f	b6c	gav	liv mix 24m25 e			no dre	P=1.	
592	M m	b6c	gav	for mix 24m25 e	.+ .		49.6mg	/ P<.0005	+
a	M m	b6c	gav	for car 24m25 e			221.mg	* P<.002	+
b	M m	b6c	gav	--- vsc 24m25 e			454.mg	* P<.009	
c	M m	b6c	gav	liv mix 24m25 e			225.mg	* P<.5	
d	M m	b6c	gav	lun mix 24m25 e			no dre	P=1.	
593	R f	f34	gav	thy mix 24m25 e	. +	.	40.6mg	* P<.009	
a	R f	f34	gav	liv nnd 24m25 e			330.mg	* P<.7	
594	R m	f34	gav	liv nnd 24m25 e	.>		no dre	P=1.	

o-BENZYL-p-CHLOROPHENOL 100ng..:..1ug....:..10.....:..100....:..1mg..:..10....:..100....:..1g.....:..10

595	M f	b6c	gav	TBA MXB 24m24	:>		6.97gm	* P<1.	-
a	M f	b6c	gav	liv MXB 24m24			684.mg	* P<.09	
b	M f	b6c	gav	lun MXB 24m24			31.5gm	* P<1.	
596	M m	b6c	gav	kid MXA 24m24	:	±	1.44gm	* P<.02	p
a	M m	b6c	gav	kid rua 24m24			2.21gm	* P<.09	p

	RefNum	LoConf	UpConf	Cntrl	1Dose	1Inc	2Dose	2Inc	Citation or Pathology / Brkly Code
c	48	23.6mg	463.mg	16/106	20.0mg	18/49			
d	48	145.mg	n.s.s.	3/88	20.0mg	1/47			
576	48	4.40mg	13.7mg	10/86	16.7mg	31/42			
a	48	5.05mg	16.5mg	10/86	16.7mg	29/42			
b	48	12.7mg	54.5mg	0/86	16.7mg	13/42			
c	48	16.4mg	n.s.s.	2/40	16.7mg	5/25			
d	48	67.2mg	n.s.s.	2/40	16.7mg	0/25			
577	1095	10.8mg	n.s.s.	8/85	10.4mg	1/25			Roe;natu,216,375-376;1967

BENZYL ACETATE   140-11-4

	RefNum	LoConf	UpConf	Cntrl	1Dose	1Inc	2Dose	2Inc	Citation or Pathology / Brkly Code
578	TR250	795.mg	7.34gm	0/50	350.mg	0/50	704.mg	9/50	for:sqp; liv:hpa. P
a	TR250	874.mg	n.s.s.	1/50	350.mg	0/50	704.mg	10/50	liv:hpa,hpc. S
b	TR250	1.02gm	n.s.s.	0/50	350.mg	0/50	704.mg	4/50	
c	TR250	1.27gm	n.s.s.	0/50	350.mg	0/50	704.mg	6/50	
d	TR250	383.mg	n.s.s.	14/50	350.mg	26/50	704.mg	25/50	
e	TR250	874.mg	n.s.s.	1/50	350.mg	0/50	704.mg	10/50	liv:hpa,hpc,nnd.
f	TR250	2.30gm	n.s.s.	1/50	350.mg	3/50	704.mg	1/50	lun:a/a,a/c.
579	TR431	21.9mg	n.s.s.	35/50	42.2mg	40/50	(128.mg	33/50	385.mg	33/50)
a	TR431	1.02gm	n.s.s.	14/50	42.2mg	15/50	128.mg	13/50	385.mg	14/50	liv:hpa,hpb,hpc.
b	TR431	1.29gm	n.s.s.	1/50	42.2mg	5/50	128.mg	2/50	385.mg	4/50	lun:a/a,a/c.
580	TR250	399.mg	1.87gm	4/50	350.mg	9/50	704.mg	21/50	for:sqc,sqp; liv:hpa. P
a	TR250	387.mg	5.38gm	10/50	350.mg	18/50	704.mg	23/50	liv:hpa,hpc. S
b	TR250	540.mg	1.96gm	0/50	350.mg	5/50	704.mg	13/50	
c	TR250	808.mg	n.s.s.	4/50	350.mg	4/50	704.mg	11/50	for:sqc,sqp.
d	TR250	877.mg	n.s.s.	3/50	350.mg	3/50	704.mg	9/50	
e	TR250	309.mg	n.s.s.	32/50	350.mg	35/50	704.mg	37/50	
f	TR250	387.mg	5.38gm	10/50	350.mg	18/50	704.mg	23/50	liv:hpa,hpc,nnd.
g	TR250	1.44gm	n.s.s.	12/50	350.mg	7/50	704.mg	7/50	lun:a/a,a/c.
581	TR431	142.mg	n.s.s.	42/50	39.0mg	40/50	118.mg	35/50	(355.mg | 23/50)
a	TR431	223.mg	n.s.s.	22/50	39.0mg	22/50	118.mg	18/50	(355.mg	9/50)	liv:hpa,hpb,hpc.
b	TR431	54.7mg	n.s.s.	14/50	39.0mg	17/50	(118.mg	8/50	355.mg	7/50)	lun:a/a,a/c.
582	TR250	116.mg	n.s.s.	43/50	175.mg	44/50	352.mg	44/50	
a	TR250	950.mg	n.s.s.	1/50	175.mg	2/50	352.mg	1/50	liv:hpa,hpc,nnd.
583	TR431	347.mg	n.s.s.	48/50	147.mg	47/50	295.mg	45/50	591.mg	49/50	
a	TR431	2.02gm	n.s.s.	1/50	147.mg	1/50	295.mg	1/50	591.mg	4/50	liv:hpa,hpb,hpc.
584	TR250	146.mg	1.86gm	22/50	174.mg	27/50	352.mg	37/50	
a	TR250	814.mg	n.s.s.	1/50	174.mg	1/50	352.mg	6/50	pre:acn,can,ccn. S
b	TR250	916.mg	n.s.s.	0/50	174.mg	1/50	352.mg	4/50	pre:acn,ccn. S
c	TR250	1.26gm	n.s.s.	0/50	174.mg	0/50	352.mg	3/50	S
d	TR250	134.mg	n.s.s.	44/50	174.mg	46/50	352.mg	45/50	
e	TR250	1.24gm	n.s.s.	1/50	174.mg	2/50	352.mg	2/50	liv:hpa,hpc,nnd.
585	TR431	379.mg	n.s.s.	39/50	118.mg	43/50	237.mg	43/50	474.mg	39/50	
a	TR431	1.91gm	n.s.s.	5/50	118.mg	6/50	237.mg	3/50	474.mg	3/50	liv:hpa,hpb,hpc.
586	1988	807.mg	n.s.s.	0/49	320.mg	3/38			Longnecker;fctx,28,665-668;1990/pers.comm.
a	1988	660.mg	n.s.s.	10/49	320.mg	8/38			

BENZYL ALCOHOL   100-51-6

	RefNum	LoConf	UpConf	Cntrl	1Dose	1Inc	2Dose	2Inc	Citation or Pathology / Brkly Code
587	TR343	120.mg	n.s.s.	27/50	70.7mg	24/50	142.mg	36/50	
a	TR343	279.mg	n.s.s.	1/50	70.7mg	6/50	142.mg	4/50	liv:hpa,hpc,nnd.
b	TR343	588.mg	n.s.s.	4/50	70.7mg	1/50	142.mg	4/50	lun:a/a,a/c.
588	TR343	489.mg	n.s.s.	0/50	70.7mg	0/50	142.mg	3/50	S
a	RefR343	134.mg	n.s.s.	34/50	70.7mg	27/50	142.mg	34/50	
b	TR343	148.mg	n.s.s.	11/50	70.7mg	16/50	142.mg	16/50	liv:hpa,hpc,nnd.
c	TR343	174.mg	n.s.s.	10/50	70.7mg	6/50	142.mg	17/50	lun:a/a,a/c.
589	TR343	133.mg	n.s.s.	46/50	142.mg	29/50	283.mg	26/50	
a	TR343	n.s.s.	n.s.s.	0/50	142.mg	0/50	283.mg	0/50	liv:hpa,hpc,nnd.
590	TR343	465.mg	n.s.s.	0/50	142.mg	3/50	283.mg	3/50	mgl:adn,fba. S
a	TR343	523.mg	n.s.s.	0/50	142.mg	2/50	283.mg	3/50	S
b	TR343	171.mg	n.s.s.	44/50	142.mg	32/50	283.mg	38/50	
c	TR343	1.47gm	n.s.s.	2/50	142.mg	0/50	283.mg	1/50	liv:hpa,hpc,nnd.

BENZYL CHLORIDE   (alpha-chloro toluene) 100-44-7

	RefNum	LoConf	UpConf	Cntrl	1Dose	1Inc	2Dose	2Inc	Citation or Pathology / Brkly Code
591	1827	49.8mg	144.mg	0/52	20.7mg	5/50	41.5mg	19/51	Lijinsky;jnci,76,1231-1236;1986
a	1827	132.mg	n.s.s.	1/52	20.7mg	2/51	41.5mg	6/51	
b	1827	169.mg	n.s.s.	0/52	20.7mg	2/50	41.5mg	3/51	
c	1827	272.mg	n.s.s.	7/52	20.7mg	5/51	41.5mg	3/51	
592	1827	33.0mg	79.2mg	0/51	20.7mg	4/52	41.5mg	32/52	
a	1827	108.mg	736.mg	0/51	20.7mg	2/52	41.5mg	8/52	
b	1827	172.mg	10.5gm	0/52	20.7mg	0/52	41.5mg	5/52	
c	1827	49.7mg	n.s.s.	17/52	20.7mg	28/52	41.5mg	20/51	
d	1827	99.1mg	n.s.s.	11/52	20.7mg	15/52	41.5mg	11/52	
593	1827	19.8mg	1.52gm	4/52	6.22mg	8/51	12.4mg	14/52	
a	1827	49.2mg	n.s.s.	2/52	6.22mg	5/52	12.4mg	3/52	
594	1827	71.9mg	n.s.s.	3/52	6.22mg	4/52	12.4mg	2/51	

o-BENZYL-p-CHLOROPHENOL   120-32-1

	RefNum	LoConf	UpConf	Cntrl	1Dose	1Inc	2Dose	2Inc		Citation or Pathology / Brkly Code
595	TR424	265.mg	n.s.s.	39/50	85.2mg	35/50	170.mg	29/50	341.mg 28/52	liv:hpa,hpb,hpc.
a	TR424	256.mg	n.s.s.	13/50	85.2mg	15/50	170.mg	17/50	341.mg 16/51	lun:a/a,a/c.
b	TR424	1.17gm	n.s.s.	5/50	85.2mg	3/50	170.mg	1/50	341.mg 4/52	kid:rua,uac.
596	TR424	680.mg	n.s.s.	0/50	85.2mg	2/50	170.mg	4/50	341.mg 3/50	
a	TR424	903.mg	n.s.s.	0/50	85.2mg	2/50	170.mg	2/50	341.mg 2/50	

```
      Spe Strain Site   Xpo+Xpt                                              TD50    2Tailpvl
          Sex  Route Hist   Notes                                               DR    AuOp

 b   M m b6c gav TBA MXB 24m24                                             3.16gm * P<.9
 c   M m b6c gav liv MXB 24m24                                             no dre  P=1.
 d   M m b6c gav lun MXB 24m24                                             no dre  P=1.
597  M m b6c gav kid MXA 24m24    with step                 . +     .      1.35gm * P<.003 p
 a   M m b6c gav kid rua 24m24                                             1.74gm * P<.007 p
598  R f f34 gav amd pob 24m24                          :     ±           615.mg * P<.05
 a   R f f34 gav kid tcc 24m24                                            4.07gm * P<.2  e
 b   R f f34 gav TBA MXB 24m24                                             867.mg * P<.8
 c   R f f34 gav liv MXB 24m24                                             no dre  P=1.
599  R m f34 gav TBA MXB 24m24                         :>                  no dre  P=1.  -
 a   R m f34 gav liv MXB 24m24                                             1.58gm * P<.5

BENZYLHYDRAZINE.2HCl   100ng..:..1ug...:..10.....:..100....:..1mg....:..10...:..100.....:..1g....:..10
600  M f swa wat lun mix 28m28 e                              .   ±        85.3mg  P<.02 +
 a   M f swa wat liv ang 28m28 e                                           252.mg  P<.2  -
 b   M f swa wat liv mix 28m28 e                                           304.mg  P<.3
601  M m swa wat liv mix 23m23 e                              .   ±        86.2mg  P<.08
 a   M m swa wat lun ade 23m23 e                                           711.mg  P<.8  -
 b   M m swa wat lun mix 23m23 e                                           no dre  P=1.

BERYLLIUM SULFATE      100ng..:..1ug...:..10.....:..100....:..1mg....:..10...:..100.....:..1g....:..10
602  M f cd1 wat lun tum 33m33 e                    .>                     no dre  P=1.  -
 a   M f cd1 wat tba mix 33m33 e                                           9.42mg  P<.4  -
 b   M f cd1 wat tba mal 33m33 e                                           36.9mg  P<.7  -
603  M m cd1 wat lun tum 41m41 e                  .>                       98.8mg  P<.9  -
 a   M m cd1 wat tba mix 41m41 e                                           17.1mg  P<.6  -
 b   M m cd1 wat tba mal 41m41 e                                           49.6mg  P<.6  -
604  R f leb wat tba mix 47m47 e         .>                                1.48mg  P<.4  -
 a   R f leb wat tba mal 47m47 e                                           3.24mg  P<.4  -
605  R m leb wat tba mix 38m38 e            .>                             2.78mg  P<.3  -
 a   R m leb wat tba mal 38m38 e                                           8.56mg  P<.6  -

BIPHENYL               100ng..:..1ug...:..10.....:..100....:..1mg....:..10...:..100.....:..1g....:..10
606  M f b6a orl liv hpt 76w76 evx                     .>                  no dre  P=1.  -
 a   M f b6a orl lun ade 76w76 evx                                         no dre  P=1.  -
 b   M f b6a orl tba mix 76w76 evx                                         187.mg  P<.4  -
607  M m b6a orl liv hpt 76w76 evx                   .>                    182.mg  P<.3  -
 a   M m b6a orl lun ade 76w76 evx                                         no dre  P=1.  -
 b   M m b6a orl tba mix 76w76 evx                                         290.mg  P<.7  -
608  M f b6c orl lun ade 76w76 evx                     .>                  467.mg  P<.3  -
 a   M f b6c orl liv hpt 76w76 evx                                         no dre  P=1.  -
 b   M f b6c orl tba mix 76w76 evx                                         227.mg  P<.2  -
609  M m b6c orl lun ade 76w76 evx               .   ±                     128.mg  P<.04 -
 a   M m b6c orl liv hpt 76w76 evx                                         199.mg  P<.1  -
 b   M m b6c orl tba mix 76w76 evx                                         71.5mg  P<.007 -

2-BIPHENYLAMINE.HCl    100ng..:..1ug...:..10.....:..100....:..1mg....:..10...:..100.....:..1g....:..10
610  M f b6c eat --- MXA 24m24                                  :  +  :    1.12gm * P<.0005c
 a   M f b6c eat --- ang 24m24                                             1.28gm * P<.002 c
 b   M f b6c eat TBA MXB 24m24                                             1.88gm * P<.7
 c   M f b6c eat liv MXB 24m24                                             2.00gm * P<.4
 d   M f b6c eat lun MXB 24m24                                             26.1gm / P<.9
611  M m b6c eat --- MXA 24m24 e                                :  ±       1.25gm * P<.02 a
 a   M m b6c eat --- MXA 24m24 e                                           1.81gm * P<.02 a
 b   M m b6c eat TBA MXB 24m24 e                                           720.mg * P<.4
 c   M m b6c eat liv MXB 24m24 e                                           1.70gm * P<.5
 d   M m b6c eat lun MXB 24m24 e                                           no dre  P=1.
612  R f f34 eat TBA MXB 24m24                         :>                  no dre  P=1.  -
 a   R f f34 eat liv MXB 24m24                                             363.mg \ P<.2
613  R m f34 eat TBA MXB 24m24                          :>                 no dre  P=1.  -
 a   R m f34 eat liv MXB 24m24                                             no dre  P=1.

BIS(2-CHLORO-1-METHYLETHYL)ETHER,  TECHNICAL GRADE..:..10.....:..100....:..1mg....:..10...:..100....:..1g.....:..10
614  M f b6c gav lun MXA 24m25                                  :  +    :  311.mg * P<.002 c
 a   M f b6c gav lun a/a 24m25                                             381.mg * P<.006 c
 b   M f b6c gav sto MXA 24m25                                  +hist 1.65gm * P<.05 a
 c   M f b6c gav TBA MXB 24m25                                             231.mg * P<.2
 d   M f b6c gav liv MXB 24m25                                             no dre  P=1.
 e   M f b6c gav lun MXB 24m25                                             311.mg * P<.002
615  M m b6c gav liv MXA 24m24                                  :  +    :  138.mg * P<.002 c
 a   M m b6c gav liv hpc 24m24                                             229.mg * P<.004 c
 b   M m b6c gav lun MXA 24m24                                             259.mg * P<.02 c
 c   M m b6c gav lun a/a 24m24                                             306.mg * P<.03 c
 d   M m b6c gav TBA MXB 24m24                                             116.mg * P<.02
 e   M m b6c gav liv MXB 24m24                                             138.mg * P<.002
 f   M m b6c gav lun MXB 24m24                                             259.mg * P<.02
616  R f f34 gav TBA MXB 24m24 s                             :  +     :     65.6mg / P<.006 -
 a   R f f34 gav liv MXB 24m24 s                                           no dre  P=1.
617  R m f34 gav TBA MXB 24m24 s                             :  ±          101.mg / P<.02 -
 a   R m f34 gav liv MXB 24m24 s                                           no dre  P=1.
```

	RefNum	LoConf	UpConf	Cntrl	1Dose	1Inc	2Dose	2Inc			Citation or Pathology
											Brkly Code
b	TR424	262.mg	n.s.s.	38/50	85.2mg	33/50	170.mg	37/50	341.mg	28/50	
c	TR424	449.mg	n.s.s.	25/50	85.2mg	25/50	170.mg	26/50	341.mg	16/50	liv:hpa,hpb,hpc.
d	TR424	1.15gm	n.s.s.	14/50	85.2mg	13/50	170.mg	7/50	341.mg	5/50	lun:a/a,a/c.
597	TR424	728.mg	6.31gm	0/50	85.2mg	2/50	170.mg	6/50	341.mg	6/50	kid:rac,rua.
a	TR424	875.mg	27.2gm	0/50	85.2mg	2/50	170.mg	4/50	341.mg	5/50	
598	TR424	298.mg	n.s.s.	0/50	42.3mg	3/50	84.6mg	3/51	169.mg	4/50	s
a	TR424	980.mg	n.s.s.	0/50	42.3mg	1/50	84.6mg	1/51	169.mg	1/50	
b	TR424	96.8mg	n.s.s.	44/50	42.3mg	40/50	84.6mg	42/51	169.mg	47/50	
c	TR424	1.76gm	n.s.s.	1/50	42.3mg	0/50	84.6mg	0/51	169.mg	0/50	liv:hpa,hpb,hpc.
599	TR424	73.6mg	n.s.s.	42/50	21.2mg	39/50	42.3mg	38/50	84.6mg	33/51	
a	TR424	305.mg	n.s.s.	0/50	21.2mg	1/50	42.3mg	0/50	84.6mg	1/51	liv:hpa,hpb,hpc.
BENZYLHYDRAZINE.2HCl		20570-96-1									
600	1056	36.8mg	n.s.s.	21/90	30.0mg	21/47					Toth;zkko,87,267-273;1976/1974
a	1056	61.4mg	n.s.s.	2/56	30.0mg	3/22					
b	1056	63.8mg	n.s.s.	3/56	30.0mg	3/22					
601	1056	24.8mg	n.s.s.	6/71	25.0mg	5/21					
a	1056	59.5mg	n.s.s.	15/99	25.0mg	8/47					
b	1056	70.9mg	n.s.s.	23/99	25.0mg	9/47					
BERYLLIUM SULFATE		13510-49-1									
602	1395	9.07mg	n.s.s.	9/47	1.00mg	5/52					Schroeder;jnut,105,452-458;1975
a	1395	2.34mg	n.s.s.	14/47	1.00mg	20/52					
b	1395	5.36mg	n.s.s.	4/47	1.00mg	6/52					
603	1395	6.34mg	n.s.s.	5/38	.833mg	7/48					
a	1395	3.29mg	n.s.s.	11/38	.833mg	17/48					
b	1395	7.70mg	n.s.s.	2/38	.833mg	4/48					
604	1456	.298mg	n.s.s.	17/24	.286mg	14/17					Schroeder;jnut,105,421-427;1975
a	1456	.689mg	n.s.s.	8/24	.286mg	8/17					
605	1456	.768mg	n.s.s.	4/26	.250mg	9/33					
a	1456	1.33mg	n.s.s.	2/26	.250mg	4/33					
BIPHENYL	92-52-4										
606	1307	137.mg	n.s.s.	0/17	73.0mg	0/17					Innes;ntis,1968/1969
a	1307	137.mg	n.s.s.	1/17	73.0mg	0/17					
b	1307	39.1mg	n.s.s.	2/17	73.0mg	4/17					
607	1307	41.6mg	n.s.s.	1/18	68.0mg	3/17					
a	1307	83.7mg	n.s.s.	2/18	68.0mg	1/17					
b	1307	39.1mg	n.s.s.	3/18	68.0mg	4/17					
608	1307	76.1mg	n.s.s.	0/16	73.0mg	1/18					
a	1307	145.mg	n.s.s.	0/16	73.0mg	0/18					
b	1307	55.7mg	n.s.s.	0/16	73.0mg	2/18					
609	1307	38.7mg	n.s.s.	0/16	68.0mg	3/17					
a	1307	48.8mg	n.s.s.	0/16	68.0mg	2/17					
b	1307	26.9mg	966.mg	0/16	68.0mg	5/17					
2-BIPHENYLAMINE.HCl		2185-92-4									
610	TR233	527.mg	3.06gm	0/50	128.mg	1/50	384.mg	8/50			---:ang,hem.
a	TR233	579.mg	4.87gm	0/50	128.mg	1/50	384.mg	7/50			
b	TR233	269.mg	n.s.s.	31/50	128.mg	30/50	384.mg	31/50			
c	TR233	475.mg	n.s.s.	7/50	128.mg	9/50	384.mg	10/50			liv:hpa,hpc,nnd.
d	TR233	1.15gm	n.s.s.	6/50	128.mg	1/50	384.mg	5/50			lun:a/a,a/c.
611	TR233	530.mg	n.s.s.	0/50	118.mg	4/50	355.mg	3/50			---:ang,hem,hes.
a	TR233	670.mg	n.s.s.	0/50	118.mg	2/50	355.mg	3/50			---:ang,hes.
b	TR233	181.mg	n.s.s.	37/50	118.mg	38/50	355.mg	27/50			
c	TR233	333.mg	n.s.s.	14/50	118.mg	19/50	355.mg	11/50			liv:hpa,hpc,nnd.
d	TR233	388.mg	n.s.s.	16/50	118.mg	6/50	(355.mg	1/50)			lun:a/a,a/c.
612	TR233	122.mg	n.s.s.	46/50	49.3mg	45/49	148.mg	46/50			
a	TR233	115.mg	n.s.s.	1/50	49.3mg	5/49	(148.mg	1/50)			liv:hpa,hpc,nnd.
613	TR233	142.mg	n.s.s.	43/50	39.4mg	40/50	118.mg	35/50			
a	TR233	n.s.s.	n.s.s.	0/50	39.4mg	0/50	118.mg	0/50			liv:hpa,hpc,nnd.
BIS(2-CHLORO-1-METHYLETHYL)ETHER, TECHNICAL GRADE				108-60-1							
614	TR239	155.mg	1.45gm	1/50	67.5mg	4/50	139.mg	10/50			lun:a/a,a/c.
a	TR239	177.mg	4.88gm	1/50	67.5mg	4/50	139.mg	8/50			
b	TR239	498.mg	n.s.s.	0/50	67.5mg	0/50	139.mg	3/50			sto:sqc,sqp.
c	TR239	75.0mg	n.s.s.	26/50	67.5mg	29/50	139.mg	29/50			
d	TR239	269.mg	n.s.s.	7/50	67.5mg	7/50	139.mg	5/50			liv:hpa,hpc,nnd.
e	TR239	155.mg	1.45gm	1/50	67.5mg	4/50	139.mg	10/50			lun:a/a,a/c.
615	TR239	73.4mg	693.mg	13/50	68.8mg	23/50	140.mg	27/50			liv:hpa,hpc.
a	TR239	117.mg	1.61gm	6/50	68.8mg	13/50	140.mg	17/50			
b	TR239	121.mg	n.s.s.	6/50	68.8mg	15/50	140.mg	13/50			lun:a/a,a/c.
c	TR239	138.mg	n.s.s.	5/50	68.8mg	13/50	140.mg	11/50			
d	TR239	55.0mg	n.s.s.	30/50	68.8mg	38/50	140.mg	40/50			
e	TR239	73.4mg	693.mg	13/50	68.8mg	23/50	140.mg	27/50			liv:hpa,hpc,nnd.
f	TR239	121.mg	n.s.s.	6/50	68.8mg	15/50	140.mg	13/50			lun:a/a,a/c.
616	TR191	30.7mg	907.mg	39/50	70.1mg	32/50	140.mg	15/50			
a	TR191	n.s.s.	n.s.s.	0/50	70.1mg	0/50	140.mg	0/50			liv:hpa,hpc,nnd.
617	TR191	45.3mg	n.s.s.	30/50	70.7mg	30/50	141.mg	18/50			
a	TR191	366.mg	n.s.s.	1/50	70.7mg	1/50	141.mg	0/50			liv:hpa,hpc,nnd.

```
        Spe Strain Site   Xpo+Xpt                                                    TD50    2Tailpvl
          Sex   Route   Hist     Notes                                                    DR      AuOp
BIS-2-CHLOROETHYLETHER        100ng..:...1ug...:...10.....:...100....:...1mg...:...10....:...100...:...1g....:...10
618 M f b6a orl liv hpt 79w79 evx                                        .>              no dre  P=1.
a   M f b6a orl lun mix 79w79 evx                                                        no dre  P=1.
b   M f b6a orl tba mix 79w79 evx                                                        no dre  P=1.
619 M m b6a orl liv hpt 79w79 evx                                   . + .                20.2mg  P<.0005+
a   M m b6a orl lun ade 79w79 evx                                                        no dre  P=1.
b   M m b6a orl tba mix 79w79 evx                                                        21.6mg  P<.01
620 M f b6c orl liv hpt 79w79 evx                                        . ±             65.0mg  P<.02
a   M f b6c orl lun ade 79w79 evx                                                        no dre  P=1.
b   M f b6c orl tba mix 79w79 evx                                                        122.mg  P<.4
621 M m b6c orl liv hpt 79w79 evx                                 . + .                  8.19mg  P<.0005+
a   M m b6c orl lun ade 79w79 evx                                                        no dre  P=1.
b   M m b6c orl tba mix 79w79 evx                                                        noTD50  P<.007

BIS-1,4-(CHLOROMETHOXY)BUTANE  100ng..:...1ug...:...10.....:...100....:...1mg...:...10....:...100...:...1g....:...10
622 M f hic ipj abd sar 81w81                               .>                           no dre  P=1.  -

BIS-1,2-(CHLOROMETHOXY)ETHANE  100ng..:...1ug...:...10.....:...100....:...1mg...:...10....:...100...:...1g....:...10
623 M f hic ipj abd mix 78w78                          . ±                               4.62mg  P<.02  +

BIS-1,6-(CHLOROMETHOXY)HEXANE  100ng..:...1ug...:...10.....:...100....:...1mg...:...10....:...100...:...1g....:...10
624 M f hic ipj abd sar 81w81                             .>                             no dre  P=1.  -

BIS-1,4-(CHLOROMETHOXY)-p-XYLENE 100ng..:...1ug...:...10.....:...100....:...1mg...:...10....:...100...:...1g....:...10
625 M f hic ipj abd sar 77w77                         . ±                                3.11mg  P<.1  +

BIS-(CHLOROMETHYL)ETHER       100ng..:...1ug...:...10.....:...100....:...1mg...:...10....:...100...:...1g....:...10
626 M f hic ipj abd sar 60w60                      . ±                                   .182mg  P<.02  +
627 M m hic inh lun mix 6m25 es                 .>                                       1.16mg * P<.4  +
a   M m hic inh lun ade 6m25 es                                                          1.26mg * P<.3  +
b   M m hic inh liv mix 6m25 es                                                          no dre  P=1.  -
c   M m hic inh liv hpc 6m25 es                                                          no dre  P=1.  -
628 R m sss inh nap ene 6m28 e          .+.                                              3.57ug Z P<.0005+
a   R m sss inh lun ade 6m28 e                                                           .196mg * P<.002 +
b   R m sss inh nap car 6m28 e                                                           .795mg * P<.2  +
c   R m sss inh liv bht 6m28 e                                                           no dre  P=1.  -
d   R m sss inh tba mix 6m28 e                                                           2.34ug * P<.0005

BIS(2,3-DIBROMOPROPYL)PHOSPHATE, MAGNESIUM SALT...:...10....:...100....:...1mg...:...10....:...100...:...1g....:...10
629 R f wis eat liv mix 52w52 ek                          . +                            11.8mg * P<.0005
a   R f wis eat liv hpa 52w52 ek                                                         33.2mg * P<.002
b   R f wis eat liv hpc 52w52 ek                                                         35.1mg * P<.009
c   R f wis eat for sqp 52w52 ek                                                         47.6mg * P<.006
630 R f wis eat liv mix 78w78 ek                               . +                       11.9mg * P<.0005
a   R f wis eat for mix 78w78 ek                                                         19.6mg * P<.0005
b   R f wis eat liv hpc 78w78 ek                                                         20.1mg * P<.0005
c   R f wis eat for sqp 78w78 ek                                                         34.5mg * P<.0005
d   R f wis eat eso sqp 78w78 ek                                                         107.mg * P<.006
e   R f wis eat smi adc 78w78 ek                                                         107.mg * P<.006
f   R f wis eat ton sqp 78w78 ek                                                         107.mg * P<.006
g   R f wis eat liv hpa 78w78 ek                                                         20.6mg Z P<.02
h   R f wis eat for sqc 78w78 ek                                                         171.mg * P<.03
631 R f wis eat liv mix 22m23 ae                               . +  .                    27.8mg * P<.0005+
a   R f wis eat liv hpc 22m23 ae                                                         37.1mg * P<.0005+
b   R f wis eat for mix 22m23 ae                                                         40.0mg * P<.0005+
c   R f wis eat for sqp 22m23 ae                                                         60.3mg * P<.0005+
d   R f wis eat smi adc 22m23 ae                                                         179.mg * P<.0005+
e   R f wis eat eso sqp 22m23 ae                                                         394.mg * P<.008 +
f   R f wis eat for sqc 22m23 ae                                                         301.mg * P<.08  +
g   R f wis eat ton mix 22m23 ae                                                         301.mg * P<.08
h   R f wis eat ton sqp 22m23 ae                                                         403.mg * P<.05
i   R f wis eat kid mix 22m23 ae                                                         608.mg * P<.03
j   R f wis eat kid rca 22m23 ae                                                         1.25gm * P<.2
k   R f wis eat kid rcc 22m23 ae                                                         1.25gm * P<.2
l   R f wis eat ton sqc 22m23 ae                                                         no dre  P=1.
m   R f wis eat liv hpa 22m23 ae                                                         no dre  P=1.
632 R m wis eat smi mix 52w52 ek                         . +      .                      38.1mg * P<.006
a   R m wis eat for mix 52w52 ek                                                         38.1mg * P<.006
b   R m wis eat smi adc 52w52 ek                                                         60.9mg * P<.03
c   R m wis eat for sqp 52w52 ek                                                         60.9mg * P<.03
d   R m wis eat for sqc 52w52 ek                                                         129.mg * P<.2
e   R m wis eat kid rcc 52w52 ek                                                         129.mg * P<.2
f   R m wis eat liv hpc 52w52 ek                                                         129.mg * P<.2
g   R m wis eat liv mix 52w52 ek                                                         129.mg * P<.2
h   R m wis eat smi ade 52w52 ek                                                         129.mg * P<.2
i   R m wis eat liv hpa 52w52 ek                                                         no dre  P=1.
j   R m wis eat eso sqp 52w52 ek                                                         no dre  P=1.
633 R m wis eat eso sqp 78w78 ek                               . ±                       16.5mg * P<.02
a   R m wis eat liv hpa 78w78 ek                                                         28.3mg * P<.5
b   R m wis eat for sqp 78w78 ek                                                         56.1mg * P<.2
c   R m wis eat kid rca 78w78 ek                                                         56.1mg * P<.2
```

	RefNum	LoConf	UpConf	Cntrl	1Dose	1Inc	2Dose	2Inc		Citation or Pathology
										Brkly Code

BIS-2-CHLOROETHYLETHER   111-44-4

	RefNum	LoConf	UpConf	Cntrl	1Dose	1Inc	2Dose	2Inc	Citation or Pathology
618	1227	88.4mg	n.s.s.	0/15	41.3mg	0/18			Innes;ntis,1968/1969
a	1227	88.4mg	n.s.s.	0/15	41.3mg	0/18			
b	1227	53.7mg	n.s.s.	1/15	41.3mg	1/18			
619	1227	9.25mg	56.1mg	0/18	38.4mg	9/17			
a	1227	40.0mg	n.s.s.	3/18	38.4mg	2/17			
b	1227	8.80mg	1.09gm	3/18	38.4mg	10/17			
620	1227	22.4mg	n.s.s.	0/18	41.3mg	4/18			
a	1227'	88.4mg	n.s.s.	1/18	41.3mg	0/18			
b	1227	25.6mg	n.s.s.	2/18	41.3mg	4/18			
621	1227	3.46mg	27.6mg	3/15	38.4mg	14/16			
a	1227	73.1mg	n.s.s.	2/15	38.4mg	0/16			
b	1227	n.s.s.	12.7mg	5/15	38.4mg	16/16			

BIS-1,4-(CHLOROMETHOXY)BUTANE   13483-19-7

	RefNum	LoConf	UpConf	Cntrl	1Dose	1Inc	Citation or Pathology
622	582	2.14mg	n.s.s.	0/30	.571mg	0/30	Van Duuren;canr,35,2553-2557;1975

BIS-1,2-(CHLOROMETHOXY)ETHANE   13483-18-6

	RefNum	LoConf	UpConf	Cntrl	1Dose	1Inc	Citation or Pathology
623	582	1.59mg	n.s.s.	0/30	1.71mg	4/30	Van Duuren;canr,35,2553-2557;1975

BIS-1,6-(CHLOROMETHOXY)HEXANE   56894-92-9

	RefNum	LoConf	UpConf	Cntrl	1Dose	1Inc	Citation or Pathology
624	582	6.43mg	n.s.s.	0/30	1.71mg	0/30	Van Duuren;canr,35,2553-2557;1975

BIS-1,4-(CHLOROMETHOXY)-p-XYLENE   56894-91-8

	RefNum	LoConf	UpConf	Cntrl	1Dose	1Inc	Citation or Pathology
625	582	.765mg	n.s.s.	0/30	.571mg	2/30	Van Duuren;canr,35,2553-2557;1975

BIS-(CHLOROMETHYL)ETHER   (BCME) 542-88-1

	RefNum	LoConf	UpConf	Cntrl	1Dose	1Inc	2Dose	2Inc			Citation or Pathology
626	582	62.8ug	n.s.s.	0/30	.114mg	4/30					Van Duuren;canr,35,2553-2557;1975
627	1086	.253mg	n.s.s.	10/157	295.ng	7/138	2.95ug	3/143	33.6ug	10/144	Leong;txap,58,269-281;1981
a	1086	.291mg	n.s.s.	6/157	295.ng	4/138	2.95ug	2/143	33.6ug	7/144	
b	1086	1.17mg	n.s.s.	4/157	295.ng	1/138	2.95ug	0/143	33.6ug	0/144	
c	1086	1.17mg	n.s.s.	3/157	295.ng	1/138	2.95ug	0/143	33.6ug	0/144	
628	1086	2.63ug	4.90ug	0/104	52.8ng	0/105	528.ng	0/103	7.78ug	96/103	
a	1086	67.7ug	1.09mg	0/104	52.8ng	0/105	528.ng	0/103	7.78ug	4/103	
b	1086	.129mg	n.s.s.	0/104	52.8ng	0/105	528.ng	0/103	7.78ug	1/103	
c	1086	.234mg	n.s.s.	0/104	52.8ng	1/105	528.ng	0/103	7.78ug	0/103	
d	1086	1.43ug	3.88ug	56/104	52.8ng	48/105	528.ng	43/103	7.78ug	102/103	

BIS(2,3-DIBROMOPROPYL)PHOSPHATE, MAGNESIUM SALT   36711-31-6

	RefNum	LoConf	UpConf	Cntrl	1Dose	1Inc	2Dose	2Inc	3Dose	3Inc	Citation or Pathology
629	2102m	4.84mg	36.2mg	0/8	4.00mg	0/8	20.0mg	1/8	100.mg	7/8	Takada;japt,11,323-331;1991/pers.comm.
a	2102m	11.2mg	185.mg	0/8	4.00mg	0/8	20.0mg	0/8	100.mg	4/8	
b	2102m	12.0mg	1.19gm	0/8	4.00mg	0/8	20.0mg	1/8	100.mg	3/8	
c	2102m	14.2mg	686.mg	0/8	4.00mg	0/8	20.0mg	0/8	100.mg	3/8	
630	2102n	4.87mg	34.9mg	0/8	4.00mg	1/8	20.0mg	3/8	100.mg	8/8	
a	2102n	7.79mg	58.3mg	0/8	4.00mg	0/8	20.0mg	1/8	100.mg	8/8	
b	2102n	8.06mg	64.1mg	0/8	4.00mg	1/8	20.0mg	0/8	100.mg	8/8	
c	2102n	14.0mg	115.mg	0/8	4.00mg	0/8	20.0mg	1/8	100.mg	6/8	
d	2102n	32.0mg	1.54gm	0/8	4.00mg	0/8	20.0mg	0/8	100.mg	3/8	
e	2102n	32.0mg	1.54gm	0/8	4.00mg	0/8	20.0mg	0/8	100.mg	3/8	
f	2102n	32.0mg	1.54gm	0/8	4.00mg	0/8	20.0mg	0/8	100.mg	3/8	
g	2102n	6.14mg	n.s.s.	0/8	4.00mg	0/8	20.0mg	3/8	(100.mg	0/8)	
h	2102n	41.8mg	n.s.s.	0/8	4.00mg	0/8	20.0mg	0/8	100.mg	2/8	
631	2102o	14.7mg	65.0mg	1/16	4.00mg	2/16	20.0mg	8/16	100.mg	14/16	
a	2102o	20.3mg	74.5mg	0/16	4.00mg	0/16	20.0mg	6/16	100.mg	13/16	
b	2102o	21.7mg	81.6mg	0/16	4.00mg	0/16	20.0mg	5/16	100.mg	13/16	
c	2102o	31.4mg	134.mg	0/16	4.00mg	0/16	20.0mg	3/16	100.mg	11/16	
d	2102o	72.7mg	644.mg	0/16	4.00mg	0/16	20.0mg	0/16	100.mg	6/16	
e	2102o	119.mg	8.01gm	0/16	4.00mg	0/16	20.0mg	0/16	100.mg	3/16	
f	2102o	104.mg	n.s.s.	0/16	4.00mg	0/16	20.0mg	2/16	100.mg	2/16	
g	2102o	104.mg	n.s.s.	0/16	4.00mg	0/16	20.0mg	2/16	100.mg	2/16	
h	2102o	122.mg	n.s.s.	0/16	4.00mg	0/16	20.0mg	1/16	100.mg	2/16	
i	2102o	149.mg	n.s.s.	0/16	4.00mg	0/16	20.0mg	0/16	100.mg	2/16	
j	2102o	203.mg	n.s.s.	0/16	4.00mg	0/16	20.0mg	0/16	100.mg	1/16	
k	2102o	203.mg	n.s.s.	0/16	4.00mg	0/16	20.0mg	0/16	100.mg	1/16	
l	2102o	227.mg	n.s.s.	0/16	4.00mg	0/16	20.0mg	1/16	100.mg	0/16	
m	2102o	189.mg	n.s.s.	1/16	4.00mg	2/16	20.0mg	2/16	100.mg	1/16	
632	2102m	11.4mg	549.mg	0/8	3.20mg	0/8	16.0mg	0/8	80.0mg	3/8	
a	2102m	11.4mg	549.mg	0/8	3.20mg	0/8	16.0mg	0/8	80.0mg	3/8	
b	2102m	14.9mg	n.s.s.	0/8	3.20mg	0/8	16.0mg	0/8	80.0mg	2/8	
c	2102m	14.9mg	n.s.s.	0/8	3.20mg	0/8	16.0mg	0/8	80.0mg	2/8	
d	2102m	21.0mg	n.s.s.	0/8	3.20mg	0/8	16.0mg	0/8	80.0mg	1/8	
e	2102m	21.0mg	n.s.s.	0/8	3.20mg	0/8	16.0mg	0/8	80.0mg	1/8	
f	2102m	21.0mg	n.s.s.	0/8	3.20mg	0/8	16.0mg	0/8	80.0mg	1/8	
g	2102m	21.0mg	n.s.s.	0/8	3.20mg	0/8	16.0mg	0/8	80.0mg	1/8	
h	2102m	21.0mg	n.s.s.	0/8	3.20mg	0/8	16.0mg	0/8	80.0mg	1/8	
i	2102m	1.06mg	n.s.s.	0/8	3.20mg	0/8	16.0mg	0/8	80.0mg	0/8	
j	2102m	1.06mg	n.s.s.	0/8	3.20mg	0/8	16.0mg	0/8	80.0mg	0/8	
633	2102n	4.92mg	n.s.s.	0/8	3.20mg	0/8	16.0mg	3/8			
a	2102n	5.00mg	n.s.s.	2/8	3.20mg	1/8	16.0mg	3/8			
b	2102n	9.11mg	n.s.s.	0/8	3.20mg	0/8	16.0mg	1/8			
c	2102n	9.11mg	n.s.s.	0/8	3.20mg	0/8	16.0mg	1/8			

	Spe	Strain	Site	Xpo+Xpt					TD50	2Tailpvl
		Sex	Route	Hist	Notes					DR	AuOp

```
         Spe Strain  Site    Xpo+Xpt                                              TD50      2Tailpvl
             Sex  Route  Hist      Notes                                                 DR    AuOp
-------------------------------------------------------------------------------------------------------
  d      R m wis eat ton sqc 78w78 ek                                            56.1mg  * P<.2
634      R m wis eat kid mix 21m24 ae                      .      +      .       35.0mg  Z P<.003
  a      R m wis eat for mix 21m24 ae                                            37.8mg  * P<.0005+
  b      R m wis eat for sqp 21m24 ae                                            43.5mg  * P<.0005+
  c      R m wis eat eso mix 21m24 ae                                            45.1mg  Z P<.006 +
  d      R m wis eat kid rca 21m24 ae                                            45.1mg  Z P<.006
  e      R m wis eat smi adc 21m24 ae                                            79.0mg  * P<.0005+
  f      R m wis eat eso sqp 21m24 ae                                            62.1mg  Z P<.02  +
  g      R m wis eat for sqc 21m24 ae                                            979.mg  * P<.5
  h      R m wis eat eso sqc 21m24 ae                                            no dre    P=1.
  i      R m wis eat kid rcc 21m24 ae                                            no dre    P=1.
  j      R m wis eat liv hpc 21m24 ae                                            no dre    P=1.
  k      R m wis eat liv mix 21m24 ae                                            no dre    P=1.
  l      R m wis eat liv hpa 21m24 ae                                            no dre    P=1.
  m      R m wis eat ton sqp 21m24 ae                                            no dre    P=1.

4-BIS(2-HYDROXYETHYL)AMINO-2-(5-NITRO-2-THIENYL)QUINAZOLINE..:..1_00...:..1mg....:..10...:..100...:..1g......:.10
635      R m sda eat smi sar 50w52 e                       .  +  .              3.14mg    P<.0005+
  a      R m sda eat smi lei 50w52 e                                            3.60mg    P<.0005
  b      R m sda eat mgl mix 50w52 e                                            6.45mg    P<.0005+
  c      R m sda eat mgl adc 50w52 e                                            11.5mg    P<.007 +
  d      R m sda eat tba mix 50w52 e                                            1.10mg    P<.0005

4-BIS(2-HYDROXYETHYL)AMINO-2-(2-THIENYL)QUINAZOLINE  ..10.....:..100.....:..1mg....:..10.....:..100...:..1g......:.10
636      R f sda eat tba mix 46w66 e                                .>          60.1mg    P<.3

BIS-2-HYDROXYETHYLDITHIOCARBAMIC ACID, POTASSIUM...:..10.....:..100.....:..1mg....:..10.....:..100...:..1g......:.10
637      M f b6a orl liv hpt 79w79 evx                   .      ±               298.mg    P<.04
  a      M f b6a orl lun ade 79w79 evx                                          850.mg    P<.6
  b      M f b6a orl tba mix 79w79 evx                                          138.mg    P<.04
638      M m b6a orl liv hpt 79w79 evx            .   +   .                     41.5mg    P<.0005+
  a      M m b6a orl lun ade 79w79 evx                                          no dre    P=1.
  b      M m b6a orl tba mix 79w79 evx                                          45.6mg    P<.0005
639      M f b6c orl liv hpt 79w79 evx              .  +  .                      56.4mg    P<.0005
  a      M f b6c orl lun ade 79w79 evx                                          1.08gm    P<.3
  b      M f b6c orl tba mix 79w79 evx                                          48.4mg    P<.0005
640      M m b6c orl liv hpt 79w79 evx             .  +  .                      34.5mg    P<.0005+
  a      M m b6c orl lun mix 79w79 evx                                          no dre    P=1.
  b      M m b6c orl tba mix 79w79 evx                                          27.8mg    P<.0005

BIS(2-HYDROXYPROPYL)AMINE         100ng..:..1ug...:..10.....:..100...:..1mg....:..10.....:..100...:..1g......:.10
641      R m wis eat liv hpc 94w94                                     .>        no dre    P=1.  -
  a      R m wis eat lun tum 94w94                                              no dre    P=1.  -
  b      R m wis eat nas tum 94w94                                              no dre    P=1.  -

BIS(TRI-N-BUTYLTIN)OXIDE, TECHNICAL GRADE ..1ug...:..10.....:..100.....:..1mg....:..10.....:..100...:..1g......:.10
642      R f wis eat amd phe 25m25 e                 .  +  .                     1.79mg  * P<.0005
  a      R f wis eat pta tum 25m25 e                                            3.73mg  * P<.03
643      R m wis eat amd phe 25m25 e               .  +  .                       1.93mg  * P<.0005
  a      R m wis eat pta tum 25m25 e                                            1.90mg  * P<.02

BISMUTH DIMETHYLDITHIOCARBAMATE   100ng..:..1ug...:..10.....:..100.....:..1mg....:..10.....:..100...:..1g......:.10
644      M f b6a orl lun ade 76w76 evx                        .>                 26.3mg    P<.6  -
  a      M f b6a orl liv hpt 76w76 evx                                          no dre    P=1.  -
  b      M f b6a orl tba mix 76w76 evx                                          no dre    P=1.  -
645      M m b6a orl lun ade 76w76 evx                          .>               no dre    P=1.  -
  a      M m b6a orl liv hpt 76w76 evx                                          no dre    P=1.  -
  b      M m b6a orl tba mix 76w76 evx                                          no dre    P=1.  -
646      M f b6c orl liv hpt 76w76 evx                        .>                 no dre    P=1.  -
  a      M f b6c orl lun mix 76w76 evx                                          no dre    P=1.  -
  b      M f b6c orl tba tum 76w76 evx                                          no dre    P=1.  -
647      M m b6c orl lun ade 76w76 evx                .      ±                   12.6mg    P<.1  -
  a      M m b6c orl liv agm 76w76 evx                                          26.0mg    P<.3  -
  b      M m b6c orl tba mix 76w76 evx                                          4.53mg    P<.007 -

BISMUTH OXYCHLORIDE               100ng..:..1ug...:..10.....:..100...:..1mg....:..10.....:..100...:..1g......:.10
648      R b bdr eat mgl fba 24m28                                              150.gm  * P<.8  -
  a      R b bdr eat mgl car 24m28                                              no dre    P=1.  -
  b      R b bdr eat hpl ade 24m28                                              no dre    P=1.  -

BISPHENOL A                       100ng..:..1ug...:..10.....:..100...:..1mg....:..10.....:..100...:..1g......:.10
649      M f b6c eat TBA MXB 24m25                                      :>       no dre    P=1.  -
  a      M f b6c eat liv MXB 24m25                                              13.1gm  * P<.05
  b      M f b6c eat lun MXB 24m25                                              41.4gm  * P<.6
650      M m b6c eat --- MXA 24m25                                 :    ±       #445.mg  \ P<.03 -
  a      M m b6c eat --- lym 24m25                                              508.mg  \ P<.05
  b      M m b6c eat pit crc 24m25                                              6.69mg  * P<.02
  c      M m b6c eat TBA MXB 24m25                                              1.32kg    P<1.
  d      M m b6c eat liv MXB 24m25                                              no dre    P=1.
  e      M m b6c eat lun MXB 24m25                                              no dre    P=1.
651      R f f34 eat TBA MXB 24m25                               :>             no dre    P=1.  -
```

	RefNum	LoConf	UpConf	Cntrl	1Dose	1Inc	2Dose	2Inc	3Dose	3Inc	Citation or Pathology / Brkly Code
d	2102n	9.11mg	n.s.s.	0/8	3.20mg	0/8	16.0mg	1/8			
634	2102o	13.2mg	178.mg	0/16	3.20mg	0/16	16.0mg	5/16	(80.0mg	0/24)	
a	2102o	22.5mg	72.1mg	0/16	3.20mg	0/16	16.0mg	8/16	80.0mg	16/24	
b	2102o	25.5mg	85.9mg	0/16	3.20mg	0/16	16.0mg	7/16	80.0mg	15/24	
c	2102o	15.5mg	492.mg	0/16	3.20mg	0/16	16.0mg	4/16	(80.0mg	2/24)	
d	2102o	15.5mg	492.mg	0/16	3.20mg	0/16	16.0mg	4/16	(80.0mg	0/24)	
e	2102o	42.0mg	175.mg	0/16	3.20mg	0/16	16.0mg	2/16	80.0mg	12/24	
f	2102o	18.7mg	n.s.s.	0/16	3.20mg	0/16	16.0mg	3/16	(80.0mg	2/24)	
g	2102o	178.mg	n.s.s.	0/16	3.20mg	0/16	16.0mg	1/16	80.0mg	1/24	
h	2102o	274.mg	n.s.s.	0/16	3.20mg	0/16	16.0mg	1/16	80.0mg	0/24	
i	2102o	274.mg	n.s.s.	0/16	3.20mg	0/16	16.0mg	1/16	80.0mg	0/24	
j	2102o	245.mg	n.s.s.	1/16	3.20mg	2/16	16.0mg	2/16	80.0mg	1/24	
k	2102o	215.mg	n.s.s.	6/16	3.20mg	4/16	16.0mg	6/16	80.0mg	3/24	
l	2102o	236.mg	n.s.s.	5/16	3.20mg	2/16	16.0mg	4/16	80.0mg	2/24	
m	2102o	274.mg	n.s.s.	0/16	3.20mg	0/16	16.0mg	1/16	80.0mg	0/24	

4-BIS(2-HYDROXYETHYL)AMINO-2-(5-NITRO-2-THIENYL)QUINAZOLINE   33372-39-3

	RefNum	LoConf	UpConf	Cntrl	1Dose	1Inc	Citation
635	1390	1.59mg	7.25mg	0/20	19.2mg	13/20	Cohen;jnci,57,277-282;1976
a	1390	1.79mg	8.58mg	0/20	19.2mg	12/20	
b	1390	2.88mg	20.4mg	0/20	19.2mg	8/20	
c	1390	4.32mg	132.mg	0/20	19.2mg	5/20	
d	1390	.452mg	2.52mg	0/20	19.2mg	19/20	

4-BIS(2-HYDROXYETHYL)AMINO-2-(2-THIENYL)QUINAZOLINE   58139-47-2

	RefNum	LoConf	UpConf	Cntrl	1Dose	1Inc	Citation
636	1390	12.7mg	n.s.s.	6/84	17.4mg	4/28	Cohen;jnci,57,277-282;1976

BIS-2-HYDROXYETHYLDITHIOCARBAMIC ACID, POTASSIUM   23746-34-1

	RefNum	LoConf	UpConf	Cntrl	1Dose	1Inc	Citation
637	1217	90.0mg	n.s.s.	0/17	157.mg	3/16	Innes;ntis,1968/1969
a	1217	124.mg	n.s.s.	1/17	157.mg	2/16	
b	1217	49.5mg	n.s.s.	2/17	157.mg	7/16	
638	1217	19.7mg	110.mg	1/18	146.mg	13/17	
a	1217	194.mg	n.s.s.	2/18	146.mg	1/17	
b	1217	20.5mg	170.mg	3/18	146.mg	13/17	
639	1217	27.7mg	136.mg	0/16	157.mg	12/18	
a	1217	176.mg	n.s.s.	0/16	157.mg	1/18	
b	1217	23.9mg	113.mg	0/16	157.mg	13/18	
640	1217	16.2mg	82.7mg	0/16	146.mg	13/16	
a	1217	278.mg	n.s.s.	0/16	146.mg	0/16	
b	1217	12.5mg	66.5mg	0/16	146.mg	14/16	

BIS(2-HYDROXYPROPYL)AMINE   110-97-4

	RefNum	LoConf	UpConf	Cntrl	1Dose	1Inc	Citation
641	1914	1.35gm	n.s.s.	0/20	400.mg	0/20	Yamamoto;carc,10,1607-1611;1989
a	1914	1.35gm	n.s.s.	0/20	400.mg	0/20	
b	1914	1.35gm	n.s.s.	0/20	400.mg	0/20	

BIS(TRI-N-BUTYLTIN)OXIDE, TECHNICAL GRADE   56-35-9

	RefNum	LoConf	UpConf	Cntrl	1Dose	1Inc	2Dose	2Inc	3Dose	3Inc	Citation
642	1975	1.15mg	3.04mg	3/50	25.0ug	3/50	.250mg	3/50	2.50mg	34/50	Wester;fctx,28,179-196;1990/pers.comm.
a	1975	1.53mg	n.s.s.	22/50	25.0ug	32/50	.250mg	22/50	2.50mg	35/50	
643	1975	1.06mg	5.03mg	16/50	20.0ug	13/50	.200mg	14/50	2.00mg	33/50	
a	1975	.774mg	n.s.s.	34/50	20.0ug	39/50	.200mg	29/50	2.00mg	43/50	

BISMUTH DIMETHYLDITHIOCARBAMATE (bismate)   21260-46-8

	RefNum	LoConf	UpConf	Cntrl	1Dose	1Inc	Citation
644	1218	3.66mg	n.s.s.	1/17	4.64mg	2/17	Innes;ntis,1968/1969
a	1218	8.68mg	n.s.s.	0/17	4.64mg	0/17	
b	1218	4.08mg	n.s.s.	2/17	4.64mg	2/17	
645	1218	4.03mg	n.s.s.	2/18	4.31mg	2/18	
a	1218	5.10mg	n.s.s.	1/18	4.31mg	1/18	
b	1218	3.39mg	n.s.s.	3/18	4.31mg	3/18	
646	1218	9.19mg	n.s.s.	0/16	4.64mg	0/18	
a	1218	9.19mg	n.s.s.	0/16	4.64mg	0/18	
b	1218	9.19mg	n.s.s.	0/16	4.64mg	0/18	
647	1218	3.10mg	n.s.s.	0/16	4.31mg	2/17	
a	1218	4.24mg	n.s.s.	0/16	4.31mg	1/17	
b	1218	1.71mg	61.2mg	0/16	4.31mg	5/17	

BISMUTH OXYCHLORIDE   7787-59-9

	RefNum	LoConf	UpConf	Cntrl	1Dose	1Inc	2Dose	2Inc	3Dose	3Inc	Citation
648	432	11.9mg	n.s.s.	2/60	391.mg	2/40	802.mg	1/40	1.98gm	2/40	Preussmann;fctx,13,543-544;1975
a	432	2.55gm	n.s.s.	1/60	391.mg	0/40	802.mg	0/40	1.98gm	0/40	
b	432	22.0gm	n.s.s.	2/60	391.mg	0/40	802.mg	0/40	1.98gm	1/40	

BISPHENOL A (4,4'-isopropylidenediphenol)   80-05-7

	RefNum	LoConf	UpConf	Cntrl	1Dose	1Inc	2Dose	2Inc	Citation/Pathology
649	TR215	2.27gm	n.s.s.	21/50	626.mg	17/50	1.25gm	19/50	
a	TR215	4.53gm	n.s.s.	0/50	626.mg	1/50	1.25gm	3/50	liv:hpa,hpc,nnd.
b	TR215	6.29gm	n.s.s.	1/50	626.mg	1/50	1.25gm	2/50	lun:a/a,a/c.
650	TR215	179.mg	n.s.s.	2/50	116.mg	9/50	(578.mg	5/50)	---:leu,lym. S
a	TR215	192.mg	n.s.s.	2/50	116.mg	8/50	(578.mg	3/50)	S
b	TR215	2.02gm	n.s.s.	0/50	116.mg	0/50	578.mg	3/50	S
c	TR215	701.mg	n.s.s.	23/50	116.mg	28/50	578.mg	24/50	
d	TR215	1.61gm	n.s.s.	16/50	116.mg	14/50	578.mg	10/50	liv:hpa,hpc,nnd.
e	TR215	2.31gm	n.s.s.	7/50	116.mg	2/50	578.mg	4/50	lun:a/a,a/c.
651	TR215	115.mg	n.s.s.	47/50	47.7mg	45/50	95.4mg	38/50	

```
     Spe Strain  Site    Xpo+Xpt                                                          TD50     2Tailpvl
         Sex  Route  Hist   Notes                                                                  DR    AuOp
 a   R f f34 eat liv MXB 24m25                                                            595.mg \ P<.6
652  R m f34 eat mgl fba 24m25                                                 :      ±   #675.mg * P<.03  -
 a   R m f34 eat TBA MXB 24m25                                                            no dre  P=1.
 b   R m f34 eat liv MXB 24m25                                                            no dre  P=1.

BLACK PN                        100ng..:..1ug...:..10......:..100....:..1mg...:..10.....:..100...:..1g...:..10
653  M f cws eat liv tum 80w80 e                                             .>          no dre  P=1.   -
 a   M f cws eat lun ade 80w80 e                                                          no dre  P=1.   -
654  M m cws eat lun ade 80w80 e                                                 .>       3.66gm * P<.3  -
 a   M m cws eat liv tum 80w80 e                                                          no dre  P=1.   -
655  R f cfe eat liv tum 24m24 es                                        .>              no dre  P=1.   -
656  R m cfe eat liv tum 24m24 es                                        .>              no dre  P=1.   -

C.I. DIRECT BLACK 38            100ng..:..1ug...:..10......:..100....:..1mg...:..10.....:..100...:..1g...:..10
657  M b icm wat liv hpt 57w57                                       . + .               71.6mg   P<.0005+
 a   M b icm wat mgl mix 57w57                                                            262.mg   P<.0005+
 b   M b icm wat tba mix 57w57                                                            32.0mg   P<.0005
658  R f f34 eat liv nnd 13w13                            :   (+)   :                     2.61mg * P<.003 c
 a   R f f34 eat TBA MXB 13w13                                                            2.61mg * P<.003
 b   R f f34 eat liv MXB 13w13                                                            2.61mg * P<.003
659  R m f34 eat liv MXA 13w13                        : (+)   :                           .945mg Z P<.0005c
 a   R m f34 eat liv hpc 13w13                                                            2.22mg * P<.004
 b   R m f34 eat TBA MXB 13w13                                                            .945mg Z P<.0005
 c   R m f34 eat liv MXB 13w13                                                            .945mg Z P<.0005

C.I. DIRECT BLUE 6              100ng..:..1ug...:..10......:..100....:..1mg...:..10.....:..100...:..1g...:..10
660  R f f34 eat liv MXA 13w13                            : (+)   :                       3.21mg Z P<.0005c
 a   R f f34 eat liv hpc 13w13                                                            5.97mg * P<.003
 b   R f f34 eat TBA MXB 13w13                                                            3.21mg   P<.0005
 c   R f f34 eat liv MXB 13w13                                                            3.21mg Z P<.0005
661  R m f34 eat liv MXA 13w13                        :(+)   :                            1.18mg Z P<.0005c
 a   R m f34 eat TBA MXB 13w13                                                            1.18mg Z P<.0005
 b   R m f34 eat liv MXB 13w13                                                            1.18mg Z P<.0005

C.I. DIRECT BLUE 15             100ng..:..1ug...:..10......:..100....:..1mg...:..10.....:..100...:..1g...:..10
662  R f f34 wat MXB MXB 96w97                                       : + :                21.3mg * P<.0005

 a   R f f34 wat cli MXA 96w97                                                            46.2mg * P<.0005c
 b   R f f34 wat mul mnl 96w97                                                            50.0mg * P<.0005c
 c   R f f34 wat MXA MXA 96w97                                                            81.0mg * P<.0005c
 d   R f f34 wat cli MXA 96w97                                                            90.1mg * P<.0005
 e   R f f34 wat zym MXA 96w97                                                            104.mg * P<.0005c
 f   R f f34 wat cli MXA 96w97                                                            107.mg * P<.0005
 g   R f f34 wat ute MXA 96w97                                                            122.mg * P<.004
 h   R f f34 wat ute esp 96w97                                                            132.mg * P<.008
 i   R f f34 wat MXA sqp 96w97                                                            132.mg * P<.0005
 j   R f f34 wat zym car 96w97                                                            150.mg * P<.0005
 k   R f f34 wat pal sqp 96w97                                                            157.mg * P<.0005
 l   R f f34 wat ski MXA 96w97                                                            158.mg * P<.0005c
 m   R f f34 wat ski sqp 96w97                                                            174.mg * P<.0005
 n   R f f34 wat MXA sqc 96w97                                                            208.mg * P<.0005
 o   R f f34 wat ton MXA 96w97                                                            238.mg * P<.002
 p   R f f34 wat zym ade 96w97                                                            321.mg * P<.002
 q   R f f34 wat ton sqc 96w97                                                            384.mg * P<.003
 r   R f f34 wat liv MXA 96w97                                                            394.mg * P<.0005c
 s   R f f34 wat liv nnd 96w97                                                            429.mg * P<.002
 t   R f f34 wat pal sqc 96w97                                                            456.mg * P<.003
 u   R f f34 wat MXA adp 96w97                                                            491.mg * P<.02  c
 v   R f f34 wat ute MXA 96w97                                                            493.mg Z P<.02  c
 w   R f f34 wat MXA MXA 96w97                                                            997.mg * P<.02  c
 x   R f f34 wat jej MXA 96w97                                                            1.18gm * P<.02
 y   R f f34 wat TBA MXB 96w97                                                            20.4mg * P<.0005
 z   R f f34 wat liv MXB 96w97                                                            394.mg * P<.0005
663  R m f34 wat MXB MXB 97w97                                        :+ :                 11.8mg * P<.0005

 a   R m f34 wat tes MXA 97w97                                                            12.9mg * P<.0005
 b   R m f34 wat ski MXA 97w97                                                            18.1mg Z P<.0005
 c   R m f34 wat ski MXA 97w97                                                            19.6mg Z P<.0005c
 d   R m f34 wat ski bca 97w97                                                            21.6mg Z P<.0005
 e   R m f34 wat mul mnl 97w97                                                            23.0mg * P<.0005e
 f   R m f34 wat amd MXA 97w97                                                            36.6mg Z P<.0005
 g   R m f34 wat MXA MXA 97w97                                                            39.3mg * P<.0005c
 h   R m f34 wat amd MXA 97w97                                                            39.8mg Z P<.0005
 i   R m f34 wat MXA sqp 97w97                                                            43.6mg * P<.0005
 j   R m f34 wat pal sqp 97w97                                                            47.0mg * P<.0005
 k   R m f34 wat thy fca 97w97                                                            49.7mg Z P<.002
 l   R m f34 wat ski MXA 97w97                                                            51.0mg Z P<.0005c
 m   R m f34 wat liv MXA 97w97                                                            55.2mg * P<.0005c
 n   R m f34 wat pre MXA 97w97                                                            55.3mg * P<.0005c
```

	RefNum	LoConf	UpConf	Cntrl	1Dose	1Inc	2Dose	2Inc							Citation or Pathology
															Brkly Code
a	TR215	103.mg	n.s.s.	4/50	47.7mg	6/50	(95.4mg	0/50)							liv:hpa,hpc,nnd.
652	TR215	231.mg	n.s.s.	0/50	38.1mg	0/50	76.3mg	4/50							s
a	TR215	54.1mg	n.s.s.	39/50	38.1mg	43/50	76.3mg	44/50							
b	TR215	256.mg	n.s.s.	4/50	38.1mg	7/50	76.3mg	2/50							liv:hpa,hpc,nnd.

BLACK PN (brilliant black BN) 2519-30-4

	RefNum	LoConf	UpConf	Cntrl	1Dose	1Inc	2Dose	2Inc	3Dose	3Inc	4Dose	4Inc	Citation
653	378	262.mg	n.s.s.	0/58	130.mg	0/28	325.mg	0/28	650.mg	0/28	1.30gm	0/29	Drake;fctx,15,503-508;1977
a	378	2.00gm	n.s.s.	14/58	130.mg	6/28	325.mg	3/28	650.mg	5/28	1.30gm	6/29	
654	378	1.02gm	n.s.s.	9/54	120.mg	1/27	300.mg	8/28	600.mg	5/26	1.20gm	7/29	
a	378	234.mg	n.s.s.	0/54	120.mg	0/27	300.mg	0/28	600.mg	0/26	1.20gm	0/29	
655	1321	177.mg	n.s.s.	0/22	50.0mg	0/22	250.mg	0/23	500.mg	0/24			Gaunt;fctx,10,17-27;1972
656	1321	144.mg	n.s.s.	0/21	40.0mg	0/23	200.mg	0/22	400.mg	0/21			

C.I. DIRECT BLACK 38   1937-37-7

	RefNum	LoConf	UpConf	Cntrl	1Dose	1Inc	2Dose	2Inc	3Dose	3Inc	4Dose	4Inc	5Dose	5Inc	Citation
657	2147	48.0mg	111.mg	0/20	526.mg	46/59									Asada;arjc,50,45-55;1981
a	2147	154.mg	603.mg	0/20	526.mg	20/59									
b	2147	18.2mg	53.3mg	0/20	526.mg	57/59									
658	TR108	.992mg	14.8mg	0/10	9.50mg	0/10	18.7mg	0/10	37.5mg	0/10	75.0mg	5/10	150.mg	0/10	
a	TR108	.992mg	14.8mg	0/10	9.50mg	0/10	18.7mg	0/10	37.5mg	0/10	75.0mg	5/10	150.mg	0/10	
b	TR108	.992mg	14.8mg	0/10	9.50mg	0/10	18.7mg	0/10	37.5mg	0/10	75.0mg	5/10	150.mg	0/10	liv:hpa,hpc,nnd.
659	TR108	.438mg	2.58mg	0/10	7.60mg	0/10	15.0mg	0/10	30.0mg	0/10	60.0mg	9/10	(120.mg	0/10)	liv:hpc,nnd.
a	TR108	.755mg	10.1mg	0/10	7.60mg	0/10	15.0mg	0/10	30.0mg	0/10	60.0mg	4/10	120.mg	0/10	s
b	TR108	.438mg	2.58mg	0/10	7.60mg	0/10	15.0mg	0/10	30.0mg	0/10	60.0mg	9/10	(120.mg	0/10)	
c	TR108	.438mg	2.58mg	0/10	7.60mg	0/10	15.0mg	0/10	30.0mg	0/10	60.0mg	9/10	(120.mg	0/10)	liv:hpa,hpc,nnd.

C.I. DIRECT BLUE 6   2602-46-2

	RefNum	LoConf	UpConf	Cntrl	1Dose	1Inc	2Dose	2Inc	3Dose	3Inc	4Dose	4Inc	5Dose	5Inc	Citation
660	TR108	1.37mg	10.2mg	0/10	9.50mg	0/10	18.7mg	0/10	37.5mg	0/10	75.0mg	0/10	150.mg	7/10	liv:hpc,nnd.
a	TR108	2.05mg	37.3mg	0/10	9.50mg	0/10	18.7mg	0/10	37.5mg	0/10	75.0mg	0/10	150.mg	4/10	s
b	TR108	1.37mg	10.2mg	0/10	9.50mg	0/10	18.7mg	0/10	37.5mg	0/10	75.0mg	0/10	150.mg	7/10	
c	TR108	1.37mg	10.2mg	0/10	9.50mg	0/10	18.7mg	0/10	37.5mg	0/10	75.0mg	0/10	150.mg	7/10	liv:hpa,hpc,nnd.
661	TR108	.532mg	3.59mg	0/10	7.60mg	0/10	15.0mg	0/10	30.0mg	0/10	60.0mg	8/10	(120.mg	1/10)	liv:hpc,nnd.
a	TR108	.532mg	3.59mg	0/10	7.60mg	0/10	15.0mg	0/10	30.0mg	0/10	60.0mg	8/10	(120.mg	1/10)	
b	TR108	.532mg	3.59mg	0/10	7.60mg	0/10	15.0mg	0/10	30.0mg	0/10	60.0mg	8/10	(120.mg	1/10)	liv:hpa,hpc,nnd.

C.I. DIRECT BLUE 15   2429-74-5

	RefNum	LoConf	UpConf	Cntrl	1Dose	1Inc	2Dose	2Inc	3Dose	3Inc	Citation or Pathology
662	TR397	15.1mg	32.5mg	16/50	35.5mg	24/35	70.3mg	59/65	141.mg	45/50	asc:adp; cli:ade,anb,car,cnb; col:adp; dsc:adp;
											duo:muc; jej:adc,muc; liv:hpc,nnd; mul:mnl; pal:sqc,sqp; ski:sqc,sqp; ton:sqc,sqp; ute:adc,ade; zym:
											ade,car. C
a	TR397	30.2mg	80.1mg	7/50	35.5mg	11/35	70.3mg	24/65	141.mg	27/50	cli:ade,anb,car,cnb.
b	TR397	31.1mg	101.mg	7/50	35.5mg	13/35	70.3mg	27/65	141.mg	15/50	
c	TR397	49.7mg	151.mg	2/50	35.5mg	4/35	70.3mg	19/65	141.mg	15/50	pal:sqc,sqp; ton:sqc,sqp.
d	TR397	50.0mg	226.mg	5/50	35.5mg	5/35	70.3mg	12/65	141.mg	12/50	cli:ade,anb. S
e	TR397	63.4mg	178.mg	0/50	35.5mg	4/35	70.3mg	11/65	141.mg	17/50	zym:ade,car.
f	TR397	62.4mg	229.mg	2/50	35.5mg	6/35	70.3mg	12/65	141.mg	15/50	cli:ade,anb. S
g	TR397	56.8mg	1.06gm	6/50	35.5mg	8/35	70.3mg	13/65	141.mg	7/50	cli:car,cnb. S
h	TR397	59.3mg	3.45gm	5/50	35.5mg	8/35	70.3mg	12/65	141.mg	5/50	ute:esp,ess. S
i	TR397	71.2mg	344.mg	2/50	35.5mg	3/35	70.3mg	12/65	141.mg	9/50	pal:sqp; ton:sqp. S
j	TR397	85.9mg	278.mg	0/50	35.5mg	4/35	70.3mg	7/65	141.mg	14/50	S
k	TR397	86.9mg	317.mg	0/50	35.5mg	2/35	70.3mg	11/65	141.mg	7/50	s
l	TR397	79.8mg	377.mg	0/50	35.5mg	2/35	70.3mg	6/65	141.mg	5/50	ski:sqc,sqp.
m	TR397	85.6mg	442.mg	0/50	35.5mg	2/35	70.3mg	5/65	141.mg	5/50	S
n	TR397	105.mg	484.mg	0/50	35.5mg	1/35	70.3mg	8/65	141.mg	6/50	pal:sqc; ton:sqc. S
o	TR397	103.mg	1.38gm	2/50	35.5mg	2/35	70.3mg	4/65	141.mg	7/50	ton:sqc,sqp. S
p	TR397	135.mg	1.48gm	0/50	35.5mg	1/35	70.3mg	5/65	141.mg	3/50	S
q	TR397	151.mg	2.35gm	0/50	35.5mg	1/35	70.3mg	3/65	141.mg	3/50	S
r	TR397	154.mg	1.43gm	0/50	35.5mg	0/35	70.3mg	2/65	141.mg	5/50	liv:hpc,nnd.
s	TR397	159.mg	1.92gm	0/50	35.5mg	0/35	70.3mg	2/65	141.mg	4/50	S
t	TR397	186.mg	2.33gm	0/50	35.5mg	0/35	70.3mg	5/65	141.mg	3/50	S
u	TR397	164.mg	n.s.s.	0/50	35.5mg	0/35	70.3mg	3/65	141.mg	1/50	asc:adp; col:adp; dsc:adp;
v	TR397	162.mg	n.s.s.	1/50	35.5mg	0/35	70.3mg	1/65	141.mg	4/50	ute:adc,ade.
w	TR397	321.mg	n.s.s.	0/50	35.5mg	0/35	70.3mg	1/65	141.mg	3/50	duo:muc; jej:adc,muc.
x	TR397	340.mg	n.s.s.	0/50	35.5mg	0/35	70.3mg	0/65	141.mg	3/50	jej:adc,muc. S
y	TR397	13.5mg	36.2mg	43/50	35.5mg	33/35	70.3mg	64/65	141.mg	49/50	
z	TR397	154.mg	1.43gm	0/50	35.5mg	0/35	70.3mg	2/65	141.mg	5/50	liv:hpa,hpc,nnd.
663	TR397	8.18mg	18.0mg	14/50	31.5mg	27/35	61.7mg	59/65	124.mg	45/50	asc:adp; col:adc,adp; dsc:adp; jej:adc,adp,muc;
											liv:hpc,nnd; pal:sqp; phr:sqc; pre:ade,anb,car,cnb; ski:bca,bcc,sea,sqc,sqp; ton:sqc,sqp; zym:ade,
											car. C
a	TR397	8.38mg	23.1mg	48/50	31.5mg	32/35	61.7mg	61/65	124.mg	43/50	tes:iab,ica. S
b	TR397	12.2mg	28.2mg	2/50	31.5mg	10/35	61.7mg	29/65	124.mg	28/50	ski:bca,bcc,sea. S
c	TR397	13.1mg	30.8mg	2/50	31.5mg	9/35	61.7mg	27/65	124.mg	28/50	ski:bca,bcc.
d	TR397	14.0mg	35.3mg	2/50	31.5mg	8/35	61.7mg	23/65	124.mg	26/50	S
e	TR397	13.7mg	46.5mg	17/50	31.5mg	19/35	61.7mg	28/65	124.mg	20/50	
f	TR397	20.7mg	84.2mg	16/50	31.5mg	5/35	61.7mg	21/65	124.mg	17/50	amd:pbb,phc,phm,pob. S
g	TR397	24.9mg	65.4mg	1/50	31.5mg	10/35	61.7mg	24/65	124.mg	17/50	pal:sqp; phr:sqc; ton:sqp. S
h	TR397	22.1mg	95.2mg	16/50	31.5mg	5/35	61.7mg	19/65	124.mg	17/50	amd:pbb,pob. S
i	TR397	27.0mg	71.6mg	0/50	31.5mg	9/35	61.7mg	18/65	124.mg	15/50	pal:sqp; ton:sqp. S
j	TR397	29.0mg	77.1mg	0/50	31.5mg	9/35	61.7mg	17/65	124.mg	15/50	S
k	TR397	15.1mg	310.mg	0/50	31.5mg	4/35	(61.7mg	1/65	124.mg	0/50)	S
l	TR397	29.3mg	97.2mg	2/50	31.5mg	4/35	61.7mg	11/65	124.mg	19/50	ski:sqc,sqp.
m	TR397	31.4mg	102.mg	0/50	31.5mg	6/35	61.7mg	9/65	124.mg	11/50	liv:hpc,nnd.
n	TR397	29.8mg	148.mg	8/50	31.5mg	5/35	61.7mg	23/65	124.mg	9/50	pre:ade,anb,car,cnb.

	Spe Strain Site Xpo+Xpt		TD50	2Tailpvl
	Sex Route Hist Notes		DR	AuOp
o	R m f34 wat liv nnd 97w97		65.3mg	* P<.0005
p	R m f34 wat pre MXA 97w97		77.1mg	* P<.0005
q	R m f34 wat zym MXA 97w97		89.2mg	* P<.0005c
r	R m f34 wat ski sqc 97w97		89.7mg	Z P<.0005
s	R m f34 wat ski sqp 97w97		92.8mg	* P<.0005
t	R m f34 wat ski bcc 97w97		101.mg	* P<.0005
u	R m f34 wat thy MXA 97w97		103.mg	Z P<.002
v	R m f34 wat ski sea 97w97		108.mg	* P<.0005c
w	R m f34 wat sub MXA 97w97		131.mg	* P<.003
x	R m f34 wat zym car 97w97		132.mg	* P<.0005
y	R m f34 wat sub fib 97w97		150.mg	* P<.007
z	R m f34 wat ski ker 97w97		152.mg	* P<.006
A	R m f34 wat MXA MXA 97w97		162.mg	* P<.0005c
B	R m f34 wat zym ade 97w97		237.mg	* P<.002
C	R m f34 wat MXA adp 97w97		310.mg	* P<.0005
D	R m f34 wat col adc 97w97		355.mg	* P<.002
E	R m f34 wat liv hpc 97w97		375.mg	* P<.002
F	R m f34 wat thy MXA 97w97		116.mg	* P<.03
G	R m f34 wat thy MXA 97w97		143.mg	* P<.04
H	R m f34 wat pre MXA 97w97		207.mg	* P<.05
I	R m f34 wat thy fcc 97w97		331.mg	* P<.04
J	R m f34 wat MXA sqc 97w97		361.mg	* P<.04
K	R m f34 wat ton MXA 97w97		371.mg	* P<.02
L	R m f34 wat phr sqc 97w97		466.mg	* P<.05
M	R m f34 wat MXA asl 97w97		600.mg	* P<.03 e
N	R m f34 wat jej MXA 97w97		1.63gm	* P<.2 c
O	R m f34 wat TBA MXB 97w97		12.0mg	* P<.0005
P	R m f34 wat liv MXB 97w97		55.2mg	* P<.0005

C.I. DIRECT BLUE 218   100ng..:..1ug....:..10......:..100.....:..1mg....:..10.....:..100....:..1g.....:..10

			TD50	2Tailpvl
664	M f b6c eat liv MXA 24m25	: + :	711.mg	Z P<.0005c
a	M f b6c eat liv hpa 24m25		774.mg	* P<.0005c
b	M f b6c eat liv MXA 24m25		4.00gm	* P<.03
c	M f b6c eat liv hpc 24m25		4.57gm	* P<.04 c
d	M f b6c eat TBA MXB 24m25		1.53gm	* P<.05
e	M f b6c eat liv MXB 24m25		711.mg	Z P<.0005
f	M f b6c eat lun MXB 24m25		no dre	P=1.
665	M m b6c eat liv MXA 24m24	: + :	1.08gm	* P<.0005c
a	M m b6c eat liv hpa 24m24		1.21gm	* P<.0005c
b	M m b6c eat liv hpc 24m24		2.99gm	* P<.004 c
c	M m b6c eat jej car 24m24		20.3gm	* P<.2 e
d	M m b6c eat kid rua 24m24		195.gm	* P<.9 e
e	M m b6c eat kid ruc 24m24		no dre	P=1. e
f	M m b6c eat kid MXA 24m24		no dre	P=1. e
g	M m b6c eat TBA MXB 24m24		1.91gm	* P<.1
h	M m b6c eat liv MXB 24m24		1.08gm	* P<.0005
i	M m b6c eat lun MXB 24m24		no dre	P=1.
666	R f f34 eat ute esp 24m24	: + :	#83.5mg	Z P<.0005-
a	R f f34 eat TBA MXB 24m24		653.mg	* P<.3
b	R f f34 eat liv MXB 24m24		no dre	P=1.
667	R m f34 eat pal MXB 24m24	: + :	1.38gm	* P<.0005
a	R m f34 eat pal MXA 24m24		1.57gm	* P<.0005p
b	R m f34 eat pal sqp 24m24		1.92gm	* P<.002 p
c	R m f34 eat for MXA 24m24		2.35gm	* P<.2 e
d	R m f34 eat for sqp 24m24		3.14gm	* P<.2 e
e	R m f34 eat pal sqc 24m24		9.51gm	* P<.2 p
f	R m f34 eat pal bsb 24m24		11.5gm	* P<.2 p
g	R m f34 eat for sqc 24m24		no dre	P=1. e
h	R m f34 eat TBA MXB 24m24		921.mg	* P<.5
i	R m f34 eat liv MXB 24m24		5.05gm	* P<.6

C.I. DISPERSE BLUE 1   100ng..:..1ug....:..10......:..100....:..1mg...:..10.....:..100....:..1g......:..10

			TD50	2Tailpvl
668	M f b6c eat liv hpa 24m25	: +	#150.mg	Z P<.01 -
a	M f b6c eat TBA MXB 24m25		no dre	P=1.
b	M f b6c eat liv MXB 24m25		no dre	P=1.
c	M f b6c eat lun MXB 24m25		9.31gm	* P<.8
669	M m b6c eat lun a/a 24m24	: ±	925.mg	* P<.05
a	M m b6c eat liv MXA 24m24		534.mg	* P<.2 e
b	M m b6c eat TBA MXB 24m24		1.29gm	Z P<.7
c	M m b6c eat liv MXB 24m24		534.mg	* P<.2
d	M m b6c eat lun MXB 24m24		701.mg	* P<.06
670	R f f34 eat ubl MXB 24m24	: +:	131.mg	Z P<.0005
a	R f f34 eat ubl MXA 24m24		198.mg	Z P<.0005c
b	R f f34 eat ubl MXA 24m24		250.mg	Z P<.0005c
c	R f f34 eat ubl lei 24m24		294.mg	Z P<.0005c
d	R f f34 eat ubl tcc 24m24		307.mg	Z P<.0005c
e	R f f34 eat ubl tpp 24m24		317.mg	Z P<.0005c
f	R f f34 eat ubl MXA 24m24		664.mg	Z P<.0005c
g	R f f34 eat ubl sqp 24m24		1.02gm	Z P<.0005c
h	R f f34 eat ubl ley 24m24		1.54gm	* P<.004 c
i	R f f34 eat ubl sqc 24m24		1.58gm	* P<.004 c

	RefNum	LoConf	UpConf	Cntrl	1Dose	1Inc	2Dose	2Inc			Citation or Pathology	Brkly Code
o	TR397	34.9mg	133.mg	0/50	31.5mg	6/35	61.7mg	8/65	124.mg	7/50		S
p	TR397	36.9mg	267.mg	6/50	31.5mg	2/35	61.7mg	12/65	124.mg	8/50	pre:ade,anb.	S
q	TR397	51.3mg	166.mg	1/50	31.5mg	5/35	61.7mg	10/65	124.mg	20/50	zym:ade,car.	
r	TR397	45.3mg	178.mg	0/50	31.5mg	1/35	61.7mg	7/65	124.mg	13/50		S
s	TR397	42.9mg	293.mg	2/50	31.5mg	3/35	61.7mg	5/65	124.mg	8/50		S
t	TR397	50.5mg	213.mg	0/50	31.5mg	2/35	61.7mg	4/65	124.mg	10/50		S
u	TR397	41.8mg	464.mg	4/50	31.5mg	4/35	61.7mg	4/65	(124.mg	0/50)	thy:fca,fcc.	S
v	TR397	50.7mg	279.mg	0/50	31.5mg	1/35	61.7mg	7/65	124.mg	3/50		S
w	TR397	53.4mg	994.mg	2/50	31.5mg	3/35	61.7mg	5/65	124.mg	4/50	sub:fib,sar.	S
x	TR397	73.7mg	267.mg	1/50	31.5mg	3/35	61.7mg	8/65	124.mg	17/50		S
y	TR397	57.0mg	2.72gm	2/50	31.5mg	2/35	61.7mg	5/65	124.mg	3/50		S
z	TR397	60.6mg	2.32gm	2/50	31.5mg	1/35	61.7mg	7/65	124.mg	2/50		S
A	TR397	84.3mg	359.mg	0/50	31.5mg	1/35	61.7mg	6/65	124.mg	8/50	asc:adp; col:adc,adp; dsc:adp.	
B	TR397	88.4mg	1.09gm	0/50	31.5mg	2/35	61.7mg	2/65	124.mg	4/50		S
C	TR397	131.mg	1.16gm	0/50	31.5mg	1/35	61.7mg	2/65	124.mg	5/50	asc:adp; col:adp; dsc:adp.	S
D	TR397	144.mg	1.66gm	0/50	31.5mg	0/35	61.7mg	4/65	124.mg	3/50		S
E	TR397	129.mg	1.87gm	0/50	31.5mg	0/35	61.7mg	1/65	124.mg	4/50		S
F	TR397	44.4mg	n.s.s.	8/50	31.5mg	9/35	61.7mg	9/65	124.mg	5/50	thy:cca,ccb,ccr,cdb.	S
G	TR397	52.4mg	n.s.s.	5/50	31.5mg	8/35	61.7mg	7/65	124.mg	4/50	thy:cca,cdb.	S
H	TR397	79.1mg	n.s.s.	2/50	31.5mg	3/35	61.7mg	11/65	124.mg	1/50	pre:car,cnb.	S
I	TR397	85.4mg	n.s.s.	0/50	31.5mg	0/35	61.7mg	3/65	124.mg	0/50		S
J	TR397	117.mg	n.s.s.	1/50	31.5mg	1/35	61.7mg	6/65	124.mg	2/50	phr:sqc; ton:sqc.	S
K	TR397	108.mg	n.s.s.	0/50	31.5mg	1/35	61.7mg	3/65	124.mg	2/50	ton:sqc,sqp.	S
L	TR397	130.mg	n.s.s.	1/50	31.5mg	0/35	61.7mg	4/65	124.mg	2/50		S
M	TR397	188.mg	n.s.s.	0/50	31.5mg	1/35	61.7mg	1/65	124.mg	2/50	bmd:asl; crb:asl.	
N	TR397	484.mg	n.s.s.	0/50	31.5mg	1/35	61.7mg	0/65	124.mg	2/50	jej:adc,adp,muc.	
O	TR397	8.00mg	20.0mg	38/50	31.5mg	34/35	61.7mg	63/65	124.mg	48/50	liv:hpa,hpc,nnd.	
P	TR397	31.4mg	102.mg	0/50	31.5mg	6/35	61.7mg	9/65	124.mg	11/50		

C.I. DIRECT BLUE 218   28407-37-6

	RefNum	LoConf	UpConf	Cntrl	1Dose	1Inc	2Dose	2Inc			Citation or Pathology	Brkly Code
664	TR430	433.mg	1.51gm	10/50	126.mg	15/50	379.mg	21/50	1.26gm	45/50	liv:hpa,hpc.	
a	TR430	480.mg	1.57gm	7/50	126.mg	12/50	379.mg	17/50	1.26gm	41/50		
b	TR430	1.65gm	n.s.s.	5/50	126.mg	5/50	379.mg	6/50	1.26gm	13/50	liv:hpb,hpc.	S
c	TR430	1.76gm	n.s.s.	5/50	126.mg	5/50	379.mg	6/50	1.26gm	12/50		
d	TR430	617.mg	n.s.s.	29/50	126.mg	35/50	379.mg	34/50	1.26gm	46/50		
e	TR430	433.mg	1.51gm	10/50	126.mg	15/50	379.mg	21/50	1.26gm	45/50	liv:hpa,hpb,hpc.	
f	TR430	8.17gm	n.s.s.	5/50	126.mg	7/50	379.mg	4/50	1.26gm	1/50	lun:a/a,a/c.	
665	TR430	571.mg	4.17gm	21/51	117.mg	20/50	352.mg	23/50	1.17gm	45/50	liv:hpa,hpc.	
a	TR430	639.mg	4.52gm	16/51	117.mg	19/50	352.mg	17/50	1.17gm	40/50		
b	TR430	1.41gm	25.6gm	7/51	117.mg	3/50	352.mg	8/50	1.17gm	17/50		
c	TR430	5.31gm	n.s.s.	1/51	117.mg	0/50	352.mg	0/50	1.17gm	3/50		
d	TR430	5.67gm	n.s.s.	0/51	117.mg	2/50	352.mg	1/50	1.17gm	1/50		
e	TR430	11.1gm	n.s.s.	0/51	117.mg	1/50	352.mg	0/50	1.17gm	0/50		
f	TR430	6.22gm	n.s.s.	0/51	117.mg	3/50	352.mg	1/50	1.17gm	1/50	kid:rua,ruc.	
g	TR430	694.mg	n.s.s.	31/51	117.mg	37/50	352.mg	28/50	1.17gm	46/50	liv:hpa,hpb,hpc.	
h	TR430	571.mg	4.17gm	21/51	117.mg	20/50	352.mg	23/50	1.17gm	45/50	lun:a/a,a/c.	
i	TR430	6.02gm	n.s.s.	14/51	117.mg	12/50	352.mg	9/50	1.17gm	5/50		
666	TR430	40.9mg	302.mg	1/51	49.2mg	12/51	(148.mg	10/50	495.mg	10/50)		S
a	TR430	192.mg	n.s.s.	45/51	49.2mg	46/51	148.mg	46/50	495.mg	45/50		
b	TR430	n.s.s.	n.s.s.	0/51	49.2mg	0/51	148.mg	0/50	495.mg	0/50	liv:hpa,hpb,hpc.	
667	TR430	592.mg	4.46gm	0/50	39.4mg	0/50	119.mg	0/50	396.mg	7/51	pal:bsb,sqc,sqp.	P
a	TR430	636.mg	5.84gm	0/50	39.4mg	0/50	119.mg	0/50	396.mg	6/51	pal:sqc,sqp.	
b	TR430	726.mg	8.98gm	0/50	39.4mg	0/50	119.mg	0/50	396.mg	5/51		
c	TR430	811.mg	n.s.s.	0/50	39.4mg	0/50	119.mg	3/50	396.mg	1/51	for:sqc,sqp.	
d	TR430	952.mg	n.s.s.	0/50	39.4mg	0/50	119.mg	2/50	396.mg	1/51		
e	TR430	1.55gm	n.s.s.	0/50	39.4mg	0/50	119.mg	0/50	396.mg	1/51		
f	TR430	1.88gm	n.s.s.	0/50	39.4mg	0/50	119.mg	0/50	396.mg	1/51		
g	TR430	1.59gm	n.s.s.	0/50	39.4mg	0/50	119.mg	1/50	396.mg	0/51		
h	TR430	192.mg	n.s.s.	44/50	39.4mg	42/50	119.mg	45/50	396.mg	44/51		
i	TR430	713.mg	n.s.s.	5/50	39.4mg	6/50	119.mg	2/50	396.mg	6/51	liv:hpa,hpb,hpc.	

C.I. DISPERSE BLUE 1   2475-45-8

	RefNum	LoConf	UpConf	Cntrl	1Dose	1Inc	2Dose	2Inc			Citation or Pathology	Brkly Code
668	TR299	67.2mg	19.8gm	2/50	76.5mg	12/50	(153.mg	3/50	319.mg	2/50)		S
a	TR299	275.mg	n.s.s.	29/50	76.5mg	33/50	153.mg	33/50	319.mg	30/50		
b	TR299	904.mg	n.s.s.	3/50	76.5mg	13/50	153.mg	3/50	319.mg	4/50	liv:hpa,hpc,nnd.	
c	TR299	791.mg	n.s.s.	1/50	76.5mg	5/50	153.mg	2/50	319.mg	3/50	lun:a/a,a/c.	
669	TR299	386.mg	n.s.s.	1/50	71.3mg	3/50	143.mg	5/50	297.mg	5/50		S
a	TR299	187.mg	n.s.s.	9/50	71.3mg	21/50	143.mg	20/50	297.mg	16/50	liv:hpa,hpc.	
b	TR299	164.mg	n.s.s.	35/50	71.3mg	44/50	143.mg	34/50	297.mg	34/50		
c	TR299	187.mg	n.s.s.	9/50	71.3mg	21/50	143.mg	20/50	297.mg	16/50	liv:hpa,hpc,nnd.	
d	TR299	275.mg	n.s.s.	4/50	71.3mg	9/50	143.mg	5/50	297.mg	11/50	lun:a/a,a/c.	
670	TR299	92.4mg	192.mg	0/50	61.3mg	0/50	124.mg	17/50	248.mg	37/50	ubl:lei,ley,sqc,sqp,tcc,tpp.	C
a	TR299	131.mg	318.mg	0/50	61.3mg	0/50	124.mg	15/50	248.mg	21/50	ubl:tcc,tpp.	
b	TR299	158.mg	424.mg	0/50	61.3mg	0/50	124.mg	3/50	248.mg	26/50	ubl:lei,ley.	
c	TR299	181.mg	520.mg	0/50	61.3mg	0/50	124.mg	2/50	248.mg	23/50		
d	TR299	186.mg	559.mg	0/50	61.3mg	0/50	124.mg	10/50	248.mg	13/50		
e	TR299	192.mg	571.mg	0/50	61.3mg	0/50	124.mg	9/50	248.mg	15/50		
f	TR299	338.mg	1.55gm	0/50	61.3mg	0/50	124.mg	1/50	248.mg	11/50	ubl:sqc,sqp.	
g	TR299	454.mg	3.15gm	0/50	61.3mg	0/50	124.mg	1/50	248.mg	7/50		
h	TR299	579.mg	10.3gm	0/50	61.3mg	0/50	124.mg	1/50	248.mg	4/50		
i	TR299	599.mg	10.9gm	0/50	61.3mg	0/50	124.mg	1/50	248.mg	4/50		

```
      Spe Strain  Site   Xpo+Xpt                                                  TD50    2Tailpvl
          Sex  Route  Hist   Notes                                                DR      AuOp
  j   R f f34 eat pni isc 24m24                                                    291.mg * P<.02
  k   R f f34 eat MXA MXA 24m24                                                    323.mg * P<.03
  l   R f f34 eat TBA MXB 24m24                                                    130.mg * P<.005
  m   R f f34 eat liv MXB 24m24                                                    no dre   P=1.
671   R m f34 eat ubl MXB 24m24                                      : + :         89.3mg Z P<.0005
  a   R m f34 eat ubl MXA 24m24                                                    129.mg Z P<.0005c
  b   R m f34 eat ubl lei 24m24                                                    131.mg Z P<.0005c
  c   R m f34 eat ubl MXA 24m24                                                    245.mg Z P<.0005c
  d   R m f34 eat ubl tpp 24m24                                                    370.mg * P<.0005c
  e   R m f34 eat pni MXA 24m24                                                    414.mg * P<.005
  f   R m f34 eat ubl tcc 24m24                                                    506.mg Z P<.0005c
  g   R m f34 eat ubl MXA 24m24                                                    803.mg * P<.002 c
  h   R m f34 eat ubl sqp 24m24                                                    1.26gm * P<.006 c
  i   R m f34 eat sub srn 24m24                                                    1.83gm Z P<.01
  j   R m f34 eat tes ict 24m24                                                    100.mg * P<.05
  k   R m f34 eat thy MXA 24m24                                                    474.mg * P<.03
  l   R m f34 eat thy cca 24m24                                                    529.mg * P<.02
  m   R m f34 eat pni isa 24m24                                                    810.mg * P<.03
  n   R m f34 eat TBA MXB 24m24                                                    39.3mg Z P<.0005
  o   R m f34 eat liv MXB 24m24                                                    no dre   P=1.

FD & C BLUE NO. 1           100ng..:..1ug....:..10.....:..100....:..1mg....:..10.....:..100....:..1g......:..10
672   M f cd1 eat lun adc 24m24 e                                                  372.gm * P<.2   -
  a   M f cd1 eat lun alc 24m24 e                                                  607.gm * P<.6   -
  b   M f cd1 eat lun bcd 24m24 e                                                  856.gm * P<.8   -
  c   M f cd1 eat lun ala 24m24 e                                                  no dre   P=1.   -
  d   M f cd1 eat liv hem 24m24 e                                                  no dre   P=1.   -
  e   M f cd1 eat lun bde 24m24 e                                                  no dre   P=1.   -
673   M m cd1 eat lun bcd 24m24 e                                                  91.1gm * P<.08  -
  a   M m cd1 eat lun adc 24m24 e                                                  293.gm * P<.4   -
  b   M m cd1 eat lun bde 24m24 e                                                  343.gm * P<.2   -
  c   M m cd1 eat liv hpc 24m24 e                                                  no dre   P=1.   -
  d   M m cd1 eat lun agc 24m24 e                                                  no dre   P=1.   -
  e   M m cd1 eat lun alc 24m24 e                                                  no dre   P=1.   -
  f   M m cd1 eat lun ala 24m24 e                                                  no dre   P=1.   -
  g   M m cd1 eat liv hes 24m24 e                                                  no dre   P=1.   -
674   R b osm eat liv hpt 24m24                                                    62.1gm * P<.2   -
  a   R b osm eat tba mix 24m24                                                    no dre   P=1.   -

FD & C BLUE NO. 2           100ng..:..1ug....:..10.....:..100....:..1mg....:..10.....:..100....:..1g......:..10
675   M f cd1 eat liv nod 84w84 es                              .>                 no dre   P=1.   -
  a   M f cd1 eat lun ade 84w84 es                                                 no dre   P=1.   -
676   M m cd1 eat liv nod 84w84 es                                     .>          20.2gm * P<.4   -
  a   M m cd1 eat lun ade 84w84 es                                                 no dre   P=1.   -
677   M f cdr eat tba mal 23m23 e                                                  .33.9gm * P<.3  -
  a   M f cdr eat tba mix 23m23 e                                                  no dre   P=1.   -
678   M m cdr eat tba mix 95w95 e                                            .>23.4gm * P<.2       -
  a   M m cdr eat tba mal 95w95 e                                                  57.1gm * P<.5   -
679   R b osm eat tba mix 24m24                                      .       ±  5.43gm * P<.06     -

HC BLUE NO. 1               100ng..:..1ug....:..10.....:..100....:..1mg....:..10.....:..100....:..1g......:..10
680   M f b6c eat liv MXA 24m24                                         :+ :       179.mg * P<.0005
  a   M f b6c eat liv hpc 24m24                                                    208.mg * P<.0005c
  b   M f b6c eat liv hpa 24m24                                                    697.mg \ P<.004
  c   M f b6c eat TBA MXB 24m24                                                    373.mg * P<.002
  d   M f b6c eat liv MXB 24m24                                                    179.mg * P<.0005
  e   M f b6c eat lun MXB 24m24                                                    no dre   P=1.
681   M f b6c eat liv mix  9m23 er                               .  +  .           41.3mg   P<.0005+
  a   M f b6c eat liv hpc  9m23 er                                                 114.mg   P<.0005+
  b   M f b6c eat liv hpa  9m23 er                                                 165.mg   P<.0005
682   M f b6c eat liv mix 15m23 er                             .  +  .             81.5mg   P<.0005+
  a   M f b6c eat liv hpc 15m23 er                                                 94.0mg   P<.0005+
  b   M f b6c eat liv hpa 15m23 er                                                 969.mg   P<.05
683   M f b6c eat liv mix 91w91 ekr                           .  +   .             85.0mg   P<.0005+
  a   M f b6c eat liv hpc 91w91 ekr                                                148.mg   P<.004 +
  b   M f b6c eat liv hpa 91w91 ekr                                                287.mg   P<.04
684   M f b6c eat liv mix 23m23 er                          .  +  .                51.4mg   P<.0005+
  a   M f b6c eat liv hpc 23m23 er                                                 163.mg   P<.0005+
  b   M f b6c eat liv hpa 23m23 er                                                 185.mg   P<.0005
685   M f b6c eat liv hpc 91w91 ekr                                <+              noTD50   P<.0005+
  a   M f b6c eat liv mix 91w91 ekr                                                noTD50   P<.0005+
  b   M f b6c eat liv hpa 91w91 ekr                                                574.mg   P<.04
686   M f b6c eat liv mix 23m23 er                                 <+              noTD50   P<.0005+
  a   M f b6c eat liv hpc 23m23 er                                                 87.1mg   P<.0005+
  b   M f b6c eat liv hpa 23m23 er                                                 1.30gm   P<.02
687   M m b6c eat liv MXA 24m24                                         : +  :     259.mg * P<.002
  a   M m b6c eat liv hpa 24m24                                                    286.mg \ P<.008
  b   M m b6c eat MXB MXB 24m24                                                    369.mg * P<.002
  c   M m b6c eat liv hpc 24m24                                                    389.mg * P<.002 c
  d   M m b6c eat thy fca 24m24                                                    2.23gm * P<.008 c
  e   M m b6c eat TBA MXB 24m24                                                    561.mg * P<.3
```

	RefNum	LoConf	UpConf	Cntrl	1Dose	1Inc	2Dose	2Inc					Citation or Pathology	Brkly Code
j	TR299	135.mg	n.s.s.	10/50	61.3mg	21/50	124.mg	20/50	248.mg	15/50				s
k	TR299	139.mg	n.s.s.	12/50	61.3mg	24/50	124.mg	20/50	248.mg	16/50			pit:can; pni:isc.	s
l	TR299	65.7mg	1.28gm	44/50	61.3mg	41/50	124.mg	45/50	248.mg	43/50				
m	TR299	1.04gm	n.s.s.	1/50	61.3mg	0/50	124.mg	2/50	248.mg	0/50			liv:hpa,hpc,nnd.	
671	TR299	63.0mg	130.mg	0/50	49.3mg	0/50	99.0mg	17/50	198.mg	45/50			ubl:lei,ley,sqc,sqp,tcc,tpp.	C
a	TR299	86.9mg	196.mg	0/50	49.3mg	0/50	99.0mg	7/50	198.mg	41/50			ubl:lei,ley.	
b	TR299	88.0mg	200.mg	0/50	49.3mg	0/50	99.0mg	6/50	198.mg	41/50				
c	TR299	141.mg	463.mg	0/50	49.3mg	0/50	99.0mg	10/50	198.mg	11/50			ubl:tcc,tpp.	
d	TR299	183.mg	910.mg	0/50	49.3mg	0/50	99.0mg	8/50	198.mg	4/50				
e	TR299	178.mg	4.62gm	1/50	49.3mg	2/50	99.0mg	5/50	198.mg	3/50			pni:isa,isc.	s
f	TR299	249.mg	1.18gm	0/50	49.3mg	0/50	99.0mg	4/50	198.mg	8/50				
g	TR299	311.mg	3.41gm	0/50	49.3mg	0/50	99.0mg	2/50	198.mg	4/50			ubl:sqc,sqp.	
h	TR299	401.mg	15.0gm	0/50	49.3mg	0/50	99.0mg	1/50	198.mg	3/50				
i	TR299	548.mg	99.7gm	0/50	49.3mg	0/50	99.0mg	0/50	198.mg	3/50				s
j	TR299	41.2mg	n.s.s.	44/50	49.3mg	44/50	99.0mg	38/50	198.mg	16/50				s
k	TR299	187.mg	n.s.s.	2/50	49.3mg	4/50	99.0mg	5/50	198.mg	3/50			thy:cca,ccr.	s
l	TR299	213.mg	n.s.s.	1/50	49.3mg	2/50	99.0mg	4/50	198.mg	3/50				s
m	TR299	278.mg	n.s.s.	1/50	49.3mg	0/50	99.0mg	4/50	198.mg	2/50				s
n	TR299	24.9mg	79.8mg	49/50	49.3mg	49/50	99.0mg	47/50	198.mg	46/50				
o	TR299	540.mg	n.s.s.	4/50	49.3mg	2/50	99.0mg	2/50	198.mg	0/50			liv:hpa,hpc,nnd.	

FD & C BLUE NO. 1 (brilliant blue FCF) 3844-45-9

	RefNum	LoConf	UpConf	Cntrl	1Dose	1Inc	2Dose	2Inc					Citation or Pathology	Brkly Code
672	1976	60.6gm	n.s.s.	0/120	650.mg	0/60	1.95gm	0/60	6.50gm	1/60			Borzelleca;fctx,28,221-234;1990	
a	1976	64.1gm	n.s.s.	1/120	650.mg	0/60	1.95gm	0/60	6.50gm	1/60				
b	1976	58.1gm	n.s.s.	1/120	650.mg	1/60	1.95gm	0/60	6.50gm	1/60				
c	1976	93.1gm	n.s.s.	5/120	650.mg	1/60	1.95gm	1/60	6.50gm	0/60				
d	1976	73.5gm	n.s.s.	1/120	650.mg	1/60	1.95gm	1/60	6.50gm	0/60				
e	1976	5.61gm	n.s.s.	2/120	650.mg	0/60	1.95gm	1/60	6.50gm	1/60				
673	1976	29.4gm	n.s.s.	0/120	600.mg	1/60	1.80gm	1/60	6.00gm	2/60				
a	1976	43.6gm	n.s.s.	0/120	600.mg	1/60	1.80gm	0/60	6.00gm	1/60				
b	1976	55.9gm	n.s.s.	0/120	600.mg	0/60	1.80gm	0/60	6.00gm	1/60				
c	1976	47.9gm	n.s.s.	8/120	600.mg	2/60	1.80gm	3/60	6.00gm	2/60				
d	1976	5.18gm	n.s.s.	1/120	600.mg	0/60	1.80gm	0/60	6.00gm	0/60				
e	1976	5.18gm	n.s.s.	1/120	600.mg	0/60	1.80gm	0/60	6.00gm	0/60				
f	1976	53.3gm	n.s.s.	4/120	600.mg	0/60	1.80gm	2/60	6.00gm	1/60				
g	1976	5.18gm	n.s.s.	3/120	600.mg	0/60	1.80gm	0/60	6.00gm	0/60				
674	415	10.1gm	n.s.s.	0/24	225.mg	0/24	450.mg	0/24	900.mg	0/24	2.25gm	1/24	Hansen;txap,8,29-36;1966	
a	415	3.75gm	n.s.s.	7/24	225.mg	10/24	450.mg	7/24	900.mg	9/24	2.25gm	6/24		

FD & C BLUE NO. 2 (indigo carmine) 860-22-0

	RefNum	LoConf	UpConf	Cntrl	1Dose	1Inc	2Dose	2Inc					Citation or Pathology	Brkly Code
675	411	432.mg	n.s.s.	0/50	260.mg	0/28	520.mg	0/19	1.04gm	0/18	2.08gm	0/25	Hooson;fctx,13,167-176;1975	
a	411	6.04gm	n.s.s.	3/50	260.mg	0/28	520.mg	0/19	1.04gm	1/18	2.08gm	1/25		
676	411	3.49gm	n.s.s.	1/42	240.mg	1/23	480.mg	1/17	960.mg	0/18	1.92gm	2/24		
a	411	4.04gm	n.s.s.	3/42	240.mg	7/23	480.mg	0/17	960.mg	1/18	1.92gm	2/24		
677	1698	9.55gm	n.s.s.	30/120	650.mg	15/57	1.95gm	15/48	6.50gm	20/60			Borzelleca;fctx,23,719-722;1985	
a	1698	14.1gm	n.s.s.	50/120	650.mg	20/57	1.95gm	20/48	6.50gm	22/60				
678	1698	7.54gm	n.s.s.	26/119	600.mg	13/55	1.80gm	19/58	6.00gm	19/60				
a	1698	11.0gm	n.s.s.	22/119	600.mg	8/55	1.80gm	16/58	6.00gm	13/60				
679	415	1.98gm	n.s.s.	6/24	225.mg	5/24	450.mg	5/24	900.mg	4/24	2.25gm	11/24	Hansen;txap,8,29-36;1966	

HC BLUE NO. 1 2784-94-3

	RefNum	LoConf	UpConf	Cntrl	1Dose	1Inc	2Dose	2Inc	Citation or Pathology	Brkly Code
680	TR271	129.mg	274.mg	3/50	379.mg	33/50	765.mg	47/50	liv:hpa,hpc.	s
a	TR271	150.mg	304.mg	1/50	379.mg	24/50	765.mg	47/50		
b	TR271	314.mg	4.97gm	2/50	379.mg	11/50	(765.mg	4/50)		s
c	TR271	199.mg	1.72gm	29/50	379.mg	35/50	765.mg	47/50		
d	TR271	129.mg	274.mg	3/50	379.mg	33/50	765.mg	47/50	liv:hpa,hpc,nnd.	
e	TR271	2.59gm	n.s.s.	4/50	379.mg	2/50	765.mg	2/50	lun:a/a,a/c.	
681	1860m	19.7mg	91.7mg	3/38	151.mg	20/22			Burnett;fctx,25,703-707;1987/pers.comm.	
a	1860m	56.4mg	302.mg	2/38	151.mg	13/22				
b	1860m	77.2mg	499.mg	1/38	151.mg	10/22				
682	1860n	41.2mg	177.mg	3/38	252.mg	20/23				
a	1860n	48.9mg	201.mg	2/38	252.mg	19/23				
b	1860n	300.mg	n.s.s.	1/38	252.mg	4/23				
683	1860o	33.8mg	291.mg	0/10	195.mg	7/10				
a	1860o	54.2mg	1.03gm	0/10	195.mg	5/10				
b	1860o	85.8mg	n.s.s.	0/10	195.mg	3/10				
684	1860r	28.5mg	95.5mg	3/38	195.mg	33/36				
a	1860r	91.4mg	358.mg	2/38	195.mg	20/36				
b	1860r	103.mg	402.mg	1/38	195.mg	18/36				
685	1860s	n.s.s.	162.mg	0/10	390.mg	10/10				
a	1860s	n.s.s.	162.mg	0/10	390.mg	10/10				
b	1860s	172.mg	n.s.s.	0/10	390.mg	3/10				
686	1860u	n.s.s.	106.mg	3/38	390.mg	36/36				
a	1860u	45.8mg	163.mg	2/38	390.mg	34/36				
b	1860u	515.mg	n.s.s.	1/38	390.mg	7/36				
687	TR271	142.mg	1.13gm	15/50	176.mg	31/50	355.mg	37/50	liv:hpa,hpc.	s
a	TR271	133.mg	6.59gm	4/50	176.mg	17/50	(355.mg	10/50)		s
b	TR271	202.mg	1.61gm	11/50	176.mg	20/50	355.mg	31/50	liv:hpc; thy:fca.	C
c	TR271	210.mg	1.95gm	11/50	176.mg	20/50	355.mg	30/50		
d	TR271	846.mg	39.1gm	0/50	176.mg	0/50	355.mg	5/50		
e	TR271	165.mg	n.s.s.	36/50	176.mg	44/50	355.mg	43/50		

Spe Strain Site	Xpo+Xpt			TD50	2Tailpvl
Sex Route Hist	Notes			DR	AuOp

f	M m b6c eat liv MXB 24m24			259.mg *	P<.002	
g	M m b6c eat lun MXB 24m24			no dre	P=1.	
688	R f f34 eat ute esp 24m24	:	±	429.mg *	P<.05	
a	R f f34 eat lun MXA 24m24			702.mg *	P<.03	p
b	R f f34 eat lun a/c 24m24			1.19gm *	P<.02	p
c	R f f34 eat TBA MXB 24m24			no dre	P=1.	
d	R f f34 eat liv MXB 24m24			3.31gm *	P<.3	
689	R m f34 eat liv MXA 24m24	:	±	938.mg *	P<.06	e
a	R m f34 eat liv nnd 24m24			1.66gm *	P<.05	
b	R m f34 eat TBA MXB 24m24			no dre	P=1.	
c	R m f34 eat liv MXB 24m24			938.mg *	P<.06	

HC BLUE NO. 1 (PURIFIED)   100ng..:..1ug....:..10....:..100....:..1mg..:..10....:..100....:..1g....:..10

690	M f b6c eat liv mix 91w91 ekr	<+	noTD50	P<.0005+	
a	M f b6c eat liv hpc 91w91 ekr		88.9mg	P<.0005+	
b	M f b6c eat liv hpa 91w91 ekr		917.mg	P<.09	
691	M f b6c eat liv mix 23m23 er	. + .	70.6mg	P<.0005+	
a	M f b6c eat liv hpc 23m23 er		115.mg	P<.0005+	
b	M f b6c eat liv hpa 23m23 er		530.mg	P<.0005	

HC BLUE NO. 2   100ng..:..1ug....:..10....:..100....:..1mg..:..10....:..100....:..1g....:..10

692	M f b6c eat TBA MXB 24m24 s	:>	42.0gm *	P<.9	-
a	M f b6c eat liv MXB 24m24 s		19.0gm /	P<.4	
b	M f b6c eat lun MXB 24m24 s		no dre	P=1.	
693	M m b6c eat MXA MXA 24m24	: ±	#970.mg \	P<.02	-
a	M m b6c eat --- MXA 24m24		4.04gm *	P<.05	
b	M m b6c eat --- mlp 24m24		5.45gm *	P<.03	
c	M m b6c eat TBA MXB 24m24		5.02gm *	P<.7	
d	M m b6c eat liv MXB 24m24		6.55gm *	P<.6	
e	M m b6c eat lun MXB 24m24		no dre	P=1.	
694	R f f34 eat liv MXA 24m24	:	#8.15gm *	P<.04	-
a	R f f34 eat TBA MXB 24m24		3.13kg	P<1.	
b	R f f34 eat liv MXB 24m24		8.15gm *	P<.04	
695	R m f34 eat thy ccr 24m24	: ±	#1.99gm *	P<.02	-
a	R m f34 eat TBA MXB 24m24		no dre	P=1.	
b	R m f34 eat liv MXB 24m24		20.3gm *	P<.7	

BORIC ACID   100ng..:..1ug....:..10....:..100....:..1mg...:..10....:..100....:..1g....:..10

696	M f b6c eat TBA MXB 24m24	:>	no dre	P=1.	-
a	M f b6c eat liv MXB 24m24		31.0gm *	P<.9	
b	M f b6c eat lun MXB 24m24		4.19gm *	P<.3	
697	M m b6c eat liv MXA 24m24 s	: +	#638.mg *	P<.009	-
a	M m b6c eat sub MXA 24m24 s		706.mg \	P<.004	
b	M m b6c eat sub MXA 24m24 s		1.07gm \	P<.009	
c	M m b6c eat liv hpc 24m24 s		1.12gm *	P<.02	
d	M m b6c eat TBA MXB 24m24 s		500.mg *	P<.03	
e	M m b6c eat liv MXB 24m24 s		638.mg *	P<.009	
f	M m b6c eat lun MXB 24m24 s		38.6mg *	P<1.	

BROMATE, POTASSIUM   100ng..:..1ug....:..10....:..100....:..1mg...:..10....:..100....:..1g....:..10

698	M f b6c wat liv ade 18m24 e	. ±	1.18gm *	P<.05	
a	M f b6c wat lun mix 18m24 e		1.05gm *	P<.2	
b	M f b6c wat liv mix 18m24 e		1.71gm *	P<.4	
c	M f b6c wat lun ade 18m24 e		2.03gm *	P<.3	
d	M f b6c wat tba mix 18m24 e		477.mg *	P<.2	
699	M f the eat liv hem 80w80 e	.>	347.mg *	P<.6	-
700	M m the eat liv tum 80w80 e	.>	no dre	P=1.	-
a	M m the eat tba mal 80w80 e		no dre	P=1.	-
701	R f f34 wat kid mix 26m26 e	. +.	13.7mg *	P<.0005+	
a	R f f34 wat kid adc 26m26 e		16.4mg *	P<.0005	
b	R f f34 wat kid ade 26m26 e		87.5mg *	P<.0005	
c	R f f34 wat thy mix 26m26 e		97.6mg *	P<.01	
d	R f f34 wat liv mix 26m26 e		no dre	P=1.	
e	R f f34 wat tba mix 26m26 e		9.05mg *	P<.003	
702	R m f34 wat kid mix 26m26 ev	.+ .	9.62mg *	P<.0005+	
a	R m f34 wat kid adc 26m26 ev		12.6mg *	P<.0005	
b	R m f34 wat per mso 26m26 ev		25.3mg *	P<.0005+	
c	R m f34 wat kid ade 26m26 ev		46.2mg \	P<.0005	
d	R m f34 wat liv mix 26m26 ev		164.mg *	P<.2	
e	R m f34 wat tba mix 26m26 ev		noTD50	P=1.	
703	R m f3d wat kid mix 6m24 e	. + .	6.67mg	P<.0005+	
a	R m f3d wat kid ade 6m24 e		6.67mg	P<.0005	
b	R m f3d wat per mso 6m24 e		7.84mg	P<.0005	
c	R m f3d wat thy fct 6m24 e		11.3mg	P<.003	
d	R m f3d wat kid adc 6m24 e		79.2mg	P<.3	
704	R m f3d wat kid mix 9m24 e	. + .	4.81mg	P<.0005+	
a	R m f3d wat kid ade 9m24 e		5.57mg	P<.0005	
b	R m f3d wat per mso 9m24 e		14.0mg	P<.002	
c	R m f3d wat kid adc 9m24 e		16.9mg	P<.003	
705	R m f3d wat kid mix 12m24 e	. + .	8.32mg	P<.0005+	
a	R m f3d wat kid ade 12m24 e		10.1mg	P<.0005	

	RefNum	LoConf	UpConf	Cntrl	1Dose	1Inc	2Dose	2Inc	Citation or Pathology
									Brkly Code
f	TR271	142.mg	1.13gm	15/50	176.mg	31/50	355.mg	37/50	liv:hpa,hpc,nnd.
g	TR271	1.12gm	n.s.s.	5/50	176.mg	6/50	355.mg	3/50	lun:a/a,a/c.
688	TR271	185.mg	n.s.s.	5/50	72.9mg	9/50	148.mg	14/50	S
a	TR271	299.mg	n.s.s.	1/50	72.9mg	3/50	148.mg	7/50	lun:a/a,a/c.
b	TR271	453.mg	n.s.s.	0/50	72.9mg	1/50	148.mg	4/50	
c	TR271	108.mg	n.s.s.	45/50	72.9mg	43/50	148.mg	45/50	liv:hpa,hpc,nnd.
d	TR271	809.mg	n.s.s.	0/50	72.9mg	1/50	148.mg	1/50	liv:hpc,nnd.
689	TR271	334.mg	n.s.s.	1/50	58.3mg	0/50	118.mg	6/50	
a	TR271	499.mg	n.s.s.	0/50	58.3mg	0/50	118.mg	3/50	S
b	TR271	117.mg	n.s.s.	41/50	58.3mg	41/50	118.mg	37/50	
c	TR271	334.mg	n.s.s.	1/50	58.3mg	0/50	118.mg	6/50	liv:hpa,hpc,nnd.
HC BLUE NO. 1 (PURIFIED)			**2784-94-3**						
690	1860m	n.s.s.	162.mg	0/10	390.mg	10/10			Burnett;fctx,25,703-707;1987/pers.comm.
a	1860m	31.2mg	277.mg	0/10	390.mg	9/10			
b	1860m	224.mg	n.s.s.	0/10	390.mg	2/10			
691	1860n	31.7mg	141.mg	3/38	390.mg	35/36			
a	1860n	66.1mg	210.mg	2/38	390.mg	32/36			
b	1860n	275.mg	1.41gm	1/38	390.mg	14/36			
HC BLUE NO. 2		**33229-34-4**							
692	TR293	2.24gm	n.s.s.	31/50	1.29gm	19/50	2.58gm	22/50	
a	TR293	4.32gm	n.s.s.	7/50	1.29gm	1/50	2.58gm	8/50	liv:hpa,hpc,nnd.
b	TR293	8.92gm	n.s.s.	1/50	1.29gm	3/50	2.58gm	0/50	lun:a/a,a/c.
693	TR293	416.mg	n.s.s.	4/50	594.mg	14/50	(1.19gm	9/50)	ski:fbs,fib; sub:fbs,fib. S
a	TR293	1.77gm	n.s.s.	1/50	594.mg	5/50	1.19gm	8/50	---:mlh,mlm,mlp. S
b	TR293	2.35gm	n.s.s.	0/50	594.mg	2/50	1.19gm	5/50	S
c	TR293	777.mg	n.s.s.	23/50	594.mg	33/50	1.19gm	37/50	
d	TR293	1.18gm	n.s.s.	10/50	594.mg	16/50	1.19gm	18/50	liv:hpa,hpc,nnd.
e	TR293	2.44gm	n.s.s.	5/50	594.mg	9/50	1.19gm	6/50	lun:a/a,a/c.
694	TR293	3.08gm	n.s.s.	0/50	495.mg	2/50	990.mg	3/50	liv:hpc,nnd. S
a	TR293	800.mg	n.s.s.	43/50	495.mg	36/50	990.mg	41/50	
b	TR293	3.08gm	n.s.s.	0/50	495.mg	2/50	990.mg	3/50	liv:hpa,hpc,nnd.
695	TR293	902.mg	n.s.s.	0/50	198.mg	3/50	396.mg	5/50	S
a	TR293	302.mg	n.s.s.	43/50	198.mg	33/50	(396.mg	32/50)	
b	TR293	2.26gm	n.s.s.	1/50	198.mg	0/50	396.mg	2/50	liv:hpa,hpc,nnd.
BORIC ACID		**10043-35-3**							
696	TR324	777.mg	n.s.s.	25/50	319.mg	27/50	638.mg	26/50	
a	TR324	1.77gm	n.s.s.	5/50	319.mg	4/50	638.mg	6/50	liv:hpa,hpc,nnd.
b	TR324	1.38gm	n.s.s.	1/50	319.mg	5/50	638.mg	4/50	lun:a/a,a/c.
697	TR324	296.mg	27.8gm	14/50	296.mg	19/50	591.mg	15/50	liv:hpa,hpc. S
a	TR324	308.mg	5.24gm	2/50	296.mg	10/50	(591.mg	2/50)	sub:fbs,fib,nfs,srn. S
b	TR324	425.mg	45.2gm	1/50	296.mg	7/50	(591.mg	2/50)	sub:fbs,nfs,srn. S
c	TR324	500.mg	n.s.s.	5/50	296.mg	12/50	591.mg	8/50	S
d	TR324	218.mg	n.s.s.	31/50	296.mg	37/50	591.mg	23/50	
e	TR324	296.mg	27.8gm	14/50	296.mg	19/50	591.mg	15/50	liv:hpa,hpc,nnd.
f	TR324	799.mg	n.s.s.	11/50	296.mg	11/50	591.mg	4/50	lun:a/a,a/c.
BROMATE, POTASSIUM			**7758-01-2**						
698	1789	480.mg	n.s.s.	0/46	75.0mg	3/48	150.mg	3/47	Kurokawa;enhp,69,221-235;1986
a	1789	369.mg	n.s.s.	3/46	75.0mg	3/48	150.mg	8/47	
b	1789	449.mg	n.s.s.	3/46	75.0mg	3/48	150.mg	6/47	
c	1789	549.mg	n.s.s.	2/46	75.0mg	1/48	150.mg	5/47	
d	1789	163.mg	n.s.s.	15/46	75.0mg	16/48	150.mg	22/47	
699	719	56.4mg	n.s.s.	0/53	6.50mg	1/54	9.75mg	0/52	Ginocchio;fctx,17,41-47;1979
700	719	21.3mg	n.s.s.	0/35	6.00mg	0/46	9.00mg	0/53	
a	719	63.9mg	n.s.s.	1/35	6.00mg	1/46	9.00mg	0/53	
701	1550	9.90mg	19.5mg	0/47	14.2mg	28/50	28.3mg	39/49	Kurokawa;jnci,71,965-972;1983
a	1550	11.7mg	23.7mg	0/47	14.2mg	21/50	28.3mg	39/49	
b	1550	49.6mg	234.mg	0/47	14.2mg	8/50	28.3mg	9/49	
c	1550	48.2mg	7.87gm	3/52	14.2mg	10/52	28.3mg	12/52	
d	1550	220.mg	n.s.s.	2/52	14.2mg	1/52	28.3mg	2/52	
e	1550	3.96mg	62.6mg	44/52	14.2mg	48/52	28.3mg	52/52	
702	1550	6.93mg	13.9mg	3/53	12.4mg	32/53	22.5mg	46/52	
a	1550	9.00mg	18.6mg	3/53	12.4mg	24/53	22.5mg	44/52	
b	1550	16.1mg	49.8mg	6/53	12.4mg	17/52	22.5mg	28/46	
c	1550	22.5mg	124.mg	0/53	12.4mg	10/53	(22.5mg	5/52)	
d	1550	62.8mg	n.s.s.	2/53	12.4mg	7/53	22.5mg	6/52	
e	1550	n.s.s.	n.s.s.	53/53	12.4mg	53/53	22.5mg	52/52	
703	1814m	3.08mg	18.8mg	0/19	6.25mg	9/19			Kurokawa;gann,78,358-364;1987
a	1814m	3.08mg	18.8mg	0/19	6.25mg	9/19			
b	1814m	3.50mg	24.7mg	0/19	6.25mg	8/19			
c	1814m	4.57mg	58.1mg	0/19	6.25mg	6/19			
d	1814m	12.9mg	n.s.s.	0/19	6.25mg	1/19			
704	1814n	2.42mg	10.9mg	0/19	9.38mg	14/19			
a	1814n	2.79mg	12.9mg	0/19	9.38mg	13/19			
b	1814n	5.97mg	52.5mg	0/19	9.38mg	7/19			
c	1814n	6.85mg	87.2mg	0/19	9.38mg	6/19			
705	1814o	3.72mg	23.3mg	0/19	12.5mg	9/14			
a	1814o	4.41mg	30.2mg	0/19	12.5mg	8/14			

```
        Spe Strain  Site    Xpo+Xpt                                                          TD50      2Tailpvl
            Sex   Route   Hist      Notes                                                            DR    AuOp
     b    R m f3d wat per mso 12m24 e                                                         19.4mg    P<.003
     c    R m f3d wat thy fct 12m24 e                                                         19.4mg    P<.003
     d    R m f3d wat kid adc 12m24 e                                                         35.5mg    P<.02
   706    R m f3d wat kid ade 52w52 e                              .  +    .                  4.98mg    P<.002
     a    R m f3d wat per mso 52w52 e                                                         25.6mg    P<.2
   707    R m f3d wat per mso 24m24 e                               .  +   .                  12.4mg    P<.0005
     a    R m f3d wat kid mix 24m24 e                                                         28.7mg    P<.0005+
     b    R m f3d wat thy fct 24m24 e                                                         39.8mg    P<.002
     c    R m f3d wat kid ade 24m24 e                                                         48.0mg    P<.004
     d    R m f3d wat thy fca 24m24 e                                                         59.5mg    P<.008
     e    R m f3d wat kid adc 24m24 e                                                         105.mg    P<.04
   708    R m f3d wat per mso 24m24                                  .  +   .                  29.1mg  * P<.0005

     a    R m f3d wat kid rct 24m24                                                           41.6mg  * P<.0005+

     b    R m f3d wat kid ade 24m24                                                           52.0mg  * P<.0005+

     c    R m f3d wat thy fct 24m24 e                                                         78.6mg  * P<.0005

     d    R m f3d wat kid adc 24m24                                                           326.mg  * P<.004

     e    R m f3d wat liv nnd 24m24                                                           68.2mg  Z P<.2


BROMOACETALDEHYDE                      100ng..:..1ug...:..10.....:..100....:..1mg....:..10.....:..100....:..1g.....:..10
   709    M f b6c wat liv hem 80w80 e                                           .>            1.02gm    P<.2    -
     a    M f b6c wat for sqp 80w80 e                                                         2.52gm    P<.8    -
     b    M f b6c wat lun tum 80w80 e                                                         no dre    P=1.    -
     c    M f b6c wat tba mix 80w80 e                                                         no dre    P=1.    -
   710    M m b6c wat for sqp 78w78 e                                           .>            1.89gm    P<.8    -
     a    M m b6c wat liv tum 78w78 e                                                         no dre    P=1.    -
     b    M m b6c wat lun ptm 78w78 e                                                         no dre    P=1.    -
     c    M m b6c wat tba mix 78w78 e                                                         no dre    P=1.    -

BROMODICHLOROMETHANE                   100ng..:..1ug...:..10.....:..100.....:..1mg....:..10.....:..100....:..1g.....:..10
   711    M f b6c gav liv MXA 24m24                                         : + :            28.9mg  * P<.0005c
     a    M f b6c gav liv hpa 24m24                                                          36.0mg  * P<.0005c
     b    M f b6c gav liv hpc 24m24                                                          144.mg  * P<.002 c
     c    M f b6c gav MXA mlh 24m24                                                          132.mg  \ P<.02
     d    M f b6c gav TBA MXB 24m24                                                          62.9mg  * P<.04
     e    M f b6c gav liv MXB 24m24                                                          28.9mg  * P<.0005
     f    M f b6c gav lun MXB 24m24                                                          644.mg  * P<.2
   712    M m b6c gav mul mlp 24m24                                      :     +        :    74.2mg  \ P<.008
     a    M m b6c gav kid MXA 24m24                                                          137.mg  * P<.02  c
     b    M m b6c gav kid tla 24m24                                                          215.mg  * P<.09  c
     c    M m b6c gav kid uac 24m24                                                          336.mg  * P<.03  c
     d    M m b6c gav TBA MXB 24m24                                                          no dre    P=1.
     e    M m b6c gav liv MXB 24m24                                                          no dre    P=1.
     f    M m b6c gav lun MXB 24m24                                                          no dre    P=1.
   713    R f f34 gav MXB MXB 24m24                                         : + :            84.8mg  / P<.0005
     a    R f f34 gav kid MXA 24m24                                                          143.mg  * P<.0005c
     b    R f f34 gav col MXA 24m24                                                          200.mg  * P<.0005c
     c    R f f34 gav kid uac 24m24                                                          272.mg  * P<.002 c
     d    R f f34 gav kid tla 24m24                                                          351.mg  * P<.008 c
     e    R f f34 gav col apn 24m24                                                          364.mg  * P<.004 c
     f    R f f34 gav col acn 24m24                                                          411.mg  * P<.007 c
     g    R f f34 gav TBA MXB 24m24                                                          no dre    P=1.
     h    R f f34 gav liv MXB 24m24                                                          no dre    P=1.
   714    R m f34 gav MXB MXB 24m24                                       :+ :               30.3mg  / P<.0005
     a    R m f34 gav MXA MXA 24m24                                                          30.7mg  / P<.0005c
     b    R m f34 gav MXA acn 24m24                                                          35.6mg  / P<.0005c
     c    R m f34 gav MXA apn 24m24                                                          55.6mg  / P<.0005c
     d    R m f34 gav kid MXA 24m24                                                          152.mg  / P<.0005c
     e    R m f34 gav kid uac 24m24                                                          213.mg  / P<.0005c
     f    R m f34 gav tnv men 24m24                                                          350.mg  * P<.05
     g    R m f34 gav lun MXA 24m24                                                          366.mg  * P<.02
     h    R m f34 gav lun a/a 24m24                                                          447.mg  * P<.05
     i    R m f34 gav kid tla 24m24                                                          583.mg  * P<.05  c
     j    R m f34 gav TBA MXB 24m24                                                          267.mg  * P<.7
     k    R m f34 gav liv MXB 24m24                                                          747.mg  * P<.2
   715    R f wis wat liv nnd 43m43 ev                                      .  +   .         544.mg    P<.002 +
   716    R m wis wat pit tum 42m42 ev                                     .   ±             911.mg    P<.06
     a    R m wis wat liv nnd 42m42 ev                                                       no dre    P=1.

BROMOETHANE                            100ng..:..1ug...:..10.....:..100....:..1mg....:..10.....:..100....:..1g.....:..10
   717    M f b6c inh ute MXA 24m24                                         : + :            535.mg  Z P<.0005c
     a    M f b6c inh ute MXA 24m24                                                          596.mg  Z P<.0005
     b    M f b6c inh ute MXA 24m24                                                          681.mg  Z P<.0005
     c    M f b6c inh ute acn 24m24                                                          822.mg  Z P<.0005
     d    M f b6c inh ute adn 24m24                                                          2.61gm  * P<.002
     e    M f b6c inh ute sqc 24m24                                                          4.44gm  * P<.03
```

	RefNum	LoConf	UpConf	Cntrl	1Dose	1Inc	2Dose	2Inc								Citation or Pathology
																Brkly Code
b	1814o	7.26mg	103.mg	0/19	12.5mg	5/14										
c	1814o	7.26mg	103.mg	0/19	12.5mg	5/14										
d	1814o	10.7mg	n.s.s.	0/19	12.5mg	3/14										
706	1814r	2.66mg	14.9mg	0/8	25.0mg	15/26										
a	1814r	8.84mg	n.s.s.	0/8	25.0mg	4/26										
707	1814s	6.33mg	27.2mg	0/19	25.0mg	15/20										
a	1814s	13.3mg	81.8mg	0/19	25.0mg	9/20										
b	1814s	17.0mg	156.mg	0/19	25.0mg	7/20										
c	1814s	19.4mg	271.mg	0/19	25.0mg	6/20										
d	1814s	22.5mg	895.mg	0/19	25.0mg	5/20										
e	1814s	31.8mg	n.s.s.	0/19	25.0mg	3/20										
708	1851	17.0mg	64.2mg	0/20	.900mg	0/20	1.70mg	3/20	3.30mg	4/24	7.30mg	2/24	16.0mg	3/20		
					43.4mg	15/20										Kurokawa;jnci,77,977-982;1986
a	1851	24.4mg	82.5mg	0/20	.900mg	0/20	1.70mg	0/20	3.30mg	1/24	7.30mg	5/24	16.0mg	5/20		
					43.4mg	9/20										
b	1851	29.4mg	133.mg	0/20	.900mg	0/20	1.70mg	0/20	3.30mg	1/24	7.30mg	5/24	16.0mg	5/20		
					43.4mg	6/20										
c	1851	39.3mg	199.mg	0/16	.900mg	0/19	1.70mg	0/20	3.30mg	1/24	7.30mg	0/24	16.0mg	3/20		
					43.4mg	7/19										
d	1851	98.7mg	2.93gm	0/20	.900mg	0/20	1.70mg	0/20	3.30mg	0/24	7.30mg	0/24	16.0mg	0/20		
					43.4mg	3/20										
e	1851	19.1mg	n.s.s.	2/20	.900mg	0/20	1.70mg	4/20	3.30mg	7/24	7.30mg	0/24	16.0mg	6/20		
					(43.4mg	5/20)										

BROMOACETALDEHYDE   17157-48-1

	RefNum	LoConf	UpConf	Cntrl	1Dose	1Inc		Citation or Pathology
709	1761	166.mg	n.s.s.	0/50	85.0mg	1/30		Van Durren;tcam,5,393-403;1985
a	1761	180.mg	n.s.s.	1/50	85.0mg	1/30		
b	1761	311.mg	n.s.s.	0/50	85.0mg	0/30		
c	1761	150.mg	n.s.s.	7/50	85.0mg	3/30		
710	1761	121.mg	n.s.s.	1/45	62.0mg	1/29		
a	1761	208.mg	n.s.s.	0/45	62.0mg	0/29		
b	1761	155.mg	n.s.s.	5/45	62.0mg	1/29		
c	1761	134.mg	n.s.s.	8/45	62.0mg	2/29		

BROMODICHLOROMETHANE   (dichlorobromomethane) 75-27-4

	RefNum	LoConf	UpConf	Cntrl	1Dose	1Inc	2Dose	2Inc		Citation or Pathology	Brkly
711	TR321	18.6mg	52.2mg	3/50	52.0mg	18/50	104.mg	29/50		liv:hpa,hpc.	
a	TR321	22.8mg	65.0mg	1/50	52.0mg	18/50	104.mg	23/50			
b	TR321	69.9mg	734.mg	2/50	52.0mg	5/50	104.mg	10/50			
c	TR321	47.5mg	n.s.s.	2/50	52.0mg	7/50	(104.mg	1/50)		mul:mlh; spl:mlh.	S
d	TR321	26.3mg	52.2mg	34/50	52.0mg	31/50	104.mg	35/50			
e	TR321	18.6mg	52.2mg	3/50	52.0mg	18/50	104.mg	29/50		liv:hpa,hpc,nnd.	
f	TR321	176.mg	n.s.s.	1/50	52.0mg	2/50	104.mg	3/50		lun:a/a,a/c.	
712	TR321	28.1mg	1.51gm	0/50	17.5mg	5/50	(35.0mg	3/50)			S
a	TR321	61.8mg	n.s.s.	1/50	17.5mg	2/50	35.0mg	9/50		kid:tla,uac.	
b	TR321	80.4mg	n.s.s.	1/50	17.5mg	2/50	35.0mg	6/50			
c	TR321	116.mg	n.s.s.	0/50	17.5mg	0/50	35.0mg	4/50			
d	TR321	42.5mg	n.s.s.	38/50	17.5mg	36/50	35.0mg	35/50			
e	TR321	51.4mg	n.s.s.	17/50	17.5mg	16/50	35.0mg	20/50		liv:hpa,hpc,nnd.	
f	TR321	158.mg	n.s.s.	12/50	17.5mg	3/50	35.0mg	7/50		lun:a/a,a/c.	
713	TR321	52.7mg	150.mg	0/50	34.7mg	1/50	70.1mg	24/50		col:acn,apn; kid:tla,uac.	C
a	TR321	79.8mg	294.mg	0/50	34.7mg	1/50	70.1mg	15/50		kid:tla,uac.	
b	TR321	103.mg	505.mg	0/50	34.7mg	0/50	70.1mg	12/50		col:acn,apn.	
c	TR321	128.mg	934.mg	0/50	34.7mg	0/50	70.1mg	9/50			
d	TR321	152.mg	5.11gm	0/50	34.7mg	1/50	70.1mg	6/50			
e	TR321	157.mg	2.04gm	0/50	34.7mg	0/50	70.1mg	7/50			
f	TR321	168.mg	4.92gm	0/50	34.7mg	0/50	70.1mg	6/50			
g	TR321	73.5mg	n.s.s.	48/50	34.7mg	43/50	70.1mg	43/50			
h	TR321	303.mg	n.s.s.	1/50	34.7mg	3/50	70.1mg	1/50		liv:hpa,hpc,nnd.	
714	TR321	21.7mg	43.8mg	0/50	35.0mg	13/50	70.1mg	46/50		col:acn,apn; kid:tla,uac; rec:acn,apn.	C
a	TR321	22.0mg	44.5mg	0/50	35.0mg	13/50	70.1mg	45/50		col:acn,apn; rec:acn,apn.	
b	TR321	24.9mg	53.3mg	0/50	35.0mg	11/50	70.1mg	38/50		col:acn; rec:acn.	
c	TR321	36.8mg	89.2mg	0/50	35.0mg	3/50	70.1mg	33/50		col:apn; rec:apn.	
d	TR321	81.6mg	345.mg	0/50	35.0mg	1/50	70.1mg	13/50		kid:tla,uac.	
e	TR321	103.mg	603.mg	0/50	35.0mg	0/50	70.1mg	10/50			
f	TR321	143.mg	n.s.s.	0/50	35.0mg	3/50	70.1mg	3/50			S
g	TR321	149.mg	n.s.s.	0/50	35.0mg	2/50	70.1mg	4/50		lun:a/a,a/c.	S
h	TR321	169.mg	n.s.s.	0/50	35.0mg	2/50	70.1mg	3/50			S
i	TR321	201.mg	n.s.s.	0/50	35.0mg	1/50	70.1mg	3/50			
j	TR321	37.0mg	n.s.s.	43/50	35.0mg	41/50	70.1mg	47/50			
k	TR321	207.mg	n.s.s.	1/50	35.0mg	0/50	70.1mg	4/50		liv:hpa,hpc,nnd.	
715	1681	307.mg	1.51gm	0/18	94.9mg	17/53				Tumasonis;eaes,9,233-240;1985	
716	1681	382.mg	n.s.s.	1/22	83.7mg	10/47					
a	1681	920.mg	n.s.s.	5/22	83.7mg	6/47					

BROMOETHANE   (ethyl bromide) 74-96-4

	RefNum	LoConf	UpConf	Cntrl	1Dose	1Inc	2Dose	2Inc	3Dose	3Inc	Citation or Pathology	Brkly
717	TR363	356.mg	855.mg	0/50	137.mg	4/50	275.mg	5/50	550.mg	27/50	ute:acn,adn,sqc.	
a	TR363	387.mg	981.mg	0/50	137.mg	3/50	275.mg	4/50	550.mg	25/50	ute:acn,adn.	S
b	TR363	434.mg	1.15gm	0/50	137.mg	3/50	275.mg	4/50	550.mg	22/50	ute:acn,sqc.	S
c	TR363	503.mg	1.47gm	0/50	137.mg	2/50	275.mg	3/50	550.mg	19/50		S
d	TR363	1.17gm	11.3gm	0/50	137.mg	1/50	275.mg	1/50	550.mg	6/50		S
e	TR363	1.67gm	n.s.s.	0/50	137.mg	1/50	275.mg	1/50	550.mg	3/50		S

```
      Spe Strain Site  Xpo+Xpt                                                TD50   2Tailpvl
          Sex  Route  Hist    Notes                                              DR  AuOp
   f   M f b6c inh TBA MXB 24m24                                              571.mg * P<.009
   g   M f b6c inh liv MXB 24m24                                              no dre   P=1.
   h   M f b6c inh lun MXB 24m24                                              5.81gm * P<.5
 718   M m b6c inh lun MXA 24m24                                   :     ±    1.10gm * P<.03  e
   a   M m b6c inh lun a/c 24m24                                              2.39gm * P<.04
   b   M m b6c inh TBA MXB 24m24                                              1.63gm * P<.5
   c   M m b6c inh liv MXB 24m24                                              3.67gm * P<.7
   d   M m b6c inh lun MXB 24m24                                              1.10gm * P<.03
 719   R f f34 inh bra gln 24m25                                   :     ±    1.02gm * P<.05  e
   a   R f f34 inh lun a/a 24m25                                              1.28gm * P<.03  e
   b   R f f34 inh TBA MXB 24m25                                              no dre   P=1.
   c   R f f34 inh liv MXB 24m25                                              670.mg Z P<.3
 720   R m f34 inh liv nnd 24m25                                   :     ±    1.14gm * P<.03
   a   R m f34 inh amd MXA 24m25                                              149.mg * P<.2   p
   b   R m f34 inh MXB MXB 24m25                                              151.mg Z P<.2
   c   R m f34 inh lun MXA 24m25                                       +hist  582.mg * P<.2   p
   d   R m f34 inh bra gcl 24m25                                       +hist  80.7mg * P<1.   p
   e   R m f34 inh bra MXA 24m25                                       +hist  no dre   P=1.   p
   f   R m f34 inh TBA MXB 24m25                                              no dre   P=1.
   g   R m f34 inh liv MXB 24m25                                              2.28gm Z P<.6

BROMOETHANOL          100ng..:..1ug....:..10......:..100.....:..1mg....:..10......:..100....:..1g......:..10
 721   M f b6c wat for sqp 79w79 e                             .   +   .                69.7mg   P<.0005+
   a   M f b6c wat liv hem 79w79 e                                               800.mg   P<.2
   b   M f b6c wat lun ptm 79w79 e                                               1.89gm   P<.7
   c   M f b6c wat tba mix 79w79 e                                               73.2mg   P<.008
 722   M m b6c wat for sqp 78w78 e                               .   +   .               83.9mg   P<.0005+
   a   M m b6c wat liv tum 78w78 e                                               no dre   P=1.
   b   M m b6c wat lun ptm 78w78 e                                               no dre   P=1.
   c   M m b6c wat tba mix 78w78 e                                               86.6mg   P<.03

C.I. DIRECT BROWN 95  100ng..:..1ug....:..10......:..100.....:..1mg....:..10......:..100....:..1g......:..10
 723   R f f34 eat liv MXA 13w13                           :  (+)  :                2.07mg Z P<.002 c
   a   R f f34 eat TBA MXB 13w13                                              2.07mg Z P<.002
   b   R f f34 eat liv MXB 13w13                                              2.07mg Z P<.002
 724   R m f34 eat TBA MXB 13w13                                              no dre   P=1.   -
   a   R m f34 eat liv MXB 13w13                                              no dre   P=1.

BUDESONIDE            100ng..:..1ug....:..10......:..100.....:..1mg....:..10......:..100....:..1g......:..10
 725   R m cdr wat liv mix 24m24 ers              .   +       .                .291mg   P<.005 +
   a   R m cdr wat liv hpa 24m24 ers                                          .491mg   P<.03
   b   R m cdr wat liv hpc 24m24 ers                                          .822mg   P<.08

1,3-BUTADIENE         100ng..:..1ug....:..10......:..100.....:..1mg....:..10......:..100....:..1g......:..10
 726   M f b6c inh MXB MXB 61w61                                :  +  :              53.0mg \ P<.0005

   a   M f b6c inh MXA hes 61w61                                              147.mg \ P<.0005
   b   M f b6c inh lun MXA 61w61                                              203.mg * P<.0005c
   c   M f b6c inh lun a/c 61w61                                              227.mg \ P<.0005c
   d   M f b6c inh lun a/a 61w61                                              259.mg \ P<.0005c
   e   M f b6c inh --- mno 61w61                                              270.mg \ P<.0005c
   f   M f b6c inh hea hes 61w61                                              283.mg * P<.0005c
   g   M f b6c inh ova MXA 61w61                                              370.mg * P<.0005c
   h   M f b6c inh for MXA 61w61                                              396.mg \ P<.002
   i   M f b6c inh for sqp 61w61                                              461.mg \ P<.005 c
   j   M f b6c inh for MXA 61w61                                              462.mg * P<.0005
   k   M f b6c inh for MXA 61w61                                              483.mg * P<.0005
   l   M f b6c inh mgl acc 61w61                                              1.03gm * P<.002 c
   m   M f b6c inh liv MXA 61w61                                              1.17gm * P<.004 c
   n   M f b6c inh liv hpa 61w61                                              1.59gm * P<.009 c
   o   M f b6c inh ova gcc 61w61                                              7.58gm * P<.2   c
   p   M f b6c inh TBA MXB 61w61                                              41.5mg \ P<.0005
   q   M f b6c inh liv MXB 61w61                                              1.17gm * P<.004
   r   M f b6c inh lun MXB 61w61                                              203.mg * P<.0005
 727   M f b6c inh MXB MXB 19m24 a                         :  +:                6.79mg Z P<.0005

   a   M f b6c inh lun MXA 19m24 a                                           12.4mg Z P<.0005c
   b   M f b6c inh lun a/a 19m24 a                                           20.3mg Z P<.0005
   c   M f b6c inh liv MXA 19m24 a                                           22.8mg * P<.0005c
   d   M f b6c inh lun MXA 19m24 a                                           23.9mg Z P<.0005
   e   M f b6c inh hag MXA 19m24 a                                           26.2mg Z P<.0005
   f   M f b6c inh hag MXA 19m24 a                                           26.2mg Z P<.0005c
   g   M f b6c inh --- MXA 19m24 a                                           27.4mg Z P<.0005

   h   M f b6c inh liv hpa 19m24 a                                           28.5mg Z P<.0005
   i   M f b6c inh --- MXA 19m24 a                                           35.0mg Z P<.0005
   j   M f b6c inh --- MXA 19m24 a                                           35.2mg Z P<.0005c

   k   M f b6c inh --- hes 19m24 a                                           44.2mg Z P<.0005
   l   M f b6c inh mgl MXA 19m24 a                                           47.4mg * P<.0005c
```

	RefNum	LoConf	UpConf	Cntrl	1Dose	1Inc	2Dose	2Inc	3Dose	3Inc	Citation or Pathology	Brkly Code
f	TR363	275.mg	16.5gm	27/50	137.mg	24/50	275.mg	29/50	550.mg	37/50		
g	TR363	1.47gm	n.s.s.	5/50	137.mg	6/50	275.mg	6/50	550.mg	3/50	liv:hpa,hpc,nnd.	
h	TR363	1.19gm	n.s.s.	6/50	137.mg	3/50	275.mg	5/50	550.mg	6/50	lun:a/a,a/c.	
718	TR363	481.mg	n.s.s.	7/50	115.mg	6/50	229.mg	12/50	458.mg	15/50		S
a	TR363	965.mg	n.s.s.	2/50	115.mg	0/50	229.mg	5/50	458.mg	6/50		
b	TR363	328.mg	n.s.s.	30/50	115.mg	30/50	229.mg	34/50	458.mg	33/50		
c	TR363	537.mg	n.s.s.	21/50	115.mg	18/50	229.mg	20/50	458.mg	22/50	liv:hpa,hpc,nnd.	
d	TR363	481.mg	n.s.s.	7/50	115.mg	6/50	229.mg	12/50	458.mg	15/50	lun:a/a,a/c.	
719	TR363	373.mg	n.s.s.	0/50	32.7mg	1/50	65.5mg	1/50	131.mg	3/50		
a	TR363	388.mg	n.s.s.	0/50	32.7mg	0/50	65.5mg	0/50	131.mg	3/50		
b	TR363	99.5mg	n.s.s.	49/50	32.7mg	46/50	65.5mg	40/50	131.mg	44/50		
c	TR363	178.mg	n.s.s.	1/50	32.7mg	0/50	65.5mg	4/50	(131.mg	0/50)	liv:hpa,hpc,nnd.	
720	TR363	339.mg	n.s.s.	0/50	22.9mg	0/50	45.8mg	0/50	91.7mg	3/50		S
a	TR363	49.7mg	n.s.s.	8/50	22.9mg	23/50	45.8mg	18/50	91.7mg	21/50	amd:phe,phm.	
b	TR363	47.4mg	n.s.s.	8/50	22.9mg	28/50	45.8mg	21/50	91.7mg	22/50	amd:phe,phm; bra:ast,gcl,gln,oli; lun:a/a,a/c.	P
c	TR363	219.mg	n.s.s.	0/50	22.9mg	0/50	45.8mg	4/50	91.7mg	1/50	lun:a/a,a/c.	
d	TR363	259.mg	n.s.s.	0/50	22.9mg	3/50	45.8mg	1/50	91.7mg	1/50		
e	TR363	557.mg	n.s.s.	0/50	22.9mg	3/50	45.8mg	0/50	91.7mg	0/50	bra:ast,gln,oli.	
f	TR363	63.2mg	n.s.s.	43/50	22.9mg	47/50	45.8mg	47/50	91.7mg	43/50		
g	TR363	358.mg	n.s.s.	2/50	22.9mg	0/50	45.8mg	0/50	91.7mg	3/50	liv:hpa,hpc,nnd.	

BROMOETHANOL   540-51-2

	RefNum	LoConf	UpConf	Cntrl	1Dose	1Inc	Citation or Pathology
721	1761	32.8mg	215.mg	1/50	71.0mg	10/29	Van Durren;tcam,5,393-403;1985
a	1761	130.mg	n.s.s.	0/50	71.0mg	1/29	
b	1761	141.mg	n.s.s.	1/50	71.0mg	1/29	
c	1761	30.8mg	1.68gm	7/50	71.0mg	12/29	
722	1761	37.9mg	315.mg	1/45	76.0mg	9/29	
a	1761	255.mg	n.s.s.	0/45	76.0mg	0/29	
b	1761	190.mg	n.s.s.	5/45	76.0mg	1/29	
c	1761	33.5mg	n.s.s.	8/45	76.0mg	12/29	

C.I. DIRECT BROWN 95   16071-86-6

	RefNum	LoConf	UpConf	Cntrl	1Dose	1Inc	2Dose	2Inc	3Dose	3Inc	4Dose	4Inc	5Dose	5Inc	Citation or Pathology
723	TR108	.765mg	9.32mg	0/10	9.50mg	0/10	18.7mg	0/10	37.5mg	0/10	75.0mg	5/10	(150.mg	0/10)	liv:hpc,nnd.
a	TR108	.765mg	9.32mg	0/10	9.50mg	0/10	18.7mg	0/10	37.5mg	0/10	75.0mg	5/10	(150.mg	0/10)	
b	TR108	.765mg	9.32mg	0/10	9.50mg	0/10	18.7mg	0/10	37.5mg	0/10	75.0mg	5/10	(150.mg	0/10)	liv:hpa,hpc,nnd.
724	TR108	n.s.s.	n.s.s.	0/10	7.60mg	0/10	15.0mg	0/10	30.0mg	0/10	60.0mg	0/10	120.mg	0/10	liv:hpc,nnd.
a	TR108	n.s.s.	n.s.s.	0/10	7.60mg	0/10	15.0mg	0/10	30.0mg	0/10	60.0mg	0/10	120.mg	0/10	liv:hpa,hpc,nnd.

BUDESONIDE   51333-22-3

	RefNum	LoConf	UpConf	Cntrl	1Dose	1Inc	Citation or Pathology
725	2059	.131mg	2.88mg	11/200	50.0ug	16/100	Ryrfeldt;txpy,20,115-117;1992
a	2059	.186mg	n.s.s.	7/200	50.0ug	10/100	
b	2059	.256mg	n.s.s.	4/200	50.0ug	6/100	

1,3-BUTADIENE   106-99-0

	RefNum	LoConf	UpConf	Cntrl	1Dose	1Inc	2Dose	2Inc	3Dose	3Inc	4Dose	4Inc	5Dose	5Inc	Citation or Pathology	Brkly
726	TR288	30.6mg	106.mg	4/50	435.mg	30/50	(869.mg	44/50)							---:mno; for:sqp; hea:hes; liv:hpa,hpc; lun:a/a,a/c; mgl:acc; ova:gcc,gct.	C
a	TR288	73.4mg	333.mg	0/50	435.mg	13/50	(869.mg	19/50)							hea:hes; liv:hes; sub:hes.	S
b	TR288	126.mg	411.mg	3/50	435.mg	12/50	869.mg	23/50							lun:a/a,a/c.	
c	TR288	91.8mg	825.mg	0/50	435.mg	6/50	(869.mg	8/50)								
d	TR288	153.mg	585.mg	3/50	435.mg	9/50	869.mg	20/50								
e	TR288	114.mg	1.03gm	1/50	435.mg	10/50	(869.mg	10/50)								
f	TR288	179.mg	483.mg	0/50	435.mg	11/50	869.mg	18/50								
g	TR288	214.mg	728.mg	0/50	435.mg	6/50	869.mg	13/50							ova:gcc,gct.	
h	TR288	133.mg	2.44gm	0/50	435.mg	4/50	(869.mg	1/50)							for:sqc,sqp.	S
i	TR288	140.mg	5.15gm	0/50	435.mg	3/50	(869.mg	1/50)								
j	TR288	252.mg	1.03gm	0/50	435.mg	5/50	869.mg	10/50							for:ppn,sqc,sqp.	S
k	TR288	259.mg	1.10gm	0/50	435.mg	4/50	869.mg	10/50							for:ppn,sqp.	S
l	TR288	462.mg	4.19gm	0/50	435.mg	2/50	869.mg	6/50								
m	TR288	496.mg	6.82gm	0/50	435.mg	2/50	869.mg	5/50							liv:hpa,hpc.	
n	TR288	596.mg	36.9gm	0/50	435.mg	1/50	869.mg	4/50								
o	TR288	1.23gm	n.s.s.	0/50	435.mg	0/50	869.mg	1/50								
p	TR288	25.1mg	75.8mg	6/50	435.mg	40/50	(869.mg	46/50)								
q	TR288	496.mg	6.82gm	0/50	435.mg	2/50	869.mg	5/50							liv:hpa,hpc,nnd.	
r	TR288	126.mg	411.mg	3/50	435.mg	12/50	869.mg	23/50							lun:a/a,a/c.	
727	TR434	4.92mg	9.63mg	28/50	4.26mg	36/50	13.7mg	39/50	42.8mg	45/50	139.mg	48/50	435.mg	71/80	---:hcs,mlm,mlp,mlu,mly; for:sqc,sqp; hag:adc,ade,anb; hea:hes; liv:hpa,hpc; lun:a/a,a/c,adc; mgl:adc,ado,mtm; ova:ade,gcb,gcm,mtb.	C
a	TR434	8.55mg	19.5mg	4/50	4.26mg	15/50	13.7mg	19/50	42.8mg	24/50	139.mg	25/50	435.mg	22/80	lun:a/a,a/c,adc.	
b	TR434	12.9mg	35.5mg	4/50	4.26mg	11/50	13.7mg	12/50	42.8mg	17/50	139.mg	14/50	435.mg	17/80		S
c	TR434	13.7mg	44.6mg	15/50	4.26mg	14/50	13.7mg	15/50	42.8mg	19/50	139.mg	16/50	435.mg	2/80	liv:hpa,hpc.	
d	TR434	15.3mg	39.1mg	0/50	4.26mg	5/50	13.7mg	11/50	42.8mg	9/50	139.mg	19/50	435.mg	8/80	lun:a/c,adc.	S
e	TR434	16.1mg	47.3mg	8/50	4.26mg	10/50	13.7mg	6/50	42.8mg	15/50	139.mg	20/50	435.mg	9/80	hag:ade,anb.	S
f	TR434	16.1mg	47.2mg	8/50	4.26mg	10/50	13.7mg	7/50	42.8mg	15/50	139.mg	20/50	435.mg	9/80	hag:adc,ade,anb.	
g	TR434	16.3mg	50.9mg	9/50	4.26mg	14/50	13.7mg	18/50	42.8mg	11/50	139.mg	16/50	435.mg	36/80	---:hcs,mlm,mlp,mlu,mly.	S
h	TR434	16.3mg	61.6mg	11/50	4.26mg	10/50	13.7mg	9/50	42.8mg	14/50	139.mg	12/50	(435.mg	1/80)		S
i	TR434	21.6mg	59.1mg	5/50	4.26mg	3/50	13.7mg	3/50	42.8mg	9/50	139.mg	24/50	435.mg	27/80	---:hem,hes.	S
j	TR434	19.6mg	74.7mg	6/50	4.26mg	12/50	13.7mg	11/50	42.8mg	7/50	139.mg	9/50	435.mg	32/80	---:mlm,mlp,mlu,mly.	
k	TR434	26.3mg	75.0mg	3/50	4.26mg	2/50	13.7mg	2/50	42.8mg	5/50	139.mg	23/50	435.mg	27/80		S
l	TR434	28.0mg	80.9mg	0/50	4.26mg	2/50	13.7mg	4/50	42.8mg	12/50	139.mg	15/50	435.mg	16/80	mgl:adc,ado,mtm.	

	Spe Strain Site Xpo+Xpt			TD50	2Tailpvl
	Sex Route Hist	Notes		DR	AuOp
m	M f b6c inh liv hpc 19m24	a		50.5mg	* P<.0005
n	M f b6c inh ova MXA 19m24	a		65.4mg	* P<.0005c
o	M f b6c inh --- mlp 19m24	a		74.9mg	* P<.0005c
p	M f b6c inh ova MXA 19m24	a		78.9mg	* P<.0005c
q	M f b6c inh for MXA 19m24	a		83.9mg	Z P<.0005c
r	M f b6c inh mgl ado 19m24	a		87.7mg	* P<.0005
s	M f b6c inh mgl adc 19m24	a		93.3mg	* P<.0005
t	M f b6c inh ova MXA 19m24	a		98.0mg	* P<.0005
u	M f b6c inh hea hes 19m24	a		104.mg	Z P<.0005c
v	M f b6c inh --- hcs 19m24	a		125.mg	* P<.0005c
w	M f b6c inh for sqp 19m24	a		126.mg	Z P<.0005
x	M f b6c inh sub MXA 19m24	a		128.mg	* P<.0005e
y	M f b6c inh sub sar 19m24	a		131.mg	* P<.0005
z	M f b6c inh ova gcm 19m24	a		198.mg	* P<.0005
A	M f b6c inh ova mtb 19m24	a		202.mg	* P<.0005
B	M f b6c inh for sqc 19m24	a		257.mg	* P<.0005
C	M f b6c inh kid rua 19m24	a		1.06gm	* P<.01 e
D	M f b6c inh mgl mtm 19m24	a		2.71gm	* P<.0005
E	M f b6c inh zym MXA 19m24	a		3.54gm	* P<.004 e
F	M f b6c inh ova ade 19m24	a		217.mg	* P<.03
G	M f b6c inh TBA MXB 19m24	a		6.28mg	Z P<.0005
H	M f b6c inh liv MXB 19m24	a		22.8mg	* P<.0005
I	M f b6c inh lun MXB 19m24	a		13.8mg	Z P<.0005
728	M m b6c inh MXB MXB 60w60	: + :		28.8mg	\ P<.0005
a	M m b6c inh lun MXA 60w60			65.9mg	\ P<.0005c
b	M m b6c inh MXA hes 60w60			96.7mg	\ P<.0005
c	M m b6c inh hea hes 60w60			96.7mg	\ P<.0005c
d	M m b6c inh lun a/a 60w60			96.9mg	* P<.0005c
e	M m b6c inh --- mno 60w60			105.mg	* P<.0005c
f	M m b6c inh for MXA 60w60			120.mg	\ P<.0005
g	M m b6c inh for MXA 60w60			140.mg	\ P<.0005c
h	M m b6c inh lun a/c 60w60			312.mg	* P<.0005c
i	M m b6c inh pre MXA 60w60			403.mg	* P<.002
j	M m b6c inh pre sqc 60w60			538.mg	* P<.004
k	M m b6c inh for MXA 60w60			557.mg	* P<.004
l	M m b6c inh TBA MXB 60w60			28.8mg	\ P<.0005
m	M m b6c inh liv MXB 60w60			541.mg	* P<.2
n	M m b6c inh lun MXB 60w60			65.9mg	* P<.0005
729	M m b6c inh MXB MXB 19m24	a	: +:	10.7mg	Z P<.0005
a	M m b6c inh lun MXA 19m24	a		19.2mg	* P<.0005c
b	M m b6c inh liv MXA 19m24	a		23.9mg	Z P<.0005c
c	M m b6c inh hag MXA 19m24	a		26.1mg	Z P<.0005c
d	M m b6c inh hag MXA 19m24	a		27.6mg	Z P<.0005
e	M m b6c inh lun MXA 19m24	a		32.4mg	* P<.0005
f	M m b6c inh lun a/a 19m24	a		33.2mg	* P<.0005
g	M m b6c inh liv hpa 19m24	a		34.1mg	Z P<.0005
h	M m b6c inh liv hpc 19m24	a		34.4mg	* P<.0005
i	M m b6c inh --- MXA 19m24	a		38.0mg	Z P<.0005
j	M m b6c inh --- hes 19m24	a		45.5mg	Z P<.0005
k	M m b6c inh --- MXA 19m24	a		63.5mg	Z P<.0005
l	M m b6c inh hea hes 19m24	a		64.8mg	Z P<.0005c
m	M m b6c inh --- MXA 19m24	a		99.8mg	Z P<.0005c
n	M m b6c inh --- mlp 19m24	a		157.mg	Z P<.0005c
o	M m b6c inh --- hcs 19m24	a		157.mg	Z P<.0005c
p	M m b6c inh for MXA 19m24	a		197.mg	Z P<.0005c
q	M m b6c inh hag adc 19m24	a		221.mg	* P<.005
r	M m b6c inh for sqp 19m24	a		230.mg	Z P<.0005
s	M m b6c inh pre car 19m24	a		384.mg	* P<.0005c
t	M m b6c inh kid rua 19m24	a		268.mg	* P<.02 c
u	M m b6c inh jej car 19m24	a		310.mg	* P<.02 e
v	M m b6c inh sub hes 19m24	a		666.mg	* P<.03
w	M m b6c inh bra onm 19m24	a		no dre	P=1.
x	M m b6c inh bra ogm 19m24	a		no dre	P=1.
y	M m b6c inh TBA MXB 19m24	a		10.5mg	Z P<.0005
z	M m b6c inh liv MXB 19m24	a		23.9mg	Z P<.0005
A	M m b6c inh lun MXB 19m24	a		19.6mg	* P<.0005
730	M m b6c inh MXB MXB 12m24	: + :		6.12mg	P<.0005
a	M m b6c inh lun MXA 12m24			7.17mg	P<.0005c
b	M m b6c inh lun a/a 12m24			9.13mg	P<.0005
c	M m b6c inh --- MXA 12m24			11.1mg	P<.0005
d	M m b6c inh --- hes 12m24			11.9mg	P<.0005
e	M m b6c inh lun MXA 12m24			13.0mg	P<.0005
f	M m b6c inh hag MXA 12m24			13.1mg	P<.0005c
g	M m b6c inh liv MXA 12m24			13.8mg	P<.0005
h	M m b6c inh hea hes 12m24			13.9mg	P<.0005c
i	M m b6c inh liv MXA 12m24			14.6mg	P<.0005
j	M m b6c inh hag MXA 12m24			14.9mg	P<.0005

	RefNum	LoConf	UpConf	Cntrl	1Dose	1Inc	2Dose	2Inc	Dose	Inc	Dose	Inc	Dose	Inc	Citation or Pathology	Brkly Code
m	TR434	26.6mg	125.mg	4/50	4.26mg	6/50	13.7mg	8/50	42.8mg	9/50	139.mg	8/50	435.mg	1/80		S
n	TR434	34.9mg	133.mg	1/50	4.26mg	0/50	13.7mg	1/50	42.8mg	9/50	139.mg	8/50	435.mg	6/80	ova:gcb,gcm.	
o	TR434	36.6mg	171.mg	1/50	4.26mg	3/50	13.7mg	6/50	42.8mg	3/50	139.mg	8/50	435.mg	31/80		
p	TR434	35.1mg	256.mg	2/50	4.26mg	4/50	13.7mg	1/50	42.8mg	4/50	139.mg	6/50	435.mg	2/80	ova:ade,mtb.	
q	TR434	39.8mg	192.mg	0/50	4.26mg	0/50	13.7mg	3/50	42.8mg	2/50	139.mg	4/50	435.mg	22/80	for:sqc,sqp.	
r	TR434	40.5mg	237.mg	0/50	4.26mg	1/50	13.7mg	2/50	42.8mg	6/50	139.mg	4/50	435.mg	0/80		S
s	TR434	49.3mg	174.mg	0/50	4.26mg	2/50	13.7mg	2/50	42.8mg	6/50	139.mg	11/50	435.mg	12/80		
t	TR434	47.4mg	235.mg	1/50	4.26mg	0/50	13.7mg	1/50	42.8mg	6/50	139.mg	6/50	435.mg	6/80	ova:gcb,glb.	S
u	TR434	60.0mg	161.mg	0/50	4.26mg	0/50	13.7mg	0/50	42.8mg	1/50	139.mg	21/50	435.mg	23/80		
v	TR434	54.2mg	431.mg	3/50	4.26mg	2/50	13.7mg	7/50	42.8mg	4/50	139.mg	7/50	435.mg	4/80		
w	TR434	51.7mg	337.mg	0/50	4.26mg	0/50	13.7mg	2/50	42.8mg	1/50	139.mg	3/50	435.mg	16/80		S
x	TR434	57.9mg	434.mg	1/50	4.26mg	2/50	13.7mg	3/50	42.8mg	5/50	139.mg	3/50	435.mg	3/80	sub:nfs,sar.	
y	TR434	58.8mg	437.mg	1/50	4.26mg	2/50	13.7mg	2/50	42.8mg	5/50	139.mg	3/50	435.mg	3/80		
z	TR434	68.7mg	790.mg	0/50	4.26mg	0/50	13.7mg	0/50	42.8mg	3/50	139.mg	2/50	435.mg	0/80		S
A	TR434	73.1mg	622.mg	0/50	4.26mg	0/50	13.7mg	0/50	42.8mg	2/50	139.mg	5/50	435.mg	1/80		S
B	TR434	77.9mg	1.04gm	0/50	4.26mg	0/50	13.7mg	1/50	42.8mg	1/50	139.mg	1/50	435.mg	6/80		S
C	TR434	199.mg	246.gm	0/50	4.26mg	0/50	13.7mg	0/50	42.8mg	0/50	139.mg	2/50	435.mg	0/80		
D	TR434	869.mg	13.2gm	0/50	4.26mg	0/50	13.7mg	0/50	42.8mg	0/50	139.mg	0/50	435.mg	4/80		S
E	TR434	808.mg	55.7gm	0/50	4.26mg	0/50	13.7mg	0/50	42.8mg	0/50	139.mg	0/50	435.mg	2/80	zym:ade,car.	
F	TR434	55.2mg	n.s.s.	2/50	4.26mg	4/50	13.7mg	1/50	42.8mg	2/50	139.mg	1/50	435.mg	1/80		
G	TR434	4.65mg	8.91mg	35/50	4.26mg	47/50	13.7mg	43/50	42.8mg	48/50	139.mg	49/50	435.mg	72/80		
H	TR434	13.7mg	44.6mg	15/50	4.26mg	14/50	13.7mg	15/50	42.8mg	19/50	139.mg	16/50	435.mg	2/80	liv:hpa,hpc,nnd.	
I	TR434	9.48mg	21.9mg	4/50	4.26mg	13/50	13.7mg	18/50	42.8mg	22/50	139.mg	24/50	435.mg	22/80	lun:a/a,a/c.	
728	TR288	17.5mg	50.0mg	2/50	362.mg	43/50	(724.mg	39/50)							---:mno; for:ppn,sqp; hea:hes; lun:a/a,a/c.	C
a	TR288	38.7mg	128.mg	2/50	362.mg	14/50	724.mg	15/50							lun:a/a,a/c.	
b	TR288	51.1mg	198.mg	0/50	362.mg	16/50	(724.mg	8/50)							hea:hes; pec:hes.	S
c	TR288	51.1mg	198.mg	0/50	362.mg	16/50	(724.mg	7/50)								
d	TR288	54.4mg	203.mg	2/50	362.mg	12/50	724.mg	11/50								
e	TR288	70.1mg	158.mg	0/50	362.mg	23/50	724.mg	29/50								
f	TR288	49.6mg	391.mg	0/50	362.mg	7/50	(724.mg	1/50)							for:ppn,sqc,sqp.	S
g	TR288	53.3mg	600.mg	0/50	362.mg	5/50	(724.mg	0/50)							for:ppn,sqp.	
h	TR288	126.mg	1.08gm	0/50	362.mg	2/50	724.mg	5/50								
i	TR288	150.mg	1.92gm	0/50	362.mg	3/50	724.mg	2/50							pre:can,sqc.	S
j	TR288	181.mg	4.83gm	0/50	362.mg	3/50	724.mg	1/50								S
k	TR288	197.mg	4.31gm	0/50	362.mg	4/50	724.mg	1/50							for:sqc,sqp.	S
l	TR288	17.3mg	52.5mg	10/50	362.mg	44/50	(724.mg	40/50)								
m	TR288	149.mg	n.s.s.	8/50	362.mg	6/50	724.mg	2/50							liv:hpa,hpc,nnd.	
n	TR288	38.7mg	128.mg	2/50	362.mg	14/50	724.mg	15/50							lun:a/a,a/c.	
729	TR434	7.80mg	15.3mg	42/50	3.55mg	38/50	11.4mg	44/50	35.6mg	47/50	114.mg	49/50	362.mg	62/73	---:hcs,mlh,mlm, mlp,mlu,mly; for:sqc,sqp; hag:adc,ade,anb; hea:hes; kid:rua; liv:hpa,hpc; lun:a/a,a/c,adc; pre:car.	C
a	TR434	12.6mg	32.9mg	21/50	3.55mg	23/50	11.4mg	19/50	35.6mg	31/50	114.mg	35/50	362.mg	3/73	lun:a/a,a/c,adc.	
b	TR434	14.9mg	43.3mg	21/50	3.55mg	23/50	11.4mg	30/50	35.6mg	25/50	114.mg	33/50	362.mg	5/73	liv:hpa,hpc.	
c	TR434	17.5mg	41.5mg	6/50	3.55mg	7/50	11.4mg	9/50	35.6mg	20/50	114.mg	31/50	362.mg	6/73	hag:adc,ade,anb.	
d	TR434	18.4mg	45.0mg	6/50	3.55mg	7/50	11.4mg	8/50	35.6mg	19/50	114.mg	30/50	362.mg	6/73	hag:ade,anb.	S
e	TR434	20.5mg	58.0mg	5/50	3.55mg	6/50	11.4mg	11/50	35.6mg	12/50	114.mg	22/50	362.mg	3/73	lun:a/a,adc.	S
f	TR434	19.7mg	69.3mg	18/50	3.55mg	20/50	11.4mg	10/50	35.6mg	25/50	114.mg	21/50	362.mg	3/73		S
g	TR434	20.5mg	65.6mg	13/50	3.55mg	13/50	11.4mg	19/50	35.6mg	16/50	114.mg	24/50	362.mg	5/73		S
h	TR434	20.5mg	68.2mg	11/50	3.55mg	16/50	11.4mg	16/50	35.6mg	17/50	114.mg	26/50	362.mg	1/73		S
i	TR434	24.5mg	63.4mg	2/50	3.55mg	5/50	11.4mg	3/50	35.6mg	11/50	114.mg	24/50	362.mg	4/73	---:hem,hes.	S
j	TR434	28.6mg	78.7mg	1/50	3.55mg	4/50	11.4mg	3/50	35.6mg	7/50	114.mg	24/50	362.mg	4/73		S
k	TR434	40.8mg	101.mg	4/50	3.55mg	2/50	11.4mg	8/50	35.6mg	11/50	114.mg	9/50	362.mg	55/73	---:hcs,mlh,mlm, mlp,mlu,mly.	S
l	TR434	38.8mg	114.mg	0/50	3.55mg	0/50	11.4mg	1/50	35.6mg	5/50	114.mg	20/50	362.mg	4/73		
m	TR434	60.0mg	174.mg	4/50	3.55mg	2/50	11.4mg	4/50	35.6mg	6/50	114.mg	2/50	362.mg	51/73	---:mlh,mlm,mlp, mlu,mly.	
n	TR434	93.7mg	248.mg	2/50	3.55mg	0/50	11.4mg	2/50	35.6mg	4/50	114.mg	2/50	362.mg	49/73		
o	TR434	82.9mg	372.mg	0/50	3.55mg	0/50	11.4mg	4/50	35.6mg	5/50	114.mg	7/50	362.mg	4/73		
p	TR434	91.0mg	494.mg	1/50	3.55mg	0/50	11.4mg	0/50	35.6mg	1/50	114.mg	8/50	362.mg	4/73	for:sqc,sqp.	
q	TR434	85.7mg	3.26gm	0/50	3.55mg	1/50	11.4mg	1/50	35.6mg	3/50	114.mg	2/50	362.mg	0/73		S
r	TR434	98.7mg	674.mg	1/50	3.55mg	0/50	11.4mg	0/50	35.6mg	1/50	114.mg	7/50	362.mg	2/73		S
s	TR434	131.mg	1.52gm	0/50	3.55mg	0/50	11.4mg	0/50	35.6mg	0/50	114.mg	5/50	362.mg	0/73		
t	TR434	92.1mg	n.s.s.	0/50	3.55mg	1/50	11.4mg	0/50	35.6mg	3/50	114.mg	1/50	362.mg	0/73		
u	TR434	97.0mg	n.s.s.	0/50	3.55mg	1/50	11.4mg	1/50	35.6mg	1/50	114.mg	2/50	362.mg	0/73		
v	TR434	164.mg	n.s.s.	0/50	3.55mg	1/50	11.4mg	1/50	35.6mg	0/50	114.mg	3/50	362.mg	0/73		S
w	TR434	n.s.s.	n.s.s.	0/50	3.55mg	0/50	11.4mg	0/50	35.6mg	0/50	114.mg	0/50	362.mg	0/73		
x	TR434	n.s.s.	n.s.s.	0/50	3.55mg	0/50	11.4mg	0/50	35.6mg	0/50	114.mg	0/50	362.mg	0/73		
y	TR434	7.76mg	15.1mg	44/50	3.55mg	40/50	11.4mg	45/50	35.6mg	48/50	114.mg	49/50	362.mg	62/73	liv:hpa,hpc,nnd.	
z	TR434	14.9mg	43.3mg	21/50	3.55mg	23/50	11.4mg	30/50	35.6mg	25/50	114.mg	33/50	362.mg	5/73	liv:hpa,hpc,nnd.	
A	TR434	12.8mg	33.8mg	21/50	3.55mg	23/50	11.4mg	30/50	35.6mg	30/50	114.mg	33/50	362.mg	5/73		
730	TR434a	3.14mg	12.8mg	38/50	90.3mg	50/50									---:hcs,mlm,mlp,mly; for:sqc,sqp; hag:adc,ade,adi,anb; hea:hes; kid:ade,rua; liv:hpa; lun:a/a,a/c,adc; pre:car,cnb.	C
a	TR434a	3.25mg	17.2mg	21/50	90.3mg	32/50									lun:a/a,a/c,adc.	
b	TR434a	3.90mg	24.2mg	18/50	90.3mg	26/50										S
c	TR434a	3.83mg	29.2mg	2/50	90.3mg	34/50									---:hem,hes.	S
d	TR434a	3.99mg	30.6mg	1/50	90.3mg	34/50										S
e	TR434a	4.19mg	45.1mg	5/50	90.3mg	16/50									lun:a/c,adc.	S
f	TR434a	5.55mg	30.1mg	6/50	90.3mg	30/50									hag:adc,ade,adi,anb.	
g	TR434a	5.73mg	38.4mg	21/50	90.3mg	25/50									liv:hpa,hpb,hpc.	S
h	TR434a	4.42mg	30.6mg	0/50	90.3mg	33/50										
i	TR434a	5.90mg	42.0mg	21/50	90.3mg	24/50									liv:hpa,hpc.	S
j	TR434a	5.97mg	35.0mg	6/50	90.3mg	28/50									hag:ade,anb.	S

	Spe	Strain	Site	Xpo+Xpt			TD50	2Tailpvl
	Sex	Route	Hist	Notes			DR	AuOp
k	M m	b6c	inh	liv	hpa	12m24	15.8mg	P<.0005c
l	M m	b6c	inh	pre	MXA	12m24	20.1mg	P<.0005c
m	M m	b6c	inh	---	MXA	12m24	27.6mg	P<.0005
n	M m	b6c	inh	liv	MXA	12m24	41.4mg	P<.0005
o	M m	b6c	inh	for	MXA	12m24	42.8mg	P<.0005c
p	M m	b6c	inh	---	MXA	12m24	46.4mg	P<.0005c
q	M m	b6c	inh	liv	hpc	12m24	47.0mg	P<.002
r	M m	b6c	inh	---	MXA	12m24	76.2mg	P<.006
s	M m	b6c	inh	for	sqp	12m24	77.7mg	P<.004
t	M m	b6c	inh	hag	MXA	12m24	88.9mg	P<.007
u	M m	b6c	inh	---	hcs	12m24	110.mg	P<.0005c
v	M m	b6c	inh	for	sqc	12m24	154.mg	P<.0005
w	M m	b6c	inh	kid	MXA	12m24	190.mg	P<.003 c
x	M m	b6c	inh	---	mlp	12m24	129.mg	P<.03 c
y	M m	b6c	inh	bra	ogm	12m24	no dre	P=1.
z	M m	b6c	inh	bra	onm	12m24	no dre	P=1.
A	M m	b6c	inh	---	hem	12m24	no dre	P=1.
B	M m	b6c	inh	zym	ade	12m24	no dre	P=1.
C	M m	b6c	inh	zym	car	12m24	no dre	P=1.
D	M m	b6c	inh	TBA	MXB	12m24	5.84mg	P<.0005
E	M m	b6c	inh	liv	MXB	12m24	14.6mg	P<.0005
F	M m	b6c	inh	lun	MXB	12m24	7.25mg	P<.0005
731	M m	b6c	inh	MXB	MXB	9m24 : + :	10.9mg	P<.0005
a	M m	b6c	inh	lun	MXA	9m24	11.0mg	P<.0005c
b	M m	b6c	inh	lun	a/c	9m24	16.3mg	P<.0005
c	M m	b6c	inh	liv	MXA	9m24	16.8mg	P<.0005
d	M m	b6c	inh	---	MXA	9m24	17.3mg	P<.0005
e	M m	b6c	inh	hag	MXA	9m24	17.5mg	P<.0005c
f	M m	b6c	inh	hag	MXA	9m24	18.0mg	P<.0005
g	M m	b6c	inh	liv	hpa	9m24	18.0mg	P<.0005c
h	M m	b6c	inh	lun	a/a	9m24	18.1mg	P<.0005
i	M m	b6c	inh	---	hes	9m24	20.8mg	P<.0005
j	M m	b6c	inh	hea	hes	9m24	21.9mg	P<.0005c
k	M m	b6c	inh	---	MXA	9m24	57.1mg	P<.002
l	M m	b6c	inh	---	hcs	9m24	128.mg	P<.002 c
m	M m	b6c	inh	kid	rua	9m24	161.mg	P<.005 c
n	M m	b6c	inh	liv	hpc	9m24	62.0mg	P<.02
o	M m	b6c	inh	---	hem	9m24	95.2mg	P<.02
p	M m	b6c	inh	---	MXA	9m24	108.mg	P<.03 c
q	M m	b6c	inh	---	mlp	9m24	142.mg	P<.04 c
r	M m	b6c	inh	for	sqp	9m24	171.mg	P<.06
s	M m	b6c	inh	for	MXA	9m24	171.mg	P<.06 c
t	M m	b6c	inh	pre	MXA	9m24	351.mg	P<.1
u	M m	b6c	inh	pre	MXA	9m24	351.mg	P<.1 c
v	M m	b6c	inh	hag	adc	9m24	531.mg	P<.07
w	M m	b6c	inh	---	MXA	9m24	487.mg	P<.5
x	M m	b6c	inh	zym	car	9m24	625.mg	P<.2
y	M m	b6c	inh	zym	MXA	9m24	647.mg	P<.5
z	M m	b6c	inh	bra	ogm	9m24	no dre	P=1.
A	M m	b6c	inh	bra	onm	9m24	no dre	P=1.
B	M m	b6c	inh	TBA	MXB	9m24	10.8mg	P<.0005
C	M m	b6c	inh	liv	MXB	9m24	16.8mg	P<.0005
D	M m	b6c	inh	lun	MXB	9m24	11.0mg	P<.0005
732	M m	b6c	inh	MXB	MXB	6m24 : + :	12.6mg	P<.0005
a	M m	b6c	inh	lun	MXA	6m24	13.4mg	P<.0005c
b	M m	b6c	inh	lun	a/a	6m24	17.3mg	P<.0005
c	M m	b6c	inh	---	hes	6m24	17.7mg	P<.0005
d	M m	b6c	inh	---	MXA	6m24	17.7mg	P<.0005
e	M m	b6c	inh	lun	a/c	6m24	17.8mg	P<.0005
f	M m	b6c	inh	hea	hes	6m24	18.5mg	P<.0005c
g	M m	b6c	inh	hag	MXA	6m24	22.7mg	P<.0005
h	M m	b6c	inh	hag	MXA	6m24	22.7mg	P<.0005c
i	M m	b6c	inh	liv	MXA	6m24	23.8mg	P<.0005
j	M m	b6c	inh	liv	MXA	6m24	23.8mg	P<.0005
k	M m	b6c	inh	pre	MXA	6m24	27.1mg	P<.0005
l	M m	b6c	inh	pre	MXA	6m24	27.1mg	P<.0005c
m	M m	b6c	inh	liv	hpa	6m24	27.8mg	P<.0005c
n	M m	b6c	inh	---	MXA	6m24	28.8mg	P<.0005
o	M m	b6c	inh	---	MXA	6m24	31.3mg	P<.0005c
p	M m	b6c	inh	---	mlp	6m24	46.0mg	P<.0005c
q	M m	b6c	inh	liv	hpc	6m24	60.0mg	P<.004
r	M m	b6c	inh	liv	MXA	6m24	60.0mg	P<.004
s	M m	b6c	inh	for	MXA	6m24	91.4mg	P<.0005c
t	M m	b6c	inh	---	MXA	6m24	111.mg	P<.002
u	M m	b6c	inh	zym	MXA	6m24	135.mg	P<.005
v	M m	b6c	inh	zym	car	6m24	135.mg	P<.005 e
w	M m	b6c	inh	for	sqc	6m24	136.mg	P<.0005
x	M m	b6c	inh	for	sqp	6m24	284.mg	P<.002
y	M m	b6c	inh	---	hcs	6m24	396.mg	P<.02 c

	RefNum	LoConf	UpConf	Cntrl	1Dose	1Inc	2Dose	2Inc	Citation or Pathology	Brkly Code
k	TR434a	5.76mg	51.1mg	13/50	90.3mg	19/50				
l	TR434a	4.83mg	145.mg	0/50	90.3mg	4/50			pre:car,cnb.	
m	TR434a	7.97mg	102.mg	4/50	90.3mg	15/50			---:hcs,mlm,mlp,mly.	S
n	TR434a	13.5mg	226.mg	11/50	90.3mg	11/50			liv:hpb,hpc.	S
o	TR434a	7.30mg	203.mg	1/50	90.3mg	9/50			for:sqc,sqp.	
p	TR434a	9.93mg	342.mg	4/50	90.3mg	8/50			---:mlm,mlp,mly.	
q	TR434a	14.4mg	341.mg	11/50	90.3mg	10/50				S
r	TR434a	14.4mg	2.13gm	2/50	90.3mg	4/50			---:mlm,mly.	S
s	TR434a	8.51mg	1.67gm	1/50	90.3mg	4/50				S
t	TR434a	15.0mg	1.69gm	0/50	90.3mg	2/50			hag:adc,adi.	S
u	TR434a	32.2mg	434.mg	0/50	90.3mg	7/50				
v	TR434a	54.0mg	670.mg	0/50	90.3mg	5/50				S
w	TR434a	44.1mg	2.09gm	0/50	90.3mg	3/50			kid:ade,rua.	
x	TR434a	12.9mg	n.s.s.	2/50	90.3mg	4/50				
y	TR434a	n.s.s.	n.s.s.	0/50	90.3mg	0/50				
z	TR434a	n.s.s.	n.s.s.	0/50	90.3mg	0/50				
A	TR434a	18.8mg	n.s.s.	1/50	90.3mg	0/50				
B	TR434a	18.8mg	n.s.s.	1/50	90.3mg	0/50				
C	TR434a	n.s.s.	n.s.s.	0/50	90.3mg	0/50				
D	TR434a	3.03mg	12.3mg	44/50	90.3mg	50/50				
E	TR434a	5.90mg	42.0mg	21/50	90.3mg	24/50			liv:hpa,hpc,nnd.	
F	TR434a	3.27mg	17.6mg	21/50	90.3mg	31/50			lun:a/a,a/c.	
731	TR434b	5.96mg	27.1mg	38/50	44.5mg	47/50			---:hcs,mlm,mlp,mly; for:sqc,sqp; hag:adc,ade,anb; hea:hes; kid:rua; liv:hpa; lun:a/a,a/c; pre:car,cnb.	C
a	TR434b	5.99mg	25.4mg	21/50	44.5mg	36/50			lun:a/a,a/c.	
b	TR434b	8.37mg	38.5mg	5/50	44.5mg	22/50				S
c	TR434b	8.75mg	44.4mg	21/50	44.5mg	33/50			liv:hpa,hpc.	S
d	TR434b	9.19mg	38.5mg	2/50	44.5mg	20/50			---:hem,hes.	S
e	TR434b	9.53mg	37.5mg	6/50	44.5mg	27/50			hag:adc,ade,anb.	
f	TR434b	9.67mg	39.1mg	6/50	44.5mg	26/50			hag:ade,anb.	S
g	TR434b	9.14mg	46.7mg	13/50	44.5mg	27/50				
h	TR434b	8.53mg	63.0mg	18/50	44.5mg	24/50				S
i	TR434b	10.8mg	47.6mg	1/50	44.5mg	17/50				S
j	TR434b	11.1mg	48.2mg	0/50	44.5mg	15/50				
k	TR434b	24.5mg	259.mg	4/50	44.5mg	13/50			---:hcs,mlm,mlp,mly.	S
l	TR434b	44.7mg	670.mg	0/50	44.5mg	5/50				
m	TR434b	49.3mg	1.60gm	0/50	44.5mg	4/50				
n	TR434b	23.9mg	n.s.s.	11/50	44.5mg	14/50				S
o	TR434b	27.8mg	n.s.s.	1/50	44.5mg	4/50				S
p	TR434b	34.5mg	n.s.s.	4/50	44.5mg	8/50			---:mlm,mlp,mly.	
q	TR434b	41.0mg	n.s.s.	2/50	44.5mg	6/50				
r	TR434b	41.5mg	n.s.s.	1/50	44.5mg	3/50				
s	TR434b	41.5mg	n.s.s.	1/50	44.5mg	3/50			for:sqc,sqp.	
t	TR434b	57.1mg	n.s.s.	0/50	44.5mg	1/50			pre:ade,car,cnb.	
u	TR434b	57.1mg	n.s.s.	0/50	44.5mg	1/50			pre:car,cnb.	
v	TR434b	130.mg	n.s.s.	0/50	44.5mg	2/50				
w	TR434b	63.3mg	n.s.s.	2/50	44.5mg	2/50			---:mlm,mly.	
x	TR434b	102.mg	n.s.s.	0/50	44.5mg	1/50				
y	TR434b	62.7mg	n.s.s.	1/50	44.5mg	1/50			zym:ade,car.	
z	TR434b	n.s.s.	n.s.s.	0/50	44.5mg	0/50				
A	TR434b	n.s.s.	n.s.s.	0/50	44.5mg	0/50				
B	TR434b	5.83mg	28.7mg	44/50	44.5mg	49/50			liv:hpa,hpc,nnd.	
C	TR434b	8.75mg	44.4mg	21/50	44.5mg	33/50			lun:a/a,a/c.	
D	TR434b	5.99mg	25.4mg	21/50	44.5mg	36/50				
732	TR434c	6.27mg	24.4mg	38/50	90.5mg	49/50			---:hcs,mlm,mlp,mly; bra:ogm; for:sqc,sqp; hag:adc,ade,adi,anb; hea:hes; kid:rua; liv:hpa; lun:a/a,a/c; pre:car,cnb.	C
a	TR434c	5.01mg	36.8mg	21/50	90.5mg	17/50			lun:a/a,a/c.	
b	TR434c	5.62mg	57.0mg	18/50	90.5mg	12/50				S
c	TR434c	4.72mg	56.1mg	1/50	90.5mg	14/50				S
d	TR434c	4.72mg	56.1mg	2/50	90.5mg	14/50			---:hem,hes.	S
e	TR434c	4.74mg	62.0mg	5/50	90.5mg	11/50				S
f	TR434c	4.76mg	64.3mg	0/50	90.5mg	13/50				
g	TR434c	6.72mg	71.9mg	6/50	90.5mg	13/50			hag:ade,anb.	S
h	TR434c	6.72mg	71.9mg	6/50	90.5mg	13/50			hag:adc,ade,adi,anb.	
i	TR434c	8.28mg	77.1mg	21/50	90.5mg	13/50			liv:hpa,hpc.	S
j	TR434c	8.28mg	77.1mg	21/50	90.5mg	13/50			liv:hpa,hpb,hpc.	S
k	TR434c	4.99mg	390.mg	0/50	90.5mg	3/50			pre:ade,car,cnb.	S
l	TR434c	4.99mg	390.mg	0/50	90.5mg	3/50			pre:car,cnb.	
m	TR434c	8.53mg	103.mg	13/50	90.5mg	11/50				
n	TR434c	13.0mg	55.5mg	4/50	90.5mg	35/50			---:hcs,mlm,mlp,mly.	S
o	TR434c	13.5mg	62.6mg	4/50	90.5mg	33/50			---:mlm,mlp,mly.	
p	TR434c	21.1mg	84.8mg	2/50	90.5mg	30/50				
q	TR434c	13.0mg	1.02gm	11/50	90.5mg	4/50				S
r	TR434c	13.0mg	1.02gm	11/50	90.5mg	4/50			liv:hpb,hpc.	S
s	TR434c	36.1mg	227.mg	1/50	90.5mg	10/50			for:sqc,sqp.	
t	TR434c	24.5mg	1.19gm	2/50	90.5mg	3/50			---:mlm,mly.	S
u	TR434c	25.1mg	3.58gm	1/50	90.5mg	2/50			zym:ade,car.	S
v	TR434c	25.1mg	3.58gm	0/50	90.5mg	2/50				
w	TR434c	41.3mg	533.mg	0/50	90.5mg	6/50				S
x	TR434c	97.1mg	1.71gm	1/50	90.5mg	4/50				S
y	TR434c	90.5mg	n.s.s.	0/50	90.5mg	2/50				

	Spe Strain Site	Xpo+Xpt			TD50	2Tailpvl
	Sex Route Hist		Notes			DR AuOp
z	M m b6c inh bra ogm	6m24			403.mg	P<.05 c
A	M m b6c inh kid rua	6m24			961.mg	P<.1 c
B	M m b6c inh --- hem	6m24			no dre	P=1.
C	M m b6c inh bra onm	6m24			no dre	P=1.
D	M m b6c inh TBA MXB	6m24			12.5mg	P<.0005
E	M m b6c inh liv MXB	6m24			23.8mg	P<.0005
F	M m b6c inh lun MXB	6m24			13.4mg	P<.0005
733	M m b6c inh MXB MXB	3m24		: (+) :	7.00mg	P<.0005
a	M m b6c inh lun MXA	3m24			8.59mg	P<.0005c
b	M m b6c inh hag MXA	3m24			11.7mg	P<.0005c
c	M m b6c inh lun MXA	3m24			14.1mg	P<.0005
d	M m b6c inh hag MXA	3m24			14.9mg	P<.0005
e	M m b6c inh liv MXA	3m24			15.1mg	P<.0005
f	M m b6c inh liv MXA	3m24			15.1mg	P<.0005
g	M m b6c inh --- MXA	3m24			16.9mg	P<.0005
h	M m b6c inh lun a/a	3m24			17.1mg	P<.0005
i	M m b6c inh liv hpa	3m24			17.9mg	P<.0005c
j	M m b6c inh --- hes	3m24			21.9mg	P<.0005
k	M m b6c inh liv MXA	3m24			23.4mg	P<.0005
l	M m b6c inh liv hpc	3m24			23.4mg	P<.0005
m	M m b6c inh --- MXA	3m24			26.2mg	P<.0005
n	M m b6c inh hea hes	3m24			32.6mg	P<.0005c
o	M m b6c inh --- MXA	3m24			34.9mg	P<.0005c
p	M m b6c inh for sqc	3m24			43.4mg	P<.0005
q	M m b6c inh for MXA	3m24			44.8mg	P<.0005c
r	M m b6c inh --- hem	3m24			45.8mg	P<.003
s	M m b6c inh hag MXA	3m24			63.5mg	P<.0005
t	M m b6c inh --- mlp	3m24			66.8mg	P<.0005c
u	M m b6c inh --- hcs	3m24			91.1mg	P<.007 c
v	M m b6c inh pre MXA	3m24			119.mg	P<.002
w	M m b6c inh pre MXA	3m24			167.mg	P<.004 c
x	M m b6c inh --- MXA	3m24			78.1mg	P<.02
y	M m b6c inh for sqp	3m24			117.mg	P<.02
z	M m b6c inh kid rua	3m24			201.mg	P<.06 c
A	M m b6c inh bra ogm	3m24			230.mg	P<.03 c
B	M m b6c inh zym car	3m24			336.mg	P<.04 e
C	M m b6c inh bra onm	3m24			394.mg	P<.05 c
D	M m b6c inh zym MXA	3m24			223.mg	P<.2
E	M m b6c inh TBA MXB	3m24			6.92mg	P<.0005
F	M m b6c inh liv MXB	3m24			15.1mg	P<.0005
G	M m b6c inh lun MXB	3m24			8.59mg	P<.0005
734	R f cdr inh mgl mix	24m24 e		. + .	133.mg \	P<.0005+
a	R f cdr inh mgl ben	24m24 e			182.mg \	P<.0005
b	R f cdr inh thy fca	24m24 e			8.27gm *	P<.0005
c	R f cdr inh mgl mal	24m24 e			7.72gm *	P<.07
d	R f cdr inh tba tum	24m24 e			no dre	P=1.
735	R m cdr inh tes ldc	26m26 e		. + 7.55gm *	P<.003 +
a	R m cdr inh pan exa	26m26 e			9.31gm *	P<.006
b	R m cdr inh tba tum	26m26 e			1.47gm *	P<.08

BUTYL BENZYL PHTHALATE   100ng..:..1ug....:..10.....:..100....:..1mg...:..10.....:..100....:..1g....:..10

	Spe Strain Site	Xpo+Xpt			TD50	2Tailpvl
736	M f b6c eat TBA MXB	24m24		:>	no dre	P=1. -
a	M f b6c eat liv MXB	24m24			8.39gm *	P<.2
b	M f b6c eat lun MXB	24m24			no dre	P=1.
737	M m b6c eat TBA MXB	24m24		:>	no dre	P=1. -
a	M m b6c eat liv MXB	24m24			20.7gm *	P<.8
b	M m b6c eat lun MXB	24m24			no dre	P=1.
738	R f f34 eat --- MXA	24m24		: +	1.41gm *	P<.03 a
a	R f f34 eat --- leu	24m24			1.54gm *	P<.04 a
b	R f f34 eat TBA MXB	24m24			no dre	P=1.
c	R f f34 eat liv MXB	24m24			5.62gm *	P<.2
739	R m f34 eat TBA MXB	29w29 s			no dre	P=1. -
a	R m f34 eat liv MXB	29w29 s			no dre	P=1. i

n-BUTYL CHLORIDE   100ng..:..1ug....:..10.....:..100....:..1mg...:..10.....:..100....:..1g....:..10

	Spe Strain Site	Xpo+Xpt			TD50	2Tailpvl
740	M f b6c gav lun a/c	24m24		: +	#1.71gm	P<.03 -
a	M f b6c gav TBA MXB	24m24			no dre	P=1.
b	M f b6c gav liv MXB	24m24			1.38gm	P<.2
c	M f b6c gav lun MXB	24m24			1.16gm	P<.08
741	M f b6c gav TBA MXB	24m24		:>	no dre	P=1.
a	M f b6c gav liv MXB	24m24			no dre	P=1.
b	M f b6c gav lun MXB	24m24			13.1gm	P<1.
742	M m b6c gav liv MXA	24m24		: +	#808.mg *	P<.02 -
a	M m b6c gav MXA hes	24m24			2.83gm *	P<.04
b	M m b6c gav TBA MXB	24m24			423.mg *	P<.01
c	M m b6c gav liv MXB	24m24			808.mg *	P<.02
d	M m b6c gav lun MXB	24m24			3.19gm *	P<.4
747	M m b6c gav TBA MXB	24m24		:>	no dre	P=1. -
a	M m b6c gav liv MXB	24m24			623.mg	P<.3
b	M m b6c gav lun MXB	24m24			no dre	P=1.

	RefNum	LoConf	UpConf	Cntrl	1Dose	1Inc	2Dose	2Inc	Citation or Pathology
									Brkly Code
z	TR434c	65.7mg	n.s.s.	0/50	90.5mg	1/50			
A	TR434c	157.mg	n.s.s.	0/50	90.5mg	1/50			
B	TR434c	n.s.s.	n.s.s.	1/50	90.5mg	0/50			
C	TR434c	n.s.s.	n.s.s.	0/50	90.5mg	0/50			
D	TR434c	6.22mg	24.0mg	44/50	90.5mg	49/50			
E	TR434c	8.28mg	77.1mg	21/50	90.5mg	13/50			liv:hpa,hpc,nnd.
F	TR434c	5.01mg	36.8mg	21/50	90.5mg	17/50			lun:a/a,a/c.
733	TR434d	3.95mg	14.6mg	38/50	45.2mg	49/50			---:hcs,mlm,mlp,mly; bra:ogm,onm; for:sqc,sqp; hag:adc,ade,adi,anb; hea:hes;
									kid:rua; liv:hpa; lun:a/a,a/c,adc; pre:car,cnb. C
a	TR434d	4.42mg	20.5mg	21/50	45.2mg	28/50			lun:a/a,a/c,adc.
b	TR434d	5.73mg	27.0mg	6/50	45.2mg	23/50			hag:adc,ade,adi,anb.
c	TR434d	6.87mg	35.0mg	5/50	45.2mg	18/50			lun:a/c,adc. S
d	TR434d	7.17mg	36.3mg	6/50	45.2mg	20/50			hag:ade,anb. S
e	TR434d	7.17mg	45.5mg	21/50	45.2mg	24/50			liv:hpa,hpb,hpc. S
f	TR434d	7.17mg	45.5mg	21/50	45.2mg	24/50			liv:hpa,hpc. S
g	TR434d	7.78mg	44.7mg	2/50	45.2mg	14/50			---:hem,hes. S
h	TR434d	7.56mg	68.5mg	18/50	45.2mg	17/50			S
i	TR434d	7.86mg	58.2mg	13/50	45.2mg	19/50			
j	TR434d	9.92mg	58.7mg	1/50	45.2mg	12/50			S
k	TR434d	9.57mg	99.0mg	11/50	45.2mg	14/50			liv:hpb,hpc. S
l	TR434d	9.57mg	99.0mg	11/50	45.2mg	14/50			S
m	TR434d	12.4mg	61.8mg	4/50	45.2mg	24/50			---:hcs,mlm,mlp,mly. S
n	TR434d	12.4mg	108.mg	0/50	45.2mg	7/50			
o	TR434d	16.2mg	86.4mg	4/50	45.2mg	22/50			---:mlm,mlp,mly.
p	TR434d	13.4mg	264.mg	0/50	45.2mg	4/50			S
q	TR434d	14.6mg	226.mg	1/50	45.2mg	7/50			for:sqc,sqp.
r	TR434d	13.7mg	449.mg	1/50	45.2mg	4/50			S
s	TR434d	19.2mg	358.mg	0/50	45.2mg	4/50			hag:adc,adi. S
t	TR434d	33.3mg	175.mg	2/50	45.2mg	17/50			
u	TR434d	21.1mg	2.84gm	0/50	45.2mg	2/50			
v	TR434d	43.9mg	550.mg	0/50	45.2mg	5/50			pre:ade,car,cnb. S
w	TR434d	56.7mg	1.32gm	0/50	45.2mg	4/50			pre:car,cnb.
x	TR434d	21.6mg	n.s.s.	2/50	45.2mg	5/50			---:mlm,mly. S
y	TR434d	27.3mg	n.s.s.	1/50	45.2mg	4/50			S
z	TR434d	32.8mg	n.s.s.	0/50	45.2mg	1/50			
A	TR434d	51.0mg	n.s.s.	0/50	45.2mg	2/50			
B	TR434d	79.6mg	n.s.s.	0/50	45.2mg	2/50			
C	TR434d	93.4mg	n.s.s.	0/50	45.2mg	2/50			
D	TR434d	34.2mg	n.s.s.	1/50	45.2mg	2/50			zym:ade,car.
E	TR434d	3.87mg	15.0mg	44/50	45.2mg	49/50			
F	TR434d	7.17mg	45.5mg	21/50	45.2mg	24/50			liv:hpa,hpc,nnd.
G	TR434d	4.42mg	20.5mg	21/50	45.2mg	28/50			lun:a/a,a/c.
734	1829	78.0mg	341.mg	50/100	166.mg	79/100	(1.32gm	81/100)	Owen;amih,48,407-413;1987
a	1829	110.mg	428.mg	32/100	166.mg	64/100	(1.32gm	55/100)	
b	1829	4.26gm	24.8mg	0/100	166.mg	2/100	1.32gm	10/100	
c	1829	2.91gm	n.s.s.	18/100	166.mg	15/100	1.32gm	26/100	
d	1829	893.mg	n.s.s.	97/100	166.mg	98/100	1.32gm	94/100	
735	1829	3.58gm	52.9gm	0/100	116.mg	3/100	927.mg	8/100	
a	1829	3.98gm	136.gm	3/100	116.mg	1/100	927.mg	10/100	
b	1829	550.mg	n.s.s.	84/100	116.mg	70/100	927.mg	87/100	

BUTYL BENZYL PHTHALATE   85-68-7

	RefNum	LoConf	UpConf	Cntrl	1Dose	1Inc	2Dose	2Inc	Citation or Pathology
736	TR213	1.64gm	n.s.s.	33/50	762.mg	27/50	1.53gm	31/50	
a	TR213	2.90gm	n.s.s.	2/50	762.mg	5/50	1.53gm	6/50	liv:hpa,hpc,nnd.
b	TR213	3.57gm	n.s.s.	8/50	762.mg	3/50	(1.53gm	3/50)	lun:a/a,a/c.
737	TR213	2.39gm	n.s.s.	38/50	710.mg	31/50	1.41gm	26/50	
a	TR213	2.49gm	n.s.s.	13/50	710.mg	12/50	1.41gm	14/50	liv:hpa,hpc,nnd.
b	TR213	5.49gm	n.s.s.	17/50	710.mg	11/50	1.41gm	8/50	lun:a/a,a/c.
738	TR213	639.mg	n.s.s.	7/49	296.mg	7/49	589.mg	19/50	---:leu,lym.
a	TR213	673.mg	n.s.s.	7/49	296.mg	7/49	589.mg	18/50	
b	TR213	402.mg	n.s.s.	44/49	296.mg	37/49	589.mg	45/50	
c	TR213	1.65gm	n.s.s.	1/49	296.mg	1/49	589.mg	4/50	liv:hpa,hpc,nnd.
739	TR213	n.s.s.	n.s.s.	0/50	228.mg	0/50	456.mg	0/50	
a	TR213	n.s.s.	n.s.s.	0/50	228.mg	0/50	456.mg	0/50	liv:hpa,hpc,nnd.

n-BUTYL CHLORIDE   109-69-3

	RefNum	LoConf	UpConf	Cntrl	1Dose	1Inc	2Dose	2Inc	Citation or Pathology
740	TR312	592.mg	n.s.s.	0/50	350.mg	4/50			S
a	TR312	295.mg	n.s.s.	31/50	350.mg	29/50			
b	TR312	442.mg	n.s.s.	3/50	350.mg	8/50			liv:hpa,hpc,nnd.
c	TR312	411.mg	n.s.s.	3/50	350.mg	9/50			lun:a/a,a/c.
741	TR312a	198.mg	n.s.s.	35/50	176.mg	35/50			
a	TR312a	521.mg	n.s.s.	9/50	176.mg	7/50			liv:hpa,hpc,nnd.
b	TR312a	366.mg	n.s.s.	6/50	176.mg	8/50			lun:a/a,a/c.
742	TR312	359.mg	n.s.s.	12/50	354.mg	13/50	707.mg	15/50	liv:hpa,hpc. S
a	TR312	1.03gm	n.s.s.	1/50	354.mg	3/50	707.mg	4/50	bom:hes; k/p:hes; liv:hes; mul:hes; omt:hes; spl:hes; thx:hes. S
b	TR312	199.mg	49.9gm	32/50	354.mg	38/50	707.mg	29/50	
c	TR312	359.mg	n.s.s.	12/50	354.mg	13/50	707.mg	15/50	liv:hpa,hpc,nnd.
d	TR312	742.mg	n.s.s.	6/50	354.mg	10/50	707.mg	4/50	lun:a/a,a/c.
747	TR312a	135.mg	n.s.s.	41/50	177.mg	39/50			
a	TR312a	177.mg	n.s.s.	15/50	177.mg	21/50			liv:hpa,hpc,nnd.
b	TR312a	354.mg	n.s.s.	14/50	177.mg	11/50			lun:a/a,a/c.

```
      Spe Strain Site   Xpo+Xpt                                               TD50   2Tailpvl
        Sex  Route  Hist    Notes                                                 DR   AuOp
748  R f f34 gav TBA MXB 24m24                            :>              100.mg * P<.3   -
a    R f f34 gav liv MXB 24m24                                           1.51gm * P<.8
749  R m f34 gav tes ict 24m24                           :  ±            #45.1mg * P<.02  -
a    R m f34 gav TBA MXB 24m24                                           40.5mg * P<.009
b    R m f34 gav liv MXB 24m24                                           no dre  P=1.

2-sec-BUTYL-4,6-DINITROPHENOL    100ng..:..1ug....:..10.......:..100....:..1mg.....:..10.......:..100....:..1g......:..10
750  M f b6a orl lun ade 76w76 evx                  .>                   no dre  P=1.   -
a    M f b6a orl liv hpt 76w76 evx                                       no dre  P=1.   -
b    M f b6a orl tba mix 76w76 evx                                       no dre  P=1.   -
751  M m b6a orl lun ade 76w76 evx                 .>                    2.17mg  P<.4   -
a    M m b6a orl liv hpt 76w76 evx                                       94.1mg  P<1.
b    M m b6a orl tba mix 76w76 evx                                       1.29mg  P<.3   -
752  M f b6c orl lun ade 76w76 evx                  .>                   6.14mg  P<.3   -
a    M f b6c orl liv hpt 76w76 evx                                       no dre  P=1.
b    M f b6c orl tba mix 76w76 evx                                       6.14mg  P<.3   -
753  M m b6c orl liv hpt 76w76 evx              .   ±                    1.30mg  P<.02  -
a    M m b6c orl lun ade 76w76 evx                                       2.77mg  P<.2   -
b    M m b6c orl tba mix 76w76 evx                                       .662mg  P<.002 -

N-n-BUTYL-N-FORMYLHYDRAZINE      100ng..:..1ug....:..10.......:..100....:..1mg.....:..10.......:..100....:..1g......:..10
754  M f swa wat lun mix 92w92 e                        . + .            19.4mg  P<.0005+
a    M f swa wat cli mix 92w92 e                                         252.mg  P<.004 +
b    M f swa wat cli sqc 92w92 e                                         321.mg  P<.01
c    M f swa wat liv mix 92w92 e                                         no dre  P=1.
755  M m swa wat lun mix 90w90 e                        . + .            19.2mg  P<.0005+
a    M m swa wat pre mix 90w90 e                                         28.2mg  P<.0005+
b    M m swa wat pre sqc 90w90 e                                         35.6mg  P<.0005
c    M m swa wat pre fbs 90w90 e                                         304.mg  P<.002
d    M m swa wat pre ang 90w90 e                                         519.mg  P<.02
e    M m swa wat liv agm 90w90 e                                         no dre  P=1.   -

BUTYL p-HYDROXYBENZOATE          100ng..:..1ug....:..10.......:..100....:..1mg.....:..10.......:..100....:..1g......:..10
756  M f icr eat lun tum 24m25                                    ± 5.10gm * P<.06  -
a    M f icr eat liv tum 24m25                                           no dre  P=1.   -
b    M f icr eat tba mix 24m25                                           2.72gm * P<.08  -
757  M m icr eat liv tum 24m25                                .> 27.4gm * P<.5   -
a    M m icr eat lun tum 24m25                                           119.gm * P<1.
b    M m icr eat tba mix 24m25                                           5.24gm * P<.4   -

N-BUTYL-N-(4-HYDROXYBUTYL)NITROSAMINE  ..:..1ug....:..10.......:..100....:..1mg.....:..10.......:..100....:..1g......:..10
758  R f f34 eat ubl car 24m24 es                  <+                    noTD50  P<.0005+
a    R f f34 eat liv hpc 24m24 es                                        no dre  P=1.
759  R m f34 wat ubl mix 25m26 ae               . + .                    .432mg * P<.0005+
a    R m f34 wat ubl pam 25m26 ae                                        .447mg  P<.0005
760  R m f34 wat ubl tcc 78w78 er            . + .                       .175mg  P<.0005+
a    R m f34 wat tes ldc 78w78 er                                        .386mg  P<.002 +
761  R m f34 eat ubl car 24m24 es               . + .                    .549mg * P<.0005+
a    R m f34 eat liv hpc 24m24 es                                        185.mg * P<.4
762  R m sda wat ubl car 24m24                     . + .                 3.30mg  P<.0005+
a    R m sda wat liv lcc 24m24                                           271.mg  P<.3
763  R m sda gav ubl mix 39w74                 .+  .                     1.17mg  P<.0005+
a    R m sda gav lun car 39w74                                           166.mg  P<.3
b    R m sda gav liv tum 39w74                                           no dre  P=1.

DI-tert-BUTYL-4-HYDROXYMETHYL PHENOL  ..:..1ug....:..10.......:..100....:..1mg.....:..10.......:..100....:..1g......:..10
764  R f nbw eat mgl fba 24m24                                      .>   3.71gm * P<.3   -
a    R f nbw eat liv tum 24m24                                           no dre  P=1.
b    R f nbw eat tba mix 24m24                                           3.71gm * P<.3   -
765  R m nbw eat liv tum 24m24                                 .>        no dre  P=1.
a    R m nbw eat tba tum 24m24                                           no dre  P=1.   -

2-tert-BUTYL-4-METHOXYPHENOL     100ng..:..1ug....:..10.......:..100....:..1mg.....:..10.......:..100....:..1g......:..10
766  R m f3d eat for tum 51w52                                     .>    no dre  P=1.
a    R m f3d eat stg tum 51w52                                           no dre  P=1.

N-BUTYL-N'-NITRO-N-NITROSOGUANIDINE  ..:..1ug....:..10.......:..100....:..1mg.....:..10.......:..100....:..1g......:..10
767  R m wis wat stg tum 52w78 er                        .>              no dre  P=1.   -

N-N-BUTYL-N-NITROSOUREA          100ng..:..1ug....:..10.......:..100....:..1mg.....:..10.......:..100....:..1g......:..10
768  R f f34 gav for tum 9m24                 .   +   .                  .267mg  P<.0005+
a    R f f34 gav lun mix 9m24                                            .277mg  P<.0005+
b    R f f34 gav ute mix 9m24                                            .551mg  P<.004 +
c    R f f34 gav mgl adc 9m24                                            .758mg  P<.0005+
d    R f f34 gav col tum 9m24                                            2.31mg  P<.04  +
e    R f f34 gav col tum 9m24                                            2.31mg  P<.04  +
f    R f f34 gav liv tum 9m24                                            no dre  P=1.
769  R m f34 gav lun mix 9m24              .   +   .                     .187mg  P<.0005+
a    R m f34 gav for tum 9m24                                            .259mg  P<.0005+
b    R m f34 gav col tum 9m24                                            1.61mg  P<.04  +
c    R m f34 gav liv tum 9m24                                            no dre  P=1.
```

	RefNum	LoConf	UpConf	Cntrl	1Dose	1Inc	2Dose	2Inc			Citation or Pathology
											Brkly Code
748	TR312	28.6mg	n.s.s.	47/50	42.0mg	46/50	84.1mg	25/50			
a	TR312	187.mg	n.s.s.	1/50	42.0mg	4/50	84.1mg	0/50			liv:hpa,hpc,nnd.
749	TR312	21.3mg	n.s.s.	46/50	42.0mg	45/50	84.5mg	39/50			S
a	TR312	19.8mg	1.01gm	48/50	42.0mg	48/50	84.5mg	42/50			
b	TR312	265.mg	n.s.s.	3/50	42.0mg	3/50	84.5mg	1/50			liv:hpa,hpc,nnd.

2-sec-BUTYL-4,6-DINITROPHENOL   88-85-7

750	1289	1.07mg	n.s.s.	1/17	.959mg	1/17					Innes;ntis,1968/1969
a	1289	1.79mg	n.s.s.	0/17	.959mg	0/17					
b	1289	.843mg	n.s.s.	2/17	.959mg	2/17					
751	1289	.475mg	n.s.s.	2/18	.891mg	4/17					
a	1289	.987mg	n.s.s.	1/18	.891mg	1/17					
b	1289	.351mg	n.s.s.	3/18	.891mg	6/17					
752	1289	.999mg	n.s.s.	0/16	.959mg	1/18					
a	1289	1.90mg	n.s.s.	0/16	.959mg	0/18					
b	1289	.999mg	n.s.s.	0/16	.959mg	1/18					
753	1289	.446mg	n.s.s.	0/16	.891mg	4/18					
a	1289	.680mg	n.s.s.	0/16	.891mg	2/18					
b	1289	.282mg	2.70mg	0/16	.891mg	7/18					

N-n-BUTYL-N-FORMYLHYDRAZINE   16120-70-0

754	1057	11.2mg	35.3mg	25/99	80.0mg	45/49					Toth;carc,1,589-593;1980
a	1057	95.7mg	1.78gm	0/40	80.0mg	5/32					
b	1057	111.mg	25.8gm	0/40	80.0mg	4/32					
c	1057	184.mg	n.s.s.	6/47	80.0mg	5/41					
755	1057	11.4mg	35.8mg	26/100	66.7mg	42/48					
a	1057	18.0mg	47.0mg	0/85	66.7mg	33/47					
b	1057	22.4mg	60.9mg	0/85	66.7mg	29/47					
c	1057	115.mg	1.44gm	0/85	66.7mg	5/47					
d	1057	157.mg	n.s.s.	0/85	66.7mg	3/47					
e	1057	123.mg	n.s.s.	7/70	66.7mg	2/23					

BUTYL p-HYDROXYBENZOATE   94-26-8

756	1688	2.00gm	n.s.s.	2/50	188.mg	4/50	375.mg	2/50	751.mg	8/50	Inai;fctx,23,575-578;1985
a	1688	1.15gm	n.s.s.	0/50	188.mg	0/50	375.mg	0/50	751.mg	0/50	
b	1688	1.04gm	n.s.s.	12/50	188.mg	14/50	375.mg	8/50	751.mg	21/50	
757	1688	5.29gm	n.s.s.	0/50	173.mg	1/50	346.mg	0/50	693.mg	1/50	
a	1688	3.87gm	n.s.s.	3/50	173.mg	2/50	346.mg	2/50	693.mg	3/50	
b	1688	1.41gm	n.s.s.	8/50	173.mg	7/50	346.mg	11/50	693.mg	11/50	

N-BUTYL-N-(4-HYDROXYBUTYL)NITROSAMINE   (butyl-butanol-nitrosamine) 3817-11-6

758	1954	n.s.s.	.471mg	0/118	1.50mg	24/24	3.00mg	23/23	6.00mg	24/24	Fears;txih,4,221-255;1988
a	1954	4.01mg	n.s.s.	0/120	1.50mg	0/23	3.00mg	0/24	6.00mg	0/22	
759	1606	.281mg	.680mg	0/50	50.0ug	0/29	.250mg	6/30	.500mg	23/30 2.50mg 30/30	Ito;zkko,108,169-173;1984
a	1606	.298mg	.691mg	0/50	50.0ug	2/29	.250mg	7/30	.500mg	23/30 2.50mg 29/30	
760	1718	.100mg	.340mg	0/30	.500mg	20/30					LaVoie;gann,76,266-271;1985/pers.comm.
a	1718	.187mg	1.55mg	2/30	.500mg	13/30					
761	1954	.347mg	.873mg	1/118	1.20mg	21/24	2.40mg	23/24	4.80mg	23/24	Fears;txih,4,221-255;1988
a	1954	22.7mg	n.s.s.	1/120	1.20mg	0/24	2.40mg	0/24	4.80mg	1/24	
762	1192	1.98mg	5.64mg	0/40	10.0mg	35/40					Schmahl;arzn,28,49-51;1978
a	1192	44.1mg	n.s.s.	0/40	10.0mg	1/40					
763	1637	.641mg	2.03mg	0/40	10.5mg	44/46					Tacchi;clet,22,89-94;1984
a	1637	27.1mg	n.s.s.	0/40	10.5mg	1/46					
b	1637	50.6mg	n.s.s.	0/40	10.5mg	0/46					

DI-tert-BUTYL-4-HYDROXYMETHYL PHENOL   88-26-6

764	127	604.mg	n.s.s.	0/20	100.mg	0/20	175.mg	1/20			Dacre;txap,17,669-678;1970
a	127	262.mg	n.s.s.	0/20	100.mg	0/20	175.mg	0/20			
b	127	604.mg	n.s.s.	0/20	100.mg	0/20	175.mg	1/20			
765	127	210.mg	n.s.s.	0/20	80.0mg	0/20	140.mg	0/20			
a	127	210.mg	n.s.s.	0/20	80.0mg	0/20	140.mg	0/20			

2-tert-BUTYL-4-METHOXYPHENOL   2409-55-4

766	1950	303.mg	n.s.s.	0/10	392.mg	0/15					Hirose;canr,48,5310-5315;1988
a	1950	303.mg	n.s.s.	0/10	392.mg	0/15					

N-BUTYL-N'-NITRO-N-NITROSOGUANIDINE   13010-08-7

767	1082	1.99mg	n.s.s.	0/9	2.14mg	0/8					Matsukura;gann,70,181-185;1979

N-N-BUTYL-N-NITROSOUREA   869-01-2

768	1915	98.3ug	.753mg	0/12	.969mg	11/12					Lijinsky;txih,5,925-935;1989
a	1915	99.5ug	.880mg	1/12	.969mg	11/12					
b	1915	.216mg	4.51mg	2/12	.969mg	9/12					
c	1915	.314mg	2.64mg	0/12	.969mg	7/12					
d	1915	.693mg	n.s.s.	0/12	.969mg	3/12					
e	1915	.693mg	n.s.s.	0/12	.969mg	3/12					
f	1915	2.40mg	n.s.s.	0/12	.969mg	0/12					
769	1915	68.8ug	.527mg	0/12	.678mg	11/12					
a	1915	.108mg	.721mg	0/12	.678mg	10/12					
b	1915	.485mg	n.s.s.	0/12	.678mg	3/12					
c	1915	1.68mg	n.s.s.	1/12	.678mg	0/12					

	Spe	Strain	Site	Xpo+Xpt				TD50	2Tailpvl
	Sex	Route	Hist	Notes				DR	AuOp
770	R f f3d wat	dgt mix	50w50 ev		. (+).			1.30mg	P<.0005+
a	R f f3d wat	for mix	50w50 ev					2.43mg	P<.0005+
b	R f f3d wat	eso mix	50w50 ev					3.21mg	P<.0005+
c	R f f3d wat	for pam	50w50 ev					4.00mg	P<.0005
d	R f f3d wat	vag mix	50w50 ev					5.05mg	P<.0005+
e	R f f3d wat	vag pam	50w50 ev					6.00mg	P<.0005
f	R f f3d wat	eso pam	50w50 ev					6.58mg	P<.0005
g	R f f3d wat	eso sqc	50w50 ev					8.05mg	P<.0005
h	R f f3d wat	phr mix	50w50 ev					8.05mg	P<.0005
i	R f f3d wat	for sqc	50w50 ev					10.2mg	P<.0005
j	R f f3d wat	phr sqc	50w50 ev					10.2mg	P<.0005
k	R f f3d wat	ton mix	50w50 ev					10.2mg	P<.0005
l	R f f3d wat	edu mix	50w50 ev					11.6mg	P<.0005+
m	R f f3d wat	ton pam	50w50 ev					13.5mg	P<.002
n	R f f3d wat	edu ssc	50w50 ev					19.4mg	P<.008
o	R f f3d wat	--- mix	50w50 ev					24.7mg	P<.02 +
p	R f f3d wat	tba mix	50w50 ev					noTD50	P<.0005
771	R m f3d wat	dgt mix	50w50 ev		.(+) .			.910mg	P<.0005+
a	R m f3d wat	for mix	50w50 ev					2.62mg	P<.0005+
b	R m f3d wat	eso mix	50w50 ev					3.02mg	P<.0005+
c	R m f3d wat	for pam	50w50 ev					3.25mg	P<.0005
d	R m f3d wat	eso pam	50w50 ev					4.08mg	P<.0005
e	R m f3d wat	--- mix	50w50 ev					10.2mg	P<.002 +
f	R m f3d wat	duo mix	50w50 ev					11.8mg	P<.002
g	R m f3d wat	edu mix	50w50 ev					11.8mg	P<.002 +
h	R m f3d wat	pls mix	50w50 ev					11.8mg	P<.002
i	R m f3d wat	eso sqc	50w50 ev					17.0mg	P<.009
j	R m f3d wat	pls sqc	50w50 ev					17.0mg	P<.009
k	R m f3d wat	tba mix	50w50 ev					noTD50	P<.0005
772	R f mrw wat	--- mye	39w85		. + .			3.27mg	P<.004 +
a	R f mrw wat	liv tum	39w85					6.63mg	P<.04
b	R f mrw wat	mgl tum	39w85					6.63mg	P<.04 -
c	R f mrw wat	tba tum	39w85					1.02mg	P<.0005
773	R m mrw wat	--- mye	39w85		. ±			2.86mg	P<.03 +
a	R m mrw wat	liv tum	39w85					26.1mg	P<.5
b	R m mrw wat	tba tum	39w85					1.64mg	P<.004

BUTYLATED HYDROXYANISOLE   100ng..:..1ug....:..10.....:..100...:..1mg....:..10......:..100....:..1g......:..10

	Spe	Strain	Site	Xpo+Xpt				TD50	2Tailpvl
774	H m syg eat	for sqc	48w72 r			.>		5.44gm	P<.4
a	H m syg eat	for pam	48w72 r					no dre	P=1.
775	H m syg eat	for pam	72w72 r			. + .		745.mg	P<.004
a	H m syg eat	for sqc	72w72 r					no dre	P=1.
776	M m b6c eat	for tum	64w64 ekr			.>		no dre	P=1.
777	M m b6c eat	for tum	72w72 ekr			.>		no dre	P=1.
778	M m b6c eat	for pam	80w80 ekr			. ±		1.02gm *	P<.03
779	M m b6c eat	for pam	88w88 ekr			.>		3.75gm *	P<.3
a	M m b6c eat	for sqc	88w88 ekr			+hist		3.75gm *	P<.3 +
780	M m b6c eat	for pam	96w96 ekr			. ±		1.53gm *	P<.02
a	M m b6c eat	for sqc	96w96 ekr			+hist		10.5gm *	P<.3 +
781	M m b6c eat	for pam	24m24 er			.>		13.1gm *	P<.4
a	M m b6c eat	for sqc	24m24 er					no dre	P=1.
782	M b swi eat	lun tum	24m24 r			.>		5.11gm	P<.2
a	M b swi eat	liv tum	24m24 r					no dre	P=1.
783	R f f34 eat	for pam	24m26 e			.+ .		490.mg /	P<.0005+
a	R f f34 eat	for sqc	24m26 e					2.75gm /	P<.0005+
b	R f f34 eat	liv hnd	24m26 e					162.gm *	P<.9
784	R m f34 eat	for pam	24m26 e			.+ .		349.mg /	P<.0005+
a	R m f34 eat	for sqc	24m26 e					1.80gm /	P<.0005+
b	R m f34 eat	liv hnd	24m26 e					no dre	P=1.
785	R m f34 eat	liv tum	60w60 er			.>		no dre	P=1.
786	R m f34 eat	for car	6m24 e				.>	no dre	P=1.
a	R m f34 eat	for pam	6m24 e					no dre	P=1.
b	R m f34 eat	liv tum	6m24 e					no dre	P=1.
787	R m f34 eat	for sqc	12m24 e			. ±		4.06gm	P<.04 +
a	R m f34 eat	fls pam	12m24 e					4.06gm	P<.04
b	R m f34 eat	fgr pam	12m24 e					12.5gm	P<.3
c	R m f34 eat	liv tum	12m24 e					no dre	P=1.
788	R m f34 eat	fgr pam	24m24 e			. + .		298.mg	P<.0005
a	R m f34 eat	fls pam	24m24 e					298.mg	P<.0005
b	R m f34 eat	for sqc	24m24 e					7.76gm	P<.04 +
c	R m f34 eat	liv tum	24m24 e					no dre	P=1.
789	R m f34 eat	ssq sqp	26m26 e			. +	+hist	907.mg	P<.0005+
a	R m f34 eat	liv hpa	26m26 e					no dre	P=1.
790	R m f3d eat	for pam	24m24 e			.+ .		598.mg Z	P<.0005+
a	R m f3d eat	for sqc	24m24 e					4.55gm Z	P<.0005+
b	R m f3d eat	liv hpc	24m24 e					no dre	P=1.
791	R m f3d eat	for mix	52w52			. ±		1.30gm *	P<.08
a	R m f3d eat	for pam	52w52					1.99gm *	P<.2 -
b	R m f3d eat	for sqc	52w52					4.04gm *	P<.4 -
c	R m f3d eat	liv tum	52w52					no dre	P=1. -

	RefNum	LoConf	UpConf	Cntrl	1Dose	1Inc	2Dose	2Inc					Citation or Pathology
													Brkly Code
770	1620m	.779mg	2.24mg	0/40	16.8mg	34/39							Takeuchi;zkko,107,32-37;1984
a	1620m	1.48mg	4.31mg	0/40	16.8mg	26/39							
b	1620m	1.91mg	5.96mg	0/40	16.8mg	22/39							
c	1620m	2.30mg	7.77mg	0/40	16.8mg	19/39							
d	1620m	2.80mg	10.5mg	0/40	16.8mg	16/39							
e	1620m	3.21mg	13.1mg	0/40	16.8mg	14/39							
f	1620m	3.45mg	14.9mg	0/40	16.8mg	13/39							
g	1620m	4.03mg	19.9mg	0/40	16.8mg	11/39							
h	1620m	4.03mg	19.9mg	0/40	16.8mg	11/39							
i	1620m	4.78mg	29.5mg	0/40	16.8mg	9/39							
j	1620m	4.78mg	29.5mg	0/40	16.8mg	9/39							
k	1620m	4.78mg	29.5mg	0/40	16.8mg	9/39							
l	1620m	5.25mg	38.3mg	0/40	16.8mg	8/39							
m	1620m	5.82mg	54.3mg	0/40	16.8mg	7/39							
n	1620m	7.38mg	282.mg	0/40	16.8mg	5/39							
o	1620m	8.51mg	n.s.s.	0/40	16.8mg	4/39							
p	1620m	n.s.s.	1.08mg	1/40	16.8mg	39/39							
771	1620n	.519mg	1.60mg	0/38	14.7mg	36/39							
a	1620n	1.57mg	4.80mg	0/38	14.7mg	23/39							
b	1620n	1.77mg	5.68mg	0/38	14.7mg	21/39							
c	1620n	1.89mg	6.21mg	0/38	14.7mg	20/39							
d	1620n	2.29mg	8.25mg	0/38	14.7mg	17/39							
e	1620n	4.60mg	34.5mg	0/38	14.7mg	8/39							
f	1620n	5.09mg	49.9mg	0/38	14.7mg	7/39							
g	1620n	5.09mg	49.9mg	0/38	14.7mg	7/39							
h	1620n	5.09mg	49.9mg	0/38	14.7mg	7/39							
i	1620n	6.45mg	362.mg	0/38	14.7mg	5/39							
j	1620n	6.45mg	362.mg	0/38	14.7mg	5/39							
k	1620n	n.s.s.	.991mg	5/38	14.7mg	39/39							
772	2038	1.38mg	18.1mg	0/10	4.49mg	7/15							Mirvish;jtxe,32,59-74;1991/pers.comm.
a	2038	2.27mg	n.s.s.	0/10	4.49mg	4/15							
b	2038	2.27mg	n.s.s.	0/10	4.49mg	4/15							
c	2038	.453mg	2.52mg	0/10	4.49mg	13/15							
773	2038	1.21mg	n.s.s.	0/5	3.93mg	7/15							
a	2038	4.25mg	n.s.s.	0/5	3.93mg	1/15							
b	2038	.756mg	9.09mg	0/5	3.93mg	10/15							

BUTYLATED HYDROXYANISOLE (BHA, 2(3)-tert-butyl-4-hydroxyanisole) 25013-16-5

	RefNum	LoConf	UpConf	Cntrl	1Dose	1Inc	2Dose	2Inc	3Dose	3Inc	4Dose	4Inc	5Dose	5Inc	Citation or Pathology
774	1997m	884.mg	n.s.s.	0/9	1.23gm	1/14									Hirose;carc,11,239-244;1990
a	1997m	1.70gm	n.s.s.	0/9	1.23gm	0/14									
775	1997n	270.mg	4.82gm	0/9	1.84gm	5/9									
a	1997n	1.64gm	n.s.s.	0/9	1.84gm	0/9									
776	1901m	312.mg	n.s.s.	0/10	600.mg	0/10	1.20gm	0/10							Masui;gann,77,1083-1090;1986/pers.comm.
777	1901n	395.mg	n.s.s.	0/10	600.mg	0/10	1.20gm	0/10							
778	1901o	413.mg	n.s.s.	0/10	600.mg	3/10	1.20gm	3/10							
779	1901r	920.mg	n.s.s.	0/10	600.mg	1/7	1.20gm	1/10							
a	1901r	920.mg	n.s.s.	0/10	600.mg	1/7	1.20gm	1/10							
780	1901s	620.mg	n.s.s.	0/13	600.mg	3/9	1.20gm	3/11							
a	1901s	1.71gm	n.s.s.	0/13	600.mg	0/9	1.20gm	1/11							
781	1901u	3.21gm	n.s.s.	0/16	600.mg	1/21	1.20gm	1/22							
a	1901u	1.76gm	n.s.s.	0/16	600.mg	0/21	1.20gm	0/22							
782	1525	1.30gm	n.s.s.	1/47	625.mg	3/30									Maru;clet,17,75-80;1982
a	1525	2.39mg	n.s.s.	7/47	625.mg	2/30									
783	1568	334.mg	744.mg	0/51	232.mg	1/51	929.mg	49/51							Ito;jnci,70,343-352;1983
a	1568	1.51gm	5.82gm	0/51	232.mg	0/51	929.mg	15/51							
b	1568	8.36gm	n.s.s.	1/51	232.mg	0/51	929.mg	1/51							
784	1568	238.mg	528.mg	0/51	186.mg	1/50	743.mg	52/52							
a	1568	1.04gm	3.56gm	0/51	186.mg	0/50	743.mg	18/52							
b	1568	4.38gm	n.s.s.	4/51	186.mg	3/50	743.mg	3/52							
785	1640	343.mg	n.s.s.	0/25	200.mg	0/25									Rao;canr,44,1072-1076;1984
786	1902m	2.06gm	n.s.s.	0/50	200.mg	0/50									Nera;txcy,53,251-268;1988
a	1902m	2.06gm	n.s.s.	0/50	200.mg	0/50									
b	1902m	2.06gm	n.s.s.	0/50	200.mg	0/50									
787	1902n	1.23gm	n.s.s.	0/50	400.mg	3/46									
a	1902n	1.23gm	n.s.s.	0/50	400.mg	3/46									
b	1902n	2.03gm	n.s.s.	0/50	400.mg	1/46									
c	1902n	3.79gm	n.s.s.	0/50	400.mg	0/46									
788	1902o	186.mg	495.mg	0/50	800.mg	37/44									
a	1902o	186.mg	495.mg	0/50	800.mg	37/44									
b	1902o	2.35gm	n.s.s.	0/50	800.mg	3/44									
c	1902o	7.25gm	n.s.s.	0/50	800.mg	0/44									
789	1973	424.mg	2.67gm	0/25	480.mg	9/27									Williams;fctx,28,799-806;1990
a	1973	1.14gm	n.s.s.	9/25	480.mg	6/27									
790	1784	431.mg	858.mg	0/50	50.0mg	0/50	100.mg	0/50	200.mg	0/50	400.mg	10/50	800.mg	50/50	Ito;jnci,77, 1261-1265;1986
a	1784	2.28gm	11.1gm	0/50	50.0mg	0/50	100.mg	0/50	200.mg	0/50	400.mg	0/50	800.mg	11/50	
b	1784	6.03gm	n.s.s.	0/50	50.0mg	0/50	100.mg	0/50	200.mg	3/50	400.mg	0/50	800.mg	0/50	
791	1883	393.mg	n.s.s.	0/10	400.mg	0/20	800.mg	3/20							Hasegawa;gann,79,320-328;1988/pers.comm.
a	1883	488.mg	n.s.s.	0/10	400.mg	0/20	800.mg	2/20							
b	1883	658.mg	n.s.s.	0/10	400.mg	0/20	800.mg	1/20							
c	1883	275.mg	n.s.s.	0/10	400.mg	0/20	800.mg	0/20							

	Spe		Strain	Site		Xpo+Xpt				TD50		2Tailpvl	
		Sex	Route	Hist	Notes							DR	AuOp
792	R m		f3d	eat eso	tum	52w52 er		.>		no dre		P=1.	-
a	R m		f3d	eat for	tum	52w52 er				no dre		P=1.	
b	R m		f3d	eat liv	tum	52w52 er				no dre		P=1.	
793	R m		f3d	eat for	mix	51w52 er		.>		907.mg		P<.3	-
a	R m		f3d	eat for	pam	51w52 er				907.mg		P<.3	
b	R m		f3d	eat for	sqc	51w52 er				907.mg		P<.3	
c	R m		f3d	eat stg	tum	51w52 er				no dre		P=1.	
794	R m		f3d	eat for	sqp	60w60 er		. ±		1.06gm		P<.04	
a	R m		f3d	eat for	sqc	60w60 er				no dre		P=1.	

BUTYLATED HYDROXYTOLUENE   100ng..:..1ug....:..10.....:..100...:..1mg....:..10.....:..100....:..1g.....:..10

795	M f		b6c	eat lun	MXA	25m25 a	: ±			#715.mg \		P<.02	-
a	M f		b6c	eat TBA	MXB	25m25 a				no dre		P=1.	
b	M f		b6c	eat liv	MXB	25m25 a				10.7gm *		P<.6	
c	M f		b6c	eat lun	MXB	25m25 a				715.mg \		P<.02	
796	M f		b6c	eat liv	hnd	22m24 e		.>		5.98gm *		P<.3	-
a	M f		b6c	eat liv	hpc	22m24 e				14.4gm *		P<.5	-
b	M f		b6c	eat lun	adc	22m24 e				32.9gm *		P<.8	-
c	M f		b6c	eat lun	ade	22m24 e				no dre		P=1.	-
797	M f		b6c	eat lun	a/c	24m28 e				71.3gm *		P<.6	-
a	M f		b6c	eat liv	hem	24m28 e				126.gm *		P<.3	-
b	M f		b6c	eat liv	hpc	24m28 e				no dre		P=1.	-
c	M f		b6c	eat liv	hpa	24m28 e				no dre		P=1.	-
d	M f		b6c	eat lun	a/a	24m28 e				no dre		P=1.	-
e	M f		b6c	eat liv	hct	24m28 e				no dre		P=1.	-
f	M f		b6c	eat tba	tum	24m28 e				no dre		P=1.	-
798	M m		b6c	eat TBA	MXB	25m25 a	:>			no dre		P=1.	-
a	M m		b6c	eat liv	MXB	25m25 a				no dre		P=1.	-
b	M m		b6c	eat lun	MXB	25m25 a				no dre		P=1.	
799	M m		b6c	eat liv	hnd	22m24 e		.>		3.00gm *		P<.3	-
a	M m		b6c	eat lun	ade	22m24 e				212.gm *		P<1.	-
b	M m		b6c	eat liv	hae	22m24 e				no dre		P=1.	-
c	M m		b6c	eat lun	adc	22m24 e				no dre		P=1.	-
d	M m		b6c	eat liv	hpc	22m24 e				no dre		P=1.	-
800	M m		b6c	eat liv	hpa	24m28 e		. +		3.45gm *		P<.003	
a	M m		b6c	eat liv	hct	24m28 e				2.90gm *		P<.02	+
b	M m		b6c	eat lun	a/a	24m28 e				18.5gm *		P<.3	
c	M m		b6c	eat liv	ang	24m28 e				128.gm *		P<.3	
d	M m		b6c	eat liv	hpc	24m28 e				no dre		P=1.	
e	M m		b6c	eat liv	hem	24m28 e				no dre		P=1.	
f	M m		b6c	eat lun	a/c	24m28 e				no dre		P=1.	
g	M m		b6c	eat tba	tum	24m28 e				no dre		P=1.	
801	M f		bal	eat lun	tum	69w69		.>		no dre		P=1.	
a	M f		bal	eat liv	tum	69w69				no dre		P=1.	
802	M m		bal	eat lun	ppa	69w69 e	. ±			368.mg		P<.03	+
a	M m		bal	eat sto	sqc	69w69 e				2.03gm		P<.09	-
b	M m		bal	eat liv	tum	69w69 e				no dre		P=1.	-
c	M m		bal	eat ---	rts	69w69 e				no dre		P=1.	-
803	M m		bal	eat lun	tum	69w69		.>		7.33gm		P<.8	-
a	M m		bal	eat liv	mix	69w69				no dre		P=1.	
804	M b		swi	eat lun	tum	24m24 r		. +		. 1.48gm		P<.002	
a	M b		swi	eat liv	tum	24m24 r				no dre		P=1.	
805	R f		f34	eat TBA	MXB	24m24	:>			no dre		P=1.	
a	R f		f34	eat liv	MXB	24m24				no dre		P=1.	
806	R m		f34	eat TBA	MXB	24m24	:>			no dre		P=1.	-
a	R m		f34	eat liv	MXB	24m24				2.30gm *		P<.4	
807	R m		f34	eat liv	hpa	26m26 e		.>		no dre		P=1.	-
a	R m		f34	eat ssq	tum	26m26 e				no dre		P=1.	-
808	R m		f3d	eat eso	tum	52w52 er		.>		no dre		P=1.	-
a	R m		f3d	eat for	tum	52w52 er				no dre		P=1.	-
b	R m		f3d	eat liv	tum	52w52 er				no dre		P=1.	-
809	R m		f3d	eat for	sqc	60w60 er		.>		no dre		P=1.	-
a	R m		f3d	eat for	sqp	60w60 er				no dre		P=1.	-
810	R f		wis	eat pit	ade	24m24 e		. ±		613.mg \		P<.02	-
a	R f		wis	eat liv	hnd	24m24 e				5.99gm *		P<.4	-
b	R f		wis	eat tba	mix	24m24 e				2.31gm *		P<.5	-
811	R m		wis	eat liv	hnd	24m24 e		.>		no dre		P=1.	-
a	R m		wis	eat tba	mix	24m24 e				no dre		P=1.	-

1,1-DI-N-BUTYLHYDRAZINE   100ng..:..1ug....:..10.....:..100...:..1mg....:..10.....:..100....:..1g.....:..10

812	M f		swa	wat lun	ade	25m25 es		. + .		51.5mg		P<.0005	
a	M f		swa	wat lun	mix	25m25 es				55.5mg		P<.0005+	
b	M f		swa	wat for	mix	25m25 es				71.5mg		P<.0005+	
c	M f		swa	wat for	sqp	25m25 es				95.9mg		P<.0005	
d	M f		swa	wat lun	adc	25m25 es				133.mg		P<.0005	
e	M f		swa	wat for	sqc	25m25 es				152.mg		P<.0005	
f	M f		swa	wat liv	agm	25m25 es				no dre		P=1.	-
813	M m		swa	wat lun	mix	25m25 es		. + .		38.1mg		P<.0005+	
a	M m		swa	wat lun	ade	25m25 es				44.9mg		P<.0005	
b	M m		swa	wat lun	adc	25m25 es				82.3mg		P<.0005	
c	M m		swa	wat for	mix	25m25 es				84.2mg		P<.0005+	

	RefNum	LoConf	UpConf	Cntrl	1Dose	1Inc	2Dose	2Inc	Citation or Pathology
									Brkly Code
792	1900	206.mg	n.s.s.	0/10	400.mg	0/15	800.mg	0/15	Hirose;carc,8,1731-1735;1987/pers.comm.
a	1900	206.mg	n.s.s.	0/10	400.mg	0/15	800.mg	0/15	
b	1900	206.mg	n.s.s.	0/10	400.mg	0/15	800.mg	0/15	
793	1921	148.mg	n.s.s.	0/10	392.mg	1/14			Hirose;carc,10,2223-2226;1989
a	1921	148.mg	n.s.s.	0/10	392.mg	1/14			
b	1921	148.mg	n.s.s.	0/10	392.mg	1/14			
c	1921	283.mg	n.s.s.	0/10	392.mg	0/14			
794	2134	321.mg	n.s.s.	0/19	800.mg	3/19			Ito;anti,183-194;1990/pers.comm.
a	2134	1.04gm	n.s.s.	0/19	800.mg	0/19			

BUTYLATED HYDROXYTOLUENE   (BHT, 2,6-DI-tert-butyl-p-cresol) 128-37-0

	RefNum	LoConf	UpConf	Cntrl	1Dose	1Inc	2Dose	2Inc	Citation or Pathology
795	TR150	357.mg	n.s.s.	1/20	390.mg	16/50	(780.mg	7/50)	lun:a/a,a/c. S
a	TR150	336.mg	n.s.s.	14/20	390.mg	32/50	(780.mg	23/50)	
b	TR150	2.18gm	n.s.s.	1/20	390.mg	4/50	780.mg	5/50	liv:hpa,hpc,nnd.
c	TR150	357.mg	n.s.s.	1/20	390.mg	16/50	(780.mg	7/50)	lun:a/a,a/c.
796	1528	1.40gm	n.s.s.	2/47	24.0mg	3/47	120.mg	5/46	600.mg	5/44	Shirai;fctx,20,861-865;1982
a	1528	2.26gm	n.s.s.	2/47	24.0mg	2/47	120.mg	1/46	600.mg	3/44	
b	1528	2.89gm	n.s.s.	3/47	24.0mg	0/47	120.mg	1/46	600.mg	2/44	
c	1528	3.81gm	n.s.s.	7/47	24.0mg	3/47	120.mg	2/46	600.mg	2/44	
797	1882	12.3gm	n.s.s.	1/41	1.13gm	2/44	2.25gm	2/40	Inai;gann,79,49-58;1988
a	1882	20.6gm	n.s.s.	0/41	1.13gm	0/44	2.25gm	1/40	
b	1882	29.7gm	n.s.s.	2/41	1.13gm	1/44	2.25gm	0/40	
c	1882	13.0gm	n.s.s.	5/41	1.13gm	7/44	2.25gm	2/40	
d	1882	16.4gm	n.s.s.	4/41	1.13gm	5/44	2.25gm	1/40	
e	1882	14.6gm	n.s.s.	7/41	1.13gm	8/44	2.25gm	2/40	
f	1882	6.04gm	n.s.s.	35/41	1.13gm	33/44	2.25gm	22/40	
798	TR150	355.mg	n.s.s.	17/20	360.mg	39/50	(720.mg	32/50)	liv:hpa,hpc,nnd.
a	TR150	531.mg	n.s.s.	11/20	360.mg	23/50	(720.mg	13/50)	lun:a/a,a/c.
b	TR150	1.23gm	n.s.s.	7/20	360.mg	21/50	720.mg	17/50	
799	1528	784.mg	n.s.s.	14/48	22.2mg	10/48	111.mg	13/50	554.mg	16/47	Shirai;fctx,20,861-865;1982
a	1528	1.48gm	n.s.s.	8/48	22.2mg	8/48	111.mg	9/50	554.mg	8/47	
b	1528	3.52gm	n.s.s.	4/48	22.2mg	5/48	111.mg	2/50	554.mg	2/47	
c	1528	4.36gm	n.s.s.	3/48	22.2mg	6/48	111.mg	2/50	554.mg	1/47	
d	1528	1.53gm	n.s.s.	11/48	22.2mg	13/48	111.mg	12/50	554.mg	10/47	
800	1882	1.90gm	17.7gm	6/32	1.04gm	16/42	2.08gm	25/47	Inai;gann,79,49-58;1988
a	1882	1.42gm	n.s.s.	12/32	1.04gm	26/42	2.08gm	31/47	
b	1882	5.38gm	n.s.s.	4/32	1.04gm	6/42	2.08gm	10/47	
c	1882	20.9gm	n.s.s.	0/32	1.04gm	0/42	2.08gm	1/47	
d	1882	7.51gm	n.s.s.	7/32	1.04gm	11/42	2.08gm	8/47	
e	1882	21.5gm	n.s.s.	4/32	1.04gm	3/42	2.08gm	4/47	
f	1882	19.0gm	n.s.s.	1/32	1.04gm	3/42	2.08gm	0/47	
g	1882	1.72gm	n.s.s.	27/32	1.04gm	36/42	2.08gm	38/47	
801	1006	1.15gm	n.s.s.	13/50	975.mg	11/50			Clapp;jnci,61,177-180;1978
a	1006	4.42gm	n.s.s.	0/50	975.mg	0/50			
802	53	122.mg	n.s.s.	6/25	900.mg	7/11			Clapp;fctx,12,367-371;1974
a	53	330.mg	n.s.s.	0/25	900.mg	1/8			
b	53	898.mg	n.s.s.	0/25	900.mg	0/11			
c	53	509.mg	n.s.s.	14/25	900.mg	1/8			
803	1006	711.mg	n.s.s.	45/100	900.mg	47/100			Clapp;jnci,61,177-180;1978
a	1006	8.16gm	n.s.s.	2/100	900.mg	0/100			
804	1525	640.mg	7.00gm	1/47	625.mg	8/30			Maru;clet,17,75-80;1982
a	1525	3.02gm	n.s.s.	7/47	625.mg	1/30			
805	TR150	374.mg	n.s.s.	12/20	150.mg	36/50	300.mg	26/50	
a	TR150	n.s.s.	n.s.s.	0/20	150.mg	0/50	300.mg	0/50	liv:hpa,hpc,nnd.
806	TR150	210.mg	n.s.s.	17/20	120.mg	36/50	240.mg	35/50	
a	TR150	796.mg	n.s.s.	0/20	120.mg	2/50	240.mg	2/50	liv:hpa,hpc,nnd.
807	1973	1.49gm	n.s.s.	9/25	480.mg	3/23			Williams;fctx,28,799-806;1990
a	1973	2.55gm	n.s.s.	0/25	480.mg	0/23			
808	1900	144.mg	n.s.s.	0/10	280.mg	0/10			Hirose;carc,8,1731-1735;1987/pers.comm.
a	1900	144.mg	n.s.s.	0/10	280.mg	0/10			
b	1900	144.mg	n.s.s.	0/10	280.mg	0/10			
809	2134	137.mg	n.s.s.	0/19	200.mg	0/10			Ito;anti,183-194;1990/pers.comm.
a	2134	137.mg	n.s.s.	0/19	200.mg	0/10			
810	1087	250.mg	171.gm	0/32	125.mg	6/46	(500.mg	3/51)	Hirose;fctx,19,147-151;1981
a	1087	1.45gm	n.s.s.	0/32	125.mg	3/46	500.mg	3/51	
b	1087	476.mg	n.s.s.	11/32	125.mg	25/46	500.mg	25/51	
811	1087	2.23gm	n.s.s.	2/26	100.mg	2/43	400.mg	1/38	
a	1087	1.12gm	n.s.s.	6/26	100.mg	13/43	400.mg	10/51	

1,1-DI-N-BUTYLHYDRAZINE   7422-80-2

	RefNum	LoConf	UpConf	Cntrl	1Dose	1Inc	Citation or Pathology
812	1438	30.3mg	107.mg	20/99	62.5mg	34/50	Toth;carc,2,651-654;1981
a	1438	31.5mg	128.mg	25/99	62.5mg	34/50	
b	1438	41.4mg	147.mg	4/71	62.5mg	23/45	
c	1438	52.5mg	227.mg	4/71	62.5mg	19/45	
d	1438	69.3mg	371.mg	6/99	62.5mg	17/50	
e	1438	77.9mg	355.mg	0/71	62.5mg	12/45	
f	1438	474.mg	n.s.s.	5/71	62.5mg	1/45	
813	1438	21.7mg	84.7mg	26/96	52.1mg	34/46	
a	1438	26.1mg	93.8mg	16/96	52.1mg	30/46	
b	1438	43.3mg	235.mg	12/96	52.1mg	21/46	
c	1438	46.7mg	174.mg	0/96	52.1mg	16/43	

```
      Spe Strain  Site   Xpo+Xpt                                                  TD50    2Tailpvl
        Sex  Route  Hist     Notes                                                   DR      AuOp
    d   M m swa wat liv mix 25m25 es                                              105.mg   P<.0005+
    e   M m swa wat for sqp 25m25 es                                              109.mg   P<.0005
    f   M m swa wat liv bhp 25m25 es                                              189.mg   P<.006
    g   M m swa wat for sqc 25m25 es                                              317.mg   P<.0005

N-BUTYLHYDRAZINE.HCl          100ng..:..1ug....:..10....:..100....:..1mg...:..10....:..100.....:..1g......:..10
  814 M f swa wat lun mix 28m28 e                             . + .               18.4mg   P<.0005+
    a M f swa wat lun ade 28m28 e                             . + .               19.3mg   P<.0005
    b M f swa wat lun adc 28m28 e                                                 50.6mg   P<.0005
    c M f swa wat liv mix 28m28 e                                                 no dre   P=1.
  815 M m swa wat lun ade 23m23 e                        . + .                    8.40mg   P<.0005
    a M m swa wat lun mix 23m23 e                        . + .                    9.03mg   P<.0005+
    b M m swa wat liv mix 23m23 e                                                 no dre   P=1.

1,2-DI-N-BUTYLHYDRAZINE.2HCl  100ng..:..1ug....:..10....:..100....:..1mg...:..10....:..100.....:..1g......:..10
  816 M f swa wat lun mix 93w93 e                             . + .               69.8mg   P<.0005+
    a M f swa wat --- mly 93w93 e                                                 198.mg   P<.003  +
    b M f swa wat liv agm 93w93 e                                                 no dre   P=1.   -
    c M f swa wat liv ang 93w93 e                                                 no dre   P=1.   -
  817 M m swa wat lun mix 87w87 e                        . + .                    34.5mg   P<.0005+
    a M m swa wat kid mix 87w87 e                                                 449.mg   P<.004  +
    b M m swa wat --- mly 87w87 e                                                 886.mg   P<.4    +
    c M m swa wat liv ang 87w87 e                                                 no dre   P=1.   -
    d M m swa wat liv agm 87w87 e                                                 no dre   P=1.   -

p-tert-BUTYLPHENOL            100ng..:..1ug....:..10....:..100....:..1mg...:..10....:..100.....:..1g......:..10
  818 R m f3d eat for pam 51w52                                          .>       1.46gm   P<.4
    a R m f3d eat stg tum 51w52                                                   no dre   P=1.

N-BUTYLUREA                   100ng..:..1ug....:..10....:..100....:..1mg...:..10....:..100.....:..1g......:..10
  819 M f cb6 eat liv hpc 52w69 e                                          .>     no dre   P=1.   -
    a M f cb6 eat lun ade 52w69 e                                                 no dre   P=1.   -
    b M f cb6 eat tba mix 52w69 e                                                 no dre   P=1.   -
  820 M m cb6 eat lun ade 52w69 e                                            .>   5.51gm   P<.4   -
    a M m cb6 eat liv hpc 52w69 e                                                 no dre   P=1.   -
    b M m cb6 eat tba mix 52w69 e                                                 699.mg   P<.02  -
  821 R f f34 eat liv tum 52w69 e                                         .>      no dre   P=1.   -
    a R f f34 eat tba mix 52w69 e                                                 no dre   P=1.   -
  822 R m f34 eat liv tum 52w69 e                                         .>      no dre   P=1.   -
    a R m f34 eat tba mix 52w69 e                                                 no dre   P=1.   -

beta-BUTYROLACTONE            100ng..:..1ug....:..10....:..100....:..1mg...:..10....:..100.....:..1g......:..10
  823 R f esd gav for sqc 70w70                               .   ±               13.8mg   P<.02  +

gamma-BUTYROLACTONE           100ng..:..1ug....:..10....:..100....:..1mg...:..10....:..100.....:..1g......:..10
  824 M f b6c gav TBA MXB 24m24                                             :>     4.30gm * P<.9  -
    a M f b6c gav liv MXB 24m24                                                   no dre   P=1.
    b M f b6c gav lun MXB 24m24                                                   no dre   P=1.
  825 M m b6c gav amd MXA 24m24 s                                              :>  2.29gm * P<.4  e
    a M m b6c gav TBA MXB 24m24 s                                                 660.mg * P<.4
    b M m b6c gav liv MXB 24m24 s                                                 no dre   P=1.
    c M m b6c gav lun MXB 24m24 s                                           :>     2.10gm * P<.5
  826 R f f34 gav TBA MXB 24m24                                             :>     no dre   P=1.
    a R f f34 gav liv MXB 24m24                                                   no dre   P=1.
  827 R m f34 gav amd phm 24m24                               :   ±               #880.mg * P<.02 +
    a R m f34 gav --- msm 24m24                                                   1.14gm * P<.03
    b R m f34 gav TBA MXB 24m24                                                   no dre   P=1.
    c R m f34 gav liv MXB 24m24                                                   no dre   P=1.

CADMIUM ACETATE               100ng..:..1ug....:..10....:..100....:..1mg...:..10....:..100.....:..1g......:..10
  828 M f cd1 wat lun car 30m30 e                                    .>            no dre   P=1.   -
    a M f cd1 wat tba tum 30m30 e                                                 no dre   P=1.   -
  829 M m cd1 wat lun car 28m28 e                                      .>          no dre   P=1.   -
    a M m cd1 wat tba tum 28m28 e                                                 no dre   P=1.   -
  830 R b leb wat liv tum 41m41 es                               .>                37.7mg   P<.8   -
    a R b leb wat tba tum 41m41 es                                               1.81mg   P<.2   -
    b R b leb wat tba mal 41m41 es                                               5.10mg   P<.2   -

CADMIUM CHLORIDE              100ng..:..1ug....:..10....:..100....:..1mg...:..10....:..100.....:..1g......:..10
  831 H f syg inh lun tum 60w76 or                                    .>           no dre   P=1.   -
  832 H f syg inh lun tum 64w76 or                         .>                      no dre   P=1.   -
  833 H m syg inh lun tum 14m24 or                               .>                no dre   P=1.   -
  834 H m syg inh lun tum 15m26 or                            .>                   no dre   P=1.   -
  835 M f nmr inh lun mix 42w82 eor                           .>                   no dre   P=1.   -
    a M f nmr inh lun ade 42w82 eor                                               no dre   P=1.   -
    b M f nmr inh lun adc 42w82 eor                                               no dre   P=1.   -
  836 M f nmr inh lun ade 69w89 eor                               .>               no dre   P=1.   -
    a M f nmr inh lun mix 69w89 eor                                               no dre   P=1.   -
    b M f nmr inh lun adc 69w89 eor                                               no dre   P=1.   -
  837 R f wis inh lun mix  6m29 eor            .   ±                               64.5ug   P<.03
    a R f wis inh lun sqa  6m29 eor                                               99.8ug   P<.08
```

	RefNum	LoConf	UpConf	Cntrl	1Dose	1Inc	2Dose	2Inc	Citation or Pathology / Brkly Code
d	1438	39.4mg	437.mg	0/40	52.1mg	5/16			
e	1438	57.2mg	246.mg	0/96	52.1mg	13/43			
f	1438	56.9mg	2.40gm	0/40	52.1mg	3/16			
g	1438	120.mg	1.37gm	0/96	52.1mg	5/43			

N-BUTYLHYDRAZINE.HCl   56795-65-4

	RefNum	LoConf	UpConf	Cntrl	1Dose	1Inc	2Dose	2Inc	Citation or Pathology
814	252	10.7mg	37.6mg	21/90	25.0mg	34/44			Nagel;ejca,11,473-478;1975/Toth 1974
a	252	11.3mg	38.5mg	18/90	25.0mg	33/44			
b	252	27.2mg	121.mg	4/90	25.0mg	17/44			
c	252	101.mg	n.s.s.	3/56	25.0mg	0/15			
815	252	4.21mg	19.9mg	15/92	20.8mg	19/23			
a	252	4.36mg	24.1mg	23/92	20.8mg	19/23			
b	252	47.6mg	n.s.s.	6/71	20.8mg	1/18			

1,2-DI-N-BUTYLHYDRAZINE.2HCl   78776-28-0

	RefNum	LoConf	UpConf	Cntrl	1Dose	1Inc	2Dose	2Inc	Citation or Pathology
816	1116	40.5mg	149.mg	25/99	125.mg	36/50			Toth;expa,37,773-775;1981
a	1116	93.2mg	1.29gm	18/100	125.mg	21/50			
b	1116	517.mg	n.s.s.	3/47	125.mg	1/36			
c	1116	379.mg	n.s.s.	3/47	125.mg	2/36			
817	1116	20.2mg	67.3mg	26/100	104.mg	38/46			
a	1116	155.mg	3.51gm	0/70	104.mg	4/38			
b	1116	186.mg	n.s.s.	8/100	104.mg	6/46			
c	1116	420.mg	n.s.s.	3/80	104.mg	1/42			
d	1116	416.mg	n.s.s.	5/70	104.mg	1/38			

p-tert-BUTYLPHENOL   98-54-4

	RefNum	LoConf	UpConf	Cntrl	1Dose	1Inc	2Dose	2Inc	Citation or Pathology
818	1950	238.mg	n.s.s.	0/10	588.mg	1/15			Hirose;canr,48,5310-5315;1988
a	1950	455.mg	n.s.s.	0/10	588.mg	0/15			

N-BUTYLUREA   592-31-4

	RefNum	LoConf	UpConf	Cntrl	1Dose	1Inc	2Dose	2Inc	Citation or Pathology
819	1361	1.24gm	n.s.s.	1/92	567.mg	0/24			Murthy;ijcn,23,253-259;1979
a	1361	1.24gm	n.s.s.	2/92	567.mg	0/24			
b	1361	501.mg	n.s.s.	17/92	567.mg	4/24			
820	1361	682.mg	n.s.s.	1/95	523.mg	1/26			
a	1361	1.23gm	n.s.s.	2/95	523.mg	0/26			
b	1361	244.mg	n.s.s.	8/95	523.mg	7/26			
821	1361	317.mg	n.s.s.	0/44	218.mg	0/16			
a	1361	127.mg	n.s.s.	20/44	218.mg	5/16			
822	1361	253.mg	n.s.s.	0/50	175.mg	0/16			
a	1361	84.4mg	n.s.s.	30/50	175.mg	7/16			

beta-BUTYROLACTONE   3068-88-0

	RefNum	LoConf	UpConf	Cntrl	1Dose	1Inc	2Dose	2Inc	Citation or Pathology
823	55	3.85mg	n.s.s.	0/5	40.8mg	3/5			Van Duuren;jnci,37,825-838;1966

gamma-BUTYROLACTONE   96-48-0

	RefNum	LoConf	UpConf	Cntrl	1Dose	1Inc	2Dose	2Inc	Citation or Pathology
824	TR406	307.mg	n.s.s.	33/50	185.mg	31/50	370.mg	35/50	
a	TR406	1.95gm	n.s.s.	8/50	185.mg	2/50	370.mg	4/50	liv:hpa,hpc,nnd.
b	TR406	1.56gm	n.s.s.	7/50	185.mg	4/50	370.mg	4/50	lun:a/a,a/c.
825	TR406	535.mg	n.s.s.	2/50	185.mg	6/50	371.mg	1/50	amd:pbb,phm,pob.
a	TR406	170.mg	n.s.s.	40/50	185.mg	31/50	371.mg	23/50	
b	TR406	690.mg	n.s.s.	24/50	185.mg	8/50	371.mg	9/50	liv:hpa,hpc,nnd.
c	TR406	408.mg	n.s.s.	10/50	185.mg	9/50	371.mg	6/50	lun:a/a,a/c.
826	TR406	226.mg	n.s.s.	46/50	158.mg	46/50	317.mg	41/50	
a	TR406	n.s.s.	n.s.s.	0/50	158.mg	0/50	317.mg	0/50	liv:hpa,hpc,nnd.
827	TR406	357.mg	n.s.s.	0/50	79.1mg	1/50	159.mg	5/50	S
a	TR406	428.mg	n.s.s.	0/50	79.1mg	1/50	159.mg	4/50	S
b	TR406	142.mg	n.s.s.	45/50	79.1mg	42/50	159.mg	47/50	
c	TR406	n.s.s.	n.s.s.	0/50	79.1mg	1/50	159.mg	0/50	liv:hpa,hpc,nnd.

CADMIUM ACETATE   543-90-8

	RefNum	LoConf	UpConf	Cntrl	1Dose	1Inc	2Dose	2Inc	Citation or Pathology
828	56	4.71mg	n.s.s.	9/60	1.00mg	5/39			Schroeder;jnut,83,239-250;1964
a	56	4.14mg	n.s.s.	22/60	1.00mg	10/39			
829	56	11.0mg	n.s.s.	8/44	.833mg	0/48			
a	56	9.55mg	n.s.s.	11/44	.833mg	1/48			
830	1036	3.29mg	n.s.s.	1/34	.265mg	2/47			Kanisawa;canr,29,892-895;1969
a	1036	.668mg	n.s.s.	10/34	.265mg	22/47			
b	1036	1.56mg	n.s.s.	2/34	.265mg	7/47			

CADMIUM CHLORIDE   10108-64-2

	RefNum	LoConf	UpConf	Cntrl	1Dose	1Inc	2Dose	2Inc	Citation or Pathology
831	1909m	.139mg	n.s.s.	0/24	52.5ug	0/24			Heinrich;expl,37,253-258;1989/pers.comm.
832	1909n	49.3ug	n.s.s.	0/24	18.7ug	0/24			
833	1909m	.164mg	n.s.s.	0/24	34.5ug	0/24			
834	1909n	64.6ug	n.s.s.	0/24	11.1ug	0/24			
835	1909m	.126mg	n.s.s.	11/43	75.0ug	8/41			
a	1909m	.139mg	n.s.s.	8/43	75.0ug	6/41			
b	1909m	.211mg	n.s.s.	3/43	75.0ug	2/41			
836	1909n	.207mg	n.s.s.	9/45	37.8ug	1/43			
a	1909n	.221mg	n.s.s.	17/45	37.8ug	1/43			
b	1909n	.246mg	n.s.s.	8/45	37.8ug	0/43			
837	1908m	19.5ug	n.s.s.	0/20	11.7ug	3/18			Oldiges;txec,19,217-222;1989/Glaser 1990
a	1908m	24.5ug	n.s.s.	0/20	11.7ug	2/18			

```
        Spe Strain Site  Xpo+Xpt                                    TD50    2Tailpvl
            Sex  Route  Hist   Notes                                    DR     AuOp
  b    R f wis inh lun adc  6m29 eor                                 .206mg   P<.3
838    R f wis inh lun mix 18m31 eor              . +  .            9.71ug   P<.0005+
  a    R f wis inh lun adc 18m31 eor                                25.3ug   P<.0005
  b    R f wis inh lun a/a 18m31 eor                                49.5ug   P<.02
  c    R f wis inh lun adq 18m31 eor                                .106mg   P<.08
839    R m wis inh lun car 18m31 eo              . +.              12.7ug * P<.0005+
  a    R m wis inh lun adc 18m31 eo                                22.0ug * P<.0005
  b    R m wis inh lun epc 18m31 eo                                58.8ug * P<.0005
  c    R m wis inh lun mec 18m31 eo                                .315mg * P<.03
  d    R m wis inh lun ade 18m31 eo                                .563mg * P<.5
  e    R m wis inh adr pbm 18m31 eo                                .591mg * P<.8
840    R m wis inh lun mix  6m30 eor             . +  .            10.6ug   P<.0005+
  a    R m wis inh lun adc  6m30 eor                                29.4ug   P<.002
  b    R m wis inh lun a/a  6m30 eor                                52.1ug   P<.01
  c    R m wis inh lun sqa  6m30 eor                                52.1ug   P<.01
841    R m wis inh lun mix 18m30 eor             . +  .            6.11ug   P<.0005+
  a    R m wis inh lun adc 18m30 eor                                9.24ug   P<.0005
  b    R m wis inh lun a/a 18m30 eor                                80.4ug   P<.04
  c    R m wis inh lun adq 18m30 eor                                .165mg   P<.2
842    R m wis eat tes icb 77w77 e '                  .  ±         31.6mg * P<.03  +

  a    R m wis eat pro ade 77w77 e                                 56.9mg * P<.3   +
  b    R m wis eat --- lle 77w77 e                                 no dre   P=1.   +

CADMIUM CHLORIDE MONOHYDRATE         100ng..:..1ug....:..10......:..100....:..1mg....:..10....:..100....:..1g......:..10
843    R f wws eat liv hpt 24m24 e                        .>       no dre   P=1.   -
  a    R f wws eat tba mix 24m24 e                                 21.6mg * P<.7   -
  b    R f wws eat tba ben 24m24 e                                 44.0mg * P<.8   -
  c    R f wws eat tba mal 24m24 e                                 124.mg * P<.9   -
844    R m wws eat adr phe 24m24 e                          .  ±   1.33mg Z P<.04  -
  a    R m wws eat liv hpt 24m24 e                                 no dre   P=1.   -
  b    R m wws eat tba mal 24m24 e                                 47.4mg * P<.5   -
  c    R m wws eat tba ben 24m24 e                                 no dre   P=1.   -
  d    R m wws eat tba mix 24m24 e                                 no dre   P=1.   -

CADMIUM DIETHYLDITHIOCARBAMATE       100ng..:..1ug....:..10......:..100....:..1mg....:..10....:..100....:..1g......:..10
845    M f b6a orl lun ade 76w76 evx                            .>    no dre   P=1.   -
  a    M f b6a orl liv hpt 76w76 evx                                 no dre   P=1.   -
  b    M f b6a orl tba mix 76w76 evx                                 47.5mg   P<.7   -
846    M m b6a orl liv hpt 76w76 evx                          .>      no dre   P=1.   -
  a    M m b6a orl lun ade 76w76 evx                                 no dre   P=1.   -
  b    M m b6a orl tba mix 76w76 evx                                 no dre   P=1.   -
847    M f b6c orl lun ade 76w76 evx                           .>     57.4mg   P<.3   -
  a    M f b6c orl liv hpt 76w76 evx                                 no dre   P=1.   -
  b    M f b6c orl tba mix 76w76 evx                                 18.0mg   P<.05  -
848    M m b6c orl liv hpt 76w76 evx                       .  ±      16.7mg   P<.05  -
  a    M m b6c orl lun ade 76w76 evx                                 25.9mg   P<.2   -
  b    M m b6c orl tba mix 76w76 evx                                 6.20mg   P<.002 -

CADMIUM SULPHATE (1:1)               100ng..:..1ug....:..10......:..100....:..1mg....:..10....:..100....:..1g......:..10
849    H f syg inh lun tum 61w76 or                         .>       no dre   P=1.   -
850    H f syg inh lun tum 64w77 or                   .>             no dre   P=1.   -
851    H m syg inh lun tum 14m24 or                     .>           no dre   P=1.   -
852    H m syg inh lun tum 15m26 or                  .>              no dre   P=1.   -
853    M f nmr inh lun adc 42w96 eor                   .>            .829mg   P<.5   -
  a    M f nmr inh lun ade 42w96 eor                                 no dre   P=1.   -
  b    M f nmr inh lun mix 42w96 eor                                 no dre   P=1.   -
854    M f nmr inh lun mix 69w94 eor                  .>             no dre   P=1.   -
  a    M f nmr inh lun ade 69w94 eor                                 no dre   P=1.   -
  b    M f nmr inh lun adc 69w94 eor                                 no dre   P=1.   -
855    R f wis inh lun mix 18m29 eor           . +  .              17.4ug   P<.0005+
  a    R f wis inh lun adc 18m29 eor                                .112mg   P<.003
  b    R f wis inh lun adq 18m29 eor                                .112mg   P<.003
  c    R f wis inh lun a/a 18m29 eor                                .180mg   P<.02
  d    R f wis inh lun sqa 18m29 eor                                .381mg   P<.1
856    R m wis inh lun mix 14m31 eor           . +  .              28.9ug   P<.0005+
  a    R m wis inh lun adc 14m31 eor                                53.7ug   P<.0005
  b    R m wis inh lun a/a 14m31 eor                                .219mg   P<.04
  c    R m wis inh lun adq 14m31 eor                                .219mg   P<.04

CADMIUM SULPHATE (1:1) HYDRATE (3:8)  ..:..1ug....:..10......:..100....:..1mg....:..10....:..100....:..1g......:..10
857    M m swi gav lun mix 78w78 e                          .>       1.78mg * P<.6   -
  a    M m swi gav liv mix 78w78 e                                  no dre   P=1.   -
858    R m cbo gav liv lca 24m26                           .  ±      1.17mg * P<.07  -

CAFFEIC ACID                         100ng..:..1ug....:..10......:..100....:..1mg....:..10....:..100....:..1g......:..10
859    M f b6c eat kid tla 96w96 e                                      .  + 4.70gm   P<.0005+
  a    M f b6c eat for sqc 96w96 e                                 43.3gm   P<.3   +
  b    M f b6c eat liv hpc 96w96 e                                 no dre   P=1.
  c    M f b6c eat lun ala 96w96 e                                 no dre   P=1.
  d    M f b6c eat tba tum 96w96 e                                 1.90gm   P<.004
```

	RefNum	LoConf	UpConf	Cntrl	1Dose	1Inc	2Dose	2Inc	3Dose	3Inc	4Dose	4Inc	Citation or Pathology / Brkly Code
b	1908m	33.5ug	n.s.s.	0/20	11.7ug	1/18							
838	1908n	4.81ug	22.7ug	0/20	10.9ug	13/18							
a	1908n	10.8ug	88.9ug	0/20	10.9ug	7/18							
b	1908n	17.0ug	n.s.s.	0/20	10.9ug	4/18							
c	1908n	25.9ug	n.s.s.	0/20	10.9ug	2/18							
839	1907	8.89ug	18.8ug	0/38	3.34ug	6/39	6.68ug	20/38	13.4ug	25/35			Takenaka;jnci,70,367-373;1983
a	1907	14.6ug	35.9ug	0/38	3.34ug	4/39	6.68ug	16/38	13.4ug	15/35			
b	1907	32.3ug	.158mg	0/38	3.34ug	2/39	6.68ug	5/38	13.4ug	8/35			
c	1907	95.5ug	n.s.s.	0/38	3.34ug	0/39	6.68ug	0/38	13.4ug	3/35			
d	1907	.118mg	n.s.s.	0/38	3.34ug	1/39	6.68ug	0/38	13.4ug	1/35			
e	1907	49.7ug	n.s.s.	0/38	3.34ug	8/39	6.68ug	4/38	13.4ug	4/35			
840	1908m	5.19ug	26.3ug	0/40	7.91ug	11/20							Oldiges;txec,19,217-222;1989/Glaser 1990
a	1908m	11.1ug	.128mg	0/40	7.91ug	5/20							
b	1908m	15.7ug	2.57mg	0/40	7.91ug	3/20							
c	1908m	15.7ug	2.57mg	0/40	7.91ug	3/20							
841	1908n	3.13ug	13.5ug	0/40	7.91ug	15/20							
a	1908n	4.61ug	22.1ug	0/40	7.91ug	12/20							
b	1908n	19.8ug	n.s.s.	0/40	7.91ug	2/20							
c	1908n	26.9ug	n.s.s.	0/40	7.91ug	1/20							
842	2027	12.1mg	n.s.s.	1/28	1.63mg	2/27	3.26mg	1/25	6.52mg	1/24	13.0mg	6/27	Waalkes;faat,19,512-520; 1992/pers.comm.
a	2027	15.3mg	n.s.s.	0/28	1.63mg	1/26	3.26mg	2/22	6.52mg	1/23	13.0mg	2/26	
b	2027	16.9mg	n.s.s.	1/28	1.63mg	4/27	3.26mg	5/24	6.52mg	5/24	13.0mg	1/27	

CADMIUM CHLORIDE MONOHYDRATE   35658-65-2

	RefNum	LoConf	UpConf	Cntrl	1Dose	1Inc	2Dose	2Inc	3Dose	3Inc	4Dose	4Inc	Citation or Pathology
843	1139	.343mg	n.s.s.	0/94	49.5ug	0/48	.149mg	0/48	.495mg	0/48	2.48mg	0/46	Loser;clet,9,191-198;1980
a	1139	2.74mg	n.s.s.	54/94	49.5ug	23/48	.149mg	17/48	.495mg	31/48	2.48mg	25/46	
b	1139	3.56mg	n.s.s.	46/94	49.5ug	19/48	.149mg	16/48	.495mg	22/48	2.48mg	21/46	
c	1139	6.66mg	n.s.s.	10/94	49.5ug	6/48	.149mg	7/48	.495mg	11/48	2.48mg	6/46	
844	1139	.432mg	n.s.s.	2/98	39.6ug	0/50	.119mg	5/50	(.396mg	2/50	1.98mg	2/50)	
a	1139	24.2mg	n.s.s.	0/98	39.6ug	1/50	.119mg	0/50	.396mg	0/50	1.98mg	0/50	
b	1139	7.94mg	n.s.s.	4/98	39.6ug	4/50	.119mg	2/50	.396mg	1/50	1.98mg	4/50	
c	1139	6.32mg	n.s.s.	40/98	39.6ug	18/50	.119mg	25/50	.396mg	23/50	1.98mg	15/50	
d	1139	5.23mg	n.s.s.	43/98	39.6ug	21/50	.119mg	26/50	.396mg	24/50	1.98mg	18/50	

CADMIUM DIETHYLDITHIOCARBAMATE   (ethyl cadmate) 14239-68-0

	RefNum	LoConf	UpConf	Cntrl	1Dose	1Inc	Citation or Pathology
845	1140	9.99mg	n.s.s.	1/17	8.96mg	1/17	Innes;ntis,1968/1969
a	1140	16.8mg	n.s.s.	0/17	8.96mg	0/17	
b	1140	5.99mg	n.s.s.	2/17	8.96mg	3/17	
846	1140	16.5mg	n.s.s.	1/18	8.34mg	0/18	
a	1140	16.5mg	n.s.s.	2/18	8.34mg	0/18	
b	1140	11.7mg	n.s.s.	3/18	8.34mg	1/18	
847	1140	9.34mg	n.s.s.	0/16	8.96mg	1/18	
a	1140	17.8mg	n.s.s.	0/16	8.96mg	0/18	
b	1140	5.43mg	n.s.s.	0/16	8.96mg	3/18	
848	1140	5.05mg	n.s.s.	0/16	8.34mg	3/18	
a	1140	6.36mg	n.s.s.	0/16	8.34mg	2/18	
b	1140	2.64mg	25.2mg	0/16	8.34mg	7/18	

CADMIUM SULPHATE (1:1)   10124-36-4

	RefNum	LoConf	UpConf	Cntrl	1Dose	1Inc	Citation or Pathology
849	1909m	.160mg	n.s.s.	0/24	60.8ug	0/24	Heinrich;expl,37,253-258;1989/pers.comm.
850	1909n	56.9ug	n.s.s.	0/24	21.0ug	0/24	
851	1909m	.191mg	n.s.s.	0/24	39.4ug	0/24	
852	1909n	73.4ug	n.s.s.	0/24	12.6ug	0/24	
853	1909m	.158mg	n.s.s.	3/43	72.9ug	5/43	
a	1909m	.232mg	n.s.s.	8/43	72.9ug	5/43	
b	1909m	.139mg	n.s.s.	11/43	72.9ug	10/43	
854	1909n	.102mg	n.s.s.	17/45	40.7ug	10/43	
a	1909n	.111mg	n.s.s.	9/45	40.7ug	6/43	
b	1909n	.145mg	n.s.s.	8/45	40.7ug	4/43	
855	1908	8.29ug	38.1ug	0/20	39.9ug	18/20	Oldiges;txec,19,217-222;1989/Glaser 1990
a	1908	45.6ug	.587mg	0/20	39.9ug	6/20	
b	1908	45.6ug	.587mg	0/20	39.9ug	6/20	
c	1908	61.9ug	n.s.s.	0/20	39.9ug	4/20	
d	1908	93.6ug	n.s.s.	0/20	39.9ug	2/20	
856	1908	14.2ug	71.8ug	0/40	20.3ug	11/20	
a	1908	23.0ug	.172mg	0/40	20.3ug	7/20	
b	1908	53.9ug	n.s.s.	0/40	20.3ug	2/20	
c	1908	53.9ug	n.s.s.	0/40	20.3ug	2/20	

CADMIUM SULPHATE (1:1) HYDRATE (3:8)   7790-84-3

	RefNum	LoConf	UpConf	Cntrl	1Dose	1Inc	2Dose	2Inc	3Dose	3Inc	Citation or Pathology
857	1044	.307mg	n.s.s.	33/149	62.9ug	14/50	.126mg	9/50	.250mg	14/49	Levy;anoh,17,213-220;1975
a	1044	.492mg	n.s.s.	50/149	62.9ug	18/50	.126mg	18/50	.250mg	11/49	
858	1045	.359mg	n.s.s.	1/90	26.2ug	2/30	52.3ug	0/30	.105mg	3/30	Levy;anoh,17,205-211;1975

CAFFEIC ACID   (3,4-dihydroxy-cinnamic acid) 331-39-5

	RefNum	LoConf	UpConf	Cntrl	1Dose	1Inc	Citation or Pathology
859	1932	2.12gm	15.4gm	0/29	2.60gm	8/29	Hagiwara;canr,51,5655-5660;1991/pers.comm.
a	1932	7.04gm	n.s.s.	0/29	2.60gm	1/29	
b	1932	13.2gm	n.s.s.	1/29	2.60gm	0/29	
c	1932	6.00gm	n.s.s.	4/29	2.60gm	3/29	
d	1932	877.mg	16.6gm	9/29	2.60gm	20/29	

```
    Spe Strain Site  Xpo+Xpt                                                      TD50    2Tailpvl
        Sex  Route  Hist     Notes                                                        DR   AuOp

860 M m b6c eat for mix 96w96 e                                        .  +     5.27gm  P<.002 +
a   M m b6c eat lun mix 96w96 e                                                  5.12gm  P<.02  +
b   M m b6c eat lun ala 96w96 e                                                  6.11gm  P<.03  +
c   M m b6c eat for pam 96w96 e                                                  9.79gm  P<.02
d   M m b6c eat for sqc 96w96 e                                                  13.3gm  P<.05
e   M m b6c eat lun alc 96w96 e                                                  41.3gm  P<.3
f   M m b6c eat liv hnd 96w96 e                                                  no dre  P=1.
g   M m b6c eat liv hpc 96w96 e                                                  no dre  P=1.
h   M m b6c eat tba tum 96w96 e                                                  3.21gm  P<.2
861 R f f3d eat for mix 24m24 e                                   .  +  .        298.mg  P<.0005+
a   R f f3d eat for pam 24m24 e                                                  426.mg  P<.0005
b   R f f3d eat for sqc 24m24 e                                                  989.mg  P<.0005
c   R f f3d eat liv hnd 24m24 e                                                  no dre  P=1.
d   R f f3d eat tba tum 24m24 e                                                  312.mg  P<.004
862 R m f3d eat for mix 24m24 e                                   .  +  .        272.mg  P<.0005+
a   R m f3d eat for pam 24m24 e                                                  377.mg  P<.0005
b   R m f3d eat for sqc 24m24 e                                                  656.mg  P<.0005
c   R m f3d eat kid tla 24m24 e                                                  3.83gm  P<.02  +
d   R m f3d eat liv hnd 24m24 e                                                  no dre  P=1.
e   R m f3d eat liv hpc 24m24 e                                                  no dre  P=1.
f   R m f3d eat tba tum 24m24 e                                                  no dre  P=1.
863 R m f3d eat for sqp 51w52 e                                     .  ±         217.mg  P<.02  +
a   R m f3d eat liv tum 51w52 e                                                  no dre  P=1.
864 R m f3d eat for sqp 60w60 er                                        +  .     542.mg  P<.004 +
a   R m f3d eat for sqc 60w60 er                                                 no dre  P=1.

CAFFEINE         100ng..:..1ug....:..10.....:..100....:..1mg....:..10.....:..100....:..1g.....:..10
865 M f cb6 eat liv tum 77w77 e                                     .>           no dre  P=1.  -
866 M m cb6 eat liv hnd 77w77 e                                     .>           775.mg  P<.3  -
867 R f sda gav mix mix 24m24 r                                       .>         no dre  P=1.  -
868 R f sda wat liv cad 24m24 e                                          .>      15.6gm * P<.9 -
a   R f sda wat liv clc 24m24 e                                                  no dre  P=1.
b   R f sda wat tba mix 24m24 e                                                  no dre  P=1.
869 R m sda gav mix mix 24m24 r                                     .>           448.mg  P<.3  -
a   R m sda gav eso ben 24m24 r                                                  734.mg  P<.3  -
b   R m sda gav for pam 24m24 r                                                  1.44gm  P<.7  -
870 R m sda eat liv tum 25m25 s                                       .>         no dre  P=1.  -
871 R m sda wat liv clc 24m24 e                                          .>      no dre  P=1.  -
a   R m sda wat liv hpa 24m24 e                                                  no dre  P=1.
b   R m sda wat tba mix 24m24 e                                                  no dre  P=1.
872 R f wis wat tba mix 18m24 e                                   .>             no dre  P=1.  -
873 R m wis wat tba mix 18m24 e                                     .>           no dre  P=1.  -

CALCIFEROL       100ng..:..1ug....:..10.....:..100....:..1mg....:..10.....:..100....:..1g.....:..10
874 M f c3h eat mgl adc 24m24 r                                  .>             39.6mg Z P<1.  +

CALCIUM ACETATE  100ng..:..1ug....:..10.....:..100....:..1mg....:..10.....:..100....:..1g.....:..10
875 R m sda eat kid tum 79w79 e                                              .>  no dre  P=1.
a   R m sda eat liv tum 79w79 e                                                  no dre  P=1.

CALCIUM CHLORIDE 100ng..:..1ug....:..10.....:..100....:..1mg....:..10.....:..100....:..1g.....:..10
876 R m wis wat sto tum 52w52 er                                 .>             no dre  P=1.

CALCIUM LACTATE  100ng..:..1ug....:..10.....:..100....:..1mg....:..10.....:..100....:..1g.....:..10
877 R f f3d wat liv hpa 24m26 e                                                 no dre  P=1.
a   R f f3d wat tba tum 24m26 e                                                  103.gm * P<1.
878 R m f3d wat liv hpa 24m26 e                                                 54.5gm * P<.4
a   R m f3d wat tba tum 24m26 e                                                  noTD50  P=1.

CAPROLACTAM      100ng..:..1ug....:..10.....:..100....:..1mg....:..10.....:..100....:..1g.....:..10
879 M f b6c eat TBA MXB 24m24                                              :>    no dre  P=1.  -
a   M f b6c eat liv MXB 24m24                                                    no dre  P=1.
b   M f b6c eat lun MXB 24m24                                                    no dre  P=1.
880 M m b6c eat TBA MXB 24m24                                               :>   no dre  P=1.  -
a   M m b6c eat liv MXB 24m24                                                    34.3gm * P<.8
b   M m b6c eat lun MXB 24m24                                                    no dre  P=1.
881 R f f34 eat TBA MXB 24m24                                            :>      no dre  P=1.  -
a   R f f34 eat liv MXB 24m24                                                    no dre  P=1.
882 R m f34 eat pit can 24m24                                              :  ±  #3.56gm * P<.05 -
a   R m f34 eat TBA MXB 24m24                                                    no dre  P=1.
b   R m f34 eat liv MXB 24m24                                                    5.69gm * P<.7

CAPSAICIN        100ng..:..1ug....:..10.....:..100....:..1mg....:..10.....:..100....:..1g.....:..10
883 M f swa eat cec pla 35m35                                          :>        167.mg  P<.2  +
a   M f swa eat lun ade 35m35                                                    225.mg  P<.7
b   M f swa eat lun mix 35m35                                                    1.08gm  P<1.
c   M f swa eat duo adc 35m35                                                    2.31gm  P<.3
d   M f swa eat liv hct 35m35                                                    no dre  P=1.
884 M m swa eat duo mix 29m29                                          :>        404.mg  P<.3
a   M m swa eat cec pla 29m29                                                    no dre  P=1.  +
```

	RefNum	LoConf	UpConf	Cntrl	1Dose	1Inc	2Dose	2Inc				Citation or Pathology
												Brkly Code
860	1932	2.27gm	22.6gm	0/28	2.40gm	7/30						
a	1932	2.13gm	n.s.s.	1/28	2.40gm	8/30						
b	1932	2.38gm	n.s.s.	1/28	2.40gm	7/30						
c	1932	3.38gm	n.s.s.	0/28	2.40gm	4/30						
d	1932	4.02gm	n.s.s.	0/28	2.40gm	3/30						
e	1932	6.73gm	n.s.s.	0/28	2.40gm	1/30						
f	1932	4.63gm	n.s.s.	6/28	2.40gm	5/30						
g	1932	7.12gm	n.s.s.	7/28	2.40gm	3/30						
h	1932	977.mg	n.s.s.	15/28	2.40gm	21/30						
861	1932	162.mg	560.mg	0/30	1.00gm	27/30						
a	1932	244.mg	795.mg	0/30	1.00gm	24/30						
b	1932	534.mg	2.11gm	0/30	1.00gm	15/30						
c	1932	4.14gm	n.s.s.	2/30	1.00gm	1/30						
d	1932	105.mg	3.07gm	21/30	1.00gm	29/30						
862	1932	152.mg	507.mg	0/30	800.mg	26/30						
a	1932	217.mg	707.mg	0/30	800.mg	23/30						
b	1932	364.mg	1.34gm	0/30	800.mg	17/30						
c	1932	1.32gm	n.s.s.	0/30	800.mg	4/30						
d	1932	3.31gm	n.s.s.	2/30	800.mg	1/30						
e	1932	4.95gm	n.s.s.	1/30	800.mg	0/30						
f	1932	138.mg	n.s.s.	30/30	800.mg	29/30						
863	2080	74.3mg	n.s.s.	0/15	392.mg	4/15						Hirose;carc,13,1825-1828;1992/pers.comm.
a	2080	303.mg	n.s.s.	0/15	392.mg	0/15						
864	2134	220.mg	3.40gm	0/19	800.mg	6/21						Ito;anti,183-194;1990/pers.comm.
a	2134	1.15gm	n.s.s.	0/19	800.mg	0/21						

CAFFEINE   58-08-2

	RefNum	LoConf	UpConf	Cntrl	1Dose	1Inc	2Dose	2Inc	3Dose	3Inc	4Dose	4Inc	Citation or Pathology
865	1028	230.mg	n.s.s.	0/36	55.0mg	0/37							Macklin;dact,3,135-163;1980
866	1028	126.mg	n.s.s.	0/35	55.0mg	1/38							
867	1326	471.mg	n.s.s.	0/32	71.4mg	0/32							Brune;zkko,102,153-157;1981
868	1615	849.mg	n.s.s.	1/100	11.4mg	2/50	24.6mg	0/50	53.1mg	1/50	114.mg	1/50	Mohr;fctx,22,377-382;1984
a	1615	1.32gm	n.s.s.	1/100	11.4mg	0/50	24.6mg	2/50	53.1mg	1/50	114.mg	0/50	
b	1615	1.21gm	n.s.s.	0/100	11.4mg	0/50	24.6mg	1/50	53.1mg	0/50	114.mg	0/50	
c	1615	281.mg	n.s.s.	84/100	11.4mg	40/50	24.6mg	40/50	53.1mg	36/50	114.mg	31/50	
869	1326	113.mg	n.s.s.	3/32	71.4mg	6/32							Brune;zkko,102,153-157;1981
a	1326	163.mg	n.s.s.	1/32	71.4mg	3/32							
b	1326	177.mg	n.s.s.	2/32	71.4mg	3/32							
870	1459	267.mg	n.s.s.	0/30	40.8mg	0/30							Johansson;ijcn,27,521-529;1981
871	1615	1.18gm	n.s.s.	1/100	10.0mg	0/50	21.5mg	0/50	46.5mg	1/50	100.mg	0/50	Mohr;fctx,22,377-382;1984
a	1615	1.65gm	n.s.s.	2/100	10.0mg	1/50	21.5mg	0/50	46.5mg	0/50	100.mg	0/50	
b	1615	366.mg	n.s.s.	69/100	10.0mg	35/50	21.5mg	29/50	46.5mg	27/50	100.mg	22/50	
872	1526	76.3mg	n.s.s.	41/50	42.9mg	44/48	85.7mg	37/50					Takayama;gann,73,365-371;1982
873	1526	126.mg	n.s.s.	24/46	37.5mg	31/48	75.0mg	18/44					

CALCIFEROL   (vitamin D2) 50-14-6

	RefNum	LoConf	UpConf	Cntrl	1Dose	1Inc	2Dose	2Inc	3Dose	3Inc	4Dose	4Inc	5Dose	5Inc	Citation or Pathology
874	1131	.889mg	n.s.s.	16/64	65.0ug	18/68	.130mg	23/46	.260mg	17/66	.520mg	9/31	(1.04mg	0/7)	Gass;ircs,5,477;1977

CALCIUM ACETATE   62-54-4

	RefNum	LoConf	UpConf	Cntrl	1Dose	1Inc	Citation or Pathology
875	1709	4.28gm	n.s.s.	0/30	1.20gm	0/30	Kasprzak;carc,6,279-282;1985
a	1709	4.28gm	n.s.s.	0/30	1.20gm	0/30	

CALCIUM CHLORIDE   10043-52-4

	RefNum	LoConf	UpConf	Cntrl	1Dose	1Inc	2Dose	2Inc	Citation or Pathology
876	2083	64.4mg	n.s.s.	0/15	100.mg	0/15	500.mg	0/15	Nishikawa;carc,13,1155-1158;1992

CALCIUM LACTATE   814-80-2

	RefNum	LoConf	UpConf	Cntrl	1Dose	1Inc	2Dose	2Inc	Citation or Pathology
877	2014	10.7gm	n.s.s.	1/49	1.31gm	0/50	2.63gm	0/50	Maekawa;fctx,29,589-594;1991
a	2014	1.84gm	n.s.s.	39/49	1.31gm	43/50	2.63gm	40/50	
878	2014	13.0gm	n.s.s.	1/49	1.15gm	1/50	2.30gm	3/50	
a	2014	n.s.s.	n.s.s.	49/49	1.15gm	50/50	2.30gm	50/50	

CAPROLACTAM   105-60-2

	RefNum	LoConf	UpConf	Cntrl	1Dose	1Inc	2Dose	2Inc	Citation or Pathology
879	TR214	1.62gm	n.s.s.	31/50	956.mg	25/50	(1.91gm	16/50)	
a	TR214	13.4gm	n.s.s.	1/50	956.mg	1/50	1.91gm	1/50	liv:hpa,hpc,nnd.
b	TR214	26.9gm	n.s.s.	3/50	956.mg	0/50	1.91gm	0/50	lun:a/a,a/c.
880	TR214	3.51gm	n.s.s.	21/50	883.mg	18/50	1.77gm	19/50	
a	TR214	3.82gm	n.s.s.	8/50	883.mg	10/50	1.77gm	10/50	liv:hpa,hpc,nnd.
b	TR214	6.47gm	n.s.s.	4/50	883.mg	5/50	1.77gm	4/50	lun:a/a,a/c.
881	TR214	361.mg	n.s.s.	45/49	184.mg	46/50	368.mg	38/50	
a	TR214	n.s.s.	n.s.s.	0/49	184.mg	0/50	368.mg	0/50	liv:hpa,hpc,nnd.
882	TR214	1.08gm	n.s.s.	0/50	147.mg	0/50	294.mg	3/50	s
a	TR214	359.mg	n.s.s.	38/50	147.mg	32/50	294.mg	35/50	
b	TR214	804.mg	n.s.s.	1/50	147.mg	5/50	294.mg	2/50	liv:hpa,hpc,nnd.

CAPSAICIN   404-86-4

	RefNum	LoConf	UpConf	Cntrl	1Dose	1Inc	Citation or Pathology
883	2039	50.5mg	n.s.s.	4/50	40.6mg	11/50	Toth;vivo,6,59-64;1992/pers.comm.
a	2039	26.3mg	n.s.s.	9/50	40.6mg	14/50	
b	2039	29.5mg	n.s.s.	14/50	40.6mg	18/50	
c	2039	377.mg	n.s.s.	0/50	40.6mg	1/50	
d	2039	n.s.s.	n.s.s.	0/50	40.6mg	0/50	
884	2039	93.0mg	n.s.s.	0/50	37.5mg	2/50	
a	2039	72.5mg	n.s.s.	4/50	37.5mg	7/50	

```
       Spe Strain Site  Xpo+Xpt                                                      TD50    2Tailpvl
           Sex  Route Hist   Notes                                                     DR     AuOp
  b     M m swa eat liv hct 29m29                                                    no dre   P=1.
  c     M m swa eat lun ade 29m29                                                    no dre   P=1.
  d     M m swa eat lun mix 29m29                                                    no dre   P=1.

CAPTAFOL                      100ng..:..1ug....:..10.....:..100....:..1mg...:..10....:..100....:..1g.....:..10
 885    M f b6c eat liv mix 22m24 ae                               . + .             89.4mg  Z P<.0005+
  a     M f b6c eat smi mix 22m24 ae                                                 359.mg  Z P<.0005+
  b     M f b6c eat liv hpc 22m24 ae                                                 398.mg  Z P<.003 +
  c     M f b6c eat smi adc 22m24 ae                                                 512.mg  Z P<.0005+
  d     M f b6c eat hea hae 22m24 ae                                                 1.40gm  * P<.0005+
  e     M f b6c eat spl hem 22m24 ae                                                 1.47gm  Z P<.02  +
  f     M f b6c eat for mix 22m24 ae                                                 2.19gm  * P<.03  +
  g     M f b6c eat for sqc 22m24 ae                                                 22.6gm  * P<.2   +
  h     M f b6c eat lun adc 22m24 ae                                                 134.gm  * P<1.
  i     M f b6c eat lun mix 22m24 ae                                                 no dre   P=1.
 886    M m b6c eat smi mix 22m24 e                          . + .                   137.mg  Z P<.0005+
  a     M m b6c eat liv hpc 22m24 e                                                  138.mg  Z P<.003 +
  b     M m b6c eat smi adc 22m24 e                                                  151.mg  Z P<.0005+
  c     M m b6c eat hea hae 22m24 e                                                  647.mg  Z P<.0005+
  d     M m b6c eat spl hem 22m24 e                                                  1.57gm  Z P<.008 +
  e     M m b6c eat liv mix 22m24 e                                                  166.mg  Z P<.03
  f     M m b6c eat for mix 22m24 e                                                  1.81gm  * P<.02  +
  g     M m b6c eat for sqc 22m24 e                                                  6.19gm  * P<.06  +
  h     M m b6c eat spl mix 22m24 e                                                  2.56gm  Z P<.3
  i     M m b6c eat spl hae 22m24 e                                                  79.1gm  * P<.9   +
  j     M m b6c eat lun adc 22m24 e                                                  no dre   P=1.
  k     M m b6c eat lun mix 22m24 e                                                  no dre   P=1.
 887    R f f34 eat kid mix 24m24 aers                       .>                      no dre   P=1.
 888    R m f34 eat kid mix 24m24 aers                              . + .            504.mg  * P<.0005+
  a     R m f34 eat kid tuc 24m24 aers                                               570.mg  * P<.0005
  b     R m f34 eat kid tua 24m24 aers                                               5.01gm  * P<.6
 889    R f f3d eat liv hnd 24m26 e                           . + .                  64.4mg  * P<.0005+
  a     R f f3d eat kid rca 24m26 e                                                  276.mg  * P<.004 +
  b     R f f3d eat thy cca 24m26 e                                                  353.mg  * P<.002
  c     R f f3d eat liv hpc 24m26 e                                                  1.01gm  * P<.02  +
 890    R m f3d eat kid rca 24m26 e                          .+ .                    30.3mg  * P<.0005+
  a     R m f3d eat liv hnd 24m26 e                                                  104.mg  * P<.0005+
  b     R m f3d eat kid rcc 24m26 e                                                  345.mg  * P<.002 +
  c     R m f3d eat liv hpc 24m26 e                                                  no dre   P=1.

CAPTAN                        100ng..:..1ug....:..10.....:..100....:..1mg...:..10....:..100....:..1g.....:..10
 891    M f b6c eat TBA MXB 80w91                                           :>       no dre   P=1.  -
  a     M f b6c eat liv MXB 80w91                                                    no dre   P=1.
  b     M f b6c eat lun MXB 80w91                                                    no dre   P=1.
 892    M f b6c eat duo aca 80w90      pool                                          : 21.6gm * P<.03  a
 893    M f b6c orl liv hpt 76w76 evx                                .>              no dre   P=1.  -
  a     M f b6c orl lun mix 76w76 evx                                                no dre   P=1.  -
  b     M f b6c orl tba mix 76w76 evx                                                244.mg    P<.2  -
 894    M m b6c eat TBA MXB 80w91                                           :>       no dre   P=1.  -
  a     M m b6c eat liv MXB 80w91                                                    no dre   P=1.
  b     M m b6c eat lun MXB 80w91                                                    no dre   P=1.
 895    M m b6c eat duo adm 80w90      pool                           : + 7.39gm * P<.003 a
  a     M m b6c eat duo aca 80w90                                                    15.1gm  * P<.03  a
 896    M m b6c orl liv hpt 76w76 evx                             .   ±              86.1mg    P<.02  -
  a     M m b6c orl lun ade 76w76 evx                                                187.mg    P<.09  -
  b     M m b6c orl tba mix 76w76 evx                                                42.5mg    P<.0005-
 897    M f b6a orl liv hpt 76w76 evx                                .>              no dre   P=1.  -
  a     M f b6a orl lun ade 76w76 evx                                                no dre   P=1.  -
  b     M f b6a orl tba mix 76w76 evx                                                no dre   P=1.  -
 898    M m b6a orl liv hpt 76w76 evx                            .>                  195.mg    P<.3  -
  a     M m b6a orl lun mix 76w76 evx                                                3.62gm    P<1.  -
  b     M m b6a orl tba mix 76w76 evx                                                161.mg    P<.4  -
 899    R f osm eat TBA MXB 19m26 v                               :>                 no dre   P=1.  -
  a     R f osm eat liv MXB 19m26 v                                                  no dre   P=1.
 900    R f osm eat adr MXA 19m25 v    pool                       :   ±              #253.mg \ P<.02  -
 901    R m osm eat TBA MXB 19m26 v                               :>                 4.78gm  * P<1.  -
  a     R m osm eat liv MXB 19m26 v                                                  no dre   P=1.

CARBAMYL HYDRAZINE.HC1        100ng..:..1ug....:..10.....:..100....:..1mg...:..10....:..100....:..1g......:..10
 902    M f swa wat lun mix 27m27 e                              . + .               223.mg    P<.002 +
  a     M f swa wat lun ade 27m27 e                                                  333.mg    P<.007
  b     M f swa wat blv mix 27m27 e                                                  395.mg    P<.003 +
  c     M f swa wat blv ang 27m27 e                                                  798.mg    P<.04
  d     M f swa wat liv mix 27m27 e                                                  798.mg    P<.04
  e     M f swa wat blv agm 27m27 e                                                  975.mg    P<.05
 903    M m swa wat liv hpt 27m27 e                            .    ±                1.42gm    P<.03  -
  a     M m swa wat lun ade 27m27 e                                                  745.mg    P<.2
  b     M m swa wat lun mix 27m27 e                                                  833.mg    P<.4   +
  c     M m swa wat liv mix 27m27 e                                                  933.mg    P<.2
```

RefNum	LoConf	UpConf	Cntrl	1Dose	1Inc	2Dose	2Inc	Citation or Pathology / Brkly Code
b 2039	n.s.s.	n.s.s.	0/50	37.5mg	0/50			
c 2039	59.8mg	n.s.s.	15/50	37.5mg	18/50			
d 2039	48.4mg	n.s.s.	19/50	37.5mg	22/50			
CAPTAFOL 2425-06-1								
885 1625	52.7mg	188.mg	4/48	90.0mg	27/50	(180.mg	22/49	390.mg 0/51) Ito;gann,75,853-865;1984
a 1625	216.mg	666.mg	0/48	90.0mg	6/50	180.mg	16/49	(390.mg 12/51)
b 1625	216.mg	2.43gm	2/48	90.0mg	13/50	180.mg	12/49	(390.mg 0/51)
c 1625	286.mg	1.09gm	0/48	90.0mg	3/50	180.mg	13/49	(390.mg 7/51)
d 1625	770.mg	3.35gm	0/48	90.0mg	2/50	180.mg	2/49	390.mg 11/51
e 1625	599.mg	n.s.s.	0/48	90.0mg	2/50	180.mg	4/49	(390.mg 0/51)
f 1625	1.07gm	n.s.s.	0/48	90.0mg	2/50	180.mg	4/49	390.mg 4/51
g 1625	3.68gm	n.s.s.	0/48	90.0mg	0/50	180.mg	0/49	390.mg 1/51
h 1625	n.s.s.	n.s.s.	1/48	90.0mg	1/50	180.mg	3/49	390.mg 1/51
i 1625	2.61gm	n.s.s.	2/48	90.0mg	1/50	180.mg	3/49	390.mg 1/51
886 1625	93.5mg	212.mg	0/47	83.1mg	10/51	166.mg	32/46	(332.mg 26/47)
a 1625	68.5mg	901.mg	8/47	83.1mg	23/51	(166.mg	15/46	332.mg 1/47)
b 1625	101.mg	237.mg	0/47	83.1mg	7/51	166.mg	32/46	(332.mg 22/47)
c 1625	401.mg	1.14gm	0/47	83.1mg	1/51	166.mg	4/46	332.mg 20/47
d 1625	596.mg	30.9gm	0/47	83.1mg	0/51	166.mg	5/46	(332.mg 0/47)
e 1625	75.7mg	n.s.s.	22/47	83.1mg	41/51	166.mg	31/46	(332.mg 8/47)
f 1625	882.mg	n.s.s.	0/47	83.1mg	2/51	166.mg	4/46	332.mg 4/47
g 1625	1.87gm	n.s.s.	0/47	83.1mg	0/51	166.mg	1/46	332.mg 2/47
h 1625	674.mg	n.s.s.	2/47	83.1mg	0/51	166.mg	5/46	(332.mg 1/47)
i 1625	3.37gm	n.s.s.	1/47	83.1mg	0/51	166.mg	0/46	332.mg 1/47
j 1625	1.68gm	n.s.s.	1/47	83.1mg	2/51	166.mg	3/46	332.mg 1/47
k 1625	2.22gm	n.s.s.	4/47	83.1mg	5/51	166.mg	4/46	332.mg 1/47
887 1953	191.mg	n.s.s.	0/50	25.0mg	0/50	100.mg	0/50	250.mg 0/49 Nyska;isjm,25,428-432;1989
888 1953	290.mg	1.05gm	0/50	20.0mg	1/49	80.0mg	5/49	200.mg 12/49
a 1953	318.mg	1.24gm	0/50	20.0mg	0/49	80.0mg	3/49	200.mg 12/49
b 1953	1.23gm	n.s.s.	0/50	20.0mg	0/49	80.0mg	2/49	200.mg 0/49
889 1941	43.4mg	106.mg	3/50	34.8mg	14/50	69.6mg	34/50	Tamano;gann,81,1222-1231;1990
a 1941	149.mg	1.48gm	0/50	34.8mg	8/50	69.6mg	6/50	
b 1941	177.mg	1.14gm	0/50	34.8mg	3/50	69.6mg	8/50	
c 1941	349.mg	n.s.s.	0/50	34.8mg	0/50	69.6mg	4/50	
890 1941	21.8mg	43.3mg	0/50	27.9mg	26/49	55.7mg	38/50	
a 1941	63.3mg	220.mg	2/50	27.9mg	8/50	55.7mg	21/50	
b 1941	163.mg	1.16gm	0/50	27.9mg	1/49	55.7mg	8/50	
c 1941	675.mg	n.s.s.	2/50	27.9mg	0/50	55.7mg	1/50	
CAPTAN 133-06-2								
891 TR15	4.11gm	n.s.s.	3/10	914.mg	5/50	1.83gm	8/50	liv:hpa,hpc,nnd. / lun:a/a,a/c.
a TR15	11.0gm	n.s.s.	1/10	914.mg	1/50	1.83gm	1/50	
b TR15	n.s.s.	n.s.s.	0/10	914.mg	0/50	1.83gm	0/50	
892 TR15	6.53gm	n.s.s.	0/79p	914.mg	0/50	1.83gm	3/50	Innes;ntis,1968/1969
893 1092	155.mg	n.s.s.	0/16	78.4mg	0/18			
a 1092	155.mg	n.s.s.	0/16	78.4mg	0/18			
b 1092	59.8mg	n.s.s.	0/16	78.4mg	2/18			
894 TR15	3.94gm	n.s.s.	5/10	844.mg	7/50	1.69gm	10/50	liv:hpa,hpc,nnd. / lun:a/a,a/c.
a TR15	8.87gm	n.s.s.	3/10	844.mg	1/50	1.69gm	3/50	
b TR15	3.04gm	n.s.s.	2/10	844.mg	3/50	(1.69gm	1/50)	
895 TR15	3.35gm	38.5gm	0/80p	844.mg	3/50	1.69gm	5/50	Innes;ntis,1968/1969
a TR15	5.22gm	n.s.s.	0/80p	844.mg	1/50	1.69gm	3/50	
896 1092	29.5mg	n.s.s.	0/16	73.0mg	4/15			
a 1092	45.8mg	n.s.s.	0/16	73.0mg	2/15			
b 1092	18.0mg	149.mg	0/16	73.0mg	7/15			
897 1092	147.mg	n.s.s.	0/17	78.4mg	0/17			
a 1092	147.mg	n.s.s.	1/17	78.4mg	0/17			
b 1092	97.0mg	n.s.s.	2/17	78.4mg	1/17			
898 1092	44.6mg	n.s.s.	1/18	73.0mg	3/17			
a 1092	63.5mg	n.s.s.	2/18	73.0mg	2/17			
b 1092	34.3mg	n.s.s.	3/18	73.0mg	5/17			
899 TR15	170.mg	n.s.s.	8/10	89.0mg	39/50	212.mg	36/50	liv:hpa,hpc,nnd. / adr:coa,coc. S
a TR15	888.mg	n.s.s.	0/10	89.0mg	3/50	212.mg	1/50	
900 TR15	109.mg	n.s.s.	4/75p	89.0mg	13/50	(212.mg	7/50)	
901 TR15	124.mg	n.s.s.	5/10	71.2mg	32/50	170.mg	25/50	liv:hpa,hpc,nnd.
a TR15	798.mg	n.s.s.	1/10	71.2mg	1/50	170.mg	1/50	
CARBAMYL HYDRAZINE.HCl 563-41-7								
902 60	111.mg	950.mg	21/90	125.mg	25/47			Toth;ejca,11,17-22;1975/1974
a 60	149.mg	5.61gm	18/90	125.mg	20/47			
b 60	164.mg	2.88gm	5/81	125.mg	9/31			
c 60	256.mg	n.s.s.	3/81	125.mg	5/31			
d 60	256.mg	n.s.s.	3/81	125.mg	5/31			
e 60	292.mg	n.s.s.	2/81	125.mg	4/31			
903 60	350.mg	n.s.s.	0/87	104.mg	2/32			
a 60	224.mg	n.s.s.	15/99	104.mg	12/48			
b 60	207.mg	n.s.s.	23/99	104.mg	15/48			
c 60	240.mg	n.s.s.	6/87	104.mg	5/32			

```
        Spe Strain Site    Xpo+Xpt                                                                    TD50    2Tailpvl
            Sex  Route  Hist   Notes                                                                          DR   AuOp
```

1-CARBAMYL-2-PHENYLHYDRAZINE   100ng..:..1ug....:..10......:..100.....:..1mg....:..10......:..100....:..1g......:..10

#	Spe	Sex	Strain	Route	Site	Hist	Xpo+Xpt	Notes	Plot	TD50	DR	2Tailpvl	AuOp
904	M	f	swa	wat	lun	ade	79w79	e	. + .	150.mg		P<.0005	
a	M	f	swa	wat	lun	mix	79w79	e		155.mg		P<.0005	+
b	M	f	swa	wat	lun	adc	79w79	e		1.25gm		P<.007	
c	M	f	swa	wat	liv	ang	79w79	e		1.98gm		P<.03	-
d	M	f	swa	wat	liv	agm	79w79	e		2.67gm		P<.06	-
905	M	m	swa	wat	lun	ade	79w79	e	. + .	158.mg		P<.0005	
a	M	m	swa	wat	lun	mix	79w79	e		176.mg		P<.0005	+
b	M	m	swa	wat	liv	mix	79w79	e		1.79gm		P<.3	
c	M	m	swa	wat	liv	ang	79w79	e		2.00gm		P<.3	-

CARBARSONE   100ng..:..1ug....:..10......:..100.....:..1mg....:..10......:..100....:..1g......:..10

#	Spe	Sex	Strain	Route	Site	Hist	Xpo+Xpt	Notes	Plot	TD50	DR	2Tailpvl	AuOp
906	R	f	fdr	eat	---	lyk	22m24	aes	.>	4.59gm	*	P<.4	-
907	R	m	fdr	eat	thm	rts	22m24	aes	.>	4.50gm	*	P<.4	-

CARBARYL   100ng..:..1ug....:..10......:..100.....:..1mg....:..10......:..100....:..1g......:..10

#	Spe	Sex	Strain	Route	Site	Hist	Xpo+Xpt	Notes	Plot	TD50	DR	2Tailpvl	AuOp
908	M	f	b6a	orl	lun	ade	76w76	evx	.>	no dre		P=1.	-
a	M	f	b6a	orl	liv	hpt	76w76	evx		no dre		P=1.	-
b	M	f	b6a	orl	tba	mix	76w76	evx		no dre		P=1.	-
909	M	m	b6a	orl	liv	hpt	76w76	evx	.>	no dre		P=1.	-
a	M	m	b6a	orl	lun	ade	76w76	evx		no dre		P=1.	-
b	M	m	b6a	orl	tba	mix	76w76	evx		no dre		P=1.	-
910	M	f	b6c	orl	liv	hpt	76w76	evx	.>	no dre		P=1.	-
a	M	f	b6c	orl	lun	mix	76w76	evx		no dre		P=1.	-
b	M	f	b6c	orl	tba	mix	76w76	evx		12.4mg		P<.3	-
911	M	m	b6c	orl	liv	hpt	76w76	evx	. ±	3.39mg		P<.04	-
a	M	m	b6c	orl	lun	ade	76w76	evx		3.39mg		P<.04	-
b	M	m	b6c	orl	tba	mix	76w76	evx		1.51mg		P<.003	-
912	R	b	mgr	gav	liv	tum	22m24	e	.>	no dre		P=1.	
a	R	b	mgr	gav	tba	mix	22m24	e		14.1mg		P<.003	+

CARBAZOLE   100ng..:..1ug....:..10......:..100.....:..1mg....:..10......:..100....:..1g......:..10

#	Spe	Sex	Strain	Route	Site	Hist	Xpo+Xpt	Notes	Plot	TD50	DR	2Tailpvl	AuOp
913	M	f	b6c	eat	liv	hpc	22m24	e	. + .	102.mg	Z	P<.0005	+
a	M	f	b6c	eat	for	pam	22m24	e		1.29gm	Z	P<.002	+
b	M	f	b6c	eat	for	mix	22m24	e		1.96gm	*	P<.009	+
c	M	f	b6c	eat	lun	mix	22m24	e		no dre		P=1.	
914	M	m	b6c	eat	liv	hpc	22m24	e	. + .	424.mg	*	P<.0005	+
a	M	m	b6c	eat	for	mix	22m24	e		2.79gm	*	P<.0005	+
b	M	m	b6c	eat	for	sqc	22m24	e		4.94gm	*	P<.002	+
c	M	m	b6c	eat	for	pam	22m24	e		7.03gm	*	P<.01	+
d	M	m	b6c	eat	lun	mix	22m24	e		no dre		P=1.	

CARBON TETRACHLORIDE   100ng..:..1ug....:..10......:..100.....:..1mg....:..10......:..100....:..1g......:..10

#	Spe	Sex	Strain	Route	Site	Hist	Xpo+Xpt	Notes	Plot	TD50	DR	2Tailpvl	AuOp
915	M	f	b6c	gav	liv	hpc	78w90	e	. + .	184.mg	*	P<.0005	+
a	M	f	b6c	gav	adr	mix	78w90	e		899.mg	\	P<.0005	+
b	M	f	b6c	gav	tba	mix	78w90	e		195.mg	*	P<.0005	
916	M	m	b6c	gav	liv	hpc	78w90	e	. + .	127.mg	*	P<.0005	+
a	M	m	b6c	gav	adr	mix	78w90	e		683.mg	*	P<.0005	+
b	M	m	b6c	gav	tba	mix	78w90	e		130.mg	*	P<.0005	
917	R	f	nss	ivj	liv	tum	47w69		. ±	.877mg		P<.1	
a	R	f	nss	ivj	mgl	mix	47w69			.765mg		P<.3	+
b	R	f	nss	ivj	tba	mix	47w69			.176mg		P<.007	
918	R	f	osm	gav	liv	nnd	18m26	e	.>	390.mg	*	P<.3	+
a	R	f	osm	gav	liv	hpc	18m26	e		no dre		P=1.	
b	R	f	osm	gav	tba	mix	18m26	e		no dre		P=1.	
919	R	m	osm	gav	liv	nnd	18m26	e	.>	529.mg	*	P<.2	+
a	R	m	osm	gav	liv	hpc	18m26	e		666.mg	*	P<.4	
b	R	m	osm	gav	tba	mix	18m26	e		166.mg	*	P<.4	

CARBOXYMETHYLNITROSOUREA   100ng..:..1ug....:..10......:..100.....:..1mg....:..10......:..100....:..1g......:..10

#	Spe	Sex	Strain	Route	Site	Hist	Xpo+Xpt	Notes	Plot	TD50	DR	2Tailpvl	AuOp
920	R	f	don	wat	mgl	mix	68w68	e	. + .	2.30mg	Z	P<.0005	+
a	R	f	don	wat	mgl	fba	68w68	e		2.92mg	Z	P<.0005	
b	R	f	don	wat	itn	mix	68w68	e		5.37mg	*	P<.0005	+
c	R	f	don	wat	smi	ade	68w68	e		8.09mg	*	P<.0005	
d	R	f	don	wat	smi	adc	68w68	e		11.6mg	*	P<.0005	
e	R	f	don	wat	tba	mix	68w68	e		1.38mg	*	P<.0005	
921	R	m	mrw	wat	epi	mix	17m30		. ±	34.8mg		P<.02	+
a	R	m	mrw	wat	liv	tum	17m30			no dre		P=1.	
b	R	m	mrw	wat	tba	mix	17m30			13.1mg		P<.09	

CARBROMAL   100ng..:..1ug....:..10......:..100.....:..1mg....:..10......:..100....:..1g......:..10

#	Spe	Sex	Strain	Route	Site	Hist	Xpo+Xpt	Plot	TD50	DR	2Tailpvl	AuOp
922	M	f	b6c	eat	lun	MXA	18w24	: ±	#553.mg	*	P<.05	-
a	M	f	b6c	eat	TBA	MXB	18m24		5.32gm	*	P<1.	
b	M	f	b6c	eat	liv	MXB	18m24		1.95gm	*	P<.09	
c	M	f	b6c	eat	lun	MXB	18m24		553.mg	*	P<.05	
923	M	m	b6c	eat	TBA	MXB	18m24	:>	no dre		P=1.	
a	M	m	b6c	eat	liv	MXB	18m24		1.58gm	*	P<.6	
b	M	m	b6c	eat	lun	MXB	18m24		no dre		P=1.	
924	R	f	f34	eat	TBA	MXB	24m24	:>	no dre		P=1.	-
a	R	f	f34	eat	liv	MXB	24m24		no dre		P=1.	

	RefNum	LoConf	UpConf	Cntrl	1Dose	1Inc	2Dose	2Inc			Citation or Pathology	Brkly Code
1-CARBAMYL-2-PHENYLHYDRAZINE 103-03-7												
904	60a	92.1mg	273.mg	18/100	500.mg	39/50					Toth;jnci,52,241-251;1974	
a	60a	93.7mg	289.mg	21/100	500.mg	39/50						
b	60a	513.mg	19.6gm	4/100	500.mg	9/50						
c	60a	682.mg	n.s.s.	0/34	500.mg	4/42						
d	60a	807.mg	n.s.s.	0/34	500.mg	3/42						
905	60a	94.8mg	305.mg	15/96	417.mg	33/47						
a	60a	101.mg	388.mg	23/96	417.mg	33/47						
b	60a	429.mg	n.s.s.	3/40	417.mg	5/32						
c	60a	477.mg	n.s.s.	2/40	417.mg	4/32						
CARBARSONE 121-59-5												
906	65	747.mg	n.s.s.	0/89	25.0mg	0/43	50.0mg	1/45	69.2mg	0/49	Oser;txap,9,528-535;1966	
907	65	732.mg	n.s.s.	0/85	25.0mg	0/39	50.0mg	1/43	69.2mg	0/50		
CARBARYL (sevin) 63-25-2												
908	1228	2.30mg	n.s.s.	1/17	1.93mg	1/18					Innes;ntis,1968/1969	
a	1228	3.82mg	n.s.s.	0/17	1.93mg	0/18						
b	1228	1.82mg	n.s.s.	2/17	1.93mg	2/18						
909	1228	2.12mg	n.s.s.	1/18	1.80mg	1/18						
a	1228	2.36mg	n.s.s.	2/18	1.80mg	1/18						
b	1228	1.86mg	n.s.s.	3/18	1.80mg	2/18						
910	1228	3.82mg	n.s.s.	0/16	1.93mg	0/18						
a	1228	3.82mg	n.s.s.	0/16	1.93mg	0/18						
b	1228	2.01mg	n.s.s.	0/16	1.93mg	1/18						
911	1228	1.02mg	n.s.s.	0/16	1.80mg	3/17						
a	1228	1.02mg	n.s.s.	0/16	1.80mg	3/17						
b	1228	.609mg	8.19mg	0/16	1.80mg	6/17						
912	1426	19.5mg	n.s.s.	0/46	7.88mg	0/12					Andrianova;vpit,29,71-74;1970	
a	1426	4.62mg	139.mg	1/46	7.88mg	4/12						
CARBAZOLE (9H-carbazole) 86-74-8												
913	1481	64.8mg	174.mg	2/45	180.mg	35/49	(360.mg	24/43	749.mg	30/46)	Tsuda;jnci,69,1383-1387;1982	
a	1481	663.mg	4.65gm	0/45	180.mg	5/49	360.mg	7/43	(749.mg	4/46)		
b	1481	1.15gm	83.3gm	0/45	180.mg	5/49	360.mg	8/43	749.mg	6/46		
c	1481	7.93gm	n.s.s.	2/45	180.mg	0/49	360.mg	0/43	749.mg	1/46		
914	1481	281.mg	763.mg	9/46	166.mg	12/42	332.mg	20/42	665.mg	37/48		
a	1481	1.44gm	6.58gm	0/46	166.mg	0/42	332.mg	1/42	665.mg	11/48		
b	1481	2.13gm	18.0gm	0/46	166.mg	0/42	332.mg	0/42	665.mg	7/48		
c	1481	2.67gm	430.gm	0/46	166.mg	0/42	332.mg	1/42	665.mg	4/48		
d	1481	4.29gm	n.s.s.	4/46	166.mg	0/42	332.mg	1/42	665.mg	3/48		
CARBON TETRACHLORIDE 56-23-5												
915	1469m	119.mg	287.mg	1/18	774.mg	40/42	1.55gm	43/45			Weisburger;enhp,21,7-16;1977/National Cancer Institute 1976	
a	1469m	490.mg	2.35gm	0/18	774.mg	15/42	(1.55gm	10/45)				
b	1469m	123.mg	324.mg	3/18	774.mg	40/42	1.55gm	43/45				
916	1469m	70.1mg	220.mg	3/18	774.mg	49/49	1.55gm	47/48				
a	1469m	487.mg	1.21gm	0/18	774.mg	28/49	1.55gm	28/48				
b	1469m	70.9mg	230.mg	4/18	774.mg	49/49	1.55gm	47/48				
917	1515	.265mg	n.s.s.	0/34	.157mg	3/57					Tourkevitch;ijcn,20,1446-1449;1964	
a	1515	.214mg	n.s.s.	1/34	.157mg	5/57						
b	1515	87.0ug	2.27mg	2/34	.157mg	16/57						
918	1469n	140.mg	n.s.s.	1/20	40.5mg	11/50	81.0mg	9/49			Weisburger;enhp,21,7-16;1977/National Cancer Institute 1976	
a	1469n	415.mg	n.s.s.	1/20	40.5mg	4/50	81.0mg	2/49				
b	1469n	144.mg	n.s.s.	10/20	40.5mg	34/50	81.0mg	20/49				
919	1469n	201.mg	n.s.s.	0/20	23.8mg	2/49	47.6mg	3/50				
a	1469n	230.mg	n.s.s.	0/20	23.8mg	2/49	47.6mg	2/50				
b	1469n	48.7mg	n.s.s.	7/20	23.8mg	21/49	47.6mg	24/50				
CARBOXYMETHYLNITROSOUREA (CMNU) 60391-92-6												
920	1547	1.44mg	4.77mg	9/36	5.71mg	28/40	11.4mg	30/38	(22.9mg	11/34)	Maekawa;zkko,106,12-16;1983	
a	1547	1.75mg	7.34mg	9/36	5.71mg	27/40	11.4mg	27/38	(22.9mg	10/34)		
b	1547	3.75mg	7.98mg	0/36	5.71mg	5/40	11.4mg	19/38	22.9mg	27/34		
c	1547	5.45mg	12.7mg	0/36	5.71mg	4/40	11.4mg	15/38	22.9mg	20/34		
d	1547	7.38mg	19.5mg	0/36	5.71mg	1/40	11.4mg	9/38	22.9mg	19/34		
e	1547	.854mg	2.48mg	14/36	5.71mg	34/40	11.4mg	35/38	22.9mg	34/34		
921	1246	14.2mg	n.s.s.	0/26	5.29mg	6/40					Bulay;jnci,62,1523-1528;1979	
a	1246	68.1mg	n.s.s.	0/26	5.29mg	0/40						
b	1246	5.06mg	n.s.s.	10/26	5.29mg	24/40						
CARBROMAL (bromodiethylacetylurea) 77-65-6												
922	TR173	314.mg	n.s.s.	0/20	122.mg	8/50	244.mg	9/49			lun:a/a,a/c.	S
a	TR173	210.mg	n.s.s.	7/20	122.mg	30/50	244.mg	25/49				
b	TR173	739.mg	n.s.s.	0/20	122.mg	1/50	244.mg	4/49			liv:hpa,hpc,nnd.	
c	TR173	314.mg	n.s.s.	0/20	122.mg	8/50	244.mg	9/49			lun:a/a,a/c.	
923	TR173	344.mg	n.s.s.	12/20	112.mg	16/49	225.mg	21/50				
a	TR173	321.mg	n.s.s.	4/20	112.mg	8/49	225.mg	13/50			liv:hpa,hpc,nnd.	
b	TR173	748.mg	n.s.s.	7/20	112.mg	4/49	225.mg	8/50			lun:a/a,a/c.	
924	TR173	138.mg	n.s.s.	10/20	61.9mg	24/50	124.mg	26/50				
a	TR173	n.s.s.	n.s.s.	0/20	61.9mg	0/50	124.mg	0/50			liv:hpa,hpc,nnd.	

```
      Spe Strain Site   Xpo+Xpt                                              TD50      2Tailpvl
        Sex  Route  Hist      Notes                                            DR       AuOp

925  R m f34 eat TBA MXB 24m24                                     :>          1.03gm * P<.8    -
  a  R m f34 eat liv MXB 24m24                                                 no dre   P=1.

CARRAGEENAN, ACID-DEGRADED       100ng..:..1ug...:..10.....:...100....:..1mg...:..10....:..100....:..1g.....:..10
926  R m f34 eat clr mix 26w78 r                                        . + .  2.43gm   P<.0005+
  a  R m f34 eat clr sqc 26w78 r                                               3.33gm   P<.003
927  R m f34 eat clr mix 39w78 r                                      . + .    1.49gm   P<.0005+
928  R b sda wat clr mix 65w65 er                                       . + .  2.20gm   P<.0005+
929  R b sda gav clr mix 65w65 er                                       . +    5.16gm * P<.0005+

CARRAGEENAN, NATIVE              100ng..:..1ug...:..10.....:...100....:..1mg...:..10....:..100....:..1g.....:..10
930  H f syg eat liv hem 78w78                                             .>  no dre   P=1.    -
  a  H f syg eat tba mix 78w78                                                 177.gm * P<.9    -
931  H m syg eat liv hem 25m26 a                                          .>   no dre   P=1.    -
  a  H m syg eat tba mix 25m26 a                                               54.1gm * P<.3    -
932  R f mrc eat pns tum 33m34 a                                              35.4gm * P<.03   -
  a  R f mrc eat ski tum 33m34 a                                               no dre   P=1.
933  R m mrc eat tba mix 32m33 a                                              .>no dre   P=1.

D-CARVONE                        100ng..:..1ug...:..10.....:...100....:..1mg...:..10....:..100....:..1g.....:..10
934  M f b6c gav TBA MXB 24m24                                          :>     no dre   P=1.    -
  a  M f b6c gav liv MXB 24m24                                                 no dre   P=1.
  b  M f b6c gav lun MXB 24m24                                                 no dre   P=1.
935  M m b6c gav TBA MXB 24m24                                        :>       no dre   P=1.
  a  M m b6c gav liv MXB 24m24                                                 63.8gm * P<1.
  b  M m b6c gav lun MXB 24m24                                                 no dre   P=1.

CATECHOL                         100ng..:..1ug...:..10.....:...100....:..1mg...:..10....:..100....:..1g.....:..10
936  M f b6c eat stg ade 96w96 e                                 . + .        471.mg   P<.0005+
  a  M f b6c eat liv hnd 96w96 e                                              17.3gm   P<.3
  b  M f b6c eat for pam 96w96 e                                              17.3gm   P<.3
  c  M f b6c eat lun ade 96w96 e                                               no dre   P=1.
  d  M f b6c eat liv hpc 96w96 e                                               no dre   P=1.
937  M m b6c eat stg ade 96w96 e                             . + .           165.mg   P<.0005+
  a  M m b6c eat --- 1km 96w96 e                                              2.46gm   P<.03
  b  M m b6c eat lun ade 96w96 e                                              16.5gm   P<.3
  c  M m b6c eat for pam 96w96 e                                              16.5gm   P<.3
  d  M m b6c eat liv hnd 96w96 e                                               no dre   P=1.
  e  M m b6c eat liv hpc 96w96 e                                               no dre   P=1.
938  R m f34 wat pro cic 78w78 er                                .>          236.mg   P<.8    -
  a  R m f34 wat ubl tum 78w78 er                                              no dre   P=1.    -
939  R f f3d eat stg ade 24m24 e                                 <+           noTD50   P<.0005
  a  R f f3d eat stg adc 24m24 e                                             490.mg   P<.0005+
  b  R f f3d eat liv hnd 24m24 e                                               no dre   P=1.
940  R m f3d eat stg ade 51w52 rv                            <+              noTD50   P<.0005
  a  R m f3d eat stg adc 51w52 rv                                            257.mg   P<.07   +
  b  R m f3d eat for pam 51w52 rv                                            833.mg   P<.4
941  R m f3d eat stg ade 24m24 e                                 <+          noTD50   P<.0005
  a  R m f3d eat stg adc 24m24 e                                             286.mg   P<.0005+
  b  R m f3d eat for pam 24m24 e                                             2.96gm   P<.09
  c  R m f3d eat liv hnd 24m24 e                                               no dre   P=1.
942  R m f3d eat stg ade 48w96 er                                <+          noTD50   P<.0005
  a  R m f3d eat stg adc 48w96 er                                            387.mg   P<.05   +
943  R m f3d eat stg ade 72w72 ekr                               <+          noTD50   P<.0005+
  a  R m f3d eat stg adc 72w72 ekr                                           206.mg   P<.02   +
944  R m f3d eat stg ade 72w96 er                                <+          noTD50   P<.0005+
  a  R m f3d eat stg adc 72w96 er                                            202.mg   P<.002  +
945  R m f3d eat stg ade 96w96 er                                <+          noTD50   P<.0005+
  a  R m f3d eat stg adc 96w96 er                                            141.mg   P<.0005+

CELIPROLOL                       100ng..:..1ug....:..10....:...100....:..1mg....:..10....:..100....:..1g.....:..10
946  M f cd1 eat lun mix 24m24 e                                           .>  8.02gm * P<.3    -
947  M m cd1 eat lun mix 24m24 e                                           .>  4.71gm * P<.2    -

CHENODEOXYCHOLIC ACID            100ng..:..1ug....:..10....:...100....:..1mg....:..10....:..100....:..1g.....:..10
948  M b cba eat itn tum 54w54 r                                    .>         no dre   P=1.    -

CHLORAL HYDRATE                  100ng..:..1ug....:..10....:...100....:..1mg....:..10....:..100....:..1g.....:..10
949  M m b6c wat liv hpc 60w60 ek                              . ±           74.1mg   P<.08
950  M m b6c wat liv mix 24m24 e                                . + .        106.mg   P<.0005+
  a  M m b6c wat liv hpc 24m24 e                                             224.mg   P<.008  +
  b  M m b6c wat liv hpa 24m24 e                                             387.mg   P<.03

CHLORAMBEN                       100ng..:..1ug....:..10....:...100....:..1mg....:..10....:..100....:..1g.....:..10
951  M f b6c eat TBA MXB 80w91                                          :>    3.62gm * P<.2    -
  a  M f b6c eat liv MXB 80w91                                               3.97gm * P<.2
  b  M f b6c eat lun MXB 80w91                                              58.0gm * P<.8
952  M f b6c eat liv hpc 80w90         pool                             :  +  5.23gm * P<.006 c
953  M m b6c eat TBA MXB 80w91                                          :>    9.80gm * P<.7    -
  a  M m b6c eat liv MXB 80w91                                                no dre   P=1.
  b  M m b6c eat lun MXB 80w91                                              18.0gm * P<.1
```

	RefNum	LoConf	UpConf	Cntrl	1Dose	1Inc	2Dose	2Inc	3Dose	3Inc	Citation or Pathology
											Brkly Code
925	TR173	125.mg	n.s.s.	8/20	49.5mg	14/50	99.0mg	24/50			
a	TR173	886.mg	n.s.s.	1/20	49.5mg	0/50	99.0mg	1/50			liv:hpa,hpc,nnd.
	CARRAGEENAN,	ACID-DEGRADED	---								
926	1517m	1.10gm	7.80gm	0/46	1.33gm	8/42					Oohashi;clet,14,267-272;1981
a	1517m	1.36gm	17.5gm	0/46	1.33gm	6/42					
927	1517n	836.mg	3.00gm	0/46	2.00gm	17/42					
928	611	1.10gm	5.92gm	0/30	2.65gm	11/40					Wakabayashi;clet,4,171-176;1978
929	611m	2.33gm	15.2gm	0/30	1.00gm	0/30	5.00gm	8/29			
	CARRAGEENAN,	NATIVE	9000-07-1								
930	1330	1.40gm	n.s.s.	1/100	523.mg	0/30	2.61gm	0/30	5.23gm	0/30	Rustia;clet,11,1-10;1980
a	1330	11.0gm	n.s.s.	5/100	523.mg	0/30	2.61gm	3/30	5.23gm	1/30	
931	1330	2.49gm	n.s.s.	1/100	460.mg	0/30	2.30gm	0/30	4.60gm	0/30	
a	1330	12.8gm	n.s.s.	6/100	460.mg	2/30	2.30gm	2/30	4.60gm	4/30	
932	1330	11.7gm	n.s.s.	2/100	250.mg	1/30	1.25gm	1/30	2.50gm	4/30	
a	1330	35.2gm	n.s.s.	1/100	250.mg	3/30	1.25gm	0/30	2.50gm	0/30	
933	1330	6.34gm	n.s.s.	56/100	200.mg	15/30	1.00gm	19/30	2.00gm	12/30	
	D-CARVONE	2244-16-8									
934	TR381	951.mg	n.s.s.	9/50	263.mg	14/50	526.mg	14/50			
a	TR381	2.73gm	n.s.s.	1/50	263.mg	2/50	526.mg	1/50			liv:hpa,hpc,nnd.
b	TR381	1.34gm	n.s.s.	1/50	263.mg	6/50	526.mg	3/50			lun:a/a,a/c.
935	TR381	1.40gm	n.s.s.	27/50	263.mg	15/50	526.mg	16/50			
a	TR381	1.27gm	n.s.s.	7/50	263.mg	7/50	526.mg	7/50			liv:hpa,hpc,nnd.
b	TR381	1.64gm	n.s.s.	7/50	263.mg	5/50	526.mg	6/50			lun:a/a,a/c.
	CATECHOL	(1,2-dihydroxybenzene)	120-80-9								
936	1971	268.mg	906.mg	0/29	1.04gm	21/29					Hirose;carc,14,525-529;1993/pers.comm.
a	1971	2.82gm	n.s.s.	0/29	1.04gm	1/29					
b	1971	2.82gm	n.s.s.	0/29	1.04gm	1/29					
c	1971	5.30gm	n.s.s.	1/29	1.04gm	0/29					
d	1971	3.22gm	n.s.s.	1/27	1.04gm	1/29					
937	1971	72.8mg	335.mg	0/28	960.mg	29/30					
a	1971	955.mg	n.s.s.	1/27	960.mg	7/30					
b	1971	2.69gm	n.s.s.	0/27	960.mg	1/30					
c	1971	2.69gm	n.s.s.	0/27	960.mg	1/30					
d	1971	2.25gm	n.s.s.	6/27	960.mg	4/30					
e	1971	2.02gm	n.s.s.	7/27	960.mg	5/30					
938	1718	23.9mg	n.s.s.	5/30	25.0mg	6/30					LaVoie;gann,76,266-271;1985/pers.comm.
a	1718	86.9mg	n.s.s.	0/30	25.0mg	0/30					
939	1971	n.s.s.	125.mg	0/30	400.mg	28/28					Hirose;carc,14,525-529;1993/pers.comm.
a	1971	249.mg	1.15gm	0/30	400.mg	12/28					
b	1971	2.31gm	n.s.s.	1/30	400.mg	0/28					
940	1845	n.s.s.	35.6mg	0/10	335.mg	15/15					Hirose;gann,78,1144-1149;1987/pers.comm.
a	1845	77.6mg	n.s.s.	0/10	335.mg	3/15					
b	1845	135.mg	n.s.s.	0/10	335.mg	1/15					
941	1971	n.s.s.	100.mg	0/30	320.mg	28/28					Hirose;carc,14,525-529;1993/pers.comm.
a	1971	154.mg	612.mg	0/30	320.mg	15/28					
b	1971	727.mg	n.s.s.	0/30	320.mg	2/28					
c	1971	1.85gm	n.s.s.	1/30	320.mg	0/28					
942	2107m	n.s.s.	60.1mg	0/12	160.mg	14/14					Hirose;canr,52,787-790;1992/pers.comm.
a	2107m	117.mg	n.s.s.	0/12	160.mg	3/14					
943	2107n	n.s.s.	83.2mg	0/10	320.mg	10/10					
a	2107n	69.6mg	n.s.s.	0/10	320.mg	4/10					
944	2107o	n.s.s.	78.7mg	0/12	240.mg	18/18					
a	2107o	93.1mg	666.mg	0/12	240.mg	9/18					
945	2107r	n.s.s.	116.mg	0/12	320.mg	15/15					
a	2107r	65.8mg	360.mg	0/12	320.mg	11/15					
	CELIPROLOL	56980-93-9									
946	1968	2.40gm	n.s.s.	13/100	100.mg	8/49	300.mg	9/50	900.mg	20/100	Markiewicz;phrm,38,421-434;1989
947	1968	1.61gm	n.s.s.	23/100	100.mg	16/51	300.mg	15/50	900.mg	34/100	
	CHENODEOXYCHOLIC ACID	474-25-9									
948	1445	17.4mg	n.s.s.	0/10	31.3mg	0/10					Martin;bjca,43,884-886;1981
	CHLORAL HYDRATE	302-17-0									
949	2072m	17.6mg	n.s.s.	0/5	166.mg	2/5					Daniel;faat,19,159-168;1992/pers.comm.
950	2072n	53.2mg	334.mg	3/20	166.mg	17/24					
a	2072n	98.7mg	4.53gm	2/20	166.mg	11/24					
b	2072n	149.mg	n.s.s.	1/20	166.mg	7/24					
	CHLORAMBEN	133-90-4									
951	TR25	1.80gm	n.s.s.	0/10	1.14gm	12/50	2.26gm	13/50			
a	TR25	2.28gm	n.s.s.	0/10	1.14gm	8/50	2.26gm	10/50			liv:hpa,hpc,nnd.
b	TR25	9.57gm	n.s.s.	0/10	1.14gm	1/50	2.26gm	1/50			lun:a/a,a/c.
952	TR25	2.58gm	57.4gm	2/69p	1.14gm	7/50	2.26gm	10/50			
953	TR25	1.53gm	n.s.s.	2/10	1.05gm	20/50	2.09gm	20/50			
a	TR25	2.26gm	n.s.s.	2/10	1.05gm	18/50	2.09gm	14/50			liv:hpa,hpc,nnd.
b	TR25	6.22gm	n.s.s.	0/10	1.05gm	0/50	2.09gm	4/50			lun:a/a,a/c.

```
       Spe Strain Site    Xpo+Xpt                                              TD50    2Tailpvl
          Sex  Route  Hist     Notes                                            DR     AuOp

954  M m b6c eat liv hpc 80w90    pool                                    :  ±    4.96gm * P<.07  a
955  R f osm eat TBA MXB 19m26                                      :>           no dre   P=1.   -
  a  R f osm eat liv MXB 19m26                                                   no dre   P=1.
956  R m osm eat TBA MXB 19m26                                         :>        1.28gm * P<.5   -
  a  R m osm eat liv MXB 19m26                                                   no dre   P=1.
957  R m osm eat --- hem 19m25    pool                                   :  ±   #3.03gm * P<.02  -

CHLORAMBUCIL                 100ng..:..1ug....:..10.....:...100....:..1mg...:..10.....:...100...:..1g.....:..10
958  M f swi ipj lun mix 26w78 e            . + .                               97.0ug * P<.0005+
  a  M f swi ipj --- lys 26w78 e                                                .709mg * P<.003
  b  M f swi ipj liv lys 26w78 e                                                no dre   P=1.
  c  M f swi ipj tba mix 26w78 e                                                30.6ug * P<.0005
  d  M f swi ipj tba mal 26w78 e                                                60.4ug * P<.0005
  e  M f swi ipj tba ben 26w78 e                                                3.60mg * P<.7
959  M m swi ipj lun mix 26w78 e              . + .                             .210mg * P<.0005+
  a  M m swi ipj --- lys 26w78 e                                                .435mg * P<.0005+
  b  M m swi ipj liv mix 26w78 e                                                no dre   P=1.
  c  M m swi ipj tba mix 26w78 e                                                .102mg * P<.0005
  d  M m swi ipj tba mal 26w78 e                                                .126mg * P<.0005
  e  M m swi ipj tba ben 26w78 e                                                no dre   P=1.
960  R m cdr ipj --- leu 26w78 e                 . + .                          1.41mg * P<.004 +
  a  R m cdr ipj --- lys 26w78 e -                                              1.73mg * P<.0005
  b  R m cdr ipj liv tum 26w78 e                                                no dre   P=1.
  c  R m cdr ipj tba mix 26w78 e                                                .603mg * P<.07
  d  R m cdr ipj tba mal 26w78 e                                                .712mg * P<.03
  e  R m cdr ipj tba ben 26w78 e                                                no dre   P=1.
961  R f sda gav mgl mix 18m24               . + .                             .657mg \ P<.0005+
  a  R f sda gav auc sqc 18m24                                                  6.41mg * P<.005 +
  b  R f sda gav ner mix 18m24                                                  3.86mg * P<.02  +
  c  R f sda gav tba mal 18m24                                                  .588mg * P<.0005
  d  R f sda gav tba ben 18m24                                                  1.47mg * P<.09

CHLORAMINATED WATER          100ng..:..1ug....:..10.....:...100....:..1mg...:..10.....:...100...:..1g.....:..10
962  M f b6c wat TBA MXB 24m24                                      :>           no dre   P=1.   -
  a  M f b6c wat liv MXB 24m24                                                   no dre   P=1.
  b  M f b6c wat lun MXB 24m24                                                   no dre   P=1.
963  M m b6c wat mul MXA 24m24                                   :  +   :       #37.2mg Z P<.007 -
  a  M m b6c wat TBA MXB 24m24                                                   8.55mg Z P<.2
  b  M m b6c wat liv MXB 24m24                                                   13.1mg Z P<.3
  c  M m b6c wat lun MXB 24m24                                                   no dre   P=1.
964  R f f34 wat pni MXA 24m24                                 :  +   :          11.7mg Z P<.008
  a  R f f34 wat mul mnl 24m24                                                   23.7mg * P<.04  e
  b  R f f34 wat TBA MXB 24m24                                                   26.9mg * P<.6
  c  R f f34 wat liv MXB 24m24                                                   116.mg * P<.4
965  R m f34 wat sub fib 24m24                                    :  ±          #23.3mg * P<.04  -
  a  R m f34 wat TBA MXB 24m24                                                   85.1mg * P<.9
  b  R m f34 wat liv MXB 24m24                                                   no dre   P=1.

CHLORAMPHENICOL              100ng..:..1ug....:..10.....:...100....:..1mg...:..10.....:...100...:..1g.....:..10
966  R f sda eat liv hpt 66w75 e                                           .>   278.mg   P<.2   -
  a  R f sda eat tba mix 66w75 e                                                 no dre   P=1.   -

CHLORANIL                    100ng..:..1ug....:..10.....:...100....:..1mg...:..10.....:...100...:..1g.....:..10
967  M f b6a orl lun ade 76w76 evx                                       .>      no dre   P=1.
  a  M f b6a orl liv hpt 76w76 evx                                               no dre   P=1.
  b  M f b6a orl tba mix 76w76 evx                                               no dre   P=1.
968  M m b6a orl lun ade 76w76 evx                                    .>         no dre   P=1.
  a  M m b6a orl liv hpt 76w76 evx                                               no dre   P=1.
  b  M m b6a orl tba mix 76w76 evx                                               440.mg   P<.7
969  M f b6c orl liv hpt 76w76 evx                                    .  ±       261.mg   P<.1
  a  M f b6c orl lun mix 76w76 evx                                               no dre   P=1.
  b  M f b6c orl tba mix 76w76 evx                                               261.mg   P<.1
970  M m b6c orl liv hpt 76w76 evx                               .  +  .         57.2mg   P<.002
  a  M m b6c orl lun ade 76w76 evx                                               113.mg   P<.02
  b  M m b6c orl tba mix 76w76 evx                                               34.2mg   P<.0005

CHLORDANE, TECHNICAL GRADE   100ng..:..1ug....:..10.....:...100....:..1mg...:..10.....:...100...:..1g.....:..10
971  M f b6c eat liv hpc 80w90 v                                   :+ :          4.89mg / P<.0005c
  a  M f b6c eat TBA MXB 80w90 v                                                 5.16mg / P<.0005
  b  M f b6c eat liv MXB 80w90 v                                                 4.89mg / P<.0005
  c  M f b6c eat lun MXB 80w90 v                                                 228.mg * P<.4
972  M f b6c eat liv hpc 80w89 v    pool                          : + :          5.28mg / P<.0005c
973  M m b6c eat liv hpc 80w90 ev                              :+ :              2.15mg / P<.0005c
  a  M m b6c eat TBA MXB 80w90 ev                                                2.31mg * P<.0005
  b  M m b6c eat liv MXB 80w90 ev                                                2.15mg / P<.0005
  c  M m b6c eat lun MXB 80w90 ev                                                656.mg * P<1.
974  M m b6c eat liv hpc 80w89 ev   pool                          :+ :           2.48mg / P<.0005c
975  M f cd1 eat liv nod 78w78 e                             . + .               1.78mg Z P<.0005
  a  M f cd1 eat liv hpt 78w78 e                                                 71.8mg * P<.2
  b  M f cd1 eat lun ade 78w78 e                                                 no dre   P=1.
976  M m cd1 eat liv nod 78w78 e                            . + .                1.45mg * P<.0005
```

	RefNum	LoConf	UpConf	Cntrl	1Dose	1Inc	2Dose	2Inc	Citation or Pathology
									Brkly Code
954	TR25	2.01gm	n.s.s.	9/70p	1.05gm	16/50	2.09gm	14/50	
955	TR25	616.mg	n.s.s.	6/10	354.mg	36/50	708.mg	34/50	
a	TR25	2.90gm	n.s.s.	0/10	354.mg	2/50	708.mg	1/50	liv:hpa,hpc,nnd.
956	TR25	339.mg	n.s.s.	2/10	286.mg	25/50	566.mg	24/50	
a	TR25	1.79gm	n.s.s.	0/10	286.mg	4/50	566.mg	1/50	liv:hpa,hpc,nnd.
957	TR25	1.31gm	n.s.s.	0/75p	286.mg	4/50	566.mg	3/50	s

CHLORAMBUCIL   305-03-3

	RefNum	LoConf	UpConf	Cntrl	1Dose	1Inc	2Dose	2Inc	Citation or Pathology
958	1336	51.4ug	.225mg	20/154	.213mg	16/19	.426mg	4/8	Skipper;srfr;1976/Weisburger;canc;1977/Prejean pers.comm.
a	1336	.229mg	8.39gm	3/154	.213mg	1/19	.426mg	3/8	
b	1336	.215mg	n.s.s.	1/154	.213mg	0/19	.426mg	0/8	
c	1336	11.8ug	77.2ug	42/154	.213mg	18/19	.426mg	8/8	
d	1336	30.9ug	.136mg	29/154	.213mg	16/19	.426mg	7/8	
e	1336	.350mg	n.s.s.	13/154	.213mg	2/19	.426mg	1/8	
959	1336	.108mg	.567mg	9/101	.213mg	10/26	.426mg	7/12	
a	1336	.194mg	1.61mg	2/101	.213mg	6/26	.426mg	3/12	
b	1336	.308mg	n.s.s.	2/101	.213mg	0/26	.426mg	0/12	
c	1336	54.3ug	.256mg	28/101	.213mg	18/26	.426mg	10/12	
d	1336	67.7ug	.305mg	19/101	.213mg	14/26	.426mg	10/12	
e	1336	.535mg	n.s.s.	9/101	.213mg	4/26	.426mg	0/12	
960	1336	.456mg	19.3mg	2/177	.314mg	2/21	.643mg	2/12	
a	1336	.524mg	12.4mg	0/177	.314mg	1/21	.643mg	2/12	
b	1336	.412mg	n.s.s.	0/177	.314mg	0/21	.643mg	0/12	
c	1336	.198mg	n.s.s.	59/177	.314mg	9/21	.643mg	7/12	
d	1336	.252mg	n.s.s.	32/177	.314mg	5/21	.643mg	6/12	
e	1336	.681mg	n.s.s.	27/177	.314mg	4/21	.643mg	1/12	
961	1770	.286mg	2.96mg	8/120	.323mg	10/30	(.645mg	5/30)	Berger;smon,13,8-13;1986
a	1770	1.94mg	65.5mg	0/120	.323mg	0/30	.645mg	3/30	
b	1770	1.37mg	n.s.s.	2/120	.323mg	3/30	.645mg	3/30	
c	1770	.343mg	1.26mg	13/120	.323mg	15/30	.645mg	15/30	
d	1770	.517mg	n.s.s.	43/120	.323mg	15/30	.645mg	15/30	

CHLORAMINATED WATER   ---

	RefNum	LoConf	UpConf	Cntrl	1Dose	1Inc	2Dose	2Inc	Citation or Pathology
962	TR392	18.1mg	n.s.s.	35/50	9.86mg	40/50	19.7mg	35/50 (39.4mg | 31/50)	
a	TR392	97.9mg	n.s.s.	20/50	9.86mg	21/50	19.7mg	24/50 39.4mg 15/50	liv:hpa,hpc,nnd.
b	TR392	207.mg	n.s.s.	5/50	9.86mg	5/50	19.7mg	3/50 39.4mg 3/50	lun:a/a,a/c.
963	TR392	13.8mg	466.mg	0/50	8.21mg	5/50	(16.4mg	3/50 32.8mg 2/51)	mul:mlm,mlp. S
a	TR392	2.95mg	n.s.s.	42/50	8.21mg	49/50	(16.4mg	41/50 32.8mg 37/51)	
b	TR392	3.85mg	n.s.s.	35/50	8.21mg	39/50	(16.4mg	33/50 32.8mg 25/51)	liv:hpa,hpc,nnd.
c	TR392	27.9mg	n.s.s.	21/50	8.21mg	19/50	16.4mg	17/50 (32.8mg 7/51)	lun:a/a,a/c.
964	TR392	4.42mg	226.mg	0/50	2.81mg	5/50	(5.63mg	1/50 11.3mg 0/50)	pni:isa,isc. S
a	TR392	10.1mg	n.s.s.	8/50	2.81mg	11/50	5.63mg	15/50 11.3mg 16/50	
b	TR392	5.25mg	n.s.s.	48/50	2.81mg	45/50	5.63mg	44/50 11.3mg 47/50	
c	TR392	26.0mg	n.s.s.	1/50	2.81mg	3/50	5.63mg	1/50 11.3mg 3/50	liv:hpa,hpc,nnd.
965	TR392	11.3mg	n.s.s.	0/51	2.47mg	4/50	4.94mg	2/51 9.87mg 5/50	s
a	TR392	4.22mg	n.s.s.	46/51	2.47mg	50/50	4.94mg	50/51 9.87mg 48/50	
b	TR392	33.6mg	n.s.s.	2/51	2.47mg	4/50	4.94mg	2/51 9.87mg 0/50	liv:hpa,hpc,nnd.

CHLORAMPHENICOL   (chloromycetin) 56-75-7

	RefNum	LoConf	UpConf	Cntrl	1Dose	1Inc	Citation or Pathology
966	200a	45.3mg	n.s.s.	0/71	22.0mg	1/36	Cohen;jnci,51,403-417;1973
a	200a	22.2mg	n.s.s.	18/71	22.0mg	9/36	

CHLORANIL   (tetrachloro-p-benzoquinone) 118-75-2

	RefNum	LoConf	UpConf	Cntrl	1Dose	1Inc	Citation or Pathology
967	1241	106.mg	n.s.s.	1/17	89.1mg	1/18	Innes;ntis,1968/1969
a	1241	177.mg	n.s.s.	0/17	89.1mg	0/18	
b	1241	84.1mg	n.s.s.	2/17	89.1mg	2/18	
968	1241	77.5mg	n.s.s.	2/18	82.9mg	2/18	
a	1241	98.0mg	n.s.s.	1/18	82.9mg	1/18	
b	1241	51.5mg	n.s.s.	3/18	82.9mg	4/18	
969	1241	64.0mg	n.s.s.	0/16	89.1mg	2/17	
a	1241	167.mg	n.s.s.	0/16	89.1mg	0/17	
b	1241	64.0mg	n.s.s.	0/16	89.1mg	2/17	
970	1241	24.3mg	220.mg	0/16	82.9mg	7/17	
a	1241	38.9mg	n.s.s.	0/16	82.9mg	4/17	
b	1241	16.1mg	89.9mg	0/16	82.9mg	10/17	

CHLORDANE, TECHNICAL GRADE   57-74-9

	RefNum	LoConf	UpConf	Cntrl	1Dose	1Inc	2Dose	2Inc	Citation or Pathology
971	TR8	3.28mg	7.88mg	0/20	3.51mg	3/50	7.41mg	34/50	
a	TR8	3.28mg	11.7mg	2/20	3.51mg	6/50	7.41mg	35/50	
b	TR8	3.28mg	7.88mg	0/20	3.51mg	3/50	7.41mg	34/50	liv:hpa,hpc,nnd.
c	TR8	37.1mg	n.s.s.	0/20	3.51mg	0/50	7.41mg	1/50	lun:a/a,a/c.
972	TR8	3.41mg	9.30mg	3/80p	3.51mg	3/50	7.41mg	34/50	
973	TR8	1.45mg	4.08mg	2/18	3.12mg	16/50	6.00mg	43/50	
a	TR8	1.47mg	5.55mg	4/18	3.12mg	19/50	6.00mg	43/50	
b	TR8	1.45mg	4.08mg	2/18	3.12mg	16/50	6.00mg	43/50	liv:hpa,hpc,nnd.
c	TR8	9.98mg	n.s.s.	1/18	3.12mg	6/50	6.00mg	3/50	lun:a/a,a/c.
974	TR8	1.61mg	4.44mg	17/95p	3.12mg	16/50	6.00mg	43/50	
975	66a	1.15mg	2.94mg	0/57	.650mg	0/62	3.25mg	32/51 (6.50mg 36/51)	Epstein(review) {irdc};stev,6,103-154;1976
a	66a	19.0mg	n.s.s.	1/57	.650mg	0/62	3.25mg	2/51 6.50mg 2/51	
b	66a	13.1mg	n.s.s.	6/57	.650mg	2/62	3.25mg	2/51 (6.50mg 0/51)	
976	66a	1.07mg	2.05mg	1/47	.600mg	6/58	3.00mg	34/52 6.00mg 38/50	

```
        Spe Strain  Site   Xpo+Xpt                                                      TD50       2Tailpvl
            Sex   Route  Hist      Notes                                                    DR       AuOp

   a    M m cd1 eat liv hpt 78w78 e                                                     no dre     P=1.
   b    M m cd1 eat lun mix 78w78 e                                                     no dre     P=1.
  977   M m cen eat liv tum 52w52 kr                                                    noTD50     P<.3
  978   M f icm eat liv mix 25m25 e                                      .     ±        13.4mg *   P<.07
   a    M f icm eat liv hpa 25m25 e                                                     19.3mg *   P<.2
   b    M f icm eat liv hpc 25m25 e                                                     60.0mg *   P<.2
   c    M f icm eat --- hem 25m25 e                                                     no dre     P=1.
  979   M m icm eat liv mix 25m25 e                                  .  +  .            1.71mg *   P<.0005
   a    M m icm eat liv hpa 25m25 e                                                     2.98mg *   P<.0005
   b    M m icm eat --- hem 25m25 e                                                     4.01mg *   P<.0005
   c    M m icm eat liv hpc 25m25 e                                                     7.38mg *   P<.03
  980   R f osm eat TBA MXB 19m25 ev                                  :>                18.4mg *   P<.5   -
   a    R f osm eat liv MXB 19m25 ev                                                    121.mg *   P<.9
  981   R f osm eat ute MXA 19m25 ev  pool                              :     ±         #101.mg *  P<.05
  982   R f osm eat thy MXA 19m25 v                                  :     ±            #41.6mg *  P<.04  -
   a    R m osm eat --- fih 19m25 v                                                     42.5mg *   P<.03
   b    R m osm eat TBA MXB 19m25 v                                                     21.6mg *   P<.4
   c    R m osm eat liv MXB 19m25 v                                                     no dre     P=1.
  983   R m osm eat --- fih 19m25 v  pool                               :     ±         #60.4mg *  P<.05
  984   R f f34 eat mgl fba 30m30 e                              .  +        .          .320mg Z   P<.01
   a    R f f34 eat liv hpa 30m30 e                                                     no dre     P=1.
  985   R m f34 eat liv hpa 30m30 e                              .     ±                15.3mg *   P<.08

CHLORENDIC ACID                    100ng..:..1ug....:..10......:...100....:...1mg..:...10.....:...100....:...1g.....:..10
  986   M f b6c eat lun MXA 24m24                                            :  +   :   #343.mg *  P<.003 -
   a    M f b6c eat lun a/a 24m24                                                       412.mg *   P<.002
   b    M f b6c eat MXA hes 24m24                                                       1.17gm *   P<.03
   c    M f b6c eat TBA MXB 24m24                                                       87.1mg /   P<.0005
   d    M f b6c eat liv MXB 24m24                                                       348.mg *   P<.02
   e    M f b6c eat lun MXB 24m24                                                       343.mg *   P<.003
  987   M m b6c eat liv MXA 24m24                                            :  +   :   141.mg *   P<.004 c
   a    M m b6c eat liv hpc 24m24                                                       206.mg *   P<.009 c
   b    M m b6c eat liv hpa 24m24                                                       387.mg *   P<.08  c
   c    M m b6c eat thy fca 24m24                                                       1.59gm *   P<.04
   d    M m b6c eat TBA MXB 24m24                                                       227.mg *   P<.2
   e    M m b6c eat liv MXB 24m24                                                       141.mg *   P<.004
   f    M m b6c eat lun MXB 24m24                                                       no dre     P=1.
  988   R f f34 eat liv MXA 24m24                                            :  +  :    98.8mg *   P<.0005c
   a    R f f34 eat liv nnd 24m24                                                       162.mg *   P<.004 c
   b    R f f34 eat liv hpc 24m24                                                       271.mg *   P<.02  c
   c    R f f34 eat TBA MXB 24m24                                                       no dre     P=1.
   d    R f f34 eat liv MXB 24m24                                                       98.8mg *   P<.0005
  989   R m f34 eat MXB MXB 24m24                                         :  +  :       25.4mg *   P<.0005
   a    R m f34 eat liv nnd 24m24                                                       25.7mg *   P<.0005c
   b    R m f34 eat liv MXA 24m24                                                       32.2mg *   P<.002
   c    R m f34 eat pan ana 24m24                                                       131.mg *   P<.005 c
   d    R m f34 eat lun MXA 24m24                                                       148.mg *   P<.01
   e    R m f34 eat lun a/a 24m24                                                       168.mg *   P<.01  e
   f    R m f34 eat pre MXA 24m24                                                       59.4mg \   P<.02
   g    R m f34 eat sub fbs 24m24                                                       380.mg *   P<.05
   h    R m f34 eat pre can 24m24                                                       189.mg *   P<.2   e
   i    R m f34 eat TBA MXB 24m24                                                       no dre     P=1.
   j    R m f34 eat liv MXB 24m24                                                       32.2mg *   P<.002

CHLORINATED PARAFFINS (C12, 60% CHLORINE) ..1ug....:..10.......:..100....:..1mg...:..10.....:...100....:..1g.....:..10
  990   M f b6c gav MXB MXB 24m24                                            :  +  :    86.8mg *   P<.0005
   a    M f b6c gav liv MXA 24m24                                                       93.8mg *   P<.0005c
   b    M f b6c gav liv hpa 24m24                                                       105.mg *   P<.0005c
   c    M f b6c gav mul mlp 24m24                                                       599.mg *   P<.002
   d    M f b6c gav thy MXA 24m24                                                       307.mg *   P<.03  c
   e    M f b6c gav hag adn 24m24                                                       347.mg \   P<.04
   f    M f b6c gav thy fca 24m24                                                       372.mg *   P<.06  c
   g    M f b6c gav liv hpc 24m24                                                       596.mg *   P<.03
   h    M f b6c gav TBA MXB 24m24                                                       170.mg *   P<.2
   i    M f b6c gav liv MXB 24m24                                                       93.8mg *   P<.0005
   j    M f b6c gav lun MXB 24m24                                                       no dre     P=1.
  991   M m b6c gav liv MXA 24m24                                            :  +     : 143.mg *   P<.006 c
   a    M m b6c gav liv hpa 24m24                                                       157.mg *   P<.002 c
   b    M m b6c gav lun a/c 24m24                                                       606.mg *   P<.004
   c    M m b6c gav TBA MXB 24m24                                                       211.mg *   P<.2
   d    M m b6c gav liv MXB 24m24                                                       143.mg *   P<.006
   e    M m b6c gav lun MXB 24m24                                                       851.mg *   P<.3
  992   R f f34 gav mul mnl 24m24 s                                          :  +       345.mg \   P<.009
   a    R f f34 gav MXB MXB 24m24 s                                                     583.mg *   P<.0005
   b    R f f34 gav thy fca 24m24 s                                                     736.mg \   P<.003
   c    R f f34 gav liv MXA 24m24 s                                                     1.14gm *   P<.002 c
   d    R f f34 gav thy MXA 24m24 s                                                     1.16gm *   P<.003 c
   e    R f f34 gav liv nnd 24m24 s                                                     1.20gm *   P<.002 c
   f    R f f34 gav ute MXA 24m24 s                                                     394.mg \   P<.02
   g    R f f34 gav pan ana 24m24 s                                                     839.mg \   P<.04
   h    R f f34 gav thy fcc 24m24 s                                                     4.06gm *   P<.04
```

	RefNum	LoConf	UpConf	Cntrl	1Dose	1Inc	2Dose	2Inc			Citation or Pathology
											Brkly Code
a	66a	14.5mg	n.s.s.	4/47	.600mg	6/58	3.00mg	10/52	6.00mg	2/50	
b	66a	32.7mg	n.s.s.	8/47	.600mg	7/58	3.00mg	2/52	6.00mg	2/50	
977	1477	n.s.s.	n.s.s.	5/8	3.00mg	8/8					Becker;canr,42,3918-3923;1982
978	1948	4.19mg	n.s.s.	0/33	.130mg	2/43	.650mg	1/42	1.62mg	4/37	Khasawinah;rtxp,10,244-254;1989/pers.comm.
a	1948	4.86mg	n.s.s.	0/33	.130mg	2/43	.650mg	1/42	1.62mg	3/37	
b	1948	9.77mg	n.s.s.	0/30	.130mg	0/32	.650mg	0/37	1.62mg	1/35	
c	1948	6.88mg	n.s.s.	6/66	.130mg	10/64	.650mg	8/68	1.62mg	8/68	
979	1948	1.02mg	4.10mg	15/61	.120mg	15/58	.600mg	21/61	1.50mg	37/60	
a	1948	1.57mg	12.0mg	12/61	.120mg	12/58	.600mg	13/61	1.50mg	28/60	
b	1948	2.18mg	13.3mg	7/68	.120mg	4/64	.600mg	11/65	1.50mg	20/66	
c	1948	3.11mg	n.s.s.	3/57	.120mg	3/54	.600mg	8/59	1.50mg	9/55	
980	TR8	4.46mg	n.s.s.	7/10	4.40mg	34/50	8.90mg	29/50			
a	TR8	11.2mg	n.s.s.	1/10	4.40mg	11/50	8.90mg	6/50			liv:hpa,hpc,nnd.
981	TR8	30.4mg	n.s.s.	0/60p	4.40mg	1/50	8.90mg	2/50			ute:adc,sar. S
982	TR8	17.9mg	n.s.s.	0/10	5.96mg	1/50	11.9mg	6/50			thy:fca,fcc. S
a	TR8	18.9mg	n.s.s.	0/10	5.96mg	11.9mg	11.9mg	7/50			S
b	TR8	6.13mg	n.s.s.	3/10	5.96mg	21/50	11.9mg	23/50			
c	TR8	40.9mg	n.s.s.	0/10	5.96mg	2/50	11.9mg	0/50			liv:hpa,hpc,nnd.
983	TR8	22.3mg	n.s.s.	2/60p	5.96mg	1/50	11.9mg	7/50			S
984	1949	.142mg	21.5mg	3/64	50.0ug	12/62	(.250mg	6/63	1.25mg	6/64)	Khasawinah;rtxp,10,95-109;1989
a	1949	27.8mg	n.s.s.	0/64	50.0ug	1/62	.250mg	0/63	1.25mg	0/64	
985	1949	4.90mg	n.s.s.	2/64	40.0ug	4/64	.200mg	2/64	1.00mg	7/64	

CHLORENDIC ACID 115-28-6

	RefNum	LoConf	UpConf	Cntrl	1Dose	1Inc	2Dose	2Inc	Pathology
986	TR304	153.mg	1.94gm	1/50	79.1mg	5/50	160.mg	6/50	lun:a/a,a/c. S
a	TR304	177.mg	1.68gm	0/50	79.1mg	4/50	160.mg	4/50	S
b	TR304	367.mg	n.s.s.	0/50	79.1mg	1/50	160.mg	3/50	mey:hes; spl:hes. S
c	TR304	46.6mg	350.mg	29/50	79.1mg	32/50	160.mg	32/50	
d	TR304	150.mg	52.5gm	3/50	79.1mg	7/50	160.mg	7/50	liv:hpa,hpc,nnd.
e	TR304	153.mg	1.94gm	1/50	79.1mg	5/50	160.mg	6/50	lun:a/a,a/c.
987	TR304	72.2mg	1.17gm	13/50	73.0mg	23/50	149.mg	27/50	liv:hpa,hpc.
a	TR304	101.mg	8.47gm	9/50	73.0mg	17/50	149.mg	20/50	
b	TR304	150.mg	n.s.s.	5/50	73.0mg	9/50	149.mg	10/50	
c	TR304	474.mg	n.s.s.	0/50	73.0mg	0/50	149.mg	3/50	S
d	TR304	74.7mg	n.s.s.	35/50	73.0mg	31/50	149.mg	39/50	
e	TR304	72.2mg	1.17gm	13/50	73.0mg	23/50	149.mg	27/50	liv:hpa,hpc,nnd.
f	TR304	379.mg	n.s.s.	15/50	73.0mg	4/50	149.mg	9/50	lun:a/a,a/c.
988	TR304	56.0mg	283.mg	1/50	30.7mg	5/50	61.9mg	16/50	liv:hpc,nnd.
a	TR304	80.3mg	1.11gm	1/50	30.7mg	3/50	61.9mg	11/50	
b	TR304	122.mg	n.s.s.	0/50	30.7mg	3/50	61.9mg	5/50	
c	TR304	41.7mg	n.s.s.	48/50	30.7mg	48/50	61.9mg	48/50	
d	TR304	56.0mg	283.mg	1/50	30.7mg	5/50	61.9mg	16/50	liv:hpa,hpc,nnd.
989	TR304	16.2mg	57.9mg	2/50	24.5mg	22/50	49.5mg	23/50	liv:nnd; pan:ana. C
a	TR304	16.3mg	58.7mg	2/50	24.5mg	21/50	49.5mg	23/50	
b	TR304	18.2mg	127.mg	5/50	24.5mg	22/50	49.5mg	23/50	liv:hpc,nnd. S
c	TR304	63.4mg	1.04gm	0/50	24.5mg	4/50	49.5mg	6/50	
d	TR304	69.7mg	10.1gm	0/50	24.5mg	4/50	49.5mg	5/50	lun:a/a,a/c. S
e	TR304	75.9mg	8.53gm	0/50	24.5mg	3/50	49.5mg	5/50	
f	TR304	26.2mg	n.s.s.	1/50	24.5mg	10/50	(49.5mg	4/50)	pre:adn,can,sqp. S
g	TR304	130.mg	n.s.s.	0/50	24.5mg	1/50	49.5mg	3/50	S
h	TR304	68.2mg	n.s.s.	1/50	24.5mg	8/50	49.5mg	4/50	
i	TR304	25.7mg	n.s.s.	50/50	24.5mg	49/50	49.5mg	50/50	
j	TR304	18.2mg	127.mg	5/50	24.5mg	22/50	49.5mg	23/50	liv:hpa,hpc,nnd.

CHLORINATED PARAFFINS (C12, 60% CHLORINE) (chlorowax 500c, avg. mol. wt. = 411) 63449-39-8

	RefNum	LoConf	UpConf	Cntrl	1Dose	1Inc	2Dose	2Inc	Pathology
990	TR308	53.3mg	200.mg	11/50	87.6mg	29/50	176.mg	34/50	liv:hpa,hpc; thy:fca,fcc. C
a	TR308	62.2mg	169.mg	3/50	87.6mg	22/50	176.mg	28/50	liv:hpa,hpc.
b	TR308	70.9mg	164.mg	0/50	87.6mg	18/50	176.mg	22/50	
c	TR308	278.mg	2.56gm	0/50	87.6mg	3/50	176.mg	6/50	S
d	TR308	136.mg	n.s.s.	8/50	87.6mg	12/50	176.mg	15/50	thy:fca,fcc.
e	TR308	126.mg	n.s.s.	1/50	87.6mg	6/50	(176.mg	2/50)	S
f	TR308	150.mg	n.s.s.	8/50	87.6mg	12/50	176.mg	13/50	
g	TR308	247.mg	n.s.s.	3/50	87.6mg	4/50	176.mg	9/50	S
h	TR308	64.1mg	n.s.s.	40/50	87.6mg	42/50	176.mg	42/50	
i	TR308	62.2mg	169.mg	3/50	87.6mg	22/50	176.mg	28/50	liv:hpa,hpc,nnd.
j	TR308	581.mg	n.s.s.	3/50	87.6mg	4/50	176.mg	1/50	lun:a/a,a/c.
991	TR308	71.9mg	1.83gm	20/50	87.6mg	34/50	177.mg	38/50	liv:hpa,hpc.
a	TR308	83.9mg	811.mg	11/50	87.6mg	20/50	177.mg	29/50	
b	TR308	284.mg	1.65gm	0/50	87.6mg	3/50	177.mg	6/50	S
c	TR308	74.9mg	n.s.s.	37/50	87.6mg	44/50	177.mg	46/50	
d	TR308	71.9mg	1.83gm	20/50	87.6mg	34/50	177.mg	38/50	liv:hpa,hpc,nnd.
e	TR308	263.mg	n.s.s.	5/50	87.6mg	6/50	177.mg	9/50	lun:a/a,a/c.
992	TR308	158.mg	11.0gm	11/50	221.mg	22/50	(442.mg	16/50)	S
a	TR308	351.mg	1.17gm	0/50	221.mg	11/50	442.mg	12/50	liv:hpc,nnd; thy:fca,fcc. C
b	TR308	290.mg	3.62gm	0/50	221.mg	6/50	(442.mg	3/50)	S
c	TR308	577.mg	4.00gm	0/50	221.mg	5/50	442.mg	7/50	liv:hpc,nnd.
d	TR308	592.mg	5.24gm	0/50	221.mg	6/50	442.mg	6/50	thy:fca,fcc.
e	TR308	597.mg	2.94gm	0/50	221.mg	4/50	442.mg	7/50	
f	TR308	167.mg	n.s.s.	6/50	221.mg	14/50	(442.mg	11/50)	ute:esp,ess. S
g	TR308	286.mg	n.s.s.	1/50	221.mg	5/50	(442.mg	2/50)	S
h	TR308	1.23gm	n.s.s.	0/50	221.mg	0/50	442.mg	3/50	S

```
        Spe Strain  Site    Xpo+Xpt                                              TD50    2Tailpvl
            Sex  Route   Hist    Notes                                              DR    AuOp
  i     R f f34 gav tes TBA MXB 24m24 s                                          1.10gm * P<.6
  j     R f f34 gav liv MXB 24m24 s                                          1.14gm * P<.002
993     R m f34 gav tes ict 24m24 s                          : +      :      75.2mg * P<.0005
  a     R m f34 gav MXB MXB 24m24 s                                          110.mg * P<.0005
  b     R m f34 gav liv MXA 24m24 s                                          123.mg * P<.0005c
  c     R m f34 gav pan MXA 24m24 s                                          144.mg * P<.0005
  d     R m f34 gav pan ana 24m24 s                                          150.mg * P<.0005
  e     R m f34 gav liv nnd 24m24 s                                          173.mg * P<.0005c
  f     R m f34 gav adr MXA 24m24 s                                          192.mg * P<.0005
  g     R m f34 gav kid MXA 24m24 s                                          196.mg \ P<.0005c
  h     R m f34 gav MXA mnl 24m24 s                                          406.mg * P<.003
  i     R m f34 gav kid tla 24m24 s                                          417.mg * P<.0005
  j     R m f34 gav liv hpc 24m24 s                                          551.mg * P<.002
  k     R m f34 gav MXA fib 24m24 s                                          332.mg \ P<.04
  l     R m f34 gav pre MXA 24m24 s                                          522.mg * P<.03
  m     R m f34 gav mgl fba 24m24 s                                          566.mg * P<.03
  n     R m f34 gav pre adn 24m24 s                                          568.mg * P<.02
  o     R m f34 gav TBA MXB 24m24 s                                          75.7mg * P<.0005
  p     R m f34 gav liv MXB 24m24 s                                          123.mg * P<.0005

CHLORINATED PARAFFINS (C23, 43% CHLORINE)  ..1ug...:..10.....:..100...:..1mg....:..10.....:..100....:..1g.....:..10
994     M f b6c gav liv MXA 24m24 s                                  :       10.5gm * P<.1   e
  a     M f b6c gav liv hpc 24m24 s                                          14.0gm * P<.05  e
  b     M f b6c gav liv hpa 24m24 s                                          16.2gm * P<.2   e
  c     M f b6c gav TBA MXB 24m24 s                                          5.83gm * P<.5
  d     M f b6c gav liv MXB 24m24 s                                          10.5gm * P<.1
  e     M f b6c gav lun MXB 24m24 s                                          1.22kg   P<1.
995     M m b6c gav MXA MXA 24m24                               : ±  6.54gm * P<.03  c
  a     M m b6c gav MXA mlp 24m24                                          22.9gm * P<.04  c
  b     M m b6c gav mul mlh 24m24                                          43.2gm * P<.05  c
  c     M m b6c gav thy fcc 24m24                                          44.0gm * P<.05
  d     M m b6c gav MXA mlm 24m24                                          12.1gm * P<.2   c
  e     M m b6c gav TBA MXB 24m24                                          36.4gm * P<.9
  f     M m b6c gav liv MXB 24m24                                          12.3gm * P<.4
  g     M m b6c gav lun MXB 24m24                                          31.1gm * P<.7
996     R f f34 gav amd MXA 24m24                                : ±  2.35gm * P<.06  e
  a     R f f34 gav TBA MXB 24m24                                          no dre   P=1.
  b     R f f34 gav liv MXB 24m24                                          24.8gm * P<.7
997     R m f34 gav liv nnd 24m24                               : #14.5gm * P<.04  -
  a     R m f34 gav pni isa 24m24                                          17.6gm * P<.02
  b     R m f34 gav TBA MXB 24m24                                          no dre   P=1.
  c     R m f34 gav liv MXB 24m24                                          14.5gm * P<.04

CHLORINATED TRISODIUM PHOSPHATE  100ng..:..1ug....:..10.....:..100...:..1mg....:..10.....:..100....:..1g......:..10
998     M f b6c gav sub srn 24m24 s                                : +  #1.49gm \ P<.009 \
  a     M f b6c gav TBA MXB 24m24 s                                          958.mg * P<.2
  b     M f b6c gav liv MXB 24m24 s                                          3.71gm * P<.5
  c     M f b6c gav lun MXB 24m24 s                                          no dre   P=1.
999     M m b6c gav adr coa 24m24                               : ±  #5.84gm * P<.04  -
  a     M m b6c gav mul mlp 24m24                                          8.42gm * P<.04
  b     M m b6c gav TBA MXB 24m24                                          11.6gm * P<.9
  c     M m b6c gav liv MXB 24m24                                          2.95gm * P<.4
  d     M m b6c gav lun MXB 24m24                                          19.9gm * P<.9

CHLORINATED WATER  100ng..:..1ug....:..10.....:..100...:..1mg....:..10.....:..100....:..1g......:..10
1000    M f b6c wat ute esp 24m24                              : ±  #657.mg * P<.03  -
  a     M f b6c wat TBA MXB 24m24                                          no dre   P=1.
  b     M f b6c wat liv MXB 24m24                                          386.mg * P<.7
  c     M f b6c wat lun MXB 24m24                                          no dre   P=1.
1001    M m b6c wat mul MXA 24m24                           : ±  #187.mg * P<.05  -
  a     M m b6c wat TBA MXB 24m24                                          161.mg * P<.7
  b     M m b6c wat liv MXB 24m24                                          no dre   P=1.
  c     M m b6c wat lun MXB 24m24                                          893.mg * P<.9
1002    R f f34 wat pni MXA 24m24                        : ±  19.3mg * P<.03
  a     R f f34 wat mul mnl 24m24                                          29.7mg * P<.06  e
  b     R f f34 wat pni isa 24m24                                          36.0mg * P<.03
  c     R f f34 wat TBA MXB 24m24                                          no dre   P=1.
  d     R f f34 wat liv MXB 24m24                                          no dre   P=1.
1003    R m f34 wat tes MXA 24m24                       : ±  #1.29mg Z P<.02  -
  a     R m f34 wat MXA MXA 24m24                                          136.mg * P<.04
  b     R m f34 wat TBA MXB 24m24                                          1.95mg Z P<.09
  c     R m f34 wat liv MXB 24m24                                          251.mg * P<.8

CHLORINE  100ng..:..1ug....:..10.....:..100...:..1mg....:..10.....:..100....:..1g......:..10
1004    R b bd2 wat tba mix 35m35 g                              .>        no dre   P=1.  -
  a     R b bd2 wat tba mal 35m35 g                                          no dre   P=1.  -
  b     R b bd2 wat tba ben 35m35 g                                          no dre   P=1.  -

CHLORMADINONE ACETATE  100ng..:..1ug....:..10.....:..100...:..1mg....:..10.....:..100....:..1g......:..10
1005    M f c3h eat mam tum 24m24 er                    .>        3.77mg * P<.5   -
1006    M f crf eat mam tum 24m24 er  pool              .>        no dre   P=1.  -
```

	RefNum	LoConf	UpConf	Cntrl	1Dose	1Inc	2Dose	2Inc		Citation or Pathology	Brkly Code
i	TR308	205.mg	n.s.s.	48/50	221.mg	45/50	442.mg	46/50			
j	TR308	577.mg	4.00gm	0/50	221.mg	5/50	442.mg	7/50		liv:hpa,hpc,nnd.	
993	TR308	41.7mg	187.mg	48/50	221.mg	49/50	442.mg	47/50			S
a	TR308	61.4mg	202.mg	0/50	221.mg	18/50	442.mg	18/50		kid:tla,uac; liv:hpc,nnd.	C
b	TR308	65.6mg	240.mg	0/50	221.mg	13/50	442.mg	16/50		liv:hpc,nnd.	
c	TR308	76.1mg	369.mg	11/50	221.mg	22/50	442.mg	17/50		pan:acc,ana.	S
d	TR308	77.5mg	409.mg	11/50	221.mg	22/50	442.mg	15/50			S
e	TR308	88.6mg	335.mg	0/50	221.mg	10/50	442.mg	16/50			
f	TR308	88.0mg	864.mg	15/50	221.mg	15/50	442.mg	15/50		adr:phe,phm.	S
g	TR308	75.8mg	591.mg	0/50	221.mg	9/50	(442.mg	3/50)		kid:tla,uac.	
h	TR308	169.mg	2.48gm	7/50	221.mg	12/50	442.mg	14/50		mul:mnl; spl:mnl.	S
i	TR308	164.mg	1.43gm	0/50	221.mg	7/50	442.mg	3/50			S
j	TR308	184.mg	3.24gm	0/50	221.mg	3/50	442.mg	2/50			S
k	TR308	90.5mg	n.s.s.	3/50	221.mg	5/50	(442.mg	1/50)		ski:fib; sub:fib.	S
l	TR308	173.mg	n.s.s.	9/50	221.mg	4/50	442.mg	9/50		pre:adn,can.	S
m	TR308	172.mg	n.s.s.	2/50	221.mg	4/50	442.mg	2/50			S
n	TR308	183.mg	n.s.s.	4/50	221.mg	3/50	442.mg	7/50			S
o	TR308	41.8mg	193.mg	50/50	221.mg	49/50	442.mg	48/50			
p	TR308	65.6mg	240.mg	0/50	221.mg	13/50	442.mg	16/50		liv:hpa,hpc,nnd.	

CHLORINATED PARAFFINS (C23, 43% CHLORINE) (chlorowax 40, avg. mol. wt. = 560) 63449-39-8

	RefNum	LoConf	UpConf	Cntrl	1Dose	1Inc	2Dose	2Inc		Citation or Pathology	Brkly Code
994	TR305	3.84gm	n.s.s.	4/50	1.75gm	3/50	3.54gm	10/50		liv:hpa,hpc.	
a	TR305	5.22gm	n.s.s.	1/50	1.75gm	1/50	3.54gm	6/50			
b	TR305	4.96gm	n.s.s.	3/50	1.75gm	2/50	3.54gm	7/50			
c	TR305	1.34gm	n.s.s.	32/50	1.75gm	29/50	3.54gm	38/50			
d	TR305	3.84gm	n.s.s.	4/50	1.75gm	3/50	3.54gm	10/50		liv:hpa,hpc,nnd.	
e	TR305	8.55gm	n.s.s.	3/50	1.75gm	3/50	3.54gm	3/50		lun:a/a,a/c.	
995	TR305	2.96gm	n.s.s.	6/50	1.77gm	12/50	3.54gm	16/50		liv:mno; mln:mlm,mlp; mul:mlh,mlm,mlp; smi:mlm; spl:mlm.	
a	TR305	8.63gm	n.s.s.	0/50	1.77gm	2/50	3.54gm	3/50		mln:mlp; mul:mlp.	
b	TR305	13.0gm	n.s.s.	0/50	1.77gm	0/50	3.54gm	3/50			
c	TR305	13.2gm	n.s.s.	0/50	1.77gm	0/50	3.54gm	3/50			S
d	TR305	4.09gm	n.s.s.	5/50	1.77gm	10/50	3.54gm	10/50		mln:mlm; mul:mlm; smi:mlm; spl:mlm.	
e	TR305	2.12gm	n.s.s.	43/50	1.77gm	41/50	3.54gm	43/50			
f	TR305	3.06gm	n.s.s.	18/50	1.77gm	21/50	3.54gm	23/50		liv:hpa,hpc,nnd.	
g	TR305	4.75gm	n.s.s.	11/50	1.77gm	10/50	3.54gm	13/50		lun:a/a,a/c.	
996	TR305	868.mg	n.s.s.	1/50	70.1mg	4/50	210.mg	6/50	631.mg 7/50	amd:phe,phm.	
a	TR305	511.mg	n.s.s.	46/50	70.1mg	45/50	210.mg	40/50	631.mg 41/50		
b	TR305	2.49gm	n.s.s.	1/50	70.1mg	2/50	210.mg	1/50	631.mg 2/50	liv:hpa,hpc,nnd.	
997	TR305	5.89gm	n.s.s.	0/50	1.31gm	3/50	2.63gm	3/50			S
a	TR305	6.56gm	n.s.s.	0/50	1.31gm	1/50	2.63gm	4/50			S
b	TR305	1.78gm	n.s.s.	46/50	1.31gm	42/50	2.63gm	41/50			
c	TR305	5.89gm	n.s.s.	0/50	1.31gm	3/50	2.63gm	3/50		liv:hpa,hpc,nnd.	

CHLORINATED TRISODIUM PHOSPHATE 32680-25-4

	RefNum	LoConf	UpConf	Cntrl	1Dose	1Inc	2Dose	2Inc	Citation or Pathology	Brkly Code
998	TR294	488.mg	48.3gm	0/50	350.mg	4/50	(701.mg	0/50)		S
a	TR294	305.mg	n.s.s.	29/50	350.mg	32/50	701.mg	29/50		
b	TR294	774.mg	n.s.s.	6/50	350.mg	8/50	701.mg	6/50	liv:hpa,hpc,nnd.	
c	TR294	2.31gm	n.s.s.	6/50	350.mg	2/50	701.mg	3/50	lun:a/a,a/c.	
999	TR294	2.02gm	n.s.s.	0/50	352.mg	1/50	707.mg	3/50		S
a	TR294	2.54gm	n.s.s.	0/50	352.mg	0/50	707.mg	3/50		S
b	TR294	577.mg	n.s.s.	35/50	352.mg	37/50	707.mg	30/50		
c	TR294	812.mg	n.s.s.	14/50	352.mg	14/50	707.mg	17/50	liv:hpa,hpc,nnd.	
d	TR294	1.22gm	n.s.s.	8/50	352.mg	12/50	707.mg	7/50	lun:a/a,a/c.	

CHLORINATED WATER   ---

	RefNum	LoConf	UpConf	Cntrl	1Dose	1Inc	2Dose	2Inc		Citation or Pathology	Brkly Code
1000	TR392	225.mg	n.s.s.	0/50	13.8mg	0/51	27.5mg	1/50	54.1mg 3/50		S
a	TR392	40.7mg	n.s.s.	35/50	13.8mg	43/51	27.5mg	37/50	54.1mg 39/50		
b	TR392	48.8mg	n.s.s.	20/50	13.8mg	22/51	27.5mg	23/50	54.1mg 24/50	liv:hpa,hpc,nnd.	
c	TR392	163.mg	n.s.s.	5/50	13.8mg	10/51	27.5mg	7/50	54.1mg 4/50	lun:a/a,a/c.	
1001	TR392	90.5mg	n.s.s.	0/50	11.5mg	3/50	22.9mg	3/50	45.1mg 4/51	mul:mlm,mlu.	S
a	TR392	25.1mg	n.s.s.	42/50	11.5mg	45/50	22.9mg	44/50	45.1mg 47/51		
b	TR392	42.5mg	n.s.s.	35/50	11.5mg	38/50	22.9mg	33/50	45.1mg 34/50	liv:hpa,hpc,nnd.	
c	TR392	57.5mg	n.s.s.	21/50	11.5mg	12/50	22.9mg	21/50	45.1mg 19/51	lun:a/a,a/c.	
1002	TR392	5.27mg	n.s.s.	0/50	3.95mg	0/50	7.89mg	3/51	15.5mg 3/50	pni:isa,isc.	S
a	TR392	9.89mg	n.s.s.	8/50	3.95mg	7/50	7.89mg	19/51	15.5mg 16/50		
b	TR392	7.45mg	n.s.s.	0/50	3.95mg	0/50	7.89mg	2/51	15.5mg 3/50		S
c	TR392	8.52mg	n.s.s.	48/50	3.95mg	46/50	7.89mg	46/51	15.5mg 43/50		
d	TR392	99.6mg	n.s.s.	1/50	3.95mg	1/50	7.89mg	2/51	15.5mg 0/50	liv:hpa,hpc,nnd.	
1003	TR392	.545mg	n.s.s.	33/51	3.46mg	41/51	(6.91mg	42/50	13.3mg 38/51)	tes:iab,ica.	S
a	TR392	36.4mg	n.s.s.	0/51	3.46mg	0/51	6.91mg	0/50	13.3mg 3/51	gnv:sqc; pal:sqp; ton:sqp.	S
b	TR392	.684mg	n.s.s.	46/51	3.46mg	48/51	(6.91mg	48/50	13.3mg 46/51)		
c	TR392	22.0mg	n.s.s.	2/51	3.46mg	2/51	6.91mg	4/50	13.3mg 3/51	liv:hpa,hpc,nnd.	

CHLORINE 7782-50-5

	RefNum	LoConf	UpConf	Cntrl	1Dose	1Inc		Citation or Pathology
1004	1421	52.6mg	n.s.s.	3/20	5.29mg	5/60		Druckrey;fctx,6,147-154;1968
a	1421	64.6mg	n.s.s.	2/20	5.29mg	3/60		
b	1421	69.1mg	n.s.s.	1/20	5.29mg	2/60		

CHLORMADINONE ACETATE 302-22-7

	RefNum	LoConf	UpConf	Cntrl	1Dose	1Inc	2Dose	2Inc	Citation or Pathology
1005	1175	.713mg	n.s.s.	54/92	.104mg	28/43	1.04mg	24/36	Rudali;jnci,49,813-819;1972
1006	1175	.966mg	n.s.s.	161/167p	.104mg	45/46	1.04mg	34/40	

```
      Spe Strain Site   Xpo+Xpt                                                          TD50    2Tailpvl
          Sex  Route  Hist     Notes                                                       DR     AuOp
1007 M f r3m eat mam tum 24m24 er                                   .>                     no dre  P=1.   -


4-CHLORO-4'-AMINODIPHENYLETHER    100ng..:..1ug....:..10.....:..100....:..1mg....:..10.....:..100....:..1g......:..10
1008 M f chi eat liv mix 68w86 av                                                 :      ±    905.mg * P<.05
 a   M f chi eat lun mix 68w86 av                                                             no dre  P=1.
 b   M f chi eat tba mix 68w86 av                                                             598.mg * P<.4
1009 M f chi eat --- vsc 68w86 av   pool                                     :    +    :      346.mg * P<.0005+
 a   M f chi eat ubl mix 68w86 av                                                             1.13gm * P<.0005+
1010 M m chi eat liv mix 51w55 v                                                              no dre  P=1.   -
 a   M m chi eat lun mix 51w55 v                                                              no dre  P=1.   -
 b   M m chi eat tba mix 51w55 v                                                              no dre  P=1.   -
1011 R m cdr eat liv hpt 77w98 v                               :     +        :              37.9mg \ P<.005 +
 a   R m cdr eat liv mix 77w98 v                                                             37.9mg \ P<.005 +
 b   R m cdr eat tba mix 77w98 v                                                             25.1mg * P<.004
1012 R m cdr eat liv hpt 77w98 v    pool                        :     +     :                37.6mg \ P<.0005+


2-CHLORO-5-(3,5-DIMETHYLPIPERIDINOSULPHONYL)BENZOIC  ACID...:..100....:..1mg....:..10.....:..100....:..1g......:..10
1013 R m f34 eat liv hpc 70w71 erv                      . + .                                4.85mg   P<.0005+


1-CHLORO-2,4-DINITROBENZENE    100ng..:..1ug....:..10.....:..100....:..1mg....:..10.....:..100....:..1g......:..10
1014 M f chi eat lun mix 77w94 v                                       :    +         :      150.mg \ P<.007 -
 a   M f chi eat liv mix 77w94 v                                                             1.52gm * P<.3   -
 b   M f chi eat tba mix 77w94 v                                                             124.mg * P<.004 -
1015 M m chi eat liv mix 77w94 v                                       :    ±                365.mg * P<.03  -
 a   M m chi eat lun mix 77w94 v                                                             541.mg * P<.06  -
 b   M m chi eat tba mix 77w94 v                                                             255.mg \ P<.09  -
1016 R m cdr eat liv mix 18m24 v                              :>                             no dre  P=1.   -
 a   R m cdr eat tba mix 18m24 v                                                             12.3mg \ P<.01  -


3-CHLORO-2-METHYLPROPENE, TECHNICAL GRADE (CONTAINING 5% DIMETHYLVINYL CHLORIDE).:..10.....:..100....:..1g......:..10
1017 M f b6c gav for MXA 24m24                                         :  +:                 82.3mg / P<.0005c
 a   M f b6c gav for sqp 24m24                                                               87.7mg / P<.0005c
 b   M f b6c gav TBA MXB 24m24                                                               149.mg / P<.03
 c   M f b6c gav liv MXB 24m24                                                               no dre  P=1.
 d   M f b6c gav lun MXB 24m24                                                               5.18gm * P<.8
1018 M m b6c gav for MXA 24m24                                         :  +  :               73.5mg * P<.0005c
 a   M m b6c gav for sqp 24m24                                                               96.2mg * P<.0005c
 b   M m b6c gav for sqc 24m24                                                               432.mg * P<.005 c
 c   M m b6c gav TBA MXB 24m24                                                               402.mg * P<.6
 d   M m b6c gav liv MXB 24m24                                                               no dre  P=1.
 e   M m b6c gav lun MXB 24m24                                                               no dre  P=1.
1019 R f f34 gav for ppn 24m24                                         :    +        :       320.mg / P<.003 c
 a   R f f34 gav sub fib 24m24                                                               524.mg * P<.02
 b   R f f34 gav TBA MXB 24m24                                                               323.mg * P<.7
 c   R f f34 gav liv MXB 24m24                                                               no dre  P=1.
1020 R m f34 gav for ppn 24m24                                         :  +  :               68.7mg / P<.0005c
 a   R m f34 gav tes ict 24m24                                                               54.0mg * P<.02
 b   R m f34 gav TBA MXB 24m24                                                               109.mg * P<.2
 c   R m f34 gav liv MXB 24m24                                                               714.mg / P<.2


1-CHLORO-2-NITROBENZENE    100ng..:..1ug....:..10.....:..100....:..1mg....:..10.....:..100....:..1g......:..10
1021 M f chi eat liv hpt 77w98 v                                      :     +        :       289.mg \ P<.007 +
 a   M f chi eat liv mix 77w98 v                                                             289.mg \ P<.007
 b   M f chi eat lun mix 77w98 v                                                             10.9gm * P<1.   -
 c   M f chi eat tba mix 77w98 v                                                             287.mg \ P<.2
1022 M f chi eat liv hpt 77w98 v    pool                              :    +     :           289.mg \ P<.0005+
1023 M m chi eat liv mix 73w81 av                                  :    +     :              113.mg * P<.003
 a   M m chi eat lun mix 73w81 av                                                            no dre  P=1.
 b   M m chi eat tba mix 73w81 av                                                            186.mg * P<.07
1024 M m chi eat liv hpt 73w81 av   pool                              :    +    :            108.mg * P<.0005+
1025 R m cdr eat liv mix 77w98 v                             :>                              no dre  P=1.   -
 a   R m cdr eat tba mix 77w98 v                                                             no dre  P=1.


1-CHLORO-4-NITROBENZENE    100ng..:..1ug....:..10.....:..100....:..1mg....:..10.....:..100....:..1g......:..10
1026 M f chi eat --- vsc 73w90 a                                      :     +    :           526.mg * P<.0005+
 a   M f chi eat lun mix 73w90 a                                                             539.mg * P<.08  -
 b   M f chi eat liv mix 73w90 a                                                             1.44gm * P<.04  -
 c   M f chi eat tba mix 73w90 a                                                             125.mg * P<.0005
1027 M f chi eat --- vsc 73w90 a    pool                             :    +    :             634.mg * P<.0005+
1028 M m chi eat --- vsc 73w86 a                                     :     +         :       430.mg * P<.006 +
 a   M m chi eat liv mix 73w86 a                                                             262.mg * P<.03
 b   M m chi eat lun mix 73w86 a                                                             1.69gm / P<.4   -
 c   M m chi eat tba mix 73w86 a                                                             215.mg * P<.02
1029 M m chi eat --- vsc 73w86 a    pool                             :     +         :       483.mg * P<.0005+
 a   M m chi eat liv hpt 73w86 a                                                             558.mg * P<.02  +
1030 R m cdr eat liv mix 77w98 v                                                             no dre  P=1.   -
 a   R m cdr eat tba mix 77w98 v                                                             60.4mg * P<.5   -


4-CHLORO-m-PHENYLENEDIAMINE    100ng..:..1ug....:..10.....:..100....:..1mg....:..10.....:..100....:..1g......:..10
1031 M f b6c eat liv MXA 78w95 v                                          :   +    :         1.23gm \ P<.0005c
 a   M f b6c eat liv hpc 78w95 v                                                             1.78gm \ P<.002 c
```

	RefNum	LoConf	UpConf	Cntrl	1Dose	1Inc	2Dose	2Inc	Citation or Pathology	Brkly Code
1007	1175	1.11mg	n.s.s.	50/73	.104mg	10/19	1.04mg	18/30		

4-CHLORO-4'-AMINODIPHENYLETHER   101-79-1

	RefNum	LoConf	UpConf	Cntrl	1Dose	1Inc	2Dose	2Inc	Citation or Pathology	Brkly Code
1008	381	225.mg	n.s.s.	1/15	287.mg	1/17	624.mg	2/18	Weisburger;jept,2,325-356;1978/pers.comm./Russfield	1973
a	381	286.mg	n.s.s.	5/15	287.mg	2/17	624.mg	0/18		
b	381	126.mg	n.s.s.	10/15	287.mg	7/17	624.mg	5/18		
1009	381	130.mg	1.49gm	8/102p	287.mg	5/17	624.mg	3/18		
a	381	389.mg	5.82gm	0/102p	287.mg	0/17	624.mg	4/18		
1010	381	n.s.s.	n.s.s.	1/14	300.mg	0/16	600.mg	0/9		
a	381	n.s.s.	n.s.s.	3/14	300.mg	0/16	600.mg	0/9		
b	381	110.mg	n.s.s.	7/14	300.mg	3/16	600.mg	0/9		
1011	381	11.8mg	381.mg	0/16	38.2mg	4/13	(84.0mg	1/23)		
a	381	11.8mg	381.mg	0/16	38.2mg	4/13	(84.0mg	1/23)		
b	381	8.14mg	248.mg	10/16	38.2mg	9/13	84.0mg	9/23		
1012	381	12.1mg	258.mg	2/111p	38.2mg	4/13	(84.0mg	1/23)		

2-CHLORO-5-(3,5-DIMETHYLPIPERIDINOSULPHONYL)BENZOIC ACID (tibric acid) 37087-94-8

	RefNum	LoConf	UpConf	Cntrl	1Dose	1Inc	2Dose	2Inc	Citation or Pathology	Brkly Code
1013	1331	2.15mg	9.78mg	0/30	52.1mg	30/31			Reddy;natu,283,397-398;1980	

1-CHLORO-2,4-DINITROBENZENE   97-00-7

	RefNum	LoConf	UpConf	Cntrl	1Dose	1Inc	2Dose	2Inc	Citation or Pathology	Brkly Code
1014	381	51.7mg	2.37gm	3/13	169.mg	5/19	(306.mg	5/16)	Weisburger;jept,2,325-356;1978/pers.comm./Russfield	1973
a	381	247.mg	n.s.s.	0/13	169.mg	0/19	306.mg	1/16		
b	381	57.4mg	1.03gm	9/13	169.mg	9/19	306.mg	11/16		
1015	381	138.mg	n.s.s.	2/16	207.mg	2/15	354.mg	4/21		
a	381	203.mg	n.s.s.	2/16	207.mg	3/15	354.mg	3/21		
b	381	79.2mg	n.s.s.	9/16	207.mg	6/15	(354.mg	10/21)		
1016	381	66.5mg	n.s.s.	1/17	17.6mg	2/18	32.2mg	0/18		
a	381	5.05mg	1.78gm	13/17	17.6mg	12/18	(32.2mg	14/18)		

3-CHLORO-2-METHYLPROPENE, TECHNICAL GRADE (CONTAINING 5% DIMETHYLVINYL CHLORIDE)   563-47-3

	RefNum	LoConf	UpConf	Cntrl	1Dose	1Inc	2Dose	2Inc	Citation or Pathology	Brkly Code
1017	TR300	57.4mg	124.mg	0/50	70.1mg	16/50	140.mg	31/50	for:sqc,sqp.	
a	TR300	60.5mg	134.mg	0/50	70.1mg	15/50	140.mg	29/50		
b	TR300	66.0mg	n.s.s.	28/50	70.1mg	32/50	140.mg	37/50		
c	TR300	707.mg	n.s.s.	4/50	70.1mg	3/50	140.mg	1/50	liv:hpa,hpc,nnd.	
d	TR300	469.mg	n.s.s.	3/50	70.1mg	2/50	140.mg	3/50	lun:a/a,a/c.	
1018	TR300	49.2mg	141.mg	3/50	70.1mg	24/50	140.mg	36/50	for:sqc,sqp.	
a	TR300	61.5mg	212.mg	3/50	70.1mg	19/50	140.mg	30/50		
b	TR300	221.mg	2.76gm	0/50	70.1mg	5/50	140.mg	7/50		
c	TR300	74.5mg	n.s.s.	33/50	70.1mg	42/50	140.mg	45/50		
d	TR300	177.mg	n.s.s.	22/50	70.1mg	16/50	(140.mg	13/50)	liv:hpa,hpc,nnd.	
e	TR300	472.mg	n.s.s.	7/50	70.1mg	10/50	140.mg	3/50	lun:a/a,a/c.	
1019	TR300	149.mg	1.82gm	1/50	52.6mg	1/50	105.mg	10/50		
a	TR300	213.mg	n.s.s.	0/50	52.6mg	2/50	105.mg	4/50		s
b	TR300	49.7mg	n.s.s.	43/50	52.6mg	42/50	105.mg	41/50		
c	TR300	714.mg	n.s.s.	2/50	52.6mg	1/50	105.mg	0/50	liv:hpa,hpc,nnd.	
1020	TR300	43.8mg	119.mg	1/50	52.6mg	5/50	105.mg	30/50		
a	TR300	25.0mg	n.s.s.	36/50	52.6mg	43/50	105.mg	43/50		s
b	TR300	37.1mg	n.s.s.	40/50	52.6mg	32/50	105.mg	40/50		
c	TR300	204.mg	n.s.s.	2/50	52.6mg	0/50	105.mg	5/50	liv:hpa,hpc,nnd.	

1-CHLORO-2-NITROBENZENE   88-73-3

	RefNum	LoConf	UpConf	Cntrl	1Dose	1Inc	2Dose	2Inc	Citation or Pathology	Brkly Code
1021	381	109.mg	3.78gm	0/20	241.mg	5/22	(461.mg	5/19)	Weisburger;jept,2,325-356;1978/pers.comm./Russfield	1973
a	381	109.mg	3.78gm	0/20	241.mg	5/22	(461.mg	7/19)		
b	381	395.mg	n.s.s.	6/20	241.mg	4/22	461.mg	4/19		
c	381	86.3mg	n.s.s.	17/20	241.mg	12/22	(461.mg	11/19)		
1022	381	109.mg	1.19gm	1/102p	241.mg	5/22	(461.mg	5/19)		
1023	381	46.0mg	885.mg	3/18	260.mg	8/17	540.mg	3/16		
a	381	246.mg	n.s.s.	5/18	260.mg	1/17	540.mg	0/16		
b	381	56.4mg	n.s.s.	12/18	260.mg	9/17	540.mg	4/16		
1024	381	45.2mg	352.mg	7/99p	260.mg	7/17	540.mg	3/16		
1025	381	46.2mg	n.s.s.	1/22	20.9mg	2/22	43.6mg	0/19		
a	381	15.3mg	n.s.s.	14/22	20.9mg	14/22	43.6mg	8/19		

1-CHLORO-4-NITROBENZENE   100-00-5

	RefNum	LoConf	UpConf	Cntrl	1Dose	1Inc	2Dose	2Inc	Citation or Pathology	Brkly Code
1026	381	208.mg	1.88gm	0/15	351.mg	2/20	780.mg	7/18	Weisburger;jept,2,325-356;1978/pers.comm./Russfield	1973
a	381	46.8mg	n.s.s.	8/15	351.mg	3/20	780.mg	4/18		
b	381	338.mg	n.s.s.	0/15	351.mg	1/20	780.mg	2/18		
c	381	31.3mg	512.mg	11/15	351.mg	10/20	780.mg	12/18		
1027	381	240.mg	2.39gm	8/102p	351.mg	2/20	780.mg	7/18		
1028	381	146.mg	4.99gm	0/14	341.mg	2/14	720.mg	4/14		
a	381	96.2mg	n.s.s.	1/14	341.mg	5/14	720.mg	1/14		
b	381	293.mg	n.s.s.	4/14	341.mg	0/14	720.mg	3/14		
c	381	83.9mg	n.s.s.	8/14	341.mg	7/14	720.mg	7/14		
1029	381	158.mg	2.62gm	5/99p	341.mg	2/14	720.mg	4/14		
a	381	153.mg	n.s.s.	7/99p	341.mg	4/14	720.mg	0/14		
1030	381	n.s.s.	n.s.s.	0/16	22.6mg	0/14	45.2mg	0/15		
a	381	12.0mg	n.s.s.	9/16	22.6mg	10/14	45.2mg	10/15		

4-CHLORO-m-PHENYLENEDIAMINE   5131-60-2

	RefNum	LoConf	UpConf	Cntrl	1Dose	1Inc	2Dose	2Inc	Citation or Pathology	Brkly Code
1031	TR85	620.mg	3.25gm	0/50	732.mg	11/50	(1.46gm	8/49)	liv:hpa,hpc.	
a	TR85	804.mg	6.61gm	0/50	732.mg	8/50	(1.46gm	5/49)		

```
     Spe Strain Site   Xpo+Xpt                                              TD50    2Tailpvl
         Sex  Route Hist     Notes                                                DR    AuOp
  b    M f b6c eat TBA MXB 78w95 v                                          15.4gm * P<.8
  c    M f b6c eat liv MXB 78w95 v                                          1.23gm \ P<.0005
  d    M f b6c eat lun MXB 78w95 v                                          no dre   P=1.
1032 M m b6c eat TBA MXB 78w94 v                                      :>    7.32gm * P<.7    -
  a    M m b6c eat liv MXB 78w94 v                                          7.05gm * P<.5
  b    M m b6c eat lun MXB 78w94 v                                          no dre   P=1.
1033 R f f34 eat ute esp 18m24                            :  +         :    #175.mg \ P<.007 -
  a    R f f34 eat TBA MXB 18m24                                            479.mg * P<.5
  b    R f f34 eat liv MXB 18m24                                            2.68gm * P<.1
1034 R m f34 eat adr phe 18m24                            :  ±              315.mg * P<.03  c
  a    R m f34 eat TBA MXB 18m24                                            422.mg / P<.5
  b    R m f34 eat liv MXB 18m24                                            1.57gm * P<.2

4-CHLORO-o-PHENYLENEDIAMINE      100ng..:..1ug....:..10.....:..100....:..1mg....:..10.....:..100....:..1g......:..10
1035 M f b6c eat liv MXA 78w95 v                                    : +  :  2.24gm * P<.0005c
  a    M f b6c eat liv hpc 78w95 v                                          5.06gm * P<.006 c
  b    M f b6c eat TBA MXB 78w95 v                                          4.09gm * P<.4
  c    M f b6c eat liv MXB 78w95 v                                          2.24gm * P<.0005
  d    M f b6c eat lun MXB 78w95 v                                          no dre   P=1.
1036 M m b6c eat liv MXA 78w95 v                                  : +    :  957.mg * P<.002 c
  a    M m b6c eat liv hpc 78w95 v                                          1.37gm * P<.002 c
  b    M m b6c eat TBA MXB 78w95 v                                          905.mg * P<.005
  c    M m b6c eat liv MXB 78w95 v                                          957.mg * P<.002
  d    M m b6c eat lun MXB 78w95 v                                          21.3gm * P<.8
1037 R f f34 eat MXB MXB 18m24                            : + :              212.mg / P<.0005
  a    R f f34 eat ubl MXA 18m24                                            212.mg / P<.0005c
  b    R f f34 eat ubl MXA 18m24                                            429.mg / P<.0005c
  c    R f f34 eat thy MXA 18m24                                            2.16gm * P<.02
  d    R f f34 eat sto sqp 18m24                                            5.27mg * P<.04  c
  e    R f f34 eat TBA MXB 18m24                                            351.mg * P<.03
  f    R f f34 eat liv MXB 18m24                                            6.50gm * P<.08
1038 R m f34 eat MXB MXB 18m24                            :+ :               197.mg * P<.0005
  a    R m f34 eat ubl MXA 18m24                                            216.mg * P<.0005c
  b    R m f34 eat ubl tcc 18m24                                            363.mg * P<.0005c
  c    R m f34 eat sto MXA 18m24                            +hist 2.50gm * P<.02  c
  d    R m f34 eat thy MXA 18m24                                            3.38gm * P<.04
  e    R m f34 eat ubl sqc 18m24                                            3.78gm * P<.04  c
  f    R m f34 eat TBA MXB 18m24                                            362.mg * P<.07
  g    R m f34 eat liv MXB 18m24                                            1.24gm * P<.02

2-CHLORO-p-PHENYLENEDIAMINE  SULFATE   ..:..1ug....:..10.....:..100....:..1mg....:..10.....:..100....:..1g......:..10
1039 M f b6c eat TBA MXB 22m24 a                                     :>     1.43gm * P<.5
  a    M f b6c eat liv MXB 22m24 a                                          1.59gm * P<.07
  b    M f b6c eat lun MXB 22m24 a                                          3.33gm * P<.2
1040 M m b6c eat TBA MXB 22m24 a                                       :>   4.38gm / P<.8    -
  a    M m b6c eat liv MXB 22m24 a                                          1.31gm * P<.2
  b    M m b6c eat lun MXB 22m24 a                                          no dre   P=1.
1041 R f f34 eat TBA MXB 24m24 a                                    :>      no dre   P=1.    -
  a    R f f34 eat liv MXB 24m24 a                                          3.60gm * P<.6
1042 R m f34 eat TBA MXB 24m24 a                                    :>      244.mg * P<.2    -
  a    R m f34 eat liv MXB 24m24 a                                          665.mg * P<.4

3-CHLORO-p-TOLUIDINE      100ng..:..1ug....:..10.....:..100....:..1mg....:..10.....:..100....:..1g......:..10
1043 M f b6c eat TBA MXB 78w90                                       :>     1.98gm * P<.8
  a    M f b6c eat liv MXB 78w90                                            919.mg * P<.6
  b    M f b6c eat lun MXB 78w90                                            5.11gm / P<.8
1044 M m b6c eat TBA MXB 78w90                                       :>     3.11gm * P<1.    -
  a    M m b6c eat liv MXB 78w90                                            no dre   P=1.
  b    M m b6c eat lun MXB 78w90                                            733.mg * P<.3
1045 R f f34 eat ute esp 18m24 v                            :  ±            #406.mg * P<.02  -
  a    R f f34 eat TBA MXB 18m24 v                                          365.mg * P<.4
  b    R f f34 eat liv MXB 18m24 v                                          5.68gm * P<.4
1046 R m f34 eat TBA MXB 18m24 v                                    :>      no dre   P=1.    -
  a    R m f34 eat liv MXB 18m24 v                                          no dre   P=1.

5-CHLORO-o-TOLUIDINE      100ng..:..1ug....:..10.....:..100....:..1mg....:..10.....:..100....:..1g......:..10
1047 M f b6c eat MXB MXB 78w91                            :+ :               143.mg * P<.0005
  a    M f b6c eat liv MXA 78w91                                            180.mg * P<.0005c
  b    M f b6c eat liv hpc 78w91                                            217.mg * P<.0005c
  c    M f b6c eat --- hes 78w91                                            373.mg / P<.0005c
  d    M f b6c eat TBA MXB 78w91                                            164.mg * P<.0005
  e    M f b6c eat liv MXB 78w91                                            180.mg * P<.0005
  f    M f b6c eat lun MXB 78w91                                            7.48gm * P<.5
1048 M m b6c eat MXB MXB 78w91                            : + :              134.mg / P<.0005
  a    M m b6c eat --- hes 78w91                                            213.mg / P<.0005c
  b    M m b6c eat liv MXA 78w91                                            249.mg / P<.0005c
  c    M m b6c eat liv hpc 78w91                                            271.mg * P<.002 c
  d    M m b6c eat TBA MXB 78w91                                            132.mg * P<.0005
  e    M m b6c eat liv MXB 78w91                                            249.mg * P<.0005
  f    M m b6c eat lun MXB 78w91                                            no dre   P=1.
1049 R f f34 eat TBA MXB 18m24                                      :>      no dre   P=1.    -
```

	RefNum	LoConf	UpConf	Cntrl	1Dose	1Inc	2Dose	2Inc	Citation or Pathology
									Brkly Code
b	TR85	1.49gm	n.s.s.	14/50	732.mg	21/50	1.46gm	16/49	
c	TR85	620.mg	3.25gm	0/50	732.mg	11/50	(1.46gm	8/49)	liv:hpa,hpc,nnd.
d	TR85	9.28gm	n.s.s.	4/50	732.mg	1/50	1.46gm	1/49	lun:a/a,a/c.
1032	TR85	1.20gm	n.s.s.	22/50	676.mg	19/50	1.35gm	25/50	
a	TR85	1.53gm	n.s.s.	15/50	676.mg	10/50	1.35gm	19/50	liv:hpa,hpc,nnd.
b	TR85	5.75gm	n.s.s.	7/50	676.mg	4/50	1.35gm	3/50	lun:a/a,a/c.
1033	TR85	79.8mg	3.15gm	2/50	75.0mg	12/50	(149.mg	5/50)	S
a	TR85	105.mg	n.s.s.	27/50	75.0mg	33/50	149.mg	30/50	
b	TR85	660.mg	n.s.s.	0/50	75.0mg	0/50	149.mg	2/50	liv:hpa,hpc,nnd.
1034	TR85	144.mg	n.s.s.	4/50	60.0mg	7/50	119.mg	14/49	
a	TR85	97.9mg	n.s.s.	28/50	60.0mg	21/50	119.mg	36/49	
b	TR85	474.mg	n.s.s.	0/50	60.0mg	1/50	119.mg	2/49	liv:hpa,hpc,nnd.

4-CHLORO-o-PHENYLENEDIAMINE   95-83-0

	RefNum	LoConf	UpConf	Cntrl	1Dose	1Inc	2Dose	2Inc	Citation or Pathology
1035	TR63	1.34gm	6.00gm	0/50	752.mg	11/50	1.50gm	10/50	liv:hpa,hpc.
a	TR63	2.47gm	42.9gm	0/50	752.mg	4/50	1.50gm	6/50	
b	TR63	1.15gm	n.s.s.	14/50	752.mg	24/50	1.50gm	21/50	
c	TR63	1.34gm	6.00gm	0/50	752.mg	11/50	1.50gm	10/50	liv:hpa,hpc,nnd.
d	TR63	6.11gm	n.s.s.	4/50	752.mg	2/50	1.50gm	3/50	lun:a/a,a/c.
1036	TR63	521.mg	4.25gm	15/50	701.mg	28/50	1.39gm	34/50	liv:hpa,hpc.
a	TR63	729.mg	7.14gm	10/50	701.mg	18/50	1.39gm	26/50	
b	TR63	460.mg	8.99gm	22/50	701.mg	37/50	1.39gm	39/50	
c	TR63	521.mg	4.25gm	15/50	701.mg	28/50	1.39gm	34/50	liv:hpa,hpc,nnd.
d	TR63	2.41gm	n.s.s.	7/50	701.mg	9/50	1.39gm	7/50	lun:a/a,a/c.
1037	TR63	148.mg	318.mg	0/50	184.mg	15/50	368.mg	32/50	sto:sqp; ubl:pas,ppc,ppn,tcc,tpp. C
a	TR63	148.mg	318.mg	0/50	184.mg	15/50	368.mg	32/50	ubl:pas,ppc,ppn,tcc,tpp.
b	TR63	270.mg	740.mg	0/50	184.mg	5/50	368.mg	22/50	ubl:ppc,tcc.
c	TR63	878.mg	n.s.s.	0/50	184.mg	2/50	368.mg	4/50	thy:fca,fcc. S
d	TR63	1.58gm	n.s.s.	0/50	184.mg	0/50	368.mg	3/50	
e	TR63	157.mg	n.s.s.	27/50	184.mg	35/50	368.mg	37/50	
f	TR63	1.60gm	n.s.s.	0/50	184.mg	0/50	368.mg	2/50	liv:hpa,hpc,nnd.
1038	TR63	136.mg	301.mg	0/50	149.mg	15/49	294.mg	29/50	sto:sqc,sqp; ubl:ppn,sqc,tcc,tpp. C
a	TR63	146.mg	337.mg	0/50	149.mg	15/49	294.mg	25/50	ubl:ppn,tcc,tpp.
b	TR63	224.mg	650.mg	0/50	149.mg	7/49	294.mg	18/50	
c	TR63	862.mg	n.s.s.	0/50	149.mg	0/49	294.mg	4/50	sto:sqc,sqp.
d	TR63	1.02gm	n.s.s.	0/50	149.mg	0/49	294.mg	3/50	thy:fca,fcc. S
e	TR63	1.13gm	n.s.s.	0/50	149.mg	0/49	294.mg	3/50	
f	TR63	147.mg	n.s.s.	28/50	149.mg	31/49	294.mg	39/50	
g	TR63	558.mg	n.s.s.	0/50	149.mg	4/49	294.mg	4/50	liv:hpa,hpc,nnd.

2-CHLORO-p-PHENYLENEDIAMINE   SULFATE   61702-44-1

	RefNum	LoConf	UpConf	Cntrl	1Dose	1Inc	2Dose	2Inc	Citation or Pathology
1039	TR113	359.mg	n.s.s.	12/20	390.mg	27/50	646.mg	22/50	
a	TR113	656.mg	n.s.s.	2/20	390.mg	5/50	646.mg	9/50	liv:hpa,hpc,nnd.
b	TR113	1.26gm	n.s.s.	0/20	390.mg	3/50	646.mg	2/50	lun:a/a,a/c.
1040	TR113	527.mg	n.s.s.	12/20	360.mg	21/50	596.mg	30/50	
a	TR113	543.mg	n.s.s.	4/20	360.mg	12/50	596.mg	20/50	liv:hpa,hpc,nnd.
b	TR113	1.55gm	n.s.s.	3/20	360.mg	7/50	596.mg	6/50	lun:a/a,a/c.
1041	TR113	112.mg	n.s.s.	17/20	75.0mg	26/50	(150.mg	19/50)	
a	TR113	886.mg	n.s.s.	0/20	75.0mg	1/50	150.mg	1/50	liv:hpa,hpc,nnd.
1042	TR113	95.0mg	n.s.s.	7/20	60.0mg	21/50	120.mg	27/50	
a	TR113	301.mg	n.s.s.	0/20	60.0mg	5/50	120.mg	3/50	liv:hpa,hpc,nnd.

3-CHLORO-p-TOLUIDINE   95-74-9

	RefNum	LoConf	UpConf	Cntrl	1Dose	1Inc	2Dose	2Inc	Citation or Pathology
1043	TR145	220.mg	n.s.s.	5/20	67.6mg	6/50	135.mg	11/50	
a	TR145	310.mg	n.s.s.	0/20	67.6mg	4/50	135.mg	2/50	liv:hpa,hpc,nnd.
b	TR145	483.mg	n.s.s.	2/20	67.6mg	0/50	135.mg	4/50	lun:a/a,a/c.
1044	TR145	148.mg	n.s.s.	5/20	62.4mg	15/50	125.mg	14/50	
a	TR145	261.mg	n.s.s.	4/20	62.4mg	10/50	125.mg	7/50	liv:hpa,hpc,nnd.
b	TR145	299.mg	n.s.s.	0/20	62.4mg	3/50	125.mg	3/50	lun:a/a,a/c.
1045	TR145	214.mg	n.s.s.	0/20	62.0mg	4/50	124.mg	9/50	S
a	TR145	102.mg	n.s.s.	7/20	62.0mg	22/50	124.mg	28/50	
b	TR145	925.mg	n.s.s.	0/20	62.0mg	0/50	124.mg	1/50	liv:hpa,hpc,nnd.
1046	TR145	196.mg	n.s.s.	9/20	49.6mg	21/50	100.mg	17/50	
a	TR145	761.mg	n.s.s.	0/20	49.6mg	1/50	100.mg	0/50	liv:hpa,hpc,nnd.

5-CHLORO-o-TOLUIDINE   95-79-4

	RefNum	LoConf	UpConf	Cntrl	1Dose	1Inc	2Dose	2Inc	Citation or Pathology
1047	TR187	104.mg	221.mg	0/20	222.mg	26/50	446.mg	35/50	---:hes; liv:hpa,hpc. C
a	TR187	128.mg	290.mg	0/20	222.mg	21/50	446.mg	31/50	liv:hpa,hpc.
b	TR187	151.mg	379.mg	0/20	222.mg	19/50	446.mg	26/50	
c	TR187	237.mg	676.mg	0/20	222.mg	6/50	446.mg	22/50	
d	TR187	103.mg	446.mg	4/20	222.mg	32/50	446.mg	38/50	
e	TR187	128.mg	290.mg	0/20	222.mg	21/50	446.mg	31/50	liv:hpa,hpc,nnd.
f	TR187	1.39gm	n.s.s.	1/20	222.mg	0/50	446.mg	3/50	lun:a/a,a/c.
1048	TR187	88.6mg	269.mg	5/20	205.mg	29/50	412.mg	45/50	---:hes; liv:hpa,hpc. C
a	TR187	145.mg	356.mg	1/20	205.mg	11/50	412.mg	37/50	
b	TR187	146.mg	876.mg	4/20	205.mg	20/50	412.mg	27/50	liv:hpa,hpc.
c	TR187	155.mg	1.16gm	4/20	205.mg	19/50	412.mg	25/50	
d	TR187	85.9mg	279.mg	6/20	205.mg	31/50	412.mg	46/50	
e	TR187	146.mg	876.mg	4/20	205.mg	20/50	412.mg	27/50	liv:hpa,hpc,nnd.
f	TR187	1.18gm	n.s.s.	2/20	205.mg	2/50	412.mg	2/50	lun:a/a,a/c.
1049	TR187	218.mg	n.s.s.	8/20	94.0mg	22/50	189.mg	23/50	

```
        Spe Strain Site   Xpo+Xpt                                          TD50    2Tailpvl
          Sex   Route Hist    Notes                                          DR    AuOp
a    R f f34 eat liv MXB 18m24                                              no dre   P=1.
1050 R m f34 eat adr phe 18m24                              :    ±          #714.mg * P<.03  -
a    R m f34 eat TBA MXB 18m24                                              1.02gm * P<.7
b    R m f34 eat liv MXB 18m24                                              3.34gm * P<.6

4-CHLORO-o-TOLUIDINE.HCl          100ng..:..1ug...:..10...:..100....:..1mg..:..10...:..100...:..1g..:..10
1051 M f b6c eat --- MXA 95w99 as                                  : +:     40.0mg / P<.0005c
a    M f b6c eat --- hes 95w99 as                                          42.3mg / P<.0005c
b    M f b6c eat TBA MXB 95w99 as                                          38.8mg / P<.0005
c    M f b6c eat liv MXB 95w99 as                                          754.mg * P<.3
d    M f b6c eat lun MXB 95w99 as                                          798.mg * P<.01
1052 M m b6c eat --- MXA 99w99 s                                     :+ :   743.mg / P<.0005c
a    M m b6c eat --- hes 99w99 s                                           911.mg / P<.0005c
b    M m b6c eat TBA MXB 99w99 s                                           537.mg / P<.0005
c    M m b6c eat liv MXB 99w99 s                                           3.26gm * P<.03
d    M m b6c eat lun MXB 99w99 s                                           no dre   P=1.
1053 M f chi eat --- vsc 51w60 a                              :   +   :     15.7mg * P<.0005+
a    M f chi eat liv mix 51w60 a                                           139.mg * P<.02   -
b    M f chi eat lun mix 51w60 a                                           153.mg * P<.06   -
c    M f chi eat tba mix 51w60 a                                           15.6mg * P<.0005
1054 M f chi eat --- vsc 51w60 a  pool                        :   +   :     18.4mg * P<.0005+
1055 M m chi eat lun mix 55w60                          :     +     :       8.85mg \ P<.0005-
a    M m chi eat --- vsc 55w60                                             15.4mg * P<.0005+
b    M m chi eat liv mix 55w60                                             95.7mg * P<.04   -
c    M m chi eat tba mix 55w60                                             7.72mg * P<.0005
1056 M m chi eat --- vsc 55w60   pool                         :   +   :     15.4mg * P<.0005+
1057 R f f34 eat TBA MXB 25m25                                      :>      no dre   P=1.   -
a    R f f34 eat liv MXB 25m25                                             no dre   P=1.
1058 R m f34 eat pit cra 25m25                              :    ±          #576.mg * P<.05  -
a    R m f34 eat TBA MXB 25m25                                             no dre   P=1.
b    R m f34 eat liv MXB 25m25                                             6.18gm * P<.8
1059 R m cdr eat liv mix 77w98 v                            :>             123.mg * P<.2   -
a    R m cdr eat tba mix 77w98 v                                           24.9mg * P<.03  -

2-CHLORO-1,1,1-TRIFLUOROETHANE    100ng..:..1ug...:..10...:..100....:..1mg..:..10...:..100...:..1g..:..10
1060 R f aap gav ute adc 12m29 e  pool                             . + .    160.mg   P<.0005+
1061 R m aap gav tes ict 12m29 e  pool                           . + .      60.0mg   P<.0005+

[4-CHLORO-6-(2,3-XYLIDINO)-2-PYRIMIDINYLTHIO]ACETIC  ACID....:..100...:..1mg..:..10...:..100...:..1g..:..10
1062 M m csb eat liv hpc 53w63 erv                                 <+       noTD50   P<.0005+
1063 R f f34 eat liv hpc 65w65 e                                   <+       noTD50   P<.0005+
a    R m f34 eat tba mix 65w65 e                                           noTD50   P<.005
1064 R m f34 eat liv thc 69w69 er                               <+         noTD50   P<.0005+

4-CHLORO-6-(2,3-XYLIDINO)-2-PYRIMIDINYLTHIO(N-beta-HYDROXYETHYL)ACETAMIDE.1 _mg...:..10...:..100...:..1g..:..10
1065 M f csb eat liv hpc 82w82 er                                  .   +   . 44.6mg   P<.0005+
1066 R m f34 eat liv hpc 75w82 aer                           .  +  .        6.49mg * P<.0005+

CHLOROACETALDEHYDE                100ng..:..1ug...:..10...:..100....:..1mg..:..10...:..100...:..1g..:..10
1067 M m b6c wat liv hpc 60w60 ek                                .>         17.4mg   P<.3
1068 M m b6c wat liv mix 24m24 e                                    . ±     36.1mg   P<.08  +
a    M m b6c wat liv hpc 24m24 e                                           44.4mg   P<.09  +
b    M m b6c wat liv hpa 24m24 e                                           405.mg   P<.8
1069 M f hic gav for tum 90w90                                   .>         13.6mg   P<.4   -
1070 M m hic gav for tum 90w90                                    .>        no dre   P=1.   -

2-CHLOROACETOPHENONE              100ng..:..1ug...:..10...:..100....:..1mg..:..10...:..100...:..1g..:..10
1071 M f b6c inh TBA MXB 24m24                                   :>         4.31mg * P<.7
a    M f b6c inh liv MXB 24m24                                             no dre   P=1.
b    M f b6c inh lun MXB 24m24                                             33.4mg * P<.8
1072 M m b6c inh TBA MXB 24m25                                   :>         2.69mg * P<.4   -
a    M m b6c inh liv MXB 24m25                                             5.24mg * P<.5
b    M m b6c inh lun MXB 24m25                                             10.4mg * P<.7
1073 R f f34 inh mgl MXA 24m24                          :   ±               .208mg * P<.07
a    R f f34 inh mgl fba 24m24                                              .211mg * P<.07 e
b    R f f34 inh mgl MXA 24m24                                              .224mg * P<.2
c    R f f34 inh TBA MXB 24m24                                              .859mg * P<.9
d    R f f34 inh liv MXB 24m24                                             no dre   P=1.
1074 R m f34 inh TBA MXB 24m24                           :>                no dre   P=1.   -
a    R m f34 inh liv MXB 24m24                                             no dre   P=1.

4'-(CHLOROACETYL)-ACETANILIDE     100ng..:..1ug...:..10...:..100....:..1mg..:..10...:..100...:..1g..:..10
1075 M f b6c eat liv hpa 21m24 v                                    :  ±    4.70mg * P<.02  a
a    M f b6c eat TBA MXB 21m24 v                                           3.50gm \ P<.7
b    M f b6c eat liv MXB 21m24 v                                           4.70gm * P<.02
c    M f b6c eat lun MXB 21m24 v                                           3.02gm \ P<.09
1076 M m b6c eat TBA MXB 21m24 v                                    :>      no dre   P=1.   -
a    M m b6c eat liv MXB 21m24 v                                           no dre   P=1.
b    M m b6c eat lun MXB 21m24 v                                           no dre   P=1.
1077 R f f34 eat TBA MXB 20m24 v                                 :>        no dre   P=1.   -
a    R f f34 eat liv MXB 20m24 v                                           no dre   P=1.
```

	RefNum	LoConf	UpConf	Cntrl	1Dose	1Inc	2Dose	2Inc	Citation or Pathology
									Brkly Code
a	TR187	1.44gm	n.s.s.	0/20	94.0mg	1/50	189.mg	0/50	liv:hpa,hpc,nnd.
1050	TR187	336.mg	n.s.s.	0/20	75.2mg	2/50	151.mg	7/50	s
a	TR187	166.mg	n.s.s.	8/20	75.2mg	17/50	151.mg	24/50	
b	TR187	821.mg	n.s.s.	0/20	75.2mg	1/50	151.mg	1/50	liv:hpa,hpc,nnd.

4-CHLORO-o-TOLUIDINE.HCl   3165-93-3

	RefNum	LoConf	UpConf	Cntrl	1Dose	1Inc	2Dose	2Inc	Citation or Pathology
1051	TR165	26.9mg	62.0mg	1/20	162.mg	43/50	650.mg	39/50	---:hem,hes.
a	TR165	28.6mg	64.0mg	0/20	162.mg	40/50	650.mg	39/50	
b	TR165	25.6mg	63.4mg	5/20	162.mg	45/50	650.mg	42/50	
c	TR165	191.mg	n.s.s.	1/20	162.mg	4/50	650.mg	0/50	liv:hpa,hpc,nnd.
d	TR165	217.mg	76.0gm	0/20	162.mg	2/50	650.mg	3/50	lun:a/a,a/c.
1052	TR165	507.mg	1.13gm	0/20	450.mg	6/50	1.80gm	41/50	---:hem,hes.
a	TR165	605.mg	1.44gm	0/20	450.mg	3/50	1.80gm	37/50	
b	TR165	336.mg	1.09gm	11/20	450.mg	28/50	1.80gm	45/50	
c	TR165	1.29gm	n.s.s.	4/20	450.mg	7/50	1.80gm	10/50	liv:hpa,hpc,nnd.
d	TR165	2.15gm	n.s.s.	4/20	450.mg	14/50	1.80gm	3/50	lun:a/a,a/c.
1053	381	6.23mg	37.4mg	0/15	260.mg	18/19	520.mg	12/16	Weisburger;jept,2,325-356;1978;pers.comm./Russfield 1973
a	381	24.8mg	n.s.s.	0/15	260.mg	1/19	520.mg	1/16	
b	381	25.0mg	n.s.s.	8/15	260.mg	1/19	520.mg	0/16	
c	381	6.21mg	36.4mg	11/15	260.mg	18/19	520.mg	13/16	
1054	381	7.46mg	41.3mg	8/102p	260.mg	18/19	520.mg	12/16	
1055	381	2.51mg	55.6mg	4/14	90.0mg	7/20	(180.mg	2/20)	
a	381	8.35mg	30.2mg	0/14	90.0mg	12/20	180.mg	13/20	
b	381	17.2mg	n.s.s.	1/14	90.0mg	0/20	180.mg	2/20	
c	381	4.02mg	15.1mg	8/14	90.0mg	18/20	180.mg	16/20	
1056	381	8.35mg	27.4mg	5/99p	90.0mg	12/20	180.mg	13/20	
1057	TR165	299.mg	n.s.s.	12/20	62.5mg	34/50	250.mg	32/50	
a	TR165	1.90gm	n.s.s.	1/20	62.5mg	0/50	250.mg	1/50	liv:hpa,hpc,nnd.
1058	TR165	239.mg	n.s.s.	2/20	50.0mg	6/50	200.mg	15/50	s
a	TR165	209.mg	n.s.s.	13/20	50.0mg	30/50	200.mg	33/50	
b	TR165	452.mg	n.s.s.	0/20	50.0mg	5/50	200.mg	4/50	liv:hpa,hpc,nnd.
1059	381	36.2mg	n.s.s.	0/16	24.6mg	2/19	49.1mg	1/13	Weisburger;jept,2,325-356;1978;pers.comm./Russfield 1973
a	381	10.7mg	n.s.s.	9/16	24.6mg	12/19	49.1mg	10/13	

2-CHLORO-1,1,1-TRIFLUOROETHANE   (fluorocarbon 133a) 75-88-7

	RefNum	LoConf	UpConf	Cntrl	1Dose	1Inc	2Dose	2Inc	Citation or Pathology
1060	1623	86.4mg	351.mg	1/104p	89.1mg	15/35			Longstaff;txap,72,15-31;1984;pers.comm.
1061	1623	34.4mg	117.mg	16/104p	89.1mg	29/36			

[4-CHLORO-6-(2,3-XYLIDINO)-2-PYRIMIDINYLTHIO]ACETIC   ACID   50892-23-4

	RefNum	LoConf	UpConf	Cntrl	1Dose	1Inc	2Dose	2Inc	Citation or Pathology
1062	1656	n.s.s.	10.8mg	0/10	76.3mg	18/18			Reddy;canr,29,152-161;1979
1063	1332	n.s.s.	13.8mg	0/12	80.0mg	14/14			Lalwani;carc,2,645-650;1981
a	1332	n.s.s.	17.8mg	3/12	80.0mg	14/14			
1064	1656	n.s.s.	7.47mg	0/10	40.0mg	15/15			Reddy;canr,29,152-161;1979

4-CHLORO-6-(2,3-XYLIDINO)-2-PYRIMIDINYLTHIO(N-beta-HYDROXYETHYL)ACETAMIDE   65089-17-0

	RefNum	LoConf	UpConf	Cntrl	1Dose	1Inc	2Dose	2Inc	Citation or Pathology
1065	1331	16.4mg	126.mg	0/15	260.mg	11/12			Reddy;natu,283,397-398;1980
1066	1331	2.77mg	14.7mg	0/30	20.0mg	7/10	80.0mg	20/20	

CHLOROACETALDEHYDE   107-20-0

	RefNum	LoConf	UpConf	Cntrl	1Dose	1Inc	2Dose	2Inc	Citation or Pathology
1067	2072m	2.80mg	n.s.s.	0/5	17.0mg	1/5			Daniel;faat,19,159-168;1992;pers.comm.
1068	2072n	13.3mg	n.s.s.	3/20	17.0mg	10/26			
a	2072n	15.8mg	n.s.s.	2/20	17.0mg	8/26			
b	2072n	40.2mg	n.s.s.	1/20	17.0mg	2/26			
1069	1011	2.43mg	n.s.s.	5/100	1.43mg	3/30			Van Duuren;jnci,63,1433-1439;1979
1070	1011	4.29mg	n.s.s.	8/60	1.19mg	1/30			

2-CHLOROACETOPHENONE   532-27-4

	RefNum	LoConf	UpConf	Cntrl	1Dose	1Inc	2Dose	2Inc	Citation or Pathology
1071	TR379	.692mg	n.s.s.	42/50	.618mg	40/50	1.22mg	38/50	
a	TR379	2.74mg	n.s.s.	12/50	.618mg	13/50	1.22mg	7/50	liv:hpa,hpc,nnd.
b	TR379	3.13mg	n.s.s.	6/50	.618mg	2/50	1.22mg	6/50	lun:a/a,a/c.
1072	TR379	.668mg	n.s.s.	33/50	.515mg	35/50	1.01mg	39/50	
a	TR379	1.23mg	n.s.s.	16/50	.515mg	19/50	1.01mg	20/50	liv:hpa,hpc,nnd.
b	TR379	1.58mg	n.s.s.	11/50	.515mg	9/50	1.01mg	13/50	lun:a/a,a/c.
1073	TR379	84.8ug	n.s.s.	12/50	73.6ug	20/50	.147mg	23/50	mgl:ade,fba.
a	TR379	85.6ug	n.s.s.	12/50	73.6ug	19/50	.147mg	23/50	
b	TR379	85.4ug	n.s.s.	13/50	73.6ug	22/50	.147mg	23/50	mgl:adc,ade,fba.
c	TR379	68.0ug	n.s.s.	45/50	73.6ug	50/50	.147mg	48/50	
d	TR379	n.s.s.	n.s.s.	0/50	73.6ug	0/50	.147mg	0/50	liv:hpa,hpc,nnd.
1074	TR379	53.4ug	n.s.s.	46/50	51.5ug	48/50	.103mg	47/50	
a	TR379	n.s.s.	n.s.s.	0/50	51.5ug	0/50	.103mg	0/50	liv:hpa,hpc,nnd.

4'-(CHLOROACETYL)-ACETANILIDE   140-49-8

	RefNum	LoConf	UpConf	Cntrl	1Dose	1Inc	2Dose	2Inc	Citation or Pathology
1075	TR177	2.29gm	n.s.s.	0/20	557.mg	2/50	1.11gm	8/50	
a	TR177	590.mg	n.s.s.	6/20	557.mg	21/50	(1.11gm	12/50)	
b	TR177	2.29gm	n.s.s.	0/20	557.mg	2/50	1.11gm	8/50	liv:hpa,hpc,nnd.
c	TR177	1.14gm	n.s.s.	0/20	557.mg	5/50	(1.11gm	1/50)	lun:a/a,a/c.
1076	TR177	877.mg	n.s.s.	9/20	514.mg	19/50	(1.03gm	5/50)	
a	TR177	1.51gm	n.s.s.	3/20	514.mg	6/50	(1.03gm	0/50)	liv:hpa,hpc,nnd.
b	TR177	3.71gm	n.s.s.	3/20	514.mg	6/50	1.03gm	4/50	lun:a/a,a/c.
1077	TR177	90.1mg	n.s.s.	11/20	42.2mg	31/50	84.5mg	29/50	
a	TR177	616.mg	n.s.s.	1/20	42.2mg	1/50	84.5mg	1/50	liv:hpa,hpc,nnd.

```
      Spe Strain Site  Xpo+Xpt                                              TD50    2Tailpvl
          Sex  Route Hist    Notes                                             DR    AuOp
1078 R m f34 eat TBA MXB 20m24 v                            :>              325.mg * P<.7   -
   a R m f34 eat liv MXB 20m24 v                                           2.47gm * P<.8


p-CHLOROANILINE                      100ng..:..1ug....:..10......:..100....:..1mg...:..10....:..100....:..1g....:..10
1079 M f b6c eat --- MXA 78w91                                       : +   1.48gm * P<.01  a
   a M f b6c eat --- hes 78w91                                             1.65gm * P<.02  a
   b M f b6c eat liv MXA 78w91                                            2.35gm * P<.02
   c M f b6c eat TBA MXB 78w91                                             889.mg * P<.05
   d M f b6c eat liv MXB 78w91                                            2.35gm * P<.02
   e M f b6c eat lun MXB 78w91                                            no dre  P=1.
1080 M m b6c eat --- MXA 78w91                                  :>    +hist 769.mg * P<.2   a
   a M m b6c eat TBA MXB 78w91                                            no dre  P=1.
   b M m b6c eat liv MXB 78w91                                            no dre  P=1.
   c M m b6c eat lun MXB 78w91                                            no dre  P=1.
1081 R f f34 eat TBA MXB 18m24                         :>                  60.7mg * P<.4   -
   a R f f34 eat liv MXB 18m24                                             867.mg * P<.4
1082 R m f34 eat spl MXA 18m24                          :   +       :+hist 85.9mg / P<.009 a
   a R m f34 eat MXA MXA 18m24                                             72.0mg / P<.03
   b R m f34 eat spl fib 18m24                                            101.mg * P<.02
   c R m f34 eat TBA MXB 18m24                                             31.4mg * P<.09
   d R m f34 eat liv MXB 18m24                                             609.mg * P<.4


p-CHLOROANILINE.HCl                  100ng..:..1ug....:..10......:..100....:..1mg...:..10....:..100....:..1g....:..10
1083 M f b6c gav TBA MXB 24m24                               :>            no dre  P=1.    -
   a M f b6c gav liv MXB 24m24                                            176.mg * P<.5
   b M f b6c gav lun MXB 24m24                                            no dre  P=1.
1084 M m b6c gav liv hpc 24m24                           : +   :          33.8mg * P<.002
   a M m b6c gav --- hes 24m24                                             89.5mg * P<.04  p
   b M m b6c gav liv MXA 24m24                                            49.3mg * P<.2   p
   c M m b6c gav MXB MXB 24m24                                             56.0mg * P<.2
   d M m b6c gav TBA MXB 24m24                                             83.7mg * P<.6
   e M m b6c gav liv MXB 24m24                                            49.3mg * P<.2
   f M m b6c gav lun MXB 24m24                                            2.78gm * P<1.
1085 R f f34 gav amd pob 24m24                               :>           105.mg * P<.2   e
   a R f f34 gav spl MXA 24m24                                            232.mg * P<.3   a
   b R f f34 gav TBA MXB 24m24                                            145.mg Z P<.9
   c R f f34 gav liv MXB 24m24                                            no dre  P=1.
1086 R m f34 gav spl MXA 24m24                     :+ :                   7.62mg Z P<.0005c
   a R m f34 gav spl MXA 24m24                                            7.90mg Z P<.0005
   b R m f34 gav amd MXA 24m24                                            13.3mg Z P<.009 e
   c R m f34 gav spl ost 24m24                                            15.5mg Z P<.0005
   d R m f34 gav spl MXA 24m24                                            16.9mg * P<.0005
   e R m f34 gav spl fbs 24m24                                            19.0mg * P<.0005
   f R m f34 gav spl hes 24m24                                           107.mg * P<.005
   g R m f34 gav amd MXA 24m24                                            14.8mg Z P<.02
   h R m f34 gav TBA MXB 24m24                                            20.8mg Z P<.4
   i R m f34 gav liv MXB 24m24                                            no dre  P=1.


o-CHLOROBENZALMALONITRILE            100ng..:..1ug....:..10......:..100....:..1mg...:..10....:..100....:..1g....:..10
1087 M f b6c inh TBA MXB 24m25                             :>              no dre  P=1.    -
   a M f b6c inh liv MXB 24m25                                            no dre  P=1.
   b M f b6c inh lun MXB 24m25                                            no dre  P=1.
1088 M m b6c inh TBA MXB 24m25                          :>                no dre  P=1.    -
   a M m b6c inh liv MXB 24m25                                            no dre  P=1.
   b M m b6c inh lun MXB 24m25                                            no dre  P=1.
1089 R f f34 inh TBA MXB 24m25                        :>                  no dre  P=1.    -
   a R f f34 inh liv MXB 24m25                                            no dre  P=1.
1090 R m f34 inh thy MXA 24m25             :   +     :                   #6.49ug Z P<.004 -
   a R m f34 inh thy cca 24m25                                            7.34ug Z P<.008
   b R m f34 inh TBA MXB 24m25                                            no dre  P=1.
   c R m f34 inh liv MXB 24m25                                            no dre  P=1.


CHLOROBENZENE                        100ng..:..1ug....:..10......:..100....:..1mg...:..10....:..100....:..1g....:..10
1091 M f b6c gav TBA MXB 24m24                               :>           373.mg * P<.5   -
   a M f b6c gav liv MXB 24m24                                            235.mg \ P<.09
   b M f b6c gav lun MXB 24m24                                            623.mg * P<.08
1092 M m b6c gav TBA MXB 24m24                              :>             60.1mg * P<.2   -
   a M m b6c gav liv MXB 24m24                                            382.mg * P<.7
   b M m b6c gav lun MXB 24m24                                            190.mg * P<.2
1093 R f f34 gav MXA MXA 24m24                              :    ±        #520.mg * P<.05  -
   a R f f34 gav TBA MXB 24m24                                            no dre  P=1.
   b R f f34 gav liv MXB 24m24                                            867.mg * P<.2
1094 R m f34 gav liv nnd 24m24                             :   ±          247.mg * P<.02  p
   a R m f34 gav TBA MXB 24m24                                            190.mg * P<.4
   b R m f34 gav liv MXB 24m24                                            328.mg * P<.08


CHLOROBENZILATE                      100ng..:..1ug....:..10......:..100....:..1mg...:..10....:..100....:..1g....:..10
1095 M f b6c eat liv hpc 78w90 dv                            : +    :     848.mg * P<.004 c
   a M f b6c eat TBA MXB 78w90 dv                                         3.93gm * P<.7
   b M f b6c eat liv MXB 78w90 dv                                         848.mg * P<.004
   c M f b6c eat lun MXB 78w90 dv                                         no dre  P=1.
```

	RefNum	LoConf	UpConf	Cntrl	1Dose	1Inc	2Dose	2Inc		Citation or Pathology Brkly Code
1078	TR177	54.6mg	n.s.s.	9/20	33.8mg	27/50	67.6mg	28/50		
a	TR177	242.mg	n.s.s.	1/20	33.8mg	2/50	67.6mg	3/50		liv:hpa,hpc,nnd.

p-CHLOROANILINE   106-47-8

	RefNum	LoConf	UpConf	Cntrl	1Dose	1Inc	2Dose	2Inc		Citation or Pathology Brkly Code
1079	TR189	744.mg	86.1gm	0/20	278.mg	3/50	558.mg	8/50		---:hem,hes.
a	TR189	802.mg	n.s.s.	0/20	278.mg	3/50	558.mg	7/50		
b	TR189	1.02gm	n.s.s.	0/20	278.mg	1/50	558.mg	6/50		liv:hpa,hpc. S
c	TR189	405.mg	n.s.s.	3/20	278.mg	11/50	558.mg	18/50		
d	TR189	1.02gm	n.s.s.	0/20	278.mg	1/50	558.mg	6/50		liv:hpa,hpc,nnd.
e	TR189	1.89gm	n.s.s.	1/20	278.mg	2/50	558.mg	2/50		lun:a/a,a/c.
1080	TR189	332.mg	n.s.s.	2/20	257.mg	10/50	275.mg	14/50		---:hem,hes.
a	TR189	374.mg	n.s.s.	9/20	257.mg	18/50	275.mg	19/50		
b	TR189	1.01gm	n.s.s.	3/20	257.mg	7/50	275.mg	2/50		liv:hpa,hpc,nnd.
c	TR189	2.25gm	n.s.s.	2/20	257.mg	1/50	275.mg	1/50		lun:a/a,a/c.
1081	TR189	16.0mg	n.s.s.	7/20	9.60mg	24/50	19.0mg	25/50		
a	TR189	141.mg	n.s.s.	0/20	9.60mg	0/50	19.0mg	1/50		liv:hpa,hpc,nnd.
1082	TR189	37.0mg	2.23gm	0/20	7.70mg	0/50	15.2mg	7/50		spl:fbs,fib.
a	TR189	31.4mg	n.s.s.	1/20	7.70mg	0/50	15.2mg	10/50	spc:srn;	spl:fbs,fib,hes,ost. S
b	TR189	41.3mg	n.s.s.	0/20	7.70mg	0/50	15.2mg	6/50		S
c	TR189	13.1mg	n.s.s.	7/20	7.70mg	14/50	15.2mg	26/50		
d	TR189	99.2mg	n.s.s.	0/20	7.70mg	0/50	15.2mg	1/50		liv:hpa,hpc,nnd.

p-CHLOROANILINE.HCl   20265-96-7

	RefNum	LoConf	UpConf	Cntrl	1Dose	1Inc	2Dose	2Inc	3Dose	3Inc	Citation or Pathology Brkly Code
1083	TR351	28.9mg	n.s.s.	36/50	2.11mg	26/50	7.04mg	21/50	21.1mg	31/50	
a	TR351	33.4mg	n.s.s.	6/50	2.11mg	9/50	7.04mg	8/50	21.1mg	11/50	liv:hpa,hpc,nnd.
b	TR351	88.7mg	n.s.s.	6/50	2.11mg	2/50	7.04mg	1/50	21.1mg	4/50	lun:a/a,a/c.
1084	TR351	17.3mg	146.mg	3/50	2.11mg	7/50	7.04mg	11/50	21.1mg	17/50	S
a	TR351	33.5mg	n.s.s.	4/50	2.11mg	4/50	7.04mg	1/50	21.1mg	10/50	
b	TR351	16.4mg	n.s.s.	11/50	2.11mg	21/50	7.04mg	20/50	21.1mg	21/50	liv:hpa,hpc.
c	TR351	16.8mg	n.s.s.	14/50	2.11mg	24/50	7.04mg	21/50	21.1mg	23/50	---:hes; liv:hpa,hpc. P
d	TR351	14.6mg	n.s.s.	40/50	2.11mg	30/50	7.04mg	35/50	21.1mg	36/50	
e	TR351	16.4mg	n.s.s.	11/50	2.11mg	21/50	7.04mg	20/50	21.1mg	21/50	liv:hpa,hpc,nnd.
f	TR351	43.1mg	n.s.s.	8/50	2.11mg	5/50	7.04mg	7/50	21.1mg	6/50	lun:a/a,a/c.
1085	TR351	29.5mg	n.s.s.	2/50	1.40mg	3/50	4.20mg	1/50	12.6mg	6/50	
a	TR351	57.0mg	n.s.s.	0/50	1.40mg	0/50	4.20mg	1/50	12.6mg	1/50	spl:fbs,ost.
b	TR351	10.8mg	n.s.s.	37/50	1.40mg	30/50	4.20mg	34/50	12.6mg	40/50	
c	TR351	130.mg	n.s.s.	1/50	1.40mg	1/50	4.20mg	0/50	12.6mg	0/50	liv:hpa,hpc,nnd.
1086	TR351	5.14mg	11.8mg	0/49	1.40mg	1/50	4.20mg	3/50	12.6mg	38/50	spl:fbs,hes,ost.
a	TR351	5.28mg	12.4mg	0/49	1.40mg	1/50	4.20mg	3/50	12.6mg	36/50	spl:fbs,ost. S
b	TR351	6.01mg	494.mg	13/49	1.40mg	14/50	4.20mg	15/50	12.6mg	26/50	amd:pbb,phm,pob.
c	TR351	9.01mg	29.7mg	0/49	1.40mg	0/50	4.20mg	1/50	12.6mg	19/50	S
d	TR351	10.1mg	31.2mg	0/49	1.40mg	1/50	4.20mg	2/50	12.6mg	19/50	spl:fbs,fib. S
e	TR351	11.0mg	36.3mg	0/49	1.40mg	1/50	4.20mg	2/50	12.6mg	17/50	S
f	TR351	36.3mg	1.04gm	0/49	1.40mg	0/50	4.20mg	0/50	12.6mg	4/50	S
g	TR351	6.47mg	n.s.s.	13/49	1.40mg	14/50	4.20mg	14/50	12.6mg	25/50	amd:pbb,pob. S
h	TR351	5.46mg	n.s.s.	42/49	1.40mg	42/50	4.20mg	39/50	12.6mg	48/50	
i	TR351	42.8mg	n.s.s.	1/49	1.40mg	6/50	4.20mg	5/50	12.6mg	0/50	liv:hpa,hpc,nnd.

o-CHLOROBENZALMALONITRILE   (CS2 [94% o-chlorobenzalmalononitrile, 5% Cab-O-Sil colloidal silica, 1% hexamethyldisilizane]) 2698-41-1

	RefNum	LoConf	UpConf	Cntrl	1Dose	1Inc	2Dose	2Inc	3Dose	3Inc	Citation or Pathology Brkly Code
1087	TR377	1.28mg	n.s.s.	46/50	.230mg	27/50	.464mg	27/50			
a	TR377	1.73mg	n.s.s.	11/50	.230mg	4/50	.464mg	9/50			liv:hpa,hpc,nnd.
b	TR377	2.71mg	n.s.s.	5/50	.230mg	2/50	.464mg	3/50			lun:a/a,a/c.
1088	TR377	.515mg	n.s.s.	31/50	.191mg	24/50	.386mg	30/50			
a	TR377	1.06mg	n.s.s.	18/50	.191mg	14/50	.386mg	13/50			liv:hpa,hpc,nnd.
b	TR377	1.22mg	n.s.s.	14/50	.191mg	8/50	.386mg	10/50			lun:a/a,a/c.
1089	TR377	37.8ug	n.s.s.	47/50	5.48ug	47/50	18.1ug	47/50	54.9ug	50/50	
a	TR377	.516mg	n.s.s.	3/50	5.48ug	1/50	18.1ug	0/50	54.9ug	0/50	liv:hpa,hpc,nnd.
1090	TR377	2.85ug	54.5ug	2/50	3.83ug	10/50	(12.7ug	9/50	38.4ug	6/50)	thy:cca,ccr. S
a	TR377	3.08ug	.178mg	2/50	3.83ug	9/50	(12.7ug	7/50	38.4ug	6/50)	S
b	TR377	28.8ug	n.s.s.	48/50	3.83ug	49/50	12.7ug	47/50	38.4ug	48/50	
c	TR377	.176mg	n.s.s.	4/50	3.83ug	3/50	12.7ug	0/50	38.4ug	2/50	liv:hpa,hpc,nnd.

CHLOROBENZENE   (monochlorobenzene) 108-90-7

	RefNum	LoConf	UpConf	Cntrl	1Dose	1Inc	2Dose	2Inc		Citation or Pathology Brkly Code
1091	TR261	82.2mg	n.s.s.	25/50	42.0mg	26/50	84.1mg	29/50		
a	TR261	79.9mg	n.s.s.	2/50	42.0mg	7/50	(84.1mg	2/50)		liv:hpa,hpc,nnd.
b	TR261	223.mg	n.s.s.	1/50	42.0mg	2/50	84.1mg	5/50		lun:a/a,a/c.
1092	TR261	22.3mg	n.s.s.	30/50	21.0mg	31/50	42.0mg	35/50		
a	TR261	48.4mg	n.s.s.	16/50	21.0mg	15/50	42.0mg	14/50		liv:hpa,hpc,nnd.
b	TR261	62.2mg	n.s.s.	6/50	21.0mg	4/50	42.0mg	10/50		lun:a/a,a/c.
1093	TR261	198.mg	n.s.s.	0/50	42.4mg	2/50	84.9mg	3/50	pty:adn;	thy:fcc; tyf:cyn. S
a	TR261	64.1mg	n.s.s.	36/50	42.4mg	30/50	84.9mg	36/50		
b	TR261	263.mg	n.s.s.	0/50	42.4mg	1/50	84.9mg	2/50		liv:hpa,hpc,nnd.
1094	TR261	109.mg	n.s.s.	2/50	42.4mg	4/50	84.9mg	8/50		
a	TR261	48.4mg	n.s.s.	33/50	42.4mg	36/50	84.9mg	28/50		
b	TR261	119.mg	n.s.s.	4/50	42.4mg	4/50	84.9mg	8/50		liv:hpa,hpc,nnd.

CHLOROBENZILATE   510-15-6

	RefNum	LoConf	UpConf	Cntrl	1Dose	1Inc	2Dose	2Inc		Citation or Pathology Brkly Code
1095	TR75	523.mg	4.51gm	0/20	356.mg	11/50	658.mg	13/50		
a	TR75	623.mg	n.s.s.	5/20	356.mg	18/50	658.mg	15/50		
b	TR75	523.mg	4.51gm	0/20	356.mg	11/50	658.mg	13/50		liv:hpa,hpc,nnd.
c	TR75	2.93gm	n.s.s.	2/20	356.mg	2/50	658.mg	2/50		lun:a/a,a/c.

```
        Spe Strain  Site   Xpo+Xpt                                                          TD50    2Tailpvl
             Sex   Route   Hist     Notes                                                       DR    AuOp
 1096 M f b6c orl lun ade 82w82 evx                                        .>               621.mg   P<.3
    a M f b6c orl liv hpt 82w82 evx                                                          no dre  P=1.
    b M f b6c orl tba mix 82w82 evx                                                          302.mg  P<.2
 1097 M m b6c eat liv hpc 78w90 dv                                              :  ±        235.mg \ P<.03  c
    a M m b6c eat TBA MXB 78w90 dv                                                           256.mg \ P<.05
    b M m b6c eat liv MXB 78w90 dv                                                           235.mg \ P<.03
    c M m b6c eat lun MXB 78w90 dv                                                           no dre  P=1.
 1098 M m b6c orl liv hpt 82w82 evx                                     .  +  .              43.8mg  P<.0005+
    a M m b6c orl lun mix 82w82 evx                                                          no dre  P=1.
    b M m b6c orl tba mix 82w82 evx                                                          31.7mg  P<.0005
 1099 M f b6a orl lun ade 82w82 evx                                           .>             621.mg  P<.6
    a M f b6a orl liv hpt 82w82 evx                                                          no dre  P=1.
    b M f b6a orl tba mix 82w82 evx                                                          621.mg  P<.7
 1100 M m b6a orl liv hpt 82w82 evx                                    .  +       .          69.8mg  P<.01  +
    a M m b6a orl lun ade 82w82 evx                                                          no dre  P=1.
    b M m b6a orl tba mix 82w82 evx                                                          72.8mg  P<.06
 1101 R f osm eat adr coa 18m26 dv                                          :  +           #490.mg * P<.01  -
    a R f osm eat TBA MXB 18m26 dv                                                           no dre  P=1.
    b R f osm eat liv MXB 18m26 dv                                                           no dre  P=1.
 1102 R m osm eat adr coa 18m26 dv                                       :  +         :    #186.mg \ P<.007 -
    a R m osm eat TBA MXB 18m26 dv                                                           no dre  P=1.
    b R m osm eat liv MXB 18m26 dv                                                           1.60gm * P<.4
 1103 R f cwf eat mgl adf 24m24                                             .>               no dre  P=1.    -

CHLORODIBROMOMETHANE              100ng..:..1ug....:..10....:..100...:..1mg...:..10....:..100...:..1g.....:..10
 1104 M f b6c gav liv MXA 24m25                                               :  ±          139.mg * P<.03  p
    a M f b6c gav liv hpa 24m25                                                              211.mg * P<.02  p
    b M f b6c gav TBA MXB 24m25                                                              no dre  P=1.
    c M f b6c gav liv MXB 24m25                                                              139.mg * P<.03
    d M f b6c gav lun MXB 24m25                                                              no dre  P=1.
 1105 M m b6c gav liv MXA 24m25 ns                                    :   +        :         20.7mg \ P<.006
    a M m b6c gav liv hpc 24m25 ns                                                           33.5mg \ P<.006 e
    b M m b6c gav TBA MXB 24m25 ns                                                           11.3mg \ P<.002
    c M m b6c gav liv MXB 24m25 ns                                                           20.7mg \ P<.006
    d M m b6c gav lun MXB 24m25 ns                                                           115.mg \ P<.4
 1106 R f f34 gav TBA MXB 24m24                                      :>                      no dre  P=1.    -
    a R f f34 gav liv MXB 24m24                                                              591.mg * P<.1
 1107 R m f34 gav TBA MXB 24m24                                     :>                       no dre  P=1.    -
    a R m f34 gav liv MXB 24m24                                                              150.mg \ P<.2

CHLORODIFLUOROMETHANE             100ng..:..1ug....:..10....:..100...:..1mg...:..10....:..100...:..1g.....:..10
 1108 M f swi inh lun ade 18m24                                                              no dre  P=1.    -
    a M f swi inh tba mal 18m24                                                              no dre  P=1.
    b M f swi inh tba mix 18m24                                                              no dre  P=1.
 1109 M m swi inh lun ade 18m24                                                              no dre  P=1.
    a M m swi inh tba mix 18m24                                                              no dre  P=1.
    b M m swi inh tba mal 18m24                                                              98.3gm * P<.8
 1110 R f sda inh liv ang 24m24                                                              no dre  P=1.    -
    a R f sda inh tba mix 24m24                                                              26.0gm * P<1.
    b R f sda inh tba mal 24m24                                                              20.1gm * P<.8   -
 1111 R m sda inh liv ang 24m24                                               .>             no dre  P=1.    -
    a R m sda inh tba mix 24m24                                                              6.54gm * P<.6   -
    b R m sda inh tba mal 24m24                                                              no dre  P=1.

CHLOROETHANE                      100ng..:..1ug....:..10....:..100...:..1mg...:..10....:..100...:..1g.....:..10
 1112 M f b6c inh ute car 23m23                                                  :  +  :     1.81gm  P<.0005c
    a M f b6c inh liv MXA 23m23                                                              7.22gm  P<.0005e
    b M f b6c inh liv hpc 23m23                                                              7.58gm  P<.0005
    c M f b6c inh --- MXA 23m23                                                              9.96gm  P<.002
    d M f b6c inh lun MXA 23m23                                                              8.64gm  P<.02
    e M f b6c inh TBA MXB 23m23                                                              1.38gm  P<.0005
    f M f b6c inh liv MXB 23m23                                                              7.22gm  P<.0005
    g M f b6c inh lun MXB 23m23                                                              8.64gm  P<.02
 1113 M m b6c inh lun MXA 23m23 s                                                 :         #9.91gm  P<.005 i
    a M m b6c inh lun a/a 23m23 s                                                            12.7gm  P<.005
    b M m b6c inh TBA MXB 23m23 s                                                            10.1gm  P<.2
    c M m b6c inh liv MXB 23m23 s                                                            34.1gm  P<.5
    d M m b6c inh lun MXB 23m23 s                                                            9.91gm  P<.005
 1114 R f f34 inh ute esp 24m24                                                        :     10.2gm  P<.05
    a R f f34 inh bra asl 24m24                                                              21.0gm  P<.04 e
    b R f f34 inh TBA MXB 24m24                                                              2.66gm  P<.2
    c R f f34 inh liv MXB 24m24                                                              no dre  P=1.
 1115 R m f34 inh ski MXA 24m24                                                  :  +        3.21gm  P<.004 e
    a R m f34 inh ski bcc 24m24                                                              4.97gm  P<.02
    b R m f34 inh TBA MXB 24m24                                                              1.60gm  P<.3
    c R m f34 inh liv MXB 24m24                                                              47.7gm  P<.8

(2-CHLOROETHYL)TRIMETHYLAMMONIUM CHLORIDE ..1ug....:..10....:..100...:..1mg...:..10....:..100...:..1g.....:..10
 1116 M f b6c eat TBA MXB 24m24                                                  :>          no dre  P=1.    -
    a M f b6c eat liv MXB 24m24                                                              no dre  P=1.
    b M f b6c eat lun MXB 24m24                                                              no dre  P=1.
```

RefNum	LoConf	UpConf	Cntrl	1Dose	1Inc	2Dose	2Inc	Citation or Pathology / Brkly Code
1096	67a	101.mg	n.s.s.	0/16	83.4mg	1/18			Innes;ntis,1968/1969
a	67a	192.mg	n.s.s.	0/16	83.4mg	0/18			
b	67a	74.1mg	n.s.s.	0/16	83.4mg	2/18			
1097	TR75	114.mg	n.s.s.	4/20	440.mg	32/50	(773.mg | 22/50)	
a	TR75	117.mg	n.s.s.	5/20	440.mg	34/50	(773.mg | 25/50)	liv:hpa,hpc,nnd.
b	TR75	114.mg	n.s.s.	4/20	440.mg	32/50	(773.mg | 22/50)	lun:a/a,a/c.
c	TR75	4.20gm	n.s.s.	1/20	440.mg	1/50	773.mg	1/50	
1098	67a	20.1mg	124.mg	0/16	77.6mg	9/17			Innes;ntis,1968/1969
a	67a	169.mg	n.s.s.	0/16	77.6mg	0/17			
b	67a	15.2mg	79.6mg	0/16	77.6mg	11/17			
1099	67a	81.7mg	n.s.s.	1/17	83.4mg	2/18			
a	67a	192.mg	n.s.s.	0/17	83.4mg	0/18			
b	67a	69.7mg	n.s.s.	2/17	83.4mg	3/18			
1100	67a	27.6mg	4.09gm	1/18	77.6mg	7/17			
a	67a	169.mg	n.s.s.	2/18	77.6mg	0/17			
b	67a	25.8mg	n.s.s.	3/18	77.6mg	8/17			
1101	TR75	211.mg	28.5gm	0/50	41.0mg	2/50	78.0mg	5/50	S
a	TR75	81.4mg	n.s.s.	36/50	41.0mg	38/50	78.0mg	34/50	
b	TR75	635.mg	n.s.s.	1/50	41.0mg	0/50	78.0mg	1/50	liv:hpa,hpc,nnd.
1102	TR75	75.1mg	2.25gm	0/50	45.6mg	6/50	(84.8mg | 3/50)	S
a	TR75	145.mg	n.s.s.	28/50	45.6mg	31/50	84.8mg	23/50	
b	TR75	395.mg	n.s.s.	0/50	45.6mg	1/50	84.8mg	1/50	liv:hpa,hpc,nnd.
1103	67	73.2mg	n.s.s.	3/20	25.0mg	1/20			Horn;agfc,3,752-756;1955

CHLORODIBROMOMETHANE   (dibromochloromethane)   124-48-1

RefNum	LoConf	UpConf	Cntrl	1Dose	1Inc	2Dose	2Inc	Citation or Pathology / Brkly Code
								liv:hpa,hpc.
1104	TR282	64.4mg	n.s.s.	6/50	35.0mg	10/50	70.4mg	19/50	
a	TR282	96.2mg	n.s.s.	2/50	35.0mg	4/50	70.4mg	11/50	
b	TR282	61.6mg	n.s.s.	33/50	35.0mg	32/50	70.4mg	37/50	
c	TR282	64.4mg	n.s.s.	6/50	35.0mg	10/50	70.4mg	19/50	liv:hpa,hpc,nnd.
d	TR282	276.mg	n.s.s.	5/50	35.0mg	0/50	70.4mg	5/50	lun:a/a,a/c.
1105	TR282	8.25mg	294.mg	23/50	35.0mg	14/50	(70.8mg | 27/50)	liv:hpa,hpc. S
a	TR282	12.5mg	491.mg	10/50	35.0mg	9/50	(70.8mg | 19/50)	
b	TR282	5.19mg	58.7mg	41/50	35.0mg	23/50	(70.8mg | 31/50)	
c	TR282	8.25mg	294.mg	23/50	35.0mg	14/50	(70.8mg | 27/50)	liv:hpa,hpc,nnd.
d	TR282	22.3mg	n.s.s.	11/50	35.0mg	5/50	(70.8mg | 4/50)	lun:a/a,a/c.
1106	TR282	23.7mg	n.s.s.	43/50	28.3mg	38/50	(56.6mg | 32/50)	
a	TR282	204.mg	n.s.s.	0/50	28.3mg	2/50	56.6mg	2/50	liv:hpa,hpc,nnd.
1107	TR282	31.5mg	n.s.s.	43/50	28.3mg	37/50	(56.6mg | 29/50)	
a	TR282	46.3mg	n.s.s.	3/50	28.3mg	8/50	(56.6mg | 3/50)	liv:hpa,hpc,nnd.

CHLORODIFLUOROMETHANE   (fluorocarbon 22)   75-45-6

RefNum	LoConf	UpConf	Cntrl	1Dose	1Inc	2Dose	2Inc	Citation or Pathology / Brkly Code
1108	BT606	30.0gm	n.s.s.	2/60	556.mg	3/60	2.78gm	0/60	Maltoni;anya,534,261-282;1988
a	BT606	8.72gm	n.s.s.	12/60	556.mg	13/60	2.78gm	11/60	
b	BT606	10.4gm	n.s.s.	14/60	556.mg	19/60	2.78gm	11/60	
1109	BT606	13.0gm	n.s.s.	6/60	463.mg	2/60	2.32gm	4/60	
a	BT606	8.02gm	n.s.s.	10/60	463.mg	10/60	2.32gm	9/60	
b	BT606	9.13gm	n.s.s.	1/60	463.mg	5/60	2.32gm	3/60	
1110	BT605	10.6gm	n.s.s.	1/60	176.mg	1/60	882.mg	0/60	
a	BT605	718.mg	n.s.s.	45/60	176.mg	44/60	882.mg	45/60	
b	BT605	2.06gm	n.s.s.	13/60	176.mg	12/60	882.mg	14/60	
1111	BT605	1.27gm	n.s.s.	0/60	124.mg	0/60	618.mg	0/60	
a	BT605	1.11gm	n.s.s.	19/60	124.mg	14/60	618.mg	20/60	
b	BT605	2.23gm	n.s.s.	9/60	124.mg	6/60	618.mg	8/60	

CHLOROETHANE   (ethyl chloride)   75-00-3

RefNum	LoConf	UpConf	Cntrl	1Dose	1Inc	2Dose	2Inc	Citation or Pathology / Brkly Code
1112	TR346	823.mg	3.55gm	1/50	12.4gm	43/50			liv:hpa,hpc.
a	TR346	2.24gm	38.5gm	3/50	12.4gm	8/50			S
b	TR346	2.29gm	46.7gm	3/50	12.4gm	7/50			
c	TR346	2.82gm	59.7gm	4/50	12.4gm	10/50			---:mlh,mlm,mlp,mlu,mly. S
d	TR346	1.97gm	n.s.s.	5/50	12.4gm	4/50			lun:a/a,a/c. S
e	TR346	721.mg	2.85gm	28/50	12.4gm	47/50			
f	TR346	2.24gm	38.5gm	3/50	12.4gm	8/50			liv:hpa,hpc,nnd.
g	TR346	1.97gm	n.s.s.	5/50	12.4gm	4/50			lun:a/a,a/c.
1113	TR346	4.00gm	107.gm	5/50	10.4gm	10/50			lun:a/a,a/c. S
a	TR346	4.89gm	159.gm	3/50	10.4gm	8/50			S
b	TR346	3.22gm	n.s.s.	27/50	10.4gm	20/50			
c	TR346	6.36gm	n.s.s.	15/50	10.4gm	10/50			liv:hpa,hpc,nnd.
d	TR346	4.00gm	107.gm	5/50	10.4gm	10/50			lun:a/a,a/c.
1114	TR346	3.58gm	n.s.s.	2/50	2.88gm	7/50			S
a	TR346	6.08gm	n.s.s.	0/50	2.88gm	3/50			
b	TR346	936.mg	n.s.s.	45/50	2.88gm	48/50			
c	TR346	13.6gm	n.s.s.	1/50	2.88gm	0/50			liv:hpa,hpc,nnd.
1115	TR346	1.15gm	20.6gm	0/50	2.01gm	5/50			ski:bcc,sea,tri.
a	TR346	1.39gm	n.s.s.	0/50	2.01gm	3/50			S
b	TR346	468.mg	n.s.s.	48/50	2.01gm	47/50			
c	TR346	2.58gm	n.s.s.	1/50	2.01gm	1/50			liv:hpa,hpc,nnd.

(2-CHLOROETHYL)TRIMETHYLAMMONIUM CHLORIDE   (CCC) 999-81-5

RefNum	LoConf	UpConf	Cntrl	1Dose	1Inc	2Dose	2Inc	Citation or Pathology / Brkly Code
1116	TR158	404.mg	n.s.s.	14/20	65.0mg	25/50	260.mg	26/50	liv:hpa,hpc,nnd.
a	TR158	1.11gm	n.s.s.	4/20	65.0mg	7/50	260.mg	4/50	lun:a/a,a/c.
b	TR158	1.15gm	n.s.s.	1/20	65.0mg	3/50	260.mg	2/50	

```
     Spe Strain Site   Xpo+Xpt                                          TD50    2Tailpvl
         Sex  Route  Hist    Notes                                         DR    AuOp
─────────────────────────────────────────────────────────────────────────────────────────
1117 M f b6c orl liv hpt 76w76 evx                      .>                no dre   P=1.
a    M f b6c orl lun mix 76w76 evx                                        no dre   P=1.
b    M f b6c orl tba mix 76w76 evx                                        57.4mg   P<.3
1118 M m b6c eat TBA MXB 24m24                                    :>      3.24gm * P<.9    -
a    M m b6c eat liv MXB 24m24                                           599.mg  * P<.2
b    M m b6c eat lun MXB 24m24                                           no dre    P=1.
1119 M m b6c orl liv hpt 76w76 evx                 .    +     .           9.38mg   P<.009
a    M m b6c orl lun mix 76w76 evx                                        no dre   P=1.
b    M m b6c orl tba mix 76w76 evx                                        5.19mg   P<.0005
1120 M f b6a orl liv hpt 76w76 evx                      .>                no dre   P=1.
a    M f b6a orl lun ade 76w76 evx                                        no dre   P=1.
b    M f b6a orl tba mix 76w76 evx                                        33.5mg   P<.6
1121 M m b6a orl liv hpt 76w76 evx                 .    ±                 11.4mg   P<.07
a    M m b6a orl lun ade 76w76 evx                                        no dre   P=1.
b    M m b6a orl tba mix 76w76 evx                                        21.3mg   P<.5
1122 R f f34 eat TBA MXB 25m25                                 :>         no dre   P=1.    -
a    R f f34 eat liv MXB 25m25                                            8.09gm * P<.4
1123 R m f34 eat pni isa 25m25                              :   ±         #458.mg* P<.02   -
a    R m f34 eat TBA MXB 25m25                                            3.36gm * P<1.
b    R m f34 eat liv MXB 25m25                                            no dre   P=1.


1-CHLOROETHYLNITROSO-3-(2-HYDROXYPROPYL)UREA  ....:..10.....:..100....:..1mg...:..10....:..100...:..1g....:..10
1124 R f f34 wat liv tum 7m24 e                    .>                     no dre   P=1.
1125 R m f34 wat lun abt 7m24 e          .    +         .                 .124mg   P<.005  +
a    R m f34 wat liv hct 7m24 e                                          .368mg   P<.09


CHLOROFLUOROMETHANE       100ng..:..1ug...:..10.....:..100....:..1mg....:..10....:..100...:..1g.....:..10
1126 R f aap gav sto mix 12m25 e  pool               .   +   .            26.5mg   P<.0005+
a    R f aap gav sto sqc 12m25 e                                          69.4mg   P<.0005+
b    R f aap gav sto fbs 12m25 e                                          119.mg   P<.0005+
c    R f aap gav sto car 12m25 e                                          2.71gm   P<.1    +
d    R f aap gav sto sar 12m25 e                                          2.71gm   P<.1    +
1127 R m aap gav sto mix 12m23 e  pool               .   +   .            28.5mg   P<.0005+
a    R m aap gav sto sqc 12m23 e                                          46.9mg   P<.0005+
b    R m aap gav sto fbs 12m23 e                                          88.1mg   P<.0005+


CHLOROFORM                100ng..:..1ug...:..10.....:..100....:..1mg....:..10....:..100...:..1g.....:..10
1128 D f beg eat liv tum 87m92 e             .>                           no dre   P=1.    -
a    D f beg eat tba mix 87m92 e                                          10.0mg \ P<.3    -
1129 D m beg eat liv tum 87m92 e             .>                           no dre   P=1.    -
a    D m beg eat tba mix 87m92 e                                          13.3mg * P<.008  -
1130 M f b6c gav liv hpc 78w92 v                   :+ :                   48.0mg * P<.0005c
a    M f b6c gav TBA MXB 78w92 v                                          54.1mg * P<.0005
b    M f b6c gav TBA MXB 78w92 v                                          48.0mg * P<.0005
c    M f b6c gav TBA MXB 78w92 v                                          no dre   P=1.
1131 M f b6c wat liv hpc 24m24 e                            .>            20.2gm * P<.5    -
a    M f b6c wat liv hpa 24m24 e                                          no dre   P=1.    -
b    M f b6c wat liv mix 24m24 e                                          no dre   P=1.    -
c    M f b6c wat tba mix 24m24 e                                          no dre   P=1.    -
1132 M m b6c gav liv hpc 78w92 v                   :  +  :                56.2mg / P<.0005c
a    M m b6c gav TBA MXB 78w92 v                                          71.1mg * P<.004
b    M m b6c gav TBA MXB 78w92 v                                          56.2mg / P<.0005
c    M m b6c gav TBA MXB 78w92 v                                          no dre   P=1.
1133 M m c51 gav lun mix 19m24 e                        .>                728.mg   P<.5    -
a    M m c51 gav liv tum 19m24 e                                          no dre   P=1.    -
b    M m c51 gav tba mix 19m24 e                                          no dre   P=1.    -
1134 M m cba gav liv mix 19m24 e                     .>                   no dre   P=1.    -
a    M m cba gav lun mix 19m24 e                                          no dre   P=1.    -
b    M m cba gav tba mix 19m24 e                                          no dre   P=1.    -
1135 M m cf1 gav lun mix 80w93 e                     .>                   376.mg   P<.6    -
a    M m cf1 gav liv tum 80w93 e                                          1.43gm   P<.9    -
b    M m cf1 gav tba mix 80w93 e                                          140.mg   P<.4    -
1136 M m cf1 gav kid mix 19m24 e                   .   +    .             153.mg   P<.0005+
a    M m cf1 gav lun tum 19m24 e                                          4.35gm   P<1.    -
b    M m cf1 gav liv tum 19m24 e                                          no dre   P=1.    -
c    M m cf1 gav tba mix 19m24 e                                          no dre   P=1.
1137 M f ici gav lun tum 80w96 e                     .>                   316.mg * P<.3    -
a    M f ici gav liv tum 80w96 e                                          no dre   P=1.    -
b    M f ici gav tba mix 80w96 e                                          no dre   P=1.    -
1138 M m ici gav kid mix 80w96 e                   .   +    .             139.mg * P<.0005+
a    M m ici gav liv tum 80w96 e                                          324.mg * P<.4    -
b    M m ici gav lun tum 80w96 e                                          9.83gm * P<1.    -
c    M m ici gav tba mix 80w96 e                                          45.4mg * P<.006
1139 M m ici gav kid mix 80w98 e                   .    ±                 278.mg   P<.08   +
a    M m ici gav lun mix 80w98 e                                          336.mg   P<.3    -
b    M m ici gav liv mix 80w98 e                                          437.mg   P<.6    -
c    M m ici gav tba mix 80w98 e                                          88.5mg   P<.2
1140 M m ici gav kid mix 80w98 e                   .   +    .             95.5mg   P<.0005+
a    M m ici gav liv mix 80w98 e                                          no dre   P=1.    -
b    M m ici gav lun mix 80w98 e                                          no dre   P=1.    -
c    M m ici gav tba mix 80w98 e                                          no dre   P=1.
```

	RefNum	LoConf	UpConf	Cntrl	1Dose	1Inc	2Dose	2Inc					Citation or Pathology
													Brkly Code
1117	1103	17.8mg	n.s.s.	0/16	8.96mg	0/18							Innes;ntis,1968/1969
a	1103	17.8mg	n.s.s.	0/16	8.96mg	0/18							
b	1103	9.34mg	n.s.s.	0/16	8.96mg	1/18							
1118	TR158	218.mg	n.s.s.	12/20	60.0mg	29/50	240.mg	29/50					
a	TR158	214.mg	n.s.s.	7/20	60.0mg	13/50	240.mg	23/50					liv:hpa,hpc,nnd.
b	TR158	806.mg	n.s.s.	4/20	60.0mg	9/50	240.mg	5/50					lun:a/a,a/c.
1119	1103	3.53mg	206.mg	0/16	8.34mg	5/18							Innes;ntis,1968/1969
a	1103	16.5mg	n.s.s.	0/16	8.34mg	0/18							
b	1103	2.31mg	17.1mg	0/16	8.34mg	8/18							
1120	1103	14.8mg	n.s.s.	0/17	8.96mg	0/15							
a	1103	14.8mg	n.s.s.	1/17	8.96mg	0/15							
b	1103	5.11mg	n.s.s.	2/17	8.96mg	3/15							
1121	1103	3.73mg	n.s.s.	1/18	8.34mg	5/18							
a	1103	10.9mg	n.s.s.	2/18	8.34mg	1/18							
b	1103	4.25mg	n.s.s.	3/18	8.34mg	5/18							
1122	TR158	135.mg	n.s.s.	13/20	75.0mg	41/50	150.mg	37/50					
a	TR158	1.32gm	n.s.s.	0/20	75.0mg	1/50	150.mg	1/50					liv:hpa,hpc,nnd.
1123	TR158	216.mg	n.s.s.	0/20	60.0mg	2/50	120.mg	7/50					S
a	TR158	83.9mg	n.s.s.	16/20	60.0mg	37/50	120.mg	36/50					
b	TR158	487.mg	n.s.s.	1/20	60.0mg	2/50	120.mg	2/50					liv:hpa,hpc,nnd.

1-CHLOROETHYLNITROSO-3-(2-HYDROXYPROPYL)UREA   ---

	RefNum	LoConf	UpConf	Cntrl	1Dose	1Inc	2Dose	2Inc					Citation or Pathology
1124	2010	.346mg	n.s.s.	0/12	.140mg	0/12							Lijinsky;vivo,4,1-6;1990
1125	2010	46.3ug	.997mg	0/12	97.8ug	5/12							
a	2010	90.0ug	n.s.s.	0/12	97.8ug	2/12							

CHLOROFLUOROMETHANE   (fluorocarbon 31) 593-70-4

	RefNum	LoConf	UpConf	Cntrl	1Dose	1Inc	2Dose	2Inc					Citation or Pathology
1126	1623	14.0mg	48.8mg	1/104p	103.mg	34/36							Longstaff;txap,72,15-31;1984/pers.comm.
a	1623	41.4mg	126.mg	0/104p	103.mg	24/36							
b	1623	66.8mg	242.mg	0/104p	103.mg	17/36							
c	1623	441.mg	n.s.s.	0/104p	103.mg	1/36							
d	1623	441.mg	n.s.s.	0/104p	103.mg	1/36							
1127	1623	16.0mg	51.3mg	1/104p	111.mg	33/36							
a	1623	28.2mg	82.9mg	0/104p	111.mg	28/36							
b	1623	50.8mg	171.mg	1/104p	111.mg	20/36							

CHLOROFORM   67-66-3

	RefNum	LoConf	UpConf	Cntrl	1Dose	1Inc	2Dose	2Inc	3Dose	3Inc	4Dose	4Inc	Citation or Pathology
1128	1003	6.52mg	n.s.s.	0/16	12.2mg	0/8	24.4mg	0/8					Heywood;jept,2,835-851;1979
a	1003	2.22mg	n.s.s.	4/16	12.2mg	4/8	(24.4mg	0/8)					
1129	1003	6.52mg	n.s.s.	0/16	12.2mg	0/8	24.4mg	0/8					
a	1003	5.39mg	230.mg	0/16	12.2mg	4/8	24.4mg	2/8					
1130	TR-A	35.2mg	68.0mg	0/20	144.mg	36/50	289.mg	39/50					
a	TR-A	37.8mg	91.8mg	2/20	144.mg	37/50	289.mg	39/50					
b	TR-A	35.2mg	68.0mg	0/20	144.mg	36/50	289.mg	39/50					
c	TR-A	n.s.s.	n.s.s.	0/20	144.mg	1/50	289.mg	0/50					
1131	1671	2.69gm	n.s.s.	2/415	40.0mg	7/410	80.0mg	1/142	180.mg	0/47	360.mg	1/44	Jorgenson;faat,5,760-769;1985
a	1671	5.18gm	n.s.s.	19/415	40.0mg	8/410	80.0mg	8/142	180.mg	0/47	360.mg	0/44	
b	1671	3.73gm	n.s.s.	21/415	40.0mg	15/410	80.0mg	9/142	180.mg	0/47	360.mg	1/44	
c	1671	664.mg	n.s.s.	225/423	40.0mg	217/415	80.0mg	90/142	180.mg	16/47	360.mg	24/44	
1132	TR-A	38.7mg	118.mg	1/20	83.6mg	19/50	168.mg	44/50					
a	TR-A	40.4mg	491.mg	4/20	83.6mg	26/50	168.mg	44/50					
b	TR-A	38.7mg	118.mg	1/20	83.6mg	19/50	168.mg	44/50					
c	TR-A	490.mg	n.s.s.	1/20	83.6mg	3/50	168.mg	2/50					
1133	710m	134.mg	n.s.s.	2/46	39.6mg	4/51							Roe;jept,2,799-819;1979
a	710m	416.mg	n.s.s.	2/46	39.6mg	0/51							
b	710m	112.mg	n.s.s.	16/46	39.6mg	13/51							
1134	710m	70.7mg	n.s.s.	37/51	39.6mg	29/51							
a	710m	155.mg	n.s.s.	13/51	39.6mg	8/51							
b	710m	64.7mg	n.s.s.	42/51	39.6mg	33/51							
1135	710m	64.2mg	n.s.s.	9/45	44.2mg	12/48							
a	710m	111.mg	n.s.s.	4/45	44.2mg	5/48							
b	710m	37.2mg	n.s.s.	16/45	44.2mg	22/48							
1136	710m	66.6mg	613.mg	6/240	39.6mg	9/49							
a	710m	65.8mg	n.s.s.	102/240	39.6mg	21/49							
b	710m	189.mg	n.s.s.	69/240	39.6mg	8/49							
c	710m	68.4mg	n.s.s.	170/240	39.6mg	30/49							
1137	710m	81.3mg	n.s.s.	5/59	12.1mg	2/35	42.9mg	6/38					
a	710m	59.2mg	n.s.s.	1/59	12.1mg	0/35	42.9mg	0/38					
b	710m	71.5mg	n.s.s.	29/59	12.1mg	10/35	42.9mg	15/38					
1138	710m	62.8mg	407.mg	0/72	12.1mg	0/37	42.9mg	8/38					
a	710m	73.5mg	n.s.s.	5/72	12.1mg	6/37	42.9mg	5/38					
b	710m	95.4mg	n.s.s.	7/72	12.1mg	7/37	42.9mg	4/38					
c	710m	21.0mg	643.mg	20/72	12.1mg	20/37	42.9mg	21/38					
1139	710n	91.2mg	n.s.s.	1/49	42.0mg	5/47							
a	710n	83.7mg	n.s.s.	4/49	42.0mg	7/47							
b	710n	78.3mg	n.s.s.	7/49	42.0mg	9/47							
c	710n	31.8mg	n.s.s.	17/49	42.0mg	24/47							
1140	710o	47.1mg	325.mg	1/50	42.0mg	12/48							
a	710o	105.mg	n.s.s.	9/50	42.0mg	8/48							
b	710o	182.mg	n.s.s.	5/50	42.0mg	3/48							
c	710o	66.8mg	n.s.s.	24/50	42.0mg	20/48							

```
         Spe Strain  Site    Xpo+Xpt                                                    TD50    2Tailpvl
           Sex  Route   Hist    Notes                                                          DR      AuOp
1141 R f osm gav thy fcc 18m26 v                                    :    +        :      #126.mg * P<.004 -
   a R f osm gav TBA MXB 18m26 v                                                         68.2mg * P<.05
   b R f osm gav TBA MXB 18m26 v                                                         1.13gm * P<.7
1142 R m osm gav kid uac 18m26                                      :    +    :          119.mg / P<.0005c
   a R m osm gav TBA MXB 18m26                                                           194.mg * P<.5
   b R m osm gav TBA MXB 18m26                                                           455.mg * P<.08
1143 R m osm wat kid mix 24m24 e                                          .    +    .    519.mg * P<.0005+
   a R m osm wat kid mix 24m24 e                                                         606.mg * P<.0005
   b R m osm wat kid tla 24m24 e                                                         972.mg * P<.002
   c R m osm wat --- nfm 24m24 e                                                         2.20gm * P<.04
   d R m osm wat --- mix 24m24 e                                                         1.36gm * P<.3
   e R m osm wat tba mix 24m24 e                                                         1.97gm * P<.9
1144 R f sda gav liv cye 80w95 e                                          .>             1.20gm * P<.3   -
   a R f sda gav tba mix 80w95 e                                                         78.3mg   P<.2   -
   b R f sda gav tba mal 80w95 e                                                         276.mg   P<.2   -
1145 R m sda gav liv tum 80w95 e                                            .>           no dre   P=1.   -
   a R m sda gav tba mix 80w95 e                                                         no dre   P=1.   -
   b R m sda gav tba mal 80w95 e                                                         no dre   P=1.   -
1146 R f wis wat liv nnd 43m43 ev                                         .    +    .    883.mg   P<.005 +
   a R f wis wat liv hpc 43m43 ev                                                        10.0gm   P<.4
   b R f wis wat kid tum 43m43 ev                                                        no dre   P=1.
1147 R m wis wat kid adc 40m40 ev                                              .>        5.30gm   P<.3   +
   a R m wis wat kid ade 40m40 ev                                                        5.30gm   P<.3   +
   b R m wis wat liv hpc 40m40 ev                                                        5.30gm   P<.3
   c R m wis wat liv nnd 40m40 ev                                                        no dre   P=1.


CHLOROMETHYL METHYL ETHER          100ng..:..1ug...:..10.....:..100....:..1mg....:..10.....:..100...:..1g.....:..10
1148 H m syg inh mix mix 26m29                             .     ±                       16.4mg   P<.1   +
   a H m syg inh lun adc 26m29                                                           32.9mg   P<.3
1149 R m sda inh mix mix 26m28                          .      ±                         5.50mg   P<.1   +


2-(CHLOROMETHYL)PYRIDINE.HCl       100ng..:..1ug...:..10.....:..100....:..1mg....:..10.....:..100...:..1g.....:..10
1150 M f b6c gav TBA MXB 23m24                                           :>              9.79gm * P<1.   -
   a M f b6c gav liv MXB 23m24                                                           no dre   P=1.
   b M f b6c gav lun MXB 23m24                                                           2.17gm * P<.6
1151 M m b6c gav TBA MXB 23m24                                           :>              no dre   P=1.   -
   a M m b6c gav liv MXB 23m24                                                           no dre   P=1.
   b M m b6c gav lun MXB 23m24                                                           2.94gm * P<.9
1152 R f f34 gav TBA MXB 23m24                                     :>                    114.mg * P<.4   -
   a R f f34 gav liv MXB 23m24                                                           no dre   P=1.
1153 R m f34 gav sub fib 23m24                                 :       ±                 #515.mg * P<.03 -
   a R m f34 gav TBA MXB 23m24                                                           671.mg * P<.9
   b R m f34 gav liv MXB 23m24                                                           no dre   P=1.


3-(CHLOROMETHYL)PYRIDINE.HCl       100ng..:..1ug...:..10.....:..100....:..1mg....:..10.....:..100...:..1g.....:..10
1154 M f b6c gav sto MXA 21m24 as                                        :   ±          +hist 394.mg * P<.04 c
   a M f b6c gav TBA MXB 21m24 as                                                        141.mg * P<.2
   b M f b6c gav liv MXB 21m24 as                                                        16.3gm * P<1.
   c M f b6c gav lun MXB 21m24 as                                                        1.97gm * P<.7
1155 M m b6c gav sto MXA 21m24 as                                        :   +    :      161.mg / P<.003 c
   a M m b6c gav TBA MXB 21m24 as                                                        64.9mg * P<.02
   b M m b6c gav liv MXB 21m24 as                                                        324.mg * P<.4
   c M m b6c gav lun MXB 21m24 as                                                        624.mg * P<.6
1156 R f f34 gav TBA MXB 22m24 as                                  :>                    160.mg * P<.6   -
   a R f f34 gav liv MXB 22m24 as                                                        no dre   P=1.
1157 R m f34 gav sto MXA 22m24 as                                        :   ±          +hist 433.mg * P<.07 c
   a R m f34 gav TBA MXB 22m24 as                                                        no dre   P=1.
   b R m f34 gav liv MXB 22m24 as                                                        no dre   P=1.


p-CHLOROPHENYL-p-CHLOROBENZENE SULFONATE.:..1ug...:..10.....:..100....:..1mg....:..10.....:..100...:..1g.....:..10
1158 M f b6a orl lun ade 76w76 evx                                    .>                 no dre   P=1.   -
   a M f b6a orl liv hpt 76w76 evx                                                       no dre   P=1.   -
   b M f b6a orl tba mix 76w76 evx                                                       no dre   P=1.   -
1159 M m b6a orl lun ade 76w76 evx                                       .>              no dre   P=1.   -
   a M m b6a orl liv hpt 76w76 evx                                                       no dre   P=1.   -
   b M m b6a orl tba mix 76w76 evx                                                       no dre   P=1.   -
1160 M f b6c orl liv hpt 76w76 evx                                        .>             931.mg   P<.3   -
   a M f b6c orl lun mix 76w76 evx                                                       no dre   P=1.
   b M f b6c orl tba mix 76w76 evx                                                       931.mg   P<.3
1161 M m b6c orl liv hpt 76w76 evx                                       .>              422.mg   P<.2   -
   a M m b6c orl lun ade 76w76 evx                                                       869.mg   P<.3   -
   b M m b6c orl tba mix 76w76 evx                                                       84.5mg   P<.0005-


3-(p-CHLOROPHENYL)-1,1-DIMETHYLUREA  ..:..1ug..:..10.....:..100....:..1mg....:..10.....:..100...:..1g.....:..10
1162 M f b6c eat TBA MXB 24m24                                          :>               no dre   P=1.   -
   a M f b6c eat liv MXB 24m24                                                           no dre   P=1.
   b M f b6c eat lun MXB 24m24                                                           no dre   P=1.
1163 M f b6c orl liv hpt 76w76 evx                                        .>             no dre   P=1.   -
   a M f b6c orl lun mix 76w76 evx                                                       no dre   P=1.
   b M f b6c orl tba tum 76w76 evx                                                       no dre   P=1.
1164 M m b6c eat sub MXA 24m24                                                :   ±      #2.10gm \ P<.02 -
```

	RefNum	LoConf	UpConf	Cntrl	1Dose	1Inc	2Dose	2Inc					Citation or Pathology / Brkly Code
1141	TR-A	65.8mg	936.mg	1/20	50.2mg	8/50	100.mg	11/50					S
a	TR-A	28.8mg	n.s.s.	12/20	50.2mg	24/50	100.mg	25/50					
b	TR-A	156.mg	n.s.s.	2/20	50.2mg	4/50	100.mg	3/50					
1142	TR-A	65.5mg	334.mg	0/20	45.2mg	4/50	90.3mg	12/50					
a	TR-A	43.5mg	n.s.s.	9/20	45.2mg	24/50	90.3mg	20/50					
b	TR-A	157.mg	n.s.s.	2/20	45.2mg	1/50	90.3mg	3/50					
1143	1671	265.mg	1.79gm	5/301	10.0mg	6/313	20.0mg	7/148	45.0mg	3/48	90.0mg	7/50	Jorgenson;faat,5,760-769;1985
a	1671	305.mg	1.97gm	4/301	10.0mg	4/313	20.0mg	4/148	45.0mg	3/48	90.0mg	7/50	
b	1671	420.mg	6.25gm	4/301	10.0mg	2/313	20.0mg	3/148	45.0mg	2/48	90.0mg	5/50	
c	1671	676.mg	n.s.s.	2/303	10.0mg	2/316	20.0mg	1/148	45.0mg	0/48	90.0mg	3/50	
d	1671	343.mg	n.s.s.	5/303	10.0mg	19/316	20.0mg	5/148	45.0mg	2/48	90.0mg	3/50	
e	1671	90.2mg	n.s.s.	212/303	10.0mg	227/316	20.0mg	105/148	45.0mg	38/48	90.0mg	34/50	
1144	711	196.mg	n.s.s.	0/50	43.3mg	1/49							Palmer;jept,2,821-833;1979
a	711	27.1mg	n.s.s.	22/50	43.3mg	29/49							
b	711	86.0mg	n.s.s.	2/50	43.3mg	6/49							
1145	711	365.mg	n.s.s.	0/48	43.3mg	0/49							
a	711	115.mg	n.s.s.	12/48	43.3mg	9/49							
b	711	195.mg	n.s.s.	6/48	43.3mg	3/49							
1146	1681	429.mg	5.42gm	0/18	115.mg	10/40							Tumasonis;eaes,9,233-240;1985
a	1681	1.63gm	n.s.s.	0/18	115.mg	1/40							
b	1681	3.06gm	n.s.s.	0/18	115.mg	0/40							
1147	1681	862.mg	n.s.s.	0/22	103.mg	1/28							
a	1681	862.mg	n.s.s.	0/22	103.mg	1/28							
b	1681	862.mg	n.s.s.	0/22	103.mg	1/28							
c	1681	574.mg	n.s.s.	5/22	103.mg	5/28							

CHLOROMETHYL METHYL ETHER (CMME) 107-30-2

	RefNum	LoConf	UpConf	Cntrl	1Dose	1Inc		Citation or Pathology
1148	348	4.02mg	n.s.s.	0/88	.377mg	2/90		Laskin;aenh,30,70-72;1975
a	348	5.36mg	n.s.s.	0/88	.377mg	1/90		
1149	348	1.35mg	n.s.s.	0/74	.160mg	2/74		

2-(CHLOROMETHYL)PYRIDINE.HCl   6959-47-3

	RefNum	LoConf	UpConf	Cntrl	1Dose	1Inc	2Dose	2Inc	Pathology
1150	TR178	174.mg	n.s.s.	5/20	51.0mg	11/50	102.mg	10/50	
a	TR178	n.s.s.	n.s.s.	0/20	51.0mg	1/50	102.mg	0/50	liv:hpa,hpc,nnd.
b	TR178	368.mg	n.s.s.	1/20	51.0mg	1/50	102.mg	3/50	lun:a/a,a/c.
1151	TR178	162.mg	n.s.s.	9/20	51.0mg	18/50	102.mg	13/50	
a	TR178	266.mg	n.s.s.	3/20	51.0mg	6/50	102.mg	4/50	liv:hpa,hpc,nnd.
b	TR178	210.mg	n.s.s.	2/20	51.0mg	5/50	102.mg	5/50	lun:a/a,a/c.
1152	TR178	32.8mg	n.s.s.	12/20	30.3mg	35/50	60.6mg	39/50	
a	TR178	n.s.s.	n.s.s.	0/20	30.3mg	0/50	60.6mg	0/50	liv:hpa,hpc,nnd.
1153	TR178	194.mg	n.s.s.	0/20	30.3mg	0/50	60.6mg	5/49	S
a	TR178	54.0mg	n.s.s.	13/20	30.3mg	28/50	60.6mg	29/49	
b	TR178	743.mg	n.s.s.	3/20	30.3mg	0/50	60.6mg	0/49	liv:hpa,hpc,nnd.

3-(CHLOROMETHYL)PYRIDINE.HCl   6959-48-4

	RefNum	LoConf	UpConf	Cntrl	1Dose	1Inc	2Dose	2Inc	Pathology
1154	TR95	161.mg	n.s.s.	0/20	42.0mg	1/50	66.8mg	5/50	sto:sqc,sqp.
a	TR95	56.8mg	n.s.s.	5/20	42.0mg	15/50	66.8mg	20/50	
b	TR95	376.mg	n.s.s.	1/20	42.0mg	0/50	66.8mg	2/50	liv:hpa,hpc,nnd.
c	TR95	269.mg	n.s.s.	1/20	42.0mg	1/50	66.8mg	3/50	lun:a/a,a/c.
1155	TR95	81.9mg	683.mg	0/20	42.0mg	2/50	66.8mg	10/50	sto:sqc,sqp.
a	TR95	32.3mg	n.s.s.	7/20	42.0mg	20/50	66.8mg	28/50	
b	TR95	93.6mg	n.s.s.	3/20	42.0mg	5/50	66.8mg	9/50	liv:hpa,hpc,nnd.
c	TR95	122.mg	n.s.s.	2/20	42.0mg	4/50	66.8mg	5/50	lun:a/a,a/c.
1156	TR95	32.3mg	n.s.s.	14/20	31.8mg	31/49	51.3mg	23/50	liv:hpa,hpc,nnd.
a	TR95	n.s.s.	n.s.s.	0/20	31.8mg	0/49	51.3mg	0/50	
1157	TR95	148.mg	n.s.s.	0/20	31.8mg	1/50	51.3mg	3/50	sto:sqc,sqp.
a	TR95	70.7mg	n.s.s.	14/20	31.8mg	21/50	51.3mg	11/50	
b	TR95	377.mg	n.s.s.	1/20	31.8mg	1/50	51.3mg	0/50	liv:hpa,hpc,nnd.

p-CHLOROPHENYL-p-CHLOROBENZENE SULFONATE (ovex) 80-33-1

	RefNum	LoConf	UpConf	Cntrl	1Dose	1Inc	Citation
1158	1287	162.mg	n.s.s.	1/17	145.mg	1/17	Innes;ntis,1968/1969
a	1287	272.mg	n.s.s.	0/17	145.mg	0/17	
b	1287	128.mg	n.s.s.	2/17	145.mg	2/17	
1159	1287	127.mg	n.s.s.	2/18	136.mg	2/18	
a	1287	160.mg	n.s.s.	1/18	136.mg	1/18	
b	1287	107.mg	n.s.s.	3/18	136.mg	3/18	
1160	1287	152.mg	n.s.s.	0/16	145.mg	1/18	
a	1287	288.mg	n.s.s.	0/16	145.mg	0/18	
b	1287	152.mg	n.s.s.	0/16	145.mg	1/18	
1161	1287	104.mg	n.s.s.	0/16	136.mg	2/18	
a	1287	141.mg	n.s.s.	0/16	136.mg	1/18	
b	1287	37.6mg	278.mg	0/16	136.mg	8/18	

3-(p-CHLOROPHENYL)-1,1-DIMETHYLUREA (telvar, monuron) 150-68-5

	RefNum	LoConf	UpConf	Cntrl	1Dose	1Inc	2Dose	2Inc	Pathology
1162	TR266	656.mg	n.s.s.	30/50	638.mg	21/50	(1.29gm	14/50)	liv:hpa,hpc,nnd.
a	TR266	7.79gm	n.s.s.	6/50	638.mg	0/50	1.29gm	3/50	lun:a/a,a/c.
b	TR266	5.33gm	n.s.s.	6/50	638.mg	7/50	1.29gm	3/50	
1163	1278	145.mg	n.s.s.	0/16	73.0mg	0/18			Innes;ntis,1968/1969
a	1278	145.mg	n.s.s.	0/16	73.0mg	0/18			
b	1278	145.mg	n.s.s.	0/16	73.0mg	0/18			
1164	TR266	864.mg	n.s.s.	1/50	589.mg	8/50	(1.19gm	1/50)	sub:fbs,fib. S

```
    Spe Strain  Site   Xpo+Xpt                                                         TD50    2Tailpvl
        Sex  Route  Hist   Notes                                                          DR      AuOp
    a   M m b6c eat sub MXA 24m24                                                      2.21gm \ P<.04
    b   M m b6c eat TBA MXB 24m24                                                      no dre   P=1.
    c   M m b6c eat liv MXB 24m24                                                      no dre   P=1.
    d   M m b6c eat lun MXB 24m24                                                      17.6gm * P<.7
 1165 M m b6c orl liv hpt 76w76 evx                                    .     ±        112.mg   P<.04
    a   M m b6c orl lun ade 76w76 evx                                                  174.mg   P<.09
    b   M m b6c orl tba mix 76w76 evx                                                  39.6mg   P<.0005
 1166 M f b6a orl lun ade 76w76 evx                                          .>        414.mg   P<.6
    a   M f b6a orl liv hpt 76w76 evx                                                  no dre   P=1.
    b   M f b6a orl tba mix 76w76 evx                                                  387.mg   P<.7
 1167 M m b6a orl lun ade 76w76 evx                                    .     ±         70.7mg   P<.07
    a   M m b6a orl liv hpt 76w76 evx                                                  no dre   P=1.
    b   M m b6a orl tba mix 76w76 evx                                                  86.6mg   P<.2
 1168 R f f34 eat TBA MXB 24m24                                         :>             no dre   P=1.   -
    a   R f f34 eat liv MXB 24m24                                                      no dre   P=1.
 1169 R m f34 eat MXB MXB 24m24                                    :  +  :             86.3mg * P<.0005
    a   R m f34 eat kid MXA 24m24                                                      131.mg * P<.0005c
    b   R m f34 eat kid tla 24m24                                                      272.mg * P<.006 c
    c   R m f34 eat kid uac 24m24                                                      281.mg * P<.003 c
    d   R m f34 eat liv MXA 24m24                                                      201.mg * P<.04  c
    e   R m f34 eat TBA MXB 24m24                                                      no dre   P=1.
    f   R m f34 eat liv MXB 24m24                                                      201.mg * P<.04

1-(4-CHLOROPHENYL)-1-PHENYL-2-PROPYNYL  CARBAMATE...:..10.......:..100.......:..1mg....:..10.......:..100....:..1g......:..10
 1170 R m hrl eat jej adc 72w94 ae                                .  +  .              8.78mg * P<.0005+
    a   R m hrl eat duo adc 72w94 ae                                                   24.9mg * P<.0005+
    b   R m hrl eat liv mix 72w94 ae                                                   58.5mg * P<.2
    c   R m hrl eat bra mix 72w94 ae                            +hist 462.mg * P<.8   +
    d   R m hrl eat tba mix 72w94 ae                                                   3.88mg * P<.0005

p-CHLOROPHENYL-2,4,5-TRICHLOROPHENYL  SULFIDE  ...:..10.......:..100.......:..1mg....:..10.......:..100....:..1g......:..10
 1171 R f wis eat liv tum 27m27                                      .>                no dre   P=1.   -
 1172 R f wis eat liv tum 24m24                              .>                        no dre   P=1.   -
 1173 R m wis eat liv hpt 27m27                                              .>        4.59gm * P<.2   -
 1174 R m wis eat liv tum 24m24                              .>                        no dre   P=1.   -

CHLOROPICRIN          100ng..:..1ug....:..10....:..100....:..1mg....:..10.......:..100....:..1g......:..10
 1175 M f b6c gav TBA MXB 78w91 v                                :>                    82.0mg * P<.2   -
    a   M f b6c gav liv MXB 78w91 v                                                    no dre   P=1.
    b   M f b6c gav lun MXB 78w91 v                                                    111.mg * P<.06
 1176 M m b6c gav TBA MXB 78w91 v                                  :>                  no dre   P=1.
    a   M m b6c gav liv MXB 78w91 v                                                    no dre   P=1.
    b   M m b6c gav lun MXB 78w91 v                                                    no dre   P=1.
 1177 R f osm gav TBA MXB 18m26 dsv                         :>                         no dre   P=1.   i
    a   R f osm gav liv MXB 18m26 dsv                                                  no dre   P=1.
 1178 R m osm gav TBA MXB 18m26 dsv                           :>                       no dre   P=1.   i
    a   R m osm gav liv MXB 18m26 dsv                                                  no dre   P=1.

2-CHLOROPROPANAL      100ng..:..1ug....:..10....:..100....:..1mg....:..10.......:..100....:..1g......:..10
 1179 M f hic gav for tum 89w89                              .  +  .                   12.9mg   P<.004 +
 1180 M m hic gav for tum 89w89                              .  ±                      22.7mg   P<.04

1-CHLOROPROPENE       100ng..:..1ug....:..10....:..100....:..1mg....:..10.......:..100....:..1g......:..10
 1181 M f hic gav for tum 89w89                             .  +  .                    5.05mg   P<.0005+
 1182 M m hic gav for tum 89w89                             .  ±                       16.7mg   P<.02

CHLOROTHALONIL        100ng..:..1ug....:..10....:..100....:..1mg....:..10.......:..100....:..1g......:..10
 1183 M f b6c eat TBA MXB 80w91 v                                         :>            no dre   P=1.   -
    a   M f b6c eat liv MXB 80w91 v                                                    no dre   P=1.
    b   M f b6c eat lun MXB 80w91 v                                                    no dre   P=1.
 1184 M m b6c eat TBA MXB 80w91 v                                    :>                no dre   P=1.
    a   M m b6c eat liv MXB 80w91 v                                                    no dre   P=1.
    b   M m b6c eat lun MXB 80w91 v                                                    no dre   P=1.
 1185 R f osm eat TBA MXB 19m26 v                                    :>                1.58gm * P<.7   -
    a   R f osm eat liv MXB 19m26 v                                                    no dre   P=1.
 1186 R f osm eat kid MXA 19m25 v    pool                                    :  +      2.51gm * P<.009 c
    a   R f osm eat kid MXA 19m25 v                                                    4.97gm * P<.04  c
 1187 R m osm eat TBA MXB 19m26 v                                   :>                 20.1gm * P<1.   -
    a   R m osm eat liv MXB 19m26 v                                                    no dre   P=1.
 1188 R m osm eat kid MXA 19m25 v    pool                                   :  +  : 1.18gm * P<.003
    a   R m osm eat kid MXA 19m25 v                                                    2.08gm * P<.02  c
    b   R m osm eat sub fih 19m25 v                                                    2.81gm * P<.02

CHLOROZOTOCIN         100ng..:..1ug....:..10....:..100....:..1mg....:..10.......:..100....:..1g......:..10
 1189 R f sda ipj pec mix 20m26 aes                      .  +  .                       84.4ug * P<.0005+
    a   R f sda ipj tba mal 20m26 aes                                                  .108mg * P<.0005+
 1190 R m sda ipj pec mix 18m23 aes                   .  +  .                          24.1ug \ P<.0005+
    a   R m sda ipj tba mal 18m23 aes                                                  52.1ug * P<.0005+

CHLORPHENIRAMINE  MALEATE     100ng..:..1ug....:..10....:..100....:..1mg....:..10.......:..100....:..1g......:..10
 1191 M f b6c gav TBA MXB 24m24                                  :>                    no dre   P=1.   -
```

	RefNum	LoConf	UpConf	Cntrl	1Dose	1Inc	2Dose	2Inc					Citation or Pathology
													Brkly Code
a	TR266	870.mg	n.s.s.	2/50	589.mg	9/50	(1.19gm	2/50)					sub:fbs,fib,srn. S
b	TR266	1.01gm	n.s.s.	30/50	589.mg	22/50	(1.19gm	19/50)					
c	TR266	1.57gm	n.s.s.	12/50	589.mg	8/50	(1.19gm	6/50)					liv:hpa,hpc,nnd.
d	TR266	2.72gm	n.s.s.	6/50	589.mg	5/50	1.19gm	10/50					lun:a/a,a/c.
1165	1278	33.6mg	n.s.s.	0/16	68.0mg	3/15							Innes;ntis,1968/1969
a	1278	42.7mg	n.s.s.	0/16	68.0mg	2/15							
b	1278	16.7mg	139.mg	0/16	68.0mg	7/15							
1166	1278	57.6mg	n.s.s.	1/17	73.0mg	2/17							
a	1278	137.mg	n.s.s.	0/17	73.0mg	0/17							
b	1278	48.8mg	n.s.s.	2/17	73.0mg	3/17							
1167	1278	23.3mg	n.s.s.	2/18	68.0mg	6/16							
a	1278	120.mg	n.s.s.	1/18	68.0mg	0/16							
b	1278	24.5mg	n.s.s.	3/18	68.0mg	6/16							
1168	TR266	79.7mg	n.s.s.	41/50	36.8mg	45/50	73.6mg	37/50					
a	TR266	527.mg	n.s.s.	4/50	36.8mg	1/50	73.6mg	2/50					liv:hpa,hpc,nnd.
1169	TR266	52.1mg	225.mg	1/50	29.4mg	8/50	58.9mg	20/50					kid:tla,uac; liv:hpc,nnd. C
a	TR266	75.2mg	286.mg	0/50	29.4mg	3/50	58.9mg	15/50					kid:tla,uac.
b	TR266	128.mg	2.64gm	0/50	29.4mg	2/50	58.9mg	7/50					
c	TR266	132.mg	764.mg	0/50	29.4mg	1/50	58.9mg	8/50					
d	TR266	93.9mg	n.s.s.	1/50	29.4mg	6/50	58.9mg	9/50					liv:hpc,nnd.
e	TR266	78.4mg	n.s.s.	36/50	29.4mg	41/50	58.9mg	36/50					
f	TR266	93.9mg	n.s.s.	1/50	29.4mg	6/50	58.9mg	9/50					liv:hpa,hpc,nnd.

1-(4-CHLOROPHENYL)-1-PHENYL-2-PROPYNYL   CARBAMATE   10473-70-8

	RefNum	LoConf	UpConf	Cntrl	1Dose	1Inc	2Dose	2Inc	3Dose	3Inc			Citation or Pathology
1170	1348	4.78mg	18.3mg	0/10	10.0mg	6/10	20.0mg	5/10	40.0mg	10/10			Harris;txap,21,414-418;1972
a	1348	12.2mg	65.5mg	0/10	10.0mg	1/10	20.0mg	2/10	40.0mg	8/10			
b	1348	23.8mg	n.s.s.	0/10	10.0mg	2/10	20.0mg	2/10	40.0mg	2/10			
c	1348	39.7mg	n.s.s.	0/10	10.0mg	2/10	20.0mg	0/10	40.0mg	1/10			
d	1348	1.73mg	10.9mg	2/10	10.0mg	9/10	20.0mg	9/10	40.0mg	10/10			

p-CHLOROPHENYL-2,4,5-TRICHLOROPHENYL   SULFIDE   (tetrasul) 2227-13-6

	RefNum	LoConf	UpConf	Cntrl	1Dose	1Inc	2Dose	2Inc	3Dose	3Inc	4Dose	4Inc	Citation or Pathology
1171	281m	15.6mg	n.s.s.	0/32	2.50mg	0/32	10.0mg	0/32	50.0mg	0/32	150.mg	0/32	Verschuuren;txcy,1,63-78;1973
1172	281n	2.36mg	n.s.s.	0/32	.500mg	0/32	1.25mg	0/32					
1173	281m	747.mg	n.s.s.	0/32	2.00mg	0/32	8.00mg	0/32	40.0mg	0/32	120.mg	1/32	
1174	281n	1.88mg	n.s.s.	0/32	.400mg	0/32	1.00mg	0/32					

CHLOROPICRIN   76-06-2

	RefNum	LoConf	UpConf	Cntrl	1Dose	1Inc	2Dose	2Inc					Citation or Pathology
1175	TR65	29.9mg	n.s.s.	4/20	20.0mg	9/50	40.0mg	10/50					
a	TR65	n.s.s.	n.s.s.	0/20	20.0mg	0/50	40.0mg	0/50					liv:hpa,hpc,nnd.
b	TR65	44.2mg	n.s.s.	1/20	20.0mg	3/50	40.0mg	6/50					lun:a/a,a/c.
1176	TR65	36.5mg	n.s.s.	5/20	20.0mg	7/50	34.0mg	9/50					
a	TR65	63.8mg	n.s.s.	2/20	20.0mg	4/50	34.0mg	2/50					liv:hpa,hpc,nnd.
b	TR65	67.7mg	n.s.s.	3/20	20.0mg	1/50	34.0mg	5/50					lun:a/a,a/c.
1177	TR65	7.12mg	n.s.s.	11/20	10.0mg	14/50	11.0mg	9/50					
a	TR65	n.s.s.	n.s.s.	0/20	10.0mg	0/50	11.0mg	0/50					liv:hpa,hpc,nnd.
1178	TR65	9.97mg	n.s.s.	5/20	12.4mg	3/50	(13.0mg	0/50)					
a	TR65	n.s.s.	n.s.s.	0/20	12.4mg	0/50	13.0mg	0/50					liv:hpa,hpc,nnd.

2-CHLOROPROPANAL   683-50-1

	RefNum	LoConf	UpConf	Cntrl	1Dose	1Inc							Citation or Pathology
1179	1011	5.23mg	73.7mg	0/30	5.71mg	6/30							Van Duuren;jnci,63,1433-1439;1979
1180	1011	6.86mg	n.s.s.	0/30	4.76mg	3/30							

1-CHLOROPROPENE   590-21-6

	RefNum	LoConf	UpConf	Cntrl	1Dose	1Inc							Citation or Pathology
1181	1011	2.63mg	11.5mg	0/30	5.71mg	13/30							Van Duuren;jnci,63,1433-1439;1979
1182	1011	5.76mg	n.s.s.	0/30	4.76mg	4/30							

CHLOROTHALONIL   1897-45-6

	RefNum	LoConf	UpConf	Cntrl	1Dose	1Inc	2Dose	2Inc					Citation or Pathology
1183	TR41	1.38gm	n.s.s.	3/10	343.mg	6/50	679.mg	8/50					
a	TR41	n.s.s.	n.s.s.	0/10	343.mg	0/50	679.mg	0/50					liv:hpa,hpc,nnd.
b	TR41	2.16gm	n.s.s.	1/10	343.mg	2/50	679.mg	3/50					lun:a/a,a/c.
1184	TR41	460.mg	n.s.s.	3/10	283.mg	9/50	(560.mg	1/50)					
a	TR41	1.42gm	n.s.s.	2/10	283.mg	1/50	(560.mg	1/50)					liv:hpa,hpc,nnd.
b	TR41	634.mg	n.s.s.	1/10	283.mg	4/50	(560.mg	0/50)					lun:a/a,a/c.
1185	TR41	240.mg	n.s.s.	5/10	184.mg	25/49	368.mg	34/50					
a	TR41	1.19gm	n.s.s.	0/10	184.mg	3/49	368.mg	2/50					liv:hpa,hpc,nnd.
1186	TR41	1.02gm	58.1gm	0/65p	184.mg	1/49	368.mg	5/50					kid:adn,can,tla,uac.
a	TR41	1.50gm	n.s.s.	0/65p	184.mg	0/49	368.mg	3/50					kid:adn,tla.
1187	TR41	212.mg	n.s.s.	4/10	147.mg	22/50	292.mg	16/49					
a	TR41	n.s.s.	n.s.s.	0/10	147.mg	0/50	292.mg	0/49					liv:hpa,hpc,nnd.
1188	TR41	500.mg	6.60gm	0/65p	147.mg	3/50	292.mg	4/49					kid:adc,can,ppa,tla,uac. S
a	TR41	701.mg	n.s.s.	0/65p	147.mg	1/50	292.mg	3/49					kid:adc,can,uac.
b	TR41	835.mg	n.s.s.	0/65p	147.mg	0/50	292.mg	3/49					S

CHLOROZOTOCIN   54749-90-5

	RefNum	LoConf	UpConf	Cntrl	1Dose	1Inc	2Dose	2Inc					Citation or Pathology
1189	1195	44.6ug	.199mg	1/17	57.1ug	10/16	.286mg	16/18					Habs;clet,8,133-137;1979
a	1195	50.6ug	.370mg	4/17	57.1ug	10/16	.286mg	16/18					
1190	1195	11.9ug	55.1ug	0/20	57.1ug	14/18	(.286mg	13/16)					
a	1195	26.9ug	.140mg	2/20	57.1ug	15/18	.286mg	14/16					

CHLORPHENIRAMINE   MALEATE   113-92-8

	RefNum	LoConf	UpConf	Cntrl	1Dose	1Inc	2Dose	2Inc					Citation or Pathology
1191	TR317	66.8mg	n.s.s.	37/50	70.7mg	35/50	(142.mg	20/50)					

	Spe/Sex	Strain	Route	Site	Hist	Xpo+Xpt	Notes	Plot	TD50	DR	2Tailpvl	AuOp
a	M f	b6c	gav	liv	MXB	24m24			no dre		P=1.	
b	M f	b6c	gav	lun	MXB	24m24			no dre		P=1.	
1192	M m	b6c	gav	arp	adn	24m24	s	: ±	#84.0mg	*	P<.03	-
a	M m	b6c	gav	sub	MXA	24m24	s		94.7mg	*	P<.02	
b	M m	b6c	gav	sub	MXA	24m24	s		95.2mg	*	P<.02	
c	M m	b6c	gav	sub	fbs	24m24	s		140.mg	*	P<.05	
d	M m	b6c	gav	TBA	MXB	24m24	s		54.1mg	*	P<.4	
e	M m	b6c	gav	liv	MXB	24m24	s		101.mg	*	P<.4	
f	M m	b6c	gav	lun	MXB	24m24	s		no dre		P=1.	
1193	R f	f34	gav	TBA	MXB	24m24	s	:>	no dre		P=1.	-
a	R f	f34	gav	liv	MXB	24m24	s		3.18gm	*	P<.9	
1194	R f	f34	eat	liv	nnd	25m30	ev	.>	no dre		P=1.	-
1195	R m	f34	gav	TBA	MXB	24m24		:>	no dre		P=1.	-
a	R m	f34	gav	liv	MXB	24m24			208.mg	*	P<.6	
1196	R m	f34	eat	liv	nnd	25m30	ev	.>	no dre		P=1.	

CHLORPROPAMIDE `100ng..:...1ug....:...10......:...100....:...1mg.....:...10.....:...100....:...1g.....:..10`

	Spe/Sex	Strain	Route	Site	Hist	Xpo+Xpt	Notes	Plot	TD50	DR	2Tailpvl	AuOp
1197	M f	b6c	eat	TBA	MXB	24m24	sv	:>	no dre		P=1.	-
a	M f	b6c	eat	liv	MXB	24m24	sv		no dre		P=1.	
b	M f	b6c	eat	lun	MXB	24m24	sv		1.44gm	*	P<.2	
1198	M m	b6c	eat	TBA	MXB	24m24	sv	:>	no dre		P=1.	
a	M m	b6c	eat	liv	MXB	24m24	sv		no dre		P=1.	
b	M m	b6c	eat	lun	MXB	24m24	sv		no dre		P=1.	
1199	R f	f34	eat	TBA	MXB	24m24	a	:>	no dre		P=1.	
a	R f	f34	eat	liv	MXB	24m24	a		no dre		P=1.	
1200	R m	f34	eat	TBA	MXB	24m24	a	:>	no dre		P=1.	
a	R m	f34	eat	liv	MXB	24m24	a		no dre		P=1.	

CHOCOLATE BROWN FB `100ng..:...1ug....:...10......:...100....:...1mg.....:...10.....:...100....:...1g.....:..10`

	Spe/Sex	Strain	Route	Site	Hist	Xpo+Xpt	Notes	Plot	TD50	DR	2Tailpvl	AuOp
1201	M f	cws	eat	liv	tum	80w80	e	.>	no dre		P=1.	-
a	M f	cws	eat	lun	ade	80w80	e		no dre		P=1.	-
1202	M m	cws	eat	lun	ade	80w80	e	.>	3.67gm	*	P<.3	-
a	M m	cws	eat	liv	tum	80w80	e		no dre		P=1.	-
1203	R f	cfe	eat	liv	tum	24m24	e	.>	no dre		P=1.	-
1204	R m	cfe	eat	liv	tum	24m24	e	.>	no dre		P=1.	-

CHOCOLATE BROWN HT `100ng..:...1ug....:...10......:...100....:...1mg.....:...10.....:...100....:...1g.....:..10`

	Spe/Sex	Strain	Route	Site	Hist	Xpo+Xpt	Notes	Plot	TD50	DR	2Tailpvl	AuOp
1205	M f	tf1	eat	---	lys	80w80	e	+ .	36.3mg	Z	P<.007	-
a	M f	tf1	eat	liv	hnd	80w80	e		8.44gm	*	P<.6	-
b	M f	tf1	eat	lun	ade	80w80	e		no dre		P=1.	-
1206	M m	tf1	eat	liv	hnd	80w80	e	.>	106.gm	*	P<1.	-
a	M m	tf1	eat	liv	lca	80w80	e		no dre		P=1.	-
b	M m	tf1	eat	lun	ade	80w80	e		no dre		P=1.	-

CHOLINE CHLORIDE `100ng..:...1ug....:...10......:...100....:...1mg.....:...10.....:...100....:...1g.....:..10`

	Spe/Sex	Strain	Route	Site	Hist	Xpo+Xpt	Notes	Plot	TD50	DR	2Tailpvl	AuOp
1207	M m	cen	eat	liv	mix	52w52	r	.>	2.77gm		P<.4	
a	M m	cen	eat	liv	hpa	52w52	r		5.65gm		P<.7	
b	M m	cen	eat	liv	hpc	52w52	r		6.06gm		P<.3	
c	M m	cen	eat	lun	mix	52w52	r		no dre		P=1.	
1208	R m	f34	eat	liv	nnd	17m24	e	.>	no dre		P=1.	-

CHROMIC OXIDE PIGMENT `100ng..:...1ug....:...10......:...100....:...1mg.....:...10.....:...100....:...1g.....:..10`

	Spe/Sex	Strain	Route	Site	Hist	Xpo+Xpt	Notes	Plot	TD50	DR	2Tailpvl	AuOp
1209	R b	bdr	eat	mgl	fba	20m43			237.gm	*	P<.8	-
a	R b	bdr	eat	mgl	car	20m43			no dre		P=1.	-
b	R b	bdr	eat	hpl	ade	20m43			no dre		P=1.	-

CHROMIUM (III) ACETATE `100ng..:...1ug....:...10......:...100....:...1mg.....:...10.....:...100....:...1g.....:..10`

	Spe/Sex	Strain	Route	Site	Hist	Xpo+Xpt	Notes	Plot	TD50	DR	2Tailpvl	AuOp
1210	M f	cd1	wat	lun	tum	34m34	e	.>	no dre		P=1.	-
a	M f	cd1	wat	tba	tum	34m34	e		no dre		P=1.	-
1211	M m	cd1	wat	lun	tum	32m32	e	.>	no dre		P=1.	-
a	M m	cd1	wat	tba	tum	32m32	e		no dre		P=1.	-
1212	R b	leb	wat	liv	tum	42m42	e	.>	no dre		P=1.	-
a	R b	leb	wat	tba	tum	42m42	e		1.61mg		P<.06	-
b	R b	leb	wat	tba	mal	42m42	e		7.62mg		P<.3	-

CHRYSAZIN `100ng..:...1ug....:...10......:...100....:...1mg.....:...10.....:...100....:...1g.....:..10`

	Spe/Sex	Strain	Route	Site	Hist	Xpo+Xpt	Notes	Plot	TD50	DR	2Tailpvl	AuOp
1213	M m	cen	eat	liv	hpc	77w77	e	±	336.mg		P<.02	+
a	M m	cen	eat	liv	hpa	77w77	e		201.mg		P<.2	+
b	M m	cen	eat	liv	mix	77w77	e		201.mg		P<.2	+
1214	R m	aci	eat	lgi	mix	69w69	r	. + .	245.mg		P<.003	+
a	R m	aci	eat	liv	tum	69w69	r		no dre		P=1.	

CIMETIDINE `100ng..:...1ug....:...10......:...100....:...1mg.....:...10.....:...100....:...1g.....:..10`

	Spe/Sex	Strain	Route	Site	Hist	Xpo+Xpt	Notes	Plot	TD50	DR	2Tailpvl	AuOp
1215	M f	cb6	wat	liv	mix	29m29	g	.>	52.2gm	*	P<1.	-
a	M f	cb6	wat	lun	tum	29m29	g		no dre		P=1.	-

CINNAMYL ANTHRANILATE `100ng..:...1ug....:...10......:...100....:...1mg.....:...10.....:...100....:...1g.....:..10`

	Spe/Sex	Strain	Route	Site	Hist	Xpo+Xpt	Notes	Plot	TD50	DR	2Tailpvl	AuOp
1216	M f	b6c	eat	liv	MXA	24m24		:+ :	2.47gm	*	P<.0005c	
a	M f	b6c	eat	liv	hpc	24m24			7.50gm	*	P<.0005c	
b	M f	b6c	eat	TBA	MXB	24m24			14.4gm	*	P<.5	
c	M f	b6c	eat	liv	MXB	24m24			2.47gm	*	P<.0005	

	RefNum	LoConf	UpConf	Cntrl	1Dose	1Inc	2Dose	2Inc	3Dose	3Inc	4Dose	4Inc	Citation or Pathology / Brkly Code
a	TR317	498.mg	n.s.s.	6/50	70.7mg	3/50	142.mg	5/50					liv:hpa,hpc,nnd.
b	TR317	480.mg	n.s.s.	6/50	70.7mg	6/50	142.mg	4/50					lun:a/a,a/c.
1192	TR317	34.7mg	n.s.s.	2/50	17.7mg	7/50	35.4mg	4/50					S
a	TR317	39.7mg	n.s.s.	4/50	17.7mg	5/50	35.4mg	8/50					sub:fbs,fib,nfs,srn. S
b	TR317	41.2mg	13.1gm	3/50	17.7mg	4/50	35.4mg	8/50					sub:fbs,nfs,srn. S
c	TR317	51.1mg	n.s.s.	3/50	17.7mg	3/50	35.4mg	6/50					S
d	TR317	14.3mg	n.s.s.	39/50	17.7mg	36/50	35.4mg	21/50					
e	TR317	25.8mg	n.s.s.	16/50	17.7mg	19/50	35.4mg	9/50					liv:hpa,hpc,nnd.
f	TR317	105.mg	n.s.s.	16/50	17.7mg	5/50	35.4mg	3/50					lun:a/a,a/c.
1193	TR317	21.5mg	n.s.s.	42/50	21.2mg	30/50	42.5mg	8/50					
a	TR317	96.6mg	n.s.s.	2/50	21.2mg	0/50	42.5mg	1/50					liv:hpa,hpc,nnd.
1194	1654	141.mg	n.s.s.	4/24	42.0mg	3/24							Lijinsky;fctx,22,715-720;1984
1195	TR317	14.2mg	n.s.s.	44/50	10.6mg	41/50	21.2mg	31/50					
a	TR317	38.5mg	n.s.s.	5/50	10.6mg	3/50	21.2mg	6/50					liv:hpa,hpc,nnd.
1196	1654	122.mg	n.s.s.	5/24	33.6mg	3/24							Lijinsky;fctx,22,715-720;1984

CHLORPROPAMIDE   94-20-2

	RefNum	LoConf	UpConf	Cntrl	1Dose	1Inc	2Dose	2Inc	Citation or Pathology / Brkly Code
1197	TR45	425.mg	n.s.s.	1/15	301.mg	9/35	616.mg	2/35	
a	TR45	693.mg	n.s.s.	0/15	301.mg	4/35	616.mg	0/35	liv:hpa,hpc,nnd.
b	TR45	546.mg	n.s.s.	0/15	301.mg	3/35	616.mg	2/35	lun:a/a,a/c.
1198	TR45	642.mg	n.s.s.	4/15	284.mg	7/35	569.mg	5/35	
a	TR45	952.mg	n.s.s.	2/15	284.mg	3/35	569.mg	2/35	liv:hpa,hpc,nnd.
b	TR45	n.s.s.	n.s.s.	0/15	284.mg	1/35	569.mg	0/35	lun:a/a,a/c.
1199	TR45	115.mg	n.s.s.	13/15	107.mg	21/35	(214.mg	8/35)	
a	TR45	n.s.s.	n.s.s.	0/15	107.mg	0/35	214.mg	0/35	liv:hpa,hpc,nnd.
1200	TR45	313.mg	n.s.s.	7/15	85.6mg	4/35	(171.mg	3/35)	
a	TR45	n.s.s.	n.s.s.	0/15	85.6mg	0/35	171.mg	0/35	liv:hpa,hpc,nnd.

CHOCOLATE BROWN FB   12236-46-3

	RefNum	LoConf	UpConf	Cntrl	1Dose	1Inc	2Dose	2Inc	3Dose	3Inc	4Dose	4Inc	Citation or Pathology / Brkly Code
1201	1335	98.7mg	n.s.s.	0/60	39.0mg	0/30	130.mg	0/30	390.mg	0/26	1.30gm	0/30	Gaunt;fctx,11,375-382;1973
a	1335	3.07gm	n.s.s.	19/60	39.0mg	13/30	130.mg	9/30	390.mg	6/26	1.30gm	4/30	
1202	1335	907.mg	n.s.s.	13/53	36.0mg	2/29	120.mg	5/29	360.mg	8/29	1.20gm	8/29	
a	1335	89.0mg	n.s.s.	0/53	36.0mg	0/29	120.mg	0/29	360.mg	0/29	1.20gm	0/29	
1203	1334	205.mg	n.s.s.	0/29	50.0mg	0/29	150.mg	0/30	500.mg	0/29	1.50gm	0/29	Gaunt;fctx,10,3-15;1972
1204	1334	158.mg	n.s.s.	0/29	40.0mg	0/28	120.mg	0/28	400.mg	0/29	1.20gm	0/30	

CHOCOLATE BROWN HT   4553-89-3

	RefNum	LoConf	UpConf	Cntrl	1Dose	1Inc	2Dose	2Inc	3Dose	3Inc	Citation or Pathology / Brkly Code
1205	1337	13.8mg	447.mg	0/39	13.0mg	5/37	(130.mg	3/41	650.mg	1/42)	Drake;txcy,10,17-27;1978
a	1337	1.21gm	n.s.s.	2/39	13.0mg	1/37	130.mg	3/41	650.mg	3/42	
b	1337	1.76gm	n.s.s.	6/39	13.0mg	3/37	130.mg	4/41	650.mg	3/42	
1206	1337	956.mg	n.s.s.	8/42	12.0mg	4/39	120.mg	6/43	600.mg	6/40	
a	1337	2.55gm	n.s.s.	1/42	12.0mg	1/39	120.mg	2/43	600.mg	0/40	
b	1337	1.35gm	n.s.s.	5/42	12.0mg	2/39	120.mg	4/43	600.mg	3/40	

CHOLINE CHLORIDE   67-48-1

	RefNum	LoConf	UpConf	Cntrl	1Dose	1Inc	Citation or Pathology / Brkly Code
1207	1992	569.mg	n.s.s.	2/30	1.20gm	4/30	Fullerton;carc,11,1301-1305;1990
a	1992	695.mg	n.s.s.	2/30	1.20gm	3/30	
b	1992	987.mg	n.s.s.	0/30	1.20gm	1/30	
c	1992	897.mg	n.s.s.	2/30	1.20gm	2/30	
1208	1803	764.mg	n.s.s.	2/28	280.mg	2/28	Shivapurkar;carc,7,547-550;1986

CHROMIC OXIDE PIGMENT   1308-38-9

	RefNum	LoConf	UpConf	Cntrl	1Dose	1Inc	2Dose	2Inc	3Dose	3Inc	Citation or Pathology / Brkly Code
1209	413	20.7gm	n.s.s.	2/60	215.mg	3/60	459.mg	1/60	1.09gm	3/60	Ivankovic;fctx,13,347-351;1975
a	413	5.16gm	n.s.s.	1/60	215.mg	0/60	459.mg	0/60	1.09gm	0/60	
b	413	45.7gm	n.s.s.	2/60	215.mg	0/60	459.mg	0/60	1.09gm	1/60	

CHROMIUM (III) ACETATE   1066-30-4

	RefNum	LoConf	UpConf	Cntrl	1Dose	1Inc	Citation or Pathology / Brkly Code
1210	56	4.60mg	n.s.s.	9/60	1.00mg	4/29	Schroeder;jnut,83,239-250;1964
a	56	3.34mg	n.s.s.	22/60	1.00mg	9/29	
1211	56	4.06mg	n.s.s.	8/44	.833mg	6/39	
a	56	4.87mg	n.s.s.	11/44	.833mg	6/39	
1212	1036	5.96mg	n.s.s.	1/34	.265mg	1/56	Kanisawa;canr,29,892-895;1969
a	1036	.681mg	n.s.s.	10/34	.265mg	28/56	
b	1036	2.09mg	n.s.s.	2/34	.265mg	7/56	

CHRYSAZIN   (danthron) 117-10-2

	RefNum	LoConf	UpConf	Cntrl	1Dose	1Inc	Citation or Pathology / Brkly Code
1213	1796	115.mg	n.s.s.	0/19	240.mg	4/17	Mori;gann,77,871-876;1986
a	1796	65.2mg	n.s.s.	5/19	240.mg	9/17	
b	1796	65.2mg	n.s.s.	5/19	240.mg	9/17	
1214	1725	104.mg	1.06gm	0/15	400.mg	7/18	Mori;bjca,52,781-783;1985
a	1725	653.mg	n.s.s.	0/15	400.mg	0/18	

CIMETIDINE   51481-61-9

	RefNum	LoConf	UpConf	Cntrl	1Dose	1Inc	2Dose	2Inc	Citation or Pathology / Brkly Code
1215	1925	541.mg	n.s.s.	3/20	22.6mg	1/15	226.mg	2/16	Anderson;canr,45,3561-3566;1985
a	1925	627.mg	n.s.s.	1/20	22.6mg	2/15	226.mg	1/16	

CINNAMYL ANTHRANILATE   87-29-6

	RefNum	LoConf	UpConf	Cntrl	1Dose	1Inc	2Dose	2Inc	Citation or Pathology / Brkly Code
1216	TR196	1.66gm	4.50gm	3/50	1.91gm	20/50	3.79gm	33/50	liv:hpa,hpc.
a	TR196	4.28gm	21.9gm	1/50	1.91gm	8/50	3.79gm	14/50	
b	TR196	2.96gm	n.s.s.	32/50	1.91gm	30/50	3.79gm	36/50	
c	TR196	1.66gm	4.50gm	3/50	1.91gm	20/50	3.79gm	33/50	liv:hpa,hpc,nnd.

```
       Spe Strain  Site    Xpo+Xpt                                                                    TD50    2Tailpvl
          Sex  Route  Hist      Notes                                                                    DR     AuOp
    d    M f b6c eat lun MXB 24m24                                                                   no dre   P=1.
1217   M m b6c eat liv MXA 24m24                                                         :  +      :2.70gm * P<.0005c
    a    M m b6c eat TBA MXB 24m24                                                                  3.08gm * P<.008
    b    M m b6c eat liv MXB 24m24                                                                  2.70gm * P<.0005
    c    M m b6c eat lun MXB 24m24                                                                  no dre   P=1.
1218   R f f34 eat ute esp 24m24                                                         :  +      :#1.46gm \ P<.002 -
    a    R f f34 eat TBA MXB 24m24                                                                  no dre   P=1.
    b    R f f34 eat liv MXB 24m24                                                                  no dre   P=1.
1219   R m f34 eat MXB MXB 24m24                                                         :  + 7.00gm * P<.003
    a    R m f34 eat MXA msm 24m24                                                                  10.9gm * P<.03
    b    R m f34 eat kcx MXA 24m24                                                                  12.1gm * P<.03   c
    c    R m f34 eat pan MXA 24m24                                                                  17.5gm * P<.05   c
    d    R m f34 eat TBA MXB 24m24                                                                  7.69gm * P<.7
    e    R m f34 eat liv MXB 24m24                                                                  9.40gm * P<.3

CIPROFIBRATE                    100ng..:..1ug....:..10......:..100....:..1mg....:..10......:..100....:..1g......:..10
1220   M m c5n eat liv mix 78w78 e                                  .    +    .                     4.17mg   P<.0005+
    a    M m c5n eat liv hpa 78w78 e                                                                5.89mg   P<.002 +
    b    M m c5n eat liv hpc 78w78 e                                                                12.3mg   P<.02  +
    c    M m c5n eat lun tum 78w78 e                                                                no dre   P=1.
1221   M m c5n eat liv mix 89w91 ev                                  .    +    .                    12.1mg   P<.0005+
    a    M m c5n eat liv hpa 89w91 ev                                                               19.2mg   P<.002 +
    b    M m c5n eat liv hpc 89w91 ev                                                               46.4mg   P<.04  +
1222   R f f34 eat stg cnd 24m24 er                                            .>                   531.mg * P<.2   +
1223   R m f34 eat liv hpc 60w60 er                        <+                                       noTD50   P<.0005+
    a    R m f34 eat liv mix 60w60 er                                                               noTD50   P<.0005+
    b    R m f34 eat liv nnd 60w60 er                                                               noTD50   P<.0005+
1224   R m f34 eat stg cnd 24m24 er                                .    +    .                      103.mg * P<.002 +

CITRININ                        100ng..:..1ug....:..10......:..100....:..1mg....:..10......:..100....:..1g......:..10
1225   R m f34 eat kid cla 60w60 kr                             <+                                  noTD50   P<.0005+
1226   R m f34 eat kid cla 80w80 r                                <+                                noTD50   P<.0005+

CLIVORINE                       100ng..:..1ug....:..10......:..100....:..1mg....:..10......:..100....:..1g......:..10
1227   R b aci wat liv mix 48w68 e                              .    +    .                         +hist .500mg P<.0005+
    a    R b aci wat liv nnd 48w68 e                                                                +hist .697mg P<.0005+
    b    R b aci wat liv hms 48w68 e                                                                +hist 3.01mg P<.06  +

CLOFIBRATE                      100ng..:..1ug....:..10......:..100....:..1mg....:..10......:..100....:..1g......:..10
1228   R m f34 eat liv hpc 28m28 e                                          .    +    .             169.mg   P<.0005+
    a    R m f34 eat pan acc 28m28 e                                                                +hist 1.30gm P<.09 +

CLOMIPHENE CITRATE              100ng..:..1ug....:..10......:..100....:..1mg....:..10......:..100....:..1g......:..10
1229   M f csc gav lun tum 69w69 ek                                 .>                              5.00mg   P<.3   -
    a    M f csc gav tba mix 69w69 ek                                                               3.62mg   P<.7   -
1230   M f csc gav tba mix 69w82 ek                         .>                                      no dre   P=1.   -

CLONITRALID                     100ng..:..1ug....:..10......:..100....:..1mg....:..10......:..100....:..1g......:..10
1231   M f b6c eat TBA MXB 78w91 sv                                                  :>             no dre   P=1.   -
    a    M f b6c eat liv MXB 78w91 sv                                                               no dre   P=1.   -
    b    M f b6c eat lun MXB 78w91 sv                                                               no dre   P=1.
1232   M m b6c eat TBA MXB 78w91 sv                                        :>                       175.mg * P<.7   i
    a    M m b6c eat liv MXB 78w91 sv                                                               no dre   P=1.
    b    M m b6c eat lun MXB 78w91 sv                                                               no dre   P=1.
1233   R f osm eat ute esp 18m26 v                                                      :  ± #5.06gm * P<.03  -
    a    R f osm eat thy MXA 18m26 v                                                                6.67gm * P<.04
    b    R f osm eat TBA MXB 18m26 v                                                                2.07gm * P<.4
    c    R f osm eat liv MXB 18m26 v                                                                13.3gm * P<.3
1234   R m osm eat TBA MXB 18m26 v                                                   :>             no dre   P=1.
    a    R m osm eat liv MXB 18m26 v                                                                no dre   P=1.

CLOPHEN A 30                    100ng..:..1ug....:..10......:..100....:..1mg....:..10......:..100....:..1g......:..10
1235   R m wis eat liv hpc 27m27 e                                          .>                      157.mg   P<.2   +

COLCEMID                        100ng..:..1ug....:..10......:..100....:..1mg....:..10......:..100....:..1g......:..10
1236   R m b46 ivj tba mix 12m24 es                   .>                                           78.7ug   P<.5   -
    a    R m b46 ivj tba mal 12m24 es                                                               .129mg   P<.6   -
    b    R m b46 ivj tba ben 12m24 es                                                               .248mg   P<.7   -

COMPOUND 50-892                 100ng..:..1ug....:..10......:..100....:..1mg....:..10......:..100....:..1g......:..10
1237   M f cd1 eat lun a/c 86w86                                                        .>          4.60gm * P<.7   -
    a    M f cd1 eat lun a/a 86w86                                                                  no dre   P=1.   -
    b    M f cd1 eat liv hes 86w86                                                                  no dre   P=1.   -
    c    M f cd1 eat lun mhb 86w86                                                                  no dre   P=1.   -
1238   M m cd1 eat lun a/c 86w86                                                         .>         8.40gm * P<.8   -
    a    M m cd1 eat lun a/a 86w86                                                                  no dre   P=1.   -
    b    M m cd1 eat liv hes 86w86                                                                  no dre   P=1.   -
1239   R f sdz eat liv nnd 24m24                                                  .>                2.22gm * P<.7   -
    a    R f sdz eat liv hpc 24m24                                                                  no dre   P=1.   -
    b    R f sdz eat mam fba 24m24                                                                  no dre   P=1.   -
```

	RefNum	LoConf	UpConf	Cntrl	1Dose	1Inc	2Dose	2Inc	Citation or Pathology	Brkly Code
d	TR196	20.2gm	n.s.s.	6/50	1.91gm	4/50	3.79gm	2/50	lun:a/a,a/c.	
1217	TR196	1.53gm	9.53gm	14/50	1.75gm	30/50	3.53gm	37/50	liv:hpa,hpc.	
a	TR196	1.53gm	73.0gm	22/50	1.75gm	39/50	3.53gm	40/50		
b	TR196	1.53gm	9.53gm	14/50	1.75gm	30/50	3.53gm	37/50	liv:hpa,hpc,nnd.	
c	TR196	12.5gm	n.s.s.	7/50	1.75gm	8/50	3.53gm	4/50	lun:a/a,a/c.	
1218	TR196	740.mg	6.34gm	2/49	736.mg	16/50	(1.46gm	9/50)		S
a	TR196	2.38gm	n.s.s.	34/49	736.mg	37/50	1.46gm	28/50		
b	TR196	13.6gm	n.s.s.	2/49	736.mg	2/50	1.46gm	0/50	liv:hpa,hpc,nnd.	
1219	TR196	3.02gm	35.3gm	0/50	583.mg	0/50	1.18gm	7/50	kcx:acn,adn; pan:acc,ana.	C
a	TR196	4.13gm	n.s.s.	0/50	583.mg	1/50	1.18gm	4/50	abc:msm; per:msm.	S
b	TR196	4.18gm	n.s.s.	0/50	583.mg	0/50	1.18gm	4/50	kcx:acn,adn.	
c	TR196	5.26gm	n.s.s.	0/50	583.mg	0/50	1.18gm	3/50	pan:acc,ana.	
d	TR196	1.12gm	n.s.s.	26/50	583.mg	30/50	1.18gm	32/50		
e	TR196	3.01gm	n.s.s.	1/50	583.mg	4/50	1.18gm	4/50	liv:hpa,hpc,nnd.	

CIPROFIBRATE   52214-84-3

	RefNum	LoConf	UpConf	Cntrl	1Dose	1Inc	2Dose	2Inc	Citation or Pathology	
1220	1895m	1.50mg	15.8mg	0/12	15.0mg	6/8			Rao;bjca,58,46-51;1988/pers.comm.	
a	1895m	2.09mg	26.4mg	0/12	15.0mg	5/8				
b	1895m	3.65mg	n.s.s.	0/12	15.0mg	3/8				
c	1895m	13.9mg	n.s.s.	0/12	15.0mg	0/8				
1221	1895n	5.16mg	37.1mg	0/12	25.4mg	8/12				
a	1895n	7.63mg	86.4mg	0/12	25.4mg	5/12				
b	1895n	13.9mg	n.s.s.	0/12	25.4mg	3/12				
1222	1952	86.5mg	n.s.s.	0/60	.500mg	0/60	2.50mg	0/60	Spencer;txpy,17,7-15;1989/pers.comm.	10.0mg 1/60
1223	1640	n.s.s.	1.09mg	0/25	10.0mg	25/25			Rao;canr,44,1072-1076;1984	
a	1640	n.s.s.	1.09mg	0/25	10.0mg	25/25				
b	1640	n.s.s.	1.09mg	0/25	10.0mg	25/25				
1224	1952	39.3mg	461.mg	0/60	.500mg	0/60	2.50mg	0/60	Spencer;txpy,17,7-15;1989/pers.comm.	10.0mg 5/60

CITRININ   (antimycin) 518-75-2

	RefNum	LoConf	UpConf	Cntrl	1Dose	1Inc			Citation or Pathology	
1225	1533m	n.s.s.	5.28mg	0/5	40.0mg	17/17			Arai;clet,17,281-287;1983	
1226	1533n	n.s.s.	12.8mg	0/10	40.0mg	10/10				

CLIVORINE   33979-15-6

	RefNum	LoConf	UpConf	Cntrl	1Dose	1Inc			Citation or Pathology	
1227	1338	.213mg	1.51mg	0/17	1.88mg	8/12			Kuhara;clet,10,117-122;1980	
a	1338	.274mg	2.56mg	0/17	1.88mg	6/11				
b	1338	.738mg	n.s.s.	0/17	1.88mg	2/12				

CLOFIBRATE   637-07-0

	RefNum	LoConf	UpConf	Cntrl	1Dose	1Inc			Citation or Pathology	
1228	1339	77.9mg	447.mg	0/15	200.mg	10/15			Reddy;bjca,40,476-482;1979	
a	1339	318.mg	n.s.s.	0/15	200.mg	2/15				

CLOMIPHENE CITRATE   43054-45-1

	RefNum	LoConf	UpConf	Cntrl	1Dose	1Inc			Citation or Pathology	
1229	585m	.813mg	n.s.s.	0/15	1.14mg	1/15			Poel;canr,28,845-859;1968	
a	585m	.413mg	n.s.s.	4/15	1.14mg	5/15				
1230	585n	.367mg	n.s.s.	8/15	.962mg	8/15				

CLONITRALID   (niclosamide) 1420-04-8

	RefNum	LoConf	UpConf	Cntrl	1Dose	1Inc	2Dose	2Inc	Citation or Pathology	Brkly Code
1231	TR91	275.mg	n.s.s.	9/20	29.9mg	5/50	61.1mg	5/50		
a	TR91	n.s.s.	n.s.s.	0/20	29.9mg	0/50	61.1mg	0/50	liv:hpa,hpc,nnd.	
b	TR91	493.mg	n.s.s.	3/20	29.9mg	0/50	61.1mg	1/50	lun:a/a,a/c.	
1232	TR91	24.8mg	n.s.s.	1/20	27.6mg	5/50	56.4mg	5/50		
a	TR91	44.1mg	n.s.s.	0/20	27.6mg	1/50	56.4mg	0/50	liv:hpa,hpc,nnd.	
b	TR91	28.5mg	n.s.s.	0/20	27.6mg	1/50	56.4mg	0/50	lun:a/a,a/c.	
1233	TR91	2.29gm	n.s.s.	0/20	504.mg	2/50	999.mg	6/50		
a	TR91	2.72gm	n.s.s.	0/20	504.mg	1/50	999.mg	5/50	thy:cca,ccr.	S
b	TR91	607.mg	n.s.s.	11/20	504.mg	33/50	999.mg	37/50		
c	TR91	4.01gm	n.s.s.	0/20	504.mg	1/50	999.mg	2/50	liv:hpa,hpc,nnd.	
1234	TR91	1.44gm	n.s.s.	10/20	403.mg	20/50	799.mg	20/50		
a	TR91	n.s.s.	n.s.s.	0/20	403.mg	0/50	799.mg	0/50	liv:hpa,hpc,nnd.	

CLOPHEN A 30   55600-34-5

	RefNum	LoConf	UpConf	Cntrl	1Dose	1Inc			Citation or Pathology	
1235	1605	43.0mg	n.s.s.	1/131	4.00mg	4/138			Schaeffer;txap,75,278-288;1984	

COLCEMID   477-30-5

	RefNum	LoConf	UpConf	Cntrl	1Dose	1Inc			Citation or Pathology	
1236	1017	14.2ug	n.s.s.	7/65	7.86ug	5/30			Schmahl;arzn,20,1461-1467;1970	
a	1017	18.4ug	n.s.s.	4/65	7.86ug	3/30				
b	1017	22.4ug	n.s.s.	3/65	7.86ug	2/30				

COMPOUND 50-892   (1-isopropyl-4-(m-methoxyphenyl)-7-methyl-2(1H)-quinazolinone)   65765-07-3

	RefNum	LoConf	UpConf	Cntrl	1Dose	1Inc	2Dose	2Inc	Citation or Pathology	
1237	1667	455.mg	n.s.s.	1/50	25.0mg	2/50	150.mg	2/50	Van Ryzin;dact,3,361-379;1980	
a	1667	96.6mg	n.s.s.	10/50	25.0mg	4/50	(150.mg	3/50)		
b	1667	1.08gm	n.s.s.	2/50	25.0mg	1/50	150.mg	0/50		
c	1667	802.mg	n.s.s.	0/50	25.0mg	1/50	150.mg	0/50		
1238	1667	463.mg	n.s.s.	0/50	25.0mg	2/50	150.mg	1/50		
a	1667	261.mg	n.s.s.	13/50	25.0mg	8/50	150.mg	11/50		
b	1667	1.08gm	n.s.s.	2/50	25.0mg	1/50	150.mg	0/50		
1239	1667	286.mg	n.s.s.	7/50	4.30mg	4/50	20.0mg	6/50	102.mg	7/50
a	1667	35.2mg	n.s.s.	1/50	4.30mg	0/50	20.0mg	0/50	102.mg	0/50
b	1667	156.mg	n.s.s.	22/50	4.30mg	32/50	20.0mg	27/50	102.mg	25/50

```
        Spe Strain  Site  Xpo+Xpt                                                                        TD50    2Tailpvl
            Sex  Route  Hist   Notes                                                                        DR    AuOp
1240 R m sdz eat adr mal 24m24                                           .    +      .                     113.mg  Z P<.003 -
   a R m sdz eat liv tum 24m24                                                                             no dre  P=1.

COMPOUND LY171883                     100ng..:..1ug....:..10.....:..100.....:..1mg....:..10.....:..100....:..1g......:..10
1241 M f b6c eat liv hpc 24m24 v                                            .  +    .                      112.mg  Z P<.0005+
   a M f b6c eat liv hpa 24m24 v                                                                           369.mg  * P<.002
1242 M m b6c eat liv hpc 24m24 v                                                        .    ±             426.mg  * P<.06
   a M m b6c eat liv hpa 24m24 v                                                                           2.74gm  * P<.6

COPPER DIMETHYLDITHIOCARBAMATE        100ng..:..1ug....:..10.....:..100.....:..1mg....:..10.....:..100....:..1g......:..10
1243 M f b6a orl lun bro 76w76 evx                                                   .>                   146.mg    P<.3   -
   a M f b6a orl liv hpt 76w76 evx                                                                         no dre    P=1.   -
   b M f b6a orl tba mix 76w76 evx                                                                         no dre    P=1.   -
1244 M m b6a orl liv hpt 76w76 evx                                                   .>                    655.mg    P<.9   -
   a M m b6a orl lun ade 76w76 evx                                                                         no dre    P=1.   -
   b M m b6a orl tba mix 76w76 evx                                                                         no dre    P=1.   -
1245 M f b6c orl liv hpt 76w76 evx                                                       .>                no dre    P=1.   -
   a M f b6c orl lun mix 76w76 evx                                                                         no dre    P=1.   -
   b M f b6c orl tba mix 76w76 evx                                                                         70.8mg    P<.2
1246 M m b6c orl liv hpt 76w76 evx                                          .    ±                         34.7mg    P<.04  -
   a M m b6c orl lun ade 76w76 evx                                                                         112.mg    P<.3   -
   b M m b6c orl tba mix 76w76 evx                                                                         19.1mg    P<.005 -

COPPER-8-HYDROXYQUINOLINE             100ng..:..1ug....:..10.....:..100.....:..1mg....:..10.....:..100....:..1g......:..10
1247 M f b6a orl liv hpt 76w76 evx                                                        .>               no dre    P=1.   -
   a M f b6a orl lun ade 76w76 evx                                                                         no dre    P=1.   -
   b M f b6a orl tba mix 76w76 evx                                                                         2.49gm    P<.7   -
1248 M m b6a orl liv hpt 76w76 evx                                                       .>                967.mg    P<.3   -
   a M m b6a orl lun ade 76w76 evx                                                                         no dre    P=1.   -
   b M m b6a orl tba mix 76w76 evx                                                                         11.2gm    P<1.
1249 M f b6c orl liv hpt 76w76 evx                                                        .>               no dre    P=1.   -
   a M f b6c orl lun mix 76w76 evx                                                                         no dre    P=1.   -
   b M f b6c orl tba mix 76w76 evx                                                                         2.49gm    P<.3   -
1250 M m b6c orl liv hpt 76w76 evx                                                       .>                no dre    P=1.   -
   a M m b6c orl lun mix 76w76 evx                                                                         no dre    P=1.   -
   b M m b6c orl tba mix 76w76 evx                                                                         2.05gm    P<.3   -

CORN OIL                              100ng..:..1ug....:..10.....:..100.....:..1mg....:..10.....:..100....:..1g.....:..10
1251 R m f34 gav pan MXA 24m24                                                                         : + 6.94gm  * P<.0005
   a R m f34 gav pan ana 24m24                                                                             7.15gm  * P<.0005
   b R m f34 gav TBA MXB 24m24                                                                             no dre    P=1.
   c R m f34 gav liv MXB 24m24                                                                             no dre    P=1.

COUMAPHOS                             100ng..:..1ug....:..10.....:..100.....:..1mg....:..10.....:..100....:..1g.....:..10
1252 M f b6c eat TBA MXB 24m24                                              :>                             10.1mg  * P<.5   -
   a M f b6c eat liv MXB 24m24                                                                             11.6mg  * P<.08
   b M f b6c eat lun MXB 24m24                                                                             42.6mg  * P<.7
1253 M m b6c eat TBA MXB 24m24                                                 :>                          no dre    P=1.   -
   a M m b6c eat liv MXB 24m24                                                                             no dre    P=1.
   b M m b6c eat lun MXB 24m24                                                                             no dre    P=1.
1254 R f f34 eat TBA MXB 24m24                                           :>                                98.1mg  * P<1.   -
   a R f f34 eat liv MXB 24m24                                                                             6.99mg  * P<.2
1255 R m f34 eat TBA MXB 24m24                                          :>                                 1.58mg  * P<.3   -
   a R m f34 eat liv MXB 24m24                                                                             12.3mg  * P<.4

COUMARIN                              100ng..:..1ug....:..10.....:..100.....:..1mg....:..10.....:..100....:..1g......:..10
1256 H f syg eat liv tum 22m24 ae                                                       .>                no dre    P=1.   -
   a H f syg eat tba mix 22m24 ae                                                                          5.52gm  * P<.9   -
1257 H m syg eat liv tum 24m24 e                                                      .>                   no dre    P=1.   -
   a H m syg eat tba mix 24m24 e                                                                           603.mg  * P<.04  -
1258 M f b6c gav liv MXA 24m24                                               :  +    :                     65.7mg  Z P<.003 c
   a M f b6c gav liv hpa 24m24                                                                             72.8mg  Z P<.006 c
   b M f b6c gav MXB MXB 24m24                                                                             85.5mg  * P<.0005
   c M f b6c gav lun MXA 24m24                                                                             135.mg  Z P<.0005c
   d M f b6c gav lun a/a 24m24                                                                             179.mg  Z P<.0005c
   e M f b6c gav lun a/c 24m24                                                                             825.mg  Z P<.0005c
   f M f b6c gav for sqp 24m24                                                                             12.2gm  * P<1.   e
   g M f b6c gav for MXA 24m24                                                                             42.0gm  * P<1.   e
   h M f b6c gav TBA MXB 24m24                                                                             198.mg  * P<.2
   i M f b6c gav liv MXB 24m24                                                                             65.7mg  Z P<.003
   j M f b6c gav lun MXB 24m24                                                                             135.mg  Z P<.0005
1259 M m b6c gav lun MXA 24m24                                                    :  +      :              241.mg  Z P<.006 p
   a M m b6c gav lun a/a 24m24                                                                             265.mg  Z P<.009 p
   b M m b6c gav liv hpb 24m24                                                                             632.mg  Z P<.009
   c M m b6c gav for MXA 24m24                                                                             152.mg  Z P<.04  e
   d M m b6c gav for sqp 24m24                                                                             179.mg  Z P<.07  e
   e M m b6c gav --- MXA 24m24                                                                             600.mg  * P<.02
   f M m b6c gav pni isa 24m24                                                                             1.31gm  * P<.03
   g M m b6c gav TBA MXB 24m24                                                                             461.mg  * P<.6
   h M m b6c gav liv MXB 24m24                                                                             no dre    P=1.
   i M m b6c gav lun MXB 24m24                                                                             241.mg  Z P<.006
```

	RefNum	LoConf	UpConf	Cntrl	1Dose	1Inc	2Dose	2Inc		Citation or Pathology
										Brkly Code
1240	1667	48.7mg	566.mg	0/50	4.30mg	1/50	20.0mg	6/50	(102.mg	1/50)
a	1667	35.2mg	n.s.s.	0/50	4.30mg	0/50	20.0mg	0/50	102.mg	0/50

COMPOUND LY171883   (1-(2-hydroxy-3-propyl-4-(4-(1H-tetrazol-5-yl)-butoxy)phenyl)ethanone)   88107-10-2

	RefNum	LoConf	UpConf	Cntrl	1Dose	1Inc	2Dose	2Inc		Citation or Pathology
1241	1969	57.7mg	267.mg	0/60	8.81mg	1/60	26.4mg	11/60	(88.1mg	11/60) Bendele;faat,15,676-682;1990
a	1969	194.mg	1.77gm	0/60	8.81mg	2/60	26.4mg	3/60	88.1mg	8/60
1242	1969	159.mg	n.s.s.	10/60	8.13mg	8/60	24.4mg	4/60	81.3mg	16/60
a	1969	372.mg	n.s.s.	1/60	8.13mg	1/60	24.4mg	4/60	81.3mg	2/60

COPPER DIMETHYLDITHIOCARBAMATE   (cumate) 137-29-1

	RefNum	LoConf	UpConf	Cntrl	1Dose	1Inc		Citation or Pathology
1243	1203	23.8mg	n.s.s.	0/17	22.8mg	1/18		Innes;ntis,1968/1969
a	1203	45.2mg	n.s.s.	0/17	22.8mg	0/18		
b	1203	30.1mg	n.s.s.	2/17	22.8mg	1/18		
1244	1203	20.4mg	n.s.s.	1/18	21.2mg	1/15		
a	1203	22.6mg	n.s.s.	2/18	21.2mg	1/15		
b	1203	17.4mg	n.s.s.	3/18	21.2mg	2/15		
1245	1203	45.2mg	n.s.s.	0/16	22.8mg	0/18		
a	1203	45.2mg	n.s.s.	0/16	22.8mg	0/18		
b	1203	17.4mg	n.s.s.	0/16	22.8mg	2/18		
1246	1203	10.5mg	n.s.s.	0/16	21.2mg	3/15		
a	1203	18.3mg	n.s.s.	0/16	21.2mg	1/15		
b	1203	7.18mg	151.mg	0/16	21.2mg	5/15		

COPPER-8-HYDROXYQUINOLINE   10380-28-6

	RefNum	LoConf	UpConf	Cntrl	1Dose	1Inc		Citation or Pathology
1247	1295	771.mg	n.s.s.	0/17	389.mg	0/18		Innes;ntis,1968/1969
a	1295	771.mg	n.s.s.	1/17	389.mg	0/18		
b	1295	279.mg	n.s.s.	2/17	389.mg	3/18		
1248	1295	221.mg	n.s.s.	1/18	362.mg	3/17		
a	1295	677.mg	n.s.s.	2/18	362.mg	0/17		
b	1295	263.mg	n.s.s.	3/18	362.mg	3/17		
1249	1295	771.mg	n.s.s.	0/16	389.mg	0/18		
a	1295	771.mg	n.s.s.	0/16	389.mg	0/18		
b	1295	405.mg	n.s.s.	0/16	389.mg	1/18		
1250	1295	638.mg	n.s.s.	0/16	362.mg	0/16		
a	1295	638.mg	n.s.s.	0/16	362.mg	0/16		
b	1295	334.mg	n.s.s.	0/16	362.mg	1/16		

CORN OIL   8001-30-7

	RefNum	LoConf	UpConf	Cntrl	1Dose	1Inc	2Dose	2Inc	3Dose	3Inc		Citation or Pathology	
1251	TR426	4.35gm	18.0gm	1/50	1.61gm	8/50	3.22gm	11/50	6.47gm	23/50		pan:acc,ana.	S
a	TR426	4.46gm	18.5gm	1/50	1.61gm	8/50	3.22gm	10/50	6.47gm	23/50			S
b	TR426	9.15gm	n.s.s.	44/50	1.61gm	42/50	3.22gm	38/50	6.47gm	42/50			
c	TR426	75.3gm	n.s.s.	4/50	1.61gm	3/50	3.22gm	0/50	6.47gm	0/50		liv:hpa,hpb,hpc.	

COUMAPHOS   56-72-4

	RefNum	LoConf	UpConf	Cntrl	1Dose	1Inc	2Dose	2Inc		Citation or Pathology
1252	TR96	2.50mg	n.s.s.	8/25	1.30mg	20/50	2.60mg	24/50		
a	TR96	5.46mg	n.s.s.	0/25	1.30mg	4/50	2.60mg	5/50		liv:hpa,hpc,nnd.
b	TR96	7.02mg	n.s.s.	1/25	1.30mg	4/50	2.60mg	4/50		lun:a/a,a/c.
1253	TR96	3.31mg	n.s.s.	15/25	1.20mg	26/50	2.40mg	22/50		
a	TR96	5.22mg	n.s.s.	7/25	1.20mg	14/50	2.40mg	9/50		liv:hpa,hpc,nnd.
b	TR96	6.82mg	n.s.s.	4/25	1.20mg	4/50	2.40mg	6/50		lun:a/a,a/c.
1254	TR96	.635mg	n.s.s.	17/23	.500mg	41/50	1.00mg	38/50		
a	TR96	2.65mg	n.s.s.	0/23	.500mg	2/50	1.00mg	3/50		liv:hpa,hpc,nnd.
1255	TR96	.471mg	n.s.s.	12/25	.400mg	32/50	.800mg	32/50		
a	TR96	3.02mg	n.s.s.	0/25	.400mg	1/50	.800mg	1/50		liv:hpa,hpc,nnd.

COUMARIN   (1,2-benzopyrone) 91-64-5

	RefNum	LoConf	UpConf	Cntrl	1Dose	1Inc	2Dose	2Inc		Citation or Pathology	
1256	1340	222.mg	n.s.s.	0/12	105.mg	0/13	523.mg	0/10		Ueno;fctx,19,353-355;1981	
a	1340	350.mg	n.s.s.	5/12	105.mg	3/13	523.mg	4/10			
1257	1340	174.mg	n.s.s.	0/12	92.0mg	0/11	460.mg	0/11			
a	1340	212.mg	n.s.s.	1/12	92.0mg	2/11	460.mg	5/11			
1258	TR422	35.7mg	438.mg	8/52	35.1mg	27/50	70.4mg	31/51	(141.mg	13/51) liv:hpa,hpc.	
a	TR422	38.0mg	966.mg	8/52	35.1mg	26/50	70.4mg	29/51	(141.mg	12/51)	
b	TR422	48.1mg	302.mg	10/52	35.1mg	28/50	70.4mg	33/51	141.mg	31/51 liv:hpa,hpc; lun:a/a,a/c.	C
c	TR422	87.4mg	252.mg	2/52	35.1mg	5/50	70.4mg	7/51	141.mg	27/51 lun:a/a,a/c.	
d	TR422	108.mg	403.mg	2/52	35.1mg	5/50	70.4mg	7/51	141.mg	20/51	
e	TR422	355.mg	2.92gm	0/52	35.1mg	0/50	70.4mg	0/51	141.mg	7/51	
f	TR422	417.mg	n.s.s.	5/52	35.1mg	5/50	70.4mg	2/51	141.mg	2/51	
g	TR422	376.mg	n.s.s.	1/52	35.1mg	6/50	70.4mg	3/51	141.mg	2/51 for:sqc,sqp.	
h	TR422	67.2mg	n.s.s.	25/52	35.1mg	41/50	70.4mg	36/51	141.mg	34/51	
i	TR422	35.7mg	438.mg	8/52	35.1mg	27/50	70.4mg	31/51	(141.mg	13/51) liv:hpa,hpb,hpc.	
j	TR422	87.4mg	252.mg	2/52	35.1mg	5/50	70.4mg	7/51	141.mg	27/51 lun:a/a,a/c.	
1259	TR422	118.mg	3.20gm	14/50	35.2mg	9/50	70.5mg	15/50	141.mg	25/51 lun:a/a,a/c.	
a	TR422	127.mg	7.96gm	14/50	35.2mg	8/50	70.5mg	14/50	141.mg	24/51	
b	TR422	240.mg	16.1gm	0/50	35.2mg	0/50	70.5mg	5/50	(141.mg	1/51)	S
c	TR422	59.2mg	n.s.s.	2/50	35.2mg	9/50	(70.5mg	4/50	141.mg	0/51) for:sqc,sqp.	
d	TR422	65.3mg	n.s.s.	2/50	35.2mg	8/50	(70.5mg	2/50	141.mg	0/51)	
e	TR422	301.mg	n.s.s.	0/50	35.2mg	3/50	70.5mg	4/50	141.mg	4/51 ---:hcs,mlh,mlm,mlp,mly.	S
f	TR422	499.mg	n.s.s.	0/50	35.2mg	0/50	70.5mg	3/50	141.mg	2/51	S
g	TR422	86.4mg	n.s.s.	42/50	35.2mg	40/50	70.5mg	42/50	141.mg	39/51	
h	TR422	137.mg	n.s.s.	35/50	35.2mg	34/50	70.5mg	32/50	141.mg	29/51 liv:hpa,hpb,hpc.	
i	TR422	118.mg	3.20gm	14/50	35.2mg	9/50	70.5mg	15/50	141.mg	25/51 lun:a/a,a/c.	

	Spe	Sex	Strain	Route	Site	Hist	Xpo+Xpt	Notes	TD50	DR	2Tailpvl	AuOp
1260	R	f	f34	gav	kid	rua	24m24		1.41gm	*	P<.07	e
a	R	f	f34	gav	TBA	MXB	24m24		456.mg	*	P<.8	
b	R	f	f34	gav	liv	MXB	24m24		no dre		P=1.	
1261	R	f	f34	gav	kid	rua	24m24	with step	930.mg	*	P<.06	e
1262	R	m	f34	gav	tes	ica	23m24	as	3.62mg	*	P<.0005	
a	R	m	f34	gav	pit	pda	23m24	as	19.7mg	Z	P<.004	
b	R	m	f34	gav	for	sqp	23m24	as	142.mg	*	P<.04	
c	R	m	f34	gav	kid	MXA	23m24	as	226.mg	*	P<.2	p
d	R	m	f34	gav	TBA	MXB	23m24	as	12.0mg	*	P<.0005	
e	R	m	f34	gav	liv	MXB	23m24	as	no dre		P=1.	
1263	R	m	f34	gav	kid	MXA	23m24	as with step	13.9mg	*	P<.0005p	
a	R	m	f34	gav	kid	MXA	23m24	as	13.9mg	*	P<.0005p	
b	R	m	f34	gav	kid	ruc	23m24	as	151.mg	*	P<.2	
1264	R	m	f34	gav	tes	ica	15m24		8.22mg		P<.003	
a	R	m	f34	gav	kid	rua	15m24		67.6mg		P<.04	
b	R	m	f34	gav	TBA	MXB	15m24		12.2mg		P<.01	
c	R	m	f34	gav	liv	MXB	15m24		171.mg		P<.4	
1265	R	m	f34	gav	kid	rua	15m24	with step	67.6mg		P<.04	
1266	R	m	f34	gav	tes	ica	9m24		19.8mg		P<.3	
a	R	m	f34	gav	kid	rua	9m24		234.mg		P<.5	
b	R	m	f34	gav	TBA	MXB	9m24		210.mg		P<.9	
c	R	m	f34	gav	liv	MXB	9m24		no dre		P=1.	
1267	R	m	f34	gav	kid	rua	9m24	with step	43.6mg		P<.01	

m-CRESIDINE   100ng..:..1ug...:..:10.....:..100....:..1mg...:..10.....:..100...:..1g..:..:10

	Spe	Sex	Strain	Route	Site	Hist	Xpo+Xpt	Notes	TD50	DR	2Tailpvl	AuOp
1268	M	f	b6c	gav	liv	MXA	53w93	sv	#206.mg	*	P<.02	-
a	M	f	b6c	gav	TBA	MXB	53w93	sv	69.2mg	*	P<.03	
b	M	f	b6c	gav	liv	MXB	53w93	sv	206.mg	*	P<.02	
c	M	f	b6c	gav	lun	MXB	53w93	sv	250.mg	*	P<.08	
1269	M	m	b6c	gav	TBA	MXB	53w93	sv	72.2mg	*	P<.6	i
a	M	m	b6c	gav	liv	MXB	53w93	sv	no dre		P=1.	
b	M	m	b6c	gav	lun	MXB	53w93	sv	no dre		P=1.	
1270	R	f	f34	gav	mgl	fba	18m25		52.7mg	\	P<.03	
a	R	f	f34	gav	ubl	tcc	18m25	+hist	769.mg	*	P<.2	c
b	R	f	f34	gav	TBA	MXB	18m25		51.5mg	\	P<.3	
c	R	f	f34	gav	liv	MXB	18m25		1.10gm	*	P<.4	
1271	R	m	f34	gav	ubl	tcc	18m25		470.mg	*	P<.02	c
a	R	m	f34	gav	TBA	MXB	18m25		368.mg	*	P<.7	
b	R	m	f34	gav	liv	MXB	18m25		506.mg	*	P<.4	

p-CRESIDINE   100ng..:..1ug...:..:10.....:..100....:..1mg...:..10.....:..100...:..1g..:..:10

	Spe	Sex	Strain	Route	Site	Hist	Xpo+Xpt	Notes	TD50	DR	2Tailpvl	AuOp
1272	M	f	b6c	eat	ubl	MXA	97w97	sv	69.0mg	/	P<.0005c	
a	M	f	b6c	eat	MXB	MXB	97w97	sv	69.0mg	/	P<.0005	
b	M	f	b6c	eat	ubl	MXA	97w97	sv	70.1mg	/	P<.0005c	
c	M	f	b6c	eat	liv	MXA	97w97	sv	301.mg	*	P<.0005c	
d	M	f	b6c	eat	liv	hpc	97w97	sv	315.mg	*	P<.0005c	
e	M	f	b6c	eat	TBA	MXB	97w97	sv	75.0mg	/	P<.0005	
f	M	f	b6c	eat	liv	MXB	97w97	sv	301.mg	/	P<.0005	
g	M	f	b6c	eat	lun	MXB	97w97	sv	3.81gm	*	P<.6	
1273	M	m	b6c	eat	ubl	MXA	97w97	sv	44.7mg	/	P<.0005c	
a	M	m	b6c	eat	TBA	MXB	97w97	sv	46.4mg	/	P<.0005	
b	M	m	b6c	eat	liv	MXB	97w97	sv	255.mg	*	P<.01	
c	M	m	b6c	eat	lun	MXB	97w97	sv	1.17gm	*	P<.4	
1274	R	f	f34	eat	ubl	MXA	24m24	s	110.mg	/	P<.0005c	
a	R	f	f34	eat	MXB	MXB	24m24	s	110.mg	/	P<.0005	
b	R	f	f34	eat	ute	esp	24m24	s	913.mg	*	P<.007	
c	R	f	f34	eat	nas	oln	24m24	s	1.19gm	/	P<.0005c	
d	R	f	f34	eat	adr	coa	24m24	s	1.49gm	*	P<.02	
e	R	f	f34	eat	TBA	MXB	24m24	s	93.8mg	/	P<.0005	
f	R	f	f34	eat	liv	MXB	24m24	s	2.18gm	*	P<.06	
1275	R	m	f34	eat	MXB	MXB	24m24	s	76.3mg	/	P<.0005	
a	R	m	f34	eat	ubl	MXA	24m24	s	88.4mg	/	P<.0005c	
b	R	m	f34	eat	tes	ict	24m24	s	165.mg	/	P<.006	
c	R	m	f34	eat	nas	MXA	24m24	s	397.mg	/	P<.0005c	
d	R	m	f34	eat	liv	MXA	24m24	s	406.mg	/	P<.0005c	
e	R	m	f34	eat	nas	MXA	24m24	s	450.mg	/	P<.0005c	
f	R	m	f34	eat	MXA	gln	24m24	s	3.51gm	/	P<.02	
g	R	m	f34	eat	MXA	MXA	24m24	s	4.62gm	*	P<.03	c
h	R	m	f34	eat	TBA	MXB	24m24	s	98.0mg	/	P<.0005	
i	R	m	f34	eat	liv	MXB	24m24	s	443.mg	*	P<.0005	

CROTONALDEHYDE   100ng..:..1ug...:..10.....:..100....:..1mg...:..10.....:..100...:..1g..:..:10

	Spe	Sex	Strain	Route	Site	Hist	Xpo+Xpt	Notes	TD50	DR	2Tailpvl	AuOp
1276	R	m	chm	wat	liv	mix	26m26		4.20mg	\	P<.0005+	
a	R	m	chm	wat	liv	nnd	26m26		4.20mg	\	P<.0005	
b	R	m	chm	wat	liv	hpc	26m26		no dre		P=1.	

CUPFERRON   100ng..:..1ug...:..10.....:..100....:..1mg...:..10.....:..100...:..1g..:..:10

	Spe	Sex	Strain	Route	Site	Hist	Xpo+Xpt	Notes	TD50	DR	2Tailpvl	AuOp
1277	M	f	b6c	eat	MXB	MXB	78w95	v	253.mg	*	P<.0005	
a	M	f	b6c	eat	liv	MXA	78w95	v	413.mg	*	P<.0005c	
b	M	f	b6c	eat	---	MXA	78w95	v	419.mg	\	P<.003	c

	RefNum	LoConf	UpConf	Cntrl	1Dose	1Inc	2Dose	2Inc		Citation or Pathology	Brkly Code
1260	TR422	347.mg	n.s.s.	0/50	17.6mg	0/50	35.2mg	0/50	70.3mg	2/50		
a	TR422	41.8mg	n.s.s.	41/50	17.6mg	44/50	35.2mg	40/50	70.3mg	43/50		
b	TR422	n.s.s.	n.s.s.	0/50	17.6mg	0/50	35.2mg	0/50	70.3mg	0/50	liv:hpa,hpb,hpc.	
1261	TR422	281.mg	n.s.s.	0/50	17.6mg	0/50	35.2mg	1/50	70.3mg	2/50		
1262	TR422	2.33mg	6.05mg	38/50	17.7mg	43/50	35.3mg	42/51	71.4mg	46/50		s
a	TR422	8.31mg	174.mg	19/50	17.7mg	12/50	35.3mg	16/51	(71.4mg | 6/50)		s
b	TR422	24.5mg	n.s.s.	0/50	17.7mg	1/50	35.3mg	0/51	71.4mg	1/50		s
c	TR422	35.2mg	n.s.s.	1/50	17.7mg	2/50	35.3mg	2/51	71.4mg	1/50	kid:ade,rua.	
d	TR422	5.71mg	47.9mg	40/50	17.7mg	36/50	35.3mg	30/51	71.4mg	15/50		
e	TR422	50.3mg	n.s.s.	2/50	17.7mg	0/50	35.3mg	0/51	71.4mg	0/50	liv:hpa,hpb,hpc.	
1263	TR422	6.49mg	37.1mg	1/50	17.7mg	6/50	35.3mg	7/51	71.4mg	5/50	kid:ade,rua.	
a	TR422	6.49mg	37.1mg	1/50	17.7mg	6/50	35.3mg	7/51	71.4mg	5/50	kid:ade,rua,ruc.	
b	TR422	24.7mg	n.s.s.	0/50	17.7mg	1/50	35.3mg	0/51	71.4mg	0/50		
1264	TR422a	3.36mg	70.0mg	38/50	44.6mg	15/20						s
a	TR422a	12.2mg	n.s.s.	1/50	44.6mg	2/20						s
b	TR422a	4.59mg	1.23gm	40/50	44.6mg	15/20						
c	TR422a	19.3mg	n.s.s.	2/50	44.6mg	1/20					liv:hpa,hpb,hpc.	
1265	TR422a	12.2mg	n.s.s.	1/50	44.6mg	2/20						s
1266	TR422b	5.48mg	n.s.s.	38/50	27.1mg	18/20						
a	TR422b	27.3mg	n.s.s.	1/50	27.1mg	1/20						
b	TR422b	9.96mg	n.s.s.	40/50	27.1mg	14/20						
c	TR422b	51.0mg	n.s.s.	2/50	27.1mg	0/20					liv:hpa,hpb,hpc.	
1267	TR422b	13.6mg	11.6gm	1/50	27.1mg	4/20						s

m-CRESIDINE   102-50-1

	RefNum	LoConf	UpConf	Cntrl	1Dose	1Inc	2Dose	2Inc	Citation or Pathology	Brkly Code
1268	TR105	83.9mg	n.s.s.	0/25	22.6mg	1/50	45.3mg	5/50	liv:hpa,hpc.	s
a	TR105	32.9mg	n.s.s.	3/25	22.6mg	10/50	45.3mg	15/50		
b	TR105	83.9mg	n.s.s.	0/25	22.6mg	1/50	45.3mg	5/50	liv:hpa,hpc,nnd.	
c	TR105	95.1mg	n.s.s.	0/25	22.6mg	2/50	45.3mg	3/50	lun:a/a,a/c.	
1269	TR105	12.5mg	n.s.s.	9/25	22.9mg	10/50	54.6mg	0/50		
a	TR105	23.8mg	n.s.s.	5/25	22.9mg	3/50	54.6mg	0/50	liv:hpa,hpc,nnd.	
b	TR105	26.9mg	n.s.s.	4/25	22.9mg	4/50	54.6mg	0/50	lun:a/a,a/c.	
1270	TR105	24.9mg	n.s.s.	3/25	40.4mg	21/49	(80.0mg	8/50)		s
a	TR105	232.mg	n.s.s.	0/25	40.4mg	1/49	80.0mg	2/50		
b	TR105	17.1mg	n.s.s.	12/25	40.4mg	38/49	(80.0mg	20/50)		
c	TR105	270.mg	n.s.s.	0/25	40.4mg	1/49	80.0mg	1/50	liv:hpa,hpc,nnd.	
1271	TR105	178.mg	n.s.s.	0/25	40.4mg	0/50	80.0mg	5/50		
a	TR105	56.4mg	n.s.s.	14/25	40.4mg	30/50	80.0mg	21/50		
b	TR105	191.mg	n.s.s.	0/25	40.4mg	4/50	80.0mg	1/50	liv:hpa,hpc,nnd.	

p-CRESIDINE   120-71-8

	RefNum	LoConf	UpConf	Cntrl	1Dose	1Inc	2Dose	2Inc	Citation or Pathology	Brkly Code
1272	TR142	49.9mg	97.7mg	0/50	281.mg	42/50	563.mg	45/50	ubl:can,nen,npm,sqc,tcc.	
a	TR142	49.9mg	97.7mg	0/50	281.mg	42/50	563.mg	45/50	liv:hpa,hpc; ubl:can,nen,npm,sqc,tcc.	C
b	TR142	50.5mg	99.5mg	0/50	281.mg	41/50	563.mg	44/50	ubl:can,sqc,tcc.	
c	TR142	172.mg	587.mg	0/50	281.mg	14/50	563.mg	6/50	liv:hpa,hpc.	
d	TR142	177.mg	627.mg	0/50	281.mg	13/50	563.mg	6/50		
e	TR142	51.5mg	117.mg	14/50	281.mg	45/50	563.mg	45/50		
f	TR142	172.mg	587.mg	0/50	281.mg	14/50	563.mg	6/50	liv:hpa,hpc,nnd.	
g	TR142	573.mg	n.s.s.	4/50	281.mg	4/50	563.mg	1/50	lun:a/a,a/c.	
1273	TR142	28.7mg	70.9mg	0/50	260.mg	40/50	552.mg	31/50	ubl:sqc,tcc.	
a	TR142	29.3mg	78.2mg	22/50	260.mg	42/50	552.mg	32/50		
b	TR142	95.5mg	32.7gm	15/50	260.mg	11/50	552.mg	3/50	liv:hpa,hpc,nnd.	
c	TR142	207.mg	n.s.s.	7/50	260.mg	2/50	552.mg	1/50	lun:a/a,a/c.	
1274	TR142	79.6mg	157.mg	0/50	245.mg	31/50	491.mg	43/50	ubl:nen,ppc,sqc,tcc.	
a	TR142	79.6mg	157.mg	0/50	245.mg	31/50	491.mg	43/50	nas:oln; ubl:nen,ppc,sqc,tcc.	C
b	TR142	401.mg	14.2gm	2/50	245.mg	8/50	491.mg	4/50		s
c	TR142	568.mg	2.89gm	0/50	245.mg	0/50	491.mg	11/50		
d	TR142	578.mg	n.s.s.	0/50	245.mg	5/50	491.mg	1/50		s
e	TR142	62.4mg	162.mg	27/50	245.mg	48/50	491.mg	44/50		
f	TR142	712.mg	n.s.s.	0/50	245.mg	4/50	491.mg	0/50	liv:hpa,hpc,nnd.	
1275	TR142	55.7mg	107.mg	0/50	198.mg	35/50	396.mg	46/50	liv:hpc,mhc,nnd; nas:adn,can,neu,npm,oln; nsp:acn; ubl:ppc,sqc, tcc,tpp,ulc.	C
a	TR142	63.9mg	126.mg	0/50	198.mg	30/50	396.mg	44/50	ubl:ppc,sqc,tcc,tpp,ulc.	
b	TR142	81.1mg	2.05gm	37/50	198.mg	45/50	396.mg	23/50		s
c	TR142	236.mg	711.mg	0/50	198.mg	2/50	396.mg	23/50	nas:adn,can,neu,npm,oln.	
d	TR142	220.mg	968.mg	0/50	198.mg	13/50	396.mg	2/50	liv:hpc,mhc,nnd.	
e	TR142	260.mg	836.mg	0/50	198.mg	1/50	396.mg	21/50	nas:neu,oln.	
f	TR142	942.mg	n.s.s.	0/50	198.mg	0/50	396.mg	3/50	bra:gln; brs:gln.	s
g	TR142	1.35gm	n.s.s.	0/50	198.mg	0/50	396.mg	3/50	nas:can,npm; nsp:acn.	
h	TR142	63.6mg	181.mg	28/50	198.mg	44/50	396.mg	46/50		
i	TR142	235.mg	1.11gm	0/50	198.mg	12/50	396.mg	2/50	liv:hpa,hpc,nnd.	

CROTONALDEHYDE   (trans-2-butenal)   123-73-9

	RefNum	LoConf	UpConf	Cntrl	1Dose	1Inc	2Dose	2Inc	Citation or Pathology	Brkly Code
1276	1774	1.96mg	12.8mg	0/23	2.10mg	9/27	(21.0mg	1/23)	Chung;canr,46,1285-1289;1986	
a	1774	1.96mg	12.8mg	0/23	2.10mg	9/27	(21.0mg	1/23)		
b	1774	94.1mg	n.s.s.	0/23	2.10mg	2/27	21.0mg	0/23		

CUPFERRON   135-20-6

	RefNum	LoConf	UpConf	Cntrl	1Dose	1Inc	2Dose	2Inc	Citation or Pathology	Brkly Code
1277	TR100	163.mg	495.mg	3/50	200.mg	18/50	405.mg	25/50	---:hem,hes; hag:adn; liv:hpa,hpc; zym:sec,sqc.	C
a	TR100	243.mg	1.05gm	2/50	200.mg	12/50	405.mg	16/50	liv:hpa,hpc.	
b	TR100	192.mg	2.20gm	1/50	200.mg	10/50	(405.mg	6/50)	---:hem,hes.	

```
       Spe Strain Site  Xpo+Xpt                                                  TD50    2Tailpvl
           Sex  Route  Hist    Notes                                               DR      AuOp
   c   M f b6c eat liv hpc 78w95 v                                              564.mg * P<.0005c
   d   M f b6c eat hag adn 78w95 v                                              1.27gm * P<.002 c
   e   M f b6c eat lun MXA 78w95 v                                              798.mg * P<.04
   f   M f b6c eat --- hes 78w95 v                                              1.33gm * P<.03
   g   M f b6c eat zym MXA 78w95 v                                              3.93gm * P<.04  c
   h   M f b6c eat TBA MXB 78w95 v                                              211.mg * P<.0005
   i   M f b6c eat liv MXB 78w95 v                                              413.mg * P<.0005
   j   M f b6c eat lun MXB 78w95 v                                              798.mg * P<.04
1278 M m b6c eat --- hes 78w95 v                                   : +        :1.00mg * P<.005 c
   a   M m b6c eat hag adn 78w95 v                                              1.04gm * P<.004
   b   M m b6c eat TBA MXB 78w95 v                                              252.mg * P<.004
   c   M m b6c eat liv MXB 78w95 v                                              5.45gm * P<.9
   d   M m b6c eat lun MXB 78w95 v                                              2.39gm * P<.6
1279 R f f34 eat MXB MXB 18m25                              : + :              14.1mg * P<.0005
   a   R f f34 eat --- hes 18m25                                               17.4mg / P<.0005c
   b   R f f34 eat liv MXA 18m25                                               17.7mg * P<.0005c
   c   R f f34 eat liv hpc 18m25                                               20.1mg * P<.0005c
   d   R f f34 eat sto MXA 18m25                                               27.8mg / P<.0005c
   e   R f f34 eat sto sqc 18m25                                               31.3mg / P<.0005c
   f   R f f34 eat zym MXA 18m25                                       +hist 171.mg * P<.002 c
   g   R f f34 eat TBA MXB 18m25                                               11.2mg / P<.0005
   h   R f f34 eat liv MXB 18m25                                               17.7mg * P<.0005
1280 R m f34 eat MXB MXB 18m24                            :  +  :              5.33mg / P<.0005
   a   R m f34 eat --- hes 18m24                                               5.49mg * P<.0005c
   b   R m f34 eat sto MXA 18m24                                               6.28mg / P<.0005c
   c   R m f34 eat sto sqc 18m24                                               6.94mg / P<.0005c
   d   R m f34 eat liv MXA 18m24                                               9.28mg * P<.0005c
   e   R m f34 eat sub fib 18m24                                               10.5mg * P<.0005
   f   R m f34 eat liv hpc 18m24                                               30.7mg * P<.0005c
   g   R m f34 eat bod MXA 18m24                                               139.mg * P<.002
   h   R m f34 eat TBA MXB 18m24                                               8.85mg / P<.0005
   i   R m f34 eat liv MXB 18m24                                               9.28mg * P<.0005

CYANAMIDE, CALCIUM      100ng..:..1ug...:..10....:..100....:..1mg...:..10.....:..100.....:..1g....:..10
1281 M f b6c eat --- MXA 23m23                                         : ±     #478.mg * P<.03  -
   a   M f b6c eat TBA MXB 23m23                                               no dre   P=1.
   b   M f b6c eat liv MXB 23m23                                               no dre   P=1.
   c   M f b6c eat lun MXB 23m23                                               4.04gm / P<.6
1282 M f b6c orl liv hpt 76w76 evx                              .>            no dre   P=1.
   a   M f b6c orl lun mix 76w76 evx                                          no dre   P=1.
   b   M f b6c orl tba mix 76w76 evx                                          68.0mg  P<.05
1283 M m b6c eat --- hes 23m23                                        : +     #766.mg * P<.009 -
   a   M m b6c eat TBA MXB 23m23                                              909.mg * P<.6
   b   M m b6c eat liv MXB 23m23                                              no dre   P=1.
   c   M m b6c eat lun MXB 23m23                                              no dre   P=1.
1284 M m b6c orl --- rts 76w76 evx                         .  +     .         30.9mg  P<.006
   a   M m b6c orl liv hpt 76w76 evx                                          55.7mg  P<.04
   b   M m b6c orl lun mix 76w76 evx                                          no dre   P=1.
   c   M m b6c orl tba mix 76w76 evx                                          16.7mg  P<.0005
1285 M f b6a orl liv hpt 76w76 evx                              .>            no dre   P=1.
   a   M f b6a orl lun ade 76w76 evx                                          no dre   P=1.
   b   M f b6a orl tba mix 76w76 evx                                          no dre   P=1.
1286 M m b6a orl liv hpt 76w76 evx                           .>               191.mg  P<.6
   a   M m b6a orl lun ade 76w76 evx                                          no dre   P=1.
   b   M m b6a orl tba mix 76w76 evx                                          no dre   P=1.
1287 R f f34 eat TBA MXB 25m25                                 :>             267.mg * P<.9
   a   R f f34 eat liv MXB 25m25                                              no dre   P=1.
1288 R m f34 eat --- MXA 25m25                              : ±               #7.45mg \ P<.05  -
   a   R m f34 eat TBA MXB 25m25                                              no dre   P=1.
   b   R m f34 eat liv MXB 25m25                                              no dre   P=1.

CYCASIN AND METHYLAZOXYMETHANOL ACETATE..:..1ug...:..10....:..100....:..1mg...:..10.....:..100....:..1g.....:..10
1289 P b cym eat liv hpc 18y24 mw                              :     ±         19.4mg  P<.03  +

   a   P b cym eat tba mal 18y24 mw                                           5.62mg  P<.3
1290 P b rhe ipj liv hpc 8y8 Ww               :    (+)     :                  32.9ug  P<.0005+
   a   P b rhe ipj MXB MXB 8y8 Ww                                             32.9ug  P<.0005
   b   P b rhe ipj eso sqc 8y8 Ww                                             47.8ug  P<.005
   c   P b rhe ipj kid rcc 8y8 Ww                                             47.8ug  P<.005 +
   d   P b rhe ipj smi adc 8y8 Ww                                             47.8ug  P<.005
   e   P b rhe ipj tba mal 8y8 Ww                                             52.0ug  P<.0005
1291 P b rhe eat asc adc 18y24 mw                              :     ±         11.3mg  P<.08
   a   P b rhe eat bil ppa 18y24 mw                                           18.8mg  P<.1
   b   P b rhe eat pan adc 18y24 mw                                           33.8mg  P<.04
   c   P b rhe eat MXB MXB 18y24 mw                                           41.3mg  P<.04
   d   P b rhe eat liv hpc 18y24 mw                                           41.3mg  P<.04  +
   e   P b rhe eat kid rcc 18y24 mw                                           41.3mg  P<.04  +
   f   P b rhe eat bil adc 18y24 mw                                           52.6mg  P<.2
   g   P b rhe eat tba mal 18y24 mw                                           10.9mg  P<.07
   h   P b rhe eat tba mix 18y24 mw                                           10.4mg  P<.2
   i   P b rhe eat tba ben 18y24 mw                                           2.92gm  P<1.
```

	RefNum	LoConf	UpConf	Cntrl	1Dose	1Inc	2Dose	2Inc	Citation or Pathology
									Brkly Code
c	TR100	309.mg	1.96gm	2/50	200.mg	9/50	405.mg	13/50	
d	TR100	576.mg	5.56gm	0/50	200.mg	2/50	405.mg	6/50	
e	TR100	344.mg	n.s.s.	4/50	200.mg	11/50	405.mg	9/50	lun:a/a,a/c. S
f	TR100	582.mg	n.s.s.	1/50	200.mg	5/50	405.mg	6/50	S
g	TR100	1.19gm	n.s.s.	0/50	200.mg	0/50	405.mg	3/50	zym:sec,sqc.
h	TR100	125.mg	591.mg	12/50	200.mg	33/50	405.mg	31/50	
i	TR100	243.mg	1.05gm	2/50	200.mg	12/50	405.mg	16/50	liv:hpa,hpc,nnd.
j	TR100	344.mg	n.s.s.	4/50	200.mg	11/50	405.mg	9/50	lun:a/a,a/c.
1278	TR100	451.mg	8.40gm	1/50	185.mg	3/50	374.mg	7/50	
a	TR100	450.mg	6.19gm	0/50	185.mg	3/50	374.mg	4/50	S
b	TR100	128.mg	1.78gm	20/50	185.mg	25/50	374.mg	26/50	
c	TR100	416.mg	n.s.s.	15/50	185.mg	9/50	374.mg	9/50	liv:hpa,hpc,nnd.
d	TR100	449.mg	n.s.s.	7/50	185.mg	8/50	374.mg	5/50	lun:a/a,a/c.
1279	TR100	9.23mg	21.8mg	2/50	55.2mg	40/50	110.mg	40/50	---:hes; liv:hpc,nnd; sto:sqc,sqp; zym:cuc,sqc. C
a	TR100	10.9mg	27.9mg	0/50	55.2mg	28/50	110.mg	37/50	
b	TR100	10.6mg	31.2mg	1/50	55.2mg	26/50	110.mg	12/50	liv:hpc,nnd.
c	TR100	11.9mg	36.0mg	1/50	55.2mg	24/50	110.mg	10/50	
d	TR100	17.0mg	45.7mg	0/50	55.2mg	19/50	110.mg	24/50	sto:sqc,sqp.
e	TR100	18.3mg	53.9mg	0/50	55.2mg	14/50	110.mg	22/50	
f	TR100	53.9mg	1.18gm	1/50	55.2mg	5/50	110.mg	4/50	zym:cuc,sqc.
g	TR100	7.18mg	18.9mg	31/50	55.2mg	42/50	110.mg	42/50	
h	TR100	10.6mg	31.2mg	1/50	55.2mg	26/50	110.mg	12/50	liv:hpa,hpc,nnd.
1280	TR100	2.06mg	9.79mg	0/50	45.0mg	42/50	96.5mg	38/49	---:hes; liv:hpc,nnd; sto:sqc,sqp. C
a	TR100	2.09mg	10.4mg	0/50	45.0mg	38/50	96.5mg	35/49	
b	TR100	2.17mg	13.1mg	0/50	45.0mg	32/50	96.5mg	24/49	sto:sqc,sqp.
c	TR100	2.24mg	16.5mg	0/50	45.0mg	19/50	96.5mg	17/49	
d	TR100	2.40mg	28.6mg	0/50	45.0mg	12/50	96.5mg	5/49	liv:hpc,nnd.
e	TR100	2.43mg	35.7mg	1/50	45.0mg	15/50	96.5mg	5/49	S
f	TR100	10.8mg	80.8mg	0/50	45.0mg	8/50	96.5mg	4/49	
g	TR100	54.4mg	571.mg	0/50	45.0mg	5/50	96.5mg	1/49	bod:men,mso. S
h	TR100	5.01mg	14.6mg	31/50	45.0mg	45/50	96.5mg	41/49	
i	TR100	2.40mg	28.6mg	0/50	45.0mg	12/50	96.5mg	5/49	liv:hpa,hpc,nnd.

CYANAMIDE, CALCIUM   156-62-7

	RefNum	LoConf	UpConf	Cntrl	1Dose	1Inc	2Dose	2Inc	Citation or Pathology
1281	TR163	213.mg	n.s.s.	1/20	65.0mg	11/50	260.mg	18/50	---:leu,lym. S
a	TR163	264.mg	n.s.s.	11/20	65.0mg	25/50	260.mg	27/50	
b	TR163	685.mg	n.s.s.	0/20	65.0mg	6/50	260.mg	3/50	liv:hpa,hpc,nnd.
c	TR163	716.mg	n.s.s.	3/20	65.0mg	1/50	260.mg	6/50	lun:a/a,a/c.
1282	1066	67.2mg	n.s.s.	0/16	33.9mg	0/18			Innes;ntis,1968/1969
a	1066	67.2mg	n.s.s.	11/16	33.9mg	0/18			
b	1066	20.5mg	n.s.s.	0/16	33.9mg	3/18			
1283	TR163	349.mg	26.9gm	1/20	60.0mg	2/50	240.mg	10/50	S
a	TR163	174.mg	n.s.s.	13/20	60.0mg	29/50	240.mg	30/50	
b	TR163	675.mg	n.s.s.	8/20	60.0mg	11/50	240.mg	7/50	liv:hpa,hpc,nnd.
c	TR163	411.mg	n.s.s.	7/20	60.0mg	11/50	240.mg	11/50	lun:a/a,a/c.
1284	1066	11.6mg	305.mg	0/16	31.6mg	5/16			Innes;ntis,1968/1969
a	1066	16.8mg	n.s.s.	0/16	31.6mg	3/16			
b	1066	55.6mg	n.s.s.	0/16	31.6mg	0/16			
c	1066	7.37mg	51.7mg	0/16	31.6mg	8/16			
1285	1066	63.4mg	n.s.s.	0/17	33.9mg	0/17			
a	1066	63.4mg	n.s.s.	1/17	33.9mg	0/17			
b	1066	29.8mg	n.s.s.	2/17	33.9mg	2/17			
1286	1066	26.5mg	n.s.s.	1/18	31.6mg	2/18			
a	1066	41.5mg	n.s.s.	2/18	31.6mg	1/18			
b	1066	24.8mg	n.s.s.	3/18	31.6mg	3/18			
1287	TR163	13.8mg	n.s.s.	18/20	5.00mg	40/50	20.0mg	42/50	
a	TR163	138.mg	n.s.s.	1/20	5.00mg	0/50	20.0mg	1/50	liv:hpa,hpc,nnd.
1288	TR163	3.39mg	n.s.s.	2/20	4.00mg	18/50	(8.00mg	7/50)	---:leu,lym,nen. S
a	TR163	5.39mg	n.s.s.	18/20	4.00mg	43/50	8.00mg	45/50	
b	TR163	40.5mg	n.s.s.	1/20	4.00mg	3/50	8.00mg	1/50	liv:hpa,hpc,nnd.

CYCASIN AND METHYLAZOXYMETHANOL ACETATE (CAS# 592-62-1 and 14901-08-7) mixture

	RefNum	LoConf	UpConf	Cntrl	1Dose	1Inc			Citation or Pathology
1289	2001	3.16mg	n.s.s.	0/73	3.61mg	1/6			Adamson;ossc,129-156;
									1982/Thorgeirsson 1994/Dalgard 1991/Thorgeirsson&Seiber pers.comm.
a	2001	.705mg	n.s.s.	2/73	3.61mg	1/6			
1290	2001	7.58ug	.218mg	0/86	.795mg	4/5			
a	2001	7.58ug	.218mg	0/86	.795mg	4/5			
b	2001	7.79ug	7.02mg	0/61	.795mg	1/1			
c	2001	7.79ug	7.02mg	0/61	.795mg	1/1			
d	2001	7.79ug	7.02mg	0/61	.795mg	1/1			
e	2001	10.7ug	.359mg	1/86	.795mg	4/5			
1291	2001m	1.83mg	n.s.s.	0/8	7.27mg	1/2			
a	2001m	3.06mg	n.s.s.	0/9	7.27mg	1/3			
b	2001m	5.50mg	n.s.s.	0/39	7.27mg	1/5			
c	2001m	6.73mg	n.s.s.	0/45	7.27mg	1/6			
d	2001m	6.73mg	n.s.s.	0/45	7.27mg	1/6			
e	2001m	6.73mg	n.s.s.	0/45	7.27mg	1/6			
f	2001m	7.67mg	n.s.s.	1/76	7.27mg	1/7			
g	2001m	2.49mg	n.s.s.	5/76	7.27mg	3/7			
h	2001m	2.45mg	n.s.s.	11/110	7.27mg	4/10			
i	2001m	5.23mg	n.s.s.	7/110	7.27mg	1/10			

```
       Spe Strain Site    Xpo+Xpt                                                              TD50      2Tailpvl
          Sex   Route   Hist      Notes                                                          DR        AuOp
CYCLAMATE, SODIUM                        100ng..:..1ug....:..10....:..100....:..1mg..:..10....:..100....:..1g......:.10
1292 M f asp eat rel lys 80w82 e                                                          .       16.7gm * P<.04   -
  a  M f asp eat lun ade 80w82 e                                                                  228.gm * P<1.    -
  b  M f asp eat liv tum 80w82 e                                                                  no dre   P=1.
1293 M m asp eat liv tum 80w82 e                                                    .>            no dre   P=1.
  a  M m asp eat lun ade 80w82 e                                                                  no dre   P=1.
1294 M f c3h wat liv tum 24m24 e                                                        .>        no dre   P=1.
  a  M f c3h wat lun tum 24m24 e                                                                  no dre   P=1.
  b  M f c3h wat tba mix 24m24 e                                                                  no dre   P=1.
1295 M m c3h wat liv mix 24m24 e                                                   .>             no dre   P=1.
  a  M m c3h wat lun mix 24m24 e                                                                  no dre   P=1.
  b  M m c3h wat tba mix 24m24 e                                                                  no dre   P=1.
1296 M f cd1 eat liv tum 24m24 e                                                                 .no dre   P=1.
  a  M f cd1 eat lun tum 24m24 e                                                                  no dre   P=1.
  b  M f cd1 eat tba mix 24m24 e                                                                  no dre   P=1.
1297 M m crf wat liv mix 24m24 e                                               .    ±            1.42gm   P<.09  +
  a  M m crf wat lun mix 24m24 e                                                                 4.38gm   P<.2
  b  M m crf wat tba mix 24m24 e                                                                 772.mg   P<.03  +
1298 M m r3m wat lun mix 24m24 e                                             .    ±              4.67gm   P<.05
  a  M m r3m wat liv tum 24m24 e                                                                  no dre   P=1.
  b  M m r3m wat tba mix 24m24 e                                                                 4.67gm   P<.05
1299 M f swa eat liv hpt 76w76 e                                                                  no dre   P=1.   -
1300 M f swi eat ubl apc 91w91 eg                                                                .no dre   P=1.
1301 M m swi eat ubl tum 91w91 e                                                                  no dre   P=1.
1302 M f xvi wat lun mix 24m24 e                                          .  +   .               587.mg   P<.0005
  a  M f xvi wat liv mix 24m24 e                                                                 16.0gm   P<.3
  b  M f xvi wat tba mix 24m24 e                                                                 587.mg   P<.0005+
1303 R b sda eat ubl pam 30m30 e                                                                  no dre   P=1.   -
  a  R b sda eat ubl tcc 30m30 e                                                                  no dre   P=1.   -
  b  R b sda eat tba mal 30m30 e                                                                 30.7gm * P<.3    -
1304 R b wis eat ubl tum 24m24 er                                                                83.9gm * P<.2

CYCLOCHLOROTINE                          100ng..:..1ug....:..10....:..100....:..1mg....:..10....:..100....:..1g......:.10
1305 M m ddn eat liv mix 37m37 e                                     .    ±                      23.6mg * P<.1    +
  a  M m ddn eat liv lca 37m37 e                                                                 48.6mg * P<.2    +
  b  M m ddn eat liv lcc 37m37 e                                                                 98.9mg * P<.4    +
  c  M m ddn eat liv hpt 37m37 e                                                                 99.5mg * P<.9    +

CYCLOHEXANONE                            100ng..:..1ug....:..10....:..100....:..1mg....:..10....:..100....:..1g......:.10
1306 M f b6c wat --- mly 24m25 ers                                            .  ±               3.69gm Z P<.03
  a  M f b6c wat liv mix 24m25 ers                                                                no dre   P=1.
  b  M f b6c wat lun mix 24m25 ers                                                                no dre   P=1.
1307 M m b6c wat liv mix 24m25 er                                                  .>             no dre   P=1.
  a  M m b6c wat lun mix 24m25 er                                                                 no dre   P=1.
1308 R f f34 wat liv nnd 24m25 e                                                 .>              6.30gm * P<.5
1309 R m f34 wat adr coa 24m25 e                                             .    ±              929.mg \ P<.03
  a  R m f34 wat liv car 24m25 e                                                                  no dre   P=1.
  b  R m f34 wat liv mix 24m25 e                                                                  no dre   P=1.

N-CYCLOHEXYL-2-BENZOTHIAZOLE SULFENAMIDE.:..1ug....:..10....:..100....:..1mg....:..10....:..100....:..1g......:.10
1310 M f b6a orl liv hpt 76w76 evx                                         .>                     no dre   P=1.   -
  a  M f b6a orl lun ade 76w76 evx                                                                no dre   P=1.   -
  b  M f b6a orl tba mix 76w76 evx                                                                no dre   P=1.   -
1311 M m b6a orl liv hpt 76w76 evx                                      .>                        no dre   P=1.   -
  a  M m b6a orl lun ade 76w76 evx                                                                no dre   P=1.   -
  b  M m b6a orl tba mix 76w76 evx                                                                no dre   P=1.   -
1312 M f b6c orl liv hpt 76w76 evx                                      .>                        no dre   P=1.   -
  a  M f b6c orl lun mix 76w76 evx                                                                no dre   P=1.   -
  b  M f b6c orl tba mix 76w76 evx                                                                243.mg   P<.09  -
1313 M m b6c orl liv hpt 76w76 evx                                  .   +   .                    92.7mg   P<.007  -
  a  M m b6c orl lun ade 76w76 evx                                                                532.mg   P<.3   -
  b  M m b6c orl tba mix 76w76 evx                                                                50.8mg   P<.0005-

CYCLOHEXYLAMINE.HCl                      100ng..:..1ug....:..10....:..100....:..1mg....:..10....:..100....:..1g......:.10
1314 M f asp eat lun ade 80w84 e                                               .>                 no dre   P=1.   -
  a  M f asp eat liv nod 80w84 e                                                                  no dre   P=1.   -
  b  M f asp eat tba mix 80w84 e                                                                  no dre   P=1.   -
1315 M m asp eat liv nod 80w84 e                                        .    ±                   231.mg Z P<.02   -
  a  M m asp eat lun adc 80w84 e                                                                  no dre   P=1.
  b  M m asp eat lun ade 80w84 e                                                                  no dre   P=1.
  c  M m asp eat tba mix 80w84 e                                                                  no dre   P=1.
1316 R b fdr eat liv mix 24m24 g                                              .>                  no dre   P=1.
  a  R b fdr eat tba mix 24m24 g                                                                  no dre   P=1.
1317 R b sda eat liv lca 30m30 e                                                .>               14.5gm   P<.2
  a  R b sda eat ubl tum 30m30 e                                                                  no dre   P=1.   -
  b  R b sda eat tba mal 30m30 e                                                                  no dre   P=1.   -
1318 R f wis eat liv hem 24m24 e                                            .    ±               1.33gm Z P<.04   -
1319 R m wis eat liv nod 24m24 e                                              .>                 47.0gm * P<1.    -
  a  R m wis eat liv hem 24m24 e                                                                  no dre   P=1.   -
```

RefNum	LoConf	UpConf	Cntrl	1Dose	1Inc	2Dose	2Inc	3Dose	3Inc	4Dose	4Inc	Citation or Pathology / Brkly Code
CYCLAMATE, SODIUM		**139-05-9**										
1292	1341	6.45gm n.s.s.	3/45	888.mg	2/19	2.22gm	3/18	4.44gm	4/21	8.88gm	6/25	Brantom;fctx,11,735-746;1973
a	1341	9.13gm n.s.s.	6/45	888.mg	2/19	2.22gm	6/18	4.44gm	6/21	8.88gm	3/25	
b	1341	1.29gm n.s.s.	0/45	888.mg	0/19	2.22gm	0/18	4.44gm	0/21	8.88gm	0/25	
1293	1341	1.39gm n.s.s.	0/46	820.mg	0/21	2.05gm	0/27	4.10gm	0/23	8.20gm	0/24	
a	1341	12.4gm n.s.s.	15/46	820.mg	5/21	2.05gm	6/27	4.10gm	1/23	8.20gm	7/24	
1294	1275	2.47gm n.s.s.	0/19	1.20gm	0/10							Rudali;adsc,269,1910-1913;1969
a	1275	2.47gm n.s.s.	0/19	1.20gm	0/10							
b	1275	201.mg n.s.s.	18/19	1.20gm	9/10							
1295	1275	652.mg n.s.s.	3/9	1.00gm	3/9							
a	1275	1.85gm n.s.s.	2/9	1.00gm	0/9							
b	1275	795.mg n.s.s.	5/9	1.00gm	3/9							
1296	1450	8.27gm n.s.s.	0/17	1.30gm	0/38	6.50gm	0/33					Homburger;ctxf,359-373;1978
a	1450	16.2gm n.s.s.	4/17	1.30gm	6/38	6.50gm	5/33					
b	1450	10.5gm n.s.s.	11/17	1.30gm	19/38	6.50gm	14/33					
1297	1275	528.mg n.s.s.	12/28	1.00gm	22/34							Rudali;adsc,269,1910-1913;1969
a	1275	1.44gm n.s.s.	2/28	1.00gm	7/34							
b	1275	314.mg n.s.s.	16/28	1.00gm	28/34							
1298	1275	1.41gm n.s.s.	0/19	1.00gm	3/22							
a	1275	4.53gm n.s.s.	0/19	1.00gm	0/22							
b	1275	1.41gm n.s.s.	0/19	1.00gm	3/22							
1299	1090	24.3gm n.s.s.	3/45	6.50gm	1/47							Roe;fctx,8,135-145;1970
1300	1349	9.41gm n.s.s.	1/41	2.60gm	0/30	6.50gm	0/39					Kroes;txcy,8,285-300;1977
1301	1349	11.1gm n.s.s.	0/40	2.40gm	0/41	6.00gm	0/41					
1302	1275	277.mg 1.95gm	3/16	1.20gm	16/20							Rudali;adsc,269,1910-1913;1969
a	1275	2.61gm n.s.s.	0/16	1.20gm	1/20							
b	1275	277.mg 1.95gm	3/16	1.20gm	16/20							
1303	1416	56.6gm n.s.s.	0/98	900.mg	1/97	2.25gm	0/101					Schmahl;arzn,23,1466-1470;1973
a	1416	56.6gm n.s.s.	0/98	900.mg	1/97	2.25gm	0/101					
b	1416	9.59gm n.s.s.	13/98	900.mg	16/97	2.25gm	20/101					
1304	1465	25.4gm n.s.s.	0/98	1.00gm	1/84	2.00gm	2/143					Hicks;carm,2,475-489;1978
CYCLOCHLOROTINE		**12663-46-6**										
1305	1346	8.15mg n.s.s.	0/11	1.33mg	1/18	2.00mg	3/19					Uraguchi;fctx,10,193-207;1972
a	1346	12.0mg n.s.s.	0/11	1.33mg	0/18	2.00mg	2/19					
b	1346	16.1mg n.s.s.	0/11	1.33mg	0/18	2.00mg	1/19					
c	1346	16.2mg n.s.s.	0/11	1.33mg	1/18	2.00mg	0/19					
CYCLOHEXANONE		**108-94-1**										
1306	1850	1.54gm n.s.s.	8/52	1.26gm	17/50	(2.53gm	4/50	4.86gm	0/41)			Lijinsky;jnci,77,941-949;1986
a	1850	24.1gm n.s.s.	3/52	1.26gm	6/50	2.53gm	3/50	4.86gm	2/41			
b	1850	36.8gm n.s.s.	3/52	1.26gm	2/50	2.53gm	2/50	4.86gm	1/41			
1307	1850	3.72gm n.s.s.	16/52	1.05gm	25/51	2.11gm	13/46					
a	1850	4.90gm n.s.s.	13/52	1.05gm	7/51	(2.11gm	3/47)					
1308	1850	1.32gm n.s.s.	3/52	183.mg	4/52	361.mg	5/52					
1309	1850	364.mg n.s.s.	1/52	160.mg	7/52	(316.mg	1/51)					
a	1850	1.20gm n.s.s.	2/52	160.mg	0/52	316.mg	0/51					
b	1850	1.66gm n.s.s.	6/52	160.mg	5/52	316.mg	4/51					
N-CYCLOHEXYL-2-BENZOTHIAZOLE SULFENAMIDE		**(durax) 95-33-0**										
1310	1299	177.mg n.s.s.	0/17	94.9mg	0/17							Innes;ntis,1968/1969
a	1299	177.mg n.s.s.	1/17	94.9mg	0/17							
b	1299	83.4mg n.s.s.	2/17	94.9mg	2/17							
1311	1299	104.mg n.s.s.	1/18	88.2mg	1/18							
a	1299	116.mg n.s.s.	2/18	88.2mg	1/18							
b	1299	69.4mg n.s.s.	3/18	88.2mg	3/18							
1312	1299	157.mg n.s.s.	0/16	94.9mg	0/15							
a	1299	157.mg n.s.s.	0/16	94.9mg	0/15							
b	1299	59.5mg n.s.s.	0/16	94.9mg	2/15							
1313	1299	34.9mg 1.25gm	0/16	88.2mg	5/17							
a	1299	86.7mg n.s.s.	0/16	88.2mg	1/17							
b	1299	22.5mg 161.mg	0/16	88.2mg	8/17							
CYCLOHEXYLAMINE.HCl		**4998-76-9**										
1314	398	875.mg n.s.s.	6/44	37.1mg	9/46	124.mg	7/42	371.mg	5/44			Hardy;fctx,14,269-276;1976
a	398	1.72gm n.s.s.	1/44	37.1mg	3/46	124.mg	0/42	371.mg	1/44			
b	398	595.mg n.s.s.	11/44	37.1mg	15/46	124.mg	15/42	371.mg	10/44			
1315	398	95.1mg n.s.s.	5/46	34.3mg	3/45	114.mg	10/31	(343.mg	3/46)			
a	398	1.54gm n.s.s.	0/46	34.3mg	0/45	114.mg	1/31	343.mg	0/46			
b	398	1.00gm n.s.s.	14/46	34.3mg	10/45	114.mg	4/31	343.mg	7/46			
c	398	665.mg n.s.s.	16/46	34.3mg	14/45	114.mg	10/31	343.mg	11/46			
1316	1458	1.91gm n.s.s.	0/60	15.0mg	2/60	50.0mg	0/60	100.mg	0/60	150.mg	1/60	Oser;txcy,6,47-65;1976
a	1458	324.mg n.s.s.	16/60	15.0mg	18/60	50.0mg	15/60	100.mg	14/60	(150.mg	3/60)	
1317	1416	2.35gm n.s.s.	0/98	200.mg	1/68							Schmahl;arzn,23,1466-1470;1973
a	1416	4.38gm n.s.s.	0/98	200.mg	0/68							
b	1416	1.81gm n.s.s.	13/98	200.mg	6/68							
1318	396	404.mg n.s.s.	0/38	30.0mg	0/43	100.mg	3/47	(300.mg	0/41)			Gaunt;fctx,14,255-267;1976
1319	396	1.43gm n.s.s.	2/34	24.0mg	1/40	80.0mg	0/39	240.mg	2/46			
a	396	2.54gm n.s.s.	3/34	24.0mg	1/40	80.0mg	1/39	240.mg	0/46			

```
        Spe Strain Site   Xpo+Xpt                                              TD50    2Tailpvl
            Sex  Route  Hist    Notes                                              DR      AuOp

CYCLOHEXYLAMINE SULFATE            100ng..:..1ug....:..10.....:..100.....:..1mg....:..10.....:..100....:..1g.....:..10
1320 M f swi eat ubl apc 91w91 eg                                                 .>    no dre    P=1.    -
1321 M m swi eat ubl tum 91w91 e                                                  .>    no dre    P=1.    -
1322 R f csa eat ubl tum 24m24 r                         .>                             no dre    P=1.
1323 R m csa eat ubl tcc 24m24 r                                       .>               280.mg  * P<.2

CYCLOPHOSPHAMIDE                   100ng..:..1ug....:..10.....:..100.....:..1mg....:..10.....:..100....:..1g.....:..10
1324 M f swi ipj --- lys 26w79 e                               .     +      .           7.09mg  * P<.003  +
   a M f swi ipj ski sqc 26w79 e                                                        11.1mg  * P<.002
   b M f swi ipj lun mix 26w79 e                                                        6.15mg  * P<.06   +
   c M f swi ipj liv lys 26w79 e                                                        no dre    P=1.
   d M f swi ipj tba mal 26w79 e                                                        1.78mg  * P<.0005
   e M f swi ipj tba mix 26w79 e                                                        1.78mg  * P<.0005
   f M f swi ipj tba ben 26w79 e                                                        no dre    P=1.
1325 M m swi ipj --- leu 26w79 e                                 .    +      .          8.69mg  * P<.004
   a M m swi ipj lun mix 26w79 e                                                        5.78mg  * P<.07   +
   b M m swi ipj liv mix 26w79 e                                                        40.9mg  * P<.5
   c M m swi ipj tba mix 26w79 e                                                        3.66mg  * P<.09
   d M m swi ipj tba mal 26w79 e                                                        3.77mg  * P<.05
   e M m swi ipj tba ben 26w79 e                                                        no dre    P=1.
1326 R m b46 ivj tba mix 12m24 es                      .     ±                          3.01mg    P<.04   +
   a R m b46 ivj tba mal 12m24 es                                                       5.36mg    P<.1    +
   b R m b46 ivj tba ben 12m24 es                                                       9.02mg    P<.3
1327 R b sda wat ubl tcc 32w36 ae                               .    +                  21.4mg  * P<.0005+
   a R b sda wat --- mix 32w36 ae                                                       34.9mg  * P<.2
   b R b sda wat vse hms 32w36 ae                                                       65.6mg  * P<.3
   c R b sda wat liv tum 32w36 ae                                                       189.mg  * P<.3
   d R b sda wat ner ngs 32w36 ae                                                       no dre    P=1.
   e R b sda wat tba mal 32w36 ae                                                       12.8mg  * P<.05   +
1328 R f sda ipj mam mal 24m24 es                      .    ±                           1.70mg    P<.05
   a R f sda ipj tba mal 24m24 es                                                       1.26mg    P<.02   +
1329 R m sda ipj liv hae 25m25 es                           .>                          no dre    P=1.
   a R m sda ipj tba mal 25m25 es                                                       2.87mg    P<.06   +
1330 R m sda ivj mix hae 12m24 e                       .    +      .                     3.38mg    P<.007
   a R m sda ivj tba mal 12m24 e                                                        1.41mg    P<.002  +

CYCLOSPORIN A                      100ng..:..1ug....:..10.....:..100.....:..1mg....:..10.....:..100....:..1g.....:..10
1331 M f of1 eat --- lcl 78w78 es                                .      +            .  16.8mg  * P<.01   -
   a M f of1 eat liv hpc 78w78 es                                                       402.mg  * P<.2    -
   b M f of1 eat lun adc 78w78 es                                                       no dre    P=1.
   c M f of1 eat lun ade 78w78 es                                                       no dre    P=1.
   d M f of1 eat tba tum 78w78 es                                                       38.2mg  * P<.7    -
1332 M m of1 eat liv hpc 78w78 e                               .>                       111.mg  * P<.6
   a M m of1 eat lun adc 78w78 e                                                        334.mg  Z P<.8
   b M m of1 eat liv hpa 78w78 e                                                        no dre    P=1.
   c M m of1 eat lun ade 78w78 e                                                        no dre    P=1.
   d M m of1 eat tba tum 78w78 e                                                        11.6mg  * P<.2
1333 R f ofs eat liv cho 24m24 e                         .>                             no dre    P=1.    -
   a R f ofs eat liv hpc 24m24 e                                                        no dre    P=1.    -
   b R f ofs eat liv hpa 24m24 e                                                        no dre    P=1.    -
   c R f ofs eat tba tum 24m24 e                                                        no dre    P=1.
1334 R m ofs eat liv hpc 96w96 es                               .>                      no dre    P=1.    -
   a R m ofs eat tba tum 96w96 es                                                       no dre    P=1.    -

CYTEMBENA                          100ng..:..1ug....:..10.....:..100.....:..1mg....:..10.....:..100....:..1g......:..10
1335 M f b6c ipj --- lhc 24m24                                         :    ±           #50.6mg * P<.02   -
   a M f b6c ipj liv hpa 24m24                                                          85.1mg  * P<.02
   b M f b6c ipj TBA MXB 24m24                                                          28.4mg  * P<.4
   c M f b6c ipj liv MXB 24m24                                                          72.6mg  * P<.3
   d M f b6c ipj lun MXB 24m24                                                          no dre    P=1.
1336 M m b6c ipj TBA MXB 24m24                                      :>                   18.9mg  * P<.5
   a M m b6c ipj liv MXB 24m24                                                          47.8mg  * P<.7
   b M m b6c ipj lun MXB 24m24                                                          161.mg  * P<.9
1337 R f f34 ipj mgl fba 24m24                               :   +    :                 4.45mg  * P<.002  c
   a R f f34 ipj liv nnd 24m24                                                          44.7mg  * P<.03
   b R f f34 ipj TBA MXB 24m24                                                          9.46mg  * P<.4
   c R f f34 ipj liv MXB 24m24                                                          44.7mg  * P<.03
1338 R m f34 ipj MXB MXB 24m24                              :  +   :                     1.05mg  \ P<.0005
   a R m f34 ipj mul msm 24m24                                                          2.01mg    P<.0005c
   b R m f34 ipj tnv men 24m24                                                          2.48mg  \ P<.0005c
   c R m f34 ipj TBA MXB 24m24                                                          1.16mg  \ P<.002
   d R m f34 ipj liv MXB 24m24                                                          56.5mg  * P<.4

DACARBAZINE                        100ng..:..1ug....:..10.....:..100.....:..1mg....:..10.....:..100....:..1g.....:..10
1339 M f swi ipj lun mix 26w61 e                               .   +    .               .595mg    P<.0005+
   a M f swi ipj ute adc 26w61 e                                                        3.20mg    P<.0005+
   b M f swi ipj --- lys 26w61 e                                                        8.00mg    P<.05   +
   c M f swi ipj spl hae 26w61 e                                                        14.5mg    P<.03   +
   d M f swi ipj liv lys 26w61 e                                                        no dre    P=1.
   e M f swi ipj tba mal 26w61 e                                                        .807mg    P<.0005
   f M f swi ipj tba ben 26w61 e                                                        7.03mg    P<.2
```

	RefNum	LoConf	UpConf	Cntrl	1Dose	1Inc	2Dose	2Inc				Citation or Pathology
												Brkly Code

CYCLOHEXYLAMINE SULFATE   19834-02-7

	RefNum	LoConf	UpConf	Cntrl	1Dose	1Inc	2Dose	2Inc				Citation
1320	1349	3.49gm	n.s.s.	1/41	650.mg	0/34						Kroes;txcy,8,285-300;1977
1321	1349	3.98gm	n.s.s.	0/40	600.mg	0/42						
1322	1350	.696mg	n.s.s.	0/25	.150mg	0/25	1.50mg	0/25	15.0mg	0/25		Price;scie,167,1131-1132;1970
1323	1350	45.6mg	n.s.s.	0/25	.150mg	0/25	1.50mg	0/25	15.0mg	1/25		

CYCLOPHOSPHAMIDE   (endoxan)   50-18-0

	RefNum	LoConf	UpConf	Cntrl	1Dose	1Inc	2Dose	2Inc				Citation
1324	1336	2.50mg	61.3mg	3/154	1.69mg	1/19	3.52mg	4/16				Skipper;srfr;1976/Weisburger;canc;1977/Prejean pers.comm.
a	1336	3.35mg	81.6mg	0/154	1.69mg	1/19	3.52mg	2/16				
b	1336	1.97mg	n.s.s.	20/154	1.69mg	4/19	3.52mg	5/16				
c	1336	2.43mg	n.s.s.	1/154	1.69mg	0/19	3.52mg	0/16				
d	1336	.858mg	5.93mg	29/154	1.69mg	7/19	3.52mg	11/16				
e	1336	.815mg	8.25mg	42/154	1.69mg	9/19	3.52mg	11/16				
f	1336	6.99mg	n.s.s.	13/154	1.69mg	2/19	3.52mg	0/16				
1325	1336	2.63mg	84.8mg	0/101	1.72mg	2/22	3.52mg	1/9				
a	1336	1.78mg	n.s.s.	9/101	1.72mg	5/22	3.52mg	2/9				
b	1336	4.60mg	n.s.s.	2/101	1.72mg	0/22	3.52mg	1/9				
c	1336	1.15mg	n.s.s.	28/101	1.72mg	8/22	3.52mg	5/9				
d	1336	1.27mg	n.s.s.	19/101	1.72mg	7/22	3.52mg	4/9				
e	1336	4.42mg	n.s.s.	9/101	1.72mg	1/22	3.52mg	1/9				
1326	1017	1.12mg	n.s.s.	7/65	.929mg	10/36						Schmahl;arzn,20,1461-1467;1970
a	1017	1.62mg	n.s.s.	4/65	.929mg	6/36						
b	1017	2.12mg	n.s.s.	3/65	.929mg	4/36						
1327	1705	12.1mg	70.4mg	0/74	.221mg	2/77	.450mg	2/78	.893mg	5/73	1.79mg	8/72	Schmahl;ijcn,23,706-712;1979
a	1705	12.1mg	n.s.s.	0/74	.221mg	3/77	.450mg	6/78	.893mg	6/73	1.79mg	4/72	
b	1705	17.1mg	n.s.s.	1/74	.221mg	4/77	.450mg	2/78	.893mg	7/73	1.79mg	3/72	
c	1705	46.4mg	n.s.s.	0/74	.221mg	0/77	.450mg	1/78	.893mg	0/73	1.79mg	1/72	
d	1705	29.6mg	n.s.s.	1/74	.221mg	7/77	.450mg	5/78	.893mg	6/73	1.79mg	1/72	
e	1705	5.20mg	n.s.s.	9/74	.221mg	22/77	.450mg	27/78	.893mg	26/73	1.79mg	22/72	
1328	1134	.660mg	n.s.s.	3/33	.571mg	10/36						Schmahl;zkko,86,77-84;1976
a	1134	.549mg	n.s.s.	3/33	.571mg	12/36						
1329	1134	3.91mg	n.s.s.	1/36	.571mg	0/32						
a	1134	.956mg	n.s.s.	1/36	.571mg	5/32						
1330	1703	1.29mg	65.1mg	1/52	.929mg	6/32						Schmahl;zkko,81,211-215;1974
a	1703	.656mg	6.75mg	6/52	.929mg	14/32						

CYCLOSPORIN A   59865-13-3

	RefNum	LoConf	UpConf	Cntrl	1Dose	1Inc	2Dose	2Inc				Citation
1331	1927	7.50mg	2.10gm	15/50	1.00mg	13/50	4.00mg	17/50	16.0mg	25/50		Ryffel;artx,5,107-141;1983/pers.comm.
a	1927	65.4mg	n.s.s.	0/50	1.00mg	0/50	4.00mg	0/50	16.0mg	1/50		
b	1927	67.6mg	n.s.s.	7/50	1.00mg	3/50	4.00mg	2/50	16.0mg	2/50		
c	1927	28.5mg	n.s.s.	8/50	1.00mg	10/50	4.00mg	7/50	16.0mg	8/50		
d	1927	4.51mg	n.s.s.	42/50	1.00mg	42/50	4.00mg	40/50	16.0mg	43/50		
1332	1927	18.6mg	n.s.s.	6/50	1.00mg	7/50	4.00mg	12/50	16.0mg	9/50		
a	1927	33.6mg	n.s.s.	9/50	1.00mg	1/50	4.00mg	2/50	16.0mg	6/50		
b	1927	33.1mg	n.s.s.	11/50	1.00mg	10/50	4.00mg	7/50	16.0mg	8/50		
c	1927	24.5mg	n.s.s.	7/50	1.00mg	13/50	4.00mg	13/50	16.0mg	9/50		
d	1927	3.47mg	n.s.s.	42/50	1.00mg	37/50	4.00mg	39/50	16.0mg	44/50		
1333	1927	4.00mg	n.s.s.	1/50	.500mg	0/50	2.00mg	0/50	8.00mg	0/50		
a	1927	63.0mg	n.s.s.	0/50	.500mg	0/50	2.00mg	1/50	8.00mg	0/50		
b	1927	89.6mg	n.s.s.	2/50	.500mg	0/50	2.00mg	1/50	8.00mg	0/50		
c	1927	13.4mg	n.s.s.	47/50	.500mg	46/50	2.00mg	41/50	8.00mg	37/50		
1334	1927	74.6mg	n.s.s.	0/50	.500mg	1/50	2.00mg	0/50	8.00mg	0/50		
a	1927	5.54mg	n.s.s.	25/50	.500mg	18/50	2.00mg	15/50	(8.00mg	11/50)		

CYTEMBENA   16170-75-5

	RefNum	LoConf	UpConf	Cntrl	1Dose	1Inc	2Dose	2Inc	Pathology	Brkly Code
1335	TR207	21.8mg	n.s.s.	0/50	5.12mg	3/50	10.2mg	4/50		S
a	TR207	29.4mg	n.s.s.	0/50	5.12mg	0/50	10.2mg	4/50		S
b	TR207	7.91mg	n.s.s.	26/50	5.12mg	23/50	10.2mg	29/50		
c	TR207	21.6mg	n.s.s.	3/50	5.12mg	3/50	10.2mg	6/50	liv:hpa,hpc,nnd.	
d	TR207	50.0mg	n.s.s.	7/50	5.12mg	4/50	10.2mg	2/50	lun:a/a,a/c.	
1336	TR207	4.29mg	n.s.s.	28/50	5.12mg	30/50	10.2mg	24/50		
a	TR207	7.03mg	n.s.s.	16/50	5.12mg	18/50	10.2mg	13/50	liv:hpa,hpc,nnd.	
b	TR207	12.7mg	n.s.s.	6/50	5.12mg	7/50	10.2mg	5/50	lun:a/a,a/c.	
1337	TR207	2.45mg	18.3mg	13/50	2.99mg	22/50	5.97mg	36/50		
a	TR207	16.8mg	n.s.s.	0/50	2.99mg	1/50	5.97mg	4/50		S
b	TR207	2.63mg	n.s.s.	38/50	2.99mg	44/50	5.97mg	47/50		
c	TR207	16.8mg	n.s.s.	0/50	2.99mg	1/50	5.97mg	4/50	liv:hpa,hpc,nnd.	
1338	TR207	.624mg	1.90mg	3/50	2.99mg	37/50	(5.97mg	36/50)	mul:msm; tnv:men. C	
a	TR207	1.15mg	4.05mg	3/50	2.99mg	26/50	(5.97mg	26/50)		
b	TR207	1.14mg	6.60mg	0/50	2.99mg	11/50	(5.97mg	10/50)		
c	TR207	.572mg	6.25mg	42/50	2.99mg	45/50	(5.97mg	48/50)		
d	TR207	11.3mg	n.s.s.	1/50	2.99mg	1/50	5.97mg	2/50	liv:hpa,hpc,nnd.	

DACARBAZINE   (DIC)   4342-03-4

	RefNum	LoConf	UpConf	Cntrl	1Dose	1Inc		Citation
1339	1336	.249mg	1.74mg	20/154	4.56mg	12/14		Skipper;srfr;1976/Weisburger;canc;1977/Prejean pers.comm.
a	1336	1.09mg	16.5mg	0/154	4.56mg	4/14		
b	1336	1.76mg	n.s.s.	3/154	4.56mg	2/14		
c	1336	2.36mg	n.s.s.	0/154	4.56mg	1/14		
d	1336	4.53mg	n.s.s.	1/154	4.56mg	0/14		
e	1336	.336mg	2.84mg	29/154	4.56mg	11/14		
f	1336	1.50mg	n.s.s.	13/154	4.56mg	3/14		

```
      Spe Strain  Site   Xpo+Xpt                                                              TD50     2Tailpvl
          Sex  Route  Hist    Notes                                                                  DR    AuOp
  g    M f swi ipj tba mix 26w61 e                                          .   +   .           noTD50    P<.2
1340  M m swi ipj lun mix 26w78 e                                          .   +   .           2.57mg    P<.0005+
  a    M m swi ipj spl mix 26w78 e                                                             5.17mg    P<.0005+
  b    M m swi ipj --- lyk 26w78 e                                                             13.0mg    P<.004 +
  c    M m swi ipj --- lys 26w78 e                                                             16.1mg    P<.07  +
  d    M m swi ipj liv mix 26w78 e                                                             98.8mg    P<.7
  e    M m swi ipj tba mix 26w78 e                                                             .813mg    P<.0005
  f    M m swi ipj tba mal 26w78 e                                                             1.54mg    P<.0005
  g    M m swi ipj tba ben 26w78 e                                                             10.6mg    P<.2
1341  R f sda eat mgl adf 46w60                             .   +   .                          .710mg    P<.0005+
  a    R f sda eat thm lys 46w60                                                               4.21mg    P<.02  +
  b    R f sda eat ute lei 46w60                                                               6.55mg    P<.05  +

DAMINOZIDE                    100ng..:..1ug....:..10.....:..100....:..1mg....:..10.....:..100....:..1g......:..10
1342  M f b6c eat TBA MXB 24m24                                                      :>       56.4gm * P<1.   -
  a    M f b6c eat liv MXB 24m24                                                               6.82gm \ P<.5
  b    M f b6c eat lun MXB 24m24                                                               2.92gm * P<.06
1343  M m b6c eat liv hpc 24m24                                                      : +      2.15gm * P<.005 a
  a    M m b6c eat liv MXA 24m24                                                               2.45gm * P<.05  a
  b    M m b6c eat TBA MXB 24m24                                                               1.50gm * P<.2
  c    M m b6c eat liv MXB 24m24                                                               2.45gm * P<.05
  d    M m b6c eat lun MXB 24m24                                                               3.40gm * P<.4
1344  M f swa wat blv mix 92w92 e                                          .   +   .           1.24gm    P<.0005+
  a    M f swa wat lun mix 92w92 e                                                             1.37gm    P<.0005+
  b    M f swa wat liv tum 92w92 e                                                             no dre    P=1.
1345  M m swa wat blv mix 81w81 e                                          .   +  .            880.mg    P<.0005+
  a    M m swa wat lun mix 81w81 e                                                             1.11gm    P<.0005+
  b    M m swa wat kid ade 81w81 e                                                             10.9gm    P<.002 +
  c    M m swa wat liv hpt 81w81 e                                                             38.8gm    P<.3   -
1346  R f f34 eat lun a/a 24m24                                                      :  ±      4.89gm * P<.05
  a    R f f34 eat MXB MXB 24m24                                                               1.84gm * P<.2
  b    R f f34 eat utm acn 24m24                                               +hist          2.50gm * P<.3   c
  c    R f f34 eat ute lei 24m24                                               +hist          5.27gm * P<.2   c
  d    R f f34 eat TBA MXB 24m24                                                               3.45gm * P<.8
  e    R f f34 eat liv MXB 24m24                                                               9.64gm * P<.5
1347  R m f34 eat TBA MXB 24m24                                                      :>       no dre    P=1.   -
  a    R m f34 eat liv MXB 24m24                                                               no dre    P=1.

·DAPSONE                      100ng..:..1ug....:..10.....:..100....:..1mg....:..10.....:..100....:..1g......:..10
1348  M f b6c eat TBA MXB 18m25                                                      :>       no dre    P=1.   -
  a    M f b6c eat liv MXB 18m25                                                               no dre    P=1.
  b    M f b6c eat lun MXB 18m25                                                               no dre    P=1.
1349  M m b6c eat TBA MXB 18m25                                                        :>     no dre    P=1.   -
  a    M m b6c eat liv MXB 18m25                                                               no dre    P=1.
  b    M m b6c eat lun MXB 18m25                                                               no dre    P=1.
1350  R f f34 eat TBA MXB 18m24                                                        :>     no dre    P=1.   -
  a    R f f34 eat liv MXB 18m24                                                               no dre    P=1.
1351  R m f34 eat spl MXA 18m24                                                  : +  :       22.4mg * P<.0005c
  a    R m f34 eat spl fib 18m24                                                               28.5mg * P<.004 c
  b    R m f34 eat TBA MXB 18m24                                                               22.4mg * P<.1
  c    R m f34 eat liv MXB 18m24                                                               no dre    P=1.
1352  R m f34 eat spl MXA 18m24    pool                                          : +  :       22.4mg * P<.0005c
  a    R m f34 eat spl fib 18m24                                                               28.5mg * P<.0005c
  b    R m f34 eat --- lym 18m24                                                               50.1mg \ P<.008
  c    R m f34 eat per MXA 18m24                                                               56.3mg * P<.002 c
  d    R m f34 eat per srn 18m24                                                               91.3mg * P<.02
  e    R m f34 eat per fbs 18m24                                                               148.mg * P<.03  c
  f    R m f34 eat spl fbs 18m24                                                               167.mg * P<.03  c
  g    R m f34 eat spl srn 18m24                                                               170.mg * P<.03  c

o,p'-DDD                      100ng..:..1ug....:..10.....:..100....:..1mg....:..10.....:..100....:..1g......:..10
1353  M f b6a orl lun ade 76w76 evx                                              .>           236.mg    P<.4
  a    M f b6a orl liv hpt 76w76 evx                                                           no dre    P=1.
  b    M f b6a orl tba mix 76w76 evx                                                           102.mg    P<.2
1354  M m b6a orl lun ade 76w76 evx                                              .>           414.mg    P<.7
  a    M m b6a orl lun car 76w76 evx                                                           467.mg    P<.3
  b    M m b6a orl liv hpt 76w76 evx                                                           no dre    P=1.
  c    M m b6a orl tba mix 76w76 evx                                                           187.mg    P<.5
1355  M f b6c orl liv hpt 76w76 evx                                                   .>      no dre    P=1.
  a    M f b6c orl lun mix 76w76 evx                                                           no dre    P=1.
  b    M f b6c orl tba tum 76w76 evx                                                           no dre    P=1.
1356  M m b6c orl --- rts 76w76 evx                                          .   +        .   82.1mg    P<.009
  a    M m b6c orl liv hpt 76w76 evx                                                           147.mg    P<.05
  b    M m b6c orl lun ade 76w76 evx                                                           467.mg    P<.3
  c    M m b6c orl tba mix 76w76 evx                                                           38.5mg    P<.0005

p,p'-DDD                      100ng..:..1ug....:..10.....:..100....:..1mg....:..10.....:..100....:..1g......:..10
1357  M f b6c eat TBA MXB 78w90 v                                                    :>       108.mg \ P<.2
  a    M f b6c eat liv MXB 78w90 v                                                             493.mg * P<.1
  b    M f b6c eat lun MXB 78w90 v                                                             no dre    P=1.
1358  M f b6c orl lun ade 76w76 evx                                                  .>       250.mg    P<.3
```

	RefNum	LoConf	UpConf	Cntrl	1Dose	1Inc	2Dose	2Inc	Citation or Pathology / Brkly Code
g	1336	n.s.s.	n.s.s.	42/154	4.56mg	14/14			
1340	1336	1.26mg	7.71mg	9/101	3.56mg	14/30			
a	1336	2.23mg	16.6mg	0/101	3.56mg	7/30			
b	1336	3.94mg	114.mg	0/101	3.56mg	3/30			
c	1336	4.16mg	n.s.s.	2/101	3.56mg	3/30			
d	1336	7.22mg	n.s.s.	2/101	3.56mg	1/30			
e	1336	.419mg	1.86mg	28/101	3.56mg	26/30			
f	1336	.796mg	4.06mg	19/101	3.56mg	20/30			
g	1336	2.97mg	n.s.s.	9/101	3.56mg	6/30			
1341	1412	.316mg	2.38mg	4/28	3.83mg	12/16			Beal;jnci,54,951-957;1975
a	1412	1.27mg	n.s.s.	0/28	3.83mg	3/16			
b	1412	1.61mg	n.s.s.	0/28	3.83mg	2/16			

DAMINOZIDE (succinic acid 2,2-dimethyl hydrazide, DMASA) 1596-84-5

	RefNum	LoConf	UpConf	Cntrl	1Dose	1Inc	2Dose	2Inc	Citation or Pathology / Brkly Code
1342	TR83	1.04gm	n.s.s.	13/20	644.mg	27/50	1.29gm	25/50	liv:hpa,hpc,nnd.
a	TR83	1.36gm	n.s.s.	1/20	644.mg	4/50	(1.29gm	0/50)	lun:a/a,a/c.
b	TR83	1.35gm	n.s.s.	1/20	644.mg	8/50	1.29gm	10/50	
1343	TR83	1.26gm	14.9gm	0/20	594.mg	7/50	1.19gm	13/50	
a	TR83	1.20gm	n.s.s.	1/20	594.mg	9/50	1.19gm	14/50	liv:hpa,hpc.
b	TR83	588.mg	n.s.s.	8/20	594.mg	28/50	1.19gm	37/50	
c	TR83	1.20gm	n.s.s.	1/20	594.mg	9/50	1.19gm	14/50	liv:hpa,hpc,nnd.
d	TR83	1.01gm	n.s.s.	4/20	594.mg	15/50	1.19gm	18/50	lun:a/a,a/c.
1344	401	757.mg	2.16gm	8/96	4.00gm	36/43			Toth;canr,37,3497-3500;1977
a	401	831.mg	2.47gm	15/99	4.00gm	37/45			
b	401	29.0gm	n.s.s.	0/71	4.00gm	0/45			
1345	401	552.mg	1.49gm	5/89	3.33gm	37/46			
a	401	649.mg	2.21gm	22/92	3.33gm	36/46			
b	401	4.15gm	51.1gm	0/77	3.33gm	5/42			
c	401	7.39gm	n.s.s.	1/77	3.33gm	2/42			
1346	TR83	1.69gm	n.s.s.	0/20	248.mg	0/50	495.mg	4/50	S
a	TR83	922.mg	n.s.s.	0/20	248.mg	6/50	495.mg	5/50	ute:lei; utm:acn. C
b	TR83	1.13gm	n.s.s.	0/20	248.mg	5/50	495.mg	3/50	
c	TR83	1.81gm	n.s.s.	0/20	248.mg	1/50	495.mg	3/50	
d	TR83	404.mg	n.s.s.	11/20	248.mg	32/50	495.mg	28/50	
e	TR83	2.37gm	n.s.s.	0/20	248.mg	1/50	495.mg	1/50	liv:hpa,hpc,nnd.
1347	TR83	771.mg	n.s.s.	10/20	198.mg	21/50	396.mg	19/50	liv:hpa,hpc,nnd.
a	TR83	3.37gm	n.s.s.	0/20	198.mg	1/50	396.mg	0/50	

DAPSONE 80-08-0

	RefNum	LoConf	UpConf	Cntrl	1Dose	1Inc	2Dose	2Inc	Citation or Pathology / Brkly Code
1348	TR20	45.8mg	n.s.s.	6/14	33.8mg	4/35	(67.6mg	2/36)	liv:hpa,hpc,nnd.
a	TR20	n.s.s.	n.s.s.	0/14	33.8mg	0/35	67.6mg	0/36	lun:a/a,a/c.
b	TR20	262.mg	n.s.s.	1/14	33.8mg	0/35	67.6mg	0/36	
1349	TR20	67.2mg	n.s.s.	0/14	31.2mg	11/35	62.4mg	5/34	liv:hpa,hpc,nnd.
a	TR20	106.mg	n.s.s.	0/14	31.2mg	6/35	62.4mg	2/34	lun:a/a,a/c.
b	TR20	133.mg	n.s.s.	0/14	31.2mg	5/35	62.4mg	1/34	
1350	TR20	35.7mg	n.s.s.	8/15	16.0mg	15/35	32.0mg	14/35	liv:hpa,hpc,nnd.
a	TR20	n.s.s.	n.s.s.	0/15	16.0mg	0/35	32.0mg	0/35	spl:fbs,fib,srn.
1351	TR20	13.1mg	61.0mg	0/15	12.8mg	6/35	25.6mg	14/35	
a	TR20	15.7mg	157.mg	0/15	12.8mg	6/35	25.6mg	10/35	
b	TR20	8.91mg	n.s.s.	6/15	12.8mg	23/35	25.6mg	24/35	
c	TR20	n.s.s.	n.s.s.	0/15	12.8mg	0/35	25.6mg	0/35	liv:hpa,hpc,nnd.
1352	TR20	13.1mg	42.8mg	0/45p	12.8mg	6/35	25.6mg	14/35	spl:fbs,fib,srn.
a	TR20	15.7mg	62.2mg	0/45p	12.8mg	6/35	25.6mg	10/35	
b	TR20	16.9mg	1.09gm	0/45p	12.8mg	4/35	(25.6mg	0/35)	S
c	TR20	27.8mg	202.mg	0/45p	12.8mg	5/35	25.6mg	6/35	per:fbs,srn.
d	TR20	38.7mg	n.s.s.	0/45p	12.8mg	4/35	25.6mg	3/35	S
e	TR20	50.0mg	n.s.s.	0/45p	12.8mg	1/35	25.6mg	3/35	
f	TR20	50.0mg	n.s.s.	0/45p	12.8mg	0/35	25.6mg	3/35	
g	TR20	50.7mg	n.s.s.	0/45p	12.8mg	0/35	25.6mg	3/35	

o,p'-DDD 53-19-0

	RefNum	LoConf	UpConf	Cntrl	1Dose	1Inc	2Dose	2Inc	Citation or Pathology / Brkly Code
1353	1202	51.4mg	n.s.s.	1/17	78.4mg	3/18			Innes;ntis,1968/1969
a	1202	155.mg	n.s.s.	0/17	78.4mg	0/18			
b	1202	31.8mg	n.s.s.	2/17	78.4mg	6/18			
1354	1202	52.0mg	n.s.s.	2/18	73.0mg	3/18			
a	1202	76.1mg	n.s.s.	0/18	73.0mg	1/18			
b	1202	86.3mg	n.s.s.	1/18	73.0mg	1/18			
c	1202	37.2mg	n.s.s.	3/18	73.0mg	5/18			
1355	1202	155.mg	n.s.s.	0/16	78.4mg	0/18			
a	1202	155.mg	n.s.s.	0/16	78.4mg	0/18			
b	1202	155.mg	n.s.s.	0/16	78.4mg	0/18			
1356	1202	30.9mg	1.80gm	0/16	73.0mg	5/18			
a	1202	44.2mg	n.s.s.	0/16	73.0mg	3/18			
b	1202	76.1mg	n.s.s.	0/16	73.0mg	1/18			
c	1202	17.8mg	111.mg	0/16	73.0mg	9/18			

p,p'-DDD (TDE) 72-54-8

	RefNum	LoConf	UpConf	Cntrl	1Dose	1Inc	2Dose	2Inc	Citation or Pathology / Brkly Code
1357	TR131	40.9mg	n.s.s.	2/20	45.5mg	13/50	(89.7mg	6/50)	liv:hpa,hpc,nnd.
a	TR131	201.mg	n.s.s.	0/20	45.5mg	2/50	89.7mg	4/50	lun:a/a,a/c.
b	TR131	240.mg	n.s.s.	0/20	45.5mg	4/50	89.7mg	1/50	
1358	79	40.7mg	n.s.s.	0/16	41.4mg	1/17			Innes;ntis,1968/1969

```
Spe Strain Site  Xpo+Xpt                                              TD50   2Tailpvl
    Sex Route Hist  Notes                                                DR    AuOp

a    M f b6c orl liv hpt 76w76 evx                                    no dre   P=1.
b    M f b6c orl tba mix 76w76 evx                                    250.mg   P<.3
1359 M m b6c eat TBA MXB 78w90 v                         :>           997.mg * P<.9  -
a    M m b6c eat liv MXB 78w90 v                                      116.mg * P<.2
b    M m b6c eat lun MXB 78w90 v                                      no dre   P=1.
1360 M m b6c orl liv hpt 76w76 evx                . +        .        43.3mg   P<.009
a    M m b6c orl lun ade 76w76 evx                                    120.mg   P<.2
b    M m b6c orl tba mix 76w76 evx                                    34.8mg   P<.004
1361 M f b6a orl lun ade 76w76 evx                      .>            73.0mg   P<.2
a    M f b6a orl liv hpt 76w76 evx                                    no dre   P=1.
b    M f b6a orl tba mix 76w76 evx                                    37.4mg   P<.05
1362 M m b6a orl lun ade 76w76 evx                     .>             82.9mg   P<.3
a    M m b6a orl liv hpt 76w76 evx                                    no dre   P=1.
b    M m b6a orl tba mix 76w76 evx                                    49.0mg   P<.2
1363 M f cf1 eat lun mix 29m29                  . +  .                39.9mg   P<.0005+
a    M f cf1 eat liv hpt 29m29                                       4.83gm   P<.8  -
b    M f cf1 eat tba mix 29m29                                       67.1mg   P<.3
1364 M m cf1 eat lun mix 29m29                   . +  .               24.9mg   P<.0005+
a    M m cf1 eat liv hpt 29m29                                       86.8mg   P<.02 +
b    M m cf1 eat tba mix 29m29                                       33.2mg   P<.2
1365 R f osm eat TBA MXB 18m26                       :>              no dre   P=1.  -
a    R f osm eat liv MXB 18m26                                       2.08gm * P<.7
1366 R m osm eat thy MXA 18m26 v                    : ±              75.8mg \ P<.05 a
a    R m osm eat TBA MXB 18m26 v                                     82.6mg \ P<.4
b    R m osm eat liv MXB 18m26 v                                     no dre   P=1.

p,p'-DDE     100ng..:..1ug....:..10......:..100....:..1mg..:..10....:..100....:..1g.....:..10
1367 H f syg eat liv nnd 28m28 e                        . +   .      354.mg * P<.003 +
a    H f syg eat adr mix 28m28 e                                    458.mg * P<.03
b    H f syg eat lun tum 28m28 e                                    no dre   P=1.
c    H f syg eat tba mix 28m28 e                                    634.mg * P<.4
1368 H m syg eat liv nnd 28m28 e                         . ±        141.mg * P<.04 +
a    H m syg eat lun tum 28m28 e                                    no dre   P=1.
b    H m syg eat tba mix 28m28 e                                    1.30gm * P<.8
1369 M f b6c eat liv hpc 78w92 dv             :+ :                  9.45mg / P<.0005c
a    M f b6c eat TBA MXB 78w92 dv                                   11.9mg / P<.0005
b    M f b6c eat liv MXB 78w92 dv                                   9.45mg / P<.0005
c    M f b6c eat lun MXB 78w92 dv                                   no dre   P=1.
1370 M m b6c eat liv hpc 78w92 dsv            : ±                   11.1mg * P<.03  c
a    M m b6c eat TBA MXB 78w92 dsv                                  7.48mg * P<.05
b    M m b6c eat liv MXB 78w92 dsv                                  11.1mg * P<.03
c    M m b6c eat lun MXB 78w92 dsv                                  119.mg * P<.5
1371 M f cf1 eat liv hpt 25m25                   . + .              10.5mg   P<.0005+
a    M f cf1 eat lun mix 25m25                                      no dre   P=1.  -
b    M f cf1 eat tba mix 25m25                                      14.8mg   P<.0005
1372 M m cf1 eat liv hpt 25m25                     . +  .           33.9mg   P<.0005+
a    M m cf1 eat lun mix 25m25                                      no dre   P=1.  -
b    M m cf1 eat tba mix 25m25                                      no dre   P=1.
1373 R f osm eat thy MXA 18m26 dev           : ±                    #33.2mg * P<.04 -
a    R f osm eat TBA MXB 18m26 dev                                  58.0mg * P<.7
b    R f osm eat liv MXB 18m26 dev                                  no dre   P=1.
1374 R m osm eat TBA MXB 18m26 dv             :>                    48.1mg * P<.4 -
a    R m osm eat liv MXB 18m26 dv                                   no dre   P=1.

DDT          100ng..:..1ug....:..10......:..100....:..1mg..:..10....:..100....:..1g.....:..10
1375 H f syg eat liv tum 28m28 e                   .>               no dre   P=1.
a    H f syg eat tba mix 28m28 e                                    639.mg * P<.4 -
1376 H f syg eat tba mix 48w88 e                 .>                 no dre   P=1.
1377 H f syg eat adr ade 28m28 e                     . +   .        345.mg   P<.005 -
a    H f syg eat liv hem 28m28 e                                    3.39gm   P<.3 -
b    H f syg eat lun tum 28m28 e                                    no dre   P=1.
c    H f syg eat tba mix 28m28 e                                    781.mg   P<.5
1378 H m syg eat liv mix 28m28 e                    . ±             311.mg Z P<.03
a    H m syg eat tba mix 28m28 e                                    145.mg * P<.02 -
1379 H m syg eat tba mix 48w85 e                 .>                 113.mg   P<.2
1380 H m syg eat liv tum 28m28 e                       .>           no dre   P=1.
a    H m syg eat lun tum 28m28 e                                    no dre   P=1.
b    H m syg eat tba mix 28m28 e                                    no dre   P=1. -
1381 M f b6c eat --- lym 78w92 v                   : ±              #61.2mg * P<.02 -
a    M f b6c eat TBA MXB 78w92 v                                    59.2mg * P<.2
b    M f b6c eat liv MXB 78w92 v                                    151.mg * P<.09
c    M f b6c eat lun MXB 78w92 v                                    no dre   P=1.
1382 M f b6c orl liv hpt 80w80 evx                . ±               31.1mg   P<.02
a    M f b6c orl lun mix 80w80 evx                                  no dre   P=1.
b    M f b6c orl tba mix 80w80 evx                                  24.0mg   P<.009
1383 M m b6c eat TBA MXB 78w91 sv           :>                      2.04mg \ P<.4 -
a    M m b6c eat liv MXB 78w91 sv                                   no dre   P=1.
b    M m b6c eat lun MXB 78w91 sv                                   no dre   P=1.
1384 M m b6c orl liv hpt 80w80 evx                . +  .            8.95mg   P<.0005+
a    M m b6c orl lun ade 80w80 evx                                  127.mg   P<.3
b    M m b6c orl lun car 80w80 evx                                  127.mg   P<.3
```

	RefNum	LoConf	UpConf	Cntrl	1Dose	1Inc	2Dose	2Inc			Citation or Pathology / Brkly Code
a	79	77.4mg	n.s.s.	0/16	41.4mg	0/17					
b	79	40.7mg	n.s.s.	0/16	41.4mg	1/17					
1359	TR131	53.0mg	n.s.s.	8/20	42.0mg	19/50	84.0mg	19/50			
a	TR131	48.5mg	n.s.s.	2/20	42.0mg	12/50	84.0mg	14/50			liv:hpa,hpc,nnd.
b	TR131	173.mg	n.s.s.	1/20	42.0mg	4/50	84.0mg	2/50			lun:a/a,a/c.
1360	79	16.3mg	950.mg	0/16	38.5mg	5/18					Innes;ntis,1968/1969
a	79	29.4mg	n.s.s.	0/16	38.5mg	2/18					
b	79	14.0mg	212.mg	0/16	38.5mg	6/18					
1361	79	20.7mg	n.s.s.	1/17	41.4mg	4/17					
a	79	77.4mg	n.s.s.	0/17	41.4mg	0/17					
b	79	13.2mg	n.s.s.	2/17	41.4mg	7/17					
1362	79	19.0mg	n.s.s.	2/18	38.5mg	4/16					
a	79	67.8mg	n.s.s.	1/18	38.5mg	0/16					
b	79	13.9mg	n.s.s.	3/18	38.5mg	6/16					
1363	80	21.8mg	115.mg	37/97	32.5mg	43/60					Tomatis;jnci,52,883-891;1974
a	80	328.mg	n.s.s.	1/97	32.5mg	1/60					
b	80	18.5mg	n.s.s.	79/97	32.5mg	53/60					
1364	80	13.5mg	68.6mg	53/101	30.0mg	51/60					
a	80	37.6mg	n.s.s.	33/101	30.0mg	31/60					
b	80	9.92mg	n.s.s.	89/101	30.0mg	57/60					
1365	TR131	59.0mg	n.s.s.	17/20	29.0mg	35/50	59.0mg	36/50			
a	TR131	295.mg	n.s.s.	1/20	29.0mg	0/50	59.0mg	3/50			´liv:hpa,hpc,nnd.
1366	TR131	35.9mg	n.s.s.	1/20	45.6mg	16/50	(91.2mg	11/50)			thy:fca,fcc.
a	TR131	25.3mg	n.s.s.	7/20	45.6mg	33/50	(91.2mg	25/50)			
b	TR131	450.mg	n.s.s.	1/20	45.6mg	1/50	91.2mg	2/50			liv:hpa,hpc,nnd.

p,p'-DDE   72-55-9

	RefNum	LoConf	UpConf	Cntrl	1Dose	1Inc	2Dose	2Inc			Citation or Pathology / Brkly Code
1367	1556	166.mg	1.76gm	0/31	52.3mg	4/26	105.mg	5/24			Rossi;canr,43,776-781;1983
a	1556	205.mg	n.s.s.	2/42	52.3mg	7/39	105.mg	8/39			
b	1556	1.13gm	n.s.s.	2/42	52.3mg	0/39	105.mg	1/39			
c	1556	163.mg	n.s.s.	13/42	52.3mg	13/39	105.mg	16/39			
1368	1556	77.0mg	n.s.s.	0/10	46.0mg	7/15	92.0mg	8/24			
a	1556	273.mg	n.s.s.	1/31	46.0mg	0/30	92.0mg	0/39			
b	1556	136.mg	n.s.s.	15/31	46.0mg	11/30	92.0mg	20/39			
1369	TR131	6.69mg	14.3mg	0/20	15.6mg	19/50	28.6mg	34/50			
a	TR131	7.31mg	32.8mg	5/20	15.6mg	25/50	28.6mg	35/50			
b	TR131	6.69mg	14.3mg	0/20	15.6mg	19/50	28.6mg	34/50			liv:hpa,hpc,nnd.
c	TR131	n.s.s.	n.s.s.	0/20	15.6mg	0/50	28.6mg	0/50			lun:a/a,a/c.
1370	TR131	6.68mg	n.s.s.	0/20	15.6mg	7/50	26.4mg	17/50			
a	TR131	4.86mg	n.s.s.	0/20	15.6mg	15/50	26.4mg	22/50			
b	TR131	6.68mg	n.s.s.	0/20	15.6mg	7/50	26.4mg	17/50			liv:hpa,hpc,nnd.
c	TR131	35.8mg	n.s.s.	0/20	15.6mg	1/50	26.4mg	2/50			lun:a/a,a/c.
1371	80	6.80mg	16.3mg	1/97	32.5mg	54/60					Tomatis;jnci,52,883-891;1974
a	80	231.mg	n.s.s.	37/97	32.5mg	9/60					
b	80	8.34mg	32.3mg	48/97	32.5mg	54/60					
1372	80	18.4mg	103.mg	33/101	30.0mg	39/60					
a	80	136.mg	n.s.s.	53/101	30.0mg	19/60					
b	80	35.9mg	n.s.s.	89/101	30.0mg	48/60					
1373	TR131	15.5mg	n.s.s.	2/20	8.40mg	9/50	14.4mg	12/50			thy:fca,fcc. S
a	TR131	9.00mg	n.s.s.	16/20	8.40mg	36/50	14.4mg	29/50			
b	TR131	n.s.s.	n.s.s.	0/20	8.40mg	0/50	14.4mg	0/50			liv:hpa,hpc,nnd.
1374	TR131	13.6mg	n.s.s.	9/20	12.0mg	21/50	23.2mg	20/50			liv:hpa,hpc,nnd.
a	TR131	n.s.s.	n.s.s.	0/20	12.0mg	0/50	23.2mg	0/50			

DDT   50-29-3

	RefNum	LoConf	UpConf	Cntrl	1Dose	1Inc	2Dose	2Inc	3Dose	3Inc	Citation or Pathology / Brkly Code
1375	1179	59.9mg	n.s.s.	0/39	13.1mg	0/28	26.1mg	0/28	52.3mg	0/40	Cabral;tumo,68,5-10;1982
a	1179	154.mg	n.s.s.	5/39	13.1mg	3/28	26.1mg	2/28	52.3mg	8/40	
1376	1400	34.8mg	n.s.s.	3/6	28.5mg	3/17					Agthe;pseb,134,113-116;1970
1377	1556	152.mg	3.14gm	2/42	105.mg	10/36					Rossi;canr,43,776-781;1983
a	1556	551.mg	n.s.s.	0/42	105.mg	1/36					
b	1556	1.03gm	n.s.s.	2/42	105.mg	0/36					
c	1556	158.mg	n.s.s.	13/42	105.mg	14/36					
1378	1179	94.1mg	n.s.s.	0/40	11.5mg	0/30	23.0mg	3/31	(46.0mg	0/39)	Cabral;tumo,68,5-10;1982
a	1179	68.7mg	n.s.s.	3/40	11.5mg	5/30	23.0mg	8/31	46.0mg	11/39	
1379	1400	34.1mg	n.s.s.	0/11	26.0mg	3/30					Agthe;pseb,134,113-116;1970
1380	1556	883.mg	n.s.s.	0/31	92.0mg	0/35					Rossi;canr,43,776-781;1983
a	1556	883.mg	n.s.s.	1/31	92.0mg	0/35					
b	1556	167.mg	n.s.s.	15/31	92.0mg	15/35					
1381	TR131	29.8mg	n.s.s.	0/20	9.50mg	3/50	19.5mg	7/50			S
a	TR131	22.4mg	n.s.s.	2/20	9.50mg	8/50	19.5mg	10/50			
b	TR131	52.1mg	n.s.s.	0/20	9.50mg	1/50	19.5mg	3/50			liv:hpa,hpc,nnd.
c	TR131	n.s.s.	n.s.s.	0/20	9.50mg	1/50	19.5mg	0/50			lun:a/a,a/c.
1382	133	10.7mg	n.s.s.	0/16	19.3mg	4/18					Innes;ntis,1968/1969
a	133	42.3mg	n.s.s.	0/16	19.3mg	0/18					
b	133	9.05mg	526.mg	0/16	19.3mg	5/18					
1383	TR131	.345mg	n.s.s.	4/20	2.30mg	6/50	(4.40mg	4/50)			
a	TR131	4.67mg	n.s.s.	2/20	2.30mg	1/50	4.40mg	1/50			liv:hpa,hpc,nnd.
b	TR131	4.67mg	n.s.s.	1/20	2.30mg	1/50	4.40mg	0/50			lun:a/a,a/c.
1384	133	4.24mg	23.7mg	0/16	17.9mg	10/18					Innes;ntis,1968/1969
a	133	20.7mg	n.s.s.	0/16	17.9mg	1/18					
b	133	20.7mg	n.s.s.	0/16	17.9mg	1/18					

```
     Spe Strain  Site   Xpo+Xpt                                           TD50    2Tailpvl
        Sex   Route  Hist     Notes                                              DR      AuOp
  c  M m b6c orl tba mix 80w80 evx                                        7.69mg   P<.0005
1385 M f b6a orl --- rts 80w80 evx            .   +      .                19.3mg   P<.003  +
a    M f b6a orl liv hpt 80w80 evx                                      137.mg    P<.3
b    M f b6a orl lun ade 80w80 evx                                       no dre   P=1.
c    M f b6a orl tba mix 80w80 evx                                       21.3mg   P<.07
1386 M m b6a orl liv hpt 80w80 evx           .   ±                       16.7mg   P<.02   +
a    M m b6a orl lun ade 80w80 evx                                       no dre   P=1.
b    M m b6a orl tba mix 80w80 evx                                       17.9mg   P<.07
1387 M f bal eat liv lct 31m31 eg                      . +   .           59.5mg * P<.0005+
a    M f bal eat lun ade 31m31 g                                         no dre   P=1.   -
b    M f bal eat tba mix 31m31 g                                         no dre   P=1.
1388 M f cf1 eat liv mix 26m26 e                  . +   .                 5.82mg * P<.0005+
a    M f cf1 eat liv lct 26m26 e                                         20.0mg * P<.0005+
b    M f cf1 eat lun tum 26m26 e                                         no dre   P=1.   -
1389 M f cf1 eat liv hpt 31m31 eg                   . +  .               43.0mg * P<.0005+
a    M f cf1 eat lun tum 31m31 eg                                       121.mg  * P<.04   -
b    M f cf1 eat ute tum 31m31 eg                                        no dre   P=1.   -
c    M f cf1 eat tba mix 31m31 eg                                       106.mg  * P<.5
1390 M f cf1 eat liv mix 26m26 e                  . +   .                 9.11mg * P<.0005+
a    M f cf1 eat liv lpb 26m26 e                                         76.5mg * P<.007
b    M f cf1 eat lun ade 26m26 e                                         no dre   P=1.   -
c    M f cf1 eat lun car 26m26 e                                         no dre   P=1.   -
1391 M f cf1 eat liv hpt 7m28                       . +   .              34.2mg   P<.0005+
a    M f cf1 eat tba tum 7m28                                            no dre   P=1.
1392 M f cf1 eat liv hpt 30w95                     . +   .               26.3mg   P<.0005+
a    M f cf1 eat tba tum 30w95                                           17.9mg   P<.4
1393 M f cf1 eat liv hpt 30w65                     .   +                  52.2mg   P<.01   +
a    M f cf1 eat tba tum 30w65                                          150.mg    P<.9
1394 M m cf1 eat liv mix 26m26 e                  . +   .                 8.04mg   P<.0005+
a    M m cf1 eat liv lct 26m26 e                                         30.3mg   P<.003  +
b    M m cf1 eat lun tum 26m26 e                                         no dre   P=1.   -
1395 M m cf1 eat liv hpt 29m31 e                    . +   .              34.7mg * P<.0005+
a    M m cf1 eat lun tum 29m31 e                                         no dre   P=1.
b    M m cf1 eat tba mix 29m31 e                                         no dre   P=1.
1396 M m cf1 eat liv mix 26m26 e                  . +   .                15.0mg * P<.0005+
a    M m cf1 eat liv lpb 26m26 e                                         72.6mg * P<.02
b    M m cf1 eat lun ade 26m26 e                                         49.3mg * P<.4
c    M m cf1 eat lun car 26m26 e                                         no dre   P=1.   -
1397 M m cf1 eat liv hpt 7m28                     . +      .             12.5mg   P<.002  +
a    M m cf1 eat tba tum 7m28                                            no dre   P=1.
1398 M m cf1 eat liv hpt 30w95                    . +   .                 6.70mg   P<.0005+
a    M m cf1 eat tba tum 30w95                                           4.59mg   P<.02
1399 M m cf1 eat liv hpt 30w65                   . +  .                   4.55mg   P<.0005+
a    M m cf1 eat tba tum 30w65                                           5.35mg   P<.02
1400 R f osm eat TBA MXB 18m26 v                         :>              no dre   P=1.   -
a    R f osm eat liv MXB 18m26 v                                         no dre   P=1.
1401 R f osm eat liv hpt 27m27                              .>          256.mg    P<.3   -
a    R f osm eat tba mix 27m27                                           no dre   P=1.
1402 R f osm eat liv mal 24m24                           .>              no dre   P=1.   -
1403 R m osm eat TBA MXB 18m26 v                      :>                 no dre   P=1.
a    R m osm eat liv MXB 18m26 v                                         no dre   P=1.
1404 R m osm eat liv tum 27m27                              .>           no dre   P=1.   -
a    R m osm eat tba mix 27m27                                           no dre   P=1.
1405 R m osm eat liv mal 24m24                           .>              no dre   P=1.   -
1406 R b alb eat liv tum 60w60 k          .>                             no dre   P=1.   -
1407 R m f34 eat liv nnd 17m24 er                        .>              56.2mg   P<.2   -
1408 R f por eat liv lct 33m33                       . +   .            140.mg  * P<.002  +
a    R f por eat tba mix 33m33                                           no dre   P=1.
1409 R m por eat liv lct 33m33                              .>          902.mg  * P<.4   -
a    R m por eat tba mix 33m33                                           90.1mg * P<.3
1410 R f wis eat liv hpt 34m34 ev                   . +   .              57.2mg   P<.0005+
a    R f wis eat tba mix 34m34 v                                        116.mg    P<.4
1411 R m wis eat liv hpt 34m34 e                    . +   .              92.6mg   P<.0005+
a    R m wis eat tba mix 34m34                                           no dre   P=1.


DECABROMODIPHENYL OXIDE       100ng..:..1ug...:..10......:.100......:..1mg...:..10......:..100....:..1g......:..10
1412 M f b6c eat TBA MXB 24m24                                      :> no dre   P=1.   -
a    M f b6c eat liv MXB 24m24                                         33.4gm * P<.5
b    M f b6c eat lun MXB 24m24                                         no dre   P=1.
1413 M m b6c eat thy MXA 24m24                                        :20.5gm * P<.09   e
a    M m b6c eat liv MXA 24m24                                         11.2gm * P<.3    e
b    M m b6c eat TBA MXB 24m24                                         9.06gm * P<.4
c    M m b6c eat liv MXB 24m24                                         11.2gm * P<.3
d    M m b6c eat lun MXB 24m24                                         no dre   P=1.
1414 R f f34 eat liv MXA 24m24                                    : + 6.58gm * P<.004
a    R f f34 eat liv nnd 24m24                                        7.74gm * P<.005 p
b    R f f34 eat TBA MXB 24m24                                        4.38gm * P<.5
c    R f f34 eat liv MXB 24m24                                        6.58gm * P<.004
1415 R m f34 eat liv nnd 24m24                                 : + 2.13gm * P<.0005p
a    R m f34 eat liv MXA 24m24                                        2.22gm * P<.0005
b    R m f34 eat pan ana 24m24                                        13.9gm * P<.02
```

	RefNum	LoConf	UpConf	Cntrl	1Dose	1Inc	2Dose	2Inc			Citation or Pathology	Brkly Code
c	133	3.72mg	19.2mg	0/16	17.9mg	11/18						
1385	133	7.78mg	106.mg	0/17	19.3mg	6/18						
a	133	22.2mg	n.s.s.	0/17	19.3mg	1/18						
b	133	42.3mg	n.s.s.	1/17	19.3mg	0/18						
c	133	7.42mg	n.s.s.	2/17	19.3mg	7/18						
1386	133	6.56mg	n.s.s.	1/18	17.9mg	7/18						
a	133	18.5mg	n.s.s.	2/18	17.9mg	2/18						
b	133	6.21mg	n.s.s.	3/18	17.9mg	8/18						
1387	88	37.8mg	101.mg	0/50	.260mg	0/58	2.60mg	1/50	32.5mg	28/57	Terracini;ijcn,11,747-764;1973	
a	88	238.mg	n.s.s.	20/62	.260mg	20/63	2.60mg	20/61	32.5mg	14/63		
b	88	73.1mg	n.s.s.	56/62	.260mg	51/63	2.60mg	48/61	32.5mg	48/63		
1388	89	3.06mg	13.0mg	10/44	12.6mg	26/30					Thorpe;fctx,11,433-442;1973	
a	89	10.2mg	47.1mg	0/44	12.6mg	12/30						
b	89	40.4mg	n.s.s.	27/44	12.6mg	9/30						
1389	90	28.1mg	71.8mg	2/56	.260mg	3/56	1.30mg	2/59	6.50mg	7/55	32.5mg 31/49	Tomatis;ijcn,10,489-506;1972
a	90	47.7mg	n.s.s.	13/56	.260mg	17/56	1.30mg	22/59	6.50mg	23/55	32.5mg 23/49	
b	90	334.mg	n.s.s.	2/56	.260mg	8/56	1.30mg	3/59	6.50mg	4/55	32.5mg 2/49	
c	90	20.1mg	n.s.s.	45/56	.260mg	50/56	1.30mg	52/59	6.50mg	48/55	32.5mg 44/49	
1390	103	5.53mg	18.6mg	8/47	6.50mg	15/30	13.0mg	24/32			Walker;fctx,11,415-432;1973	
a	103	31.2mg	895.mg	0/47	6.50mg	2/30	13.0mg	4/32				
b	103	52.1mg	n.s.s.	19/47	6.50mg	6/30	13.0mg	7/32				
c	103	48.8mg	n.s.s.	3/47	6.50mg	5/30	13.0mg	1/32				
1391	1012	16.7mg	99.9mg	1/90	8.13mg	11/54					Tomatis;zkko,82,25-35;1974	
a	1012	13.0mg	n.s.s.	77/90	8.13mg	41/54						
1392	1012m	13.2mg	64.3mg	0/72	10.3mg	11/55						
a	1012m	4.65mg	n.s.s.	52/72	10.3mg	44/55						
1393	1012n	18.0mg	5.13gm	0/69	15.0mg	4/54						
a	1012n	9.20mg	n.s.s.	27/69	15.0mg	22/54						
1394	89	4.17mg	21.3mg	11/45	11.7mg	23/30					Thorpe;fctx,11,433-442;1973	
a	89	13.1mg	195.mg	2/45	11.7mg	9/30						
b	89	29.5mg	n.s.s.	27/45	11.7mg	11/30						
1395	90	19.4mg	86.6mg	12/55	.240mg	25/58	1.20mg	28/53	6.00mg	24/53	30.0mg 38/50	Tomatis;ijcn,10,489-506;1972
a	90	151.mg	n.s.s.	23/55	.240mg	38/58	1.20mg	29/53	6.00mg	27/53	30.0mg 19/50	
b	90	24.9mg	n.s.s.	46/55	.240mg	53/58	1.20mg	48/53	6.00mg	49/53	30.0mg 44/50	
1396	103	8.40mg	43.3mg	6/47	6.00mg	12/32	12.0mg	17/32			Walker;fctx,11,415-432;1973	
a	103	29.6mg	n.s.s.	0/47	6.00mg	3/32	12.0mg	3/32				
b	103	12.9mg	n.s.s.	18/47	6.00mg	13/32	12.0mg	16/32				
c	103	30.6mg	n.s.s.	0/47	6.00mg	0/32	12.0mg	0/32				
1397	1012	6.40mg	53.1mg	33/98	7.50mg	37/60					Tomatis;zkko,82,25-35;1974	
a	1012	8.22mg	n.s.s.	83/98	7.50mg	49/60						
1398	1012m	3.86mg	15.9mg	24/83	9.47mg	41/60						
a	1012m	1.90mg	n.s.s.	65/83	9.47mg	56/60						
1399	1012n	2.76mg	9.13mg	12/70	13.8mg	38/60						
a	1012n	2.39mg	n.s.s.	42/70	13.8mg	48/60						
1400	TR131	16.8mg	n.s.s.	16/20	7.40mg	38/50	15.0mg	27/50				
a	TR131	n.s.s.	n.s.s.	0/20	7.40mg	0/50	15.0mg	0/50			liv:hpa,hpc,nnd.	
1401	21	41.7mg	n.s.s.	0/30	10.0mg	1/30					Deichmann;txap,11,88-103;1967	
a	21	24.3mg	n.s.s.	13/30	10.0mg	9/30						
1402	84a	24.7mg	n.s.s.	1/30	4.00mg	0/30					Radomski;txap,7,652-656;1965	
1403	TR131	14.2mg	n.s.s.	10/20	8.80mg	29/50	18.4mg	32/50				
a	TR131	n.s.s.	n.s.s.	0/20	8.80mg	0/50	18.4mg	0/50			liv:hpa,hpc,nnd.	
1404	21	62.6mg	n.s.s.	0/30	8.00mg	0/30					Deichmann;txap,11,88-103;1967	
a	21	37.8mg	n.s.s.	1/30	8.00mg	1/30						
1405	84a	19.8mg	n.s.s.	0/30	3.20mg	0/30					Radomski;txap,7,652-656;1965	
1406	1457	.148mg	n.s.s.	0/6	.360mg	0/6					Cameron;bmjl,2,819-821;1951	
1407	1803	17.6mg	n.s.s.	2/28	14.0mg	6/28					Shivapurkar;carc,7,547-550;1986	
1408	1178	74.0mg	552.mg	0/38	6.25mg	2/30	12.5mg	4/30	25.0mg	7/38	Cabral;tumo,68,11-17;1982	
a	1178	47.0mg	n.s.s.	32/38	6.25mg	26/30	12.5mg	27/30	25.0mg	27/38		
1409	1178	174.mg	n.s.s.	1/38	5.00mg	0/30	10.0mg	1/30	20.0mg	2/38		
a	1178	24.5mg	n.s.s.	20/38	5.00mg	19/30	10.0mg	18/30	20.0mg	25/38		
1410	85	31.1mg	122.mg	0/32	25.0mg	15/34					Rossi;ijcn,19,179-185;1977	
a	85	29.5mg	n.s.s.	19/35	25.0mg	23/35						
1411	85	43.5mg	272.mg	0/35	20.0mg	9/36						
a	85	42.6mg	n.s.s.	19/36	20.0mg	19/37						

DECABROMODIPHENYL OXIDE   1163-19-5

	RefNum	LoConf	UpConf	Cntrl	1Dose	1Inc	2Dose	2Inc	Citation or Pathology	Brkly Code
1412	TR309	5.88gm	n.s.s.	35/50	3.22gm	37/50	6.44gm	35/50		
a	TR309	6.93gm	n.s.s.	8/50	3.22gm	13/50	6.44gm	13/50	liv:hpa,hpc,nnd.	
b	TR309	22.4gm	n.s.s.	6/50	3.22gm	4/50	6.44gm	4/50	lun:a/a,a/c.	
1413	TR309	8.82gm	n.s.s.	0/50	2.97gm	4/50	6.00gm	3/50	thy:fca,fcc.	
a	TR309	3.50gm	n.s.s.	8/50	2.97gm	22/50	6.00gm	18/50	liv:hpa,hpc.	
b	TR309	2.36gm	n.s.s.	21/50	2.97gm	37/50	6.00gm	36/50		
c	TR309	3.50gm	n.s.s.	8/50	2.97gm	22/50	6.00gm	18/50	liv:hpa,hpc,nnd.	
d	TR309	12.7gm	n.s.s.	5/50	2.97gm	4/50	6.00gm	5/50	lun:a/a,a/c.	
1414	TR309	3.25gm	48.3gm	1/50	1.24gm	5/50	2.48gm	9/50	liv:hpc,nnd.	S
a	TR309	3.68gm	61.9gm	1/50	1.24gm	3/50	2.48gm	9/50		
b	TR309	1.06gm	n.s.s.	49/50	1.24gm	49/50	2.48gm	50/50		
c	TR309	3.25gm	48.3gm	1/50	1.24gm	5/50	2.48gm	9/50	liv:hpa,hpc,nnd.	
1415	TR309	1.22gm	5.14gm	1/50	990.mg	7/50	1.98gm	15/50		
a	TR309	1.23gm	6.55gm	2/50	990.mg	8/50	1.98gm	15/50	liv:hpc,nnd.	S
b	TR309	4.77gm	n.s.s.	0/50	990.mg	0/50	1.98gm	4/50		S

```
         Spe Strain  Site    Xpo+Xpt                                                      TD50      2Tailpvl
             Sex  Route   Hist    Notes                                                         DR    AuOp
   c    R m f34 eat TBA MXB 24m24                                                          2.23gm * P<.3
   d    R m f34 eat liv MXB 24m24                                                          2.22gm * P<.0005

DEHYDROEPIANDROSTERONE ACETATE  100ng..:..1ug....:..10....:...100....:..1mg....:..10....:...100....:...1g.....:..10
1416 R m f34 eat liv mix 84w84 er                                        .   +   .        31.4mg   P<.0005+
   a R m f34 eat liv hpc 84w84 er                                                         43.0mg   P<.0005+
   b R m f34 eat tes ldc 84w84 er                                                         no dre   P=1.   -

DELTAMETHRIN                    100ng..:..1ug....:..10....:...100....:..1mg....:..10....:...100....:...1g.....:..10
1417 M f cb6 gav --- lym 24m26                                        .>                  no dre   P=1.   -
   a M f cb6 gav liv hpt 24m26                                                            no dre   P=1.   -
   b M f cb6 gav lun ade 24m26                                                            no dre   P=1.   -
   c M f cb6 gav tba tum 24m26                                                            no dre   P=1.   -
1418 M m cb6 gav lun ade 24m26                                           .>               no dre   P=1.   -
   a M m cb6 gav liv hpt 24m26                                                            no dre   P=1.   -
   b M m cb6 gav tba tum 24m26                                                            no dre   P=1.   -
1419 R f bdf gav thy ade 24m26                                .   +            .          16.9mg * P<.01 -
   a R f bdf gav liv hpt 24m26                                                            no dre   P=1.   -
   b R f bdf gav tba tum 24m26                                                            36.1mg * P<.7  -
1420 R m bdf gav thy ade 24m26                             .   +    .                     4.60mg \ P<.003 -
   a R m bdf gav liv hpt 24m26                                                            no dre   P=1.   -
   b R m bdf gav tba tum 24m26                                                            no dre   P=1.   -

DESERPIDINE                     100ng..:..1ug....:..10....:...100....:..1mg....:..10....:...100....:...1g.....:..10
1421 R b wis eat mix mix 78w78 r                    .>                                    1.28mg   P<.6   -

DEXTRAN                         100ng..:..1ug....:..10....:...100....:..1mg....:..10....:...100....:...1g.....:..10
1422 R f aci eat ubl tum 68w68                                                .   ±       1.64gm   P<.02  -
1423 R m aci eat ubl tum 68w68                                                   .>       3.19gm   P<.4   -

DEXTRAN SULFATE SODIUM (DS-M-1) 100ng..:..1ug....:..10....:...100....:..1mg....:..10....:...100....:...1g.....:..10
1424 R b aci eat itn mix 94w94 e                                          .   +   .       191.mg   P<.0005+
   a R b aci eat clr pam 94w94 e                                                          331.mg   P<.0005
   b R b aci eat clr sqc 94w94 e                                                          1.76gm   P<.04  +
1425 R f aci eat clr pam 68w68                                            .   +   .        219.mg   P<.0005+
   a R f aci eat clr adc 68w68                                                            373.mg   P<.0005+
   b R f aci eat clr ade 68w68                                                            977.mg   P<.004 +
   c R f aci eat cec ade 68w68                                                            1.27gm   P<.009
   d R f aci eat clr sqc 68w68                                                            1.76gm   P<.03  +
1426 R m aci eat clr adc 68w68                                            .   +   .        182.mg   P<.0005+
   a R m aci eat clr pam 68w68                                                            466.mg   P<.0005+
   b R m aci eat clr sqc 68w68                                                            2.05gm   P<.06  +
   c R m aci eat clr ade 68w68                                                            1.70gm   P<.2   +

DEXTRAN SULFATE SODIUM (DST-H)  100ng..:..1ug....:..10....:...100....:..1mg....:..10....:...100....:...1g.....:..10
1427 R f aci eat ubl tum 68w68                                                .   ±       2.56gm   P<.06  -
1428 R m aci eat cec ade 68w68                                                   .>       4.25gm   P<.2   -

DEXTRAN SULFATE SODIUM (KMDS-H) 100ng..:..1ug....:..10....:...100....:..1mg....:..10....:...100....:...1g.....:..10
1429 R f aci eat ubl tum 68w68                                                .   ±       1.64gm   P<.02  -
1430 R m aci eat clr pam 68w68                                                   .>       4.25gm   P<.2   -

N-1-DIACETAMIDOFLUORENE         100ng..:..1ug....:..10....:...100....:..1mg....:..10....:...100....:...1g.....:..10
1431 R f buf eat pit ade 53w93 e                              .   ±                       8.63mg   P<.03
   a R f buf eat mgl adc 53w93 e                                                          19.0mg   P<.03  +
   b R f buf eat liv hem 53w93 e                                                          61.2mg   P<.3
   c R f buf eat edu sqc 53w93 e                                                          61.2mg   P<.3   +

DIACETYL HYDRAZINE              100ng..:..1ug....:..10....:...100....:..1mg....:..10....:...100....:...1g.....:..10
1432 M f swi gav lun mix 95w95 rs                                        .>               no dre   P=1.   -
1433 M f swi gav lun mix 34w95 rs                                     .>                  no dre   P=1.   -
1434 M m swi gav lun mix 95w95 rs                                        .>               411.mg * P<.7  -
1435 M m swi gav lun mix 34w95 rs                                  .>                     144.mg   P<.8   -

DIALLATE                        100ng..:..1ug....:..10....:...100....:..1mg....:..10....:...100....:...1g.....:..10
1436 M f b6a orl liv hpt 85w85 evx                                               .>       516.mg   P<.3
   a M f b6a orl lun ade 85w85 evx                                                        4.26gm   P<1.
   b M f b6a orl tba mix 85w85 evx                                                        1.99gm   P<.9
1437 M m b6a orl liv hpt 84w84 evx                                        .   +   .        43.0mg   P<.002 +
   a M m b6a orl lun ade 84w84 evx                                                        243.mg   P<.4
   b M m b6a orl tba mix 84w84 evx                                                        35.4mg   P<.003
1438 M f b6c orl liv mix 83w83 evx                                        .   ±           164.mg   P<.04
   a M f b6c orl lun ade 83w83 evx                                                        255.mg   P<.09
   b M f b6c orl tba mix 83w83 evx                                                        90.8mg   P<.006
1439 M m b6c orl liv hpt 84w84 evx                                    .   +   .           19.4mg   P<.0005+
   a M m b6c orl lun ade 84w84 evx                                                        113.mg   P<.02
   b M m b6c orl tba mix 84w84 evx                                                        15.6mg   P<.0005

DIALLYL PHTHALATE               100ng..:..1ug....:..10....:...100....:..1mg....:..10....:...100....:...1g.....:..10
1440 M f b6c gav for ppn 24m25                                                   :   +hist 2.73gm * P<.1  a
   a M f b6c gav TBA MXB 24m25                                                            10.1gm * P<1.
```

	RefNum	LoConf	UpConf	Cntrl	1Dose	1Inc	2Dose	2Inc	Citation or Pathology
									Brkly Code
c	TR309	681.mg	n.s.s.	49/50	990.mg	50/50	1.98gm	49/50	
d	TR309	1.23gm	6.55gm	2/50	990.mg	8/50	1.98gm	15/50	liv:hpa,hpc,nnd.

DEHYDROEPIANDROSTERONE ACETATE   853-23-6

	RefNum	LoConf	UpConf	Cntrl	1Dose	1Inc			Citation
1416	2057	11.8mg	85.1mg	0/8	180.mg	12/13			Rao;canr,52,2977-2979;1992a/Rao 1992b/pers.comm.
a	2057	18.4mg	114.mg	0/8	180.mg	11/13			
b	2057	315.mg	n.s.s.	8/8	180.mg	0/13			

DELTAMETHRIN   52918-63-5

	RefNum	LoConf	UpConf	Cntrl	1Dose	1Inc	2Dose	2Inc	3Dose	3Inc	Citation
1417	1965	13.7mg	n.s.s.	19/50	.652mg	12/30	2.61mg	17/30	5.21mg	15/50	Cabral;clet,49,147-152;1990/pers.comm.
a	1965	53.4mg	n.s.s.	2/50	.652mg	1/30	2.61mg	2/30	5.21mg	0/50	
b	1965	53.4mg	n.s.s.	2/50	.652mg	1/30	2.61mg	2/30	5.21mg	0/50	
c	1965	13.0mg	n.s.s.	22/50	.652mg	14/30	2.61mg	20/30	5.21mg	17/50	
1418	1965	29.6mg	n.s.s.	2/50	.652mg	3/30	2.61mg	3/30	5.21mg	2/50	
a	1965	32.0mg	n.s.s.	4/50	.652mg	2/30	2.61mg	0/30	5.21mg	4/50	
b	1965	28.3mg	n.s.s.	25/50	.652mg	10/30	2.61mg	11/30	5.21mg	11/50	
1419	1965	7.94mg	1.16gm	4/50	1.95mg	4/50	3.91mg	14/50			
a	1965	16.1mg	n.s.s.	0/50	1.95mg	0/50	3.91mg	0/50			
b	1965	4.68mg	n.s.s.	30/50	1.95mg	25/50	3.91mg	32/50			
1420	1965	2.25mg	28.9mg	6/50	1.95mg	19/50	(3.91mg	10/50)			
a	1965	16.1mg	n.s.s.	0/50	1.95mg	0/50	3.91mg	0/50			
b	1965	14.7mg	n.s.s.	23/50	1.95mg	22/50	3.91mg	14/50			

DESERPIDINE   131-01-1

	RefNum	LoConf	UpConf	Cntrl	1Dose	1Inc			Citation
1421	1188	.272mg	n.s.s.	17/130	.100mg	36/230			Tuchmann-Duplessis;adsc,254,1535-1537;1962

DEXTRAN   9004-54-0

	RefNum	LoConf	UpConf	Cntrl	1Dose	1Inc			Citation
1422	1540	494.mg	n.s.s.	0/20	1.25gm	3/15			Hirono;clet,18,29-34;1983
1423	1540	540.mg	n.s.s.	1/20	1.00gm	2/15			

DEXTRAN SULFATE SODIUM (DS-M-1)   (mol. wt. = 54,000) 9011-18-1

	RefNum	LoConf	UpConf	Cntrl	1Dose	1Inc			Citation
1424	1482	110.mg	361.mg	0/20	450.mg	22/30			Hirono;carc,3,353-355;1982
a	1482	181.mg	691.mg	0/20	450.mg	16/30			
b	1482	607.mg	n.s.s.	0/20	450.mg	4/30			
1425	1540	103.mg	525.mg	0/20	1.25gm	13/16			Hirono;clet,18,29-34;1983
a	1540	174.mg	982.mg	0/20	1.25gm	10/16			
b	1540	368.mg	5.89gm	0/20	1.25gm	5/16			
c	1540	437.mg	31.1gm	0/20	1.25gm	4/16			
d	1540	532.mg	n.s.s.	0/20	1.25gm	3/16			
1426	1540	84.0mg	452.mg	0/20	1.00gm	12/15			
a	1540	197.mg	1.53gm	0/20	1.00gm	7/15			
b	1540	502.mg	n.s.s.	0/20	1.00gm	2/15			
c	1540	419.mg	n.s.s.	1/20	1.00gm	3/15			

DEXTRAN SULFATE SODIUM (DST-H)   (mol. wt. = 9500) 9011-18-1

	RefNum	LoConf	UpConf	Cntrl	1Dose	1Inc			Citation
1427	1540	628.mg	n.s.s.	0/20	1.25gm	2/15			Hirono;clet,18,29-34;1983
1428	1540	691.mg	n.s.s.	0/20	1.00gm	1/15			

DEXTRAN SULFATE SODIUM (KMDS-H)   (mol. wt. = 520,000) 9011-18-1

	RefNum	LoConf	UpConf	Cntrl	1Dose	1Inc			Citation
1429	1540	494.mg	n.s.s.	0/20	1.25gm	3/15			Hirono;clet,18,29-34;1983
1430	1540	691.mg	n.s.s.	0/20	1.00gm	1/15			

N-1-DIACETAMIDOFLUORENE   63019-65-8

	RefNum	LoConf	UpConf	Cntrl	1Dose	1Inc			Citation
1431	144	3.14mg	n.s.s.	2/18	7.21mg	7/16			Morris;jnci,24,149-180;1960
a	144	5.74mg	n.s.s.	0/18	7.21mg	3/16			
b	144	9.96mg	n.s.s.	0/18	7.21mg	1/16			
c	144	9.96mg	n.s.s.	0/18	7.21mg	1/16			

DIACETYL HYDRAZINE   3148-73-0

	RefNum	LoConf	UpConf	Cntrl	1Dose	1Inc	2Dose	2Inc	Citation
1432	1661m	59.0mg	n.s.s.	3/20	22.9mg	0/15			Bhide;clet,23,235-240;1984/pers.comm.
1433	1661n	29.5mg	n.s.s.	3/20	11.4mg	0/15			
1434	1661m	43.8mg	n.s.s.	2/20	19.0mg	0/15	26.2mg	3/15	
1435	1661n	10.9mg	n.s.s.	2/20	9.52mg	2/15			

DIALLATE   (avadex) 2303-16-4

	RefNum	LoConf	UpConf	Cntrl	1Dose	1Inc			Citation
1436	1219	84.0mg	n.s.s.	0/17	77.8mg	1/15			Innes;ntis,1968/1969
a	1219	94.1mg	n.s.s.	1/17	77.8mg	1/15			
b	1219	73.2mg	n.s.s.	2/17	77.8mg	2/15			
1437	1219	19.5mg	172.mg	1/18	72.5mg	10/18			
a	1219	50.7mg	n.s.s.	2/18	72.5mg	4/18			
b	1219	15.6mg	210.mg	3/18	72.5mg	12/18			
1438	1219	49.4mg	n.s.s.	0/16	77.9mg	3/16			
a	1219	62.5mg	n.s.s.	0/16	77.9mg	2/16			
b	1219	34.1mg	898.mg	0/16	77.9mg	5/16			
1439	1219	9.11mg	46.4mg	0/16	72.5mg	13/16			
a	1219	38.7mg	n.s.s.	0/16	72.5mg	4/16			
b	1219	7.03mg	37.3mg	0/16	72.5mg	14/16			

DIALLYL PHTHALATE   131-17-9

	RefNum	LoConf	UpConf	Cntrl	1Dose	1Inc	2Dose	2Inc	Citation
1440	TR242	826.mg	n.s.s.	0/50	104.mg	1/50	208.mg	2/50	
a	TR242	215.mg	n.s.s.	31/50	104.mg	28/50	208.mg	32/50	

```
      Spe Strain  Site   Xpo+Xpt                                                                    TD50    2Tailpvl
          Sex  Route  Hist   Notes                                                                          DR   AuOp
  b   M f b6c gav liv MXB 24m25                                                                     2.93gm * P<.4
  c   M f b6c gav lun MXB 24m25                                                                     4.18gm * P<.4
1441  M m b6c gav --- mno 24m25                                               :       ±            816.mg * P<.09  e
  a   M m b6c gav liv hpa 24m25                                                                     2.45gm * P<.04
  b   M m b6c gav for ppn 24m25                                                             +hist  2.46gm * P<.08  a
  c   M m b6c gav TBA MXB 24m25                                                                     717.mg * P<.4
  d   M m b6c gav liv MXB 24m25                                                                     6.68gm * P<.9
  e   M m b6c gav lun MXB 24m25                                                                     8.52gm * P<.9
1442  R f f34 gav --- mnl 24m24                                               :  ·  ±             138.mg * P<.05  e
  a   R f f34 gav TBA MXB 24m24                                                                     163.mg * P<.5
  b   R f f34 gav liv MXB 24m24                                                                     no dre  P=1.
1443  R m f34 gav TBA MXB 24m24                                                      :>           2.99gm * P<1.   -
  a   R m f34 gav liv MXB 24m24                                                                     no dre  P=1.

1,1-DIALLYLHYDRAZINE               100ng..:..1ug....:..10......:..100...:..1mg....:..10......:..100....:..1g.....:..10
1444  M f swa wat lun mix 23m23 es                                         . + .                   34.9mg  P<.0005+
  a   M f swa wat lun ade 23m23 es                                                                  37.8mg  P<.0005
  b   M f swa wat lun adc 23m23 es                                                                  112.mg  P<.0005
  c   M f swa wat for sqp 23m23 es                                                                  332.mg  P<.1    +
  d   M f swa wat liv ang 23m23 es                                                                  no dre  P=1.
  e   M f swa wat liv agm 23m23 es                                                                  no dre  P=1.    -
1445  M m swa wat lun mix 93w93 es                                       . + .                      25.7mg  P<.0005+
  a   M m swa wat lun ade 93w93 es                                                                  29.8mg  P<.0005
  b   M m swa wat lun adc 93w93 es                                                                  38.9mg  P<.0005
  c   M m swa wat for sqp 93w93 es                                                                  48.1mg  P<.0005+
  d   M m swa wat liv ang 93w93 es                                                                  723.mg  P<.5    -
  e   M m swa wat liv agm 93w93 es                                                                  no dre  P=1.    -

1,2-DIALLYLHYDRAZINE.2HCl          100ng..:..1ug....:..10......:..100...:..1mg....:..10......:..100....:..1g.....:..10
1446  M f swa wat lun mix 83w83 es                                         . + .                   33.8mg  P<.0005+
  a   M f swa wat lun ade 83w83 es                                                                  47.9mg  P<.0005
  b   M f swa wat lun adc 83w83 es                                                                  78.3mg  P<.0005
  c   M f swa wat liv hpt 83w83 es                                                                  409.mg  P<.4    -
  d   M f swa wat liv ang 83w83 es                                                                  no dre  P=1.    -
  e   M f swa wat liv agm 83w83 es                                                                  no dre  P=1.    -
1447  M m swa wat lun mix 82w82 es                                         . + .                   33.9mg  P<.0005+
  a   M m swa wat lun ade 82w82 es                                                                  35.4mg  P<.0005
  b   M m swa wat lun adc 82w82 es                                                                  60.0mg  P<.0005
  c   M m swa wat liv mix 82w82 es                                                                  no dre  P=1.
  d   M m swa wat liv agm 82w82 es                                                                  no dre  P=1.    -

DIALLYLNITROSAMINE                 100ng..:..1ug....:..10......:..100...:..1mg....:..10......:..100....:..1g.....:..10
1448  H b syg gav trh ppp 53w59 ae                                     . + .                       2.84mg * P<.0005
  a   H b syg gav ncp adc 53w59 ae                                                                  2.86mg * P<.0005+
  b   H b syg gav lar ppp 53w59 ae                                                                  3.69mg Z P<.0005
  c   H b syg gav phr pam 53w59 ae                                                                  25.3mg * P<.2
  d   H b syg gav liv cho 53w59 ae                                                                  743.mg * P<1.
  e   H b syg gav lun ade 53w59 ae                                                                  no dre  P=1.
  f   H b syg gav tba mix 53w59 ae                                                                  1.54mg * P<.0005+
1449  R f bd9 wat npc mix 24m24 r                                          . + .                   32.1mg * P<.0005+
  a   R f bd9 wat npc adc 24m24 r                                           . + .                   33.6mg * P<.0005+
1450  R m bd9 wat npc mix 24m24 r                                           . + .                   36.0mg Z P<.0005+
  a   R m bd9 wat npc adc 24m24 r                                                                   46.9mg Z P<.0005+

4,6-DIAMINO-2-(5-NITRO-2-FURYL)-s-TRIAZINE..1_ug....:..10......:..100...:..1mg....:..10......:..100....:..1g.....:..10
1451  R f sda eat mgl mix 46w66 e                                      . + .                       1.71mg  P<.0005+
  a   R f sda eat mgl adc 46w66 e                                                                   4.49mg  P<.0005
  b   R f sda eat liv tum 46w66 e                                                                   no dre  P=1.
  c   R f sda eat tba mix 46w66 e                                                                   1.71mg  P<.0005

4,4'-DIAMINO-2,2'-STILBENEDISULFONIC  ACID, DISODIUM SALT....:..100...:..1mg....:..10......:..100.....:..1g.....:..10
1452  M f b6c eat TBA MXB 24m24                                                             :>     no dre  P=1.    -
  a   M f b6c eat liv MXB 24m24                                                                     no dre  P=1.
  b   M f b6c eat lun MXB 24m24                                                                     no dre  P=1.
1453  M m b6c eat liv hpa 24m24                                               :       ±           #2.57gm \ P<.03  -
  a   M m b6c eat TBA MXB 24m24                                                                     12.9gm * P<.8
  b   M m b6c eat liv MXB 24m24                                                                     9.72gm * P<.4
  c   M m b6c eat lun MXB 24m24                                                                     no dre  P=1.
1454  R f f34 eat TBA MXB 24m24                                                         :>        17.6gm * P<1.   -
  a   R f f34 eat liv MXB 24m24                                                                     no dre  P=1.
1455  R m f34 eat TBA MXB 24m24                                                         :>        5.81gm * P<.9   -
  a   R m f34 eat liv MXB 24m24                                                                     5.02gm * P<.2

2,4-DIAMINOANISOLE SULFATE         100ng..:..1ug....:..10......:..100...:..1mg....:..10......:..100....:..1g.....:..10
1456  M f b6c eat --- lym 78w96                                                          :  +  :   262.mg \ P<.005
  a   M f b6c eat thy MXA 78w96                                                                     1.06gm / P<.0005c
  b   M f b6c eat thy fca 78w96                                                                     1.45gm * P<.002  c
  c   M f b6c eat TBA MXB 78w96                                                                     311.mg * P<.02
  d   M f b6c eat liv MXB 78w96                                                                     7.35gm * P<.6
  e   M f b6c eat lun MXB 78w96                                                                     10.9gm * P<.9
1457  M m b6c eat thy fca 78w96                                                          :  +  :   791.mg / P<.0005c
```

	RefNum	LoConf	UpConf	Cntrl	1Dose	1Inc	2Dose	2Inc			Citation or Pathology
											Brkly Code
b	TR242	736.mg	n.s.s.	1/50	104.mg	2/50	208.mg	3/50			liv:hpa,hpc,nnd.
c	TR242	911.mg	n.s.s.	1/50	104.mg	0/50	208.mg	3/50			lun:a/a,a/c.
1441	TR242	304.mg	n.s.s.	6/50	104.mg	5/50	208.mg	12/50			s
a	TR242	740.mg	n.s.s.	0/50	104.mg	0/50	208.mg	3/50			
b	TR242	744.mg	n.s.s.	0/50	104.mg	1/50	208.mg	2/50			
c	TR242	180.mg	n.s.s.	23/50	104.mg	23/50	208.mg	26/50			
d	TR242	517.mg	n.s.s.	7/50	104.mg	5/50	208.mg	7/50			liv:hpa,hpc,nnd.
e	TR242	587.mg	n.s.s.	5/50	104.mg	4/50	208.mg	5/50			lun:a/a,a/c.
1442	TR284	58.3mg	n.s.s.	15/50	35.5mg	15/50	71.1mg	25/50			
a	TR284	37.9mg	n.s.s.	41/50	35.5mg	37/50	71.1mg	42/50			
b	TR284	371.mg	n.s.s.	5/50	35.5mg	0/50	71.1mg	3/50			liv:hpa,hpc,nnd.
1443	TR284	61.9mg	n.s.s.	36/50	35.5mg	37/50	71.1mg	29/50			
a	TR284	542.mg	n.s.s.	2/50	35.5mg	0/50	71.1mg	1/50			liv:hpa,hpc,nnd.

1,1-DIALLYLHYDRAZINE   5164-11-4

	RefNum	LoConf	UpConf	Cntrl	1Dose	1Inc					Citation or Pathology
1444	1676	20.6mg	70.2mg	25/99	62.5mg	38/50					Toth;acnr,1,259-262;1981
a	1676	22.6mg	74.3mg	20/99	62.5mg	36/50					
b	1676	58.3mg	312.mg	6/99	62.5mg	17/50					
c	1676	105.mg	n.s.s.	4/61	62.5mg	7/41					
d	1676	336.mg	n.s.s.	3/61	62.5mg	1/41					
e	1676	421.mg	n.s.s.	5/67	62.5mg	1/47					
1445	1676	15.0mg	53.2mg	26/96	52.1mg	38/50					
a	1676	17.9mg	58.3mg	16/96	52.1mg	34/50					
b	1676	22.8mg	80.1mg	12/96	52.1mg	29/50					
c	1676	27.0mg	97.4mg	0/63	52.1mg	17/38					
d	1676	120.mg	n.s.s.	4/68	52.1mg	4/42					
e	1676	254.mg	n.s.s.	7/63	52.1mg	1/38					

1,2-DIALLYLHYDRAZINE.2HCl   26072-78-6

	RefNum	LoConf	UpConf	Cntrl	1Dose	1Inc					Citation or Pathology
1446	1531	20.0mg	63.9mg	25/99	125.mg	40/47					Toth;onco,39,104-108;1982
a	1531	28.4mg	93.7mg	20/99	125.mg	35/47					
b	1531	46.2mg	154.mg	6/99	125.mg	25/47					
c	1531	66.3mg	n.s.s.	0/5	125.mg	1/8					
d	1531	389.mg	n.s.s.	3/32	125.mg	1/33					
e	1531	328.mg	n.s.s.	1/32	125.mg	1/33					
1447	1531	20.2mg	65.8mg	26/100	104.mg	40/50					
a	1531	21.8mg	64.0mg	16/100	104.mg	38/50					
b	1531	35.3mg	122.mg	12/100	104.mg	29/50					
c	1531	290.mg	n.s.s.	6/52	104.mg	2/36					
d	1531	202.mg	n.s.s.	4/47	104.mg	2/29					

DIALLYLNITROSAMINE   16338-97-9

	RefNum	LoConf	UpConf	Cntrl	1Dose	1Inc	2Dose	2Inc	3Dose	3Inc	Citation or Pathology
1448	1826	1.86mg	5.49mg	0/30	3.11mg	11/28	6.21mg	8/27	12.4mg	15/27	Grandjean;jnci,74,1043-1046;1985/pers.comm.
a	1826	1.85mg	4.74mg	0/30	3.11mg	1/28	6.21mg	13/27	12.4mg	18/27	
b	1826	2.33mg	11.8mg	0/30	3.11mg	13/28	6.21mg	1/27	12.4mg	14/27	
c	1826	9.60mg	n.s.s.	0/30	3.11mg	1/28	6.21mg	3/27	12.4mg	1/27	
d	1826	n.s.s.	n.s.s.	1/30	3.11mg	1/28	6.21mg	1/27	12.4mg	1/27	
e	1826	23.6mg	n.s.s.	0/30	3.11mg	1/28	6.21mg	0/27	12.4mg	0/27	
f	1826	.933mg	3.37mg	7/30	3.11mg	18/28	6.21mg	16/27	12.4mg	24/27	
1449	1714	20.9mg	53.2mg	0/20	20.0mg	6/20	40.0mg	17/20	80.0mg	13/20	Pour;zkko,109,5-8;1985
a	1714	21.8mg	56.3mg	0/20	20.0mg	6/20	40.0mg	16/20	80.0mg	13/20	
1450	1714	19.7mg	75.0mg	0/20	20.0mg	0/20	40.0mg	16/20	(80.0mg	15/20)	
a	1714	29.0mg	81.8mg	0/20	20.0mg	0/20	40.0mg	13/20	80.0mg	14/20	

4,6-DIAMINO-2-(5-NITRO-2-FURYL)-s-TRIAZINE   720-69-4

	RefNum	LoConf	UpConf	Cntrl	1Dose	1Inc					Citation or Pathology
1451	200a	.895mg	3.23mg	2/39	17.4mg	33/35					Cohen;jnci,51,403-417;1973
a	200a	2.66mg	8.28mg	0/39	17.4mg	23/35					
b	200a	50.6mg	n.s.s.	0/39	17.4mg	0/35					
c	200a	.895mg	3.23mg	2/39	17.4mg	33/35					

4,4'-DIAMINO-2,2'-STILBENEDISULFONIC ACID, DISODIUM SALT   7336-20-1

	RefNum	LoConf	UpConf	Cntrl	1Dose	1Inc	2Dose	2Inc			Citation or Pathology
1452	TR412	3.18gm	n.s.s.	34/50	803.mg	17/50	1.61gm	24/50			
a	TR412	12.0gm	n.s.s.	5/50	803.mg	0/50	1.61gm	2/50			liv:hpa,hpc,nnd.
b	TR412	8.73gm	n.s.s.	13/50	803.mg	3/50	1.61gm	5/50			lun:a/a,a/c.
1453	TR412	1.03gm	n.s.s.	2/50	741.mg	9/50	(1.48gm	6/50)			s
a	TR412	1.57gm	n.s.s.	23/50	741.mg	24/50	1.48gm	25/50			
b	TR412	2.47gm	n.s.s.	5/50	741.mg	11/50	1.48gm	8/50			liv:hpa,hpc,nnd.
c	TR412	5.18gm	n.s.s.	15/50	741.mg	4/50	1.48gm	10/50			lun:a/a,a/c.
1454	TR412	706.mg	n.s.s.	45/50	615.mg	44/50	1.23gm	50/50			
a	TR412	9.09gm	n.s.s.	3/50	615.mg	0/50	1.23gm	1/50			liv:hpa,hpc,nnd.
1455	TR412	499.mg	n.s.s.	44/50	492.mg	46/50	984.mg	49/50			
a	TR412	1.68gm	n.s.s.	1/50	492.mg	1/50	984.mg	5/50			liv:hpa,hpc,nnd.

2,4-DIAMINOANISOLE SULFATE   39156-41-7

	RefNum	LoConf	UpConf	Cntrl	1Dose	1Inc	2Dose	2Inc			Citation or Pathology
1456	TR84	116.mg	2.51gm	7/100	126.mg	14/50	(253.mg	9/50)			s
a	TR84	480.mg	3.18gm	0/100	126.mg	0/50	253.mg	8/50			thy:fca,fcc.
b	TR84	591.mg	5.87gm	0/100	126.mg	0/50	253.mg	6/50			
c	TR84	145.mg	n.s.s.	30/100	126.mg	25/50	253.mg	30/50			
d	TR84	1.03gm	n.s.s.	2/100	126.mg	1/50	253.mg	2/50			liv:hpa,hpc,nnd.
e	TR84	716.mg	n.s.s.	7/100	126.mg	5/50	253.mg	4/50			lun:a/a,a/c.
1457	TR84	383.mg	2.47gm	1/100	116.mg	0/50	234.mg	11/50			

```
       Spe Strain  Site    Xpo+Xpt                                                                    TD50    2Tailpvl
           Sex   Route  Hist    Notes                                                                        DR    AuOp
  a    M m b6c eat TBA MXB 78w96                                                                      354.mg * P<.04
  b    M m b6c eat liv MXB 78w96                                                                      926.mg * P<.2
  c    M m b6c eat lun MXB 78w96                                                                      2.56gm / P<.5
1458 R f f34 eat MXB MXB 18m25 v                                              :    +    :             301.mg * P<.0005
  a    R f f34 eat MXA MXA 18m25 v                                                                    408.mg * P<.0005
  b    R f f34 eat cli MXA 18m25 v                                                                    416.mg * P<.003
  c    R f f34 eat thy MXA 18m25 v                                                                    425.mg * P<.0005c
  d    R f f34 eat zym sec 18m25 v                                                                    665.mg * P<.0005c
  e    R f f34 eat TBA MXB 18m25 v                                                                    181.mg * P<.09
  f    R f f34 eat liv MXB 18m25 v                                                                    14.5gm * P<.8
1459 R f f34 eat cli mix 83w86 ae                                                  .   +   .          113.mg Z P<.0005+
  a    R f f34 eat cli ssc 83w86 ae                                                                   162.mg Z P<.0005+
  b    R f f34 eat thy mix 83w86 ae                                                                   162.mg Z P<.0005+
  c    R f f34 eat thy fca 83w86 ae                                                                   296.mg Z P<.0005
  d    R f f34 eat mgl mix 83w86 ae                                                                   1.29gm * P<.5    +
  e    R f f34 eat tba mix 83w86 ae                                                                   67.3mg * P<.0005
1460 R m f34 eat MXB MXB 18m25 v                                                   :   +   :          72.6mg * P<.0005

  a    R m f34 eat thy MXA 18m25 v                                                                    192.mg * P<.0005c
  b    R m f34 eat thy acn 18m25 v                                                                    252.mg * P<.0005c
  c    R m f34 eat thy MXA 18m25 v                                                                    319.mg * P<.0005c
  d    R m f34 eat pre MXA 18m25 v                                                                    351.mg * P<.0005c
  e    R m f34 eat ski MXA 18m25 v                                                                    358.mg * P<.0005c
  f    R m f34 eat pre MXA 18m25 v                                                                    530.mg * P<.002 c
  g    R m f34 eat MXA MXA 18m25 v                                                                    709.mg * P<.0005c
  h    R m f34 eat TBA MXB 18m25 v                                                                    135.mg * P<.03
  i    R m f34 eat liv MXB 18m25 v                                                                    1.61gm * P<.3
1461 R m f3d eat liv tum 52w52 r                                          .>                          no dre   P=1.
  a    R m f3d eat thy tum 52w52 r                                                                    no dre   P=1.


4,4'-DIAMINOAZOBENZENE            100ng..:..1ug...:..10.....:..100...:..1mg...:..10...:..100...:..1g.....:..10
1462 M f bld eat lun mix 14m31 e                                              .>                      240.mg * P<.2   -
  a    M f bld eat liv mix 14m31 e                                                                    no dre   P=1.   -
1463 M m bld eat lun mix 14m31 e                                                  .>                  213.mg * P<.2   -
  a    M m bld eat liv mix 14m31 e                                                                    no dre   P=1.   -


4,4'-DIAMINOBENZANILIDE           100ng..:..1ug...:..10.....:..100...:..1mg...:..10...:..100...:..1g.....:..10
1464 M f bld eat lun mix 14m31 e                                                  .>                  360.mg * P<.5   -
  a    M f bld eat liv mix 14m31 e                                                                    no dre   P=1.   -
1465 M m bld eat lun mix 14m31 e                                                       .>             no dre   P=1.   -
  a    M m bld eat liv mix 14m31 e                                                                    no dre   P=1.   -


2,4-DIAMINOPHENOL.2HCl            100ng..:..1ug...:..10.....:..100...:..1mg...:..10...:..100...:..1g.....:..10
1466 M f b6c gav TBA MXB 24m24                                                  :>                    no dre   P=1.   -
  a    M f b6c gav liv MXB 24m24                                                                      2.89gm * P<1.
  b    M f b6c gav lun MXB 24m24                                                                      239.mg * P<.5
1467 M m b6c gav kid rua 24m24                                                     :         ±        320.mg * P<.06  p
  a    M m b6c gav TBA MXB 24m24                                                                      no dre   P=1.
  b    M m b6c gav liv MXB 24m24                                                                      164.mg \ P<.8
  c    M m b6c gav lun MXB 24m24                                                                      194.mg * P<.4
1468 M m b6c gav kid rua 24m24    with step                                        .    +    .        143.mg   P<.004 p
1469 R f f34 gav TBA MXB 24m24                                                 :>                      46.3mg * P<.6   -
  a    R f f34 gav liv MXB 24m24                                                                      no dre   P=1.
1470 R m f34 gav TBA MXB 24m24                                                :>                       543.mg * P<1.   -
  a    R m f34 gav liv MXB 24m24                                                                      no dre   P=1.


2,4-DIAMINOTOLUENE                100ng..:..1ug...:..10.....:..100...:..1mg...:..10...:..100...:..1g.....:..10
1471 M f b6c eat --- MXA 23m23                                            :    +    :                 10.6mg \ P<.004 a
  a    M f b6c eat liv hpc 23m23                                                                      26.7mg * P<.002 c
  b    M f b6c eat TBA MXB 23m23                                                                      8.64mg \ P<.01
  c    M f b6c eat liv MXB 23m23                                                                      26.7mg * P<.002
  d    M f b6c eat lun MXB 23m23                                                                      no dre   P=1.
1472 M m b6c eat lun a/c 23m23                                               :    ±                   #35.8mg \ P<.02  -
  a    M m b6c eat TBA MXB 23m23                                                                      105.mg * P<.7
  b    M m b6c eat liv MXB 23m23                                                                      no dre   P=1.
  c    M m b6c eat lun MXB 23m23                                                                      35.8mg \ P<.02
1473 R f f34 eat MXB MXB 22m24 asv                                       :    +    :                  1.43mg / P<.0005
  a    R f f34 eat mgl ade 22m24 asv                                                                  1.46mg * P<.0005c
  b    R f f34 eat mgl MXA 22m24 asv                                                                  5.54mg * P<.0005c
  c    R f f34 eat mgl car 22m24 asv                                                                  8.98mg * P<.002
  d    R f f34 eat sub fib 22m24 asv                                                                  16.4mg * P<.002
  e    R f f34 eat liv MXA 22m24 asv                                                                  58.0mg * P<.02  c
  f    R f f34 eat TBA MXB 22m24 asv                                                                  1.46mg / P<.0005
  g    R f f34 eat liv MXB 22m24 asv                                                                  58.0mg * P<.02
1474 R m f34 eat sub fib 21m24 asv                                       :    +    :                  2.52mg / P<.0005
  a    R m f34 eat --- mso 21m24 asv                                                                  6.81mg / P<.0005
  b    R m f34 eat liv MXA 21m24 asv                                                                  8.11mg / P<.0005c
  c    R m f34 eat sub lip 21m24 asv                                                                  13.2mg / P<.002
  d    R m f34 eat TBA MXB 21m24 asv                                                                  .991mg / P<.0005
  e    R m f34 eat liv MXB 21m24 asv                                                                  8.11mg / P<.0005
```

	RefNum	LoConf	UpConf	Cntrl	1Dose	1Inc	2Dose	2Inc			Citation or Pathology
											Brkly Code
a	TR84	148.mg	n.s.s.	39/100	116.mg	24/50	234.mg	33/50			
b	TR84	296.mg	n.s.s.	15/100	116.mg	14/50	234.mg	12/50			liv:hpa,hpc,nnd.
c	TR84	480.mg	n.s.s.	16/100	116.mg	3/50	234.mg	12/50			lun:a/a,a/c.
1458	TR84	155.mg	785.mg	3/100	44.0mg	1/50	182.mg	14/50			thy:acn,fcc,pcn; zym:sec. C
a	TR84	191.mg	1.11gm	0/100	44.0mg	4/50	182.mg	5/50			cli:can; utm:acn,can,ppc. S
b	TR84	184.mg	2.88gm	3/100	44.0mg	5/50	182.mg	8/50			cli:adn,can,cyn,sqc,sqp. S
c	TR84	195.mg	1.62gm	3/100	44.0mg	1/50	182.mg	10/50			thy:acn,fcc,pcn.
d	TR84	284.mg	2.08gm	0/100	44.0mg	0/50	182.mg	7/50			
e	TR84	65.1mg	n.s.s.	70/100	44.0mg	44/50	182.mg	37/50			
f	TR84	773.mg	n.s.s.	2/100	44.0mg	0/50	182.mg	1/50			liv:hpa,hpc,nnd.
1459	1027	68.8mg	206.mg	0/37	60.0mg	8/47	120.mg	15/33	(250.mg	9/40)	Evarts;jnci,65,197-204;1980
a	1027	91.4mg	336.mg	0/37	60.0mg	5/47	120.mg	12/33	(250.mg	9/40)	
b	1027	107.mg	267.mg	1/37	60.0mg	2/47	120.mg	3/33	250.mg	31/40	
c	1027	178.mg	545.mg	0/37	60.0mg	0/47	120.mg	1/33	250.mg	21/40	
d	1027	297.mg	n.s.s.	2/37	60.0mg	7/47	120.mg	8/33	250.mg	5/40	
e	1027	47.3mg	103.mg	3/37	60.0mg	14/47	120.mg	19/33	250.mg	36/40	
1460	TR84	48.0mg	123.mg	4/99	35.2mg	11/50	146.mg	36/50			eac:sec,sqc; ear:sqc; pre:adn,can,cyn,ppa; ski:bcc,sec; thy:acn,
											ccr,fcc,pac,pcn; zym:sec. C
a	TR84	107.mg	430.mg	2/99	35.2mg	2/50	146.mg	17/50			thy:acn,fcc,pac,pcn.
b	TR84	131.mg	648.mg	2/99	35.2mg	1/50	146.mg	14/50			
c	TR84	154.mg	1.27gm	2/99	35.2mg	4/50	146.mg	10/50			thy:cca,ccr.
d	TR84	171.mg	954.mg	0/99	35.2mg	2/50	146.mg	8/50			pre:adn,can,cyn,ppa.
e	TR84	177.mg	875.mg	0/99	35.2mg	2/50	146.mg	9/50			ski:bcc,sec,sqc.
f	TR84	228.mg	2.44gm	0/99	35.2mg	2/50	146.mg	5/50			pre:adn,cyn,ppa.
g	TR84	305.mg	2.51gm	0/99	35.2mg	1/50	146.mg	6/50			eac:sec,sqc; ear:sqc; zym:sec.
h	TR84	57.5mg	n.s.s.	52/99	35.2mg	35/50	146.mg	41/50			
i	TR84	356.mg	n.s.s.	1/99	35.2mg	2/50	146.mg	2/50			liv:hpa,hpc,nnd.
1461	2024	26.4mg	n.s.s.	0/20	24.4mg	0/21					Hasegawa;carc,12,1515-1518;1991/pers.comm.
a	2024	26.4mg	n.s.s.	0/20	24.4mg	0/21					

4,4'-DIAMINOAZOBENZENE   (DAAB) 538-41-0

	RefNum	LoConf	UpConf	Cntrl	1Dose	1Inc	2Dose	2Inc	3Dose	3Inc	Citation or Pathology
1462	1368	75.0mg	n.s.s.	11/40	5.91mg	7/40	17.7mg	11/40	35.5mg	14/39	Della Porta;clet,14,329-336;1981
a	1368	52.2mg	n.s.s.	1/40	5.91mg	0/40	17.7mg	0/40	35.5mg	0/39	
1463	1368	70.9mg	n.s.s.	10/39	5.45mg	6/39	16.4mg	10/40	32.7mg	14/40	
a	1368	492.mg	n.s.s.	1/39	5.45mg	0/39	16.4mg	1/40	32.7mg	0/40	

4,4'-DIAMINOBENZANILIDE   (DABA) 785-30-8

	RefNum	LoConf	UpConf	Cntrl	1Dose	1Inc	2Dose	2Inc	3Dose	3Inc	Citation or Pathology
1464	1368	77.8mg	n.s.s.	11/40	5.91mg	13/40	17.7mg	10/38	35.5mg	15/40	Della Porta;clet,14,329-336;1981
a	1368	51.7mg	n.s.s.	1/40	5.91mg	0/40	17.7mg	0/38	35.5mg	0/40	
1465	1368	90.6mg	n.s.s.	10/39	5.45mg	17/39	16.4mg	10/40	32.7mg	13/39	
a	1368	596.mg	n.s.s.	1/39	5.45mg	1/39	16.4mg	0/40	32.7mg	0/39	

2,4-DIAMINOPHENOL.2HCl   137-09-7

	RefNum	LoConf	UpConf	Cntrl	1Dose	1Inc	2Dose	2Inc			Citation or Pathology
1466	TR401	29.6mg	n.s.s.	24/50	13.4mg	29/50	26.8mg	17/50			
a	TR401	60.1mg	n.s.s.	4/50	13.4mg	7/50	26.8mg	3/50			liv:hpa,hpc,nnd.
b	TR401	53.3mg	n.s.s.	2/50	13.4mg	5/50	26.8mg	3/50			lun:a/a,a/c.
1467	TR401	97.0mg	n.s.s.	0/50	13.3mg	0/50	26.7mg	3/50			
a	TR401	30.8mg	n.s.s.	30/50	13.3mg	27/50	26.7mg	31/50			
b	TR401	15.5mg	n.s.s.	15/50	13.3mg	18/50	(26.7mg	8/50)			liv:hpa,hpc,nnd.
c	TR401	53.0mg	n.s.s.	3/50	13.3mg	4/50	26.7mg	7/50			lun:a/a,a/c
1468	TR401	58.3mg	892.mg	0/50	26.7mg	6/50					
1469	TR401	9.06mg	n.s.s.	40/50	8.78mg	41/50	17.6mg	47/50			
a	TR401	n.s.s.	n.s.s.	0/50	8.78mg	0/50	17.6mg	0/50			liv:hpa,hpc,nnd.
1470	TR401	9.45mg	n.s.s.	44/51	8.80mg	34/50	17.6mg	36/50			liv:hpa,hpc,nnd.
a	TR401	n.s.s.	n.s.s.	0/51	8.80mg	0/50	17.6mg	0/50			liv:hpa,hpc,nnd.

2,4-DIAMINOTOLUENE   95-80-7

	RefNum	LoConf	UpConf	Cntrl	1Dose	1Inc	2Dose	2Inc			Citation or Pathology
1471	TR162	5.93mg	75.5mg	2/20	13.0mg	29/50	(26.0mg	11/50)			---:leu,lym.
a	TR162	17.3mg	99.8mg	0/20	13.0mg	13/50	26.0mg	18/50			
b	TR162	4.52mg	629.mg	5/20	13.0mg	40/50	(26.0mg	26/50)			
c	TR162	17.3mg	99.8mg	0/20	13.0mg	13/50	26.0mg	18/50			liv:hpa,hpc,nnd.
d	TR162	134.mg	n.s.s.	0/20	13.0mg	2/50	26.0mg	0/50			lun:a/a,a/c.
1472	TR162	16.9mg	n.s.s.	0/20	12.0mg	9/50	(24.0mg	6/50)			S
a	TR162	17.4mg	n.s.s.	10/20	12.0mg	31/50	24.0mg	30/50			
b	TR162	34.9mg	n.s.s.	5/20	12.0mg	17/50	24.0mg	13/50			liv:hpa,hpc,nnd.
c	TR162	16.9mg	n.s.s.	0/20	12.0mg	9/50	(24.0mg	6/50)			lun:a/a,a/c.
1473	TR162	.984mg	2.24mg	1/20	3.95mg	34/50	8.55mg	41/50			liv:hpc,nnd; mgl:ade,ade,car. C
a	TR162	1.00mg	2.33mg	1/20	3.95mg	34/50	8.55mg	38/50			
b	TR162	3.14mg	11.5mg	0/20	3.95mg	11/50	8.55mg	14/50			mgl:ade,car.
c	TR162	4.53mg	32.3mg	0/20	3.95mg	9/50	8.55mg	8/50			S
d	TR162	7.85mg	55.9mg	0/20	3.95mg	4/50	8.55mg	10/50			S
e	TR162	23.6mg	n.s.s.	0/20	3.95mg	0/50	8.55mg	6/50			liv:hpc,nnd.
f	TR162	.918mg	2.96mg	13/20	3.95mg	49/50	8.55mg	49/50			
g	TR162	23.6mg	n.s.s.	0/20	3.95mg	0/50	8.55mg	6/50			liv:hpa,hpc,nnd.
1474	TR162	1.50mg	4.48mg	0/20	3.20mg	15/50	7.00mg	19/50			S
a	TR162	3.13mg	18.9mg	0/20	3.20mg	5/50	7.00mg	8/50			S
b	TR162	3.81mg	21.2mg	0/20	3.20mg	5/50	7.00mg	10/50			liv:hpc,nnd.
c	TR162	5.48mg	49.3mg	0/20	3.20mg	3/50	7.00mg	8/50			S
d	TR162	.627mg	1.85mg	13/20	3.20mg	42/50	7.00mg	44/50			
e	TR162	3.81mg	21.2mg	0/20	3.20mg	5/50	7.00mg	10/50			liv:hpa,hpc,nnd.

	Spe	Strain	Site	Xpo+Xpt		Plot	TD50	DR	2Tailpvl	AuOp
	Sex	Route	Hist	Notes						

2,4-DIAMINOTOLUENE.2HCl   `100ng..:..1ug...:..10.....:..100.....:..1mg.....:..10.....:..100.....:..1g......:..10`

ID	Sex/Strain	Route	Site	Hist	Xpo+Xpt	Notes	Plot	TD50	DR	2Tailpvl	AuOp
1475	M f chi	eat	lun	mix	77w90		: + :	72.0mg	\	P<.007	-
a	M f chi	eat	liv	mix	77w90			180.mg	*	P<.09	
b	M f chi	eat	tba	mix	77w90			100.mg	\	P<.4	
1476	M f chi	eat	liv	hpt	77w90	pool	: + :	206.mg	*	P<.006	+
1477	M m chi	eat	liv	mix	77w90		:>	191.mg	*	P<.2	
a	M m chi	eat	lun	mix	77w90			no dre		P=1.	-
b	M m chi	eat	tba	mix	77w90			no dre		P=1.	-
1478	M m chi	eat	liv	hpt	77w90	pool	: ±	201.mg	*	P<.1	+
a	M m chi	eat	---	vsc	77w90			317.mg	*	P<.2	+
1479	R m cdr	eat	sub	fib	64w73 av		: + :	5.42mg	/	P<.0005	+
a	R m cdr	eat	liv	mix	64w73 av			26.2mg	*	P<.02	
b	R m cdr	eat	tba	mix	64w73 av			2.98mg	/	P<.0005	
1480	R m cdr	eat	sub	fib	64w73 av	pool	: + :	4.42mg	/	P<.0005	+
a	R m cdr	eat	liv	hpt	64w73 av			22.2mg	*	P<.003	+

2,6-DIAMINOTOLUENE.2HCl   `100ng..:..1ug...:..10.....:..100.....:..1mg.....:..10.....:..100.....:..1g......:..10`

ID	Sex/Strain	Route	Site	Hist	Xpo+Xpt	Notes	Plot	TD50	DR	2Tailpvl	AuOp
1481	M f b6c	eat	liv	hpc	24m24		: ±	#181.mg	*	P<.05	-
a	M f b6c	eat	TBA	MXB	24m24			75.9mg	*	P<.7	
b	M f b6c	eat	liv	MXB	24m24			129.mg	*	P<.4	
c	M f b6c	eat	lun	MXB	24m24			no dre		P=1.	
1482	M m b6c	eat	---	lym	24m24		: ±	#20.5mg	\	P<.05	-
a	M m b6c	eat	TBA	MXB	24m24			no dre		P=1.	
b	M m b6c	eat	liv	MXB	24m24			no dre		P=1.	
c	M m b6c	eat	lun	MXB	24m24			no dre		P=1.	
1483	R f f34	eat	TBA	MXB	24m24		:>	no dre		P=1.	-
a	R f f34	eat	liv	MXB	24m24			455.mg	*	P<.1	
1484	R m f34	eat	liv	MXA	24m24		: ±	#117.mg	*	P<.03	-
a	R m f34	eat	pni	isa	24m24			140.mg	*	P<.03	
b	R m f34	eat	TBA	MXB	24m24			no dre		P=1.	
c	R m f34	eat	liv	MXB	24m24			117.mg	*	P<.03	

2,5-DIAMINOTOLUENE SULFATE   `100ng..:..1ug...:..10.....:..100.....:..1mg.....:..10.....:..100.....:..1g......:..10`

ID	Sex/Strain	Route	Site	Hist	Xpo+Xpt	Notes	Plot	TD50	DR	2Tailpvl	AuOp
1485	M f b6c	eat	lun	MXA	78w94		: ±	#364.mg	*	P<.03	-
a	M f b6c	eat	TBA	MXB	78w94			33.2gm	*	P<1.	
b	M f b6c	eat	liv	MXB	78w94			1.96gm	*	P<.6	
c	M f b6c	eat	lun	MXB	78w94			364.mg	*	P<.03	
1486	M m b6c	eat	TBA	MXB	78w94		:>	272.mg	*	P<.3	
a	M m b6c	eat	liv	MXB	78w94			288.mg	/	P<.06	
b	M m b6c	eat	lun	MXB	78w94			3.56gm	*	P<.9	
1487	R f f34	eat	TBA	MXB	18m25 v		:>	no dre		P=1.	-
a	R f f34	eat	liv	MXB	18m25 v			no dre		P=1.	
1488	R m f34	eat	TBA	MXB	18m25 v		:>	no dre		P=1.	
a	R m f34	eat	liv	MXB	18m25 v			233.mg	\	P<.09	

DIAZEPAM   `100ng..:..1ug...:..10.....:..100.....:..1mg.....:..10.....:..100.....:..1g......:..10`

ID	Sex/Strain	Route	Site	Hist	Xpo+Xpt	Notes	Plot	TD50	DR	2Tailpvl	AuOp
1489	M f cf1	eat	bon	ost	80w80 e		. ±	247.mg		P<.03	-
a	M f cf1	eat	liv	hpc	80w80 e			988.mg		P<.3	-
b	M f cf1	eat	liv	hpa	80w80 e			no dre		P=1.	-
c	M f cf1	eat	lun	agt	80w80 e			no dre		P=1.	-
d	M f cf1	eat	tba	ben	80w80 e			371.mg		P<.3	-
e	M f cf1	eat	tba	mal	80w80 e			no dre		P=1.	-
f	M f cf1	eat	tba	mix	80w80 e			no dre		P=1.	-
1490	M m cf1	eat	liv	hpc	80w80 e		. ±	242.mg		P<.05	-
a	M m cf1	eat	liv	hpa	80w80 e			988.mg		P<.3	-
b	M m cf1	eat	lun	agt	80w80 e			no dre		P=1.	-
c	M m cf1	eat	tba	mal	80w80 e			204.mg		P<.3	-
d	M m cf1	eat	tba	mix	80w80 e			220.mg		P<.4	-
e	M m cf1	eat	tba	ben	80w80 e			no dre		P=1.	-
1491	R f wal	eat	liv	hpa	24m24 e		.>	no dre		P=1.	-
a	R f wal	eat	tba	mix	24m24 e			no dre		P=1.	-
b	R f wal	eat	tba	mal	24m24 e			457.mg		P<.2	-
c	R f wal	eat	tba	ben	24m24 e			no dre		P=1.	-
1492	R m wal	eat	liv	hpc	24m24 e		. ±	1.09gm		P<.02	-
a	R m wal	eat	liv	hpa	24m24 e			no dre		P=1.	-
b	R m wal	eat	tba	mal	24m24 e			865.mg		P<.5	-
c	R m wal	eat	tba	ben	24m24 e			no dre		P=1.	-
d	R m wal	eat	tba	mix	24m24 e			no dre		P=1.	-

DIAZINON   `100ng..:..1ug...:..10.....:..100.....:..1mg.....:..10.....:..100.....:..1g......:..10`

ID	Sex/Strain	Route	Site	Hist	Xpo+Xpt	Notes	Plot	TD50	DR	2Tailpvl	AuOp
1493	M f b6c	eat	TBA	MXB	24m24		:>	230.mg	*	P<.7	-
a	M f b6c	eat	liv	MXB	24m24			no dre		P=1.	
b	M f b6c	eat	lun	MXB	24m24			6.23gm	*	P<1.	
1494	M m b6c	eat	TBA	MXB	24m24		:>	no dre		P=1.	-
a	M m b6c	eat	liv	MXB	24m24			26.8mg	\	P<.2	
b	M m b6c	eat	lun	MXB	24m24			no dre		P=1.	
1495	R f f34	eat	TBA	MXB	24m24		:>	51.3mg	*	P<.4	-
a	R f f34	eat	liv	MXB	24m24			no dre		P=1.	
1496	R m f34	eat	---	MXA	24m24		: ±	#20.4mg	\	P<.03	-
a	R m f34	eat	TBA	MXB	24m24			53.4mg	\	P<.7	
b	R m f34	eat	liv	MXB	24m24			no dre		P=1.	

RefNum	LoConf	UpConf	Cntrl	1Dose	1Inc	2Dose	2Inc	Citation or Pathology	Brkly Code
2,4-DIAMINOTOLUENE.2HCl (2,4-toluenediamine.2HCl) 636-23-7									
1475	381	29.3mg 894.mg	5/22	58.5mg	6/19	(111.mg	2/17)	Weisburger;jept,2,325-356;1978/pers.comm./Russfield 1973	
a	381	58.7mg n.s.s.	1/22	58.5mg	1/19	111.mg	5/17		
b	381	24.6mg n.s.s.	18/22	58.5mg	12/19	(111.mg	10/17)		
1476	381	65.9mg 3.37gm	1/102p	58.5mg	1/19	111.mg	3/17		
1477	381	54.2mg n.s.s.	1/18	54.0mg	2/22	103.mg	4/17		
a	381	118.mg n.s.s.	8/18	54.0mg	1/22	(103.mg	2/17)		
b	381	45.1mg n.s.s.	13/18	54.0mg	7/22	103.mg	11/17		
1478	381	54.8mg n.s.s.	7/99p	54.0mg	2/22	103.mg	4/17		
a	381	76.7mg n.s.s.	5/99p	54.0mg	0/22	103.mg	4/17		
1479	381	2.81mg 10.7mg	0/24	12.4mg	6/19	25.7mg	15/24		
a	381	6.43mg n.s.s.	0/24	12.4mg	1/19	25.7mg	3/24		
b	381	1.58mg 7.35mg	7/24	12.4mg	12/19	25.7mg	20/24		
1480	381	2.27mg 10.2mg	14/111p	12.4mg	6/19	25.7mg	15/24		
a	381	5.70mg 263.mg	2/111p	12.4mg	1/19	25.7mg	3/24		
2,6-DIAMINOTOLUENE.2HCl (2,6-toluenediamine.2HCl) 15481-70-6									
1481	TR200	54.8mg n.s.s.	0/50	6.40mg	0/50	12.9mg	3/50		S
a	TR200	12.6mg n.s.s.	21/50	6.40mg	30/50	12.9mg	24/50		
b	TR200	32.0mg n.s.s.	4/50	6.40mg	3/50	12.9mg	7/50	liv:hpa,hpc,nnd.	
c	TR200	38.9mg n.s.s.	4/50	6.40mg	8/50	12.9mg	3/50	lun:a/a,a/c.	
1482	TR200	7.63mg n.s.s.	2/50	5.90mg	8/50	(11.9mg	2/50)		S
a	TR200	12.2mg n.s.s.	31/50	5.90mg	36/50	11.9mg	26/50		
b	TR200	17.1mg n.s.s.	21/50	5.90mg	17/50	11.9mg	18/50	liv:hpa,hpc,nnd.	
c	TR200	25.7mg n.s.s.	11/50	5.90mg	13/50	11.9mg	7/50	lun:a/a,a/c.	
1483	TR200	22.7mg n.s.s.	42/50	12.4mg	38/50	24.8mg	39/50		
a	TR200	112.mg n.s.s.	0/50	12.4mg	0/50	24.8mg	2/50	liv:hpa,hpc,nnd.	
1484	TR200	47.6mg n.s.s.	0/50	9.90mg	2/50	19.8mg	4/50	liv:hpc,nnd.	S
a	TR200	53.0mg n.s.s.	0/50	9.90mg	1/50	19.8mg	4/50		S
b	TR200	16.3mg n.s.s.	32/50	9.90mg	38/50	19.8mg	36/50		
c	TR200	47.6mg n.s.s.	0/50	9.90mg	2/50	19.8mg	4/50	liv:hpa,hpc,nnd.	
2,5-DIAMINOTOLUENE SULFATE (2,5-toluenediamine sulfate) 6369-59-1									
1485	TR126	152.mg n.s.s.	5/100	64.7mg	6/50	104.mg	8/50	lun:a/a,a/c.	S
a	TR126	128.mg n.s.s.	42/100	64.7mg	15/50	104.mg	22/50		
b	TR126	307.mg n.s.s.	5/100	64.7mg	2/50	104.mg	4/50	liv:hpa,hpc,nnd.	
c	TR126	152.mg n.s.s.	5/100	64.7mg	6/50	104.mg	8/50	lun:a/a,a/c.	
1486	TR126	83.9mg n.s.s.	38/100	59.8mg	19/50	96.5mg	27/50		
a	TR126	113.mg n.s.s.	17/100	59.8mg	8/50	96.5mg	18/50	liv:hpa,hpc,nnd.	
b	TR126	211.mg n.s.s.	17/100	59.8mg	6/50	96.5mg	10/50	lun:a/a,a/c.	
1487	TR126	27.0mg n.s.s.	51/75	21.9mg	34/50	(71.6mg	33/50)		
a	TR126	520.mg n.s.s.	4/75	21.9mg	1/50	71.6mg	1/50	liv:hpa,hpc,nnd.	
1488	TR126	102.mg n.s.s.	36/75	17.1mg	20/50	57.8mg	25/50		
a	TR126	57.1mg n.s.s.	0/75	17.1mg	2/50	(57.8mg	0/50)	liv:hpa,hpc,nnd.	
DIAZEPAM 439-14-5									
1489	2139	89.5mg n.s.s.	5/100	75.0mg	8/50			de la Iglesia;txap,57,39-54;1981/pers.comm.	
a	2139	193.mg n.s.s.	1/100	75.0mg	2/50				
b	2139	457.mg n.s.s.	1/100	75.0mg	0/50				
c	2139	247.mg n.s.s.	16/100	75.0mg	4/50				
d	2139	95.7mg n.s.s.	11/100	75.0mg	9/50				
e	2139	121.mg n.s.s.	41/100	75.0mg	16/50				
f	2139	113.mg n.s.s.	49/100	75.0mg	19/50				
1490	2139	85.2mg n.s.s.	7/100	75.0mg	9/50				
a	2139	193.mg n.s.s.	1/100	75.0mg	2/50				
b	2139	173.mg n.s.s.	19/100	75.0mg	7/50				
c	2139	53.2mg n.s.s.	35/100	75.0mg	22/50				
d	2139	52.5mg n.s.s.	38/100	75.0mg	23/50				
e	2139	201.mg n.s.s.	6/100	75.0mg	3/50				
1491	2139	1.00gm n.s.s.	1/115	75.0mg	0/65				
a	2139	110.mg n.s.s.	80/115	75.0mg	42/65				
b	2139	145.mg n.s.s.	18/115	75.0mg	16/65				
c	2139	185.mg n.s.s.	72/115	75.0mg	33/65				
1492	2139	329.mg n.s.s.	0/115	75.0mg	3/65				
a	2139	1.00gm n.s.s.	1/115	75.0mg	0/65				
b	2139	168.mg n.s.s.	23/115	75.0mg	16/65				
c	2139	277.mg n.s.s.	37/115	75.0mg	16/65				
d	2139	193.mg n.s.s.	57/115	75.0mg	27/65				
DIAZINON 333-41-5									
1493	TR137	33.4mg n.s.s.	9/25	12.7mg	14/50	25.5mg	21/50		
a	TR137	143.mg n.s.s.	2/25	12.7mg	0/50	25.5mg	3/50	liv:hpa,hpc,nnd.	
b	TR137	139.mg n.s.s.	1/25	12.7mg	1/50	25.5mg	2/50	lun:a/a,a/c.	
1494	TR137	29.6mg n.s.s.	12/25	11.8mg	30/50	23.5mg	24/50		
a	TR137	9.56mg n.s.s.	5/25	11.8mg	20/50	(23.5mg	13/50)	liv:hpa,hpc,nnd.	
b	TR137	126.mg n.s.s.	2/25	11.8mg	3/50	23.5mg	1/50	lun:a/a,a/c.	
1495	TR137	14.7mg n.s.s.	20/25	19.8mg	44/50	39.6mg	41/50		
a	TR137	231.mg n.s.s.	1/25	19.8mg	1/50	39.6mg	0/50	liv:hpa,hpc,nnd.	
1496	TR137	9.49mg n.s.s.	5/25	15.7mg	25/50	(31.4mg	12/50)	---:leu,lym.	S
a	TR137	7.64mg n.s.s.	20/25	15.7mg	41/50	(31.4mg	33/50)		
b	TR137	203.mg n.s.s.	1/25	15.7mg	2/50	31.4mg	0/50	liv:hpa,hpc,nnd.	

```
        Spe Strain  Site    Xpo+Xpt                                                                    TD50    2Tailpvl
            Sex  Route   Hist    Notes                                                                         DR    AuOp

3-DIAZOTYRAMINE.HCl             100ng..:..1ug....:..10......:..100.....:..1mg....:..10..:..100....:..1g.....:..10
1497 R m f3d wat orc sqc 27m27 e                                              . + .                    37.6mg   P<.0005+

DIBENZ(a,h)ANTHRACENE           100ng..:..1ug....:..10......:..100.....:..1mg....:..10..:..100....:..1g.....:..10
1498 M m dba wat lun alc 60w60                                             . + .                       5.88mg   P<.0005+

DIBENZO-p-DIOXIN                100ng..:..1ug....:..10......:..100.....:..1mg....:..10..:..100....:..1g.....:..10
1499 M f b6c eat TBA MXB 90w91 s                                                          :>           no dre   P=1.    -
   a M f b6c eat liv MXB 90w91 s                                                                       no dre   P=1.
   b M f b6c eat lun MXB 90w91 s                                                                       8.09gm * P<.2
1500 M m b6c eat lun a/c 88w92 as                                                      :  +            #2.84gm \ P<.01  -
   a M m b6c eat TBA MXB 88w92 as                                                                      no dre   P=1.
   b M m b6c eat liv MXB 88w92 as                                                                      no dre   P=1.
   c M m b6c eat lun MXB 88w92 as                                                                      no dre   P=1.
1501 R f osm eat TBA MXB 26m26 s                                                     :>                no dre   P=1.    -
   a R f osm eat liv MXB 26m26 s                                                                       5.37gm * P<.08
1502 R m osm eat TBA MXB 26m26                                                          :>             no dre   P=1.
   a R m osm eat liv MXB 26m26                                                                         no dre   P=1.

3-DIBENZOFURANAMINE             100ng..:..1ug....:..10......:..100.....:..1mg....:..10..:..100....:..1g.....:..10
1503 R m wis eat liv hpt 89w89 e                                        .>                             9.44mg   P<.3
   a R m wis eat tba ben 89w89 e                                                                       1.57mg   P<.06
   b R m wis eat tba mal 89w89 e                                                                       2.48mg   P<.02  +

O,S-DIBENZOYL THIAMINE.HCl      100ng..:..1ug....:..10......:..100.....:..1mg....:..10..:..100....:..1g.....:..10
1504 R f sda eat liv hga 24m24 e                                                             .>  no dre P=1.    -
   a R f sda eat tba mix 24m24 e                                                                       no dre   P=1.    -
   b R f sda eat tba mal 24m24 e                                                                       14.3gm * P<.9    -
1505 R m sda eat liv lcm 24m24 e                                                           .>          6.77gm * P<.2    -
   a R m sda eat liv lcb 24m24 e                                                                       6.98gm * P<.2    -
   b R m sda eat tba mal 24m24 e                                                                       2.34gm / P<.4    -
   c R m sda eat tba mix 24m24 e                                                                       no dre   P=1.    -

1,2-DIBROMO-3-CHLOROPROPANE     100ng..:..1ug....:..10......:..100.....:..1mg....:..10..:..100....:..1g.....:..10
1506 M f b6c gav sto sqc 53w59 av                                            : + :                     4.29mg / P<.0005c
   a M f b6c gav TBA MXB 53w59 av                                                                      4.29mg / P<.0005
   b M f b6c gav liv MXB 53w59 av                                                                      1.42gm * P<.3
   c M f b6c gav lun MXB 53w59 av                                                                      150.mg * P<.04
1507 M f b6c inh MXB MXB 21m24 a                                         : + :                         1.28mg / P<.0005

   a M f b6c inh MXA MXA 21m24 a                                                                       2.13mg / P<.0005c
   b M f b6c inh MXA MXA 21m24 a                                                                       2.27mg / P<.0005c
   c M f b6c inh lun MXA 21m24 a                                                                       5.21mg / P<.0005c
   d M f b6c inh nas can 21m24 a                                                                       6.49mg / P<.0005c
   e M f b6c inh lun a/a 21m24 a                                                                       7.89mg / P<.0005c
   f M f b6c inh hag adn 21m24 a                                                                       8.21mg * P<.003
   g M f b6c inh nas MXA 21m24 a                                                                       10.2mg * P<.0005c
   h M f b6c inh nas acn 21m24 a                                                                       14.5mg * P<.0005c
   i M f b6c inh nas sqc 21m24 a                                                                       19.0mg * P<.0005c
   j M f b6c inh nas cas 21m24 a                                                                       40.0mg * P<.003 c
   k M f b6c inh nas fbs 21m24 a                                                                       46.8mg * P<.003 c
   l M f b6c inh l/b ppc 21m24 a                                                                       72.2mg * P<.02  a
   m M f b6c inh TBA MXB 21m24 a                                                                       .945mg / P<.0005
   n M f b6c inh liv MXB 21m24 a                                                                       9.84mg * P<.02
   o M f b6c inh lun MXB 21m24 a                                                                       5.21mg / P<.0005
1508 M m b6c gav sto sqc 53w59 av                                          : + :                       4.62mg / P<.0005c
   a M m b6c gav TBA MXB 53w59 av                                                                      4.64mg / P<.0005
   b M m b6c gav liv MXB 53w59 av                                                                      no dre   P=1.
   c M m b6c gav lun MXB 53w59 av                                                                      1.11gm * P<.3
1509 M m b6c inh MXB MXB 76w76                                            : + :                        1.44mg * P<.0005
   a M m b6c inh MXA MXA 76w76                                                                         1.82mg * P<.0005c
   b M m b6c inh lun --- 76w76                                                                         2.48mg * P<.0005a
   c M m b6c inh lun MXA 76w76                                                                         3.95mg * P<.002 c
   d M m b6c inh nas apn 76w76                                                                         4.75mg * P<.002 c
   e M m b6c inh nas can 76w76                                                                         6.09mg * P<.0005c
   f M m b6c inh nas sqc 76w76                                                                         7.05mg * P<.002 c
   g M m b6c inh lun a/a 76w76                                                                         7.38mg * P<.003 c
   h M m b6c inh nas hes 76w76                                                                         16.3mg * P<.02  a
   i M m b6c inh sto MXA 76w76                                                                         16.8mg * P<.02  a
   j M m b6c inh sto sqp 76w76                                                                         21.5mg * P<.05  a
   k M m b6c inh TBA MXB 76w76                                                                         1.32mg * P<.0005
   l M m b6c inh liv MXB 76w76                                                                         32.4mg * P<.7
   m M m b6c inh lun MXB 76w76                                                                         3.95mg * P<.002
1510 R f osm gav MXB MXB 68w72 av                                        : + :                         .855mg / P<.0005
   a R f osm gav sto sqc 68w72 av                                                                      .909mg / P<.0005c
   b R f osm gav mgl MXA 68w72 av                                                                      2.33mg / P<.0005c
   c R f osm gav mgl acn 68w72 av                                                                      2.37mg / P<.0005c
   d R f osm gav sto sqp 68w72 av                                                                      30.1mg * P<.002 c
   e R f osm gav TBA MXB 68w72 av                                                                      .892mg / P<.0005
   f R f osm gav liv MXB 68w72 av                                                                      90.3mg * P<.4
```

```
        RefNum  LoConf UpConf   Cntrl  1Dose   1Inc  2Dose   2Inc                        Citation or Pathology
                                                                                                              Brkly Code
3-DIAZOTYRAMINE.HCl   (4-(2-aminoethyl)-6-diazo-2,4-cyclohexadienone.HCl)   ---
1497    1825   21.1mg 74.3mg   0/16  50.0mg  19/28                                         Fujita;carc,8,527-529;1987

DIBENZ(a,h)ANTHRACENE    53-70-3
1498    1128   3.03mg 13.2mg   0/25  28.3mg  14/21                                         Snell;jnci,28,1043-1051;1962

DIBENZO-p-DIOXIN    262-12-4
1499    TR122  1.74gm n.s.s.  15/50  636.mg  12/50  1.29gm   9/50
a       TR122  n.s.s. n.s.s.   0/50  636.mg   1/50  1.29gm   0/50                          liv:hpa,hpc,nnd.
b       TR122  2.28gm n.s.s.   3/50  636.mg   2/50  1.29gm   5/50                          lun:a/a,a/c.
1500    TR122  1.08gm 289.gm   0/50  557.mg   5/50 (1.12gm   1/50)                                        S
a       TR122  1.67gm n.s.s.  19/50  557.mg  19/50  1.12gm  16/50
b       TR122  3.33gm n.s.s.   8/50  557.mg   8/50  1.12gm   5/50                          liv:hpa,hpc,nnd.
c       TR122  2.58gm n.s.s.   8/50  557.mg  11/50  1.12gm   6/50                          lun:a/a,a/c.
1501    TR122  500.mg n.s.s.  26/35  250.mg  19/35  500.mg  13/35
a       TR122  1.32gm n.s.s.   0/35  250.mg   0/35  500.mg   2/35                          liv:hpa,hpc,nnd.
1502    TR122  1.13gm n.s.s.  21/35  200.mg  11/35  400.mg  11/35
a       TR122  3.52gm n.s.s.   1/35  200.mg   0/35  400.mg   0/35                          liv:hpa,hpc,nnd.

3-DIBENZOFURANAMINE    4106-66-5
1503    1420   1.53mg n.s.s.   0/6   3.43mg   1/6                                          Hackmann;zkko,61,45-54;1956
a       1420   .329mg n.s.s.   0/3   3.43mg   2/3
b       1420   .825mg n.s.s.   0/6   3.43mg   4/8

O,S-DIBENZOYL THIAMINE.HCl   35660-60-7
1504    1728   4.65gm n.s.s.   0/55  50.0mg   1/55  500.mg   0/55                          Heywood;txlt,26,53-58;1985
a       1728   212.mg n.s.s.  53/55  50.0mg  50/55  500.mg  51/55
b       1728   998.mg n.s.s.  13/55  50.0mg  18/55  500.mg  16/55
1505    1728   1.75gm n.s.s.   1/55  40.0mg   0/55  400.mg   3/55
a       1728   1.14gm n.s.s.   0/21  40.0mg   0/25  400.mg   1/23
b       1728   544.mg n.s.s.  25/55  40.0mg  13/55  400.mg  24/55
c       1728   429.mg n.s.s.  48/55  40.0mg  42/55  400.mg  42/55

1,2-DIBROMO-3-CHLOROPROPANE   (DBCP)  96-12-8
1506    TR28   2.79mg 6.57mg   0/20  79.0mg  50/50  149.mg  47/50
a       TR28   2.79mg 6.57mg   0/20  79.0mg  50/50  149.mg  47/50
b       TR28   231.mg n.s.s.   0/20  79.0mg   0/50  149.mg   1/50                          liv:hpa,hpc,nnd.
c       TR28   29.4mg n.s.s.   0/20  79.0mg   1/50  149.mg   2/50                          lun:a/a,a/c.
1507    TR206  .851mg 2.03mg   4/50  1.80mg  20/50  9.00mg  43/50   b/l:ppa,ppc; l/b:ppc,sqc; lun:a/a,a/c,sqc; nas:---,acn,can,cas,
                                                                                          fbs,sqc,sqp; ntu:apn. C
a       TR206  1.35mg 3.43mg   0/50  1.80mg  11/50  9.00mg  38/50                          nas:---; ntu:apn.
b       TR206  1.33mg 4.37mg   4/50  1.80mg  13/50  9.00mg  20/50   b/l:ppa,ppc; l/b:ppc,sqc; lun:a/a,a/c,sqc.
c       TR206  2.54mg 13.8mg   4/50  1.80mg   5/50  9.00mg  13/50                          lun:a/a,a/c.
d       TR206  3.28mg 12.7mg   0/50  1.80mg   3/50  9.00mg  17/50
e       TR206  3.39mg 25.9mg   3/50  1.80mg   3/50  9.00mg  10/50
f       TR206  3.19mg 50.0mg   0/50  1.80mg   5/50  9.00mg   1/50                                         S
g       TR206  3.99mg 33.2mg   0/50  1.80mg   3/50  9.00mg   6/50   nas:sqc,sqp.
h       TR206  5.14mg 51.6mg   0/50  1.80mg   2/50  9.00mg   6/50
i       TR206  6.40mg 69.0mg   0/50  1.80mg   1/50  9.00mg   6/50
j       TR206  11.3mg 346.mg   0/50  1.80mg   0/50  9.00mg   3/50
k       TR206  14.1mg 395.mg   0/50  1.80mg   0/50  9.00mg   3/50
l       TR206  17.6mg n.s.s.   0/50  1.80mg   0/50  9.00mg   2/50
m       TR206  .629mg 1.53mg  23/50  1.80mg  32/50  9.00mg  45/50
n       TR206  3.41mg n.s.s.   2/50  1.80mg   5/50  9.00mg   2/50                          liv:hpa,hpc,nnd.
o       TR206  2.54mg 13.8mg   4/50  1.80mg   5/50  9.00mg  13/50                          lun:a/a,a/c.
1508    TR28   2.96mg 7.21mg   0/20  81.0mg  43/50  156.mg  47/50
a       TR28   2.98mg 7.51mg   2/20  81.0mg  43/50  156.mg  47/50
b       TR28   50.6mg n.s.s.   2/20  81.0mg   1/50  156.mg   0/50                          liv:hpa,hpc,nnd.
c       TR28   181.mg n.s.s.   0/20  81.0mg   0/50  156.mg   1/50                          lun:a/a,a/c.
1509    TR206  .847mg 2.61mg   0/50  1.50mg   4/50  7.50mg  22/50   l/b:sqc; lun:a/a,a/c; nas:---,apn,can,sqc; trh:---. C
a       TR206  1.03mg 3.42mg   0/50  1.50mg   1/50  7.50mg  21/50   l/b:sqc; nas:---; trh:---.
b       TR206  1.26mg 6.41mg   0/50  1.50mg   3/50  7.50mg  11/50
c       TR206  1.81mg 20.5mg   0/50  1.50mg   3/50  7.50mg   7/50                          lun:a/a,a/c.
d       TR206  1.85mg 26.7mg   0/50  1.50mg   1/50  7.50mg   5/50
e       TR206  2.48mg 20.9mg   0/50  1.50mg   0/50  7.50mg   7/50
f       TR206  2.70mg 28.8mg   0/50  1.50mg   0/50  7.50mg   6/50
g       TR206  3.06mg 42.1mg   0/50  1.50mg   1/50  7.50mg   6/50
h       TR206  4.77mg n.s.s.   0/50  1.50mg   0/50  7.50mg   3/50
i       TR206  4.85mg n.s.s.   0/50  1.50mg   0/50  7.50mg   3/50                          sto:sqc,sqp.
j       TR206  5.20mg n.s.s.   0/50  1.50mg   0/50  7.50mg   2/50
k       TR206  .735mg 3.11mg   3/50  1.50mg   7/50  7.50mg  23/50
l       TR206  3.68mg n.s.s.   0/50  1.50mg   2/50  7.50mg   3/50                          liv:hpa,hpc,nnd.
m       TR206  1.81mg 20.5mg   0/50  1.50mg   3/50  7.50mg   7/50                          lun:a/a,a/c.
1510    TR28   .544mg 1.38mg   1/20  10.0mg  43/50  20.0mg  42/50   mgl:acn,fba; sto:sqc,sqp. C
a       TR28   .568mg 1.47mg   0/20  10.0mg  38/50  20.0mg  29/50
b       TR28   1.35mg 4.15mg   1/20  10.0mg  24/50  20.0mg  30/50   mgl:acn,fba.
c       TR28   1.39mg 3.84mg   0/20  10.0mg  24/50  20.0mg  30/50
d       TR28   13.5mg 117.mg   0/20  10.0mg   1/50  20.0mg   9/50
e       TR28   .560mg 1.49mg   4/20  10.0mg  44/50  20.0mg  43/50
f       TR28   14.7mg n.s.s.   0/20  10.0mg   1/50  20.0mg   0/50                          liv:hpa,hpc,nnd.
```

```
     Spe Strain Site   Xpo+Xpt                                                    TD50    2Tailpvl
         Sex  Route  Hist     Notes                                                  DR      AuOp
1511 R f f34 inh MXB MXB 22m24 a                      :+ :                        .199mg / P<.0005

   a  R f f34 inh MXA MXA 22m24 a                                                 .252mg / P<.0005c
   b  R f f34 inh mgl --- 22m24 a                                                 .727mg / P<.0005
   c  R f f34 inh mgl fba 22m24 a                                                 .896mg * P<.0005
   d  R f f34 inh nas sqp 22m24 a                                                 .948mg * P<.0005c
   e  R f f34 inh adr coa 22m24 a                                                 1.20mg * P<.0005c
   f  R f f34 inh nas adn 22m24 a                                                 1.51mg * P<.0005c
   g  R f f34 inh nas apn 22m24 a                                                 1.70mg * P<.0005c
   h  R f f34 inh ton MXA 22m24 a                                                 1.80mg * P<.0005c
   i  R f f34 inh nas can 22m24 a                                                 1.83mg / P<.0005c
   j  R f f34 inh ntu apn 22m24 a                                                 2.08mg * P<.002
   k  R f f34 inh ton sqp 22m24 a                                                 2.46mg / P<.0005c
   l  R f f34 inh nas sqc 22m24 a                                                 2.90mg / P<.0005c
   m  R f f34 inh nas acn 22m24 a                                                 3.55mg * P<.0005c
   n  R f f34 inh phr MXA 22m24 a                                                 4.70mg * P<.0005c
   o  R f f34 inh phr sqp 22m24 a                                                 5.86mg / P<.0005c
   p  R f f34 inh ton sqc 22m24 a                                                 6.75mg * P<.003 c
   q  R f f34 inh TBA MXB 22m24 a                                                 .164mg / P<.0005
   r  R f f34 inh liv MXB 22m24 a                                                 5.52mg * P<.4
1512 R m osm gav sto sqc 71w83 av               : + :                            .967mg / P<.0005c
   a  R m osm gav·--- MXA 71w83 av                                                4.23mg * P<.02
   b  R m osm gav TBA MXB 71w83 av                                                .988mg / P<.0005
   c  R m osm gav liv MXB 71w83 av                                               no dre   P=1.
1513 R m f34 inh MXB MXB 22m24 a                : +:                              .106mg / P<.0005
   a  R m f34 inh MXA MXA 22m24 a                                                 .107mg / P<.0005c
   b  R m f34 inh nas apn 22m24 a                                                 .567mg * P<.0005c
   c  R m f34 inh nas adn 22m24 a                                                 .837mg * P<.0005c
   d  R m f34 inh nas acn 22m24 a                                                 .864mg * P<.0005c
   e  R m f34 inh ntu apn 22m24 a                                                 .976mg * P<.002
   f  R m f34 inh nas sqp 22m24 a                                                 .978mg * P<.0005c
   g  R m f34 inh nas can 22m24 a                                                 1.15mg / P<.0005c
   h  R m f34 inh nas sqc 22m24 a                                                 1.44mg / P<.0005c
   i  R m f34 inh ton MXA 22m24 a                                                 2.08mg / P<.0005c
   j  R m f34 inh tnv mso 22m24 a                                                 2.73mg * P<.002
   k  R m f34 inh ton sqp 22m24 a                                                 2.75mg * P<.0005c
   l  R m f34 inh ski tri 22m24 a                                                 3.25mg * P<.0005
   m  R m f34 inh tnv men 22m24 a                                                 3.70mg * P<.0005
   n  R m f34 inh tnv men 22m24 a                                                 3.70mg * P<.0005
   o  R m f34 inh ton sqc 22m24 a                                                 9.14mg * P<.005 c
   p  R m f34 inh TBA MXB 22m24 a                                                 .112mg / P<.0005
   q  R m f34 inh liv MXB 22m24 a                                                 5.65mg * P<.03


DIBROMODULCITOL        100ng..:..1ug....:..10....:...100....:..1mg....:...10...:....100....:..1g.....:..10
1514 M f swi ipj lun mix 26w78 e                          .      +     .          9.23mg * P<.003 +
   a  M f swi ipj --- lys 26w78 e                                                 12.2mg * P<.0005+
   b  M f swi ipj --- leu 26w78 e                                                 46.3mg * P<.006
   c  M f swi ipj liv lys 26w78 e                                                no dre   P=1.
   d  M f swi ipj tba mix 26w78 e                                                 2.45mg * P<.0005
   e  M f swi ipj tba mal 26w78 e                                                 4.75mg * P<.0005
   f  M f swi ipj tba ben 26w78 e                                                 24.5mg * P<.09
1515 M m swi ipj --- lys 26w78 e                                 .     +     .    17.6mg * P<.002 +
   a  M m swi ipj --- lyk 26w78 e                                                 32.4mg * P<.002 +
   b  M m swi ipj lun mix 26w78 e                                                 13.5mg \ P<.05  +
   c  M m swi ipj liv mix 26w78 e                                                no dre   P=1.
   d  M m swi ipj tba mix 26w78 e                                                 12.7mg * P<.06
   e  M m swi ipj tba mal 26w78 e                                                 15.2mg * P<.04
   f  M m swi ipj tba ben 26w78 e                                                no dre   P=1.
1516 R m cdr ipj ski mix 26w78 e                          .    +    .             8.37mg * P<.0005+
   a  R m cdr ipj per sar 26w78 e                                                 59.2mg * P<.002
   b  R m cdr ipj --- lys 26w78 e                                                 182.mg * P<.04
   c  R m cdr ipj mgl sqc 26w78 e                                                 182.mg * P<.04
   d  R m cdr ipj liv tum 26w78 e                                                no dre   P=1.
   e  R m cdr ipj tba mix 26w78 e                                                 3.13mg * P<.0005
   f  R m cdr ipj tba mal 26w78 e                                                 5.05mg * P<.0005
   g  R m cdr ipj tba ben 26w78 e                                                no dre   P=1.


   1,2-DIBROMOETHANE   100ng..:..1ug....:..10....:...100....:..1mg....:...10...:....100....:..1g.....:..10
1517 M f b6c gav MXB MXB 53w78 sv                         :+ :                    3.74mg * P<.0005
   a  M f b6c gav sto sqc 53w78 sv                                                4.07mg * P<.0005
   b  M f b6c gav lun MXA 53w78 sv                                                15.4mg * P<.04  c
   c  M f b6c gav lun a/a 53w78 sv                                                17.3mg * P<.03  c
   d  M f b6c gav TBA MXB 53w78 sv                                                3.52mg * P<.0005
   e  M f b6c gav liv MXB 53w78 sv                                               no dre   P=1.
   f  M f b6c gav lun MXB 53w78 sv                                                15.4mg * P<.04
1518 M f b6c inh MXB MXB 22m24 s                       :+ :                       9.60mg / P<.0005

   a  M f b6c inh lun MXA 22m24 s                                                 18.4mg / P<.0005c
   b  M f b6c inh lun MXA 22m24 s                                                 18.7mg / P<.0005c
   c  M f b6c inh --- MXA 22m24 s                                                 24.1mg / P<.0005c
   d  M f b6c inh --- hes 22m24 s                                                 25.7mg / P<.0005c
```

	RefNum	LoConf	UpConf	Cntrl	1Dose	1Inc	2Dose	2Inc	Citation or Pathology
									Brkly Code
1511	TR206	.139mg	.295mg	1/50	.429mg	32/50	2.14mg	45/51	adr:coa; nas:acn,adn,apn,can,sqc,sqp; ntu:---; phr:sqc,sqp; ton:
									sqc,sqp. C
a	TR206	.172mg	.382mg	1/50	.429mg	26/50	2.14mg	42/51	nas:acn,adn,apn,can,sqc,sqp; ntu:---.
b	TR206	.392mg	1.79mg	5/50	.429mg	14/50	2.14mg	9/51	S
c	TR206	.444mg	3.03mg	4/50	.429mg	13/50	2.14mg	4/51	S
d	TR206	.486mg	2.19mg	0/50	.429mg	10/50	2.14mg	3/51	
e	TR206	.586mg	2.93mg	0/50	.429mg	7/50	2.14mg	5/51	
f	TR206	.683mg	4.25mg	0/50	.429mg	6/50	2.14mg	5/51	
g	TR206	.750mg	4.88mg	0/50	.429mg	5/50	2.14mg	5/51	
h	TR206	.828mg	4.33mg	0/50	.429mg	4/50	2.14mg	9/51	ton:sqc,sqp.
i	TR206	1.05mg	3.36mg	0/50	.429mg	0/50	2.14mg	23/51	
j	TR206	.830mg	10.1mg	0/50	.429mg	5/50	2.14mg	1/51	S
k	TR206	.985mg	7.58mg	0/50	.429mg	3/50	2.14mg	6/51	
l	TR206	1.13mg	11.4mg	1/50	.429mg	2/50	2.14mg	7/51	
m	TR206	1.27mg	12.9mg	0/50	.429mg	2/50	2.14mg	6/51	
n	TR206	1.71mg	22.1mg	1/50	.429mg	0/50	2.14mg	6/51	phr:sqc,sqp.
o	TR206	2.14mg	24.2mg	0/50	.429mg	0/50	2.14mg	5/51	
p	TR206	1.96mg	54.8mg	0/50	.429mg	1/50	2.14mg	3/51	
q	TR206	.110mg	.264mg	33/50	.429mg	44/50	2.14mg	47/51	
r	TR206	1.17mg	n.s.s.	2/50	.429mg	0/50	2.14mg	0/51	liv:hpa,hpc,nnd.
1512	TR28	.584mg	1.55mg	0/20	10.0mg	47/50	21.0mg	47/50	
a	TR28	1.71mg	n.s.s.	0/20	10.0mg	13/50	21.0mg	2/50	---:hem,hes. S
b	TR28	.600mg	1.61mg	1/20	10.0mg	49/50	21.0mg	48/50	
c	TR28	n.s.s.	n.s.s.	0/20	10.0mg	0/50	21.0mg	0/50	liv:hpa,hpc,nnd.
1513	TR206	75.6ug	.154mg	0/50	.300mg	41/50	1.50mg	40/49	nas:acn,adn,apn,can,cas,sqc,sqp; nsm:sqc; ntu:---; ton:sqc,sqp. C
a	TR206	76.4ug	.156mg	0/50	.300mg	40/50	1.50mg	39/49	nas:acn,adn,apn,can,cas,sqc,sqp; nsm:sqc; ntu:---.
b	TR206	.301mg	1.29mg	0/50	.300mg	13/50	1.50mg	1/49	
c	TR206	.399mg	2.46mg	0/50	.300mg	9/50	1.50mg	1/49	
d	TR206	.424mg	2.09mg	0/50	.300mg	8/50	1.50mg	6/49	
e	TR206	.442mg	3.71mg	0/50	.300mg	8/50	1.50mg	0/49	S
f	TR206	.456mg	2.72mg	0/50	.300mg	7/50	1.50mg	3/49	
g	TR206	.625mg	2.01mg	0/50	.300mg	2/50	1.50mg	22/49	
h	TR206	.684mg	3.18mg	0/50	.300mg	2/50	1.50mg	11/49	
i	TR206	1.01mg	4.85mg	0/50	.300mg	1/50	1.50mg	11/49	ton:sqc,sqp.
j	TR206	.947mg	18.0mg	1/50	.300mg	2/50	1.50mg	5/49	S
k	TR206	1.23mg	7.53mg	0/50	.300mg	1/50	1.50mg	8/49	
l	TR206	1.07mg	18.8mg	0/50	.300mg	1/50	1.50mg	3/49	S
m	TR206	1.19mg	16.3mg	0/50	.300mg	1/50	1.50mg	5/49	S
n	TR206	1.19mg	16.3mg	0/50	.300mg	1/50	1.50mg	5/49	S
o	TR206	2.47mg	104.mg	0/50	.300mg	0/50	1.50mg	3/49	
p	TR206	75.9ug	.179mg	30/50	.300mg	48/50	1.50mg	47/49	
q	TR206	1.30mg	n.s.s.	0/50	.300mg	1/50	1.50mg	1/49	liv:hpa,hpc,nnd.

DIBROMODULCITOL   10318-26-0

	RefNum	LoConf	UpConf	Cntrl	1Dose	1Inc	2Dose	2Inc	Citation or Pathology
1514	1336	3.83mg	79.3mg	20/154	6.40mg	9/21	12.8mg	3/9	Skipper;srfr;1976/Weisburger;canc;1977/Prejean pers.comm.
a	1336	5.05mg	46.7mg	3/154	6.40mg	2/21	12.8mg	5/9	
b	1336	11.4mg	1.06gm	0/154	6.40mg	1/21	12.8mg	1/9	
c	1336	7.19mg	n.s.s.	1/154	6.40mg	0/21	12.8mg	0/9	
d	1336	1.23mg	6.20mg	42/154	6.40mg	14/21	12.8mg	9/9	
e	1336	2.28mg	15.2mg	29/154	6.40mg	10/21	12.8mg	7/9	
f	1336	6.95mg	n.s.s.	13/154	6.40mg	4/21	12.8mg	2/9	
1515	1336	7.37mg	97.0mg	2/101	6.40mg	6/29	15.8mg	2/11	
a	1336	11.2mg	179.mg	0/101	6.40mg	2/29	15.8mg	2/11	
b	1336	4.45mg	n.s.s.	9/101	6.40mg	7/29	(15.8mg	0/11)	
c	1336	10.4mg	n.s.s.	2/101	6.40mg	0/29	15.8mg	0/11	
d	1336	4.53mg	n.s.s.	28/101	6.40mg	14/29	15.8mg	5/11	
e	1336	5.56mg	n.s.s.	19/101	6.40mg	9/29	15.8mg	5/11	
f	1336	16.1mg	n.s.s.	9/101	6.40mg	5/29	15.8mg	0/11	
1516	1336	4.50mg	18.2mg	4/177	10.7mg	8/23	21.5mg	8/11	
a	1336	17.9mg	435.mg	0/177	10.7mg	2/23	21.5mg	1/11	
b	1336	29.6mg	n.s.s.	0/177	10.7mg	0/23	21.5mg	1/11	
c	1336	29.6mg	n.s.s.	0/177	10.7mg	0/23	21.5mg	1/11	
d	1336	14.0mg	n.s.s.	0/177	10.7mg	0/23	21.5mg	0/11	
e	1336	1.58mg	7.35mg	59/177	10.7mg	18/23	21.5mg	11/11	
f	1336	2.69mg	11.4mg	32/177	10.7mg	12/23	21.5mg	11/11	
g	1336	21.6mg	n.s.s.	27/177	10.7mg	6/23	21.5mg	0/11	

1,2-DIBROMOETHANE   (ethylene dibromide, EDB) 106-93-4

	RefNum	LoConf	UpConf	Cntrl	1Dose	1Inc	2Dose	2Inc	Citation or Pathology
1517	TR86	2.58mg	6.93mg	0/20	26.0mg	46/50	52.0mg	30/50	lun:a/a,a/c; sto:sqc. C
a	TR86	2.81mg	8.09mg	0/20	26.0mg	46/50	52.0mg	28/50	
b	TR86	8.06mg	n.s.s.	0/20	26.0mg	11/50	52.0mg	6/50	lun:a/a,a/c.
c	TR86	8.85mg	n.s.s.	0/20	26.0mg	10/50	52.0mg	6/50	
d	TR86	2.42mg	6.31mg	0/20	26.0mg	47/50	52.0mg	31/50	
e	TR86	n.s.s.	n.s.s.	0/20	26.0mg	1/50	52.0mg	0/50	liv:hpa,hpc,nnd.
f	TR86	8.06mg	n.s.s.	0/20	26.0mg	11/50	52.0mg	6/50	lun:a/a,a/c.
1518	TR210	6.41mg	14.8mg	6/50	23.9mg	31/50	95.6mg	47/50	---:hem,hes; lun:a/a,a/c,adn,apn,can; mgl:acn; nas:---,adn,apn,
									can; sub:fbs. C
a	TR210	11.3mg	30.9mg	4/50	23.9mg	11/50	95.6mg	42/50	lun:a/a,a/c,adn,apn,can.
b	TR210	11.4mg	31.6mg	4/50	23.9mg	11/50	95.6mg	41/50	lun:a/a,a/c.
c	TR210	14.1mg	41.6mg	0/50	23.9mg	12/50	95.6mg	25/50	---:hem,hes.
d	TR210	14.7mg	45.6mg	0/50	23.9mg	11/50	95.6mg	23/50	

	Spe Strain Site Xpo+Xpt		TD50	2Tailpvl
	Sex Route Hist Notes		DR	AuOp
e	M f b6c inh mgl acn 22m24 s		26.2mg *	P<.0005c
f	M f b6c inh lun a/c 22m24 s		30.3mg /	P<.0005c
g	M f b6c inh lun a/a 22m24 s		38.6mg /	P<.0005c
h	M f b6c inh sub fbs 22m24 s		90.7mg *	P<.0005c
i	M f b6c inh l/b MXA 22m24 s		169.mg *	P<.0005
j	M f b6c inh nas --- 22m24 s		187.mg /	P<.0005c
k	M f b6c inh --- hem 22m24 s		245.mg *	P<.0005
l	M f b6c inh l/b can 22m24 s		259.mg *	P<.002
m	M f b6c inh l/b MXA 22m24 s		272.mg /	P<.0005
n	M f b6c inh nas MXA 22m24 s		337.mg *	P<.0005c
o	M f b6c inh nas MXA 22m24 s		353.mg *	P<.0005c
p	M f b6c inh l/b adn 22m24 s		361.mg *	P<.0005
q	M f b6c inh nas can 22m24 s		443.mg *	P<.0005c
r	M f b6c inh nas apn 22m24 s		470.mg *	P<.004 c
s	M f b6c inh TBA MXB 22m24 s		6.83mg /	P<.0005
t	M f b6c inh liv MXB 22m24 s		107.mg *	P<.07
u	M f b6c inh lun MXB 22m24 s		18.7mg *	P<.0005
1519	M f b6c wat for mix 73w73 e	. + .	13.1mg	P<.0005+
a	M f b6c wat for sqc 73w73 e		24.5mg	P<.0005+
b	M f b6c wat eso sqp 73w73 e		318.mg	P<.02 +
c	M f b6c wat lun ptm 73w73 e		2.34gm	P<.7 -
d	M f b6c wat liv tum 73w73 e		no dre	P=1. -
e	M f b6c wat tba mix 73w73 e		13.8mg	P<.0005
1520	M f b6c wat for pam 78w78 e	. + .	23.2mg	P<.0005
a	M f b6c wat for sqc 78w78 e		35.3mg	P<.0005+
b	M f b6c wat eso pam 78w78 e		217.mg	P<.004
c	M f b6c wat liv hpt 78w78 e		no dre	P=1.
d	M f b6c wat lun ptm 78w78 e		no dre	P=1.
e	M f b6c wat tba mix 78w78 e		noTD50	P<.2
1521	M m b6c gav MXB MXB 53w78 esv	: + :	2.34mg *	P<.0005
a	M m b6c gav sto MXA 53w78 esv		2.36mg *	P<.0005c
b	M m b6c gav sto sqc 53w78 esv		2.38mg *	P<.0005c
c	M m b6c gav lun a/a 53w78 esv		13.4mg *	P<.003 c
d	M m b6c gav TBA MXB 53w78 esv		2.66mg *	P<.0005
e	M m b6c gav liv MXB 53w78 esv		no dre	P=1.
f	M m b6c gav lun MXB 53w78 esv		13.4mg *	P<.003
1522	M m b6c inh MXB MXB 78w78	: + :	18.0mg *	P<.0005
a	M m b6c inh lun --- 78w78		18.2mg *	P<.0005c
b	M m b6c inh lun MXA 78w78		21.0mg *	P<.0005c
c	M m b6c inh lun a/c 78w78		24.7mg *	P<.0005c
d	M m b6c inh lun a/a 78w78		60.1mg *	P<.0005c
e	M m b6c inh MXA apn 78w78		124.mg *	P<.02
f	M m b6c inh l/b MXA 78w78		124.mg *	P<.02
g	M m b6c inh --- MXA 78w78		200.mg *	P<.03 c
h	M m b6c inh l/b apn 78w78		217.mg *	P<.05
i	M m b6c inh TBA MXB 78w78		21.2mg *	P<.0005
j	M m b6c inh liv MXB 78w78		no dre	P=1.
k	M m b6c inh lun MXB 78w78		21.0mg *	P<.0005
1523	M m b6c wat for sqc 65w65 e	. + .	11.8mg	P<.0005+
a	M m b6c wat for mix 65w65 e		11.9mg	P<.0005+
b	M m b6c wat liv tum 65w65 e		no dre	P=1.
c	M m b6c wat lun ptm 65w65 e		no dre	P=1. -
d	M m b6c wat tba mix 65w65 e		12.7mg	P<.0005
1524	M m b6c wat for sqc 78w78 e	. + .	9.44mg	P<.0005+
a	M m b6c wat eso sqc 78w78 e		207.mg	P<.003
b	M m b6c wat eso pam 78w78 e		207.mg	P<.003
c	M m b6c wat liv hpt 78w78 e		no dre	P=1.
d	M m b6c wat tba mix 78w78 e		noTD50	P<.05
1525	R f osm gav MXB MXB 50w61 ades	: + :	1.26mg *	P<.0005
a	R f osm gav sto sqc 50w61 ades		1.26mg *	P<.0005c
b	R f osm gav liv MXA 50w61 ades		5.38mg *	P<.0005c
c	R f osm gav liv hpc 50w61 ades		5.52mg *	P<.0005c
d	R f osm gav adr MXA 50w61 ades		20.9mg /	P<.02
e	R f osm gav TBA MXB 50w61 ades		1.19mg *	P<.0005
f	R f osm gav liv MXB 50w61 ades		5.38mg *	P<.0005
1526	R f f34 inh MXB MXB 23m24 as	:+ :	1.81mg /	P<.0005
a	R f f34 inh nas --- 23m24 as		2.33mg /	P<.0005c
b	R f f34 inh mgl fba 23m24 as		3.60mg /	P<.0005c
c	R f f34 inh nas acn 23m24 as		4.28mg /	P<.0005c
d	R f f34 inh pit adn 23m24 as		7.46mg *	P<.0005c
e	R f f34 inh nas adn 23m24 as		10.8mg *	P<.0005c
f	R f f34 inh nas can 23m24 as		21.6mg *	P<.0005c
g	R f f34 inh nas apn 23m24 as		22.9mg *	P<.0005c
h	R f f34 inh nas sqc 23m24 as		54.2mg /	P<.002 c
i	R f f34 inh liv hpc 23m24 as		66.4mg *	P<.002
j	R f f34 inh lun MXA 23m24 as		83.5mg *	P<.0005c
k	R f f34 inh --- hes 23m24 as		87.3mg *	P<.0005c
l	R f f34 inh spl hes 23m24 as		87.3mg *	P<.0005c
m	R f f34 inh lun a/c 23m24 as		99.9mg *	P<.002 c
n	R f f34 inh sub fib 23m24 as		170.mg *	P<.008

	RefNum	LoConf	UpConf	Cntrl	1Dose	1Inc	2Dose	2Inc	Citation or Pathology Brkly Code
e	TR210	13.7mg	61.7mg	2/50	23.9mg	14/50	95.6mg	8/50	
f	TR210	18.0mg	50.3mg	1/50	23.9mg	5/50	95.6mg	37/50	
g	TR210	18.6mg	98.3mg	3/50	23.9mg	7/50	95.6mg	13/50	
h	TR210	41.5mg	201.mg	0/50	23.9mg	4/50	95.6mg	11/50	
i	TR210	64.5mg	470.mg	0/50	23.9mg	1/50	95.6mg	8/50	l/b:adn,can. S
j	TR210	89.0mg	425.mg	0/50	23.9mg	0/50	95.6mg	12/50	
k	TR210	81.6mg	1.18gm	0/50	23.9mg	1/50	95.6mg	4/50	S
l	TR210	74.7mg	1.51gm	0/50	23.9mg	1/50	95.6mg	4/50	S
m	TR210	105.mg	968.mg	0/50	23.9mg	0/50	95.6mg	6/50	l/b:adn,apn. S
n	TR210	148.mg	978.mg	0/50	23.9mg	0/50	95.6mg	8/50	nas:adn,can.
o	TR210	122.mg	1.57gm	0/50	23.9mg	0/50	95.6mg	5/50	nas:adn,apn.
p	TR210	129.mg	1.55gm	0/50	23.9mg	0/50	95.6mg	5/50	S
q	TR210	173.mg	1.67gm	0/50	23.9mg	0/50	95.6mg	6/50	
r	TR210	133.mg	4.80gm	0/50	23.9mg	0/50	95.6mg	3/50	
s	TR210	4.55mg	10.9mg	23/50	23.9mg	45/50	95.6mg	49/50	
t	TR210	31.4mg	n.s.s.	2/50	23.9mg	6/50	95.6mg	1/50	liv:hpa,hpc,nnd.
u	TR210	11.4mg	31.6mg	4/50	23.9mg	11/50	95.6mg	41/50	lun:a/a,a/c.
1519	1761	6.68mg	25.7mg	1/50	103.mg	27/29			Van Durren;tcam,5,393-403;1985
a	1761	13.9mg	46.5mg	0/50	103.mg	22/29			
b	1761	96.3mg	n.s.s.	0/50	103.mg	3/29			
c	1761	175.mg	n.s.s.	1750	103.mg	1/29			
d	1761	303.mg	n.s.s.	0/50	103.mg	0/29			
e	1761	6.84mg	28.6mg	7/50	103.mg	27/29			
1520	1806	13.9mg	44.8mg	9/96	48.0mg	29/49			Van Duuren;enhp,69,109-117;1986
a	1806	20.7mg	67.1mg	0/96	48.0mg	20/49			
b	1806	75.0mg	1.58gm	0/96	48.0mg	4/49			
c	1806	273.mg	n.s.s.	2/96	48.0mg	0/49			
d	1806	192.mg	n.s.s.	13/96	48.0mg	2/49			
e	1806	n.s.s.	n.s.s.	84/96	48.0mg	49/49			
1521	TR86	1.07mg	4.23mg	0/20	30.0mg	45/50	53.0mg	33/50	lun:a/a; sto:sqc,sqp. C
a	TR86	1.07mg	4.32mg	0/20	30.0mg	45/50	53.0mg	31/50	sto:sqc,sqp.
b	TR86	1.08mg	4.43mg	0/20	30.0mg	45/50	53.0mg	29/50	
c	TR86	4.58mg	55.8mg	0/20	30.0mg	4/50	53.0mg	10/50	
d	TR86	1.19mg	5.83mg	2/20	30.0mg	45/50	53.0mg	33/50	
e	TR86	51.5mg	n.s.s.	2/20	30.0mg	1/50	53.0mg	1/50	liv:hpa,hpc,nnd.
f	TR86	4.58mg	55.8mg	0/20	30.0mg	4/50	53.0mg	10/50	lun:a/a,a/c.
1522	TR210	11.3mg	32.5mg	0/50	19.9mg	3/50	79.5mg	26/50	---:hem,hes; lun:---,a/a,a/c. C
a	TR210	11.4mg	33.6mg	0/50	19.9mg	3/50	79.5mg	25/50	
b	TR210	13.0mg	40.2mg	0/50	19.9mg	3/50	79.5mg	23/50	lun:a/a,a/c.
c	TR210	14.7mg	53.7mg	0/50	19.9mg	3/50	79.5mg	19/50	
d	TR210	29.9mg	146.mg	0/50	19.9mg	0/50	79.5mg	11/50	
e	TR210	47.1mg	n.s.s.	0/50	19.9mg	0/50	79.5mg	5/50	b/l:apn; l/b:apn. S
f	TR210	47.1mg	n.s.s.	0/50	19.9mg	0/50	79.5mg	5/50	l/b:adn,apn. S
g	TR210	67.2mg	n.s.s.	0/50	19.9mg	0/50	79.5mg	4/50	---:hem,hes.
h	TR210	65.7mg	n.s.s.	0/50	19.9mg	0/50	79.5mg	3/50	S
i	TR210	11.7mg	65.5mg	3/50	19.9mg	5/50	79.5mg	27/50	
j	TR210	87.5mg	n.s.s.	3/50	19.9mg	1/50	79.5mg	3/50	liv:hpa,hpc,nnd.
k	TR210	13.0mg	40.2mg	0/50	19.9mg	3/50	79.5mg	23/50	lun:a/a,a/c.
1523	1761	5.98mg	23.1mg	0/45	116.mg	26/28			Van Durren;tcam,5,393-403;1985
a	1761	6.01mg	23.5mg	1/45	116.mg	26/28			
b	1761	261.mg	n.s.s.	0/45	116.mg	0/28			
c	1761	146.mg	n.s.s.	5/45	116.mg	2/28			
d	1761	6.21mg	27.4mg	8/45	116.mg	26/28			
1524	1806	5.97mg	15.5mg	2/99	46.7mg	41/48			Van Duuren;enhp,69,109-117;1986
a	1806	71.4mg	1.40gm	0/99	46.7mg	4/48			
b	1806	71.4mg	1.40gm	0/99	46.7mg	4/48			
c	1806	215.mg	n.s.s.	12/99	46.7mg	1/48			
d	1806	n.s.s.	n.s.s.	76/99	46.7mg	48/48			
1525	TR86	.665mg	2.14mg	0/20	26.7mg	40/50	28.1mg	29/50	liv:hpc,nnd; sto:sqc. C
a	TR86	.665mg	2.14mg	0/20	26.7mg	40/50	28.1mg	29/50	
b	TR86	1.51mg	32.8mg	0/20	26.7mg	1/50	28.1mg	6/50	liv:hpc,nnd.
c	TR86	1.52mg	38.6mg	0/20	26.7mg	1/50	28.1mg	5/50	
d	TR86	5.40mg	n.s.s.	0/20	26.7mg	0/50	28.1mg	4/50	adr:coa,coc. S
e	TR86	.628mg	2.23mg	2/20	26.7mg	40/50	28.1mg	29/50	
f	TR86	1.51mg	32.8mg	0/20	26.7mg	1/50	28.1mg	6/50	liv:hpa,hpc,nnd.
1526	TR210	1.27mg	2.71mg	6/50	5.71mg	44/50	22.8mg	45/50	---:hes; lun:a/a,a/c; mgl:fba; nas:---,acn,adn,apn,can,sqc; pit: adn; spl:hes. C
a	TR210	1.64mg	3.43mg	1/50	5.71mg	34/50	22.8mg	43/50	
b	TR210	2.37mg	5.91mg	4/50	5.71mg	29/50	22.8mg	24/50	
c	TR210	2.81mg	6.78mg	0/50	5.71mg	20/50	22.8mg	29/50	
d	TR210	4.25mg	16.0mg	1/50	5.71mg	18/50	22.8mg	4/50	
e	TR210	5.67mg	24.0mg	0/50	5.71mg	11/50	22.8mg	5/50	
f	TR210	13.0mg	38.3mg	0/50	5.71mg	0/50	22.8mg	25/50	
g	TR210	10.0mg	62.4mg	0/50	5.71mg	5/50	22.8mg	5/50	
h	TR210	18.6mg	358.mg	1/50	5.71mg	1/50	22.8mg	5/50	
i	TR210	20.4mg	461.mg	0/50	5.71mg	1/50	22.8mg	3/50	S
j	TR210	30.9mg	350.mg	0/50	5.71mg	0/50	22.8mg	5/50	lun:a/a,a/c.
k	TR210	31.8mg	319.mg	0/50	5.71mg	0/50	22.8mg	5/50	
l	TR210	31.8mg	319.mg	0/50	5.71mg	0/50	22.8mg	5/50	
m	TR210	33.6mg	555.mg	0/50	5.71mg	0/50	22.8mg	4/50	
n	TR210	47.7mg	4.00gm	0/50	5.71mg	0/50	22.8mg	3/50	S

	Spe		Strain	Site		Xpo+Xpt		Notes		TD50		2Tailpvl	
		Sex		Route	Hist						DR	AuOp	
o	R	f	f34	inh	sub	MXA	23m24	as			170.mg	*	P<.008	
p	R	f	f34	inh	mgl	acn	23m24	as			98.8mg	/	P<.02	
q	R	f	f34	inh	TBA	MXB	23m24	as			2.02mg	/	P<.0005	
r	R	f	f34	inh	liv	MXB	23m24	as			45.6mg	/	P<.003	
1527	R	m	osm	gav	MXB	MXB	40w49	adsv	: (+):		1.64mg	/	P<.0005	
a	R	m	osm	gav	sto	sqc	40w49	adsv			1.65mg	/	P<.0005c	
b	R	m	osm	gav	---	hes	40w49	adsv			9.62mg	*	P<.002	c
c	R	m	osm	gav	thy	MXA	40w49	adsv			10.4mg	/	P<.002	
d	R	m	osm	gav	TBA	MXB	40w49	adsv			1.56mg	/	P<.0005	
e	R	m	osm	gav	liv	MXB	40w49	adsv			23.8mg	*	P<.03	
1528	R	m	f34	inh	MXB	MXB	22m24	as	:+ :		1.10mg	/	P<.0005	
a	R	m	f34	inh	MXA	MXA	22m24	as			1.23mg	/	P<.0005c	
b	R	m	f34	inh	nas	acn	22m24	as			2.73mg	/	P<.0005c	
c	R	m	f34	inh	nas	apn	22m24	as			3.97mg	*	P<.0005c	
d	R	m	f34	inh	tnv	mso	22m24	as			5.63mg	/	P<.0005c	
e	R	m	f34	inh	tnv	men	22m24	as			6.06mg	/	P<.0005c	
f	R	m	f34	inh	nas	adn	22m24	as			7.66mg	*	P<.0005c	
g	R	m	f34	inh	pit	adn	22m24	as			13.8mg	*	P<.0005c	
h	R	m	f34	inh	spl	hes	22m24	as			16.5mg	*	P<.0005c	
i	R	m	f34	inh	nas	can	22m24	as			18.0mg	*	P<.0005c	
j	R	m	f34	inh	mul	msm	22m24	as			18.5mg	*	P<.003	
k	R	m	f34	inh	slg	sar	22m24	as			58.8mg	*	P<.004	
l	R	m	f34	inh	thy	MXA	22m24	as			86.1mg	*	P<.004	
m	R	m	f34	inh	TBA	MXB	22m24	as			1.34mg	/	P<.0005	
n	R	m	f34	inh	liv	MXB	22m24	as			78.5mg	*	P<.06	
1529	R	f	cdr	inh	nas	tum	78w78	e	. + .		2.20mg		P<.0005+	
a	R	f	cdr	inh	nas	ben	78w78	e			5.02mg		P<.0005+	
b	R	f	cdr	inh	nas	mal	78w78	e			8.59mg		P<.0005+	
c	R	f	cdr	inh	mgl	car	78w78	e			19.9mg		P<.0005+	
d	R	f	cdr	inh	liv	hpc	78w78	e			59.4mg		P<.02	
e	R	f	cdr	inh	spl	hes	78w78	e			122.mg		P<.1	
1530	R	m	cdr	inh	nas	tum	78w78	e	.+ .		1.19mg		P<.0005+	
a	R	m	cdr	inh	nas	ben	78w78	e			2.52mg		P<.0005+	
b	R	m	cdr	inh	nas	mal	78w78	e			3.52mg		P<.0005+	
c	R	m	cdr	inh	spl	hes	78w78	e			12.6mg		P<.0005+	
d	R	m	cdr	inh	bra	mng	78w78	e			56.1mg		P<.04	
e	R	m	cdr	inh	liv	hpc	78w78	e			85.1mg		P<.1	

DIBROMOMANNITOL   100ng..:..1ug....:..10......:..100....:..1mg....:..10......:..100....:..1g......:..10

	Spe		Strain	Site		Xpo+Xpt		Notes	TD50		2Tailpvl	
1531	M	f	swi	ipj	lun	mix	26w78	e	. + .	21.4mg	*	P<.002 +
a	M	f	swi	ipj	---	lys	26w78	e		38.4mg	*	P<.0005+
b	M	f	swi	ipj	sub	car	26w78	e		66.0mg	*	P<.0005
c	M	f	swi	ipj	---	lyk	26w78	e		89.8mg	*	P<.002
d	M	f	swi	ipj	adr	coa	26w78	e		136.mg	*	P<.006
e	M	f	swi	ipj	liv	lys	26w78	e		no dre		P=1.
f	M	f	swi	ipj	tba	mix	26w78	e		7.67mg	*	P<.0005
g	M	f	swi	ipj	tba	mal	26w78	e		11.2mg	*	P<.0005
h	M	f	swi	ipj	tba	ben	26w78	e		130.mg	*	P<.4
1532	M	m	swi	ipj	lun	mix	26w78	e	. + .	11.4mg		P<.002 +
a	M	m	swi	ipj	liv	mix	26w78	e		65.7mg		P<.2
b	M	m	swi	ipj	tba	mix	26w78	e		6.03mg		P<.0005
c	M	m	swi	ipj	tba	mal	26w78	e		8.53mg		P<.002
d	M	m	swi	ipj	tba	ben	26w78	e		92.8mg		P<.6
1533	R	f	cdr	ipj	mgl	adc	26w78	e	. + .	24.9mg		P<.003 +
a	R	f	cdr	ipj	per	sar	26w78	e		79.2mg		P<.004 +
b	R	f	cdr	ipj	lun	car	26w78	e		84.6mg		P<.03
c	R	f	cdr	ipj	liv	tum	26w78	e		no dre		P=1.
d	R	f	cdr	ipj	tba	mal	26w78	e		11.5mg		P<.002
e	R	f	cdr	ipj	tba	mix	26w78	e		12.5mg		P<.08
f	R	f	cdr	ipj	tba	ben	26w78	e		no dre		P=1.
1534	R	m	cdr	ipj	per	sar	26w78	e	. + .	30.9mg	\	P<.0005+
a	R	m	cdr	ipj	ski	mix	26w78	e		103.mg	*	P<.03 +
b	R	m	cdr	ipj	mgl	fba	26w78	e		255.mg	*	P<.2
c	R	m	cdr	ipj	liv	tum	26w78	e		no dre		P=1.
d	R	m	cdr	ipj	tba	mix	26w78	e		7.95mg	\	P<.0005
e	R	m	cdr	ipj	tba	ben	26w78	e		16.6mg	\	P<.003
f	R	m	cdr	ipj	tba	mal	26w78	e		no dre		P=1.

DIBROMONEOPENTYL GLYCOL   100ng..:..1ug....:..10......:..100....:..1mg....:..10......:..100....:..1g....:..10

	Spe		Strain	Site		Xpo+Xpt		Notes	TD50		2Tailpvl	
1535	R	f	sss	eat	liv	hph	24m24	e	.>	2.55gm	*	P<.5	-
a	R	f	sss	eat	liv	hpc	24m24	e		no dre		P=1.	-
b	R	f	sss	eat	tba	mix	24m24	e		1.73gm	*	P<1.	-
1536	R	m	sss	eat	liv	hph	24m24	e	.>	no dre		P=1.	-
a	R	m	sss	eat	liv	hpc	24m24	e		no dre		P=1.	-
b	R	m	sss	eat	tba	mix	24m24	e		no dre		P=1.	-

5,7-DIBROMOQUINOLINE   100ng..:..1ug....:..10......:..100....:..1mg....:..10......:..100....:..1g......:..10

	Spe		Strain	Site		Xpo+Xpt		Notes	TD50		2Tailpvl	
1537	R	f	f34	eat	liv	hnd	24m24	e	.>	no dre		P=1.	-
1538	R	m	f34	eat	tes	ict	24m24	e	±	69.4mg		P<.06	-
a	R	m	f34	eat	liv	hnd	24m24	e		754.mg		P<.3	-

	RefNum	LoConf	UpConf	Cntrl	1Dose	1Inc	2Dose	2Inc	Citation or Pathology	Brkly Code
o	TR210	47.7mg	4.00gm	0/50	5.71mg	0/50	22.8mg	3/50	sub:fbs,fib.	S
p	TR210	26.5mg	27.1gm	1/50	5.71mg	0/50	22.8mg	4/50		S
q	TR210	1.36mg	3.30mg	33/50	5.71mg	49/50	22.8mg	48/50		
r	TR210	16.0mg	364.mg	2/50	5.71mg	1/50	22.8mg	5/50	liv:hpa,hpc,nnd.	
1527	TR86	1.18mg	2.33mg	0/20	27.4mg	45/50	29.2mg	34/50	---:hes; sto:sqc.	C
a	TR86	1.18mg	2.35mg	0/20	27.4mg	45/50	29.2mg	33/50		
b	TR86	4.98mg	21.1mg	0/20	27.4mg	11/50	29.2mg	4/50		
c	TR86	5.20mg	34.2mg	0/20	27.4mg	5/50	29.2mg	8/50	thy:fca,fcc.	S
d	TR86	1.12mg	2.22mg	0/20	27.4mg	46/50	29.2mg	35/50		
e	TR86	8.97mg	n.s.s.	0/20	27.4mg	2/50	29.2mg	2/50	liv:hpa,hpc,nnd.	
1528	TR210	.787mg	1.60mg	1/50	4.00mg	43/50	15.9mg	48/50	nas:---,acn,adn,apn,can; nse:sqp; pit:adn; spl:hes; tnv:men,mso.	C
a	TR210	.872mg	1.79mg	0/50	4.00mg	39/50	15.9mg	41/50	nas:---; nse:sqp.	
b	TR210	1.78mg	4.36mg	0/50	4.00mg	20/50	15.9mg	28/50		
c	TR210	2.34mg	7.45mg	0/50	4.00mg	18/50	15.9mg	5/50		
d	TR210	3.36mg	10.1mg	1/50	4.00mg	8/50	15.9mg	25/50		
e	TR210	3.61mg	10.6mg	0/50	4.00mg	7/50	15.9mg	25/50		
f	TR210	3.84mg	19.7mg	0/50	4.00mg	11/50	15.9mg	0/50		
g	TR210	6.07mg	46.3mg	0/50	4.00mg	7/50	15.9mg	2/50		
h	TR210	8.48mg	33.7mg	0/50	4.00mg	1/50	15.9mg	15/50		
i	TR210	10.4mg	33.2mg	0/50	4.00mg	0/50	15.9mg	21/50		
j	TR210	7.21mg	109.mg	0/50	4.00mg	5/50	15.9mg	1/50		S
k	TR210	17.3mg	528.mg	0/50	4.00mg	1/50	15.9mg	3/50		S
l	TR210	24.6mg	856.mg	0/50	4.00mg	0/50	15.9mg	3/50	thy:fca,fcc.	S
m	TR210	.904mg	2.14mg	30/50	4.00mg	48/50	15.9mg	49/50		
n	TR210	15.3mg	n.s.s.	0/50	4.00mg	1/50	15.9mg	1/50	liv:hpa,hpc,nnd.	
1529	1032	1.31mg	3.75mg	0/47	13.4mg	38/42			Wong;txap,63,155-165;1982/Plotnick pers.comm.	
a	1032	3.09mg	8.79mg	0/47	13.4mg	27/42				
b	1032	4.96mg	16.7mg	0/47	13.4mg	19/42				
c	1032	9.95mg	49.5mg	0/48	13.4mg	11/48				
d	1032	20.5mg	n.s.s.	0/48	13.4mg	4/48				
e	1032	29.9mg	n.s.s.	0/48	13.4mg	2/48				
1530	1032	.645mg	2.11mg	0/47	9.39mg	40/42				
a	1032	1.57mg	4.27mg	0/47	9.39mg	32/42				
b	1032	2.17mg	6.15mg	0/47	9.39mg	27/42				
c	1032	6.47mg	29.7mg	0/48	9.39mg	12/48				
d	1032	17.0mg	n.s.s.	0/48	9.39mg	3/48				
e	1032	20.9mg	n.s.s.	0/48	9.39mg	2/48				

DIBROMOMANNITOL   (DBM) 488-41-5

	RefNum	LoConf	UpConf	Cntrl	1Dose	1Inc	2Dose	2Inc	Citation or Pathology	Brkly Code
1531	1336m	9.53mg	136.mg	20/154	12.8mg	11/23	25.7mg	5/17	Skipper;srfr;1976/Weisburger;canc;1977/Prejean pers.comm.	
a	1336m	15.8mg	162.mg	3/154	12.8mg	1/23	25.7mg	6/17		
b	1336m	22.8mg	340.mg	0/154	12.8mg	1/23	25.7mg	3/17		
c	1336m	27.2mg	684.mg	0/154	12.8mg	1/23	25.7mg	2/17		
d	1336m	33.4mg	2.86gm	0/154	12.8mg	0/23	25.7mg	2/17		
e	1336m	20.4mg	n.s.s.	1/154	12.8mg	0/23	25.7mg	0/17		
f	1336m	4.05mg	19.2mg	42/154	12.8mg	12/23	25.7mg	15/17		
g	1336m	5.76mg	30.7mg	29/154	12.8mg	10/23	25.7mg	12/17		
h	1336m	25.3mg	n.s.s.	13/154	12.8mg	2/23	25.7mg	3/17		
1532	1336m	4.75mg	60.4mg	9/101	12.8mg	9/22				
a	1336m	13.4mg	n.s.s.	2/101	12.8mg	2/22				
b	1336m	2.64mg	28.7mg	28/101	12.8mg	15/22				
c	1336m	3.62mg	50.3mg	19/101	12.8mg	12/22				
d	1336m	12.4mg	n.s.s.	9/101	12.8mg	3/22				
1533	1336n	9.35mg	232.mg	12/182	17.9mg	7/24				
a	1336n	19.5mg	1.19gm	0/182	17.9mg	2/24				
b	1336n	19.8mg	n.s.s.	1/182	17.9mg	2/24				
c	1336n	49.8mg	n.s.s.	0/182	17.9mg	0/24				
d	1336n	4.92mg	68.1mg	44/182	17.9mg	14/24				
e	1336n	3.95mg	n.s.s.	103/182	17.9mg	18/24				
f	1336n	27.8mg	n.s.s.	59/182	17.9mg	4/24				
1534	1336n	11.7mg	129.mg	0/177	17.9mg	5/25	(53.0mg	0/12)		
a	1336n	32.5mg	n.s.s.	4/177	17.9mg	4/25	53.0mg	1/12		
b	1336n	52.7mg	n.s.s.	1/177	17.9mg	2/25	53.0mg	0/12		
c	1336n	30.4mg	n.s.s.	0/177	17.9mg	0/25	53.0mg	0/12		
d	1336n	3.57mg	34.3mg	59/177	17.9mg	18/25	(53.0mg	2/12)		
e	1336n	6.83mg	125.mg	27/177	17.9mg	11/25	(53.0mg	1/12)		
f	1336n	44.3mg	n.s.s.	32/177	17.9mg	7/25	53.0mg	1/12		

DIBROMONEOPENTYL GLYCOL   3296-90-0

	RefNum	LoConf	UpConf	Cntrl	1Dose	1Inc	2Dose	2Inc	Citation or Pathology	Brkly Code
1535	1642	409.mg	n.s.s.	3/48	5.00mg	0/50	100.mg	3/50	Keyes;jctx,7,77-98;1980	
a	1642	49.1mg	n.s.s.	1/48	5.00mg	0/50	100.mg	0/50		
b	1642	23.6mg	n.s.s.	47/50	5.00mg	49/50	100.mg	48/50		
1536	1642	49.1mg	n.s.s.	1/50	5.00mg	0/50	100.mg	0/50		
a	1642	1.01gm	n.s.s.	1/50	5.00mg	1/50	100.mg	0/50		
b	1642	74.9mg	n.s.s.	39/50	5.00mg	40/50	100.mg	39/50		

5,7-DIBROMOQUINOLINE   34522-69-5

	RefNum	LoConf	UpConf	Cntrl	1Dose	1Inc	2Dose	2Inc	Citation or Pathology	Brkly Code
1537	1529	199.mg	n.s.s.	3/44	50.0mg	2/37			Fukushima;clet,14,115-123;1981	
1538	1529	26.1mg	n.s.s.	8/31	40.0mg	14/28				
a	1529	123.mg	n.s.s.	0/31	40.0mg	1/28				

```
          Spe Strain  Site    Xpo+Xpt                                                              TD50    2Tailpvl
             Sex  Route   Hist     Notes                                                             DR    AuOp
1,3-DIBUTYL-1-NITROSOUREA            100ng..:..1ug...:..10.....:..100....:..1mg....:..10.....:..100....:..1g......:..10
1539 R f don wat mam fba 75w75 e                                    .  +  .                          2.23mg  Z P<.0005
   a R f don wat mam mix 75w75 e                                                                     4.28mg  * P<.002  +
   b R f don wat mgl adc 75w75 e                                                                     22.7mg  * P<.002
   c R f don wat ute mix 75w75 e                                                                     15.3mg  Z P<.03
   d R f don wat --- leu 75w75 e                                                                     32.0mg  * P<.06  +
   e R f don wat liv ade 75w75 e                                                                     303.mg  * P<.6
   f R f don wat tba mix 75w75 e                                                                     2.67mg  * P<.0005

DIBUTYLTIN DIACETATE                 100ng..:..1ug...:..10.....:..100....:..1mg....:..10.....:..100....:..1g......:..10
1540 M f b6c eat liv hpa 78w92 ev                                      :  +        :                 #31.6mg * P<.006  -
   a M f b6c eat TBA MXB 78w92 ev                                                                    29.6mg  * P<.07
   b M f b6c eat liv MXB 78w92 ev                                                                    31.6mg  * P<.006
   c M f b6c eat lun MXB 78w92 ev                                                                    no dre   P=1.
1541 M m b6c eat TBA MXB 78w92 v                                           :>                        178.mg  * P<.9
   a M m b6c eat liv MXB 78w92 v                                                                     29.2mg  * P<.06
   b M m b6c eat lun MXB 78w92 v                                                                     no dre   P=1.
1542 R f f34 eat TBA MXB 18m24 v                                  :>                                 no dre   P=1.    i
   a R f f34 eat liv MXB 18m24 v                                                                     94.4mg  * P<.5
1543 R m f34 eat TBA MXB 18m24 v                              :>                                     no dre   P=1.
   a R m f34 eat liv MXB 18m24 v                                                                     no dre   P=1.

3,5-DICHLORO(N-1,1-DIMETHYL-2-PROPYNYL)BENZAMIDE...:..1_0.....:..100....:..1mg....:..10.....:..100....:..1g......:..10
1544 M m b6c eat liv hnd 52w52 r                                         .  +     .                  43.4mg  Z P<.002
   a M m b6c eat liv hpa 52w52 r                                                                     969.mg  * P<.3
   b M m b6c eat liv hpc 52w52 r                                                                     56.6gm  * P<1.
1545 M m b6c eat liv hnd 78w78 r                                              .  +      .            286.mg  * P<.003
   a M m b6c eat liv hpc 78w78 r                                                                     1.02gm  * P<.06
   b M m b6c eat liv hpa 78w78 r                                                                     1.30gm  * P<.2
1546 M m b6c eat liv hnd 24m24 r                                          .  +     .                 113.mg  Z P<.002
   a M m b6c eat liv hpc 24m24 r                                                                     119.mg  Z P<.0005+
   b M m b6c eat liv hpa 24m24 r                                                                     400.mg  * P<.0005

2,3-DICHLORO-p-DIOXANE               100ng..:..1ug...:..10.....:..100....:..1mg....:..10.....:..100....:..1g......:..10
1547 M f hic ipj lun ptm 64w64                          .>                                          7.04mg    P<.6

alpha,beta-DICHLORO-beta-FORMYLACRYLIC ACID.1ug...:..10.....:..100....:..1mg....:..10.....:..100....:..1g......:..10
1548 M f b6a orl liv hpt 76w76 evx                               .>                                 no dre   P=1.    -
   a M f b6a orl lun ade 76w76 evx                                                                  no dre   P=1.    -
   b M f b6a orl tba mix 76w76 evx                                                                  no dre   P=1.    -
1549 M m b6a orl lun mix 76w76 evx                           .>                                     17.8mg   P<.4    -
   a M m b6a orl liv hpt 76w76 evx                                                                  771.mg   P<1.    -
   b M m b6a orl tba mix 76w76 evx                                                                  16.1mg   P<.4    -
1550 M f b6c orl liv hpt 76w92 evx                              .>                                  no dre   P=1.    -
   a M f b6c orl lun mix 76w76 evx                                                                  no dre   P=1.    -
   b M f b6c orl tba mix 76w76 evx                                                                  24.4mg   P<.2    -
1551 M m b6c orl liv hpt 76w76 evx                           .>                                     44.1mg   P<.3    -
   a M m b6c orl lun ade 76w76 evx                                                                  44.1mg   P<.3    -
   b M m b6c orl tba mix 76w76 evx                                                                  5.03mg   P<.002  -

3,4'-DICHLORO-2-METHYLACRYLANILIDE    ..:..1ug...:..10.....:..100....:..1mg....:..10.....:..100....:..1g......:..10
1552 M f b6a orl liv hpt 76w76 evx                             .>                                   no dre   P=1.    -
   a M f b6a orl lun ade 76w76 evx                                                                  no dre   P=1.    -
   b M f b6a orl tba mix 76w76 evx                                                                  no dre   P=1.    -
1553 M m b6a orl liv hpt 76w76 evx                        .>                                        27.1mg   P<.3    -
   a M m b6a orl lun ade 76w76 evx                                                                  no dre   P=1.    -
   b M m b6a orl tba mix 76w76 evx                                                                  49.1mg   P<.7    -
1554 M f b6c orl liv hpt 76w76 evx                             .>                                   no dre   P=1.    -
   a M f b6c orl lun mix 76w76 evx                                                                  no dre   P=1.    -
   b M f b6c orl tba tum 76w76 evx                                                                  no dre   P=1.    -
1555 M m b6c orl liv hpt 76w76 evx                        .  ±                                      15.2mg   P<.04   -
   a M m b6c orl lun ade 76w76 evx                                                                  23.7mg   P<.09   -
   b M m b6c orl tba mix 76w76 evx                                                                  5.39mg   P<.0005 -

2,3-DICHLORO-1,4-NAPHTHOQUINONE      100ng..:..1ug...:..10.....:..100....:..1mg....:..10.....:..100....:..1g......:..10
1556 M f b6a orl liv hpt 76w76 evx                            .>                                    26.5mg   P<.3    -
   a M f b6a orl lun ade 76w76 evx                                                                  no dre   P=1.    -
   b M f b6a orl tba mix 76w76 evx                                                                  no dre   P=1.    -
1557 M m b6a orl liv hpt 76w76 evx                         .>                                       no dre   P=1.    -
   a M m b6a orl lun ade 76w76 evx                                                                  no dre   P=1.    -
   b M m b6a orl tba mix 76w76 evx                                                                  26.5mg   P<.3    -
1558 M f b6c orl lun ade 76w76 evx                            .>                                    26.5mg   P<.3    -
   a M f b6c orl liv hpt 76w76 evx                                                                  no dre   P=1.    -
   b M f b6c orl tba mix 76w76 evx                                                                  8.31mg   P<.05   -
1559 M m b6c orl lun ade 76w76 evx                     .  ±                                         7.26mg   P<.04   -
   a M m b6c orl liv hpt 76w76 evx                                                                  23.2mg   P<.3    -
   b M m b6c orl tba mix 76w76 evx                                                                  3.24mg   P<.003  -

2,6-DICHLORO-4-NITROANILINE          100ng..:..1ug...:..10.....:..100....:..1mg....:..10.....:..100....:..1g......:..10
1560 M f b6a orl lun ade 76w76 evx                                 .>                               no dre   P=1.    -
   a M f b6a orl liv hpt 76w76 evx                                                                  no dre   P=1.    -
```

```
        RefNum LoConf UpConf   Cntrl  1Dose   1Inc  2Dose    2Inc              Citation or Pathology
                                                                                                Brkly Code

1,3-DIBUTYL-1-NITROSOUREA      56654-52-5
1539    1216  1.31mg 4.93mg    5/25   5.71mg  21/26 11.4mg  19/24 (22.9mg 11/19)       Ogiu;zkko,96,35-41;1980
a       1216  2.26mg 18.1mg    8/25   5.71mg  21/26 11.4mg  20/24  22.9mg 15/19
b       1216  11.7mg 86.6mg    0/25   5.71mg   1/26 11.4mg   7/24  22.9mg  4/19
c       1216  7.22mg n.s.s.    0/25   5.71mg   6/26 11.4mg   3/24 (22.9mg  2/19)
d       1216  13.0mg n.s.s.    1/25   5.71mg   3/26 11.4mg   5/24  22.9mg  4/19
e       1216  49.4mg n.s.s.    0/25   5.71mg   0/26 11.4mg   1/24  22.9mg  0/19
f       1216  1.45mg 7.88mg   10/25   5.71mg  23/26 11.4mg  22/24  22.9mg 17/19


DIBUTYLTIN DIACETATE    1067-33-0
1540    TR183 16.1mg 313.mg    1/20   8.30mg   4/50 16.8mg  12/50                                            S
a       TR183 12.8mg n.s.s.    3/20   8.30mg  12/50 16.8mg  13/50              liv:hpa,hpc,nnd.
b       TR183 16.1mg 313.mg    1/20   8.30mg   4/50 16.8mg  12/50              lun:a/a,a/c.
c       TR183 69.4mg n.s.s.    2/20   8.30mg   4/50 16.8mg   0/50
1541    TR183 15.4mg n.s.s.    6/20   7.70mg  19/50 15.5mg  16/50
a       TR183 13.3mg n.s.s.    2/20   7.70mg  11/50 15.5mg  15/50              liv:hpa,hpc,nnd.
b       TR183 47.6mg n.s.s.    2/20   7.70mg   8/50 15.5mg   2/50              lun:a/a,a/c.
1542    TR183 3.35mg n.s.s.   12/19   2.50mg  26/50 (5.00mg 13/50)
a       TR183 23.2mg n.s.s.    0/19   2.50mg   1/50  5.00mg   1/50             liv:hpa,hpc,nnd.
1543    TR183 5.12mg n.s.s.    8/20   2.00mg  20/50  4.00mg  11/50
a       TR183 n.s.s. n.s.s.    0/20   2.00mg   2/50  4.00mg   0/50             liv:hpa,hpc,nnd.


3,5-DICHLORO(N-1,1-DIMETHYL-2-PROPYNYL)BENZAMIDE     23950-58-5
1544    1473m 19.5mg 263.mg    9/84   2.40mg   3/42 12.0mg   2/42 60.0mg  13/42 (300.mg 13/42)  Essigmann;canr,41,2823-2831;1981
a       1473m 214.mg n.s.s.    2/84   2.40mg   1/42 12.0mg   2/42 60.0mg   3/42  300.mg  3/42
b       1473m 361.mg n.s.s.    2/84   2.40mg   0/42 12.0mg   2/42 60.0mg   2/42  300.mg  1/42
1545    1473n 133.mg 1.86gm   13/84   2.40mg  14/42 12.0mg   9/42 60.0mg  13/42  300.mg 19/41
a       1473n 329.mg n.s.s.    3/84   2.40mg   3/42 12.0mg   3/42 60.0mg   4/42  300.mg  6/41
b       1473n 375.mg n.s.s.    3/84   2.40mg   4/42 12.0mg   4/42 60.0mg   2/42  300.mg  6/41
1546    1473o 56.1mg 544.mg   22/126  2.40mg  14/63 12.0mg  24/63 60.0mg  26/63 (300.mg 19/63)
a       1473o 64.5mg 323.mg    6/126  2.40mg   9/63 12.0mg  12/63 60.0mg  20/63 (300.mg 14/63)
b       1473o 247.mg 756.mg    5/126  2.40mg   6/63 12.0mg   7/63 60.0mg   8/63  300.mg 28/63


2,3-DICHLORO-p-DIOXANE    3883-43-0
1547    1143  1.11mg n.s.s.   10/30   2.86mg  12/30                             Van Duuren;jnci,53,695-700;1974


alpha,beta-DICHLORO-beta-FORMYLACRYLIC  ACID  (mucochloric acid) 87-56-9
1548    1296  15.5mg n.s.s.    0/17   7.84mg   0/18                             Innes;ntis,1968/1969
a       1296  15.5mg n.s.s.    1/17   7.84mg   0/18
b       1296  10.3mg n.s.s.    2/17   7.84mg   1/18
1549    1296  3.89mg n.s.s.    2/18   7.30mg   4/17
a       1296  8.09mg n.s.s.    1/18   7.30mg   1/17
b       1296  3.43mg n.s.s.    3/18   7.30mg   5/17
1550    1296  15.5mg n.s.s.    0/16   7.84mg   0/18
a       1296  15.5mg n.s.s.    0/16   7.84mg   0/18
b       1296  5.98mg n.s.s.    0/16   7.84mg   2/18
1551    1296  7.17mg n.s.s.    0/16   7.30mg   1/17
a       1296  7.17mg n.s.s.    0/16   7.30mg   1/17
b       1296  2.14mg 19.3mg    0/16   7.30mg   7/17


3,4'-DICHLORO-2-METHYLACRYLANILIDE     (dicryl) 2164-09-2
1552    1279  18.6mg n.s.s.    0/17   9.96mg   0/17                             Innes;ntis,1968/1969
a       1279  18.6mg n.s.s.    1/17   9.96mg   0/17
b       1279  18.6mg n.s.s.    2/17   9.96mg   0/17
1553    1279  6.05mg n.s.s.    1/18   9.26mg   3/18
a       1279  18.3mg n.s.s.    2/18   9.26mg   0/18
b       1279  5.75mg n.s.s.    3/18   9.26mg   4/18
1554    1279  19.7mg n.s.s.    0/16   9.96mg   0/18
a       1279  19.7mg n.s.s.    0/16   9.96mg   0/18
b       1279  19.7mg n.s.s.    0/16   9.96mg   0/18
1555    1279  4.58mg n.s.s.    0/16   9.26mg   3/15
a       1279  5.81mg n.s.s.    0/16   9.26mg   2/15
b       1279  2.28mg 18.9mg    0/16   9.26mg   7/15


2,3-DICHLORO-1,4-NAPHTHOQUINONE    (dichlone) 117-80-6
1556    1240  4.31mg n.s.s.    0/17   4.14mg   1/18                             Innes;ntis,1968/1969
a       1240  4.92mg n.s.s.    1/17   4.14mg   1/18
b       1240  3.91mg n.s.s.    2/17   4.14mg   2/18
1557    1240  7.63mg n.s.s.    1/18   3.85mg   0/18
a       1240  7.63mg n.s.s.    2/18   3.85mg   0/18
b       1240  7.63mg n.s.s.    3/18   3.85mg   0/18
1558    1240  4.31mg n.s.s.    0/16   4.14mg   1/18
a       1240  8.20mg n.s.s.    0/16   4.14mg   0/18
b       1240  2.51mg n.s.s.    0/16   4.14mg   3/18
1559    1240  2.19mg n.s.s.    0/16   3.85mg   3/17
a       1240  3.78mg n.s.s.    0/16   3.85mg   1/17
b       1240  1.31mg 17.6mg    0/16   3.85mg   6/17


2,6-DICHLORO-4-NITROANILINE    (botran) 99-30-9
1560    1291  119.mg n.s.s.    1/17   99.7mg   1/18                             Innes;ntis,1968/1969
a       1291  198.mg n.s.s.    0/17   99.7mg   0/18
```

```
       Spe Strain  Site    Xpo+Xpt                                           TD50     2Tailpvl
           Sex     Route   Hist    Notes                                       DR       AuOp

b      M f b6a orl tba mix 76w76 evx                                          182.mg   P<.3    -
1561   M f b6a orl liv hpt 76w76 evx                         .>               no dre   P=1.    -
a      M f b6a orl lun ade 76w76 evx                                          no dre   P=1.    -
b      M f b6a orl tba mix 76w76 evx                                          no dre   P=1.    -
1562   M m b6a orl liv hpt 76w76 evx                            .>            no dre   P=1.    -
a      M m b6a orl lun ade 76w76 evx                                          no dre   P=1.    -
b      M m b6a orl tba mix 76w76 evx                                          no dre   P=1.    -
1563   M m b6a orl liv hpt 76w76 evx                               .>         471.mg   P<.6    -
a      M m b6a orl lun ade 76w76 evx                                          no dre   P=1.    -
b      M m b6a orl tba mix 76w76 evx                                          no dre   P=1.    -
1564   M f b6c orl lun ade 76w76 evx                              .>          602.mg   P<.3    -
a      M f b6c orl liv hpt 76w76 evx                                          no dre   P=1.    -
b      M f b6c orl tba mix 76w76 evx                                          188.mg   P<.04   -
1565   M f b6c orl lun ade 76w76 evx                             .>           506.mg   P<.3    -
a      M f b6c orl liv hpt 76w76 evx                                          no dre   P=1.    -
b      M f b6c orl tba mix 76w76 evx                                          114.mg   P<.02   -
1566   M m b6c orl liv mix 76w76 evx                 .    +            .      104.mg   P<.009  -
a      M m b6c orl lun ade 76w76 evx                                          594.mg   P<.3    -
b      M m b6c orl tba mix 76w76 evx                                          57.7mg   P<.0005-
1567   M m b6c orl liv agm 76w76 evx                            .>            499.mg   P<.3    -
a      M m b6c orl liv hpt 76w76 evx                                          499.mg   P<.3    -
b      M m b6c orl lun ade 76w76 evx                                          499.mg   P<.3    -
c      M m b6c orl tba mix 76w76 evx                                          156.mg   P<.05   -


2,6-DICHLORO-p-PHENYLENEDIAMINE   100ng..:..1ug....:..10......:..100....:..1mg....:..10......:..100....:..1g......:..10
1568   M f b6c eat liv MXA 24m26                                      :   +   883.mg * P<.008  c
a      M f b6c eat liv hpc 24m26                                              2.03gm * P<.04   c
b      M f b6c eat TBA MXB 24m26                                              801.mg * P<.2
c      M f b6c eat liv MXB 24m26                                              883.mg * P<.008
d      M f b6c eat lun MXB 24m26                                              42.1gm * P<.9
1569   M m b6c eat liv MXA 24m26                                          :  ± 737.mg * P<.07  c
a      M m b6c eat liv hpa 24m26                                              933.mg * P<.02   c
b      M m b6c eat TBA MXB 24m26                                              1.73gm * P<.6
c      M m b6c eat liv MXB 24m26                                              737.mg * P<.07
d      M m b6c eat lun MXB 24m26                                              no dre   P=1.
1570   R f f34 eat TBA MXB 24m26                              :>              no dre   P=1.    -
a      R f f34 eat liv MXB 24m26                                              2.96gm * P<.3
1571   R m f34 eat TBA MXB 24m26                           :>                 ?14.mg * P<.9    -
a      R m f34 eat liv MXB 24m26                                              340.mg * P<.05


DICHLOROACETIC ACID   100ng..:..1ug....:..10......:..100....:..1mg....:..10......:..100....:..1g......:..10
1572   M m b6c wat liv mix 24m24 e                   .  +    .                49.3mg   P<.0005+
a      M m b6c wat liv hpc 24m24 e                                            68.9mg   P<.0005+
b      M m b6c wat liv hpa 24m24 e                                            124.mg   P<.004


DICHLOROACETYLENE   100ng..:..1ug....:..10......:..100....:..1mg....:..10......:..100....:..1g......:..10
1573   M f nmr inh hag cye 18m28 es              .  +    .                    .550mg   P<.002  +
a      M f nmr inh kid cye 18m28 es                                           1.08mg   P<.002  +
b      M f nmr inh lun ade 18m28 es                                           2.73mg   P<.04
c      M f nmr inh lun car 18m28 es                                           no dre   P=1.
d      M f nmr inh lun mix 18m28 es                                           no dre   P=1.
1574   M f nmr inh kid cye 52w98 es                  . + .                    1.03mg   P<.0005+
a      M f nmr inh hag cye 52w98 es                                           3.02mg   P<.07   +
b      M f nmr inh lun ade 52w98 es                                           no dre   P=1.
c      M f nmr inh lun car 52w98 es                                           no dre   P=1.
1575   M f nmr inh hag cye 18m30 es                . + .                      .499mg   P<.0005+
a      M f nmr inh kid cye 18m30 es                                           7.80mg   P<.5    +
b      M f nmr inh lun ade 18m30 es                                           no dre   P=1.
c      M f nmr inh lun car 18m30 es                                           no dre   P=1.
1576   M m nmr inh kid cye 18m30 es                .  +   .                   .466mg   P<.003  +
a      M m nmr inh kid cyc 18m30 es                                           .501mg   P<.0005+
b      M m nmr inh hag cye 18m30 es                                           1.01mg   P<.2    +
c      M m nmr inh lun ade 18m30 es                                           no dre   P=1.
d      M m nmr inh lun car 18m30 es                                           no dre   P=1.
1577   M m nmr inh kid cye 52w92 es                .  +   .                   .486mg   P<.0005+
a      M m nmr inh kid cyc 52w92 es                                           3.89mg   P<.02   +
b      M m nmr inh hag cye 52w92 es                                           no dre   P=1.    +
c      M m nmr inh lun ade 52w92 es                                           no dre   P=1.
1578   M m nmr inh hag cye 18m27 es                 .  +   .                  .687mg   P<.0005+
a      M m nmr inh kid cye 18m27 es                                           .769mg   P<.002  +
b      M m nmr inh kid cyc 18m27 es                                           4.52mg   P<.04   +
c      M m nmr inh lun ade 18m27 es                                           no dre   P=1.
d      M m nmr inh lun car 18m27 es                                           no dre   P=1.
1579   R f wis inh --- mly 18m35 es                      . ±                  3.86mg   P<.04   +
a      R f wis inh liv cho 18m35 es                                           3.86mg   P<.04   +
b      R f wis inh kid cye 18m35 es                                           11.5mg   P<.04   +
c      R f wis inh liv hpa 18m35 es                                           17.5mg   P<.1    +
1580   R m wis inh kid cye 18m36 es                       .  +   .            3.34mg   P<.002  +
a      R m wis inh liv cho 18m36 es                                           3.97mg   P<.004  +
b      R m wis inh liv hpa 18m36 es                                           12.9mg   P<.1    +
c      R m wis inh kid cyc 18m36 es                                           26.2mg   P<.3    +
```

	RefNum	LoConf	UpConf	Cntrl	1Dose	1Inc	2Dose	2Inc	Citation or Pathology
									Brkly Code
b	1291	47.7mg	n.s.s.	2/17	99.7mg	5/18			
1561	1292	157.mg	n.s.s.	0/17	83.8mg	0/17			
a	1292	157.mg	n.s.s.	1/17	83.8mg	0/17			
b	1292	157.mg	n.s.s.	2/17	83.8mg	0/17			
1562	1291	184.mg	n.s.s.	1/18	92.7mg	0/18			
a	1291	184.mg	n.s.s.	2/18	92.7mg	0/18			
b	1291	130.mg	n.s.s.	3/18	92.7mg	1/18			
1563	1292	65.3mg	n.s.s.	1/18	78.0mg	2/18			
a	1292	102.mg	n.s.s.	2/18	78.0mg	1/18			
b	1292	61.3mg	n.s.s.	3/18	78.0mg	3/18			
1564	1291	98.0mg	n.s.s.	0/16	99.7mg	1/17			
a	1291	187.mg	n.s.s.	0/16	99.7mg	0/17			
b	1291	56.7mg	n.s.s.	0/16	99.7mg	3/17			
1565	1292	82.3mg	n.s.s.	0/16	83.8mg	1/17			
a	1292	157.mg	n.s.s.	0/16	83.8mg	0/17			
b	1292	39.3mg	n.s.s.	0/16	83.8mg	4/17			
1566	1291	39.3mg	2.29gm	0/16	92.7mg	5/18			
a	1291	96.6mg	n.s.s.	0/16	92.7mg	1/18			
b	1291	25.7mg	190.mg	0/16	92.7mg	8/18			
1567	1292	81.2mg	n.s.s.	0/16	78.0mg	1/18			
a	1292	81.2mg	n.s.s.	0/16	78.0mg	1/18			
b	1292	81.2mg	n.s.s.	0/16	78.0mg	1/18			
c	1292	47.2mg	n.s.s.	0/16	78.0mg	3/18			

2,6-DICHLORO-p-PHENYLENEDIAMINE   609-20-1

	RefNum	LoConf	UpConf	Cntrl	1Dose	1Inc	2Dose	2Inc	Citation or Pathology
1568	TR219	409.mg	18.5gm	6/50	121.mg	6/50	362.mg	16/50	liv:hpa,hpc.
a	TR219	770.mg	n.s.s.	2/50	121.mg	2/50	362.mg	7/50	
b	TR219	273.mg	n.s.s.	31/50	121.mg	26/50	362.mg	37/50	
c	TR219	409.mg	18.5gm	6/50	121.mg	6/50	362.mg	16/50	liv:hpa,hpc,nnd.
d	TR219	1.67gm	n.s.s.	2/50	121.mg	2/50	362.mg	2/50	lun:a/a,a/c.
1569	TR219	284.mg	n.s.s.	16/50	111.mg	19/50	334.mg	29/50	liv:hpa,hpc.
a	TR219	431.mg	n.s.s.	4/50	111.mg	7/50	334.mg	15/50	
b	TR219	299.mg	n.s.s.	32/50	111.mg	30/50	334.mg	38/50	
c	TR219	284.mg	n.s.s.	16/50	111.mg	19/50	334.mg	29/50	liv:hpa,hpc,nnd.
d	TR219	559.mg	n.s.s.	13/50	111.mg	5/50	(334.mg	4/50)	lun:a/a,a/c.
1570	TR219	66.1mg	n.s.s.	49/50	92.8mg	45/50	(278.mg	38/50)	
a	TR219	757.mg	n.s.s.	3/50	92.8mg	2/50	278.mg	6/50	liv:hpa,hpc,nnd.
1571	TR219	58.3mg	n.s.s.	42/50	37.1mg	37/50	74.2mg	34/50	
a	TR219	132.mg	n.s.s.	1/50	37.1mg	3/50	74.2mg	5/50	liv:hpa,hpc,nnd.

DICHLOROACETIC ACID   79-43-6

	RefNum	LoConf	UpConf	Cntrl	1Dose	1Inc	2Dose	2Inc	Citation or Pathology
1572	2072	24.9mg	139.mg	3/20	88.0mg	18/24			Daniel;faat,19,159-168;1992/pers.comm.
a	2072	34.3mg	223.mg	2/20	88.0mg	15/24			
b	2072	56.0mg	817.mg	1/20	88.0mg	10/24			

DICHLOROACETYLENE   7572-29-4

	RefNum	LoConf	UpConf	Cntrl	1Dose	1Inc	2Dose	2Inc	Citation or Pathology
1573	1651m	.264mg	2.65mg	3/30	.321mg	14/30			Reichert;carc,5,1411-1420;1984
a	1651m	.467mg	4.34mg	0/30	.321mg	7/30			
b	1651m	.827mg	n.s.s.	0/30	.321mg	3/30			
c	1651m	2.60mg	n.s.s.	4/30	.321mg	0/30			
d	1651m	1.18mg	n.s.s.	4/30	.321mg	3/30			
1574	1651n	.554mg	2.19mg	0/30	1.17mg	15/30			
a	1651n	1.10mg	n.s.s.	2/24	1.17mg	8/29			
b	1651n	4.98mg	n.s.s.	5/30	1.17mg	1/30			
c	1651n	6.42mg	n.s.s.	1/30	1.17mg	0/30			
1575	1651o	.280mg	1.01mg	1/28	.592mg	21/29			
a	1651o	1.44mg	n.s.s.	4/30	.592mg	6/30			
b	1651o	3.40mg	n.s.s.	1/30	.592mg	1/30			
c	1651o	3.40mg	n.s.s.	1/30	.592mg	1/30			
1576	1651m	.221mg	2.66mg	4/30	.251mg	15/30			
a	1651m	.256mg	1.18mg	0/30	.251mg	12/30			
b	1651m	.332mg	n.s.s.	6/30	.251mg	11/29			
c	1651m	.842mg	n.s.s.	4/30	.251mg	4/30			
d	1651m	1.79mg	n.s.s.	5/30	.251mg	1/30			
1577	1651n	.248mg	1.50mg	8/30	1.04mg	23/30			
a	1651n	1.34mg	n.s.s.	0/30	1.04mg	4/30			
b	1651n	1.18mg	n.s.s.	6/26	1.04mg	5/22			
c	1651n	3.08mg	n.s.s.	5/30	1.04mg	2/30			
1578	1651o	.345mg	2.45mg	4/30	.540mg	17/30			
a	1651o	.376mg	3.31mg	4/30	.540mg	16/30			
b	1651o	1.37mg	n.s.s.	0/30	.540mg	3/30			
c	1651o	2.49mg	n.s.s.	8/30	.540mg	3/30			
d	1651o	4.30mg	n.s.s.	3/30	.540mg	0/30			
1579	1651n	1.52mg	n.s.s.	4/30	.849mg	11/30			
a	1651n	1.52mg	n.s.s.	4/30	.849mg	11/30			
b	1651n	3.47mg	n.s.s.	0/30	.849mg	3/30			
c	1651n	4.31mg	n.s.s.	0/30	.849mg	2/30			
1580	1651n	1.44mg	13.4mg	0/30	.568mg	7/30			
a	1651n	1.62mg	22.8mg	0/30	.568mg	6/30			
b	1651n	3.16mg	n.s.s.	0/30	.568mg	2/30			
c	1651n	4.26mg	n.s.s.	0/30	.568mg	1/30			

```
      Spe Strain Site   Xpo+Xpt                                                              TD50    2Tailpvl
          Sex  Route Hist   Notes                                                            DR      AuOp
  d    R m wis inh liv cvh 18m36 es                                                          no dre  P=1.
  e    R m wis inh liv hpc 18m36 es                                                          no dre  P=1.

1,2-DICHLOROBENZENE                  100ng..:..1ug....:..10......:..100....:..1mg....:..10.....:..100....:..1g.....:..10
1581 M f b6c gav --- mlh 24m24                                                  :      ±     #1.11gm * P<.05  -
  a    M f b6c gav TBA MXB 24m24                                                              no dre  P=1.
  b    M f b6c gav liv MXB 24m24                                                              no dre  P=1.
  c    M f b6c gav lun MXB 24m24                                                              no dre  P=1.
1582 M m b6c gav --- mlh 24m24                                                 :      ±      #648.mg * P<.04  -
  a    M m b6c gav TBA MXB 24m24                                                              no dre  P=1.
  b    M m b6c gav liv MXB 24m24                                                              no dre  P=1.
  c    M m b6c gav lun MXB 24m24                                                              767.mg  * P<.6
1583 R f f34 gav TBA MXB 24m24                                      :>                        1.16gm  * P<.9  -
  a    R f f34 gav liv MXB 24m24                                                              1.13gm  * P<.4
1584 R m f34 gav tes ict 24m24 s                                        :     ±              #48.5mg * P<.03  -
  a    R m f34 gav TBA MXB 24m24 s                                                            81.4mg  * P<.05
  b    R m f34 gav liv MXB 24m24 s                                                            696.mg  * P<.09

1,4-DICHLOROBENZENE                  100ng..:..1ug....:..10......:..100....:..1mg....:..10.....:..100....:..1g.....:..10
1585 M f b6c gav liv MXA 24m24                                                 :      +    :  483.mg  / P<.005 c
  a    M f b6c gav liv hpc 24m24                                                              852.mg  / P<.005 c
  b    M f b6c gav liv hpa 24m24                                                              1.03gm  / P<.06  c
  c    M f b6c gav thy fca 24m24                                                              5.09gm  * P<.05
  d    M f b6c gav TBA MXB 24m24                                                              870.mg  * P<.4
  e    M f b6c gav liv MXB 24m24                                                              483.mg  / P<.005
  f    M f b6c gav lun MXB 24m24                                                              no dre  P=1.
1586 M m b6c gav liv MXA 24m24                                                 :      +    :  339.mg  / P<.005 c
  a    M m b6c gav liv hpc 24m24                                                              559.mg  / P<.02  c
  b    M m b6c gav liv hpa 24m24                                                              700.mg  * P<.03  c
  c    M m b6c gav amd MXA 24m24                                                              2.24gm  * P<.03
  d    M m b6c gav liv hpb 24m24                                                              3.51gm  * P<.03
  e    M m b6c gav TBA MXB 24m24                                                              1.31gm  * P<.6
  f    M m b6c gav liv MXB 24m24                                                              339.mg  / P<.005
  g    M m b6c gav lun MXB 24m24                                                              666.mg  \ P<.2
1587 R f f34 gav TBA MXB 24m24                                                  :>            no dre  P=1.  -
  a    R f f34 gav liv MXB 24m24                                                              9.29gm  * P<.5
1588 R m f34 gav kid MXA 24m24                                                 :      +    :  586.mg  * P<.005
  a    R m f34 gav kid uac 24m24                                                              644.mg  * P<.009 c
  b    R m f34 gav mul mnl 24m24                                                              615.mg  * P<.05
  c    R m f34 gav TBA MXB 24m24                                                              224.mg  * P<.2
  d    R m f34 gav liv MXB 24m24                                                              no dre  P=1.

3,3'-DICHLOROBENZIDINE               100ng..:..1ug....:..10......:..100....:..1mg....:..10.....:..100....:..1g.....:..10
1589 D f beg cap liv mix 86m86 ev                                  <+                         noTD50  P<.008 +
  a    D f beg cap ubl mix 86m86 ev                                                           noTD50  P<.008 +
  b    D f beg cap tba mix 86m86 ev                                                           noTD50  P<.6
1590 R f cdr eat mgl adc 69w69                                 .    +    .                    18.3mg  P<.0005+
  a    R f cdr eat liv tum 69w69                                                              no dre  P=1.  -
1591 R m cdr eat zym sqc 69w69                                      .    +    .               60.1mg  P<.002 +
  a    R m cdr eat mgl adc 69w69                                                              69.6mg  P<.002 +
  b    R m cdr eat --- grl 69w69                                                              66.2mg  P<.03  +
  c    R m cdr eat liv tum 69w69                                                              no dre  P=1.  -

trans-1,4-DICHLOROBUTENE-2           100ng..:..1ug....:..10......:..100....:..1mg....:..10.....:..100....:..1g.....:..10
1592 M f hic ipj abd sar 76w76                                 .    ±                         1.52mg  P<.1   +

2,7-DICHLORODIBENZO-p-DIOXIN         100ng..:..1ug....:..10......:..100....:..1mg....:..10.....:..100....:..1g.....:..10
1593 M f b6c eat TBA MXB 90w91                                                        :>      no dre  P=1.  -
  a    M f b6c eat liv MXB 90w91                                                              no dre  P=1.
  b    M f b6c eat lun MXB 90w91                                                              no dre  P=1.
1594 M m b6c eat --- MXA 90w92                                                     :   +      :1.92gm  \ P<.003 a
  a    M m b6c eat --- MXA 90w92                                                              2.75gm  \ P<.01  a
  b    M m b6c eat liv MXA 90w92                                                              2.12gm  * P<.07  a
  c    M m b6c eat TBA MXB 90w92                                                              2.91gm  * P<.4
  d    M m b6c eat liv MXB 90w92                                                              2.12gm  * P<.07
  e    M m b6c eat lun MXB 90w92                                                              no dre  P=1.
1595 R f osm eat TBA MXB 26m26                                                  :>            no dre  P=1.  -
  a    R f osm eat liv MXB 26m26                                                              no dre  P=1.
1596 R m osm eat TBA MXB 26m26                                              :>                no dre  P=1.  -
  a    R m osm eat liv MXB 26m26                                                              12.9gm  * P<.8

DICHLORODIFLUOROMETHANE              100ng..:..1ug....:..10......:..100....:..1mg....:..10.....:..100....:..1g.....:..10
1597 M f swi inh --- leu 18m24                                                         .  ±   4.10gm  \ P<.06
  a    M f swi inh lun ade 18m24                                                              91.9gm  * P<.4
  b    M f swi inh tba mal 18m24                                                              3.32gm  \ P<.03
  c    M f swi inh tba mix 18m24                                                              59.0gm  * P<.7
1598 M m swi inh lun ade 18m24                                                               41.9gm  * P<.2
  a    M m swi inh tba mix 18m24                                                              12.1gm  * P<.02
  b    M m swi inh tba mal 18m24                                                              33.1gm  * P<.2
1599 R f sda inh liv ang 24m24                                                               .31.3gm  * P<.2
  a    R f sda inh tba mal 24m24                                                              no dre  P=1.  -
```

	RefNum	LoConf	UpConf	Cntrl	1Dose	1Inc	2Dose	2Inc	Citation or Pathology	Brkly Code
d	1651n	8.00mg	n.s.s.	1/30	.568mg	0/30				
e	1651n	8.00mg	n.s.s.	1/30	.568mg	0/30				

1,2-DICHLOROBENZENE   (o-dichlorobenzene) 95-50-1

	RefNum	LoConf	UpConf	Cntrl	1Dose	1Inc	2Dose	2Inc	Citation or Pathology	Brkly Code
1581	TR255	335.mg	n.s.s.	0/50	42.0mg	0/50	84.1mg	3/50		s
a	TR255	120.mg	n.s.s.	32/50	42.0mg	28/50	84.1mg	29/50		
b	TR255	319.mg	n.s.s.	4/50	42.0mg	5/50	84.1mg	3/50	liv:hpa,hpc,nnd.	
c	TR255	306.mg	n.s.s.	3/50	42.0mg	4/50	84.1mg	3/50	lun:a/a,a/c.	
1582	TR255	245.mg	n.s.s.	0/50	42.0mg	1/50	84.1mg	4/50		s
a	TR255	109.mg	n.s.s.	38/50	42.0mg	32/50	84.1mg	36/50		
b	TR255	93.2mg	n.s.s.	19/50	42.0mg	14/50	(84.1mg	11/50)	liv:hpa,hpc,nnd.	
c	TR255	130.mg	n.s.s.	8/50	42.0mg	8/50	84.1mg	13/50	lun:a/a,a/c.	
1583	TR255	57.8mg	n.s.s.	39/50	42.0mg	40/50	84.9mg	40/50		
a	TR255	269.mg	n.s.s.	1/50	42.0mg	1/50	84.9mg	3/50	liv:hpa,hpc,nnd.	
1584	TR255	22.1mg	n.s.s.	47/50	42.2mg	49/50	84.9mg	41/50		s
a	TR255	34.0mg	n.s.s.	33/50	42.2mg	38/50	84.9mg	29/50		
b	TR255	211.mg	n.s.s.	0/50	42.2mg	2/50	84.9mg	1/50	liv:hpa,hpc,nnd.	

1,4-DICHLOROBENZENE   106-46-7

	RefNum	LoConf	UpConf	Cntrl	1Dose	1Inc	2Dose	2Inc	Citation or Pathology	Brkly Code
1585	TR319	244.mg	4.20gm	15/50	212.mg	10/50	424.mg	36/50	liv:hpa,hpc.	
a	TR319	427.mg	7.32gm	5/50	212.mg	5/50	424.mg	19/50		
b	TR319	422.mg	n.s.s.	10/50	212.mg	6/50	424.mg	21/50		
c	TR319	1.54gm	n.s.s.	0/50	212.mg	0/50	424.mg	3/50		s
d	TR319	220.mg	n.s.s.	37/50	212.mg	36/50	424.mg	46/50		
e	TR319	244.mg	4.20gm	15/50	212.mg	10/50	424.mg	36/50	liv:hpa,hpc,nnd.	
f	TR319	2.27gm	n.s.s.	5/50	212.mg	5/50	424.mg	1/50	lun:a/a,a/c.	
1586	TR319	172.mg	3.15gm	17/50	212.mg	22/50	424.mg	40/50	liv:hpa,hpc.	
a	TR319	267.mg	n.s.s.	14/50	212.mg	11/50	424.mg	32/50		
b	TR319	324.mg	n.s.s.	5/50	212.mg	13/50	424.mg	16/50		
c	TR319	916.mg	n.s.s.	0/50	212.mg	2/50	424.mg	4/50	amd:phe,phm.	s
d	TR319	1.21gm	n.s.s.	0/50	212.mg	0/50	424.mg	4/50		s
e	TR319	230.mg	n.s.s.	35/50	212.mg	33/50	424.mg	42/50		
f	TR319	172.mg	3.15gm	17/50	212.mg	22/50	424.mg	40/50	liv:hpa,hpc,nnd.	
g	TR319	213.mg	n.s.s.	6/50	212.mg	13/50	(424.mg	2/50)	lun:a/a,a/c.	
1587	TR319	289.mg	n.s.s.	43/50	212.mg	45/50	424.mg	36/50		
a	TR319	1.62gm	n.s.s.	1/50	212.mg	1/50	424.mg	2/50	liv:hpa,hpc,nnd.	
1588	TR319	270.mg	5.06gm	1/50	106.mg	3/50	212.mg	8/50	kid:tla,uac.	s
a	TR319	285.mg	20.0gm	1/50	106.mg	3/50	212.mg	7/50		
b	TR319	250.mg	n.s.s.	5/50	106.mg	7/50	212.mg	11/50		s
c	TR319	78.2mg	n.s.s.	38/50	106.mg	38/50	212.mg	38/50		
d	TR319	933.mg	n.s.s.	2/50	106.mg	2/50	212.mg	0/50	liv:hpa,hpc,nnd.	

3,3'-DICHLOROBENZIDINE   (DCB) 91-94-1

	RefNum	LoConf	UpConf	Cntrl	1Dose	1Inc	2Dose	2Inc	Citation or Pathology
1589	1379	n.s.s.	1.78mg	0/6	4.42mg	5/5			Stula;jept,1,475-490;1978
a	1379	n.s.s.	1.78mg	0/6	4.42mg	5/5			
b	1379	n.s.s.	n.s.s.	5/6	4.42mg	5/5			
1590	192	10.8mg	36.6mg	3/44	50.0mg	26/44			Stula;txap,31,159-176;1975/pers.comm.
a	192	200.mg	n.s.s.	0/44	50.0mg	0/44			
1591	192	27.2mg	203.mg	0/44	40.0mg	8/44			
a	192	30.0mg	290.mg	0/44	40.0mg	7/44			
b	192	26.9mg	n.s.s.	2/44	40.0mg	9/44			
c	192	160.mg	n.s.s.	0/44	40.0mg	0/44			

trans-1,4-DICHLOROBUTENE-2   110-57-6

	RefNum	LoConf	UpConf	Cntrl	1Dose	1Inc	Citation or Pathology
1592	582	.373mg	n.s.s.	0/30	.286mg	2/30	Van Duuren;canr,35,2553-2557;1975

2,7-DICHLORODIBENZO-p-DIOXIN   (DCDD) 33857-26-0

	RefNum	LoConf	UpConf	Cntrl	1Dose	1Inc	2Dose	2Inc	Citation or Pathology
1593	TR123	2.30gm	n.s.s.	15/50	636.mg	18/50	1.19gm	9/50	
a	TR123	n.s.s.	n.s.s.	0/50	636.mg	1/50	1.19gm	0/50	liv:hpa,hpc,nnd.
b	TR123	8.39gm	n.s.s.	3/50	636.mg	2/50	1.19gm	0/50	lun:a/a,a/c.
1594	TR123	831.mg	9.58gm	0/50	540.mg	7/50	(1.09gm	3/50)	---:leu,lym.
a	TR123	1.04gm	281.gm	0/50	540.mg	5/50	(1.09gm	1/50)	---:hem,hes.
b	TR123	885.mg	n.s.s.	8/50	540.mg	20/50	1.09gm	17/50	liv:hpa,hpc.
c	TR123	787.mg	n.s.s.	19/50	540.mg	30/50	1.09gm	25/50	
d	TR123	885.mg	n.s.s.	8/50	540.mg	20/50	1.09gm	17/50	liv:hpa,hpc,nnd.
e	TR123	3.95gm	n.s.s.	8/50	540.mg	5/50	1.09gm	5/50	lun:a/a,a/c.
1595	TR123	456.mg	n.s.s.	26/35	250.mg	22/35	500.mg	16/35	
a	TR123	n.s.s.	n.s.s.	0/35	250.mg	0/35	500.mg	0/35	liv:hpa,hpc,nnd.
1596	TR123	131.mg	n.s.s.	21/35	198.mg	14/35	(396.mg	6/35)	
a	TR123	716.mg	n.s.s.	1/35	198.mg	0/35	396.mg	1/35	liv:hpa,hpc,nnd.

DICHLORODIFLUOROMETHANE   (fluorocarbon 12) 75-71-8

	RefNum	LoConf	UpConf	Cntrl	1Dose	1Inc	2Dose	2Inc	Citation or Pathology
1597	BT602	1.51gm	n.s.s.	8/90	777.mg	12/60	(3.89gm	6/60)	Maltoni;anya,534,261-282;1988
a	BT602	19.1gm	n.s.s.	2/90	777.mg	1/60	3.89gm	3/60	
b	BT602	1.33gm	n.s.s.	9/90	777.mg	14/60	(3.89gm	6/60)	
c	BT602	8.68gm	n.s.s.	15/90	777.mg	15/60	3.89gm	13/60	
1598	BT602	11.1gm	n.s.s.	3/90	648.mg	3/60	3.24gm	5/60	
a	BT602	5.11gm	n.s.s.	9/90	648.mg	9/60	3.24gm	15/60	
b	BT602	9.46gm	n.s.s.	5/90	648.mg	4/60	3.24gm	7/60	
1599	BT601	8.46gm	n.s.s.	1/150	247.mg	1/90	1.23gm	3/90	
a	BT601	6.91gm	n.s.s.	43/150	247.mg	24/90	1.23gm	18/90	

```
      Spe Strain  Site  Xpo+Xpt                                                          TD50    2Tailpvl
          Sex  Route  Hist   Notes                                                          DR       AuOp
b      R f sda inh tba mix 24m24                                                          no dre  P=1.   -
1600   R m sda inh adr phe 24m24                                          .     ±         1.32gm \ P<.02  -
a      R m sda inh liv ang 24m24                                                          no dre  P=1.   -
b      R m sda inh tba mix 24m24                                                          555.mg \ P<.06  -
c      R m sda inh tba mal 24m24                                                          no dre  P=1.   -

1,1-DICHLOROETHANE            100ng..:..1ug....:..10.....:..100....:..1mg....:..10.....:..100....:..1g.....:..10
1601  M f b6c gav TBA MXB 78w90 sv                                                 :>    8.40gm / P<.5   -
a     M f b6c gav liv MXB 78w90 sv                                                       no dre  P=1.
b     M f b6c gav lun MXB 78w90 sv                                                       no dre  P=1.
1602  M f b6c gav ute esp 78w90 sv  pool                                          :      11.2gm * P<.004  a
1603  M m b6c gav TBA MXB 78w90 esv                                        :  ±          1.37gm * P<.07  -
a     M m b6c gav liv MXB 78w90 esv                                                      2.04gm * P<.05
b     M m b6c gav lun MXB 78w90 esv                                                      7.32gm * P<.04
1604  R f osm gav mgl adc 18m26 dsv                                 :     ±              986.mg * P<.04  a
a     R f osm gav --- hes 18m26 dsv                                                      1.35gm * P<.05  a
b     R f osm gav TBA MXB 18m26 dsv                                                      336.mg * P<.2
c     R f osm gav liv MXB 18m26 dsv                                                      4.65gm * P<.4
1605  R f osm gav --- hes 18m26 dsv  pool                           :     ±              1.36gm * P<.02  a
1606  R m osm gav TBA MXB 18m26 dsv                                             :>       no dre  P=1.   -
a     R m osm gav liv MXB 18m26 dsv                                                      no dre  P=1.

1,2-DICHLOROETHANE            100ng..:..1ug....:..10.....:..100....:..1mg....:..10.....:..100....:..1g.....:..10
1607  M f b6c gav MXB MXB 78w90 v                                   :  +  :              61.2mg / P<.0005
a     M f b6c gav lun a/a 78w90 v                                                        118.mg / P<.0005c
b     M f b6c gav mgl acn 78w90 v                                                        133.mg / P<.0005c
c     M f b6c gav TBA MXB 78w90 v                                                        36.0mg / P<.0005
d     M f b6c gav liv MXB 78w90 v                                                        4.74gm / P<.8
e     M f b6c gav lun MXB 78w90 v                                                        112.mg / P<.0005
1608  M f b6c gav lun a/a 78w90 v  pool                                  :  +  :         116.mg / P<.0005c
a     M f b6c gav mgl acn 78w90 v                                                        133.mg / P<.0005c
b     M f b6c gav ute MXA 78w90 v                                                        230.mg / P<.0005c
c     M f b6c gav sto sqc 78w90 v                                                        438.mg / P<.003
1609  M m b6c gav lun a/a 78w90 v                                       :  +  :          89.7mg * P<.0005c
a     M m b6c gav liv hpc 78w90 v                                                        133.mg * P<.04
b     M m b6c gav TBA MXB 78w90 v                                                        57.7mg * P<.02
c     M m b6c gav liv MXB 78w90 v                                                        133.mg * P<.04
d     M m b6c gav lun MXB 78w90 v                                                        89.7mg * P<.0005
1610  M m b6c gav lun a/a 78w90 v  pool                                  :  +  :         89.7mg * P<.0005c
a     M m b6c gav liv hpc 78w90 v                                                        148.mg * P<.02
1611  M f swi inh liv mix 18m25 ev                                               .>      16.9gm * P<.7   -
a     M f swi inh lun ade 18m25 ev                                                       55.4gm * P<1.   -
b     M f swi inh liv hpt 18m25 ev                                                       no dre  P=1.   -
c     M f swi inh liv ang 18m25 ev                                                       no dre  P=1.   -
d     M f swi inh tba mix 18m25 ev                                                       no dre  P=1.   -
1612  M m swi inh liv ang 18m25 ev                                      .>               no dre  P=1.   -
a     M m swi inh liv hpt 18m25 ev                                                       no dre  P=1.   -
b     M m swi inh liv mix 18m25 ev                                                       no dre  P=1.   -
c     M m swi inh lun ade 18m25 ev                                                       no dre  P=1.   -
d     M m swi inh tba mix 18m25 ev                                                       no dre  P=1.   -
1613  R f osm gav mgl MXA 18m26 dsv                          :  +  :                     5.49mg / P<.0005c
a     R f osm gav mgl fba 18m26 dsv                                                      5.84mg / P<.0005
b     R f osm gav mgl acn 18m26 dsv                                                      74.8mg / P<.0005c
c     R f osm gav TBA MXB 18m26 dsv                                                      3.19mg / P<.0005
d     R f osm gav liv MXB 18m26 dsv                                                      60.6mg * P<.2
1614  R f osm gav mgl MXA 18m26 dsv  pool                    :  +  :                     5.63mg / P<.0005c
a     R f osm gav mgl fba 18m26 dsv                                                      6.38mg / P<.0005
b     R f osm gav --- hes 18m26 dsv                                                      19.5mg * P<.0005
c     R f osm gav mgl acn 18m26 dsv                                                      54.0mg / P<.0005c
1615  R m osm gav MXB MXB 18m26 dsv                             :  +  :                  11.5mg / P<.0005
a     R m osm gav --- hes 18m26 dsv                                                      15.0mg * P<.0005c
b     R m osm gav sto sqc 18m26 dsv                                                      46.3mg / P<.0005c
c     R m osm gav TBA MXB 18m26 dsv                                                      3.18mg / P<.0005
d     R m osm gav liv MXB 18m26 dsv                                                      132.mg / P<.07
1616  R m osm gav --- hes 18m26 dsv  pool                    :     +  :                  18.4mg * P<.0005c
a     R m osm gav sub fib 18m26 dsv                                                      43.2mg / P<.0005c
b     R m osm gav sto sqc 18m26 dsv                                                      46.3mg / P<.0005c
1617  R f cdr inh liv hpc 24m24 e                                        .>              594.mg  P<.3   -
a     R f cdr inh liv nnd 24m24 e                                                        no dre  P=1.   -
b     R f cdr inh tba tum 24m24 e                                                        no dre  P=1.   -
1618  R m cdr inh liv nnd 24m24 e                                         .     ±        206.mg  P<.1   -
a     R m cdr inh liv hpc 24m24 e                                                        no dre  P=1.   -
b     R m cdr inh tba tum 24m24 e                                                        17.9mg  P<.4   -
1619  R f sda inh liv mix 18m31 ev                                       .>              no dre  P=1.   -
a     R f sda inh mam mix 18m31 ev                                                       no dre  P=1.   -
b     R f sda inh tba mix 18m31 ev                                                       no dre  P=1.   -
1620  R f sda inh mam mix 18m31 ev                                        .     ±        38.1mg Z P<.07  -
a     R f sda inh liv mix 18m31 ev                                                       no dre  P=1.   -
b     R f sda inh tba mix 18m31 ev                                                       48.8mg Z P<.2   -
1621  R m sda inh liv mix 18m33 ev                                       .>              no dre  P=1.   -
a     R m sda inh tba mix 18m33 ev                                                       2.40gm Z P<1.   -
```

	RefNum	LoConf	UpConf	Cntrl	1Dose	1Inc	2Dose	2Inc					Citation or Pathology
													Brkly Code
b	BT601	1.34gm	n.s.s.	124/150	247.mg	71/90	1.23gm	72/90					
1600	BT601	529.mg	n.s.s.	6/150	173.mg	11/90	(864.mg	5/90)					
a	BT601	2.67gm	n.s.s.	1/150	173.mg	0/90	864.mg	0/90					
b	BT601	221.mg	n.s.s.	51/150	173.mg	42/90	(864.mg	27/90)					
c	BT601	6.54gm	n.s.s.	25/150	173.mg	16/90	864.mg	9/90					

1,1-DICHLOROETHANE   75-34-3

	RefNum	LoConf	UpConf	Cntrl	1Dose	1Inc	2Dose	2Inc					Citation or Pathology
1601	TR66	2.03gm	n.s.s.	6/20	1.02gm	6/50	2.04gm	12/50					
a	TR66	10.3gm	n.s.s.	1/20	1.02gm	1/50	2.04gm	0/50					liv:hpa,hpc,nnd.
b	TR66	8.16gm	n.s.s.	1/20	1.02gm	2/50	2.04gm	0/50					lun:a/a,a/c.
1602	TR66	3.89gm	86.3gm	0/80p	1.02gm	0/50	2.04gm	4/50					
1603	TR66	588.mg	n.s.s.	4/20	883.mg	19/50	1.77gm	15/50					
a	TR66	936.mg	n.s.s.	1/20	883.mg	8/50	1.77gm	8/50					liv:hpa,hpc,nnd.
b	TR66	2.67gm	n.s.s.	0/20	883.mg	1/50	1.77gm	4/50					lun:a/a,a/c.
1604	TR66	377.mg	n.s.s.	0/20	238.mg	1/50	477.mg	5/50					
a	TR66	450.mg	n.s.s.	0/20	238.mg	0/50	477.mg	4/50					
b	TR66	119.mg	n.s.s.	4/20	238.mg	12/50	477.mg	18/50					
c	TR66	758.mg	n.s.s.	0/20	238.mg	0/50	477.mg	1/50					liv:hpa,hpc,nnd.
1605	TR66	456.mg	n.s.s.	0/40p	238.mg	0/50	477.mg	4/50					
1606	TR66	258.mg	n.s.s.	6/20	192.mg	6/50	383.mg	5/50					
a	TR66	926.mg	n.s.s.	1/20	192.mg	0/50	383.mg	0/50					liv:hpa,hpc,nnd.

1,2-DICHLOROETHANE   (ethylene dichloride, EDC) 107-06-2

	RefNum	LoConf	UpConf	Cntrl	1Dose	1Inc	2Dose	2Inc	3Dose	3Inc	4Dose	4Inc	Citation or Pathology
1607	TR55	38.2mg	113.mg	1/20	92.5mg	16/50	183.mg	19/50					lun:a/a; mgl:acn. C
a	TR55	65.2mg	280.mg	1/20	92.5mg	7/50	183.mg	15/50					
b	TR55	71.4mg	312.mg	0/20	92.5mg	9/50	183.mg	7/50					
c	TR55	23.5mg	65.4mg	6/20	92.5mg	33/50	183.mg	29/50					
d	TR55	349.mg	n.s.s.	1/20	92.5mg	0/50	183.mg	1/50					liv:hpa,hpc,nnd.
e	TR55	63.2mg	253.mg	1/20	92.5mg	7/50	183.mg	16/50					lun:a/a,a/c.
1608	TR55	64.5mg	244.mg	2/60p	92.5mg	7/50	183.mg	15/50					
a	TR55	71.4mg	280.mg	0/60p	92.5mg	9/50	183.mg	7/50					
b	TR55	108.mg	598.mg	0/60p	92.5mg	5/50	183.mg	5/50					ute:esp,ess.
c	TR55	170.mg	3.27gm	1/60p	92.5mg	2/50	183.mg	5/50					s
1609	TR55	49.7mg	230.mg	0/20	60.0mg	1/50	120.mg	15/50					
a	TR55	62.1mg	n.s.s.	1/20	60.0mg	6/50	120.mg	12/50					s
b	TR55	29.3mg	n.s.s.	4/20	60.0mg	15/50	120.mg	28/50					
c	TR55	62.1mg	n.s.s.	1/20	60.0mg	6/50	120.mg	12/50					liv:hpa,hpc,nnd.
d	TR55	49.7mg	230.mg	0/20	60.0mg	1/50	120.mg	15/50					lun:a/a,a/c.
1610	TR55	49.7mg	186.mg	0/60p	60.0mg	1/50	120.mg	15/50					
a	TR55	67.0mg	n.s.s.	4/60p	60.0mg	6/50	120.mg	12/50					s
1611	1001m	2.76gm	n.s.s.	0/133	5.31mg	0/89	11.9mg	0/88	53.1mg	1/87	195.mg	0/84	Maltoni;banb,5,3-33;1980
a	1001m	1.80gm	n.s.s.	4/133	5.31mg	4/89	11.9mg	2/88	53.1mg	2/87	195.mg	3/84	
b	1001m	67.7mg	n.s.s.	0/133	5.31mg	0/89	11.9mg	0/88	53.1mg	0/87	195.mg	0/84	
c	1001m	67.7mg	n.s.s.	0/133	5.31mg	0/89	11.9mg	0/88	53.1mg	0/87	195.mg	0/84	
d	1001m	1.52gm	n.s.s.	27/133	5.31mg	19/89	11.9mg	17/88	53.1mg	15/87	195.mg	11/84	
1612	1001m	47.1mg	n.s.s.	0/111	4.43mg	0/69	9.19mg	0/89	49.7mg	0/87	156.mg	0/81	
a	1001m	47.1mg	n.s.s.	4/111	4.43mg	0/69	9.19mg	0/89	49.7mg	0/87	156.mg	0/81	
b	1001m	47.1mg	n.s.s.	1/111	4.43mg	0/69	9.19mg	0/89	49.7mg	0/87	156.mg	0/81	
c	1001m	2.69gm	n.s.s.	4/111	4.43mg	1/69	9.19mg	4/89	49.7mg	3/87	156.mg	0/81	
d	1001m	285.mg	n.s.s.	14/111	4.43mg	12/69	9.19mg	12/89	49.7mg	9/87	(156.mg	2/81)	
1613	TR55	1.98mg	13.7mg	0/20	24.0mg	15/50	48.0mg	24/50					mgl:acn,fba.
a	TR55	2.02mg	17.2mg	0/20	24.0mg	14/50	48.0mg	8/50					s
b	TR55	40.8mg	147.mg	0/20	24.0mg	1/50	48.0mg	18/50					
c	TR55	1.47mg	7.60mg	7/20	24.0mg	24/50	48.0mg	33/50					
d	TR55	10.4mg	n.s.s.	0/20	24.0mg	2/50	48.0mg	0/50					liv:hpa,hpc,nnd.
1614	TR55	2.27mg	13.8mg	6/60p	24.0mg	15/50	48.0mg	24/50					mgl:acn,fba.
a	TR55	2.42mg	18.9mg	5/60p	24.0mg	14/50	48.0mg	8/50					s
b	TR55	4.35mg	99.1mg	0/60p	24.0mg	4/50	48.0mg	4/50					s
c	TR55	20.3mg	125.mg	1/60p	24.0mg	1/50	48.0mg	18/50					
1615	TR55	3.91mg	27.3mg	0/20	24.0mg	12/50	48.0mg	14/50					---:hes; sto:sqc. C
a	TR55	4.16mg	56.2mg	0/20	24.0mg	9/50	48.0mg	7/50					
b	TR55	17.4mg	153.mg	0/20	24.0mg	9/50	48.0mg	9/50					
c	TR55	1.12mg	10.7mg	4/20	24.0mg	20/50	48.0mg	20/50					
d	TR55	23.8mg	n.s.s.	0/20	24.0mg	0/50	48.0mg	2/50					liv:hpa,hpc,nnd.
1616	TR55	4.98mg	61.0mg	1/60p	24.0mg	9/50	48.0mg	7/50					
a	TR55	17.6mg	108.mg	0/60p	24.0mg	5/50	48.0mg	6/50					
b	TR55	17.4mg	121.mg	0/60p	24.0mg	3/50	48.0mg	9/50					
1617	1962	96.7mg	n.s.s.	0/50	17.5mg	1/50							Cheever;faat,14,243-261;1990
a	1962	110.mg	n.s.s.	1/50	17.5mg	1/50							
b	1962	5.32mg	n.s.s.	47/50	17.5mg	47/50							
1618	1962	50.6mg	n.s.s.	0/50	12.3mg	2/50							
a	1962	76.8mg	n.s.s.	1/50	12.3mg	1/50							
b	1962	4.15mg	n.s.s.	42/50	12.3mg	45/50							
1619	1001m	19.7mg	n.s.s.	0/90	1.08mg	0/90	2.15mg	0/90	10.4mg	0/90	39.8mg	0/90	Maltoni;banb,5,3-33;1980
a	1001m	117.mg	n.s.s.	52/90	1.08mg	65/90	2.15mg	43/90	10.4mg	58/90	39.8mg	52/90	
b	1001m	111.mg	n.s.s.	56/90	1.08mg	65/90	2.15mg	43/90	10.4mg	56/90	39.8mg	54/90	
1620	1001n	14.6mg	n.s.s.	38/90	1.08mg	65/90	2.15mg	43/90	10.4mg	58/90	(39.8mg	52/90)	
a	1001m	19.7mg	n.s.s.	0/90	1.08mg	0/90	2.15mg	0/90	10.4mg	0/90	39.8mg	0/90	
b	1001m	16.4mg	n.s.s.	38/90	1.08mg	65/90	2.15mg	43/90	10.4mg	56/90	(39.8mg	54/90)	
1621	1001m	15.9mg	n.s.s.	0/90	.670mg	0/90	1.79mg	0/89	7.09mg	0/90	26.8mg	0/89	
a	1001m	45.6mg	n.s.s.	14/90	.670mg	30/90	1.79mg	13/89	7.09mg	20/90	(26.8mg	15/89)	

```
        Spe Strain Site  Xpo+Xpt                                                          TD50    2Tailpvl
          Sex  Route  Hist   Notes                                                           DR      AuOp
 1622 R m sda inh liv mix 18m33 ev                                    .>                   no dre   P=1.    -
    a R m sda inh tba mix 18m33 ev                                                          no dre   P=1.    -


2,4-DICHLOROPHENOL              100ng..:..1ug....:..10......:..100....:..1mg....:..10......:..100....:..1g......:..10
 1623 M f b6c eat TBA MXB 24m24                                                       :>    no dre   P=1.
    a M f b6c eat liv MXB 24m24                                                             no dre   P=1.
    b M f b6c eat lun MXB 24m24                                                             no dre   P=1.
 1624 M m b6c eat for MXA 24m24                                                        :    #12.4gm * P<.04   -
    a M m b6c eat TBA MXB 24m24                                                             no dre   P=1.
    b M m b6c eat liv MXB 24m24                                                             458.gm  * P<1.
    c M m b6c eat lun MXB 24m24                                                             13.3gm  * P<.5
 1625 R f f34 eat TBA MXB 24m24                                                 :>          no dre   P=1.    -
    a R f f34 eat liv MXB 24m24                                                             no dre   P=1.    -
 1626 R m f34 eat TBA MXB 24m24                                                    :>       no dre   P=1.    -
    a R m f34 eat liv MXB 24m24                                                             no dre   P=1.


alpha-(2,4-DICHLOROPHENOXY)PROPIONIC  ACID ..1ug....:..10......:..100....:..1mg....:..10......:..100....:..1g......:..10
 1627 M f b6a orl liv agm 76w76 evx                                   .>                   233.mg   P<.3    -
    a M f b6a orl lun ade 76w76 evx                                                         no dre   P=1.    -
    b M f b6a orl tba mix 76w76 evx                                                         66.5mg   P<.3    -
 1628 M f b6a orl lun ade 76w76 evx                            .>                          364.mg   P<1.    -
    a M f b6a orl liv hpt 76w76 evx                                                         no dre   P=1.    -
    b M f b6a orl tba mix 76w76 evx                                                         no dre   P=1.    -
 1629 M m b6a orl liv hpt 76w76 evx                                       .>                3.58gm   P<1.    -
    a M m b6a orl lun ade 76w76 evx                                                         no dre   P=1.    -
    b M m b6a orl tba mix 76w76 evx                                                         no dre   P=1.    -
 1630 M m b6a orl lun mix 76w76 evx                            .>                          no dre   P=1.    -
    a M m b6a orl liv hpt 76w76 evx                                                         no dre   P=1.    -
    b M m b6a orl tba mix 76w76 evx                                                         9.26mg   P<.5    -
 1631 M f b6c orl lun ade 76w76 evx                               .>                       220.mg   P<.3    -
    a M f b6c orl liv hpt 76w76 evx                                                         no dre   P=1.    -
    b M f b6c orl tba mix 76w76 evx                                                         106.mg   P<.1    -
 1632 M f b6c orl liv hpt 76w76 evx                      .>                                no dre   P=1.    -
    a M f b6c orl lun mix 76w76 evx                                                         no dre   P=1.    -
    b M f b6c orl tba tum 76w76 evx                                                         no dre   P=1.    -
 1633 M m b6c orl liv hpt 76w76 evx                   .       ±                            99.1mg   P<.1    -
    a M m b6c orl lun ade 76w76 evx                                                         no dre   P=1.    -
    b M m b6c orl tba mix 76w76 evx                                                         99.1mg   P<.1    -
 1634 M m b6c orl liv hpt 76w76 evx                .       ±                               4.27mg   P<.02   -
    a M m b6c orl lun mix 76w76 evx                                                         no dre   P=1.    -
    b M m b6c orl tba mix 76w76 evx                                                         2.59mg   P<.002  -


alpha-(2,5-DICHLOROPHENOXY)PROPIONIC  ACID ..1ug....:..10......:..100....:..1mg....:..10......:..100....:..1g......:..10
 1635 M f b6a orl lun ade 76w76 evx                               .>                       no dre   P=1.    -
    a M f b6a orl liv hpt 76w76 evx                                                         no dre   P=1.    -
    b M f b6a orl tba mix 76w76 evx                                                         no dre   P=1.    -
 1636 M m b6a orl lun ade 76w76 evx                             .>                         no dre   P=1.    -
    a M m b6a orl liv hpt 76w76 evx                                                         no dre   P=1.    -
    b M m b6a orl tba mix 76w76 evx                                                         no dre   P=1.    -
 1637 M f b6c orl liv hpt 76w76 evx                               .>                       no dre   P=1.    -
    a M f b6c orl lun mix 76w76 evx                                                         no dre   P=1.    -
    b M f b6c orl tba tum 76w76 evx                                                         no dre   P=1.    -
 1638 M m b6c orl liv hpt 76w76 evx                      .       ±                         36.7mg   P<.1    -
    a M m b6c orl lun ade 76w76 evx                                                         36.7mg   P<.1    -
    b M m b6c orl tba mix 76w76 evx                                                         10.5mg   P<.003  -


2,4-DICHLOROPHENOXYACETIC  ACID      100ng..:..1ug....:..10......:..100....:..1mg....:..10......:..100....:..1g......:..10
 1639 M f b6a orl lun ade 76w76 evx                               .>                       2.75gm * P<1.    -
    a M f b6a orl liv hpt 76w76 evx                                                         no dre   P=1.    -
    b M f b6a orl tba mix 76w76 evx                                                         1.07gm * P<.9    -
 1640 M m b6a orl liv hpt 76w76 evx                            .>                          no dre   P=1.    -
    a M m b6a orl lun ade 76w76 evx                                                         no dre   P=1.    -
    b M m b6a orl tba mix 76w76 evx                                                         no dre   P=1.    -
 1641 M f b6c orl liv hpt 76w76 evx                               .>                       123.mg   P<.3    -
    a M f b6c orl lun mix 76w76 evx                                                         no dre   P=1.    -
    b M f b6c orl tba mix 76w76 evx                                                         123.mg   P<.3    -
 1642 M m b6c orl liv hpt 76w76 evx                            .>                          no dre   P=1.    -
    a M m b6c orl lun mix 76w76 evx                                                         no dre   P=1.    -
    b M m b6c orl tba mix 76w76 evx                                                         55.5mg   P<.1    -


2,4-DICHLOROPHENOXYACETIC  ACID, N-BUTYL ESTER  ....:..10......:..100....:..1mg....:..10......:..100....:..1g......:..10
 1643 M f b6a orl liv hpt 76w76 evx                               .> .                     no dre   P=1.    -
    a M f b6a orl lun ade 76w76 evx                                                         no dre   P=1.    -
    b M f b6a orl tba mix 76w76 evx                                                         no dre   P=1.    -
 1644 M m b6a orl lun ade 76w76 evx                            .>                          no dre   P=1.    -
    a M m b6a orl liv hpt 76w76 evx                                                         no dre   P=1.    -
    b M m b6a orl tba mix 76w76 evx                                                         no dre   P=1.    -
 1645 M f b6c orl liv hpt 76w76 evx                               .>                       no dre   P=1.    -
    a M f b6c orl lun mix 76w76 evx                                                         no dre   P=1.    -
    b M f b6c orl tba mix 76w76 evx                                                         29.7mg   P<.02   -
 1646 M m b6c orl lun ade 76w76 evx                         .>                             59.0mg   P<.2    -
```

RefNum	LoConf	UpConf	Cntrl	1Dose	1Inc	2Dose	2Inc			Citation or Pathology
										Brkly Code

	RefNum	LoConf	UpConf	Cntrl	1Dose	1Inc	2Dose	2Inc				Citation or Pathology
1622	1001n	15.9mg	n.s.s.	0/90	.670mg	0/90	1.79mg	0/89	7.09mg	0/90	26.8mg	0/89
a	1001n	51.5mg	n.s.s.	17/90	.670mg	30/90	1.79mg	13/89	7.09mg	20/90	(26.8mg	15/89)

2,4-DICHLOROPHENOL   120-83-2

	RefNum	LoConf	UpConf	Cntrl	1Dose	1Inc	2Dose	2Inc	Citation or Pathology
1623	TR353	2.31gm	n.s.s.	26/50	638.mg	18/50	1.29gm	21/50	
a	TR353	8.98gm	n.s.s.	2/50	638.mg	3/50	1.29gm	0/50	liv:hpa,hpc,nnd.
b	TR353	7.95gm	n.s.s.	3/50	638.mg	1/50	1.29gm	2/50	lun:a/a,a/c.
1624	TR353	3.77gm	n.s.s.	0/50	589.mg	0/50	1.19gm	3/50	for:sqc,sqp. S
a	TR353	1.20gm	n.s.s.	29/50	589.mg	29/50	1.19gm	24/50	
b	TR353	1.93gm	n.s.s.	10/50	589.mg	12/50	1.19gm	9/50	liv:hpa,hpc,nnd.
c	TR353	2.85gm	n.s.s.	3/50	589.mg	2/50	1.19gm	5/50	lun:a/a,a/c.
1625	TR353	254.mg	n.s.s.	47/50	123.mg	36/50	248.mg	44/50	
a	TR353	n.s.s.	n.s.s.	0/50	123.mg	1/50	248.mg	0/50	liv:hpa,hpc,nnd.
1626	TR353	343.mg	n.s.s.	48/50	196.mg	38/50	396.mg	41/50	
a	TR353	748.mg	n.s.s.	5/50	196.mg	1/50	(396.mg	1/50)	liv:hpa,hpc,nnd.

alpha-(2,4-DICHLOROPHENOXY)PROPIONIC ACID   (2-(2,4-dichlorophenoxy)propionic acid) 120-36-5

	RefNum	LoConf	UpConf	Cntrl	1Dose	1Inc	Citation or Pathology
1627	1230	37.9mg	n.s.s.	0/17	36.4mg	1/18	Innes;ntis,1968/1969
a	1230	43.3mg	n.s.s.	1/17	36.4mg	1/18	
b	1230	17.4mg	n.s.s.	2/17	36.4mg	5/18	
1628	1238	4.05mg	n.s.s.	1/17	3.89mg	1/16	
a	1238	6.85mg	n.s.s.	0/17	3.89mg	0/16	
b	1238	4.49mg	n.s.s.	2/17	3.89mg	1/16	
1629	1230	37.6mg	n.s.s.	1/18	33.9mg	1/17	
a	1230	41.7mg	n.s.s.	2/18	33.9mg	1/17	
b	1230	32.7mg	n.s.s.	3/18	33.9mg	2/17	
1630	1238	3.38mg	n.s.s.	2/18	3.62mg	2/18	
a	1238	4.28mg	n.s.s.	1/18	3.62mg	1/18	
b	1238	1.84mg	n.s.s.	3/18	3.62mg	5/18	
1631	1230	35.8mg	n.s.s.	0/16	36.4mg	1/17	
a	1230	68.1mg	n.s.s.	0/16	36.4mg	0/17	
b	1230	26.1mg	n.s.s.	0/16	36.4mg	2/17	
1632	1238	7.71mg	n.s.s.	0/16	3.89mg	0/18	
a	1238	7.71mg	n.s.s.	0/16	3.89mg	0/18	
b	1238	7.71mg	n.s.s.	0/16	3.89mg	0/18	
1633	1230	24.3mg	n.s.s.	0/16	33.9mg	2/17	
a	1230	63.4mg	n.s.s.	0/16	33.9mg	0/17	
b	1230	24.3mg	n.s.s.	0/16	33.9mg	2/17	
1634	1238	1.46mg	n.s.s.	0/16	3.62mg	4/15	
a	1238	5.98mg	n.s.s.	0/16	3.62mg	0/15	
b	1238	1.04mg	11.7mg	0/16	3.62mg	6/15	

alpha-(2,5-DICHLOROPHENOXY)PROPIONIC ACID   6965-71-5

	RefNum	LoConf	UpConf	Cntrl	1Dose	1Inc	Citation or Pathology
1635	1231	16.0mg	n.s.s.	1/17	13.4mg	1/18	Innes;ntis,1968/1969
a	1231	26.6mg	n.s.s.	0/17	13.4mg	0/18	
b	1231	17.7mg	n.s.s.	2/17	13.4mg	1/18	
1636	1231	15.4mg	n.s.s.	2/18	12.5mg	1/17	
a	1231	23.5mg	n.s.s.	1/18	12.5mg	0/17	
b	1231	16.5mg	n.s.s.	3/18	12.5mg	1/17	
1637	1231	25.1mg	n.s.s.	0/16	13.4mg	0/17	
a	1231	25.1mg	n.s.s.	0/16	13.4mg	0/17	
b	1231	25.1mg	n.s.s.	0/16	13.4mg	0/17	
1638	1231	9.00mg	n.s.s.	0/16	12.5mg	2/17	
a	1231	9.00mg	n.s.s.	0/16	12.5mg	2/17	
b	1231	4.25mg	57.2mg	0/16	12.5mg	6/17	

2,4-DICHLOROPHENOXYACETIC ACID   (2,4-D) 94-75-7

	RefNum	LoConf	UpConf	Cntrl	1Dose	1Inc	2Dose	2Inc	Citation or Pathology
1639	1232	64.6mg	n.s.s.	1/17	20.4mg	0/16	44.3mg	1/15	Innes;ntis,1968/1969
a	1232	24.1mg	n.s.s.	0/17	20.4mg	0/16	44.3mg	0/15	
b	1232	43.4mg	n.s.s.	2/17	20.4mg	1/16	44.3mg	2/15	
1640	1232	22.2mg	n.s.s.	1/18	19.0mg	0/18	41.2mg	0/12	
a	1232	40.0mg	n.s.s.	2/18	19.0mg	2/18	41.2mg	1/12	
b	1232	45.0mg	n.s.s.	3/18	19.0mg	2/18	41.2mg	1/12	
1641	1232	20.1mg	n.s.s.	0/16	20.4mg	1/17			
a	1232	38.2mg	n.s.s.	0/16	20.4mg	0/17			
b	1232	20.1mg	n.s.s.	0/16	20.4mg	1/17			
1642	1232	35.5mg	n.s.s.	0/16	19.0mg	0/17			
a	1232	35.5mg	n.s.s.	0/16	19.0mg	0/17			
b	1232	13.6mg	n.s.s.	0/16	19.0mg	2/17			

2,4-DICHLOROPHENOXYACETIC ACID, N-BUTYL ESTER   94-80-4

	RefNum	LoConf	UpConf	Cntrl	1Dose	1Inc	Citation or Pathology
1643	1237	36.0mg	n.s.s.	0/17	20.4mg	0/16	Innes;ntis,1968/1969
a	1237	36.0mg	n.s.s.	1/17	20.4mg	0/16	
b	1237	36.0mg	n.s.s.	2/17	20.4mg	0/16	
1644	1237	24.9mg	n.s.s.	2/18	19.0mg	1/18	
a	1237	37.6mg	n.s.s.	1/18	19.0mg	0/18	
b	1237	14.9mg	n.s.s.	3/18	19.0mg	3/18	
1645	1237	40.5mg	n.s.s.	0/16	20.4mg	0/18	
a	1237	40.5mg	n.s.s.	0/16	20.4mg	0/18	
b	1237	10.2mg	n.s.s.	0/16	20.4mg	4/18	
1646	1237	14.5mg	n.s.s.	0/16	19.0mg	2/18	

Spe	Sex	Strain	Route	Site	Hist	Xpo+Xpt	Notes	plot	TD50	DR	2Tailpvl	AuOp
a	M m	b6c	orl	liv	hpt	76w76	evx		122.mg		P<.3	-
b	M m	b6c	orl	tba	mix	76w76	evx		38.1mg		P<.05	-

2,4-DICHLOROPHENOXYACETIC ACID, ISOOCTYL ESTER ...:..10....:..100...:..1mg...:..10...:..100...:..1g...:..10

Spe	Sex	Strain	Route	Site	Hist	Xpo+Xpt	Notes	plot	TD50	DR	2Tailpvl	AuOp
1647	M f	b6a	orl	liv	hpt	76w76	evx	.>	no dre		P=1.	-
a	M f	b6a	orl	lun	ade	76w76	evx		no dre		P=1.	-
b	M f	b6a	orl	tba	mix	76w76	evx		no dre		P=1.	-
1648	M m	b6a	orl	lun	mix	76w76	evx	.>	no dre		P=1.	-
a	M m	b6a	orl	liv	hpt	76w76	evx		no dre		P=1.	-
b	M m	b6a	orl	tba	mix	76w76	evx		no dre		P=1.	-
1649	M f	b6c	orl	lun	ade	76w76	evx	. ±	34.0mg		P<.04	-
a	M f	b6c	orl	liv	hpt	76w76	evx		no dre		P=1.	-
b	M f	b6c	orl	tba	mix	76w76	evx		24.6mg		P<.02	-
1650	M m	b6c	orl	liv	hpt	76w76	evx	. ±	31.7mg		P<.04	-
a	M m	b6c	orl	lun	mix	76w76	evx		no dre		P=1.	-
b	M m	b6c	orl	tba	mix	76w76	evx		17.7mg		P<.007	-

2,4-DICHLOROPHENOXYACETIC ACID, ISOPROPYL ESTER...:..10....:..100...:..1mg...:..10...:..100...:..1g...:..10

Spe	Sex	Strain	Route	Site	Hist	Xpo+Xpt	Notes	plot	TD50	DR	2Tailpvl	AuOp
1651	M f	b6a	orl	liv	hpc	76w76	evx	.>	94.7mg		P<.3	-
a	M f	b6a	orl	lun	ade	76w76	evx		no dre		P=1.	-
b	M f	b6a	orl	tba	mix	76w76	evx		83.2mg		P<.7	-
1652	M m	b6a	orl	lun	ade	76w76	evx	.>	no dre		P=1.	-
a	M m	b6a	orl	liv	hpt	76w76	evx		no dre		P=1.	-
b	M m	b6a	orl	tba	mix	76w76	evx		no dre		P=1.	-
1653	M f	b6c	orl	liv	hpt	76w76	evx	.>	100.mg		P<.3	-
a	M f	b6c	orl	lun	ade	76w76	evx		100.mg		P<.3	-
b	M f	b6c	orl	tba	mix	76w76	evx		48.7mg		P<.2	-
1654	M m	b6c	orl	lun	ade	76w76	evx	. ±	21.3mg		P<.02	-
a	M m	b6c	orl	liv	hpt	76w76	evx		93.6mg		P<.3	-
b	M m	b6c	orl	tba	mix	76w76	evx		16.4mg		P<.009	-

3-(3,4-DICHLOROPHENYL)-1,1-DIMETHYLUREA..:..1_ug...:..10....:..100...:..1mg...:..10....:..100...:..1g......:..10

Spe	Sex	Strain	Route	Site	Hist	Xpo+Xpt	Notes	plot	TD50	DR	2Tailpvl	AuOp
1655	M f	b6a	orl	lun	ade	76w76	evx	.>	no dre		P=1.	-
a	M f	b6a	orl	liv	hpt	76w76	evx		no dre		P=1.	-
b	M f	b6a	orl	tba	mix	76w76	evx		no dre		P=1.	-
1656	M m	b6a	orl	lun	ade	76w76	evx	.>	437.mg		P<.4	-
a	M m	b6a	orl	liv	hpt	76w76	evx		no dre		P=1.	-
b	M m	b6a	orl	tba	mix	76w76	evx		260.mg		P<.3	-
1657	M f	b6c	orl	lun	ade	76w76	evx	.>	600.mg		P<.2	-
a	M f	b6c	orl	liv	hpt	76w76	evx		no dre		P=1.	-
b	M f	b6c	orl	tba	mix	76w76	evx		388.mg		P<.05	-
1658	M m	b6c	orl	liv	hpt	76w76	evx	.>	1.15gm		P<.3	-
a	M m	b6c	orl	lun	ade	76w76	evx		1.15gm		P<.3	-
b	M m	b6c	orl	tba	mix	76w76	evx		361.mg		P<.05	-

2,4-DICHLOROPHENYLBENZENE SULFONATE ..:..1ug...:..10....:..100...:..1mg...:..10....:..100...:..1g......:..10

Spe	Sex	Strain	Route	Site	Hist	Xpo+Xpt	Notes	plot	TD50	DR	2Tailpvl	AuOp
1659	M f	b6a	orl	lun	mix	76w76	evx	.>	1.92gm		P<.6	-
a	M f	b6a	orl	liv	hpt	76w76	evx		no dre		P=1.	-
b	M f	b6a	orl	tba	mix	76w76	evx		867.mg		P<.4	-
1660	M m	b6a	orl	lun	ade	76w76	evx	.>	866.mg		P<.4	-
a	M m	b6a	orl	liv	hpt	76w76	evx		no dre		P=1.	-
b	M m	b6a	orl	tba	mix	76w76	evx		1.68gm		P<.7	-
1661	M f	b6c	orl	lun	ade	76w76	evx	.>	1.92gm		P<.3	-
a	M f	b6c	orl	liv	hpt	76w76	evx		no dre		P=1.	-
b	M f	b6c	orl	tba	mix	76w76	evx		929.mg		P<.09	-
1662	M m	b6c	orl	liv	hpt	76w76	evx	. + .	266.mg		P<.003	-
a	M m	b6c	orl	lun	ade	76w76	evx		924.mg		P<.1	-
b	M m	b6c	orl	tba	mix	76w76	evx		218.mg		P<.002	-

1,2-DICHLOROPROPANE 100ng..:..1ug...:..10....:..100...:..1mg...:..10....:..100...:..1g......:..10

Spe	Sex	Strain	Route	Site	Hist	Xpo+Xpt	Notes	plot	TD50	DR	2Tailpvl	AuOp
1663	M f	b6c	gav	liv	MXA	24m24	s	: +	: 346.mg	*	P<.008	p
a	M f	b6c	gav	liv	hpa	24m24	s		562.mg	*	P<.04	p
b	M f	b6c	gav	for	MXA	24m24	s		1.02gm	*	P<.03	
c	M f	b6c	gav	TBA	MXB	24m24	s		341.mg	*	P<.4	
d	M f	b6c	gav	liv	MXB	24m24	s		346.mg	*	P<.008	
e	M f	b6c	gav	lun	MXB	24m24	s		no dre		P=1.	
1664	M m	b6c	gav	liv	MXA	24m24		: ±	229.mg	*	P<.05	p
a	M m	b6c	gav	liv	hpa	24m24			384.mg	*	P<.05	p
b	M m	b6c	gav	for	sqp	24m24			1.52gm	*	P<.05	
c	M m	b6c	gav	TBA	MXB	24m24			397.mg	*	P<.4	
d	M m	b6c	gav	liv	MXB	24m24			229.mg	*	P<.05	
e	M m	b6c	gav	lun	MXB	24m24			5.07gm	*	P<.9	
1665	R f	f34	gav	mgl	acn	24m24		: ±	767.mg	*	P<.03	e
a	R f	f34	gav	TBA	MXB	24m24			240.mg	*	P<.3	
b	R f	f34	gav	liv	MXB	24m24			no dre		P=1.	
1666	R m	f34	gav	TBA	MXB	24m25		: ±	69.7mg	/	P<.04	-
a	R m	f34	gav	liv	MXB	24m25			2.57gm	*	P<.8	

DICHLORVOS 100ng..:..1ug...:..10....:..100...:..1mg...:..10....:..100...:..1g......:..10

Spe	Sex	Strain	Route	Site	Hist	Xpo+Xpt	Notes	plot	TD50	DR	2Tailpvl	AuOp
1667	M f	b6c	eat	TBA	MXB	80w92	v	:>	7.85gm	*	P<1.	-
a	M f	b6c	eat	liv	MXB	80w92	v		no dre		P=1.	

	RefNum	LoConf	UpConf	Cntrl	1Dose	1Inc	2Dose	2Inc	Citation or Pathology
									Brkly Code
a	1237	19.8mg	n.s.s.	0/16	19.0mg	1/18			
b	1237	11.5mg	n.s.s.	0/16	19.0mg	3/18			

2,4-DICHLOROPHENOXYACETIC ACID, ISOOCTYL ESTER   25168-26-7

									Innes;ntis,1968/1969
1647	1236	33.8mg	n.s.s.	0/17	18.1mg	0/17			
a	1236	33.8mg	n.s.s.	1/17	18.1mg	0/17			
b	1236	22.3mg	n.s.s.	2/17	18.1mg	1/17			
1648	1236	15.7mg	n.s.s.	2/18	16.8mg	2/18			
a	1236	19.9mg	n.s.s.	1/18	16.8mg	1/18			
b	1236	13.2mg	n.s.s.	3/18	16.8mg	3/18			
1649	1236	10.3mg	n.s.s.	0/16	18.1mg	3/17			
a	1236	33.8mg	n.s.s.	0/16	18.1mg	0/17			
b	1236	8.46mg	n.s.s.	0/16	18.1mg	4/17			
1650	1236	9.56mg	n.s.s.	0/16	16.8mg	3/17			
a	1236	31.4mg	n.s.s.	0/16	16.8mg	0/17			
b	1236	6.65mg	239.mg	0/16	16.8mg	5/17			

2,4-DICHLOROPHENOXYACETIC ACID, ISOPROPYL ESTER   94-11-1

									Innes;ntis,1968/1969
1651	1235	15.4mg	n.s.s.	0/17	15.7mg	1/17			
a	1235	29.3mg	n.s.s.	1/17	15.7mg	0/17			
b	1235	10.5mg	n.s.s.	2/17	15.7mg	3/17			
1652	1235	13.7mg	n.s.s.	2/18	14.6mg	2/18			
a	1235	29.0mg	n.s.s.	1/18	14.6mg	0/18			
b	1235	11.5mg	n.s.s.	3/18	14.6mg	3/18			
1653	1235	16.3mg	n.s.s.	0/16	15.7mg	1/18			
a	1235	16.3mg	n.s.s.	0/16	15.7mg	1/18			
b	1235	12.0mg	n.s.s.	0/16	15.7mg	2/18			
1654	1235	7.32mg	n.s.s.	0/16	14.6mg	4/18			
a	1235	15.2mg	n.s.s.	0/16	14.6mg	1/18			
b	1235	6.20mg	361.mg	0/16	14.6mg	5/18			

3-(3,4-DICHLOROPHENYL)-1,1-DIMETHYLUREA   (karmex, diuron) 330-54-1

									Innes;ntis,1968/1969
1655	1276	215.mg	n.s.s.	1/17	193.mg	1/17			
a	1276	361.mg	n.s.s.	0/17	193.mg	0/17			
b	1276	170.mg	n.s.s.	2/17	193.mg	2/17			
1656	1276	95.6mg	n.s.s.	2/18	180.mg	4/17			
a	1276	336.mg	n.s.s.	1/18	180.mg	0/17			
b	1276	70.7mg	n.s.s.	3/18	180.mg	6/17			
1657	1276	147.mg	n.s.s.	0/16	193.mg	2/18			
a	1276	382.mg	n.s.s.	0/16	193.mg	0/18			
b	1276	117.mg	n.s.s.	0/16	193.mg	3/18			
1658	1276	187.mg	n.s.s.	0/16	180.mg	1/18			
a	1276	187.mg	n.s.s.	0/16	180.mg	1/18			
b	1276	109.mg	n.s.s.	0/16	180.mg	3/18			

2,4-DICHLOROPHENYLBENZENE SULFONATE   (genite-R99) 97-16-5

									Innes;ntis,1968/1969
1659	1288	267.mg	n.s.s.	1/17	339.mg	2/17			
a	1288	634.mg	n.s.s.	0/17	339.mg	0/17			
b	1288	182.mg	n.s.s.	2/17	339.mg	4/17			
1660	1288	181.mg	n.s.s.	2/18	316.mg	4/18			
a	1288	626.mg	n.s.s.	1/18	316.mg	0/18			
b	1288	196.mg	n.s.s.	3/18	316.mg	4/18			
1661	1288	313.mg	n.s.s.	0/16	339.mg	1/16			
a	1288	597.mg	n.s.s.	0/16	339.mg	0/16			
b	1288	228.mg	n.s.s.	0/16	339.mg	2/16			
1662	1288	107.mg	1.44gm	0/16	316.mg	6/17			
a	1288	227.mg	n.s.s.	0/16	316.mg	2/17			
b	1288	92.7mg	837.mg	0/16	316.mg	7/17			

1,2-DICHLOROPROPANE   (propylene dichloride) 78-87-5

1663	TR263	168.mg	6.75gm	2/50	86.8mg	8/50	175.mg	9/50	liv:hpa,hpc.	
a	TR263	235.mg	n.s.s.	1/50	86.8mg	5/50	175.mg	5/50		
b	TR263	387.mg	n.s.s.	0/50	86.8mg	2/50	175.mg	3/50	for:sqc,sqp.	S
c	TR263	92.1mg	n.s.s.	35/50	86.8mg	29/50	175.mg	34/50		
d	TR263	168.mg	6.75gm	2/50	86.8mg	8/50	175.mg	9/50	liv:hpa,hpc,nnd.	
e	TR263	1.07gm	n.s.s.	6/50	86.8mg	1/50	175.mg	1/50	lun:a/a,a/c.	
1664	TR263	98.6mg	n.s.s.	18/50	86.8mg	26/50	175.mg	33/50	liv:hpa,hpc.	
a	TR263	165.mg	n.s.s.	7/50	86.8mg	10/50	175.mg	17/50		
b	TR263	524.mg	n.s.s.	0/50	86.8mg	1/50	175.mg	3/50		S
c	TR263	102.mg	n.s.s.	33/50	86.8mg	36/50	175.mg	42/50		
d	TR263	98.6mg	n.s.s.	18/50	86.8mg	26/50	175.mg	33/50	liv:hpa,hpc,nnd.	
e	TR263	312.mg	n.s.s.	11/50	86.8mg	8/50	175.mg	12/50	lun:a/a,a/c.	
1665	TR263	294.mg	n.s.s.	1/50	86.8mg	2/50	175.mg	5/50		
a	TR263	72.5mg	n.s.s.	42/50	86.8mg	46/50	175.mg	30/50		
b	TR263	2.18gm	n.s.s.	1/50	86.8mg	1/50	175.mg	0/50	liv:hpa,hpc,nnd.	
1666	TR263	29.8mg	n.s.s.	45/50	42.8mg	37/50	86.4mg	43/50	liv:hpa,hpc,nnd.	
a	TR263	208.mg	n.s.s.	3/50	42.8mg	3/50	86.4mg	2/50	liv:hpa,hpc,nnd.	

DICHLORVOS   (DDVP, Vapona) 62-73-7

1667	TR10	86.5mg	n.s.s.	1/10	35.1mg	11/50	70.2mg	9/50		
a	TR10	n.s.s.	n.s.s.	0/10	35.1mg	0/50	70.2mg	0/50	liv:hpa,hpc,nnd.	

```
        Spe Strain  Site   Xpo+Xpt                                                          TD50     2Tailpvl
          Sex  Route  Hist     Notes                                                           DR    AuOp
  b    M f b6c eat lun MXB 80w92 v                                                          1.53gm * P<.8
1668   M f b6c gav for MXA 24m24                                          :   ±            56.3mg * P<.02
  a    M f b6c gav for sqp 24m24                                                          61.3mg * P<.03  c
  b    M f b6c gav TBA MXB 24m24                                                          no dre   P=1.
  c    M f b6c gav liv MXB 24m24                                                          no dre   P=1.
  d    M f b6c gav lun MXB 24m24                                                          321.mg * P<.5
1669   M m b6c eat TBA MXB 80w92 v                                          :   ±          40.3mg \ P<.08  -
  a    M m b6c eat liv MXB 80w92 v                                                        no dre   P=1.
  b    M m b6c eat lun MXB 80w92 v                                                        725.mg * P<.8
1670   M m b6c gav for sqp 24m24                                          :      ±        82.8mg * P<.07  p
  a    M m b6c gav TBA MXB 24m24                                                          38.2mg * P<.6
  b    M m b6c gav liv MXB 24m24                                                          43.4mg * P<.3
  c    M m b6c gav lun MXB 24m24                                                          121.mg * P<.7
1671   R f osm eat TBA MXB 19m26 v                              :>                        no dre   P=1.   -
  a    R f osm eat liv MXB 19m26 v                                                        no dre   P=1.
1672   R f f34 gav mgl fba 24m24                                    :   ±                  3.97mg \ P<.03  e
  a    R f f34 gav mgl MXA 24m24                                                          8.51mg * P<.05
  b    R f f34 gav sub MXA 24m24                                                          27.3mg * P<.03
  c    R f f34 gav mgl MXA 24m24                                                          10.4mg * P<.2
  d    R f f34 gav pan ade 24m24                                        '                  44.6mg * P<.2   e
  e    R f f34 gav TBA MXB 24m24              -                                            13.9mg * P<.6
  f    R f f34 gav liv MXB 24m24                                                          83.9mg * P<.3
1673   R m osm eat TBA MXB 19m26 v                               :>                       no dre   P=1.   -
  a    R m osm eat liv MXB 19m26 v                                                        no dre   P=1.
1674   R m f34 gav MXB MXB 24m24                                    :   +        :         3.21mg * P<.005
  a    R m f34 gav pan ade 24m24                                                          4.16mg * P<.005 p
  b    R m f34 gav --- mnl 24m24                                                          6.89mg * P<.02  p
  c    R m f34 gav lun a/a 24m24                                                          52.2mg * P<.04
  d    R m f34 gav TBA MXB 24m24                                                          12.3mg * P<.6
  e    R m f34 gav liv MXB 24m24                                                          56.2mg * P<.4
1675   R f cfe inh pit tum 24m24 e                          .   +         .                4.65mg * P<.009 -
  a    R f cfe inh liv tum 24m24 e                                                        no dre   P=1.   -
  b    R f cfe inh tba mix 24m24 e                                                        24.5mg * P<.9
1676   R m cfe inh pit tum 24m24 e                                .>                       11.3mg * P<.3   -
  a    R m cfe inh liv tum 24m24 e                                                        50.0mg * P<.2   -
  b    R m cfe inh thy tum 24m24 e                                                        no dre   P=1.   -
  c    R m cfe inh tba mix 24m24 e                                                        no dre   P=1.   -

DICOFOL                100ng..:..1ug....:..10....:..100....:..1mg....:..10....:..100....:..1g......:..10
1677   M f b6c eat TBA MXB 78w91 v                                 :>                      no dre   P=1.   -
  a    M f b6c eat liv MXB 78w91 v                                                        no dre   P=1.
  b    M f b6c eat lun MXB 78w91 v                                                        no dre   P=1.
1678   M m b6c eat liv MXA 78w91 v                               :   ±                     32.9mg * P<.04  c
  a    M m b6c eat liv hpc 78w91 v                                                        34.8mg * P<.04  c
  b    M m b6c eat TBA MXB 78w91 v                                                        59.2mg * P<.4
  c    M m b6c eat liv MXB 78w91 v                                                        32.9mg * P<.04
  d    M m b6c eat lun MXB 78w91 v                                                        616.mg * P<.6
1679   R f osm eat TBA MXB 18m26                                :>                         no dre   P=1.   -
  a    R f osm eat liv MXB 18m26                                                          no dre   P=1.
1680   R m osm eat TBA MXB 18m26 v                           :>                            no dre   P=1.   -
  a    R m osm eat liv MXB 18m26 v                                                        517.mg * P<.7

N,N'-DICYCLOHEXYLTHIOUREA   100ng..:..1ug....:..10....:..100....:..1mg....:..10....:..100....:..1g......:..10
1681   M f b6c eat TBA MXB 25m24                                                   :>     122.gm * P<1.    -
  a    M f b6c eat liv MXB 25m24                                                          no dre   P=1.
  b    M f b6c eat lun MXB 25m24                                                          no dre   P=1.
1682   M m b6c eat TBA MXB 25m24                                                   :>     no dre   P=1.    -
  a    M m b6c eat liv MXB 25m24                                                          no dre   P=1.
  b    M m b6c eat lun MXB 25m24                                                          no dre   P=1.
1683   R f f34 eat TBA MXB 25m25                                                  :>      no dre   P=1.    -
  a    R f f34 eat liv MXB 25m25                                                          no dre   P=1.
1684   R m f34 eat thy cca 25m25                                               :   ± #5.48gm * P<.03  -
  a    R m f34 eat thy MXA 25m25                                                          7.41gm * P<.04
  b    R m f34 eat ski bcc 25m25                                                          26.1gm * P<.04
  c    R m f34 eat TBA MXB 25m25                                                          3.91gm * P<.3
  d    R m f34 eat liv MXB 25m25                                                          15.6gm * P<.09

DICYCLOPENTADIENE DIOXIDE   100ng..:..1ug....:..10....:..100....:..1mg....:..10....:..100....:..1g......:..10
1685   M f chi eat lun mix 64w86 a                                           :>           1.66gm * P<.2   -
  a    M f chi eat liv mix 64w86 a                                                        no dre   P=1.
  b    M f chi eat tba mix 64w86 a                                                        789.mg * P<.2
1686   M m chi eat lun mix 77w86                                       :    +             373.mg \ P<.009 -
  a    M m chi eat liv mix 77w86                                                          930.mg * P<.05
  b    M m chi eat tba mix 77w86                                                          149.mg \ P<.003 -
1687   R m cdr eat liv mix 18m25                                             :>           3.91gm * P<.8   -
  a    R m cdr eat tba mix 18m25                                                          176.mg * P<.03  -

DIELDRIN                100ng..:..1ug....:..10....:..100....:..1mg....:..10....:..100....:..1g......:..10
1688   H f syg eat liv hpt 23m26 e                                       .>               176.mg * P<.3   -
  a    H f syg eat lun tum 23m26 e                                                        no dre   P=1.   -
  b    H f syg eat tba mix 23m26 e                                                        355.mg * P<.6   -
```

	RefNum	LoConf	UpConf	Cntrl	1Dose	1Inc	2Dose	2Inc			Citation or Pathology
											Brkly Code
b	TR10	277.mg	n.s.s.	0/10	35.1mg	1/50	70.2mg	1/50			lun:a/a,a/c.
1668	TR342	26.8mg	n.s.s.	5/50	14.0mg	6/50	28.0mg	19/50			for:sqc,sqp. S
a	TR342	28.2mg	n.s.s.	5/50	14.0mg	6/50	28.0mg	18/50			
b	TR342	31.8mg	n.s.s.	37/50	14.0mg	26/50	28.0mg	37/50			
c	TR342	75.7mg	n.s.s.	6/50	14.0mg	4/50	28.0mg	7/50			liv:hpa,hpc,nnd.
d	TR342	63.6mg	n.s.s.	3/50	14.0mg	3/50	28.0mg	6/50			lun:a/a,a/c.
1669	TR10	20.1mg	n.s.s.	1/10	32.4mg	22/50	(64.8mg	14/50)			
a	TR10	76.1mg	n.s.s.	0/10	32.4mg	12/50	64.8mg	7/50			liv:hpa,hpc,nnd.
b	TR10	103.mg	n.s.s.	0/10	32.4mg	7/50	64.8mg	5/50			lun:a/a,a/c.
1670	TR342	28.9mg	n.s.s.	1/50	7.01mg	1/50	14.0mg	5/50			
a	TR342	7.52mg	n.s.s.	37/50	7.01mg	41/50	14.0mg	37/50			
b	TR342	12.5mg	n.s.s.	16/50	7.01mg	18/50	14.0mg	20/50			liv:hpa,hpc,nnd.
c	TR342	17.2mg	n.s.s.	10/50	7.01mg	15/50	14.0mg	10/50			lun:a/a,a/c.
1671	TR10	5.73mg	n.s.s.	8/10	5.50mg	34/50	(11.9mg	30/50)			
a	TR10	75.7mg	n.s.s.	1/10	5.50mg	3/50	11.9mg	1/50			liv:hpa,hpc,nnd.
1672	TR342	1.65mg	n.s.s.	9/50	2.80mg	19/50	(5.61mg	16/50)			
a	TR342	3.66mg	n.s.s.	9/50	2.80mg	19/50	5.61mg	17/50			mgl:ade,fba. S
b	TR342	11.1mg	n.s.s.	0/50	2.80mg	3/50	5.61mg	3/50			sub:fbs,fib. S
c	TR342	3.91mg	n.s.s.	11/50	2.80mg	20/50	5.61mg	17/50			mgl:ade,car,fba.
d	TR342	13.6mg	n.s.s.	1/50	2.80mg	1/50	5.61mg	4/50			
e	TR342	2.62mg	n.s.s.	47/50	2.80mg	46/50	5.61mg	46/50			
f	TR342	20.5mg	n.s.s.	0/50	2.80mg	1/50	5.61mg	1/50			liv:hpa,hpc,nnd.
1673	TR10	8.10mg	n.s.s.	6/10	4.40mg	21/50	9.50mg	33/50			
a	TR10	n.s.s.	n.s.s.	0/10	4.40mg	0/50	9.50mg	0/50			liv:hpa,hpc,nnd.
1674	TR342	1.62mg	31.1mg	25/50	2.80mg	37/50	5.61mg	41/50			---:mnl; pan:ade. P
a	TR342	2.09mg	44.6mg	16/50	2.80mg	25/50	5.61mg	30/50			
b	TR342	3.24mg	n.s.s.	11/50	2.80mg	20/50	5.61mg	21/50			
c	TR342	15.5mg	n.s.s.	0/50	2.80mg	0/50	5.61mg	3/50			S
d	TR342	2.34mg	n.s.s.	48/50	2.80mg	45/50	5.61mg	45/50			
e	TR342	12.7mg	n.s.s.	1/50	2.80mg	3/50	5.61mg	2/50			liv:hpa,hpc,nnd.
1675	96	1.97mg	172.mg	7/47	18.9mg	5/47	.189mg	12/47	1.89mg	16/46	Blair;artx,35,281-294;1976
a	96	.165mg	n.s.s.	0/47	18.9ug	0/47	.189mg	0/47	1.89mg	0/46	
b	96	1.48mg	n.s.s.	36/47	18.9ug	28/47	.189mg	36/47	1.89mg	33/46	
1676	96	2.76mg	n.s.s.	4/50	13.3ug	10/50	.133mg	6/50	1.33mg	10/50	
a	96	8.14mg	n.s.s.	0/50	13.3ug	0/50	.133mg	0/50	1.33mg	1/50	
b	96	5.04mg	n.s.s.	2/50	13.3ug	9/50	.133mg	5/50	1.33mg	5/50	
c	96	1.97mg	n.s.s.	18/50	13.3ug	30/50	.133mg	26/50	1.33mg	24/50	

DICOFOL (kelthane) 115-32-2

	RefNum	LoConf	UpConf	Cntrl	1Dose	1Inc	2Dose	2Inc			Citation or Pathology
1677	TR90	65.1mg	n.s.s.	5/20	13.0mg	6/50	26.0mg	8/50			
a	TR90	309.mg	n.s.s.	1/20	13.0mg	0/50	26.0mg	0/50			liv:hpa,hpc,nnd.
b	TR90	296.mg	n.s.s.	1/20	13.0mg	0/50	26.0mg	0/50			lun:a/a,a/c.
1678	TR90	15.9mg	n.s.s.	3/20	26.4mg	23/50	54.0mg	36/50			liv:hpa,hpc.
a	TR90	16.8mg	n.s.s.	3/20	26.4mg	22/50	54.0mg	35/50			
b	TR90	16.1mg	n.s.s.	5/20	26.4mg	34/50	54.0mg	38/50			
c	TR90	15.9mg	n.s.s.	3/20	26.4mg	23/50	54.0mg	36/50			liv:hpa,hpc,nnd.
d	TR90	114.mg	n.s.s.	1/20	26.4mg	2/50	54.0mg	5/50			lun:a/a,a/c.
1679	TR90	46.3mg	n.s.s.	16/20	13.0mg	27/50	26.0mg	28/50			
a	TR90	n.s.s.	n.s.s.	0/20	13.0mg	0/50	26.0mg	0/50			liv:hpa,hpc,nnd.
1680	TR90	13.4mg	n.s.s.	10/20	12.8mg	23/50	(26.4mg	16/50)			
a	TR90	103.mg	n.s.s.	0/20	12.8mg	1/50	26.4mg	1/50			liv:hpa,hpc,nnd.

N,N'-DICYCLOHEXYLTHIOUREA   1212-29-9

	RefNum	LoConf	UpConf	Cntrl	1Dose	1Inc	2Dose	2Inc			Citation or Pathology
1681	TR56	4.84gm	n.s.s.	36/49	3.25gm	37/50	6.50gm	37/50			
a	TR56	30.1gm	n.s.s.	5/49	3.25gm	2/50	6.50gm	3/50			liv:hpa,hpc,nnd.
b	TR56	23.9gm	n.s.s.	4/49	3.25gm	2/50	6.50gm	4/50			lun:a/a,a/c.
1682	TR56	4.43gm	n.s.s.	39/50	3.00gm	37/50	6.00gm	39/50			
a	TR56	6.61gm	n.s.s.	26/50	3.00gm	18/50	6.00gm	25/50			liv:hpa,hpc,nnd.
b	TR56	13.6gm	n.s.s.	9/50	3.00gm	5/50	6.00gm	9/50			lun:a/a,a/c.
1683	TR56	3.02gm	n.s.s.	48/50	1.25gm	31/50	2.50gm	43/50			
a	TR56	15.2gm	n.s.s.	1/50	1.25gm	1/50	2.50gm	1/50			liv:hpa,hpc,nnd.
1684	TR56	2.45gm	n.s.s.	4/50	1.00gm	9/50	2.00gm	12/50			S
a	TR56	3.32gm	n.s.s.	1/50	1.00gm	7/50	2.00gm	6/50			thy:fca,fcc. S
b	TR56	7.90gm	n.s.s.	0/50	1.00gm	0/50	2.00gm	3/50			S
c	TR56	1.18gm	n.s.s.	36/50	1.00gm	36/50	2.00gm	43/50			
d	TR56	5.92gm	n.s.s.	0/50	1.00gm	3/50	2.00gm	2/50			liv:hpa,hpc,nnd.

DICYCLOPENTADIENE DIOXIDE   81-21-0

	RefNum	LoConf	UpConf	Cntrl	1Dose	1Inc	2Dose	2Inc			Citation or Pathology
1685	381	407.mg	n.s.s.	3/13	493.mg	2/19	1.04gm	0/15			Weisburger;jept,2,325-356;1978/pers.comm./Russfield 1973
a	381	n.s.s.	n.s.s.	0/13	493.mg	0/19	1.04gm	0/15			
b	381	232.mg	n.s.s.	9/13	493.mg	7/19	1.04gm	1/15			
1686	381	140.mg	12.1gm	2/16	455.mg	5/15	(909.mg	2/16)			
a	381	380.mg	n.s.s.	2/16	455.mg	3/15	909.mg	3/16			
b	381	66.5mg	962.mg	9/16	455.mg	13/15	(909.mg	8/16)			
1687	381	309.mg	n.s.s.	1/17	120.mg	1/19	240.mg	1/13			
a	381	74.5mg	n.s.s.	13/17	120.mg	10/19	240.mg	9/13			

DIELDRIN   60-57-1

	RefNum	LoConf	UpConf	Cntrl	1Dose	1Inc	2Dose	2Inc	3Dose	3Inc	Citation or Pathology
1688	1000	28.7mg	n.s.s.	0/6	2.09mg	0/2	6.27mg	0/8	18.8mg	1/9	Cabral;clet,6,241-246;1979
a	1000	11.8mg	n.s.s.	0/39	2.09mg	0/32	6.27mg	0/34	18.8mg	0/38	
b	1000	54.3mg	n.s.s.	5/39	2.09mg	1/32	6.27mg	5/34	18.8mg	5/38	

ID	Spe	Sex	Strain	Route	Site	Hist	Xpo+Xpt	Notes	DRplot	TD50	DR	2Tailpvl	AuOp
1689	H	m	syg	eat	liv	hpt	26m26	e	.>	198.mg	*	P<.4	-
a	H	m	syg	eat	lun	ade	26m26	e		732.mg	*	P<.2	-
b	H	m	syg	eat	tba	mix	26m26	e		69.9mg	*	P<.05	-
1690	M	f	b6c	eat	TBA	MXB	80w91	s	:>	no dre		P=1.	-
a	M	f	b6c	eat	liv	MXB	80w91	s		1.09mg	\	P<.06	
b	M	f	b6c	eat	lun	MXB	80w91	s		5.05mg	*	P<.4	
1691	M	m	b6c	eat	TBA	MXB	80w91	s	:>	1.06mg	*	P<.2	-
a	M	m	b6c	eat	liv	MXB	80w91	s		1.04mg	*	P<.2	
b	M	m	b6c	eat	lun	MXB	80w91	s		24.7mg	*	P<.9	
1692	M	m	b6c	eat	liv	hpc	80w89	s pool	: ±	1.35mg	*	P<.08	a
1693	M	b	c3e	eat	liv	hpa	24m24	e	. + .	4.09mg			
a	M	b	c3e	eat	lun	ade	24m24	e		no dre		P=1.	
b	M	b	c3e	eat	lun	car	24m24	e		no dre		P=1.	
1694	M	b	c3h	eat	liv	hpt	24m24		. + .	3.08mg		P<.0005	
a	M	b	c3h	eat	liv	hpc	24m24			167.mg		P<.8	
b	M	b	c3h	eat	tba	ben	24m24			3.10mg		P<.0005	
c	M	b	c3h	eat	tba	mal	24m24			no dre		P=1.	
1695	M	f	cf1	eat	liv	mix	26m26	e	. + .	.567mg		P<.0005	+
a	M	f	cf1	eat	liv	lct	26m26	e		1.59mg		P<.0005	+
b	M	f	cf1	eat	lun	tum	26m26	e		no dre		P=1.	-
1696	M	f	cf1	eat	lun	ade	28m32	ae	. + .	.606mg	Z	P<.002	-
a	M	f	cf1	eat	liv	mix	28m32	ae		.642mg	*	P<.0005	+
b	M	f	cf1	eat	liv	lpb	28m32	ae		1.96mg	*	P<.0005	
c	M	f	cf1	eat	lun	car	28m32	ae		1.78mg	Z	P<.06	
1697	M	f	cf1	eat	liv	mix	25m30	aes	. + .	1.49mg	*	P<.0005	+
a	M	f	cf1	eat	liv	lpb	25m30	aes		9.72mg	*	P<.0005	
b	M	f	cf1	eat	lun	ade	25m30	aes		no dre		P=1.	-
c	M	f	cf1	eat	lun	car	25m30	aes		no dre		P=1.	-
1698	M	f	cf1	eat	liv	lpb	90w90	e	. + .	2.59mg		P<.007	
a	M	f	cf1	eat	liv	mix	90w90	e		1.05mg		P<.02	+
b	M	f	cf1	eat	lun	ade	90w90	e		no dre		P=1.	-
c	M	f	cf1	eat	lun	car	90w90	e		no dre		P=1.	-
1699	M	f	cf1	eat	liv	mix	30m30	e	. ±	3.11mg		P<.02	+
a	M	f	cf1	eat	liv	lpb	30m30	e		12.1mg		P<.06	
b	M	f	cf1	eat	lun	ade	30m30	e		no dre		P=1.	-
c	M	f	cf1	eat	lun	car	30m30	e		no dre		P=1.	-
1700	M	f	cf1	eat	liv	mix	25m25	e	. + .	1.16mg		P<.002	+
a	M	f	cf1	eat	liv	lpb	25m25	e		3.96mg		P<.007	
b	M	f	cf1	eat	lun	ade	25m25	e		5.69mg		P<.4	-
c	M	f	cf1	eat	lun	car	25m25	e		no dre		P=1.	-
1701	M	m	cf1	eat	liv	mix	26m26	e	<+	noTD50		P<.0005	+
a	M	m	cf1	eat	liv	lct	26m26	e		1.28mg		P<.0005	+
b	M	m	cf1	eat	lun	tum	26m26	e		no dre		P=1.	-
1702	M	m	cf1	eat	liv	mix	28m31	ae	.+.	.547mg	*	P<.0005	+
a	M	m	cf1	eat	liv	lpb	28m31	ae		1.68mg	*	P<.0005	
b	M	m	cf1	eat	lun	ade	28m31	ae		2.13mg	Z	P<.5	-
c	M	m	cf1	eat	lun	car	28m31	ae		3.71mg	Z	P<.4	-
1703	M	m	cf1	eat	liv	mix	26m30	aes	.+.	1.12mg		P<.0005	+
a	M	m	cf1	eat	liv	lpb	26m30	aes		4.66mg	*	P<.0005	
b	M	m	cf1	eat	lun	ade	26m30	aes		no dre		P=1.	-
c	M	m	cf1	eat	lun	car	26m30	aes		no dre		P=1.	-
1704	M	m	cf1	eat	liv	mix	30m30	e	. + .	.913mg		P<.002	+
a	M	m	cf1	eat	liv	lpb	30m30	e		5.12mg		P<.04	
b	M	m	cf1	eat	lun	ade	30m30	e		no dre		P=1.	-
c	M	m	cf1	eat	lun	car	30m30	e		no dre		P=1.	-
1705	M	m	cf1	eat	liv	mix	30m30	e	. ±	1.91mg		P<.04	+
a	M	m	cf1	eat	liv	lpb	30m30	e		6.58mg		P<.2	
b	M	m	cf1	eat	lun	ade	30m30	e		no dre		P=1.	-
c	M	m	cf1	eat	lun	car	30m30	e		no dre		P=1.	-
1706	M	m	cf1	eat	liv	mix	26m29	e	. + .	.713mg		P<.002	+
a	M	m	cf1	eat	liv	lpb	26m29	e		4.02mg		P<.006	
b	M	m	cf1	eat	lun	ade	26m29	e		no dre		P=1.	-
c	M	m	cf1	eat	lun	car	26m29	e		no dre		P=1.	-
1707	R	b	osm	eat	lun	lys	24m24	es	.>	no dre		P=1.	
a	R	b	osm	eat	tba	mix	24m24	es		no dre		P=1.	+
1708	R	f	osm	eat	TBA	MXB	21m26	sv	:>	no dre		P=1.	-
a	R	f	osm	eat	liv	MXB	21m26	sv		38.4mg	*	P<.8	
1709	R	f	osm	eat	adr	MXA	21m25	sv pool	: + :	#4.42mg	\	P<.003	-
1710	R	f	f34	eat	TBA	MXB	24m24	a	:>	489.mg	*	P<1.	-
a	R	f	f34	eat	liv	MXB	24m24	a		no dre		P=1.	
1711	R	f	osm	eat	liv	tum	28m29	ev	.>	no dre		P=1.	
a	R	f	osm	eat	tba	mix	28m29	ev		no dre		P=1.	
1712	R	m	osm	eat	TBA	MXB	21m26	sv	:>	3.34mg	*	P<.6	-
a	R	m	osm	eat	liv	MXB	21m26	sv		no dre		P=1.	
1713	R	m	f34	eat	TBA	MXB	24m24	a	:>	7.28mg	*	P<.4	-
a	R	m	f34	eat	liv	MXB	24m24	a		no dre		P=1.	
1714	R	m	osm	eat	liv	hem	29m29	ev	.>	no dre		P=1.	
a	R	m	osm	eat	tba	mix	29m29	ev		no dre		P=1.	
1715	R	f	cfe	eat	tba	mix	24m24	e	.>	no dre		P=1.	-

RefNum		LoConf	UpConf	Cntrl	1Dose	1Inc	2Dose	2Inc					Citation or Pathology
													Brkly Code
1689	1000	32.2mg	n.s.s.	0/2	1.84mg	0/3	5.52mg	0/5	16.6mg	1/13			
a	1000	119.mg	n.s.s.	0/40	1.84mg	0/32	5.52mg	0/32	16.6mg	1/40			
b	1000	26.5mg	n.s.s.	3/40	1.84mg	5/32	5.52mg	5/32	16.6mg	10/40			
1690	TR21	.878mg	n.s.s.	3/20	.290mg	13/50	.560mg	9/50					
a	TR21	.446mg	n.s.s.	0/20	.290mg	6/50	(.560mg	2/50)					liv:hpa,hpc,nnd.
b	TR21	1.75mg	n.s.s.	0/20	.290mg	2/50	.560mg	2/50					lun:a/a,a/c.
1691	TR21	.399mg	n.s.s.	4/20	.260mg	14/50	.520mg	18/50					
a	TR21	.422mg	n.s.s.	3/20	.260mg	12/50	.520mg	16/50					liv:hpa,hpc,nnd.
b	TR21	1.30mg	n.s.s.	1/20	.260mg	4/50	.520mg	3/50					lun:a/a,a/c.
1692	TR21	.508mg	n.s.s.	17/95p	.260mg	12/50	.520mg	16/50					
1693	20a	2.40mg	10.6mg	9/134	1.25mg	36/148							Davis;txap,4,187-189;1962
a	20a	27.8mg	n.s.s.	3/134	1.25mg	1/148							
b	20a	38.1mg	n.s.s.	0/134	1.25mg	0/148							
1694	22a	1.95mg	6.43mg	27/200	1.25mg	69/200							Epstein(review) {H J Davis};stev,4,1-52;1975
a	22a	16.8mg	n.s.s.	4/200	1.25mg	5/200							
b	22a	1.94mg	6.79mg	30/200	1.25mg	71/200							
c	22a	24.7mg	n.s.s.	21/200	1.25mg	9/200							
1695	89	.298mg	1.27mg	10/44	1.30mg	26/30							Thorpe;fctx,11,433-442;1973
a	89	.842mg	3.48mg	0/44	1.30mg	14/30							
b	89	4.36mg	n.s.s.	27/44	1.30mg	8/30							
1696	103a	.300mg	2.84mg	48/297	12.8ug	23/90	.124mg	30/87	(1.30mg	15/148)			Walker;fctx,11,415-432;1973
a	103a	.481mg	.870mg	39/297	12.8ug	24/90	.124mg	32/87	1.30mg	136/148			
b	103a	1.49mg	2.66mg	0/297	12.8ug	4/90	.124mg	5/87	1.30mg	81/148			
c	103a	.624mg	n.s.s.	18/297	12.8ug	12/90	.124mg	12/87	(1.30mg	0/148)			
1697	103b	.959mg	2.69mg	8/78	.162mg	5/30	.325mg	12/28	.650mg	18/30	1.30mg	9/17	2.60mg 16/21
a	103b	4.87mg	29.5mg	0/78	.162mg	0/30	.325mg	1/28	.650mg	5/30	1.30mg	2/17	2.60mg 3/21
b	103b	2.03mg	n.s.s.	24/78	.162mg	7/30	.325mg	3/28	(.650mg	3/30	1.30mg	1/17	2.60mg 0/21)
c	103b	30.6mg	n.s.s.	8/78	.162mg	0/30	.325mg	0/28	.650mg	1/30	1.30mg	0/17	2.60mg 0/21
1698	103c	.978mg	31.8mg	0/22	1.30mg	5/22							
a	103c	.438mg	n.s.s.	5/22	1.30mg	13/22							
b	103c	2.93mg	n.s.s.	7/22	1.30mg	2/22							
c	103c	4.41mg	n.s.s.	0/22	1.30mg	0/22							
1699	103d	1.20mg	n.s.s.	3/28	1.30mg	8/19							
a	103d	2.98mg	n.s.s.	0/28	1.30mg	2/19							
b	103d	7.71mg	n.s.s.	5/28	1.30mg	0/19							
c	103d	7.71mg	n.s.s.	0/28	1.30mg	0/19							
1700	103e	.548mg	5.32mg	4/24	1.30mg	15/24							
a	103e	1.50mg	51.4mg	0/24	1.30mg	5/24							
b	103e	1.41mg	n.s.s.	4/24	1.30mg	7/24							
c	103e	6.68mg	n.s.s.	3/24	1.30mg	0/24							
1701	89	n.s.s.	.469mg	11/45	1.20mg	30/30							Thorpe;fctx,11,433-442;1973
a	89	.682mg	3.03mg	2/45	1.20mg	16/30							
b	89	2.87mg	n.s.s.	27/45	1.20mg	11/30							
1702	103a	.416mg	.729mg	58/288	11.8ug	32/124	.118mg	34/111	1.15mg	165/176			Walker;fctx,11,415-432;1973
a	103a	1.29mg	2.25mg	12/288	11.8ug	5/124	.118mg	9/111	1.15mg	100/176			
b	103a	.439mg	n.s.s.	95/288	11.8ug	47/124	.118mg	42/111	(1.15mg	32/176)			
c	103a	.844mg	n.s.s.	23/288	11.8ug	14/124	.118mg	13/111	(1.15mg	2/176)			
1703	103b	.714mg	2.09mg	9/78	.150mg	6/30	.300mg	13/30	.600mg	26/30	1.20mg	5/11	2.40mg 12/17
a	103b	2.59mg	9.94mg	0/78	.150mg	2/30	.300mg	1/30	.600mg	3/30	1.20mg	1/11	2.40mg 9/17
b	103b	3.96mg	n.s.s.	45/78	.150mg	17/30	.300mg	11/30	.600mg	14/30	1.20mg	2/11	(2.40mg 1/17)
c	103b	14.7mg	n.s.s.	1/78	.150mg	1/30	.300mg	1/30	.600mg	1/30	1.20mg	0/11	2.40mg 0/17
1704	103c	.420mg	3.87mg	8/23	1.20mg	20/24							
a	103c	1.85mg	n.s.s.	1/23	1.20mg	6/24							
b	103c	4.13mg	n.s.s.	7/23	1.20mg	4/24							
c	103c	8.99mg	n.s.s.	0/23	1.20mg	0/24							
1705	103d	.589mg	n.s.s.	7/30	1.20mg	6/10							
a	103d	1.41mg	n.s.s.	1/30	1.20mg	2/10							
b	103d	2.93mg	n.s.s.	13/30	1.20mg	1/10							
c	103d	3.75mg	n.s.s.	1/30	1.20mg	0/10							
1706	103e	.305mg	3.96mg	10/24	1.06mg	19/22							
a	103e	1.52mg	35.7mg	0/24	1.06mg	5/22							
b	103e	1.93mg	n.s.s.	11/24	1.06mg	8/22							
c	103e	6.86mg	n.s.s.	1/24	1.06mg	0/22							
1707	23	35.4mg	n.s.s.	1/17	22.5ug	4/22	90.0ug	2/23	.450mg	2/18	2.25mg	1/20	4.50mg 0/18
				6.75mg	0/11								Fitzhugh;fctx,2,551-562;1964
a	23	23.3mg	n.s.s.	3/17	22.5ug	8/22	90.0ug	8/23	.450mg	4/18	2.25mg	4/20	4.50mg 3/18
				6.75mg	0/11								
1708	TR21	.635mg	n.s.s.	7/10	1.10mg	39/50	(1.70mg	27/50)					
a	TR21	9.44mg	n.s.s.	0/10	1.10mg	1/50	1.70mg	1/50					liv:hpa,hpc,nnd.
1709	TR21	1.80mg	22.2mg	0/60p	1.10mg	6/50	(1.70mg	2/50)					adr:coa,coc. S
1710	TR22	1.62mg	n.s.s.	17/24	.100mg	17/24	.500mg	16/24	2.50mg	14/24			
a	TR22	n.s.s.	n.s.s.	0/24	.100mg	0/24	.500mg	0/24	2.50mg	0/24			liv:hpa,hpc,nnd.
1711	1004	6.19mg	n.s.s.	0/88	.960mg	0/48	1.44mg	0/41	2.40mg	0/41			Deichmann;imed,39,426-434;1970
a	1004	15.1mg	n.s.s.	60/88	.960mg	23/48	1.44mg	16/41	2.40mg	16/41			
1712	TR21	.774mg	n.s.s.	5/10	.880mg	24/50	1.36mg	22/50					
a	TR21	7.38mg	n.s.s.	1/10	.880mg	0/50	1.36mg	1/50					liv:hpa,hpc,nnd.
1713	TR22	1.64mg	n.s.s.	5/24	80.0ug	9/24	.400mg	7/24	2.00mg	9/24			
a	TR22	n.s.s.	n.s.s.	0/24	80.0ug	0/24	.400mg	0/24	2.00mg	0/24			liv:hpa,hpc,nnd.
1714	1004	4.90mg	n.s.s.	1/75	.768mg	0/48	1.15mg	0/38	1.92mg	0/44			Deichmann;imed,39,426-434;1970
a	1004	8.93mg	n.s.s.	19/75	.768mg	4/48	1.15mg	7/38	(1.92mg	1/44)			
1715	100	.388mg	n.s.s.	18/43	5.00ug	18/23	50.0ug	16/23	.500mg	13/23			Stevenson;txap,36,247-254;1976

```
        Spe Strain  Site   Xpo+Xpt                                                         TD50    2Tailpvl
            Sex   Route  Hist     Notes                                                      DR      AuOp
1716 R m cfe eat tba mix 24m24 e                              .>                           2.01mg * P<.5   -
1717 R f nss eat tba tum 24m24                                      .>                      57.2mg * P<.9   -
1718 R m nss eat tba tum 24m24                                     .>                       40.5mg * P<.8   -

DIELDRIN, PHOTO-              100ng..:..1ug....:..10......:..100....:..1mg...:..10....:..100...:..1g.....:..10
1719 M f b6c eat TBA MXB 80w92                          :>                                  .743mg * P<.6   -
a    M f b6c eat liv MXB 80w92                                                              2.53mg * P<.4
b    M f b6c eat lun MXB 80w92                                                              6.78mg * P<.9
1720 M m b6c eat TBA MXB 80w93                      :>                                      .494mg * P<.7   -
a    M m b6c eat liv MXB 80w93                                                              1.96mg * P<1.
b    M m b6c eat lun MXB 80w93                                                              .465mg * P<.09
1721 R f osm eat mgl MXA 16m26 av                             :  ±                        #.612mg * P<.04
a    R f osm eat TBA MXB 16m26 av                                                           .327mg * P<.4
b    R f osm eat liv MXB 16m26 av                                                           no dre   P=1.
1722 R m osm eat TBA MXB 19m26                            :>                                no dre   P=1.
a    R m osm eat liv MXB 19m26                                                              7.72mg * P<.4
1723 R m osm eat --- hem 19m25     pool                          :    ±                    #2.89mg * P<.02  -

D,L-DIEPOXYBUTANE            100ng..:..1ug....:..10......:..100....:..1mg...:..10....:..100...:..1g.....:..10
1724 R f esd gav sto tum 52w52                             .>                               no dre   P=1.

DIETHYL-beta,gamma-EPOXYPROPYLPHOSPHONATE  ..1ug....:..10......:..100....:..1mg...:..10....:..100...:..1g.....:..10
1725 M f hic ipj lun ptm 64w64                                   .  ±                       12.4mg   P<.02

N,N-DIETHYL-4-(4'-[PYRIDYL-1'-OXIDE]AZO)ANILINE...:..1_0....:..100....:..1mg...:..10....:..100...:..1g.....:..10
1726 R m sda eat liv tum 52w52 bfr                           <+                             noTD50   P<.0005+

O,O-DIETHYL-O-(3,5,6-TRICHLORO-2-PYRIDYL)PHOSPHOROTHIOATE....:..1_00...:..1mg...:..10....:..100...:..1g.....:..10
1727 R f she eat liv tum 24m24 e                       .>                                   no dre   P=1.

a    R f she eat tba mix 24m24 e                                                            no dre   P=1.
1728 R m she eat liv bda 24m24 e                                         .>                 70.6mg * P<.4   -
a    R m she eat tba mix 24m24 e                                                            203.mg * P<1.

DIETHYLACETAMIDE            100ng..:..1ug....:..10......:..100....:..1mg...:..10....:..100...:..1g.....:..10
1729 R m wis gav kid ptc 73w73 e                             .>                             8.85mg   P<.5   +
a    R m wis gav liv tum 73w73 e                                                            no dre   P=1.
b    R m wis gav tba mix 73w73 e                                                            11.8mg   P<.9

DIETHYLACETYLUREA           100ng..:..1ug....:..10......:..100....:..1mg...:..10....:..100...:..1g.....:..10
1730 R m f34 eat kid tla 71w71 e                                          ±                 30.4mg   P<.09
a    R m f34 eat liv hpt 71w71 e                                                            no dre   P=1.

DIETHYLENE GLYCOL           100ng..:..1ug....:..10......:..100....:..1mg...:..10....:..100...:..1g.....:..10
1731 R m osm eat ubl mix 24m24 r                                              .  +  .       1.66gm * P<.002 +

DIETHYLFORMAMIDE            100ng..:..1ug....:..10......:..100....:..1mg...:..10....:..100...:..1g.....:..10
1732 R m wis gav liv tum 73w73 e                                  .>                        no dre   P=1.
a    R m wis gav tba mix 73w73 e                                                            no dre   P=1.

DIETHYLMALEATE              100ng..:..1ug....:..10......:..100....:..1mg...:..10....:..100...:..1g.....:..10
1733 R m f3d eat for tum 51w52 er                                       .>                  no dre   P=1.    -
a    R m f3d eat stg tum 51w52 er                                                           no dre   P=1.    -

DIETHYLSTILBESTROL          100ng..:..1ug....:..10......:..100....:..1mg...:..10....:..100...:..1g.....:..10
1734 M f b62 eat pit ade 33m36 aes              .+.                                         45.6ug * P<.0005+

a    M f b62 eat thy fca 33m36 aes                                                          .186mg Z P<.003 +

b    M f b62 eat cvx sqc 33m36 aes                                                          1.03mg * P<.0005+

c    M f b62 eat ute mso 33m36 aes                                                          1.72mg * P<.0005+

d    M f b62 eat ova gct 33m36 aes                                                          1.74mg * P<.03

e    M f b62 eat pit adc 33m36 aes                                                          17.0mg * P<.8

1735 M m b62 eat tes ict 34m37 aes              .+.                                         66.1ug Z P<.0005+

a    M m b62 eat pit ade 34m37 aes                                                          79.4ug * P<.0005+

b    M m b62 eat pit adc 34m37 aes                                                          4.36mg * P<.4

1736 M f bcn eat ova gct 31m31 ae                  . + .                                    .166mg Z P<.0005+

a    M f bcn eat pit ade 31m31 ae                                                           .200mg Z P<.0005+

b    M f bcn eat cvx sqc 31m31 ae                                                           .311mg * P<.0005+

c    M f bcn eat mgl adb 31m31 ae                                                           .476mg * P<.0005+
```

```
      RefNum LoConf UpConf  Cntrl  1Dose  1Inc  2Dose  2Inc                              Citation or Pathology
                                                                                                    Brkly Code
1716   100  .384mg n.s.s.  12/43  4.00ug  9/23  40.0ug  5/23  .400mg  9/23
1717  1002  3.29mg n.s.s.   6/60  .125mg  7/40  .625mg  7/40  1.25mg  5/40              Cleveland;aenh,13,195-198;1966
1718  1002  4.09mg n.s.s.   3/60  .100mg  1/40  .500mg  3/40  1.00mg  2/40

DIELDRIN, PHOTO-   13366-73-9
1719  TR17  .147mg n.s.s.   3/20  38.0ug  3/50  71.5ug  8/50
a     TR17  .412mg n.s.s.   0/20  38.0ug  0/50  71.5ug  1/50                            liv:hpa,hpc,nnd.
b     TR17  .368mg n.s.s.   1/20  38.0ug  0/50  71.5ug  2/50                            lun:a/a,a/c.
1720  TR17  73.1ug n.s.s.   3/20  34.0ug 14/50  66.0ug 13/50
a     TR17  97.7ug n.s.s.   3/20  34.0ug 10/50  66.0ug 10/50                            liv:hpa,hpc,nnd.
b     TR17  .176mg n.s.s.   0/20  34.0ug  1/50  66.0ug  4/50                            lun:a/a,a/c.
1721  TR17  .327mg n.s.s.   0/10  .120mg  5/50  .200mg  9/50                            mgl:ade,fba.  S
a     TR17  .106mg n.s.s.   4/10  .120mg 29/50  .200mg 29/50
b     TR17  1.30mg n.s.s.   0/10  .120mg  1/50  .200mg  0/50                            liv:hpa,hpc,nnd.
1722  TR17  .263mg n.s.s.   5/10  .140mg 23/50  .290mg 21/50
a     TR17  1.26mg n.s.s.   0/10  .140mg  0/50  .290mg  1/50                            liv:hpa,hpc,nnd.
1723  TR17  .868mg n.s.s.  0/75p  .140mg  0/50  .290mg  3/50                                         S

D,L-DIEPOXYBUTANE   298-18-0
1724   55  .506mg n.s.s.   0/5   2.04mg  0/5                                            Van Duuren;jnci,37,825-838;1966

DIETHYL-beta,gamma-EPOXYPROPYLPHOSPHONATE   7316-37-2
1725  1143  5.23mg n.s.s.  10/30  28.6mg 19/30                                          Van Duuren;jnci,53,695-700;1974

N,N-DIETHYL-4-(4'-[PYRIDYL-1'-OXIDE]AZO)ANILINE   7347-49-1
1726  1176  n.s.s. 1.63mg   0/10  12.0mg 10/10                                          Brown;jnci,37,365-367;1966

O,O-DIETHYL-O-(3,5,6-TRICHLORO-2-PYRIDYL)PHOSPHOROTHIOATE   2921-88-2
1727  1333  35.6ug n.s.s.   0/25  10.0ug  0/25  30.0ug  0/25  .100mg  0/25  1.00mg  0/25  3.00mg  0/25  McCollister;fctx,12,
                                                                                                           45-61;1974
a     1333  4.60mg n.s.s.  12/25  10.0ug  6/25  30.0ug 11/25  .100mg  4/25  1.00mg  8/25  3.00mg  8/25
1728  1333  11.5mg n.s.s.   0/25  10.0ug  0/25  30.0ug  0/25  .100mg  0/25  1.00mg  1/25  3.00mg  0/25
a     1333  6.32mg n.s.s.   4/25  10.0ug  2/25  30.0ug  3/25  .100mg  6/25  1.00mg  3/25  3.00mg  4/25

DIETHYLACETAMIDE   685-91-6
1729   104  1.44mg n.s.s.   0/9   .889mg  1/30                                          Argus;jnci,35,949-958;1965
a      104  2.71mg n.s.s.   0/9   .889mg  0/30
b      104  .854mg n.s.s.   1/9   .889mg  4/30

DIETHYLACETYLUREA   ---
1730  1951  7.47mg n.s.s.   0/15  13.6mg  2/15                                          Diwan;carc,10,189-194;1989
a     1951  19.7mg n.s.s.   0/15  13.6mg  0/15

DIETHYLENE GLYCOL   111-46-6
1731   105  824.mg 5.28gm   0/12  400.mg  0/12  800.mg  6/12  1.60gm  5/12              Fitzhugh;jiht,28,40-43;1946

DIETHYLFORMAMIDE   617-84-5
1732   104  2.14mg n.s.s.   0/9   .780mg  0/27                                          Argus;jnci,35,949-958;1965
a      104  1.40mg n.s.s.   1/9   .780mg  1/27

DIETHYLMALEATE   141-05-9
1733  1921  60.6mg n.s.s.   0/10  78.5mg  0/15                                          Hirose;carc,10,2223-2226;1989
a     1921  60.6mg n.s.s.   0/10  78.5mg  0/15

DIETHYLSTILBESTROL   (DES) 56-53-1
1734  1936r 34.6ug 62.2ug   7/57  650.ng  8/56  1.30ug  9/61  2.60ug 10/53  5.20ug 16/53  20.8ug 41/61
                                  41.6ug 47/62  83.2ug 54/60
                                                                                       Greenman;jtxe,29,269-278;1990/1988
a     1936r 82.4ug 1.49mg   4/58  650.ng  1/45  1.30ug  3/50  2.60ug  2/50  5.20ug  5/55  20.8ug 10/51
                                 (41.6ug  2/49  83.2ug  1/53)
b     1936r .567mg 2.19mg   0/72  650.ng  0/71  1.30ug  0/72  2.60ug  0/72  5.20ug  0/71  20.8ug  1/72
                                  41.6ug  2/68  83.2ug 12/72
c     1936r .810mg 4.69mg   0/67  650.ng  0/71  1.30ug  0/72  2.60ug  0/70  5.20ug  0/70  20.8ug  0/69
                                  41.6ug  2/66  83.2ug  7/71
d     1936r .839mg n.s.s.   0/57  650.ng  1/64  1.30ug  0/63  2.60ug  2/66  5.20ug  0/57  20.8ug  5/62
                                  41.6ug  2/48  83.2ug  3/61
e     1936r 1.24mg n.s.s.   1/57  650.ng  0/56  1.30ug  0/61  2.60ug  0/53  5.20ug  3/53  20.8ug  3/61
                                  41.6ug  1/62  83.2ug  1/60
1735  1936r 52.3ug 84.8ug   1/71  600.ng  1/70  1.20ug  2/70  2.40ug  1/72  4.80ug  1/70  19.2ug 15/68
                                  38.4ug 56/68  76.8ug 60/71
a     1936r 61.2ug .107mg   0/51  600.ng  1/55  1.20ug  2/52  2.40ug  2/53  4.80ug  6/51  19.2ug 25/56
                                  38.4ug 34/59  76.8ug 38/57
b     1936r 1.01mg n.s.s.   0/51  600.ng  0/55  1.20ug  0/52  2.40ug  0/53  4.80ug  1/51  19.2ug  2/56
                                  38.4ug  2/59  76.8ug  0/57
1736  1936n .104mg .291mg   1/36  650.ng  0/65  1.30ug  0/55  2.60ug  0/56  5.20ug  0/63  20.8ug 10/69
                                  41.6ug 17/57 (83.2ug 14/57)
a     1936n .137mg .307mg   0/48  650.ng  0/53  1.30ug  1/51  2.60ug  0/48  5.20ug  0/50  20.8ug  3/53
                                  41.6ug  3/52  83.2ug 35/65
b     1936n .205mg .512mg   1/68  650.ng  0/72  1.30ug  1/67  2.60ug  0/68  5.20ug  1/68  20.8ug  2/72
                                  41.6ug  9/64  83.2ug 23/72
c     1936n .259mg 1.29mg   1/58  650.ng  1/62  1.30ug  4/62  2.60ug  4/57  5.20ug  3/57  20.8ug  6/62
                                  41.6ug  4/59  83.2ug 16/67
```

	Spe	Sex	Strain	Route	Site	Hist	Xpo+Xpt	Notes	DR	TD50	2Tailpvl
d	M	f	bcn	eat	cvx	ado	31m31	ae		1.28mg *	P<.0005+
e	M	f	bcn	eat	vag	sqc	31m31	ae		1.31mg *	P<.003
f	M	f	bcn	eat	mgl	ado	31m31	ae		1.88mg *	P<.004 +
g	M	f	bcn	eat	pit	adc	31m31	ae		5.10mg *	P<.2
h	M	f	bcn	eat	ute	adc	31m31	ae		5.32mg *	P<.3
1737	M	m	bcn	eat	tes	ict	28m29	aes	.+.	21.5ug Z	P<.0005+
a	M	m	bcn	eat	pit	ade	28m29	aes		.171mg Z	P<.0005+
1738	M	m	c3c	eat	mgl	car	24m24	er	.+.	35.9ug *	P<.0005+
1739	M	m	c3c	eat	mgl	car	24m24	er	. + .	42.1ug \	P<.0005+
1740	M	m	c3c	eat	mgl	car	24m24	er	.+ .	78.9ug *	P<.0005+
1741	M	f	c3h	eat	mgl	car	85w85	r	. + .	29.2ug *	P<.0005+
1742	M	f	c3h	eat	mgl	adc	24m24	r	. + .	26.0ug	P<.0005+
1743	M	f	c3j	eat	mgl	adc	52w52	ek	. ±	82.5ug *	P<.03
1744	M	f	c3j	eat	ova	tua	78w78	ek	.>	22.2ug *	P<.3
a	M	f	c3j	eat	mgl	adc	78w78	ek		no dre	P=1.
1745	M	f	c3j	eat	mgl	adc	24m24	ek	.>	14.2mg *	P<1.
a	M	f	c3j	eat	ova	tua	24m24	ek		no dre	P=1.
1746	M	f	c3v	eat	mgl	mix	25m25	e	.+ .	24.7ug	P<.0005+
a	M	f	c3v	eat	mgl	adb	25m25	e		30.2ug	P<.0005
b	M	f	c3v	eat	mgl	ada	25m25	e		2.63mg	P<.5
c	M	f	c3v	eat	lun	act	25m25	e		no dre	P=1.
d	M	f	c3v	eat	liv	ade	25m25	e		no dre	P=1.
e	M	f	c3v	eat	liv	adc	25m25	e		no dre	P=1.
1747	M	f	c3v	eat	mgl	mix	26m26	e	.+.	42.7ug	P<.0005+
a	M	f	c3v	eat	mgl	adb	26m26	e		43.4ug	P<.0005
b	M	f	c3v	eat	---	mso	26m26	e		.928mg	P<.003
c	M	f	c3v	eat	ute	ena	26m26	e		2.05mg	P<.05
d	M	f	c3v	eat	cvu	adc	26m26	e		2.47mg	P<.07
e	M	f	c3v	eat	liv	ade	26m26	e		no dre	P=1.
f	M	f	c3v	eat	mgl	ada	26m26	e		no dre	P=1.
g	M	f	c3v	eat	liv	adc	26m26	e		no dre	P=1.
h	M	f	c3v	eat	lun	act	26m26	e		no dre	P=1.
1748	M	f	c3v	eat	mgl	mix	29m29	e	.+.	87.2ug	P<.0005+
a	M	f	c3v	eat	mgl	adb	29m29	e		95.6ug	P<.0005
b	M	f	c3v	eat	---	mso	29m29	e		.520mg	P<.0005
c	M	f	c3v	eat	cvu	adc	29m29	e		1.17mg	P<.004
d	M	f	c3v	eat	ute	ena	29m29	e		1.71mg	P<.02
e	M	f	c3v	eat	mgl	ada	29m29	e		2.10mg	P<.3
f	M	f	c3v	eat	liv	ade	29m29	e		no dre	P=1.
g	M	f	c3v	eat	liv	adc	29m29	e		no dre	P=1.
h	M	f	c3v	eat	lun	act	29m29	e		no dre	P=1.
1749	M	f	c3v	eat	mgl	mix	29m29	e	.+ .	.115mg	P<.0005+
a	M	f	c3v	eat	mgl	adb	29m29	e		.118mg	P<.0005
b	M	f	c3v	eat	---	mso	29m29	e		.487mg	P<.0005
c	M	f	c3v	eat	cvu	adc	29m29	e		1.66mg	P<.02
d	M	f	c3v	eat	ute	ena	29m29	e		1.87mg	P<.02
e	M	f	c3v	eat	mgl	ada	29m29	e		4.39mg	P<.5
f	M	f	c3v	eat	liv	ade	29m29	e		no dre	P=1.
g	M	f	c3v	eat	liv	adc	29m29	e		no dre	P=1.
h	M	f	c3v	eat	lun	act	29m29	e		no dre	P=1.
1750	M	f	c7b	eat	pit	ade	34m37	aes	.+ .	38.7ug *	P<.0005+
a	M	f	c7b	eat	cvx	sqc	34m37	aes		1.98mg *	P<.0005+
b	M	f	c7b	eat	pit	adc	34m37	aes		2.01mg *	P<.03
c	M	f	c7b	eat	mgl	adb	34m37	aes		no dre	P=1. +
d	M	f	c7b	eat	thy	fca	34m37	aes		no dre	P=1.
1751	M	m	c7b	eat	tes	ict	35m39	aes	.+.	40.8ug Z	P<.0005+
a	M	m	c7b	eat	pit	ade	35m39	aes		72.2ug Z	P<.0005+
b	M	m	c7b	eat	pit	adc	35m39	aes		no dre	P=1.
1752	M	f	cb6	eat	pit	ade	27m33	aes	. + .	25.9ug *	P<.0005+
a	M	f	cb6	eat	pit	adc	27m33	aes		no dre	P=1.
b	M	f	cb6	eat	thy	fct	27m33	aes		no dre	P=1. +

```
      RefNum  LoConf UpConf   Cntrl  1Dose   1Inc  2Dose   2Inc                               Citation or Pathology
                                                                                                                      Brkly Code
 d    1936n  .605mg 3.50mg   0/63  650.ng  0/71  1.30ug  0/66  2.60ug  0/66  5.20ug  0/68  20.8ug  1/71
                                    41.6ug  2/64  83.2ug  6/70
 e    1936n  .550mg 10.6mg   0/65  650.ng  0/71  1.30ug  0/64  2.60ug  2/65  5.20ug  1/68  20.8ug  1/71
                                    41.6ug  4/64  83.2ug  4/70
 f    1936n  .718mg 19.3mg   0/58  650.ng  0/62  1.30ug  1/62  2.60ug  0/57  5.20ug  0/57  20.8ug  1/62
                                    41.6ug  1/59  83.2ug  4/67
 g    1936n  1.26mg n.s.s.   0/48  650.ng  0/53  1.30ug  0/51  2.60ug  0/48  5.20ug  0/50  20.8ug  1/53
                                    41.6ug  0/52  83.2ug  1/65
 h    1936n  1.19mg n.s.s.   0/63  650.ng  0/71  1.30ug  0/66  2.60ug  2/66  5.20ug  0/68  20.8ug  0/71
                                    41.6ug  2/64  83.2ug  1/70
1737  1936n  17.0ug 27.2ug   0/72  600.ng  2/70  1.20ug  2/64  2.40ug  3/68  4.80ug  2/71  19.2ug  40/69
                                    38.4ug  65/70  76.8ug  69/69
 a    1936n  95.3ug .373mg   0/50  600.ng  0/52  1.20ug  1/47  2.40ug  0/46  4.80ug  0/46  19.2ug  10/50
                                    38.4ug  5/40  (76.8ug  5/56)
1738  109m   27.9ug 46.9ug   0/78  30.0ug  48/92  60.0ug  58/94                              Okey;jnci,40,225-230;1968
1739  109n   27.7ug 68.1ug   0/78  30.0ug  34/88  (60.0ug  35/92)
1740  109o   57.1ug .113mg   0/78  30.0ug  26/93  60.0ug  32/89
1741  106a   19.4ug 48.1ug   40/121  813.ng  27/56  1.63ug  26/60  3.25ug  26/60  6.50ug  36/68  13.0ug  42/64
                                    65.0ug  50/59  .130mg  50/58                             Gass;jnci,33,971-977;1964
1742  1131   15.7ug 55.6ug   16/64  32.5ug  45/66                                            Gass;ircs,5,477;1977
1743  1468m  30.2ug n.s.s.   2/43  1.30ug  0/29  13.0ug  3/35  65.0ug  6/41                  Highman;jept,4,81-95;1980/pers.comm.
1744  1468n  5.68ug n.s.s.   2/14  1.30ug  5/24  13.0ug  6/18
 a    1468n  11.8ug n.s.s.   1/13  1.30ug  7/22  13.0ug  2/16
1745  1468o  35.5ug n.s.s.   4/24  1.30ug  10/38  13.0ug  3/9  65.0ug  1/5
 a    1468o  35.7ug n.s.s.   12/24  1.30ug  19/40  13.0ug  7/11  65.0ug  2/6
1746  1852m  19.0ug 32.3ug   4/73  83.2ug  167/182                                           Greenman;jnci,77,891-898;1986/pers.comm.
 a    1852m  23.7ug 38.9ug   2/73  83.2ug  158/182
 b    1852m  .673mg n.s.s.   2/73  83.2ug  9/182
 c    1852m  1.97mg n.s.s.   5/75  83.2ug  3/182
 d    1852m  3.29mg n.s.s.   3/77  83.2ug  0/181
 e    1852m  2.93mg n.s.s.   10/77  83.2ug  1/181
1747  1852n  33.7ug 55.3ug   4/73  83.2ug  151/189
 a    1852n  34.3ug 55.6ug   2/73  83.2ug  149/189
 b    1852n  .490mg 4.06mg   0/77  83.2ug  13/189
 c    1852n  .836mg n.s.s.   0/77  83.2ug  6/189
 d    1852n  .936mg n.s.s.   0/77  83.2ug  5/189
 e    1852n  3.74mg n.s.s.   3/77  83.2ug  0/188
 f    1852n  2.13mg n.s.s.   2/73  83.2ug  2/189
 g    1852n  2.24mg n.s.s.   10/77  83.2ug  5/188
 h    1852n  1.71mg n.s.s.   5/75  83.2ug  5/189
1748  1852o  67.3ug .118mg   4/73  83.2ug  117/185
 a    1852o  73.9ug .128mg   2/73  83.2ug  109/185
 b    1852o  .331mg .932mg   0/77  83.2ug  28/191
 c    1852o  .616mg 5.24mg   0/77  83.2ug  13/191
 d    1852o  .804mg n.s.s.   0/77  83.2ug  9/191
 e    1852o  .737mg n.s.s.   2/73  83.2ug  12/185
 f    1852o  2.05mg n.s.s.   3/77  83.2ug  4/191
 g    1852o  1.85mg n.s.s.   10/77  83.2ug  10/191
 h    1852o  1.70mg n.s.s.   5/75  83.2ug  7/191
1749  1852r  86.3ug .164mg   4/73  83.2ug  96/182
 a    1852r  89.3ug .163mg   2/73  83.2ug  92/182
 b    1852r  .312mg .860mg   0/77  83.2ug  29/192
 c    1852r  .783mg n.s.s.   0/77  83.2ug  9/192
 d    1852r  .849mg n.s.s.   0/77  83.2ug  8/192
 e    1852r  1.04mg n.s.s.   2/96  83.2ug  7/182
 f    1852r  2.00mg n.s.s.   3/77  83.2ug  4/192
 g    1852r  1.81mg n.s.s.   10/77  83.2ug  10/192
 h    1852r  1.00mg n.s.s.   5/75  83.2ug  12/192
1750  1936o  29.6ug 51.9ug   4/55  650.ng  6/55  1.30ug  10/54  2.60ug  14/55  5.20ug  19/63  20.8ug  50/67
                                    41.6ug  54/64  83.2ug  59/64                             Greenman;jtxe,29,269-278;1990/1988
 a    1936o  .933mg 5.40mg   0/71  650.ng  0/68  1.30ug  0/71  2.60ug  0/67  5.20ug  0/68  20.8ug  0/72
                                    41.6ug  4/72  83.2ug  5/72
 b    1936o  .765mg n.s.s.   0/55  650.ng  0/55  1.30ug  0/54  2.60ug  1/55  5.20ug  2/63  20.8ug  2/67
                                    41.6ug  5/64  83.2ug  2/64
 c    1936o  .647mg n.s.s.   0/56  650.ng  0/54  1.30ug  4/58  2.60ug  4/61  5.20ug  5/64  20.8ug  5/61
                                    41.6ug  1/65  (83.2ug  0/61)
 d    1936o  1.26mg n.s.s.   1/43  650.ng  1/45  1.30ug  1/50  2.60ug  0/49  5.20ug  1/50  20.8ug  4/54
                                    41.6ug  2/61  83.2ug  0/42
1751  1936o  32.4ug 51.9ug   0/72  600.ng  1/66  1.20ug  1/70  2.40ug  1/65  4.80ug  1/65  19.2ug  37/67
                                    38.4ug  70/71  76.8ug  65/67
 a    1936o  53.5ug .101mg   1/52  600.ng  1/49  1.20ug  1/58  2.40ug  2/53  4.80ug  3/47  19.2ug  32/60
                                    38.4ug  36/64  (76.8ug  24/57)
 b    1936o  1.04mg n.s.s.   0/52  600.ng  0/49  1.20ug  0/58  2.40ug  1/53  4.80ug  0/47  19.2ug  8/60
                                    38.4ug  0/64  76.8ug  0/57
1752  1936m  17.9ug 39.3ug   21/57  650.ng  22/60  1.30ug  21/65  2.60ug  32/63  5.20ug  49/65  20.8ug  46/53
                                    41.6ug  59/62  83.2ug  58/62
 a    1936m  1.65mg n.s.s.   1/57  650.ng  1/60  1.30ug  4/65  2.60ug  0/63  5.20ug  3/65  20.8ug  1/53
                                    41.6ug  0/62  83.2ug  2/62
 b    1936m  .504mg n.s.s.   13/64  650.ng  11/56  1.30ug  10/51  2.60ug  10/55  5.20ug  16/61  20.8ug  14/48
                                    41.6ug  4/60  (83.2ug  0/50)
```

```
      Spe Strain  Site   Xpo+Xpt                                              TD50    2Tailpvl
          Sex   Route   Hist     Notes                                        DR      AuOp
───────────────────────────────────────────────────────────────────────────────────────────
1753 M m cb6 eat pit ade 29m36 aes                .+.                         17.6ug Z P<.0005+

  a   M m cb6 eat thy fct 29m36 aes                                           50.0ug Z P<.0005+

  b   M m cb6 eat tes ict 29m36 aes                                           .257mg Z P<.0005+

  c   M m cb6 eat pit adc 29m36 aes                                           1.42mg * P<.0005

1754 M f cbj eat mgl tum 52w52 ek           .>                                no dre   P=1.
1755 M f cbj eat mgl adc 78w78 ek                        .    ±               .295mg * P<.04
  a   M f cbj eat ova tua 78w78 ek                                            .329mg * P<.3
1756 M f cbj eat ova tua 24m24 ek                    .>                       .597mg * P<.8
1757 M f cbj eat ova tua 30m30 ek           .>                                no dre   P=1.
  a   M f cbj eat mgl adc 30m30 ek                                            no dre   P=1.
1758 R f cdr eat liv hpt 21m24 aes                 .   ±                      .130mg \ P<.1   -
1759 R m cdr eat liv tum 21m24 aes                   .>                       no dre   P=1.   -
1760 R m cdr eat pit cra 66w66                   . + .                        .114mg   P<.002 +
  a   R m cdr eat adr coa 66w66                                               .390mg   P<.07  +
  b   R m cdr eat liv nod 66w66                                               .390mg   P<.07
  c   R m cdr eat liv car 66w66                                               1.21mg   P<.3

N,N'-DIETHYLTHIOUREA        100ng..:..1ug....:..10......:..100...:..1mg....:..10....:..100....:..1g......:..10
1761 M f b6c eat TBA MXB 24m24                                     :>         no dre   P=1.   -
  a   M f b6c eat liv MXB 24m24                                               687.mg * P<.3
  b   M f b6c eat lun MXB 24m24                                               no dre   P=1.
1762 M m b6c eat TBA MXB 24m24                                        :>      no dre   P=1.   -
  a   M m b6c eat liv MXB 24m24                                               no dre   P=1.
  b   M m b6c eat lun MXB 24m24                                               no dre   P=1.
1763 R f f34 eat thy MXA 24m24                               :  +  :          23.8mg * P<.0005c
  a   R f f34 eat thy fcc 24m24                                               57.9mg * P<.008 c
  b   R f f34 eat TBA MXB 24m24                                               21.5mg * P<.3
  c   R f f34 eat liv MXB 24m24                                               no dre   P=1.
1764 R m f34 eat thy MXA 24m24                               :  +  :          24.3mg / P<.0005c
  a   R m f34 eat thy fcc 24m24                                               33.6mg * P<.003 c
  b   R m f34 eat TBA MXB 24m24                                               12.9mg * P<.04
  c   R m f34 eat liv MXB 24m24                                               no dre   P=1.
1765 R m f3d eat thy fcc 52w52 r                    .>                        28.1mg   P<.3
  a   R m f3d eat liv tum 52w52 r                                             no dre   P=1.

1,2-DIFORMYLHYDRAZINE       100ng..:..1ug....:..10......:..100...:..1mg....:..10....:..100....:..1g......:..10
1766 M f swa wat lun ade 83w83 e                            . + .             561.mg   P<.0005
  a   M f swa wat lun mix 83w83 e                                             571.mg   P<.0005+
  b   M f swa wat lun adc 83w83 e                                             2.17gm   P<.0005
  c   M f swa wat liv hpt 83w83 e                                             7.82gm   P<.07  -
  d   M f swa wat liv mix 83w83 e                                             89.2gm   P<.8
1767 M m swa wat lun ade 83w83 e                               . + .          779.mg   P<.0005
  a   M m swa wat lun mix 83w83 e                                             806.mg   P<.0005+
  b   M m swa wat lun adc 83w83 e                                             4.53gm   P<.0005
  c   M m swa wat liv mix 83w83 e                                             8.14gm   P<.02
  d   M m swa wat liv hpt 83w83 e                                             11.3gm   P<.02  -

DIFTALONE                   100ng..:..1ug....:..10......:..100...:..1mg....:..10....:..100....:..1g......:..10
1768 M f bld eat liv ang 19m28 e                          . + .              852.mg * P<.002 +
  a   M f bld eat liv hpa 19m28 e                                            1.15gm * P<.009 +
  b   M f bld eat lun mix 19m28 e                                            269.mg * P<.04  -
  c   M f bld eat liv hpc 19m28 e                                            3.54gm * P<.08  +
  d   M f bld eat liv agm 19m28 e                                            7.13gm * P<.3
1769 M m bld eat liv ang 19m28 e                           . + .             879.mg Z P<.0005+
  a   M m bld eat liv agm 19m28 e                                            1.37gm Z P<.03  +
  b   M m bld eat lun mix 19m28 e                                            412.mg * P<.3   -
  c   M m bld eat liv hpc 19m28 e                                            no dre   P=1.
  d   M m bld eat liv hpa 19m28 e                                            no dre   P=1.

DIGLYCIDYL RESORCINOL ETHER, TECHNICAL GRADE  ....:..10......:..100...:..1mg....:..10......:..100....:..1g......:..10
1770 M f b6c gav sto MXA 24m24 s                       :  +  :               17.9mg * P<.0005c
  a   M f b6c gav sto sqc 24m24 s                                            36.3mg * P<.0005c
  b   M f b6c gav sto MXA 24m24 s                                            48.2mg * P<.0005c
  c   M f b6c gav liv hpc 24m24 s                                            221.mg * P<.02
  d   M f b6c gav TBA MXB 24m24 s                                            32.3mg * P<.007
  e   M f b6c gav liv MXB 24m24 s                                            169.mg * P<.06
  f   M f b6c gav lun MXB 24m24 s                                            4.79gm * P<1.
1771 M m b6c gav sto MXA 24m24 s                             :+ :            37.8mg * P<.0005c
  a   M m b6c gav sto sqc 24m24 s                                            55.5mg * P<.0005c
  b   M m b6c gav sto MXA 24m24 s                                            145.mg * P<.0005c
  c   M m b6c gav TBA MXB 24m24 s                                            36.6mg \ P<.2
  d   M m b6c gav liv MXB 24m24 s                                            84.7mg \ P<.3
  e   M m b6c gav lun MXB 24m24 s                                            1.35gm * P<.8
1772 R f f34 gav sto MXB 24m24 s                         :  +  :             4.47mg * P<.0005
  a   R f f34 gav sto sqc 24m24 s                                            5.10mg * P<.0005c
  b   R f f34 gav sto sqp 24m24 s                                            45.5mg * P<.0005c
  c   R f f34 gav TBA MXB 24m24 s                                            8.53mg * P<.004
```

	RefNum	LoConf	UpConf	Cntrl	1Dose	1Inc	2Dose	2Inc							Citation or Pathology
															Brkly Code
1753	1936m	13.3ug	23.7ug	0/52	600.ng	2/57	1.20ug	3/57	2.40ug	13/59	4.80ug	21/59	19.2ug	51/64	
					(38.4ug	39/62	76.8ug	22/43)							
a	1936m	32.7ug	84.1ug	2/48	600.ng	0/51	1.20ug	1/47	2.40ug	3/48	4.80ug	6/51	19.2ug	27/54	
					(38.4ug	3/58	76.8ug	0/43)							
b	1936m	.155mg	.489mg	0/70	600.ng	1/71	1.20ug	0/66	2.40ug	0/69	4.80ug	1/69	19.2ug	3/64	
					38.4ug	17/61	(76.8ug	3/52)							
c	1936m	.592mg	6.74mg	0/52	600.ng	1/57	1.20ug	1/57	2.40ug	0/59	4.80ug	1/59	19.2ug	0/64	
					38.4ug	2/62	76.8ug	5/43							
1754	1468r	2.06ug	n.s.s.	0/17	1.30ug	0/38	13.0ug	0/18	65.0ug	0/30					Highman;jept,4,81-95;1980/pers.comm.
1755	1468s	89.2ug	n.s.s.	0/31	1.30ug	0/40	13.0ug	1/33	65.0ug	2/29					
a	1468s	70.9ug	n.s.s.	0/34	1.30ug	5/42	13.0ug	3/34	65.0ug	4/31					
1756	1468t	48.1ug	n.s.s.	9/29	1.30ug	10/31	13.0ug	12/36	65.0ug	4/11					
1757	1468u	2.14ug	n.s.s.	4/11	1.30ug	3/12									
a	1468u	3.54ug	n.s.s.	3/11	1.30ug	1/12									
1758	108	32.0ug	n.s.s.	0/20	20.0ug	2/20	(.200mg	0/20)							Gibson;txap,11,489-510;1967
1759	108	74.9ug	n.s.s.	0/20	20.0ug	0/20	.200mg	0/20							
1760	333	53.3ug	.391mg	0/20	.160mg	9/28									Newberne;aenh,19,489-498;1969
a	333	.118mg	n.s.s.	0/20	.160mg	3/28									
b	333	.118mg	n.s.s.	0/20	.160mg	3/28									
c	333	.198mg	n.s.s.	0/20	.160mg	1/28									

N,N'-DIETHYLTHIOUREA　　105-55-5

	RefNum	LoConf	UpConf	Cntrl	1Dose	1Inc	2Dose	2Inc	Pathology
1761	TR149	102.mg	n.s.s.	11/20	32.2mg	23/50	64.3mg	15/50	
a	TR149	208.mg	n.s.s.	0/20	32.2mg	1/50	64.3mg	2/50	liv:hpa,hpc,nnd.
b	TR149	472.mg	n.s.s.	1/20	32.2mg	1/50	64.3mg	0/50	lun:a/a,a/c.
1762	TR149	77.7mg	n.s.s.	8/19	29.8mg	21/50	59.4mg	23/50	
a	TR149	94.1mg	n.s.s.	5/19	29.8mg	7/50	(59.4mg	3/50)	liv:hpa,hpc,nnd.
b	TR149	163.mg	n.s.s.	2/19	29.8mg	4/50	59.4mg	6/50	lun:a/a,a/c.
1763	TR149	14.2mg	55.0mg	0/20	6.20mg	4/50	12.4mg	17/50	thy:fca,fcc.
a	TR149	27.3mg	774.mg	0/20	6.20mg	1/50	12.4mg	8/50	
b	TR149	6.90mg	n.s.s.	13/20	6.20mg	33/50	12.4mg	41/50	
c	TR149	n.s.s.	n.s.s.	0/20	6.20mg	0/50	12.4mg	0/50	liv:hpa,hpc,nnd.
1764	TR149	13.5mg	50.0mg	0/20	5.00mg	1/50	9.90mg	15/50	thy:fca,fcc.
a	TR149	17.3mg	136.mg	0/20	5.00mg	1/50	9.90mg	11/50	
b	TR149	6.15mg	n.s.s.	5/20	5.00mg	23/50	9.90mg	31/50	
c	TR149	n.s.s.	n.s.s.	0/20	5.00mg	0/50	9.90mg	0/50	liv:hpa,hpc,nnd.
1765	2024	4.57mg	n.s.s.	0/20	8.00mg	1/21			Hasegawa;carc,12,1515-1518;1991/pers.comm.
a	2024	8.66mg	n.s.s.	0/20	8.00mg	0/21			

1,2-DIFORMYLHYDRAZINE　　628-36-4

	RefNum	LoConf	UpConf	Cntrl	1Dose	1Inc	Citation
1766	491	307.mg	987.mg	10/100	4.00gm	48/50	Toth;zkko,92,11-16;1978
a	491	310.mg	1.02gm	15/100	4.00gm	48/50	
b	491	1.33gm	3.98gm	6/100	4.00gm	29/50	
c	491	1.26gm	n.s.s.	0/19	4.00gm	1/5	
d	491	7.47gm	n.s.s.	3/46	4.00gm	3/36	
1767	491	469.mg	1.38gm	17/98	3.33gm	41/47	
a	491	479.mg	1.48gm	22/98	3.33gm	41/47	
b	491	2.25gm	14.5gm	6/98	3.33gm	15/47	
c	491	2.86gm	n.s.s.	3/69	3.33gm	6/30	
d	491	3.62gm	n.s.s.	1/69	3.33gm	4/30	

DIFTALONE　　21626-89-1

	RefNum	LoConf	UpConf	Cntrl	1Dose	1Inc	2Dose	2Inc	3Dose	3Inc	Citation
1768	1638	386.mg	2.95gm	0/38	26.2mg	0/43	52.4mg	1/45	105.mg	7/43	Della Porta;zkko,108,308-311;1984
a	1638	471.mg	39.9gm	0/38	26.2mg	1/43	52.4mg	0/45	105.mg	5/43	
b	1638	118.mg	n.s.s.	8/38	26.2mg	13/43	52.4mg	19/45	105.mg	18/43	
c	1638	871.mg	n.s.s.	0/38	26.2mg	0/43	52.4mg	0/45	105.mg	2/43	
d	1638	1.16gm	n.s.s.	0/38	26.2mg	0/43	52.4mg	0/45	105.mg	1/43	
1769	1638	398.mg	2.78gm	0/40	24.2mg	0/48	48.4mg	0/50	96.8mg	8/48	
a	1638	507.mg	n.s.s.	1/40	24.2mg	0/48	48.4mg	0/50	96.8mg	6/48	
b	1638	122.mg	n.s.s.	13/40	24.2mg	19/48	48.4mg	26/50	96.8mg	21/48	
c	1638	847.mg	n.s.s.	2/40	24.2mg	1/48	48.4mg	3/50	96.8mg	1/48	
d	1638	563.mg	n.s.s.	3/40	24.2mg	5/48	48.4mg	2/50	96.8mg	4/48	

DIGLYCIDYL RESORCINOL ETHER, TECHNICAL GRADE　　101-90-6

	RefNum	LoConf	UpConf	Cntrl	1Dose	1Inc	2Dose	2Inc	Pathology	Code
1770	TR257	11.9mg	27.9mg	0/50	35.0mg	17/50	70.1mg	33/50	sto:pas,sqc,sqp.	
a	TR257	23.1mg	59.8mg	0/50	35.0mg	12/50	70.1mg	23/50		
b	TR257	25.4mg	113.mg	0/50	35.0mg	5/50	70.1mg	10/50	sto:pas,sqp.	
c	TR257	74.9mg	n.s.s.	0/50	35.0mg	1/50	70.1mg	3/50		S
d	TR257	15.5mg	484.mg	28/50	35.0mg	27/50	70.1mg	38/50		
e	TR257	58.0mg	n.s.s.	3/50	35.0mg	1/50	70.1mg	7/50	liv:hpa,hpc,nnd.	
f	TR257	112.mg	n.s.s.	3/50	35.0mg	3/50	70.1mg	2/50	lun:a/a,a/c.	
1771	TR257	26.5mg	56.5mg	0/50	35.0mg	17/50	70.4mg	33/50	sto:pas,sqc,sqp.	
a	TR257	37.4mg	88.8mg	0/50	35.0mg	14/50	70.4mg	25/50		
b	TR257	78.0mg	423.mg	0/50	35.0mg	4/50	70.4mg	10/50	sto:pas,sqp.	
c	TR257	12.6mg	n.s.s.	31/50	35.0mg	40/50	(70.4mg	41/50)		
d	TR257	24.7mg	n.s.s.	13/50	35.0mg	18/50	(70.4mg	11/50)	liv:hpa,hpc,nnd.	
e	TR257	156.mg	n.s.s.	6/50	35.0mg	2/50	70.4mg	8/50	lun:a/a,a/c.	
1772	TR257	2.88mg	7.26mg	0/50	17.7mg	38/50	35.4mg	4/50	sto:sqc,sqp.	C
a	TR257	3.23mg	8.51mg	0/50	17.7mg	34/50	35.4mg	3/50		
b	TR257	19.4mg	141.mg	0/50	17.7mg	7/50	35.4mg	1/50		
c	TR257	4.09mg	64.0mg	46/50	17.7mg	41/50	35.4mg	6/50		

```
     Spe Strain  Site   Xpo+Xpt                                          TD50      2Tailpvl
         Sex  Route  Hist       Notes                                        DR      AuOp

  d   R f f34 gav liv MXB 24m24 s                                        no dre   P=1.
1773  R f f34 gav for MXB 24m24                        : + :            3.53mg   P<.0005
  a   R f f34 gav for sqc 24m24                                         5.93mg   P<.0005c
  b   R f f34 gav for sqp 24m24                                         8.95mg   P<.0005c
  c   R f f34 gav TBA MXB 24m24                                         16.0mg   P<.4
  d   R f f34 gav liv MXB 24m24                                         174.mg   P<.6
1774  R m f34 gav tes ict 23m24 a                    : + :             2.35mg / P<.0005
  a   R m f34 gav sto MXB 23m24 a                                      2.89mg * P<.0005
  b   R m f34 gav sto sqc 23m24 a                                      3.06mg * P<.0005c
  c   R m f34 gav pit adn 23m24 a                                      12.2mg * P<.008
  d   R m f34 gav sto sqp 23m24 a                                      24.7mg * P<.0005c
  e   R m f34 gav TBA MXB 23m24 a                                      2.43mg * P<.0005
  f   R m f34 gav liv MXB 23m24 a                                      114.mg * P<.3
1775  R m f34 gav for MXB 24m24                        : + :            2.33mg   P<.0005
  a   R m f34 gav for sqc 24m24                                         2.73mg   P<.0005c
  b   R m f34 gav for sqp 24m24                                         8.16mg   P<.0005c
  c   R m f34 gav TBA MXB 24m24                                         4.61mg   P<.006
  d   R m f34 gav liv MXB 24m24                                         640.mg   P<.9

5,6-DIHYDRO-5-AZACYTIDINE        100ng..:..1ug....:..10.....:..100....:..1mg....:..10.....:..100....:..1g.....:..10
1776  R m f34 ipj liv tum 52w52 e                                .>         no dre   P=1.
  a   R m f34 ipj tba tum 52w52 e                                          20.7mg   P<.5

1,2-DIHYDRO-2-(5-NITRO-2-THIENYL)QUINAZOLIN-4(3H)-ONE.1_0....:..100....:..1mg....:..10.....:..100....:..1g.....:..10
1777  R f sda eat mgl adc 46w66 e                           .  +  .        13.3mg   P<.0005
  a   R f sda eat tba mix 46w66 e                                          1.53mg   P<.0005+

3,6-DIHYDRO-2-NITROSO-2H-1,2-OXAZINE    ..:..1ug...:..10.....:..100....:..1mg....:..10.....:..100....:..1g.....:..10
1778  R b sda wat liv hpt 20m23                                  .>        417.mg   P<.3    -
  a   R b sda wat tba mal 20m23                                           90.6mg   P<.07   +
  b   R b sda wat tba mix 20m23                                           110.mg   P<.3

3,4-DIHYDROCOUMARIN              100ng..:..1ug....:..10.....:..100....:..1mg....:..10.....:..100....:..1g.....:..10
1779  M f b6c gav --- MXA 24m24                                      :   +  4.81gm * P<.007
  a   M f b6c gav --- hes 24m24                                            6.14gm * P<.009
  b   M f b6c gav liv MXA 24m24                                           723.mg * P<.02   p
  c   M f b6c gav liv hpa 24m24                                           790.mg * P<.03   p
  d   M f b6c gav TBA MXB 24m24                                           541.mg * P<.02
  e   M f b6c gav liv MXB 24m24                                           723.mg * P<.02
  f   M f b6c gav lun MXB 24m24                                           96.1gm * P<1.
1780  M m b6c gav --- MXA 24m24                                      :   + #3.78gm * P<.006 -
  a   M m b6c gav lun a/a 24m24                                           596.mg   Z P<.05
  b   M m b6c gav --- MXA 24m24                                           3.32gm * P<.02
  c   M m b6c gav TBA MXB 24m24                                           1.40gm * P<.5
  d   M m b6c gav liv MXB 24m24                                           2.89gm * P<.7
  e   M m b6c gav lun MXB 24m24                                           3.26gm * P<.5
1781  R f f34 gav amd MXA 24m24                                  :   ±   #538.mg   Z P<.03 -
  a   R f f34 gav TBA MXB 24m24                                           3.18gm * P<.9
  b   R f f34 gav liv MXB 24m24                                           no dre   P=1.
1782  R m f34 gav tes ica 24m24 s                              : + :      70.6mg * P<.0005
  a   R m f34 gav kid rua 24m24 s                                         3.87gm * P<.08   p
  b   R m f34 gav kid tcc 24m24 s                                         11.8gm * P<.07   e
  c   R m f34 gav TBA MXB 24m24 s                                         143.mg * P<.005
  d   R m f34 gav liv MXB 24m24 s                                         1.38gm * P<.02
1783  R m f34 gav kid rua 24m24 s   with step                     .   ±   2.97gm * P<.02   p
1784  R m f34 gav tes ica 15m24                         :   +   :        51.7mg   P<.006
  a   R m f34 gav ski bca 15m24                                           378.mg   P<.002
  b   R m f34 gav ski MXA 15m24                                           258.mg   P<.02
  c   R m f34 gav ski fib 15m24                                           403.mg   P<.04
  d   R m f34 gav kid rua 15m24                                           no dre   P=1.
  e   R m f34 gav TBA MXB 15m24                                           186.mg   P<.3
  f   R m f34 gav liv MXB 15m24                                           no dre   P=1.
1785  R m f34 gav kid rua 15m24   with step                  :     ±     408.mg   P<.04
1786  R m f34 gav kid rua 9m24                                           no dre   P=1.
  a   R m f34 gav ski bca 9m24                                            no dre   P=1.
  b   R m f34 gav TBA MXB 9m24                                            1.37gm   P<1.
  c   R m f34 gav liv MXB 9m24                                            819.mg   P<.09
1787  R m f34 gav kid rua 9m24    with step                  :     ±     310.mg   P<.03

DIHYDROSAFROLE                   100ng..:..1ug....:..10.....:..100....:..1mg....:..10.....:..100....:..1g.....:..10
1788  M f b6a orl sto pam 81w81 evx                              .  +  .   84.7mg   P<.0005
  a   M f b6a orl lun ade 81w81 evx                                       739.mg   P<.6
  b   M f b6a orl liv agm 81w81 evx                                       1.40gm   P<.3
  c   M f b6a orl liv hpt 81w81 evx                                       no dre   P=1.
  d   M f b6a orl tba mix 81w81 evx                                       305.mg   P<.2
1789  M m b6a orl liv hpt 81w81 evx                              .  +  .   117.mg   P<.0005+
  a   M m b6a orl lun ade 81w81 evx                                       323.mg   P<.2
  b   M m b6a orl lun mix 81w81 evx                                       448.mg   P<.4
  c   M m b6a orl tba mix 81w81 evx                                       86.6mg   P<.004
1790  M f b6c orl lun ade 81w81 evx                               .  +   . 230.mg   P<.006 +
  a   M f b6c orl liv hpt 81w81 evx                                       no dre   P=1.
```

	RefNum	LoConf	UpConf	Cntrl	1Dose	1Inc	2Dose	2Inc			Citation or Pathology
											Brkly Code
d	TR257	66.5mg	n.s.s.	1/50	17.7mg	0/50	35.4mg	0/50			liv:hpa,hpc,nnd.
1773	TR257a	2.36mg	5.59mg	0/50	8.45mg	38/50					for:sqc,sqp. C
a	TR257a	3.72mg	10.3mg	0/50	8.45mg	27/50					
b	TR257a	5.20mg	17.3mg	0/50	8.45mg	19/50					
c	TR257a	3.84mg	n.s.s.	45/50	8.45mg	49/50					
d	TR257a	24.5mg	n.s.s.	2/50	8.45mg	3/50					liv:hpa,hpc,nnd.
1774	TR257	1.33mg	4.84mg	47/50	17.7mg	39/50	35.7mg	11/50			S
a	TR257	1.61mg	5.01mg	0/50	17.7mg	45/50	35.7mg	10/50			sto:sqc,sqp. C
b	TR257	1.67mg	5.54mg	0/50	17.7mg	38/50	35.7mg	4/50			
c	TR257	4.25mg	446.mg	17/50	17.7mg	8/50	35.7mg	2/50			S
d	TR257	14.5mg	45.0mg	0/50	17.7mg	17/50	35.7mg	6/50			
e	TR257	1.41mg	4.70mg	39/50	17.7mg	47/50	35.7mg	11/50			
f	TR257	14.0mg	n.s.s.	1/50	17.7mg	1/50	35.7mg	0/50			liv:hpa,hpc,nnd.
1775	TR257a	1.55mg	3.66mg	0/50	8.49mg	44/50					for:sqc,sqp. C
a	TR257a	1.79mg	4.38mg	0/50	8.49mg	39/50					
b	TR257a	4.42mg	17.1mg	0/50	8.49mg	16/50					
c	TR257a	2.19mg	61.9mg	40/50	8.49mg	49/50					
d	TR257a	29.7mg	n.s.s.	1/50	8.49mg	1/50					liv:hpa,hpc,nnd.

5,6-DIHYDRO-5-AZACYTIDINE   (DHAC) 62488-57-7

	RefNum	LoConf	UpConf	Cntrl	1Dose	1Inc			Citation or Pathology
1776	1906	9.94mg	n.s.s.	0/49	21.4mg	0/9			Carr;bjca,57,395-402;1988
a	1906	3.21mg	n.s.s.	10/49	21.4mg	3/9			

1,2-DIHYDRO-2-(5-NITRO-2-THIENYL)QUINAZOLIN-4(3H)-ONE   33389-33-2

	RefNum	LoConf	UpConf	Cntrl	1Dose	1Inc			Citation or Pathology
1777	1390	5.76mg	48.5mg	2/84	17.4mg	8/25			Cohen;jnci,57,277-282;1976
a	1390	.647mg	3.36mg	6/84	17.4mg	24/25			

3,6-DIHYDRO-2-NITROSO-2H-1,2-OXAZINE   (N-nitroso-3,6-dihydrooxazine-1,2) 3276-41-3

	RefNum	LoConf	UpConf	Cntrl	1Dose	1Inc			Citation or Pathology
1778	1417	67.9mg	n.s.s.	0/20	33.1mg	1/20			Wiessler;zkko,79,114-117;1973
a	1417	29.7mg	n.s.s.	1/20	33.1mg	5/20			
b	1417	28.5mg	n.s.s.	3/20	33.1mg	6/20			

3,4-DIHYDROCOUMARIN   119-84-6

	RefNum	LoConf	UpConf	Cntrl	1Dose	1Inc	2Dose	2Inc	3Dose	3Inc	Citation or Pathology
1779	TR423	1.82gm	67.5gm	0/51	141.mg	0/50	282.mg	1/51	564.mg	4/52	---:hem,hes. S
a	TR423	2.11gm	169.gm	0/51	141.mg	0/50	282.mg	0/51	564.mg	4/52	S
b	TR423	334.mg	n.s.s.	13/51	141.mg	21/50	282.mg	25/51	564.mg	24/52	liv:hpa,hpc.
c	TR423	352.mg	n.s.s.	10/51	141.mg	20/50	282.mg	22/51	564.mg	20/52	
d	TR423	249.mg	n.s.s.	24/51	141.mg	36/50	282.mg	36/51	564.mg	38/52	
e	TR423	334.mg	n.s.s.	13/51	141.mg	21/50	282.mg	25/51	564.mg	24/52	liv:hpa,hpb,hpc.
f	TR423	1.83gm	n.s.s.	2/51	141.mg	6/50	282.mg	1/51	564.mg	3/52	lun:a/a,a/c.
1780	TR423	1.63gm	39.0gm	0/50	141.mg	0/51	282.mg	3/51	565.mg	4/50	---:mlh,mlm,mlp,mly. S
a	TR423	252.mg	n.s.s.	8/50	141.mg	15/51	282.mg	15/51	(565.mg	10/50)	S
b	TR423	1.50gm	451.gm	0/50	141.mg	1/51	282.mg	3/51	565.mg	4/50	---:hcs,mlh,mlp,mly. S
c	TR423	314.mg	n.s.s.	42/50	141.mg	40/51	282.mg	44/51	565.mg	44/50	
d	TR423	423.mg	n.s.s.	36/50	141.mg	30/51	282.mg	40/51	565.mg	34/50	liv:hpa,hpb,hpc.
e	TR423	653.mg	n.s.s.	9/50	141.mg	18/51	282.mg	16/51	565.mg	13/50	lun:a/a,a/c.
1781	TR423	242.mg	n.s.s.	1/50	106.mg	7/51	210.mg	6/50	(422.mg	4/51)	amd:pbb,pob. S
a	TR423	214.mg	n.s.s.	46/50	106.mg	43/51	210.mg	48/50	422.mg	39/51	
b	TR423	n.s.s.	n.s.s.	0/50	106.mg	0/51	210.mg	0/50	422.mg	0/51	liv:hpa,hpb,hpc.
1782	TR423	41.0mg	164.mg	43/51	106.mg	39/50	212.mg	39/50	423.mg	42/50	S
a	TR423	1.08gm	n.s.s.	0/51	106.mg	1/50	212.mg	0/50	423.mg	2/50	
b	TR423	2.90gm	n.s.s.	0/51	106.mg	0/50	212.mg	0/50	423.mg	2/50	
c	TR423	66.8mg	1.53gm	43/51	106.mg	39/50	212.mg	32/50	423.mg	29/50	
d	TR423	405.mg	n.s.s.	0/51	106.mg	1/50	212.mg	2/50	423.mg	2/50	liv:hpa,hpb,hpc.
1783	TR423	1.26gm	n.s.s.	1/51	106.mg	1/50	212.mg	3/50	423.mg	6/50	
1784	TR423a	19.8mg	730.mg	43/51	266.mg	16/20					S
a	TR423a	109.mg	3.03gm	0/51	266.mg	3/20					S
b	TR423a	60.8mg	n.s.s.	2/51	266.mg	3/20					ski:bca,ker,sqp. S
c	TR423a	73.8mg	n.s.s.	1/51	266.mg	2/20					S
d	TR423a	n.s.s.	n.s.s.	0/51	266.mg	0/20					
e	TR423a	40.2mg	n.s.s.	43/51	266.mg	10/20					
f	TR423a	n.s.s.	n.s.s.	0/51	266.mg	0/20					liv:hpa,hpb,hpc.
1785	TR423a	63.9mg	n.s.s.	1/51	266.mg	2/20					S
1786	TR423b	n.s.s.	n.s.s.	0/51	159.mg	0/20					
a	TR423b	n.s.s.	n.s.s.	0/51	159.mg	0/20					
b	TR423b	49.7mg	n.s.s.	43/51	159.mg	14/20					
c	TR423b	133.mg	n.s.s.	0/51	159.mg	1/20					liv:hpa,hpb,hpc.
1787	TR423b	86.2mg	n.s.s.	1/51	159.mg	3/20					S

DIHYDROSAFROLE   94-58-6

	RefNum	LoConf	UpConf	Cntrl	1Dose	1Inc			Citation or Pathology
1788	111	41.0mg	212.mg	0/15	192.mg	11/18			Innes;ntis,1968/1969
a	111	128.mg	n.s.s.	2/15	192.mg	4/18			
b	111	228.mg	n.s.s.	0/15	192.mg	1/18			
c	111	263.mg	n.s.s.	1/15	192.mg	1/18			
d	111	89.7mg	n.s.s.	2/15	192.mg	6/18			
1789	111	51.9mg	358.mg	0/18	179.mg	8/17			
a	111	89.7mg	n.s.s.	2/18	179.mg	5/17			
b	111	95.6mg	n.s.s.	3/18	179.mg	5/17			
c	111	37.0mg	685.mg	3/18	179.mg	11/17			
1790	111	86.5mg	2.37gm	0/17	192.mg	5/17			
a	111	409.mg	n.s.s.	0/17	192.mg	0/17			

```
        Spe Strain  Site    Xpo+Xpt                                                         TD50     2Tailpvl
            Sex  Route  Hist       Notes                                                         DR       AuOp
     b   M f b6c orl tba mix 81w81 evx                                                      170.mg    P<.02
1791 M m b6c orl liv hpt 81w81 evx                                          .    +      .   90.0mg    P<.0005+
     a   M m b6c orl lun ade 81w81 evx                                                      358.mg    P<.2
     b   M m b6c orl tba mix 81w81 evx                                                      107.mg    P<.09
1792 R b osm eat eso mix 24m24                                                  .  +  .     143.mg  * P<.0005+


(R,R)-DILEVALOL.HCl           100ng..:..1ug....:..10....:...100....:...1mg..:...10.....:...100....:...1g.....:..10
1793 R f sda eat hea ens 24m24 e                                                    .>     no dre    P=1.    -
1794 R m sda eat hea ens 24m24 e                                                       .>  no dre    P=1.    -


DIMETHADIONE                  100ng..:..1ug....:..10....:...100....:...1mg..:...10.....:...100....:...1g.....:..10
1795 R m f34 eat liv hct 71w71 e                                             .>            no dre    P=1.


DIMETHOATE                    100ng..:..1ug....:..10....:...100....:...1mg..:...10.....:...100....:...1g.....:..10
1796 M f b6c eat TBA MXB 80w93                                          :>                  no dre    P=1.    -
     a   M f b6c eat liv MXB 80w93                                                          no dre    P=1.
     b   M f b6c eat lun MXB 80w93                                                          222.mg  * P<.4
1797 M m b6c eat TBA MXB 69w93                                     :>                       no dre    P=1.    -
     a   M m b6c eat liv MXB 69w93                                                          no dre    P=1.
     b   M m b6c eat lun MXB 69w93                                                          no dre    P=1.
1798 R f osm eat TBA MXB 19m26 sv                          :>                               92.4mg  * P<.8    -
     a   R f osm eat liv MXB 19m26 sv                                                       126.mg  * P<.3
1799 R m osm eat TBA MXB 19m26 sv                          :>                               no dre    P=1.    -
     a   R m osm eat liv MXB 19m26 sv                                                       215.mg  * P<.8


DIMETHOXANE                   100ng..:..1ug....:..10....:...100....:...1mg....:...10.....:...100....:...1g.....:..10
1800 R m wis wat liv hpt 20m23 e                                                .   +     . 716.mg    P<.005 +
     a   R m wis wat --- mlk 20m23 e                                                        6.77gm    P<.4   +
     b   R m wis wat kid ptc 20m23 e                                                        6.77gm    P<.4   +
     c   R m wis wat sub fbs 20m23 e                                                        6.77gm    P<.4   +
     d   R m wis wat ski epc 20m23 e                                                        6.77gm    P<.4   +
     e   R m wis wat --- lys 20m23 e                                                        29.8gm    P<1.   +
     f   R m wis wat lun tum 20m23 e                                                        no dre    P=1.
     g   R m wis wat tba mix 20m23 e                                                        419.mg    P<.004


DIMETHOXANE, COMMERCIAL GRADE 100ng..:..1ug....:..10....:...100....:...1mg....:...10.....:...100....:...1g.....:..10
1801 M f b6c gav TBA MXB 24m24                                                        :>    3.38gm  * P<.8    -
     a   M f b6c gav liv MXB 24m24                                                          no dre    P=1.
     b   M f b6c gav lun MXB 24m24                                                          10.4gm  * P<.8
1802 M m b6c gav for MXA 24m24                                                   :     ±    1.44gm  * P<.04  e
     a   M m b6c gav TBA MXB 24m24                                                          1.82gm  * P<.7
     b   M m b6c gav liv MXB 24m24                                                          1.18gm  * P<.3
     c   M m b6c gav lun MXB 24m24                                                          1.45gm  * P<.2
1803 R f f34 gav TBA MXB 24m24                                              :>              no dre    P=1.    -
     a   R f f34 gav liv MXB 24m24                                                          2.43gm  * P<.2
1804 R m f34 gav TBA MXB 24m24                                         :>                   no dre    P=1.    -
     a   R m f34 gav liv MXB 24m24                                                          no dre    P=1.


2,5-DIMETHOXY-4'-AMINOSTILBENE 100ng..:..1ug....:..10....:...100....:...1mg....:...10.....:...100....:...1g.....:..10
1805 M f chi eat lun mix 77w90                                             :     ±          152.mg  \ P<.09   -
     a   M f chi eat liv mix 77w90                                                          no dre    P=1.    -
     b   M f chi eat tba mix 77w90                                                          118.mg  \ P<.06   -
1806 M m chi eat lun mix 77w90                                         :   ±                95.9mg  \ P<.02   +
     a   M m chi eat liv mix 77w90                                                          329.mg  \ P<.04
     b   M m chi eat tba mix 77w90                                                          126.mg  \ P<.09
1807 M m chi eat lun mix 77w90   pool                                      :   +   :        115.mg  \ P<.0005+
     a   M m chi eat liv hpt 77w90                                                          376.mg  \ P<.005 +
1808 R m cdr eat sto mix 47w51 v              :  (+)   :                                    .721mg  * P<.0005+
     a   R m cdr eat smi mix 47w51 v                                                        .866mg  * P<.0005+
     b   R m cdr eat ear mix 47w51 v                                                        1.45mg  * P<.0005+
     c   R m cdr eat liv mix 47w51 v                                                        2.61mg  * P<.02
     d   R m cdr eat tba mix 47w51 v                                                        .269mg  * P<.0005
1809 R m cdr eat sto mix 47w51 v   pool       :  (+)   :                                    .721mg  * P<.0005+
     a   R m cdr eat smi mix 47w51 v                                                        .866mg  * P<.0005+
     b   R m cdr eat ski mix 47w51 v                                                        1.48mg  * P<.0005+
     c   R m cdr eat ear mix 47w51 v                                                        1.54mg  * P<.0005+


2,4-DIMETHOXYANILINE.HCl      100ng..:..1ug....:..10....:...100....:...1mg....:...10.....:...100....:...1g.....:..10
1810 M f b6c eat TBA MXB 24m24                                                   :   ±      986.mg  * P<.08   -
     a   M f b6c eat liv MXB 24m24                                                          2.79gm  * P<.4
     b   M f b6c eat lun MXB 24m24                                                          50.6gm  * P<.9
1811 M m b6c eat TBA MXB 24m24                                                        :>    1.65gm  / P<.3    -
     a   M m b6c eat liv MXB 24m24                                                          1.39gm  / P<.09
     b   M m b6c eat lun MXB 24m24                                                          no dre    P=1.
1812 R f f34 eat TBA MXB 24m24                                                  :>          no dre    P=1.    -
     a   R f f34 eat liv MXB 24m24                                                          no dre    P=1.
1813 R m f34 eat TBA MXB 24m24                                                  :>          no dre    P=1.    -
     a   R m f34 eat liv MXB 24m24                                                          no dre    P=1.


3,3'-DIMETHOXYBENZIDINE-4,4'-DIISOCYANATE  ..1ug..:..10....:...100....:...1mg....:...10.....:...100....:...1g.....:..10
1814 M f b6c eat TBA MXB 18m24                                                        :> 32.8gm  * P<.8    -
```

	RefNum	LoConf	UpConf	Cntrl	1Dose	1Inc	2Dose	2Inc	Citation or Pathology
									Brkly Code
b	111	66.8mg	n.s.s.	1/17	192.mg	7/17			
1791	111	40.6mg	352.mg	1/17	179.mg	10/17			
a	111	101.mg	n.s.s.	1/17	179.mg	4/17			
b	111	36.1mg	n.s.s.	7/17	179.mg	12/17			
1792	110	98.5mg	215.mg	0/20	225.mg	37/50	450.mg	15/20	Hagan;txap,7,18-24;1965

(R,R)-DILEVALOL.HCl   75659-08-4

| 1793 | 2174 | 294.mg | n.s.s. | 1/97 | 50.0mg | 0/50 | 100.mg | 0/50 | 200.mg 0/50 | Selan;faat,18,471-476;1992 |
| 1794 | 2174 | 1.25gm | n.s.s. | 0/100 | 50.0mg | 4/49 | 100.mg | 0/50 | 200.mg 0/49 | |

DIMETHADIONE   (5,5-dimethyl-2,4-oxazolidinedione)   695-53-4

| 1795 | 2109 | 16.1mg | n.s.s. | 0/27 | 11.2mg | 0/15 | | | Diwan;artx,66,413-422;1992 |

DIMETHOATE   60-51-5

1796	TR4	85.1mg	n.s.s.	3/10	28.0mg	15/50	55.4mg	12/50	
a	TR4	632.mg	n.s.s.	1/10	28.0mg	0/50	55.4mg	0/50	liv:hpa,hpc,nnd.
b	TR4	102.mg	n.s.s.	0/10	28.0mg	4/50	55.4mg	5/50	lun:a/a,a/c.
1797	TR4	48.0mg	n.s.s.	6/7	22.3mg	11/50	(38.3mg	11/50)	
a	TR4	51.0mg	n.s.s.	4/7	22.3mg	8/50	(38.3mg	6/50)	liv:hpa,hpc,nnd.
b	TR4	101.mg	n.s.s.	3/7	22.3mg	2/50	(38.3mg	1/50)	lun:a/a,a/c.
1798	TR4	9.18mg	n.s.s.	7/10	6.70mg	30/50	13.4mg	21/50	
a	TR4	35.8mg	n.s.s.	1/10	6.70mg	1/50	13.4mg	5/50	liv:hpa,hpc,nnd.
1799	TR4	9.62mg	n.s.s.	7/10	4.30mg	23/50	8.60mg	24/50	
a	TR4	39.5mg	n.s.s.	0/10	4.30mg	1/50	8.60mg	1/50	liv:hpa,hpc,nnd.

DIMETHOXANE   828-00-2

1800	112	322.mg	4.88gm	0/14	436.mg	8/25			Hoch-Ligeti;jnci,53,791-793;1974
a	112	1.10gm	n.s.s.	0/14	436.mg	1/25			
b	112	1.10gm	n.s.s.	0/14	436.mg	1/25			
c	112	1.10gm	n.s.s.	0/14	436.mg	1/25			
d	112	1.10gm	n.s.s.	0/14	436.mg	1/25			
e	112	946.mg	n.s.s.	1/14	436.mg	2/25			
f	112	2.08gm	n.s.s.	0/14	436.mg	0/25			
g	112	203.mg	2.46gm	1/14	436.mg	13/25			

DIMETHOXANE, COMMERCIAL GRADE   828-00-2

1801	TR354	322.mg	n.s.s.	32/50	176.mg	26/50	352.mg	33/50	
a	TR354	1.39gm	n.s.s.	8/50	176.mg	5/50	352.mg	4/50	liv:hpa,hpc,nnd.
b	TR354	1.15gm	n.s.s.	4/50	176.mg	1/50	352.mg	5/50	lun:a/a,a/c.
1802	TR354	584.mg	n.s.s.	2/50	176.mg	3/50	352.mg	8/50	for:sqc,sqp.
a	TR354	237.mg	n.s.s.	38/50	176.mg	31/50	352.mg	37/50	
b	TR354	355.mg	n.s.s.	14/50	176.mg	12/50	352.mg	19/50	liv:hpa,hpc,nnd.
c	TR354	462.mg	n.s.s.	8/50	176.mg	4/50	352.mg	13/50	lun:a/a,a/c.
1803	TR354	135.mg	n.s.s.	41/50	88.1mg	37/50	176.mg	32/50	
a	TR354	598.mg	n.s.s.	0/50	88.1mg	1/50	176.mg	1/50	liv:hpa,hpc,nnd.
1804	TR354	58.7mg	n.s.s.	39/50	44.0mg	36/50	88.1mg	32/50	
a	TR354	n.s.s.	n.s.s.	0/50	44.0mg	1/50	88.1mg	0/50	liv:hpa,hpc,nnd.

2,5-DIMETHOXY-4'-AMINOSTILBENE   5803-51-0

1805	381	39.1mg	n.s.s.	5/15	234.mg	6/16	(468.mg	3/17)	Weisburger;jept,2,325-356;1978/pers.comm./Russfield 1973
a	381	n.s.s.	n.s.s.	1/15	234.mg	0/16	468.mg	0/17	
b	381	39.2mg	n.s.s.	10/15	234.mg	12/16	(468.mg	10/17)	
1806	381	41.6mg	n.s.s.	3/14	227.mg	11/17	(432.mg	7/20)	
a	381	111.mg	n.s.s.	1/14	227.mg	4/17	(432.mg	3/20)	
b	381	44.5mg	n.s.s.	7/14	227.mg	14/17	(432.mg	10/20)	
1807	381	50.2mg	389.mg	23/99p	227.mg	11/17	(432.mg	7/20)	
a	381	119.mg	5.61gm	7/99p	227.mg	4/17	(432.mg	2/20)	
1808	381	.361mg	1.69mg	0/16	3.20mg	6/23	6.40mg	8/24	
a	381	.369mg	2.82mg	0/16	3.20mg	3/23	6.40mg	6/24	
b	381	.712mg	4.51mg	0/16	3.20mg	8/23	6.40mg	8/24	
c	381	.777mg	n.s.s.	0/16	3.20mg	2/23	6.40mg	2/24	
d	381	.162mg	.458mg	10/16	3.20mg	18/23	6.40mg	22/24	
1809	381	.361mg	1.61mg	2/111p	3.20mg	6/23	6.40mg	8/24	
a	381	.369mg	2.54mg	0/111p	3.20mg	3/23	6.40mg	6/24	
b	381	.601mg	4.01mg	0/111p	3.20mg	4/23	6.40mg	5/24	
c	381	.759mg	3.35mg	1/111p	3.20mg	8/23	6.40mg	8/24	

2,4-DIMETHOXYANILINE.HCl   54150-69-5

1810	TR171	427.mg	n.s.s.	5/20	325.mg	25/50	650.mg	24/50	
a	TR171	805.mg	n.s.s.	3/20	325.mg	12/50	650.mg	11/50	liv:hpa,hpc,nnd.
b	TR171	3.01gm	n.s.s.	1/20	325.mg	1/50	650.mg	2/50	lun:a/a,a/c.
1811	TR171	521.mg	n.s.s.	11/20	300.mg	15/50	600.mg	33/50	
a	TR171	573.mg	n.s.s.	7/20	300.mg	9/50	600.mg	27/50	liv:hpa,hpc,nnd.
b	TR171	1.03gm	n.s.s.	4/20	300.mg	6/50	(600.mg	2/50)	lun:a/a,a/c.
1812	TR171	124.mg	n.s.s.	16/20	75.0mg	31/50	(150.mg	21/50)	
a	TR171	n.s.s.	n.s.s.	0/20	75.0mg	0/50	150.mg	0/50	liv:hpa,hpc,nnd.
1813	TR171	143.mg	n.s.s.	11/20	60.0mg	28/50	120.mg	25/50	
a	TR171	n.s.s.	n.s.s.	0/20	60.0mg	0/50	120.mg	0/50	liv:hpa,hpc,nnd.

3,3'-DIMETHOXYBENZIDINE-4,4'-DIISOCYANATE   91-93-0

| 1814 | TR128 | 4.03gm | n.s.s. | 6/20 | 1.67gm | 14/50 | 3.33gm | 13/50 | |

```
      Spe Strain Site   Xpo+Xpt                                                        TD50   2Tailpvl
          Sex  Route  Hist    Notes                                                          DR    AuOp

   a   M f b6c eat liv MXB 18m24                                                        no dre   P=1.
   b   M f b6c eat lun MXB 18m24                                                        no dre   P=1.
1815 M m b6c eat TBA MXB 18m24                                              :>          24.0gm * P<.7   -
   a   M m b6c eat liv MXB 18m24                                                        12.1gm * P<.2
   b   M m b6c eat lun MXB 18m24                                                        87.9gm * P<.9
1816 R f f34 orl MXB MXB 18m24 v                                    : +   :            1.45gm * P<.0005
   a   R f f34 orl --- MXA 18m24 v                                                      2.28gm * P<.004 c
   b   R f f34 orl ute esp 18m24 v                                                      2.74gm * P<.003 c
   c   R f f34 orl MXA MXA 18m24 v                                                      2.59gm \ P<.02  a
   d   R f f34 orl TBA MXB 18m24 v                                                      773.mg * P<.006
   e   R f f34 orl liv MXB 18m24 v                                                      17.1gm * P<.2
1817 R m f34 orl MXB MXB 18m24 ev                                  : +   :             742.mg * P<.0005

   a   R m f34 orl --- MXA 18m24 ev                                                     1.27gm * P<.0005c
   b   R m f34 orl MXA MXA 18m24 ev                                                     1.33gm * P<.002 c
   c   R m f34 orl MXA MXA 18m24 ev                                                     1.77gm * P<.006 c
   d   R m f34 orl MXA sqc 18m24 ev                                                     7.01gm * P<.02  a
   e   R m f34 orl TBA MXB 18m24 ev                                                     809.mg * P<.005
   f   R m f34 orl liv MXB 18m24 ev                                                     6.60gm * P<.5


3,3'-DIMETHOXYBENZIDINE.2HCl          100ng..:..1ug....:..10.....:..100....:..1mg...:..10......:..100...:..1g......:..10
1818 M f bcn wat hag ade 78w78 kr                                        .>            no dre   P=1.  -

   a   M f bcn wat hag car 78w78 kr                                                     no dre   P=1.  -

1819 M f bcn wat hag ade 26m26 r                                     .>                no dre   P=1.  -

1820 M m bcn wat hag ade 78w78 kr                                   .>                 3.56gm * P<.9  -

1821 M m bcn wat hag ade 26m26 r                                      .>               no dre   P=1.  -

1822 R f f34 wat MXB MXB 91w93                           :+  :                         1.19mg * P<.0005


   a   R f f34 wat cli MXA 91w93                                                        1.60mg * P<.0005c
   b   R f f34 wat cli MXA 91w93                                                        2.62mg * P<.0005
   c   R f f34 wat mgl MXA 91w93                                                        3.24mg Z P<.0005
   d   R f f34 wat cli MXA 91w93                                                        3.84mg * P<.0005
   e   R f f34 wat mgl MXA 91w93                                                        6.22mg * P<.0005
   f   R f f34 wat --- mnl 91w93                                                        6.71mg * P<.003
   g   R f f34 wat zym MXA 91w93                                                        6.74mg * P<.0005c
   h   R f f34 wat mgl fba 91w93                                                        7.05mg * P<.002
   i   R f f34 wat mgl adc 91w93                                                        7.70mg Z P<.0005c
   j   R f f34 wat zym MXA 91w93                                                        8.41mg * P<.0005
   k   R f f34 wat pit MXA 91w93                                                        8.84mg Z P<.002
   l   R f f34 wat pit MXA 91w93                                                        9.40mg Z P<.003
   m   R f f34 wat ski MXA 91w93                                                        10.2mg * P<.0005c
   n   R f f34 wat ute esp 91w93                                                        11.6mg * P<.002
   o   R f f34 wat ski bca 91w93                                                        12.3mg * P<.0005
   p   R f f34 wat MXA MXA 91w93                                                        13.8mg * P<.0005c
   q   R f f34 wat MXA MXA 91w93                                                        14.0mg * P<.0005c
   r   R f f34 wat MXA MXA 91w93                                                        15.8mg * P<.0005
   s   R f f34 wat MXA sqp 91w93                                                        16.6mg * P<.002
   t   R f f34 wat ton MXA 91w93                                                        19.9mg * P<.0005
   u   R f f34 wat zym ade 91w93                                                        27.5mg * P<.0005
   v   R f f34 wat liv MXA 91w93                                                        33.3mg Z P<.002 c
   w   R f f34 wat MXA MXA 91w93                                                        44.3mg * P<.003 c
   x   R f f34 wat MXA sqc 91w93                                                        113.mg * P<.01
   y   R f f34 wat pal MXA 91w93                                                        45.3mg * P<.04
   z   R f f34 wat pal sqp 91w93                                                        47.8mg * P<.04
   A   R f f34 wat TBA MXB 91w93                                                        1.09mg Z P<.0005
   B   R f f34 wat liv MXB 91w93                                                        33.3mg Z P<.002
1823 R m f34 wat tes MXA 88w92 a                         :+ :                          .612mg Z P<.0005
   a   R m f34 wat MXB MXB 88w92 a                                                      .677mg * P<.0005


   b   R m f34 wat ski MXA 88w92 a                                                      .767mg * P<.0005c
   c   R m f34 wat ski MXA 88w92 a                                                      .770mg * P<.0005
   d   R m f34 wat ski MXA 88w92 a                                                      .785mg * P<.0005
   e   R m f34 wat ski bca 88w92 a                                                      .789mg * P<.0005
   f   R m f34 wat pre MXA 88w92 a                                                      1.78mg Z P<.0005c
   g   R m f34 wat ski MXA 88w92 a                                                      2.01mg * P<.0005c
   h   R m f34 wat amd MXA 88w92 a                                                      2.06mg * P<.0005
   i   R m f34 wat amd MXA 88w92 a                                                      2.11mg * P<.0005
   j   R m f34 wat ski sqc 88w92 a                                                      2.35mg Z P<.0005
   k   R m f34 wat --- mnl 88w92 a                                                      2.50mg * P<.0005
   l   R m f34 wat zym MXA 88w92 a                                                      3.02mg * P<.0005c
   m   R m f34 wat pre MXA 88w92 a                                                      3.28mg Z P<.0005
   n   R m f34 wat MXA MXA 88w92 a                                                      4.22mg * P<.0005c
   o   R m f34 wat MXA sqp 88w92 a                                                      4.49mg * P<.0005
   p   R m f34 wat pre MXA 88w92 a                                                      4.65mg * P<.0005
```

	RefNum	LoConf	UpConf	Cntrl	1Dose	1Inc	2Dose	2Inc	(add'l doses) / Citation or Pathology — Brkly Code
a	TR128	24.2gm	n.s.s.	1/20	1.67gm	1/50	3.33gm	0/50	liv:hpa,hpc,nnd.
b	TR128	18.6gm	n.s.s.	3/20	1.67gm	2/50	3.33gm	1/50	lun:a/a,a/c.
1815	TR128	3.71gm	n.s.s.	7/20	1.67gm	17/50	3.33gm	21/50	
a	TR128	4.87gm	n.s.s.	1/20	1.67gm	8/50	3.33gm	10/50	liv:hpa,hpc,nnd.
b	TR128	6.79gm	n.s.s.	2/20	1.67gm	8/50	3.33gm	7/50	lun:a/a,a/c.
1816	TR128	869.mg	4.77gm	1/20	819.mg	13/50	1.64gm	20/49	---:leu,lym; ute:esp. C
a	TR128	1.25gm	16.1gm	1/20	819.mg	8/50	1.64gm	15/49	---:leu,lym.
b	TR128	1.47gm	11.9gm	0/20	819.mg	5/50	1.64gm	10/49	
c	TR128	1.14gm	n.s.s.	0/20	819.mg	8/50	(1.64gm	6/49)	ear:sec,sqc,tri; zym:sec,sqc,tri.
d	TR128	405.mg	8.45gm	10/20	819.mg	40/50	1.64gm	44/49	
e	TR128	4.21gm	n.s.s.	0/20	819.mg	0/50	1.64gm	2/49	liv:hpa,hpc,nnd.
1817	TR128	498.mg	1.56gm	1/20	819.mg	29/50	1.64gm	27/50	---:leu,lym; sft:ker; skb:bct,ker; skf:bct; ski:bct,ker,ppn,sea, seb,sqc,tri. C
a	TR128	826.mg	2.97gm	0/20	819.mg	18/50	1.64gm	16/50	---:leu,lym.
b	TR128	775.mg	6.21gm	1/20	819.mg	17/50	1.64gm	15/50	sft:ker; skb:bct,ker; skf:bct; ski:bct,ker,ppn,seb,sqc,tri.
c	TR128	954.mg	17.9gm	1/20	819.mg	12/50	1.64gm	12/50	sft:ker; skb:bct,ker; skf:bct; ski:bct,ker,ppn,sea,tri.
d	TR128	3.07gm	n.s.s.	0/20	819.mg	2/50	1.64gm	6/50	ear:sqc; zym:sqc.
e	TR128	430.mg	7.28gm	10/20	819.mg	42/50	1.64gm	38/50	
f	TR128	1.76gm	n.s.s.	1/20	819.mg	8/50	1.64gm	3/50	liv:hpa,hpc,nnd.

3,3'-DIMETHOXYBENZIDINE.2HCl   20325-40-0

	RefNum	LoConf	UpConf	Cntrl	1Dose	1Inc	2Dose	2Inc	(add'l doses) / Citation or Pathology — Brkly Code
1818	1919m	243.mg	n.s.s.	4/24	4.00mg	3/21	8.00mg	1/19	16.0mg 6/24 32.0mg 0/20 63.0mg 2/21 126.mg 2/22 Schieferstein;jact,9,71-77;1990/pers.comm.
a	1919m	551.mg	n.s.s.	0/24	4.00mg	1/21	8.00mg	0/19	16.0mg 0/24 32.0mg 0/20 63.0mg 0/21 126.mg 0/22
1819	1919n	352.mg	n.s.s.	1/25	4.00mg	3/27	8.00mg	3/26	16.0mg 3/22 32.0mg 6/28 63.0mg 3/24 126.mg 2/26
1820	1919m	151.mg	n.s.s.	2/20	3.33mg	3/22	6.67mg	2/21	13.3mg 3/19 26.7mg 0/23 52.5mg 1/20 105.mg 3/20
1821	1919n	615.mg	n.s.s.	2/26	3.33mg	4/25	6.67mg	4/25	13.3mg 3/27 26.7mg 4/26 52.5mg 4/30 105.mg 0/28
1822	TR372	.844mg	1.72mg	9/60	4.57mg	37/45	9.71mg	67/75	18.9mg 53/60 cli:ade,anb,car,cnb; cvu:car,ppa; dsc:adc,adp; liv:hpc,nnd; mgl:adc; pal:sqc,sqp; phr:sqp; rec:adc,adp; ski:bca,bcc; ton:sqc,sqp; ute:ade,car; zym:ade,car,cnb. C
a	TR372	1.12mg	2.39mg	7/60	4.57mg	27/45	9.71mg	48/75	18.9mg 41/60 cli:ade,anb,car,cnb.
b	TR372	1.73mg	4.02mg	2/60	4.57mg	17/45	9.71mg	41/75	18.9mg 30/60 cli:car,cnb. S
c	TR372	1.93mg	6.12mg	15/60	4.57mg	13/45	9.71mg	21/75	18.9mg 22/60 mgl:adc,ade,fba. S
d	TR372	2.26mg	7.35mg	5/60	4.57mg	15/45	9.71mg	13/75	18.9mg 16/60 cli:ade,anb. S
e	TR372	2.91mg	24.8mg	14/60	4.57mg	11/45	9.71mg	9/75	18.9mg 6/60 mgl:ade,fba. S
f	TR372	2.98mg	51.2mg	21/60	4.57mg	15/45	9.71mg	12/75	18.9mg 4/60 S
g	TR372	4.19mg	11.2mg	1/60	4.57mg	12/45	9.71mg	21/75	18.9mg 16/60 zym:ade,car,cnb.
h	TR372	3.09mg	44.0mg	14/60	4.57mg	11/45	9.71mg	9/75	18.9mg 4/60 S
i	TR372	4.23mg	14.0mg	1/60	4.57mg	2/45	9.71mg	14/75	18.9mg 20/60
j	TR372	4.92mg	15.3mg	1/60	4.57mg	10/45	9.71mg	17/75	18.9mg 13/60 zym:car,cnb. S
k	TR372	3.83mg	47.8mg	15/60	4.57mg	9/45	9.71mg	5/75	18.9mg 8/60 pit:ade,pda. S
l	TR372	3.97mg	66.1mg	17/60	4.57mg	9/45	9.71mg	5/75	18.9mg 8/60 pit:ade,pda,pdc. S
m	TR372	4.47mg	29.6mg	0/60	4.57mg	4/45	9.71mg	3/75	18.9mg 2/60 ski:bca,bcc.
n	TR372	4.90mg	62.8mg	6/60	4.57mg	8/45	9.71mg	7/75	18.9mg 5/60 S
o	TR372	5.10mg	38.6mg	0/60	4.57mg	3/45	9.71mg	3/75	18.9mg 2/60 S
p	TR372	5.69mg	50.3mg	2/60	4.57mg	2/45	9.71mg	6/75	18.9mg 5/60 pal:sqc,sqp; phr:sqp; ton:sqc,sqp.
q	TR372	5.69mg	47.3mg	0/60	4.57mg	4/45	9.71mg	2/75	18.9mg 2/60 cvu:car,ppa; ute:ade,car.
r	TR372	6.01mg	63.7mg	0/60	4.57mg	3/45	9.71mg	1/75	18.9mg 2/60 cvu:ppa; ute:ade. S
s	TR372	6.07mg	107.mg	2/60	4.57mg	2/45	9.71mg	3/75	18.9mg 3/60 pal:sqp; ton:sqp. S
t	TR372	7.17mg	106.mg	1/60	4.57mg	2/45	9.71mg	2/75	18.9mg 4/60 ton:sqc,sqp. S
u	TR372	11.6mg	94.6mg	0/60	4.57mg	3/45	9.71mg	4/75	18.9mg 3/60 S
v	TR372	9.60mg	237.mg	0/60	4.57mg	1/45	9.71mg	0/75	18.9mg 3/60 liv:hpc,nnd.
w	TR372	10.7mg	416.mg	0/60	4.57mg	1/45	9.71mg	1/75	18.9mg 3/60 dsc:adc,adp; rec:adc,adp.
x	TR372	40.3mg	8.26gm	0/60	4.57mg	0/45	9.71mg	3/75	18.9mg 2/60 pal:sqc; ton:sqc. S
y	TR372	10.6mg	n.s.s.	1/60	4.57mg	0/45	9.71mg	4/75	18.9mg 1/60 pal:sqc,sqp. S
z	TR372	10.8mg	n.s.s.	1/60	4.57mg	0/45	9.71mg	3/75	18.9mg 1/60 S
A	TR372	.764mg	1.64mg	48/60	4.57mg	42/45	9.71mg	73/75	18.9mg 57/60
B	TR372	9.60mg	237.mg	0/60	4.57mg	1/45	9.71mg	0/75	18.9mg 3/60 liv:hpa,hpc,nnd.
1823	TR372	.412mg	.961mg	57/60	4.00mg	39/45	8.50mg	68/75	16.5mg 42/60 tes:iab,ica. S
a	TR372	.461mg	1.03mg	22/60	4.00mg	41/45	8.50mg	71/75	16.5mg 60/60 ---:msb,msm; asc:adp; cec:muc; col:adc; dsc:adc, adp; duo:adc,muc; ilm:adc; jej:adc,muc; liv:hpc,nnd; pal:sqc; phr:sqp; pre:ade,anb,car,cnb; rec:adc, adp; ski:bca,bcc,sbr,sea,sqc,sqp; ton:sqc,sqp; zym:ade,anb,car,cnb. C
b	TR372	.507mg	1.16mg	2/60	4.00mg	33/45	8.50mg	56/75	16.5mg 41/60 ski:bca,bcc,sbr,sea.
c	TR372	.508mg	1.17mg	2/60	4.00mg	32/45	8.50mg	54/75	16.5mg 40/60 ski:bca,bcc. S
d	TR372	.515mg	1.21mg	1/60	4.00mg	32/45	8.50mg	49/75	16.5mg 35/60 ski:bca,sea. S
e	TR372	.516mg	1.22mg	1/60	4.00mg	31/45	8.50mg	47/75	16.5mg 35/60 S
f	TR372	1.06mg	3.19mg	16/60	4.00mg	12/45	8.50mg	33/75	16.5mg 29/60 pre:ade,anb,car,cnb.
g	TR372	1.20mg	3.29mg	1/60	4.00mg	13/45	8.50mg	28/75	16.5mg 22/60 ski:sqc,sqp.
h	TR372	1.18mg	4.09mg	14/60	4.00mg	17/45	8.50mg	23/75	16.5mg 9/60 amd:pbb,pob. S
i	TR372	1.22mg	4.24mg	15/60	4.00mg	18/45	8.50mg	23/75	16.5mg 9/60 amd:pbb,phm,pob. S
j	TR372	1.38mg	3.92mg	0/60	4.00mg	9/45	8.50mg	24/75	16.5mg 21/60 S
k	TR372	1.23mg	7.45mg	19/60	4.00mg	17/45	8.50mg	17/75	16.5mg 4/60 S
l	TR372	1.65mg	5.21mg	0/60	4.00mg	10/45	8.50mg	25/75	16.5mg 30/60 zym:ade,anb,car,cnb.
m	TR372	1.67mg	7.70mg	14/60	4.00mg	6/45	8.50mg	19/75	16.5mg 12/60 pre:ade,anb. S
n	TR372	2.04mg	9.43mg	1/60	4.00mg	8/45	8.50mg	10/75	16.5mg 11/60 pal:sqc; phr:sqp; ton:sqc,sqp.
o	TR372	2.11mg	10.6mg	1/60	4.00mg	7/45	8.50mg	10/75	16.5mg 9/60 phr:sqp; ton:sqp. S
p	TR372	2.28mg	9.77mg	2/60	4.00mg	6/45	8.50mg	15/75	16.5mg 19/60 pre:car,cnb. S

```
      Spe Strain  Site   Xpo+Xpt                                                          TD50    2Tailpvl
          Sex  Route  Hist      Notes                                                         DR    AuOp
  q   R m f34 wat thy MXA 88w92 a                                                         4.70mg  * P<.0005
  r   R m f34 wat sub MXA 88w92 a                                                         5.19mg  * P<.0005
  s   R m f34 wat zym MXA 88w92 a                                                         5.84mg  * P<.0005
  t   R m f34 wat sub MXA 88w92 a                                                         6.49mg  * P<.0005
  u   R m f34 wat ski MXA 88w92 a                                                         6.50mg  * P<.0005
  v   R m f34 wat ski bcc 88w92 a                                                         6.55mg  * P<.0005
  w   R m f34 wat ski sqp 88w92 a                                                         6.60mg  * P<.0005
  x   R m f34 wat zym MXA 88w92 a                                                         6.71mg  * P<.0005
  y   R m f34 wat thy cca 88w92 a                                                         6.92mg  * P<.003
  z   R m f34 wat liv MXA 88w92 a                                                         7.80mg  * P<.0005c
  A   R m f34 wat pal MXA 88w92 a                                                         8.21mg  * P<.0005
  B   R m f34 wat ton MXA 88w92 a                                                         8.25mg  * P<.0005
  C   R m f34 wat pal sqp 88w92 a                                                         8.35mg  * P<.0005
  D   R m f34 wat MXA MXA 88w92 a                                                         8.44mg  * P<.0005c
  E   R m f34 wat ski ker 88w92 a                                                         9.13mg  * P<.0005
  F   R m f34 wat ton sqp 88w92 a                                                         9.16mg  * P<.0005
  G   R m f34 wat MXA MXA 88w92 a                                                         9.91mg  * P<.0005c

  H   R m f34 wat MXA asl 88w92 a                                                         10.9mg  * P<.0005e
  I   R m f34 wat liv nnd 88w92 a                                                         12.4mg  * P<.0005
  J   R m f34 wat MXA adp 88w92 a                                                         13.6mg  * P<.0005
  K   R m f34 wat sub MXA 88w92 a                                                         13.6mg  * P<.0005
  L   R m f34 wat sub fib 88w92 a                                                         13.9mg  * P<.0005
  M   R m f34 wat ski MXA 88w92 a                                                         24.6mg  * P<.009
  N   R m f34 wat --- MXA 88w92 a                                                         27.4mg  * P<.002 c
  O   R m f34 wat MXA MXA 88w92 a                                                         38.0mg  * P<.002
  P   R m f34 wat TBA MXB 88w92 a                                                         .565mg  Z P<.0005
  Q   R m f34 wat liv MXB 88w92 a                                                         7.80mg  * P<.0005
```

```
5,7-DIMETHOXYCYCLOPENTENE[c]COUMARIN    ..:...1ug...:...10......:...100.....:...1mg....:...10......:...100...:...1g.....:..10
1824 R m fis gav liv tum 52w87 r                                        .>                           no dre   P=1.   -
```

```
5,7-DIMETHOXYCYCLOPENTENONE[2,3-c]COUMARIN..1_ug...:...10......:...100.....:...1mg....:...10......:...100.....:...1g.....:..10
1825 R m fis gav liv tum 52w74 r                                        .>                           no dre   P=1.   -
```

```
5,7-DIMETHOXYCYCLOPENTENONE[3,2-c]COUMARIN..1_ug...:...10......:...100.....:...1mg....:...10......:...100.....:...1g.....:..10
1826 R m fis gav liv tum 52w87 r                                         .>                          no dre   P=1.   -
```

```
5,6-DIMETHOXYSTERIGMATOCYSTIN   100ng..:...1ug...:...10......:...100.....:...1mg....:...10......:...100.....:...1g.....:..10
1827 R m ain eat liv nnd 38w80 e                                 <+                                  noTD50   P<.002
  a  R m ain eat liv mix 38w80 e                                                                     noTD50   P<.002 +
  b  R m ain eat liv hpc 38w80 e                                                                     .400mg   P<.002
  c  R m ain eat liv hms 38w80 e                                                                     .566mg   P<.006
  d  R m ain eat bon ost 38w80 e                                                                     1.36mg   P<.06
```

```
N,N-DIMETHYL-4-AMINOAZOBENZENE   100ng..:...1ug...:...10......:...100.....:...1mg....:...10......:...100.....:...1g.....:..10
1828 P b rhe eat lun a/c  5y20 Ww                             :        ±                             4.84mg   P<.02

  a  P b rhe eat tba mal  5y20 Ww                                                                    17.0mg   P<.2
1829 R f wal eat liv hpt 33w56 fv                          .        +        .                       3.31mg   P<.002 +
  a  R f wal eat liv bdt 33w56 fv                                                                    4.90mg   P<.006 +
  b  R f wal eat liv lcc 33w56 fv                                                                    7.42mg   P<.03  +
```

```
N,N'-DIMETHYL-N,N'-DINITROSOPHTHALAMIDE..:...1_ug...:...10......:...100.....:...1mg....:...10......:...100.....:...1g.....:..10
1830 R m wis gav liv tum 67w67 ev                                   .>                               no dre   P=1.
  a  R m wis gav tba mix 67w67 ev                                                                    no dre   P=1.   -
```

```
DIMETHYL HYDROGEN PHOSPHITE   100ng..:...1ug...:...10......:...100.....:...1mg....:...10......:...100.....:...1g.....:..10
1831 M f b6c gav liv hpa 24m24                                                  :    +    :         #286.mg \ P<.005 -
  a  M f b6c gav TBA MXB 24m24                                                                       no dre   P=1.
  b  M f b6c gav liv MXB 24m24                                                                       1.71gm   * P<.6
  c  M f b6c gav lun MXB 24m24                                                                       no dre   P=1.
1832 M m b6c gav TBA MXB 24m24                                               :>                      864.mg  * P<.7   -
  a  M m b6c gav liv MXB 24m24                                                                       no dre   P=1.
  b  M m b6c gav lun MXB 24m24                                                                       2.43gm  * P<.8
1833 R f f34 gav lun a/c 24m24                                             :       ±                 600.mg  * P<.04  e
  a  R f f34 gav for MXA 24m24                                               +hist 1.27gm  * P<.09  e
  b  R f f34 gav TBA MXB 24m24                                                                       183.mg  * P<.5
  c  R f f34 gav liv MXB 24m24                                                                       2.46gm  * P<.3
1834 R m f34 gav MXB MXB 24m24 s                                             :  +  :                 105.mg  / P<.0005
  a  R m f34 gav lun MXA 24m24 s                                                                     139.mg  / P<.0005c
  b  R m f34 gav lun a/c 24m24 s                                                                     167.mg  / P<.0005c
  c  R m f34 gav for MXA 24m24 s                                                                     577.mg  * P<.002 c
  d  R m f34 gav lun a/a 24m24 s                                                                     903.mg  * P<.006 c
  e  R m f34 gav lun sqc 24m24 s                                                                     951.mg  * P<.006 c
  f  R m f34 gav for sqp 24m24 s                                                                     990.mg  * P<.03  c
  g  R m f34 gav for sqc 24m24 s                                                                     1.42gm  * P<.03  c
  h  R m f34 gav TBA MXB 24m24 s                                                                     119.mg  * P<.04
  i  R m f34 gav liv MXB 24m24 s                                                                     no dre   P=1.
```

	RefNum	LoConf	UpConf	Cntrl	1Dose	1Inc	2Dose	2Inc			Citation or Pathology / Brkly Code
q	TR372	2.01mg	17.8mg	6/60	4.00mg	7/45	8.50mg	7/75	16.5mg	2/60	thy:cca,ccr. S
r	TR372	2.13mg	14.9mg	0/60	4.00mg	6/45	8.50mg	6/75	16.5mg	4/60	sub:fib,nfm. S
s	TR372	2.51mg	13.8mg	0/60	4.00mg	4/45	8.50mg	11/75	16.5mg	9/60	zym:ade,anb. S
t	TR372	2.51mg	25.8mg	2/60	4.00mg	6/45	8.50mg	6/75	16.5mg	5/60	sub:fbs,fib,nfm,sar. S
u	TR372	3.44mg	11.5mg	1/60	4.00mg	4/45	8.50mg	18/75	16.5mg	18/60	ski:bcc,sbr. S
v	TR372	3.46mg	11.7mg	1/60	4.00mg	4/45	8.50mg	18/75	16.5mg	17/60	S
w	TR372	2.81mg	16.4mg	0/60	4.00mg	5/45	8.50mg	7/75	16.5mg	5/60	S
x	TR372	3.25mg	11.7mg	0/60	4.00mg	7/45	8.50mg	14/75	16.5mg	21/60	zym:car,cnb. S
y	TR372	2.61mg	57.2mg	6/60	4.00mg	6/45	8.50mg	5/75	16.5mg	1/60	S
z	TR372	3.05mg	21.9mg	1/60	4.00mg	4/45	8.50mg	7/75	16.5mg	8/60	liv:hpc,nnd.
A	TR372	3.05mg	25.6mg	0/60	4.00mg	4/45	8.50mg	5/75	16.5mg	4/60	pal:sqc,sqp. S
B	TR372	3.39mg	23.6mg	1/60	4.00mg	4/45	8.50mg	5/75	16.5mg	8/60	ton:sqc,sqp. S
C	TR372	3.06mg	27.2mg	0/60	4.00mg	4/45	8.50mg	5/75	16.5mg	3/60	S
D	TR372	3.04mg	27.0mg	0/60	4.00mg	4/45	8.50mg	7/75	16.5mg	5/60	duo:adc,muc; ilm:adc; jej:adc,muc.
E	TR372	3.33mg	42.4mg	1/60	4.00mg	5/45	8.50mg	7/75	16.5mg	1/60	S
F	TR372	3.57mg	28.6mg	1/60	4.00mg	3/45	8.50mg	5/75	16.5mg	7/60	S
G	TR372	3.90mg	23.1mg	0/60	4.00mg	1/45	8.50mg	8/75	16.5mg	8/60	asc:adp; cec:muc; col:adc; dsc:adc,adp; rec:adc, adp.
H	TR372	3.35mg	57.4mg	0/60	4.00mg	2/45	8.50mg	3/75	16.5mg	1/60	bra:asl; clb:asl; crb:asl; crl:asl.
I	TR372	4.43mg	30.2mg	0/60	4.00mg	3/45	8.50mg	7/75	16.5mg	6/60	S
J	TR372	4.44mg	42.8mg	0/60	4.00mg	1/45	8.50mg	4/75	16.5mg	5/60	asc:adp; dsc:adp; rec:adp. S
K	TR372	4.44mg	50.4mg	0/60	4.00mg	4/45	8.50mg	4/75	16.5mg	3/60	sub:fbs,fib. S
L	TR372	4.46mg	57.2mg	0/60	4.00mg	4/45	8.50mg	4/75	16.5mg	2/60	S
M	TR372	5.01mg	1.67gm	0/60	4.00mg	2/45	8.50mg	3/75	16.5mg	2/60	ski:sbr,sea. S
N	TR372	7.62mg	165.mg	2/60	4.00mg	1/45	8.50mg	7/75	16.5mg	6/60	---:msb,msm.
O	TR372	10.5mg	194.mg	0/60	4.00mg	0/45	8.50mg	4/75	16.5mg	3/60	cec:muc; col:adc; dsc:adc; rec:adc. S
P	TR372	.386mg	.868mg	59/60	4.00mg	45/45	8.50mg	75/75	16.5mg	60/60	
Q	TR372	3.05mg	21.9mg	1/60	4.00mg	4/45	8.50mg	7/75	16.5mg	8/60	liv:hpa,hpc,nnd.

5,7-DIMETHOXYCYCLOPENTENE[c]COUMARIN   1146-71-0

1824	1455	.591mg	n.s.s.	0/9	.512mg	0/8					Wogan;canr,31,1936-1942;1971

5,7-DIMETHOXYCYCLOPENTENONE[2,3-c]COUMARIN   1150-37-4

1825	1455	.566mg	n.s.s.	0/9	.602mg	0/9	Wogan;canr,31,1936-1942;1971

5,7-DIMETHOXYCYCLOPENTENONE[3,2-c]COUMARIN   1150-42-1

1826	1455	.665mg	n.s.s.	0/9	.512mg	0/9	Wogan;canr,31,1936-1942;1971

5,6-DIMETHOXYSTERIGMATOCYSTIN   65176-75-2

1827	1889	n.s.s.	.364mg	0/10	.967mg	8/8	Mori;carc,9,1039-1042;1988
a	1889	n.s.s.	.364mg	0/10	.967mg	8/8	
b	1889	.142mg	1.94mg	0/10	.967mg	5/8	
c	1889	.188mg	5.90mg	0/10	.967mg	4/8	
d	1889	.332mg	n.s.s.	0/10	.967mg	2/8	

N,N-DIMETHYL-4-AMINOAZOBENZENE   (butter yellow) 60-11-7

1828	2001	.788mg	n.s.s.	0/23	14.3mg	1/1		Adamson;ossc,129-156; 1982/Thorgeirsson 1994/Dalgard 1991/Thorgeirsson&Seiber pers.comm.
a	2001	1.95mg	n.s.s.	4/76	14.3mg	1/3		
1829	27	1.12mg	16.3mg	0/8	20.9mg	5/7		Kirby;jpat,59,1-18;1947
a	27	1.60mg	56.0mg	0/8	20.9mg	4/7		
b	27	2.18mg	n.s.s.	0/8	20.9mg	3/7		

N,N'-DIMETHYL-N,N'-DINITROSOPHTHALAMIDE   3851-16-9

1830	104	8.28mg	n.s.s.	0/9	3.46mg	0/28	Argus;jnci,35,949-958;1965
a	104	5.45mg	n.s.s.	1/9	3.46mg	1/28	

DIMETHYL HYDROGEN PHOSPHITE   (DMHP) 868-85-9

	RefNum	LoConf	UpConf	Cntrl	1Dose	1Inc	2Dose	2Inc	Pathology/Code
1831	TR287	117.mg	2.00gm	0/50	70.1mg	6/50	(140.mg	3/50)	S
a	TR287	145.mg	n.s.s.	37/50	70.1mg	30/50	140.mg	32/50	
b	TR287	331.mg	n.s.s.	2/50	70.1mg	6/50	140.mg	3/50	liv:hpa,hpc,nnd.
c	TR287	839.mg	n.s.s.	4/50	70.1mg	3/50	140.mg	1/50	lun:a/a,a/c.
1832	TR287	124.mg	n.s.s.	34/50	70.1mg	21/50	140.mg	31/50	
a	TR287	277.mg	n.s.s.	19/50	70.1mg	10/50	140.mg	13/50	liv:hpa,hpc,nnd.
b	TR287	251.mg	n.s.s.	12/50	70.1mg	7/50	140.mg	11/50	lun:a/a,a/c.
1833	TR287	207.mg	n.s.s.	0/50	35.2mg	1/50	70.4mg	3/50	
a	TR287	312.mg	n.s.s.	0/50	35.2mg	0/50	70.4mg	2/50	for:sqc,sqp.
b	TR287	42.8mg	n.s.s.	36/50	35.2mg	38/50	70.4mg	38/50	
c	TR287	401.mg	n.s.s.	0/50	35.2mg	0/50	70.4mg	1/50	liv:hpa,hpc,nnd.
1834	TR287	68.4mg	172.mg	0/50	70.4mg	2/50	141.mg	31/50	for:sqc,sqp; lun:a/a,a/c,sqc. C
a	TR287	85.7mg	246.mg	0/50	70.4mg	1/50	141.mg	24/50	lun:a/a,a/c.
b	TR287	98.8mg	311.mg	0/50	70.4mg	1/50	141.mg	20/50	
c	TR287	248.mg	2.58gm	0/50	70.4mg	1/50	141.mg	6/50	for:sqc,sqp.
d	TR287	339.mg	8.99gm	0/50	70.4mg	0/50	141.mg	5/50	
e	TR287	358.mg	3.71gm	0/50	70.4mg	0/50	141.mg	5/50	
f	TR287	342.mg	n.s.s.	0/50	70.4mg	1/50	141.mg	3/50	
g	TR287	426.mg	n.s.s.	0/50	70.4mg	0/50	141.mg	3/50	
h	TR287	52.0mg	n.s.s.	38/50	70.4mg	36/50	141.mg	45/50	
i	TR287	1.34gm	n.s.s.	3/50	70.4mg	0/50	141.mg	0/50	liv:hpa,hpc,nnd.

```
        Spe Strain  Site   Xpo+Xpt                                                      TD50   2Tailpvl
            Sex  Route  Hist   Notes                                                        DR    AuOp
DIMETHYL METHYLPHOSPHONATE        100ng..:..1ug....:..10......:..100....:..1mg....:..10......:..100....:..1g......:..10
1835 M f b6c gav lun a/a 24m24 as                                                :       ±    #4.09gm * P<.04  -
a    M f b6c gav TBA MXB 24m24 as                                                             851.mg * P<.02
b    M f b6c gav liv MXB 24m24 as                                                             6.56gm * P<.4
c    M f b6c gav lun MXB 24m24 as                                                             7.25gm * P<.3
1836 M m b6c gav liv MXA 23m24 as                                   :    +       :       #458.mg * P<.002 i
a    M m b6c gav liv hpa 23m24 as                                                             563.mg * P<.003
b    M m b6c gav TBA MXB 23m24 as                                                             370.mg / P<.002
c    M m b6c gav liv MXB 23m24 as                                                             458.mg / P<.002
d    M m b6c gav lun MXB 23m24 as                                                             7.27gm / P<.5
1837 R f f34 gav TBA MXB 24m24                                           :>                   868.mg * P<.3  -
a    R f f34 gav liv MXB 24m24                                                                no dre  P=1.
1838 R m f34 gav tes ict 24m24                                      :  +   :                  169.mg * P<.0005
a    R m f34 gav amd MXA 24m24                                                                306.mg * P<.0005
b    R m f34 gav MXA mnl 24m24                                                                453.mg / P<.0005
c    R m f34 gav MXB MXB 24m24                                                                520.mg * P<.0005
d    R m f34 gav k/p MXA 24m24                                                                608.mg * P<.0005
e    R m f34 gav k/p tpp 24m24                                                                700.mg * P<.0005p
f    R m f34 gav thy ccr 24m24                                                                983.mg * P<.005
g    R m f34 gav tnv men 24m24                                                                1.01gm * P<.0005
h    R m f34 gav MXA MXA 24m24                                                                1.02gm * P<.005
i    R m f34 gav thy MXA 24m24                                                                1.66gm * P<.004
j    R m f34 gav kid uac 24m24                                                                2.24gm * P<.008 p
k    R m f34 gav thy MXA 24m24                                                                1.14gm * P<.04
l    R m f34 gav amd phm 24m24                                                                2.02gm * P<.05
m    R m f34 gav TBA MXB 24m24                                                                178.mg * P<.0005
n    R m f34 gav liv MXB 24m24                                                                1.63gm * P<.09

DIMETHYL MORPHOLINOPHOSPHORAMIDATE        ..:..1ug....:..10......:..100....:..1mg....:..10......:..100....:..1g......:..10
1839 M f b6c gav TBA MXB 24m24 s                                            :>                no dre  P=1.  -
a    M f b6c gav liv MXB 24m24 s                                                              no dre  P=1.
b    M f b6c gav lun MXB 24m24 s                                                              17.6gm * P<.9
1840 M m b6c gav TBA MXB 24m24                                          :>                    no dre  P=1.
a    M m b6c gav liv MXB 24m24                                                                1.27gm * P<.4
b    M m b6c gav lun MXB 24m24                                                                3.90gm * P<.8
1841 R f f34 gav --- mnl 24m24 s                                    :  ±                      788.mg * P<.02 p
a    R f f34 gav TBA MXB 24m24 s                                                              774.mg * P<.3
b    R f f34 gav liv MXB 24m24 s                                                              8.24gm * P<.2
1842 R m f34 gav --- mnl 24m24                                      :  +       : 503.mg * P<.006 p
a    R m f34 gav tes ict 24m24                                                                283.mg * P<.04
b    R m f34 gav TBA MXB 24m24                                                                283.mg * P<.04
c    R m f34 gav liv MXB 24m24                                                                2.82gm * P<.2

4,6-DIMETHYL-2-(5-NITRO-2-FURYL)PYRIMIDINE..1_ug....:..10......:..100....:..1mg....:..10......:..100....:..1g......:..10
1843 R f sda eat for sqc 42w59 e                        <+                                    noTD50  P<.0005+
a    R f sda eat itn sar 42w59 e                                                              2.61mg  P<.0005+
b    R f sda eat itn hms 42w59 e                                                              3.43mg  P<.0005+
c    R f sda eat mgl adc 42w59 e                                                              6.45mg  P<.0005+
d    R f sda eat kid tcc 42w59 e                                                              45.5mg  P<.03  +
e    R f sda eat tba mix 42w59 e                                                              noTD50  P<.0005

1,2-DIMETHYL-5-NITROIMIDAZOLE        100ng..:..1ug....:..10......:..100....:..1mg....:..10......:..100....:..1g......:..10
1844 R f sda eat mgl fba 46w66 e                                      .  +  .                 17.0mg  P<.0005+
a    R f sda eat liv tum 46w66 e                                                              no dre  P=1.
b    R f sda eat tba mix 46w66 e                                                              17.0mg  P<.0005

DIMETHYL TEREPHTHALATE        100ng..:..1ug....:..10......:..100....:..1mg....:..10......:..100....:..1g......:..10
1845 M f b6c eat TBA MXB 24m24                                              :>                4.34gm / P<.7  -
a    M f b6c eat liv MXB 24m24                                                                no dre  P=1.
b    M f b6c eat lun MXB 24m24                                                                44.0gm \ P<1.
1846 M m b6c eat lun MXA 24m24                                                 : + : #1.16gm * P<.002 -
a    M m b6c eat TBA MXB 24m24                                                                no dre  P=1.
b    M m b6c eat liv MXB 24m24                                                                no dre  P=1.
c    M m b6c eat lun MXB 24m24                                                                1.16gm * P<.002
1847 R f f34 eat TBA MXB 24m24                                          :>                    342.mg \ P<.5  -
a    R f f34 eat liv MXB 24m24                                                                no dre  P=1.
1848 R m f34 eat TBA MXB 24m24                                      :>                        no dre  P=1.  -
a    R m f34 eat liv MXB 24m24                                                                no dre  P=1.

6-DIMETHYLAMINO-4,4-DIPHENYL-3-HEPTANOL  ACETATE.HCl  ..10......:..100....:..1mg....:..10......:..100....:..1g......:..10
1849 M f b6c eat liv hct 25m25 e                                         .>                   1.85gm * P<.9
a    M f b6c eat lun mix 25m25 e                                                              no dre  P=1.
1850 M m b6c eat liv hct 25m25 e                                    .>                        no dre  P=1.
a    M m b6c eat lun mix 25m25 e                                                              no dre  P=1.
1851 R f f34 eat ute esp 25m25 e                                    .  +       .              43.3mg * P<.006
a    R f f34 eat liv nnd 25m25 e                                                              60.7mg * P<.007 +
b    R f f34 eat liv hpc 25m25 e                                                       +hist 260.mg * P<.1  +
1852 R m f34 eat liv nnd 25m25 e                                    .  ±                      77.3mg * P<.09 +
a    R m f34 eat liv hpc 25m25 e                                                       +hist 110.mg * P<.02 +
```

	RefNum	LoConf	UpConf	Cntrl	1Dose	1Inc	2Dose	2Inc			Citation or Pathology	Brkly Code
DIMETHYL METHYLPHOSPHONATE (DMMP) 756-79-6												
1835	TR323	1.41gm	n.s.s.	1/50	701.mg	5/50	1.40gm	1/50				s
a	TR323	382.mg	n.s.s.	27/50	701.mg	31/50	1.40gm	8/50				
b	TR323	1.40gm	n.s.s.	3/50	701.mg	5/50	1.40gm	0/50			liv:hpa,hpc,nnd.	
c	TR323	1.69gm	n.s.s.	3/50	701.mg	5/50	1.40gm	1/50			lun:a/a,a/c.	
1836	TR323	211.mg	2.42gm	17/50	701.mg	21/50	1.43gm	4/50			liv:hpa,hpc.	s
a	TR323	242.mg	4.34gm	12/50	701.mg	15/50	1.43gm	3/50				s
b	TR323	178.mg	1.95gm	34/50	701.mg	27/50	1.43gm	10/50				
c	TR323	211.mg	2.42gm	17/50	701.mg	21/50	1.43gm	4/50			liv:hpa,hpc,nnd.	
d	TR323	1.05gm	n.s.s.	6/50	701.mg	0/50	1.43gm	3/50			lun:a/a,a/c.	
1837	TR323	280.mg	n.s.s.	40/50	350.mg	42/50	701.mg	40/50				
a	TR323	n.s.s.	n.s.s.	0/50	350.mg	2/50	701.mg	0/50			liv:hpa,hpc,nnd.	
1838	TR323	93.5mg	500.mg	41/50	350.mg	39/50	701.mg	39/50				s
a	TR323	163.mg	899.mg	12/50	350.mg	18/50	701.mg	18/50			amd:phe,phm.	s
b	TR323	227.mg	1.58gm	10/50	350.mg	11/50	701.mg	17/50			mul:mnl; spl:mnl.	s
c	TR323	267.mg	1.18gm	0/50	350.mg	9/50	701.mg	6/50			k/p:tpp; kid:uac.	P
d	TR323	292.mg	1.66gm	0/50	350.mg	8/50	701.mg	3/50			k/p:tcc,tpp.	s
e	TR323	324.mg	2.07gm	0/50	350.mg	7/50	701.mg	3/50				
f	TR323	384.mg	9.77gm	1/50	350.mg	4/50	701.mg	4/50				s
g	TR323	438.mg	3.03gm	0/50	350.mg	4/50	701.mg	6/50				s
h	TR323	415.mg	10.2gm	2/50	350.mg	5/50	701.mg	6/50			mul:men,msm; tnv:men.	s
i	TR323	591.mg	11.5gm	0/50	350.mg	2/50	701.mg	3/50			thy:fca,fcc.	s
j	TR323	737.mg	44.7gm	0/50	350.mg	2/50	701.mg	3/50				
k	TR323	398.mg	n.s.s.	4/50	350.mg	4/50	701.mg	5/50			thy:cca,ccr.	s
l	TR323	647.mg	n.s.s.	0/50	350.mg	4/50	701.mg	0/50				s
m	TR323	97.5mg	549.mg	40/50	350.mg	36/50	701.mg	37/50				
n	TR323	498.mg	n.s.s.	1/50	350.mg	4/50	701.mg	1/50			liv:hpa,hpc,nnd.	
DIMETHYL MORPHOLINOPHOSPHORAMIDATE 597-25-1												
1839	TR298	493.mg	n.s.s.	31/50	210.mg	28/50	420.mg	24/50				
a	TR298	2.18gm	n.s.s.	7/50	210.mg	4/50	420.mg	2/50			liv:hpa,hpc,nnd.	
b	TR298	1.34gm	n.s.s.	4/50	210.mg	3/50	420.mg	4/50			lun:a/a,a/c.	
1840	TR298	247.mg	n.s.s.	26/50	105.mg	29/50	210.mg	25/50				
a	TR298	310.mg	n.s.s.	11/50	105.mg	13/50	210.mg	15/50			liv:hpa,hpc,nnd.	
b	TR298	458.mg	n.s.s.	6/50	105.mg	8/50	210.mg	7/50			lun:a/a,a/c.	
1841	TR298	362.mg	n.s.s.	9/50	105.mg	13/50	210.mg	12/50	420.mg	18/50		
a	TR298	229.mg	n.s.s.	35/50	105.mg	37/50	210.mg	35/50	420.mg	34/50		
b	TR298	2.01gm	n.s.s.	0/50	105.mg	0/50	210.mg	1/50	420.mg	1/50	liv:hpa,hpc,nnd.	
1842	TR298	247.mg	6.51gm	14/50	105.mg	21/50	210.mg	19/50	420.mg	25/50		
a	TR298	124.mg	n.s.s.	45/50	105.mg	45/50	210.mg	44/50	420.mg	46/50		s
b	TR298	122.mg	n.s.s.	49/50	105.mg	47/50	210.mg	48/50	420.mg	49/50		
c	TR298	880.mg	n.s.s.	1/50	105.mg	3/50	210.mg	3/50	420.mg	3/50	liv:hpa,hpc,nnd.	
4,6-DIMETHYL-2-(5-NITRO-2-FURYL)PYRIMIDINE 59-35-8												
1843	1390	n.s.s.	1.39mg	0/84	14.2mg	30/30					Cohen;jnci,57,277-282;1976	
a	1390	1.49mg	4.99mg	0/84	14.2mg	21/30						
b	1390	1.92mg	6.85mg	0/84	14.2mg	18/30						
c	1390	3.21mg	16.8mg	2/84	14.2mg	12/30						
d	1390	11.2mg	n.s.s.	0/84	14.2mg	2/30						
e	1390	n.s.s.	1.44mg	6/84	14.2mg	30/30						
1,2-DIMETHYL-5-NITROIMIDAZOLE 551-92-8												
1844	200a	9.67mg	35.9mg	4/35	69.7mg	25/35					Cohen;jnci,51,403-417;1973	
a	200a	202.mg	n.s.s.	0/35	69.7mg	0/35						
b	200a	9.67mg	35.9mg	4/35	69.7mg	25/35						
DIMETHYL TEREPHTHALATE (DMT) 120-61-6												
1845	TR121	660.mg	n.s.s.	29/50	319.mg	20/50	638.mg	32/50				
a	TR121	2.21gm	n.s.s.	5/50	319.mg	1/50	(638.mg	0/50)			liv:hpa,hpc,nnd.	
b	TR121	926.mg	n.s.s.	4/50	319.mg	5/50	(638.mg	0/50)			lun:a/a,a/c.	
1846	TR121	640.mg	5.49gm	1/50	294.mg	8/50	589.mg	13/50			lun:a/a,a/c.	s
a	TR121	731.mg	n.s.s.	36/50	294.mg	33/50	589.mg	34/50				
b	TR121	1.33gm	n.s.s.	19/50	294.mg	13/50	589.mg	16/50			liv:hpa,hpc,nnd.	
c	TR121	640.mg	5.49gm	1/50	294.mg	8/50	589.mg	13/50			lun:a/a,a/c.	
1847	TR121	75.7mg	n.s.s.	38/50	123.mg	38/50	(245.mg	29/50)				
a	TR121	2.03gm	n.s.s.	1/50	123.mg	1/50	245.mg	0/50			liv:hpa,hpc,nnd.	
1848	TR121	168.mg	n.s.s.	40/50	98.2mg	28/50	(196.mg	26/50)				
a	TR121	1.91gm	n.s.s.	4/50	98.2mg	0/50	196.mg	1/50			liv:hpa,hpc,nnd.	
6-DIMETHYLAMINO-4,4-DIPHENYL-3-HEPTANOL ACETATE.HCl (L-alpha-acetylmethadol.HCl, LAAM) 43033-72-3												
1849	1894	85.3mg	n.s.s.	6/50	7.60mg	8/50	30.0mg	7/50			Rosenkrantz;faat,11,626-639;1988/pers.comm.	
a	1894	186.mg	n.s.s.	5/50	7.60mg	0/50	30.0mg	3/50				
1850	1894	13.4mg	n.s.s.	18/50	7.60mg	18/50	(30.0mg	5/50)				
a	1894	182.mg	n.s.s.	11/50	7.60mg	8/50	30.0mg	4/50				
1851	1894	21.6mg	520.mg	2/50	5.70mg	10/50	16.6mg	12/50				
a	1894	28.2mg	999.mg	3/50	5.70mg	4/50	16.6mg	12/50				
b	1894	78.7mg	n.s.s.	0/50	5.70mg	1/50	16.6mg	2/50				
1852	1894	27.2mg	n.s.s.	1/50	3.10mg	3/50	9.70mg	5/50				
a	1894	38.1mg	19.3gm	0/50	3.10mg	0/50	9.70mg	4/50				

```
       Spe Strain  Site    Xpo+Xpt
       Sex    Route   Hist     Notes                                                            TD50    2Tailpvl
                                                                                                        DR  AuOp
6-DIMETHYLAMINO-4,4-DIPHENYL-3-HEPTANONE.HCl        ....:..10......:..100...:..1mg...:..10.....:..100....:..1g......:..10
 1853 M f b6c eat pta ade 25m25 e                                              ·         ±              42.5mg \ P<.02  -
    a  M f b6c eat lun a/a 25m25 e                                                                      371.mg * P<.09  -
    b  M f b6c eat liv hpa 25m25 e                                                                      385.mg * P<.2   -
    c  M f b6c eat liv hpc 25m25 e                                                                      no dre   P=1.   -
    d  M f b6c eat lun a/c 25m25 e                                                                      no dre   P=1.   -
 1854 M m b6c eat liv hpc 25m25 e                                                      ·>              281.mg * P<.2   -
    a  M m b6c eat liv hpa 25m25 e                                                                      2.13gm / P<.8   -
    b  M m b6c eat lun a/a 25m25 e                                                                      no dre   P=1.   -
    c  M m b6c eat lun a/c 25m25 e                                                                      no dre   P=1.   -
 1855 R f f34 eat liv nnd 25m25 e                                                         ·>           1.42gm * P<.4   -
 1856 R m f34 eat liv hpc 25m25 e                                                         ·>           1.09gm * P<.3   -
    a  R m f34 eat liv nnd 25m25 e                                                                      no dre   P=1.   -

trans-2-[(DIMETHYLAMINO)METHYLIMINO]-5-[2-(5-NITRO-2-FURYL)VINYL]-1,3,4-OXADIAZOLE..1 _0......:..100...:..1g.....:..10
 1857 R f sda eat mgl adc 46w66 e                                            ·    +   ·               20.4mg   P<.0005
    a  R f sda eat mgl mix 46w66 e                                                                      22.4mg   P<.0005+
    b  R f sda eat for sqp 46w66 e                                                                      221.mg   P<.08  +
    c  R f sda eat duo adp 46w66 e                                                                      336.mg   P<.2   +
    d  R f sda eat duo adc 46w66 e                                                                      336.mg   P<.2   +
    e  R f sda eat lun alc 46w66 e                                                               +hist 683.mg   P<.4   +
    f  R f sda eat liv tum 46w66 e                                                                      no dre   P=1.
    g  R f sda eat tba mix 46w66 e                                                                      20.6mg   P<.0005

4-DIMETHYLAMINO-3,5-XYLENOL          100ng..:..1ug....:..10......:..100.....:..1mg...:..10......:..100...:..1g......:..10
 1858 M f b6a orl lun ade 76w76 evx                                             ·>                     no dre   P=1.   -
    a  M f b6a orl liv hpt 76w76 evx                                                                    no dre   P=1.   -
    b  M f b6a orl tba mix 76w76 evx                                                                    220.mg   P<.7   -
 1859 M m b6a orl liv hpt 76w76 evx                                            ·>                      207.mg   P<.6   -
    a  M m b6a orl lun ade 76w76 evx                                                                    no dre   P=1.   -
    b  M m b6a orl tba mix 76w76 evx                                                                    1.19gm   P<1.   -
 1860 M f b6c orl lun ade 76w76 evx                                             ·>                     265.mg   P<.3   -
    a  M f b6c orl liv hpt 76w76 evx                                                                    no dre   P=1.   -
    b  M f b6c orl tba mix 76w76 evx                                                                    265.mg   P<.3   -
 1861 M m b6c orl lun ade 76w76 evx                                     ·         ±                    56.1mg   P<.02  -
    a  M m b6c orl liv hpt 76w76 evx                                                                    247.mg   P<.3   -
    b  M m b6c orl tba mix 76w76 evx                                                                    34.8mg   P<.004 -

4-DIMETHYLAMINOANTIPYRINE          100ng..:..1ug....:..10......:..100.....:..1mg...:..10......:..100...:..1g......:..10
 1862 M f b6c wat liv hpa 23m24 e                                                      ·>              6.15gm * P<.6   -
    a  M f b6c wat lun a/c 23m24 e                                                                      6.54gm * P<.3   -
    b  M f b6c wat liv hpc 23m24 e                                                                      no dre   P=1.   -
    c  M f b6c wat lun a/a 23m24 e                                                                      no dre   P=1.   -
    d  M f b6c wat tba tum 23m24 e                                                                      no dre   P=1.   -
 1863 M m b6c wat lun a/a 23m24 e                                                   ·         ±         1.48gm * P<.09  -
    a  M m b6c wat liv hpa 23m24 e                                                                      683.mg * P<.3   -
    b  M m b6c wat liv hpc 23m24 e                                                                      5.39gm * P<.9   -
    c  M m b6c wat lun a/c 23m24 e                                                                      no dre   P=1.   -
    d  M m b6c wat tba tum 23m24 e                                                                      156.mg * P<.02  -

2-DIMETHYLAMINOETHANOL          100ng..:..1ug....:..10......:..100.....:..1mg...:..10......:..100...:..1g......:..10
 1864 M f c3j wat liv tum 29m29 e                                                      ·>              6.72gm   P<.7
    a  M f c3j wat tba tum 29m29 e                                                                      no dre   P=1.
 1865 M f cen wat liv tum 24m24 e                                                  ·>                  2.99gm   P<.6
    a  M f cen wat tba tum 24m24 e                                                                      1.66gm   P<.8   -

DIMETHYLAMINOETHYLNITROSOETHYLUREA,  NITRITE SALT...:..10......:..100....:..1mg...:..10......:..100...:..1g......:..10
 1866 R f f34 gav mgl fba 9m24 e                                          ·    +   ·                   .704mg   P<.0005+
    a  R f f34 gav ute emp 9m24 e                                                                       1.41mg   P<.002 +
    b  R f f34 gav mgl adc 9m24 e                                                                       3.39mg   P<.04  +
    c  R f f34 gav lun tum 9m24 e                                                                       5.35mg   P<.09
    d  R f f34 gav cvx adc 9m24 e                                                                       5.35mg   P<.09
    e  R f f34 gav ute ade 9m24 e                                                                       11.2mg   P<.3
    f  R f f34 gav mgl ade 9m24 e                                                                       11.2mg   P<.3
    g  R f f34 gav liv tum 9m24 e                                                                       no dre   P=1.
 1867 R m f34 gav lun tum 9m24 e                                          ·      ±                      2.37mg   P<.04
    a  R m f34 gav ski ker 9m24 e                                                                       1.92mg   P<.2
    b  R m f34 gav liv tum 9m24 e                                                                       no dre   P=1.

N,N-DIMETHYLANILINE          100ng..:..1ug....:..10......:..100....:..1mg...:..10......:..100...:..1g......:..10
 1868 M f b6c gav for sqp 24m24                                                 :     ±                99.3mg * P<.05  e
    a  M f b6c gav TBA MXB 24m24                                                                        108.mg / P<.7
    b  M f b6c gav liv MXB 24m24                                                                        163.mg * P<.4
    c  M f b6c gav lun MXB 24m24                                                                        276.mg * P<.5
 1869 M m b6c gav TBA MXB 24m24                                              :>                        489.mg * P<1.   -
    a  M m b6c gav liv MXB 24m24                                                                        216.mg * P<.7
    b  M m b6c gav lun MXB 24m24                                                                        102.mg * P<.4
 1870 R f f34 gav TBA MXB 24m24                                         :>                             no dre   P=1.
    a  R f f34 gav liv MXB 24m24                                                                        no dre   P=1.
 1871 R m f34 gav srp MXA 24m24                                              :    +        :           125.mg * P<.005 p
    a  R m f34 gav srp srn 24m24                                                                        175.mg * P<.02
```

RefNum	LoConf	UpConf	Cntrl	1Dose	1Inc	2Dose	2Inc	Citation or Pathology
								Brkly Code

6-DIMETHYLAMINO-4,4-DIPHENYL-3-HEPTANONE.HCl   (DL-methadone.HCl) 1095-90-5

1853	1893	19.0mg n.s.s.	5/50	15.0mg	15/50	(60.0mg	4/50)	Rosenkrantz;faat,11,640-651;1988/pers.comm.
a	1893	132.mg n.s.s.	1/50	15.0mg	5/50	60.0mg	6/50	
b	1893	124.mg n.s.s.	3/50	15.0mg	6/50	60.0mg	8/50	
c	1893	217.mg n.s.s.	5/50	15.0mg	6/50	60.0mg	5/50	
d	1893	128.mg n.s.s.	1/50	15.0mg	0/50	60.0mg	0/50	
1854	1893	86.0mg n.s.s.	12/50	15.0mg	11/50	60.0mg	17/50	
a	1893	196.mg n.s.s.	9/50	15.0mg	2/50	60.0mg	8/50	
b	1893	384.mg n.s.s.	8/50	15.0mg	5/50	60.0mg	3/50	
c	1893	607.mg n.s.s.	5/50	15.0mg	0/50	60.0mg	1/50	
1855	1893	328.mg n.s.s.	3/50	28.0mg	1/50	88.0mg	5/50	
1856	1893	269.mg n.s.s.	0/50	16.0mg	1/50	46.0mg	1/50	
a	1893	155.mg n.s.s.	10/50	16.0mg	9/50	46.0mg	8/50	

trans-2-[(DIMETHYLAMINO)METHYLIMINO]-5-[2-(5-NITRO-2-FURYL)VINYL]-1,3,4-OXADIAZOLE   55738-54-0

1857	1126	12.0mg 37.9mg	0/24	69.7mg	22/36			Cohen;jnci,54,841-850;1975
a	1126	12.6mg 53.5mg	2/24	69.7mg	22/36			
b	1126	66.9mg n.s.s.	0/24	69.7mg	3/36			
c	1126	82.7mg n.s.s.	0/24	69.7mg	2/36			
d	1126	82.7mg n.s.s.	0/24	69.7mg	2/36			
e	1126	111.mg n.s.s.	0/24	69.7mg	1/36			
f	1126	208.mg n.s.s.	0/24	69.7mg	0/36			
g	1126	11.7mg 46.8mg	2/24	69.7mg	23/36			

4-DIMETHYLAMINO-3,5-XYLENOL   6120-10-1

1858	1248	46.1mg n.s.s.	1/17	41.4mg	1/17			Innes;ntis,1968/1969
a	1248	77.4mg n.s.s.	0/17	41.4mg	0/17			
b	1248	27.7mg n.s.s.	2/17	41.4mg	3/17			
1859	1248	30.2mg n.s.s.	1/18	38.5mg	2/17			
a	1248	47.4mg n.s.s.	2/18	38.5mg	1/17			
b	1248	28.0mg n.s.s.	3/18	38.5mg	3/17			
1860	1248	43.1mg n.s.s.	0/16	41.4mg	1/18			
a	1248	82.0mg n.s.s.	0/16	41.4mg	0/18			
b	1248	43.1mg n.s.s.	0/16	41.4mg	1/18			
1861	1248	19.3mg n.s.s.	0/16	38.5mg	4/18			
a	1248	40.1mg n.s.s.	0/16	38.5mg	1/18			
b	1248	14.0mg 212.mg	0/16	38.5mg	6/18			

4-DIMETHYLAMINOANTIPYRINE   (aminopyrine) 58-15-1

1862	1991	874.mg n.s.s.	1/45	76.9mg	0/41	154.mg	2/42	Inai;gann,81,122-128;1990
a	1991	1.06gm n.s.s.	0/45	76.9mg	0/41	154.mg	1/42	
b	1991	437.mg n.s.s.	1/45	76.9mg	0/41	154.mg	0/42	
c	1991	932.mg n.s.s.	2/45	76.9mg	2/41	154.mg	1/42	
d	1991	196.mg n.s.s.	22/45	76.9mg	21/41	154.mg	20/42	
1863	1991	478.mg n.s.s.	1/46	64.1mg	0/44	128.mg	5/48	
a	1991	213.mg n.s.s.	10/46	64.1mg	6/44	128.mg	16/48	
b	1991	406.mg n.s.s.	6/46	64.1mg	4/44	128.mg	7/48	
c	1991	902.mg n.s.s.	1/46	64.1mg	1/44	128.mg	1/48	
d	1991	74.5mg n.s.s.	22/46	64.1mg	19/44	128.mg	36/48	

2-DIMETHYLAMINOETHANOL   108-01-0

1864	1877	931.mg n.s.s.	4/44	268.mg	5/40			Stenback;made,42,129-138;1988
a	1877	166.mg n.s.s.	39/44	268.mg	35/40			
1865	1877	478.mg n.s.s.	6/58	178.mg	7/50			
a	1877	177.mg n.s.s.	33/58	178.mg	30/50			

DIMETHYLAMINOETHYLNITROSOETHYLUREA,   NITRITE SALT   ---

1866	2111	.301mg 2.02mg	0/12	1.48mg	9/12			Lijinsky;clet,63,101-107;1992
a	2111	.558mg 6.32mg	0/12	1.48mg	6/12			
b	2111	1.02mg n.s.s.	0/12	1.48mg	3/12			
c	2111	1.31mg n.s.s.	0/12	1.48mg	2/12			
d	2111	1.31mg n.s.s.	0/12	1.48mg	2/12			
e	2111	1.82mg n.s.s.	0/12	1.48mg	1/12			
f	2111	1.82mg n.s.s.	0/12	1.48mg	1/12			
g	2111	3.52mg n.s.s.	0/12	1.48mg	0/12			
1867	2111	.713mg n.s.s.	0/12	1.04mg	3/12			
a	2111	.526mg n.s.s.	2/12	1.04mg	5/12			
b	2111	2.47mg n.s.s.	0/12	1.04mg	0/12			

N,N-DIMETHYLANILINE   121-69-7

1868	TR360	39.3mg n.s.s.	2/50	10.6mg	2/50	21.2mg	8/50	
a	TR360	15.7mg n.s.s.	39/50	10.6mg	27/50	21.2mg	41/50	
b	TR360	40.2mg n.s.s.	5/50	10.6mg	5/50	21.2mg	8/50	liv:hpa,hpc,nnd.
c	TR360	54.8mg n.s.s.	4/50	10.6mg	3/50	21.2mg	6/50	lun:a/a,a/c.
1869	TR360	17.4mg n.s.s.	32/50	10.6mg	37/50	21.2mg	33/50	
a	TR360	28.3mg n.s.s.	11/50	10.6mg	16/50	21.2mg	13/50	liv:hpa,hpc,nnd.
b	TR360	26.7mg n.s.s.	7/50	10.6mg	12/50	21.2mg	11/50	lun:a/a,a/c.
1870	TR360	2.41mg n.s.s.	48/50	2.12mg	42/50	(21.2mg	42/50)	
a	TR360	n.s.s. n.s.s.	0/50	2.12mg	0/50	21.2mg	0/50	liv:hpa,hpc,nnd.
1871	TR360	42.5mg 1.09gm	0/50	2.12mg	0/50	21.2mg	4/50	srp:ost,srn.
a	TR360	52.3mg n.s.s.	0/50	2.12mg	0/50	21.2mg	3/50	

S

```
     Spe Strain  Site    Xpo+Xpt
         Sex  Route  Hist   Notes                                        TD50    2Tailpvl
                                                                         DR      AuOp
b    R m f34 gav TBA MXB 24m24                                           93.2mg * P<.7
c    R m f34 gav liv MXB 24m24                                           no dre   P=1.


DIMETHYLARSINIC ACID            100ng..:..1ug....:..:.10......:..100....:..1mg....:..10.....:..100....:.1g.....:..10
1872 M f b6a orl lun ade 76w76 evx                            .>         108.mg  P<.3   -
a    M f b6a orl liv hpt 76w76 evx                                       no dre  P=1.
b    M f b6a orl tba mix 76w76 evx                                       34.0mg  P<.04
1873 M m b6a orl lun ade 76w76 evx                           .>          95.2mg  P<.3   -
a    M m b6a orl liv hpt 76w76 evx                                       no dre  P=1.
b    M m b6a orl tba mix 76w76 evx                                       no dre  P=1.
1874 M f b6c orl liv hpt 76w76 evx                              .>       no dre  P=1.   -
a    M f b6c orl lun ade 76w76 evx                                       no dre  P=1.   -
b    M f b6c orl tba mix 76w76 evx                                       no dre  P=1.   -
1875 M m b6c orl liv hpt 76w76 evx                           .>          no dre  P=1.   -
a    M m b6c orl lun mix 76w76 evx                                       no dre  P=1.   -
b    M m b6c orl tba mix 76w76 evx                                       no dre  P=1.   -


5,5-DIMETHYLBARBITURIC ACID     100ng..:..1ug....:..:.10......:..100....:..1mg....:..10.....:..100....:.1g.....:..10
1876 R m f34 eat liv hct 71w71 e                                .>       no dre  P=1.


7,12-DIMETHYLBENZ(a)ANTHRACENE  100ng..:..1ug....:..:.10......:..100....:..1mg....:..10.....:..100....:.1g.....:..10
1877 M f alb eat mei ane 60w60 er           . + .                        84.0ug  P<.0005+
a    M f alb eat for pam 60w60 er                                        .287mg  P<.0005


3,3'-DIMETHYLBENZIDINE.2HCl     100ng..:..1ug....:..:.10......:..100....:..1mg....:..10.....:..100....:.1g.....:..10
1878 M f bcn wat lun mix 52w52 ekr                       .>              28.1mg * P<.2

a    M f bcn wat lun adc 52w52 ekr                                       147.mg * P<.2

1879 M f bcn wat lun adc 78w78 ekr                          .>           168.mg * P<.2

a    M f bcn wat lun mix 78w78 ekr                                       295.mg * P<.8

1880 M f bcn wat lun mix 26m26 er                           .>           no dre   P=1.

a    M f bcn wat lun adc 26m26 er                                        no dre   P=1.

1881 M m bcn wat lun adc 52w52 ekr                          .>           123.mg * P<.3

a    M m bcn wat lun mix 52w52 ekr                                       684.mg * P<1.

1882 M m bcn wat lun mix 78w78 ekr                          .>           423.mg * P<1.

a    M m bcn wat lun adc 78w78 ekr                                       no dre   P=1.

1883 M m bcn wat lun mix 26m26 er                  . +       .           28.6mg * P<.002 +

a    M m bcn wat lun adc 26m26 er                                        50.1mg * P<.002 +

1884 R f f34 wat MXB MXB 60w60             :+:                           .418mg * P<.0005


a    R f f34 wat cli MXA 60w60                                           .847mg * P<.0005c
b    R f f34 wat zym MXA 60w60                                           1.27mg * P<.0005c
c    R f f34 wat cli MXA 60w60                                           1.31mg * P<.0005
d    R f f34 wat zym MXA 60w60                                           2.00mg Z P<.0005
e    R f f34 wat cli MXA 60w60                                           2.58mg * P<.0005
f    R f f34 wat zym MXA 60w60                                           3.08mg * P<.0005
g    R f f34 wat ski MXA 60w60                                           3.17mg * P<.0005c
h    R f f34 wat MXA MXA 60w60                                           3.20mg * P<.0005c
i    R f f34 wat ski MXA 60w60                                           3.80mg * P<.0005c
j    R f f34 wat MXA sqp 60w60                                           4.26mg * P<.0005
k    R f f34 wat ski sqc 60w60                                           5.56mg * P<.0005
l    R f f34 wat mgl MXA 60w60                                           5.77mg * P<.0005
m    R f f34 wat ski bca 60w60                                           5.77mg * P<.0005
n    R f f34 wat --- mnl 60w60                                           6.46mg * P<.002 e
o    R f f34 wat MXA MXA 60w60                                           6.65mg * P<.0005c
p    R f f34 wat liv MXA 60w60                                           6.97mg * P<.0005c
q    R f f34 wat ski sqp 60w60                                           7.06mg * P<.0005
r    R f f34 wat MXA adp 60w60                                           7.11mg * P<.0005
s    R f f34 wat liv nnd 60w60                                           7.62mg * P<.0005
t    R f f34 wat mgl adc 60w60                                           8.33mg * P<.0005c
u    R f f34 wat MXA MXA 60w60                                           11.0mg * P<.0005c
v    R f f34 wat ski bcc 60w60                                           11.6mg * P<.0005
w    R f f34 wat MXA sqc 60w60                                           11.8mg * P<.002
x    R f f34 wat lun MXA 60w60                                           11.9mg * P<.006 c
y    R f f34 wat MXA MXA 60w60                                           15.4mg * P<.0005
z    R f f34 wat lun a/a 60w60                                           13.6mg * P<.02
A    R f f34 wat bra MXA 60w60                                           15.4mg * P<.06  e
B    R f f34 wat TBA MXB 60w60                                           .462mg * P<.0005
C    R f f34 wat liv MXB 60w60                                           6.97mg * P<.0005
```

	RefNum	LoConf	UpConf	Cntrl	1Dose	1Inc	2Dose	2Inc	3Dose	3Inc	4Dose	4Inc	5Dose	5Inc	6Dose	6Inc	Citation or Pathology / Brkly Code
b	TR360	12.8mg	n.s.s.	42/50	2.12mg	35/50	21.2mg	39/50									
c	TR360	126.mg	n.s.s.	1/50	2.12mg	1/50	21.2mg	0/50									liv:hpa,hpc,nnd.

DIMETHYLARSINIC ACID (cacodylic acid) 75-60-5

	RefNum	LoConf	UpConf	Cntrl	1Dose	1Inc	2Dose	2Inc	Citation or Pathology / Brkly Code
1872	1198	17.6mg	n.s.s.	0/18	16.9mg	1/18			Innes;ntis,1968/1969
a	1198	33.5mg	n.s.s.	0/18	16.9mg	0/18			
b	1198	10.3mg	n.s.s.	0/18	16.9mg	3/18			
1873	1198	15.5mg	n.s.s.	0/18	15.8mg	1/17			
a	1198	29.5mg	n.s.s.	1/18	15.8mg	0/17			
b	1198	19.4mg	n.s.s.	2/18	15.8mg	1/17			
1874	1198	33.5mg	n.s.s.	0/18	16.9mg	0/18			
a	1198	33.5mg	n.s.s.	1/18	16.9mg	0/18			
b	1198	33.5mg	n.s.s.	2/18	16.9mg	0/18			
1875	1198	17.1mg	n.s.s.	3/14	15.8mg	1/14			
a	1198	24.3mg	n.s.s.	0/14	15.8mg	0/14			
b	1198	10.3mg	n.s.s.	4/14	15.8mg	3/14			

5,5-DIMETHYLBARBITURIC ACID ---

	RefNum	LoConf	UpConf	Cntrl	1Dose	1Inc	Citation or Pathology / Brkly Code
1876	2109	19.6mg	n.s.s.	0/27	13.6mg	0/15	Diwan;artx,66,413-422;1992

7,12-DIMETHYLBENZ(a)ANTHRACENE 57-97-6

	RefNum	LoConf	UpConf	Cntrl	1Dose	1Inc	Citation or Pathology / Brkly Code
1877	1274	58.2ug	.126mg	0/40	.390mg	49/75	Chouroulinkov;bdca,54,67-78;1967
a	1274	.169mg	.563mg	0/40	.390mg	20/75	

3,3'-DIMETHYLBENZIDINE.2HCl 612-82-8

	RefNum	LoConf	UpConf	Cntrl	1Dose	1Inc	2Dose	2Inc	3Dose	3Inc	4Dose	4Inc	5Dose	5Inc	6Dose	6Inc	Citation or Pathology / Brkly Code
1878	1913m	8.34mg	n.s.s.	0/16	1.00mg	1/15	1.80mg	2/16	3.60mg	1/13	7.00mg	3/16	14.0mg	0/16	28.0mg	4/16	Schieferstein;fctx,27,801-806;1989
a	1913m	24.0mg	n.s.s.	0/16	1.00mg	0/15	1.80mg	0/16	3.60mg	0/13	7.00mg	0/16	14.0mg	0/16	28.0mg	1/16	
1879	1913n	36.9mg	n.s.s.	0/21	1.00mg	0/23	1.80mg	1/20	3.60mg	1/21	7.00mg	0/20	14.0mg	0/21	28.0mg	2/18	
a	1913n	23.8mg	n.s.s.	4/21	1.00mg	1/23	1.80mg	8/20	3.60mg	5/21	7.00mg	4/20	14.0mg	2/21	28.0mg	5/18	
1880	1913o	55.2mg	n.s.s.	8/26	1.00mg	6/25	1.80mg	7/28	3.60mg	8/25	7.00mg	10/28	14.0mg	9/25	28.0mg	7/29	
a	1913o	115.mg	n.s.s.	5/26	1.00mg	3/25	1.80mg	3/28	3.60mg	4/25	7.00mg	5/28	14.0mg	2/25	28.0mg	3/29	
1881	1913m	20.0mg	n.s.s.	0/15	.833mg	0/16	1.50mg	0/14	3.00mg	0/14	5.83mg	0/15	11.7mg	1/16	23.3mg	0/16	
a	1913m	8.25mg	n.s.s.	1/15	.833mg	3/16	1.50mg	1/14	3.00mg	5/14	5.83mg	2/15	11.7mg	4/16	23.3mg	2/16	
1882	1913n	15.7mg	n.s.s.	11/23	.833mg	4/20	1.50mg	8/18	3.00mg	8/23	5.83mg	5/18	11.7mg	7/21	23.3mg	8/20	
a	1913n	60.0mg	n.s.s.	0/23	.833mg	0/20	1.50mg	0/18	3.00mg	2/23	5.83mg	0/18	11.7mg	0/21	23.3mg	0/20	
1883	1913o	14.0mg	168.mg	8/26	.833mg	12/26	1.50mg	5/29	3.00mg	11/28	5.83mg	10/32	11.7mg	15/27	23.3mg	17/27	
a	1913o	24.3mg	278.mg	3/26	.833mg	5/26	1.50mg	2/29	3.00mg	2/28	5.83mg	7/32	11.7mg	6/27	23.3mg	11/27	

	RefNum	LoConf	UpConf	Cntrl	1Dose	1Inc	2Dose	2Inc	3Dose	3Inc	Citation or Pathology / Brkly Code
1884	TR390	.330mg	.542mg	1/60	1.71mg	25/45	4.00mg	70/75	8.57mg	58/60	cli:ade,anb,car,cnb; col:adc,adp; duo:adc; ilm: adp,mua; jej:adc,adp,muc; liv:hpc,nnd; lun:a/a,a/c; mgl:adc; phr:sqc,sqp; rec:adc,adp; ski:bca,bcc, sqc,sqp; ton:sqc,sqp; zym:ade,anb,car,cnb. C
a	TR390	.634mg	1.16mg	0/60	1.71mg	14/45	4.00mg	42/75	8.57mg	32/60	cli:ade,anb,car,cnb.
b	TR390	.936mg	1.76mg	0/60	1.71mg	6/45	4.00mg	32/75	8.57mg	42/60	zym:ade,anb,car,cnb.
c	TR390	.926mg	1.93mg	0/60	1.71mg	9/45	4.00mg	32/75	8.57mg	17/60	cli:ade,anb. S
d	TR390	1.42mg	2.90mg	0/60	1.71mg	2/45	4.00mg	22/75	8.57mg	35/60	zym:car,cnb. S
e	TR390	1.66mg	4.23mg	0/60	1.71mg	5/45	4.00mg	11/75	8.57mg	18/60	cli:car,cnb. S
f	TR390	1.88mg	5.43mg	0/60	1.71mg	4/45	4.00mg	11/75	8.57mg	12/60	zym:ade,anb. S
g	TR390	1.91mg	5.75mg	0/60	1.71mg	3/45	4.00mg	9/75	8.57mg	12/60	ski:sqc,sqp.
h	TR390	1.92mg	5.76mg	0/60	1.71mg	3/45	4.00mg	9/75	8.57mg	13/60	phr:sqc,sqp; ton:sqc,sqp.
i	TR390	2.22mg	7.20mg	0/60	1.71mg	3/45	4.00mg	10/75	8.57mg	9/60	ski:bca,bcc.
j	TR390	2.40mg	8.54mg	0/60	1.71mg	3/45	4.00mg	7/75	8.57mg	9/60	phr:sqp; ton:sqp. S
k	TR390	2.87mg	13.1mg	0/60	1.71mg	2/45	4.00mg	4/75	8.57mg	7/60	S
l	TR390	2.81mg	22.3mg	2/60	1.71mg	2/45	4.00mg	7/75	8.57mg	6/60	mgl:adc,fba. S
m	TR390	2.95mg	15.5mg	0/60	1.71mg	3/45	4.00mg	5/75	8.57mg	5/60	S
n	TR390	3.09mg	33.1mg	1/60	1.71mg	3/45	4.00mg	6/75	8.57mg	4/60	
o	TR390	3.31mg	17.3mg	0/60	1.71mg	1/45	4.00mg	7/75	8.57mg	4/60	col:adc,adp; rec:adc,adp.
p	TR390	3.41mg	17.7mg	0/60	1.71mg	0/45	4.00mg	7/75	8.57mg	4/60	liv:hpc,nnd.
q	TR390	3.51mg	17.9mg	0/60	1.71mg	1/45	4.00mg	6/75	8.57mg	5/60	S
r	TR390	3.45mg	19.6mg	0/60	1.71mg	1/45	4.00mg	6/75	8.57mg	4/60	col:adp; rec:adp. S
s	TR390	3.60mg	21.2mg	0/60	1.71mg	0/45	4.00mg	7/75	8.57mg	3/60	S
t	TR390	3.87mg	23.4mg	0/60	1.71mg	1/45	4.00mg	3/75	8.57mg	6/60	
u	TR390	5.01mg	37.5mg	0/60	1.71mg	1/45	4.00mg	3/75	8.57mg	5/60	duo:adc; ilm:adp,mua; jej:adc,adp,muc.
v	TR390	5.21mg	37.6mg	0/60	1.71mg	0/45	4.00mg	5/75	8.57mg	4/60	S
w	TR390	4.82mg	52.6mg	0/60	1.71mg	1/45	4.00mg	2/75	8.57mg	4/60	phr:sqc; ton:sqc. S
x	TR390	4.75mg	174.mg	1/60	1.71mg	1/45	4.00mg	3/75	8.57mg	4/60	lun:a/a,a/c.
y	TR390	6.37mg	56.1mg	0/60	1.71mg	0/45	4.00mg	2/75	8.57mg	5/60	duo:adc; ilm:mua; jej:adc,muc. S
z	TR390	5.07mg	n.s.s.	1/60	1.71mg	1/45	4.00mg	3/75	8.57mg	3/60	S
A	TR390	5.67mg	n.s.s.	0/60	1.71mg	2/45	4.00mg	2/75	8.57mg	1/60	bra:asl,mag,mnm.
B	TR390	.344mg	.660mg	18/60	1.71mg	33/45	4.00mg	71/75	8.57mg	58/60	
C	TR390	3.41mg	17.7mg	0/60	1.71mg	0/45	4.00mg	7/75	8.57mg	4/60	liv:hpa,hpc,nnd.

```
      Spe Strain Site   Xpo+Xpt                                              TD50    2Tailpvl
          Sex  Route Hist   Notes                                             DR      AuOp
1885 R m f34 wat MXB MXB 58w60 a                      :+:                    .346mg Z P<.0005

 a   R m f34 wat ski MXA 58w60 a                                             .500mg Z P<.0005c
 b   R m f34 wat ski bca 58w60 a                                             .534mg Z P<.0005
 c   R m f34 wat liv MXA 58w60 a                                             .898mg Z P<.0005c
 d   R m f34 wat zym MXA 58w60 a                                            1.12mg  Z P<.0005c
 e   R m f34 wat liv nnd 58w60 a                                            1.12mg  Z P<.0005
 f   R m f34 wat ski MXA 58w60 a                                            1.55mg  Z P<.0005c
 g   R m f34 wat zym MXA 58w60 a                                            1.93mg  Z P<.0005
 h   R m f34 wat zym MXA 58w60 a                                            2.38mg  Z P<.0005
 i   R m f34 wat liv hpc 58w60 a                                            2.76mg  Z P<.0005
 j   R m f34 wat ski sqc 58w60 a                                            2.80mg  Z P<.0005
 k   R m f34 wat ski sqp 58w60 a                                            3.28mg  Z P<.0005
 l   R m f34 wat MXA MXA 58w60 a                                            3.76mg  Z P<.0005c
 m   R m f34 wat pre MXA 58w60 a                                            3.96mg  Z P<.0005c
 n   R m f34 wat ski ker 58w60 a                                            4.58mg  Z P<.0005c
 o   R m f34 wat MXA adp 58w60 a                                            4.87mg  Z P<.0005
 p   R m f34 wat lun MXA 58w60 a                                            4.93mg  Z P<.0005c
 q   R m f34 wat pre ade 58w60 a                                            4.94mg  Z P<.0005
 r   R m f34 wat lun a/a 58w60 a                                            5.47mg  Z P<.0005
 s   R m f34 wat ski sea 58w60 a                                            5.62mg  Z P<.0005c
 t   R m f34 wat MXA MXA 58w60 a                                            6.99mg  Z P<.0005c
 u   R m f34 wat MXA MXA 58w60 a                                            7.79mg  Z P<.0005
 v   R m f34 wat MXA MXA 58w60 a                                            8.47mg  Z P<.0005c
 w   R m f34 wat ski bcc 58w60 a                                            8.85mg  * P<.002
 x   R m f34 wat mul msm 58w60 a                                            9.45mg  Z P<.0005c
 y   R m f34 wat amd pob 58w60 a                                           10.6mg   Z P<.003
 z   R m f34 wat MXA MXA 58w60 a                                           14.4mg   Z P<.0005
 A   R m f34 wat MXA sqc 58w60 a                                           26.7mg   * P<.007
 B   R m f34 wat bra MXA 58w60 a                                           26.1mg   * P<.02  e
 C   R m f34 wat TBA MXB 58w60 a                                            .334mg  Z P<.0005
 D   R m f34 wat liv MXB 58w60 a                                            .898mg  Z P<.0005

DIMETHYLCARBAMYL CHLORIDE         100ng..:...1ug....:..10....:...100....:...1mg....:..10....:...100....:...1g......:...10
1886 H m syg inh nas sqc 26m26 er                      .+ .                +hist .625mg   P<.0005+
1887 M f hic ipj abd mix 64w64                      .    +    .                   4.59mg   P<.004
 a   M f hic ipj abd sar 64w64                                                    5.37mg   P<.008 +
 b   M f hic ipj lun ptm 64w64                                                    6.65mg   P<.3

DIMETHYLDITHIOCARBAMIC ACID, DIMETHYLAMINE..1ug....:...10....:...100....:...1mg....:..10....:...100....:...1g....:...10
1888 M f b6a orl liv agm 76w76 evx                                 .>               233.mg   P<.3   -
 a   M f b6a orl lun ade 76w76 evx                                                 no dre   P=1.   .
 b   M f b6a orl tba mix 76w76 evx                                                 no dre   P=1.   .
1889 M m b6a orl liv hpt 76w76 evx                             .>                   99.1mg   P<.3   .
 a   M m b6a orl lun ade 76w76 evx                                                 no dre   P=1.   .
 b   M m b6a orl tba mix 76w76 evx                                                  180.mg   P<.7   .
1890 M f b6c orl liv hpt 76w76 evx                             .>                   233.mg   P<.3   .
 a   M f b6c orl lun mix 76w76 evx                                                 no dre   P=1.   .
 b   M f b6c orl tba mix 76w76 evx                                                  113.mg   P<.2   .
1891 M m b6c orl lun ade 76w76 evx                         .    ±                   63.9mg   P<.04  .
 a   M m b6c orl liv hpt 76w76 evx                                                  99.1mg   P<.1   .
 b   M m b6c orl tba mix 76w76 evx                                                  19.5mg   P<.0005-

N,N-DIMETHYLDODECYLAMINE-N-OXIDE  100ng..:...1ug....:..10....:...100....:...1mg..:...10....:...100....:...1g......:...10
1892 R f f34 wat liv nnd 22m30 e                                  .>               no dre   P=1.   -
1893 R m f34 wat liv mix 22m30 e                             .>                     231.mg   P<.5   -
 a   R m f34 wat liv nnd 22m30 e                                                    496.mg   P<.7   -
 b   R m f34 wat liv hpc 22m30 e                                                    519.mg   P<.6   -

1,1-DIMETHYLHYDRAZINE             100ng..:...1ug....:..10....:...100....:...1mg....:..10....:...100....:...1g......:...10
1894 H f syg wat cec mix 79w79 e                                       .+    .      104.mg   P<.0005+
 a   H f syg wat blv mix 79w79 e                                                    620.mg   P<.04
 b   H f syg wat liv mix 79w79 e                                                    620.mg   P<.04
 c   H f syg wat lun tum 79w79 e                                                   no dre   P=1.
1895 H m syg wat cec mix 91w91 e                                     .   +    .     155.mg   P<.0005+
 a   H m syg wat liv mix 91w91 e                                                    183.mg   P<.0005
 b   H m syg wat blv mix 91w91 e                                                    183.mg   P<.0005
 c   H m syg wat liv ang 91w91 e                                                    242.mg   P<.0005
 d   H m syg wat cec adc 91w91 e                                                    322.mg   P<.0005
 e   H m syg wat cec pla 91w91 e                                                    372.mg   P<.0005
 f   H m syg wat liv agm 91w91 e                                                    975.mg   P<.02
 g   H m syg wat lun tum 91w91 e                                                   no dre   P=1.
1896 M f cd1 wat lun a/a 52w52 ek                          .>                       28.5mg   * P<.8   -
 a   M f cd1 wat liv tum 52w52 ek                                                  no dre   P=1.   .
 b   M f cd1 wat --- vsc 52w52 ek                                                  no dre   P=1.   .
1897 M f cd1 wat lun mix 24m24 e                       .    +    .                  4.12mg   * P<.0005
 a   M f cd1 wat lun a/a 24m24 e                                                    6.46mg   * P<.0005
 b   M f cd1 wat lun a/c 24m24 e                                                   23.3mg   * P<.005
 c   M f cd1 wat liv hes 24m24 e                                                   54.9mg   * P<.3
 d   M f cd1 wat liv vsc 24m24 e                                                   68.2mg   * P<.4
```

	RefNum	LoConf	UpConf	Cntrl	1Dose	1Inc	2Dose	2Inc			Citation or Pathology
											Brkly Code
1885	TR390	.270mg	.454mg	5/60	1.48mg	19/45	3.45mg	73/75	7.50mg	58/60	cec:adp; col:adc,adp,muc; duo:muc; ilm:adc,adp;
											jej:muc; lgi:adc; liv:hpc,nnd; lun:a/a,a/c; mul:msm; phr:sqc,sqp; pre:ade,car; rec:adc,adp; ski:bca,
											bcc,ker,sea,sqc,sqp; ton:sqc,sqp; zym:ade,anb,car,cnb. C
a	TR390	.380mg	.672mg	0/60	1.48mg	11/45	3.45mg	54/75	7.50mg	30/60	ski:bca,bcc. S
b	TR390	.405mg	.723mg	0/60	1.48mg	10/45	3.45mg	52/75	7.50mg	29/60	
c	TR390	.650mg	1.28mg	0/60	1.48mg	0/45	3.45mg	35/75	7.50mg	33/60	liv:hpc,nnd.
d	TR390	.806mg	1.59mg	1/60	1.48mg	3/45	3.45mg	32/75	7.50mg	36/60	zym:ade,anb,car,cnb. S
e	TR390	.789mg	1.66mg	0/60	1.48mg	0/45	3.45mg	29/75	7.50mg	26/60	
f	TR390	1.05mg	2.38mg	0/60	1.48mg	2/45	3.45mg	17/75	7.50mg	27/60	ski:sqc,sqp.
g	TR390	1.30mg	2.98mg	0/60	1.48mg	2/45	3.45mg	21/75	7.50mg	23/60	zym:car,cnb. S
h	TR390	1.48mg	4.22mg	1/60	1.48mg	1/45	3.45mg	13/75	7.50mg	16/60	zym:ade,anb. S
i	TR390	1.65mg	5.02mg	0/60	1.48mg	0/45	3.45mg	12/75	7.50mg	12/60	S
j	TR390	1.68mg	5.05mg	0/60	1.48mg	2/45	3.45mg	10/75	7.50mg	13/60	S
k	TR390	1.93mg	6.04mg	0/60	1.48mg	0/45	3.45mg	8/75	7.50mg	15/60	S
l	TR390	2.18mg	7.05mg	0/60	1.48mg	0/45	3.45mg	6/75	7.50mg	15/60	cec:adp; col:adc,adp,muc; lgi:adc; rec:adc,adp;
m	TR390	2.09mg	10.8mg	2/60	1.48mg	4/45	3.45mg	6/75	7.50mg	9/60	pre:ade,car.
n	TR390	2.34mg	12.4mg	1/60	1.48mg	1/45	3.45mg	8/75	7.50mg	5/60	
o	TR390	2.58mg	10.4mg	0/60	1.48mg	0/45	3.45mg	6/75	7.50mg	9/60	cec:adp; col:adp; rec:adp. S
p	TR390	2.47mg	13.2mg	1/60	1.48mg	0/45	3.45mg	8/75	7.50mg	6/60	lun:a/a,a/c.
q	TR390	2.43mg	16.8mg	2/60	1.48mg	4/45	3.45mg	4/75	7.50mg	8/60	S
r	TR390	2.67mg	15.5mg	1/60	1.48mg	0/45	3.45mg	7/75	7.50mg	6/60	S
s	TR390	2.79mg	13.4mg	0/60	1.48mg	0/45	3.45mg	7/75	7.50mg	5/60	
t	TR390	3.39mg	16.6mg	0/60	1.48mg	0/45	3.45mg	4/75	7.50mg	8/60	duo:muc; ilm:adc,adp; jej:muc.
u	TR390	3.64mg	19.3mg	0/60	1.48mg	0/45	3.45mg	3/75	7.50mg	8/60	duo:muc; ilm:adc; jej:muc. S
v	TR390	3.80mg	23.9mg	0/60	1.48mg	0/45	3.45mg	4/75	7.50mg	5/60	phr:sqc,sqp; ton:sqc,sqp.
w	TR390	3.67mg	43.0mg	0/60	1.48mg	1/45	3.45mg	4/75	7.50mg	2/60	S
x	TR390	3.95mg	31.6mg	0/60	1.48mg	0/45	3.45mg	3/75	7.50mg	4/60	
y	TR390	4.22mg	78.6mg	0/60	1.48mg	2/45	3.45mg	1/75	7.50mg	3/60	S
z	TR390	6.16mg	45.9mg	0/60	1.48mg	0/45	3.45mg	0/75	7.50mg	7/60	col:adc,muc; lgi:adc; rec:adc. S
A	TR390	9.18mg	129.mg	0/60	1.48mg	0/45	3.45mg	1/75	7.50mg	3/60	phr:sqc; ton:sqc. S
B	TR390	7.38mg	n.s.s.	0/60	1.48mg	0/45	3.45mg	1/75	7.50mg	2/60	bra:gmf,mag,mnm.
C	TR390	.260mg	.445mg	9/60	1.48mg	25/45	3.45mg	73/75	7.50mg	58/60	
D	TR390	.650mg	1.28mg	0/60	1.48mg	0/45	3.45mg	35/75	7.50mg	33/60	liv:hpa,hpc,nnd.

DIMETHYLCARBAMYL CHLORIDE   79-44-7

	RefNum	LoConf	UpConf	Cntrl	1Dose	1Inc		Citation or Pathology
1886	1142	.439mg	.928mg	0/50	.553mg	50/99		Sellakumar;jept,4,107-115;1980
1887	1143	2.02mg	34.7mg	1/30	5.71mg	9/30		Van Duuren;jnci,53,695-700;1974
a	1143	2.25mg	134.mg	1/30	5.71mg	8/30		
b	1143	1.76mg	n.s.s.	10/30	5.71mg	14/30		

DIMETHYLDITHIOCARBAMIC ACID, DIMETHYLAMINE   598-64-1

	RefNum	LoConf	UpConf	Cntrl	1Dose	1Inc		Citation or Pathology
1888	1222	37.9mg	n.s.s.	0/17	36.4mg	1/18		Innes;ntis,1968/1969
a	1222	43.3mg	n.s.s.	1/17	36.4mg	1/18		
b	1222	34.4mg	n.s.s.	2/17	36.4mg	2/18		
1889	1222	22.1mg	n.s.s.	1/18	33.9mg	3/18		
a	1222	44.5mg	n.s.s.	2/18	33.9mg	1/18		
b	1222	21.1mg	n.s.s.	3/18	33.9mg	4/18		
1890	1222	37.9mg	n.s.s.	0/16	36.4mg	1/18		
a	1222	72.1mg	n.s.s.	0/16	36.4mg	0/18		
b	1222	27.8mg	n.s.s.	0/16	36.4mg	2/18		
1891	1222	19.3mg	n.s.s.	0/16	33.9mg	3/17		
a	1222	24.3mg	n.s.s.	0/16	33.9mg	2/17		
b	1222	8.66mg	62.1mg	0/16	33.9mg	8/17		

N,N-DIMETHYLDODECYLAMINE-N-OXIDE   1643-20-5

	RefNum	LoConf	UpConf	Cntrl	1Dose	1Inc		Citation or Pathology
1892	1654	107.mg	n.s.s.	4/24	31.8mg	3/24		Lijinsky;fctx,22,715-720;1984
1893	1654	45.1mg	n.s.s.	3/24	22.2mg	5/24		
a	1654	61.5mg	n.s.s.	2/24	22.2mg	3/24		
b	1654	71.6mg	n.s.s.	1/24	22.2mg	2/24		

1,1-DIMETHYLHYDRAZINE   (UDMH) 57-14-7

	RefNum	LoConf	UpConf	Cntrl	1Dose	1Inc	2Dose	2Inc		Citation or Pathology
1894	367	49.0mg	301.mg	1/50	136.mg	10/24				Toth;canc,40,2427-2431;1977
a	367	152.mg	n.s.s.	0/50	136.mg	2/24				
b	367	152.mg	n.s.s.	0/50	136.mg	2/24				
c	367	519.mg	n.s.s.	0/100	136.mg	0/32				
1895	367	84.8mg	329.mg	0/64	120.mg	15/45				
a	367	98.0mg	399.mg	0/85	120.mg	14/48				
b	367	98.0mg	399.mg	0/85	120.mg	14/48				
c	367	121.mg	592.mg	0/85	120.mg	11/48				
d	367	145.mg	957.mg	0/64	120.mg	8/45				
e	367	161.mg	1.25gm	0/64	120.mg	7/45				
f	367	295.mg	n.s.s.	0/85	120.mg	3/48				
g	367	909.mg	n.s.s.	0/88	120.mg	0/48				
1896	1916m	1.93mg	n.s.s.	2/20	.200mg	1/20	1.00mg	2/20	4.00mg	2/20	Goldenthal;irdc,399-063;1989/pers.comm.
a	1916m	.165mg	n.s.s.	0/20	.200mg	0/20	1.00mg	0/20	4.00mg	0/20	
b	1916m	.165mg	n.s.s.	0/20	.200mg	0/20	1.00mg	0/20	4.00mg	0/20	
1897	1916n	2.41mg	9.38mg	6/50	.200mg	10/49	1.00mg	13/50	4.00mg	28/50	
a	1916n	3.37mg	23.0mg	5/50	.200mg	9/49	1.00mg	12/50	4.00mg	21/50	
b	1916n	9.61mg	236.mg	1/50	.200mg	1/49	1.00mg	1/50	4.00mg	7/50	
c	1916n	13.6mg	n.s.s.	3/50	.200mg	2/50	1.00mg	1/50	4.00mg	5/50	
d	1916n	14.3mg	n.s.s.	4/50	.200mg	2/50	1.00mg	1/50	4.00mg	5/50	

	Spe	Sex	Strain	Route	Site	Hist	Xpo+Xpt	Notes	TD50	DR	2Tailpvl	AuOp
e	M	f	cd1	wat	liv	mix	24m24	e	74.7mg	Z	P<.4	
f	M	f	cd1	wat	liv	hpa	24m24	e	98.7mg	*	P<.5	
g	M	f	cd1	wat	---	hes	24m24	e	328.mg	*	P<.9	
h	M	f	cd1	wat	liv	hpc	24m24	e	340.mg	*	P<.7	
i	M	f	cd1	wat	liv	hem	24m24	e	no dre		P=1.	
j	M	f	cd1	wat	---	vsc	24m24	e	no dre		P=1.	
k	M	f	cd1	wat	---	hem	24m24	e	no dre		P=1.	
1898	M	f	cd1	wat	lun	mix	52w52	eks	3.68mg	*	P<.006	+
a	M	f	cd1	wat	lun	a/a	52w52	eks	3.93mg	*	P<.007	+
b	M	f	cd1	wat	---	vsc	52w52	eks	26.0mg	*	P<.04	+
c	M	f	cd1	wat	liv	vsc	52w52	eks	26.0mg	*	P<.04	+
d	M	f	cd1	wat	liv	hem	52w52	eks	39.7mg	*	P<.1	
e	M	f	cd1	wat	---	hem	52w52	eks	39.7mg	*	P<.1	
f	M	f	cd1	wat	---	hes	52w52	eks	80.8mg	*	P<.3	
g	M	f	cd1	wat	liv	hes	52w52	eks	80.8mg	*	P<.3	
h	M	f	cd1	wat	lun	a/c	52w52	eks	81.5mg	*	P=1.	
1899	M	f	cd1	wat	---	vsc	24m24	es	11.1mg	*	P<.0005	+
a	M	f	cd1	wat	---	hes	24m24	es	11.4mg	*	P<.0005	+
b	M	f	cd1	wat	liv	vsc	24m24	es	11.8mg	/	P<.0005	+
c	M	f	cd1	wat	liv	hes	24m24	es	12.2mg	/	P<.0005	+
d	M	f	cd1	wat	lun	mix	24m24	es	27.0mg	*	P<.03	+
e	M	f	cd1	wat	lun	a/a	24m24	es	37.1mg	*	P<.07	+
f	M	f	cd1	wat	liv	hpa	24m24	es	63.6mg	\	P<.2	
g	M	f	cd1	wat	lun	a/c	24m24	es	199.mg	*	P<.4	
h	M	f	cd1	wat	liv	hem	24m24	es	no dre		P=1.	
i	M	f	cd1	wat	---	hem	24m24	es	no dre		P=1.	
j	M	f	cd1	wat	lun	msm	24m24	es	no dre		P=1.	
1900	M	m	cd1	wat	liv	hpa	52w52	ek	26.7mg	*	P<.8	-
a	M	m	cd1	wat	---	vsc	52w52	ek	no dre		P=1.	-
b	M	m	cd1	wat	liv	hpc	52w52	ek	no dre		P=1.	-
c	M	m	cd1	wat	liv	mix	52w52	ek	no dre		P=1.	-
d	M	m	cd1	wat	lun	a/a	52w52	ek	no dre		P=1.	-
1901	M	m	cd1	wat	liv	hpc	24m24	e	11.9mg	*	P<.09	
a	M	m	cd1	wat	---	hes	24m24	e	26.7mg	*	P<.09	
b	M	m	cd1	wat	liv	mix	24m24	e	7.22mg	Z	P<.2	
c	M	m	cd1	wat	liv	hpa	24m24	e	29.7mg	*	P<.6	
d	M	m	cd1	wat	lun	mix	24m24	e	39.6mg	*	P<.9	
e	M	m	cd1	wat	liv	hes	24m24	e	44.8mg	*	P<.3	
f	M	m	cd1	wat	lun	a/c	24m24	e	89.0mg	*	P<.9	
g	M	m	cd1	wat	lun	a/a	24m24	e	116.mg	*	P<1.	
1902	M	m	cd1	wat	lun	mix	52w52	eks	4.33mg	*	P<.04	+
a	M	m	cd1	wat	lun	a/a	52w52	eks	5.18mg	*	P<.08	+
b	M	m	cd1	wat	liv	hpc	52w52	eks	10.8mg	\	P<.1	
c	M	m	cd1	wat	liv	vsc	52w52	eks	12.6mg	*	P<.02	+
d	M	m	cd1	wat	---	vsc	52w52	eks	12.6mg	*	P<.02	+
e	M	m	cd1	wat	liv	hes	52w52	eks	21.7mg	*	P<.04	
f	M	m	cd1	wat	---	hes	52w52	eks	21.7mg	*	P<.04	
g	M	m	cd1	wat	liv	hem	52w52	eks	33.4mg	*	P<.3	
h	M	m	cd1	wat	---	hem	52w52	eks	33.4mg	*	P<.3	
i	M	m	cd1	wat	lun	a/c	52w52	eks	67.4mg	*	P<.3	
j	M	m	cd1	wat	liv	mix	52w52	eks	6.28kg		P=1.	
k	M	m	cd1	wat	liv	hpa	52w52	eks	no dre		P=1.	
1903	M	m	cd1	wat	---	vsc	24m24	es	6.13mg	*	P<.0005	+
a	M	m	cd1	wat	liv	vsc	24m24	es	6.47mg	*	P<.0005	+
b	M	m	cd1	wat	---	hes	24m24	es	6.53mg	*	P<.0005	+
c	M	m	cd1	wat	liv	hes	24m24	es	6.88mg	*	P<.0005	+
d	M	m	cd1	wat	lun	a/c	24m24	es	33.4mg	\	P<.07	
e	M	m	cd1	wat	liv	hpc	24m24	es	33.4mg	\	P<.07	
f	M	m	cd1	wat	lun	mix	24m24	es	26.4mg	*	P<.2	
g	M	m	cd1	wat	lun	a/a	24m24	es	34.3mg	*	P<.2	
h	M	m	cd1	wat	liv	hpa	24m24	es	140.mg	*	P<.5	
i	M	m	cd1	wat	liv	hem	24m24	es	339.mg	*	P<.3	
j	M	m	cd1	wat	---	hem	24m24	es	339.mg	*	P<.3	
k	M	m	cd1	wat	liv	mix	24m24	es	24.3kg		P=1.	
1904	M	f	swa	wat	blv	ang	72w72	e	3.57mg		P<.0005	+
a	M	f	swa	wat	blv	mix	72w72	e	3.65mg		P<.0005	+
b	M	f	swa	wat	lun	ade	72w72	e	5.58mg		P<.0005	
c	M	f	swa	wat	lun	mix	72w72	e	5.69mg		P<.0005	+
d	M	f	swa	wat	lun	adc	72w72	e	51.6mg		P<.007	
e	M	f	swa	wat	kid	ade	72w72	e	148.mg		P<.2	
f	M	f	swa	wat	liv	tum	72w72	e	no dre		P=1.	
1905	M	m	swa	wat	blv	ang	62w62	e	2.09mg		P<.0005	+
a	M	m	swa	wat	blv	mix	62w62	e	2.11mg		P<.0005	+
b	M	m	swa	wat	lun	mix	62w62	e	2.62mg		P<.0005	+
c	M	m	swa	wat	lun	ade	62w62	e	2.62mg		P<.0005	
d	M	m	swa	wat	kid	ade	62w62	e	19.5mg		P<.0005	
e	M	m	swa	wat	liv	hpt	62w62	e	30.4mg		P<.005	+
f	M	m	swa	wat	lun	adc	62w62	e	46.6mg		P<.005	
1906	M	f	swi	gav	lun	tum	40w55		16.0mg		P<.2	+
1907	R	f	f34	wat	liv	hem	52w52	ek	no dre		P=1.	-
1908	R	f	f34	wat	liv	mix	24m24	e	30.9mg	*	P<.005	-

	RefNum	LoConf	UpConf	Cntrl	1Dose	1Inc	2Dose	2Inc			Citation or Pathology
											Brkly Code
e	1916n	15.6mg	n.s.s.	5/50	.200mg	1/50	1.00mg	0/50	4.00mg	5/50	
f	1916n	17.6mg	n.s.s.	4/50	.200mg	1/50	1.00mg	0/50	4.00mg	4/50	
g	1916n	15.5mg	n.s.s.	4/50	.200mg	6/50	1.00mg	2/50	4.00mg	5/50	
h	1916n	30.8mg	n.s.s.	1/50	.200mg	0/50	1.00mg	0/50	4.00mg	1/50	
i	1916n	1.65mg	n.s.s.	1/50	.200mg	0/50	1.00mg	0/50	4.00mg	0/50	
j	1916n	15.6mg	n.s.s.	7/50	.200mg	11/50	1.00mg	3/50	4.00mg	7/50	
k	1916n	27.5mg	n.s.s.	3/50	.200mg	5/50	1.00mg	1/50	4.00mg	2/50	
1898	2013m	1.82mg	43.3mg	3/20	8.00mg	10/20	16.0mg	11/20			Goldenthal;irdc,399-065;1990/pers.comm.
a	2013m	1.92mg	54.2mg	3/20	8.00mg	9/20	16.0mg	11/20			
b	2013m	7.87mg	n.s.s.	0/20	8.00mg	0/20	16.0mg	3/20			
c	2013m	7.87mg	n.s.s.	0/20	8.00mg	0/20	16.0mg	3/20			
d	2013m	9.77mg	n.s.s.	0/20	8.00mg	0/20	16.0mg	2/20			
e	2013m	9.77mg	n.s.s.	0/20	8.00mg	0/20	16.0mg	2/20			
f	2013m	13.2mg	n.s.s.	0/20	8.00mg	0/20	16.0mg	1/20			
g	2013m	13.2mg	n.s.s.	0/20	8.00mg	0/20	16.0mg	1/20			
h	2013m	13.3mg	n.s.s.	0/20	8.00mg	1/20	16.0mg	0/20			
1899	2013n	7.41mg	18.8mg	6/50	8.00mg	18/50	16.0mg	37/50			
a	2013n	7.72mg	18.6mg	4/50	8.00mg	15/50	16.0mg	37/50			
b	2013n	8.07mg	18.5mg	2/49	8.00mg	12/50	16.0mg	37/50			
c	2013n	8.36mg	18.8mg	1/49	8.00mg	10/50	16.0mg	37/50			
d	2013n	12.4mg	n.s.s.	14/49	8.00mg	25/50	16.0mg	25/50			
e	2013n	15.3mg	n.s.s.	13/49	8.00mg	21/50	16.0mg	22/50			
f	2013n	19.5mg	n.s.s.	2/49	8.00mg	6/50	(16.0mg	1/50)			
g	2013n	55.4mg	n.s.s.	1/49	8.00mg	4/50	16.0mg	3/50			
h	2013n	138.mg	n.s.s.	1/49	8.00mg	2/50	16.0mg	0/50			
i	2013n	134.mg	n.s.s.	2/50	8.00mg	3/50	16.0mg	0/50			
j	2013n	134.mg	n.s.s.	0/49	8.00mg	1/50	16.0mg	0/50			
1900	1916m	1.60mg	n.s.s.	1/20	.167mg	0/20	.833mg	0/20	1.67mg	1/20	Goldenthal;irdc,399-063;1989/pers.comm.
a	1916m	.132mg	n.s.s.	0/20	.167mg	0/20	.833mg	0/20	1.67mg	0/20	
b	1916m	2.06mg	n.s.s.	0/20	.167mg	1/20	.833mg	0/20	1.67mg	0/20	
c	1916m	1.55mg	n.s.s.	1/20	.167mg	1/20	.833mg	0/20	1.67mg	1/20	
d	1916m	1.02mg	n.s.s.	3/20	.167mg	1/20	.833mg	2/20	1.67mg	2/20	
1901	1916m	4.23mg	n.s.s.	5/50	.167mg	1/50	.833mg	5/50	1.67mg	8/50	
a	1916m	7.82mg	n.s.s.	0/50	.167mg	1/50	.833mg	0/50	1.67mg	3/50	
b	1916m	2.55mg	n.s.s.	16/50	.167mg	6/50	.833mg	10/50	1.67mg	19/50	
c	1916m	4.60mg	n.s.s.	11/50	.167mg	5/50	.833mg	5/50	1.67mg	11/50	
d	1916m	2.67mg	n.s.s.	16/50	.167mg	17/50	.833mg	24/50	1.67mg	16/50	
e	1916m	9.43mg	n.s.s.	0/50	.167mg	1/50	.833mg	0/50	1.67mg	2/50	
f	1916m	5.75mg	n.s.s.	4/50	.167mg	4/50	.833mg	7/50	1.67mg	4/50	
g	1916m	3.47mg	n.s.s.	12/50	.167mg	13/50	.833mg	17/50	1.67mg	12/50	
1902	2013m	1.84mg	n.s.s.	4/20	6.67mg	9/20	13.3mg	10/20			Goldenthal;irdc,399-065;1990/pers.comm.
a	2013m	2.02mg	n.s.s.	4/20	6.67mg	9/20	13.3mg	9/20			
b	2013m	2.66mg	n.s.s.	0/20	6.67mg	2/20	(13.3mg	0/20)			
c	2013m	4.79mg	n.s.s.	0/20	6.67mg	1/20	13.3mg	4/20			
d	2013m	4.79mg	n.s.s.	0/20	6.67mg	1/20	13.3mg	4/20			
e	2013m	6.56mg	n.s.s.	0/20	6.67mg	0/20	13.3mg	3/20			
f	2013m	6.56mg	n.s.s.	0/20	6.67mg	0/20	13.3mg	3/20			
g	2013m	8.21mg	n.s.s.	0/20	6.67mg	1/20	13.3mg	1/20			
h	2013m	8.21mg	n.s.s.	0/20	6.67mg	1/20	13.3mg	1/20			
i	2013m	11.0mg	n.s.s.	0/20	6.67mg	0/20	13.3mg	1/20			
j	2013m	7.02mg	n.s.s.	1/20	6.67mg	3/20	13.3mg	1/20			
k	2013m	9.85mg	n.s.s.	1/20	6.67mg	1/20	13.3mg	1/20			
1903	2013n	4.28mg	9.51mg	5/50	6.67mg	30/50	13.3mg	39/50			
a	2013n	4.54mg	10.0mg	4/50	6.67mg	30/50	13.3mg	37/50			
b	2013n	4.54mg	10.3mg	5/50	6.67mg	29/50	13.3mg	38/50			
c	2013n	4.80mg	10.8mg	4/50	6.67mg	29/50	13.3mg	36/50			
d	2013n	12.3mg	n.s.s.	3/50	6.67mg	9/50	(13.3mg	3/50)			
e	2013n	12.3mg	n.s.s.	3/50	6.67mg	9/50	(13.3mg	0/50)			
f	2013n	9.70mg	n.s.s.	19/50	6.67mg	34/50	13.3mg	26/50			
g	2013n	12.3mg	n.s.s.	16/50	6.67mg	25/50	13.3mg	23/50			
h	2013n	32.9mg	n.s.s.	6/50	6.67mg	6/50	13.3mg	9/50			
i	2013n	83.4mg	n.s.s.	0/50	6.67mg	1/50	13.3mg	1/50			
j	2013n	83.4mg	n.s.s.	0/50	6.67mg	1/50	13.3mg	1/50			
k	2013n	29.8mg	n.s.s.	9/50	6.67mg	15/50	13.3mg	9/50			
1904	117	2.23mg	5.94mg	0/47	20.0mg	37/44					Toth;jnci,50,181-194;1973
a	117	2.26mg	6.17mg	4/104	20.0mg	37/44					
b	117	3.38mg	10.3mg	12/104	20.0mg	32/44					
c	117	3.41mg	10.7mg	14/104	20.0mg	32/44					
d	117	19.2mg	1.11gm	2/104	20.0mg	6/44					
e	117	24.1mg	n.s.s.	0/32	20.0mg	1/23					
f	117	94.8mg	n.s.s.	0/109	20.0mg	0/48					
1905	117	1.33mg	3.37mg	0/50	16.7mg	42/49					
a	117	1.34mg	3.44mg	2/91	16.7mg	42/49					
b	117	1.63mg	4.53mg	10/86	16.7mg	39/48					
c	117	1.63mg	4.53mg	10/86	16.7mg	39/48					
d	117	9.19mg	59.3mg	0/45	16.7mg	9/48					
e	117	12.4mg	218.mg	0/45	16.7mg	6/48					
f	117	16.1mg	391.mg	0/86	16.7mg	4/48					
1906	1095	4.00mg	n.s.s.	8/85	10.4mg	5/25					Roe;natu,216,375-376;1967
1907	2012m	57.2ug	n.s.s.	1/20	57.1ug	0/20	2.86mg	0/20	5.71mg	0/20	Goldenthal;irdc,399-062;1989/pers.comm.
1908	2012n	14.5mg	310.mg	0/50	57.1ug	1/50	2.86mg	5/50	5.71mg	5/50	

```
      Spe Strain  Site   Xpo+Xpt
        Sex  Route  Hist    Notes                          .                                              TD50    2Tailpvl
                                                                                                          DR      AuOp
  a   R f f34 wat liv hpc 24m24 e                                                                        40.7mg  * P<.003 -
  b   R f f34 wat pit ade 24m24 e                                                                        14.8mg  * P<.06  -
  c   R f f34 wat liv hpa 24m24 e                                                                        184.mg  * P<.5   -
1909  R m f34 wat liv tum 52w52 ek                          .>                                          no dre    P=1.   -
1910  R m f34 wat liv mix 24m24 e                                              .>                        155.mg  * P<.5   -
  a   R m f34 wat liv hpa 24m24 e                                                                        205.mg  * P<.6   -
  b   R m f34 wat liv hpc 24m24 e                                                                        650.mg  * P<.8   -

1,2-DIMETHYLHYDRAZINE.2HCl        100ng..:..1ug....:..10......:..100....:..1mg....:..10.....:..100....:..1g.....:..10
1911  H f syg wat blv ang 67w67 e                                   . + .                               .156mg    P<.0005+
  a   H f syg wat liv ang 67w67 e                                                                        .156mg    P<.0005+
  b   H f syg wat lun ang 67w67 e                                                                        .395mg    P<.0005+
  c   H f syg wat cec mix 67w67 e                                                                        .864mg    P<.0005+
  d   H f syg wat liv mix 67w67 e                                                                        .917mg    P<.0005+
  e   H f syg wat mus ang 67w67 e                                                                        1.03mg    P<.0005+
  f   H f syg wat liv hpt 67w67 e                                                                        1.40mg    P<.003
  g   H f syg wat cec pla 67w67 e                                                                        2.08mg    P<.004
  h   H f syg wat hea ang 67w67 e                                                                        2.13mg    P<.004 +
  i   H f syg wat cec adc 67w67 e                                                                        2.84mg    P<.02
  j   H f syg wat pan ang 67w67 e                                                                        18.4mg    P<.4   +
1912  H m syg wat blv ang 71w71 e                                 . + .                                  .211mg    P<.0005+
  a   H m syg wat liv ang 71w71 e                                                                        .211mg    P<.0005+
  b   H m syg wat lun ang 71w71 e                                                                        .685mg    P<.0005+
  c   H m syg wat mus ang 71w71 e                                                                        1.36mg    P<.0005+
  d   H m syg wat liv mix 71w71 e                                                                        2.05mg    P<.006 +
  e   H m syg wat hea ang 71w71 e                                                                        2.49mg    P<.008 +
  f   H m syg wat cec pla 71w71 e                                                                        2.42mg    P<.02  +
  g   H m syg wat pan ang 71w71 e                                                                        18.6mg    P<.4   +
1913  M f swa wat blv ang 52w52 e                             <+                                         noTD50    P<.0005+
  a   M f swa wat lun ade 52w52 e                                                                        .559mg    P<.0005+
  b   M f swa wat liv hpt 52w52 e                                                                        13.9mg    P<.5
1914  M m swa wat blv ang 52w52 e                            . + .                                       .102mg    P<.0005+
  a   M m swa wat prn ang 52w52 e                                                                        .180mg    P<.0005
  b   M m swa wat mus ang 52w52 e                                                                        .203mg    P<.0005
  c   M m swa wat fat ang 52w52 e                                                                        .215mg    P<.0005
  d   M m swa wat pep ang 52w52 e                                                                        .337mg    P<.0005
  e   M m swa wat liv ang 52w52 e                                                                        .377mg    P<.0005
  f   M m swa wat sub ang 52w52 e                                                                        .781mg    P<.0005
  g   M m swa wat lyd ang 52w52 e                                                                        1.02mg    P<.0005
  h   M m swa wat lun mix 52w52 e                                                                        1.81mg    P<.06  +

2-(2,2-DIMETHYLHYDRAZINO)-4-(5-NITRO-2-FURYL)THIAZOLE.1_0....:..100....:..1mg....:..10.....:..100....:..1g.....:..10
1915  R f sda eat mgl mix 46w66 e                                  <+                                    noTD50    P<.0005+
  a   R f sda eat mgl adc 46w66 e                                                                        .391mg    P<.0005
  b   R f sda eat --- lbl 46w66 e                                               +hist 16.3mg  P<.09  +
  c   R f sda eat liv tum 46w66 e                                                                        no dre    P=1.
  d   R f sda eat tba mix 46w66 e                                                                        noTD50    P<.0005

DIMETHYLNITRAMINE                 100ng..:..1ug....:..10......:..100....:..1mg....:..10.....:..100....:..1g.....:..10
1916  R f bdf gav nas ene 24m28 a                                         . + .                          10.4mg  * P<.0005+
1917  R m bdf gav nas ene 23m28 a                                         . + .                          4.74mg  * P<.0005+
1918  R f nzd wat liv hpc 12m24                          . + .                                           .256mg  * P<.0005+
  a   R f nzd wat tba tum 12m24                                                                          .321mg    P<.002
1919  R m nzd wat liv hpc 12m24                            . + .                                         .322mg  * P<.0005+
  a   R m nzd wat tba tum 12m24                                                                          .403mg    P<.002

DIMETHYLVINYL CHLORIDE            100ng..:..1ug....:..10......:..100....:..1mg....:..10.....:..100....:..1g.....:..10
1920  M f b6c gav for MXA 24m24                                              : + :                       14.3mg  \ P<.0005c
  a   M f b6c gav for MXA 24m24                                                                          14.3mg  \ P<.0005c
  b   M f b6c gav for sqc 24m24                                                                          14.4mg  \ P<.0005c
  c   M f b6c gav for sqc 24m24                                                                          14.4mg  \ P<.0005c
  d   M f b6c gav MXA MXA 24m24                                                                          124.mg  / P<.004
  e   M f b6c gav hag ppa 24m24                                                                          146.mg    P<.0005
  f   M f b6c gav liv hpa 24m24                                                                          150.mg  * P<.007
  g   M f b6c gav lun MXA 24m24                                                                          186.mg  * P<.003
  h   M f b6c gav lun a/a 24m24                                                                          235.mg  * P<.004
  i   M f b6c gav for sqp 24m24                                                                          373.mg  * P<.005 c
  j   M f b6c gav MXA mlm 24m24                                                                          229.mg  * P<.04
  k   M f b6c gav TBA MXB 24m24                                                                          14.6mg  * P<.0005
  l   M f b6c gav liv MXB 24m24                                                                          150.mg  * P<.007
  m   M f b6c gav lun MXB 24m24                                                                          186.mg  * P<.003
1921  M m b6c gav MXB MXB 24m24                                              : + :                       15.2mg  * P<.0005
  a   M m b6c gav for MXA 24m24                                                                          15.5mg  * P<.0005c
  b   M m b6c gav for sqc 24m24                                                                          17.2mg  * P<.0005c
  c   M m b6c gav liv MXA 24m24                                                                          48.8mg  * P<.0005
  d   M m b6c gav liv hpa 24m24                                                                          87.1mg  * P<.0005
  e   M m b6c gav lun MXA 24m24                                                                          87.7mg  * P<.0005
  f   M m b6c gav pre sqc 24m24                                                                          90.7mg  / P<.0005c
  g   M m b6c gav liv hpc 24m24                                                                          104.mg  * P<.0005
  h   M m b6c gav lun a/a 24m24                                                                          117.mg  * P<.002
  i   M m b6c gav MXA MXA 24m24                                                                          148.mg  * P<.008
```

	RefNum	LoConf	UpConf	Cntrl	1Dose	1Inc	2Dose	2Inc			Citation or Pathology / Brkly Code
a	2012n	17.6mg	197.mg	0/50	57.1ug	0/50	2.86mg	3/50	5.71mg	4/50	
b	2012n	5.82mg	n.s.s.	15/50	57.1ug	22/50	2.86mg	16/50	5.71mg	28/50	
c	2012n	38.9mg	n.s.s.	0/50	57.1ug	1/50	2.86mg	2/50	5.71mg	1/50	
1909	2012m	50.0ug	n.s.s.	0/20	50.0ug	0/20	2.50mg	0/20	5.00mg	0/20	
1910	2012n	26.3mg	n.s.s.	3/50	50.0ug	0/50	2.50mg	1/50	5.00mg	3/50	
a	2012n	30.0mg	n.s.s.	2/50	50.0ug	0/50	2.50mg	1/50	5.00mg	2/50	
b	2012n	45.5mg	n.s.s.	1/50	50.0ug	0/50	2.50mg	0/50	5.00mg	1/50	

1,2-DIMETHYLHYDRAZINE.2HCl   306-37-6

	RefNum	LoConf	UpConf	Cntrl	1Dose	1Inc	Citation or Pathology
1911	1108	94.9ug	.258mg	0/32	1.36mg	44/48	Toth;canr,32,804-807;1972/1967a
a	1108	94.9ug	.258mg	0/32	1.36mg	44/48	
b	1108	.250mg	.670mg	0/32	1.36mg	30/48	
c	1108	.487mg	1.76mg	0/32	1.36mg	17/47	
d	1108	.444mg	2.51mg	0/25	1.36mg	10/29	
e	1108	.566mg	2.27mg	0/32	1.36mg	15/48	
f	1108	.604mg	6.57mg	0/25	1.36mg	7/29	
g	1108	.940mg	11.0mg	0/32	1.36mg	8/47	
h	1108	.962mg	11.8mg	0/32	1.36mg	8/48	
i	1108	1.16mg	n.s.s.	0/32	1.36mg	6/47	
j	1108	3.00mg	n.s.s.	0/32	1.36mg	1/48	
1912	1108	.135mg	.342mg	0/31	1.20mg	41/49	
a	1108	.135mg	.342mg	0/31	1.20mg	41/49	
b	1108	.406mg	1.28mg	0/31	1.20mg	21/49	
c	1108	.701mg	3.69mg	0/31	1.20mg	12/49	
d	1108	.883mg	18.3mg	0/28	1.20mg	7/41	
e	1108	1.07mg	37.5mg	0/31	1.20mg	7/49	
f	1108	.987mg	n.s.s.	0/28	1.20mg	6/41	
g	1108	3.03mg	n.s.s.	0/31	1.20mg	1/49	
1913	119	n.s.s.	.128mg	4/109	2.00mg	49/49	Toth;ajpa,64,585-600;1971
a	119	.334mg	1.11mg	0/18	2.00mg	22/48	
b	119	2.26mg	n.s.s.	0/15	2.00mg	1/41	
1914	119	60.3ug	.171mg	0/40	1.67mg	46/49	
a	119	.116mg	.291mg	0/40	1.67mg	39/49	
b	119	.131mg	.330mg	0/40	1.67mg	37/49	
c	119	.139mg	.351mg	0/40	1.67mg	36/49	
d	119	.211mg	.580mg	0/40	1.67mg	28/49	
e	119	.233mg	.662mg	0/40	1.67mg	26/49	
f	119	.427mg	1.66mg	0/40	1.67mg	15/49	
g	119	.522mg	2.49mg	0/40	1.67mg	12/49	
h	119	.655mg	n.s.s.	11/95	1.67mg	12/49	

2-(2,2-DIMETHYLHYDRAZINO)-4-(5-NITRO-2-FURYL)THIAZOLE   26049-69-4

	RefNum	LoConf	UpConf	Cntrl	1Dose	1Inc	Citation or Pathology
1915	200a	n.s.s.	.410mg	2/39	3.48mg	35/35	Cohen;jnci,51,403-417;1973
a	200a	.219mg	.707mg	0/39	3.48mg	32/35	
b	200a	4.02mg	n.s.s.	0/39	3.48mg	2/35	
c	200a	10.1mg	n.s.s.	0/39	3.48mg	0/35	
d	200a	n.s.s.	.410mg	2/39	3.48mg	35/35	

DIMETHYLNITRAMINE   4164-28-7

	RefNum	LoConf	UpConf	Cntrl	1Dose	1Inc	2Dose	2Inc	Citation or Pathology
1916	1931	4.98mg	27.4mg	0/10	6.43mg	3/10	12.9mg	8/10	Scherf;carc,10,1977-1981;1989/pers.comm.
1917	1931	2.26mg	10.7mg	0/10	6.43mg	6/10	12.9mg	10/10	
1918	119a	89.8ug	.796mg	0/107	.860mg	9/10			Goodall;clet,1,295-298;1976
a	119a	96.7ug	2.35mg	40/107	.860mg	9/10			
1919	119a	.126mg	1.02mg	1/107	.753mg	8/10			
a	119a	.137mg	3.12mg	30/107	.753mg	8/10			

DIMETHYLVINYL CHLORIDE   513-37-1

	RefNum	LoConf	UpConf	Cntrl	1Dose	1Inc	2Dose	2Inc	Citation or Pathology / Brkly Code
1920	TR316	8.29mg	24.3mg	0/50	71.4mg	41/50	(142.mg	36/50)	for:adq,can,sqc.
a	TR316	8.29mg	24.3mg	0/50	71.4mg	41/50	(142.mg	38/50)	for:adq,can,sqc,sqp.
b	TR316	8.34mg	24.8mg	0/50	71.4mg	40/50	(142.mg	36/50)	
c	TR316	8.34mg	24.8mg	0/50	71.4mg	40/50	(142.mg	38/50)	for:sqc,sqp.
d	TR316	45.7mg	1.32gm	10/50	71.4mg	2/50	142.mg	10/50	jej:mlp; liv:mlm; mul:mlh,mlm,mlp; spl:mlm. S
e	TR316	52.9mg	489.mg	0/50	71.4mg	3/50	142.mg	5/50	S
f	TR316	50.0mg	3.68gm	4/50	71.4mg	4/50	142.mg	4/50	S
g	TR316	58.7mg	1.62gm	3/50	71.4mg	1/50	142.mg	7/50	lun:a/a,a/c. S
h	TR316	68.5mg	2.72gm	2/50	71.4mg	1/50	142.mg	6/50	S
i	TR316	83.4mg	5.10gm	0/50	71.4mg	1/50	142.mg	3/50	
j	TR316	64.1mg	n.s.s.	6/50	71.4mg	1/50	142.mg	5/50	liv:mlm; mul:mlm; spl:mlm. S
k	TR316	9.20mg	25.3mg	32/50	71.4mg	45/50	142.mg	43/50	
l	TR316	50.0mg	3.68gm	4/50	71.4mg	4/50	142.mg	4/50	liv:hpa,hpc,nnd.
m	TR316	58.7mg	1.62gm	3/50	71.4mg	1/50	142.mg	7/50	lun:a/a,a/c.
1921	TR316	9.93mg	23.6mg	2/50	71.4mg	44/50	142.mg	42/50	for:sqc,sqp; pre:sqc. C
a	TR316	10.1mg	23.7mg	1/50	71.4mg	43/50	142.mg	41/50	for:sqc,sqp.
b	TR316	11.2mg	26.2mg	0/50	71.4mg	42/50	142.mg	35/50	
c	TR316	24.0mg	133.mg	11/50	71.4mg	12/50	142.mg	13/50	liv:hpa,hpc. S
d	TR316	36.4mg	407.mg	8/50	71.4mg	7/50	142.mg	8/50	S
e	TR316	37.2mg	355.mg	6/50	71.4mg	9/50	142.mg	8/50	lun:a/a,a/c. S
f	TR316	41.8mg	213.mg	1/50	71.4mg	3/50	142.mg	16/50	
g	TR316	44.3mg	375.mg	3/50	71.4mg	6/50	142.mg	7/50	S
h	TR316	43.2mg	778.mg	3/50	71.4mg	6/50	142.mg	4/50	S
i	TR316	51.1mg	4.32gm	6/50	71.4mg	6/50	142.mg	6/50	duo:mlm; lyd:mno; mul:grl,mlh,mlm,mlp,mlu; spl:mlm. S

```
     Spe Strain  Site   Xpo+Xpt                                                                        TD50    2Tailpvl
         Sex  Route   Hist    Notes                                                                      DR      AuOp
  j    M m b6c gav hag ppa 24m24                                                                       149.mg  * P<.004
  k    M m b6c gav MXA MXA 24m24                                                                       154.mg  * P<.009
  l    M m b6c gav for sqp 24m24                                                                       195.mg  * P<.0005c
  m    M m b6c gav lun a/c 24m24                                                                       199.mg  * P<.007
  n    M m b6c gav MXA MXA 24m24                                                                       221.mg  * P<.008
  o    M m b6c gav MXA hes 24m24                                                                       375.mg  * P<.03
  p    M m b6c gav TBA MXB 24m24                                                                       13.9mg  * P<.0005
  q    M m b6c gav liv MXB 24m24                                                                       48.8mg  * P<.0005
  r    M m b6c gav lun MXB 24m24                                                                       87.7mg  * P<.0005
1922 R f f34 gav MXB MXB 23m24 a                                       :  +  :                          25.3mg  / P<.0005

  a    R f f34 gav nas MXA 23m24 a                                                                     34.3mg  / P<.0005c
  b    R f f34 gav nas MXA 23m24 a                                                                     35.2mg  / P<.0005c
  c    R f f34 gav mgl MXA 23m24 a                                                                     36.7mg  / P<.0005
  d    R f f34 gav mgl fba 23m24 a                                                                     38.3mg  / P<.0005
  e    R f f34 gav nas MXA 23m24 a                                                                     46.8mg  / P<.0005c
  f    R f f34 gav nas can 23m24 a                                                                     52.7mg  / P<.0005c
  g    R f f34 gav pta MXA 23m24 a                                                                     68.0mg  * P<.004
  h    R f f34 gav pta MXA 23m24 a                                                                     71.7mg  * P<.006
  i    R f f34 gav ute MXA 23m24 a                                                                     84.9mg  * P<.002
  j    R f f34 gav ute esp 23m24 a                                                                     91.9mg  * P<.004
  k    R f f34 gav for MXA 23m24 a                                                                     95.2mg  * P<.0005c
  l    R f f34 gav thy MXA 23m24 a                                                                     131.mg  * P<.002
  m    R f f34 gav thy fcc 23m24 a                                                                     136.mg  * P<.004
  n    R f f34 gav nas MXA 23m24 a                                                                     143.mg  / P<.0005c
  o    R f f34 gav nas adc 23m24 a                                                                     159.mg  / P<.0005c
  p    R f f34 gav for sqc 23m24 a                                                                     191.mg  * P<.0005c
  q    R f f34 gav for sqp 23m24 a                                                                     209.mg  * P<.008 c
  r    R f f34 gav lun MXA 23m24 a                                                                     218.mg  * P<.002
  s    R f f34 gav eso sqc 23m24 a                                                                     250.mg  * P<.003 c
  t    R f f34 gav MXA MXA 23m24 a                                                                     318.mg  / P<.0005c
  u    R f f34 gav MXA sqp 23m24 a                                                                     1.01gm  * P<.006 c
  v    R f f34 gav pta MXA 23m24 a                                                                     101.mg  * P<.03
  w    R f f34 gav pta cra 23m24 a                                                                     109.mg  * P<.04
  x    R f f34 gav mul MXA 23m24 a                                                                     170.mg  * P<.02
  y    R f f34 gav mul mnl 23m24 a                                                                     202.mg  * P<.04
  z    R f f34 gav pta crc 23m24 a                                                                     222.mg  * P<.02
  A    R f f34 gav thy cca 23m24 a                                                                     225.mg  * P<.04
  B    R f f34 gav orm sqp 23m24 a                                                                     1.22gm  / P<.02  c
  C    R f f34 gav TBA MXB 23m24 a                                                                     11.9mg  / P<.0005
  D    R f f34 gav liv MXB 23m24 a                                                                     no dre    P=1.
1923 R m f34 gav tes ict 22m24 a                                      :  +  :                          13.9mg  / P<.0005
  a    R m f34 gav MXB MXB 22m24 a                                                                     17.5mg  / P<.0005

  b    R m f34 gav nas MXA 22m24 a                                                                     29.7mg  / P<.0005c
  c    R m f34 gav for MXA 22m24 a                                                                     41.6mg  * P<.0005c
  d    R m f34 gav nas MXA 22m24 a                                                                     52.1mg  / P<.0005c
  e    R m f34 gav nas can 22m24 a                                                                     63.4mg  / P<.0005c
  f    R m f34 gav for sqp 22m24 a                                                                     67.1mg  * P<.0005c
  g    R m f34 gav nas acn 22m24 a                                                                     70.1mg  / P<.0005
  h    R m f34 gav eso MXA 22m24 a                                                                     81.6mg  / P<.0005c
  i    R m f34 gav mul mnl 22m24 a                                                                     128.mg  * P<.006
  j    R m f34 gav for sqc 22m24 a                                                                     135.mg  * P<.0005c
  k    R m f34 gav eso sqc 22m24 a                                                                     143.mg  * P<.0005c
  l    R m f34 gav MXA MXA 22m24 a                                                                     163.mg  * P<.0005c
  m    R m f34 gav MXA sqc 22m24 a                                                                     178.mg  * P<.0005c
  n    R m f34 gav eso sqp 22m24 a                                                                     218.mg  * P<.002 c
  o    R m f34 gav nas sqc 22m24 a                                                                     297.mg  * P<.009 c
  p    R m f34 gav ton MXA 22m24 a                                                                     415.mg  * P<.002 c
  q    R m f34 gav ton sqc 22m24 a                                                                     469.mg  * P<.005 c
  r    R m f34 gav MXA MXA 22m24 a                                                                     86.7mg  / P<.02
  s    R m f34 gav TBA MXB 22m24 a                                                                     10.2mg  / P<.0005
  t    R m f34 gav liv MXB 22m24 a                                                                     1.03gm  / P<.3
```

```
DINITRO(1-METHYLHEPTYL)PHENYL  CROTONATE..:..1ug....:..10......:..100....:..1mg....:..10.....:..100....:..1g.....:..10
1924 M f b6a orl liv hpt 76w76 evx                                                .>                   no dre    P=1.   -
  a    M f b6a orl lun ade 76w76 evx                                                                   no dre    P=1.   -
  b    M f b6a orl tba mix 76w76 evx                                                                   no dre    P=1.   -
1925 M m b6a orl liv hpt 76w76 evx                                             .>                       2.07mg    P<.6   -
  a    M m b6a orl lun ade 76w76 evx                                                                   19.1mg    P<1.   -
  b    M m b6a orl tba mix 76w76 evx                                                                   11.9mg    P<1.   -
1926 M f b6c orl liv hpt 76w76 evx                                                .>                   2.50mg    P<.3   -
  a    M f b6c orl lun mix 76w76 evx                                                                   no dre    P=1.   -
  b    M f b6c orl tba mix 76w76 evx                                                                   1.21mg    P<.1   -
1927 M m b6c orl liv hpt 76w76 evx                                                .>                   2.32mg    P<.3   -
  a    M m b6c orl lun mix 76w76 evx                                                                   no dre    P=1.   -
  b    M m b6c orl tba mix 76w76 evx                                                                   .405mg    P<.007 -
```

```
2,4-DINITROPHENOL                   100ng..:..1ug....:..10....:..100....:..1mg....:..10.....:..100....:..1g.....:..10
1928 M b c51 eat pit ade 73w73 er                                                      .>              no dre    P=1.
```

	RefNum	LoConf	UpConf	Cntrl	1Dose	1Inc	2Dose	2Inc	Citation or Pathology
									Brkly Code
j	TR316	50.4mg	1.51gm	2/50	71.4mg	3/50	142.mg	3/50	S
k	TR316	52.0mg	7.57gm	6/50	71.4mg	5/50	142.mg	6/50	duo:mlm; lyd:mno; mul:mlh,mlm,mlp,mlu; spl:mlm. S
l	TR316	66.8mg	827.mg	1/50	71.4mg	3/50	142.mg	8/50	
m	TR316	65.5mg	3.88gm	3/50	71.4mg	4/50	142.mg	5/50	S
n	TR316	68.2mg	5.79gm	2/50	71.4mg	3/50	142.mg	4/50	liv:hem,hes; mln:hem; pre:hem; spl:hes. S
o	TR316	83.9mg	n.s.s.	1/50	71.4mg	1/50	142.mg	3/50	liv:hes; spl:hes. S
p	TR316	8.92mg	23.2mg	26/50	71.4mg	47/50	142.mg	44/50	
q	TR316	24.0mg	133.mg	11/50	71.4mg	12/50	142.mg	13/50	liv:hpa,hpc,nnd.
r	TR316	37.2mg	355.mg	6/50	71.4mg	9/50	142.mg	8/50	lun:a/a,a/c.
1922	TR316	15.9mg	40.5mg	1/50	70.7mg	25/50	143.mg	36/50	eso:sqc; for:sqc,sqp; gnv:sqc; nas:adc,adn,can,sqc; orm:sqp;
									pal:sqc,sqp; ton:sqc. C
a	TR316	20.8mg	55.6mg	0/50	70.7mg	17/50	143.mg	35/50	nas:adc,adn,can,sqc.
b	TR316	21.1mg	57.5mg	0/50	70.7mg	16/50	143.mg	35/50	nas:adc,can,sqc.
c	TR316	19.1mg	89.9mg	11/50	70.7mg	21/50	143.mg	5/50	mgl:adc,fba. S
d	TR316	19.5mg	98.1mg	10/50	70.7mg	18/50	143.mg	5/50	S
e	TR316	26.9mg	77.7mg	0/50	70.7mg	13/50	143.mg	29/50	nas:can,sqc.
f	TR316	29.2mg	89.1mg	0/50	70.7mg	11/50	143.mg	28/50	
g	TR316	29.5mg	616.mg	17/50	70.7mg	17/50	143.mg	2/50	pta:adn,cra,crc. S
h	TR316	30.3mg	940.mg	17/50	70.7mg	16/50	143.mg	2/50	pta:cra,crc. S
i	TR316	35.3mg	543.mg	9/50	70.7mg	12/50	143.mg	2/50	ute:esp,ess. S
j	TR316	36.6mg	801.mg	8/50	70.7mg	11/50	143.mg	2/50	S
k	TR316	40.2mg	313.mg	1/50	70.7mg	9/50	143.mg	2/50	for:sqc,sqp.
l	TR316	44.3mg	994.mg	1/50	70.7mg	5/50	143.mg	1/50	thy:fca,fcc. S
m	TR316	44.9mg	1.53gm	1/50	70.7mg	5/50	143.mg	0/50	S
n	TR316	54.6mg	412.mg	0/50	70.7mg	4/50	143.mg	6/50	nas:adc,adn.
o	TR316	56.8mg	504.mg	0/50	70.7mg	3/50	143.mg	6/50	
p	TR316	65.9mg	832.mg	0/50	70.7mg	5/50	143.mg	1/50	
q	TR316	62.0mg	7.92gm	0/50	70.7mg	4/50	143.mg	1/50	
r	TR316	70.5mg	1.11gm	0/50	70.7mg	4/50	143.mg	1/50	lun:a/a,a/c. S
s	TR316	72.5mg	1.89gm	0/50	70.7mg	3/50	143.mg	1/50	
t	TR316	117.mg	1.13gm	0/50	70.7mg	2/50	143.mg	5/50	gnv:sqc; orm:sqp; pal:sqc,sqp; ton:sqc.
u	TR316	343.mg	10.5gm	0/50	70.7mg	0/50	143.mg	4/50	orm:sqp; pal:sqp.
v	TR316	37.4mg	n.s.s.	16/50	70.7mg	13/50	143.mg	2/50	pta:adn,cra.
w	TR316	38.9mg	n.s.s.	16/50	70.7mg	12/50	143.mg	2/50	S
x	TR316	59.6mg	n.s.s.	5/50	70.7mg	8/50	143.mg	1/50	mul:lle,mnl. S
y	TR316	64.5mg	n.s.s.	5/50	70.7mg	7/50	143.mg	1/50	S
z	TR316	64.6mg	n.s.s.	1/50	70.7mg	4/50	143.mg	0/50	S
A	TR316	64.4mg	n.s.s.	3/50	70.7mg	4/50	143.mg	1/50	S
B	TR316	369.mg	n.s.s.	0/50	70.7mg	0/50	143.mg	3/50	
C	TR316	7.69mg	19.4mg	33/50	70.7mg	48/50	143.mg	42/50	
D	TR316	n.s.s.	n.s.s.	0/50	70.7mg	0/50	143.mg	0/50	liv:hpa,hpc,nnd.
1923	TR316	7.93mg	30.9mg	40/50	70.7mg	41/50	143.mg	6/50	S
a	TR316	10.8mg	28.5mg	0/50	70.7mg	33/50	143.mg	29/50	eso:sqc,sqp; for:sqc,sqp; lpp:sqc; nas:adc,can,sqc; pal:sqc,sqp;
									ton:sqc,sqp. C
b	TR316	18.2mg	47.1mg	0/50	70.7mg	23/50	143.mg	28/50	nas:adc,can,sqc.
c	TR316	19.8mg	98.0mg	0/50	70.7mg	14/50	143.mg	0/50	for:sqc,sqp.
d	TR316	32.0mg	84.8mg	0/50	70.7mg	15/50	143.mg	24/50	nas:can,sqc.
e	TR316	38.2mg	101.mg	0/50	70.7mg	12/50	143.mg	24/50	
f	TR316	26.7mg	230.mg	0/50	70.7mg	7/50	143.mg	0/50	
g	TR316	29.7mg	185.mg	0/50	70.7mg	8/50	143.mg	4/50	S
h	TR316	32.0mg	249.mg	0/50	70.7mg	6/50	143.mg	4/50	eso:sqc,sqp.
i	TR316	44.2mg	2.22gm	3/50	70.7mg	6/50	143.mg	1/50	S
j	TR316	50.3mg	463.mg	0/50	70.7mg	7/50	143.mg	0/50	
k	TR316	44.3mg	770.mg	0/50	70.7mg	4/50	143.mg	1/50	
l	TR316	56.5mg	511.mg	0/50	70.7mg	5/50	143.mg	4/50	lpp:sqc; pal:sqc,sqp; ton:sqc,sqp.
m	TR316	58.0mg	708.mg	0/50	70.7mg	5/50	143.mg	2/50	lpp:sqc; pal:sqc; ton:sqc.
n	TR316	59.8mg	1.31gm	0/50	70.7mg	2/50	143.mg	3/50	
o	TR316	87.7mg	13.1gm	0/50	70.7mg	3/50	143.mg	0/50	
p	TR316	157.mg	2.06gm	0/50	70.7mg	3/50	143.mg	3/50	ton:sqc,sqp.
q	TR316	165.mg	4.25gm	0/50	70.7mg	3/50	143.mg	2/50	
r	TR316	29.2mg	n.s.s.	13/50	70.7mg	7/50	143.mg	4/50	adr:phm; amd:phe. S
s	TR316	6.56mg	17.2mg	40/50	70.7mg	48/50	143.mg	37/50	
t	TR316	107.mg	n.s.s.	1/50	70.7mg	0/50	143.mg	1/50	liv:hpa,hpc,nnd.

DINITRO(1-METHYLHEPTYL)PHENYL CROTONATE (karathane) 6119-92-2

1924	1290	.774mg	n.s.s.	0/17	.414mg	0/17			Innes;ntis,1968/1969
a	1290	.774mg	n.s.s.	1/17	.414mg	0/17			
b	1290	.774mg	n.s.s.	2/17	.414mg	0/17			
1925	1290	.302mg	n.s.s.	1/18	.385mg	2/17			
a	1290	.335mg	n.s.s.	2/18	.385mg	2/17			
b	1290	.280mg	n.s.s.	3/18	.385mg	3/17			
1926	1290	.407mg	n.s.s.	0/16	.414mg	1/17			
a	1290	.774mg	n.s.s.	0/16	.414mg	0/17			
b	1290	.297mg	n.s.s.	0/16	.414mg	2/17			
1927	1290	.378mg	n.s.s.	0/16	.385mg	1/17			
a	1290	.721mg	n.s.s.	0/16	.385mg	0/17			
b	1290	.152mg	5.47mg	0/16	.385mg	5/17			

2,4-DINITROPHENOL   51-28-5

1928	257	241.mg	n.s.s.	0/28	62.5mg	0/38			King;pseb,112,365-366;1963

```
     Spe Strain Site   Xpo+Xpt                                                              TD50    2Tailpvl
       Sex   Route  Hist    Notes                                                             DR     AuOp

2,4-DINITROPHENOL, SODIUM        100ng..:..1ug....:..10.....:..100....:..1mg....:..10.....:..100....:..1g.....:..10
1929 M f dbx eat mgl car 90w90 er                                                             .>     no dre   P=1.  -

1,4-DINITROSO-2,6-DIMETHYLPIPERAZINE  ..:..1ug....:..10.....:..100....:..1mg....:..10.....:..100....:..1g.....:..10
1930 H m syg gav for pam 35w76                              .      ±                          3.10mg   P<.04  +
   a H m syg gav lun ade 35w76                                                                6.04mg   P<.02
   b H m syg gav liv ang 35w76                                                                8.30mg   P<.04
   c H m syg gav tba mix 35w76                                                                .932mg   P<.0005

DINITROSOHOMOPIPERAZINE          100ng..:..1ug....:..10.....:..100....:..1mg....:..10.....:..100....:..1g.....:..10
1931 R f f34 wat ugi car  7m31 es                      .  +  .                               .240mg Z P<.0005+

   a R f f34 wat eso mix  7m31 es                                                             .253mg Z P<.0005+

   b R f f34 wat ugi mix  7m31 es                                                             .708mg Z P<.0005+

   c R f f34 wat liv mix  7m31 es                                                             no dre   P=1.  +

   d R f f34 wat nas olc  7m31 es                                                             no dre   P=1.  +

1932 R f f34 wat ugi mix 60w82 es                     .    +   .                             29.7ug   P<.0005+
   a R f f34 wat ugi car 60w82 es                                                             46.9ug   P<.0005+
   b R f f34 wat eso mix 60w82 es                                                             55.3ug   P<.0005+
   c R f f34 wat nas olc 60w82 es                                                             .207mg   P<.002 +
1933 R f f34 wat ugi mix 28m31 es                          .  +  .                            91.4ug * P<.0005+
   a R f f34 wat liv mix 28m31 es                                                             97.3ug \ P<.006 +
   b R f f34 wat eso mix 28m31 es                                                             .105mg * P<.0005+
   c R f f34 wat ton bcp 28m31 es                                                             .630mg * P<.004 +

N,N-DINITROSOPENTAMETHYLENETETRAMINE  ..:..1ug....:..10.....:..100....:..1mg....:..10.....:..100....:..1g.....:..10
1934 R m cbr ipj liv hpt  6m23 e                                        .>                    no dre   P=1.  -

DINITROSOPIPERAZINE              100ng..:..1ug....:..10.....:..100....:..1mg....:..10.....:..100....:..1g.....:..10
1935 M f c17 gav for sqc  9m24 e                                 .    +    .                  4.66mg   P<.003 +
   a M f c17 gav ute rna  9m24 e                                                              4.13mg   P<.02
1936 M m c17 gav for sqc  9m24 e                              .    +    .                     2.01mg   P<.0005+
1937 M f swi wat lun ade 12m23 e                            .>                                no dre   P=1.
   a M f swi wat liv hpc 12m23 e                                                              no dre   P=1.
   b M f swi wat tba mix 12m23 e                                                              no dre   P=1.  +
1938 M m swi wat lun ade 12m23 e                                 .    +    .                  8.70mg   P<.0005
   a M m swi wat liv hpc 12m23 e                                                              51.9mg   P<.2
   b M m swi wat tba mix 12m23 e                                                              8.24mg   P<.02 +

2,4-DINITROTOLUENE               100ng..:..1ug....:..10.....:..100....:..1mg....:..10.....:..100....:..1g.....:..10
1939 M f b6c eat TBA MXB 78w93                                         :>                     no dre   P=1.  -
   a M f b6c eat liv MXB 78w93                                                                no dre   P=1.
   b M f b6c eat lun MXB 78w93                                                                68.0mg \ P<.2
1940 M m b6c eat TBA MXB 78w92                                            :>                   no dre   P=1.  -
   a M m b6c eat liv MXB 78w92                                                                no dre   P=1.
   b M m b6c eat lun MXB 78w92                                                                no dre   P=1.
1941 R f f34 eat mgl fba 18m25 v                                  :    ±                      12.7mg * P<.02  a
   a R f f34 eat TBA MXB 18m25 v                                                              no dre   P=1.
   b R f f34 eat liv MXB 18m25 v                                                              no dre   P=1.
1942 R m f34 eat MXA fib 18m25 v                                   :  +  :                    9.35mg * P<.0005a
   a R m f34 eat TBA MXB 18m25 v                                                              43.7mg * P<.8
   b R m f34 eat liv MXB 18m25 v                                                              31.1mg * P<.04

2,4-DINITROTOLUENE (PURIFIED)    100ng..:..1ug....:..10.....:..100....:..1mg....:..10.....:..100....:..1g.....:..10
1943 R m f34 eat liv mix 52w52 er                                       .>                    90.2mg   P<.3  -
   a R m f34 eat liv nnd 52w52 er                                                             90.2mg   P<.3  -

2,6-DINITROTOLUENE               100ng..:..1ug....:..10.....:..100....:..1mg....:..10.....:..100....:..1g.....:..10
1944 R m f34 eat liv thc 52w52 er                          .  +  .                            .574mg * P<.0005
   a R m f34 eat liv hpc 52w52 er                                                             .574mg * P<.0005+
   b R m f34 eat liv nnd 52w52 er                                                             .964mg * P<.0005+
   c R m f34 eat liv clc 52w52 er                                                             34.2mg * P<.8
   d R m f34 eat liv hpd 52w52 er                                                             68.9mg * P<.9

DINITROTOLUENE, TECHNICAL GRADE (2,4 (77%)- and 2,6 (19%)-)..:..100....:..1mg....:..10.....:..100....:..1g.....:..10
1945 R m f34 eat liv nnd 52w52 er                          .    +    .                        8.02mg   P<.0005+
   a R m f34 eat liv thc 52w52 er                                                             9.34mg   P<.0005+
   b R m f34 eat liv clc 52w52 er                                                             53.9mg   P<.09

1,4-DIOXANE                      100ng..:..1ug....:..10.....:..100....:..1mg....:..10.....:..100....:..1g.....:..10
1946 M f b6c wat liv MXA 90w90                                                       :+ :      594.mg * P<.0005c
   a M f b6c wat liv hpc 90w90                                                                938.mg / P<.0005c
   b M f b6c wat TBA MXB 90w90                                                                847.mg * P<.0005
   c M f b6c wat liv MXB 90w90                                                                594.mg * P<.0005
   d M f b6c wat lun MXB 90w90                                                                71.9gm   P<.8
1947 M m b6c wat liv hpc 90w91                                                       :  +  :   1.42gm * P<.0005c
   a M m b6c wat liv MXA 90w91                                                                1.46gm * P<.0005c
```

	RefNum	LoConf	UpConf	Cntrl	1Dose	1Inc	2Dose	2Inc				Citation or Pathology
												Brkly Code

2,4-DINITROPHENOL, SODIUM   1011-73-0

	RefNum	LoConf	UpConf	Cntrl	1Dose	1Inc	2Dose	2Inc				Citation
1929	1652	2.38gm	n.s.s.	37/50	325.mg	1/50						Tannenbaum;canr,9,403-410;1949

1,4-DINITROSO-2,6-DIMETHYLPIPERAZINE   55380-34-2

									Citation
1930	1570	1.17mg	n.s.s.	3/20	3.68mg	9/20			Lijinsky;carc,4,1165-1167;1983
a	1570	2.08mg	n.s.s.	0/20	3.68mg	4/20			
b	1570	2.50mg	n.s.s.	0/20	3.68mg	3/20			
c	1570	.448mg	2.63mg	3/20	3.68mg	16/20			

DINITROSOHOMOPIPERAZINE   55557-00-1

	RefNum	LoConf	UpConf	Cntrl	1Dose	1Inc	2Dose	2Inc	3Dose	3Inc	4Dose	4Inc	5Dose	5Inc
1931	1375m	.144mg	.501mg	0/20	10.1ug	1/20	26.4ug	3/20	72.0ug	7/20	.269mg	13/20	(1.18mg	10/20
					2.93mg	14/20)									
a	1375m	.151mg	.538mg	0/20	10.1ug	1/20	26.4ug	3/20	72.0ug	6/20	.269mg	13/20	(1.18mg	10/20
					2.93mg	14/20)									
b	1375m	.398mg	1.43mg	0/20	10.1ug	1/20	26.4ug	4/20	72.0ug	9/20	.269mg	16/20	1.18mg	15/20
					2.93mg	19/20									
c	1375m	29.3mg	n.s.s.	1/20	10.1ug	6/20	26.4ug	4/20	72.0ug	1/20	.269mg	0/20	1.18mg	0/20
					2.93mg	0/20									
d	1375m	26.8mg	n.s.s.	0/20	10.1ug	0/20	26.4ug	0/20	72.0ug	7/20	.269mg	0/20	1.18mg	0/20
					2.93mg	0/20									

Lijinsky;eaes,6,513-527;1982/pers.comm.

1932	1375n	12.2ug	68.1ug	0/20	.209mg	19/20			
a	1375n	23.4ug	.102mg	0/20	.209mg	17/20			
b	1375n	28.2ug	.120mg	0/20	.209mg	16/20			
c	1375n	88.4ug	.783mg	0/20	.209mg	7/20			
1933	1375o	52.8ug	.173mg	0/20	39.9ug	5/20	.116mg	17/20	
a	1375o	40.6ug	1.20mg	1/20	39.9ug	8/20	(.116mg	6/20)	
b	1375o	59.4ug	.203mg	0/20	39.9ug	3/20	.116mg	17/20	
c	1375o	.239mg	4.43mg	0/20	39.9ug	0/20	.116mg	5/20	

N,N-DINITROSOPENTAMETHYLENETETRAMINE   101-25-7

									Citation
1934	1258	4.93mg	n.s.s.	1/22	1.86mg	1/23			Boyland;ejca,4,233-239;1968

DINITROSOPIPERAZINE   140-79-4

									Citation
1935	1347	1.59mg	28.1mg	0/22	3.08mg	4/11			Pai;carc,2,175-177;1981
a	1347	1.36mg	n.s.s.	2/22	3.08mg	5/11			
1936	1347	.831mg	7.00mg	0/12	2.56mg	7/12			
1937	1250	11.7mg	n.s.s.	7/29	10.4mg	3/14			Borzsonyi;canr,40,2925-2927;1980
a	1250	27.8mg	n.s.s.	4/29	10.4mg	0/14			
b	1250	5.26mg	n.s.s.	25/29	10.4mg	10/14			
1938	1250	4.07mg	27.6mg	3/50	8.67mg	11/22			
a	1250	12.2mg	n.s.s.	2/50	8.67mg	3/22			
b	1250	3.17mg	n.s.s.	19/50	8.67mg	15/22			

2,4-DINITROTOLUENE   121-14-2

									Pathology
1939	TR54	18.6mg	n.s.s.	39/100	8.80mg	14/50	(44.2mg	11/50)	liv:hpa,hpc,nnd.
a	TR54	234.mg	n.s.s.	8/100	8.80mg	1/50	44.2mg	1/50	lun:a/a,a/c.
b	TR54	18.6mg	n.s.s.	2/100	8.80mg	4/50	(44.2mg	0/50)	
1940	TR54	86.7mg	n.s.s.	40/100	8.20mg	13/50	40.8mg	13/50	
a	TR54	86.7mg	n.s.s.	22/100	8.20mg	6/50	40.8mg	9/50	liv:hpa,hpc,nnd.
b	TR54	39.4mg	n.s.s.	18/100	8.20mg	3/50	(40.8mg	2/50)	lun:a/a,a/c.
1941	TR54	5.78mg	n.s.s.	13/75	3.00mg	12/50	7.50mg	23/50	
a	TR54	5.97mg	n.s.s.	57/75	3.00mg	41/50	7.50mg	40/50	
b	TR54	35.0mg	n.s.s.	3/75	3.00mg	0/50	7.50mg	2/50	liv:hpa,hpc,nnd.
1942	TR54	5.49mg	19.1mg	0/75	2.40mg	7/50	6.00mg	13/50	ski:fib; sub:fib.
a	TR54	4.84mg	n.s.s.	40/75	2.40mg	29/50	6.00mg	35/50	
b	TR54	12.6mg	n.s.s.	0/75	2.40mg	3/50	6.00mg	3/50	liv:hpa,hpc,nnd.

2,4-DINITROTOLUENE (PURIFIED)   121-14-2

									Citation
1943	1834	14.7mg	n.s.s.	0/20	27.0mg	1/20			Leonard;jnci,79,1313-1319;1987/pers.comm.
a	1834	14.7mg	n.s.s.	0/20	27.0mg	1/20			

2,6-DINITROTOLUENE   606-20-2

									Citation
1944	1834	.305mg	1.06mg	0/20	7.00mg	17/20	14.0mg	19/19	Leonard;jnci,79,1313-1319;1987/pers.comm.
a	1834	.305mg	1.06mg	0/20	7.00mg	17/20	14.0mg	19/19	
b	1834	.587mg	1.65mg	0/20	7.00mg	18/20	14.0mg	15/19	
c	1834	8.41mg	n.s.s.	0/20	7.00mg	2/20	14.0mg	0/19	
d	1834	11.2mg	n.s.s.	0/20	7.00mg	1/20	14.0mg	0/19	

DINITROTOLUENE, TECHNICAL GRADE (2,4 (77%)- and 2,6 (19%)-)   25321-14-6

									Citation
1945	1834	3.82mg	20.9mg	0/20	35.0mg	10/19			Leonard;jnci,79,1313-1319;1987/pers.comm.
a	1834	4.32mg	26.1mg	0/20	35.0mg	9/19			
b	1834	13.2mg	n.s.s.	0/20	35.0mg	2/19			

1,4-DIOXANE (p-dioxane) 123-91-1

									Pathology
1946	TR80	426.mg	861.mg	0/50	984.mg	21/50	1.99gm	35/50	liv:hpa,hpc.
a	TR80	639.mg	1.45gm	0/50	984.mg	12/50	1.99gm	29/50	
b	TR80	501.mg	2.19gm	15/50	984.mg	31/50	1.99gm	35/50	
c	TR80	426.mg	861.mg	0/50	984.mg	21/50	1.99gm	35/50	
d	TR80	6.34gm	n.s.s.	3/50	984.mg	0/50	1.99gm	3/50	liv:hpa,hpc,nnd.
									lun:a/a,a/c.
1947	TR80	863.mg	3.66gm	4/50	820.mg	18/50	1.65gm	24/50	
a	TR80	833.mg	5.11gm	8/50	820.mg	19/50	1.65gm	28/50	liv:hpa,hpc.

```
      Spe Strain Site   Xpo+Xpt                                                          TD50    2Tailpvl
          Sex  Route  Hist    Notes                                                           DR     AuOp
  b   M m b6c wat --- MXA 90w91                                                          3.15gm \ P<.005
  c   M m b6c wat --- lym 90w91                                                          3.93gm \ P<.009
  d   M m b6c wat TBA MXB 90w91                                                          1.88gm * P<.03
  e   M m b6c wat liv MXB 90w91                                                          1.46gm * P<.0005
  f   M m b6c wat lun MXB 90w91                                                          no dre   P=1.
1948  R f osm wat MXB MXB 26m26                                  :   +  :                126.mg  * P<.0005
  a   R f osm wat liv hpa 26m26                                                          160.mg  * P<.0005c
  b   R f osm wat ntu sqc 26m26                                                          476.mg  * P<.0005c
  c   R f osm wat TBA MXB 26m26                                                          40.5mg  \ P<.0005
  d   R f osm wat liv MXB 26m26                                                          124.mg  * P<.0005
1949  R m osm wat ntu sqc 26m26                                  :   +  :                168.mg  * P<.0005c
  a   R m osm wat TBA MXB 26m26                                                          114.mg  * P<.0005
  b   R m osm wat liv MXB 26m26                                                          796.mg  * P<.2
1950  R m cdr wat nas sqc 56w69 r                                      . +hist          2.95gm  * P<.07 +
1951  R b she wat liv mix 24m24 e                                      . +  .           2.12gm  * P<.0005+
  a   R b she wat liv hpc 24m24 e                                                        2.50gm  * P<.0005+
  b   R b she wat ntu sqc 24m24 e                                                        15.9gm  * P<.006 +
  c   R b she wat tba mix 24m24 e                                                        no dre   P=1.
1952  R b wis wat liv hpt 63w64                                      .  ±               515.mg    P<.05 +
  a   R b wis wat tba mix 63w64                                                          690.mg    P<.4

DIOXATHION                      100ng..:..1ug...:..10.....:..100....:..1mg.....:..10....:..100....:..1g.....:..10
1953  M f b6c eat TBA MXB 78w90 v                                           :>           2.59gm    P<.9   -
  a   M f b6c eat liv MXB 78w90 v                                                        no dre   P=1.
  b   M f b6c eat lun MXB 78w90 v                                                        1.82gm    P<.5
1954  M m b6c eat TBA MXB 78w90 v                                        :>              no dre   P=1.   -
  a   M m b6c eat liv MXB 78w90 v                                                        no dre   P=1.
  b   M m b6c eat lun MXB 78w90 v                                                        628.mg  * P<.5
1955  R f osm eat TBA MXB 18m26 v                               :>                       no dre   P=1.
  a   R f osm eat liv MXB 18m26 v                                                        no dre   P=1.
1956  R m osm eat TBA MXB 18m26 v                                 :>                     no dre   P=1.
  a   R m osm eat liv MXB 18m26 v                                                        no dre   P=1.

DIPENTAMETHYLENETHIURAM HEXASULFIDE   ..:..1ug...:..10.....:..100....:..1mg.....:..10....:..100....:..1g.....:..10
1957  M f b6a orl lun ade 76w76 evx                                        .>            no dre   P=1.
  a   M f b6a orl liv hpt 76w76 evx                                                      no dre   P=1.
  b   M f b6a orl tba mix 76w76 evx                                                      120.mg    P<.5
1958  M m b6a orl lun ade 76w76 evx                                         .>           no dre   P=1.
  a   M m b6a orl liv hpt 76w76 evx                                                      no dre   P=1.
  b   M m b6a orl tba mix 76w76 evx                                                      no dre   P=1.
1959  M f b6c orl lun ade 76w76 evx                                 .  ±                 83.1mg    P<.05
  a   M f b6c orl liv hpt 76w76 evx                                                      no dre   P=1.
  b   M f b6c orl tba mix 76w76 evx                                                      83.1mg    P<.05
1960  M m b6c orl liv hpt 76w76 evx                              .   ±                   113.mg    P<.1
  a   M m b6c orl lun ade 76w76 evx                                                      113.mg    P<.1
  b   M m b6c orl tba mix 76w76 evx                                                      52.5mg    P<.02

DIPENTYLNITROSAMINE             100ng..:..1ug...:..10.....:..100....:..1mg.....:..10....:..100....:..1g.....:..10
1961  R f f34 eat liv hpc 23m24 er                         . + .                         7.57mg  * P<.0005+
1962  R m f34 eat liv hpc 21m24 er                         . + .                         2.75mg  * P<.0005+

DIPHENHYDRAMINE.HCl             100ng..:..1ug...:..10.....:..100....:..1mg.....:..10....:..100....:..1g.....:..10
1963  M f b6c eat TBA MXB 24m24                                        :>                no dre   P=1.   -
  a   M f b6c eat liv MXB 24m24                                                          363.mg  * P<.5
  b   M f b6c eat lun MXB 24m24                                                          357.mg  * P<.3
1964  M m b6c eat liv hpc 24m25                                 :  ±                     #37.3mg \ P<.02  -
  a   M m b6c eat TBA MXB 24m25                                                          no dre   P=1.
  b   M m b6c eat liv MXB 24m25                                                          201.mg  * P<.6
  c   M m b6c eat lun MXB 24m25                                                          447.mg  \ P<.9
1965  R f f34 eat pta adn 24m24                                 :   ±                    20.5mg  * P<.06 e
  a   R f f34 eat TBA MXB 24m24                                                          65.1mg  * P<.8
  b   R f f34 eat liv MXB 24m24                                                          no dre   P=1.
1966  R f f34 eat liv nnd 25m30 e                                           .>           no dre   P=1.   -
1967  R m f34 eat lun a/a 24m24                             :   +    :                   66.1mg  * P<.009
  a   R m f34 eat bra ast 24m24                                                          140.mg  / P<.01
  b   R m f34 eat lun MXA 24m24                                                          57.3mg  * P<.02 e
  c   R m f34 eat bra MXA 24m24                                                          143.mg  / P<.05 e
  d   R m f34 eat TBA MXB 24m24                                                          17.8mg  * P<.07
  e   R m f34 eat liv MXB 24m24                                                          243.mg  * P<.2
1968  R m f34 eat liv nnd 25m30 e                                        .>              no dre   P=1.   -

DIPHENYL-p-PHENYLENEDIAMINE     100ng..:..1ug...:..10.....:..100....:..1mg.....:..10....:..100....:..1g.....:..10
1969  M f b6a orl lun ade 76w76 evx                                        .>            2.79gm    P<.3   -
  a   M f b6a orl liv hpt 76w76 evx                                                      no dre   P=1.
  b   M f b6a orl tba mix 76w76 evx                                                      2.79gm    P<.3
1970  M m b6a orl lun ade 76w76 evx                                     .  ±             1.33gm    P<.09  -
  a   M m b6a orl liv hpt 76w76 evx                                                      no dre   P=1.
  b   M m b6a orl tba mix 76w76 evx                                                      no dre   P=1.
1971  M f b6c orl liv hpt 76w76 evx                                         .>           no dre   P=1.   -
  a   M f b6c orl lun ade 76w76 evx                                                      no dre   P=1.
  b   M f b6c orl tba mix 76w76 evx                                                      no dre   P=1.
```

	RefNum	LoConf	UpConf	Cntrl	1Dose	1Inc	2Dose	2Inc					Citation or Pathology	
														Brkly Code
b	TR80	1.29gm	22.1gm	0/50	820.mg	6/50	(1.65gm	3/50)					---:hem,hes.	S
c	TR80	1.49gm	16.3gm	0/50	820.mg	5/50	(1.65gm	2/50)						S
d	TR80	841.mg	n.s.s.	19/50	820.mg	28/50	1.65gm	33/50						
e	TR80	833.mg	5.11gm	8/50	820.mg	19/50	1.65gm	28/50					liv:hpa,hpc,nnd.	
f	TR80	8.27gm	n.s.s.	8/50	820.mg	3/50	1.65gm	3/50					lun:a/a,a/c.	
1948	TR80	69.7mg	218.mg	0/35	284.mg	18/35	569.mg	18/35					liv:hpa; ntu:sqc. C	
a	TR80	79.1mg	323.mg	0/35	284.mg	10/35	569.mg	11/35						
b	TR80	249.mg	970.mg	0/35	284.mg	10/35	569.mg	8/35						
c	TR80	20.0mg	108.mg	26/35	284.mg	28/35	(569.mg	22/35)						
d	TR80	60.1mg	263.mg	0/35	284.mg	10/35	569.mg	12/35					liv:hpa,hpc,nnd.	
1949	TR80	94.3mg	304.mg	0/35	250.mg	12/35	500.mg	16/35						
a	TR80	59.3mg	339.mg	21/35	250.mg	18/35	500.mg	27/35						
b	TR80	190.mg	n.s.s.	1/35	250.mg	2/35	500.mg	1/35					liv:hpa,hpc,nnd.	
1950	124	1.20gm	n.s.s.	0/30	305.mg	1/30	406.mg	1/30	569.mg	2/30	731.mg	2/30	Hoch-Ligeti;bjca,24,164-167;1969	
1951	125	1.08gm	5.36gm	2/106	5.29mg	0/110	52.9mg	1/106	529.mg	12/66			Kociba;txap,30,275-286;1974	
a	125	1.22gm	6.73gm	1/106	5.29mg	0/110	52.9mg	1/106	529.mg	10/66				
b	125	4.82gm	231.gm	0/120	5.29mg	0/120	52.9mg	0/120	529.mg	3/120				
c	125	4.33gm	n.s.s.	31/120	5.29mg	34/120	52.9mg	28/120	529.mg	21/120				
1952	104	209.mg	n.s.s.	0/9	521.mg	6/26							Argus;jnci,35,949-958;1965	
a	104	206.mg	n.s.s.	1/9	521.mg	7/26								

DIOXATHION   78-34-2

	RefNum	LoConf	UpConf	Cntrl	1Dose	1Inc	2Dose	2Inc					Citation or Pathology	
1953	TR125	207.mg	n.s.s.	2/20	52.5mg	6/50	105.mg	6/50						
a	TR125	623.mg	n.s.s.	0/20	52.5mg	1/50	105.mg	0/50					liv:hpa,hpc,nnd.	
b	TR125	447.mg	n.s.s.	0/20	52.5mg	1/50	105.mg	1/50					lun:a/a,a/c.	
1954	TR125	77.0mg	n.s.s.	7/20	29.5mg	10/50	58.8mg	13/50						
a	TR125	128.mg	n.s.s.	4/20	29.5mg	4/50	58.8mg	6/50					liv:hpa,hpc,nnd.	
b	TR125	141.mg	n.s.s.	1/20	29.5mg	1/50	58.8mg	4/50					lun:a/a,a/c.	
1955	TR125	3.98mg	n.s.s.	36/50	1.60mg	27/50	3.00mg	29/50						
a	TR125	36.3mg	n.s.s.	1/50	1.60mg	0/50	3.00mg	0/50					liv:hpa,hpc,nnd.	
1956	TR125	6.81mg	n.s.s.	28/50	2.40mg	23/50	4.80mg	21/50						
a	TR125	n.s.s.	n.s.s.	0/50	2.40mg	0/50	4.80mg	0/50					liv:hpa,hpc,nnd.	

DIPENTAMETHYLENETHIURAM HEXASULFIDE   (sulfads) 971-15-3

	RefNum	LoConf	UpConf	Cntrl	1Dose	1Inc							Citation or Pathology	
1957	1221	49.2mg	n.s.s.	1/17	41.4mg	1/18							Innes;ntis,1968/1969	
a	1221	82.0mg	n.s.s.	0/17	41.4mg	0/18								
b	1221	23.9mg	n.s.s.	2/17	41.4mg	4/18								
1958	1221	47.4mg	n.s.s.	2/18	38.5mg	1/17								
a	1221	72.1mg	n.s.s.	1/18	38.5mg	0/17								
b	1221	50.7mg	n.s.s.	3/18	38.5mg	1/17								
1959	1221	25.1mg	n.s.s.	0/16	41.4mg	3/18								
a	1221	82.0mg	n.s.s.	0/16	41.4mg	0/18								
b	1221	25.1mg	n.s.s.	0/16	41.4mg	3/18								
1960	1221	27.6mg	n.s.s.	0/16	38.5mg	2/17								
a	1221	27.6mg	n.s.s.	0/16	38.5mg	2/17								
b	1221	18.1mg	n.s.s.	0/16	38.5mg	4/17								

DIPENTYLNITROSAMINE   13256-06-9

	RefNum	LoConf	UpConf	Cntrl	1Dose	1Inc	2Dose	2Inc	3Dose	3Inc			Citation or Pathology	
1961	1824	4.75mg	12.7mg	0/144	2.50mg	3/24	7.50mg	8/24	22.5mg	24/24			Elashoff;jnci,79,509-526;1987	
1962	1824	1.74mg	4.46mg	1/144	2.00mg	7/24	6.00mg	22/24	18.0mg	23/24				

DIPHENHYDRAMINE.HCl   (benadryl) 147-24-0

	RefNum	LoConf	UpConf	Cntrl	1Dose	1Inc	2Dose	2Inc					Citation or Pathology	
1963	TR355	33.1mg	n.s.s.	37/50	20.1mg	39/50	40.3mg	32/50						
a	TR355	80.6mg	n.s.s.	5/50	20.1mg	5/50	40.3mg	7/50					liv:hpa,hpc,nnd.	
b	TR355	98.7mg	n.s.s.	3/50	20.1mg	2/50	40.3mg	6/50					lun:a/a,a/c.	
1964	TR355	16.0mg	n.s.s.	4/50	18.5mg	14/50	(37.2mg	5/50)						
a	TR355	31.3mg	n.s.s.	30/50	18.5mg	32/50	37.2mg	22/50						S
b	TR355	34.1mg	n.s.s.	12/50	18.5mg	18/50	37.2mg	12/50					liv:hpa,hpc,nnd.	
c	TR355	31.8mg	n.s.s.	6/50	18.5mg	7/50	(37.2mg	0/50)					lun:a/a,a/c.	
1965	TR355	8.55mg	n.s.s.	23/50	7.62mg	26/50	15.4mg	35/50						S
a	TR355	8.16mg	n.s.s.	47/50	7.62mg	46/50	15.4mg	46/50						
b	TR355	n.s.s.	n.s.s.	0/50	7.62mg	0/50	15.4mg	0/50					liv:hpa,hpc,nnd.	
1966	1654	277.mg	n.s.s.	4/24	82.5mg	3/24							Lijinsky;fctx,22,715-720;1984	
1967	TR355	29.3mg	1.71gm	0/50	12.3mg	5/50	24.6mg	3/50						S
a	TR355	46.0mg	6.01gm	0/50	12.3mg	0/50	24.6mg	4/50						S
b	TR355	25.0mg	n.s.s.	0/50	12.3mg	6/50	24.6mg	5/50					lun:a/a,a/c.	
c	TR355	45.5mg	n.s.s.	1/50	12.3mg	0/50	24.6mg	5/50					bra:ast,gln.	
d	TR355	7.16mg	n.s.s.	47/50	12.3mg	49/50	24.6mg	47/50						
e	TR355	59.7mg	n.s.s.	0/50	12.3mg	1/50	24.6mg	1/50					liv:hpa,hpc,nnd.	
1968	1654	193.mg	n.s.s.	5/24	66.0mg	4/24							Lijinsky;fctx,22,715-720;1984	

DIPHENYL-p-PHENYLENEDIAMINE   (agerite DPPD) 74-31-7

	RefNum	LoConf	UpConf	Cntrl	1Dose	1Inc							Citation or Pathology	
1969	1158	454.mg	n.s.s.	0/18	462.mg	1/17							Innes;ntis,1968/1969	
a	1158	864.mg	n.s.s.	0/18	462.mg	0/17								
b	1158	454.mg	n.s.s.	0/18	462.mg	1/17								
1970	1158	328.mg	n.s.s.	0/18	430.mg	2/18								
a	1158	851.mg	n.s.s.	1/18	430.mg	0/18								
b	1158	401.mg	n.s.s.	2/18	430.mg	2/18								
1971	1158	915.mg	n.s.s.	0/18	462.mg	0/18								
a	1158	915.mg	n.s.s.	1/18	462.mg	0/18								
b	1158	915.mg	n.s.s.	2/18	462.mg	0/18								

```
          Spe Strain Site   Xpo+Xpt                                                          TD50    2Tailpvl
          Sex  Route  Hist   Notes                                                             DR      AuOp
1972 M m b6c orl liv hpt 76w76 evx                                      .>                   no dre   P=1.    -
a    M m b6c orl lun mix 76w76 evx                                                           no dre   P=1.    -
b    M m b6c orl tba mix 76w76 evx                                                           4.11gm   P<.9    -
1973 R f f3d eat liv clc 24m26 e                                                             .>44.8gm * P<.2  -
1974 R m f3d eat liv hnd 24m26 e                                        .>                   19.5gm * P<.4    -

DIPHENYLACETONITRILE          100ng..:..1ug...:..10.....:..100...:..1mg....:..10.....:..100...:..1g.....:..10
1975 M f b6a orl lun ade 76w76 evx                                              .>           no dre   P=1.    -
a    M f b6a orl liv hpt 76w76 evx                                                           no dre   P=1.    -
b    M f b6a orl tba mix 76w76 evx                                                           no dre   P=1.    -
1976 M m b6a orl lun ade 76w76 evx                                         .>                3.62gm   P<1.    -
a    M m b6a orl liv hpt 76w76 evx                                                           no dre   P=1.    -
b    M m b6a orl tba mix 76w76 evx                                                           no dre   P=1.    -
1977 M f b6c orl lun mix 76w76 evx                                  .        ±               229.mg   P<.1    -
a    M f b6c orl liv hpt 76w76 evx                                                           no dre   P=1.    -
b    M f b6c orl tba mix 76w76 evx                                                           148.mg   P<.04   -
1978 M m b6c orl lun ade 76w76 evx                              .        ±                   129.mg   P<.04   -
a    M m b6c orl liv hpt 76w76 evx                                                           200.mg   P<.09   -
b    M m b6c orl tba mix 76w76 evx                                                           46.4mg   P<.002  -

DIPHENYLCARBONATE             100ng..:..1ug...:..10.....:..100...:..1mg....:..10.....:..100...:..1g.....:..10
1979 M f b6a orl lun ade 76w76 evx                                         .>                206.mg   P<.6    -
a    M f b6a orl liv hpt 76w76 evx                                                           no dre   P=1.    -
b    M f b6a orl tba mix 76w76 evx                                                           no dre   P=1.    -
1980 M m b6a orl lun ade 76w76 evx                                       .>                  no dre   P=1.    -
a    M m b6a orl liv hpt 76w76 evx                                                           no dre   P=1.    -
b    M m b6a orl tba mix 76w76 evx                                                           no dre   P=1.    -
1981 M f b6c orl liv hem 76w76 evx                                       .>                  233.mg   P<.3    -
a    M f b6c orl lun ade 76w76 evx                                                           233.mg   P<.3    -
b    M f b6c orl tba mix 76w76 evx                                                           73.1mg   P<.05   -
1982 M m b6c orl liv hpt 76w76 evx                                .        ±                 68.0mg   P<.05   -
a    M m b6c orl lun ade 76w76 evx                                                           217.mg   P<.3    -
b    M m b6c orl tba mix 76w76 evx                                                           25.2mg   P<.002  -

5,5-DIPHENYLHYDANTOIN         100ng..:..1ug...:..10.....:..100...:..1mg....:..10.....:..100...:..1g.....:..10
1983 M f b6c eat liv MXA 25m25                                  :  +  :                      59.1mg * P<.0005c
a    M f b6c eat liv hpa 25m25                                                               92.4mg * P<.002 c
b    M f b6c eat liv MXA 25m25                                                               148.mg * P<.0005c
c    M f b6c eat liv hpc 25m25                                                               165.mg * P<.0005c
d    M f b6c eat TBA MXB 25m25                                                               2.39gm * P<1.
e    M f b6c eat liv MXB 25m25                                                               59.1mg * P<.0005
f    M f b6c eat lun MXB 25m25                                                               no dre   P=1.
1984 M f b6c eat liv hct 78w86 e                                       .>                    151.mg * P<.3    -
a    M f b6c eat liv hpa 78w86 e                                                             229.mg * P<.8    -
b    M f b6c eat liv hpc 78w86 e                                                             457.mg * P<.3    -
c    M f b6c eat lun act 78w86 e                                                             2.06gm * P<1.   -
d    M f b6c eat liv hem 78w86 e                                                             no dre   P=1.    -
1985 M m b6c eat TBA MXB 24m24                                       :>                       no dre   P=1.    -
a    M m b6c eat liv MXB 24m24                                                               724.mg * P<.9
b    M m b6c eat lun MXB 24m24                                                               no dre   P=1.    -
1986 M m b6c eat liv hem 78w86 e                                       .>                    321.mg * P<.7    -
a    M m b6c eat liv hct 78w86 e                                                             no dre   P=1.    -
b    M m b6c eat liv hpa 78w86 e                                                             no dre   P=1.    -
c    M m b6c eat liv hpc 78w86 e                                                             no dre   P=1.    -
d    M m b6c eat lun act 78w86 e                                                             no dre   P=1.    -
1987 R f f34 eat TBA MXB 24m24                                      :>                        no dre   P=1.    -
a    R f f34 eat liv MXB 24m24                                                               2.08gm * P<.4    -
1988 R m f34 eat liv MXA 24m24                                      :        ±               351.mg * P<.02   e
a    R m f34 eat liv hpa 24m24                                                               429.mg * P<.02   e
b    R m f34 eat TBA MXB 24m24                                                               no dre   P=1.
c    R m f34 eat liv MXB 24m24                                                               351.mg * P<.02
1989 R f f3d eat liv nnd 24m26 e                                          .>                 47.8gm * P<1.   -
a    R f f3d eat liv hpc 24m26 e                                                             no dre   P=1.    -
b    R f f3d eat tba mal 24m26 e                                                             110.mg * P<.2    -
1990 R m f3d eat liv hpc 24m26 e                                          .>                 1.08gm * P<.3    -
a    R m f3d eat liv nnd 24m26 e                                                             no dre   P=1.    -
b    R m f3d eat tba mal 24m26 e                                                             98.7mg * P<.5    -
1991 R f hza eat --- mly 60w60 es                                        .>                  422.mg   P<.3
a    R f hza eat liv tum 60w60 es                                                            no dre   P=1.
b    R f hza eat mam tum 60w60 es                                                            no dre   P=1.

N,N-DIPROPYL-4-(4'-[PYRIDYL-1'-OXIDE]AZO)ANILINE...:..1_0...:..100...:..1mg....:..10.....:..100...:..1g.....:..10
1992 R m sda eat liv tum 52w52 bfr                                       .>                  no dre   P=1.

DIPYRONE                      100ng..:..1ug...:..10.....:..100...:..1mg....:..10.....:..100...:..1g.....:..10
1993 M f b6c wat liv mix 75w86 ae                                           .  +  .          742.mg / P<.0005+
a    M f b6c wat liv hpc 75w86 ae                                                            7.12gm * P<.02
1994 M m b6c wat liv mix 77w86 ae                                        .  +  .             547.mg * P<.0005+
1995 R m f3d wat liv nnd 72w83 e                                               ±             400.mg   P<.08
a    R m f3d wat liv hpc 72w83 e                                                             no dre   P=1.
b    R m f3d wat kid tum 72w83 e                                                             no dre   P=1.
```

	RefNum	LoConf	UpConf	Cntrl	1Dose	1Inc	2Dose	2Inc	Citation or Pathology
									Brkly Code
1972	1158	306.mg	n.s.s.	3/14	430.mg	3/16			
a	1158	756.mg	n.s.s.	0/14	430.mg	0/16			
b	1158	213.mg	n.s.s.	4/14	430.mg	5/16			
1973	1943	7.30gm	n.s.s.	0/47	232.mg	0/49	929.mg	1/49	Hasegawa;txcy,54,69-78;1989
1974	1943	4.40gm	n.s.s.	0/50	186.mg	1/49	743.mg	1/49	

DIPHENYLACETONITRILE   86-29-3

	RefNum	LoConf	UpConf	Cntrl	1Dose	1Inc	2Dose	2Inc	Citation or Pathology
1975	1311	93.2mg	n.s.s.	1/17	78.4mg	1/18			Innes;ntis,1968/1969
a	1311	155.mg	n.s.s.	0/17	78.4mg	0/18			
b	1311	74.0mg	n.s.s.	2/17	78.4mg	2/18			
1976	1311	63.5mg	n.s.s.	2/18	73.0mg	2/17			
a	1311	137.mg	n.s.s.	1/18	73.0mg	0/17			
b	1311	70.3mg	n.s.s.	3/18	73.0mg	2/17			
1977	1311	56.3mg	n.s.s.	0/16	78.4mg	2/17			
a	1311	147.mg	n.s.s.	0/16	78.4mg	0/17			
b	1311	44.6mg	n.s.s.	0/16	78.4mg	3/17			
1978	1311	38.8mg	n.s.s.	0/16	73.0mg	3/16			
a	1311	49.1mg	n.s.s.	0/16	73.0mg	2/16			
b	1311	19.7mg	170.mg	0/16	73.0mg	7/16			

DIPHENYLCARBONATE   102-09-0

	RefNum	LoConf	UpConf	Cntrl	1Dose	1Inc	2Dose	2Inc	Citation or Pathology
1979	1313	28.7mg	n.s.s.	1/17	36.4mg	2/17			Innes;ntis,1968/1969
a	1313	68.1mg	n.s.s.	0/17	36.4mg	0/17			
b	1313	32.0mg	n.s.s.	2/17	36.4mg	2/17			
1980	1313	38.9mg	n.s.s.	2/18	33.9mg	1/16			
a	1313	59.7mg	n.s.s.	1/18	33.9mg	0/16			
b	1313	41.6mg	n.s.s.	3/18	33.9mg	1/16			
1981	1313	37.9mg	n.s.s.	0/16	36.4mg	1/18			
a	1313	37.9mg	n.s.s.	0/16	36.4mg	1/18			
b	1313	22.0mg	n.s.s.	0/16	36.4mg	3/18			
1982	1313	20.5mg	n.s.s.	0/16	33.9mg	3/18			
a	1313	35.3mg	n.s.s.	0/16	33.9mg	1/18			
b	1313	10.7mg	103.mg	0/16	33.9mg	7/18			

5,5-DIPHENYLHYDANTOIN   (phenytoin) 57-41-0

	RefNum	LoConf	UpConf	Cntrl	1Dose	1Inc	2Dose	2Inc	Citation or Pathology
1983	TR404	34.8mg	151.mg	5/50	26.0mg	14/50	78.0mg	30/50	liv:hpa,hpb,hpc.
a	TR404	48.2mg	496.mg	5/50	26.0mg	13/50	78.0mg	22/50	
b	TR404	76.5mg	354.mg	0/50	26.0mg	1/50	78.0mg	12/50	liv:hpb,hpc.
c	TR404	83.1mg	418.mg	0/50	26.0mg	1/50	78.0mg	11/50	
d	TR404	52.8mg	n.s.s.	40/50	26.0mg	36/50	78.0mg	42/50	
e	TR404	34.8mg	151.mg	5/50	26.0mg	14/50	78.0mg	30/50	liv:hpa,hpb,hpc.
f	TR404	385.mg	n.s.s.	5/50	26.0mg	3/50	78.0mg	2/50	lun:a/a,a/c.
1984	1887	45.8mg	n.s.s.	0/49	7.07mg	2/49	14.1mg	1/45	Maeda;jtxe,24,111-119;1988
a	1887	56.3mg	n.s.s.	0/49	7.07mg	2/49	14.1mg	0/45	
b	1887	74.5mg	n.s.s.	0/49	7.07mg	0/49	14.1mg	1/45	
c	1887	55.4mg	n.s.s.	1/49	7.07mg	2/49	14.1mg	1/45	
d	1887	97.3mg	n.s.s.	1/49	7.07mg	1/49	14.1mg	0/45	
1985	TR404	26.7mg	n.s.s.	43/50	12.0mg	39/50	36.0mg	36/50	
a	TR404	33.2mg	n.s.s.	29/50	12.0mg	29/50	36.0mg	26/50	liv:hpa,hpb,hpc.
b	TR404	91.7mg	n.s.s.	10/50	12.0mg	11/50	36.0mg	6/50	lun:a/a,a/c.
1986	1887	41.0mg	n.s.s.	3/47	6.53mg	0/44	13.1mg	4/43	Maeda;jtxe,24,111-119;1988
a	1887	22.3mg	n.s.s.	26/47	6.53mg	20/44	13.1mg	16/43	
b	1887	28.9mg	n.s.s.	19/47	6.53mg	12/44	13.1mg	11/43	
c	1887	28.7mg	n.s.s.	7/47	6.53mg	8/44	13.1mg	5/43	
d	1887	58.2mg	n.s.s.	6/47	6.53mg	3/44	13.1mg	2/43	
1987	TR404	37.9mg	n.s.s.	45/50	40.0mg	38/50	(120.mg	39/50)	liv:hpa,hpb,hpc.
a	TR404	512.mg	n.s.s.	0/50	40.0mg	1/50	120.mg	1/50	liv:hpa,hpc.
1988	TR404	142.mg	n.s.s.	0/50	32.0mg	2/50	96.0mg	4/50	
a	TR404	162.mg	n.s.s.	0/50	32.0mg	1/50	96.0mg	4/50	
b	TR404	58.9mg	n.s.s.	49/50	32.0mg	46/50	96.0mg	42/50	
c	TR404	142.mg	n.s.s.	0/50	32.0mg	2/50	96.0mg	4/50	liv:hpa,hpb,hpc.
1989	1855	241.mg	n.s.s.	1/50	11.6mg	0/47	23.2mg	1/48	Jang;fctx,25,697-702;1987
a	1855	87.5mg	n.s.s.	0/50	11.6mg	0/47	23.2mg	0/48	
b	1855	37.0mg	n.s.s.	13/50	11.6mg	11/50	23.2mg	19/48	
1990	1855	177.mg	n.s.s.	0/50	9.29mg	0/48	18.6mg	1/50	
a	1855	133.mg	n.s.s.	1/50	9.29mg	2/48	18.6mg	1/50	
b	1855	22.8mg	n.s.s.	22/50	9.29mg	22/48	18.6mg	26/50	
1991	1063	68.7mg	n.s.s.	0/16	100.mg	1/19			Morris;canr,29,2145-2156;1969
a	1063	130.mg	n.s.s.	0/16	100.mg	0/19			
b	1063	93.9mg	n.s.s.	3/16	100.mg	1/19			

N,N-DIPROPYL-4-(4'-[PYRIDYL-1'-OXIDE]AZO)ANILINE   ---

	RefNum	LoConf	UpConf	Cntrl	1Dose	1Inc	2Dose	2Inc	Citation or Pathology
1992	1176	6.18mg	n.s.s.	0/10	12.0mg	0/10			Brown;jnci,37,365-367;1966

DIPYRONE   (sulpyrin) 68-89-3

	RefNum	LoConf	UpConf	Cntrl	1Dose	1Inc	2Dose	2Inc	Citation or Pathology
1993	1545	484.mg	1.22gm	3/51	448.mg	3/47	1.71gm	36/44	Kumagai;jnci,71,1295-1297;1983
a	1545	2.77gm	n.s.s.	1/51	448.mg	1/47	1.71gm	6/44	
1994	1545	300.mg	1.79gm	8/44	189.mg	16/46	746.mg	27/48	
1995	2029	130.mg	n.s.s.	1/23	173.mg	5/24			Izumi;carc,12,1221-1225;1991
a	2029	547.mg	n.s.s.	0/23	173.mg	0/24			
b	2029	547.mg	n.s.s.	0/23	173.mg	0/24			

```
      Spe Strain  Site   Xpo+Xpt                                                                        TD50    2Tailpvl
          Sex  Route   Hist    Notes                                                                       DR      AuOp
1996 R f wis eat mgl ade 24m30 e                                               .   +     .             328.mg \ P<.004 -
   a R f wis eat liv lcc 24m30 e                                                                       8.33gm * P<.3   -
   b R f wis eat liv ade 24m30 e                                                                       33.1gm * P<1.   -
   c R f wis eat liv ang 24m30 e                                                                       41.4gm * P<.9   -
1997 R m wis eat liv ade 24m30 e                                                         .>            3.34gm * P<.4   -
   a R m wis eat liv apc 24m30 e                                                                       no dre  P=1.    -

2,5-DITHIOBIUREA              100ng..:..1ug....:..10......:..100.....:...1mg......:..10......:...100.....:..1g......:..10
1998 M f b6c eat liv hpc 94w94                                                                  :  ±   5.67gm * P<.03  a
   a M f b6c eat TBA MXB 94w94                                                                         4.35gm * P<.2
   b M f b6c eat liv MXB 94w94                                                                         5.67gm * P<.03
   c M f b6c eat lun MXB 94w94                                                                         22.9gm * P<.6
1999 M m b6c eat TBA MXB 94w94                                                            :>           20.0gm * P<.8   -
   a M m b6c eat liv MXB 94w94                                                                         no dre  P=1.
   b M m b6c eat lun MXB 94w94                                                                         no dre  P=1.
2000 R f f34 eat TBA MXB 25m25                                                      :>                 842.mg * P<.3   -
   a R f f34 eat liv MXB 25m25                                                                         131.gm * P<1.
2001 R m f34 eat pit can 25m25                                                        :  ±            #2.62gm * P<.02  -
   a R m f34 eat TBA MXB 25m25                                                                         960.mg / P<.4
   b R m f34 eat liv MXB 25m25                                                                         5.01gm * P<.09

DITHIOOXAMIDE                 100ng..:..1ug....:..10......:..100.....:...1mg......:..10......:...100.....:..1g......:..10
2002 R f cdr eat mgl tum 18m24 ev                                             .   ±                    12.6mg \ P<.02

N-DODECYLGUANIDINE ACETATE    100ng..:..1ug....:..10......:..100.....:...1mg......:..10......:...100.....:..1g......:..10
2003 M f b6a orl liv hpt 76w76 evx                                                  .>                 no dre  P=1.    -
   a M f b6a orl lun ade 76w76 evx                                                                     no dre  P=1.    -
   b M f b6a orl tba mix 76w76 evx                                                                     no dre  P=1.    -
2004 M m b6a orl lun mix 76w76 evx                                                   .>                41.9mg   P<.6   -
   a M m b6a orl liv hpt 76w76 evx                                                                     no dre  P=1.    -
   b M m b6a orl tba mix 76w76 evx                                                                     149.mg   P<.9   -
2005 M f b6c orl liv hpt 76w76 evx                                                  .>                 54.7mg   P<.3   -
   a M f b6c orl lun mix 76w76 evx                                                                     no dre  P=1.    -
   b M f b6c orl tba mix 76w76 evx                                                                     54.7mg   P<.3   -
2006 M m b6c orl liv hpt 76w76 evx                                             .   ±                   19.4mg   P<.04  -
   a M m b6c orl lun ade 76w76 evx                                                                     30.1mg   P<.1   -
   b M m b6c orl tba mix 76w76 evx                                                                     10.8mg   P<.007 -

DL-DOPA                       100ng..:..1ug....:..10......:..100.....:...1mg......:..10......:...100.....:..1g......:..10
2007 R m f3d eat for tum 51w52 e                                                         .>            no dre  P=1.    -
   a R m f3d eat liv tum 51w52 e                                                                       no dre  P=1.

DOPAMINE.HCl                  100ng..:..1ug....:..10......:..100.....:...1mg......:..10......:...100.....:..1g......:..10
2008 R m f3d eat for tum 51w52 e                                                         .>            no dre  P=1.
   a R m f3d eat liv tum 51w52 e                                                                       no dre  P=1.

DOXEFAZEPAM                   100ng..:..1ug....:..10......:..100.....:...1mg......:..10......:...100.....:..1g......:..10
2009 R f cdr eat adr phe 24m24 e                                                    .   +     .        216.mg * P<.003 -
   a R f cdr eat liv hpa 24m24 e                                                                       213.mg * P<.02  -
   b R f cdr eat cns gli 24m24 e                                                                       543.mg * P<.4   -
   c R f cdr eat liv hpc 24m24 e                                                                       1.28gm * P<.6   -
   d R f cdr eat pit ade 24m24 e                                                                       no dre  P=1.    -
   e R f cdr eat tba mal 24m24 e                                                                       no dre  P=1.    -
2010 R m cdr eat sub mix 24m24 e                                              .   +     .              9.04mg Z P<.002 -
   a R m cdr eat sub fib 24m24 e                                                                       11.8mg Z P<.002 -
   b R m cdr eat liv hpa 24m24 e                                                                       355.mg * P<.02  -
   c R m cdr eat cns gli 24m24 e                                                                       366.mg * P<.3   -
   d R m cdr eat liv cho 24m24 e                                                                       1.44gm * P<.2   -
   e R m cdr eat liv hpc 24m24 e                                                                       12.6gm * P<1.   -
   f R m cdr eat tba mal 24m24 e                                                                       181.mg * P<.5   -

EDIFAS A                      100ng..:..1ug....:..10......:..100.....:...1mg......:..10......:...100.....:..1g......:..10
2011 M f aps eat liv tum 23m23 e                                                               .> no dre  P=1.    -
   a M f aps eat lun ade 23m23 e                                                                       no dre  P=1.    -
   b M f aps eat tba mix 23m23 e                                                                       no dre  P=1.    -
2012 M m aps eat liv hpt 23m23 e                                                                       254.gm * P<.9   -
   a M m aps eat lun ade 23m23 e                                                                       no dre  P=1.    -
   b M m aps eat tba mix 23m23 e                                                                       19.4gm * P<.4   -
2013 R f aps eat lun ade 24m24 e                                                                       161.gm * P<.2   -
   a R f aps eat liv mix 24m24 e                                                                       375.gm * P<.7   -
   b R f aps eat tba mix 24m24 e                                                                       113.gm * P<.9   -
2014 R m aps eat liv tum 24m24 e                                                               .> no dre  P=1.    -
   a R m aps eat lun ade 24m24 e                                                                       no dre  P=1.    -
   b R m aps eat tba mix 24m24 e                                                                       27.3gm * P<.4   -

EDIFAS B                      100ng..:..1ug....:..10......:..100.....:...1mg......:..10......:...100.....:..1g......:..10
2015 M f aps eat liv hpt 23m23 e                                                                       538.gm * P<.7   -
   a M f aps eat lun ade 23m23 e                                                                       no dre  P=1.    -
   b M f aps eat tba mix 23m23 e                                                                       no dre  P=1.    -
2016 M m aps eat lun ade 23m23 e                                                                       no dre  P=1.    -
   a M m aps eat liv hpt 23m23 e                                                                       no dre  P=1.    -
```

	RefNum	LoConf	UpConf	Cntrl	1Dose	1Inc	2Dose	2Inc	3Dose	3Inc	Citation or Pathology Brkly Code
1996	1673	134.mg	2.04gm	0/49	40.0mg	6/49	(120.mg	2/49)			Donaubauer;txap,81,443-451;1985
a	1673	1.36gm	n.s.s.	0/49	40.0mg	0/49	120.mg	1/49			
b	1673	1.07gm	n.s.s.	2/49	40.0mg	1/49	120.mg	2/49			
c	1673	1.51gm	n.s.s.	1/49	40.0mg	0/49	120.mg	1/49			
1997	1673	822.mg	n.s.s.	0/47	32.0mg	1/50	96.0mg	1/49			
a	1673	384.mg	n.s.s.	1/47	32.0mg	0/50	96.0mg	0/49			

2,5-DITHIOBIUREA   142-46-1

	RefNum	LoConf	UpConf	Cntrl	1Dose	1Inc	2Dose	2Inc			Citation or Pathology Brkly Code
1998	TR132	2.60gm	n.s.s.	2/50	1.08gm	8/50	2.16gm	9/50			
a	TR132	1.58gm	n.s.s.	12/50	1.08gm	24/50	2.16gm	20/50			
b	TR132	2.60gm	n.s.s.	2/50	1.08gm	8/50	2.16gm	9/50			liv:hpa,hpc,nnd.
c	TR132	4.47gm	n.s.s.	4/50	1.08gm	5/50	2.16gm	6/50			lun:a/a,a/c.
1999	TR132	1.94gm	n.s.s.	20/50	996.mg	23/50	1.99gm	18/50			
a	TR132	5.70gm	n.s.s.	15/50	996.mg	9/50	1.99gm	7/50			liv:hpa,hpc,nnd.
b	TR132	3.98gm	n.s.s.	7/50	996.mg	13/50	1.99gm	4/50			lun:a/a,a/c.
2000	TR132	262.mg	n.s.s.	31/50	215.mg	36/50	429.mg	32/50			
a	TR132	3.07gm	n.s.s.	1/50	215.mg	0/50	429.mg	1/50			liv:hpa,hpc,nnd.
2001	TR132	905.mg	n.s.s.	0/50	172.mg	0/50	343.mg	4/50			S
a	TR132	251.mg	n.s.s.	31/50	172.mg	28/50	343.mg	30/50			
b	TR132	1.46gm	n.s.s.	0/50	172.mg	1/50	343.mg	2/50			liv:hpa,hpc,nnd.

DITHIOOXAMIDE   79-40-3

2002	1112	5.72mg	n.s.s.	3/16	12.0mg	15/26	(24.0mg	9/27)	Weisburger;jnci,67,75-88;1981

N-DODECYLGUANIDINE ACETATE   2439-10-3

	RefNum	LoConf	UpConf	Cntrl	1Dose	1Inc	Citation or Pathology Brkly Code
2003	1293	22.0mg	n.s.s.	0/17	11.1mg	0/18	Innes;ntis,1968/1969
a	1293	22.0mg	n.s.s.	1/17	11.1mg	0/18	
b	1293	10.5mg	n.s.s.	2/17	11.1mg	2/18	
2004	1293	6.32mg	n.s.s.	2/18	10.3mg	3/16	
a	1293	18.1mg	n.s.s.	1/18	10.3mg	0/16	
b	1293	6.90mg	n.s.s.	3/18	10.3mg	3/16	
2005	1293	8.90mg	n.s.s.	0/16	11.1mg	1/14	
a	1293	17.1mg	n.s.s.	0/16	11.1mg	0/14	
b	1293	8.90mg	n.s.s.	0/16	11.1mg	1/14	
2006	1293	5.85mg	n.s.s.	0/16	10.3mg	3/17	
a	1293	7.39mg	n.s.s.	0/16	10.3mg	2/17	
b	1293	4.07mg	146.mg	0/16	10.3mg	5/17	

DL-DOPA   (dl-3,4-dihydroxyphenylalanine) 63-84-3

2007	2080	455.mg	n.s.s.	0/15	588.mg	0/15	Hirose;carc,13,1825-1828;1992/pers.comm.
a	2080	455.mg	n.s.s.	0/15	588.mg	0/15	

DOPAMINE.HCl   62-31-7

2008	2080	455.mg	n.s.s.	0/15	588.mg	0/15	Hirose;carc,13,1825-1828;1992/pers.comm.
a	2080	455.mg	n.s.s.	0/15	588.mg	0/15	

DOXEFAZEPAM   40762-15-0

	RefNum	LoConf	UpConf	Cntrl	1Dose	1Inc	2Dose	2Inc	3Dose	3Inc	Citation or Pathology Brkly Code
2009	1983	88.5mg	1.38gm	1/49	3.00mg	0/50	10.0mg	0/49	30.0mg	7/50	Borelli;faat,15,82-92;1990
a	1983	84.6mg	n.s.s.	1/49	3.00mg	0/50	10.0mg	3/49	30.0mg	5/50	
b	1983	124.mg	n.s.s.	1/49	3.00mg	0/50	10.0mg	3/49	30.0mg	2/50	
c	1983	158.mg	n.s.s.	0/49	3.00mg	1/50	10.0mg	1/49	30.0mg	1/50	
d	1983	28.9mg	n.s.s.	32/49	3.00mg	41/50	10.0mg	33/49	30.0mg	35/50	
e	1983	55.4mg	n.s.s.	19/49	3.00mg	21/50	10.0mg	18/49	30.0mg	19/50	
2010	1983	4.30mg	41.2mg	1/48	3.00mg	11/50	(10.0mg	8/47	30.0mg	6/50)	
a	1983	5.33mg	41.1mg	0/48	3.00mg	8/50	(10.0mg	5/47	30.0mg	4/50)	
b	1983	122.mg	n.s.s.	0/48	3.00mg	0/50	10.0mg	1/47	30.0mg	3/50	
c	1983	86.2mg	n.s.s.	1/48	3.00mg	3/50	10.0mg	4/47	30.0mg	4/50	
d	1983	235.mg	n.s.s.	0/48	3.00mg	0/50	10.0mg	0/47	30.0mg	1/50	
e	1983	133.mg	n.s.s.	1/48	3.00mg	3/50	10.0mg	3/47	30.0mg	2/50	
f	1983	39.4mg	n.s.s.	17/48	3.00mg	15/50	10.0mg	18/47	30.0mg	20/50	

EDIFAS A   (methyl ethyl cellulose) 9004-59-5

	RefNum	LoConf	UpConf	Cntrl	1Dose	1Inc	2Dose	2Inc			Citation or Pathology Brkly Code
2011	127a	5.78gm	n.s.s.	0/35	1.30gm	0/25	13.0gm	0/35			McElligott;fctx,6,449-460;1968
a	127a	38.9gm	n.s.s.	1/35	1.30gm	3/25	13.0gm	2/35			
b	127a	12.6gm	n.s.s.	16/35	1.30gm	16/25	13.0gm	18/35			
2012	127a	14.8gm	n.s.s.	4/29	1.20gm	6/24	12.0gm	4/20			
a	127a	18.3gm	n.s.s.	8/29	1.20gm	2/24	12.0gm	4/20			
b	127a	4.17gm	n.s.s.	18/29	1.20gm	16/24	12.0gm	15/20			
2013	127a	26.2gm	n.s.s.	0/43	500.mg	0/44	5.00gm	1/43			
a	127a	28.3gm	n.s.s.	1/43	500.mg	0/44	5.00gm	1/43			
b	127a	5.18gm	n.s.s.	22/43	500.mg	26/44	5.00gm	24/43			
2014	127a	3.26gm	n.s.s.	0/46	400.mg	0/43	4.00gm	0/49			
a	127a	3.26gm	n.s.s.	1/46	400.mg	0/43	4.00gm	0/49			
b	127a	6.71gm	n.s.s.	6/46	400.mg	10/43	4.00gm	12/49			

EDIFAS B   (cellulose carboxymethyl ether, sodium) 9004-32-4

	RefNum	LoConf	UpConf	Cntrl	1Dose	1Inc	2Dose	2Inc			Citation or Pathology Brkly Code
2015	127a	42.9gm	n.s.s.	0/23	1.30gm	1/32	13.0gm	2/35			McElligott;fctx,6,449-460;1968
a	127a	44.7gm	n.s.s.	1/23	1.30gm	5/32	13.0gm	2/35			
b	127a	17.3gm	n.s.s.	12/23	1.30gm	25/32	13.0gm	18/35			
2016	127a	18.9gm	n.s.s.	5/31	1.20gm	9/27	12.0gm	7/31			
a	127a	51.1gm	n.s.s.	3/31	1.20gm	1/27	12.0gm	1/31			

```
      Spe Strain  Site   Xpo+Xpt                                              TD50    2Tailpvl
          Sex  Route  Hist    Notes                                              DR     AuOp
b    M m aps eat tba mix 23m23 e                                              132.gm * P<.9   -
2017 R f aps eat liv tum 24m24 e                                       .>     no dre   P=1.  -
a    R f aps eat tba mix 24m24 e                                              7.79gm * P<.2   -
2018 R m aps eat liv hpt 24m24 e                                              133.gm * P<.2   -
a    R m aps eat tba mix 24m24 e                                              no dre   P=1.  -

EDTA, TRISODIUM SALT TRIHYDRATE   100ng..:..1ug....:..10......:..100....:..1mg....:..10......:..100....:..1g......:..10
2019 M f b6c eat TBA MXB 24m24                                        :>      no dre   P=1.  -
a    M f b6c eat liv MXB 24m24                                                18.8gm * P<.5
b    M f b6c eat lun MXB 24m24                                                5.29gm * P<.2
2020 M m b6c eat TBA MXB 24m24                                  :>            4.92gm * P<.7
a    M m b6c eat liv MXB 24m24                                                12.3gm * P<.8
b    M m b6c eat lun MXB 24m24                                                3.19gm * P<.2
2021 R f f34 eat TBA MXB 24m24                                       :>       no dre   P=1.  -
a    R f f34 eat liv MXB 24m24                                                no dre   P=1.  -
2022 R m f34 eat TBA MXB 24m24                                     :>         1.86gm * P<.7
a    R m f34 eat liv MXB 24m24                                                6.59gm * P<.6

EMETINE.2HCl                      100ng..:..1ug....:..10......:..100....:..1mg....:..10......:..100....:..1g......:..10
2023 M f b6c ipj TBA MXB 40w78 as              :     ±                        1.20mg * P<.03  i
a    M f b6c ipj liv MXB 40w78 as                                            no dre   P=1.
b    M f b6c ipj lun MXB 40w78 as                                            49.4mg * P<.2
2024 M m b6c ipj TBA MXB 44w78 as               :     ±                       .552mg * P<.09  i
a    M m b6c ipj liv MXB 44w78 as                                            1.30mg * P<.3
b    M m b6c ipj lun MXB 44w78 as                                            1.19mg * P<.2
2025 R f sda ipj TBA MXB 52w83 s        :>                                    .235mg * P<.3   i
a    R f sda ipj liv MXB 52w83 s                                             no dre   P=1.
2026 R m sda ipj TBA MXB 52w83 s         :>                                   .382mg * P<.5   i
a    R m sda ipj liv MXB 52w83 s                                             no dre   P=1.

EMULSIFIER YN                     100ng..:..1ug....:..10......:..100....:..1mg....:..10......:..100....:..1g......:..10
2027 M f nss eat liv tum 80w80 e                                             no dre   P=1.  -
a    M f nss eat lun ade 80w80 e                                             no dre   P=1.  -
2028 M m nss eat lun ade 80w80 e                                             no dre   P=1.  -
a    M m nss eat liv 80w80 e                                                 no dre   P=1.  -
2029 R f wis eat pit cra 24m24 e                                      . +    5.54gm * P<.009 -
a    R f wis eat liv nod 24m24 e                                             511.gm * P<.9   -
2030 R m wis eat liv cho 24m24 e                                             no dre   P=1.  -

ENDOSULFAN                        100ng..:..1ug....:..10......:..100....:..1mg....:..10......:..100....:..1g......:..10
2031 M f b6c eat TBA MXB 78w91 v                  :>                         no dre   P=1.  -
a    M f b6c eat liv MXB 78w91 v                                             15.9mg * P<.4
b    M f b6c eat lun MXB 78w91 v                                             no dre   P=1.
2032 M f b6c orl liv hpt 76w76 evx          .>                               no dre   P=1.
a    M f b6c orl lun ade 76w76 evx                                           no dre   P=1.
b    M f b6c orl tba mix 76w76 evx                                           .425mg \ P<.02  -
2033 M m b6c eat TBA MXB 78w91 av                 :>                         6.36mg * P<.9   i
a    M m b6c eat liv MXB 78w91 av                                            no dre   P=1.
b    M m b6c eat lun MXB 78w91 av                                            1.70mg * P<.2
2034 M m b6c orl liv hpt 76w76 evx            .    ±                         .584mg \ P<.03  -
a    M m b6c orl lun ade 76w76 evx                                           .914mg \ P<.08  -
b    M m b6c orl tba mix 76w76 evx                                           .319mg \ P<.004 -
2035 M f b6a orl liv hpt 76w76 evx              .>                           no dre   P=1.  -
a    M f b6a orl lun ade 76w76 evx                                           no dre   P=1.  -
b    M f b6a orl tba mix 76w76 evx                                           no dre   P=1.  -
2036 M m b6a orl liv hpt 76w76 evx              .>                           no dre   P=1.  -
a    M m b6a orl lun ade 76w76 evx                                           no dre   P=1.  -
b    M m b6a orl tba mix 76w76 evx                                           no dre   P=1.  -
2037 R f osm eat TBA MXB 18m26 dv                       :>                   no dre   P=1.  -
a    R f osm eat liv MXB 18m26 dv                                            224.mg * P<.4
2038 R m osm eat TBA MXB 70w74 adsv                      :>                  no dre   P=1.  i
a    R m osm eat liv MXB 70w74 adsv                                          no dre   P=1.

ENDRIN                            100ng..:..1ug....:..10......:..100....:..1mg....:..10......:..100....:..1g......:..10
2039 M f b6c eat TBA MXB 80w90                    :>                         6.88mg * P<.8   -
a    M f b6c eat liv MXB 80w90                                               no dre   P=1.
b    M f b6c eat lun MXB 80w90                                               67.0mg * P<1.
2040 M m b6c eat TBA MXB 80w90 dsv          :     ±                          .546mg * P<.04  -
a    M m b6c eat liv MXB 80w90 dsv                                           .854mg * P<.2
b    M m b6c eat lun MXB 80w90 dsv                                           1.27mg * P<.08
2041 R f osm eat TBA MXB 19m26 sv                 :>                         no dre   P=1.  -
a    R f osm eat liv MXB 19m26 sv                                            7.44mg * P<.8
2042 R f osm eat adr MXA 19m25 sv pool            :  +    :                  #.227mg \ P<.003 -
a    R f osm eat pit adn 19m25 sv                                            .579mg * P<.02
2043 R f osm eat liv hem 27m28 ev                      .>                    42.3mg * P<.4   -
a    R f osm eat tba mix 27m28 ev                                            no dre   P=1.  -
2044 R m osm eat TBA MXB 19m26                    :>                         .776mg * P<.8   -
a    R m osm eat liv MXB 19m26                                               no dre   P=1.
2045 R m osm eat adr MXA 19m26 pool               :  ±                       #.553mg * P<.04  -
a    R m osm eat adr adn 19m26                                               .660mg * P<.04
b    R m osm eat --- hem 19m26                                               .661mg * P<.03
```

	RefNum	LoConf	UpConf	Cntrl	1Dose	1Inc	2Dose	2Inc			Citation or Pathology
											Brkly Code
b	127a	9.17gm	n.s.s.	18/31	1.20gm	14/27	12.0gm	18/31			
2017	127a	4.46gm	n.s.s.	0/43	500.mg	0/48	5.00gm	0/44			
a	127a	2.57gm	n.s.s.	27/43	500.mg	29/48	5.00gm	33/44			
2018	127a	21.7gm	n.s.s.	0/49	400.mg	0/42	4.00gm	1/45			
a	127a	8.70gm	n.s.s.	21/49	400.mg	14/42	4.00gm	15/45			
EDTA, TRISODIUM SALT TRIHYDRATE (EDTA) 150-38-9											
2019	TR11	1.22gm	n.s.s.	12/20	483.mg	24/50	966.mg	23/50			
a	TR11	4.62gm	n.s.s.	0/20	483.mg	1/50	966.mg	1/50			liv:hpa,hpc,nnd.
b	TR11	2.29gm	n.s.s.	0/20	483.mg	3/50	966.mg	4/50			lun:a/a,a/c.
2020	TR11	739.mg	n.s.s.	9/20	446.mg	24/50	891.mg	27/50			
a	TR11	1.41gm	n.s.s.	3/20	446.mg	10/50	891.mg	10/50			liv:hpa,hpc,nnd.
b	TR11	1.16gm	n.s.s.	2/20	446.mg	8/50	891.mg	12/50			lun:a/a,a/c.
2021	TR11	563.mg	n.s.s.	14/20	186.mg	28/50	371.mg	27/50			
a	TR11	2.96gm	n.s.s.	0/20	186.mg	1/50	371.mg	0/50			liv:hpa,hpc,nnd.
2022	TR11	263.mg	n.s.s.	9/20	149.mg	25/50	297.mg	30/50			
a	TR11	1.62gm	n.s.s.	0/20	149.mg	1/50	297.mg	1/50			liv:hpa,hpc,nnd.
EMETINE.2HCl 316-42-7											
2023	TR43	.298mg	n.s.s.	0/25	.460mg	2/35	1.40mg	0/35	2.70mg	1/35	
a	TR43	n.s.s.	n.s.s.	0/25	.460mg	0/35	1.40mg	0/35	2.70mg	0/35	liv:hpa,hpc,nnd.
b	TR43	8.05mg	n.s.s.	0/25	.460mg	0/35	1.40mg	0/35	2.70mg	1/35	lun:a/a,a/c.
2024	TR43	.175mg	n.s.s.	3/25	.460mg	7/35	1.00mg	0/35	2.70mg	0/35	
a	TR43	.293mg	n.s.s.	1/25	.460mg	3/35	1.00mg	0/35	2.70mg	0/35	liv:hpa,hpc,nnd.
b	TR43	.279mg	n.s.s.	1/25	.460mg	3/35	1.00mg	0/35	2.70mg	0/35	lun:a/a,a/c.
2025	TR43	71.5ug	n.s.s.	6/10	.130mg	26/35	.270mg	22/35			
a	TR43	n.s.s.	n.s.s.	0/10	.130mg	0/35	.270mg	0/35			liv:hpa,hpc,nnd.
2026	TR43	96.7ug	n.s.s.	3/10	.130mg	13/35	.270mg	5/35			
a	TR43	n.s.s.	n.s.s.	0/10	.130mg	2/35	.270mg	0/35			liv:hpa,hpc,nnd.
EMULSIFIER YN 55965-13-4											
2027	425	10.1gm	n.s.s.	0/41	2.60gm	0/41	7.80gm	0/47			Gaunt;fctx,15,1-5;1977
a	425	n.s.s.	n.s.s.	1/41	2.60gm	0/41	7.80gm	1/47			
2028	425	15.8gm	n.s.s.	2/25	2.40gm	4/34	7.20gm	2/37			
a	425	32.5gm	n.s.s.	1/25	2.40gm	1/34	7.20gm	0/37			
2029	1359	2.58gm	214.gm	13/47	1.00gm	9/45	3.00gm	23/44			Brantom;fctx,11,755-769;1973
a	1359	21.7gm	n.s.s.	1/47	1.00gm	0/45	3.00gm	1/44			
2030	1359	14.7gm	n.s.s.	0/41	800.mg	1/39	2.40gm	0/39			
ENDOSULFAN (thiodan) 115-29-7											
2031	TR62	.250mg	n.s.s.	7/20	.210mg	18/50	(.430mg	8/50)			
a	TR62	2.59mg	n.s.s.	0/20	.210mg	0/50	.430mg	1/50			liv:hpa,hpc,nnd.
b	TR62	.497mg	n.s.s.	2/20	.210mg	5/50	(.430mg	0/50)			lun:a/a,a/c.
2032	283	.353mg	n.s.s.	0/16	.414mg	0/10	.834mg	0/17			Innes;ntis,1968/1969
a	283	1.09mg	n.s.s.	0/16	.414mg	1/10	.834mg	0/17			
b	283	.127mg	n.s.s.	0/16	.414mg	3/10	(.834mg	0/17)			
2033	TR62	.364mg	n.s.s.	4/20	.350mg	8/50	.700mg	7/50			
a	TR62	.408mg	n.s.s.	1/20	.350mg	6/50	.700mg	2/50			liv:hpa,hpc,nnd.
b	TR62	.576mg	n.s.s.	0/20	.350mg	2/50	.700mg	2/50			lun:a/a,a/c.
2034	283	.176mg	n.s.s.	0/16	.385mg	3/14	(.776mg	0/16)			Innes;ntis,1968/1969
a	283	.224mg	n.s.s.	0/16	.385mg	2/14	(.776mg	0/16)			
b	283	.120mg	2.11mg	0/16	.385mg	5/14	(.776mg	0/16)			
2035	283	.506mg	n.s.s.	0/17	.414mg	0/16	.834mg	0/18			
a	283	1.68mg	n.s.s.	1/17	.414mg	1/16	.834mg	0/18			
b	283	1.29mg	n.s.s.	2/17	.414mg	3/16	.834mg	0/18			
2036	283	.433mg	n.s.s.	1/18	.385mg	0/16	.776mg	0/14			
a	283	.620mg	n.s.s.	2/18	.385mg	4/16	.776mg	1/14			
b	283	.637mg	n.s.s.	3/18	.385mg	5/16	.776mg	1/14			
2037	TR62	10.5mg	n.s.s.	15/20	6.60mg	30/50	15.6mg	24/50			
a	TR62	55.0mg	n.s.s.	0/20	6.60mg	1/50	15.6mg	1/50			liv:hpa,hpc,nnd.
2038	TR62	15.3mg	n.s.s.	12/20	15.7mg	8/50	(32.9mg	3/50)			
a	TR62	n.s.s.	n.s.s.	0/20	15.7mg	0/50	32.9mg	0/50			liv:hpa,hpc,nnd.
ENDRIN 72-20-8											
2039	TR12	.956mg	n.s.s.	2/10	.290mg	5/50	.570mg	8/50			
a	TR12	.973mg	n.s.s.	2/10	.290mg	3/50	(.570mg	1/50)			liv:hpa,hpc,nnd.
b	TR12	1.83mg	n.s.s.	0/10	.290mg	2/50	.570mg	1/50			lun:a/a,a/c.
2040	TR12	.261mg	n.s.s.	1/10	.170mg	3/50	.340mg	14/50			
a	TR12	.335mg	n.s.s.	1/10	.170mg	3/50	.340mg	10/50			liv:hpa,hpc,nnd.
b	TR12	.520mg	n.s.s.	0/10	.170mg	1/50	.340mg	1/50			lun:a/a,a/c.
2041	TR12	.214mg	n.s.s.	8/10	.110mg	37/50	.210mg	33/50			
a	TR12	1.21mg	n.s.s.	0/10	.110mg	1/50	.210mg	1/50			liv:hpa,hpc,nnd.
2042	TR12	.110mg	1.27mg	4/60p	.110mg	16/50	(.210mg	6/50)			adr:adn,can. S
a	TR12	.277mg	n.s.s.	4/60p	.110mg	11/50	.210mg	13/50			S
2043	1004	6.88mg	n.s.s.	0/88	95.6ug	0/48	.287mg	1/45	.575mg	0/49	Deichmann;imed,39,426-434;1970
a	1004	1.77mg	n.s.s.	60/88	95.6ug	37/48	.287mg	26/45	.575mg	27/49	
2044	TR12	.101mg	n.s.s.	5/10	70.0ug	25/50	.140mg	27/50			
a	TR12	n.s.s.	n.s.s.	0/10	70.0ug	0/50	.140mg	0/50			liv:hpa,hpc,nnd.
2045	TR12	.233mg	n.s.s.	2/60p	70.0ug	4/50	.140mg	8/50			adr:adn,can. S
a	TR12	.265mg	n.s.s.	2/60p	70.0ug	2/50	.140mg	8/50			S
b	TR12	.298mg	n.s.s.	0/60p	70.0ug	5/50	.140mg	3/50			S

```
      Spe Strain  Site   Xpo+Xpt                                                                          TD50    2Tailpvl
          Sex Route  Hist      Notes                                                                           DR   AuOp
  c   R m osm eat pni isc 19m26                                                                            1.75mg  * P<.04
2046  R m osm eat liv hem 28m29 ev                                      .>                                no dre    P=1.   -
  a   R m osm eat tba mix 28m29 ev                                                                         no dre    P=1.   -

ENFLURANE                       100ng..:..1ug....:..10.....:..100....:..1mg..:..10.....:..100..:..1g.....:..10
2047  M f sic inh lun ade 52w52 ek                                                               .>       8.93gm    P<.4   -
  a   M f sic inh liv tum 52w52 ek                                                                        no dre    P=1.   -
2048  M f sic inh lun ade 78w82 e                                                                   .>   >18.9gm    P<.2   -
  a   M f sic inh liv mix 78w82 e                                                                         no dre    P=1.   -
2049  M m sic inh liv mix 52w52 ek                                                              .        ±8.12gm    P<.1   -
  a   M m sic inh lun ade 52w52 ek                                                                        no dre    P=1.   -
2050  M m sic inh liv mix 78w82 e                                                              .         11.7gm    P<.1   -
  a   M m sic inh lun ade 78w82 e                                                                         60.9gm    P<.8   -

ENOVID                          100ng..:..1ug....:..10.....:..100....:..1mg..:..10.....:..100..:..1g.....:..10
2051  M f aah eat lun tum 24m24 g                                         .>                              4.62mg    P<.6   -
  a   M f aah eat liv hpt 24m24 g                                                                         25.2mg    P<.3   -
2052  M f aah eat lun tum 24m24                                               .>                          4.31mg    P<.2   -
  a   M f aah eat liv hpt 24m24                                                                           no dre    P=1.   -
2053  M f aah eat lun tum 24m24                                                  .>                       no dre    P=1.   -
2054  M f bal eat mix sqc 86w86 r                                       .     ±                           2.00mg    P<.1   -
2055  M f bce eat lun tum 24m24 g                                             .>                          no dre    P=1.   -
2056  M f bce eat lun tum 24m24 g                                          .>                             no dre    P=1.   -
2057  M f bce eat lun tum 24m24                                                     .>                    no dre    P=1.   -
2058  M f c71 gav pit tum 89w89 e                  .    +       .                                        .151mg  * P<.0005+ -
2059  M f cf1 eat liv hct 78w78 er                                             .>                         no dre    P=1.   -
2060  M m cf1 eat liv hct 78w78 er                                             .>                         no dre    P=1.   -
2061  M f che eat lun tum 24m24 g                                          .>                             20.7mg    P<.3   -
2062  M f che eat hpl cra 24m24                                 .   +           .                         1.75mg    P<.008 -
2063  M f che eat hpl cra 24m24                                 .   +  .                                  1.82mg    P<.0005+ -
  a   M f che eat liv hpt 24m24                                                                           no dre    P=1.   -
2064  M f chf eat --- fbs 24m24 g                                     .   +        .                      3.93mg    P<.005 -
  a   M f chf eat lun tum 24m24 g                                                                         27.6mg    P<.8   -
  b   M f chf eat liv hpt 24m24 g                                                                         no dre    P=1.   -
2065  M f chf eat lun tum 24m24 g                                            .>                           no dre    P=1.   -
  a   M f chf eat liv hpt 24m24 g                                                                         no dre    P=1.   -
2066  M f chf eat lun tum 24m24                                                  .>                       no dre    P=1.   -
  a   M f chf eat liv hpt 24m24                                                                           no dre    P=1.   -
2067  M f chh eat liv hpt 24m24 g                                              .>                         no dre    P=1.   -
2068  M f csc gav lun tum 69w69 ek                                  .>                                    no dre    P=1.   -
  a   M f csc gav tba mix 69w69 ek                                                                        no dre    P=1.   -
2069  M f csc gav tba mix 69w82 ek                            .>                                          .332mg    P<.8   -
2070  M f r3m eat mam tum 24m24 r                                      .>                                 no dre    P=1.   -

ENOVID-E                        100ng..:..1ug....:..10.....:..100....:..1mg..:..10.....:..100..:..1g.....:..10
2071  M f cf1 eat liv hct 78w78 er                                             .>                         143.mg  * P<.8   -
2072  M m cf1 eat liv hct 78w78 er                                             .>                         no dre    P=1.   -
2073  M f sww gav liv hct 65w80 er                                          .>                            20.8mg    P<.6   -

EPHEDRINE SULPHATE              100ng..:..1ug....:..10.....:..100....:..1mg..:..10.....:..100..:..1g.....:..10
2074  M f b6c eat TBA MXB 24m24                                                  :>                        259.mg  * P<.9   -
  a   M f b6c eat liv MXB 24m24                                                                           no dre    P=1.
  b   M f b6c eat lun MXB 24m24                                                                           789.mg    P<.6
2075  M m b6c eat MXA mlm 24m24                                                    :     ±                #194.mg  * P<.02  -
  a   M m b6c eat TBA MXB 24m24                                                                           no dre    P=1.
  b   M m b6c eat liv MXB 24m24                                                                            264.mg    P<.7
  c   M m b6c eat lun MXB 24m24                                                                           no dre    P=1.
2076  R f f34 eat TBA MXB 24m24                                               :>                          no dre    P=1.
  a   R f f34 eat liv MXB 24m24                                                                           no dre    P=1.
2077  R m f34 eat TBA MXB 24m24                                               :>                          no dre    P=1.
  a   R m f34 eat liv MXB 24m24                                                                            166.mg  * P<.7

EPICHLOROHYDRIN                 100ng..:..1ug....:..10.....:..100....:..1mg..:..10.....:..100..:..1g.....:..10
2078  M f hic ipj lun ptm 64w64                                                  .>                       28.9mg    P<.8
2079  R m sda inh nas sqc 13m32 as                                                    .>                   421.mg  * P<.2
  a   R m sda inh liv tum 13m32 as                                                                        no dre    P=1.
2080  R f wis gav for sqc 24m24 e                                       .   +  .                          2.55mg  / P<.0005+
  a   R f wis gav liv nnd 24m24 e                                                                         no dre    P=1.
  b   R f wis gav tba mix 24m24 e                                                                         no dre    P=1.
2081  R m wis gav for sqc 24m24 e                                         .  +  .                         3.53mg  * P<.0005+
  a   R m wis gav liv tum 24m24 e                                                                          218.mg    P<.3   -
  b   R m wis gav liv hpc 24m24 e                                                                         no dre    P=1.   -
  c   R m wis gav liv nnd 24m24 e                                                                         no dre    P=1.   -
  d   R m wis gav tba mix 24m24 e                                                                         4.83mg    P<.4

L-EPINEPHRINE.HCl               100ng..:..1ug....:..10.....:..100....:..1mg..:..10.....:..100..:..1g.....:..10
2082  M f b6c inh ute MXA 24m25                                                  :     ±                  #6.86mg  * P<.03  i
  a   M f b6c inh ute esp 24m25                                                                           8.52mg    P<.05
  b   M f b6c inh TBA MXB 24m25                                                                           no dre    P=1.
  c   M f b6c inh liv MXB 24m25                                                                           9.74mg  * P<.4
  d   M f b6c inh lun MXB 24m25                                                                           no dre    P=1.
```

	RefNum	LoConf	UpConf	Cntrl	1Dose	1Inc	2Dose	2Inc		Citation or Pathology
										Brkly Code
c	TR12	.529mg	n.s.s.	0/60p	70.0ug	0/50	.140mg	3/50		S
2046	1004	.706mg	n.s.s.	1/75	76.5ug	0/47	.230mg	0/44	.461mg 0/42	Deichmann;imed,39,426-434;1970
a	1004	2.08mg	n.s.s.	19/75	76.5ug	11/47	.230mg	6/44	.461mg 10/42	

ENFLURANE   13838-16-9

	RefNum	LoConf	UpConf	Cntrl	1Dose	1Inc	Citation or Pathology
2047	1677m	1.84gm	n.s.s.	2/25	4.74gm	4/25	Baden;anes,56,9-13;1982
a	1677m	6.11gm	n.s.s.	0/25	4.74gm	0/25	
2048	1677n	6.69gm	n.s.s.	13/98	4.51gm	21/97	
a	1677n	56.1gm	n.s.s.	2/98	4.51gm	0/97	
2049	1677m	2.00gm	n.s.s.	0/25	3.95gm	2/25	
a	1677m	3.10gm	n.s.s.	5/25	3.95gm	2/25	
2050	1677n	4.41gm	n.s.s.	18/96	3.76gm	28/96	
a	1677n	7.07gm	n.s.s.	19/96	3.76gm	21/96	

ENOVID   (norethynodrel/mestranol [66:1]) 8015-30-3

	RefNum	LoConf	UpConf	Cntrl	1Dose	1Inc	2Dose	2Inc	3Dose	3Inc	Citation or Pathology
2051	1367m	.763mg	n.s.s.	28/61	.650mg	29/57					Heston;jnci,51,209-224;1973
a	1367m	4.10mg	n.s.s.	0/61	.650mg	1/57					
2052	1367n	1.42mg	n.s.s.	21/59	1.30mg	30/63					
a	1367n	16.9mg	n.s.s.	1/59	1.30mg	0/63					
2053	1367o	6.12mg	n.s.s.	17/49	2.60mg	16/54					
2054	1472	.492mg	n.s.s.	0/20	.450mg	2/20					Munoz;canr,38,1504-1508;1973/Dunn 1969
2055	1367m	4.21mg	n.s.s.	18/52	.650mg	6/55					Heston;jnci,51,209-224;1973
2056	1367n	3.77mg	n.s.s.	15/53	1.30mg	11/50					
2057	1367o	19.8mg	n.s.s.	12/55	2.60mg	3/55					
2058	230	52.7ug	.995mg	1/8	.200mg	6/11	2.00mg	7/7			Poel;scie,154,402-403;1966
2059	1453	23.8mg	n.s.s.	0/39	.125mg	1/40	1.50mg	1/40	5.00mg	0/38	Barrows;jtxe,3,219-230;1977
2060	1453	17.8mg	n.s.s.	6/39	.125mg	4/40	1.50mg	5/39	5.00mg	1/40	
2061	1367m	3.37mg	n.s.s.	0/61	.650mg	1/47					Heston;jnci,51,209-224;1973
2062	1367n	.830mg	40.3mg	17/50	1.30mg	32/53					
2063	1367o	1.03mg	4.55mg	15/51	2.60mg	36/49					
a	1367o	26.3mg	n.s.s.	1/51	2.60mg	0/49					
2064	1367m	1.60mg	26.9mg	0/54	.650mg	6/56					
a	1367m	2.21mg	n.s.s.	5/54	.650mg	6/56					
b	1367m	4.36mg	n.s.s.	46/54	.650mg	13/56					
2065	1367n	10.2mg	n.s.s.	3/52	1.30mg	1/53					
a	1367n	11.4mg	n.s.s.	34/52	1.30mg	4/53					
2066	1367o	12.5mg	n.s.s.	6/50	2.60mg	4/52					
a	1367o	24.0mg	n.s.s.	37/50	2.60mg	3/52					
2067	1367o	28.4mg	n.s.s.	2/55	2.60mg	0/53					
2068	585m	.194mg	n.s.s.	0/15	.143mg	0/15					Poel;canr,28,845-859;1968
a	585m	83.4ug	n.s.s.	4/15	.143mg	3/15					
2069	585n	36.7ug	n.s.s.	8/15	.120mg	9/15					
2070	1463	1.93mg	n.s.s.	50/73	1.95mg	12/21					Rudali;gmcr,17,243-252;1975

ENOVID-E (norethynodrel/mestranol [25:1]) 8015-30-3

	RefNum	LoConf	UpConf	Cntrl	1Dose	1Inc	2Dose	2Inc	3Dose	3Inc	Citation or Pathology
2071	1453m	9.95mg	n.s.s.	0/39	.130mg	1/39	1.56mg	0/38	2.60mg	1/39	Barrows;jtxe,3,219-230;1977
2072	1453m	8.87mg	n.s.s.	6/39	.130mg	8/40	1.56mg	7/40	2.60mg	1/39	
2073	1453n	2.84mg	n.s.s.	1/50	1.06mg	2/50					

EPHEDRINE SULPHATE   134-72-5

	RefNum	LoConf	UpConf	Cntrl	1Dose	1Inc	2Dose	2Inc	Citation or Pathology
2074	TR307	19.8mg	n.s.s.	43/50	16.1mg	45/50	32.2mg	41/50	
a	TR307	67.6mg	n.s.s.	9/50	16.1mg	3/50	(32.2mg	3/50)	liv:hpa,hpc,nnd.
b	TR307	132.mg	n.s.s.	1/50	16.1mg	1/50	32.2mg	2/50	lun:a/a,a/c.
2075	TR307	79.1mg	n.s.s.	0/50	14.9mg	2/50	29.7mg	4/50	mey:mlm; mul:mlm; spl:mlm. S
a	TR307	27.8mg	n.s.s.	38/50	14.9mg	31/50	29.7mg	36/50	
b	TR307	37.9mg	n.s.s.	19/50	14.9mg	18/50	29.7mg	21/50	liv:hpa,hpc,nnd.
c	TR307	98.7mg	n.s.s.	9/50	14.9mg	11/50	29.7mg	4/50	lun:a/a,a/c.
2076	TR307	13.6mg	n.s.s.	45/50	6.19mg	43/50	12.4mg	45/50	
a	TR307	98.6mg	n.s.s.	2/50	6.19mg	2/50	12.4mg	0/50	liv:hpa,hpc,nnd.
2077	TR307	8.33mg	n.s.s.	43/50	4.95mg	47/50	9.90mg	45/50	
a	TR307	21.5mg	n.s.s.	3/50	4.95mg	5/50	9.90mg	5/50	liv:hpa,hpc,nnd.

EPICHLOROHYDRIN   106-89-8

	RefNum	LoConf	UpConf	Cntrl	1Dose	1Inc	2Dose	2Inc	Citation or Pathology
2078	1143	2.54mg	n.s.s.	10/30	5.71mg	11/30			Van Duuren;jnci,53,695-700;1974
2079	1167m	68.5mg	n.s.s.	0/100	.729mg	0/100	2.88mg	1/100	Laskin;jnci,65,751-757;1980/Kuschner pers.comm.
a	1167m	20.5mg	n.s.s.	0/100	.729mg	0/100	2.88mg	0/100	
2080	1727	1.46mg	4.71mg	0/38	1.43mg	2/27	7.14mg	24/24	Wester;txcy,36,325-339;1985
a	1727	72.1mg	n.s.s.	1/50	7.14mg	0/49			
b	1727	10.3mg	n.s.s.	33/50	7.14mg	28/49			
2081	1727	2.34mg	5.56mg	0/49	1.43mg	6/43	7.14mg	35/43	
a	1727	35.5mg	n.s.s.	0/49	7.14mg	1/45			
b	1727	66.2mg	n.s.s.	1/49	7.14mg	0/45			
c	1727	66.2mg	n.s.s.	1/49	7.14mg	0/45			
d	1727	.898mg	n.s.s.	46/49	7.14mg	44/45			

L-EPINEPHRINE.HCl   55-31-2

	RefNum	LoConf	UpConf	Cntrl	1Dose	1Inc	2Dose	2Inc	Citation or Pathology
2082	TR380	2.61mg	n.s.s.	0/50	.458mg	1/50	.920mg	4/50	ute:esp,ess. S
a	TR380	2.94mg	n.s.s.	0/50	.458mg	1/50	.920mg	3/50	S
b	TR380	1.19mg	n.s.s.	36/50	.458mg	35/50	.920mg	30/50	
c	TR380	2.42mg	n.s.s.	3/50	.458mg	2/50	.920mg	6/50	liv:hpa,hpc,nnd.
d	TR380	4.44mg	n.s.s.	5/50	.458mg	5/50	.920mg	2/50	lun:a/a,a/c.

Spe Strain Site Xpo+Xpt				Notes		TD50	2Tailpvl
	Sex	Route	Hist				DR	AuOp

2083	M m b6c	inh	TBA MXB	24m25		:>	no dre	P=1.	i
a	M m b6c	inh	liv MXB	24m25			no dre	P=1.	
b	M m b6c	inh	lun MXB	24m25			no dre	P=1.	
2084	R f f34	inh	TBA MXB	24m24		:>	no dre	P=1.	i
a	R f f34	inh	liv MXB	24m24			no dre	P=1.	
2085	R m f34	inh	TBA MXB	24m24		:>	4.70mg *	P<1.	i
a	R m f34	inh	liv MXB	24m24			no dre	P=1.	

1,2-EPOXYBUTANE   100ng..:..1ug....:..10.....:..100....:..1mg....:..10.....:..100...:..1g.....:..10

2086	M f b6c	inh	TBA MXB	24m24 s		:>	141.mg *	P<.4	-
a	M f b6c	inh	liv MXB	24m24 s			445.mg *	P<.3	
b	M f b6c	inh	lun MXB	24m24 s			834.mg *	P<.6	
2087	M m b6c	inh	TBA MXB	24m24		:>	2.47gm *	P<1.	
a	M m b6c	inh	liv MXB	24m24			8.60gm *	P<1.	
b	M m b6c	inh	lun MXB	24m24			no dre	P=1.	
2088	R f f34	inh	pta adn	24m24		: ±	83.8mg *	P<.05	
a	R f f34	inh	thy MXA	24m24			509.mg *	P<.03	
b	R f f34	inh	ova MXA	24m24			545.mg *	P<.03	
c	R f f34	inh	nas ppa	24m24			1.73gm *	P<.09	e
d	R f f34	inh	TBA MXB	24m24			78.0mg *	P<.2	
e	R f f34	inh	liv MXB	24m24			no dre	P=1.	
2089	R m f34	inh	MXB MXB	24m24		: + :	106.mg *	P<.0005	
a	R m f34	inh	lun MXA	24m24			220.mg *	P<.006	c
b	R m f34	inh	nas ppa	24m24			220.mg *	P<.002	c
c	R m f34	inh	MXA mnl	24m24			30.7mg \	P<.04	
d	R m f34	inh	lun a/c	24m24			314.mg *	P<.02	c
e	R m f34	inh	TBA MXB	24m24			79.6mg *	P<.3	
f	R m f34	inh	liv MXB	24m24			622.mg *	P<.5	

ERYTHORBATE, SODIUM   100ng..:..1ug....:..10.....:..100....:..1mg....:..10.....:..100...:..1g.....:..10

2090	M f b6c	wat	liv hem	22m26 e			149.gm *	P<.3	-
a	M f b6c	wat	lun a/c	22m26 e			225.gm *	P<.3	
b	M f b6c	wat	liv hpa	22m26 e			225.gm *	P<.3	
c	M f b6c	wat	liv hpc	22m26 e			452.gm *	P<.6	
d	M f b6c	wat	lun a/a	22m26 e			no dre	P=1.	
e	M f b6c	wat	--- lkm	22m26 e			no dre	P=1.	
f	M f b6c	wat	tba mix	22m26 e			40.9gm *	P<.3	
2091	M m b6c	wat	liv hem	22m26 e		.	16.9gm \	P<.04	-
a	M m b6c	wat	liv hpa	22m26 e			25.1gm *	P<.2	
b	M m b6c	wat	liv hpc	22m26 e			57.2gm *	P<.5	
c	M m b6c	wat	lun a/c	22m26 e			85.4gm *	P<.3	
d	M m b6c	wat	lun a/a	22m26 e			no dre	P=1.	
e	M m b6c	wat	tba mix	22m26 e			8.69gm *	P<.06	-
2092	R f f3d	wat	liv hpa	24m26 e			no dre	P=1.	
a	R f f3d	wat	tba mix	24m26 e			no dre	P=1.	
2093	R m f3d	wat	liv hpa	24m26 e			no dre	P=1.	
a	R m f3d	wat	tba mix	24m26 e			no dre	P=1.	

ERYTHROMYCIN STEARATE   100ng..:..1ug....:..10.....:..100....:..1mg....:..10.....:..100...:..1g.....:..10

2094	M f b6c	eat	TBA MXB	24m24		:>	no dre	P=1.	
a	M f b6c	eat	liv MXB	24m24			no dre	P=1.	
b	M f b6c	eat	lun MXB	24m24			no dre	P=1.	
2095	M m b6c	eat	TBA MXB	24m24		:>	no dre	P=1.	
a	M m b6c	eat	liv MXB	24m24			no dre	P=1.	
b	M m b6c	eat	lun MXB	24m24			no dre	P=1.	
2096	R f f34	eat	TBA MXB	24m24		:>	no dre	P=1.	
a	R f f34	eat	liv MXB	24m24			5.85gm *	P<.2	
2097	R m f34	eat	TBA MXB	24m24		:>	2.02gm *	P<.8	
a	R m f34	eat	liv MXB	24m24			3.95gm *	P<.3	

ESTAZOLAM   100ng..:..1ug....:..10.....:..100....:..1mg....:..10.....:..100...:..1g.....:..10

2098	M f b6c	eat	liv agm	24m24 e		.>	756.mg *	P<.6	-
a	M f b6c	eat	liv hpc	24m24 e			820.mg *	P<.9	
b	M f b6c	eat	lun a/a	24m24 e			1.96gm *	P<.9	
c	M f b6c	eat	liv ang	24m24 e			no dre	P=1.	
d	M f b6c	eat	lun a/c	24m24 e			no dre	P=1.	
e	M f b6c	eat	tba ben	24m24 e			70.8mg *	P<.3	
f	M f b6c	eat	tba mal	24m24 e			no dre	P=1.	
g	M f b6c	eat	tba mix	24m24 e			no dre	P=1.	
2099	M m b6c	eat	lun a/a	24m24 e		. ±	89.5mg *	P<.09	
a	M m b6c	eat	liv agm	24m24 e			471.mg *	P<.5	
b	M m b6c	eat	liv hpc	24m24 e			no dre	P=1.	
c	M m b6c	eat	lun a/c	24m24 e			no dre	P=1.	
d	M m b6c	eat	tba ben	24m24 e			271.mg *	P<.8	
e	M m b6c	eat	tba mal	24m24 e			no dre	P=1.	
f	M m b6c	eat	tba mix	24m24 e			no dre	P=1.	
2100	R f cdr	eat	liv nnd	24m24 e		.>	no dre	P=1.	
a	R f cdr	eat	liv hpc	24m24 e			no dre	P=1.	
b	R f cdr	eat	tba mix	24m24 e			5.81mg *	P<.3	-
c	R f cdr	eat	tba ben	24m24 e			476.mg *	P<1.	-
d	R f cdr	eat	tba mal	24m24 e			no dre	P=1.	-

	RefNum	LoConf	UpConf	Cntrl	1Dose	1Inc	2Dose	2Inc			Citation or Pathology
											Brkly Code
2083	TR380	.993mg	n.s.s.	38/50	.382mg	33/50	.774mg	33/50			
a	TR380	1.93mg	n.s.s.	20/50	.382mg	14/50	.774mg	15/50			liv:hpa,hpc,nnd.
b	TR380	1.34mg	n.s.s.	15/50	.382mg	11/50	.774mg	16/50			lun:a/a,a/c.
2084	TR380	.281mg	n.s.s.	49/50	.111mg	41/50	.368mg	43/50			
a	TR380	1.96mg	n.s.s.	2/50	.111mg	0/50	.368mg	1/50			liv:hpa,hpc,nnd.
2085	TR380	.162mg	n.s.s.	48/50	77.1ug	40/50	.258mg	46/50			
a	TR380	1.82mg	n.s.s.	1/50	77.1ug	1/50	.258mg	0/50			liv:hpa,hpc,nnd.
1,2-EPOXYBUTANE	**106-88-7**										
2086	TR329	35.3mg	n.s.s.	35/50	45.5mg	29/50	90.9mg	20/50			
a	TR329	118.mg	n.s.s.	4/50	45.5mg	3/50	90.9mg	5/50			liv:hpa,hpc,nnd.
b	TR329	135.mg	n.s.s.	4/50	45.5mg	3/50	90.9mg	3/50			lun:a/a,a/c.
2087	TR329	81.5mg	n.s.s.	28/50	37.9mg	26/50	75.8mg	24/50			
a	TR329	137.mg	n.s.s.	14/50	37.9mg	13/50	75.8mg	12/50			liv:hpa,hpc,nnd.
b	TR329	228.mg	n.s.s.	11/50	37.9mg	9/50	75.8mg	6/50			lun:a/a,a/c.
2088	TR329	35.2mg	n.s.s.	25/50	43.3mg	26/50	86.6mg	32/50			S
a	TR329	174.mg	n.s.s.	0/50	43.3mg	1/50	86.6mg	3/50			thy:fca,fcc. S
b	TR329	181.mg	n.s.s.	0/50	43.3mg	1/50	86.6mg	3/50			ova:gcc,gct,tcm. S
c	TR329	422.mg	n.s.s.	0/50	43.3mg	0/50	86.6mg	2/50			
d	TR329	27.9mg	n.s.s.	45/50	43.3mg	47/50	86.6mg	47/50			
e	TR329	601.mg	n.s.s.	1/50	43.3mg	0/50	86.6mg	0/50			liv:hpa,hpc,nnd.
2089	TR329	56.3mg	241.mg	0/50	30.3mg	2/50	60.6mg	12/50			lun:a/a,a/c; nas:ppa. C
a	TR329	92.8mg	684.mg	0/50	30.3mg	2/50	60.6mg	5/50			lun:a/a,a/c.
b	TR329	93.7mg	872.mg	0/50	30.3mg	0/50	60.6mg	1/50			
c	TR329	12.5mg	n.s.s.	25/50	30.3mg	31/50	(60.6mg	22/50)			mul:mnl; spl:mnl. S
d	TR329	116.mg	n.s.s.	0/50	30.3mg	1/50	60.6mg	4/50			
e	TR329	22.4mg	n.s.s.	46/50	30.3mg	48/50	60.6mg	47/50			
f	TR329	120.mg	n.s.s.	2/50	30.3mg	2/50	60.6mg	3/50			liv:hpa,hpc,nnd.
ERYTHORBATE, SODIUM	**(isoascorbate)**	**6381-77-7**									
2090	1897	45.2gm	n.s.s.	0/45	4.36gm	2/44	8.73gm	1/46			Inai;hijm,38,135-139;1989
a	1897	55.3gm	n.s.s.	0/45	4.36gm	1/44	8.73gm	1/46			
b	1897	55.3gm	n.s.s.	0/45	4.36gm	1/44	8.73gm	1/46			
c	1897	60.7gm	n.s.s.	1/45	4.36gm	0/44	8.73gm	2/46			
d	1897	29.9gm	n.s.s.	1/45	4.36gm	0/44	8.73gm	0/46			
e	1897	30.0gm	n.s.s.	5/45	4.36gm	7/44	8.73gm	5/46			
f	1897	12.3gm	n.s.s.	10/45	4.36gm	16/44	8.73gm	15/46			
2091	1897	5.13gm	n.s.s.	0/38	1.82gm	3/38	(3.64gm	0/43)			
a	1897	9.33gm	n.s.s.	1/38	1.82gm	4/38	3.64gm	5/43			
b	1897	11.6gm	n.s.s.	2/38	1.82gm	4/38	3.64gm	4/43			
c	1897	21.0gm	n.s.s.	0/38	1.82gm	1/38	3.64gm	1/43			
d	1897	20.2gm	n.s.s.	4/38	1.82gm	0/38	3.64gm	4/43			
e	1897	3.68gm	n.s.s.	9/38	1.82gm	14/38	3.64gm	19/43			
2092	1736	15.5gm	n.s.s.	1/48	663.mg	1/49	1.33gm	0/45			Abe;eamp,41,35-43;1984
a	1736	1.69gm	n.s.s.	45/48	663.mg	43/49	1.33gm	35/45			
2093	1736	12.5gm	n.s.s.	1/49	580.mg	0/48	1.16gm	1/50			
a	1736	n.s.s.	n.s.s.	49/49	580.mg	48/48	1.16gm	50/50			
ERYTHROMYCIN STEARATE	**643-22-1**										
2094	TR338	634.mg	n.s.s.	44/50	322.mg	41/50	644.mg	39/50			
a	TR338	2.62gm	n.s.s.	4/50	322.mg	6/50	644.mg	2/50			liv:hpa,hpc,nnd.
b	TR338	3.34gm	n.s.s.	4/50	322.mg	3/50	644.mg	2/50			lun:a/a,a/c.
2095	TR338	1.02gm	n.s.s.	37/50	297.mg	26/50	594.mg	29/50			
a	TR338	1.77gm	n.s.s.	15/50	297.mg	8/50	594.mg	11/50			liv:hpa,hpc,nnd.
b	TR338	2.30gm	n.s.s.	6/50	297.mg	5/50	594.mg	4/50			lun:a/a,a/c.
2096	TR338	153.mg	n.s.s.	49/50	248.mg	48/50	(495.mg	42/50)			
a	TR338	1.77gm	n.s.s.	0/50	248.mg	0/50	495.mg	2/50			liv:hpa,hpc,nnd.
2097	TR338	197.mg	n.s.s.	45/50	198.mg	46/50	396.mg	47/50			
a	TR338	957.mg	n.s.s.	1/50	198.mg	1/50	396.mg	3/50			liv:hpa,hpc,nnd.
ESTAZOLAM	**29975-16-4**										
2098	1984	80.8mg	n.s.s.	1/100	.800mg	0/50	3.00mg	0/50	10.0mg	1/50	Kimura;faat,4,827-842;1984
a	1984	40.6mg	n.s.s.	2/100	.800mg	4/50	3.00mg	4/50	10.0mg	2/50	
b	1984	85.7mg	n.s.s.	2/100	.800mg	0/50	3.00mg	0/50	10.0mg	1/50	
c	1984	102.mg	n.s.s.	0/100	.800mg	1/50	3.00mg	0/50	10.0mg	0/50	
d	1984	64.4mg	n.s.s.	4/100	.800mg	0/50	3.00mg	3/50	10.0mg	1/50	
e	1984	18.1mg	n.s.s.	19/100	.800mg	6/50	3.00mg	13/50	10.0mg	12/50	
f	1984	21.0mg	n.s.s.	32/100	.800mg	21/50	3.00mg	20/50	10.0mg	16/50	
g	1984	15.1mg	n.s.s.	47/100	.800mg	24/50	3.00mg	28/50	10.0mg	23/50	
2099	1984	28.5mg	n.s.s.	7/100	.800mg	1/50	3.00mg	2/50	10.0mg	7/50	
a	1984	64.2mg	n.s.s.	1/100	.800mg	0/50	3.00mg	1/50	10.0mg	1/50	
b	1984	33.9mg	n.s.s.	23/100	.800mg	11/50	3.00mg	12/50	10.0mg	9/50	
c	1984	35.6mg	n.s.s.	13/100	.800mg	6/50	3.00mg	7/50	10.0mg	6/50	
d	1984	24.9mg	n.s.s.	21/100	.800mg	5/50	3.00mg	11/50	10.0mg	10/50	
e	1984	25.2mg	n.s.s.	42/100	.800mg	20/50	3.00mg	28/50	10.0mg	16/50	
f	1984	21.4mg	n.s.s.	57/100	3.00mg	57/100	10.0mg	23/50			
2100	1984	73.8mg	n.s.s.	5/100	.500mg	0/50	2.00mg	2/50	10.0mg	1/50	
a	1984	77.5mg	n.s.s.	0/100	.500mg	1/50	2.00mg	1/50	10.0mg	0/50	
b	1984	1.21mg	n.s.s.	95/100	.500mg	46/50	2.00mg	49/50	10.0mg	49/50	
c	1984	4.35mg	n.s.s.	91/100	.500mg	43/50	2.00mg	47/50	10.0mg	45/50	
d	1984	27.3mg	n.s.s.	26/100	.500mg	16/50	2.00mg	17/50	10.0mg	12/50	

```
       Spe Strain  Site    Xpo+Xpt                                                    TD50      2Tailpvl
         Sex  Route   Hist   Notes                                                       DR       AuOp
2101 R m cdr eat ski fib 24m24 e                                          .>                   234.mg  * P<.6    -
   a R m cdr eat liv nnd 24m24 e                                                               854.mg  * P<.8    -
   b R m cdr eat liv hpc 24m24 e                                                               no dre    P=1.    -
   c R m cdr eat tba mal 24m24 e                                                               75.7mg  * P<.3    -
   d R m cdr eat tba ben 24m24 e                                                               no dre    P=1.    -
   e R m cdr eat tba mix 24m24 e                                                               no dre    P=1.    -

ESTRADIOL                       100ng..:..1ug....:..10.....:..100....:..1mg....:..10.....:..100....:..1g.....:..10
2102 M f c3h eat mgl adc 24m24 r                           .>                                  .282mg    P<.7    +
2103 M f c3j eat mgl adc 52w52 ek                   .  ±                                        .904mg  * P<.02
2104 M f c3j eat ova tua 78w78 ek                     .>                                        no dre    P=1.
2105 M f c3j eat mgl adc 24m24 ek                     .>                                        2.75mg    P<.6

ESTRADIOL MUSTARD               100ng..:..1ug....:..10.....:..100....:..1mg....:..10.....:..100....:..1g.....:..10
2106 M f b6c gav MXB MXB 52w84 s                 :  +  :                                        .682mg  * P<.0005
   a M f b6c gav lun MXA 52w84 s                                                                1.14mg  * P<.0005c
   b M f b6c gav --- lym 52w84 s                                                                2.34mg  * P<.0005c
   c M f b6c gav myc srn 52w84 s                                                                3.01mg  * P<.006 c
   d M f b6c gav sto sqc 52w84 s                                                          +hist 5.17mg  * P<.009 c
   e M f b6c gav mgl MXA 52w84 s                                                                13.4mg  * P<.02
   f M f b6c gav TBA MXB 52w84 s                                                                .562mg  * P<.0005
   g M f b6c gav liv MXB 52w84 s                                                                no dre    P=1.
   h M f b6c gav lun MXB 52w84 s                                                                1.14mg  * P<.0005
2107 M f b6c gav lun MXA 52w82 s     pool                :    +    :                            1.28mg  * P<.0005c
   a M f b6c gav --- lym 52w82 s                                                                2.23mg  * P<.0005c
   b M f b6c gav lun a/c 52w82 s                                                                2.87mg  * P<.002
   c M f b6c gav myc srn 52w82 s                                                                3.22mg  * P<.0005c
   d M f b6c gav mgl MXA 52w82 s                                                                12.8mg  * P<.007
2108 M m b6c gav MXB MXB 52w83 s                             :  +  :                            4.10mg  * P<.0005
   a M m b6c gav --- MXA 52w83 s                                                                4.39mg  * P<.0005c
   b M m b6c gav sto sqc 52w83 s                                                          +hist 29.0mg  * P<.08  c
   c M m b6c gav TBA MXB 52w83 s                                                                1.96mg  * P<.0005
   d M m b6c gav liv MXB 52w83 s                                                                no dre    P=1.
   e M m b6c gav lun MXB 52w83 s                                                                2.34mg  \ P<.02
2109 M m b6c gav lun MXA 52w81 s     pool                :    +    :                            1.99mg  \ P<.0005c
   a M m b6c gav lun a/c 52w81 s                                                                3.61mg  \ P<.0005c
   b M m b6c gav --- MXA 52w81 s                                                                4.31mg  \ P<.0005c
   c M m b6c gav myc srn 52w81 s                                                                5.55mg  \ P<.003 c
2110 R f sda gav TBA MXB 52w85                   :>                                             .189mg  * P<.2    -
   a R f sda gav liv MXB 52w85                                                                  no dre    P=1.
2111 R m sda gav TBA MXB 52w85                       :>                                         3.85mg  * P<.9    -
   a R m sda gav liv MXB 52w85                                                                  no dre    P=1.
2112 R m sda gav mgl fba 52w85       pool             :   ±                                    #.756mg  * P<.02   -

ESTRAGOLE                       100ng..:..1ug....:..10.....:..100....:..1mg....:..10.....:..100....:..1g.....:..10
2113 M f cd1 eat liv hpt 51w86 v                                    .  +.                       51.8mg  * P<.0005+
   a M f cd1 eat lun ade 51w86 v                                                                no dre    P=1.

ETHIONAMIDE                     100ng..:..1ug....:..10.....:..100....:..1mg....:..10.....:..100....:..1g.....:..10
2114 M f b6c eat TBA MXB 18m24 s                                           :>                   450.mg  * P<.5    -
   a M f b6c eat liv MXB 18m24 s                                                                no dre    P=1.
   b M f b6c eat lun MXB 18m24 s                                                                no dre    P=1.
2115 M m b6c eat TBA MXB 18m24 s                                              :>                no dre    P=1.
   a M m b6c eat liv MXB 18m24 s                                                                no dre    P=1.
   b M m b6c eat lun MXB 18m24 s                                                                no dre    P=1.
2116 M f bal gav thy car 50w69 e                                     .  +                      . 69.3mg    P<.01   +
   a M f bal gav lun ade 50w69 e                                                                124.mg    P<.4    -
   b M f bal gav liv ppc 50w69 e                                                                240.mg    P<.2    -
2117 R f f34 eat TBA MXB 18m24 s                                        :>                      no dre    P=1.    -
   a R f f34 eat liv MXB 18m24 s                                                                508.mg  * P<.2
2118 R m f34 eat TBA MXB 18m24                                       :>                         no dre    P=1.    -
   a R m f34 eat liv MXB 18m24                                                                  623.mg  * P<.5

ETHIONINE                       100ng..:..1ug....:..10.....:..100....:..1mg....:..10.....:..100....:..1g.....:..10
2119 R m wis eat liv car 34w52 e                         <+                                     noTD50    P<.0005+

DL-ETHIONINE                    100ng..:..1ug....:..10.....:..100....:..1mg....:..10.....:..100....:..1g.....:..10
2120 M f bal eat liv mix 24m24 e                                       .  +  .                  59.7mg  * P<.0005+
   a M f bal eat liv hpc 24m24 e                                                                112.mg  * P<.0005+
   b M f bal eat lun adc 24m24 e                                                                no dre    P=1.
   c M f bal eat lun ade 24m24 e                                                                no dre    P=1.
2121 M m bal eat lun adc 24m24 e                                         .  ±                   222.mg  \ P<.04
   a M m bal eat liv mix 24m24 e                                                                791.mg  * P<.02   +
   b M m bal eat liv hpc 24m24 e                                                                1.54gm  * P<.02   +
   c M m bal eat lun ade 24m24 e                                                                no dre    P=1.
2122 M f cen gav liv mix  7m24                                      .  ±                        33.8mg  \ P<.05   +
2123 M f cen eat liv mix 68w68 esv                                  .  +  .                     62.5mg  * P<.0005+
   a M f cen eat liv hpc 68w68 esv                                                              128.mg  * P<.0005+
   b M f cen eat lun ade 68w68 esv                                                              1.13gm  * P<.06
2124 M m cen eat liv hpc 68w68 e                                        .  +    .               182.mg    P<.003  +
   a M m cen eat liv mix 68w68 e                                                                181.mg  * P<.08   +
```

	RefNum	LoConf	UpConf	Cntrl	1Dose	1Inc	2Dose	2Inc			Citation or Pathology
											Brkly Code
2101	1984	33.4mg	n.s.s.	3/99	.500mg	6/49	2.00mg	3/49	10.0mg	4/49	
a	1984	53.3mg	n.s.s.	3/100	.500mg	2/50	2.00mg	1/50	10.0mg	2/50	
b	1984	76.1mg	n.s.s.	4/100	.500mg	0/50	2.00mg	1/50	10.0mg	1/50	
c	1984	18.3mg	n.s.s.	18/100	.500mg	11/50	2.00mg	9/50	10.0mg	13/50	
d	1984	14.5mg	n.s.s.	64/100	.500mg	36/50	2.00mg	42/50	10.0mg	30/50	
e	1984	15.2mg	n.s.s.	76/100	.500mg	44/50	2.00mg	43/50	10.0mg	34/50	
ESTRADIOL	(estradiol-17beta)	50-28-2									
2102	1131	38.1ug	n.s.s.	11/23	56.8ug	12/22					Gass;ircs,5,477;1977
2103	1468	.338mg	n.s.s.	2/43	13.0ug	1/34	.130mg	1/34	.650mg	7/45	Highman;jept,4,81-95;1980/pers.comm.
2104	1468m	.525mg	n.s.s.	2/14	13.0ug	0/5	.130mg	2/16	.650mg	0/7	
2105	1468n	.368mg	n.s.s.	4/24	13.0ug	5/23	.130mg	5/18	.650mg	2/7	
ESTRADIOL MUSTARD		22966-79-6									
2106	TR59	.330mg	1.38mg	0/16	3.98mg	21/36	8.80mg	14/35			---:lym; lun:a/a,a/c; myc:srn; sto:sqc. C
a	TR59	.409mg	5.29mg	0/16	3.98mg	7/36	8.80mg	1/35			lun:a/a,a/c.
b	TR59	.991mg	5.45mg	0/16	3.98mg	9/36	8.80mg	11/35			
c	TR59	.925mg	35.0mg	0/16	3.98mg	8/36	8.80mg	1/35			
d	TR59	1.43mg	178.mg	0/16	3.98mg	2/36	8.80mg	2/35			
e	TR59	2.93mg	n.s.s.	0/16	3.98mg	0/36	8.80mg	3/35			mgl:acn,can. S
f	TR59	.298mg	.998mg	0/16	3.98mg	25/36	8.80mg	20/35			
g	TR59	n.s.s.	n.s.s.	0/16	3.98mg	0/36	8.80mg	0/35			liv:hpa,hpc,nnd.
h	TR59	.409mg	5.29mg	0/16	3.98mg	7/36	8.80mg	1/35			lun:a/a,a/c.
2107	TR59	.442mg	6.82mg	1/31p	3.98mg	7/36	8.80mg	1/35			lun:a/a,a/c.
a	TR59	.944mg	4.65mg	0/31p	3.98mg	9/36	8.80mg	11/35			
b	TR59	.807mg	22.7mg	0/31p	3.98mg	4/36	8.80mg	0/35			S
c	TR59	1.09mg	11.0mg	0/31p	3.98mg	8/36	8.80mg	1/35			
d	TR59	2.79mg	272.mg	0/31p	3.98mg	0/36	8.80mg	3/35			mgl:acn,can. S
2108	TR59	2.41mg	8.00mg	0/14	3.98mg	6/34	8.20mg	18/35			---:leu,lym; sto:sqc. C
a	TR59	2.56mg	8.90mg	0/14	3.98mg	6/34	8.20mg	17/35			---:leu,lym.
b	TR59	7.13mg	n.s.s.	0/14	3.98mg	0/34	8.20mg	2/35			
c	TR59	1.12mg	6.77mg	4/14	3.98mg	24/34	8.20mg	23/35			
d	TR59	5.23mg	n.s.s.	2/14	3.98mg	5/34	8.20mg	1/35			liv:hpa,hpc,nnd.
e	TR59	1.01mg	n.s.s.	2/14	3.98mg	12/34	(8.20mg	5/35)			lun:a/a,a/c.
2109	TR59	.929mg	7.16mg	2/29p	3.98mg	12/34	(8.20mg	5/35)			lun:a/a,a/c.
a	TR59	1.46mg	12.7mg	0/29p	3.98mg	6/34	(8.20mg	1/35)			
b	TR59	2.54mg	7.89mg	0/29p	3.98mg	6/34	8.20mg	17/35			---:leu,lym.
c	TR59	2.18mg	27.4mg	0/29p	3.98mg	6/34	(8.20mg	2/35)			
2110	TR59	65.7ug	n.s.s.	5/10	.160mg	27/35	.330mg	26/35			
a	TR59	n.s.s.	n.s.s.	0/10	.160mg	0/35	.330mg	0/35			liv:hpa,hpc,nnd.
2111	TR59	.234mg	n.s.s.	4/10	.160mg	10/35	.330mg	10/35			
a	TR59	n.s.s.	n.s.s.	0/10	.160mg	0/35	.330mg	0/35			liv:hpa,hpc,nnd.
2112	TR59	.327mg	n.s.s.	0/20p	.160mg	2/35	.330mg	5/35			S
ESTRAGOLE	140-67-0										
2113	1582	37.3mg	74.3mg	0/50	74.2mg	27/50	148.mg	35/50			Miller;canr,43,1124-1134;1983
a	1582	666.mg	n.s.s.	2/50	74.2mg	1/50	148.mg	2/50			
ETHIONAMIDE	536-33-4										
2114	TR46	113.mg	n.s.s.	4/15	70.2mg	7/35	139.mg	14/35			
a	TR46	n.s.s.	n.s.s.	0/15	70.2mg	0/35	139.mg	0/35			liv:hpa,hpc,nnd.
b	TR46	650.mg	n.s.s.	1/15	70.2mg	0/35	139.mg	1/35			lun:a/a,a/c.
2115	TR46	101.mg	n.s.s.	7/15	64.8mg	16/35	128.mg	10/34			
a	TR46	159.mg	n.s.s.	3/15	64.8mg	6/35	128.mg	4/34			liv:hpa,hpc,nnd.
b	TR46	192.mg	n.s.s.	2/15	64.8mg	5/35	128.mg	2/34			lun:a/a,a/c.
2116	1014	29.9mg	7.42gm	0/20	49.7mg	7/36					Biancifiori;lapp,24,145-165;1964
a	1014	31.8mg	n.s.s.	2/18	49.7mg	7/33					
b	1014	58.9mg	n.s.s.	0/18	49.7mg	2/33					
2117	TR46	49.5mg	n.s.s.	12/15	40.0mg	26/35	81.0mg	25/35			
a	TR46	175.mg	n.s.s.	0/15	40.0mg	1/35	81.0mg	3/35			liv:hpa,hpc,nnd.
2118	TR46	33.4mg	n.s.s.	10/15	32.0mg	16/35	(64.8mg	4/34)			
a	TR46	153.mg	n.s.s.	0/15	32.0mg	1/35	64.8mg	1/34			liv:hpa,hpc,nnd.
ETHIONINE	13073-35-3										
2119	1373	n.s.s.	4.97mg	0/30	65.4mg	30/30					Argus;zkko,75,201-208;1971
DL-ETHIONINE	67-21-0										
2120	1782n	36.2mg	106.mg	0/30	65.0mg	14/29	130.mg	12/14			Hoover;carc,7,1143-1148;1986
a	1782n	62.6mg	227.mg	0/30	65.0mg	7/29	130.mg	10/14			
b	1782n	342.mg	n.s.s.	7/30	65.0mg	5/29	130.mg	1/14			
c	1782n	318.mg	n.s.s.	6/30	65.0mg	1/29	130.mg	3/14			
2121	1782n	88.4mg	n.s.s.	4/24	120.mg	12/28	(300.mg	1/23)			
a	1782n	306.mg	n.s.s.	1/26	120.mg	2/24	300.mg	5/16			
b	1782n	584.mg	n.s.s.	0/26	120.mg	1/30	300.mg	4/27			
c	1782n	1.20gm	n.s.s.	2/24	120.mg	1/28	300.mg	1/23			
2122	1782	13.5mg	n.s.s.	12/41	18.7mg	20/39	(62.4mg	7/37)			
2123	1782m	36.4mg	127.mg	1/26	130.mg	12/24	308.mg	12/16			
a	1782m	66.5mg	377.mg	1/26	130.mg	6/24	308.mg	9/16			
b	1782m	278.mg	n.s.s.	0/26	130.mg	0/24	308.mg	2/16			
2124	1782m	79.6mg	1.38gm	9/56	300.mg	13/27					
a	1782m	69.7mg	n.s.s.	8/26	120.mg	16/30	300.mg	15/27			

```
      Spe Strain  Site   Xpo+Xpt                                                            TD50     2Tailpvl
         Sex   Route   Hist     Notes                                                          DR      AuOp
  b   M m cen eat lun ade 68w68 e                                                           no dre   P=1.
2125 M f scd eat liv mix 24m24 esv                                     .  +  .             67.3mg \ P<.0005+
  a   M f scd eat liv hpc 24m24 esv                                                        151.mg * P<.0005+
  b   M f scd eat lun adc 24m24 esv                                                        no dre   P=1.
  c   M f scd eat lun ade 24m24 esv                                                        no dre   P=1.
2126 R m f3d eat liv hpc 24m24 e                                          .  +  .          46.0mg Z P<.0005+
  a   R m f3d eat liv thc 24m24 e                                                          46.0mg Z P<.0005+
  b   R m f3d eat liv ghc 24m24 e                                                          no dre   P=1.
  c   R m f3d eat liv clc 24m24 e                                                          no dre   P=1.
2127 R m fis eat liv hpc 69w69 e                                   .  +  .                 5.24mg   P<.0005+
  a   R m fis eat liv clc 69w69 e                                                          235.mg   P<.3
2128 R m fis eat liv hpc 52w52                              .  +  .                         12.4mg   P<.0005+
2129 R m fis eat liv hpc 39w52                           <+                                noTD50   P<.0005+
  a   R m fis eat liv clc 39w52                                                            250.mg   P<.3

o-ETHOXYBENZAMIDE                    100ng..:..1ug....:..10.....:..100....:..1mg....:..10.....:..100....:..1g......:..10
2130 M f b6c eat liv mix 22m23 er                                                    .>    3.31gm * P<.4
2131 M m b6c eat liv mix 22m23 er                                           .  +    .      513.mg * P<.005 +

ETHOXYQUIN                           100ng..:..1ug....:..10.....:..100....:..1mg....:..10.....:..100....:..1g......:..10
2132 R m f34 eat liv tum 60w60 er                                                    .>    no dre   P=1.
2133 R m f3d eat eso tum 52w52 er                                            .>            no dre   P=1.  -
  a   R m f3d eat for tum 52w52 er                                                         no dre   P=1.  -
  b   R m f3d eat liv tum 52w52 er                                                         no dre   P=1.  -

ETHYL ACRYLATE                       100ng..:..1ug....:..10.....:..100....:..1mg....:..10.....:..100....:..1g......:..10
2134 M f b6c gav for MXA 24m24                                                 :  ±        430.mg * P<.03  c
  a   M f b6c gav TBA MXB 24m24                                                            no dre   P=1.
  b   M f b6c gav liv MXB 24m24                                                            26.3gm * P<1.
  c   M f b6c gav lun MXB 24m24                                                            2.81gm * P<.6
2135 M f b6c inh nac tum 24m24 r                                         .>                no dre   P=1.  -
2136 M f b6c inh lun mix  6m27 e                                                .>         4.19gm   P<.8  -
  a   M f b6c inh liv mix  6m27 e                                                          no dre   P=1.  -
  b   M f b6c inh tba ben  6m27 e                                                          no dre   P=1.  -
  c   M f b6c inh tba mal  6m27 e                                                          no dre   P=1.  -
  d   M f b6c inh tba mix  6m27 e                                                          no dre   P=1.  -
2137 M f b6c inh lun mix 27m27 e                                                .>         1.26gm * P<.2  -
  a   M f b6c inh liv vsc 27m27 e                                                          no dre   P=1.  -
  b   M f b6c inh liv mix 27m27 e                                                          no dre   P=1.  -
  c   M f b6c inh tba mix 27m27 e                                                          11.7gm * P<1.  -
  d   M f b6c inh tba mal 27m27 e                                                          no dre   P=1.  -
  e   M f b6c inh tba ben 27m27 e                                                          no dre   P=1.  -
2138 M m b6c gav for MXA 24m24                                                 :  +  :      260.mg * P<.0005c
  a   M m b6c gav for sqp 24m24                                                            339.mg * P<.002 c
  b   M m b6c gav for sqc 24m24                                                            687.mg * P<.01  c
  c   M m b6c gav TBA MXB 24m24                                                            no dre   P=1.
  d   M m b6c gav liv MXB 24m24                                                            no dre   P=1.
  e   M m b6c gav lun MXB 24m24                                                            no dre   P=1.
2139 M m b6c inh nac tum 24m24 r                                         .>                no dre   P=1.  -
2140 M m b6c inh thy ade  6m27 e                                               .  +        515.mg   P<.01  -
  a   M m b6c inh lun mix  6m27 e                                                          no dre   P=1.  -
  b   M m b6c inh liv mix  6m27 e                                                          no dre   P=1.  -
  c   M m b6c inh tba ben  6m27 e                                                          no dre   P=1.  -
  d   M m b6c inh tba mal  6m27 e                                                          no dre   P=1.  -
  e   M m b6c inh tba mix  6m27 e                                                          no dre   P=1.  -
2141 M m b6c inh lun mix 27m27 e                                                .>         no dre   P=1.  -
  a   M m b6c inh liv mix 27m27 e                                                          no dre   P=1.  -
  b   M m b6c inh liv vsc 27m27 e                                                          no dre   P=1.  -
  c   M m b6c inh tba mix 27m27 e                                                          2.17gm * P<.9  -
  d   M m b6c inh tba mal 27m27 e                                                          300.mg * P<.2  -
  e   M m b6c inh tba ben 27m27 e                                                          no dre   P=1.  -
2142 R f f34 gav for MXA 24m24                                                 :  +    :    362.mg * P<.004 c
  a   R f f34 gav for sqp 24m24                                                            420.mg * P<.01  c
  b   R f f34 gav TBA MXB 24m24                                                            534.mg * P<.7
  c   R f f34 gav liv MXB 24m24                                                            4.66gm * P<.3
2143 R f f34 inh nac tum 24m24 r                                      .>                   no dre   P=1.  -
2144 R f f34 inh liv bht  6m27 e                                                .>         no dre   P=1.  -
  a   R f f34 inh liv sar  6m27 e                                                          no dre   P=1.  -
  b   R f f34 inh tyf mix  6m27 e                                                          no dre   P=1.  -
  c   R f f34 inh tba mix  6m27 e                                                          no dre   P=1.  -
  d   R f f34 inh tba ben  6m27 e                                                          no dre   P=1.  -
  e   R f f34 inh tba mal  6m27 e                                                          no dre   P=1.  -
2145 R f f34 inh tyf mix 27m27 e                                                .>         1.03gm * P<.2  -
  a   R f f34 inh liv sar 27m27 e                                                          no dre   P=1.  -
  b   R f f34 inh liv bht 27m27 e                                                          no dre   P=1.  -
  c   R f f34 inh tba ben 27m27 e                                                          no dre   P=1.  -
  d   R f f34 inh tba mal 27m27 e                                                          no dre   P=1.  -
  e   R f f34 inh tba mix 27m27 e                                                          no dre   P=1.  -
2146 R m f34 gav for MXA 24m24                                               :  +  :        71.9mg * P<.0005c
  a   R m f34 gav for sqp 24m24                                                            92.9mg * P<.0005c
  b   R m f34 gav for sqc 24m24                                                            234.mg * P<.0005c
```

	RefNum	LoConf	UpConf	Cntrl	1Dose	1Inc	2Dose	2Inc			Citation or Pathology / Brkly Code
b	1782m	573.mg	n.s.s.	0/26	120.mg	1/30	300.mg	0/27			
2125	1782n	37.6mg	132.mg	0/29	130.mg	20/27	(313.mg	16/25)			
a	1782n	97.3mg	251.mg	0/29	130.mg	18/27	313.mg	15/25			
b	1782n	328.mg	n.s.s.	13/29	130.mg	6/27	(313.mg	4/25)			
c	1782n	1.87gm	n.s.s.	2/29	130.mg	1/27	313.mg	0/25			
2126	1756	27.2mg	83.0mg	0/30	4.00mg	0/29	20.0mg	0/27	100.mg	26/27	Ogiso;txpy,13,257-265;1985
a	1756	27.2mg	83.0mg	0/30	4.00mg	0/29	20.0mg	0/27	100.mg	26/27	
b	1756	19.0mg	n.s.s.	0/30	4.00mg	0/29	20.0mg	0/27	100.mg	0/27	
c	1756	19.0mg	n.s.s.	0/30	4.00mg	0/29	20.0mg	0/27	100.mg	0/27	
2127	1491m	2.49mg	11.5mg	0/20	40.0mg	18/20					Leopold;canr,42,4364-4374;1982
a	1491m	38.3mg	n.s.s.	0/20	40.0mg	1/20					
2128	1491n	6.33mg	27.2mg	0/20	100.mg	15/20					
2129	1491o	n.s.s.	6.85mg	0/20	75.0mg	20/20					
a	1491o	40.8mg	n.s.s.	0/20	75.0mg	1/20					

o-ETHOXYBENZAMIDE   (ethenzamide) 938-73-8

	RefNum	LoConf	UpConf	Cntrl	1Dose	1Inc	2Dose	2Inc	Citation or Pathology / Brkly Code
2130	1797	787.mg	n.s.s.	5/19	499.mg	9/18	1.50gm	7/16	Naito;jnci,76,115-118;1986
2131	1797	226.mg	5.10gm	5/10	461.mg	8/15	1.38gm	17/18	

ETHOXYQUIN   91-53-2

	RefNum	LoConf	UpConf	Cntrl	1Dose	1Inc	Citation or Pathology / Brkly Code
2132	1640	343.mg	n.s.s.	0/25	200.mg	0/25	Rao;canr,44,1072-1076;1984
2133	1900	51.5mg	n.s.s.	0/10	100.mg	0/10	Hirose;carc,8,1731-1735;1987/pers.comm.
a	1900	51.5mg	n.s.s.	0/10	100.mg	0/10	
b	1900	51.5mg	n.s.s.	0/10	100.mg	0/10	

ETHYL ACRYLATE   140-88-5

	RefNum	LoConf	UpConf	Cntrl	1Dose	1Inc	2Dose	2Inc	Citation or Pathology / Brkly Code
2134	TR259	193.mg	n.s.s.	1/50	70.4mg	5/50	142.mg	7/50	for:sqc,sqp.
a	TR259	134.mg	n.s.s.	32/50	70.4mg	31/50	142.mg	28/50	
b	TR259	392.mg	n.s.s.	3/50	70.4mg	3/50	142.mg	3/50	liv:hpa,hpc,nnd.
c	TR259	420.mg	n.s.s.	2/50	70.4mg	2/50	142.mg	3/50	lun:a/a,a/c.
2135	1754i	119.mg	n.s.s.	0/80	6.44mg	0/90			Miller;dact,8,1-42;1985
2136	1754j	403.mg	n.s.s.	6/125	64.4mg	4/66			
a	1754j	312.mg	n.s.s.	27/125	64.4mg	12/66			
b	1754j	155.mg	n.s.s.	73/125	64.4mg	35/66			
c	1754j	199.mg	n.s.s.	90/125	64.4mg	38/66			
d	1754j	93.0mg	n.s.s.	109/125	64.4mg	53/66			
2137	1754k	405.mg	n.s.s.	6/125	32.2mg	7/78	96.5mg	8/76	
a	1754k	2.20gm	n.s.s.	4/125	32.2mg	1/78	96.5mg	0/76	
b	1754k	614.mg	n.s.s.	27/125	32.2mg	9/78	96.5mg	13/76	
c	1754k	89.5mg	n.s.s.	109/125	32.2mg	61/78	96.5mg	66/76	
d	1754k	182.mg	n.s.s.	90/125	32.2mg	43/78	96.5mg	52/76	
e	1754k	308.mg	n.s.s.	73/125	32.2mg	38/78	96.5mg	38/76	
2138	TR259	147.mg	523.mg	0/50	70.4mg	5/50	142.mg	12/50	for:sqc,sqp.
a	TR259	178.mg	1.04gm	0/50	70.4mg	4/50	142.mg	9/50	
b	TR259	296.mg	48.3gm	0/50	70.4mg	2/50	142.mg	5/50	
c	TR259	134.mg	n.s.s.	30/50	70.4mg	32/50	142.mg	29/50	
d	TR259	189.mg	n.s.s.	17/50	70.4mg	12/50	(142.mg	6/50)	liv:hpa,hpc,nnd.
e	TR259	428.mg	n.s.s.	8/50	70.4mg	6/50	142.mg	5/50	lun:a/a,a/c.
2139	1754i	99.5mg	n.s.s.	0/80	5.36mg	0/90			Miller;dact,8,1-42;1985
2140	1754j	199.mg	53.5gm	2/121	53.6mg	7/69			
a	1754j	302.mg	n.s.s.	28/121	53.6mg	12/69			
b	1754j	254.mg	n.s.s.	44/121	53.6mg	19/69			
c	1754j	208.mg	n.s.s.	60/121	53.6mg	27/69			
d	1754j	160.mg	n.s.s.	62/121	53.6mg	31/69			
e	1754j	132.mg	n.s.s.	90/121	53.6mg	44/69			
2141	1754k	215.mg	n.s.s.	28/121	26.8mg	10/75	(80.4mg	9/76)	
a	1754k	351.mg	n.s.s.	44/121	26.8mg	22/75	80.4mg	23/76	
b	1754k	1.17gm	n.s.s.	5/121	26.8mg	3/75	80.4mg	1/76	
c	1754k	100.mg	n.s.s.	90/121	26.8mg	53/75	80.4mg	57/76	
d	1754k	104.mg	n.s.s.	62/121	26.8mg	34/75	80.4mg	47/76	
e	1754k	466.mg	n.s.s.	60/121	26.8mg	30/75	80.4mg	25/76	
2142	TR259	188.mg	2.29gm	1/50	70.1mg	6/50	141.mg	11/50	for:sqc,sqp.
a	TR259	205.mg	37.9gm	1/50	70.1mg	6/50	141.mg	9/50	
b	TR259	81.4mg	n.s.s.	38/50	70.1mg	44/50	141.mg	44/50	
c	TR259	760.mg	n.s.s.	0/50	70.1mg	0/50	141.mg	1/50	liv:hpa,hpc,nnd.
2143	1754m	28.4mg	n.s.s.	0/80	1.53mg	0/90			Miller;dact,8,1-42;1985
2144	1754n	199.mg	n.s.s.	4/121	15.3mg	1/70			
a	1754n	280.mg	n.s.s.	1/121	15.3mg	0/70			
b	1754n	280.mg	n.s.s.	0/121	15.3mg	0/70			
c	1754n	11.5mg	n.s.s.	114/121	15.3mg	64/70			
d	1754n	20.3mg	n.s.s.	97/121	15.3mg	54/70			
e	1754n	59.0mg	n.s.s.	72/121	15.3mg	32/70			
2145	1754o	253.mg	n.s.s.	0/121	7.66mg	1/77	23.0mg	1/78	
a	1754o	116.mg	n.s.s.	1/121	7.66mg	0/77	23.0mg	0/78	
b	1754o	299.mg	n.s.s.	4/121	7.66mg	1/77	23.0mg	2/78	
c	1754o	48.7mg	n.s.s.	97/121	7.66mg	51/77	23.0mg	56/78	
d	1754o	81.1mg	n.s.s.	72/121	7.66mg	41/77	23.0mg	38/78	
e	1754o	13.7mg	n.s.s.	114/121	7.66mg	71/77	23.0mg	73/78	
2146	TR259	50.0mg	111.mg	1/50	70.1mg	18/50	141.mg	36/50	for:sqc,sqp.
a	TR259	62.2mg	152.mg	1/50	70.1mg	15/50	141.mg	29/50	
b	TR259	130.mg	488.mg	0/50	70.1mg	5/50	141.mg	12/50	

```
     Spe Strain  Site   Xpo+Xpt                                                    TD50    2Tailpvl
         Sex  Route  Hist     Notes                                                    DR     AuOp
  c   R m f34 gav mul mle 24m24                                                   330.mg \ P<.03
  d   R m f34 gav pan MXA 24m24                                                   402.mg \ P<.02
  e   R m f34 gav TBA MXB 24m24                                                   94.8mg * P<.01
  f   R m f34 gav liv MXB 24m24                                                   3.29gm * P<.2
2147 R m f34 inh nac tum 24m24 r                        .>                        no dre  P=1.  -
2148 R m f34 inh tyf mix 6m27 e                                .>                 267.mg  P<.2  -
  a   R m f34 inh liv hpc 6m27 e                                                  no dre  P=1.  -
  b   R m f34 inh liv bht 6m27 e                                                  no dre  P=1.  -
  c   R m f34 inh tba ben 6m27 e                                                  12.8mg  P<.4  -
  d   R m f34 inh tba mal 6m27 e                                                  no dre  P=1.  -
  e   R m f34 inh tba mix 6m27 e                                                  noTD50  P<.4  -
2149 R m f34 inh tyf mix 27m27 e                             .  ±                 77.9mg \ P<.03 -
  a   R m f34 inh liv bht 27m27 e                                                 599.mg * P<.6  -
  b   R m f34 inh liv hpc 27m27 e                                                 1.67gm * P<.6  -
  c   R m f34 inh tba mix 27m27 e                                                 5.76mg * P<.08 -
  d   R m f34 inh tba ben 27m27 e                                                 6.27mg * P<.03 -
  e   R m f34 inh tba mal 27m27 e                                                 39.0mg * P<.04 -

ETHYL ALCOHOL              100ng..:..1ug...:..10....:...100...:..1mg...:..10....:..100...:..1g....:..10
2150 M m amm wat liv hpa 72w72 e                                                 127.gm  P<.4
  a   M m amm wat lun ala 72w72 e                                                no dre  P=1.
2151 M f c3s wat mam adc 86w86 r                                                 no dre  P=1.
2152 R m aci wat for pam 52w52 er                               .    ±           4.98gm  P<.03
  a   R m aci wat stg adc 52w52 er                                               no dre  P=1.
2153 R f cdr wat liv nnd 24m24 e                                                 no dre  P=1.  -
  a   R f cdr wat tba tum 24m24 e                                                no dre  P=1.  -
2154 R m cdr wat liv nnd 24m24 e                                                 84.8gm  P<.3  -
  a   R m cdr wat liv hpc 24m24 e                                                no dre  P=1.  -
  b   R m cdr wat tba tum 24m24 e                                                no dre  P=1.  -
2155 R m cdr wat stg adc 52w52 er                                    .>          no dre  P=1.
2156 R f nbr wat liv tum 64w64                                                   .no dre P=1.
2157 R m nbr wat liv tum 64w64                                                   .>no dre P=1.
2158 R b sda wat tba mal 26m26 e                                                 no dre  P=1.  -
2159 R m sda wat liv hnd 30m30 e                               .                 +8.26gm P<.0005
  a   R m sda wat pit tum 30m30 e                                                9.11gm  P<.0005+
  b   R m sda wat adr tum 30m30 e                                                13.7gm  P<.0005+
  c   R m sda wat pan tum 30m30 e                                                13.7gm  P<.0005+
  d   R m sda wat liv hpc 30m30 e                                                28.4gm  P<.02  +
  e   R m sda wat tba mix 30m30 e                                                2.13gm  P<.0005
2160 R f wis wat liv mix 52w52 r                                    .>           no dre  P=1.
2161 R m wis wat for sqc 91w91 r                                    .>           no dre  P=1.

Z-ETHYL-O,N,N-AZOXYETHANE  100ng..:..1ug...:..10....:...100...:..1mg...:..10....:..100...:..1g....:..10
2162 R m f34 wat liv mix 30w55            <+                                     noTD50  P<.0005+
  a   R m f34 wat nas mix 30w55                                                  22.0ug  P<.0005+
  b   R m f34 wat eso mix 30w55                                                  30.2ug  P<.0005+
  c   R m f34 wat liv hes 30w55                                                  34.7ug  P<.0005+
  d   R m f34 wat liv hpc 30w55                                                  34.7ug  P<.0005+
  e   R m f34 wat liv nsc 30w55                                                  97.1ug  P<.002
  f   R m f34 wat liv nnd 30w55                                                  no dre  P=1.

Z-ETHYL-O,N,N-AZOXYMETHANE 100ng..:..1ug...:..10....:...100...:..1mg...:..10....:..100...:..1g....:..10
2163 R m f34 wat liv mix 30w55            <+                                     noTD50  P<.0005+
  a   R m f34 wat liv hpc 30w55                                                  18.9ug  P<.0005+
  b   R m f34 wat liv hes 30w55                                                  39.1ug  P<.0005+
  c   R m f34 wat liv nsc 30w55                                                  83.2ug  P<.002
  d   R m f34 wat col mix 30w55                                                  .125mg  P<.007 +
  e   R m f34 wat ilm tum 30w55                                                  .221mg  P<.04  +
  f   R m f34 wat liv nnd 30w55                                                  no dre  P=1.

ETHYL BENZENE             100ng..:..1ug...:..10....:...100...:..1mg...:..10....:..100...:..1g....:..10
2164 R f sda gav tba mal 24m33 e                               .    ±            1.21gm  P<.08 +
2165 R m sda gav tba mal 24m33 e                                    .>           1.98gm  P<.3  +

ETHYL BROMOACETATE        100ng..:..1ug...:..10....:...100...:..1mg...:..10....:..100...:..1g....:..10
2166 M f hic ipj lun ptm 64w64                  .>                               no dre  P=1.

S-ETHYL-L-CYSTEINE        100ng..:..1ug...:..10....:...100...:..1mg...:..10....:..100...:..1g....:..10
2167 R m wis eat abd lps 58w89 aev                                  .>           3.63gm * P<.4  -
  a   R m wis eat liv tum 58w89 aev                                              no dre  P=1.  -
  b   R m wis eat tba mix 58w89 aev                                             3.63gm * P<.4  -

p,p'-ETHYL-DDD            100ng..:..1ug...:..10....:...100...:..1mg...:..10....:..100...:..1g....:..10
2168 M f b6c eat liv MXA 24m24 v                               :    ±            2.67gm * P<.04 a
  a   M f b6c eat liv hpc 24m24 v                                                3.16gm * P<.04 a
  b   M f b6c eat TBA MXB 24m24 v                                                2.36gm * P<.4
  c   M f b6c eat liv MXB 24m24 v                                                2.67gm * P<.04
  d   M f b6c eat lun MXB 24m24 v                                                no dre  P=1.
2169 M f b6c orl liv hpt 76w76 evx                  .>                           665.mg  P<.3
  a   M f b6c orl lun mix 76w76 evx                                              no dre  P=1.
  b   M f b6c orl tba mix 76w76 evx                                              322.mg  P<.1
```

	RefNum	LoConf	UpConf	Cntrl	1Dose	1Inc	2Dose	2Inc	Citation or Pathology
									Brkly Code
c	TR259	120.mg	n.s.s.	1/50	70.1mg	6/50	(141.mg	1/50)	S
d	TR259	138.mg	n.s.s.	0/50	70.1mg	4/50	(141.mg	0/50)	pan:acc,ana. S
e	TR259	45.7mg	9.25gm	38/50	70.1mg	40/50	141.mg	45/50	
f	TR259	535.mg	n.s.s.	0/50	70.1mg	0/50	141.mg	1/50	liv:hpa,hpc,nnd.
2147	1754m	19.9mg	n.s.s.	0/80	1.07mg	0/90			Miller;dact,8,1-42;1985
2148	1754n	68.6mg	n.s.s.	1/120	10.7mg	3/71			
a	1754n	199.mg	n.s.s.	2/120	10.7mg	0/71			
b	1754n	199.mg	n.s.s.	7/120	10.7mg	0/71			
c	1754n	2.72mg	n.s.s.	113/120	10.7mg	69/71			
d	1754n	28.0mg	n.s.s.	60/120	10.7mg	33/71			
e	1754n	n.s.s.	n.s.s.	116/120	10.7mg	71/71			
2149	1754o	27.0mg	n.s.s.	1/120	5.36mg	5/76	(16.1mg	2/75)	
a	1754o	94.8mg	n.s.s.	7/120	5.36mg	5/76	16.1mg	6/75	
b	1754o	194.mg	n.s.s.	2/120	5.36mg	0/76	16.1mg	2/75	
c	1754o	1.43mg	n.s.s.	116/120	5.36mg	74/76	16.1mg	75/75	
d	1754o	2.09mg	n.s.s.	113/120	5.36mg	72/76	16.1mg	75/75	
e	1754o	16.5mg	n.s.s.	60/120	5.36mg	40/76	16.1mg	49/75	

ETHYL ALCOHOL   64-17-5

	RefNum	LoConf	UpConf	Cntrl	1Dose	1Inc	2Dose	2Inc	Citation or Pathology
2150	2081	27.8gm	n.s.s.	1/47	16.7gm	3/48			Anderson;carc,13,2107-2111;1992
a	2081	10.4gm	n.s.s.	39/47	16.7gm	33/48			
2151	1632	15.3gm	n.s.s.	23/30	24.0gm	8/15			Schrauzer;jsac,40,240-246;1979
2152	2076	1.50gm	n.s.s.	0/22	5.00gm	3/19			Watanabe;gann,83,588-593;1992/pers.comm.
a	2076	4.89gm	n.s.s.	0/22	5.00gm	0/19			
2153	1962	28.9gm	n.s.s.	1/50	2.86gm	0/49			Cheever;faat,14,243-261;1990
a	1962	2.70gm	n.s.s.	47/50	2.86gm	41/50			
2154	1962	13.8gm	n.s.s.	0/50	2.50gm	1/50			
a	1962	25.8gm	n.s.s.	1/50	2.50gm	0/50			
b	1962	1.38gm	n.s.s.	42/50	2.50gm	42/50			
2155	2074	3.86gm	n.s.s.	0/12	5.00gm	0/15			Watanabe;gann,83,1267-1272;1992/pers.comm.
2156	162a	8.92gm	n.s.s.	0/20	5.71gm	0/20			Yamamoto;ijcn,2,337-343;1967
2157	162a	7.80gm	n.s.s.	0/20	5.00gm	0/20			
2158	1264	39.9gm	n.s.s.	4/48	9.45gm	4/48			Schmahl;clet,1,215-218;1976
2159	1440	4.47gm	29.0gm	10/80	2.50gm	29/79			Radike;enhp,41,59-62;1981
a	1440	4.89gm	32.1gm	8/80	2.50gm	26/79			
b	1440	7.38gm	29.9gm	0/80	2.50gm	14/79			
c	1440	7.38gm	29.9gm	0/80	2.50gm	14/79			
d	1440	11.9mg	n.s.s.	1/80	2.50gm	8/79			
e	1440	1.42gm	3.51gm	16/80	2.50gm	61/79			
2160	2028	2.10gm	n.s.s.	0/5	4.08gm	0/10			Yamagiwa;gann,82,771-778;1991
2161	1800	3.94gm	n.s.s.	0/10	2.50gm	0/10			Salmon;carc,7,1447-1450;1986

Z-ETHYL-O,N,N-AZOXYETHANE   16301-26-1

	RefNum	LoConf	UpConf	Cntrl	1Dose	1Inc	2Dose	2Inc	Citation or Pathology
2162	1723	n.s.s.	23.8ug	2/20	.218mg	20/20			Lijinsky;canr,45,76-79;1985
a	1723	11.0ug	47.7ug	0/20	.218mg	17/20			
b	1723	15.4ug	66.5ug	0/20	.218mg	15/20			
c	1723	17.7ug	78.1ug	0/20	.218mg	14/20			
d	1723	17.7ug	78.1ug	0/20	.218mg	14/20			
e	1723	41.5ug	.368mg	0/20	.218mg	7/20			
f	1723	.252mg	n.s.s.	2/20	.218mg	0/20			

Z-ETHYL-O,N,N-AZOXYMETHANE   57497-29-7

	RefNum	LoConf	UpConf	Cntrl	1Dose	1Inc	2Dose	2Inc	Citation or Pathology
2163	1723	n.s.s.	20.4ug	2/20	.187mg	20/20			Lijinsky;canr,45,76-79;1985
a	1723	9.44ug	40.9ug	0/20	.187mg	17/20			
b	1723	19.5ug	93.3ug	0/20	.187mg	12/20			
c	1723	35.6ug	.315mg	0/20	.187mg	7/20			
d	1723	47.0ug	1.44mg	0/20	.187mg	5/20			
e	1723	66.6ug	n.s.s.	0/20	.187mg	3/20			
f	1723	.216mg	n.s.s.	2/20	.187mg	0/20			

ETHYL BENZENE   100-41-4

	RefNum	LoConf	UpConf	Cntrl	1Dose	1Inc	2Dose	2Inc	Citation or Pathology
2164	BT905	433.mg	n.s.s.	10/49	237.mg	14/37			Maltoni;ajim,7,415-446;1985
2165	BT905	528.mg	n.s.s.	11/45	237.mg	14/40			

ETHYL BROMOACETATE   105-36-2

	RefNum	LoConf	UpConf	Cntrl	1Dose	1Inc	2Dose	2Inc	Citation or Pathology
2166	1143	.390mg	n.s.s.	12/30	.571mg	9/30			Van Duuren;jnci,53,695-700;1974

S-ETHYL-L-CYSTEINE   2629-59-6

	RefNum	LoConf	UpConf	Cntrl	1Dose	1Inc	2Dose	2Inc	Citation or Pathology
2167	1373	592.mg	n.s.s.	0/30	116.mg	0/30	127.mg	1/30	Argus;zkko,75,201-208;1971
a	1373	275.mg	n.s.s.	0/30	116.mg	0/30	127.mg	0/30	
b	1373	592.mg	n.s.s.	0/30	116.mg	0/30	127.mg	1/30	

p,p'-ETHYL-DDD   (perthane) 72-56-0

	RefNum	LoConf	UpConf	Cntrl	1Dose	1Inc	2Dose	2Inc	Citation or Pathology
2168	TR156	1.23gm	n.s.s.	1/20	368.mg	3/50	806.mg	11/50	liv:hpa,hpc.
a	TR156	1.39gm	n.s.s.	1/20	368.mg	2/50	806.mg	10/50	
b	TR156	617.mg	n.s.s.	7/20	368.mg	24/50	806.mg	26/50	
c	TR156	1.23gm	n.s.s.	1/20	368.mg	3/50	806.mg	11/50	liv:hpa,hpc,nnd.
d	TR156	2.94gm	n.s.s.	0/20	368.mg	3/50	806.mg	1/50	lun:a/a,a/c.
2169	1098	108.mg	n.s.s.	0/16	110.mg	1/17			Innes;ntis,1968/1969
a	1098	206.mg	n.s.s.	0/16	110.mg	0/17			
b	1098	79.1mg	n.s.s.	0/16	110.mg	2/17			

```
       Spe Strain Site   Xpo+Xpt                                                              TD50    2Tailpvl
          Sex Route Hist Notes                                                                  DR     AuOp
2170 M m b6c eat lun a/a 24m24                                              :    ±          #1.64gm * P<.02  -
   a M m b6c eat TBA MXB 24m24                                                              no dre    P=1.
   b M m b6c eat liv MXB 24m24                                                              9.16gm  * P<.9
   c M m b6c eat lun MXB 24m24                                                              5.07gm  * P<.6
2171 M m b6c orl liv hpt 76w76 evx                                  .    +    .             65.1mg    P<.002
   a M m b6c orl lun ade 76w76 evx                                                          581.mg    P<.3
   b M m b6c orl tba mix 76w76 evx                                                          54.1mg    P<.0005
2172 M f b6a orl liv hpt 76w76 evx                                              .>          no dre    P=1.
   a M f b6a orl lun ade 76w76 evx                                                          no dre    P=1.
   b M f b6a orl tba mix 76w76 evx                                                          no dre    P=1.
2173 M m b6a orl lun ade 76w76 evx                                        .>                581.mg    P<.7
   a M m b6a orl liv hpt 76w76 evx                                                          no dre    P=1.
   b M m b6a orl tba mix 76w76 evx                                                          543.mg    P<.7
2174 R f f34 eat TBA MXB 24m24                                    :>                        no dre    P=1.    -
   a R f f34 eat liv MXB 24m24                                                              8.54gm  * P<.8
2175 R m f34 eat TBA MXB 24m24                                         :>                   no dre    P=1.    -
   a R m f34 eat liv MXB 24m24                                                              no dre    P=1.

N-ETHYL-N-FORMYLHYDRAZINE        100ng..:..1ug...:..10.....:..100....:..1mg...:..10.....:..100...:..1g.....:..10
2176 M f swa wat blv mix 61w61 e                                 .   +   .                  2.49mg    P<.0005+
   a M f swa wat lun mix 61w61 e                                                            2.52mg    P<.0005+
   b M f swa wat liv tum 61w61 e                                                            no dre    P=1.
2177 M m swa wat lun mix 60w60 e                              .   +   .                      3.19mg    P<.0005+
   a M m swa wat blv mix 60w60 e                                                            3.55mg    P<.0005+
   b M m swa wat liv mix 60w60 e                                                            17.7mg    P<.0005+
   c M m swa wat gal mix 60w60 e                                                            47.4mg    P<.005 +
   d M m swa wat pre mix 60w60 e                                                            50.8mg    P<.003 +

ETHYL METHYLPHENYLGLYCIDATE      100ng..:..1ug...:..10.....:..100....:..1mg...:..10.....:..100...:..1g.....:..10
2178 R f wis eat pit ade 24m24 e                                   .    ±                   331.mg  * P<.05  -
   a R f wis eat liv hae 24m24 e                                                            9.36gm  * P<.2   -
   b R f wis eat tba ben 24m24 e                                                            253.mg  * P<.06  -
   c R f wis eat tba mal 24m24 e                                                            2.23gm  * P<.2   -
2179 R m wis eat tes ict 24m24 e                                           .>               1.15gm  Z P<.2   -
   a R m wis eat liv tum 24m24 e                                                            no dre    P=1.    -
   b R m wis eat tba ben 24m24 e                                                            1.15gm  * P<.5   -
   c R m wis eat tba mal 24m24 e                                                            4.11gm  * P<.7   -

N-ETHYL-N'-NITRO-N-NITROSOGUANIDINE   ..:..1ug...:..10.....:..100....:..1mg...:..10.....:..100...:..1g.....:..10
2180 M b cbh wat duo mix 43w69 e                                   .   +   .                2.84mg    P<.0005+
   a M b cbh wat eso mix 43w69 e                                                            3.85mg    P<.0005+
   b M b cbh wat duo adc 43w69 e                                                            4.20mg    P<.0005+
   c M b cbh wat eso sqc 43w69 e                                                            5.60mg    P<.0005+
   d M b cbh wat sto mix 43w69 e                                                            8.03mg    P<.002
   e M b cbh wat liv hms 43w69 e                                                            70.3mg    P<.3
   f M b cbh wat liv hpt 43w69 e                                                            no dre    P=1.
   g M b cbh wat lun ade 43w69 e                                                            no dre    P=1.
   h M b cbh wat tba mix 43w69 e                                                            284.mg    P<1.

1-ETHYL-1-NITROSOUREA            100ng..:..1ug...:..10.....:..100....:..1mg...:..10.....:..100....:..1g...:..10
2181 R f f3d wat --- mnl 24m26 ae                           .   +   .                       .151mg  Z P<.0005
   a R f f3d wat ute esp 24m26 ae                                                           .883mg  * P<.003
   b R f f3d wat bra mix 24m26 ae                                                           .904mg  * P<.0005+
   c R f f3d wat mgl adc 24m26 ae                                                           2.54mg  * P<.002
   d R f f3d wat utm sar 24m26 ae                                                           2.56mg  * P<.003
   e R f f3d wat dgt mix 24m26 ae                                                           2.56mg  * P<.003 +
   f R f f3d wat duo mix 24m26 ae                                                           7.68mg  * P<.006 +
   g R f f3d wat ute ade 24m26 ae                                                           4.08mg  * P<.1
   h R f f3d wat tyf ppa 24m26 ae                                                           8.70mg  * P<.08
   i R f f3d wat liv nnd 24m26 ae                                                           no dre    P=1.
   j R f f3d wat tba mix 24m26 ae                                                           68.7ug  * P<.0005
2182 R m f3d wat per mso 23m26 ae                     .   +   .                             .109mg  Z P<.0005
   a R m f3d wat lun ade 23m26 ae                                                           .518mg  * P<.0005
   b R m f3d wat mgl fib 23m26 ae                                                           .889mg  * P<.0005
   c R m f3d wat bra mix 23m26 ae                                                           .997mg  * P<.0005+
   d R m f3d wat dgt mix 23m26 ae                                                           1.15mg  * P<.0005+
   e R m f3d wat lun adc 23m26 ae                                                           2.00mg  * P<.0005
   f R m f3d wat sub fib 23m26 ae                                                           2.42mg  * P<.02
   g R m f3d wat tyf ppa 23m26 ae                                                           3.90mg  * P<.05  +
   h R m f3d wat tyf pac 23m26 ae                                                           8.53mg  * P<.09  +
   i R m f3d wat pns mix 23m26 ae                                                           10.2mg  * P<.4   +
   j R m f3d wat liv nnd 23m26 ae                                                           16.4mg  * P<.7
   k R m f3d wat duo mix 23m26 ae                                                           31.2mg  * P<.5   +
   l R m f3d wat tba mix 23m26 ae                                                           25.0ug  * P<.3

ETHYL TELLURAC                   100ng..:..1ug...:..10.....:..100....:..1mg...:..10.....:..100...:..1g.....:..10
2183 M f b6c eat TBA MXB 25m25 v                                        :>                  1.03gm  * P<.3   -
   a M f b6c eat liv MXB 25m25 v                                                            2.27gm  * P<.2
   b M f b6c eat lun MXB 25m25 v                                                            3.40gm  * P<.5
2184 M f b6c orl lun ade 76w76 evx                              .>                          131.mg    P<.3
   a M f b6c orl liv hpt 76w76 evx                                                          no dre    P=1.
```

	RefNum	LoConf	UpConf	Cntrl	1Dose	1Inc	2Dose	2Inc	Citation or Pathology / Brkly Code
2170	TR156	866.mg	n.s.s.	0/20	300.mg	4/50	600.mg	9/50	S
a	TR156	430.mg	n.s.s.	11/20	300.mg	38/50	600.mg	33/50	
b	TR156	508.mg	n.s.s.	8/20	300.mg	27/50	600.mg	25/50	liv:hpa,hpc,nnd.
c	TR156	988.mg	n.s.s.	4/20	300.mg	5/50	600.mg	12/50	lun:a/a,a/c.
2171	1098	27.6mg	238.mg	0/16	102.mg	7/16			Innes;ntis,1968/1969
a	1098	94.5mg	n.s.s.	0/16	102.mg	1/16			
b	1098	23.9mg	167.mg	0/16	102.mg	8/16			
2172	1098	218.mg	n.s.s.	0/17	110.mg	0/18			
a	1098	218.mg	n.s.s.	1/17	110.mg	0/18			
b	1098	104.mg	n.s.s.	2/17	110.mg	2/18			
2173	1098	72.9mg	n.s.s.	2/18	102.mg	3/18			
a	1098	121.mg	n.s.s.	1/18	102.mg	1/18			
b	1098	63.6mg	n.s.s.	3/18	102.mg	4/18			
2174	TR156	166.mg	n.s.s.	15/20	175.mg	32/50	(350.mg	21/50)	
a	TR156	1.59gm	n.s.s.	0/20	175.mg	2/50	350.mg	1/50	liv:hpa,hpc,nnd.
2175	TR156	286.mg	n.s.s.	13/20	140.mg	36/50	280.mg	30/50	
a	TR156	1.59gm	n.s.s.	0/20	140.mg	2/50	280.mg	0/50	liv:hpa,hpc,nnd.

N-ETHYL-N-FORMYLHYDRAZINE   74920-78-8

	RefNum	LoConf	UpConf	Cntrl	1Dose	1Inc	2Dose	2Inc	Citation or Pathology / Brkly Code
2176	1052	1.17mg	4.67mg	8/96	40.0mg	47/48			Toth;carc,1,61-65;1980
a	1052	1.17mg	4.76mg	15/99	40.0mg	49/50			
b	1052	142.mg	n.s.s.	0/21	40.0mg	0/50			
2177	1052	1.76mg	6.03mg	22/98	33.3mg	39/42			
a	1052	2.04mg	6.45mg	5/88	33.3mg	32/36			
b	1052	9.07mg	45.6mg	1/52	33.3mg	13/36			
c	1052	16.4mg	452.mg	0/44	33.3mg	4/27			
d	1052	19.3mg	290.mg	0/52	33.3mg	5/36			

ETHYL METHYLPHENYLGLYCIDATE   77-83-8

	RefNum	LoConf	UpConf	Cntrl	1Dose	1Inc	2Dose	2Inc			Citation or Pathology / Brkly Code
2178	1383	126.mg	n.s.s.	18/43	10.0mg	30/44	50.0mg	24/41	250.mg	31/43	Dunnington;fctx,19,691-699;1981
a	1383	1.52gm	n.s.s.	0/44	10.0mg	0/44	50.0mg	0/42	250.mg	1/45	
b	1383	92.7mg	n.s.s.	24/44	10.0mg	35/44	50.0mg	33/42	250.mg	37/45	
c	1383	601.mg	n.s.s.	1/44	10.0mg	3/44	50.0mg	3/42	250.mg	5/45	
2179	1383	335.mg	n.s.s.	2/38	8.00mg	8/35	40.0mg	1/35	200.mg	8/36	
a	1383	50.8mg	n.s.s.	0/37	8.00mg	0/38	40.0mg	0/39	200.mg	0/39	
b	1383	230.mg	n.s.s.	10/38	8.00mg	20/39	40.0mg	11/39	200.mg	17/39	
c	1383	491.mg	n.s.s.	2/38	8.00mg	6/39	40.0mg	4/39	200.mg	5/39	

N-ETHYL-N'-NITRO-N-NITROSOGUANIDINE   63885-23-4

	RefNum	LoConf	UpConf	Cntrl	1Dose	1Inc	Citation or Pathology / Brkly Code
2180	600	1.64mg	5.50mg	0/38	5.48mg	19/43	Nakamura;jnci,52,519-522;1974
a	600	2.10mg	8.18mg	0/38	5.48mg	15/43	
b	600	2.25mg	9.20mg	0/38	5.48mg	14/43	
c	600	2.80mg	14.3mg	0/38	5.48mg	11/43	
d	600	3.63mg	29.5mg	0/38	5.48mg	8/43	
e	600	11.4mg	n.s.s.	0/38	5.48mg	1/43	
f	600	13.8mg	n.s.s.	19/38	5.48mg	5/43	
g	600	21.4mg	n.s.s.	3/38	5.48mg	0/43	
h	600	2.37mg	n.s.s.	22/38	5.48mg	25/43	

1-ETHYL-1-NITROSOUREA   (ENU, N-ethyl-N-nitrosourea) 759-73-9

	RefNum	LoConf	UpConf	Cntrl	1Dose	1Inc	2Dose	2Inc	3Dose	3Inc	4Dose	4Inc	Citation or Pathology / Brkly Code
2181	1614	90.2ug	.351mg	5/52	12.0ug	8/51	40.2ug	18/52	.120mg	24/50	(.429mg	27/50)	Maekawa;gann,75,117-125;1984
a	1614	.419mg	6.46mg	9/52	12.0ug	10/51	40.2ug	16/52	.120mg	16/50	.429mg	22/50	
b	1614	.544mg	1.66mg	0/52	12.0ug	0/51	40.2ug	0/52	.120mg	2/50	.429mg	20/50	
c	1614	1.12mg	12.3mg	1/52	12.0ug	1/51	40.2ug	1/52	.120mg	1/50	.429mg	8/50	
d	1614	1.10mg	18.2mg	2/52	12.0ug	1/51	40.2ug	1/52	.120mg	2/50	.429mg	8/50	
e	1614	1.10mg	18.2mg	2/52	12.0ug	1/51	40.2ug	1/52	.120mg	2/50	.429mg	8/50	
f	1614	2.32mg	106.mg	0/52	12.0ug	0/51	40.2ug	0/52	.120mg	0/50	.429mg	3/50	
g	1614	1.24mg	n.s.s.	0/52	12.0ug	5/51	40.2ug	3/52	.120mg	3/50	.429mg	6/50	
h	1614	2.35mg	n.s.s.	0/52	12.0ug	0/51	40.2ug	1/52	.120mg	0/50	.429mg	2/50	
i	1614	4.88mg	n.s.s.	0/52	12.0ug	0/51	40.2ug	1/52	.120mg	0/50	.429mg	2/50	
j	1614	32.6ug	.202mg	38/52	12.0ug	39/51	40.2ug	44/52	.120mg	46/50	.429mg	50/50	
2182	1614	71.0ug	.200mg	1/51	11.2ug	10/52	37.5ug	13/52	.112mg	28/51	(.400mg	32/52)	
a	1614	.326mg	.976mg	5/51	11.2ug	0/52	37.5ug	6/52	.112mg	11/51	.400mg	25/52	
b	1614	.460mg	3.25mg	6/51	11.2ug	5/52	37.5ug	6/52	.112mg	13/51	.400mg	18/52	
c	1614	.588mg	1.90mg	0/51	11.2ug	0/52	37.5ug	2/52	.112mg	2/51	.400mg	16/52	
d	1614	.629mg	2.84mg	1/51	11.2ug	2/52	37.5ug	1/52	.112mg	4/51	.400mg	14/52	
e	1614	.942mg	7.80mg	1/51	11.2ug	1/52	37.5ug	1/52	.112mg	2/51	.400mg	9/52	
f	1614	.961mg	n.s.s.	0/51	11.2ug	1/52	37.5ug	2/52	.112mg	5/51	.400mg	5/52	
g	1614	1.29mg	n.s.s.	0/51	11.2ug	1/52	37.5ug	4/52	.112mg	1/51	.400mg	5/52	
h	1614	2.27mg	n.s.s.	0/51	11.2ug	0/52	37.5ug	1/52	.112mg	0/51	.400mg	2/52	
i	1614	2.34mg	n.s.s.	0/51	11.2ug	2/52	37.5ug	0/52	.112mg	2/51	.400mg	2/52	
j	1614	1.73mg	n.s.s.	4/51	11.2ug	0/52	37.5ug	6/52	.112mg	3/51	.400mg	4/52	
k	1614	3.55mg	n.s.s.	0/51	11.2ug	1/52	37.5ug	0/52	.112mg	0/51	.400mg	1/52	
l	1614	n.s.s.	n.s.s.	51/51	11.2ug	51/52	37.5ug	52/52	.112mg	51/51	.400mg	52/52	

ETHYL TELLURAC   20941-65-5

	RefNum	LoConf	UpConf	Cntrl	1Dose	1Inc	2Dose	2Inc	Citation or Pathology / Brkly Code
2183	TR152	325.mg	n.s.s.	8/20	277.mg	36/50	639.mg	36/50	
a	TR152	908.mg	n.s.s.	1/20	277.mg	7/50	639.mg	10/50	liv:hpa,hpc,nnd.
b	TR152	856.mg	n.s.s.	3/20	277.mg	9/50	639.mg	12/50	lun:a/a,a/c.
2184	132	21.3mg	n.s.s.	0/16	20.4mg	1/18			Innes;ntis,1968/1969
a	132	40.5mg	n.s.s.	0/16	20.4mg	0/18			

	Spe	Strain	Site	Xpo+Xpt					TD50	2Tailpvl
	Sex	Route	Hist	Notes					DR	AuOp
b	M f	b6c	orl	tba mix	76w76	evx			63.5mg	P<.2
2185	M f	b6c	eat	eye adn	25m25	v	: + :		201.mg	\ P<.002 a
a	M m	b6c	eat	lun MXA	25m25	v			228.mg	\ P<.002
b	M m	b6c	eat	lun a/c	25m25	v			260.mg	\ P<.003
c	M m	b6c	eat	TBA MXB	25m25	v			902.mg	* P<.5
d	M m	b6c	eat	liv MXB	25m25	v			no dre	P=1.
e	M m	b6c	eat	lun MXB	25m25	v			228.mg	\ P<.002
2186	M m	b6c	orl	liv hpt	76w76	evx	. ±		24.2mg	P<.02
a	M m	b6c	orl	lun mix	76w76	evx			52.1mg	P<.09
b	M m	b6c	orl	lun ade	76w76	evx			108.mg	P<.3
c	M m	b6c	orl	lun car	76w76	evx			108.mg	P<.3
d	M m	b6c	orl	tba mix	76w76	evx			8.41mg	P<.0005
2187	M f	b6a	orl	lun ade	76w76	evx	. ±		28.2mg	P<.08
a	M f	b6a	orl	liv hpt	76w76	evx			no dre	P=1.
b	M f	b6a	orl	tba mix	76w76	evx			26.7mg	P<.2
2188	M m	b6a	orl	liv hpt	76w76	evx	.>		55.5mg	P<.3
a	M m	b6a	orl	lun ade	76w76	evx			no dre	P=1.
b	M m	b6a	orl	tba mix	76w76	evx			31.2mg	P<.3
2189	R f	f34	eat	pit adn	24m24		: ±		#12.3mg	\ P<.04 -
a	R f	f34	eat	TBA MXB	24m24				no dre	P=1.
b	R f	f34	eat	liv MXB	24m24				650.mg	* P<.4
2190	R m	f34	eat	--- men	24m24		: +	:	77.6mg	* P<.009 a
a	R m	f34	eat	bod men	24m24				181.mg	* P<.05 a
b	R m	f34	eat	TBA MXB	24m24				no dre	P=1.
c	R m	f34	eat	liv MXB	24m24				no dre	P=1.

ETHYLENE GLYCOL   100ng..:..1ug....:..10......:..100....:..1mg....:..10......:..100....:..1g......:..10

2191	M f	b6c	eat	TBA MXB	24m24			:no dre	P=1. -
a	M f	b6c	eat	liv MXB	24m24			no dre	P=1.
b	M f	b6c	eat	lun MXB	24m24			11.3gm	Z P<.06
2192	M m	b6c	eat	hag MXA	24m24			:#19.2gm	* P<.02 -
a	M m	b6c	eat	TBA MXB	24m24			9.98gm	* P<.6
b	M m	b6c	eat	liv MXB	24m24			48.1gm	* P<.9
c	M m	b6c	eat	lun MXB	24m24			no dre	P=1.

ETHYLENE GLYCOL, CYCLIC SULFATE   100ng..:..1ug....:..10......:..100....:..1mg....:..10......:..100....:..1g......:..10

| 2193 | M f | hic | ipj | lun ptm | 64w64 | | .> | no dre | P=1. |

ETHYLENE IMINE   100ng..:..1ug....:..10......:..100....:..1mg....:..10......:..100....:..1g......:..10

2194	M f	b6a	orl	lun ade	76w76	evx	. + .	.283mg	P<.0005+
a	M f	b6a	orl	liv hpt	76w76	evx		3.29mg	P<.05
b	M f	b6a	orl	tba mix	76w76	evx		noTD50	P<.003
2195	M m	b6a	orl	lun ade	76w76	evx	. + .	.485mg	P<.0005
a	M m	b6a	orl	liv hpt	76w76	evx		.800mg	P<.002 +
b	M m	b6a	orl	lun bro	76w76	evx		2.14mg	P<.02
c	M m	b6a	orl	tba mix	76w76	evx		noTD50	P<.002
2196	M f	b6c	orl	lun ade	76w76	evx	<+	noTD50	P<.0005+
a	M f	b6c	orl	liv hpt	76w76	evx		.500mg	P<.0005
b	M f	b6c	orl	tba mix	76w76	evx		noTD50	P<.0005
2197	M m	b6c	orl	liv hpt	77w77	evx	. + .	.295mg	P<.0005+
a	M m	b6c	orl	lun mix	77w77	evx		.295mg	P<.0005+
b	M m	b6c	orl	lun ade	77w77	evx		.364mg	P<.0005
c	M m	b6c	orl	tba mix	77w77	evx		.223mg	P<.0005

ETHYLENE OXIDE   100ng..:..1ug....:..10......:..100....:..1mg....:..10......:..100....:..1g......:..10

2198	M f	b6c	inh	MXB MXB	24m24	: + :	39.2mg	* P<.002
a	M f	b6c	inh	MXB MXB	24m24		45.8mg	* P<.0005
b	M f	b6c	inh	lun MXA	24m24		61.8mg	* P<.0005c
c	M f	b6c	inh	lun a/a	24m24		87.5mg	* P<.002 c
d	M f	b6c	inh	lun a/c	24m24		200.mg	* P<.004 c
e	M f	b6c	inh	liv hpa	24m24		75.8mg	\ P<.02
f	M f	b6c	inh	mgl MXA	24m24		76.3mg	\ P<.02 p
g	M f	b6c	inh	MXB MXB	24m24		79.2mg	* P<.02
h	M f	b6c	inh	MXA MXA	24m24		139.mg	* P<.06 p
i	M f	b6c	inh	hag pcy	24m24		142.mg	* P<.04 c
j	M f	b6c	inh	ute MXA	24m24		250.mg	* P<.02 p
k	M f	b6c	inh	ute acn	24m24		297.mg	* P<.02 p
l	M f	b6c	inh	TBA MXB	24m24		103.mg	* P<.4
m	M f	b6c	inh	liv MXB	24m24		149.mg	* P<.4
n	M f	b6c	inh	lun MXB	24m24		61.8mg	* P<.0005
2199	M m	b6c	inh	MXB MXB	24m24	: ±	51.2mg	* P<.04
a	M m	b6c	inh	lun MXA	24m24		65.7mg	* P<.06 c
b	M m	b6c	inh	lun a/c	24m24		110.mg	* P<.08 c
c	M m	b6c	inh	hag pcy	24m24		110.mg	* P<.04 c
d	M m	b6c	inh	lun a/a	24m24		181.mg	* P<.3 c
e	M m	b6c	inh	TBA MXB	24m24		108.mg	* P<.5
f	M m	b6c	inh	liv MXB	24m24		273.mg	* P<.6
g	M m	b6c	inh	lun MXB	24m24		65.7mg	* P<.06

	RefNum	LoConf	UpConf	Cntrl	1Dose	1Inc	2Dose	2Inc	Citation or Pathology
									Brkly Code
b	132	15.6mg	n.s.s.	0/16	20.4mg	2/18			
2185	TR152	112.mg	617.mg	0/20	151.mg	16/50	(376.mg	10/50)	
a	TR152	127.mg	706.mg	0/20	151.mg	16/50	(376.mg	11/50)	lun:a/a,a/c. S
b	TR152	140.mg	1.02gm	0/20	151.mg	14/50	(376.mg	11/50)	S
c	TR152	198.mg	n.s.s.	10/20	151.mg	39/50	376.mg	35/50	
d	TR152	518.mg	n.s.s.	8/20	151.mg	15/50	376.mg	16/50	liv:hpa,hpc,nnd.
e	TR152	127.mg	706.mg	0/20	151.mg	16/50	376.mg	11/50	lun:a/a,a/c.
2186	132	8.30mg	n.s.s.	0/16	19.0mg	4/16			Innes;ntis,1968/1969
a	132	12.8mg	n.s.s.	0/16	19.0mg	2/16			
b	132	17.5mg	n.s.s.	0/16	19.0mg	1/16			
c	132	17.5mg	n.s.s.	0/16	19.0mg	1/16			
d	132	3.84mg	23.6mg	0/16	19.0mg	9/16			
2187	132	9.17mg	n.s.s.	1/17	20.4mg	5/18			
a	132	40.5mg	n.s.s.	0/17	20.4mg	0/18			
b	132	8.28mg	n.s.s.	2/17	20.4mg	6/18			
2188	132	12.4mg	n.s.s.	1/18	19.0mg	3/18			
a	132	17.8mg	n.s.s.	2/18	19.0mg	2/18			
b	132	8.12mg	n.s.s.	3/18	19.0mg	6/18			
2189	TR152	5.78mg	n.s.s.	2/20	7.50mg	19/50	(15.0mg	9/50)	S
a	TR152	11.7mg	n.s.s.	15/20	7.50mg	40/50	15.0mg	39/50	
b	TR152	106.mg	n.s.s.	0/20	7.50mg	0/50	15.0mg	1/50	liv:hpa,hpc,nnd.
2190	TR152	37.1mg	2.43gm	0/20	12.0mg	2/50	24.0mg	8/50	
a	TR152	61.8mg	n.s.s.	0/20	12.0mg	0/50	24.0mg	4/50	
b	TR152	12.7mg	n.s.s.	17/20	12.0mg	43/50	24.0mg	39/50	
c	TR152	119.mg	n.s.s.	1/20	12.0mg	2/50	24.0mg	0/50	liv:hpa,hpc,nnd.

ETHYLENE GLYCOL   107-21-1

	RefNum	LoConf	UpConf	Cntrl	1Dose	1Inc	2Dose	2Inc	3Dose	3Inc	Citation or Pathology
2191	TR413	8.40gm	n.s.s.	39/50	1.59gm	33/50	3.19gm	31/51	6.37gm	36/50	
a	TR413	12.7gm	n.s.s.	10/50	1.59gm	9/50	3.19gm	14/51	6.37gm	10/50	liv:hpa,hpc,nnd.
b	TR413	4.83gm	n.s.s.	1/50	1.59gm	6/50	3.19gm	6/51	(6.37gm	1/50)	lun:a/a,a/c.
2192	TR413	7.26gm	n.s.s.	0/54	740.mg	0/54	1.48gm	2/54	2.96gm	3/54	hag:ade,car. S
a	TR413	1.83gm	n.s.s.	34/54	740.mg	36/54	1.48gm	31/54	2.96gm	33/54	
b	TR413	3.13gm	n.s.s.	19/54	740.mg	23/54	1.48gm	16/54	2.96gm	18/54	liv:hpa,hpc,nnd.
c	TR413	6.61gm	n.s.s.	7/54	740.mg	3/54	1.48gm	9/54	2.96gm	3/54	lun:a/a,a/c.

ETHYLENE GLYCOL, CYCLIC SULFATE   (glycol sulfate) 1072-53-3

	RefNum	LoConf	UpConf	Cntrl	1Dose	1Inc	Citation or Pathology
2193	1143	.169mg	n.s.s.	10/30	.286mg	9/30	Van Duuren;jnci,53,695-700;1974

ETHYLENE IMINE   151-56-4

	RefNum	LoConf	UpConf	Cntrl	1Dose	1Inc	Citation or Pathology
2194	1245	.100mg	.900mg	1/17	1.81mg	10/11	Innes;ntis,1968/1969
a	1245	.806mg	n.s.s.	0/17	1.81mg	2/11	
b	1245	n.s.s.	.551mg	2/17	1.81mg	11/11	
2195	1245	.219mg	1.59mg	2/18	1.68mg	12/16	
a	1245	.349mg	3.45mg	1/18	1.68mg	9/16	
b	1245	.734mg	n.s.s.	0/18	1.68mg	4/16	
c	1245	n.s.s.	.419mg	3/18	1.68mg	16/16	
2196	1245	n.s.s.	.409mg	15/15	1.81mg	15/15	
a	1245	.233mg	1.27mg	0/16	1.81mg	11/15	
b	1245	n.s.s.	.409mg	0/15	1.81mg	15/15	
2197	1245	.135mg	.688mg	0/16	1.68mg	15/17	
a	1245	.135mg	.688mg	0/16	1.68mg	15/17	
b	1245	.174mg	.847mg	0/16	1.68mg	14/17	
c	1245	88.7ug	.540mg	0/16	1.68mg	16/17	

ETHYLENE OXIDE   (EO) 75-21-8

	RefNum	LoConf	UpConf	Cntrl	1Dose	1Inc	2Dose	2Inc	Citation or Pathology
2198	TR326	21.7mg	163.mg	13/50	27.8mg	22/50	55.6mg	43/50	duo:mno; hag:pcy; kid:mno; lun:a/a,a/c; mds:mno; mgl:acn,adq; mln:mno; mul:mlh,mno; spl:mno; ute:acn,adn,mno. M
a	TR326	27.3mg	120.mg	3/50	27.8mg	9/50	55.6mg	26/50	hag:pcy; lun:a/a,a/c. C
b	TR326	36.0mg	163.mg	2/50	27.8mg	5/50	55.6mg	22/50	lun:a/a,a/c.
c	TR326	47.0mg	362.mg	2/50	27.8mg	4/50	55.6mg	17/50	
d	TR326	90.3mg	1.25gm	0/50	27.8mg	1/50	55.6mg	7/50	
e	TR326	31.2mg	41.7gm	1/50	27.8mg	8/50	(55.6mg	3/50)	S
f	TR326	31.5mg	n.s.s.	1/50	27.8mg	8/50	(55.6mg	6/50)	mgl:acn,adq.
g	TR326	38.7mg	n.s.s.	10/50	27.8mg	14/50	55.6mg	29/50	duo:mno; kid:mno; mds:mno; mgl:acn,adq; mln:mno; mul:mlh,mno; spl:mno; ute:acn,adn,mno. P
h	TR326	57.8mg	n.s.s.	9/50	27.8mg	6/50	55.6mg	22/50	duo:mno; kid:mno; mds:mno; mln:mno; mul:mlh,mno; spl:mno; ute: mno.
i	TR326	64.9mg	n.s.s.	1/50	27.8mg	6/50	55.6mg	8/50	
j	TR326	107.mg	n.s.s.	0/50	27.8mg	2/50	55.6mg	5/50	ute:acn,adn.
k	TR326	120.mg	n.s.s.	0/50	27.8mg	1/50	55.6mg	5/50	
l	TR326	27.6mg	n.s.s.	30/50	27.8mg	33/50	55.6mg	46/50	
m	TR326	34.7mg	n.s.s.	6/50	27.8mg	9/50	(55.6mg	3/50)	liv:hpa,hpc,nnd.
n	TR326	36.0mg	163.mg	2/50	27.8mg	5/50	55.6mg	22/50	lun:a/a,a/c.
2199	TR326	22.5mg	n.s.s.	12/50	23.1mg	26/50	46.3mg	29/50	hag:pcy; lun:a/a,a/c. C
a	TR326	27.8mg	n.s.s.	11/50	23.1mg	19/50	46.3mg	26/50	lun:a/a,a/c.
b	TR326	44.0mg	n.s.s.	6/50	23.1mg	10/50	46.3mg	16/50	
c	TR326	52.1mg	n.s.s.	1/50	23.1mg	9/50	46.3mg	8/50	
d	TR326	54.5mg	n.s.s.	5/50	23.1mg	11/50	46.3mg	11/50	
e	TR326	24.4mg	n.s.s.	29/50	23.1mg	38/50	46.3mg	41/50	
f	TR326	53.7mg	n.s.s.	15/50	23.1mg	17/50	46.3mg	21/50	liv:hpa,hpc,nnd.
g	TR326	27.8mg	n.s.s.	11/50	23.1mg	19/50	46.3mg	26/50	lun:a/a,a/c.

```
      Spe Strain  Site    Xpo+Xpt                                              TD50    2Tailpvl
          Sex   Route  Hist     Notes                                          DR      AuOp

2200 R m f34 inh per mso 23m24 eis pool                        .(+) .          30.8mg * P<.0005+
a    R m f34 inh bra gli 23m24 eis                                             121.mg * P<.01 +
b    R m f34 inh spl mnl 23m24 eis                                             47.3mg * P<.3  +
c    R m f34 inh liv nnd 23m24 eis                                             no dre   P=1.
2201 R f fmf inh pit ade 76w78 ikr pool                    .  (±)             29.8mg   P<.08 -
a    R f fmf inh bra mix 76w78 ikr                                             237.mg   P<.6  +
2202 R f fmf inh bra mix 23m24 ir  pool                          . (±)         143.mg   P<.02 +
2203 R m fmf inh adr phe 76w78 ikr pool                    .  (±)             26.1mg   P<.08 -
a    R m fmf inh bra ast 76w78 ikr                                             102.mg * P<.7  +
b    R m fmf inh pit ade 76w78 ikr                                             no dre   P=1.  -
c    R m fmf inh tes ict 76w78 ikr                                             no dre   P=1.  -
2204 R m fmf inh bra mix 24m25 ir  pool                        . (+) .         70.7mg * P<.002 +
2205 R f sda gav sto mix 25m35 e                        .+ .                   7.43mg * P<.0005+
a    R f sda gav for sqc 25m35 e                                               10.6mg * P<.0005+
b    R f sda gav mgl adf 25m35 e                                               10.0mg \ P<.02 -

ETHYLENE THIOUREA         100ng..:..1ug....:..10.....:..100....:..1mg....:..10.....:..100....:..1g......:..10
2206 M f b6c eat liv MXA 25m25                                        :+ :     19.9mg \ P<.0005c
a    M f b6c eat MXB MXB 25m25                                                 26.6mg \ P<.0005
b    M f b6c eat liv hpa 25m25                                                 30.5mg \ P<.0005c
c    M f b6c eat liv hpc 25m25                                                 39.7mg * P<.0005c
d    M f b6c eat thy MXA 25m25                                                 82.0mg / P<.0005c
e    M f b6c eat thy fca 25m25                                                 90.9mg * P<.0005c
f    M f b6c eat pit pda 25m25                                                 147.mg * P<.003 c
g    M f b6c eat pit MXA 25m25                                                 155.mg * P<.005 c
h    M f b6c eat thy fcc 25m25                                                 546.mg * P<.0005c
i    M f b6c eat TBA MXB 25m25                                                 120.mg * P<.08
j    M f b6c eat liv MXB 25m25                                                 19.9mg \ P<.0005
k    M f b6c eat lun MXB 25m25                                                 641.mg \ P<.7
2207 M f b6c orl liv hpt 81w81 evx                             <+              noTD50   P<.0005
a    M f b6c orl lun ade 81w81 evx                                             203.mg   P<.05
b    M f b6c orl tba mix 81w81 evx                                             noTD50   P<.0005
2208 M m b6c eat MXB MXB 25m25                                   :  +  :       54.1mg * P<.0005
a    M m b6c eat liv MXA 25m25                                                 54.1mg * P<.0005c
b    M m b6c eat liv hpc 25m25                                                 54.6mg / P<.0005c
c    M m b6c eat thy MXA 25m25                                                 97.1mg / P<.0005c
d    M m b6c eat thy fca 25m25                                                 112.mg / P<.0005c
e    M m b6c eat pit pda 25m25                                                 387.mg * P<.0005c
f    M m b6c eat thy fcc 25m25                                                 730.mg * P<.04 c
g    M m b6c eat MXA hem 25m25                                                 840.mg * P<.03
h    M m b6c eat TBA MXB 25m25                                                 89.2mg * P<.03
i    M m b6c eat liv MXB 25m25                                                 54.1mg * P<.0005
j    M m b6c eat lun MXB 25m25                                                 532.mg * P<.3
2209 M m b6c orl liv hpt 82w82 evx                           . + .             16.9mg   P<.0005+
a    M m b6c orl lun ade 82w82 evx                                             545.mg   P<.3
b    M m b6c orl tba mix 82w82 evx                                             16.9mg   P<.0005
2210 M f b6a orl liv hpt 81w81 evx                             . + .           44.7mg   P<.0005
a    M f b6a orl lun ade 81w81 evx                                             no dre   P=1.
b    M f b6a orl tba mix 81w81 evx                                             29.3mg   P<.0005
2211 M m b6a orl liv hpt 81w81 evx                             <+              noTD50   P<.0005
a    M m b6a orl lun ade 81w81 evx                                             no dre   P=1.
b    M m b6a orl tba mix 81w81 evx                                             noTD50   P<.0005
2212 R f f34 eat thy MXA 25m25                               :+ :              7.42mg / P<.0005c
a    R f f34 eat thy MXA 25m25                                                 7.42mg / P<.0005c
b    R f f34 eat thy MXA 25m25                                                 39.6mg / P<.02 c
c    R f f34 eat TBA MXB 25m25                                                 24.6mg * P<.5
d    R f f34 eat liv MXB 25m25                                                 no dre   P=1.
2213 R m f34 eat thy MXA 25m25                             :+ :                3.16mg * P<.0005c
a    R m f34 eat thy MXA 25m25                                                 5.25mg * P<.0005c
b    R m f34 eat thy MXA 25m25                                                 5.50mg * P<.0005c
c    R m f34 eat TBA MXB 25m25                                                 17.8mg * P<.5
d    R m f34 eat liv MXB 25m25                                                 186.mg * P<.2
2214 R b cdr eat --- fbs 24m24 e                              .  +  .          13.2mg Z P<.003

a    R b cdr eat thy mix 24m24 e                                               16.6mg Z P<.0005+
b    R b cdr eat thy ade 24m24 e                                               33.9mg Z P<.0005+
c    R b cdr eat liv hpt 24m24 e                                               330.mg * P<.04 +
d    R b cdr eat mgl adc 24m24 e                                               no dre   P=1.
e    R b cdr eat tba mix 24m24 e                                               7.23mg * P<.0005
f    R b cdr eat tba mal 24m24 e                                               13.4mg Z P<.0005
2215 R f cdr eat thy car 18m24                                .  +  .          34.4mg * P<.002 +
a    R f cdr eat thy fcc 18m24 e                                               39.8mg * P<.03 +
b    R f cdr eat thy sca 18m24 e                                               98.9mg * P<.2  +
c    R f cdr eat liv hnd 18m24                                                 304.mg * P<.6
2216 R m cdr eat thy car 18m24                               .  +  .           10.8mg / P<.0005+
a    R m cdr eat thy fcc 18m24 e                                               12.8mg / P<.0005+
b    R m cdr eat liv hnd 18m24                                                 79.1mg * P<.2

ETHYLENE UREA            100ng..:..1ug....:..10.....:..100....:..1mg....:..10.....:..100....:..1g......:..10
2217 M f b6a orl lun ade 76w76 evx                                 .>          505.mg   P<.6  -
a    M f b6a orl liv hpt 76w76 evx                                             no dre   P=1.  -
```

	RefNum	LoConf	UpConf	Cntrl	1Dose	1Inc	2Dose	2Inc	3Dose	3Inc	Citation or Pathology / Brkly Code
2200	1624	18.3mg	77.8mg	3/78p	5.42mg	9/79	10.8mg	21/79			Lynch;txap,76,69-84;1984
a	1624	52.3mg	4.81gm	0/76p	5.42mg	2/77	10.8mg	5/79			
b	1624	14.3mg	n.s.s.	24/77p	5.42mg	38/79	10.8mg	30/76			
c	1624	88.2mg	n.s.s.	2/78p	5.42mg	3/78	10.8mg	2/79			
2201	1666m	8.48mg	n.s.s.	2/40p	13.3mg	4/20					Snellings;txap,75,105-117;1984/Garman 1985
a	1666m	24.7mg	n.s.s.	1/40p	1.31mg	0/20	4.34mg	0/20	13.3mg	1/20	
2202	1666o	61.7mg	n.s.s.	0/154p	1.31mg	1/78	4.32mg	3/79	13.2mg	3/79	
2203	1666m	6.94mg	n.s.s.	1/40p	9.29mg	3/20					
a	1666m	16.5mg	n.s.s.	0/40p	.920mg	0/20	3.04mg	1/20	9.29mg	0/20	
b	1666m	9.53mg	n.s.s.	7/40p	9.29mg	3/20					
c	1666m	2.56mg	n.s.s.	35/40p	9.29mg	16/20					
2204	1666o	33.5mg	313.mg	1/156p	.916mg	1/79	3.02mg	4/78	9.25mg	7/79	
2205	1486	5.12mg	11.3mg	0/50	1.53mg	12/50	6.11mg	35/50			Dunkelberg;bjca,46,924-933;1982
a	1486	7.06mg	16.9mg	0/50	1.53mg	8/50	6.11mg	29/50			
b	1486	4.33mg	n.s.s.	4/50	1.53mg	13/50	(6.11mg	1/50)			

ETHYLENE THIOUREA   (ETU) 96-45-7

	RefNum	LoConf	UpConf	Cntrl	1Dose	1Inc	2Dose	2Inc	Citation or Pathology / Brkly Code
2206	TR388	12.7mg	38.1mg	4/50	42.9mg	44/50	(130.mg	48/50)	liv:hpa,hpc.
a	TR388	14.5mg	96.7mg	14/50	42.9mg	47/50	(130.mg	48/50)	liv:hpa,hpc; pit:pda,pdc; thy:fca,fcc. C
b	TR388	19.1mg	59.9mg	2/50	42.9mg	33/50	(130.mg	14/50)	
c	TR388	28.4mg	62.9mg	2/50	42.9mg	29/50	130.mg	47/50	
d	TR388	55.5mg	128.mg	0/50	42.9mg	2/50	130.mg	38/50	
e	TR388	60.6mg	144.mg	0/50	42.9mg	2/50	130.mg	35/50	thy:fca,fcc.
f	TR388	74.6mg	973.mg	10/50	42.9mg	19/50	130.mg	26/50	
g	TR388	76.7mg	1.57gm	11/50	42.9mg	19/50	130.mg	26/50	pit:pda,pdc.
h	TR388	246.mg	1.69gm	0/50	42.9mg	0/50	130.mg	8/50	
i	TR388	46.2mg	n.s.s.	36/50	42.9mg	49/50	130.mg	48/50	
j	TR388	12.7mg	38.1mg	4/50	42.9mg	44/50	(130.mg	48/50)	liv:hpa,hpc,nnd.
k	TR388	93.4mg	n.s.s.	5/50	42.9mg	8/50	(130.mg	0/50)	lun:a/a,a/c.
2207	1141	n.s.s.	20.7mg	0/16	88.8mg	18/18			Innes;ntis,1968/1969
a	1141	61.1mg	n.s.s.	0/16	88.8mg	3/18			
b	1141	n.s.s.	20.7mg	0/16	88.8mg	18/18			
2208	TR388	30.6mg	159.mg	21/50	39.6mg	32/50	120.mg	47/50	liv:hpa,hpc; pit:pda; thy:fca,fcc. C
a	TR388	30.6mg	158.mg	20/50	39.6mg	32/50	120.mg	46/50	liv:hpa,hpc.
b	TR388	33.5mg	116.mg	13/50	39.6mg	19/50	120.mg	45/50	
c	TR388	60.7mg	172.mg	1/50	39.6mg	1/50	120.mg	29/50	thy:fca,fcc.
d	TR388	69.6mg	194.mg	0/50	39.6mg	1/50	120.mg	26/50	
e	TR388	173.mg	1.20gm	0/50	39.6mg	0/50	120.mg	8/50	
f	TR388	246.mg	n.s.s.	1/50	39.6mg	0/50	120.mg	5/50	
g	TR388	284.mg	n.s.s.	0/50	39.6mg	1/50	120.mg	3/50	ear:hem; hea:hem; liv:hem. S
h	TR388	38.6mg	n.s.s.	35/50	39.6mg	40/50	120.mg	47/50	
i	TR388	30.6mg	158.mg	20/50	39.6mg	32/50	120.mg	46/50	liv:hpa,hpc,nnd.
j	TR388	156.mg	n.s.s.	5/50	39.6mg	8/50	120.mg	8/50	lun:a/a,a/c.
2209	1141	7.63mg	40.5mg	0/16	82.5mg	14/16			Innes;ntis,1968/1969
a	1141	88.7mg	n.s.s.	0/16	82.5mg	1/16			
b	1141	7.63mg	40.5mg	0/16	82.5mg	14/16			
2210	1141	20.4mg	124.mg	0/17	88.8mg	9/16			
a	1141	178.mg	n.s.s.	1/17	88.8mg	0/16			
b	1141	13.2mg	99.9mg	2/17	88.8mg	12/16			
2211	1141	n.s.s.	20.0mg	1/18	82.6mg	18/18			
a	1141	87.7mg	n.s.s.	2/18	82.6mg	2/18			
b	1141	n.s.s.	21.8mg	3/18	82.6mg	18/18			
2212	TR388	4.74mg	13.2mg	1/50	4.15mg	6/50	12.5mg	28/50	thy:fab,fca.
a	TR388	4.64mg	14.2mg	3/50	4.15mg	7/50	12.5mg	30/50	thy:fab,fca,fcy,fdc.
b	TR388	15.9mg	n.s.s.	2/50	4.15mg	1/50	12.5mg	8/50	thy:fcy,fdc.
c	TR388	5.13mg	n.s.s.	48/50	4.15mg	46/50	12.5mg	49/50	
d	TR388	82.4mg	n.s.s.	0/50	4.15mg	1/50	12.5mg	0/50	liv:hpa,hpc,nnd.
2213	TR388	2.08mg	5.23mg	1/50	3.32mg	12/50	10.0mg	37/50	thy:fab,fca,fcy,fdc.
a	TR388	3.28mg	9.00mg	0/50	3.32mg	9/50	10.0mg	23/50	thy:fab,fca.
b	TR388	3.29mg	10.4mg	1/50	3.32mg	3/50	10.0mg	26/50	thy:fcy,fdc.
c	TR388	3.49mg	n.s.s.	43/50	3.32mg	47/50	10.0mg	43/50	
d	TR388	30.3mg	n.s.s.	0/50	3.32mg	0/50	10.0mg	1/50	liv:hpa,hpc,nnd.

	RefNum	LoConf	UpConf	Cntrl	1Dose	1Inc	2Dose	2Inc	3Dose	3Inc	4Dose	4Inc	5Dose	5Inc	Citation / Brkly Code
2214	139	5.00mg	75.7mg	0/72	.225mg	0/75	1.13mg	5/73	(5.63mg	4/73	11.3mg	5/69	22.5mg	3/70)	Graham;fctx,13, 493-499;1975
a	139	12.5mg	22.9mg	2/72	.225mg	2/75	1.13mg	1/73	5.63mg	2/73	11.3mg	16/69	22.5mg	62/70	
b	139	20.0mg	70.6mg	2/72	.225mg	0/75	1.13mg	5/73	5.63mg	1/73	11.3mg	21/69	(22.5mg	3/70)	
c	139	117.mg	n.s.s.	1/72	.225mg	1/75	1.13mg	1/73	5.63mg	2/73	11.3mg	1/69	22.5mg	5/70	
d	139	100.mg	n.s.s.	2/72	.225mg	14/75	1.13mg	5/69	5.63mg	5/69	11.3mg	5/69	22.5mg	(0/70)	
e	139	4.58mg	13.5mg	51/72	.225mg	49/75	1.13mg	45/73	5.63mg	54/73	11.3mg	61/69	22.5mg	68/70	
f	139	9.62mg	19.9mg	15/72	.225mg	23/75	1.13mg	13/73	5.63mg	16/73	11.3mg	31/69	22.5mg	63/70	

	RefNum	LoConf	UpConf	Cntrl	1Dose	1Inc	2Dose	2Inc	Citation / Brkly Code
2215	141m	15.6mg	149.mg	0/32	6.56mg	3/26	13.1mg	5/21	Ulland;jnci,49,583-584;1972/Weisburger 1981
a	141m	18.0mg	n.s.s.	0/10	6.56mg	2/26	13.1mg	6/26	
b	141m	29.9mg	n.s.s.	0/32	6.56mg	2/26	13.1mg	1/21	
c	141m	49.4mg	n.s.s.	0/32	6.56mg	1/26	13.1mg	0/21	
2216	141m	6.03mg	21.8mg	0/32	5.25mg	3/26	10.5mg	14/21	
a	141m	7.21mg	28.7mg	0/10	5.25mg	2/26	10.5mg	15/26	
b	141m	23.9mg	n.s.s.	0/32	5.25mg	2/26	10.5mg	1/21	

ETHYLENE UREA (2-imidazolidinone) 120-93-4

	RefNum	LoConf	UpConf	Cntrl	1Dose	1Inc	Citation / Brkly Code
2217	1197	70.3mg	n.s.s.	1/17	89.1mg	2/17	Innes;ntis,1968/1969
a	1197	167.mg	n.s.s.	0/17	89.1mg	0/17	

Spe		Strain	Site		Xpo+Xpt			TD50	2Tailpvl
	Sex	Route	Hist	Notes				DR	AuOp

										TD50	2Tailpvl
b	M f	b6a	orl tba	mix 76w76	evx			473.mg	P<.7	-
2218	M m	b6a	orl liv	hpt 76w76	evx	.>		500.mg	P<.6	-
a	M m	b6a	orl lun	ade 76w76	evx			no dre	P=1.	-
b	M m	b6a	orl tba	mix 76w76	evx			no dre	P=1.	-
2219	M f	b6c	orl liv	hpt 76w76	evx	.>		no dre	P=1.	-
a	M f	b6c	orl lun	mix 76w76	evx			no dre	P=1.	-
b	M f	b6c	orl tba	tum 76w76	evx			no dre	P=1.	-
2220	M m	b6c	orl liv	hpt 76w76	evx	. ±		242.mg	P<.1	-
a	M m	b6c	orl lun	ade 76w76	evx			242.mg	P<.1	-
b	M m	b6c	orl tba	mix 76w76	evx			87.1mg	P<.007	-

ETHYLENEBISDITHIOCARBAMATE, DISODIUM ..:..1ug....:..10......:..100....:..1mg..:..10......:..100....:..1g......:..10

2221	M f	b6a	orl liv	hpt 76w76	evx	.>		no dre	P=1.	-
a	M f	b6a	orl lun	mix 76w76	evx			no dre	P=1.	-
b	M f	b6a	orl tba	mix 76w76	evx			no dre	P=1.	-
2222	M m	b6a	orl lun	ade 76w76	evx	.>		no dre	P=1.	-
a	M m	b6a	orl liv	hpt 76w76	evx			no dre	P=1.	-
b	M m	b6a	orl tba	mix 76w76	evx			no dre	P=1.	-
2223	M f	b6c	orl liv	hpt 76w76	evx	.>		no dre	P=1.	-
a	M f	b6c	orl lun	ade 76w76	evx			no dre	P=1.	-
b	M f	b6c	orl tba	mix 76w76	evx			no dre	P=1.	-
2224	M m	b6c	orl liv	hpt 76w76	evx	.>		866.mg	P<1.	-
a	M m	b6c	orl lun	ade 76w76	evx			no dre	P=1.	-
b	M m	b6c	orl tba	mix 76w76	evx			no dre	P=1.	-

1-ETHYLENEOXY-3,4-EPOXYCYCLOHEXANE ..:..1ug....:..10......:..100....:..1mg..:..10......:..100....:..1g......:..10

| 2225 | R f | esd | gav sto | tum 84w84 | | .> | | no dre | P=1. | - |

DI(2-ETHYLHEXYL)ADIPATE 100ng..:..1ug....:..10......:..100....:..1mg....:..10......:..100....:..1g......:..10

2226	M f	b6c	eat liv	hpc 24m24		: +	:3.05gm \	P<.0005c
a	M f	b6c	eat liv	MXA 24m24			3.84gm *	P<.0005c
b	M f	b6c	eat TBA	MXB 24m24			no dre	P=1.
c	M f	b6c	eat liv	MXB 24m24			3.84gm *	P<.0005
d	M f	b6c	eat lun	MXB 24m24			no dre	P=1.
2227	M m	b6c	eat liv	MXA 24m24		: ±	5.33gm *	P<.06 c
a	M m	b6c	eat liv	hpa 24m24			9.04gm *	P<.07 c
b	M m	b6c	eat TBA	MXB 24m24			no dre	P=1.
c	M m	b6c	eat liv	MXB 24m24			5.33gm *	P<.06
d	M m	b6c	eat lun	MXB 24m24			11.7gm \	P<.6
2228	R f	f34	eat TBA	MXB 24m24		:>	no dre	P=1.	-
a	R f	f34	eat liv	MXB 24m24			14.5gm *	P<.6
2229	R m	f34	eat TBA	MXB 24m24		:>	no dre	P=1.	-
a	R m	f34	eat liv	MXB 24m24			no dre	P=1.

DI(2-ETHYLHEXYL)PHTHALATE 100ng..:..1ug....:..10......:..100....:..1mg....:..10......:..100....:..1g...:..10

2230	M f	b6c	eat liv	MXA 24m24		: + :	825.mg *	P<.0005c
a	M f	b6c	eat liv	hpc 24m24			1.05gm *	P<.0005c
b	M f	b6c	eat TBA	MXB 24m24			350.mg \	P<.003
c	M f	b6c	eat liv	MXB 24m24			825.mg *	P<.0005
d	M f	b6c	eat lun	MXB 24m24			7.26gm *	P<.07
2231	M m	b6c	eat liv	MXA 24m24		: ±	976.mg *	P<.03 c
a	M m	b6c	eat liv	hpc 24m24			1.82gm *	P<.08 c
b	M m	b6c	eat TBA	MXB 24m24			1.54gm *	P<.4
c	M m	b6c	eat liv	MXB 24m24			976.mg *	P<.03
d	M m	b6c	eat lun	MXB 24m24			no dre	P=1.
2232	R f	f34	eat liv	MXA 24m24		: + :	1.11gm *	P<.0005c
a	R f	f34	eat liv	hpc 24m24			2.30gm *	P<.002 c
b	R f	f34	eat liv	nnd 24m24			2.35gm *	P<.02
c	R f	f34	eat TBA	MXB 24m24			1.67gm *	P<.6
d	R f	f34	eat liv	MXB 24m24			1.11gm *	P<.0005
2233	R m	f34	eat liv	MXA 24m24		: ±	1.17gm *	P<.03 c
a	R m	f34	eat liv	hpc 24m24			3.36gm *	P<.2 c
b	R m	f34	eat TBA	MXB 24m24			no dre	P=1.
c	R m	f34	eat liv	MXB 24m24			1.17gm *	P<.03
2234	R m	f34	eat liv	mix 95w95	er	. +	499.mg	P<.003 +
a	R m	f34	eat liv	hpc 95w95	er		895.mg	P<.02
b	R m	f34	eat liv	hpn 95w95	er		2.05gm	P<.2
2235	R m	f34	eat liv	mix 25m25	er	. + .	412.mg	P<.0005+

ETHYLHYDRAZINE.HCl 100ng..:..1ug....:..10......:..100....:..1mg..:..10..:..100....:..1g....:..10

2236	M f	swa	wat lun	mix 82w82	e	. + .	5.22mg	P<.0005+
a	M f	swa	wat lun	ade 82w82	e		5.62mg	P<.0005
b	M f	swa	wat blv	mix 82w82	e		11.6mg	P<.0005+
c	M f	swa	wat liv	mix 82w82	e		13.0mg	P<.0005
d	M f	swa	wat lun	adc 82w82	e		14.0mg	P<.0005
e	M f	swa	wat blv	ang 82w82	e		27.6mg	P<.0005
f	M f	swa	wat liv	ang 82w82	e		27.6mg	P<.0005
g	M f	swa	wat liv	agm 82w82	e		42.9mg	P<.0005
h	M f	swa	wat blv	agm 82w82	e		42.9mg	P<.0005
2237	M m	swa	wat lun	ade 72w72	e	. + .	7.82mg	P<.0005
a	M m	swa	wat lun	mix 72w72	e		8.81mg	P<.0005+

	RefNum	LoConf	UpConf	Cntrl	1Dose	1Inc	2Dose	2Inc	Citation or Pathology
									Brkly Code
b	1197	59.5mg	n.s.s.	2/17	89.1mg	3/17			
2218	1197	69.5mg	n.s.s.	1/18	82.9mg	2/18			
a	1197	109.mg	n.s.s.	2/18	82.9mg	1/18			
b	1197	65.2mg	n.s.s.	3/18	82.9mg	3/18			
2219	1197	167.mg	n.s.s.	0/16	89.1mg	0/17			
a	1197	167.mg	n.s.s.	0/16	89.1mg	0/17			
b	1197	167.mg	n.s.s.	0/16	89.1mg	0/17			
2220	1197	59.5mg	n.s.s.	0/16	82.9mg	2/17			
a	1197	59.5mg	n.s.s.	0/16	82.9mg	2/17			
b	1197	32.8mg	1.18gm	0/16	82.9mg	5/17			

ETHYLENEBISDITHIOCARBAMATE, DISODIUM (dithane, nabam) 142-59-6

	RefNum	LoConf	UpConf	Cntrl	1Dose	1Inc	2Dose	2Inc	Citation or Pathology
2221	1215	19.7mg	n.s.s.	0/17	9.96mg	0/18			Innes;ntis,1968/1969
a	1215	19.7mg	n.s.s.	0/17	9.96mg	0/18			
b	1215	19.7mg	n.s.s.	2/17	9.96mg	0/18			
2222	1215	12.2mg	n.s.s.	2/18	9.26mg	1/18			
a	1215	18.3mg	n.s.s.	3/18	9.26mg	0/18			
b	1215	14.1mg	n.s.s.	5/18	9.26mg	1/18			
2223	1215	17.5mg	n.s.s.	0/18	9.96mg	0/16			
a	1215	17.5mg	n.s.s.	1/18	9.96mg	0/16			
b	1215	17.5mg	n.s.s.	3/18	9.96mg	0/16			
2224	1215	9.63mg	n.s.s.	1/17	9.26mg	1/16			
a	1215	10.7mg	n.s.s.	2/17	9.26mg	1/16			
b	1215	10.1mg	n.s.s.	6/17	9.26mg	2/16			

1-ETHYLENEOXY-3,4-EPOXYCYCLOHEXANE 106-87-6

	RefNum	LoConf	UpConf	Cntrl	1Dose	1Inc	2Dose	2Inc	Citation or Pathology
2225	55	27.4mg	n.s.s.	0/5	40.8mg	0/5			Van Duuren;jnci,37,825-838;1966

DI(2-ETHYLHEXYL)ADIPATE 103-23-1

	RefNum	LoConf	UpConf	Cntrl	1Dose	1Inc	2Dose	2Inc	Citation or Pathology
2226	TR212	1.57gm	9.40gm	1/50	1.52gm	14/50	(3.19gm	12/50)	
a	TR212	2.29gm	11.3gm	3/50	1.52gm	19/50	3.19gm	18/50	
b	TR212	3.44gm	n.s.s.	37/50	1.52gm	33/50	3.19gm	30/50	liv:hpa,hpc.
c	TR212	2.29gm	11.3gm	3/50	1.52gm	19/50	3.19gm	18/50	liv:hpa,hpc,nnd.
d	TR212	17.8mg	n.s.s.	6/50	1.52gm	1/50	3.19gm	3/50	lun:a/a,a/c.
2227	TR212	2.25gm	n.s.s.	13/50	1.41gm	20/50	2.96gm	27/50	liv:hpa,hpc.
a	TR212	3.62gm	n.s.s.	6/50	1.41gm	8/50	2.96gm	15/50	
b	TR212	2.99gm	n.s.s.	33/50	1.41gm	32/50	2.96gm	34/50	
c	TR212	2.25gm	n.s.s.	13/50	1.41gm	20/50	2.96gm	27/50	liv:hpa,hpc,nnd.
d	TR212	2.04gm	n.s.s.	8/50	1.41gm	9/50	(2.96gm	3/50)	lun:a/a,a/c.
2228	TR212	699.mg	n.s.s.	44/50	589.mg	41/50	(1.23gm	34/50)	
a	TR212	4.76gm	n.s.s.	0/50	589.mg	3/50	1.23gm	1/50	liv:hpa,hpc,nnd.
2229	TR212	1.47gm	n.s.s.	30/49	466.mg	32/50	986.mg	26/50	
a	TR212	4.46gm	n.s.s.	2/49	466.mg	2/50	986.mg	2/50	liv:hpa,hpc,nnd.

DI(2-ETHYLHEXYL)PHTHALATE (di-sec-octyl phthalate) 117-81-7

	RefNum	LoConf	UpConf	Cntrl	1Dose	1Inc	2Dose	2Inc	Citation or Pathology
2230	TR217	513.mg	1.67gm	1/50	383.mg	12/50	765.mg	18/50	liv:hpa,hpc.
a	TR217	638.mg	1.89gm	0/50	383.mg	7/50	765.mg	17/50	
b	TR217	175.mg	2.07gm	20/50	383.mg	35/50	(765.mg	35/50)	
c	TR217	513.mg	1.67gm	1/50	383.mg	12/50	765.mg	18/50	liv:hpa,hpc,nnd.
d	TR217	2.20gm	n.s.s.	0/50	383.mg	1/50	765.mg	2/50	lun:a/a,a/c.
2231	TR217	440.mg	n.s.s.	14/50	353.mg	25/49	713.mg	29/50	liv:hpa,hpc.
a	TR217	736.mg	n.s.s.	9/50	353.mg	14/49	713.mg	18/50	
b	TR217	413.mg	n.s.s.	29/50	353.mg	37/49	713.mg	38/50	
c	TR217	440.mg	n.s.s.	14/50	353.mg	25/49	713.mg	29/50	liv:hpa,hpc,nnd.
d	TR217	1.97gm	n.s.s.	10/50	353.mg	9/49	713.mg	7/50	lun:a/a,a/c.
2232	TR217	644.mg	2.40gm	0/50	294.mg	6/50	589.mg	13/50	liv:hpc,nnd.
a	TR217	1.12gm	5.90gm	0/50	294.mg	2/50	589.mg	8/50	
b	TR217	1.11gm	n.s.s.	0/50	294.mg	4/50	589.mg	5/50	
c	TR217	321.mg	n.s.s.	41/50	294.mg	43/50	589.mg	49/50	s
d	TR217	644.mg	2.40gm	0/50	294.mg	6/50	589.mg	13/50	liv:hpa,hpc,nnd.
2233	TR217	526.mg	n.s.s.	3/50	235.mg	6/50	475.mg	12/50	liv:hpc,nnd.
a	TR217	1.13gm	n.s.s.	1/50	235.mg	1/50	475.mg	5/50	
b	TR217	455.mg	n.s.s.	36/50	235.mg	35/50	475.mg	34/50	
c	TR217	526.mg	n.s.s.	3/50	235.mg	6/50	475.mg	12/50	liv:hpa,hpc,nnd.
2234	1823	193.mg	2.61gm	0/8	800.mg	6/10			Rao;carc,8,1347-1350;1987
a	1823	303.mg	n.s.s.	0/8	800.mg	4/10			
b	1823	501.mg	n.s.s.	0/8	800.mg	2/10			
2235	1982	178.mg	1.59gm	1/10	800.mg	11/14			Rao;jtxe,30,85-89;1990

ETHYLHYDRAZINE.HCl 18413-14-4

	RefNum	LoConf	UpConf	Cntrl	1Dose	1Inc	2Dose	2Inc	Citation or Pathology
2236	142	3.11mg	9.34mg	21/97	25.0mg	44/49			Shimizu;ijcn,13,500-505;1974
a	142	3.40mg	9.90mg	18/97	25.0mg	43/49			
b	142	7.37mg	19.6mg	0/43	25.0mg	30/50			
c	142	8.13mg	22.3mg	0/43	25.0mg	28/50			
d	142	8.54mg	25.8mg	4/97	25.0mg	27/49			
e	142	15.4mg	57.0mg	0/43	25.0mg	16/50			
f	142	15.4mg	57.0mg	0/43	25.0mg	16/50			
g	142	21.5mg	111.mg	0/43	25.0mg	11/50			
h	142	21.5mg	111.mg	0/43	25.0mg	11/50			
2237	142	4.62mg	15.7mg	15/100	20.8mg	31/48			
a	142	4.93mg	21.1mg	23/100	20.8mg	31/48			

```
      Spe Strain  Site   Xpo+Xpt                                                              TD50    2Tailpvl
          Sex   Route  Hist   Notes                                                            DR      AuOp

  b   M m swa wat blv mix 72w72 e                                                             38.6mg   P<.09  +
  c   M m swa wat liv ang 72w72 e                                                             46.9mg   P<.08
  d   M m swa wat blv ang 72w72 e                                                             46.9mg   P<.08
  e   M m swa wat liv mix 72w72 e                                                             48.1mg   P<.2

1-ETHYLNITROSO-3-(2-HYDROXYETHYL)UREA   ..:..1ug....:..10......:..100.....:..1mg....:..10......:..100.....:..1g.....:..10
2238 R f f34 wat ner tum 7m24 e                                . + .                          .714mg * P<.0005
  a   R f f34 wat mam car 7m24 e                                                              .748mg * P<.004 +
  b   R f f34 wat ski tum 7m24 e                                                              .956mg * P<.004 +
  c   R f f34 wat ute tum 7m24 e                                                              .972mg * P<.07 +
  d   R f f34 wat liv tum 7m24 e                                                              no dre   P=1.
2239 R m f34 wat tnv mso 7m24 e                           . + .                               .226mg * P<.0005
  a   R m f34 wat lun tum 7m24 e                                                              .390mg * P<.0005
  b   R m f34 wat ski tum 7m24 e                                                              .401mg * P<.0005+
  c   R m f34 wat ner tum 7m24 e                                                              .769mg * P<.005
  d   R m f34 wat liv tum 7m24 e                                                              no dre   P=1.

1-ETHYLNITROSO-3-(2-OXOPROPYL)UREA      ..:..1ug....:..10......:..100.....:..1mg....:..10......:..100.....:..1g.....:..10
2240 R f f34 gav mam car 7m24 e                                . + .                          .192mg * P<.0005+
  a   R f f34 gav lun tum 7m24 e                                                              .873mg * P<.02  +
  b   R f f34 gav ner tum 7m24 e                                                              1.66mg \ P<.09
  c   R f f34 gav ute tum 7m24 e                                                              1.84mg * P<.4  +
  d   R f f34 gav liv tum 7m24 e                                                              no dre   P=1.
2241 R f f34 wat ner tum 7m24 e                               . + .                           .605mg \ P<.005 +
  a   R f f34 wat mam car 7m24 e                                                              .748mg * P<.0005+
  b   R f f34 wat lun tum 7m24 e                                                              .885mg * P<.002 +
  c   R f f34 wat thy fct 7m24 e                                                              .804mg \ P<.02  +
  d   R f f34 wat col tum 7m24 e                                                              1.13mg \ P<.04  +
  e   R f f34 wat ute tum 7m24 e                                                              1.49mg * P<.09  +
  f   R f f34 wat ski tum 7m24 e                                                              1.79mg \ P<.09  +
  g   R f f34 wat liv tum 7m24 e                                                              5.53mg * P<.09
2242 R m f34 gav lun tum 7m24 e                          <+                                   noTD50   P<.0005+
  a   R m f34 gav tnv mso 7m24 e                                                              .225mg * P<.0005+
  b   R m f34 gav zym tum 7m24 e                                                              .499mg * P<.0005
  c   R m f34 gav col tum 7m24 e                                                              .562mg * P<.0005+
  d   R m f34 gav thy fct 7m24 e                                                              .767mg * P<.005
  e   R m f34 gav for tum 7m24 e                                                              .861mg * P<.002
  f   R m f34 gav ner tum 7m24 e                                                              1.37mg * P<.08
  g   R m f34 gav liv tum 7m24 e                                                              3.39mg * P<.3
2243 R m f34 wat ner tum 7m24 e                             . + .                             .127mg \ P<.0005+
  a   R m f34 wat lun tum 7m24 e                                                              .231mg * P<.0005+
  b   R m f34 wat tnv mso 7m24 e                                                              .264mg * P<.0005+
  c   R m f34 wat col tum 7m24 e                                                              .524mg * P<.0005+
  d   R m f34 wat zym tum 7m24 e                                                              1.13mg * P<.003
  e   R m f34 wat ski tum 7m24 e                                                              .982mg * P<.02  +
  f   R m f34 wat liv tum 7m24 e                                                              no dre   P=1.

ETHYLNITROSOCYANAMIDE   100ng..:..1ug....:..10......:..100.....:..1mg...:..10......:..100.....:..1g...:..10
2244 R f mrw wat res tum 52w90                                 . + .                          2.91mg   P<.0005+
  a   R f mrw wat liv tum 52w90                                                               no dre   P=1.
  b   R f mrw wat tba mix 52w90                                                               3.17mg   P<.2
2245 R m mrw wat res tum 52w90                                   . + .                        4.99mg   P<.005 +
  a   R m mrw wat liv tum 52w90                                                               no dre   P=1.
  b   R m mrw wat tba mix 52w90                                                               no dre   P=1.

ETHYLPHENYLACETYLUREA   100ng..:..1ug....:..10......:..100.....:..1mg...:..10......:..100.....:..1g...:..10
2246 R m f34 eat liv hpt 71w71 e                                      .>                      no dre   P=1.

4-ETHYLSULPHONYLNAPHTHALENE-1-SULFONAMIDE  ..1ug....:..10......:..100.....:..1mg...:..10......:..100.....:..1g...:..10
2247 M f aif eat ubl car 23m23 e                                  . + .                       34.2mg   P<.0005+
  a   M f aif eat lun ade 23m23 e                                                             53.4mg   P<.005
2248 M f ifc eat ubl car 92w92 r                              . + .                           15.3mg   P<.0005+

ETHYNODIOL DIACETATE   100ng..:..1ug....:..10......:..100.....:..1mg...:..10......:..100.....:..1g...:..10
2249 M f crf eat mam tum 24m24 er pool                 .>                                     no dre   P=1.  -
2250 R f win gav liv tum 60w60 er                               .>                            no dre   P=1.  -

EUCALYPTOL   100ng..:..1ug....:..10......:..100.....:..1mg...:..10......:..100.....:..1g...:..10
2251 M m cfl gav lun tum 19m24 e                                      .>                      no dre   P=1.
  a   M m cfl gav liv tum 19m24 e                                                             no dre   P=1.
  b   M m cfl gav tba mix 19m24 e                                                             119.mg * P<.7
  c   M m cfl gav tba mal 19m24 e                                                             no dre   P=1.

EUGENOL   100ng..:..1ug....:..10......:..100.....:..1mg...:..10......:..100.....:..1g...:..10
2252 M f b6c eat liv MXA 24m24                                              : ±               2.91gm * P<.04  e
  a   M f b6c eat TBA MXB 24m24                                                               no dre   P=1.
  b   M f b6c eat liv MXB 24m24                                                               2.91gm * P<.04
  c   M f b6c eat lun MXB 24m24                                                               25.3gm * P<.8
2253 M m b6c eat liv MXA 24m24                                        : ±                     451.mg \ P<.02  e
  a   M m b6c eat liv hpc 24m24                                                               710.mg \ P<.03  e
  b   M m b6c eat liv hpa 24m24                                                               781.mg \ P<.02  e
```

	RefNum	LoConf	UpConf	Cntrl	1Dose	1Inc	2Dose	2Inc	Citation or Pathology
									Brkly Code
b	142	13.7mg	n.s.s.	2/28	20.8mg	8/36			
c	142	16.3mg	n.s.s.	1/28	20.8mg	6/36			
d	142	16.3mg	n.s.s.	1/28	20.8mg	6/36			
e	142	15.4mg	n.s.s.	2/28	20.8mg	7/36			

1-ETHYLNITROSO-3-(2-HYDROXYETHYL)UREA   ---

	RefNum	LoConf	UpConf	Cntrl	1Dose	1Inc	2Dose	2Inc	Citation
2238	2010	.351mg	2.15mg	0/12	.443mg	4/12	.885mg	7/12	Lijinsky;vivo,4,1-6;1990
a	2010	.370mg	3.97mg	0/12	.443mg	6/12	.885mg	5/12	
b	2010	.445mg	6.16mg	0/12	.443mg	4/12	.885mg	5/12	
c	2010	.376mg	n.s.s.	2/12	.443mg	6/12	.885mg	6/12	
d	2010	.730mg	n.s.s.	0/12	.443mg	0/12	.885mg	0/12	
2239	2010	.121mg	.473mg	0/12	.310mg	9/12	.620mg	9/12	
a	2010	.199mg	.979mg	0/12	.310mg	5/12	.620mg	8/12	
b	2010	.206mg	1.10mg	0/12	.310mg	6/12	.620mg	7/12	
c	2010	.345mg	5.53mg	0/12	.310mg	3/12	.620mg	5/12	
d	2010	.511mg	n.s.s.	0/12	.310mg	0/12	.620mg	0/12	

1-ETHYLNITROSO-3-(2-OXOPROPYL)UREA   ---

	RefNum	LoConf	UpConf	Cntrl	1Dose	1Inc	2Dose	2Inc	Citation
2240	2010m	98.9ug	.388mg	0/12	.442mg	12/12	.883mg	10/12	Lijinsky;vivo,4,1-6;1990
a	2010m	.383mg	n.s.s.	1/12	.442mg	5/12	.883mg	6/12	
b	2010m	.407mg	n.s.s.	0/12	.442mg	2/12	(.883mg	0/12)	
c	2010m	.501mg	n.s.s.	2/12	.442mg	6/12	.883mg	4/12	
d	2010m	.728mg	n.s.s.	0/12	.442mg	0/12	.883mg	0/12	
2241	2010n	.225mg	4.85mg	0/12	.476mg	5/12	(.951mg	0/12)	
a	2010n	.366mg	2.03mg	0/12	.476mg	3/12	.951mg	8/12	
b	2010n	.424mg	3.41mg	0/12	.476mg	4/12	.951mg	6/12	
c	2010n	.274mg	n.s.s.	0/12	.476mg	4/12	(.951mg	0/12)	
d	2010n	.340mg	n.s.s.	0/12	.476mg	3/12	(.951mg	0/12)	
e	2010n	.529mg	n.s.s.	2/12	.476mg	3/12	.951mg	6/12	
f	2010n	.438mg	n.s.s.	0/12	.476mg	2/12	(.951mg	0/12)	
g	2010n	1.36mg	n.s.s.	0/12	.476mg	0/12	.951mg	2/12	
2242	2010m	n.s.s.	.135mg	0/12	.309mg	12/12	.618mg	12/12	
a	2010m	.121mg	.472mg	0/12	.309mg	9/12	.618mg	9/12	
b	2010m	.245mg	1.50mg	0/12	.309mg	4/12	.618mg	7/12	
c	2010m	.269mg	1.77mg	0/12	.309mg	3/12	.618mg	7/12	
d	2010m	.344mg	5.51mg	0/12	.309mg	3/12	.618mg	5/12	
e	2010m	.367mg	3.21mg	0/12	.309mg	0/12	.618mg	7/12	
f	2010m	.519mg	n.s.s.	0/12	.309mg	3/12	.618mg	2/12	
g	2010m	.757mg	n.s.s.	1/12	.309mg	0/12	.618mg	3/12	
2243	2010n	53.1ug	.354mg	0/12	.333mg	10/12	(.666mg	5/12)	
a	2010n	.123mg	.480mg	0/12	.333mg	8/12	.666mg	10/12	
b	2010n	.144mg	.605mg	0/12	.333mg	11/12	.666mg	7/12	
c	2010n	.256mg	1.42mg	0/12	.333mg	3/12	.666mg	8/12	
d	2010n	.455mg	5.71mg	0/12	.333mg	0/12	.666mg	6/12	
e	2010n	.421mg	n.s.s.	0/12	.333mg	3/12	.666mg	4/12	
f	2010n	.549mg	n.s.s.	0/12	.333mg	0/12	.666mg	0/12	

ETHYLNITROSOCYANAMIDE   (nitrosoethanecarbamonitrile) 38434-77-4

	RefNum	LoConf	UpConf	Cntrl	1Dose	1Inc	2Dose	2Inc	Citation
2244	1246	1.35mg	8.02mg	0/25	3.40mg	9/20			Bulay;jnci,62,1523-1528;1979
a	1246	10.5mg	n.s.s.	0/25	3.40mg	0/20			
b	1246	.981mg	n.s.s.	12/25	3.40mg	14/20			
2245	1246	1.88mg	36.2mg	0/22	2.97mg	5/19			
a	1246	8.71mg	n.s.s.	0/22	2.97mg	0/19			
b	1246	1.67mg	n.s.s.	12/22	2.97mg	10/19			

ETHYLPHENYLACETYLUREA   (pheneturide) 90-49-3

	RefNum	LoConf	UpConf	Cntrl	1Dose	1Inc	2Dose	2Inc	Citation
2246	1951	25.6mg	n.s.s.	0/15	17.8mg	0/15			Diwan;carc,10,189-194;1989

4-ETHYLSULPHONYLNAPHTHALENE-1-SULFONAMIDE   842-00-2

	RefNum	LoConf	UpConf	Cntrl	1Dose	1Inc	2Dose	2Inc	Citation
2247	142c	16.1mg	102.mg	0/40	13.0mg	9/42			Flaks;bjca,28,227-231;1973
a	142c	21.8mg	360.mg	0/40	13.0mg	6/42			
2248	142a	8.89mg	31.0mg	0/26	13.0mg	19/52			Flaks;bjca,31,585-587;1975

ETHYNODIOL DIACETATE   (ovulen-50) 297-76-7

	RefNum	LoConf	UpConf	Cntrl	1Dose	1Inc	2Dose	2Inc	Citation
2249	1175	.190mg	n.s.s.	161/167p	9.75ug	30/32	.130mg	46/56	Rudali;jnci,49,813-819;1972
2250	1905	1.68mg	n.s.s.	0/6	4.08mg	0/6			Annapurna;ijbb,25,708-713;1988

EUCALYPTOL   470-82-6

	RefNum	LoConf	UpConf	Cntrl	1Dose	1Inc	2Dose	2Inc	Citation
2251	710	38.5mg	n.s.s.	102/240	5.27mg	30/52	21.1mg	18/47	Roe;jept,2,799-819;1979
a	710	72.8mg	n.s.s.	69/240	5.27mg	5/52	21.1mg	12/47	
b	710	16.2mg	n.s.s.	170/240	5.27mg	36/52	21.1mg	35/47	
c	710	67.8mg	n.s.s.	75/240	5.27mg	18/52	21.1mg	11/47	

EUGENOL   (1-allyl-3-methoxy-4-hydroxybenzene) 97-53-0

	RefNum	LoConf	UpConf	Cntrl	1Dose	1Inc	2Dose	2Inc	Pathology
2252	TR223	1.29gm	n.s.s.	2/50	383.mg	7/50	773.mg	9/50	liv:hpa,hpc.
a	TR223	1.12gm	n.s.s.	27/50	383.mg	22/50	773.mg	26/50	
b	TR223	1.29gm	n.s.s.	2/50	383.mg	7/50	773.mg	9/50	liv:hpa,hpc,nnd.
c	TR223	2.22gm	n.s.s.	4/50	383.mg	6/50	773.mg	5/50	lun:a/a,a/c.
2253	TR223	208.mg	124.gm	14/50	353.mg	28/50	(713.mg	18/50)	liv:hpa,hpc.
a	TR223	298.mg	n.s.s.	10/50	353.mg	20/50	(713.mg	9/50)	
b	TR223	336.mg	n.s.s.	4/50	353.mg	13/50	(713.mg	10/50)	

```
     Spe Strain  Site   Xpo+Xpt                                                        TD50    2Tailpvl
         Sex  Route  Hist   Notes                                                          DR    AuOp

   c   M m b6c eat thy fca 24m24                                                        8.52gm * P<.04
   d   M m b6c eat TBA MXB 24m24                                                        1.55gm * P<.3
   e   M m b6c eat liv MXB 24m24                                                        451.mg \ P<.02
   f   M m b6c eat lun MXB 24m24                                                        no dre   P=1.
 2254 M f cd1 eat lun ade 50w78 v                                      .>              no dre   P=1.  -
   a   M f cd1 eat liv tum 50w78 v                                                      no dre   P=1.  -
 2255 R f f34 eat TBA MXB 24m24                                              :>         no dre   P=1.  -
   a   R f f34 eat liv MXB 24m24                                                        no dre   P=1.
 2256 R m f34 eat lun MXA 24m24                                        :    ±         #424.mg \ P<.02  -
   a   R m f34 eat TBA MXB 24m24                                                        293.mg \ P<.5
   b   R m f34 eat liv MXB 24m24                                                        no dre   P=1.

FENAMINOSULF, FORMULATED    100ng..:..1ug....:..10......:..100....:..1mg....:..10......:..100.....:..1g......:..10
 2257 M f b6c eat liv MXA 78w95 v                                       :    ±         #152.mg \ P<.02  -
   a   M f b6c eat TBA MXB 78w95 v                                                      259.mg * P<.2
   b   M f b6c eat liv MXB 78w95 v                                                      152.mg \ P<.02
   c   M f b6c eat lun MXB 78w95 v                                                      1.75gm * P<.7
 2258 M m b6c eat TBA MXB 78w94 v                                            :>         no dre   P=1.  -
   a   M m b6c eat liv MXB 78w94 v                                                      no dre   P=1.
   b   M m b6c eat lun MXB 78w94 v                                                      no dre   P=1.
 2259 R f f34 eat TBA MXB 18m24 v                                    :>                 no dre   P=1.  -
   a   R f f34 eat liv MXB 18m24 v                                                      no dre   P=1.
 2260 R m f34 eat TBA MXB 18m24 v                                   :>                  no dre   P=1.  -
   a   R m f34 eat liv MXB 18m24 v                                                      no dre   P=1.

FENTHION    100ng..:..1ug....:..10......:..100....:..1mg....:..10......:..100.....:..1g......:..10
 2261 M f b6c eat TBA MXB 24m24                                 :>                      17.8mg * P<.7  -
   a   M f b6c eat liv MXB 24m24                                                        no dre   P=1.
   b   M f b6c eat lun MXB 24m24                                                        no dre   P=1.
 2262 M m b6c eat MXA MXA 24m24                             :    ±                      6.12mg * P<.02 a
   a   M m b6c eat TBA MXB 24m24                                                        1.03mg \ P<.02
   b   M m b6c eat liv MXB 24m24                                                        9.44mg * P<.5
   c   M m b6c eat lun MXB 24m24                                                        13.3mg * P<.3
 2263 R f f34 eat TBA MXB 24m24                             :>                          29.4mg * P<1.  -
   a   R f f34 eat liv MXB 24m24                                                        no dre   P=1.
 2264 R m f34 eat TBA MXB 24m24                         :>                              no dre   P=1.  -
   a   R m f34 eat liv MXB 24m24                                                        10.3mg * P<.7

FENVALERATE    100ng..:..1ug....:..10......:..100....:..1mg....:..10......:..100.....:..1g......:..10
 2265 M f cb6 gav --- lym 24m26 e                                      .>              384.mg * P<.6  -
   a   M f cb6 gav liv lct 24m26 e                                                      no dre   P=1.  -
   b   M f cb6 gav lun ade 24m26 e                                                      no dre   P=1.  -
   c   M f cb6 gav tba tum 24m26 e                                                      435.mg * P<.7  -
 2266 M m cb6 gav liv lct 24m26 e                                      .    ±          328.mg * P<.09 -
   a   M m cb6 gav --- lym 24m26 e                                                      659.mg * P<.7  -
   b   M m cb6 gav lun ade 24m26 e                                                      no dre   P=1.  -
   c   M m cb6 gav tba tum 24m26 e                                                      no dre   P=1.  -
 2267 R f sda eat mgl fib 24m24 e                         .    ±                        2.55mg Z P<.02 -
   a   R f sda eat mam ben 24m24 e                                                      43.7mg * P<.2  -
   b   R f sda eat mgl fba 24m24 e                                                      58.0mg * P<.3  -
   c   R f sda eat mam mal 24m24 e                                                      150.mg * P<.5  -
 2268 R f sda eat mam ben 24m24 e                                      .>              247.mg   P<.4  -
   a   R f sda eat liv rcs 24m24 e                                                      no dre   P=1.  -
   b   R f sda eat mam mal 24m24 e                                                      no dre   P=1.  -
 2269 R m sda eat sub fbs 24m24 e                       .    +       .                  5.04mg Z P<.003 -
 2270 R m sda eat sub scs 24m24 e                                   .    +      .       266.mg   P<.009 -

FERRIC CHLORIDE    100ng..:..1ug....:..10......:..100....:..1mg....:..10......:..100....:..1g.......:..10
 2271 R f f3d wat liv nnd 24m26 e                                               .>      no dre   P=1.  -
   a   R f f3d wat tba tum 24m26 e                                                      633.mg * P<.4  -
 2272 R m f3d wat liv nnd 24m26 e                                               .>      11.5gm * P<.7  -
   a   R m f3d wat tba tum 24m26 e                                                      no dre   P=1.  -

FERRIC DIMETHYLDITHIOCARBAMATE    100ng..:..1ug....:..10......:..100....:..1mg....:..10......:..100.....:..1g......:..10
 2273 M f b6a orl liv hpt 76w76 evx                           .>                        no dre   P=1.  -
   a   M f b6a orl lun ade 76w76 evx                                                    no dre   P=1.
   b   M f b6a orl tba mix 76w76 evx                                                    no dre   P=1.
 2274 M m b6a orl lun ade 76w76 evx                         .>                          11.2mg   P<.4  -
   a   M m b6a orl liv hpt 76w76 evx                                                    no dre   P=1.
   b   M m b6a orl tba mix 76w76 evx                                                    21.6mg   P<.7  -
 2275 M f b6c orl lun ade 76w76 evx                          .>                         24.9mg   P<.3  -
   a   M f b6c orl liv hpt 76w76 evx                                                    no dre   P=1.
   b   M f b6c orl tba mix 76w76 evx                                                    12.0mg   P<.09 -
 2276 M m b6c orl liv hpt 76w76 evx                      .    ±                         5.57mg   P<.02 -
   a   M m b6c orl lun ade 76w76 evx                                                    11.9mg   P<.1  -
   b   M m b6c orl tba mix 76w76 evx                                                    2.81mg   P<.002 -

FLECAINIDE ACETATE    100ng..:..1ug....:..10......:..100....:..1mg....:..10......:..100....:..1g......:..10
 2277 M f cd1 eat lun ade 78w78 e                                      .>              938.mg * P<.7  -
   a   M f cd1 eat liv ade 78w78 e                                                      no dre   P=1.  -
 2278 M m cd1 eat lun ade 78w78 e                                  .>                   no dre   P=1.  -
```

	RefNum	LoConf	UpConf	Cntrl	1Dose	1Inc	2Dose	2Inc				Citation or Pathology
												Brkly Code
c	TR223	2.58gm	n.s.s.	0/50	353.mg	0/50	713.mg	3/50				S
d	TR223	464.mg	n.s.s.	27/50	353.mg	36/50	713.mg	32/50				
e	TR223	208.mg	124.gm	14/50	353.mg	28/50	(713.mg	18/50)				liv:hpa,hpc,nnd.
f	TR223	1.77gm	n.s.s.	13/50	353.mg	8/50	713.mg	9/50				lun:a/a,a/c.
2254	1582	370.mg	n.s.s.	1/30	176.mg	1/30						Miller;canr,43,1124-1134;1983
a	1582	612.mg	n.s.s.	0/30	176.mg	0/30						
2255	TR223	723.mg	n.s.s.	33/40	294.mg	41/50	619.mg	38/50				
a	TR223	n.s.s.	n.s.s.	0/40	294.mg	0/50	619.mg	0/50				liv:hpa,hpc,nnd.
2256	TR223	158.mg	n.s.s.	0/40	118.mg	5/50	(238.mg	2/50)				lun:a/a,a/c. S
a	TR223	62.8mg	n.s.s.	30/40	118.mg	40/50	(238.mg	33/50)				
b	TR223	1.75gm	n.s.s.	2/40	118.mg	0/50	238.mg	1/50				liv:hpa,hpc,nnd.

FENAMINOSULF, FORMULATED (methyl orange B) 140-56-7

	RefNum	LoConf	UpConf	Cntrl	1Dose	1Inc	2Dose	2Inc				Citation or Pathology
2257	TR101	47.6mg	n.s.s.	2/100	54.3mg	4/50	(103.mg	3/49)				liv:hpa,hpc. S
a	TR101	82.6mg	n.s.s.	26/100	54.3mg	10/50	103.mg	16/49				
b	TR101	47.6mg	n.s.s.	2/100	54.3mg	4/50	(103.mg	3/49)				liv:hpa,hpc,nnd.
c	TR101	210.mg	n.s.s.	8/100	54.3mg	3/50	103.mg	4/49				lun:a/a,a/c.
2258	TR101	137.mg	n.s.s.	20/50	94.7mg	20/50	189.mg	3/50				
a	TR101	265.mg	n.s.s.	15/50	94.7mg	9/50	189.mg	2/50				liv:hpa,hpc,nnd.
b	TR101	249.mg	n.s.s.	7/50	94.7mg	7/50	189.mg	1/50				lun:a/a,a/c.
2259	TR101	36.7mg	n.s.s.	31/50	19.3mg	33/50	34.4mg	30/50				
a	TR101	325.mg	n.s.s.	1/50	19.3mg	1/50	34.4mg	0/50				liv:hpa,hpc,nnd.
2260	TR101	19.3mg	n.s.s.	31/50	15.4mg	18/50	(27.5mg	17/50)				
a	TR101	n.s.s.	n.s.s.	0/50	15.4mg	1/50	27.5mg	0/50				liv:hpa,hpc,nnd.

FENTHION 55-38-9

	RefNum	LoConf	UpConf	Cntrl	1Dose	1Inc	2Dose	2Inc				Citation or Pathology
2261	TR103	2.57mg	n.s.s.	12/25	1.30mg	22/50	2.60mg	24/50				
a	TR103	11.0mg	n.s.s.	2/25	1.30mg	4/50	2.60mg	2/50				liv:hpa,hpc,nnd.
b	TR103	11.2mg	n.s.s.	3/25	1.30mg	3/50	2.60mg	3/50				lun:a/a,a/c.
2262	TR103	3.34mg	n.s.s.	0/25	1.20mg	7/50	2.40mg	8/50				ski:fbs,srn; sub:fbs,rhb,srn.
a	TR103	.500mg	n.s.s.	10/25	1.20mg	35/50	(2.40mg	26/50)				
b	TR103	2.37mg	n.s.s.	6/25	1.20mg	17/50	2.40mg	17/50				liv:hpa,hpc,nnd.
c	TR103	4.15mg	n.s.s.	2/25	1.20mg	5/50	2.40mg	8/50				lun:a/a,a/c.
2263	TR103	.581mg	n.s.s.	18/25	.500mg	44/50	1.00mg	39/50				
a	TR103	9.77mg	n.s.s.	1/25	.500mg	0/50	1.00mg	0/50				liv:hpa,hpc,nnd.
2264	TR103	1.04mg	n.s.s.	17/25	.400mg	32/50	.800mg	26/50				
a	TR103	2.82mg	n.s.s.	0/25	.400mg	2/50	.800mg	1/50				liv:hpa,hpc,nnd.

FENVALERATE (cyano-(3-phenoxyphenyl)methyl-4-chloro-alpha-(1-methylethyl)benzene acetate) 51630-58-1

	RefNum	LoConf	UpConf	Cntrl	1Dose	1Inc	2Dose	2Inc	3Dose	3Inc	4Dose	4Inc	Citation or Pathology
2265	1963	67.5mg	n.s.s.	19/49	26.1mg	24/47	52.1mg	18/41					Cabral;clet,49,13-18;1990/pers.comm.
a	1963	555.mg	n.s.s.	2/49	26.1mg	0/47	52.1mg	1/41					
b	1963	193.mg	n.s.s.	2/49	26.1mg	0/47	52.1mg	0/41					
c	1963	62.6mg	n.s.s.	22/49	26.1mg	27/47	52.1mg	20/41					
2266	1963	125.mg	n.s.s.	4/48	26.1mg	5/47	52.1mg	10/48					
a	1963	93.6mg	n.s.s.	15/48	26.1mg	17/47	52.1mg	17/48					
b	1963	385.mg	n.s.s.	2/48	26.1mg	3/47	52.1mg	1/48					
c	1963	88.7mg	n.s.s.	25/48	26.1mg	22/47	52.1mg	24/48					
2267	1650m	.882mg	n.s.s.	0/102	50.0ug	1/51	.250mg	3/51	(1.25mg	1/51	12.5mg	0/48)	Parker;jtxe,13,83-97;1984
a	1650m	14.0mg	n.s.s.	25/102	50.0ug	16/51	.250mg	18/51	1.25mg	21/51	12.5mg	20/48	
b	1650m	15.8mg	n.s.s.	25/102	50.0ug	18/51	.250mg	18/51	1.25mg	19/51	12.5mg	19/48	
c	1650m	26.2mg	n.s.s.	18/102	50.0ug	9/51	.250mg	11/51	1.25mg	9/51	12.5mg	11/48	
2268	1650n	60.6mg	n.s.s.	16/50	50.0mg	20/49							
a	1650n	505.mg	n.s.s.	1/50	50.0mg	0/49							
b	1650n	179.mg	n.s.s.	10/50	50.0mg	7/49							
2269	1650m	2.18mg	38.3mg	0/103	40.0ug	1/51	.200mg	3/51	1.00mg	5/51	(10.0mg	0/51)	
2270	1650n	101.mg	6.25gm	0/50	40.0mg	5/51							

FERRIC CHLORIDE 7705-08-0

	RefNum	LoConf	UpConf	Cntrl	1Dose	1Inc	2Dose	2Inc				Citation or Pathology
2271	2090	3.82gm	n.s.s.	2/50	174.mg	0/49	312.mg	1/48				Sato;fctx,30,837-842;1992
a	2090	161.mg	n.s.s.	40/50	174.mg	39/49	312.mg	42/48				
2272	2090	1.60gm	n.s.s.	2/50	158.mg	2/48	297.mg	3/49				
a	2090	28.6mg	n.s.s.	50/50	158.mg	47/48	297.mg	49/49				

FERRIC DIMETHYLDITHIOCARBAMATE (ferbam) 14484-64-1

	RefNum	LoConf	UpConf	Cntrl	1Dose	1Inc						Citation or Pathology
2273	1209	8.21mg	n.s.s.	0/17	4.39mg	0/17						Innes;ntis,1968/1969
a	1209	8.21mg	n.s.s.	1/17	4.39mg	0/17						
b	1209	5.43mg	n.s.s.	2/17	4.39mg	1/17						
2274	1209	2.34mg	n.s.s.	2/18	4.08mg	4/18						
a	1209	4.82mg	n.s.s.	1/18	4.08mg	1/18						
b	1209	2.54mg	n.s.s.	3/18	4.08mg	4/18						
2275	1209	4.05mg	n.s.s.	0/16	4.39mg	1/16						
a	1209	7.73mg	n.s.s.	0/16	4.39mg	0/16						
b	1209	2.95mg	n.s.s.	0/16	4.39mg	2/16						
2276	1209	1.91mg	n.s.s.	0/16	4.08mg	4/17						
a	1209	2.93mg	n.s.s.	0/16	4.08mg	2/17						
b	1209	1.20mg	10.8mg	0/16	4.08mg	7/17						

FLECAINIDE ACETATE (2,5-bis(2,2,2-trifluorethoxy)-N-(2-piperidylmethyl)benzamide acetate) 54143-56-5

	RefNum	LoConf	UpConf	Cntrl	1Dose	1Inc	2Dose	2Inc	3Dose	3Inc		Citation or Pathology
2277	1649n	136.mg	n.s.s.	5/69	15.0mg	5/69	30.0mg	7/68	60.0mg	6/67		Case;txap,73,232-242;1984
a	1649n	68.0mg	n.s.s.	1/69	15.0mg	0/69	30.0mg	0/68	60.0mg	0/67		
2278	1649n	165.mg	n.s.s.	10/67	15.0mg	7/69	30.0mg	7/69	60.0mg	7/60		

```
    Spe Strain Site   Xpo+Xpt                                                    TD50    2Tailpvl
       Sex  Route  Hist    Notes                                                         DR    AuOp
a     M m cd1 eat liv ade 78w78 e                                              no dre   P=1.  -
b     M m cd1 eat liv hpc 78w78 e                                              no dre   P=1.  -
2279  R f crw eat liv ade 24m24 e                                   .>         no dre   P=1.  -
2280  R m crw eat tes ica 24m24 e                        . +    .              79.1mg * P<.004 -
a     R m crw eat liv ade 24m24 e                                              3.58gm * P<.2  -

FLUOMETURON                    100ng..:..1ug....:..10.....:..100....:..1mg....:..10......:..100.....:..1g......:..10
2281  M f b6c eat TBA MXB 24m24                                       :>       369.mg * P<.2
a     M f b6c eat liv MXB 24m24                                               1.23gm * P<.4
b     M f b6c eat lun MXB 24m24                                                no dre   P=1.
2282  M m b6c eat liv MXA 24m24                              :  ±              229.mg * P<.06 a
a     M m b6c eat TBA MXB 24m24                                               265.mg * P<.3
b     M m b6c eat liv MXB 24m24                                               229.mg * P<.06
c     M m b6c eat lun MXB 24m24                                              1.42gm * P<.6
2283  R f f34 eat TBA MXB 24m24 v                         :>                   no dre   P=1.  -
a     R f f34 eat liv MXB 24m24 v                                              no dre   P=1.
2284  R m f34 eat liv nnd 24m24 v                           :  ±              #55.4mg * P<.04 -
a     R m f34 eat TBA MXB 24m24 v                                              no dre   P=1.
b     R m f34 eat liv MXB 24m24 v                                             55.4mg * P<.04

N-(2-FLUORENYL)-2,2,2-TRIFLUOROACETAMIDE.:...1_ug....:..10.....:..100....:..1mg....:..10......:..100.....:..1g......:..10
2285  R f buf eat mgl adc 53w54 e                    . +   .                  1.62mg   P<.0005+
a     R f buf eat liv hpt 53w54 e                                             1.90mg   P<.0005+
b     R f buf eat edu sqc 53w54 e                                             4.94mg   P<.002 +

FLUORIDE, SODIUM               100ng..:..1ug....:..10.....:..100....:..1mg....:..10......:..100.....:..1g......:..10
2286  M f b6c wat TBA MXB 24m24                                 :>             no dre   P=1.  -
a     M f b6c wat liv MXB 24m24                                                no dre   P=1.
b     M f b6c wat lun MXB 24m24                                                no dre   P=1.
2287  M m b6c wat TBA MXB 24m24                                   :>           no dre   P=1.  -
a     M m b6c wat liv MXB 24m24                                                no dre   P=1.
b     M m b6c wat lun MXB 24m24                                                no dre   P=1.
2288  M b cd1 wat lun tum 30m30 e                        .>                    no dre   P=1.  -
a     M b cd1 wat liv tum 30m30 e                                              no dre   P=1.
b     M b cd1 wat tba tum 30m30 e                                              no dre   P=1.
c     M b cd1 wat tba mal 30m30 e                                              no dre   P=1.
2289  M f dbx eat mgl car 90w90 er                             .>             no dre   P=1.  -
2290  R f f34 wat TBA MXB 24m24                               :>              73.9mg * P<.8
a     R f f34 wat liv MXB 24m24                                                no dre   P=1.
2291  R m f34 wat MXA ost 24m24                            :  +         :     82.7mg * P<.009
a     R m f34 wat MXA ost 24m24                                              109.mg * P<.03 e
b     R m f34 wat TBA MXB 24m24                                              147.mg * P<.1
c     R m f34 wat liv MXB 24m24                                               no dre   P=1.

4'-FLUORO-4-AMINODIPHENYL      100ng..:..1ug....:..10.....:..100....:..1mg....:..10......:..100.....:..1g......:..10
2292  M f cba gav liv hpt  6m26 e                    .+  .                    1.09mg   P<.0005+
2293  M m cba gav liv hpt  6m24 e                    . +     .                1.19mg   P<.005 +
a     M m cba gav liv mix  6m24 e                                            1.19mg   P<.005

N-4-(4'-FLUOROBIPHENYL)ACETAMIDE  100ng..:..1ug....:..10.....:..100....:..1mg....:..10......:..100.....:..1g......:..10
2294  R m f34 eat kid adc 52w52 ekr                  . +   .                 1.01mg   P<.0005+

2-FLUOROETHYL-NITROSOUREA      100ng..:..1ug....:..10.....:..100....:..1mg....:..10......:..100.....:..1g......:..10
2295  R m f34 gav for tum  9m24                  .  ±                         .125mg   P<.02 +
a     R m f34 gav lun mix  9m24                                               .279mg   P<.09
b     R m f34 gav liv tum  9m24                                               no dre   P=1.

5-FLUOROURACIL                 100ng..:..1ug....:..10.....:..100....:..1mg....:..10......:..100.....:..1g......:..10
2296  M f bal ipj lmr mix 50w87 e                        . +   .             2.83mg   P<.002 +
a     M f bal ipj lun mix 50w87 e                                            7.46mg   P<.2  +
b     M f bal ipj lun sqc 50w87 e                                           58.5mg   P<.3
c     M f bal ipj tba tum 50w87 e                                            2.19mg   P<.003
2297  M m bal ipj lun ade 50w87 e                        . ±                 3.11mg   P<.02 +
a     M m bal ipj tba tum 50w87 e                                            2.73mg   P<.008
2298  R m b46 ivj liv ade 12m24 es                           .     ±         34.7mg   P<.1  -
a     R m b46 ivj tba ben 12m24 es                                           16.3mg   P<.2  -
b     R m b46 ivj tba mix 12m24 es                                           18.6mg   P<.4  -
c     R m b46 ivj tba mal 12m24 es                                           no dre   P=1.  -

FLUOXETINE.HCl                 100ng..:..1ug....:..10.....:..100....:..1mg....:..10......:..100.....:..1g......:..10
2299  M f b6c eat lun a/a 24m24 e                              .>             no dre   P=1.  -
a     M f b6c eat liv mix 24m24 e                                             no dre   P=1.  -
2300  M f b6c eat liv mix 24m24 e                             .>             239.mg * P<.4  -
a     M f b6c eat lun mix 24m24 e                                             no dre   P=1.  -
2301  M m b6c eat lun mix 24m24 e                          . ±               19.2mg Z P<.07 -
a     M m b6c eat liv mix 24m24 e                                             no dre   P=1.  -
2302  M m b6c eat lun mix 24m24 e                             .>             37.1mg Z P<.3  -
a     M m b6c eat liv mix 24m24 e                                             no dre   P=1.  -
2303  R f sda eat liv hpa 24m24 e                         .>                  no dre   P=1.  -
2304  R m sda eat liv hpa 24m24 e                          .     ±           207.mg * P<.04 -
```

	RefNum	LoConf	UpConf	Cntrl	1Dose	1Inc	2Dose	2Inc			Citation or Pathology
											Brkly Code
a	1649n	269.mg	n.s.s.	10/67	15.0mg	4/69	30.0mg	5/69	60.0mg	4/60	
b	1649n	584.mg	n.s.s.	2/67	15.0mg	0/69	30.0mg	1/69	60.0mg	0/60	
2279	1649m	820.mg	n.s.s.	2/49	15.0mg	0/50	30.0mg	1/50	60.0mg	0/50	
2280	1649m	40.8mg	561.mg	19/50	15.0mg	16/50	30.0mg	23/50	60.0mg	32/50	
a	1649m	583.mg	n.s.s.	0/50	15.0mg	0/50	30.0mg	0/50	60.0mg	1/50	

FLUOMETURON   2164-17-2

	RefNum	LoConf	UpConf	Cntrl	1Dose	1Inc	2Dose	2Inc			Citation or Pathology
2281	TR195	127.mg	n.s.s.	9/25	64.4mg	15/50	129.mg	23/50			
a	TR195	324.mg	n.s.s.	1/25	64.4mg	3/50	129.mg	4/50			liv:hpa,hpc,nnd.
b	TR195	636.mg	n.s.s.	1/25	64.4mg	2/50	129.mg	1/50			lun:a/a,a/c.
2282	TR195	103.mg	n.s.s.	4/25	59.4mg	13/50	118.mg	21/50			liv:hpa,hpc.
a	TR195	80.2mg	n.s.s.	9/25	59.4mg	27/50	118.mg	29/50			
b	TR195	103.mg	n.s.s.	4/25	59.4mg	13/50	118.mg	21/50			liv:hpa,hpc,nnd.
c	TR195	262.mg	n.s.s.	2/25	59.4mg	4/50	118.mg	6/50			lun:a/a,a/c.
2283	TR195	16.0mg	n.s.s.	45/50	6.10mg	41/50	12.4mg	42/50			
a	TR195	77.6mg	n.s.s.	3/50	6.10mg	3/50	12.4mg	1/50			liv:hpa,hpc,nnd.
2284	TR195	21.1mg	n.s.s.	0/50	5.00mg	1/50	9.90mg	4/50			S
a	TR195	9.84mg	n.s.s.	30/50	5.00mg	24/50	9.90mg	35/50			
b	TR195	21.1mg	n.s.s.	0/50	5.00mg	1/50	9.90mg	4/50			liv:hpa,hpc,nnd.

N-(2-FLUORENYL)-2,2,2-TRIFLUOROACETAMIDE   363-17-7

	RefNum	LoConf	UpConf	Cntrl	1Dose	1Inc					Citation or Pathology
2285	144	.800mg	3.71mg	0/18	13.2mg	14/18					Morris;jnci,24,149-180;1960
a	144	.940mg	4.44mg	0/18	13.2mg	13/18					
b	144	2.11mg	18.4mg	0/18	13.2mg	7/18					

FLUORIDE, SODIUM   7681-49-4

	RefNum	LoConf	UpConf	Cntrl	1Dose	1Inc	2Dose	2Inc			Citation or Pathology
2286	TR393	38.0mg	n.s.s.	72/80	4.90mg	41/52	19.6mg	37/50	34.3mg	61/80	
a	TR393	59.1mg	n.s.s.	55/80	4.90mg	33/52	19.6mg	26/50	34.3mg	41/80	liv:hpa,hpc,nnd.
b	TR393	138.mg	n.s.s.	8/80	4.90mg	6/52	19.6mg	4/50	34.3mg	7/80	lun:a/a,a/c.
2287	TR393	33.2mg	n.s.s.	71/79	4.08mg	45/50	16.3mg	43/51	28.6mg	69/80	
a	TR393	36.2mg	n.s.s.	62/79	4.08mg	39/50	16.3mg	37/51	28.6mg	61/80	liv:hpa,hpc,nnd.
b	TR393	96.2mg	n.s.s.	23/79	4.08mg	13/50	16.3mg	13/51	28.6mg	18/80	lun:a/a,a/c.
2288	1036	11.0mg	n.s.s.	15/71	1.75mg	12/72					Kanisawa;canr,29,892-895;1969
a	1036	15.3mg	n.s.s.	4/71	1.75mg	4/72					
b	1036	7.10mg	n.s.s.	24/71	1.75mg	22/72					
c	1036	17.4mg	n.s.s.	8/71	1.75mg	5/72					
2289	1652	276.mg	n.s.s.	37/50	117.mg	20/48					Tannenbaum;canr,9,403-410;1949
2290	TR393	7.56mg	n.s.s.	73/80	1.41mg	41/50	5.63mg	48/50	9.86mg	71/81	
a	TR393	83.0mg	n.s.s.	2/80	1.41mg	0/50	5.63mg	1/50	9.86mg	1/81	liv:hpa,hpc,nnd.
2291	TR393	30.6mg	2.86gm	0/80	1.23mg	0/51	4.92mg	1/49	8.63mg	4/80	cyx:ost; hum:ost; sub:ost; ver:ost. S
a	TR393	36.4mg	n.s.s.	0/80	1.23mg	0/51	4.92mg	1/49	8.63mg	3/80	cyx:ost; hum:ost; ver:ost.
b	TR393	6.37mg	n.s.s.	71/80	1.23mg	48/51	4.92mg	42/49	8.63mg	74/80	
c	TR393	53.0mg	n.s.s.	1/80	1.23mg	1/51	4.92mg	1/49	8.63mg	1/80	liv:hpa,hpc,nnd.

4'-FLUORO-4-AMINODIPHENYL   324-93-6

	RefNum	LoConf	UpConf	Cntrl	1Dose	1Inc					Citation or Pathology
2292	1165	.663mg	1.96mg	0/18	1.30mg	25/40					Clayson;bjca,19,297-310;1965/Williams 1962
2293	1165	.601mg	8.97mg	2/15	1.19mg	18/32					
a	1165	.601mg	8.97mg	2/15	1.19mg	18/32					

N-4-(4'-FLUOROBIPHENYL)ACETAMIDE   398-32-3

	RefNum	LoConf	UpConf	Cntrl	1Dose	1Inc					Citation or Pathology
2294	1150	.392mg	2.58mg	0/15	16.0mg	14/15					Hinton;bect,23,464-469;1979

2-FLUOROETHYL-NITROSOUREA   69112-98-7

	RefNum	LoConf	UpConf	Cntrl	1Dose	1Inc					Citation or Pathology
2295	1915	42.8ug	n.s.s.	0/12	74.2ug	4/12					Lijinsky;txih,5,925-935;1989
a	1915	68.3ug	n.s.s.	0/12	74.2ug	2/12					
b	1915	.184mg	n.s.s.	1/12	74.2ug	0/12					

5-FLUOROURACIL   (fluracil) 51-21-8

	RefNum	LoConf	UpConf	Cntrl	1Dose	1Inc					Citation or Pathology
2296	1964	1.46mg	11.2mg	6/50	2.46mg	21/50					Cavaliere;tumo,76,179-181;1990
a	1964	2.43mg	n.s.s.	9/50	2.46mg	15/50					
b	1964	9.52mg	n.s.s.	0/50	2.46mg	1/50					
c	1964	1.09mg	14.2mg	14/50	2.46mg	29/50					
2297	1964	1.42mg	n.s.s.	12/50	2.46mg	24/50					
a	1964	1.28mg	69.1mg	13/50	2.46mg	26/50					
2298	1017	5.65mg	n.s.s.	0/65	2.36mg	1/22					Schmahl;arzn,20,1461-1467;1970
a	1017	3.65mg	n.s.s.	3/65	2.36mg	3/22					
b	1017	3.37mg	n.s.s.	7/65	2.36mg	4/22					
c	1017	6.96mg	n.s.s.	4/65	2.36mg	1/22					

FLUOXETINE.HCl   (prozac) 59333-67-4

	RefNum	LoConf	UpConf	Cntrl	1Dose	1Inc	2Dose	2Inc			Citation or Pathology
2299	2045m	86.9mg	n.s.s.	2/59	1.30mg	2/59	5.20mg	2/60	13.0mg	2/60	Bendele;canr,52,6931-6935;1992/pers.comm.
a	2045n	101.mg	n.s.s.	2/59	1.30mg	2/59	5.20mg	3/60	13.0mg	1/60	
2300	2045n	54.9mg	n.s.s.	4/59	1.30mg	1/60	5.20mg	3/61	13.0mg	5/59	
a	2045n	157.mg	n.s.s.	3/59	1.30mg	3/60	5.20mg	0/61	13.0mg	1/59	
2301	2045m	7.34mg	n.s.s.	9/60	1.20mg	11/60	4.80mg	17/60	(12.0mg	3/59)	
a	2045m	51.1mg	n.s.s.	16/60	1.20mg	12/60	4.80mg	17/60	12.0mg	9/59	
2302	2045n	10.2mg	n.s.s.	5/60	1.20mg	10/60	4.80mg	10/59	(12.0mg	2/61)	
a	2045n	66.9mg	n.s.s.	16/60	1.20mg	11/60	4.80mg	11/59	12.0mg	8/61	
2303	2045m	4.86mg	n.s.s.	1/60	.500mg	0/60	2.25mg	0/60	10.0mg	0/60	
2304	2045m	50.9mg	n.s.s.	0/60	.400mg	0/60	1.80mg	0/60	8.00mg	2/60	

```
     Spe Strain Site   Xpo+Xpt                                                                TD50    2Tailpvl
        Sex  Route  Hist     Notes                                                                  DR    AuOp
FORMALDEHYDE                       100ng..:..1ug....:..10.....:..100....:..1mg...:..10....:..100....:..1g......:..10
2305 H m syg inh res tum 94w94 r                                           .>                        no dre  P=1.    -
2306 H m syg inh res tum 25m25 rs                                               .>                   no dre  P=1.    -
2307 M f b6c inh nas tum 52w52 ek                      .>                                            no dre  P=1.
  a  M f b6c inh lun tum 52w52 ek                                                                    no dre  P=1.
  b  M f b6c inh liv tum 52w52 ek                                                                    no dre  P=1.
2308 M f b6c inh nas tum 78w78 ek                          .>                                        no dre  P=1.
  a  M f b6c inh liv mix 78w78 ek                                                                    no dre  P=1.
  b  M f b6c inh lun tum 78w78 ek                                                                    no dre  P=1.
2309 M f b6c inh lun mix 24m24 ek                                .>                                  47.3mg  P<.3
  a  M f b6c inh liv mix 24m24 ek                                    .>                              98.7mg  P<.6
  b  M f b6c inh nas tum 24m24 ek                                                                    no dre  P=1.
2310 M f b6c inh nas tum 24m27 e                                 .>                                  no dre  P=1.
  a  M f b6c inh lun mix 24m27 e                                                                     no dre  P=1.
  b  M f b6c inh liv tum 24m27 e                                                                     no dre  P=1.
2311 M m b6c inh liv hpc 52w52 ek                              .>                                    7.85mg  P<.3
  a  M m b6c inh lun tum 52w52 ek                                                                    no dre  P=1.
  b  M m b6c inh nas tum 52w52 ek                                                                    no dre  P=1.
2312 M m b6c inh ntu sqc 24m24 es                                   .   ±                            43.9mg * P<.04  +
  a  M m b6c inh liv mix 24m24 es                                                                    34.5mg  P<.2
  b  M m b6c inh lun mix 24m24 es                                                                    no dre  P=1.
2313 R f f34 inh nas tum 52w52 ek                   .>                                               no dre  P=1.
  a  R f f34 inh liv tum 52w52 ek                                                                    no dre  P=1.
2314 R f f34 inh ntu sqc 78w78 ek                             .   +      .                           3.67mg * P<.005 +
  a  R f f34 inh liv nnd 78w78 ek                                                                    192.mg  P<1.
  b  R f f34 inh ntu pla 78w78 ek                                                                    no dre  P=1.
2315 R f f34 inh ntu sqc 25m25 s                         :  +  :                                     1.37mg Z P<.0005+
  a  R f f34 inh ntu pla 25m25 s                                                                     7.95mg Z P<.4
  b  R f f34 inh liv nnd 25m25 es                                                                    no dre  P=1.
2316 R m f34 inh nas tum 52w52 ek                    .>                                              no dre  P=1.
  a  R m f34 inh liv tum 52w52 ek                                                                    no dre  P=1.
2317 R m f34 inh ntu sqc 78w78 ek                             .   +      .                           2.57mg * P<.005 +
  a  R m f34 inh ntu pla 78w78 ek                                                                    5.39mg * P<.2
  b  R m f34 inh liv tum 78w78 ek                                                                     no dre  P=1.
2318 R m f34 inh ntu sqc 24m24 s                         :  +:                                       .798mg Z P<.0005+
  a  R m f34 inh ntu pla 24m24 s                                                                     3.01mg * P<.03
  b  R m f34 inh liv mix 24m24 es                                                                    no dre  P=1.
2319 R f sda wat --- leu 24m34 er                                          .    ±                    815.mg * P<.04  +

  a  R f sda wat --- lls 24m34 er                                                                    996.mg * P<.03  +

  b  R f sda wat git mix 24m34 er                                                                    2.96gm * P<.2   +

2320 R m sda wat --- lls 24m34 er                                          .  +       .              424.mg * P<.0005+

  a  R m sda wat --- leu 24m34 er                                                                    480.mg * P<.01  +

  b  R m sda wat git mix 24m34 er                                                                    1.41gm * P<.02  +

2321 R m sda inh nac sqc 28m28 e                               .  +  .                               1.82mg  P<.0005+
  a  R m sda inh nac car 28m28 e                                                                     86.4mg  P<.3    +
  b  R m sda inh nac fbs 28m28 e                                                                     86.4mg  P<.3    +
  c  R m sda inh liv hpc 28m28 e                                                                     no dre  P=1.
2322 R f wal wat for exp 24m24 e                                       .>                            no dre  P=1.    -
  a  R f wal wat stg tum 24m24 e                                                                     no dre  P=1.    -
  b  R f wal wat liv hpc 24m24                                                                       no dre  P=1.    -
  c  R f wal wat pit het 24m24                                                                       no dre  P=1.    -
  d  R f wal wat tba mix 24m24                                                                       no dre  P=1.    -
  e  R f wal wat tba ben 24m24                                                                       no dre  P=1.    -
  f  R f wal wat tba mal 24m24                                                                       no dre  P=1.    -
2323 R m wal wat stg tum 24m24 e                                   .>                                no dre  P=1.    -
  a  R m wal wat for exp 24m24 e                                                                     no dre  P=1.    -
  b  R m wal wat liv hpc 24m24                                                                       no dre  P=1.    -
  c  R m wal wat tba mix 24m24                                                                       no dre  P=1.    -
  d  R m wal wat tba ben 24m24                                                                       no dre  P=1.    -
  e  R m wal wat tba mal 24m24                                                                       no dre  P=1.    -
2324 R m wal inh nac sqc 28m29 r                                       .>                            42.9mg * P<.7   -


FORMIC ACID 2-[4-(2-FURYL)-2-THIAZOLYL]HYDRAZIDE...:..1_0......:..100...:..1mg...:..10....:..100....:..1g......:..10
2325 R f sda eat liv tum 46w64 ev                                          .>                        no dre  P=1.
  a  R f sda eat tba mix 46w64 ev                                                                    no dre  P=1.    -


FORMIC ACID 2-(4-METHYL-2-THIAZOLYL)HYDRAZIDE     ...:..10.........:..100...:..1mg...:..10....:..100....:..1g......:..10
2326 R f sda eat mgl adf 46w64 ev                              .   ±                                 14.4mg  P<.03   +
  a  R f sda eat liv tum 46w64 ev                                                                    no dre  P=1.
  b  R f sda eat tba mix 46w64 ev                                                                    14.4mg  P<.03


FORMIC ACID 2-[4-(5-NITRO-2-FURYL)-2-THIAZOLYL]HYDRAZIDE...:..1_00...:..1mg...:..10....:..100....:..1g......:..10
2327 H m syg eat ubl tcc 48w70 e                              .  +  .                                16.6mg  P<.0005+
  a  H m syg eat for sqp 48w70 e                                                                     25.1mg  P<.0005+
  b  H m syg eat liv tum 48w70 e                                                                     no dre  P=1.
```

RefNum	LoConf	UpConf	Cntrl	1Dose	1Inc	2Dose	2Inc						Citation or Pathology
													Brkly Code
FORMALDEHYDE	50-00-0												
2305	1414m	6.50mg n.s.s.	0/50	.772mg	0/50								Dalbey;txcy,24,9-14;1982
2306	1414n	25.6mg n.s.s.	0/132	1.29mg	0/88								
2307	1566a	.271mg n.s.s.	0/10	.772mg	0/10	2.32mg	0/10	5.79mg	0/10				Pavkov;ciit;1981/Kerns 1983
a	1566a	2.98mg n.s.s.	0/10	5.79mg	0/10								
b	1566a	2.98mg n.s.s.	0/10	5.79mg	0/10								
2308	1566b	1.21mg n.s.s.	0/20	.772mg	0/20	2.32mg	0/20	5.79mg	0/19				
a	1566b	8.42mg n.s.s.	2/20	5.79mg	1/19								
b	1566b	12.8mg n.s.s.	0/20	5.79mg	0/19								
2309	1566c	10.9mg n.s.s.	1/30	5.79mg	3/27								
a	1566c	14.4mg n.s.s.	1/30	5.79mg	2/28								
b	1566c	3.10mg n.s.s.	0/30	.772mg	0/26	2.32mg	0/41	5.79mg	0/28				
2310	1566m	6.06mg n.s.s.	0/50	.686mg	0/54	2.06mg	0/39	5.15mg	0/54				
a	1566m	44.5mg n.s.s.	1/50	5.15mg	1/54								
b	1566m	72.5mg n.s.s.	0/50	5.15mg	0/54								
2311	1566a	1.28mg n.s.s.	0/10	4.83mg	0/10								
a	1566a	2.49mg n.s.s.	0/10	4.83mg	0/10								
b	1566a	.226mg n.s.s.	0/10	.644mg	0/10	1.93mg	0/10	4.83mg	0/10				
2312	1566c	10.8mg n.s.s.	0/20	.644mg	0/22	1.93mg	0/19	4.83mg	2/17				
a	1566c	9.67mg n.s.s.	4/62	4.83mg	6/40								
b	1566c	6.53mg n.s.s.	6/25	4.83mg	4/18								
2313	1566n	64.6ug n.s.s.	0/10	.184mg	0/10	.552mg	0/10	1.38mg	0/10				
a	1566n	.710mg n.s.s.	0/10	1.38mg	0/10								
2314	1566o	1.27mg 33.9mg	0/20	.184mg	0/20	.552mg	0/20	1.38mg	4/19				
a	1566o	1.81mg n.s.s.	1/20	1.38mg	1/19								
b	1566o	3.23mg n.s.s.	0/20	.184mg	1/20	.552mg	0/20	1.38mg	0/19				
2315	1566r	.952mg 2.06mg	0/78	.184mg	0/79	.552mg	1/76	1.38mg	48/73				
a	1566r	1.25mg n.s.s.	0/78	.184mg	3/79	.552mg	0/76	1.38mg	1/73				
b	1566r	4.13mg n.s.s.	1/67	1.38mg	0/14								
2316	1566n	45.2ug n.s.s.	0/10	.129mg	0/10	.386mg	0/10	.965mg	0/10				
a	1566n	.497mg n.s.s.	0/10	.965mg	0/10								
2317	1566o	.886mg 23.8mg	0/20	.129mg	0/20	.386mg	0/20	.965mg	4/19				
a	1566o	1.32mg n.s.s.	0/20	.129mg	0/20	.386mg	1/20	.965mg	1/19				
b	1566o	2.24mg n.s.s.	0/20	.965mg	0/20								
2318	1566r	.510mg 1.23mg	0/79	.129mg	0/80	.386mg	1/79	.965mg	47/73				
a	1566r	1.10mg n.s.s.	1/79	.129mg	4/80	.386mg	5/79	.965mg	3/73				
b	1566r	2.85mg n.s.s.	4/74	.965mg	1/22								
2319	BT7001	316.mg n.s.s.	3/50	.410mg	2/50	2.05mg	4/50	4.10mg	4/50	20.5mg	4/50	41.0mg	7/50
			61.5mg	7/50									Soffritti;txih,5,699-730;1989
a	BT7001	386.mg n.s.s.	1/50	.410mg	1/50	2.05mg	3/50	4.10mg	2/50	20.5mg	2/50	41.0mg	5/50
			61.5mg	5/50									
b	BT7001	753.mg n.s.s.	0/50	.410mg	1/50	2.05mg	2/50	4.10mg	0/50	20.5mg	0/50	41.0mg	1/50
			61.5mg	3/50									
2320	BT7001	213.mg 1.73gm	4/50	.359mg	0/50	1.79mg	2/50	3.59mg	4/50	17.9mg	4/50	35.9mg	6/50
			53.8mg	11/50									
a	BT7001	213.mg 63.8gm	5/50	.359mg	1/50	1.79mg	5/50	3.59mg	5/50	17.9mg	8/50	35.9mg	6/50
			53.8mg	11/50									
b	BT7001	517.mg n.s.s.	0/50	.359mg	2/50	1.79mg	0/50	3.59mg	0/50	17.9mg	0/50	35.9mg	1/50
			53.8mg	5/50									
2321	1674	1.22mg 2.86mg	0/99	.952mg	38/100								Sellakumar;txap,81,401-406;1985
a	1674	14.1mg n.s.s.	0/99	.952mg	1/100								
b	1674	14.1mg n.s.s.	0/99	.952mg	1/100								
c	1674	16.0mg n.s.s.	1/99	.952mg	1/100								
2322	1918	16.7mg n.s.s.	1/48	1.80mg	0/49	21.0mg	0/47	109.mg	0/48				Til;fctx,27,77-87;1989
a	1918	16.7mg n.s.s.	0/48	1.80mg	0/49	21.0mg	0/47	109.mg	0/48				
b	1918	17.2mg n.s.s.	0/50	1.80mg	0/50	21.0mg	0/50	109.mg	0/50				
c	1918	410.mg n.s.s.	6/50	1.80mg	11/50	21.0mg	3/50	109.mg	7/50				
d	1918	136.mg n.s.s.	37/50	1.80mg	38/50	21.0mg	32/50	109.mg	34/50				
e	1918	144.mg n.s.s.	34/50	1.80mg	37/50	21.0mg	25/50	109.mg	32/50				
f	1918	418.mg n.s.s.	9/50	1.80mg	9/50	21.0mg	7/50	109.mg	7/50				
2323	1918	10.4mg n.s.s.	0/47	1.20mg	0/45	15.0mg	0/44	82.0mg	0/47				
a	1918	913.mg n.s.s.	0/47	1.20mg	1/45	15.0mg	0/44	82.0mg	0/47				
b	1918	981.mg n.s.s.	0/50	1.20mg	1/50	15.0mg	0/50	82.0mg	0/47				
c	1918	182.mg n.s.s.	33/50	1.20mg	27/50	15.0mg	30/50	82.0mg	24/50				
d	1918	242.mg n.s.s.	28/50	1.20mg	22/50	15.0mg	26/50	82.0mg	18/50				
e	1918	263.mg n.s.s.	8/50	1.20mg	6/50	15.0mg	8/50	82.0mg	7/50				
2324	1926	4.83mg n.s.s.	0/30	6.21ug	1/30	62.1ug	1/30	.621mg	1/30				Woutersen;japt,9,39-46;1989
FORMIC ACID 2-[4-(2-FURYL)-2-THIAZOLYL]HYDRAZIDE				31873-81-1									
2325	1073	52.9mg n.s.s.	0/29	29.5mg	0/23								Erturk;jnci,47,437-445;1971/1970a
a	1073	52.9mg n.s.s.	2/29	29.5mg	0/23								
FORMIC ACID 2-(4-METHYL-2-THIAZOLYL)HYDRAZIDE				32852-21-4									
2326	1073	5.54mg n.s.s.	2/29	14.7mg	8/28								Erturk;jnci,47,437-445;1971/1970a
a	1073	32.1mg n.s.s.	0/29	14.7mg	0/28								
b	1073	5.54mg n.s.s.	2/29	14.7mg	8/28								
FORMIC ACID 2-[4-(5-NITRO-2-FURYL)-2-THIAZOLYL]HYDRAZIDE		(FNT) 3570-75-0											
2327	1077	7.32mg 46.9mg	0/17	63.1mg	9/13								Croft;jnci,51,941-949;1973
a	1077	12.9mg 57.3mg	0/24	63.1mg	13/24								
b	1077	141.mg n.s.s.	0/24	63.1mg	0/24								

	Spe	Sex	Strain	Route	Site	Hist	Xpo+Xpt	Notes	DR	TD50	2Tailpvl AuOp
c	H	m	syg	eat	lun	tum	48w70	e		no dre	P=1.
2328	M	f	swi	eat	for	sqp	33w52	e	. + .	8.85mg	P<.0005+
a	M	f	swi	eat	---	lle	33w52	e		11.4mg	P<.0005+
b	M	f	swi	eat	lun	alc	33w52	e		67.1mg	P<.06
c	M	f	swi	eat	liv	tum	33w52	e		no dre	P=1.
d	M	f	swi	eat	tba	mix	33w52	e		2.42mg	P<.0005
2329	M	f	swi	eat	for	pam	33w52		. + .	13.8mg	P<.0005+
a	M	f	swi	eat	---	leu	33w52			25.6mg	P<.008
b	M	f	swi	eat	liv	tum	33w52			no dre	P=1.
c	M	f	swi	eat	lun	tum	33w52			no dre	P=1.
2330	R	f	buf	eat	mgl	mix	46w64	e	. + .	5.54mg	P<.0005+
a	R	f	buf	eat	mgl	adc	46w64	e		13.1mg	P<.0005+
b	R	f	buf	eat	kid	mix	46w64	e		17.5mg	P<.0005+
c	R	f	buf	eat	liv	cye	46w64	e		39.1mg	P<.0005+
d	R	f	buf	eat	kid	uac	46w64	e		44.1mg	P<.0005+
e	R	f	buf	eat	mgl	adf	46w64	e		80.5mg	P<.003 +
f	R	f	buf	eat	kid	tua	46w64	e		98.6mg	P<.007 +
g	R	f	buf	eat	tba	mix	46w64	e		noTD50	P<.0005
2331	R	f	hza	eat	mam	tum	44w60	es	. + .	5.50mg	P<.0005+
a	R	f	hza	eat	kid	ade	44w60	es		17.6mg	P<.0005+
b	R	f	hza	eat	kid	car	44w60	es		78.5mg	P<.03 +
c	R	f	hza	eat	smi	adc	44w60	es		78.5mg	P<.03 +
d	R	f	hza	eat	eac	sqc	44w60	es		+hist 100.mg	P<.05 +
e	R	f	hza	eat	cec	adc	44w60	es		210.mg	P<.2 +
f	R	f	hza	eat	liv	hms	44w60	es		428.mg	P<.4 -
2332	R	f	sda	eat	mgl	adc	45w75	ev	. + .	3.54mg	P<.0005+
a	R	f	sda	eat	for	sqp	45w75	ev		65.4mg	P<.0005
b	R	f	sda	eat	kid	rcc	45w75	ev		65.4mg	P<.0005
c	R	f	sda	eat	k/p	tcc	45w75	ev		+hist 137.mg	P<.008 +
d	R	f	sda	eat	---	lbl	45w75	ev		+hist 184.mg	P<.03 +
e	R	f	sda	eat	liv	tum	45w75	ev		no dre	P=1. -
f	R	f	sda	eat	tba	mix	45w75	ev		3.79mg	P<.0005
2333	R	f	sda	eat	mgl	mix	46w64	e	. + .	5.85mg	P<.0005+
a	R	f	sda	eat	kid	mix	46w64	e		9.96mg	P<.0005+
b	R	f	sda	eat	mgl	adc	46w64	e		19.5mg	P<.0005+
c	R	f	sda	eat	liv	mix	46w64	e		30.1mg	P<.0005+
d	R	f	sda	eat	kid	uac	46w64	e		33.9mg	P<.0005+
e	R	f	sda	eat	liv	cye	46w64	e		43.9mg	P<.0005+
f	R	f	sda	eat	kid	tua	46w64	e		50.7mg	P<.0005+
g	R	f	sda	eat	mgl	adf	46w64	e		52.8mg	P<.009 +
h	R	f	sda	eat	tba	mix	46w64	e		noTD50	P<.0005
2334	R	m	sda	eat	kid	mix	46w64	er	. + .	5.82mg	P<.0005+
a	R	m	sda	eat	kid	uac	46w64	er		19.3mg	P<.0005+
b	R	m	sda	eat	liv	cye	46w64	e		35.3mg	P<.0005+
c	R	m	sda	eat	mgl	adf	46w64	e		+hist 46.2mg	P<.0005+
d	R	m	sda	eat	kid	tua	46w64	er		47.6mg	P<.002 +
e	R	m	sda	eat	k/p	tcc	46w64	er		69.9mg	P<.006 +
f	R	m	sda	eat	tba	mix	46w64	e		noTD50	P<.0005

1-FORMYL-3-THIOSEMICARBAZIDE   `100ng..:..1ug...:..10....:..100...:..1mg...:..10.....:..100...:..1g.....:..10`

	Spe	Sex	Strain	Route	Site	Hist	Xpo+Xpt	Notes	DR	TD50	2Tailpvl AuOp
2335	R	f	sda	eat	liv	tum	46w64	e	.>	no dre	P=1.
a	R	f	sda	eat	tba	mix	46w64	e		140.mg	P<.2 -

FORMYLHYDRAZINE   `100ng..:..1ug...:..10....:..100...:..1mg...:..10.....:..100...:..1g.....:..10`

	Spe	Sex	Strain	Route	Site	Hist	Xpo+Xpt	Notes	DR	TD50	2Tailpvl AuOp
2336	M	f	swa	wat	lun	mix	83w83	e	. + .	36.0mg	P<.0005+
a	M	f	swa	wat	liv	mix	83w83	e		6.86gm	P<.9
2337	M	m	swa	wat	lun	mix	83w83	e	<+	noTD50	P<.0005+
a	M	m	swa	wat	liv	mix	83w83	e		1.15gm	P<.2

FOSETYL Al   `100ng..:..1ug...:..10....:..100...:..1mg...:..10.....:..100...:..1g.....:..10`

	Spe	Sex	Strain	Route	Site	Hist	Xpo+Xpt	Notes	DR	TD50	2Tailpvl AuOp
2338	R	f	cd1	eat	k/p	mix	24m24	v	.	14.6gm	* P<.005
a	R	f	cd1	eat	k/p	tcc	24m24	v		17.8gm	* P<.0005
b	R	f	cd1	eat	ubl	mix	24m24	v		17.6gm	Z P<.02
c	R	f	cd1	eat	ubl	tcc	24m24	v		23.9gm	* P<.02
d	R	f	cd1	eat	k/p	tpp	24m24	v		80.4gm	* P<.5
e	R	f	cd1	eat	ubl	adq	24m24	v		110.gm	* P<.2
f	R	f	cd1	eat	ubl	tpp	24m24	v		452.gm	* P<.9
2339	R	m	cd1	eat	ubl	mix	24m24	ev	. +	.3.66gm	* P<.0005+
a	R	m	cd1	eat	ubl	tcc	24m24	ev		5.00gm	* P<.0005
b	R	m	cd1	eat	ubl	tpp	24m24	ev		17.2gm	* P<.04
c	R	m	cd1	eat	k/p	tpp	24m24	v		31.8gm	* P<.3
d	R	m	cd1	eat	k/p	mix	24m24	v		36.4gm	* P<.4
e	R	m	cd1	eat	k/p	tcc	24m24	v		no dre	P=1.
f	R	m	cd1	eat	amd	mix	24m24	ev		no dre	P=1.

FUMONISIN B1   `100ng..:..1ug...:..10....:..100...:..1mg...:..10.....:..100...:..1g.....:..10`

	Spe	Sex	Strain	Route	Site	Hist	Xpo+Xpt	Notes	DR	TD50	2Tailpvl AuOp
2340	R	m	bd9	eat	liv	hpc	52w52	k	.>	no dre	P=1.
2341	R	m	bd9	eat	liv	hpc	86w86	k	±	1.02mg	P<.02 +
2342	R	m	bd9	eat	liv	hpc	26m26		. + .	1.34mg	P<.005 +

	RefNum	LoConf	UpConf	Cntrl	1Dose	1Inc	2Dose	2Inc	Citation or Pathology
									Brkly Code
c	1077	141.mg	n.s.s.	0/24	63.1mg	0/24			
2328	1076	4.33mg	21.9mg	0/29	41.3mg	11/20			Cohen;canr,33,1593-1597;1973
a	1076	5.09mg	47.4mg	2/29	41.3mg	10/20			
b	1076	16.5mg	n.s.s.	0/29	41.3mg	2/20			
c	1076	42.5mg	n.s.s.	0/29	41.3mg	0/20			
d	1076	.980mg	5.76mg	2/29	41.3mg	19/20			
2329	1118	7.06mg	32.6mg	0/30	41.3mg	12/30			Cohen;canr,38,1398-1405;1978
a	1118	10.7mg	639.mg	1/30	41.3mg	8/30			
b	1118	63.8mg	n.s.s.	0/30	41.3mg	0/30			
c	1118	63.8mg	n.s.s.	0/30	41.3mg	0/30			
2330	1073	2.43mg	11.4mg	0/30	71.9mg	28/29			Erturk;jnci,47,437-445;1971/1970a
a	1073	7.48mg	25.0mg	0/30	71.9mg	22/29			
b	1073	9.87mg	34.5mg	0/30	71.9mg	19/29			
c	1073	19.5mg	96.4mg	0/30	71.9mg	11/29			
d	1073	21.3mg	116.mg	0/30	71.9mg	10/29			
e	1073	32.7mg	434.mg	0/30	71.9mg	6/29			
f	1073	37.3mg	1.20gm	0/30	71.9mg	5/29			
g	1073	n.s.s.	8.39mg	0/30	71.9mg	29/29			
2331	1063	2.28mg	13.1mg	3/16	73.6mg	25/26			Morris;canr,29,2145-2156;1969
a	1063	9.51mg	37.0mg	0/16	73.6mg	16/26			
b	1063	29.7mg	n.s.s.	0/16	73.6mg	5/26			
c	1063	29.7mg	n.s.s.	0/16	73.6mg	5/26			
d	1063	34.6mg	n.s.s.	0/16	73.6mg	4/26			
e	1063	51.5mg	n.s.s.	0/16	73.6mg	2/26			
f	1063	69.6mg	n.s.s.	0/16	73.6mg	1/26			
2332	200a	1.95mg	6.19mg	6/71	31.3mg	49/51			Cohen;jnci,51,403-417;1973
a	200a	29.6mg	196.mg	0/71	31.3mg	8/51			
b	200a	29.6mg	196.mg	0/71	31.3mg	8/51			
c	200a	47.2mg	3.07gm	0/71	31.3mg	4/51			
d	200a	55.7mg	n.s.s.	0/71	31.3mg	3/51			
e	200a	171.mg	n.s.s.	0/71	31.3mg	0/51			
f	200a	2.02mg	7.04mg	18/71	31.3mg	49/51			
2333	1073	2.49mg	12.8mg	2/29	71.9mg	25/26			Erturk;jnci,47,437-445;1971/1970a
a	1073	5.42mg	19.5mg	0/29	71.9mg	22/26			
b	1073	10.6mg	41.0mg	0/29	71.9mg	16/26			
c	1073	15.3mg	71.1mg	0/29	71.9mg	12/26			
d	1073	16.8mg	83.4mg	0/29	71.9mg	11/26			
e	1073	20.5mg	123.mg	0/29	71.9mg	9/26			
f	1073	22.8mg	157.mg	0/29	71.9mg	8/26			
g	1073	22.0mg	2.37gm	2/29	71.9mg	9/26			
h	1073	n.s.s.	9.13mg	2/29	71.9mg	26/26			
2334	1073	2.92mg	11.7mg	0/28	57.5mg	24/26			
a	1073	10.2mg	42.6mg	0/28	57.5mg	14/26			
b	1073	17.1mg	92.9mg	0/29	57.5mg	10/29			
c	1073	20.8mg	151.mg	0/29	57.5mg	8/29			
d	1073	20.4mg	177.mg	0/28	57.5mg	7/26			
e	1073	26.4mg	690.mg	0/28	57.5mg	5/26			
f	1073	n.s.s.	6.71mg	0/29	57.5mg	29/29			

1-FORMYL-3-THIOSEMICARBAZIDE   2302-84-3

	RefNum	LoConf	UpConf	Cntrl	1Dose	1Inc	2Dose	2Inc	Citation or Pathology
2335	1073	151.mg	n.s.s.	0/29	71.9mg	0/27			Erturk;jnci,47,437-445;1971/1970a
a	1073	38.6mg	n.s.s.	2/29	71.9mg	5/27			

FORMYLHYDRAZINE   624-84-0

	RefNum	LoConf	UpConf	Cntrl	1Dose	1Inc	2Dose	2Inc	Citation or Pathology
2336	397	19.5mg	64.9mg	15/96	250.mg	47/49			Toth;bjca,37,960-964;1978
a	397	292.mg	n.s.s.	3/33	250.mg	2/19			
2337	397	n.s.s.	36.9mg	22/92	208.mg	50/50			
a	397	316.mg	n.s.s.	3/69	208.mg	5/43			

FOSETYL Al   (aliette) 39148-24-8

	RefNum	LoConf	UpConf	Cntrl	1Dose	1Inc	2Dose	2Inc	Citation or Pathology
2338	2138	6.04gm	163.gm	1/80	100.mg	1/80	400.mg	1/80	1.51gm	7/80	Quest;rtxp,14,3-11;
									1991/Environmental Protection Agency 1993/pers.comm.
a	2138	7.28gm	65.1gm	0/80	100.mg	0/80	400.mg	0/80	1.51gm	6/80	
b	2138	6.25gm	n.s.s.	0/80	100.mg	3/80	400.mg	0/80	1.51gm	6/80	
c	2138	8.23gm	n.s.s.	0/80	100.mg	1/78	400.mg	0/79	1.51gm	4/80	
d	2138	12.1gm	n.s.s.	1/80	100.mg	1/80	400.mg	1/80	1.51gm	2/80	
e	2138	17.9gm	n.s.s.	0/80	100.mg	0/80	400.mg	0/80	1.51gm	1/80	
f	2138	16.9gm	n.s.s.	0/80	100.mg	2/78	400.mg	0/79	1.51gm	1/80	
2339	2138	2.10gm	8.15gm	3/80	80.0mg	3/78	320.mg	2/79	1.21gm	21/80	
a	2138	2.69gm	12.7gm	2/80	80.0mg	2/78	320.mg	1/79	1.21gm	16/80	
b	2138	5.94gm	n.s.s.	1/80	80.0mg	1/78	320.mg	1/79	1.21gm	5/80	
c	2138	7.40gm	n.s.s.	1/80	80.0mg	3/80	320.mg	0/80	1.21gm	4/80	
d	2138	7.07gm	n.s.s.	1/80	80.0mg	4/80	320.mg	1/80	1.21gm	4/80	
e	2138	19.5gm	n.s.s.	1/80	80.0mg	1/80	320.mg	1/80	1.21gm	0/80	
f	2138	5.89gm	n.s.s.	6/80	80.0mg	5/78	320.mg	10/79	1.21gm	6/80	

FUMONISIN B1   116355-83-0

	RefNum	LoConf	UpConf	Cntrl	1Dose	1Inc	2Dose	2Inc	Citation or Pathology
2340	2018m	.515mg	n.s.s.	0/5	2.00mg	0/5			Gelderblom;carc,12,1247-1251;1991
2341	2018n	.285mg	n.s.s.	0/5	2.00mg	3/5			
2342	2018o	.534mg	8.70mg	0/5	2.00mg	7/10			

```
     Spe Strain  Site   Xpo+Xpt                                                                    TD50    2Tailpvl
        Sex  Route  Hist     Notes                                                                       DR    AuOp
2-FURALDEHYDE SEMICARBAZONE          100ng..:..1ug...:..10.....:..100...:..1mg....:..10....:..100...:..1g.....:..10
2343 R f sda eat mgl fba 46w66 er                                               .>                        48.8mg    P<.2    -

FURAN                               100ng..:..1ug...:..10.....:..100...:..1mg....:..10....:..100...:..1g.....:..10
2344 M f b6c gav liv MXA 24m24                                             : + :                          2.19mg  / P<.0005c
   a M f b6c gav MXB MXB 24m24                                                                            2.26mg  / P<.0005
   b M f b6c gav liv hpa 24m24                                                                            2.26mg  / P<.0005
   c M f b6c gav liv MXA 24m24                                                                            6.85mg  / P<.0005
   d M f b6c gav for sqp 24m24                                                                            23.8mg  * P<.04
   e M f b6c gav amd pob 24m24                                                                            36.0mg  / P<.03   c
   f M f b6c gav TBA MXB 24m24                                                                            2.81mg  / P<.0005
   g M f b6c gav liv MXB 24m24                                                                            2.19mg  / P<.0005
   h M f b6c gav lun MXB 24m24                                                                            38.5mg  * P<.2
2345 M m b6c gav MXB MXB 24m24                                                      : + :                 3.60mg  * P<.0005
   a M m b6c gav liv MXA 24m24                                                                            3.60mg  * P<.0005c
   b M m b6c gav liv hpa 24m24                                                                            4.02mg  * P<.0005
   c M m b6c gav liv hpc 24m24                                                                            5.13mg  * P<.0005
   d M m b6c gav amd pob 24m24                                                                            20.1mg  * P<.002  c
   e M m b6c gav mul MXA 24m24                                                                            11.0mg  \ P<.02
   f M m b6c gav lun a/a 24m24                                                                            18.2mg  * P<.02
   g M m b6c gav for sqp 24m24                                                                            64.6mg  * P<.03
   h M m b6c gav TBA MXB 24m24                                                                            5.43mg  * P<.008
   i M m b6c gav liv MXB 24m24                                                                            3.60mg  * P<.0005
   j M m b6c gav lun MXB 24m24                                                                            21.5mg  * P<.08
2346 R f f34 gav liv clc 24m24                                     :+ :                                   .395mg  Z P<.0005c
   a R f f34 gav MXB MXB 24m24                                                                            .543mg  Z P<.0005
   b R f f34 gav mul mnl 24m24                                                                            5.84mg  * P<.0005c
   c R f f34 gav liv MXA 24m24                                                                            11.2mg  * P<.0005c
   d R f f34 gav liv hpa 24m24                                                                            11.8mg  * P<.0005
   e R f f34 gav ubl tpp 24m24                                                                            50.7mg  * P<.02
   f R f f34 gav TBA MXB 24m24                                                                            2.83mg  * P<.008
   g R f f34 gav liv MXB 24m24                                                                            11.2mg  * P<.0005
2347 R m f34 gav liv clc 24m24                                     :+ :                                   .398mg  * P<.0005c
   a R m f34 gav MXB MXB 24m24                                                                            .529mg  Z P<.0005
   b R m f34 gav liv MXA 24m24                                                                            2.31mg  * P<.0005c
   c R m f34 gav liv hpa 24m24                                                                            3.10mg  * P<.0005
   d R m f34 gav mul mnl 24m24                                                                            4.64mg  * P<.0005c
   e R m f34 gav liv hpc 24m24                                                                            6.18mg  Z P<.0005
   f R m f34 gav tes MXA 24m24                                                                            3.14mg  * P<.03
   g R m f34 gav amd pob 24m24                                                                            12.7mg  * P<.05
   h R m f34 gav ski ker 24m24                                                                            16.4mg  * P<.05
   i R m f34 gav sub fib 24m24                                                                            18.1mg  * P<.03
   j R m f34 gav pre MXA 24m24                                                                            21.5mg  * P<.04
   k R m f34 gav TBA MXB 24m24                                                                            1.96mg  * P<.0005
   l R m f34 gav liv MXB 24m24                                                                            2.31mg  * P<.0005

FURFURAL                            100ng..:..1ug...:..10.....:..100....:..1mg....:..10....:..100...:..1g.....:..10
2348 H f syg inh liv tum 52w81 ev                                             .>                          no dre    P=1.    -
   a H f syg inh lun tum 52w81 ev                                                                         no dre    P=1.    -
2349 H m syg inh liv tum 52w81 ev                                          .>                             no dre    P=1.    -
   a H m syg inh lun tum 52w81 ev                                                                         no dre    P=1.    -
2350 M f b6c gav liv hpa 24m24                                                      : +                   364.mg  * P<.01   p
   a M f b6c gav liv MXA 24m24                                                                            377.mg  * P<.04
   b M f b6c gav for sqp 24m24                                                                            920.mg  * P<.03   e
   c M f b6c gav TBA MXB 24m24                                                                            no dre    P=1.
   d M f b6c gav liv MXB 24m24                                                                            377.mg  * P<.04
   e M f b6c gav lun MXB 24m24                                                                            no dre    P=1.
2351 M m b6c gav liv MXA 24m24                                                         : +            :   135.mg  * P<.008  c
   a M m b6c gav liv hpc 24m24                                                                            219.mg  * P<.007  c
   b M m b6c gav sub MXA 24m24                                                                            169.mg  Z P<.02
   c M m b6c gav sub MXA 24m24                                                                            187.mg  Z P<.02
   d M m b6c gav sub MXA 24m24                                                                            190.mg  Z P<.04
   e M m b6c gav liv hpa 24m24                                                                            207.mg  * P<.03   c
   f M m b6c gav sub fbs 24m24                                                                            212.mg  Z P<.03
   g M m b6c gav kcx MXA 24m24                                                                            1.40gm  * P<.3    e
   h M m b6c gav TBA MXB 24m24                                                                            165.mg  * P<.3
   i M m b6c gav liv MXB 24m24                                                                            135.mg  * P<.008
   j M m b6c gav lun MXB 24m24                                                                            662.mg  * P<.4
2352 R f f34 gav TBA MXB 24m24                                                 :>                         194.mg  * P<.8    -
   a R f f34 gav liv MXB 24m24                                                                            no dre    P=1.
2353 R m f34 gav liv clc 24m24                                                        :      ±            683.mg  * P<.09   p
   a R m f34 gav TBA MXB 24m24                                                                            150.mg  * P<.7
   b R m f34 gav liv MXB 24m24                                                                            no dre    P=1.

FUROSEMIDE                          100ng..:..1ug...:..10.....:..100...:..1mg....:..10....:..100...:..1g.....:..10
2354 M f b6c eat mgl MXA 24m24                                                        : +            :    732.mg  * P<.004  p
   a M f b6c eat mgl mtm 24m24                                                                            845.mg  * P<.004
   b M f b6c eat thy fca 24m24                                                                            703.mg  * P<.02
   c M f b6c eat TBA MXB 24m24                                                                            136.mg  * P<.02
   d M f b6c eat liv MXB 24m24                                                                            3.97gm  * P<.8
   e M f b6c eat lun MXB 24m24                                                                            2.71gm  * P<.6
```

	RefNum	LoConf	UpConf	Cntrl	1Dose	1Inc	2Dose	2Inc	Citation or Pathology
									Brkly Code

2-FURALDEHYDE SEMICARBAZONE 2411-74-7

	RefNum	LoConf	UpConf	Cntrl	1Dose	1Inc	2Dose	2Inc	Citation or Pathology	Brkly
2343	1120	15.1mg	n.s.s.	2/29	26.8mg	6/30			Erturk;canr,30,1409-1412;1970	

FURAN 110-00-9

	RefNum	LoConf	UpConf	Cntrl	1Dose	1Inc	2Dose	2Inc	3Dose	3Inc	Citation or Pathology	Brkly
2344	TR402	1.52mg	3.36mg	7/50	5.64mg	34/50	10.6mg	50/50			liv:hpa,hpc.	
a	TR402	1.56mg	3.53mg	9/50	5.64mg	34/50	10.6mg	50/50			amd:pob; liv:hpa,hpc.	C
b	TR402	1.57mg	3.44mg	5/50	5.64mg	31/50	10.6mg	48/50				S
c	TR402	4.20mg	12.4mg	2/50	5.64mg	7/50	10.6mg	27/50			liv:car,hpc.	S
d	TR402	8.90mg	n.s.s.	2/50	5.64mg	6/50	10.6mg	3/50				S
e	TR402	12.6mg	n.s.s.	2/50	5.64mg	1/50	10.6mg	6/50				
f	TR402	1.65mg	6.99mg	43/50	5.64mg	43/50	10.6mg	50/50				
g	TR402	1.52mg	3.36mg	7/50	5.64mg	34/50	10.6mg	50/50			liv:hpa,hpc,nnd.	
h	TR402	11.3mg	n.s.s.	4/50	5.64mg	6/50	10.6mg	4/50			lun:a/a,a/c.	
2345	TR402	2.17mg	8.50mg	26/50	5.65mg	44/50	10.6mg	50/50			amd:pob; liv:hpa,hpc.	C
a	TR402	2.17mg	8.50mg	26/50	5.65mg	44/50	10.6mg	50/50			liv:hpa,hpc.	
b	TR402	2.40mg	9.61mg	20/50	5.65mg	33/50	10.6mg	42/50				S
c	TR402	3.37mg	9.50mg	7/50	5.65mg	32/50	10.6mg	34/50				S
d	TR402	10.3mg	72.7mg	1/50	5.65mg	6/50	10.6mg	10/50				
e	TR402	4.46mg	n.s.s.	5/50	5.65mg	12/50	(10.6mg	1/50)			mul:mlh,mlm.	S
f	TR402	7.89mg	n.s.s.	9/50	5.65mg	7/50	10.6mg	16/50				S
g	TR402	21.4mg	n.s.s.	0/50	5.65mg	1/50	10.6mg	3/50				S
h	TR402	2.65mg	106.mg	43/50	5.65mg	45/50	10.6mg	50/50				
i	TR402	2.17mg	8.50mg	26/50	5.65mg	44/50	10.6mg	50/50			liv:hpa,hpc,nnd.	
j	TR402	8.12mg	n.s.s.	12/50	5.65mg	9/50	10.6mg	16/50			lun:a/a,a/c.	
2346	TR402	.273mg	.598mg	0/50	1.40mg	49/50	(2.81mg	50/50	5.64mg	48/50)		
a	TR402	.342mg	1.04mg	8/50	1.40mg	49/50	(2.81mg	50/50	5.64mg	50/50)	liv:clc,hpa,hpc; mul:mnl.	C
b	TR402	3.31mg	18.0mg	8/50	1.40mg	9/50	2.81mg	17/50	5.64mg	21/50		
c	TR402	6.00mg	28.1mg	0/50	1.40mg	2/50	2.81mg	4/50	5.64mg	8/50	liv:hpa,hpc.	
d	TR402	6.20mg	32.8mg	0/50	1.40mg	2/50	2.81mg	4/50	5.64mg	7/50		S
e	TR402	15.4mg	n.s.s.	0/50	1.40mg	0/50	2.81mg	0/50	5.64mg	3/50		S
f	TR402	1.38mg	67.0mg	45/50	1.40mg	50/50	2.81mg	50/50	5.64mg	50/50		
g	TR402	6.00mg	28.1mg	0/50	1.40mg	2/50	2.81mg	4/50	5.64mg	8/50	liv:hpa,hpc,nnd.	
2347	TR402	.268mg	.620mg	0/50	1.41mg	43/50	(2.82mg	48/50	5.64mg	49/50)		
a	TR402	.323mg	1.06mg	8/50	1.41mg	44/50	(2.82mg	48/50	5.64mg	49/50)	liv:clc,hpa,hpc; mul:mnl.	C
b	TR402	1.65mg	3.45mg	1/50	1.41mg	5/50	2.82mg	22/50	5.64mg	35/50	liv:hpa,hpc.	
c	TR402	2.13mg	4.94mg	1/50	1.41mg	4/50	2.82mg	18/50	5.64mg	27/50		S
d	TR402	2.71mg	12.3mg	8/50	1.41mg	11/50	2.82mg	17/50	5.64mg	25/50		
e	TR402	3.79mg	10.9mg	0/50	1.41mg	1/50	2.82mg	6/50	5.64mg	18/50		S
f	TR402	1.42mg	n.s.s.	45/50	1.41mg	38/50	2.82mg	40/50	5.64mg	45/50	tes:ica,ica.	S
g	TR402	4.99mg	n.s.s.	8/50	1.41mg	4/50	2.82mg	9/50	5.64mg	10/50		S
h	TR402	7.84mg	n.s.s.	0/50	1.41mg	3/50	2.82mg	5/50	5.64mg	2/50		S
i	TR402	8.52mg	n.s.s.	0/50	1.41mg	3/50	2.82mg	4/50	5.64mg	3/50		S
j	TR402	8.50mg	n.s.s.	1/50	1.41mg	2/50	2.82mg	4/50	5.64mg	4/50	pre:ade,anb.	S
k	TR402	1.11mg	6.22mg	31/50	1.41mg	45/50	2.82mg	48/50	5.64mg	49/50		
l	TR402	1.65mg	3.45mg	1/50	1.41mg	5/50	2.82mg	22/50	5.64mg	35/50	liv:hpa,hpc,nnd.	

FURFURAL 98-01-1

	RefNum	LoConf	UpConf	Cntrl	1Dose	1Inc	2Dose	2Inc	3Dose	3Inc	Citation or Pathology	Brkly
2348	1078	92.4mg	n.s.s.	0/6	123.mg	0/6					Feron;txcy,11,127-144;1978	
a	1078	216.mg	n.s.s.	0/14	123.mg	0/14						
2349	1078	81.3mg	n.s.s.	0/6	108.mg	0/6						
a	1078	203.mg	n.s.s.	0/15	108.mg	0/15						
2350	TR382	177.mg	26.3gm	1/50	35.1mg	2/50	70.5mg	5/50	123.mg	8/50		
a	TR382	160.mg	n.s.s.	5/50	35.1mg	3/50	70.5mg	7/50	123.mg	12/50	liv:hpa,hpc.	S
b	TR382	355.mg	n.s.s.	1/50	35.1mg	0/50	70.5mg	1/50	123.mg	6/50		
c	TR382	118.mg	n.s.s.	33/50	35.1mg	26/50	70.5mg	24/50	123.mg	31/50		
d	TR382	160.mg	n.s.s.	5/50	35.1mg	3/50	70.5mg	7/50	123.mg	12/50	liv:hpa,hpc,nnd.	
e	TR382	585.mg	n.s.s.	5/50	35.1mg	2/50	70.5mg	1/50	123.mg	3/50	lun:a/a,a/c.	
2351	TR382	65.9mg	3.47gm	16/50	35.1mg	22/50	70.7mg	17/50	123.mg	32/50	liv:hpa,hpc.	
a	TR382	107.mg	3.10gm	7/50	35.1mg	12/50	70.7mg	6/50	123.mg	21/50		
b	TR382	75.6mg	n.s.s.	5/50	35.1mg	7/50	70.7mg	13/50	(123.mg	3/50)	sub:fbs,fib,sar.	S
c	TR382	86.4mg	n.s.s.	3/50	35.1mg	6/50	70.7mg	11/50	(123.mg	2/50)	sub:fbs,sar.	S
d	TR382	80.9mg	n.s.s.	5/50	35.1mg	7/50	70.7mg	12/50	(123.mg	3/50)	sub:fbs,fib.	S
e	TR382	91.8mg	n.s.s.	9/50	35.1mg	13/50	70.7mg	11/50	123.mg	19/50		
f	TR382	93.6mg	n.s.s.	3/50	35.1mg	6/50	70.7mg	10/50	(123.mg	2/50)		S
g	TR382	425.mg	n.s.s.	0/50	35.1mg	1/50	70.7mg	1/50	123.mg	1/50	kcx:ade,car.	
h	TR382	52.4mg	n.s.s.	38/50	35.1mg	41/50	70.7mg	42/50	123.mg	42/50		
i	TR382	65.9mg	3.47gm	16/50	35.1mg	22/50	70.7mg	17/50	123.mg	32/50	liv:hpa,hpc,nnd.	
j	TR382	166.mg	n.s.s.	10/50	35.1mg	8/50	70.7mg	5/50	123.mg	13/50	lun:a/a,a/c.	
2352	TR382	20.4mg	n.s.s.	43/50	21.1mg	44/50	42.4mg	33/50				
a	TR382	n.s.s.	n.s.s.	0/50	21.1mg	0/50	42.4mg	0/50			liv:hpa,hpc,nnd.	
2353	TR382	162.mg	n.s.s.	0/50	21.1mg	0/50	42.3mg	2/50				
a	TR382	20.9mg	n.s.s.	43/50	21.1mg	39/50	42.3mg	40/50				
b	TR382	345.mg	n.s.s.	1/50	21.1mg	0/50	42.3mg	0/50			liv:hpa,hpc,nnd.	

FUROSEMIDE 54-31-9

	RefNum	LoConf	UpConf	Cntrl	1Dose	1Inc	2Dose	2Inc	Citation or Pathology	Brkly
2354	TR356	310.mg	4.35gm	0/50	89.3mg	2/50	180.mg	5/50	mgl:acc,mtm.	
a	TR356	337.mg	5.57gm	0/50	89.3mg	1/50	180.mg	5/50		S
b	TR356	292.mg	n.s.s.	0/50	89.3mg	4/50	180.mg	3/50		S
c	TR356	62.7mg	n.s.s.	34/50	89.3mg	35/50	180.mg	39/50		
d	TR356	380.mg	n.s.s.	6/50	89.3mg	4/50	180.mg	5/50	liv:hpa,hpc,nnd.	
e	TR356	416.mg	n.s.s.	3/50	89.3mg	2/50	180.mg	3/50	lun:a/a,a/c.	

```
         Spe Strain  Site   Xpo+Xpt                                                    TD50     2Tailpvl
              Sex  Route  Hist   Notes                                                       DR      AuOp
```

2355	M m b6c eat TBA MXB 24m24			:>	393.mg * P<.5	-
a	M m b6c eat liv MXB 24m24				405.mg * P<.2	
b	M m b6c eat lun MXB 24m24				no dre P=1.	
2356	R f f34 eat thy MXA 24m24		: + :		#47.1mg * P<.003	-
a	R f f34 eat thy cca 24m24				51.4mg * P<.004	
b	R f f34 eat TBA MXB 24m24				20.9mg / P<.04	
c	R f f34 eat liv MXB 24m24				no dre P=1.	
2357	R m f34 eat pta adn 24m24		: ±		27.0mg \ P<.05	
a	R m f34 eat kid MXA 24m24				370.mg * P<.7	e
b	R m f34 eat brm mng 24m24				no dre P=1.	e
c	R m f34 eat TBA MXB 24m24				no dre P=1.	
d	R m f34 eat liv MXB 24m24				no dre P=1.	
2358	R m f34 eat kid MXA 24m24	with step		.>	277.mg * P<.3	e

FUSARENON-X   100ng..:..1ug....:..10......:..100...:..1mg....:..10......:..100....:..1g......:..10

2359	R m don eat liv hnd 24m24 e	.>	no dre P=1.
a	R m don eat tba mix 24m24 e		no dre P=1.
2360	R m don eat liv hnd 18m24 ae	.>	no dre P=1.
a	R m don eat tba mix 18m24 ae		no dre P=1.

GALLIC ACID   100ng..:..1ug....:..10......:..100...:..1mg....:..10......:..100....:..1g......:..10

2361	R m f3d eat for sqc 60w60 er	.>	no dre P=1.	-
a	R m f3d eat for sqp 60w60 er		no dre P=1.	-

GEMCADIOL   100ng..:..1ug....:..10......:..100...:..1mg....:..10......:..100....:..1g......:..10

2362	R f cdr eat liv hpa 52w52 k	.>	no dre P=1.
2363	R m cdr eat liv hpa 52w52 k	.>	no dre P=1.

GEMFIBROZIL   100ng..:..1ug....:..10......:..100...:..1mg....:..10......:..100....:..1g......:..10

2364	M f cd1 eat tba mix 78w78 e	.>	220.mg * P<.6	-
a	M f cd1 eat tba ben 78w78 e		173.mg * P<.3	-
b	M f cd1 eat tba mal 78w78 e		no dre P=1.	-
2365	M m cd1 eat tba mix 78w78 e	.>	no dre P=1.	-
a	M m cd1 eat tba ben 78w78 e		no dre P=1.	-
b	M m cd1 eat tba mal 78w78 e		no dre P=1.	-
2366	R f cdr eat tba mix 24m24 e	.>	no dre P=1.	-
a	R f cdr eat tba ben 24m24 e		no dre P=1.	
b	R f cdr eat tba mal 24m24 e		587.mg * P<.9	
2367	R m cdr eat tba mix 24m24 e	. ±	8.07mg * P<.08	
a	R m cdr eat tba ben 24m24 e		7.85mg * P<.02	
b	R m cdr eat tba mal 24m24 e		41.6mg * P<.2	

GENTIAN VIOLET   100ng..:..1ug....:..10......:..100...:..1mg....:..10......:..100....:..1g......:..10

2368	M f b6c eat liv tum 52w52	.>	no dre P=1.	-
2369	M f b6c eat liv hpa 78w78 e	. + .	115.mg * P<.003	+
a	M f b6c eat liv hpc 78w78 e		343.mg * P<.07	+
2370	M f b6c eat liv hpc 25m25 e	.+.	57.9mg Z P<.0005+
a	M f b6c eat liv hpa 25m25 e		75.5mg Z P<.0005+
b	M f b6c eat hag ade 25m25 e		305.mg * P<.0005+
c	M f b6c eat ute rta 25m25 e		413.mg * P<.0005+
d	M f b6c eat vag rta 25m25 e		668.mg * P<.0005+
e	M f b6c eat ubl rta 25m25 e		834.mg * P<.002 +
f	M f b6c eat ova rta 25m25 e		912.mg * P<.0005+
2371	M m b6c eat liv hpa 52w52	.>	no dre P=1.	-
2372	M m b6c eat liv hpa 78w78 e	.>	636.mg * P<.5	-
a	M m b6c eat liv hpc 78w78 e		no dre P=1.	-
2373	M m b6c eat liv hpa 25m25 e	. + .	139.mg * P<.0005+
a	M m b6c eat liv hpc 25m25 e		222.mg * P<.0005+
b	M m b6c eat hag ade 25m25 e		615.mg * P<.03 +

GERANYL ACETATE, FOOD GRADE (71% GERANYL ACETATE, 29% CITRONELLYL ACETATE)   ...:..10......:..100...:..1g......:..10

2374	M f b6c gav TBA MXB 23m24 as	: ±	353.mg * P<.02	-
a	M f b6c gav liv MXB 23m24 as		2.73gm * P<.4	
b	M f b6c gav lun MXB 23m24 as		5.30gm * P<.4	
2375	M m b6c gav TBA MXB 23m24 as	: ±	485.mg / P<.04	-
a	M m b6c gav liv MXB 23m24 as		647.mg / P<.005	
b	M m b6c gav lun MXB 23m24 as		5.26gm * P<.6	
2376	R f f34 gav TBA MXB 24m24	:>	no dre P=1.	-
a	R f f34 gav liv MXB 24m24		45.1gm * P<.3	
2377	R m f34 gav tes ict 24m24 s	: ±	772.mg * P<.03	
a	R m f34 gav ski MXA 24m24 s		5.05gm * P<.05	
b	R m f34 gav ski sqp 24m24 s		6.12gm * P<.06	e
c	R m f34 gav kid tla 24m24 s		20.2gm * P<.5	e
d	R m f34 gav TBA MXB 24m24 s		9.26gm * P<.8	
e	R m f34 gav liv MXB 24m24 s		15.4gm * P<.2	

GERMANATE, SODIUM   100ng..:..1ug....:..10......:..100...:..1mg....:..10......:..100....:..1g......:..10

2378	M b cd1 wat lun mix 24m24 e	.>	no dre P=1.	-
a	M b cd1 wat liv mix 24m24 e		no dre P=1.	-
b	M b cd1 wat tba mix 24m24 e		no dre P=1.	-
c	M b cd1 wat tba ben 24m24 e		no dre P=1.	-

	RefNum	LoConf	UpConf	Cntrl	1Dose	1Inc	2Dose	2Inc	Citation or Pathology
									Brkly Code
2355	TR356	89.7mg	n.s.s.	33/50	82.4mg	34/50	166.mg	34/50	
a	TR356	133.mg	n.s.s.	15/50	82.4mg	16/50	166.mg	20/50	liv:hpa,hpc,nnd.
b	TR356	471.mg	n.s.s.	5/50	82.4mg	3/50	166.mg	4/50	lun:a/a,a/c.
2356	TR356	22.2mg	296.mg	4/50	17.2mg	7/50	34.5mg	11/50	thy:cca,ccr. S
a	TR356	23.6mg	375.mg	4/50	17.2mg	6/50	34.5mg	11/50	S
b	TR356	8.91mg	n.s.s.	48/50	17.2mg	45/50	34.5mg	46/50	
c	TR356	n.s.s.	n.s.s.	0/50	17.2mg	0/50	34.5mg	0/50	liv:hpa,hpc,nnd.
2357	TR356	10.0mg	n.s.s.	4/50	13.7mg	11/50	(27.7mg	8/50)	S
a	TR356	48.3mg	n.s.s.	1/50	13.7mg	4/50	27.7mg	2/50	kid:tla,uac.
b	TR356	n.s.s.	n.s.s.	0/50	13.7mg	3/50	27.7mg	0/50	
c	TR356	15.6mg	n.s.s.	47/50	13.7mg	44/50	27.7mg	46/50	
d	TR356	n.s.s.	n.s.s.	0/50	13.7mg	0/50	27.7mg	0/50	liv:hpa,hpc,nnd.
2358	TR356	76.9mg	n.s.s.	3/50	13.7mg	5/50	27.7mg	6/50	kid:tla,uac.
FUSARENON-X	**23255-69-8**								
2359	1748m	.319mg	n.s.s.	1/19	.100mg	0/22	.210mg	0/25	Saito;jjem,50,293-302;1980/Ohtsubo pers.comm.
a	1748m	.684mg	n.s.s.	4/19	.100mg	3/22	.210mg	2/25	
2360	1748n	.218mg	n.s.s.	0/23	.100mg	0/24	.105mg	0/18	
a	1748n	.464mg	n.s.s.	4/23	.100mg	2/24	.105mg	1/18	
GALLIC ACID	**149-91-7**								
2361	2134	1.10gm	n.s.s.	0/19	800.mg	0/20			Ito;anti,183-194;1990/pers.comm.
a	2134	1.10gm	n.s.s.	0/19	800.mg	0/20			
GEMCADIOL	**(2,2,9,9-tetramethyl-1,10-decanediol)**		**35449-36-6**						
2362	1768	186.mg	n.s.s.	0/10	30.0mg	1/10	150.mg	0/10	300.mg 0/10	Fitzgerald;faat,6,520-531;1986
2363	1768	185.mg	n.s.s.	0/10	30.0mg	2/10	150.mg	0/10	300.mg 0/10	
GEMFIBROZIL	**25812-30-0**								
2364	1518n	40.2mg	n.s.s.	21/72	3.90mg	23/72	39.0mg	25/72	Fitzgerald;jnci,67,1105-1115;1981
a	1518n	44.7mg	n.s.s.	14/72	3.90mg	15/72	39.0mg	19/72	
b	1518n	83.1mg	n.s.s.	10/72	3.90mg	10/72	39.0mg	10/72	
2365	1518n	28.5mg	n.s.s.	47/72	3.60mg	46/72	36.0mg	45/72	
a	1518n	48.8mg	n.s.s.	39/72	3.60mg	37/72	36.0mg	33/72	
b	1518n	62.9mg	n.s.s.	15/72	3.60mg	20/72	36.0mg	16/72	
2366	1518m	11.2mg	n.s.s.	47/50	1.50mg	50/50	15.0mg	44/50	
a	1518m	15.0mg	n.s.s.	44/50	1.50mg	50/50	15.0mg	41/50	
b	1518m	35.3mg	n.s.s.	12/50	1.50mg	8/50	15.0mg	11/50	
2367	1518m	2.62mg	n.s.s.	41/50	1.20mg	44/50	12.0mg	47/50	
a	1518m	3.21mg	n.s.s.	35/50	1.20mg	39/50	12.0mg	45/50	
b	1518m	12.9mg	n.s.s.	15/50	1.20mg	18/50	12.0mg	22/50	
GENTIAN VIOLET	**548-62-9**								
2368	1680m	10.7mg	n.s.s.	0/48	13.0mg	0/24	39.0mg	0/24	78.0mg 0/24	Littlefield;faat,5,902-912;1985
2369	1680n	51.5mg	734.mg	3/47	13.0mg	0/22	39.0mg	3/24	78.0mg 8/24	
a	1680n	103.mg	n.s.s.	1/47	13.0mg	0/22	39.0mg	1/24	78.0mg 3/24	
2370	1680o	44.7mg	77.0mg	7/185	13.0mg	5/93	39.0mg	30/93	78.0mg 73/95	
a	1680o	49.8mg	128.mg	8/185	13.0mg	8/93	39.0mg	36/93	(78.0mg 20/95)	
b	1680o	166.mg	1.13gm	8/186	13.0mg	11/93	39.0mg	18/89	78.0mg 15/94	
c	1680o	244.mg	786.mg	0/188	13.0mg	2/95	39.0mg	6/90	78.0mg 12/93	
d	1680o	335.mg	2.14gm	1/182	13.0mg	1/90	39.0mg	4/88	78.0mg 8/87	
e	1680o	407.mg	3.57gm	0/188	13.0mg	2/92	39.0mg	3/89	78.0mg 5/91	
f	1680o	430.mg	3.27gm	0/178	13.0mg	1/90	39.0mg	3/89	78.0mg 5/89	
2371	1680m	79.4mg	n.s.s.	0/48	12.0mg	2/24	36.0mg	0/24	72.0mg 0/24	
2372	1680n	106.mg	n.s.s.	3/48	12.0mg	0/24	36.0mg	2/24	72.0mg 2/22	
a	1680n	119.mg	n.s.s.	5/48	12.0mg	1/24	36.0mg	2/24	72.0mg 2/22	
2373	1680o	90.3mg	260.mg	17/183	12.0mg	14/92	36.0mg	20/93	72.0mg 37/93	
a	1680o	123.mg	742.mg	27/183	12.0mg	15/92	36.0mg	17/93	72.0mg 33/93	
b	1680o	255.mg	n.s.s.	7/187	12.0mg	7/92	36.0mg	10/94	72.0mg 9/89	
GERANYL ACETATE, FOOD GRADE (71% GERANYL ACETATE, 29% CITRONELLYL ACETATE)				**(CAS# 105-87-3 and 150-84-5) mixture**					
2374	TR252	146.mg	n.s.s.	20/50	354.mg	20/50	714.mg	6/50	
a	TR252	520.mg	n.s.s.	5/50	354.mg	4/50	714.mg	2/50	liv:hpa,hpc,nnd.
b	TR252	779.mg	n.s.s.	1/50	354.mg	1/50	714.mg	1/50	lun:a/a,a/c.
2375	TR252	197.mg	n.s.s.	33/50	354.mg	33/50	714.mg	21/50	
a	TR252	296.mg	7.59gm	13/50	354.mg	17/50	714.mg	15/50	liv:hpa,hpc,nnd.
b	TR252	766.mg	n.s.s.	6/50	354.mg	6/50	714.mg	3/50	lun:a/a,a/c.
2376	TR252	1.61gm	n.s.s.	38/50	707.mg	31/50	1.41gm	28/50	
a	TR252	7.34gm	n.s.s.	0/50	707.mg	0/50	1.41gm	1/50	liv:hpa,hpc,nnd.
2377	TR252	353.mg	n.s.s.	43/50	707.mg	44/50	1.41gm	44/50	S
a	TR252	2.06gm	n.s.s.	0/50	707.mg	5/50	1.41gm	1/50	ski:sqc,sqp. S
b	TR252	2.32gm	n.s.s.	0/50	707.mg	4/50	1.41gm	1/50	
c	TR252	4.71gm	n.s.s.	0/50	707.mg	2/50	1.41gm	0/50	
d	TR252	867.mg	n.s.s.	36/50	707.mg	33/50	1.41gm	23/50	
e	TR252	3.79gm	n.s.s.	0/50	707.mg	1/50	1.41gm	1/50	liv:hpa,hpc,nnd.
GERMANATE, SODIUM	**12025-19-3**								
2378	1512	7.48mg	n.s.s.	26/170	.877mg	14/131			Kanisawa;canr,27,1192-1195;1967
a	1512	10.5mg	n.s.s.	7/170	.877mg	4/131			
b	1512	7.84mg	n.s.s.	55/170	.877mg	25/131			
c	1512	8.97mg	n.s.s.	29/170	.877mg	13/131			

```
     Spe Strain Site   Xpo+Xpt                                        TD50   2Tailpvl
       Sex  Route  Hist   Notes                                         DR     AuOp

d      M b cd1 wat tba mal 24m24 e                                    no dre  P=1.   -
2379   R b leb wat liv tum 35m35 e                 .>                 no dre  P=1.   -
a      R b leb wat tba tum 35m35 e                                    no dre  P=1.   -
b      R b leb wat tba mal 35m35 e                                    no dre  P=1.   -

GIBBERELLIC ACID       100ng..:..1ug....:..10......:..100...:..1mg....:..10......:..100....:..1g......:..10
2380  M f b6a orl liv hpt 76w76 evx                          .>       no dre  P=1.   -
a     M f b6a orl lun ade 76w76 evx                                   no dre  P=1.   -
b     M f b6a orl tba mix 76w76 evx                                   no dre  P=1.   -
2381  M m b6a orl liv hpt 76w76 evx                              .>   316.mg  P<.2   -
a     M m b6a orl lun mix 76w76 evx                                   460.mg  P<.4   -
b     M m b6a orl tba mix 76w76 evx                                   151.mg  P<.07  -
2382  M f b6c orl liv hpt 76w76 evx                              .>   no dre  P=1.   -
a     M f b6c orl lun mix 76w76 evx                                   no dre  P=1.   -
b     M f b6c orl tba mix 76w76 evx                                   1.15gm  P<.3   -
2383  M m b6c orl liv hpt 76w76 evx                         .    ±    460.mg  P<.09  -
a     M m b6c orl lun ade 76w76 evx                                   952.mg  P<.3   -
b     M m b6c orl tba mix 76w76 evx                                   214.mg  P<.02  -

Glu-P-1       100ng..:..1ug....:..10......:..100...:..1mg....:..10......:..100....:..1g......:..10
2384  M f cdf eat liv mix 67w67 e                  . + .              5.08mg  P<.0005+
a     M f cdf eat blv mix 67w67 e                                     10.9mg  P<.0005+
b     M f cdf eat blv hms 67w67 e                                     13.8mg  P<.0005
c     M f cdf eat liv hpc 67w67 e                                     18.5mg  P<.0005
d     M f cdf eat liv hpa 67w67 e                                     44.1mg  P<.0005
e     M f cdf eat lun ade 67w67 e                                     166.mg  P<.02
f     M f cdf eat lun mix 67w67 e                                     308.mg  P<.4
2385  M m cdf eat blv mix 57w57 e                 . + .               5.77mg  P<.0005+
a     M m cdf eat blv hms 57w57 e                                     7.81mg  P<.0005
b     M m cdf eat blv hae 57w57 e                                     98.7mg  P<.02
c     M m cdf eat liv hpa 57w57 e                                     98.7mg  P<.02  +
d     M m cdf eat lun ade 57w57 e                                     1.00gm  P<.9
e     M m cdf eat lun mix 57w57 e                                     no dre  P=1.
2386  R f f3d eat liv mix 67w67 e                   . + .             8.39mg  P<.0005+
a     R f f3d eat zym sqc 67w67 e                                     12.7mg  P<.0005+
b     R f f3d eat smi mix 67w67 e                                     26.1mg  P<.0005+
c     R f f3d eat col mix 67w67 e                                     39.0mg  P<.002 +
d     R f f3d eat smi adc 67w67 e                                     56.1mg  P<.005 +
e     R f f3d eat cli tum 67w67 e                                     56.1mg  P<.005 +
f     R f f3d eat col ade 67w67 e                                     71.0mg  P<.02  +
g     R f f3d eat smi mcc 67w67 e                                     71.0mg  P<.02  +
h     R f f3d eat col adc 67w67 e                                     95.9mg  P<.03  +
i     R f f3d eat bra ast 67w67 e                                     95.9mg  P<.03  +
j     R f f3d eat smi ade 67w67 e                                     146.mg  P<.08  +
2387  R m f3d eat liv mix 67w67 e                 . + .               3.25mg  P<.0005+
a     R m f3d eat smi mix 67w67 e                                     5.89mg  P<.0005+
b     R m f3d eat smi adc 67w67 e                                     6.71mg  P<.0005+
c     R m f3d eat smi ade 67w67 e                                     8.21mg  P<.0005+
d     R m f3d eat col mix 67w67 e                                     9.45mg  P<.0005+
e     R m f3d eat zym sqc 67w67 e                                     10.2mg  P<.0005+
f     R m f3d eat col ade 67w67 e                                     15.4mg  P<.0005+
g     R m f3d eat col adc 67w67 e                                     16.9mg  P<.0005+
h     R m f3d eat smi mcc 67w67 e                                     236.mg  P<.3   +

Glu-P-2       100ng..:..1ug....:..10......:..100...:..1mg....:..10......:..100....:..1g......:..10
2388  M f cdf eat liv mix 82w82                    . + .              12.0mg  P<.0005+
a     M f cdf eat liv hpc 82w82                                       20.0mg  P<.0005
b     M f cdf eat blv mix 82w82                                       39.9mg  P<.0005+
c     M f cdf eat blv hms 82w82                                       43.0mg  P<.0005
d     M f cdf eat liv hpa 82w82                                       170.mg  P<.004
e     M f cdf eat lun ade 82w82                                       1.09gm  P<.3
f     M f cdf eat lun mix 82w82                                       no dre  P=1.
2389  M m cdf eat blv mix 84w84                       . + .           23.9mg  P<.0005+
a     M m cdf eat blv hms 84w84                                       27.3mg  P<.0005
b     M m cdf eat liv mix 84w84                                       93.2mg  P<.0005+
c     M m cdf eat liv hpa 84w84                                       201.mg  P<.008
d     M m cdf eat liv hpc 84w84                                       255.mg  P<.02
e     M m cdf eat lun ade 84w84                                       483.mg  P<.5
f     M m cdf eat lun mix 84w84                                       no dre  P=1.
2390  R f f3d eat cli tum 24m24                        . + .          56.4mg  P<.0005+
a     R f f3d eat smi mix 24m24                                       81.1mg  P<.0005+
b     R f f3d eat col mix 24m24                                       81.1mg  P<.0005+
c     R f f3d eat zym sqc 24m24                                       94.0mg  P<.002 +
d     R f f3d eat col ade 24m24                                       111.mg  P<.003 +
e     R f f3d eat smi adc 24m24                                       135.mg  P<.005 +
f     R f f3d eat smi ade 24m24                                       171.mg  P<.02  +
g     R f f3d eat liv mix 24m24                                       351.mg  P<.08  +
h     R f f3d eat col adc 24m24                                       351.mg  P<.08  +
i     R f f3d eat smi mcc 24m24                                       711.mg  P<.3   +
2391  R m f3d eat smi mix 24m24                     . + .             33.8mg  P<.0005+
a     R m f3d eat smi adc 24m24                                       45.1mg  P<.0005+
```

	RefNum	LoConf	UpConf	Cntrl	1Dose	1Inc	2Dose	2Inc	Citation or Pathology
									Brkly Code
d	1512	10.0mg	n.s.s.	15/170	.877mg	7/131			
2379	1036	6.82mg	n.s.s.	1/82	.265mg	1/98			Kanisawa;canr,29,892-895;1969
a	1036	2.70mg	n.s.s.	31/82	.265mg	25/98			
b	1036	8.11mg	n.s.s.	9/82	.265mg	2/98			

GIBBERELLIC ACID   77-06-5

	RefNum	LoConf	UpConf	Cntrl	1Dose	1Inc	2Dose	2Inc	Citation or Pathology
2380	1281	318.mg	n.s.s.	0/17	180.mg	0/16			Innes;ntis,1968/1969
a	1281	318.mg	n.s.s.	1/17	180.mg	0/16			
b	1281	208.mg	n.s.s.	2/17	180.mg	1/16			
2381	1281	89.6mg	n.s.s.	1/18	168.mg	4/18			
a	1281	96.2mg	n.s.s.	2/18	168.mg	4/18			
b	1281	52.6mg	n.s.s.	3/18	168.mg	8/18			
2382	1281	357.mg	n.s.s.	0/16	180.mg	0/18			
a	1281	357.mg	n.s.s.	0/16	180.mg	0/18			
b	1281	188.mg	n.s.s.	0/16	180.mg	1/18			
2383	1281	113.mg	n.s.s.	0/16	168.mg	2/16			
a	1281	155.mg	n.s.s.	0/16	168.mg	1/16			
b	1281	73.3mg	n.s.s.	0/16	168.mg	4/16			

Glu-P-1   (2-amino-6-methyldipyrido[1,2-a:3',2'-d]imidazole)   67730-11-4

	RefNum	LoConf	UpConf	Cntrl	1Dose	1Inc	2Dose	2Inc	Citation or Pathology
2384	1616	2.33mg	9.78mg	0/40	65.0mg	37/38			Ohgaki;carc,5,815-819;1984
a	1616	6.63mg	18.9mg	0/40	65.0mg	31/38			
b	1616	8.44mg	24.3mg	0/40	65.0mg	28/38			
c	1616	11.1mg	33.6mg	0/40	65.0mg	24/38			
d	1616	23.1mg	99.8mg	0/40	65.0mg	13/38			
e	1616	57.4mg	n.s.s.	0/40	65.0mg	4/38			
f	1616	65.6mg	n.s.s.	2/40	65.0mg	4/38			
2385	1616	3.31mg	10.4mg	0/39	60.0mg	30/34			
a	1616	4.63mg	14.0mg	0/39	60.0mg	27/34			
b	1616	34.1mg	n.s.s.	0/39	60.0mg	4/34			
c	1616	34.1mg	n.s.s.	0/39	60.0mg	4/34			
d	1616	50.9mg	n.s.s.	3/39	60.0mg	3/34			
e	1616	50.5mg	n.s.s.	11/39	60.0mg	6/34			
2386	1619m	5.07mg	15.1mg	0/50	25.0mg	24/42			Takayama;gann,75,207-213;1984/Masuda;expl,26,123-129;1984
a	1619m	7.24mg	25.1mg	0/50	25.0mg	18/42			
b	1619m	12.7mg	67.7mg	0/50	25.0mg	10/42			
c	1619m	16.8mg	142.mg	0/50	25.0mg	7/42			
d	1619m	21.3mg	463.mg	0/50	25.0mg	5/42			
e	1619m	21.3mg	463.mg	0/50	25.0mg	5/42			
f	1619m	24.5mg	n.s.s.	0/50	25.0mg	4/42			
g	1619m	24.5mg	n.s.s.	0/50	25.0mg	4/42			
h	1619m	29.0mg	n.s.s.	0/50	25.0mg	3/42			
i	1619m	29.0mg	n.s.s.	0/50	25.0mg	3/42			
j	1619m	35.8mg	n.s.s.	0/50	25.0mg	2/42			
2387	1619m	1.99mg	5.58mg	2/50	20.0mg	35/42			
a	1619m	3.61mg	10.4mg	0/50	20.0mg	26/42			
b	1619m	4.06mg	12.1mg	0/50	20.0mg	24/42			
c	1619m	4.84mg	15.4mg	0/50	20.0mg	21/42			
d	1619m	5.45mg	18.3mg	0/50	20.0mg	19/42			
e	1619m	5.80mg	20.1mg	0/50	20.0mg	18/42			
f	1619m	8.06mg	34.7mg	0/50	20.0mg	13/42			
g	1619m	8.68mg	39.6mg	0/50	20.0mg	12/42			
h	1619m	38.4mg	n.s.s.	0/50	20.0mg	1/42			

Glu-P-2   (2-aminodipyrido[1,2-a:3',2'-d]imidazole)   67730-10-3

	RefNum	LoConf	UpConf	Cntrl	1Dose	1Inc	2Dose	2Inc	Citation or Pathology
2388	1616	7.10mg	20.7mg	0/40	65.0mg	36/40			Ohgaki;carc,5,815-819;1984
a	1616	12.3mg	34.4mg	0/40	65.0mg	30/40			
b	1616	23.3mg	76.3mg	0/40	65.0mg	20/40			
c	1616	24.8mg	83.5mg	0/40	65.0mg	19/40			
d	1616	69.4mg	1.03gm	0/40	65.0mg	6/40			
e	1616	178.mg	n.s.s.	0/40	65.0mg	1/40			
f	1616	224.mg	n.s.s.	2/40	65.0mg	1/40			
2389	1616	14.6mg	41.9mg	0/40	60.0mg	27/40			
a	1616	16.6mg	48.9mg	0/40	60.0mg	25/40			
b	1616	45.3mg	248.mg	0/40	60.0mg	10/40			
c	1616	76.2mg	3.49gm	0/40	60.0mg	5/40			
d	1616	87.9mg	n.s.s.	0/40	60.0mg	4/40			
e	1616	92.2mg	n.s.s.	3/40	60.0mg	5/40			
f	1616	105.mg	n.s.s.	11/40	60.0mg	8/40			
2390	1619	28.2mg	138.mg	0/50	25.0mg	11/42			Takayama;gann,75,207-213;1984
a	1619	36.6mg	252.mg	0/50	25.0mg	8/42			
b	1619	36.6mg	252.mg	0/50	25.0mg	8/42			
c	1619	40.5mg	341.mg	0/50	25.0mg	7/42			
d	1619	45.3mg	525.mg	0/50	25.0mg	6/42			
e	1619	51.3mg	1.12gm	0/50	25.0mg	5/42			
f	1619	59.1mg	n.s.s.	0/50	25.0mg	4/42			
g	1619	86.3mg	n.s.s.	0/50	25.0mg	2/42			
h	1619	86.3mg	n.s.s.	0/50	25.0mg	2/42			
i	1619	116.mg	n.s.s.	0/50	25.0mg	1/42			
2391	1619	18.1mg	73.9mg	0/50	20.0mg	14/42			
a	1619	22.6mg	111.mg	0/50	20.0mg	11/42			

	Spe	Strain	Site	Xpo+Xpt			TD50	2Tailpvl
	Sex	Route	Hist	Notes			DR	AuOp
b	R m	f3d	eat	liv mix 24m24			52.1mg	P<.003 +
c	R m	f3d	eat	col mix 24m24			88.9mg	P<.003 +
d	R m	f3d	eat	smi ade 24m24			137.mg	P<.02 +
e	R m	f3d	eat	col ade 24m24			185.mg	P<.03 +
f	R m	f3d	eat	col adc 24m24			185.mg	P<.03 +
g	R m	f3d	eat	smi mcc 24m24			281.mg	P<.08 +
h	R m	f3d	eat	bra ast 24m24			281.mg	P<.08 +
i	R m	f3d	eat	zym sqc 24m24			569.mg	P<.3 +

L-GLUTAMIC ACID   100ng..:..1ug....:..10.....:..100....:..1mg..:..10.....:..100...:..1g.....:..10

	Spe	Strain	Site	Xpo+Xpt			TD50	2Tailpvl
2392	M m	cbl	eat	liv tum 24m24 e		.>	no dre	P=1. -
a	M m	cbl	eat	lun tum 24m24 e			no dre	P=1. -
b	M m	cbl	eat	tba tum 24m24 e			no dre	P=1. -

N2-gamma-GLUTAMYL-p-HYDRAZINOBENZOIC ACID ..1ug....:..10.....:..100..:..1mg.....:..10.....:..100...:..1g.....:..10

	Spe	Strain	Site	Xpo+Xpt			TD50	2Tailpvl
2393	M f	swa	gav	liv mix 12m31 e		.>	41.4gm	P<1.
a	M f	swa	gav	lun mix 12m31 e			no dre	P=1.
b	M f	swa	gav	sub mix 12m31 e			no dre	P=1.
2394	M m	swa	gav	sub fbs 12m31 e		. + .	277.mg	P<.0005+
a	M m	swa	gav	lun mix 12m31 e			250.mg	P<.05
b	M m	swa	gav	liv mix 12m31 e			no dre	P=1.

beta-N-[gamma-L(+)-GLUTAMYL]-4-HYDROXYMETHYLPHENYLHYDRAZINE..:..1 _00...:..1mg....:..10.....:..100...:..1g.....:..10

	Spe	Strain	Site	Xpo+Xpt			TD50	2Tailpvl
2395	M f	swa	wat	liv ang 28m28 e		.>	2.72gm	P<.4
a	M f	swa	wat	lun ade 28m28 e			3.29gm	P<.8
b	M f	swa	wat	lun mix 28m28 e			no dre	P=1. -
2396	M m	swa	wat	liv ang 28m28 aes		.>	2.04gm	P<.4
a	M m	swa	wat	liv hpt 28m28 aes			3.31gm	P<.5 -
b	M m	swa	wat	lun ade 28m28 aes			no dre	P=1. -
c	M m	swa	wat	lun mix 28m28 aes			no dre	P=1. -

GLYCEROL alpha-MONOCHLOROHYDRIN   100ng..:..1ug....:..10....:..100....:..1mg....:..10.....:..100....:..1g.....:..10

	Spe	Strain	Site	Xpo+Xpt			TD50	2Tailpvl
2397	R f	cdr	gav	pty ade 17m24 ev		.>	no dre	P=1. -
2398	R m	cdr	gav	pty ade 17m24 ev		. ±	116.mg *	P<.06

GLYCIDALDEHYDE   100ng..:..1ug....:..10....:..100....:..1mg....:..10.....:..100....:..1g.....:..10

	Spe	Strain	Site	Xpo+Xpt			TD50	2Tailpvl
2399	R f	esd	gav	sto tum 70w70		.>	no dre	P=1. -

GLYCIDOL   100ng..:..1ug....:..10....:..100....:..1mg....:..10.....:..100..:..1g.....:..10

	Spe	Strain	Site	Xpo+Xpt			TD50	2Tailpvl
2400	H f	syg	gav	spl hes 14m24 e		. ±	+hist 56.1mg	P<.05 +
a	H f	syg	gav	liv hcs 14m24 e			244.mg	P<.4
b	H f	syg	gav	liv cho 14m24 e			244.mg	P<.4
c	H f	syg	gav	lun adc 14m24 e			no dre	P=1.
d	H f	syg	gav	tba tum 14m24 e			71.8mg	P<.8
2401	H m	syg	gav	for mal 14m24 e		. ±	64.1mg	P<.08
a	H m	syg	gav	spl hes 14m24 e			+hist 99.1mg	P<.2 +
b	H m	syg	gav	liv hes 14m24 e			204.mg	P<.4
c	H m	syg	gav	tba tum 14m24 e			no dre	P=1.
2402	M f	b6c	gav	MXB MXB 24m24		:+ :	15.3mg *	P<.0005
a	M f	b6c	gav	hag MXA 24m24			32.9mg *	P<.0005c
b	M f	b6c	gav	hag adn 24m24			36.0mg *	P<.0005
c	M f	b6c	gav	mgl acn 24m24			56.6mg *	P<.0005
d	M f	b6c	gav	mgl MXA 24m24			56.9mg *	P<.0005
e	M f	b6c	gav	sub MXA 24m24			86.4mg *	P<.0005c
f	M f	b6c	gav	sub srn 24m24			145.mg *	P<.003
g	M f	b6c	gav	ute MXA 24m24			127.mg *	P<.02 c
h	M f	b6c	gav	MXA MXA 24m24			137.mg *	P<.04
i	M f	b6c	gav	ute acn 24m24			144.mg *	P<.04
j	M f	b6c	gav	sub fbs 24m24			217.mg *	P<.03
k	M f	b6c	gav	ski MXA 24m24			417.mg *	P<.07 c
l	M f	b6c	gav	TBA MXB 24m24			27.5mg *	P<.08
m	M f	b6c	gav	liv MXB 24m24			76.0mg *	P<.07
n	M f	b6c	gav	lun MXB 24m24			96.7mg *	P<.2
2403	M m	b6c	gav	hag MXA 24m24		: + :	41.7mg *	P<.003 c
a	M m	b6c	gav	for sqp 24m24			97.5mg *	P<.0005
b	M m	b6c	gav	for MXA 24m24			100.mg *	P<.003 c
c	M m	b6c	gav	MXB MXB 24m24			29.2mg *	P<.05
d	M m	b6c	gav	liv MXA 24m24			36.6mg *	P<.04 c
e	M m	b6c	gav	MXA MXA 24m24			38.8mg \	P<.03
f	M m	b6c	gav	liv hpa 24m24			41.2mg *	P<.03
g	M m	b6c	gav	hag adn 24m24			60.4mg *	P<.03
h	M m	b6c	gav	lun MXA 24m24			66.2mg *	P<.07 c
i	M m	b6c	gav	hag acn 24m24			148.mg *	P<.02
j	M m	b6c	gav	ski sqp 24m24			253.mg *	P<.02 c
k	M m	b6c	gav	mul mlh 24m24			297.mg *	P<.04
l	M m	b6c	gav	epy srn 24m24			473.mg *	P<.09 e
m	M m	b6c	gav	ubl tcc 24m24			761.mg *	P<.3 e
n	M m	b6c	gav	TBA MXB 24m24			34.6mg *	P<.2
o	M m	b6c	gav	liv MXB 24m24			36.6mg *	P<.04
p	M m	b6c	gav	lun MXB 24m24			66.2mg *	P<.07

	RefNum	LoConf	UpConf	Cntrl	1Dose	1Inc	2Dose	2Inc	Citation or Pathology
									Brkly Code
b	1619	24.0mg	290.mg	2/50	20.0mg	11/42			
c	1619	36.2mg	420.mg	0/50	20.0mg	6/42			
d	1619	47.3mg	n.s.s.	0/50	20.0mg	4/42			
e	1619	56.0mg	n.s.s.	0/50	20.0mg	3/42			
f	1619	56.0mg	n.s.s.	0/50	20.0mg	3/42			
g	1619	69.1mg	n.s.s.	0/50	20.0mg	2/42			
h	1619	69.1mg	n.s.s.	0/50	20.0mg	2/42			
i	1619	92.6mg	n.s.s.	0/50	20.0mg	1/42			

L-GLUTAMIC ACID   56-86-0

	RefNum	LoConf	UpConf	Cntrl	1Dose	1Inc	2Dose	2Inc	Citation or Pathology
2392	1631	4.06gm	n.s.s.	0/55	1.20gm	0/20	4.80gm	0/29	Ebert;txlt,3,65-70;1979
a	1631	4.06gm	n.s.s.	0/55	1.20gm	0/20	4.80gm	0/29	
b	1631	4.06gm	n.s.s.	0/55	1.20gm	0/20	4.80gm	0/29	

N2-gamma-GLUTAMYL-p-HYDRAZINOBENZOIC ACID (N2-[gamma-L(+)-GLUTAMYL]-4-CARBOXYPHENYLHYDRAZINE) 69644-85-5

	RefNum	LoConf	UpConf	Cntrl	1Dose	1Inc	Citation or Pathology
2393	1832	666.mg	n.s.s.	1/46	78.2mg	1/42	Toth;acnr,6,917-920;1986
a	1832	264.mg	n.s.s.	12/47	78.2mg	12/47	
b	1832	585.mg	n.s.s.	3/43	78.2mg	1/32	
2394	1832	146.mg	632.mg	0/43	78.2mg	13/48	
a	1832	101.mg	n.s.s.	15/49	78.2mg	23/45	
b	1832	443.mg	n.s.s.	7/40	78.2mg	3/32	

beta-N-[gamma-L(+)-GLUTAMYL]-4-HYDROXYMETHYLPHENYLHYDRAZINE   (agaritine) 2757-90-6

	RefNum	LoConf	UpConf	Cntrl	1Dose	1Inc	Citation or Pathology
2395	1584	500.mg	n.s.s.	1/56	125.mg	2/34	Toth;acnr,1,255-258;1981/1982a
a	1584	362.mg	n.s.s.	20/94	125.mg	12/50	
b	1584	458.mg	n.s.s.	29/94	125.mg	13/50	
2396	1584	311.mg	n.s.s.	3/83	104.mg	2/25	
a	1584	396.mg	n.s.s.	1/83	104.mg	1/25	
b	1584	643.mg	n.s.s.	13/100	104.mg	4/47	
c	1584	567.mg	n.s.s.	19/100	104.mg	6/47	

GLYCEROL alpha-MONOCHLOROHYDRIN   96-24-2

	RefNum	LoConf	UpConf	Cntrl	1Dose	1Inc	2Dose	2Inc	Citation or Pathology
2397	1112	24.2mg	n.s.s.	0/20	6.79mg	0/26	13.6mg	0/26	Weisburger;jnci,67,75-88;1981
2398	1112	35.1mg	n.s.s.	0/20	6.79mg	0/26	13.6mg	3/26	

GLYCIDALDEHYDE   765-34-4

	RefNum	LoConf	UpConf	Cntrl	1Dose	1Inc	Citation or Pathology
2399	55	6.29mg	n.s.s.	0/5	13.5mg	0/5	Van Duuren;jnci,37,825-838;1966

GLYCIDOL   556-52-5

	RefNum	LoConf	UpConf	Cntrl	1Dose	1Inc	2Dose	2Inc	Citation or Pathology / Brkly Code
2400	2042	19.3mg	n.s.s.	0/12	17.9mg	4/20			Lijinsky;txih,8,267-271;1992
a	2042	39.8mg	n.s.s.	0/12	17.9mg	1/20			
b	2042	39.8mg	n.s.s.	0/12	17.9mg	1/20			
c	2042	75.3mg	n.s.s.	1/12	17.9mg	0/20			
d	2042	8.94mg	n.s.s.	7/12	17.9mg	13/20			
2401	2042	19.4mg	n.s.s.	0/12	15.8mg	3/19			
a	2042	24.3mg	n.s.s.	0/12	15.8mg	2/19			
b	2042	33.2mg	n.s.s.	0/12	15.8mg	1/19			
c	2042	14.0mg	n.s.s.	7/12	15.8mg	9/19			
2402	TR374	10.2mg	26.6mg	5/50	17.7mg	21/50	35.4mg	37/50	hag:acn,adn; mgl:acn,adn,fba; ski:sqc,sqp; sub:fbs,srn; ute:acn, can. C
a	TR374	18.2mg	103.mg	4/50	17.7mg	11/50	35.4mg	17/50	hag:acn,adn.
b	TR374	19.4mg	128.mg	4/50	17.7mg	10/50	35.4mg	16/50	S
c	TR374	31.5mg	138.mg	1/50	17.7mg	5/50	35.4mg	15/50	S
d	TR374	31.0mg	170.mg	2/50	17.7mg	6/50	35.4mg	15/50	mgl:acn,adn,fba.
e	TR374	43.1mg	229.mg	0/50	17.7mg	3/50	35.4mg	9/50	sub:fbs,srn.
f	TR374	60.9mg	702.mg	0/50	17.7mg	1/50	35.4mg	6/50	S
g	TR374	51.3mg	n.s.s.	0/50	17.7mg	3/50	35.4mg	3/50	ute:acn,can.
h	TR374	52.6mg	n.s.s.	1/50	17.7mg	3/50	35.4mg	5/50	liv:hem,hes; ova:hem; pec:hes; sub:hes; ute:hem,hes. S
i	TR374	54.8mg	n.s.s.	0/50	17.7mg	3/50	35.4mg	2/50	S
j	TR374	78.1mg	n.s.s.	0/50	17.7mg	2/50	35.4mg	3/50	S
k	TR374	102.mg	n.s.s.	0/50	17.7mg	0/50	35.4mg	2/50	ski:sqc,sqp.
l	TR374	10.9mg	n.s.s.	45/50	17.7mg	46/50	35.4mg	47/50	
m	TR374	29.1mg	n.s.s.	9/50	17.7mg	7/50	35.4mg	14/50	liv:hpa,hpc,nnd.
n	TR374	32.3mg	n.s.s.	6/50	17.7mg	10/50	35.4mg	8/50	lun:a/a,a/c.
2403	TR374	21.5mg	270.mg	8/50	17.7mg	12/50	35.4mg	22/50	hag:acn,adn.
a	TR374	48.3mg	274.mg	0/50	17.7mg	2/50	35.4mg	9/50	S
b	TR374	47.7mg	533.mg	1/50	17.7mg	2/50	35.4mg	10/50	for:sqc,sqp.
c	TR374	12.3mg	n.s.s.	35/50	17.7mg	41/50	35.4mg	46/50	for:sqc,sqp; hag:acn,adn; liv:hpa,hpc; lun:a/a,a/c; ski:sqp. C
d	TR374	15.8mg	n.s.s.	24/50	17.7mg	31/50	35.4mg	35/50	liv:hpa,hpc.
e	TR374	15.4mg	n.s.s.	5/50	17.7mg	12/50	(35.4mg	7/50)	mul:mlh,mlm,mlp; thm:mlm. S
f	TR374	18.4mg	n.s.s.	18/50	17.7mg	16/50	35.4mg	30/50	S
g	TR374	27.2mg	n.s.s.	7/50	17.7mg	10/50	35.4mg	16/50	S
h	TR374	26.2mg	n.s.s.	13/50	17.7mg	11/50	35.4mg	21/50	lun:a/a,a/c.
i	TR374	61.5mg	n.s.s.	1/50	17.7mg	2/50	35.4mg	7/50	S
j	TR374	86.5mg	n.s.s.	0/50	17.7mg	0/50	35.4mg	4/50	
k	TR374	100.mg	n.s.s.	0/50	17.7mg	1/50	35.4mg	3/50	S
l	TR374	116.mg	n.s.s.	0/50	17.7mg	0/50	35.4mg	2/50	
m	TR374	186.mg	n.s.s.	0/50	17.7mg	1/50	35.4mg	2/50	
n	TR374	12.6mg	n.s.s.	42/50	17.7mg	47/50	35.4mg	49/50	
o	TR374	15.8mg	n.s.s.	24/50	17.7mg	31/50	35.4mg	35/50	liv:hpa,hpc,nnd.
p	TR374	26.2mg	n.s.s.	13/50	17.7mg	11/50	35.4mg	21/50	lun:a/a,a/c.

```
       Spe Strain  Site   Xpo+Xpt                                              TD50      2Tailpvl
           Sex   Route   Hist     Notes                                                DR     AuOp

2404 R f f34 gav MXB MXB 23m24 a                         :   +   :             4.13mg / P<.0005

  a  R f f34 gav mgl MXA 23m24 a                                               4.15mg / P<.0005c
  b  R f f34 gav mgl fba 23m24 a                                               4.55mg * P<.0005c
  c  R f f34 gav cli MXA 23m24 a                                               12.5mg * P<.0005c
  d  R f f34 gav ute esp 23m24 a                                               13.5mg * P<.0005
  e  R f f34 gav mul mnl 23m24 a                                               14.0mg * P<.0005c
  f  R f f34 gav cli adn 23m24 a                                               16.7mg * P<.0005c
  g  R f f34 gav mgl acn 23m24 a                                               23.0mg * P<.0005c
  h  R f f34 gav for MXA 23m24 a                                               28.5mg / P<.0005c
  i  R f f34 gav for sqp 23m24 a                                               31.4mg / P<.0005
  j  R f f34 gav bra gln 23m24 a                                               43.6mg * P<.0005c
  k  R f f34 gav MXA MXA 23m24 a                                               44.7mg * P<.0005c
  l  R f f34 gav MXA sqp 23m24 a                                               46.7mg * P<.0005
  m  R f f34 gav ton sqp 23m24 a                                               53.7mg * P<.0005
  n  R f f34 gav cli MXA 23m24 a                                               62.7mg * P<.008
  o  R f f34 gav cli can 23m24 a                                               64.1mg * P<.008
  p  R f f34 gav thy MXA 23m24 a                                               68.2mg * P<.003  c
  q  R f f34 gav pta adn 23m24 a                                               30.6mg / P<.03
  r  R f f34 gav sub MXA 23m24 a                                               71.1mg * P<.02
  s  R f f34 gav adr MXA 23m24 a                                               91.1mg / P<.04
  t  R f f34 gav ute MXA 23m24 a                                               130.mg / P<.02
  u  R f f34 gav for sqc 23m24 a                                               317.mg / P<.02
  v  R f f34 gav thy fcc 23m24 a                                               411.mg / P<.02
  w  R f f34 gav stg fbs 23m24 a                                               752.mg * P<.07  e
  x  R f f34 gav TBA MXB 23m24 a                                               3.88mg / P<.0005
  y  R f f34 gav liv MXB 23m24 a                                               no dre   P=1.

2405 R m f34 gav tes ict 93w98 a                         :   +   :             1.49mg / P<.0005
  a  R m f34 gav MXB MXB 93w98 a                                               2.30mg / P<.0005

  b  R m f34 gav mul mnl 93w98 a                                               3.14mg / P<.0005
  c  R m f34 gav tnv MXA 93w98 a                                               4.42mg / P<.0005c
  d  R m f34 gav tnv msm 93w98 a                                               5.68mg / P<.0005
  e  R m f34 gav ski sqp 93w98 a                                               7.00mg * P<.0005
  f  R m f34 gav bra gln 93w98 a                                               7.32mg * P<.0005c
  g  R m f34 gav mgl fba 93w98 a                                               7.97mg * P<.0005c
  h  R m f34 gav lun MXA 93w98 a                                               10.7mg * P<.0005
  i  R m f34 gav pre adn 93w98 a                                               10.8mg * P<.002
  j  R m f34 gav pre MXA 93w98 a                                               11.2mg / P<.0005
  k  R m f34 gav pta MXA 93w98 a                                               11.9mg * P<.005
  l  R m f34 gav pta adn 93w98 a                                               13.1mg * P<.009
  m  R m f34 gav thy MXA 93w98 a                                               16.5mg / P<.0005c
  n  R m f34 gav lun a/a 93w98 a                                               17.0mg * P<.003
  o  R m f34 gav thy fcc 93w98 a                                               19.7mg / P<.0005
  p  R m f34 gav ski ker 93w98 a                                               21.6mg * P<.003
  q  R m f34 gav sub MXA 93w98 a                                               25.3mg / P<.0005
  r  R m f34 gav for MXA 93w98 a                                               25.6mg * P<.0005c
  s  R m f34 gav tnv men 93w98 a                                               25.8mg * P<.0005
  t  R m f34 gav ski MXA 93w98 a                                               29.5mg * P<.0005c
  u  R m f34 gav sub MXA 93w98 a                                               30.0mg / P<.002
  v  R m f34 gav ski MXA 93w98 a                                               30.6mg * P<.0005
  w  R m f34 gav ton MXA 93w98 a                                               37.8mg * P<.006
  x  R m f34 gav ski bct 93w98 a                                               38.8mg * P<.0005
  y  R m f34 gav thy cca 93w98 a                                               42.8mg / P<.005
  z  R m f34 gav sub MXA 93w98 a                                               42.9mg / P<.002
  A  R m f34 gav zym can 93w98 a                                               49.8mg / P<.0005c
  B  R m f34 gav sub fib 93w98 a                                               55.9mg / P<.005
  C  R m f34 gav for sqp 93w98 a                                               101.mg / P<.0005
  D  R m f34 gav MXA MXA 93w98 a                                               244.mg * P<.004  c
  E  R m f34 gav MXA MXA 93w98 a                                               72.3mg * P<.03
  F  R m f34 gav TBA MXB 93w98 a                                               1.66mg / P<.0005
  G  R m f34 gav liv MXB 93w98 a                                               101.mg * P<.2

GLYCOL SULFITE                   100ng..:..1ug...:..10......:..100...:..1mg....:..10.....:..100...:..1g.....:..10
2406 M f hic ipj lun ptm 64w64                               .   ±                           1.03mg   P<.07

GLYCYRRHETINIC ACID             100ng..:..1ug...:..10......:..100...:..1mg....:..10.....:..100...:..1g.....:..10
2407 R m f34 eat col mix 54w54 er                                        .>                  no dre   P=1.

GLYCYRRHIZINATE, DISODIUM       100ng..:..1ug...:..10......:..100...:..1mg....:..10.....:..100...:..1g.....:..10
2408 M f b6c wat lun ade 22m26 e                                              .>             14.4gm * P<.2    -
  a  M f b6c wat liv hpa 22m26 e                                                             28.9gm * P<.2    -
  b  M f b6c wat liv hpc 22m26 e                                                             29.0gm * P<.6    -
  c  M f b6c wat lun adc 22m26 e                                                             no dre   P=1.    -
  d  M f b6c wat liv hem 22m26 e                                                             no dre   P=1.    -
  e  M f b6c wat tba mix 22m26 e                                                             no dre   P=1.    -
2409 M m b6c wat liv hem 22m26 e                                              .>             no dre   P=1.    -
  a  M m b6c wat liv hpc 22m26 e                                                             no dre   P=1.    -
  b  M m b6c wat liv hpa 22m26 e                                                             no dre   P=1.    -
  c  M m b6c wat lun ade 22m26 e                                                             no dre   P=1.    -
```

	RefNum	LoConf	UpConf	Cntrl	1Dose	1Inc	2Dose	2Inc	Citation or Pathology
									Brkly Code
2404	TR374	2.43mg	7.45mg	25/50	26.5mg	39/50	53.6mg	42/50	bra:gln; cli:acn,adn,can; for:sqc,sqp; mgl:acn,fba; mth:sqc,sqp;
									mul:mnl; thy:fca,fcc; ton:sqp. C
a	TR374	2.40mg	7.58mg	14/50	26.5mg	34/50	53.6mg	37/50	mgl:acn,fba.
b	TR374	2.52mg	8.70mg	14/50	26.5mg	32/50	53.6mg	29/50	
c	TR374	5.27mg	36.5mg	5/50	26.5mg	9/50	53.6mg	12/50	cli:acn,adn,can.
d	TR374	6.57mg	39.2mg	19/50	26.5mg	21/50	53.6mg	14/50	S
e	TR374	7.12mg	32.9mg	13/50	26.5mg	14/50	53.6mg	20/50	
f	TR374	6.73mg	56.7mg	3/50	26.5mg	7/50	53.6mg	7/50	S
g	TR374	11.9mg	46.8mg	1/50	26.5mg	11/50	53.6mg	16/50	
h	TR374	11.0mg	62.5mg	0/50	26.5mg	4/50	53.6mg	11/50	for:sqc,sqp.
i	TR374	11.3mg	77.8mg	0/50	26.5mg	4/50	53.6mg	8/50	S
j	TR374	12.3mg	174.mg	0/50	26.5mg	4/50	53.6mg	4/50	
k	TR374	14.1mg	181.mg	1/50	26.5mg	3/50	53.6mg	7/50	mth:sqc,sqp; ton:sqp.
l	TR374	14.4mg	212.mg	1/50	26.5mg	3/50	53.6mg	6/50	mth:sqp; ton:sqp. S
m	TR374	15.1mg	319.mg	1/50	26.5mg	3/50	53.6mg	5/50	S
n	TR374	14.0mg	2.21gm	2/50	26.5mg	2/50	53.6mg	5/50	cli:acn,can. S
o	TR374	14.1mg	2.38gm	2/50	26.5mg	1/50	53.6mg	5/50	S
p	TR374	13.1mg	878.mg	0/50	26.5mg	1/50	53.6mg	3/50	thy:fca,fcc.
q	TR374	10.0mg	n.s.s.	18/50	26.5mg	14/50	53.6mg	6/50	S
r	TR374	15.8mg	n.s.s.	1/50	26.5mg	4/50	53.6mg	2/50	sub:fbs,fib. S
s	TR374	17.7mg	n.s.s.	2/50	26.5mg	0/50	53.6mg	3/50	adr:con,crn. S
t	TR374	20.1mg	n.s.s.	1/50	26.5mg	0/50	53.6mg	4/50	ute:acn,adn. S
u	TR374	94.6mg	n.s.s.	0/50	26.5mg	0/50	53.6mg	3/50	S
v	TR374	109.mg	n.s.s.	0/50	26.5mg	0/50	53.6mg	3/50	S
w	TR374	164.mg	n.s.s.	0/50	26.5mg	0/50	53.6mg	2/50	
x	TR374	2.29mg	7.16mg	43/50	26.5mg	43/50	53.6mg	47/50	
y	TR374	n.s.s.	n.s.s.	0/50	26.5mg	0/50	53.6mg	0/50	liv:hpa,hpc,nnd.
2405	TR374	.818mg	2.82mg	46/50	26.8mg	50/50	53.6mg	49/50	S
a	TR374	1.04mg	4.74mg	8/50	26.8mg	40/50	53.6mg	46/50	bra:gln; col:adc,adp; for:sqc,sqp; mgl:fba; ski:bct,sea,sec;
									smi:mua; thy:fca,fcc; tnv:men,msm; zym:can. C
b	TR374	1.49mg	6.95mg	25/50	26.8mg	33/50	53.6mg	21/50	S
c	TR374	1.75mg	8.36mg	3/50	26.8mg	34/50	53.6mg	39/50	tnv:men,msm.
d	TR374	1.94mg	12.1mg	3/50	26.8mg	24/50	53.6mg	31/50	S
e	TR374	1.31mg	80.1mg	0/50	26.8mg	3/50	53.6mg	3/50	S
f	TR374	1.31mg	74.5mg	0/50	26.8mg	5/50	53.6mg	6/50	
g	TR374	2.65mg	27.8mg	3/50	26.8mg	8/50	53.6mg	7/50	
h	TR374	2.61mg	70.1mg	2/50	26.8mg	5/50	53.6mg	4/50	lun:a/a,a/c. S
i	TR374	2.87mg	85.3mg	5/50	26.8mg	7/50	53.6mg	1/50	S
j	TR374	3.43mg	68.2mg	10/50	26.8mg	7/50	53.6mg	5/50	pre:acn,adn. S
k	TR374	2.94mg	205.mg	8/50	26.8mg	8/50	53.6mg	2/50	pta:adn,can. S
l	TR374	3.08mg	767.mg	8/50	26.8mg	7/50	53.6mg	2/50	S
m	TR374	3.45mg	81.4mg	1/50	26.8mg	4/50	53.6mg	6/50	thy:fca,fcc.
n	TR374	3.07mg	344.mg	1/50	26.8mg	3/50	53.6mg	2/50	S
o	TR374	3.46mg	147.mg	1/50	26.8mg	2/50	53.6mg	5/50	S
p	TR374	3.55mg	333.mg	1/50	26.8mg	4/50	53.6mg	1/50	S
q	TR374	6.08mg	160.mg	4/50	26.8mg	5/50	53.6mg	5/50	sub:fbs,fib,nfm. S
r	TR374	3.64mg	194.mg	1/50	26.8mg	2/50	53.6mg	6/50	for:sqc,sqp.
s	TR374	10.8mg	61.5mg	0/50	26.8mg	10/50	53.6mg	8/50	S
t	TR374	10.2mg	87.2mg	0/50	26.8mg	5/50	53.6mg	4/50	ski:bct,sea,sec.
u	TR374	6.33mg	248.mg	4/50	26.8mg	4/50	53.6mg	5/50	sub:fbs,fib. S
v	TR374	10.3mg	98.1mg	0/50	26.8mg	5/50	53.6mg	3/50	ski:bct,sea. S
w	TR374	3.94mg	1.42gm	1/50	26.8mg	1/50	53.6mg	4/50	ton:sqc,sqp. S
x	TR374	11.1mg	176.mg	0/50	26.8mg	4/50	53.6mg	2/50	S
y	TR374	7.41mg	651.mg	2/50	26.8mg	3/50	53.6mg	3/50	S
z	TR374	8.90mg	377.mg	2/50	26.8mg	3/50	53.6mg	4/50	sub:fib,nfm. S
A	TR374	14.7mg	235.mg	1/50	26.8mg	3/50	53.6mg	6/50	
B	TR374	9.83mg	942.mg	2/50	26.8mg	2/50	53.6mg	4/50	S
C	TR374	35.5mg	395.mg	0/50	26.8mg	1/50	53.6mg	5/50	S
D	TR374	89.0mg	1.88gm	0/50	26.8mg	1/50	53.6mg	4/50	col:adc,adp; smi:mua.
E	TR374	8.91mg	n.s.s.	3/50	26.8mg	2/50	53.6mg	5/50	mth:sqc,sqp; ton:sqc,sqp. S
F	TR374	.903mg	3.17mg	43/50	26.8mg	50/50	53.6mg	48/50	
G	TR374	16.6mg	n.s.s.	1/50	26.8mg	1/50	53.6mg	1/50	liv:hpa,hpc,nnd.

GLYCOL SULFITE 3741-38-6

	RefNum	LoConf	UpConf	Cntrl	1Dose	1Inc			Citation or Pathology
2406	1143	.385mg	n.s.s.	10/30	1.71mg	17/30			Van Duuren;jnci,53,695-700;1974

GLYCYRRHETINIC ACID 471-53-4

	RefNum	LoConf	UpConf	Cntrl	1Dose	1Inc			Citation or Pathology
2407	2078	64.0mg	n.s.s.	0/12	96.0mg	0/12			Reddy;carc,13,1019-1023;1992

GLYCYRRHIZINATE, DISODIUM 71277-79-7

	RefNum	LoConf	UpConf	Cntrl	1Dose	1Inc	2Dose	2Inc	3Dose	3Inc	Citation or Pathology
2408	1685	3.54gm	n.s.s.	0/44	140.mg	0/47	262.mg	1/46	524.mg	1/37	Kobuke;fctx,23,979-983;1985
a	1685	4.71gm	n.s.s.	0/44	140.mg	0/47	262.mg	0/46	524.mg	1/37	
b	1685	4.73gm	n.s.s.	0/44	140.mg	1/47	262.mg	1/46	524.mg	0/37	
c	1685	3.72gm	n.s.s.	0/44	140.mg	1/47	262.mg	1/46	524.mg	0/37	
d	1685	5.96gm	n.s.s.	1/44	140.mg	0/47	262.mg	1/46	524.mg	0/37	
e	1685	1.33gm	n.s.s.	12/44	140.mg	12/47	262.mg	11/46	524.mg	9/37	
2409	1685	1.22gm	n.s.s.	3/37	58.2mg	0/38	116.mg	3/37	218.mg	2/40	
a	1685	1.23gm	n.s.s.	5/37	58.2mg	0/38	116.mg	3/37	218.mg	3/40	
b	1685	1.41gm	n.s.s.	4/37	58.2mg	3/38	116.mg	2/37	218.mg	2/40	
c	1685	1.53gm	n.s.s.	0/37	58.2mg	1/38	116.mg	1/37	218.mg	0/40	

```
        Spe Strain  Site   Xpo+Xpt                                                                        TD50    2Tailpvl
            Sex  Route  Hist    Notes                                                                       DR     AuOp

     d   M m b6c wat lun adc 22m26 e                                                                      no dre    P=1.    -
     e   M m b6c wat tba mix 22m26 e                                                                      3.17gm  * P<.8    -

FD & C GREEN NO. 1              100ng..:..1ug....:..10.....:..100....:..1mg....:..10.....:..100....:..1g.....:..10
  2410 M f cbj eat liv mix 24m24                                                                          146.gm  * P<.6    -
     a   M f cbj eat lun ade 24m24                                                                        146.gm  * P<.6    -
     b   M f cbj eat tba mal 24m24                                                                        18.4gm  \ P<.1    -
     c   M f cbj eat tba mix 24m24                                                                        no dre    P=1.    -
  2411 M m cbj eat liv mix 24m24                                                                     .> 26.1gm  * P<.5    -
     a   M m cbj eat lun mix 24m24                                                                        59.2gm  * P<.3    -
     b   M m cbj eat tba mix 24m24                                                                        20.8gm  * P<.4    -
     c   M m cbj eat tba mal 24m24                                                                        no dre    P=1.    -
  2412 R f osm eat liv mix 24m24 e                                                                     . 13.8gm  * P<.003  +
     a   R f osm eat tba mix 24m24 e                                                                      97.5gm  * P<1.    -
     b   R f osm eat tba mal 24m24 e                                                                      no dre    P=1.    -
  2413 R m osm eat liv mix 24m24 e                                                                   . + 5.98gm  * P<.003  +
     a   R m osm eat tba mix 24m24 e                                                                      11.4gm  * P<.4    -
     b   R m osm eat tba mal 24m24 e                                                                      no dre    P=1.    -
  2414 R b wis eat mix mly 93w93 er                                                               . + 3.92gm    P<.006  +
     a   R b wis eat mds mly 93w93 er                                                                     8.37gm    P<.05   +
     b   R b wis eat abd mly 93w93 er                                                                     8.37gm    P<.05   +

FD & C GREEN NO. 2              100ng..:..1ug....:..10.....:..100....:..1mg....:..10.....:..100....:..1g.....:..10
  2415 M f cbj eat lun mix 24m24                                                                          43.8gm  * P<.04
     a   M f cbj eat liv mix 24m24                                                                        no dre    P=1.
     b   M f cbj eat tba mix 24m24                                                                        148.gm  * P<.9
     c   M f cbj eat tba mal 24m24                                                                        152.gm  * P<.8    -
  2416 M m cbj eat liv mix 24m24                                                                     .> 18.9gm  \ P<.4    -
     a   M m cbj eat lun mix 24m24                                                                        58.2gm  * P<.3    -
     b   M m cbj eat tba mal 24m24                                                                        49.3gm  * P<.5    -
     c   M m cbj eat tba mix 24m24                                                                        318.gm  * P<1.
  2417 R m nss eat liv tum 65w65 e                               .>                                       no dre    P=1.    -
     a   R m nss eat tba tum 65w65 e                                                                      no dre    P=1.    -
  2418 R f osm eat liv tum 24m24 e                                             .>                         no dre    P=1.    -
     a   R f osm eat tba mix 24m24 e                                                                      no dre    P=1.    -
     b   R f osm eat tba mal 24m24 e                                                                      no dre    P=1.    -
  2419 R m osm eat liv hpa 24m24 e                                                .>                      no dre    P=1.    -
     a   R m osm eat tba mix 24m24 e                                                                      no dre    P=1.    -
     b   R m osm eat tba mal 24m24 e                                                                      no dre    P=1.    -
  2420 R b wis eat cec mly 95w95 er                                                         .  ± 5.64gm    P<.04   +

FD & C GREEN NO. 3              100ng..:..1ug....:..10.....:..100....:..1mg....:..10.....:..100....:..1g.....:..10
  2421 M f cbj eat lun ade 24m24                                                                          133.gm  * P<.5    -
     a   M f cbj eat liv hpa 24m24                                                                        no dre    P=1.    -
     b   M f cbj eat tba mal 24m24                                                                        no dre    P=1.    -
     c   M f cbj eat tba mix 24m24                                                                        no dre    P=1.    -
  2422 M m cbj eat liv mix 24m24                                                                          no dre    P=1.    -
     a   M m cbj eat lun ade 24m24                                                                        no dre    P=1.    -
     b   M m cbj eat tba mix 24m24                                                                        18.1gm  \ P<.5    -
     c   M m cbj eat tba mal 24m24                                                                        no dre    P=1.    -
  2423 R f osm eat liv tum 24m24 e                                                     .>                 no dre    P=1.    -
     a   R f osm eat mgl fba 24m24 e                                                                      no dre    P=1.    -
     b   R f osm eat tba mix 24m24 e                                                                      18.5gm  * P<.8    -
     c   R f osm eat tba mal 24m24 e                                                                      46.7gm  * P<.9    -
  2424 R m osm eat liv hpa 24m24 e                                                                        no dre    P=1.    -
     a   R m osm eat tba mix 24m24 e                                                                      no dre    P=1.    -
     b   R m osm eat tba mal 24m24 e                                                                      no dre    P=1.    -

GRISEOFULVIN                    100ng..:..1ug....:..10.....:..100....:..1mg....:..10.....:..100....:..1g.....:..10
  2425 H f syg eat lun tum 23m23 ae                                                           .>          no dre    P=1.
     a   H f syg eat liv clc 23m23 ae                                                                     no dre    P=1.
     b   H f syg eat liv mix 23m23 ae                                                                     no dre    P=1.
  2426 H m syg eat liv tum 29m30 ae                                                             .>        no dre    P=1.
     a   H m syg eat lun tum 29m30 ae                                                                     no dre    P=1.
  2427 M f swi eat lun tum 27m27 e                                                     .>                 no dre    P=1.
     a   M f swi eat liv hem 27m27 e                                                                      no dre    P=1.
  2428 M m swi eat liv hpt 27m27 e                                                       .>               1.66gm    P<.2    +
     a   M m swi eat liv hem 27m27 e                                                                      no dre    P=1.
     b   M m swi eat lun tum 27m27 e                                                                      no dre    P=1.    -

GUAR GUM                        100ng..:..1ug....:..10.....:..100....:..1mg....:..10.....:..100....:..1g.....:..10
  2429 M f b6c eat TBA MXB 24m25                                                                        :no dre    P=1.
     a   M f b6c eat liv MXB 24m25                                                                        no dre    P=1.
     b   M f b6c eat lun MXB 24m25                                                                        no dre    P=1.
  2430 M m b6c eat TBA MXB 24m24                                                                        :no dre    P=1.
     a   M m b6c eat liv MXB 24m24                                                                        no dre    P=1.
     b   M m b6c eat lun MXB 24m24                                                                        no dre    P=1.
  2431 R f f34 eat TBA MXB 24m24                                                                   :>     no dre    P=1.
     a   R f f34 eat liv MXB 24m24                                                                        no dre    P=1.
  2432 R m f34 eat sub fib 24m24                                                                      : #13.4gm  * P<.03   -
```

	RefNum	LoConf	UpConf	Cntrl	1Dose	1Inc	2Dose	2Inc					Citation or Pathology
													Brkly Code
d	1685	1.88gm	n.s.s.	2/37	58.2mg	0/38	116.mg	1/37	218.mg	1/40			
e	1685	377.mg	n.s.s.	14/37	58.2mg	7/38	116.mg	14/37	218.mg	14/40			
FD & C GREEN NO. 1 (guinea green B) 4680-78-8													
2410	143	18.2gm	n.s.s.	1/101	1.30gm	1/53	2.60gm	1/49 ·					Hansen;fctx,4,389-410;1966
a	143	18.2gm	n.s.s.	1/101	1.30gm	1/53	2.60gm	1/49					
b	143	4.84gm	n.s.s.	1/101	1.30gm	3/53	(2.60gm	0/49)					
c	143	16.5gm	n.s.s.	6/101	1.30gm	3/53	2.60gm	2/49					
2411	143	5.50gm	n.s.s.	12/101	1.20gm	13/50	2.40gm	7/51					
a	143	13.4gm	n.s.s.	1/101	1.20gm	1/50	2.40gm	2/51					
b	143	4.84gm	n.s.s.	15/101	1.20gm	15/50	2.40gm	9/51					
c	143	9.44gm	n.s.s.	11/101	1.20gm	9/50	2.40gm	4/51					
2412	143	5.23gm	85.1gm	0/25	250.mg	0/25	500.mg	0/25	1.00gm	1/25	2.50gm	4/25	
a	143	2.54gm	n.s.s.	12/25	250.mg	10/25	500.mg	13/25	1.00gm	9/25	2.50gm	12/25	
b	143	5.09gm	n.s.s.	6/25	250.mg	5/25	500.mg	8/25	1.00gm	4/25	2.50gm	5/25	
2413	143	2.82gm	31.6gm	0/25	200.mg	0/25	400.mg	2/25	800.mg	2/25	2.00gm	5/25	
a	143	2.59gm	n.s.s.	4/25	200.mg	7/25	400.mg	2/25	800.mg	5/25	2.00gm	7/25	
b	143	4.77gm	n.s.s.	2/25	200.mg	5/25	400.mg	1/25	800.mg	5/25	2.00gm	2/25	
2414	1136	958.mg	76.8gm	0/50	1.80gm	2/9							Willheim;gaga,23,1-19;1953
a	1136	1.36gm	n.s.s.	0/50	1.80gm	1/9							
b	1136	1.36gm	n.s.s.	0/50	1.80gm	1/9							
FD & C GREEN NO. 2 (light green SF yellowish) 5141-20-8													
2415	143	13.3gm	n.s.s.	0/100	1.30gm	1/50	2.60gm	2/50					Hansen;fctx,4,389-410;1966
a	143	22.3gm	n.s.s.	3/100	1.30gm	1/50	2.60gm	1/50					
b	143	11.0gm	n.s.s.	7/100	1.30gm	4/50	2.60gm	4/50					
c	143	14.5gm	n.s.s.	3/100	1.30gm	2/50	2.60gm	2/50					
2416	143	3.68gm	n.s.s.	6/100	1.20gm	5/50	(2.40gm	0/50)					
a	143	13.2gm	n.s.s.	1/100	1.20gm	1/50	2.40gm	2/50					
b	143	9.39gm	n.s.s.	3/100	1.20gm	5/50	2.40gm	2/50					
c	143	9.31gm	n.s.s.	7/100	1.20gm	7/50	2.40gm	3/50					
2417	176a	5.75mg	n.s.s.	0/5	12.0mg	0/6	1.20gm	0/8					Allmark;jphp,8,417-424;1956
a	176a	5.75mg	n.s.s.	0/5	12.0mg	0/6	1.20gm	0/8					
2418	143	696.mg	n.s.s.	0/25	250.mg	0/25	500.mg	0/25	1.00gm	0/25	2.50gm	0/25	Hansen;fctx,4,389-410;1966
a	143	4.73gm	n.s.s.	10/25	250.mg	15/25	500.mg	16/25	1.00gm	10/25	2.50gm	8/25	
b	143	4.81gm	n.s.s.	6/25	250.mg	6/25	500.mg	9/25	1.00gm	6/25	2.50gm	5/25	
2419	143	557.mg	n.s.s.	1/25	200.mg	0/25	400.mg	0/25	800.mg	0/25	2.00gm	0/25	
a	143	4.57gm	n.s.s.	8/25	200.mg	7/25	400.mg	10/25	800.mg	6/25	2.00gm	5/25	
b	143	4.99gm	n.s.s.	4/25	200.mg	3/25	400.mg	4/25	800.mg	3/25	2.00gm	3/25	
2420	1136	913.mg	n.s.s.	0/50	1.80gm	1/6							Willheim;gaga,23,1-19;1953
FD & C GREEN NO. 3 (fast green FCF) 2353-45-9													
2421	143	21.7gm	n.s.s.	0/100	1.30gm	2/50	2.60gm	0/50					Hansen;fctx,4,389-410;1966
a	143	25.6gm	n.s.s.	2/100	1.30gm	0/50	2.60gm	1/50					
b	143	25.4gm	n.s.s.	3/100	1.30gm	2/50	2.60gm	0/50					
c	143	19.2gm	n.s.s.	5/100	1.30gm	3/50	2.60gm	1/50					
2422	143	19.1gm	n.s.s.	8/100	1.20gm	4/50	2.40gm	0/50					
a	143	21.3gm	n.s.s.	2/100	1.20gm	2/50	2.40gm	0/50					
b	143	3.21gm	n.s.s.	10/100	1.20gm	7/50	(2.40gm	1/50)					
c	143	23.5gm	n.s.s.	8/100	1.20gm	2/50	2.40gm	1/50					
2423	143	696.mg	n.s.s.	0/25	250.mg	0/25	500.mg	0/25	1.00gm	0/25	2.50gm	0/25	
a	143	2.70gm	n.s.s.	6/25	250.mg	6/25	500.mg	10/25	1.00gm	15/25	2.50gm	6/25	
b	143	1.90gm	n.s.s.	13/25	250.mg	10/25	500.mg	11/25	1.00gm	18/25	2.50gm	12/25	
c	143	3.30gm	n.s.s.	7/25	250.mg	6/25	500.mg	3/25	1.00gm	11/25	2.50gm	6/25	
2424	143	10.6gm	n.s.s.	0/25	200.mg	0/25	400.mg	0/25	800.mg	0/25	2.00gm	0/25	
a	143	5.03gm	n.s.s.	2/25	200.mg	3/25	400.mg	5/25	800.mg	3/25	2.00gm	2/25	
b	143	5.37gm	n.s.s.	1/25	200.mg	2/25	400.mg	3/25	800.mg	3/25	2.00gm	1/25	
GRISEOFULVIN 126-07-8													
2425	402	1.01gm	n.s.s.	0/29	314.mg	0/21	1.57gm	0/24	3.14gm	0/22			Rustia;bjca,38,237-249;1978
a	402	3.34gm	n.s.s.	0/16	314.mg	1/9	1.57gm	0/5	3.14gm	0/5			
b	402	4.25gm	n.s.s.	1/16	314.mg	1/9	1.57gm	0/5	3.14gm	0/5			
2426	402	1.13gm	n.s.s.	0/22	276.mg	0/16	1.38gm	0/23	2.76gm	0/24			
a	402	1.13gm	n.s.s.	0/22	276.mg	0/16	1.38gm	0/23	2.76gm	0/24			
2427	402	471.mg	n.s.s.	19/89	130.mg	6/37							
a	402	845.mg	n.s.s.	5/76	130.mg	1/36							
2428	402	270.mg	n.s.s.	0/32	120.mg	1/17							
a	402	525.mg	n.s.s.	6/86	120.mg	2/33							
b	402	326.mg	n.s.s.	16/79	120.mg	6/32							
GUAR GUM 9000-30-0													
2429	TR229	8.80gm	n.s.s.	32/50	3.16gm	26/50	6.32gm	27/50					
a	TR229	27.4gm	n.s.s.	5/50	3.16gm	2/50	6.32gm	4/50					liv:hpa,hpc,nnd.
b	TR229	33.9gm	n.s.s.	5/50	3.16gm	1/50	6.32gm	3/50					lun:a/a,a/c.
2430	TR229	8.14gm	n.s.s.	32/50	2.92gm	33/50	5.89gm	32/50					
a	TR229	7.71gm	n.s.s.	16/50	2.92gm	12/50	(5.89gm	7/50)					liv:hpa,hpc,nnd.
b	TR229	21.6gm	n.s.s.	12/50	2.92gm	9/50	5.89gm	8/50					lun:a/a,a/c.
2431	TR229	2.53gm	n.s.s.	46/50	1.24gm	47/50	2.48gm	46/50					
a	TR229	18.5gm	n.s.s.	2/50	1.24gm	1/50	2.48gm	1/50					liv:hpa,hpc,nnd.
2432	TR229	5.10gm	n.s.s.	0/50	990.mg	1/50	1.98gm	4/50					

S

```
       Spe Strain  Site    Xpo+Xpt                                              TD50     2Tailpvl
            Sex  Route  Hist       Notes                                            DR     AuOp
     a   R m f34 eat TBA MXB 24m24                                              16.5gm * P<.9
     b   R m f34 eat liv MXB 24m24                                              no dre   P=1.

  GUM ARABIC                        100ng..:..1ug...:..10.....:..100....:..1mg...:..10.....:..100...:..1g.....:..10
  2433 M f b6c eat TBA MXB 24m24                                                :>no dre P=1.   -
     a   M f b6c eat liv MXB 24m24                                              31.1gm * P<.06
     b   M f b6c eat lun MXB 24m24                                              24.7gm \ P<.3
  2434 M m b6c eat --- hem 24m24                                                #80.8gm * P<.05 -
     a   M m b6c eat TBA MXB 24m24                                              46.0gm / P<.8
     b   M m b6c eat liv MXB 24m24                                              no dre   P=1.
     c   M m b6c eat lun MXB 24m24                                              no dre   P=1.
  2435 R f f34 eat TBA MXB 24m24                                          :>    8.31gm * P<.7   -
     a   R f f34 eat liv MXB 24m24                                              no dre   P=1.
  2436 R m f34 eat TBA MXB 24m24                                          :>    7.97gm * P<.7   -
     a   R m f34 eat liv MXB 24m24                                              33.5gm * P<.7

  HCDD MIXTURE                      100ng..:..1ug...:..10.....:..100....:..1mg...:..10.....:..100...:..1g.....:..10
  2437 M f b6c gav liv MXA 24m24              : + :                            3.87ug * P<.005 c
     a   M f b6c gav liv hpa 24m24                                             4.31ug * P<.005 c
     b   M f b6c gav TBA MXB 24m24                                             6.60ug * P<.5
     c   M f b6c gav liv MXB 24m24                                             3.87ug * P<.005
     d   M f b6c gav lun MXB 24m24                    .                        63.0ug * P<.9
  2438 M m b6c gav liv MXA 24m24              : +      :                       876.ng * P<.007 c
     a   M m b6c gav liv hpa 24m24                                             1.26ug * P<.005 c
     b   M m b6c gav TBA MXB 24m24                                             1.02ug * P<.2
     c   M m b6c gav liv MXB 24m24                                             876.ng * P<.007
     d   M m b6c gav lun MXB 24m24                                             40.1ug * P<1.
  2439 R f osm gav liv MXA 24m24           : + :                               596.ng * P<.0005c
     a   R f osm gav TBA MXB 24m24                                             no dre   P=1.
     b   R f osm gav liv MXB 24m24                                             596.ng * P<.0005
  2440 R m osm gav liv MXA 24m25           :  ±                                #2.30ug * P<.02  -
     a   R m osm gav TBA MXB 24m25                                             no dre   P=1.
     b   R m osm gav liv MXB 24m25                                             2.30ug * P<.02

  HEMATOXYLIN                       100ng..:..1ug...:..10.....:..100....:..1mg...:..10.....:..100...:..1g.....:..10
  2441 R b wis eat mix mly 78w78 er                                   .    +    1.00gm   P<.002 +
     a   R b wis eat fhd mly 78w78 er                                          2.41gm   P<.02  +
     b   R b wis eat abd mly 78w78 er                                          2.41gm   P<.02  +

  HEPTACHLOR                        100ng..:..1ug...:..10.....:..100....:..1mg...:..10.....:..100...:..1g.....:..10
  2442 M f b6c eat liv hpc 80w89 v                           : + :             1.47mg / P<.0005c
     a   M f b6c eat TBA MXB 80w89 v                                           1.46mg / P<.007
     b   M f b6c eat liv MXB 80w89 v                                           1.47mg / P<.0005
     c   M f b6c eat lun MXB 80w89 v                                           26.5mg * P<.6
  2443 M f b6c eat liv hpc 80w89 v  pool                     : + :             1.37mg / P<.0005c
  2444 M m b6c eat liv hpc 73w89 av                        : +      :          1.09mg / P<.006 c
     a   M m b6c eat TBA MXB 73w89 av                                          1.35mg / P<.03
     b   M m b6c eat liv MXB 73w89 av                                          1.09mg / P<.006
     c   M m b6c eat lun MXB 73w89 av                                          7.15mg * P<.2
  2445 M m b6c eat liv hpc 73w89 av pool                     : + :             1.09mg / P<.0005c
  2446 R f osm eat thy MXA 19m26 v                          : ±                #4.11mg / P<.02  -
     a   R f osm eat TBA MXB 19m26 v                                           125.mg * P<1.
     b   R f osm eat liv MXB 19m26 v                                           no dre   P=1.
  2447 R f osm eat thy MXA 19m25 v  pool                     : + :             #4.05mg / P<.002 -
     a   R f osm eat thy fcc 19m25 v                                           10.5mg * P<.04
  2448 R m osm eat TBA MXB 19m26 v                    :>                       no dre   P=1.
     a   R m osm eat liv MXB 19m26 v                                           18.1mg * P<.5
  2449 R f cfr eat tba tum 26m26 r                 .>                          1.25mg * P<.2   -

     a   R f cfr eat tba mal 26m26 r                                           1.87mg * P<.2   -
     b   R f cfr eat tba ben 26m26 r                                           3.07mg * P<.4   -
  2450 R m cfr eat tba tum 26m26 r                    .>                       no dre   P=1.   -
     a   R m cfr eat tba mal 26m26 r                                           no dre   P=1.   -
     b   R m cfr eat tba ben 26m26 r                                           no dre   P=1.   -

  HEPTAMETHYLENEIMINE               100ng..:..1ug...:..10.....:..100.....:..1mg...:..10.....:..100...:..1g.....:..10
  2451 R f mrc wat liv tum 17m24 e                                .>           no dre   P=1.
     a   R f mrc wat tba mix 17m24 e                                           15.8mg   P<.3
  2452 R m mrc wat liv tum 17m24 e                                  .>         no dre   P=1.
     a   R m mrc wat tba mix 17m24 e                                           6.37mg   P<.07

  HEPTYLAMINE                       100ng..:..1ug...:..10.....:..100.....:..1mg...:..10.....:..100...:..1g.....:..10
  2453 R m wis wat liv tum 61w61 e                             .>              no dre   P=1.
     a   R m wis wat tba mix 61w61 e                                           no dre   P=1.   -

  HEXACHLOROBENZENE                 100ng..:..1ug...:..10.....:..100....:..1mg...:..10.....:..100...:..1g.....:..10
  2454 H f syg eat liv hpt 24m24                          .+ .                 7.42mg * P<.0005+
     a   H f syg eat liv hae 24m24                                             125.mg * P<.003 +
     b   H f syg eat tba mix 24m24                                             7.68mg * P<.0005+
  2455 H m syg eat liv hpt 24m24                        .+.                    5.48mg * P<.0005+
     a   H m syg eat liv hae 24m24                                             32.3mg * P<.0005+
```

	RefNum	LoConf	UpConf	Cntrl	1Dose	1Inc	2Dose	2Inc							Citation or Pathology	Brkly Code
a	TR229	1.29gm	n.s.s.	39/50	990.mg	41/50	1.98gm	42/50								
b	TR229	16.4gm	n.s.s.	3/50	990.mg	0/50	1.98gm	1/50							liv:hpa,hpc,nnd.	

GUM ARABIC   (gum acacia) 9000-01-5

	RefNum	LoConf	UpConf	Cntrl	1Dose	1Inc	2Dose	2Inc							Citation or Pathology	Brkly Code
2433	TR227	6.45gm	n.s.s.	30/50	3.19gm	33/50	6.38gm	31/50								
a	TR227	12.2gm	n.s.s.	3/50	3.19gm	2/50	6.38gm	10/50							liv:hpa,hpc,nnd.	
b	TR227	6.59gm	n.s.s.	3/50	3.19gm	7/50	(6.38gm	1/50)							lun:a/a,a/c.	
2434	TR227	24.5gm	n.s.s.	0/50	2.94gm	0/50	5.89gm	3/50								S
a	TR227	5.01gm	n.s.s.	36/50	2.94gm	28/50	5.89gm	40/50								
b	TR227	12.4gm	n.s.s.	16/50	2.94gm	11/50	5.89gm	15/50							liv:hpa,hpc,nnd.	
c	TR227	12.1gm	n.s.s.	12/50	2.94gm	10/50	5.89gm	12/50							lun:a/a,a/c.	
2435	TR227	1.29gm	n.s.s.	45/50	1.23gm	46/50	2.45gm	47/50								
a	TR227	10.2gm	n.s.s.	3/50	1.23gm	3/50	2.45gm	2/50							liv:hpa,hpc,nnd.	
2436	TR227	1.08gm	n.s.s.	40/50	981.mg	45/50	1.96gm	42/50								
a	TR227	4.15gm	n.s.s.	4/50	981.mg	5/50	1.96gm	5/50							liv:hpa,hpc,nnd.	

HCDD MIXTURE   (1,2,3,7,8,9-hexachlorodibenzo-p-dioxin and 1,2,3,6,7,8-isomer. CAS# 19408-74-3 and 57653-85-7) mixture

	RefNum	LoConf	UpConf	Cntrl	1Dose	1Inc	2Dose	2Inc	3Dose	3Inc					Citation or Pathology	Brkly Code
2437	TR198	1.86ug	36.0ug	3/75	344.ng	4/50	688.ng	6/50	1.39ug	10/50					liv:hpa,hpc.	
a	TR198	2.05ug	42.4ug	2/75	344.ng	4/50	688.ng	4/50	1.39ug	9/50						
b	TR198	1.42ug	n.s.s.	36/75	344.ng	24/50	688.ng	22/50	1.39ug	28/50						
c	TR198	1.86ug	36.0ug	3/75	344.ng	4/50	688.ng	6/50	1.39ug	10/50					liv:hpa,hpc,nnd.	
d	TR198	4.57ug	n.s.s.	2/75	344.ng	2/50	688.ng	5/50	1.39ug	1/50					lun:a/a,a/c.	
2438	TR198	420.ng	16.6ug	15/75	172.ng	14/50	347.ng	14/50	688.ng	24/50					liv:hpa,hpc.	
a	TR198	612.ng	11.9ug	7/75	172.ng	5/50	347.ng	9/50	688.ng	15/50						
b	TR198	350.ng	n.s.s.	40/75	172.ng	33/50	347.ng	29/50	688.ng	38/50						
c	TR198	420.ng	16.6ug	15/75	172.ng	14/50	347.ng	14/50	688.ng	24/50					liv:hpa,hpc,nnd.	
d	TR198	1.04ug	n.s.s.	10/75	172.ng	11/50	347.ng	10/50	688.ng	7/50					lun:a/a,a/c.	
2439	TR198	369.ng	1.33ug	5/75	174.ng	10/50	347.ng	12/50	694.ng	30/50					liv:hpc,nnd.	
a	TR198	756.ng	n.s.s.	54/75	174.ng	36/50	347.ng	33/50	694.ng	41/50						
b	TR198	369.ng	1.33ug	5/75	174.ng	10/50	347.ng	12/50	694.ng	30/50					liv:hpa,hpc,nnd.	
2440	TR198	842.ng	n.s.s.	0/75	175.ng	0/50	347.ng	1/50	694.ng	4/50					liv:hpc,nnd.	S
a	TR198	807.ng	n.s.s.	40/75	175.ng	22/50	347.ng	22/50	694.ng	25/50						
b	TR198	842.ng	n.s.s.	0/75	175.ng	0/50	347.ng	1/50	694.ng	4/50					liv:hpa,hpc,nnd.	

HEMATOXYLIN   517-28-2

	RefNum	LoConf	UpConf	Cntrl	1Dose	1Inc									Citation or Pathology	Brkly Code
2441	1136	231.mg	13.6gm	0/50	1.80gm	2/4									Willheim;gaga,23,1-19;1953	
a	1136	386.mg	n.s.s.	0/50	1.80gm	1/4										
b	1136	386.mg	n.s.s.	0/50	1.80gm	1/4										

HEPTACHLOR   76-44-8

	RefNum	LoConf	UpConf	Cntrl	1Dose	1Inc	2Dose	2Inc	3Dose	3Inc	4Dose	4Inc	5Dose	5Inc	Citation or Pathology	Brkly Code
2442	TR9	.887mg	3.94mg	2/10	1.00mg	3/50	2.10mg	30/50								
a	TR9	.779mg	24.9mg	4/10	1.00mg	12/50	2.10mg	31/50								
b	TR9	.887mg	3.94mg	2/10	1.00mg	3/50	2.10mg	30/50							liv:hpa,hpc,nnd.	
c	TR9	6.52mg	n.s.s.	0/10	1.00mg	1/50	2.10mg	1/50							lun:a/a,a/c.	
2443	TR9	.866mg	2.47mg	3/80p	1.00mg	3/50	2.10mg	30/50								
2444	TR9	.586mg	12.2mg	5/20	.650mg	11/50	1.30mg	34/48								
a	TR9	.645mg	n.s.s.	7/20	.650mg	11/50	1.30mg	34/48								
b	TR9	.586mg	12.2mg	5/20	.650mg	11/50	1.30mg	34/48							liv:hpa,hpc,nnd.	
c	TR9	2.51mg	n.s.s.	1/20	.650mg	1/50	1.30mg	7/48							lun:a/a,a/c.	
2445	TR9	.628mg	2.83mg	17/95p	.650mg	11/50	1.30mg	34/48							thy:fca,fcc.	S
2446	TR9	2.03mg	n.s.s.	1/10	.920mg	3/49	1.85mg	14/50								
a	TR9	.998mg	n.s.s.	9/10	.920mg	37/49	1.85mg	33/50								
b	TR9	3.28mg	n.s.s.	1/10	.920mg	9/49	1.85mg	5/50							liv:hpa,hpc,nnd.	
2447	TR9	2.05mg	16.1mg	3/60p	.920mg	3/49	1.85mg	14/50							thy:fca,fcc.	S
a	TR9	4.01mg	n.s.s.	1/60p	.920mg	2/49	1.85mg	5/50								S
2448	TR9	2.33mg	n.s.s.	7/10	1.10mg	24/50	2.20mg	20/50								
a	TR9	4.46mg	n.s.s.	1/10	1.10mg	3/50	2.20mg	6/50							liv:hpa,hpc,nnd.	
2449	66b	.426mg	n.s.s.	6/20	75.0ug	7/20	.150mg	7/20	.250mg	6/20	.350mg	11/20	.500mg	9/20	Epstein(review) {S Witherup};stev,6,103-154;1976	
a	66b	.695mg	n.s.s.	1/20	75.0ug	3/20	.150mg	2/20	.250mg	4/20	.350mg	6/20	.500mg	3/20		
b	66b	.728mg	n.s.s.	5/20	75.0ug	4/20	.150mg	5/20	.250mg	2/20	.350mg	6/20	.500mg	7/20		
2450	66b	1.34mg	n.s.s.	10/20	60.0ug	2/20	.120mg	6/20	.200mg	6/20	.280mg	4/20	.400mg	3/20		
a	66b	1.46mg	n.s.s.	5/20	60.0ug	1/20	.120mg	5/20	.200mg	2/20	.280mg	1/20	.400mg	3/20		
b	66b	1.82mg	n.s.s.	5/20	60.0ug	1/20	.120mg	1/20	.200mg	4/20	.280mg	3/20	.400mg	0/20		

HEPTAMETHYLENEIMINE   1121-92-2

	RefNum	LoConf	UpConf	Cntrl	1Dose	1Inc									Citation or Pathology	Brkly Code
2451	216	22.7mg	n.s.s.	0/15	7.36mg	0/15									Garcia;zkko,79,141-144;1973	
a	216	4.09mg	n.s.s.	4/15	7.36mg	7/15										
2452	216	19.9mg	n.s.s.	0/15	6.44mg	0/15										
a	216	2.17mg	n.s.s.	5/15	6.44mg	10/15										

HEPTYLAMINE   1241-27-6

	RefNum	LoConf	UpConf	Cntrl	1Dose	1Inc									Citation or Pathology	Brkly Code
2453	104	2.72mg	n.s.s.	0/9	1.92mg	0/20									Argus;jnci,35,949-958;1965	
a	104	1.74mg	n.s.s.	1/9	1.92mg	1/20										

HEXACHLOROBENZENE   118-74-1

	RefNum	LoConf	UpConf	Cntrl	1Dose	1Inc	2Dose	2Inc	3Dose	3Inc					Citation or Pathology	Brkly Code
2454	151a	5.47mg	10.3mg	0/40	5.23mg	14/30	10.5mg	17/30	20.9mg	51/60					Cabral;natu,269,510-511;1977	
a	151a	58.8mg	577.mg	0/40	5.23mg	0/30	10.5mg	2/30	20.9mg	7/60						
b	151a	5.41mg	11.9mg	5/40	5.23mg	16/30	10.5mg	18/30	20.9mg	52/60						
2455	151a	4.05mg	7.55mg	0/40	4.60mg	14/30	9.20mg	26/30	18.4mg	49/59						
a	151a	20.4mg	56.1mg	0/40	4.60mg	1/30	9.20mg	6/30	18.4mg	20/59						

```
       Spe Strain Site   Xpo+Xpt
       Sex  Route  Hist  Notes                                              TD50    2Tailpvl
                                                                              DR     AuOp
   b   H m syg eat thy ald 24m24                                           108.mg * P<.002 +
   c   H m syg eat tba mix 24m24                                           3.72mg * P<.0005+
2456 M m c5c eat liv hnd 52w52 ekr                             .>          17.5mg   P<.3
   a   M m c5c eat liv hpc 52w52 ekr                                       no dre   P=1.
2457 M m c5c eat liv tum 78w78 er                                 .>       no dre   P=1.   -
2458 M f swi eat liv lct 86w93 aes                           . + .         46.4mg * P<.0005+
   a   M f swi eat lun tum 86w93 aes                                       no dre   P=1.
   b   M f swi eat tba mix 86w93 aes                                       no dre   P=1.
2459 M m swi eat liv lct 23m25 aes                             . + .       109.mg * P<.0005+
   a   M m swi eat lun tum 23m25 aes                                       no dre   P=1.
   b   M m swi eat tba mix 23m25 aes                                       no dre   P=1.
2460 R f agu eat liv lct 90w90 r                        <+                 noTD50   P<.0005+
2461 R m cdr eat liv nnd 28m28 be                     .     ±              27.3mg   P<.1
   a   R m cdr eat liv hpc 28m28 be                                        55.3mg   P<.3
   b   R m cdr eat liv blc 28m28 be                                        55.3mg   P<.3
2462 R m cdr eat liv nnd 28m28 e                        .   ±              17.9mg   P<.05
   a   R m cdr eat liv kcs 28m28 e                                         55.3mg   P<.3
2463 R f f34 eat liv mix 90w90 r                    . + .                  4.67mg   P<.0005+
   a   R f f34 eat liv hpc 90w90 r                                         12.7mg   P<.006 +
   b   R f f34 eat liv nnd 90w90 r                                         12.7mg   P<.006 +
2464 R m f34 eat liv mix 90w90 r                      .    ±               28.7mg   P<.09  +

HEXACHLOROBUTADIENE              100ng..:..1ug....:..10.....:..100.....:..1mg....:..10.....:..100....:..1g.....:..10
2465 R f sss eat kid mix 24m24 e                           . + .          94.4mg * P<.0005+
   a   R f sss eat liv mht 24m24 e                                         no dre   P=1.   -
   b   R f sss eat tba mix 24m24 e                                         9.84mg * P<.1
2466 R m sss eat kid mix 95w95 es                          . + .          50.5mg * P<.0005+
   a   R m sss eat liv hpc 95w95 es                                        no dre   P=1.   -
   b   R m sss eat tba mix 95w95 es                                        no dre   P=1.

alpha-1,2,3,4,5,6-HEXACHLOROCYCLOHEXANE..:..1_ug..:..10.....:..100.....:..1mg....:..10.....:..100....:..1g.....:..10
2467 M m ddy eat liv mix 36w72 ekr                           <+            noTD50   P<.0005+
2468 R m buf eat liv mix 35w65 ekr                        .>               no dre   P=1.   -
2469 R m wis eat liv nod 72w72 ekr                      . + .              11.2mg * P<.0005+
   a   R m wis eat liv hpc 72w72 ekr                                       107.mg * P<.09  +

beta-1,2,3,4,5,6-HEXACHLOROCYCLOHEXANE ..:..1ug....:..10.....:..100.....:..1mg....:..10.....:..100....:..1g......:..10
2470 M f cf1 eat liv lct 26m26 e                           . + .   .       139.mg   P<.007 +
   a   M f cf1 eat liv mix 26m26 e                                         64.3mg   P<.07  +
   b   M f cf1 eat lun tum 26m26 e                                         no dre   P=1.   -
2471 M m cf1 eat liv mix 26m26 e                        . + .              17.7mg   P<.0005+
   a   M m cf1 eat liv lct 26m26 e                                         51.1mg   P<.002 +
   b   M m cf1 eat lun tum 26m26 e                                         no dre   P=1.   -

gamma-1,2,3,4,5,6-HEXACHLOROCYCLOHEXANE..:..1_ug....:..10.....:..100.....:..1mg....:..10.....:..100....:..1g......:..10
2472 M f b6c eat TBA MXB 80w90                              :>             no dre   P=1.   -
   a   M f b6c eat liv MXB 80w90                                           no dre   P=1.
   b   M f b6c eat lun MXB 80w90                                           no dre   P=1.
2473 M m b6c eat TBA MXB 80w90                           :>                no dre   P=1.   -
   a   M m b6c eat liv MXB 80w90                                           no dre   P=1.
   b   M m b6c eat lun MXB 80w90                                           no dre   P=1.
2474 M m b6c eat liv hpc 80w90     pool              :  +    :            #12.0mg \ P<.004 -
   a   M m b6c eat liv MXA 80w90                                           15.3mg \ P<.04
2475 M f baa eat lun tum 26w52 er                         .>               1.26gm   P<1.
   a   M f baa eat liv hpc 26w52 er                                        no dre   P=1.   -
   b   M f baa eat liv hpa 26w52 er                                        no dre   P=1.
2476 M f baa eat lun tum  6m24 er                            .>            15.9gm   P<1.
   a   M f baa eat liv hpc  6m24 er                                        no dre   P=1.   -
   b   M f baa eat liv hpa  6m24 er                                        no dre   P=1.
2477 M f baa eat liv hpa 52w52 er                            .>            no dre   P=1.   -
   a   M f baa eat liv hpc 52w52 er                                        no dre   P=1.   -
   b   M f baa eat lun tum 52w52 er                                        no dre   P=1.
2478 M f baa eat lun tum 24m24 er                             .>           1.33gm   P<.7   -
   a   M f baa eat liv hpc 24m24 er                                        no dre   P=1.   -
   b   M f baa eat liv hpa 24m24 er                                        no dre   P=1.   -
   c   M f baa eat liv mix 24m24 er                                        no dre   P=1.   -
2479 M f cf1 eat liv mix 26m26 e                        . + .              43.7mg   P<.0005+
   a   M f cf1 eat liv lct 26m26 e                                         94.3mg   P<.0005+
   b   M f cf1 eat lun tum 26m26 e                                         no dre   P=1.   -
2480 M m cf1 eat liv mix 26m26 e                      . + .                15.4mg   P<.0005+
   a   M m cf1 eat liv lct 26m26 e                                         48.6mg   P<.0005+
   b   M m cf1 eat lun tum 26m26 e                                         no dre   P=1.   -
2481 M f nmr eat liv mix 80w80 r                          .>               803.mg * P<1.   -
2482 M m nmr eat liv mix 80w80 r                          .>               no dre   P=1.   -
   a   M m nmr eat liv ret 80w80 r                                         no dre   P=1.   -
2483 M f pva eat lun tum 52w52 er                            .>            no dre   P=1.   -
   a   M f pva eat liv hpc 52w52 er                                        no dre   P=1.   -
   b   M f pva eat liv hpa 52w52 er                                        no dre   P=1.   -
2484 M f pva eat liv mix 24m24 er                          .   ±           132.mg   P<.05  +
   a   M f pva eat lun tum 24m24 er                                        170.mg   P<.09  +
   b   M f pva eat liv hpa 24m24 er                                        207.mg   P<.2
```

	RefNum	LoConf	UpConf	Cntrl	1Dose	1Inc	2Dose	2Inc			Citation or Pathology
											Brkly Code
b	151a	50.9mg	382.mg	0/40	4.60mg	0/30	9.20mg	1/30	18.4mg	8/59	
c	151a	2.62mg	5.40mg	3/40	4.60mg	18/30	9.20mg	27/30	18.4mg	56/59	
2456	1938m	2.83mg	n.s.s.	0/9	12.0mg	1/9					Smith;ijcn,43,492-496;1989/pers.comm.
a	1938m	5.56mg	n.s.s.	0/9	12.0mg	0/9					
2457	1938n	15.3mg	n.s.s.	0/11	12.0mg	0/11					
2458	384	26.3mg	93.4mg	0/49	6.50mg	0/30	13.0mg	3/30	26.0mg	14/41	Cabral;ijcn,23,47-51;1979
a	384	32.2mg	n.s.s.	14/49	6.50mg	4/30	13.0mg	6/30	(26.0mg	2/41)	
b	384	63.5mg	n.s.s.	39/49	6.50mg	21/30	13.0mg	13/30	26.0mg	19/41	
2459	384	53.3mg	329.mg	0/47	6.00mg	0/30	12.0mg	3/29	24.0mg	7/44	
a	384	189.mg	n.s.s.	13/47	6.00mg	4/30	12.0mg	0/29	24.0mg	4/44	
b	384	86.7mg	n.s.s.	22/47	6.00mg	15/30	12.0mg	10/29	24.0mg	12/44	
2460	1180	n.s.s.	1.65mg	0/12	5.00mg	14/14					Smith;clet,11,169-172;1980
2461	1833m	6.71mg	n.s.s.	0/39	1.60mg	2/39					Arnold;fctx,23,779-793;1985
a	1833m	9.00mg	n.s.s.	0/39	1.60mg	1/39					
b	1833m	9.00mg	n.s.s.	0/39	1.60mg	1/39					
2462	1833n	5.43mg	n.s.s.	0/38	1.60mg	3/39					
a	1833n	9.00mg	n.s.s.	0/38	1.60mg	1/39					
2463	1708	2.16mg	12.4mg	0/15	10.0mg	10/15					Smith;carc,6,631-636;1985
a	1708	4.75mg	120.mg	0/15	10.0mg	5/15					
b	1708	4.75mg	120.mg	0/15	10.0mg	5/15					
2464	1708	7.04mg	n.s.s.	0/15	8.00mg	2/15					
	HEXACHLOROBUTADIENE		87-68-3								
2465	373	38.5mg	338.mg	0/90	.200mg	0/40	2.00mg	0/40	20.0mg	6/40	Kociba;amih,38,589-602;1977
a	373	1.49mg	n.s.s.	1/90	.200mg	0/40	2.00mg	0/40	20.0mg	0/40	
b	373	2.42mg	n.s.s.	82/90	.200mg	35/40	2.00mg	37/40	20.0mg	39/40	
2466	373	23.4mg	148.mg	1/90	.200mg	0/40	2.00mg	0/40	20.0mg	9/39	
a	373	1.24mg	n.s.s.	1/90	.200mg	0/40	2.00mg	0/40	20.0mg	0/39	
b	373	34.8mg	n.s.s.	39/90	.200mg	24/40	2.00mg	13/40	20.0mg	15/39	
	alpha-1,2,3,4,5,6-HEXACHLOROCYCLOHEXANE		(alpha-lindane)	319-84-6							
2467	1149n	n.s.s.	6.62mg	0/18	30.0mg	13/13					Ito;canr,36,2227-2234;1976
2468	1071	6.07mg	n.s.s.	0/8	10.8mg	0/7					Angsubhakorn;bjca,43,881-883;1981
2469	45a	6.37mg	23.3mg	0/8	40.0mg	12/16	60.0mg	10/13			Ito;jnci,54,801-804;1975
a	45a	37.0mg	n.s.s.	0/8	40.0mg	1/16	60.0mg	3/13			
	beta-1,2,3,4,5,6-HEXACHLOROCYCLOHEXANE		(beta-lindane)	319-85-7							
2470	89	48.1mg	1.92gm	0/44	26.0mg	4/30					Thorpe;fctx,11,433-442;1973
a	89	23.2mg	n.s.s.	10/44	26.0mg	13/30					
b	89	116.mg	n.s.s.	27/44	26.0mg	5/30					
2471	89	9.03mg	51.0mg	11/45	24.0mg	22/30					
a	89	23.1mg	231.mg	2/45	24.0mg	10/30					
b	89	79.6mg	n.s.s.	27/45	24.0mg	8/30					
	gamma-1,2,3,4,5,6-HEXACHLOROCYCLOHEXANE		(lindane)	58-89-9							
2472	TR14	46.4mg	n.s.s.	4/10	9.20mg	7/50	18.5mg	8/50			
a	TR14	56.9mg	n.s.s.	1/10	9.20mg	4/50	18.5mg	3/50			liv:hpa,hpc,nnd.
b	TR14	86.0mg	n.s.s.	1/10	9.20mg	1/50	18.5mg	2/50			lun:a/a,a/c.
2473	TR14	19.9mg	n.s.s.	4/10	8.50mg	21/50	17.0mg	15/50			
a	TR14	25.8mg	n.s.s.	3/10	8.50mg	19/50	17.0mg	10/50			liv:hpa,hpc,nnd.
b	TR14	65.5mg	n.s.s.	2/10	8.50mg	2/50	17.0mg	3/50			lun:a/a,a/c.
2474	TR14	5.84mg	95.4mg	5/50p	8.50mg	19/50	(17.0mg	9/50)			S
a	TR14	6.39mg	n.s.s.	8/50p	8.50mg	19/50	(17.0mg	10/50)			liv:hpc,nnd. S
2475	1828m	14.6mg	n.s.s.	1/48	10.4mg	1/45					Wolff;carc,8,1889-1897;1987
a	1828m	24.1mg	n.s.s.	0/48	10.4mg	0/45					
b	1828m	24.1mg	n.s.s.	0/48	10.4mg	0/45					
2476	1828n	50.4mg	n.s.s.	2/96	5.20mg	2/95					
a	1828n	55.5mg	n.s.s.	3/96	5.20mg	2/95					
b	1828n	54.8mg	n.s.s.	6/96	5.20mg	3/95					
2477	1828o	51.4mg	n.s.s.	0/48	20.8mg	0/48					
a	1828o	51.4mg	n.s.s.	0/48	20.8mg	0/48					
b	1828o	51.4mg	n.s.s.	1/48	20.8mg	0/48					
2478	1828r	160.mg	n.s.s.	2/96	20.8mg	3/96					
a	1828r	297.mg	n.s.s.	3/96	20.8mg	1/96					
b	1828r	222.mg	n.s.s.	6/96	20.8mg	3/96					
c	1828r	219.mg	n.s.s.	9/96	20.8mg	4/96					
2479	89	21.9mg	140.mg	10/44	52.0mg	20/29					Thorpe;fctx,11,433-442;1973
a	89	45.6mg	243.mg	0/44	52.0mg	10/29					
b	89	133.mg	n.s.s.	27/44	52.0mg	10/29					
2480	89	7.43mg	34.4mg	11/45	48.0mg	27/29					
a	89	25.8mg	114.mg	2/45	48.0mg	16/29					
b	89	95.0mg	n.s.s.	27/45	48.0mg	12/29					
2481	1119	16.7mg	n.s.s.	6/100	1.63mg	7/50	3.25mg	2/50	6.50mg	4/50	Herbst;txcy,4,91-96;1975
2482	1119	22.9mg	n.s.s.	7/100	1.50mg	5/50	3.00mg	1/50	6.00mg	3/50	
a	1119	25.2mg	n.s.s.	0/100	1.50mg	3/50	3.00mg	0/50	6.00mg	0/50	
2483	1828o	49.3mg	n.s.s.	1/46	20.8mg	0/46					Wolff;carc,8,1889-1897;1987
a	1828o	51.4mg	n.s.s.	0/48	20.8mg	0/48					
b	1828o	51.4mg	n.s.s.	0/46	20.8mg	0/48					
2484	1828r	53.5mg	n.s.s.	7/95	20.8mg	16/95					
a	1828r	62.5mg	n.s.s.	6/95	20.8mg	13/94					
b	1828r	71.0mg	n.s.s.	5/95	20.8mg	11/95					

```
     Spe Strain Site    Xpo+Xpt
     Sex  Route  Hist    Notes                                          TD50    2Tailpvl
                                                                                DR      AuOp
c    M f pva eat liv hpc 24m24 er                                       435.mg  P<.3
2485 M f yva eat lun tum 26w52 er                        .>             84.6mg  P<.3
a    M f yva eat liv hpc 26w52 er                                       no dre  P=1.   -
b    M f yva eat liv hpa 26w52 er                                       no dre  P=1.   -
2486 M f yva eat lun tum 6m24  er                   .    ±              52.2mg  P<.1
a    M f yva eat liv hpa 6m24  er                                       62.3mg  P<.3   -
b    M f yva eat liv hpc 6m24  er                                       396.mg  P<.9   -
2487 M f yva eat liv hpa 52w52 er                        .>             162.mg  P<.7
a    M f yva eat lun tum 52w52 er                                       169.mg  P<.3
b    M f yva eat liv hpc 52w52 er                                       no dre  P=1.
2488 M f yva eat liv mix 24m24 er                     .  +  .           28.8mg  P<.0005+
a    M f yva eat liv hpa 24m24 er                                       41.6mg  P<.0005+
b    M f yva eat lun tum 24m24 er                                       85.3mg  P<.002 +
c    M f yva eat liv hpc 24m24 er                                       294.mg  P<.5   -
2489 R f osm eat TBA MXB 19m25 sv            :>                         no dre  P=1.   -
a    R f osm eat liv MXB 19m25 sv                                       no dre  P=1.
2490 R m osm eat TBA MXB 19m25 v                 :>                     38.4mg * P<.6   -
a    R m osm eat liv MXB 19m25 v                                        143.mg * P<.8
2491 R m osm eat --- hem 19m25 v  pool               :    ±           #131.mg * P<.03  -


HEXACHLOROCYCLOHEXANE, TECHNICAL GRADE ..:...1ug....:..10......:..100....:..1mg....:..10......:..100....:..1g......:..10
2492 M f swi eat liv mix 66w66 r          <+                            noTD50  P<.0005+
2493 M m swi eat liv hpc 52w52 r              .    ±                    25.3mg  P<.05  +
2494 M m swi eat liv mix 66w66 r          <+                            noTD50  P<.002 +


HEXACHLOROCYCLOPENTADIENE          100ng..:...1ug....:..10......:..100....:..1mg....:..10......:..100....:..1g......:..10
2495 M f b6c inh TBA MXB 24m24             :>                           no dre  P=1.   -
a    M f b6c inh liv MXB 24m24                                          no dre  P=1.
b    M f b6c inh lun MXB 24m24                                          10.3mg * P<.7
2496 M m b6c inh TBA MXB 24m24            :>                            no dre  P=1.   -
a    M m b6c inh liv MXB 24m24                                          no dre  P=1.
b    M m b6c inh lun MXB 24m24                                          2.45mg * P<.3
2497 M m b6c inh lun MXA 15m24           :>                             1.05mg  P<.2
a    M m b6c inh lun MXB 15m24                                          1.05mg  P<.2
2498 M m b6c inh lun a/c 10m24           :  +  :                        2.21mg  P<.004
a    M m b6c inh lun MXB 10m24                                          3.11mg  P<.5
2499 M m b6c inh lun a/c 8m24         :  ±                              1.15mg  P<.02
a    M m b6c inh lun MXB 8m24                                           2.04mg  P<.7
2500 M m b6c inh lun a/c 6m24         :  ±                              1.98mg  P<.02
a    M m b6c inh lun MXB 6m24                                           10.9mg  P<.9
2501 R f f34 inh thy ccr 24m24        :  ±                            #.838mg * P<.05  -
a    R f f34 inh TBA MXB 24m24                                          1.81mg * P<.9
b    R f f34 inh liv MXB 24m24                                          no dre  P=1.
2502 R m f34 inh pit pda 24m24      :  ±                              #.121mg * P<.05  -
a    R m f34 inh TBA MXB 24m24                                          .188mg * P<.4
b    R m f34 inh liv MXB 24m24                                          .932mg * P<.2


HEXACHLOROETHANE          100ng..:...1ug....:..10......:..100....:..1mg....:..10......:..100....:..1g......:..10
2503 M f b6c gav liv hpc 78w90 v                         :>             873.mg * P<.2   c
a    M f b6c gav TBA MXB 78w90 v                                        1.19gm * P<.4
b    M f b6c gav liv MXB 78w90 v                                        873.mg * P<.2
c    M f b6c gav lun MXB 78w90 v                                        6.95gm * P<.4
2504 M f b6c gav liv hpc 78w90 v  pool               :  +  :            319.mg \ P<.0005c
2505 M m b6c gav liv hpc 78w90 v                       :>               585.mg * P<.2   c
a    M m b6c gav TBA MXB 78w90 v                                        675.mg * P<.4
b    M m b6c gav liv MXB 78w90 v                                        585.mg * P<.2
c    M m b6c gav lun MXB 78w90 v                                        23.6gm * P<1.
2506 M m b6c gav liv hpc 78w90 v  pool               :  +  :            359.mg * P<.0005c
2507 R f osm gav TBA MXB 18m26 dv                   :>                  no dre  P=1.   -
a    R f osm gav liv MXB 18m26 dv                                       no dre  P=1.
2508 R f f34 gav TBA MXB 24m24                   :>                     no dre  P=1.   -
a    R f f34 gav liv MXB 24m24                                          276.gm * P<1.
2509 R m osm gav TBA MXB 18m26 dv                  :>                   no dre  P=1.   -
a    R m osm gav liv MXB 18m26 dv                                       no dre  P=1.
2510 R m f34 gav MXA MXA 24m24          :    ±                          8.02mg \ P<.03  e
a    R m f34 gav MXA MXA 24m24                                          9.05mg \ P<.04
b    R m f34 gav kid MXA 24m24                                          55.4mg * P<.02  c
c    R m f34 gav kid ruc 24m24                                          159.mg * P<.04
d    R m f34 gav TBA MXB 24m24                                          28.0mg * P<.5
e    R m f34 gav liv MXB 24m24                                          1.34gm * P<.9


HEXACHLOROPHENE          100ng..:...1ug....:..10......:..100....:..1mg....:..10......:..100....:..1g......:..10
2511 M f c51 eat liv hpt 24m24                        .>                434.mg  P<.3   -
a    M f c51 eat tba mix 24m24                                          no dre  P=1.   -
2512 M m c51 eat liv tum 24m24                          .>              no dre  P=1.
a    M m c51 eat lun tum 24m24                                          no dre  P=1.
b    M m c51 eat tba mix 24m24                                          900.mg  P<1.
2513 M f xvi eat lun tum 24m24                       .>                 33.4mg  P<.2
a    M f xvi eat liv tum 24m24                                          no dre  P=1.   -
b    M f xvi eat tba mix 24m24                                          39.4mg  P<.3   -
2514 M m xvi eat lun tum 24m24                     .    ±               17.8mg  P<.07  -
```

	RefNum	LoConf	UpConf	Cntrl	1Dose	1Inc	2Dose	2Inc	Citation or Pathology
									Brkly Code
c	1828r	113.mg	n.s.s.	2/95	20.8mg	5/95			
2485	1828m	13.8mg	n.s.s.	0/48	10.4mg	1/48			
a	1828m	25.7mg	n.s.s.	0/48	10.4mg	0/48			
b	1828m	25.7mg	n.s.s.	2/48	10.4mg	0/48			
2486	1828n	18.4mg	n.s.s.	4/95	5.20mg	10/95			
a	1828n	17.4mg	n.s.s.	8/93	5.20mg	13/95			
b	1828n	21.5mg	n.s.s.	12/93	5.20mg	13/95			
2487	1828o	19.7mg	n.s.s.	2/48	20.8mg	3/48			
a	1828o	27.6mg	n.s.s.	0/48	20.8mg	1/48			
b	1828o	51.4mg	n.s.s.	0/48	20.8mg	0/48			
2488	1828r	17.1mg	71.0mg	20/93	20.8mg	49/94			
a	1828r	24.6mg	97.7mg	8/93	20.8mg	33/94			
b	1828r	43.3mg	383.mg	4/95	20.8mg	18/95			
c	1828r	65.1mg	n.s.s.	12/93	20.8mg	16/94			
2489	TR14	3.27mg	n.s.s.	9/10	4.90mg	44/50	(9.80mg	34/50)	
a	TR14	24.7mg	n.s.s.	0/10	4.90mg	4/50	9.80mg	2/50	liv:hpa,hpc,nnd.
2490	TR14	7.29mg	n.s.s.	3/10	6.90mg	21/50	13.6mg	19/50	
a	TR14	27.8mg	n.s.s.	0/10	6.90mg	3/50	13.6mg	2/50	liv:hpa,hpc,nnd.
2491	TR14	39.5mg	n.s.s.	0/55p	6.90mg	0/50	13.6mg	3/50	
									S

HEXACHLOROCYCLOHEXANE, TECHNICAL GRADE   608-73-1

	RefNum	LoConf	UpConf	Cntrl	1Dose	1Inc	2Dose	2Inc	Citation or Pathology
2492	2164	n.s.s.	12.0mg	1/20	65.0mg	14/14			Munir;tumo,69,383-386;1983
2493	1744	10.9mg	n.s.s.	0/6	60.0mg	7/21			Kandarkar;ijmr,78,155-161;1983
2494	2164	n.s.s.	12.6mg	2/22	60.0mg	12/12			Munir;tumo,69,383-386;1983

HEXACHLOROCYCLOPENTADIENE   77-47-4

	RefNum	LoConf	UpConf	Cntrl	1Dose	1Inc	2Dose	2Inc	Citation or Pathology
2495	TR437	.796mg	n.s.s.	34/50	34.5ug	37/50	.173mg	33/50	.691mg	20/50
a	TR437	1.18mg	n.s.s.	9/50	34.5ug	12/50	.173mg	10/50	.691mg	6/50
b	TR437	1.17mg	n.s.s.	7/50	34.5ug	4/50	.173mg	5/50	.691mg	5/50	liv:hpa,hpb,hpc. lun:a/a,a/c.
2496	TR437	.535mg	n.s.s.	35/50	28.9ug	32/50	.145mg	39/50	.580mg	33/50
a	TR437	.866mg	n.s.s.	24/50	28.9ug	21/50	.145mg	28/50	.580mg	20/50
b	TR437	.649mg	n.s.s.	11/50	28.9ug	11/50	.145mg	14/50	.580mg	16/50	liv:hpa,hpb,hpc. lun:a/a,a/c.
2497	TR437a	.323mg	n.s.s.	11/50	.371mg	17/50			lun:a/a,a/c.
a	TR437a	.323mg	n.s.s.	11/50	.371mg	17/50			lun:a/a,a/c.
2498	TR437b	.899mg	13.4mg	0/50	.590mg	6/50			
a	TR437b	.668mg	n.s.s.	11/50	.590mg	14/50			lun:a/a,a/c. S
2499	TR437c	.392mg	n.s.s.	0/50	.185mg	4/50			S
a	TR437c	.262mg	n.s.s.	11/50	.185mg	13/50			lun:a/a,a/c.
2500	TR437d	.750mg	n.s.s.	0/50	.365mg	5/50			S
a	TR437d	.577mg	n.s.s.	11/50	.365mg	14/50			lun:a/a,a/c.
2501	TR437	.291mg	n.s.s.	0/50	8.26ug	1/50	41.4ug	3/50	.166mg	4/50
a	TR437	.101mg	n.s.s.	47/50	8.26ug	48/50	41.4ug	41/50	.166mg	49/50	S
b	TR437	.725mg	n.s.s.	1/50	8.26ug	1/50	41.4ug	1/50	.166mg	1/50	liv:hpa,hpb,hpc.
2502	TR437	47.1ug	n.s.s.	23/50	5.78ug	23/50	29.0ug	23/50	.116mg	33/50	S
a	TR437	44.4ug	n.s.s.	47/50	5.78ug	45/50	29.0ug	43/50	.116mg	49/50	
b	TR437	.225mg	n.s.s.	1/50	5.78ug	1/50	29.0ug	1/50	.116mg	3/50	liv:hpa,hpb,hpc.

HEXACHLOROETHANE   67-72-1

	RefNum	LoConf	UpConf	Cntrl	1Dose	1Inc	2Dose	2Inc	Citation or Pathology
2503	TR68	362.mg	n.s.s.	2/20	361.mg	20/50	722.mg	15/50	
a	TR68	313.mg	n.s.s.	8/20	361.mg	32/50	722.mg	26/50	
b	TR68	362.mg	n.s.s.	2/20	361.mg	20/50	722.mg	15/50	
c	TR68	1.65gm	n.s.s.	1/20	361.mg	1/50	722.mg	4/50	liv:hpa,hpc,nnd. lun:a/a,a/c.
2504	TR68	178.mg	760.mg	2/60p	361.mg	20/50	(722.mg	15/50)	
2505	TR68	236.mg	n.s.s.	3/20	361.mg	15/50	722.mg	31/50	
a	TR68	216.mg	n.s.s.	4/20	361.mg	17/50	722.mg	34/50	
b	TR68	236.mg	n.s.s.	3/20	361.mg	15/50	722.mg	31/50	
c	TR68	883.mg	n.s.s.	0/20	361.mg	2/50	722.mg	3/50	liv:hpa,hpc,nnd. lun:a/a,a/c.
2506	TR68	213.mg	958.mg	6/60p	361.mg	15/50	722.mg	31/50	
2507	TR68	169.mg	n.s.s.	14/20	105.mg	33/50	210.mg	20/50	
a	TR68	n.s.s.	n.s.s.	0/20	105.mg	0/50	210.mg	0/50	liv:hpa,hpc,nnd.
2508	TR361	89.8mg	n.s.s.	50/50	56.1mg	44/50	111.mg	43/50	
a	TR361	384.mg	n.s.s.	1/50	56.1mg	3/50	111.mg	1/50	liv:hpa,hpc,nnd.
2509	TR68	118.mg	n.s.s.	9/20	105.mg	17/50	210.mg	11/50	
a	TR68	n.s.s.	n.s.s.	0/20	105.mg	0/50	210.mg	0/50	liv:hpa,hpc,nnd.
2510	TR361	3.35mg	n.s.s.	15/50	7.01mg	28/50	(14.0mg	21/50)	adr:pob; amd:pbb,phc,phm,pob.
a	TR361	3.70mg	n.s.s.	14/50	7.01mg	26/50	(14.0mg	19/50)	adr:pob; amd:pbb,pob. S
b	TR361	23.3mg	n.s.s.	1/50	7.01mg	2/50	14.0mg	7/50	kid:ade,rua,ruc.
c	TR361	47.3mg	n.s.s.	0/50	7.01mg	0/50	14.0mg	3/50	S
d	TR361	6.04mg	n.s.s.	45/50	7.01mg	48/50	14.0mg	45/50	
e	TR361	53.9mg	n.s.s.	2/50	7.01mg	1/50	14.0mg	2/50	liv:hpa,hpc,nnd.

HEXACHLOROPHENE   70-30-4

	RefNum	LoConf	UpConf	Cntrl	1Dose	1Inc	2Dose	2Inc	Citation or Pathology
2511	706	70.7mg	n.s.s.	0/25	19.5mg	1/33			Rudali;clet,5,325-332;1978
a	706	31.2mg	n.s.s.	7/25	19.5mg	9/33			
2512	706	130.mg	n.s.s.	0/25	18.0mg	0/35			
a	706	130.mg	n.s.s.	1/25	18.0mg	0/35			
b	706	37.0mg	n.s.s.	4/25	18.0mg	6/35			
2513	706	10.7mg	n.s.s.	21/38	19.5mg	28/40			
a	706	161.mg	n.s.s.	0/38	19.5mg	0/40			
b	706	11.1mg	n.s.s.	22/38	19.5mg	28/40			
2514	706	6.56mg	n.s.s.	23/37	18.0mg	30/37			

```
        Spe Strain  Site  Xpo+Xpt                                                    TD50     2Tailpvl
            Sex Route   Hist    Notes                                                    DR    AuOp
   a    M m xvi eat liv ade 24m24                                                     450.mg   P<.3   -
   b    M m xvi eat tba mix 24m24                                                     17.8mg   P<.07  -
 2515 R f f34 eat TBA MXB 25m25                          :>                           no dre   P=1.   -
   a    R f f34 eat liv MXB 25m25                                                     no dre   P=1.   -
 2516 R m f34 eat sub fbs 25m25                          :      ±                     #41.3mg * P<.02  -
   a    R m f34 eat tnv men 25m25                                                     57.5mg * P<.05  -
   b    R m f34 eat TBA MXB 25m25                                                     30.4mg * P<.5   -
   c    R m f34 eat liv MXB 25m25                                                     no dre   P=1.   -
```

3-(HEXAHYDRO-4,7-METHANOINDAN-5-YL)-1,1-DIMETHYLUREA..1_0....:..100....:...1mg..:..10....:...100....:...1g....:..10

```
 2517 M f b6a orl liv hpt 76w76 evx                                           .>      no dre   P=1.   -
   a    M f b6a orl lun ade 76w76 evx                                                 no dre   P=1.   -
   b    M f b6a orl tba mix 76w76 evx                                                 no dre   P=1.   -
 2518 M m b6a orl lun ade 76w76 evx                                       .>          no dre   P=1.   -
   a    M m b6a orl liv hpt 76w76 evx                                                 no dre   P=1.   -
   b    M m b6a orl tba mix 76w76 evx                                                 no dre   P=1.   -
 2519 M f b6c orl lun ade 76w76 evx                                     .>            1.23gm   P<.3   -
   a    M f b6c orl liv hpt 76w76 evx                                                 no dre   P=1.   -
   b    M f b6c orl tba mix 76w76 evx                                                 1.23gm   P<.3   -
 2520 M m b6c orl liv hpt 76w76 evx                                .      ±           359.mg   P<.04  -
   a    M m b6c orl lun ade 76w76 evx                                                 no dre   P=1.   -
   b    M m b6c orl tba mix 76w76 evx                                                 259.mg   P<.02  -
```

HEXAMETHYLENETETRAMINE   100ng..:..1ug....:..10.......:..100....:..1mg......:..10....:...100....:...1g.....:..10

```
 2521 M f c3d wat liv agm 14m24 e                                              .>     51.9gm   P<.7   -
   a    M f c3d wat liv hpt 14m24 e                                                   no dre   P=1.   -
   b    M f c3d wat lun ade 14m24 e                                                   no dre   P=1.   -
   c    M f c3d wat tba mix 14m24 e                                                   no dre   P=1.   -
 2522 M m c3d wat liv agm 14m24 e                                             .> 32.3gm   P<.4   -
   a    M m c3d wat lun ade 14m24 e                                                   293.gm   P<1.   -
   b    M m c3d wat liv hpt 14m24 e                                                   no dre   P=1.   -
   c    M m c3d wat tba mix 14m24 e                                                   no dre   P=1.   -
 2523 M f ctn wat mix tum 12m23 ae                                                    .20.9gm * P<.02  -
   a    M f ctn wat hag tum 12m23 ae                                                  26.3gm  Z P<.05  -
   b    M f ctn wat liv agm 12m23 ae                                                  no dre   P=1.   -
   c    M f ctn wat lun ade 12m23 ae                                                  no dre   P=1.   -
   d    M f ctn wat tba mix 12m23 ae                                                  11.6gm * P<.6   -
 2524 M m ctn wat lun ade 12m23 ae                                                    no dre   P=1.   -
   a    M m ctn wat liv agm 12m23 ae                                                  no dre   P=1.   -
   b    M m ctn wat tba mix 12m23 ae                                                  no dre   P=1.   -
 2525 M f swr wat lun ade 14m24 e                                      .>             no dre   P=1.   -
   a    M f swr wat liv tum 14m24 e                                                   no dre   P=1.   -
   b    M f swr wat tba mix 14m24 e                                                   no dre   P=1.   -
 2526 M m swr wat lun ade 14m24 e                                       .>            no dre   P=1.   -
   a    M m swr wat liv agm 14m24 e                                                   no dre   P=1.   -
   b    M m swr wat tba mix 14m24 e                                                   no dre   P=1.   -
 2527 R f wis wat liv tum 24m34 e                                                     .>no dre  P=1.   -
   a    R f wis wat tba mix 24m34 e                                                   no dre   P=1.   -
 2528 R m wis wat liv lcc 24m34 e                                                     .>no dre  P=1.   -
   a    R m wis wat tba mix 24m34 e                                                   no dre   P=1.   -
```

HEXAMETHYLMELAMINE   100ng..:..1ug....:..10....:...100....:...1mg......:..10...:...100....:...1g......:..10

```
 2529 R f sda eat mgl adc 44w66 ev                      .    +      .                 10.2mg   P<.003 +
   a    R f sda eat mgl mix 44w66 ev                                                  11.9mg   P<.03  +
   b    R f sda eat k/p tcc 44w66 ev                                        +hist 33.6mg   P<.09  +
   c    R f sda eat liv tum 44w66 ev                                                  no dre   P=1.   -
   d    R f sda eat tba mix 44w66 ev                                                  6.82mg   P<.003
```

HEXANAL METHYLFORMYLHYDRAZONE   100ng..:..1ug....:..10....:...100....:...1mg......:..10....:...100....:...1g......:..10

```
 2530 M f swi gav lun mix 12m28 e                            .  +  .                  2.96mg * P<.0005+
   a    M f swi gav lun ade 12m28 e                                                   4.68mg * P<.0005
   b    M f swi gav lun adc 12m28 e                                                   8.73mg * P<.0005
   c    M f swi gav liv mix 12m28 e                                                   19.9mg * P<.0005+
   d    M f swi gav liv bhp 12m28 e                                                   23.3mg * P<.0005
   e    M f swi gav liv hpc 12m28 e                                                   180.mg * P<.1
 2531 M m swi gav lun mix 12m30 es                      .    +                        1.92mg \ P<.0005+
   a    M m swi gav lun adc 12m30 es                                                  5.52mg \ P<.0005
   b    M m swi gav pre mix 12m30 es                                                  6.51mg / P<.0005+
   c    M m swi gav lun ade 12m30 es                                                  6.63mg * P<.0005
   d    M m swi gav pre sqc 12m30 es                                                  8.03mg / P<.0005
   e    M m swi gav liv mix 12m30 es                                                  31.4mg * P<.007 +
   f    M m swi gav liv bhp 12m30 es                                                  45.4mg * P<.08
   g    M m swi gav pre sqp 12m30 es                                                  93.3mg * P<.05
   h    M m swi gav liv hpc 12m30 es                                                  115.mg * P<.05
   i    M m swi gav pre ade 12m30 es                                                  no dre   P=1.
```

HEXANAMIDE   100ng..:..1ug....:..10....:...100....:...1mg......:..10....:...100....:...1g......:..10

```
 2532 M f cb6 eat sto sqp 52w69 e    pool                              .              ±8.90gm * P<.07
   a    M f cb6 eat lun a/a 52w69 e                                                   37.2gm * P<.7   -
   b    M f cb6 eat liv hpc 52w69 e                                                   no dre   P=1.   -
 2533 M m cb6 eat --- mix 52w69 e    pool                              .  +  .        1.95gm * P<.0005+
```

	RefNum	LoConf	UpConf	Cntrl	1Dose	1Inc	2Dose	2Inc			Citation or Pathology
											Brkly Code
a	706	73.3mg	n.s.s.	0/37	18.0mg	1/37					
b	706	6.56mg	n.s.s.	23/37	18.0mg	30/37					
2515	TR40	9.82mg	n.s.s.	21/24	.850mg	17/24	2.50mg	21/24	7.50mg	13/24	
a	TR40	n.s.s.	n.s.s.	0/24	.850mg	0/24	2.50mg	0/24	7.50mg	0/24	liv:hpa,hpc,nnd.
2516	TR40	12.4mg	n.s.s.	0/24	.680mg	0/24	2.00mg	0/24	6.00mg	3/24	
a	TR40	14.1mg	n.s.s.	0/24	.680mg	0/24	2.00mg	0/24	6.00mg	2/24	s
b	TR40	5.68mg	n.s.s.	11/24	.680mg	5/24	2.00mg	7/24	6.00mg	11/24	s
c	TR40	n.s.s.	n.s.s.	0/24	.680mg	0/24	2.00mg	0/24	6.00mg	0/24	liv:hpa,hpc,nnd.

3-(HEXAHYDRO-4,7-METHANOINDAN-5-YL)-1,1-DIMETHYLUREA   (hercules-7531) 2163-79-3

	RefNum	LoConf	UpConf	Cntrl	1Dose	1Inc	2Dose	2Inc			Citation or Pathology
2517	1277	383.mg	n.s.s.	0/17	205.mg	0/17					Innes;ntis,1968/1969
a	1277	383.mg	n.s.s.	1/17	205.mg	0/17					
b	1277	253.mg	n.s.s.	2/17	205.mg	1/17					
2518	1277	250.mg	n.s.s.	2/18	190.mg	1/18					
a	1277	377.mg	n.s.s.	1/18	190.mg	0/18					
b	1277	197.mg	n.s.s.	3/18	190.mg	2/18					
2519	1277	201.mg	n.s.s.	0/16	205.mg	1/17					
a	1277	383.mg	n.s.s.	0/16	205.mg	0/17					
b	1277	201.mg	n.s.s.	0/16	205.mg	1/17					
2520	1277	108.mg	n.s.s.	0/16	190.mg	3/17					
a	1277	356.mg	n.s.s.	0/16	190.mg	0/17					
b	1277	89.2mg	n.s.s.	0/16	190.mg	4/17					

HEXAMETHYLENETETRAMINE   100-97-0

	RefNum	LoConf	UpConf	Cntrl	1Dose	1Inc	2Dose	2Inc			Citation or Pathology
2521	155m	4.76gm	n.s.s.	2/63	1.14gm	2/43					Della Porta;fctx,6,707-715;1968
a	155m	5.98gm	n.s.s.	15/63	1.14gm	4/43					
b	155m	6.38gm	n.s.s.	12/63	1.14gm	3/43					
c	155m	3.01gm	n.s.s.	46/63	1.14gm	20/43					
2522	155m	5.25gm	n.s.s.	0/30	952.mg	1/49					
a	155m	3.16gm	n.s.s.	3/30	952.mg	5/49					
b	155m	3.24gm	n.s.s.	20/30	952.mg	16/49					
c	155m	2.21gm	n.s.s.	21/30	952.mg	22/49					
2523	155m	9.06gm	n.s.s.	1/99	600.mg	1/48	1.20gm	6/102	3.00gm	4/50	
a	155m	9.65gm	n.s.s.	3/99	600.mg	4/48	1.20gm	1/102	3.00gm	7/50	
b	155m	14.7gm	n.s.s.	3/99	600.mg	2/48	1.20gm	7/102	3.00gm	1/50	
c	155m	8.34gm	n.s.s.	17/99	600.mg	10/48	1.20gm	21/102	3.00gm	8/50	
d	155m	1.97gm	n.s.s.	67/99	600.mg	36/48	1.20gm	77/102	3.00gm	36/50	
2524	155m	10.3gm	n.s.s.	19/99	500.mg	10/50	1.00gm	19/94	2.50gm	1/29	
a	155m	12.6gm	n.s.s.	9/99	500.mg	3/50	1.00gm	10/94	2.50gm	0/29	
b	155m	5.51gm	n.s.s.	55/99	500.mg	27/50	1.00gm	57/94	2.50gm	8/29	
2525	155m	1.77gm	n.s.s.	15/30	1.17gm	10/27					
a	155m	6.36gm	n.s.s.	0/30	1.17gm	0/27					
b	155m	1.61gm	n.s.s.	18/30	1.17gm	12/27					
2526	155m	1.33gm	n.s.s.	19/45	971.mg	11/29					
a	155m	5.69gm	n.s.s.	1/45	971.mg	0/29					
b	155m	1.34gm	n.s.s.	22/45	971.mg	12/29					
2527	155n	7.94gm	n.s.s.	0/48	407.mg	0/48					
a	155n	1.56gm	n.s.s.	37/48	407.mg	27/48					
2528	155n	6.94gm	n.s.s.	1/48	356.mg	0/48					
a	155n	627.mg	n.s.s.	39/48	356.mg	36/48					

HEXAMETHYLMELAMINE   531-18-0

	RefNum	LoConf	UpConf	Cntrl	1Dose	1Inc	2Dose	2Inc			Citation or Pathology
2529	200a	4.13mg	52.4mg	0/25	10.6mg	6/24					Cohen;jnci,51,403-417;1973
a	200a	4.33mg	n.s.s.	1/25	10.6mg	6/24					
b	200a	8.27mg	n.s.s.	0/25	10.6mg	2/24					
c	200a	21.1mg	n.s.s.	0/25	10.6mg	0/24					
d	200a	3.01mg	40.7mg	1/25	10.6mg	9/24					

HEXANAL METHYLFORMYLHYDRAZONE   ---

	RefNum	LoConf	UpConf	Cntrl	1Dose	1Inc	2Dose	2Inc			Citation or Pathology
2530	2019	1.97mg	4.96mg	13/48	3.03mg	37/50	6.52mg	45/50			Toth;myco,115,65-71;1991
a	2019	3.17mg	7.81mg	7/48	3.03mg	26/50	6.52mg	39/50			
b	2019	5.41mg	20.0mg	6/48	3.03mg	21/50	6.52mg	27/50			
c	2019	11.1mg	41.5mg	0/41	3.03mg	2/36	6.52mg	14/43			
d	2019	12.5mg	52.5mg	0/41	3.03mg	2/36	6.52mg	12/43			
e	2019	44.4mg	n.s.s.	0/41	3.03mg	0/36	6.52mg	2/43			
2531	2019	1.12mg	3.86mg	13/48	2.90mg	39/46	(6.69mg	35/50)			
a	2019	3.09mg	13.0mg	3/48	2.90mg	21/46	(6.69mg	15/50)			
b	2019	4.47mg	9.92mg	0/47	2.90mg	3/44	6.69mg	43/50			
c	2019	4.02mg	16.1mg	12/48	2.90mg	31/46	6.69mg	34/50			
d	2019	5.40mg	12.6mg	0/47	2.90mg	0/44	6.69mg	40/50			
e	2019	13.5mg	413.mg	0/36	2.90mg	4/37	6.69mg	3/18			
f	2019	17.2mg	n.s.s.	0/36	2.90mg	4/37	6.69mg	1/18			
g	2019	35.4mg	n.s.s.	0/47	2.90mg	2/44	6.69mg	3/50			
h	2019	28.2mg	n.s.s.	0/36	2.90mg	0/37	6.69mg	2/18			
i	2019	79.8mg	n.s.s.	0/47	2.90mg	1/44	6.69mg	0/50			

HEXANAMIDE   628-02-4

	RefNum	LoConf	UpConf	Cntrl	1Dose	1Inc	2Dose	2Inc			Citation or Pathology
2532	1343	2.69mg	n.s.s.	0/86p	978.mg	3/41	1.47gm	0/34			Fleischman;jept,3,149-170;1980
a	1343	3.94mg	n.s.s.	2/89p	978.mg	0/41	1.47gm	2/37			
b	1343	2.09mg	n.s.s.	1/89p	978.mg	0/41	1.47gm	0/37			
2533	1343	1.00gm	4.56gm	0/95p	903.mg	6/35	1.35gm	6/39			

```
     Spe Strain Site   Xpo+Xpt                                          TD50    2Tailpvl
         Sex  Route Hist   Notes                                          DR      AuOp

a    M m cb6 eat --- mlh 52w69 e                                       2.14gm * P<.0005
b    M m cb6 eat mul mlh 52w69 e                                       2.66gm * P<.0005
c    M m cb6 eat lun a/a 52w69 e                                       3.49gm \ P<.06
d    M m cb6 eat liv hes 52w69 e                                       25.3gm * P<.2   -
e    M m cb6 eat liv mlh 52w69 e                                       25.3gm * P<.2   -
2534 R f f34 eat liv tum 52w69 e                              .>       no dre  P=1.    -
2535 R m f34 eat liv tum 52w69 e                         .>           no dre  P=1.    -

N-HEXYLNITROSOUREA           100ng..:..1ug....:..10......:..100....:..1mg...:..10.....:..100....:..1g.....:..10
2536 R f f34 gav ute mix 12m24                    .  +   .            .579mg  P<.002 +
a    R f f34 gav for tum 12m24                                       .848mg  P<.005+
b    R f f34 gav mgl adc 12m24                                       1.73mg  P<.005 +
c    R f f34 gav lun mix 12m24                                       2.06mg  P<.06  +
d    R f f34 gav liv tum 12m24                                       no dre  P=1.
2537 R m f34 gav for tum 12m24         <+                            noTD50  P<.0005+
a    R m f34 gav lun mix 12m24                                       noTD50  P<.0005+
b    R m f34 gav --- mso 12m24                                       .745mg  P<.0005
c    R m f34 gav col tum 12m24                                       2.27mg  P<.04  +
d    R m f34 gav liv tum 12m24                                       no dre  P=1.

4-HEXYLRESORCINOL            100ng..:..1ug....:..10......:..100....:..1mg...:..10.....:..100....:..1g.....:..10
2538 M f b6c gav TBA MXB 24m24                         :>            no dre  P=1.    -
a    M f b6c gav liv MXB 24m24                                       no dre  P=1.
b    M f b6c gav lun MXB 24m24                                       no dre  P=1.
2539 M m b6c gav hag MXA 24m24                            :   ±      368.mg * P<.03 e
a    M m b6c gav amd phe 24m24                                       519.mg * P<.07 e
b    M m b6c gav TBA MXB 24m24                                       no dre  P=1.
c    M m b6c gav liv MXB 24m24                                       no dre  P=1.
d    M m b6c gav lun MXB 24m24                                       no dre  P=1.
2540 R f f34 gav TBA MXB 24m24                         :>            no dre  P=1.    -
a    R f f34 gav liv MXB 24m24      .                                128.gm * P<1.
2541 R m f34 gav TBA MXB 24m24                    :>                 615.mg \ P<1.
a    R m f34 gav liv MXB 24m24                                       1.53gm * P<.3

HUMIC ACIDS, COMMERCIAL GRADE 100ng..:..1ug....:..10......:..100....:..1mg...:..10.....:..100....:..1g.....:..10
2542 M f b6c wat liv hem 24m24 e                          .>         1.24gm  P<.2
a    M f b6c wat liv hnd 24m24 e                                     3.19gm  P<.5
b    M f b6c wat lun ptm 24m24 e                                     5.65gm  P<.9
c    M f b6c wat tba mix 24m24 e                                     169.mg  P<.5
2543 M m b6c wat lun ptm 24m24 e                     .>              348.mg  P<.2
a    M m b6c wat liv hpt 24m24 e                                     2.64gm  P<.8
b    M m b6c wat tba mix 24m24 e                                     153.mg  P<.3

HYDRAZINE                    100ng..:..1ug....:..10......:..100....:..1mg...:..10.....:..100....:..1g.....:..10
2544 H m syg inh nas adp 12m24 es                .  +  .            4.16mg * P<.0005+
a    H m syg inh thy pfa 12m24 es                                   17.5mg * P<.003 +
b    H m syg inh col adc 12m24 es                                   14.0mg * P<.03  +
c    H m syg inh sto bcc 12m24 es                                   37.6mg * P<.4   +
d    H m syg inh col ley 12m24 es                                   70.5mg * P<.2   +
e    H m syg inh col pam 12m24 es                                   70.5mg * P<.2   +
2545 M f cb6 inh lun ade 12m27 e                  .    ±            7.78mg  P<.04
2546 M f nmb wat tba ben 24m24                 .>                   58.5mg * P<.5   -
a    M f nmb wat tba mal 24m24                                      no dre  P=1.    -
2547 M m nmb wat tba mal 24m24                   .>                 no dre  P=1.    -
a    M m nmb wat tba ben 24m24                                      no dre  P=1.    -
2548 M f swi gav lun tum 40w55                .    ±                5.67mg  P<.08 +
2549 M f swi wat lun mix 26m26 e              .  +  .               2.54mg  P<.0005+
a    M f swi wat lun ade 26m26 e                                    2.81mg  P<.0005
b    M f swi wat lun adc 26m26 e                                    9.00mg  P<.0005
c    M f swi wat liv mix 26m26 e                                    no dre  P=1.
2550 M m swi wat lun mix 26m26 e             .  +  .                2.20mg  P<.0005+
a    M m swi wat lun ade 26m26 e                                    2.77mg  P<.0005
b    M m swi wat lun adc 26m26 e                                    6.19mg  P<.0005
c    M m swi wat --- mly 26m26 e                                    9.52mg  P<.007
d    M m swi wat liv mix 26m26 e                                    29.8mg  P<.5
2551 R f f34 inh nas adp 12m30 e           .  + .                   .758mg Z P<.0005+
a    R f f34 inh nas sqp 12m30 e                                    8.72mg * P<.004 +
b    R f f34 inh nas vlp 12m30 e                                    6.56mg * P<.03  +
c    R f f34 inh nas adc 12m30 e                                    9.20mg * P<.03  +
d    R f f34 inh nas sqc 12m30 e                                    13.1mg * P<.02  +
e    R f f34 inh lun bcd 12m30 e                                    26.4mg * P<.09  +
2552 R m f34 inh nas adp 12m30 e          .+ .                      .194mg * P<.0005+
a    R m f34 inh nas vlp 12m30 e                                    1.36mg * P<.0005+
b    R m f34 inh lun bcd 12m30 e                                    6.10mg * P<.004 +
c    R m f34 inh nas sqp 12m30 e                                    6.10mg * P<.004 +
d    R m f34 inh thy car 12m30 e                                    1.60mg * P<.02
e    R m f34 inh nas sqc 12m30 e                                    6.12mg * P<.02  +
f    R m f34 inh nas adc 12m30 e                                    no dre  P=1.    +

HYDRAZINE SULFATE            100ng..:..1ug....:..10......:..100....:..1mg...:..10.....:..100....:..1g.....:..10
2553 H f syg wat cec tum 25m25 e                 .  +  .            115.mg  P<.004 -
```

	RefNum	LoConf	UpConf	Cntrl	1Dose	1Inc	2Dose	2Inc				Citation or Pathology
												Brkly Code
a	1343	1.08gm	5.28gm	0/95p	903.mg	6/35	1.35gm	5/39				
b	1343	1.25gm	7.72gm	0/91p	903.mg	5/35	1.35gm	4/39				
c	1343	952.mg	n.s.s.	1/87p	903.mg	3/35	(1.35gm	0/39)				
d	1343	4.11gm	n.s.s.	0/91p	903.mg	0/35	1.35gm	1/39				
e	1343	4.11gm	n.s.s.	0/91p	903.mg	0/35	1.35gm	1/39				
2534	1343	1.89gm	n.s.s.	0/49	564.mg	0/37						
2535	1343	655.mg	n.s.s.	0/50	452.mg	0/16						

N-HEXYLNITROSOUREA   18774-85-1

	RefNum	LoConf	UpConf	Cntrl	1Dose	1Inc	2Dose	2Inc				Citation or Pathology
2536	1915	.226mg	2.72mg	2/12	1.36mg	10/12						Lijinsky;txih,5,925-935;1989
a	1915	.361mg	2.59mg	0/12	1.36mg	8/12						
b	1915	.644mg	13.9mg	0/12	1.36mg	5/12						
c	1915	.677mg	n.s.s.	1/12	1.36mg	5/12						
d	1915	3.36mg	n.s.s.	0/12	1.36mg	0/12						
2537	1915	n.s.s.	.460mg	0/12	.952mg	12/12						
a	1915	n.s.s.	.460mg	0/12	.952mg	12/12						
b	1915	.308mg	2.60mg	0/12	.952mg	7/12						
c	1915	.681mg	n.s.s.	0/12	.952mg	3/12						
d	1915	1.38mg	n.s.s.	1/12	.952mg	1/12						

4-HEXYLRESORCINOL   136-77-6

	RefNum	LoConf	UpConf	Cntrl	1Dose	1Inc	2Dose	2Inc				Citation or Pathology
2538	TR330	124.mg	n.s.s.	39/50	43.8mg	22/50	87.6mg	31/50				
a	TR330	699.mg	n.s.s.	3/50	43.8mg	0/50	87.6mg	1/50				
b	TR330	585.mg	n.s.s.	5/50	43.8mg	0/50	87.6mg	2/50				liv:hpa,hpc,nnd.
2539	TR330	158.mg	n.s.s.	0/50	43.8mg	4/50	87.6mg	3/50				lun:a/a,a/c.
a	TR330	187.mg	n.s.s.	1/50	43.8mg	2/50	87.6mg	5/50				hag:adn,can.
b	TR330	72.4mg	n.s.s.	36/50	43.8mg	34/50	87.6mg	30/50				
c	TR330	115.mg	n.s.s.	21/50	43.8mg	9/50	(87.6mg	9/50)				liv:hpa,hpc,nnd.
d	TR330	225.mg	n.s.s.	10/50	43.8mg	9/50	87.6mg	5/50				lun:a/a,a/c.
2540	TR330	76.1mg	n.s.s.	44/50	44.2mg	39/50	88.4mg	35/50				
a	TR330	548.mg	n.s.s.	1/50	44.2mg	0/50	88.4mg	1/50				liv:hpa,hpc,nnd.
2541	TR330	22.8mg	n.s.s.	44/50	44.2mg	44/50	(88.4mg	36/50)				
a	TR330	376.mg	n.s.s.	0/50	44.2mg	1/50	88.4mg	1/50				liv:hpa,hpc,nnd.

HUMIC ACIDS, COMMERCIAL GRADE   1415-93-6

	RefNum	LoConf	UpConf	Cntrl	1Dose	1Inc	2Dose	2Inc				Citation or Pathology
2542	1806	305.mg	n.s.s.	3/96	100.mg	4/48						Van Duuren;enhp,69,109-117;1986
a	1806	442.mg	n.s.s.	2/96	100.mg	2/48						
b	1806	289.mg	n.s.s.	13/96	100.mg	7/48						
c	1806	32.4mg	n.s.s.	84/96	100.mg	44/48						
2543	1806	113.mg	n.s.s.	22/99	83.3mg	17/50						
a	1806	242.mg	n.s.s.	12/99	83.3mg	7/50						
b	1806	38.8mg	n.s.s.	76/99	83.3mg	42/50						

HYDRAZINE   302-01-2

	RefNum	LoConf	UpConf	Cntrl	1Dose	1Inc	2Dose	2Inc	3Dose	3Inc	4Dose	4Inc	Citation or Pathology
2544	1679n	2.32mg	8.95mg	1/200	20.6ug	0/200	82.4ug	1/200	.412mg	16/200			Vernot;faat,5,1050-1064;1985
a	1679n	6.05mg	115.mg	0/200	20.6ug	0/200	82.4ug	0/200	.412mg	4/200			
b	1679n	5.33mg	n.s.s.	0/200	20.6ug	0/200	82.4ug	2/200	.412mg	3/200			
c	1679n	7.19mg	n.s.s.	0/200	20.6ug	0/200	82.4ug	2/200	.412mg	1/200			
d	1679n	11.5mg	n.s.s.	0/200	20.6ug	0/200	82.4ug	0/200	.412mg	1/200			
e	1679n	11.5mg	n.s.s.	0/200	20.6ug	0/200	82.4ug	0/200	.412mg	1/200			
2545	1679n	3.12mg	n.s.s.	4/400	.183mg	12/400							
2546	1960	12.0mg	n.s.s.	21/50	.400mg	18/50	2.00mg	21/50	10.0mg	23/50			Steinhoff;expl,39,1-9;1990
a	1960	17.1mg	n.s.s.	27/50	.400mg	37/50	2.00mg	27/50	10.0mg	27/50			
2547	1960	20.7mg	n.s.s.	17/50	.333mg	26/50	1.67mg	22/50	8.33mg	17/50			
a	1960	28.3mg	n.s.s.	26/50	.333mg	24/50	1.67mg	22/50	8.33mg	16/50			
2548	1095	1.71mg	n.s.s.	8/85	5.19mg	6/25							
2549	1117	1.45mg	5.59mg	14/108	2.00mg	27/50							Roe;natu,216,375-376;1967
a	1117	1.59mg	6.34mg	12/108	2.00mg	25/50							Toth;ijcn,9,109-118;1972/1969a
b	1117	4.00mg	37.2mg	2/108	2.00mg	9/50							
c	1117	13.1mg	n.s.s.	3/88	2.00mg	1/41							
2550	1117	1.23mg	5.11mg	10/86	1.67mg	24/46							
a	1117	1.47mg	7.69mg	10/86	1.67mg	21/46							
b	1117	2.91mg	16.9mg	0/86	1.67mg	9/46							
c	1117	3.73mg	180.mg	2/86	1.67mg	7/46							
d	1117	4.90mg	n.s.s.	2/41	1.67mg	3/33							
2551	1679m	.485mg	1.29mg	0/150	1.96ug	2/100	9.81ug	0/100	39.3ug	2/100	.196mg	28/100	Vernot;faat,5,1050-1064;1985
a	1679m	2.64mg	79.3mg	0/150	1.96ug	0/100	9.81ug	0/100	39.3ug	0/100	.196mg	3/100	
b	1679m	2.27mg	n.s.s.	0/150	1.96ug	0/100	9.81ug	0/100	39.3ug	2/100	.196mg	2/100	
c	1679m	2.72mg	n.s.s.	0/150	1.96ug	1/100	9.81ug	0/100	39.3ug	0/100	.196mg	3/100	
d	1679m	3.23mg	n.s.s.	0/150	1.96ug	0/100	9.81ug	0/100	39.3ug	0/100	.196mg	2/100	
e	1679m	4.30mg	n.s.s.	0/150	1.96ug	0/100	9.81ug	0/100	39.3ug	0/100	.196mg	1/100	
2552	1679m	.144mg	.271mg	0/150	1.37ug	2/100	6.87ug	1/100	27.5ug	9/100	.137mg	58/100	
a	1679m	.715mg	3.05mg	0/150	1.37ug	0/100	6.87ug	0/100	27.5ug	1/100	.137mg	12/100	
b	1679m	1.85mg	55.5mg	0/150	1.37ug	0/100	6.87ug	0/100	27.5ug	0/100	.137mg	3/100	
c	1679m	1.85mg	55.5mg	0/150	1.37ug	0/100	6.87ug	0/100	27.5ug	0/100	.137mg	3/100	
d	1679m	.658mg	n.s.s.	7/150	1.37ug	6/100	6.87ug	5/100	27.5ug	9/100	.137mg	13/100	
e	1679m	1.85mg	n.s.s.	0/150	1.37ug	0/100	6.87ug	0/100	27.5ug	1/100	.137mg	2/100	
f	1679m	5.33mg	n.s.s.	0/150	1.37ug	1/100	6.87ug	0/100	27.5ug	0/100	.137mg	0/100	

HYDRAZINE SULFATE   10034-93-2

	RefNum	LoConf	UpConf	Cntrl	1Dose	1Inc	2Dose	2Inc				Citation or Pathology
2553	1108	39.6mg	813.mg	0/79	16.4mg	4/40						Toth;canr,32,804-807;1972/1967a

	Spe	Strain	Site	Xpo+Xpt				TD50	2Tailpvl
		Sex	Route	Hist	Notes			DR	AuOp
a	H	f syg wat	thy ade	25m25 e				124.mg	P<.004 -
b	H	f syg wat	liv tum	25m25 e				no dre	P=1.
c	H	f syg wat	lun tum	25m25 e				no dre	P=1.
2554	H	m syg wat	cec tum	28m28 e	.	±		136.mg	P<.02 -
a	H	m syg wat	liv hem	28m28 e				no dre	P=1. -
b	H	m syg wat	lun tum	28m28 e				no dre	P=1.
2555	H	m syg wat	liv hpc	24m24 er	. +			181.mg *	P<.0005+
a	H	m syg wat	liv mhs	24m24 er				2.58gm *	P<.3
b	H	m syg wat	liv hpa	24m24 er				2.59gm *	P<.5
c	H	m syg wat	liv rts	24m24 er	.>			2.59gm *	P<.5
2556	M	f akr wat	liv tum	69w69 e		.>		no dre	P=1.
a	M	f akr wat	lun tum	69w69 e				no dre	P=1. -
2557	M	m akr wat	lun ade	69w69 e		.>		no dre	P=1. -
a	M	m akr wat	liv hem	69w69 e				no dre	P=1. -
2558	M	f c3h wat	lun ade	92w92 e			±	109.mg	P<.05 +
2559	M	f cbc gav	lun mix	36w74 e	. + .			3.35mg	P<.0005+
a	M	f cbc gav	liv hpt	36w74 e				6.27mg	P<.0005+
2560	M	m cbc gav	lun mix	36w84 e	. + .			5.10mg	P<.0005+
a	M	m cbc gav	liv hpt	36w84 e				8.43mg	P<.0005+
2561	M	f ic3 gav	lun adc	73w73	. + .			10.3mg	P<.0005+
a	M	f ic3 gav	liv tum	73w73				no dre	P=1. -
2562	M	m ic3 gav	lun adc	73w73	.>			79.5mg	P<.2 +
a	M	m ic3 gav	liv tum	73w73				no dre	P=1. -
2563	M	f swa wat	lun mix	95w95 e	. + .			23.8mg	P<.0005+
a	M	f swa wat	lun ade	95w95 e				26.7mg	P<.0005
b	M	f swa wat	lun adc	95w95 e				146.mg	P<.03
c	M	f swa wat	liv hem	95w95 e				no dre	P=1. -
2564	M	m swa wat	lun mix	94w94 e	. + .			18.7mg	P<.0005+
a	M	m swa wat	lun ade	94w94 e				23.1mg	P<.0005
b	M	m swa wat	lun adc	94w94 e				74.2mg	P<.0005
c	M	m swa wat	liv mix	94w94 e				78.4mg	P<.2
2565	M	b swi gav	lun tum	54w54 r	. + .			3.92mg	P<.0005+
a	M	b swi gav	liv tum	54w54 r				no dre	P=1.
2566	M	b swi gav	lun adc	95w95 r	.+ .			9.81mg	P<.0005+
2567	R	f cbs gav	lun mix	68w94 e	. + .			42.4mg	P<.004 +
a	R	f cbs gav	liv mix	68w94 e				no dre	P=1.
2568	R	m cbs gav	liv mix	68w94 e	. + .		+hist	39.4mg	P<.002 +
a	R	m cbs gav	lun mix	68w94 e				60.1mg	P<.009 +

2-HYDRAZINO-4-(p-AMINOPHENYL)THIAZOLE   ..:...1ug....:...10.......:...100....:...1mg....:...10.......:...100....:...1g......:...10

	Spe	Strain	Site	Xpo+Xpt				TD50	2Tailpvl
2569	M	f swi eat	--- lle	46w66 e		±		11.3mg	P<.04 +
a	M	f swi eat	liv tum	46w66 e				no dre	P=1.
b	M	f swi eat	lun tum	46w66 e				no dre	P=1.
c	M	f swi eat	tba mix	46w66 e				11.3mg	P<.04
2570	R	f sda eat	mgl adc	46w75 e	. + .			1.02mg	P<.0005
a	R	f sda eat	mgl mix	46w75 e				1.03mg	P<.0005+
b	R	f sda eat	--- lbl	46w75 e			+hist	7.09mg	P<.002 +
c	R	f sda eat	liv tum	46w75 e				no dre	P=1.
d	R	f sda eat	tba mix	46w75 e				.582mg	P<.0005

2-HYDRAZINO-4-(5-NITRO-2-FURYL)THIAZOLE..:...1_ug....:...10.......:...100....:...1mg....:...10.......:...100....:...1g......:...10

	Spe	Strain	Site	Xpo+Xpt				TD50	2Tailpvl
2571	M	f swi eat	for sqp	46w55 e	. +			16.4mg	P<.0005+
a	M	f swi eat	liv tum	46w55 e				no dre	P=1.
b	M	f swi eat	lun tum	46w55 e				no dre	P=1.
c	M	f swi eat	tba mix	46w55 e				7.57mg	P<.0005
2572	R	f sda eat	mgl mix	46w57 ev	. + .			2.83mg	P<.0005+
a	R	f sda eat	mgl adc	46w57 ev				7.79mg	P<.0005
b	R	f sda eat	mgl adf	46w57 ev				25.2mg	P<.04
c	R	f sda eat	kid mix	46w57 ev				36.8mg	P<.02 +
d	R	f sda eat	liv tum	46w57 ev				no dre	P=1.
e	R	f sda eat	tba mix	46w57 ev				2.83mg	P<.0005
2573	R	f sda eat	mgl mix	46w66 e	. + .			3.66mg	P<.0005+
a	R	f sda eat	mgl adc	46w66 e				10.9mg	P<.0005+
b	R	f sda eat	slg adc	46w66 e				145.mg	P<.1
c	R	f sda eat	liv cye	46w66 e				295.mg	P<.3
d	R	f sda eat	tba mix	46w66 e				noTD50	P<.0005

2-HYDRAZINO-4-(p-NITROPHENYL)THIAZOLE   ..:...1ug....:...10.......:...100....:...1mg....:...10.......:...100....:...1g......:...10

	Spe	Strain	Site	Xpo+Xpt				TD50	2Tailpvl
2574	M	f swi eat	--- lle	46w66 e		±		10.6mg	P<.03 +
a	M	f swi eat	lun alc	46w66 e				25.0mg	P<.07
b	M	f swi eat	liv tum	46w66 e				no dre	P=1.
c	M	f swi eat	tba mix	46w66 e				8.33mg	P<.02
2575	R	f sda eat	mgl mix	46w75 e	. +			1.97mg	P<.002 +
a	R	f sda eat	mgl adc	46w75 e				3.78mg	P<.004
b	R	f sda eat	for sqp	46w75 e				9.00mg	P<.003
c	R	f sda eat	liv tum	46w75 e				no dre	P=1.
d	R	f sda eat	tba mix	46w75 e				.743mg	P<.0005
2576	R	f sda eat	mgl mix	45w52 ev	. + .			8.66mg	P<.0005+
a	R	f sda eat	mgl adc	45w52 ev				9.51mg	P<.0005+
b	R	f sda eat	slg adc	45w52 ev				21.7mg	P<.0005

	RefNum	LoConf	UpConf	Cntrl	1Dose	1Inc	2Dose	2Inc	Citation or Pathology
									Brkly Code
a	1108	42.8mg	895.mg	0/84	16.4mg	4/43			
b	1108	156.mg	n.s.s.	0/84	16.4mg	0/43			
c	1108	156.mg	n.s.s.	0/84	16.4mg	0/43			
2554	1108	41.0mg	241.gm	0/64	14.4mg	3/33			
a	1108	23.3mg	n.s.s.	1/22	14.4mg	0/6			
b	1108	194.mg	n.s.s.	0/86	14.4mg	0/50			
2555	1821	95.1mg	431.mg	0/31	20.4mg	0/31	40.8mg	4/31	61.2mg	9/31	Bosan;carc,8,439-444;1987
a	1821	420.mg	n.s.s.	0/31	20.4mg	0/31	40.8mg	0/31	61.2mg	1/31	
b	1821	421.mg	n.s.s.	0/31	20.4mg	0/31	40.8mg	1/31	61.2mg	1/31	
c	1821	421.mg	n.s.s.	0/31	20.4mg	0/31	40.8mg	1/31	61.2mg	0/31	
2556	158	87.1mg	n.s.s.	0/30	24.0mg	0/40			Toth;jnci,42,469-475;1969a/1966a
a	158	87.1mg	n.s.s.	0/30	24.0mg	0/40			
2557	158	22.9mg	n.s.s.	1/11	20.0mg	1/20			
a	158	52.6mg	n.s.s.	1/16	20.0mg	0/29			
2558	158	37.7mg	n.s.s.	0/22	24.0mg	4/36			
2559	1074	1.58mg	7.60mg	4/47	21.8mg	19/21			Biancifiori;bjca,18,543-550;1964
a	1074	3.19mg	14.5mg	2/47	21.8mg	15/21			
2560	1074	2.62mg	11.3mg	1/37	16.0mg	16/21			
a	1074	4.01mg	26.7mg	4/37	16.0mg	13/21			
2561	156	5.58mg	21.6mg	0/12	37.7mg	17/24			Bhide;ijcn,18,530-535;1976
a	156	91.9mg	n.s.s.	0/12	37.7mg	0/24			
2562	156	19.5mg	n.s.s.	0/12	31.4mg	2/16			
a	156	51.1mg	n.s.s.	0/12	31.4mg	0/16			
2563	158	13.1mg	57.1mg	14/109	24.0mg	24/47			Toth;jnci,42,469-475;1969a/1966a
a	158	14.5mg	66.4mg	12/109	24.0mg	22/47			
b	158	49.0mg	n.s.s.	2/109	24.0mg	5/47			
c	158	97.1mg	n.s.s.	3/78	24.0mg	1/36			
2564	158	10.8mg	39.8mg	10/110	20.0mg	25/50			
a	158	12.7mg	55.2mg	10/110	20.0mg	22/50			
b	158	32.0mg	237.mg	0/110	20.0mg	7/50			
c	158	18.7mg	n.s.s.	2/40	20.0mg	3/17			
2565	1525	2.23mg	7.57mg	1/47	27.6mg	22/30			Maru;clet,17,75-80;1982
a	1525	28.4mg	n.s.s.	7/47	27.6mg	2/30			
2566	1552	6.55mg	15.7mg	1/20	27.6mg	51/63			Menon;zkko,105,258-261;1983
2567	157	16.0mg	274.mg	0/22	24.7mg	5/18			Severi;jnci,41,331-349;1968
a	157	54.0mg	n.s.s.	0/22	24.7mg	0/13			
2568	157	13.5mg	231.mg	0/28	25.9mg	4/13			
a	157	18.1mg	1.95gm	0/28	25.9mg	3/14			

2-HYDRAZINO-4-(p-AMINOPHENYL)THIAZOLE   26049-71-8

	RefNum	LoConf	UpConf	Cntrl	1Dose	1Inc	2Dose	2Inc	Citation or Pathology
2569	1076	3.84mg	n.s.s.	1/28	9.06mg	5/22			Cohen;canr,33,1593-1597;1973
a	1076	16.5mg	n.s.s.	0/28	9.06mg	0/22			
b	1076	16.5mg	n.s.s.	0/28	9.06mg	0/22			
c	1076	3.84mg	n.s.s.	1/28	9.06mg	5/22			
2570	200a	.587mg	2.05mg	6/71	3.07mg	24/35			Cohen;jnci,51,403-417;1973
a	200a	.548mg	2.56mg	18/71	3.07mg	26/35			
b	200a	2.69mg	31.4mg	0/71	3.07mg	5/35			
c	200a	11.5mg	n.s.s.	0/71	3.07mg	0/35			
d	200a	.315mg	1.21mg	18/71	3.07mg	31/35			

2-HYDRAZINO-4-(5-NITRO-2-FURYL)THIAZOLE   (HNT) 26049-68-3

	RefNum	LoConf	UpConf	Cntrl	1Dose	1Inc	2Dose	2Inc	Citation or Pathology
2571	1076	7.27mg	48.5mg	0/29	54.4mg	8/17			Cohen;canr,33,1593-1597;1973
a	1076	53.3mg	n.s.s.	0/29	54.4mg	0/17			
b	1076	53.3mg	n.s.s.	0/29	54.4mg	0/17			
c	1076	3.58mg	20.0mg	2/29	54.4mg	13/17			
2572	1073	1.09mg	7.37mg	2/29	37.1mg	15/16			Erturk;jnci,47,437-445;1971/1970a
a	1073	3.64mg	20.5mg	0/29	37.1mg	10/16			
b	1073	8.18mg	n.s.s.	2/29	37.1mg	5/16			
c	1073	11.1mg	n.s.s.	0/29	37.1mg	3/16			
d	1073	36.8mg	n.s.s.	0/29	37.1mg	0/16			
e	1073	1.09mg	7.37mg	2/29	37.1mg	15/16			
2573	1121	2.00mg	7.00mg	4/35	31.0mg	32/35			Cohen;canr,30,897-901;1970
a	1121	6.26mg	21.3mg	0/35	31.0mg	19/35			
b	1121	35.8mg	n.s.s.	0/35	31.0mg	2/35			
c	1121	48.1mg	n.s.s.	0/35	31.0mg	1/35			
d	1121	n.s.s.	3.77mg	4/35	31.0mg	35/35			

2-HYDRAZINO-4-(p-NITROPHENYL)THIAZOLE   26049-70-7

	RefNum	LoConf	UpConf	Cntrl	1Dose	1Inc	2Dose	2Inc	Citation or Pathology
2574	1076	3.63mg	n.s.s.	1/28	9.06mg	5/21			Cohen;canr,33,1593-1597;1973
a	1076	6.14mg	n.s.s.	0/28	9.06mg	2/21			
b	1076	15.8mg	n.s.s.	0/28	9.06mg	0/21			
c	1076	3.13mg	n.s.s.	1/28	9.06mg	6/21			
2575	200a	.917mg	11.6mg	18/71	3.07mg	20/35			Cohen;jnci,51,403-417;1973
a	200a	1.62mg	34.6mg	6/71	3.07mg	11/35			
b	200a	3.11mg	61.2mg	0/71	3.07mg	4/35			
c	200a	11.5mg	n.s.s.	0/71	3.07mg	0/35			
d	200a	.406mg	1.62mg	18/71	3.07mg	29/35			
2576	1121	4.95mg	17.3mg	4/35	73.8mg	27/34			Cohen;canr,30,897-901;1970
a	1121	5.64mg	17.3mg	0/35	73.8mg	25/34			
b	1121	11.8mg	46.3mg	0/35	73.8mg	15/34			

```
     Spe Strain  Site    Xpo+Xpt                                                      TD50    2Tailpvl
         Sex  Route  Hist      Notes                                                       DR    AuOp
   c   R f sda eat liv tum 45w52 ev                                                   no dre   P=1.
   d   R f sda eat tba mix 45w52 ev                                                   noTD50   P<.0005

2-HYDRAZINO-4-PHENYLTHIAZOLE          100ng..:..1ug....:..10.....:..100.....:..1mg....:..10.....:..100....:..1g.....:..10
2577 R f sda eat liv tum 46w66 ev                                         .>                  no dre   P=1.
   a   R f sda eat tba mix 46w66 ev                                                   311.mg   P<.8    -

p-HYDRAZINOBENZOIC ACID               100ng..:..1ug....:..10.....:..100.....:..1mg....:..10.....:..100....:..1g.....:..10
2578 M f swi gav lun tum 40w55                                                .>             no dre   P=1.

p-HYDRAZINOBENZOIC ACID.HCl           100ng..:..1ug....:..10.....:..100.....:..1mg....:..10.....:..100....:..1g.....:..10
2579 M f swa wat aol mix 28m28 e                                            .  +  .          1.07gm   P<.0005+
   a   M f swa wat aol lei 28m28 e                                                    1.97gm   P<.01  +
   b   M f swa wat lun ade 28m28 e                                                    1.17gm   P<.04
   c   M f swa wat aol ley 28m28 e                                                    2.66gm   P<.03  +
   d   M f swa wat lun mix 28m28 e                                                    1.14gm   P<.2
   e   M f swa wat lun adc 28m28 e                                                    11.2mg   P<.9
2580 M m swa wat aol mix 28m28 e                                        .  +  .              380.mg   P<.0005+
   a   M m swa wat aol lei 28m28 e                                                    609.mg   P<.002 +
   b   M m swa wat aol ley 28m28 e                                                    1.49gm   P<.005 +
   c   M m swa wat lun ade 28m28 e                                                    597.mg   P<.07
   d   M m swa wat lun mix 28m28 e                                                    675.mg   P<.3
   e   M m swa wat lun adc 28m28 e                                                    no dre   P=1.

HYDRAZOBENZENE                        100ng..:..1ug....:..10.....:..100.....:..1mg....:..10.....:..100....:..1g.....:..10
2581 M f b6c eat liv MXA 78w95                                    :  +  :                    26.0mg * P<.0005c
   a   M f b6c eat liv hpc 78w95                                                      29.7mg * P<.0005c
   b   M f b6c eat TBA MXB 78w95                                                      31.5mg * P<.0005
   c   M f b6c eat liv MXB 78w95                                                      26.0mg * P<.0005
   d   M f b6c eat lun MXB 78w95                                                      8.76gm * P<1.
2582 M m b6c eat TBA MXB 78w95 v                                          :>                 no dre   P=1.    -
   a   M m b6c eat liv MXB 78w95 v                                                    1.09gm * P<.9
   b   M m b6c eat lun MXB 78w95 v                                                    no dre   P=1.
2583 R f f34 eat MXB MXB 18m25                                      :  +  :                  6.74mg * P<.0005
   a   R f f34 eat mgl acn 18m25                                                      11.4mg * P<.002 c
   b   R f f34 eat liv nnd 18m25                                                      18.1mg * P<.0005c
   c   R f f34 eat TBA MXB 18m25                                                      4.50mg * P<.2
   d   R f f34 eat liv MXB 18m25                                                      18.1mg * P<.0005
2584 R m f34 eat MXB MXB 18m25 v                                 :  +  :                     3.55mg * P<.0005
   a   R m f34 eat liv MXA 18m25 v                                                    3.70mg * P<.0005c
   b   R m f34 eat liv hpc 18m25 v                                                    4.67mg / P<.0005c
   c   R m f34 eat adr phe 18m25 v                                                    16.2mg * P<.007
   d   R m f34 eat MXA MXA 18m25 v                                                    35.8mg * P<.003
   e   R m f34 eat zym sqc 18m25 v                                                    38.5mg * P<.0005c
   f   R m f34 eat TBA MXB 18m25 v                                                    6.85mg * P<.01
   g   R m f34 eat liv MXB 18m25 v                                                    3.70mg * P<.0005

HYDROCHLORIC ACID                     100ng..:..1ug....:..10.....:..100.....:..1mg....:..10.....:..100....:..1g.....:..10
2585 R m sda inh liv hpc 30m30 e                                      .>                     no dre   P=1.    -
   a   R m sda inh nac tum 30m30 e                                                    no dre   P=1.    -

HYDROCHLOROTHIAZIDE                   100ng..:..1ug....:..10.....:..100.....:..1mg....:..10.....:..100....:..1g.....:..10
2586 M f b6c eat TBA MXB 24m24                                                :>             no dre   P=1.    c
   a   M f b6c eat liv MXB 24m24                                                      no dre   P=1.
   b   M f b6c eat lun MXB 24m24                                                      no dre   P=1.
2587 M m b6c eat liv MXA 24m24                                                    :  +        1.23gm * P<.008 e
   a   M m b6c eat liv hpa 24m24                                                      1.49gm * P<.007
   b   M m b6c eat TBA MXB 24m24                                                      3.82gm / P<.7
   c   M m b6c eat liv MXB 24m24                                                      1.23gm * P<.008
   d   M m b6c eat lun MXB 24m24                                                      19.8gm * P<.9
2588 R f f34 eat TBA MXB 24m25                                           :>                  no dre   P=1.    -
   a   R f f34 eat liv MXB 24m25                                                      no dre   P=1.
2589 R f f34 eat pit mix 24m30                                         .  +  .               33.9mg   P<.0005
   a   R f f34 eat adr phe 24m30                                                      100.mg   P<.0005
   b   R f f34 eat liv hpc 24m30                                                      1.11gm   P<.3
   c   R f f34 eat liv nnd 24m30                                                      no dre   P=1.
   d   R f f34 eat liv mix 24m30                                                      no dre   P=1.
2590 R m f34 eat TBA MXB 24m25                                            :>                 214.mg * P<.7    -
   a   R m f34 eat liv MXB 24m25                                                      771.mg * P<.3
2591 R m f34 eat tes car 24m30                                         .  +  .               35.6mg   P<.0005
   a   R m f34 eat pit mix 24m30                                                      71.3mg   P<.0005
   b   R m f34 eat adr phe 24m30                                                      271.mg   P<.4
   c   R m f34 eat liv mix 24m30                                                      no dre   P=1.
   d   R m f34 eat liv hpc 24m30                                                      no dre   P=1.
   e   R m f34 eat liv nnd 24m30                                                      no dre   P=1.

HYDROCORTISONE                        100ng..:..1ug....:..10.....:..100.....:..1mg....:..10.....:..100....:..1g.....:..10
2592 R f sda gav tba mal 26m26 e                                             .>             no dre   P=1.    -
2593 R m sda gav liv hae 24m24 e                                             .>             no dre   P=1.    -
   a   R m sda gav tba mal 24m24 e                                                    no dre   P=1.    -
```

	RefNum	LoConf	UpConf	Cntrl	1Dose	1Inc	2Dose	2Inc	Citation or Pathology	Brkly Code
c	1121	129.mg	n.s.s.	0/35	73.8mg	0/34				
d	1121	n.s.s.	5.63mg	4/35	73.8mg	34/34				

2-HYDRAZINO-4-PHENYLTHIAZOLE   34176-52-8

	RefNum	LoConf	UpConf	Cntrl	1Dose	1Inc	2Dose	2Inc	Citation or Pathology	Brkly Code
2577	200a	70.9mg	n.s.s.	0/25	26.7mg	0/32			Cohen;jnci,51,403-417;1973	
a	200a	31.5mg	n.s.s.	1/25	26.7mg	2/32				

p-HYDRAZINOBENZOIC ACID   619-67-0

	RefNum	LoConf	UpConf	Cntrl	1Dose	1Inc	2Dose	2Inc	Citation or Pathology	Brkly Code
2578	1095	21.7mg	n.s.s.	8/85	20.8mg	1/25			Roe;natu,216,375-376;1967	

p-HYDRAZINOBENZOIC ACID.HCl   24589-77-3

	RefNum	LoConf	UpConf	Cntrl	1Dose	1Inc	2Dose	2Inc	Citation or Pathology	Brkly Code
2579	1741	462.mg	3.77gm	0/45	250.mg	7/36			McManus;livt,57,78-85;1987	
a	1741	680.mg	212.gm	0/45	250.mg	4/36				
b	1741	469.mg	n.s.s.	5/48	250.mg	13/49				
c	1741	806.mg	n.s.s.	0/45	250.mg	3/36				
d	1741	385.mg	n.s.s.	12/48	250.mg	19/49				
e	1741	865.mg	n.s.s.	7/48	250.mg	8/49				
2580	1741	216.mg	862.mg	2/46	208.mg	21/50				
a	1741	309.mg	2.24gm	2/46	208.mg	15/50				
b	1741	606.mg	11.3gm	0/46	208.mg	6/50				
c	1741	219.mg	n.s.s.	10/40	208.mg	15/33				
d	1741	191.mg	n.s.s.	15/39	208.mg	15/28				
e	1741	652.mg	n.s.s.	9/40	208.mg	6/33				

HYDRAZOBENZENE   122-66-7

	RefNum	LoConf	UpConf	Cntrl	1Dose	1Inc	2Dose	2Inc	Citation or Pathology	Brkly Code
2581	TR92	15.2mg	53.2mg	3/100	4.30mg	4/47	42.2mg	22/50	liv:hpa,hpc.	
a	TR92	16.9mg	63.7mg	3/100	4.30mg	4/47	42.2mg	20/50		
b	TR92	15.7mg	126.mg	29/100	4.30mg	19/47	42.2mg	29/50		
c	TR92	15.2mg	53.2mg	3/100	4.30mg	4/47	42.2mg	22/50	liv:hpa,hpc,nnd.	
d	TR92	100.mg	n.s.s.	5/100	4.30mg	3/47	42.2mg	2/50	lun:a/a,a/c.	
2582	TR92	49.6mg	n.s.s.	45/100	7.90mg	18/50	39.4mg	17/50		
a	TR92	61.0mg	n.s.s.	20/100	7.90mg	11/50	39.4mg	9/50	liv:hpa,hpc,nnd.	
b	TR92	148.mg	n.s.s.	15/100	7.90mg	2/50	39.4mg	3/50	lun:a/a,a/c.	
2583	TR92	3.58mg	15.9mg	1/100	1.40mg	3/50	3.60mg	12/50	liv:nnd; mgl:acn. C	
a	TR92	5.11mg	50.7mg	1/100	1.40mg	3/50	3.60mg	6/50		
b	TR92	7.27mg	64.9mg	0/100	1.40mg	0/50	3.60mg	6/50		
c	TR92	1.61mg	n.s.s.	80/100	1.40mg	39/50	3.60mg	36/50		
d	TR92	7.27mg	64.9mg	0/100	1.40mg	0/50	3.60mg	6/50	liv:hpa,hpc,nnd.	
2584	TR92	2.37mg	5.97mg	6/99	2.30mg	14/50	8.80mg	38/50	liv:hpc,nnd; zym:sqc. C	
a	TR92	2.44mg	6.30mg	6/99	2.30mg	13/50	8.80mg	37/50	liv:hpc,nnd.	
b	TR92	3.05mg	7.77mg	1/99	2.30mg	5/50	8.80mg	31/50		
c	TR92	7.08mg	323.mg	15/99	2.30mg	7/50	8.80mg	16/50		S
d	TR92	14.8mg	258.mg	1/99	2.30mg	2/50	8.80mg	6/50	eac:sqc,sqp; zym:sqc. S	
e	TR92	15.5mg	157.mg	0/99	2.30mg	1/50	8.80mg	5/50		
f	TR92	3.13mg	342.mg	62/99	2.30mg	36/50	8.80mg	43/50		
g	TR92	2.44mg	6.30mg	6/99	2.30mg	13/50	8.80mg	37/50	liv:hpa,hpc,nnd.	

HYDROCHLORIC ACID   (hydrogen chloride) 7647-01-0

	RefNum	LoConf	UpConf	Cntrl	1Dose	1Inc	2Dose	2Inc	Citation or Pathology	Brkly Code
2585	1674	14.8mg	n.s.s.	1/99	.781mg	1/99			Sellakumar;txap,81,401-406;1985	
a	1674	24.1mg	n.s.s.	0/99	.781mg	0/99				

HYDROCHLOROTHIAZIDE   58-93-5

	RefNum	LoConf	UpConf	Cntrl	1Dose	1Inc	2Dose	2Inc	Citation or Pathology	Brkly Code
2586	TR357	846.mg	n.s.s.	36/50	320.mg	35/50	638.mg	27/50		
a	TR357	2.74gm	n.s.s.	3/50	320.mg	5/50	638.mg	1/50	liv:hpa,hpc,nnd.	
b	TR357	3.18gm	n.s.s.	4/50	320.mg	3/50	638.mg	2/50	lun:a/a,a/c.	
2587	TR357	612.mg	24.4gm	7/50	296.mg	10/50	589.mg	21/50	liv:hpa,hpc.	
a	TR357	747.mg	20.3gm	3/50	296.mg	8/50	589.mg	14/50		S
b	TR357	610.mg	n.s.s.	30/50	296.mg	20/50	589.mg	34/50		
c	TR357	612.mg	24.4gm	7/50	296.mg	10/50	589.mg	21/50	liv:hpa,hpc,nnd.	
d	TR357	1.59gm	n.s.s.	8/50	296.mg	4/50	589.mg	9/50	lun:a/a,a/c.	
2588	TR357	88.0mg	n.s.s.	49/50	12.4mg	39/50	24.6mg	39/50	99.1mg	37/50		
a	TR357	n.s.s.	n.s.s.	0/50	12.4mg	0/50	24.6mg	0/50	99.1mg	0/50	liv:hpa,hpc,nnd.	
2589	1854	18.4mg	69.4mg	0/24	44.0mg	18/24			Lijinsky;txih,3,413-422;1987/pers.comm.	
a	1854	46.7mg	285.mg	0/24	44.0mg	9/24				
b	1854	180.mg	n.s.s.	0/24	44.0mg	1/24				
c	1854	194.mg	n.s.s.	4/24	44.0mg	2/24				
d	1854	152.mg	n.s.s.	4/24	44.0mg	3/24				
2590	TR357	29.2mg	n.s.s.	50/50	9.91mg	43/50	19.8mg	44/50	79.2mg	45/50		
a	TR357	143.mg	n.s.s.	1/50	9.91mg	0/50	19.8mg	1/50	79.2mg	2/50	liv:hpa,hpc,nnd.	
2591	1854	19.3mg	72.9mg	0/24	46.2mg	18/24			Lijinsky;txih,3,413-422;1987/pers.comm.	
a	1854	36.1mg	169.mg	0/24	46.2mg	12/24				
b	1854	64.1mg	n.s.s.	6/24	46.2mg	9/24				
c	1854	357.mg	n.s.s.	3/24	46.2mg	0/24				
d	1854	357.mg	n.s.s.	1/24	46.2mg	0/24				
e	1854	357.mg	n.s.s.	2/24	46.2mg	0/24				

HYDROCORTISONE   (cortisol) 50-23-7

	RefNum	LoConf	UpConf	Cntrl	1Dose	1Inc	2Dose	2Inc	Citation or Pathology	Brkly Code
2592	1134	27.5mg	n.s.s.	3/33	5.36mg	2/40			Schmahl;zkko,86,77-84;1976	
2593	1134	32.0mg	n.s.s.	1/36	5.36mg	0/29				
a	1134	32.0mg	n.s.s.	1/36	5.36mg	0/29				

```
     Spe Strain  Site    Xpo+Xpt                                                      TD50    2Tailpvl
         Sex  Route  Hist     Notes                                                        DR   AuOp
HYDROGEN PEROXIDE        100ng..:..1ug....:..10.....:..100....:..1mg.....:..10......:..100.....:..1g......:..10
2594 M f c56 wat duo car 23m23 e                                                   .  + 7.54gm * P<.008 +
a    M f c56 wat duo ade 23m23 e                                                      2.01gm \ P<.2
b    M f c56 wat duo mix 23m23 e                                                      7.83gm \ P<.3
c    M f c56 wat liv tum 23m23 e                                                      no dre   P=1.
2595 M m c56 wat duo mix 23m23 e                                                .>    4.80gm * P<.2
a    M m c56 wat duo ade 23m23 e                                                      7.35gm * P<.3
b    M m c56 wat duo car 23m23 e                                                      15.0gm * P<.4
c    M m c56 wat liv hem 23m23 e                                                      no dre   P=1.

HYDROQUINONE             100ng..:..1ug....:..10.....:..100....:..1mg.....:..10......:..100.....:..1g......:..10
2596 M f b6c gav liv hpa 24m24                          :   +     :                   65.0mg \ P<.002
a    M f b6c gav liv MXA 24m24                                                        122.mg * P<.009 p
b    M f b6c gav TBA MXB 24m24                                                        no dre   P=1.
c    M f b6c gav liv MXB 24m24                                                        122.mg * P<.009
d    M f b6c gav lun MXB 24m24                                                        19.2gm * P<1.
2597 M f b6c eat liv hpa 96w96 e                                                .>    17.9gm   P<.3
a    M f b6c eat kid ade 96w96 e                                                      no dre   P=1.
b    M f b6c eat liv hpc 96w96 e                                                      no dre   P=1.
2598 M m b6c gav TBA MXB 24m24                                   :>                   497.mg   P<.8  -
a    M m b6c gav liv MXB 24m24                                                        383.mg   P<.7
b    M m b6c gav lun MXB 24m24                                                        no dre   P=1.
2599 M m b6c eat liv hpa 96w96 e                                            .   ±     1.45gm   P<.05 +
a    M m b6c eat kid ade 96w96 e                                                      5.32gm   P<.05 +
b    M m b6c eat liv hpc 96w96 e                                                      no dre   P=1.
2600 R f f34 gav --- mnl 24m24                              :   +        :            55.8mg * P<.006 p
a    R f f34 gav TBA MXB 24m24                                                        51.6mg * P<.3
b    R f f34 gav liv MXB 24m24                                                        no dre   P=1.
2601 R m f34 gav kid rua 24m24                              :   +     :               64.7mg * P<.0005p
a    R m f34 gav amd MXA 24m24                                                        40.4mg * P<.03
b    R m f34 gav amd MXA 24m24                                                        48.2mg * P<.05
c    R m f34 gav TBA MXB 24m24                                                        41.8mg * P<.3
d    R m f34 gav liv MXB 24m24                                                        5.71gm * P<1.
2602 R f f3d eat liv hpc 24m24 e                                              .>      8.08gm   P<.3
a    R f f3d eat liv hnd 24m24 e                                                      no dre   P=1.
b    R f f3d eat kid ade 24m24 e                                                      no dre   P=1.
2603 R m f3d eat kid ade 24m24 e                                      .  +  .         349.mg   P<.0005+
a    R m f3d eat liv hnd 24m24 e                                                      no dre   P=1.
b    R m f3d eat liv hpc 24m24 e                                                      no dre   P=1.

HYDROQUINONE MONOBENZYL ETHER   100ng..:..1ug....:..10.....:..100....:..1mg.....:..10......:..100...:..1g......:..10
2604 M f b6a orl liv hpt 76w76 evx                                          .>        no dre   P=1.  -
a    M f b6a orl lun ade 76w76 evx                                                    no dre   P=1.  -
b    M f b6a orl tba mix 76w76 evx                                                    no dre   P=1.  -
2605 M m b6a orl lun ade 76w76 evx                                       .>           no dre   P=1.  -
a    M m b6a orl liv hpt 76w76 evx                                                    no dre   P=1.  -
b    M m b6a orl tba mix 76w76 evx                                                    no dre   P=1.  -
2606 M f b6c orl liv hpt 76w76 evx                                         .>         no dre   P=1.  -
a    M f b6c orl lun mix 76w76 evx                                                    no dre   P=1.  -
b    M f b6c orl tba mix 76w76 evx                                                    1.23gm   P<.3
2607 M m b6c orl liv hpt 76w76 evx                                       .   ±        382.mg   P<.05 -
a    M m b6c orl lun ade 76w76 evx                                                    382.mg   P<.05 -
b    M m b6c orl tba mix 76w76 evx                                                    141.mg   P<.002 -

3-HYDROXY-4-ACETYLAMINOBIPHENYL   100ng..:..1ug....:..10.....:..100....:..1mg.....:..10......:..100...:..1g......:..10
2608 R f nss eat liv tum 43w65                              .>                        no dre   P=1.  -

N-HYDROXY-2-ACETYLAMINOFLUORENE   100ng..:..1ug....:..10.....:..100....:..1mg.....:..10......:..100...:..1g......:..10
2609 H m nss eat for mix 32w56                            .  +  .                     2.10mg   P<.0005+
a    H m nss eat liv bda 32w56                                                        20.8mg   P<.05  +
2610 H m nss ipj per fbs 34w69                         .   ±                          2.20mg   P<.03
a    H m nss ipj for pam 34w69                                                        no dre   P=1.
b    H m nss ipj smi adc 34w69                                                        no dre   P=1.
2611 M f ddd eat for mix 30w76 e                               .  +  .                6.23mg   P<.0005+
a    M f ddd eat ubl mix 30w76 e                                                      7.32mg   P<.0005+
b    M f ddd eat liv hpt 30w76 e                                                      16.0mg   P<.0005+
c    M f ddd eat ubl epc 30w76 e                                                      16.0mg   P<.0005
d    M f ddd eat eso ept 30w76 e                                                      28.8mg   P<.009 +
e    M f ddd eat k/p ept 30w76 e                                                      164.mg   P<.3   +
f    M f ddd eat lun ade 30w76 e                                                      no dre   P=1.   -
2612 R f nbr eat liv mix 64w64                .  +  .                                 1.74ug   P<.0005+
a    R f nbr eat liv cho 64w64                                                        2.03ug   P<.0005+
2613 R m nbr eat liv mix 64w64             .  +  .                                    690.ng   P<.0005+
a    R m nbr eat liv cho 64w64                                                        16.2ug   P<.3

3-HYDROXY-4-AMINOBIPHENYL   100ng..:..1ug....:..10.....:..100....:..1mg.....:..10......:..100...:..1g......:..10
2614 R f nss eat liv tum 43w65                              .>                        no dre   P=1.  -

3-HYDROXY-p-BUTYROPHENETIDIDE   100ng..:..1ug....:..10.....:..100....:..1mg.....:..10......:..100...:..1g......:..10
2615 M f b6c eat liv hpa 72w84 ae                                                .  ± 4.00gm \ P<.02
a    M f b6c eat lun act 72w84 ae                                                     11.3gm * P<.4
```

RefNum	LoConf	UpConf	Cntrl	1Dose	1Inc	2Dose	2Inc	Citation or Pathology	Brkly Code

HYDROGEN PEROXIDE   7722-84-1

2594	1381m	2.60gm 151.gm	0/50	200.mg	0/50	800.mg	4/49	Ito;gann,72,174-175;1981/1981a/pers.comm.	
a	1381m	561.mg n.s.s.	1/50	200.mg	4/50	(800.mg	0/49)		
b	1381m	2.02gm n.s.s.	1/50	200.mg	4/50	800.mg	4/49		
c	1381m	1.52gm n.s.s.	0/50	200.mg	0/50	800.mg	0/49		
2595	1381m	1.75gm n.s.s.	0/48	167.mg	3/51	667.mg	3/50		
a	1381m	2.25gm n.s.s.	0/48	167.mg	2/51	667.mg	2/50		
b	1381m	3.23gm n.s.s.	0/48	167.mg	1/51	667.mg	1/50		
c	1381m	4.80gm n.s.s.	0/48	167.mg	1/51	667.mg	0/50		

HYDROQUINONE   123-31-9

2596	TR366	32.3mg 305.mg	2/55	35.0mg	15/55	(70.0mg	12/55)		S
a	TR366	62.8mg 4.53gm	3/55	35.0mg	16/55	70.0mg	13/55	liv:hpa,hpc.	
b	TR366	63.5mg n.s.s.	43/55	35.0mg	42/55	70.0mg	39/55		
c	TR366	62.8mg 4.53gm	3/55	35.0mg	16/55	70.0mg	13/55	liv:hpa,hpc,nnd.	
d	TR366	203.mg n.s.s.	4/55	35.0mg	6/55	70.0mg	4/55	lun:a/a,a/c.	
2597	1929	2.92gm n.s.s.	0/29	1.04gm	1/30			Shibata;gann,82,1211-1219;1991	
a	1929	5.48gm n.s.s.	0/29	1.04gm	0/30				
b	1929	5.48gm n.s.s.	1/29	1.04gm	0/30				
2598	TR366	45.4mg n.s.s.	39/55	35.3mg	46/55	70.5mg	44/55		
a	TR366	63.4mg n.s.s.	20/55	35.3mg	29/55	70.5mg	25/55	liv:hpa,hpc,nnd.	
b	TR366	183.mg n.s.s.	14/55	35.3mg	11/55	70.5mg	10/55	lun:a/a,a/c.	
2599	1929	572.mg n.s.s.	6/28	960.mg	14/30			Shibata;gann,82,1211-1219;1991	
a	1929	1.61gm n.s.s.	0/28	960.mg	3/30				
b	1929	1.69gm n.s.s.	7/28	960.mg	6/30				
2600	TR366	27.9mg 747.mg	9/55	17.6mg	15/55	35.3mg	22/55		
a	TR366	15.6mg n.s.s.	47/55	17.6mg	49/55	35.3mg	50/55		
b	TR366	n.s.s. n.s.s.	0/55	17.6mg	0/55	35.3mg	0/55	liv:hpa,hpc,nnd.	
2601	TR366	32.1mg 181.mg	0/55	17.6mg	4/55	35.2mg	8/55		
a	TR366	17.7mg n.s.s.	14/55	17.6mg	19/55	35.2mg	21/55	amd:pbb,phm,pob.	S
b	TR366	19.9mg n.s.s.	13/55	17.6mg	17/55	35.2mg	19/55	amd:pbb,pob.	S
c	TR366	12.7mg n.s.s.	49/55	17.6mg	46/55	35.2mg	48/55		
d	TR366	82.6mg n.s.s.	3/55	17.6mg	2/55	35.2mg	2/55	liv:hpa,hpc,nnd.	
2602	1929	1.32gm n.s.s.	0/30	400.mg	1/30			Shibata;gann,82,1211-1219;1991	
a	1929	1.20gm n.s.s.	2/30	400.mg	2/30				
b	1929	2.47gm n.s.s.	0/30	400.mg	0/30				
2603	1929	185.mg 766.mg	0/30	320.mg	14/30				
a	1929	1.32gm n.s.s.	2/30	320.mg	1/30				
b	1929	1.19gm n.s.s.	1/30	320.mg	1/30				

HYDROQUINONE MONOBENZYL ETHER   (agerite alba) 103-16-2

2604	1308	383.mg n.s.s.	0/17	205.mg	0/17			Innes;ntis,1968/1969	
a	1308	383.mg n.s.s.	1/17	205.mg	0/17				
b	1308	180.mg n.s.s.	2/17	205.mg	2/17				
2605	1308	178.mg n.s.s.	2/18	190.mg	2/18				
a	1308	225.mg n.s.s.	1/18	190.mg	1/18				
b	1308	150.mg n.s.s.	3/18	190.mg	3/18				
2606	1308	383.mg n.s.s.	0/16	205.mg	0/17				
a	1308	383.mg n.s.s.	0/16	205.mg	0/17				
b	1308	201.mg n.s.s.	0/16	205.mg	1/17				
2607	1308	115.mg n.s.s.	0/16	190.mg	3/18				
a	1308	115.mg n.s.s.	0/16	190.mg	3/18				
b	1308	60.3mg 576.mg	0/16	190.mg	7/18				

3-HYDROXY-4-ACETYLAMINOBIPHENYL   4463-22-3

2608	1424	7.94mg n.s.s.	0/15	12.3mg	0/8			Miller;jnci,15,1571-1590;1955	

N-HYDROXY-2-ACETYLAMINOFLUORENE   (hydroxy-N-2-fluorenylacetamide) 53-95-2

2609	308m	1.07mg 4.56mg	0/17	17.0mg	16/20			Miller;canr,24,2018-2026;1964	
a	308m	6.27mg n.s.s.	0/17	17.0mg	3/20				
2610	308n	.652mg n.s.s.	0/8	3.43mg	3/8				
a	308n	2.49mg n.s.s.	0/8	3.43mg	0/8				
b	308n	2.49mg n.s.s.	0/8	3.43mg	0/8				
2611	1628	3.08mg 14.3mg	0/16	25.6mg	14/18			Enomoto;jjem,44,37-54;1974	
a	1628	3.62mg 17.1mg	0/16	25.6mg	13/18				
b	1628	7.10mg 52.5mg	0/16	25.6mg	8/18				
c	1628	7.10mg 52.5mg	0/16	25.6mg	8/18				
d	1628	10.9mg 632.mg	0/16	25.6mg	5/18				
e	1628	26.7mg n.s.s.	0/16	25.6mg	1/18				
f	1628	34.0mg n.s.s.	2/16	25.6mg	1/18				
2612	162a	805.ng 4.90ug	0/20	4.00ug	9/20			Yamamoto;ijcn,2,337-343;1967	
a	162a	908.ng 6.43ug	0/20	4.00ug	8/20				
2613	162a	352.ng 1.55ug	0/20	3.20ug	14/20				
a	162a	2.64ug n.s.s.	0/20	3.20ug	1/20				

3-HYDROXY-4-AMINOBIPHENYL   (4-amino-3-hydroxybiphenyl) 4363-03-5

2614	1424	5.45mg n.s.s.	0/15	9.67mg	0/7			Miller;jnci,15,1571-1590;1955	

3-HYDROXY-p-BUTYROPHENETIDIDE   (betadid, bucetin) 1083-57-4

2615	1835	1.38gm n.s.s.	0/46	795.mg	4/47	(1.76gm	0/46)	Togei;jnci,79,1151-1158;1987	
a	1835	3.06gm n.s.s.	1/46	795.mg	6/47	1.76gm	3/46		

```
      Spe Strain Site   Xpo+Xpt                                                        TD50     2Tailpvl
      Sex  Route  Hist     Notes                                                         DR      AuOp
  b   M f b6c eat liv hpc 72w84 ae                                                    no dre    P=1.
2616 M m b6c eat kid rcc 70w84 ae                                             .   +   5.53gm  * P<.004 +
  a   M m b6c eat --- lkm 70w84 ae                                                    2.74gm  \ P<.04
  b   M m b6c eat kid rca 70w84 ae                                                    5.58gm  * P<.02  +
  c   M m b6c eat lun act 70w84 ae                                                    5.28gm  * P<.3
  d   M m b6c eat liv hpa 70w84 ae                                                    17.5gm  \ P<.9
  e   M m b6c eat ubl pam 70w84 ae                                                    46.4gm  * P<.3
  f   M m b6c eat liv hem 70w84 ae                                                    406.gm  * P<1.
  g   M m b6c eat liv hpc 70w84 ae                                                    no dre    P=1.

1-HYDROXYANTHRAQUINONE          100ng..:..1ug...:..10.....:..100....:..1mg....:..10.....:..100...:..1g.....:.10
2617 R m ain eat lgi mix 68w68 e                                       .   +   .       59.2mg    P<.0005+
  a   R m ain eat col ade 68w68 e                                                     219.mg     P<.0005+
  b   R m ain eat liv nnd 68w68 e                                                     219.mg     P<.0005
  c   R m ain eat liv mix 68w68 e                                                     219.mg     P<.0005+
  d   R m ain eat col adc 68w68 e                                                     246.mg     P<.0005+
  e   R m ain eat cec ade 68w68 e                                                     277.mg     P<.0005+
  f   R m ain eat cec adc 68w68 e                                                     619.mg     P<.007 +
  g   R m ain eat sto mix 68w68 e                                                     619.mg     P<.007 +
  h   R m ain eat liv hpc 68w68 e                                                     789.mg     P<.02
  i   R m ain eat for pam 68w68 e                                                     789.mg     P<.02
  j   R m ain eat stg ade 68w68 e                                                     3.34gm     P<.3

1'-HYDROXYESTRAGOLE             100ng..:..1ug...:..10.....:..100....:..1mg....:..10.....:..100...:..1g.....:.10
2618 M f cd1 eat liv hpt 51w86 v                                       .   +   .       57.8mg    P<.0005+
  a   M f cd1 eat lun ade 51w86 v                                                     no dre     P=1.

1-(2-HYDROXYETHYL)-3-[(5-NITROFURFURYLIDENE)AMINO]-2-IMIDAZOLIDINONE...:..1 _mg....:..10.....:..100....:..1g.....:.10
2619 R f sda eat mgl adc 46w66 e                                    .   +   .          16.7mg    P<.0005+
  a   R f sda eat liv tum 46w66 e                                                     no dre     P=1.
  b   R f sda eat tba mix 46w66 e                                                     18.0mg     P<.0005

1-(2-HYDROXYETHYL)-NITROSO-3-CHLOROETHYLUREA   ...:..10.....:..100....:..1mg....:..10.....:..100...:..1g.....:.10
2620 R f f34 wat ute tum 7m24 e                              .    ±                   .669mg     P<.08
  a   R f f34 wat liv tum 7m24 e                                                     1.19mg      P<.04  +
2621 R m f34 wat liv tum 7m24 e                           .    ±                      .590mg     P<.02  +

1-(2-HYDROXYETHYL)-NITROSO-3-ETHYLUREA  ..:..1ug....:..10.....:..100....:..1mg....:..10.....:..100....:..1g.....:.10
2622 R f f34 wat mam car 7m24 e                           .   +   .                   .501mg   * P<.0005+
  a   R f f34 wat liv tum 7m24 e                                                     no dre      P=1.
2623 R m f34 wat lun tum 7m24 e                        .   +   .                      .446mg   * P<.0005
  a   R m f34 wat tnv mso 7m24 e                                                      .576mg   * P<.0005
  b   R m f34 wat ski tum 7m24 e                                                      .641mg   * P<.002 +
  c   R m f34 wat ner tum 7m24 e                                                      .754mg   * P<.003
  d   R m f34 wat liv tum 7m24 e                                                     1.09mg    * P<.02

1-(2-HYDROXYETHYL)-1-NITROSOUREA 100ng..:..1ug...:..10.....:..100....:..1mg....:..10.....:..100...:..1g.....:.10
2624 M f swi wat --- lym 50w51 e                             .(+)  .                  .603mg   * P<.0005+
  a   M f swi wat lun ade 50w51 e                                                    1.97mg    * P<.0005
  b   M f swi wat tba tum 50w51 e                                                    noTD50
2625 M m swi wat lun ade 50w51 e                          . (+)  .                    .382mg   \ P<.0005
  a   M m swi wat --- lym 50w51 e                                                    1.27mg    * P<.0005+
  b   M m swi wat tba tum 50w51 e                                                     .353mg   * P<.0005
2626 R f f34 gav mgl adc 37w60                            .    +                      .452mg     P<.003 +
  a   R f f34 gav col ade 37w60                                                      1.53mg      P<.1
  b   R f f34 gav for sqc 37w60                                                      3.15mg      P<.3
  c   R f f34 gav zym car 37w60                                                      3.15mg      P<.3
  d   R f f34 gav for mix 37w60                                                      no dre      P=1.
  e   R f f34 gav lun a/a 37w60                                                      no dre      P=1.
2627 R m f34 gav lun abt 55w55 e                  .   +   .                           46.0ug     P<.0005+
  a   R m f34 gav for mix 55w55 e                                                     .173mg     P<.0005+
  b   R m f34 gav liv nnd 55w55 e                                                    no dre      P=1.
2628 R m f34 gav lun mix 37w55                   .   +   .                            77.6ug     P<.0005+
  a   R m f34 gav for mix 37w55                                                       .203mg     P<.0005+
  b   R m f34 gav col mix 37w55                                                       .240mg     P<.002 +
  c   R m f34 gav for sqc 37w55                                                       .637mg     P<.04
  d   R m f34 gav col adc 37w55                                                       .983mg     P<.1
  e   R m f34 gav zym mix 37w55                                                       .568mg     P<.3
  f   R m f34 gav lun mal 37w55                                                       .602mg     P<.2
  g   R m f34 gav zym car 37w55                                                       .879mg     P<.4
2629 R f mrw wat --- bly 39w85                                .>                      3.16mg     P<.2  +
  a   R f mrw wat --- mye 39w85                                                      3.16mg      P<.2  -
  b   R f mrw wat liv tum 39w85                                                      10.0mg      P<.4
  c   R f mrw wat tba tum 39w85                                                       .320mg     P<.0005
2630 R m mrw wat bon ost 52w70                            .   +   .                  1.52mg      P<.002 +
  a   R m mrw wat --- mly 52w70                                                      2.03mg      P<.0005+
  b   R m mrw wat --- lcl 52w70                                                      2.03mg      P<.0005
  c   R m mrw wat tba mix 52w70                                                       .360mg     P<.0005
2631 R m mrw wat --- bly 39w85                                .>                     1.56mg      P<.2  +
  a   R m mrw wat bon ost 39w85                                                      1.56mg      P<.2  +
  b   R m mrw wat --- mye 39w85                                                      2.02mg      P<.2  -
```

	RefNum	LoConf	UpConf	Cntrl	1Dose	1Inc	2Dose	2Inc	Citation or Pathology
									Brkly Code
b	1835	8.72gm	n.s.s.	0/46	795.mg	1/47	1.76gm	0/46	
2616	1835	2.50gm	32.0gm	0/47	745.mg	2/45	1.54gm	6/46	
a	1835	997.mg	n.s.s.	1/47	745.mg	6/45	(1.54gm	1/46)	
b	1835	2.53gm	n.s.s.	0/47	745.mg	4/45	1.54gm	4/46	
c	1835	1.60gm	n.s.s.	6/47	745.mg	12/45	1.54gm	10/46	
d	1835	833.mg	n.s.s.	14/47	745.mg	14/45	(1.54gm	0/46)	
e	1835	7.55gm	n.s.s.	0/47	745.mg	0/45	1.54gm	1/46	
f	1835	5.29gm	n.s.s.	3/47	745.mg	1/45	1.54gm	3/46	
g	1835	6.91gm	n.s.s.	7/47	745.mg	4/45	1.54gm	2/46	
1-HYDROXYANTHRAQUINONE		129-43-1							
2617	1937	32.9mg	112.mg	0/30	400.mg	25/29			Mori;carc,11,799-802;1990
a	1937	112.mg	517.mg	0/30	400.mg	12/29			
b	1937	112.mg	517.mg	0/30	400.mg	12/29			
c	1937	112.mg	517.mg	0/30	400.mg	12/29			
d	1937	122.mg	605.mg	0/30	400.mg	11/29			
e	1937	134.mg	726.mg	0/30	400.mg	10/29			
f	1937	235.mg	7.54gm	0/30	400.mg	5/29			
g	1937	235.mg	7.54gm	0/30	400.mg	5/29			
h	1937	272.mg	n.s.s.	0/30	400.mg	4/29			
i	1937	272.mg	n.s.s.	0/30	400.mg	4/29			
j	1937	544.mg	n.s.s.	0/30	400.mg	1/29			
1'-HYDROXYESTRAGOLE		51410-44-7							
2618	1582	35.2mg	104.mg	0/50	80.7mg	24/50			Miller;canr,43,1124-1134;1983
a	1582	383.mg	n.s.s.	2/50	80.7mg	1/50			
1-(2-HYDROXYETHYL)-3-[(5-NITROFURFURYLIDENE)AMINO]-2-IMIDAZOLIDINONE							5036-03-3		
2619	200a	8.90mg	36.9mg	0/25	34.8mg	14/32			Cohen;jnci,51,403-417;1973
a	200a	92.6mg	n.s.s.	0/25	34.8mg	0/32			
b	200a	9.19mg	56.5mg	1/25	34.8mg	14/32			
1-(2-HYDROXYETHYL)-NITROSO-3-CHLOROETHYLUREA		---							
2620	2010	.214mg	n.s.s.	2/12	.499mg	6/12			Lijinsky;vivo,4,1-6;1990
a	2010	.357mg	n.s.s.	0/12	.499mg	3/12			
2621	2010	.201mg	n.s.s.	0/12	.349mg	4/12			
1-(2-HYDROXYETHYL)-NITROSO-3-ETHYLUREA		---							
2622	2010	.261mg	1.22mg	0/12	.443mg	6/12	.885mg	8/12	Lijinsky;vivo,4,1-6;1990
a	2010	.730mg	n.s.s.	0/12	.443mg	0/12	.885mg	0/12	
2623	2010	.224mg	1.28mg	0/12	.310mg	5/12	.620mg	7/12	
a	2010	.276mg	2.22mg	0/12	.310mg	4/12	.620mg	6/12	
b	2010	.297mg	2.13mg	0/12	.310mg	2/12	.620mg	7/12	
c	2010	.337mg	3.30mg	0/12	.310mg	2/12	.620mg	6/12	
d	2010	.440mg	n.s.s.	0/12	.310mg	2/12	.620mg	4/12	
1-(2-HYDROXYETHYL)-1-NITROSOUREA		(N-nitroso-2-hydroxethylurea)		13743-07-2					
2624	1967	.327mg	1.28mg	1/15	4.29mg	11/14	8.57mg	13/15	Mirvish;clet,54,101-106;1990/pers.comm.
a	1967	1.00mg	6.03mg	0/15	4.29mg	5/14	8.57mg	7/15	
b	1967	n.s.s.	.453mg	2/15	4.29mg	14/14	8.57mg	15/15	
2625	1967	.173mg	.988mg	0/12	3.57mg	11/14	(7.14mg	6/15)	
a	1967	.671mg	3.06mg	0/12	3.57mg	4/14	7.14mg	10/15	
b	1967	.170mg	.832mg	2/12	3.57mg	11/14	7.14mg	15/15	
2626	1792m	.183mg	2.36mg	0/20	.707mg	6/20			Lijinsky;gann,79,181-186;1988/1986
a	1792m	.376mg	n.s.s.	0/20	.707mg	2/20			
b	1792m	.512mg	n.s.s.	0/20	.707mg	1/20			
c	1792m	.512mg	n.s.s.	0/20	.707mg	1/20			
d	1792m	.461mg	n.s.s.	2/20	.707mg	2/20			
e	1792m	.581mg	n.s.s.	1/20	.707mg	1/20			
2627	1792	21.6ug	.105mg	1/20	.540mg	18/20			Lijinsky;zkko,112,221-228;1986
a	1792	80.3ug	.489mg	0/20	.540mg	9/20			
b	1792	.623mg	n.s.s.	2/20	.540mg	0/20			
2628	1792m	38.9ug	.188mg	1/20	.540mg	15/20			Lijinsky;gann,79,181-186;1988/1986
a	1792m	90.6ug	.641mg	0/20	.540mg	8/20			
b	1792m	.103mg	.910mg	0/20	.540mg	7/20			
c	1792m	.192mg	n.s.s.	0/20	.540mg	3/20			
d	1792m	.241mg	n.s.s.	0/20	.540mg	2/20			
e	1792m	.153mg	n.s.s.	2/20	.540mg	5/20			
f	1792m	.170mg	n.s.s.	1/20	.540mg	4/20			
g	1792m	.183mg	n.s.s.	0/20	.540mg	4/20			
2629	2038m	.956mg	n.s.s.	0/10	1.12mg	3/20			Mirvish;jtxe,32,59-74;1991/pers.comm.
a	2038m	.956mg	n.s.s.	0/10	1.12mg	3/20			
b	2038m	1.63mg	n.s.s.	0/10	1.12mg	1/20			
c	2038m	.163mg	.696mg	0/10	1.12mg	16/20			
2630	1246	.786mg	5.40mg	1/26	2.10mg	15/40			Bulay;jnci,62,1523-1528;1979
a	1246	1.02mg	5.89mg	0/26	2.10mg	11/40			
b	1246	1.02mg	5.89mg	0/26	2.10mg	11/40			
c	1246	.190mg	.903mg	10/26	2.10mg	36/40			
2631	2038m	.591mg	n.s.s.	0/5	.983mg	5/20			Mirvish;jtxe,32,59-74;1991/pers.comm.
a	2038m	.591mg	n.s.s.	0/5	.983mg	5/20			
b	2038m	.694mg	n.s.s.	0/5	.983mg	4/20			

```
       Spe Strain Site   Xpo+Xpt                                                      TD50    2Tailpvl
           Sex  Route  Hist      Notes                                                   DR      AuOp
   c   R m mrw wat liv tum 39w85                                                      2.77mg   P<.3
   d   R m mrw wat tba tum 39w85                                                      .195mg   P<.0005
2632 R m wis wat lun tum 90w90                              .  +  .                   1.59mg   P<.002 +
   a   R m wis wat mgl adc 90w90                                                      4.14mg   P<.04  +
   b   R m wis wat kid tum 90w90                                                      4.93mg   P<.06  +
   c   R m wis wat --- mye 90w90                                                      3.66mg   P<.2   +
   d   R m wis wat --- bly 90w90                                                      15.9mg   P<.3
   e   R m wis wat liv tum 90w90                                                      no dre   P=1.
   f   R m wis wat tba tum 90w90                                                      .576mg   P<.0005

4-(2-HYDROXYETHYLAMINO)-2-(5-NITRO-2-THIENYL)QUINAZOLINE....:..1 _00....:..1mg...:..10......:..100...:..1g.....:..10
2633 R f sda eat itn lei 46w66 e                                      .  +  .          3.84mg   P<.0005
   a   R f sda eat liv bda 46w66 e                                                    132.mg   P<.1
   b   R f sda eat tba mix 46w66 e                                                    1.87mg   P<.0005+

2-HYDROXYETHYLHYDRAZINE           100ng..:..1ug...:..10....:..100.....:..1mg...:..10......:..100....:..1g.....:..10
2634 H f syg wat liv hpt 84w84 s                                            .  ±       148.mg   P<.02  -
   a   H f syg wat lun tum 84w84 s                                                    no dre   P=1.
2635 H m syg wat liv hpt 84w84 s                                         .  ±          84.4mg   P<.02  -
   a   H m syg wat lun tum 84w84 s                                                    no dre   P=1.
2636 M f b6a orl lun ade 78w78 evx                          .>                        no dre   P=1.
   a   M f b6a orl liv hpt 78w78 evx                                                  no dre   P=1.
   b   M f b6a orl tba mix 78w78 evx                                                  5.34mg   P<.8
2637 M m b6a orl liv hpt 78w78 evx                 .  +  .                            .314mg   P<.0005+
   a   M m b6a orl lun ade 78w78 evx                                                  no dre   P=1.
   b   M m b6a orl lun mix 78w78 evx                                                  no dre   P=1.
   c   M m b6a orl tba mix 78w78 evx                                                  .404mg   P<.02
2638 M f b6c orl liv hpt 78w78 evx                          .>                        4.77mg   P<.3
   a   M f b6c orl lun ade 78w78 evx                                                  4.77mg   P<.3
   b   M f b6c orl tba mix 78w78 evx                                                  4.77mg   P<.6
2639 M m b6c orl liv hpt 78w78 evx                    .  ±                            .541mg   P<.02  +
   a   M m b6c orl lun ade 78w78 evx                                                  no dre   P=1.
   b   M m b6c orl tba mix 78w78 evx                                                  .713mg   P<.4
2640 M f swi wat lun tum 75w75 s                                    .>                no dre   P=1.   -
   a   M f swi wat liv mix 75w75 s                                                    no dre   P=1.
2641 M m swi wat lun tum 75w75 s                                         .>           no dre   P=1.   -
   a   M m swi wat liv mix 75w75 s                                                    no dre   P=1.

HYDROXYPROPYL DISTARCH GLYCEROL   100ng..:..1ug...:..10......:..100.....:..1mg...:..10......:..100...:..1g.....:..10
2642 R f wis eat liv tum 24m24 e                                                     no dre   P=1.   -
   a   R f wis eat tba mix 24m24 e                                                    930.gm   P<1.
2643 R m wis eat liv hpt 24m24 e                                                     243.gm   P<.3
   a   R m wis eat tba mix 24m24 e                                                    no dre   P=1.   -

1-(2-HYDROXYPROPYL)-NITROSO-3-CHLOROETHYLUREA  ....:..10....:..100.....:..1mg...:..10......:..100...:..1g.....:..10
2644 R f f34 wat liv tum 7m24 e                              .>                       no dre   P=1.
2645 R m f34 wat liv tum 7m24 e                         .>                            no dre   P=1.

1-(3-HYDROXYPROPYL)-1-NITROSOUREA 100ng..:..1ug...:..10....:..100.....:..1mg...:..10......:..100...:..1g.....:..10
2646 R f f34 gav for tum 7m24                          .  ±                           1.27mg   P<.02  +
   a   R f f34 gav liv tum 7m24                                                       1.27mg   P<.02
   b   R f f34 gav lun mix 7m24                                                       no dre   P=1.
2647 R f f34 gav liv tum 9m24                               .>                        no dre   P=1.
2648 R m f34 gav ski mix 7m24                      .  +  .                            .668mg   P<.005
   a   R m f34 gav stg tum 7m24                                                       .888mg   P<.02  +
   b   R m f34 gav for tum 7m24                                                       1.98mg   P<.09  +
   c   R m f34 gav lun mix 7m24                                                       1.79mg   P<.3
   d   R m f34 gav liv tum 7m24                                                       3.78mg   P<.6
2649 R m f34 gav for tum 9m24                          .  ±                           .866mg   P<.04  +
   a   R m f34 gav liv tum 9m24                                                       1.24mg   P<.3

8-HYDROXYQUINOLINE                100ng..:..1ug...:..10....:..100.....:..1mg...:..10......:..100...:..1g.....:..10
2650 M f b6c eat --- MXA 24m24                                           :  ±         #819.mg \ P<.02  -
   a   M f b6c eat TBA MXB 24m24                                                      no dre   P=1.
   b   M f b6c eat liv MXB 24m24                                                      no dre   P=1.
   c   M f b6c eat lun MXB 24m24                                                      2.98gm * P<.5
2651 M m b6c eat TBA MXB 24m24                                              :>        no dre   P=1.
   a   M m b6c eat liv MXB 24m24                                                      17.7gm * P<1.
   b   M m b6c eat lun MXB 24m24                                                      2.77gm * P<.6
2652 R f f34 eat TBA MXB 24m24                                          :>            434.mg * P<.6   -
   a   R f f34 eat liv MXB 24m24                                                      no dre   P=1.
2653 R f f34 eat liv hnd 24m24 e                                    .>                no dre   P=1.   -
2654 R m f34 eat lun MXA 24m24                                           :  ±         #651.mg * P<.03  -
   a   R m f34 eat thy ccr 24m24                                                      1.06gm * P<.03
   b   R m f34 eat TBA MXB 24m24                                                      no dre   P=1.
   c   R m f34 eat liv MXB 24m24                                                      no dre   P=1.
2655 R m f34 eat liv tum 52w52 e                                    .>                no dre   P=1.   -
2656 R m f34 eat tes ldc 78w78                         .  ±                           398.mg   P<.02  -
   a   R m f34 eat liv tum 78w78                                                      no dre   P=1.
2657 R m f34 eat liv hnd 24m24 e                       .  ±                           269.mg   P<.04  -
```

	RefNum	LoConf	UpConf	Cntrl	1Dose	1Inc	2Dose	2Inc	Citation or Pathology
									Brkly Code
c	2038m	.836mg	n.s.s.	0/5	.983mg	3/20			
d	2038m	92.9ug	.450mg	0/5	.983mg	18/20			
2632	2038n	.857mg	4.69mg	0/10	2.14mg	15/30			
a	2038n	1.78mg	n.s.s.	0/10	2.14mg	7/30			
b	2038n	2.00mg	n.s.s.	0/10	2.14mg	6/30			
c	2038n	1.43mg	n.s.s.	1/10	2.14mg	10/30			
d	2038n	3.92mg	n.s.s.	0/10	2.14mg	2/30			
e	2038n	9.92mg	n.s.s.	0/10	2.14mg	0/30			
f	2038n	.314mg	1.27mg	1/10	2.14mg	26/30			

4-(2-HYDROXYETHYLAMINO)-2-(5-NITRO-2-THIENYL)QUINAZOLINE   33389-36-5

	RefNum	LoConf	UpConf	Cntrl	1Dose	1Inc	2Dose	2Inc	Citation or Pathology
2633	1390	2.16mg	7.49mg	0/84	17.4mg	20/28			Cohen;jnci,57,277-282;1976
a	1390	21.5mg	n.s.s.	0/84	17.4mg	1/28			
b	1390	.939mg	3.79mg	6/84	17.4mg	26/28			

2-HYDROXYETHYLHYDRAZINE   (BOH) 109-84-2

	RefNum	LoConf	UpConf	Cntrl	1Dose	1Inc	2Dose	2Inc	Citation or Pathology
2634	160	44.7mg	28.1gm	0/100	20.5mg	3/50			Shimizu;jnci,52,903-906;1974
a	160	137.mg	n.s.s.	0/100	20.5mg	0/50			
2635	160	29.9mg	49.0gm	1/100	18.0mg	5/50			
a	160	121.mg	n.s.s.	0/100	18.0mg	0/50			
2636	159	.670mg	n.s.s.	2/15	.708mg	2/17			Innes;ntis,1968/1969
a	159	1.39mg	n.s.s.	1/15	.708mg	0/17			
b	159	.508mg	n.s.s.	2/15	.708mg	3/17			
2637	159	.148mg	.820mg	0/18	.660mg	10/18			
a	159	.912mg	n.s.s.	2/18	.660mg	1/18			
b	159	.975mg	n.s.s.	3/18	.660mg	1/18			
c	159	.163mg	n.s.s.	3/18	.660mg	10/18			
2638	159	.777mg	n.s.s.	0/17	.708mg	1/18			
a	159	.777mg	n.s.s.	0/17	.708mg	1/18			
b	159	.628mg	n.s.s.	1/17	.708mg	2/18			
2639	159	.212mg	n.s.s.	1/17	.660mg	7/17			
a	159	.774mg	n.s.s.	1/17	.660mg	1/17			
b	159	.177mg	n.s.s.	7/17	.660mg	10/17			
2640	160	65.1mg	n.s.s.	21/100	30.0mg	7/50			Shimizu;jnci,52,903-906;1974
a	160	107.mg	n.s.s.	3/100	30.0mg	1/50			
2641	160	84.2mg	n.s.s.	23/100	25.0mg	4/50			
a	160	100.mg	n.s.s.	6/100	25.0mg	1/50			

HYDROXYPROPYL DISTARCH GLYCEROL   ---

	RefNum	LoConf	UpConf	Cntrl	1Dose	1Inc	2Dose	2Inc	Citation or Pathology
2642	1407	80.4gm	n.s.s.	0/30	15.0gm	0/26			de Groot;fctx,12,651-663;1974
a	1407	7.41gm	n.s.s.	23/30	15.0gm	20/26			
2643	1407	39.5gm	n.s.s.	0/23	12.0gm	1/30			
a	1407	14.8gm	n.s.s.	21/23	12.0gm	19/30			

1-(2-HYDROXYPROPYL)-NITROSO-3-CHLOROETHYLUREA   ---

	RefNum	LoConf	UpConf	Cntrl	1Dose	1Inc	2Dose	2Inc	Citation or Pathology
2644	2010	1.32mg	n.s.s.	0/12	.535mg	0/12			Lijinsky;vivo,4,1-6;1990
2645	2010	.926mg	n.s.s.	0/12	.374mg	0/12			

1-(3-HYDROXYPROPYL)-1-NITROSOUREA   71752-70-0

	RefNum	LoConf	UpConf	Cntrl	1Dose	1Inc	2Dose	2Inc	Citation or Pathology
2646	1915m	.433mg	n.s.s.	0/12	.751mg	4/12			Lijinsky;txih,5,925-935;1989
a	1915m	.433mg	n.s.s.	0/12	.751mg	4/12			
b	1915m	1.86mg	n.s.s.	1/12	.751mg	0/12			
2647	1915n	1.29mg	n.s.s.	0/12	.520mg	0/12			
2648	1915m	.249mg	5.36mg	0/12	.526mg	5/12			
a	1915m	.303mg	n.s.s.	0/12	.526mg	4/12			
b	1915m	.484mg	n.s.s.	0/12	.526mg	2/12			
c	1915m	.405mg	n.s.s.	1/12	.526mg	3/12			
d	1915m	.531mg	n.s.s.	1/12	.526mg	2/12			
2649	1915n	.260mg	n.s.s.	0/12	.364mg	3/12			
a	1915n	.280mg	n.s.s.	1/12	.364mg	3/12			

8-HYDROXYQUINOLINE   (8-quinolinol) 148-24-3

	RefNum	LoConf	UpConf	Cntrl	1Dose	1Inc	2Dose	2Inc	Citation or Pathology
2650	TR276	305.mg	n.s.s.	0/50	193.mg	5/50	(386.mg	1/50)	---:hem,hes. S
a	TR276	172.mg	n.s.s.	33/50	193.mg	30/50	(386.mg	25/50)	
b	TR276	1.36gm	n.s.s.	5/50	193.mg	2/50	386.mg	4/50	liv:hpa,hpc,nnd.
c	TR276	713.mg	n.s.s.	2/50	193.mg	5/50	386.mg	5/50	lun:a/a,a/c.
2651	TR276	373.mg	n.s.s.	35/50	178.mg	35/50	357.mg	35/50	
a	TR276	478.mg	n.s.s.	14/50	178.mg	15/50	357.mg	17/50	liv:hpa,hpc,nnd.
b	TR276	550.mg	n.s.s.	6/50	178.mg	10/50	357.mg	10/50	lun:a/a,a/c.
2652	TR276	84.9mg	n.s.s.	40/50	74.3mg	44/50	149.mg	45/50	
a	TR276	552.mg	n.s.s.	4/50	74.3mg	2/50	149.mg	4/50	liv:hpa,hpc,nnd.
2653	1529	285.mg	n.s.s.	3/44	50.0mg	1/39			Fukushima;clet,14,115-123;1981
2654	TR276	277.mg	n.s.s.	0/50	59.4mg	3/50	119.mg	4/50	lun:a/a,a/c. S
a	TR276	362.mg	n.s.s.	0/50	59.4mg	0/50	119.mg	4/50	S
b	TR276	141.mg	n.s.s.	43/50	59.4mg	42/50	119.mg	36/50	
c	TR276	693.mg	n.s.s.	7/50	59.4mg	1/50	119.mg	3/50	liv:hpa,hpc,nnd.
2655	166m	165.mg	n.s.s.	0/10	320.mg	0/10			Yamamoto;txap,19,687-698;1971
2656	166n	136.mg	n.s.s.	0/15	320.mg	4/15			
a	166n	556.mg	n.s.s.	0/15	320.mg	0/15			
2657	1529	81.4mg	n.s.s.	0/31	40.0mg	3/31			Fukushima;clet,14,115-123;1981

```
      Spe Strain  Site   Xpo+Xpt                                                          TD50   2Tailpvl
          Sex  Route  Hist    Notes                                                         DR     AuOp

1'-HYDROXYSAFROLE              100ng..:..1ug....:..10.....:..100...:..1mg....:..10....:..100...:..1g.....:..10
2658 M f b6b eat liv hpt 52w69 ev                                              .>         1.53gm * P<.3
2659 M f b6n eat liv hpt 52w69 ev                               . + .                     53.3mg * P<.0005+
2660 M f cd1 eat liv hpc 51w73 ev                                 . + .                   68.0mg * P<.0005+
  a  M f cd1 eat sub ang 51w73 ev                                                         1.55gm   P<.007
  b  M f cd1 eat lun tum 51w73 ev                                                         no dre   P=1.
2661 M f cd1 eat liv hpt 52w69 ev                              .+ .                        49.1mg * P<.0005+
  a  M f cd1 eat lun ade 52w69 ev                                                         3.67gm * P<.9
2662 M m cd1 eat liv car 51w73 ev                                                 .>      no dre   P=1.
  a  M m cd1 eat lun tum 51w73 ev                                                         no dre   P=1.
2663 M m cd1 eat isp ang 56w69                                          . + .             429.mg * P<.0005+
  a  M m cd1 eat lun ade 56w69                                                            5.47gm * P<.5
  b  M m cd1 eat liv car 56w69                                                            18.9gm * P<.9
2664 R m cdr eat liv hpc 58w95 ae                                .  + .                    47.8mg * P<.0005+
2665 R m cdr eat liv car 43w69                               .   +  .                      16.7mg * P<.0005+
  a  R m cdr eat for pam 43w69                                                            59.8mg * P<.002 +
2666 R m cdr eat liv car 36w52                               .  + .                        12.1mg * P<.0005+
  a  R m cdr eat for pam 36w52                                                            65.8mg * P<.003 +


IBUPROFEN                      100ng..:..1ug....:..10.....:..100...:..1mg....:..10....:..100...:..1g.....:..10
2667 R m f34 eat col mix 54w54 er                                        .>               no dre   P=1.


ICRF-159                       100ng..:..1ug....:..10.....:..100...:..1mg....:..10....:..100...:..1g.....:..10
2668 M f b6c ipj --- MXA 52w86                                     : +    :               23.7mg * P<.005 c
  a  M f b6c ipj --- MXB 52w86                                                            23.7mg * P<.005
  b  M f b6c ipj --- lhc 52w86                                                            53.7mg * P<.03  c
  c  M f b6c ipj TBA MXB 52w86                                                            16.8mg * P<.003
  d  M f b6c ipj liv MXB 52w86                                                            no dre   P=1.
  e  M f b6c ipj lun MXB 52w86                                                            no dre   P=1.
2669 M f b6c ipj --- MXA 52w86      pool                           :  +  :                25.9mg * P<.0005
  a  M f b6c ipj --- lhc 52w86                                                            53.7mg * P<.003 c
2670 M m b6c ipj TBA MXB 52w86                                        :>                  262.mg * P<.8   -
  a  M m b6c ipj liv MXB 52w86                                                            no dre   P=1.
  b  M m b6c ipj lun MXB 52w86                                                            212.mg * P<.3
2671 R f sda ipj ute acn 52w81                                 :  + :                      11.3mg * P<.002 c
  a  R f sda ipj TBA MXB 52w81                                                            6.48mg * P<.005
  b  R f sda ipj liv MXB 52w81                                                            no dre   P=1.
2672 R f sda ipj ute acn 52w79      pool                       :  + :                      10.7mg * P<.0005c
2673 R m sda ipj TBA MXB 52w81 s                            :  ±                           9.55mg * P<.03  -
  a  R m sda ipj liv MXB 52w81 s                                                          105.mg * P<.5


3,3'-IMINOBIS-1-PROPANOL DIMETHANESULFONATE(ESTER).HCl  ....:..100...:..1mg....:..10....:..100...:..1g.....:..10
2674 M f b6c ipj TBA MXB 52w86 e                      : +    :                            1.17mg * P<.0005-
  a  M f b6c ipj liv MXB 52w86 e                                                          no dre   P=1.
  b  M f b6c ipj lun MXB 52w86 e                                                          6.88mg * P<.04
2675 M m b6c ipj --- lym 52w78                             :    ±                          #37.7mg * P<.04  -
  a  M m b6c ipj TBA MXB 52w78                                                            2.07mg / P<.0005
  b  M m b6c ipj liv MXB 52w78                                                            no dre   P=1.
  c  M m b6c ipj lun MXB 52w78                                                            no dre   P=1.
2676 R f sda ipj per MXA 36w80 as                 :    +    :                             2.74mg Z P<.003 a
  a  R f sda ipj TBA MXB 36w80 as                                                         .616mg * P<.006
  b  R f sda ipj liv MXB 36w80 as                                                         5.01mg * P<.05
2677 R m sda ipj per MXA 36w80 as             :    +    :                                 .915mg * P<.003 a
  a  R m sda ipj TBA MXB 36w80 as                                                         1.30mg * P<.0005
  b  R m sda ipj liv MXB 36w80 as                                                         no dre   P=1.


IMINODIACETIC ACID, MONOSODIUM 100ng..:..1ug....:..10.....:..100...:..1mg....:..10....:..100...:..1g.....:..10
2678 R f mrc wat tba mix 20m24 e                                          .  ±             143.mg   P<.03  -
2679 R m mrc wat tba mix 20m24 e                                             .>           no dre   P=1.   -


INDOLE-3-ACETIC ACID           100ng..:..1ug....:..10.....:..100...:..1mg....:..10....:..100...:..1g.....:..10
2680 M f b6a orl lun ade 76w76 evx                                        .>              no dre   P=1.   -
  a  M f b6a orl liv hpt 76w76 evx                                                        no dre   P=1.   -
  b  M f b6a orl tba mix 76w76 evx                                                        no dre   P=1.   -
2681 M m b6a orl liv hpt 76w76 evx                                        .>              no dre   P=1.   -
  a  M m b6a orl lun ade 76w76 evx                                                        no dre   P=1.   -
  b  M m b6a orl tba mix 76w76 evx                                                        no dre   P=1.   -
2682 M f b6c orl lun ade 76w76 evx                                          .>            538.mg   P<.3   -
  a  M f b6c orl liv hpt 76w76 evx                                                        no dre   P=1.   -
  b  M f b6c orl tba mix 76w76 evx                                                        538.mg   P<.3   -
2683 M m b6c orl liv hpt 76w76 evx                                     .  ±                113.mg   P<.02  -
  a  M m b6c orl lun ade 76w76 evx                                                        500.mg   P<.3   -
  b  M m b6c orl tba mix 76w76 evx                                                        40.3mg   P<.0005-


INDOLIDAN                      100ng..:..1ug....:..10.....:..100...:..1mg....:..10....:..100...:..1g.....:..10
2684 R f f34 eat amd mix 24m24 e                                 . +   .                  4.82mg * P<.0005+
  a  R f f34 eat amd pob 24m24 e                                                          5.05mg * P<.0005
  b  R f f34 eat amd phm 24m24 e                                                          no dre   P=1.
2685 R m f34 eat amd mix 95w95 es                           . + .                         1.27mg * P<.0005+
  a  R m f34 eat amd pob 95w95 es                                                         1.38mg * P<.0005
  b  R m f34 eat amd phm 95w95 es                                                         23.8mg * P<.2
```

RefNum	LoConf	UpConf	Cntrl	1Dose	1Inc	2Dose	2Inc			Citation or Pathology

										Brkly Code

1'-HYDROXYSAFROLE   5208-87-7

RefNum	LoConf	UpConf	Cntrl	1Dose	1Inc	2Dose	2Inc	3Dose	3Inc	Citation or Pathology / Brkly Code
2658	1581n	404.mg	n.s.s.	1/44	135.mg	2/41	260.mg	3/42			Boberg;canr,43,5163-5173;1983
2659	1581n	36.6mg	81.6mg	1/49	135.mg	24/38	260.mg	27/38			
2660	1035c	37.7mg	125.mg	0/53	483.mg	30/33					Wislocki;canr,37,1883-1891;1977
a	1035c	587.mg	17.2gm	0/55	483.mg	5/50					
b	1035c	2.45gm	n.s.s.	0/55	483.mg	0/50					
2661	1581m	32.7mg	77.2mg	0/32	135.mg	17/29	260.mg	25/32			Boberg;canr,43,5163-5173;1983
a	1581m	597.mg	n.s.s.	0/32	135.mg	1/29	260.mg	0/32			
2662	1035c	1.13gm	n.s.s.	0/44	446.mg	0/25					Wislocki;canr,37,1883-1891;1977
a	1035c	2.04gm	n.s.s.	0/55	446.mg	0/45					
2663	1042a	245.mg	972.mg	1/50	429.mg	7/25	536.mg	13/40			Borchert;canr,33,590-600;1973
a	1042a	1.05gm	n.s.s.	1/50	429.mg	1/25	536.mg	2/40			
b	1042a	872.mg	n.s.s.	4/50	429.mg	3/25	536.mg	3/40			
2664	1035c	28.2mg	88.3mg	0/18	77.3mg	7/18	100.mg	16/18			Wislocki;canr,37,1883-1891;1977
2665	1042a	6.14mg	47.1mg	0/12	138.mg	11/12					Borchert;canr,33,590-600;1973
a	1042a	23.7mg	269.mg	0/12	138.mg	6/12					
2666	1042c	5.64mg	27.7mg	0/18	156.mg	16/18					
a	1042c	26.6mg	334.mg	0/18	156.mg	6/18					

IBUPROFEN   15687-27-1

RefNum	LoConf	UpConf	Cntrl	1Dose	1Inc					Citation or Pathology
2667	2078	10.7mg	n.s.s.	0/12	16.0mg	0/12				Reddy;carc,13,1019-1023;1992

ICRF-159   21416-87-5

RefNum	LoConf	UpConf	Cntrl	1Dose	1Inc	2Dose	2Inc	Citation or Pathology / Brkly Code
2668	TR78	12.7mg	187.mg	0/15	10.0mg	5/35	21.0mg	9/35	---:leu,lym.
a	TR78	12.7mg	187.mg	0/15	10.0mg	5/35	21.0mg	9/35	---:leu,lhc,lym. C
b	TR78	21.9mg	n.s.s.	0/15	10.0mg	1/35	21.0mg	5/35	
c	TR78	9.72mg	86.4mg	0/15	10.0mg	8/35	21.0mg	11/35	
d	TR78	n.s.s.	n.s.s.	0/15	10.0mg	0/35	21.0mg	0/35	liv:hpa,hpc,nnd.
e	TR78	n.s.s.	n.s.s.	0/15	10.0mg	0/35	21.0mg	0/35	lun:a/a,a/c.
2669	TR78	13.2mg	90.5mg	1/45p	10.0mg	5/35	21.0mg	9/35	---:leu,lym. S
a	TR78	21.9mg	268.mg	0/45p	10.0mg	1/35	21.0mg	5/35	
2670	TR78	23.0mg	n.s.s.	2/15	10.0mg	2/35	21.0mg	3/35	
a	TR78	65.8mg	n.s.s.	2/15	10.0mg	0/35	21.0mg	0/35	liv:hpa,hpc,nnd.
b	TR78	34.5mg	n.s.s.	0/15	10.0mg	0/35	21.0mg	1/35	lun:a/a,a/c.
2671	TR78	6.67mg	34.9mg	0/10	13.0mg	10/34	26.0mg	11/35	
a	TR78	3.48mg	66.8mg	4/10	13.0mg	24/34	26.0mg	22/35	
b	TR78	n.s.s.	n.s.s.	0/10	13.0mg	0/34	26.0mg	0/35	liv:hpa,hpc,nnd.
2672	TR78	6.34mg	20.1mg	0/40p	13.0mg	10/34	26.0mg	11/35	
2673	TR78	4.34mg	n.s.s.	1/10	13.0mg	11/36	26.0mg	4/35	liv:hpa,hpc,nnd.
a	TR78	17.2mg	n.s.s.	0/10	13.0mg	1/36	26.0mg	0/35	

3,3'-IMINOBIS-1-PROPANOL DIMETHANESULFONATE(ESTER).HCl   (IPD) 3458-22-8

RefNum	LoConf	UpConf	Cntrl	1Dose	1Inc	2Dose	2Inc	3Dose	3Inc	Citation or Pathology / Brkly Code
2674	TR18	.541mg	3.20mg	3/15	5.20mg	13/36	11.6mg	14/35			
a	TR18	n.s.s.	n.s.s.	0/15	5.20mg	0/36	11.6mg	0/35			liv:hpa,hpc,nnd.
b	TR18	1.77mg	n.s.s.	1/15	5.20mg	2/36	11.6mg	2/35			lun:a/a,a/c.
2675	TR18	9.01mg	n.s.s.	0/15	5.80mg	0/34	14.0mg	3/35			S
a	TR18	.728mg	7.51mg	1/15	5.80mg	6/34	14.0mg	4/35			
b	TR18	n.s.s.	n.s.s.	0/15	5.80mg	0/34	14.0mg	0/35			liv:hpa,hpc,nnd.
c	TR18	n.s.s.	n.s.s.	0/15	5.80mg	0/34	14.0mg	0/35			lun:a/a,a/c.
2676	TR18	.722mg	27.7mg	0/20	3.30mg	0/35	4.90mg	3/35	21.0mg	0/35	per:fbs,srn.
a	TR18	.228mg	9.06mg	12/20	3.30mg	8/35	4.90mg	3/35	21.0mg	0/35	
b	TR18	.816mg	n.s.s.	0/20	3.30mg	0/35	4.90mg	1/35	21.0mg	0/35	liv:hpa,hpc,nnd.
2677	TR18	.160mg	23.9mg	0/20	4.90mg	3/70	21.0mg	0/35			per:fbs,fib,srn.
a	TR18	.226mg	11.1mg	5/20	4.90mg	7/70	21.0mg	1/35			
b	TR18	n.s.s.	n.s.s.	0/20	4.90mg	0/70	21.0mg	0/35			liv:hpa,hpc,nnd.

IMINODIACETIC ACID, MONOSODIUM   32607-00-4

RefNum	LoConf	UpConf	Cntrl	1Dose	1Inc					Citation or Pathology
2678	213	53.5mg	n.s.s.	4/15	165.mg	10/15				Lijinsky;jnci,50,1061-1063;1973
2679	213	134.mg	n.s.s.	8/15	115.mg	5/15				

INDOLE-3-ACETIC ACID   (heteroauxin) 87-51-4

RefNum	LoConf	UpConf	Cntrl	1Dose	1Inc					Citation or Pathology
2680	1280	99.3mg	n.s.s.	1/17	89.1mg	1/17				Innes;ntis,1968/1969
a	1280	167.mg	n.s.s.	0/17	89.1mg	0/17				
b	1280	78.3mg	n.s.s.	2/17	89.1mg	2/17				
2681	1280	98.0mg	n.s.s.	1/18	82.9mg	1/18				
a	1280	109.mg	n.s.s.	2/18	82.9mg	1/18				
b	1280	65.2mg	n.s.s.	3/18	82.9mg	3/18				
2682	1280	87.5mg	n.s.s.	0/16	89.1mg	1/17				
a	1280	167.mg	n.s.s.	0/16	89.1mg	0/17				
b	1280	87.5mg	n.s.s.	0/16	89.1mg	1/17				
2683	1280	38.9mg	n.s.s.	0/16	82.9mg	4/17				
a	1280	81.5mg	n.s.s.	0/16	82.9mg	1/17				
b	1280	18.5mg	114.mg	0/16	82.9mg	9/17				

INDOLIDAN   100643-96-7

RefNum	LoConf	UpConf	Cntrl	1Dose	1Inc	2Dose	2Inc	3Dose	3Inc	Citation or Pathology
2684	2033	2.38mg	19.9mg	3/60	.125mg	2/59	.375mg	3/58	1.25mg	13/60	Sandusky;faat,16,198-209;1991/pers.comm.
a	2033	2.48mg	20.6mg	3/60	.125mg	2/59	.375mg	2/58	1.25mg	13/60	
b	2033	12.0mg	n.s.s.	0/60	.125mg	0/59	.375mg	1/58	1.25mg	0/60	
2685	2033	.737mg	3.15mg	4/60	.100mg	9/60	.300mg	13/60	1.00mg	24/60	
a	2033	.784mg	3.64mg	4/60	.100mg	9/60	.300mg	12/60	1.00mg	23/60	
b	2033	5.86mg	n.s.s.	0/60	.100mg	0/60	.300mg	1/60	1.00mg	1/60	

```
      Spe Strain  Site   Xpo+Xpt                                                   TD50   2Tailpvl
         Sex  Route  Hist    Notes                                                         DR    AuOp
IODINATED GLYCEROL            100ng..:..1ug....:..10......:..100.....:..1mg....:..10......:..100....:..1g.....:..10
2686 M f b6c gav MXB MXB 24m24                                       :  +        :         92.4mg * P<.005
   a M f b6c gav pta adn 24m24                                                             138.mg * P<.02   p
   b M f b6c gav hag MXA 24m24                                                             263.mg * P<.06   p
   c M f b6c gav liv MXA 24m24                                                             373.mg * P<.02
   d M f b6c gav liv hpa 24m24                                                             471.mg * P<.04
   e M f b6c gav for sqp 24m24                                                             608.mg * P<.08   e
   f M f b6c gav TBA MXB 24m24                                                             172.mg * P<.3
   g M f b6c gav liv MXB 24m24                                                             373.mg * P<.02
   h M f b6c gav lun MXB 24m24                                                             2.74gm / P<.7
2687 M m b6c gav TBA MXB 24m24                                          :>                 no dre   P=1.   -
   a M m b6c gav liv MXB 24m24                                                             1.10gm * P<.4
   b M m b6c gav lun MXB 24m24                                                             2.41gm * P<.7
2688 R f f34 gav TBA MXB 24m24                                      :>                     214.mg * P<.6    -
   a R f f34 gav liv MXB 24m24                                                             2.49gm * P<.3
2689 R m f34 gav tes ict 24m24                                       :  +     :            40.6mg * P<.0005
   a R m f34 gav MXB MXB 24m24                                                             84.1mg * P<.0005
   b R m f34 gav mul mnl 24m24                                                             101.mg * P<.0005p
   c R m f34 gav thy fcc 24m24                                                             280.mg * P<.005  p
   d R m f34 gav nas adn 24m24                                                             1.21gm * P<.04   e
   e R m f34 gav TBA MXB 24m24                                                             52.1mg * P<.0005
   f R m f34 gav liv MXB 24m24                                                             571.mg * P<.02


IODOACETAMIDE                 100ng..:..1ug....:..10......:..100.....:..1mg....:..10......:..100....:..1g.....:..10
2690 R m wis wat sto car 74w75 r                                              .>           no dre   P=1.   -


IODOFORM                      100ng..:..1ug....:..10......:..100.....:..1mg....:..10......:..100....:..1g.....:..10
2691 M f b6c gav TBA MXB 78w90 v                                         :>                no dre   P=1.   -
   a M f b6c gav liv MXB 78w90 v                                                           no dre   P=1.
   b M f b6c gav lun MXB 78w90 v                                                           no dre   P=1.
2692 M m b6c gav TBA MXB 78w90 v                                            :>             73.8mg * P<.2    -
   a M m b6c gav liv MXB 78w90 v                                                           1.66gm * P<.9
   b M m b6c gav lun MXB 78w90 v                                                           630.mg * P<.7
2693 R f osm gav TBA MXB 18m26 v                            :   ±                          11.3mg \ P<.05   -
   a R f osm gav liv MXB 18m26 v                                                           no dre   P=1.
2694 R m osm gav thy MXA 18m26 v                                 :   ±                     #82.2mg * P<.02  -
   a R m osm gav TBA MXB 18m26 v                                                           113.mg * P<.4
   b R m osm gav liv MXB 18m26 v                                                           no dre   P=1.
2695 R m osm gav thy MXA 18m26 v  pool                               :   ±                 #116.mg * P<.05  -


IQ                            100ng..:..1ug....:..10......:..100.....:..1mg....:..10......:..100....:..1g.....:..10
2696 M f cdf eat liv mix 96w96 e                                    . + .                  17.5mg   P<.0005+
   a M f cdf eat liv hpc 96w96 e                                                           24.1mg   P<.0005
   b M f cdf eat for mix 96w96 e                                                           62.4mg   P<.0005+
   c M f cdf eat lun mix 96w96 e                                                           67.9mg   P<.03   +
2697 M m cdf eat lun mix 96w96 e                                 . + .                     22.4mg   P<.0005+
   a M m cdf eat for mix 96w96 e                                                           42.3mg   P<.0005+
   b M m cdf eat liv mix 96w96 e                                                           48.6mg   P<.002  +
   c M m cdf eat lun adc 96w96 e                                                           60.2mg   P<.007
   d M m cdf eat liv hpc 96w96 e                                                           91.6mg   P<.002
   e M m cdf eat for sqc 96w96 e                                                           153.mg   P<.02
2698 P b cym gav liv hpc  8y8  u                        .(+).                              .577mg * P<.0005+
   a P b cym gav tba mal  8y8  u                                                           .577mg * P<.0005
2699 R f f3d eat zym sqc 72w72                               . + .                         4.38mg   P<.0005+
   a R f f3d eat cli sqc 72w72                                                             7.11mg   P<.0005+
   b R f f3d eat liv hpc 72w72                                                             8.24mg   P<.0005+
   c R f f3d eat lgi adc 72w72                                                             19.3mg   P<.0005+
   d R f f3d eat ski sqc 72w72                                                             63.2mg   P<.03   +
   e R f f3d eat smi adc 72w72                                                             195.mg   P<.3    +
   f R f f3d eat orc sqc 72w72                                                             195.mg   P<.3    +
2700 R m f3d eat zym sqc 55w55                               . + .                         .999mg   P<.0005+
   a R m f3d eat liv hpc 55w55                                                             2.08mg   P<.0005+
   b R m f3d eat lgi adc 55w55                                                             2.34mg   P<.0005+
   c R m f3d eat ski sqc 55w55                                                             4.16mg   P<.0005+
   d R m f3d eat smi adc 55w55                                                             6.45mg   P<.0005+
   e R m f3d eat orc sqc 55w55                                                             44.8mg   P<.07   +


IQ.HCl                        100ng..:..1ug....:..10......:..100.....:..1mg....:..10......:..100....:..1g.....:..10
2701 R f sda gav mgl car 31w52 ev                               . + .                      3.29mg   P<.0005+
   a R f sda gav mgl mix 31w52 ev                                                          3.80mg   P<.002  +
   b R f sda gav zym epc 31w52 ev                                                          4.50mg   P<.0005+
   c R f sda gav liv mix 31w52 ev                                                          9.12mg   P<.006  +
   d R f sda gav liv nnd 31w52 ev                                                          19.2mg   P<.06   +
   e R f sda gav liv hpc 31w52 ev                                                          29.3mg   P<.2    +
   f R f sda gav liv hae 31w52 ev                                                          29.3mg   P<.2    +
   g R f sda gav unt tpp 31w52 ev                                                          59.7mg   P<.3    +
   h R f sda gav pan ana 31w52 ev                                                          59.7mg   P<.3    +
   i R f sda gav tba mix 31w52 ev                                                          1.59mg   P<.0005


ISOBUTYL p-HYDROXYBENZOATE    100ng..:..1ug....:..10......:..100.....:..1mg....:..10......:..100....:..1g.....:..10
2702 M f icr eat lun tum 24m25                                                    .>       37.2gm * P<.8    -
```

	RefNum	LoConf	UpConf	Cntrl	1Dose	1Inc	2Dose	2Inc	Citation or Pathology	Brkly Code
IODINATED GLYCEROL		(organidin)	5634-39-9							
2686	TR340	47.4mg	806.mg	15/50	43.4mg	21/50	87.6mg	34/50	hag:adn,can; pta:adn.	P
a	TR340	67.0mg	n.s.s.	10/50	43.4mg	15/50	87.6mg	24/50		
b	TR340	107.mg	n.s.s.	6/50	43.4mg	8/50	87.6mg	14/50	hag:adn,can.	
c	TR340	175.mg	n.s.s.	0/50	43.4mg	5/50	87.6mg	4/50	liv:hpa,hpc.	S
d	TR340	203.mg	n.s.s.	0/50	43.4mg	4/50	87.6mg	3/50		S
e	TR340	218.mg	n.s.s.	1/50	43.4mg	2/50	87.6mg	5/50		
f	TR340	50.0mg	n.s.s.	37/50	43.4mg	30/50	87.6mg	45/50		
g	TR340	175.mg	n.s.s.	0/50	43.4mg	5/50	87.6mg	4/50	liv:hpa,hpc,nnd.	
h	TR340	315.mg	n.s.s.	4/50	43.4mg	0/50	87.6mg	5/50	lun:a/a,a/c.	
2687	TR340	184.mg	n.s.s.	33/50	87.6mg	29/50	176.mg	29/50		
a	TR340	273.mg	n.s.s.	10/50	87.6mg	9/50	176.mg	13/50	liv:hpa,hpc,nnd.	
b	TR340	341.mg	n.s.s.	9/50	87.6mg	6/50	176.mg	10/50	lun:a/a,a/c.	
2688	TR340	41.1mg	n.s.s.	44/50	43.9mg	46/50	87.6mg	47/50		
a	TR340	406.mg	n.s.s.	0/50	43.9mg	0/50	87.6mg	1/50	liv:hpa,hpc,nnd.	
2689	TR340	23.1mg	110.mg	46/50	88.4mg	49/50	175.mg	48/50		S
a	TR340	47.3mg	223.mg	14/50	88.4mg	31/50	175.mg	25/50	mul:mnl; thy:fcc.	P
b	TR340	55.2mg	303.mg	14/50	88.4mg	29/50	175.mg	24/50		
c	TR340	111.mg	2.52gm	0/50	88.4mg	5/50	175.mg	1/50		
d	TR340	259.mg	n.s.s.	0/50	88.4mg	0/50	175.mg	2/50		
e	TR340	27.5mg	198.mg	45/50	88.4mg	46/50	175.mg	42/50		
f	TR340	182.mg	n.s.s.	0/50	88.4mg	2/50	175.mg	2/50	liv:hpa,hpc,nnd.	
IODOACETAMIDE	144-48-9									
2690	1759	21.1mg	n.s.s.	0/20	9.87mg	0/20			Shirai;acpj,35,35-43;1985	
IODOFORM	(triiodomethane)	75-47-8								
2691	TR110	40.2mg	n.s.s.	7/20	28.5mg	14/50	(56.3mg	5/50)		
a	TR110	418.mg	n.s.s.	1/20	28.5mg	1/50	56.3mg	0/50	liv:hpa,hpc,nnd.	
b	TR110	418.mg	n.s.s.	1/20	28.5mg	1/50	56.3mg	0/50	lun:a/a,a/c.	
2692	TR110	30.0mg	n.s.s.	5/20	28.5mg	14/50	56.9mg	23/50		
a	TR110	81.1mg	n.s.s.	3/20	28.5mg	5/50	56.9mg	7/50	liv:hpa,hpc,nnd.	
b	TR110	91.4mg	n.s.s.	1/20	28.5mg	4/50	56.9mg	4/50	lun:a/a,a/c.	
2693	TR110	4.66mg	n.s.s.	10/20	13.6mg	27/50	(27.1mg	23/50)		
a	TR110	241.mg	n.s.s.	1/20	13.6mg	0/50	27.1mg	0/50	liv:hpa,hpc,nnd.	
2694	TR110	40.8mg	n.s.s.	0/20	35.2mg	8/50	70.4mg	4/50	thy:fca,fcc.	S
a	TR110	28.4mg	n.s.s.	7/20	35.2mg	17/50	70.4mg	10/50		
b	TR110	n.s.s.	n.s.s.	0/20	35.2mg	0/50	70.4mg	0/50	liv:hpa,hpc,nnd.	
2695	TR110	46.2mg	n.s.s.	2/40p	35.2mg	8/50	70.4mg	4/50	thy:fca,fcc.	S
IQ	(2-amino-3-methylimidazo[4,5-f]quinoline)	76180-96-6								
2696	1617	10.2mg	33.6mg	3/38	39.0mg	27/36			Ohgaki;carc,5,921-924;1984	
a	1617	14.2mg	44.9mg	0/38	39.0mg	22/36				
b	1617	31.2mg	154.mg	0/38	39.0mg	11/36				
c	1617	27.9mg	n.s.s.	7/38	39.0mg	15/36				
2697	1617	12.2mg	60.4mg	7/33	36.0mg	27/39				
a	1617	22.6mg	109.mg	1/33	36.0mg	16/39				
b	1617	24.0mg	241.mg	3/33	36.0mg	16/39				
c	1617	27.9mg	811.mg	3/33	36.0mg	14/39				
d	1617	41.4mg	347.mg	0/33	36.0mg	8/39				
e	1617	58.1mg	n.s.s.	0/33	36.0mg	5/39				
2698	2002	.331mg	1.03mg	0/9	7.14mg	14/20	14.3mg	18/18	Adamson;enhp,102,190-193;1994/Thorgeirsson pers.comm.	
a	2002	.331mg	1.03mg	0/9	7.14mg	14/20	14.3mg	18/18		
2699	1767m	2.68mg	7.70mg	0/50	15.0mg	27/40			Ohgaki;enhp,67,129-134;1986/1991	
a	1767m	4.14mg	13.6mg	0/50	15.0mg	20/40				
b	1767m	4.69mg	16.3mg	0/50	15.0mg	18/40				
c	1767m	9.08mg	53.5mg	0/50	15.0mg	9/40				
d	1767m	19.1mg	n.s.s.	0/50	15.0mg	3/40				
e	1767m	31.7mg	n.s.s.	0/50	15.0mg	1/40				
f	1767m	31.7mg	n.s.s.	0/50	15.0mg	1/40				
2700	1767m	.590mg	1.72mg	0/50	12.0mg	36/40				
a	1767m	1.27mg	3.74mg	1/50	12.0mg	27/40				
b	1767m	1.42mg	4.19mg	0/50	12.0mg	25/40				
c	1767m	2.34mg	8.41mg	0/50	12.0mg	17/40				
d	1767m	3.31mg	15.1mg	0/50	12.0mg	12/40				
e	1767m	11.0mg	n.s.s.	0/50	12.0mg	2/40				
IQ.HCl	(2-amino-3-methylimidazo[4,5-f]quinoline.HCl)	---								
2701	1721	1.75mg	7.23mg	0/27	11.1mg	14/32			Tanaka;gann,76,570-576;1985/pers.comm.	
a	1721	1.86mg	16.6mg	2/27	11.1mg	14/32				
b	1721	2.24mg	11.5mg	0/27	11.1mg	11/32				
c	1721	3.71mg	81.0mg	0/27	11.1mg	6/32				
d	1721	5.82mg	n.s.s.	0/27	11.1mg	3/32				
e	1721	7.22mg	n.s.s.	0/27	11.1mg	2/32				
f	1721	7.22mg	n.s.s.	0/27	11.1mg	2/32				
g	1721	9.71mg	n.s.s.	0/27	11.1mg	1/32				
h	1721	9.71mg	n.s.s.	0/27	11.1mg	1/32				
i	1721	.900mg	3.30mg	2/27	11.1mg	23/32				
ISOBUTYL p-HYDROXYBENZOATE	4247-02-3									
2702	1688	3.62gm	n.s.s.	2/50	188.mg	3/50	375.mg	2/50	751.mg 3/50	Inai;fctx,23,575-578;1985

```
     Spe Strain  Site   Xpo+Xpt                                                      TD50    2Tailpvl
        Sex  Route   Hist    Notes                                                      DR      AuOp
─────────────────────────────────────────────────────────────────────────────────────────────────────
   a   M f icr eat liv tum 24m25                                                      no dre  P=1.  -
   b   M f icr eat tba mix 24m25                                                      12.3gm * P<.7  -
2703 M m icr eat liv tum 24m24                                                 .>    26.3gm Z P<.7  -
   a   M m icr eat lun tum 24m24                                                      no dre  P=1.  -
   b   M m icr eat tba mix 24m24                                                      6.57gm * P<.5  -

N-ISOBUTYL-N'-NITRO-N-NITROSOGUANIDINE  ..:..1ug...:.10.......:..100...:..1mg....:..10....:..100....:..1g.....:.10
2704 R m wis wat stg tum 52w78 er                                    .>              no dre  P=1.  -

ISOFLURANE                  100ng..:..1ug...:.10.......:..100...:..1mg....:..10....:..100....:..1g.....:.10
2705 M f sww inh lun ala 78w81                                                       .30.0gm * P<.3  -
   a   M f sww inh liv bsa 78w81                                                      no dre  P=1.  -
2706 M m sww inh lun ala 78w81                                                       no dre  P=1.  -
   a   M m sww inh liv bsa 78w81                                                      no dre  P=1.  -

ISOMALT                     100ng..:..1ug...:.10.......:..100...:..1mg....:..10....:..100....:..1g.....:.10
2707 M f scp eat lun act 24m24 e                                                     92.6gm * P<.4  -
   a   M f scp eat liv tum 24m24 e                                                   no dre  P=1.  -
   b   M f scp eat tba tum 24m24 e                                                   83.4gm * P<.7  -
2708 M m scp eat lun act 94w94 e                                                     .26.2gm Z P<.3  -
   a   M m scp eat liv nnd 94w94 e                                                   no dre  P=1.  -
   b   M m scp eat liv hpc 94w94 e                                                   no dre  P=1.  -
   c   M m scp eat tba tum 94w94 e                                                   no dre  P=1.  -

ISOMAZOLE                   100ng..:..1ug...:.10.......:..100...:..1mg....:..10....:..100....:..1g.....:.10
2709 R f f34 eat amd mix 24m24 e                                          .  ±       385.mg * P<.02  +
   a   R f f34 eat amd pob 24m24 e                                                   450.mg * P<.02
   b   R f f34 eat amd phm 24m24 e                                                   no dre  P=1.
2710 R m f34 eat amd mix 24m24 es                              .+  .                 38.8mg * P<.0005+
   a   R m f34 eat amd pob 24m24 es                                                  45.4mg * P<.0005
   b   R m f34 eat amd phm 24m24 es                                                  481.mg * P<.04

ISONIAZID                   100ng..:..1ug...:.10.......:..100...:..1mg....:..10....:..100....:..1g.....:.10
2711 H f syg wat for pam 28m28 e                                          .  ±       575.mg  P<.02  -
   a   H f syg wat liv tum 28m28 e                                                   no dre  P=1.
   b   H f syg wat lun tum 28m28 e                                                   no dre  P=1.
2712 H m syg wat liv hem 30m30 e                                          .>         no dre  P=1.  -
   a   H m syg wat lun tum 30m30 e                                                   no dre  P=1.
2713 M f akr wat lun tum 70w70 e                              .>                     89.6mg  P<.4  -
   a   M f akr wat liv tum 70w70 e                                                   no dre  P=1.
2714 M m akr wat lun tum 59w59 e                                 .>                  202.mg  P<.2  -
   a   M m akr wat liv hem 59w59 e                                                   no dre  P=1.
2715 M m amm gav liv car 79w79 r                           .>                        159.mg \ P<.2  -
   a   M m amm gav lun car 79w79 r                                                   206.mg * P<.4
   b   M m amm gav mix car 79w79 r                                                   219.mg * P<.5  +
2716 M f c3h wat lun ade 72w72 e                                    .>               755.mg  P<.5
2717 M m c3h wat lun ade 72w72 e                                    .>               219.mg  P<.2
2718 M f cbc gav lun mix 36w84 e                        .  +  .                      11.2mg  P<.0005+
   a   M f cbc gav liv hpt 36w84 e                                                   187.mg  P<.4  -
2719 M m cbc gav lun mix 36w74 e                        .  +  .                      12.2mg  P<.0005+
   a   M m cbc gav liv hpt 36w74 e                                                   165.mg  P<.6  -
2720 M b nss gav lun ade 52w52 e                        .  +  .                      19.3mg  P<.0005+
   a   M b nss gav mgl adc 52w52 e                                                   80.9mg  P<.06  +
   b   M b nss gav tba mix 52w52 e                                                   14.1mg  P<.0005+
2721 M f swa wat lun ade 23m25 ae                             .  +  .                386.mg * P<.0005+
   a   M f swa wat liv hem 23m25 ae                                                  no dre  P=1.  -
2722 M f swa wat lun mix 80w80 e                           .  +  .                   153.mg  P<.0005+
   a   M f swa wat lun ade 80w80 e                                                   189.mg  P<.0005+
   b   M f swa wat lun adc 80w80 e                                                   435.mg  P<.01
   c   M f swa wat liv hem 80w80 e                                                   no dre  P=1.  -
2723 M m swa wat lun ade 86w87 aes                         .  +  .                   104.mg \ P<.0005+
   a   M m swa wat liv mix 86w87 aes                                                 no dre  P=1.
2724 M m swa wat lun ade 80w80 e                           .  +  .                   124.mg  P<.0005+
   a   M m swa wat liv hem 80w80 e                                                   no dre  P=1.  -
2725 M b swi gav lun tum 97w97 r                     .  +  .                         24.5mg  P<.0005+
   a   M b swi gav liv tum 97w97 r                                                   no dre  P=1.
2726 M b swi gav lun adc 95w95 r                        .  +  .                      28.8mg  P<.0005+
2727 M f swi gav lun car 83w83 r                        .  +  .                      53.1mg  P<.0005
   a   M f swi gav mix car 83w83 r                                                   53.1mg  P<.0005+
   b   M f swi gav liv car 83w83 r                                                   no dre  P=1.
2728 M f swi gav mix car 81w83 gr                       .  +  .                      39.9mg  P<.0005+
   a   M f swi gav lun car 81w83 gr                                                  88.6mg  P<.008
   b   M f swi gav liv car 81w83 gr                                                  88.6mg  P<.008
2729 M m swi gav mix car 85w85 r                     .  +  .                         24.0mg * P<.0005+
   a   M m swi gav lun car 85w85 r                                                   28.9mg * P<.0005
   b   M m swi gav liv car 85w85 r                                                   211.mg * P<.3
2730 M m swi gav mix car 79w79 fr                    .  +  .                         25.7mg * P<.003 +
   a   M m swi gav lun car 79w79 fr                                                  41.1mg * P<.007
   b   M m swi gav liv car 79w79 fr                                                  88.8mg * P<.3
2731 M m swi gav mix car 79w79 br                    .  +  .                         21.2mg  P<.005 +
   a   M m swi gav lun car 79w79 br                                                  39.2mg  P<.05
```

	RefNum	LoConf	UpConf	Cntrl	1Dose	1Inc	2Dose	2Inc					Citation or Pathology
													Brkly Code
a	1688	1.15gm	n.s.s.	0/50	188.mg	0/50	375.mg	0/50	751.mg	0/50			
b	1688	1.61gm	n.s.s.	12/50	188.mg	10/50	375.mg	12/50	751.mg	13/50			
2703	1688	3.08gm	n.s.s.	0/50	177.mg	4/50	353.mg	0/50	706.mg	2/50			
a	1688	2.84gm	n.s.s.	3/50	177.mg	6/50	353.mg	3/50	706.mg	4/50			
b	1688	1.38gm	n.s.s.	8/50	177.mg	12/50	353.mg	9/50	706.mg	12/50			

N-ISOBUTYL-N'-NITRO-N-NITROSOGUANIDINE   5461-85-8

	RefNum	LoConf	UpConf	Cntrl	1Dose	1Inc	2Dose	2Inc					Citation or Pathology
2704	1082	2.24mg	n.s.s.	0/9	2.14mg	0/9							Matsukura;gann,70,181-185;1979

ISOFLURANE   26675-46-7

	RefNum	LoConf	UpConf	Cntrl	1Dose	1Inc	2Dose	2Inc					Citation or Pathology
2705	1879	8.67gm	n.s.s.	15/92	1.52gm	14/83	6.09gm	19/83					Baden;anes,69,750-753;1988
a	1879	36.0gm	n.s.s.	1/92	1.52gm	2/83	6.09gm	1/83					
2706	1879	16.1gm	n.s.s.	18/89	1.27gm	23/84	5.08gm	12/82					
a	1879	27.2gm	n.s.s.	4/89	1.27gm	7/84	5.08gm	2/82					

ISOMALT   64519-82-0

	RefNum	LoConf	UpConf	Cntrl	1Dose	1Inc	2Dose	2Inc	3Dose	3Inc			Citation or Pathology
2707	1979	22.5gm	n.s.s.	3/44	3.25gm	10/43	6.50gm	9/49	13.0gm	7/42			Smits-van Prooije;fctx,28,243-251;1990
a	1979	16.9gm	n.s.s.	0/44	3.25gm	0/44	6.50gm	0/45	13.0gm	0/44			
b	1979	11.8gm	n.s.s.	28/46	3.25gm	28/48	6.50gm	27/50	13.0gm	30/46			
2708	1979	8.30gm	n.s.s.	5/49	3.00gm	13/49	6.00gm	9/49	(12.0gm	3/50)			
a	1979	48.6gm	n.s.s.	4/48	3.00gm	4/49	6.00gm	4/47	12.0gm	3/50			
b	1979	90.4gm	n.s.s.	1/48	3.00gm	1/49	6.00gm	0/47	12.0gm	1/50			
c	1979	22.2gm	n.s.s.	20/50	3.00gm	22/50	6.00gm	19/50	12.0gm	17/50			

ISOMAZOLE   86315-52-8

	RefNum	LoConf	UpConf	Cntrl	1Dose	1Inc	2Dose	2Inc	3Dose	3Inc			Citation or Pathology
2709	2033	169.mg	n.s.s.	1/60	12.5mg	1/60	25.0mg	4/60	50.0mg	6/60			Sandusky;faat,16,198-209;1991/pers.comm.
a	2033	192.mg	n.s.s.	1/60	12.5mg	0/60	25.0mg	3/60	50.0mg	6/60			
b	2033	474.mg	n.s.s.	0/60	12.5mg	1/60	25.0mg	1/60	50.0mg	0/60			
2710	2033	25.7mg	70.8mg	10/60	10.0mg	13/60	20.0mg	18/60	40.0mg	40/60			
a	2033	29.3mg	89.6mg	9/60	10.0mg	13/60	20.0mg	16/60	40.0mg	36/60			
b	2033	183.mg	n.s.s.	1/60	10.0mg	0/60	20.0mg	2/60	40.0mg	5/60			

ISONIAZID (INH) 54-85-3

	RefNum	LoConf	UpConf	Cntrl	1Dose	1Inc	2Dose	2Inc					Citation or Pathology
2711	170	195.mg	n.s.s.	2/54	136.mg	5/22							Toth;ejca,5,165-171;1969
a	170	1.37gm	n.s.s.	0/72	136.mg	0/36							
b	170	1.37gm	n.s.s.	0/72	136.mg	0/36							
2712	170	425.mg	n.s.s.	2/25	120.mg	0/11							
a	170	1.51gm	n.s.s.	0/67	120.mg	0/39							
2713	169	13.2mg	n.s.s.	0/1	200.mg	1/2							Toth;ijcn,2,413-420;1967
a	169	560.mg	n.s.s.	0/30	200.mg	0/30							
2714	169	32.6mg	n.s.s.	0/9	167.mg	1/6							
a	169	155.mg	n.s.s.	1/14	167.mg	0/14							
2715	584m	48.2mg	n.s.s.	0/20	31.4mg	3/40	(62.9mg	0/40)					Bhide;ijcn,21,381-386;1978
a	584m	64.0mg	n.s.s.	1/20	31.4mg	7/40	62.9mg	6/40					
b	584m	55.8mg	n.s.s.	1/20	31.4mg	10/40	62.9mg	6/40					
2716	1115	123.mg	n.s.s.	0/4	200.mg	1/12							Toth;scie,152,1376-1377;1966a
2717	1115	71.3mg	n.s.s.	1/12	167.mg	6/21							
2718	1074	5.26mg	29.8mg	4/47	34.1mg	13/17							Biancifiori;bjca,18,543-550;1964
a	1074	32.3mg	n.s.s.	2/47	34.1mg	2/17							
2719	1074	5.80mg	32.7mg	1/37	32.2mg	11/18							
a	1074	22.1mg	n.s.s.	4/37	32.2mg	3/18							
2720	1425	11.1mg	47.4mg	5/94	30.1mg	27/98							Pershin;vopr,XVIII,50-53;1972
a	1425	30.1mg	n.s.s.	2/94	30.1mg	8/98							
b	1425	8.62mg	29.9mg	7/94	30.1mg	35/98							
2721	169	201.mg	1.37gm	14/108	100.mg	17/39	300.mg	13/32					Toth;ijcn,2,413-420;1967
a	169	413.mg	n.s.s.	3/88	100.mg	0/30	300.mg	0/18					
2722	1127	82.3mg	426.mg	9/68	200.mg	23/47							Toth;canr,26,1473-1475;1966
a	1127	97.3mg	635.mg	8/68	200.mg	20/47							
b	1127	165.mg	27.1gm	1/48	200.mg	6/32							
c	1127	534.mg	n.s.s.	3/48	200.mg	1/32							
2723	169	49.2mg	437.mg	10/90	83.3mg	15/38	(250.mg	3/14)					Toth;ijcn,2,413-420;1967
a	169	555.mg	n.s.s.	2/85	83.3mg	1/38	250.mg	0/10					
2724	1127	67.5mg	311.mg	8/80	167.mg	21/44							Toth;canr,26,1473-1475;1966
a	1127	203.mg	n.s.s.	2/35	167.mg	0/10							
2725	1525	13.0mg	55.8mg	1/47	27.6mg	15/30							Maru;clet,17,75-80;1982
a	1525	116.mg	n.s.s.	7/47	27.6mg	1/30							
2726	1552	17.2mg	80.7mg	1/20	27.6mg	27/60							Menon;zkko,105,258-261;1983
2727	584m	23.9mg	174.mg	0/30	37.7mg	8/30							Bhide;ijcn,21,381-386;1978
a	584m	23.9mg	174.mg	0/30	37.7mg	8/30							
b	584m	149.mg	n.s.s.	0/30	37.7mg	0/30							
2728	584n	19.3mg	105.mg	0/30	37.0mg	10/30							
a	584n	33.6mg	1.32gm	0/30	37.0mg	5/30							
b	584n	33.6mg	1.32gm	0/30	37.0mg	5/30							
2729	584m	13.9mg	53.4mg	1/30	15.7mg	6/30	31.4mg	16/30					
a	584m	16.2mg	70.0mg	1/30	15.7mg	4/30	31.4mg	15/30					
b	584m	63.8mg	n.s.s.	0/30	15.7mg	2/30	31.4mg	1/30					
2730	584o	14.1mg	107.mg	0/15	15.7mg	6/25	31.4mg	9/25					
a	584o	20.0mg	420.mg	0/15	15.7mg	3/25	31.4mg	7/25					
b	584o	33.7mg	n.s.s.	0/15	15.7mg	3/25	31.4mg	2/25					
2731	584r	10.1mg	158.mg	1/15	31.4mg	12/25							
a	584r	15.4mg	n.s.s.	1/15	31.4mg	8/25							

```
        Spe Strain  Site    Xpo+Xpt                                                        TD50    2Tailpvl
             Sex  Route  Hist    Notes                                                         DR    AuOp
b     M m swi gav liv car 79w79 br                                                        71.3mg   P<.05
2732  R f cbs wat mgl fba 48w64 s                             . +     .                    120.mg   P<.002 +
a     R f cbs wat lun tum 48w64 s                                                         no dre   P=1.
b     R f cbs wat liv tum 48w64 s                                                         no dre   P=1.
2733  R m cbs wat liv tum 48w84 es                         .       ±                      199.mg   P<.07 +
a     R m cbs wat lun mix 48w84 s                                                         1.07gm   P<.2  +

ISONICOTINAMIDE              100ng..:..1ug...:..10....:...100....:...1mg...:..10....:..100...:..1g..:..10
2734  M f swa wat liv hpt 28m28 e                                                    .>   21.3gm   P<.2  -
a     M f swa wat lun mix 28m28 e                                                         no dre   P=1.  -
2735  M m swa wat lun mix 27m27 e                                                     .> no dre   P=1.  -
a     M m swa wat liv mix 27m27 e                                                         no dre   P=1.  -

ISONICOTINIC ACID           100ng..:..1ug...:..10....:...100....:...1mg...:..10....:..100...:..1g....:..10
2736  H f syg wat cec tum 28m28 e                                                  . ±8.08gm   P<.03 -
a     H f syg wat liv tum 28m28 e                                                         no dre   P=1.
b     H f syg wat lun tum 28m28 e                                                         no dre   P=1.
2737  H m syg wat liv lcc 28m28 e                                                  . ± 6.95gm   P<.1  -
a     H m syg wat liv mix 28m28 e                                                         10.3gm   P<.5
b     H m syg wat lun tum 28m28 e                                                         no dre   P=1.
2738  M f cbc gav liv hpt 36w89 e                                         .>              612.mg   P<.8  -
a     M f cbc gav lun ade 36w89 e                                                         no dre   P=1.  -
2739  M m cbc gav lun ade 36w89 e                                      .>                 101.mg   P<.3  -
a     M m cbc gav liv hpt 36w89 e                                                         no dre   P=1.  -

ISONICOTINIC ACID VANILLYLIDENEHYDRAZIDE.:..1ug....:..10....:..100....:..1mg....:..10....:..100....:..1g....:..10
2740  M b nss gav lun adc 52w52 e                                    . ±                  45.0mg   P<.03 +
a     M b nss gav mgl adc 52w52 e                                                         86.6mg   P<.07 +
b     M b nss gav tba mix 52w52 e                                                         27.4mg   P<.003 +

ISOPHORONE                  100ng..:..1ug...:..10....:...100....:...1mg...:..10....:..100...:..1g....:..10
2741  M f b6c gav TBA MXB 24m24                                           :>              no dre   P=1.  -
a     M f b6c gav liv MXB 24m24                                                           2.74gm * P<.5
b     M f b6c gav lun MXB 24m24                                                           no dre   P=1.
2742  M m b6c gav --- mlh 24m24 s                                    :  +  :              244.mg \ P<.002 e
a     M m b6c gav --- MXA 24m24 s                                                         203.mg \ P<.06 e
b     M m b6c gav --- MXA 24m24 s                                                         236.mg \ P<.1  e
c     M m b6c gav sub MXA 24m24 s                                                         654.mg * P<.03 e
d     M m b6c gav sub MXA 24m24 s                                                         703.mg * P<.04 e
e     M m b6c gav sub fbs 24m24 s                                                         1.17gm * P<.09 e
f     M m b6c gav sub fib 24m24 s                                                         1.21gm * P<.06 e
g     M m b6c gav liv MXA 24m24 s                                                         573.mg * P<.3  e
h     M m b6c gav MXA MXA 24m24 s                                                         840.mg * P<.2  e
i     M m b6c gav MXA MXA 24m24 s                                                         882.mg * P<.2  e
j     M m b6c gav MXA MXA 24m24 s                                                         1.21gm * P<.2  e
k     M m b6c gav sub MXA 24m24 s                                                         1.26gm * P<.2  e
l     M m b6c gav TBA MXB 24m24 s                                                         no dre   P=1.
m     M m b6c gav liv MXB 24m24 s                                                         573.mg * P<.3
n     M m b6c gav lun MXB 24m24 s                                                         no dre   P=1.
2743  R f f34 gav TBA MXB 24m24 s                                          :>             16.6mg * P<1.  -
a     R f f34 gav liv MXB 24m24 s                                                         no dre   P=1.
2744  R m f34 gav MXB MXB 24m24                                            :  +  :        774.mg * P<.0005
a     R m f34 gav kid MXA 24m24                                                           1.21gm * P<.008 p
b     R m f34 gav pre can 24m24                                                           2.25gm * P<.005 p
c     R m f34 gav TBA MXB 24m24                                                           237.mg * P<.05
d     R m f34 gav liv MXB 24m24                                                           9.08gm * P<.9

ISOPHOSPHAMIDE              100ng..:..1ug...:..10....:...100....:...1mg...:..10....:..100....:..1g....:..10
2745  M f b6c ipj --- lhc 52w79 s                          :  +  :                        5.06mg * P<.0005c
a     M f b6c ipj TBA MXB 52w79 s                                                         3.63mg * P<.0005
b     M f b6c ipj liv MXB 52w79 s                                                         no dre   P=1.
c     M f b6c ipj lun MXB 52w79 s                                                         no dre   P=1.
2746  M f b6c ipj --- lhc 52w79 s  pool                    :  +  :                        5.06mg * P<.0005c
2747  M m b6c ipj TBA MXB 52w79 s                                  :>                     no dre   P=1.  -
a     M m b6c ipj liv MXB 52w79 s                                                         no dre   P=1.
b     M m b6c ipj lun MXB 52w79 s                                                         37.6mg * P<.5
2748  M m b6c ipj liv MXA 52w79 s  pool                          :  ±                     #2.42mg \ P<.02 -
2749  R f sda ipj mgl fba 52w83 s                     :  +  :                             .301mg * P<.0005a
a     R f sda ipj ute lei 52w83 s                                                         .739mg * P<.003 c
b     R f sda ipj TBA MXB 52w83 s                                                         .383mg / P<.009
c     R f sda ipj liv MXB 52w83 s                                                         17.2mg * P<.5
2750  R f sda ipj mgl fba 52w80 s  pool               :  +  :                            .358mg * P<.0005a
a     R f sda ipj ute lei 52w80 s                                                         .814mg * P<.0005c
b     R f sda ipj mgl MXA 52w80 s                                                         2.96mg * P<.007
2751  R m sda ipj TBA MXB 52w79 s                           :  ±                          1.01mg / P<.02 -
a     R m sda ipj liv MXB 52w79 s                                                         no dre   P=1.
2752  R m sda ipj --- MXA 52w79 s  pool                        :  +  :                    #3.23mg * P<.0005-

p-ISOPROPOXYDIPHENYLAMINE   100ng..:..1ug...:..10....:...100....:...1mg...:..10....:..100....:..1g....:..10
2753  M f b6a orl liv agm 76w76 evx                                           .>          2.65gm   P<.3  -
a     M f b6a orl lun ade 76w76 evx                                                       no dre   P=1.  -
```

	RefNum	LoConf	UpConf	Cntrl	1Dose	1Inc	2Dose	2Inc	Citation or Pathology
									Brkly Code
b	584r	24.6mg	n.s.s.	0/15	31.4mg	4/25			
2732	157	60.1mg	399.mg	0/22	149.mg	11/40			Severi;jnci,41,331-349;1968
a	157	465.mg	n.s.s.	0/22	149.mg	0/40			
b	157	465.mg	n.s.s.	0/22	149.mg	0/40			
2733	157	32.1mg	n.s.s.	0/21	99.4mg	1/5			
a	157	262.mg	n.s.s.	0/28	99.4mg	2/49			

ISONICOTINAMIDE   1453-82-3

	RefNum	LoConf	UpConf	Cntrl	1Dose	1Inc	2Dose	2Inc	Citation or Pathology
2734	1732	3.47gm	n.s.s.	0/28	2.00gm	1/12			Toth;onco,40,72-75;1983/1979
a	1732	6.83gm	n.s.s.	15/91	2.00gm	5/33			
2735	1732	5.77gm	n.s.s.	22/88	1.67gm	9/43			
a	1732	13.9gm	n.s.s.	2/62	1.67gm	0/32			

ISONICOTINIC ACID   55-22-1

	RefNum	LoConf	UpConf	Cntrl	1Dose	1Inc	2Dose	2Inc	Citation or Pathology
2736	1108	1.99gm	n.s.s.	0/66	682.mg	2/27			Toth;canr,32,804-807;1972/1967a
a	1108	6.36gm	n.s.s.	0/72	682.mg	0/34			
b	1108	6.36gm	n.s.s.	0/72	682.mg	0/34			
2737	1108	1.13gm	n.s.s.	0/39	600.mg	1/13			
a	1108	1.19gm	n.s.s.	1/39	600.mg	1/13			
b	1108	4.85gm	n.s.s.	0/64	600.mg	0/29			
2738	1074	31.4mg	n.s.s.	2/47	20.9mg	1/17			Biancifiori;bjca,18,543-550;1964
a	1074	32.6mg	n.s.s.	4/47	20.9mg	1/16			
2739	1074	16.3mg	n.s.s.	0/13	17.4mg	1/12			
a	1074	24.6mg	n.s.s.	4/27	17.4mg	2/18			

ISONICOTINIC ACID VANILLYLIDENEHYDRAZIDE   (phthivazid) 149-17-7

	RefNum	LoConf	UpConf	Cntrl	1Dose	1Inc	2Dose	2Inc	Citation or Pathology
2740	1425	19.2mg	n.s.s.	5/94	30.1mg	14/90			Pershin;vopr,XVIII,50-53;1972
a	1425	30.3mg	n.s.s.	2/94	30.1mg	7/90			
b	1425	13.6mg	174.mg	7/94	30.1mg	21/90			

ISOPHORONE   78-59-1

	RefNum	LoConf	UpConf	Cntrl	1Dose	1Inc	2Dose	2Inc	Citation or Pathology
2741	TR291	136.mg	n.s.s.	36/50	175.mg	41/50	(354.mg	28/50)	
a	TR291	604.mg	n.s.s.	4/50	175.mg	6/50	354.mg	8/50	liv:hpa,hpc,nnd.
b	TR291	1.88gm	n.s.s.	3/50	175.mg	1/50	354.mg	2/50	lun:a/a,a/c.
2742	TR291	109.mg	928.mg	0/50	177.mg	9/50	(354.mg	4/50)	
a	TR291	80.2mg	n.s.s.	7/50	177.mg	19/50	(354.mg	5/50)	---:mlh,mlm,mlp.
b	TR291	85.7mg	n.s.s.	8/50	177.mg	19/50	(354.mg	5/50)	---:lkn,mlh,mlm,mlp.
c	TR291	293.mg	n.s.s.	3/50	177.mg	6/50	354.mg	13/50	sub:fbs,fib.
d	TR291	301.mg	n.s.s.	4/50	177.mg	6/50	354.mg	14/50	sub:fbs,fib,nfs,sar.
e	TR291	444.mg	n.s.s.	3/50	177.mg	4/50	354.mg	10/50	
f	TR291	460.mg	n.s.s.	0/50	177.mg	2/50	354.mg	3/50	
g	TR291	162.mg	n.s.s.	18/50	177.mg	18/50	354.mg	29/50	liv:hpa,hpc.
h	TR291	308.mg	n.s.s.	5/50	177.mg	7/50	354.mg	13/50	ski:fib; sub:fbs,fib.
i	TR291	309.mg	n.s.s.	6/50	177.mg	8/50	354.mg	14/50	ski:fib,nfs; sub:fbs,fib,nfs,sar.
j	TR291	434.mg	n.s.s.	4/50	177.mg	5/50	354.mg	11/50	ski:nfs; sub:fbs,nfs,sar.
k	TR291	451.mg	n.s.s.	4/50	177.mg	4/50	354.mg	11/50	sub:fbs,nfs,sar.
l	TR291	174.mg	n.s.s.	35/50	177.mg	40/50	354.mg	40/50	
m	TR291	162.mg	n.s.s.	18/50	177.mg	18/50	354.mg	29/50	liv:hpa,hpc,nnd.
n	TR291	1.39gm	n.s.s.	7/50	177.mg	1/50	354.mg	3/50	lun:a/a,a/c.
2743	TR291	192.mg	n.s.s.	43/50	175.mg	36/50	350.mg	30/50	
a	TR291	1.45gm	n.s.s.	3/50	175.mg	1/50	350.mg	1/50	liv:hpa,hpc,nnd.
2744	TR291	378.mg	2.07gm	0/50	175.mg	3/50	350.mg	8/50	kid:tla,uac; pre:can. P
a	TR291	492.mg	21.4gm	0/50	175.mg	3/50	350.mg	3/50	kid:tla,uac.
b	TR291	818.mg	19.3gm	0/50	175.mg	0/50	350.mg	5/50	
c	TR291	99.5mg	n.s.s.	38/50	175.mg	42/50	350.mg	33/50	
d	TR291	522.mg	n.s.s.	5/50	175.mg	9/50	350.mg	2/50	liv:hpa,hpc,nnd.

ISOPHOSPHAMIDE   3778-73-2

	RefNum	LoConf	UpConf	Cntrl	1Dose	1Inc	2Dose	2Inc	Citation or Pathology
2745	TR32	2.81mg	14.1mg	0/15	2.80mg	3/35	5.60mg	13/35	
a	TR32	2.18mg	9.48mg	0/15	2.80mg	7/35	5.60mg	15/35	
b	TR32	n.s.s.	n.s.s.	0/15	2.80mg	1/35	5.60mg	0/35	liv:hpa,hpc,nnd.
c	TR32	n.s.s.	n.s.s.	0/15	2.80mg	0/35	5.60mg	0/35	lun:a/a,a/c.
2746	TR32	2.81mg	10.9mg	0/30p	2.80mg	3/35	5.60mg	13/35	
2747	TR32	1.85mg	n.s.s.	0/15	2.80mg	6/35	5.60mg	4/35	liv:hpa,hpc,nnd.
a	TR32	2.70mg	n.s.s.	0/15	2.80mg	5/35	5.60mg	2/35	lun:a/a,a/c.
b	TR32	6.13mg	n.s.s.	0/15	2.80mg	0/35	5.60mg	1/35	
2748	TR32	.900mg	n.s.s.	0/30p	2.80mg	5/35	(5.60mg	2/35)	liv:hpa,hpc. S
2749	TR32	.158mg	1.01mg	3/10	1.60mg	28/35	3.30mg	6/35	
a	TR32	.375mg	3.07mg	0/10	1.60mg	15/35	3.30mg	1/35	
b	TR32	.175mg	17.4mg	9/10	1.60mg	32/35	3.30mg	9/35	
c	TR32	2.79mg	n.s.s.	0/10	1.60mg	1/35	3.30mg	0/35	liv:hpa,hpc,nnd.
2750	TR32	.199mg	.841mg	8/30p	1.60mg	28/35	3.30mg	6/35	
a	TR32	.429mg	1.76mg	0/30p	1.60mg	15/35	3.30mg	1/35	
b	TR32	.977mg	49.3mg	0/30p	1.60mg	3/35	3.30mg	1/35	mgl:ccn,cyn. S
2751	TR32	.476mg	n.s.s.	5/10	1.60mg	17/35	3.30mg	12/35	
a	TR32	n.s.s.	n.s.s.	0/10	1.60mg	0/35	3.30mg	0/35	liv:hpa,hpc,nnd.
2752	TR32	1.37mg	11.2mg	0/30p	1.60mg	3/35	3.30mg	5/35	---:leu,lym. S

p-ISOPROPOXYDIPHENYLAMINE   (agerite 150) 101-73-5

	RefNum	LoConf	UpConf	Cntrl	1Dose	1Inc	2Dose	2Inc	Citation or Pathology
2753	1306	431.mg	n.s.s.	0/17	414.mg	1/18			Innes;ntis,1968/1969
a	1306	492.mg	n.s.s.	1/17	414.mg	1/18			

Spe Strain Site Xpo+Xpt				TD50	2Tailpvl
Sex Route Hist Notes				DR	AuOp
b M f b6a orl tba mix 76w76 evx				1.20gm	P<.5 −
2754 M m b6a orl liv hpt 76w76 evx		.>		1.13gm	P<.3 −
a M m b6a orl lun ade 76w76 evx				2.18gm	P<.7 −
b M m b6a orl tba mix 76w76 evx				985.mg	P<.5 −
2755 M f b6c orl liv hpt 76w76 evx			.>	2.65gm	P<.3 −
a M f b6c orl lun mix 76w76 evx				no dre	P=1. −
b M f b6c orl tba mix 76w76 evx				1.29gm	P<.2 −
2756 M f b6c orl liv hpt 76w76 evx			. ±	679.mg	P<.04 −
a M m b6c orl lun ade 76w76 evx				2.18gm	P<.3 −
b M m b6c orl tba mix 76w76 evx				300.mg	P<.003 −

ISOPROPYL-N-(3-CHLOROPHENYL)CARBAMATE ..:..1ug....:..10.......:..100......:..1mg.....:..10.....:..100....:..1g.....:..10

				TD50	2Tailpvl
2757 H f syg eat liv tum 33m33 e			.>	no dre	P=1.
a H f syg eat lun sqc 33m33 e				no dre	P=1. −
2758 H m syg eat liv hpc 33m33 e			.>	no dre	P=1.
2759 M f b6a orl lun ade 76w76 evx		.>		no dre	P=1. −
a M f b6a orl liv hpt 76w76 evx				no dre	P=1. −
b M f b6a orl tba mix 76w76 evx				no dre	P=1. −
2760 M m b6a orl liv hpt 76w76 evx		.>		no dre	P=1. −
a M m b6a orl lun ade 76w76 evx				no dre	P=1. −
b M m b6a orl tba mix 76w76 evx				no dre	P=1. −
2761 M f b6c orl liv hpt 76w76 evx		.>		no dre	P=1. −
a M f b6c orl lun mix 76w76 evx				no dre	P=1. −
b M f b6c orl tba tum 76w76 evx				no dre	P=1. −
2762 M m b6c orl liv hpt 76w76 evx		. ±		258.mg	P<.04 −
a M m b6c orl lun ade 76w76 evx				830.mg	P<.3 −
b M m b6c orl tba mix 76w76 evx				114.mg	P<.003 −
2763 M b swi eat lun ade 27m27 e		.>		682.mg	P<.2 −
a M b swi eat lun adc 27m27 e				4.95gm	P<.3 −
b M b swi eat liv tum 27m27 e				no dre	P=1. −
2764 R f alb eat mix mly 24m24			.>	no dre	P=1.
2765 R m alb eat liv tum 24m24 s	.>			no dre	P=1.
a R m alb eat mds mly 24m24 s				no dre	P=1.

1-ISOPROPYL-3-METHYL-s-PYRAZOLYLDIMETHYL CARBAMATE.:..10......:..100....:..1mg....:..10....:..100....:..1g.....:..10

				TD50	2Tailpvl
2766 M f b6a orl liv hpt 76w76 evx	.>			no dre	P=1. −
a M f b6a orl lun mix 76w76 evx				no dre	P=1. −
b M f b6a orl tba mix 76w76 evx				23.0ug	P<.08 −
2767 M m b6a orl lun ade 76w76 evx	. ±			14.7ug	P<.04 −
a M m b6a orl liv hpt 76w76 evx				41.9ug	P<.6 −
b M m b6a orl tba mix 76w76 evx				12.4ug	P<.2 −
2768 M f b6c orl liv hpt 76w76 evx	.>			no dre	P=1. −
a M f b6c orl lun ade 76w76 evx				no dre	P=1. −
b M f b6c orl tba mix 76w76 evx				no dre	P=1. −
2769 M m b6c orl lun ade 76w76 evx	. ±			15.6ug	P<.06 −
a M m b6c orl liv hpt 76w76 evx				no dre	P=1. −
b M m b6c orl tba mix 76w76 evx				41.3ug	P<.8 −

ISOPROPYL-N-PHENYL CARBAMATE   100ng..:..1ug....:..10......:..100......:..1mg.....:..10....:..100....:..1g.....:..10

				TD50	2Tailpvl
2770 H f syg eat liv tum 33m33 e			.>	no dre	P=1.
a H f syg eat lun sqc 33m33 e				no dre	P=1. −
2771 H m syg eat liv hpc 33m33 e			.>	no dre	P=1. −
2772 M f b6a orl lun ade 76w76 evx		.>		444.mg	P<.6 −
a M f b6a orl liv hpt 76w76 evx				no dre	P=1. −
b M f b6a orl tba mix 76w76 evx				416.mg	P<.7 −
2773 M m b6a orl liv hpt 76w76 evx		.>		350.mg	P<.5 −
a M m b6a orl lun ade 76w76 evx				1.70gm	P<1. −
b M m b6a orl tba mix 76w76 evx				254.mg	P<.6 −
2774 M f b6c orl lun ade 76w76 evx		.>		244.mg	P<.2 −
a M f b6c orl liv hpt 76w76 evx				502.mg	P<.3 −
b M f b6c orl tba mix 76w76 evx				88.1mg	P<.009 −
2775 M m b6c orl liv hpt 76w76 evx		. ±		120.mg	P<.04 −
a M m b6c orl lun ade 76w76 evx				387.mg	P<.3 −
b M m b6c orl tba mix 76w76 evx				52.3mg	P<.002 −

ISOSAFROLE   100ng..:..1ug....:..10......:..100......:..1mg.....:..10......:..100....:..1g.....:..10

				TD50	2Tailpvl
2776 M f b6a orl liv hpt 81w81 evx			.>	no dre	P=1.
a M f b6a orl lun ade 81w81 evx				no dre	P=1.
b M f b6a orl tba mix 81w81 evx				no dre	P=1.
2777 M m b6a orl liv hpt 81w81 evx		. ±		225.mg	P<.09
a M m b6a orl lun mix 81w81 evx				no dre	P=1.
b M m b6a orl tba mix 81w81 evx				327.mg	P<.7
2778 M f b6c orl liv hpt 81w81 evx		.>		468.mg	P<.3
a M f b6c orl lun ade 81w81 evx				468.mg	P<.3
b M f b6c orl tba mix 81w81 evx				414.mg	P<.6
2779 M m b6c orl liv hpt 81w81 evx		. ±		106.mg	P<.08
a M m b6c orl lun ade 81w81 evx				231.mg	P<.4
b M m b6c orl tba mix 81w81 evx				492.mg	P<.9

KAEMPFEROL   100ng..:..1ug......:..10......:..100......:..1mg.....:..10......:..100....:..1g.......:..10

				TD50	2Tailpvl
2780 R f aci eat tba mix 77w77 e		.>		55.3mg	P<.4 −

	RefNum	LoConf	UpConf	Cntrl	1Dose	1Inc	2Dose	2Inc	Citation or Pathology
									Brkly Code
b	1306	239.mg	n.s.s.	2/17	414.mg	4/18			
2754	1306	252.mg	n.s.s.	1/18	385.mg	3/18			
a	1306	274.mg	n.s.s.	2/18	385.mg	3/18			
b	1306	196.mg	n.s.s.	3/18	385.mg	5/18			
2755	1306	431.mg	n.s.s.	0/16	414.mg	1/18			
a	1306	820.mg	n.s.s.	0/16	414.mg	0/18			
b	1306	316.mg	n.s.s.	0/16	414.mg	2/18			
2756	1306	205.mg	n.s.s.	0/16	385.mg	3/16			
a	1306	355.mg	n.s.s.	0/16	385.mg	1/16			
b	1306	121.mg	1.47gm	0/16	385.mg	6/16			

ISOPROPYL-N-(3-CHLOROPHENYL)CARBAMATE   (CIPC, chlorpropham) 101-21-3

	RefNum	LoConf	UpConf	Cntrl	1Dose	1Inc	2Dose	2Inc	3Dose	3Inc	Citation or Pathology
2757	171a	2.12gm	n.s.s.	0/27	209.mg	0/26					Van Esch;fctx,10,373-381;1972
a	171a	2.12gm	n.s.s.	1/27	209.mg	0/26					
2758	171a	1.65gm	n.s.s.	1/22	184.mg	0/23					
2759	1196	175.mg	n.s.s.	1/17	157.mg	1/17					Innes;ntis,1968/1969
a	1196	294.mg	n.s.s.	0/17	157.mg	0/17					
b	1196	138.mg	n.s.s.	2/17	157.mg	2/17					
2760	1196	258.mg	n.s.s.	1/18	146.mg	0/16					
a	1196	258.mg	n.s.s.	2/18	146.mg	0/16					
b	1196	180.mg	n.s.s.	3/18	146.mg	1/16					
2761	1196	311.mg	n.s.s.	0/16	157.mg	0/18					
a	1196	311.mg	n.s.s.	0/16	157.mg	0/18					
b	1196	311.mg	n.s.s.	0/16	157.mg	0/18					
2762	1196	77.8mg	n.s.s.	0/16	146.mg	3/16					
a	1196	135.mg	n.s.s.	0/16	146.mg	1/16					
b	1196	45.9mg	559.mg	0/16	146.mg	6/16					
2763	171a	209.mg	n.s.s.	10/49	125.mg	15/47					Van Esch;fctx,10,373-381;1972
a	171a	807.mg	n.s.s.	0/49	125.mg	1/47					
b	171a	1.51gm	n.s.s.	0/49	125.mg	0/47					
2764	170a	4.86gm	n.s.s.	1/25	10.0mg	0/25	100.mg	1/25	1.00gm	0/25	Larson;txap,2,659-673;1960
2765	170a	37.1mg	n.s.s.	0/25	8.00mg	0/25	80.0mg	0/25	800.mg	0/25	
a	170a	3.11gm	n.s.s.	0/25	8.00mg	0/25	80.0mg	1/25	800.mg	0/25	

1-ISOPROPYL-3-METHYL-s-PYRAZOLYLDIMETHYL CARBAMATE   (isolan) 119-38-0

	RefNum	LoConf	UpConf	Cntrl	1Dose	1Inc	2Dose	2Inc	Citation or Pathology
2766	1247	14.7ug	n.s.s.	0/18	8.38ug	0/16			Innes;ntis,1968/1969
a	1247	14.7ug	n.s.s.	0/18	8.38ug	0/16			
b	1247	5.63ug	n.s.s.	0/18	8.38ug	2/16			
2767	1247	4.43ug	n.s.s.	0/18	7.80ug	3/17			
a	1247	6.12ug	n.s.s.	1/18	7.80ug	2/17			
b	1247	3.44ug	n.s.s.	2/18	7.80ug	5/17			
2768	1247	14.7ug	n.s.s.	0/18	8.38ug	0/16			
a	1247	14.7ug	n.s.s.	1/18	8.38ug	0/16			
b	1247	14.7ug	n.s.s.	2/18	8.38ug	0/16			
2769	1247	4.72ug	n.s.s.	0/14	7.80ug	3/18			
a	1247	11.2ug	n.s.s.	3/14	7.80ug	1/18			
b	1247	3.81ug	n.s.s.	4/14	7.80ug	6/18			

ISOPROPYL-N-PHENYL CARBAMATE   (IPC) 122-42-9

	RefNum	LoConf	UpConf	Cntrl	1Dose	1Inc	2Dose	2Inc	Citation or Pathology
2770	171a	2.12gm	n.s.s.	0/27	209.mg	0/26			Van Esch;fctx,10,373-381;1972
a	171a	2.12gm	n.s.s.	1/27	209.mg	0/26			
2771	171a	995.mg	n.s.s.	1/22	184.mg	1/23			
2772	171	61.8mg	n.s.s.	1/17	78.4mg	2/17			Innes;ntis,1968/1969
a	171	147.mg	n.s.s.	0/17	78.4mg	0/17			
b	171	52.4mg	n.s.s.	2/17	78.4mg	3/17			
2773	171	53.4mg	n.s.s.	1/18	73.0mg	2/16			
a	171	58.9mg	n.s.s.	2/18	73.0mg	2/16			
b	171	38.6mg	n.s.s.	3/18	73.0mg	4/16			
2774	171	59.8mg	n.s.s.	0/16	78.4mg	2/18			
a	171	81.7mg	n.s.s.	0/16	78.4mg	1/18			
b	171	33.2mg	1.93gm	0/16	78.4mg	5/18			
2775	171	36.1mg	n.s.s.	0/16	73.0mg	3/15			
a	171	63.0mg	n.s.s.	0/16	73.0mg	1/15			
b	171	21.0mg	236.mg	0/16	73.0mg	6/15			

ISOSAFROLE   120-58-1

	RefNum	LoConf	UpConf	Cntrl	1Dose	1Inc	2Dose	2Inc	Citation or Pathology
2776	172	145.mg	n.s.s.	1/15	72.7mg	0/16			Innes;ntis,1968/1969
a	172	145.mg	n.s.s.	2/15	72.7mg	0/16			
b	172	145.mg	n.s.s.	2/15	72.7mg	0/16			
2777	172	55.2mg	n.s.s.	0/18	67.7mg	2/17			
a	172	144.mg	n.s.s.	3/18	67.7mg	0/17			
b	172	44.2mg	n.s.s.	3/18	67.7mg	4/17			
2778	172	76.2mg	n.s.s.	0/17	72.7mg	1/16			
a	172	76.2mg	n.s.s.	0/17	72.7mg	1/16			
b	172	60.7mg	n.s.s.	1/17	72.7mg	2/16			
2779	172	34.5mg	n.s.s.	1/17	67.7mg	5/18			
a	172	50.4mg	n.s.s.	1/17	67.7mg	3/18			
b	172	31.4mg	n.s.s.	7/17	67.7mg	8/18			

KAEMPFEROL   520-18-3

	RefNum	LoConf	UpConf	Cntrl	1Dose	1Inc	2Dose	2Inc	Citation or Pathology
2780	1662	6.92mg	n.s.s.	1/22	20.0mg	1/6			Takanashi;jfds,5,55-60;1983

```
       Spe Strain  Site   Xpo+Xpt                                                                    TD50    2Tailpvl
          Sex  Route  Hist      Notes                                                                   DR    AuOp
2781 R m aci eat adr coa 77w77 e                               .            ±                        17.9mg   P<.1   -

KANECHLOR 400                  100ng..:..1ug....:..10.....:..100....:..1mg....:..10......:..100.....:..1g......:..10
2782 R f don eat liv tum 26w84 e                                                    .>               no dre   P=1.   -

KEPONE                         100ng..:..1ug....:..10.....:..100....:..1mg....:..10......:..100.....:..1g......:..10
2783 M f b6c eat liv hpc 80w90 v                                      :  +  :                        1.62mg \ P<.003 c
a    M f b6c eat TBA MXB 80w90 v                                                                     1.57mg \ P<.003
b    M f b6c eat TBA MXB 80w90 v                                                                     1.62mg \ P<.003
c    M f b6c eat TBA MXB 80w90 v                                                                     no dre   P=1.
2784 M m b6c eat liv hpc 80w89 dv                                  :  +  :                           .705mg * P<.0005c
a    M m b6c eat TBA MXB 80w89 dv                                                                    .705mg * P<.0005
b    M m b6c eat TBA MXB 80w89 dv                                                                    .705mg * P<.0005
c    M m b6c eat TBA MXB 80w89 dv                                                                    29.3mg * P<.3
2785 R f osm eat TBA MXB 19m25 v                                 :>                                  1.83mg * P<.5   -
a    R f osm eat TBA MXB 19m25 v                                                                     2.96mg / P<.08
2786 R m osm eat TBA MXB 19m25 v                                    :>                               1.42mg * P<.4   -
a    R m osm eat TBA MXB 19m25 v                                                                     2.49mg * P<.2
2787 R m sda eat liv nnd 91w91                          .>                                           .199mg   P<.3

KETOPROFEN                     100ng..:..1ug....:..10.....:..100....:..1mg....:..10......:..100.....:..1g......:..10
2788 R m f34 eat col mix 54w54 er                                                   .>               no dre   P=1.

LASIOCARPINE                   100ng..:..1ug....:..10.....:..100....:..1mg....:..10......:..100.....:..1g......:..10
2789 R f f34 eat MXB MXB 21m24 as                              :  +  :                               .141mg Z P<.0005
a    R f f34 eat mul MXA 21m24 as                                                                    .325mg * P<.0005c
b    R f f34 eat liv ang 21m24 as                                                                    .355mg * P<.0005c
c    R f f34 eat mul MXA 21m24 as                                                                    .387mg * P<.0005c
d    R f f34 eat liv MXA 21m24 as                                                                    .453mg Z P<.0005c
e    R f f34 eat TBA MXB 21m24 as                                                                    .102mg Z P<.0005
f    R f f34 eat liv MXB 21m24 as                                                                    22.7mg * P<.2
2790 R f f34 eat liv hpc 21m24 er                                      .  +  .                       .938mg   P<.0005+
2791 R m f34 eat liv MXB 23m24 as                                   :  +  :                          .250mg Z P<.0005
a    R m f34 eat liv ang 23m24 as                                                                    .355mg Z P<.0005c
b    R m f34 eat --- MXA 23m24 as                                                                    .514mg Z P<.0005
c    R m f34 eat liv MXA 23m24 as                                                                    .908mg Z P<.0005c
d    R m f34 eat TBA MXB 23m24 as                                                                    .152mg Z P<.0005
e    R m f34 eat liv MXB 23m24 as                                                                    7.43mg Z P<.03
2792 R m f34 eat liv ang 55w59 e                                       .  +  .                        .688mg   P<.004 +
a    R m f34 eat liv car 55w59 e                                                                     .954mg   P<.02  +
b    R m f34 eat tba mix 55w59 e                                                                     .217mg   P<.0005
2793 R m f34 ipj liv mix 56w76 ev                              .  +  .                               .341mg   P<.0005+
a    R m f34 ipj liv hpc 56w76 ev                                                                    .397mg   P<.0005+
b    R m f34 ipj ski sqc 56w76 ev                                                                    .794mg   P<.002 +
c    R m f34 ipj lun ade 56w76 ev                                                                    1.33mg   P<.09
d    R m f34 ipj smi adc 56w76 ev                                            +hist                   2.73mg   P<.06  +
e    R m f34 ipj tba mix 56w76 ev                                                                    .152mg   P<.0005
2794 R m f34 eat liv hpc 23m24 er                              .  +  .                               .800mg * P<.0005+

LEAD ACETATE                   100ng..:..1ug....:..10.....:..100....:..1mg....:..10......:..100....:..1g......:..10
2795 M f cd1 wat lun tum 30m30 e                                              .>                     no dre   P=1.   -
a    M f cd1 wat tba tum 30m30 e                                                                     51.5mg   P<1.   -
2796 M f cd1 wat lun tum 28m28 e                                         .   ±                        7.56mg   P<.05  -
a    M f cd1 wat tba mix 28m28 e                                                                     45.3mg   P<.9   -
b    M f cd1 wat tba mal 28m28 e                                                                     no dre   P=1.   -
2797 M m cd1 wat lun tum 30m30 e                                             .>                      no dre   P=1.   -
a    M m cd1 wat liv sar 30m30 e                                                                     no dre   P=1.   -
b    M m cd1 wat tba mix 30m30 e                                                                     no dre   P=1.   -
2798 M m cd1 wat lun tum 25m25 e                                           .>                        7.78mg   P<.4   -
a    M m cd1 wat tba mal 25m25 e                                                                     4.04mg   P<.04  -
b    M m cd1 wat tba mix 25m25 e                                                                     6.22mg   P<.5   -
2799 R f f34 eat kid mal 24m24 e                                                    .   +            1.58gm * P<.0005-
a    R f f34 eat liv hpc 24m24 e                                                                     no dre   P=1.
2800 R m f34 eat kid adc 52w52 ekr                                                .>                 993.mg   P<.3
a    R m f34 eat liv hpc 52w52 ekr                                                                   no dre   P=1.
2801 R m f34 eat kid mal 24m24 e                                                     .  +.           129.mg * P<.0005+
a    R m f34 eat liv hpc 24m24 e                                                                     no dre   P=1.
2802 R b leb wat liv tum 41m41 es                                             .>                     no dre   P=1.   -
a    R b leb wat tba tum 41m41 es                                                                    no dre   P=1.   -
b    R b leb wat tba mal 41m41 es                                                                    133.mg   P<1.   -
2803 R m sda wat kid tcc 76w76 e                                              .  +  .                28.4mg   P<.0005+
a    R m sda wat liv tum 76w76 e                                                                     no dre   P=1.
b    R m sda wat tba mix 76w76 e                                                                     28.4mg   P<.0005+

LEAD ACETATE, BASIC            100ng..:..1ug....:..10.....:..100....:..1mg....:..10......:..100....:..1g......:..10
2804 H f syg eat liv tum 23m25 as                                                     .>             no dre   P=1.
2805 H m syg eat liv tum 24m24 s                                                      .>             no dre   P=1.
2806 M f swi eat kid car 24m24 sv                                                       .>           14.8gm * P<.2   +
a    M f swi eat kid ade 24m24 sv                                                                    no dre   P=1.   +
b    M f swi eat liv hpt 24m24 sv                                                                    no dre   P=1.
c    M f swi eat lun ade 24m24 sv                                                                    no dre   P=1.
```

	RefNum	LoConf	UpConf	Cntrl	1Dose	1Inc	2Dose	2Inc	Citation or Pathology
									Brkly Code
2781	1662	3.71mg	n.s.s.	2/30	16.0mg	2/6			

KANECHLOR 400 (PCBs, polychlorinated biphenyls) 12737-87-0

	RefNum	LoConf	UpConf	Cntrl	1Dose	1Inc	2Dose	2Inc	Citation or Pathology
2782	1762	20.8mg	n.s.s.	0/10	6.19mg	0/25			Kimura;zkko,87,257-266;1976

KEPONE (chlordecone) 143-50-0

	RefNum	LoConf	UpConf	Cntrl	1Dose	1Inc	2Dose	2Inc	Citation or Pathology
2783	TR-B	1.01mg	6.36mg	0/10	2.30mg	26/50	(4.70mg	23/50)	
a	TR-B	.992mg	5.65mg	0/10	2.30mg	27/50	(4.70mg	25/50)	
b	TR-B	1.01mg	6.36mg	0/10	2.30mg	26/50	(4.70mg	23/50)	
c	TR-B	n.s.s.	n.s.s.	0/10	2.30mg	0/50	4.70mg	0/50	
2784	TR-B	.462mg	1.59mg	6/20	2.40mg	43/50	2.20mg	39/50	
a	TR-B	.462mg	1.59mg	6/20	2.40mg	43/50	2.20mg	39/50	
b	TR-B	.462mg	1.59mg	6/20	2.40mg	43/50	2.20mg	39/50	
c	TR-B	7.21mg	n.s.s.	0/20	2.40mg	2/50	2.20mg	0/50	
2785	TR-B	.469mg	n.s.s.	7/10	.640mg	29/50	.930mg	29/50	
a	TR-B	1.30mg	n.s.s.	1/10	.640mg	1/50	.930mg	12/50	
2786	TR-B	.459mg	n.s.s.	3/15	.230mg	24/50	.690mg	16/50	
a	TR-B	1.22mg	n.s.s.	0/15	.230mg	3/50	.690mg	3/50	
2787	1160	32.4ug	n.s.s.	0/10	40.0ug	1/10			Chu;txap,59,268-278;1981

KETOPROFEN 22071-15-4

	RefNum	LoConf	UpConf	Cntrl	1Dose	1Inc	2Dose	2Inc	Citation or Pathology
2788	2078	5.33mg	n.s.s.	0/12	8.00mg	0/12			Reddy;carc,13,1019-1023;1992

LASIOCARPINE 303-34-4

	RefNum	LoConf	UpConf	Cntrl	1Dose	1Inc	2Dose	2Inc	Citation or Pathology
2789	TR39	85.9ug	.247mg	2/24	.350mg	17/24	.750mg	19/24	1.50mg	8/24	liv:adn,ang,hpc; mul:grl,leu,lle,lym,ule. C
a	TR39	.165mg	.822mg	2/24	.350mg	9/24	.750mg	11/24	1.50mg	1/24	mul:leu,lym.
b	TR39	.183mg	.741mg	0/24	.350mg	8/24	.750mg	7/24	1.50mg	2/24	
c	TR39	.186mg	1.04mg	1/24	.350mg	7/24	.750mg	8/24	1.50mg	1/24	mul:grl,lle,ule.
d	TR39	.195mg	1.26mg	0/24	.350mg	5/24	.750mg	1/24	1.50mg	7/24	liv:adn,hpc.
e	TR39	60.1ug	.205mg	18/24	.350mg	22/24	.750mg	22/24	1.50mg	8/24	
f	TR39	3.70mg	n.s.s.	0/24	.350mg	0/24	.750mg	0/24	1.50mg	0/24	liv:hpa,hpc,nnd.
2790	1824	.521mg	1.94mg	0/144	.350mg	9/24	.750mg	7/24	(1.50mg	3/23)	Elashoff;jnci,79,509-526;1987
2791	TR39	.149mg	.428mg	0/24	.280mg	5/24	.600mg	14/24	1.20mg	17/24	liv:adn,ang,hpc. C
a	TR39	.208mg	.621mg	0/24	.280mg	5/24	.600mg	11/24	1.20mg	13/24	
b	TR39	.256mg	1.45mg	4/24	.280mg	3/24	.600mg	11/24	1.20mg	7/24	
c	TR39	.329mg	3.21mg	0/24	.280mg	0/24	.600mg	3/24	1.20mg	5/24	---:leu,lym. S
d	TR39	97.2ug	.257mg	7/24	.280mg	14/24	.600mg	22/24	1.20mg	23/24	liv:adn,hpc.
e	TR39	1.82mg	n.s.s.	0/24	.280mg	0/24	.600mg	0/24	1.20mg	2/24	liv:hpa,hpc,nnd.
2792	390	.319mg	3.32mg	0/10	1.86mg	9/20			Rao;bjca,37,289-293;1978
a	390	.408mg	170.mg	0/10	1.86mg	7/20			
b	390	.108mg	.469mg	0/10	1.86mg	17/20			
2793	1013	.165mg	.851mg	0/25	.880mg	11/18			Svoboda;canr,32,908-911;1972
a	1013	.188mg	1.04mg	0/25	.880mg	10/18			
b	1013	.321mg	3.10mg	0/25	.880mg	6/18			
c	1013	.406mg	n.s.s.	2/25	.880mg	5/18			
d	1013	.671mg	n.s.s.	0/25	.880mg	2/18			
e	1013	69.2ug	.372mg	2/25	.880mg	16/18			
2794	1824	.507mg	1.36mg	1/144	.280mg	5/24	.600mg	11/24	1.20mg	14/23	Elashoff;jnci,79,509-526;1987

LEAD ACETATE 301-04-2

	RefNum	LoConf	UpConf	Cntrl	1Dose	1Inc	2Dose	2Inc	Citation or Pathology
2795	56	3.58mg	n.s.s.	9/60	1.00mg	4/29			Schroeder;jnut,83,239-250;1964
a	56	1.90mg	n.s.s.	22/60	1.00mg	11/29			
2796	1395	2.54mg	n.s.s.	1/45	1.00mg	5/37			Schroeder;jnut,105,452-458;1975
a	1395	2.62mg	n.s.s.	9/45	1.00mg	8/37			
b	1395	3.06mg	n.s.s.	9/45	1.00mg	7/37			
2797	56	8.23mg	n.s.s.	8/44	.833mg	1/39			Schroeder;jnut,83,239-250;1964
a	56	9.99mg	n.s.s.	1/44	.833mg	0/39			
b	56	8.57mg	n.s.s.	11/44	.833mg	1/39			
2798	1395	1.98mg	n.s.s.	4/43	.833mg	8/49			Schroeder;jnut,105,452-458;1975
a	1395	1.59mg	n.s.s.	2/43	.833mg	9/49			
b	1395	1.39mg	n.s.s.	10/43	.833mg	15/49			
2799	1947	598.mg	6.55gm	0/214	25.0mg	0/24	100.mg	1/24	400.mg	4/24	Fears;txih,5,1-23;1989
a	1947	92.4mg	n.s.s.	0/214	25.0mg	0/24	100.mg	0/24	400.mg	0/24	
2800	1150	162.mg	n.s.s.	0/15	400.mg	1/15			Hinton;bect,23,464-469;1979
a	1150	309.mg	n.s.s.	0/15	400.mg	0/15			
2801	1947	80.1mg	222.mg	0/213	20.0mg	0/24	80.0mg	11/24	320.mg	19/24	Fears;txih,5,1-23;1989
a	1947	73.8mg	n.s.s.	0/214	20.0mg	0/24	80.0mg	0/24	320.mg	0/23	
2802	1036	5.00mg	n.s.s.	1/34	.265mg	0/32			Kanisawa;canr,29,892-895;1969
a	1036	1.67mg	n.s.s.	10/34	.265mg	7/32			
b	1036	2.40mg	n.s.s.	2/34	.265mg	2/32			
2803	1755	13.4mg	68.2mg	0/10	130.mg	13/16			Koller;txpy,13,50-57;1985
a	1755	229.mg	n.s.s.	0/10	130.mg	0/16			
b	1755	13.4mg	68.2mg	0/10	130.mg	13/16			

LEAD ACETATE, BASIC 1335-32-6

	RefNum	LoConf	UpConf	Cntrl	1Dose	1Inc	2Dose	2Inc	Citation or Pathology
2804	176	456.mg	n.s.s.	0/23	105.mg	0/24	523.mg	0/24	Van Esch;bjca,23,765-771;1969
2805	176	348.mg	n.s.s.	0/22	92.0mg	0/22	460.mg	0/22	
2806	176	2.42gm	n.s.s.	0/25	130.mg	0/25	751.mg	1/25	
a	176	2.93gm	n.s.s.	0/25	130.mg	1/25	751.mg	0/25	
b	176	2.93gm	n.s.s.	0/25	130.mg	1/25	751.mg	0/25	
c	176	2.58gm	n.s.s.	3/25	130.mg	5/25	751.mg	1/25	

```
        Spe Strain Site  Xpo+Xpt                                              TD50      2Tailpvl
            Sex Route Hist      Notes                                              DR       AuOp
2807 M m swi eat kid car 20m24 asv                          . ±       472.mg \ P<.02  +
a    M m swi eat kid ade 20m24 asv                                    986.mg \ P<.1   +
b    M m swi eat liv hpt 20m24 asv                                    no dre   P=1.
c    M m swi eat lun ade 20m24 asv                                    no dre   P=1.
2808 R m sda eat kid mix 79w79 e                        . + .         266.mg   P<.0005+
a    R m sda eat liv tum 79w79 e                                      no dre   P=1.    -
2809 R f wis eat kid mix 29m29 e                          . + .       107.mg   P<.003  +
2810 R f wis eat kid mix 24m24 e                          . + .       339.mg   P<.0005+
2811 R m wis eat kid mix 29m29 e                        . + .         107.mg   P<.008  +
2812 R m wis eat kid mix 24m24 e                        . + .         443.mg   P<.002  +

LEAD DIMETHYLDITHIOCARBAMATE    100ng..:..1ug...:..10......:..100...:..1mg....:..10......:..100......:..1g......:..10
2813 M f b6c eat TBA MXB 24m24              :>                        no dre   P=1.    -
a    M f b6c eat liv MXB 24m24                                        149.mg * P<.6
b    M f b6c eat lun MXB 24m24                                        no dre   P=1.
2814 M f b6c orl liv hpt 76w76 evx                   .>               no dre   P=1.
a    M f b6c orl lun mix 76w76 evx                                    no dre   P=1.
b    M f b6c orl tba mix 76w76 evx                                    116.mg   P<.3
2815 M m b6c eat TBA MXB 24m24               :>                       no dre   P=1.
a    M m b6c eat liv MXB 24m24                                        no dre   P=1.
b    M m b6c eat lun MXB 24m24                                        no dre   P=1.
2816 M m b6c orl --- rts 76w76 evx       . + .                        16.4mg   P<.006
a    M m b6c orl liv hpt 76w76 evx                                    95.3mg   P<.3
b    M m b6c orl lun mix 76w76 evx                                    no dre   P=1.
c    M m b6c orl tba mix 76w76 evx                                    10.7mg   P<.002
2817 M f b6a orl liv hpt 76w76 evx                   .>               no dre   P=1.
a    M f b6a orl lun ade 76w76 evx                                    no dre   P=1.
b    M f b6a orl tba mix 76w76 evx                                    no dre   P=1.
2818 M m b6a orl lun ade 76w76 evx               .>                   833.mg   P<1.
a    M m b6a orl liv hpt 76w76 evx                                    no dre   P=1.
b    M m b6a orl tba mix 76w76 evx                                    no dre   P=1.
2819 R f f34 eat TBA MXB 24m24          :>                            27.8mg * P<.9    -
a    R f f34 eat liv MXB 24m24                                        no dre   P=1.
2820 R m f34 eat TBA MXB 24m24          :>                            no dre   P=1.    -
a    R m f34 eat liv MXB 24m24                                        72.6mg * P<.4

LEUPEPTIN                       100ng..:..1ug...:..10......:..100...:..1mg....:..10......:..100......:..1g......:..10
2821 M f ajj eat lun mix 68w68 e                         .>           229.mg   P<.3    -
a    M f ajj eat liv lpb 68w68 e                                      933.mg   P<.3    -
2822 M m ajj eat liv mix 68w68 e                    . +               55.8mg   P<.006  +
a    M m ajj eat lun mix 68w68 e                                      422.mg   P<.9    -

LEVOBUNOLOL                     100ng..:..1ug...:..10......:..100...:..1mg....:..10......:..100......:..1g......:..10
2823 M f cf1 eat ute ley 78w78 e                          . +        1.22gm * P<.002
a    M f cf1 eat lun agc 78w78 e                                      2.50gm * P<.7
b    M f cf1 eat liv hpt 78w78 e                                      5.01gm * P<.2
2824 M m cf1 eat lun agc 78w78 e                        .>            no dre   P=1.
a    M m cf1 eat liv bhp 78w78 e                                      no dre   P=1.
2825 R f wal eat liv hct 78w78 ek                  .>                 no dre   P=1.
2826 R f wal eat mgl fba 24m24 e                    . ±               71.0mg Z P<.05
a    R f wal eat liv bhp 24m24 e                                      2.36gm * P<.3
2827 R f wal eat liv bhp 52w52 ek               .>                    no dre   P=1.
2828 R f wal eat liv tum 78w78 ek             .>                      no dre   P=1.
2829 R f wal eat liv bhp 24m24 e                .>                    no dre   P=1.
a    R f wal eat --- lyt 24m24 e                                      no dre \ P=1.
b    R f wal eat liv mhp 24m24 e                                      no dre   P=1.
2830 R m wal eat liv hct 78w78 ek                  .>                 no dre   P=1.
2831 R m wal eat liv bhp 24m24 e                          . +         982.mg * P<.005
a    R m wal eat liv mhp 24m24 e                                      no dre   P=1.
2832 R m wal eat liv tum 52w52 ek            .>                       no dre   P=1.
2833 R m wal eat liv tum 78w78 ek             .>                      no dre   P=1.
2834 R m wal eat pit ade 24m24 e                .>                    71.9mg * P<.7
a    R m wal eat liv bhp 24m24 e                                      no dre   P=1.

D-LIMONENE                      100ng..:..1ug...:..10......:..100...:..1mg....:..10......:..100......:..1g......:..10
2835 M f b6c gav liv hpc 24m24                                :      #10.5gm * P<.05   -
a    M f b6c gav TBA MXB 24m24                                        no dre   P=1.
b    M f b6c gav liv MXB 24m24                                        6.44gm * P<.3
c    M f b6c gav lun MXB 24m24                                        no dre   P=1.
2836 M m b6c gav TBA MXB 24m24                        :>              no dre   P=1.    -
a    M m b6c gav liv MXB 24m24                                        no dre   P=1.
b    M m b6c gav lun MXB 24m24                                        no dre   P=1.
2837 R f f34 gav ute esp 24m24                       :  +           #533.mg \ P<.009   -
a    R f f34 gav TBA MXB 24m24                                        1.33gm * P<.6
b    R f f34 gav liv MXB 24m24                                        no dre   P=1.
2838 R m f34 gav kid MXA 24m24                      :  +  :           204.mg * P<.002  c
a    R m f34 gav kid tla 24m24                                        325.mg * P<.004  c
b    R m f34 gav kid uac 24m24                                        595.mg * P<.1    c
c    R m f34 gav TBA MXB 24m24                                        no dre   P=1.
d    R m f34 gav liv MXB 24m24                                        no dre   P=1.
```

	RefNum	LoConf	UpConf	Cntrl	1Dose	1Inc	2Dose	2Inc	Citation or Pathology	Brkly Code
2807	176	163.mg	n.s.s.	0/25	120.mg	4/25	(713.mg	0/25)		
a	176	242.mg	n.s.s.	0/25	120.mg	2/25	(713.mg	0/25)		
b	176	529.mg	n.s.s.	0/25	120.mg	0/25	713.mg	0/25		
c	176	2.75gm	n.s.s.	3/25	120.mg	1/25	713.mg	1/25		
2808	1709	138.mg	604.mg	0/30	400.mg	13/29			Kasprzak;carc,6,279-282;1985	
a	1709	1.38gm	n.s.s.	0/30	400.mg	0/29				
2809	175m	43.1mg	571.mg	0/15	50.0mg	6/16			Van Esch;bjca,16,289-297;1962	
2810	175n	138.mg	1.13gm	0/13	500.mg	7/11				
2811	175m	40.4mg	2.21gm	0/14	40.0mg	5/16				
2812	175n	176.mg	2.04gm	0/13	400.mg	6/13				

LEAD DIMETHYLDITHIOCARBAMATE (ledate) 19010-66-3

	RefNum	LoConf	UpConf	Cntrl	1Dose	1Inc	2Dose	2Inc	Citation or Pathology	Brkly Code
2813	TR151	11.9mg	n.s.s.	9/20	3.25mg	18/50	6.50mg	18/50		
a	TR151	36.6mg	n.s.s.	0/20	3.25mg	1/50	6.50mg	1/50	liv:hpa,hpc,nnd.	
b	TR151	23.0mg	n.s.s.	3/20	3.25mg	4/50	6.50mg	5/50	lun:a/a,a/c.	
2814	1093	35.8mg	n.s.s.	0/16	18.1mg	0/18			Innes;ntis,1968/1969	
a	1093	35.8mg	n.s.s.	0/16	18.1mg	0/18				
b	1093	18.8mg	n.s.s.	0/16	18.1mg	1/18				
2815	TR151	5.92mg	n.s.s.	12/20	3.00mg	32/50	6.00mg	28/50		
a	TR151	16.0mg	n.s.s.	4/20	3.00mg	11/50	6.00mg	7/50	liv:hpa,hpc,nnd.	
b	TR151	12.0mg	n.s.s.	9/20	3.00mg	14/50	6.00mg	15/50	lun:a/a,a/c.	
2816	1093	6.17mg	162.mg	0/16	16.8mg	5/16			Innes;ntis,1968/1969	
a	1093	15.5mg	n.s.s.	0/16	16.8mg	1/16				
b	1093	29.6mg	n.s.s.	0/16	16.8mg	0/16				
c	1093	4.54mg	39.0mg	0/16	16.8mg	7/16				
2817	1093	35.8mg	n.s.s.	0/17	18.1mg	0/18				
a	1093	35.8mg	n.s.s.	1/17	18.1mg	0/18				
b	1093	35.8mg	n.s.s.	2/17	18.1mg	0/18				
2818	1093	14.6mg	n.s.s.	2/18	16.8mg	2/17				
a	1093	31.4mg	n.s.s.	1/18	16.8mg	0/17				
b	1093	16.2mg	n.s.s.	3/18	16.8mg	2/17				
2819	TR151	1.82mg	n.s.s.	15/20	1.25mg	33/50	2.50mg	36/50		
a	TR151	16.3mg	n.s.s.	1/20	1.25mg	1/50	2.50mg	1/50	liv:hpa,hpc,nnd.	
2820	TR151	1.40mg	n.s.s.	17/20	1.00mg	40/50	2.00mg	37/50		
a	TR151	11.8mg	n.s.s.	0/20	1.00mg	0/50	2.00mg	1/50	liv:hpa,hpc,nnd.	

LEUPEPTIN 24365-47-7

	RefNum	LoConf	UpConf	Cntrl	1Dose	1Inc	2Dose	2Inc	Citation or Pathology	Brkly Code
2821	1432	59.6mg	n.s.s.	3/20	130.mg	7/25			Hosaka;gann,71,913-917;1980	
a	1432	152.mg	n.s.s.	0/20	130.mg	1/25				
2822	1432	25.2mg	641.mg	4/23	120.mg	14/25				
a	1432	29.0mg	n.s.s.	15/23	120.mg	17/25				

LEVOBUNOLOL 47141-42-4

	RefNum	LoConf	UpConf	Cntrl	1Dose	1Inc	2Dose	2Inc	Citation or Pathology	Brkly Code
2823	2067	423.mg	6.69gm	0/100	12.0mg	0/50	50.0mg	0/50	200.mg 4/50 — Rothwell;faat,18,353-359;1992/pers.comm.	
a	2067m	288.mg	n.s.s.	18/100	12.0mg	8/50	50.0mg	7/50	200.mg 10/50	
b	2067m	816.mg	n.s.s.	0/100	12.0mg	0/50	50.0mg	0/50	200.mg 1/50	
2824	2067m	495.mg	n.s.s.	22/100	12.0mg	6/50	50.0mg	6/50	200.mg 7/50	
a	2067m	594.mg	n.s.s.	9/100	12.0mg	1/50	50.0mg	3/50	200.mg 3/50	
2825	2067m	4.85mg	n.s.s.	0/10	5.00mg	0/10	30.0mg	0/10	180.mg 0/10	
2826	2067n	27.0mg	n.s.s.	27/100	5.00mg	24/50	30.0mg	23/50	(180.mg 15/50)	
a	2067n	554.mg	n.s.s.	4/100	5.00mg	4/50	30.0mg	2/50	180.mg 5/50	
2827	2067o	2.85mg	n.s.s.	0/10	.500mg	1/10	2.00mg	0/10	5.00mg 0/10	
2828	2067r	.429mg	n.s.s.	0/10	.500mg	0/10	2.00mg	0/10	5.00mg 0/10	
2829	2067s	3.05mg	n.s.s.	2/80	.500mg	0/40	2.00mg	0/40	5.00mg 0/40	
a	2067s	26.5mg	n.s.s.	1/80	.500mg	4/40	2.00mg	1/40	5.00mg 1/40	
b	2067s	29.1mg	n.s.s.	0/80	.500mg	1/40	2.00mg	1/40	5.00mg 0/40	
2830	2067m	4.85mg	n.s.s.	0/10	5.00mg	0/10	30.0mg	0/10	180.mg 0/10	
2831	2067n	388.mg	12.9gm	2/100	5.00mg	2/50	30.0mg	2/50	180.mg 7/50	
a	2067n	43.1mg	n.s.s.	1/100	5.00mg	0/50	30.0mg	0/50	180.mg 0/50	
2832	2067o	.191mg	n.s.s.	0/10	.500mg	0/10	2.00mg	0/10	5.00mg 0/10	
2833	2067r	.429mg	n.s.s.	0/10	.500mg	0/10	2.00mg	0/10	5.00mg 0/10	
2834	2067s	9.99mg	n.s.s.	10/80	.500mg	7/40	2.00mg	11/40	5.00mg 6/40	
a	2067s	41.1mg	n.s.s.	1/80	.500mg	0/40	2.00mg	1/40	5.00mg 0/40	

D-LIMONENE 5989-27-5

	RefNum	LoConf	UpConf	Cntrl	1Dose	1Inc	2Dose	2Inc	Citation or Pathology	Brkly Code
2835	TR347	3.18gm	n.s.s.	0/50	354.mg	0/50	707.mg	3/50		S
a	TR347	532.mg	n.s.s.	41/50	354.mg	31/50	(707.mg	26/50)		
b	TR347	1.84gm	n.s.s.	4/50	354.mg	2/50	707.mg	8/50	liv:hpa,hpc,nnd.	
c	TR347	4.74gm	n.s.s.	5/50	354.mg	2/50	707.mg	2/50	lun:a/a,a/c.	
2836	TR347	134.mg	n.s.s.	43/50	177.mg	29/50	(354.mg	35/50)		
a	TR347	862.mg	n.s.s.	22/50	177.mg	14/50	354.mg	15/50	liv:hpa,hpc,nnd.	
b	TR347	1.01gm	n.s.s.	15/50	177.mg	9/50	354.mg	10/50	lun:a/a,a/c.	
2837	TR347	240.mg	14.6gm	3/50	211.mg	13/50	(424.mg	5/50)		S
a	TR347	224.mg	n.s.s.	37/50	211.mg	44/50	424.mg	26/50		
b	TR347	4.07gm	n.s.s.	2/50	211.mg	0/50	424.mg	0/50	liv:hpa,hpc,nnd.	
2838	TR347	119.mg	644.mg	0/50	53.1mg	8/50	106.mg	11/50		
a	TR347	167.mg	1.90gm	0/50	53.1mg	4/50	106.mg	8/50	kid:tla,uac.	
b	TR347	257.mg	n.s.s.	0/50	53.1mg	4/50	106.mg	3/50		
c	TR347	102.mg	n.s.s.	43/50	53.1mg	40/50	106.mg	45/50		
d	TR347	504.mg	n.s.s.	2/50	53.1mg	2/50	106.mg	2/50	liv:hpa,hpc,nnd.	

```
        Spe Strain  Site    Xpo+Xpt                                                              TD50     2Tailpvl
            Sex   Route   Hist       Notes                                                              DR       AuOp
LITHOCHOLIC ACID               100ng..:..1ug....:..10.....:..100....:..1mg....:..10.....:..100....:..1g.....:..10
2839 M f b6c gav TBA  MXB 24m24                                              :>                          960.mg  * P<.8   -
a    M f b6c gav liv  MXB 24m24                                                                          1.14gm  * P<.2
b    M f b6c gav lun  MXB 24m24                                                                          1.03gm  * P<.2
2840 M m b6c gav TBA  MXB 24m24                                                 :>                        no dre   P=1.   -
a    M m b6c gav liv  MXB 24m24                                                                           no dre   P=1.
b    M m b6c gav lun  MXB 24m24                                                                          834.mg  * P<.6
2841 R f f34 gav TBA  MXB 24m24                                                 :>                       686.mg  * P<.6   -
a    R f f34 gav liv  MXB 24m24                                                                          1.89gm  * P<.3
2842 R m f34 gav TBA  MXB 24m24                                              :>                          501.mg  * P<.3   -
a    R m f34 gav liv  MXB 24m24                                                                          2.25gm  * P<.2

LOCUST BEAN GUM                100ng..:..1ug....:..10.....:..100....:..1mg....:..10.....:..100....:..1g.....:..10
2843 M f b6c eat pit  adn 24m24                                                                        :#20.2gm \ P<.03   -
a    M f b6c eat ute  esp 24m24                                                                          86.5gm  * P<.05
b    M f b6c eat TBA  MXB 24m24                                                                           no dre   P=1.
c    M f b6c eat liv  MXB 24m24                                                                           no dre   P=1.
d    M f b6c eat lun  MXB 24m24                                                                           no dre   P=1.
2844 M m b6c eat lun  a/a 24m24                                                                   :     ±#7.16gm \ P<.05   -
a    M m b6c eat TBA  MXB 24m24                                                                           no dre   P=1.
b    M m b6c eat liv  MXB 24m24                                                                           no dre   P=1.
c    M m b6c eat lun  MXB 24m24                                                                           no dre   P=1.
2845 R f f34 eat adr  coa 24m24                                                                  :      #11.9gm  * P<.04   -
a    R f f34 eat TBA  MXB 24m24                                                                          29.5gm  * P<.9
b    R f f34 eat liv  MXB 24m24                                                                           no dre   P=1.
2846 R m f34 eat TBA  MXB 24m24                                                      :>                   no dre   P=1.    -
a    R m f34 eat liv  MXB 24m24                                                                         516.gm  * P<1.

LOFEXIDINE.HCl                 100ng..:..1ug....:..10.....:..100....:..1mg....:..10.....:..100....:..1g.....:..10
2847 R f lev eat tba  tum 52w52 e                           .     ±                                      2.30mg  \ P<.09   -
2848 R m lev eat tba  tum 52w52 e                           .     ±                                      2.30mg  \ P<.09   -

LONIDAMINE                     100ng..:..1ug....:..10.....:..100....:..1mg....:..10.....:..100....:..1g.....:..10
2849 R f sda eat tba  ben 24m24                                                    .>                     no dre   P=1.
a    R f sda eat tba  mix 24m24                                                                           no dre   P=1.
b    R f sda eat tba  mal 24m24                                                                           no dre   P=1.
2850 R m sda eat tba  mix 24m24                                               .>                         190.mg  * P<.2
a    R m sda eat tba  ben 24m24                                                                         866.mg  * P<.7
b    R m sda eat tba  mal 24m24                                                                           no dre   P=1.

LUTEOSKYRIN                    100ng..:..1ug....:..10.....:..100....:..1mg....:..10.....:..100....:..1g.....:..10
2851 M f ddn eat liv  mix 37m37 e                                         .>                             79.4mg    P<.3   +
a    M f ddn eat liv  lca 37m37 e                                                                       122.mg    P<.4   +
b    M f ddn eat liv  lcc 37m37 e                                                                       248.mg    P<.5   +
2852 M m ddn eat liv  mix 37m37 e                                    .  +  .                             14.8mg  Z P<.0005+
a    M m ddn eat liv  lca 37m37 e                                                                       83.7mg  * P<.03   +
b    M m ddn eat liv  hpt 37m37 e                                                                       89.5mg  * P<.08   +
c    M m ddn eat liv  lcc 37m37 e                                                                       332.mg  * P<.2   +
d    M m ddn eat liv  akt 37m37 e                                                                        no dre   P=1.   +
2853 M m ddn eat liv  lca 37m37 e                                    .  ±                                25.1mg    P<.03   +
a    M m ddn eat liv  hpt 37m37 e                                                                       74.3mg    P<.2   +
b    M m ddn eat liv  akt 37m37 e                                                                       113.mg    P<.3   +
2854 M m ddn eat liv  hpt 37m37 e                                           .>                          113.mg    P<.3   +
a    M m ddn eat liv  lca 37m37 e                                                                       231.mg    P<.5   +

LUTESTRAL                      100ng..:..1ug....:..10.....:..100....:..1mg....:..10.....:..100....:..1g.....:..10
2855 M f c3h eat mam  tum 24m24 er                                 .>                                    5.98mg    P<.7   -
2856 M f crf eat mam  tum 24m24 er  pool                        .>                                       no dre   P=1.   -
2857 M m crf eat mam  tum 24m24 r                            .  +  .                                     1.76mg    P<.0005
2858 M f r3m eat mam  tum 24m24 er                              .>                                       2.12mg    P<.4   -

LYNESTRENOL                    100ng..:..1ug....:..10.....:..100....:..1mg....:..10.....:..100....:..1g.....:..10
2859 D f beg eat mgl  car 85m85 er                          .  +  .                                      .580mg  * P<.002  +

MAGNESIUM CHLORIDE HEXAHYDRATE 100ng..:..1ug....:..10.....:..100....:..1mg....:..10.....:..100....:..1g.....:..10
2860 M f b6c eat liv  hes 22m24 e                                                                       57.5gm  * P<.4   -
a    M f b6c eat liv  hnd 22m24 e                                                                       60.5gm  * P<.7   -
b    M f b6c eat lun  ade 22m24 e                                                                       100.gm  * P<.2   -
c    M f b6c eat lun  adc 22m24 e                                                                       260.gm  * P<.9   -
d    M f b6c eat liv  hpc 22m24 e                                                                        no dre   P=1.   -
e    M f b6c eat liv  hem 22m24 e                                                                        no dre   P=1.   -
f    M f b6c eat ---  lkm 22m24 e                                                                        no dre   P=1.   -
2861 M m b6c eat ---  lkm 22m24 e                                                                .      10.7gm  * P<.1   -
a    M m b6c eat liv  hes 22m24 e                                                                       46.7gm  * P<.07  -
b    M m b6c eat liv  hem 22m24 e                                                                       24.4gm  * P<.3   -
c    M m b6c eat lun  ade 22m24 e                                                                       87.6gm  * P<.7   -
d    M m b6c eat liv  hpc 22m24 e                                                                        no dre   P=1.   -
e    M m b6c eat lun  adc 22m24 e                                                                        no dre   P=1.   -
f    M m b6c eat liv  hnd 22m24 e                                                                        no dre   P=1.   -
```

RefNum	LoConf	UpConf	Cntrl	1Dose	1Inc	2Dose	2Inc			Citation or Pathology
										Brkly Code

LITHOCHOLIC ACID   434-13-9

2839	TR175	111.mg n.s.s.	6/20	52.5mg	20/50	105.mg	23/50			
a	TR175	392.mg n.s.s.	0/20	52.5mg	1/50	105.mg	3/50			liv:hpa,hpc,nnd.
b	TR175	357.mg n.s.s.	0/20	52.5mg	1/50	105.mg	3/50			lun:a/a,a/c.
2840	TR175	132.mg n.s.s.	12/20	52.5mg	26/50	105.mg	20/50			
a	TR175	187.mg n.s.s.	6/20	52.5mg	17/50	105.mg	9/50			liv:hpa,hpc,nnd.
b	TR175	180.mg n.s.s.	1/20	52.5mg	7/50	105.mg	5/50			lun:a/a,a/c.
2841	TR175	129.mg n.s.s.	14/20	106.mg	35/49	212.mg	35/50			
a	TR175	653.mg n.s.s.	0/20	106.mg	2/49	212.mg	2/50			liv:hpa,hpc,nnd.
2842	TR175	150.mg n.s.s.	9/20	106.mg	26/50	212.mg	30/50			
a	TR175	778.mg n.s.s.	0/20	106.mg	1/50	212.mg	3/50			liv:hpa,hpc,nnd.

LOCUST BEAN GUM (carob seed gum) 9000-40-2

2843	TR221	6.98gm n.s.s.	0/50	3.19gm	4/50	(6.44gm	1/50)			S
a	TR221	26.2gm n.s.s.	0/50	3.19gm	0/50	6.44gm	3/50			S
b	TR221	4.02gm n.s.s.	45/50	3.19gm	36/50	(6.44gm	30/50)			
c	TR221	34.8gm n.s.s.	3/50	3.19gm	2/50	6.44gm	2/50			liv:hpa,hpc,nnd.
d	TR221	29.4gm n.s.s.	5/50	3.19gm	2/50	6.44gm	4/50			lun:a/a,a/c.
2844	TR221	2.90gm n.s.s.	7/50	2.94gm	17/50	(5.94gm	11/50)			S
a	TR221	5.06gm n.s.s.	36/50	2.94gm	41/50	5.94gm	38/50			
b	TR221	14.0gm n.s.s.	18/50	2.94gm	16/50	5.94gm	14/50			liv:hpa,hpc,nnd.
c	TR221	9.89gm n.s.s.	14/50	2.94gm	21/50	5.94gm	14/50			lun:a/a,a/c.
2845	TR221	4.97gm n.s.s.	1/50	1.23gm	4/50	2.45gm	6/50			S
a	TR221	1.74gm n.s.s.	44/50	1.23gm	43/50	2.45gm	42/50			
b	TR221	n.s.s. n.s.s.	0/50	1.23gm	0/50	2.45gm	0/50			liv:hpa,hpc,nnd.
2846	TR221	1.78gm n.s.s.	37/50	981.mg	35/50	1.96gm	35/50			
a	TR221	9.23gm n.s.s.	1/50	981.mg	2/50	1.96gm	1/50			liv:hpa,hpc,nnd.

LOFEXIDINE.HCl (2-[1-(2,6-dichlorphenoxy)-ethyl]-2-imidazoline.HCl) 21498-08-8

| 2847 | 1668 | .563mg n.s.s. | 0/10 | 3.00mg | 2/10 | (8.00mg | 0/20) | | | Tsai;arzn,31,955-962;1982 |
| 2848 | 1668 | .563mg n.s.s. | 0/10 | 3.00mg | 2/10 | (8.00mg | 0/20) | | | |

LONIDAMINE 50264-69-2

2849	2055	228.mg n.s.s.	87/100	20.0mg	44/50	60.0mg	36/50	180.mg	39/50	Patton;txlt,62,209-214;1992/pers.comm.
a	2055	239.mg n.s.s.	92/100	20.0mg	45/50	60.0mg	38/50	180.mg	40/50	
b	2055	902.mg n.s.s.	14/100	20.0mg	7/50	60.0mg	8/50	180.mg	4/50	
2850	2055	54.0mg n.s.s.	90/100	20.0mg	42/50	60.0mg	40/50	180.mg	48/50	
a	2055	108.mg n.s.s.	83/100	20.0mg	37/50	60.0mg	38/50	180.mg	42/50	
b	2055	677.mg n.s.s.	18/100	20.0mg	11/50	60.0mg	6/50	180.mg	8/50	

LUTEOSKYRIN 21884-44-6

2851	1346m	24.0mg n.s.s.	0/7	6.00mg	3/26					Uraguchi;fctx,10,193-207;1972
a	1346m	29.9mg n.s.s.	0/7	6.00mg	2/26					
b	1346m	40.4mg n.s.s.	0/7	6.00mg	1/26					
2852	1346m	7.71mg 45.5mg	0/10	1.67mg	2/18	5.00mg	11/24	(16.7mg	11/29)	
a	1346m	42.0mg n.s.s.	0/10	1.67mg	0/18	5.00mg	5/24	16.7mg	6/29	
b	1346m	34.0mg n.s.s.	0/10	1.67mg	1/18	5.00mg	7/24	16.7mg	7/29	
c	1346m	100.mg n.s.s.	0/10	1.67mg	0/18	5.00mg	1/24	16.7mg	2/29	
d	1346m	108.mg n.s.s.	0/10	1.67mg	1/18	5.00mg	3/24	16.7mg	1/29	
2853	1346n	11.3mg n.s.s.	0/9	5.00mg	8/29					
a	1346n	22.5mg n.s.s.	0/9	5.00mg	3/29					
b	1346n	27.9mg n.s.s.	0/9	5.00mg	2/29					
2854	1346o	27.9mg n.s.s.	0/10	5.00mg	2/29					
a	1346o	37.6mg n.s.s.	0/10	5.00mg	1/29					

LUTESTRAL 8065-91-6

2855	1175	.759mg n.s.s.	54/92	1.04mg	19/30					Rudali;jnci,49,813-819;1972
2856	1175	.487mg n.s.s.	161/167p	1.04mg	28/31					
2857	1463	.851mg 4.52mg	0/76	.960mg	10/32					Rudali;gmcr,17,243-252;1975
2858	1175	.531mg n.s.s.	50/73	1.04mg	31/40					Rudali;jnci,49,813-819;1972

LYNESTRENOL 52-76-6

| 2859 | 2168 | .265mg 2.68mg | 1/16 | 78.1ug | 0/16 | .391mg | 3/16 | .977mg | 7/16 | Misdorp;aenc,125,27-31;1991/pers.comm. |

MAGNESIUM CHLORIDE HEXAHYDRATE   7791-18-6

2860	1945	12.5gm n.s.s.	0/49	600.mg	1/50	2.40gm	1/50			Kurata;fctx,27,559-563;1989
a	1945	7.51gm n.s.s.	1/50	600.mg	5/50	2.40gm	3/50			
b	1945	16.3gm n.s.s.	0/49	600.mg	0/50	2.40gm	1/49			
c	1945	9.45gm n.s.s.	0/49	600.mg	3/50	2.40gm	1/49			
d	1945	18.6gm n.s.s.	0/49	600.mg	1/50	2.40gm	0/50			
e	1945	15.8gm n.s.s.	1/49	600.mg	1/50	2.40gm	1/50			
f	1945	5.08gm n.s.s.	9/49	600.mg	17/50	2.40gm	11/50			
2861	1945	3.73gm n.s.s.	5/50	554.mg	7/50	2.22gm	11/50			
a	1945	11.5gm n.s.s.	0/50	554.mg	0/50	2.22gm	2/50			
b	1945	6.05gm n.s.s.	2/50	554.mg	4/50	2.22gm	5/50			
c	1945	9.95gm n.s.s.	1/49	554.mg	2/50	2.22gm	2/50			
d	1945	2.72gm n.s.s.	13/50	554.mg	6/50	(2.22gm	4/50)			
e	1945	11.5gm n.s.s.	2/49	554.mg	2/50	2.22gm	2/50			
f	1945	8.24gm n.s.s.	17/50	554.mg	11/50	2.22gm	10/50			

```
              Spe Strain Site  Xpo+Xpt                                                                    TD50   2Tailpvl
              Sex    Route   Hist     Notes                                                               DR     AuOp
```

MALAOXON   `100ng..:..1ug....:..10.....:..100.....:..1mg....:..10.....:..100....:..1g.....:..10`

ID	Spe Sex	Strain	Route	Site	Hist	Xpo+Xpt	Notes	plot	TD50	DR	2Tailpvl	AuOp
2862	M f	b6c	eat	TBA	MXB	24m24		:>	no dre		P=1.	-
a	M f	b6c	eat	liv	MXB	24m24			no dre		P=1.	
b	M f	b6c	eat	lun	MXB	24m24			2.76gm	*	P<.3	
2863	M m	b6c	eat	TBA	MXB	24m24		:>	472.mg	*	P<.4	-
a	M m	b6c	eat	liv	MXB	24m24			579.mg	/	P<.2	
b	M m	b6c	eat	lun	MXB	24m24			3.23gm	*	P<.8	
2864	R f	f34	eat	thy	MXA	24m24		: ±	#296.mg	*	P<.02	-
a	R f	f34	eat	TBA	MXB	24m24			no dre		P=1.	
b	R f	f34	eat	liv	MXB	24m24			no dre		P=1.	
2865	R m	f34	eat	TBA	MXB	24m24		:>	no dre		P=1.	-
a	R m	f34	eat	liv	MXB	24m24			1.56gm	*	P<.3	

MALATHION   `100ng..:..1ug....:..10.....:..100.....:..1mg....:..10.....:..100....:..1g.....:..10`

ID	Spe Sex	Strain	Route	Site	Hist	Xpo+Xpt	Notes	plot	TD50	DR	2Tailpvl	AuOp
2866	M f	b6c	eat	TBA	MXB	80w94		:>	no dre		P=1.	-
a	M f	b6c	eat	liv	MXB	80w94			31.7gm	*	P<.3	
b	M f	b6c	eat	lun	MXB	80w94			no dre		P=1.	
2867	M m	b6c	eat	liv	nnd	80w94		:	#10.2gm	*	P<.04	-
a	M m	b6c	eat	TBA	MXB	80w94			5.19gm	*	P<.3	
b	M m	b6c	eat	liv	MXB	80w94			4.40gm	*	P<.2	
c	M m	b6c	eat	lun	MXB	80w94			44.7gm	*	P<.8	
2868	R f	osm	eat	TBA	MXB	19m25	v	:>	434.mg	*	P<.2	-
a	R f	osm	eat	liv	MXB	19m25	v		3.87gm	*	P<.2	
2869	R f	osm	eat	thy	MXA	19m25	v pool	: ±	#2.96gm	*	P<.03	-
2870	R f	f34	eat	TBA	MXB	24m24		:>	293.mg	*	P<.3	-
a	R f	f34	eat	liv	MXB	24m24			no dre		P=1.	
2871	R m	osm	eat	TBA	MXB	19m25	v	:>	335.mg	*	P<.2	-
a	R m	osm	eat	liv	MXB	19m25	v		3.07gm	*	P<.7	
2872	R m	f34	eat	adr	phe	24m24		: + :	#145.mg	*	P<.0005	-
a	R m	f34	eat	TBA	MXB	24m24			66.6mg	*	P<.002	
b	R m	f34	eat	liv	MXB	24m24			411.mg	*	P<.05	

MALEIC HYDRAZIDE   `100ng..:..1ug....:..10.....:..100.....:..1mg....:..10.....:..100....:..1g.....:..10`

ID	Spe Sex	Strain	Route	Site	Hist	Xpo+Xpt	Notes	plot	TD50	DR	2Tailpvl	AuOp
2873	M f	b6a	orl	liv	hpt	76w76	evx	.>	2.35gm		P<.3	-
a	M f	b6a	orl	lun	ade	76w76	evx		38.7gm		P<1.	
b	M f	b6a	orl	tba	mix	76w76	evx		932.mg		P<.4	
2874	M m	b6a	orl	liv	hpt	76w76	evx	.>	2.07gm		P<.6	-
a	M m	b6a	orl	lun	ade	76w76	evx		no dre		P=1.	
b	M m	b6a	orl	tba	mix	76w76	evx		11.9gm		P<1.	
2875	M f	b6c	orl	liv	hpt	76w76	evx	.>	no dre		P=1.	-
a	M f	b6c	orl	lun	mix	76w76	evx		no dre		P=1.	
b	M f	b6c	orl	tba	mix	76w76	evx		no dre		P=1.	
2876	M m	b6c	orl	liv	hpt	76w76	evx	. ±	1.06gm		P<.09	-
a	M m	b6c	orl	lun	mix	76w76	evx		no dre		P=1.	
b	M m	b6c	orl	tba	mix	76w76	evx		679.mg		P<.04	
2877	M f	cb6	gav	liv	tum	28m28	e	.>	no dre		P=1.	-
a	M f	cb6	gav	lun	tum	28m28	e		no dre		P=1.	
b	M f	cb6	gav	tba	tum	28m28	e		147.mg		P<.3	
2878	M m	cb6	gav	lun	tum	28m28	e	.>	2.43gm		P<.5	-
a	M m	cb6	gav	liv	tum	28m28	e		3.48gm		P<.9	
b	M m	cb6	gav	tba	tum	28m28	e		no dre		P=1.	
2879	R f	nss	eat	liv	hpt	23m23	e	.>	no dre		P=1.	-
a	R f	nss	eat	tba	mix	23m23	e		3.72gm		P<.8	
2880	R m	nss	eat	liv	hpt	23m23	e	.>	no dre		P=1.	-
a	R m	nss	eat	tba	mix	23m23	e		12.3gm		P<1.	

MALONALDEHYDE, SODIUM SALT   `100ng..:..1ug....:..10.....:..100.....:..1mg....:..10.....:..100....:..1g.....:..10`

ID	Spe Sex	Strain	Route	Site	Hist	Xpo+Xpt	Notes	plot	TD50	DR	2Tailpvl	AuOp
2881	M f	b6c	gav	TBA	MXB	24m24		:>	274.mg	*	P<.5	-
a	M f	b6c	gav	liv	MXB	24m24			568.mg	*	P<.2	
b	M f	b6c	gav	lun	MXB	24m24			4.41gm	*	P<.9	
2882	M m	b6c	gav	TBA	MXB	24m24		:>	217.mg	*	P<.6	-
a	M m	b6c	gav	liv	MXB	24m24			230.mg	*	P<.4	
b	M m	b6c	gav	lun	MXB	24m24			1.13gm	*	P<.8	
2883	M f	swi	wat	liv	hpt	52w52	e pool	. ±	4.62gm	Z	P<.03	-
a	M f	swi	wat	liv	mix	52w52	e		14.1mg	*	P<.02	+
b	M f	swi	wat	liv	hem	52w52	e		24.8mg	*	P<.02	
c	M f	swi	wat	liv	hnd	52w52	e		44.3mg	*	P<.3	
d	M f	swi	wat	lun	ade	52w52	e		no dre		P=1.	
2884	R f	f34	gav	thy	MXA	24m24		: ±	252.mg	/	P<.02	c
a	R f	f34	gav	TBA	MXB	24m24			164.mg	*	P<.5	
b	R f	f34	gav	liv	MXB	24m24			1.95gm	*	P<.2	
2885	R m	f34	gav	MXB	MXB	24m24		: + :	67.7mg	*	P<.0005	-
a	R m	f34	gav	pni	isa	24m24			80.5mg	\	P<.0005	c
b	R m	f34	gav	pni	MXA	24m24			90.3mg	\	P<.005	
c	R m	f34	gav	thy	MXA	24m24			113.mg	*	P<.002	c
d	R m	f34	gav	tes	ict	24m24			39.8mg	*	P<.02	
e	R m	f34	gav	amd	phe	24m24			184.mg	*	P<.05	
f	R m	f34	gav	sub	MXA	24m24			195.mg	*	P<.04	
g	R m	f34	gav	thy	fca	24m24			206.mg	*	P<.02	
h	R m	f34	gav	thy	fcc	24m24			236.mg	*	P<.02	
i	R m	f34	gav	adr	coa	24m24			566.mg	*	P<.03	

	RefNum	LoConf	UpConf	Cntrl	1Dose	1Inc	2Dose	2Inc			Citation or Pathology
											Brkly Code

MALAOXON (malathion-O-analog) 1634-78-2

	RefNum	LoConf	UpConf	Cntrl	1Dose	1Inc	2Dose	2Inc			Pathology / Brkly Code
2862	TR135	265.mg	n.s.s.	15/50	65.0mg	15/50	130.mg	15/50			
a	TR135	263.mg	n.s.s.	6/50	65.0mg	3/50	(130.mg	1/50)			
b	TR135	680.mg	n.s.s.	0/50	65.0mg	1/50	130.mg	1/50			liv:hpa,hpc,nnd.
2863	TR135	127.mg	n.s.s.	23/50	59.4mg	18/50	119.mg	26/50			lun:a/a,a/c.
a	TR135	183.mg	n.s.s.	12/50	59.4mg	5/50	119.mg	17/50			
b	TR135	315.mg	n.s.s.	6/50	59.4mg	5/50	119.mg	6/50			liv:hpa,hpc,nnd.
2864	TR135	120.mg	n.s.s.	0/50	25.0mg	1/50	49.5mg	5/50			lun:a/a,a/c.
a	TR135	41.9mg	n.s.s.	45/50	25.0mg	44/50	49.5mg	44/50			thy:cca,ccr. S
b	TR135	n.s.s.	n.s.s.	0/50	25.0mg	3/50	49.5mg	0/50			
2865	TR135	27.7mg	n.s.s.	44/50	20.0mg	43/50	40.0mg	40/50			liv:hpa,hpc,nnd.
a	TR135	254.mg	n.s.s.	0/50	20.0mg	0/50	40.0mg	1/50			liv:hpa,hpc,nnd.

MALATHION 121-75-5

	RefNum	LoConf	UpConf	Cntrl	1Dose	1Inc	2Dose	2Inc			Pathology / Brkly Code
2866	TR24	4.12gm	n.s.s.	1/10	884.mg	5/50	1.75gm	5/50			
a	TR24	7.80gm	n.s.s.	0/10	884.mg	0/50	1.75gm	2/50			
b	TR24	n.s.s.	n.s.s.	0/10	884.mg	0/50	1.75gm	0/50			liv:hpa,hpc,nnd.
2867	TR24	4.17gm	n.s.s.	0/10	816.mg	0/50	1.62gm	6/50			lun:a/a,a/c.
a	TR24	1.66gm	n.s.s.	3/10	816.mg	9/50	1.62gm	19/50			S
b	TR24	1.74gm	n.s.s.	2/10	816.mg	7/50	1.62gm	17/50			
c	TR24	7.80gm	n.s.s.	0/10	816.mg	1/50	1.62gm	1/50			liv:hpa,hpc,nnd.
2868	TR24	175.mg	n.s.s.	7/15	166.mg	28/50	299.mg	32/50			lun:a/a,a/c.
a	TR24	1.17gm	n.s.s.	0/15	166.mg	0/50	299.mg	3/50			
2869	TR24	1.02gm	n.s.s.	0/55p	166.mg	0/50	299.mg	4/50			liv:hpa,hpc,nnd.
2870	TR192	82.4mg	n.s.s.	43/50	97.2mg	45/50	194.mg	45/50			thy:fca,fcc. S
a	TR192	1.16gm	n.s.s.	2/50	97.2mg	0/50	194.mg	1/50			liv:hpa,hpc,nnd.
2871	TR24	139.mg	n.s.s.	4/15	133.mg	23/50	239.mg	25/50			
a	TR24	851.mg	n.s.s.	0/15	133.mg	2/50	239.mg	1/50			liv:hpa,hpc,nnd.
2872	TR192	64.8mg	587.mg	2/50	78.5mg	11/50	157.mg	6/49			
a	TR192	32.0mg	340.mg	32/50	78.5mg	30/50	157.mg	26/49			S
b	TR192	101.mg	n.s.s.	0/50	78.5mg	2/50	157.mg	0/49			liv:hpa,hpc,nnd.

MALEIC HYDRAZIDE (1,2-dihydro-3,6-pyridazinedione) 123-33-1

	RefNum	LoConf	UpConf	Cntrl	1Dose	1Inc	2Dose	2Inc			Citation
2873	1193	382.mg	n.s.s.	0/17	414.mg	1/16					
a	1193	431.mg	n.s.s.	1/17	414.mg	1/16					Innes;ntis,1968/1969
b	1193	205.mg	n.s.s.	2/17	414.mg	4/16					
2874	1193	302.mg	n.s.s.	1/18	385.mg	2/17					
a	1193	721.mg	n.s.s.	2/18	385.mg	0/17					
b	1193	280.mg	n.s.s.	3/18	385.mg	3/17					
2875	1193	774.mg	n.s.s.	0/16	414.mg	0/17					
a	1193	774.mg	n.s.s.	0/16	414.mg	0/17					
b	1193	774.mg	n.s.s.	0/16	414.mg	0/17					
2876	1193	259.mg	n.s.s.	0/16	385.mg	2/16					
a	1193	678.mg	n.s.s.	0/16	385.mg	0/16					
b	1193	205.mg	n.s.s.	0/16	385.mg	3/16					
2877	1520	340.mg	n.s.s.	1/12	72.9mg	2/35					
a	1520	394.mg	n.s.s.	2/12	72.9mg	2/35					Cabral;txcy,24,169-173;1982
b	1520	51.4mg	n.s.s.	5/12	72.9mg	22/35					
2878	1520	395.mg	n.s.s.	0/11	72.9mg	1/37					
a	1520	237.mg	n.s.s.	1/11	72.9mg	4/37					
b	1520	112.mg	n.s.s.	7/11	72.9mg	18/37					
2879	1109	1.33gm	n.s.s.	0/9	500.mg	1/14					
a	1109	393.mg	n.s.s.	2/9	500.mg	4/14					Barnes;natu,180,62-64;1957
2880	1109	1.07gm	n.s.s.	0/8	400.mg	0/14					
a	1109	467.mg	n.s.s.	1/8	400.mg	2/14					

MALONALDEHYDE, SODIUM SALT (3-hydroxy-2-propenal, sodium salt) 24382-04-5

	RefNum	LoConf	UpConf	Cntrl	1Dose	1Inc	2Dose	2Inc	(col)	(col)	Citation / Pathology / Brkly Code
2881	TR331	64.5mg	n.s.s.	27/50	42.0mg	31/50	84.9mg	26/50			
a	TR331	184.mg	n.s.s.	2/50	42.0mg	3/50	84.9mg	5/50			liv:hpa,hpc,nnd.
b	TR331	200.mg	n.s.s.	5/50	42.0mg	7/50	84.9mg	4/50			lun:a/a,a/c.
2882	TR331	36.0mg	n.s.s.	39/50	42.4mg	37/50	84.9mg	31/50			
a	TR331	55.8mg	n.s.s.	17/50	42.4mg	21/50	84.9mg	17/50			liv:hpa,hpc,nnd.
b	TR331	102.mg	n.s.s.	10/50	42.4mg	5/50	84.9mg	8/50			lun:a/a,a/c.
2883	1521	1.14mg	n.s.s.	0/97p	.100mg	0/49	1.00mg	2/50	(10.0mg	0/48)	Bird;jtxe,10,897-905;1982
a	1521	5.07mg	n.s.s.	1/97p	.100mg	2/49	1.00mg	4/50	10.0mg	6/48	
b	1521	7.78mg	n.s.s.	1/97p	.100mg	1/49	1.00mg	0/50	10.0mg	4/48	
c	1521	10.3mg	n.s.s.	0/97p	.100mg	1/49	1.00mg	2/50	10.0mg	2/48	
d	1521	14.3mg	n.s.s.	6/97p	.100mg	4/49	1.00mg	5/50	10.0mg	2/48	
2884	TR331	98.7mg	n.s.s.	2/50	35.0mg	1/50	70.7mg	7/50			
a	TR331	34.5mg	n.s.s.	37/50	35.0mg	37/50	70.7mg	21/50			thy:fca,fcc.
b	TR331	318.mg	n.s.s.	0/50	35.0mg	0/50	70.7mg	1/50			liv:hpa,hpc,nnd.
2885	TR331	37.4mg	216.mg	4/50	35.4mg	17/50	70.7mg	13/50			pni:isa; thy:fca,fcc. C
a	TR331	37.9mg	219.mg	0/50	35.4mg	9/50	(70.7mg	1/50)			
b	TR331	39.6mg	876.mg	1/50	35.4mg	9/50	(70.7mg	1/50)			pni:isa,isc. S
c	TR331	56.2mg	545.mg	4/50	35.4mg	8/50	70.7mg	13/50			thy:fca,fcc.
d	TR331	18.5mg	n.s.s.	40/50	35.4mg	45/50	70.7mg	36/50			S
e	TR331	70.8mg	n.s.s.	5/50	35.4mg	6/50	70.7mg	8/50			S
f	TR331	81.3mg	n.s.s.	1/50	35.4mg	7/50	70.7mg	8/50			S
g	TR331	86.5mg	n.s.s.	3/50	35.4mg	3/50	70.7mg	9/50			sub:fbs,fib. S
h	TR331	98.2mg	n.s.s.	1/50	35.4mg	5/50	70.7mg	5/50			S
i	TR331	186.mg	n.s.s.	0/50	35.4mg	1/50	70.7mg	3/50			S
											S

```
        Spe Strain Site    Xpo+Xpt                                                          TD50    2Tailpvl
           Sex  Route  Hist      Notes                                                        DR      AuOp
```

```
  j   R m f34 gav TBA MXB 24m24                                                            41.8mg  / P<.008
  k   R m f34 gav liv MXB 24m24                                                            665.mg  * P<.4
```

MANGANESE ETHYLENEBISTHIOCARBAMATE   ..:..1ug....:..10......:..100...:..1mg....:..10....:..100....:..1g......:..10

```
2886 M f b6a orl lun ade 76w76 evx                              .>                         no dre    P=1.   -
  a  M f b6a orl liv hpt 76w76 evx                                                         no dre    P=1.   -
  b  M f b6a orl tba mix 76w76 evx                                                         138.mg    P<.7   -
2887 M m b6a orl lun ade 76w76 evx                          .>                             54.9mg    P<.4   -
  a  M m b6a orl liv hpt 76w76 evx                                                         no dre    P=1.   -
  b  M m b6a orl tba mix 76w76 evx                                                         106.mg    P<.7   -
2888 M f b6c orl liv hpt 76w76 evx                               .>                        no dre    P=1.   -
  a  M f b6c orl lun mix 76w76 evx                                                         no dre    P=1.   -
  b  M f b6c orl tba tum 76w76 evx                                                         no dre    P=1.   -
2889 M m b6c orl liv hpt 76w76 evx                             .>                          128.mg    P<.3   -
  a  M m b6c orl lun ade 76w76 evx                                                         128.mg    P<.3   -
  b  M m b6c orl tba mix 76w76 evx                                                         62.3mg    P<.2   -
2890 R b mgr gav liv tum 22m24 e                                   .>                      no dre    P=1.
  a  R b mgr gav tba mix 22m24 e                                                           157.mg    P<.02  +
```

MANGANESE (II) SULFATE MONOHYDRATE   ..:..1ug....:..10......:..100...:..1mg....:..10....:..100....:..1g......:..10

```
2891 M f b6c eat thy fca 24m24                                                           : 20.5gm * P<.1   e
  a  M f b6c eat TBA MXB 24m24                                                             no dre    P=1.
  b  M f b6c eat liv MXB 24m24                                                             no dre    P=1.
  c  M f b6c eat lun MXB 24m24                                                           : 23.2gm * P<.05
2892 M m b6c eat hag adc 24m24                                                             26.2gm * P<.02  e
  a  M m b6c eat thy fca 24m24                                                             no dre    P=1.
  b  M m b6c eat TBA MXB 24m24                                                             no dre    P=1.
  c  M m b6c eat liv MXB 24m24                                                             no dre    P=1.
  d  M m b6c eat lun MXB 24m24                                                             no dre    P=1.
2893 R f f34 eat TBA MXB 24m24                                         :>                  no dre    P=1.   -
  a  R f f34 eat liv MXB 24m24                                                             no dre    P=1.
2894 R m f34 eat pni MXA 24m24                                      :   ±                 #1.46gm * P<.04  -
  a  R m f34 eat TBA MXB 24m24                                                             386.mg  * P<.07
  b  R m f34 eat liv MXB 24m24                                                             no dre    P=1.
```

D-MANNITOL   100ng..:..1ug....:..10......:..100...:..1mg....:..10......:..100....:..1g......:..10

```
2895 M f b6c eat --- hes 24m24                                                           #46.7gm * P<.04  -
  a  M f b6c eat --- lle 24m24                                                             48.0gm * P<.02
  b  M f b6c eat TBA MXB 24m24                                                             no dre    P=1.
  c  M f b6c eat liv MXB 24m24                                                             no dre    P=1.
  d  M f b6c eat lun MXB 24m24                                                             no dre    P=1.   -
2896 M m b6c eat TBA MXB 24m24                                                             no dre    P=1.
  a  M m b6c eat liv MXB 24m24                                                             98.1gm * P<.8
  b  M m b6c eat lun MXB 24m24                                           :>                5.00gm \ P<.7   -
2897 R f f34 eat TBA MXB 24m24                                                             no dre    P=1.
  a  R f f34 eat liv MXB 24m24                                              :>             no dre    P=1.   -
2898 R m f34 eat TBA MXB 24m24                                                             18.0gm * P<.1
  a  R m f34 eat liv MXB 24m24
```

MANNITOL NITROGEN MUSTARD   100ng..:..1ug....:..10......:..100...:..1mg....:..10......:..100....:..1g......:..10

```
2899 R m b46 ivj tba mix 12m24 es                            .>                            3.11mg    P<.5
  a  R m b46 ivj tba mal 12m24 es                                                          3.85mg    P<.5
  b  R m b46 ivj tba ben 12m24 es                                                          23.5mg    P<.9
```

MeA-alpha-C   100ng..:..1ug....:..10......:..100...:..1mg....:..10......:..100....:..1g......:..10

```
2900 M f cdf eat blv hms 84w84                                      . + .                  38.6mg    P<.0005+
  a  M f cdf eat liv mix 84w84                                                             38.6mg    P<.0005+
  b  M f cdf eat liv hpc 84w84                                                             98.9mg    P<.0005
  c  M f cdf eat liv hpa 84w84                                                             118.mg    P<.0005
  d  M f cdf eat lun ade 84w84                                                             1.84gm    P<.3
  e  M f cdf eat lun mix 84w84                                                             no dre    P=1.
2901 M m cdf eat blv hms 73w73                                      . + .                  15.6mg    P<.0005+
  a  M m cdf eat liv mix 73w73                                                             43.5mg    P<.0005+
  b  M m cdf eat liv hpa 73w73                                                             90.9mg    P<.0005
  c  M m cdf eat liv hpc 73w73                                                             127.mg    P<.0005
  d  M m cdf eat lun ade 73w73                                                             1.18gm    P<.7
  e  M m cdf eat lun mix 73w73                                                             no dre    P=1.
```

MeIQ   100ng..:..1ug....:..10......:..100...:..1mg....:..10......:..100....:..1g......:..10

```
2902 M f cdf eat for mix 91w91 e                                    . + .                  10.7mg  * P<.0005+
  a  M f cdf eat for sqc 91w91 e                                                            24.5mg  * P<.0005+
  b  M f cdf eat liv mix 91w91 e                                                            27.1mg  * P<.0005+
  c  M f cdf eat liv hpc 91w91 e                                                            65.6mg  / P<.0005+
  d  M f cdf eat lun mix 91w91 e                                                            25.1mg  \ P<.03   -
2903 M m cdf eat for mix 91w91 e                                    . + .                   14.5mg  * P<.0005+
  a  M m cdf eat for sqc 91w91 e                                                            22.4mg  / P<.0005+
  b  M m cdf eat lun mix 91w91 e                                                            148.mg  * P<.2   -
  c  M m cdf eat liv mix 91w91 e                                                            no dre    P=1.
  d  M m cdf eat liv hpc 91w91 e                                                            no dre    P=1.
```

	RefNum	LoConf	UpConf	Cntrl	1Dose	1Inc	2Dose	2Inc	Citation or Pathology
									Brkly Code
j	TR331	20.2mg	889.mg	40/50	35.4mg	40/50	70.7mg	41/50	
k	TR331	133.mg	n.s.s.	3/50	35.4mg	2/50	70.7mg	3/50	liv:hpa,hpc,nnd.

MANGANESE ETHYLENEBISTHIOCARBAMATE   (maneb) 12427-38-2

2886	1214	25.6mg	n.s.s.	1/17	21.6mg	1/18			Innes;ntis,1968/1969
a	1214	42.7mg	n.s.s.	0/17	21.6mg	0/18			
b	1214	15.5mg	n.s.s.	2/17	21.6mg	3/18			
2887	1214	11.5mg	n.s.s.	2/18	20.0mg	4/18			
a	1214	39.7mg	n.s.s.	1/18	20.0mg	0/18			
b	1214	12.4mg	n.s.s.	3/18	20.0mg	4/18			
2888	1214	40.3mg	n.s.s.	0/16	21.6mg	0/17			
a	1214	40.3mg	n.s.s.	0/16	21.6mg	0/17			
b	1214	40.3mg	n.s.s.	0/16	21.6mg	0/17			
2889	1214	20.9mg	n.s.s.	0/16	20.0mg	1/18			
a	1214	20.9mg	n.s.s.	0/16	20.0mg	1/18			
b	1214	15.3mg	n.s.s.	0/16	20.0mg	2/18			
2890	1426	109.mg	n.s.s.	0/46	88.0mg	0/6			Andrianova;vpit,29,71-74;1970
a	1426	36.3mg	n.s.s.	1/46	88.0mg	2/6			

MANGANESE (II) SULFATE MONOHYDRATE   10034-96-5

2891	TR428	6.11gm	n.s.s.	2/51	192.mg	1/50	641.mg	0/51	1.92gm	5/51
a	TR428	1.89gm	n.s.s.	37/51	192.mg	37/50	641.mg	37/51	1.92gm	36/51
b	TR428	4.20gm	n.s.s.	13/51	192.mg	15/50	641.mg	7/51	1.92gm	13/51
c	TR428	7.11gm	n.s.s.	6/51	192.mg	4/50	641.mg	6/51	1.92gm	4/51
2892	TR428	7.20gm	n.s.s.	1/50	178.mg	0/50	592.mg	0/51	1.78gm	4/51
a	TR428	7.93gm	n.s.s.	0/50	178.mg	0/50	592.mg	0/51	1.78gm	3/51
b	TR428	2.01gm	n.s.s.	46/50	178.mg	44/50	592.mg	40/51	1.78gm	39/51
c	TR428	182.mg	n.s.s.	34/50	178.mg	31/50	(592.mg	24/51	1.78gm	22/51)
d	TR428	4.17gm	n.s.s.	12/50	178.mg	13/50	592.mg	9/51	1.78gm	12/51
2893	TR428	557.mg	n.s.s.	45/50	74.0mg	47/50	247.mg	49/51	740.mg	42/50
a	TR428	4.29gm	n.s.s.	0/50	74.0mg	1/50	247.mg	1/51	740.mg	0/50
2894	TR428	513.mg	n.s.s.	0/52	59.2mg	3/51	197.mg	4/51	591.mg	4/52
a	TR428	147.mg	n.s.s.	44/52	59.2mg	46/51	197.mg	46/51	591.mg	44/52
b	TR428	1.46gm	n.s.s.	0/52	59.2mg	1/51	197.mg	1/51	591.mg	0/52

Citation/Pathology column (Manganese sulfate):
- 2891c: liv:hpa,hpb,hpc. lun:a/a,a/c.   s
- 2892d: liv:hpa,hpb,hpc. lun:a/a,a/c.
- 2894a: liv:hpa,hpb,hpc. pni:isa,isc. s
- 2894b: liv:hpa,hpb,hpc.

D-MANNITOL   69-65-8

2895	TR236	17.7gm	n.s.s.	0/50	3.19gm	2/50	6.38gm	3/50	s
a	TR236	19.5gm	n.s.s.	0/50	3.19gm	2/50	6.38gm	4/50	s
b	TR236	11.2gm	n.s.s.	26/50	3.19gm	24/50	6.38gm	18/50	
c	TR236	25.7gm	n.s.s.	3/50	3.19gm	3/50	6.38gm	2/50	
d	TR236	36.9gm	n.s.s.	3/50	3.19gm	2/50	6.38gm	1/50	liv:hpa,hpc,nnd. lun:a/a,a/c.
2896	TR236	10.6gm	n.s.s.	33/50	2.94gm	26/50	5.89gm	26/50	
a	TR236	14.5gm	n.s.s.	14/50	2.94gm	14/50	5.89gm	11/50	
b	TR236	10.7gm	n.s.s.	9/50	2.94gm	12/50	5.89gm	11/50	liv:hpa,hpc,nnd. lun:a/a,a/c.
2897	TR236	686.mg	n.s.s.	45/50	1.23gm	44/50	(2.45gm	36/50)	
a	TR236	n.s.s.	n.s.s.	0/50	1.23gm	1/50	2.45gm	0/50	liv:hpa,hpc,nnd.
2898	TR236	1.77gm	n.s.s.	40/50	981.mg	40/50	1.96gm	36/50	
a	TR236	6.18gm	n.s.s.	0/50	981.mg	2/50	1.96gm	2/50	liv:hpa,hpc,nnd.

MANNITOL NITROGEN MUSTARD   (degranol) 576-68-1

2899	1017	.583mg	n.s.s.	7/65	.286mg	6/37		Schmahl;arzn,20,1461-1467;1970
a	1017	.706mg	n.s.s.	4/65	.286mg	4/37		
b	1017	1.05mg	n.s.s.	3/65	.286mg	2/37		

MeA-alpha-C (2-amino-3-methyl-9H-pyrido-[2,3-b]-indole)   68006-83-7

2900	1616	23.7mg	67.3mg	0/40	104.mg	28/40		Ohgaki;carc,5,815-819;1984
a	1616	23.7mg	67.3mg	0/40	104.mg	28/40		
b	1616	53.9mg	210.mg	0/40	104.mg	15/40		
c	1616	62.1mg	267.mg	0/40	104.mg	13/40		
d	1616	299.mg	n.s.s.	0/40	104.mg	1/40		
e	1616	273.mg	n.s.s.	2/40	104.mg	2/40		
2901	1616	9.38mg	26.7mg	0/40	96.0mg	35/40		
a	1616	25.6mg	81.9mg	0/40	96.0mg	21/40		
b	1616	46.6mg	214.mg	0/40	96.0mg	12/40		
c	1616	59.7mg	372.mg	0/40	96.0mg	9/40		
d	1616	132.mg	n.s.s.	3/40	96.0mg	4/40		
e	1616	192.mg	n.s.s.	11/40	96.0mg	5/40		

MeIQ (2-amino-3,4-dimethylimidazo[4,5-f]quinoline)   77094-11-2

2902	1798	7.18mg	16.5mg	0/40	13.0mg	19/36	52.0mg	34/38	Ohgaki;carc,7,1889-1893;1986
a	1798	16.0mg	39.7mg	0/40	13.0mg	11/36	52.0mg	24/38	
b	1798	17.2mg	45.4mg	0/40	13.0mg	4/36	52.0mg	27/38	
c	1798	36.4mg	135.mg	0/40	13.0mg	0/36	52.0mg	16/38	
d	1798	10.1mg	n.s.s.	5/40	13.0mg	12/36	(52.0mg	7/38)	
2903	1798	9.53mg	23.1mg	0/29	12.0mg	7/38	48.0mg	35/38	
a	1798	14.4mg	37.1mg	0/29	12.0mg	3/38	48.0mg	30/38	
b	1798	46.8mg	n.s.s.	6/29	12.0mg	6/38	48.0mg	12/38	
c	1798	70.6mg	n.s.s.	4/29	12.0mg	11/38	48.0mg	7/38	
d	1798	163.mg	n.s.s.	1/29	12.0mg	3/38	48.0mg	1/38	

```
     Spe Strain Site    Xpo+Xpt                                                          TD50     2Tailpvl
         Sex  Route Hist      Notes                                                        DR       AuOp
```

```
MeIQx                              100ng..:..1ug....:..10.....:..100...:..1mg....:..10.....:..100...:..1g......:..10
2904 M f cdf eat liv mix 84w84 e                                    . + .                 14.2mg   P<.0005+
  a  M f cdf eat liv hpa 84w84 e                                    . + .                 14.2mg   P<.0005
  b  M f cdf eat liv hpc 84w84 e                                                          27.8mg   P<.0005
  c  M f cdf eat lun mix 84w84 e                                                          77.2mg   P<.002 +
  d  M f cdf eat lun ade 84w84 e                                                          143.mg   P<.02
  e  M f cdf eat lun adc 84w84 e                                                          257.mg   P<.1
  f  M f cdf eat --- lkm 84w84 e                                                          6.82gm   P<1.
2905 M m cdf eat liv hpa 84w84 e                                           . +            86.9mg   P<.01
  a  M m cdf eat liv hpc 84w84 e                                                          102.mg   P<.0005
  b  M m cdf eat --- lkm 84w84 e                                                          109.mg   P<.006 +
  c  M m cdf eat liv mix 84w84 e                                                          83.8mg   P<.02  +
  d  M m cdf eat lun mix 84w84 e                                                          134.mg   P<.2
  e  M m cdf eat lun adc 84w84 e                                                          236.mg   P<.4
  f  M m cdf eat lun ade 84w84 e                                                          262.mg   P<.2
2906 R f f3d eat cli sqc 61w61 e                              . + . .                     4.72mg   P<.0005+
  a  R f f3d eat liv nnd 61w61 e                                                          6.31mg   P<.0005+
  b  R f f3d eat zym sqc 61w61 e                                                          6.31mg   P<.0005+
  c  R f f3d eat ski sqc 61w61 e                                                          87.2mg   P<.3
2907 R m f3d eat liv mix 61w61 e                          <+                              noTD50   P<.0005+
  a  R m f3d eat liv hpc 61w61 e                                                          1.26mg   P<.0005+
  b  R m f3d eat zym mix 61w61 e                                                          2.72mg   P<.0005+
  c  R m f3d eat zym sqc 61w61 e                                                          3.59mg   P<.0005+
  d  R m f3d eat ski mix 61w61 e                                                          8.76mg   P<.002 +
  e  R m f3d eat ski sqc 61w61 e                                                          13.1mg   P<.008 +
  f  R m f3d eat zym sqp 61w61 e                                                          35.8mg   P<.1
  g  R m f3d eat ski sqp 61w61 e                                                          73.5mg   P<.3
  h  R m f3d eat ski bcc 61w61 e                                                          73.5mg   P<.3
  i  R m f3d eat liv nnd 61w61 e                                                          73.5mg   P<.3

MELAMINE                           100ng..:..1ug....:..10.....:..100...:..1mg....:..10.....:..100...:..1g......:..10
2908 M f b6c eat TBA MXB 24m24                                                       :>   no dre   P=1.    -
  a  M f b6c eat liv MXB 24m24                                                            no dre   P=1.
  b  M f b6c eat lun MXB 24m24                                                            no dre   P=1.
2909 M m b6c eat TBA MXB 24m24                                                  :>        3.20gm * P<.7    -
  a  M m b6c eat liv MXB 24m24                                                            4.37gm * P<.6
  b  M m b6c eat lun MXB 24m24                                                            no dre   P=1.
2910 R f f34 eat thy MXA 24m24                                             :   ±          #2.47gm* P<.03   -
  a  R f f34 eat thy ccr 24m24                                                            4.16gm * P<.04
  b  R f f34 eat TBA MXB 24m24                                                            3.93gm * P<.9
  c  R f f34 eat liv MXB 24m24                                                            4.41gm * P<.2
2911 R m f34 eat ubl MXA 24m24                                            :   +   :       679.mg / P<.0005
  a  R m f34 eat ubl tcc 24m24                                                            735.mg / P<.002 c
  b  R m f34 eat TBA MXB 24m24                                                            149.mg * P<.04
  c  R m f34 eat liv MXB 24m24                                                            no dre   P=1.

MELPHALAN                          100ng..:..1ug....:..10.....:..100...:..1mg....:..10.....:..100...:..1g......:..10
2912 M f swi ipj lun mix 26w79 e                          .    +        .                 .165mg * P<.007 +
  a  M f swi ipj mgl ade 26w79 e                                                          .617mg * P<.004
  b  M f swi ipj --- lys 26w79 e                                                          .702mg * P<.06
  c  M f swi ipj adr fbs 26w79 e                                                          1.26mg * P<.03
  d  M f swi ipj liv lys 26w79 e                                                          no dre   P=1.
  e  M f swi ipj tba mix 26w79 e                                                          51.0ug * P<.0005
  f  M f swi ipj tba mal 26w79 e                                                          .104mg * P<.002
  g  M f swi ipj tba ben 26w79 e                                                          .365mg * P<.08
2913 M m swi ipj lun mix 26w77 e                       .   +        .                     .137mg * P<.005 +
  a  M m swi ipj --- lys 26w77 e                                                          .302mg   P<.02  +
  b  M m swi ipj liv mix 26w77 e                                                          no dre   P=1.
  c  M m swi ipj tba mix 26w77 e                                                          91.4ug * P<.02
  d  M m swi ipj tba mal 26w77 e                                                          .115mg   P<.02
  e  M m swi ipj tba ben 26w77 e                                                          2.02mg   P<.8
2914 R f cdr ipj per sar 26w78 e                       .   +                              .135mg * P<.0005+
  a  R f cdr ipj lun a/a 26w78 e                                                          1.44mg * P<.02
  b  R f cdr ipj pan adc 26w78 e                                                          1.44mg * P<.02
  c  R f cdr ipj --- lys 26w78 e                                                          1.54mg * P<.09
  d  R f cdr ipj liv tum 26w78 e                                                          no dre   P=1.
  e  R f cdr ipj tba mal 26w78 e                                                          78.1ug * P<.0005
  f  R f cdr ipj tba mix 26w78 e                                                          .106mg * P<.07
  g  R f cdr ipj tba ben 26w78 e                                                          no dre   P=1.
2915 R m cdr ipj per mix 26w67 e                      .   +                               71.9ug   P<.0005+
  a  R m cdr ipj --- lys 26w67 e                                                          .359mg   P<.003
  b  R m cdr ipj liv tum 26w67 e                                                          no dre   P=1.
  c  R m cdr ipj tba mix 26w67 e                                                          23.6ug   P<.0005
  d  R m cdr ipj tba mal 26w67 e                                                          47.0ug   P<.0005
  e  R m cdr ipj tba ben 26w67 e                                                          .493mg   P<.5

DL-MENTHOL                         100ng..:..1ug....:..10.....:..100...:..1mg....:..10.....:..100...:..1g......:..10
2916 M f b6c eat TBA MXB 24m24                                                       :>   no dre   P=1.    -
  a  M f b6c eat liv MXB 24m24                                                            5.19gm * P<.3
  b  M f b6c eat lun MXB 24m24                                                            3.17gm * P<.06
2917 M m b6c eat TBA MXB 24m24                                                  :>        2.75gm * P<.7    -
```

RefNum	LoConf	UpConf	Cntrl	1Dose	1Inc	2Dose	2Inc	Citation or Pathology	Brkly Code
MeIQx (2-amino-3,8-dimethylimidazo[4,5-f]quinoxaline) 77500-04-0									
2904	1820m	7.95mg	25.6mg	0/39	78.0mg	32/35			Ohgaki;carc,8,665-668;1987/Wakabayashi 1992
a	1820m	7.95mg	25.6mg	0/39	78.0mg	32/35			
b	1820m	16.6mg	50.3mg	0/39	78.0mg	25/35			
c	1820m	37.2mg	370.mg	4/39	78.0mg	15/35			
d	1820m	59.2mg	n.s.s.	2/39	78.0mg	9/35			
e	1820m	83.2mg	n.s.s.	2/39	78.0mg	6/35			
f	1820m	83.2mg	n.s.s.	11/39	78.0mg	10/35			
2905	1820m	38.8mg	10.9gm	5/36	72.0mg	15/37			
a	1820m	49.6mg	273.mg	0/36	72.0mg	10/37			
b	1820m	48.9mg	1.23gm	2/36	72.0mg	11/37			
c	1820m	37.0mg	n.s.s.	6/36	72.0mg	16/37			
d	1820m	42.9mg	n.s.s.	10/36	72.0mg	16/37			
e	1820m	60.9mg	n.s.s.	7/36	72.0mg	11/37			
f	1820m	77.2mg	n.s.s.	3/36	72.0mg	7/37			
2906	1867	2.34mg	11.3mg	0/20	20.0mg	12/19			Kato;carc,9,71-73;1988/pers.comm.
a	1867	3.00mg	16.4mg	0/20	20.0mg	10/19			
b	1867	3.00mg	16.4mg	0/20	20.0mg	10/19			
c	1867	14.2mg	n.s.s.	0/20	20.0mg	1/19			
2907	1867	n.s.s.	2.01mg	0/19	16.0mg	20/20			
a	1867	.518mg	2.88mg	0/19	16.0mg	19/20			
b	1867	1.39mg	6.00mg	0/19	16.0mg	15/20			
c	1867	1.82mg	8.30mg	0/19	16.0mg	13/20			
d	1867	3.75mg	34.5mg	0/19	16.0mg	7/20			
e	1867	4.95mg	197.mg	0/19	16.0mg	5/20			
f	1867	8.79mg	n.s.s.	0/19	16.0mg	2/20			
g	1867	12.0mg	n.s.s.	0/19	16.0mg	1/20			
h	1867	12.0mg	n.s.s.	0/19	16.0mg	1/20			
i	1867	12.0mg	n.s.s.	0/19	16.0mg	1/20			
MELAMINE 108-78-1									
2908	TR245	804.mg	n.s.s.	28/50	287.mg	33/50	574.mg	26/50	
a	TR245	2.61gm	n.s.s.	4/50	287.mg	6/50	574.mg	2/50	liv:hpa,hpc,nnd.
b	TR245	3.53gm	n.s.s.	5/50	287.mg	1/50	574.mg	3/50	lun:a/a,a/c.
2909	TR245	525.mg	n.s.s.	23/50	265.mg	22/50	530.mg	20/50	
a	TR245	820.mg	n.s.s.	12/50	265.mg	8/50	530.mg	12/50	liv:hpa,hpc,nnd.
b	TR245	2.37gm	n.s.s.	5/50	265.mg	4/50	530.mg	1/50	lun:a/a,a/c.
2910	TR245	937.mg	n.s.s.	0/50	221.mg	2/50	441.mg	3/50	thy:cca,ccr. S
a	TR245	1.26gm	n.s.s.	0/50	221.mg	0/50	441.mg	3/50	S
b	TR245	273.mg	n.s.s.	42/50	221.mg	43/50	441.mg	37/50	
c	TR245	1.33gm	n.s.s.	0/50	221.mg	2/50	441.mg	1/50	liv:hpa,hpc,nnd.
2911	TR245	308.mg	2.04gm	0/50	88.3mg	0/50	177.mg	9/50	ubl:tcc,tpp. S
a	TR245	322.mg	2.51gm	0/50	88.3mg	0/50	177.mg	8/50	
b	TR245	65.3mg	n.s.s.	30/50	88.3mg	36/50	177.mg	38/50	
c	TR245	n.s.s.	n.s.s.	0/50	88.3mg	1/50	177.mg	0/50	liv:hpa,hpc,nnd.
MELPHALAN (L-sarcolysin) 148-82-3									
2912	1336m	61.7ug	4.11mg	20/154	.106mg	6/15	.212mg	3/8	Skipper;srfr;1976/Weisburger;canc;1977/Prejean pers.comm.
a	1336m	.152mg	10.1mg	0/154	.106mg	1/15	.212mg	1/8	
b	1336m	.154mg	n.s.s.	3/154	.106mg	0/15	.212mg	2/8	
c	1336m	.204mg	n.s.s.	0/154	.106mg	0/15	.212mg	1/8	
d	1336m	97.5ug	n.s.s.	1/154	.106mg	0/15	.212mg	0/8	
e	1336m	23.3ug	.167mg	42/154	.106mg	10/15	.212mg	7/8	
f	1336m	43.4ug	.617mg	29/154	.106mg	7/15	.212mg	5/8	
g	1336m	.100mg	n.s.s.	13/154	.106mg	3/15	.212mg	2/8	
2913	1336m	54.6ug	1.57mg	9/101	.108mg	9/28			
a	1336m	94.4ug	n.s.s.	2/101	.108mg	4/28			
b	1336m	.341mg	n.s.s.	2/101	.108mg	0/28			
c	1336m	36.2ug	n.s.s.	28/101	.108mg	15/28			
d	1336m	44.9ug	n.s.s.	19/101	.108mg	12/28			
e	1336m	.140mg	n.s.s.	9/101	.108mg	3/28			
2914	1336n	63.2ug	.371mg	0/182	.129mg	8/16	.304mg	1/6	
a	1336n	.234mg	n.s.s.	0/182	.129mg	0/16	.304mg	1/6	
b	1336n	.234mg	n.s.s.	0/182	.129mg	0/16	.304mg	1/6	
c	1336n	.236mg	n.s.s.	1/182	.129mg	0/16	.304mg	1/6	
d	1336n	.112mg	n.s.s.	0/182	.129mg	0/16	.304mg	0/6	
e	1336n	35.2ug	.310mg	44/182	.129mg	13/16	.304mg	3/6	
f	1336n	33.1ug	n.s.s.	103/182	.129mg	14/16	.304mg	4/6	
g	1336n	.319mg	n.s.s.	59/182	.129mg	1/16	.304mg	1/6	
2915	1336n	32.0ug	.212mg	0/177	.149mg	8/18			
a	1336n	88.1ug	4.84mg	0/177	.149mg	2/18			
b	1336n	.229mg	n.s.s.	0/177	.149mg	0/18			
c	1336n	9.76ug	77.1ug	59/177	.149mg	16/18			
d	1336n	20.7ug	.171mg	32/177	.149mg	12/18			
e	1336n	74.1ug	n.s.s.	27/177	.149mg	4/18			
DL-MENTHOL 15356-70-4									
2916	TR98	723.mg	n.s.s.	30/50	257.mg	20/50	515.mg	24/50	
a	TR98	1.54gm	n.s.s.	1/50	257.mg	3/50	515.mg	3/50	liv:hpa,hpc,nnd.
b	TR98	1.22gm	n.s.s.	1/50	257.mg	3/50	515.mg	5/50	lun:a/a,a/c.
2917	TR98	357.mg	n.s.s.	29/50	238.mg	28/50	475.mg	33/50	

```
     Spe Strain Site   Xpo+Xpt                                                      TD50    2Tailpvl
         Sex  Route  Hist     Notes                                                    DR    AuOp

 a    M m b6c eat liv MXB 24m24                                                     1.96gm * P<.3
 b    M m b6c eat lun MXB 24m24                                                     no dre   P=1.
2918  R f f34 eat TBA MXB 24m24                                           :>        no dre   P=1.   -
 a    R f f34 eat liv MXB 24m24                                                     no dre   P=1.
2919  R m f34 eat TBA MXB 24m24                                     :>              1.41gm * P<.7   -
 a    R m f34 eat liv MXB 24m24                                                     5.69gm * P<.6


MER-25                             100ng..:..1ug....:..10......:..100...:..1mg...:..10...:..100....:..1g.....:..10
2920  R f nss gav mam tum 52w52 er                                 .>              no dre   P=1.   -


2-MERCAPTOBENZOTHIAZOLE            100ng..:..1ug....:..10......:..100...:..1mg...:..10...:..100....:..1g.....:..10
2921  M f b6c gav liv MXA 24m24                                              :>     2.16gm * P<.3   e
 a    M f b6c gav TBA MXB 24m24                                                     no dre   P=1.
 b    M f b6c gav liv MXB 24m24                                                     2.16gm * P<.3
 c    M f b6c gav lun MXB 24m24                                                     no dre   P=1.
2922  M f b6c orl lun ade 76w76 evx                         .>                      283.mg   P<.3   -
 a    M f b6c orl liv hpt 76w76 evx                                                 no dre   P=1.   -
 b    M f b6c orl tba mix 76w76 evx                                                 283.mg   P<.3   -
2923  M m b6c gav TBA MXB 24m24                                          :>         3.33gm * P<.8   -
 a    M m b6c gav liv MXB 24m24                                                     3.70gm * P<.7   -
 b    M m b6c gav lun MXB 24m24                                                     no dre   P=1.
2924  M m b6c orl liv hpt 76w76 evx                    .    ±                       56.2mg   P<.02  -
 a    M m b6c orl lun ade 76w76 evx                                                 120.mg   P<.1   -
 b    M m b6c orl tba mix 76w76 evx                                                 17.0mg   P<.0005-
2925  M f b6a orl lun ade 76w76 evx                         .>                      no dre   P=1.   -
 a    M f b6a orl liv hpt 76w76 evx                                                 no dre   P=1.   -
 b    M f b6a orl tba mix 76w76 evx                                                 no dre   P=1.   -
2926  M m b6a orl lun ade 76w76 evx                         .>                      no dre   P=1.   -
 a    M m b6a orl liv hpt 76w76 evx                                                 no dre   P=1.   -
 b    M m b6a orl tba mix 76w76 evx                                                 105.mg   P<.5   -
2927  R f f34 gav MXB MXB 24m24                                     :    ±          247.mg * P<.03
 a    R f f34 gav pta adn 24m24                                                     343.mg * P<.07  p
 b    R f f34 gav amd phe 24m24                                                     805.mg * P<.04  p
 c    R f f34 gav TBA MXB 24m24                                                     576.mg * P<.6
 d    R f f34 gav liv MXB 24m24                                                     no dre   P=1.
2928  R m f34 gav MXB MXB 24m24                                     :  +     :      157.mg \ P<.0005
 a    R m f34 gav tes ict 24m24                                                     240.mg * P<.005
 b    R m f34 gav pta adn 24m24                                                     333.mg \ P<.009
 c    R m f34 gav pan ana 24m24                                                     345.mg \ P<.0005p
 d    R m f34 gav amd MXA 24m24                                                     394.mg * P<.003  p
 e    R m f34 gav mul mnl 24m24                                                     401.mg \ P<.004  p
 f    R m f34 gav pre adn 24m24                                                     1.71gm * P<.004
 g    R m f34 gav pre MXA 24m24                                                     1.58gm * P<.02   p
 h    R m f34 gav sub MXA 24m24                                                     1.73gm * P<.04
 i    R m f34 gav sub MXA 24m24                                                     1.90gm * P<.03
 j    R m f34 gav sub fib 24m24                                                     2.20gm * P<.04
 k    R m f34 gav MXA MXA 24m24                                                     3.20gm * P<.02
 l    R m f34 gav TBA MXB 24m24                                                     136.mg \ P<.002
 m    R m f34 gav liv MXB 24m24                                                     no dre   P=1.


2-MERCAPTOBENZOTHIAZOLE,  ZINC     100ng..:..1ug....:..10......:..100...:..1mg...:..10...:..100....:..1g.....:..10
2929  M f b6a orl lun ade 76w76 evx                                .>               no dre   P=1.   -
 a    M f b6a orl liv hpt 76w76 evx                                                 no dre   P=1.   -
 b    M f b6a orl tba mix 76w76 evx                                                 no dre   P=1.   -
2930  M m b6a orl lun ade 76w76 evx                                .>               no dre   P=1.   -
 a    M m b6a orl liv hpt 76w76 evx                                                 no dre   P=1.   -
 b    M m b6a orl tba mix 76w76 evx                                                 no dre   P=1.   -
2931  M f b6c orl liv hpt 76w76 evx                                   .>            no dre   P=1.   -
 a    M f b6c orl lun mix 76w76 evx                                                 no dre   P=1.   -
 b    M f b6c orl tba mix 76w76 evx                                                 2.79gm   P<.3   -
2932  M m b6c orl liv hpt 76w76 evx                    .    +       .               388.mg   P<.004 -
 a    M m b6c orl lun ade 76w76 evx                                                 2.75gm   P<.3   -
 b    M m b6c orl tba mix 76w76 evx                                                 319.mg   P<.002 -


2-MERCAPTOETHANESULFONATE,  SODIUM 100ng..:..1ug....:..10......:..100...:..1mg...:..10...:..100....:..1g.....:..10
2933  R m sda wat liv tum 9m24                                           .>         no dre   P=1.
 a    R m sda wat lun tum 9m24                                                      no dre   P=1.
 b    R m sda wat ubl tum 9m24                                                      no dre   P=1.


6-MERCAPTOPURINE                   100ng..:..1ug....:..10......:..100...:..1mg...:..10...:..100....:..1g.....:..10
2934  R m b46 ivj tba mix 12m24 es                          .>                      58.1mg   P<.9   -
2935  R f f3d eat adr phe 24m26                        .    ±                       22.5mg * P<.02
 a    R f f3d eat cns mix 24m26                                                     34.1mg * P<.04
 b    R f f3d eat ute adc 24m26                                                     34.1mg * P<.04
 c    R f f3d eat liv hpc 24m26                                                     139.mg * P<.3
 d    R f f3d eat liv hpa 24m26                                                     no dre   P=1.
 e    R f f3d eat tba mix 24m26                                                     4.74mg * P<.4
 f    R f f3d eat tba mal 24m26                                                     10.3mg * P<.3
2936  R m f3d eat liv hpa 24m26                             .>                      no dre   P=1.
 a    R m f3d eat tba mix 24m26                                                     no dre   P=1.
 b    R m f3d eat tba mal 24m26                                                     no dre   P=1.
```

	RefNum	LoConf	UpConf	Cntrl	1Dose	1Inc	2Dose	2Inc	Citation or Pathology
									Brkly Code
a	TR98	601.mg	n.s.s.	8/50	238.mg	8/50	475.mg	14/50	liv:hpa,hpc,nnd.
b	TR98	1.10gm	n.s.s.	6/50	238.mg	7/50	475.mg	6/50	lun:a/a,a/c.
2918	TR98	462.mg	n.s.s.	41/50	184.mg	36/50	368.mg	32/50	
a	TR98	2.11gm	n.s.s.	1/50	184.mg	1/50	368.mg	1/50	liv:hpa,hpc,nnd.
2919	TR98	200.mg	n.s.s.	32/50	147.mg	41/50	294.mg	37/50	
a	TR98	932.mg	n.s.s.	1/50	147.mg	3/50	294.mg	2/50	liv:hpa,hpc,nnd.

MER-25   67-98-1

2920	481	13.9mg	n.s.s.	0/9	12.2mg	0/22			Shay;mpoc,305-318;1962

2-MERCAPTOBENZOTHIAZOLE   (captax, rotax) 149-30-4

2921	TR332	629.mg	n.s.s.	4/50	265.mg	12/50	531.mg	4/50	liv:hpa,hpc.
a	TR332	697.mg	n.s.s.	38/50	265.mg	33/50	531.mg	15/50	
b	TR332	629.mg	n.s.s.	4/50	265.mg	12/50	531.mg	4/50	liv:hpa,hpc,nnd.
c	TR332	2.14gm	n.s.s.	3/50	265.mg	1/50	531.mg	2/50	lun:a/a,a/c.
2922	1302	46.1mg	n.s.s.	0/16	44.3mg	1/18			Innes;ntis,1968/1969
a	1302	87.7mg	n.s.s.	0/16	44.3mg	0/18			
b	1302	46.1mg	n.s.s.	0/16	44.3mg	1/18			
2923	TR332	359.mg	n.s.s.	31/50	265.mg	39/50	531.mg	25/50	
a	TR332	557.mg	n.s.s.	16/50	265.mg	21/50	531.mg	14/50	liv:hpa,hpc,nnd.
b	TR332	1.06gm	n.s.s.	7/50	265.mg	9/50	531.mg	5/50	lun:a/a,a/c.
2924	1302	19.3mg	n.s.s.	0/16	41.2mg	4/17			Innes;ntis,1968/1969
a	1302	29.5mg	n.s.s.	0/16	41.2mg	2/17			
b	1302	7.99mg	44.6mg	0/16	41.2mg	10/17			
2925	1302	49.3mg	n.s.s.	1/17	44.3mg	1/17			
a	1302	82.8mg	n.s.s.	0/17	44.3mg	0/17			
b	1302	54.8mg	n.s.s.	2/17	44.3mg	1/17			
2926	1302	38.5mg	n.s.s.	2/18	41.2mg	2/18			
a	1302	48.7mg	n.s.s.	1/18	41.2mg	1/18			
b	1302	21.0mg	n.s.s.	3/18	41.2mg	5/18			
2927	TR332	112.mg	n.s.s.	16/50	133.mg	28/50	265.mg	29/50	amd:phe; pta:adn. P
a	TR332	140.mg	n.s.s.	15/50	133.mg	24/50	265.mg	25/50	
b	TR332	347.mg	n.s.s.	1/50	133.mg	5/50	265.mg	6/50	
c	TR332	116.mg	n.s.s.	37/50	133.mg	46/50	265.mg	40/50	
d	TR332	3.30gm	n.s.s.	1/50	133.mg	0/50	265.mg	0/50	liv:hpa,hpc,nnd.
2928	TR332	82.2mg	543.mg	24/50	265.mg	38/50	(531.mg	31/50)	amd:phe,phm; mul:mnl; pan:ana; pre:adn,can. P
a	TR332	121.mg	2.27gm	48/50	265.mg	48/50	531.mg	48/50	S
b	TR332	146.mg	13.6gm	14/50	265.mg	21/50	(531.mg	12/50)	S
c	TR332	167.mg	1.10gm	2/50	265.mg	13/50	(531.mg	6/50)	
d	TR332	203.mg	2.31gm	18/50	265.mg	27/50	531.mg	24/50	
e	TR332	183.mg	2.99gm	7/50	265.mg	16/50	(531.mg	3/50)	amd:phe,phm.
f	TR332	749.mg	10.5gm	0/50	265.mg	4/50	531.mg	4/50	S
g	TR332	684.mg	n.s.s.	1/50	265.mg	6/50	531.mg	5/50	pre:adn,can.
h	TR332	686.mg	n.s.s.	3/50	265.mg	6/50	531.mg	7/50	sub:fbs,fib,nfm,srn. S
i	TR332	744.mg	n.s.s.	2/50	265.mg	4/50	531.mg	6/50	sub:fib,nfm. S
j	TR332	822.mg	n.s.s.	2/50	265.mg	3/50	531.mg	6/50	S
k	TR332	1.14gm	n.s.s.	0/50	265.mg	2/50	531.mg	3/50	S
l	TR332	69.9mg	617.mg	37/50	265.mg	47/50	(531.mg	41/50)	mul:msm; tna:men. S
m	TR332	1.94gm	n.s.s.	3/50	265.mg	2/50	531.mg	1/50	liv:hpa,hpc,nnd.

2-MERCAPTOBENZOTHIAZOLE, ZINC   (zetax) 155-04-4

2929	1300	515.mg	n.s.s.	1/17	462.mg	1/17			Innes;ntis,1968/1969
a	1300	864.mg	n.s.s.	0/17	462.mg	0/17			
b	1300	572.mg	n.s.s.	2/17	462.mg	1/17			
2930	1300	564.mg	n.s.s.	2/18	430.mg	1/18			
a	1300	851.mg	n.s.s.	1/18	430.mg	0/18			
b	1300	603.mg	n.s.s.	3/18	430.mg	1/18			
2931	1300	864.mg	n.s.s.	0/16	462.mg	0/17			
a	1300	864.mg	n.s.s.	0/16	462.mg	0/17			
b	1300	454.mg	n.s.s.	0/16	462.mg	1/17			
2932	1300	157.mg	2.37gm	0/16	430.mg	6/18			
a	1300	448.mg	n.s.s.	0/16	430.mg	1/18			
b	1300	136.mg	1.30gm	0/16	430.mg	7/18			

2-MERCAPTOETHANESULFONATE, SODIUM   19767-45-4

2933	1637	717.mg	n.s.s.	0/40	87.0mg	0/40			Tacchi;clet,22,89-94;1984
a	1637	717.mg	n.s.s.	0/40	87.0mg	0/40			
b	1637	717.mg	n.s.s.	0/40	87.0mg	0/40			

6-MERCAPTOPURINE   50-44-2

2934	1017	2.51mg	n.s.s.	7/65	1.18mg	3/25			Schmahl;arzn,20,1461-1467;1970
2935	1980	9.17mg	n.s.s.	0/50	1.15mg	2/50	2.30mg	4/50	Maekawa;zkko,116,245-250;1990/pers.comm.
a	1980	11.8mg	n.s.s.	0/50	1.15mg	1/50	2.30mg	3/50	
b	1980	11.8mg	n.s.s.	0/50	1.15mg	1/50	2.30mg	3/50	
c	1980	22.6mg	n.s.s.	0/50	1.15mg	0/50	2.30mg	1/50	
d	1980	9.33mg	n.s.s.	1/50	1.15mg	0/50	2.30mg	0/50	
e	1980	1.27mg	n.s.s.	39/50	1.15mg	38/50	2.30mg	43/50	
f	1980	3.31mg	n.s.s.	14/50	1.15mg	17/50	2.30mg	20/50	
2936	1980	20.6mg	n.s.s.	3/50	.920mg	1/50	1.84mg	1/50	
a	1980	.616mg	n.s.s.	50/50	.920mg	50/50	1.84mg	48/50	
b	1980	4.44mg	n.s.s.	20/50	.920mg	17/50	1.84mg	18/50	

```
     Spe Strain  Site    Xpo+Xpt                                                          TD50    2Tailpvl
        Sex  Route   Hist    Notes                                                               DR     AuOp
MERCURIC CHLORIDE              100ng..:..1ug....:..10.....:..100....:..1mg....:..10.....:..100....:..1g.....:..10
2937 M f b6c gav TBA MXB 24m24                                             :    ±                   5.83mg \ P<.09  -
  a  M f b6c gav liv MXB 24m24                                                                     no dre   P=1.
  b  M f b6c gav lun MXB 24m24                                                                      40.8mg * P<.3
2938 M m b6c gav kid MXA 24m24                                                  :      ±            81.6mg * P<.04  e
  a  M m b6c gav kid rua 24m24                                                                     125.mg * P<.09  e
  b  M m b6c gav kid uac 24m24                                                                     240.mg * P<.3   e
  c  M m b6c gav TBA MXB 24m24                                                                     no dre   P=1.
  d  M m b6c gav liv MXB 24m24                                                                     no dre   P=1.
  e  M m b6c gav lun MXB 24m24                                                                     no dre   P=1.
2939 M f cd1 wat --- lkm 27m27 e                                .    +         .                    3.58mg   P<.009 -
  a  M f cd1 wat lun tum 27m27 e                                                                   no dre   P=1.   -
  b  M f cd1 wat tba mix 27m27 e                                                                    2.42mg   P<.05  -
  c  M f cd1 wat tba mal 27m27 e                                                                    5.55mg   P<.08  -
2940 M m cd1 wat lun tum 28m28 e                                            .>                      11.2mg   P<.5   -
  a  M m cd1 wat tba mix 28m28 e                                                                    3.20mg   P<.2   -
  b  M m cd1 wat tba mal 28m28 e                                                                   no dre   P=1.   -
2941 R f f34 gav for sqp 24m24                                                  :      ±            51.8mg * P<.09  e
  a  R f f34 gav TBA MXB 24m24                                                                     no dre   P=1.
  b  R f f34 gav liv MXB 24m24                                                                    105.mg * P<.8
2942 R m f34 gav tes MXA 24m24                                         :    +    :                  .825mg * P<.0005
  a  R m f34 gav amd MXA 24m24                                                                      1.56mg * P<.0005
  b  R m f34 gav amd MXA 24m24                                                                      1.60mg * P<.002
  c  R m f34 gav for sqp 24m24                                                                      3.12mg * P<.0005p
  d  R m f34 gav thy MXA 24m24                                                                      5.25mg * P<.006  e
  e  R m f34 gav thy fcc 24m24                                                                      7.32mg * P<.005  e
  f  R m f34 gav thy fca 24m24                                                                      5.98mg \ P<.05   e
  g  R m f34 gav TBA MXB 24m24                                                                      1.04mg * P<.002
  h  R m f34 gav liv MXB 24m24                                                                     10.3mg * P<.02

MERCURYMETHYL CHLORIDE         100ng..:..1ug....:..10.....:..100....:..1mg....:..10.....:..100....:..1g.....:..10
2943 M f b6c eat kid rca 24m24 er                                                   .>             65.8mg * P<.2
2944 M m b6c eat kid mix 24m24 ers                              .    +   .                          3.19mg * P<.0005+
  a  M m b6c eat kid rcc 24m24 ers                                                                  4.28mg * P<.0005
  b  M m b6c eat kid rca 24m24 ers                                                                  9.83mg * P<.002
2945 M f cd1 wat lun tum 33m35 esv                                       .>                        no dre   P=1.   -
  a  M f cd1 wat tba mix 33m35 esv                                                                  2.13mg \ P<.5   -
  b  M f cd1 wat tba mal 33m35 esv                                                                 no dre   P=1.   -
2946 M m cd1 wat lun tum 31m37 esv                                   .>                            no dre   P=1.   -
  a  M m cd1 wat tba mix 31m37 esv                                                                  5.63mg * P<.8   -
  b  M m cd1 wat tba mal 31m37 esv                                                                  9.64mg * P<.6   -
2947 M f icm eat lun adc 80w80 e                                        .>                          11.6mg   P<.6   -
  a  M f icm eat liv hem 80w80 e                                                                    26.1mg   P<.3   -
  b  M f icm eat kid ade 80w80 e                                                                   no dre   P=1.   -
  c  M f icm eat liv hpa 80w80 e                                                                   no dre   P=1.   -
  d  M f icm eat liv hpc 80w80 e                                                                   no dre   P=1.   -
  e  M f icm eat lun ade 80w80 e                                                                   no dre   P=1.   -
2948 M m icm eat kid mix 80w80 e                                 .   +   .                          1.36mg   P<.0005+
  a  M m icm eat kid adc 80w80 e                                                                    1.48mg   P<.0005
  b  M m icm eat lun adc 80w80 e                                                                    5.10mg   P<.3
  c  M m icm eat kid ade 80w80 e                                                                    24.1mg   P<.3
  d  M m icm eat liv hpa 80w80 e                                                                   no dre   P=1.
  e  M m icm eat liv hpc 80w80 e                                                                   no dre   P=1.
  f  M m icm eat liv hpb 80w80 e                                                                   no dre   P=1.
  g  M m icm eat lun ade 80w80 e                                                                   no dre   P=1.

MESTRANOL                      100ng..:..1ug....:..10.....:..100....:..1mg....:..10.....:..100....:..1g.....:..10
2949 M f cfl eat liv hct 78w78 er                                       .>                          6.27mg * P<.7   -
2950 M m cfl eat liv hct 78w78 er                                    .>                             1.81mg * P<.5   -
2951 M f crf eat mam tum 24m24 r                                .>                                 no dre   P=1.   -
2952 M f r3m eat mam tum 24m24 r                              .>                                   no dre   P=1.   -

METEPA                         100ng..:..1ug....:..10.....:..100....:..1mg....:..10.....:..100....:..1g.....:..10
2953 R m she gav --- lyk 60w60 e                                .    ±                              4.46mg * P<.02  +

METHAFURYLENE                  100ng..:..1ug....:..10.....:..100....:..1mg....:..10.....:..100....:..1g.....:..10
2954 R f f34 wat tyf mix 22m30 e                                                      .>           559.mg   P<.6   -
  a  R f f34 wat liv mix 22m30 e                                                                   no dre   P=1.   -
2955 R m f34 wat liv mix 22m27 e                                                   .>              no dre   P=1.   -

METHAPHENILENE                 100ng..:..1ug....:..10.....:..100....:..1mg....:..10.....:..100....:..1g.....:..10
2956 R f f34 wat tyf mix 19m30 e                                                      .>           487.mg   P<.6   -
  a  R f f34 wat liv mix 19m30 e                                                                   no dre   P=1.   -
2957 R m f34 wat liv mix 19m30 e                                                   .>              no dre   P=1.   -
  a  R m f34 wat tyf mix 19m30 e                                                                   no dre   P=1.   -

METHAPYRILENE.HCl              100ng..:..1ug....:..10.....:..100....:..1mg....:..10.....:..100....:..1g.....:..10
2958 H m syg gav liv nnd 58w61                                             .>                      150.mg   P<.4
  a  H m syg gav tba tum 58w61                                                                     no dre   P=1.
2959 R f f34 eat liv mix 26m31                                    .    +   .                        7.65mg * P<.0005+
  a  R f f34 eat liv nnd 26m31                                                                      11.8mg * P<.0005
```

	RefNum	LoConf	UpConf	Cntrl	1Dose	1Inc	2Dose	2Inc					Citation or Pathology
													Brkly Code

	RefNum	LoConf	UpConf	Cntrl	1Dose	1Inc	2Dose	2Inc	3Dose	3Inc	4Dose	4Inc	Citation or Pathology
MERCURIC CHLORIDE		7487-94-7											
2937	TR408	2.18mg	n.s.s.	23/50	3.53mg	32/50	(7.07mg	19/50)					
a	TR408	33.4mg	n.s.s.	3/50	3.53mg	1/50	7.07mg	2/50					liv:hpa,hpc,nnd.
b	TR408	12.4mg	n.s.s.	3/50	3.53mg	8/50	7.07mg	5/50					lun:a/a,a/c.
2938	TR408	24.7mg	n.s.s.	0/50	3.53mg	0/50	7.07mg	3/50					kid:rua,uac.
a	TR408	30.7mg	n.s.s.	0/50	3.53mg	0/50	7.07mg	2/50					
b	TR408	39.0mg	n.s.s.	0/50	3.53mg	0/50	7.07mg	1/50					
c	TR408	12.2mg	n.s.s.	28/50	3.53mg	20/50	7.07mg	17/50					
d	TR408	18.5mg	n.s.s.	7/50	3.53mg	4/50	7.07mg	6/50					liv:hpa,hpc,nnd.
e	TR408	18.4mg	n.s.s.	10/50	3.53mg	8/50	7.07mg	6/50					lun:a/a,a/c.
2939	1395	1.55mg	105.mg	3/47	1.00mg	11/41							Schroeder;jnut,105,452-458;1975
a	1395	3.36mg	n.s.s.	9/47	1.00mg	7/41							
b	1395	.973mg	n.s.s.	14/47	1.00mg	21/41							
c	1395	1.95mg	n.s.s.	4/47	1.00mg	9/41							
2940	1395	2.25mg	n.s.s.	5/38	.833mg	9/48							
a	1395	1.09mg	n.s.s.	11/38	.833mg	21/48							
b	1395	7.43mg	n.s.s.	2/38	.833mg	1/48							
2941	TR408	12.7mg	n.s.s.	0/50	1.76mg	0/50	3.51mg	2/50					
a	TR408	2.12mg	n.s.s.	46/50	1.76mg	42/50	3.51mg	39/50					
b	TR408	17.2mg	n.s.s.	0/50	1.76mg	1/50	3.51mg	0/50					liv:hpa,hpc,nnd.
2942	TR408	.437mg	2.88mg	45/50	1.75mg	37/50	3.51mg	39/50					tes:iab,ica. S
a	TR408	.770mg	6.42mg	21/50	1.75mg	18/50	3.51mg	23/50					amd:pbb,pob. S
b	TR408	.767mg	8.98mg	24/50	1.75mg	18/50	3.51mg	23/50					amd:pbb,phc,phm,pmb,pob. S
c	TR408	1.48mg	7.20mg	0/50	1.75mg	3/50	3.51mg	12/50					
d	TR408	2.14mg	68.4mg	2/50	1.75mg	6/50	3.51mg	6/50					thy:fca,fcc.
e	TR408	2.71mg	71.8mg	1/50	1.75mg	2/50	3.51mg	6/50					
f	TR408	1.65mg	n.s.s.	1/50	1.75mg	4/50	(3.51mg	0/50)					
g	TR408	.521mg	4.77mg	45/50	1.75mg	34/50	3.51mg	39/50					
h	TR408	2.91mg	n.s.s.	1/50	1.75mg	1/50	3.51mg	2/50					liv:hpa,hpc,nnd.
MERCURYMETHYL CHLORIDE		(methylmercuric chloride) 115-09-3											
2943	1961	10.7mg	n.s.s.	0/60	52.0ug	0/60	.260mg	1/60	1.30mg	1/60			Mitsumori;faat,14,179-190;1990
2944	1961	1.81mg	6.42mg	0/60	48.0ug	0/60	.240mg	1/60	1.20mg	16/60			
a	1961	2.26mg	9.65mg	0/60	48.0ug	0/60	.240mg	0/60	1.20mg	13/60			
b	1961	4.01mg	45.6mg	0/60	48.0ug	0/60	.240mg	1/60	1.20mg	5/60			
2945	1395	3.28mg	n.s.s.	9/47	.200mg	3/31	.253mg	3/40					Schroeder;jnut,105,452-458;1975
a	1395	.446mg	n.s.s.	14/47	.200mg	12/31	(.253mg	4/40)					
b	1395	4.64mg	n.s.s.	4/47	.200mg	2/31	.253mg	0/40					
2946	1395	1.57mg	n.s.s.	5/38	.167mg	4/25	.208mg	3/32					
a	1395	.727mg	n.s.s.	11/38	.167mg	9/25	.208mg	10/32					
b	1395	1.53mg	n.s.s.	2/38	.167mg	3/25	.208mg	2/32					
2947	2041	1.99mg	n.s.s.	5/50	1.30mg	7/50							Hirano;jjvs,50,886-893;1988
a	2041	4.25mg	n.s.s.	0/50	1.30mg	1/50							
b	2041	7.93mg	n.s.s.	1/50	1.30mg	0/50							
c	2041	7.93mg	n.s.s.	5/50	1.30mg	0/50							
d	2041	7.93mg	n.s.s.	1/50	1.30mg	0/50							
e	2041	5.70mg	n.s.s.	3/50	1.30mg	1/50							
2948	2041	.746mg	2.89mg	0/50	1.20mg	15/50							
a	2041	.795mg	3.23mg	0/50	1.20mg	14/50							
b	2041	1.37mg	n.s.s.	6/50	1.20mg	10/50							
c	2041	3.92mg	n.s.s.	0/50	1.20mg	1/50							
d	2041	3.33mg	n.s.s.	12/50	1.20mg	6/50							
e	2041	3.45mg	n.s.s.	7/50	1.20mg	4/50							
f	2041	7.32mg	n.s.s.	2/50	1.20mg	0/50							
g	2041	2.48mg	n.s.s.	7/50	1.20mg	6/50							
MESTRANOL		72-33-3											
2949	1453	.551mg	n.s.s.	1/40	5.00ug	0/40	30.0ug	1/40	60.0ug	1/39	.200mg	1/39	Barrows;jtxe,3,219-230;1977
2950	1453	.335mg	n.s.s.	2/40	5.00ug	4/40	30.0ug	4/40	60.0ug	0/40	.200mg	5/40	
2951	1174	60.3ug	n.s.s.	145/160	.130mg	30/34							Rudali;reec,16,425-429;1971
2952	1174	12.2ug	n.s.s.	12/15	13.0ug	26/40							
METEPA		57-39-6											
2953	179	1.35mg	n.s.s.	0/20	.625mg	0/20	2.50mg	3/20					Gaines;bwho,34,317-320;1966
METHAFURYLENE		531-06-6											
2954	1790	80.6mg	n.s.s.	1/21	29.5mg	2/20							Lijinsky;zkko,112,57-60;1986
a	1790	77.7mg	n.s.s.	6/21	29.5mg	4/20							
2955	1790	88.0mg	n.s.s.	4/19	23.4mg	1/20							
METHAPHENILENE		493-78-7											
2956	1790	70.3mg	n.s.s.	1/21	25.7mg	2/20							Lijinsky;zkko,112,57-60;1986
a	1790	83.9mg	n.s.s.	6/21	25.7mg	3/20							
2957	1790	66.5mg	n.s.s.	4/19	18.0mg	2/20							
a	1790	77.5mg	n.s.s.	2/19	18.0mg	1/20							
METHAPYRILENE.HCl		135-23-9											
2958	1572	24.4mg	n.s.s.	0/12	32.6mg	1/20							Lijinsky;jtxe,12,653-657;1983
a	1572	23.8mg	n.s.s.	2/12	32.6mg	2/20							
2959	1644	4.54mg	15.6mg	3/20	5.17mg	12/30	10.3mg	20/20					Lijinsky;fctx,22,27-30;1984
a	1644	6.62mg	35.1mg	3/20	5.17mg	12/30	10.3mg	15/20					

```
      Spe Strain  Site   Xpo+Xpt                                                                              TD50    2Tailpvl
          Sex  Route  Hist   Notes                                                                                    DR    AuOp
  b     R f f34  eat liv hpc 26m31                                                                            61.6mg / P<.004 +
2960 R m f34  eat liv bsa 73w73 ekr                                                <+                         noTD50   P<.009 +
  a     R m f34  eat liv esa 73w73 ekr                                                                        noTD50   P<.009 +
  b     R m f34  eat liv hpc 73w73 ekr                                                                        14.7mg   P<.02  +
  c     R m f34  eat liv hps 73w73 ekr                                                                        26.4mg   P<.08  +
  d     R m f34  eat liv thc 73w73 ekr                                                                        26.4mg   P<.08  +
2961 R m f34  eat liv mix 26m31                                         .  +    .                             7.70mg * P<.0005+
  a     R m f34  eat liv nnd 26m31                                                                            11.1mg * P<.003
  b     R m f34  eat liv hpc 26m31                                                                            77.0mg * P<.04  +
  c     R m f34  eat liv clc 26m31                                                                            400.mg * P<.3

METHIDATHION                           100ng..:..1ug....:..10......:..100...:..1mg.....:..10....:..100.....:..1g.....:.10
2962 M f cd1  eat liv hpc 23m23 e                                                    .>                       184.mg * P<.3   -
  a     M f cd1  eat liv mix 23m23 e                                                                          16.2gm * P<1.   -
  b     M f cd1  eat liv hpa 23m23 e                                                                          no dre   P=1.   -
2963 M m cd1  eat liv mix 23m23 es                                            .  +   .                        6.04mg * P<.0005+
  a     M m cd1  eat liv hpa 23m23 es                                                                         17.2mg * P<.0005+
  b     M m cd1  eat liv hpc 23m23 es                                                                         21.8mg * P<.0005+

METHIMAZOLE                            100ng..:..1ug....:..10......:..100...:..1mg.....:..10....:..100.....:..1g.....:.10
2964 R f hrl  eat thy fca 24m24 es                                           .  +   .                         1.57mg Z P<.0005+
  a     R f hrl  eat thy fdc 24m24 es                                                                         9.47mg Z P<.007 +
2965 R m hrl  eat thy fca 24m24 es                                       .  +   .                             .900mg Z P<.0005+
  a     R m hrl  eat thy fdc 24m24 es                                                                         37.6mg * P<.05  +

DL-METHIONINE                          100ng..:..1ug....:..10......:..100...:..1mg.....:..10....:..100.....:..1g.....:.10
2966 M m cen  eat liv hpc 52w52 r                                                                 .     ±     4.47gm   P<.1
  a     M m cen  eat liv mix 52w52 r                                                                          2.00gm   P<.2
  b     M m cen  eat liv hpa 52w52 r                                                                          4.16gm   P<.4
  c     M m cen  eat lun mix 52w52 r                                                                          no dre   P=1.
2967 R m f34  eat liv nnd 17m24 e                                                              .>             no dre   P=1.   -

METHOTREXATE                           100ng..:..1ug....:..10......:..100...:..1mg.....:..10....:..100.....:..1g.....:.10
2968 H f syg  eat liv hpa 23m24 as                                                  .>                        51.1mg * P<.2   -
  a     H f syg  eat liv mix 23m24 as                                                                         167.mg * P<.8
  b     H f syg  eat lun tum 23m24 as                                                                         no dre   P=1.
  c     H f syg  eat tba mix 23m24 as                                                                         no dre   P=1.
2969 H m syg  eat liv tum 25m27 as                                        .>                                  no dre   P=1.
  a     H m syg  eat lun tum 25m27 as                                                                         no dre   P=1.
  b     H m syg  eat tba mix 25m27 as                                                                         no dre   P=1.
2970 M f swi  eat liv hpc 28m28                                                      .>                       65.2mg * P<.4   -
  a     M f swi  eat lun mix 28m28                                                                            no dre   P=1.
  b     M f swi  eat tba mix 28m28                                                                            48.0mg * P<1.   -
2971 M m swi  eat liv hem 27m28 a                                              .>                             no dre   P=1.   -
  a     M m swi  eat lun mix 27m28 a                                                                          no dre   P=1.   -
  b     M m swi  eat tba mix 27m28 a                                                                          no dre   P=1.   -
2972 R m b46  ivj tba ben 12m24 es                                    .    ±                                  .409mg   P<.1   -
  a     R m b46  ivj tba mix 12m24 es                                                                         .330mg   P<.2   -
  b     R m b46  ivj tba mal 12m24 es                                                                         2.96mg   P<.8   -
2973 R f sda  ipj tba mal 24m24 e                                          .>                                 no dre   P=1.   -
2974 R m sda  ipj liv hae 24m24 e                                          .>                                 no dre   P=1.
  a     R m sda  ipj tba mal 24m24 e                                                                          no dre   P=1.

2-METHOXY-4-AMINOAZOBENZENE            100ng..:..1ug....:.10....:..100...:..1mg.....:..10....:..100.....:..1g....:.10
2975 M f b6c  eat liv tum 56w56 r                                                         .>                  no dre   P=1.   -
2976 M m b6c  eat liv tum 56w56 r                                                         .>                  no dre   P=1.   -

3-METHOXY-4-AMINOAZOBENZENE            100ng..:..1ug....:.10....:..100...:..1mg.....:..10....:..100.....:..1g....:.10
2977 M f b6c  eat liv hpa 56w56 r                                                     .  +       .            60.2mg * P<.005 +
2978 M m b6c  eat liv tum 56w56 r                                                         .>                  no dre   P=1.   -

2-METHOXY-3-AMINODIBENZOFURAN          100ng..:..1ug....:.10....:..100...:..1mg.....:..10....:..100.....:..1g....:.10
2979 R f fis  eat ubl mix 75w75 e                                                   .  +    .                 25.7mg   P<.002 +
  a     R f fis  eat ubl tcc 75w75 e                                                                          43.9mg   P<.02  +
  b     R f fis  eat auc sqc 75w75 e                                                                          43.9mg   P<.02  +
  c     R f fis  eat mgl adc 75w75 e                                                                          205.mg   P<.3   +
  d     R f fis  eat liv tum 75w75 e                                                                          no dre   P=1.
  e     R f fis  eat tba mal 75w75 e                                                                          33.1mg   P<.005
2980 R m fis  eat ubl mix 75w75 e                                              .   +       .                  26.4mg   P<.005 +
  a     R m fis  eat auc sqc 75w75 e                                                                          78.2mg   P<.09  +
  b     R m fis  eat ubl tcc 75w75 e                                                                          78.2mg   P<.09  +
  c     R m fis  eat liv tum 75w75 e                                                                          no dre   P=1.
  d     R m fis  eat tba mal 75w75 e                                                                          78.2mg   P<.09
2981 R f sda  eat ubl mix 75w75 e                                                   .      ±                  43.9mg   P<.02  +
  a     R f sda  eat auc sqc 75w75 e                                                                          61.9mg   P<.04  +
  b     R f sda  eat mgl adc 75w75 e                                                                          61.9mg   P<.04  +
  c     R f sda  eat ubl tcc 75w75 e                                                                          205.mg   P<.3   +
  d     R f sda  eat liv tum 75w75 e                                                                          no dre   P=1.
  e     R f sda  eat tba mal 75w75 e                                                                          25.7mg   P<.002
2982 R m sda  eat kid sqc 75w75 e                                                 .>                          164.mg   P<.3   +
  a     R m sda  eat ubl tpp 75w75 e                                                                          164.mg   P<.3   +
```

	RefNum	LoConf	UpConf	Cntrl	1Dose	1Inc	2Dose	2Inc	3Dose	3Inc	4Dose	4Inc	5Dose	5Inc	Citation or Pathology / Brkly Code
b	1644	25.1mg	363.mg	0/20	5.17mg	0/30	10.3mg	6/20							
2960	1618	n.s.s.	18.7mg	0/5	40.0mg	5/5									Ohshima;jnci,72,759-768;1984
a	1618	n.s.s.	18.7mg	0/5	40.0mg	5/5									
b	1618	4.11mg	n.s.s.	0/5	40.0mg	3/5									
c	1618	6.29mg	n.s.s.	0/5	40.0mg	2/5									
d	1618	6.29mg	n.s.s.	0/5	40.0mg	2/5									
2961	1644	4.27mg	22.2mg	5/20	4.14mg	13/30	8.27mg	18/20							Lijinsky;fctx,22,27-30;1984
a	1644	5.69mg	68.0mg	5/20	4.14mg	12/30	8.27mg	15/20							
b	1644	26.6mg	n.s.s.	0/20	4.14mg	1/30	8.27mg	3/20							
c	1644	65.1mg	n.s.s.	0/20	5.17mg	0/30	10.3mg	1/20							

METHIDATHION   950-37-8

	RefNum	LoConf	UpConf	Cntrl	1Dose	1Inc	2Dose	2Inc	3Dose	3Inc	4Dose	4Inc	5Dose	5Inc	Citation or Pathology / Brkly Code
2962	2130	45.1mg	n.s.s.	2/50	.390mg	3/50	1.30mg	2/50	6.50mg	2/50	13.0mg	5/50			Quest;rtxp,12,117-126;1990
a	2130	48.8mg	n.s.s.	5/50	.390mg	10/50	1.30mg	5/50	6.50mg	2/50	13.0mg	8/50			
b	2130	97.6mg	n.s.s.	3/50	.390mg	9/50	1.30mg	3/50	6.50mg	1/50	13.0mg	3/50			
2963	2130	3.92mg	10.6mg	9/46	.360mg	15/45	1.20mg	11/47	6.00mg	21/43	12.0mg	38/45			
a	2130	9.67mg	43.9mg	1/46	.360mg	9/45	1.20mg	7/47	6.00mg	8/43	12.0mg	21/45			
b	2130	11.4mg	83.6mg	8/46	.360mg	6/45	1.20mg	4/47	6.00mg	13/43	12.0mg	17/45			

METHIMAZOLE   60-56-0

	RefNum	LoConf	UpConf	Cntrl	1Dose	1Inc	2Dose	2Inc	3Dose	3Inc	4Dose	4Inc	5Dose	5Inc	Citation or Pathology / Brkly Code
2964	180	.834mg	3.46mg	0/50	.250mg	0/25	1.50mg	14/25	(9.00mg	12/25)					Owen;fctx,11,649-653;1973
a	180	2.86mg	150.mg	0/50	.250mg	0/25	1.50mg	3/25	(9.00mg	2/25)					
2965	180	.500mg	1.85mg	1/50	.200mg	1/25	1.20mg	17/25	(7.20mg	5/25)					
a	180	11.5mg	n.s.s.	1/50	.200mg	0/25	1.20mg	2/25	7.20mg	3/25					

DL-METHIONINE   59-51-8

	RefNum	LoConf	UpConf	Cntrl	1Dose	1Inc	2Dose	2Inc	3Dose	3Inc	4Dose	4Inc	5Dose	5Inc	Citation or Pathology / Brkly Code
2966	1992	1.10gm	n.s.s.	0/30	1.80gm	2/30									Fullerton;carc,11,1301-1305;1990
a	1992	625.mg	n.s.s.	2/30	1.80gm	6/30									
b	1992	854.mg	n.s.s.	2/30	1.80gm	4/30									
c	1992	1.86gm	n.s.s.	2/30	1.80gm	1/30									
2967	1803	1.15gm	n.s.s.	2/28	419.mg	2/28									Shivapurkar;carc,7,547-550;1986

METHOTREXATE   59-05-2

	RefNum	LoConf	UpConf	Cntrl	1Dose	1Inc	2Dose	2Inc	3Dose	3Inc	4Dose	4Inc	Citation or Pathology / Brkly Code
2968	1324	8.32mg	n.s.s.	0/49	.261mg	0/42	.523mg	0/42	1.05mg	1/39			Rustia;txap,26,329-338;1973
a	1324	9.14mg	n.s.s.	1/49	.261mg	0/42	.523mg	0/42	1.05mg	1/39			
b	1324	1.30mg	n.s.s.	0/49	.261mg	0/42	.523mg	0/42	1.05mg	0/39			
c	1324	6.01mg	n.s.s.	7/49	.261mg	4/42	.523mg	2/42	1.05mg	3/39			
2969	1324	1.38mg	n.s.s.	0/49	.230mg	0/42	.460mg	0/42	.920mg	0/40			
a	1324	1.38mg	n.s.s.	0/49	.230mg	0/42	.460mg	0/42	.920mg	0/40			
b	1324	12.8mg	n.s.s.	8/49	.230mg	1/42	.460mg	0/42	.920mg	1/40			
2970	1324	10.6mg	n.s.s.	0/70	.195mg	0/36	.325mg	0/36	.520mg	1/42	.650mg	0/48	
a	1324	3.57mg	n.s.s.	14/70	.195mg	8/36	.325mg	8/36	.520mg	6/42	.650mg	8/48	
b	1324	1.35mg	n.s.s.	34/70	.195mg	21/36	.325mg	13/36	.520mg	19/42	.650mg	26/48	
2971	1324	3.20mg	n.s.s.	6/70	.180mg	2/36	.300mg	6/36	.480mg	1/42	(.600mg	0/48)	
a	1324	3.45mg	n.s.s.	13/70	.180mg	3/36	.300mg	3/36	.480mg	9/42	.600mg	6/48	
b	1324	1.99mg	n.s.s.	29/70	.180mg	15/36	.300mg	17/36	.480mg	18/42	.600mg	15/48	
2972	1017	.111mg	n.s.s.	3/65	71.4ug	4/26							Schmahl;arzn,20,1461-1467;1970
a	1017	90.2ug	n.s.s.	7/65	71.4ug	6/26							
b	1017	.181mg	n.s.s.	4/65	71.4ug	2/26							
2973	1134	.406mg	n.s.s.	3/33	89.3ug	1/31							Schmahl;zkko,86,77-84;1976
2974	1134	.552mg	n.s.s.	1/36	89.3ug	0/30							
a	1134	.552mg	n.s.s.	1/36	89.3ug	0/30							

2-METHOXY-4-AMINOAZOBENZENE   80830-39-3

	RefNum	LoConf	UpConf	Cntrl	1Dose	1Inc	Citation or Pathology / Brkly Code
2975	1500	90.9mg	n.s.s.	0/13	117.mg	0/13	Watanabe;gann,73,136-140;1982
2976	1500	83.9mg	n.s.s.	0/13	108.mg	0/13	

3-METHOXY-4-AMINOAZOBENZENE   3544-23-8

	RefNum	LoConf	UpConf	Cntrl	1Dose	1Inc	2Dose	2Inc	Citation or Pathology / Brkly Code
2977	1500	25.8mg	388.mg	0/13	78.0mg	1/13	117.mg	6/13	Watanabe;gann,73,136-140;1982
2978	1500	27.7mg	n.s.s.	0/13	72.0mg	0/10	108.mg	0/12	

2-METHOXY-3-AMINODIBENZOFURAN   5834-17-3

	RefNum	LoConf	UpConf	Cntrl	1Dose	1Inc	Citation or Pathology / Brkly Code
2979	183n	10.2mg	115.mg	0/12	50.0mg	6/12	Radomski;jnci,39,1069-1080;1967
a	183n	15.0mg	n.s.s.	0/12	50.0mg	4/12	
b	183n	15.0mg	n.s.s.	0/12	50.0mg	4/12	
c	183n	33.3mg	n.s.s.	0/12	50.0mg	1/12	
d	183n	64.3mg	n.s.s.	0/12	50.0mg	0/12	
e	183n	12.3mg	265.mg	0/12	50.0mg	5/12	
2980	183n	9.84mg	212.mg	0/12	40.0mg	5/12	
a	183n	19.2mg	n.s.s.	0/12	40.0mg	2/12	
b	183n	19.2mg	n.s.s.	0/12	40.0mg	2/12	
c	183n	51.4mg	n.s.s.	0/12	40.0mg	0/12	
d	183n	19.2mg	n.s.s.	0/12	40.0mg	2/12	
2981	183m	15.0mg	n.s.s.	0/12	50.0mg	4/12	
a	183m	18.6mg	n.s.s.	0/12	50.0mg	3/12	
b	183m	18.6mg	n.s.s.	0/12	50.0mg	3/12	
c	183m	33.3mg	n.s.s.	0/12	50.0mg	1/12	
d	183m	64.3mg	n.s.s.	0/12	50.0mg	0/12	
e	183m	10.2mg	115.mg	0/12	50.0mg	6/12	
2982	183m	26.6mg	n.s.s.	0/12	40.0mg	1/12	
a	183m	26.6mg	n.s.s.	0/12	40.0mg	1/12	

	Spe	Strain	Site	Xpo+Xpt					TD50		2Tailpvl
	Sex	Route	Hist	Notes						DR	AuOp
b	R m	sda	eat	liv tum 75w75 e					no dre	P=1.	
c	R m	sda	eat	tba mal 75w75 e					78.2mg	P<.09	
2983	R f	wis	eat	ubl mix 75w75 e		.	±		43.9mg	P<.02	+
a	R f	wis	eat	mgl adc 75w75 e					97.7mg	P<.09	+
b	R f	wis	eat	ubl tcc 75w75 e					97.7mg	P<.09	+
c	R f	wis	eat	liv cye 75w75 e					205.mg	P<.3	
d	R f	wis	eat	tba mal 75w75 e					43.9mg	P<.02	
2984	R m	wis	eat	ubl mix 75w75 e		.	±		35.2mg	P<.02	+
a	R m	wis	eat	auc sqc 75w75 e					78.2mg	P<.09	+
b	R m	wis	eat	ubl tcc 75w75 e					78.2mg	P<.09	+
c	R m	wis	eat	kid sqc 75w75 e					164.mg	P<.3	+
d	R m	wis	eat	liv tum 75w75 e					no dre	P=1.	
e	R m	wis	eat	tba mal 75w75 e					26.4mg	P<.005	
2985	R m	wis	eat	ubl mix 84w84 e			<+		noTD50	P<.004	+
a	R m	wis	eat	tba mal 84w84 e					noTD50	P<.004	+
b	R m	wis	eat	tba ben 84w84 e					53.3mg	P<.07	

METHOXYCHLOR   100ng..:..1ug...:..10.....:..100....:..1mg....:..10.....:..100....:..1g.....:..10

	Spe	Strain	Site	Xpo+Xpt		TD50		2Tailpvl
2986	M f	b6c	eat	TBA MXB 78w92 v	:>	no dre	P=1.	-
a	M f	b6c	eat	liv MXB 78w92 v		no dre	P=1.	
b	M f	b6c	eat	lun MXB 78w92 v		no dre	P=1.	
2987	M m	b6c	eat	TBA MXB 78w91 v	:>	no dre	P=1.	-
a	M m	b6c	eat	liv MXB 78w91 v		no dre	P=1.	
b	M m	b6c	eat	lun MXB 78w91 v		no dre	P=1.	
2988	R f	osm	eat	TBA MXB 18m26 dv	:>	no dre	P=1.	-
a	R f	osm	eat	liv MXB 18m26 dv		no dre	P=1.	
2989	R f	osm	eat	liv tum 27m27	.>	no dre	P=1.	
a	R f	osm	eat	tba mix 27m27		no dre	P=1.	
2990	R f	osm	eat	liv tum 24m24	.>	no dre	P=1.	
a	R f	osm	eat	tba ben 24m24		no dre	P=1.	
b	R f	osm	eat	tba mal 24m24		no dre	P=1.	
2991	R m	osm	eat	TBA MXB 18m26 dv	:>	no dre	P=1.	-
a	R m	osm	eat	liv MXB 18m26 dv		no dre	P=1.	
2992	R m	osm	eat	liv tum 27m27	.>	no dre	P=1.	
a	R m	osm	eat	tba mix 27m27		234.mg	P<.08	
2993	R m	osm	eat	liv tum 24m24	.>	no dre	P=1.	-
a	R m	osm	eat	tba ben 24m24		31.8mg	P<.1	
b	R m	osm	eat	tba mal 24m24		no dre	P=1.	-

4-METHOXYPHENOL   100ng..:..1ug...:..10.....:..100....:..1mg....:..10.....:..100....:..1g.....:..10

	Spe	Strain	Site	Xpo+Xpt		TD50		2Tailpvl
2994	R m	f3d	eat	for pam 51w52	.>	1.46gm	P<.4	
a	R m	f3d	eat	stg tum 51w52		no dre	P=1.	
2995	R m	f3d	eat	for tum 51w52 er	.>	no dre	P=1.	
a	R m	f3d	eat	stg tum 51w52 er		no dre	P=1.	

METHOXYPHENYLACETIC ACID   100ng..:..1ug...:..10.....:..100....:..1mg....:..10.....:..100....:..1g.....:..10

	Spe	Strain	Site	Xpo+Xpt		TD50		2Tailpvl
2996	M f	b6a	orl	lun ade 76w76 evx	.>	236.mg	P<.4	-
a	M f	b6a	orl	liv hpt 76w76 evx		no dre	P=1.	-
b	M f	b6a	orl	tba mix 76w76 evx		227.mg	P<.5	
2997	M m	b6a	orl	lun ade 76w76 evx	.>	157.mg	P<.3	-
a	M m	b6a	orl	liv hpt 76w76 evx		3.62gm	P<1.	
b	M m	b6a	orl	tba mix 76w76 evx		139.mg	P<.4	
2998	M f	b6c	orl	lun ade 76w76 evx	.>	244.mg	P<.2	-
a	M f	b6c	orl	liv hpt 76w76 evx		no dre	P=1.	-
b	M f	b6c	orl	tba mix 76w76 evx		157.mg	P<.05	-
2999	M m	b6c	orl	--- rts 76w76 evx	. + .	82.1mg	P<.009	-
a	M m	b6c	orl	liv hpt 76w76 evx		227.mg	P<.2	
b	M m	b6c	orl	lun mix 76w76 evx		no dre	P=1.	-
c	M m	b6c	orl	tba mix 76w76 evx		54.2mg	P<.002	-

8-METHOXYPSORALEN   100ng..:..1ug...:..10.....:..100....:..1mg....:..10.....:..100....:..1g.....:..10

	Spe	Strain	Site	Xpo+Xpt		TD50		2Tailpvl
3000	M f	hra	eat	ski sqc 52w80 r	.>	48.9mg *	P<.3	-
a	M f	hra	eat	ski mix 52w80 r		56.8mg *	P<.4	-
b	M f	hra	eat	ski sqp 52w80 r		151.mg *	P<.7	-
3001	M m	hra	eat	ski sqp 52w80 r	.>	137.mg *	P<.6	-
a	M m	hra	eat	ski sqc 52w80 r		no dre	P=1.	-
b	M m	hra	eat	ski mix 52w80 r		no dre	P=1.	-
c	M m	hra	eat	ski sqn 52w80 r		no dre	P=1.	-
3002	R f	f34	gav	MXA sqp 24m24	: ±	#333.mg *	P<.02	-
a	R f	f34	gav	TBA MXB 24m24		45.6mg *	P<.09	
b	R f	f34	gav	liv MXB 24m24		5.14gm *	P<.9	
3003	R m	f34	gav	tes ict 24m24	: + :	14.6mg *	P<.0005	
a	R m	f34	gav	MXB MXB 24m24		27.3mg *	P<.0005	
b	R m	f34	gav	kid MXA 24m24		32.4mg *	P<.0005c	
c	R m	f34	gav	kid tla 24m24		43.2mg *	P<.0005c	
d	R m	f34	gav	lun MXA 24m24		57.1mg *	P<.003	
e	R m	f34	gav	lun a/a 24m24		57.1mg *	P<.003	e
f	R m	f34	gav	sub MXA 24m24		70.2mg *	P<.0005	
g	R m	f34	gav	sub fib 24m24		72.6mg *	P<.002	e
h	R m	f34	gav	kid uac 24m24		167.mg *	P<.008	c
i	R m	f34	gav	zym MXA 24m24		101.mg *	P<.02	c

	RefNum	LoConf	UpConf	Cntrl	1Dose	1Inc	2Dose	2Inc				Citation or Pathology
												Brkly Code
b	183m	51.4mg	n.s.s.	0/12	40.0mg	0/12						
c	183m	19.2mg	n.s.s.	0/12	40.0mg	2/12						
2983	183n	15.0mg	n.s.s.	0/12	50.0mg	4/12						
a	183n	23.9mg	n.s.s.	0/12	50.0mg	2/12						
b	183n	23.9mg	n.s.s.	0/12	50.0mg	2/12						
c	183n	33.3mg	n.s.s.	0/12	50.0mg	1/12						
d	183n	15.0mg	n.s.s.	0/12	50.0mg	4/12						
2984	183n	12.0mg	n.s.s.	0/12	40.0mg	4/12						
a	183n	19.2mg	n.s.s.	0/12	40.0mg	2/12						
b	183n	19.2mg	n.s.s.	0/12	40.0mg	2/12						
c	183n	26.6mg	n.s.s.	0/12	40.0mg	1/12						
d	183n	51.4mg	n.s.s.	0/12	40.0mg	0/12						
e	183n	9.84mg	212.mg	0/12	40.0mg	5/12						
2985	1420	n.s.s.	17.9mg	0/7	34.3mg	6/6						Hackmann;zkko,61,45-54;1956
a	1420	n.s.s.	17.9mg	0/7	34.3mg	6/6						
b	1420	13.0mg	n.s.s.	0/9	34.3mg	2/8						
METHOXYCHLOR	**72-43-5**											
2986	TR35	828.mg	n.s.s.	3/20	109.mg	6/50	217.mg	3/50				
a	TR35	1.41gm	n.s.s.	0/20	109.mg	1/50	217.mg	0/50				liv:hpa,hpc,nnd.
b	TR35	n.s.s.	n.s.s.	0/20	109.mg	0/50	217.mg	0/50				lun:a/a,a/c.
2987	TR35	466.mg	n.s.s.	3/20	178.mg	5/50	352.mg	9/50				
a	TR35	664.mg	n.s.s.	3/20	178.mg	3/50	352.mg	6/50				liv:hpa,hpc,nnd.
b	TR35	n.s.s.	n.s.s.	0/20	178.mg	0/50	352.mg	0/50				lun:a/a,a/c.
2988	TR35	63.8mg	n.s.s.	12/20	26.4mg	30/50	48.2mg	30/50				
a	TR35	329.mg	n.s.s.	1/20	26.4mg	1/50	48.2mg	2/50				liv:hpa,hpc,nnd.
2989	21	391.mg	n.s.s.	0/30	50.0mg	0/30						Deichmann;txap,11,88-103;1967
a	21	181.mg	n.s.s.	13/30	50.0mg	6/30						
2990	84a	24.7mg	n.s.s.	1/30	4.00mg	0/30						Radomski;txap,7,652-656;1965
a	84a	7.36mg	n.s.s.	6/30	4.00mg	6/30						
b	84a	8.82mg	n.s.s.	6/30	4.00mg	5/30						
2991	TR35	44.6mg	n.s.s.	11/20	12.6mg	23/50	23.8mg	21/50				
a	TR35	145.mg	n.s.s.	0/20	12.6mg	2/50	23.8mg	0/50				liv:hpa,hpc,nnd.
2992	21	313.mg	n.s.s.	0/30	40.0mg	0/30						Deichmann;txap,11,88-103;1967
a	21	76.3mg	n.s.s.	1/30	40.0mg	5/30						
2993	84a	19.8mg	n.s.s.	0/30	3.20mg	0/30						Radomski;txap,7,652-656;1965
a	84a	7.81mg	n.s.s.	0/30	3.20mg	2/30						
b	84a	19.8mg	n.s.s.	3/30	3.20mg	0/30						
4-METHOXYPHENOL	**150-76-5**											
2994	1950	238.mg	n.s.s.	0/10	588.mg	1/15						Hirose;canr,48,5310-5315;1988
a	1950	455.mg	n.s.s.	0/10	588.mg	0/15						
2995	1990	29.6mg	n.s.s.	0/10	98.1mg	0/11	196.mg	0/11	392.mg	0/11	785.mg	0/11	Wada;carc,11,1891-1894;1990
a	1990	29.6mg	n.s.s.	0/10	98.1mg	0/11	196.mg	0/11	392.mg	0/11	785.mg	0/11	
METHOXYPHENYLACETIC ACID	**1701-77-5**											
2996	1312	51.4mg	n.s.s.	1/17	78.4mg	3/18						Innes;ntis,1968/1969
a	1312	155.mg	n.s.s.	0/17	78.4mg	0/18						
b	1312	45.2mg	n.s.s.	2/17	78.4mg	4/18						
2997	1312	35.9mg	n.s.s.	2/18	73.0mg	4/16						
a	1312	n.s.s.	n.s.s.	1/18	73.0mg	1/16						
b	1312	31.5mg	n.s.s.	3/18	73.0mg	5/16						
2998	1312	59.8mg	n.s.s.	0/16	78.4mg	2/18						
a	1312	155.mg	n.s.s.	0/16	78.4mg	0/18						
b	1312	47.5mg	n.s.s.	0/16	78.4mg	3/18						
2999	1312	30.9mg	1.80gm	0/16	73.0mg	5/18						
a	1312	55.7mg	n.s.s.	0/16	73.0mg	2/18						
b	1312	145.mg	n.s.s.	0/16	73.0mg	0/18						
c	1312	23.1mg	221.mg	0/16	73.0mg	7/18						
8-METHOXYPSORALEN (8-MOP)	**298-81-7**											
3000	2032	14.8mg	n.s.s.	0/36	.836mg	0/36	1.95mg	2/36	7.43mg	1/36		Dunnick;faat,16,92-102;1991/pers.comm.
a	2032	10.8mg	n.s.s.	1/36	.836mg	2/36	1.95mg	3/36	7.43mg	3/36		
b	2032	15.4mg	n.s.s.	1/36	.836mg	2/36	1.95mg	1/36	7.43mg	2/36		
3001	2032	15.9mg	n.s.s.	0/36	.836mg	1/36	1.95mg	1/36	7.43mg	1/36		
a	2032	31.2mg	n.s.s.	0/36	.836mg	1/36	1.95mg	0/36	7.43mg	0/36		
b	2032	20.9mg	n.s.s.	1/36	.836mg	2/36	1.95mg	1/36	7.43mg	1/36		
c	2032	2.38mg	n.s.s.	1/36	.836mg	0/36	1.95mg	0/36	7.43mg	0/36		
3002	TR359	108.mg	n.s.s.	0/50	26.3mg	1/50	52.8mg	3/50				pal:sqp; ton:sqp. S
a	TR359	17.5mg	n.s.s.	46/50	26.3mg	43/50	52.8mg	37/50				
b	TR359	240.mg	n.s.s.	1/50	26.3mg	0/50	52.8mg	1/50				liv:hpa,hpc,nnd.
3003	TR359	8.08mg	43.1mg	38/50	26.3mg	44/50	52.8mg	43/50				S
a	TR359	15.5mg	59.4mg	2/50	26.3mg	16/50	52.8mg	14/50				kid:tla,uac; zym:can,sqc. C
b	TR359	17.6mg	73.7mg	1/50	26.3mg	12/50	52.8mg	11/50				kid:tla,uac.
c	TR359	22.2mg	120.mg	1/50	26.3mg	11/50	52.8mg	8/50				
d	TR359	25.5mg	397.mg	4/50	26.3mg	9/50	52.8mg	9/50				lun:a/a,a/c. S
e	TR359	25.5mg	397.mg	4/50	26.3mg	9/50	52.8mg	9/50				
f	TR359	32.1mg	265.mg	1/50	26.3mg	5/50	52.8mg	8/50				sub:fib,srn. S
g	TR359	32.6mg	316.mg	1/50	26.3mg	5/50	52.8mg	7/50				
h	TR359	52.4mg	3.38gm	0/50	26.3mg	1/50	52.8mg	3/50				
i	TR359	41.2mg	n.s.s.	1/50	26.3mg	7/50	52.8mg	4/50				zym:can,sqc.

```
      Spe Strain Site  Xpo+Xpt                                               TD50   2Tailpvl
        Sex  Route  Hist    Notes                                              DR     AuOp
  j   R m f34 gav pan ana 24m24                                             113.mg * P<.04
  k   R m f34 gav lun a/c 24m24                                             488.mg * P<.3
  l   R m f34 gav TBA MXB 24m24                                             16.1mg * P<.002
  m   R m f34 gav liv MXB 24m24                                             no dre  P=1.

Z-METHYL-O,N,N-AZOXYETHANE      100ng..:..1ug...:..10...:..100...:..1mg...:..10...:..100...:..1g...:..10
3004 R m f34 wat liv mix 30w95                          . + .              11.5mg \ P<.002 +
  a   R m f34 wat liv hpc 30w95                                            18.2mg \ P<.0005+
  b   R m f34 wat kid mix 30w95                                            123.mg * P<.0005+
  c   R m f34 wat liv hes 30w95                                            302.mg * P<.3
  d   R m f34 wat liv nnd 30w95                                            664.mg * P<.8

METHYL BROMIDE                  100ng..:..1ug...:..10...:..100...:..1mg...:..10...:..100...:..1g...:..10
3005 M f b6c inh TBA MXB 17m24 as                               :>         no dre  P=1.  -
  a   M f b6c inh liv MXB 17m24 as                                         no dre  P=1.  -
  b   M f b6c inh lun MXB 17m24 as                                         no dre  P=1.  -
3006 M m b6c inh TBA MXB 17m24 as                            :>            no dre  P=1.  -
  a   M m b6c inh liv MXB 17m24 as                                         no dre  P=1.  -
  b   M m b6c inh lun MXB 17m24 as                                         no dre  P=1.  -
3007 R f wsr inh liv nnd 30m30 e                              .>           2.24gm * P<.2  -
  a   R f wsr inh liv clc 30m30 e                                          no dre  P=1.  -
  b   R f wsr inh nof tum 30m30 e                                          no dre  P=1.  -
  c   R f wsr inh nre tum 30m30 e                                          no dre  P=1.  -
  d   R f wsr inh tba mal 30m30 e                                          no dre  P=1.  -
  e   R f wsr inh tba tum 30m30 e                                          no dre  P=1.  -
3008 R m wsr inh nof tum 30m30 e                          .>               no dre  P=1.  -
  a   R m wsr inh pit het 30m30 e                                          no dre  P=1.  -
  b   R m wsr inh nre fbs 30m30 e                                          no dre  P=1.  -
  c   R m wsr inh nre ost 30m30 e                                          no dre  P=1.  -
  d   R m wsr inh liv hpc 30m30 e                                          no dre  P=1.  -
  e   R m wsr inh liv nnd 30m30 e                                          no dre  P=1.  -
  f   R m wsr inh liv clc 30m30 e                                          no dre  P=1.  -
  g   R m wsr inh nre sqc 30m30 e                                          no dre  P=1.  -
  h   R m wsr inh tba tum 30m30 e                                          3.82mg \ P<.6  -
  i   R m wsr inh tba mal 30m30 e                                          no dre  P=1.  -

METHYL CARBAMATE                100ng..:..1ug...:..10...:..100...:..1mg...:..10...:..100...:..1g...:..10
3009 M f b6c gav TBA MXB 24m24                                  :>         no dre  P=1.  -
  a   M f b6c gav liv MXB 24m24                                            4.70gm * P<.4  -
  b   M f b6c gav lun MXB 24m24                                            no dre  P=1.
3010 M m b6c gav TBA MXB 24m24                               :>            5.22gm * P<.8  -
  a   M m b6c gav liv MXB 24m24                                            5.08gm * P<.7  -
  b   M m b6c gav lun MXB 24m24                                            no dre  P=1.
3011 R f f34 gav liv MXA 24m24                               :  +          :839.mg * P<.006 c
  a   R f f34 gav liv nnd 24m24                                            979.mg * P<.02
  b   R f f34 gav TBA MXB 24m24                                            no dre  P=1.
  c   R f f34 gav liv MXB 24m24                                            839.mg * P<.006
3012 R m f34 gav liv MXA 24m24                               :>            2.03gm / P<.6  c
  a   R m f34 gav TBA MXB 24m24                                            no dre  P=1.
  b   R m f34 gav liv MXB 24m24                                            2.03gm / P<.6

METHYL CARBAZATE                100ng..:..1ug...:..10...:..100...:..1mg...:..10...:..100...:..1g...:..10
3013 R f wis eat tba tum 24m24                        .>                   no dre  P=1.  -
3014 R m wis eat tba tum 24m24                        .>                   no dre  P=1.  -

METHYL CLOFENAPATE              100ng..:..1ug...:..10...:..100...:..1mg...:..10...:..100...:..1g...:..10
3015 R m f34 eat liv hpc 75w75 er                     <+                   noTD50   P<.0005+

1-METHYL-1,4-DIHYDRO-7-[2-(5-NITROFURYL)VINYL]-4-OXO-1,8-NAPHTHYRIDINE-3-CARBOXYLATE,    POTASSIUM ...:..1g...:..10
3016 M b icr eat lun ade 54w54 s                        . + .              8.03mg   P<.0005+
  a   M b icr eat for sqc 54w54 s                                          10.3mg   P<.002 +
  b   M b icr eat lun car 54w54 s                                          21.9mg   P<.03  +
  c   M b icr eat thm lyk 54w54 s                                          26.5mg   P<.05  +

3'-METHYL-4-DIMETHYLAMINOAZOBENZENE      ..:..1ug...:..10...:..100...:..1mg...:..10...:..100...:..1g...:..10
3017 P b rhe eat liv hpc  5y24 w                        :        ±         67.1mg   P<.06

  a   P b rhe eat --- hem  5y24 w                                          48.9mg   P<.3
  b   P b rhe eat tba mix  5y24 w                                          109.mg   P<.7
  c   P b rhe eat tba mal  5y24 w                                          260.mg   P<.8
  d   P b rhe eat tba ben  5y24 w                                          464.mg   P<1.
3018 R m aap gav liv hpc 52w52 bekr                   .>                   3.64mg   P<.4   +
3019 R m f3d eat liv hpc 24m24 e                        . + .              3.28mg * P<.0005+
  a   R m f3d eat liv thc 24m24 e                                          4.94mg * P<.0005
  b   R m f3d eat liv ghc 24m24 e                                          10.1mg Z P<.0005
  c   R m f3d eat liv clc 24m24 e                                          46.7mg * P<.0005

N-METHYL-N,4-DINITROSOANILINE   100ng..:..1ug...:..10...:..100...:..1mg...:..10...:..100...:..1g...:..10
3020 R m cbr ipj pec mix 26w86 e                      .>                   1.30mg   P<.2   +
  a   R m cbr ipj liv hpt 26w86 e                                          1.50mg   P<.3   -
```

	RefNum	LoConf	UpConf	Cntrl	1Dose	1Inc	2Dose	2Inc		Citation or Pathology	Brkly Code
j	TR359	39.2mg	n.s.s.	2/50	26.3mg	3/50	52.8mg	4/50			s
k	TR359	79.4mg	n.s.s.	0/50	26.3mg	1/50	52.8mg	0/50			
l	TR359	8.23mg	76.0mg	44/50	26.3mg	41/50	52.8mg	41/50			
m	TR359	n.s.s.	n.s.s.	0/50	26.3mg	0/50	52.8mg	0/50		liv:hpa,hpc,nnd.	

Z-METHYL-O,N,N-AZOXYETHANE   57497-34-4

	RefNum	LoConf	UpConf	Cntrl	1Dose	1Inc	2Dose	2Inc		Citation or Pathology	Brkly Code
3004	1723	5.32mg	46.7mg	2/20	16.2mg	12/20	(72.6mg	10/20)		Lijinsky;canr,45,76-79;1985	
a	1723	8.13mg	57.5mg	0/20	16.2mg	8/20	(72.6mg	6/20)			
b	1723	52.9mg	405.mg	0/20	16.2mg	0/20	72.6mg	7/20			
c	1723	83.6mg	n.s.s.	0/20	16.2mg	2/20	72.6mg	2/20			
d	1723	67.4mg	n.s.s.	2/20	16.2mg	6/20	72.6mg	4/20			

METHYL BROMIDE   74-83-9

	RefNum	LoConf	UpConf	Cntrl	1Dose	1Inc	2Dose	2Inc	3Dose	3Inc	Citation or Pathology
3005	TR385	65.4mg	n.s.s.	27/50	12.0mg	29/50	39.7mg	27/50	23.4mg	27/60	
a	TR385	149.mg	n.s.s.	10/50	12.0mg	11/50	39.7mg	8/50	23.4mg	6/60	liv:hpa,hpc,nnd.
b	TR385	202.mg	n.s.s.	4/50	12.0mg	4/50	39.7mg	1/50	23.4mg	7/60	lun:a/a,a/c.
3006	TR385	27.7mg	n.s.s.	37/50	10.0mg	41/49	33.1mg	38/50	19.5mg	16/70	
a	TR385	45.0mg	n.s.s.	28/50	10.0mg	31/49	33.1mg	26/50	19.5mg	9/70	liv:hpa,hpc,nnd.
b	TR385	60.1mg	n.s.s.	14/50	10.0mg	14/49	33.1mg	14/50	19.5mg	5/70	lun:a/a,a/c.
3007	2017	365.mg	n.s.s.	0/60	.872mg	0/60	8.72mg	0/59	26.2mg	1/60	Reuzel;fctx,29,31-39;1991/pers.comm.
a	2017	611.mg	n.s.s.	0/60	.872mg	1/60	8.72mg	0/59	26.2mg	0/60	
b	2017	14.2mg	n.s.s.	0/58	.872mg	0/58	8.72mg	0/59	26.2mg	0/59	
c	2017	14.2mg	n.s.s.	0/58	.872mg	0/58	8.72mg	0/59	26.2mg	0/59	
d	2017	141.mg	n.s.s.	17/60	.872mg	11/60	8.72mg	13/60	26.2mg	12/60	
e	2017	56.5mg	n.s.s.	52/60	.872mg	54/60	8.72mg	52/60	26.2mg	46/60	
3008	2017	8.08mg	n.s.s.	0/46	.610mg	0/48	6.10mg	0/49	18.3mg	0/48	
a	2017	19.8mg	n.s.s.	10/47	.610mg	21/50	6.10mg	14/50	(18.3mg	7/44)	
b	2017	205.mg	n.s.s.	0/46	.610mg	0/48	6.10mg	1/49	18.3mg	0/48	
c	2017	341.mg	n.s.s.	0/46	.610mg	1/48	6.10mg	0/49	18.3mg	0/48	
d	2017	346.mg	n.s.s.	0/48	.610mg	1/48	6.10mg	0/49	18.3mg	0/49	
e	2017	346.mg	n.s.s.	0/48	.610mg	1/48	6.10mg	0/49	18.3mg	0/49	
f	2017	346.mg	n.s.s.	0/48	.610mg	1/48	6.10mg	0/49	18.3mg	0/49	
g	2017	353.mg	n.s.s.	1/46	.610mg	2/48	6.10mg	0/49	18.3mg	0/48	
h	2017	.807mg	n.s.s.	31/48	.610mg	70/100	(18.3mg	21/49)			
i	2017	154.mg	n.s.s.	11/48	.610mg	16/50	6.10mg	14/50	18.3mg	5/49	

METHYL CARBAMATE   598-55-0

	RefNum	LoConf	UpConf	Cntrl	1Dose	1Inc	2Dose	2Inc		Citation or Pathology	Brkly Code
3009	TR328	700.mg	n.s.s.	32/50	354.mg	28/50	707.mg	27/50			
a	TR328	1.20gm	n.s.s.	4/50	354.mg	7/50	707.mg	6/50		liv:hpa,hpc,nnd.	
b	TR328	2.30gm	n.s.s.	7/50	354.mg	5/50	707.mg	4/50		lun:a/a,a/c.	
3010	TR328	495.mg	n.s.s.	27/50	354.mg	35/50	707.mg	28/50			
a	TR328	780.mg	n.s.s.	14/50	354.mg	17/50	707.mg	16/50		liv:hpa,hpc,nnd.	
b	TR328	1.61gm	n.s.s.	11/50	354.mg	8/50	707.mg	8/50		lun:a/a,a/c.	
3011	TR328	342.mg	8.57gm	0/50	70.7mg	0/50	142.mg	6/50		liv:hpc,nnd.	
a	TR328	372.mg	n.s.s.	0/50	70.7mg	0/50	142.mg	5/50			s
b	TR328	146.mg	n.s.s.	47/50	70.7mg	43/50	142.mg	42/50			
c	TR328	342.mg	8.57gm	0/50	70.7mg	0/50	142.mg	6/50		liv:hpa,hpc,nnd.	
3012	TR328	357.mg	n.s.s.	4/50	70.7mg	0/50	142.mg	7/50		liv:hpc,nnd.	
a	TR328	56.6mg	n.s.s.	47/50	70.7mg	45/50	(142.mg	40/50)			
b	TR328	357.mg	n.s.s.	4/50	70.7mg	0/50	142.mg	7/50		liv:hpa,hpc,nnd.	

METHYL CARBAZATE   6294-89-9

	RefNum	LoConf	UpConf	Cntrl	1Dose	1Inc	2Dose	2Inc	3Dose	3Inc	Citation or Pathology
3013	1389	8.26mg	n.s.s.	16/24	2.50mg	11/24	5.00mg	15/24	10.0mg	14/24	Truhaut;txcy,22,219-221;1981
3014	1389	17.5mg	n.s.s.	7/24	2.50mg	6/24	5.00mg	8/24	10.0mg	5/24	

METHYL CLOFENAPATE   21340-68-1

	RefNum	LoConf	UpConf	Cntrl	1Dose	1Inc		Citation or Pathology
3015	1478	n.s.s.	9.17mg	0/10	40.0mg	14/14		Reddy;canr,42,259-266;1982

1-METHYL-1,4-DIHYDRO-7-[2-(5-NITROFURYL)VINYL]-4-OXO-1,8-NAPHTHYRIDINE-3-CARBOXYLATE,   POTASSIUM   ---

	RefNum	LoConf	UpConf	Cntrl	1Dose	1Inc		Citation or Pathology
3016	211	4.40mg	19.9mg	0/30	12.5mg	15/60		Matsuzaki;gann,66,259-267;1975
a	211	5.33mg	34.8mg	0/30	12.5mg	12/60		
b	211	8.94mg	n.s.s.	0/30	12.5mg	6/60		
c	211	10.1mg	n.s.s.	0/30	12.5mg	5/60		

3'-METHYL-4-DIMETHYLAMINOAZOBENZENE   55-80-1

	RefNum	LoConf	UpConf	Cntrl	1Dose	1Inc	2Dose	2Inc	3Dose	3Inc	Citation or Pathology
3017	2001	10.9mg	n.s.s.	0/23	18.5mg	1/4					Adamson;ossc,129-156; 1982/Thorgeirsson 1994/Dalgard 1991/Thorgeirsson&Seiber pers.comm.
a	2001	3.35mg	n.s.s.	2/110	18.5mg	1/12					
b	2001	7.82mg	n.s.s.	11/110	18.5mg	2/12					
c	2001	14.8mg	n.s.s.	5/76	18.5mg	1/8					
d	2001	8.09mg	n.s.s.	7/110	18.5mg	1/12					
3018	1710	.590mg	n.s.s.	0/5	2.50mg	1/9					Styles;carc,6,21-28;1985
3019	1756	2.04mg	5.42mg	0/30	.960mg	4/27	4.80mg	21/23	24.0mg	26/28	Ogiso;txpy,13,257-265;1985
a	1756	3.06mg	8.18mg	0/30	.960mg	3/27	4.80mg	14/23	24.0mg	26/28	
b	1756	4.56mg	32.2mg	0/30	.960mg	1/27	4.80mg	7/23	(24.0mg	0/28)	
c	1756	22.6mg	120.mg	0/30	.960mg	0/27	4.80mg	0/23	24.0mg	10/28	

N-METHYL-N,4-DINITROSOANILINE   99-80-9

	RefNum	LoConf	UpConf	Cntrl	1Dose	1Inc		Citation or Pathology
3020	1258	.319mg	n.s.s.	0/10	.429mg	2/14		Boyland;ejca,4,233-239;1968
a	1258	.244mg	n.s.s.	0/7	.429mg	1/8		

```
         Spe Strain  Site    Xpo+Xpt                                                  TD50      2Tailpvl
            Sex   Route   Hist      Notes                                                DR     AuOp

N-METHYL-N-FORMYLHYDRAZINE          100ng..:..1ug....:..10.....:..100....:..1mg...:..10.....:..100..:..1g.....:..10
3021 H f syg wat --- mhs 24m24 e                                        . + .                     4.71mg    P<.0005+
  a  H f syg wat liv mix 24m24 e                                                                  11.3mg    P<.0005+
  b  H f syg wat liv lcc 24m24 e                                                                  25.1mg    P<.0005
  c  H f syg wat liv hpt 24m24 e                                                                  28.3mg    P<.0005
  d  H f syg wat gal mix 24m24 e                                                                  29.1mg    P<.5   +
  e  H f syg wat lun tum 24m24 e                                                                  no dre    P=1.
  f  H f syg wat tba mix 24m24 e                                                                  6.04mg    P<.0005
3022 H m syg wat liv mix 24m24 e                                            . + .                 7.55mg    P<.0005+
  a  H m syg wat --- mhs 24m24 e                                                                  12.7mg    P<.002 +
  b  H m syg wat liv hpt 24m24 e                                                                  13.3mg    P<.0005
  c  H m syg wat bil mix 24m24 e                                                                  19.7mg    P<.007 +
  d  H m syg wat gal mix 24m24 e                                                                  25.3mg    P<.002 +
  e  H m syg wat liv lcc 24m24 e                                                                  30.0mg    P<.0005
  f  H m syg wat gal ppa 24m24 e                                                                  42.5mg    P<.01
  g  H m syg wat lun tum 24m24 e                                                                  no dre    P=1.
  h  H m syg wat tba mix 24m24 e                                                                  4.89mg    P<.0005
3023 M f swa wat lun mix 65w73 aes                                     . + .                      6.44mg \  P<.0005+
  a  M f swa wat liv mix 65w73 aes                                                                10.4mg *  P<.0005+
  b  M f swa wat liv hpt 65w73 aes                                                                19.5mg *  P<.0005+
  c  M f swa wat liv lcc 65w73 aes                                                                19.9mg \  P<.0005+
  d  M f swa wat gal ade 65w73 aes                                                                31.5mg *  P<.008 +
3024 M f swa wat lun mix 23m23 aes                                .+ .                            .921mg *  P<.0005+
  a  M f swa wat blv mix 23m23 aes                                                                6.04mg *  P<.0005+
  b  M f swa wat gal mix 23m23 aes                                                                57.9mg *  P<.02
  c  M f swa wat liv bhp 23m23 aes                                                                24.7mg *  P<.2
3025 M f swa wat lun mix 23m24 aes                                  . + .                         .745mg *  P<.0005+
  a  M f swa wat lun adc 23m24 aes                                                                1.12mg *  P<.0005
  b  M f swa wat lun ade 23m24 aes                                                                1.50mg *  P<.002
  c  M f swa wat liv hpt 23m24 aes                                                                9.23mg *  P<.2   -
  d  M f swa wat liv agm 23m24 aes                                                                23.8mg *  P<.5   -
3026 M f swa wat lun mix 92w92 es                                       . + .                     2.16mg    P<.0005+
  a  M f swa wat liv mix 92w92 es                                                                 8.66mg    P<.0005+
  b  M f swa wat blv mix 92w92 es                                                                 12.2mg    P<.0005+
  c  M f swa wat liv bhp 92w92 es                                                                 17.5mg    P<.0005
  d  M f swa wat liv lcc 92w92 es                                                                 22.4mg    P<.0005
  e  M f swa wat gal mix 92w92 es                                                                 31.3mg    P<.003 +
  f  M f swa wat gal ade 92w92 es                                                                 39.7mg    P<.006
3027 M m swa wat liv mix 65w72 aes                                          . + .                 16.8mg \  P<.0005+
  a  M m swa wat bil mix 65w72 aes                                                                36.2mg *  P<.003 +
  b  M m swa wat gal ade 65w72 aes                                                                38.0mg \  P<.002 +
  c  M m swa wat lun mix 65w72 aes                                                                14.2mg \  P<.02  +
3028 M m swa wat lun mix 22m23 ae                                    . + .                        1.72mg *  P<.0005+
  a  M m swa wat blv mix 22m23 ae                                                                 3.47mg *  P<.0005+
  b  M m swa wat liv mix 22m23 ae                                                                 5.47mg *  P<.0005+
  c  M m swa wat gal mix 22m23 ae                                                                 9.10mg *  P<.0005+
  d  M m swa wat gal ade 22m23 ae                                                                 9.88mg *  P<.0005
3029 M m swa wat lun mix 23m24 aes                                  . + .                         .865mg *  P<.0005+
  a  M m swa wat lun ade 23m24 aes                                                                1.30mg *  P<.002
  b  M m swa wat stg pla 23m24 aes                                                                3.43mg \  P<.01  -
  c  M m swa wat lun adc 23m24 aes                                                                2.48mg \  P<.03
  d  M m swa wat sub fbs 23m24 aes                                                                2.60mg \  P<.02  -
  e  M m swa wat for sqp 23m24 aes                                                                11.2mg *  P<.02  -
  f  M m swa wat liv mix 23m24 aes                                                                no dre    P=1.
3030 M m swa wat liv mix 82w82 es                                    . + .                        1.03mg    P<.0005+
  a  M m swa wat lun mix 82w82 es                                                                 1.87mg    P<.0005+
  b  M m swa wat bil cho 82w82 es                                                                 12.4mg    P<.0005+
  c  M m swa wat gal ade 82w82 es                                                                 15.2mg    P<.0005+
  d  M m swa wat sev ade 82w82 es                                                                 20.7mg    P<.009 -

METHYL HESPERIDIN                  100ng..:..1ug....:..10.....:..100....:..1mg...:..10.....:..100...:..1g.....:..10
3031 M f b6c eat liv hpc 22m24 e                                                                 194.gm *  P<.6   -
  a  M f b6c eat liv hnd 22m24 e                                                                 379.gm *  P<.8   -
  b  M f b6c eat liv hes 22m24 e                                                                 887.gm *  P<.9   -
  c  M f b6c eat liv hem 22m24 e                                                                 no dre    P=1.
  d  M f b6c eat lun ade 22m24 e                                                                 no dre    P=1.  -
3032 M m b6c eat liv hpc 22m24 e                                                                 146.gm *  P<.9   -
  a  M m b6c eat liv hnd 22m24 e                                                                 416.gm *  P<1.
  b  M m b6c eat liv hem 22m24 e                                                                 no dre    P=1.
  c  M m b6c eat lun adc 22m24 e                                                                 no dre    P=1.
  d  M m b6c eat lun ade 22m24 e                                                                 no dre    P=1.  -

METHYL LINOLEATE HYDROPEROXIDE     100ng..:..1ug....:..10.....:..100....:..1mg...:..10.....:..100....:..1g.....:..10
3033 R m wis gav git mix 30w87 e                                           .>                    no dre    P=1.
  a  R m wis gav liv tum 30w87 e                                                                 no dre    P=1.

METHYL LINOLEATE, NATIVE           100ng..:..1ug....:..10.....:..100....:..1mg...:..10.....:..100....:..1g.....:..10
3034 R m wis gav git mix 30w87 e                                           .>                    no dre    P=1.
  a  R m wis gav liv tum 30w87 e                                                                 no dre    P=1.  -
```

	RefNum	LoConf	UpConf	Cntrl	1Dose	1Inc	2Dose	2Inc	Citation or Pathology
									Brkly Code
N-METHYL-N-FORMYLHYDRAZINE			758-17-8						
3021	716	2.73mg	8.73mg	0/13	10.6mg	24/31			Toth;zkko,93,109-121;1979
a	716	6.50mg	21.9mg	0/79	10.6mg	19/41			
b	716	12.2mg	64.5mg	0/79	10.6mg	10/41			
c	716	13.3mg	77.2mg	0/79	10.6mg	9/41			
d	716	8.75mg	n.s.s.	0/1	10.6mg	3/14			
e	716	92.8mg	n.s.s.	0/100	10.6mg	0/44			
f	716	3.63mg	11.3mg	13/100	10.6mg	32/44			
3022	716	4.57mg	13.6mg	0/65	9.36mg	24/43			
a	716	6.13mg	41.1mg	0/16	9.36mg	10/26			
b	716	7.35mg	27.4mg	0/65	9.36mg	16/43			
c	716	8.45mg	218.mg	0/16	9.36mg	7/26			
d	716	11.4mg	89.1mg	0/34	9.36mg	8/37			
e	716	13.5mg	88.4mg	0/65	9.36mg	8/43			
f	716	16.1mg	1.95gm	0/34	9.36mg	5/37			
g	716	85.4mg	n.s.s.	0/88	9.36mg	0/46			
h	716	3.04mg	8.58mg	7/88	9.36mg	34/46			
3023	1110	3.77mg	13.3mg	15/100	15.6mg	30/48	(31.2mg	9/37)	Toth;jnci,60,201-204;1978
a	1110	6.44mg	18.8mg	0/33	15.6mg	22/43	31.2mg	3/11	
b	1110	10.7mg	47.2mg	0/33	15.6mg	12/43	31.2mg	3/11	
c	1110	9.67mg	59.0mg	0/33	15.6mg	10/43	(31.2mg	0/11)	
d	1110	11.9mg	510.mg	0/29	15.6mg	4/23	31.2mg	1/5	
3024	1265	.636mg	1.37mg	15/99	2.00mg	39/49	4.00mg	47/50	Toth;nplm,27,25-31;1980
a	1265	3.49mg	14.5mg	8/95	2.00mg	10/44	4.00mg	20/48	
b	1265	17.5mg	n.s.s.	0/82	2.00mg	0/44	4.00mg	3/48	
c	1265	4.02mg	n.s.s.	0/32	2.00mg	1/10	4.00mg	0/5	
3025	1266	.444mg	1.77mg	29/94	.500mg	31/48	1.00mg	32/48	Toth;myco,78,11-16;1982a
a	1266	.665mg	2.72mg	14/94	.500mg	23/48	1.00mg	22/48	
b	1266	.779mg	7.82mg	20/94	.500mg	23/48	1.00mg	21/48	
c	1266	2.27mg	n.s.s.	0/50	.500mg	2/23	1.00mg	0/17	
d	1266	4.03mg	n.s.s.	3/77	.500mg	1/45	1.00mg	3/38	
3026	1314	1.32mg	3.73mg	15/99	7.80mg	43/49			Toth;myco,68,121-128;1979
a	1314	4.95mg	17.1mg	0/68	7.80mg	18/47			
b	1314	6.17mg	39.7mg	8/99	7.80mg	17/49			
c	1314	8.50mg	45.0mg	0/68	7.80mg	10/47			
d	1314	10.1mg	66.6mg	0/68	7.80mg	8/47			
e	1314	11.9mg	164.mg	0/63	7.80mg	5/40			
f	1314	13.7mg	449.mg	0/63	7.80mg	4/40			
3027	1110	8.19mg	48.2mg	1/81	13.0mg	11/47	(26.0mg	3/26)	Toth;jnci,60,201-204;1978
a	1110	13.8mg	203.mg	0/75	13.0mg	5/29	26.0mg	0/8	
b	1110	14.4mg	186.mg	0/81	13.0mg	5/47	(26.0mg	0/26)	
c	1110	6.00mg	n.s.s.	22/98	13.0mg	20/47	(26.0mg	4/26)	
3028	1265	1.12mg	3.09mg	22/98	1.67mg	36/49	3.33mg	31/46	Toth;nplm,27,25-31;1980
a	1265	1.64mg	14.5mg	5/78	1.67mg	13/42	(3.33mg	7/36)	
b	1265	3.10mg	12.2mg	2/78	1.67mg	6/42	3.33mg	14/36	
c	1265	4.90mg	20.0mg	0/90	1.67mg	6/46	3.33mg	8/43	
d	1265	5.21mg	22.8mg	0/90	1.67mg	6/46	3.33mg	7/43	
3029	1266	.488mg	2.69mg	19/83	.417mg	27/45	.833mg	24/47	Toth;myco,78,11-16;1982a
a	1266	.686mg	5.95mg	13/83	.417mg	20/45	.833mg	18/47	
b	1266	1.04mg	215.mg	0/80	.417mg	3/39	(.833mg	0/43)	
c	1266	1.08mg	n.s.s.	9/83	.417mg	14/45	.833mg	11/47	
d	1266	.916mg	n.s.s.	1/83	.417mg	5/45	(.833mg	0/47)	
e	1266	3.38mg	n.s.s.	0/80	.417mg	0/39	.833mg	3/43	
f	1266	3.97mg	n.s.s.	5/80	.417mg	3/39	.833mg	2/43	
3030	1314	.527mg	1.99mg	1/69	6.50mg	28/30			Toth;myco,68,121-128;1979
a	1314	1.07mg	3.78mg	22/88	6.50mg	34/41			
b	1314	5.05mg	44.5mg	0/69	6.50mg	6/30			
c	1314	5.75mg	64.6mg	0/69	6.50mg	5/30			
d	1314	6.27mg	618.mg	0/51	6.50mg	3/24			
METHYL HESPERIDIN			11013-97-1						
3031	1987	29.5gm	n.s.s.	1/50	1.50gm	1/50	6.00gm	2/50	Kurata;fctx,28,613-618;1990
a	1987	28.8gm	n.s.s.	3/50	1.50gm	1/50	6.00gm	3/50	
b	1987	45.7gm	n.s.s.	1/50	1.50gm	0/50	6.00gm	1/50	
c	1987	65.1gm	n.s.s.	2/50	1.50gm	1/50	6.00gm	0/50	
d	1987	45.5gm	n.s.s.	3/50	1.50gm	2/50	6.00gm	1/50	
3032	1987	10.8gm	n.s.s.	10/50	1.38gm	14/50	5.54gm	12/50	
a	1987	9.74gm	n.s.s.	17/50	1.38gm	16/50	5.54gm	17/50	
b	1987	32.1gm	n.s.s.	3/50	1.38gm	2/50	5.54gm	2/50	
c	1987	45.2gm	n.s.s.	5/50	1.38gm	3/50	5.54gm	1/50	
d	1987	32.6gm	n.s.s.	6/50	1.38gm	2/50	5.54gm	3/50	
METHYL LINOLEATE HYDROPEROXIDE			27323-65-5						
3033	1475	41.6mg	n.s.s.	0/30	9.61mg	0/30			Arffmann;jnci,67,1071-1075;1981
a	1475	41.6mg	n.s.s.	0/30	9.61mg	0/30			
METHYL LINOLEATE, NATIVE			---						
3034	1475	42.0mg	n.s.s.	0/30	9.71mg	0/30			Arffmann;jnci,67,1071-1075;1981
a	1475	42.0mg	n.s.s.	0/30	9.71mg	0/30			

```
      Spe Strain  Site    Xpo+Xpt                                                    TD50    2Tailpvl
         Sex  Route  Hist     Notes                                                   DR      AuOp

METHYL METHACRYLATE            100ng..:..1ug...:..10....:..100...:..1mg...:..10....:..100...:..1g....:..10
3035 M f b6c inh TBA MXB 24m24                                           :>           no dre  P=1.    -
 a   M f b6c inh liv MXB 24m24                                                        no dre  P=1.
 b   M f b6c inh lun MXB 24m24                                                        4.01gm \ P<.3
3036 M m b6c inh TBA MXB 24m24                                    :>                  no dre  P=1.    -
 a   M m b6c inh liv MXB 24m24                                                        no dre  P=1.
 b   M m b6c inh lun MXB 24m24                                                        no dre  P=1.
3037 R f f34 inh TBA MXB 24m24                              :>                        500.mg  * P<.7  -
 a   R f f34 inh liv MXB 24m24                                                        2.14gm  * P<.1
3038 R m f34 inh TBA MXB 24m24                           :>                           no dre  P=1.    -
 a   R m f34 inh liv MXB 24m24                                                        1.27gm  * P<.2

METHYL METHANESULFONATE        100ng..:..1ug...:..10....:..100...:..1mg...:..10....:..100...:..1g....:..10
3039 M m rfm wat lun ade 24m24 e                                    . +  .           31.8mg  P<.0005+
 a   M m rfm wat thm lym 24m24 e                                                      185.mg  P<.02 +
 b   M m rfm wat liv tum 24m24 e                                                      3.98gm  P<.9  -

N-METHYL-N'-NITRO-N-NITROSOGUANIDINE ..:..1ug...:..10....:..100...:..1mg...:..10....:..100...:..1g....:..10
3040 P b rhe eat liv hpa 23y23 Ww                      :      ±                       3.03mg  P<.06

 a   P b rhe eat tba ben 23y23 Ww                                                     no dre  P=1.
 b   P b rhe eat tba mal 23y23 Ww                                                     no dre  P=1.
 c   P b rhe eat tba mix 23y23 Ww                                                     no dre  P=1.
3041 R b alb wat for pam 78w78 r                             .     ±                  22.9mg  P<.02
 a   R b alb wat stg adc 78w78 r                                                      22.9mg  P<.02
3042 R m alb wat gam adc 34w52 er                   .   +   .                         .403mg  P<.003 +
3043 R m bfm wat stg ivc 52w52 er                              .>                     no dre  P=1.    -
 a   R m bfm wat stg pvc 52w52 er                                                     no dre  P=1.    -
3044 R f f34 eat sto car 24m24 es                       . +  .                        1.44mg  * P<.0005+
 a   R f f34 eat liv hpc 24m24 es                                                     17.8mg  * P<.0005
3045 R m f34 wat stg mix 12m29                          . +  .                        1.42mg  P<.0005+
 a   R m f34 wat stg ade 12m29                                                        2.38mg  P<.003 +
 b   R m f34 wat stg adc 12m29                                                        5.22mg  P<.04 +
 c   R m f34 wat stg sar 12m29                                                        16.5mg  P<.3  +
 d   R m f34 wat liv nnd 12m29                                                        no dre  P=1.
3046 R m f34 wat stg mix 12m24                       . +  .                           .592mg  P<.0005+
 a   R m f34 wat stg ade 12m24                                                        1.65mg  P<.002 +
 b   R m f34 wat stg sar 12m24                                                        2.48mg  P<.007 +
 c   R m f34 wat stg adc 12m24                                                        3.19mg  P<.02 +
 d   R m f34 wat liv nnd 12m24                                                        no dre  P=1.
3047 R m f34 eat sto car 24m24 es                       . +.                          .724mg  * P<.0005+
 a   R m f34 eat liv hpc 24m24 es                                                     no dre  P=1.
3048 R m sda wat stg adc 30w52 er                         . ±                         1.61mg  P<.03 +
3049 R m sda wat stg adc 52w52 er                         . +  .                      1.12mg  P<.0005+
3050 R f wis wat bil cye 35w60 er                   . +   .                           .178mg  P<.0005
 a   R f wis wat stg cnd 35w60 er                                                     2.38mg  P<.2  +
3051 R f wis wat duo adc 32w57 r                            .>                        6.38mg  P<.5  +
 a   R f wis wat stg adc 32w57 r                                                      6.38mg  P<.5  +
3052 R m wis wat stg mix 77w77 e                       . +   .                        .693mg  P<.0005+
 a   R m wis wat stg ade 77w77 e                                                      1.13mg  P<.003 +
 b   R m wis wat stg adc 77w77 e                                                      3.06mg  P<.06 +
 c   R m wis wat liv hpa 77w77 e                                                      14.3mg  P<.4
3053 R m wis wat stg adc 30w65 er                     . +    .                        .523mg  P<.002 +
3054 R m wis wat stg cnd 35w60 er                          .>                         2.09mg  P<.2  +
 a   R m wis wat bil cye 35w60 er                                                     no dre  P=1.
3055 R m wis wat git mix 32w87 e                        . + .                         .581mg  * P<.0005+
 a   R m wis wat liv tum 32w87 e                                                      no dre  P=1.
3056 R m wis wat git adc 28w58 r                      . + .                           .810mg  P<.0005+
 a   R m wis wat smi adc 28w58 r                                                      1.21mg  P<.003
 b   R m wis wat stg adc 28w58 r                                                      4.31mg  P<.1
3057 R m wis wat duo tum 30w54 er                        . ±                          3.44mg  P<.1  +
3058 R m wis wat git mix 26w52 er                        . ±                          1.60mg  P<.04 +
 a   R m wis wat duo tum 26w52 er                                                     2.21mg  P<.07 +
 b   R m wis wat stg tum 26w52 er                                                     7.06mg  P<.3  +
3059 R m wis wat git mix 26w52 er                       . ±                           1.92mg  P<.04 +
 a   R m wis wat stg mix 26w52 er                                                     2.28mg  P<.06 +
 b   R m wis wat duo tum 26w52 er                                                     7.28mg  P<.3
3060 R m wis wat duo adc 32w57 r                            .>                        2.72mg  P<.4  +
3061 R m wis wat duo adc 75w75 e                       . + .                          .910mg  P<.0005+
 a   R m wis wat gam adc 75w75 e                                                      5.31mg  P<.02 +
 b   R m wis wat eso sqc 75w75 e                                                      11.0mg  P<.1  +
 c   R m wis wat liv tum 75w75 e                                                      no dre  P=1.
 d   R m wis wat tba mix 75w75 e                                                      no dre  P=1.
3062 R m wmf wat stg ivc 52w52 er                          .>                         6.75mg  P<.2
3063 R m wsr wat stg ivc 52w52 er                      .+  .                          1.03mg  P<.0005+
 a   R m wsr wat stg pvc 52w52 er                                                     2.72mg  P<.0005+

2-METHYL-1-NITROANTHRAQUINONE  100ng..:..1ug...:..10....:..100...:..1mg...:..10....:..100...:..1g....:..10
3064 M f b6c eat sub hes 49w49                            : (+) :                     1.88mg  * P<.0005c
 a   M f b6c eat TBA MXB 49w49                                                        1.76mg  * P<.0005
 b   M f b6c eat liv MXB 49w49                                                        no dre  P=1.
```

	RefNum	LoConf	UpConf	Cntrl	1Dose	1Inc	2Dose	2Inc	Citation or Pathology
									Brkly Code

METHYL METHACRYLATE   80-62-6

	RefNum	LoConf	UpConf	Cntrl	1Dose	1Inc	2Dose	2Inc	Citation or Pathology
3035	TR314	1.08gm	n.s.s.	34/50	631.mg	19/50	(1.26gm	19/50)	
a	TR314	1.82gm	n.s.s.	7/50	631.mg	4/50	(1.26gm	2/50)	liv:hpa,hpc,nnd.
b	TR314	1.04gm	n.s.s.	2/50	631.mg	5/50	(1.26gm	0/50)	lun:a/a,a/c.
3036	TR314	1.32gm	n.s.s.	30/50	526.mg	17/50	(1.05gm	17/50)	
a	TR314	2.07gm	n.s.s.	16/50	526.mg	7/50	(1.05gm	7/50)	liv:hpa,hpc,nnd.
b	TR314	7.86gm	n.s.s.	11/50	526.mg	1/50	1.05gm	4/50	lun:a/a,a/c.
3037	TR314	78.2mg	n.s.s.	40/50	75.1mg	46/50	150.mg	44/50	
a	TR314	526.mg	n.s.s.	0/50	75.1mg	0/50	150.mg	2/50	liv:hpa,hpc,nnd.
3038	TR314	139.mg	n.s.s.	44/50	105.mg	46/50	210.mg	45/50	
a	TR314	477.mg	n.s.s.	0/50	105.mg	3/50	210.mg	2/50	liv:hpa,hpc,nnd.

METHYL METHANESULFONATE   (MMS) 66-27-3

	RefNum	LoConf	UpConf	Cntrl	1Dose	1Inc	2Dose	2Inc	Citation or Pathology
3039	195	16.0mg	113.mg	63/162	33.3mg	33/47			Clapp;scie,161,913-914;1968
a	195	67.5mg	n.s.s.	6/162	33.3mg	7/47			
b	195	160.mg	n.s.s.	6/162	33.3mg	2/47			

N-METHYL-N'-NITRO-N-NITROSOGUANIDINE   (MNNG) 70-25-7

	RefNum	LoConf	UpConf	Cntrl	1Dose	1Inc	2Dose	2Inc	Citation or Pathology
3040	2001	.493mg	n.s.s.	0/23	.964mg	1/4			Adamson;ossc,129-156;
									1982/Thorgeirsson 1994/Dalgard 1991/Thorgeirsson&Seiber pers.comm.
a	2001	.880mg	n.s.s.	7/110	.964mg	1/18			
b	2001	1.97mg	n.s.s.	5/76	.964mg	0/18			
c	2001	1.14mg	n.s.s.	11/110	.964mg	1/18			
3041	1571	7.87mg	n.s.s.	0/23	13.2mg	4/20			Tsung-Hsien;jnci,70,1067-1069;1983
a	1571	7.87mg	n.s.s.	0/23	13.2mg	4/20			
3042	1778	.160mg	1.90mg	0/6	2.83mg	7/10			Gurkalo;bexb,101,833-837;1986
3043	196	10.7mg	n.s.s.	0/20	4.15mg	0/50			Bralow;onco,27,168-180;1973
a	196	10.7mg	n.s.s.	0/20	4.15mg	0/50			
3044	1954	.961mg	2.25mg	1/119	1.00mg	8/24	2.00mg	16/24	4.00mg 20/24	Fears;txih,4,221-255;1988
a	1954	7.25mg	64.3mg	0/120	1.00mg	0/24	2.00mg	3/24	4.00mg 3/24	
3045	1611m	.658mg	4.00mg	0/20	.857mg	9/20			Lijinsky;canr,44,447-449;1984
a	1611m	.963mg	12.4mg	0/20	.857mg	6/20			
b	1611m	1.58mg	n.s.s.	0/20	.857mg	3/20			
c	1611m	2.69mg	n.s.s.	0/20	.857mg	1/20			
d	1611m	3.94mg	n.s.s.	5/20	.857mg	1/20			
3046	1611n	.302mg	1.33mg	0/20	1.02mg	14/20			
a	1611n	.708mg	6.27mg	0/20	1.02mg	7/20			
b	1611n	.936mg	28.6mg	0/20	1.02mg	5/20			
c	1611n	1.10mg	n.s.s.	0/20	1.02mg	4/20			
d	1611n	3.31mg	n.s.s.	5/20	1.02mg	1/20			
3047	1954	.486mg	1.11mg	0/119	.800mg	14/24	1.60mg	18/23	3.20mg 22/24	Fears;txih,4,221-255;1988
a	1954	2.19mg	n.s.s.	1/120	.800mg	0/24	1.60mg	0/23	3.20mg 0/24	
3048	2141m	.551mg	n.s.s.	0/10	3.46mg	4/13			Basso;zkko,118,441-446;1992
3049	2141n	.508mg	3.53mg	0/10	6.00mg	9/15			
3050	1105	73.1ug	.411mg	0/5	2.33mg	19/20			Tahara;zkko,100,1-12;1981
a	1105	.821mg	n.s.s.	0/5	2.33mg	4/20			
3051	1726	1.04mg	n.s.s.	0/5	1.59mg	1/20			Yasui;canr,45,4763-4767;1985
a	1726	1.04mg	n.s.s.	0/5	1.59mg	1/20			
3052	199	.289mg	2.05mg	0/6	3.30mg	10/12			Sugimura;canr,30,455-465;1970
a	199	.480mg	5.31mg	0/6	3.30mg	8/12			
b	199	1.04mg	n.s.s.	0/6	3.30mg	4/12			
c	199	2.32mg	n.s.s.	0/6	3.30mg	1/12			
3053	435	.186mg	2.54mg	0/10	1.92mg	5/8			Matsukura;jnci,61,141-143;1978
3054	1105	.510mg	n.s.s.	0/5	2.04mg	2/10			Tahara;zkko,100,1-12;1981
a	1105	1.40mg	n.s.s.	0/5	2.04mg	0/10			
3055	1475	.366mg	1.00mg	0/30	.366mg	10/30	1.52mg	20/30	Arffmann;jnci,67,1071-1075;1981
a	1475	1.28mg	n.s.s.	0/30	.366mg	0/30	1.52mg	0/30	
3056	1678	.444mg	1.92mg	0/12	2.90mg	16/30			Domellof;ajsu,142,551-554;1981
a	1678	.617mg	4.60mg	0/12	2.90mg	12/30			
b	1678	1.49mg	n.s.s.	0/12	2.90mg	4/30			
3057	1724m	.839mg	n.s.s.	0/8	4.68mg	2/9			Morishita;clet,17,347-352;1983/pers.comm.
3058	1724n	.549mg	n.s.s.	0/12	2.50mg	4/17			
a	1724n	.665mg	n.s.s.	0/12	2.50mg	3/17			
b	1724n	1.15mg	n.s.s.	0/12	2.50mg	1/17			
3059	1724o	.827mg	n.s.s.	0/12	2.50mg	7/35			
a	1724o	.927mg	n.s.s.	0/12	2.50mg	6/35			
b	1724o	1.79mg	n.s.s.	0/12	2.50mg	2/35			
3060	1726	.668mg	n.s.s.	0/5	1.39mg	2/20			Yasui;canr,45,4763-4767;1985
3061	1822	.505mg	1.85mg	0/30	2.13mg	17/30			Fujii;nutc,9,185-193;1987
a	1822	1.83mg	n.s.s.	0/30	2.13mg	4/30			
b	1822	2.71mg	n.s.s.	0/30	2.13mg	2/30			
c	1822	6.86mg	n.s.s.	0/30	2.13mg	0/30			
d	1822	.764mg	n.s.s.	20/30	2.13mg	20/30			
3062	196	1.66mg	n.s.s.	0/16	4.15mg	2/20			Bralow;onco,27,168-180;1973
3063	196	.684mg	1.63mg	0/40	4.15mg	37/74			
a	196	1.54mg	5.91mg	0/40	4.15mg	17/74			

2-METHYL-1-NITROANTHRAQUINONE   129-15-7

	RefNum	LoConf	UpConf	Cntrl	1Dose	1Inc	2Dose	2Inc	Citation or Pathology
3064	TR29	1.12mg	3.17mg	0/50	32.0mg	29/50	64.0mg	38/50	
a	TR29	1.07mg	2.85mg	10/50	32.0mg	34/50	64.0mg	44/50	
b	TR29	n.s.s.	n.s.s.	1/50	32.0mg	0/50	64.0mg	0/50	liv:hpa,hpc,nnd.

```
          Spe Strain  Site    Xpo+Xpt                                                          TD50     2Tailpvl
              Sex  Route  Hist    Notes                                                             DR       AuOp
      c   M f b6c eat lun MXB 49w49                                                             97.3mg * P<.06
3065  M m b6c eat sub hes 46w46                               : (+):                           1.34mg / P<.0005c
   a  M m b6c eat TBA MXB 46w46                                                                1.02mg / P<.0005
   b  M m b6c eat liv MXB 46w46                                                                no dre   P=1.
   c  M m b6c eat lun MXB 46w46                                                                45.5mg * P<.08
3066  R f f34 eat sub fib 18m25 s                                   :  +    :                  131.mg / P<.0005a
   a  R f f34 eat ubl MXA 18m25 s                                                              224.mg * P<.01  a
   b  R f f34 eat pit ade 18m25 s                                                              63.4mg * P<.04
   c  R f f34 eat liv MXA 18m25 s                                                              324.mg * P<.02  c
   d  R f f34 eat sto sqc 18m25 s                                                              507.mg * P<.03  a
   e  R f f34 eat liv hpc 18m25 s                                                              547.mg * P<.04  c
   f  R f f34 eat TBA MXB 18m25 s                                                              47.1mg * P<.08
   g  R f f34 eat liv MXB 18m25 s                                                              324.mg * P<.02
3067  R m f34 eat sub fib 18m25                                   : + :                        25.8mg / P<.0005a
   a  R m f34 eat liv MXA 18m25                                                                48.8mg * P<.0005c
   b  R m f34 eat liv hpc 18m25                                                                89.2mg * P<.005 c
   c  R m f34 eat liv nnd 18m25                                                                133.mg * P<.004
   d  R m f34 eat TBA MXB 18m25                                                                31.9mg * P<.05
   e  R m f34 eat liv MXB 18m25                                                                48.8mg * P<.0005


4-METHYL-1-[(5-NITROFURFURYLIDENE)AMINO]-2-IMIDAZOLIDINONE...:..1 _00....:..1mg....:....10.......:...100......:...1g......:..10
3068  R f sda eat mgl adc 46w66 e                                        .  +   .             5.34mg   P<.0005+
   a  R f sda eat liv tum 46w66 e                                                             no dre   P=1.
   b  R f sda eat tba mix 46w66 e                                                             5.59mg   P<.0005


4-(4-N-METHYL-N-NITROSAMINOSTYRYL)QUINOLINE.1_ug....:..10....:..100.....:..1mg.....:....10.......:...100......:...1g......:..10
3069  R f fis eat liv hnd 34w52                                  .  +   .                     .468mg   P<.0005+
3070  R m fis eat liv hnd 34w52                                  .   ±                        1.38mg   P<.04  +
   a  R m fis eat liv car 34w52                                                               4.37mg   P<.3   +


N-METHYL-N-NITROSOBENZAMIDE          100ng..:..1ug....:...10....:...100....:..1mg....:....10.......:...100......:...1g......:..10
3071  R f mrw wat liv tum 12m30                                             .>               no dre   P=1.   -
   a  R f mrw wat tba mix 12m30                                                               11.2mg   P<.4   +
3072  R m mrw wat for mix 12m30                                      .  +   .                 3.23mg   P<.0005+
   a  R m mrw wat for sqp 12m30                                                               7.73mg   P<.002
   b  R m mrw wat for sqc 12m30                                                               9.66mg   P<.003
   c  R m mrw wat liv tum 12m30                                                               no dre   P=1.
   d  R m mrw wat tba mix 12m30                                                               2.49mg   P<.02


N-(N-METHYL-N-NITROSOCARBAMOYL)-L-ORNITHINE.1_ug....:...10....:...100.....:..1mg.....:....10.......:...100......:...1g......:..10
3073  R f wal ipj kid epn 26w52 es                               .  +   .                    .633mg   P<.0005+
   a  R f wal ipj kid mnp 26w52 es                                                            .633mg   P<.0005+
   b  R f wal ipj ski neo 26w52 es                                                            1.42mg   P<.002 +
   c  R f wal ipj pan aca 26w52 es                                                            11.0mg   P<.3   +
   d  R f wal ipj mam mix 26w52 es                                                            no dre   P=1.   +
3074  R m wal ipj kid epn 26w52 es                               .  ±                        1.04mg   P<.02  +
   a  R m wal ipj ski neo 26w52 es                                                            1.04mg   P<.02  +
   b  R m wal ipj kid mnp 26w52 es                                                            1.57mg   P<.06  +
   c  R m wal ipj pan mix 26w52 es                                                            5.25mg   P<.3   +
   d  R m wal ipj pan adc 26w52 es                                                            11.0mg   P<.5   +
   e  R m wal ipj zym car 26w52 es                                                            11.8mg   P<.5   +


R(-)-2-METHYL-N-NITROSOPIPERIDINE   100ng..:..1ug....:...10....:...100....:..1mg....:....10.......:...100......:...1g......:..10
3075  R b sda wat nol epd 24m24                                  .  +   .                     22.1mg   P<.0005+
   a  R b sda wat liv hpa 24m24                                                               108.mg   P<.04
   b  R b sda wat tba mal 24m24                                                               20.4mg   P<.0005+


S(+)-2-METHYL-N-NITROSOPIPERIDINE   100ng..:..1ug....:...10....:...100....:..1mg....:....10.......:...100......:...1g......:..10
3076  R b sda wat mix epd 24m24                                     .  +   .                  34.5mg   P<.0005+
   a  R b sda wat liv mix 24m24                                                               49.4mg   P<.003
   b  R b sda wat liv lcc 24m24                                                               61.2mg   P<.007
   c  R b sda wat tba mal 24m24                                                               13.2mg   P<.0005+


METHYL 12-OXO-trans-10-OCTADECENOATE  ..:..1ug....:...10....:...100....:..1mg....:....10.......:...100......:...1g......:..10
3077  M b stm eat --- mly 42w84 e                                             .>             2.94gm   P<.7   -


METHYL PARATHION                     100ng..:..1ug....:...10....:...100....:..1mg....:....10.......:...100......:...1g......:..10
3078  M f b6c eat TBA MXB 24m24                                         :>                    no dre   P=1.   -
   a  M f b6c eat liv MXB 24m24                                                               no dre   P=1.
   b  M f b6c eat lun MXB 24m24                                                               125.mg * P<.5
3079  M m b6c eat TBA MXB 24m24 v                                          :>                 no dre   P=1.   -
   a  M m b6c eat liv MXB 24m24 v                                                             no dre   P=1.
   b  M m b6c eat lun MXB 24m24 v                                                             58.1mg * P<.5
3080  R f f34 eat TBA MXB 24m24 s                                    :>                       no dre   P=1.   -
   a  R f f34 eat liv MXB 24m24 s                                                             no dre   P=1.
3081  R m f34 eat TBA MXB 24m24                                      :>                       no dre   P=1.   -
   a  R m f34 eat liv MXB 24m24                                                               no dre   P=1.


N-METHYL-2-PYRROLIDONE               100ng..:..1ug....:...10....:...100....:..1mg....:....10.......:...100......:...1g......:..10
3082  R f cdr inh liv tum 24m24 e                                               .>           no dre   P=1.   -
3083  R m cdr inh liv tum 24m24 e                                               .>           no dre   P=1.   -
```

	RefNum	LoConf	UpConf	Cntrl	1Dose	1Inc	2Dose	2Inc	Citation or Pathology	Brkly Code
c	TR29	15.9mg	n.s.s.	3/50	32.0mg	0/50	64.0mg	1/50		lun:a/a,a/c.
3065	TR29	.780mg	2.31mg	0/50	29.5mg	25/50	59.0mg	36/50		
a	TR29	.621mg	1.65mg	22/50	29.5mg	42/50	59.0mg	46/50		
b	TR29	n.s.s.	n.s.s.	8/50	29.5mg	0/50	59.0mg	0/50	liv:hpa,hpc,nnd.	
c	TR29	7.41mg	n.s.s.	10/50	29.5mg	1/50	59.0mg	0/50	lun:a/a,a/c.	
3066	TR29	66.1mg	403.mg	1/50	21.0mg	0/50	43.0mg	13/50		
a	TR29	96.4mg	28.3gm	0/50	21.0mg	3/50	43.0mg	4/50	ubl:ppn,srn,tpp.	s
b	TR29	27.4mg	n.s.s.	17/50	21.0mg	26/50	43.0mg	25/50		
c	TR29	123.mg	n.s.s.	0/50	21.0mg	1/50	43.0mg	4/50	liv:hpc,nnd.	
d	TR29	154.mg	n.s.s.	0/50	21.0mg	0/50	43.0mg	3/50		
e	TR29	165.mg	n.s.s.	0/50	21.0mg	0/50	43.0mg	3/50		
f	TR29	18.6mg	n.s.s.	38/50	21.0mg	43/50	43.0mg	44/50		
g	TR29	123.mg	n.s.s.	0/50	21.0mg	1/50	43.0mg	4/50	liv:hpa,hpc,nnd.	
3067	TR29	16.8mg	47.4mg	3/49	16.8mg	10/50	34.4mg	34/50		
a	TR29	28.0mg	130.mg	1/49	16.8mg	7/50	34.4mg	15/50	liv:hpc,nnd.	
b	TR29	43.8mg	879.mg	1/49	16.8mg	5/50	34.4mg	9/50		
c	TR29	60.2mg	779.mg	0/49	16.8mg	2/50	34.4mg	6/50		s
d	TR29	13.6mg	n.s.s.	32/49	16.8mg	41/50	34.4mg	47/50		
e	TR29	28.0mg	130.mg	1/49	16.8mg	7/50	34.4mg	15/50	liv:hpa,hpc,nnd.	

4-METHYL-1-[(5-NITROFURFURYLIDENE)AMINO]-2-IMIDAZOLIDINONE   21638-36-8

	RefNum	LoConf	UpConf	Cntrl	1Dose	1Inc	Citation or Pathology
3068	200a	3.04mg	10.5mg	0/25	17.4mg	19/32	Cohen;jnci,51,403-417;1973
a	200a	46.3mg	n.s.s.	0/25	17.4mg	0/32	
b	200a	3.11mg	12.4mg	1/25	17.4mg	19/32	

4-(4-N-METHYL-N-NITROSAMINOSTYRYL)QUINOLINE   16699-10-8

	RefNum	LoConf	UpConf	Cntrl	1Dose	1Inc	Citation or Pathology
3069	1162	.217mg	1.32mg	0/20	1.63mg	9/20	Yamamoto;jnci,51,1313-1315;1973
3070	1162	.416mg	n.s.s.	0/20	1.31mg	3/20	
a	1162	.711mg	n.s.s.	0/20	1.31mg	1/20	

N-METHYL-N-NITROSOBENZAMIDE   63412-06-6

	RefNum	LoConf	UpConf	Cntrl	1Dose	1Inc	Citation or Pathology
3071	1246	22.0mg	n.s.s.	0/25	3.59mg	0/19	Bulay;jnci,62,1523-1528;1979
a	1246	2.67mg	n.s.s.	12/25	3.59mg	12/19	
3072	1246	1.55mg	8.10mg	0/22	3.14mg	11/17	
a	1246	3.12mg	31.1mg	0/22	3.14mg	6/17	
b	1246	3.64mg	56.4mg	0/22	3.14mg	5/17	
c	1246	17.2mg	n.s.s.	0/22	3.14mg	0/17	
d	1246	.856mg	n.s.s.	12/22	3.14mg	15/17	

N-(N-METHYL-N-NITROSOCARBAMOYL)-L-ORNITHINE   (nitrosourea amino acid) 63642-17-1

	RefNum	LoConf	UpConf	Cntrl	1Dose	1Inc	Citation or Pathology
3073	1317	.271mg	1.81mg	0/14	5.12mg	9/12	Longnecker;jept,4,117-129;1980
a	1317	.271mg	1.81mg	0/14	5.12mg	9/12	
b	1317	.565mg	6.13mg	0/14	5.12mg	6/13	
c	1317	1.78mg	n.s.s.	0/14	5.12mg	1/13	
d	1317	1.88mg	n.s.s.	5/14	5.12mg	3/15	
3074	1317	.452mg	n.s.s.	0/4	5.12mg	8/14	
a	1317	.452mg	n.s.s.	0/4	5.12mg	8/14	
b	1317	.628mg	n.s.s.	0/4	5.12mg	6/14	
c	1317	1.29mg	n.s.s.	0/4	5.12mg	2/13	
d	1317	1.78mg	n.s.s.	0/4	5.12mg	1/13	
e	1317	1.93mg	n.s.s.	0/4	5.12mg	1/14	

R(-)-2-METHYL-N-NITROSOPIPERIDINE   14026-03-0

	RefNum	LoConf	UpConf	Cntrl	1Dose	1Inc	Citation or Pathology
3075	1418	10.8mg	54.7mg	0/20	25.7mg	11/20	Wiessler;zkko,79,118-122;1973
a	1418	32.7mg	n.s.s.	0/20	25.7mg	3/20	
b	1418	9.83mg	61.1mg	1/20	25.7mg	12/20	

S(+)-2-METHYL-N-NITROSOPIPERIDINE   36702-44-0

	RefNum	LoConf	UpConf	Cntrl	1Dose	1Inc	Citation or Pathology
3076	1418	15.4mg	109.mg	0/20	25.7mg	8/20	Wiessler;zkko,79,118-122;1973
a	1418	20.0mg	258.mg	0/20	25.7mg	6/20	
b	1418	23.1mg	706.mg	0/20	25.7mg	5/20	
c	1418	6.62mg	31.9mg	1/20	25.7mg	15/20	

METHYL 12-OXO-trans-10-OCTADECENOATE   21308-79-2

	RefNum	LoConf	UpConf	Cntrl	1Dose	1Inc	Citation or Pathology
3077	1627	361.mg	n.s.s.	2/30	239.mg	3/30	Kiaer;apms,83,550-558;1975

METHYL PARATHION   298-00-0

	RefNum	LoConf	UpConf	Cntrl	1Dose	1Inc	2Dose	2Inc	Citation or Pathology
3078	TR157	7.79mg	n.s.s.	14/20	8.10mg	30/50	(16.2mg	20/50)	
a	TR157	58.5mg	n.s.s.	1/20	8.10mg	4/50	16.2mg	2/50	liv:hpa,hpc,nnd.
b	TR157	47.3mg	n.s.s.	0/20	8.10mg	3/50	16.2mg	2/50	lun:a/a,a/c.
3079	TR157	10.0mg	n.s.s.	13/20	4.20mg	31/50	9.20mg	28/50	
a	TR157	8.86mg	n.s.s.	10/20	4.20mg	14/50	(9.20mg	12/50)	liv:hpa,hpc,nnd.
b	TR157	13.9mg	n.s.s.	1/20	4.20mg	10/50	9.20mg	8/50	lun:a/a,a/c.
3080	TR157	1.49mg	n.s.s.	18/20	1.00mg	33/50	2.00mg	20/50	
a	TR157	7.80mg	n.s.s.	2/20	1.00mg	5/50	2.00mg	0/50	liv:hpa,hpc,nnd.
3081	TR157	1.52mg	n.s.s.	16/20	.800mg	37/50	1.60mg	32/50	
a	TR157	19.7mg	n.s.s.	2/20	.800mg	0/50	1.60mg	0/50	liv:hpa,hpc,nnd.

N-METHYL-2-PYRROLIDONE   872-50-4

	RefNum	LoConf	UpConf	Cntrl	1Dose	1Inc	Citation or Pathology
3082	1818	513.mg	n.s.s.	0/83	30.4mg	0/82	Lee;faat,9,222-235;1987/pers.comm.
3083	1818	372.mg	n.s.s.	0/82	21.2mg	0/85	

```
         Spe Strain Site   Xpo+Xpt                                                          TD50    2Tailpvl
         Sex  Route  Hist    Notes                                                            DR     AuOp

(N-6)-METHYLADENINE          100ng..:..1ug....:..10.....:..100....:..1mg....:..10.....:..100....:..1g.....:..10
3084 M f scd wat lun tum  6m24 s                                        .>                   no dre  P=1.    -
  a  M f scd wat liv hct  6m24 s                                                             no dre  P=1.    -
3085 M m scd wat liv hct  6m25 s                                  .>                         no dre  P=1.    -
  a  M m scd wat lun tum  6m25 s                                                             no dre  P=1.    -

(N-6)-METHYLADENOSINE        100ng..:..1ug....:..10.....:..100....:..1mg....:..10.....:..100....:..1g.....:..10
3086 M f scd wat liv mix  8m27 s                                             .>              364.mg  P<.9
  a  M f scd wat lun tum  8m27 s                                                             no dre  P=1.
3087 M m scd wat lun tum  8m27 s                                          .>                 55.6mg  P<.6
  a  M m scd wat liv mix  8m27 s                                                             108.mg  P<.8

alpha-METHYLBENZYL ALCOHOL   100ng..:..1ug....:..10.....:..100....:..1mg....:..10.....:..100....:..1g.....:..10
3088 M f b6c gav TBA MXB 24m24                                                   :>          no dre  P=1.
  a  M f b6c gav liv MXB 24m24                                                               no dre  P=1.
  b  M f b6c gav lun MXB 24m24                                                               no dre  P=1.
3089 M m b6c gav TBA MXB 24m24                                                :>             2.23gm * P<.7   -
  a  M m b6c gav liv MXB 24m24                                                               no dre  P=1.
  b  M m b6c gav lun MXB 24m24                                                               no dre  P=1.
3090 R f f34 gav pni isa 24m24 s                                          :   +              #2.30gm * P<.009 -
  a  R f f34 gav TBA MXB 24m24 s                                                             826.mg * P<.5
  b  R f f34 gav liv MXB 24m24 s                                                             8.63gm * P<.4
3091 R m f34 gav tes ict 24m24 s                                 :  + :                      51.5mg / P<.0005
  a  R m f34 gav amd MXA 24m24 s                                                             135.mg * P<.0005
  b  R m f34 gav kid MXA 24m24 s                                                             458.mg / P<.0005p
  c  R m f34 gav kid tla 24m24 s                                                             564.mg / P<.0005
  d  R m f34 gav lun MXA 24m24 s                                                             1.05gm * P<.04
  e  R m f34 gav pre can 24m24 s                                                             1.27gm * P<.03
  f  R m f34 gav TBA MXB 24m24 s                                                             88.6mg * P<.0005
  g  R m f34 gav liv MXB 24m24 s                                                             4.80gm * P<.7
3092 R m f34 gav kid uac 24m24    with step                             . +  .               944.mg * P<.0005p

3-METHYLBUTANAL METHYLFORMYLHYDRAZONE ..:..1ug....:..10.....:..100....:..1mg....:..10.....:..100....:..1g.....:..10
3093 M f swa gav lun mix 12m30 e                                  . + .                      1.99mg  P<.0005+
  a  M f swa gav lun ade 12m30 e                                                             2.39mg  P<.0005
  b  M f swa gav liv mix 12m30 e                                                             7.90mg  P<.0005+
  c  M f swa gav lun adc 12m30 e                                                             8.70mg  P<.003
  d  M f swa gav liv hpt 12m30 e                                                             9.27mg  P<.0005
  e  M f swa gav gal mix 12m30 e                                                             28.9mg  P<.008 +
  f  M f swa gav gal ade 12m30 e                                                             36.5mg  P<.02
  g  M f swa gav liv hpc 12m30 e                                                             74.6mg  P<.1
  h  M f swa gav gal adc 12m30 e                                                             151.mg  P<.3
3094 M m swa gav lun mix 12m28 e                            . + .                            2.53mg  P<.0005+
  a  M m swa gav lun ade 12m28 e                                                             3.30mg  P<.0005
  b  M m swa gav pre mix 12m28 e                                                             4.36mg  P<.0005+
  c  M m swa gav pre sqc 12m28 e                                                             5.59mg  P<.0005
  d  M m swa gav liv mix 12m28 e                                                             5.97mg  P<.0005+
  e  M m swa gav liv hpt 12m28 e                                                             8.00mg  P<.0005
  f  M m swa gav lun adc 12m28 e                                                             10.7mg  P<.003
  g  M m swa gav liv hpc 12m28 e                                                             34.2mg  P<.02
  h  M m swa gav pre sqp 12m28 e                                                             34.2mg  P<.02
  i  M m swa gav gal ade 12m28 e                                                             46.1mg  P<.04  +

3-METHYLCHOLANTHRENE         100ng..:..1ug....:..10.....:..100....:..1mg....:..10.....:..100....:..1g.....:..10
3095 P b cym eat tba mal  5y27 w                                        :>                   no dre  P=1.
3096 P b rhe eat tba mal  5y26 w                                        :>                   no dre  P=1.    -
3097 R m lee eat liv tum 26w65                              .>                               no dre  P=1.    -
3098 R m lee eat liv tum 39w65                                 .>                            no dre  P=1.    -
3099 R m lee eat liv tum 65w65                                    .>                         no dre  P=1.    -
3100 R f nss gav mam tum 52w52 er                             . + .                          .506mg  P<.0005+
3101 R f wis gav mgl adc 52w52 r                              . + .                          .714mg  P<.0005+
3102 R f wis gav mgl adc 26w52 r                           . + .                             .202mg  P<.0005+
3103 R f wis gav mgl adc 39w52 r                            . + .                            .304mg  P<.0005+
3104 R f wis gav mgl adc 52w52 r                              . + .                          .764mg  P<.0005+
3105 R f wis gav mgl adc 52w52 r                             . + .                           .506mg  P<.0005+
3106 R f wis gav mgl adc 52w52 r                           <+                                noTD50  P<.0005+
3107 R f wis gav mgl adc 52w52 r                              . + .                          .689mg  P<.0005+
3108 R f wis gav mgl adc 52w52 r                             . + .                           .590mg  P<.0005+
3109 R f wis gav mgl adc 52w52 m                              .+  .                          1.25mg  P<.0005+
3110 R f wis gav mam tum 52w52 r                              . + .                          .669mg  P<.0005+

alpha-METHYLDOPA SESQUIHYDRATE 100ng..:..1ug....:..10.....:..100....:..1mg....:..10.....:..100....:..1g......:..10
3111 M f b6c eat TBA MXB 24m24                                                   :>          no dre  P=1.    -
  a  M f b6c eat liv MXB 24m24                                                               no dre  P=1.    -
  b  M f b6c eat lun MXB 24m24                                                               25.0gm * P<.5
3112 M m b6c eat kid MXA 24m24                                                   :> 20.1gm * P<.2   e
  a  M m b6c eat TBA MXB 24m24                                                               no dre  P=1.
  b  M m b6c eat liv MXB 24m24                                                               no dre  P=1.
  c  M m b6c eat lun MXB 24m24                                                               no dre  P=1.
3113 R f f34 eat TBA MXB 24m24                                              :>               592.mg * P<.5   -
```

	RefNum	LoConf	UpConf	Cntrl	1Dose	1Inc	2Dose	2Inc	Citation or Pathology	Brkly Code
(N-6)-METHYLADENINE	**443-72-1**									
3084	1255	6.08mg	n.s.s.	3/16	4.33mg	3/17			Anderson;ijcn,24,319-322;1979	
a	1255	15.2mg	n.s.s.	1/16	4.33mg	0/17				
3085	1255	5.70mg	n.s.s.	5/21	3.52mg	3/17				
a	1255	6.92mg	n.s.s.	4/21	3.52mg	2/17				
(N-6)-METHYLADENOSINE	**1867-73-8**									
3086	1255	16.9mg	n.s.s.	1/16	9.76mg	1/12			Anderson;ijcn,24,319-322;1979	
a	1255	14.1mg	n.s.s.	3/16	9.76mg	2/12				
3087	1255	8.65mg	n.s.s.	4/21	7.89mg	4/14				
a	1255	9.22mg	n.s.s.	5/21	7.89mg	4/14				
alpha-METHYLBENZYL ALCOHOL	**98-85-1**									
3088	TR369	870.mg	n.s.s.	33/50	265.mg	23/50	531.mg	24/50		
a	TR369	4.97gm	n.s.s.	2/50	265.mg	1/50	531.mg	0/50	liv:hpa,hpc,nnd.	
b	TR369	2.76gm	n.s.s.	7/50	265.mg	1/50	531.mg	4/50	lun:a/a,a/c.	
3089	TR369	321.mg	n.s.s.	39/50	265.mg	40/50	531.mg	32/50		
a	TR369	711.mg	n.s.s.	20/50	265.mg	22/50	531.mg	13/50	liv:hpa,hpc,nnd.	
b	TR369	1.07gm	n.s.s.	17/50	265.mg	14/50	531.mg	9/50	lun:a/a,a/c.	
3090	TR369	791.mg	60.5gm	0/50	265.mg	1/50	531.mg	3/50		S
a	TR369	190.mg	n.s.s.	46/50	265.mg	40/50	531.mg	22/50		
b	TR369	1.41gm	n.s.s.	0/50	265.mg	1/50	531.mg	0/50	liv:hpa,hpc,nnd.	
3091	TR369	31.1mg	98.0mg	46/50	265.mg	36/50	531.mg	35/50		S
a	TR369	62.4mg	508.mg	16/50	265.mg	20/50	531.mg	4/50	amd:phe,phm.	S
b	TR369	161.mg	1.70gm	0/50	265.mg	2/50	531.mg	5/50	kid:tla,uac.	
c	TR369	176.mg	2.50gm	0/50	265.mg	1/50	531.mg	5/50		S
d	TR369	243.mg	n.s.s.	1/50	265.mg	3/50	531.mg	1/50	lun:a/a,a/c.	S
e	TR369	276.mg	n.s.s.	0/50	265.mg	3/50	531.mg	4/50		S
f	TR369	46.3mg	249.mg	43/50	265.mg	32/50	531.mg	18/50		
g	TR369	456.mg	n.s.s.	2/50	265.mg	2/50	531.mg	0/50	liv:hpa,hpc,nnd.	
3092	TR369	567.mg	2.38gm	1/50	265.mg	13/50	531.mg	14/50		
3-METHYLBUTANAL METHYLFORMYLHYDRAZONE	**---**									
3093	1966	1.19mg	3.85mg	13/50	2.98mg	42/50			Toth;vivo,4,283-288;1990	
a	1966	1.48mg	4.33mg	7/50	2.98mg	38/50				
b	1966	4.39mg	16.3mg	0/50	2.98mg	16/50				
c	1966	4.26mg	54.6mg	6/50	2.98mg	19/50				
d	1966	4.98mg	20.2mg	0/50	2.98mg	14/50				
e	1966	11.0mg	559.mg	0/50	2.98mg	5/50				
f	1966	12.6mg	n.s.s.	0/50	2.98mg	4/50				
g	1966	18.4mg	n.s.s.	0/50	2.98mg	2/50				
h	1966	24.6mg	n.s.s.	0/50	2.98mg	1/50				
3094	1966	1.48mg	5.47mg	13/50	3.18mg	38/50				
a	1966	1.88mg	8.11mg	12/50	3.18mg	34/50				
b	1966	2.66mg	7.83mg	0/50	3.18mg	24/50				
c	1966	3.28mg	10.6mg	0/50	3.18mg	20/50				
d	1966	3.46mg	11.6mg	0/50	3.18mg	19/50				
e	1966	4.37mg	17.0mg	0/50	3.18mg	15/50				
f	1966	5.10mg	69.2mg	3/50	3.18mg	14/50				
g	1966	11.8mg	n.s.s.	0/50	3.18mg	4/50				
h	1966	11.8mg	n.s.s.	0/50	3.18mg	4/50				
i	1966	14.0mg	n.s.s.	0/50	3.18mg	3/50				
3-METHYLCHOLANTHRENE	**56-49-5**									
3095	2001	9.28mg	n.s.s.	2/10	8.37mg	0/3			Adamson;ossc,129-156;	
3096	2001	8.96mg	n.s.s.	5/76	10.7mg	0/7			1982/Thorgeirsson 1994/Dalgard 1991/Thorgeirsson&Seiber pers.comm.	
3097	1484m	.518mg	n.s.s.	0/40	1.07mg	0/6			Flaks;carc,3,981-991;1982	
3098	1484n	1.94mg	n.s.s.	0/40	1.61mg	0/15				
3099	1484o	3.24mg	n.s.s.	0/40	2.68mg	0/15				
3100	481	.261mg	1.08mg	0/9	4.90mg	17/21			Shay;mpoc,305-318;1962	
3101	188BT	.343mg	1.53mg	0/20	9.80mg	19/21			Shay;jnci,27,503-513;1961	
3102	188bm	91.4ug	.485mg	0/20	2.46mg	14/16				
3103	188bn	.137mg	.727mg	0/20	3.69mg	14/16				
3104	188bo	.375mg	1.84mg	0/20	4.90mg	12/18				
3105	188br	.261mg	1.08mg	0/20	4.90mg	17/21				
3106	188bs	n.s.s.	.486mg	0/20	4.90mg	17/17				
3107	188bu	.286mg	1.55mg	0/20	12.2mg	20/21				
3108	188bv	.336mg	1.12mg	0/20	4.90mg	22/29				
3109	1130m	.808mg	2.04mg	1/54	8.57mg	37/53			Gruenstein;canr,24,1656-1658;1964	
3110	1130n	.363mg	1.39mg	1/54	4.90mg	18/25				
alpha-METHYLDOPA SESQUIHYDRATE	**41372-08-1**									
3111	TR348	3.22gm	n.s.s.	33/50	811.mg	22/50	1.61gm	21/50		
a	TR348	5.13gm	n.s.s.	4/50	811.mg	1/50	(1.61gm	0/50)	liv:hpa,hpc,nnd.	
b	TR348	4.97gm	n.s.s.	4/50	811.mg	1/50	1.61gm	6/50	lun:a/a,a/c.	
3112	TR348	6.08gm	n.s.s.	0/50	749.mg	2/50	1.49gm	1/50	kid:tla,uac.	
a	TR348	4.11gm	n.s.s.	32/50	749.mg	15/50	1.49gm	17/50		
b	TR348	7.29gm	n.s.s.	15/50	749.mg	5/50	1.49gm	6/50	liv:hpa,hpc,nnd.	
c	TR348	4.94gm	n.s.s.	10/50	749.mg	5/50	1.49gm	7/50	lun:a/a,a/c.	
3113	TR348	140.mg	n.s.s.	45/50	154.mg	47/50	312.mg	47/50		

```
    Spe Strain Site   Xpo+Xpt                                              TD50    2Tailpvl
        Sex  Route  Hist     Notes                                           DR     AuOp
───────────────────────────────────────────────────────────────────────────────────────────
  a   R f f34 eat liv MXB 24m24                                            no dre  P=1.
3114  R m f34 eat TBA MXB 24m24                                    :>      no dre  P=1.   -
  a   R m f34 eat liv MXB 24m24                                            no dre  P=1.


N-METHYLDOPAMINE,O,O'-DIISOBUTYROYL  ESTER.HCl  ...:..10......:..100.....:..1mg...:..10......:..100...:..1g......:..10
3115  R f cdr gav adr coa 24m24                                      .>    no dre  P=1.   -
  a   R f cdr gav pit ade 24m24                                           no dre  P=1.   -
  b   R f cdr gav mgl adc 24m24                                           no dre  P=1.   -
  c   R f cdr gav mgl ade 24m24                                           no dre  P=1.   -
  d   R f cdr gav adr phm 24m24                                           no dre  P=1.   -
3116  R m cdr gav pit ade 24m24              .  +       .                 30.2mg Z P<.002 -
  a   R m cdr gav adr phm 24m24                                           no dre  P=1.   -
  b   R m cdr gav adr coa 24m24                                           no dre  P=1.   -


4,4'-METHYLENE-BIS(2-CHLOROANILINE)   ..:..1ug...:..10.....:..100....:..1mg...:..10......:..100......:..1g.....:..10
3117  D f beg eat ubl ptc 9y9  emv              .    +    .              2.12mg  P<.003 +
  a   D f beg eat liv hnd 9y9  emv              3.72mg  P<.02  +
  b   D f beg eat mgl mix 9y9  emv                                        no dre  P=1.   -
3118  R f cdr eat lun adc 67w67 ef                         .  +   .       42.3mg  P<.003 +
  a   R f cdr eat mgl adc 67w67 ef                                        42.3mg  P<.003 +
  b   R f cdr eat liv hpa 67w67 ef                                        142.mg  P<.1   -
  c   R f cdr eat liv hpc 67w67 ef                                        291.mg  P<.3   -
3119  R f cdr eat lun adc 24m24 e                            .  +  .      34.7mg  P<.0005+
  a   R f cdr eat liv hpc 24m24 e                                         467.mg  P<.04  -
  b   R f cdr eat liv hpa 24m24 e                                         708.mg  P<.1   -
  c   R f cdr eat liv cho 24m24 e                                         1.43gm  P<.3   -
3120  R m cdr eat liv hpc 67w67 ef                         .  +   .       15.3mg  P<.0005+
  a   R m cdr eat liv hpa 67w67 ef                                        41.8mg  P<.007 +
  b   R m cdr eat lun adc 67w67 ef                                        41.8mg  P<.007 +
3121  R m cdr eat lun adc 24m24 e                            .  +  .      42.3mg  P<.0005+
  a   R m cdr eat liv hpc 24m24 e                                         388.mg  P<.04  -
  b   R m cdr eat liv hpa 24m24 e                                         388.mg  P<.04  -
3122  R m cdr eat lun neo 18m24 es                       .  +.            20.8mg * P<.0005+
  a   R m cdr eat lun mix 18m24 es                                        29.2mg * P<.0005+
  b   R m cdr eat mgl adc 18m24 es                                        87.6mg * P<.0005+
  c   R m cdr eat liv hpc 18m24 es                                        92.2mg Z P<.0005+
  d   R m cdr eat zym car 18m24 es                                        107.mg * P<.0005+
  e   R m cdr eat --- ker 18m24 es                                        144.mg Z P<.01
  f   R m cdr eat ski sqc 18m24 es                                        286.mg * P<.2
  g   R m cdr eat tba mix 18m24 es                                        10.4mg * P<.0005
3123  R m cdr eat lun neo 18m24 efs                          .  +  .      35.1mg * P<.0005+
  a   R m cdr eat lun mix 18m24 efs                                       60.9mg * P<.0005+
  b   R m cdr eat zym car 18m24 efs                                       112.mg * P<.0005+
  c   R m cdr eat liv hpc 18m24 efs                                       124.mg * P<.0005+
  d   R m cdr eat mgl adc 18m24 efs                                       162.mg * P<.006 +
  e   R m cdr eat --- hes 18m24 efs                                       138.mg * P<.02  +
  f   R m cdr eat ski sqc 18m24 efs                                       263.mg * P<.03
  g   R m cdr eat tba mix 18m24 efs                                       14.8mg * P<.0005
3124  R f wi2 eat liv mix 16m28 e                        .  +   .         15.8mg  P<.0005+
  a   R f wi2 eat mgl mix 16m28 e                                         51.4mg  P<.003
  b   R f wi2 eat lun mix 16m28 e                                         104.mg  P<.005 +
  c   R f wi2 eat tba mix 16m28 e                                         11.6mg  P<.0005+
3125  R m wi2 eat liv mix 16m30 e                        .  +   .         10.8mg  P<.0005+
  a   R m wi2 eat lun mix 16m30 e                                         59.5mg  P<.0005+
  b   R m wi2 eat tba mix 16m30 e                                         9.09mg  P<.0005+


4,4'-METHYLENE-BIS(2-CHLOROANILINE).2HCl.:..1_ug...:..10.....:..100....:..1mg...:..10......:..100.....:..1g.....:..10
3126  M f chi eat liv hpt 77w90                           :  +  :         66.6mg  / P<.0005+
  a   M f chi eat liv mix 77w90                                           66.6mg  / P<.0005+
  b   M f chi eat lun mix 77w90                                           556.mg * P<.4   -
  c   M f chi eat tba mix 77w90                                           52.8mg  / P<.0005
3127  M m chi eat lun mix 77w94                               :>          342.mg * P<.2   -
  a   M m chi eat liv mix 77w94                                           471.mg * P<.4
  b   M m chi eat tba mix 77w94                                           97.9mg * P<.02
3128  R m cdr eat liv mix 77w98                         :>                70.1mg * P<.2
  a   R m cdr eat tba mix 77w98                                           33.4mg  P<.5


4,4'-METHYLENE-BIS(2-METHYLANILINE)  ..:..1ug...:..10......:..100.....:..1mg...:..10......:..100...:..1g.....:..10
3129  R f cdr eat liv hpc 82w82 v                       .  +  .           7.92mg  P<.0005+
  a   R f cdr eat liv hpa 82w82 v                                         30.5mg  P<.004
  b   R f cdr eat liv clc 82w82 v                                         96.3mg  P<.1
3130  R m cdr eat mgl fba 68w68 v                       .  +   .          6.91mg  P<.0005+
  a   R m cdr eat ski fib 68w68 v                                         8.19mg  P<.002 +
  b   R m cdr eat liv hpc 68w68 v                                         12.4mg  P<.002 +
  c   R m cdr eat liv hpa 68w68 v                                         17.0mg  P<.004
  d   R m cdr eat liv hem 68w68 v                                         35.2mg  P<.04


METHYLENE CHLORIDE          100ng..:..1ug....:..10.....:..100....:..1mg...:..10......:..100......:..1g.......:..10
3131  M f b6c inh MXB MXB 24m24                                 : + :     817.mg * P<.0005
  a   M f b6c inh lun MXA 24m24                                           917.mg * P<.0005c
  b   M f b6c inh lun a/a 24m24                                           1.42gm * P<.0005c
```

	RefNum	LoConf	UpConf	Cntrl	1Dose	1Inc	2Dose	2Inc	Citation or Pathology
									Brkly Code
a	TR348	n.s.s.	n.s.s.	0/50	154.mg	1/50	312.mg	0/50	liv:hpa,hpc,nnd.
3114	TR348	216.mg	n.s.s.	46/50	123.mg	31/50	250.mg	42/50	
a	TR348	351.mg	n.s.s.	6/50	123.mg	3/50	(250.mg	1/50)	liv:hpa,hpc,nnd.

N-METHYLDOPAMINE,O,O'-DIISOBUTYROYL   ESTER.HCl   (ibopamine.HCl) 75011-65-3

	RefNum	LoConf	UpConf	Cntrl	1Dose	1Inc	2Dose	2Inc	Citation or Pathology
3115	1875	164.mg	n.s.s.	71/200	30.0mg	26/100	(90.0mg	22/100 180.mg 19/100)	Walker;neag,9,291-301;1988/pers.comm.
a	1875	249.mg	n.s.s.	163/200	30.0mg	83/100	90.0mg	92/100 180.mg 75/100	
b	1875	626.mg	n.s.s.	31/200	30.0mg	17/100	90.0mg	11/100 (180.mg 6/100)	
c	1875	1.22gm	n.s.s.	67/200	30.0mg	30/100	90.0mg	27/100 180.mg 23/100	
d	1875	2.59gm	n.s.s.	11/200	30.0mg	1/100	90.0mg	3/100 180.mg 3/100	
3116	1875	15.5mg	142.mg	125/200	30.0mg	81/100	(90.0mg	59/100 180.mg 41/100)	
a	1875	850.mg	n.s.s.	45/200	30.0mg	11/100	90.0mg	13/100 (180.mg 7/100)	
b	1875	145.mg	n.s.s.	50/200	30.0mg	21/100	(90.0mg	11/100 180.mg 10/100)	

4,4'-METHYLENE-BIS(2-CHLOROANILINE)   (3,3'-dichloro-4,4'-diaminodiphenylmethane,   MOCA) 101-14-4

	RefNum	LoConf	UpConf	Cntrl	1Dose	1Inc	2Dose	2Inc	Citation or Pathology
3117	1030	.586mg	14.8mg	0/6	7.31mg	4/5			Stula;jept,1,31-50;1977
a	1030	1.04mg	n.s.s.	0/6	7.31mg	3/5			
b	1030	2.29mg	n.s.s.	4/6	7.31mg	2/5			
3118	192m	17.1mg	223.mg	0/21	50.0mg	6/21			Stula;txap,31,159-176;1975/pers.comm.
a	192m	17.1mg	223.mg	0/21	50.0mg	6/21			
b	192m	34.9mg	n.s.s.	0/21	50.0mg	2/21			
c	192m	47.4mg	n.s.s.	0/21	50.0mg	1/21			
3119	192n	21.4mg	60.5mg	0/44	50.0mg	27/44			
a	192n	141.mg	n.s.s.	0/44	50.0mg	3/44			
b	192n	174.mg	n.s.s.	0/44	50.0mg	2/44			
c	192n	233.mg	n.s.s.	0/44	50.0mg	1/44			
3120	192m	7.53mg	38.0mg	0/21	40.0mg	11/21			
a	192m	15.8mg	498.mg	0/21	40.0mg	5/21			
b	192m	15.8mg	498.mg	0/21	40.0mg	5/21			
3121	192n	25.0mg	79.2mg	0/44	40.0mg	21/44			
a	192n	117.mg	n.s.s.	0/44	40.0mg	3/44			
b	192n	117.mg	n.s.s.	0/44	40.0mg	3/44			
3122	1031m	15.7mg	28.6mg	1/100	7.53mg	23/100	15.1mg	28/75 34.0mg 35/50	Kommineni;jept,2,149-171;1979
a	1031m	21.5mg	41.1mg	0/100	7.53mg	14/100	15.1mg	20/75 34.0mg 31/50	
b	1031m	53.7mg	174.mg	1/100	7.53mg	5/100	15.1mg	8/75 34.0mg 14/50	
c	1031m	56.8mg	164.mg	0/100	7.53mg	3/100	15.1mg	3/75 34.0mg 18/50	
d	1031m	61.4mg	288.mg	1/100	7.53mg	8/100	15.1mg	5/75 34.0mg 11/50	
e	1031m	63.9mg	6.58gm	1/100	7.53mg	3/100	15.1mg	7/75 (34.0mg 1/50)	
f	1031m	109.mg	n.s.s.	1/100	7.53mg	4/100	15.1mg	7/75 34.0mg 2/50	
g	1031m	6.48mg	21.1mg	58/100	7.53mg	80/100	15.1mg	61/75 34.0mg 48/50	
3123	1031n	22.7mg	58.9mg	0/100	3.76mg	6/100	7.53mg	11/75 15.1mg 13/50	
a	1031n	35.0mg	127.mg	0/100	3.76mg	3/100	7.53mg	7/75 15.1mg 8/50	
b	1031n	54.5mg	298.mg	0/100	3.76mg	0/100	7.53mg	4/75 15.1mg 6/50	
c	1031n	58.3mg	337.mg	0/100	3.76mg	0/100	7.53mg	0/75 15.1mg 9/50	
d	1031n	70.1mg	1.51gm	0/100	3.76mg	1/100	7.53mg	3/75 15.1mg 3/50	
e	1031n	58.8mg	n.s.s.	1/100	3.76mg	2/100	7.53mg	4/75 15.1mg 4/50	
f	1031n	90.3mg	n.s.s.	1/100	3.76mg	0/100	7.53mg	1/75 15.1mg 4/50	
g	1031n	8.87mg	37.5mg	37/100	3.76mg	34/100	7.53mg	40/75 15.1mg 36/50	
3124	1018	8.23mg	32.9mg	0/25	29.9mg	18/22			Grundmann;zkko,74,28-39;1970
a	1018	22.5mg	352.mg	2/25	29.9mg	10/22			
b	1018	39.4mg	827.mg	0/25	29.9mg	5/22			
c	1018	5.53mg	25.9mg	2/25	29.9mg	20/22			
3125	1018	5.69mg	21.6mg	0/25	22.5mg	22/25			
a	1018	26.8mg	192.mg	0/25	22.5mg	8/25			
b	1018	4.52mg	18.4mg	0/25	22.5mg	23/25			

4,4'-METHYLENE-BIS(2-CHLOROANILINE).2HCl   64049-29-2

	RefNum	LoConf	UpConf	Cntrl	1Dose	1Inc	2Dose	2Inc	Citation or Pathology
3126	191	35.3mg	147.mg	0/20	117.mg	9/21	246.mg	7/14	Russfield;txap,31,47-54;1975
a	191	35.3mg	147.mg	0/20	117.mg	9/21	246.mg	7/14	
b	191	108.mg	n.s.s.	6/20	117.mg	3/21	246.mg	2/14	
c	191	26.1mg	196.mg	17/20	117.mg	16/21	246.mg	12/14	
3127	191	86.4mg	n.s.s.	5/18	108.mg	3/13	206.mg	4/20	
a	191	112.mg	n.s.s.	3/18	108.mg	4/13	206.mg	4/20	
b	191	28.0mg	n.s.s.	12/18	108.mg	7/13	206.mg	16/20	
3128	191	18.6mg	n.s.s.	1/22	18.0mg	1/22	31.3mg	4/19	
a	191	6.57mg	n.s.s.	14/22	18.0mg	13/22	31.3mg	14/19	

4,4'-METHYLENE-BIS(2-METHYLANILINE)   838-88-0

	RefNum	LoConf	UpConf	Cntrl	1Dose	1Inc	2Dose	2Inc	Citation or Pathology
3129	192	4.58mg	15.4mg	0/44	10.5mg	19/44			Stula;txap,31,159-176;1975/pers.comm.
a	192	12.4mg	187.mg	0/44	10.5mg	6/44			
b	192	23.7mg	n.s.s.	0/44	10.5mg	2/44			
3130	192	3.58mg	19.2mg	1/44	8.49mg	14/44			
a	192	3.97mg	37.4mg	2/44	8.49mg	13/44			
b	192	5.61mg	41.8mg	0/44	8.49mg	8/44			
c	192	6.92mg	104.mg	0/44	8.49mg	6/44			
d	192	10.7mg	n.s.s.	0/44	8.49mg	3/44			

METHYLENE CHLORIDE   (dichloromethane, Freon 30) 75-09-2

	RefNum	LoConf	UpConf	Cntrl	1Dose	1Inc	2Dose	2Inc	Citation or Pathology
3131	TR306	568.mg	1.28gm	5/50	2.14gm	36/50	4.28gm	46/50	liv:hpa,hpc; lun:a/a,a/c. C
a	TR306	635.mg	1.43gm	3/50	2.14gm	30/50	4.28gm	41/50	lun:a/a,a/c.
b	TR306	938.mg	2.39gm	2/50	2.14gm	23/50	4.28gm	28/50	

	Spe	Sex	Strain	Route	Site	Hist	Xpo+Xpt	Notes	Plot	TD50	DR	2Tailpvl	AuOp
c	M	f	b6c	inh	liv	MXA	24m24			1.43gm	/	P<.0005c	
d	M	f	b6c	inh	lun	a/c	24m24			1.69gm	/	P<.0005c	
e	M	f	b6c	inh	liv	hpc	24m24			1.85gm	/	P<.0005c	
f	M	f	b6c	inh	liv	hpa	24m24			3.17gm	/	P<.0005c	
g	M	f	b6c	inh	thy	fca	24m24			15.8gm	*	P<.05	
h	M	f	b6c	inh	TBA	MXB	24m24			1.02gm	*	P<.0005	
i	M	f	b6c	inh	liv	MXB	24m24			1.43gm	*	P<.0005	
j	M	f	b6c	inh	lun	MXB	24m24			917.mg	*	P<.0005	
3132	M	m	b6c	inh	lun	MXA	24m24		: + :	920.mg	*	P<.0005c	
a	M	m	b6c	inh	MXB	MXB	24m24			1.09gm	*	P<.0005	
b	M	m	b6c	inh	lun	a/a	24m24			1.68gm	/	P<.0005c	
c	M	m	b6c	inh	lun	a/c	24m24			1.71gm	/	P<.0005c	
d	M	m	b6c	inh	liv	MXA	24m24			1.80gm	*	P<.0005c	
e	M	m	b6c	inh	liv	hpc	24m24			2.69gm	*	P<.0005c	
f	M	m	b6c	inh	liv	hpa	24m24			3.48gm	*	P<.003 c	
g	M	m	b6c	inh	---	hes	24m24			12.9gm	*	P<.02	
h	M	m	b6c	inh	---	MXA	24m24			13.2gm	*	P<.03	
i	M	m	b6c	inh	TBA	MXB	24m24			1.16gm	*	P<.0005	
j	M	m	b6c	inh	liv	MXB	24m24			1.80gm	*	P<.0005	
k	M	m	b6c	inh	lun	MXB	24m24			920.mg	*	P<.0005	
3133	M	m	b6c	wat	liv	mix	24m24	e	. ±	1.31gm	*	P<.07	-
a	M	m	b6c	wat	liv	hpc	24m24	e		2.35gm	*	P<.2	-
b	M	m	b6c	wat	liv	hpa	24m24	e		2.90gm	*	P<.2	-
c	M	m	b6c	wat	thy	tum	24m24	e		no dre		P=1.	
3134	M	f	swi	gav	lun	ade	15m24	s	.>	7.05gm	*	P<.9	
a	M	f	swi	gav	liv	hpt	15m24	s		no dre		P=1.	
b	M	f	swi	gav	tba	mal	15m24	s		no dre		P=1.	
c	M	f	swi	gav	tba	mix	15m24	s		no dre		P=1.	
3135	M	m	swi	gav	lun	ade	15m24	s	. ±	916.mg	*	P<.05	
a	M	m	swi	gav	liv	hpt	15m24	s		no dre		P=1.	
b	M	m	swi	gav	tba	mal	15m24	s		no dre		P=1.	
c	M	m	swi	gav	tba	mix	15m24	s		no dre		P=1.	
3136	R	f	f34	inh	mgl	MXA	24m24		: + :	598.mg	*	P<.0005c	
a	R	f	f34	inh	mgl	MXA	24m24			630.mg	*	P<.0005	
b	R	f	f34	inh	mgl	fba	24m24			632.mg	*	P<.0005c	
c	R	f	f34	inh	mgl	MXA	24m24			670.mg	*	P<.0005	
d	R	f	f34	inh	---	mnl	24m24			1.28gm	*	P<.02	
e	R	f	f34	inh	liv	MXA	24m24			4.27gm	*	P<.05	
f	R	f	f34	inh	TBA	MXB	24m24			501.mg	*	P<.007	
g	R	f	f34	inh	liv	MXB	24m24			4.27gm	*	P<.05	
3137	R	f	f34	wat	liv	nnd	78w78	ek	.>	no dre		P=1.	-
3138	R	f	f34	wat	liv	mix	18m24	e	. ±	1.54gm		P<.02	-
a	R	f	f34	wat	liv	nnd	18m24	e		1.54gm		P<.02	-
b	R	f	f34	wat	liv	hpc	18m24	e		no dre		P=1.	-
3139	R	f	f34	wat	liv	mix	24m24	e	. +	1.43gm	*	P<.003	-
a	R	f	f34	wat	liv	nnd	24m24	e		2.34gm	*	P<.02	-
b	R	f	f34	wat	liv	hpc	24m24	e		3.63gm	*	P<.06	-
3140	R	m	f34	inh	MXA	MXA	24m24		: + :	917.mg	*	P<.0005p	
a	R	m	f34	inh	MXA	MXA	24m24			1.44gm	*	P<.006	
b	R	m	f34	inh	mgl	MXA	24m24			1.64gm	*	P<.002 p	
c	R	m	f34	inh	tnv	MXA	24m24			1.64gm	*	P<.005	
d	R	m	f34	inh	mgl	fba	24m24			1.77gm	*	P<.004 p	
e	R	m	f34	inh	sub	MXA	24m24			1.99gm	*	P<.003	
f	R	m	f34	inh	sub	fib	24m24			2.18gm	*	P<.007	
g	R	m	f34	inh	tnv	msm	24m24			4.31gm	*	P<.04	
h	R	m	f34	inh	TBA	MXB	24m24			539.mg	*	P<.2	
i	R	m	f34	inh	liv	MXB	24m24			no dre		P=1.	
3141	R	m	f34	wat	liv	nnd	78w78	ek	.>	no dre		P=1.	-
3142	R	m	f34	wat	liv	nnd	18m24	e	.>	1.61gm		P<.4	-
a	R	m	f34	wat	liv	mix	18m24	e		3.66gm		P<.8	-
b	R	m	f34	wat	liv	hpc	18m24	e		no dre		P=1.	-
3143	R	m	f34	wat	liv	mix	24m24	e	.>	no dre		P=1.	-
a	R	m	f34	wat	liv	nnd	24m24	e		no dre		P=1.	-
b	R	m	f34	wat	liv	hpc	24m24	e		no dre		P=1.	-
3144	R	f	sda	gav	mam	mal	15m24	s	. ±	1.13gm	*	P<.07	
a	R	f	sda	gav	mam	mix	15m24	s		1.03gm	*	P<.7	
b	R	f	sda	gav	tba	mal	15m24	s		1.74gm	*	P<.6	
c	R	f	sda	gav	tba	mix	15m24	s		no dre		P=1.	
3145	R	f	sda	inh	mam	mix	24m24	gv	. + .	37.8mg	*	P<.009	
a	R	f	sda	inh	tba	mix	24m24	gv		32.1mg		P<.03	
b	R	f	sda	inh	tba	mal	24m24	gv		477.mg		P<.7	
3146	R	m	sda	gav	tba	mix	15m24	s	.>	no dre		P=1.	
a	R	m	sda	gav	tba	mal	15m24	s		935.mg	\	P<.9	
3147	R	f	sss	inh	mgl	ben	24m24	e	.>	4.43gm	*	P<.7	
a	R	f	sss	inh	pit	ade	24m24	e		no dre		P=1.	
3148	R	f	sss	inh	mgl	fba	24m24	e	.>	631.mg	*	P<.7	
a	R	f	sss	inh	mgl	ben	24m24	e		1.55gm	*	P<.9	
b	R	f	sss	inh	liv	kcs	24m24	e		9.34gm	*	P<.8	-
c	R	f	sss	inh	liv	hpc	24m24	e		9.52gm	*	P<.7	-
d	R	f	sss	inh	liv	nnd	24m24	e		no dre		P=1.	-
e	R	f	sss	inh	mgl	adc	24m24	e		no dre		P=1.	-

	RefNum	LoConf	UpConf	Cntrl	1Dose	1Inc	2Dose	2Inc	3Dose	3Inc	4Dose	4Inc	Citation or Pathology	Brkly Code
c	TR306	956.mg	2.31gm	3/50	2.14gm	16/50	4.28gm	40/50					liv:hpa,hpc.	
d	TR306	1.10gm	2.81gm	1/50	2.14gm	13/50	4.28gm	29/50						
e	TR306	1.20gm	3.04gm	1/50	2.14gm	11/50	4.28gm	32/50						
f	TR306	1.85gm	6.50gm	2/50	2.14gm	6/50	4.28gm	22/50						
g	TR306	5.11gm	n.s.s.	1/50	2.14gm	1/50	4.28gm	4/50						S
h	TR306	652.mg	1.93gm	18/50	2.14gm	41/50	4.28gm	47/50						
i	TR306	956.mg	2.31gm	3/50	2.14gm	16/50	4.28gm	40/50					liv:hpa,hpc,nnd.	
j	TR306	635.mg	1.43gm	3/50	2.14gm	30/50	4.28gm	41/50					lun:a/a,a/c.	
3132	TR306	630.mg	1.46gm	5/50	1.79gm	27/50	3.57gm	40/50					lun:a/a,a/c.	
a	TR306	672.mg	2.27gm	27/50	1.79gm	35/50	3.57gm	45/50					liv:hpa,hpc; lun:a/a,a/c.	C
b	TR306	1.08gm	2.98gm	3/50	1.79gm	19/50	3.57gm	24/50						
c	TR306	1.09gm	2.95gm	2/50	1.79gm	20/50	3.57gm	28/50						
d	TR306	1.01gm	4.99gm	22/50	1.79gm	24/50	3.57gm	33/50					liv:hpa,hpc.	
e	TR306	1.49gm	7.70gm	13/50	1.79gm	15/50	3.57gm	26/50						
f	TR306	1.68gm	23.5gm	10/50	1.79gm	14/50	3.57gm	14/50						
g	TR306	4.78gm	n.s.s.	1/50	1.79gm	2/50	3.57gm	5/50						S
h	TR306	4.82gm	n.s.s.	2/50	1.79gm	2/50	3.57gm	6/50					---:hem,hes.	S
i	TR306	688.mg	2.76gm	34/50	1.79gm	37/50	3.57gm	46/50					liv:hpa,hpc,nnd.	
j	TR306	1.01gm	4.99gm	22/50	1.79gm	24/50	3.57gm	33/50					liv:hpa,hpc.	
k	TR306	630.mg	1.46gm	5/50	1.79gm	27/50	3.57gm	40/50					lun:a/a,a/c.	
3133	1802	540.mg	n.s.s.	24/125	60.0mg	51/200	125.mg	30/100	185.mg	31/99	250.mg	35/125	Serota;fctx,24,959-963;1986	
a	1802	815.mg	n.s.s.	14/125	60.0mg	33/200	125.mg	18/100	185.mg	17/99	250.mg	23/125		
b	1802	1.01gm	n.s.s.	10/125	60.0mg	20/200	125.mg	14/100	185.mg	14/99	250.mg	15/125		
c	1802	825.mg	n.s.s.	0/125	60.0mg	0/200	125.mg	0/100	185.mg	0/99	250.mg	0/125		
3134	BT3003	502.mg	n.s.s.	10/60	39.6mg	8/50	198.mg	9/50					Maltoni;anya,534,352-366;1988	
a	BT3003	340.mg	n.s.s.	0/60	39.6mg	0/50	198.mg	0/50						
b	BT3003	586.mg	n.s.s.	15/60	39.6mg	8/50	198.mg	10/50						
c	BT3003	427.mg	n.s.s.	31/60	39.6mg	17/50	198.mg	20/50						
3135	BT3003	345.mg	n.s.s.	3/60	39.6mg	6/50	198.mg	9/50						
a	BT3003	1.27gm	n.s.s.	5/60	39.6mg	2/50	198.mg	2/50						
b	BT3003	1.18gm	n.s.s.	12/60	39.6mg	4/50	198.mg	4/50						
c	BT3003	477.mg	n.s.s.	18/60	39.6mg	10/50	198.mg	13/50						
3136	TR306	361.mg	1.36gm	5/50	255.mg	11/50	510.mg	13/50	1.02gm	23/50			mgl:adn,fba.	
a	TR306	371.mg	1.57gm	6/50	255.mg	13/50	510.mg	14/50	1.02gm	23/50			mgl:acn,adn,fba.	S
b	TR306	378.mg	1.49gm	5/50	255.mg	11/50	510.mg	13/50	1.02gm	22/50				
c	TR306	385.mg	1.80gm	7/50	255.mg	13/50	510.mg	14/50	1.02gm	23/50			mgl:acn,adn,fba,mtm.	S
d	TR306	578.mg	n.s.s.	17/50	255.mg	17/50	510.mg	23/50	1.02gm	23/50				S
e	TR306	1.56gm	n.s.s.	2/50	255.mg	1/50	510.mg	4/50	1.02gm	5/50			liv:hpc,nnd.	S
f	TR306	244.mg	8.34gm	41/50	255.mg	46/50	510.mg	48/50	1.02gm	48/50				
g	TR306	1.56gm	n.s.s.	2/50	255.mg	1/50	510.mg	4/50	1.02gm	5/50			liv:hpa,hpc,nnd.	
3137	1801m	9.99mg	n.s.s.	0/20	5.00mg	0/20	50.0mg	0/20	125.mg	0/20	250.mg	0/20	Serota;fctx,24,951-958;1986	
3138	1801n	379.mg	n.s.s.	0/100	188.mg	2/25								
a	1801n	379.mg	n.s.s.	0/100	188.mg	2/25								
b	1801n	966.mg	n.s.s.	0/100	188.mg	0/25								
3139	1801o	638.mg	10.1gm	0/100	5.00mg	1/50	50.0mg	4/50	125.mg	1/50	250.mg	6/50		
a	1801o	880.mg	n.s.s.	0/100	5.00mg	1/50	50.0mg	2/50	125.mg	1/50	250.mg	4/50		
b	1801o	1.25gm	n.s.s.	0/100	5.00mg	0/50	50.0mg	2/50	125.mg	0/50	250.mg	2/50		
3140	TR306	477.mg	2.17gm	0/50	179.mg	1/50	357.mg	4/50	714.mg	9/50			mgl:adn,fba; sub:fib.	
a	TR306	686.mg	16.1gm	0/50	179.mg	2/50	357.mg	5/50	714.mg	4/50			mul:men,msm; tnv:men,msm.	S
b	TR306	693.mg	7.00gm	0/50	179.mg	0/50	357.mg	2/50	714.mg	5/50			mgl:adn,fba.	
c	TR306	739.mg	12.2gm	0/50	179.mg	1/50	357.mg	4/50	714.mg	4/50			tnv:men,msm.	S
d	TR306	716.mg	11.2gm	0/50	179.mg	0/50	357.mg	2/50	714.mg	4/50				
e	TR306	853.mg	11.1gm	0/50	179.mg	1/50	357.mg	2/50	714.mg	5/50			sub:fib,srn.	S
f	TR306	892.mg	28.3gm	0/50	179.mg	1/50	357.mg	2/50	714.mg	4/50				S
g	TR306	1.34gm	n.s.s.	0/50	179.mg	1/50	357.mg	0/50	714.mg	3/50				S
h	TR306	186.mg	n.s.s.	46/50	179.mg	48/50	357.mg	49/50	714.mg	48/50				
i	TR306	1.43gm	n.s.s.	2/50	179.mg	2/50	357.mg	4/50	714.mg	1/50			liv:hpa,hpc,nnd.	
3141	1801m	489.mg	n.s.s.	0/20	5.00mg	1/20	50.0mg	2/20	125.mg	1/20	250.mg	0/20	Serota;fctx,24,951-958;1986	
3142	1801n	307.mg	n.s.s.	9/100	188.mg	4/25								
a	1801n	341.mg	n.s.s.	13/100	188.mg	4/25								
b	1801n	966.mg	n.s.s.	4/100	188.mg	0/25								
3143	1801o	2.41gm	n.s.s.	13/100	5.00mg	1/50	50.0mg	1/50	125.mg	2/50	250.mg	2/50		
a	1801o	2.58gm	n.s.s.	9/100	5.00mg	1/50	50.0mg	1/50	125.mg	2/50	250.mg	1/50		
b	1801o	3.01gm	n.s.s.	4/100	5.00mg	0/50	50.0mg	0/50	125.mg	0/50	250.mg	1/50		
3144	BT3002	403.mg	n.s.s.	4/50	39.6mg	3/50	198.mg	9/50					Maltoni;anya,534,352-366;1988	
a	BT3002	151.mg	n.s.s.	28/50	39.6mg	37/50	198.mg	33/50						
b	BT3002	320.mg	n.s.s.	14/50	39.6mg	12/50	198.mg	16/50						
c	BT3002	141.mg	n.s.s.	39/50	39.6mg	41/50	198.mg	39/50						
3145	BT4005	17.7mg	1.23gm	24/60	29.4mg	35/54								
a	BT4005	13.6mg	n.s.s.	35/60	29.4mg	42/54								
b	BT4005	70.6mg	n.s.s.	9/60	29.4mg	10/54								
3146	BT3002	47.0mg	n.s.s.	37/50	39.6mg	33/50	(198.mg	22/50)						
a	BT3002	68.6mg	n.s.s.	15/50	39.6mg	16/50	(198.mg	6/50)						
3147	1600	627.mg	n.s.s.	79/96	130.mg	81/95	390.mg	80/96	910.mg	83/97			Burek;faat,4,30-47;1984/pers.comm.	
a	1600	1.25gm	n.s.s.	34/96	130.mg	24/95	390.mg	30/96	(910.mg	16/97)				
3148	1890m	84.1mg	n.s.s.	51/69	13.0mg	57/69	52.0mg	60/69	130.mg	55/69			Nitschke;faat,11,48-59;1988/pers.comm.	
a	1890m	93.8mg	n.s.s.	52/70	13.0mg	58/70	52.0mg	61/70	130.mg	55/70				
b	1890m	1.52gm	n.s.s.	0/70	13.0mg	0/70	52.0mg	1/70	130.mg	0/70				
c	1890m	1.09gm	n.s.s.	1/70	13.0mg	0/70	52.0mg	2/70	130.mg	1/70				
d	1890m	813.mg	n.s.s.	4/70	13.0mg	4/70	52.0mg	3/70	130.mg	4/70				
e	1890m	859.mg	n.s.s.	3/69	13.0mg	5/69	52.0mg	4/69	130.mg	3/69				

```
      Spe Strain Site   Xpo+Xpt                                                         TD50     2Tailpvl
         Sex   Route  Hist    Notes                                                        DR      AuOp
3149 R f sss inh mgl fba 12m24 e                                    .     ±              37.7mg   P<.05
  a  R f sss inh mgl ben 12m24 e                                                         38.1mg   P<.05
  b  R f sss inh mgl adc 12m24 e                                                         1.14gm   P<.6    -
  c  R f sss inh liv nnd 12m24 e                                                         no dre   P=1.    -
  d  R f sss inh liv hpc 12m24 e                                                         no dre   P=1.    -
3150 R m sss inh slg mix 24m24 e                                         .   +           4.26gm * P<.0005
  a  R m sss inh mgl ben 24m24 e                                                         5.08gm * P<.03
3151 R m sss inh liv tum 86w86 e                                             .>          no dre   P=1.    -


4,4'-METHYLENEBIS(N,N-DIMETHYL)BENZENAMINE..1 ug...:..10.....:...100...:..1mg...:..10.....:...100...:..1g....:..10
3152 M f b6c eat liv MXA 78w91                                        :  +   :           207.mg * P<.003 c
  a  M f b6c eat TBA MXB 78w91                                                           280.mg * P<.07
  b  M f b6c eat liv MXB 78w91                                                           207.mg * P<.003
  c  M f b6c eat lun MXB 78w91                                              .>            1.08gm \ P<.2
3153 M m b6c eat TBA MXB 78w91                                              :>            367.mg * P<.2    -
  a  M m b6c eat liv MXB 78w91                                                           558.mg * P<.2
  b  M m b6c eat lun MXB 78w91                                                           1.41gm * P<.4
3154 R f f34 eat thy MXA 14m24                                   :  +  :                 16.4mg / P<.0005c
  a  R f f34 eat thy fcc 14m24                                                           27.7mg / P<.0005c
  b  R f f34 eat liv nnd 14m24                                                           203.mg * P<.05
  c  R f f34 eat TBA MXB 14m24                                                           41.7mg * P<.4
  d  R f f34 eat liv MXB 14m24                                                           203.mg * P<.05
3155 R m f34 eat thy MXA 14m24                                   :  +  :                 16.5mg / P<.0005c
  a  R m f34 eat thy fcc 14m24                                                           28.6mg / P<.002 c
  b  R m f34 eat TBA MXB 14m24                                                           42.9mg * P<.5
  c  R m f34 eat liv MXB 14m24                                                           no dre   P=1.


4,4'-METHYLENEDIANILINE.2HCl        100ng..:..1ug...:..10....:..100....:..1mg...:..10.....:...100...:..1g.....:..10
3156 M f b6c wat MXB MXB 24m24                                        :  +   :           32.6mg * P<.0005
  a  M f b6c wat liv MXA 24m24                                                           59.3mg * P<.0005c
  b  M f b6c wat --- MXA 24m24                                                           64.7mg * P<.007 c
  c  M f b6c wat thy MXA 24m24                                                           128.mg * P<.0005
  d  M f b6c wat liv hpc 24m24                                                           137.mg * P<.002 c
  e  M f b6c wat thy fca 24m24                                                           150.mg * P<.0005c
  f  M f b6c wat liv hpa 24m24                                                           135.mg * P<.02  c
  g  M f b6c wat lun MXA 24m24                                                           268.mg * P<.04
  h  M f b6c wat lun a/a 24m24                                                           327.mg * P<.04
  i  M f b6c wat TBA MXB 24m24                                                           52.1mg * P<.05
  j  M f b6c wat liv MXB 24m24                                                           59.3mg * P<.0005
  k  M f b6c wat lun MXB 24m24                                                           268.mg * P<.04
3157 M m b6c wat liv MXA 24m24                                   :  +    :               17.0mg \ P<.0005
  a  M m b6c wat liv hpc 24m24                                                           22.3mg \ P<.0005c
  b  M m b6c wat MXB MXB 24m24                                                           28.7mg * P<.0005
  c  M m b6c wat adr phe 24m24                                                           70.0mg * P<.0005c
  d  M m b6c wat thy fca 24m24                                                           88.4mg / P<.0005c
  e  M m b6c wat TBA MXB 24m24                                                           47.4mg * P<.07
  f  M m b6c wat liv MXB 24m24                                                           17.0mg \ P<.0005
  g  M m b6c wat lun MXB 24m24                                                           no dre   P=1.
3158 R f f34 wat thy MXB 24m24                                   :  +  :                 22.0mg * P<.0005
  a  R f f34 wat thy MXA 24m24                                                           27.2mg * P<.0005
  b  R f f34 wat thy fca 24m24                                                           33.5mg / P<.0005c
  c  R f f34 wat thy cca 24m24                                                           73.7mg * P<.007 c
  d  R f f34 wat thy MXA 24m24                                                           70.0mg * P<.04
  e  R f f34 wat TBA MXB 24m24                                                           277.mg * P<1.
  f  R f f34 wat liv MXB 24m24                                                           118.mg * P<.4
3159 R m f34 wat MXB MXB 24m24                                   :+   :                  12.5mg * P<.0005
  a  R m f34 wat liv nnd 24m24                                                           14.3mg * P<.0005c
  b  R m f34 wat thy MXA 24m24                                                           49.5mg * P<.006
  c  R m f34 wat thy fcc 24m24                                                           85.4mg * P<.003 c
  d  R m f34 wat TBA MXB 24m24                                                           no dre   P=1.
  e  R m f34 wat liv MXB 24m24                                                           14.1mg * P<.0005


METHYLGUANIDINE                     100ng..:..1ug...:..10....:..100....:..1mg...:..10.....:...100...:..1g.....:..10
3160 R m wis eat liv hem 68w68 e                                           .>            84.0mg   P<.2


7-METHYLGUANINE                     100ng..:..1ug...:..10....:..100....:..1mg...:..10.....:...100...:..1g.....:..10
3161 R b b46 ivj tba mal 25m29                                       .>                  8.24mg   P<.2    -


METHYLHYDRAZINE                     100ng..:..1ug...:..10....:..100....:..1mg...:..10.....:...100...:..1g.....:..10
3162 H f syg wat liv mhs 86w86 e                                      .  +  .            15.4mg   P<.0005+
  a  H f syg wat cec mix 86w86 e                                                         25.8mg   P<.0005+
  b  H f syg wat cec pla 86w86 e                                                         29.8mg   P<.002
  c  H f syg wat liv mix 86w86 e                                                         30.8mg   P<.006
  d  H f syg wat lun ang 86w86 e                                                         297.mg   P<.2    -
3163 H m syg wat liv mhs 23m23 e                                  .  +  .                9.20mg   P<.0005+
  a  H m syg wat cec mix 23m23 e                                                         42.1mg   P<.004 +
  b  H m syg wat cec pla 23m23 e                                                         50.7mg   P<.01
  c  H m syg wat lun tum 23m23 e                                                         no dre   P=1.
3164 M f swi wat lun ade 62w62 es                                     .  +        .      19.6mg   P<.008
  a  M f swi wat bil mix 62w62 es                                                        24.6mg   P<.003 +
  b  M f swi wat bil cho 62w62 es                                                        29.2mg   P<.006
```

	RefNum	LoConf	UpConf	Cntrl	1Dose	1Inc	2Dose	2Inc	Citation or Pathology
									Brkly Code
3149	1890n	11.8mg	n.s.s.	51/69	65.0mg	23/25			
a	1890n	11.9mg	n.s.s.	52/70	65.0mg	23/25			
b	1890n	149.mg	n.s.s.	3/69	65.0mg	2/25			
c	1890n	220.mg	n.s.s.	4/70	65.0mg	1/25			
d	1890n	335.mg	n.s.s.		65.0mg	0/25			
3150	1600	2.28gm	12.1gm	1/92	91.0mg	0/95	273.mg	5/95	637.mg 11/97	Burek;faat,4,30-47;1984/pers.comm.
a	1600	2.17gm	n.s.s.	7/92	91.0mg	3/95	273.mg	7/95	637.mg 14/97	
3151	1890m	66.5mg	n.s.s.	0/70	9.10mg	0/70	36.4mg	0/70	91.0mg 0/70	Nitschke;faat,11,48-59;1988/pers.comm.

4,4'-METHYLENEBIS(N,N-DIMETHYL)BENZENAMINE   101-61-1

	RefNum	LoConf	UpConf	Cntrl	1Dose	1Inc	2Dose	2Inc	
3152	TR186	126.mg	1.14gm	1/20	139.mg	19/50	278.mg	23/50	liv:hpa,hpc.
a	TR186	122.mg	n.s.s.	6/20	139.mg	26/50	278.mg	30/50	
b	TR186	126.mg	1.14gm	1/20	139.mg	19/50	278.mg	23/50	liv:hpa,hpc,nnd.
c	TR186	328.mg	n.s.s.	0/20	139.mg	3/50	(278.mg	0/50)	lun:a/a,a/c.
3153	TR186	140.mg	n.s.s.	7/20	128.mg	24/50	257.mg	31/50	
a	TR186	216.mg	n.s.s.	5/20	128.mg	12/50	257.mg	22/50	liv:hpa,hpc,nnd.
b	TR186	437.mg	n.s.s.	1/20	128.mg	6/50	257.mg	7/50	lun:a/a,a/c.
3154	TR186	11.2mg	25.8mg	0/20	11.0mg	4/50	21.0mg	36/50	thy:fca,fcc.
a	TR186	17.4mg	51.4mg	0/20	11.0mg	3/50	21.0mg	23/50	
b	TR186	70.2mg	n.s.s.	0/20	11.0mg	0/50	21.0mg	4/50	S
c	TR186	10.8mg	n.s.s.	17/20	11.0mg	33/50	21.0mg	44/50	
d	TR186	70.2mg	n.s.s.	0/20	11.0mg	0/50	21.0mg	4/50	liv:hpa,hpc,nnd.
3155	TR186	10.7mg	33.5mg	1/20	8.80mg	4/50	16.8mg	34/50	thy:fca,fcc.
a	TR186	16.5mg	117.mg	1/20	8.80mg	4/50	16.8mg	21/50	
b	TR186	10.2mg	n.s.s.	14/20	8.80mg	29/50	16.8mg	44/50	
c	TR186	76.0mg	n.s.s.	3/20	8.80mg	5/50	16.8mg	3/50	liv:hpa,hpc,nnd.

4,4'-METHYLENEDIANILINE.2HCl   13552-44-8

	RefNum	LoConf	UpConf	Cntrl	1Dose	1Inc	2Dose	2Inc	
3156	TR248	19.1mg	95.5mg	16/50	29.4mg	36/50	59.1mg	44/50	---:mlh,mlm,mlu,mno; liv:hpa,hpc; thy:fca. C
a	TR248	35.2mg	162.mg	4/50	29.4mg	15/50	59.1mg	23/50	liv:hpa,hpc.
b	TR248	32.7mg	1.16gm	13/50	29.4mg	28/50	59.1mg	29/50	---:mlh,mlm,mlu,mno.
c	TR248	71.6mg	265.mg	0/50	29.4mg	1/50	59.1mg	15/50	thy:fca,fcc. S
d	TR248	72.0mg	611.mg	1/50	29.4mg	6/50	59.1mg	11/50	
e	TR248	80.7mg	338.mg	0/50	29.4mg	1/50	59.1mg	13/50	
f	TR248	65.1mg	n.s.s.	3/50	29.4mg	9/50	59.1mg	12/50	
g	TR248	109.mg	n.s.s.	2/50	29.4mg	3/50	59.1mg	8/50	lun:a/a,a/c. S
h	TR248	129.mg	n.s.s.	1/50	29.4mg	2/50	59.1mg	6/50	S
i	TR248	22.1mg	n.s.s.	32/50	29.4mg	47/50	59.1mg	47/50	
j	TR248	35.2mg	162.mg	4/50	29.4mg	15/50	59.1mg	23/50	liv:hpa,hpc,nnd.
k	TR248	109.mg	n.s.s.	2/50	29.4mg	3/50	59.1mg	8/50	lun:a/a,a/c.
3157	TR248	9.07mg	65.8mg	17/50	24.5mg	43/50	(49.5mg	37/50)	liv:hpa,hpc. S
a	TR248	12.1mg	75.5mg	10/50	24.5mg	33/50	(49.5mg	29/50)	
b	TR248	17.1mg	78.7mg	12/50	24.5mg	36/50	49.5mg	34/50	adr:phe; liv:hpc; thy:fca. C
c	TR248	39.7mg	231.mg	2/50	24.5mg	12/50	49.5mg	14/50	
d	TR248	51.4mg	172.mg	0/50	24.5mg	3/50	49.5mg	16/50	
e	TR248	19.1mg	n.s.s.	37/50	24.5mg	47/50	49.5mg	45/50	
f	TR248	9.07mg	65.8mg	17/50	24.5mg	43/50	(49.5mg	37/50)	liv:hpa,hpc,nnd.
g	TR248	168.mg	n.s.s.	13/50	24.5mg	12/50	49.5mg	4/50	lun:a/a,a/c.
3158	TR248	13.9mg	38.4mg	0/50	8.41mg	5/50	16.7mg	22/50	thy:cca,fca. C
a	TR248	16.6mg	50.3mg	0/50	8.41mg	4/50	16.7mg	19/50	thy:fca,fcc. S
b	TR248	19.5mg	66.1mg	0/50	8.41mg	2/50	16.7mg	17/50	
c	TR248	34.7mg	873.mg	0/50	8.41mg	3/50	16.7mg	6/50	
d	TR248	30.6mg	n.s.s.	1/50	8.41mg	5/50	16.7mg	7/50	thy:cca,ccr. S
e	TR248	10.6mg	n.s.s.	43/50	8.41mg	48/50	16.7mg	49/50	
f	TR248	31.5mg	n.s.s.	4/50	8.41mg	8/50	16.7mg	8/50	liv:hpa,hpc,nnd.
3159	TR248	8.33mg	22.0mg	1/50	7.36mg	12/50	14.7mg	29/50	liv:nnd; thy:fcc. C
a	TR248	9.35mg	27.1mg	1/50	7.36mg	12/50	14.7mg	25/50	
b	TR248	24.4mg	535.mg	1/50	7.36mg	4/50	14.7mg	10/50	thy:fca,fcc. S
c	TR248	36.9mg	422.mg	0/50	7.36mg	0/50	14.7mg	7/50	
d	TR248	15.3mg	n.s.s.	45/50	7.36mg	42/50	14.7mg	40/50	
e	TR248	9.22mg	26.6mg	1/50	7.36mg	13/50	14.7mg	25/50	liv:hpa,hpc,nnd.

METHYLGUANIDINE   471-29-4

	RefNum	LoConf	UpConf	Cntrl	1Dose	1Inc		
3160	1471	13.5mg	n.s.s.	0/10	64.0mg	1/5		Matsukura;zkko,90,87-94;1977

7-METHYLGUANINE   578-76-7

	RefNum	LoConf	UpConf	Cntrl	1Dose	1Inc		
3161	1443	2.03mg	n.s.s.	0/20	.565mg	2/30		Kruger;zkko,75,253-254;1971

METHYLHYDRAZINE   60-34-4

	RefNum	LoConf	UpConf	Cntrl	1Dose	1Inc		
3162	194	8.53mg	31.7mg	0/85	13.6mg	16/47		Toth;canr,33,2744-2753;1973
a	194	11.7mg	93.3mg	1/67	13.6mg	9/39		
b	194	12.9mg	132.mg	1/67	13.6mg	8/39		
c	194	9.28mg	418.mg	0/39	13.6mg	3/16		
d	194	48.4mg	n.s.s.	0/85	13.6mg	1/47		
3163	194	5.72mg	16.0mg	0/72	12.0mg	27/48		
a	194	17.1mg	359.mg	1/59	12.0mg	7/39		
b	194	19.2mg	4.28gm	1/59	12.0mg	6/39		
c	194	110.mg	n.s.s.	0/87	12.0mg	0/48		
3164	1117	8.03mg	558.mg	12/107	20.0mg	12/39		Toth;ijcn,9,109-118;1972/1969a
a	1117	10.6mg	119.mg	0/34	20.0mg	7/39		
b	1117	11.9mg	247.mg	0/34	20.0mg	6/39		

```
      Spe Strain  Site   Xpo+Xpt                                                      TD50    2Tailpvl
         Sex  Route  Hist     Notes                                                     DR      AuOp
```

```
  c   M f swi wat lun mix 62w62 es                                                     21.4mg   P<.02  +
  d   M f swi wat liv hpt 62w62 es                                                     51.1mg   P<.05  +
3165  M m swi wat liv mix 69w69 es              .        .      +      .               4.58mg   P<.0005+
  a   M m swi wat liv hpt 69w69 es                                                     7.25mg   P<.0005
  b   M m swi wat lun ade 69w69 es                                                     10.1mg   P<.0005
  c   M m swi wat bil mix 69w69 es                                                     22.5mg   P<.002 +
```

METHYLHYDRAZINE SULFATE   100ng..:..1ug....:..10....:..100...:..1mg....:..10...:..100....:..1g.....:..10
```
3166  M f swi gav lun tum 40w55                                     .>                no dre   P=1.   -
3167  M f swi wat lun mix 26m26 e                              . + .                   2.98mg   P<.0005+
  a   M f swi wat lun ade 26m26 e                                                      4.02mg   P<.0005
  b   M f swi wat lun adc 26m26 e                                      .               5.63mg   P<.0005
  c   M f swi wat liv mix 26m26 e                                                      47.2mg   P<.5
3168  M m swi wat lun mix 26m26 e                            . +    .                  2.51mg   P<.0005+
  a   M m swi wat lun ade 26m26 e                                                      2.93mg   P<.0005
  b   M m swi wat lun adc 26m26 e                                                      12.0mg   P<.002
  c   M m swi wat --- mly 26m26 e                                                      7.85mg   P<.02
  d   M m swi wat liv mix 26m26 e                                                      no dre   P=1.
```

METHYLHYDROQUINONE   100ng..:..1ug....:..10....:..100...:..1mg....:..10...:..100....:..1g.....:..10
```
3169  R m f3d eat for tum 51w52                                            .>          no dre   P=1.
  a   R m f3d eat stg tum 51w52                                                        no dre   P=1.
```

METHYLNITRAMINE   100ng..:..1ug....:..10....:..100...:..1mg....:..10...:..100....:..1g.....:..10
```
3170  R f bdf gav spd nim 27m30 a                                 .   +    .           28.7mg * P<.006 +
  a   R f bdf gav spn nim 27m30 a                                    +                 39.8mg * P<.3   +
  b   R f bdf gav pnr nim 27m30 a                                                      171.mg * P<1.   +
3171  R m bdf gav spd nim 26m29 a                             .   +    .               12.5mg * P<.0005+
  a   R m bdf gav spn nim 26m29 a                                                      13.6mg * P<.008 +
  b   R m bdf gav pnr nim 26m29 a                                                      50.1mg * P<.3   +
```

METHYLNITROSAMINO-N,N-DIMETHYLETHYLAMINE.:..1_ug....:..10....:..100...:..1mg....:..10...:..100....:..1g.....:..10
```
3172  H f syg gav nac mix 14m25 e                            . +    .                  2.76mg   P<.0005+
  a   H f syg gav lun a/a 14m25 e                                                      9.45mg   P<.02  +
  b   H f syg gav nac mal 14m25 e                                                      9.45mg   P<.02
  c   H f syg gav col mix 14m25 e                                                      21.0mg   P<.09
  d   H f syg gav liv hct 14m25 e                                                      44.0mg   P<.3
3173  H m syg gav nac mix 14m25 e                                .   +     .           6.25mg   P<.005 +
  a   H m syg gav liv hct 14m25 e                                                      11.7mg   P<.04  +
  b   H m syg gav col adc 14m25 e                                                      38.7mg   P<.3
  c   H m syg gav nac mal 14m25 e                                                      38.7mg   P<.3
  d   H m syg gav lun tum 14m25 e                                                      no dre   P=1.
```

4-(METHYLNITROSAMINO)-1-(3-PYRRIDYL)-1-BUTANOL ...:..10....:..100...:..1mg....:..10...:..100....:..1g.....:..10
```
3174  R m f34 wat lun mix 26m26              .   + .                                   .103mg   P<.0005+
  a   R m f34 wat lun adc 26m26                                                        .409mg   P<.0005
  b   R m f34 wat lun adq 26m26                                                        .577mg   P<.0005
  c   R m f34 wat pae mix 26m26                                                        .668mg   P<.0005+
  d   R m f34 wat pae aod 26m26                                                        1.09mg   P<.0005
  e   R m f34 wat amd tum 26m26                                                        1.39mg   P<.002
  f   R m f34 wat lun ade 26m26                                                        1.38mg   P<.04
  g   R m f34 wat pae ana 26m26                                                        2.14mg   P<.05
  h   R m f34 wat liv hpt 26m26                                                        5.86mg   P<.2
  i   R m f34 wat liv mix 26m26                                                        7.25mg   P<.7
  j   R m f34 wat liv ade 26m26                                                        no dre   P=1.
```

4-(METHYLNITROSAMINO)-1-(3-PYRRIDYL)-1-(BUTANONE)..:..1_0....:..100...:..1mg....:..10...:..100....:..1g.....:..10
```
3175  R m f34 wat lun mix 27m30               .  + .                                   .182mg * P<.0005+
  a   R m f34 wat lun ade 27m30                                                        .343mg Z P<.002
  b   R m f34 wat pae mix 27m30                                                        .490mg Z P<.006 +
  c   R m f34 wat liv mix 27m30                                                        .651mg * P<.0005+
  d   R m f34 wat lun adc 27m30                                                        .672mg * P<.0005
  e   R m f34 wat liv ade 27m30                                                        .871mg * P<.0005
  f   R m f34 wat lun adq 27m30                                                        1.46mg Z P<.0005
  g   R m f34 wat nas mix 27m30                                                        1.66mg * P<.0005
  h   R m f34 wat lun sqc 27m30                                                        3.40mg * P<.005
  i   R m f34 wat --- mix 27m30                                                        .140mg Z P<.02
  j   R m f34 wat liv hpt 27m30                                                        2.74mg * P<.03
  k   R m f34 wat pae aod 27m30                                                        6.93mg * P<.08
  l   R m f34 wat pae ana 27m30                                                        29.2mg * P<1.
```

(N-6)-(METHYLNITROSO)ADENINE   100ng..:..1ug....:..10....:..100...:..1mg....:..10...:..100....:..1g.....:..10
```
3176  M f scd wat liv mix 24m24                                       .>               63.2mg   P<.2
  a   M f scd wat lun tum 24m24                                                        909.mg   P<1.
3177  M m scd wat lun tum 24m24                             .    ±                     18.0mg   P<.02  +
  a   M m scd wat liv mix 24m24                                                        62.6mg   P<.4
```

(N-6)-(METHYLNITROSO)ADENOSINE   100ng..:..1ug....:..10....:..100...:..1mg....:..10...:..100....:..1g.....:..10
```
3178  M f scd wat rep mix 24m24                                  . +   .               15.8mg   P<.0005+
  a   M f scd wat lun tum 24m24                                                        17.4mg   P<.0005+
  b   M f scd wat mgl car 24m24                                                        28.7mg   P<.0005+
```

	RefNum	LoConf	UpConf	Cntrl	1Dose	1Inc	2Dose	2Inc	Citation or Pathology
									Brkly Code
c	1117	8.30mg	n.s.s.	14/107	20.0mg	12/39			
d	1117	15.5mg	n.s.s.	0/31	20.0mg	3/33			
3165	1117	1.43mg	24.5mg	0/45	16.7mg	4/6			
a	1117	2.10mg	52.8mg	0/45	16.7mg	3/6			
b	1117	4.45mg	44.8mg	10/91	16.7mg	11/24			
c	1117	6.79mg	167.mg	0/64	16.7mg	3/15			

METHYLHYDRAZINE SULFATE   302-15-8

	RefNum	LoConf	UpConf	Cntrl	1Dose	1Inc	Citation or Pathology
3166	1095	10.8mg	n.s.s.	8/85	10.4mg	1/25	Roe;natu,216,375-376;1967
3167	1117	1.58mg	8.60mg	14/88	2.00mg	23/45	Toth;ijcn,9,109-118;1972/1969a
a	1117	1.99mg	15.7mg	12/88	2.00mg	19/45	
b	1117	2.78mg	16.3mg	2/88	2.00mg	12/45	
c	1117	8.16mg	n.s.s.	3/88	2.00mg	3/45	
3168	1117	1.37mg	6.32mg	10/86	1.67mg	23/48	
a	1117	1.54mg	8.52mg	10/86	1.67mg	21/48	
b	1117	4.57mg	57.4mg	0/86	1.67mg	5/48	
c	1117	3.12mg	n.s.s.	2/55	1.67mg	8/43	
d	1117	7.98mg	n.s.s.	2/41	1.67mg	2/41	

METHYLHYDROQUINONE   95-71-6

	RefNum	LoConf	UpConf	Cntrl	1Dose	1Inc	Citation or Pathology
3169	1950	455.mg	n.s.s.	0/10	588.mg	0/15	Hirose;canr,48,5310-5315;1988
a	1950	455.mg	n.s.s.	0/10	588.mg	0/15	

METHYLNITRAMINE   (N-nitromethylamine) 598-57-2

	RefNum	LoConf	UpConf	Cntrl	1Dose	1Inc	2Dose	2Inc	Citation or Pathology
3170	1931	10.8mg	271.mg	0/10	5.43mg	0/10	10.9mg	5/10	Scherf;carc,10,1977-1981;1989/pers.comm.
a	1931	13.7mg	n.s.s.	0/10	5.43mg	3/10	10.9mg	1/10	
b	1931	27.9mg	n.s.s.	0/10	5.43mg	1/10	10.9mg	0/10	
3171	1931	5.72mg	39.8mg	0/10	5.43mg	2/10	10.9mg	7/10	
a	1931	6.31mg	265.mg	0/10	5.43mg	5/10	10.9mg	4/10	
b	1931	15.1mg	n.s.s.	0/10	5.43mg	2/10	10.9mg	1/10	

METHYLNITROSAMINO-N,N-DIMETHYLETHYLAMINE   ---

	RefNum	LoConf	UpConf	Cntrl	1Dose	1Inc	Citation or Pathology
3172	2111	1.18mg	7.91mg	0/12	5.38mg	9/12	Lijinsky;clet,63,101-107;1992
a	2111	3.22mg	n.s.s.	0/12	5.38mg	4/12	
b	2111	3.22mg	n.s.s.	0/12	5.38mg	4/12	
c	2111	5.15mg	n.s.s.	0/12	5.38mg	2/12	
d	2111	7.16mg	n.s.s.	0/12	5.38mg	1/12	
3173	2111	2.33mg	50.2mg	0/12	4.74mg	5/12	
a	2111	3.52mg	n.s.s.	0/12	4.74mg	3/12	
b	2111	6.30mg	n.s.s.	0/12	4.74mg	1/12	
c	2111	6.30mg	n.s.s.	0/12	4.74mg	1/12	
d	2111	12.2mg	n.s.s.	0/12	4.74mg	0/12	

4-(METHYLNITROSAMINO)-1-(3-PYRRIDYL)-1-BUTANOL   ---

	RefNum	LoConf	UpConf	Cntrl	1Dose	1Inc	Citation or Pathology
3174	1866	56.5ug	.199mg	6/80	.250mg	26/30	Rivenson;canr,48,6912-6917;1988/pers.comm.
a	1866	.203mg	1.07mg	2/80	.250mg	12/30	
b	1866	.265mg	1.76mg	1/80	.250mg	9/30	
c	1866	.294mg	2.29mg	1/80	.250mg	8/30	
d	1866	.413mg	4.54mg	0/80	.250mg	5/30	
e	1866	.479mg	7.65mg	0/80	.250mg	4/30	
f	1866	.445mg	n.s.s.	3/80	.250mg	5/30	
g	1866	.589mg	n.s.s.	1/80	.250mg	3/30	
h	1866	.954mg	n.s.s.	0/80	.250mg	1/30	
i	1866	.712mg	n.s.s.	6/80	.250mg	3/30	
j	1866	.935mg	n.s.s.	6/80	.250mg	2/30	

4-(METHYLNITROSAMINO)-1-(3-PYRRIDYL)-1-(BUTANONE)   64091-91-4

	RefNum	LoConf	UpConf	Cntrl	1Dose	1Inc	2Dose	2Inc			Citation or Pathology
3175	1866	.123mg	.297mg	6/80	25.0mg	9/80	50.0ug	20/80	.250mg	27/30	Rivenson;canr,48,6912-6917;1988/pers.comm.
a	1866	.181mg	1.52mg	3/80	25.0mg	5/80	50.0ug	16/80	(.250mg	2/30)	
b	1866	.241mg	5.70mg	1/80	25.0mg	5/80	50.0ug	9/80	(.250mg	2/30)	
c	1866	.345mg	1.90mg	6/80	25.0mg	3/80	50.0ug	11/80	.250mg	12/30	
d	1866	.375mg	1.53mg	2/80	25.0mg	3/80	50.0ug	4/80	.250mg	13/30	
e	1866	.427mg	3.37mg	6/80	25.0mg	2/80	50.0ug	9/80	.250mg	10/30	
f	1866	.677mg	4.26mg	1/80	25.0mg	0/80	50.0ug	0/80	.250mg	9/30	
g	1866	.751mg	5.90mg	0/80	25.0mg	1/80	50.0ug	2/80	.250mg	5/30	
h	1866	1.17mg	42.2mg	0/80	25.0mg	1/80	50.0ug	0/80	.250mg	3/30	
i	1866	63.5ug	n.s.s.	9/80	25.0mg	21/80	(50.0ug	10/80	.250mg	2/30)	
j	1866	1.04mg	n.s.s.	0/80	25.0mg	1/80	50.0ug	2/80	.250mg	1/30	
k	1866	1.70mg	n.s.s.	0/80	25.0mg	0/80	50.0ug	1/80	.250mg	1/30	
l	1866	.859mg	n.s.s.	1/80	25.0mg	5/80	50.0ug	8/80	.250mg	1/30	

(N-6)-(METHYLNITROSO)ADENINE   ---

	RefNum	LoConf	UpConf	Cntrl	1Dose	1Inc	Citation or Pathology
3176	1255	19.8mg	n.s.s.	1/16	20.6mg	5/20	Anderson;ijcn,24,319-322;1979
a	1255	28.5mg	n.s.s.	3/16	20.6mg	4/20	
3177	1255	7.37mg	n.s.s.	4/21	17.1mg	11/19	
a	1255	13.8mg	n.s.s.	5/21	17.1mg	7/19	

(N-6)-(METHYLNITROSO)ADENOSINE   21928-82-5

	RefNum	LoConf	UpConf	Cntrl	1Dose	1Inc	Citation or Pathology
3178	1255	7.88mg	38.3mg	1/16	35.7mg	16/20	Anderson;ijcn,24,319-322;1979
a	1255	8.22mg	58.0mg	3/16	35.7mg	16/20	
b	1255	13.7mg	101.mg	1/16	35.7mg	12/20	

```
     Spe Strain Site   Xpo+Xpt                                                        TD50    2Tailpvl
         Sex   Route  Hist      Notes                                                       DR     AuOp
──────────────────────────────────────────────────────────────────────────────────────────────────────
     c   M f scd wat ute tum 24m24                                                     84.9mg   P<.02  +
     d   M f scd wat liv mix 24m24                                                     no dre   P=1.
  3179   M m scd wat lun tum 24m24                              .     +        .       21.6mg   P<.002 +
     a   M m scd wat liv mix 24m24                                                     no dre   P=1.

METHYLNITROSOCYANAMIDE         100ng..:..1ug...:..10....:..100....:..1mg...:..10....:..100....:..1g......:..10
  3180   R f sda wat for mix 60w95 er                              .  +    .           .480mg * P<.0005+
     a   R f sda wat for car 60w95 er                                                  1.17mg * P<.0005+

N-METHYLOLACRYLAMIDE           100ng..:..1ug...:..10....:..100....:..1mg...:..10....:..100....:..1g......:..10
  3181   M f b6c gav MXB MXB 24m24                                       :  +   :      27.8mg * P<.0005
     a   M f b6c gav hag MXA 24m24                                                     44.6mg * P<.0005
     b   M f b6c gav hag MXA 24m24                                                     55.5mg * P<.002  c
     c   M f b6c gav liv hpa 24m24                                                     66.9mg / P<.0005c
     d   M f b6c gav liv MXA 24m24                                                     74.8mg * P<.009
     e   M f b6c gav ova gcb 24m24                                                     119.mg * P<.006  c
     f   M f b6c gav lun MXA 24m24                                                     111.mg * P<.06   c
     g   M f b6c gav TBA MXB 24m24                                                     30.9mg * P<.03
     h   M f b6c gav liv MXB 24m24                                                     74.8mg * P<.009
     i   M f b6c gav lun MXB 24m24                                                     111.mg * P<.06
  3182   M m b6c gav MXB MXB 24m24                                       :  +   :      13.3mg * P<.0005
     a   M m b6c gav hag MXA 24m24                                                     17.4mg * P<.0005
     b   M m b6c gav hag ade 24m24                                                     17.5mg * P<.0005c
     c   M m b6c gav liv MXA 24m24                                                     29.4mg * P<.003  c
     d   M m b6c gav lun MXA 24m24                                                     38.5mg * P<.002  c
     e   M m b6c gav liv hpa 24m24                                                     50.2mg / P<.008
     f   M m b6c gav lun a/a 24m24                                                     66.0mg * P<.009
     g   M m b6c gav lun a/c 24m24                                                     75.1mg * P<.006
     h   M m b6c gav liv hpc 24m24                                                     63.4mg * P<.05
     i   M m b6c gav TBA MXB 24m24                                                     22.3mg * P<.02
     j   M m b6c gav liv MXB 24m24                                                     29.4mg * P<.003
     k   M m b6c gav lun MXB 24m24                                                     38.5mg * P<.002
  3183   R f f34 gav TBA MXB 24m24                                   :>                no dre   P=1.   -
     a   R f f34 gav liv MXB 24m24                                                     no dre   P=1.
  3184   R m f34 gav ski ker 24m24                               :    ±                #13.0mg \ P<.03  -
     a   R m f34 gav TBA MXB 24m24                                                     3.12gm * P<1.
     b   R m f34 gav liv MXB 24m24                                                     no dre   P=1.

6-METHYLQUINOLINE              100ng..:..1ug...:..10....:..100....:..1mg...:..10....:..100....:..1g......:..10
  3185   R f f34 eat liv hnd 24m24 e                                          .>       230.mg   P<.4   -
  3186   R m f34 eat liv hnd 24m24 e                                       .  ±        123.mg   P<.03  -

8-METHYLQUINOLINE              100ng..:..1ug...:..10....:..100....:..1mg...:..10....:..100....:..1g......:..10
  3187   R f f34 eat liv hnd 24m24 e                                           .>      no dre   P=1.   -
  3188   R m f34 eat liv hnd 24m24 e                                          .>       no dre   P=1.   -

p-METHYLSTYRENE                100ng..:..1ug...:..10....:..100....:..1mg...:..10....:..100....:..1g......:..10
  3189   M f swi gav lun ade 78w83                                              .>     616.mg * P<.2   -
     a   M f swi gav liv hpt 78w83                                                     no dre   P=1.
     b   M f swi gav tba mix 78w83                                                     343.mg * P<.2   -
     c   M f swi gav tba mal 78w83                                                     no dre   P=1.
  3190   M m swi gav lun ade 78w83                                                .>   962.mg * P<.3   -
     a   M m swi gav liv hpt 78w83                                                     no dre   P=1.
     b   M m swi gav tba mal 78w83                                                     no dre   P=1.
     c   M m swi gav tba mix 78w83                                                     no dre   P=1.

METHYLTHIOURACIL               100ng..:..1ug...:..10....:..100....:..1mg...:..10....:..100....:..1g......:..10
  3191   H f nss wat thy mix 52w52 ekr                            .     +        .     53.4mg   P<.006 +
  3192   M b cfi eat liv hpt 51w78 erv                                           .>    1.37gm   P<.3   -
     a   M b cfi eat thy tum 51w78 erv                                                 no dre   P=1.

METIAPINE                      100ng..:..1ug...:..10....:..100....:..1mg...:..10....:..100....:..1g......:..10
  3193   R f alb eat mgl fba 78w78 e                                          .>       no dre   P=1.   -
  3194   R m alb eat pit ade 78w78 e                                       .>          no dre   P=1.   -

METRONIDAZOLE                  100ng..:..1ug...:..10....:..100....:..1mg...:..10....:..100....:..1g......:..10
  3195   M f swi eat lun mix 28m30 e                                            .  +   .   937.mg * P<.0005+
     a   M f swi eat --- mly 28m30 e                                                   1.30gm * P<.002 +
     b   M f swi eat liv mix 28m30 e                                                   24.2gm   P<.6   -
  3196   M m swi eat lun mix 27m29 e                                          .  +  .   347.mg * P<.0005+
     a   M m swi eat liv mix 27m29 e                                                   20.7gm   P<.8   -
  3197   R f sda eat ute amy 66w75 e                              .     +      .        180.mg   P<.004 -
     a   R f sda eat liv tum 66w75 e                                                   no dre   P=1.
     b   R f sda eat tba mix 66w75 e                                                   61.1mg   P<.03  -
  3198   R f smw eat mgl fba 34m35 ae                                      .  +  .      431.mg * P<.0005+
     a   R f smw eat mgl mix 34m35 ae                                                  446.mg * P<.0005+
     b   R f smw eat liv hpc 34m35 ae                                                  1.86gm * P<.0005+
     c   R f smw eat sub mix 34m35 a                                                   4.05gm * P<.03
     d   R f smw eat ute mix 34m35 a                                                   2.97gm * P<.2
     e   R f smw eat pit mix 34m35 a                                                   no dre   P=1.   -
     f   R f smw eat tba mix 34m35 a                                                   139.mg * P<.0005
```

	RefNum	LoConf	UpConf	Cntrl	1Dose	1Inc	2Dose	2Inc			Citation or Pathology
											Brkly Code
c	1255	32.1mg	n.s.s.	0/16	35.7mg	5/20					
d	1255	89.9mg	n.s.s.	1/16	35.7mg	1/20					
3179	1255	9.65mg	115.mg	4/21	29.7mg	13/19					
a	1255	54.0mg	n.s.s.	5/21	29.7mg	3/19					

METHYLNITROSOCYANAMIDE   33868-17-6

	RefNum	LoConf	UpConf	Cntrl	1Dose	1Inc	2Dose	2Inc			Citation or Pathology
3180	1733	.283mg	.828mg	0/9	.573mg	2/12	1.08mg	12/15	2.23mg 15/15	4.39mg 14/14	Endo;pjpa,50,497-502;1974/pers.comm.
a	1733	.713mg	2.02mg	0/9	.573mg	0/12	1.08mg	5/15	2.23mg 12/15	4.39mg 13/14	

N-METHYLOLACRYLAMIDE   (N-(hydroxymethyl)-acrylamide)   924-42-5

	RefNum	LoConf	UpConf	Cntrl	1Dose	1Inc	2Dose	2Inc	Citation or Pathology	Brkly Code
3181	TR352	15.8mg	89.4mg	14/50	17.5mg	20/50	35.0mg	37/50	hag:ade,anb; liv:hpa; lun:a/a,a/c; ova:gcb.	C
a	TR352	25.3mg	141.mg	5/50	17.5mg	11/50	35.0mg	22/50	hag:ade,anb,car.	S
b	TR352	29.9mg	222.mg	5/50	17.5mg	8/50	35.0mg	20/50	hag:ade,anb.	
c	TR352	35.6mg	241.mg	3/50	17.5mg	4/50	35.0mg	17/50		
d	TR352	35.7mg	2.08gm	6/50	17.5mg	7/50	35.0mg	17/50		
e	TR352	58.1mg	996.mg	0/50	17.5mg	5/50	35.0mg	5/50	liv:hpa,hpc.	S
f	TR352	45.2mg	n.s.s.	6/50	17.5mg	8/50	35.0mg	13/50	lun:a/a,a/c.	
g	TR352	13.9mg	n.s.s.	33/50	17.5mg	41/50	35.0mg	47/50		
h	TR352	35.7mg	2.08gm	6/50	17.5mg	7/50	35.0mg	17/50	liv:hpa,hpc,nnd.	
i	TR352	45.2mg	n.s.s.	6/50	17.5mg	8/50	35.0mg	13/50	lun:a/a,a/c.	
3182	TR352	8.07mg	32.0mg	16/50	17.5mg	30/50	35.0mg	43/50	hag:ade; liv:hpa,hpc; lun:a/a,a/c.	C
a	TR352	11.4mg	30.6mg	2/50	17.5mg	14/50	35.0mg	30/50	hag:ade,car.	S
b	TR352	11.7mg	29.4mg	1/50	17.5mg	14/50	35.0mg	29/50		
c	TR352	15.2mg	186.mg	12/50	17.5mg	17/50	35.0mg	26/50	liv:hpa,hpc.	
d	TR352	20.4mg	163.mg	5/50	17.5mg	10/50	35.0mg	18/50	lun:a/a,a/c.	
e	TR352	23.3mg	1.15gm	8/50	17.5mg	4/50	35.0mg	19/50		
f	TR352	30.7mg	2.27gm	3/50	17.5mg	6/50	35.0mg	11/50		S
g	TR352	35.1mg	897.mg	2/50	17.5mg	4/50	35.0mg	10/50		S
h	TR352	27.1mg	n.s.s.	6/50	17.5mg	13/50	35.0mg	12/50		S
i	TR352	10.2mg	n.s.s.	35/50	17.5mg	39/50	35.0mg	47/50		S
j	TR352	15.2mg	186.mg	12/50	17.5mg	17/50	35.0mg	26/50	liv:hpa,hpc,nnd.	
k	TR352	20.4mg	163.mg	5/50	17.5mg	10/50	35.0mg	18/50	lun:a/a,a/c.	
3183	TR352	6.33mg	n.s.s.	45/50	4.20mg	36/50	8.41mg	42/50		
a	TR352	n.s.s.	n.s.s.	0/50	4.20mg	0/50	8.41mg	0/50	liv:hpa,hpc,nnd.	
3184	TR352	4.72mg	n.s.s.	1/50	4.20mg	6/50	(8.41mg	3/50)		S
a	TR352	5.39mg	n.s.s.	45/50	4.20mg	40/50	8.41mg	45/50		
b	TR352	11.0mg	n.s.s.	4/50	4.20mg	2/50	(8.41mg	0/50)	liv:hpa,hpc,nnd.	

6-METHYLQUINOLINE   91-62-3

	RefNum	LoConf	UpConf	Cntrl	1Dose	1Inc			Citation or Pathology
3185	1529	52.6mg	n.s.s.	3/44	25.0mg	5/37			
3186	1529	42.5mg	n.s.s.	0/31	20.0mg	4/38			Fukushima;clet,14,115-123;1981

8-METHYLQUINOLINE   611-32-5

	RefNum	LoConf	UpConf	Cntrl	1Dose	1Inc			Citation or Pathology
3187	1529	175.mg	n.s.s.	3/44	25.0mg	0/34			
3188	1529	78.3mg	n.s.s.	0/31	20.0mg	0/19			Fukushima;clet,14,115-123;1981

p-METHYLSTYRENE   622-97-9

	RefNum	LoConf	UpConf	Cntrl	1Dose	1Inc	2Dose	2Inc			Citation or Pathology
3189	BT107	200.mg	n.s.s.	13/60	6.71mg	5/60	33.6mg	10/60	168.mg 15/60		
a	BT107	42.6mg	n.s.s.	0/60	6.71mg	0/60	33.6mg	0/60	168.mg 0/60		
b	BT107	104.mg	n.s.s.	33/60	6.71mg	18/60	33.6mg	33/60	168.mg 33/60		
c	BT107	343.mg	n.s.s.	19/60	6.71mg	10/60	33.6mg	20/60	168.mg 13/60		Conti;anya,534,203-234;1988
3190	BT107	262.mg	n.s.s.	7/60	6.71mg	5/60	33.6mg	7/60	168.mg 10/60		
a	BT107	793.mg	n.s.s.	5/60	6.71mg	4/60	33.6mg	6/60	168.mg 2/60		
b	BT107	583.mg	n.s.s.	12/60	6.71mg	7/60	33.6mg	9/60	168.mg 6/60		
c	BT107	320.mg	n.s.s.	20/60	6.71mg	14/60	33.6mg	18/60	168.mg 15/60		

METHYLTHIOURACIL   56-04-2

	RefNum	LoConf	UpConf	Cntrl	1Dose	1Inc			Citation or Pathology
3191	1385	22.1mg	507.mg	0/6	273.mg	7/12			Christov;zkko,77,171-179;1972
3192	1630	337.mg	n.s.s.	0/9	375.mg	2/20			Jemec;canc,40,2188-2202;1977
a	1630	869.mg	n.s.s.	0/9	375.mg	0/20			

METIAPINE   5800-19-1

	RefNum	LoConf	UpConf	Cntrl	1Dose	1Inc	2Dose	2Inc	3Dose	3Inc	Citation or Pathology
3193	200	38.9mg	n.s.s.	1/22	3.00mg	6/20	10.0mg	1/21	30.0mg	2/20	Gibson;txap,25,220-229;1973
3194	200	8.00mg	n.s.s.	3/18	3.00mg	7/20	10.0mg	3/19	(30.0mg	0/20)	

METRONIDAZOLE   443-48-1

	RefNum	LoConf	UpConf	Cntrl	1Dose	1Inc	2Dose	2Inc	3Dose	3Inc	4Dose	4Inc	Citation or Pathology
3195	201	498.mg	3.54gm	14/65	78.0mg	4/10	195.mg	10/20	390.mg	14/19	650.mg	16/32	Rustia;jnci,48,721-729;1972
a	201	654.mg	6.48gm	16/68	78.0mg	3/10	195.mg	5/20	390.mg	10/19	650.mg	18/34	
b	201	3.34gm	n.s.s.	2/53	78.0mg	0/10	195.mg	0/18	390.mg	2/19	650.mg	1/25	
3196	201	217.mg	635.mg	13/67	72.0mg	3/8	180.mg	11/15	360.mg	12/16	600.mg	27/34	
a	201	2.06gm	n.s.s.	6/64	72.0mg	0/8	180.mg	3/14	360.mg	3/16	600.mg	3/32	
3197	200a	62.0mg	1.27gm	0/71	59.4mg	4/36							Cohen;jnci,51,403-417;1973
a	200a	229.mg	n.s.s.	0/71	59.4mg	0/36							
b	200a	24.3mg	n.s.s.	18/71	59.4mg	17/36							
3198	1064	231.mg	1.25gm	30/95	30.0mg	17/29	150.mg	14/28	300.mg	23/29			Rustia;jnci,63,863-868;1979
a	1064	232.mg	1.46gm	34/95	30.0mg	17/29	150.mg	15/28	300.mg	23/29			
b	1064	842.mg	5.46gm	0/58	30.0mg	0/28	150.mg	1/21	300.mg	7/25			
c	1064	1.54gm	n.s.s.	0/100	30.0mg	1/30	150.mg	3/30	300.mg	1/30			
d	1064	855.mg	n.s.s.	13/100	30.0mg	9/30	150.mg	8/30	300.mg	7/30			
e	1064	761.mg	n.s.s.	61/100	30.0mg	24/30	150.mg	19/30	300.mg	17/30			
f	1064	53.2mg	720.mg	80/100	30.0mg	27/30	150.mg	28/30	300.mg	30/30			

```
        Spe Strain  Site    Xpo+Xpt                                          TD50     2Tailpvl
          Sex  Route   Hist      Notes                                         DR      AuOp
```

```
3199 R m smw eat pit mix 33m33 a                              . +        . 731.mg * P<.004 +
 a   R m smw eat tes mix 33m33 a                                          993.mg * P<.01  +
 b   R m smw eat sub mix 33m33 a                                          2.36gm * P<.06
 c   R m smw eat mgl fba 33m33 a                                          2.49gm Z P<.02
 d   R m smw eat liv hpc 33m33 a                                          45.9gm * P<.8
 e   R m smw eat tba mix 33m33 a                                          211.mg * P<.0005
```

MEXACARBATE   100ng..:..1ug....:..10.....:..100....:..1mg....:..10.....:..100....:..1g.....:..10
```
3200 M f b6c eat TBA MXB 78w92 v                                :>        no dre   P=1.   -
 a   M f b6c eat liv MXB 78w92 v                                          no dre   P=1.
 b   M f b6c eat lun MXB 78w92 v                                          no dre   P=1.
3201 M f b6c orl lun ade 76w76 evx                      .     ±           2.93mg   P<.04
 a   M f b6c orl liv hpt 76w76 evx                                        no dre   P=1.
 b   M f b6c orl tba mix 76w76 evx                                        2.12mg   P<.02
3202 M m b6c eat liv MXA 78w91 esv                             : ±        #76.4mg * P<.03  -
 a   M m b6c eat liv hpc 78w91 esv                                        85.5mg * P<.02
 b   M m b6c eat TBA MXB 78w91 esv                                        53.2mg * P<.3
 c   M m b6c eat liv MXB 78w91 esv                                        76.4mg * P<.03
 d   M m b6c eat lun MXB 78w91 esv                                        391.mg * P<.6
3203 M m b6c eat sub fbs 78w90 esv pool                 :   ±             #137.mg * P<.05  -
 a   M m b6c eat ski fib 78w90 esv                                        242.mg * P<.03
3204 M m b6c orl liv hpt 76w76 evx                   .   +       .        1.42mg   P<.006
 a   M m b6c orl lun ade 76w76 evx                                        1.84mg   P<.02
 b   M m b6c orl tba mix 76w76 evx                                        .642mg   P<.0005
3205 M f b6a orl liv hpt 76w76 evx                         .>             no dre   P=1.
 a   M f b6a orl lun ade 76w76 evx                                        no dre   P=1.
 b   M f b6a orl tba mix 76w76 evx                                        no dre   P=1.
3206 M m b6a orl liv hpt 76w76 evx                       .>               7.80mg   P<.6
 a   M m b6a orl lun ade 76w76 evx                                        no dre   P=1.
 b   M m b6a orl tba mix 76w76 evx                                        44.8mg   P<1.
3207 R f osm eat TBA MXB 18m26 v                            :>            no dre   P=1.   -
 a   R f osm eat liv MXB 18m26 v                                          1.06gm * P<.4
3208 R m osm eat TBA MXB 18m26 v                        :>                24.7mg \ P<.8   -
 a   R m osm eat liv MXB 18m26 v                                          no dre   P=1.
```

MICHLER'S KETONE   100ng..:..1ug....:..10.....:..100....:..1mg....:..10.....:..100....:..1g.....:..10
```
3209 M f b6c eat liv MXA 78w91                                 :+:        53.0mg * P<.0005c
 a   M f b6c eat liv hpc 78w91                                            111.mg / P<.0005c
 b   M f b6c eat TBA MXB 78w91                                            69.1mg * P<.0005
 c   M f b6c eat liv MXB 78w91                                            53.0mg * P<.0005
 d   M f b6c eat lun MXB 78w91                                            no dre   P=1.
3210 M m b6c eat --- MXA 78w91                                  : + :     203.mg / P<.0005c
 a   M m b6c eat --- hes 78w91                                            233.mg / P<.0005c
 b   M m b6c eat MXA MXA 78w91                                            753.mg * P<.008
 c   M m b6c eat TBA MXB 78w91                                            99.3mg / P<.0005
 d   M m b6c eat liv MXB 78w91                                            459.mg * P<.07
 e   M m b6c eat lun MXB 78w91                                            1.14gm / P<.3
3211 R f f34 eat liv MXA 18m25                          :+ :              4.87mg / P<.0005c
 a   R f f34 eat liv hpc 18m25                                            5.47mg / P<.0005c
 b   R f f34 eat TBA MXB 18m25                                            5.97mg / P<.0005
 c   R f f34 eat liv MXB 18m25                                            4.87mg / P<.0005
3212 R m f34 eat liv MXA 18m25                            : +:            6.69mg / P<.0005c
 a   R m f34 eat liv hpc 18m25                                            8.86mg / P<.0005c
 b   R m f34 eat TBA MXB 18m25                                            7.29mg / P<.0005
 c   R m f34 eat liv MXB 18m25                                            6.69mg / P<.0005
```

MIREX   100ng..:..1ug....:..10.....:..100....:..1mg....:..10.....:..100....:..1g.....:..10
```
3213 M f b6a orl liv hpt 68w68 evx                   . +     .           1.10mg   P<.0005+
 a   M f b6a orl lun ade 68w68 evx                                        no dre   P=1.
 b   M f b6a orl tba mix 68w68 evx                                        1.26mg   P<.003
3214 M m b6a orl liv hpt 58w58 evx                   .   ±               2.13mg   P<.04   +
 a   M m b6a orl lun ade 58w58 evx                                        no dre   P=1.
 b   M m b6a orl tba mix 58w58 evx                                        3.32mg   P<.3
3215 M f b6c orl liv hpt 69w69 evx                   .  +   .            1.60mg   P<.0005
 a   M f b6c orl lun mix 69w69 evx                                        no dre   P=1.
 b   M f b6c orl tba mix 69w69 evx                                        1.60mg   P<.0005
3216 M m b6c orl liv mix 58w58 evx                   .  +   .            1.50mg   P<.002
 a   M m b6c orl liv hpt 58w58 evx                                        1.83mg   P<.004
 b   M m b6c orl lun mix 58w58 evx                                        no dre   P=1.
 c   M m b6c orl tba mix 58w58 evx                                        1.50mg   P<.002
3217 R f f34 eat MXA mnl 24m25                          : ±              5.97mg * P<.02   c
 a   R f f34 eat TBA MXB 24m25                                            22.4mg * P<.9
 b   R f f34 eat liv MXB 24m25                                            16.5mg * P<.3
3218 R f f34 eat liv MXA 24m24                          : + :            3.17mg * P<.0005c
 a   R f f34 eat liv nnd 24m24                                            3.22mg * P<.0005c
 b   R f f34 eat MXB MXB 24m24                                            3.34mg * P<.0005
 c   R f f34 eat MXA mnl 24m24                                            15.7mg * P<.05   c
 d   R f f34 eat TBA MXB 24m24                                            21.9mg * P<.7
 e   R f f34 eat liv MXB 24m24                                            3.17mg * P<.0005
3219 R m f34 eat MXB MXB 24m24                          : +:             .669mg * P<.0005
```

RefNum		LoConf	UpConf	Cntrl	1Dose	1Inc	2Dose	2Inc						Citation or Pathology	Brkly Code
3199	1064	338.mg	6.36gm	20/100	24.0mg	11/30	120.mg	10/30	240.mg	15/30					
a	1064	424.mg	178.gm	18/100	24.0mg	9/30	120.mg	6/30	240.mg	14/30					
b	1064	816.mg	n.s.s.	5/100	24.0mg	2/30	120.mg	5/30	240.mg	4/30					
c	1064	989.mg	n.s.s.	0/100	24.0mg	2/30	120.mg	1/30	240.mg	3/30					
d	1064	2.70gm	n.s.s.	2/100	24.0mg	0/30	120.mg	0/30	240.mg	1/30					
e	1064	112.mg	592.mg	47/100	24.0mg	22/30	120.mg	21/30	240.mg	27/30					

MEXACARBATE   (zectran) 315-18-4

RefNum		LoConf	UpConf	Cntrl	1Dose	1Inc	2Dose	2Inc	Citation or Pathology
3200	TR147	11.2mg	n.s.s.	5/20	7.50mg	16/50	(14.9mg	13/50)	
a	TR147	57.7mg	n.s.s.	1/20	7.50mg	1/50	14.9mg	3/50	liv:hpa,hpc,nnd.
b	TR147	28.4mg	n.s.s.	1/20	7.50mg	2/50	(14.9mg	1/50)	lun:a/a,a/c.
3201	1102	.885mg	n.s.s.	0/16	1.56mg	3/17			Innes;ntis,1968/1969
a	1102	2.91mg	n.s.s.	0/16	1.56mg	0/17			
b	1102	.729mg	n.s.s.	0/16	1.56mg	4/17			
3202	TR147	45.5mg	n.s.s.	0/20	33.6mg	6/50	66.0mg	15/50	liv:hpa,hpc. S
a	TR147	49.8mg	n.s.s.	0/20	33.6mg	4/50	66.0mg	15/50	S
b	TR147	20.2mg	n.s.s.	0/20	33.6mg	24/50	66.0mg	28/50	
c	TR147	45.5mg	n.s.s.	0/20	33.6mg	6/50	66.0mg	15/50	liv:hpa,hpc,nnd.
d	TR147	109.mg	n.s.s.	0/20	33.6mg	3/50	66.0mg	4/50	lun:a/a,a/c.
3203	TR147	72.2mg	n.s.s.	0/40p	33.6mg	6/50	66.0mg	7/50	
a	TR147	105.mg	n.s.s.	0/40p	33.6mg	1/50	66.0mg	6/50	S
3204	1102	.533mg	14.0mg	0/16	1.45mg	5/16			Innes;ntis,1968/1969
a	1102	.633mg	n.s.s.	0/16	1.45mg	4/16			
b	1102	.293mg	1.80mg	0/16	1.45mg	9/16			
3205	1102	2.91mg	n.s.s.	0/17	1.56mg	0/17			
a	1102	2.91mg	n.s.s.	1/17	1.56mg	0/17			
b	1102	1.93mg	n.s.s.	2/17	1.56mg	1/17			
3206	1102	1.14mg	n.s.s.	1/18	1.45mg	2/17			
a	1102	1.78mg	n.s.s.	2/18	1.45mg	1/17			
b	1102	1.06mg	n.s.s.	3/18	1.45mg	3/17			
3207	TR147	23.7mg	n.s.s.	12/20	11.9mg	33/50	23.6mg	28/50	
a	TR147	173.mg	n.s.s.	0/20	11.9mg	0/50	23.6mg	1/50	liv:hpa,hpc,nnd.
3208	TR147	3.26mg	n.s.s.	6/20	5.80mg	21/50	(11.8mg	16/50)	
a	TR147	n.s.s.	n.s.s.	0/20	5.80mg	0/50	11.8mg	0/50	liv:hpa,hpc,nnd.

MICHLER'S KETONE   90-94-8

RefNum		LoConf	UpConf	Cntrl	1Dose	1Inc	2Dose	2Inc	Citation or Pathology
3209	TR181	40.5mg	77.0mg	0/20	139.mg	41/50	278.mg	49/50	liv:hpa,hpc.
a	TR181	79.2mg	168.mg	0/20	139.mg	16/50	278.mg	38/50	
b	TR181	45.6mg	159.mg	4/20	139.mg	43/50	278.mg	49/50	
c	TR181	40.5mg	77.0mg	0/20	139.mg	41/50	278.mg	49/50	liv:hpa,hpc,nnd.
d	TR181	n.s.s.	n.s.s.	0/20	139.mg	0/50	278.mg	0/50	lun:a/a,a/c.
3210	TR181	126.mg	349.mg	0/20	128.mg	5/50	257.mg	23/50	---:hem,hes.
a	TR181	141.mg	425.mg	0/20	128.mg	5/50	257.mg	20/50	
b	TR181	331.mg	13.1gm	0/20	128.mg	2/50	257.mg	6/50	ski:fbs; sub:fbs,srn. S
c	TR181	63.1mg	207.mg	7/20	128.mg	21/50	257.mg	39/50	
d	TR181	183.mg	n.s.s.	3/20	128.mg	8/50	257.mg	9/50	liv:hpa,hpc,nnd.
e	TR181	310.mg	n.s.s.	2/20	128.mg	2/50	257.mg	4/50	lun:a/a,a/c.
3211	TR181	3.53mg	6.86mg	0/20	18.0mg	46/50	37.0mg	48/50	liv:hpc,nnd.
a	TR181	3.93mg	7.81mg	0/20	18.0mg	41/50	37.0mg	44/50	
b	TR181	4.06mg	10.1mg	7/20	18.0mg	47/50	37.0mg	48/50	
c	TR181	3.53mg	6.86mg	0/20	18.0mg	46/50	37.0mg	48/50	liv:hpa,hpc,nnd.
3212	TR181	4.85mg	9.78mg	0/20	7.40mg	17/50	14.4mg	43/50	liv:hpc,nnd.
a	TR181	6.24mg	13.3mg	0/20	7.40mg	9/50	14.4mg	40/50	
b	TR181	4.74mg	16.3mg	5/20	7.40mg	26/50	14.4mg	48/50	
c	TR181	4.85mg	9.78mg	0/20	7.40mg	17/50	14.4mg	43/50	liv:hpa,hpc,nnd.

MIREX   (1,1a,2,2,3,3a,4,5,5,5a,5b,6-dodecachlorooctahydro-1,3,4-metheno-1H-cyclobuta[cd]pentalene)   2385-85-5

RefNum		LoConf	UpConf	Cntrl	1Dose	1Inc	2Dose	2Inc	3Dose	3Inc	4Dose	4Inc	Citation or Pathology
3213	202	.512mg	2.89mg	0/17	3.67mg	10/16							Innes;ntis,1968/1969
a	202	5.18mg	n.s.s.	1/17	3.67mg	0/16							
b	202	.540mg	7.45mg	2/17	3.67mg	10/16							
3214	202	.718mg	n.s.s.	1/18	3.48mg	5/15							
a	202	3.34mg	n.s.s.	2/18	3.48mg	0/15							
b	202	.797mg	n.s.s.	3/18	3.48mg	5/15							
3215	202	.705mg	4.95mg	0/16	3.67mg	8/16							
a	202	5.32mg	n.s.s.	0/16	3.67mg	0/16							
b	202	.705mg	4.95mg	0/16	3.67mg	8/16							
3216	202	.641mg	6.13mg	0/16	3.48mg	7/18							
a	202	.738mg	11.2mg	0/16	3.48mg	6/18							
b	202	4.01mg	n.s.s.	0/16	3.48mg	0/18							
c	202	.641mg	6.13mg	0/16	3.48mg	7/18							
3217	TR313	2.69mg	n.s.s.	8/52	4.77ug	8/52	47.9ug	11/52	.481mg	14/52	1.21mg	18/52	2.41mg 18/52 mul:mnl; spl:mnl.
a	TR313	1.75mg	n.s.s.	48/52	4.77ug	48/52	47.9ug	49/52	.481mg	47/52	1.21mg	50/52	2.41mg 49/52
b	TR313	4.54mg	n.s.s.	10/52	4.77ug	5/52	47.9ug	4/52	.481mg	5/52	1.21mg	10/52	2.41mg 9/52 liv:hpa,hpc,nnd.
3218	TR313a	2.17mg	5.50mg	2/52	2.45mg	23/52	4.95mg	31/52					liv:hpc,nnd.
a	TR313a	2.19mg	5.67mg	2/52	2.45mg	23/52	4.95mg	30/52					
b	TR313a	2.11mg	7.24mg	8/52	2.45mg	25/52	4.95mg	37/52					liv:hpc,nnd; mul:mnl; spl:mnl. C
c	TR313a	6.61mg	n.s.s.	6/52	2.45mg	9/52	4.95mg	14/52					mul:mnl; spl:mnl.
d	TR313a	2.93mg	n.s.s.	49/52	2.45mg	47/52	4.95mg	48/52					
e	TR313a	2.17mg	5.50mg	2/52	2.45mg	23/52	4.95mg	31/52					liv:hpa,hpc,nnd.
3219	TR313	.443mg	1.14mg	16/52	3.89ug	11/52	39.1ug	17/52	.392mg	23/52	1.00mg	27/52	1.96mg 37/52 adr:phe,phm; k/p: tpp; liv:hpc,nnd. C

```
        Spe Strain  Site     Xpo+Xpt                                                            TD50    2Tailpvl
            Sex  Route  Hist      Notes                                                           DR     AuOp
   a    R m f34 eat tes ict 24m24                                                                .708mg  * P<.0005
   b    R m f34 eat liv MXA 24m24                                                                .825mg  * P<.0005c
   c    R m f34 eat liv nnd 24m24                                                                .870mg  * P<.0005c
   d    R m f34 eat adr MXA 24m24                                                                1.43mg  * P<.0005c
   e    R m f34 eat liv hpc 24m24                                                                7.09mg  * P<.009
   f    R m f34 eat thy MXA 24m24                                                                10.4mg  * P<.003
   g    R m f34 eat thy fca 24m24                                                                12.7mg  * P<.003
   h    R m f34 eat k/p tpp 24m24                                                                13.5mg  * P<.002 c
   i    R m f34 eat TBA MXB 24m24                                                                .618mg  * P<.0005
   j    R m f34 eat liv MXB 24m24                                                                .825mg  * P<.0005
3220    R m sda eat liv nnd 91w91                        .>                                      no dre    P=1.
```

MIREX, PHOTO-   100ng..:..1ug....:..10.....:..100....:..1mg....:..10.....:..100.....:..1g...:..10
```
3221    R m sda eat thy ade 91w91                                 .       +         .              1.46mg  * P<.007 +
   a    R m sda eat liv nnd 91w91                                                                  6.68mg  * P<.5
```

MISOPROSTOL   100ng..:..1ug....:..10.....:..100....:..1mg....:..10.....:..100.....:..1g...:..10
```
3222    M f cd1 gav lun ala 91w91 e                                                    .>         2.35gm  * P<.9   -
   a    M f cd1 gav liv hpa 91w91 e                                                               no dre    P=1.   -
   b    M f cd1 gav liv hem 91w91 e                                                               no dre    P=1.   -
   c    M f cd1 gav lun alc 91w91 e                                                               no dre    P=1.   -
   d    M f cd1 gav tba mix 91w91 e                                                               .130mg  Z P<.0005-
3223    M m cd1 gav liv hpa 91w91 e                                        .>                     no dre    P=1.   -
   a    M m cd1 gav liv hpc 91w91 e                                                               no dre    P=1.   -
   b    M m cd1 gav lun ala 91w91 e                                                               no dre    P=1.   -
   c    M m cd1 gav lun alc 91w91 e                                                               no dre    P=1.   -
   d    M m cd1 gav liv hem 91w91 e                                                               no dre    P=1.   -
   e    M m cd1 gav tba mix 91w91 e                                                               no dre    P=1.   -
3224    R f cdr gav liv hcs 24m24 e                                             .>                9.53gm  * P<1.   -
   a    R f cdr gav liv hpc 24m24 e                                                               no dre    P=1.   -
   b    R f cdr gav mgl fba 24m24 e                                                               no dre    P=1.   -
   c    R f cdr gav tba mix 24m24 e                                                               no dre    P=1.   -
3225    R m cdr gav liv hcs 24m24 e                                             .>                no dre    P=1.   -
   a    R m cdr gav tba mix 24m24 e                                                               no dre    P=1.   -
```

MITOMYCIN-C   100ng..:..1ug....:..10......:..100....:..1mg......:..10......:..100....:..1g......:..10
```
3226    R f cdr ipj per sar 26w78 e          . + .                                                981.ng    P<.0005+
   a    R f cdr ipj liv tum 26w78 e                                                               no dre    P=1.
   b    R f cdr ipj tba mal 26w78 e                                                               1.13ug    P<.0005
   c    R f cdr ipj tba mix 26w78 e                                                               1.23ug    P<.0005
   d    R f cdr ipj tba ben 26w78 e                                                               no dre    P=1.
3227    R m cdr ipj per sar 26w63 e          . + .                                                1.07ug    P<.0005+
   a    R m cdr ipj liv lys 26w63 e                                                               39.3ug    P<.04
   b    R m cdr ipj tba mal 26w63 e                                                               732.ng    P<.0005
   c    R m cdr ipj tba mix 26w63 e                                                               805.ng    P<.0005
   d    R m cdr ipj tba ben 26w63 e                                                               no dre    P=1.
```

MONOACETYL HYDRAZINE   100ng..:..1ug....:..10......:..100....:..1mg......:..10......:..100....:..1g......:..10
```
3228    M f swi gav lun mix 95w95 rs                                           .>                 53.8mg    P<.3   +
3229    M f swi gav lun mix 34w95 rs                                        .  ±                   18.8mg    P<.1   +
3230    M m swi gav lun mix 95w95 rs                                              ±                35.6mg  * P<.02  +
3231    M m swi gav lun mix 34w95 rs                                   .  +                        4.48mg    P<.0005+
```

MONOCHLOROACETIC ACID   100ng..:..1ug....:..10......:..100....:..1mg......:..10......:..100....:..1g......:..10
```
3232    M f b6c gav TBA MXB 24m24                                                 :>              no dre    P=1.   -
   a    M f b6c gav liv MXB 24m24                                                                 2.05gm  * P<.7   -
   b    M f b6c gav lun MXB 24m24                                                                 640.mg  * P<.2   -
3233    M f b6c orl liv hpt 76w76 evx                                      .>                     no dre    P=1.   -
   a    M f b6c orl lun ade 76w76 evx                                                             no dre    P=1.   -
   b    M f b6c orl tba mix 76w76 evx                                                             1.01gm    P<1.   -
3234    M m b6c gav TBA MXB 24m24                                                 :>              159.mg  * P<.3   -
   a    M m b6c gav liv MXB 24m24                                                                 no dre    P=1.
   b    M m b6c gav lun MXB 24m24                                                                 no dre    P=1.
3235    M m b6c orl lun ade 76w76 evx                                         .>                  122.mg    P<.3   -
   a    M m b6c orl liv hpt 76w76 evx                                                             no dre    P=1.   -
   b    M m b6c orl tba mix 76w76 evx                                                             no dre    P=1.   -
3236    M f b6a orl lun ade 76w76 evx                                       .  ±                  59.7mg    P<.09  -
   a    M f b6a orl liv hpt 76w76 evx                                                             no dre    P=1.   -
   b    M f b6a orl tba mix 76w76 evx                                                             38.5mg    P<.04  -
3237    M m b6a orl lun ade 76w76 evx                                         .>                  115.mg    P<.3   -
   a    M m b6a orl liv hpt 76w76 evx                                                             2.01gm    P<1.   -
   b    M m b6a orl tba mix 76w76 evx                                                             942.mg    P<1.   -
3238    R f f34 gav ute esp 24m24                                            :  +   :             #41.7mg * P<.002 -
   a    R f f34 gav ute MXA 24m24                                                                 47.3mg  * P<.007 -
   b    R f f34 gav TBA MXB 24m24                                                                 85.4mg  * P<.8   -
   c    R f f34 gav liv MXB 24m24                                                                 no dre    P=1.   -
3239    R m f34 gav TBA MXB 24m24                                          :>                     53.3mg  * P<.7   -
   a    R m f34 gav liv MXB 24m24                                                                 623.mg  * P<.7
```

MONOCROTALINE   100ng..:..1ug....:..10.....:..100....:..1mg....:..10.....:..100.....:..1g...:..10
```
3240    R m cdr gav liv hpc 42w71 erv                              .  ±                           1.16mg    P<.02  +
```

	RefNum	LoConf	UpConf	Cntrl	1Dose	1Inc	2Dose	2Inc							Citation or Pathology	Brkly Code
a	TR313	.392mg	2.25mg	50/52	3.89ug	51/52	39.1ug	43/52	.392mg	48/52	1.00mg	39/52	1.96mg	42/52		S
b	TR313	.545mg	1.39mg	6/52	3.89ug	5/52	39.1ug	6/52	.392mg	15/52	1.00mg	16/52	1.96mg	28/52	liv:hpc,nnd.	
c	TR313	.574mg	1.47mg	3/52	3.89ug	5/52	39.1ug	5/52	.392mg	14/52	1.00mg	15/52	1.96mg	26/52		
d	TR313	.840mg	3.18mg	3/52	3.89ug	7/52	39.1ug	13/52	.392mg	12/52	1.00mg	18/52	1.96mg	20/52	adr:phe,phm.	
e	TR313	2.70mg	356.mg	3/52	3.89ug	0/52	39.1ug	2/52	.392mg	2/52	1.00mg	3/52	1.96mg	4/52		S
f	TR313	3.67mg	100.mg	0/52	3.89ug	1/52	39.1ug	0/52	.392mg	1/52	1.00mg	0/52	1.96mg	4/52	thy:fca,fcc.	S
g	TR313	4.33mg	88.7mg	0/52	3.89ug	0/52	39.1ug	0/52	.392mg	1/52	1.00mg	0/52	1.96mg	3/52		S
h	TR313	4.58mg	72.4mg	0/52	3.89ug	0/52	39.1ug	0/52	.392mg	0/52	1.00mg	1/52	1.96mg	3/52		
i	TR313	.374mg	1.39mg	45/52	3.89ug	46/52	39.1ug	40/52	.392mg	45/52	1.00mg	48/52	1.96mg	45/52		
j	TR313	.545mg	1.39mg	6/52	3.89ug	5/52	39.1ug	6/52	.392mg	15/52	1.00mg	16/52	1.96mg	28/52	liv:hpa,hpc,nnd.	
3220	1160	52.6ug	n.s.s.	0/10	40.0ug	0/10	.200mg	0/10							Chu;txap,59,268-278;1981	

MIREX, PHOTO- 39801-14-4

	RefNum	LoConf	UpConf	Cntrl	1Dose	1Inc	2Dose	2Inc	3Dose	3Inc	4Dose	4Inc	Citation or Pathology
3221	1160	.474mg	28.7mg	1/10	8.00ug	0/10	40.0ug	0/10	.200mg	0/10	1.00mg	4/10	Chu;txap,59,268-278;1981
a	1160	1.06mg	n.s.s.	0/10	8.00ug	1/10	40.0ug	0/10	.200mg	1/10	1.00mg	1/10	

MISOPROSTOL 59122-46-2

	RefNum	LoConf	UpConf	Cntrl	1Dose	1Inc	2Dose	2Inc	3Dose	3Inc	Citation or Pathology
3222	1841	106.mg	n.s.s.	1/64	.160mg	1/64	1.60mg	0/64	16.0mg	1/64	Port;txpy,15,134-142;1987/pers.comm.
a	1841	110.mg	n.s.s.	0/64	.160mg	3/64	1.60mg	0/64	16.0mg	1/64	
b	1841	176.mg	n.s.s.	2/64	.160mg	2/64	1.60mg	0/64	16.0mg	0/64	
c	1841	96.2mg	n.s.s.	0/64	.160mg	2/64	1.60mg	2/64	16.0mg	1/64	
d	1841	69.7ug	.481mg	22/64	.160mg	42/64	(1.60mg	21/64	16.0mg	18/64)	
3223	1841	5.97mg	n.s.s.	15/64	.160mg	11/64	1.60mg	9/64	(16.0mg	2/64)	
a	1841	109.mg	n.s.s.	2/64	.160mg	1/64	1.60mg	1/64	16.0mg	1/64	
b	1841	148.mg	n.s.s.	0/64	.160mg	1/64	1.60mg	1/64	16.0mg	0/64	
c	1841	167.mg	n.s.s.	2/64	.160mg	4/64	1.60mg	1/64	16.0mg	0/64	
d	1841	176.mg	n.s.s.	1/64	.160mg	1/64	1.60mg	0/64	16.0mg	0/64	
e	1841	2.86mg	n.s.s.	34/64	.160mg	27/64	1.60mg	27/64	(16.0mg	15/64)	
3224	1840	18.1mg	n.s.s.	0/60	24.0ug	2/60	.240mg	1/59	2.40mg	1/60	Dodd;txpy,15,125-133;1987/pers.comm.
a	1840	22.4mg	n.s.s.	0/60	24.0ug	0/60	.240mg	1/59	2.40mg	0/60	
b	1840	5.70mg	n.s.s.	23/60	24.0ug	27/60	.240mg	37/59	2.40mg	24/60	
c	1840	2.28mg	n.s.s.	56/60	24.0ug	58/60	.240mg	54/59	2.40mg	52/60	
3225	1840	18.8mg	n.s.s.	2/60	24.0ug	0/60	.240mg	1/60	2.40mg	1/60	
a	1840	.350mg	n.s.s.	47/60	24.0ug	49/60	.240mg	42/60	(2.40mg	24/60)	

MITOMYCIN-C 50-07-7

	RefNum	LoConf	UpConf	Cntrl	1Dose	1Inc	Citation or Pathology
3226	1336	516.ng	1.95ug	0/182	5.40ug	22/25	Skipper;srfr;1976/Weisburger;canc;1977/Prejean pers.comm.
a	1336	15.6ug	n.s.s.	0/182	5.40ug	0/25	
b	1336	553.ng	2.67ug	44/182	5.40ug	22/25	
c	1336	490.ng	5.41ug	103/182	5.40ug	23/25	
d	1336	13.8ug	n.s.s.	59/182	5.40ug	1/25	
3227	1336	575.ng	2.16ug	0/177	6.65ug	19/24	
a	1336	6.40ug	n.s.s.	0/177	6.65ug	1/24	
b	1336	346.ng	1.66ug	32/177	6.65ug	22/24	
c	1336	361.ng	2.11ug	59/177	6.65ug	22/24	
d	1336	12.1ug	n.s.s.	27/177	6.65ug	0/24	

MONOACETYL HYDRAZINE 1068-57-1

	RefNum	LoConf	UpConf	Cntrl	1Dose	1Inc	2Dose	2Inc	Citation or Pathology
3228	1661m	13.9mg	n.s.s.	3/20	22.9mg	5/15			Bhide;clet,23,235-240;1984/pers.comm.
3229	1661n	5.77mg	n.s.s.	3/20	11.4mg	6/15			
3230	1661m	15.2mg	n.s.s.	2/20	19.0mg	4/15	26.2mg	7/15	
3231	1661n	1.98mg	15.0mg	2/20	9.52mg	11/15			

MONOCHLOROACETIC ACID 79-11-8

	RefNum	LoConf	UpConf	Cntrl	1Dose	1Inc	2Dose	2Inc	Citation or Pathology	Brkly Code
3232	TR396	49.7mg	n.s.s.	41/60	35.6mg	33/60	(71.2mg	27/60)		
a	TR396	312.mg	n.s.s.	1/60	35.6mg	2/60	71.2mg	2/60	liv:hpa,hpc,nnd.	
b	TR396	243.mg	n.s.s.	0/60	35.6mg	3/60	71.2mg	2/60	lun:a/a,a/c.	
3233	1263	38.2mg	n.s.s.	0/18	20.4mg	0/17			Innes;ntis,1968/1969	
a	1263	38.2mg	n.s.s.	1/18	20.4mg	0/17				
b	1263	17.8mg	n.s.s.	2/18	20.4mg	2/17				
3234	TR396	46.4mg	n.s.s.	31/60	35.6mg	32/60	71.2mg	21/60		
a	TR396	150.mg	n.s.s.	12/60	35.6mg	8/60	71.2mg	6/60	liv:hpa,hpc,nnd.	
b	TR396	135.mg	n.s.s.	12/60	35.6mg	10/60	71.2mg	5/60	lun:a/a,a/c.	
3235	1263	19.8mg	n.s.s.	0/14	19.0mg	1/18			Innes;ntis,1968/1969	
a	1263	20.5mg	n.s.s.	3/14	19.0mg	2/18				
b	1263	11.3mg	n.s.s.	4/14	19.0mg	5/18				
3236	1263	14.7mg	n.s.s.	0/18	20.4mg	2/17				
a	1263	38.2mg	n.s.s.	0/18	20.4mg	0/17				
b	1263	11.6mg	n.s.s.	0/18	20.4mg	3/17				
3237	1263	18.7mg	n.s.s.	0/17	19.0mg	1/17				
a	1263	21.0mg	n.s.s.	1/18	19.0mg	1/17				
b	1263	16.5mg	n.s.s.	2/18	19.0mg	2/17				
3238	TR396	21.4mg	193.mg	1/53	10.7mg	7/53	21.3mg	9/53		S
a	TR396	22.5mg	816.mg	2/53	10.7mg	7/53	21.3mg	9/53	ute:esp,ess.	S
b	TR396	10.4mg	n.s.s.	50/53	10.7mg	45/53	21.3mg	39/53		
c	TR396	161.mg	n.s.s.	1/53	10.7mg	0/53	21.3mg	0/53	liv:hpa,hpc,nnd.	
3239	TR396	8.54mg	n.s.s.	49/53	10.7mg	45/53	21.3mg	34/53	liv:hpa,hpc,nnd.	
a	TR396	60.9mg	n.s.s.	1/53	10.7mg	1/53	21.3mg	1/53	liv:hpa,hpc,nnd.	

MONOCROTALINE 315-22-0

	RefNum	LoConf	UpConf	Cntrl	1Dose	1Inc	Citation or Pathology
3240	1719m	.566mg	n.s.s.	0/15	.813mg	10/50	Newberne;pffm,1,23-31;1973

```
      Spe Strain  Site   Xpo+Xpt                                              TD50    2Tailpvl
          Sex   Route   Hist    Notes                                            DR     AuOp
 3241 R m cdr gav liv hpc 42w71 cerv              .  +     .                    .790mg  P<.005 +

 DL-MONOSODIUM GLUTAMATE          100ng..:..1ug....:..10.....:..100....:..1mg..:..10....:..100....:..1g.....:..10
 3242 M m cbl eat liv tum 24m24 e                                         .>   no dre   P=1.   -
   a  M m cbl eat lun ade 24m24 e                                              no dre   P=1.   -
   b  M m cbl eat tba mix 24m24 e                                              no dre   P=1.   -

 L-MONOSODIUM GLUTAMATE           100ng..:..1ug....:..10.....:..100....:..1mg..:..10....:..100....:..1g.....:..10
 3243 M m cbl eat liv tum 24m24 e                                         .>   no dre   P=1.   -
   a  M m cbl eat lun tum 24m24 e                                              no dre   P=1.   -
   b  M m cbl eat tba tum 24m24 e                                              no dre   P=1.   -

 MONOSODIUM SUCCINATE             100ng..:..1ug....:..10.....:..100....:..1mg..:..10....:..100....:..1g.....:..10
 3244 R f f3d wat liv tum 24m26 e                                         .>   no dre   P=1.   -
   a  R f f3d wat tba tum 24m26 e                                              4.82gm * P<.7   -
 3245 R m f3d wat liv hpc 24m26 e                                              .54.7gm * P<.3  -
   a  R m f3d wat liv hpa 24m26 e                                              no dre   P=1.   -
   b  R m f3d wat tba tum 24m26 e                                              noTD50   P=1.   -

 4-MORPHOLINO-2-(5-NITRO-2-THIENYL)QUINAZOLINE   ....:..10.....:..100....:..1mg...:..10....:..100....:..1g.....:..10
 3246 R f sda eat tba mix 46w66 e                          .  +   .             5.03mg  P<.0005+

 L-5-MORPHOLINOMETHYL-3-[(5-NITROFURFURYLIDENE)AMINO]-2-OXAZOLIDINONE.HC1..1 _mg...:..10.....:..100....:..1g.....:..10
 3247 R f sda eat mgl mix 46w66 e                       .  +   .                2.81mg  P<.0005
   a  R f sda eat mgl adc 46w66 e                                              6.33mg  P<.0005+
   b  R f sda eat --- lbl 46w66 e                                        +hist 39.0mg  P<.004 +
   c  R f sda eat k/p tcc 46w66 e                                        +hist 149.mg  P<.2   +
   d  R f sda eat liv tum 46w66 e                                              no dre   P=1.
   e  R f sda eat tba mix 46w66 e                                              2.81mg  P<.0005

 MYLERAN                          100ng..:..1ug....:..10.....:..100....:..1mg..:..10....:..100....:..1g.....:..10
 3248 R m b46 gav tba mix 12m24 es              .>                             93.1ug  P<.6
   a  R m b46 gav tba mal 12m24 es                                             .117mg  P<.5
   b  R m b46 gav tba ben 12m24 es                                             .642mg  P<.9

 NAFENOPIN                        100ng..:..1ug....:..10.....:..100....:..1mg..:..10....:..100....:..1g.....:..10
 3249 R m wis eat liv mix 59w59 e                          .  +   .             22.1mg  P<.0005+
   a  R m wis eat liv hpa 59w59 e                                              48.0mg  P<.002
   b  R m wis eat liv hpc 59w59 e                                              58.1mg  P<.004 +
   c  R m wis eat tba tum 59w59 e                                              12.3mg  P<.0005

 NALIDIXIC ACID                   100ng..:..1ug....:..10.....:..100....:..1mg..:..10....:..100....:..1g.....:..10
 3250 M f b6c eat amd phe 24m24                                          :   ±  #3.63gm * P<.03  -
   a  M f b6c eat TBA MXB 24m24                                                1.69gm * P<.6
   b  M f b6c eat liv MXB 24m24                                                3.16gm * P<.3
   c  M f b6c eat lun MXB 24m24                                                4.90gm * P<.4
 3251 M m b6c eat sub MXA 24m24                                          :   ±  1.21gm * P<.04  e
   a  M m b6c eat TBA MXB 24m24                                                1.33gm * P<.4
   b  M m b6c eat liv MXB 24m24                                                no dre   P=1.
   c  M m b6c eat lun MXB 24m24                                                4.60gm * P<.6
 3252 R f f34 eat cli MXA 24m24                                         :   ±   372.mg * P<.05  c
   a  R f f34 eat TBA MXB 24m24                                                no dre   P=1.
   b  R f f34 eat liv MXB 24m24                                                no dre   P=1.
 3253 R m f34 eat pre MXA 24m24                                       :  +  :   138.mg * P<.0005c
   a  R m f34 eat pre can 24m24                                                211.mg * P<.0005
   b  R m f34 eat pre adn 24m24                                                301.mg * P<.02
   c  R m f34 eat pre MXA 24m24                                                358.mg * P<.05
   d  R m f34 eat TBA MXB 24m24                                                7.60gm * P<1.
   e  R m f34 eat liv MXB 24m24                                                no dre   P=1.

 NAPHTHALENE                      100ng..:..1ug....:..10.....:..100....:..1mg..:..10....:..100....:..1g.....:..10
 3254 M f b6c inh lun MXA 24m24                                        :  +  :   163.mg / P<.0005p
   a  M f b6c inh lun a/a 24m24                                                171.mg / P<.002 p
   b  M f b6c inh lun a/c 24m24                                                4.02gm * P<.3   p
   c  M f b6c inh TBA MXB 24m24                                                128.mg * P<.04
   d  M f b6c inh liv MXB 24m24                                                no dre   P=1.
   e  M f b6c inh lun MXB 24m24                                                163.mg / P<.0005
 3255 M m b6c inh TBA MXB 24m24                              :>                 no dre   P=1.   -
   a  M m b6c inh liv MXB 24m24                                                no dre   P=1.
   b  M m b6c inh lun MXB 24m24                                                no dre   P=1.

 1-NAPHTHALENE ACETAMIDE          100ng..:..1ug....:..10.....:..100....:..1mg..:..10....:..100....:..1g.....:..10
 3256 M f b6a orl liv hpt 76w76 evx                                  .>        no dre   P=1.   -
   a  M f b6a orl lun ade 76w76 evx                                            no dre   P=1.   -
   b  M f b6a orl tba mix 76w76 evx                                            956.mg  P<.7   -
 3257 M m b6a orl liv hpt 76w76 evx                                  .>        no dre   P=1.   -
   a  M m b6a orl lun ade 76w76 evx                                            no dre   P=1.   -
   b  M m b6a orl tba mix 76w76 evx                                            no dre   P=1.   -
 3258 M f b6c orl lun ade 76w76 evx                              .    ±        461.mg  P<.09  -
   a  M f b6c orl liv hpt 76w76 evx                                            956.mg  P<.3   -
   b  M f b6c orl tba mix 76w76 evx                                            213.mg  P<.02  -
```

	RefNum	LoConf	UpConf	Cntrl	1Dose	1Inc	2Dose	2Inc	Citation or Pathology
									Brkly Code
3241	1719n	.424mg	4.36mg	0/15	.813mg	14/50			

DL-MONOSODIUM GLUTAMATE   32221-81-1

	RefNum	LoConf	UpConf	Cntrl	1Dose	1Inc	2Dose	2Inc	Citation or Pathology
3242	1631	4.51gm	n.s.s.	0/55	1.20gm	0/23	4.80gm	0/27	Ebert;txlt,3,65-70;1979
a	1631	17.6gm	n.s.s.	0/55	1.20gm	1/23	4.80gm	0/27	
b	1631	17.6gm	n.s.s.	0/55	1.20gm	1/23	4.80gm	0/27	

L-MONOSODIUM GLUTAMATE   142-47-2

	RefNum	LoConf	UpConf	Cntrl	1Dose	1Inc	2Dose	2Inc	Citation or Pathology
3243	1631	4.68gm	n.s.s.	0/55	1.20gm	0/25	4.80gm	0/23	Ebert;txlt,3,65-70;1979
a	1631	4.68gm	n.s.s.	0/55	1.20gm	0/25	4.80gm	0/23	
b	1631	4.68gm	n.s.s.	0/55	1.20gm	0/25	4.80gm	0/23	

MONOSODIUM SUCCINATE   2922-54-5

	RefNum	LoConf	UpConf	Cntrl	1Dose	1Inc	2Dose	2Inc	Citation or Pathology
3244	1977	4.12gm	n.s.s.	0/49	526.mg	0/48	1.05gm	0/49	Maekawa;fctx,28,235-241;1990
a	1977	655.mg	n.s.s.	40/49	526.mg	37/48	1.05gm	40/47	
3245	1977	8.91gm	n.s.s.	0/50	460.mg	0/48	920.mg	1/50	
a	1977	6.14gm	n.s.s.	2/50	460.mg	4/48	920.mg	1/50	
b	1977	n.s.s.	n.s.s.	50/50	460.mg	48/48	920.mg	50/50	

4-MORPHOLINO-2-(5-NITRO-2-THIENYL)QUINAZOLINE   58139-48-3

	RefNum	LoConf	UpConf	Cntrl	1Dose	1Inc	2Dose	2Inc	Citation or Pathology
3246	1390	2.70mg	11.2mg	6/84	17.4mg	18/28			Cohen;jnci,57,277-282;1976

L-5-MORPHOLINOMETHYL-3-[(5-NITROFURFURYLIDENE)AMINO]-2-OXAZOLIDINONE.HCl   3031-51-4

	RefNum	LoConf	UpConf	Cntrl	1Dose	1Inc	2Dose	2Inc	Citation or Pathology
3247	200a	1.25mg	5.71mg	1/25	34.8mg	31/32			Cohen;jnci,51,403-417;1973
a	200a	3.70mg	11.6mg	0/25	34.8mg	25/32			
b	200a	16.8mg	219.mg	0/25	34.8mg	7/32			
c	200a	36.6mg	n.s.s.	0/25	34.8mg	2/32			
d	200a	92.6mg	n.s.s.	0/25	34.8mg	0/32			
e	200a	1.25mg	5.71mg	1/25	34.8mg	31/32			

MYLERAN   (busulfan)   55-98-1

	RefNum	LoConf	UpConf	Cntrl	1Dose	1Inc	2Dose	2Inc	Citation or Pathology
3248	1017	12.7ug	n.s.s.	7/65	9.29ug	3/18			Schmahl;arzn,20,1461-1467;1970
a	1017	15.1ug	n.s.s.	4/65	9.29ug	2/18			
b	1017	20.6ug	n.s.s.	3/65	9.29ug	1/18			

NAFENOPIN   3771-19-5

	RefNum	LoConf	UpConf	Cntrl	1Dose	1Inc	2Dose	2Inc	Citation or Pathology
3249	2155	10.9mg	52.9mg	0/18	100.mg	12/19			Kraupp-Grasl;canr,51,666-671;1991/pers.comm.
a	2155	20.5mg	187.mg	0/18	100.mg	7/19			
b	2155	23.5mg	324.mg	0/18	100.mg	6/19			
c	2155	5.98mg	29.2mg	1/18	100.mg	16/19			

NALIDIXIC ACID   389-08-2

	RefNum	LoConf	UpConf	Cntrl	1Dose	1Inc	2Dose	2Inc	Citation or Pathology / Brkly Code
3250	TR368	1.38gm	n.s.s.	0/50	258.mg	2/50	515.mg	3/50	s
a	TR368	344.mg	n.s.s.	36/50	258.mg	39/50	515.mg	35/50	
b	TR368	963.mg	n.s.s.	4/50	258.mg	6/50	515.mg	7/50	liv:hpa,hpc,nnd.
c	TR368	1.27gm	n.s.s.	1/50	258.mg	5/50	515.mg	2/50	lun:a/a,a/c.
3251	TR368	536.mg	n.s.s.	5/50	238.mg	9/50	475.mg	14/50	sub:fbs,fib.
a	TR368	342.mg	n.s.s.	26/50	238.mg	24/50	475.mg	32/50	
b	TR368	1.07gm	n.s.s.	10/50	238.mg	12/50	475.mg	7/50	liv:hpa,hpc,nnd.
c	TR368	848.mg	n.s.s.	6/50	238.mg	5/50	475.mg	8/50	lun:a/a,a/c.
3252	TR368	165.mg	n.s.s.	5/50	99.0mg	15/50	198.mg	16/50	cli:adn,can,ppc,ppn.
a	TR368	102.mg	n.s.s.	47/50	99.0mg	43/50	(198.mg	41/50)	
b	TR368	1.14gm	n.s.s.	1/50	99.0mg	0/50	198.mg	1/50	liv:hpa,hpc,nnd.
3253	TR368	82.0mg	416.mg	3/50	79.2mg	19/50	159.mg	20/50	pre:adn,can,ppn.
a	TR368	126.mg	458.mg	0/50	79.2mg	10/50	159.mg	12/50	s
b	TR368	145.mg	n.s.s.	2/50	79.2mg	10/50	159.mg	10/50	s
c	TR368	155.mg	n.s.s.	3/50	79.2mg	10/50	159.mg	10/50	pre:adn,ppn. s
d	TR368	94.9mg	n.s.s.	44/50	79.2mg	45/50	159.mg	44/50	
e	TR368	746.mg	n.s.s.	2/50	79.2mg	1/50	159.mg	1/50	liv:hpa,hpc,nnd.

NAPHTHALENE   91-20-3

	RefNum	LoConf	UpConf	Cntrl	1Dose	1Inc	2Dose	2Inc	Citation or Pathology
3254	TR410	91.2mg	572.mg	5/70	16.0mg	2/68	48.3mg	29/138	lun:a/a,a/c.
a	TR410	94.4mg	662.mg	5/70	16.0mg	2/68	48.3mg	28/138	
b	TR410	654.mg	n.s.s.	0/70	16.0mg	0/68	48.3mg	1/138	
c	TR410	58.4mg	n.s.s.	23/70	16.0mg	22/68	48.3mg	63/138	
d	TR410	457.mg	n.s.s.	2/70	16.0mg	2/68	48.3mg	3/138	liv:hpa,hpc,nnd.
e	TR410	91.2mg	572.mg	5/70	16.0mg	2/68	48.3mg	29/138	lun:a/a,a/c.
3255	TR410	30.0mg	n.s.s.	36/70	13.3mg	36/70	(40.3mg	66/137)	
a	TR410	182.mg	n.s.s.	8/70	13.3mg	14/70	40.3mg	23/137	liv:hpa,hpc,nnd.
b	TR410	116.mg	n.s.s.	7/70	13.3mg	17/70	40.3mg	31/137	lun:a/a,a/c.

1-NAPHTHALENE ACETAMIDE   86-86-2

	RefNum	LoConf	UpConf	Cntrl	1Dose	1Inc	2Dose	2Inc	Citation or Pathology
3256	1156	337.mg	n.s.s.	0/17	180.mg	0/17			Innes;ntis,1968/1969
a	1156	337.mg	n.s.s.	1/17	180.mg	0/17			
b	1156	120.mg	n.s.s.	2/17	180.mg	3/17			
3257	1156	314.mg	n.s.s.	1/18	168.mg	0/17			
a	1156	314.mg	n.s.s.	2/18	168.mg	0/17			
b	1156	314.mg	n.s.s.	3/18	168.mg	0/17			
3258	1156	113.mg	n.s.s.	0/16	180.mg	2/15			
a	1156	156.mg	n.s.s.	0/16	180.mg	1/15			
b	1156	73.0mg	n.s.s.	0/16	180.mg	4/15			

```
      Spe Strain  Site   Xpo+Xpt                                                      TD50    2Tailpvl
          Sex Route  Hist    Notes                                                       DR    AuOp.
3259 M m b6c orl liv hpt 76w76 evx                                    .       ±       316.mg  P<.04  -
   a M m b6c orl lun ade 76w76 evx                                                    1.01gm  P<.3   -
   b M m b6c orl tba mix 76w76 evx                                                    141.mg  P<.003 -

1-NAPHTHALENE ACETIC ACID       100ng..:..1ug...:..10.....:..100....:..1mg....:..10....:..100...:..1g.....:.10
3260 M f b6a orl liv agm 76w76 evx                                    .       ±       213.mg  P<.09  -
   a M f b6a orl lun ade 76w76 evx                                                    no dre  P=1.   -
   b M f b6a orl tba mix 76w76 evx                                                    387.mg  P<.7   -
3261 M m b6a orl liv hpt 76w76 evx                               .>                   no dre  P=1.   -
   a M m b6a orl lun ade 76w76 evx                                                    no dre  P=1.   -
   b M m b6a orl tba mix 76w76 evx                                                    no dre  P=1.   -
3262 M f b6c orl liv hpt 76w76 evx                                   .>               no dre  P=1.   -
   a M f b6c orl lun mix 76w76 evx                                                    no dre  P=1.   -
   b M f b6c orl tba tum 76w76 evx                                                    no dre  P=1.   -
3263 M m b6c orl liv hpt 76w76 evx                              .       ±             199.mg  P<.1   -
   a M m b6c orl lun mix 76w76 evx                                                    no dre  P=1.   -
   b M m b6c orl tba mix 76w76 evx                                                    71.5mg  P<.007 -

1,5-NAPHTHALENEDIAMINE          100ng..:..1ug...:..10.....:..100....:..1mg....:..10....:..100...:..1g.....:.10
3264 M f b6c eat MXB MXB 24m24                              :  +  :                    66.6mg  \ P<.0005
   a M f b6c eat liv hpc 24m24                                                        115.mg  \ P<.0005c
   b M f b6c eat liv MXA 24m24                                                        129.mg  * P<.0005c
   c M f b6c eat thy MXA 24m24                                                        327.mg  * P<.002 c
   d M f b6c eat lun MXA 24m24                                                        331.mg  \ P<.0005c
   e M f b6c eat thy MXA 24m24                                                        936.mg  * P<.002 c
   f M f b6c eat thy ccr 24m24                                                        1.32gm  * P<.005 c
   g M f b6c eat TBA MXB 24m24                                                        223.mg  * P<.03
   h M f b6c eat liv MXB 24m24                                                        129.mg  \ P<.0005
   i M f b6c eat lun MXB 24m24                                                        331.mg  \ P<.0005
3265 M m b6c eat thy MXB 24m24                                 :  +  :                 217.mg  * P<.0005
   a M m b6c eat thy MXA 24m24                                                        276.mg  * P<.0005c
   b M m b6c eat thy MXA 24m24                                                        1.26gm  * P<.02  c
   c M m b6c eat thy ccr 24m24                                                        1.91gm  * P<.02  c
   d M m b6c eat TBA MXB 24m24                                                        583.mg  * P<.5
   e M m b6c eat liv MXB 24m24                                                        1.64gm  * P<.6
   f M m b6c eat lun MXB 24m24                                                        no dre  P=1.
3266 R f f34 eat MXB MXB 24m25                             :  +     :                  50.8mg  * P<.002
   a R f f34 eat ute esp 24m25                                                        69.6mg  * P<.009 c
   b R f f34 eat cli MXA 24m25                                                        137.mg  * P<.009 c
   c R f f34 eat thy MXA 24m25                                                        57.6mg  \ P<.03
   d R f f34 eat TBA MXB 24m25                                                        55.8mg  * P<.2
   e R f f34 eat liv MXB 24m25                                                        236.mg  * P<.2
3267 R m f34 eat TBA MXB 24m25                               :    ±                   48.1mg  * P<.2   -
   a R m f34 eat liv MXB 24m25                                                        1.04gm  * P<.9

N-(1-NAPHTHYL)ETHYLENEDIAMINE.2HCl     ..:..1ug...:..10.....:..100....:..1mg....:..10....:..100...:..1g.....:.10
3268 M f b6c eat TBA MXB 24m25 sv                                          :>         no dre  P=1.   -
   a M f b6c eat liv MXB 24m25 sv                                                     25.6gm  * P<.8
   b M f b6c eat lun MXB 24m25 sv                                                     4.39gm  * P<.2
3269 M m b6c eat TBA MXB 24m24                                       :>               no dre  P=1.   -
   a M m b6c eat liv MXB 24m24                                                        no dre  P=1.
   b M m b6c eat lun MXB 24m24                                                        no dre  P=1.
3270 R f f34 eat liv MXA 24m25                                 :    ±                 #93.5mg \ P<.02  -
   a R f f34 eat TBA MXB 24m25                                                        no dre  P=1.
   b R f f34 eat liv MXB 24m25                                                        93.5mg  \ P<.02
3271 R m f34 eat TBA MXB 24m25                                    :>                   41.6mg  \ P<.3   -
   a R m f34 eat liv MXB 24m25                                                        no dre  P=1.

sym.-dibeta-NAPHTHYL-p-PHENYLENEDIAMINE..:..1_ug...:..10.....:..100....:..1mg...:..10....:..100...:..1g.....:.10
3272 M f b6a orl lun ade 76w76 evx                                  .>                221.mg  P<.6   -
   a M f b6a orl liv hpt 76w76 evx                                                    no dre  P=1.   -
   b M f b6a orl tba mix 76w76 evx                                                    63.8mg  P<.2   -
3273 M m b6a orl lun ade 76w76 evx                                .>                  no dre  P=1.   -
   a M m b6a orl liv hpt 76w76 evx                                                    no dre  P=1.   -
   b M m b6a orl tba mix 76w76 evx                                                    no dre  P=1.   -
3274 M f b6c orl liv hpt 76w76 evx                                   .>               no dre  P=1.   -
   a M f b6c orl lun mix 76w76 evx                                                    no dre  P=1.   -
   b M f b6c orl tba mix 76w76 evx                                                    235.mg  P<.3   -
3275 M m b6c orl liv hpt 76w76 evx                              .       ±             59.4mg  P<.04  -
   a M m b6c orl lun ade 76w76 evx                                                    192.mg  P<.3   -
   b M m b6c orl tba mix 76w76 evx                                                    25.9mg  P<.002 -

1-(1-NAPHTHYL)-2-THIOUREA       100ng..:..1ug...:..10.....:..100....:..1mg....:..10....:..100...:..1g.....:.10
3276 M f b6a orl lun ade 76w76 evx                               .>                   no dre  P=1.   -
   a M f b6a orl liv hpt 76w76 evx                                                    no dre  P=1.   -
   b M f b6a orl tba mix 76w76 evx                                                    no dre  P=1.   -
3277 M m b6a orl lun ade 76w76 evx                              .>                    no dre  P=1.   -
   a M m b6a orl liv hpt 76w76 evx                                                    no dre  P=1.   -
   b M m b6a orl tba mix 76w76 evx                                                    no dre  P=1.   -
3278 M f b6c orl liv hpt 76w76 evx                                 .>                 no dre  P=1.   -
   a M f b6c orl lun mix 76w76 evx                                                    no dre  P=1.   -
```

	RefNum	LoConf	UpConf	Cntrl	1Dose	1Inc	2Dose	2Inc	Citation or Pathology
									Brkly Code
3259	1156	95.4mg	n.s.s.	0/16	168.mg	3/17			
a	1156	165.mg	n.s.s.	0/16	168.mg	1/17			
b	1156	56.9mg	765.mg	0/16	168.mg	6/17			

1-NAPHTHALENE ACETIC ACID   (planofix) 86-87-3

	RefNum	LoConf	UpConf	Cntrl	1Dose	1Inc	2Dose	2Inc	Citation or Pathology
3260	1157	52.4mg	n.s.s.	0/17	73.0mg	2/17			Innes;ntis,1968/1969
a	1157	81.4mg	n.s.s.	1/17	73.0mg	1/17			
b	1157	48.8mg	n.s.s.	2/17	73.0mg	3/17			
3261	1157	80.4mg	n.s.s.	1/18	68.0mg	1/18			
a	1157	89.3mg	n.s.s.	2/18	68.0mg	1/18			
b	1157	53.5mg	n.s.s.	3/18	68.0mg	3/18			
3262	1157	129.mg	n.s.s.	0/16	73.0mg	0/16			
a	1157	129.mg	n.s.s.	0/16	73.0mg	0/16			
b	1157	129.mg	n.s.s.	0/16	73.0mg	0/16			
3263	1157	48.8mg	n.s.s.	0/16	68.0mg	2/17			
a	1157	127.mg	n.s.s.	0/16	68.0mg	0/17			
b	1157	26.9mg	966.mg	0/16	68.0mg	5/17			

1,5-NAPHTHALENEDIAMINE   (1,5-diaminonaphthalene) 2243-62-1

	RefNum	LoConf	UpConf	Cntrl	1Dose	1Inc	2Dose	2Inc	Citation or Pathology
3264	TR143	42.1mg	129.mg	3/50	127.mg	38/50	(253.mg	34/50)	liv:hpa,hpc; lun:a/a,a/c; thy:cca,ccr,fca,pcy,ppa. C
a	TR143	69.5mg	236.mg	1/50	127.mg	25/50	253.mg	16/50)	
b	TR143	89.8mg	223.mg	1/50	127.mg	28/50	253.mg	27/50	liv:hpa, hpc.
c	TR143	186.mg	1.46gm	2/50	127.mg	17/50	253.mg	14/50	thy:fca,pcy,ppa.
d	TR143	161.mg	1.00gm	0/50	127.mg	10/50	(253.mg	5/50)	lun:a/a,a/c.
e	TR143	455.mg	3.26gm	0/50	127.mg	2/50	253.mg	8/50	thy:cca,ccr.
f	TR143	571.mg	9.23gm	0/50	127.mg	1/50	253.mg	6/50	
g	TR143	102.mg	n.s.s.	21/50	127.mg	41/50	253.mg	37/50	
h	TR143	89.8mg	223.mg	1/50	127.mg	28/50	253.mg	27/50	liv:hpa,hpc,nnd.
i	TR143	161.mg	1.00gm	0/50	127.mg	10/50	(253.mg	5/50)	lun:a/a,a/c.
3265	TR143	139.mg	373.mg	0/50	118.mg	10/50	235.mg	19/50	thy:cca,ccr,fca,pcy,ppa. C
a	TR143	170.mg	510.mg	0/50	118.mg	8/50	235.mg	16/50	thy:fca,pcy,ppa.
b	TR143	515.mg	n.s.s.	0/50	118.mg	2/50	235.mg	4/50	thy:cca,ccr.
c	TR143	661.mg	n.s.s.	0/50	118.mg	0/50	235.mg	4/50	
d	TR143	131.mg	n.s.s.	24/50	118.mg	40/50	235.mg	25/50	
e	TR143	291.mg	n.s.s.	12/50	118.mg	13/50	235.mg	13/50	liv:hpa,hpc,nnd.
f	TR143	607.mg	n.s.s.	4/50	118.mg	9/50	235.mg	2/50	lun:a/a,a/c.
3266	TR143	29.5mg	242.mg	3/25	24.3mg	16/50	48.1mg	28/50	cli:adn,can; ute:esp. C
a	TR143	37.7mg	2.04gm	2/25	24.3mg	14/50	48.1mg	20/50	
b	TR143	69.1mg	4.53gm	1/25	24.3mg	3/50	48.1mg	13/50	cli:adn,can.
c	TR143	26.2mg	n.s.s.	1/25	24.3mg	12/50	(48.1mg	4/50)	thy:cca,ccr. S
d	TR143	19.7mg	n.s.s.	17/25	24.3mg	41/50	48.1mg	49/50	
e	TR143	107.mg	n.s.s.	0/25	24.3mg	4/50	48.1mg	4/50	liv:hpa,hpc,nnd.
3267	TR143	19.2mg	n.s.s.	10/25	19.4mg	32/50	38.9mg	38/50	
a	TR143	86.4mg	n.s.s.	1/25	19.4mg	7/50	38.9mg	4/50	liv:hpa,hpc,nnd.

N-(1-NAPHTHYL)ETHYLENEDIAMINE.2HCl   1465-25-4

	RefNum	LoConf	UpConf	Cntrl	1Dose	1Inc	2Dose	2Inc	Citation or Pathology
3268	TR168	851.mg	n.s.s.	21/50	253.mg	15/50	392.mg	9/50	
a	TR168	1.94gm	n.s.s.	1/50	253.mg	1/50	392.mg	1/50	liv:hpa,hpc,nnd.
b	TR168	1.33gm	n.s.s.	0/50	253.mg	2/50	392.mg	1/50	lun:a/a,a/c.
3269	TR168	119.mg	n.s.s.	24/50	60.0mg	19/50	(120.mg	16/50)	
a	TR168	368.mg	n.s.s.	12/50	60.0mg	5/50	120.mg	9/50	liv:hpa,hpc,nnd.
b	TR168	384.mg	n.s.s.	4/50	60.0mg	7/50	120.mg	3/50	lun:a/a,a/c.
3270	TR168	42.3mg	n.s.s.	0/25	25.0mg	8/50	(50.0mg	1/50)	liv:hpc,nnd. S
a	TR168	69.2mg	n.s.s.	17/25	25.0mg	36/50	50.0mg	30/50	
b	TR168	42.3mg	n.s.s.	0/25	25.0mg	8/50	(50.0mg	1/50)	liv:hpa,hpc,nnd.
3271	TR168	12.5mg	n.s.s.	10/25	20.0mg	35/50	(40.0mg	27/50)	
a	TR168	265.mg	n.s.s.	1/25	20.0mg	2/50	40.0mg	1/50	liv:hpa,hpc,nnd.

sym.-dibeta-NAPHTHYL-p-PHENYLENEDIAMINE   (agerite white) 93-46-9

	RefNum	LoConf	UpConf	Cntrl	1Dose	1Inc	2Dose	2Inc	Citation or Pathology
3272	1305	30.7mg	n.s.s.	1/17	38.9mg	2/17			Innes;ntis,1968/1969
a	1305	72.8mg	n.s.s.	0/17	38.9mg	0/17			
b	1305	17.3mg	n.s.s.	2/17	38.9mg	5/17			
3273	1305	47.5mg	n.s.s.	2/18	36.2mg	1/18			
a	1305	71.7mg	n.s.s.	1/18	36.2mg	0/18			
b	1305	37.5mg	n.s.s.	3/18	36.2mg	2/18			
3274	1305	72.8mg	n.s.s.	0/16	38.9mg	0/17			
a	1305	72.8mg	n.s.s.	0/17	38.9mg	0/17			
b	1305	38.2mg	n.s.s.	0/16	38.9mg	1/17			
3275	1305	17.9mg	n.s.s.	0/16	36.2mg	3/15			
a	1305	31.2mg	n.s.s.	0/16	36.2mg	1/15			
b	1305	10.4mg	117.mg	0/16	36.2mg	6/15			

1-(1-NAPHTHYL)-2-THIOUREA   (ANTU) 86-88-4

	RefNum	LoConf	UpConf	Cntrl	1Dose	1Inc	2Dose	2Inc	Citation or Pathology
3276	1155	.991mg	n.s.s.	1/17	.834mg	1/18			Innes;ntis,1968/1969
a	1155	1.65mg	n.s.s.	0/17	.834mg	0/18			
b	1155	1.10mg	n.s.s.	2/17	.834mg	1/18			
3277	1155	1.02mg	n.s.s.	2/18	.776mg	1/18			
a	1155	1.54mg	n.s.s.	1/18	.776mg	0/18			
b	1155	.803mg	n.s.s.	3/18	.776mg	2/18			
3278	1155	1.65mg	n.s.s.	0/16	.834mg	0/18			
a	1155	1.65mg	n.s.s.	0/16	.834mg	0/18			

```
       Spe Strain  Site   Xpo+Xpt                                                          TD50    2Tailpvl
          Sex  Route  Hist      Notes                                                         DR   AuOp
  b    M f b6c orl tba mix 76w76 evx                                                        5.34mg   P<.3   -
  3279 M m b6c orl lun ade 76w76 evx                                  .         ±           1.27mg   P<.04  -
  a    M m b6c orl liv hpt 76w76 evx                                                        no dre   P=1.   -
  b    M m b6c orl tba mix 76w76 evx                                                        .700mg   P<.005 -

2-NAPHTHYLAMINE                           100ng..:..1ug....:..10.....:..100.....:..1mg....:..10.....:..100....:..1g......:..10
  3280 M f bal eat liv hnd 40w55 er                                   .    +    .           17.4mg   P<.0005
  a    M f bal eat liv ade 40w55 er                                                         36.9mg   P<.0005
  b    M f bal eat liv mix 40w55 er                                                         63.0mg   P<.0005
  c    M f bal eat liv hpt 40w55 er                                                         175.mg   P<.04  +
  d    M f bal eat ubl tum 40w55 er                                                         no dre   P=1.   -
  3281 M b cba gav liv hpt 79w79 e                           .    +    .                     20.5mg   P<.002  +
  3282 P f rhe mix ubl mix 60m60 emrv                   .    (±)                            5.74mg   P<.06   +
  a    P f rhe mix ubl ppa 60m60 emrv                                                       25.8mg   P<.4    +
  b    P f rhe mix ubl pam 60m60 emrv                                                       25.8mg   P<.4    +
  c    P f rhe mix ubl tcc 60m60 emrv                                                       25.8mg   P<.4    +
  d    P f rhe mix ubl car 60m60 emrv                                                       53.7mg   P<.6    +
  e    P f rhe mix ubl ppc 60m60 emrv                                                       53.7mg   P<.6    +
  3283 R b alb eat ubl pam 24m24 r                                        .>                no dre   P=1.
  3284 R b alb eat ubl pam 24m24 er                                  .>                     9.68mg   P<.2
  3285 R b alb eat ubl pam 24m24 efr                            .>                          35.4mg   P<.4
  3286 R f wis gav ubl mix 13m23 er                           .    ±                        61.6mg   P<.02   +

2-NAPHTHYLAMINO,1-SULFONIC ACID           100ng..:..1ug....:..10.....:..100.....:..1mg....:..10.....:..100....:..1g......:..10
  3287 M f bld eat lun mix 15m33 e                                                    .>    9.96gm   P<.8   -
  a    M f bld eat liv mix 15m33 e                                                         876.gm    P<1.
  3288 M m bld eat lun mix 15m33 e                                               .>         4.92gm   P<.7   -
  a    M m bld eat liv mix 15m33 e                                                         379.gm    P<1.

NEOSUGAR                                  100ng..:..1ug....:..10.....:..100.....:..1mg....:..10.....:..100....:..1g......:..10
  3289 R f f3d eat spl leu 24m24                                                      .> 25.5gm * P<.5  -
  a    R f f3d eat pit ade 24m24                                                           no dre   P=1.
  3290 R m f3d eat pit ade 24m24                                                .    +  3.55gm * P<.007 -

NICKEL                                    100ng..:..1ug....:..10.....:..100.....:..1mg....:..10.....:..100....:..1g......:..10
  3291 R f leb wat tba mix 43m43 e                                        .>                no dre   P=1.
  a    R f leb wat tba mal 43m43 e                                                          no dre   P=1.
  3292 R m leb wat tba mix 37m37 e                                  .>                       no dre   P=1.
  a    R m leb wat tba mal 37m37 e                                                          no dre   P=1.

NICKEL (II) ACETATE                       100ng..:..1ug....:..10.....:..100.....:..1mg....:..10.....:..100....:..1g......:..10
  3293 M f cd1 wat lun tum 33m33 e                                            .>             no dre   P=1.
  a    M f cd1 wat liv car 33m33 e                                                          no dre   P=1.
  b    M f cd1 wat tba mix 33m33 e                                                          no dre   P=1.
  3294 M f cd1 wat tba mix 29m29 e                                   .>                      29.0mg   P<.8   -
  3295 M m cd1 wat lun tum 32m32 e                                        .>                 no dre   P=1.
  a    M m cd1 wat tba tum 32m32 e                                                          no dre   P=1.
  3296 M m cd1 wat tba mix 32m32 e                                        .>                 no dre   P=1.

NICKEL DIBUTYLDITHIOCARBAMATE             100ng..:..1ug....:..10.....:..100.....:..1mg....:..10.....:..100....:..1g......:..10
  3297 M f b6a orl lun ade 76w76 evx                         .>                             86.9ug   P<.4   -
  a    M f b6a orl liv hpt 76w76 evx                                                        no dre   P=1.   -
  b    M f b6a orl tba mix 76w76 evx                                                        .185mg   P<.7   -
  3298 M m b6a orl lun ade 76w76 evx                         .>                             .153mg   P<.7   -
  a    M m b6a orl liv hpt 76w76 evx                                                        no dre   P=1.   -
  b    M m b6a orl tba mix 76w76 evx                                                        no dre   P=1.   -
  3299 M f b6c orl lun ade 76w76 evx                    .>                                  .174mg   P<.3   -
  a    M f b6c orl liv hpt 76w76 evx                                                        no dre   P=1.   -
  b    M f b6c orl tba mix 76w76 evx                                                        .174mg   P<.3   -
  3300 M m b6c orl liv hpt 76w76 evx                 .    ±                                 50.9ug   P<.04  -
  a    M m b6c orl lun ade 76w76 evx                                                        .163mg   P<.3   -
  b    M m b6c orl tba mix 76w76 evx                                                        28.3ug   P<.007 -

NICOTINAMIDE                              100ng..:..1ug....:..10.....:..100.....:..1mg....:..10.....:..100....:..1g......:..10
  3301 M f swa wat lun mix 26m26 e                   .                                    .>no dre   P=1.
  a    M f swa wat liv tum 26m26 e                                                          no dre   P=1.
  3302 M m swa wat lun mix 25m25 e                                                         .no dre   P=1.   -
  a    M m swa wat liv mix 25m25 e                                                          no dre   P=1.

NICOTINE                                  100ng..:..1ug....:..10.....:..100.....:..1mg....:..10.....:..100....:..1g......:..10
  3303 R f sda ipj tba mal 24m24 es                                 .>                       no dre   P=1.   -
  3304 R m sda ipj liv hae 24m24 es                            .>                            no dre   P=1.   -
  a    R m sda ipj tba mal 24m24 es                                                         6.38mg   P<.6   -

NICOTINE.HCl                              100ng..:..1ug....:..10.....:..100.....:..1mg....:..10.....:..100....:..1g......:..10
  3305 M f swa wat lun tum 28m29 e                                                    .>    11.1gm * P<.9   -
  a    M f swa wat liv mix 28m29 e                                                          no dre   P=1.   -
  3306 M m swa wat liv mix 26m28 e                                               .>          no dre   P=1.   -
  a    M m swa wat lun tum 26m28 e                                                          no dre   P=1.   -
```

	RefNum	LoConf	UpConf	Cntrl	1Dose	1Inc	2Dose	2Inc	Citation or Pathology
									Brkly Code
b	1155	.869mg	n.s.s.	0/16	.834mg	1/18			
3279	1155	.383mg	n.s.s.	0/16	.776mg	3/15			
a	1155	1.28mg	n.s.s.	0/16	.776mg	0/15			
b	1155	.263mg	5.52mg	0/16	.776mg	5/15			

2-NAPHTHYLAMINE   91-59-8

	RefNum	LoConf	UpConf	Cntrl	1Dose	1Inc	2Dose	2Inc	Citation or Pathology
3280	1446	7.87mg	41.7mg	0/17	189.mg	14/16			Yoshida;gann,70,645-652;1979
a	1446	17.2mg	97.2mg	0/17	189.mg	10/16			
b	1446	26.7mg	223.mg	0/17	189.mg	7/16			
c	1446	52.6mg	n.s.s.	0/17	189.mg	3/16			
d	1446	174.mg	n.s.s.	0/17	189.mg	0/16			
3281	207	10.5mg	67.5mg	0/7	50.1mg	13/21			Bonser;bjca,6,412-424;1952
3282	1470	2.42mg	n.s.s.	0/3	90.5mg	7/14			Conzelman;jnci,42,825-831;1969
a	1470	6.34mg	n.s.s.	0/3	90.5mg	2/14			
b	1470	6.34mg	n.s.s.	0/3	90.5mg	2/14			
c	1470	6.34mg	n.s.s.	0/3	90.5mg	2/14			
d	1470	8.74mg	n.s.s.	0/3	90.5mg	1/14			
e	1470	8.74mg	n.s.s.	0/3	90.5mg	1/14			
3283	207m	15.8mg	n.s.s.	0/17	4.50mg	0/17			Bonser;bjca,6,412-424;1952
3284	207n	2.90mg	n.s.s.	0/5	4.50mg	3/11			
3285	207o	5.76mg	n.s.s.	0/5	4.50mg	1/12			
3286	1564	21.2mg	n.s.s.	0/20	24.4mg	4/18			Hicks;bjca,46,646-661;1982

2-NAPHTHYLAMINO,1-SULFONIC ACID   81-16-3

	RefNum	LoConf	UpConf	Cntrl	1Dose	1Inc	2Dose	2Inc	Citation or Pathology
3287	1488	942.mg	n.s.s.	14/49	306.mg	15/48			Della Porta;carc,3,647-649;1982
a	1488	3.33gm	n.s.s.	1/49	306.mg	1/48			
3288	1488	765.mg	n.s.s.	14/48	283.mg	16/47			
a	1488	2.42gm	n.s.s.	2/48	283.mg	2/47			

NEOSUGAR   88385-81-3

	RefNum	LoConf	UpConf	Cntrl	1Dose	1Inc	2Dose	2Inc	3Dose	3Inc	Citation or Pathology
3289	1880	5.05gm	n.s.s.	4/50	400.mg	7/50	1.00gm	12/50	2.50gm	7/50	Clevenger;jact,7,643-662;1988
a	1880	8.67gm	n.s.s.	24/50	400.mg	19/50	1.00gm	19/50	2.50gm	14/50	
3290	1880	1.71gm	59.6gm	10/50	320.mg	13/50	800.mg	19/50	2.00gm	22/50	

NICKEL   7440-02-0

	RefNum	LoConf	UpConf	Cntrl	1Dose	1Inc	2Dose	2Inc	Citation or Pathology
3291	1464	1.29mg	n.s.s.	10/26	.286mg	9/27			Schroeder;jnut,104,239-243;1974
a	1464	2.83mg	n.s.s.	7/26	.286mg	3/27			
3292	1464	.919mg	n.s.s.	9/23	.250mg	4/17			
a	1464	1.10mg	n.s.s.	4/23	.250mg	2/17			

NICKEL (II) ACETATE   373-02-4

	RefNum	LoConf	UpConf	Cntrl	1Dose	1Inc	2Dose	2Inc	Citation or Pathology
3293	56	6.58mg	n.s.s.	9/60	1.00mg	3/33			Schroeder;jnut,83,239-250;1964
a	56	12.7mg	n.s.s.	1/60	1.00mg	0/33			
b	56	8.98mg	n.s.s.	22/60	1.00mg	3/33			
3294	1395	2.89mg	n.s.s.	9/45	1.00mg	10/44			Schroeder;jnut,105,452-458;1975
3295	56	4.94mg	n.s.s.	8/44	.833mg	5/41			Schroeder;jnut,83,239-250;1964
a	56	4.36mg	n.s.s.	11/44	.833mg	7/41			
3296	1395	5.82mg	n.s.s.	10/43	.833mg	4/37			Schroeder;jnut,105,452-458;1975

NICKEL DIBUTYLDITHIOCARBAMATE   (vanguard N) 13927-77-0

	RefNum	LoConf	UpConf	Cntrl	1Dose	1Inc	2Dose	2Inc	Citation or Pathology
3297	1357	19.0ug	n.s.s.	1/17	28.9ug	3/18			Innes;ntis,1968/1969
a	1357	57.3ug	n.s.s.	0/17	28.9ug	0/18			
b	1357	20.8ug	n.s.s.	2/17	28.9ug	3/18			
3298	1357	19.2ug	n.s.s.	2/18	27.0ug	3/18			
a	1357	53.5ug	n.s.s.	1/18	27.0ug	0/18			
b	1357	21.2ug	n.s.s.	3/18	27.0ug	3/18			
3299	1357	28.4ug	n.s.s.	0/16	28.9ug	1/17			
a	1357	54.1ug	n.s.s.	0/16	28.9ug	0/17			
b	1357	28.4ug	n.s.s.	0/16	28.9ug	1/17			
3300	1357	15.3ug	n.s.s.	0/16	27.0ug	3/17			
a	1357	26.5ug	n.s.s.	0/16	27.0ug	1/17			
b	1357	10.7ug	.383mg	0/16	27.0ug	5/17			

NICOTINAMIDE   98-92-0

	RefNum	LoConf	UpConf	Cntrl	1Dose	1Inc	2Dose	2Inc	Citation or Pathology
3301	1732	7.03gm	n.s.s.	15/89	2.00gm	7/45			Toth;onco,40,72-75;1983/1979
a	1732	23.8gm	n.s.s.	0/98	2.00gm	0/49			
3302	1732	9.34gm	n.s.s.	21/87	1.67gm	6/50			
a	1732	15.8gm	n.s.s.	2/61	1.67gm	0/42			

NICOTINE   54-11-5

	RefNum	LoConf	UpConf	Cntrl	1Dose	1Inc	2Dose	2Inc	Citation or Pathology
3303	1134	1.00mg	n.s.s.	3/33	.286mg	2/32			Schmahl;zkko,86,77-84;1976
3304	1134	2.06mg	n.s.s.	1/36	.286mg	0/35			
a	1134	.898mg	n.s.s.	1/36	.286mg	2/35			

NICOTINE.HCl   636-79-3

	RefNum	LoConf	UpConf	Cntrl	1Dose	1Inc	2Dose	2Inc	Citation or Pathology
3305	1530	834.mg	n.s.s.	15/95	125.mg	6/46	188.mg	9/48	Toth;acnr,2,71-74;1982/1979
a	1530	1.06gm	n.s.s.	0/99	125.mg	0/50	188.mg	0/48	
3306	1530	606.mg	n.s.s.	2/62	104.mg	0/38	156.mg	0/32	
a	1530	1.41gm	n.s.s.	22/88	104.mg	6/50	156.mg	6/48	

```
         Spe Strain  Site   Xpo+Xpt                                                 TD50      2Tailpvl
            Sex   Route  Hist      Notes                                                 DR     AuOp
NICOTINIC ACID               100ng..:..1ug...:..10....:..100...:..1mg..:..10....:..100...:..1g.....:.10
3307 M f swa wat lun tum 32m32 e                                                  .> 17.0gm   P<.2   -
   a M f swa wat liv mix 32m32 e                                                     no dre   P=1.   -
3308 M m swa wat lun tum 28m28 e                                                  .>no dre   P=1.   -
   a M m swa wat liv mix 28m28 e                                                     no dre   P=1.   -

NICOTINIC ACID HYDRAZIDE     100ng..:..1ug...:..10....:..100...:..1mg..:..10....:..100...:..1g.....:.10
3309 M f swa wat lun mix 24m24 e                             . + .                   145.mg   P<.0005+
   a M f swa wat lun ade 24m24 e                                                     181.mg   P<.0005
   b M f swa wat lun adc 24m24 e                                                     511.mg   P<.0005
   c M f swa wat liv ang 24m24 e                                                     no dre   P=1.   -
   d M f swa wat liv agm 24m24 e                                                     no dre   P=1.   -
3310 M m swa wat lun ade 23m23 e                                     . +    .        428.mg   P<.004
   a M m swa wat lun mix 23m23 e                                                     528.mg   P<.05  +
   b M m swa wat liv hpt 23m23 e                                                    1.41gm    P<.08  -
   c M m swa wat liv ang 23m23 e                                                     no dre   P=1.   -
   d M m swa wat liv agm 23m23 e                                                     no dre   P=1.   -

NIGROSINE                    100ng..:..1ug...:..10....:..100...:..1mg..:..10....:..100...:..1g.....:.10
3311 R f nss eat liv tum 64w64 e                                      .>             no dre   P=1.   -
   a R f nss eat tba mix 64w64 e                                                     no dre   P=1.   -
3312 R m nss eat liv tum 64w64 e                                  .>                 no dre   P=1.   -
   a R m nss eat tba mix 64w64 e                                                     no dre   P=1.   -

NIOBATE, SODIUM              100ng..:..1ug...:..10....:..100...:..1mg..:..10....:..100...:..1g.....:.10
3313 M b cd1 wat lun tum 31m31 e                                    .>               no dre   P=1.   -
   a M b cd1 wat liv tum 31m31 e                                                     no dre   P=1.   -
   b M b cd1 wat tba tum 31m31 e                                                     no dre   P=1.   -
   c M b cd1 wat tba mal 31m31 e                                                     no dre   P=1.   -

NITHIAZIDE                   100ng..:..1ug...:..10....:..100...:..1mg..:..10....:..100...:..1g.....:.10
3314 M f b6c eat TBA MXB 22m24 v                                         :>         1.40gm  * P<.2
   a M f b6c eat liv MXB 22m24 v                                                    2.72gm  * P<.3
   b M f b6c eat lun MXB 22m24 v                                                     no dre   P=1.
3315 M m b6c eat liv MXA 22m24 v                                          : +        758.mg  * P<.05  c
   a M m b6c eat liv hpc 22m24 v                                                    1.90gm  * P<.2   c
   b M m b6c eat TBA MXB 22m24 v                                                    4.75gm  * P<.9
   c M m b6c eat liv MXB 22m24 v                                                     758.mg  * P<.05
   d M m b6c eat lun MXB 22m24 v                                                     no dre   P=1.
3316 R f f34 eat MXA MXA 22m24 v                                      : +            131.mg  * P<.02  c
   a R f f34 eat TBA MXB 22m24 v                                                     175.mg  * P<.6
   b R f f34 eat liv MXB 22m24 v                                                     no dre   P=1.
3317 R m f34 eat TBA MXB 22m24 v                                 :>                  no dre   P=1.   -
   a R m f34 eat liv MXB 22m24 v                                                     508.mg  * P<.3

NITRATE, SODIUM              100ng..:..1ug...:..10....:..100...:..1mg..:..10....:..100...:..1g.....:.10
3318 R f f34 eat liv mix 24m29 e                                                     no dre   P=1.   -
   a R f f34 eat tba mix 24m29 e                                                     no dre   P=1.   -
3319 R m f34 eat liv mix 24m29 e                                                     no dre   P=1.   -
   a R m f34 eat tba mix 24m29 e                                                    2.62gm  * P<.6   -
3320 R f mrc wat pit tum 20m24 e                             . +    .                103.mg   P<.004 -
   a R f mrc wat tba mix 20m24 e                                                    86.9mg   P<.004 -
3321 R m mrc wat tba mix 20m24 e                                      .>             750.mg   P<.8   -

NITRIC OXIDE                 100ng..:..1ug...:..10....:..100...:..1mg..:..10....:..100...:..1g.....:.10
3322 M f jic inh lun ade 29m29                                         .>            no dre   P=1.   -

NITRILOTRIACETIC ACID        100ng..:..1ug...:..10....:..100...:..1mg..:..10....:..100...:..1g.....:.10
3323 M f b6c eat kid uac 77w91                                              :       13.8gm  * P<.06  c
   a M f b6c eat TBA MXB 77w91                                                      3.95gm  * P<.3
   b M f b6c eat liv MXB 77w91                                                       no dre   P=1.
   c M f b6c eat lun MXB 77w91                                                      9.31gm  * P<.2
3324 M m b6c eat MXA MXA 77w91                                          :+ :        1.47gm  / P<.0005c
   a M m b6c eat TBA MXB 77w91                                                      2.21gm  / P<.09
   b M m b6c eat liv MXB 77w91                                                       no dre   P=1.
   c M m b6c eat lun MXB 77w91                                                       no dre   P=1.
3325 R f f34 eat liv nnd 77w99                                        : + :          810.mg  * P<.004
   a R f f34 eat MXA MXA 77w99                                                      1.45gm  * P<.002 c
   b R f f34 eat ubl MXA 77w99                                                      1.57gm  * P<.002 c
   c R f f34 eat adr phe 77w99                                                      1.71gm  / P<.005
   d R f f34 eat lun a/c 77w99                                                      2.25gm  * P<.03
   e R f f34 eat TBA MXB 77w99                                                       631.mg  * P<.2
   f R f f34 eat liv MXB 77w99                                                       810.mg  * P<.004
3326 R f f34 eat kur mal 24m24 e                                          . +       2.70gm  * P<.0005+
   a R f f34 eat liv hpc 24m24 e                                                     no dre   P=1.
3327 R m f34 eat MXA MXA 18m24                                        : +            446.mg  \ P<.05
   a R m f34 eat MXA MXA 18m24                                                      2.26gm  * P<.02  c
   b R m f34 eat TBA MXB 18m24                                                      7.12gm  * P<.9
   c R m f34 eat liv MXB 18m24                                                       no dre   P=1.
3328 R m f34 eat kur mal 24m24 es                                         . +    .  1.30gm  * P<.0005+
   a R m f34 eat liv hpc 24m24 es                                                    no dre   P=1.
```

	RefNum	LoConf	UpConf	Cntrl	1Dose	1Inc	2Dose	2Inc	Citation or Pathology
									Brkly Code

NICOTINIC ACID (niacin) 59-67-6

	RefNum	LoConf	UpConf	Cntrl	1Dose	1Inc			
3307	1530	5.50gm	n.s.s.	15/95	2.00gm	13/48			Toth;acnr,2,71-74;1982/1979
a	1530	36.8gm	n.s.s.	0/99	2.00gm	0/50			
3308	1530	7.84gm	n.s.s.	22/88	1.67gm	9/50			
a	1530	9.24gm	n.s.s.	2/62	1.67gm	1/33			

NICOTINIC ACID HYDRAZIDE (3-pyridoyl hydrazine) 553-53-7

	RefNum	LoConf	UpConf	Cntrl	1Dose	1Inc			
3309	1084	85.8mg	294.mg	25/98	250.mg	38/50			Toth;onco,38,106-109;1981
a	1084	106.mg	376.mg	20/98	250.mg	34/50			
b	1084	259.mg	1.57gm	6/98	250.mg	16/50			
c	1084	1.41gm	n.s.s.	3/71	250.mg	1/42			
d	1084	1.54gm	n.s.s.	5/71	250.mg	1/42			
3310	1084	197.mg	3.49gm	16/99	208.mg	19/49			
a	1084	203.mg	n.s.s.	26/99	208.mg	21/49			
b	1084	230.mg	n.s.s.	0/43	208.mg	1/11			
c	1084	1.30gm	n.s.s.	4/90	208.mg	0/32			
d	1084	1.01gm	n.s.s.	8/80	208.mg	0/25			

NIGROSINE ---

	RefNum	LoConf	UpConf	Cntrl	1Dose	1Inc	2Dose	2Inc	
3311	1372	89.3mg	n.s.s.	0/7	150.mg	0/9	1.50gm	0/5	Allmark;jphp,9,622-628;1957
a	1372	586.mg	n.s.s.	1/7	150.mg	1/9	1.50gm	0/5	
3312	1372	4.26mg	n.s.s.	0/5	12.0mg	0/5	120.mg	0/5	
a	1372	4.26mg	n.s.s.	0/5	12.0mg	0/5	120.mg	0/5	

NIOBATE, SODIUM 12034-09-2

	RefNum	LoConf	UpConf	Cntrl	1Dose	1Inc			
3313	1036	7.82mg	n.s.s.	15/71	.877mg	11/79			Kanisawa;canr,29,892-895;1969
a	1036	24.1mg	n.s.s.	4/71	.877mg	0/79			
b	1036	6.56mg	n.s.s.	24/71	.877mg	18/79			
c	1036	10.8mg	n.s.s.	8/71	.877mg	5/79			

NITHIAZIDE 139-94-6

	RefNum	LoConf	UpConf	Cntrl	1Dose	1Inc	2Dose	2Inc	
3314	TR146	522.mg	n.s.s.	6/20	294.mg	13/50	587.mg	23/50	liv:hpa,hpc,nnd.
a	TR146	924.mg	n.s.s.	3/20	294.mg	4/50	587.mg	12/50	lun:a/a,a/c.
b	TR146	3.71gm	n.s.s.	3/20	294.mg	1/50	587.mg	2/50	liv:hpa,hpc.
3315	TR146	349.mg	n.s.s.	4/20	271.mg	15/50	542.mg	25/50	
a	TR146	721.mg	n.s.s.	2/20	271.mg	6/50	542.mg	12/50	
b	TR146	415.mg	n.s.s.	10/20	271.mg	27/50	542.mg	28/50	
c	TR146	349.mg	n.s.s.	4/20	271.mg	15/50	542.mg	25/50	liv:hpa,hpc,nnd.
d	TR146	874.mg	n.s.s.	3/20	271.mg	4/50	(542.mg	1/50)	lun:a/a,a/c.
3316	TR146	68.2mg	n.s.s.	1/20	28.2mg	5/50	56.5mg	15/50	mgl:cyn,fba; ski:fba; sub:fba.
a	TR146	36.5mg	n.s.s.	13/20	28.2mg	28/50	56.5mg	38/50	
b	TR146	n.s.s.	n.s.s.	0/20	28.2mg	0/50	56.5mg	0/50	liv:hpa,hpc,nnd.
3317	TR146	48.5mg	n.s.s.	11/20	22.6mg	25/50	45.2mg	24/50	
a	TR146	154.mg	n.s.s.	0/20	22.6mg	1/50	45.2mg	2/50	liv:hpa,hpc,nnd.

NITRATE, SODIUM 7631-99-4

	RefNum	LoConf	UpConf	Cntrl	1Dose	1Inc	2Dose	2Inc	
3318	1490	10.1gm	n.s.s.	2/50	1.06gm	0/50	2.11gm	0/49	Maekawa;fctx,20,25-33;1982
a	1490	2.99gm	n.s.s.	46/50	1.06gm	43/50	2.11gm	39/49	
3319	1490	10.2gm	n.s.s.	6/50	846.mg	7/50	1.69gm	4/50	
a	1490	409.mg	n.s.s.	47/50	846.mg	50/50	1.69gm	48/50	
3320	213	42.8mg	764.mg	3/15	165.mg	11/15			Lijinsky;jnci,50,1061-1063;1973
a	213	35.3mg	665.mg	4/15	165.mg	12/15			
3321	213	83.5mg	n.s.s.	5/15	115.mg	6/15			

NITRIC OXIDE 10102-43-9

	RefNum	LoConf	UpConf	Cntrl	1Dose	1Inc			
3322	1388	38.4mg	n.s.s.	6/64	5.19mg	5/65			Oda;envr,22,254-263;1980

NITRILOTRIACETIC ACID 139-13-9

	RefNum	LoConf	UpConf	Cntrl	1Dose	1Inc	2Dose	2Inc	3Dose	3Inc	Citation/Pathology	Brkly
3323	TR6	4.77gm	n.s.s.	0/20	836.mg	0/50	1.67gm	4/50			liv:hpa,hpc,nnd.	
a	TR6	1.33gm	n.s.s.	4/20	836.mg	13/50	1.67gm	19/50			lun:a/a,a/c.	
b	TR6	9.71gm	n.s.s.	0/20	836.mg	1/50	1.67gm	0/50			k/p:pam; kid:tla,uac.	
c	TR6	3.80gm	n.s.s.	0/20	836.mg	2/50	1.67gm	4/50				
3324	TR6	1.04gm	2.17gm	0/20	771.mg	5/50	1.54gm	24/50				
a	TR6	1.06gm	n.s.s.	10/20	771.mg	14/50	1.54gm	31/50			liv:hpa,hpc,nnd.	
b	TR6	9.66gm	n.s.s.	3/20	771.mg	3/50	1.54gm	2/50			lun:a/a,a/c.	
c	TR6	4.68gm	n.s.s.	4/20	771.mg	5/50	1.54gm	6/50				S
3325	TR6	508.mg	2.19gm	2/20	281.mg	8/50	562.mg	22/50			kid:tla,tpp; ubl:sqc,tcc.	
a	TR6	908.mg	2.72gm	0/20	281.mg	2/50	562.mg	13/50			ubl:sqc,tcc.	
b	TR6	967.mg	3.08gm	0/20	281.mg	2/50	562.mg	12/50				S
c	TR6	984.mg	4.86gm	1/20	281.mg	0/50	562.mg	14/50				S
d	TR6	1.28gm	10.1gm	0/20	281.mg	3/50	562.mg	7/50				
e	TR6	278.mg	n.s.s.	14/20	281.mg	33/50	562.mg	47/50			liv:hpa,hpc,nnd.	
f	TR6	508.mg	2.19gm	2/20	281.mg	8/50	562.mg	22/50				
3326	1954	1.08gm	10.6gm	1/120	10.0mg	0/24	100.mg	0/24	1.00gm	6/24	Fears;txih,4,221-255;1988	
a	1954	43.7mg	n.s.s.	0/120	10.0mg	0/24	100.mg	0/24	1.00gm	0/24		
3327	TR6	234.mg	10.6gm	2/20	225.mg	16/50	(450.mg	8/50)			adr:phe; pit:bsa,cra; pni:isa; thy:fca,fcc. S	
a	TR6	1.21gm	7.43gm	0/20	225.mg	1/50	450.mg	7/50			kid:tla,uac; ure:pam,ppa.	
b	TR6	442.mg	n.s.s.	8/20	225.mg	26/50	450.mg	22/50				
c	TR6	3.30gm	n.s.s.	3/20	225.mg	5/50	450.mg	2/50			liv:hpa,hpc,nnd.	
3328	1954	610.mg	3.57gm	0/120	8.00mg	0/24	80.0mg	0/24	800.mg	9/24	Fears;txih,4,221-255;1988	
a	1954	33.6mg	n.s.s.	1/120	8.00mg	0/23	80.0mg	0/24	800.mg	0/24		

```
          Spe Strain Site   Xpo+Xpt                                                    TD50    2Tailpvl
              Sex   Route  Hist    Notes                                                  DR    AuOp
────────────────────────────────────────────────────────────────────────────────────────────────────
3329 R f mrc wat tba mix 20m24                                      .>                  221.mg  P<.2   -
3330 R m mrc wat tba mix 20m24                                      .>                  248.mg  P<.3   -

NITRILOTRIACETIC ACID, TRISODIUM SALT, MONOHYDRATE.:..10.....:..100....:..1mg....:..10....:..100...:..1g....:..10
3331 M f b6c eat TBA MXB 78w91                                         :>               5.31gm * P<.8   -
  a  M f b6c eat liv MXB 78w91                                                          no dre   P=1.
  b  M f b6c eat lun MXB 78w91                                                          61.8mg * P<.9
3332 M m b6c eat --- MXA 78w91                                           :  ±           #1.55gm * P<.02  -
  a  M m b6c eat TBA MXB 78w91                                                          no dre   P=1.
  b  M m b6c eat liv MXB 78w91                                                          no dre   P=1.
  c  M m b6c eat lun MXB 78w91                                                          1.99gm \ P<.6
3333 R f f34 eat kid MXA 24m24                                        :    +            2.88gm * P<.002 c
3334 R f f34 eat liv MXA 18m25                                             :>           12.0mg * P<.5
3335 R f f34 eat TBA MXB 24m24                                    :>                    2.99gm * P<.6   -
  a  R f f34 eat liv MXB 24m24                                                          no dre   P=1.
3336 R f f34 eat TBA MXB 18m25                                       :>                 no dre   P=1.   -
3337 R f f34 eat MXA MXA 24m24                                    :  +  :               783.mg * P<.0005c
  a  R f f34 eat ubl MXA 24m24                                                          1.99gm * P<.002 c
  b  R f f34 eat ure tcc 24m24                                                          2.28gm * P<.0005c
3338 R m f34 eat kid MXA 24m24                                    :  +  :               826.mg * P<.0005c
  a  R m f34 eat kid tcc 24m24                                                          2.53gm * P<.002 c
  b  R m f34 eat liv MXB 24m24                                                          2.20gm * P<.05
3339 R m f34 eat liv MXB 18m25                                             :>           no dre   P=1.
3340 R m f34 eat TBA MXB 24m24                                  :   +  :                329.mg Z P<.0005-
3341 R m f34 eat TBA MXB 18m25                                       :>                 2.63gm * P<.7   -
3342 R m f34 eat MXA MXA 24m24                                    :  +  :               511.mg * P<.0005c
  a  R m f34 eat ure tcc 24m24                                                          965.mg * P<.0005c
3343 R m cdr wat kid ade 24m24 e                                 .   ±                  320.mg   P<.02  +
  a  R m cdr wat tba mix 24m24 e                                                        88.4mg   P<.07
3344 R m cdr wat kid ade 24m24 e                                 .  +   .               224.mg   P<.0005+
  a  R m cdr wat tba mix 24m24 e                                                        105.mg   P<.2

NITRITE, SODIUM                      100ng..:..1ug....:..10.....:..100....:..1mg....:..10....:..100...:..1g.....:..10
3345 H f syg eat liv cgf 97w97 e                                    .>                  no dre   P=1.
  a  H f syg eat tba tum 97w97 e                                                        477.mg   P<.6
3346 H m syg eat liv tum 24m24 e                                 .>                     no dre   P=1.
  a  H m syg eat tba tum 24m24 e                                                        no dre   P=1.
3347 M f cb6 eat liv hpc 52w69 e                                    .>                  no dre   P=1.   -
  a  M f cb6 eat lun ade 52w69 e                                                        no dre   P=1.   -
  b  M f cb6 eat tba mix 52w69 e                                                        no dre   P=1.   -
3348 M f cb6 wat lun tum 27m29 g                                       .>               5.20gm * P<.5   -
  a  M f cb6 wat liv mix 27m29 g                                                        no dre   P=1.
3349 M m cb6 eat lun ade 52w69 e                                 .>                     1.61gm   P<.2
  a  M m cb6 eat liv hpc 52w69 e                                                        no dre   P=1.
  b  M m cb6 eat tba mix 52w69 e                                                        1.21gm   P<.4
3350 M f scd wat liv hct  8m29 s                              .>                        744.mg   P<.7   -
  a  M f scd wat lun tum  8m29 s                                                        no dre   P=1.   -
3351 M m scd wat lun tum  8m27 s                            .>                          103.mg   P<.3   -
  a  M m scd wat liv hct  8m27 s                                                        no dre   P=1.
3352 R b cdr eat --- mly 29m29                                   .   ±                  539.mg   P<.03  +
  a  R b cdr eat spl lym 29m29                                                          601.mg   P<.04
  b  R b cdr eat liv ang 29m29                                                          5.84gm   P<.3
  c  R b cdr eat liv mix 29m29                                                          no dre   P=1.
3353 R f f34 eat liv tum 52w69 e                                 .>                     no dre   P=1.   -
  a  R f f34 eat tba mix 52w69 e                                                        no dre   P=1.   -
3354 R f f34 wat liv mix 24m28 e                                    .>                  no dre   P=1.   -
  a  R f f34 wat tba mix 24m28 e                                                        no dre   P=1.
3355 R f f34 eat liv mix 24m30 e                              .  +                      124.mg   P<.003 +
  a  R f f34 eat liv nnd 24m30 e                                                        199.mg   P<.03
  b  R f f34 eat liv car 24m30 e                                                        470.mg   P<.02
3356 R f f34 eat liv mix 24m30 e                              .  +                      141.mg   P<.007 +
  a  R f f34 eat liv nnd 24m30 e                                                        164.mg   P<.02
  b  R f f34 eat liv hpc 24m30 e                                                        629.mg   P<.04
3357 R f f34 wat adr mdt 24m30 e                              .  +   .                  163.mg   P<.0005
  a  R f f34 wat thy car 24m30 e                                                        163.mg   P<.0005
  b  R f f34 wat mam tum 24m30 e                                                        115.mg   P<.05
  c  R f f34 wat liv mix 24m30 e                                                        153.mg   P<.03  +
  d  R f f34 wat liv nnd 24m30 e                                                        185.mg   P<.06
  e  R f f34 wat ute pol 24m30 e                                                        218.mg   P<.02
  f  R f f34 wat liv hpc 24m30 e                                                        1.55gm   P<.3
3358 R f f34 eat liv mix 24m30 e                              .  +   .                  136.mg   P<.003
  a  R f f34 eat liv nnd 24m30 e                                                        218.mg   P<.03
  b  R f f34 eat liv hpc 24m30 e                                                        516.mg   P<.02
3359 R m f34 eat liv tum 52w69 e                                 .>                     no dre   P=1.   -
  a  R m f34 eat tba mix 52w69 e                                                        no dre   P=1.   -
3360 R m f34 wat liv mix 24m28 e                                    .>                  1.31gm * P<.5   -
  a  R m f34 wat tba mix 24m28 e                                                        noTD50   P=1.   -
3361 R m f34 eat liv mix 24m30 e                                 .>                     445.mg   P<.3
  a  R m f34 eat liv nnd 24m30 e                                                        467.mg   P<.3
  b  R m f34 eat liv car 24m30 e                                                        no dre   P=1.
3362 R m f34 eat liv nnd 24m30 e                                 .>                     no dre   P=1.
```

	RefNum	LoConf	UpConf	Cntrl	1Dose	1Inc	2Dose	2Inc	Citation or Pathology	Brkly Code
3329	213	67.2mg	n.s.s.	5/15	165.mg	9/15			Lijinsky;jnci,50,1061-1063;1973	
3330	213	64.2mg	n.s.s.	4/15	115.mg	7/15				

NITRILOTRIACETIC ACID, TRISODIUM SALT, MONOHYDRATE   18662-53-8

	RefNum	LoConf	UpConf	Cntrl	1Dose	1Inc	2Dose	2Inc	Citation or Pathology	Brkly Code
3331	TR6	667.mg	n.s.s.	6/20	279.mg	11/50	557.mg	16/50		
a	TR6	2.31gm	n.s.s.	2/20	279.mg	3/50	557.mg	2/50	liv:hpa,hpc,nnd.	
b	TR6	2.72gm	n.s.s.	1/20	279.mg	0/50	557.mg	2/50	lun:a/a,a/c.	
3332	TR6	778.mg	n.s.s.	0/20	257.mg	3/49	514.mg	8/50	---:leu,lym. S	
a	TR6	730.mg	n.s.s.	7/20	257.mg	17/49	514.mg	14/50		
b	TR6	2.12gm	n.s.s.	4/20	257.mg	4/49	514.mg	3/50	liv:hpa,hpc,nnd.	
c	TR6	409.mg	n.s.s.	2/20	257.mg	7/49	(514.mg	0/50)	lun:a/a,a/c.	
3333	TR6	994.mg	17.1gm	0/24	10.0mg	0/24	100.mg	0/24	1.00gm	4/24	kid:tla,uac.	
3334	TR6	2.96gm	n.s.s.	0/20	276.mg	1/50	552.mg	1/49	liv:hpa,hpc,nnd.	
3335	TR6	481.mg	n.s.s.	18/24	10.0mg	19/24	100.mg	17/24	1.00gm	20/24		
a	TR6	2.37gm	n.s.s.	1/24	10.0mg	3/24	100.mg	1/24	1.00gm	1/24	liv:hpa,hpc,nnd.	
3336	TR6	559.mg	n.s.s.	13/20	276.mg	36/50	552.mg	26/49		
3337	TR6	417.mg	1.73gm	0/24	10.0mg	0/24	100.mg	1/24	1.00gm	13/24	kid:tla,uac; ubl:pam,tcc; ure:tcc.	
a	TR6	808.mg	10.9gm	0/24	10.0mg	0/24	100.mg	1/24	1.00gm	5/24	ubl:pam,tcc.	
b	TR6	919.mg	8.10gm	0/24	10.0mg	0/24	100.mg	0/24	1.00gm	6/24		
3338	TR6	367.mg	2.13gm	0/24	8.00mg	0/24	80.0mg	0/24	800.mg	9/24	kid:hem,tcc,tla,uac.	
a	TR6	830.mg	13.7gm	0/24	8.00mg	0/24	80.0mg	0/24	800.mg	4/24		
b	TR6	579.mg	n.s.s.	2/24	8.00mg	0/24	80.0mg	1/24	800.mg	3/24	liv:hpa,hpc,nnd.	
3339	TR6	3.59gm	n.s.s.	1/20	221.mg	0/50	441.mg	1/50	liv:hpa,hpc,nnd.	
3340	TR6	157.mg	1.21gm	8/24	8.00mg	15/24	80.0mg	10/24	800.mg	18/24		
3341	TR6	409.mg	n.s.s.	10/20	221.mg	26/50	441.mg	24/50		
3342	TR6	261.mg	1.10gm	0/24	8.00mg	0/24	80.0mg	0/24	800.mg	14/24	kid:hem,tcc,tla,uac; ubl:tcc; ure:tcc.	
a	TR6	412.mg	2.66gm	0/24	8.00mg	0/24	80.0mg	0/24	800.mg	8/24		
3343	1089m	139.mg	n.s.s.	4/101	50.0mg	13/96			Goyer;jnci,66,869-880;1981/pers.comm.	
a	1089m	35.5mg	n.s.s.	61/101	50.0mg	70/96				
3344	1089n	113.mg	748.mg	1/85	50.0mg	13/87				
a	1089n	37.9mg	n.s.s.	50/85	50.0mg	61/87				

NITRITE, SODIUM   7632-00-0

	RefNum	LoConf	UpConf	Cntrl	1Dose	1Inc	2Dose	2Inc	Citation or Pathology	Brkly Code
3345	1831	402.mg	n.s.s.	1/14	149.mg	0/15			Ernst;carc,8,1843-1845;1987/pers.comm.	
a	1831	78.7mg	n.s.s.	5/14	149.mg	7/15				
3346	1831	391.mg	n.s.s.	0/16	131.mg	0/15				
a	1831	134.mg	n.s.s.	10/16	131.mg	6/15				
3347	1361	532.mg	n.s.s.	1/92	489.mg	0/12			Murthy;ijcn,23,253-259;1979	
a	1361	532.mg	n.s.s.	2/92	489.mg	0/12				
b	1361	268.mg	n.s.s.	17/92	489.mg	2/12				
3348	1925	788.mg	n.s.s.	1/20	36.8mg	1/15	368.mg	2/17	Anderson;canr,45,3561-3566;1985	
a	1925	1.33gm	n.s.s.	3/20	36.8mg	1/15	368.mg	1/17		
3349	1361	236.mg	n.s.s.	1/95	452.mg	1/11			Murthy;ijcn,23,253-259;1979	
a	1361	451.mg	n.s.s.	2/95	452.mg	0/11				
b	1361	185.mg	n.s.s.	8/95	452.mg	2/11				
3350	1255	79.3mg	n.s.s.	1/16	30.7mg	2/20			Anderson;ijcn,24,319-322;1979	
a	1255	132.mg	n.s.s.	3/16	30.7mg	1/20				
3351	1255	25.7mg	n.s.s.	4/21	27.7mg	6/17				
a	1255	52.8mg	n.s.s.	5/21	27.7mg	3/17				
3352	471	231.mg	n.s.s.	6/136	45.0mg	16/136			Newberne;scie,204,1079-1081;1979	
a	471	246.mg	n.s.s.	6/136	45.0mg	15/136				
b	471	952.mg	n.s.s.	0/136	45.0mg	1/136				
c	471	1.08gm	n.s.s.	1/136	45.0mg	1/136				
3353	1361	273.mg	n.s.s.	0/44	188.mg	0/16			Murthy;ijcn,23,253-259;1979	
a	1361	88.4mg	n.s.s.	20/44	188.mg	6/16				
3354	1490	420.mg	n.s.s.	1/49	51.0mg	0/48	85.0mg	0/48	Maekawa;fctx,20,25-33;1982	
a	1490	154.mg	n.s.s.	45/49	51.0mg	41/48	85.0mg	35/48		
3355	1645m	56.7mg	817.mg	4/24	80.0mg	14/24			Lijinsky;jtxe,13,609-614;1984/1983a	
a	1645m	78.9mg	n.s.s.	4/24	80.0mg	11/24				
b	1645m	162.mg	n.s.s.	0/24	80.0mg	4/24				
3356	1654m	62.0mg	2.04gm	4/24	80.9mg	13/24			Lijinsky;fctx,22,715-720;1984/1983a	
a	1654m	69.2mg	n.s.s.	4/24	80.9mg	12/24				
b	1654m	190.mg	n.s.s.	0/24	80.9mg	3/24				
3357	1654n	73.1mg	523.mg	0/24	63.5mg	8/24				
a	1654n	73.1mg	523.mg	0/24	63.5mg	8/24				
b	1654n	44.1mg	n.s.s.	8/24	63.5mg	15/24				
c	1654n	60.7mg	n.s.s.	4/24	63.5mg	11/24				
d	1654n	68.2mg	n.s.s.	4/24	63.5mg	10/24				
e	1654n	85.8mg	n.s.s.	1/24	63.5mg	7/24				
f	1654n	252.mg	n.s.s.	0/24	63.5mg	1/24				
3358	1854	62.3mg	897.mg	4/24	87.9mg	14/24			Lijinsky;txih,3,413-422;1987/pers.comm.	
a	1854	86.7mg	n.s.s.	4/24	87.9mg	11/24				
b	1854	178.mg	n.s.s.	0/24	87.9mg	4/24				
3359	1361	218.mg	n.s.s.	0/50	151.mg	0/16			Murthy;ijcn,23,253-259;1979	
a	1361	58.8mg	n.s.s.	30/50	151.mg	8/16				
3360	1490	306.mg	n.s.s.	4/46	45.2mg	5/49	81.0mg	7/50	Maekawa;fctx,20,25-33;1982	
a	1490	n.s.s.	n.s.s.	46/46	45.2mg	49/49	81.0mg	50/50		
3361	1645m	114.mg	n.s.s.	3/24	64.0mg	6/24			Lijinsky;jtxe,13,609-614;1984/1983a	
a	1645m	125.mg	n.s.s.	2/24	64.0mg	5/24				
b	1645m	297.mg	n.s.s.	1/24	64.0mg	1/24				
3362	1654m	235.mg	n.s.s.	5/24	64.7mg	3/24			Lijinsky;fctx,22,715-720;1984/1983a	

	Spe	Strain	Site	Xpo+Xpt				TD50		2Tailpvl
	Sex	Route	Hist	Notes					DR	AuOp
3363	R m f34 wat liv mix 24m30 e				.>			946.mg	P<.7	
a	R m f34 wat liv hpc 24m30 e							1.04gm	P<.6	
b	R m f34 wat liv nnd 24m30 e							no dre	P=1.	
3364	R m f34 eat liv mix 24m30					.>		641.mg	P<.3	
a	R m f34 eat liv nnd 24m30							674.mg	P<.3	
b	R m f34 eat liv hpc 24m30							no dre	P=1.	
3365	R m f34 eat tes ict 27m27 ef					.>		no dre	P=1.	-
a	R m f34 eat --- mnl 27m27 ef							no dre	P=1.	-
b	R m f34 eat mix lym 27m27 ef							no dre	P=1.	-
c	R m f34 eat --- grl 27m27 ef							no dre	P=1.	-
d	R m f34 eat tba tum 27m27 ef							no dre	P=1.	-
3366	R f mrc wat tba mix 20m24 e				. ±			57.3mg	P<.03	-
3367	R f mrc wat liv tum 16m28 e				.>			no dre	P=1.	-
a	R f mrc wat tba tum 16m28 e							no dre	P=1.	-
3368	R m mrc wat tba mix 20m24 e			.>				61.9mg	P<.2	-
3369	R m mrc wat liv tum 16m28 e			.>				no dre	P=1.	-
a	R m mrc wat tba tum 16m28 e							no dre	P=1.	-
3370	R m wis eat liv mix 92w92 e			. +	.			155.mg *	P<.007	+
3371	R m wis wat liv tum 24m24 r				.>			no dre	P=1.	-
3372	R m wis wat liv hpc 94w94				.>			no dre	P=1.	-
a	R m wis wat lun tum 94w94							no dre	P=1.	-
b	R m wis wat nas tum 94w94							no dre	P=1.	-

3-NITRO-p-ACETOPHENETIDE   100ng..:..1ug...:..10.....:..100...:..1mg....:..10.....:..100...:.1g.....:.10

								TD50	DR	AuOp
3373	M f b6c eat TBA MXB 78w97 v				:>			no dre	P=1.	
a	M f b6c eat liv MXB 78w97 v							no dre	P=1.	
b	M f b6c eat lun MXB 78w97 v							30.8gm *	P<.7	
3374	M m b6c eat liv MXA 78w98 v				: ±			2.27gm *	P<.03	c
a	M m b6c eat TBA MXB 78w98 v							3.49gm *	P<.3	
b	M m b6c eat liv MXB 78w98 v							2.27gm *	P<.03	
c	M m b6c eat lun MXB 78w98 v							no dre	P=1.	
3375	R f f34 eat TBA MXB 18m25 v				:>			no dre	P=1.	-
a	R f f34 eat liv MXB 18m25 v							6.24gm *	P<.7	
3376	R m f34 eat TBA MXB 18m24 v				:>			no dre	P=1.	-
a	R m f34 eat liv MXB 18m24 v							no dre	P=1.	

5-NITRO-o-ANISIDINE   100ng..:..1ug...:..10.....:..100...:..1mg....:..10.....:..100...:.1g.....:.10

								TD50	DR	AuOp
3377	M f b6c eat --- MXA 78w96 v				: ±			2.33gm *	P<.04	
a	M f b6c eat liv hpc 78w96 v							3.72gm /	P<.06	c
b	M f b6c eat TBA MXB 78w96 v							1.82gm *	P<.2	
c	M f b6c eat liv MXB 78w96 v							3.72gm /	P<.06	
d	M f b6c eat lun MXB 78w96 v							no dre	P=1.	
3378	M m b6c eat liv hpc 78w95 v				: +	:		#1.05gm \	P<.003	-
a	M m b6c eat liv MXA 78w95 v							1.12gm \	P<.005	
b	M m b6c eat TBA MXB 78w95 v							1.69gm \	P<.2	
c	M m b6c eat liv MXB 78w95 v							1.12gm \	P<.005	
d	M m b6c eat lun MXB 78w95 v							no dre	P=1.	
3379	R f f34 eat MXB MXB 18m25				: + :			170.mg *	P<.0005	
a	R f f34 eat mgl MXA 18m25							222.mg *	P<.0005	
b	R f f34 eat cli MXA 18m25							230.mg *	P<.0005c	
c	R f f34 eat mgl acn 18m25							254.mg \	P<.0005	
d	R f f34 eat lun MXA 18m25							280.mg \	P<.0005	
e	R f f34 eat MXA MXA 18m25							602.mg *	P<.0005c	
f	R f f34 eat cli MXA 18m25							966.mg *	P<.0005c	
g	R f f34 eat cli can 18m25							1.26gm /	P<.0005c	
h	R f f34 eat TBA MXB 18m25							73.4mg *	P<.0005	
i	R f f34 eat liv MXB 18m25							3.20gm *	P<.2	
3380	R m f34 eat MXB MXB 18m25				: + :			28.1mg *	P<.0005	
a	R m f34 eat ski MXA 18m25							30.5mg *	P<.0005c	
b	R m f34 eat ski tri 18m25							43.7mg *	P<.0005c	
c	R m f34 eat ski bcc 18m25							52.0mg /	P<.0005c	
d	R m f34 eat ski MXA 18m25							66.0mg *	P<.0005c	
e	R m f34 eat MXA MXA 18m25							105.mg /	P<.0005c	
f	R m f34 eat pit adn 18m25							115.mg *	P<.0005	
g	R m f34 eat ski sec 18m25							162.mg *	P<.0005c	
h	R m f34 eat pre MXA 18m25							196.mg *	P<.0005	
i	R m f34 eat ski sqc 18m25							252.mg /	P<.0005c	
j	R m f34 eat adr MXA 18m25							345.mg *	P<.0005	
k	R m f34 eat TBA MXB 18m25							12.8mg *	P<.0005	
l	R m f34 eat liv MXB 18m25							194.mg *	P<.002	

5-NITRO-2-FURALDEHYDE SEMICARBAZONE   ..:..1ug....:..10.......:..100....:..1mg....:..10.....:..100....:..1g......:.10

								TD50	DR	AuOp
3381	M f b6c eat ova MXB 24m24				: + :			22.4mg *	P<.0005	
a	M f b6c eat ova MXA 24m24							29.8mg *	P<.0005	
b	M f b6c eat ova mtb 24m24							30.8mg *	P<.0005c	
c	M f b6c eat ova gct 24m24							124.mg *	P<.005	c
d	M f b6c eat TBA MXB 24m24							66.2mg *	P<.3	
e	M f b6c eat liv MXB 24m24							no dre	P=1.	
f	M f b6c eat lun MXB 24m24							241.mg *	P<.3	
3382	M m b6c eat sub MXA 24m24				: ±			#154.mg *	P<.05	-
a	M m b6c eat TBA MXB 24m24							89.3mg *	P<.4	

	RefNum	LoConf	UpConf	Cntrl	1Dose	1Inc	2Dose	2Inc	Citation or Pathology
									Brkly Code
3363	1654n	109.mg	n.s.s.	3/24	44.5mg	4/24			
a	1654n	143.mg	n.s.s.	1/24	44.5mg	2/24			
b	1654n	160.mg	n.s.s.	2/24	44.5mg	2/24			
3364	1854	164.mg	n.s.s.	3/24	92.3mg	6/24			Lijinsky;txih,3,413-422;1987/pers.comm.
a	1854	180.mg	n.s.s.	2/24	92.3mg	5/24			
b	1854	429.mg	n.s.s.	1/24	92.3mg	1/24			
3365	1920	109.mg	n.s.s.	16/20	80.0mg	31/48	(200.mg	7/50)	Grant;fctx,27,565-571;1989
a	1920	348.mg	n.s.s.	9/18	80.0mg	7/39	(200.mg	3/40)	
b	1920	371.mg	n.s.s.	15/20	80.0mg	14/50	(200.mg	10/49)	
c	1920	1.63gm	n.s.s.	0/18	80.0mg	1/39	200.mg	0/40	
d	1920	389.mg	n.s.s.	20/20	80.0mg	45/50	200.mg	34/50	
3366	213	21.4mg	n.s.s.	4/15	65.9mg	10/15			Lijinsky;jnci,50,1061-1063;1973
3367	1189	41.3mg	n.s.s.	0/20	11.2mg	0/13			Greenblatt;jnci,50,799-802;1973
a	1189	6.48mg	n.s.s.	14/20	11.2mg	9/13			
3368	213	18.8mg	n.s.s.	5/15	46.2mg	9/15			Lijinsky;jnci,50,1061-1063;1973
3369	1189	33.4mg	n.s.s.	0/20	7.85mg	0/15			Greenblatt;jnci,50,799-802;1973
a	1189	12.9mg	n.s.s.	7/20	7.85mg	4/15			
3370	1352	63.2mg	2.04gm	0/19	32.0mg	1/22	64.0mg	5/19	Aoyagi;jnci,65,411-414;1980
3371	1910	216.mg	n.s.s.	0/21	75.0mg	0/21	150.mg	0/21	Yamamoto;clet,45,221-225;1989
3372	1914	168.mg	n.s.s.	0/20	75.0mg	0/20	150.mg	0/20	Yamamoto;carc,10,1607-1611;1989
a	1914	168.mg	n.s.s.	0/20	75.0mg	0/20	150.mg	0/20	
b	1914	168.mg	n.s.s.	0/20	75.0mg	0/20	150.mg	0/20	

3-NITRO-p-ACETOPHENETIDE   1777-84-0

	RefNum	LoConf	UpConf	Cntrl	1Dose	1Inc	2Dose	2Inc	Citation or Pathology
3373	TR133	1.44gm	n.s.s.	22/50	763.mg	17/50	(1.53gm	15/50)	liv:hpa,hpc,nnd.
a	TR133	8.86gm	n.s.s.	4/50	763.mg	3/50	1.53gm	1/50	
b	TR133	4.33gm	n.s.s.	1/50	763.mg	4/50	1.53gm	2/50	lun:a/a,a/c.
3374	TR133	1.02gm	n.s.s.	10/50	697.mg	13/49	1.39gm	23/50	liv:hpa,hpc.
a	TR133	989.mg	n.s.s.	21/50	697.mg	20/49	1.39gm	20/50	
b	TR133	1.02gm	n.s.s.	10/50	697.mg	13/49	1.39gm	23/50	liv:hpa,hpc,nnd.
c	TR133	5.73gm	n.s.s.	11/50	697.mg	2/49	1.39gm	6/50	lun:a/a,a/c.
3375	TR133	649.mg	n.s.s.	45/50	66.0mg	12/50	130.mg	15/50	
a	TR133	692.mg	n.s.s.	2/50	66.0mg	0/50	130.mg	2/50	liv:hpa,hpc,nnd.
3376	TR133	118.mg	n.s.s.	26/50	52.8mg	19/50	(105.mg	13/50)	
a	TR133	1.38gm	n.s.s.	3/50	52.8mg	0/50	105.mg	0/50	liv:hpa,hpc,nnd.

5-NITRO-o-ANISIDINE   99-59-2

	RefNum	LoConf	UpConf	Cntrl	1Dose	1Inc	2Dose	2Inc	Citation or Pathology
3377	TR127	917.mg	n.s.s.	9/100	836.mg	5/50	666.mg	10/50	---:leu,lym. S
a	TR127	1.32gm	n.s.s.	3/100	836.mg	0/50	666.mg	8/50	
b	TR127	632.mg	n.s.s.	29/100	836.mg	9/50	666.mg	22/50	
c	TR127	1.32gm	n.s.s.	3/100	836.mg	0/50	666.mg	8/50	liv:hpa,hpc,nnd.
d	TR127	3.87gm	n.s.s.	5/100	836.mg	0/50	666.mg	2/50	lun:a/a,a/c.
3378	TR127	504.mg	6.19gm	18/100	780.mg	25/50	(615.mg	3/50)	S
a	TR127	518.mg	11.3gm	20/100	780.mg	25/50	(615.mg	3/50)	liv:hpa,hpc. S
b	TR127	516.mg	n.s.s.	45/100	780.mg	32/50	(615.mg	17/50)	
c	TR127	518.mg	11.3gm	20/100	780.mg	25/50	(615.mg	3/50)	liv:hpa,hpc,nnd.
d	TR127	2.53gm	n.s.s.	15/100	780.mg	5/50	(615.mg	2/50)	lun:a/a,a/c.
3379	TR127	107.mg	292.mg	3/100	147.mg	17/50	294.mg	22/50	cli:adn,can,ppa,sqc; ski:sec,sqc; zym:sec,sqc. C
a	TR127	102.mg	642.mg	2/100	147.mg	11/50	(294.mg	4/50)	mgl:acn,adn,pac. S
b	TR127	133.mg	454.mg	3/100	147.mg	12/50	294.mg	14/50	cli:adn,can,ppa,sqc.
c	TR127	117.mg	609.mg	0/100	147.mg	10/50	(294.mg	4/50)	S
d	TR127	99.1mg	1.63gm	1/100	147.mg	5/50	(294.mg	1/50)	lun:a/a,a/c. S
e	TR127	306.mg	1.23gm	0/100	147.mg	5/50	294.mg	10/50	ski:sec,sqc; zym:sec,sqc.
f	TR127	458.mg	2.42gm	0/100	147.mg	1/50	294.mg	9/50	cli:can,sqc.
g	TR127	529.mg	3.95gm	0/100	147.mg	0/50	294.mg	7/50	
h	TR127	44.8mg	154.mg	83/100	147.mg	45/50	294.mg	41/50	
i	TR127	589.mg	n.s.s.	2/100	147.mg	0/50	294.mg	2/50	liv:hpa,hpc,nnd.
3380	TR127	16.4mg	43.8mg	2/99	118.mg	36/50	235.mg	45/50	ski:bcc,cuc,sea,sec,sgc,sqc,tri; zym:cuc,sec,sqc. C
a	TR127	17.3mg	49.6mg	2/99	118.mg	30/50	235.mg	42/50	ski:bcc,sec,sgc,sqc,tri.
b	TR127	21.5mg	78.8mg	0/99	118.mg	20/50	235.mg	9/50	
c	TR127	22.7mg	105.mg	1/99	118.mg	7/50	235.mg	30/50	
d	TR127	27.1mg	113.mg	0/99	118.mg	14/50	235.mg	26/50	
e	TR127	61.3mg	176.mg	1/99	118.mg	10/50	235.mg	29/50	ski:sea,sec.
f	TR127	38.0mg	528.mg	10/99	118.mg	8/50	235.mg	5/50	ski:cuc,sec,sqc; zym:cuc,sec,sqc.
g	TR127	87.7mg	286.mg	0/99	118.mg	5/50	235.mg	21/50	S
h	TR127	45.7mg	1.17gm	2/99	118.mg	2/50	235.mg	5/50	
i	TR127	107.mg	627.mg	1/99	118.mg	3/50	235.mg	12/50	pre:adn,can. S
j	TR127	55.5mg	2.47gm	1/99	118.mg	1/50	235.mg	5/50	adr:coa,coc. S
k	TR127	7.94mg	21.3mg	58/99	118.mg	44/50	235.mg	48/50	
l	TR127	49.9mg	1.68gm	4/99	118.mg	3/50	235.mg	3/50	liv:hpa,hpc,nnd.

5-NITRO-2-FURALDEHYDE SEMICARBAZONE   (nitrofurazone) 59-87-0

	RefNum	LoConf	UpConf	Cntrl	1Dose	1Inc	2Dose	2Inc	Citation or Pathology
3381	TR337	15.4mg	37.7mg	1/50	19.3mg	20/50	39.9mg	29/50	ova:gct,mtb. C
a	TR337	20.1mg	48.8mg	0/50	19.3mg	18/50	39.9mg	20/50	ova:mtb,tua. S
b	TR337	20.7mg	50.6mg	0/50	19.3mg	17/50	39.9mg	20/50	
c	TR337	60.2mg	1.14gm	1/50	19.3mg	4/50	39.9mg	9/50	
d	TR337	19.9mg	n.s.s.	36/50	19.3mg	43/50	39.9mg	42/50	
e	TR337	214.mg	n.s.s.	3/50	19.3mg	3/50	39.9mg	1/50	liv:hpa,hpc,nnd.
f	TR337	73.0mg	n.s.s.	3/50	19.3mg	7/50	39.9mg	6/50	lun:a/a,a/c.
3382	TR337	60.1mg	n.s.s.	3/50	17.8mg	3/50	36.8mg	8/50	sub:fbs,fib,nfs,srn. S
a	TR337	23.9mg	n.s.s.	28/50	17.8mg	26/50	36.8mg	26/50	

	Spe	Sex	Strain	Route	Site	Hist	Xpo+Xpt	Notes	TD50 DR	2Tailpvl AuOp
b	M	m	b6c	eat	liv	MXB	24m24		152.mg \	P<.7
c	M	m	b6c	eat	lun	MXB	24m24		1.38gm *	P<.9
3383	R	f	f34	eat	mgl	fba	24m24	: + :	15.7mg *	P<.0005c
a	R	f	f34	eat	TBA	MXB	24m24		no dre	P=1.
b	R	f	f34	eat	liv	MXB	24m24		253.mg *	P<.2
3384	R	m	f34	eat	tnv	MXA	24m24	: + :	44.0mg \	P<.003 e
a	R	m	f34	eat	ski	MXA	24m24		151.mg *	P<.006 e
b	R	m	f34	eat	pre	can	24m24		65.5mg *	P<.03 e
c	R	m	f34	eat	ski	sea	24m24		177.mg *	P<.02
d	R	m	f34	eat	TBA	MXB	24m24		33.8mg *	P<.3
e	R	m	f34	eat	liv	MXB	24m24		no dre	P=1.
3385	R	f	hza	eat	mgl	fba	36w54	es	. + .	6.52mg	P<.006 +
a	R	f	hza	eat	liv	tum	36w54	es	no dre	P=1.
3386	R	f	hza	eat	mam	tum	44w60	es	<+	noTD50	P<.0005+
a	R	f	hza	eat	liv	tum	44w60	es	no dre	P=1.
3387	R	f	sda	eat	mam	tum	46w66	er	. + .	7.12mg	P<.0005+

5-NITRO-2-FURAMIDOXIME `100ng..:..1ug....:..:..10......:..:..100...:..:..1mg..:..:..10......:..:..100....:..:..1g......:..:..10`

	Spe	Sex	Strain	Route	Site	Hist	Xpo+Xpt	Notes	TD50 DR	2Tailpvl AuOp
3388	R	f	sda	eat	liv	tum	46w66	e	.>	no dre	P=1.
a	R	f	sda	eat	tba	mix	46w66	e		42.0mg	P<.5 -

5-NITRO-2-FURANMETHANEDIOL DIACETATE `..:..1ug....:..:..10......:..:..100...:..:..1mg..:..:..10......:..:..100....:..:..1g......:..:..10`

	Spe	Sex	Strain	Route	Site	Hist	Xpo+Xpt	Notes	TD50 DR	2Tailpvl AuOp
3389	R	f	hza	eat	liv	tum	36w54	es	.>	no dre	P=1.
a	R	f	hza	eat	mam	tum	36w54	es		no dre	P=1. -
3390	R	f	hza	eat	mam	tum	44w60	es	.>	105.mg	P<.5 -
a	R	f	hza	eat	liv	tum	44w60	es		no dre	P=1.

3-(5-NITRO-2-FURYL)-IMIDAZO(1,2-alpha)PYRIDINE `....:..10......:..:..100...:..:..1mg..:..:..10......:..:..100....:..:..1g......:..:..10`

	Spe	Sex	Strain	Route	Site	Hist	Xpo+Xpt	Notes	TD50 DR	2Tailpvl AuOp
3391	M	f	ctn	eat	mix	car	40w79	es	. + .	21.6mg Z	P<.0005+
a	M	f	ctn	eat	mix	pam	40w79	es		117.mg Z	P<.0005+
b	M	f	ctn	eat	thy	lys	40w79	es		781.mg *	P<.0005
c	M	f	ctn	eat	liv	hpt	40w79	es		no dre	P=1. -
d	M	f	ctn	eat	lun	tum	40w79	es		no dre	P=1. -
e	M	f	ctn	eat	tba	mix	40w79	es		noTD50	P<.0005+
3392	M	m	ctn	eat	mix	car	40w73	es	. + .	40.1mg Z	P<.0005+
a	M	m	ctn	eat	mix	pam	40w73	es		65.8mg Z	P<.0005+
b	M	m	ctn	eat	lun	tum	40w73	es		763.mg *	P<.2
c	M	m	ctn	eat	liv	hpt	40w73	es		no dre	P=1. -
d	M	m	ctn	eat	tba	mix	40w73	es		57.1mg *	P<.002 +
3393	R	f	wis	eat	mix	pam	40w89	es	. + .	27.5mg	P<.0005+
a	R	f	wis	eat	mam	tum	40w89	es		33.3mg	P<.0005+
b	R	f	wis	eat	kid	tum	40w89	es		116.mg	P<.0005+
c	R	f	wis	eat	mix	car	40w89	es		443.mg	P<.04 +
d	R	f	wis	eat	liv	hpt	40w89	es		644.mg	P<.3
e	R	f	wis	eat	tba	mix	40w89	es		15.8mg	P<.0005+
3394	R	m	wis	eat	kid	tum	40w83	es	. + .	21.4mg	P<.0005+
a	R	m	wis	eat	mix	pam	40w83	es		43.7mg	P<.0005+
b	R	m	wis	eat	mix	car	40w83	es		104.mg	P<.0005+
c	R	m	wis	eat	liv	hpt	40w83	es		959.mg	P<.3
d	R	m	wis	eat	tba	mix	40w83	es		11.9mg	P<.0005+

5-(5-NITRO-2-FURYL)-1,3,4-OXADIAZOLE-2-OL `..1ug....:..:..10......:..:..100...:..:..1mg..:..:..10......:..:..100....:..:..1g......:..:..10`

	Spe	Sex	Strain	Route	Site	Hist	Xpo+Xpt	Notes	TD50 DR	2Tailpvl AuOp
3395	R	f	sda	eat	mgl	mix	46w66	e	. + .	8.61mg	P<.0005
a	R	f	sda	eat	mgl	adc	46w66	e		31.2mg	P<.006
b	R	f	sda	eat	liv	tum	46w66	e		no dre	P=1.
c	R	f	sda	eat	tba	mix	46w66	e		8.61mg	P<.0005+

N-{[3-(5-NITRO-2-FURYL)-1,2,4-OXADIAZOLE-5-YL]-METHYL}ACETAMIDE `.1_00...:..:..1mg..:..:..10......:..:..100....:..:..1g......:..:..10`

	Spe	Sex	Strain	Route	Site	Hist	Xpo+Xpt	Notes	TD50 DR	2Tailpvl AuOp
3396	R	f	sda	eat	mgl	mix	46w66	e	. + .	6.35mg	P<.0005
a	R	f	sda	eat	mgl	adc	46w66	e		10.3mg	P<.0005
b	R	f	sda	eat	k/p	tcc	46w66	e		+hist 59.6mg	P<.2 +
c	R	f	sda	eat	lun	alc	46w66	e		59.6mg	P<.2 +
d	R	f	sda	eat	liv	tum	46w66	e		no dre	P=1.
e	R	f	sda	eat	tba	mix	46w66	e		5.20mg	P<.0005

N-[5-(5-NITRO-2-FURYL)-1,3,4-THIADIAZOL-2-YL]ACETAMIDE `......:..100...:..:..1mg..:..:..10......:..:..100....:..:..1g......:..:..10`

	Spe	Sex	Strain	Route	Site	Hist	Xpo+Xpt	Notes	TD50 DR	2Tailpvl AuOp
3397	M	f	swi	eat	for	mix	46w55	e	. + .	6.74mg	P<.0005+
a	M	f	swi	eat	for	sqp	46w55	e		14.1mg	P<.0005
b	M	f	swi	eat	for	sqc	46w55	e		104.mg	P<.008
c	M	f	swi	eat	liv	tum	46w55	e		no dre	P=1.
d	M	f	swi	eat	lun	tum	46w55	e		no dre	P=1.
e	M	f	swi	eat	tba	mix	46w55	e		6.90mg	P<.0005
3398	R	f	sda	eat	for	mix	40w59	ev	. + .	8.84mg	P<.0005+
a	R	f	sda	eat	mix	hms	40w59	ev		10.6mg	P<.0005+
b	R	f	sda	eat	for	sqp	40w59	ev		11.6mg	P<.0005
c	R	f	sda	eat	lun	alc	40w59	ev		15.3mg	P<.0005+
d	R	f	sda	eat	k/p	tcc	40w59	ev		+hist 16.9mg	P<.0005+
e	R	f	sda	eat	itn	hms	40w59	ev		18.6mg	P<.0005+
f	R	f	sda	eat	liv	hms	40w59	ev		22.9mg	P<.0005+
g	R	f	sda	eat	mgl	adc	40w59	ev		37.7mg	P<.002
h	R	f	sda	eat	lun	hms	40w59	ev		52.3mg	P<.007 +

	RefNum	LoConf	UpConf	Cntrl	1Dose	1Inc	2Dose	2Inc	Citation or Pathology
									Brkly Code
b	TR337	20.2mg	n.s.s.	16/50	17.8mg	15/50	(36.8mg	5/50)	liv:hpa,hpc,nnd.
c	TR337	69.1mg	n.s.s.	8/50	17.8mg	7/50	36.8mg	6/50	lun:a/a,a/c.
3383	TR337	9.47mg	44.2mg	8/50	15.4mg	36/50	30.7mg	36/50	
a	TR337	20.3mg	n.s.s.	44/50	15.4mg	47/50	30.7mg	46/50	
b	TR337	87.3mg	n.s.s.	0/50	15.4mg	2/50	30.7mg	2/50	liv:hpa,hpc,nnd.
3384	TR337	18.8mg	195.mg	0/50	12.3mg	7/50	(24.6mg	2/50)	tnv:men,msm.
a	TR337	55.2mg	548.mg	0/50	12.3mg	0/50	24.6mg	5/50	ski:sea,tri.
b	TR337	29.4mg	n.s.s.	1/50	12.3mg	8/50	24.6mg	5/50	
c	TR337	59.1mg	n.s.s.	0/50	12.3mg	0/50	24.6mg	4/50	s
d	TR337	9.75mg	n.s.s.	47/50	12.3mg	46/50	24.6mg	40/50	
e	TR337	73.2mg	n.s.s.	7/50	12.3mg	4/50	24.6mg	3/50	liv:hpa,hpc,nnd.
3385	1063m	3.15mg	54.3mg	0/5	33.3mg	11/18			Morris;canr,29,2145-2156;1969
a	1063m	33.3mg	n.s.s.	0/5	33.3mg	0/18			
3386	1063n	n.s.s.	4.64mg	3/16	36.8mg	24/24			
a	1063n	60.5mg	n.s.s.	0/16	36.8mg	0/24			
3387	1120	3.96mg	14.7mg	2/29	34.8mg	22/29			Erturk;canr,30,1409-1412;1970

5-NITRO-2-FURAMIDOXIME   772-43-0

3388	200a	18.5mg	n.s.s.	0/39	6.97mg	0/32			Cohen;jnci,51,403-417;1973
a	200a	6.78mg	n.s.s.	2/39	6.97mg	3/32			

5-NITRO-2-FURANMETHANEDIOL DIACETATE   92-55-7

3389	1063m	18.5mg	n.s.s.	0/5	66.7mg	0/5			Morris;canr,29,2145-2156;1969
a	1063m	18.5mg	n.s.s.	0/5	66.7mg	0/5			
3390	1063n	18.5mg	n.s.s.	3/16	73.6mg	4/13			
a	1063n	65.6mg	n.s.s.	0/16	73.6mg	0/13			

3-(5-NITRO-2-FURYL)-IMIDAZO(1,2-alpha)PYRIDINE   (NFIP) 75198-31-1

3391	1163	13.3mg	37.7mg	0/39	65.8mg	28/40	(160.mg	23/36	371.mg 35/44)	Cabral;tumo,66,131-144;1980
a	1163	68.0mg	329.mg	2/39	65.8mg	12/40	160.mg	16/39	(371.mg 15/50)	
b	1163	412.mg	2.15gm	0/39	65.8mg	0/40	160.mg	5/40	371.mg 8/50	
c	1163	138.mg	n.s.s.	1/37	65.8mg	0/36	160.mg	0/23	371.mg 0/16	
d	1163	704.mg	n.s.s.	6/39	65.8mg	6/40	160.mg	3/35	371.mg 6/43	
e	1163	n.s.s.	20.3mg	28/39	65.8mg	40/40	160.mg	40/40	371.mg 50/50	
3392	1163	26.8mg	63.3mg	0/39	65.8mg	19/35	143.mg	20/33	(409.mg 18/30)	
a	1163	39.4mg	174.mg	3/39	65.8mg	19/38	143.mg	17/37	(409.mg 18/30)	
b	1163	248.mg	n.s.s.	5/39	65.8mg	5/36	143.mg	8/35	409.mg 8/31	
c	1163	765.mg	n.s.s.	7/37	65.8mg	1/33	143.mg	1/28	409.mg 1/15	
d	1163	26.0mg	338.mg	21/39	65.8mg	38/38	143.mg	37/37	409.mg 38/42	
3393	1163	15.9mg	50.7mg	0/32	89.9mg	25/31			
a	1163	17.5mg	85.0mg	8/32	89.9mg	25/31			
b	1163	56.1mg	304.mg	0/32	89.9mg	10/31			
c	1163	134.mg	n.s.s.	0/32	89.9mg	3/31			
d	1163	145.mg	n.s.s.	1/32	89.9mg	3/31			
e	1163	6.29mg	43.0mg	14/32	89.9mg	30/31			
3394	1163	12.1mg	40.4mg	0/31	77.1mg	23/29			
a	1163	23.4mg	101.mg	1/31	77.1mg	16/29			
b	1163	47.0mg	331.mg	0/31	77.1mg	8/29			
c	1163	156.mg	n.s.s.	0/31	77.1mg	1/29			
d	1163	4.72mg	32.9mg	13/31	77.1mg	28/29			

5-(5-NITRO-2-FURYL)-1,3,4-OXADIAZOLE-2-OL   2122-86-3

3395	1126	4.53mg	22.7mg	2/39	17.4mg	16/35			Cohen;jnci,54,841-850;1975
a	1126	11.8mg	302.mg	0/39	17.4mg	5/35			
b	1126	50.6mg	n.s.s.	0/39	17.4mg	0/35			
c	1126	4.53mg	22.7mg	2/39	17.4mg	16/35			

N-{[3-(5-NITRO-2-FURYL)-1,2,4-OXADIAZOLE-5-YL]-METHYL}ACETAMIDE   36133-88-7

3396	1126	3.24mg	22.9mg	2/24	13.9mg	16/32			Cohen;jnci,54,841-850;1975
a	1126	4.98mg	30.0mg	0/24	13.9mg	10/32			
b	1126	14.7mg	n.s.s.	0/24	13.9mg	2/32			
c	1126	14.7mg	n.s.s.	0/24	13.9mg	2/32			
d	1126	37.0mg	n.s.s.	0/24	13.9mg	0/32			
e	1126	2.76mg	14.7mg	2/24	13.9mg	18/32			

N-[5-(5-NITRO-2-FURYL)-1,3,4-THIADIAZOL-2-YL]ACETAMIDE   2578-75-8

3397	1076	2.82mg	15.0mg	0/29	109.mg	21/22			Cohen;canr,33,1593-1597;1973
a	1076	7.41mg	29.6mg	0/29	109.mg	17/22			
b	1076	35.8mg	2.28gm	0/29	109.mg	4/22			
c	1076	138.mg	n.s.s.	0/29	109.mg	0/22			
d	1076	138.mg	n.s.s.	0/29	109.mg	0/22			
e	1076	2.85mg	15.9mg	2/29	109.mg	21/22			
3398	1126	5.08mg	16.7mg	0/24	53.0mg	22/30			Cohen;jnci,54,841-850;1975
a	1126	6.06mg	20.6mg	0/24	53.0mg	20/30			
b	1126	6.59mg	22.9mg	0/24	53.0mg	19/30			
c	1126	8.40mg	31.9mg	0/24	53.0mg	16/30			
d	1126	9.10mg	36.0mg	0/24	53.0mg	15/30			
e	1126	9.87mg	40.9mg	0/24	53.0mg	14/30			
f	1126	11.7mg	55.1mg	0/24	53.0mg	12/30			
g	1126	17.0mg	145.mg	0/24	53.0mg	8/30			
h	1126	21.3mg	561.mg	0/24	53.0mg	6/30			

```
        Spe Strain  Site    Xpo+Xpt                                                           TD50      2Tailpvl
            Sex   Route   Hist      Notes                                                       DR      AuOp
     i   R f sda eat mgl mix 40w59 ev                                                         43.3mg   P<.05
     j   R f sda eat pan hms 40w59 ev                                                         81.6mg   P<.03  +
     k   R f sda eat --- lbl 40w59 ev                                                         111.mg   P<.06  +
     l   R f sda eat smi adc 40w59 ev                                                         111.mg   P<.06  +
     m   R f sda eat smi lei 40w59 ev                                                         111.mg   P<.06  +
     n   R f sda eat tba mix 40w59 ev                                                         noTD50   P<.0005
```

4-(5-NITRO-2-FURYL)THIAZOLE   100ng..:..1ug...:..10....:..100...:..1mg....:..10....:..100...:..1g.....:..10
```
3399 R f asd eat mgl fba 46w68 e                                            . + .             15.6mg   P<.0005+
   a R f asd eat for sqc 46w68 e                                                               19.5mg   P<.0005+
   b R f asd eat liv tum 46w68 e                                                               no dre   P=1.
   c R f asd eat tba mix 46w68 e                                                               7.68mg   P<.0005+
```

N-[4-(5-NITRO-2-FURYL)-2-THIAZOLYL]ACETAMIDE   ...:..10.....:..100....:..1mg....:..10.....:..100...:..1g.....:..10
```
3400 H m syg eat ubl mix 48w70 e                                        . + .                  8.91mg   P<.0005+
   a H m syg eat ubl sqc 48w70 e                                                               60.2mg   P<.004 +
   b H m syg eat for sqp 48w70 e                                                               68.1mg   P<.003 +
   c H m syg eat liv tum 48w70 e                                                               no dre   P=1.
   d H m syg eat lun tum 48w70 e                                                               no dre   P=1.
3401 R f sda eat mam mix 46w66 e                                      . + .                    10.5mg   P<.0005+
   a R f sda eat mgl adc 46w66 e                                                               36.2mg   P<.0005+
   b R f sda eat mgl fba 46w66 e                                                               56.9mg   P<.0005
   c R f sda eat liv tum 46w66 e                                                               no dre   P=1.
   d R f sda eat tba mix 46w66 e                                                               7.25mg   P<.0005
3402 R f sda eat mgl fba 46w66 e                                            . + .              25.0mg   P<.0005+
   a R f sda eat k/p tcc 46w66 e                                                               316.mg   P<.09
   b R f sda eat liv tum 46w66 e                                                               no dre   P=1.
   c R f sda eat tba mix 46w66 e                                                               22.8mg   P<.0005
3403 R f sda eat mgl fba 46w66 e                                            . + .              30.2mg   P<.0005+
   a R f sda eat k/p tcc 46w66 e                                                               660.mg   P<.3
   b R f sda eat liv tum 46w66 e                                                               no dre   P=1.
   c R f sda eat tba mix 46w66 e                                                               27.7mg   P<.0005
```

N-[4-(5-NITRO-2-FURYL)-2-THIAZOLYL]FORMAMIDE   ...:..10.....:..100....:..1mg....:..10.....:..100...:..1g.....:..10
```
3404 H m syg eat ubl mix 48w70 e                                        <+                     noTD50   P<.0005+
   a H m syg eat ubl sqc 48w70 e                                                               54.0mg   P<.003 +
   b H m syg eat for sqp 48w70 e                                                               83.8mg   P<.007 +
   c H m syg eat liv tum 48w70 e                                                               no dre   P=1.
   d H m syg eat lun tum 48w70 e                                                               no dre   P=1.
3405 M f cd1 eat ubl tcc 36w68 e                                          . + .                33.3mg   P<.0005+
   a M f cd1 eat ubl spt 36w68 e                                                               128.mg   P<.002 +
   b M f cd1 eat ubl sqc 36w68 e                                                               128.mg   P<.002 +
   c M f cd1 eat lun ade 36w68 e                                                               1.12gm   P<.6
   d M f cd1 eat liv tum 36w68 e                                                               no dre   P=1.
   e M f cd1 eat tba mix 36w68 e                                                               18.7mg   P<.0005
3406 M m cd1 eat ubl tcc 40w77 e                                            . + .              30.5mg   P<.0005+
   a M m cd1 eat ubl spt 40w77 e                                                               97.1mg   P<.0005+
   b M m cd1 eat ubl sqc 40w77 e                                                               250.mg   P<.009 +
   c M m cd1 eat lun ade 40w77 e                                                               no dre   P=1.
   d M m cd1 eat liv hpc 40w77 e                                                               no dre   P=1.
   e M m cd1 eat tba mix 40w77 e                                                               21.5mg   P<.0005
3407 M f nmr gav for sqc 68w75 esv                                              . + .          139.mg * P<.0005+
   a M f nmr gav for pam 68w75 esv                                                             164.mg * P<.0005+
   b M f nmr gav stg mal 68w75 esv                                                             944.mg * P<.02
   c M f nmr gav liv mal 68w75 esv                                                             2.39gm * P<.2
   d M f nmr gav lun adc 68w75 esv                                                             no dre   P=1.
   e M f nmr gav tba mal 68w75 esv                                                             59.5mg * P<.0005
   f M f nmr gav tba ben 68w75 esv                                                             69.4mg * P<.0005
3408 M f swi eat ubl tum 46w53                                              . + .              18.2mg   P<.0005+
   a M f swi eat --- leu 46w53                                                                 55.4mg   P<.03  +
   b M f swi eat lun car 46w53                                                                 163.mg   P<.03  +
3409 M f swi eat ubl tcc 46w66 e                                          . + .                7.72mg   P<.0005+
   a M f swi eat lun alc 46w66 e                                                               86.4mg   P<.003
   b M f swi eat --- lle 46w66 e                                                               78.3mg   P<.02  +
   c M f swi eat for sqp 46w66 e                                                               235.mg   P<.07  +
   d M f swi eat liv tum 46w66 e                                                               no dre   P=1.
   e M f swi eat tba mix 46w66 e                                                               noTD50   P<.0005
3410 M f swi eat ubl car 33w52                                            . + .                19.8mg   P<.0005+
   a M f swi eat --- leu 33w52                                                                 47.6mg   P<.08
   b M f swi eat lun ala 33w52                                                                 208.mg   P<.3
   c M f swi eat liv tum 33w52                                                                 no dre   P=1.
3411 R m f34 eat ubl mix 30w52 e                                      . + .                    1.31mg * P<.0005+

   a R m f34 eat ubl ppc 30w52 e                                                               5.92mg * P<.0005

   b R m f34 eat liv tum 30w52 e                                                               no dre   P=1.

3412 R m f34 eat ubl car  7m24                                        . +.                     3.18mg Z P<.0005+

3413 R m fis eat ubl car 72w72                                              <+                 noTD50   P<.0005+
3414 R m fis eat liv tum 24m24                                                .>               no dre   P=1.
```

	RefNum	LoConf	UpConf	Cntrl	1Dose	1Inc	2Dose	2Inc				Citation or Pathology
												Brkly Code
i	1126	16.9mg	n.s.s.	2/24	53.0mg	9/30						
j	1126	28.2mg	n.s.s.	0/24	53.0mg	4/30						
k	1126	33.5mg	n.s.s.	0/24	53.0mg	3/30						
l	1126	33.5mg	n.s.s.	0/24	53.0mg	3/30						
m	1126	33.5mg	n.s.s.	0/24	53.0mg	3/30						
n	1126	n.s.s.	5.41mg	2/24	53.0mg	30/30						

4-(5-NITRO-2-FURYL)THIAZOLE   53757-28-1

3399	1411	8.56mg	38.6mg	6/36	52.1mg	24/35						Swaminathan;canr,41,2648-2653;1981
a	1411	11.2mg	38.0mg	0/36	52.1mg	19/35						
b	1411	161.mg	n.s.s.	0/36	52.1mg	0/35						
c	1411	4.26mg	15.2mg	6/36	52.1mg	31/35						

N-[4-(5-NITRO-2-FURYL)-2-THIAZOLYL]ACETAMIDE   (NFTA) 531-82-8

3400	1077	4.14mg	20.3mg	0/22	63.1mg	16/18						Croft;jnci,51,941-949;1973
a	1077	22.7mg	388.mg	0/22	63.1mg	5/18						
b	1077	27.6mg	372.mg	0/24	63.1mg	6/24						
c	1077	141.mg	n.s.s.	0/24	63.1mg	0/24						
d	1077	141.mg	n.s.s.	0/24	63.1mg	0/24						
3401	1122	6.89mg	16.4mg	0/28	69.3mg	47/56						Erturk;canr,30,936-941;1970
a	1122	21.9mg	66.4mg	0/28	69.3mg	23/56						
b	1122	31.7mg	132.mg	0/28	69.3mg	16/56						
c	1122	322.mg	n.s.s.	0/28	69.3mg	0/56						
d	1122	4.51mg	11.6mg	0/28	69.3mg	52/56						
3402	1126m	13.4mg	66.1mg	4/35	69.3mg	20/34						Cohen;jnci,54,841-850;1975
a	1126m	77.6mg	n.s.s.	0/35	69.3mg	2/34						
b	1126m	196.mg	n.s.s.	0/35	69.3mg	0/34						
c	1126m	12.4mg	56.4mg	4/35	69.3mg	21/34						
3403	1126n	16.5mg	70.4mg	1/32	69.3mg	17/35						
a	1126n	108.mg	n.s.s.	0/32	69.3mg	1/35						
b	1126n	201.mg	n.s.s.	0/32	69.3mg	0/35						
c	1126n	15.3mg	61.8mg	1/32	69.3mg	18/35						

N-[4-(5-NITRO-2-FURYL)-2-THIAZOLYL]FORMAMIDE   (FANFT) 24554-26-5

3404	1077	n.s.s.	9.77mg	0/20	63.1mg	23/23						Croft;jnci,51,941-949;1973
a	1077	23.2mg	238.mg	0/20	63.1mg	7/23						
b	1077	31.7mg	1.09gm	0/24	63.1mg	5/24						
c	1077	141.mg	n.s.s.	0/24	63.1mg	0/24						
d	1077	141.mg	n.s.s.	0/24	63.1mg	0/24						
3405	1601	20.5mg	58.9mg	0/53	68.8mg	25/55						Dunsford;jnci,73,151-160;1984
a	1601	58.0mg	449.mg	0/53	68.8mg	8/55						
b	1601	58.0mg	449.mg	0/53	68.8mg	8/55						
c	1601	148.mg	n.s.s.	1/53	68.8mg	2/55						
d	1601	333.mg	n.s.s.	0/53	68.8mg	0/55						
e	1601	10.6mg	45.9mg	19/53	68.8mg	43/55						
3406	1601	19.5mg	51.4mg	0/55	62.3mg	30/56						
a	1601	49.9mg	230.mg	0/55	62.3mg	12/56						
b	1601	95.0mg	6.08gm	0/55	62.3mg	5/56						
c	1601	298.mg	n.s.s.	4/55	62.3mg	1/56						
d	1601	298.mg	n.s.s.	4/55	62.3mg	1/56						
e	1601	12.6mg	47.6mg	17/55	62.3mg	43/56						
3407	1771	89.7mg	232.mg	0/100	114.mg	14/60	115.mg	4/13	118.mg	12/44		Berger;clet,31,311-318;1986
a	1771	103.mg	285.mg	0/100	114.mg	18/60	115.mg	1/13	118.mg	7/44		
b	1771	358.mg	n.s.s.	0/100	114.mg	5/60	115.mg	0/13	118.mg	0/44		
c	1771	588.mg	n.s.s.	0/100	114.mg	0/60	115.mg	0/13	118.mg	2/44		
d	1771	106.mg	n.s.s.	1/100	114.mg	0/60	115.mg	0/13	118.mg	0/44		
e	1771	40.0mg	103.mg	18/100	114.mg	35/60	115.mg	7/13	118.mg	27/44		
f	1771	48.4mg	107.mg	4/100	114.mg	32/60	115.mg	5/13	118.mg	18/44		
3408	1068	11.5mg	30.6mg	0/56	106.mg	31/48						Erturk;canr,30,1309-1311;1970
a	1068	23.4mg	n.s.s.	15/56	106.mg	23/48						
b	1068	60.3mg	n.s.s.	1/56	106.mg	6/48						
3409	1076	3.21mg	17.4mg	0/28	85.2mg	20/21						Cohen;canr,33,1593-1597;1973
a	1076	32.7mg	508.mg	0/28	85.2mg	5/21						
b	1076	29.4mg	n.s.s.	1/28	85.2mg	6/21						
c	1076	57.7mg	n.s.s.	0/28	85.2mg	2/21						
d	1076	148.mg	n.s.s.	0/28	85.2mg	0/21						
e	1076	n.s.s.	12.5mg	1/28	85.2mg	21/21						
3410	1118	9.28mg	57.1mg	0/30	41.3mg	9/30						Cohen;canr,38,1398-1405;1978
a	1118	15.6mg	n.s.s.	1/30	41.3mg	5/30						
b	1118	33.9mg	n.s.s.	0/30	41.3mg	1/30						
c	1118	63.8mg	n.s.s.	0/30	41.3mg	0/30						
3411	727	.739mg	2.38mg	0/20	.115mg	0/16	.231mg	0/15	1.15mg	0/16	2.31mg 1/15 11.5mg 13/14	
					23.1mg	16/16						Arai;jnci,62,1013-1016;1979
a	727	3.09mg	13.4mg	0/20	.115mg	0/16	.231mg	0/15	1.15mg	0/16	2.31mg 0/15 11.5mg 4/14	
					23.1mg	9/16						
b	727	55.8ug	n.s.s.	0/20	.115mg	0/16	.231mg	0/15	1.15mg	0/16	2.31mg 0/15 11.5mg 0/14	
					23.1mg	0/16						
3412	1575	2.25mg	4.56mg	0/40	.115mg	0/40	.577mg	0/40	1.15mg	0/40	5.77mg 34/40 11.5mg 39/40	
					(23.1mg	36/40)						Arai;clet,18,261-269;1983
3413	1430	n.s.s.	24.4mg	0/27	80.0mg	8/8						Fukushima;canr,41,3100-3103;1981
3414	1574	8.24mg	n.s.s.	0/30	2.00mg	0/20						Murasaki;carc,4,97-99;1983

```
     Spe Strain Site   Xpo+Xpt                                          TD50     2Tailpvl
         Sex   Route Hist    Notes                                        DR       AuOp

  a   R m fis eat ubl tum 24m24                                        no dre    P=1.
3415  R m fis eat ubl car 36w74              .   +   .                 4.51mg    P<.0005+
  a   R m fis eat ubl ivc 36w74                                        9.74mg    P<.0005
  b   R m fis eat ubl nvc 36w74                                        60.5mg    P<.003
  c   R m fis eat liv hnd 36w74                                        263.mg    P<.2
3416  R m fis eat ubl car 77w77                   .   +   .            9.87mg    P<.0005+
  a   R m fis eat ubl ivc 77w77                                        18.1mg    P<.0005
  b   R m fis eat ubl nvc 77w77                                        195.mg    P<.004
  c   R m fis eat liv tum 77w77                                        no dre    P=1.
3417  R f sda eat ubl mix 46w63              .   +   .                 5.07mg    P<.0005+
3418  R f sda eat ubl mix 26w70 er                       <+           noTD50    P<.0005+
  a   R f sda eat ubl tcc 26w70 er                                    noTD50    P<.0005+
  b   R f sda eat ubl ppc 26w70 er                                    48.6mg    P<.003 +
3419  R f sda eat ubl mix 46w70 er                       <+           noTD50    P<.0005+
  a   R f sda eat ubl tcc 46w70 er                                    noTD50    P<.0005+
  b   R f sda eat ubl ppc 46w70 er                                    82.7mg    P<.003 +
3420  R m sda eat ubl tcc 26w52 er                   <+               noTD50    P<.0005+
3421  R m sda eat ubl tcc 46w52 er                    <+              noTD50    P<.0005+
3422  R f wis eat ubl tcc 34w52 er                    <+              noTD50    P<.0005+


N,N'-[6-(5-NITRO-2-FURYL)-s-TRIAZINE-2,4-DIYL]BISACETAMIDE...:..1_00....:..1mg....:..10....:..100....:..1g.....:..10
3423  R f sda eat mgl mix 46w66 e            .   +   .                 14.1mg    P<.0005+
  a   R f sda eat mgl adc 46w66 e                                      29.0mg    P<.0005
  b   R f sda eat liv tum 46w66 e                                      no dre    P=1.
  c   R f sda eat tba mix 46w66 e                                      12.8mg    P<.0005


3-NITRO-3-HEXENE                     100ng..:..1ug....:..10....:..100....:..1mg....:..10....:..100....:..1g.....:..10
3424  M b swi inh lun adc 62w62                    .   ±              .346mg    P<.02  +
3425  M f sww inh lun adc 62w62 r                  .   ±              .363mg    P<.09
  a   M f sww inh lun mix 62w62 r                                     .687mg    P<.6
3426  M f sww inh lun adc 62w62 r                    .>               .768mg    P<.3
3427  R b cfn inh lun ulc 37m37 e                  .   +   .          8.66mg  * P<.0005+


3-NITRO-4-HYDROXYPHENYLARSONIC ACID     ..:..1ug....:..10....:..100....:..1mg....:..10....:..100....:..1g.....:..10
3428  M f b6c eat TBA MXB 24m24                            :>         no dre    P=1.    -
  a   M f b6c eat liv MXB 24m24                                       no dre    P=1.
  b   M f b6c eat lun MXB 24m24                                       193.mg  * P<.6
3429  M m b6c eat adr MXA 24m24                       :   ±           #138.mg * P<.03  -
  a   M m b6c eat TBA MXB 24m24                                       no dre    P=1.
  b   M m b6c eat liv MXB 24m24                                       no dre    P=1.
  c   M m b6c eat lun MXB 24m24                                       no dre    P=1.
3430  M f asw eat tba tum 24m24 e                  .>                 no dre    P=1.    -
  a   M f asw eat tba mal 24m24 e                                     93.6mg  * P<.8   -
3431  M m asw eat liv hpt 24m24 e                      .>             185.mg  * P<.3   -
  a   M m asw eat tba mal 24m24 e                                     85.6mg  * P<.7   -
  b   M m asw eat tba tum 24m24 e                                     no dre    P=1.
3432  R f f34 eat TBA MXB 24m24                    :>                 no dre    P=1.    -
  a   R f f34 eat liv MXB 24m24                                       no dre    P=1.
3433  R m f34 eat pit pda 24m24                :   ±                  3.40mg  \ P<.05
  a   R m f34 eat pni isa 24m24                                       16.5mg  * P<.02
  b   R m f34 eat pan ade 24m24                                       19.0mg  * P<.07  e
  c   R m f34 eat TBA MXB 24m24                                       7.20mg  * P<.5
  d   R m f34 eat liv MXB 24m24                                       no dre    P=1.
3434  R f asd eat tba tum 24m24 e                  .>                 no dre    P=1.    -
  a   R f asd eat tba mal 24m24 e                                     no dre    P=1.
3435  R m asd eat tba tum 24m24 e                      .>             no dre    P=1.    -
  a   R m asd eat tba mal 24m24 e                                     1.30gm  * P<1.


2-NITRO-p-PHENYLENEDIAMINE          100ng..:..1ug....:..10....:..100....:..1mg....:..10....:..100....:..1g.....:..10
3436  M f b6c eat liv MXA 78w90                            :   +      :  614.mg * P<.006 c
  a   M f b6c eat TBA MXB 78w90                                          468.mg * P<.02
  b   M f b6c eat liv MXB 78w90                                          614.mg * P<.006
  c   M f b6c eat lun MXB 78w90                                          1.98gm * P<.3
3437  M m b6c eat TBA MXB 78w90                                :>        no dre   P=1.   -
  a   M m b6c eat liv MXB 78w90                                          no dre   P=1.
  b   M m b6c eat lun MXB 78w90                                          945.mg \ P<.3
3438  R f f34 eat --- MXA 18m24                            :   ±         #892.mg * P<.05  -
  a   R f f34 eat TBA MXB 18m24                                          no dre   P=1.
  b   R f f34 eat liv MXB 18m24                                          no dre   P=1.
3439  R m f34 eat thy MXA 18m24                           :   ±         #224.mg * P<.03  -
  a   R m f34 eat TBA MXB 18m24                                          no dre   P=1.
  b   R m f34 eat liv MXB 18m24                                          no dre   P=1.


4-NITRO-o-PHENYLENEDIAMINE          100ng..:..1ug....:..10....:..100....:..1mg....:..10....:..100....:..1g.....:..10
3440  M f b6c eat TBA MXB 24m24                                  :>     no dre   P=1.   -
  a   M f b6c eat liv MXB 24m24                                         no dre   P=1.
  b   M f b6c eat lun MXB 24m24                                         no dre   P=1.
3441  M m b6c eat TBA MXB 24m24                             :>          no dre   P=1.   -
  a   M m b6c eat liv MXB 24m24                                         no dre   P=1.
  b   M m b6c eat lun MXB 24m24                                         no dre   P=1.
3442  R f f34 eat TBA MXB 24m24                      :>                 no dre   P=1.   -
```

	RefNum	LoConf	UpConf	Cntrl	1Dose	1Inc	2Dose	2Inc	Citation or Pathology
									Brkly Code
a	1574	8.24mg	n.s.s.	0/30	2.00mg	0/20			
3415	1657m	1.85mg	10.3mg	0/42	38.9mg	19/20			Cohen;canr,39,1207-1217;1979
a	1657m	4.99mg	21.5mg	0/42	38.9mg	15/20			
b	1657m	20.8mg	378.mg	0/42	38.9mg	4/20			
c	1657m	42.8mg	n.s.s.	0/42	38.9mg	1/20			
3416	1657n	5.36mg	17.5mg	0/42	80.0mg	40/42			
a	1657n	11.3mg	30.5mg	0/42	80.0mg	34/42			
b	1657n	79.4mg	1.19gm	0/42	80.0mg	6/42			
c	1657n	380.mg	n.s.s.	0/42	80.0mg	0/42			
3417	1124	2.24mg	10.3mg	0/30	68.6mg	29/30			Erturk;canr,27,1998-2002;1967
3418	1125m	n.s.s.	4.81mg	0/34	34.9mg	30/30			Erturk;canr,29,2219-2228;1969
a	1125m	n.s.s.	5.30mg	0/27	34.9mg	24/24			
b	1125m	19.8mg	234.mg	0/34	34.9mg	6/30			
3419	1125n	n.s.s.	8.63mg	0/34	61.8mg	29/29			
a	1125n	n.s.s.	9.56mg	0/27	61.8mg	23/23			
b	1125n	33.6mg	383.mg	0/34	61.8mg	6/29			
3420	1125m	n.s.s.	3.99mg	0/30	37.6mg	15/15			
3421	1125n	n.s.s.	6.07mg	0/30	66.5mg	20/20			
3422	1123	n.s.s.	4.76mg	0/15	62.7mg	30/30			Adolphs;urre,6,19-27;1978

N,N'-[6-(5-NITRO-2-FURYL)-s-TRIAZINE-2,4-DIYL]BISACETAMIDE   51325-35-0

	RefNum	LoConf	UpConf	Cntrl	1Dose	1Inc	2Dose	2Inc	Citation or Pathology
3423	200a	8.17mg	27.0mg	2/39	69.7mg	25/33			Cohen;jnci,51,403-417;1973
a	200a	16.0mg	60.3mg	0/39	69.7mg	16/33			
b	200a	191.mg	n.s.s.	0/39	69.7mg	0/33			
c	200a	7.44mg	24.3mg	2/39	69.7mg	26/33			

3-NITRO-3-HEXENE   4812-22-0

	RefNum	LoConf	UpConf	Cntrl	1Dose	1Inc	2Dose	2Inc	Citation or Pathology
3424	1360	.131mg	n.s.s.	0/21	.291mg	5/27			Deichmann;imed,34,800-807;1965
3425	1366m	88.6ug	n.s.s.	0/10	.332mg	2/10			Deichmann;txap,5,445-456;1963
a	1366m	97.1ug	n.s.s.	1/10	.332mg	2/10			
3426	1366n	.125mg	n.s.s.	0/10	.332mg	1/10			
3427	1360	4.91mg	19.8mg	0/100	.326mg	6/100	.651mg	11/100	Deichmann;imed,34,800-807;1965

3-NITRO-4-HYDROXYPHENYLARSONIC ACID (roxarsone) 121-19-7

	RefNum	LoConf	UpConf	Cntrl	1Dose	1Inc	2Dose	2Inc	Citation or Pathology
3428	TR345	32.7mg	n.s.s.	24/50	12.8mg	11/50	25.5mg	18/50	
a	TR345	196.mg	n.s.s.	3/50	12.8mg	0/50	25.5mg	0/50	liv:hpa,hpc,nnd.
b	TR345	34.8mg	n.s.s.	3/50	12.8mg	4/50	25.5mg	5/50	lun:a/a,a/c.
3429	TR345	56.5mg	n.s.s.	0/50	11.8mg	2/50	23.5mg	4/50	adr:csa,coa. S
a	TR345	29.5mg	n.s.s.	35/50	11.8mg	30/50	23.5mg	34/50	
b	TR345	71.9mg	n.s.s.	12/50	11.8mg	15/50	23.5mg	7/50	liv:hpa,hpc,nnd.
c	TR345	63.0mg	n.s.s.	11/50	11.8mg	5/50	23.5mg	10/50	lun:a/a,a/c.
3430	213b	6.88mg	n.s.s.	14/16	6.50mg	20/30	13.0mg	27/33	Prier;txap,5,526-542;1963
a	213b	10.9mg	n.s.s.	6/16	6.50mg	16/30	13.0mg	15/33	
3431	213b	30.1mg	n.s.s.	0/10	6.00mg	0/22	12.0mg	1/12	
a	213b	12.5mg	n.s.s.	1/10	6.00mg	5/22	12.0mg	2/12	
b	213b	12.6mg	n.s.s.	4/10	6.00mg	6/22	12.0mg	4/12	
3432	TR345	4.18mg	n.s.s.	48/50	2.45mg	47/50	4.93mg	46/50	
a	TR345	n.s.s.	n.s.s.	0/50	2.45mg	1/50	4.93mg	0/50	liv:hpa,hpc,nnd.
3433	TR345	1.28mg	n.s.s.	6/50	1.96mg	13/50	(3.92mg	8/50)	S
a	TR345	6.87mg	n.s.s.	0/50	1.96mg	3/50	3.92mg	4/50	S
b	TR345	6.46mg	n.s.s.	1/50	1.96mg	1/50	3.92mg	5/50	
c	TR345	1.58mg	n.s.s.	46/50	1.96mg	48/50	3.92mg	47/50	
d	TR345	18.5mg	n.s.s.	2/50	1.96mg	2/50	3.92mg	1/50	liv:hpa,hpc,nnd.
3434	213b	15.2mg	n.s.s.	14/27	2.50mg	18/28	10.0mg	11/28	Prier;txap,5,526-542;1963
a	213b	45.4mg	n.s.s.	4/27	2.50mg	2/28	10.0mg	1/28	
3435	213b	21.7mg	n.s.s.	4/24	2.00mg	2/24	8.00mg	2/22	
a	213b	22.9mg	n.s.s.	1/24	2.00mg	1/24	8.00mg	1/22	

2-NITRO-p-PHENYLENEDIAMINE   5307-14-2

	RefNum	LoConf	UpConf	Cntrl	1Dose	1Inc	2Dose	2Inc	Citation or Pathology
3436	TR169	346.mg	6.17gm	1/20	248.mg	10/50	490.mg	17/50	liv:hpa,hpc.
a	TR169	244.mg	n.s.s.	4/20	248.mg	18/50	490.mg	27/50	
b	TR169	346.mg	6.17gm	1/20	248.mg	10/50	490.mg	17/50	liv:hpa,hpc,nnd.
c	TR169	898.mg	n.s.s.	0/20	248.mg	5/50	490.mg	3/50	lun:a/a,a/c.
3437	TR169	326.mg	n.s.s.	9/20	229.mg	17/50	(457.mg	7/50)	
a	TR169	1.60gm	n.s.s.	3/20	229.mg	7/50	457.mg	3/50	liv:hpa,hpc,nnd.
b	TR169	318.mg	n.s.s.	1/20	229.mg	8/50	(457.mg	2/50)	lun:a/a,a/c.
3438	TR169	308.mg	n.s.s.	0/20	41.0mg	0/50	82.0mg	4/50	---:leu,lym. S
a	TR169	153.mg	n.s.s.	11/20	41.0mg	12/50	82.0mg	17/50	
b	TR169	666.mg	n.s.s.	1/20	41.0mg	0/50	82.0mg	1/50	liv:hpa,hpc,nnd.
3439	TR169	96.6mg	n.s.s.	0/20	16.0mg	1/50	32.8mg	6/50	thy:cca,ccr. S
a	TR169	90.8mg	n.s.s.	9/20	16.0mg	15/50	32.8mg	13/50	
b	TR169	489.mg	n.s.s.	1/20	16.0mg	0/50	32.8mg	0/50	liv:hpa,hpc,nnd.

4-NITRO-o-PHENYLENEDIAMINE   99-56-9

	RefNum	LoConf	UpConf	Cntrl	1Dose	1Inc	2Dose	2Inc	Citation or Pathology
3440	TR180	924.mg	n.s.s.	8/20	478.mg	18/50	(956.mg	11/50)	
a	TR180	2.64gm	n.s.s.	1/20	478.mg	9/50	956.mg	3/50	liv:hpa,hpc,nnd.
b	TR180	3.79gm	n.s.s.	3/20	478.mg	5/50	956.mg	4/50	lun:a/a,a/c.
3441	TR180	539.mg	n.s.s.	12/20	441.mg	25/50	(883.mg	12/50)	
a	TR180	3.51gm	n.s.s.	4/20	441.mg	5/50	883.mg	4/50	liv:hpa,hpc,nnd.
b	TR180	1.96gm	n.s.s.	3/20	441.mg	13/50	883.mg	6/50	lun:a/a,a/c.
3442	TR180	36.0mg	n.s.s.	15/20	18.4mg	31/50	36.8mg	33/50	

```
      Spe Strain Site   Xpo+Xpt                                                          TD50      2Tailpvl
          Sex Route Hist      Notes                                                         DR        AuOp
 a    R f f34 eat liv MXB 24m24                                                           no dre   P=1.
3443  R m f34 eat TBA MXB 24m24                                          :>               no dre   P=1.      -
 a    R m f34 eat liv MXB 24m24                                                           no dre   P=1.
```

5-NITRO-o-TOLUIDINE   100ng..:..1ug....:..10.....:..100....:..1mg....:..10......:..100....:..1g.....:..10
```
3444  M f b6c eat MXB MXB 78w96 v                               : + :                     242.mg  * P<.0005
 a    M f b6c eat liv hpc 78w96 v                                                         288.mg  / P<.0005c
 b    M f b6c eat --- hes 78w96 v                                             +hist 1.86gm * P<.4     c
 c    M f b6c eat TBA MXB 78w96 v                                                         989.mg  * P<.5
 d    M f b6c eat liv MXB 78w96 v                                                         288.mg  / P<.0005
 e    M f b6c eat lun MXB 78w96 v                                                         no dre   P=1.
3445  M m b6c eat liv hpc 78w95 v                                 : + :                    266.mg  * P<.005 c
 a    M m b6c eat MXB MXB 78w95 v                                                         269.mg  * P<.005
 b    M m b6c eat --- MXA 78w95 v                                             +hist 2.55gm * P<.2     c
 c    M m b6c eat TBA MXB 78w95 v                                                         431.mg  * P<.2
 d    M m b6c eat liv MXB 78w95 v                                                         266.mg  * P<.005
 e    M m b6c eat lun MXB 78w95 v                                                         no dre   P=1.
3446  R f f34 eat TBA MXB 18m25 v                        :>                               no dre   P=1.      -
 a    R f f34 eat liv MXB 18m25 v                                                         43.0mg  * P<.3
3447  R m f34 eat liv hpc 18m25 v                  :     ±                                #27.7mg  * P<.04     -
 a    R m f34 eat TBA MXB 18m25 v                                                         43.7mg  * P<.9
 b    R m f34 eat liv MXB 18m25 v                                                         no dre   P=1.
```

5-NITROACENAPHTHENE   100ng..:..1ug....:..10.....:..100....:..1mg....:..10......:..100....:..1g.....:..10
```
3448  H f syg eat bil cho 26w71                            .    +    .                     354.mg    P<.003
3449  M f b6c eat MXB MXB 78w96 ev                      : + :                              45.3mg  * P<.0005
 a    M f b6c eat liv hpc 78w96 ev                                                        45.8mg  * P<.0005c
 b    M f b6c eat ova MXA 78w96 ev                                                        217.mg  * P<.0005c
 c    M f b6c eat ova tua 78w96 ev                                                        395.mg  * P<.005 c
 d    M f b6c eat TBA MXB 78w96 ev                                                        90.4mg  * P<.02
 e    M f b6c eat liv MXB 78w96 ev                                                        45.8mg  * P<.0005
 f    M f b6c eat lun MXB 78w96 ev                                                        1.32gm  * P<.5
3450  M m b6c eat TBA MXB 78w95 es                              :>                         no dre   P=1.      -
 a    M m b6c eat liv MXB 78w95 es                                                        no dre   P=1.
 b    M m b6c eat lun MXB 78w95 es                                                        no dre   P=1.
3451  R f f34 eat MXB MXB 18m23 e                 :   +  :                                 5.98mg  / P<.0005
 a    R f f34 eat lun MXA 18m23 e                                                         9.63mg  * P<.0005c
 b    R f f34 eat eac MXA 18m23 e                                                         14.9mg  / P<.0005c
 c    R f f34 eat eac cuc 18m23 e                                                         17.4mg  * P<.0005c
 d    R f f34 eat cli can 18m23 e                                                         52.2mg  * P<.0005c
 e    R f f34 eat mgl MXA 18m23 e                                                         77.9mg  * P<.0005c
 f    R f f34 eat mgl acn 18m23 e                                                         81.7mg  * P<.0005c
 g    R f f34 eat eac sqc 18m23 e                                                         90.5mg  / P<.0005c
 h    R f f34 eat TBA MXB 18m23 e                                                         6.44mg  / P<.0005
 i    R f f34 eat liv MXB 18m23 e                                                         no dre   P=1.
3452  R m f34 eat MXB MXB 17m23 ae               :    +  :                                 6.45mg  / P<.0005
 a    R m f34 eat eac MXA 17m23 ae                                                        7.88mg  * P<.0005c
 b    R m f34 eat eac cuc 17m23 ae                                                        8.92mg  * P<.0005c
 c    R m f34 eat lun MXA 17m23 ae                                                        29.9mg  * P<.0005c
 d    R m f34 eat lun a/c 17m23 ae                                                        35.1mg  * P<.0005c
 e    R m f34 eat thy MXA 17m23 ae                                                        37.7mg  / P<.0005
 f    R m f34 eat kid MXA 17m23 ae                                                        42.9mg  * P<.0005
 g    R m f34 eat thy MXA 17m23 ae                                                        58.5mg  * P<.0005
 h    R m f34 eat eac sqc 17m23 ae                                                        68.2mg  * P<.0005
 i    R m f34 eat TBA MXB 17m23 ae                                                        5.58mg  / P<.0005
 j    R m f34 eat liv MXB 17m23 ae                                                        71.4mg  * P<.08
```

p-NITROANILINE   100ng..:..1ug....:..10.....:..100....:..1mg....:..10......:..100....:..1g.....:..10
```
3453  M f b6c gav TBA MXB 24m24                                        :>                  487.mg  * P<.8      -
 a    M f b6c gav liv MXB 24m24                                                           no dre   P=1.
 b    M f b6c gav lun MXB 24m24                                                           2.37gm  * P<.8
3454  M m b6c gav --- MXA 24m24                              :   ±                         374.mg  * P<.1      e
 a    M m b6c gav liv hes 24m24                                                           501.mg  * P<.05     e
 b    M m b6c gav TBA MXB 24m24                                                           no dre   P=1.
 c    M m b6c gav liv MXB 24m24                                                           no dre   P=1.
 d    M m b6c gav lun MXB 24m24                                                           no dre   P=1.
3455  R f cdr gav liv nnd 24m24 e                                .>                        no dre   P=1.      -
 a    R f cdr gav liv hpc 24m24 e                                                         no dre   P=1.      -
 b    R f cdr gav tba mix 24m24 e                                                         9.05mg  * P<.2      -
 c    R f cdr gav tba ben 24m24 e                                                         9.72mg  * P<.2      -
 d    R f cdr gav tba mal 24m24 e                                                         47.0mg  * P<.3      -
3456  R m cdr gav liv hpc 24m24 e                                .>                        129.mg  * P<.3      -
 a    R m cdr gav liv nnd 24m24 e                                                         no dre   P=1.      -
 b    R m cdr gav tba mix 24m24 e                                                         12.2mg  * P<.08     -
 c    R m cdr gav tba mal 24m24 e                                                         41.6mg  * P<.09     -
 d    R m cdr gav tba ben 24m24 e                                                         16.2mg  * P<.2      -
```

o-NITROANISOLE   100ng..:..1ug....:..10.....:..100....:..1mg....:..10......:..100....:..1g.....:..10
```
3457  M f b6c eat --- MXA 24m24                             :  +                 :         153.mg  Z P<.008
 a    M f b6c eat liv hpa 24m24                                                           179.mg  Z P<.0005p
 b    M f b6c eat liv MXA 24m24                                                           194.mg  Z P<.003 p
```

	RefNum	LoConf	UpConf	Cntrl	1Dose	1Inc	2Dose	2Inc			Citation or Pathology
											Brkly Code
a	TR180	217.mg	n.s.s.	0/20	18.4mg	2/50	36.8mg	0/50			liv:hpa,hpc,nnd.
3443	TR180	47.2mg	n.s.s.	9/20	14.7mg	26/50	29.4mg	19/50			
a	TR180	147.mg	n.s.s.	1/20	14.7mg	2/50	29.4mg	2/50			liv:hpa,hpc,nnd.

5-NITRO-o-TOLUIDINE   99-55-8

	RefNum	LoConf	UpConf	Cntrl	1Dose	1Inc	2Dose	2Inc			Citation or Pathology
3444	TR107	142.mg	698.mg	3/50	125.mg	12/50	240.mg	22/50			---:hes; liv:hpc. C
a	TR107	168.mg	787.mg	2/50	125.mg	7/50	240.mg	20/50			
b	TR107	529.mg	n.s.s.	1/50	125.mg	5/50	240.mg	3/50			
c	TR107	205.mg	n.s.s.	19/50	125.mg	19/50	240.mg	25/50			
d	TR107	168.mg	787.mg	2/50	125.mg	7/50	240.mg	20/50			liv:hpa,hpc,nnd.
e	TR107	1.36gm	n.s.s.	2/50	125.mg	3/50	240.mg	0/50			lun:a/a,a/c.
3445	TR107	135.mg	2.35gm	12/49	115.mg	12/49	224.mg	29/50			
a	TR107	135.mg	2.90gm	13/50	115.mg	12/49	224.mg	30/50			---:hem,hes; liv:hpc. C
b	TR107	718.mg	n.s.s.	1/50	115.mg	0/49	224.mg	4/50			---:hem,hes.
c	TR107	146.mg	n.s.s.	23/50	115.mg	20/49	224.mg	31/50			
d	TR107	135.mg	2.35gm	12/49	115.mg	12/49	224.mg	29/50			liv:hpa,hpc,nnd.
e	TR107	1.03gm	n.s.s.	5/50	115.mg	3/49	224.mg	2/50			lun:a/a,a/c.
3446	TR107	3.72mg	n.s.s.	42/50	1.80mg	33/50	3.60mg	35/50			
a	TR107	13.0mg	n.s.s.	0/50	1.80mg	2/50	3.60mg	1/50			liv:hpa,hpc,nnd.
3447	TR107	8.39mg	n.s.s.	0/50	1.40mg	0/50	2.90mg	3/50			S
a	TR107	2.67mg	n.s.s.	30/50	1.40mg	20/50	2.90mg	27/50			
b	TR107	9.23mg	n.s.s.	5/50	1.40mg	1/50	2.90mg	4/50			liv:hpa,hpc,nnd.

5-NITROACENAPHTHENE   602-87-9

	RefNum	LoConf	UpConf	Cntrl	1Dose	1Inc	2Dose	2Inc			Citation or Pathology
3448	1091	152.mg	1.66gm	0/20	382.mg	7/24					Takemura;bjca,30,481-483;1974
3449	TR118	29.3mg	87.4mg	2/50	53.3mg	23/50	127.mg	19/50			liv:hpc; ova:gct,lut,tua. C
a	TR118	29.5mg	90.3mg	2/50	53.3mg	23/50	127.mg	18/50			
b	TR118	108.mg	623.mg	0/50	53.3mg	4/50	127.mg	7/50			ova:gct,lut,tua.
c	TR118	161.mg	3.08gm	0/50	53.3mg	2/50	127.mg	4/50			
d	TR118	40.3mg	n.s.s.	19/50	53.3mg	29/50	127.mg	21/50			
e	TR118	29.5mg	90.3mg	2/50	53.3mg	23/50	127.mg	18/50			liv:hpa,hpc,nnd.
f	TR118	232.mg	n.s.s.	2/50	53.3mg	3/50	127.mg	2/50			lun:a/a,a/c.
3450	TR118	74.6mg	n.s.s.	23/50	58.8mg	15/50	140.mg	1/50			
a	TR118	132.mg	n.s.s.	12/50	58.8mg	7/50	140.mg	0/50			liv:hpa,hpc,nnd.
b	TR118	152.mg	n.s.s.	5/50	58.8mg	4/50	140.mg	0/50			lun:a/a,a/c.
3451	TR118	1.96mg	11.3mg	1/100	43.0mg	37/50	87.0mg	38/50		cli:can; eac:cas,cuc,sqc; lun:a/a,a/c; mgl:acn,pcn. C	
a	TR118	2.18mg	44.5mg	1/100	43.0mg	8/50	87.0mg	3/50			lun:a/a,a/c.
b	TR118	7.28mg	24.8mg	1/100	43.0mg	27/50	87.0mg	35/50			eac:cas,cuc,sqc.
c	TR118	7.85mg	30.9mg	0/100	43.0mg	25/50	87.0mg	26/50			
d	TR118	18.7mg	143.mg	0/100	43.0mg	6/50	87.0mg	5/50			
e	TR118	31.5mg	198.mg	0/100	43.0mg	6/50	87.0mg	5/50			mgl:acn,pcn.
f	TR118	32.1mg	220.mg	0/100	43.0mg	5/50	87.0mg	5/50			
g	TR118	29.9mg	230.mg	0/100	43.0mg	3/50	87.0mg	9/50			
h	TR118	3.78mg	10.5mg	83/100	43.0mg	46/50	87.0mg	45/50			liv:hpa,hpc,nnd.
i	TR118	n.s.s.	n.s.s.	2/100	43.0mg	0/50	87.0mg	0/50			
3452	TR118	1.67mg	17.7mg	1/99	34.4mg	25/50	62.4mg	21/50			eac:cuc,sqc; lun:a/a,a/c. C
a	TR118	1.71mg	27.8mg	0/99	34.4mg	21/50	62.4mg	20/50			eac:cuc,sqc.
b	TR118	1.75mg	45.1mg	0/99	34.4mg	14/50	62.4mg	18/50			
c	TR118	9.96mg	93.8mg	1/99	34.4mg	7/50	62.4mg	3/50			lun:a/a,a/c.
d	TR118	10.4mg	171.mg	1/99	34.4mg	5/50	62.4mg	0/50			
e	TR118	10.6mg	210.mg	0/99	34.4mg	4/50	62.4mg	1/50			thy:fcc,pac. S
f	TR118	13.5mg	204.mg	0/99	34.4mg	4/50	62.4mg	0/50			kid:tla,uac. S
g	TR118	15.4mg	323.mg	0/99	34.4mg	4/50	62.4mg	0/50			thy:fca,ppa. S
h	TR118	27.4mg	188.mg	0/99	34.4mg	7/50	62.4mg	2/50			
i	TR118	2.91mg	10.1mg	58/99	34.4mg	39/50	62.4mg	30/50			
j	TR118	5.92mg	n.s.s.	4/99	34.4mg	1/50	62.4mg	0/50			liv:hpa,hpc,nnd.

p-NITROANILINE   100-01-6

	RefNum	LoConf	UpConf	Cntrl	1Dose	1Inc	2Dose	2Inc	3Dose	3Inc	Citation or Pathology
3453	TR418	57.7mg	n.s.s.	35/50	2.11mg	33/50	21.1mg	34/49	70.3mg	36/50	
a	TR418	104.mg	n.s.s.	17/50	2.11mg	17/50	21.1mg	21/49	70.3mg	16/50	liv:hpa,hpc,nnd.
b	TR418	178.mg	n.s.s.	2/50	2.11mg	5/50	21.1mg	5/49	70.3mg	4/50	lun:a/a,a/c.
3454	TR418	126.mg	n.s.s.	5/50	2.11mg	3/50	21.1mg	4/50	70.5mg	10/50	---:hem,hes.
a	TR418	171.mg	n.s.s.	0/50	2.11mg	1/50	21.1mg	2/50	70.5mg	4/50	
b	TR418	135.mg	n.s.s.	33/50	2.11mg	38/50	21.1mg	36/50	70.5mg	28/50	
c	TR418	28.9mg	n.s.s.	25/50	2.11mg	26/50	21.1mg	25/50	(70.5mg	13/50)	liv:hpa,hpc,nnd.
d	TR418	193.mg	n.s.s.	9/50	2.11mg	10/50	21.1mg	9/50	70.5mg	9/50	lun:a/a,a/c.
3455	1985	29.9mg	n.s.s.	17/60	.250mg	14/59	1.50mg	10/60	9.00mg	13/60	Nair;faat,15,607-621;1990
a	1985	115.mg	n.s.s.	2/60	.250mg	2/59	1.50mg	1/60	9.00mg	0/60	
b	1985	2.90mg	n.s.s.	55/60	.250mg	50/60	1.50mg	45/60	9.00mg	56/60	
c	1985	3.15mg	n.s.s.	53/60	.250mg	49/60	1.50mg	45/60	9.00mg	55/60	
d	1985	13.6mg	n.s.s.	22/60	.250mg	15/60	1.50mg	13/60	9.00mg	23/60	
3456	1985	30.7mg	n.s.s.	0/60	.250mg	1/60	1.50mg	4/60	9.00mg	3/60	
a	1985	49.9mg	n.s.s.	16/60	.250mg	7/60	1.50mg	3/60	9.00mg	7/60	
b	1985	4.41mg	n.s.s.	51/60	.250mg	36/60	1.50mg	39/60	9.00mg	50/60	
c	1985	14.6mg	n.s.s.	12/60	.250mg	8/60	1.50mg	12/60	9.00mg	17/60	
d	1985	5.47mg	n.s.s.	50/60	.250mg	33/60	1.50mg	34/60	9.00mg	47/60	

o-NITROANISOLE   91-23-6

	RefNum	LoConf	UpConf	Cntrl	1Dose	1Inc	2Dose	2Inc	3Dose	3Inc	Citation or Pathology
3457	TR416	67.1mg	3.07gm	5/50	85.7mg	14/50	(255.mg	14/50	767.mg	5/50)	---:mlh,mlm,mlp,mlu. S
a	TR416	96.6mg	695.mg	14/50	85.7mg	20/50	255.mg	36/50	(767.mg	18/50)	
b	TR416	99.4mg	1.09gm	17/50	85.7mg	21/50	255.mg	37/50	(767.mg	20/50)	liv:hpa,hpc.

```
        Spe Strain  Site    Xpo+Xpt                                                         TD50     2Tailpvl
           Sex    Route   Hist      Notes                                                      DR     AuOp

    c   M f b6c eat TBA MXB 24m24                                                            350.mg Z P<.3
    d   M f b6c eat liv MXB 24m24                                                            194.mg Z P<.003
    e   M f b6c eat lun MXB 24m24                                                            no dre   P=1.
 3458 M m b6c eat liv hpa 24m24                                             :   +   :        177.mg Z P<.002 c
    a   M m b6c eat liv MXB 24m24                                                            198.mg Z P<.006
    b   M m b6c eat liv MXA 24m24                                                            208.mg Z P<.007 c
    c   M m b6c eat liv MXA 24m24                                                            391.mg Z P<.005 c
    d   M m b6c eat liv hpb 24m24                                                            428.mg Z P<.0005c
    e   M m b6c eat liv hpc 24m24                                                            no dre   P=1.
    f   M m b6c eat TBA MXB 24m24                                                            43.5gm * P<1.
    g   M m b6c eat liv MXB 24m24                                                            198.mg Z P<.006
    h   M m b6c eat lun MXB 24m24                                                            no dre   P=1.
 3459 R f f34 eat --- mnl 24m24                                               :   +   :      152.mg * P<.006 p
    a   R f f34 eat lun a/a 24m24                                                            666.mg * P<.02
    b   R f f34 eat liv hpa 24m24                                                            1.01gm * P<.02
    c   R f f34 eat for MXA 24m24                                                            1.27gm * P<.2  e
    d   R f f34 eat for sqp 24m24                                                            3.27gm * P<.5  e
    e   R f f34 eat TBA MXB 24m24                                                            336.mg * P<.7
    f   R f f34 eat liv MXB 24m24                                                            1.01gm * P<.02
 3460 R f f34 eat MXB MXB 6m24                                            :   + :            12.9mg / P<.0005
    a   R f f34 eat ubl MXA 6m24                                                             13.0mg / P<.0005c
    b   R f f34 eat ubl tcc 6m24                                                             13.2mg / P<.0005c
    c   R f f34 eat lgi MXA 6m24                                                             55.0mg / P<.0005c
    d   R f f34 eat lgi adp 6m24                                                             55.0mg / P<.0005c
    e   R f f34 eat col adp 6m24                                                             60.9mg / P<.0005
    f   R f f34 eat ubl sar 6m24                                                             192.mg / P<.0005c
    g   R f f34 eat ubl MXA 6m24                                                             588.mg * P<.003 c
    h   R f f34 eat ubl sqp 6m24                                                             751.mg * P<.006 c
    i   R f f34 eat ubl tpp 6m24                                                             543.mg * P<.07 c
    j   R f f34 eat kid MXA 6m24                                                             1.24gm * P<.04 c
    k   R f f34 eat lgi car 6m24                                                             1.30gm * P<.04 c
    l   R f f34 eat col car 6m24                                                             1.30gm * P<.04
    m   R f f34 eat rec adp 6m24                                                             560.mg * P<.2
    n   R f f34 eat kid tpp 6m24                                                             1.89gm * P<.2  c
    o   R f f34 eat ubl sqc 6m24                                                             2.70gm * P<.2  c
    p   R f f34 eat kid tcc 6m24                                                             3.55gm * P<.2  c
    q   R f f34 eat TBA MXB 6m24                                                             10.7mg / P<.0005
    r   R f f34 eat liv MXB 6m24                                                             no dre   P=1.
 3461 R m f34 eat tes MXA 24m24                                           :   +   :          21.3mg * P<.0005
    a   R m f34 eat --- mnl 24m24                                                            25.9mg * P<.0005p
    b   R m f34 eat amd MXA 24m24                                                            89.6mg * P<.004
    c   R m f34 eat kid MXA 24m24                                                            251.mg * P<.005 e
    d   R m f34 eat for MXA 24m24                                                            304.mg * P<.007 e
    e   R m f34 eat amd MXA 24m24                                                            109.mg * P<.04
    f   R m f34 eat pre MXA 24m24                                                            195.mg * P<.04
    g   R m f34 eat for sqp 24m24                                                            543.mg * P<.05 e
    h   R m f34 eat kid ruc 24m24                                                            612.mg * P<.02 e
    i   R m f34 eat for sqc 24m24                                                            715.mg * P<.07 e
    j   R m f34 eat kid rua 24m24                                                            435.mg * P<.2  e
    k   R m f34 eat TBA MXB 24m24                                                            21.4mg * P<.0005
    l   R m f34 eat liv MXB 24m24                                                            330.mg * P<.2
 3462 R m f34 eat MXB MXB 6m24                                            :   + :            7.47mg / P<.0005
    a   R m f34 eat ubl MXA 6m24                                                             7.48mg / P<.0005c
    b   R m f34 eat ubl tcc 6m24                                                             8.28mg / P<.0005c
    c   R m f34 eat lgi MXA 6m24                                                             8.54mg / P<.0005c
    d   R m f34 eat lgi adp 6m24                                                             8.58mg / P<.0005c
    e   R m f34 eat col adp 6m24                                                             17.2mg / P<.0005
    f   R m f34 eat rec adp 6m24                                                             51.8mg * P<.0005
    g   R m f34 eat ubl tpp 6m24                                                             75.6mg * P<.004 c
    h   R m f34 eat kid MXA 6m24                                                             83.6mg * P<.0005c
    i   R m f34 eat kid tcc 6m24                                                             90.9mg * P<.0005c
    j   R m f34 eat ubl MXA 6m24                                                             273.mg / P<.0005c
    k   R m f34 eat ubl sar 6m24                                                             292.mg * P<.0005c
    l   R m f34 eat ubl sqc 6m24                                                             422.mg / P<.002 c
    m   R m f34 eat ubl sqp 6m24                                                             482.mg / P<.003 c
    n   R m f34 eat lgi car 6m24                                                             573.mg * P<.004 c
    o   R m f34 eat col car 6m24                                                             573.mg * P<.004
    p   R m f34 eat cec adp 6m24                                                             21.3mg * P<.03
    q   R m f34 eat kid tpp 6m24                                                             1.04gm * P<.03  c
    r   R m f34 eat TBA MXB 6m24                                                             3.97mg / P<.0005
    s   R m f34 eat liv MXB 6m24                                                             no dre   P=1.


4-NITROANTHRANILIC ACID            100ng..:..1ug....:..10.....:..100....:..1mg....:..10.....:..100....:..1g......:..10
 3463 M f b6c eat TBA MXB 78w93                                                     :>      no dre   P=1.  -
    a   M f b6c eat liv MXB 78w93                                                            no dre   P=1.
    b   M f b6c eat lun MXB 78w93                                                            2.56gm \ P<.05
 3464 M m b6c eat --- MXA 78w93                                                     :       #12.6gm * P<.02  -
    a   M m b6c eat TBA MXB 78w93                                                            1.66gm * P<.4
    b   M m b6c eat liv MXB 78w93                                                            no dre   P=1.
    c   M m b6c eat lun MXB 78w93                                                            no dre   P=1.
 3465 R f f34 eat TBA MXB 18m24 v                                               :>          no dre   P=1.  -
```

	RefNum	LoConf	UpConf	Cntrl	1Dose	1Inc	2Dose	2Inc			Citation or Pathology	Brkly Code
c	TR416	104.mg	n.s.s.	39/50	85.7mg	44/50	255.mg	47/50	(767.mg	28/50)		
d	TR416	99.1mg	1.13gm	17/50	85.7mg	22/50	255.mg	37/50	(767.mg	20/50)	liv:hpa,hpb,hpc.	
e	TR416	1.02gm	n.s.s.	6/50	85.7mg	3/50	255.mg	0/50	(767.mg	0/50)	lun:a/a,a/c.	
3458	TR416	95.4mg	724.mg	14/50	79.2mg	26/50	237.mg	41/50	(710.mg	29/50)		
a	TR416	97.0mg	2.44gm	21/50	79.2mg	33/50	237.mg	46/50	(710.mg	34/50)	liv:hpa,hpb,hpc.	C
b	TR416	101.mg	3.82gm	21/50	79.2mg	32/50	237.mg	45/50	(710.mg	32/50)	liv:hpa,hpc.	
c	TR416	194.mg	4.17mg	7/50	79.2mg	14/50	237.mg	23/50	(710.mg	15/50)	liv:hpb,hpc.	
d	TR416	251.mg	825.mg	0/50	79.2mg	3/50	237.mg	17/50	(710.mg	9/50)		
e	TR416	1.99gm	n.s.s.	7/50	79.2mg	12/50	237.mg	11/50	710.mg	7/50		
f	TR416	531.mg	n.s.s.	38/50	79.2mg	41/50	237.mg	49/50	710.mg	41/50		
g	TR416	97.0mg	2.44gm	21/50	79.2mg	33/50	237.mg	46/50	(710.mg	34/50)	liv:hpa,hpb,hpc.	
h	TR416	3.36gm	n.s.s.	6/50	79.2mg	12/50	237.mg	4/50	710.mg	4/50	lun:a/a,a/c.	
3459	TR416	71.8mg	1.85mg	14/50	11.0mg	11/50	32.9mg	14/50	98.9mg	26/50		
a	TR416	251.mg	n.s.s.	0/50	11.0mg	0/50	32.9mg	2/50	98.9mg	3/50		S
b	TR416	307.mg	n.s.s.	0/50	11.0mg	0/50	32.9mg	0/50	98.9mg	3/50		S
c	TR416	313.mg	n.s.s.	0/50	11.0mg	1/50	32.9mg	0/50	98.9mg	2/50	for:sqc,sqp.	
d	TR416	404.mg	n.s.s.	0/50	11.0mg	1/50	32.9mg	0/50	98.9mg	1/50		
e	TR416	52.3mg	n.s.s.	46/50	11.0mg	46/50	32.9mg	46/50	98.9mg	46/50		
f	TR416	307.mg	n.s.s.	0/50	11.0mg	0/50	32.9mg	0/50	98.9mg	3/50	liv:hpa,hpb,hpc.	
3460	TR416a	6.72mg	23.8mg	0/22	77.9mg	20/20	234.mg	32/34			kid:tcc,tpp; lgi:adp,car; ubl:sar,sqc,sqp,tcc,tpp.	C
a	TR416a	6.74mg	24.0mg	0/22	77.9mg	19/20	234.mg	32/34			ubl:tcc,tpp.	
b	TR416a	6.81mg	24.8mg	0/22	77.9mg	18/20	234.mg	32/34				
c	TR416a	24.6mg	114.mg	0/22	77.9mg	5/20	234.mg	17/34			lgi:adp,car.	
d	TR416a	24.6mg	114.mg	0/22	77.9mg	5/20	234.mg	17/34				
e	TR416a	26.0mg	130.mg	0/22	77.9mg	4/20	234.mg	17/34				S
f	TR416a	73.8mg	544.mg	0/22	77.9mg	1/20	234.mg	9/34				
g	TR416a	208.mg	3.24gm	0/22	77.9mg	0/20	234.mg	5/34			ubl:sqc,sqp.	
h	TR416a	236.mg	9.40gm	0/22	77.9mg	0/20	234.mg	4/34				
i	TR416a	99.4mg	n.s.s.	0/22	77.9mg	1/20	234.mg	1/34				
j	TR416a	287.mg	n.s.s.	0/22	77.9mg	0/20	234.mg	2/34			kid:tcc,tpp.	
k	TR416a	317.mg	n.s.s.	0/22	77.9mg	0/20	234.mg	2/34				
l	TR416a	317.mg	n.s.s.	0/22	77.9mg	0/20	234.mg	2/34				S
m	TR416a	91.3mg	n.s.s.	0/22	77.9mg	1/20	234.mg	0/34				
n	TR416a	309.mg	n.s.s.	0/22	77.9mg	0/20	234.mg	1/34				
o	TR416a	439.mg	n.s.s.	0/22	77.9mg	0/20	234.mg	1/34				
p	TR416a	578.mg	n.s.s.	0/22	77.9mg	0/20	234.mg	1/34				
q	TR416a	5.84mg	22.2mg	19/22	77.9mg	20/20	234.mg	32/34				
r	TR416a	n.s.s.	n.s.s.	0/22	77.9mg	0/20	234.mg	0/34			liv:hpa,hpb,hpc.	
3461	TR416	11.8mg	63.4mg	48/50	8.75mg	45/50	26.2mg	45/50	79.0mg	45/51	tes:iab,ica.	S
a	TR416	15.1mg	62.1mg	26/50	8.75mg	25/50	26.2mg	42/50	79.0mg	34/51		
b	TR416	39.4mg	767.mg	7/50	8.75mg	7/50	26.2mg	8/50	79.0mg	10/51	amd:pbb,pob.	S
c	TR416	93.2mg	3.11gm	0/50	8.75mg	1/50	26.2mg	1/50	79.0mg	3/51	kid:rua,ruc.	
d	TR416	102.mg	5.23gm	0/50	8.75mg	0/50	26.2mg	2/50	79.0mg	2/51	for:sqc,sqp.	
e	TR416	40.9mg	n.s.s.	12/50	8.75mg	10/50	26.2mg	10/50	79.0mg	10/51	amd:pbb,phm,pob.	S
f	TR416	68.6mg	n.s.s.	7/50	8.75mg	4/50	26.2mg	3/50	79.0mg	8/51	pre:car,cnb.	S
g	TR416	134.mg	n.s.s.	0/50	8.75mg	0/50	26.2mg	1/50	79.0mg	1/51		
h	TR416	149.mg	n.s.s.	0/50	8.75mg	0/50	26.2mg	0/50	79.0mg	2/51		
i	TR416	167.mg	n.s.s.	0/50	8.75mg	0/50	26.2mg	1/50	79.0mg	1/51		
j	TR416	127.mg	n.s.s.	0/50	8.75mg	1/50	26.2mg	1/50	79.0mg	1/51		
k	TR416	12.1mg	59.4mg	48/50	8.75mg	44/50	26.2mg	48/50	79.0mg	47/51		
l	TR416	78.5mg	n.s.s.	0/50	8.75mg	4/50	26.2mg	1/50	79.0mg	2/51	liv:hpa,hpb,hpc.	
3462	TR416a	2.84mg	14.0mg	0/21	62.3mg	26/27	187.mg	34/34			kid:tcc,tpp; lgi:adp,car; ubl:sar,sqc,sqp,tcc,tpp.	C
a	TR416a	2.84mg	14.1mg	0/21	62.3mg	27/27	187.mg	33/34			ubl:tcc,tpp.	
b	TR416a	2.93mg	16.4mg	0/21	62.3mg	23/27	187.mg	33/34				
c	TR416a	2.96mg	17.9mg	0/21	62.3mg	21/27	187.mg	25/34			lgi:adp,car.	
d	TR416a	2.96mg	18.0mg	0/21	62.3mg	21/27	187.mg	24/34				
e	TR416a	7.65mg	29.1mg	0/21	62.3mg	18/27	187.mg	24/34				S
f	TR416a	15.7mg	212.mg	0/21	62.3mg	6/27	187.mg	1/34				S
g	TR416a	17.0mg	965.mg	0/21	62.3mg	3/27	187.mg	1/34				
h	TR416a	16.8mg	468.mg	0/21	62.3mg	1/27	187.mg	9/34			kid:tcc,tpp.	
i	TR416a	17.0mg	806.mg	0/21	62.3mg	1/27	187.mg	6/34				
j	TR416a	117.mg	801.mg	0/21	62.3mg	0/27	187.mg	8/34			ubl:sqc,sqp.	
k	TR416a	117.mg	994.mg	0/21	62.3mg	1/27	187.mg	7/34				
l	TR416a	147.mg	2.04gm	0/21	62.3mg	0/27	187.mg	5/34				
m	TR416a	155.mg	3.27gm	0/21	62.3mg	0/27	187.mg	4/34				
n	TR416a	194.mg	4.36gm	0/21	62.3mg	0/27	187.mg	4/34				
o	TR416a	194.mg	4.36gm	0/21	62.3mg	0/27	187.mg	4/34				S
p	TR416a	3.48mg	n.s.s.	0/21	62.3mg	1/27	187.mg	0/34				S
q	TR416a	291.mg	n.s.s.	0/21	62.3mg	0/27	187.mg	3/34				
r	TR416a	1.92mg	8.76mg	19/21	62.3mg	27/27	187.mg	34/34				
s	TR416a	12.8mg	n.s.s.	1/21	62.3mg	0/27	187.mg	0/34			liv:hpa,hpb,hpc.	

4-NITROANTHRANILIC ACID   619-17-0

	RefNum	LoConf	UpConf	Cntrl	1Dose	1Inc	2Dose	2Inc			Citation or Pathology	Brkly Code
3463	TR109	1.35gm	n.s.s.	39/100	502.mg	18/50	1.09gm	18/50				
a	TR109	6.86gm	n.s.s.	8/100	502.mg	1/50	1.09gm	1/50			liv:hpa,hpc,nnd.	
b	TR109	824.mg	n.s.s.	2/100	502.mg	5/50	(1.09gm	1/50)			lun:a/a,a/c.	
3464	TR109	3.81gm	n.s.s.	0/100	463.mg	0/50	1.01gm	0/50			---:hem,hes.	S
a	TR109	381.mg	n.s.s.	40/100	463.mg	30/50	(1.01gm	15/50)				
b	TR109	2.25gm	n.s.s.	22/100	463.mg	16/50	1.01gm	9/50			liv:hpa,hpc,nnd.	
c	TR109	4.11gm	n.s.s.	18/100	463.mg	10/50	1.01gm	4/50			lun:a/a,a/c.	
3465	TR109	411.mg	n.s.s.	57/75	170.mg	35/50	557.mg	22/50				

```
        Spe Strain Site   Xpo+Xpt                                             TD50    2Tailpvl
            Sex  Route Hist    Notes                                                  DR    AuOp
 a   R f f34  eat liv MXB 18m24 v                                             11.8gm  * P<.6
 3466 R m f34 eat TBA MXB 18m24 v                                :>           no dre    P=1.   -
 a   R m f34  eat liv MXB 18m24 v                                             3.71gm  * P<.09

6-NITROBENZIMIDAZOLE                100ng..:..1ug...:..10.....:...100...:..1mg...:..10.....:..100...:..1g.....:..10
 3467 M f b6c eat liv MXA 78w95 v                              : + :          354.mg  / P<.0005c
 a   M f b6c  eat liv hpc 78w95 v                                             762.mg  * P<.002 c
 b   M f b6c  eat pit MXA 78w95 v                                             335.mg  \ P<.03
 c   M f b6c  eat TBA MXB 78w95 v                                             429.mg  * P<.05
 d   M f b6c  eat liv MXB 78w95 v                                             354.mg  / P<.0005
 e   M f b6c  eat lun MXB 78w95 v                                             no dre    P=1.
 3468 M m b6c eat liv MXA 78w95 v                              :  ±           392.mg  * P<.02  c
 a   M m b6c  eat liv hpc 78w95 v                                             423.mg  * P<.02
 b   M m b6c  eat TBA MXB 78w95 v                                             1.98gm  * P<.8
 c   M m b6c  eat liv MXB 78w95 v                                             392.mg  * P<.02
 d   M m b6c  eat lun MXB 78w95 v                                             no dre    P=1.
 3469 R f f34 eat TBA MXB 18m24 v                             :>              no dre    P=1.   -
 a   R f f34  eat liv MXB 18m24 v                                             no dre    P=1.
 3470 R m f34 eat TBA MXB 18m24 v                                :>           no dre    P=1.   -
 a   R m f34  eat liv MXB 18m24 v                                             8.83gm  * P<.7

p-NITROBENZOIC ACID                 100ng..:..1ug...:..10.....:...100...:..1mg...:..10.....:..100...:..1g.....:..10
 3471 M f b6c eat TBA MXB 24m24                                  :>           3.16gm  * P<.7   -
 a   M f b6c  eat liv MXB 24m24                                               no dre    P=1.
 b   M f b6c  eat lun MXB 24m24                                               2.33gm  * P<.1
 3472 M m b6c eat TBA MXB 24m24                                    :>         no dre    P=1.   -
 a   M m b6c  eat liv MXB 24m24                                               no dre    P=1.
 b   M m b6c  eat lun MXB 24m24                                               4.35gm  * P<.5
 3473 R f f34 eat cli MXA 24m24                              :  +       :     287.mg  * P<.007 p
 a   R f f34  eat ute MXA 24m24                                               197.mg  Z P<.02
 b   R f f34  eat ute esp 24m24                                               220.mg  Z P<.03
 c   R f f34  eat cli MXA 24m24                                               439.mg  * P<.04  p
 d   R f f34  eat cli MXA 24m24                                               1.08gm  * P<.08  p
 e   R f f34  eat TBA MXB 24m24                                               373.mg  * P<.4
 f   R f f34  eat liv MXB 24m24                                               no dre    P=1.
 3474 R m f34 eat pre MXA 24m24                               :  ±            #926.mg  * P<.04  -
 a   R m f34  eat TBA MXB 24m24                                               no dre    P=1.
 b   R m f34  eat liv MXB 24m24                                               no dre    P=1.

NITROFEN                            100ng..:..1ug...:..10.....:...100...:..1mg...:..10.....:..100...:..1g...:..10
 3475 M f b6c eat liv hpc 78w92 v                              : + :          64.2mg  \ P<.0005c
 a   M f b6c  eat TBA MXB 78w92 v                                             74.6mg  \ P<.0005
 b   M f b6c  eat liv MXB 78w92 v                                             64.2mg  \ P<.0005
 c   M f b6c  eat lun MXB 78w92 v                                             no dre    P=1.
 3476 M f b6c eat liv hpc 78w90 v       pool                   : + :          67.7mg  \ P<.0005c
 a   M f b6c  eat --- hes 78w90 v                                             2.84gm  * P<.02  a
 3477 M f b6c eat liv MXA 78w91                                  : + :        406.mg  * P<.0005c
 a   M f b6c  eat liv hpc 78w91                                               1.20gm  * P<.008 c
 b   M f b6c  eat TBA MXB 78w91                                               606.mg  * P<.05
 c   M f b6c  eat liv MXB 78w91                                               406.mg  * P<.0005
 d   M f b6c  eat lun MXB 78w91                                               4.73gm  * P<.3
 3478 M m b6c eat liv hpc 78w92 v                             :  +        :   133.mg  * P<.007 c
 a   M m b6c  eat TBA MXB 78w92 v                                             167.mg  * P<.04
 b   M m b6c  eat liv MXB 78w92 v                                             133.mg  * P<.007
 c   M m b6c  eat lun MXB 78w92 v                                             no dre    P=1.
 3479 M m b6c eat liv hpc 78w90 v       pool                   : + :          85.3mg  * P<.0005c
 a   M m b6c  eat --- hes 78w90 v                                             1.49gm  * P<.005 c
 b   M m b6c  eat sub MXA 78w90 v                                             481.mg  \ P<.03
 3480 M m b6c eat liv MXA 78w91                                : + :          204.mg  * P<.0005c
 a   M m b6c  eat liv hpc 78w91                                               540.mg  * P<.0005c
 b   M m b6c  eat TBA MXB 78w91                                               310.mg  * P<.003
 c   M m b6c  eat liv MXB 78w91                                               204.mg  * P<.0005
 d   M m b6c  eat lun MXB 78w91                                               no dre    P=1.
 3481 R f osm eat pan can 18m26                              :  ±             #459.mg  * P<.02  -
 a   R f osm  eat ova gct 18m26                                               827.mg  * P<.04
 b   R f osm  eat --- lym 18m26                                               934.mg  * P<.05
 c   R f osm  eat TBA MXB 18m26                                               100.mg  / P<.07
 d   R f osm  eat liv MXB 18m26                                               no dre    P=1.
 3482 R f osm eat pan can 18m24         pool                  :  +  :         420.mg  * P<.0005c
 a   R f osm  eat ova MXA 18m24                                               820.mg  * P<.03
 b   R f osm  eat ova gct 18m24                                               1.02gm  * P<.04
 c   R f osm  eat --- lym 18m24                                               1.06gm  * P<.04
 3483 R f f34 eat TBA MXB 18m24                                   :>          no dre    P=1.   -
 a   R f f34  eat liv MXB 18m24                                               no dre    P=1.
 3484 R m osm eat TBA MXB 18m26 sv                         :>                 202.mg  * P<.5   i
 a   R m osm  eat liv MXB 18m26 sv                                            no dre    P=1.
 3485 R m f34 eat TBA MXB 18m24                                    :>         no dre    P=1.   -
 a   R m f34  eat liv MXB 18m24                                               no dre    P=1.

1-[(5-NITROFURFURYLIDENE)AMINO]HYDANTOIN.:...1_ug...:..10.....:..100.....:..1mg....:..10.....:..100...:..1g.....:..10
 3486 M f b6c eat ova MXB 24m24 s                            :  +       :     866.mg  * P<.003
```

	RefNum	LoConf	UpConf	Cntrl	1Dose	1Inc	2Dose	2Inc			Citation or Pathology
											Brkly Code
a	TR109	1.31gm	n.s.s.	3/75	170.mg	0/50	557.mg	2/50			liv:hpa,hpc,nnd.
3466	TR109	353.mg	n.s.s.	40/75	136.mg	24/50	446.mg	15/50			
a	TR109	911.mg	n.s.s.	0/75	136.mg	1/50	446.mg	1/50			liv:hpa,hpc,nnd.

6-NITROBENZIMIDAZOLE   94-52-0

	RefNum	LoConf	UpConf	Cntrl	1Dose	1Inc	2Dose	2Inc			Citation or Pathology
3467	TR117	206.mg	769.mg	3/100	127.mg	4/50	253.mg	20/50			liv:hpa,hpc.
a	TR117	359.mg	3.92gm	3/100	127.mg	2/50	253.mg	11/50			
b	TR117	131.mg	n.s.s.	8/100	127.mg	12/50	(253.mg	0/50)			pit:adn,cra. S
c	TR117	176.mg	n.s.s.	29/100	127.mg	22/50	253.mg	25/50			
d	TR117	206.mg	769.mg	3/100	127.mg	4/50	253.mg	20/50			liv:hpa,hpc,nnd.
e	TR117	870.mg	n.s.s.	5/100	127.mg	4/50	253.mg	2/50			lun:a/a,a/c.
3468	TR117	180.mg	n.s.s.	20/100	118.mg	19/50	234.mg	22/50			liv:hpa,hpc.
a	TR117	194.mg	n.s.s.	18/100	118.mg	16/50	234.mg	21/50			S
b	TR117	229.mg	n.s.s.	45/100	118.mg	29/50	234.mg	25/50			
c	TR117	180.mg	n.s.s.	20/100	118.mg	19/50	234.mg	22/50			liv:hpa,hpc,nnd.
d	TR117	901.mg	n.s.s.	15/100	118.mg	8/50	234.mg	4/50			lun:a/a,a/c.
3469	TR117	55.8mg	n.s.s.	70/100	45.0mg	31/50	(182.mg	26/50)			
a	TR117	1.92gm	n.s.s.	2/100	45.0mg	0/50	182.mg	0/50			liv:hpa,hpc,nnd.
3470	TR117	247.mg	n.s.s.	54/102	36.0mg	22/50	146.mg	22/50			
a	TR117	797.mg	n.s.s.	1/102	36.0mg	0/50	146.mg	1/50			liv:hpa,hpc,nnd.

p-NITROBENZOIC ACID   62-23-7

	RefNum	LoConf	UpConf	Cntrl	1Dose	1Inc	2Dose	2Inc	3Dose	3Inc	Citation or Pathology
3471	TR442	445.mg	n.s.s.	36/50	160.mg	38/50	321.mg	35/50	641.mg	32/50	
a	TR442	1.01gm	n.s.s.	15/50	160.mg	16/50	321.mg	15/50	641.mg	11/50	liv:hpa,hpb,hpc.
b	TR442	839.mg	n.s.s.	3/50	160.mg	10/50	321.mg	4/50	641.mg	9/50	lun:a/a,a/c.
3472	TR442	787.mg	n.s.s.	33/50	148.mg	36/50	297.mg	35/50	594.mg	30/50	
a	TR442	1.14gm	n.s.s.	22/50	148.mg	26/50	297.mg	23/50	594.mg	19/50	liv:hpa,hpb,hpc.
b	TR442	887.mg	n.s.s.	7/50	148.mg	14/50	297.mg	10/50	594.mg	13/50	lun:a/a,a/c.
3473	TR442	141.mg	5.60gm	4/50	61.7mg	14/50	123.mg	15/50	247.mg	15/50	cli:ade,anb,car,cnb.
a	TR442	92.4mg	n.s.s.	5/50	61.7mg	12/50	123.mg	13/50	(247.mg	5/50)	ute:esp,ess. S
b	TR442	97.5mg	n.s.s.	5/50	61.7mg	11/50	123.mg	12/50	(247.mg	5/50)	S
c	TR442	182.mg	n.s.s.	4/50	61.7mg	12/50	123.mg	10/50	247.mg	12/50	cli:ade,anb.
d	TR442	411.mg	n.s.s.	1/50	61.7mg	2/50	123.mg	5/50	247.mg	4/50	cli:car,cnb.
e	TR442	95.8mg	n.s.s.	44/50	61.7mg	48/50	123.mg	50/50	247.mg	45/50	
f	TR442	1.93gm	n.s.s.	2/50	61.7mg	0/50	123.mg	0/50	247.mg	0/50	liv:hpa,hpb,hpc.
3474	TR442	386.mg	n.s.s.	1/50	49.5mg	1/50	98.9mg	4/50	198.mg	6/50	pre:car,cnb. S
a	TR442	215.mg	n.s.s.	46/50	49.5mg	48/50	98.9mg	47/50	198.mg	41/50	
b	TR442	509.mg	n.s.s.	4/50	49.5mg	4/50	98.9mg	1/50	198.mg	4/50	liv:hpa,hpb,hpc.

NITROFEN   1836-75-5

	RefNum	LoConf	UpConf	Cntrl	1Dose	1Inc	2Dose	2Inc			Citation or Pathology
3475	TR26	42.8mg	102.mg	0/20	264.mg	42/50	(529.mg	44/50)			
a	TR26	46.4mg	157.mg	3/20	264.mg	43/50	(529.mg	44/50)			
b	TR26	42.8mg	102.mg	0/20	264.mg	42/50	(529.mg	44/50)			liv:hpa,hpc,nnd.
c	TR26	n.s.s.	n.s.s.	0/20	264.mg	1/50	529.mg	0/50			lun:a/a,a/c.
3476	TR26	45.2mg	108.mg	1/80p	264.mg	42/50	(529.mg	44/50)			
a	TR26	993.mg	n.s.s.	1/80p	264.mg	0/50	529.mg	5/50			
3477	TR184	282.mg	942.mg	0/20	334.mg	14/50	668.mg	30/50			liv:hpa,hpc.
a	TR184	692.mg	20.6gm	0/20	334.mg	5/50	668.mg	13/50			
b	TR184	287.mg	n.s.s.	3/20	334.mg	23/50	668.mg	33/50			
c	TR184	282.mg	942.mg	0/20	334.mg	14/50	668.mg	30/50			liv:hpa,hpc,nnd.
d	TR184	1.80gm	n.s.s.	0/20	334.mg	2/50	668.mg	3/50			lun:a/a,a/c.
3478	TR26	73.0mg	1.86gm	4/20	244.mg	38/50	488.mg	46/50			
a	TR26	78.7mg	n.s.s.	6/20	244.mg	40/50	488.mg	46/50			
b	TR26	73.0mg	1.86gm	4/20	244.mg	38/50	488.mg	46/50			liv:hpa,hpc,nnd.
c	TR26	1.40gm	n.s.s.	1/20	244.mg	1/50	488.mg	1/50			lun:a/a,a/c.
3479	TR26	60.9mg	129.mg	5/80p	244.mg	38/50	488.mg	46/50			
a	TR26	566.mg	13.6gm	0/80p	244.mg	1/50	488.mg	4/50			
b	TR26	193.mg	n.s.s.	3/80p	244.mg	10/50	(488.mg	0/50)			sub:fbs,fib. S
3480	TR184	144.mg	401.mg	1/20	308.mg	31/50	617.mg	40/50			liv:hpa,hpc.
a	TR184	355.mg	1.20gm	0/20	308.mg	13/50	617.mg	20/50			
b	TR184	175.mg	1.70gm	7/20	308.mg	31/50	617.mg	43/50			
c	TR184	144.mg	401.mg	1/20	308.mg	31/50	617.mg	40/50			liv:hpa,hpc,nnd.
d	TR184	2.90gm	n.s.s.	3/20	308.mg	0/50	617.mg	3/50			lun:a/a,a/c.
3481	TR26	215.mg	n.s.s.	0/20	46.1mg	2/50	92.0mg	7/50			S
a	TR26	286.mg	n.s.s.	0/20	46.1mg	0/50	92.0mg	4/50			S
b	TR26	320.mg	n.s.s.	0/20	46.1mg	0/50	92.0mg	4/50			S
c	TR26	42.4mg	n.s.s.	14/20	46.1mg	31/50	92.0mg	42/50			
d	TR26	n.s.s.	n.s.s.	0/20	46.1mg	0/50	92.0mg	0/50			liv:hpa,hpc,nnd.
3482	TR26	197.mg	1.17gm	0/110p	46.1mg	2/50	92.0mg	7/50			
a	TR26	281.mg	n.s.s.	1/110p	46.1mg	1/50	92.0mg	4/50			ova:gcc,gct. S
b	TR26	321.mg	n.s.s.	1/110p	46.1mg	0/50	92.0mg	4/50			S
c	TR26	333.mg	n.s.s.	1/110p	46.1mg	0/50	92.0mg	4/50			S
3483	TR184	206.mg	n.s.s.	15/20	113.mg	21/50	(225.mg	15/50)			
a	TR184	n.s.s.	n.s.s.	0/20	113.mg	0/50	225.mg	0/50			liv:hpa,hpc,nnd.
3484	TR26	45.9mg	n.s.s.	10/20	65.3mg	30/50	136.mg	4/50			
a	TR26	n.s.s.	n.s.s.	0/20	65.3mg	0/50	136.mg	0/50			liv:hpa,hpc,nnd.
3485	TR184	441.mg	n.s.s.	10/20	90.4mg	20/50	180.mg	14/50			
a	TR184	n.s.s.	n.s.s.	0/20	90.4mg	0/50	180.mg	0/50			liv:hpa,hpc,nnd.

1-[(5-NITROFURFURYLIDENE)AMINO]HYDANTOIN   (macrodantin, nitrofurantoin) 67-20-9

	RefNum	LoConf	UpConf	Cntrl	1Dose	1Inc	2Dose	2Inc			Citation or Pathology
3486	TR341	466.mg	3.57gm	0/50	166.mg	3/50	319.mg	11/50			ova:gcb,gcm,mtb,tua. C

```
    Spe Strain Site  Xpo+Xpt                                              TD50    2Tailpvl
       Sex  Route Hist  Notes                                               DR      AuOp
    a   M f b6c eat ova MXA 24m24 s                                              1.40gm /  P<.004 c
    b   M f b6c eat ova tua 24m24 s                                             2.53gm *  P<.03  c
    c   M f b6c eat ova mtb 24m24 s                                             3.38gm *  P<.05  c
    d   M f b6c eat ova MXA 24m24 s                                      +hist  2.56gm *  P<.4   c
    e   M f b6c eat ova gcb 24m24 s                                      +hist  8.47gm *  P<.9   c
    f   M f b6c eat TBA MXB 24m24 s                                             no dre    P=1.
    g   M f b6c eat liv MXB 24m24 s                                             2.33gm *  P<.2
    h   M f b6c eat lun MXB 24m24 s                                             no dre    P=1.
 3487  M f b6c eat cvu lei 64w64 er                            .>                4.00gm *  P<.3   -
    a   M f b6c eat ova gct 64w64 er                                            4.02gm *  P<.5
 3488  M m b6c eat TBA MXB 24m24                                   :>            no dre    P=1.   -
    a   M m b6c eat liv MXB 24m24                                                no dre    P=1.
    b   M m b6c eat lun MXB 24m24                                                no dre    P=1.
 3489  M f bd1 eat lun adc 24m24                              .>                 no dre    P=1.   -
    a   M f bd1 eat liv hem 24m24                                                no dre    P=1.
    b   M f bd1 eat tba mix 24m24                                                no dre    P=1.
 3490  M m bd1 eat lun ade 24m24                              .>                53.9gm *  P<.9   -
    a   M m bd1 eat liv hem 24m24                                                no dre    P=1.
    b   M m bd1 eat liv ade 24m24                                                no dre    P=1.
    c   M m bd1 eat liv mix 24m24                                                no dre    P=1.
    d   M m bd1 eat tba mix 24m24                                                no dre    P=1.
 3491  M f cd1 eat liv hem 22m23 e                       .    ±                 4.46gm *  P<.07  -
    a   M f cd1 eat liv hes 22m23 e                                             no dre    P=1.
    b   M f cd1 eat liv hpa 22m23 e                                             no dre    P=1.
    c   M f cd1 eat lun ala 22m23 e                                             no dre    P=1.
    d   M f cd1 eat lun alc 22m23 e                                             no dre    P=1.
 3492  M m cd1 eat --- myl 22m23 e                            .    ±            588.mg *  P<.02  -
    a   M m cd1 eat lun ala 22m23 e                                             833.mg *  P<.05  -
    b   M m cd1 eat liv hpa 22m23 e                                             1.20gm *  P<.2   -
    c   M m cd1 eat lun alc 22m23 e                                             17.0gm *  P<.9   -
    d   M m cd1 eat liv hes 22m23 e                                             no dre    P=1.   -
    e   M m cd1 eat liv hpc 22m23 e                                             no dre    P=1.   -
 3493  R f f34 eat cli ade 24m24                              :    ±           #93.7mg \  P<.03  -
    a   R f f34 eat TBA MXB 24m24                                               no dre    P=1.
    b   R f f34 eat liv MXB 24m24                                               no dre    P=1.
 3494  R m f34 eat sub fib 24m24                              :    ±            303.mg *  P<.02
    a   R m f34 eat kid MXA 24m24                                        +hist  698.mg *  P<.05  p
    b   R m f34 eat TBA MXB 24m24                                               no dre    P=1.
    c   R m f34 eat liv MXB 24m24                                               332.mg \  P<.2
 3495  R m f34 eat kid MXA 24m24     with step                .   +    .        163.mg *  P<.0005p
 3496  R m f34 eat ubl tum 24m24 r                                 .>           no dre    P=1.   -
 3497  R f cdr eat mgl mix 24m24 e                                 .>           no dre    P=1.   -
    a   R f cdr eat tba tum 24m24 e                                             no dre    P=1.
    b   R f cdr eat tba mal 24m24 e                                             no dre    P=1.
 3498  R m cdr eat mgl fba 24m24 e                                  .>          878.mg *  P<.2   -
    a   R m cdr eat tba mal 24m24 e                                             1.20gm *  P<.8   -
    b   R m cdr eat tba tum 24m24 e                                             no dre    P=1.
 3499  R f hza eat mgl fba 36w54 es                       .>                    120.mg    P<.3   -
    a   R f hza eat liv tum 36w54 es                                            no dre    P=1.
 3500  R f hza eat mam tum 44w60 es                        .>                   214.mg    P<.6   -
    a   R f hza eat liv tum 44w60 es                                            no dre    P=1.
 3501  R f sda eat liv tum 75w80 ev                      .>                     no dre    P=1.
    a   R f sda eat tba mix 75w80 ev                                            185.mg    P<.7   -
 3502  R f sda eat mgl fba 24m24                          .   +    .            54.3mg    P<.006
    a   R f sda eat tba mix 24m24                                               64.2mg    P<.2
```

```
1-[(5-NITROFURFURYLIDENE)AMINO]-2-IMIDAZOLIDINONE..:..1_0.......:..100....:...1mg....:...10.....:...100....:...1g....:...10
 3503  R f sda eat mgl mix 46w66 e                       <+                     noTD50    P<.0005
    a   R f sda eat mgl adc 46w66 e                                             5.26mg    P<.0005+
    b   R f sda eat --- lbl 46w66 e                                      +hist  82.0mg    P<.02  +
    c   R f sda eat liv tum 46w66 e                                             no dre    P=1.
    d   R f sda eat tba mix 46w66 e                                             noTD50    P<.0005
```

```
NITROGEN MUSTARD                100ng..:..1ug....:..10.......:...100....:...1mg....:...10.....:...100....:...1g.....:..10
 3504  R m b46 ivj tba mix 12m24 es             .   +    .                      11.4ug    P<.0005+
    a   R m b46 ivj tba mal 12m24 es                                            22.8ug    P<.02  +
    b   R m b46 ivj tba ben 12m24 es                                            34.2ug    P<.05
```

```
NITROGEN MUSTARD N-OXIDE        100ng..:..1ug....:..10.......:...100....:...1mg....:...10.....:...100....:...1g.....:..10
 3505  R m b46 ivj --- mix 12m24 es               .   +    .                    1.40mg    P<.002
    a   R m b46 ivj tba mix 12m24 es                                            .764mg    P<.008 +
    b   R m b46 ivj tba mal 12m24 es                                            .806mg    P<.003 +
    c   R m b46 ivj tba ben 12m24 es                                            no dre    P=1.
```

```
1-NITRONAPHTHALENE              100ng..:..1ug....:..10.......:...100....:...1mg....:...10.....:...100....:...1g.....:..10
 3506  M f b6c eat TBA MXB 78w96                                  :>            no dre    P=1.   -
    a   M f b6c eat liv MXB 78w96                                               no dre    P=1.
    b   M f b6c eat lun MXB 78w96                                               682.mg *  P<.2
 3507  M m b6c eat TBA MXB 78w96                                :>              no dre    P=1.   -
    a   M m b6c eat liv MXB 78w96                                               no dre    P=1.
    b   M m b6c eat lun MXB 78w96                                               no dre    P=1.
```

	RefNum	LoConf	UpConf	Cntrl	1Dose	1Inc	2Dose	2Inc	Citation or Pathology / Brkly Code
a	TR341	660.mg	3.82gm	0/50	166.mg	0/50	319.mg	9/50	ova:mtb,tua.
b	TR341	960.mg	n.s.s.	0/50	166.mg	0/50	319.mg	5/50	
c	TR341	1.16gm	n.s.s.	0/50	166.mg	0/50	319.mg	4/50	
d	TR341	972.mg	n.s.s.	0/50	166.mg	3/50	319.mg	2/50	ova:gcb,gcm.
e	TR341	1.12gm	n.s.s.	0/50	166.mg	3/50	319.mg	1/50	
f	TR341	368.mg	n.s.s.	30/50	166.mg	35/50	319.mg	41/50	
g	TR341	764.mg	n.s.s.	2/50	166.mg	2/50	319.mg	8/50	liv:hpa,hpc,nnd.
h	TR341	822.mg	n.s.s.	3/50	166.mg	2/50	(319.mg	0/50)	lun:a/a,a/c.
3487	1923	651.mg	n.s.s.	0/20	350.mg	0/19	500.mg	1/18	Stitzel;txpy,17,774-781;1989
a	1923	654.mg	n.s.s.	0/20	350.mg	1/19	500.mg	0/18	
3488	TR341	357.mg	n.s.s.	31/50	153.mg	27/50	294.mg	33/50	
a	TR341	567.mg	n.s.s.	10/50	153.mg	12/50	294.mg	11/50	liv:hpa,hpc,nnd.
b	TR341	787.mg	n.s.s.	6/50	153.mg	4/50	294.mg	7/50	lun:a/a,a/c.
3489	1747	3.26gm	n.s.s.	0/54	97.5mg	1/54	390.mg	0/54	Ito;hijm,32,99-102;1983
a	1747	3.65gm	n.s.s.	1/54	97.5mg	2/54	390.mg	0/54	
b	1747	956.mg	n.s.s.	16/54	97.5mg	9/54	390.mg	14/54	
3490	1747	2.85gm	n.s.s.	1/53	90.0mg	0/52	360.mg	1/52	
a	1747	2.45gm	n.s.s.	3/53	90.0mg	0/52	360.mg	2/52	
b	1747	771.mg	n.s.s.	6/53	90.0mg	1/52	(360.mg	0/52)	
c	1747	3.17gm	n.s.s.	9/53	90.0mg	1/52	360.mg	2/52	
d	1747	2.54gm	n.s.s.	12/53	90.0mg	1/52	360.mg	4/52	
3491	1972	1.10gm	n.s.s.	0/50	40.2mg	0/49	78.4mg	0/50	167.mg	2/50	Butler;fctx,28,49-54;1990

	RefNum	LoConf	UpConf	Cntrl	1Dose	1Inc	2Dose	2Inc	3Dose	3Inc	Citation or Pathology / Brkly Code
3491	1972	1.10gm	n.s.s.	0/50	40.2mg	0/49	78.4mg	0/50	167.mg	2/50	Butler;fctx,28,49-54;1990
a	1972	1.16gm	n.s.s.	2/50	40.2mg	1/49	78.4mg	2/50	167.mg	1/50	
b	1972	1.66gm	n.s.s.	0/50	40.2mg	1/49	78.4mg	0/50	167.mg	0/50	
c	1972	869.mg	n.s.s.	1/50	40.2mg	5/50	78.4mg	0/50	167.mg	2/50	
d	1972	1.09gm	n.s.s.	1/50	40.2mg	2/50	78.4mg	1/50	167.mg	1/50	
3492	1972	263.mg	n.s.s.	2/50	37.1mg	6/50	72.4mg	4/49	152.mg	10/50	
a	1972	350.mg	n.s.s.	1/50	37.1mg	2/50	72.4mg	6/49	152.mg	5/50	
b	1972	412.mg	n.s.s.	3/50	37.1mg	2/50	72.4mg	3/49	152.mg	7/50	
c	1972	945.mg	n.s.s.	1/50	37.1mg	1/50	72.4mg	0/49	152.mg	3/50	
d	1972	955.mg	n.s.s.	1/50	37.1mg	3/50	72.4mg	3/49	152.mg	0/50	
e	1972	446.mg	n.s.s.	4/50	37.1mg	7/50	72.4mg	9/49	152.mg	4/50	

	RefNum	LoConf	UpConf	Cntrl	1Dose	1Inc	2Dose	2Inc	Citation or Pathology / Brkly Code
3493	TR341	36.2mg	n.s.s.	1/50	29.4mg	7/50	(63.8mg	4/50)	S
a	TR341	45.2mg	n.s.s.	49/50	29.4mg	48/50	63.8mg	46/50	
b	TR341	n.s.s.	n.s.s.	0/50	29.4mg	1/50	63.8mg	0/50	liv:hpa,hpc,nnd.
3494	TR341	142.mg	n.s.s.	0/50	51.3mg	5/50	98.1mg	4/50	S
a	TR341	240.mg	n.s.s.	0/50	51.3mg	1/50	98.1mg	3/50	kid:rua,ruc.
b	TR341	62.1mg	n.s.s.	48/50	51.3mg	47/50	98.1mg	45/50	
c	TR341	90.0mg	n.s.s.	1/50	51.3mg	4/50	(98.1mg	0/50)	liv:hpa,hpc,nnd.
3495	TR341	97.1mg	401.mg	3/50	51.3mg	11/50	98.1mg	20/50	kid:rua,ruc.
3496	1986	370.mg	n.s.s.	0/30	74.8mg	0/24			Hasegawa;txcy,62,333-347;1990
3497	1978	137.mg	n.s.s.	17/50	16.8mg	17/50	31.0mg	17/50	55.8mg	14/50	Butler;fctx,28,269-277;1990

Rows 3497-3498 have an additional 3Dose/3Inc column:

	RefNum	LoConf	UpConf	Cntrl	1Dose	1Inc	2Dose	2Inc	3Dose	3Inc	Citation or Pathology / Brkly Code
3497	1978	137.mg	n.s.s.	17/50	16.8mg	17/50	31.0mg	17/50	55.8mg	14/50	Butler;fctx,28,269-277;1990
a	1978	35.4mg	n.s.s.	48/50	16.8mg	45/50	31.0mg	47/50	55.8mg	44/50	
b	1978	205.mg	n.s.s.	5/50	16.8mg	7/50	31.0mg	8/50	55.8mg	4/50	
3498	1978	266.mg	n.s.s.	0/50	12.4mg	0/50	22.3mg	2/50	43.1mg	1/50	
a	1978	132.mg	n.s.s.	7/50	12.4mg	8/50	22.3mg	3/50	43.1mg	9/50	
b	1978	97.2mg	n.s.s.	39/50	12.4mg	38/50	22.3mg	32/50	43.1mg	28/50	

	RefNum	LoConf	UpConf	Cntrl	1Dose	1Inc	2Dose	2Inc	Citation or Pathology / Brkly Code
3499	1063m	19.4mg	n.s.s.	0/5	100.mg	1/7			Morris;canr,29,2145-2156;1969
a	1063m	38.9mg	n.s.s.	0/5	100.mg	0/7			
3500	1063n	35.8mg	n.s.s.	3/16	110.mg	5/18			
a	1063n	136.mg	n.s.s.	0/16	110.mg	0/18			
3501	200a	244.mg	n.s.s.	0/30	55.6mg	0/36			Cohen;jnci,51,403-417;1973
a	200a	28.6mg	n.s.s.	14/30	55.6mg	19/36			
3502	1622	21.1mg	675.mg	2/11	94.0mg	9/12			Wang;clet,21,303-308;1984
a	1622	17.9mg	n.s.s.	6/11	94.0mg	10/12			

1-[(5-NITROFURFURYLIDENE)AMINO]-2-IMIDAZOLIDINONE   555-84-0

	RefNum	LoConf	UpConf	Cntrl	1Dose	1Inc	2Dose	2Inc	Citation or Pathology / Brkly Code
3503	200a	n.s.s.	6.44mg	1/25	52.3mg	31/31			Cohen;jnci,51,403-417;1973
a	200a	2.72mg	10.0mg	0/25	52.3mg	29/31			
b	200a	31.1mg	n.s.s.	0/25	52.3mg	5/31			
c	200a	134.mg	n.s.s.	0/25	52.3mg	0/31			
d	200a	n.s.s.	6.44mg	1/25	52.3mg	31/31			

NITROGEN MUSTARD   (2,2'-dichloro-N-methyldiethylamine)   51-75-2

	RefNum	LoConf	UpConf	Cntrl	1Dose	1Inc	2Dose	2Inc	Citation or Pathology / Brkly Code
3504	1017	5.13ug	51.5ug	7/65	7.86ug	12/27			Schmahl;arzn,20,1461-1467;1970
a	1017	8.41ug	n.s.s.	4/65	7.86ug	7/27			
b	1017	10.8ug	n.s.s.	3/65	7.86ug	5/27			

NITROGEN MUSTARD N-OXIDE   (mitomen) 126-85-2

	RefNum	LoConf	UpConf	Cntrl	1Dose	1Inc	2Dose	2Inc	Citation or Pathology / Brkly Code
3505	1017	.571mg	5.62mg	0/65	.300mg	6/44			Schmahl;arzn,20,1461-1467;1970
a	1017	.335mg	15.9mg	7/65	.300mg	14/44			
b	1017	.364mg	5.46mg	4/65	.300mg	12/44			
c	1017	1.36mg	n.s.s.	3/65	.300mg	2/44			

1-NITRONAPHTHALENE   86-57-7

	RefNum	LoConf	UpConf	Cntrl	1Dose	1Inc	2Dose	2Inc	Citation or Pathology / Brkly Code
3506	TR64	230.mg	n.s.s.	42/100	63.7mg	12/50	127.mg	20/50	
a	TR64	1.01gm	n.s.s.	5/100	63.7mg	0/50	127.mg	1/50	liv:hpa,hpc,nnd.
b	TR64	231.mg	n.s.s.	5/100	63.7mg	4/50	127.mg	7/50	lun:a/a,a/c.
3507	TR64	168.mg	n.s.s.	38/100	58.8mg	19/50	118.mg	21/50	
a	TR64	323.mg	n.s.s.	17/100	58.8mg	8/50	118.mg	8/50	liv:hpa,hpc,nnd.
b	TR64	290.mg	n.s.s.	17/100	58.8mg	8/50	118.mg	9/50	lun:a/a,a/c.

```
      Spe Strain Site    Xpo+Xpt                                                                      TD50    2Tailpvl
          Sex  Route Hist      Notes                                                                    DR      AuOp
3508 R f f34 eat TBA MXB 18m25 v                                          :>                 no dre   P=1.    -
a    R f f34 eat liv MXB 18m25 v                                            :>               no dre   P=1.    -
3509 R m f34 eat TBA MXB 18m25 v                                            :>               no dre   P=1.    -
a    R m f34 eat liv MXB 18m25 v                                                           509.mg * P<.2

1-NITROPROPANE                      100ng..:..1ug....:..10.....:..100....:..1mg....:..10.....:..100....:..1g.....:..10
3510 R f leb inh liv hpc 52w52 ekr                                       .>                  no dre   P=1.    -
3511 R f leb inh liv hpc 78w78 ekr                                         .>                no dre   P=1.    -
3512 R f leb inh liv hpc 93w93 ekr                                            .>             no dre   P=1.    -
3513 R f leb inh liv hpc 52w93 ekr                                         .>                no dre   P=1.    -
3514 R m leb inh liv hpc 52w52 ekr                                      .>                   no dre   P=1.    -
3515 R m leb inh liv hpc 78w78 ekr                                         .>                no dre   P=1.    -
3516 R m leb inh liv hpc 93w93 ekr                                            .>             no dre   P=1.    -
3517 R m leb inh liv hpc 52w93 ekr                                         .>                no dre   P=1.    -
3518 R m sda gav liv hpa 26w77 ev                 .>                                        .872mg   P<1.    -

2-NITROPROPANE                      100ng..:..1ug....:..10.....:..100....:..1mg....:..10.....:..100....:..1g.....:..10
3519 R f sda inh liv tum 95w95 e                                           .>              293.mg   P<.2    -
a    R f sda inh tba ben 95w95 e                                                             no dre   P=1.    -
b    R f sda inh tba mal 95w95 e                                                             no dre   P=1.    -
3520 R f sda inh liv tum 52w52 ek                             .>                             no dre   P=1.    -
a    R f sda inh tba ben 52w52 ek                                                            no dre   P=1.    -
b    R f sda inh tba mal 52w52 ek                                                            no dre   P=1.    -
3521 R f sda inh liv tum 52w95 ek                             .>                             no dre   P=1.    -
a    R f sda inh tba ben 52w95 ek                                                            no dre   P=1.    -
b    R f sda inh tba mal 52w95 ek                                                            no dre   P=1.    -
3522 R m sda inh liv agm 95w95 e                                      .>                     no dre   P=1.    -
a    R m sda inh tba ben 95w95 e                                                             no dre   P=1.    -
b    R m sda inh tba mal 95w95 e                                                             no dre   P=1.    -
3523 R m sda inh liv tum 52w52 ek                         .>                                 no dre   P=1.    -
a    R m sda inh tba ben 52w52 ek                                                          9.06mg   P<.3    -
b    R m sda inh tba mal 52w52 ek                                                            no dre   P=1.    -
3524 R m sda inh liv agm 52w95 ek                         .>                                 no dre   P=1.    -
a    R m sda inh tba ben 52w95 ek                                                          7.78mg   P<.6    -
b    R m sda inh tba mal 52w95 ek                                                          23.6mg   P<.4    -

3-NITROPROPIONIC ACID               100ng..:..1ug....:..10.....:..100....:..1mg....:..10.....:..100....:..1g.....:..10
3525 M f b6c gav TBA MXB 24m24                                   :>                          no dre   P=1.    -
a    M f b6c gav liv MXB 24m24                                                             337.mg * P<.5
b    M f b6c gav lun MXB 24m24                                                             499.mg * P<.8
3526 M m b6c gav TBA MXB 24m24                                   :>                          no dre   P=1.    -
a    M m b6c gav liv MXB 24m24                                                               no dre   P=1.    -
b    M m b6c gav lun MXB 24m24                                                               no dre   P=1.    -
3527 R f f34 gav TBA MXB 26m26                              :>                             43.8mg * P<1.    -
a    R f f34 gav liv MXB 26m26                                                             29.0mg * P<.3
3528 R m f34 gav liv MXA 26m26                         :   +   :                           4.41mg * P<.004 a
a    R m f34 gav liv nnd 26m26                                                             5.03mg * P<.008 a
b    R m f34 gav pni isa 26m26                                                             4.53mg * P<.06  a
c    R m f34 gav TBA MXB 26m26                                                             7.02mg * P<.7
d    R m f34 gav liv MXB 26m26                                                             4.41mg * P<.004

6-NITROQUINOLINE                    100ng..:..1ug....:..10.....:..100....:..1mg....:..10.....:..100....:..1g.....:..10
3529 R f f34 eat liv hnd 24m24 e                                              .>             no dre   P=1.    -
3530 R m f34 eat liv hnd 24m24 e                                            .>             267.mg   P<.2    -

8-NITROQUINOLINE                    100ng..:..1ug....:..10.....:..100....:..1mg....:..10.....:..100....:..1g.....:..10
3531 R f f34 eat for sqp 24m24 e                                    .  +  .                9.55mg   P<.0005+
a    R f f34 eat for sqc 24m24 e                                                          32.8mg   P<.0005+
b    R f f34 eat liv hnd 24m24 e                                                             no dre   P=1.    -
3532 R m f34 eat for sqp 24m24 e                                  .  +  .                  10.1mg   P<.0005+
a    R m f34 eat for sqc 24m24 e                                                          24.9mg   P<.0005+
b    R m f34 eat liv hnd 24m24 e                                                             no dre   P=1.    -

NITROSO-BAYGON                      100ng..:..1ug....:..10.....:..100....:..1mg....:..10.....:..100....:..1g.....:..10
3533 R m sda gav for tum 31w90 erv                .  +  .                                  .364mg   P<.0005+
a    R m sda gav for car 31w90 erv                                                         .434mg   P<.0005+

N-NITROSO-BIS-(4,4,4-TRIFLUORO-N-BUTYL)AMINE    ...:..10.....:..100....:..1mg....:..10.....:..100....:..1g.....:..10
3534 R f sda gav liv hpc 7m23                                   .  +  .                     .707mg   P<.0005+
a    R f sda gav lun mix 7m23                                                             1.12mg   P<.0005+
b    R f sda gav tba mal 7m23                                                              .291mg   P<.0005
3535 R m sda gav liv hpc 7m23                                 .  +  .                       .793mg   P<.0005+
a    R m sda gav lun mix 7m23                                                              .793mg   P<.0005+
b    R m sda gav tba mal 7m23                                                              .363mg   P<.0005

1-NITROSO-5,6-DIHYDROTHYMINE         100ng..:..1ug....:..10.....:..100....:..1mg....:..10.....:..100....:..1g.....:..10
3536 R f mrw wat liv mix 12m30 e                                    ±                      31.2mg   P<.1
a    R f mrw wat tba mix 12m30 e                                                             no dre   P=1.    -
3537 R m mrw wat liv tum 12m30 e                                        .>                   no dre   P=1.    -
a    R m mrw wat tba mix 12m30 e                                                             no dre   P=1.    -
```

	RefNum	LoConf	UpConf	Cntrl	1Dose	1Inc	2Dose	2Inc	Citation or Pathology
									Brkly Code
3508	TR64	72.0mg	n.s.s.	51/75	21.0mg	32/50	66.0mg	34/50	
a	TR64	346.mg	n.s.s.	4/75	21.0mg	0/50	66.0mg	2/50	
3509	TR64	23.4mg	n.s.s.	36/75	16.8mg	21/50	(52.8mg	19/50)	liv:hpa,hpc,nnd.
a	TR64	175.mg	n.s.s.	0/75	16.8mg	2/50	52.8mg	2/50	
									liv:hpa,hpc,nnd.

1-NITROPROPANE   108-03-2

	RefNum	LoConf	UpConf	Cntrl	1Dose	1Inc	Citation or Pathology
3510	1670m	16.4mg	n.s.s.	0/10	31.8mg	0/10	
3511	1670n	36.9mg	n.s.s.	0/10	31.8mg	0/10	Griffin;eaes,6,268-282;1982
3512	1670o	147.mg	n.s.s.	0/59	31.8mg	0/28	
3513	1670r	23.4mg	n.s.s.	0/59	17.8mg	0/8	
3514	1670m	11.5mg	n.s.s.	0/10	22.3mg	0/10	
3515	1670n	25.8mg	n.s.s.	0/10	22.3mg	0/10	
3516	1670o	99.1mg	n.s.s.	0/60	22.3mg	0/27	
3517	1670r	12.3mg	n.s.s.	0/60	12.4mg	0/6	
3518	1837	16.8ug	n.s.s.	1/29	9.59ug	1/26	Fiala;carc,8,1947-1949;1987/pers.comm.

2-NITROPROPANE   79-46-9

	RefNum	LoConf	UpConf	Cntrl	1Dose	1Inc	Citation or Pathology
3519	1444	47.8mg	n.s.s.	0/85	7.96mg	1/65	
a	1444	24.1mg	n.s.s.	66/85	7.96mg	32/65	Griffin;eaes,5,194-201;1981
b	1444	45.9mg	n.s.s.	18/85	7.96mg	6/65	
3520	1444m	4.10mg	n.s.s.	0/10	7.96mg	0/10	
a	1444m	2.38mg	n.s.s.	1/10	7.96mg	1/10	
b	1444m	4.10mg	n.s.s.	0/10	7.96mg	0/10	
3521	1444n	7.46mg	n.s.s.	0/48	4.34mg	0/10	
a	1444n	1.43mg	n.s.s.	36/48	4.34mg	7/10	
b	1444n	4.83mg	n.s.s.	7/48	4.34mg	1/10	
3522	1444	62.2mg	n.s.s.	1/85	5.57mg	0/65	
a	1444	12.2mg	n.s.s.	42/85	5.57mg	26/65	
b	1444	21.1mg	n.s.s.	11/85	5.57mg	7/65	
3523	1444m	2.87mg	n.s.s.	0/10	5.57mg	0/10	
a	1444m	1.47mg	n.s.s.	0/10	5.57mg	1/10	
b	1444m	2.87mg	n.s.s.	0/10	5.57mg	0/10	
3524	1444n	5.22mg	n.s.s.	1/64	3.04mg	0/10	
a	1444n	1.02mg	n.s.s.	32/64	3.04mg	6/10	
b	1444n	2.81mg	n.s.s.	2/64	3.04mg	1/10	

3-NITROPROPIONIC ACID   504-88-1

	RefNum	LoConf	UpConf	Cntrl	1Dose	1Inc	2Dose	2Inc	Citation or Pathology
3525	TR52	22.1mg	n.s.s.	26/50	8.80mg	32/50	17.7mg	26/50	
a	TR52	65.8mg	n.s.s.	2/50	8.80mg	1/50	17.7mg	4/50	
b	TR52	49.0mg	n.s.s.	2/50	8.80mg	6/50	17.7mg	3/50	liv:hpa,hpc,nnd.
3526	TR52	17.9mg	n.s.s.	39/50	8.80mg	28/50	17.7mg	38/50	lun:a/a,a/c.
a	TR52	39.7mg	n.s.s.	20/50	8.80mg	10/50	17.7mg	16/50	
b	TR52	49.1mg	n.s.s.	14/50	8.80mg	8/50	17.7mg	10/50	liv:hpa,hpc,nnd.
3527	TR52	1.43mg	n.s.s.	40/50	.850mg	30/50	1.70mg	39/50	lun:a/a,a/c.
a	TR52	7.14mg	n.s.s.	0/50	.850mg	1/50	1.70mg	1/50	
3528	TR52	2.08mg	27.1mg	0/50	.600mg	3/50	1.20mg	6/50	liv:hpa,hpc,nnd.
a	TR52	2.28mg	87.3mg	0/50	.600mg	3/50	1.20mg	5/50	liv:hpc,nnd.
b	TR52	1.85mg	n.s.s.	4/50	.600mg	6/50	1.20mg	11/50	
c	TR52	.998mg	n.s.s.	36/50	.600mg	31/50	1.20mg	38/50	
d	TR52	2.08mg	27.1mg	0/50	.600mg	3/50	1.20mg	6/50	liv:hpa,hpc,nnd.

6-NITROQUINOLINE   613-50-3

	RefNum	LoConf	UpConf	Cntrl	1Dose	1Inc	Citation or Pathology
3529	1529	185.mg	n.s.s.	3/44	25.0mg	0/36	
3530	1529	65.7mg	n.s.s.	0/31	20.0mg	2/40	Fukushima;clet,14,115-123;1981

8-NITROQUINOLINE   607-35-2

	RefNum	LoConf	UpConf	Cntrl	1Dose	1Inc	Citation or Pathology
3531	1529	4.34mg	18.6mg	1/44	50.0mg	36/37	
a	1529	19.6mg	59.5mg	0/44	50.0mg	24/37	Fukushima;clet,14,115-123;1981
b	1529	381.mg	n.s.s.	3/44	50.0mg	0/37	
3532	1529	5.21mg	19.5mg	0/31	40.0mg	28/30	
a	1529	14.2mg	48.3mg	0/31	40.0mg	20/30	
b	1529	247.mg	n.s.s.	0/31	40.0mg	0/30	

NITROSO-BAYGON   38777-13-8

	RefNum	LoConf	UpConf	Cntrl	1Dose	1Inc	Citation or Pathology
3533	1409	.173mg	.891mg	0/40	.984mg	12/16	
a	1409	.206mg	1.10mg	0/40	.984mg	11/16	Lijinsky;eaes,2,413-419;1978

N-NITROSO-BIS-(4,4,4-TRIFLUORO-N-BUTYL)AMINE   83335-32-4

	RefNum	LoConf	UpConf	Cntrl	1Dose	1Inc	Citation or Pathology
3534	1489	.382mg	1.47mg	0/24	1.35mg	17/24	
a	1489	.576mg	2.55mg	0/24	1.35mg	13/24	Preussmann;carc,3,1219-1222;1982
b	1489	.120mg	.693mg	4/24	1.35mg	23/24	
3535	1489	.425mg	1.68mg	0/24	1.35mg	16/24	
a	1489	.425mg	1.68mg	0/24	1.35mg	16/24	
b	1489	.176mg	.788mg	2/24	1.35mg	22/24	

1-NITROSO-5,6-DIHYDROTHYMINE   62641-67-2

	RefNum	LoConf	UpConf	Cntrl	1Dose	1Inc	Citation or Pathology
3536	1607	7.66mg	n.s.s.	0/20	3.21mg	2/20	
a	1607	2.67mg	n.s.s.	15/20	3.21mg	14/20	Lawson;jnci,73,515-519;1984
3537	1607	13.1mg	n.s.s.	0/23	2.25mg	0/19	
a	1607	7.43mg	n.s.s.	9/23	2.25mg	3/19	

Spe	Sex	Strain	Route	Site	Hist	Xpo+Xpt	Notes	TD50 DR	2Tailpvl AuOp

1-NITROSO-5,6-DIHYDROURACIL `100ng..:..1ug....:..10.....:..100...:..1mg..:..10..:..100....:..1g.....:..10`

ID	Spe	Sex	Strain	Route	Site	Hist	Xpt	Notes	TD50	DR	AuOp
3538	R	f	mrw	wat	liv	hpc	41w70		.104mg		P<.0005+
a	R	f	mrw	wat	liv	mix	41w70		.104mg		P<.0005+
b	R	f	mrw	wat	tba	mix	41w70		.130mg		P<.0005
3539	R	m	mrw	wat	liv	hpc	41w70		93.2ug		P<.0005+
a	R	m	mrw	wat	liv	mix	41w70		93.2ug		P<.0005+
b	R	m	mrw	wat	tba	mix	41w70		.125mg		P<.002

N-NITROSO-2,3-DIHYDROXYPROPYL-2-HYDROXYPROPYLAMINE `.:..1_0....:..100...:..1mg..:..10..:..100....:..1g.....:..10`

ID	Spe	Sex	Strain	Route	Site	Hist	Xpt	Notes	TD50	DR	AuOp
3540	H	f	syg	gav	for	mix	41w75	e	1.59mg		P<.0005+
a	H	f	syg	gav	for	sqp	41w75	e	1.59mg		P<.0005+
b	H	f	syg	gav	liv	cad	41w75	e	32.6mg		P<.3
c	H	f	syg	gav	lun	tum	41w75	e	no dre		P=1.
3541	R	f	f34	wat	eso	mix	37w85		53.5ug		P<.0005+
a	R	f	f34	wat	ton	mix	37w85		88.8ug		P<.0005+
b	R	f	f34	wat	eso	car	37w85		.117mg		P<.0005
c	R	f	f34	wat	for	mix	37w85		.178mg		P<.0005+
d	R	f	f34	wat	ton	sqc	37w85		.241mg		P<.0005
e	R	f	f34	wat	liv	hpc	37w85		2.40mg		P<.3
f	R	f	f34	wat	liv	mix	37w85		no dre		P=1.

NITROSO-2,3-DIHYDROXYPROPYL-2-OXOPROPYLAMINE `...:..10.....:..100...:..1mg..:..10..:..100....:..1g.....:..10`

ID	Spe	Sex	Strain	Route	Site	Hist	Xpt	Notes	TD50	DR	AuOp
3542	H	f	syg	gav	for	mix	35w55	e	.754mg		P<.0005+
a	H	f	syg	gav	for	sqp	35w55	e	.868mg		P<.0005+
b	H	f	syg	gav	pdu	mix	35w55	e	3.63mg		P<.007 +
c	H	f	syg	gav	liv	hpa	35w55	e	9.92mg		P<.1
d	H	f	syg	gav	liv	mix	35w55	e	9.92mg		P<.1
e	H	f	syg	gav	lun	car	35w55	e	20.4mg		P<.3
f	H	f	syg	gav	liv	hpc	35w55	e	20.4mg		P<.3
3543	R	f	f34	wat	eso	mix	31w55		35.2ug		P<.0005+
a	R	f	f34	wat	for	mix	31w55		48.2ug		P<.0005+
b	R	f	f34	wat	ton	mix	31w55		55.5ug		P<.0005+
c	R	f	f34	wat	eso	car	31w55		63.7ug		P<.0005+
d	R	f	f34	wat	ton	car	31w55		96.4ug		P<.0005+
e	R	f	f34	wat	for	car	31w55		.155mg		P<.002 +
f	R	f	f34	wat	liv	hpc	31w55		1.30mg		P<.3 -

N-NITROSO-2,3-DIHYDROXYPROPYLETHANOLAMINE `..1ug....:..10.....:..100...:..1mg..:..10..:..100....:..1g.....:..10`

ID	Spe	Sex	Strain	Route	Site	Hist	Xpt	Notes	TD50	DR	AuOp
3544	H	f	syg	gav	trh	ade	12m24	e	52.7mg		P<.04
a	H	f	syg	gav	lun	tum	12m24	e	no dre		P=1.
b	H	f	syg	gav	liv	tum	12m24	e	no dre		P=1.
3545	R	f	f34	wat	liv	mix	17m28		5.98mg	*	P<.02 +
a	R	f	f34	wat	liv	nnd	17m28		8.69mg	*	P<.04 +
b	R	f	f34	wat	liv	hpc	17m28		11.5mg	*	P<.02 +

1-NITROSO-3,5-DIMETHYL-4-BENZOYLPIPERAZINE `.1_ug....:..10.....:..100...:..1mg..:..10..:..100....:..1g.....:..10`

ID	Spe	Sex	Strain	Route	Site	Hist	Xpt	Notes	TD50	DR	AuOp
3546	R	f	f34	wat	for	pam	12m29	e	9.66mg		P<.007 +
a	R	f	f34	wat	liv	tum	12m29	e	9.10mg		P<.04
b	R	f	f34	wat	for	bcc	12m29	e	26.4mg		P<.1 +
c	R	f	f34	wat	tba	mix	12m29	e	no dre		P=1. +

NITROSO-2,6-DIMETHYLMORPHOLINE `100ng..:..1ug....:..10.....:..100...:..1mg..:..10..:..100....:..1g.....:..10`

ID	Spe	Sex	Strain	Route	Site	Hist	Xpt	Notes	TD50	DR	AuOp
3547	H	f	syg	gav	trh	ppp	59w83	a	6.40mg	*	P<.0005+
a	H	f	syg	gav	mix	mix	59w83	a	9.17mg	*	P<.0005
b	H	f	syg	gav	liv	mix	59w83	a	11.0mg	*	P<.0005
c	H	f	syg	gav	mix	mix	59w83	a	16.9mg	*	P<.008
d	H	f	syg	gav	pan	mix	59w83	a	5.75mg	*	P<.02 +
e	H	f	syg	gav	pdu	ade	59w83	a	6.99mg	*	P<.02
f	H	f	syg	gav	for	mix	59w83	a	9.45mg	*	P<.03
g	H	f	syg	gav	gal	mix	59w83	a	12.9mg	Z	P<.04
h	H	f	syg	gav	pdu	car	59w83	a	32.1mg	*	P<.3
i	H	f	syg	gav	lun	mix	59w83	a	34.6mg	*	P<.6 +
j	H	f	syg	gav	lar	sqp	59w83	a	no dre		P=1. +
k	H	f	syg	gav	tba	mix	59w83	a	1.55mg	*	P<.0005
3548	H	m	syg	gav	lun	mix	66w87	a	1.21mg	Z	P<.0005+
a	H	m	syg	gav	pan	mix	66w87	a	2.78mg	Z	P<.0005+
b	H	m	syg	gav	for	mix	66w87	a	2.91mg	Z	P<.002
c	H	m	syg	gav	pdu	car	66w87	a	6.67mg	*	P<.0005+
d	H	m	syg	gav	liv	mix	66w87	a	10.1mg	*	P<.002
e	H	m	syg	gav	trh	ppp	66w87	a	10.5mg	*	P<.03 +
f	H	m	syg	gav	pdu	ade	66w87	a	21.2mg	Z	P<.3
g	H	m	syg	gav	lar	sqp	66w87	a	27.3mg	*	P<.3 +
h	H	m	syg	gav	tba	mix	66w87	a	1.68mg	*	P<.0005

N-NITROSO-ETHYL-2-OXOPROPYLUREA `100ng..:..1ug....:..10.....:..100...:..1mg..:..10..:..100....:..1g.....:..10`

ID	Spe	Sex	Strain	Route	Site	Hist	Xpt	Notes	TD50	DR	AuOp
3549	H	f	syg	gav	spl	vsc	30w85	e	.394mg		P<.003 +
a	H	f	syg	gav	for	mix	30w85	e	.983mg		P<.02 +
b	H	f	syg	gav	for	car	30w85	e	2.19mg		P<.09
c	H	f	syg	gav	liv	cho	30w85	e	no dre		P=1.
3550	H	m	syg	gav	for	mix	30w85	e	.506mg		P<.002 +
a	H	m	syg	gav	spl	hes	30w85	e	.865mg		P<.02 +

	RefNum	LoConf	UpConf	Cntrl	1Dose	1Inc	2Dose	2Inc	Citation or Pathology

1-NITROSO-5,6-DIHYDROURACIL   16813-36-8

	RefNum	LoConf	UpConf	Cntrl	1Dose	1Inc			Citation or Pathology
3538	1246	44.5ug	.222mg	0/25	1.08mg	24/25			Bulay;jnci,62,1523-1528;1979
a	1246	44.5ug	.222mg	0/25	1.08mg	24/25			
b	1246	48.6ug	.458mg	12/25	1.08mg	24/25			
3539	1246	39.4ug	.204mg	0/22	.941mg	22/23			
a	1246	39.4ug	.204mg	0/22	.941mg	22/23			
b	1246	43.9ug	.639mg	12/22	.941mg	22/23			

N-NITROSO-2,3-DIHYDROXYPROPYL-2-HYDROXYPROPYLAMINE   89911-79-5

	RefNum	LoConf	UpConf	Cntrl	1Dose	1Inc			Citation or Pathology
3540	1682	.804mg	3.67mg	0/20	4.69mg	13/20			Lijinsky;jnci,74,923-926;1985
a	1682	.804mg	3.67mg	0/20	4.69mg	13/20			
b	1682	5.30mg	n.s.s.	0/20	4.69mg	1/20			
c	1682	10.0mg	n.s.s.	0/20	4.69mg	0/20			
3541	1612	25.4ug	.117mg	0/20	.269mg	18/20			Lijinsky;carc,5,167-170;1984
a	1612	45.5ug	.196mg	0/20	.269mg	15/20			
b	1612	59.3ug	.271mg	0/20	.269mg	13/20			
c	1612	84.8ug	.464mg	0/20	.269mg	10/20			
d	1612	.108mg	.762mg	0/20	.269mg	8/20			
e	1612	.391mg	n.s.s.	0/20	.269mg	1/20			
f	1612	.526mg	n.s.s.	3/20	.269mg	1/20			

NITROSO-2,3-DIHYDROXYPROPYL-2-OXOPROPYLAMINE   92177-50-9

	RefNum	LoConf	UpConf	Cntrl	1Dose	1Inc			Citation or Pathology
3542	1682	.386mg	1.66mg	0/20	5.45mg	15/20			Lijinsky;jnci,74,923-926;1985
a	1682	.443mg	1.95mg	0/20	5.45mg	14/20			
b	1682	1.37mg	41.9mg	0/20	5.45mg	5/20			
c	1682	2.44mg	n.s.s.	0/20	5.45mg	2/20			
d	1682	2.44mg	n.s.s.	0/20	5.45mg	2/20			
e	1682	3.32mg	n.s.s.	0/20	5.45mg	1/20			
f	1682	3.32mg	n.s.s.	0/20	5.45mg	1/20			
3543	1639	17.6ug	76.3ug	0/20	.349mg	17/20			Lijinsky;zkko,107,178-182;1984
a	1639	24.7ug	.106mg	0/20	.349mg	15/20			
b	1639	28.3ug	.125mg	0/20	.349mg	14/20			
c	1639	32.2ug	.147mg	0/20	.349mg	13/20			
d	1639	46.0ug	.252mg	0/20	.349mg	10/20			
e	1639	66.4ug	.588mg	0/20	.349mg	7/20			
f	1639	.212mg	n.s.s.	0/20	.349mg	1/20			

N-NITROSO-2,3-DIHYDROXYPROPYLETHANOLAMINE   89911-78-4

	RefNum	LoConf	UpConf	Cntrl	1Dose	1Inc	2Dose	2Inc	Citation or Pathology
3544	1682	15.9mg	n.s.s.	0/20	12.5mg	3/20			Lijinsky;jnci,74,923-926;1985
a	1682	51.5mg	n.s.s.	0/20	12.5mg	0/20			
b	1682	51.5mg	n.s.s.	0/20	12.5mg	0/20			
3545	1612	2.75mg	n.s.s.	3/20	1.87mg	8/20	3.74mg	10/20	Lijinsky;carc,5,167-170;1984
a	1612	3.59mg	n.s.s.	3/20	1.87mg	5/20	3.74mg	9/20	
b	1612	5.18mg	n.s.s.	4/20	1.87mg	4/20	3.74mg	4/20	

1-NITROSO-3,5-DIMETHYL-4-BENZOYLPIPERAZINE   61034-40-0

	RefNum	LoConf	UpConf	Cntrl	1Dose	1Inc			Citation or Pathology
3546	1208	3.65mg	111.mg	0/20	2.81mg	5/20			Singer;canr,41,1034-1038;1981
a	1208	3.31mg	n.s.s.	1/20	2.81mg	6/20			
b	1208	6.48mg	n.s.s.	0/20	2.81mg	2/20			
c	1208	.701mg	n.s.s.	20/20	2.81mg	19/20			

NITROSO-2,6-DIMETHYLMORPHOLINE   1456-28-6

	RefNum	LoConf	UpConf	Cntrl	1Dose	1Inc	2Dose	2Inc	3Dose	3Inc	4Dose	4Inc	Citation or Pathology
3547	430	3.54mg	21.8mg	0/15	1.31mg	2/15	2.63mg	3/15	5.24mg	3/15	10.5mg	8/15	Reznik;jnci,60,371-378;1978
a	430	4.70mg	27.7mg	0/15	1.31mg	0/15	2.63mg	1/15	5.24mg	7/15	10.5mg	4/15	
b	430	5.34mg	32.0mg	0/15	1.31mg	0/15	2.63mg	1/15	5.24mg	3/15	10.5mg	6/15	
c	430	7.29mg	279.mg	0/15	1.31mg	0/15	2.63mg	0/15	5.24mg	5/15	10.5mg	2/15	
d	430	2.77mg	n.s.s.	0/15	1.31mg	3/15	2.63mg	7/15	5.24mg	6/15	10.5mg	6/15	
e	430	3.39mg	n.s.s.	0/15	1.31mg	2/15	2.63mg	6/15	5.24mg	5/15	10.5mg	5/15	
f	430	4.86mg	n.s.s.	0/15	1.31mg	1/15	2.63mg	3/15	5.24mg	5/15	10.5mg	3/15	
g	430	6.07mg	n.s.s.	0/15	1.31mg	0/15	2.63mg	5/15	5.24mg	0/15	10.5mg	4/15	
h	430	8.48mg	n.s.s.	0/15	1.31mg	1/15	2.63mg	2/15	5.24mg	1/15	10.5mg	2/15	
i	430	5.06mg	n.s.s.	0/15	1.31mg	3/15	2.63mg	5/15	5.24mg	5/15	10.5mg	2/15	
j	430	16.5mg	n.s.s.	0/15	1.31mg	3/15	2.63mg	2/15	5.24mg	0/15	10.5mg	1/15	
k	430	.987mg	4.99mg	0/15	1.31mg	8/15	2.63mg	12/15	5.24mg	14/15	10.5mg	10/15	
3548	430	.659mg	2.72mg	0/15	1.31mg	7/15	2.63mg	9/15	(5.24mg	5/15	10.5mg	5/15)	
a	430	1.58mg	7.41mg	0/15	1.31mg	2/15	2.63mg	10/15	5.24mg	6/15	(10.5mg	9/15)	
b	430	1.31mg	10.6mg	0/15	1.31mg	1/15	2.63mg	7/15	(5.24mg	3/15	10.5mg	4/15)	
c	430	3.75mg	18.8mg	0/15	1.31mg	0/15	2.63mg	5/15	5.24mg	6/15	10.5mg	6/15	
d	430	5.16mg	38.4mg	0/15	1.31mg	0/15	2.63mg	3/15	5.24mg	4/15	10.5mg	5/15	
e	430	4.25mg	n.s.s.	0/15	1.31mg	4/15	2.63mg	3/15	5.24mg	3/15	10.5mg	6/15	
f	430	5.29mg	n.s.s.	0/15	1.31mg	2/15	2.63mg	7/15	5.24mg	1/15	10.5mg	4/15	
g	430	7.74mg	n.s.s.	0/15	1.31mg	2/15	2.63mg	1/15	5.24mg	3/15	10.5mg	2/15	
h	430	1.07mg	2.90mg	0/15	1.31mg	5/15	2.63mg	10/15	5.24mg	12/15	10.5mg	13/15	

N-NITROSO-ETHYL-2-OXOPROPYLUREA   ---

	RefNum	LoConf	UpConf	Cntrl	1Dose	1Inc			Citation or Pathology
3549	1956	.159mg	2.42mg	1/12	.871mg	8/12			Lijinsky;bmes,2,167-173;1989
a	1956	.335mg	n.s.s.	0/12	.871mg	4/12			
b	1956	.536mg	n.s.s.	0/12	.871mg	2/12			
c	1956	.845mg	n.s.s.	1/12	.871mg	1/12			
3550	1956	.201mg	2.27mg	0/12	.766mg	6/12			
a	1956	.295mg	n.s.s.	0/12	.766mg	4/12			

Brkly Code

```
          Spe Strain Site   Xpo+Xpt                                              TD50      2Tailpvl
              Sex   Route  Hist      Notes                                        DR        AuOp
      b    H m syg gav for car 30w85 e                                           1.92mg    P<.09
      c    H m syg gav liv cho 30w85 e                                           1.10mg    P<.2

  N-NITROSO-ETHYLHYDROXYETHYLUREA     100ng..:..1ug...:.10.....:..100....:..1mg....:.10.....:..100...:..1g.....:.10
  3551 H f syg gav cvu tum 30w75 e                               .   +    .      1.04mg    P<.002 +
      a    H f syg gav for ben 30w75 e                                           1.76mg    P<.01  +
      b    H f syg gav spl vsc 30w75 e                                           3.05mg    P<.2   +
      c    H f syg gav liv cho 30w75 e                                           27.5mg    P<.9

  1-NITROSO-1-HYDROXYETHYL-3-CHLOROETHYLUREA..1_ug...:..10.....:..100...:..1mg....:..10.....:..100...:..1g...:.10
  3552 R f f34 gav liv mix 9m27 e                          .   +   .             .488mg    P<.0005+
      a    R f f34 gav liv nnd 9m27 e                                            .564mg    P<.0005+
      b    R f f34 gav liv hpc 9m27 e                                            .653mg    P<.0005+
      c    R f f34 gav --- mnl 9m27 e                                            .760mg    P<.02
      d    R f f34 gav liv clc 9m27 e                                            4.38mg    P<.09  +
      e    R f f34 gav ktu mix 9m27 e                                            4.38mg    P<.09  +
  3553 R m f34 gav liv mix 8m28 e                       .   +   .                .150mg *  P<.0005+
      a    R m f34 gav liv hpc 8m28 e                                            .233mg *  P<.0005+
      b    R m f34 gav ktu mix 8m28 e                                            .663mg *  P<.0005+
      c    R m f34 gav liv clc 8m28 e                                            .926mg *  P<.0005+
      d    R m f34 gav liv nnd 8m28 e                                            1.19mg *  P<.02  +

  N-NITROSO-2-HYDROXYMORPHOLINE      100ng..:..1ug....:.10.....:..100.....:.1mg....:..10.....:.100...:..1g.....:.10
  3554 R f f34 wat liv hpc 12m28 e                                  .>           14.9mg *  P<1.   -

  1-NITROSO-1-(2-HYDROXYPROPYL)-3-CHLOROETHYLUREA...:..1_0....:..100....:..1mg...:..10....:..100...:..1g.....:.10
  3555 R f f34 gav liv mix 40w95 e                          .   +    .           .885mg    P<.0005+
      a    R f f34 gav liv nnd 40w95 e                                           .885mg    P<.0005+
      b    R f f34 gav --- mnl 40w95 e                                           .962mg    P<.05
      c    R f f34 gav liv hpc 40w95 e                                           8.81mg    P<.3   +
  3556 R m f34 gav liv mix 8m28 e                         .   +   .              .861mg *  P<.0005+
      a    R m f34 gav liv nnd 8m28 e                                            1.17mg *  P<.002 +
      b    R m f34 gav liv hpc 8m28 e                                            2.71mg *  P<.004 +
      c    R m f34 gav lun abt 8m28 e                                            2.39mg *  P<.02

  N-NITROSO-(2-HYDROXYPROPYL)-(2-HYDROXYETHYL)AMINE..:..1_0.....:..100....:..1mg...:..10....:..100...:..1g....:.10
  3557 R f f34 wat liv hpc 50w75                            .   +    .           1.02mg    P<.0005+
      a    R f f34 wat liv mix 50w75                                             1.12mg    P<.0005+
      b    R f f34 wat liv ang 50w75                                             4.50mg    P<.002 +
      c    R f f34 wat nas olc 50w75                                             5.44mg    P<.003
      d    R f f34 wat eso mix 50w75                                             6.74mg    P<.007 +
      e    R f f34 wat liv nnd 50w75                                             no dre    P=1.

  N-NITROSO-3-HYDROXYPYRROLIDINE     100ng..:..1ug....:.10.....:..100.....:.1mg....:..10.....:.100...:..1g.....:.10
  3558 R b sda wat liv hpc 26m26                                  .   +          8.11mg    P<.006
      a    R b sda wat tba mal 26m26                                             7.65mg    P<.03  +

  N-NITROSO-N-ISOBUTYLUREA      100ng..:..1ug....:.10.....:..100.....:.1mg....:..10.....:.100...:..1g.....:.10
  3559 R f don wat dgt mix 76w76 e                            .+   .             4.73mg *  P<.0005+
      a    R f don wat duo mix 76w76 e                                           7.22mg *  P<.0005+
      b    R f don wat stg mix 76w76 e                                           23.1mg *  P<.05
      c    R f don wat liv mix 76w76 e                                           1.22gm *  P<1.
      d    R f don wat tba mix 76w76 e                                           4.70mg *  P<.02

  N-NITROSO-N-METHYL-N-DODECYLAMINE   100ng..:..1ug....:.10.....:..100.....:.1mg....:..10.....:.100...:..1g.....:.10
  3560 R f f34 gav ubl tum 30w65                            .   +   .            .656mg    P<.0005+
      a    R f f34 gav lun tum 30w65                                             1.32mg    P<.0005
      b    R f f34 gav liv tum 30w65                                             no dre    P=1.
      c    R f f34 gav tba tum 30w65                                             noTD50    P=1.
  3561 R m f34 gav ubl tcc 7m26 e                             .   +   .          .487mg    P<.0005+
      a    R m f34 gav for car 7m26 e                                            5.07mg    P<.007 +
      b    R m f34 gav --- leu 7m26 e                                            .970mg    P<.02
      c    R m f34 gav pni isc 7m26 e                                            2.75mg    P<.02
      d    R m f34 gav lun adc 7m26 e                                            6.54mg    P<.02  +
      e    R m f34 gav liv hpc 7m26 e                                            8.98mg    P<.04  +
      f    R m f34 gav for pam 7m26 e                                            13.9mg    P<.1   +
      g    R m f34 gav lun ade 7m26 e                                            28.5mg    P<.3   +
      h    R m f34 gav tba mix 7m26 e                                            noTD50    P<.6   +
  3562 R m f34 gav ubl tcc 28w66 er                     <+                       noTD50    P<.0005+

  N-NITROSO-N-METHYL-4-FLUOROANILINE   ..:..1ug...:..10....:..100....:..1mg...:..10.....:..100....:..1g......:.10
  3563 R m f34 wat eso mix 12m26                           .   +   .             .255mg    P<.0005+
      a    R m f34 wat liv hnd 12m26                                             10.3mg    P<.7
      b    R m f34 wat tba mix 12m26                                             no dre    P=1.

  N-NITROSO-N-METHYL-4-NITROANILINE   100ng..:..1ug....:.10.....:..100.....:.1mg....:..10.....:.100...:..1g.....:.10
  3564 R m f34 wat liv hnd 12m26                                  .>             5.87mg    P<.4   -
      a    R m f34 wat tba mix 12m26                                             no dre    P=1.   -
```

	RefNum	LoConf	UpConf	Cntrl	1Dose	1Inc	2Dose	2Inc	Citation or Pathology
									Brkly Code
b	1956	.471mg	n.s.s.	0/12	.766mg	2/12			
c	1956	.313mg	n.s.s.	1/12	.766mg	4/12			

N-NITROSO-ETHYLHYDROXYETHYLUREA   ---

	RefNum	LoConf	UpConf	Cntrl	1Dose	1Inc	2Dose	2Inc	Citation or Pathology
3551	1956	.480mg	3.62mg	0/12	1.87mg	9/19			Lijinsky;bmes,2,167-173;1989
a	1956	.710mg	141.mg	0/12	1.87mg	6/19			
b	1956	.893mg	n.s.s.	1/12	1.87mg	5/19			
c	1956	1.69mg	n.s.s.	1/12	1.87mg	2/19			

1-NITROSO-1-HYDROXYETHYL-3-CHLOROETHYLUREA   96806-34-7

	RefNum	LoConf	UpConf	Cntrl	1Dose	1Inc	2Dose	2Inc	Citation or Pathology
3552	1792	.242mg	1.17mg	0/20	.582mg	12/19			Lijinsky;zkko,112,221-228;1986
a	1792	.274mg	1.40mg	0/20	.582mg	11/19			
b	1792	.310mg	1.70mg	0/20	.582mg	10/19			
c	1792	.308mg	n.s.s.	4/20	.582mg	11/19			
d	1792	1.08mg	n.s.s.	0/20	.582mg	2/19			
e	1792	1.08mg	n.s.s.	0/20	.582mg	2/19			
3553	1792	73.4ug	.298mg	2/20	.390mg	18/20	.758mg	20/20	
a	1792	.135mg	.412mg	0/20	.390mg	14/20	.758mg	20/20	
b	1792	.387mg	1.25mg	0/20	.390mg	7/20	.758mg	14/20	
c	1792	.519mg	2.01mg	0/20	.390mg	7/20	.758mg	10/20	
d	1792	.564mg	n.s.s.	2/20	.390mg	10/20	.758mg	8/20	

N-NITROSO-2-HYDROXYMORPHOLINE   67587-52-4

	RefNum	LoConf	UpConf	Cntrl	1Dose	1Inc	2Dose	2Inc	Citation or Pathology
3554	1928	2.42mg	n.s.s.	0/20	.265mg	1/20	.530mg	0/20	Hecht;carc,10,1475-1477;1989/pers.comm.

1-NITROSO-1-(2-HYDROXYPROPYL)-3-CHLOROETHYLUREA   96806-35-8

	RefNum	LoConf	UpConf	Cntrl	1Dose	1Inc	2Dose	2Inc	Citation or Pathology
3555	1792	.396mg	2.80mg	0/20	.791mg	8/20			Lijinsky;zkko,112,221-228;1986
a	1792	.396mg	2.80mg	0/20	.791mg	8/20			
b	1792	.358mg	n.s.s.	4/20	.791mg	10/20			
c	1792	1.43mg	n.s.s.	0/20	.791mg	1/20			
3556	1792	.462mg	2.43mg	2/20	.438mg	4/19	.751mg	15/20	
a	1792	.587mg	5.55mg	2/20	.438mg	4/19	.751mg	12/20	
b	1792	1.17mg	16.8mg	0/20	.438mg	1/19	.751mg	6/20	
c	1792	1.01mg	n.s.s.	1/20	.438mg	2/19	.751mg	7/20	

N-NITROSO-(2-HYDROXYPROPYL)-(2-HYDROXYETHYL)AMINE   75896-33-2

	RefNum	LoConf	UpConf	Cntrl	1Dose	1Inc	2Dose	2Inc	Citation or Pathology
3557	1612	.511mg	2.21mg	0/20	5.44mg	17/20			Lijinsky;carc,5,167-170;1984
a	1612	.532mg	2.94mg	3/20	5.44mg	17/20			
b	1612	1.93mg	17.0mg	0/20	5.44mg	7/20			
c	1612	2.20mg	28.4mg	0/20	5.44mg	6/20			
d	1612	2.55mg	77.7mg	0/20	5.44mg	5/20			
e	1612	11.7mg	n.s.s.	3/20	5.44mg	0/20			

N-NITROSO-3-HYDROXYPYRROLIDINE   56222-35-6

	RefNum	LoConf	UpConf	Cntrl	1Dose	1Inc	2Dose	2Inc	Citation or Pathology
3558	1634	3.07mg	85.1mg	0/24	2.50mg	5/23			Eisenbrand;iarc,657-663;1980
a	1634	2.80mg	n.s.s.	1/24	2.50mg	6/23			

N-NITROSO-N-ISOBUTYLUREA   (N-isobutyl-N-nitrosourea)   760-60-1

	RefNum	LoConf	UpConf	Cntrl	1Dose	1Inc	2Dose	2Inc	3Dose	3Inc	Citation or Pathology
3559	1549	3.18mg	7.32mg	0/17	5.71mg	6/25	11.4mg	14/24	22.9mg	25/28	Ogiu;gann,74,342-350;1983
a	1549	4.72mg	11.7mg	0/17	5.71mg	2/25	11.4mg	12/24	22.9mg	21/28	
b	1549	12.7mg	n.s.s.	0/17	5.71mg	3/25	11.4mg	7/24	22.9mg	5/28	
c	1549	30.4mg	n.s.s.	1/17	5.71mg	2/25	11.4mg	3/24	22.9mg	2/28	
d	1549	2.04mg	n.s.s.	12/17	5.71mg	20/25	11.4mg	20/24	22.9mg	27/28	

N-NITROSO-N-METHYL-N-DODECYLAMINE   55090-44-3

	RefNum	LoConf	UpConf	Cntrl	1Dose	1Inc	2Dose	2Inc	Citation or Pathology
3560	1561	.323mg	1.50mg	1/20	4.52mg	17/20			Lijinsky;fctx,21,601-605;1983
a	1561	.658mg	3.15mg	0/20	4.52mg	12/20			
b	1561	7.28mg	n.s.s.	0/20	4.52mg	0/20			
c	1561	n.s.s.	n.s.s.	20/20	4.52mg	20/20			
3561	1206	.200mg	1.12mg	0/20	1.84mg	19/20			Lijinsky;canr,41,1288-1292;1981
a	1206	1.92mg	58.5mg	0/20	1.84mg	5/20			
b	1206	.356mg	n.s.s.	11/20	1.84mg	18/20			
c	1206	1.11mg	n.s.s.	3/20	1.84mg	10/20			
d	1206	2.25mg	n.s.s.	0/20	1.84mg	4/20			
e	1206	2.71mg	n.s.s.	0/20	1.84mg	3/20			
f	1206	3.40mg	n.s.s.	0/20	1.84mg	2/20			
g	1206	4.63mg	n.s.s.	0/20	1.84mg	1/20			
h	1206	n.s.s.	n.s.s.	19/20	1.84mg	20/20			
3562	1912	n.s.s.	.497mg	0/5	2.91mg	15/15			Reznik;acnr,1,389-392;1981

N-NITROSO-N-METHYL-4-FLUOROANILINE   937-25-7

	RefNum	LoConf	UpConf	Cntrl	1Dose	1Inc	2Dose	2Inc	Citation or Pathology
3563	1544	.121mg	.558mg	0/20	.714mg	18/20			Kroeger-Koepke;carc,4,157-160;1983
a	1544	1.29mg	n.s.s.	2/20	.714mg	3/20			
b	1544	.234mg	n.s.s.	19/20	.714mg	18/20			

N-NITROSO-N-METHYL-4-NITROANILINE   943-41-9

	RefNum	LoConf	UpConf	Cntrl	1Dose	1Inc	2Dose	2Inc	Citation or Pathology
3564	1544	1.22mg	n.s.s.	2/20	.840mg	4/20			Kroeger-Koepke;carc,4,157-160;1983
a	1544	.149mg	n.s.s.	19/20	.840mg	19/20			

```
      Spe Strain  Site   Xpo+Xpt                                                    TD50    2Tailpvl
          Sex  Route  Hist    Notes                                                  DR      AuOp
```

NITROSO-N-METHYL-N-(2-PHENYL)ETHYLAMINE..:..1_ug...:..10.....:..100..:..1mg..:..10....:..100...:..1g....:..10

```
3565 R m f34 wat ugi mix  7m30 e                      . + .                         7.88ug  Z P<.0005+

  a  R m f34 wat eso mix  7m30 e                                                    40.1ug  Z P<.0005+
  b  R m f34 wat for bcp  7m30 e                                                    2.21mg  Z P<.4   +
  c  R m f34 wat for bcc  7m30 e                                                    2.57mg  Z P<.3   +
  d  R m f34 wat mul mle  7m30 e                                                    no dre    P=1.   +
  e  R m f34 wat ton bcc  7m30 e                                                    no dre    P=1.   +
  f  R m f34 wat liv hnd  7m30 e                                                    no dre    P=1.   +
3566 R m f34 wat ugi mix 24m30 e                     . + .                          13.6ug    P<.0005+
  a  R m f34 wat for bcc 24m30 e                                                    42.2ug    P<.02  +
  b  R m f34 wat eso mix 24m30 e                                                    42.2ug    P<.02  +
  c  R m f34 wat mul mle 24m30 e                                                    98.9ug    P<.8   +
  d  R m f34 wat liv hnd 24m30 e                                                    .184mg    P<.3   +
  e  R m f34 wat ton bcc 24m30 e                                                    .184mg    P<.3   +
```

N-NITROSO-N-METHYL-N-TETRADECYLAMINE ..:..1ug....:..10.....:..100...:..1mg..:..10....:..100...:..1g.....:..10

```
3567 R m f34 gav ubl tcc  7m24 e                                      <+           noTD50    P<.0005+
  a  R m f34 gav lun ade  7m24 e                                                   29.4mg    P<.1   +
  b  R m f34 gav lun adc  7m24 e                                                   60.5mg    P<.3   +
  c  R m f34 gav liv tum  7m24 e                                                   no dre    P=1.
  d  R m f34 gav tba mix  7m24 e                                                   noTD50    P<.6   +
```

N-NITROSO-N-METHYLDECYLAMINE 100ng..:..1ug....:..10....:..100....:..1mg....:..10.....:..100...:..1g.....:..10

```
3568 R m f34 gav ubl tcc  7m24 e                          . + .                    1.26mg    P<.0005+
  a  R m f34 gav lun adc  7m24 e                                                   8.32mg    P<.007 +
  b  R m f34 gav lun ade  7m24 e                                                   14.7mg    P<.04  +
  c  R m f34 gav liv hpc  7m24 e                                                   22.7mg    P<.1   +
  d  R m f34 gav for pam  7m24 e                                                   46.6mg    P<.3   +
  e  R m f34 gav nas mix  7m24 e                                                   46.6mg    P<.3   +
  f  R m f34 gav for car  7m24 e                                                   46.6mg    P<.3   +
  g  R m f34 gav tba mix  7m24 e                                                   no dre    P=1.   +
```

NITROSO-5-METHYLOXAZOLIDONE 100ng..:..1ug....:..10.....:..100....:..1mg....:..10.....:..100...:..1g.....:..10

```
3569 H f syg gav for mix 28w60 e                        . + .                      .172mg    P<.0005+
```

N-NITROSO-N-METHYLUREA 100ng..:..1ug....:..10.....:..100....:..1mg....:..10.....:..100...:..1g.....:..10

```
3570 P b cym eat ugi sqc 18y22 w                            :    +    :            4.52mg    P<.0005+

  a  P b cym eat eso sqc 18y22 w                                                   10.5mg    P<.0005+
  b  P b cym eat orc mix 18y22 w                                                   2.65mg    P<.03
  c  P b cym eat hea rhm 18y22 w                                                   66.3mg    P<.09
  d  P b cym eat sto adc 18y22 w                                                   66.3mg    P<.09
  e  P b cym eat phr sqc 18y22 w                                                   82.2mg    P<.06
  f  P b cym eat pls sqc 18y22 w                                                   7.95mg    P<.2
  g  P b cym eat ton sqp 18y22 w                                                   7.95mg    P<.2
  h  P b cym eat sku rhb 18y22 w                                                   23.9mg    P<.2
  i  P b cym eat eso sqp 18y22 w                                                   29.2mg    P<.2
  j  P b cym eat col lei 18y22 w                                                   34.5mg    P<.2
  k  P b cym eat tba mix 18y22 w                                                   2.25mg    P<.0005
  l  P b cym eat tba mal 18y22 w                                                   3.39mg    P<.002
  m  P b cym eat tba ben 18y22 w                                                   11.2mg    P<.2
3571 P b rhe eat ugi sqc 18y22 w                       :    +    :                 7.46mg    P<.0005+
  a  P b rhe eat eso sqc 18y22 w                                                   8.20mg    P<.0005+
  b  P b rhe eat orc mix 18y22 w                                                   16.8mg    P<.0005
  c  P b rhe eat phr sqc 18y22 w                                                   44.2mg    P<.008
  d  P b rhe eat epg sqc 18y22 w                                                   42.8mg    P<.06
  e  P b rhe eat sto adc 18y22 w                                                   42.8mg    P<.06
  f  P b rhe eat gnv sqc 18y22 w                                                   62.6mg    P<.07
  g  P b rhe eat lar sqc 18y22 w                                                   95.6mg    P<.06
  h  P b rhe eat chp sqc 18y22 w                                                   95.6mg    P<.06
  i  P b rhe eat ton sqc 18y22 w                                                   102.mg    P<.07
  j  P b rhe eat pls sqc 18y22 w                                                   102.mg    P<.07
  k  P b rhe eat tba mal 18y22 w                                                   6.72mg    P<.0005
3572 R m f34 gav for tum  6m24                      . + .                          92.7ug    P<.0005+
  a  R m f34 gav lun mix  6m24                                                     .357mg    P<.002 +
  b  R m f34 gav ner mix  6m24                                                     2.03mg    P<.2   +
  c  R m f34 gav liv tum  6m24                                                     11.6mg    P<.9
```

N-NITROSO-N-METHYLURETHAN 100ng..:..1ug....:..10......:..100....:..1mg..:..10.....:..100....:..1g.....:..10

```
3573 H b syg gav for epc 27w73 ev                  <+                              noTD50    P<.0005+
  a  H b syg gav eso epc 27w73 ev                                                  .127mg    P<.0005+
```

N-NITROSO-1,3-OXAZOLIDINE 100ng..:..1ug....:..10.....:..100....:..1mg....:..10.....:..100....:..1g.....:..10

```
3574 H f syg gav for mix  7m24 e                          . ±                     2.61mg    P<.07  +
  a  H f syg gav liv tum  7m24 e                                                   no dre    P=1.
3575 H m syg gav liv clc 35w65                      . + .                          .471mg    P<.0005+
  a  H m syg gav liv hpa 35w65                                                     .949mg    P<.0005+
  b  H m syg gav liv hpc 35w65                                                     2.63mg    P<.007 +
  c  H m syg gav lun car 35w65                                                     14.8mg    P<.3
  d  H m syg gav tba mix 35w65                                                     .267mg    P<.0005
```

	RefNum	LoConf	UpConf	Cntrl	1Dose	1Inc	2Dose	2Inc					Citation or Pathology
													Brkly Code

NITROSO-N-METHYL-N-(2-PHENYL)ETHYLAMINE   13256-11-6

	RefNum	LoConf	UpConf	Cntrl	1Dose	1Inc	2Dose	2Inc	3Dose	3Inc	4Dose	4Inc	Citation
3565	1374m	4.57ug	17.1ug	0/20	2.92ug	9/20	8.10ug	11/20	(44.6ug	16/20	.197mg	19/20	.600mg 19/20) Lijinsky;fctx,20, 393-399;1982
a	1374m	22.3ug	80.6ug	0/20	2.92ug	1/20	8.10ug	2/20	44.6ug	15/20	(.197mg	13/20	.600mg 16/20)
b	1374m	.424mg	n.s.s.	0/20	2.92ug	6/20	8.10ug	1/20	44.6ug	1/20	.197mg	4/20	(.600mg 5/20)
c	1374m	.666mg	n.s.s.	0/20	2.92ug	4/20	8.10ug	7/20	44.6ug	15/20	.197mg	4/20	.600mg 9/20
d	1374m	4.81mg	n.s.s.	9/20	2.92ug	16/20	8.10ug	13/20	44.6ug	9/20	.197mg	0/20	.600mg 0/20
e	1374m	3.60mg	n.s.s.	0/20	2.92ug	0/20	8.10ug	0/20	44.6ug	6/20	.197mg	0/20	.600mg 0/20
f	1374m	4.94mg	n.s.s.	0/20	2.92ug	0/20	8.10ug	5/20	44.6ug	0/20	.197mg	0/20	.600mg 0/20
3566	1374n	6.49ug	35.5ug	0/20	9.22ug	10/20							
a	1374n	14.5ug	n.s.s.	0/20	9.22ug	4/20							
b	1374n	14.5ug	n.s.s.	0/20	9.22ug	4/20							
c	1374n	9.81ug	n.s.s.	9/20	9.22ug	10/20							
d	1374n	29.9ug	n.s.s.	0/20	9.22ug	1/20							
e	1374n	29.9ug	n.s.s.	0/20	9.22ug	1/20							

N-NITROSO-N-METHYL-N-TETRADECYLAMINE   75881-20-8

	RefNum	LoConf	UpConf	Cntrl	1Dose	1Inc	Citation
3567	1206	n.s.s.	1.65mg	0/20	4.71mg	20/20	Lijinsky;canr,41,1288-1292;1981
a	1206	7.23mg	n.s.s.	0/20	4.71mg	2/20	
b	1206	9.84mg	n.s.s.	0/20	4.71mg	1/20	
c	1206	18.7mg	n.s.s.	0/20	4.71mg	0/20	
d	1206	n.s.s.	n.s.s.	19/20	4.71mg	20/20	

N-NITROSO-N-METHYLDECYLAMINE   75881-22-0

	RefNum	LoConf	UpConf	Cntrl	1Dose	1Inc	Citation
3568	1206	.630mg	2.73mg	0/20	3.63mg	17/20	Lijinsky;canr,41,1288-1292;1981
a	1206	3.14mg	95.9mg	0/20	3.63mg	5/20	
b	1206	4.45mg	n.s.s.	0/20	3.63mg	3/20	
c	1206	5.58mg	n.s.s.	0/20	3.63mg	2/20	
d	1206	7.59mg	n.s.s.	0/20	3.63mg	1/20	
e	1206	7.59mg	n.s.s.	0/20	3.63mg	1/20	
f	1206	7.59mg	n.s.s.	0/20	3.63mg	1/20	
g	1206	.953mg	n.s.s.	19/20	3.63mg	18/20	

NITROSO-5-METHYLOXAZOLIDONE   79624-33-2

	RefNum	LoConf	UpConf	Cntrl	1Dose	1Inc	Citation
3569	1672	80.7ug	.391mg	1/20	1.70mg	18/20	Lijinsky;canr,45,542-545;1985

N-NITROSO-N-METHYLUREA   684-93-5

	RefNum	LoConf	UpConf	Cntrl	1Dose	1Inc	Citation
3570	2001	1.22mg	17.0mg	0/90	6.32mg	6/16	Adamson;ossc,129-156; 1982/Thorgeirsson 1994/Dalgard 1991/Thorgeirsson&Seiber pers.comm.
a	2001	3.75mg	41.3mg	0/90	6.32mg	5/16	
b	2001	.652mg	n.s.s.	0/5	6.32mg	2/2	
c	2001	10.8mg	n.s.s.	0/45	6.32mg	1/13	
d	2001	10.8mg	n.s.s.	0/45	6.32mg	1/13	
e	2001	13.4mg	n.s.s.	0/90	6.32mg	1/16	
f	2001	1.29mg	n.s.s.	0/5	6.32mg	1/2	
g	2001	1.29mg	n.s.s.	0/5	6.32mg	1/2	
h	2001	3.88mg	n.s.s.	0/12	6.32mg	1/5	
i	2001	4.75mg	n.s.s.	0/14	6.32mg	1/6	
j	2001	5.61mg	n.s.s.	0/15	6.32mg	1/7	
k	2001	.755mg	11.8mg	3/90	6.32mg	9/16	
l	2001	1.14mg	20.4mg	2/90	6.32mg	7/16	
m	2001	1.70mg	n.s.s.	1/19	6.32mg	2/8	
3571	2001	3.40mg	19.3mg	0/74	7.80mg	9/16	
a	2001	3.58mg	22.4mg	0/74	7.80mg	8/16	
b	2001	5.42mg	81.5mg	0/74	7.80mg	4/16	
c	2001	10.8mg	1.35gm	0/74	7.80mg	2/15	
d	2001	6.98mg	n.s.s.	0/39	7.80mg	1/7	
e	2001	6.98mg	n.s.s.	0/39	7.80mg	1/7	
f	2001	10.2mg	n.s.s.	0/49	7.80mg	1/10	
g	2001	15.6mg	n.s.s.	0/74	7.80mg	1/15	
h	2001	15.6mg	n.s.s.	0/74	7.80mg	1/15	
i	2001	16.6mg	n.s.s.	0/74	7.80mg	1/16	
j	2001	16.6mg	n.s.s.	0/74	7.80mg	1/16	
k	2001	2.78mg	25.7mg	5/76	7.80mg	9/16	
3572	1915	44.1ug	.203mg	0/12	.312mg	18/20	Lijinsky;txih,5,925-935;1989
a	1915	.165mg	1.33mg	0/12	.312mg	9/20	
b	1915	.498mg	n.s.s.	0/12	.312mg	2/20	
c	1915	.574mg	n.s.s.	1/12	.312mg	2/20	

N-NITROSO-N-METHYLURETHAN   615-53-2

	RefNum	LoConf	UpConf	Cntrl	1Dose	1Inc	Citation
3573	1181	n.s.s.	.127mg	0/16	.629mg	16/16	Herrold;jnci,37,389-394;1966
a	1181	59.7ug	.305mg	0/16	.629mg	13/16	

N-NITROSO-1,3-OXAZOLIDINE   (nitrosooxazolidone) 39884-52-1

	RefNum	LoConf	UpConf	Cntrl	1Dose	1Inc	Citation
3574	1672	.855mg	n.s.s.	1/20	.899mg	5/20	Lijinsky;canr,45,542-545;1985
a	1672	3.71mg	n.s.s.	0/20	.899mg	0/20	
3575	1608	.240mg	1.02mg	0/20	2.83mg	16/20	Lijinsky;carc,5,875-878;1984
a	1608	.464mg	2.35mg	0/20	2.83mg	11/20	
b	1608	.995mg	30.4mg	0/20	2.83mg	5/20	
c	1608	2.40mg	n.s.s.	0/20	2.83mg	1/20	
d	1608	.106mg	.681mg	3/20	2.83mg	19/20	

```
          Spe Strain  Site    Xpo+Xpt                                                      TD50    2Tailpvl
          Sex  Route   Hist    Notes                                                       DR      AuOp

3-NITROSO-2-OXAZOLIDINONE          100ng..:..1ug....:..10.....:..100...:..1mg.....:..10....:..100...:..1g.....:..10
3576 R m mrw wat itn mix 64w64 e                                <+                         noTD50  P<.0005+
   a R m mrw wat liv mix 64w64 e                                                           .582mg  P<.0005+
   b R m mrw wat tba tum 64w64 e                                                           noTD50  P<.0005
3577 R m mrw wat itn mix 70w70 ev                          .  +  .                         .729mg  P<.0005+
   a R m mrw wat liv mix 70w70 ev                                                          1.15mg  P<.0005+
   b R m mrw wat tba tum 70w70 ev                                                          .335mg  P<.0005

N-NITROSO-OXOPROPYLCHLOROETHYLUREA    ..:..1ug....:..10.....:..100...:..1mg.....:..10....:..100...:..1g.....:..10
3578 H f syg gav spl vsc 43w75 e                                  .  +     .               .774mg  P<.008 +
   a H f syg gav liv cho 43w75 e                                                           no dre  P=1.
3579 H m syg gav spl hes 43w75 e                             .  +  .                        .216mg  P<.0005+
   a H m syg gav liv cho 43w75 e                                                           no dre  P=1.

NITROSO-2-OXOPROPYLETHANOLAMINE    100ng..:..1ug....:..10.....:..100...:..1mg.....:..10....:..100...:..1g.....:..10
3580 H f syg gav liv mix 50w65 e                                     .  +   .              .997mg  P<.0005+
   a H f syg gav pdu mix 50w65 e                                                           2.31mg  P<.0005+
   b H f syg gav liv caa 50w65 e                                                           2.68mg  P<.0005+
   c H f syg gav liv hca 50w65 e                                                           3.72mg  P<.002 +
   d H f syg gav pdu car 50w65 e                                                           3.72mg  P<.002
   e H f syg gav liv hpa 50w65 e                                                           5.58mg  P<.007
   f H f syg gav liv clc 50w65 e                                                           5.58mg  P<.007
   g H f syg gav lun ade 50w65 e                                                           31.3mg  P<.3
3581 R f f34 wat liv mix 12m28                                  .  +   .                    1.80mg  P<.002 +
   a R f f34 wat liv hpc 12m28                                                             2.70mg  P<.007 +
   b R f f34 wat liv nnd 12m28                                                             2.70mg  P<.007 +

N-NITROSO-OXOPROPYLUREA            100ng..:..1ug....:..10.....:..100...:..1mg.....:..10....:..100...:..1g.....:..10
3582 H f syg gav spl vsc 40w55 e                             .  +   .                       .121mg  P<.0005+
   a H f syg gav liv cho 40w55 e                                                           no dre  P=1.
3583 H m syg gav spl hes 40w75 e                             .  +   .                       .140mg  P<.0005+
   a H m syg gav liv cho 40w75 e                                                           no dre  P=1.

DI(N-NITROSO)-PERHYDROPYRIMIDINE   100ng..:..1ug....:..10.....:..100...:..1mg.....:..10....:..100...:..1g.....:..10
3584 R m sda ipj pnl sqc 24m24 rs                            .  +   .                       .166mg * P<.003 +

N-NITROSO-2-PHENYLETHYLUREA        100ng..:..1ug....:..10.....:..100...:..1mg.....:..10....:..100...:..1g.....:..10
3585 H f syg gav spl vsc 30w75 e                                  .  +    .                .801mg  P<.005 +
   a H f syg gav for mix 30w75 e                                                           .939mg  P<.002 +
   b H f syg gav liv cho 30w75 e                                                           6.51mg  P<.5
   c H f syg gav for car 30w75 e                                                           7.77mg  P<.3
3586 H m syg gav for mix 30w75 e                                  .  +   .                 1.21mg  P<.005 +
   a H m syg gav for car 30w75 e                                                           2.27mg  P<.04
   b H m syg gav spl hes 30w75 e                                                           2.27mg  P<.04  +
   c H m syg gav liv cho 30w75 e                                                           3.25mg  P<.3

NITROSO-1,2,3,6-TETRAHYDROPYRIDINE    ..:..1ug....:..10.....:..100...:..1mg.....:..10....:..100...:..1g.....:..10
3587 R f f34 wat ugi mix  6m30 esy                           .  (+)  .                     .142mg  Z P<.0005+

   a R f f34 wat eso mix  6m30 esy                                                         .216mg  Z P<.0005+

   b R f f34 wat ugi car  6m30 esy                                                         .271mg  Z P<.0005+

   c R f f34 wat liv hes  6m30 esy                                                         4.04mg  * P<.0005+

   d R f f34 wat liv hpc  6m30 esy                                                         no dre  P=1.  +

   e R f f34 wat stn bcp  6m30 esy                                                         no dre  P=1.  +

   f R f f34 wat liv mix  6m30 esy                                                         no dre  P=1.  +

3588 R f f34 wat ugi mix 16m23 es                               <+                         noTD50  P<.0005+
   a R f f34 wat ugi car 16m23 es                                                          96.6ug  P<.0005+
   b R f f34 wat eso mix 16m23 es                                                          .127mg  P<.0005+
   c R f f34 wat liv hnd 16m23 es                                                          2.15mg  P<.6  +
3589 R f f34 wat ugi mix 23m30 e                              .  +  .                       37.4ug  * P<.0005+
   a R f f34 wat stn bcp 23m30 e                                                           55.1ug  \ P<.0005+
   b R f f34 wat eso mix 23m30 e                                                           66.9ug  / P<.0005+
   c R f f34 wat ugi car 23m30 e                                                           85.6ug  / P<.0005+
   d R f f34 wat liv mix 23m30 e                                                           .132mg  * P<.003 +
   e R f f34 wat eso sqc 23m30 e                                                           .430mg  * P<.005 +
   f R f f34 wat liv hnd 23m30 e                                                           .284mg  * P<.03  +

N-NITROSO(2,2,2-TRIFLUOROETHYL)ETHYLAMINE  ..1ug....:..10.....:..100...:..1mg.....:..10....:..100...:..1g.....:..10
3590 R m sda gav eso ppc 88w88                                       .  +  .               2.52mg  * P<.0005+
   a R m sda gav nas mix 88w88                                                             7.34mg  Z P<.007 +
   b R m sda gav liv hes 88w88                                                             no dre  P=1.

N-NITROSO-2,2,4-TRIMETHYL-1,2-DIHYDROQUINOLINE POLYMER  ....:..100...:..1mg.....:..10....:..100...:..1g.....:..10
3591 R m cbr ipj pec sar  6m23 e                                  .  +   .                 3.31mg  P<.003 +
```

	RefNum	LoConf	UpConf	Cntrl	1Dose	1Inc	2Dose	2Inc					Citation or Pathology
													Brkly Code

3-NITROSO-2-OXAZOLIDINONE 38347-74-9

3576	1813m	n.s.s.	.262mg	0/17	2.14mg	26/26							Mirvish;jnci,78,387-393;1987/pers.comm.
a	1813m	.315mg	1.22mg	0/17	2.14mg	16/26							
b	1813m	n.s.s.	.281mg	2/17	2.14mg	26/26							
3577	1813n	.389mg	1.57mg	0/17	2.15mg	15/25							
a	1813n	.571mg	3.09mg	0/17	2.15mg	11/25							
b	1813n	.171mg	.750mg	2/17	2.15mg	22/25							

N-NITROSO-OXOPROPYLCHLOROETHYLUREA ---

3578	1956	.299mg	15.1mg	1/12	1.71mg	7/12							Lijinsky;bmes,2,167-173;1989
a	1956	2.20mg	n.s.s.	1/12	1.71mg	0/12							
3579	1956	79.5ug	.610mg	0/12	1.51mg	11/12							
a	1956	1.94mg	n.s.s.	1/12	1.51mg	0/12							

NITROSO-2-OXOPROPYLETHANOLAMINE 92177-49-6

3580	1682	.508mg	2.17mg	0/20	5.99mg	16/20							Lijinsky;jnci,74,923-926;1985
a	1682	1.10mg	6.05mg	0/20	5.99mg	10/20							
b	1682	1.24mg	7.57mg	0/20	5.99mg	9/20							
c	1682	1.59mg	14.1mg	0/20	5.99mg	7/20							
d	1682	1.59mg	14.1mg	0/20	5.99mg	7/20							
e	1682	2.11mg	64.3mg	0/20	5.99mg	5/20							
f	1682	2.11mg	64.3mg	0/20	5.99mg	5/20							
g	1682	5.09mg	n.s.s.	0/20	5.99mg	1/20							
3581	1639	.770mg	6.82mg	0/20	.850mg	7/20							Lijinsky;zkko,107,178-182;1984
a	1639	1.02mg	31.1mg	0/20	.850mg	5/20							
b	1639	1.02mg	31.1mg	0/20	.850mg	5/20							

N-NITROSO-OXOPROPYLUREA ---

3582	1956	43.4ug	.384mg	1/12	1.51mg	11/12							Lijinsky;bmes,2,167-173;1989
a	1956	1.05mg	n.s.s.	1/12	1.51mg	0/12							
3583	1956	51.5ug	.394mg	0/12	.975mg	11/12							
a	1956	.737mg	n.s.s.	1/12	.975mg	1/12							

DI(N-NITROSO)-PERHYDROPYRIMIDINE 15973-99-6

3584	1538	67.8ug	.778mg	0/40	10.0ug	0/40	28.6ug	6/40					Habs;zkko,105,191-193;1983

N-NITROSO-2-PHENYLETHYLUREA ---

3585	1956	.308mg	7.43mg	1/12	2.08mg	7/11							Lijinsky;bmes,2,167-173;1989
a	1956	.369mg	3.87mg	0/12	2.08mg	6/11							
b	1956	.984mg	n.s.s.	1/12	2.08mg	2/11							
c	1956	1.26mg	n.s.s.	0/12	2.08mg	1/11							
3586	1956	.450mg	9.70mg	0/12	1.83mg	5/12							
a	1956	.680mg	n.s.s.	0/12	1.83mg	3/12							
b	1956	.680mg	n.s.s.	0/12	1.83mg	3/12							
c	1956	.732mg	n.s.s.	1/12	1.83mg	3/12							

NITROSO-1,2,3,6-TETRAHYDROPYRIDINE 55556-92-8

	RefNum	LoConf	UpConf	Cntrl	1Dose	1Inc	2Dose	2Inc	3Dose	3Inc	4Dose	4Inc	5Dose	5Inc	Citation
3587	1375m	76.3ug	.350mg	0/20	7.94ug	6/20	20.6ug	1/20	50.8ug	4/20	.140mg	15/20	(.785mg	17/20	
					2.83mg	15/20)										Lijinsky;eaes,6,513-527;1982/pers.comm.
a	1375m	.129mg	.378mg	0/20	7.94ug	0/20	20.6ug	0/20	50.8ug	1/20	.140mg	16/20	.785mg	19/20	
					(2.83mg	15/20)										
b	1375m	.141mg	.614mg	0/20	7.94ug	0/20	20.6ug	0/20	50.8ug	0/20	.140mg	13/20	(.785mg	10/20	
					2.83mg	8/20)										
c	1375m	2.14mg	8.91mg	0/20	7.94ug	0/20	20.6ug	0/20	50.8ug	0/20	.140mg	0/20	.785mg	0/20	
					2.83mg	14/20										
d	1375m	12.8mg	n.s.s.	0/20	7.94ug	2/20	20.6ug	6/20	50.8ug	4/20	.140mg	2/20	.785mg	0/20	
					2.83mg	2/20										
e	1375m	14.4mg	n.s.s.	0/20	7.94ug	6/20	20.6ug	0/20	50.8ug	2/20	.140mg	4/20	.785mg	1/20	
					2.83mg	1/20										
f	1375m	8.09mg	n.s.s.	1/20	7.94ug	4/20	20.6ug	9/20	50.8ug	6/20	.140mg	5/20	.785mg	0/20	
					2.83mg	5/20										
3588	1375n	n.s.s.	62.0ug	0/20	.184mg	20/20									
a	1375n	49.3ug	.217mg	0/20	.184mg	14/20									
b	1375n	63.3ug	.303mg	0/20	.184mg	12/20									
c	1375n	.298mg	n.s.s.	1/20	.184mg	2/20									
3589	1375o	22.0ug	66.7ug	0/20	31.8ug	10/20	82.6ug	19/20							
a	1375o	25.6ug	.156mg	0/20	31.8ug	9/20	(82.6ug	3/20)							
b	1375o	37.6ug	.131mg	0/20	31.8ug	0/20	82.6ug	20/20							
c	1375o	47.9ug	.171mg	0/20	31.8ug	2/20	82.6ug	16/20							
d	1375o	66.5ug	.852mg	0/20	31.8ug	7/20	82.6ug	9/20							
e	1375o	.163mg	3.43mg	0/20	31.8ug	0/20	82.6ug	5/20							
f	1375o	.115mg	n.s.s.	1/20	31.8ug	3/20	82.6ug	6/20							

N-NITROSO(2,2,2-TRIFLUOROETHYL)ETHYLAMINE 82018-90-4

3590	1554	1.73mg	4.45mg	0/20	2.14mg	9/20	4.29mg	18/20	8.57mg	14/20	17.1mg	17/20	Preussmann;carc,4,755-757;1983
a	1554	4.14mg	120.mg	0/20	2.14mg	4/20	4.29mg	8/20	8.57mg	5/20	(17.1mg	3/20)	
b	1554	65.7mg	n.s.s.	0/20	2.14mg	1/20	4.29mg	0/20	8.57mg	0/20	17.1mg	0/20	

N-NITROSO-2,2,4-TRIMETHYL-1,2-DIHYDROQUINOLINE POLYMER 29929-77-9

3591	1258	1.34mg	15.4mg	0/22	1.86mg	6/20							Boyland;ejca,4,233-239;1968

	Spe	Strain	Site	Xpo+Xpt		TD50	2Tailpvl
	Sex	Route	Hist	Notes		DR	AuOp

1-NITROSO-3,4,5-TRIMETHYLPIPERAZINE ..:...1ug....:...10.....:...100.....:...1mg....:...10.....:...100....:..1g.....:..10

ID	Spe	Sex	Strain	Route	Site	Hist	Xpo+Xpt	Notes	TD50	DR	2Tailpvl	AuOp
3592	H	m	syg	gav	for	pam	46w69		1.32mg		P<.0005	+
a	H	m	syg	gav	lun	ade	46w69		1.73mg		P<.0005	+
b	H	m	syg	gav	lun	car	46w69		3.20mg		P<.002	+
c	H	m	syg	gav	liv	ang	46w69		6.18mg		P<.02	
d	H	m	syg	gav	tba	mix	46w69		noTD50		P<.0005	
3593	R	f	f34	wat	nas	olc	30w85		.151mg	*	P<.0005	+
a	R	f	f34	wat	liv	nnd	30w85		no dre		P=1.	
b	R	f	f34	wat	tba	mix	30w85		no dre		P=1.	

N-NITROSOALLYL-2,3-DIHYDROXYPROPYLAMINE ..:...1_ug....:...10.....:...100.....:...1mg....:...10.....:...100....:..1g.....:..10

ID	Spe	Sex	Strain	Route	Site	Hist	Xpo+Xpt	Notes	TD50	DR	2Tailpvl	AuOp
3594	R	f	f34	wat	eso	mix	50w55		.825mg		P<.0005	+
a	R	f	f34	wat	nas	mix	50w55		.972mg		P<.0005	+
b	R	f	f34	wat	eso	car	50w55		1.13mg		P<.0005	+
c	R	f	f34	wat	liv	nnd	50w55		no dre		P=1.	-

N-NITROSOALLYL-2-HYDROXYPROPYLAMINE ..:...1ug....:...10.....:...100.....:...1mg....:...10.....:...100....:..1g.....:..10

ID	Spe	Sex	Strain	Route	Site	Hist	Xpo+Xpt	Notes	TD50	DR	2Tailpvl	AuOp
3595	R	f	f34	wat	nas	mix	50w65		.877mg		P<.0005	+
a	R	f	f34	wat	liv	hpc	50w65		1.38mg		P<.0005	+
b	R	f	f34	wat	eso	mix	50w65		4.67mg		P<.003	+

N-NITROSOALLYL-2-OXOPROPYLAMINE 100ng..:...1ug....:...10.....:...100.....:...1mg....:...10.....:...100....:..1g.....:..10

ID	Spe	Sex	Strain	Route	Site	Hist	Xpo+Xpt	Notes	TD50	DR	2Tailpvl	AuOp
3596	H	f	syg	gav	nac	mix	40w55	e	1.19mg		P<.0005	+
a	H	f	syg	gav	nac	car	40w55	e	1.36mg		P<.0005	
b	H	f	syg	gav	liv	mix	40w55	e	3.78mg		P<.007	
c	H	f	syg	gav	liv	hpa	40w55	e	4.87mg		P<.02	+
d	H	f	syg	gav	liv	clc	40w55	e	21.2mg		P<.3	
e	H	f	syg	gav	lun	tum	40w55	e	no dre		P=1.	
3597	R	f	f34	wat	liv	hpc	50w85		.335mg		P<.0005	+
a	R	f	f34	wat	eso	mix	50w85		1.25mg		P<.002	+

N-NITROSOALLYLETHANOLAMINE 100ng..:...1ug....:...10.....:...100.....:...1mg....:...10.....:...100....:..1g.....:..10

ID	Spe	Sex	Strain	Route	Site	Hist	Xpo+Xpt	Notes	TD50	DR	2Tailpvl	AuOp
3598	R	f	f34	wat	nas	mix	50w65		.491mg		P<.0005	+
a	R	f	f34	wat	liv	hpc	50w65		4.12mg		P<.003	+

NITROSOAMYLURETHAN 100ng..:...1ug....:...10.....:...100.....:...1mg....:...10.....:...100....:..1g.....:..10

ID	Spe	Sex	Strain	Route	Site	Hist	Xpo+Xpt	Notes	TD50	DR	2Tailpvl	AuOp
3599	R	f	don	wat	eso	sqc	52w60	ae	1.01mg	Z	P<.0005	+
a	R	f	don	wat	mix	sqc	52w60	ae	1.09mg	*	P<.0005	+
b	R	f	don	wat	eso	pam	52w60	ae	1.46mg	*	P<.0005	+
c	R	f	don	wat	for	pam	52w60	ae	1.54mg	Z	P<.0005	
d	R	f	don	wat	mix	pam	52w60	ae	1.70mg	Z	P<.0005	
e	R	f	don	wat	for	sqc	52w60	ae	2.81mg	Z	P<.002	
f	R	f	don	wat	tba	mix	52w60	ae	.336mg	*	P<.0005	

NITROSOANABASINE 100ng..:...1ug....:...10.....:...100.....:...1mg....:...10.....:...100....:..1g.....:..10

ID	Spe	Sex	Strain	Route	Site	Hist	Xpo+Xpt	Notes	TD50	DR	2Tailpvl	AuOp
3600	R	f	cbt	wat	eso	mix	73w73	e	12.9mg		P<.0005	+
a	R	f	cbt	wat	eso	ben	73w73	e	17.7mg		P<.0005	+
b	R	f	cbt	wat	eso	sqc	73w73	e	413.mg		P<.5	+
c	R	f	cbt	wat	liv	tum	73w73	e	no dre		P=1.	
3601	R	m	cbt	wat	eso	mix	73w73	e	11.0mg		P<.002	+
a	R	m	cbt	wat	eso	ben	73w73	e	28.1mg		P<.03	+
b	R	m	cbt	wat	eso	sqc	73w73	e	86.0mg		P<.2	+
c	R	m	cbt	wat	liv	tum	73w73	e	no dre		P=1.	

N-NITROSOAZETIDINE 100ng..:...1ug....:...10.....:...100.....:...1mg....:...10.....:...100....:..1g.....:..10

ID	Spe	Sex	Strain	Route	Site	Hist	Xpo+Xpt	Notes	TD50	DR	2Tailpvl	AuOp
3602	H	m	syg	gav	liv	mix	50w85		7.14mg		P<.002	+
a	H	m	syg	gav	liv	hpa	50w85		10.7mg		P<.007	
b	H	m	syg	gav	lun	tum	50w85		no dre		P=1.	
c	H	m	syg	gav	tba	mix	50w85		8.83mg		P<.08	

N-NITROSOBENZTHIAZURON 100ng..:...1ug....:...10.....:...100.....:...1mg....:...10.....:...100....:..1g.....:..10

ID	Spe	Sex	Strain	Route	Site	Hist	Xpo+Xpt	Notes	TD50	DR	2Tailpvl	AuOp
3603	R	b	wis	gav	for	mix	51w58		1.13mg	*	P<.0005	+
a	R	b	wis	gav	for	sqc	51w58		1.48mg	*	P<.0005	+
b	R	b	wis	gav	for	pam	51w58		3.75mg	\	P<.0005	
c	R	b	wis	gav	liv	mix	51w58		11.3mg	*	P<.03	

N-NITROSOBIS(2-HYDROXYPROPYL)AMINE ..:...1ug....:...10.....:...100.....:...1mg....:...10.....:...100....:..1g.....:..10

ID	Spe	Sex	Strain	Route	Site	Hist	Xpo+Xpt	Notes	TD50	DR	2Tailpvl	AuOp
3604	R	f	f34	wat	eso	mix	42w55		.813mg		P<.0005	+
a	R	f	f34	wat	eso	car	42w55		1.25mg		P<.0005	
b	R	f	f34	wat	nas	olc	42w55		3.04mg		P<.002	+
c	R	f	f34	wat	liv	nnd	42w55		no dre		P=1.	
3605	R	m	mrw	gav	pro	car	74w74	er	437.mg	*	P<.7	+
3606	R	m	wis	wat	lun	ade	45w52	es	.881mg	*	P<.0005	+
a	R	m	wis	wat	lun	adc	45w52	es	3.36mg	*	P<.0005	
b	R	m	wis	wat	liv	hem	45w52	es	4.55mg	*	P<.0005	
c	R	m	wis	wat	lun	crt	45w52	es	9.48mg	*	P<.003	
d	R	m	wis	wat	liv	hpc	45w52	es	2.40mg	\	P<.03	

N-NITROSOBIS(2-OXOPROPYL)AMINE 100ng..:...1ug....:...10.....:...100.....:...1mg....:...10.....:...100....:..1g.....:..10

ID	Spe	Sex	Strain	Route	Site	Hist	Xpo+Xpt	Notes	TD50	DR	2Tailpvl	AuOp
3607	R	f	f34	gav	liv	hpc	30w95		.684mg		P<.002	+
a	R	f	f34	gav	tba	mix	30w95		noTD50		P<.6	

	RefNum	LoConf	UpConf	Cntrl	1Dose	1Inc	2Dose	2Inc			Citation or Pathology
											Brkly Code

1-NITROSO-3,4,5-TRIMETHYLPIPERAZINE   75881-18-4

3592	1570	.623mg	4.85mg	3/20	4.57mg	14/20					Lijinsky;carc,4,1165-1167;1983
a	1570	.844mg	4.28mg	0/20	4.57mg	11/20					
b	1570	1.37mg	12.1mg	0/20	4.57mg	7/20					
c	1570	2.13mg	n.s.s.	0/20	4.57mg	4/20					
d	1570	n.s.s.	.813mg	3/20	4.57mg	20/20					
3593	1570	88.2ug	.270mg	0/20	.259mg	13/20	.980mg	18/20			
a	1570	.564mg	n.s.s.	2/20	.259mg	0/20	.980mg	0/20			
b	1570	.184mg	n.s.s.	19/20	.259mg	17/20	.980mg	18/20			

N-NITROSOALLYL-2,3-DIHYDROXYPROPYLAMINE   88208-16-6

3594	1613	.412mg	1.78mg	0/20	8.16mg	17/20					Lijinsky;clet,22,281-288;1984
a	1613	.495mg	2.11mg	0/20	8.16mg	16/20					
b	1613	.578mg	2.49mg	0/20	8.16mg	15/20					
c	1613	9.41mg	n.s.s.	2/20	8.16mg	0/20					

N-NITROSOALLYL-2-HYDROXYPROPYLAMINE   91308-70-2

3595	1613	.438mg	1.90mg	0/20	6.22mg	17/20					Lijinsky;clet,22,281-288;1984
a	1613	.705mg	3.11mg	0/20	6.22mg	14/20					
b	1613	1.89mg	24.4mg	0/20	6.22mg	6/20					

N-NITROSOALLYL-2-OXOPROPYLAMINE   91308-71-3

3596	1682	.591mg	2.83mg	0/20	5.67mg	12/20					Lijinsky;jnci,74,923-926;1985
a	1682	.665mg	3.37mg	0/20	5.67mg	11/20					
b	1682	1.43mg	43.5mg	0/20	5.67mg	5/20					
c	1682	1.68mg	n.s.s.	0/20	5.67mg	4/20					
d	1682	3.45mg	n.s.s.	0/20	5.67mg	1/20					
e	1682	6.53mg	n.s.s.	0/20	5.67mg	0/20					
3597	1613	.170mg	.728mg	0/20	1.18mg	16/20					Lijinsky;clet,22,281-288;1984
a	1613	.535mg	4.73mg	0/20	1.18mg	7/20					

N-NITROSOALLYLETHANOLAMINE   91308-69-9

3598	1613	.202mg	1.12mg	0/20	5.49mg	19/20					Lijinsky;clet,22,281-288;1984
a	1613	1.67mg	21.5mg	0/20	5.49mg	6/20					

NITROSOAMYLURETHAN   (1-amyl-1-nitrosourethan) 64005-62-5

3599	1494	.664mg	1.63mg	0/37	2.86mg	19/29	5.71mg	17/29	(11.4mg	18/27)	Onodera;gann,73,48-54;1982
a	1494	.769mg	1.59mg	0/37	2.86mg	21/29	5.71mg	21/27	11.4mg	21/27	
b	1494	1.02mg	2.18mg	0/37	2.86mg	16/29	5.71mg	19/29	11.4mg	18/27	
c	1494	.746mg	3.98mg	0/37	2.86mg	10/29	(5.71mg	3/29	11.4mg	2/27)	
d	1494	1.05mg	3.02mg	0/37	2.86mg	11/29	5.71mg	14/29	(11.4mg	8/27)	
e	1494	1.14mg	12.1mg	0/37	2.86mg	6/29	(5.71mg	1/29	11.4mg	0/27)	
f	1494	.187mg	.626mg	11/37	2.86mg	29/29	5.71mg	29/29	11.4mg	26/27	

NITROSOANABASINE   1133-64-8

3600	1260	4.84mg	35.9mg	0/5	98.0mg	12/13					Boyland;bjca,18,265-270;1964
a	1260	7.54mg	53.0mg	0/5	98.0mg	11/13					
b	1260	67.2mg	n.s.s.	0/5	98.0mg	1/13					
c	1260	129.mg	n.s.s.	0/5	98.0mg	0/13					
3601	1260	4.19mg	37.0mg	0/3	85.7mg	13/14					
a	1260	12.6mg	n.s.s.	0/3	85.7mg	9/14					
b	1260	29.5mg	n.s.s.	0/3	85.7mg	4/14					
c	1260	122.mg	n.s.s.	0/3	85.7mg	0/14					

N-NITROSOAZETIDINE   15216-10-1

3602	1608	3.06mg	27.1mg	0/20	6.72mg	7/20					Lijinsky;carc,5,875-878;1984
a	1608	4.04mg	123.mg	0/20	6.72mg	5/20					
b	1608	18.5mg	n.s.s.	0/20	6.72mg	0/20					
c	1608	3.05mg	n.s.s.	3/20	6.72mg	8/20					

N-NITROSOBENZTHIAZURON   (1-(2'-benzothiazolyl)-3-methyl-3-nitrosourea) 51542-33-7

3603	1422	.746mg	1.74mg	0/22	7.14mg	27/30	14.3mg	25/30			Ungerer;zkko,81,217-224;1974
a	1422	.984mg	2.31mg	0/22	7.14mg	21/30	14.3mg	25/30			
b	1422	1.82mg	11.0mg	0/22	7.14mg	10/30	(14.3mg	4/30)			
c	1422	5.67mg	n.s.s.	0/22	7.14mg	6/30	14.3mg	5/30			

N-NITROSOBIS(2-HYDROXYPROPYL)AMINE   (N-bis(2-hydroxypropyl)nitrosamine) 53609-64-6

3604	1612	.414mg	1.77mg	0/20	6.83mg	16/20					Lijinsky;carc,5,167-170;1984
a	1612	.630mg	2.88mg	0/20	6.83mg	13/20					
b	1612	1.30mg	11.5mg	0/20	6.83mg	7/20					
c	1612	7.87mg	n.s.s.	3/20	6.83mg	0/20					
3605	1553	55.4mg	n.s.s.	0/15	17.9mg	2/14	35.7mg	1/15	71.4mg	1/15	Pour;carc,4,49-55;1983
3606	1191	.358mg	2.34mg	0/12	5.00mg	6/10	25.0mg	9/9			Konishi;gann,69,573-577;1978
a	1191	1.51mg	12.6mg	0/12	5.00mg	3/10	25.0mg	6/9			
b	1191	1.88mg	15.8mg	0/12	5.00mg	1/10	25.0mg	6/9			
c	1191	3.22mg	61.8mg	0/12	5.00mg	0/10	25.0mg	4/9			
d	1191	.718mg	n.s.s.	0/12	5.00mg	3/10	(25.0mg	1/9)			

N-NITROSOBIS(2-OXOPROPYL)AMINE   60599-38-4

3607	1551	.254mg	3.08mg	0/20	.644mg	5/12					Lijinsky;clet,19,207-213;1983
a	1551	n.s.s.	n.s.s.	18/20	.644mg	12/12					

```
       Spe Strain Site   Xpo+Xpt                                                       TD50    2Tailpvl
       Sex  Route  Hist  Notes                                                           DR      AuOp

3608 R f f34 wat liv mix 50w75                               .  +  .                   .232mg   P<.0005+
a    R f f34 wat lun mix 50w75                                                         .399mg   P<.0005+
b    R f f34 wat liv hpc 50w75                                                         .582mg   P<.0005+
c    R f f34 wat liv hes 50w75                                                         .892mg   P<.0005+
3609 R f mrc gav liv mix 73w77 as                              .  +  .                 .283mg Z P<.0005+
a    R f mrc gav thy mix 73w77 as                                                      1.39mg * P<.02
b    R f mrc gav lun ade 73w77 as                                                      14.0mg * P<.7
c    R f mrc gav npc mix 73w77 as                                                      14.0mg * P<.7
d    R f mrc gav tba tum 73w77 as                                                      .171mg * P<.0005
3610 R m mrc gav urt pam 72w80 as                                .  +  .               .878mg * P<.0005+
a    R m mrc gav liv mix 72w80 as                                                      .989mg * P<.005 +
b    R m mrc gav lun mix 72w80 as                                                      1.10mg * P<.0005+
c    R m mrc gav npc mix 72w80 as                                                      1.23mg * P<.0005+
d    R m mrc gav col mix 72w80 as                                                      1.60mg * P<.002 +
e    R m mrc gav pro sqc 72w80 as                                                      2.21mg * P<.003
f    R m mrc gav thy mix 72w80 as                                                      .968mg * P<.04
g    R m mrc gav tba tum 72w80 as                                                      99.5ug * P<.0005
3611 R m mrw gav urt mix 24m24                                   .  +  .               .891mg   P<.0005+
a    R m mrw gav liv tum 24m24                                                         1.56mg   P<.002 +
b    R m mrw gav nas tum 24m24                                                         1.56mg   P<.002 +
c    R m mrw gav clr tum 24m24                                                         1.56mg   P<.002 +
d    R m mrw gav lun tum 24m24                                                         1.92mg   P<.003 +
e    R m mrw gav pro sqk 24m24                                                         2.41mg   P<.006 +
f    R m mrw gav thy tum 24m24                                                         3.16mg   P<.02  +
3612 R m mrw gav pro car 74w74 er                                 .  +  .              1.53mg * P<.003 +

N-NITROSOBIS(2,2,2-TRIFLUOROETHYL)AMINE..:..1_ug....:.10.......:.100...:..1mg...:..10....:.100....:.1g......:.10
3613 R f f34 wat liv mal  7m30 e                                    .>                 4.98mg   P<.3  -
3614 R m sda wat mix mal  7m26 e                                        .>              no dre   P=1.  -
a    R m sda wat liv tum  7m26 e                                                       no dre   P=1.

NITROSOCHLORDIAZEPOXIDE                 100ng..:..1ug....:.10......:.100...:..1mg...:..10.....:..100...:..1g.....:.10
3615 M f c7b wat lun tum 17m25                                         .>              no dre   P=1.  -
a    M f c7b wat liv tum 17m25                                                         no dre   P=1.  -
b    M f c7b wat tba mix 17m25                                                         no dre   P=1.  -
3616 M m c7b wat lun tum 17m25                                      .>                 no dre   P=1.  -
a    M m c7b wat tba mix 17m25                                                         no dre   P=1.  -

N-NITROSOCIMETIDINE                     100ng..:..1ug....:.10......:.100...:..1mg...:..10.....:..100...:..1g.....:.10
3617 M f cb6 wat liv mix 24m24 g                                              .>       no dre   P=1.  -
a    M f cb6 wat lun tum 24m24 g                                                       no dre   P=1.  -
3618 R f f34 wat thy tum 25m30                                 .   +                   18.8mg   P<.007 -
a    R f f34 wat pit tum 25m30                                                         3.60mg   P<.02  -
b    R f f34 wat liv nnd 25m30                                                         43.3mg   P<.5  -
3619 R m f34 wat liv nnd 25m30                                      .>                 no dre   P=1.  -
3620 R f sda gav tba mix 12m24                                           .>            no dre   P=1.  -
a    R f sda gav tba mal 12m24                                                         no dre   P=1.  -
3621 R m sda gav tba mix 12m24                                           .>            no dre   P=1.  -
a    R m sda gav tba mal 12m24                                                         4.82gm * P<.9  -

NITROSODIBUTYLAMINE                     100ng..:..1ug....:.10......:.100...:..1mg...:..10.....:..100...:..1g.....:.10
3622 M m icr eat for mix 52w65                                 .  +.                   .686mg   P<.0005
a    M m icr eat for sqc 52w65                                                         1.09mg   P<.0005+
b    M m icr eat liv mix 52w65                                                         2.65mg   P<.0005+
c    M m icr eat liv ade 52w65                                                         4.34mg   P<.0005
d    M m icr eat for sqp 52w65                                                         7.69mg   P<.008
e    M m icr eat lun ade 52w65                                                         8.00mg   P<.1  +
f    M m icr eat eso pam 52w65                                                         11.9mg   P<.03  +
3623 R m f34 gav liv hpc  7m25                                 .  +  .                 .691mg   P<.0005+
a    R m f34 gav for car  7m25                                                         .914mg   P<.0005+
b    R m f34 gav lun car  7m25                                                         1.16mg   P<.003 +
c    R m f34 gav ubl tcc  7m25                                                         1.47mg   P<.002 +
d    R m f34 gav tba mix  7m25                                                         no dre   P=1.

N-NITROSODIETHANOLAMINE                 100ng..:..1ug....:.10......:.100...:..1mg...:..10.....:..100...:..1g.....:.10
3624 R f f34 wat liv hpc 13m24 a                                  .  +  .              3.27mg * P<.0005+

a    R f f34 wat nas adc 13m24 a                                                       44.1mg Z P<.0005+
b    R f f34 wat nas olc 13m24 a                                                       128.mg Z P<.005 +
c    R f f34 wat liv clc 13m24 a                                                       197.mg * P<.007 +
d    R f f34 wat kid tum 13m24 a                                                       327.mg * P<.009 +
e    R f f34 wat eso pam 13m24 a                                                       2.05gm * P<.2  +
f    R f f34 wat nas sqc 13m24 a                                                       4.32gm * P<.9  +
3625 R f f34 wat liv mix 17m30 a                                  .+  .                1.90mg * P<.0005+
a    R f f34 wat liv nnd 17m30 a                                                       2.93mg * P<.0005+
b    R f f34 wat liv hpc 17m30 a                                                       4.05mg * P<.0005+
c    R f f34 wat kid mix 17m30 a                                                       42.2mg * P<.03  +
d    R f f34 wat liv caa 17m30 a                                                       56.8mg * P<.08
3626 R f f34 wat liv hpc 12m28 e                                  .  +  .              1.94mg   P<.0005+
a    R f f34 wat liv hes 12m28 e                                                       45.4mg   P<.3
```

	RefNum	LoConf	UpConf	Cntrl	1Dose	1Inc	2Dose	2Inc			Citation or Pathology
											Brkly Code
3608	1639	.110mg	.506mg	0/20	1.50mg	18/20					
a	1639	.200mg	.967mg	1/20	1.50mg	15/20					Lijinsky;zkko,107,178-182;1984
b	1639	.290mg	1.39mg	0/20	1.50mg	12/20					
c	1639	.413mg	2.52mg	0/20	1.50mg	9/20					
3609	1911	.167mg	.511mg	0/15	.357mg	0/15	.714mg	12/15	1.43mg	14/15	Pour;clet,45,49-57;1989
a	1911	.651mg	n.s.s.	0/15	.357mg	1/15	.714mg	5/15	1.43mg	3/15	
b	1911	2.27mg	n.s.s.	0/15	.357mg	0/15	.714mg	1/15	1.43mg	0/15	
c	1911	2.27mg	n.s.s.	0/15	.357mg	0/15	.714mg	1/15	1.43mg	0/15	
d	1911	90.5ug	.399mg	5/15	.357mg	5/15	.714mg	15/15	1.43mg	15/15	
3610	1911	.458mg	2.02mg	0/15	.357mg	0/15	.714mg	4/14	1.43mg	9/15	
a	1911	.476mg	9.48mg	1/15	.357mg	2/15	.714mg	5/14	1.43mg	7/15	
b	1911	.548mg	2.94mg	0/15	.357mg	0/15	.714mg	4/14	1.43mg	7/15	
c	1911	.594mg	3.45mg	0/15	.357mg	0/15	.714mg	3/14	1.43mg	7/15	
d	1911	.719mg	5.63mg	0/15	.357mg	0/15	.714mg	2/14	1.43mg	6/15	
e	1911	.898mg	12.1mg	0/15	.357mg	0/15	.714mg	1/14	1.43mg	5/15	
f	1911	.508mg	n.s.s.	0/15	.357mg	4/15	.714mg	5/14	1.43mg	4/15	
g	1911	49.8ug	.217mg	3/15	.357mg	11/15	.714mg	14/14	1.43mg	15/15	
3611	1393	.411mg	2.36mg	0/15	1.43mg	10/15					Pour;clet,13,303-308;1981
a	1393	.658mg	5.63mg	0/15	1.43mg	7/15					
b	1393	.658mg	5.63mg	0/15	1.43mg	7/15					
c	1393	.658mg	5.63mg	0/15	1.43mg	7/15					
d	1393	.770mg	9.22mg	0/15	1.43mg	6/15					
e	1393	.907mg	22.8mg	0/15	1.43mg	5/15					
f	1393	1.08mg	n.s.s.	0/15	1.43mg	4/15					
3612	1553	.658mg	7.19mg	0/15	.357mg	0/15	.714mg	2/12	1.43mg	5/15	Pour;carc,4,49-55;1983

N-NITROSOBIS(2,2,2-TRIFLUOROETHYL)AMINE   (6-F-DEN) 625-89-8

	RefNum	LoConf	UpConf	Cntrl	1Dose	1Inc	2Dose	2Inc			Citation or Pathology
3613	1342	.931mg	n.s.s.	1/20	.610mg	2/12					Preussmann;carc,2,753-756;1981
3614	1342	2.36mg	n.s.s.	2/35	.561mg	2/35					
a	1342	4.86mg	n.s.s.	0/35	.561mg	0/35					

NITROSOCHLORDIAZEPOXIDE   51715-17-4

	RefNum	LoConf	UpConf	Cntrl	1Dose	1Inc					Citation or Pathology
3615	1257	5.85mg	n.s.s.	14/43	4.02mg	6/21					Giner-Sorolla;fctx,18,81-83;1980
a	1257	18.1mg	n.s.s.	0/43	4.02mg	0/21					
b	1257	8.43mg	n.s.s.	24/43	4.02mg	6/21					
3616	1257	3.32mg	n.s.s.	15/18	3.35mg	14/23					
a	1257	3.12mg	n.s.s.	16/18	3.35mg	15/23					

N-NITROSOCIMETIDINE   73785-40-7

	RefNum	LoConf	UpConf	Cntrl	1Dose	1Inc	2Dose	2Inc			Citation or Pathology
3617	1925	558.mg	n.s.s.	3/20	22.6mg	1/17	226.mg	1/17			Anderson;canr,45,3561-3566;1985
a	1925	714.mg	n.s.s.	1/20	22.6mg	2/17	226.mg	0/17			
3618	1611	7.11mg	217.mg	0/20	4.99mg	5/20					Lijinsky;canr,44,447-449;1984
a	1611	1.32mg	n.s.s.	11/20	4.99mg	18/20					
b	1611	8.56mg	n.s.s.	3/20	4.99mg	5/20					
3619	1611	7.07mg	n.s.s.	5/20	3.49mg	5/20					
3620	1699	142.mg	n.s.s.	10/50	7.17mg	1/20	71.7mg	3/20			Habs;hepg,29,265-266;1982
a	1699	232.mg	n.s.s.	7/50	7.17mg	0/20	71.7mg	1/20			
3621	1699	136.mg	n.s.s.	9/50	7.17mg	1/20	71.7mg	3/20			
a	1699	149.mg	n.s.s.	5/50	7.17mg	1/20	71.7mg	2/20			

NITROSODIBUTYLAMINE   924-16-3

	RefNum	LoConf	UpConf	Cntrl	1Dose	1Inc					Citation or Pathology
3622	1434	.416mg	1.18mg	0/30	4.80mg	33/39					Takayama;gann,60,353;1969
a	1434	.666mg	1.92mg	0/30	4.80mg	27/39					
b	1434	1.44mg	5.65mg	0/30	4.80mg	15/39					
c	1434	2.11mg	12.7mg	0/30	4.80mg	10/39					
d	1434	3.13mg	122.mg	0/30	4.80mg	6/39					
e	1434	2.82mg	n.s.s.	2/30	4.80mg	8/39					
f	1434	4.10mg	n.s.s.	0/30	4.80mg	4/39					
3623	1551	.345mg	1.65mg	0/20	.857mg	12/20					Lijinsky;clet,19,207-213;1983
a	1551	.436mg	2.39mg	0/20	.857mg	10/20					
b	1551	.508mg	6.79mg	1/20	.857mg	9/20					
c	1551	.629mg	5.57mg	0/20	.857mg	7/20					
d	1551	.136mg	n.s.s.	19/20	.857mg	19/20					

N-NITROSODIETHANOLAMINE   1116-54-7

	RefNum	LoConf	UpConf	Cntrl	1Dose	1Inc	2Dose	2Inc	3Dose	3Inc	Citation or Pathology
3624	1610	1.86mg	5.75mg	0/40	7.77mg	15/16	14.4mg	16/16	31.4mg	19/20	99.8mg 20/20	Lijinsky;fctx,22,23-26;
a	1610	23.1mg	106.mg	0/40	7.77mg	3/16	14.4mg	2/16	31.4mg	8/20	(99.8mg 1/20)	1984/pers.comm.
b	1610	48.4mg	1.12gm	0/40	7.77mg	1/16	14.4mg	0/16	31.4mg	4/20	(99.8mg 1/20)	
c	1610	81.3mg	3.82gm	0/40	7.77mg	3/16	14.4mg	1/16	31.4mg	3/20	99.8mg 5/20	
d	1610	133.mg	13.1gm	0/40	7.77mg	0/16	14.4mg	0/16	31.4mg	4/20	99.8mg 2/20	
e	1610	333.mg	n.s.s.	0/40	7.77mg	0/16	14.4mg	0/16	31.4mg	0/20	99.8mg 1/20	
f	1610	190.mg	n.s.s.	0/40	7.77mg	3/16	14.4mg	1/16	31.4mg	2/20	99.8mg 1/20	
3625	1707	1.32mg	2.92mg	1/20	.879mg	10/39	1.00mg	5/20	2.01mg	14/20	2.51mg 27/27	Lijinsky;carc,6,1697-1681;1985
a	1707	1.96mg	5.05mg	1/20	.879mg	8/39	1.00mg	4/20	2.01mg	13/20	2.51mg 19/27	
b	1707	2.65mg	6.65mg	0/20	.879mg	5/39	1.00mg	2/20	2.01mg	7/20	2.51mg 19/27	
c	1707	14.6mg	n.s.s.	0/20	.879mg	0/39	1.00mg	0/20	2.01mg	1/20	2.51mg 3/27	
d	1707	17.2mg	n.s.s.	0/20	.879mg	0/39	1.00mg	0/20	2.01mg	2/20	2.51mg 1/27	
3626	1928	.988mg	4.36mg	0/20	2.56mg	14/20					Hecht;carc,10,1475-1477;1989/pers.comm.
a	1928	7.40mg	n.s.s.	0/20	2.56mg	1/20					

```
        Spe Strain  Site   Xpo+Xpt                                                                      TD50      2Tailpvl
            Sex   Route   Hist    Notes                                                                    DR      AuOp
3627 R m f34 wat liv hpc 48w95 a                                        . + .                           2.77mg * P<.0005+

a    R m f34 wat nas adc 48w95 a                                                                        15.5mg * P<.0005+
b    R m f34 wat liv clc 48w95 a                                                                        23.9mg Z P<.002 +
c    R m f34 wat nas olc 48w95 a                                                                        50.4mg Z P<.003 +
d    R m f34 wat kid tum 48w95 a                                                                        99.7mg * P<.002 +
e    R m f34 wat eso pam 48w95 a                                                                        207.mg * P<.003 +
f    R m f34 wat nas sqc 48w95 a                                                                        280.mg * P<.009 +
g    R m f34 wat eso car 48w95 a                                                                        432.mg * P<.2  +
3628 R m f34 wat liv mix 17m29 a                                       . + .                            2.84mg Z P<.0005+
a    R m f34 wat liv nnd 17m29 a                                                                        3.93mg * P<.002 +
b    R m f34 wat liv hpc 17m29 a                                                                        9.03mg * P<.0005+
c    R m f34 wat kid mix 17m29 a                                                                        19.0mg * P<.07 +
d    R m f34 wat liv caa 17m29 a                                                                        38.6mg * P<.06
3629 R m sda wat liv mix 27m34 ae                                         . +.                          8.23mg Z P<.0005+

a    R m sda wat liv hpd 27m34 ae                                                                       9.51mg Z P<.0005+
b    R m sda wat nas mix 27m34 ae                                                                       168.mg Z P<.0005+
c    R m sda wat nas olp 27m34 ae                                                                       679.mg Z P<.002
d    R m sda wat nas sqc 27m34 ae                                                                       954.mg Z P<.004
e    R m sda wat liv cgd 27m34 ae                                                                       10.7gm * P<.2
3630 R m sda wat liv mix 38m39 a                                            . +    .                    38.4mg * P<.0005+
a    R m sda wat liv hpc 38m39 a                                                                        70.4mg * P<.0005+
b    R m sda wat liv hmb 38m39 a                                                                        126.mg * P<.06 +
c    R m sda wat ner tum 38m39 a                                                                        42.5mg * P<.3  +
d    R m sda wat --- tum 38m39 a                                                                        55.6mg * P<.2  +
e    R m sda wat git mix 38m39 a                                                                        143.mg * P<.6  +
f    R m sda wat liv hpa 38m39 a                                                                        347.mg * P<.3  +
g    R m sda wat tba ben 38m39 a                                                                        7.93mg * P<.2
h    R m sda wat tba mal 38m39 a                                                                        23.5mg * P<.3  +

N-NITROSODIETHYLAMINE               100ng..:..1ug...:..10....:..100....:..1mg...:..10....:..100...:..1g.....:..10
3631 N b bbb ipj nac mec 32m32 Ww                :    (+)    :                                          12.2ug   P<.0005+

a    N b bbb ipj liv hpc 32m32 Ww                                                                       18.4ug   P<.008
b    N b bbb ipj tba mal 32m32 Ww                                                                       12.2ug   P<.0005
3632 P b cym ipj liv hpc 13y13 Wuw                  :    +   :                                          27.7ug * P<.0005+
a    P b cym ipj tba mal 13y13 Wuw                                                                      27.7ug * P<.0005
3633 P b cym eat liv hpc 16y16 Ww                             :    +  :                                 2.08mg   P<.0005+
a    P b cym eat tba mal 16y16 Ww                                                                       2.08mg   P<.0005
3634 P b rhe ipj liv hpc 20y20 Wuw                :    +  :                                             57.8ug * P<.0005+

a    P b rhe ipj tba mal 20y20 Wuw                                                                      37.4ug * P<.0005

3635 P b rhe eat liv hpc 22y22 Ww                            :    +  :                                  2.62mg   P<.0005+
a    P b rhe eat tba mal 22y22 Ww                                                                       1.59mg   P<.0005
3636 R f f34 wat eso mix  7m30                        . + .                                             25.5ug Z P<.0005+

a    R f f34 wat liv hpc  7m30                                                                          no dre   P=1.  +
b    R f f34 wat ton bcc  7m30                                                                          no dre   P=1.  +
3637 R f f34 wat liv mix 14m30                    . + .                                                 7.87ug \ P<.0005+
a    R f f34 wat eso mix 14m30                                                                          20.7ug / P<.0005+
3638 R f f34 wat mix mix 24m30                     . + .                                                13.1ug   P<.0005+
a    R f f34 wat eso mix 24m30                                                                          15.0ug   P<.0005+
b    R f f34 wat for bcp 24m30                                                                          54.9ug   P<.007 +
c    R f f34 wat liv mix 24m30                                                                          41.6ug   P<.02 +
3639 R f f34 wat eso tum 30w65                    . + .                                                 21.9ug   P<.0005+
a    R f f34 wat liv tum 30w65                                                                          no dre   P=1.
b    R f f34 wat tba tum 30w65                                                                          no dre   P=1.
3640 R f fis wat liv hpt 86w86 e                          .>                                            no dre   P=1.  -
a    R f fis wat tba mix 86w86 e                                                                        no dre   P=1.  -
3641 R m fis wat liv hpt 86w86 e                           .>                                           no dre   P=1.  -
a    R m fis wat tba tum 86w86 e                                                                        no dre   P=1.  -
3642 R m sda wat mix tum 27m27                         . + .                                            70.6ug   P<.0005+
a    R m sda wat liv tum 27m27                                                                          .119mg   P<.0005+
b    R m sda wat eso tum 27m27                                                                          .133mg   P<.0005+
3643 R m sda wat liv mix 35m37 a                              . + .                                     .270mg Z P<.0005+
a    R m sda wat git mix 35m37 a                                                                        .401mg * P<.0005+
b    R m sda wat liv hpc 35m37 a                                                                        .541mg Z P<.0005+
c    R m sda wat eso pam 35m37 a                                                                        .715mg Z P<.0005+
d    R m sda wat liv hmm 35m37 a                                                                        .719mg * P<.0005+
e    R m sda wat eso sqc 35m37 a                                                                        3.38mg   P<.002 +
f    R m sda wat unt tum 35m37 a                                                                        6.51mg * P<.3  +
g    R m sda wat tba mal 35m37 a                                                                        .188mg * P<.0005+
h    R m sda wat tba ben 35m37 a                                                                        no dre   P=1.
3644 R f wio wat liv hpt 60w63 e                  . +  .                                                49.3ug   P<.0005+
a    R f wio wat tba mix 60w63 e                                                                        53.7ug   P<.002
3645 R m wio wat liv hpt 60w63 e                .  ±                                                    .104mg   P<.02 +
a    R m wio wat tba mix 60w63 e                                                                        .104mg   P<.02
3646 R f wis gav mgl adc 28m28 e                 .  ±                                                   .127mg   P<.08 -
a    R f wis gav liv hnd 28m28 e                                                                        31.3mg   P<1.  -
```

	RefNum	LoConf	UpConf	Cntrl	1Dose	1Inc	2Dose	2Inc					Citation or Pathology
													Brkly Code
3627	1610	1.72mg	4.60mg	1/28	6.02mg	14/16	9.02mg	14/16	16.8mg	18/20	48.4mg	20/20	Lijinsky;fctx,22,23-26;
													1984/pers.comm.
a	1610	10.1mg	32.1mg	0/28	6.02mg	5/16	9.02mg	9/16	16.8mg	8/20	48.4mg	14/20	
b	1610	12.3mg	89.9mg	0/28	6.02mg	4/16	9.02mg	2/16	16.8mg	6/20	(48.4mg	4/20)	
c	1610	20.5mg	265.mg	0/28	6.02mg	0/16	9.02mg	1/16	16.8mg	5/20	(48.4mg	2/20)	
d	1610	45.1mg	409.mg	0/28	6.02mg	0/16	9.02mg	1/16	16.8mg	2/20	48.4mg	5/20	
e	1610	71.3mg	1.32gm	0/28	6.02mg	0/16	9.02mg	0/16	16.8mg	0/20	48.4mg	4/20	
f	1610	84.7mg	9.40gm	0/28	6.02mg	0/16	9.02mg	0/16	16.8mg	1/20	48.4mg	3/20	
g	1610	106.mg	n.s.s.	0/28	6.02mg	0/16	9.02mg	0/16	16.8mg	1/20	48.4mg	1/20	
3628	1707	1.72mg	6.81mg	4/20	.640mg	6/39	.795mg	2/20	1.46mg	11/20	1.83mg	19/27	Lijinsky;carc,6,1697-1681;1985
a	1707	2.19mg	15.1mg	4/20	.640mg	6/39	.795mg	2/20	1.46mg	7/20	1.83mg	17/27	
b	1707	4.65mg	24.9mg	0/20	.640mg	0/39	.795mg	1/20	1.46mg	4/20	1.83mg	7/27	
c	1707	7.77mg	n.s.s.	0/20	.640mg	1/39	.795mg	1/20	1.46mg	2/20	1.83mg	2/27	
d	1707	11.7mg	n.s.s.	0/20	.640mg	0/39	.795mg	0/20	1.46mg	1/20	1.83mg	2/27	
3629	1483	6.04mg	11.5mg	0/88	1.07mg	7/72	4.29mg	43/72	17.9mg	33/36	(71.4mg	32/36	286.mg	31/36)	Preussmann;canr,42,
														5167-5171;1982
a	1483	6.94mg	13.3mg	0/88	1.07mg	5/72	4.29mg	40/72	17.9mg	32/36	(71.4mg	31/36	286.mg	31/36)	
b	1483	75.3mg	795.mg	0/88	1.07mg	2/72	4.29mg	0/72	17.9mg	6/36	(71.4mg	6/36	286.mg	1/36)	
c	1483	276.mg	4.62gm	0/88	1.07mg	1/72	4.29mg	0/72	17.9mg	4/36	71.4mg	3/36	(286.mg	0/36)	
d	1483	339.mg	11.0gm	0/88	1.07mg	1/72	4.29mg	0/72	17.9mg	2/36	71.4mg	3/36	(286.mg	1/36)	
e	1483	2.17gm	n.s.s.	0/88	1.07mg	0/72	4.29mg	0/72	17.9mg	0/36	71.4mg	1/36	286.mg	1/36	
3630	1838	16.0mg	201.mg	3/500	.143mg	2/80	.450mg	1/80	1.43mg	6/80			Berger;carc,8,1635-1643;1987/pers.comm.
a	1838	24.3mg	368.mg	0/500	.143mg	0/80	.450mg	1/80	1.43mg	3/80			
b	1838	30.8mg	n.s.s.	1/500	.143mg	1/80	.450mg	0/80	1.43mg	2/80			
c	1838	11.5mg	n.s.s.	54/500	.143mg	8/80	.450mg	16/80	1.43mg	11/80			
d	1838	15.3mg	n.s.s.	23/500	.143mg	8/80	.450mg	5/80	1.43mg	7/80			
e	1838	20.7mg	n.s.s.	26/500	.143mg	7/80	.450mg	3/80	1.43mg	6/80			
f	1838	47.6mg	n.s.s.	1/500	.143mg	0/80	.450mg	0/80	1.43mg	1/80			
g	1838	2.58mg	n.s.s.	362/500	.143mg	61/80	.450mg	60/80	1.43mg	64/80			
h	1838	6.85mg	n.s.s.	144/500	.143mg	29/80	.450mg	29/80	1.43mg	28/80			

N-NITROSODIETHYLAMINE   (DEN) 55-18-5

	RefNum	LoConf	UpConf	Cntrl	1Dose	1Inc	2Dose	2Inc					Citation or Pathology
3631	2001s	3.36ug	46.0ug	0/9	.897mg	10/13							Adamson;ossc,129-156;
													1982/Thorgeirsson 1994/Dalgard 1991/Thorgeirsson&Seiber pers.comm.
a	2001s	3.57ug	1.03mg	0/9	.897mg	2/3							
b	2001s	3.36ug	46.0ug	0/9	.897mg	10/13							
3632	2001m	8.71ug	76.8ug	0/132	6.30ug	0/7	.322mg	3/3	.910mg	5/5	1.35mg	5/5	2.29mg	37/40	
a	2001m	8.71ug	76.8ug	0/132	6.30ug	0/7	.322mg	3/3	.910mg	5/5	1.35mg	5/5	2.29mg	37/40	
3633	2001n	1.01mg	4.44mg	0/131	8.07mg	13/16							
a	2001n	1.01mg	4.44mg	0/131	8.07mg	13/16							
3634	2001m	24.9ug	.118mg	0/122	7.40ug	0/4	80.0ug	2/8	.340mg	6/6	.650mg	5/5	1.59mg	6/6	
					2.09mg	52/53							
a	2001m	12.8ug	92.0ug	4/122	7.40ug	0/4	80.0ug	2/8	.340mg	6/6	.650mg	5/5	1.59mg	6/6	
					2.09mg	52/53							
3635	2001n	1.04mg	5.89mg	0/104	6.87mg	11/14							
a	2001n	.620mg	4.89mg	5/104	6.87mg	11/14							
3636	1173m	14.7ug	48.3ug	0/20	4.40ug	0/20	10.4ug	3/20	26.4ug	18/19	(.119mg	13/20	.538mg	10/12)	Lijinsky;canr,41,
														4997-5003;1981/pers.comm.
a	1173m	1.53mg	n.s.s.	0/20	4.40ug	1/20	10.4ug	5/20	26.4ug	5/19	.119mg	1/20	.538mg	1/12	
b	1173m	2.24mg	n.s.s.	0/20	4.40ug	0/20	10.4ug	6/19	26.4ug	6/19	.119mg	0/20	.538mg	0/12	
3637	1173n	3.92ug	20.0ug	1/20	8.48ug	14/20	(20.7ug	5/20)					
a	1173n	11.7ug	40.8ug	0/20	8.48ug	2/20	20.7ug	17/20					
3638	1173o	6.70ug	29.5ug	0/20	14.8ug	14/20							
a	1173o	7.61ug	34.8ug	0/20	14.8ug	13/20							
b	1173o	20.7ug	.633mg	0/20	14.8ug	5/20							
c	1173o	16.4ug	n.s.s.	1/20	14.8ug	7/20							
3639	1561	11.2ug	47.7ug	0/20	.132mg	16/20							Lijinsky;fctx,21,601-605;1983
a	1561	.127mg	n.s.s.	1/20	.132mg	1/20							
b	1561	16.1ug	n.s.s.	20/20	.132mg	18/20							
3640	1041	.109mg	n.s.s.	0/15	51.6ug	0/15							Nixon;jnci,53,453-458;1974
a	1041	.109mg	n.s.s.	3/15	51.6ug	0/15							
3641	1041	84.3ug	n.s.s.	0/16	46.0ug	0/13							
a	1041	84.3ug	n.s.s.	0/16	46.0ug	0/13							
3642	1303	49.8ug	.104mg	0/90	71.4ug	52/90							Habs;onco,37,259-265;1980
a	1303	79.4ug	.190mg	0/90	71.4ug	36/90							
b	1303	87.4ug	.217mg	0/90	71.4ug	33/90							
3643	1838	.183mg	.422mg	3/500	7.14ug	2/80	22.9ug	3/80	71.4ug	36/80			Berger;carc,8,1635-1643;1987/pers.comm.
a	1838	.241mg	.807mg	26/500	7.14ug	9/80	22.9ug	7/80	71.4ug	25/80			
b	1838	.326mg	.991mg	0/500	7.14ug	1/80	22.9ug	0/80	71.4ug	21/80			
c	1838	.405mg	1.44mg	0/500	7.14ug	0/80	22.9ug	0/80	71.4ug	17/80			
d	1838	.408mg	1.45mg	0/500	7.14ug	0/80	22.9ug	2/80	71.4ug	15/80			
e	1838	1.13mg	26.3mg	1/500	7.14ug	0/80	22.9ug	0/80	71.4ug	4/80			
f	1838	1.32mg	n.s.s.	1/500	7.14ug	2/80	22.9ug	1/80	71.4ug	1/80			
g	1838	.118mg	.370mg	144/500	7.14ug	23/80	22.9ug	27/80	71.4ug	52/80			
h	1838	.368mg	n.s.s.	362/500	7.14ug	58/80	22.9ug	54/80	71.4ug	53/80			
3644	1041	23.5ug	.131mg	0/18	.136mg	10/20							Nixon;jnci,53,453-458;1974
a	1041	24.3ug	.262mg	1/18	.136mg	10/20							
3645	1041	35.9ug	n.s.s.	0/17	.104mg	4/18							
a	1041	35.9ug	n.s.s.	0/17	.104mg	4/18							
3646	1399	41.5ug	n.s.s.	1/59	10.2ug	5/58							Kroes;fctx,12,671-679;1974
a	1399	98.7ug	n.s.s.	1/59	10.2ug	1/58							

```
        Spe Strain  Site   Xpo+Xpt                                                              TD50    2Tailpvl
          Sex  Route  Hist        Notes                                                             DR     AuOp
    b   R f wis gav tba ben 28m28 e                                                             35.5ug  P<.08  -
    c   R f wis gav tba mal 28m28 e                                                             4.01mg  P<1.   -
 3647 R m wis gav liv tum 28m28 e                        .>                                     no dre  P=1.   -
    a   R m wis gav tba mal 28m28 e                                                             77.4ug  P<.4   -
    b   R m wis gav tba ben 28m28 e                                                             .257mg  P<.9   -

N-NITROSODIMETHYLAMINE          100ng..:..1ug...:..10....:..100....:..1mg...:..10....:..100....:..1g....:..10
 3648 M m amm wat lun ala 72w72 e                                  .>                           .177mg  P<.6
    a   M m amm wat liv hpa 72w72 e                                                             no dre  P=1.
 3649 M f bal wat lun ade 85w85 gs                            . + .                             .324mg  P<.0005+
    a   M f bal wat tba mix 85w85 gs                                                            .396mg  P<.3
 3650 M f cbl gav frb olp 50w72 e                         . + .                                 .153mg  P<.0005+
    a   M f cbl gav liv ben 50w72 e                                                             .350mg  P<.003
    b   M f cbl gav liv mal 50w72 e                                                             .429mg  P<.006
    c   M f cbl gav lun tum 50w72 e                                                             no dre  P=1.
    d   M f cbl gav tba mix 50w72 e                                                             96.4ug  P<.002
 3651 M m cbl gav frb olp 50w72 e                           . + .                               .161mg  P<.0005+
    a   M m cbl gav liv mal 50w72 e                                                             .179mg  P<.0005+
    b   M m cbl gav liv ben 50w72 e                                                             .508mg  P<.2   +
    c   M m cbl gav lun ade 50w72 e                                                             21.1mg  P<1.
    d   M m cbl gav tba mix 50w72 e                                                             60.0ug  P<.0005
 3652 P b rhe ipj tba mal 10y10 Ww                  (:>)                                        no dre  P=1.

 3653 R m buf eat liv hpc 26w52 ekr                              .>                             1.76mg  P<.4   +
 3654 R f f34 wat liv mix  7m26                 . + .                                           58.7ug * P<.0005+
    a   R f f34 wat liv hpc  7m26                                                               .135mg * P<.0005+
    b   R f f34 wat liv hes  7m26                                                               .438mg * P<.002 +
    c   R f f34 wat liv nnd  7m26                                                               no dre  P=1.
 3655 R m f34 gav lun mix 30w65 e               . + .                                           76.5ug  P<.0005+
    a   R m f34 gav lun a/a 30w65 e                                                             .106mg  P<.0005+
    b   R m f34 gav kid mnp 30w65 e                                                             .189mg  P<.0005+
    c   R m f34 gav liv mix 30w65 e                                                             .189mg  P<.0005+
    d   R m f34 gav liv bht 30w65 e                                                             .372mg  P<.003
    e   R m f34 gav lun sqc 30w65 e                                                             .372mg  P<.003
    f   R m f34 gav liv cab 30w65 e                                                             .822mg  P<.04
    g   R m f34 gav liv hes 30w65 e                                                             .822mg  P<.04
 3656 R f por eat liv tum 25m28 ar                          . + .                              1.40mg * P<.0005+

 3657 R f por eat liv tum 12m28 r                      . +      .                               .443mg  P<.01
 3658 R m por eat liv tum 28m28 r                        .>                                    1.18mg * P<.2   +
 3659 R f wis eat liv nod 96w96                 . + .                                           .113mg Z P<.002 +
    a   R f wis eat --- leu 96w96                                                               1.46mg * P<.005
    b   R f wis eat liv hae 96w96                                                               1.83mg * P<.03  +
    c   R f wis eat liv hpc 96w96                                                               4.10mg * P<.4   +
 3660 R m wis eat tes ldc 54w69 er              . + .                                           .136mg  P<.0005+
    a   R m wis eat liv car 54w69 er                                                            no dre  P=1.
 3661 R m wis eat liv nod 96w96                      . +     .                                  .782mg * P<.0005+
    a   R m wis eat liv hae 96w96                                                               1.95mg * P<.007 +
    b   R m wis eat liv hpc 96w96                                                               5.52mg * P<.4   +

N-NITROSODIPHENYLAMINE          100ng..:..1ug...:..10....:..100....:..1mg...:..10....:..100....:..1g....:..10
 3662 M f b6c eat TBA MXB 23m23 v                                                 :>           no dre  P=1.   -
    a   M f b6c eat liv MXB 23m23 v                                                             no dre  P=1.
    b   M f b6c eat lun MXB 23m23 v                                                             no dre  P=1.
 3663 M f b6c orl liv hpt 76w76 evx                                           .>               no dre  P=1.
    a   M f b6c orl lun ade 76w76 evx                                                           no dre  P=1.
    b   M f b6c orl tba mix 76w76 evx                                                           no dre  P=1.
 3664 M m b6c eat TBA MXB 23m23                                                     :>         no dre  P=1.   -
    a   M m b6c eat liv MXB 23m23                                                               no dre  P=1.
    b   M m b6c eat lun MXB 23m23                                                               no dre  P=1.
 3665 M m b6c orl liv hpt 76w76 evx                                      . +                    385.mg  P<.02
    a   M m b6c orl lun ade 76w76 evx                                                           9.67gm  P<.9
    b   M m b6c orl tba mix 76w76 evx                                                           897.mg  P<.6
 3666 M f b6a orl lun ade 76w76 evx                                           . +             1.02gm  P<.04
    a   M f b6a orl liv hpt 76w76 evx                                                           no dre  P=1.
    b   M f b6a orl tba mix 76w76 evx                                                           3.26gm  P<.7
 3667 M m b6a orl lun ade 76w76 evx                                        .>                  no dre  P=1.
    a   M m b6a orl liv hpt 76w76 evx                                                           no dre  P=1.
    b   M m b6a orl tba mix 76w76 evx                                                           no dre  P=1.
 3668 R f f34 eat ubl tcc 23m23                                        : + :                   116.mg  / P<.0005c
    a   R f f34 eat TBA MXB 23m23                                                               159.mg  / P<.008
    b   R f f34 eat liv MXB 23m23                                                               no dre  P=1.
 3669 R m f34 eat ubl tcc 23m23                                            : + :               299.mg  / P<.0005c
    a   R m f34 eat --- fib 23m23                                                               537.mg * P<.01
    b   R m f34 eat TBA MXB 23m23                                                               136.gm * P<1.
    c   R m f34 eat liv MXB 23m23                                                               5.45gm * P<.3
 3670 R m cbr ipj liv hpt  6m23 e                                .>                            18.9mg  P<1.   -

p-NITROSODIPHENYLAMINE          100ng..:..1ug...:..10....:..100....:..1mg...:..10....:..100....:..1g....:..10
 3671 M f b6c eat TBA MXB 53w92 asv                                                 :>         2.17gm * P<.5   -
    a   M f b6c eat liv MXB 53w92 asv                                                           2.25gm * P<.2
```

	RefNum	LoConf	UpConf	Cntrl	1Dose	1Inc	2Dose	2Inc					Citation or Pathology
													Brkly Code
b	1399	13.6ug	n.s.s.	18/59	10.2ug	27/58							
c	1399	47.3ug	n.s.s.	7/59	10.2ug	7/58							
3647	1399	78.4ug	n.s.s.	0/39	7.14ug	0/40							
a	1399	18.1ug	n.s.s.	4/39	7.14ug	7/40							
b	1399	18.2ug	n.s.s.	8/39	7.14ug	9/40							

N-NITROSODIMETHYLAMINE   (DMN) 62-75-9

	RefNum	LoConf	UpConf	Cntrl	1Dose	1Inc	2Dose	2Inc	3Dose	3Inc	4Dose	4Inc	Citation or Pathology
3648	2081	31.8ug	n.s.s.	39/47	.167mg	42/48							Anderson;carc,13,2107-2111;1992
a	2081	.790mg	n.s.s.	1/47	.167mg	0/48							
3649	88	.186mg	.824mg	20/62	.600mg	44/62							Terracini;ijcn,11,747-764;1973
a	88	97.8ug	n.s.s.	56/62	.600mg	59/62							
3650	1522	78.1ug	.360mg	0/32	.238mg	12/30							Griciute;clet,13,345-351;1981
a	1522	.143mg	1.82mg	0/32	.238mg	6/30							
b	1522	.163mg	4.64mg	0/32	.238mg	5/30							
c	1522	.706mg	n.s.s.	0/32	.238mg	0/30							
d	1522	46.8ug	.441mg	8/32	.238mg	20/30							
3651	1522	82.3ug	.377mg	0/38	.198mg	12/36							
a	1522	89.3ug	.440mg	0/38	.198mg	11/36							
b	1522	.161mg	n.s.s.	2/38	.198mg	6/36							
c	1522	.340mg	n.s.s.	2/38	.198mg	2/36							
d	1522	30.8ug	.196mg	13/38	.198mg	28/36							
3652	2001	.108mg	n.s.s.	1/76	1.01mg	0/2							Adamson;ossc,129-156;
													1982/Thorgeirsson 1994/Dalgard 1991/Thorgeirsson&Seiber pers.comm.
3653	1664	.286mg	n.s.s.	0/14	.500mg	1/21							Angsubhakorn;ijcn,28,621-626;1981
3654	1609	33.8ug	.123mg	2/20	63.1ug	14/20	.168mg	17/20					Lijinsky;clet,22,83-88;1984
a	1609	77.5ug	.316mg	0/20	63.1ug	9/20	.168mg	10/20					
b	1609	.188mg	1.56mg	0/20	63.1ug	0/20	.168mg	7/20					
c	1609	.309mg	n.s.s.	2/20	63.1ug	5/20	.168mg	2/20					
3655	1864	37.7ug	.169mg	0/19	.527mg	16/19							Lijinsky;canr,47,3968-3972;1987/pers.comm.
a	1864	53.3ug	.240mg	0/19	.527mg	14/19							
b	1864	89.9ug	.494mg	0/19	.527mg	10/19							
c	1864	89.9ug	.494mg	0/19	.527mg	10/19							
d	1864	.151mg	1.92mg	0/19	.527mg	6/19							
e	1864	.151mg	1.92mg	0/19	.527mg	6/19							
f	1864	.248mg	n.s.s.	0/19	.527mg	3/19							
g	1864	.248mg	n.s.s.	0/19	.527mg	3/19							
3656	122m	.893mg	2.36mg	0/29	.100mg	0/18	.250mg	4/62	.500mg	2/5	1.00mg	15/23	2.50mg 10/12 Terracini;bjca,20,
													559-565;1967
3657	122n	.133mg	23.7mg	0/29	.108mg	3/15							
3658	122m	.289mg	n.s.s.	0/12	80.0ug	1/19	.200mg	1/6					
3659	1201	46.0ug	.439mg	0/18	5.00ug	0/24	50.0ug	6/24	(.500mg	4/24)			Arai;gann,70,549-558;1979
a	1201	.546mg	16.3mg	0/18	5.00ug	1/24	50.0ug	0/24	.500mg	5/24			
b	1201	.631mg	n.s.s.	0/18	5.00ug	0/24	50.0ug	1/24	.500mg	3/24			
c	1201	.666mg	n.s.s.	0/18	5.00ug	0/24	50.0ug	3/24	.500mg	2/24			
3660	707	57.3ug	.441mg	0/29	.313mg	7/14							Terao;fctx,16,591-596;1978
a	707	.398mg	n.s.s.	0/29	.313mg	0/14							
3661	1201	.337mg	3.05mg	0/18	4.00ug	0/24	40.0ug	1/24	.400mg	6/24			Arai;gann,70,549-558;1979
a	1201	.591mg	32.7mg	0/18	4.00ug	0/24	40.0ug	0/24	.400mg	3/24			
b	1201	.780mg	n.s.s.	0/18	4.00ug	0/24	40.0ug	1/24	.400mg	1/24			

N-NITROSODIPHENYLAMINE   (redax, diphenylnitrosamine) 86-30-6

	RefNum	LoConf	UpConf	Cntrl	1Dose	1Inc	2Dose	2Inc	Citation or Pathology
3662	TR164	912.mg	n.s.s.	9/20	322.mg	24/50	798.mg	15/50	
a	TR164	1.89gm	n.s.s.	3/20	322.mg	7/50	798.mg	4/50	liv:hpa,hpc,nnd.
b	TR164	1.48gm	n.s.s.	3/20	322.mg	11/50	798.mg	5/50	lun:a/a,a/c.
3663	1094	842.mg	n.s.s.	0/18	510.mg	0/15			Innes;ntis,1968/1969
a	1094	842.mg	n.s.s.	1/18	510.mg	0/15			
b	1094	582.mg	n.s.s.	3/18	510.mg	1/15			
3664	TR164	2.44gm	n.s.s.	10/20	1.20gm	29/50	2.40gm	22/50	
a	TR164	6.54gm	n.s.s.	6/20	1.20gm	12/50	2.40gm	7/50	liv:hpa,hpc,nnd.
b	TR164	5.78gm	n.s.s.	4/20	1.20gm	9/50	2.40gm	7/50	lun:a/a,a/c.
3665	1094	142.mg	n.s.s.	1/17	474.mg	6/15			Innes;ntis,1968/1969
a	1094	356.mg	n.s.s.	2/17	474.mg	2/15			
b	1094	154.mg	n.s.s.	6/17	474.mg	7/15			
3666	1094	309.mg	n.s.s.	0/17	510.mg	3/18			
a	1094	1.01gm	n.s.s.	0/17	510.mg	0/18			
b	1094	366.mg	n.s.s.	2/17	510.mg	3/18			
3667	1094	443.mg	n.s.s.	2/18	474.mg	2/18			
a	1094	939.mg	n.s.s.	3/18	474.mg	0/18			
b	1094	445.mg	n.s.s.	5/18	474.mg	3/18			
3668	TR164	78.4mg	181.mg	0/20	50.0mg	40/50	200.mg	40/50	
a	TR164	76.7mg	3.27gm	13/20	50.0mg	23/50	200.mg	43/50	
b	TR164	n.s.s.	n.s.s.	0/20	50.0mg	0/50	200.mg	0/50	liv:hpa,hpc,nnd.
3669	TR164	167.mg	616.mg	0/20	40.0mg	0/50	160.mg	16/50	
a	TR164	242.mg	24.5gm	1/20	40.0mg	10/50	160.mg	10/50	
b	TR164	137.mg	n.s.s.	16/20	40.0mg	29/50	160.mg	33/50	s
c	TR164	887.mg	n.s.s.	0/20	40.0mg	0/50	160.mg	1/50	liv:hpa,hpc,nnd.
3670	1258	.441mg	n.s.s.	1/24	.186mg	1/21			Boyland;ejca,4,233-239;1968

p-NITROSODIPHENYLAMINE   156-10-5

	RefNum	LoConf	UpConf	Cntrl	1Dose	1Inc	2Dose	2Inc	Citation or Pathology
3671	TR190	493.mg	n.s.s.	4/20	343.mg	16/50	636.mg	9/50	
a	TR190	974.mg	n.s.s.	0/20	343.mg	5/50	636.mg	2/50	liv:hpa,hpc,nnd.

```
       Spe Strain Site    Xpo+Xpt                                                                   TD50    2Tailpvl
       Sex  Route  Hist      Notes                                                                          DR   AuOp
     b  M f b6c eat lun MXB 53w92 asv                                                              5.41gm * P<.4
3672 M m b6c eat liv MXA 53w92 asv                                          : ±                    340.mg \ P<.02  c
   a  M m b6c eat liv hpc 53w92 asv                                                                661.mg \ P<.02  c
   b  M m b6c eat TBA MXB 53w92 asv                                                                260.mg \ P<.03
   c  M m b6c eat liv MXB 53w92 asv                                                                340.mg \ P<.02
   d  M m b6c eat lun MXB 53w92 asv                                                                3.52gm * P<.7
3673 R f f34 eat TBA MXB 18m24                                                    :>               no dre   P=1.  -
   a  R f f34 eat liv MXB 18m24                                                                    1.21gm * P<.06
3674 R m f34 eat liv MXA 18m24                                              : + :                  201.mg * P<.0005c
   a  R m f34 eat TBA MXB 18m24                                                                    no dre   P=1.
   b  R m f34 eat liv MXB 18m24                                                                    201.mg * P<.0005

N-NITROSODIPROPYLAMINE          100ng..:..1ug....:..10......:..100....:..1mg....:..10......:..100....:..1g......:..10
3675 P b rhe ipj liv hpc 31m31 Ww            :     (+)     :                                       12.1ug   P<.0005+

   a  P b rhe ipj tba mal 31m31 Ww                                                                 12.1ug   P<.0005
3676 R f f34 gav liv hpc 30w60           .    +    .                                               .186mg   P<.0005+
   a  R f f34 gav nas car 30w60                                                                    .186mg   P<.0005+
   b  R f f34 gav eso mix 30w60                                                                    .505mg   P<.004 +
   c  R f f34 gav tba mix 30w60                                                                    noTD50   P<.6

N-NITROSODITHIAZINE             100ng..:..1ug....:..10......:..100....:..1mg....:..10......:..100....:..1g......:..10
3677 R f f34 gav nac car 11m26 e                                 .         ±                       5.11mg   P<.1
   a  R f f34 gav liv hpc 11m26 e                                                                  10.5mg   P<.3  -

NITROSODODECAMETHYLENEIMINE     100ng..:..1ug....:..10......:..100....:..1mg....:..10......:..100....:..1g......:..10
3678 R f nzd gav liv hct 12m40 e                                         .         +          .    10.9mg   P<.007 +
   a  R f nzd gav lun mix 12m40 e                                                                  7.96mg   P<.03
3679 R m nzd gav hea tum 12m35 e                                 .         ±                       2.15mg   P<.03
   a  R m nzd gav lun mix 12m35 e                                                                  14.6mg   P<.5
   b  R m nzd gav liv hct 12m35 e                                                                  16.8mg   P<.3  +

N-NITROSOEPHEDRINE              100ng..:..1ug....:..10......:..100....:..1mg....:..10......:..100....:..1g......:..10
3680 R m sda gav liv mix 24m24                                                     .    +     .    95.2mg   P<.0005+
   a  R m sda gav liv hpc 24m24                                                                    138.mg   P<.004
   b  R m sda gav lun mix 24m24                                                                    217.mg   P<.1  +
   c  R m sda gav for pam 24m24                                                                    239.mg   P<.03  +
   d  R m sda gav for sqc 24m24                                                                    740.mg   P<.3  +
   e  R m sda gav tba mix 24m24                                                                    40.0mg   P<.0005

NITROSOETHYLMETHYLAMINE         100ng..:..1ug....:..10......:..100....:..1mg....:..10......:..100....:..1g......:..10
3681 R f f34 wat --- tum  7m27                .    ±                                               38.6ug   P<.03
   a  R f f34 wat liv tum  7m27                                                                    no dre   P=1.
   b  R f f34 wat tba tum  7m27                                                                    no dre   P=1.
3682 R m f34 gav liv bht 30w55 e                      .    +    .                                  50.3ug   P<.0005
   a  R m f34 gav liv mix 30w55 e                                                                  50.3ug   P<.0005+
   b  R m f34 gav liv hes 30w55 e                                                                  .120mg   P<.0005
   c  R m f34 gav nac tum 30w55 e                                                                  .201mg   P<.0005+
   d  R m f34 gav lun a/a 30w55 e                                                                  .242mg   P<.0005+

NITROSOETHYLURETHAN             100ng..:..1ug....:..10......:..100....:..1mg....:..10......:..100....:..1g......:..10
3683 R f don wat mix sqc 51w60 ae                           .  +  .                                .164mg Z P<.0005
   a  R f don wat for sqc 51w60 ae                                                                 .248mg * P<.0005+
   b  R f don wat mix pam 51w60 ae                                                                 .339mg Z P<.0005
   c  R f don wat eso pam 51w60 ae                                                                 .473mg Z P<.0005
   d  R f don wat for pam 51w60 ae                                                                 .508mg * P<.0005+
   e  R f don wat eso sqc 51w60 ae                                                                 .540mg Z P<.0005
   f  R f don wat duo adc 51w60 ae                                                                 .555mg * P<.0005+
   g  R f don wat tba mix 51w60 ae                                                                 78.4ug * P<.0005
3684 R f f3d wat ugi mix 23m26 es                          . + .                                   55.3ug * P<.0005+

   a  R f f3d wat for pam 23m26 es                                                                 79.6ug Z P<.0005+
   b  R f f3d wat eso sqc 23m26 es                                                                 .139mg Z P<.0005+
   c  R f f3d wat for sqc 23m26 es                                                                 .213mg Z P<.0005+
   d  R f f3d wat eso pam 23m26 es                                                                 .514mg Z P<.0005+
   e  R f f3d wat orc sqc 23m26 es                                                                 .784mg * P<.0005+
   f  R f f3d wat orc pam 23m26 es                                                                 3.02mg * P<.0005+
   g  R f f3d wat tba tum 23m26 es                                                                 .135mg * P<.002

N-NITROSOGUVACOLINE             100ng..:..1ug....:..10......:..100....:..1mg....:..10......:..100....:..1g......:..10
3685 R m f34 wat pae ana 30m30                                   .         ±                       7.95mg   P<.02
   a  R m f34 wat liv ade 30m30                                                                    no dre   P=1.

NITROSOHEPTAMETHYLENEIMINE      100ng..:..1ug....:..10......:..100....:..1mg....:..10......:..100....:..1g......:..10
3686 R m f34 wat ugi car  6m27 esy           .   (+)   .                                           36.3ug Z P<.0005+
   a  R m f34 wat ugi mix  6m27 esy                                                                51.2ug Z P<.0005+
   b  R m f34 wat eso mix  6m27 esy                                                                .118mg Z P<.0005+
   c  R m f34 wat liv hem  6m27 esy                                                                no dre   P=1.
3687 R m f34 wat ugi mix 12m29 es                          . + .                                   57.1ug * P<.0005+
   a  R m f34 wat eso bcc 12m29 es                                                                 .159mg * P<.0005+
   b  R m f34 wat ton bcc 12m29 es                                                                 .171mg * P<.0005+
```

	RefNum	LoConf	UpConf	Cntrl	1Dose	1Inc	2Dose	2Inc			Citation or Pathology
											Brkly Code
b	TR190	1.64gm	n.s.s.	0/20	343.mg	2/50	636.mg	1/50			lun:a/a,a/c.
3672	TR190	172.mg	n.s.s.	2/20	316.mg	22/50	(587.mg	12/50)			liv:hpa,hpc.
a	TR190	323.mg	n.s.s.	0/20	316.mg	10/50	(587.mg	1/50)			
b	TR190	128.mg	n.s.s.	4/20	316.mg	30/50	(587.mg	17/50)			
c	TR190	172.mg	n.s.s.	2/20	316.mg	22/50	(587.mg	12/50)			liv:hpa,hpc,nnd.
d	TR190	711.mg	n.s.s.	1/20	316.mg	10/50	587.mg	4/50			lun:a/a,a/c.
3673	TR190	195.mg	n.s.s.	11/20	92.9mg	36/50	186.mg	28/50			
a	TR190	522.mg	n.s.s.	0/20	92.9mg	2/50	186.mg	5/50			liv:hpa,hpc,nnd.
3674	TR190	129.mg	547.mg	0/20	74.3mg	10/50	149.mg	19/50			liv:hpc,nnd.
a	TR190	129.mg	n.s.s.	13/20	74.3mg	34/50	149.mg	33/50			
b	TR190	129.mg	547.mg	0/20	74.3mg	10/50	149.mg	19/50			liv:hpa,hpc,nnd.

N-NITROSODIPROPYLAMINE   621-64-7

	RefNum	LoConf	UpConf	Cntrl	1Dose	1Inc	2Dose	2Inc			Citation
3675	2001	3.14ug	59.9ug	0/104	3.52mg	4/4					Adamson;ossc,129-156;
											1982/Thorgeirsson 1994/Dalgard 1991/Thorgeirsson&Seiber pers.comm.
a	2001	3.14ug	59.9ug	0/104	3.52mg	4/4					
3676	1551	79.3ug	.564mg	0/20	.898mg	8/12					Lijinsky;clet,19,207-213;1983
a	1551	79.3ug	.564mg	0/20	.898mg	8/12					
b	1551	.172mg	3.75mg	0/20	.898mg	4/12					
c	1551	n.s.s.	n.s.s.	18/20	.898mg	12/12					

N-NITROSODITHIAZINE   114282-83-6

	RefNum	LoConf	UpConf	Cntrl	1Dose	1Inc					Citation
3677	1884	1.25mg	n.s.s.	0/20	.702mg	2/20					Lijinsky;fctx,26,3-7;1988
a	1884	1.71mg	n.s.s.	0/20	.702mg	1/20					

NITROSODODECAMETHYLENEIMINE   40580-89-0

	RefNum	LoConf	UpConf	Cntrl	1Dose	1Inc					Citation
3678	1603	2.67mg	255.mg	0/109	.619mg	2/20					Goodall;carc,5,537-540;1984
a	1603	2.20mg	n.s.s.	2/109	.619mg	3/20					
3679	1603	.730mg	n.s.s.	18/108	.495mg	8/20					
a	1603	1.88mg	n.s.s.	6/108	.495mg	2/20					
b	1603	2.30mg	n.s.s.	1/108	.495mg	1/20					

N-NITROSOEPHEDRINE   17608-59-2

	RefNum	LoConf	UpConf	Cntrl	1Dose	1Inc					Citation
3680	1259	41.0mg	332.mg	0/40	34.3mg	7/32					Eisenbrand;clet,5,103-106;1978
a	1259	52.4mg	974.mg	0/40	34.3mg	5/32					
b	1259	64.1mg	n.s.s.	1/40	34.3mg	4/32					
c	1259	72.2mg	n.s.s.	0/40	34.3mg	3/32					
d	1259	120.mg	n.s.s.	0/40	34.3mg	1/32					
e	1259	20.1mg	129.mg	4/40	34.3mg	16/32					

NITROSOETHYLMETHYLAMINE   (N-nitrosomethylethylamine) 10595-95-6

	RefNum	LoConf	UpConf	Cntrl	1Dose	1Inc					Citation
3681	1561	13.4ug	n.s.s.	12/20	63.9ug	18/20					Lijinsky;fctx,21,601-605;1983
a	1561	.322mg	n.s.s.	1/20	63.9ug	0/20					
b	1561	13.5ug	n.s.s.	20/20	63.9ug	19/20					
3682	1864	19.8ug	.125mg	0/19	.727mg	15/16					Lijinsky;canr,47,3968-3972;1987/pers.comm.
a	1864	19.8ug	.125mg	0/19	.727mg	15/16					
b	1864	56.8ug	.302mg	0/19	.727mg	11/16					
c	1864	88.9ug	.602mg	0/19	.727mg	8/16					
d	1864	.103mg	.821mg	0/19	.727mg	7/16					

NITROSOETHYLURETHAN   (1-ethyl-1-nitrosourethan) 614-95-9

	RefNum	LoConf	UpConf	Cntrl	1Dose	1Inc	2Dose	2Inc	3Dose	3Inc	Citation
3683	1494	.108mg	.260mg	0/37	.714mg	21/26	1.43mg	21/28	(2.86mg	11/24)	Onodera;gann,73,48-54;1982
a	1494	.170mg	.375mg	0/37	.714mg	11/26	1.43mg	22/28	2.86mg	22/24	
b	1494	.213mg	.581mg	0/37	.714mg	14/26	1.43mg	14/28	(2.86mg	9/24)	
c	1494	.284mg	.884mg	0/37	.714mg	11/26	1.43mg	11/28	(2.86mg	7/24)	
d	1494	.338mg	.807mg	0/37	.714mg	6/26	1.43mg	16/28	2.86mg	16/24	
e	1494	.317mg	1.08mg	0/37	.714mg	11/26	1.43mg	9/28	(2.86mg	4/24)	
f	1494	.364mg	.900mg	0/37	.714mg	5/26	1.43mg	12/28	2.86mg	18/24	
g	1494	41.3ug	.151mg	11/37	.714mg	25/26	1.43mg	27/28	2.86mg	24/24	
3684	1939	37.8ug	83.4ug	1/40	7.96ug	0/39	31.8ug	16/38	.133mg	38/39 .571mg 37/38	Maekawa;gann,80,632-636;
											1989/pers.comm.
a	1939	41.9ug	.180mg	0/40	7.96ug	0/39	31.8ug	13/38	(.133mg	14/39 .571mg 19/38)	
b	1939	87.0ug	.242mg	0/40	7.96ug	0/39	31.8ug	1/38	.133mg	26/39 (.571mg 23/38)	
c	1939	.125mg	.405mg	0/40	7.96ug	0/39	31.8ug	2/38	.133mg	18/39 (.571mg 9/38)	
d	1939	.241mg	1.72mg	1/40	7.96ug	0/39	31.8ug	1/38	.133mg	9/39 (.571mg 6/38)	
e	1939	.475mg	1.42mg	0/40	7.96ug	0/39	31.8ug	0/38	.133mg	6/39 .571mg 17/38	
f	1939	1.30mg	9.75mg	0/40	7.96ug	0/39	31.8ug	0/38	.133mg	1/39 .571mg 6/38	
g	1939	49.0ug	.868mg	33/40	7.96ug	30/39	31.8ug	31/38	.133mg	39/39 .571mg 37/38	

N-NITROSOGUVACOLINE   55557-02-3

	RefNum	LoConf	UpConf	Cntrl	1Dose	1Inc					Citation
3685	1866	2.57mg	n.s.s.	1/80	1.00mg	4/30					Rivenson;canr,48,6912-6917;1988/pers.comm.
a	1866	9.37mg	n.s.s.	6/80	1.00mg	0/30					

NITROSOHEPTAMETHYLENEIMINE   20917-49-1

	RefNum	LoConf	UpConf	Cntrl	1Dose	1Inc	2Dose	2Inc	3Dose	3Inc	Citation
3686	1376m	18.1ug	86.7ug	0/20	39.8ug	12/20	(.197mg	11/20	.733mg	10/20)	Lijinsky;jnci,69,1127-1133;1982/pers.comm.
a	1376m	30.3ug	96.3ug	0/20	39.8ug	15/20	.197mg	17/20	(.733mg	15/20)	
b	1376m	66.9ug	.233mg	0/20	39.8ug	5/20	.197mg	15/20	(.733mg	12/20)	
c	1376m	.160mg	n.s.s.	1/20	39.8ug	0/20	.197mg	0/20	.733mg	0/20	
3687	1376n	35.3ug	99.6ug	0/20	11.4ug	2/20	31.5ug	13/20	.125mg	16/20	
a	1376n	85.8ug	.340mg	0/20	11.4ug	0/20	31.5ug	2/20	.125mg	13/20	
b	1376n	92.7ug	.410mg	0/20	11.4ug	1/20	31.5ug	4/20	.125mg	10/20	

```
      Spe Strain  Site    Xpo+Xpt                                                                        TD50    2Tailpvl
      Sex  Route  Hist      Notes                                                                          DR      AuOp

 c  R m f34 wat liv hpc 12m29 es                                                                          no dre   P=1.   +
3688 R m f34 wat ugi mix 23m27 e                          . + .                                           29.2ug * P<.0005+
 a  R m f34 wat eso mix 23m27 e                                                                           73.5ug * P<.0005+
 b  R m f34 wat ton bcc 23m27 e                                                                           .105mg * P<.004 +
 c  R m f34 wat liv mix 23m27 e                                                                           .198mg \ P<.1   +

N-NITROSOHEXAMETHYLENEIMINE        100ng..:..1ug....:...10......:...100....:...1mg....:...10.......:..100....:...1g......:..10
3689 M f nzb wat ssq mix 32w74                            .   +   .                                       .415mg   P<.0005+
 a  M f nzb wat eso mix 32w74                                                                             .725mg   P<.0005+
 b  M f nzb wat liv hpc 32w74                                                                             3.21mg   P<.004 +
 c  M f nzb wat stg mix 32w74                                                                             4.05mg   P<.0005+
 d  M f nzb wat opx mix 32w74                                                                             6.37mg   P<.003 +
 e  M f nzb wat lun mix 32w74                                                                             no dre   P=1.
3690 M m nzb wat eso mix 32w60                         .   +   .                                          .313mg   P<.0005+
 a  M m nzb wat ssq mix 32w60                                                                             .448mg   P<.0005+
 b  M m nzb wat opx mix 32w60                                                                             1.41mg   P<.0005+
 c  M m nzb wat stg mix 32w60                                                                             3.23mg   P<.0005+
 d  M m nzb wat lun mix 32w60                                                                             1.93mg   P<.2
 e  M m nzb wat liv hpc 32w60                                                                             8.03mg   P<.2   +

1-NITROSOHYDANTOIN                 100ng..:..1ug....:...10......:...100....:...1mg....:...10.......:..100....:...1g......:..10
3691 R f mrw wat liv tum 12m26                                                   .>                       no dre   P=1.
 a  R f mrw wat tba mix 12m26                                                                             19.6mg   P<.2   +
3692 R m mrw wat for sqp 12m26                                              . + .                          27.8mg   P<.005
 a  R m mrw wat phr sqc 12m26                                                                             43.8mg   P<.03  +
 b  R m mrw wat liv tum 12m26                                                                             no dre   P=1.
 c  R m mrw wat tba mix 12m26                                                                             7.31mg   P<.03

NITROSOHYDROXYPROLINE              100ng..:..1ug....:...10......:...100....:...1mg....:...10.......:..100....:...1g......:..10
3693 R f mrc wat liv tum 17m24                                                   .>                       no dre   P=1.
 a  R f mrc wat tba mix 17m24                                                                             3.84mg   P<.03  -
3694 R m mrc wat liv tum 17m24                                              .>                             no dre   P=1.
 a  R m mrc wat tba mix 17m24                                                                             no dre   P=1.   -

NITROSOIMINODIACETIC ACID          100ng..:..1ug....:...10......:...100....:...1mg....:...10.......:..100....:...1g......:..10
3695 R f mrc wat tba mix 17m24 e                                      .    ±                               63.9mg   P<.07  -
3696 R m mrc wat pit tum 17m24 e                                      .    +    .                          66.9mg   P<.006 -
 a  R m mrc wat tba mix 17m24 e                                                                           76.0mg   P<.3   -

N-NITROSOMETHYL-2,3-DIHYDROXYPROPYLAMINE.:..1_ug....:...10......:...100......:...1mg....:...10.......:..100....:...1g......:..10
3697 H f syg gav nac mix 41w81                                     . + .                                   .940mg   P<.0005+
 a  H f syg gav nac car 41w81                                                                             1.51mg   P<.002
 b  H f syg gav liv vsc 41w81                                                                             12.7mg   P<.3   -
 c  H f syg gav tba mix 41w81                                                                             .509mg   P<.0005
3698 R f f34 wat lun mix 9m26                              . + .                                           .646mg   P<.0005+
 a  R f f34 wat liv mix 9m26                                                                              .670mg   P<.02
 b  R f f34 wat liv nnd 9m26                                                                              1.01mg   P<.06  +
 c  R f f34 wat eso mix 9m26                                                                              1.48mg   P<.02  +
 d  R f f34 wat nas mix 9m26                                                                              1.48mg   P<.02  +
 e  R f f34 wat liv hpc 9m26                                                                              3.13mg   P<.1

N-NITROSOMETHYL-(2-HYDROXYETHYL)AMINE   .:..1ug....:...10......:...100......:...1mg....:...10....:...100....:...1g......:..10
3699 R m f34 gav liv hpc 69w69 es                            . + .                                        1.29mg   P<.003 +
 a  R m f34 gav nas sqc 69w69 es                                                                          2.06mg   P<.02  +
 b  R m f34 gav liv hes 69w69 es                                                                          8.97mg   P<.3
 c  R m f34 gav nas sqp 69w69 es                                                                          8.97mg   P<.3
 d  R m f34 gav liv nnd 69w69 es                                                                          no dre   P=1.

N-NITROSOMETHYL-(3-HYDROXYPROPYL)AMINE  .:..1ug....:...10......:...100......:...1mg....:...10.:...100....:...1g......:..10
3700 R f f34 gav liv nnd 30m30 e                                       . + .                              3.48mg   P<.004 +
 a  R f f34 gav lun mix 30m30 e                                                                           8.06mg   P<.06  +
 b  R f f34 gav lun a/c 30m30 e                                                                           24.9mg   P<.1   +
 c  R f f34 gav lun a/a 30m30 e                                                                           14.4mg   P<.3
3701 R m f34 gav lun mix 30m30 e                           . + .                                          1.09mg   P<.0005+
 a  R m f34 gav tes ict 30m30 e                                                                           2.26mg   P<.0005
 b  R m f34 gav lun a/c 30m30 e                                                                           2.46mg   P<.0005
 c  R m f34 gav liv nnd 30m30 e                                                                           4.78mg   P<.06  +
 d  R m f34 gav lun a/a 30m30 e                                                                           8.46mg   P<.3

N-NITROSOMETHYL-2-HYDROXYPROPYLAMINE    .:..1ug....:...10......:...100......:...1mg....:...10....:...100....:...1g......:..10
3702 R f f34 wat nas mix 30w75                          . + .                                             48.6ug   P<.0005+
 a  R f f34 wat eso mix 30w75                                                                             .105mg   P<.0005+
 b  R f f34 wat eso car 30w75                                                                             .408mg   P<.003
 c  R f f34 wat lun mix 30w75                                                                             .408mg   P<.003
 d  R f f34 wat liv hpc 30w75                                                                             .895mg   P<.04
 e  R f f34 wat liv mix 30w75                                                                             .798mg   P<.3
 f  R f f34 wat liv nnd 30w75                                                                             2.54mg   P<.7
3703 R m f34 wat nas mix 30w75                          . + .                                             44.2ug   P<.0005+
 a  R m f34 wat eso mix 30w75                                                                             .170mg   P<.0005+
 b  R m f34 wat liv mix 30w75                                                                             no dre   P=1.
 c  R m f34 wat liv nnd 30w75                                                                             no dre   P=1.
```

	RefNum	LoConf	UpConf	Cntrl	1Dose	1Inc	2Dose	2Inc	Citation or Pathology
									Brkly Code
c	1376n	.729mg	n.s.s.	0/20	11.4ug	5/20	31.5ug	0/20	.125mg 0/20
3688	1376o	17.4ug	52.1ug	0/20	24.8ug	10/20	64.6ug	17/20	
a	1376o	39.7ug	.158mg	0/20	24.8ug	3/20	64.6ug	12/20	
b	1376o	53.7ug	.736mg	0/20	24.8ug	6/20	64.6ug	6/20	
c	1376o	48.5ug	n.s.s.	0/20	24.8ug	2/20	(64.6ug	0/20)	

N-NITROSOHEXAMETHYLENEIMINE   932-83-2

	RefNum	LoConf	UpConf	Cntrl	1Dose	1Inc	Citation or Pathology
3689	1738	.156mg	1.13mg	0/113	3.07mg	12/13	Goodall;txcy,33,251-259;1984
a	1738	.320mg	1.96mg	0/113	3.07mg	10/13	
b	1738	1.02mg	40.6mg	4/113	3.07mg	4/13	
c	1738	1.22mg	29.1mg	0/113	3.07mg	3/13	
d	1738	1.56mg	88.1mg	0/113	3.07mg	2/13	
e	1738	1.58mg	n.s.s.	39/113	3.07mg	4/13	
3690	1738	.110mg	.974mg	0/194	3.16mg	9/10	
a	1738	.176mg	1.40mg	0/194	3.16mg	8/10	
b	1738	.477mg	7.31mg	0/194	3.16mg	4/10	
c	1738	.789mg	43.0mg	0/194	3.16mg	2/10	
d	1738	.458mg	n.s.s.	53/194	3.16mg	5/10	
e	1738	1.14mg	n.s.s.	3/194	3.16mg	1/10	

1-NITROSOHYDANTOIN   42579-28-2

	RefNum	LoConf	UpConf	Cntrl	1Dose	1Inc	Citation or Pathology
3691	1246	63.0mg	n.s.s.	0/25	11.4mg	0/24	Bulay;jnci,62,1523-1528;1979
a	1246	5.90mg	n.s.s.	12/25	11.4mg	16/24	
3692	1246	11.3mg	198.mg	0/22	9.96mg	6/25	
a	1246	15.1mg	n.s.s.	0/22	9.96mg	4/25	
b	1246	57.4mg	n.s.s.	0/22	9.96mg	0/25	
c	1246	2.84mg	n.s.s.	12/22	9.96mg	21/25	

NITROSOHYDROXYPROLINE   30310-80-6

	RefNum	LoConf	UpConf	Cntrl	1Dose	1Inc	Citation or Pathology
3693	216	13.6mg	n.s.s.	0/15	4.42mg	0/15	Garcia;zkko,79,141-144;1973
a	216	1.43mg	n.s.s.	4/15	4.42mg	10/15	
3694	216	11.9mg	n.s.s.	0/15	3.86mg	0/15	
a	216	3.47mg	n.s.s.	5/15	3.86mg	5/15	

NITROSOIMINODIACETIC ACID   25081-31-6

	RefNum	LoConf	UpConf	Cntrl	1Dose	1Inc	Citation or Pathology
3695	213	22.0mg	n.s.s.	4/15	56.5mg	9/15	Lijinsky;jnci,50,1061-1063;1973
3696	213	25.1mg	631.mg	0/15	39.6mg	5/15	
a	213	19.4mg	n.s.s.	5/15	39.6mg	8/15	

N-NITROSOMETHYL-2,3-DIHYDROXYPROPYLAMINE   86451-37-8

	RefNum	LoConf	UpConf	Cntrl	1Dose	1Inc	Citation or Pathology
3697	1715	.449mg	2.46mg	0/20	1.57mg	10/20	Lijinsky;zkko,109,1-4;1985
a	1715	.647mg	5.73mg	0/20	1.57mg	7/20	
b	1715	2.07mg	n.s.s.	0/20	1.57mg	1/20	
c	1715	.250mg	1.38mg	2/20	1.57mg	15/20	
3698	1609	.289mg	2.04mg	0/20	.430mg	8/20	Lijinsky;clet,22,83-88;1984
a	1609	.275mg	n.s.s.	2/20	.430mg	9/20	
b	1609	.358mg	n.s.s.	2/20	.430mg	7/20	
c	1609	.509mg	n.s.s.	0/20	.430mg	4/20	
d	1609	.509mg	n.s.s.	0/20	.430mg	4/20	
e	1609	.769mg	n.s.s.	0/20	.430mg	2/20	

N-NITROSOMETHYL-(2-HYDROXYETHYL)AMINE   26921-68-6

	RefNum	LoConf	UpConf	Cntrl	1Dose	1Inc	Citation or Pathology
3699	1904	.522mg	6.74mg	0/20	1.53mg	6/20	Koepke;canr,48,1533-1536;1988/pers.comm.
a	1904	.710mg	n.s.s.	0/20	1.53mg	4/20	
b	1904	1.46mg	n.s.s.	0/20	1.53mg	1/20	
c	1904	1.46mg	n.s.s.	0/20	1.53mg	1/20	
d	1904	1.45mg	n.s.s.	3/20	1.53mg	2/20	

N-NITROSOMETHYL-(3-HYDROXYPROPYL)AMINE   70415-59-7

	RefNum	LoConf	UpConf	Cntrl	1Dose	1Inc	Citation or Pathology
3700	1904	1.55mg	24.4mg	3/20	2.45mg	12/20	Koepke;canr,48,1533-1536;1988/pers.comm.
a	1904	2.85mg	n.s.s.	2/20	2.45mg	7/20	
b	1904	6.11mg	n.s.s.	0/20	2.45mg	2/20	
c	1904	3.87mg	n.s.s.	2/20	2.45mg	5/20	
3701	1904	.511mg	2.98mg	3/20	1.71mg	16/19	
a	1904	1.06mg	7.37mg	1/20	1.71mg	11/19	
b	1904	1.17mg	6.40mg	0/20	1.71mg	10/19	
c	1904	1.69mg	n.s.s.	3/20	1.71mg	8/19	
d	1904	2.28mg	n.s.s.	3/20	1.71mg	6/19	

N-NITROSOMETHYL-2-HYDROXYPROPYLAMINE   75411-83-5

	RefNum	LoConf	UpConf	Cntrl	1Dose	1Inc	Citation or Pathology
3702	1609	20.0ug	.111mg	0/20	.408mg	19/20	Lijinsky;clet,22,83-88;1984
a	1609	53.7ug	.231mg	0/20	.408mg	15/20	
b	1609	.165mg	2.13mg	0/20	.408mg	6/20	
c	1609	.165mg	2.13mg	0/20	.408mg	6/20	
d	1609	.270mg	n.s.s.	0/20	.408mg	3/20	
e	1609	.215mg	n.s.s.	2/20	.408mg	5/20	
f	1609	.318mg	n.s.s.	2/20	.408mg	3/20	
3703	1609	21.0ug	96.7ug	0/20	.286mg	18/20	
a	1609	78.9ug	.480mg	0/20	.286mg	9/20	
b	1609	.291mg	n.s.s.	2/20	.286mg	2/20	
c	1609	.407mg	n.s.s.	2/20	.286mg	1/20	

```
      Spe Strain  Site    Xpo+Xpt                                                                      TD50    2Tailpvl
      Sex  Route  Hist    Notes                                                                         DR      AuOp
N-NITROSOMETHYL(2-OXOPROPYL)AMINE  100ng..:..1ug...:..10....:..100....:..1mg....:..10....:..100....:..1g......:..10
3704  R f mrc gav npc mix 67w76 as                    .   +    .                                       16.7ug  Z P<.0005+
  a   R f mrc gav phr mix 67w76 as                                                                     75.8ug  * P<.0005+
  b   R f mrc gav eso mix 67w76 as                                                                     .160mg  * P<.0005+
  c   R f mrc gav liv mix 67w76 as                                                                     .268mg  * P<.002 +
  d   R f mrc gav lun ade 67w76 as                                                                     no dre    P=1.
  e   R f mrc gav nsp pam 67w76 as                                                                     no dre    P=1.
  f   R f mrc gav tba tum 67w76 as                                                                     noTD50    P<.0005
3705  R m mrc gav npc mix 51w58 as                        .   +    .                                   17.8ug  Z P<.0005+
  a   R m mrc gav phr mix 51w58 as                                                                     23.0ug  Z P<.0005+
  b   R m mrc gav eso mix 51w58 as                                                                     64.2ug  * P<.0005+
  c   R m mrc gav liv mix 51w58 as                                                                     .252mg  Z P<.07  +
  d   R m mrc gav nsp pam 51w58 as                                                                     .942mg  * P<.4
  e   R m mrc gav lun ade 51w58 as                                                                     no dre    P=1.
  f   R m mrc gav tba tum 51w58 as                                                                     noTD50    P<.0005

N-NITROSOMETHYL-(2-TOSYLOXYETHYL)AMINE  ..:..1ug...:..10....:..100....:..1mg....:..10....:..100....:..1g....:..10
3706  R f f34 gav liv mix 28m28 e                                             .   +    .               3.47mg    P<.0005+
  a   R f f34 gav liv hpc 28m28 e                                                                      7.55mg    P<.0005+
  b   R f f34 gav liv hes 28m28 e                                                                      12.8mg    P<.003 +
  c   R f f34 gav liv nnd 28m28 e                                                                      22.3mg    P<.3
3707  R m f34 gav liv hes 28m28 e                                                 .   +    .           11.8mg    P<.007 +
  a   R m f34 gav liv mix 28m28 e                                                                      7.79mg    P<.04  +
  b   R m f34 gav tes ict 28m28 e                                                                      11.1mg    P<.04
  c   R m f34 gav pit ade 28m28 e                                                                      11.1mg    P<.04
  d   R m f34 gav liv hpc 28m28 e                                                                      32.2mg    P<.1   +
  e   R m f34 gav liv nnd 28m28 e                                                                      12.6mg    P<.2

2-NITROSOMETHYLAMINOPYRIDINE  100ng..:..1ug...:..10....:..100....:..1mg....:..10....:..100....:..1g......:..10
3708  R f bdf gav eso mix 72w72 e                          .   +      .                                .214mg    P<.003 +
  a   R f bdf gav liv mix 72w72 e                                                                      no dre    P=1.
  b   R f bdf gav tba mal 72w72 e                                                                       .668mg    P<.08
  c   R f bdf gav tba mix 72w72 e                                                                      .280mg    P<.2
  d   R f bdf gav tba ben 72w72 e                                                                      3.05mg    P<.9

3-NITROSOMETHYLAMINOPYRIDINE  100ng..:..1ug...:..10....:..100....:..1mg....:..10....:..100....:..1g......:..10
3709  R f bdf gav liv mix 27m27 e                                           .>                         no dre    P=1.   -
  a   R f bdf gav tba mix 27m27 e                                                                      no dre    P=1.   -
  b   R f bdf gav tba mal 27m27 e                                                                      6.09mg    P<.4   -
  c   R f bdf gav tba ben 27m27 e                                                                      no dre    P=1.   -

4-NITROSOMETHYLAMINOPYRIDINE  100ng..:..1ug...:..10....:..100....:..1mg....:..10....:..100....:..1g......:..10
3710  R f bdf gav liv mix 23m23 e                                              .>                      no dre    P=1.   -
  a   R f bdf gav tba mix 23m23 e                                                                      no dre    P=1.   -
  b   R f bdf gav tba ben 23m23 e                                                                      no dre    P=1.   -
  c   R f bdf gav tba mal 23m23 e                                                                      26.8mg    P<.5

NITROSOMETHYLANILINE  100ng..:..1ug...:..10....:..100....:..1mg....:..10....:..100....:..1g......:..10
3711  R f cbt wat eso mix 67w67 ev                                             <+                      noTD50    P<.0005+
  a   R f cbt wat eso sqc 67w67 ev                                                                     26.4mg    P<.02  +
  b   R f cbt wat eso ben 67w67 ev                                                                     32.0mg    P<.03  +
  c   R f cbt wat liv tum 67w67 ev                                                                     no dre    P=1.   -
3712  R m cbt wat eso mix 60w60 ev                                   .   +    .                        5.27mg    P<.002 +
  a   R m cbt wat eso sqc 60w60 ev                                                                     10.5mg    P<.009 +
  b   R m cbt wat eso ben 60w60 ev                                                                     70.4mg    P<.3   +
  c   R m cbt wat liv tum 60w60 ev                                                                     no dre    P=1.   -
3713  R f f34 wat eso car 12m26                                 .   +    .                             .458mg    P<.0005+
  a   R f f34 wat liv car 12m26                                                                        no dre    P=1.
  b   R f f34 wat tba mix 12m26                                                                        no dre    P=1.
3714  R m f34 wat eso mix 12m26                                .   +    .                              .272mg    P<.0005+
  a   R m f34 wat liv car 12m26                                                                        no dre    P=1.
  b   R m f34 wat tba mix 12m26                                                                        no dre    P=1.
3715  R b sda wat eso mix 24m24 e                        .   +   .                                     34.3ug  \ P<.0005+

NITROSOMETHYLPHENIDATE  100ng..:..1ug...:..10....:..100....:..1mg....:..10....:..100....:..1g......:..10
3716  M f c7b wat lun tum 17m25                                                  .>                    no dre    P=1.   -
  a   M f c7b wat liv tum 17m25                                                                        no dre    P=1.   -
  b   M f c7b wat tba mix 17m25                                                                        no dre    P=1.   -
3717  M m c7b wat lun tum 17m25                                               .>                       no dre    P=1.   -
  a   M m c7b wat tba mix 17m25                                                                        no dre    P=1.   -

NITROSOMETHYLUNDECYLAMINE  100ng..:..1ug...:..10....:..100....:..1mg....:..10....:..100....:..1g......:..10
3718  R m f34 gav lun sqc 30w65                                            .   +    .                  2.37mg    P<.0005+
  a   R m f34 gav liv hpc 30w65                                                                        2.38mg    P<.0005+
  b   R m f34 gav lun ald 30w65                                                                        2.41mg    P<.0005+
  c   R m f34 gav liv clc 30w65                                                                        3.18mg    P<.0005+
  d   R m f34 gav lun ala 30w65                                                                        4.99mg    P<.0005
  e   R m f34 gav eso sqc 30w65                                                                        15.4mg    P<.002
  f   R m f34 gav stn pam 30w65                                                                        15.4mg    P<.002
```

	RefNum	LoConf	UpConf	Cntrl	1Dose	1Inc	2Dose	2Inc		Citation or Pathology
										Brkly Code

N-NITROSOMETHYL(2-OXOPROPYL)AMINE   55984-51-5

	RefNum	LoConf	UpConf	Cntrl	1Dose	1Inc	2Dose	2Inc			Citation or Pathology
3704	1911	6.68ug	36.8ug	0/15	.129mg	14/15	.257mg	15/15	(.500mg	9/14)	Pour;clet,45,49-57;1989
a	1911	47.6ug	.149mg	0/15	.129mg	10/15	.257mg	14/15	.500mg	9/14	
b	1911	90.8ug	.314mg	0/15	.129mg	1/15	.257mg	6/15	.500mg	12/14	
c	1911	.143mg	1.05mg	0/15	.129mg	3/15	.257mg	5/15	.500mg	6/14	
d	1911	1.09mg	n.s.s.	6/15	.129mg	0/15	.257mg	0/15	.500mg	1/14	
e	1911	.618mg	n.s.s.	0/15	.129mg	3/15	.257mg	0/15	.500mg	0/14	
f	1911	n.s.s.	34.9ug	5/15	.129mg	15/15	.257mg	15/15	.500mg	14/14	
3705	1911	9.79ug	32.5ug	0/15	.129mg	12/13	.257mg	15/15	.500mg	14/15	
a	1911	12.7ug	44.4ug	0/15	.129mg	10/13	.257mg	13/15	(.500mg	8/15)	
b	1911	38.9ug	.120mg	0/15	.129mg	4/13	.257mg	12/15	.500mg	10/15	
c	1911	82.7ug	n.s.s.	1/15	.129mg	0/13	.257mg	5/15	(.500mg	1/15)	
d	1911	.270mg	n.s.s.	0/15	.129mg	1/13	.257mg	1/15	.500mg	1/15	
e	1911	64.7ug	n.s.s.	0/15	.129mg	0/13	.257mg	0/15	.500mg	0/15	
f	1911	n.s.s.	19.0ug	3/15	.129mg	13/13	.257mg	15/15	.500mg	15/15	

N-NITROSOMETHYL-(2-TOSYLOXYETHYL)AMINE   ---

	RefNum	LoConf	UpConf	Cntrl	1Dose	1Inc	Citation or Pathology
3706	1904	1.64mg	10.3mg	3/20	5.22mg	15/19	Koepke;canr,48,1533-1536;1988/pers.comm.
a	1904	3.49mg	21.1mg	0/20	5.22mg	9/19	
b	1904	5.17mg	61.8mg	0/20	5.22mg	6/19	
c	1904	6.01mg	n.s.s.	3/20	5.22mg	6/19	
3707	1904	4.45mg	136.mg	0/20	3.66mg	5/20	
a	1904	2.94mg	n.s.s.	3/20	3.66mg	9/20	
b	1904	4.04mg	n.s.s.	1/20	3.66mg	6/20	
c	1904	4.04mg	n.s.s.	1/20	3.66mg	6/20	
d	1904	7.91mg	n.s.s.	0/20	3.66mg	2/20	
e	1904	3.88mg	n.s.s.	3/20	3.66mg	7/20	

2-NITROSOMETHYLAMINOPYRIDINE   16219-98-0

	RefNum	LoConf	UpConf	Cntrl	1Dose	1Inc	Citation or Pathology
3708	1261	.118mg	.826mg	0/5	.714mg	18/27	Preussmann;jnci,62,153-156;1979
a	1261	.964mg	n.s.s.	1/5	.714mg	2/27	
b	1261	.301mg	n.s.s.	0/5	.714mg	8/27	
c	1261	.112mg	n.s.s.	2/5	.714mg	20/27	
d	1261	.263mg	n.s.s.	2/5	.714mg	12/27	

3-NITROSOMETHYLAMINOPYRIDINE   69658-91-9

	RefNum	LoConf	UpConf	Cntrl	1Dose	1Inc	Citation or Pathology
3709	1261	1.92mg	n.s.s.	1/5	.571mg	2/26	Preussmann;jnci,62,153-156;1979
a	1261	1.10mg	n.s.s.	2/5	.571mg	6/26	
b	1261	1.50mg	n.s.s.	0/5	.571mg	2/26	
c	1261	1.51mg	n.s.s.	2/5	.571mg	4/26	

4-NITROSOMETHYLAMINOPYRIDINE   16219-99-1

	RefNum	LoConf	UpConf	Cntrl	1Dose	1Inc	Citation or Pathology
3710	1261	8.33mg	n.s.s.	1/5	2.86mg	0/15	Preussmann;jnci,62,153-156;1979
a	1261	2.85mg	n.s.s.	2/5	2.86mg	4/15	
b	1261	3.56mg	n.s.s.	2/5	2.86mg	3/15	
c	1261	4.35mg	n.s.s.	0/5	2.86mg	1/15	

NITROSOMETHYLANILINE   614-00-6

	RefNum	LoConf	UpConf	Cntrl	1Dose	1Inc	2Dose	2Inc		Citation or Pathology
3711	1260	n.s.s.	12.4mg	0/5	70.7mg	15/15				Boyland;bjca,18,265-270;1964
a	1260	11.6mg	n.s.s.	0/5	70.7mg	8/15				
b	1260	13.5mg	n.s.s.	0/5	70.7mg	7/15				
c	1260	90.7mg	n.s.s.	0/5	70.7mg	0/15				
3712	1260	2.07mg	16.2mg	0/3	64.1mg	15/16				
a	1260	5.02mg	241.mg	0/3	64.1mg	12/16				
b	1260	21.2mg	n.s.s.	0/3	64.1mg	3/16				
c	1260	70.4mg	n.s.s.	0/3	64.1mg	0/16				
3713	1544	.233mg	.996mg	0/20	.895mg	16/20				Kroeger-Koepke;carc,4,157-160;1983
a	1544	3.30mg	n.s.s.	4/20	.895mg	1/20				
b	1544	.258mg	n.s.s.	18/20	.895mg	18/20				
3714	1544	.136mg	.588mg	0/20	.627mg	17/20				
a	1544	2.20mg	n.s.s.	3/20	.627mg	1/20				
b	1544	.205mg	n.s.s.	19/20	.627mg	18/20				
3715	1264	22.0ug	55.7ug	0/48	83.8ug	39/48	(.319mg	42/48)		Schmahl;clet,1,215-218;1976

NITROSOMETHYLPHENIDATE   55557-03-4

	RefNum	LoConf	UpConf	Cntrl	1Dose	1Inc	Citation or Pathology
3716	1257	19.3mg	n.s.s.	14/43	8.05mg	7/31	Giner-Sorolla;fctx,18,81-83;1980
a	1257	53.4mg	n.s.s.	0/43	8.05mg	0/31	
b	1257	10.6mg	n.s.s.	24/43	8.05mg	15/31	
3717	1257	4.69mg	n.s.s.	15/18	6.70mg	17/24	
a	1257	3.59mg	n.s.s.	16/18	6.70mg	19/24	

NITROSOMETHYLUNDECYLAMINE   68107-26-6

	RefNum	LoConf	UpConf	Cntrl	1Dose	1Inc	Citation or Pathology
3718	1207	1.12mg	6.28mg	2/260	6.07mg	10/20	Lijinsky;clet,5,209-213;1978
a	1207	1.13mg	6.38mg	3/260	6.07mg	10/20	
b	1207	1.13mg	6.59mg	5/260	6.07mg	10/20	
c	1207	1.42mg	9.37mg	0/260	6.07mg	8/20	
d	1207	1.91mg	24.9mg	8/260	6.07mg	6/20	
e	1207	3.79mg	205.mg	0/260	6.07mg	2/20	
f	1207	3.79mg	205.mg	0/260	6.07mg	2/20	

```
       Spe Strain Site    Xpo+Xpt                                                       TD50     2Tailpvl
          Sex   Route  Hist    Notes                                                       DR      AuOp

N-NITROSOMORPHOLINE         100ng..:..1ug...:..10.....:..100....:..1mg....:..10......:..100....:..1g.....:..10
3719 H f syg wat res mix 24m24 e                              . + .                      1.35mg  Z P<.0005+
a    H f syg wat liv mix 24m24 e                                                         51.6mg  * P<.0005+
b    H f syg wat tba mix 24m24 e                                                         1.48mg  Z P<.0005
3720 H f syg wat trh mem 24m24 er                                   . + .                11.3mg  * P<.0005+
a    H f syg wat lar mix 24m24 er                                                        15.4mg  * P<.0005+
3721 H m syg wat res mix 24m24 e                                  .+ .                    5.98mg  * P<.0005+
a    H m syg wat liv mix 24m24 e                                                         9.14mg  * P<.0005+
b    H m syg wat tba mix 24m24 e                                                         5.56mg  * P<.0005
3722 H m syg wat trh mem 24m24 er                                   . + .                7.98mg  * P<.0005+
a    H m syg wat lar mix 24m24 er                                                        11.5mg  * P<.0005+
3723 R f f34 wat liv mix 10m29 es                     . + .                              .140mg  * P<.0005+

a    R f f34 wat amd phe 10m29 es                                                        .325mg  Z P<.007

b    R f f34 wat thy ccr 10m29 es                                                        .363mg  Z P<.0005

c    R f f34 wat liv hpc 10m29 es                                                        .564mg  Z P<.0005+

d    R f f34 wat liv hes 10m29 es                                                        .666mg  * P<.0005+

e    R f f34 wat liv nnd 10m29 es                                                        1.28mg  Z P<.0005+

f    R f f34 wat eso mix 10m29 es                                                        1.97mg  Z P<.0005

g    R f f34 wat eso sqp 10m29 es                                                        2.53mg  Z P<.0005

h    R f f34 wat ton mix 10m29 es                                                        6.54mg  Z P<.003

i    R f f34 wat ton sqp 10m29 es                                                        50.2mg  * P<.3

j    R f f34 wat thy fca 10m29 es                                                        no dre    P=1.

3724 R f f34 wat liv mix 23m29 e                      . +.                               .127mg  * P<.0005+

a    R f f34 wat liv nnd 23m29 e                                                         .239mg  * P<.0005+

b    R f f34 wat liv hpc 23m29 e                                                         .431mg  * P<.0005+

c    R f f34 wat thy fca 23m29 e                                                         .581mg  Z P<.0005

d    R f f34 wat liv hes 23m29 e                                                         .630mg  * P<.0005+

e    R f f34 wat amd phe 23m29 e                                                         1.07mg  * P<.06

f    R f f34 wat ton sqp 23m29 e                                                         3.49mg  * P<.02

3725 R f f34 wat liv mix 50w73 e                            <+                           noTD50    P<.0005+
a    R f f34 wat liv hpc 50w73 e                                                         80.5ug    P<.0005+
b    R f f34 wat liv hes 50w73 e                                                         .348mg    P<.0005+

N'-NITROSONORNICOTINE        100ng..:..1ug...:..10.....:..100....:..1mg....:..10......:..100....:..1g.....:..10
3726 H f syg wat res pam 31w96                                     .    ±                11.5mg    P<.04  +
a    H f syg wat liv tum 31w96                                                           no dre    P=1.
3727 H m syg wat res pam 31w96                                   .    ±                  10.1mg    P<.04  +
a    H m syg wat liv ang 31w96                                                           34.4mg    P<.3

N'-NITROSONORNICOTINE-1-N-OXIDE  100ng..:..1ug...:..10.....:..100....:..1mg....:..10......:..100....:..1g.....:..10
3728 H f syg wat liv ang 31w96                                       .>                  39.0mg    P<.3   -
3729 H m syg wat liv tum 31w96                                          .>                no dre    P=1.
3730 R f f34 wat nas mix  8m24                               .    +    .                  1.86mg    P<.0005+
a    R f f34 wat eso mix  8m24                                                           5.65mg    P<.04  +
3731 R m f34 wat nas mix  8m24                             .    +    .                    .573mg    P<.0005+
a    R m f34 wat eso mix  8m24                                                           1.63mg    P<.0005+
b    R m f34 wat nas mal  8m24                                                           1.63mg    P<.0005
c    R m f34 wat eso pam  8m24                                                           2.64mg    P<.005
d    R m f34 wat nas pam  8m24                                                           2.64mg    P<.005

NITROSOPIPECOLIC ACID        100ng..:..1ug...:..10.....:..100....:..1mg....:..10......:..100....:..1g.....:..10
3732 R f mrc wat liv tum 17m24 e                                    .>                   no dre    P=1.
a    R f mrc wat tba mix 17m24 e                                                         4.99mg    P<.07  -
3733 R m mrc wat liv tum 17m24 e                                      .>                 no dre    P=1.
a    R m mrc wat tba mix 17m24 e                                                         3.82mg    P<.07  -

N-NITROSOPIPERAZINE         100ng..:..1ug...:..10.....:..100....:..1mg....:..10......:..100....:..1g.....:..10
3734 R f mrc wat liv tum 14m30 e                                      .>                 no dre    P=1.
a    R f mrc wat tba mix 14m30 e                                                         5.51mg  * P<.2   +
3735 R m mrc wat liv hpt 14m30 e                            .        ±                    6.52mg  \ P<.05  +
a    R m mrc wat tba mix 14m30 e                                                         21.6mg  * P<.5   +
```

	RefNum	LoConf	UpConf	Cntrl	1Dose	1Inc	2Dose	2Inc	3Dose	3Inc	4Dose	4Inc	5Dose	5Inc	6Dose	6Inc	7Dose	7Inc	Citation or Pathology / Brkly Code

N-NITROSOMORPHOLINE   59-89-2

	RefNum	LoConf	UpConf	Cntrl	1Dose	1Inc	2Dose	2Inc	3Dose	3Inc	4Dose	4Inc	5Dose	5Inc	6Dose	6Inc	7Dose	7Inc	Citation
3719	1543	.714mg	2.97mg	0/50	1.36mg	14/28	(6.82mg	16/30	13.6mg	22/30)									Ketkar;clet,17,333-338;1983/pers.comm.
a	1543	23.4mg	156.mg	0/50	1.36mg	0/28	6.82mg	2/30	13.6mg	6/30									
b	1543	.746mg	4.02mg	3/50	1.36mg	14/28	(6.82mg	17/30	13.6mg	23/30)									
3720	1998	6.50mg	29.0mg	0/25	1.36mg	7/28	6.82mg	11/30	13.6mg	16/30									Cardesa;expl,401,267-281;1990
a	1998	8.95mg	44.7mg	0/25	1.36mg	4/28	6.82mg	9/30	13.6mg	12/30									
3721	1543	4.03mg	9.71mg	0/50	1.20mg	8/29	6.00mg	13/29	12.0mg	21/30									Ketkar;clet,17,333-338;1983/pers.comm.
a	1543	5.87mg	15.4mg	0/50	1.20mg	4/29	6.00mg	9/29	12.0mg	18/30									
b	1543	3.40mg	10.8mg	8/50	1.20mg	12/29	6.00mg	14/29	12.0mg	26/30									
3722	1998	5.21mg	14.9mg	0/25	1.20mg	5/29	6.00mg	9/29	12.0mg	20/30									Cardesa;expl,401,267-281;1990
a	1998	7.14mg	20.2mg	0/25	1.20mg	1/29	6.00mg	10/29	12.0mg	15/30									
3723	1886m	95.6ug	.222mg	1/80	7.29ug	6/48	17.8ug	7/48	42.1ug	15/48	.104mg	14/24	.338mg	22/23	1.06mg	23/24	2.83mg	24/24	Lijinsky;canr,48,2089-2095;1988
a	1886m	.146mg	6.40mg	8/80	7.29ug	10/48	17.8ug	3/48	42.1ug	15/48	.104mg	7/24	(.338mg	5/24	1.06mg	2/24	2.83mg	0/24)	
b	1886m	.185mg	1.38mg	2/80	7.29ug	1/48	17.8ug	4/48	42.1ug	8/48	.104mg	5/24	(.338mg	2/24	1.06mg	2/24	2.83mg	0/24)	
c	1886m	.382mg	.871mg	0/80	7.29ug	0/48	17.8ug	1/48	42.1ug	5/48	.104mg	7/24	.338mg	15/23	1.06mg	16/24	(2.83mg	15/24)	
d	1886m	.432mg	1.05mg	0/80	7.29ug	0/48	17.8ug	0/48	42.1ug	1/48	.104mg	0/24	.338mg	8/23	1.06mg	23/24	2.83mg	24/24	
e	1886m	.793mg	2.31mg	1/80	7.29ug	6/48	17.8ug	6/48	42.1ug	11/48	.104mg	9/24	.338mg	15/23	1.06mg	11/24	2.83mg	20/24	
f	1886m	1.09mg	4.07mg	0/80	7.29ug	0/48	17.8ug	0/48	42.1ug	0/48	.104mg	0/24	.338mg	3/24	1.06mg	13/24	(2.83mg	5/24)	
g	1886m	1.33mg	5.72mg	0/80	7.29ug	0/48	17.8ug	0/48	42.1ug	0/48	.104mg	0/24	.338mg	2/24	1.06mg	11/24	(2.83mg	5/24)	
h	1886m	2.44mg	63.6mg	2/80	7.29ug	2/48	17.8ug	0/48	42.1ug	0/48	.104mg	1/24	.338mg	2/24	1.06mg	4/24	(2.83mg	0/24)	
i	1886m	14.1mg	n.s.s.	0/80	7.29ug	1/48	17.8ug	0/48	42.1ug	0/48	.104mg	0/24	.338mg	1/24	1.06mg	2/24	2.83mg	0/24	
j	1886m	20.0mg	n.s.s.	0/80	7.29ug	1/48	17.8ug	0/48	42.1ug	4/48	.104mg	1/24	.338mg	3/24	1.06mg	0/24	2.83mg	0/24	
3724	1886n	89.3ug	.192mg	1/80	2.27ug	6/100	5.83ug	5/99	14.6ug	7/47	35.6ug	9/48	84.2ug	22/48	.249mg	23/24			
a	1886n	.156mg	.411mg	1/80	2.27ug	5/100	5.83ug	5/99	14.6ug	6/47	35.6ug	8/48	84.2ug	15/48	.249mg	15/24			
b	1886n	.266mg	.761mg	0/80	2.27ug	1/100	5.83ug	0/99	14.6ug	0/47	35.6ug	1/48	84.2ug	7/48	.249mg	16/24			
c	1886n	.299mg	1.95mg	0/80	2.27ug	0/100	5.83ug	2/100	14.6ug	3/48	35.6ug	0/48	84.2ug	7/48	(.249mg	2/24)			
d	1886n	.361mg	1.24mg	0/80	2.27ug	0/100	5.83ug	0/99	14.6ug	0/47	35.6ug	0/48	84.2ug	5/48	.249mg	13/24			
e	1886n	.377mg	n.s.s.	8/80	2.27ug	22/100	5.83ug	18/100	14.6ug	8/48	35.6ug	12/48	84.2ug	13/48	.249mg	7/24			
f	1886n	1.08mg	n.s.s.	0/80	2.27ug	1/100	5.83ug	1/100	14.6ug	0/48	35.6ug	1/48	84.2ug	1/48	.249mg	2/24			
3725	1928	n.s.s.	.129mg	0/20	.714mg	20/20													Hecht;carc,10,1475-1477;1989/pers.comm.
a	1928	33.1ug	.184mg	0/20	.714mg	19/20													
b	1928	.166mg	.909mg	0/20	.714mg	10/20													

N'-NITROSONORNICOTINE   16543-55-8

	RefNum	LoConf	UpConf	Cntrl	1Dose	1Inc	Citation
3726	1539	3.45mg	n.s.s.	0/10	7.05mg	3/10	Hecht;clet,20,333-340;1983
a	1539	12.4mg	n.s.s.	0/10	7.05mg	0/10	
3727	1539	3.03mg	n.s.s.	0/10	6.20mg	3/10	
a	1539	5.58mg	n.s.s.	0/10	6.20mg	1/10	

N'-NITROSONORNICOTINE-1-N-OXIDE   78246-24-9

	RefNum	LoConf	UpConf	Cntrl	1Dose	1Inc	Citation
3728	1539n	6.34mg	n.s.s.	0/10	7.05mg	1/10	Hecht;clet,20,333-340;1983
3729	1539n	10.9mg	n.s.s.	0/10	6.20mg	0/10	
3730	1539m	.769mg	6.48mg	0/12	2.37mg	7/12	
a	1539m	1.70mg	n.s.s.	0/12	2.37mg	3/12	
3731	1539m	.211mg	1.62mg	0/12	2.08mg	11/12	
a	1539m	.673mg	5.67mg	0/12	2.08mg	7/12	
b	1539m	.673mg	5.67mg	0/12	2.08mg	7/12	
c	1539m	.983mg	21.2mg	0/12	2.08mg	5/12	
d	1539m	.983mg	21.2mg	0/12	2.08mg	5/12	

NITROSOPIPECOLIC ACID   4515-18-8

	RefNum	LoConf	UpConf	Cntrl	1Dose	1Inc	Citation
3732	216	13.6mg	n.s.s.	0/15	4.42mg	0/15	Garcia;zkko,79,141-144;1973
a	216	1.72mg	n.s.s.	4/15	4.42mg	9/15	
3733	216	11.9mg	n.s.s.	0/15	3.86mg	0/15	
a	216	1.30mg	n.s.s.	5/15	3.86mg	10/15	

N-NITROSOPIPERAZINE   5632-47-3

	RefNum	LoConf	UpConf	Cntrl	1Dose	1Inc	2Dose	2Inc	Citation
3734	229	2.48mg	n.s.s.	0/13	.964mg	0/10	4.90mg	0/10	Garcia;zkko,74,179-184;1970
a	229	1.55mg	n.s.s.	0/13	.964mg	8/10	4.90mg	7/10	
3735	229	1.06mg	n.s.s.	0/57	.672mg	1/10	(3.30mg	0/10)	
a	229	3.16mg	n.s.s.	12/57	.672mg	4/10	3.30mg	3/10	

```
        Spe Strain  Site    Xpo+Xpt                                              TD50    2Tailpvl
         Sex  Route  Hist       Notes                                             DR      AuOp
N-NITROSOPIPERIDINE              100ng..:..1ug....:..10.....:..100.....:..1mg....:..10.....:..100....:..1g.....:..10
3736 H f syg wat res mix 24m24                                              . + .         96.7mg * P<.0005+
  a  H f syg wat dgt mix 24m24                                                            132.mg * P<.002 +
  b  H f syg wat liv mix 24m24                                                            307.mg * P<.005 +
  c  H f syg wat tba mix 24m24                                                            9.89mg Z P<.0005
3737 H f syg wat trh mem 24m24 er                                              . + .      205.mg * P<.005 +
  a  H f syg wat lar mix 24m24 er                                                         258.mg * P<.002 +
3738 H m syg wat res mix 24m24                                           . + .            50.9mg * P<.0005+
  a  H m syg wat dgt mix 24m24                                                            98.2mg * P<.0005+
  b  H m syg wat liv mix 24m24                                                            135.mg * P<.0005+
  c  H m syg wat tba mix 24m24                                                            52.9mg * P<.0005
3739 H m syg wat lar mix 24m24 er                                         . + .           76.1mg * P<.0005+
  a  H m syg wat trh mem 24m24 er                                                         180.mg * P<.0005+
3740 M m icm eat for sqc 52w52                              . + .                         1.30mg   P<.0005+
  a  M m icm eat liv mix 52w52                                                            2.53mg   P<.0005+
  b  M m icm eat liv ade 52w52                                                            5.12mg   P<.005
  c  M m icm eat lun ade 52w52                                                            3.52mg   P<.02  +
3741 P b cym eat liv hpc 12y12 Ww                      :     +     :                      12.1mg   P<.0005+

  a  P b cym eat liv hes 12y12 Ww                                                         21.7mg   P<.006
  b  P b cym eat tba mal 12y12 Ww                                                         12.1mg   P<.0005
3742 P b rhe eat liv hpc 12y12 Ww                         :     +     :                   8.36mg   P<.0005+
  a  P b rhe eat tba mal 12y12 Ww                                                         10.3mg   P<.0005
3743 P b rhe ipj liv hpc 20y20 Ww                :     +     :                            1.65mg   P<.0005+
  a  P b rhe ipj bil hmt 20y20 Ww                                                         1.07mg   P<.02
  b  P b rhe ipj tba mix 20y20 Ww                                                         1.04mg   P<.002
  c  P b rhe ipj tba mal 20y20 Ww                                                         1.39mg   P<.003
  d  P b rhe ipj tba ben 20y20 Ww                                                         4.36mg   P<.3
3744 R b clw wat eso mix 30m30 e                       . +.                               1.31mg * P<.0005+


  a  R b clw wat liv hct 30m30 e                                                          1.41mg * P<.0005+


  b  R b clw wat liv hpc 30m30 e                                                          1.51mg * P<.0005+


  c  R b clw wat eso mal 30m30 e                                                          10.3mg * P<.0005+


  d  R b clw wat nsp mix 30m30 e                                                          27.4mg Z P<.0005+


  e  R b clw wat nsp mal 30m30 e                                                          77.1mg Z P<.02   +


  f  R b clw wat ski mal 30m30 e                                                          222.mg * P<.07


  g  R b clw wat lgi mal 30m30 e                                                          224.mg Z P<.8


3745 R b sda wat liv mix 27m33 a                    . + .                                 .963mg Z P<.0005
  a  R b sda wat mix mix 27m33 a                                                          2.32mg * P<.0005
  b  R b sda wat tba mal 27m33 a                                                          1.57mg * P<.0005+

NITROSOPROLINE                  100ng..:..1ug....:..10.....:..100.....:..1mg....:..10.....:..100....:..1g.....:..10
3746 R f mrc wat liv tum 17m24 e                                     .>                   no dre   P=1.
  a  R f mrc wat tba mix 17m24 e                                                          6.69mg   P<.2    -
3747 R m mrc wat liv tum 17m24 e                                  .>                      no dre   P=1.
  a  R m mrc wat tba mix 17m24 e                                                          no dre   P=1.    -

N-NITROSOPYRROLIDINE            100ng..:..1ug....:..10.....:..100.....:..1mg....:..10.....:..100....:..1g.....:..10
3748 H f syg wat liv hct 24m24                                             . + .    .     35.9mg * P<.008 +
  a  H f syg wat tba mix 24m24                                                            8.44mg * P<.002
3749 H m syg wat liv hct 24m24                                      . + .                 8.88mg * P<.0005+
  a  H m syg wat liv hae 24m24                                                            130.mg * P<.2
  b  H m syg wat tba mix 24m24                                                            9.26mg * P<.007
3750 M m swi gav tba mix 76w76 e                       . + .                              .679mg   P<.002 +
3751 R m chm wat liv hpc 84w84                         . + .                              .659mg   P<.0005+
  a  R m chm wat liv nnd 84w84                                                            1.13mg   P<.0005+
  b  R m chm wat liv clc 84w84                                                            14.8mg   P<.1   +
3752 R b clw wat liv hct 30m30 e                         . +.                             1.70mg * P<.0005+


  a  R b clw wat liv hpc 30m30 e                                                          2.20mg * P<.0005+


  b  R b clw wat bil ben 30m30 e                                                          99.7mg * P<.003 +
```

```
       RefNum  LoConf UpConf    Cntrl  1Dose  1Inc   2Dose  2Inc              Citation or Pathology
                                                                                       Brkly Code

N-NITROSOPIPERIDINE      100-75-4
3736   1541  56.9mg 240.mg   0/50  8.18mg  4/30  34.1mg  6/30  68.2mg 10/30      Ketkar;clet,21,219-224;1983
a      1541  70.6mg 624.mg   0/50  8.18mg  4/30  34.1mg  5/30  68.2mg  7/30
b      1541  133.mg 3.08gm   0/50  8.18mg  1/30  34.1mg  2/30  68.2mg  4/30
c      1541  4.97mg 27.9mg   3/50  8.18mg 14/30 (34.1mg 12/30  68.2mg 16/30)
3737   1998  99.9mg 1.93gm   0/25  8.18mg  1/30  34.1mg  4/28  68.2mg  5/30      Cardesa;expl,401,267-281;1990
a      1998  117.mg 1.10gm   0/25  8.18mg  0/30  34.1mg  3/28  68.2mg  5/30
3738   1541  32.6mg 90.0mg   0/50  7.20mg  5/30  30.0mg 10/30  60.0mg 15/30      Ketkar;clet,21,219-224;1983
a      1541  55.5mg 198.mg   0/50  7.20mg  0/30  30.0mg  4/30  60.0mg 13/30
b      1541  71.2mg 312.mg   0/50  7.20mg  1/30  30.0mg  2/30  60.0mg 10/30
c      1541  29.6mg 142.mg   8/50  7.20mg  9/30  30.0mg 12/30  60.0mg 19/30
3739   1998  45.2mg 146.mg   0/25  7.20mg  1/28  30.0mg  7/29  60.0mg 13/30      Cardesa;expl,401,267-281;1990
a      1998  87.5mg 569.mg   0/25  7.20mg  0/28  30.0mg  4/29  60.0mg  6/30
3740    225  .736mg 2.60mg   0/30  6.00mg 18/33                                  Takayama;nawi,56,142;1969
a       225  1.26mg 6.36mg   0/30  6.00mg 11/33
b       225  2.09mg 36.6mg   0/30  6.00mg  6/33
c       225  1.49mg n.s.s.   2/30  6.00mg 10/33
3741   2001  3.30mg 47.3mg   0/89  176.mg  5/5                                   Adamson;ossc,129-156;
                                            1982/Thorgeirsson 1994/Dalgard 1991/Thorgeirsson&Seiber pers.comm.
a      2001  3.53mg 3.31gm   0/55  176.mg  1/1
b      2001  3.30mg 47.3mg   0/89  176.mg  5/5
3742   2001  2.35mg 28.4mg   0/96  123.mg  6/7
a      2001  3.05mg 46.5mg   2/96  123.mg  6/7
3743  2001m  .454mg 11.7mg   0/71  3.10mg  3/5
a     2001m  .174mg n.s.s.   0/23  3.10mg  1/1
b     2001m  .285mg 10.5mg  8/110  3.10mg  4/6
c     2001m  .351mg 19.8mg   4/76  3.10mg  3/6
d     2001m  .457mg n.s.s.  5/110  3.10mg  1/6
3744   1958  .970mg 1.82mg  0/246  11.0ug  0/12  22.0ug  0/12  44.0ug  1/12  87.0ug  0/12  .174mg  1/12
                                   .349mg  3/12  .523mg  8/12  .698mg  4/12  .872mg 10/12  1.05mg 10/12
                                   1.39mg  8/12  1.74mg  8/12  2.09mg 10/12  2.79mg 12/12  5.58mg  9/12 Gray;canr,51,6470-6491;1991
a      1958  1.04mg 2.00mg  8/246  11.0ug  0/12  22.0ug  1/12  44.0ug  1/12  87.0ug  3/12  .174mg  3/12
                                   .349mg  5/12  .523mg  9/12  .698mg  7/12  .872mg  9/12  1.05mg 11/12
                                   1.39mg  8/12  1.74mg  8/12  2.09mg 10/12  2.79mg  9/12  5.58mg  8/12
b      1958  1.11mg 2.10mg  2/246  11.0ug  0/12  22.0ug  1/12  44.0ug  1/12  87.0ug  3/12  .174mg  2/12
                                   .349mg  4/12  .523mg  9/12  .698mg  5/12  .872mg  7/12  1.05mg 10/12
                                   1.39mg  8/12  1.74mg  8/12  2.09mg 10/12  2.79mg  9/12  5.58mg  8/12
c      1958  6.07mg 19.5mg  0/246  11.0ug  0/12  22.0ug  0/12  44.0ug  0/12  87.0ug  0/12  .174mg  1/12
                                   .349mg  1/12  .523mg  2/12  .698mg  0/12  .872mg  3/12  1.05mg  3/12
                                   1.39mg  1/12  1.74mg  1/12  2.09mg  4/12  2.79mg  3/12  5.58mg
d      1958  12.0mg 110.mg  0/246  11.0ug  0/12  22.0ug  1/12  44.0ug  0/12  87.0ug  0/12  .174mg  0/12
                                   .349mg  0/12  .523mg  0/12  .698mg  0/12  .872mg  2/12  1.05mg  1/12
                                   1.39mg  1/12  1.74mg  2/12  2.09mg  1/12  2.79mg  0/12  5.58mg  1/12
e      1958  22.7mg n.s.s.  0/246  11.0ug  0/12  22.0ug  0/12  44.0ug  0/12  87.0ug  0/12  .174mg  0/12
                                   .349mg  0/12  .523mg  0/12  .698mg  0/12  .872mg  0/12  1.05mg  0/12
                                   1.39mg  1/12  1.74mg  1/12  2.09mg  0/12  2.79mg  0/12  5.58mg  1/12
f      1958  36.2mg n.s.s.  0/246  11.0ug  0/12  22.0ug  0/12  44.0ug  0/12  87.0ug  0/12  .174mg  0/12
                                   .349mg  0/12  .523mg  0/12  .698mg  0/12  .872mg  0/12  1.05mg  0/12
                                   1.39mg  0/12  1.74mg  0/12  2.09mg  0/12  2.79mg  1/12  5.58mg  0/12
g      1958  36.4mg n.s.s.  0/246  11.0ug  0/12  22.0ug  0/12  44.0ug  0/12  87.0ug  0/12  .174mg  0/12
                                   .349mg  0/12  .523mg  1/12  .698mg  0/12  .872mg  0/12  1.05mg  0/12
                                   1.39mg  0/12  1.74mg  0/12  2.09mg  0/12  2.79mg  0/12  5.58mg  0/12
3745   1634  .590mg 1.96mg   0/40  17.1ug  3/78  85.7ug  5/75  .429mg 16/34 (2.14mg 11/34)  Eisenbrand;iarc,657-663;1980
a      1634  1.51mg 3.80mg   0/40  17.1ug  0/78  85.7ug  2/75  .429mg  8/34  2.14mg 24/34
b      1634  .998mg 2.69mg   4/40  17.1ug  7/78  85.7ug  7/75  .429mg 14/34  2.14mg 28/34


NITROSOPROLINE      7519-36-0
3746    216  13.6mg n.s.s.   0/15  4.42mg  0/15                                  Garcia;zkko,79,141-144;1973
a       216  2.05mg n.s.s.   4/15  4.42mg  8/15
3747    216  11.9mg n.s.s.   0/15  3.86mg  0/15
a       216  4.39mg n.s.s.   5/15  3.86mg  4/15


N-NITROSOPYRROLIDINE     930-55-2
3748   1503  12.4mg 770.mg   0/50  .573mg  0/30  2.18mg  1/30  4.50mg  3/30      Ketkar;zkko,104,75-79;1982
a      1503  4.27mg 39.7mg   3/50  .573mg  2/30  2.18mg 10/30  4.50mg  8/30
3749   1503  4.67mg 20.4mg   0/50  .504mg  1/30  1.92mg  2/30  3.96mg 10/30
a      1503  21.2mg n.s.s.   0/50  .504mg  0/30  1.92mg  0/30  3.96mg  1/30
b      1503  4.14mg 159.mg   8/50  .504mg  3/30  1.92mg  4/30  3.96mg 13/30
3750   1713  .309mg 2.84mg   1/20  1.19mg 10/20                                  Shah;zkko,109,203-207;1985
3751   1774  .340mg 1.35mg   0/23  3.00mg 20/23                                  Chung;canr,46,1285-1289;1986
a      1774  .601mg 2.40mg   0/23  3.00mg 16/23
b      1774  3.63mg n.s.s.   0/23  3.00mg  2/23
3752   1958  1.24mg 2.36mg  8/246  34.0ug  0/12  69.0ug  0/12  .137mg  1/12  .275mg  5/12  .550mg  6/12
                                   1.10mg  9/12  1.65mg 12/12  2.20mg 12/12  2.75mg 11/12  3.30mg 10/12
                                   4.40mg 12/12  5.50mg  9/12  6.60mg 10/12  8.80mg 11/12  17.6mg 12/12 Gray;canr,51,6470-6491;1991
a      1958  1.63mg 3.03mg  2/246  34.0ug  0/12  69.0ug  0/12  .137mg  1/12  .275mg  5/12  .550mg  4/12
                                   1.10mg  7/12  1.65mg 10/12  2.20mg 12/12  2.75mg 10/12  3.30mg  9/12
                                   4.40mg 11/12  5.50mg  9/12  6.60mg  9/12  8.80mg 11/12  17.6mg 12/12
b      1958  40.1mg 814.mg  2/246  34.0ug  0/12  69.0ug  0/12  .137mg  0/12  .275mg  1/12  .550mg  0/12
                                   1.10mg  1/12  1.65mg  0/12  2.20mg  0/12  2.75mg  0/12  3.30mg  1/12
                                   4.40mg  0/12  5.50mg  2/12  6.60mg  2/12  8.80mg  2/12  17.6mg  0/12
```

	Spe Strain Site Xpo+Xpt Sex Route Hist Notes		TD50 DR 2Tailpvl AuOp
c	R b clw wat nsp tum 30m30 e		350.mg * P<.04
3753	R f f34 wat liv hpc 7m26 e	<+	noTD50 P<.0005+
a	R f f34 wat liv nnd 7m26 e		2.51mg P<.04 +
b	R f f34 wat liv hes 7m26 e		15.0mg P<.3 +
c	R f f34 wat liv clc 7m26 e		15.0mg P<.3 +
d	R f f34 wat tba mix 7m26 e		noTD50 P<.6 +
3754	R f f34 wat liv hpc 12m26 e	. + .	.439mg P<.0005+
a	R f f34 wat liv mix 12m26 e		.439mg P<.0005+
b	R f f34 wat liv clc 12m26 e		4.49mg P<.007 +
c	R f f34 wat liv cho 12m26 e		7.95mg P<.04
d	R f f34 wat liv lyp 12m26 e		7.95mg P<.04
e	R f f34 wat liv sar 12m26 e		25.2mg P<.3
f	R f f34 wat tba tum 12m26 e		no dre P=1.
3755	R f mrw wat liv hpc 16m24 e	. + .	1.81mg P<.0005+
a	R f mrw wat tba mix 16m24 e		4.49mg P<.3
3756	R m mrw wat liv hpc 16m24 e	. + .	1.58mg P<.0005+
a	R m mrw wat tes pms 16m24 e		8.21mg P<.007
b	R m mrw wat tba mix 16m24 e		2.03mg P<.004
3757	R b sda wat liv hpc 23m30 as	. +.	3.26mg Z P<.0005
a	R b sda wat liv hpa 23m30 as		55.5mg * P<.005
b	R b sda wat tba mal 23m30 as		2.37mg Z P<.0005+
c	R b sda wat tba ben 23m30 as		71.1mg * P<.2
3758	R m sda wat liv hpc 20m30	. +.	.648mg P<.0005+
a	R m sda wat tba mal 20m30		.631mg P<.0005
3759	R m sda wat liv hpc 10m30	.+ .	.489mg P<.0005+
a	R m sda wat tba mal 10m30		.501mg P<.0005
3760	R m sda wat liv mix 37m38 a	. + .	2.43mg * P<.0005+
a	R m sda wat liv hpc 37m38 a		4.06mg * P<.0005+
b	R m sda wat liv hpa 37m38 a		8.13mg * P<.0005+
c	R m sda wat unt tum 37m38 a		7.03mg Z P<.1 +
d	R m sda wat git mix 37m38 a		14.9mg * P<.4 +
e	R m sda wat liv hmm 37m38 a		55.7mg * P<.3
f	R m sda wat tba mal 37m38 a		2.03mg * P<.007 +
g	R m sda wat tba ben 37m38 a		no dre P=1.

N-NITROSOTHIALDINE `100ng..:..1ug...:..10......:..100.....:..1mg....:..10......:..100....:..1g.....:..10`

			TD50 DR 2Tailpvl AuOp
3761	R f f34 gav eso sqa 43w90 e	. + .	.483mg P<.0005+
a	R f f34 gav eso sqc 43w90 e		1.63mg P<.003
b	R f f34 gav ton sqa 43w90 e		1.53mg P<.02 +
c	R f f34 gav liv hpa 43w90 e		2.61mg P<.02 +
d	R f f34 gav liv cho 43w90 e		3.58mg P<.04 +
e	R f f34 gav ton sqc 43w90 e		no dre P=1.

N-NITROSOTHIOMORPHOLINE `100ng..:..1ug...:..10......:..100.....:..1mg....:..10......:..100....:..1g.....:..10`

			TD50 DR 2Tailpvl AuOp
3762	R f mrc wat eso mix 9m24 e	. + .	7.69mg * P<.002 +
a	R f mrc wat ton sqc 9m24 e		22.0mg * P<.3 +
b	R f mrc wat liv tum 9m24 e		no dre P=1.
c	R f mrc wat tba mix 9m24 e		7.38mg * P<.2
3763	R m mrc wat eso mix 38w95 e	. + .	4.15mg P<.0005+
a	R m mrc wat liv tum 38w95 e		no dre P=1.
b	R m mrc wat tba mix 38w95 e		2.29mg * P<.02

o-NITROSOTOLUENE `100ng..:..1ug...:..10......:..100.....:..1mg....:..10......:..100....:..1g.....:..10`

			TD50 DR 2Tailpvl AuOp
3764	R m f34 eat liv mix 72w93 e	. + .	50.7mg P<.0005+
a	R m f34 eat ski fib 72w93 e		55.8mg P<.0005+
b	R m f34 eat liv hpt 72w93 e		59.2mg P<.0005+
c	R m f34 eat ubl mix 72w93 e		71.5mg P<.0005+
d	R m f34 eat ubl pam 72w93 e		78.8mg P<.0005+
e	R m f34 eat spl fib 72w93 e		87.0mg P<.0005+
f	R m f34 eat pec scs 72w93 e		303.mg P<.009
g	R m f34 eat pec mso 72w93 e		195.mg P<.03

NITROUS OXIDE `100ng..:..1ug...:..10......:..100.....:..1mg....:..10......:..100....:..1g.....:..10`

			TD50 DR 2Tailpvl AuOp
3765	M f sww inh lun ade 78w83		309.gm * P<.08 -
a	M f sww inh liv tum 78w83		no dre P=1. -
b	M f sww inh mix tum 78w83		no dre P=1. -
3766	M m sww inh lun ade 78w83 s		no dre P=1. -
a	M m sww inh mix tum 78w83 s		no dre P=1. -
b	M m sww inh liv tum 78w83 s		no dre P=1. -

NIVALENOL `100ng..:..1ug...:..10......:..100.....:..1mg....:..10......:..100....:..1g.....:..10`

			TD50 DR 2Tailpvl AuOp
3767	M f c6s eat liv mfh 24m24 e	.>	no dre P=1.
a	M f c6s eat lun mix 24m24 e		no dre P=1.
b	M f c6s eat tba tum 24m24 e		no dre P=1.

NORETHYNODREL `100ng..:..1ug...:..10......:..100.....:..1mg....:..10......:..100....:..1g.....:..10`

			TD50 DR 2Tailpvl AuOp
3768	M f c3h eat mam tum 24m24 er	.>	no dre P=1. -
3769	M f cf1 eat liv hct 78w78 er	. ±	49.1mg * P<.04 -
3770	M m cf1 eat liv hct 78w78 er	.>	no dre P=1. -

	RefNum	LoConf	UpConf	Cntrl	1Dose	1Inc	2Dose	2Inc	(additional doses)	Citation or Pathology / Brkly Code
c	1958	86.1mg	n.s.s.	0/246	34.0ug	0/12	69.0ug	0/12	.137mg 0/12 .275mg 0/12 .550mg 0/12	
									1.10mg 0/12 1.65mg 0/12 2.20mg 0/12 2.75mg 1/12 3.30mg 0/12	
									4.40mg 0/12 5.50mg 0/12 6.60mg 0/12 8.80mg 1/12 17.6mg 0/12	
3753	1795	n.s.s.	.409mg	0/20	1.00mg	20/20				Michejda;canr,46,2252-2256;1986/pers.comm.
a	1795	.915mg	n.s.s.	1/20	1.00mg	6/20				
b	1795	2.44mg	n.s.s.	0/20	1.00mg	1/20				
c	1795	2.44mg	n.s.s.	0/20	1.00mg	1/20				
d	1795	n.s.s.	n.s.s.	19/20	1.00mg	20/20				
3754	2022	.179mg	1.04mg	1/20	1.65mg	19/20				Lijinsky;clet,12,99-103;1981
a	2022	.179mg	1.04mg	1/20	1.65mg	19/20				
b	2022	1.70mg	51.8mg	0/20	1.65mg	5/20				
c	2022	2.40mg	n.s.s.	0/20	1.65mg	3/20				
d	2022	2.40mg	n.s.s.	0/20	1.65mg	3/20				
e	2022	4.10mg	n.s.s.	0/20	1.65mg	1/20				
f	2022	.326mg	n.s.s.	20/20	1.65mg	19/20				
3755	228	.802mg	4.45mg	0/20	5.21mg	13/15				Greenblatt;jnci,48,1687-1696;1972/1973
a	228	1.06mg	n.s.s.	14/20	5.21mg	13/15				
3756	228	.731mg	3.93mg	0/20	3.65mg	12/15				
a	228	2.82mg	123.mg	0/20	3.65mg	4/15				
b	228	.807mg	16.5mg	6/20	3.65mg	12/15				
3757	371	2.23mg	5.01mg	0/61	.300mg	0/60	1.00mg	13/62	3.00mg 30/38 (10.0mg 9/24)	Preussmann;zkko,90,161-166;1977
a	371	20.9mg	857.mg	0/61	.300mg	3/60	1.00mg	4/62	3.00mg 1/38 10.0mg 5/24	
b	371	1.60mg	3.90mg	6/61	.300mg	12/60	1.00mg	20/62	3.00mg 32/38 (10.0mg 11/24)	
c	371	20.9mg	n.s.s.	5/61	.300mg	7/60	1.00mg	9/62	3.00mg 3/38 10.0mg 6/24	
3758	1717m	.454mg	.958mg	0/80	.657mg	53/80				Hoos;clet,26,77-82;1985
a	1717m	.412mg	1.10mg	19/80	.657mg	60/80				
3759	1717n	.346mg	.712mg	0/80	.657mg	61/80				
a	1717n	.333mg	.826mg	19/80	.657mg	65/80				
3760	1838	1.44mg	4.68mg	3/500	28.6ug	1/80	95.0ug	4/80	.286mg 17/80	Berger;carc,8,1635-1643;1987/pers.comm.
a	1838	2.14mg	9.15mg	0/500	28.6ug	1/80	95.0ug	0/80	.286mg 12/80	
b	1838	3.42mg	31.2mg	1/500	28.6ug	0/80	95.0ug	2/80	.286mg 5/80	
c	1838	1.78mg	n.s.s.	1/500	28.6ug	2/80	95.0ug	1/80	(.286mg 0/80)	
d	1838	3.25mg	n.s.s.	26/500	28.6ug	6/80	95.0ug	7/80	.286mg 6/80	
e	1838	9.07mg	n.s.s.	0/500	28.6ug	0/80	95.0ug	1/80	.286mg 0/80	
f	1838	.933mg	38.1mg	144/500	28.6ug	23/80	95.0ug	28/80	.286mg 35/80	
g	1838	1.18mg	n.s.s.	362/500	28.6ug	63/80	95.0ug	59/80	.286mg 55/80	

N-NITROSOTHIALDINE   81795-07-5

	RefNum	LoConf	UpConf	Cntrl	1Dose	1Inc	2Dose	2Inc	(additional doses)	Citation or Pathology
3761	1884	.247mg	1.09mg	0/20	1.13mg	14/20				Lijinsky;fctx,26,3-7;1988
a	1884	.661mg	8.52mg	0/20	1.13mg	6/20				
b	1884	.603mg	n.s.s.	1/20	1.13mg	7/20				
c	1884	.897mg	n.s.s.	0/20	1.13mg	4/20				
d	1884	1.08mg	n.s.s.	0/20	1.13mg	3/20				
e	1884	2.09mg	n.s.s.	1/20	1.13mg	1/20				

N-NITROSOTHIOMORPHOLINE   26541-51-5

	RefNum	LoConf	UpConf	Cntrl	1Dose	1Inc	2Dose	2Inc	Citation or Pathology
3762	229	3.11mg	34.4mg	0/13	.739mg	0/10	4.25mg	6/17	Garcia;zkko,74,179-184;1970
a	229	5.31mg	n.s.s.	0/13	.739mg	1/10	4.25mg	2/17	
b	229	1.41mg	n.s.s.	0/13	.739mg	0/10	4.25mg	0/17	
c	229	2.22mg	n.s.s.	4/13	.739mg	3/10	4.25mg	9/17	
3763	229	1.42mg	21.5mg	0/57	.571mg	0/11	2.90mg	4/10	
a	229	.888mg	n.s.s.	0/57	.571mg	0/11	2.90mg	0/10	
b	229	.794mg	n.s.s.	12/57	.571mg	4/11	2.90mg	6/10	

o-NITROSOTOLUENE   611-23-4

	RefNum	LoConf	UpConf	Cntrl	1Dose	1Inc	Citation or Pathology
3764	1487	28.3mg	105.mg	1/27	105.mg	20/29	Hecht;clet,16,103-108;1982
a	1487	30.9mg	119.mg	1/27	105.mg	19/29	
b	1487	33.1mg	119.mg	0/27	105.mg	18/29	
c	1487	39.1mg	149.mg	0/27	105.mg	16/29	
d	1487	42.5mg	168.mg	0/27	105.mg	15/29	
e	1487	46.1mg	191.mg	0/27	105.mg	14/29	
f	1487	115.mg	8.01gm	0/27	105.mg	5/29	
g	1487	78.5mg	n.s.s.	2/27	105.mg	9/29	

NITROUS OXIDE   (nitrogen oxide) 10024-97-2

	RefNum	LoConf	UpConf	Cntrl	1Dose	1Inc	2Dose	2Inc	Citation or Pathology
3765	1769	118.gm	n.s.s.	25/88	35.5gm	24/77	142.gm	31/75	Baden;anes,64,747-750;1986
a	1769	841.gm	n.s.s.	4/88	35.5gm	2/77	142.gm	2/75	
b	1769	465.gm	n.s.s.	19/88	35.5gm	11/77	142.gm	11/75	
3766	1769	39.7gm	n.s.s.	34/91	29.5gm	28/75	(118.gm	16/76)	
a	1769	320.gm	n.s.s.	9/91	29.5gm	10/75	118.gm	8/76	
b	1769	483.gm	n.s.s.	11/91	29.5gm	8/75	118.gm	6/76	

NIVALENOL   23282-20-4

	RefNum	LoConf	UpConf	Cntrl	1Dose	1Inc	2Dose	2Inc	(additional doses)	Citation or Pathology
3767	1946	28.3mg	n.s.s.	0/38	.780mg	2/38	1.56mg	0/41	3.90mg 0/41	Ohtsubo;fctx,27,591-598;1989/pers.comm.
a	1946	45.3mg	n.s.s.	2/38	.780mg	1/38	1.56mg	0/41	3.90mg 0/41	
b	1946	4.58mg	n.s.s.	22/38	.780mg	25/38	1.56mg	23/41	3.90mg 23/41	

NORETHYNODREL   68-23-5

	RefNum	LoConf	UpConf	Cntrl	1Dose	1Inc	2Dose	2Inc	(additional doses)	Citation or Pathology
3768	1175	2.48mg	n.s.s.	54/92	1.76mg	45/77				Rudali;jnci,49,813-819;1972
3769	1453m	12.1mg	n.s.s.	0/39	.125mg	0/39	1.50mg	0/40	5.00mg 2/39	Barrows;jtxe,3,219-230;1977
3770	1453m	16.3mg	n.s.s.	6/39	.125mg	7/40	1.50mg	1/40	5.00mg 3/40	

```
         Spe Strain  Site   Xpo+Xpt                                                        TD50    2Tailpvl
            Sex   Route   Hist      Notes                                                      DR     AuOp
3771 M f crf eat mam tum 24m24 er  pool                    .>                               2.13mg   P<.6   -
3772 M f r3m eat mam tum 24m24 er                                    .>                      no dre  P=1.   -
3773 M f sww gav liv hct 65w80 er                                          .>                no dre  P=1.   -

NORHARMAN                          100ng..:..1ug....:..10......:..100....:..1mg.....:..10......:..100....:..1g......:..10
3774 R m wis eat for pam 80w80                                             .>               191.mg   P<.3   -
   a R m wis eat ubl tum 80w80                                                               no dre  P=1.   -

NORLESTRIN                         100ng..:..1ug....:..10......:..100....:..1mg.....:..10......:..100....:..1g......:..10
3775 M f c71 gav pit tum 89w89 e                           .>                               1.34mg * P<.2   +
3776 M f csc gav lun tum 69w69 ek                    .>                                      no dre  P=1.   -
   a M f csc gav tba mix 69w69 ek                                                            no dre  P=1.   -
3777 M f csc gav tba mix 69w82 ek              .>                                            no dre  P=1.   -
3778 P f rhe eat ute ley 8y10 e                                .>                            6.42mg * P<.2
   a P f rhe eat ski pam 8y10 e                                                             10.1mg * P<.6
   b P f rhe eat lun tum 8y10 e                                                              no dre  P=1.
   c P f rhe eat liv tum 8y10 e                                                              no dre  P=1.
   d P f rhe eat pdu ade 8y10 e                                                              no dre  P=1.   -
3779 R b asd eat liv hnd 24m24 e                               .  +  .                        1.94mg \ P<.0005+
   a R b asd eat pit cra 24m24 e                                                             5.31mg / P<.0005+
   b R b asd eat mgl mix 24m24 e                                                             6.66mg * P<.0005+
   c R b asd eat mgl ade 24m24 e                                                            14.8mg * P<.0005+
   d R b asd eat ute aep 24m24 e                                                            27.4mg * P<.002 +
   e R b asd eat liv hpa 24m24 e                                                             7.59mg \ P<.02 +
   f R b asd eat mgl adc 24m24 e                                                            76.8mg * P<.05 +
   g R b asd eat mgl fep 24m24 e                                                            21.0mg * P<.2 +
   h R b asd eat tba mix 24m24 e                                                            14.2mg * P<.6

NOVADELOX                          100ng..:..1ug....:..10......:..100....:..1mg.....:..10......:..100....:..1g......:..10
3780 M b alb eat lun ade 80w80 e                                                            .no dre  P=1.   -
   a M b alb eat lun mix 80w80 e                                                             no dre  P=1.   -
   b M b alb eat liv hpt 80w80 e                                                             no dre  P=1.   -
3781 R b alb eat liv mhp 28m28 e                                                            .>no dre  P=1.   -
   a R b alb eat tba ben 28m28 e                                                             6.48gm * P<.04  -
   b R b alb eat tba mal 28m28 e                                                            15.6gm * P<.5   -

OCHRATOXIN A                       100ng..:..1ug....:..10......:..100....:..1mg.....:..10......:..100....:..1g......:..10
3782 M f b6c eat liv hpc 24m24 er                                   .     ±                 34.8mg * P<.02  +
   a M f b6c eat liv hpa 24m24 er                                                           109.mg * P<.3
   b M f b6c eat kid mix 24m24 er                                                            no dre  P=1.
3783 M m b6c eat kid mix 24m24 er                             . + .                          3.53mg * P<.0005+
   a M m b6c eat kid ade 24m24 er                                                            4.64mg * P<.0005
   b M m b6c eat kid car 24m24 er                                                           10.3mg * P<.0005
   c M m b6c eat liv hpa 24m24 er                                                           48.1mg * P<.3
   d M m b6c eat liv hpc 24m24 er                                                           58.7mg * P<.2
3784 R f f34 gav kid MXA 24m24                         :  +  :                               .485mg * P<.0005c
   a R f f34 gav kid rua 24m24                                                               .813mg * P<.003 c
   b R f f34 gav kid ruc 24m24                                                              1.31mg * P<.02  c
   c R f f34 gav MXB MXB 24m24                                                               .386mg * P<.3
   d R f f34 gav mgl fba 24m24                                                               .550mg * P<.5   c
   e R f f34 gav TBA MXB 24m24                                                               no dre  P=1.
   f R f f34 gav liv MXB 24m24                                                               no dre  P=1.
3785 R m f34 gav kid MXA 24m24                         :+ :                                 57.9ug * P<.0005c
   a R m f34 gav kid MXA 24m24                                                              75.6ug * P<.0005c
   b R m f34 gav kid MXA 24m24                                                               .241mg * P<.0005c
   c R m f34 gav TBA MXB 24m24                                                              86.3ug * P<.007
   d R m f34 gav liv MXB 24m24                                                              2.50mg * P<.3
3786 R b wis gav liv tum 12m26 e                           .>                                no dre  P=1.
   a R b wis gav tba tum 12m26 e                                                             no dre  P=1.   -

OCTACHLOROSTYRENE                  100ng..:..1ug....:..10......:..100....:..1mg.....:..10......:..100....:..1g......:..10
3787 R f sda eat liv tum 52w52                                                               no dre  P=1.
3788 R m sda eat liv tum 52w52                                                               no dre  P=1.

OLEATE, SODIUM                     100ng..:..1ug....:..10......:..100....:..1mg.....:..10......:..100....:..1g......:..10
3789 R f f3d wat liv hnd 25m25 e                                                             no dre  P=1.   -
   a R f f3d wat tba mix 25m25 e                                                             no dre  P=1.   -
3790 R m f3d wat pan mix 25m25 e                                                            .10.1gm * P<.003 -
   a R m f3d wat pan ism 25m25 e                                                            12.6gm * P<.005 -
   b R m f3d wat liv hnd 25m25 e                                                            29.1gm * P<.03  -
   c R m f3d wat liv mix 25m25 e                                                            39.3gm * P<.2   -
   d R m f3d wat tba mix 25m25 e                                                             no dre  P=1.   -

OMEPRAZOLE                         100ng..:..1ug....:..10......:..100....:..1mg.....:..10......:..100....:..1g......:..10
3791 M f cd1 gav gam cnd 78w78 er                                   .>                       no dre  P=1.
3792 M m cd1 gav gam cnd 78w78 er                                   .>                       no dre  P=1.
3793 R f sda gav gam cnd 24m24 er                                .+  .                       63.0mg Z P<.0005
3794 R m sda gav gam cnd 24m24 er                                         .  +  .            1.12gm * P<.0005

C.I. ACID ORANGE 3                 100ng..:..1ug....:..10......:..100....:..1mg.....:..10......:..100....:..1g......:..10
3795 M f b6c gav lun a/a 24m24                                          :     ±             #2.59gm * P<.05  -
```

	RefNum	LoConf	UpConf	Cntrl	1Dose	1Inc	2Dose	2Inc					Citation or Pathology
													Brkly Code
3771	1175	.243mg	n.s.s.	161/167p	1.76mg	48/49							Rudali;jnci,49,813-819;1972
3772	1175	1.86mg	n.s.s.	50/73	1.76mg	19/31							
3773	1453	6.19mg	n.s.s.	1/50	1.02mg	0/50							Barrows;jtxe,3,219-230;1977
NORHARMAN	(9H-pyrido(3,4-b)indole)		244-63-3										
3774	1460	31.0mg	n.s.s.	0/28	20.0mg	1/24							Hagiwara;txlt,6,71-75;1980
a	1460	58.5mg	n.s.s.	0/28	20.0mg	0/24							
NORLESTRIN	8015-12-1												
3775	230	.347mg	n.s.s.	1/8	.200mg	7/15	2.00mg	5/8					Poel;scie,154,402-403;1966
3776	585m	.194mg	n.s.s.	0/15	.143mg	0/15							Poel;canr,28,845-859;1968
a	585m	83.4ug	n.s.s.	4/15	.143mg	3/15							
3777	585n	57.2ug	n.s.s.	8/15	.120mg	7/15							
3778	1441	1.05mg	n.s.s.	0/16	37.5ug	0/16	.383mg	0/16	1.91mg	1/16			Fitzgerald;jtxe,10,879-896;1982
a	1441	1.13mg	n.s.s.	0/16	37.5ug	1/16	.383mg	0/16	1.91mg	1/16			
b	1441	28.5ug	n.s.s.	0/16	37.5ug	0/16	.383mg	0/16	1.91mg	0/16			
c	1441	28.5ug	n.s.s.	0/16	37.5ug	0/16	.383mg	0/16	1.91mg	0/16			
d	1441	1.86mg	n.s.s.	0/16	37.5ug	1/16	.383mg	0/16	1.91mg	0/16			
3779	1403	.941mg	6.88mg	4/200	.338mg	13/100	(3.38mg	10/100)					Schardein;txap,16,10-23;1970
a	1403	3.07mg	13.3mg	64/200	.338mg	19/100	3.38mg	55/100					
b	1403	3.47mg	27.9mg	61/200	.338mg	34/100	3.38mg	51/100					
c	1403	7.91mg	37.3mg	5/200	.338mg	2/100	3.38mg	17/100					
d	1403	12.0mg	164.mg	4/200	.338mg	2/100	3.38mg	10/100					
e	1403	2.30mg	37.2gm	0/200	.338mg	3/100	(3.38mg	4/100)					
f	1403	22.9mg	n.s.s.	2/200	.338mg	0/100	3.38mg	4/100					
g	1403	6.38mg	n.s.s.	58/200	.338mg	33/100	3.38mg	37/100					
h	1403	2.59mg	n.s.s.	159/200	.338mg	82/100	3.38mg	83/100					
NOVADELOX	94-36-0												
3780	1345	9.23gm	n.s.s.	0/41	19.6mg	1/42	196.mg	1/42	1.96gm	0/42			Sharratt;fctx,2,527-538;1964
a	1345	9.85gm	n.s.s.	1/41	19.6mg	1/42	196.mg	1/42	1.96gm	0/42			
b	1345	11.0mg	n.s.s.	1/41	19.6mg	1/42	196.mg	0/42	1.96gm	0/42			
3781	1345	6.57gm	n.s.s.	0/46	7.07mg	0/47	70.7mg	1/47	707.mg	0/45			
a	1345	2.16gm	n.s.s.	3/46	7.07mg	1/47	70.7mg	1/47	707.mg	6/45			
b	1345	2.84gm	n.s.s.	6/46	7.07mg	2/47	70.7mg	1/47	707.mg	5/45			
OCHRATOXIN A	303-47-9												
3782	1691	12.3mg	n.s.s.	0/47	.130mg	1/45	5.20mg	5/49					Bendele;jnci,75,733-739;1985
a	1691	21.7mg	n.s.s.	0/47	.130mg	1/45	5.20mg	2/49					
b	1691	1.18mg	n.s.s.	0/47	.130mg	0/45	5.20mg	0/49					
3783	1691	2.25mg	5.92mg	0/50	.120mg	0/47	4.80mg	31/50					
a	1691	2.87mg	8.13mg	0/50	.120mg	0/47	4.80mg	26/50					
b	1691	5.52mg	22.5mg	0/50	.120mg	0/47	4.80mg	14/50					
c	1691	12.2mg	n.s.s.	1/50	.120mg	5/47	4.80mg	6/50					
d	1691	14.5mg	n.s.s.	0/50	.120mg	3/47	4.80mg	4/50					
3784	TR358	.235mg	1.40mg	0/50	14.8ug	0/51	49.4ug	2/50	.148mg	8/50			kid:rua,ruc.
a	TR358	.331mg	4.82mg	0/50	14.8ug	0/51	49.4ug	1/50	.148mg	5/50			
b	TR358	.448mg	n.s.s.	0/50	14.8ug	0/51	49.4ug	1/50	.148mg	3/50			
c	TR358	.103mg	n.s.s.	17/50	14.8ug	23/51	49.4ug	24/50	.148mg	30/50			kid:rua,ruc; mgl:fba. C
d	TR358	.116mg	n.s.s.	17/50	14.8ug	23/51	49.4ug	22/50	.148mg	28/50			
e	TR358	.117mg	n.s.s.	39/50	14.8ug	44/51	49.4ug	40/50	.148mg	45/50			
f	TR358	n.s.s.	n.s.s.	0/50	14.8ug	0/51	49.4ug	0/50	.148mg	0/50			liv:hpa,hpc,nnd.
3785	TR358	40.6ug	87.9ug	1/50	14.8ug	1/51	49.3ug	20/51	.148mg	36/50			kid:rua,ruc,rue,tcb.
a	TR358	51.8ug	.115mg	0/50	14.8ug	0/51	49.3ug	16/51	.148mg	30/50			kid:ruc,tcb.
b	TR358	.127mg	.729mg	1/50	14.8ug	1/51	49.3ug	6/51	.148mg	10/50			kid:rua,rue.
c	TR358	40.9ug	1.42mg	43/50	14.8ug	38/51	49.3ug	44/51	.148mg	50/50			
d	TR358	.495mg	n.s.s.	1/50	14.8ug	0/51	49.3ug	0/51	.148mg	2/50			liv:hpa,hpc,nnd.
3786	1365	.132mg	n.s.s.	0/10	76.4ug	0/10	.229mg	0/10					Purchase;fctx,9,681-682;1971
a	1365	.132mg	n.s.s.	0/10	76.4ug	0/10	.229mg	0/10					
OCTACHLOROSTYRENE	29082-74-4												
3787	1773	n.s.s.	n.s.s.	0/20	250.ng	0/20	2.50ug	0/20	25.0ug	0/20	.250mg	0/20	2.50mg 0/20 Chu;faat,6,69-77;1986
3788	1773	n.s.s.	n.s.s.	0/20	200.ng	0/20	2.00ug	0/20	20.0ug	0/20	.200mg	0/20	2.00mg 0/20
OLEATE, SODIUM	143-19-1												
3789	1687	22.2gm	n.s.s.	0/43	1.43gm	1/39	2.86gm	0/45					Hiasa;fctx,23,619-623;1985
a	1687	2.99gm	n.s.s.	24/43	1.43gm	28/39	2.86gm	25/45					
3790	1687	5.09gm	45.3gm	0/41	1.25gm	4/40	2.50gm	7/45					
a	1687	5.91gm	94.5gm	0/41	1.25gm	3/40	2.50gm	6/45					
b	1687	10.0gm	n.s.s.	0/41	1.25gm	0/40	2.50gm	4/45					
c	1687	10.8mg	n.s.s.	1/41	1.25gm	0/40	2.50gm	4/45					
d	1687	625.mg	n.s.s.	40/41	1.25gm	40/40	2.50gm	43/45					
OMEPRAZOLE	73590-58-6												
3791	1781m	63.3mg	n.s.s.	0/109	14.1mg	0/55	44.0mg	0/55	141.mg	0/55			Havu;dgsn,35,42-55;1986/pers.comm.
3792	1781m	63.3mg	n.s.s.	0/109	14.1mg	0/55	44.0mg	0/55	141.mg	0/55			
3793	1781n	41.1mg	103.mg	0/120	14.1mg	13/60	44.0mg	19/60	(141.mg	24/60)			
3794	1781n	486.mg	3.65gm	0/119	14.1mg	0/60	44.0mg	1/60	141.mg	6/60			
C.I. ACID ORANGE 3	6373-74-6												
3795	TR335	868.mg	n.s.s.	0/50	177.mg	1/50	354.mg	3/50					S

```
     Spe Strain  Site   Xpo+Xpt                                           TD50    2Tailpvl
         Sex  Route  Hist      Notes                                        DR      AuOp

     a   M f b6c gav TBA MXB 24m24                                        no dre   P=1.
     b   M f b6c gav liv MXB 24m24                                        7.55gm * P<.8
     c   M f b6c gav lun MXB 24m24                                        11.6gm * P<.8
  3796 M m b6c gav liv hpc 24m24                          :    ±          #184.mg \ P<.02  -
     a   M m b6c gav TBA MXB 24m24                                        372.mg * P<.4
     b   M m b6c gav liv MXB 24m24                                        no dre   P=1.
     c   M m b6c gav lun MXB 24m24                                        5.86gm * P<.9
  3797 R f f34 gav kid tcc 24m24 s                            :   +    :  1.71gm / P<.0005c
     a   R f f34 gav TBA MXB 24m24 s                                      361.mg * P<.05
     b   R f f34 gav liv MXB 24m24 s                                      5.09gm * P<.5
  3798 R m f34 gav TBA MXB 23m24 ans                          :    ±      270.mg / P<.03  -
     a   R m f34 gav liv MXB 23m24 ans                                    no dre   P=1.


C.I. ACID ORANGE 10               100ng..:..1ug....:..10......:..100....:..1mg...:..10....:..100....:..1g......:.10
  3799 M f b6c eat TBA MXB 24m24                                   :>     no dre   P=1.
     a   M f b6c eat liv MXB 24m24                                        no dre   P=1.
     b   M f b6c eat lun MXB 24m24                                        no dre   P=1.
  3800 M m b6c eat TBA MXB 24m24                                      :>  no dre   P=1.
     a   M m b6c eat liv MXB 24m24                                        no dre   P=1.
     b   M m b6c eat lun MXB 24m24                                        22.0gm * P<.8
  3801 R f f34 eat TBA MXB 24m24                                :>        no dre   P=1.
     a   R f f34 eat liv MXB 24m24                                        no dre   P=1.
  3802 R m f34 eat tnv MXA 24m24                              :    ±      #915.mg * P<.05
     a   R m f34 eat TBA MXB 24m24                                        2.71gm * P<.9
     b   R m f34 eat liv MXB 24m24                                        620.mg * P<.06


OVULEN                            100ng..:..1ug....:..10......:..100....:..1mg...:..10....:..100....:..1g....:..10
  3803 M f crf eat mam tum 24m24 r              .>                        .858mg   P<.8   -
  3804 M m crf eat mam tum 24m24 r         .  +  .                        .300mg   P<.0005


OXAMYL                            100ng..:..1ug....:..10......:..100....:..1mg...:..10....:..100....:..1g....:..10
  3805 M f cd1 eat tba mix 24m24 v                      .>                no dre   P=1.  -
  3806 M m cd1 eat tba mix 24m24 v                    .>                  no dre   P=1.  -
  3807 R f cdr eat liv hph 24m24 eg                    .>                 no dre   P=1.  -
  3808 R m cdr eat liv tum 24m24 eg                    .>                 no dre   P=1.  -


OXAZEPAM                          100ng..:..1ug....:..10......:..100....:..1mg...:..10....:..100....:..1g......:.10
  3809 M f b6c eat MXB MXB 23m24 a                      :+  :             19.8mg Z P<.0005
     a   M f b6c eat liv MXA 23m24 a                                      19.8mg Z P<.0005c
     b   M f b6c eat liv hpa 23m24 a                                      27.2mg Z P<.0005c
     c   M f b6c eat liv hpc 23m24 a                                      47.2mg Z P<.0005c
     d   M f b6c eat thy fca 23m24 a                                      160.mg Z P<.0005p
     e   M f b6c eat liv hpb 23m24 a                                      476.mg Z P<.0005c
     f   M f b6c eat TBA MXB 23m24 a                                      20.2mg Z P<.0005
     g   M f b6c eat liv MXB 23m24 a                                      19.8mg Z P<.0005
     h   M f b6c eat lun MXB 23m24 a                                      811.mg * P<.2
  3810 M f sww eat liv MXA 57w57 s                          : +:          64.8mg * P<.0005c
     a   M f sww eat liv hpa 57w57 s                                      65.8mg * P<.0005c
     b   M f sww eat liv hpc 57w57 s                                      360.mg / P<.0005c
     c   M f sww eat TBA MXB 57w57 s                                      79.9mg * P<.0005
     d   M f sww eat liv MXB 57w57 s                                      64.8mg * P<.0005
     e   M f sww eat lun MXB 57w57 s                                      1.48gm * P<.5
  3811 M m b6c eat liv MXA 23m24 a                          :+ :          59.0mg Z P<.0005c
     a   M m b6c eat liv MXA 23m24 a                                      68.3mg Z P<.0005c
     b   M m b6c eat liv hpc 23m24 a                                      73.5mg Z P<.0005c
     c   M m b6c eat liv hpa 23m24 a                                      93.7mg Z P<.0005c
     d   M m b6c eat liv hpb 23m24 a                                      205.mg * P<.0005c
     e   M m b6c eat TBA MXB 23m24 a                                      54.0mg Z P<.0005
     f   M m b6c eat liv MXB 23m24 a                                      59.0mg Z P<.0005
     g   M m b6c eat lun MXB 23m24 a                                      no dre   P=1.
  3812 M m sww eat liv MXA 57w57 s                          : +:          34.6mg * P<.0005c
     a   M m sww eat liv hpa 57w57 s                                      35.0mg * P<.0005c
     b   M m sww eat liv hpc 57w57 s                                      139.mg / P<.0005c
     c   M m sww eat TBA MXB 57w57 s                                      47.0mg * P<.0005
     d   M m sww eat liv MXB 57w57 s                                      34.6mg * P<.0005
     e   M m sww eat lun MXB 57w57 s                                      755.mg * P<.4


N-(9-OXO-2-FLUORENYL)ACETAMIDE    100ng..:..1ug....:..10....:..100....:..1mg...:..10....:..100....:..1g....:..10
  3813 R f buf eat mgl adc 86w86 e                      .  +  .           6.17mg   P<.0005+
     a   R f buf eat edu sqc 86w86 e                                      32.0mg   P<.04  +
     b   R f buf eat liv hpt 86w86 e                                      102.mg   P<.3   +


1-(2-OXOPROPYL)NITROSO-3-(2-CHLOROETHYL)UREA  ...:..10....:..100....:..1mg...:..10....:..100....:..1g....:..10
  3814 R f f34 gav liv tum 7m24 e                    .   ±                1.36mg   P<.04
     a   R f f34 gav ner tum 7m24 e                                       4.49mg   P<.3
  3815 R f f34 wat lun tum 7m24 e                  .>                      4.49mg   P<.3
     a   R f f34 wat ner tum 7m24 e                                       4.49mg   P<.3
     b   R f f34 wat liv tum 7m24 e                                       no dre   P=1.
  3816 R m f34 gav ner tum 7m24 e                .   ±                     .951mg   P<.04
     a   R m f34 gav lun tum 7m24 e                                       3.14mg   P<.3
     b   R m f34 gav liv tum 7m24 e                                       no dre   P=1.
```

	RefNum	LoConf	UpConf	Cntrl	1Dose	1Inc	2Dose	2Inc	Citation or Pathology	Brkly Code
a	TR335	165.mg	n.s.s.	35/50	177.mg	27/50	(354.mg	23/50)		
b	TR335	707.mg	n.s.s.	3/50	177.mg	4/50	354.mg	4/50	liv:hpa,hpc,nnd.	
c	TR335	1.10gm	n.s.s.	2/50	177.mg	2/50	354.mg	3/50	lun:a/a,a/c.	
3796	TR335	79.4mg	n.s.s.	7/50	88.4mg	16/50	(177.mg	10/50)		S
a	TR335	96.8mg	n.s.s.	42/50	88.4mg	36/50	177.mg	39/50		
b	TR335	224.mg	n.s.s.	21/50	88.4mg	20/50	177.mg	15/50	liv:hpa,hpc,nnd.	
c	TR335	266.mg	n.s.s.	13/50	88.4mg	9/50	177.mg	10/50	lun:a/a,a/c.	
3797	TR335	683.mg	6.26gm	0/50	264.mg	0/50	531.mg	6/50		
a	TR335	149.mg	n.s.s.	45/50	264.mg	44/50	531.mg	19/50		
b	TR335	945.mg	n.s.s.	1/50	264.mg	3/50	531.mg	0/50	liv:hpa,hpc,nnd.	
3798	TR335	113.mg	n.s.s.	42/50	265.mg	33/50	536.mg	13/50		
a	TR335	1.65gm	n.s.s.	2/50	265.mg	0/50	536.mg	0/50	liv:hpa,hpc,nnd.	

C.I. ACID ORANGE 10   1936-15-8

	RefNum	LoConf	UpConf	Cntrl	1Dose	1Inc	2Dose	2Inc	Citation or Pathology	Brkly Code
3799	TR211	709.mg	n.s.s.	25/50	386.mg	27/50	(773.mg	19/50)		
a	TR211	3.72gm	n.s.s.	3/50	386.mg	3/50	773.mg	3/50		
b	TR211	2.15gm	n.s.s.	2/50	386.mg	1/50	(773.mg	0/50)	liv:hpa,hpc,nnd.	
3800	TR211	1.45gm	n.s.s.	30/50	360.mg	19/50	720.mg	25/50	lun:a/a,a/c.	
a	TR211	2.09gm	n.s.s.	15/50	360.mg	7/50	720.mg	12/50		
b	TR211	2.56gm	n.s.s.	1/50	360.mg	3/50	720.mg	2/50	liv:hpa,hpc,nnd.	
3801	TR211	89.9mg	n.s.s.	68/90	49.5mg	33/50	(149.mg	31/50)	lun:a/a,a/c.	
a	TR211	967.mg	n.s.s.	3/90	49.5mg	2/50	149.mg	1/50		
3802	TR211	346.mg	n.s.s.	0/90	39.6mg	3/50	119.mg	2/50	liv:hpa,hpc,nnd.	
a	TR211	125.mg	n.s.s.	61/90	39.6mg	31/50	119.mg	34/50	tnv:men,msm.	S
b	TR211	223.mg	n.s.s.	5/90	39.6mg	3/50	119.mg	8/50		
									liv:hpa,hpc,nnd.	

OVULEN (ethynodiol diacetate/ethinyl estradiol [10:1]) 8056-92-6

	RefNum	LoConf	UpConf	Cntrl	1Dose	1Inc	Citation or Pathology
3803	1463	57.0ug	n.s.s.	161/167	.390mg	37/38	
3804	1463	.158mg	.664mg	0/76	.360mg	14/25	Rudali;gmcr,17,243-252;1975

OXAMYL   23135-22-0

	RefNum	LoConf	UpConf	Cntrl	1Dose	1Inc	2Dose	2Inc	3Dose	3Inc	Citation or Pathology
3805	1788	35.8mg	n.s.s.	32/80	3.25mg	42/79	6.50mg	28/79	9.94mg	27/80	
3806	1788	30.7mg	n.s.s.	35/80	3.00mg	36/79	6.00mg	28/80	9.17mg	29/78	Kennedy;faat,7,106-118;1986/pers.comm.
3807	1788	46.4mg	n.s.s.	1/60	7.50mg	0/30					
3808	1788	37.1mg	n.s.s.	0/60	6.00mg	0/30					

OXAZEPAM   604-75-1

	RefNum	LoConf	UpConf	Cntrl	1Dose	1Inc	2Dose	2Inc	3Dose	3Inc	Citation or Pathology
3809	TR443	12.7mg	32.6mg	28/50	16.2mg	36/50	325.mg	50/50	650.mg	47/50	liv:hpa,hpb,hpc; thy:fca. M
a	TR443	12.7mg	32.6mg	28/50	16.2mg	36/50	325.mg	50/50	650.mg	47/50	liv:hpa,hpb,hpc.
b	TR443	16.8mg	47.7mg	25/50	16.2mg	35/50	325.mg	35/50	650.mg	36/50	
c	TR443	30.0mg	76.0mg	9/50	16.2mg	5/50	325.mg	49/50	650.mg	44/50	
d	TR443	68.3mg	573.mg	0/50	16.2mg	4/50	325.mg	5/50	650.mg	6/50	
e	TR443	221.mg	1.07gm	0/50	16.2mg	1/50	325.mg	8/50	650.mg	8/50	
f	TR443	12.9mg	33.4mg	37/50	16.2mg	43/50	325.mg	50/50	650.mg	47/50	
g	TR443	12.7mg	32.6mg	28/50	16.2mg	36/50	325.mg	50/50	650.mg	47/50	liv:hpa,hpb,hpc.
h	TR443	121.mg	n.s.s.	5/50	16.2mg	4/50	325.mg	1/50	650.mg	1/50	lun:a/a,a/c.
3810	TR443	46.4mg	94.0mg	1/60	325.mg	23/60	650.mg	47/60			liv:hpa,hpc.
a	TR443	47.3mg	94.1mg	0/60	325.mg	22/60	650.mg	47/60			
b	TR443	176.mg	1.06gm	1/60	325.mg	11/60	650.mg	11/60			
c	TR443	50.9mg	155.mg	24/60	325.mg	37/60	650.mg	49/60			
d	TR443	46.4mg	94.0mg	1/60	325.mg	23/60	650.mg	47/60			liv:hpa,hpb,hpc.
e	TR443	273.mg	n.s.s.	11/60	325.mg	6/60	650.mg	7/60			lun:a/a,a/c.
3811	TR443	40.5mg	89.4mg	23/50	15.0mg	19/50	300.mg	50/50	600.mg	50/50	liv:hpa,hpb,hpc.
a	TR443	47.6mg	101.mg	9/50	15.0mg	6/50	300.mg	47/50	600.mg	50/50	liv:hpb,hpc.
b	TR443	51.1mg	109.mg	9/50	15.0mg	5/50	300.mg	45/50	600.mg	50/50	
c	TR443	60.0mg	157.mg	17/50	15.0mg	18/50	300.mg	34/50	600.mg	32/50	
d	TR443	126.mg	347.mg	0/50	15.0mg	2/50	300.mg	21/50	600.mg	13/50	
e	TR443	37.0mg	83.5mg	37/50	15.0mg	35/50	300.mg	50/50	600.mg	50/50	
f	TR443	40.5mg	89.4mg	23/50	15.0mg	19/50	300.mg	50/50	600.mg	50/50	
g	TR443	407.mg	n.s.s.	13/50	15.0mg	18/50	300.mg	5/50	600.mg	1/50	liv:hpa,hpb,hpc.
3812	TR443	24.7mg	49.6mg	1/60	300.mg	35/60	600.mg	52/60			lun:a/a,a/c.
a	TR443	25.0mg	50.6mg	1/60	300.mg	35/60	600.mg	50/60			liv:hpa,hpc.
b	TR443	80.4mg	257.mg	0/60	300.mg	5/60	600.mg	19/60			
c	TR443	30.3mg	83.5mg	26/60	300.mg	42/60	600.mg	54/60			
d	TR443	24.7mg	49.6mg	1/60	300.mg	35/60	600.mg	52/60			
e	TR443	163.mg	n.s.s.	14/60	300.mg	9/60	600.mg	7/60			liv:hpa,hpb,hpc.
											lun:a/a,a/c.

N-(9-OXO-2-FLUORENYL)ACETAMIDE   3096-50-2

	RefNum	LoConf	UpConf	Cntrl	1Dose	1Inc	Citation or Pathology
3813	144	2.98mg	15.4mg	0/18	12.4mg	11/18	
a	144	9.65mg	n.s.s.	0/18	12.4mg	3/18	Morris;jnci,24,149-180;1960
b	144	16.6mg	n.s.s.	0/18	12.4mg	1/18	

1-(2-OXOPROPYL)NITROSO-3-(2-CHLOROETHYL)UREA   110559-85-8

	RefNum	LoConf	UpConf	Cntrl	1Dose	1Inc	Citation or Pathology
3814	2010m	.408mg	n.s.s.	0/12	.570mg	3/12	
a	2010m	.730mg	n.s.s.	0/12	.570mg	1/12	Lijinsky;vivo,4,1-6;1990
3815	2010n	.730mg	n.s.s.	0/12	.570mg	1/12	
a	2010n	.730mg	n.s.s.	0/12	.570mg	1/12	
b	2010n	1.41mg	n.s.s.	0/12	.570mg	0/12	
3816	2010m	.286mg	n.s.s.	0/12	.399mg	3/12	
a	2010m	.511mg	n.s.s.	0/12	.399mg	1/12	
b	2010m	.987mg	n.s.s.	1/12	.399mg	0/12	

	Spe Strain Site Xpo+Xpt		TD50	2Tailpvl
	Sex Route Hist Notes		DR	AuOp
3817	R m f34 wat lun tum 7m24 e	±	1.50mg	P<.09
a	R m f34 wat liv tum 7m24 e		3.14mg	P<.3
b	R m f34 wat ner tum 7m24 e		3.14mg	P<.3

2-OXOPROPYLNITROSOUREA 100ng..:..1ug....:..10....:..:..100....:..1mg..:..:..10....:..:..100....:..1g......:..10

3818	R f f34 gav kid mnp 7m24 e	. + .	.320mg	P<.0005
a	R f f34 gav liv tum 7m24 e		4.04mg	P<.3
3819	R m f34 gav kid mnp 7m24 e	. + .	.456mg	P<.005
a	R m f34 gav liv tum 7m24 e		no dre	P=1.

1'-OXOSAFROLE 100ng..:..1ug....:..10....:..:..100....:..1mg..:..:..10....:..:..100....:..1g......:..10

| 3820 | R m cdr eat liv car 73w95 e | .> | no dre | P=1. | - |

OXPRENOLOL.HCl 100ng..:..1ug....:..10....:..:..100....:..1mg..:..:..10....:..:..100....:..1g......:..10

3821	M f cf1 eat liv car 78w91 e	.>	2.40gm *	P<.2	-
a	M f cf1 eat lun ade 78w91 e		no dre	P=1.	-
b	M f cf1 eat lun nfs 78w91 e		no dre	P=1.	-
3822	M f cf1 eat lun car 78w78 e	.>	425.mg	P<.2	-
a	M f cf1 eat liv tum 78w78 e		no dre	P=1.	-
3823	M m cf1 eat liv pca 78w91 e	.>	2.12gm *	P<.5	-
a	M m cf1 eat lun ade 78w91 e		no dre	P=1.	-
3824	M m cf1 eat liv car 78w78 e	.>	2.62gm *	P<.7	-
a	M m cf1 eat lun ade 78w78 e		no dre	P=1.	-
b	M m cf1 eat lun car 78w78 e		no dre	P=1.	-
3825	R f cdr eat liv tum 78w78 e	.>	1.68gm *	P<.3	-
3826	R m cdr eat liv hpa 78w78 e	.>			

4,4'-OXYDIANILINE 100ng..:..1ug....:..10....:..:..100....:..1mg..:..:..10....:..:..100....:..1g......:..10

3827	M f b6c eat MXB MXB 24m24 s	: + :	19.7mg Z	P<.002	
a	M f b6c eat hag adn 24m24 s		46.8mg Z	P<.0005c	
b	M f b6c eat liv MXA 24m24 s		108.mg *	P<.0005c	
c	M f b6c eat liv hpc 24m24 s		252.mg *	P<.01 c	
d	M f b6c eat thy fca 24m24 s		598.mg *	P<.0005c	
e	M f b6c eat liv hpa 24m24 s		244.mg *	P<.02 c	
f	M f b6c eat TBA MXB 24m24 s		32.9mg Z	P<.02	
g	M f b6c eat liv MXB 24m24 s		108.mg *	P<.0005	
h	M f b6c eat lun MXB 24m24 s		156.mg Z	P<.08	
3828	M m b6c eat hag adn 24m24	: + :	26.2mg Z	P<.0005c	
a	M m b6c eat --- hem 24m24		379.mg *	P<.003	
b	M m b6c eat pit adn 24m24		568.mg *	P<.004	
c	M m b6c eat MXB MXB 24m24		167.mg	P<.3	
d	M m b6c eat liv MXA 24m24		225.mg *	P<.5 c	
e	M m b6c eat TBA MXB 24m24		318.mg *	P<.7	
f	M m b6c eat liv MXB 24m24		225.mg *	P<.5	
g	M m b6c eat lun MXB 24m24		no dre	P=1.	
3829	R f f34 eat MXB MXB 24m24 s	:+ :	12.1mg Z	P<.0005	
a	R f f34 eat thy MXA 24m24 s		14.3mg Z	P<.0005c	
b	R f f34 eat liv MXA 24m24 s		20.1mg Z	P<.0005c	
c	R f f34 eat thy fca 24m24 s		27.5mg Z	P<.0005c	
d	R f f34 eat liv nnd 24m24 s		29.5mg Z	P<.0005c	
e	R f f34 eat thy fcc 24m24 s		41.2mg *	P<.0005c	
f	R f f34 eat liv hpc 24m24 s		93.7mg Z	P<.0005c	
g	R f f34 eat TBA MXB 24m24 s		20.6mg Z	P<.02	
h	R f f34 eat liv MXB 24m24 s		20.1mg Z	P<.0005	
3830	R m f34 eat MXB MXB 24m24	: +:	6.65mg *	P<.0005	
a	R m f34 eat liv MXA 24m24		7.12mg Z	P<.0005c	
b	R m f34 eat liv hpc 24m24		15.7mg Z	P<.0005c	
c	R m f34 eat thy MXA 24m24		17.7mg Z	P<.0005c	
d	R m f34 eat liv nnd 24m24		22.5mg *	P<.0005c	
e	R m f34 eat thy fcc 24m24		32.1mg *	P<.0005c	
f	R m f34 eat thy fca 24m24		47.4mg *	P<.0005c	
g	R m f34 eat TBA MXB 24m24		no dre	P=1.	
h	R m f34 eat liv MXB 24m24		7.12mg Z	P<.0005	

N-OXYDIETHYLENE THIOCARBAMYL-N-OXYDIETHYLENE SULFENAMIDE....:..100....:..1mg....:..:..10....:..:..100....:..1g......:..10

3831	R f sda eat unt tum 52w52 ek	.>	no dre	P=1.	
3832	R f sda eat unt mix 26m26 e	. + .	96.9mg Z	P<.0005+	
a	R f sda eat ubl mix 26m26 e		206.mg *	P<.0005+	
b	R f sda eat kid mix 26m26 e		338.mg *	P<.0005+	
c	R f sda eat kid sqc 26m26 e		571.mg *	P<.007	
d	R f sda eat kid utc 26m26 e		571.mg *	P<.007	
e	R f sda eat ubl utp 26m26 e		571.mg *	P<.007 +	
f	R f sda eat ubl sqp 26m26 e		862.mg *	P<.03 +	
g	R f sda eat ubl utc 26m26 e		862.mg *	P<.03 +	
h	R f sda eat kid utp 26m26 e		1.74gm *	P<.2	
i	R f sda eat ubl sqc 26m26 e		1.74gm *	P<.2 +	
3833	R m sda eat unt tum 52w52 ek	.>	no dre	P=1.	
3834	R m sda eat unt mix 26m26 e	. + .	85.5mg Z	P<.0005+	
a	R m sda eat ubl mix 26m26 e		130.mg *	P<.0005+	
b	R m sda eat ubl utc 26m26 e		340.mg *	P<.003 +	

RefNum	LoConf	UpConf	Cntrl	1Dose	1Inc	2Dose	2Inc	Citation or Pathology / Brkly Code
3817	2010n	.368mg	n.s.s.	0/12	.399mg	2/12		
a	2010n	.511mg	n.s.s.	0/12	.399mg	1/12		
b	2010n	.511mg	n.s.s.	0/12	.399mg	1/12		

2-OXOPROPYLNITROSOUREA   ---

RefNum	LoConf	UpConf	Cntrl	1Dose	1Inc	2Dose	2Inc	Citation or Pathology / Brkly Code
3818	2010	.136mg	.977mg	0/12	.513mg	8/12			Lijinsky;vivo,4,1-6;1990
a	2010	.656mg	n.s.s.	0/12	.513mg	1/12		
3819	2010	.170mg	3.66mg	0/12	.359mg	5/12		
a	2010	.887mg	n.s.s.	1/12	.359mg	0/12		

1'-OXOSAFROLE   30418-53-2

RefNum	LoConf	UpConf	Cntrl	1Dose	1Inc	Citation or Pathology / Brkly Code
3820	1035d	239.mg	n.s.s.	0/18	77.3mg	0/18 Wislocki;canr,37,1883-1891;1977

OXPRENOLOL.HCl   6452-73-9

RefNum	LoConf	UpConf	Cntrl	1Dose	1Inc	2Dose	2Inc	3Dose	3Inc	Citation or Pathology / Brkly Code
3821	469m	391.mg	n.s.s.	0/24	12.9mg	0/28	42.9mg	0/28	129.mg	1/24 Newberne;txap,41,535-546;1977
a	469m	464.mg	n.s.s.	5/49	12.9mg	6/49	42.9mg	4/49	129.mg	4/50
b	469m	529.mg	n.s.s.	0/24	12.9mg	1/28	42.9mg	0/28	129.mg	0/24
3822	469n	143.mg	n.s.s.	1/25	150.mg	6/37				
a	469n	939.mg	n.s.s.	0/40	150.mg	0/54				
3823	469m	366.mg	n.s.s.	0/49	12.9mg	4/49	42.9mg	1/49	129.mg	3/49
a	469m	453.mg	n.s.s.	4/49	12.9mg	7/49	42.9mg	3/49	129.mg	4/49
3824	469n	279.mg	n.s.s.	1/29	150.mg	2/36				
a	469n	316.mg	n.s.s.	2/29	150.mg	2/36				
b	469n	291.mg	n.s.s.	9/44	150.mg	7/48				
3825	469m	37.3mg	n.s.s.	0/30	15.0mg	0/30	50.0mg	0/30	150.mg	0/30
3826	469m	317.mg	n.s.s.	1/29	15.0mg	0/30	50.0mg	0/30	150.mg	2/30

4,4'-OXYDIANILINE   101-80-4

RefNum	LoConf	UpConf	Cntrl	1Dose	1Inc	2Dose	2Inc	3Dose	3Inc	Citation or Pathology / Brkly Code
3827	TR205	9.78mg	97.5mg	10/50	19.1mg	25/50	(38.3mg	23/50	102.mg	35/50) hag:adn; liv:hpa,hpc; thy:fca. C
a	TR205	27.4mg	135.mg	2/50	19.1mg	15/50	38.3mg	14/50	(102.mg	12/50)
b	TR205	58.1mg	413.mg	8/50	19.1mg	13/50	38.3mg	15/50	102.mg	29/50 liv:hpa,hpc.
c	TR205	115.mg	17.8gm	4/50	19.1mg	7/50	38.3mg	6/50	102.mg	15/50
d	TR205	258.mg	2.04gm	0/50	19.1mg	0/50	38.3mg	0/50	102.mg	7/50
e	TR205	109.mg	n.s.s.	4/50	19.1mg	6/50	38.3mg	9/50	102.mg	14/50
f	TR205	15.9mg	4.65gm	28/50	19.1mg	37/50	38.3mg	40/50	(102.mg	42/50)
g	TR205	58.1mg	413.mg	8/50	19.1mg	13/50	38.3mg	15/50	102.mg	29/50 liv:hpa,hpc,nnd.
h	TR205	58.4mg	n.s.s.	5/50	19.1mg	5/50	38.3mg	10/50	(102.mg	3/50) lun:a/a,a/c.
3828	TR205	14.2mg	73.2mg	1/50	17.7mg	17/50	(35.3mg	13/50	94.2mg	17/50)
a	TR205	183.mg	2.05gm	0/50	17.7mg	0/50	35.3mg	5/50	94.2mg	5/50
b	TR205	231.mg	4.96gm	1/50	17.7mg	0/50	35.3mg	0/50	94.2mg	7/50 s / s
c	TR205	44.8mg	n.s.s.	30/50	17.7mg	42/50	35.3mg	36/50	94.2mg	39/50 hag:adn; liv:hpa,hpc. C
d	TR205	50.7mg	n.s.s.	29/50	17.7mg	40/50	35.3mg	34/50	94.2mg	36/50 liv:hpa,hpc.
e	TR205	50.1mg	n.s.s.	39/50	17.7mg	45/50	35.3mg	40/50	94.2mg	42/50
f	TR205	50.7mg	n.s.s.	29/50	17.7mg	40/50	35.3mg	34/50	94.2mg	36/50 liv:hpa,hpc,nnd.
g	TR205	381.mg	n.s.s.	13/50	17.7mg	10/50	35.3mg	8/50	94.2mg	4/50 lun:a/a,a/c.
3829	TR205	8.62mg	18.3mg	3/50	9.80mg	4/50	19.6mg	37/50	24.5mg	26/50 liv:hpc,nnd; thy:fca,fcc. C
a	TR205	10.2mg	20.8mg	0/50	9.80mg	4/50	19.6mg	29/50	24.5mg	23/50 thy:fca,fcc.
b	TR205	13.2mg	35.5mg	3/50	9.80mg	0/50	19.6mg	24/50	24.5mg	17/50 liv:hpc,nnd.
c	TR205	18.1mg	44.4mg	0/50	9.80mg	2/50	19.6mg	17/50	24.5mg	16/50
d	TR205	18.1mg	61.1mg	3/50	9.80mg	0/50	19.6mg	20/50	24.5mg	11/50
e	TR205	24.6mg	77.8mg	2/50	9.80mg	2/50	19.6mg	12/50	24.5mg	7/50
f	TR205	45.6mg	260.mg	0/50	9.80mg	0/50	19.6mg	4/50	24.5mg	6/50
g	TR205	9.60mg	n.s.s.	42/50	9.80mg	36/50	19.6mg	45/50	24.5mg	31/50
h	TR205	13.2mg	35.5mg	3/50	9.80mg	0/50	19.6mg	24/50	24.5mg	17/50 liv:hpa,hpc,nnd.
3830	TR205	5.00mg	9.96mg	2/50	7.80mg	18/50	15.7mg	43/50	19.6mg	43/50 liv:hpc,nnd; thy:fca,fcc. C
a	TR205	5.38mg	10.3mg	1/50	7.80mg	13/50	15.7mg	41/50	19.6mg	39/50 liv:hpc,nnd.
b	TR205	11.1mg	23.7mg	0/50	7.80mg	4/50	15.7mg	23/50	19.6mg	22/50
c	TR205	12.2mg	30.1mg	1/50	7.80mg	6/50	15.7mg	17/50	19.6mg	28/50
d	TR205	14.7mg	50.7mg	1/50	7.80mg	9/50	15.7mg	18/50	19.6mg	17/50 thy:fca,fcc.
e	TR205	20.6mg	63.2mg	0/50	7.80mg	5/50	15.7mg	9/50	19.6mg	15/50
f	TR205	27.0mg	143.mg	1/50	7.80mg	1/50	15.7mg	8/50	19.6mg	13/50
g	TR205	15.2mg	n.s.s.	45/50	7.80mg	38/50	15.7mg	48/50	19.6mg	46/50
h	TR205	5.38mg	10.3mg	1/50	7.80mg	13/50	15.7mg	41/50	19.6mg	39/50 liv:hpa,hpc,nnd.

N-OXYDIETHYLENE THIOCARBAMYL-N-OXYDIETHYLENE SULFENAMIDE   (OTOS) 13752-51-7

RefNum	LoConf	UpConf	Cntrl	1Dose	1Inc	2Dose	2Inc	3Dose	3Inc	4Dose	4Inc	Citation or Pathology / Brkly Code
3831	1765m	.351mg	n.s.s.	0/10	1.00mg	0/10	3.00mg	0/10	10.0mg	0/10	30.0mg	0/10 Hinderer;txap,82,521-531; 1986/pers.comm.
3832	1765n	54.0mg	200.mg	0/50	1.00mg	0/50	3.00mg	0/50	10.0mg	0/50	30.0mg	16/50
a	1765n	93.4mg	604.mg	0/50	1.00mg	0/50	3.00mg	0/50	10.0mg	0/50	30.0mg	8/50
b	1765n	128.mg	1.44gm	0/50	1.00mg	0/50	3.00mg	0/50	10.0mg	0/50	30.0mg	5/50
c	1765n	173.mg	10.6gm	0/50	1.00mg	0/50	3.00mg	0/50	10.0mg	0/50	30.0mg	3/50
d	1765n	173.mg	10.6gm	0/50	1.00mg	0/50	3.00mg	0/50	10.0mg	0/50	30.0mg	3/50
e	1765n	173.mg	10.6gm	0/50	1.00mg	0/50	3.00mg	0/50	10.0mg	0/50	30.0mg	3/50
f	1765n	212.mg	n.s.s.	0/50	1.00mg	0/50	3.00mg	0/50	10.0mg	0/50	30.0mg	2/50
g	1765n	212.mg	n.s.s.	0/50	1.00mg	0/50	3.00mg	0/50	10.0mg	0/50	30.0mg	2/50
h	1765n	283.mg	n.s.s.	0/50	1.00mg	0/50	3.00mg	0/50	10.0mg	0/50	30.0mg	1/50
i	1765n	283.mg	n.s.s.	0/50	1.00mg	0/50	3.00mg	0/50	10.0mg	0/50	30.0mg	1/50
3833	1765m	.281mg	n.s.s.	0/10	.800mg	0/10	2.40mg	0/10	8.00mg	0/10	24.0mg	0/10
3834	1765n	46.3mg	189.mg	1/50	.800mg	0/50	2.40mg	0/50	8.00mg	0/50	24.0mg	15/50
a	1765n	63.4mg	334.mg	0/50	.800mg	0/50	2.40mg	0/50	8.00mg	0/50	24.0mg	10/50
b	1765n	117.mg	2.03gm	0/50	.800mg	0/50	2.40mg	0/50	8.00mg	0/50	24.0mg	4/50

Spe Sex	Strain	Site Route	Xpo+Xpt Hist	Notes	TD50 DR	2Tailpvl AuOp
c	R m sda	eat ubl	utp 26m26	e	457.mg *	P<.007 +
d	R m sda	eat kid	mix 26m26	e	528.mg *	P<.06
e	R m sda	eat kid	utc 26m26	e	690.mg *	P<.03
f	R m sda	eat ubl	sqc 26m26	e	690.mg *	P<.03 +
g	R m sda	eat ubl	sqp 26m26	e	1.39gm *	P<.2 +
h	R m sda	eat kid	sqc 26m26	e	1.39gm *	P<.2
i	R m sda	eat kid	utp 26m26	e	no dre	P=1.

N-OXYDIETHYLENEBENZOTHIAZOLE-2-SULFENAMIDE

					TD50 DR	2Tailpvl AuOp
3835 M f b6a	orl lun	ade 76w76	evx		no dre	P=1. -
a M f b6a	orl liv	hpt 76w76	evx		no dre	P=1. -
b M f b6a	orl tba	mix 76w76	evx		1.31gm	P<.7 -
3836 M m b6a	orl liv	hpt 76w76	evx		20.1gm	P<1. -
a M m b6a	orl lun	ade 76w76	evx		no dre	P=1. -
b M m b6a	orl tba	mix 76w76	evx		598.mg	P<.1 -
3837 M f b6c	orl lun	ade 76w76	evx		1.23gm	P<.3 -
a M f b6c	orl liv	hpt 76w76	evx		386.mg	P<.04 -
b M f b6c	orl tba	mix 76w76	evx		359.mg	P<.04 -
3838 M m b6c	orl liv	hpt 76w76	evx		556.mg	P<.1 -
a M m b6c	orl lun	ade 76w76	evx		200.mg	P<.007 -
b M m b6c	orl tba	mix 76w76	evx			

OXYTETRACYCLINE.HCl

					TD50 DR	2Tailpvl AuOp
3839 M f b6c	eat TBA	MXB 24m24			no dre	P=1. -
a M f b6c	eat liv	MXB 24m24			no dre	P=1.
b M f b6c	eat lun	MXB 24m24			no dre	P=1.
3840 M m b6c	eat TBA	MXB 24m24			no dre	P=1.
a M m b6c	eat liv	MXB 24m24			no dre	P=1.
b M m b6c	eat lun	MXB 24m24			24.0gm *	P<.05
3841 R f f34	eat ute	ess 24m24			5.23gm *	P<.2 e
a R f f34	eat pta	adn 24m24			no dre	P=1.
b R f f34	eat TBA	MXB 24m24			76.2gm *	P<.9
c R f f34	eat liv	MXB 24m24			10.9gm *	P<.6 e
3842 R m f34	eat amd	MXA 24m24	e		no dre	P=1.
a R m f34	eat TBA	MXB 24m24			no dre	P=1.
b R m f34	eat liv	MXB 24m24				

OZONE

					TD50 DR	2Tailpvl AuOp
3843 M f b6c	inh lun	a/a 30m30			1.78mg *	P<.04
a M f b6c	inh lun	MXA 30m30			2.09mg *	P<.2 p
b M f b6c	inh TBA	MXB 30m30			no dre	P=1.
c M f b6c	inh liv	MXB 30m30			2.09mg *	P<.2
d M f b6c	inh lun	MXB 30m30			1.88mg *	P<.06 p
3844 M f b6c	inh lun	MXA 24m24			2.75mg *	P<.05
a M f b6c	inh lun	a/c 24m24			6.49mg *	P<.05
b M f b6c	inh ute	esp 24m24			no dre	P=1.
c M f b6c	inh TBA	MXB 24m24			no dre	P=1.
d M f b6c	inh liv	MXB 24m24			1.88mg *	P<.06
e M f b6c	inh lun	MXB 24m24			.633mg *	P<.02
3845 M m b6c	inh lun	a/c 30m30			.961mg *	P<.2 e
a M m b6c	inh lun	MXA 30m30			1.54mg *	P<.6
b M m b6c	inh TBA	MXB 30m30			2.02mg *	P<.6
c M m b6c	inh liv	MXB 30m30			.961mg *	P<.2
d M m b6c	inh lun	MXB 30m30			.314mg Z	P<.02
3846 M m b6c	inh MXA	hes 24m24			.894mg *	P<.07 e
a M m b6c	inh lun	MXA 24m24			.997mg *	P<.4
b M m b6c	inh TBA	MXB 24m24			1.68mg *	P<.5
c M m b6c	inh liv	MXB 24m24			.894mg *	P<.07
d M m b6c	inh lun	MXB 24m24			no dre	P=1. -
3847 R f f34	inh TBA	MXB 29m29			no dre	P=1.
a R f f34	inh liv	MXB 29m29			no dre	P=1. -
3848 R f f34	inh TBA	MXB 24m24			no dre	P=1.
a R f f34	inh liv	MXB 24m24			.138mg \	P<.7
3849 R m f34	inh TBA	MXB 28m29	a		2.41mg *	P<.8
a R m f34	inh liv	MXB 28m29	a		#15.6ug Z	P<.009 -
3850 R m f34	inh thy	MXA 24m24			19.0ug Z	P<.02
a R m f34	inh thy	cca 24m24			1.20mg *	P<1.
b R m f34	inh TBA	MXB 24m24			no dre	P=1.
c R m f34	inh liv	MXB 24m24				

PARATHION

					TD50 DR	2Tailpvl AuOp
3851 M f b6c	eat TBA	MXB 80w89			no dre	P=1.
a M f b6c	eat liv	MXB 80w89			no dre	P=1.
b M f b6c	eat lun	MXB 80w89			no dre	P=1.
3852 M m b6c	eat TBA	MXB 67w89	a		75.1mg *	P<.7 -
a M m b6c	eat liv	MXB 67w89	a		111.mg *	P<.7
b M m b6c	eat lun	MXB 67w89	a		52.5mg *	P<.2
3853 R f osm	eat TBA	MXB 19m26	v		no dre	P=1.
a R f osm	eat liv	MXB 19m26	v		16.4mg *	P<.2
3854 R f osm	eat mgl	fba 19m25	v	pool	2.07mg \	P<.004
a R f osm	eat adr	MXA 19m25	v		3.93mg *	P<.0005a

	RefNum	LoConf	UpConf	Cntrl	1Dose	1Inc	2Dose	2Inc				Citation or Pathology
												Brkly Code
c	1765n	138.mg	8.46gm	0/50	.800mg	0/50	2.40mg	0/50	8.00mg	0/50	24.0mg 3/50	
d	1765n	143.mg	n.s.s.	1/50	.800mg	0/50	2.40mg	0/50	8.00mg	0/50	24.0mg 3/50	
e	1765n	170.mg	n.s.s.	0/50	.800mg	0/50	2.40mg	0/50	8.00mg	0/50	24.0mg 2/50	
f	1765n	170.mg	n.s.s.	0/50	.800mg	0/50	2.40mg	0/50	8.00mg	0/50	24.0mg 2/50	
g	1765n	226.mg	n.s.s.	0/50	.800mg	0/50	2.40mg	0/50	8.00mg	0/50	24.0mg 1/50	
h	1765n	226.mg	n.s.s.	0/50	.800mg	0/50	2.40mg	0/50	8.00mg	0/50	24.0mg 1/50	
i	1765n	6.52mg	n.s.s.	1/50	.800mg	0/50	2.40mg	0/50	8.00mg	0/50	24.0mg 0/50	

N-OXYDIETHYLENEBENZOTHIAZOLE-2-SULFENAMIDE   (amax) 102-77-2

	RefNum	LoConf	UpConf	Cntrl	1Dose	1Inc		Citation or Pathology
3835	1298	243.mg	n.s.s.	1/17	205.mg	1/18		Innes;ntis,1968/1969
a	1298	405.mg	n.s.s.	0/17	205.mg	0/18		
b	1298	147.mg	n.s.s.	2/17	205.mg	3/18		
3836	1298	211.mg	n.s.s.	1/18	190.mg	1/17		
a	1298	356.mg	n.s.s.	2/18	190.mg	0/17		
b	1298	250.mg	n.s.s.	3/18	190.mg	1/17		
3837	1298	147.mg	n.s.s.	0/16	205.mg	2/17		
a	1298	201.mg	n.s.s.	0/16	205.mg	1/17		
b	1298	116.mg	n.s.s.	0/16	205.mg	3/17		
3838	1298	108.mg	n.s.s.	0/16	190.mg	3/17		
a	1298	137.mg	n.s.s.	0/16	190.mg	2/17		
b	1298	75.3mg	2.70gm	0/16	190.mg	5/17		

OXYTETRACYCLINE.HCl   2058-46-0

	RefNum	LoConf	UpConf	Cntrl	1Dose	1Inc	2Dose	2Inc	Citation or Pathology
3839	TR315	2.12gm	n.s.s.	43/50	811.mg	34/50	1.61gm	36/50	
a	TR315	12.4gm	n.s.s.	6/50	811.mg	0/50	1.61gm	2/50	liv:hpa,hpc,nnd.
b	TR315	6.10gm	n.s.s.	3/50	811.mg	3/50	1.61gm	3/50	lun:a/a,a/c.
3840	TR315	1.58gm	n.s.s.	36/50	749.mg	32/50	1.49gm	33/50	
a	TR315	2.37gm	n.s.s.	18/50	749.mg	15/50	1.49gm	17/50	liv:hpa,hpc,nnd.
b	TR315	4.32gm	n.s.s.	10/50	749.mg	9/50	1.49gm	6/50	lun:a/a,a/c.
3841	TR315	8.10gm	n.s.s.	0/50	1.24gm	1/50	2.48gm	3/50	S
a	TR315	1.79gm	n.s.s.	19/50	1.24gm	17/50	2.48gm	30/50	
b	TR315	1.60gm	n.s.s.	49/50	1.24gm	44/50	2.48gm	49/50	
c	TR315	6.17gm	n.s.s.	5/50	1.24gm	4/50	2.48gm	6/50	liv:hpa,hpc,nnd.
3842	TR315	1.91gm	n.s.s.	12/50	990.mg	19/50	1.98gm	24/50	amd:phe,phm.
a	TR315	723.mg	n.s.s.	48/50	990.mg	48/50	(1.98gm	42/50)	
b	TR315	4.52gm	n.s.s.	6/50	990.mg	5/50	1.98gm	9/50	liv:hpa,hpc,nnd.

OZONE   10028-15-6

	RefNum	LoConf	UpConf	Cntrl	1Dose	1Inc	2Dose	2Inc	3Dose	3Inc	Citation or Pathology
3843	TR440a	.702mg	n.s.s.	3/50	.308mg	3/50	.616mg	11/50			S
a	TR440a	.655mg	n.s.s.	6/50	.308mg	8/50	.616mg	12/50			lun:a/a,a/c.
b	TR440a	.375mg	n.s.s.	48/50	.308mg	49/50	.616mg	47/50			
c	TR440a	.668mg	n.s.s.	27/50	.308mg	28/50	.616mg	21/50			liv:hpa,hpb,hpc.
d	TR440a	.655mg	n.s.s.	6/50	.308mg	8/50	.616mg	12/50			lun:a/a,a/c.
3844	TR440b	.741mg	n.s.s.	6/50	74.0ug	7/50	.308mg	9/50	.617mg	16/50	lun:a/a,a/c.
a	TR440b	1.11mg	n.s.s.	2/50	74.0ug	2/50	.308mg	5/50	.617mg	8/50	S
b	TR440b	2.20mg	n.s.s.	1/50	74.0ug	0/50	.308mg	0/50	.617mg	5/50	S
c	TR440b	.835mg	n.s.s.	47/50	74.0ug	44/50	.308mg	42/50	.617mg	40/50	
d	TR440b	.132mg	n.s.s.	27/50	74.0ug	22/50	(.308mg	20/50	.617mg	11/50)	liv:hpa,hpb,hpc.
e	TR440b	.741mg	n.s.s.	6/50	74.0ug	7/50	.308mg	9/50	.617mg	16/50	lun:a/a,a/c.
3845	TR440a	.293mg	n.s.s.	8/50	.257mg	15/50	.513mg	18/50			S
a	TR440a	.318mg	n.s.s.	16/50	.257mg	22/50	.513mg	21/50			lun:a/a,a/c.
b	TR440a	.282mg	n.s.s.	43/50	.257mg	50/50	.513mg	42/50			
c	TR440a	.365mg	n.s.s.	31/50	.257mg	34/50	.513mg	31/50			liv:hpa,hpb,hpc.
d	TR440a	.318mg	n.s.s.	16/50	.257mg	22/50	.513mg	21/50			lun:a/a,a/c.
3846	TR440b	.118mg	n.s.s.	0/50	61.7ug	5/50	(.257mg	1/50	.514mg	0/50)	liv:hes; mey:hes; spl:hes. S
a	TR440b	.340mg	n.s.s.	14/50	61.7ug	13/50	.257mg	18/50	.514mg	19/50	lun:a/a,a/c.
b	TR440b	.246mg	n.s.s.	39/50	61.7ug	43/50	.257mg	42/50	.514mg	40/50	
c	TR440b	.376mg	n.s.s.	30/50	61.7ug	27/50	.257mg	29/50	.514mg	29/50	liv:hpa,hpb,hpc.
d	TR440b	.340mg	n.s.s.	14/50	61.7ug	13/50	.257mg	18/50	.514mg	19/50	lun:a/a,a/c.
3847	TR440a	86.8ug	n.s.s.	49/56	73.5ug	48/56	.147mg	50/56			
a	TR440a	n.s.s.	n.s.s.	0/56	73.5ug	0/56	.147mg	0/56			liv:hpa,hpb,hpc.
3848	TR440b	92.5ug	n.s.s.	48/56	17.6ug	48/56	73.5ug	48/56	.147mg	46/56	
a	TR440b	1.47mg	n.s.s.	0/56	17.6ug	1/56	73.5ug	0/56	.147mg	0/56	liv:hpa,hpb,hpc.
3849	TR440a	15.9ug	n.s.s.	50/56	51.4ug	47/56	(.103mg	49/56)			
a	TR440a	.164mg	n.s.s.	1/56	51.4ug	1/56	.103mg	2/56			liv:hpa,hpb,hpc.
3850	TR440b	5.62ug	.729mg	2/56	12.3ug	11/56	(51.4ug	4/56	.103mg	2/56)	thy:cca,ccr. S
a	TR440b	5.78ug	n.s.s.	1/56	12.3ug	8/56	(51.4ug	2/56	.103mg	1/56)	S
b	TR440b	35.5ug	n.s.s.	49/56	12.3ug	49/56	51.4ug	49/56	.103mg	49/56	
c	TR440b	.354mg	n.s.s.	2/56	12.3ug	1/56	51.4ug	1/56	.103mg	1/56	liv:hpa,hpb,hpc.

PARATHION   56-38-2

	RefNum	LoConf	UpConf	Cntrl	1Dose	1Inc	2Dose	2Inc	Citation or Pathology
3851	TR70	33.0mg	n.s.s.	3/10	9.40mg	12/50	18.5mg	9/50	
a	TR70	107.mg	n.s.s.	1/10	9.40mg	1/50	18.5mg	1/50	liv:hpa,hpc,nnd.
b	TR70	94.0mg	n.s.s.	1/10	9.40mg	0/50	18.5mg	2/50	lun:a/a,a/c.
3852	TR70	12.6mg	n.s.s.	2/10	7.70mg	14/50	13.2mg	12/50	
a	TR70	19.6mg	n.s.s.	2/10	7.70mg	6/50	13.2mg	9/50	liv:hpa,hpc,nnd.
b	TR70	23.8mg	n.s.s.	0/10	7.70mg	3/50	13.2mg	5/50	lun:a/a,a/c.
3853	TR70	1.57mg	n.s.s.	10/10	.820mg	34/50	1.60mg	32/50	
a	TR70	5.67mg	n.s.s.	0/10	.820mg	1/50	1.60mg	3/50	liv:hpa,hpc,nnd.
3854	TR70	.947mg	16.4mg	9/90p	.820mg	16/50	(1.60mg	8/50)	S
a	TR70	2.01mg	15.0mg	4/90p	.820mg	6/50	1.60mg	13/50	adr:coa,coc.

```
    Spe Strain  Site    Xpo+Xpt                                              TD50    2Tailpvl
      Sex  Route  Hist     Notes                                               DR      AuOp
  b   R f osm eat adr coa 19m25 v                                           5.31mg * P<.004 a
3855 R m osm eat TBA MXB 19m26 v                          :>                3.65mg * P<.4   -
  a   R m osm eat liv MXB 19m26 v                                           17.7mg * P<.09
3856 R m osm eat adr MXA 19m25 v   pool             :  +        :           5.22mg * P<.005 a
  a   R m osm eat adr coa 19m25 v                                           6.39mg * P<.007 a
  b   R m osm eat pni isc 19m25 v                                           19.2mg * P<.03

PATULIN                            100ng..:..1ug....:..10.....:...100.....:...1mg.....:...10.....:...100.....:...1g.....:...10
3857 R f sda gav liv mhs 17m37 v                          .>                no dre  P=1.   -
  a   R f sda gav tba mix 17m37 v                                           no dre  P=1.   -
  b   R f sda gav tba ben 17m37 v                                           no dre  P=1.   -
  c   R f sda gav tba mal 17m37 v                                           no dre  P=1.   -

PENICILLIN VK                      100ng..:..1ug....:..10.....:...100.....:...1mg.....:...10.....:...100.....:...1g.....:...10
3858 M f b6c gav TBA MXB 24m24                                     :>        no dre  P=1.   -
  a   M f b6c gav liv MXB 24m24                                              10.2gm * P<.6
  b   M f b6c gav lun MXB 24m24                                              16.6gm * P<.7
3859 M m b6c gav TBA MXB 24m24                                     :>        no dre  P=1.   -
  a   M m b6c gav liv MXB 24m24                                              no dre  P=1.
  b   M m b6c gav lun MXB 24m24                                              no dre  P=1.
3860 R f f34 gav thy MXA 24m24 s                                :   ±       #1.36gm * P<.03  -
  a   R f f34 gav thy cca 24m24 s                                           1.62gm * P<.05
  b   R f f34 gav cli adn 24m24 s                                           3.65gm * P<.02
  c   R f f34 gav TBA MXB 24m24 s                                           503.mg * P<.08
  d   R f f34 gav liv MXB 24m24 s                                           no dre  P=1.
3861 R m f34 gav pta MXA 24m24 s                             :   ±         #874.mg * P<.02  -
  a   R m f34 gav pta adn 24m24 s                                           949.mg * P<.03
  b   R m f34 gav TBA MXB 24m24 s                                           544.mg * P<.08
  c   R m f34 gav liv MXB 24m24 s                                           12.2gm * P<.5

PENTACHLOROANISOLE                 100ng..:..1ug....:..10.....:...100.....:...1mg....:...10.....:...100....:...1g.....:...10
3862 M f b6c gav TBA MXB 24m24                                     :>        47.5mg * P<.4
  a   M f b6c gav liv MXB 24m24                                              63.0mg * P<.2
  b   M f b6c gav lun MXB 24m24                                              776.mg * P<.8
3863 M m b6c gav MXB MXB 24m24                                  :  +  :      38.3mg * P<.0005
  a   M m b6c gav amd pob 24m24                                              77.4mg * P<.002 p
  b   M m b6c gav liv hes 24m24                                              68.0mg * P<.02  p
  c   M m b6c gav TBA MXB 24m24                                              40.2mg * P<.3
  d   M m b6c gav liv MXB 24m24                                              no dre  P=1.
  e   M m b6c gav lun MXB 24m24                                              no dre  P=1.
3864 R f f34 gav amd pob 24m24                                        :>     174.mg * P<.4   e
  a   R f f34 gav TBA MXB 24m24                                              no dre  P=1.
  b   R f f34 gav liv MXB 24m24                                              1.18gm * P<.3
3865 R m f34 gav amd pob 24m24 s                             :   ±          24.8mg * P<.03  p
  a   R m f34 gav amd MXA 24m24 s                                            30.0mg * P<.08  p
  b   R m f34 gav TBA MXB 24m24 s                                            no dre  P=1.
  c   R m f34 gav liv MXB 24m24 s                                            6.46gm * P<1.

PENTACHLOROETHANE                  100ng..:..1ug....:..10.....:...100.....:...1mg....:...10.....:...100....:...1g.....:...10
3866 M f b6c gav liv MXA 24m24 as                               :  +  :      39.8mg / P<.0005c
  a   M f b6c gav liv hpc 24m24 as                                           66.6mg * P<.0005c
  b   M f b6c gav liv hpa 24m24 as                                           153.mg / P<.0005c
  c   M f b6c gav TBA MXB 24m24 as                                           39.5mg / P<.0005
  d   M f b6c gav liv MXB 24m24 as                                           39.8mg / P<.0005
  e   M f b6c gav lun MXB 24m24 as                                           1.03gm / P<.2
3867 M m b6c gav liv hpc 24m24 as                               :  +  :      102.mg / P<.0005c
  a   M m b6c gav liv MXA 24m24 as                                           107.mg / P<.0005
  b   M m b6c gav TBA MXB 24m24 as                                           85.6mg / P<.0005
  c   M m b6c gav liv MXB 24m24 as                                           107.mg / P<.0005
  d   M m b6c gav lun MXB 24m24 as                                           no dre  P=1.
3868 R f f34 gav TBA MXB 24m24 s                                   :>        1.66gm * P<.9
  a   R f f34 gav liv MXB 24m24 s                                            no dre  P=1.
3869 R m f34 gav kid tla 24m24 s                             :   +     +hist 635.mg * P<.009 a
  a   R m f34 gav kid MXA 24m24 s                                      +hist 662.mg * P<.07  a
  b   R m f34 gav TBA MXB 24m24 s                                            no dre  P=1.
  c   R m f34 gav liv MXB 24m24 s                                            no dre  P=1.

PENTACHLORONITROBENZENE            100ng..:..1ug....:..10.....:...100....:...1mg....:...10.....:...100....:...1g.....:...10
3870 M f b6c eat TBA MXB 78w91 v                                      :>     3.32gm * P<.2
  a   M f b6c eat liv MXB 78w91 v                                            9.69gm * P<.09
  b   M f b6c eat lun MXB 78w91 v                                            30.7gm * P<.4
3871 M f b6c eat lun a/a 24m24 s                                   :   ±    #2.00gm * P<.04  -
  a   M f b6c eat MXA mlm 24m24 s                                            2.34gm * P<.03
  b   M f b6c eat TBA MXB 24m24 s                                            1.01gm * P<.3
  c   M f b6c eat liv MXB 24m24 s                                            2.50gm * P<.2
  d   M f b6c eat lun MXB 24m24 s                                            3.44gm * P<.3
3872 M f b6c orl liv hpt 77w77 evx                              .   ±        252.mg * P<.02
  a   M f b6c orl lun ade 77w77 evx                                          1.11gm   P<.3
  b   M f b6c orl tba mix 77w77 evx                                          195.mg   P<.009
3873 M m b6c eat TBA MXB 78w91 v                                      :>     no dre  P=1.   -
  a   M m b6c eat liv MXB 78w91 v                                            no dre  P=1.
```

	RefNum	LoConf	UpConf	Cntrl	1Dose	1Inc	2Dose	2Inc		Citation or Pathology
										Brkly Code
b	TR70	2.48mg	38.8mg	4/90p	.820mg	4/50	1.60mg	11/50		
3855	TR70	1.03mg	n.s.s.	5/10	.900mg	25/50	1.80mg	35/50		
a	TR70	6.12mg	n.s.s.	0/10	.900mg	0/50	1.80mg	4/50		liv:hpa,hpc,nnd.
3856	TR70	2.54mg	54.3mg	3/90p	.900mg	7/50	1.80mg	11/50		adr:coa,coc.
a	TR70	2.99mg	111.mg	2/90p	.900mg	5/50	1.80mg	9/50		
b	TR70	6.62mg	n.s.s.	0/90p	.900mg	1/50	1.80mg	3/50		S
PATULIN	149-29-1									
3857	393	7.80mg	n.s.s.	1/50	.316mg	0/50				Osswald;fctx,16,243-247;1978
a	393	.809mg	n.s.s.	42/50	.316mg	37/50				
b	393	1.23mg	n.s.s.	29/50	.316mg	25/50				
c	393	1.80mg	n.s.s.	13/50	.316mg	12/50				
PENICILLIN VK	132-98-9									
3858	TR336	754.mg	n.s.s.	38/50	357.mg	26/50	714.mg	30/50		
a	TR336	1.66gm	n.s.s.	3/50	357.mg	4/50	714.mg	4/50		liv:hpa,hpc,nnd.
b	TR336	2.19gm	n.s.s.	4/50	357.mg	1/50	714.mg	5/50		lun:a/a,a/c.
3859	TR336	719.mg	n.s.s.	36/50	357.mg	35/50	714.mg	29/50		
a	TR336	657.mg	n.s.s.	19/50	357.mg	18/50	(714.mg	8/50)		liv:hpa,hpc,nnd.
b	TR336	1.81gm	n.s.s.	10/50	357.mg	9/50	714.mg	6/50		lun:a/a,a/c.
3860	TR336	576.mg	n.s.s.	-6/50	354.mg	7/50	707.mg	11/50		thy:cca,ccr. S
a	TR336	638.mg	n.s.s.	6/50	354.mg	6/50	707.mg	10/50		S
b	TR336	1.25gm	n.s.s.	0/50	354.mg	1/50	707.mg	3/50		S
c	TR336	197.mg	n.s.s.	44/50	354.mg	45/50	707.mg	38/50		
d	TR336	n.s.s.	n.s.s.	0/50	354.mg	1/50	707.mg	0/50		liv:hpa,hpc,nnd.
3861	TR336	390.mg	n.s.s.	10/50	354.mg	11/50	707.mg	14/50		pta:adn,can. S
a	TR336	406.mg	n.s.s.	10/50	354.mg	11/50	707.mg	13/50		S
b	TR336	210.mg	n.s.s.	42/50	354.mg	40/50	707.mg	31/50		
c	TR336	1.99gm	n.s.s.	0/50	354.mg	1/50	707.mg	0/50		liv:hpa,hpc,nnd.
PENTACHLOROANISOLE	1825-21-4									
3862	TR414	11.7mg	n.s.s.	32/50	14.1mg	31/50	28.2mg	30/50		
a	TR414	19.8mg	n.s.s.	11/50	14.1mg	10/50	28.2mg	14/50		liv:hpa,hpc,nnd.
b	TR414	66.1mg	n.s.s.	2/50	14.1mg	2/50	28.2mg	2/50		lun:a/a,a/c.
3863	TR414	21.8mg	124.mg	2/50	14.1mg	11/50	28.3mg	16/50		amd:pob; liv:hes. P
a	TR414	38.5mg	189.mg	0/50	14.1mg	4/50	28.3mg	7/50		
b	TR414	32.6mg	n.s.s.	2/50	14.1mg	8/50	28.3mg	10/50		
c	TR414	11.9mg	n.s.s.	38/50	14.1mg	47/50	28.3mg	46/50		
d	TR414	25.8mg	n.s.s.	26/50	14.1mg	34/50	28.3mg	24/50		liv:hpa,hpc,nnd.
e	TR414	50.9mg	n.s.s.	11/50	14.1mg	12/50	28.3mg	8/50		lun:a/a,a/c.
3864	TR414	50.8mg	n.s.s.	3/50	14.0mg	7/50	28.2mg	9/50		
a	TR414	14.7mg	n.s.s.	45/50	14.0mg	41/50	(28.2mg	41/50)		
b	TR414	192.mg	n.s.s.	0/50	14.0mg	0/50	28.2mg	1/50		liv:hpa,hpc,nnd.
3865	TR414	10.8mg	n.s.s.	12/50	6.99mg	17/50	14.0mg	23/50	28.2mg 15/50	
a	TR414	11.5mg	n.s.s.	15/50	6.99mg	18/50	14.0mg	25/50	28.2mg 15/50	amd:phm,pob.
b	TR414	15.0mg	n.s.s.	47/50	6.99mg	49/50	14.0mg	44/50	28.2mg 23/50	
c	TR414	86.3mg	n.s.s.	0/50	6.99mg	1/50	14.0mg	1/50	28.2mg 0/50	liv:hpa,hpc,nnd.
PENTACHLOROETHANE	(pentalin) 76-01-7									
3866	TR232	24.0mg	68.0mg	3/50	177.mg	36/50	357.mg	32/50		liv:hpa,hpc.
a	TR232	38.3mg	118.mg	1/50	177.mg	28/50	357.mg	13/50		
b	TR232	69.1mg	367.mg	2/50	177.mg	8/50	357.mg	19/50		
c	TR232	24.1mg	69.1mg	22/50	177.mg	38/50	357.mg	33/50		
d	TR232	24.0mg	68.0mg	3/50	177.mg	36/50	357.mg	32/50		liv:hpa,hpc,nnd.
e	TR232	193.mg	n.s.s.	4/50	177.mg	0/50	357.mg	3/50		lun:a/a,a/c.
3867	TR232	59.1mg	234.mg	4/50	177.mg	26/50	357.mg	7/50		
a	TR232	58.4mg	304.mg	14/50	177.mg	30/50	357.mg	14/50		liv:hpa,hpc. S
b	TR232	46.5mg	278.mg	20/50	177.mg	38/50	357.mg	14/50		
c	TR232	58.4mg	304.mg	14/50	177.mg	30/50	357.mg	14/50		liv:hpa,hpc,nnd.
d	TR232	389.mg	n.s.s.	6/50	177.mg	5/50	357.mg	0/50		lun:a/a,a/c.
3868	TR232	79.4mg	n.s.s.	37/50	53.1mg	36/50	106.mg	26/50		
a	TR232	409.mg	n.s.s.	1/50	53.1mg	3/50	106.mg	0/50		liv:hpa,hpc,nnd.
3869	TR232	240.mg	14.6gm	0/50	53.1mg	1/50	106.mg	4/50		
a	TR232	229.mg	n.s.s.	1/50	53.1mg	2/50	106.mg	4/50		kid:tla,uac.
b	TR232	107.mg	n.s.s.	36/50	53.1mg	26/50	106.mg	22/50		
c	TR232	500.mg	n.s.s.	2/50	53.1mg	1/50	106.mg	1/50		liv:hpa,hpc,nnd.
PENTACHLORONITROBENZENE	(PCNB) 82-68-8									
3870	TR61	1.22gm	n.s.s.	2/20	446.mg	5/50	893.mg	10/50		
a	TR61	2.93gm	n.s.s.	0/20	446.mg	0/50	893.mg	3/50		liv:hpa,hpc,nnd.
b	TR61	5.00gm	n.s.s.	0/20	446.mg	0/50	893.mg	1/50		lun:a/a,a/c.
3871	TR325	770.mg	n.s.s.	1/50	322.mg	4/50	644.mg	4/50		S
a	TR325	881.mg	n.s.s.	0/50	322.mg	3/50	644.mg	2/50		mln:mlm; mul:mlm. S
b	TR325	284.mg	n.s.s.	27/50	322.mg	24/50	644.mg	21/50		
c	TR325	720.mg	n.s.s.	3/50	322.mg	4/50	644.mg	4/50		liv:hpa,hpc,nnd.
d	TR325	896.mg	n.s.s.	3/50	322.mg	4/50	644.mg	4/50		lun:a/a,a/c.
3872	1101	86.7mg	n.s.s.	0/16	169.mg	4/18				Innes;ntis,1968/1969
a	1101	180.mg	n.s.s.	0/16	169.mg	1/18				
b	1101	73.4mg	4.27gm	0/16	169.mg	5/18				
3873	TR61	498.mg	n.s.s.	6/20	265.mg	11/50	530.mg	12/50		
a	TR61	669.mg	n.s.s.	2/20	265.mg	8/50	530.mg	4/50		liv:hpa,hpc,nnd.

```
        Spe Strain  Site   Xpo+Xpt                                                      TD50    2Tailpvl
            Sex  Route  Hist   Notes                                                              DR    AuOp
b      M m b6c eat lun MXB 78w91 v                                                      no dre   P=1.
3874   M m b6c eat TBA MXB 24m24                                              :>        13.0gm * P<1.    -
a      M m b6c eat liv MXB 24m24                                                        23.1gm * P<1.
b      M m b6c eat lun MXB 24m24                                                        no dre   P=1.
3875   M m b6c orl liv hpt 77w77 evx                              .>                    501.mg   P<.2
a      M m b6c orl lun ade 77w77 evx                                                    501.mg   P<.2
b      M m b6c orl tba mix 77w77 evx                                                    181.mg   P<.009
3876   M f b6a orl liv hpt 77w77 evx                                 .>                 1.05gm   P<.3
a      M f b6a orl lun ade 77w77 evx                                                    no dre   P=1.
b      M f b6a orl tba mix 77w77 evx                                                    no dre   P=1.
3877   M m b6a orl liv hpt 77w77 evx                          .  +   .                  71.1mg   P<.0005+
a      M m b6a orl lun ade 77w77 evx                                                    no dre   P=1.
b      M m b6a orl tba mix 77w77 evx                                                    68.7mg   P<.004
3878   R f osm eat TBA MXB 18m26 dv                              :>                     no dre   P=1.    -
a      R f osm eat liv MXB 18m26 dv                                                     no dre   P=1.
3879   R m osm eat TBA MXB 18m26 dv                              :>                     no dre   P=1.    -
a      R m osm eat liv MXB 18m26 dv                                                     no dre   P=1.


2,3,4,5,6-PENTACHLOROPHENOL          100ng..:..1ug....:..10......:..100...:..1mg....:..10......:..100....:..1g.....:..10
3880   M f cd1 eat lun ade 52w69 e                               .>                     448.mg   P<.3
a      M f cd1 eat liv tum 52w69 e                                                      no dre   P=1.


2,3,4,5,6-PENTACHLOROPHENOL  (Dowicide EC-7).1ug....:..10......:..100.....:..1mg....:..10......:..100.....:..1g.....:..10
3881   M f b6c eat MXB MXB 24m24                                      : + :             28.7mg * P<.0005
a      M f b6c eat liv MXA 24m24                                                        38.3mg * P<.0005c
b      M f b6c eat amd MXA 24m24                                                        38.8mg Z P<.0005c
c      M f b6c eat amd MXA 24m24                                                        40.3mg Z P<.0005
d      M f b6c eat liv hpa 24m24                                                        40.6mg * P<.0005
e      M f b6c eat --- MXA 24m24                                                        187.mg * P<.0005
f      M f b6c eat --- hes 24m24                                                        200.mg * P<.002 c
g      M f b6c eat TBA MXB 24m24                                                        86.1mg Z P<.2
h      M f b6c eat liv MXB 24m24                                                        38.3mg * P<.0005
i      M f b6c eat lun MXB 24m24                                                        12.8gm * P<1.
3882   M f b6c orl lun ade 76w76 evx                             .>                     56.1mg   P<.2    -
a      M f b6c orl liv hpt 76w76 evx                                                    no dre   P=1.    -
b      M f b6c orl tba mix 76w76 evx                                                    26.3mg   P<.02   -
3883   M m b6c eat amd MXA 24m24                                      : + :             17.4mg * P<.0005
a      M m b6c eat amd MXA 24m24                                                        17.5mg * P<.0005c
b      M m b6c eat MXB MXB 24m24                                                        24.9mg Z P<.0005
c      M m b6c eat liv hpa 24m24                                                        36.2mg * P<.0005
d      M m b6c eat liv MXA 24m24                                                        38.3mg Z P<.003 c
e      M m b6c eat TBA MXB 24m24                                                        57.8mg * P<.09
f      M m b6c eat liv MXB 24m24                                                        38.3mg Z P<.003
g      M m b6c eat lun MXB 24m24                                                        1.94gm * P<.9
3884   M m b6c orl liv hpt 76w76 evx                             .>                     108.mg   P<.3    -
a      M m b6c orl lun ade 76w76 evx                                                    108.mg   P<.3    -
b      M m b6c orl tba mix 76w76 evx                                                    33.7mg   P<.05   -
3885   M f b6a orl liv hpt 76w76 evx                                .>                  no dre   P=1.    -
a      M f b6a orl lun ade 76w76 evx                                                    no dre   P=1.    -
b      M f b6a orl tba mix 76w76 evx                                                    no dre   P=1.    -
3886   M m b6a orl liv mix 76w76 evx                              .>                    90.4mg   P<.6    -
a      M m b6a orl lun ade 76w76 evx                                                    no dre   P=1.    -
b      M m b6a orl tba mix 76w76 evx                                                    520.mg   P<1.    -
3887   R f sss eat tba tum 24m24 e                     .>                               no dre   P=1.    -
3888   R m sss eat tba tum 95w95 e                        .>                            no dre   P=1.    -


2,3,4,5,6-PENTACHLOROPHENOL, TECHNICAL GRADE      ...:..10......:..100...:..1mg....:..10......:..100....:..1g......:..10
3889   M f b6c eat --- hes 24m24                                      :   +      :      106.mg * P<.007 p
a      M f b6c eat MXB MXB 24m24                                                        50.4mg * P<.02
b      M f b6c eat liv MXA 24m24                                                        90.1mg * P<.2    p
c      M f b6c eat TBA MXB 24m24                                                        79.3mg / P<.5
d      M f b6c eat liv MXB 24m24                                                        90.1mg * P<.2
e      M f b6c eat lun MXB 24m24                                                        395.mg * P<.4
3890   M m b6c eat MXB MXB 24m24                                      :  +     :        10.5mg * P<.002
a      M m b6c eat amd MXA 24m24                                                        12.9mg * P<.0005c
b      M m b6c eat liv hpa 24m24                                                        13.5mg * P<.002
c      M m b6c eat liv MXA 24m24                                                        13.7mg * P<.005 c
d      M m b6c eat TBA MXB 24m24                                                        22.6mg * P<.2
e      M m b6c eat liv MXB 24m24                                                        13.7mg * P<.005
f      M m b6c eat lun MXB 24m24                                                        154.mg * P<.5
3891   R f mrw eat liv hpa 87w87                              .  +    .                 13.1mg   P<.002 +
a      R f mrw eat tba tum 87w87                                                        9.96mg   P<.0005
3892   R m mrw eat tba tum 87w87                                 .>                     43.0mg   P<.3


PENTAERYTHRITOL TETRANITRATE WITH 80% D-LACTOSE MONOHYDRATE..:..100...:..1mg....:..10......:..100...:..1g.....:..10
3893   M f b6c eat TBA MXB 24m24                                                        :no dre  P=1.    -
a      M f b6c eat liv MXB 24m24                                                        no dre   P=1.
b      M f b6c eat lun MXB 24m24                                                        82.0gm * P<.5
3894   M m b6c eat TBA MXB 24m24                                                        no dre   P=1.    -
a      M m b6c eat liv MXB 24m24                                                        no dre   P=1.
b      M m b6c eat lun MXB 24m24                                                        no dre   P=1.
```

	RefNum	LoConf	UpConf	Cntrl	1Dose	1Inc	2Dose	2Inc	3Dose	3Inc	4Dose	4Inc	Citation or Pathology	Brkly Code
b	TR61	n.s.s.	n.s.s.	0/20	265.mg	0/50	530.mg	0/50					lun:a/a,a/c.	
3874	TR325	527.mg	n.s.s.	31/50	297.mg	34/50	594.mg	29/50						
a	TR325	804.mg	n.s.s.	17/50	297.mg	18/50	594.mg	16/50					liv:hpa,hpc,nnd.	
b	TR325	1.90gm	n.s.s.	9/50	297.mg	4/50	594.mg	6/50					lun:a/a,a/c.	
3875	1101	123.mg	n.s.s.	0/16	157.mg	2/18							Innes;ntis,1968/1969	
a	1101	123.mg	n.s.s.	0/16	157.mg	2/18								
b	1101	68.4mg	3.98gm	0/16	157.mg	5/18								
3876	1101	170.mg	n.s.s.	0/17	169.mg	1/17								
a	1101	324.mg	n.s.s.	1/17	169.mg	0/17								
b	1101	152.mg	n.s.s.	2/17	169.mg	2/17								
3877	1101	32.2mg	264.mg	1/18	157.mg	10/17								
a	1101	198.mg	n.s.s.	2/18	157.mg	1/17								
b	1101	29.4mg	544.mg	3/18	157.mg	11/17								
3878	TR61	322.mg	n.s.s.	14/20	272.mg	34/50	(514.mg	20/50)						
a	TR61	n.s.s.	n.s.s.	0/20	272.mg	0/50	514.mg	0/50					liv:hpa,hpc,nnd.	
3879	TR61	409.mg	n.s.s.	8/20	152.mg	20/50	280.mg	18/50						
a	TR61	1.67gm	n.s.s.	0/20	152.mg	1/50	280.mg	0/50					liv:hpa,hpc,nnd.	

2,3,4,5,6-PENTACHLOROPHENOL   (penta, PCP) 87-86-5

	RefNum	LoConf	UpConf	Cntrl	1Dose	1Inc	Citation or Pathology
3880	1581	73.0mg	n.s.s.	0/32	48.8mg	1/31	Boberg;canr,43,5163-5173;1983
a	1581	137.mg	n.s.s.	0/32	48.8mg	0/31	

2,3,4,5,6-PENTACHLOROPHENOL   (Dowicide EC-7)   (dowicide 7, penta, PCP) 87-86-5

	RefNum	LoConf	UpConf	Cntrl	1Dose	1Inc	2Dose	2Inc	3Dose	3Inc	4Dose	4Inc	Citation or Pathology	Brkly Code
3881	TR349	19.4mg	47.8mg	1/35	12.8mg	6/50	25.6mg	9/50	76.9mg	42/50			---:hes; amd:pbb,phm,pob; liv:hpa,hpc.	C
a	TR349	24.4mg	71.6mg	1/35	12.8mg	4/50	25.6mg	6/50	76.9mg	31/50			liv:hpa,hpc.	
b	TR349	25.7mg	61.2mg	0/35	12.8mg	2/50	25.6mg	2/50	76.9mg	38/50			amd:pbb,phm,pob.	
c	TR349	26.6mg	64.0mg	0/35	12.8mg	1/50	25.6mg	2/50	76.9mg	38/50			amd:pbb,pob.	S
d	TR349	25.8mg	75.3mg	1/35	12.8mg	3/50	25.6mg	6/50	76.9mg	30/50				S
e	TR349	93.8mg	646.mg	0/35	12.8mg	1/50	25.6mg	3/50	76.9mg	9/50			---:hem,hes.	S
f	TR349	97.8mg	878.mg	0/35	12.8mg	1/50	25.6mg	3/50	76.9mg	8/50				
g	TR349	29.6mg	n.s.s.	24/35	12.8mg	29/50	25.6mg	25/50	76.9mg	46/50				
h	TR349	24.4mg	71.6mg	1/35	12.8mg	4/50	25.6mg	6/50	76.9mg	31/50			liv:hpa,hpc,nnd.	
i	TR349	221.mg	n.s.s.	2/35	12.8mg	3/50	25.6mg	1/50	76.9mg	3/50			lun:a/a,a/c.	
3882	1285	13.8mg	n.s.s.	0/16	18.1mg	2/18							Innes;ntis,1968/1969	
a	1285	35.8mg	n.s.s.	0/16	18.1mg	0/18								
b	1285	9.04mg	n.s.s.	0/16	18.1mg	4/18								
3883	TR349	12.4mg	25.6mg	0/35	11.8mg	4/50	23.7mg	21/50	71.0mg	44/50			amd:pbb,pob.	S
a	TR349	12.2mg	28.2mg	1/35	11.8mg	4/50	23.7mg	21/50	71.0mg	45/50			amd:pbb,phm,pmb,pob.	
b	TR349	13.9mg	72.0mg	7/35	11.8mg	21/50	23.7mg	27/50	71.0mg	45/50			amd:pbb,phm,pmb,pob; liv:hpa,hpc.	C
c	TR349	19.2mg	128.mg	5/35	11.8mg	13/50	23.7mg	17/50	71.0mg	32/50				S
d	TR349	18.8mg	224.mg	6/35	11.8mg	19/50	23.7mg	21/50	71.0mg	34/50			liv:hpa,hpc.	
e	TR349	21.2mg	n.s.s.	20/35	11.8mg	32/50	23.7mg	33/50	71.0mg	46/50				
f	TR349	18.8mg	224.mg	6/35	11.8mg	19/50	23.7mg	21/50	71.0mg	34/50			liv:hpa,hpc,nnd.	
g	TR349	143.mg	n.s.s.	5/35	11.8mg	1/50	23.7mg	2/50	71.0mg	5/50			lun:a/a,a/c.	
3884	1285	17.5mg	n.s.s.	0/16	16.8mg	1/18							Innes;ntis,1968/1969	
a	1285	17.5mg	n.s.s.	0/16	16.8mg	1/18								
b	1285	10.2mg	n.s.s.	0/16	16.8mg	3/18								
3885	1285	35.8mg	n.s.s.	0/17	18.1mg	0/18								
a	1285	35.8mg	n.s.s.	1/17	18.1mg	0/18								
b	1285	17.0mg	n.s.s.	2/17	18.1mg	2/18								
3886	1285	13.2mg	n.s.s.	1/18	16.8mg	2/17								
a	1285	20.7mg	n.s.s.	2/18	16.8mg	1/17								
b	1285	12.2mg	n.s.s.	3/18	16.8mg	3/17								
3887	1401	10.0mg	n.s.s.	27/27	1.00mg	26/27	3.00mg	25/27	10.0mg	25/27	30.0mg	25/27	Schwetz;pcpl,301-309;1978	
3888	1401	33.4mg	n.s.s.	11/27	1.00mg	13/26	3.00mg	13/27	10.0mg	12/27	30.0mg	11/27		

2,3,4,5,6-PENTACHLOROPHENOL,   TECHNICAL GRADE   (penta, PCP) 87-86-5

	RefNum	LoConf	UpConf	Cntrl	1Dose	1Inc	2Dose	2Inc	Citation or Pathology	Brkly Code
3889	TR349	49.9mg	1.32gm	0/35	12.8mg	3/50	25.5mg	6/50		
a	TR349	25.0mg	n.s.s.	3/35	12.8mg	11/50	25.5mg	14/50	---:hes; liv:hpa,hpc.	P
b	TR349	34.1mg	n.s.s.	3/35	12.8mg	9/50	25.5mg	9/50	liv:hpa,hpc.	
c	TR349	17.5mg	n.s.s.	27/35	12.8mg	30/50	25.5mg	37/50		
d	TR349	34.1mg	n.s.s.	3/35	12.8mg	9/50	25.5mg	9/50	liv:hpa,hpc,nnd.	
e	TR349	89.9mg	n.s.s.	1/35	12.8mg	1/50	25.5mg	3/50	lun:a/a,a/c.	
3890	TR349	5.89mg	51.5mg	7/35	11.8mg	29/50	23.5mg	41/50	amd:pbb,pob; liv:hpa,hpc.	C
a	TR349	8.42mg	24.2mg	0/35	11.8mg	10/50	23.5mg	23/50	amd:pbb,pob.	
b	TR349	7.57mg	66.9mg	5/35	11.8mg	20/50	23.5mg	33/50		S
c	TR349	7.30mg	123.mg	7/35	11.8mg	26/50	23.5mg	37/50	liv:hpa,hpc.	
d	TR349	7.67mg	n.s.s.	17/35	11.8mg	38/50	23.5mg	45/50		
e	TR349	7.30mg	123.mg	7/35	11.8mg	26/50	23.5mg	37/50	liv:hpa,hpc,nnd.	
f	TR349	36.5mg	n.s.s.	2/35	11.8mg	2/50	23.5mg	6/50	lun:a/a,a/c.	
3891	2038	5.07mg	55.4mg	0/10	25.0mg	6/10			Mirvish;jtxe,32,59-74;1991/pers.comm.	
a	2038	3.96mg	34.1mg	0/10	25.0mg	7/10				
3892	2038	6.93mg	n.s.s.	0/5	20.0mg	1/5				

PENTAERYTHRITOL TETRANITRATE WITH 80% D-LACTOSE MONOHYDRATE   (PETN, NF) 78-11-5

	RefNum	LoConf	UpConf	Cntrl	1Dose	1Inc	2Dose	2Inc	Citation or Pathology
3893	TR365	9.43gm	n.s.s.	37/50	3.19gm	29/50	6.38gm	28/50	
a	TR365	13.6gm	n.s.s.	6/50	3.19gm	2/50	(6.38gm	1/50)	liv:hpa,hpc,nnd.
b	TR365	17.0gm	n.s.s.	3/50	3.19gm	3/50	6.38gm	5/50	lun:a/a,a/c.
3894	TR365	10.4gm	n.s.s.	36/50	2.94gm	28/50	5.89gm	32/50	
a	TR365	14.5gm	n.s.s.	11/50	2.94gm	11/50	5.89gm	11/50	liv:hpa,hpc,nnd.
b	TR365	9.00gm	n.s.s.	11/50	2.94gm	8/50	(5.89gm	7/50)	lun:a/a,a/c.

```
      Spe Strain Site   Xpo+Xpt                                                          TD50    2Tailpvl
          Sex   Route Hist      Notes                                                      DR       AuOp
3895 R f f34 eat zym MXA 24m25                                                    :  ±  5.03gm * P<.04  e
   a R f f34 eat thy MXA 24m25                                                          5.80gm * P<.03
   b R f f34 eat zym ade 24m25                                                          11.1gm * P<.1   e
   c R f f34 eat zym car 24m25                                                          9.20gm * P<.2   e
   d R f f34 eat TBA MXB 24m25                                                          416.mg * P<.04
   e R f f34 eat liv MXB 24m25                                                          no dre   P=1.
3896 R m f34 eat amd phm 24m25                                                    :     12.2gm * P<.04
   a R m f34 eat zym MXA 24m25                                                          13.6gm * P<.2   e
   b R m f34 eat zym car 24m25                                                          20.6gm * P<.2   e
   c R m f34 eat zym ade 24m25                                                          no dre   P=1.   e
   d R m f34 eat TBA MXB 24m25                                                          24.0gm / P<1.
   e R m f34 eat liv MXB 24m25                                                          no dre   P=1.

PENTANAL METHYLFORMYLHYDRAZONE    100ng..:..1ug...:..10......:..100....:..1mg...:..10......:..100.....:..1g.....:..10
3897 M f swa gav lun mix 12m33 e                                     . + .                 3.12mg   P<.0005+
   a M f swa gav lun ade 12m33 e                                                           5.64mg   P<.0005
   b M f swa gav lun adc 12m33 e                                                           9.29mg   P<.0005
   c M f swa gav liv bhp 12m33 e                                                           11.8mg   P<.003 +
3898 M m swa gav lun mix 12m31 e                                  . + .                    3.79mg   P<.0005+
   a M m swa gav lun ade 12m31 e                                                           5.52mg   P<.002
   b M m swa gav pre mix 12m31 e                                                           16.2mg   P<.007 +
   c M m swa gav lun adc 12m31 e                                                           11.0mg   P<.02
   d M m swa gav pre sqc 12m31 e                                                           19.8mg   P<.02
   e M m swa gav liv hpt 12m31 e                                                           33.1mg   P<.3   +
   f M m swa gav pre sqp 12m31 e                                                           106.mg   P<.3

N-PENTYL-N'-NITRO-N-NITROSOGUANIDINE    ..:..1ug....:..10.....:..100....:..1mg...:..10....:..100.....:..1g......:..10
3899 R m wis wat stg tum 52w78 er                                      .>                   no dre   P=1.   -

n-PENTYLHYDRAZINE.HCl              100ng..:..1ug...:..10.....:..100....:..1mg...:..10......:..100.....:..1g......:..10
3900 M f swa wat lun mix 94w94                                       . + .                  5.87mg   P<.0005+
   a M f swa wat lun ade 94w94                                                              7.95mg   P<.0005
   b M f swa wat lun adc 94w94                                                              18.7mg   P<.0005
   c M f swa wat blv mix 94w94                                                              35.5mg   P<.003 +
   d M f swa wat blv agm 94w94                                                              71.9mg   P<.04
   e M f swa wat blv ang 94w94                                                              82.2mg   P<.04
   f M f swa wat liv mix 94w94                                                              223.mg   P<.4
3901 M m swa wat lun ade 94w94                                        . ±                   30.0mg   P<.04
   a M m swa wat lun mix 94w94                                                              46.9mg   P<.3
   b M m swa wat blv mix 94w94                                                              65.6mg   P<.2
   c M m swa wat liv mix 94w94                                                              271.mg   P<.7

PEPPERMINT OIL                     100ng..:..1ug...:..10.....:..100....:..1mg...:..10......:..100.....:..1g......:..10
3902 M m cfl gav lun tum 19m24 e                                           .>               no dre   P=1.
   a M m cfl gav liv tum 19m24 e                                                            no dre   P=1.
   b M m cfl gav tba mix 19m24 e                                                            no dre   P=1.
   c M m cfl gav tba mal 19m24 e                                                            112.mg * P<.6

PETASITENINE                       100ng..:..1ug...:..10.....:..100....:..1mg...:..10......:..100.....:..1g......:..10
3903 R f aci wat liv hms 68w68 er                          . + .               +hist 1.52mg   P<.003 +
   a R f aci wat liv mix 68w68 er                                               +hist 1.52mg   P<.003 +
   b R f aci wat liv lca 68w68 er                                               +hist 4.13mg   P<.05  +
3904 R m aci wat liv mix 58w58 er                       . + .                   +hist .662mg  P<.0005+
   a R m aci wat liv lca 58w58 er                                               +hist 1.16mg   P<.005 +
   b R m aci wat liv hms 58w58 er                                               +hist 4.78mg   P<.2   +

PHENACETIN                         100ng..:..1ug...:..10.....:..100....:..1mg...:..10......:..100.....:..1g......:..10
3905 M f b6c eat lun ade 22m24 e                                             . ± 6.89gm * P<.1   -
   a M f b6c eat ubl pam 22m24 e                                                  37.0gm * P<.1   +
   b M f b6c eat ubl tcc 22m24 e                                                  37.0gm * P<.1   +
   c M f b6c eat lun adc 22m24 e                                                  8.83gm * P<.2   -
   d M f b6c eat liv hnd 22m24 e                                                  14.0gm * P<.3   -
   e M f b6c eat liv hem 22m24 e                                                  29.2gm * P<.4   -
   f M f b6c eat liv hpc 22m24 e                                                  47.7gm * P<.7   -
3906 M m b6c eat kid rca 22m24 e                                      . + .      1.10gm * P<.0005+
   a M m b6c eat lun adc 22m24 e                                                  3.80gm * P<.006 +
   b M m b6c eat kid rcc 22m24 e                                                  4.02gm / P<.0005
   c M m b6c eat lun ade 22m24 e                                                  5.28gm * P<.2   -
   d M m b6c eat liv hpc 22m24 e                                                  17.1gm \ P<.9   -
   e M m b6c eat liv hem 22m24 e                                                  no dre   P=1.   -
   f M m b6c eat liv hnd 22m24 e                                                  no dre   P=1.   -
3907 M f cb6 eat liv tum 77w77 e                                       .>        no dre   P=1.   -
3908 M m cb6 eat liv hnd 77w77 e                                        .>       12.9gm * P<.2   -
3909 R f sda eat --- myl 18m24                                        . + .      2.56gm * P<.0005-
   a R f sda eat mix mix 18m24                                                    2.73gm * P<.0005+
   b R f sda eat nas adc 18m24                                                    7.73gm * P<.005 +
   c R f sda eat ubl tcc 18m24                                                    11.7gm * P<.02  +
   d R f sda eat ubl pam 18m24                                                    23.8gm * P<.08  +
   e R f sda eat nas tcc 18m24                                                    23.8gm * P<.2   +
   f R f sda eat nas sqc 18m24                                                    48.0gm * P<.7   +
   g R f sda eat liv tum 18m24                                                    no dre   P=1.
```

	RefNum	LoConf	UpConf	Cntrl	1Dose	1Inc	2Dose	2Inc	Citation or Pathology
									Brkly Code
3895	TR365	1.66gm	n.s.s.	0/50	304.mg	1/50	616.mg	3/50	zym:ade,car.
a	TR365	1.71gm	n.s.s.	0/50	304.mg	0/50	616.mg	3/50	thy:fca,fcc. S
b	TR365	2.72gm	n.s.s.	0/50	304.mg	0/50	616.mg	2/50	
c	TR365	2.10gm	n.s.s.	0/50	304.mg	1/50	616.mg	1/50	
d	TR365	182.mg	n.s.s.	46/50	304.mg	47/50	616.mg	50/50	
e	TR365	n.s.s.	n.s.s.	0/50	304.mg	0/50	616.mg	0/50	liv:hpa,hpc,nnd.
3896	TR365	4.24gm	n.s.s.	0/50	981.mg	2/50	1.97gm	3/50	
									S
a	TR365	4.84gm	n.s.s.	0/50	981.mg	3/50	1.97gm	2/50	zym:ade,car.
b	TR365	7.00gm	n.s.s.	0/50	981.mg	2/50	1.97gm	2/50	
c	TR365	n.s.s.	n.s.s.	0/50	981.mg	1/50	1.97gm	0/50	
d	TR365	1.02gm	n.s.s.	47/50	981.mg	43/50	1.97gm	48/50	
e	TR365	11.4gm	n.s.s.	3/50	981.mg	0/50	1.97gm	1/50	liv:hpa,hpc,nnd.

PENTANAL METHYLFORMYLHYDRAZONE   57590-20-2

	RefNum	LoConf	UpConf	Cntrl	1Dose	1Inc	2Dose	2Inc	Citation or Pathology
3897	1859	1.80mg	6.98mg	13/50	2.58mg	36/48			Toth;myco,98,83-89;1987/pers.comm.
a	1859	2.94mg	22.1mg	12/50	2.58mg	28/48			
b	1859	4.88mg	26.7mg	2/50	2.58mg	16/48			
c	1859	5.31mg	52.3mg	0/23	2.58mg	8/32			
3898	1859	2.04mg	11.7mg	11/41	2.77mg	30/44			
a	1859	2.84mg	23.2mg	8/41	2.77mg	24/44			
b	1859	6.61mg	193.mg	0/27	2.77mg	6/34			
c	1859	4.82mg	n.s.s.	5/41	2.77mg	15/44			
d	1859	7.52mg	n.s.s.	0/27	2.77mg	5/34			
e	1859	5.38mg	n.s.s.	0/8	2.77mg	1/11			
f	1859	17.2mg	n.s.s.	0/27	2.77mg	1/34			

N-PENTYL-N'-NITRO-N-NITROSOGUANIDINE   13010-10-1

	RefNum	LoConf	UpConf	Cntrl	1Dose	1Inc	2Dose	2Inc	Citation or Pathology
3899	1082	1.60mg	n.s.s.	0/9	2.30mg	0/6			Matsukura;gann,70,181-185;1979

n-PENTYLHYDRAZINE.HCl   1119-68-2

	RefNum	LoConf	UpConf	Cntrl	1Dose	1Inc	2Dose	2Inc	Citation or Pathology
3900	36	3.54mg	11.2mg	21/100	12.5mg	38/50			Shimizu;bjca,31,492-496;1975
a	36	4.69mg	16.3mg	18/100	12.5mg	33/50			
b	36	10.0mg	45.4mg	4/100	12.5mg	17/50			
c	36	15.6mg	245.mg	5/100	12.5mg	11/50			
d	36	24.4mg	n.s.s.	3/100	12.5mg	6/50			
e	36	26.9mg	n.s.s.	2/100	12.5mg	5/50			
f	36	39.3mg	n.s.s.	3/100	12.5mg	3/50			
3901	36	11.6mg	n.s.s.	15/100	10.4mg	15/50			
a	36	12.8mg	n.s.s.	23/100	10.4mg	16/50			
b	36	19.8mg	n.s.s.	6/100	10.4mg	7/50			
c	36	31.1mg	n.s.s.	6/100	10.4mg	4/50			

PEPPERMINT OIL   8006-90-4

	RefNum	LoConf	UpConf	Cntrl	1Dose	1Inc	2Dose	2Inc	Citation or Pathology
3902	710	21.1mg	n.s.s.	102/240	2.64mg	19/51	10.5mg	20/49	Roe;jept,2,799-819;1979
a	710	25.3mg	n.s.s.	69/240	2.64mg	13/51	10.5mg	14/49	
b	710	11.0mg	n.s.s.	170/240	2.64mg	37/51	10.5mg	34/49	
c	710	17.3mg	n.s.s.	75/240	2.64mg	20/51	10.5mg	17/49	

PETASITENINE   60102-37-6

	RefNum	LoConf	UpConf	Cntrl	1Dose	1Inc	2Dose	2Inc	Citation or Pathology
3903	427	.477mg	9.69mg	0/9	5.71mg	4/6			Hirono;jnci,58,1155-1157;1977
a	427	.477mg	9.69mg	0/9	5.71mg	4/6			
b	427	.994mg	n.s.s.	0/9	5.71mg	2/6			
3904	427	.183mg	3.71mg	0/10	5.00mg	4/5			
a	427	.324mg	13.1mg	0/10	5.00mg	3/5			
b	427	.770mg	n.s.s.	0/10	5.00mg	1/5			

PHENACETIN   62-44-2

	RefNum	LoConf	UpConf	Cntrl	1Dose	1Inc	2Dose	2Inc	Citation or Pathology
3905	1501	2.60gm	n.s.s.	7/48	720.mg	6/50	1.50gm	14/49	Nakanishi;ijcn,29,439-444;1982
a	1501	9.10gm	n.s.s.	0/48	720.mg	0/50	1.50gm	2/49	
b	1501	9.10gm	n.s.s.	0/48	720.mg	0/50	1.50gm	2/49	
c	1501	3.23gm	n.s.s.	3/48	720.mg	6/50	1.50gm	8/49	
d	1501	4.15gm	n.s.s.	2/48	720.mg	5/50	1.50gm	5/49	
e	1501	6.97gm	n.s.s.	1/48	720.mg	1/50	1.50gm	3/49	
f	1501	6.28gm	n.s.s.	2/48	720.mg	3/50	1.50gm	3/49	
3906	1501	754.mg	1.69gm	0/48	665.mg	11/48	1.38gm	32/48	
a	1501	1.90gm	41.0gm	3/48	665.mg	8/48	1.38gm	13/48	
b	1501	2.20gm	8.53gm	0/48	665.mg	1/48	1.38gm	14/48	
c	1501	1.87gm	n.s.s.	8/48	665.mg	14/48	1.38gm	14/48	
d	1501	1.39gm	n.s.s.	10/48	665.mg	11/48	(1.38gm	3/48)	
e	1501	10.5gm	n.s.s.	4/48	665.mg	1/48	1.38gm	2/48	
f	1501	4.58gm	n.s.s.	14/48	665.mg	10/48	1.38gm	10/48	
3907	1028	848.mg	n.s.s.	0/36	268.mg	0/37	754.mg	0/41	Macklin;dact,3,135-163;1980
3908	1028	2.11gm	n.s.s.	0/35	268.mg	0/40	754.mg	1/32	
3909	708	1.45gm	5.32gm	0/65	469.mg	6/50	938.mg	11/50	Isaka;gann,70,29-36;1979
a	708	1.52gm	5.79gm	0/65	469.mg	5/50	938.mg	11/50	
b	708	3.15gm	58.5gm	0/65	469.mg	1/50	938.mg	5/50	
c	708	4.05gm	n.s.s.	0/65	469.mg	0/50	938.mg	4/50	
d	708	5.85gm	n.s.s.	0/65	469.mg	0/50	938.mg	2/50	
e	708	5.87gm	n.s.s.	0/65	469.mg	1/50	938.mg	1/50	
f	708	7.82gm	n.s.s.	0/65	469.mg	1/50	938.mg	0/50	
g	708	3.22gm	n.s.s.	0/65	469.mg	0/50	938.mg	0/50	

```
       Spe Strain  Site   Xpo+Xpt                                                                    TD50    2Tailpvl
           Sex  Route  Hist     Notes                                                                  DR    AuOp
  h     R f sda eat tba mix 18m24                                                                    1.21gm  * P<.0005
3910  R f sda eat mgl adc 26m26 e                                              .       ±            1.38gm  P<.08  +
  a     R f sda eat edu sqc 26m26 e                                                                 1.43gm  P<.02  +
3911  R m sda eat mix mix 18m24                                                         .+ .        741.mg  * P<.0005+
  a     R m sda eat --- myl 18m24                                                                   2.38gm  * P<.0005-
  b     R m sda eat nas adc 18m24                                                                   2.38gm  * P<.0005+
  c     R m sda eat ubl tcc 18m24                                                                   2.71gm  / P<.0005+
  d     R m sda eat nas tcc 18m24                                                                   4.05gm  * P<.002 +
  e     R m sda eat nas sqc 18m24                                                                   7.53gm  * P<.06  +
  f     R m sda eat k/p rcc 18m24                                                                   38.3gm  * P<.3  +
  g     R m sda eat liv tum 18m24                                                                   no dre  P=1.
  h     R m sda eat tba mix 18m24                                                                   631.mg  * P<.0005
3912  R m sda eat kcx rct 27m27                                              .       ±   +hist     1.30gm  P<.02  +
  a     R m sda eat for sqp 27m27                                                        +hist     2.69gm  P<.1   +
  b     R m sda eat ubl tum 27m27                                                                  1.64gm  P<.3   +
  c     R m sda eat k/p tum 27m27                                                        +hist     5.47gm  P<.3   +
  d     R m sda eat liv hem 27m27                                                                  5.47gm  P<.3
3913  R f wis eat unt tum 69w69 er                                                      .>         no dre  P=1.   -

PHENAZONE                      100ng..:..1ug....:...10......:...100....:..1mg..:...10....:...100.....:..1g.....:..10
3914  R m sda eat k/p mix 26m26                                                 .       ±          1.23gm  P<.02  +
  a     R m sda eat liv ccy 26m26                                                                  2.55gm  P<.1
  b     R m sda eat ubl tum 26m26                                                                  1.55gm  P<.3   +

PHENAZOPYRIDINE.HCl            100ng..:..1ug....:...10......:...100....:..1mg..:...10....:...100.....:..1g.....:..10
3915  M f b6c eat liv MXA 19m24                                             :   +          :       71.1mg  * P<.007 c
  a     M f b6c eat liv hpc 19m24                                                                  132.mg  * P<.03 c
  b     M f b6c eat TBA MXB 19m24                                                                  89.7mg  * P<.2
  c     M f b6c eat liv MXB 19m24                                                                  71.1mg  * P<.007
  d     M f b6c eat lun MXB 19m24                                                                  4.65gm  * P<.9
3916  M m b6c eat TBA MXB 19m24                                                 :>                 496.mg  * P<.8   -
  a     M m b6c eat liv MXB 19m24                                                                  811.mg  * P<.9
  b     M m b6c eat lun MXB 19m24                                                                  1.30gm  * P<.8
3917  R f f34 eat col acn 18m24                                                     :>   +hist     687.mg  * P<.2  c
  a     R f f34 eat TBA MXB 18m24                                                                  no dre  P=1.
  b     R f f34 eat liv MXB 18m24                                                                  no dre  P=1.
3918  R m f34 eat MXA MXA 18m24                                             :   ±                  303.mg  * P<.02 c
  a     R m f34 eat MXA acn 18m24                                                                  333.mg  * P<.02 c
  b     R m f34 eat TBA MXB 18m24                                                                  no dre  P=1.
  c     R m f34 eat liv MXB 18m24                                                                  2.17gm  * P<.5

PHENESTERIN                    100ng..:..1ug....:...10......:...100....:..1mg..:...10....:...100.....:..1g.....:..10
3919  M f b6c gav MXB MXB 70w83 aes                              :   +   :                         .211mg  Z P<.0005
  a     M f b6c gav lun MXA 70w83 aes                                                              .416mg  * P<.0005c
  b     M f b6c gav --- MXA 70w83 aes                                                              .652mg  Z P<.0005c
  c     M f b6c gav ova tua 70w83 aes                                                              .912mg  * P<.0005
  d     M f b6c gav myc srn 70w83 aes                                                              .951mg  * P<.0005c
  e     M f b6c gav --- hes 70w83 aes                                                              1.52mg  * P<.0005
  f     M f b6c gav TBA MXB 70w83 aes                                                              .183mg  Z P<.0005
  g     M f b6c gav liv MXB 70w83 aes                                                              no dre  P=1.
  h     M f b6c gav lun MXB 70w83 aes                                                              .416mg  * P<.0005
3920  M f b6c gav --- MXA 70w82 aes pool                                 :   + :                   2.07mg  * P<.0005c
3921  M m b6c gav MXB MXB 64w81 a                                       :   + :                    .757mg  * P<.0005
  a     M m b6c gav lun MXA 64w81 a                                                                1.19mg  * P<.0005c
  b     M m b6c gav lun a/c 64w81 a                                                                2.05mg  Z P<.0005c
  c     M m b6c gav --- MXA 64w81 a                                                                2.13mg  * P<.0005c
  d     M m b6c gav myc srn 64w81 a                                                                3.87mg  * P<.0005c
  e     M m b6c gav TBA MXB 64w81 a                                                                .728mg  * P<.0005
  f     M m b6c gav liv MXB 64w81 a                                                                32.9mg  * P<.9
  g     M m b6c gav lun MXB 64w81 a                                                                1.19mg  * P<.0005
3922  M m b6c gav --- MXA 64w81 a    pool                               :   +   :                  1.89mg  * P<.0005c
  a     M m b6c gav --- hes 64w81 a                                                                6.15mg  * P<.002
  b     M m b6c gav myc srn 64w81 a                                                                6.28mg  * P<.003 c
3923  R f sda gav TBA MXB 52w85                              :   +   :                             .271mg  * P<.0005-
  a     R f sda gav liv MXB 52w85                                                                  no dre  P=1.
3924  R f sda gav mgl acn 52w85       pool                          :   +   :                      .523mg  * P<.0005c
3925  R m sda gav TBA MXB 52w84                               :   +   :                            .481mg  * P<.002 -
  a     R m sda gav liv MXB 52w84                                                                  no dre  P=1.
3926  R m sda gav --- MXA 52w84       pool                          :   +   :                      #2.73mg * P<.003 -

PHENFORMIN.HCl                 100ng..:..1ug....:...10......:...100....:..1mg..:...10....:...100.....:..1g.....:..10
3927  M f b6c eat TBA MXB 18m24                                                     :>             no dre  P=1.   -
  a     M f b6c eat liv MXB 18m24                                                                  no dre  P=1.
  b     M f b6c eat lun MXB 18m24                                                                  no dre  P=1.
3928  M m b6c eat TBA MXB 18m24 s                                                      :>          no dre  P=1.   -
  a     M m b6c eat liv MXB 18m24 s                                                                no dre  P=1.
  b     M m b6c eat lun MXB 18m24 s                                                                no dre  P=1.
3929  R f f34 eat TBA MXB 18m24                                                :>                  no dre  P=1.   -
  a     R f f34 eat liv MXB 18m24                                                                  no dre  P=1.
3930  R m f34 eat TBA MXB 18m24                                             :>                     36.4mg  / P<.2  -
  a     R m f34 eat liv MXB 18m24                                                                  no dre  P=1.
```

	RefNum	LoConf	UpConf	Cntrl	1Dose	1Inc	2Dose	2Inc	Citation or Pathology
									Brkly Code
h	708	728.mg	2.89gm	6/65	469.mg	19/50	938.mg	21/50	
3910	1449	451.mg	n.s.s.	1/30	268.mg	5/30			Johansson;apms,84,375-383;1976
a	1449	494.mg	n.s.s.	0/30	268.mg	4/30			
3911	708	503.mg	1.15gm	0/65	375.mg	17/50	750.mg	23/50	Isaka;gann,70,29-36;1979
a	708	1.30gm	6.32gm	0/65	375.mg	8/50	750.mg	7/50	
b	708	1.30gm	6.32gm	0/65	375.mg	8/50	750.mg	7/50	
c	708	1.43gm	6.12gm	0/65	375.mg	1/50	750.mg	12/50	
d	708	1.91gm	13.5gm	0/65	375.mg	2/50	750.mg	7/50	
e	708	2.86gm	n.s.s.	0/65	375.mg	3/50	750.mg	2/50	
f	708	6.24gm	n.s.s.	0/65	375.mg	0/50	750.mg	1/50	
g	708	2.58gm	n.s.s.	0/65	375.mg	0/50	750.mg	0/50	
h	708	432.mg	990.mg	1/65	375.mg	20/50	750.mg	26/50	
3912	1459	447.mg	n.s.s.	0/30	214.mg	4/30			Johansson;ijcn,27,521-529;1981
a	1459	661.mg	n.s.s.	0/30	214.mg	2/30			
b	1459	435.mg	n.s.s.	2/30	214.mg	5/30			
c	1459	891.mg	n.s.s.	0/30	214.mg	1/30			
d	1459	891.mg	n.s.s.	0/30	214.mg	1/30			
3913	1740	1.86gm	n.s.s.	0/30	500.mg	0/41			Kunze;zkko,105,38-47;1983

PHENAZONE   60-80-0

3914	1459	425.mg	n.s.s.	0/30	214.mg	4/30			Johansson;ijcn,27,521-529;1981
a	1459	628.mg	n.s.s.	0/30	214.mg	2/30			
b	1459	413.mg	n.s.s.	2/30	214.mg	5/30			

PHENAZOPYRIDINE.HCl   136-40-3

3915	TR99	38.0mg	991.mg	2/15	41.6mg	11/35	84.5mg	19/35	liv:hpa,hpc.
a	TR99	61.6mg	n.s.s.	2/15	41.6mg	6/35	84.5mg	14/35	
b	TR99	34.1mg	n.s.s.	8/15	41.6mg	20/35	84.5mg	27/35	
c	TR99	38.0mg	991.mg	2/15	41.6mg	11/35	84.5mg	19/35	liv:hpa,hpc,nnd.
d	TR99	252.mg	n.s.s.	1/15	41.6mg	1/35	84.5mg	2/35	lun:a/a,a/c.
3916	TR99	51.8mg	n.s.s.	7/15	39.6mg	17/35	78.0mg	20/35	
a	TR99	62.1mg	n.s.s.	5/15	39.6mg	15/35	78.0mg	15/35	liv:hpa,hpc,nnd.
b	TR99	144.mg	n.s.s.	1/15	39.6mg	4/35	78.0mg	4/35	lun:a/a,a/c.
3917	TR99	311.mg	n.s.s.	0/15	98.0mg	3/34	199.mg	5/35	
a	TR99	232.mg	n.s.s.	5/15	98.0mg	10/34	199.mg	13/35	
b	TR99	n.s.s.	n.s.s.	0/15	98.0mg	0/34	199.mg	0/35	liv:hpa,hpc,nnd.
3918	TR99	160.mg	43.4gm	0/15	78.4mg	4/35	161.mg	9/35	col:acn,adn; rec:acn.
a	TR99	172.mg	n.s.s.	0/15	78.4mg	4/35	161.mg	8/35	col:acn; rec:acn.
b	TR99	210.mg	n.s.s.	9/15	78.4mg	19/35	161.mg	14/35	
c	TR99	535.mg	n.s.s.	0/15	78.4mg	1/35	161.mg	1/35	liv:hpa,hpc,nnd.

PHENESTERIN   3546-10-9

3919	TR60	.103mg	.440mg	7/35	3.00mg	30/40	5.40mg	16/35	10.8mg	21/35	---:leu,lym; lun:a/a,a/c; myc:srn. C
a	TR60	.152mg	1.25mg	2/35	3.00mg	15/40	5.40mg	1/35	10.8mg	1/35	lun:a/a,a/c.
b	TR60	.288mg	1.65mg	5/35	3.00mg	12/40	5.40mg	14/35	10.8mg	17/35	---:leu,lym.
c	TR60	.273mg	3.43mg	0/35	3.00mg	8/40	5.40mg	0/35	10.8mg	0/35	S
d	TR60	.288mg	2.78mg	0/35	3.00mg	8/40	5.40mg	2/35	10.8mg	3/35	
e	TR60	.465mg	6.27mg	0/35	3.00mg	4/40	5.40mg	1/35	10.8mg	2/35	S
f	TR60	96.1ug	.368mg	11/35	3.00mg	33/40	5.40mg	17/35	10.8mg	21/35	
g	TR60	n.s.s.	n.s.s.	3/35	3.00mg	1/40	5.40mg	0/35	10.8mg	0/35	liv:hpa,hpc,nnd.
h	TR60	.152mg	1.25mg	2/35	3.00mg	15/40	5.40mg	1/35	10.8mg	1/35	lun:a/a,a/c.
3920	TR60	.862mg	3.97mg	0/31p	5.40mg	14/35	10.8mg	17/35			---:leu,lym.
3921	TR60	.491mg	1.30mg	3/35	3.00mg	25/40	4.10mg	19/35	9.00mg	15/35	---:leu,lym; lun:a/a,a/c; myc:srn. C
a	TR60	.684mg	2.84mg	2/35	3.00mg	18/40	4.10mg	6/35	9.00mg	2/35	lun:a/a,a/c.
b	TR60	1.08mg	5.57mg	0/35	3.00mg	14/40	4.10mg	0/35	9.00mg	0/35	
c	TR60	1.20mg	4.46mg	1/35	3.00mg	11/40	4.10mg	9/35	9.00mg	11/35	---:leu,lym.
d	TR60	1.85mg	11.2mg	0/35	3.00mg	5/40	4.10mg	7/35	9.00mg	2/35	
e	TR60	.446mg	1.43mg	11/35	3.00mg	29/40	4.10mg	20/35	9.00mg	17/35	
f	TR60	2.07mg	n.s.s.	7/35	3.00mg	10/40	4.10mg	2/35	9.00mg	0/35	liv:hpa,hpc,nnd.
g	TR60	.684mg	2.84mg	2/35	3.00mg	18/40	4.10mg	6/35	9.00mg	2/35	lun:a/a,a/c.
3922	TR60	.784mg	4.46mg	0/29p	4.10mg	9/35	9.00mg	11/35			---:leu,lym.
a	TR60	1.99mg	36.3mg	0/29p	4.10mg	1/35	9.00mg	3/35			S
b	TR60	2.64mg	30.5mg	0/29p	4.10mg	7/35	9.00mg	2/35			
3923	TR60	.137mg	.880mg	7/10	1.30mg	28/35	2.60mg	22/35			
a	TR60	n.s.s.	n.s.s.	0/10	1.30mg	0/35	2.60mg	0/35			liv:hpa,hpc,nnd.
3924	TR60	.251mg	1.30mg	1/20p	1.30mg	12/35	2.60mg	12/35			
3925	TR60	.258mg	2.10mg	1/10	1.30mg	18/35	2.60mg	11/35			
a	TR60	n.s.s.	n.s.s.	0/10	1.30mg	0/35	2.60mg	0/35			liv:hpa,hpc,nnd.
3926	TR60	1.07mg	8.49mg	0/20p	1.30mg	2/35	2.60mg	5/35			---:leu,lym. S

PHENFORMIN.HCl   834-28-6

3927	TR7	203.mg	n.s.s.	7/15	117.mg	13/35	(244.mg	9/35)	
a	TR7	701.mg	n.s.s.	1/15	117.mg	3/35	244.mg	2/35	
b	TR7	n.s.s.	n.s.s.	0/15	117.mg	0/35	244.mg	0/35	liv:hpa,hpc,nnd.
3928	TR7	266.mg	n.s.s.	8/15	108.mg	10/35	226.mg	11/35	lun:a/a,a/c.
a	TR7	262.mg	n.s.s.	1/15	108.mg	6/35	226.mg	4/35	liv:hpa,hpc,nnd.
b	TR7	376.mg	n.s.s.	1/15	108.mg	4/35	226.mg	1/35	lun:a/a,a/c.
3929	TR7	39.3mg	n.s.s.	14/15	15.0mg	28/35	30.0mg	26/35	
a	TR7	n.s.s.	n.s.s.	0/15	15.0mg	0/35	30.0mg	0/35	liv:hpa,hpc,nnd.
3930	TR7	13.0mg	n.s.s.	7/15	12.0mg	14/35	24.0mg	24/35	
a	TR7	n.s.s.	n.s.s.	0/15	12.0mg	0/35	24.0mg	0/35	liv:hpa,hpc,nnd.

```
     Spe Strain Site    Xpo+Xpt                                                              TD50    2Tailpvl
     Sex  Route  Hist   Notes                                                                 DR      AuOp
PHENOBARBITAL              100ng..:..1ug....:.10.....:..100.....:.1mg....:..10......:..100....:.1g......:.10
3931 H m syg eat liv mix 52w52 er                                          .>                 no dre   P=1.  -
3932 M m b6c wat liv mix 52w52 r                              <+                               noTD50   P<.006 +
3933 M f c31 eat liv tum 52w52 r                              <+                               noTD50   P<.0005+
3934 M f c31 eat liv tum 52w52 r                         .  +  .                               12.2mg   P<.0005+
3935 M m c31 eat liv tum 52w52 r                       .  +  .                                 4.18mg   P<.0005+
3936 M m c31 eat liv tum 52w52 r                       .  +  .                                 4.46mg   P<.0005+
3937 M m c5n wat liv tum 78w78 r                                     .>                        no dre   P=1.  -
3938 M m cb6 eat liv hct 53w53 r                                  .>                           no dre   P=1.  -
3939 M f cd1 wat liv hpt 78w78                                        .   ±                     559.mg   P<.1
   a M f cd1 wat lun ade 78w78                                                                  no dre   P=1.
3940 M m cen wat liv mix 52w52 kr                                                              noTD50   P<.3
3941 M m cen wat liv mix 52w52 r                                                               noTD50   P<.09  +
3942 M m cen eat liv mix 52w52 r                              .  +  .                           6.67mg   P<.0005+
   a M m cen eat liv hpc 52w52 r                              .  +  .                           7.78mg   P<.0005+
   b M m cen eat lun mix 52w52 r                                                                139.mg   P<.4
   c M m cen eat liv hpa 52w52 r                                                                no dre   P=1.
3943 M m chg eat liv tum 52w52 r                            .  +  .                             5.56mg   P<.002 +
3944 M m chh eat liv esn 55w55 ekr                           <+                                noTD50   P<.009
   a M m chh eat liv bsn 55w55 ekr                                                              56.6mg   P<.5
3945 M m chh eat liv esn 65w65 ekr                              .  +  .                         49.8mg   P<.0005
   a M m chh eat liv bsn 65w65 ekr                                                              no dre   P=1.
3946 M m chh eat liv tum 52w52 r                                                               noTD50   P<.09
3947 M m dba eat liv hpa 53w53 r                                 .   ±                          61.2mg   P<.02
3948 R f aci eat liv tum 67w67                                    .>                            no dre   P=1.  -
   a R f aci eat tba mix 67w67                                                                  81.7mg   P<.3
3949 R m aci eat liv tum 67w67                                    .>                            no dre   P=1.  -
   a R m aci eat tba mix 67w67                                                                  no dre   P=1.
3950 R f f34 wat liv hpa 76w76 er                                   .>                          99.2mg   P<.3
3951 R m f34 wat liv hpa 76w76 er                                   .>                          86.8mg   P<.3
3952 R m f34 wat liv hpa 72w72 e                                .   ±                           45.0mg   P<.08
3953 R m f34 eat liv nnd 17m24 e                                       .>                        no dre   P=1.  -
3954 R m f34 eat liv mix 52w52 er                                     .>                         no dre   P=1.  -
3955 R m f34 eat liv hpt 71w71 e                                      .>                         no dre   P=1.  -
3956 R m f34 eat liv hpa 71w71                                    .   ±                          76.6mg   P<.1
3957 R m f34 eat liv hpa 71w71 e                                  .   ±                          44.6mg   P<.04
3958 R m fis eat liv tum 39w78 e                                  .>                            no dre   P=1.  -
   a R m fis eat tba tum 39w78 e                                                                no dre   P=1.
3959 R f sda ipj tba mal 24m24 es                        .>                                    no dre   P=1.  -
3960 R m sda ipj liv hae 24m24 es                          .>                                  no dre   P=1.  -
   a R m sda ipj tba mal 24m24 es                                                               54.7mg   P<1.
3961 R m wis eat liv mix 59w59 e                               .>                               303.mg   P<.4
   a R m wis eat tba mix 59w59 e                                                                9.22mg   P<.0005


PHENOBARBITAL, SODIUM      100ng..:..1ug....:.10.....:..100.....:.1mg....:..10......:..100....:.1g......:.10
3962 M f bal wat lun ade 27m28                                             .>                  1.21gm   P<.5  -
   a M f bal wat liv hpc 27m28                                                                  no dre   P=1.  -
   b M f bal wat tba mix 27m28                                                                  no dre   P=1.  -
3963 M m bal wat lun ade 27m27                                             .>                  no dre   P=1.  -
   a M m bal wat liv hpc 27m27                                                                  no dre   P=1.  -
   b M m bal wat tba mix 27m27                                                                  no dre   P=1.  -
3964 M f cf1 eat liv mix 26m26 e                              .  +  .                           44.2mg   P<.0005+
   a M f cf1 eat liv lct 26m26 e                                                                128.mg   P<.0005+
   b M f cf1 eat lun tum 26m26 e                                                                no dre   P=1.  -
3965 M f cf1 wat liv hpt 28m28 e                                   .+  .                        95.2mg   P<.0005+
   a M f cf1 wat lun ade 28m28 e                                                                no dre   P=1.  -
   b M f cf1 wat tba mix 28m28 e                                                                no dre   P=1.
3966 M m cf1 eat liv mix 26m26 e                             .  +  .                            34.6mg   P<.0005+
   a M m cf1 eat liv lct 26m26 e                                                                174.mg   P<.006 +
   b M m cf1 eat lun tum 26m26 e                                                                no dre   P=1.  -
3967 M m cf1 wat liv hpt 28m28 e                                  .+  .                         62.2mg   P<.0005+
   a M m cf1 wat lun ade 28m28 e                                                                no dre   P=1.  -
   b M m cf1 wat tba mix 28m28 e                                                                132.mg   P<.4
3968 R m cdr wat liv sad 95w95 e                              .>                                500.mg   P<.3
3969 R m f34 eat liv nnd 95w95 e                             .  +  .                            120.mg * P<.004
   a R m f34 eat thy cca 95w95                                                                  31.7mg \ P<.03
3970 R m f34 eat ubl tum 56w56 e                            .>                                  no dre   P=1.  -
3971 R m fis eat liv hnd 24m24 ev                           .  +  .                             67.3mg   P<.0005
3972 R f wis wat liv hpt 34m34 e                            .  +  .                             102.mg   P<.0005+
   a R f wis wat tba tum 34m34                                                                  73.1mg   P<.2
3973 R m wis wat liv hpt 34m34 e                            .  +  .                             74.3mg   P<.0005+
   a R m wis wat tba tum 34m34 e                                                                206.mg   P<.6


PHENOL                     100ng..:..1ug....:.10.....:..100.....:.1mg....:..10......:..100....:.1g......:.10
3974 M f b6c wat TBA MXB 24m24                                             :>                   no dre   P=1.  -
   a M f b6c wat liv MXB 24m24                                                                  no dre   P=1.
   b M f b6c wat lun MXB 24m24                                                                  18.5gm * P<.6
3975 M m b6c wat TBA MXB 24m24                                                :>                no dre   P=1.  -
   a M m b6c wat liv MXB 24m24                                                                  2.45gm \ P<.5
   b M m b6c wat lun MXB 24m24                                                                  8.29gm * P<.5
3976 R f f34 wat TBA MXB 24m24                                          :>                      no dre   P=1.  -
```

RefNum	LoConf	UpConf	Cntrl	1Dose	1Inc	2Dose	2Inc	Citation or Pathology
								Brkly Code

PHENOBARBITAL (phenobarbitone) 50-06-6

RefNum	LoConf	UpConf	Cntrl	1Dose	1Inc	2Dose	2Inc	Citation or Pathology
3931	1804	23.7mg n.s.s.	0/5	46.0mg	0/10			Stenback;jnci,76,327-333;1986
3932	1477m	n.s.s. 11.5mg	5/16	83.3mg	16/16			Becker;canr,42,3918-3923;1982
3933	235m	n.s.s. 5.36mg	5/39	65.0mg	29/29			Peraino;jnci,51,1349-1350;1973
3934	235n	5.45mg 46.2mg	1/16	65.0mg	10/16			
3935	235m	1.52mg 19.0mg	25/37	60.0mg	35/36			
3936	235n	1.56mg 21.4mg	7/17	60.0mg	16/17			
3937	1477n	155.mg n.s.s.	0/16	83.3mg	0/16			Becker;canr,42,3918-3923;1982
3938	2084m	80.3mg n.s.s.	0/25	60.0mg	0/25			Diwan;carc,13,1893-1901;1992
3939	1582	137.mg n.s.s.	0/30	100.mg	2/30			Miller;canr,43,1124-1134;1983
a	1582	210.mg n.s.s.	1/30	100.mg	1/30			
3940	1477o	n.s.s. n.s.s.	5/8	83.3mg	8/8			Becker;canr,42,3918-3923;1982
3941	1477r	n.s.s. n.s.s.	10/16	83.3mg	16/16			
3942	1992	3.75mg 13.3mg	2/30	60.0mg	24/30			Fullerton;carc,11,1301-1305;1990
a	1992	4.47mg 14.7mg	0/30	60.0mg	22/30			
b	1992	28.5mg n.s.s.	2/30	60.0mg	4/30			
c	1992	44.9mg n.s.s.	2/30	60.0mg	2/30			
3943	1891	2.31mg 38.5mg	42/139	24.0mg	14/21			Mizutani;clet,39,233-237;1988
3944	1776m	n.s.s. 22.6mg	0/5	85.0mg	5/5			Evans;carc,7,627-631;1986
a	1776m	8.26mg n.s.s.	1/5	85.0mg	2/5			
3945	1776n	24.8mg 123.mg	0/30	85.0mg	11/30			
a	1776n	68.8mg n.s.s.	12/30	85.0mg	8/30			
3946	1891	n.s.s. n.s.s.	42/56	24.0mg	31/31			Mizutani;clet,39,233-237;1988
3947	2084n	21.1mg n.s.s.	0/25	60.0mg	4/25			Diwan;carc,13,1893-1901;1992
3948	1448	25.7mg n.s.s.	0/10	25.0mg	0/12			Uchida;zkko,100,231-238;1981
a	1448	13.3mg n.s.s.	0/10	25.0mg	1/12			
3949	1448	20.5mg n.s.s.	0/10	20.0mg	0/12			
a	1448	15.3mg n.s.s.	4/10	20.0mg	1/12			
3950	1689	16.1mg n.s.s.	0/10	28.6mg	1/10			Diwan;jnci,75,1099-1105;1985
3951	1689	14.1mg n.s.s.	0/10	25.0mg	1/10			
3952	1690	11.0mg n.s.s.	0/13	25.0mg	2/12			Diwan;jnci,74,509-516;1985
3953	1803	38.2mg n.s.s.	2/28	14.0mg	2/28			Shivapurkar;carc,7,547-550;1986
3954	1834	20.6mg n.s.s.	0/20	20.0mg	0/20			Leonard;jnci,79,1313-1319;1987/pers.comm.
3955	1951	28.8mg n.s.s.	0/15	20.0mg	0/15			Diwan;carc,10,189-194;1989
3956	2084m	18.8mg n.s.s.	0/25	20.0mg	2/25			Diwan;carc,13,1893-1901;1992
3957	2109	11.0mg n.s.s.	0/27	20.0mg	2/15			Diwan;artx,66,413-422;1992
3958	1653	6.96mg n.s.s.	0/12	10.0mg	0/6			Carr;carc,5,1583-1590;1984
a	1653	6.96mg n.s.s.	0/12	10.0mg	0/6			
3959	1134	1.25mg n.s.s.	3/33	.286mg	1/30			Schmahl;zkko,86,77-84;1976
3960	1134	1.88mg n.s.s.	1/36	.286mg	0/32			
a	1134	n.s.s. n.s.s.	1/36	.286mg	1/32			
3961	2155	49.4mg n.s.s.	0/18	50.0mg	1/28			Kraupp-Grasl;canr,51,666-671;1991/pers.comm.
a	2155	5.07mg 20.4mg	1/18	50.0mg	20/28			

PHENOBARBITAL, SODIUM (phenobarbitone, sodium) 57-30-7

RefNum	LoConf	UpConf	Cntrl	1Dose	1Inc	2Dose	2Inc	Citation or Pathology
3962	1772	209.mg n.s.s.	7/50	95.8mg	6/30			Cavaliere;tumo,72,125-128;1986
a	1772	789.mg n.s.s.	0/50	95.8mg	0/30			
b	1772	274.mg n.s.s.	12/50	95.8mg	6/30			
3963	1772	219.mg n.s.s.	19/50	83.3mg	8/30			
a	1772	630.mg n.s.s.	0/50	83.3mg	0/30			
b	1772	246.mg n.s.s.	22/50	83.3mg	8/30			
3964	89	22.5mg 122.mg	10/44	65.0mg	21/28			Thorpe;fctx,11,433-442;1973
a	89	60.1mg 352.mg	0/44	65.0mg	9/28			
b	89	175.mg n.s.s.	27/44	65.0mg	9/28			
3965	237	65.2mg 145.mg	0/47	100.mg	45/73			Ponomarkov;clet,1,165-172;1976
a	237	338.mg n.s.s.	15/47	100.mg	21/73			
b	237	93.4mg n.s.s.	45/47	100.mg	65/73			
3966	89	18.1mg 86.1mg	11/45	60.0mg	24/30			Thorpe;fctx,11,433-442;1973
a	89	70.9mg 2.46gm	2/45	60.0mg	8/30			
b	89	86.7mg n.s.s.	27/45	60.0mg	15/30			
3967	237	41.6mg 109.mg	12/44	83.3mg	77/98			Ponomarkov;clet,1,165-172;1976
a	237	244.mg n.s.s.	20/44	83.3mg	40/98			
b	237	33.0mg n.s.s.	40/44	83.3mg	93/98			
3968	1035	81.4mg n.s.s.	0/15	50.0mg	1/18			Wislocki;canr,37,1883-1891;1977
3969	1783	60.5mg 908.mg	1/42	20.0mg	7/42	60.0mg	10/42	Imaida;canr,46,6160-6164;1986
a	1783	13.7mg n.s.s.	9/42	20.0mg	19/42	(60.0mg	11/42)	
3970	1942	35.9mg n.s.s.	0/15	40.0mg	0/15			Diwan;txap,98,269-277;1989
3971	389	33.6mg 177.mg	0/25	39.8mg	11/33			Butler;bjca,37,418-423;1978
3972	85	48.0mg 287.mg	0/32	28.6mg	9/29			Rossi;ijcn,19,179-185;1977
a	85	22.5mg n.s.s.	19/32	28.6mg	22/29			
3973	85	38.9mg 168.mg	0/35	25.0mg	13/36			
a	85	35.8mg n.s.s.	19/35	25.0mg	22/36			

PHENOL   108-95-2

RefNum	LoConf	UpConf	Cntrl	1Dose	1Inc	2Dose	2Inc	Citation or Pathology
3974	TR203	1.71gm n.s.s.	27/50	491.mg	21/50	981.mg	21/50	
a	TR203	7.62gm n.s.s.	3/50	491.mg	1/50	981.mg	1/50	liv:hpa,hpc,nnd.
b	TR203	3.60gm n.s.s.	1/50	491.mg	3/50	981.mg	2/50	lun:a/a,a/c.
3975	TR203	1.48gm n.s.s.	30/50	409.mg	28/50	818.mg	25/50	
a	TR203	523.mg n.s.s.	14/50	409.mg	19/50	(818.mg	9/50)	liv:hpa,hpc,nnd.
b	TR203	1.91gm n.s.s.	6/50	409.mg	5/50	818.mg	10/50	lun:a/a,a/c.
3976	TR203	257.mg n.s.s.	45/50	140.mg	45/50	280.mg	38/50	

```
      Spe Strain Site  Xpo+Xpt                                              TD50      2Tailpvl
         Sex   Route  Hist     Notes                                          DR       AuOp
 a    R f f34 wat liv MXB 24m24                                            no dre   P=1.
3977  R m f34 wat thy ccr 24m24                               :     +      : #420.mg \ P<.007 -
 a    R m f34 wat --- MXA 24m24                                            133.mg \ P<.03
 b    R m f34 wat --- leu 24m24                                            143.mg \ P<.04
 c    R m f34 wat --- mle 24m24                                            143.mg \ P<.04
 d    R m f34 wat TBA MXB 24m24                                            no dre   P=1.
 e    R m f34 wat liv MXB 24m24                                            no dre   P=1.

PHENOTHIAZINE                    100ng..:..1ug...:..10....:...100...:..1mg..:..10....:..100...:..1g.....:..10
3978  M f b6a orl lun ade 76w76 evx                              .>       no dre   P=1.   -
 a    M f b6a orl liv hpt 76w76 evx                                       no dre   P=1.   -
 b    M f b6a orl tba mix 76w76 evx                                       no dre   P=1.   -
3979  M m b6a orl liv hpt 76w76 evx                          .>          393.mg   P<.6   -
 a    M m b6a orl lun ade 76w76 evx                                       no dre   P=1.   -
 b    M m b6a orl tba mix 76w76 evx                                       311.mg   P<.7   -
3980  M f b6c orl liv hpt 76w76 evx                            .>        no dre   P=1.   -
 a    M f b6c orl lun mix 76w76 evx                                       no dre   P=1.   -
 b    M f b6c orl tba tum 76w76 evx                                       no dre   P=1.   -
3981  M m b6c orl liv hpt 76w76 evx                          .>          414.mg   P<.3   -
 a    M m b6c orl lun mix 76w76 evx                                       no dre   P=1.   -
 b    M m b6c orl tba mix 76w76 evx                                       200.mg   P<.09  -

PHENOXYBENZAMINE.HC1            100ng..:..1ug...:..10....:...100...:..1mg..:..10....:..100...:..1g.....:..10
3982  M f b6c ipj abc srn 52w84 s                      :   +   :          5.85mg / P<.0005c
 a    M f b6c ipj TBA MXB 52w84 s                                         3.08mg / P<.0005
 b    M f b6c ipj liv MXB 52w84 s                                         no dre   P=1.
 c    M f b6c ipj lun MXB 52w84 s                                         no dre   P=1.
3983  M m b6c ipj abc srn 51w83 as                     :   +   :          4.95mg / P<.0005c
 a    M m b6c ipj TBA MXB 51w83 as                                        2.17mg / P<.0005
 b    M m b6c ipj liv MXB 51w83 as                                        5.84mg * P<.2
 c    M m b6c ipj lun MXB 51w83 as                                        63.0mg * P<.9
3984  R f sda ipj per srn 52w83                        :   +   :          2.36mg / P<.0005c
 a    R f sda ipj TBA MXB 52w83                                           1.34mg * P<.05
 b    R f sda ipj liv MXB 52w83                                           no dre   P=1.
3985  R m sda ipj per srn 52w83 s                    :   +   :            .710mg / P<.0005c
 a    R m sda ipj TBA MXB 52w83 s                                         .555mg / P<.0005
 b    R m sda ipj liv MXB 52w83 s                                         no dre   P=1.

1-PHENYL-3,3-DIMETHYLTRIAZENE   100ng..:..1ug...:..10....:...100...:..1mg..:..10....:..100...:..1g.....:..10
3986  R b bdr gav bra mix 24m24                       .   +   .           5.81mg * P<.0005+
 a    R b bdr gav tba mal 24m24                                           2.31mg * P<.0005+

PHENYL ISOTHIOCYANATE           100ng..:..1ug...:..10....:...100...:..1mg..:..10....:..100...:..1g.....:..10
3987  M f b6a orl liv hem 76w76 evx                          .>          138.mg   P<.3   -
 a    M f b6a orl lun ade 76w76 evx                                       no dre   P=1.   -
 b    M f b6a orl tba mix 76w76 evx                                       39.4mg   P<.3   -
3988  M m b6a orl lun ade 76w76 evx                           .>         no dre   P=1.   -
 a    M m b6a orl liv hpt 76w76 evx                                       no dre   P=1.   -
 b    M m b6a orl tba mix 76w76 evx                                       no dre   P=1.   -
3989  M f b6c orl liv hpt 76w76 evx                            .>        no dre   P=1.   -
 a    M f b6c orl lun mix 76w76 evx                                       no dre   P=1.   -
 b    M f b6c orl tba tum 76w76 evx                                       no dre   P=1.   -
3990  M m b6c orl liv hpt 76w76 evx                           .>         128.mg   P<.3   -
 a    M m b6c orl lun ade 76w76 evx                                       128.mg   P<.3   -
 b    M m b6c orl tba mix 76w76 evx                                       40.2mg   P<.05  -

1-PHENYL-3-METHYL-5-PYRAZOLONE  100ng..:..1ug...:..10....:...100...:..1mg..:..10....:..100...:..1g.....:..10
3991  M f b6c eat TBA MXB 24m24                                  :>      no dre   P=1.   -
 a    M f b6c eat liv MXB 24m24                                           no dre   P=1.
 b    M f b6c eat lun MXB 24m24                                           no dre   P=1.
3992  M m b6c eat TBA MXB 24m24                                   :>     no dre   P=1.   -
 a    M m b6c eat liv MXB 24m24                                           no dre   P=1.
 b    M m b6c eat lun MXB 24m24                                           no dre   P=1.
3993  R f f34 eat TBA MXB 24m24                                :>        no dre   P=1.   -
 a    R f f34 eat liv MXB 24m24                                           no dre   P=1.
3994  R m f34 eat TBA MXB 24m24                                :>        no dre   P=1.   -
 a    R m f34 eat liv MXB 24m24                                           3.70gm * P<.5

PHENYL-beta-NAPHTHYLAMINE       100ng..:..1ug...:..10....:...100...:..1mg..:..10....:..100...:..1g.....:..10
3995  H f syg gav liv tum 84w84 e                             .>         no dre   P=1.
 a    H f syg gav lun tum 84w84 e                                         no dre   P=1.
 b    H f syg gav tba mix 84w84 e                                         no dre   P=1.   -
3996  H m syg gav liv tum 84w84 e                              .>        no dre   P=1.
 a    H m syg gav lun tum 84w84 e                                         no dre   P=1.
 b    H m syg gav tba mix 84w84 e                                         no dre   P=1.   -
3997  M f b6c eat kid MXA 24m24                                      :   10.8gm * P<.1   e
 a    M f b6c eat TBA MXB 24m24                                           no dre   P=1.
 b    M f b6c eat liv MXB 24m24                                           5.63gm * P<.4
 c    M f b6c eat lun MXB 24m24                                           no dre   P=1.
3998  M f b6c orl liv hpt 76w76 evx                          .>          1.08gm   P<.3
 a    M f b6c orl lun mix 76w76 evx                                       no dre   P=1.
```

	RefNum	LoConf	UpConf	Cntrl	1Dose	1Inc	2Dose	2Inc	Citation or Pathology
									Brkly Code
a	TR203	866.mg	n.s.s.	4/50	140.mg	1/50	(280.mg	0/50)	liv:hpa,hpc,nnd.
3977	TR203	158.mg	5.60gm	0/50	123.mg	5/50	(245.mg	1/50)	S
a	TR203	57.0mg	n.s.s.	18/50	123.mg	31/50	(245.mg	25/50)	---:leu,lym. S
b	TR203	59.1mg	n.s.s.	18/50	123.mg	30/50	(245.mg	25/50)	S
c	TR203	59.1mg	n.s.s.	18/50	123.mg	30/50	(245.mg	24/50)	S
d	TR203	189.mg	n.s.s.	40/50	123.mg	44/50	245.mg	38/50	
e	TR203	716.mg	n.s.s.	5/50	123.mg	4/50	245.mg	4/50	liv:hpa,hpc,nnd.

PHENOTHIAZINE   92-84-2

	RefNum	LoConf	UpConf	Cntrl	1Dose	1Inc	Citation
3978	1254	87.3mg	n.s.s.	1/17	78.4mg	1/17	Innes;ntis,1968/1969
a	1254	147.mg	n.s.s.	0/17	78.4mg	0/17	
b	1254	68.9mg	n.s.s.	2/17	78.4mg	2/17	
3979	1254	57.3mg	n.s.s.	1/18	73.0mg	2/17	
a	1254	89.8mg	n.s.s.	2/18	73.0mg	1/17	
b	1254	41.9mg	n.s.s.	3/18	73.0mg	4/17	
3980	1254	155.mg	n.s.s.	0/16	78.4mg	0/18	
a	1254	155.mg	n.s.s.	0/16	78.4mg	0/18	
b	1254	155.mg	n.s.s.	0/16	78.4mg	0/18	
3981	1254	67.4mg	n.s.s.	0/16	73.0mg	1/16	
a	1254	129.mg	n.s.s.	0/16	73.0mg	0/16	
b	1254	49.1mg	n.s.s.	0/16	73.0mg	2/16	

PHENOXYBENZAMINE.HCl   63-92-3

	RefNum	LoConf	UpConf	Cntrl	1Dose	1Inc	2Dose	2Inc	Pathology
3982	TR72	3.11mg	12.0mg	0/15	3.30mg	0/35	10.3mg	16/35	
a	TR72	1.59mg	8.85mg	5/15	3.30mg	7/35	10.3mg	16/35	
b	TR72	7.21mg	n.s.s.	1/15	3.30mg	1/35	10.3mg	0/35	liv:hpa,hpc,nnd.
c	TR72	7.21mg	n.s.s.	1/15	3.30mg	1/35	10.3mg	0/35	lun:a/a,a/c.
3983	TR72	2.63mg	9.91mg	0/15	3.40mg	0/35	10.7mg	17/35	
a	TR72	1.17mg	4.88mg	3/15	3.40mg	7/35	10.7mg	18/35	
b	TR72	1.98mg	n.s.s.	1/15	3.40mg	6/35	10.7mg	0/35	liv:hpa,hpc,nnd.
c	TR72	3.24mg	n.s.s.	2/15	3.40mg	3/35	10.7mg	0/35	lun:a/a,a/c.
3984	TR72	1.30mg	5.16mg	0/10	1.30mg	0/35	2.70mg	16/35	
a	TR72	.610mg	n.s.s.	5/10	1.30mg	20/35	2.70mg	20/35	
b	TR72	n.s.s.	n.s.s.	0/10	1.30mg	0/35	2.70mg	0/35	liv:hpa,hpc,nnd.
3985	TR72	.403mg	1.31mg	0/10	1.30mg	11/35	3.80mg	16/35	
a	TR72	.303mg	1.35mg	3/10	1.30mg	15/35	3.80mg	16/35	
b	TR72	n.s.s.	n.s.s.	0/10	1.30mg	0/35	3.80mg	0/35	liv:hpa,hpc,nnd.

1-PHENYL-3,3-DIMETHYLTRIAZENE   7227-91-0

	RefNum	LoConf	UpConf	Cntrl	1Dose	1Inc	2Dose	2Inc	Citation
3986	1467	2.65mg	16.1mg	0/120	3.57mg	1/7	7.14mg	8/12	Preussmann;zkko,81,285-310;1974
a	1467	1.10mg	5.34mg	2/120	3.57mg	4/7	7.14mg	11/12	

PHENYL ISOTHIOCYANATE   103-72-0

	RefNum	LoConf	UpConf	Cntrl	1Dose	1Inc	Citation
3987	1304	22.5mg	n.s.s.	0/17	21.6mg	1/18	Innes;ntis,1968/1969
a	1304	25.6mg	n.s.s.	1/17	21.6mg	1/18	
b	1304	10.3mg	n.s.s.	2/17	21.6mg	5/18	
3988	1304	21.4mg	n.s.s.	2/18	20.0mg	1/15	
a	1304	33.1mg	n.s.s.	1/18	20.0mg	0/15	
b	1304	22.9mg	n.s.s.	3/18	20.0mg	1/15	
3989	1304	42.7mg	n.s.s.	0/16	21.6mg	0/18	
a	1304	42.7mg	n.s.s.	0/16	21.6mg	0/18	
b	1304	42.7mg	n.s.s.	0/16	21.6mg	0/18	
3990	1304	20.9mg	n.s.s.	0/16	20.0mg	1/18	
a	1304	20.9mg	n.s.s.	0/16	20.0mg	1/18	
b	1304	12.1mg	n.s.s.	0/16	20.0mg	3/18	

1-PHENYL-3-METHYL-5-PYRAZOLONE   89-25-8

	RefNum	LoConf	UpConf	Cntrl	1Dose	1Inc	2Dose	2Inc	Pathology
3991	TR141	2.85gm	n.s.s.	9/20	956.mg	19/50	1.91gm	15/50	
a	TR141	14.2gm	n.s.s.	2/20	956.mg	2/50	1.91gm	0/50	liv:hpa,hpc,nnd.
b	TR141	8.43gm	n.s.s.	1/20	956.mg	3/50	1.91gm	1/50	lun:a/a,a/c.
3992	TR141	1.38gm	n.s.s.	11/20	883.mg	20/50	(1.77gm	14/50)	
a	TR141	2.85gm	n.s.s.	8/20	883.mg	8/50	(1.77gm	6/50)	liv:hpa,hpc,nnd.
b	TR141	3.62gm	n.s.s.	3/20	883.mg	3/50	(1.77gm	0/50)	lun:a/a,a/c.
3993	TR141	115.mg	n.s.s.	12/20	123.mg	36/50	(245.mg	25/50)	
a	TR141	n.s.s.	n.s.s.	0/20	123.mg	0/50	245.mg	0/50	liv:hpa,hpc,nnd.
3994	TR141	192.mg	n.s.s.	14/20	98.1mg	32/49	196.mg	31/50	
a	TR141	907.mg	n.s.s.	0/20	98.1mg	1/49	196.mg	1/50	liv:hpa,hpc,nnd.

PHENYL-beta-NAPHTHYLAMINE   (agerite powder, N-phenyl-2-naphthylamine)   135-88-6

	RefNum	LoConf	UpConf	Cntrl	1Dose	1Inc	2Dose	2Inc	Citation or Pathology
3995	1151	202.mg	n.s.s.	0/40	37.5mg	0/40			Green;zkko,95,51-55;1979
a	1151	202.mg	n.s.s.	0/40	37.5mg	0/40			
b	1151	72.3mg	n.s.s.	8/40	37.5mg	6/40			
3996	1151	202.mg	n.s.s.	0/40	37.5mg	0/40			
a	1151	202.mg	n.s.s.	0/40	37.5mg	0/40			
b	1151	84.5mg	n.s.s.	3/40	37.5mg	3/40			
3997	TR333	2.66gm	n.s.s.	0/50	322.mg	0/50	644.mg	2/50	kid:tla,uac.
a	TR333	830.mg	n.s.s.	42/50	322.mg	23/50	644.mg	33/50	
b	TR333	1.41gm	n.s.s.	4/50	322.mg	3/50	644.mg	7/50	liv:hpa,hpc,nnd.
c	TR333	3.31gm	n.s.s.	5/50	322.mg	1/50	644.mg	3/50	lun:a/a,a/c.
3998	1153	176.mg	n.s.s.	0/16	169.mg	1/18			Innes;ntis,1968/1969
a	1153	334.mg	n.s.s.	0/16	169.mg	0/18			

Spe Strain Site	Xpo+Xpt		TD50	2Tailpvl
Sex Route Hist	Notes		DR	AuOp
b M f b6c orl tba mix 76w76	evx		1.08gm	P<.3
3999 M m b6c eat sub MXA 24m24		: ±	#2.38gm *	P<.04 -
a M m b6c eat TBA MXB 24m24			4.51gm *	P<.8
b M m b6c eat liv MXB 24m24			1.35gm *	P<.2
c M m b6c eat lun MXB 24m24			no dre	P=1.
4000 M m b6c orl liv hpt 76w76	evx	. + .	165.mg	P<.007
a M m b6c orl lun mix 76w76	evx		no dre	P=1.
b M m b6c orl tba mix 76w76	evx		108.mg	P<.002
4001 M f b6a orl lun ade 76w76	evx	.>	no dre	P=1.
a M f b6a orl liv hpt 76w76	evx		no dre	P=1.
b M f b6a orl tba mix 76w76	evx		1.08gm	P<.7
4002 M m b6a orl lun ade 76w76	evx	.>	431.mg	P<.4
a M m b6a orl liv hpt 76w76	evx		460.mg	P<.3
b M m b6a orl tba mix 76w76	evx		186.mg	P<.2
4003 R f f34 eat TBA MXB 24m24		:>	no dre	P=1. -
a R f f34 eat liv MXB 24m24			no dre	P=1.
4004 R m f34 eat TBA MXB 24m24		:>	no dre	P=1.
a R m f34 eat liv MXB 24m24			no dre	P=1.
4005 R f sda gav liv tum 32m32	e	.>	no dre	P=1. -
a R f sda gav tba mix 32m32	e		no dre	P=1. -
4006 R m sda gav liv tum 37m37	e	.>	no dre	P=1.
a R m sda gav tba mix 37m37	e		no dre	P=1.

N-PHENYL-p-PHENYLENEDIAMINE.HCl   100ng..:..1ug....:..10.......:..100.....:..1mg...:..10.....:..100.....:..1g.....:..10

4007 M f b6c eat TBA MXB 48w91 sv		:>	no dre	P=1. -
a M f b6c eat liv MXB 48w91 sv			no dre	P=1.
b M f b6c eat lun MXB 48w91 sv			no dre	P=1.
4008 M m b6c eat TBA MXB 48w91 sv		:>	no dre	P=1. -
a M m b6c eat liv MXB 48w91 sv			213.mg \	P<.06
b M m b6c eat lun MXB 48w91 sv			no dre	P=1.
4009 R f f34 eat TBA MXB 18m24		:>	no dre	P=1. -
a R f f34 eat liv MXB 18m24			no dre	P=1.
4010 R m f34 eat TBA MXB 18m24		:>	161.mg *	P<.5 -
a R m f34 eat liv MXB 18m24			no dre	P=1.

1-PHENYL-2-THIOUREA   100ng..:..1ug....:..10.......:..100.....:..1mg...:..10.....:..100.....:..1g.....:..10

4011 M f b6c eat TBA MXB 78w91		:>	133.mg *	P<.4 -
a M f b6c eat liv MXB 78w91			268.mg *	P<.06
b M f b6c eat lun MXB 78w91			134.mg *	P<.07
4012 M m b6c eat TBA MXB 78w91		:>	no dre	P=1. -
a M m b6c eat liv MXB 78w91			no dre	P=1.
b M m b6c eat lun MXB 78w91			no dre	P=1.
4013 R f f34 eat TBA MXB 18m24		:>	no dre	P=1. -
a R f f34 eat liv MXB 18m24			no dre	P=1.
4014 R m f34 eat TBA MXB 18m24		:>	no dre	P=1. -
a R m f34 eat liv MXB 18m24			no dre	P=1.

1-PHENYLAZO-2-NAPHTHOL   100ng..:..1ug....:..10.......:..100.....:..1mg...:..10.....:..100.....:..1g.....:..10

4015 M f b6c eat --- lym 24m24		: ±	#128.mg \	P<.03 -
a M f b6c eat TBA MXB 24m24			391.mg *	P<.4
b M f b6c eat liv MXB 24m24			833.mg *	P<.2
c M f b6c eat lun MXB 24m24			3.03gm *	P<.8
4016 M m b6c eat TBA MXB 24m24		: ±	154.mg *	P<.04 -
a M m b6c eat liv MXB 24m24			587.mg *	P<.4
b M m b6c eat lun MXB 24m24			1.08gm *	P<.5
4017 M f cba eat liv hpt 12m24		.>	no dre	P=1. -
4018 M m cba eat liv hpt 12m24		.>	no dre	P=1. -
4019 R f f34 eat liv MXA 24m24		: + :	86.5mg *	P<.01 c
a R f f34 eat liv nnd 24m24			96.6mg *	P<.02 c
b R f f34 eat sub fib 24m24			346.mg *	P<.05
c R f f34 eat TBA MXB 24m24			no dre	P=1.
d R f f34 eat liv MXB 24m24			86.5mg *	P<.01
4020 R m f34 eat liv nnd 24m24		: + :	17.7mg /	P<.0005c
a R m f34 eat liv MXA 24m24			17.9mg /	P<.0005c
b R m f34 eat TBA MXB 24m24			106.mg *	P<.8
c R m f34 eat liv MXB 24m24			17.9mg /	P<.0005

PHENYLBUTAZONE   100ng..:..1ug....:..10.......:..100.....:..1mg...:..10.....:..100.....:..1g.....:..10

4021 M f b6c gav TBA MXB 24m24		:>	no dre	P=1. -
a M f b6c gav liv MXB 24m24			6.01gm *	P<.9
b M f b6c gav lun MXB 24m24			no dre	P=1.
4022 M m b6c gav liv hpa 24m24		: + :	323.mg *	P<.007
a M m b6c gav liv MXA 24m24			353.mg /	P<.03 p
b M m b6c gav TBA MXB 24m24			434.mg *	P<.3
c M m b6c gav liv MXB 24m24			353.mg /	P<.03
d M m b6c gav lun MXB 24m24			no dre	P=1.
4023 R f f34 gav kid tcc 24m24		: ± +hist	1.16gm *	P<.09 p
a R f f34 gav TBA MXB 24m24			1.58gm *	P<1.
b R f f34 gav liv MXB 24m24			no dre	P=1.
4024 R m f34 gav kid MXA 24m24		: ±	458.mg *	P<.04 e
a R m f34 gav kid MXA 24m24			621.mg *	P<.04

	RefNum	LoConf	UpConf	Cntrl	1Dose	1Inc	2Dose	2Inc	Citation or Pathology
									Brkly Code
b	1153	176.mg	n.s.s.	0/16	169.mg	1/18			
3999	TR333	992.mg	n.s.s.	2/50	297.mg	4/50	594.mg	8/50	sub:fbs,nfs,srn. S
a	TR333	437.mg	n.s.s.	34/50	297.mg	30/50	594.mg	32/50	
b	TR333	487.mg	n.s.s.	11/50	297.mg	16/50	594.mg	17/50	liv:hpa,hpc,nnd.
c	TR333	1.36gm	n.s.s.	11/50	297.mg	9/50	594.mg	7/50	lun:a/a,a/c.
4000	1153	62.2mg	2.23gm	0/16	157.mg	5/17			Innes;ntis,1968/1969
a	1153	294.mg	n.s.s.	0/16	157.mg	0/17			
b	1153	46.1mg	417.mg	0/16	157.mg	7/17			
4001	1153	201.mg	n.s.s.	1/17	169.mg	1/18			
a	1153	334.mg	n.s.s.	0/17	169.mg	0/18			
b	1153	121.mg	n.s.s.	2/17	169.mg	3/18			
4002	1153	90.1mg	n.s.s.	2/18	157.mg	4/18			
a	1153	103.mg	n.s.s.	1/18	157.mg	3/18			
b	1153	57.2mg	n.s.s.	3/18	157.mg	7/18			
4003	TR333	175.mg	n.s.s.	46/50	124.mg	38/50	(248.mg	33/50)	
a	TR333	3.10gm	n.s.s.	2/50	124.mg	0/50	248.mg	0/50	liv:hpa,hpc,nnd.
4004	TR333	108.mg	n.s.s.	48/50	99.0mg	40/50	(198.mg	42/50)	
a	TR333	1.11gm	n.s.s.	2/50	99.0mg	3/50	198.mg	0/50	liv:hpa,hpc,nnd.
4005	1524	2.42gm	n.s.s.	0/40	171.mg	0/40			Ketkar;clet,16,203-206;1982
a	1524	2.00gm	n.s.s.	30/40	171.mg	3/40			
4006	1524	3.34gm	n.s.s.	0/40	171.mg	0/40			
a	1524	1.80gm	n.s.s.	24/40	171.mg	8/40			

N-PHENYL-p-PHENYLENEDIAMINE.HCl   2198-59-6

	RefNum	LoConf	UpConf	Cntrl	1Dose	1Inc	2Dose	2Inc	Citation or Pathology
4007	TR82	648.mg	n.s.s.	5/20	252.mg	4/50	(478.mg	2/50)	
a	TR82	1.53gm	n.s.s.	1/20	252.mg	2/50	478.mg	1/50	liv:hpa,hpc,nnd.
b	TR82	n.s.s.	n.s.s.	0/20	252.mg	0/50	478.mg	0/50	lun:a/a,a/c.
4008	TR82	404.mg	n.s.s.	10/20	131.mg	25/49	260.mg	15/50	
a	TR82	96.5mg	n.s.s.	2/20	131.mg	18/49	(260.mg	10/50)	liv:hpa,hpc,nnd.
b	TR82	818.mg	n.s.s.	4/20	131.mg	4/49	260.mg	5/50	lun:a/a,a/c.
4009	TR82	68.6mg	n.s.s.	13/20	22.5mg	26/50	45.0mg	24/50	
a	TR82	n.s.s.	n.s.s.	0/20	22.5mg	0/50	45.0mg	0/50	liv:hpa,hpc,nnd.
4010	TR82	40.5mg	n.s.s.	6/20	18.0mg	17/50	36.0mg	20/50	
a	TR82	n.s.s.	n.s.s.	0/20	18.0mg	0/50	36.0mg	0/50	liv:hpa,hpc,nnd.

1-PHENYL-2-THIOUREA   103-85-5

	RefNum	LoConf	UpConf	Cntrl	1Dose	1Inc	2Dose	2Inc	Citation or Pathology
4011	TR148	35.3mg	n.s.s.	4/20	16.9mg	11/50	33.8mg	15/50	
a	TR148	92.7mg	n.s.s.	0/20	16.9mg	0/50	33.8mg	4/50	liv:hpa,hpc,nnd.
b	TR148	60.5mg	n.s.s.	0/20	16.9mg	3/50	33.8mg	5/50	lun:a/a,a/c.
4012	TR148	49.0mg	n.s.s.	6/20	15.6mg	16/50	31.2mg	12/50	
a	TR148	74.6mg	n.s.s.	1/20	15.6mg	7/50	31.2mg	3/50	liv:hpa,hpc,nnd.
b	TR148	73.7mg	n.s.s.	3/20	15.6mg	6/50	31.2mg	6/50	lun:a/a,a/c.
4013	TR148	2.46mg	n.s.s.	12/20	2.30mg	37/50	(4.50mg	30/50)	
a	TR148	n.s.s.	n.s.s.	0/20	2.30mg	0/50	4.50mg	0/50	liv:hpa,hpc,nnd.
4014	TR148	2.71mg	n.s.s.	7/20	1.80mg	22/50	(3.60mg	22/50)	
a	TR148	17.3mg	n.s.s.	0/20	1.80mg	2/50	3.60mg	1/50	liv:hpa,hpc,nnd.

1-PHENYLAZO-2-NAPHTHOL   (C.I. solvent yellow 14) 842-07-9

	RefNum	LoConf	UpConf	Cntrl	1Dose	1Inc	2Dose	2Inc	Citation or Pathology
4015	TR226	55.6mg	n.s.s.	9/50	63.8mg	23/50	(128.mg	17/50)	S
a	TR226	96.1mg	n.s.s.	28/50	63.8mg	34/50	128.mg	36/50	
b	TR226	279.mg	n.s.s.	2/50	63.8mg	4/50	128.mg	6/50	liv:hpa,hpc,nnd.
c	TR226	342.mg	n.s.s.	3/50	63.8mg	6/50	128.mg	4/50	lun:a/a,a/c.
4016	TR226	68.5mg	n.s.s.	24/50	58.9mg	30/50	118.mg	37/50	
a	TR226	159.mg	n.s.s.	15/50	58.9mg	11/50	118.mg	19/50	liv:hpa,hpc,nnd.
b	TR226	250.mg	n.s.s.	5/50	58.9mg	7/50	118.mg	7/50	lun:a/a,a/c.
4017	1165	348.mg	n.s.s.	0/18	64.9mg	0/26			Clayson;bjca,19,297-310;1965/Williams 1962
4018	1165	150.mg	n.s.s.	2/15	59.9mg	1/18			
4019	TR226	40.2mg	4.20gm	2/50	12.4mg	3/49	24.8mg	11/50	liv:hpc,nnd.
a	TR226	43.2mg	n.s.s.	2/50	12.4mg	3/49	24.8mg	10/50	
b	TR226	104.mg	n.s.s.	0/50	12.4mg	0/49	24.8mg	3/50	
c	TR226	24.4mg	n.s.s.	44/50	12.4mg	41/49	24.8mg	38/50	S
d	TR226	40.2mg	4.20gm	2/50	12.4mg	3/49	24.8mg	11/50	liv:hpa,hpc,nnd.
4020	TR226	10.4mg	50.0mg	5/50	9.90mg	10/50	19.8mg	30/50	
a	TR226	10.3mg	55.3mg	6/50	9.90mg	10/50	19.8mg	31/50	liv:hpc,nnd.
b	TR226	13.5mg	n.s.s.	36/50	9.90mg	33/50	19.8mg	45/50	
c	TR226	10.3mg	55.3mg	6/50	9.90mg	10/50	19.8mg	31/50	liv:hpa,hpc,nnd.

PHENYLBUTAZONE   50-33-9

	RefNum	LoConf	UpConf	Cntrl	1Dose	1Inc	2Dose	2Inc	Citation or Pathology
4021	TR367	323.mg	n.s.s.	31/50	105.mg	20/50	210.mg	25/50	
a	TR367	405.mg	n.s.s.	5/50	105.mg	5/50	210.mg	7/50	liv:hpa,hpc,nnd.
b	TR367	704.mg	n.s.s.	3/50	105.mg	3/50	210.mg	3/50	lun:a/a,a/c.
4022	TR367	160.mg	4.75gm	8/50	105.mg	12/50	210.mg	23/50	S
a	TR367	157.mg	n.s.s.	16/50	105.mg	14/50	210.mg	31/50	liv:hpa,hpc.
b	TR367	136.mg	n.s.s.	30/50	105.mg	28/50	210.mg	40/50	
c	TR367	157.mg	n.s.s.	16/50	105.mg	14/50	210.mg	31/50	liv:hpa,hpc,nnd.
d	TR367	616.mg	n.s.s.	6/50	105.mg	6/50	210.mg	5/50	lun:a/a,a/c.
4023	TR367	276.mg	n.s.s.	0/50	34.9mg	0/50	69.7mg	2/50	
a	TR367	40.3mg	n.s.s.	46/50	34.9mg	46/50	69.7mg	36/50	
b	TR367	339.mg	n.s.s.	1/50	34.9mg	2/50	69.7mg	0/50	liv:hpa,hpc,nnd.
4024	TR367	158.mg	n.s.s.	0/50	35.0mg	1/50	70.1mg	3/50	kid:rab,rua,ruc.
a	TR367	187.mg	n.s.s.	0/50	35.0mg	0/50	70.1mg	3/50	kid:rab,rua. S

```
      Spe Strain  Site   Xpo+Xpt                                                      TD50    2Tailpvl
          Sex  Route   Hist    Notes                                                     DR      AuOp
   b  R m f34 gav TBA MXB 24m24                                                       121.mg * P<.4
   c  R m f34 gav liv MXB 24m24                                                       1.09gm * P<.3
4025 R m f34 gav kid MXA 24m24    with step                               .  ±        382.mg * P<.02  e
4026 R f don eat adr phe 24m26 e                                          .    ±      707.mg * P<.03  -
   a  R f don eat liv nnd 24m26 e                                                     1.14gm * P<.06  -
   b  R f don eat kid tum 24m26 e                                                     no dre  P=1.
   c  R f don eat tba mix 24m26 e                                                     no dre  P=1.
4027 R m don eat kid tum 24m26 e                                                .>    no dre  P=1.
   a  R m don eat liv hpc 24m26 e                                                     no dre  P=1.
   b  R m don eat liv nnd 24m26 e                                                     no dre  P=1.
   c  R m don eat tba mix 24m26 e                                                     no dre  P=1.

m-PHENYLENEDIAMINE        100ng..:..1ug...:..10......:..100....:..1mg..:..:..10......:..100....:..1g.....:..10
4028 M f b6c wat liv hem 78w84 e                                             .>       916.mg * P<.2
   a  M f b6c wat liv hct 78w84 e                                                     no dre  P=1.
   b  M f b6c wat liv hnd 78w84 e                                                     no dre  P=1.
   c  M f b6c wat lun ade 78w84 e                                                     no dre  P=1.
4029 M m b6c wat liv hem 78w84 e                                                .>    2.73gm * P<.9
   a  M m b6c wat liv hct 78w84 e                                                     no dre  P=1.
   b  M m b6c wat liv hnd 78w84 e                                                     no dre  P=1.
   c  M m b6c wat lun ade 78w84 e                                                     no dre  P=1.

p-PHENYLENEDIAMINE        100ng..:..1ug...:..10......:..100....:..1mg..:..:..10......:..100....:..1g.....:..10
4030 R f f34 eat adr phe 80w80 e                                                .>    4.28gm * P<1.  -
   a  R f f34 eat liv tum 80w80 e                                                     no dre  P=1.
4031 R m f34 eat adr phe 80w80 e                                          .>          no dre  P=1.
   a  R m f34 eat liv tum 80w80 e                                                     no dre  P=1.

m-PHENYLENEDIAMINE.2HCl   100ng..:..1ug...:..10......:..100....:..1mg..:..:..10......:..100....:..1g.....:..10
4032 M f chi eat lun mix 64w77 a                                     :     ±          253.mg * P<.03  -
   a  M f chi eat liv mix 64w77 a                                                     1.17gm * P<.3
   b  M f chi eat tba mix 64w77 a                                                     370.mg * P<.3
4033 M m chi eat lun mix 64w77 a                                       :       ±      421.mg * P<.08  -
   a  M m chi eat liv mix 64w77 a                                                     889.mg * P<.3
   b  M m chi eat tba mix 64w77 a                                                     307.mg * P<.3
4034 R m cdr eat liv mix 18m25                                            :>          1.69gm * P<.9  -
   a  R m cdr eat tba mix 18m25                                                       114.mg * P<.3  -

o-PHENYLENEDIAMINE.2HCl   100ng..:..1ug...:..10......:..100....:..1mg..:..:..10......:..100....:..1g.....:..10
4035 M f chi eat liv mix 77w90 v                                         :  +  :      507.mg * P<.002
   a  M f chi eat liv hpt 77w90 v                                                     611.mg * P<.002 +
   b  M f chi eat lun mix 77w90 v                                                     2.26gm * P<.3  -
   c  M f chi eat tba mix 77w90 v                                                     676.mg * P<.07
4036 M f chi eat liv hpt 77w90 v   pool                                   :  +  :     611.mg * P<.0005+
4037 M m chi eat liv hpt 77w86 v                                            :   ±     921.mg * P<.04  +
   a  M m chi eat liv mix 77w86 v                                                     921.mg * P<.04  +
   b  M m chi eat lun mix 77w86 v                                                     3.21gm * P<.6
   c  M m chi eat tba mix 77w86 v                                                     1.97gm * P<.5
4038 M m chi eat liv hpt 77w86 v   pool                                   :   +   :   1.11gm * P<.0005+
4039 R m cdr eat liv hpt 18m25                                        :    ±          248.mg * P<.02  +
   a  R m cdr eat liv mix 18m25                                        :    ±          248.mg * P<.02  +
   b  R m cdr eat tba mix 18m25                                                       no dre  P=1.
4040 R m cdr eat liv hpt 18m25     pool                               :    ±          315.mg * P<.02  +

p-PHENYLENEDIAMINE.2HCl   100ng..:..1ug...:..10......:..100....:..1mg..:..:..10......:..100....:..1g.....:..10
4041 M f b6c eat TBA MXB 24m24                                              :>        1.50gm * P<.8  -
   a  M f b6c eat liv MXB 24m24                                                       1.37gm * P<.5
   b  M f b6c eat lun MXB 24m24                                                       1.39gm * P<.2
4042 M m b6c eat TBA MXB 24m24                                              :>        no dre  P=1.  -
   a  M m b6c eat liv MXB 24m24                                                       no dre  P=1.
   b  M m b6c eat lun MXB 24m24                                                       no dre  P=1.
4043 R f f34 eat TBA MXB 24m27                                        :>              247.mg * P<.6  -
   a  R f f34 eat liv MXB 24m27                                                       no dre  P=1.
4044 R m f34 eat TBA MXB 24m27                                        :>              no dre  P=1.  -
   a  R m f34 eat liv MXB 24m27                                                       2.23gm * P<.4

PHENYLEPHRINE.HCl         100ng..:..1ug...:..10......:..100....:..1mg..:..:..10......:..100....:..1g.....:..10
4045 M f b6c eat TBA MXB 24m24                                              :>        911.mg * P<.4  -
   a  M f b6c eat liv MXB 24m24                                                       7.85gm * P<.7
   b  M f b6c eat lun MXB 24m24                                                       1.61gm * P<.2
4046 M m b6c eat sub fbs 24m24                                            :    ±      #1.64gm * P<.04  -
   a  M m b6c eat MXA mlm 24m24                                                       1.72gm * P<.02
   b  M m b6c eat TBA MXB 24m24                                                       6.33gm * P<1.
   c  M m b6c eat liv MXB 24m24                                                       4.27gm * P<.8
   d  M m b6c eat lun MXB 24m24                                                       2.13gm * P<.5
4047 R f f34 eat TBA MXB 24m24                                        :>              272.mg * P<.7  -
   a  R f f34 eat liv MXB 24m24                                                       no dre  P=1.
4048 R m f34 eat TBA MXB 24m24                                        :>              no dre  P=1.  -
   a  R m f34 eat liv MXB 24m24                                                       no dre  P=1.
```

	RefNum	LoConf	UpConf	Cntrl	1Dose	1Inc	2Dose	2Inc	Citation or Pathology
									Brkly Code
b	TR367	31.1mg	n.s.s.	41/50	35.0mg	42/50	70.1mg	44/50	
c	TR367	261.mg	n.s.s.	0/50	35.0mg	1/50	70.1mg	1/50	liv:hpa,hpc,nnd.
4025	TR367	180.mg	n.s.s.	0/50	35.0mg	5/50	70.1mg	4/50	kid:rab,rua,ruc.
4026	1815	283.mg	n.s.s.	4/96	59.1mg	6/50	118.mg	6/42	Maekawa;jnci,79,577-584;1987/pers.comm.
a	1815	396.mg	n.s.s.	2/96	59.1mg	3/50	118.mg	4/42	
b	1815	427.mg	n.s.s.	0/96	59.1mg	0/50	118.mg	0/42	
c	1815	151.mg	n.s.s.	85/96	59.1mg	43/50	118.mg	32/42	
4027	1815	335.mg	n.s.s.	0/93	47.3mg	0/47	95.4mg	0/44	
a	1815	335.mg	n.s.s.	1/93	47.3mg	0/47	95.4mg	0/44	
b	1815	778.mg	n.s.s.	4/93	47.3mg	0/47	95.4mg	2/44	
c	1815	159.mg	n.s.s.	69/93	47.3mg	36/47	95.4mg	27/44	
m-PHENYLENEDIAMINE		**108-45-2**							
4028	1888	277.mg	n.s.s.	0/50	37.1mg	1/50	74.3mg	2/59	Amo;fctx,26,893-897;1988
a	1888	175.mg	n.s.s.	3/50	37.1mg	0/50	74.3mg	0/59	
b	1888	175.mg	n.s.s.	0/50	37.1mg	0/50	74.3mg	0/59	
c	1888	406.mg	n.s.s.	4/50	37.1mg	2/50	74.3mg	2/59	
4029	1888	165.mg	n.s.s.	4/50	31.0mg	4/50	61.9mg	5/56	
a	1888	112.mg	n.s.s.	20/50	31.0mg	7/50	(61.9mg	3/56)	
b	1888	98.2mg	n.s.s.	7/50	31.0mg	4/50	(61.9mg	1/56)	
c	1888	123.mg	n.s.s.	8/50	31.0mg	3/50	(61.9mg	1/56)	
p-PHENYLENEDIAMINE		**106-50-3**							
4030	1542	146.mg	n.s.s.	1/21	25.0mg	1/37	50.0mg	2/42	Imaida;txlt,16,259-269;1983
a	1542	78.3mg	n.s.s.	0/21	25.0mg	0/37	50.0mg	0/42	
4031	1542	42.8mg	n.s.s.	6/19	20.0mg	8/35	40.0mg	10/36	
a	1542	57.4mg	n.s.s.	0/19	20.0mg	0/35	40.0mg	0/36	
m-PHENYLENEDIAMINE.2HCl		**541-69-5**							
4032	381	87.6mg	n.s.s.	3/13	260.mg	4/13	520.mg	0/16	Weisburger;jept,2,325-356;1978/pers.comm./Russfield 1973
a	381	190.mg	n.s.s.	0/13	260.mg	1/13	520.mg	0/16	
b	381	93.9mg	n.s.s.	9/13	260.mg	5/13	520.mg	1/16	
4033	381	104.mg	n.s.s.	2/16	240.mg	2/15	480.mg	0/10	
a	381	145.mg	n.s.s.	2/16	240.mg	1/15	480.mg	0/10	
b	381	77.9mg	n.s.s.	9/16	240.mg	5/15	480.mg	0/10	
4034	381	87.6mg	n.s.s.	1/17	30.0mg	1/17	60.0mg	1/15	
a	381	31.5mg	n.s.s.	13/17	30.0mg	7/17	60.0mg	6/15	
o-PHENYLENEDIAMINE.2HCl		**615-28-1**							
4035	381	236.mg	1.83gm	1/15	806.mg	7/18	1.70gm	6/15	Weisburger;jept,2,325-356;1978/pers.comm./Russfield 1973
a	381	276.mg	2.46gm	1/15	806.mg	6/18	1.70gm	6/15	
b	381	522.mg	n.s.s.	5/15	806.mg	3/18	1.70gm	4/15	
c	381	233.mg	n.s.s.	10/15	806.mg	13/18	1.70gm	10/15	
4036	381	276.mg	1.49gm	1/102p	806.mg	6/18	1.70gm	6/15	
4037	381	406.mg	n.s.s.	0/14	783.mg	5/17	1.57gm	3/14	
a	381	406.mg	n.s.s.	1/14	783.mg	5/17	1.57gm	3/14	
b	381	643.mg	n.s.s.	3/14	783.mg	4/17	1.57gm	2/14	
c	381	381.mg	n.s.s.	7/14	783.mg	10/17	1.57gm	6/14	
4038	381	456.mg	4.81gm	7/99p	783.mg	5/17	1.57gm	3/14	
4039	381	92.5mg	n.s.s.	0/16	57.6mg	0/14	115.mg	5/16	
a	381	92.5mg	n.s.s.	0/16	57.6mg	0/14	115.mg	5/16	
b	381	67.7mg	n.s.s.	10/16	57.6mg	9/14	115.mg	12/16	
4040	381	105.mg	n.s.s.	2/111p	57.6mg	0/14	115.mg	5/16	
p-PHENYLENEDIAMINE.2HCl		**624-18-0**							
4041	TR174	198.mg	n.s.s.	8/20	80.5mg	17/50	161.mg	21/49	
a	TR174	325.mg	n.s.s.	2/20	80.5mg	6/50	161.mg	8/49	liv:hpa,hpc,nnd.
b	TR174	527.mg	n.s.s.	0/20	80.5mg	2/50	161.mg	3/49	lun:a/a,a/c.
4042	TR174	167.mg	n.s.s.	15/20	74.3mg	35/50	149.mg	31/50	
a	TR174	193.mg	n.s.s.	9/20	74.3mg	18/50	149.mg	21/50	liv:hpa,hpc,nnd.
b	TR174	346.mg	n.s.s.	4/20	74.3mg	10/50	149.mg	8/50	lun:a/a,a/c.
4043	TR174	44.0mg	n.s.s.	10/20	30.6mg	37/50	61.3mg	36/50	
a	TR174	n.s.s.	n.s.s.	0/20	30.6mg	0/50	61.3mg	0/50	liv:hpa,hpc,nnd.
4044	TR174	65.7mg	n.s.s.	14/20	24.5mg	34/50	49.0mg	29/50	
a	TR174	364.mg	n.s.s.	0/20	24.5mg	0/50	49.0mg	1/50	liv:hpa,hpc,nnd.
PHENYLEPHRINE.HCl		**61-76-7**							
4045	TR322	236.mg	n.s.s.	22/50	161.mg	31/50	322.mg	27/50	
a	TR322	1.01gm	n.s.s.	3/50	161.mg	2/50	322.mg	4/50	liv:hpa,hpc,nnd.
b	TR322	579.mg	n.s.s.	2/50	161.mg	6/50	322.mg	6/50	lun:a/a,a/c.
4046	TR322	744.mg	n.s.s.	0/50	149.mg	4/50	297.mg	4/50	S
a	TR322	742.mg	n.s.s.	0/50	149.mg	2/50	297.mg	5/50	mul:mlm; spl:mlm. S
b	TR322	274.mg	n.s.s.	27/50	149.mg	30/50	297.mg	34/50	
c	TR322	409.mg	n.s.s.	15/50	149.mg	15/50	297.mg	20/50	liv:hpa,hpc,nnd.
d	TR322	485.mg	n.s.s.	6/50	149.mg	11/50	297.mg	11/50	lun:a/a,a/c.
4047	TR322	37.9mg	n.s.s.	44/50	30.7mg	45/50	61.9mg	42/50	
a	TR322	714.mg	n.s.s.	2/50	30.7mg	0/50	61.9mg	0/50	liv:hpa,hpc,nnd.
4048	TR322	20.1mg	n.s.s.	45/50	24.6mg	42/50	(49.5mg	29/50)	
a	TR322	348.mg	n.s.s.	5/50	24.6mg	1/50	49.5mg	2/50	liv:hpa,hpc,nnd.

```
        Spe Strain  Site    Xpo+Xpt                                                                    TD50    2Tailpvl
         Sex  Route  Hist   Notes                                                                         DR    AuOp
```

Dose-response scale (applies to each chemical block):
`100ng..:..1ug....:..10.....:..100.....:..1mg....:..10....:..100....:..1g.....:..10`

PHENYLETHYLHYDRAZINE SULFATE

Ref	Spe	Sex	Strain	Route	Site	Hist	Xpo+Xpt	Notes	TD50	DR	2Tailpvl	AuOp
4049	M	f	swa	wat	blv	mix	93w93	e	14.6mg		P<.0005	+
a	M	f	swa	wat	sub	ang	93w93	e	19.9mg		P<.0005	
b	M	f	swa	wat	blv	ang	93w93	e	20.4mg		P<.0005	
c	M	f	swa	wat	mus	ang	93w93	e	21.7mg		P<.0005	
d	M	f	swa	wat	fat	ang	93w93	e	23.7mg		P<.0005	
e	M	f	swa	wat	lun	ade	93w93	e	25.8mg		P<.0005	
f	M	f	swa	wat	lun	mix	93w93	e	27.5mg		P<.0005	+
g	M	f	swa	wat	lun	ang	93w93	e	123.mg		P<.006	
h	M	f	swa	wat	liv	mix	93w93	e	278.mg		P<.4	
4050	M	m	swa	wat	lun	ade	99w99	e	30.0mg		P<.0005	-
a	M	m	swa	wat	lun	mix	99w99	e	33.1mg		P<.004	-
b	M	m	swa	wat	liv	mix	99w99	e	no dre		P=1.	-

PHENYLGLYCIDYL ETHER

Ref	Spe	Sex	Strain	Route	Site	Hist	Xpo+Xpt	Notes	TD50	DR	2Tailpvl	AuOp
4051	R	f	cdr	inh	nas	epc	24m24	r	90.2mg	*	P<.005	+
4052	R	m	cdr	inh	nas	mix	24m24	r	29.1mg	*	P<.0005	+

PHENYLHYDRAZINE

Ref	Spe	Sex	Strain	Route	Site	Hist	Xpo+Xpt	Notes	TD50	DR	2Tailpvl	AuOp
4053	M	f	swi	gav	lun	tum	39w55	v	no dre		P=1.	-

PHENYLHYDRAZINE.HCl

Ref	Spe	Sex	Strain	Route	Site	Hist	Xpo+Xpt	Notes	TD50	DR	2Tailpvl	AuOp
4054	M	f	swa	wat	liv	mix	26m26	e	68.6mg		P<.0005	
a	M	f	swa	wat	blv	mix	26m26	e	72.1mg		P<.0005	+
b	M	f	swa	wat	blv	agm	26m26	e	147.mg		P<.02	
c	M	f	swa	wat	blv	ang	26m26	e	165.mg		P<.02	
d	M	f	swa	wat	liv	ang	26m26	e	165.mg		P<.02	
e	M	f	swa	wat	liv	agm	26m26	e	165.mg		P<.02	
f	M	f	swa	wat	lun	mix	26m26	e	no dre		P=1.	-
4055	M	m	swa	wat	blv	mix	26m26	e	70.6mg		P<.02	+
a	M	m	swa	wat	liv	mix	26m26	e	83.9mg		P<.04	
b	M	m	swa	wat	lun	mix	26m26	e	no dre		P=1.	-

beta-PHENYLISOPROPYLHYDRAZINE.HCl

Ref	Spe	Sex	Strain	Route	Site	Hist	Xpo+Xpt	Notes	TD50	DR	2Tailpvl	AuOp
4056	M	f	swa	wat	liv	hpt	28m30	aes	no dre		P=1.	-
a	M	f	swa	wat	lun	tum	28m30	aes	no dre		P=1.	-
4057	M	m	swa	wat	liv	hpt	23m26	aes	no dre		P=1.	-
a	M	m	swa	wat	lun	tum	23m26	aes	no dre		P=1.	-

PHENYLMERCURIC ACETATE

Ref	Spe	Sex	Strain	Route	Site	Hist	Xpo+Xpt	Notes	TD50	DR	2Tailpvl	AuOp
4058	M	f	b6a	orl	liv	hpt	76w76	evx	no dre		P=1.	-
a	M	f	b6a	orl	lun	ade	76w76	evx	no dre		P=1.	-
b	M	f	b6a	orl	tba	mix	76w76	evx	no dre		P=1.	-
4059	M	m	b6a	orl	lun	ade	76w76	evx	no dre		P=1.	-
a	M	m	b6a	orl	liv	hpt	76w76	evx	no dre		P=1.	-
b	M	m	b6a	orl	tba	mix	76w76	evx	no dre		P=1.	-
4060	M	f	b6c	orl	liv	hpt	76w76	evx	21.7mg		P<.3	-
a	M	f	b6c	orl	lun	ade	76w76	evx	21.7mg		P<.3	-
b	M	f	b6c	orl	tba	mix	76w76	evx	10.5mg		P<.2	-
4061	M	m	b6c	orl	lun	ade	76w76	evx	9.24mg		P<.1	-
a	M	m	b6c	orl	liv	hpt	76w76	evx	no dre		P=1.	-
b	M	m	b6c	orl	tba	mix	76w76	evx	9.24mg		P<.1	-

o-PHENYLPHENATE, SODIUM

Ref	Spe	Sex	Strain	Route	Site	Hist	Xpo+Xpt	Notes	TD50	DR	2Tailpvl	AuOp
4062	M	f	b6c	eat	liv	hpc	22m24	e	13.0gm	Z	P<.4	-
a	M	f	b6c	eat	lun	ade	22m24	e	143.gm	*	P<.3	-
b	M	f	b6c	eat	liv	hnd	22m24	e	no dre		P=1.	-
4063	M	m	b6c	eat	liv	hpc	22m24	e	5.42gm	*	P<.009	-
a	M	m	b6c	eat	liv	hes	22m24	e	11.6gm	*	P<.09	-
b	M	m	b6c	eat	liv	fbs	22m24	e	132.gm	*	P<.7	-
c	M	m	b6c	eat	liv	hem	22m24	e	no dre		P=1.	-
d	M	m	b6c	eat	liv	hnd	22m24	e	no dre		P=1.	-
e	M	m	b6c	eat	lun	adc	22m24	e	no dre		P=1.	-
f	M	m	b6c	eat	lun	ade	22m24	e	no dre		P=1.	-
4064	R	f	f3d	eat	ubl	mix	65w65	er	2.88gm	*	P<.1	+
a	R	f	f3d	eat	ubl	tcc	65w65	er	5.89gm	*	P<.3	+
b	R	f	f3d	eat	ubl	tpp	65w65	er	5.89gm	*	P<.3	+
4065	R	m	f3d	eat	unt	mix	91w91	er	414.mg	*	P<.0005	+
a	R	m	f3d	eat	ubl	mix	91w91	er	470.mg	Z	P<.0005	
b	R	m	f3d	eat	kpp	tcc	91w91	er	2.42gm	*	P<.0005	
4066	R	m	f3d	eat	ubl	can	24m24	er	3.18gm	*	P<.04	+
a	R	m	f3d	eat	ubl	pam	24m24	er	3.18gm	*	P<.04	+
4067	R	m	f3d	eat	ubl	mix	65w65	er	195.mg	/	P<.0005	+
a	R	m	f3d	eat	ubl	tcc	65w65	er	364.mg	/	P<.0005	+
b	R	m	f3d	eat	ubl	tpp	65w65	er	852.mg	*	P<.008	+
4068	R	m	f3d	eat	ubl	car	24m24	er	1.03gm		P<.0005	+
a	R	m	f3d	eat	ubl	pam	24m24	er	5.02gm		P<.05	

RefNum		LoConf	UpConf	Cntrl	1Dose	1Inc	2Dose	2Inc				Citation or Pathology
												Brkly Code

PHENYLETHYLHYDRAZINE SULFATE   156-51-4

4049	239	8.37mg	28.7mg	2/52	30.0mg	22/32			Toth;canr,36,917-921;1976
a	239	11.2mg	39.6mg	0/52	30.0mg	18/32			
b	239	11.3mg	42.2mg	1/52	30.0mg	18/32			
c	239	12.1mg	44.1mg	0/52	30.0mg	17/32			
d	239	13.0mg	49.3mg	0/52	30.0mg	16/32			
e	239	14.2mg	65.5mg	18/95	30.0mg	28/49			
f	239	14.7mg	80.1mg	21/95	30.0mg	28/49			
g	239	42.5mg	1.22gm	0/52	30.0mg	4/32			
h	239	56.2mg	n.s.s.	2/52	30.0mg	3/32			
4050	239	14.5mg	113.mg	15/93	25.0mg	17/34			
a	239	14.6mg	304.mg	23/93	25.0mg	18/34			
b	239	136.mg	n.s.s.	6/88	25.0mg	2/50			

PHENYLGLYCIDYL ETHER   122-60-1

| 4051 | 1737 | 31.2mg | 812.mg | 0/90 | .460mg | 0/90 | 5.52mg | 4/90 | Lee;ajpa,111,140-148;1983 |
| 4052 | 1737 | 13.2mg | 108.mg | 1/90 | .322mg | 0/90 | 3.86mg | 9/90 | |

PHENYLHYDRAZINE   100-63-0

| 4053 | 1095 | 8.51mg | n.s.s. | 8/85 | 5.91mg | 0/25 | | | Roe;natu,216,375-376;1967 |

PHENYLHYDRAZINE.HCl   59-88-1

4054	1056	32.6mg	229.mg	2/90	20.0mg	11/49			Toth;zkko,87,267-273;1976/1974
a	1056	33.3mg	301.mg	3/90	20.0mg	11/49			
b	1056	52.8mg	n.s.s.	2/90	20.0mg	6/49			
c	1056	58.0mg	n.s.s.	1/90	20.0mg	5/49			
d	1056	58.0mg	n.s.s.	1/90	20.0mg	5/49			
e	1056	58.0mg	n.s.s.	1/90	20.0mg	5/49			
f	1056	56.9mg	n.s.s.	21/87	20.0mg	8/38			
4055	1056	27.8mg	n.s.s.	6/81	16.7mg	10/43			
a	1056	30.7mg	n.s.s.	6/81	16.7mg	9/43			
b	1056	94.2mg	n.s.s.	23/87	16.7mg	5/43			

beta-PHENYLISOPROPYLHYDRAZINE.HCl   66-05-7

4056	1675	23.0mg	n.s.s.	1/15	31.2mg	0/4	62.4mg	0/3	Toth;faat,2,173-176;1982/1980
a	1675	145.mg	n.s.s.	25/99	31.2mg	10/49	(62.4mg	2/45)	
4057	1675	65.2mg	n.s.s.	0/99	26.0mg	0/42	52.0mg	0/7	
a	1675	143.mg	n.s.s.	26/100	26.0mg	5/42	(52.0mg	0/7)	

PHENYLMERCURIC ACETATE   62-38-4

4058	1286	6.34mg	n.s.s.	0/17	3.39mg	0/17			Innes;ntis,1968/1969
a	1286	6.34mg	n.s.s.	1/17	3.39mg	0/17			
b	1286	6.34mg	n.s.s.	2/17	3.39mg	0/17			
4059	1286	3.89mg	n.s.s.	2/18	3.16mg	1/17			
a	1286	5.91mg	n.s.s.	1/18	3.16mg	0/17			
b	1286	3.04mg	n.s.s.	3/18	3.16mg	2/17			
4060	1286	3.53mg	n.s.s.	0/16	3.39mg	1/18			
a	1286	3.53mg	n.s.s.	0/16	3.39mg	1/18			
b	1286	2.59mg	n.s.s.	0/16	3.39mg	2/18			
4061	1286	2.27mg	n.s.s.	0/16	3.16mg	2/17			
a	1286	5.91mg	n.s.s.	0/16	3.16mg	0/17			
b	1286	2.27mg	n.s.s.	0/16	3.16mg	2/17			

o-PHENYLPHENATE, SODIUM   132-27-4

4062	1647	3.28gm	n.s.s.	4/49	600.mg	5/50	1.20gm	7/50	(2.40gm	0/50)		Hagiwara;fctx,22,809-814;1984
a	1647	23.2gm	n.s.s.	0/49	600.mg	0/49	1.20gm	0/50	2.40gm	1/50		
b	1647	9.85gm	n.s.s.	5/49	600.mg	6/50	1.20gm	9/50	2.40gm	3/50		
4063	1647	2.68gm	188.gm	4/49	554.mg	9/50	1.11gm	13/50	2.22gm	14/50		
a	1647	5.84gm	n.s.s.	0/49	554.mg	3/50	1.11gm	5/50	2.22gm	3/50		
b	1647	21.6gm	n.s.s.	0/49	554.mg	0/50	1.11gm	1/50	2.22gm	0/50		
c	1647	13.3gm	n.s.s.	1/49	554.mg	3/50	1.11gm	2/50	2.22gm	1/50		
d	1647	5.00gm	n.s.s.	16/49	554.mg	19/50	1.11gm	16/50	2.22gm	15/50		
e	1647	14.4gm	n.s.s.	3/49	554.mg	3/50	1.11gm	2/50	2.22gm	2/50		
f	1647	10.3gm	n.s.s.	4/49	554.mg	4/50	1.11gm	5/50	2.22gm	3/50		
4064	1777	707.mg	n.s.s.	0/15	500.mg	0/15	1.00gm	2/15				Fujii;fctx,24,207-211;1986
a	1777	958.mg	n.s.s.	0/15	500.mg	0/15	1.00gm	1/15				
b	1777	958.mg	n.s.s.	0/15	500.mg	0/15	1.00gm	1/15				
4065	1378	279.mg	642.mg	0/20	50.0mg	0/20	100.mg	0/20	200.mg	1/21	400.mg	7/21	800.mg	20/21
					1.60gm	17/20							Hiraga;fctx,19,303-310;1981
a	1378	287.mg	836.mg	0/20	50.0mg	0/20	100.mg	0/20	200.mg	0/21	400.mg	6/21	800.mg	19/21
					(1.60gm | 8/20)							
b	1378	1.24gm	5.68gm	0/20	50.0mg	0/20	100.mg	0/20	200.mg	1/21	400.mg	0/21	800.mg	1/21
					1.60gm	10/20							
4066	1752	779.mg	n.s.s.	0/7	100.mg	0/9	200.mg	0/8	400.mg	0/9	800.mg	2/5	Fukushima;onco,42,304-311;1985
a	1752	779.mg	n.s.s.	0/7	100.mg	0/9	200.mg	0/8	400.mg	0/9	800.mg	2/5	
4067	1777	103.mg	423.mg	0/14	400.mg	0/15	800.mg	15/15				Fujii;fctx,24,207-211;1986
a	1777	175.mg	959.mg	0/14	400.mg	0/15	800.mg	10/15				
b	1777	322.mg	13.7gm	0/14	400.mg	0/15	800.mg	5/15				
4068	1930	523.mg	2.42gm	0/27	800.mg	12/29						Fukushima;carc,10,1635-1640;1989
a	1930	1.52gm	n.s.s.	0/27	800.mg	3/29						

	Spe	Sex	Strain	Route	Site	Hist	Xpo+Xpt	Notes	DR plot	TD50	DR	AuOp
o-PHENYLPHENOL									100ng..:..1ug....:..10......:..100.....:..1mg....:..10......:..100....:..1g.....:..10			
4069	M	f	b6a	orl	liv	hpt	76w76	evx	.>	no dre	P=1.	-
a	M	f	b6a	orl	lun	ade	76w76	evx		no dre	P=1.	-
b	M	f	b6a	orl	tba	mix	76w76	evx		no dre	P=1.	-
4070	M	m	b6a	orl	liv	hpt	76w76	evx		1.80gm	P<1.	-
a	M	m	b6a	orl	lun	ade	76w76	evx		no dre	P=1.	-
b	M	m	b6a	orl	tba	mix	76w76	evx		no dre	P=1.	-
4071	M	f	b6c	orl	lun	ade	76w76	evx	.>	221.mg	P<.3	-
a	M	f	b6c	orl	liv	hpt	76w76	evx		no dre	P=1.	-
b	M	f	b6c	orl	tba	mix	76w76	evx		107.mg	P<.09	-
4072	M	m	b6c	orl	liv	hpt	76w76	evx	. ±	46.1mg	P<.02	-
a	M	m	b6c	orl	lun	ade	76w76	evx		205.mg	P<.3	-
b	M	m	b6c	orl	tba	mix	76w76	evx		35.4mg	P<.006	-
4073	R	m	f3d	eat	ubl	mix	91w91	ers	. + .	232.mg	Z P<.0005+	
a	R	m	f3d	eat	ubl	tcc	91w91	ers		296.mg	Z P<.0005+	
b	R	m	f3d	eat	ubl	nvt	91w91	ers		451.mg	Z P<.0005+	
c	R	m	f3d	eat	ubl	ivt	91w91	ers		1.65gm	Z P<.008 +	
4074	R	m	f3d	eat	ubl	tum	24m24	r	.>	no dre	P=1.	
p-PHENYLPHENOL									100ng..:..1ug....:..10......:..100.....:..1mg....:..10......:..100....:..1g.....:..10			
4075	M	f	b6a	orl	lun	ade	76w76	evx	.>	no dre	P=1.	-
a	M	f	b6a	orl	liv	hpt	76w76	evx		no dre	P=1.	-
b	M	f	b6a	orl	tba	mix	76w76	evx		no dre	P=1.	-
4076	M	m	b6a	orl	liv	hpt	76w76	evx	.>	no dre	P=1.	-
a	M	m	b6a	orl	lun	ade	76w76	evx		no dre	P=1.	-
b	M	m	b6a	orl	tba	mix	76w76	evx		953.mg	P<.7	-
4077	M	f	b6c	orl	liv	hpt	76w76	evx	.>	no dre	P=1.	-
a	M	f	b6c	orl	lun	mix	76w76	evx		no dre	P=1.	-
b	M	f	b6c	orl	tba	tum	76w76	evx		no dre	P=1.	-
4078	M	m	b6c	orl	lun	ade	76w76	evx	. ±	426.mg	P<.08	-
a	M	m	b6c	orl	liv	hpt	76w76	evx		no dre	P=1.	-
b	M	m	b6c	orl	tba	mix	76w76	evx		195.mg	P<.01	-
PhIP.HCl									100ng..:..1ug....:..10......:..100.....:..1mg....:..10......:..100....:..1g.....:..10			
4079	M	f	cdf	eat	---	lym	82w82	e	. + .	22.4mg	P<.0005+	
4080	M	m	cdf	eat	---	lym	82w82	e	. + .	63.9mg	P<.004 +	
4081	R	f	f3d	eat	mgl	adc	52w52	e	. + .	5.45mg	P<.0005+	
a	R	f	f3d	eat	col	adc	52w52	e		49.7mg	P<.07 +	
b	R	f	f3d	eat	liv	tum	52w52	e		no dre	P=1.	
4082	R	f	f3d	eat	mgl	adc	24m24	e	. + .	6.34mg	* P<.0005+	
a	R	f	f3d	eat	col	adc	24m24	e		30.4mg	* P<.008 +	
b	R	f	f3d	eat	liv	nnd	24m24	e		no dre	P=1.	
c	R	f	f3d	eat	mgl	ade	24m24	e		no dre	P=1.	
d	R	f	f3d	eat	mgl	fba	24m24	e		no dre	P=1.	
4083	R	m	f3d	eat	col	adc	52w52	e	. + .	3.42mg	P<.0005+	
a	R	m	f3d	eat	liv	tum	52w52	e		no dre	P=1.	
4084	R	m	f3d	eat	---	lle	24m24	e	. + .	5.92mg	* P<.003 +	
a	R	m	f3d	eat	col	adc	24m24	e		6.44mg	* P<.0005+	
b	R	m	f3d	eat	liv	nnd	24m24	e		4.62mg	\ P<.08	
c	R	m	f3d	eat	mgl	fba	24m24	e		no dre	P=1.	
PHORBOL									100ng..:..1ug....:..10......:..100.....:..1mg....:..10......:..100....:..1g.....:..10			
4085	M	f	aks	ipj	lmr	lyk	24m24	er	.>	no dre	P=1.	-
4086	M	m	aks	ipj	lmr	lyk	24m24	er	.>	8.26mg	P<.7	-
4087	M	f	bal	ipj	lmr	tum	24m24	er	.>	no dre	P=1.	-
4088	M	m	bal	ipj	lmr	tum	24m24	er	.>	no dre	P=1.	-
4089	M	f	c3p	ipj	lmr	tum	24m24	er	.>	no dre	P=1.	-
4090	M	m	c3p	ipj	lmr	tum	24m24	er	.>	no dre	P=1.	-
4091	M	f	c51	ipj	lmr	tum	24m24	er	.>	no dre	P=1.	-
4092	M	f	sjs	ipj	lmr	rtb	24m24	er	.>	92.6mg	P<1.	-
4093	M	f	swr	ipj	lmr	lyk	24m24	er	. + .	2.21mg	P<.0005+	
PHOSPHAMIDON									100ng..:..1ug....:..10......:..100.....:..1mg....:..10......:..100....:..1g.....:..10			
4094	M	f	b6c	eat	TBA	MXB	80w90		:>	215.mg	* P<.8	-
a	M	f	b6c	eat	liv	MXB	80w90			34.2mg	\ P<.2	
b	M	f	b6c	eat	lun	MXB	80w90			no dre	P=1.	-
4095	M	m	b6c	eat	TBA	MXB	66w90		:>	no dre	P=1.	-
a	M	m	b6c	eat	liv	MXB	66w90			no dre	P=1.	
b	M	m	b6c	eat	lun	MXB	66w90			no dre	P=1.	
4096	R	f	osm	eat	TBA	MXB	19m26		:>	no dre	P=1.	i
a	R	f	osm	eat	liv	MXB	19m26			no dre	P=1.	
4097	R	f	osm	eat	thy	MXA	19m25	pool	: + :	#19.0mg	* P<.008	i
4098	R	m	osm	eat	TBA	MXB	19m26		:>	no dre	P=1.	i
a	R	m	osm	eat	liv	MXB	19m26			185.mg	* P<.5	
4099	R	m	osm	eat	spl	MXA	19m25	pool	: ±	#26.9mg	* P<.02	i
PHOSPHATED DISTARCH PHOSPHATE									100ng..:..1ug....:..10......:..100.....:..1mg....:..10......:..100....:..1g.....:..10			
4100	R	f	wis	eat	liv	tum	24m24	e		no dre	P=1.	-
a	R	f	wis	eat	tba	mix	24m24	e		no dre	P=1.	-
4101	R	m	wis	eat	liv	tum	24m24	e		no dre	P=1.	-
a	R	m	wis	eat	tba	mix	24m24	e		24.4gm	P<.6	-

	RefNum	LoConf	UpConf	Cntrl	1Dose	1Inc	2Dose	2Inc	Citation or Pathology
									Brkly Code

o-PHENYLPHENOL   (orthoxenol, Dowicide-1) 90-43-7

	RefNum	LoConf	UpConf	Cntrl	1Dose	1Inc	2Dose	2Inc	Citation or Pathology
4069	1310	64.2mg	n.s.s.	0/17	38.9mg	0/15			Innes;ntis,1968/1969
a	1310	64.2mg	n.s.s.	1/17	38.9mg	0/15			
b	1310	41.7mg	n.s.s.	2/17	38.9mg	1/15			
4070	1310	n.s.s.	n.s.s.	1/18	36.2mg	1/16			
a	1310	41.5mg	n.s.s.	2/18	36.2mg	1/16			
b	1310	32.3mg	n.s.s.	3/18	36.2mg	2/16			
4071	1310	35.9mg	n.s.s.	0/16	38.9mg	1/16			
a	1310	68.5mg	n.s.s.	0/16	38.9mg	0/16			
b	1310	26.2mg	n.s.s.	0/16	38.9mg	2/16			
4072	1310	15.8mg	n.s.s.	0/16	36.2mg	4/16			
a	1310	33.4mg	n.s.s.	0/16	36.2mg	1/16			
b	1310	13.3mg	350.mg	0/16	36.2mg	5/16			
4073	1646	137.mg	430.mg	0/24	250.mg	0/20	500.mg	23/24 (1.00gm 4/23)	Hiraga;fctx,22,865-870;1984
a	1646	171.mg	568.mg	0/24	250.mg	0/20	500.mg	20/24 (1.00gm 2/23)	
b	1646	245.mg	960.mg	0/24	250.mg	0/20	500.mg	15/24 (1.00gm 2/23)	
c	1646	625.mg	26.3gm	0/24	250.mg	0/20	500.mg	5/24 (1.00gm 0/23)	
4074	1930	3.19gm	n.s.s.	0/30	500.mg	0/31			Fukushima;carc,10,1635-1640;1989

p-PHENYLPHENOL   (paraxenol) 92-69-3

	RefNum	LoConf	UpConf	Cntrl	1Dose	1Inc	2Dose	2Inc	Citation or Pathology
4075	1309	215.mg	n.s.s.	1/17	193.mg	1/17			Innes;ntis,1968/1969
a	1309	361.mg	n.s.s.	0/17	193.mg	0/17			
b	1309	170.mg	n.s.s.	2/17	193.mg	2/17			
4076	1309	212.mg	n.s.s.	1/18	180.mg	1/18			
a	1309	356.mg	n.s.s.	2/18	180.mg	0/18			
b	1309	112.mg	n.s.s.	3/18	180.mg	4/18			
4077	1309	361.mg	n.s.s.	0/16	193.mg	0/17			
a	1309	361.mg	n.s.s.	0/16	193.mg	0/17			
b	1309	361.mg	n.s.s.	0/16	193.mg	0/17			
4078	1309	105.mg	n.s.s.	0/16	180.mg	2/14			
a	1309	277.mg	n.s.s.	0/16	180.mg	0/14			
b	1309	66.9mg	13.2gm	0/16	180.mg	4/14			

PhIP.HCl   (2-amino-1-methyl-6-phenylimidazo[4,5-b]-pyridine.HCl)   ---

	RefNum	LoConf	UpConf	Cntrl	1Dose	1Inc	2Dose	2Inc	Citation or Pathology
4079	1872	12.6mg	50.4mg	6/40	52.0mg	26/38			Esumi;gann,80,1176-1178;1989/pers.comm.
4080	1872	28.9mg	513.mg	2/36	48.0mg	11/35			
4081	1922	2.90mg	12.0mg	0/40	20.0mg	14/30			Ito;carc,12,1503-1506;1991/pers.comm.
a	1922	12.2mg	n.s.s.	0/40	20.0mg	2/30			
b	1922	30.9mg	n.s.s.	0/40	20.0mg	0/30			
4082	2062	3.51mg	13.1mg	0/30	1.25mg	2/30	5.00mg	14/30	Hasegawa;carc,14,2553-2557;1993/pers.comm.
a	2062	10.5mg	584.mg	0/30	1.25mg	0/30	5.00mg	4/30	
b	2062	28.8mg	n.s.s.	2/30	1.25mg	2/30	5.00mg	0/30	
c	2062	20.3mg	n.s.s.	3/30	1.25mg	4/30	5.00mg	1/30	
d	2062	3.19mg	n.s.s.	8/30	1.25mg	5/30	(5.00mg	1/30)	
4083	1922	1.87mg	7.13mg	0/40	16.0mg	16/29			Ito;carc,12,1503-1506;1991/pers.comm.
a	1922	23.9mg	n.s.s.	0/40	16.0mg	0/29			
4084	2062	2.85mg	38.2mg	3/30	1.00mg	6/30	4.00mg	13/30	Hasegawa;carc,14,2553-2557;1993/pers.comm.
a	2062	3.37mg	14.6mg	0/30	1.00mg	0/30	4.00mg	13/30	
b	2062	1.51mg	n.s.s.	1/30	1.00mg	5/30	(4.00mg	1/30)	
c	2062	12.2mg	n.s.s.	2/30	1.00mg	2/30	4.00mg	2/30	

PHORBOL   17673-25-5

	RefNum	LoConf	UpConf	Cntrl	1Dose	1Inc	2Dose	2Inc	Citation or Pathology
4085	1415	1.35mg	n.s.s.	15/15	4.16mg	11/13			Armuth;bjca,34,516-522;1976
4086	1415	1.01mg	n.s.s.	10/14	3.47mg	11/14			
4087	1415	58.3mg	n.s.s.	0/53	4.16mg	0/68			
4088	1415	29.3mg	n.s.s.	0/58	3.47mg	0/41			
4089	1415	39.4mg	n.s.s.	0/61	4.16mg	0/46			
4090	1415	42.2mg	n.s.s.	0/41	3.47mg	0/59			
4091	1415	25.7mg	n.s.s.	0/20	4.16mg	0/30			
4092	1415	1.95mg	n.s.s.	26/32	4.16mg	27/33			
4093	1415	1.32mg	4.07mg	3/45	4.16mg	29/39			

PHOSPHAMIDON   13171-21-6

	RefNum	LoConf	UpConf	Cntrl	1Dose	1Inc	2Dose	2Inc	Citation or Pathology
4094	TR16	22.4mg	n.s.s.	1/10	9.10mg	11/50	18.3mg	9/50	
a	TR16	14.0mg	n.s.s.	0/10	9.10mg	6/50	(18.3mg	1/50)	liv:hpa,hpc,nnd.
b	TR16	81.7mg	n.s.s.	1/10	9.10mg	3/50	18.3mg	1/50	lun:a/a,a/c.
4095	TR16	18.4mg	n.s.s.	3/10	7.60mg	14/50	13.2mg	11/50	
a	TR16	23.5mg	n.s.s.	2/10	7.60mg	9/50	13.2mg	7/50	liv:hpa,hpc,nnd.
b	TR16	47.9mg	n.s.s.	1/10	7.60mg	3/50	13.2mg	2/50	lun:a/a,a/c.
4096	TR16	6.09mg	n.s.s.	7/10	2.90mg	29/50	5.80mg	26/50	
a	TR16	47.0mg	n.s.s.	0/10	2.90mg	1/50	5.80mg	0/50	liv:hpa,hpc,nnd.
4097	TR16	9.26mg	361.mg	2/95p	2.90mg	9/50	5.80mg	8/50	thy:cca,ccr. S
4098	TR16	5.55mg	n.s.s.	4/10	2.30mg	26/50	4.60mg	20/50	
a	TR16	30.2mg	n.s.s.	0/10	2.30mg	0/50	4.60mg	1/50	liv:hpa,hpc,nnd.
4099	TR16	10.9mg	n.s.s.	1/95p	2.30mg	3/50	4.60mg	5/50	spl:hem,hes. S

PHOSPHATED DISTARCH PHOSPHATE   ---

	RefNum	LoConf	UpConf	Cntrl	1Dose	1Inc	2Dose	2Inc	Citation or Pathology
4100	1407	89.6gm	n.s.s.	0/29	15.0gm	0/29			de Groot;fctx,12,651-663;1974
a	1407	10.4gm	n.s.s.	25/29	15.0gm	22/29			
4101	1407	69.2gm	n.s.s.	0/28	12.0gm	0/28			
a	1407	4.43gm	n.s.s.	21/28	12.0gm	23/28			

```
           Spe Strain  Site    Xpo+Xpt                                                            TD50     2Tailpvl
               Sex Route  Hist       Notes                                                           DR     AuOp

PHOSPHINE                          100ng..:..1ug...:..10.....:..100.....:..1mg...:..10.....:..100...:..1g.....:..10
4102 R f sda eat tba tum 52w52 ek    .>                                                          no dre   P=1.    -
4103 R f sda eat tba mix 24m24 e                                                                 no dre   P=1.    -
4104 R m sda eat tba tum 52w52 ek    .>                                                          no dre   P=1.    -
4105 R m sda eat tba mix 24m24 e      .>                                                         no dre   P=1.    -

PHTHALAMIDE                        100ng..:..1ug...:..10.....:..100.....:..1mg...:..10.....:..100...:..1g.....:..10
4106 M f b6c eat TBA MXB 24m24 as                                                           :>   no dre   P=1.    -
a    M f b6c eat liv MXB 24m24 as                                                                no dre   P=1.
b    M f b6c eat lun MXB 24m24 as                                                                no dre   P=1.
4107 M m b6c eat TBA MXB 24m24                                                               :>   1.09kg   P<1.    -
a    M m b6c eat liv MXB 24m24                                                                   no dre   P=1.
b    M m b6c eat lun MXB 24m24                                                                   35.8gm * P<.5
4108 R f f34 eat TBA MXB 25m25                                                  :>               722.mg * P<.3    -
a    R f f34 eat liv MXB 25m25                                                                   5.39gm * P<.5
4109 R m f34 eat TBA MXB 25m25                                                         :>        no dre   P=1.    -
a    R m f34 eat liv MXB 25m25                                                                   16.6gm * P<.3

PHTHALIC ANHYDRIDE                 100ng..:..1ug...:..10.....:..100.....:..1mg...:..10.....:..100...:..1g.....:..10
4110 M f b6c eat TBA MXB 24m24 v                                                            :>   no dre   P=1.    -
a    M f b6c eat liv MXB 24m24 v                                                                 no dre   P=1.
b    M f b6c eat lun MXB 24m24 v                                                                 no dre   P=1.
4111 M m b6c eat TBA MXB 24m24 v                                                            :>no dre      P=1.    -
a    M m b6c eat liv MXB 24m24 v                                                                 7.46gm \ P<.3
b    M m b6c eat lun MXB 24m24 v                                                                 no dre   P=1.
4112 R f f34 eat lun a/a 24m24                                                              :  ± #6.23gm * P<.03  -
a    R f f34 eat TBA MXB 24m24                                                                   4.50gm * P<.8
b    R f f34 eat liv MXB 24m24                                                                   no dre   P=1.
4113 R m f34 eat TBA MXB 24m24                                                         :>        no dre   P=1.    -
a    R m f34 eat liv MXB 24m24                                                                   no dre   P=1.

PICLORAM, TECHNICAL GRADE          100ng..:..1ug...:..10.....:..100.....:..1mg...:..10.....:..100......:..1g.....:..10
4114 M f b6c eat TBA MXB 80w90 v                                                       :>        no dre   P=1.    -
a    M f b6c eat liv MXB 80w90 v                                                                 20.1gm * P<.4
b    M f b6c eat lun MXB 80w90 v                                                                 no dre   P=1.
4115 M m b6c eat TBA MXB 80w90 v                                               :>                no dre   P=1.    -
a    M m b6c eat liv MXB 80w90 v                                                                 no dre   P=1.
b    M m b6c eat lun MXB 80w90 v                                                                 no dre   P=1.
4116 R f osm eat TBA MXB 19m26 v                                               :>                no dre   P=1.    -
a    R f osm eat liv MXB 19m26 v                                                                 1.64gm * P<.09
4117 R f osm eat liv nnd 19m26 v   pool                                             :  +         1.70gm * P<.007 a
4118 R m osm eat TBA MXB 19m26 v                                                   :>            2.93gm * P<.8    -
a    R m osm eat liv MXB 19m26 v                                                                 1.32gm \ P<.3
4119 R f f34 eat liv hpa 24m24                                                          .>       15.8gm * P<.8    -
a    R f f34 eat thy pfa 24m24                                                                   no dre   P=1.
4120 R m f34 eat liv hpa 24m24                                                          .>       5.56gm * P<.6    -
a    R m f34 eat liv hpc 24m24                                                                   8.71gm * P<.6

PILDRALAZINE                       100ng..:..1ug...:..10.....:..100.....:..1mg...:..10.....:..100...:..1g.....:..10
4121 M f b6c wat liv mix 19m29 e                                                  .>             503.mg * P<.2    -
a    M f b6c wat lun mix 19m29 e                                                                 no dre   P=1.
4122 M m b6c wat liv mix 19m29 e                                                  .>             922.mg Z P<.6    -
a    M m b6c wat lun mix 19m29 e                                                                 no dre   P=1.
4123 M f bld wat --- lym 19m29 e                                                .>               408.mg * P<.6    -
a    M f bld wat liv hpa 19m29 e                                                                 no dre   P=1.
b    M f bld wat lun mix 19m29 e                                                                 no dre   P=1.

PILOCARPINE                        100ng..:..1ug...:..10.....:..100.....:..1mg...:..10.....:..100...:..1g.....:..10
4124 R f sda ipj tba mal 24m24 e                                             .>                  98.4mg   P<.8    -
4125 R m sda ipj liv hae 24m24 e                                          .>                     no dre   P=1.
a    R m sda ipj tba mal 24m24 e                                                                 1.75gm   P<1.

PIMARICIN                          100ng..:..1ug...:..10.....:..100.....:..1mg...:..10.....:..100...:..1g.....:..10
4126 R f cfn eat liv mhp 24m24                                                          .>       no dre   P=1.    -
4127 R m cfn eat lun lys 24m24                                                         .>        1.97gm * P<.5    -
a    R m cfn eat thm tma 24m24                                                                   no dre   P=1.

PIPERAZINE                         100ng..:..1ug...:..10.....:..100.....:..1mg...:..10.....:..100...:..1g.....:..10
4128 R f mrc wat liv tum 17m24 e                                              .>                 no dre   P=1.
a    R f mrc wat tba mix 17m24 e                                                                 11.2mg   P<.2
4129 R m mrc wat liv tum 17m24 e                                              .>                 no dre   P=1.
a    R m mrc wat tba mix 17m24 e                                                                 no dre   P=1.

PIPERIDINE                         100ng..:..1ug...:..10.....:..100.....:..1mg...:..10.....:..100...:..1g.....:..10
4130 R f mrc wat liv tum 17m24 e                                                  .>             no dre   P=1.
a    R f mrc wat tba mix 17m24 e                                                                 63.3mg   P<.3
4131 R m mrc wat liv tum 17m24 e                                                .>               no dre   P=1.
a    R m mrc wat tba mix 17m24 e                                                                 no dre   P=1.

PIPERONYL BUTOXIDE                 100ng..:..1ug...:..10.....:..100.....:..1mg...:..10.....:..100...:..1g.....:..10
4132 M f b6c eat TBA MXB 26m26 v                                                   :>            no dre   P=1.    -
```

	RefNum	LoConf	UpConf	Cntrl	1Dose	1Inc	2Dose	2Inc					Citation or Pathology
													Brkly Code

	RefNum	LoConf	UpConf	Cntrl	1Dose	1Inc	2Dose	2Inc			Citation or Pathology
PHOSPHINE	7803-51-2										
4102	1692m	245.ng	n.s.s.	0/20	250.ng	0/19					Cabrol Telle;fctx,23,1001-1009;1985
4103	1692n	43.1ng	n.s.s.	10/10	250.ng	10/11					
4104	1692m	206.ng	n.s.s.	0/19	200.ng	0/20					
4105	1692n	264.ng	n.s.s.	2/11	200.ng	1/10					
PHTHALAMIDE	88-96-0										
4106	TR161	3.07gm	n.s.s.	19/40	806.mg	29/50	1.62gm	27/50	3.25gm	8/50	
a	TR161	17.3gm	n.s.s.	4/40	806.mg	2/50	1.62gm	2/50	3.25gm	0/50	liv:hpa,hpc,nnd.
b	TR161	15.8gm	n.s.s.	3/40	806.mg	5/50	1.62gm	1/50	3.25gm	0/50	lun:a/a,a/c.
4107	TR161	4.34gm	n.s.s.	15/20	3.00gm	33/50	6.00gm	32/50			
a	TR161	10.8gm	n.s.s.	9/20	3.00gm	17/50	6.00gm	13/50			liv:hpa,hpc,nnd.
b	TR161	9.09gm	n.s.s.	3/20	3.00gm	7/50	6.00gm	10/50			lun:a/a,a/c.
4108	TR161	217.mg	n.s.s.	14/20	250.mg	39/50	500.mg	39/50			
a	TR161	1.25gm	n.s.s.	2/20	250.mg	2/50	500.mg	6/50			liv:hpa,hpc,nnd.
4109	TR161	1.06gm	n.s.s.	17/20	600.mg	43/50	1.20gm	34/50			
a	TR161	5.02gm	n.s.s.	0/20	600.mg	1/50	1.20gm	2/50			liv:hpa,hpc,nnd.
PHTHALIC ANHYDRIDE	85-44-9										
4110	TR159	5.42gm	n.s.s.	10/20	1.56gm	21/50	3.12gm	17/50			
a	TR159	25.9gm	n.s.s.	1/20	1.56gm	0/50	3.12gm	1/50			liv:hpa,hpc,nnd.
b	TR159	10.7gm	n.s.s.	1/20	1.56gm	6/50	3.12gm	2/50			lun:a/a,a/c.
4111	TR159	7.76gm	n.s.s.	11/20	1.96gm	22/50	3.92gm	18/50			
a	TR159	2.27gm	n.s.s.	3/20	1.96gm	12/50	(3.92gm	7/50)			liv:hpa,hpc,nnd.
b	TR159	13.3gm	n.s.s.	7/20	1.96gm	6/50	3.92gm	9/50			lun:a/a,a/c.
4112	TR159	2.37gm	n.s.s.	0/20	375.mg	0/50	750.mg	5/50			s
a	TR159	511.mg	n.s.s.	13/20	375.mg	37/50	750.mg	36/50			
b	TR159	5.50gm	n.s.s.	0/20	375.mg	1/50	750.mg	0/50			liv:hpa,hpc,nnd.
4113	TR159	451.mg	n.s.s.	15/20	300.mg	37/50	600.mg	36/50			
a	TR159	4.79gm	n.s.s.	1/20	300.mg	2/50	600.mg	0/50			liv:hpa,hpc,nnd.
PICLORAM, TECHNICAL GRADE	1918-02-1										
4114	TR23	1.57gm	n.s.s.	1/10	292.mg	6/50	585.mg	3/50			
a	TR23	3.28gm	n.s.s.	0/10	292.mg	0/50	585.mg	1/50			liv:hpa,hpc,nnd.
b	TR23	1.25gm	n.s.s.	1/10	292.mg	1/50	(585.mg	0/50)			lun:a/a,a/c.
4115	TR23	688.mg	n.s.s.	5/10	270.mg	18/50	540.mg	14/50			
a	TR23	398.mg	n.s.s.	5/10	270.mg	13/50	(540.mg	8/50)			liv:hpa,hpc,nnd.
b	TR23	1.33gm	n.s.s.	1/10	270.mg	5/50	540.mg	3/50			lun:a/a,a/c.
4116	TR23	397.mg	n.s.s.	7/10	263.mg	32/50	527.mg	33/50			
a	TR23	864.mg	n.s.s.	0/10	263.mg	5/50	527.mg	8/50			liv:hpa,hpc,nnd.
4117	TR23	873.mg	23.4gm	0/40p	263.mg	5/50	527.mg	7/50			
4118	TR23	290.mg	n.s.s.	3/10	210.mg	28/50	422.mg	22/50			
a	TR23	456.mg	n.s.s.	0/10	210.mg	4/50	(422.mg	0/50)			liv:hpa,hpc,nnd.
4119	1981	1.27gm	n.s.s.	2/50	20.0mg	0/50	60.0mg	0/50	200.mg	2/50	Stott;jtxe,30,91-104;1990
a	1981	909.mg	n.s.s.	3/50	20.0mg	10/50	60.0mg	4/50	200.mg	4/50	
4120	1981	759.mg	n.s.s.	1/50	20.0mg	1/50	60.0mg	4/50	200.mg	2/50	
a	1981	1.04gm	n.s.s.	0/50	20.0mg	1/50	60.0mg	1/50	200.mg	1/50	
PILDRALAZINE	56393-22-7										
4121	1567	160.mg	n.s.s.	6/49	13.0mg	5/48	25.9mg	5/49	51.8mg	10/46	Della Porta;zkko,106,97-101;1983
a	1567	475.mg	n.s.s.	4/49	13.0mg	4/48	25.9mg	2/49	51.8mg	2/46	
4122	1567	171.mg	n.s.s.	11/50	10.8mg	8/50	21.6mg	8/50	43.2mg	10/49	
a	1567	232.mg	n.s.s.	5/50	10.8mg	10/48	21.6mg	5/50	43.2mg	5/49	
4123	1567	72.9mg	n.s.s.	16/46	13.0mg	31/49	25.9mg	23/50	51.8mg	23/47	
a	1567	765.mg	n.s.s.	0/46	13.0mg	1/49	25.9mg	0/50	51.8mg	0/47	
b	1567	149.mg	n.s.s.	14/46	13.0mg	12/49	25.9mg	16/50	51.8mg	13/47	
PILOCARPINE	92-13-7										
4124	1134	10.1mg	n.s.s.	3/33	4.29mg	4/34					Schmahl;zkko,86,77-84;1976
4125	1134	30.0mg	n.s.s.	1/36	4.29mg	0/34					
a	1134	18.1mg	n.s.s.	1/36	4.29mg	1/34					
PIMARICIN	7681-93-8										
4126	240	639.mg	n.s.s.	1/40	6.25mg	1/35	12.5mg	0/35	25.0mg	0/35	50.0mg	0/40	Levinskas;txap,8,97-109;1966
4127	240	320.mg	n.s.s.	0/40	5.00mg	0/35	10.0mg	0/35	20.0mg	1/35	40.0mg	0/40	
a	240	350.mg	n.s.s.	0/40	5.00mg	0/35	10.0mg	1/35	20.0mg	0/35	40.0mg	0/40	
PIPERAZINE	110-85-0										
4128	216	22.7mg	n.s.s.	0/15	7.36mg	0/15					Garcia;zkko,79,141-144;1973
a	216	3.42mg	n.s.s.	4/15	7.36mg	8/15					
4129	216	19.9mg	n.s.s.	0/15	6.44mg	0/15					
a	216	5.79mg	n.s.s.	5/15	6.44mg	5/15					
PIPERIDINE	110-89-4										
4130	216	91.0mg	n.s.s.	0/15	29.4mg	0/15					Garcia;zkko,79,141-144;1973
a	216	16.4mg	n.s.s.	4/15	29.4mg	7/15					
4131	216	79.6mg	n.s.s.	0/15	25.8mg	0/15					
a	216	29.2mg	n.s.s.	5/15	25.8mg	4/15					
PIPERONYL BUTOXIDE	51-03-6										
4132	TR120	241.mg	n.s.s.	15/20	135.mg	22/50	(364.mg	20/50)			

Spe		Strain	Site		Xpo+Xpt			TD50		2Tailpvl
	Sex	Route	Hist	Notes				DR		AuOp
a	M f b6c	eat	liv	MXB 26m26	v			3.26gm	*	P<.3
b	M f b6c	eat	lun	MXB 26m26	v			no dre		P=1.
4133	M f b6c	orl	liv	hpt 76w76	evx	.>		no dre		P=1.
a	M f b6c	orl	lun	mix 76w76	evx			no dre		P=1.
b	M f b6c	orl	tba	mix 76w76	evx			83.1mg		P<.05
4134	M m b6c	eat	TBA	MXB 26m26	v	:>		no dre		P=1.	-
a	M m b6c	eat	liv	MXB 26m26	v			no dre		P=1.
b	M m b6c	eat	lun	MXB 26m26	v			no dre		P=1.
4135	M m b6c	orl	---	rts 76w76	evx	. + .		34.8mg		P<.005
a	M m b6c	orl	liv	hpt 76w76	evx			98.5mg		P<.09
b	M m b6c	orl	lun	mix 76w76	evx			no dre		P=1.
c	M m b6c	orl	tba	mix 76w76	evx			22.4mg		P<.0005
4136	M f b6a	orl	lun	ade 76w76	evx	.>		265.mg		P<.6
a	M f b6a	orl	liv	hpt 76w76	evx			no dre		P=1.
b	M f b6a	orl	tba	mix 76w76	evx			265.mg		P<.7
4137	M m b6a	orl	lun	ade 76w76	evx	.>		no dre		P=1.
a	M m b6a	orl	liv	hpt 76w76	evx			no dre		P=1.
b	M m b6a	orl	tba	mix 76w76	evx			no dre		P=1.
4138	R f f34	eat	---	lym 25m25		: +		#1.18gm	*	P<.008	-
a	R f f34	eat	TBA	MXB 25m25				7.82gm	*	P<.9
b	R f f34	eat	liv	MXB 25m25				no dre		P=1.
4139	R m f34	eat	TBA	MXB 25m25		:>		no dre		P=1.	-
a	R m f34	eat	liv	MXB 25m25				no dre		P=1.
4140	R f f3d	eat	liv	nnd 24m26	e		.>	no dre		P=1.
a	R f f3d	eat	tba	mix 24m26	e			no dre		P=1.
4141	R m f3d	eat	liv	nnd 24m26	e		.>	7.69gm	*	P<.3
a	R m f3d	eat	liv	hpc 24m26	e			20.1gm	*	P<.3
b	R m f3d	eat	tba	mix 24m26	e			no dre		P=1.
4142	R b wis	eat	liv	hpt 22m24	sv		.>	22.5gm	*	P<.5	-
a	R b wis	eat	liv	hpc 22m24	sv			26.1gm	*	P<.2	-

PIPERONYL BUTOXIDE IN SOLVENT   100ng..:..1ug....:..10......:..100....:..1mg....:..10.....:..100....:..1g.....:..10

4143	M f b6a	orl	lun	ade 76w76	evx	.>		no dre		P=1.	-
a	M f b6a	orl	liv	hpt 76w76	evx			no dre		P=1.
b	M f b6a	orl	tba	mix 76w76	evx			1.01gm		P<.7
4144	M m b6a	orl	liv	hpt 76w76	evx	.>		no dre		P=1.
a	M m b6a	orl	lun	ade 76w76	evx			no dre		P=1.
b	M m b6a	orl	tba	mix 76w76	evx			no dre		P=1.
4145	M f b6c	orl	liv	hpt 76w76	evx	.>		no dre		P=1.
a	M f b6c	orl	lun	mix 76w76	evx			no dre		P=1.
b	M f b6c	orl	tba	tum 76w76	evx			no dre		P=1.
4146	M m b6c	orl	liv	hpt 76w76	evx	. ±		186.mg		P<.02	-
a	M m b6c	orl	lun	mix 76w76	evx			no dre		P=1.
b	M m b6c	orl	tba	mix 76w76	evx			93.1mg		P<.002	-

PIPERONYL SULFOXIDE   100ng..:..1ug....:..10......:..100....:..1mg....:..10.....:..100....:..1g.....:..10

4147	M f b6c	eat	TBA	MXB 24m24	av	:>		no dre		P=1.	-
a	M f b6c	eat	liv	MXB 24m24	av			725.mg	*	P<.3
b	M f b6c	eat	lun	MXB 24m24	av			no dre		P=1.
4148	M f b6c	orl	liv	hpt 76w76	evx	.>		no dre		P=1.
a	M f b6c	orl	lun	mix 76w76	evx			no dre		P=1.
b	M f b6c	orl	tba	mix 76w76	evx			100.mg		P<.3
4149	M m b6c	eat	liv	hpc 24m24	a	: ±		62.2mg	*	P<.02	c
a	M m b6c	eat	TBA	MXB 24m24	a			423.mg	*	P<.8
b	M m b6c	eat	liv	MXB 24m24	a			62.2mg	*	P<.02
c	M m b6c	eat	lun	MXB 24m24	a			no dre		P=1.
4150	M m b6c	orl	---	rts 76w76	evx	. + .		9.10mg		P<.0005
a	M m b6c	orl	liv	hpt 76w76	evx			45.4mg		P<.2
b	M m b6c	orl	lun	mix 76w76	evx			no dre		P=1.
c	M m b6c	orl	tba	mix 76w76	evx			7.72mg		P<.0005
4151	M f b6a	orl	lun	ade 76w76	evx	.>		no dre		P=1.
a	M f b6a	orl	liv	hpt 76w76	evx			no dre		P=1.
b	M f b6a	orl	tba	mix 76w76	evx			no dre		P=1.
4152	M m b6a	orl	liv	hpt 76w76	evx	.>		88.2mg		P<.6
a	M m b6a	orl	lun	ade 76w76	evx			no dre		P=1.
b	M m b6a	orl	tba	mix 76w76	evx			no dre		P=1.
4153	R f f34	eat	TBA	MXB 24m24		:>		2.30gm	*	P<.8	-
a	R f f34	eat	liv	MXB 24m24				no dre		P=1.
4154	R m f34	eat	TBA	MXB 24m24		:>		no dre		P=1.	-
a	R m f34	eat	liv	MXB 24m24				2.65gm	*	P<.6

PIROXICAM   100ng..:..1ug....:..10......:..100....:..1mg....:..10.....:..100....:..1g.....:..10

| 4155 | R m f34 | eat | col | mix 54w54 | er | .> | | no dre | | P=1. |

PIVALOLACTONE   100ng..:..1ug....:..10......:..100....:..1mg....:..10.....:..100....:..1g.....:..10

4156	M f b6c	gav	TBA	MXB 24m24		:>		179.mg	*	P<.3	-
a	M f b6c	gav	liv	MXB 24m24				no dre		P=1.
b	M f b6c	gav	lun	MXB 24m24				482.mg	*	P<.6
4157	M m b6c	gav	TBA	MXB 24m24		:>		no dre		P=1.	-
a	M m b6c	gav	liv	MXB 24m24				no dre		P=1.
b	M m b6c	gav	lun	MXB 24m24				435.mg	*	P<.5

	RefNum	LoConf	UpConf	Cntrl	1Dose	1Inc	2Dose	2Inc					Citation or Pathology
													Brkly Code
a	TR120	929.mg	n.s.s.	1/20	135.mg	2/50	364.mg	5/50					liv:hpa,hpc,nnd.
b	TR120	1.49gm	n.s.s.	2/20	135.mg	6/50	364.mg	2/50					lun:a/a,a/c.
4133	1096	82.0mg	n.s.s.	0/16	41.4mg	0/18							Innes;ntis,1968/1969
a	1096	82.0mg	n.s.s.	0/16	41.4mg	0/18							
b	1096	25.1mg	n.s.s.	0/16	41.4mg	3/18							
4134	TR120	406.mg	n.s.s.	16/20	124.mg	32/50	336.mg	31/50					
a	TR120	528.mg	n.s.s.	10/20	124.mg	17/50	336.mg	20/50					liv:hpa,hpc,nnd.
b	TR120	998.mg	n.s.s.	5/20	124.mg	6/50	336.mg	8/50					lun:a/a,a/c.
4135	1096	13.1mg	274.mg	0/16	38.5mg	5/15							Innes;ntis,1968/1969
a	1096	24.2mg	n.s.s.	0/16	38.5mg	2/15							
b	1096	63.6mg	n.s.s.	0/16	38.5mg	0/15							
c	1096	9.48mg	78.6mg	0/16	38.5mg	7/15							
4136	1096	34.8mg	n.s.s.	1/17	41.4mg	2/18							
a	1096	82.0mg	n.s.s.	0/17	41.4mg	0/18							
b	1096	29.7mg	n.s.s.	2/17	41.4mg	3/18							
4137	1096	50.5mg	n.s.s.	2/18	38.5mg	1/18							
a	1096	76.3mg	n.s.s.	1/18	38.5mg	0/18							
b	1096	54.0mg	n.s.s.	3/18	38.5mg	1/18							
4138	TR120	631.mg	27.1gm	1/20	250.mg	7/50	500.mg	15/50					s
a	TR120	422.mg	n.s.s.	15/20	250.mg	33/50	500.mg	33/50					
b	TR120	n.s.s.	n.s.s.	0/20	250.mg	0/50	500.mg	0/50					liv:hpa,hpc,nnd.
4139	TR120	384.mg	n.s.s.	14/20	200.mg	34/50	400.mg	32/50					
a	TR120	5.20gm	n.s.s.	2/20	200.mg	0/50	400.mg	0/50					liv:hpa,hpc,nnd.
4140	1686	4.33gm	n.s.s.	0/47	236.mg	1/49	473.mg	0/49					Maekawa;fctx,23,675-682;1985
a	1686	726.mg	n.s.s.	35/47	236.mg	39/49	473.mg	30/49					
4141	1686	1.88gm	n.s.s.	1/48	189.mg	1/48	378.mg	3/46					
a	1686	3.28gm	n.s.s.	0/48	189.mg	0/48	378.mg	1/46					
b	1686	267.mg	n.s.s.	48/48	189.mg	48/48	378.mg	41/46					
4142	241	3.48gm	n.s.s.	1/24	4.50mg	0/24	45.0mg	0/24	450.mg	1/24	1.11gm	1/24	Sarles;atmh,1,862-883;1952
a	241	4.25gm	n.s.s.	0/24	4.50mg	0/24	45.0mg	0/24	450.mg	0/24	1.11gm	1/24	

PIPERONYL BUTOXIDE IN SOLVENT (butacide) 51-03-6

	RefNum	LoConf	UpConf	Cntrl	1Dose	1Inc							Citation or Pathology
4143	1097	187.mg	n.s.s.	1/17	157.mg	1/18							Innes;ntis,1968/1969
a	1097	311.mg	n.s.s.	0/17	157.mg	0/18							
b	1097	113.mg	n.s.s.	2/17	157.mg	3/18							
4144	1097	173.mg	n.s.s.	1/18	146.mg	1/18							
a	1097	192.mg	n.s.s.	2/18	146.mg	1/18							
b	1097	115.mg	n.s.s.	3/18	146.mg	3/18							
4145	1097	311.mg	n.s.s.	0/16	157.mg	0/18							
a	1097	311.mg	n.s.s.	0/16	157.mg	0/18							
b	1097	311.mg	n.s.s.	0/16	157.mg	0/18							
4146	1097	63.9mg	n.s.s.	0/16	146.mg	4/16							
a	1097	258.mg	n.s.s.	0/16	146.mg	0/16							
b	1097	39.5mg	340.mg	0/16	146.mg	7/16							

PIPERONYL SULFOXIDE   120-62-7

	RefNum	LoConf	UpConf	Cntrl	1Dose	1Inc	2Dose	2Inc					Citation or Pathology
4147	TR124	110.mg	n.s.s.	10/20	38.3mg	23/50	98.0mg	25/50					liv:hpa,hpc,nnd.
a	TR124	215.mg	n.s.s.	1/20	38.3mg	3/50	98.0mg	6/50					lun:a/a,a/c.
b	TR124	523.mg	n.s.s.	2/20	38.3mg	2/50	98.0mg	2/50					Innes;ntis,1968/1969
4148	1099	31.1mg	n.s.s.	0/16	15.7mg	0/18							
a	1099	31.1mg	n.s.s.	0/16	15.7mg	0/18							
b	1099	16.3mg	n.s.s.	0/16	15.7mg	1/18							
4149	TR124	31.0mg	n.s.s.	6/20	42.2mg	31/50	84.0mg	46/50					
a	TR124	46.4mg	n.s.s.	14/20	42.2mg	39/50	84.0mg	47/50					liv:hpa,hpc,nnd.
b	TR124	31.0mg	n.s.s.	6/20	42.2mg	31/50	84.0mg	46/50					lun:a/a,a/c.
c	TR124	106.mg	n.s.s.	6/20	42.2mg	10/50	(84.0mg	7/50)					Innes;ntis,1968/1969
4150	1099	4.05mg	29.9mg	0/16	14.6mg	8/18							
a	1099	11.2mg	n.s.s.	0/16	14.6mg	2/18							
b	1099	29.0mg	n.s.s.	0/16	14.6mg	0/18							
c	1099	3.56mg	22.2mg	0/16	14.6mg	9/18							
4151	1099	17.5mg	n.s.s.	1/17	15.7mg	1/17							
a	1099	29.3mg	n.s.s.	0/17	15.7mg	0/17							
b	1099	13.8mg	n.s.s.	2/17	15.7mg	2/17							
4152	1099	12.3mg	n.s.s.	1/18	14.6mg	2/18							
a	1099	29.0mg	n.s.s.	2/18	14.6mg	0/18							
b	1099	15.1mg	n.s.s.	3/18	14.6mg	2/18							
4153	TR124	283.mg	n.s.s.	12/20	150.mg	24/50	300.mg	27/50					
a	TR124	n.s.s.	n.s.s.	0/20	150.mg	0/50	300.mg	0/50					liv:hpa,hpc,nnd.
4154	TR124	60.1mg	n.s.s.	15/20	60.0mg	32/50	(120.mg	22/50)					
a	TR124	651.mg	n.s.s.	0/20	60.0mg	1/50	120.mg	1/50					liv:hpa,hpc,nnd.

PIROXICAM   36322-90-4

	RefNum	LoConf	UpConf	Cntrl	1Dose	1Inc							Citation or Pathology
4155	2078	10.7mg	n.s.s.	0/12	16.0mg	0/12							Reddy;carc,13,1019-1023;1992

PIVALOLACTONE   1955-45-9

	RefNum	LoConf	UpConf	Cntrl	1Dose	1Inc	2Dose	2Inc					Citation or Pathology
4156	TR140	55.7mg	n.s.s.	6/20	31.8mg	20/50	63.7mg	22/50					
a	TR140	340.mg	n.s.s.	0/20	31.8mg	2/50	63.7mg	0/50					liv:hpa,hpc,nnd.
b	TR140	161.mg	n.s.s.	0/20	31.8mg	4/50	63.7mg	2/50					lun:a/a,a/c.
4157	TR140	97.0mg	n.s.s.	12/20	31.8mg	24/50	63.7mg	25/50					
a	TR140	64.1mg	n.s.s.	4/20	31.8mg	10/50	(63.7mg	4/50)					liv:hpa,hpc,nnd.
b	TR140	114.mg	n.s.s.	2/20	31.8mg	6/50	63.7mg	10/50					lun:a/a,a/c.

```
      Spe Strain  Site   Xpo+Xpt                                                          TD50    2Tailpvl
          Sex   Route   Hist       Notes                                                      DR    AuOp
4158 R f f34 gav sto MXA 24m24                                      :  +   :               333.mg / P<.002 c
   a R f f34 gav TBA MXB 24m24                                                             200.mg * P<.3
   b R f f34 gav liv MXB 24m24                                                             no dre   P=1.
4159 R m f34 gav sto MXA 24m24                                       : + :                 154.mg * P<.0005c
   a R m f34 gav sto sqc 24m24                                                             532.mg * P<.009 c
   b R m f34 gav TBA MXB 24m24                                                             181.mg * P<.2
   c R m f34 gav liv MXB 24m24                                                             850.mg * P<.3

POLYBROMINATED BIPHENYL MIXTURE   100ng..:..1ug...:..10....:..100....:..1mg...:..10......:..100....:..1g....:..10
4160 M f b6c gav liv hpc  6m30                                     :   +   :               .534mg * P<.0005c
   a M f b6c gav TBA MXB  6m30                                                             .338mg * P<.003
   b M f b6c gav liv MXB  6m30                                                             .402mg * P<.0005
   c M f b6c gav lun MXB  6m30                                                             .915mg * P<.02
4161 M f b6c eat liv MXA 24m24                                   :+ :                       .307mg * P<.0005c
   a M f b6c eat liv hpa 24m24                                                             .367mg * P<.0005c
   b M f b6c eat liv hpc 24m24                                                             .588mg * P<.0005c
   c M f b6c eat TBA MXB 24m24                                                             .362mg * P<.0005
   d M f b6c eat liv MXB 24m24                                                             .307mg * P<.0005
   e M f b6c eat lun MXB 24m24                                                             19.1mg * P<.3
4162 M m b6c gav liv hpc  6m30                                     :  +  :                  .381mg * P<.0005c
   a M m b6c gav TBA MXB  6m30                                                             .418mg * P<.0005
   b M m b6c gav liv MXB  6m30                                                             .285mg * P<.0005
   c M m b6c gav lun MXB  6m30                                                             no dre   P=1.
4163 M m b6c eat liv hpa 24m24                                    :  +  :                   .233mg * P<.0005c
   a M m b6c eat liv MXA 24m24                                                             .262mg * P<.0005c
   b M m b6c eat liv hpc 24m24                                                             .540mg * P<.0005
   c M m b6c eat TBA MXB 24m24                                                             .349mg * P<.0005
   d M m b6c eat liv MXB 24m24                                                             .262mg * P<.0005
   e M m b6c eat lun MXB 24m24                                                             54.9mg * P<.9
4164 R f f34 gav liv MXA  6m29 y                                  :(+):                     .596mg * P<.0005c
   a R f f34 gav liv nnd  6m29 y                                                           1.10mg * P<.0005c
   b R f f34 gav liv hpc  6m29 y                                                           1.65mg * P<.0005c
   c R f f34 gav liv clc  6m29 y                                                           2.10mg * P<.0005c
   d R f f34 gav TBA MXB  6m29 y                                                           .365mg * P<.009
   e R f f34 gav liv MXB  6m29 y                                                           .645mg * P<.0005
4165 R f f34 eat liv MXA 24m24                                      : +:                    .656mg * P<.0005c
   a R f f34 eat liv hpa 24m24                                                             .700mg * P<.0005c
   b R f f34 eat mgl fba 24m24                                                             1.59mg \ P<.05
   c R f f34 eat liv hpc 24m24                                                             8.22mg * P<.02  c
   d R f f34 eat TBA MXB 24m24                                                             2.38mg * P<.3
   e R f f34 eat liv MXB 24m24                                                             .656mg * P<.0005
4166 R m f34 gav liv MXA  6m27 y                               :   (+)   :                  .148mg * P<.0005c
   a R m f34 gav liv nnd  6m27 y                                                           .168mg * P<.0005c
   b R m f34 gav liv hpc  6m27 y                                                           1.30mg * P<.0005c
   c R m f34 gav liv clc  6m27 y                                                           14.4mg * P<.006 c
   d R m f34 gav TBA MXB  6m27 y                                                           64.5ug * P<.0005
   e R m f34 gav liv MXB  6m27 y                                                           .151mg * P<.0005
4167 R m f34 eat liv MXA 24m24                                     : + :                    .408mg * P<.0005c
   a R m f34 eat liv hpa 24m24                                                             .461mg * P<.0005c
   b R m f34 eat liv hpc 24m24                                                             1.19mg * P<.0005c
   c R m f34 eat TBA MXB 24m24                                                             1.56mg * P<.3
   d R m f34 eat liv MXB 24m24                                                             .408mg * P<.0005

POLYBROMINATED BIPHENYLS   100ng..:..1ug...:..10....:..100....:..1mg...:..10......:..100....:..1g....:..10
4168 R f sss eat liv tum 57w57                                .>                           no dre   P=1.   -
   a R f sss eat tba mix 57w57                                                             5.91mg   P<.4

POLYSORBATE 80   100ng..:..1ug...:..10....:..100....:..1mg...:..10......:..100....:..1g....:..10
4169 M f b6c eat TBA MXB 24m24                                                           :> 42.9gm / P<.8   -
   a M f b6c eat liv MXB 24m24                                                             no dre   P=1.
   b M f b6c eat lun MXB 24m24                                                             332.gm * P<.9
4170 M m b6c eat TBA MXB 24m24                                                           :> no dre   P=1.   -
   a M m b6c eat liv MXB 24m24                                                             no dre   P=1.
   b M m b6c eat lun MXB 24m24                                                             no dre   P=1.
4171 R f f34 eat liv hpa 24m24                                                           : #15.8gm * P<.05  -
   a R f f34 eat TBA MXB 24m24                                                             no dre   P=1.
   b R f f34 eat liv MXB 24m24                                                             15.8gm * P<.05
4172 R m f34 eat amd MXA 99w99                                        :   ±                 2.58gm * P<.07  e
   a R m f34 eat TBA MXB 99w99                                                             3.29gm * P<.5
   b R m f34 eat liv MXB 99w99                                                             217.gm * P<.9

POLYVINYLPYRIDINE-N-OXIDE   100ng..:..1ug...:..10....:..100....:..1mg...:..10......:..100....:..1g....:..10
4173 M f ici ivj lun ade 17m24                                        .>                   no dre   P=1.   -
   a M f ici ivj tba mal 17m24                                                             no dre   P=1.   -
   b M f ici ivj tba ben 17m24                                                             1.69gm   P<1.   -
4174 R f wis ivj tba mal 17m24                                  .>                         175.mg   P<.8   -
   a R f wis ivj tba ben 17m24                                                             no dre   P=1.   -

POTASSIUM CHLORIDE   100ng..:..1ug...:..10....:..100....:..1mg...:..10......:..100....:..1g....:..10
4175 R m f34 eat amd phe 24m24                                             .>              6.21gm * P<.2   -
   a R m f34 eat mgl fba 24m24                                                             33.5gm Z P<.5   -
```

	RefNum	LoConf	UpConf	Cntrl	1Dose	1Inc	2Dose	2Inc							Citation or Pathology
															Brkly Code
4158	TR140	175.mg	1.03gm	0/20	63.1mg	2/50	126.mg	11/50							sto:sqc,sqp.
a	TR140	62.5mg	n.s.s.	14/20	63.1mg	35/50	126.mg	34/50							
b	TR140	n.s.s.	n.s.s.	0/20	63.1mg	0/50	126.mg	0/50							liv:hpa,hpc,nnd.
4159	TR140	96.7mg	293.mg	0/20	63.1mg	6/50	126.mg	21/50							sto:sqc,sqp.
a	TR140	241.mg	12.5gm	0/20	63.1mg	1/50	126.mg	7/50							
b	TR140	66.4mg	n.s.s.	11/20	63.1mg	29/50	126.mg	34/50							
c	TR140	323.mg	n.s.s.	0/20	63.1mg	3/50	126.mg	2/50							liv:hpa,hpc,nnd.

POLYBROMINATED BIPHENYL MIXTURE (firemaster FF-1) 67774-32-7

	RefNum	LoConf	UpConf	Cntrl	1Dose	1Inc	2Dose	2Inc	3Dose	3Inc	4Dose	4Inc	5Dose	5Inc	Citation or Pathology
4160	TR244	.250mg	1.66mg	0/13	14.2ug	0/19	42.5ug	2/15	.142mg	2/11	.425mg	3/17	1.45mg	7/8	
a	TR244	.141mg	2.97mg	12/13	14.2ug	13/19	42.5ug	12/15	.142mg	9/11	.425mg	16/17	1.45mg	8/8	
b	TR244	.180mg	1.22mg	0/13	14.2ug	2/19	42.5ug	2/15	.142mg	3/11	.425mg	4/17	1.45mg	8/8	liv:hpa,hpc,nnd.
c	TR244	.285mg	n.s.s.	0/13	14.2ug	1/19	42.5ug	1/15	.142mg	1/11	.425mg	3/17	1.45mg	1/8	lun:a/a,a/c.
4161	TR398	.204mg	.477mg	5/50	1.30mg	42/50	3.90mg	47/50							liv:hpa,hpc.
a	TR398	.245mg	.562mg	4/50	1.30mg	39/50	3.90mg	46/50							
b	TR398	.364mg	.975mg	1/50	1.30mg	22/50	3.90mg	35/50							
c	TR398	.234mg	.601mg	25/50	1.30mg	46/50	3.90mg	47/50							
d	TR398	.204mg	.477mg	5/50	1.30mg	42/50	3.90mg	47/50							liv:hpa,hpb,hpc.
e	TR398	3.06mg	n.s.s.	1/50	1.30mg	2/50	3.90mg	1/50							lun:a/a,a/c.
4162	TR244	.200mg	.899mg	12/25	14.2ug	8/27	42.5ug	8/24	.142mg	12/25	.425mg	15/23	1.47mg	21/22	
a	TR244	.190mg	1.87mg	20/25	14.2ug	23/27	42.5ug	22/24	.142mg	23/25	.425mg	21/23	1.47mg	22/22	
b	TR244	.147mg	.685mg	14/25	14.2ug	9/27	42.5ug	12/24	.142mg	14/25	.425mg	17/23	1.47mg	22/22	liv:hpa,hpc,nnd.
c	TR244	.471mg	n.s.s.	1/25	14.2ug	1/27	42.5ug	3/24	.142mg	4/25	.425mg	3/23	1.47mg	0/22	lun:a/a,a/c.
4163	TR398	.139mg	.401mg	9/50	1.20mg	48/50	3.60mg	42/50							
a	TR398	.156mg	.453mg	16/50	1.20mg	48/50	3.60mg	48/50							liv:hpa,hpc.
b	TR398	.319mg	.922mg	8/50	1.20mg	30/50	3.60mg	36/50							
c	TR398	.203mg	.660mg	37/50	1.20mg	48/50	3.60mg	48/50							
d	TR398	.156mg	.453mg	16/50	1.20mg	48/50	3.60mg	48/50							liv:hpa,hpb,hpc.
e	TR398	1.92mg	n.s.s.	13/50	1.20mg	10/50	3.60mg	2/50							lun:a/a,a/c.
4164	TR244	.323mg	1.22mg	0/20	14.4ug	2/21	43.2ug	0/21	.144mg	2/11	.432mg	7/19	1.51mg	16/20	liv:clc,hpc,nnd.
a	TR244	.505mg	3.47mg	0/20	14.4ug	2/21	43.2ug	0/21	.144mg	2/11	.432mg	5/19	1.51mg	8/20	
b	TR244	.619mg	4.21mg	0/20	14.4ug	0/21	43.2ug	0/21	.144mg	0/11	.432mg	3/19	1.51mg	7/20	
c	TR244	.800mg	6.40mg	0/20	14.4ug	0/21	43.2ug	0/21	.144mg	0/11	.432mg	0/19	1.51mg	7/20	
d	TR244	.136mg	13.7mg	18/20	14.4ug	18/21	43.2ug	17/21	.144mg	11/11	.432mg	19/19	1.51mg	17/20	
e	TR244	.343mg	1.35mg	0/20	14.4ug	2/21	43.2ug	0/21	.144mg	2/11	.432mg	7/19	1.51mg	15/20	liv:hpa,hpc,nnd.
4165	TR398	.464mg	.970mg	0/50	.500mg	12/50	1.50mg	39/50							liv:hpa,hpc.
a	TR398	.490mg	1.05mg	0/50	.500mg	10/50	1.50mg	38/50							
b	TR398	.616mg	n.s.s.	4/50	.500mg	11/50	(1.50mg	2/50)							S
c	TR398	3.34mg	n.s.s.	0/50	.500mg	2/50	1.50mg	4/50							
d	TR398	.699mg	n.s.s.	40/50	.500mg	42/50	1.50mg	47/50							
e	TR398	.464mg	.970mg	0/50	.500mg	12/50	1.50mg	39/50							liv:hpa,hpb,hpc.
4166	TR244	45.7ug	.491mg	0/33	15.1ug	2/39	46.2ug	1/40	.151mg	5/31	.478mg	10/33	1.73mg	10/32	liv:clc,hpc,nnd.
a	TR244	47.2ug	.608mg	0/33	15.1ug	0/39	46.2ug	1/40	.151mg	4/31	.478mg	4/33	1.73mg	1/32	
b	TR244	.484mg	3.70mg	0/33	15.1ug	2/39	46.2ug	0/40	.151mg	1/31	.478mg	7/33	1.73mg	7/32	
c	TR244	3.45mg	286.mg	0/33	15.1ug	0/39	46.2ug	0/40	.151mg	0/31	.478mg	0/33	1.73mg	2/32	
d	TR244	29.8ug	.163mg	30/33	15.1ug	34/39	46.2ug	34/40	.151mg	24/31	.478mg	25/33	1.73mg	18/32	
e	TR244	45.9ug	.619mg	0/33	15.1ug	2/39	46.2ug	1/40	.151mg	5/31	.478mg	10/33	1.73mg	8/32	liv:hpa,hpc,nnd.
4167	TR398	.282mg	.646mg	1/50	.400mg	12/50	1.20mg	41/50							liv:hpa,hpc.
a	TR398	.313mg	.745mg	1/50	.400mg	10/50	1.20mg	38/50							
b	TR398	.700mg	2.25mg	0/50	.400mg	2/50	1.20mg	19/50							
c	TR398	.429mg	n.s.s.	45/50	.400mg	48/50	1.20mg	50/50							
d	TR398	.282mg	.646mg	1/50	.400mg	12/50	1.20mg	41/50							liv:hpa,hpb,hpc.

POLYBROMINATED BIPHENYLS (firemaster BP-6) 59536-65-1

	RefNum	LoConf	UpConf	Cntrl	1Dose	1Inc	Citation or Pathology
4168	1061	1.86mg	n.s.s.	0/8	2.50mg	0/12	Schwartz;jnci,64,63-67;1980
a	1061	.962mg	n.s.s.	0/8	2.50mg	1/12	

POLYSORBATE 80   9005-65-6

	RefNum	LoConf	UpConf	Cntrl	1Dose	1Inc	2Dose	2Inc	Citation or Pathology
4169	TR415	5.34gm	n.s.s.	30/51	3.19gm	16/50	6.42gm	29/50	
a	TR415	23.6gm	n.s.s.	3/51	3.19gm	2/50	6.42gm	2/50	liv:hpa,hpc,nnd.
b	TR415	18.1gm	n.s.s.	3/51	3.19gm	2/50	6.42gm	3/50	lun:a/a,a/c.
4170	TR415	5.33gm	n.s.s.	32/53	2.95gm	34/53	5.92gm	31/50	
a	TR415	12.5gm	n.s.s.	15/53	2.95gm	13/53	5.92gm	11/50	liv:hpa,hpc,nnd.
b	TR415	15.7gm	n.s.s.	6/53	2.95gm	6/53	5.92gm	5/50	lun:a/a,a/c.
4171	TR415	5.46gm	n.s.s.	0/50	1.23gm	1/50	2.47gm	3/50	S
a	TR415	1.41gm	n.s.s.	49/50	1.23gm	48/50	2.47gm	48/50	
b	TR415	5.46gm	n.s.s.	0/50	1.23gm	1/50	2.47gm	5/50	liv:hpa,hpc,nnd.
4172	TR415	1.02gm	n.s.s.	21/50	1.00gm	19/50	2.00gm	29/50	amd:pbb,phm,pob.
a	TR415	781.mg	n.s.s.	47/50	1.00gm	46/50	2.00gm	46/50	
b	TR415	8.20gm	n.s.s.	2/50	1.00gm	0/50	2.00gm	2/50	liv:hpa,hpc,nnd.

POLYVINYLPYRIDINE-N-OXIDE   ---

	RefNum	LoConf	UpConf	Cntrl	1Dose	1Inc	Citation or Pathology
4173	1419	48.0mg	n.s.s.	3/39	8.24mg	2/50	Schmahl;arzn,19,1313-1314;1969
a	1419	14.9mg	n.s.s.	14/39	8.24mg	17/50	
b	1419	30.9mg	n.s.s.	3/39	8.24mg	4/50	
4174	1419	18.2mg	n.s.s.	4/48	5.89mg	5/48	
a	1419	21.8mg	n.s.s.	4/48	5.89mg	4/48	

POLYASSIUM CHLORIDE   7447-40-7

	RefNum	LoConf	UpConf	Cntrl	1Dose	1Inc	2Dose	2Inc	3Dose	3Inc	Citation or Pathology
4175	1807	1.84gm	n.s.s.	11/50	100.mg	16/50	400.mg	20/50	1.60gm	19/50	Imai;jnma,37,115-127;1986
a	1807	5.60gm	n.s.s.	0/50	100.mg	6/50	400.mg	1/50	1.60gm	4/50	

```
     Spe Strain Site  Xpo+Xpt                                                          TD50    2Tailpvl
        Sex Route  Hist   Notes                                                        DR      AuOp

PRAZEPAM                   100ng..:..1ug....:..10......:..100.....:..1mg....:..10.....:..100....:..1g....:..10
4176 M f cf1 eat bon otm 80w80 e                                                       264.mg * P<.01  -
a    M f cf1 eat --- lyt 80w80 e                                             .   +     146.mg * P<.07  -
b    M f cf1 eat lun agt 80w80 e                                                       15.4gm * P<1.   -
c    M f cf1 eat liv hpc 80w80 e                                                       no dre   P=1.   -
d    M f cf1 eat liv hpa 80w80 e                                                       no dre   P=1.   -
e    M f cf1 eat tba mal 80w80 e                                                       108.mg * P<.07  -
f    M f cf1 eat tba ben 80w80 e                                                       251.mg * P<.09  -
g    M f cf1 eat tba mix 80w80 e                                                       139.mg * P<.3   -
4177 M m cf1 eat liv hpa 80w80 e                                              .>       2.15gm * P<.5   -
a    M m cf1 eat liv hpc 80w80 e                                                       8.96gm * P<1.   -
b    M m cf1 eat --- lyt 80w80 e                                                       no dre   P=1.   -
c    M m cf1 eat lun agt 80w80 e                                                       no dre   P=1.   -
d    M m cf1 eat tba ben 80w80 e                                                       639.mg * P<.3   -
e    M m cf1 eat tba mix 80w80 e                                                       6.41gm * P<1.   -
f    M m cf1 eat tba mal 80w80 e                                                       no dre   P=1.   -
4178 R f wal eat mgl mix 24m24 e                                          .   ±        207.mg * P<.06  -
a    R f wal eat liv hpa 24m24 e                                                       1.45gm * P<.06  -
b    R f wal eat pit cra 24m24 e                                                       no dre   P=1.   -
c    R f wal eat tba mal 24m24 e                                                       415.mg * P<.1   -
d    R f wal eat tba mix 24m24 e                                                       230.mg * P<.4   -
e    R f wal eat tba ben 24m24 e                                                       67.5gm * P<1.   -
4179 R m wal eat liv hpc 24m24 e                                                .>     2.39gm * P<.2   -
a    R m wal eat liv hpa 24m24 e                                                       3.54gm * P<.5   -
b    R m wal eat pit cra 24m24 e                                                       no dre   P=1.   -
c    R m wal eat tba ben 24m24 e                                                       1.51gm Z P<.7   -
d    R m wal eat tba mix 24m24 e                                                       3.08gm * P<.9   -
e    R m wal eat tba mal 24m24 e                                                       no dre   P=1.   -

PRAZIQUANTEL               100ng..:..1ug....:..10......:..100.....:..1mg....:..10......:..100....:..1g......:..10
4180 H f syg gav tba mix 80w80 e                                         .>            250.mg * P<.5   -
4181 H m syg gav tba mix 80w80 e                                           .>          no dre   P=1.   -
4182 R f sda gav tba mix 24m30 e                                         .>            no dre   P=1.   -
4183 R m sda gav tba mix 24m30 e                                          .>           no dre   P=1.   -

PREDNIMUSTINE              100ng..:..1ug....:..10......:..100.....:..1mg....:..10......:..100....:..1g......:..10
4184 R f sda gav auc sqc 18m24                                 .   +      .            19.2mg * P<.005 +
a    R f sda gav tba ben 18m24                                                         14.0mg * P<.5   -
b    R f sda gav tba mal 18m24                                                         21.6mg * P<.3   -

PREDNISOLONE               100ng..:..1ug....:..10......:..100.....:..1mg....:..10......:..100....:..1g......:..10
4185 R m cdr wat liv mix 24m24 ers                         .   +    .                  1.53mg   P<.0005+
a    R m cdr wat liv hpa 24m24 ers                                                     2.16mg   P<.002
b    R m cdr wat liv hpc 24m24 ers                                                     6.58mg   P<.08
4186 R f sda gav tba ben 18m24                                  .>                     1.96mg * P<.2   -
a    R f sda gav tba mal 18m24                                                         no dre   P=1.   -

PREDNISONE                 100ng..:..1ug....:..10......:..100.....:..1mg....:..10......:..100....:..1g......:..10
4187 M f cd1 eat lun abt 78w78 es                                    .>                no dre   P=1.   -
a    M f cd1 eat lun ulc 78w78 es                                                      no dre   P=1.   -
b    M f cd1 eat liv mix 78w78 es                                                      no dre   P=1.   -
4188 M m cd1 eat liv mix 78w78 es                                 .>                   no dre   P=1.   -
a    M m cd1 eat lun abt 78w78 es                                                      no dre   P=1.   -

PREMARIN                   100ng..:..1ug....:..10......:..100.....:..1mg....:..10......:..100....:..1g......:..10
4189 R f cdr eat liv tum 24m24 e                                .>                     no dre   P=1.   -
4190 R m cdr eat liv hpt 24m24 e                                   .>                  no dre   P=1.   -

PRIMIDOLOL.HCl             100ng..:..1ug....:..10......:..100.....:..1mg....:..10......:..100....:..1g......:..10
4191 M f cd1 eat lun ade 78w78 e                                     .>                449.mg * P<.4
a    M f cd1 eat lun adc 78w78 e                                                       1.67gm * P<.4
b    M f cd1 eat liv hpc 78w78 e                                                       1.67gm * P<.2
c    M f cd1 eat tba mix 78w78 e                                                       52.3mg * P<.005
4192 M m cd1 eat liv hpc 78w78 e                                     .>                935.mg * P<.5
a    M m cd1 eat liv cho 78w78 e                                                       no dre   P=1.
b    M m cd1 eat liv ade 78w78 e                                                       no dre   P=1.
c    M m cd1 eat lun adc 78w78 e                                                       no dre   P=1.
d    M m cd1 eat lun ade 78w78 e                                                       no dre   P=1.
e    M m cd1 eat tba mix 78w78 e                                                       no dre   P=1.

PROBENECID                 100ng..:..1ug....:..10......:..100.....:..1mg....:..10......:..100....:..1g......:..10
4193 M f b6c gav liv MXA 24m24                                              :   +    : 540.mg * P<.003 p
a    M f b6c gav liv hpa 24m24                                                         561.mg * P<.002 p
b    M f b6c gav TBA MXB 24m24                                                         498.mg * P<.2
c    M f b6c gav liv MXB 24m24                                                         540.mg * P<.003
d    M f b6c gav lun MXB 24m24                                                         1.98gm * P<.3
4194 M m b6c gav sub fib 24m24                                              :   ±     #2.13gm * P<.02
a    M m b6c gav TBA MXB 24m24                                                         276.mg * P<.02
b    M m b6c gav liv MXB 24m24                                                         731.mg * P<.2
c    M m b6c gav lun MXB 24m24                                                         2.32gm * P<.5
4195 R f f34 gav TBA MXB 24m24                                        :>                1.54gm * P<.8  -
```

RefNum	LoConf	UpConf	Cntrl	1Dose	1Inc	2Dose	2Inc				Citation or Pathology

Brkly Code

PRAZEPAM   2955-38-6

RefNum	LoConf	UpConf	Cntrl	1Dose	1Inc	2Dose	2Inc	3Dose	3Inc	Citation or Pathology
4176	2139	108.mg 15.3gm	5/100	8.00mg	0/50	25.0mg	3/50	75.0mg	8/50	de la Iglesia;txap,57,39-54;1981/pers.comm.
a	2139	54.5mg n.s.s.	21/100	8.00mg	11/50	25.0mg	17/50	75.0mg	17/50	
b	2139	142.mg n.s.s.	16/100	8.00mg	4/50	25.0mg	9/50	75.0mg	7/50	
c	2139	452.mg n.s.s.	1/100	8.00mg	0/50	25.0mg	1/50	75.0mg	0/50	
d	2139	338.mg n.s.s.	1/100	8.00mg	1/50	25.0mg	2/50	75.0mg	0/50	
e	2139	40.6mg n.s.s.	41/100	8.00mg	16/50	25.0mg	22/50	75.0mg	27/50	
f	2139	86.2mg n.s.s.	11/100	8.00mg	4/50	25.0mg	8/50	75.0mg	10/50	
g	2139	41.5mg n.s.s.	49/100	8.00mg	18/50	25.0mg	27/50	75.0mg	28/50	
4177	2139	296.mg n.s.s.	1/100	8.00mg	0/50	25.0mg	1/50	75.0mg	1/50	
a	2139	179.mg n.s.s.	7/100	8.00mg	5/50	25.0mg	4/50	75.0mg	4/50	
b	2139	135.mg n.s.s.	10/100	8.00mg	7/50	25.0mg	11/50	75.0mg	5/50	
c	2139	175.mg n.s.s.	19/100	8.00mg	4/50	25.0mg	7/50	75.0mg	7/50	
d	2139	164.mg n.s.s.	6/100	8.00mg	0/50	25.0mg	2/50	75.0mg	5/50	
e	2139	80.2mg n.s.s.	38/100	8.00mg	16/50	25.0mg	21/50	75.0mg	18/50	
f	2139	89.6mg n.s.s.	35/100	8.00mg	15/50	25.0mg	21/50	75.0mg	16/50	
4178	2139	80.4mg n.s.s.	41/115	8.00mg	26/65	25.0mg	25/65	75.0mg	33/65	
a	2139	431.mg n.s.s.	1/115	8.00mg	0/65	25.0mg	1/65	75.0mg	3/65	
b	2139	314.mg n.s.s.	34/115	8.00mg	18/65	25.0mg	23/65	75.0mg	13/65	
c	2139	145.mg n.s.s.	18/115	8.00mg	9/65	25.0mg	13/65	75.0mg	16/65	
d	2139	55.1mg n.s.s.	80/115	8.00mg	42/65	25.0mg	50/65	75.0mg	48/65	
e	2139	91.5mg n.s.s.	72/115	8.00mg	39/65	25.0mg	46/65	75.0mg	40/65	
4179	2139	587.mg n.s.s.	0/115	8.00mg	0/65	25.0mg	1/65	75.0mg	1/65	
a	2139	548.mg n.s.s.	1/115	8.00mg	0/65	25.0mg	2/65	75.0mg	1/65	
b	2139	385.mg n.s.s.	17/115	8.00mg	2/65	25.0mg	6/65	75.0mg	7/65	
c	2139	183.mg n.s.s.	37/115	8.00mg	7/65	25.0mg	16/65	75.0mg	19/65	
d	2139	137.mg n.s.s.	57/115	8.00mg	21/65	25.0mg	26/65	75.0mg	30/65	
e	2139	254.mg n.s.s.	23/115	8.00mg	16/65	25.0mg	12/65	75.0mg	13/65	

PRAZIQUANTEL   (embay 8440, droncit) 55268-74-1

RefNum	LoConf	UpConf	Cntrl	1Dose	1Inc	2Dose	2Inc	Citation or Pathology
4180	1519m	50.3mg n.s.s.	12/99	14.3mg	9/49	35.7mg	8/49	Ketkar;txcy,24,345-350;1982
4181	1519m	79.5mg n.s.s.	23/99	14.3mg	16/49	35.7mg	7/50	
4182	1519n	57.7mg n.s.s.	76/97	11.4mg	37/50	28.6mg	35/50	
4183	1519n	106.mg n.s.s.	61/103	11.4mg	27/50	28.6mg	23/50	

PREDNIMUSTINE   29069-24-7

RefNum	LoConf	UpConf	Cntrl	1Dose	1Inc	2Dose	2Inc	Citation or Pathology
4184	1770	6.64mg 167.mg	0/120	1.29mg	2/30	2.58mg	2/30	Berger;smon,13,8-13;1986
a	1770	2.92mg n.s.s.	43/120	1.29mg	12/30	2.58mg	13/30	
b	1770	5.23mg n.s.s.	13/120	1.29mg	3/30	2.58mg	6/30	

PREDNISOLONE   50-24-8

RefNum	LoConf	UpConf	Cntrl	1Dose	1Inc	2Dose	2Inc	Citation or Pathology
4185	2059	.799mg 4.81mg	11/200	.400mg	21/100			Ryrfeldt;txpy,20,115-117;1992
a	2059	1.04mg 9.22mg	7/200	.400mg	15/100			
b	2059	2.05mg n.s.s.	4/200	.400mg	6/100			
4186	1770	.599mg n.s.s.	43/120	.323mg	12/30	.645mg	15/30	Berger;smon,13,8-13;1986
a	1770	2.37mg n.s.s.	13/120	.323mg	4/30	.645mg	2/30	

PREDNISONE   53-03-2

RefNum	LoConf	UpConf	Cntrl	1Dose	1Inc	2Dose	2Inc	3Dose	3Inc	4Dose	4Inc	Citation or Pathology
4187	2104	19.7mg n.s.s.	6/48	.250mg	4/50	.500mg	3/50	1.00mg	2/50	5.00mg	2/49	Dillberger;txpy,20,18-26;1992
a	2104	22.1mg n.s.s.	0/48	.250mg	0/50	.500mg	0/50	1.00mg	1/50	5.00mg	0/49	
b	2104	31.0mg n.s.s.	3/47	.250mg	3/50	.500mg	1/50	1.00mg	1/50	5.00mg	0/49	
4188	2104	3.52mg n.s.s.	12/50	.250mg	7/50	.500mg	8/50	1.00mg	4/50	(5.00mg	1/50)	
a	2104	17.7mg n.s.s.	9/50	.250mg	6/50	.500mg	4/50	1.00mg	6/50	5.00mg	3/50	

PREMARIN   (conjugated equine estrogens) 12126-59-9

RefNum	LoConf	UpConf	Cntrl	1Dose	1Inc	2Dose	2Inc	Citation or Pathology
4189	108	.262mg n.s.s.	0/20	70.0ug	0/20	.700mg	0/20	Gibson;txap,11,489-510;1967
4190	108	2.32mg n.s.s.	0/20	70.0ug	1/20	.700mg	0/20	

PRIMIDOLOL.HCl   40778-40-3

RefNum	LoConf	UpConf	Cntrl	1Dose	1Inc	2Dose	2Inc	3Dose	3Inc	Citation or Pathology
4191	1669	95.9mg n.s.s.	8/100	12.5mg	5/50	25.0mg	2/49	50.0mg	7/50	Faccini;txcy,21,279-290;1981
a	1669	272.mg n.s.s.	0/100	12.5mg	0/50	25.0mg	1/49	50.0mg	0/50	
b	1669	271.mg n.s.s.	0/100	12.5mg	0/50	25.0mg	0/49	50.0mg	1/50	
c	1669	25.5mg 595.mg	24/100	12.5mg	19/50	25.0mg	17/49	50.0mg	24/50	
4192	1669	130.mg n.s.s.	2/100	50.0mg	2/50					
a	1669	290.mg n.s.s.	1/100	50.0mg	0/50					
b	1669	290.mg n.s.s.	8/100	50.0mg	0/50					
c	1669	290.mg n.s.s.	1/100	50.0mg	0/50					
d	1669	121.mg n.s.s.	22/100	50.0mg	7/50					
e	1669	101.mg n.s.s.	48/100	50.0mg	15/50					

PROBENECID   57-66-9

RefNum	LoConf	UpConf	Cntrl	1Dose	1Inc	2Dose	2Inc	Citation or Pathology
4193	TR395	256.mg 3.80gm	5/50	70.4mg	3/50	282.mg	16/50	liv:hpa,hpc.
a	TR395	273.mg 2.65gm	3/50	70.4mg	2/50	282.mg	14/50	
b	TR395	161.mg n.s.s.	23/50	70.4mg	21/50	282.mg	31/50	
c	TR395	256.mg 3.80gm	5/50	70.4mg	3/50	282.mg	16/50	liv:hpa,hpc,nnd.
d	TR395	488.mg n.s.s.	3/50	70.4mg	4/50	282.mg	6/50	lun:a/a,a/c.
4194	TR395	643.mg n.s.s.	0/50	70.5mg	0/50	282.mg	3/50	s
a	TR395	122.mg n.s.s.	25/50	70.5mg	26/50	282.mg	36/50	
b	TR395	241.mg n.s.s.	15/50	70.5mg	12/50	282.mg	19/50	liv:hpa,hpc,nnd.
c	TR395	466.mg n.s.s.	5/50	70.5mg	4/50	282.mg	6/50	lun:a/a,a/c.
4195	TR395	129.mg n.s.s.	48/50	70.1mg	46/50	280.mg	37/50	

Spe Strain Site Xpo+Xpt		TD50	2Tailpvl
Sex Route Hist Notes		DR	AuOp

```
     Spe Strain Site  Xpo+Xpt                                                          TD50     2Tailpvl
         Sex Route Hist   Notes                                                          DR       AuOp
 a    R f f34 gav liv MXB 24m24                                                         1.79gm * P<.07
4196  R m f34 gav tes MXA 24m24                                      :     ±           #212.mg * P<.05  -
 a    R m f34 gav TBA MXB 24m24                                                         296.mg * P<.1
 b    R m f34 gav liv MXB 24m24                                                         no dre   P=1.

PROCARBAZINE                         100ng..:..1ug....:..10......:..100....:..1mg....:..10......:..100....:..1g.....:..10
4197  R m b46 ivj tba mix 12m24                                      .     +     .       4.01mg   P<.002 +
 a    R m b46 ivj tba mal 12m24                                                         3.93mg   P<.0005
 b    R m b46 ivj tba ben 12m24                                                         no dre   P=1.

PROCARBAZINE.HCl                     100ng..:..1ug....:..10......:..100....:..1mg....:..10......:..100....:..1g.....:..10
4198  M f b6c ipj MXB MXB 52w85 s                              :     +     :             .194mg / P<.0005
 a    M f b6c ipj ute acn 52w85 s                                                       .253mg / P<.0005c
 b    M f b6c ipj --- MXA 52w85 s                                                       .758mg * P<.0005c
 c    M f b6c ipj --- lym 52w85 s                                                       .908mg * P<.002 c
 d    M f b6c ipj lun a/a 52w85 s                                                      3.83mg / P<.002 c
 e    M f b6c ipj bra oln 52w85 s                                                      6.34mg / P<.0005c
 f    M f b6c ipj TBA MXB 52w85 s                                                       .193mg / P<.0005
 g    M f b6c ipj liv MXB 52w85 s                                                       no dre   P=1.
 h    M f b6c ipj lun MXB 52w85 s                                                      3.83mg / P<.002
4199  M f b6c ipj ute acn 52w80 s         pool                 :     +     :             .411mg / P<.0005c
 a    M f b6c ipj --- MXA 52w80 s                                                       .690mg * P<.0005c
 b    M f b6c ipj --- lym 52w80 s                                                       .855mg * P<.0005c
 c    M f b6c ipj lun a/a 52w80 s                                                      3.39mg / P<.0005c
 d    M f b6c ipj bra oln 52w80 s                                                      5.62mg / P<.0005c
4200  M m b6c ipj MXB MXB 52w85 s                              :     +     :             .511mg / P<.0005
 a    M m b6c ipj lun MXA 52w85 s                                                       .623mg / P<.0005c
 b    M m b6c ipj lun a/a 52w85 s                                                       .721mg / P<.0005c
 c    M m b6c ipj --- MXA 52w85 s                                                      3.34mg * P<.009 c
 d    M m b6c ipj bra MXA 52w85 s                                                      4.51mg / P<.0005c
 e    M m b6c ipj bra oln 52w85 s                                                      5.19mg / P<.0005c
 f    M m b6c ipj TBA MXB 52w85 s                                                       .464mg / P<.0005
 g    M m b6c ipj liv MXB 52w85 s                                                      3.28mg / P<.01
 h    M m b6c ipj lun MXB 52w85 s                                                       .623mg / P<.0005
4201  M m b6c ipj lun MXA 52w85 s         pool                 :     +     :             .623mg / P<.0005c
 a    M m b6c ipj lun a/a 52w85 s                                                       .721mg / P<.0005c
 b    M m b6c ipj --- MXA 52w85 s                                                      3.34mg * P<.002 c
 c    M m b6c ipj bra MXA 52w85 s                                                      4.51mg / P<.0005c
 d    M m b6c ipj bra oln 52w85 s                                                      5.19mg / P<.0005c
4202  M f swi ipj lun mix 26w76 e                           .     +     .              1.05mg   P<.0005+
 a    M f swi ipj --- leu 26w76 e                                                      3.72mg   P<.0005
 b    M f swi ipj ute car 26w76 e                                                      5.75mg   P<.004
 c    M f swi ipj ute sar 26w76 e                                                      6.11mg   P<.03
 d    M f swi ipj --- lys 26w76 e                                                      6.99mg   P<.09  +
 e    M f swi ipj liv lys 26w76 e                                                      no dre   P=1.
 f    M f swi ipj tba mix 26w76 e                                                       .244mg   P<.0005
 g    M f swi ipj tba mal 26w76 e                                                       .391mg   P<.0005
 h    M f swi ipj tba ben 26w76 e                                                      27.8mg   P<.8
4203  M m swi ipj lun mix 26w78 e                           .     +     .              1.53mg   P<.0005+
 a    M m swi ipj liv mix 26w78 e                                                      no dre   P=1.
 b    M m swi ipj tba mix 26w78 e                                                      1.15mg   P<.004
 c    M m swi ipj tba mal 26w78 e                                                      1.74mg   P<.009
 d    M m swi ipj tba ben 26w78 e                                                      9.84mg   P<.4
4204  P b cym mix --- ost 16y19 Wmw                               :     +     :        5.91mg   P<.003

 a    P b cym mix --- lna 16y19 Wmw                                                    14.8mg   P<.005 +
 b    P b cym mix bra ast 16y19 Wmw                                                    6.19mg   P<.06
 c    P b cym mix --- lcl 16y19 Wmw                                                    21.1mg   P<.04
 d    P b cym mix kid hes 16y19 Wmw                                                    23.5mg   P<.05
 e    P b cym mix tba mal 16y19 Wmw                                                    2.05mg   P<.0005
4205  P b rhe mix --- lna 16y23 mw                                :     +     :        4.21mg   P<.0005+
 a    P b rhe mix cec adc 16y23 mw                                                     5.16mg   P<.04
 b    P b rhe mix hum ost 16y23 mw                                                     32.7mg   P<.05
 c    P b rhe mix kid hes 16y23 mw                                                     39.6mg   P<.06
 d    P b rhe mix tba mal 16y23 mw                                                     1.89mg   P<.0005
4206  R f sda ipj MXB MXB 26w86 s                           :     +     :               .337mg / P<.0005
 a    R f sda ipj MXA MXA 26w86 s                                                       .416mg * P<.0005c
 b    R f sda ipj bra oln 26w86 s                                                       .418mg * P<.0005c
 c    R f sda ipj mgl MXA 26w86 s                                                       .458mg * P<.0005c
 d    R f sda ipj mgl acn 26w86 s                                                       .459mg * P<.0005c
 e    R f sda ipj --- lym 26w86 s                                                      5.18mg / P<.0005c
 f    R f sda ipj TBA MXB 26w86 s                                                       .329mg / P<.0005
 g    R f sda ipj liv MXB 26w86 s                                                      no dre   P=1.
4207  R f sda ipj MXA MXA 26w80 s         pool              :     +     :               .360mg * P<.0005c
 a    R f sda ipj bra oln 26w80 s                                                       .361mg * P<.0005c
 b    R f sda ipj mgl MXA 26w80 s                                                       .396mg * P<.0005c
 c    R f sda ipj mgl acn 26w80 s                                                       .397mg * P<.0005c
 d    R f sda ipj --- lym 26w80 s                                                      4.48mg / P<.0005c
4208  R m sda ipj MXB MXB 26w61 s                           :     +     :               .284mg * P<.0005
 a    R m sda ipj MXA MXA 26w61 s                                                       .343mg * P<.0005c
 b    R m sda ipj --- MXA 26w61 s                                                      2.10mg * P<.0005c
```

	RefNum	LoConf	UpConf	Cntrl	1Dose	1Inc	2Dose	2Inc	Citation or Pathology
									Brkly Code
a	TR395	543.mg	n.s.s.	0/50	70.1mg	1/50	280.mg	2/50	
4196	TR395	85.3mg	n.s.s.	45/50	70.0mg	39/50	281.mg	44/50	liv:hpa,hpc,nnd.
a	TR395	107.mg	n.s.s.	42/50	70.0mg	41/50	281.mg	41/50	tes:iab,ica. S
b	TR395	1.13gm	n.s.s.	1/50	70.0mg	2/50	281.mg	0/50	liv:hpa,hpc,nnd.
PROCARBAZINE	671-16-9								
4197	1017	1.92mg	17.2mg	7/89	1.71mg	15/48			Schmahl;arzn,20,1461-1467;1970
a	1017	1.95mg	12.4mg	4/89	1.71mg	14/48			
b	1017	11.4mg	n.s.s.	3/89	1.71mg	1/48			
PROCARBAZINE.HCl	366-70-1								
4198	TR19	65.1ug	.432mg	0/15	2.00mg	19/35	5.20mg	18/35	---:leu,lym; bra:oln; lun:a/a; ute:acn. C
a	TR19	68.9ug	.728mg	0/15	2.00mg	14/35	5.20mg	8/35	
b	TR19	.289mg	2.96mg	0/15	2.00mg	8/35	5.20mg	2/35	---:leu,lym.
c	TR19	.308mg	5.34mg	0/15	2.00mg	6/35	5.20mg	2/35	
d	TR19	1.05mg	19.5mg	0/15	2.00mg	1/35	5.20mg	6/35	
e	TR19	2.93mg	16.6mg	0/15	2.00mg	0/35	5.20mg	11/35	
f	TR19	65.0ug	.422mg	0/15	2.00mg	19/35	5.20mg	20/35	
g	TR19	n.s.s.	n.s.s.	0/15	2.00mg	0/35	5.20mg	0/35	liv:hpa,hpc,nnd.
h	TR19	1.05mg	19.5mg	0/15	2.00mg	1/35	5.20mg	6/35	lun:a/a,a/c.
4199	TR19	.203mg	.834mg	0/45p	2.00mg	14/35	5.20mg	8/35	
a	TR19	.258mg	2.69mg	1/45p	2.00mg	8/35	5.20mg	2/35	---:leu,lym.
b	TR19	.283mg	4.57mg	1/45p	2.00mg	6/35	5.20mg	2/35	
c	TR19	.927mg	13.6mg	0/45p	2.00mg	1/35	5.20mg	6/35	
d	TR19	2.60mg	13.2mg	0/45p	2.00mg	0/35	5.20mg	11/35	
4200	TR19	.289mg	.947mg	0/15	1.90mg	14/35	4.80mg	19/35	---:leu,lym; bra:oln,ulc; lun:a/a,a/c. C
a	TR19	.323mg	1.39mg	0/15	1.90mg	11/35	4.80mg	10/35	lun:a/a,a/c.
b	TR19	.369mg	1.63mg	0/15	1.90mg	10/35	4.80mg	10/35	
c	TR19	1.36mg	135.mg	0/15	1.90mg	4/35	4.80mg	4/35	---:leu,lym.
d	TR19	2.07mg	13.1mg	0/15	1.90mg	0/35	4.80mg	10/35	bra:oln,ulc.
e	TR19	2.28mg	17.1mg	0/15	1.90mg	0/35	4.80mg	9/35	
f	TR19	.271mg	.821mg	0/15	1.90mg	18/35	4.80mg	20/35	
g	TR19	1.27mg	222.mg	0/15	1.90mg	3/35	4.80mg	3/35	liv:hpa,hpc,nnd.
h	TR19	.323mg	1.39mg	0/15	1.90mg	11/35	4.80mg	10/35	lun:a/a,a/c.
4201	TR19	.323mg	1.31mg	0/45p	1.90mg	11/35	4.80mg	10/35	lun:a/a,a/c.
a	TR19	.369mg	1.53mg	0/45p	1.90mg	10/35	4.80mg	10/35	
b	TR19	1.36mg	13.8mg	0/45p	1.90mg	4/35	4.80mg	4/35	---:leu,lym.
c	TR19	2.07mg	11.7mg	0/45p	1.90mg	0/35	4.80mg	10/35	bra:oln,ulc.
d	TR19	2.28mg	14.6mg	0/45p	1.90mg	0/35	4.80mg	9/35	
4202	1336	.446mg	4.68mg	20/154	1.75mg	10/19			Skipper;srfr;1976/Weisburger;canc;1977/Prejean pers.comm.
a	1336	1.12mg	26.7mg	0/154	1.75mg	3/19			
b	1336	1.41mg	82.8mg	0/154	1.75mg	2/19			
c	1336	1.43mg	n.s.s.	1/154	1.75mg	2/19			
d	1336	1.47mg	n.s.s.	3/154	1.75mg	2/19			
e	1336	3.66mg	n.s.s.	1/154	1.75mg	0/19			
f	1336	92.6ug	.683mg	42/154	1.75mg	18/19			
g	1336	.181mg	1.03mg	29/154	1.75mg	16/19			
h	1336	1.74mg	n.s.s.	13/154	1.75mg	2/19			
4203	1336	.683mg	6.43mg	9/101	1.71mg	11/27			
a	1336	5.35mg	n.s.s.	2/101	1.71mg	0/27			
b	1336	.492mg	9.99mg	28/101	1.71mg	16/27			
c	1336	.689mg	94.8mg	19/101	1.71mg	12/27			
d	1336	1.74mg	n.s.s.	9/101	1.71mg	4/27			
4204	2001	1.34mg	86.3mg	0/71	3.96mg	2/8			Adamson;ossc,129-156;
									1982/Thorgeirsson 1994/Dalgard 1991/Thorgeirsson&Seiber pers.comm.
a	2001	3.65mg	248.mg	0/94	3.96mg	2/13			
b	2001	1.01mg	n.s.s.	0/16	3.96mg	1/3			
c	2001	3.43mg	n.s.s.	0/71	3.96mg	1/9			
d	2001	3.83mg	n.s.s.	0/72	3.96mg	1/10			
e	2001	.749mg	5.70mg	0/94	3.96mg	7/13			
4205	2001	1.48mg	14.4mg	0/107	3.78mg	6/22			
a	2001	.840mg	n.s.s.	0/17	3.78mg	1/2			
b	2001	5.32mg	n.s.s.	0/66	3.78mg	1/10			
c	2001	6.44mg	n.s.s.	0/70	3.78mg	1/12			
d	2001	.715mg	6.99mg	5/107	3.78mg	9/22			
4206	TR19	.106mg	.701mg	0/10	3.15mg	25/36	10.8mg	30/35	---:lym; bra:oln; crb:mua; mgl:acn,adn. C
a	TR19	.112mg	1.47mg	0/10	3.15mg	17/36	10.8mg	3/35	bra:oln; crb:mua.
b	TR19	.112mg	1.51mg	0/10	3.15mg	17/36	10.8mg	2/35	
c	TR19	.112mg	1.37mg	0/10	3.15mg	17/36	10.8mg	25/35	mgl:acn,adn.
d	TR19	.112mg	1.37mg	0/10	3.15mg	16/36	10.8mg	25/35	
e	TR19	2.79mg	9.68mg	0/10	3.15mg	0/36	10.8mg	20/35	
f	TR19	.105mg	.668mg	4/10	3.15mg	27/36	10.8mg	30/35	
g	TR19	n.s.s.	n.s.s.	0/10	3.15mg	0/36	10.8mg	0/35	liv:hpa,hpc,nnd.
4207	TR19	96.8ug	1.10mg	0/40p	3.15mg	17/36	10.8mg	3/35	bra:oln; crb:mua.
a	TR19	96.9ug	1.11mg	0/40p	3.15mg	17/36	10.8mg	2/35	
b	TR19	96.7ug	1.14mg	3/40p	3.15mg	17/36	10.8mg	25/35	mgl:acn,adn.
c	TR19	96.8ug	1.15mg	2/40p	3.15mg	16/36	10.8mg	25/35	
d	TR19	2.41mg	8.16mg	0/40p	3.15mg	0/36	10.8mg	20/35	
4208	TR19	.104mg	.627mg	1/10	2.20mg	16/34	6.20mg	24/35	---:leu,lym; bra:can,oln; mgl:acn,adn; olb:acn. C
a	TR19	.110mg	1.13mg	0/10	2.20mg	12/34	6.20mg	9/35	bra:can,oln; olb:acn.
b	TR19	1.00mg	5.59mg	1/10	2.20mg	3/34	6.20mg	12/35	---:leu,lym.

```
      Spe Strain  Site   Xpo+Xpt                                                  TD50      2Tailpvl
          Sex  Route  Hist     Notes                                             DR        AuOp

c     R m sda ipj mgl MXA 26w61 s                                                4.50mg * P<.003 c
d     R m sda ipj mgl acn 26w61 s                                                4.85mg * P<.004 c
e     R m sda ipj TBA MXB 26w61 s                                                .186mg / P<.0005
f     R m sda ipj liv MXB 26w61 s                                                no dre   P=1.
4209  R m sda ipj MXA MXA 26w61 s     pool              :      +      :          .343mg * P<.0005c
a     R m sda ipj --- MXA 26w61 s                                                2.10mg * P<.0005c
b     R m sda ipj --- lym 26w61 s                                                2.51mg * P<.0005c
c     R m sda ipj mgl MXA 26w61 s                                                4.50mg * P<.0005c
d     R m sda ipj olb acn 26w61 s                                                4.67mg * P<.003
e     R m sda ipj mgl acn 26w61 s                                                4.85mg * P<.0005c
f     R m sda ipj --- leu 26w61 s                                                13.3mg * P<.02

PROFLAVINE.HCl HEMIHYDRATE  100ng..:..1ug....:..10......:..100....:..1mg....:..10......:..100....:..1g......:..10
4210  M f b6c eat liv MXA 24m24                        :   +   :                 #26.4mg \ P<.0005i
a     M f b6c eat TBA MXB 24m24                                                  86.2mg * P<.3
b     M f b6c eat liv MXB 24m24                                                  26.4mg \ P<.0005
c     M f b6c eat lun MXB 24m24                                                  487.mg * P<.5
4211  M m b6c eat TBA MXB 24m24                                    :>            no dre   P=1.    i
a     M m b6c eat liv MXB 24m24                                                  248.mg * P<.7
b     M m b6c eat lun MXB 24m24                                                  no dre   P=1.
4212  R f f34 eat TBA MXB 25m25                               :>                 no dre   P=1.    i
a     R f f34 eat liv MXB 25m25                                                  no dre   P=1.
4213  R m f34 eat MXA MXA 25m25                         :    ±                   #213.mg * P<.02  i
a     R m f34 eat liv MXA 25m25                                                  254.mg * P<.05
b     R m f34 eat TBA MXB 25m25                                                  no dre   P=1.
c     R m f34 eat liv MXB 25m25                                                  254.mg * P<.05

PROMETHAZINE.HCl            100ng..:..1ug....:..10......:..100....:..1mg....:..10......:..100....:..1g......:..10
4214  M f b6c gav TBA MXB 24m24                                  :>              no dre   P=1.    -
a     M f b6c gav liv MXB 24m24                                                  47.8mg * P<.08
b     M f b6c gav lun MXB 24m24                                                  no dre   P=1.
4215  M m b6c gav TBA MXB 24m24                                      :>          393.mg * P<.9    -
a     M m b6c gav liv MXB 24m24                                                  157.mg * P<.4
b     M m b6c gav lun MXB 24m24                                                  no dre   P=1.
4216  R f f34 gav TBA MXB 24m24                                  :>              no dre   P=1.    -
a     R f f34 gav liv MXB 24m24                                                  418.mg * P<.2
4217  R m f34 gav tes MXA 24m24                              :   ±               #10.3mg * P<.02  -
a     R m f34 gav TBA MXB 24m24                                                  24.6mg * P<.3
b     R m f34 gav liv MXB 24m24                                                  no dre   P=1.

PRONETHALOL                 100ng..:..1ug....:..10......:..100....:..1mg....:..10......:..100....:..1g......:..10
4218  M f nss eat thm tum 64w64 r                                      .>        408.mg * P<.3
4219  M m nss eat thm tum 64w64 r                                           .>   1.93gm * P<.7

PRONETHALOL.HCl             100ng..:..1ug....:..10......:..100....:..1mg....:..10......:..100....:..1g......:..10
4220  M f cf1 eat lun ade 78w78 e                                           .>   1.30gm   P<.6    -
a     M f cf1 eat liv tum 78w78 e                                                no dre   P=1.    -
4221  M m cf1 eat lun ade 78w78 e                                           .>   931.mg   P<.5    -
a     M m cf1 eat liv ade 78w78 e                                                1.82gm   P<.3    -
b     M m cf1 eat liv car 78w78 e                                                no dre   P=1.    -

PROPANE SULTONE             100ng..:..1ug....:..10......:..100....:..1mg....:..10......:..100....:..1g......:..10
4222  R f cdr gav crb mag 46w60 aes                     .   ±                    4.06mg * P<.1    +
a     R f cdr gav mgl adc 46w60 aes                                              4.13mg * P<.02   +
b     R f cdr gav crl mag 46w60 aes                                              7.20mg * P<.08   +
c     R f cdr gav --- grl 46w60 aes                                        +hist 18.6mg * P<.3    +
d     R f cdr gav smi adc 46w60 aes                                        +hist 31.7mg * P<.5    +
4223  R m cdr gav crb mag 46w60 aes                         .  ±                 3.64mg * P<.02   +
a     R m cdr gav crl mag 46w60 aes                                              4.75mg * P<.03   +
b     R m cdr gav smi adc 46w60 aes                                        +hist 15.4mg * P<.3    +
c     R m cdr gav --- grl 46w60 aes                                        +hist 23.5mg * P<.3    +

PROPAZINE                   100ng..:..1ug....:..10......:..100....:..1mg....:..10......:..100....:..1g......:..10
4224  M f b6a orl lun ade 76w76 evx                                    .>        82.6mg   P<.6    -
a     M f b6a orl liv hpt 76w76 evx                                              no dre   P=1.    -
b     M f b6a orl tba mix 76w76 evx                                              37.2mg   P<.4    -
4225  M m b6a orl liv hpt 76w76 evx                                       .>     73.1mg   P<.6    -
a     M m b6a orl lun ade 76w76 evx                                              673.mg   P<1.    -
b     M m b6a orl tba mix 76w76 evx                                              57.8mg   P<.7    -
4226  M f b6c orl liv hpt 76w76 evx                                   .>         no dre   P=1.    -
a     M f b6c orl lun mix 76w76 evx                                              no dre   P=1.    -
b     M f b6c orl tba mix 76w76 evx                                              93.2mg   P<.3    -
4227  M m b6c orl liv hpt 76w76 evx                          .      ±            39.7mg   P<.1    -
a     M m b6c orl lun ade 76w76 evx                                              82.0mg   P<.3    -
b     M m b6c orl tba mix 76w76 evx                                              11.4mg   P<.003  -

beta-PROPIOLACTONE          100ng..:..1ug....:..10......:..100....:..1mg....:..10......:..100....:..1g......:..10
4228  M f hic gav for tum 77w77                         .   +   .                1.33mg   P<.0005+
a     M f hic gav for sqc 77w77                                                  1.47mg   P<.0005
4229  M m hic gav for tum 83w83                     .+   .                       1.16mg   P<.0005+
a     M m hic gav for sqc 83w83                                                  1.43mg   P<.0005
```

	RefNum	LoConf	UpConf	Cntrl	1Dose	1Inc	2Dose	2Inc			Citation or Pathology
											Brkly Code
c	TR19	1.93mg	20.1mg	0/10	2.20mg	1/34	6.20mg	8/35			mgl:acn,adn.
d	TR19	2.00mg	29.1mg	0/10	2.20mg	1/34	6.20mg	7/35			
e	TR19	84.2ug	.361mg	4/10	2.20mg	19/34	6.20mg	30/35			
f	TR19	n.s.s.	n.s.s.	0/10	2.20mg	0/34	6.20mg	0/35			liv:hpa,hpc,nnd.
4209	TR19	.110mg	.885mg	0/40p	2.20mg	12/34	6.20mg	9/35			bra:can,oln; olb:acn.
a	TR19	1.00mg	4.63mg	1/40p	2.20mg	3/34	6.20mg	12/35			---:leu,lym.
b	TR19	1.10mg	6.37mg	1/40p	2.20mg	3/34	6.20mg	9/35			
c	TR19	1.93mg	12.8mg	1/40p	2.20mg	1/34	6.20mg	8/35			mgl:acn,adn.
d	TR19	1.22mg	50.7mg	0/40p	2.20mg	0/34	6.20mg	3/35			S
e	TR19	2.00mg	15.3mg	1/40p	2.20mg	1/34	6.20mg	7/35			
f	TR19	3.78mg	n.s.s.	0/40p	2.20mg	0/34	6.20mg	3/35			S

PROFLAVINE.HCl HEMIHYDRATE   952-23-8

	RefNum	LoConf	UpConf	Cntrl	1Dose	1Inc	2Dose	2Inc			Citation or Pathology
4210	TR5	13.6mg	89.3mg	5/49	26.0mg	20/50	(52.0mg	22/50)			liv:hpa,hpc. S
a	TR5	26.0mg	n.s.s.	36/49	26.0mg	37/50	52.0mg	41/50			
b	TR5	13.6mg	89.3mg	5/49	26.0mg	20/50	(52.0mg	22/50)			liv:hpa,hpc,nnd.
c	TR5	107.mg	n.s.s.	4/49	26.0mg	3/50	52.0mg	6/50			lun:a/a,a/c.
4211	TR5	35.5mg	n.s.s.	39/50	24.0mg	38/50	48.0mg	39/50			
a	TR5	34.1mg	n.s.s.	26/50	24.0mg	28/50	48.0mg	30/50			liv:hpa,hpc,nnd.
b	TR5	95.5mg	n.s.s.	9/50	24.0mg	11/50	48.0mg	8/50			lun:a/a,a/c.
4212	TR5	21.2mg	n.s.s.	48/50	15.0mg	42/50	(30.0mg	35/50)			
a	TR5	157.mg	n.s.s.	1/50	15.0mg	2/50	30.0mg	1/50			liv:hpa,hpc,nnd.
4213	TR5	80.8mg	25.0gm	0/50	12.0mg	0/50	24.0mg	5/50			lgi:srn; smi:acn,lei. S
a	TR5	87.8mg	n.s.s.	0/50	12.0mg	1/50	24.0mg	3/50			liv:hpc,nnd. S
b	TR5	35.6mg	n.s.s.	36/50	12.0mg	30/50	24.0mg	32/50			
c	TR5	87.8mg	n.s.s.	0/50	12.0mg	1/50	24.0mg	3/50			liv:hpa,hpc,nnd.

PROMETHAZINE.HCl   58-33-3

	RefNum	LoConf	UpConf	Cntrl	1Dose	1Inc	2Dose	2Inc	3Dose	3Inc	Citation or Pathology
4214	TR425	17.0mg	n.s.s.	31/50	2.63mg	33/50	5.28mg	29/51	10.6mg	26/51	
a	TR425	18.6mg	n.s.s.	4/50	2.63mg	4/50	5.28mg	7/51	10.6mg	10/51	liv:hpa,hpb,hpc.
b	TR425	102.mg	n.s.s.	8/50	2.63mg	2/50	5.28mg	0/51	10.6mg	2/51	lun:a/a,a/c.
4215	TR425	31.1mg	n.s.s.	31/50	7.90mg	35/50	15.9mg	29/50	31.8mg	37/50	
a	TR425	37.8mg	n.s.s.	18/50	7.90mg	18/50	15.9mg	17/50	31.8mg	25/50	liv:hpa,hpb,hpc.
b	TR425	111.mg	n.s.s.	11/50	7.90mg	9/50	15.9mg	7/50	31.8mg	8/50	lun:a/a,a/c.
4216	TR425	16.7mg	n.s.s.	46/50	5.81mg	42/50	11.6mg	39/50	23.4mg	32/53	
a	TR425	102.mg	n.s.s.	0/50	5.81mg	0/50	11.6mg	1/50	23.4mg	1/53	liv:hpa,hpb,hpc.
4217	TR425	4.68mg	n.s.s.	47/50	5.85mg	44/50	11.8mg	44/50	23.7mg	43/51	tes:iab,ica. S
a	TR425	6.85mg	n.s.s.	44/50	5.85mg	46/50	11.8mg	47/50	23.7mg	31/51	
b	TR425	49.7mg	n.s.s.	4/50	5.85mg	2/50	11.8mg	3/50	23.7mg	1/51	liv:hpa,hpb,hpc.

PRONETHALOL   (alderlin) 54-80-8

	RefNum	LoConf	UpConf	Cntrl	1Dose	1Inc	2Dose	2Inc	3Dose	3Inc	Citation or Pathology
4218	20	147.mg	n.s.s.	0/25	65.0mg	3/25	130.mg	4/25	260.mg	2/25	Howe;natu,207,594-595;1965
4219	20	266.mg	n.s.s.	1/25	60.0mg	0/25	120.mg	3/25	240.mg	1/25	

PRONETHALOL.HCl   (alderlin.HCl) 51-02-5

	RefNum	LoConf	UpConf	Cntrl	1Dose	1Inc	Citation or Pathology
4220	469	239.mg	n.s.s.	3/40	150.mg	6/52	Newberne;txap,41,535-546;1977
a	469	904.mg	n.s.s.	0/40	150.mg	0/52	
4221	469	174.mg	n.s.s.	2/29	150.mg	4/32	
a	469	297.mg	n.s.s.	0/29	150.mg	1/32	
b	469	340.mg	n.s.s.	1/29	150.mg	1/32	

PROPANE SULTONE   1120-71-4

	RefNum	LoConf	UpConf	Cntrl	1Dose	1Inc	2Dose	2Inc	Citation or Pathology
4222	1112	2.02mg	n.s.s.	1/7	8.00mg	12/26	8.53mg	12/26	Weisburger;jnci,67,75-88;1981
a	1112	2.40mg	n.s.s.	0/7	8.00mg	6/26	8.53mg	13/26	
b	1112	3.70mg	n.s.s.	0/7	8.00mg	8/26	8.53mg	4/26	
c	1112	7.07mg	n.s.s.	0/7	8.00mg	2/26	8.53mg	2/26	
d	1112	9.60mg	n.s.s.	0/7	8.00mg	2/26	8.53mg	1/26	
4223	1112	2.16mg	n.s.s.	0/6	8.00mg	10/26	8.53mg	11/26	
a	1112	2.68mg	n.s.s.	0/6	8.00mg	6/26	8.53mg	11/26	
b	1112	6.27mg	n.s.s.	0/6	8.00mg	3/26	8.53mg	3/26	
c	1112	8.12mg	n.s.s.	0/6	8.00mg	0/26	8.53mg	4/26	

PROPAZINE   (gesamil) 139-40-2

	RefNum	LoConf	UpConf	Cntrl	1Dose	1Inc	Citation or Pathology
4224	1242	11.5mg	n.s.s.	1/17	14.6mg	2/17	Innes;ntis,1968/1969
a	1242	27.2mg	n.s.s.	0/17	14.6mg	0/17	
b	1242	7.81mg	n.s.s.	2/17	14.6mg	4/17	
4225	1242	10.7mg	n.s.s.	1/18	13.6mg	2/17	
a	1242	11.8mg	n.s.s.	2/18	13.6mg	2/17	
b	1242	7.80mg	n.s.s.	3/18	13.6mg	4/17	
4226	1242	28.8mg	n.s.s.	0/16	14.6mg	0/18	
a	1242	28.8mg	n.s.s.	0/16	14.6mg	0/18	
b	1242	15.2mg	n.s.s.	0/16	14.6mg	1/18	
4227	1242	9.75mg	n.s.s.	0/16	13.6mg	2/17	
a	1242	13.3mg	n.s.s.	0/16	13.6mg	1/17	
b	1242	4.61mg	61.9mg	0/16	13.6mg	6/17	

beta-PROPIOLACTONE   57-57-8

	RefNum	LoConf	UpConf	Cntrl	1Dose	1Inc	Citation or Pathology
4228	1011	.765mg	2.49mg	0/30	5.71mg	24/30	Van Duuren;jnci,63,1433-1439;1979
a	1011	.848mg	2.77mg	0/30	5.71mg	23/30	
4229	1011	.660mg	2.16mg	0/30	4.76mg	25/30	
a	1011	.821mg	2.68mg	0/30	4.76mg	23/30	

	Spe	Strain	Site	Xpo+Xpt						TD50	2Tailpvl
	Sex	Route	Hist	Notes						DR	AuOp
4230	R f	esd	gav	sto	sqc	69w69		·	±		1.34mg	P<.02 +
4231	R f	sda	gav	sto	tum	12m35 e		· + ·			1.61mg	P<.0005+

PROPRANOLOL.HCl   100ng..:..1ug....:..10.....:..100....:..1mg....:..10.....:..100...:..1g.....:..10

	Spe	Strain	Site	Xpo+Xpt	Notes		TD50	2Tailpvl
4232	M f	cd1	eat	lun	ade 78w92 e	.>	2.12gm	P<.7 -
a	M f	cd1	eat	liv	tum 78w92 e		no dre	P=1. -
b	M f	cd1	eat	tba	ben 78w92 e		1.99gm	P<.7 -
c	M f	cd1	eat	tba	mal 78w92 e		no dre	P=1. -
d	M f	cd1	eat	tba	mix 78w92 e		no dre	P=1. -
4233	M m	cd1	eat	liv	car 78w92 e	.>	2.91gm	P<.6 -
a	M m	cd1	eat	liv	ade 78w92 e		no dre	P=1. -
b	M m	cd1	eat	lun	ade 78w92 e		no dre	P=1. -
c	M m	cd1	eat	lun	car 78w92 e		no dre	P=1. -
d	M m	cd1	eat	tba	mal 78w92 e		670.mg	P<.4 -
e	M m	cd1	eat	tba	ben 78w92 e		no dre	P=1. -
f	M m	cd1	eat	tba	mix 78w92 e		no dre	P=1. -
4234	R f	lev	eat	liv	tum 18m24 e	.>	no dre	P=1. -
a	R f	lev	eat	tba	mix 18m24 e		144.mg	P<.7 -
b	R f	lev	eat	tba	ben 18m24 e		no dre	P=1. -
c	R f	lev	eat	tba	mal 18m24 e		342.mg	P<.4 -
4235	R m	lev	eat	liv	hpa 18m24 e	.>	no dre	P=1. -
a	R m	lev	eat	tba	mix 18m24 e		no dre	P=1. -
b	R m	lev	eat	tba	ben 18m24 e		no dre	P=1. -
c	R m	lev	eat	tba	mal 18m24 e		no dre	P=1. -

PROPYL N-ETHYL-N-BUTYLTHIOCARBAMATE   ..:..1ug....:..10.....:..100....:..1mg....:..10.....:..100...:..1g.....:..10

	Spe	Strain	Site	Xpo+Xpt	Notes		TD50	2Tailpvl
4236	M f	b6a	orl	lun	ade 76w76 evx	· ±	113.mg	P<.09 -
a	M f	b6a	orl	liv	hpt 76w76 evx		no dre	P=1. -
b	M f	b6a	orl	tba	mix 76w76 evx		219.mg	P<.6 -
4237	M m	b6a	orl	lun	ade 76w76 evx	.>	no dre	P=1. -
a	M m	b6a	orl	liv	hpt 76w76 evx		no dre	P=1. -
b	M m	b6a	orl	tba	mix 76w76 evx		no dre	P=1. -
4238	M f	b6c	orl	lun	ade 76w76 evx	.>	267.mg	P<.6 -
a	M f	b6c	orl	liv	hpt 76w76 evx		no dre	P=1. -
b	M f	b6c	orl	tba	mix 76w76 evx		251.mg	P<.7 -
4239	M m	b6c	orl	lun	ade 76w76 evx	.>	no dre	P=1. -
a	M m	b6c	orl	liv	hpt 76w76 evx		no dre	P=1. -
b	M m	b6c	orl	tba	mix 76w76 evx		no dre	P=1. -

N-N'-PROPYL-N-FORMYLHYDRAZINE   100ng..:..1ug....:..10.....:..100....:..1mg....:..10.....:..100...:..1g....:..10

	Spe	Strain	Site	Xpo+Xpt	Notes		TD50	2Tailpvl
4240	M f	swa	wat	lun	mix 75w79 ae	· + ·	8.74mg \	P<.0005+
a	M f	swa	wat	liv	mix 75w79 ae		215.mg *	P<.0005+
b	M f	swa	wat	gal	mix 75w79 ae		444.mg *	P<.006 +
c	M f	swa	wat	gal	ade 75w79 ae		497.mg *	P<.02
4241	M m	swa	wat	lun	mix 51w61 aes	· + ·	8.84mg \	P<.0005+
a	M m	swa	wat	pre	mix 51w61 aes		60.4mg \	P<.0005+
b	M m	swa	wat	pre	sqc 51w61 aes		75.7mg \	P<.002
c	M m	swa	wat	liv	hpt 51w61 aes		no dre	P=1.

PROPYL GALLATE   100ng..:..1ug....:..10.....:..100....:..1mg....:..10.....:..100...:..1g.....:..10

	Spe	Strain	Site	Xpo+Xpt	Notes		TD50	2Tailpvl
4242	M f	b6c	eat	liv	hpa 24m24	:	#8.26gm *	P<.01 -
a	M f	b6c	eat	TBA	MXB 24m24		no dre	P=1.
b	M f	b6c	eat	liv	MXB 24m24		21.4gm *	P<.5
c	M f	b6c	eat	lun	MXB 24m24		44.3gm *	P<.6
4243	M m	b6c	eat	---	lym 24m24	:	±#6.54gm *	P<.02 -
a	M m	b6c	eat	---	mly 24m24		12.1gm *	P<.03
b	M m	b6c	eat	---	lhc 24m24		15.1gm *	P<.03
c	M m	b6c	eat	TBA	MXB 24m24		3.64gm \	P<.6
d	M m	b6c	eat	liv	MXB 24m24		no dre	P=1.
e	M m	b6c	eat	lun	MXB 24m24		55.6gm *	P<.9
4244	R f	f34	eat	mgl	ade 24m24	:	#8.08gm *	P<.05 -
a	R f	f34	eat	TBA	MXB 24m24		no dre	P=1.
b	R f	f34	eat	liv	MXB 24m24		no dre	P=1.
4245	R m	f34	eat	pni	isa 24m24	: + :	#749.mg \	P<.002 -
a	R m	f34	eat	adr	MXA 24m24		613.mg \	P<.03
b	R m	f34	eat	adr	phe 24m24		613.mg \	P<.03
c	R m	f34	eat	pni	MXA 24m24		840.mg \	P<.03
d	R m	f34	eat	pre	MXA 24m24		1.04gm \	P<.03
e	R m	f34	eat	thy	MXA 24m24		6.87gm *	P<.05
f	R m	f34	eat	TBA	MXB 24m24		no dre	P=1.
g	R m	f34	eat	liv	MXB 24m24		no dre	P=1.
4246	R m	f3d	eat	eso	tum 52w52 er	.>	no dre	P=1. -
a	R m	f3d	eat	for	tum 52w52 er		no dre	P=1. -
b	R m	f3d	eat	liv	tum 52w52 er		no dre	P=1. -

N-PROPYL ISOME   100ng..:..1ug....:..10.....:..100....:..1mg....:..10.....:..100...:..1g.....:..10

	Spe	Strain	Site	Xpo+Xpt	Notes		TD50	2Tailpvl
4247	M f	b6a	orl	lun	ade 76w76 evx	.>	4.69gm	P<.6 -
a	M f	b6a	orl	liv	hpt 76w76 evx		no dre	P=1. -
b	M f	b6a	orl	tba	mix 76w76 evx		4.39gm	P<.7 -
4248	M m	b6a	orl	lun	ade 76w76 evx	.>	no dre	P=1. -
a	M m	b6a	orl	liv	hpt 76w76 evx		no dre	P=1. -

	RefNum	LoConf	UpConf	Cntrl	1Dose	1Inc	2Dose	2Inc	Citation or Pathology
									Brkly Code
4230	55	.374mg	n.s.s.	0/5	4.08mg	3/5			Van Duuren;jnci,37,825-838;1966
4231	1486	.986mg	2.65mg	0/50	2.86mg	46/50			Dunkelberg;bjca,46,924-933;1982
PROPRANOLOL.HCl		318-98-9							
4232	1635	267.mg	n.s.s.	3/60	84.8mg	4/57			Weikel;jcph,19,591-604;1979
a	1635	779.mg	n.s.s.	0/60	84.8mg	0/57			
b	1635	243.mg	n.s.s.	4/60	84.8mg	5/57			
c	1635	323.mg	n.s.s.	13/60	84.8mg	7/57			
d	1635	211.mg	n.s.s.	16/60	84.8mg	12/57			
4233	1635	371.mg	n.s.s.	1/57	84.8mg	2/61			
a	1635	672.mg	n.s.s.	6/57	84.8mg	1/61			
b	1635	228.mg	n.s.s.	11/57	84.8mg	10/61			
c	1635	512.mg	n.s.s.	1/57	84.8mg	1/61			
d	1635	171.mg	n.s.s.	5/57	84.8mg	9/61			
e	1635	314.mg	n.s.s.	19/57	84.8mg	11/61			
f	1635	203.mg	n.s.s.	23/57	84.8mg	18/61			
4234	1635	347.mg	n.s.s.	0/60	28.0mg	0/60			
a	1635	19.4mg	n.s.s.	44/60	28.0mg	46/60			
b	1635	33.5mg	n.s.s.	43/60	28.0mg	40/60			
c	1635	78.0mg	n.s.s.	5/60	28.0mg	8/60			
4235	1635	347.mg	n.s.s.	1/60	28.0mg	0/60			
a	1635	39.0mg	n.s.s.	44/60	28.0mg	39/60			
b	1635	59.6mg	n.s.s.	41/60	28.0mg	32/60			
c	1635	76.7mg	n.s.s.	12/60	28.0mg	12/60			
PROPYL N-ETHYL-N-BUTYLTHIOCARBAMATE			(tillam-6-E)	1114-71-2					
4236	1223	27.8mg	n.s.s.	0/15	44.3mg	2/15			Innes;ntis,1968/1969
a	1223	73.1mg	n.s.s.	0/15	44.3mg	0/15			
b	1223	30.5mg	n.s.s.	1/15	44.3mg	2/15			
4237	1223	33.8mg	n.s.s.	3/18	41.2mg	2/15			
a	1223	68.0mg	n.s.s.	0/18	41.2mg	0/15			
b	1223	33.8mg	n.s.s.	3/18	41.2mg	2/15			
4238	1223	37.1mg	n.s.s.	1/18	44.3mg	2/18			
a	1223	87.7mg	n.s.s.	0/18	44.3mg	0/18			
b	1223	31.5mg	n.s.s.	2/18	44.3mg	3/18			
4239	1223	48.2mg	n.s.s.	2/15	41.2mg	1/16			
a	1223	51.5mg	n.s.s.	3/15	41.2mg	1/16			
b	1223	43.8mg	n.s.s.	5/15	41.2mg	2/16			
N-N'-PROPYL-N-FORMYLHYDRAZINE			77337-54-3						
4240	1053	3.99mg	17.1mg	25/99	80.0mg	49/50	(160.mg	22/49)	Toth;bjca,42,922-928;1980
a	1053	108.mg	653.mg	0/38	80.0mg	6/45	160.mg	5/19	
b	1053	216.mg	3.77gm	0/54	80.0mg	5/50	160.mg	5/49	
c	1053	234.mg	n.s.s.	0/54	80.0mg	5/50	160.mg	4/49	
4241	1053	5.21mg	16.5mg	26/100	66.7mg	42/48	(133.mg	6/37)	
a	1053	30.2mg	172.mg	0/33	66.7mg	11/48	(133.mg	1/37)	
b	1053	35.6mg	294.mg	0/33	66.7mg	9/48	(133.mg	1/37)	
c	1053	304.mg	n.s.s.	0/28	66.7mg	2/50	133.mg	0/50	
PROPYL GALLATE		121-79-9							
4242	TR240	3.56gm	370.gm	0/50	758.mg	2/50	1.52gm	5/50	S
a	TR240	2.54gm	n.s.s.	25/50	758.mg	17/50	1.52gm	22/50	
b	TR240	4.17gm	n.s.s.	3/50	758.mg	3/50	1.52gm	5/50	liv:hpa,hpc,nnd.
c	TR240	6.71gm	n.s.s.	1/50	758.mg	1/50	1.52gm	2/50	lun:a/a,a/c.
4243	TR240	2.92gm	n.s.s.	1/50	700.mg	3/50	1.40gm	8/50	S
a	TR240	4.60gm	n.s.s.	0/50	700.mg	1/50	1.40gm	4/50	S
b	TR240	5.22gm	n.s.s.	0/50	700.mg	0/50	1.40gm	4/50	S
c	TR240	604.mg	n.s.s.	29/50	700.mg	31/50	(1.40gm	22/50)	
d	TR240	4.54gm	n.s.s.	17/50	700.mg	15/50	1.40gm	10/50	liv:hpa,hpc,nnd.
e	TR240	4.19gm	n.s.s.	4/50	700.mg	5/50	1.40gm	5/50	lun:a/a,a/c.
4244	TR240	2.44gm	n.s.s.	0/50	292.mg	0/50	583.mg	3/50	
a	TR240	663.mg	n.s.s.	38/50	292.mg	34/50	583.mg	36/50	S
b	TR240	n.s.s.	n.s.s.	0/50	292.mg	1/50	583.mg	0/50	liv:hpa,hpc,nnd.
4245	TR240	338.mg	2.74gm	0/50	233.mg	8/50	(466.mg	2/50)	S
a	TR240	256.mg	n.s.s.	4/50	233.mg	13/50	(466.mg	8/50)	adr:phe,phm. S
b	TR240	256.mg	n.s.s.	4/50	233.mg	13/50	(466.mg	8/50)	S
c	TR240	335.mg	n.s.s.	2/50	233.mg	9/50	(466.mg	4/50)	pni:isa,isc. S
d	TR240	404.mg	n.s.s.	1/50	233.mg	7/50	(466.mg	0/50)	pre:adc,ade,car. S
e	TR240	2.08gm	n.s.s.	0/50	233.mg	0/50	466.mg	3/50	thy:fca,fcc. S
f	TR240	581.mg	n.s.s.	33/50	233.mg	37/50	466.mg	32/50	
g	TR240	3.51gm	n.s.s.	2/50	233.mg	1/50	466.mg	1/50	liv:hpa,hpc,nnd.
4246	1900	206.mg	n.s.s.	0/10	400.mg	0/10			Hirose;carc,8,1731-1735;1987/pers.comm.
a	1900	206.mg	n.s.s.	0/10	400.mg	0/10			
b	1900	206.mg	n.s.s.	0/10	400.mg	0/10			
N-PROPYL ISOME		83-59-0							
4247	1252	653.mg	n.s.s.	1/17	828.mg	2/17			Innes;ntis,1968/1969
a	1252	1.55gm	n.s.s.	0/17	828.mg	0/17			
b	1252	553.mg	n.s.s.	2/17	828.mg	3/17			
4248	1252	720.mg	n.s.s.	2/18	770.mg	2/18			
a	1252	1.53gm	n.s.s.	1/18	770.mg	0/18			

```
     Spe Strain Site    Xpo+Xpt                                              TD50    2Tailpvl
         Sex  Route Hist      Notes                                            DR     AuOp
```

						TD50	DR	2Tailpvl AuOp

```
 b    M m b6a orl tba mix 76w76 evx                                   no dre    P=1.  -
4249  M f b6c orl liv hpt 76w76 evx                           .>      no dre    P=1.  -
 a    M f b6c orl lun mix 76w76 evx                                   no dre    P=1.  -
 b    M f b6c orl tba tum 76w76 evx                                   no dre    P=1.  -
4250  M m b6c orl liv hpt 76w76 evx                    .      ±       1.05gm    P<.02 -
 a    M m b6c orl lun ade 76w76 evx                                   4.65gm    P<.3  -
 b    M m b6c orl tba mix 76w76 evx                                   374.mg    P<.0005-
```

N-PROPYL-N'-NITRO-N-NITROSOGUANIDINE   `..:..1ug...:....10...:....100...:..1mg...:...10...:...100...:..1g...:..10`

```
4251  R m wis wat stg ade 34w65 er                    .      ±        1.90mg    P<.04
4252  R m wis wat stg ade 52w65 er              .     +      .        .919mg    P<.002 +
 a    R m wis wat stg adc 52w65 er                                    2.85mg    P<.05  +
4253  R m wis wat stg adc 52w78 er                    .      ±        2.27mg    P<.06  +
 a    R m wis wat stg ade 52w78 er                                    2.27mg    P<.06
```

N-PROPYL-N-NITROSOUREA   `100ng..:..1ug...:....10...:....100...:..1mg...:...10...:...100...:..1g...:..10`

```
4254  R f f3d wat thm lym 70w70 e                           <+       noTD50    P<.0005+
 a    R f f3d wat tba tum 70w70 e                                    noTD50    P<.02
4255  R f f3d wat thm lym 75w75 e                             <+     noTD50    P<.0005+
 a    R f f3d wat tba tum 75w75 e                                    noTD50    P<.02
4256  R m f3d wat thm lym 70w70 e                             <+     noTD50    P<.0005+
 a    R m f3d wat tba tum 70w70 e                                    noTD50    P<.4
4257  R m f3d wat thm lym 75w75 e                             <+     noTD50    P<.0005+
 a    R m f3d wat tba tum 75w75 e                                    noTD50    P<.4
```

PROPYLENE   `100ng..:..1ug...:....10...:....100...:..1mg...:...10...:...100...:..1g...:..10`

```
4258  M f b6c inh TBA MXB 24m24                                      :>no dre  P=1.  -
 a    M f b6c inh liv MXB 24m24                                      no dre    P=1.
 b    M f b6c inh lun MXB 24m24                                      no dre    P=1.
4259  M m b6c inh TBA MXB 24m24                                      no dre    P=1.  -
 a    M m b6c inh liv MXB 24m24                                      no dre    P=1.
 b    M m b6c inh lun MXB 24m24                                      no dre    P=1.
4260  M f swi inh lun ade 18m24                              .        19.8gm * P<.09 -
 a    M f swi inh liv hpt 18m24                                      286.gm  * P<.6  -
 b    M f swi inh lun adc 18m24                                      1.17kg    P<.9  -
 c    M f swi inh tba mix 18m24                                      10.1gm  * P<.08 -
 d    M f swi inh tba mal 18m24                                      20.1gm  * P<.3  -
4261  M m swi inh lun ade 18m24                                      no dre    P=1.  -
 a    M m swi inh lun adc 18m24                                      no dre    P=1.
 b    M m swi inh liv hpt 18m24                                      no dre    P=1.
 c    M m swi inh tba mix 18m24                                      no dre    P=1.
 d    M m swi inh tba mal 18m24                                      no dre    P=1.
4262  R f f34 inh TBA MXB 24m24                             :>       no dre    P=1.  -
 a    R f f34 inh liv MXB 24m24                                      20.5gm  * P<.2
4263  R m f34 inh TBA MXB 24m24                            :>        15.5gm  * P<1.  -
 a    R m f34 inh liv MXB 24m24                                      4.61gm  * P<.06
```

PROPYLENE GLYCOL   `100ng..:..1ug...:....10...:....100...:..1mg...:...10...:...100...:..1g...:..10`

```
4264  R f cdr eat liv nod 24m24 e                                   .no dre   P=1.  -
4265  R m cdr eat liv nod 24m24 e                                   .no dre   P=1.  -
```

1,2-PROPYLENE OXIDE   `100ng..:..1ug...:....10...:....100...:..1mg...:...10...:...100...:..1g...:..10`

```
4266  M f b6c inh --- MXA 24m24                             :   +   :  863.mg / P<.003
 a    M f b6c inh nas MXA 24m24                                        1.21gm / P<.002 c
 b    M f b6c inh mgl MXA 24m24                                        1.26gm * P<.01
 c    M f b6c inh lun a/a 24m24                                        588.mg * P<.02
 d    M f b6c inh --- hes 24m24                                        1.61gm * P<.02
 e    M f b6c inh --- hem 24m24                                        1.81gm / P<.05
 f    M f b6c inh nas hem 24m24                                        2.02gm * P<.02
 g    M f b6c inh mgl adq 24m24                                        2.77gm * P<.03
 h    M f b6c inh TBA MXB 24m24                                        226.mg * P<.01
 i    M f b6c inh liv MXB 24m24                                        979.mg * P<.2
 j    M f b6c inh lun MXB 24m24                                        588.mg * P<.02
4267  M m b6c inh nas MXA 24m24                             :   +   :  732.mg / P<.0005c
 a    M m b6c inh --- MXA 24m24                                        753.mg * P<.006
 b    M m b6c inh nas hem 24m24                                        1.47gm * P<.005
 c    M m b6c inh nas hes 24m24                                        1.61gm * P<.006
 d    M m b6c inh TBA MXB 24m24                                        722.mg * P<.4
 e    M m b6c inh liv MXB 24m24                                        no dre   P=1.
 f    M m b6c inh lun MXB 24m24                                        no dre   P=1.
4268  R f f34 inh MXA MXA 24m24                             :   ±      156.mg * P<.04
 a    R f f34 inh nas ppa 24m24                                        732.mg * P<.04 p
 b    R f f34 inh TBA MXB 24m24                                        124.mg * P<.4
 c    R f f34 inh liv MXB 24m24                                        no dre   P=1.
4269  R m f34 inh nas ppa 24m24                             :   ±      826.mg * P<.1  p
 a    R m f34 inh TBA MXB 24m24                                        220.mg * P<.8
 b    R m f34 inh liv MXB 24m24                                        528.mg * P<.4
4270  R m f34 inh adr phe 23m24 eis pool          .  (+)   .          35.1mg \ P<.002 +
 a    R m f34 inh liv hpc 23m24 eis                                    2.96gm   P<.3
 b    R m f34 inh liv nnd 23m24 eis                                    no dre   P=1.
4271  R f sda gav sto mix 25m35 e                         .  +  .      39.5mg * P<.0005+
```

	RefNum	LoConf	UpConf	Cntrl	1Dose	1Inc	2Dose	2Inc					Citation or Pathology
													Brkly Code
b	1252	797.mg	n.s.s.	3/18	770.mg	2/18							
4249	1252	1.64gm	n.s.s.	0/16	828.mg	0/18							
a	1252	1.64gm	n.s.s.	0/16	828.mg	0/18							
b	1252	1.64gm	n.s.s.	0/16	828.mg	0/18							
4250	1252	361.mg	n.s.s.	0/16	770.mg	4/17							
a	1252	757.mg	n.s.s.	0/16	770.mg	1/17							
b	1252	172.mg	1.06gm	0/16	770.mg	9/17							

N-PROPYL-N'-NITRO-N-NITROSOGUANIDINE   13010-07-6

	RefNum	LoConf	UpConf	Cntrl	1Dose	1Inc	2Dose	2Inc					Citation or Pathology
4251	1081m	.573mg	n.s.s.	0/15	1.59mg	3/15							Sasajima;zkko,94,201-206;1979
4252	1081n	.338mg	4.24mg	0/15	2.38mg	5/10							
a	1081n	.698mg	n.s.s.	0/15	2.38mg	2/10							
4253	1082	.551mg	n.s.s.	0/9	1.98mg	2/7							Matsukura;gann,70,181-185;1979
a	1082	.551mg	n.s.s.	0/9	1.98mg	2/7							

N-PROPYL-N-NITROSOUREA   816-57-9

	RefNum	LoConf	UpConf	Cntrl	1Dose	1Inc	2Dose	2Inc					Citation or Pathology
4254	2112m	n.s.s.	3.78mg	0/20	22.9mg	20/20							Ogiu;thym,20,249-258;1992
a	2112m	n.s.s.	n.s.s.	10/20	22.9mg	20/20							
4255	2112n	n.s.s.	4.34mg	0/20	22.9mg	20/20							
a	2112n	n.s.s.	n.s.s.	10/20	22.9mg	20/20							
4256	2112m	n.s.s.	3.31mg	0/20	20.0mg	20/20							
a	2112m	n.s.s.	n.s.s.	18/20	20.0mg	20/20							
4257	2112n	n.s.s.	3.80mg	0/20	20.0mg	20/20							
a	2112n	n.s.s.	n.s.s.	18/20	20.0mg	20/20							

PROPYLENE   115-07-1

	RefNum	LoConf	UpConf	Cntrl	1Dose	1Inc	2Dose	2Inc	3Dose	3Inc			Citation or Pathology
4258	TR272	7.47gm	n.s.s.	31/50	2.65gm	35/50	5.31gm	39/50					
a	TR272	15.3gm	n.s.s.	2/50	2.65gm	3/50	5.31gm	5/50					liv:hpa,hpc,nnd.
b	TR272	17.1gm	n.s.s.	6/50	2.65gm	4/50	5.31gm	7/50					lun:a/a,a/c.
4259	TR272	12.7gm	n.s.s.	37/50	2.21gm	25/50	4.42gm	28/50					
a	TR272	13.2gm	n.s.s.	14/50	2.21gm	11/50	4.42gm	14/50					liv:hpa,hpc,nnd.
b	TR272	25.7gm	n.s.s.	16/50	2.21gm	4/50	4.42gm	7/50					lun:a/a,a/c.
4260	BT702	6.88gm	n.s.s.	6/100	94.7mg	10/100	474.mg	13/100	2.37gm	15/100			Ciliberti;anya,534,235-245;1988
a	BT702	30.2gm	n.s.s.	1/100	94.7mg	1/100	474.mg	0/100	2.37gm	1/100			
b	BT702	34.8gm	n.s.s.	1/100	94.7mg	1/100	474.mg	0/100	2.37gm	1/100			
c	BT702	3.79gm	n.s.s.	24/100	94.7mg	34/100	474.mg	31/100	2.37gm	39/100			
d	BT702	5.89gm	n.s.s.	18/100	94.7mg	20/100	474.mg	19/100	2.37gm	25/100			
4261	BT702	26.0gm	n.s.s.	10/100	78.9mg	8/100	395.mg	9/100	1.97gm	3/100			
a	BT702	29.9gm	n.s.s.	0/100	78.9mg	0/100	395.mg	1/100	1.97gm	0/100			
b	BT702	44.5gm	n.s.s.	4/100	78.9mg	2/100	395.mg	1/100	1.97gm	0/100			
c	BT702	1.99gm	n.s.s.	26/100	78.9mg	17/100	395.mg	19/100	(1.97gm	6/100)			
d	BT702	2.63gm	n.s.s.	14/100	78.9mg	11/100	395.mg	10/100	(1.97gm	2/100)			
4262	TR272	915.mg	n.s.s.	37/50	638.mg	42/50	1.28gm	40/50					
a	TR272	5.05gm	n.s.s.	0/50	638.mg	0/50	1.28gm	2/50					liv:hpa,hpc,nnd.
4263	TR272	704.mg	n.s.s.	36/50	447.mg	38/50	893.mg	41/50					
a	TR272	1.99gm	n.s.s.	0/50	447.mg	4/50	893.mg	3/50					liv:hpa,hpc,nnd.

PROPYLENE GLYCOL   57-55-6

	RefNum	LoConf	UpConf	Cntrl	1Dose	1Inc	2Dose	2Inc	3Dose	3Inc	4Dose	4Inc	Citation or Pathology
4264	1351	8.74gm	n.s.s.	4/28	313.mg	5/25	625.mg	2/27	1.25gm	4/28	2.50gm	2/27	Gaunt;fctx,10,151-162;1972
4265	1351	9.11gm	n.s.s.	2/26	250.mg	4/27	500.mg	3/27	1.00gm	0/23	2.00gm	1/24	

1,2-PROPYLENE OXIDE (1,2-epoxypropane) 75-56-9

	RefNum	LoConf	UpConf	Cntrl	1Dose	1Inc	2Dose	2Inc					Citation or Pathology	
4266	TR267	360.mg	5.43gm	1/50	148.mg	1/50	296.mg	7/50					---:hem,hes.	S
a	TR267	451.mg	4.63gm	0/50	148.mg	0/50	296.mg	5/50					nas:hem,hes.	
b	TR267	475.mg	80.3gm	0/50	148.mg	3/50	296.mg	3/50					mgl:adc,adq.	S
c	TR267	243.mg	n.s.s.	4/50	148.mg	7/50	296.mg	6/50						S
d	TR267	522.mg	n.s.s.	0/50	148.mg	1/50	296.mg	3/50						S
e	TR267	554.mg	n.s.s.	1/50	148.mg	0/50	296.mg	4/50						S
f	TR267	606.mg	n.s.s.	0/50	148.mg	0/50	296.mg	3/50						S
g	TR267	732.mg	n.s.s.	0/50	148.mg	0/50	296.mg	3/50						S
h	TR267	107.mg	15.0gm	28/50	148.mg	35/50	296.mg	23/50						
i	TR267	320.mg	n.s.s.	3/50	148.mg	7/50	296.mg	3/50					liv:hpa,hpc,nnd.	
j	TR267	243.mg	n.s.s.	4/50	148.mg	7/50	296.mg	6/50					lun:a/a,a/c.	
4267	TR267	355.mg	1.95gm	0/50	123.mg	0/50	247.mg	10/50					nas:hem,hes.	
a	TR267	342.mg	8.20gm	2/50	123.mg	2/50	247.mg	10/50					---:hem,hes.	S
b	TR267	558.mg	12.1gm	0/50	123.mg	0/50	247.mg	5/50						S
c	TR267	604.mg	15.1gm	0/50	123.mg	0/50	247.mg	5/50						S
d	TR267	173.mg	n.s.s.	29/50	123.mg	31/50	247.mg	27/50						
e	TR267	425.mg	n.s.s.	14/50	123.mg	16/50	247.mg	9/50					liv:hpa,hpc,nnd.	
f	TR267	469.mg	n.s.s.	15/50	123.mg	14/50	247.mg	8/50					lun:a/a,a/c.	
4268	TR267	71.3mg	n.s.s.	3/50	35.2mg	12/50	70.4mg	10/50					cvu:esp; ute:esp,ess.	S
a	TR267	221.mg	n.s.s.	0/50	35.2mg	0/50	70.4mg	3/50						
b	TR267	32.8mg	n.s.s.	38/50	35.2mg	46/50	70.4mg	44/50						
c	TR267	1.01gm	n.s.s.	1/50	35.2mg	0/50	70.4mg	0/50					liv:hpa,hpc,nnd.	
4269	TR267	202.mg	n.s.s.	0/50	24.7mg	0/50	49.3mg	2/50						
a	TR267	27.8mg	n.s.s.	40/50	24.7mg	41/50	49.3mg	45/50						
b	TR267	133.mg	n.s.s.	1/50	24.7mg	2/50	49.3mg	3/50					liv:hpa,hpc,nnd.	
4270	1624	18.4mg	144.mg	8/78p	14.2mg	25/78	(42.4mg	22/80)					Lynch;txap,76,69-84;1984	
a	1624	482.mg	n.s.s.	0/78p	14.2mg	0/78	42.4mg	1/76						
b	1624	509.mg	n.s.s.	2/78p	14.2mg	1/78	42.4mg	1/76						
4271	1486	24.0mg	71.6mg	0/50	3.13mg	2/50	12.5mg	21/50					Dunkelberg;bjca,46,924-933;1982	

```
        Spe Strain Site  Xpo+Xpt                                              TD50    2Tailpvl
            Sex  Route  Hist   Notes                                              DR    AuOp

 a   R f sda gav for sqc 25m35 e                                            44.3mg * P<.0005
4272 R f wsr inh mgl fba 29m29 e                        . +      .          92.6mg * P<.003 +
 a   R f wsr inh mgl adc 29m29 e                                            813.mg * P<.2
 b   R f wsr inh liv cho 29m29                                              2.46gm * P<.5
 c   R f wsr inh res car 29m29 e                                            no dre   P=1.
 d   R f wsr inh tba tum 29m29 e                                            25.7mg * P<.0005
 e   R f wsr inh tba mal 29m29 e                                            150.mg * P<.0005
4273 R m wsr inh res car 29m29 e                        .    +   +hist      896.mg * P<.005
 a   R m wsr inh liv nnd 29m29                                              no dre   P=1.
 b   R m wsr inh liv clc 29m29                                              no dre   P=1.
 c   R m wsr inh tba mal 29m29 e                                            101.mg * P<.002
 d   R m wsr inh tba tum 29m29 e                                            79.3mg Z P<.02

PROPYLHYDRAZINE.HCl        100ng..:..1ug...:..10.....:..100....:..1mg..:..10.....:..100...:..1g.....:..10
4274 M f swa wat lun mix 99w99 e                             . +   .        41.4mg   P<.0005+
 a   M f swa wat lun ade 99w99 e                                            42.7mg   P<.0005
 b   M f swa wat lun adc 99w99 e                                            109.mg   P<.0005
 c   M f swa wat liv mix 99w99 e                                            no dre   P=1.
4275 M m swa wat lun ade 93w93 e                             . +   .        41.3mg   P<.0005
 a   M m swa wat lun mix 93w93 e                                            50.5mg   P<.002 +
 b   M m swa wat liv lcc 93w93 e                                            102.mg   P<.006 -
 c   M m swa wat liv mix 93w93 e                                            80.8mg   P<.06

PROPYLTHIOURACIL           100ng..:..1ug...:..10.....:..100....:..1mg..:..10.....:..100...:..1g.....:..10
4276 M b c51 eat pit ade 73w73 er                                 . +  .    409.mg * P<.0005+
4277 R m lev eat thy ade 52w52 ers                  . +   .                 10.3mg   P<.0005+
4278 R f wal wat thy ade 78w78 erv                    . +   .               15.8mg   P<.0005+
 a   R f wal wat thy car 78w78 erv                                          115.mg   P<.04  +
4279 R m wal wat thy ade 78w78 erv                  . +   .                 17.2mg   P<.002 +
 a   R m wal wat thy car 78w78 erv                                          79.3mg   P<.03  +
4280 R m wis wat thy ade  6m24 rs                                           noTD50   P<.2   -

PROQUAZONE                 100ng..:..1ug...:..10.....:..100....:..1mg..:..10.....:..100...:..1g.....:..10
4281 M f cd1 eat liv hem 86w86                                       .>     48.7gm * P<.9   -
 a   M f cd1 eat liv hms 86w86                                              no dre   P=1.
 b   M f cd1 eat liv hpc 86w86                                              no dre   P=1.
 c   M f cd1 eat lun a/a 86w86                                              no dre   P=1.
 d   M f cd1 eat lun a/c 86w86                                              no dre   P=1.
4282 M m cd1 eat liv hpc 86w86                                   .>         4.35gm * P<.6   -
 a   M m cd1 eat lun a/a 86w86                                              5.45gm * P<.7   -
 b   M m cd1 eat liv hem 86w86                                              14.0gm * P<.2   -
 c   M m cd1 eat lun a/c 86w86                                              no dre   P=1.   -

PRORESID                   100ng..:..1ug...:..10.....:..100....:..1mg..:..10.....:..100...:..1g.....:..10
4283 R m b46 ivj tba mix 12m24 es               .>                         no dre   P=1.   -

PROTOCATECHUIC ACID        100ng..:..1ug...:..10.....:..100....:..1mg..:..10.....:..100...:..1g.....:..10
4284 R m f3d eat for tum 51w52 e                                  .>        no dre   P=1.
 a   R m f3d eat liv tum 51w52 e                                            no dre   P=1.

SX PURPLE                  100ng..:..1ug...:..10.....:..100....:..1mg..:..10.....:..100...:..1g.....:..10
4285 R b non eat liv tum 33m33                                              no dre   P=1.
 a   R b non eat tba mix 33m33                                              24.5gm   P<.1   +
4286 R f nss eat liv tum 64w64 e                                 .>         no dre   P=1.   -
4287 R m nss eat liv tum 64w64 e                                 .>         no dre   P=1.   -
 a   R m nss eat tba tum 64w64 e                                            no dre   P=1.   -

PYRAZINAMIDE               100ng..:..1ug...:..10.....:..100....:..1mg..:..10.....:..100...:..1g.....:..10
4288 M f b6c eat TBA MXB 18m24 s                                     :>     3.43gm * P<.7   i
 a   M f b6c eat liv MXB 18m24 s                                            10.7gm * P<.5
 b   M f b6c eat lun MXB 18m24 s                                            no dre   P=1.
4289 M m b6c eat TBA MXB 18m24 s                                 :>         546.mg * P<.2
 a   M m b6c eat liv MXB 18m24 s                                            5.18gm * P<.4
 b   M m b6c eat lun MXB 18m24 s                                            no dre   P=1.
4290 R f f34 eat pit MXA 18m24                                 :  ±         #178.mg \ P<.03 -
 a   R f f34 eat TBA MXB 18m24                                              182.mg \ P<.4
 b   R f f34 eat liv MXB 18m24                                              no dre   P=1.
4291 R m f34 eat TBA MXB 18m24                                   :>         284.mg \ P<.4   -
 a   R m f34 eat liv MXB 18m24                                              no dre   P=1.

PYRILAMINE MALEATE         100ng..:..1ug...:..10.....:..100....:..1mg..:..10.....:..100...:..1g.....:..10
4292 R f f34 eat liv mix 26m31                                 .   ±        175.mg   P<.02  +
 a   R f f34 eat liv nnd 26m31                                              266.mg   P<.08
 b   R f f34 eat liv hpc 26m31                                              880.mg   P<.1   +
4293 R f f34 wat liv nnd 26m31                                   .>         1.25gm   P<.7
4294 R m f34 eat liv hpc 26m31                                 .    ±       704.mg   P<.1   +
 a   R m f34 eat liv mix 26m31                                              518.mg   P<.5
 b   R m f34 eat liv nnd 26m31                                              no dre   P=1.
4295 R m f34 wat liv nnd 26m31                                   .>         no dre   P=1.
```

	RefNum	LoConf	UpConf	Cntrl	1Dose	1Inc	2Dose	2Inc		Citation or Pathology
a	1486	26.3mg	82.6mg	0/50	3.13mg	2/50	12.5mg	19/50		
4272	1830	46.7mg	602.mg	32/69	5.33mg	30/71	17.8mg	39/69	53.3mg	47/70	Kuper;fctx,29,159-167;1988
a	1830	235.mg	n.s.s.	3/69	5.33mg	6/71	17.8mg	5/69	53.3mg	8/70	
b	1830	447.mg	n.s.s.	2/69	5.33mg	1/71	17.8mg	2/69	53.3mg	3/70	
c	1830	77.4mg	n.s.s.	0/69	5.33mg	0/71	17.8mg	0/69	53.3mg	0/70	
d	1830	14.0mg	69.4mg	52/69	5.33mg	49/71	17.8mg	61/69	53.3mg	67/70	
e	1830	81.1mg	510.mg	6/69	5.33mg	15/71	17.8mg	14/69	53.3mg	26/70	
4273	1830	310.mg	7.93gm	0/70	3.73mg	0/69	12.4mg	0/71	37.3mg	4/70	
a	1830	604.mg	n.s.s.	0/70	3.73mg	0/69	12.4mg	1/71	37.3mg	0/70	
b	1830	630.mg	n.s.s.	1/70	3.73mg	0/69	12.4mg	2/71	37.3mg	0/70	
c	1830	51.3mg	569.mg	19/70	3.73mg	17/69	12.4mg	22/71	37.3mg	34/70	
d	1830	35.9mg	n.s.s.	49/70	3.73mg	28/69	12.4mg	34/71	37.3mg	53/70	

PROPYLHYDRAZINE.HCl   56795-66-5

	RefNum	LoConf	UpConf	Cntrl	1Dose	1Inc	2Dose	2Inc	Citation or Pathology
4274	252	22.6mg	104.mg	21/99	50.0mg	27/43			Nagel;ejca,11,473-478;1975/Toth 1974
a	252	23.5mg	103.mg	18/99	50.0mg	26/43			
b	252	51.8mg	367.mg	4/99	50.0mg	12/43			
c	252	18.7mg	n.s.s.	2/43	50.0mg	0/2			
4275	252	21.8mg	116.mg	15/99	41.7mg	22/43			
a	252	23.9mg	281.mg	23/99	41.7mg	22/43			
b	252	25.0mg	2.02gm	0/56	41.7mg	2/10			
c	252	20.3mg	n.s.s.	4/56	41.7mg	3/10			

PROPYLTHIOURACIL   51-52-5

	RefNum	LoConf	UpConf	Cntrl	1Dose	1Inc	2Dose	2Inc	Citation or Pathology
4276	257	266.mg	663.mg	0/28	1.25gm	15/24	1.50gm	21/29	King;pseb,112,365-366;1963
4277	254	5.69mg	21.5mg	0/68	40.0mg	16/33			Lindsay;arpa,81,308-316;1966
4278	256	8.77mg	34.7mg	1/20	42.9mg	20/30			Willis;jpat,82,23-27;1961
a	256	39.8mg	n.s.s.	0/20	42.9mg	4/30			
4279	256	7.77mg	73.3mg	2/20	37.5mg	11/18			
a	256	23.9mg	n.s.s.	0/20	37.5mg	3/18			
4280	1273	n.s.s.	n.s.s.	3/8	7.50mg	5/5			Stoll;bdca,50,389-398;1963

PROQUAZONE   (1-isopropyl-7-methyl-4-phenyl-2(IH)-quinazolinone)   22760-18-5

	RefNum	LoConf	UpConf	Cntrl	1Dose	1Inc	2Dose	2Inc			Citation or Pathology
4281	1667	2.50gm	n.s.s.	1/50	60.0mg	0/50	180.mg	1/50	360.mg	1/50	Van Ryzin;dact,3,361-379;1980
a	1667	2.88gm	n.s.s.	1/50	60.0mg	0/50	180.mg	1/50	360.mg	0/50	
b	1667	1.65gm	n.s.s.	0/50	60.0mg	3/50	180.mg	0/50	360.mg	1/50	
c	1667	866.mg	n.s.s.	4/50	60.0mg	6/50	180.mg	10/50	360.mg	3/50	
d	1667	284.mg	n.s.s.	12/50	60.0mg	3/50	(180.mg	2/50	360.mg	1/50)	
4282	1667	762.mg	n.s.s.	4/50	60.0mg	2/50	180.mg	8/50	360.mg	4/50	
a	1667	666.mg	n.s.s.	5/50	60.0mg	6/50	180.mg	9/50	360.mg	6/50	
b	1667	2.28gm	n.s.s.	0/50	60.0mg	0/50	180.mg	0/50	360.mg	1/50	
c	1667	861.mg	n.s.s.	12/50	60.0mg	11/50	180.mg	7/50	360.mg	10/50	

PRORESID   1508-45-8

	RefNum	LoConf	UpConf	Cntrl	1Dose	1Inc	Citation or Pathology
4283	1017	2.58mg	n.s.s.	7/65	.714mg	1/24	Schmahl;arzn,20,1461-1467;1970

PROTOCATECHUIC ACID   (3,4-dihydroxybenzoic acid) 99-50-3

	RefNum	LoConf	UpConf	Cntrl	1Dose	1Inc	Citation or Pathology
4284	2080	455.mg	n.s.s.	0/15	588.mg	0/15	Hirose;carc,13,1825-1828;1992/pers.comm.
a	2080	455.mg	n.s.s.	0/15	588.mg	0/15	

SX PURPLE   (ponceau 4R) 2611-82-7

	RefNum	LoConf	UpConf	Cntrl	1Dose	1Inc	Citation or Pathology
4285	1429	15.0gm	n.s.s.	0/50	771.mg	0/50	Andrianova;vpit,29,61-65;1970
a	1429	6.02gm	n.s.s.	0/50	771.mg	2/50	
4286	1372	819.mg	n.s.s.	0/7	1.50gm	0/7	Allmark;jphp,9,622-628;1957
4287	1372	562.mg	n.s.s.	0/5	1.20gm	0/6	
a	1372	562.mg	n.s.s.	0/5	1.20gm	0/6	

PYRAZINAMIDE   98-96-4

	RefNum	LoConf	UpConf	Cntrl	1Dose	1Inc	2Dose	2Inc	Citation or Pathology / Brkly Code
4288	TR48	575.mg	n.s.s.	1/15	348.mg	5/35	697.mg	9/35	
a	TR48	1.75gm	n.s.s.	0/15	348.mg	0/35	697.mg	1/35	liv:hpa,hpc,nnd.
b	TR48	2.62gm	n.s.s.	1/15	348.mg	0/35	697.mg	1/35	lun:a/a,a/c.
4289	TR48	105.mg	n.s.s.	2/15	322.mg	5/35	726.mg	6/35	
a	TR48	956.mg	n.s.s.	1/15	322.mg	0/35	726.mg	3/35	liv:hpa,hpc,nnd.
b	TR48	n.s.s.	n.s.s.	0/15	322.mg	0/35	726.mg	0/35	lun:a/a,a/c.
4290	TR48	78.9mg	n.s.s.	2/15	134.mg	14/35	(268.mg	7/34)	pit:cra,crc. S
a	TR48	50.4mg	n.s.s.	11/15	134.mg	26/35	(268.mg	19/34)	
b	TR48	n.s.s.	n.s.s.	0/15	134.mg	0/35	268.mg	0/34	liv:hpa,hpc,nnd.
4291	TR48	79.4mg	n.s.s.	4/15	107.mg	17/35	(214.mg	8/36)	
a	TR48	n.s.s.	n.s.s.	0/15	107.mg	0/35	214.mg	0/36	liv:hpa,hpc,nnd.

PYRILAMINE MALEATE   59-33-6

	RefNum	LoConf	UpConf	Cntrl	1Dose	1Inc	Citation or Pathology
4292	1644	70.5mg	n.s.s.	3/20	82.7mg	10/20	Lijinsky;fctx,22,27-30;1984
a	1644	91.9mg	n.s.s.	3/20	82.7mg	8/20	
b	1644	216.mg	n.s.s.	0/20	82.7mg	2/20	
4293	1644m	145.mg	n.s.s.	3/20	67.5mg	4/20	
4294	1644	173.mg	n.s.s.	0/20	66.2mg	2/20	
a	1644	95.7mg	n.s.s.	5/20	66.2mg	7/20	
b	1644	138.mg	n.s.s.	5/20	66.2mg	5/20	
4295	1644m	98.7mg	n.s.s.	5/20	47.3mg	5/20	

```
        Spe Strain Site  Xpo+Xpt                                                 TD50    2Tailpvl
            Sex  Route  Hist      Notes                                           DR       AuOp
PYRIMETHAMINE       100ng..:..1ug....:..10.....:..100....:..1mg...:..10.....:..100....:..1g.....:..10
4296 M f b6c eat TBA MXB 18m24                              :>                    315.mg * P<.8   -
a    M f b6c eat liv MXB 18m24                                                    no dre   P=1.
b    M f b6c eat lun MXB 18m24                                                    405.mg * P<.4
4297 M m b6c eat TBA MXB 18m24 s                              :>                  419.mg * P<.9   i
a    M m b6c eat liv MXB 18m24 s                                                  no dre   P=1.
b    M m b6c eat lun MXB 18m24 s                                                  no dre   P=1.
4298 R f f34 eat TBA MXB 18m24                         :>                         27.8mg * P<.5   -
a    R f f34 eat liv MXB 18m24                                                    no dre   P=1.
4299 R m f34 eat TBA MXB 18m24                            :>                      no dre   P=1.
a    R m f34 eat liv MXB 18m24                                                    no dre   P=1.

QUERCETIN           100ng..:..1ug....:..10.....:..100....:..1mg...:..10.....:..100....:..1g.....:..10
4300 M f ddy eat lun mix 28m28 e                                      .   ±       4.24gm   P<.05  -
a    M f ddy eat liv tum 28m28 e                                                  no dre   P=1.
4301 M m ddy eat liv mix 27m27 e                                      .          15.1gm   P<.09  -
a    M m ddy eat lun mix 27m27 e                                                  5.46gm   P<.2
b    M m ddy eat hea scs 27m27 e                                                  73.4gm   P<.4
4302 R f f34 eat ton sqc 24m24                                                    #36.5gm * P<.04  -
a    R f f34 eat TBA MXB 24m24                                                    no dre   P=1.
b    R f f34 eat liv MXB 24m24                                                    75.7gm * P<.6
4303 R f f34 eat tba mix 64w64 fg                       .>                        no dre   P=1.  -
4304 R m f34 eat kid MXA 24m24                                        :          +8.99gm * P<.003 p
a    R m f34 eat kid rua 24m24                                                   12.5gm * P<.009 p
b    R m f34 eat TBA MXB 24m24                                                    no dre   P=1.
c    R m f34 eat liv MXB 24m24                                                    no dre   P=1.
4305 R m f34 eat kid MXA 24m24  with step                     .   +              5.25gm * P<.004
a    R m f34 eat kid rua 24m24                                                   6.03gm * P<.009 p
4306 R m f34 eat tba mix 64w64                          .>                        no dre   P=1.  -
4307 R f f3d eat --- lkm 24m26 e                                      .   ±       7.86gm * P<.06  -
a    R f f3d eat ute esp 24m26 e                                                 19.3gm * P<.4   -
b    R f f3d eat liv hnd 24m26 e                                                  no dre   P=1.  -
4308 R m f3d eat --- lkm 24m26 e                                   .   +          1.16gm \ P<.005 -
a    R m f3d eat liv hnd 24m26 e                                                 66.8gm * P<.7   -
4309 R f nra eat ilm mix 58w58 er            .   +                                5.12mg   P<.0005+
a    R f nra eat ilm adc 58w58 er                                                15.4mg   P<.003
b    R f nra eat ubl mix 58w58 er                                                51.3mg   P<.09  +
4310 R m nra eat ilm mix 58w58 er                <+                               noTD50   P<.004 +
a    R m nra eat ubl tcc 58w58 er                                                21.0mg   P<.06  +

QUERCETIN DIHYDRATE 100ng..:..1ug....:..10.....:..100....:..1mg...:..10.....:..100....:..1g.....:..10
4311 H f syg eat ilm adc 24m24                                                    140.gm   P<.3   -
a    H f syg eat for pam 24m24                                                    no dre   P=1.
4312 H f syg eat for pam 23m23                                          .>        18.9gm   P<.2   -
a    H f syg eat ute ley 23m23                                                    39.2gm   P<.4   -
4313 H f syg eat for pam 12m23                                             .>     no dre   P=1.   -
4314 H m syg eat for pam 24m24                                                    56.7gm   P<.3   -
a    H m syg eat adr coa 24m24                                                    no dre   P=1.   -
4315 H m syg eat for pam 23m23                                                    no dre   P=1.   -
a    H m syg eat adr coa 23m23                                                    no dre   P=1.   -
4316 H m syg eat for pam 12m23                                        .>          1.44gm   P<.5   -
4317 R f aci eat adr coa 77w77                                           .>       no dre   P=1.   -
a    R f aci eat pit ade 77w77                                                    no dre   P=1.   -
4318 R f aci eat thi sar 28m28                                                    90.4gm   P<.2   -
a    R f aci eat adr coa 28m28                                                    226.gm   P<.8   -
b    R f aci eat pit ade 28m28                                                    461.gm   P<1.   -
4319 R m aci eat tes ict 77w77                                        .   ±       2.29gm * P<.09  -
a    R m aci eat cec ade 77w77                                                    7.13gm * P<.08  -
b    R m aci eat adr phe 77w77                                                    7.13gm * P<.08  -
4320 R m aci eat cec adc 28m28                                                    .35.2gm  P<.05  -
a    R m aci eat cec ade 28m28                                                    72.3gm   P<.2   -
b    R m aci eat pan exa 28m28                                                    72.3gm   P<.2   -
4321 R f f34 eat adr coa 77w77 e                            .>                    118.mg   P<.3   -
4322 R m f34 eat tba mix 77w77 e                                                  noTD50   P<.6   -

QUILLAIA EXTRACT    100ng..:..1ug....:..10.....:..100....:..1mg...:..10.....:..100....:..1g.....:..10
4323 M f the eat lun ppa 84w84 e                                         .>       no dre   P=1.   -
a    M f the eat liv hnd 84w84 e                                                  no dre   P=1.
4324 M m the eat lun ppa 84w84 e                                      .   ±       4.09gm * P<.03  -
a    M m the eat liv mhp 84w84 e                                                  no dre   P=1.
b    M m the eat liv hnd 84w84 e                                                  no dre   P=1.
4325 R f wis eat thy ade 25m25 e                                      .          ±8.70gm * P<.09  -
a    R f wis eat liv ade 25m25 e                                                 72.4gm * P<.2   -
4326 R m wis eat liv ade 25m25 e                                         .>       no dre   P=1.   -

p-QUINONE DIOXIME   100ng..:..1ug....:..10.....:..100....:..1mg...:..10.....:..100....:..1g.....:..10
4327 M f b6c eat TBA MXB 24m24                              :>                    6.71gm * P<1.   -
a    M f b6c eat liv MXB 24m24                                                    656.mg * P<.02
b    M f b6c eat lun MXB 24m24                                                    6.66gm * P<.8
4328 M m b6c eat TBA MXB 24m24                              :>                    936.mg * P<.6
a    M m b6c eat liv MXB 24m24                                                    1.23kg   P<1.
```

	RefNum	LoConf	UpConf	Cntrl	1Dose	1Inc	2Dose	2Inc		Citation or Pathology
										Brkly Code
PYRIMETHAMINE	(daraprin)	58-14-0								
4296	TR77	44.2mg	n.s.s.	1/15	34.7mg	5/35	69.5mg	9/35		
a	TR77	n.s.s.	n.s.s.	0/15	34.7mg	0/35	69.5mg	0/35		liv:hpa,hpc,nnd.
b	TR77	99.6mg	n.s.s.	0/15	34.7mg	0/35	69.5mg	2/35		lun:a/a,a/c.
4297	TR77	30.1mg	n.s.s.	0/15	32.0mg	1/35	64.2mg	1/35		
a	TR77	n.s.s.	n.s.s.	0/15	32.0mg	0/35	64.2mg	0/35		liv:hpa,hpc,nnd.
b	TR77	n.s.s.	n.s.s.	0/15	32.0mg	0/35	64.2mg	0/35		lun:a/a,a/c.
4298	TR77	7.01mg	n.s.s.	7/15	5.30mg	16/35	10.7mg	26/35		
a	TR77	n.s.s.	n.s.s.	0/15	5.30mg	0/35	10.7mg	0/35		liv:hpa,hpc,nnd.
4299	TR77	14.9mg	n.s.s.	8/15	4.20mg	17/35	8.60mg	11/35		
a	TR77	n.s.s.	n.s.s.	0/15	4.20mg	0/35	8.60mg	0/35		liv:hpa,hpc,nnd.
QUERCETIN	117-39-5									
4300	1146	1.82gm	n.s.s.	4/15	2.60gm	18/31				Saito;tcam,1,213-221;1980
a	1146	23.5gm	n.s.s.	0/15	2.60gm	0/33				
4301	1146	5.20gm	n.s.s.	0/14	2.40gm	4/32				
a	1146	1.87gm	n.s.s.	6/16	2.40gm	21/37				
b	1146	12.0gm	n.s.s.	0/16	2.40gm	1/37				
4302	TR409	8.92gm	n.s.s.	0/50	50.0mg	0/50	500.mg	0/50	2.00gm 2/50	S
a	TR409	1.27gm	n.s.s.	49/50	50.0mg	48/50	500.mg	43/50	2.00gm 44/50	
b	TR409	7.97gm	n.s.s.	0/50	50.0mg	1/50	500.mg	0/50	2.00gm 1/50	liv:hpa,hpc,nnd.
4303	1648	69.1mg	n.s.s.	0/10	50.0mg	1/11	100.mg	0/11		Stoewsand;jtxe,14,105-114;1984
4304	TR409	3.06gm	58.8gm	0/50	40.0mg	0/50	400.mg	0/50	1.60gm 4/50	kid:rua,uac.
a	TR409	3.72gm	393.gm	0/50	40.0mg	0/50	400.mg	0/50	1.60gm 3/50	
b	TR409	1.19gm	n.s.s.	46/50	40.0mg	40/50	400.mg	46/50	1.60gm 39/50	
c	TR409	5.34gm	n.s.s.	3/50	40.0mg	4/50	400.mg	4/50	1.60gm 1/50	liv:hpa,hpc,nnd.
4305	TR409	2.40gm	45.6gm	1/50	40.0mg	2/50	400.mg	7/50	1.60gm 9/50	kid:rua,uac. S
a	TR409	2.59gm	305.gm	1/50	40.0mg	2/50	400.mg	7/50	1.60gm 8/50	
4306	1648	59.8mg	n.s.s.	1/10	40.0mg	0/10	80.0mg	1/11		Stoewsand;jtxe,14,105-114;1984
4307	1934	3.04gm	n.s.s.	8/50	580.mg	14/50	2.32gm	17/50		Ito;gann,80,317-325;1989
a	1934	4.71gm	n.s.s.	8/50	580.mg	10/50	2.32gm	12/50		
b	1934	5.55gm	n.s.s.	2/50	580.mg	0/50	2.32gm	0/50		
4308	1934	550.mg	11.6gm	6/50	464.mg	18/50	(1.86gm	14/50)		
a	1934	8.86gm	n.s.s.	2/50	464.mg	2/50	1.86gm	3/50		
4309	1392	2.31mg	12.3mg	0/9	50.0mg	14/16				Pamukcu;canr,40,3468-3472;1980
a	1392	6.79mg	76.3mg	0/9	50.0mg	8/16				
b	1392	15.5mg	n.s.s.	0/9	50.0mg	3/16				
4310	1392	n.s.s.	9.97mg	0/8	40.0mg	6/6				
a	1392	n.s.s.	5.06mg	0/8	40.0mg	2/6				
QUERCETIN DIHYDRATE	6151-25-3									
4311	1144m	22.7gm	n.s.s.	0/20	10.5gm	1/20				Morino;carc,3,93-97;1982
a	1144m	20.5gm	n.s.s.	2/20	10.5gm	2/20				
4312	1144n	4.63gm	n.s.s.	0/8	4.18gm	2/15				
a	1144n	6.37gm	n.s.s.	0/8	4.18gm	1/15				
4313	1144o	698.mg	n.s.s.	0/8	523.mg	0/7				
4314	1144m	12.6gm	n.s.s.	1/20	9.20gm	3/20				
a	1144m	18.0gm	n.s.s.	2/20	9.20gm	2/20				
4315	1144n	10.7gm	n.s.s.	1/8	3.68gm	0/15				
a	1144n	12.2gm	n.s.s.	1/8	4.18gm	0/15				
4316	1144o	228.mg	n.s.s.	1/8	461.mg	2/7				
4317	1145m	462.mg	n.s.s.	0/22	500.mg	0/10	2.50gm	0/9		Hirono;clet,13,15-21;1981
a	1145m	462.mg	n.s.s.	0/22	500.mg	0/10	2.50gm	0/9		
4318	1145n	14.7gm	n.s.s.	0/33	5.00gm	1/20				
a	1145n	16.0gm	n.s.s.	1/33	5.00gm	1/20				
b	1145n	13.2gm	n.s.s.	3/33	5.00gm	2/20				
4319	1145m	593.mg	n.s.s.	3/30	400.mg	1/10	2.00gm	3/8		
a	1145m	1.16gm	n.s.s.	0/30	400.mg	0/10	2.00gm	1/8		
b	1145m	1.16gm	n.s.s.	0/30	400.mg	0/10	2.00gm	1/8		
4320	1145n	8.65gm	n.s.s.	0/33	4.00gm	2/20				
a	1145n	11.8gm	n.s.s.	0/33	4.00gm	1/20				
b	1145n	11.8gm	n.s.s.	0/33	4.00gm	1/20				
4321	1662	27.2mg	n.s.s.	1/16	50.0mg	3/15				Takanashi;jfds,5,55-60;1983
4322	1662	n.s.s.	n.s.s.	15/16	40.0mg	15/15				
QUILLAIA EXTRACT	(spray-dried aqueous extract of quillaia bark) ---									
4323	1404	5.84gm	n.s.s.	9/44	130.mg	5/43	650.mg	4/43	1.95gm 5/46	Phillips;fctx,17,23-27;1979
a	1404	12.1gm	n.s.s.	1/44	130.mg	2/43	650.mg	1/43	1.95gm 0/46	
4324	1404	1.64gm	n.s.s.	7/45	120.mg	3/42	600.mg	7/41	1.80gm 12/43	
a	1404	8.58gm	n.s.s.	5/45	120.mg	1/42	600.mg	2/41	1.80gm 1/43	
b	1404	3.93gm	n.s.s.	8/45	120.mg	4/42	600.mg	3/41	1.80gm 6/43	
4325	1527	3.06gm	n.s.s.	0/39	150.mg	2/40	500.mg	5/45	1.50gm 4/42	Drake;fctx,20,15-23;1982
a	1527	11.8gm	n.s.s.	0/42	150.mg	0/45	500.mg	0/46	1.50gm 1/46	
4326	1527	605.mg	n.s.s.	0/40	120.mg	0/33	400.mg	0/26	1.20gm 0/44	
p-QUINONE DIOXIME	105-11-3									
4327	TR179	252.mg	n.s.s.	7/20	96.6mg	14/50	193.mg	17/50		
a	TR179	329.mg	n.s.s.	0/20	96.6mg	3/50	193.mg	8/50		liv:hpa,hpc,nnd.
b	TR179	664.mg	n.s.s.	1/20	96.6mg	2/50	193.mg	3/50		lun:a/a,a/c.
4328	TR179	178.mg	n.s.s.	5/18	89.2mg	20/50	178.mg	22/50		
a	TR179	398.mg	n.s.s.	5/18	89.2mg	6/50	178.mg	12/50		liv:hpa,hpc,nnd.

```
        Spe Strain Site    Xpo+Xpt                                                      TD50     2Tailpvl
            Sex  Route  Hist      Notes                                                   DR     AuOp
   b    M m b6c eat lun MXB 24m24                                                       no dre   P=1.
4329  R f f34 eat ubl MXA 24m24                               :    +       :          106.mg * P<.003 c
   a  R f f34 eat ubl tcc 24m24                                                       182.mg * P<.02  c
   b  R f f34 eat TBA MXB 24m24                                                       no dre   P=1.
   c  R f f34 eat liv MXB 24m24                                                       no dre   P=1.
4330  R m f34 eat TBA MXB 24m24                                        :>             349.mg * P<.9   -
   a  R m f34 eat liv MXB 24m24                                                       1.13gm * P<.4

C.I. ACID RED 114                      100ng..:..1ug....:..10....:..100....:..1mg....:..10.....:..100....:..1g.....:..10
4331  R f f34 wat MXB MXB 23m24 a                                  :+ :               3.11mg Z P<.0005

   a  R f f34 wat cli MXA 23m24 a                                                     4.75mg Z P<.0005c
   b  R f f34 wat cli MXA 23m24 a                                                     8.97mg Z P<.0005
   c  R f f34 wat mul MXA 23m24 a                                                     9.44mg * P<.0005e
   d  R f f34 wat pit pda 23m24 a                                                     11.1mg * P<.003
   e  R f f34 wat mgl fba 23m24 a                                                     11.4mg * P<.008
   f  R f f34 wat cli MXA 23m24 a                                                     11.9mg Z P<.0005
   g  R f f34 wat liv MXA 23m24 a                                                     14.2mg Z P<.0005c
   h  R f f34 wat zym MXA 23m24 a                                                     15.6mg Z P<.0005c
   i  R f f34 wat ute MXA 23m24 a                                                     16.4mg Z P<.0005
   j  R f f34 wat liv nnd 23m24 a                                                     18.0mg Z P<.0005
   k  R f f34 wat zym car 23m24 a                                                     18.0mg * P<.0005
   l  R f f34 wat ute esp 23m24 a                                                     20.1mg * P<.002
   m  R f f34 wat ski MXA 23m24 a                                                     21.9mg * P<.0005c
   n  R f f34 wat lun MXA 23m24 a                                                     23.6mg * P<.0005c
   o  R f f34 wat lun a/a 23m24 a                                                     26.3mg * P<.0005
   p  R f f34 wat ski bca 23m24 a                                                     28.2mg * P<.0005
   q  R f f34 wat thy MXA 23m24 a                                                     28.8mg * P<.0005
   r  R f f34 wat MXA MXA 23m24 a                                                     31.4mg Z P<.0005c
   s  R f f34 wat amd MXA 23m24 a                                                     31.6mg * P<.003 e
   t  R f f34 wat MXA sqp 23m24 a                                                     36.5mg * P<.0005
   u  R f f34 wat amd pob 23m24 a                                                     38.3mg * P<.007
   v  R f f34 wat mgl adc 23m24 a                                                     40.6mg * P<.0005e
   w  R f f34 wat liv hpc 23m24 a                                                     42.5mg Z P<.0005
   x  R f f34 wat ski MXA 23m24 a                                                     49.7mg * P<.002
   y  R f f34 wat thy fca 23m24 a                                                     53.5mg * P<.007
   z  R f f34 wat ski sqc 23m24 a                                                     57.3mg * P<.005
   A  R f f34 wat zym ade 23m24 a                                                     65.4mg Z P<.0005
   B  R f f34 wat ton MXA 23m24 a                                                     65.7mg * P<.002
   C  R f f34 wat phr sqp 23m24 a                                                     72.4mg * P<.004
   D  R f f34 wat ton sqp 23m24 a                                                     75.0mg * P<.003
   E  R f f34 wat MXA MXA 23m24 a                                                     136.mg Z P<.007 c
   F  R f f34 wat thy cca 23m24 a                                                     48.7mg Z P<.05
   G  R f f34 wat mgl ade 23m24 a                                                     67.0mg * P<.04
   H  R f f34 wat thy fcc 23m24 a                                                     68.9mg * P<.03
   I  R f f34 wat sub MXA 23m24 a                                                     78.2mg * P<.05
   J  R f f34 wat MXA MXA 23m24 a                                                     213.mg * P<.02  c
   K  R f f34 wat MXA sqc 23m24 a                                                     227.mg * P<.02
   L  R f f34 wat TBA MXB 23m24 a                                                     2.31mg Z P<.0005
   M  R f f34 wat liv MXB 23m24 a                                                     14.2mg Z P<.0005
4332  R m f34 wat MXB MXB 24m24                                   : + :               2.80mg Z P<.0005

   a  R m f34 wat ski MXA 24m24                                                       3.30mg Z P<.0005c
   b  R m f34 wat tes MXA 24m24                                                       3.39mg Z P<.0005
   c  R m f34 wat ski bca 24m24                                                       3.58mg Z P<.0005
   d  R m f34 wat amd MXA 24m24                                                       5.94mg Z P<.0005e
   e  R m f34 wat amd pob 24m24                                                       6.69mg Z P<.0005
   f  R m f34 wat liv MXA 24m24                                                       7.02mg Z P<.0005c
   g  R m f34 wat liv nnd 24m24                                                       10.4mg Z P<.0005
   h  R m f34 wat ski MXA 24m24                                                       12.7mg * P<.0005c
   i  R m f34 wat ski sqc 24m24                                                       16.2mg * P<.0005
   j  R m f34 wat liv hpc 24m24                                                       20.2mg * P<.0005
   k  R m f34 wat ski ker 24m24                                                       20.4mg * P<.0005c
   l  R m f34 wat ski bcc 24m24                                                       22.8mg * P<.0005
   m  R m f34 wat ski MXA 24m24                                                       23.4mg * P<.002 c
   n  R m f34 wat zym MXA 24m24                                                       27.0mg * P<.0005c
   o  R m f34 wat zym car 24m24                                                       29.8mg * P<.0005
   p  R m f34 wat lun a/a 24m24                                                       39.0mg Z P<.008
   q  R m f34 wat mul mnl 24m24                                                       12.3mg * P<.05
   r  R m f34 wat lun MXA 24m24                                                       47.2mg Z P<.1   e
   s  R m f34 wat MXA sqp 24m24                                                       71.0mg Z P<.02  e
   t  R m f34 wat TBA MXB 24m24                                                       3.23mg Z P<.0005
   u  R m f34 wat liv MXB 24m24                                                       7.02mg Z P<.0005

C.I. FOOD RED 3                        100ng..:..1ug....:..10....:..100....:..1mg....:..10....:..100....:..1g......:..10
4333  M f b6c eat TBA MXB 24m24                                               :>      no dre   P=1.   -
   a  M f b6c eat liv MXB 24m24                                                       no dre   P=1.
   b  M f b6c eat lun MXB 24m24                                                       no dre   P=1.
4334  M m b6c eat TBA MXB 24m24                                            :>         no dre   P=1.   -
   a  M m b6c eat liv MXB 24m24                                                       no dre   P=1.
   b  M m b6c eat lun MXB 24m24                                                       no dre   P=1.
```

	RefNum	LoConf	UpConf	Cntrl	1Dose	1Inc	2Dose	2Inc		Citation or Pathology
										Brkly Code
b	TR179	466.mg	n.s.s.	2/18	89.2mg	7/50	178.mg	6/50		lun:a/a,a/c.
4329	TR179	57.2mg	514.mg	0/20	18.6mg	3/49	37.1mg	11/50		ubl:sqc,tcc,tpp.
a	TR179	82.6mg	n.s.s.	0/20	18.6mg	1/49	37.1mg	7/50		
b	TR179	39.7mg	n.s.s.	14/20	18.6mg	30/49	37.1mg	28/50		
c	TR179	204.mg	n.s.s.	0/20	18.6mg	2/49	37.1mg	0/50		liv:hpa,hpc,nnd.
4330	TR179	31.3mg	n.s.s.	11/20	14.9mg	21/50	29.7mg	24/50		
a	TR179	184.mg	n.s.s.	0/20	14.9mg	0/50	29.7mg	1/50		liv:hpa,hpc,nnd.

C.I. ACID RED 114   6459-94-5

	RefNum	LoConf	UpConf	Cntrl	1Dose	1Inc	2Dose	2Inc		Citation or Pathology
4331	TR405	2.14mg	4.68mg	12/50	8.42mg	22/35	16.8mg	54/65	34.3mg 39/50	cli:ade,anb,car,cnb; col:adp,muc; duo:adp,muc;
					jej:adp; liv:hpc,nnd; lun:a/a,a/c; phr:sqc,sqp; rec:adp; ski:bca,bcc; ton:sqc,sqp; zym:ade,car. C
a	TR405	2.95mg	8.36mg	11/50	8.42mg	17/35	16.8mg	28/65	34.3mg 23/50	cli:ade,anb,car,cnb.
b	TR405	5.19mg	17.0mg	4/50	8.42mg	9/35	16.8mg	19/65	34.3mg 15/50	cli:car,cnb. S
c	TR405	5.10mg	24.3mg	12/50	8.42mg	13/35	16.8mg	18/65	34.3mg 5/50	mul:leu,mnl.
d	TR405	5.14mg	73.3mg	25/50	8.42mg	17/35	16.8mg	17/65	34.3mg 5/50	S
e	TR405	4.86mg	282.mg	19/50	8.42mg	13/35	16.8mg	12/65	34.3mg 1/50	S
f	TR405	6.08mg	30.5mg	7/50	8.42mg	10/35	16.8mg	10/65	34.3mg 10/50	cli:ade,anb. S
g	TR405	8.08mg	26.0mg	0/50	8.42mg	0/35	16.8mg	19/65	34.3mg 8/50	liv:hpc,nnd.
h	TR405	8.95mg	25.7mg	0/50	8.42mg	3/35	16.8mg	18/65	34.3mg 19/50	zym:ade,car.
i	TR405	7.79mg	63.7mg	5/50	8.42mg	9/35	16.8mg	10/65	34.3mg 2/50	ute:esp,ess. S
j	TR405	9.57mg	35.4mg	0/50	8.42mg	0/35	16.8mg	15/65	34.3mg 6/50	S
k	TR405	9.72mg	31.6mg	0/50	8.42mg	3/35	16.8mg	17/65	34.3mg 13/50	S
l	TR405	9.03mg	98.6mg	4/50	8.42mg	8/35	16.8mg	8/65	34.3mg 2/50	S
m	TR405	10.2mg	51.0mg	0/50	8.42mg	4/35	16.8mg	7/65	34.3mg 5/50	ski:bca,bcc.
n	TR405	11.2mg	62.9mg	1/50	8.42mg	2/35	16.8mg	9/65	34.3mg 4/50	lun:a/a,a/c.
o	TR405	12.1mg	75.0mg	1/50	8.42mg	2/35	16.8mg	8/65	34.3mg 4/50	S
p	TR405	11.6mg	82.7mg	0/50	8.42mg	3/35	16.8mg	5/65	34.3mg 3/50	S
q	TR405	11.0mg	125.mg	0/50	8.42mg	3/35	16.8mg	3/65	34.3mg 1/50	thy:fca,fcc. S
r	TR405	13.7mg	71.1mg	0/50	8.42mg	3/35	16.8mg	9/65	34.3mg 6/50	phr:sqc,sqp; ton:sqc,sqp.
s	TR405	12.3mg	220.mg	1/50	8.42mg	3/35	16.8mg	5/65	34.3mg 1/50	amd:phm,pob.
t	TR405	14.6mg	102.mg	0/50	8.42mg	3/35	16.8mg	6/65	34.3mg 4/50	phr:sqp; ton:sqp. S
u	TR405	13.4mg	763.mg	1/50	8.42mg	3/35	16.8mg	4/65	34.3mg 1/50	S
v	TR405	16.7mg	124.mg	0/50	8.42mg	3/35	16.8mg	6/65	34.3mg 3/50	
w	TR405	17.7mg	121.mg	0/50	8.42mg	0/35	16.8mg	6/65	34.3mg 3/50	S
x	TR405	16.5mg	264.mg	0/50	8.42mg	0/35	16.8mg	4/65	34.3mg 1/50	ski:sqc,sqp. S
y	TR405	16.2mg	878.mg	0/50	8.42mg	1/35	16.8mg	3/65	34.3mg 0/50	S
z	TR405	17.2mg	705.mg	0/50	8.42mg	0/35	16.8mg	3/65	34.3mg 0/50	S
A	TR405	20.9mg	210.mg	0/50	8.42mg	0/35	16.8mg	2/65	34.3mg 6/50	S
B	TR405	21.3mg	300.mg	0/50	8.42mg	2/35	16.8mg	4/65	34.3mg 3/50	ton:sqc,sqp. S
C	TR405	21.7mg	566.mg	0/50	8.42mg	1/35	16.8mg	4/65	34.3mg 1/50	S
D	TR405	22.0mg	509.mg	0/50	8.42mg	2/35	16.8mg	2/65	34.3mg 3/50	S
E	TR405	31.5mg	3.20gm	0/50	8.42mg	1/35	16.8mg	0/65	34.3mg 3/50	col:adp,muc; rec:adp.
F	TR405	13.9mg	n.s.s.	6/50	8.42mg	2/35	16.8mg	2/65	34.3mg 3/50	S
G	TR405	19.9mg	n.s.s.	1/50	8.42mg	1/35	16.8mg	4/65	34.3mg 0/50	S
H	TR405	17.6mg	n.s.s.	2/50	8.42mg	2/35	16.8mg	0/65	34.3mg 1/50	S
I	TR405	20.9mg	n.s.s.	3/50	8.42mg	0/35	16.8mg	3/65	34.3mg 2/50	sub:fbs,fib. S
J	TR405	46.1mg	n.s.s.	0/50	8.42mg	0/35	16.8mg	1/65	34.3mg 2/50	duo:adp,muc; jej:adp.
K	TR405	81.3mg	n.s.s.	0/50	8.42mg	0/35	16.8mg	3/65	34.3mg 2/50	phr:sqc; ton:sqc. S
L	TR405	1.58mg	3.59mg	47/50	8.42mg	35/35	16.8mg	61/65	34.3mg 47/50	
M	TR405	8.08mg	26.0mg	0/50	8.42mg	0/35	16.8mg	19/65	34.3mg 8/50	liv:hpa,hpc,nnd.
4332	TR405	1.96mg	4.26mg	6/50	3.45mg	9/35	7.39mg	40/65	14.8mg 40/50	liv:hpc,nnd; ski:bca,bcc,ker,sbr,sea,sqc,sqp;
										zym:ade,car. C
a	TR405	2.29mg	5.00mg	1/50	3.45mg	5/35	7.39mg	28/65	14.8mg 32/50	ski:bca,bcc.
b	TR405	1.97mg	8.88mg	44/50	3.45mg	35/35	7.39mg	59/65	14.8mg 43/50	tes:iab,ica. S
c	TR405	2.45mg	5.50mg	1/50	3.45mg	4/35	7.39mg	26/65	14.8mg 30/50	S
d	TR405	3.39mg	15.5mg	17/50	3.45mg	11/35	7.39mg	27/65	14.8mg 22/50	amd:pbb,phm,pob.
e	TR405	3.71mg	19.5mg	16/50	3.45mg	11/35	7.39mg	24/65	14.8mg 20/50	S
f	TR405	4.34mg	12.7mg	2/50	3.45mg	2/35	7.39mg	15/65	14.8mg 20/50	liv:hpc,nnd.
g	TR405	5.97mg	21.8mg	2/50	3.45mg	1/35	7.39mg	10/65	14.8mg 15/50	S
h	TR405	6.90mg	30.1mg	1/50	3.45mg	2/35	7.39mg	11/65	14.8mg 9/50	ski:sqc,sqp.
i	TR405	8.33mg	37.2mg	0/50	3.45mg	2/35	7.39mg	8/65	14.8mg 7/50	S
j	TR405	9.95mg	48.9mg	0/50	3.45mg	1/35	7.39mg	6/65	14.8mg 7/50	S
k	TR405	9.42mg	73.0mg	1/50	3.45mg	1/35	7.39mg	4/65	14.8mg 7/50	
l	TR405	10.8mg	61.0mg	0/50	3.45mg	1/35	7.39mg	5/65	14.8mg 6/50	S
m	TR405	10.5mg	100.mg	1/50	3.45mg	1/35	7.39mg	5/65	14.8mg 6/50	ski:sbr,sea.
n	TR405	13.3mg	62.3mg	0/50	3.45mg	0/35	7.39mg	8/65	14.8mg 7/50	zym:ade,car.
o	TR405	14.1mg	76.6mg	0/50	3.45mg	0/35	7.39mg	7/65	14.8mg 6/50	S
p	TR405	15.1mg	1.09gm	0/50	3.45mg	2/35	7.39mg	1/65	14.8mg 3/50	S
q	TR405	4.95mg	n.s.s.	20/50	3.45mg	20/35	7.39mg	37/65	14.8mg 12/50	S
r	TR405	14.5mg	n.s.s.	2/50	3.45mg	2/35	7.39mg	2/65	14.8mg 3/50	lun:a/a,a/c.
s	TR405	20.0mg	n.s.s.	0/50	3.45mg	0/35	7.39mg	1/65	14.8mg 2/50	phr:sqp; ton:sqp.
t	TR405	1.95mg	7.30mg	43/50	3.45mg	33/35	7.39mg	62/65	14.8mg 48/50	
u	TR405	4.34mg	12.7mg	2/50	3.45mg	2/35	7.39mg	15/65	14.8mg 20/50	liv:hpa,hpc,nnd.

C.I. FOOD RED 3 (carmoisine, C.I. acid red 14, disodium salt) 3567-69-9

	RefNum	LoConf	UpConf	Cntrl	1Dose	1Inc	2Dose	2Inc		Citation or Pathology
4333	TR220	1.45gm	n.s.s.	28/50	386.mg	29/50	773.mg	23/49		
a	TR220	3.32gm	n.s.s.	3/50	386.mg	5/50	773.mg	2/49		liv:hpa,hpc,nnd.
b	TR220	2.93gm	n.s.s.	4/50	386.mg	4/50	773.mg	4/49		lun:a/a,a/c.
4334	TR220	680.mg	n.s.s.	31/50	357.mg	28/50	713.mg	31/50		
a	TR220	1.28gm	n.s.s.	15/50	357.mg	9/50	713.mg	14/50		liv:hpa,hpc,nnd.
b	TR220	2.18gm	n.s.s.	4/50	357.mg	4/50	713.mg	4/50		lun:a/a,a/c.

	Spe	Sex	Strain	Route	Site	Hist	Xpo+Xpt	Notes	DR	TD50	2Tailpvl	AuOp
4335	M	f	asp	eat	lun	ppa	80w80	e	.>	32.9gm	* P<.9	-
a	M	f	asp	eat	liv	tum	80w80	e		no dre	P=1.	-
4336	M	m	asp	eat	lun	ppa	80w80	e	.>	12.3gm	* P<.7	-
a	M	m	asp	eat	liv	nod	80w80	e		no dre	P=1.	-
b	M	m	asp	eat	liv	ade	80w80	e		no dre	P=1.	-
4337	R	f	f34	eat	ute	esp	24m24		: ±	#3.27gm	* P<.02	-
a	R	f	f34	eat	cli	sea	24m24			16.4gm	* P<.02	
b	R	f	f34	eat	TBA	MXB	24m24			7.18gm	* P<.7	
c	R	f	f34	eat	liv	MXB	24m24			no dre	P=1.	
4338	R	m	f34	eat	TBA	MXB	24m24		:>	1.86gm	* P<.5	-
a	R	m	f34	eat	liv	MXB	24m24			9.85gm	* P<.7	

C.I. PIGMENT RED 3   100ng..:..1ug...:..10.....:..100....:..1mg...:..10.....:..100...:..1g.....:..10

	Spe	Sex	Strain	Route	Site	Hist	Xpo+Xpt	Notes	DR	TD50	2Tailpvl	AuOp
4339	M	f	b6c	eat	lyd	MXA	24m24			#104.gm	* P<.03	-
a	M	f	b6c	eat	TBA	MXB	24m24			13.8gm	* P<.3	
b	M	f	b6c	eat	liv	MXB	24m24			204.gm	* P<.9	
c	M	f	b6c	eat	lun	MXB	24m24			252.gm	* P<.8	
4340	M	m	b6c	eat	MXB	MXB	24m24			:19.5gm	* P<.0005	
a	M	m	b6c	eat	kcx	ade	24m24			35.5gm	* P<.003	p
b	M	m	b6c	eat	thy	fca	24m24			38.3gm	* P<.005	p
c	M	m	b6c	eat	---	MXA	24m24			22.7gm	* P<.05	
d	M	m	b6c	eat	TBA	MXB	24m24			9.42gm	* P<.2	
e	M	m	b6c	eat	liv	MXB	24m24			17.9gm	* P<.3	
f	M	m	b6c	eat	lun	MXB	24m24			36.9gm	* P<.4	
4341	R	f	f34	eat	liv	hpa	24m24		: +	5.24gm	* P<.0005p	
a	R	f	f34	eat	TBA	MXB	24m24			no dre	P=1.	
b	R	f	f34	eat	liv	MXB	24m24			5.24gm	* P<.0005	
4342	R	m	f34	eat	amd	MXA	24m24		: +	: 661.mg	* P<.005	p
a	R	m	f34	eat	amd	pob	24m24			674.mg	* P<.004	
b	R	m	f34	eat	ski	sqp	24m24			2.54gm	* P<.004	e
c	R	m	f34	eat	ski	MXA	24m24			3.66gm	* P<.04	
d	R	m	f34	eat	zym	car	24m24			9.01gm	* P<.02	e
e	R	m	f34	eat	TBA	MXB	24m24			906.mg	* P<.2	
f	R	m	f34	eat	liv	MXB	24m24			5.17gm	* P<.08	

C.I. PIGMENT RED 23   100ng..:..1ug...:..10.....:..100....:..1mg...:..10.....:..100...:..1g.....:..10

	Spe	Sex	Strain	Route	Site	Hist	Xpo+Xpt	Notes	DR	TD50	2Tailpvl	AuOp
4343	M	f	b6c	eat	TBA	MXB	24m24		:>	29.7gm	* P<.6	-
a	M	f	b6c	eat	liv	MXB	24m24			no dre	P=1.	
b	M	f	b6c	eat	lun	MXB	24m24			45.2gm	* P<.09	
4344	M	m	b6c	eat	TBA	MXB	24m24		:>	20.2gm	* P<.6	-
a	M	m	b6c	eat	liv	MXB	24m24			15.1gm	* P<.2	
b	M	m	b6c	eat	lun	MXB	24m24			112.gm	* P<.8	
4345	R	f	f34	eat	TBA	MXB	24m24		:>	no dre	P=1.	-
a	R	f	f34	eat	liv	MXB	24m24			no dre	P=1.	
4346	R	m	f34	eat	kid	MXA	24m24		:	19.8gm	* P<.04	e
a	R	m	f34	eat	TBA	MXB	24m24			no dre	P=1.	
b	R	m	f34	eat	liv	MXB	24m24			no dre	P=1.	
4347	R	m	f34	eat	kid	MXA	24m24	with step	.	16.3gm	P<.09	e

D & C RED NO. 5   100ng..:..1ug...:..10.....:..100....:..1mg...:..10.....:..100...:..1g.....:..10

	Spe	Sex	Strain	Route	Site	Hist	Xpo+Xpt	Notes	DR	TD50	2Tailpvl	AuOp
4348	M	f	ddy	eat	liv	hpc	81w81	erv	. + .	784.mg	Z P<.0005+	
a	M	f	ddy	eat	liv	hpa	81w81	erv		21.5gm	* P<.2	+
4349	M	m	ddy	eat	liv	hpc	81w81	erv	. + .	659.mg	Z P<.0005+	
a	M	m	ddy	eat	liv	hpa	81w81	erv		8.79gm	* P<.06	+
4350	R	f	cfe	eat	liv	nod	24m24	e	. + .	704.mg	* P<.0005	
4351	R	m	cfe	eat	liv	nod	24m24	e	. + ±	2.91gm	* P<.02	
4352	R	f	wal	eat	liv	mix	52w91	er	. + .	233.mg	P<.0005+	
a	R	f	wal	eat	liv	hpc	52w91	er		2.62gm	P<.3	+
4353	R	f	wal	eat	liv	mix	91w91	er	<+	noTD50	P<.0005+	
a	R	f	wal	eat	liv	hpc	91w91	er		3.27gm	P<.2	+
4354	R	m	wis	eat	liv	lca	65w65	es	. ±	1.13gm	* P<.06	+

D & C RED NO. 9   100ng..:..1ug...:..10.....:..100....:..1mg...:..10.....:..100...:..1g.....:..10

	Spe	Sex	Strain	Route	Site	Hist	Xpo+Xpt	Notes	DR	TD50	2Tailpvl	AuOp
4355	M	f	b6c	eat	TBA	MXB	24m24		:>	18.2gm	* P<1.	-
a	M	f	b6c	eat	liv	MXB	24m24			9.24gm	* P<.8	
b	M	f	b6c	eat	lun	MXB	24m24			9.20gm	* P<.7	
4356	M	m	b6c	eat	TBA	MXB	24m24		:>	2.12gm	* P<.8	-
a	M	m	b6c	eat	liv	MXB	24m24			781.mg	* P<.2	
b	M	m	b6c	eat	lun	MXB	24m24			5.39gm	* P<.7	
4357	R	f	f34	eat	liv	nnd	24m24		: ±	1.14gm	* P<.08	a
a	R	f	f34	eat	TBA	MXB	24m24			no dre	P=1.	
b	R	f	f34	eat	liv	MXB	24m24			1.14gm	* P<.08	
4358	R	m	f34	eat	MXB	MXB	24m24		: + :	104.mg	/ P<.0005	
a	R	m	f34	eat	spl	MXA	24m24			146.mg	/ P<.0005c	
b	R	m	f34	eat	spl	fbs	24m24			211.mg	/ P<.0005c	
c	R	m	f34	eat	liv	nnd	24m24			265.mg	* P<.004	c
d	R	m	f34	eat	spl	ost	24m24			728.mg	* P<.005	c
e	R	m	f34	eat	liv	MXA	24m24			357.mg	* P<.03	c
f	R	m	f34	eat	TBA	MXB	24m24			331.mg	/ P<.5	
g	R	m	f34	eat	liv	MXB	24m24			357.mg	* P<.03	

	RefNum	LoConf	UpConf	Cntrl	1Dose	1Inc	2Dose	2Inc					Citation or Pathology	Brkly Code
4335	1325	1.65gm	n.s.s.	4/55	13.0mg	2/23	65.0mg	4/26	325.mg	4/27	1.63gm	2/19	Mason;fctx,12,601-607;1974	
a	1325	29.9mg	n.s.s.	0/55	13.0mg	0/23	65.0mg	0/26	325.mg	0/27	1.63gm	0/19		
4336	1325	1.33gm	n.s.s.	13/48	12.0mg	5/26	60.0mg	7/26	300.mg	4/24	1.50gm	8/28		
a	1325	3.17gm	n.s.s.	6/48	12.0mg	1/26	60.0mg	0/26	300.mg	0/24	1.50gm	2/28		
b	1325	30.4mg	n.s.s.	1/48	12.0mg	0/26	60.0mg	0/26	300.mg	0/24	1.50gm	0/28		
4337	TR220	1.47gm	n.s.s.	9/90	613.mg	11/50	1.23gm	14/50						S
a	TR220	4.96gm	n.s.s.	0/90	613.mg	0/50	1.23gm	3/50						S
b	TR220	1.01gm	n.s.s.	68/90	613.mg	34/50	1.23gm	44/50						
c	TR220	9.34gm	n.s.s.	3/90	613.mg	1/50	1.23gm	1/50					liv:hpa,hpc,nnd.	
4338	TR220	407.mg	n.s.s.	61/90	238.mg	22/50	495.mg	34/50						
a	TR220	1.34gm	n.s.s.	5/90	238.mg	3/50	495.mg	3/50					liv:hpa,hpc,nnd.	

C.I. PIGMENT RED 3   2425-85-6

	RefNum	LoConf	UpConf	Cntrl	1Dose	1Inc	2Dose	2Inc	3Dose	3Inc			Citation or Pathology	Brkly Code
4339	TR407	31.0gm	n.s.s.	0/51	1.60gm	0/50	3.20gm	0/50	6.42gm	3/50			lyd:ahs,dhs,ihs,nhs,rhs,shs.	S
a	TR407	4.15gm	n.s.s.	36/51	1.60gm	29/50	3.20gm	23/50	6.42gm	33/50				
b	TR407	10.9gm	n.s.s.	10/51	1.60gm	14/50	3.20gm	4/50	6.42gm	9/50			liv:hpa,hpc,nnd.	
c	TR407	24.6gm	n.s.s.	4/51	1.60gm	0/50	3.20gm	1/50	6.42gm	3/50			lun:a/a,a/c.	
4340	TR407	9.79gm	49.6gm	0/50	1.47gm	0/50	2.96gm	1/51	5.92gm	10/52			kcx:ade; thy:fca.	P
a	TR407	14.5gm	176.gm	0/50	1.47gm	0/50	2.96gm	0/51	5.92gm	6/52				
b	TR407	15.6gm	271.gm	0/50	1.47gm	0/50	2.96gm	1/51	5.92gm	5/52				
c	TR407	9.42gm	n.s.s.	2/50	1.47gm	2/50	2.96gm	6/51	5.92gm	7/52			---:mlh,mlm.	S
d	TR407	3.15gm	n.s.s.	25/50	1.47gm	27/50	2.96gm	33/51	5.92gm	35/52				
e	TR407	5.12gm	n.s.s.	12/50	1.47gm	16/50	2.96gm	16/51	5.92gm	19/52			liv:hpa,hpc,nnd.	
f	TR407	9.86gm	n.s.s.	2/50	1.47gm	5/50	2.96gm	7/51	5.92gm	5/52			lun:a/a,a/c.	
4341	TR407	2.63gm	13.4gm	0/50	295.mg	0/50	617.mg	1/50	1.23gm	10/50				
a	TR407	1.41gm	n.s.s.	46/50	295.mg	35/50	617.mg	38/50	1.23gm	41/50				
b	TR407	2.63gm	13.4gm	0/50	295.mg	0/50	617.mg	1/50	1.23gm	10/50			liv:hpa,hpc,nnd.	
4342	TR407	332.mg	6.55gm	24/50	236.mg	32/51	493.mg	37/51	988.mg	36/51			amd:phm,pob.	
a	TR407	343.mg	5.48gm	22/50	236.mg	29/51	493.mg	35/51	988.mg	34/51				S
b	TR407	1.30gm	21.7gm	0/50	236.mg	4/51	493.mg	2/51	988.mg	6/51				
c	TR407	1.49gm	n.s.s.	1/50	236.mg	4/51	493.mg	2/51	988.mg	6/51			ski:car,sqp.	S
d	TR407	3.34gm	n.s.s.	0/50	236.mg	0/51	493.mg	2/51	988.mg	3/51				
e	TR407	332.mg	n.s.s.	47/50	236.mg	47/51	493.mg	46/51	988.mg	48/51				
f	TR407	2.08gm	n.s.s.	0/50	236.mg	1/51	493.mg	4/51	988.mg	1/51			liv:hpa,hpc,nnd.	

C.I. PIGMENT RED 23   6471-49-4

	RefNum	LoConf	UpConf	Cntrl	1Dose	1Inc	2Dose	2Inc	3Dose	3Inc			Citation or Pathology	Brkly Code
4343	TR411	5.16gm	n.s.s.	27/50	1.29gm	31/50	3.23gm	31/50	6.46gm	32/50				
a	TR411	16.2gm	n.s.s.	5/50	1.29gm	8/50	3.23gm	10/50	6.46gm	4/50			liv:hpa,hpc,nnd.	
b	TR411	15.9gm	n.s.s.	1/50	1.29gm	2/50	3.23gm	2/50	6.46gm	5/50			lun:a/a,a/c.	
4344	TR411	3.71gm	n.s.s.	35/52	1.19gm	27/53	2.98gm	33/52	5.96gm	41/51				
a	TR411	5.46gm	n.s.s.	13/52	1.19gm	8/53	2.98gm	14/52	5.96gm	21/51			liv:hpa,hpc,nnd.	
b	TR411	10.0gm	n.s.s.	5/52	1.19gm	7/53	2.98gm	8/52	5.96gm	7/51			lun:a/a,a/c.	
4345	TR411	2.99gm	n.s.s.	48/50	490.mg	48/50	1.23gm	44/50	2.47gm	44/50				
a	TR411	n.s.s.	n.s.s.	0/50	490.mg	0/50	1.23gm	0/50	2.47gm	0/50			liv:hpa,hpc,nnd.	
4346	TR411	6.85gm	n.s.s.	0/50	392.mg	0/50	983.mg	1/50	1.97gm	3/51			kid:rua,ruc.	
a	TR411	2.64gm	n.s.s.	46/50	392.mg	45/50	983.mg	40/50	1.97gm	45/51				
b	TR411	9.92gm	n.s.s.	3/50	392.mg	1/50	983.mg	1/50	1.97gm	3/51			liv:hpa,hpc,nnd.	
4347	TR411	5.28gm	n.s.s.	1/50	1.97gm	5/51							kid:rua,ruc.	

D & C RED NO. 5   (ponceau MX) 3761-53-3

	RefNum	LoConf	UpConf	Cntrl	1Dose	1Inc	2Dose	2Inc	3Dose	3Inc	4Dose	4Inc	Citation or Pathology	Brkly Code
4348	244	365.mg	2.43gm	0/34	260.mg	4/27	1.27gm	5/11	(6.22gm	5/12)			Ikeda;fctx,6,591-598;1968	
a	244	4.19gm	n.s.s.	1/34	260.mg	2/27	1.27gm	0/11	6.22gm	2/12				
4349	244	218.mg	3.45gm	0/31	240.mg	0/8	1.18gm	4/6	(5.74gm	6/12)				
a	244	2.47gm	n.s.s.	1/31	240.mg	1/8	1.18gm	1/6	6.42gm	3/12				
4350	242	414.mg	1.34gm	0/30	62.5mg	1/30	125.mg	0/29	250.mg	5/28	500.mg	14/25	Grasso;fctx,7,425-442;1969	
4351	242	1.10gm	n.s.s.	0/30	50.0mg	0/30	100.mg	1/29	200.mg	1/30	400.mg	3/29		
4352	1730m	107.mg	641.mg	0/6	714.mg	12/15							Grasso;txcy,7,327-347;1977/pers.comm.	
a	1730m	643.mg	n.s.s.	0/6	714.mg	2/15								
4353	1730n	n.s.s.	488.mg	0/6	1.25gm	11/11								
a	1730n	800.mg	n.s.s.	0/6	1.25gm	2/11								
4354	243	368.mg	n.s.s.	0/12	80.0mg	4/15	400.mg	4/12	2.00gm	6/14			Ikeda;fctx,4,485-492;1966	

D & C RED NO. 9   (brilliant red) 5160-02-1

	RefNum	LoConf	UpConf	Cntrl	1Dose	1Inc	2Dose	2Inc					Citation or Pathology	Brkly Code
4355	TR225	319.mg	n.s.s.	26/50	128.mg	25/50	255.mg	27/50						
a	TR225	803.mg	n.s.s.	5/50	128.mg	3/50	255.mg	6/50					liv:hpa,hpc,nnd.	
b	TR225	1.13gm	n.s.s.	2/50	128.mg	1/50	255.mg	3/50					lun:a/a,a/c.	
4356	TR225	258.mg	n.s.s.	23/50	118.mg	28/50	235.mg	24/50						
a	TR225	293.mg	n.s.s.	8/50	118.mg	13/50	235.mg	15/50					liv:hpa,hpc,nnd.	
b	TR225	703.mg	n.s.s.	4/50	118.mg	4/50	235.mg	5/50					lun:a/a,a/c.	
4357	TR225	386.mg	n.s.s.	1/50	49.5mg	1/50	149.mg	5/50						
a	TR225	149.mg	n.s.s.	44/50	49.5mg	42/50	149.mg	40/50						
b	TR225	386.mg	n.s.s.	1/50	49.5mg	1/50	149.mg	5/50					liv:hpa,hpc,nnd.	
4358	TR225	66.0mg	190.mg	1/50	39.6mg	6/50	119.mg	26/50					liv:hpc,nnd; spl:fbs,lei,ost,srn.	C
a	TR225	88.7mg	265.mg	0/50	39.6mg	0/50	119.mg	23/50					spl:fbs,lei,ost,srn.	
b	TR225	119.mg	425.mg	0/50	39.6mg	0/50	119.mg	17/50						
c	TR225	140.mg	1.80gm	0/50	39.6mg	6/50	119.mg	7/50						
d	TR225	276.mg	5.35gm	0/50	39.6mg	0/50	119.mg	5/50						
e	TR225	154.mg	n.s.s.	1/50	39.6mg	6/50	119.mg	7/50					liv:hpc,nnd.	
f	TR225	69.0mg	n.s.s.	43/50	39.6mg	34/50	119.mg	44/50						
g	TR225	154.mg	n.s.s.	1/50	39.6mg	6/50	119.mg	7/50					liv:hpa,hpc,nnd.	

```
     Spe Strain  Site    Xpo+Xpt                                         TD50    2Tailpvl
         Sex  Route  Hist    Notes                                           DR      AuOp

4359 R f osm eat pit ade 25m25 e                               .>       no dre   P=1.   -
4360 R m osm eat tes ica 25m25 e                                  .>    28.6mg * P<.8   -
   a R m osm eat --- fbs 25m25 e                                       no dre   P=1.   -


D & C RED NO. 10         100ng..:..1ug...:..10.....:..100....:..1mg...:..10.....:..100...:..1g.....:..10
4361 R f osm eat pit ade 24m24 e                                  .>   3.56gm * P<.5   -
   a R f osm eat liv ade 24m24 e                                       no dre   P=1.
4362 R m osm eat liv hpa 24m24 e                               .>      no dre   P=1.


FD & C RED NO. 1         100ng..:..1ug...:..10.....:..100....:..1mg...:..10.....:..100...:..1g.....:..10
4363 R b bbl eat liv mix 24m24                          . + .          746.mg   P<.0005
   a R b bbl eat liv car 24m24                                         2.48gm   P<.0005+
   b R b bbl eat liv nod 24m24                                         3.04gm   P<.003
   c R b bbl eat liv bht 24m24                                         3.54gm   P<.002 +
4364 R f nss eat bil ade 65w65 r                             . +      .1.57gm * P<.002
   a R f nss eat liv car 65w65 r                                       1.99gm * P<.003 +
4365 R m nss eat bil ade 65w65 r                             . + .     755.mg * P<.0005
   a R m nss eat liv car 65w65 r                                       1.24gm * P<.0005+
4366 R b osm eat liv nod 24m24                          . + .          639.mg Z P<.0005
   a R b osm eat liv bht 24m24                                         909.mg Z P<.0005+
   b R b osm eat liv car 24m24                                         7.54gm * P<.0005+
   c R b osm eat tba mal 24m24                                         12.9gm   P<.2
4367 R b wis eat liv tum 65w65 ekr                   . + .            225.mg   P<.0005+
4368 R b wis eat liv tum 65w65 ekr                   . +    .         394.mg   P<.006 +
4369 R b wis eat liv tum 70w70 er                    . + .           314.mg   P<.0005+
4370 R b wis eat liv tum 70w70 er                    . + .           420.mg   P<.0005+


FD & C RED NO. 2         100ng..:..1ug...:..10.....:..100....:..1mg...:..10.....:..100...:..1g.....:..10
4371 R b mgr eat liv tum 25m25 e                              .>      no dre   P=1.
   a R b mgr eat tba mix 25m25 e                                       632.mg   P<.0005+
4372 R b non eat liv hpt 33m33                                        .49.5gm   P<.3
   a R b non eat tba mix 33m33                                        3.32gm   P<.0005+
4373 R f nss eat liv tum 64w64 e                 .>                  no dre   P=1.   -
   a R f nss eat tba mix 64w64 e                                       1.64gm * P<.6   -
4374 R m nss eat liv tum 64w64 e                 .>                  no dre   P=1.   -
   a R m nss eat tba mix 64w64 e                                       no dre   P=1.   -
4375 R b wis eat abd mly 91w91 er                          .    ±    6.13gm   P<.04 +


FD & C RED NO. 3         100ng..:..1ug...:..10.....:..100....:..1mg...:..10.....:..100...:..1g.....:..10
4376 M f cd1 eat tba mix 24m24 e                                      no dre   P=1.   -
   a M f cd1 eat tba ben 24m24 e                                       no dre   P=1.   -
   b M f cd1 eat tba mal 24m24 e                                       no dre   P=1.   -
4377 M m cd1 eat --- lcl 24m24 e                          . ±         3.14gm Z P<.02   -
   a M m cd1 eat tba mix 24m24 e                                       17.0gm   P<.3
   b M m cd1 eat tba ben 24m24 e                                       25.9gm * P<.3
   c M m cd1 eat tba mal 24m24 e                                       87.6gm * P<.8
4378 R b osm gav --- lys 20m24 e                      . +    .        122.mg Z P<.004 -
   a R b osm gav liv tum 20m24 e                                       no dre   P=1.
   b R b osm gav tba mal 20m24 e                                       no dre   P=1.
4379 R b osm eat liv hpt 20m24 e                                      84.9gm * P<.4
   a R b osm eat tba mal 20m24 e                                       14.1gm * P<.5
4380 R f osm eat liv tum 24m24                                .>      no dre   P=1.
   a R f osm eat tba mix 24m24                                         no dre   P=1.
4381 R m osm eat liv tum 24m24                              .>        no dre   P=1.
   a R m osm eat tba mix 24m24                                         4.81gm * P<.2   -


FD & C RED NO. 4         100ng..:..1ug...:..10.....:..100....:..1mg...:..10.....:..100...:..1g.....:..10
4382 D f beg eat adr mda 85m85 e                          .>         316.mg   P<.2
   a D f beg eat mgl adc 85m85 e                                       316.mg   P<.2   -
   b D f beg eat liv nod 85m85 e                                       no dre   P=1.
   c D f beg eat tba mix 85m85 e                                       670.mg   P<.9
4383 M b che eat liv hpt 24m24 e                                     .>57.1gm * P<.7   -
   a M b che eat tba mix 24m24 e                                       20.3gm * P<.3   -
4384 M b chj eat liv hpt 24m24 e                                     .>no dre   P=1.   -
   a M b chj eat tba mix 24m24 e                                       no dre   P=1.   -
4385 R b non eat liv hpt 33m33                                        .49.5gm   P<.3
   a R b non eat tba mix 33m33                                        12.0gm   P<.02 +
4386 R b osm eat liv ade 24m24 e                                     .59.8gm * P<.2   -
   a R b osm eat tba mix 24m24 e                                       no dre   P=1.   -
4387 R b osm eat --- fbs 24m24 e                              .       20.3gm * P<.006 -
   a R b osm eat tba mix 24m24 e                                       no dre   P=1.
4388 R b sda eat tba mix 24m24 e                          .>         no dre   P=1.   -
4389 R b wis eat abd mly 91w91 er                          .    ±    6.13gm   P<.04 +


HC RED NO. 3             100ng..:..1ug...:..10.....:..100....:..1mg...:..10.....:..100...:..1g.....:..10
4390 M f b6c gav for sqp 24m24 s                             :  ±    #721.mg * P<.04  i
   a M f b6c gav TBA MXB 24m24 s                                       292.mg * P<.5
   b M f b6c gav liv MXB 24m24 s                                       no dre   P=1.
   c M f b6c gav lun MXB 24m24 s                                       4.67gm * P<.9
4391 M m b6c gav liv MXA 24m24                               :>       427.mg / P<.2   e
   a M m b6c gav TBA MXB 24m24                                         no dre   P=1.
```

	RefNum	LoConf	UpConf	Cntrl	1Dose	1Inc	2Dose	2Inc					Citation or Pathology
4359	262a	998.mg	n.s.s.	4/25	5.00mg	1/25	25.0mg	9/25	125.mg	6/25	500.mg	4/25	Davis;txap,4,200-205;1962
4360	262a	1.65gm	n.s.s.	0/25	4.00mg	2/25	20.0mg	0/25	100.mg	0/25	400.mg	1/25	
a	262a	1.68gm	n.s.s.	0/25	4.00mg	1/25	20.0mg	1/25	100.mg	2/25	400.mg	0/25	

D & C RED NO. 10   1248-18-6

	RefNum	LoConf	UpConf	Cntrl	1Dose	1Inc	2Dose	2Inc					Citation or Pathology
4361	263	666.mg	n.s.s.	2/25	5.00mg	5/25	25.0mg	9/25	125.mg	3/25	500.mg	7/25	Davis;txap,5,728-734;1963
a	263	21.0mg	n.s.s.	0/25	5.00mg	0/25	25.0mg	0/25	125.mg	0/25	500.mg	0/25	
4362	263	2.70gm	n.s.s.	1/25	4.00mg	1/25	20.0mg	0/25	100.mg	0/25	400.mg	0/25	

FD & C RED NO. 1 (ponceau 3R) 3564-09-8

	RefNum	LoConf	UpConf	Cntrl	1Dose	1Inc	2Dose	2Inc	3Dose	3Inc	4Dose	4Inc	Citation or Pathology
4363	246	460.mg	1.36gm	2/50	900.mg	29/50							Hansen;txap,5,105-118;1963
a	246	1.24gm	6.19gm	0/50	900.mg	11/50							
b	246	1.40gm	17.3gm	1/50	900.mg	10/50							
c	246	1.60gm	12.0gm	0/50	900.mg	8/50							
4364	245	596.mg	8.32gm	0/45	150.mg	0/15	500.mg	2/15	1.50gm	3/15			Grice;txap,3,509-520;1961
a	245	684.mg	13.2gm	0/45	150.mg	0/15	500.mg	1/15	1.50gm	3/15			
4365	245	340.mg	2.92gm	0/45	120.mg	1/15	400.mg	3/15	1.20gm	4/15			
a	245	468.mg	5.35gm	0/45	120.mg	0/15	400.mg	1/15	1.20gm	4/15			
4366	246	411.mg	1.08gm	0/49	225.mg	12/50	450.mg	18/50	(900.mg	9/50	2.25gm	4/49)	Hansen;txap,5,105-118;1963
a	246	616.mg	1.69gm	1/49	225.mg	12/50	450.mg	17/50	900.mg	22/50	(2.25gm	25/49)	
b	246	4.21gm	19.9gm	0/49	225.mg	1/50	450.mg	2/50	900.mg	4/50	2.25gm	9/49	
c	246	4.09gm	n.s.s.	10/49	225.mg	8/50	450.mg	7/50	900.mg	7/50	2.25gm	14/49	
4367	247m	88.2mg	785.mg	0/6	1.35gm	8/10							Mannell;fctx,2,169-174;1964
4368	247n	153.mg	4.09gm	0/6	1.35gm	6/10							
4369	247o	158.mg	711.mg	0/23	1.35gm	14/19							
4370	247r	208.mg	1.00gm	0/24	1.35gm	12/19							

FD & C RED NO. 2 (amaranth) 915-67-3

	RefNum	LoConf	UpConf	Cntrl	1Dose	1Inc	2Dose	2Inc	3Dose	3Inc	Citation or Pathology
4371	1427	3.03gm	n.s.s.	0/35	758.mg	0/18					Baigusheva;vpit,27,46-50;1968
a	1427	295.mg	1.87gm	2/35	758.mg	11/18					
4372	1429	8.06gm	n.s.s.	0/50	771.mg	1/50					Andrianova;vpit,29,61-65;1970
a	1429	1.75gm	7.50gm	0/50	771.mg	13/50					
4373	1358	14.4mg	n.s.s.	0/10	15.0mg	0/14	150.mg	0/13	750.mg	0/9	Mannell;jphp,10,625-634;1958
a	1358	192.mg	n.s.s.	2/10	15.0mg	1/14	150.mg	4/13	750.mg	2/9	
4374	1358	8.41mg	n.s.s.	0/11	12.0mg	0/10	120.mg	0/11	600.mg	0/9	
a	1358	8.41mg	n.s.s.	1/11	12.0mg	0/10	120.mg	0/11	600.mg	0/9	
4375	1136	993.mg	n.s.s.	0/50	1.80gm	1/7					Willheim;gaga,23,1-19;1953

FD & C RED NO. 3 (erythrosine) 16423-68-0

	RefNum	LoConf	UpConf	Cntrl	1Dose	1Inc	2Dose	2Inc	3Dose	3Inc	4Dose	4Inc	Citation or Pathology
4376	1811	11.7gm	n.s.s.	67/120	3.90mg	25/60							Borzelleca;fctx,25,735-737;1987
a	1811	14.4gm	n.s.s.	40/120	390.mg	13/60	1.30gm	9/60	3.90gm	16/60			
b	1811	16.0gm	n.s.s.	35/120	390.mg	14/60	1.30gm	17/60	3.90gm	12/60			
4377	1811	1.11gm	n.s.s.	1/120	360.mg	5/60	(1.20gm	2/60	3.60gm	2/60)			
a	1811	4.50gm	n.s.s.	46/120	3.60gm	28/60							
b	1811	7.34gm	n.s.s.	29/120	360.mg	5/60	1.20gm	12/60	3.60gm	17/60			
c	1811	9.15gm	n.s.s.	17/120	360.mg	18/60	1.20gm	12/60	3.60gm	12/60			
4378	130m	49.8mg	759.mg	0/49	23.3mg	6/49	(54.7mg	3/47	175.mg	2/46	349.mg	1/43)	Hansen;fctx,11,535-545;1973
a	130m	141.mg	n.s.s.	0/49	23.3mg	0/49	54.7mg	0/47	175.mg	0/46	349.mg	0/43	
b	130m	1.36gm	n.s.s.	5/49	23.3mg	12/49	54.7mg	6/47	175.mg	9/46	349.mg	4/43	
4379	130n	13.8gm	n.s.s.	0/89	185.mg	0/41	371.mg	1/45	742.mg	1/45	1.48gm	0/45	
a	130n	3.08gm	n.s.s.	20/89	185.mg	6/48	371.mg	8/41	742.mg	9/45	1.48gm	12/45	
4380	1371	334.mg	n.s.s.	0/12	250.mg	0/12	500.mg	0/12	1.00gm	0/12	2.50gm	0/12	Hansen;fctx,11,527-534;1973
a	1371	7.16gm	n.s.s.	5/12	250.mg	4/12	500.mg	1/12	1.00gm	2/12	2.50gm	0/12	
4381	1371	267.mg	n.s.s.	0/12	200.mg	0/12	400.mg	0/12	800.mg	0/12	2.00gm	0/12	
a	1371	1.42gm	n.s.s.	2/12	200.mg	0/12	400.mg	2/12	800.mg	5/12	2.00gm	3/12	

FD & C RED NO. 4 (ponceau SX) 4548-53-2

	RefNum	LoConf	UpConf	Cntrl	1Dose	1Inc	2Dose	2Inc	3Dose	3Inc	4Dose	4Inc	Citation or Pathology
4382	248m	51.0mg	n.s.s.	0/9	250.mg	1/5							Davis;txap,8,306-317;1966
a	248m	51.0mg	n.s.s.	0/9	250.mg	1/5							
b	248m	106.mg	n.s.s.	1/9	250.mg	0/5							
c	248m	39.1mg	n.s.s.	3/9	250.mg	2/5							
4383	248m	7.69gm	n.s.s.	12/91	1.25gm	6/47	2.50gm	9/56					
a	248m	5.60gm	n.s.s.	13/91	1.25gm	8/47	2.50gm	12/56					
4384	248m	6.80gm	n.s.s.	0/66	1.25gm	0/50	2.50gm	0/28					
a	248m	6.80gm	n.s.s.	5/66	1.25gm	0/50	2.50gm	0/28					
4385	1429	8.06gm	n.s.s.	0/50	771.mg	1/50							Andrianova;vpit,29,61-65;1970
a	1429	4.14gm	n.s.s.	0/50	771.mg	4/50							
4386	248m	9.74gm	n.s.s.	0/16	225.mg	0/19	450.mg	0/23	900.mg	0/22	2.25gm	1/24	Davis;txap,8,306-317;1966
a	248m	5.14gm	n.s.s.	7/16	225.mg	10/19	450.mg	8/23	900.mg	10/22	2.25gm	5/24	
4387	248n	7.00gm	239.gm	0/171	450.mg	0/89	900.mg	4/89					
a	248n	3.48gm	n.s.s.	67/171	450.mg	23/89	900.mg	32/89					
4388	248m	4.95gm	n.s.s.	38/147	450.mg	16/83	900.mg	14/74					
4389	1136	993.mg	n.s.s.	0/50	1.80gm	1/7							Willheim;gaga,23,1-19;1953

HC RED NO. 3   2871-01-4

	RefNum	LoConf	UpConf	Cntrl	1Dose	1Inc	2Dose	2Inc	Citation or Pathology
4390	TR281	208.mg	n.s.s.	0/50	88.4mg	0/50	177.mg	3/50	s
a	TR281	67.0mg	n.s.s.	16/50	88.4mg	15/50	177.mg	18/50	
b	TR281	273.mg	n.s.s.	4/50	88.4mg	1/50	177.mg	2/50	liv:hpa,hpc,nnd.
c	TR281	241.mg	n.s.s.	1/50	88.4mg	2/50	177.mg	1/50	lun:a/a,a/c.
4391	TR281	141.mg	n.s.s.	25/50	88.4mg	15/50	177.mg	35/50	liv:hpa,hpc.
a	TR281	142.mg	n.s.s.	44/50	88.4mg	36/50	177.mg	42/50	

```
      Spe Strain  Site   Xpo+Xpt                                                          TD50    2Tailpvl
          Sex  Route Hist      Notes                                                        DR     AuOp
  b    M m b6c gav liv MXB 24m24                                                          427.mg / P<.2
  c    M m b6c gav lun MXB 24m24                                                          1.75gm * P<.7
  4392 R f f34 gav TBA MXB 24m25                                          :>             4.87gm * P<.9   -
  a    R f f34 gav liv MXB 24m25                                                          no dre  P=1.
  4393 R m f34 gav TBA MXB 24m25                                                :>        500.mg * P<.3   -
  a    R m f34 gav liv MXB 24m25                                                          1.51kg   P<1.
```

RESERPINE   100ng..:..1ug...:..10.....:..100....:..1mg....:..10.....:..100...:..1g.....:..10
```
  4394 M f b6c eat mgl --- 24m24                         :  +  :                          3.58mg * P<.002 c
  a    M f b6c eat TBA MXB 24m24                                                          6.42mg * P<.5
  b    M f b6c eat liv MXB 24m24                                                          no dre  P=1.
  c    M f b6c eat lun MXB 24m24                                                          54.5mg * P<.9
  4395 M m b6c eat sev ulc 24m24                               :  +        :             8.39mg * P<.01  c
  a    M m b6c eat TBA MXB 24m24                                                          no dre  P=1.
  b    M m b6c eat liv MXB 24m24                                                          5.46mg \ P<.6
  c    M m b6c eat lun MXB 24m24                                                          no dre  P=1.
  4396 M f c3h eat mgl car 24m24            .>                                            34.2ug * P<.6   +
  4397 R f f34 eat TBA MXB 24m24                      :>                                  .727mg * P<.3
  a    R f f34 eat liv MXB 24m24                                                          no dre  P=1.
  4398 R m f34 eat adr MXA 24m24                  :  +   :                                .306mg * P<.0005c
  a    R m f34 eat TBA MXB 24m24                                                          no dre  P=1.
  b    R m f34 eat liv MXB 24m24                                                          6.77mg * P<.4
  4399 R f wis eat pit mix 75w75 e                           .>                           no dre  P=1.   -
  a    R f wis eat mgl fba 75w75 e                                                        no dre  P=1.   -
  4400 R m wis eat pit ade 75w75 e                         .>                             38.3mg * P<.8   -
```

RESORCINOL   100ng..:..1ug...:..10.....:..100....:..1mg....:..10.....:..100...:..1g.....:..10
```
  4401 M f b6c gav TBA MXB 24m24                                      :>                  1.88gm / P<.9   -
  a    M f b6c gav liv MXB 24m24                                                          1.56gm * P<.3
  b    M f b6c gav lun MXB 24m24                                                          1.35gm * P<.2
  4402 M m b6c gav TBA MXB 24m24                                    :>                    no dre  P=1.   -
  a    M m b6c gav liv MXB 24m24                                                          no dre  P=1.
  b    M m b6c gav lun MXB 24m24                                                          6.80gm * P<.9
  4403 R f f34 gav TBA MXB 24m24 s                       :>                               219.mg * P<.5   -
  a    R f f34 gav liv MXB 24m24 s                                                        no dre  P=1.
  4404 R m f34 gav tes ica 24m24 s                   :  ±                                 #84.0mg * P<.05 -
  a    R m f34 gav TBA MXB 24m24 s                                                        213.mg * P<.4
  b    R m f34 gav liv MXB 24m24 s                                                        no dre  P=1.
```

RETINOIC ACID   100ng..:..1ug...:..10.....:..100....:..1mg....:..10.....:..100...:..1g.....:..10
```
  4405 R f sda ipj tba mal 24m24 e                              .>                        no dre  P=1.
  4406 R m sda ipj liv hae 24m24 e                            .>                          no dre  P=1.
  a    R m sda ipj tba mal 24m24 e                                                        188.mg   P<1.   -
```

RETINOL ACETATE   100ng..:..1ug...:..10.....:..100....:..1mg....:..10.....:..100...:..1g.....:..10
```
  4407 R f f3d wat amd phe 24m25 e                                   .  +  .             227.mg * P<.0005
  a    R f f3d wat cli ade 24m25 e                                                        1.02gm * P<.02
  b    R f f3d wat liv nnd 24m25 e                                                        7.34gm * P<.9
  c    R f f3d wat tba mix 24m25 e                                                        no dre  P=1.
  4408 R m f3d wat amd phe 24m25 e                                   .  +  .             86.4mg * P<.0005
  a    R m f3d wat spl mnl 24m25 e                                                        548.mg * P<.04
  b    R m f3d wat amd phm 24m25 e                                                        552.mg * P<.02
  c    R m f3d wat liv nnd 24m25 e                                                        2.06gm * P<.3
  d    R m f3d wat liv mix 24m25 e                                                        3.67gm * P<.7
  e    R m f3d wat liv cho 24m25 e                                                        6.47gm * P<1.
  f    R m f3d wat liv hpc 24m25 e                                                        no dre  P=1.
  g    R m f3d wat tba mix 24m25 e                                                        noTD50  P=1.
  4409 R m f3d wat for neo 52w52                               .>                         no dre  P=1.   -
  a    R m f3d wat liv tum 52w52                                                          no dre  P=1.
```

RETINOL PALMITATE   100ng..:..1ug...:..10.....:..100....:..1mg....:..10.....:..100...:..1g.....:..10
```
  4410 R m cdr eat liv tum 28m28 e                              .>                        no dre  P=1.
```

RHODAMINE 6G   100ng..:..1ug...:..10.....:..100....:..1mg....:..10.....:..100...:..1g.....:..10
```
  4411 M f b6c eat TBA MXB 24m24                                       :>                 no dre  P=1.   -
  a    M f b6c eat liv MXB 24m24                                                          no dre  P=1.
  b    M f b6c eat lun MXB 24m24                                                          no dre  P=1.
  4412 M m b6c eat thy MXA 24m24                                          :  ±           #1.21gm * P<.05 -
  a    M m b6c eat TBA MXB 24m24                                                          no dre  P=1.
  b    M m b6c eat liv MXB 24m24                                                          no dre  P=1.
  c    M m b6c eat lun MXB 24m24                                                          no dre  P=1.
  4413 R f f34 eat amd MXA 24m24                                  :  ±                    47.1mg * P<.06  e
  a    R f f34 eat TBA MXB 24m24                                                          no dre  P=1.
  b    R f f34 eat liv MXB 24m24                                                          no dre  P=1.
  4414 R m f34 eat ski ker 24m24                                 :  ±                     34.5mg * P<.02  e
  a    R m f34 eat TBA MXB 24m24                                                          no dre  P=1.
  b    R m f34 eat liv MXB 24m24                                                          553.mg / P<.9
```

RIFAMPICIN   100ng..:..1ug...:..10.....:..100....:..1mg....:..10.....:..100...:..1g.....:..10
```
  4415 M f bal wat lun tum 14m26 e                                    .>                  1.01gm * P<.8   -
  a    M f bal wat liv hpt 14m26 e                                                        3.18gm * P<.6   -
```

	RefNum	LoConf	UpConf	Cntrl	1Dose	1Inc	2Dose	2Inc			Citation or Pathology
											Brkly Code
b	TR281	141.mg	n.s.s.	25/50	88.4mg	15/50	177.mg	35/50			liv:hpa,hpc,nnd.
c	TR281	251.mg	n.s.s.	11/50	88.4mg	13/50	177.mg	13/50			lun:a/a,a/c.
4392	TR281	238.mg	n.s.s.	45/50	177.mg	44/50	354.mg	41/50			
a	TR281	n.s.s.	n.s.s.	0/50	177.mg	0/50	354.mg	0/50			liv:hpa,hpc,nnd.
4393	TR281	154.mg	n.s.s.	44/50	177.mg	38/50	356.mg	38/50			
a	TR281	1.18gm	n.s.s.	4/50	177.mg	1/50	356.mg	3/50			liv:hpa,hpc,nnd.

RESERPINE 50-55-5

	RefNum	LoConf	UpConf	Cntrl	1Dose	1Inc	2Dose	2Inc			Citation or Pathology
4394	TR193	1.92mg	12.2mg	0/50	.650mg	7/50	1.30mg	7/50			
a	TR193	1.44mg	n.s.s.	21/50	.650mg	19/50	1.30mg	23/50			
b	TR193	10.1mg	n.s.s.	2/50	.650mg	0/50	1.30mg	1/50			liv:hpa,hpc,nnd.
c	TR193	4.03mg	n.s.s.	4/50	.650mg	4/50	1.30mg	4/50			lun:a/a,a/c.
4395	TR193	3.42mg	947.mg	0/50	.600mg	1/50	1.20mg	5/50			
a	TR193	1.78mg	n.s.s.	30/50	.600mg	34/50	1.20mg	25/50			
b	TR193	.868mg	n.s.s.	12/50	.600mg	14/50	(1.20mg	4/50)			liv:hpa,hpc,nnd.
c	TR193	3.89mg	n.s.s.	9/50	.600mg	9/50	1.20mg	6/50			lun:a/a,a/c.
4396	1187	5.54ug	n.s.s.	12/22	9.60ug	15/24					Lacassagne;adsc,810-812;1959
4397	TR193	.216mg	n.s.s.	42/50	.250mg	45/50	.500mg	45/50			
a	TR193	6.05mg	n.s.s.	1/50	.250mg	0/50	.500mg	0/50			liv:hpa,hpc,nnd.
4398	TR193	.181mg	1.01mg	3/50	.200mg	18/50	.400mg	24/50			adr:phe,phm.
a	TR193	.367mg	n.s.s.	43/50	.200mg	42/50	.400mg	42/50			
b	TR193	1.64mg	n.s.s.	0/50	.200mg	1/50	.400mg	1/50			liv:hpa,hpc,nnd.
4399	1008	5.14mg	n.s.s.	1/22	1.50mg	1/18	3.00mg	1/24			Tatematsu;txlt,1,201-205;1978
a	1008	7.52mg	n.s.s.	1/22	1.50mg	1/18	3.00mg	0/24			
4400	1008	2.81mg	n.s.s.	1/22	1.20mg	1/18	2.40mg	1/15			

RESORCINOL 108-46-3

	RefNum	LoConf	UpConf	Cntrl	1Dose	1Inc	2Dose	2Inc			Citation or Pathology
4401	TR403	140.mg	n.s.s.	35/60	79.5mg	21/60	160.mg	33/60			
a	TR403	441.mg	n.s.s.	2/60	79.5mg	1/60	160.mg	5/60			liv:hpa,hpc,nnd.
b	TR403	465.mg	n.s.s.	0/60	79.5mg	3/60	160.mg	1/60			lun:a/a,a/c.
4402	TR403	251.mg	n.s.s.	33/60	79.5mg	29/60	160.mg	22/60			
a	TR403	447.mg	n.s.s.	12/60	79.5mg	11/60	160.mg	7/60			liv:hpa,hpc,nnd.
b	TR403	412.mg	n.s.s.	6/60	79.5mg	6/60	160.mg	6/60			lun:a/a,a/c.
4403	TR403	49.5mg	n.s.s.	49/50	35.3mg	46/50	70.5mg	44/50	106.mg	42/50	
a	TR403	1.20gm	n.s.s.	1/50	35.3mg	0/50	70.5mg	0/50	106.mg	0/50	liv:hpa,hpc,nnd.
4404	TR403	35.2mg	n.s.s.	45/50	79.1mg	44/50	159.mg	31/60			S
a	TR403	51.0mg	n.s.s.	42/50	79.1mg	45/50	159.mg	20/60			
b	TR403	951.mg	n.s.s.	2/50	79.1mg	0/50	159.mg	0/60			liv:hpa,hpc,nnd.

RETINOIC ACID (vitamin A acid) 302-79-4

	RefNum	LoConf	UpConf	Cntrl	1Dose	1Inc	2Dose	2Inc			Citation or Pathology
4405	1134	2.80mg	n.s.s.	3/33	.714mg	2/35					Schmahl;zkko,86,77-84;1976
4406	1134	4.86mg	n.s.s.	1/36	.714mg	0/33					
a	1134	2.91mg	n.s.s.	1/36	.714mg	1/33					

RETINOL ACETATE (vitamin A, acetate) 127-47-9

	RefNum	LoConf	UpConf	Cntrl	1Dose	1Inc	2Dose	2Inc			Citation or Pathology
4407	1693	135.mg	539.mg	3/50	68.8mg	11/49	138.mg	20/48			Kurokawa;jnci,74,715-723;1985
a	1693	416.mg	n.s.s.	1/50	68.8mg	1/49	138.mg	7/48			
b	1693	1.20gm	n.s.s.	0/50	68.8mg	1/49	138.mg	0/48			
c	1693	115.mg	n.s.s.	39/50	68.8mg	40/49	138.mg	36/48			
4408	1693	51.8mg	212.mg	18/49	60.2mg	27/50	120.mg	39/48			
a	1693	251.mg	n.s.s.	1/49	60.2mg	8/50	120.mg	6/48			
b	1693	246.mg	n.s.s.	3/49	60.2mg	4/50	120.mg	11/48			
c	1693	533.mg	n.s.s.	1/49	60.2mg	2/50	120.mg	3/48			
d	1693	537.mg	n.s.s.	2/49	60.2mg	3/50	120.mg	3/48			
e	1693	1.05gm	n.s.s.	0/49	60.2mg	1/50	120.mg	0/48			
f	1693	440.mg	n.s.s.	1/49	60.2mg	0/50	120.mg	0/48			
g	1693	n.s.s.	n.s.s.	49/49	60.2mg	50/50	120.mg	48/48			
4409	1883	64.4mg	n.s.s.	0/10	125.mg	0/10					Hasegawa;gann,79,320-328;1988/pers.comm.
a	1883	64.4mg	n.s.s.	0/10	125.mg	0/10					

RETINOL PALMITATE (vitamin A, palmitate) 79-81-2

	RefNum	LoConf	UpConf	Cntrl	1Dose	1Inc	2Dose	2Inc			Citation or Pathology
4410	1833	16.8mg	n.s.s.	0/38	1.60mg	0/39					Arnold;fctx,23,779-793;1985

RHODAMINE 6G (C.I. basic red 1.HCl) 989-38-8

	RefNum	LoConf	UpConf	Cntrl	1Dose	1Inc	2Dose	2Inc			Citation or Pathology
4411	TR364	144.mg	n.s.s.	29/50	63.8mg	33/50	128.mg	25/50			
a	TR364	481.mg	n.s.s.	8/50	63.8mg	4/50	128.mg	5/50			liv:hpa,hpc,nnd.
b	TR364	397.mg	n.s.s.	4/50	63.8mg	6/50	128.mg	3/50			lun:a/a,a/c.
4412	TR364	523.mg	n.s.s.	0/50	118.mg	4/50	235.mg	3/50			thy:fca,fcc. S
a	TR364	351.mg	n.s.s.	39/50	118.mg	33/50	235.mg	29/50			
b	TR364	527.mg	n.s.s.	13/50	118.mg	12/50	235.mg	11/50			liv:hpa,hpc,nnd.
c	TR364	831.mg	n.s.s.	9/50	118.mg	7/50	235.mg	5/50			lun:a/a,a/c.
4413	TR364	18.8mg	n.s.s.	3/50	5.89mg	3/50	12.3mg	10/50			amd:phm,pob.
a	TR364	7.23mg	n.s.s.	48/50	5.89mg	47/50	12.3mg	49/50			
b	TR364	n.s.s.	n.s.s.	0/50	5.89mg	1/50	12.3mg	0/50			liv:hpa,hpc,nnd.
4414	TR364	14.9mg	n.s.s.	1/50	4.71mg	2/50	9.81mg	8/50			
a	TR364	6.86mg	n.s.s.	48/50	4.71mg	49/50	9.81mg	46/50			
b	TR364	23.7mg	n.s.s.	5/50	4.71mg	0/50	9.81mg	6/50			liv:hpa,hpc,nnd.

RIFAMPICIN 13292-46-1

	RefNum	LoConf	UpConf	Cntrl	1Dose	1Inc	2Dose	2Inc	3Dose	3Inc	Citation or Pathology
4415	400	115.mg	n.s.s.	13/37	10.7mg	10/36	32.1mg	10/37	64.3mg	14/38	Della Porta;txap,43,293-302;1978
a	400	518.mg	n.s.s.	0/37	10.7mg	0/36	32.1mg	1/37	64.3mg	0/38	

```
          Spe Strain Site  Xpo+Xpt                                                              TD50   2Tailpvl
              Sex   Route  Hist   Notes                                                           DR    AuOp
      b   M f bal wat tba mix 14m26 e                                                           no dre    P=1.
  4416 M m bal wat lun tum 14m26 es                                               .>            320.mg * P<.3    -
      a   M m bal wat liv hpt 14m26 es                                                          3.00gm * P<.6    -
      b   M m bal wat tba mix 14m26 es                                                          509.mg * P<.6    -
  4417 M f c3d wat liv hpt 14m25 e                                          .  +  .             33.6mg * P<.0005+
      a   M f c3d wat lun tum 14m25 e                                                           no dre    P=1.
      b   M f c3d wat tba mix 14m25 e                                                           39.3mg * P<.0005
  4418 M m c3d wat lun tum 14m25 e                                                .>            356.mg * P<.2    -
      a   M m c3d wat liv hpt 14m25 e                                                           1.05gm * P<.9    -
      b   M m c3d wat tba mix 14m25 e                                                           474.mg * P<.9    -
  4419 R f wis wat liv hpt 24m32 e                                             .   ±            436.mg * P<.09   -
      a   R f wis wat tba mix 24m32 e                                                           no dre    P=1.   -
  4420 R m wis wat liv hpt 24m32 e                                                 .>           no dre    P=1.   -
      a   R m wis wat tba mix 24m32 e                                                           1.38gm * P<1.

  RIPAZEPAM            100ng..:..1ug....:..10.....:..100....:..1mg....:..10.....:..100...:..1g.....:..10
  4421 M f cd1 eat liv hpa 78w78 e                                             .  ±             352.mg * P<.02   +
      a   M f cd1 eat lun car 78w78 e                                                           1.03gm * P<.2
      b   M f cd1 eat lun ade 78w78 e                                                           no dre    P=1.
      c   M f cd1 eat tba ben 78w78 e                                                           287.mg * P<.04
      d   M f cd1 eat tba mal 78w78 e                                                           1.19gm * P<.5
      e   M f cd1 eat tba mix 78w78 e                                                           219.gm \ P=1.
  4422 M m cd1 eat liv hpa 78w78 e                                         .  +  .              67.8mg * P<.0005+
      a   M m cd1 eat liv hpc 78w78 e                                                           3.15gm * P<.2
      b   M m cd1 eat lun car 78w78 e                                                           no dre    P=1.
      c   M m cd1 eat tba mix 78w78 e                                                           72.2mg * P<.004
      d   M m cd1 eat tba ben 78w78 e                                                           79.4mg * P<.002
      e   M m cd1 eat tba mal 78w78 e                                                           35.0mg \ P<.05
  4423 R f cdr eat liv hpc 24m24 e                                             .  ±             2.77gm * P<.05   -
      a   R f cdr eat liv hpa 24m24 e                                                           3.67gm * P<.7    -
      b   R f cdr eat tba mal 24m24 e                                                           1.44gm * P<.5    -
      c   R f cdr eat tba ben 24m24 e                                                           no dre    P=1.   -
      d   R f cdr eat tba mix 24m24 e                                                           no dre    P=1.   -
  4424 R m cdr eat pit ade 24m24 e                                             .  +  .          147.mg * P<.004  -
      a   R m cdr eat liv hpa 24m24 e                                                           988.mg * P<.2    -
      b   R m cdr eat tba mix 24m24 e                                                           89.7mg * P<.04   -
      c   R m cdr eat tba ben 24m24 e                                                           106.mg * P<.03   -
      d   R m cdr eat tba mal 24m24 e                                                           2.25gm * P<.7    -

  ROSANILINE.HCl      100ng..:..1ug....:..10.....:..100....:..1mg....:..10.....:..100...:..1g.....:..10
  4425 H f syg gav liv tum 68w68 e                                                  .>          no dre    P=1.
      a   H f syg gav lun tum 68w68 e                                                           no dre    P=1.
      b   H f syg gav tba mix 68w68 e                                                           no dre    P=1.   -
  4426 H m syg gav lun bcd 84w84 e                                                 .>           2.02gm    P<.3   -
      a   H m syg gav liv tum 84w84 e                                                           no dre    P=1.
      b   H m syg gav tba mix 84w84 e                                                           no dre    P=1.
  4427 R f sda gav liv tum 25m25 ev                                              .>             no dre    P=1.
      a   R f sda gav tba mix 25m25 ev                                                          no dre    P=1.
  4428 R m sda gav liv tum 26m26 ev                                             .>              no dre    P=1.
      a   R m sda gav tba mix 26m26 ev                                                          no dre    P=1.

  p-ROSANILINE.HCl    100ng..:..1ug....:..10.....:..100....:..1mg....:..10.....:..100...:..1g.....:..10
  4429 H f syg gav liv tum 84w84 e                                                  .>          no dre    P=1.
      a   H f syg gav lun tum 84w84 e                                                           no dre    P=1.
      b   H f syg gav tba mix 84w84 e                                                           no dre    P=1.   -
  4430 H m syg gav liv tum 84w84 e                                                  .>          no dre    P=1.
      a   H m syg gav lun tum 84w84 e                                                           no dre    P=1.
      b   H m syg gav tba mix 84w84 e                                                           no dre    P=1.   -
  4431 M f b6c eat liv MXA 24m24                                       :  +  :                  20.3mg * P<.0005
      a   M f b6c eat MXB MXB 24m24                                                             28.8mg * P<.0005
      b   M f b6c eat liv hpc 24m24                                                             32.3mg * P<.0005c
      c   M f b6c eat liv hpa 24m24                                                             35.9mg \ P<.0005
      d   M f b6c eat --- MXA 24m24                                                             62.2mg * P<.0005e
      e   M f b6c eat adr MXA 24m24                                                             90.6mg * P<.0005c
      f   M f b6c eat --- mlp 24m24                                                             151.mg \ P<.002 e
      g   M f b6c eat lun MXA 24m24                                                             302.mg * P<.002
      h   M f b6c eat hag MXA 24m24                                                             316.mg / P<.002
      i   M f b6c eat lun a/a 24m24                                                             326.mg * P<.003
      j   M f b6c eat --- mlh 24m24                                                             380.mg * P<.004 e
      k   M f b6c eat hag ade 24m24                                                             548.mg * P<.01
      l   M f b6c eat hag MXA 24m24                                                             388.mg / P<.02
      m   M f b6c eat TBA MXB 24m24                                                             23.4mg * P<.0005
      n   M f b6c eat liv MXB 24m24                                                             20.3mg * P<.0005
      o   M f b6c eat lun MXB 24m24                                                             302.mg * P<.002
  4432 M m b6c eat liv hpc 24m24                                           :  +  :              127.mg * P<.002 c
      a   M m b6c eat liv MXA 24m24                                                             116.mg * P<.04
      b   M m b6c eat --- mlm 24m24                                                             610.mg * P<.02
      c   M m b6c eat TBA MXB 24m24                                                             222.mg * P<.4
      d   M m b6c eat liv MXB 24m24                                                             116.mg * P<.04
      e   M m b6c eat lun MXB 24m24                                                             no dre    P=1.
  4433 R f f34 eat mgl MXA 24m24                                       :  +  :                  25.4mg * P<.0005e
```

	RefNum	LoConf	UpConf	Cntrl	1Dose	1Inc	2Dose	2Inc			Citation or Pathology
											Brkly Code
b	400	84.1mg	n.s.s.	26/37	10.7mg	30/36	32.1mg	25/37	64.3mg	25/38	
4416	400	88.4mg	n.s.s.	6/42	8.93mg	9/41	26.8mg	14/42	53.6mg	10/43	
a	400	489.mg	n.s.s.	0/42	8.93mg	0/41	26.8mg	1/42	53.6mg	0/43	
b	400	88.5mg	n.s.s.	11/42	8.93mg	13/41	26.8mg	15/42	53.6mg	14/43	
4417	400	21.6mg	63.7mg	8/40	11.3mg	16/40	34.0mg	27/44	67.9mg	29/36	
a	400	244.mg	n.s.s.	3/40	11.3mg	6/40	34.0mg	2/44	67.9mg	4/36	
b	400	21.0mg	148.mg	18/40	11.3mg	27/40	34.0mg	30/44	67.9mg	31/36	
4418	400	114.mg	n.s.s.	3/30	9.43mg	5/40	28.3mg	3/40	56.6mg	9/40	
a	400	59.8mg	n.s.s.	18/30	9.43mg	20/40	28.3mg	21/40	56.6mg	23/40	
b	400	38.5mg	n.s.s.	20/30	9.43mg	30/40	28.3mg	29/40	56.6mg	29/40	
4419	400	107.mg	n.s.s.	0/20	12.9mg	0/20	25.8mg	2/19			
a	400	22.0mg	n.s.s.	17/20	12.9mg	19/20	25.8mg	14/19			
4420	400	130.mg	n.s.s.	1/19	11.3mg	1/21	22.6mg	1/22			
a	400	26.2mg	n.s.s.	12/19	11.3mg	12/21	22.6mg	14/22			

RIPAZEPAM (pyrazapon) 26308-28-1

	RefNum	LoConf	UpConf	Cntrl	1Dose	1Inc	2Dose	2Inc	Citation or Pathology
4421	1602	142.mg	n.s.s.	5/50	15.0mg	2/50	150.mg	11/50	Fitzgerald;faat,4,178-190;1984
a	1602	252.mg	n.s.s.	0/50	15.0mg	2/50	150.mg	3/50	
b	1602	633.mg	n.s.s.	3/50	15.0mg	1/50	150.mg	1/50	
c	1602	111.mg	n.s.s.	10/50	15.0mg	6/50	150.mg	16/50	
d	1602	214.mg	n.s.s.	2/50	15.0mg	7/50	150.mg	6/50	
e	1602	17.8mg	n.s.s.	13/50	15.0mg	13/50	(150.mg	2/50)	
4422	1602	38.9mg	157.mg	10/50	15.0mg	15/50	150.mg	33/50	
a	1602	513.mg	n.s.s.	0/50	15.0mg	0/50	150.mg	1/50	
b	1602	496.mg	n.s.s.	3/50	15.0mg	6/50	150.mg	2/50	
c	1602	33.7mg	618.mg	24/50	15.0mg	31/50	150.mg	39/50	
d	1602	39.2mg	394.mg	21/50	15.0mg	23/50	150.mg	36/50	
e	1602	13.5mg	n.s.s.	4/50	15.0mg	11/50	(150.mg	5/50)	
4423	1602	683.mg	n.s.s.	0/50	15.0mg	0/50	150.mg	2/50	
a	1602	429.mg	n.s.s.	4/50	15.0mg	6/50	150.mg	6/50	
b	1602	301.mg	n.s.s.	9/50	15.0mg	7/50	150.mg	11/50	
c	1602	74.0mg	n.s.s.	45/50	15.0mg	43/50	150.mg	44/50	
d	1602	68.2mg	n.s.s.	48/50	15.0mg	47/50	150.mg	46/50	
4424	1602	69.5mg	1.14gm	20/50	15.0mg	28/50	150.mg	36/50	
a	1602	304.mg	n.s.s.	5/50	15.0mg	4/50	150.mg	9/50	
b	1602	31.6mg	n.s.s.	40/50	15.0mg	43/50	150.mg	47/50	
c	1602	41.3mg	n.s.s.	36/50	15.0mg	40/50	150.mg	45/50	
d	1602	324.mg	n.s.s.	7/50	15.0mg	10/50	150.mg	10/50	

ROSANILINE.HCl   (magenta I) 632-99-5

	RefNum	LoConf	UpConf	Cntrl	1Dose	1Inc	Citation or Pathology
4425	1151	403.mg	n.s.s.	0/40	114.mg	0/40	Green;zkko,95,51-55;1979
a	1151	403.mg	n.s.s.	0/40	114.mg	0/40	
b	1151	248.mg	n.s.s.	5/40	114.mg	2/40	
4426	1151	329.mg	n.s.s.	0/40	114.mg	1/40	
a	1151	615.mg	n.s.s.	0/40	114.mg	0/40	
b	1151	331.mg	n.s.s.	3/40	114.mg	2/40	
4427	1524	266.mg	n.s.s.	0/40	30.4mg	0/40	Ketkar;clet,16,203-206;1982
a	1524	210.mg	n.s.s.	23/40	30.4mg	3/40	
4428	1524	285.mg	n.s.s.	0/40	30.4mg	0/40	
a	1524	244.mg	n.s.s.	10/40	30.4mg	1/40	

p-ROSANILINE.HCl   (p-magenta, C.I. basic red 9.HCl) 569-61-9

	RefNum	LoConf	UpConf	Cntrl	1Dose	1Inc	2Dose	2Inc	Citation or Pathology
4429	1151	461.mg	n.s.s.	0/40	85.7mg	0/40			Green;zkko,95,51-55;1979
a	1151	461.mg	n.s.s.	0/40	85.7mg	0/40			
b	1151	187.mg	n.s.s.	5/40	85.7mg	4/40			
4430	1151	461.mg	n.s.s.	0/40	85.7mg	0/40			
a	1151	461.mg	n.s.s.	0/40	85.7mg	0/40			
b	1151	331.mg	n.s.s.	3/40	85.7mg	1/40			
4431	TR285	13.5mg	32.3mg	5/50	63.8mg	35/50	128.mg	41/50	liv:hpa,hpc. S
a	TR285	18.9mg	46.5mg	4/50	63.8mg	25/50	128.mg	37/50	adr:phe,phm; liv:hpc. C
b	TR285	20.9mg	52.5mg	3/50	63.8mg	19/50	128.mg	37/50	
c	TR285	18.6mg	85.9mg	2/50	63.8mg	18/50	(128.mg	4/50)	S
d	TR285	33.7mg	179.mg	17/50	63.8mg	24/50	128.mg	25/50	---:mlh,mlm,mlp,mlu,mno.
e	TR285	45.0mg	245.mg	1/50	63.8mg	8/50	128.mg	8/50	adr:phe,phm.
f	TR285	51.6mg	810.mg	0/50	63.8mg	5/50	(128.mg	1/50)	
g	TR285	108.mg	1.35gm	0/50	63.8mg	2/50	128.mg	5/50	lun:a/a,a/c. S
h	TR285	106.mg	1.58gm	0/50	63.8mg	0/50	128.mg	5/50	hag:adn,cyn. S
i	TR285	111.mg	1.95gm	0/50	63.8mg	2/50	128.mg	4/50	S
j	TR285	163.mg	2.64gm	1/50	63.8mg	3/50	128.mg	8/50	
k	TR285	147.mg	45.2gm	0/50	63.8mg	0/50	128.mg	3/50	S
l	TR285	121.mg	n.s.s.	1/50	63.8mg	0/50	128.mg	5/50	hag:adn,can,cyn. S
m	TR285	14.7mg	44.0mg	31/50	63.8mg	50/50	128.mg	49/50	
n	TR285	13.5mg	32.3mg	5/50	63.8mg	35/50	128.mg	41/50	liv:hpa,hpc,nnd.
o	TR285	108.mg	1.35gm	0/50	63.8mg	2/50	128.mg	5/50	lun:a/a,a/c.
4432	TR285	69.0mg	582.mg	10/50	58.9mg	20/50	118.mg	27/50	
a	TR285	50.1mg	n.s.s.	29/50	58.9mg	37/50	118.mg	41/50	liv:hpa,hpc. S
b	TR285	263.mg	n.s.s.	0/50	58.9mg	3/50	118.mg	4/50	S
c	TR285	56.8mg	n.s.s.	44/50	58.9mg	43/50	118.mg	46/50	
d	TR285	50.1mg	n.s.s.	29/50	58.9mg	37/50	118.mg	41/50	liv:hpa,hpc,nnd.
e	TR285	284.mg	n.s.s.	11/50	58.9mg	7/50	118.mg	8/50	lun:a/a,a/c.
4433	TR285	14.1mg	83.6mg	23/50	24.6mg	32/50	49.3mg	32/50	mgl:acn,adn,fba.

```
     Spe Strain  Site    Xpo+Xpt                                                              TD50      2Tailpvl
         Sex   Route  Hist    Notes                                                            DR        AuOp

a    R f f34 eat mgl fba 24m24                                                                 27.9mg  * P<.0005e
b    R f f34 eat MXB MXB 24m24                                                                 32.2mg  * P<.0005
c    R f f34 eat sub MXA 24m24                                                                 40.9mg  * P<.0005
d    R f f34 eat sub MXA 24m24                                                                 42.8mg  * P<.0005
e    R f f34 eat sub fib 24m24                                                                 44.5mg  * P<.0005c
f    R f f34 eat ute ess 24m24                                                                 133.mg  * P<.006
g    R f f34 eat thy MXA 24m24                                                                 154.mg  * P<.0005c
h    R f f34 eat zym can 24m24                                                                 190.mg  * P<.002 c
i    R f f34 eat thy fca 24m24                                                                 318.mg  / P<.007
j    R f f34 eat mgl MXA 24m24                                                                 156.mg  * P<.03  e
k    R f f34 eat liv MXA 24m24                                                                 189.mg  * P<.04
l    R f f34 eat mgl acn 24m24                                                                 232.mg  * P<.06  e
m    R f f34 eat TBA MXB 24m24                                                                 17.8mg  * P<.0005
n    R f f34 eat liv MXB 24m24                                                                 189.mg  * P<.04
4434 R m f34 eat tes ict 24m24                              :  +  :                            15.6mg  / P<.0005
a    R m f34 eat MXB MXB 24m24                                                                 21.2mg  / P<.0005
b    R m f34 eat sub MXA 24m24                                                                 33.0mg  / P<.0005
c    R m f34 eat sub MXA 24m24                                                                 33.4mg  / P<.0005
d    R m f34 eat sub fib 24m24                                                                 35.3mg  / P<.0005c
e    R m f34 eat liv MXA 24m24                                                                 41.5mg  / P<.0005
f    R m f34 eat thy MXA 24m24                                                                 57.0mg  / P<.0005c
g    R m f34 eat liv nnd 24m24                                                                 62.2mg  * P<.0005
h    R m f34 eat thy fcc 24m24                                                                 73.9mg  / P<.0005c
i    R m f34 eat ski MXA 24m24                                                                 93.8mg  / P<.0005
j    R m f34 eat zym can 24m24                                                                 124.mg  / P<.0005c
k    R m f34 eat ski sqc 24m24                                                                 130.mg  / P<.0005c
l    R m f34 eat liv hpc 24m24                                                                 140.mg  / P<.0005c
m    R m f34 eat pni MXA 24m24                                                                 152.mg  * P<.006
n    R m f34 eat thy fca 24m24                                                                 176.mg  / P<.0005c
o    R m f34 eat ski tri 24m24                                                                 221.mg  / P<.0005c
p    R m f34 eat lun a/a 24m24                                                                 281.mg  * P<.006
q    R m f34 eat ski sea 24m24                                                                 424.mg  / P<.003 c
r    R m f34 eat MXA MXA 24m24                                                                 184.mg  * P<.02
s    R m f34 eat pni isc 24m24                                                                 252.mg  * P<.02
t    R m f34 eat lun MXA 24m24                                                                 254.mg  * P<.02
u    R m f34 eat ski bcc 24m24                                                                 385.mg  / P<.03
v    R m f34 eat liv MXA 24m24                                                                 557.mg  / P<.02
w    R m f34 eat TBA MXB 24m24                                                                 15.7mg  / P<.0005
x    R m f34 eat liv MXB 24m24                                                                 41.5mg  / P<.0005
4435 R f sda gav liv tum 30m30 ev                                                 .>          no dre   P=1.
a    R f sda gav tba mix 30m30 ev                                                             no dre   P=1.  -
4436 R m sda gav liv tum 29m29 ev                                                 .>          no dre   P=1.
a    R m sda gav tba mix 29m29 ev                                                             no dre   P=1.  -

ROTENONE                       100ng..:..1ug...:..10.....:..100...:..1mg...:..10.....:..100...:..1g.....:..10
4437 M f b6c eat TBA MXB 24m24                                                        :>      no dre   P=1.  -
a    M f b6c eat liv MXB 24m24                                                                no dre   P=1.
b    M f b6c eat lun MXB 24m24                                                                15.0gm  * P<1.
4438 M f b6c orl liv hpt 76w76 evx                      .>                                    2.65mg    P<.3  -
a    M f b6c orl lun car 76w76 evx                                                            2.65mg    P<.3  -
b    M f b6c orl tba mix 76w76 evx                                                            .831mg    P<.05 -
4439 M m b6c eat TBA MXB 24m24                                               :>               no dre   P=1.  -
a    M m b6c eat liv MXB 24m24                                                                no dre   P=1.
b    M m b6c eat lun MXB 24m24                                                                no dre   P=1.
4440 M m b6c orl liv mix 76w76 evx                  .    ±                                    .773mg    P<.05 -
a    M m b6c orl lun ade 76w76 evx                                                            2.47mg    P<.3  -
b    M m b6c orl tba mix 76w76 evx                                                            .433mg    P<.009 -
4441 M f b6a orl lun ade 76w76 evx                    .>                                      2.65mg    P<.6  -
a    M f b6a orl liv hpt 76w76 evx                                                            no dre   P=1.  -
b    M f b6a orl tba mix 76w76 evx                                                            no dre   P=1.  -
4442 M m b6a orl liv hpt 76w76 evx                    .>                                      no dre   P=1.  -
a    M m b6a orl lun ade 76w76 evx                                                            no dre   P=1.  -
b    M m b6a orl tba mix 76w76 evx                                                            no dre   P=1.  -
4443 R f f34 eat TBA MXB 24m24                                        :>                       no dre   P=1.  -
a    R f f34 eat liv MXB 24m24                                                                no dre   P=1.
4444 R m f34 eat pty adn 24m24                                            :>                   35.6mg  * P<.3  e
a    R m f34 eat TBA MXB 24m24                                                                no dre   P=1.
b    R m f34 eat liv MXB 24m24                                                                no dre   P=1.

RUTIN SULFATE                  100ng..:..1ug...:..10.....:..100...:..1mg...:..10.....:..100...:..1g.....:..10
4445 R f sda gav liv hpc 24m32 e                                                  .>          44.8gm  * P<1.  -

4446 R m sda gav liv hpc 24m31 e                                     .>                        no dre   P=1.  -

RUTIN TRIHYDRATE               100ng..:..1ug...:..10.....:..100...:..1mg...:..10.....:..100...:..1g.....:..10
4447 H f syg eat adr coa 24m24                                                                132.gm    P<.6  -
a    H f syg eat ute ley 24m24                                                                140.gm    P<.3  -
4448 H m syg eat adr coa 24m24                                                                no dre   P=1.  -
a    H m syg eat for pam 24m24                                                                no dre   P=1.
4449 R f aci eat pit ade 77w77                                                    .>          8.91gm    P<.2  -
a    R f aci eat ubl pam 77w77                                                                8.91gm    P<.2  -
```

	RefNum	LoConf	UpConf	Cntrl	1Dose	1Inc	2Dose	2Inc							Citation or Pathology
															Brkly Code
a	TR285	15.0mg	108.mg	22/50	24.6mg	31/50	49.3mg	29/50							
b	TR285	21.2mg	51.7mg	0/50	24.6mg	16/50	49.3mg	21/50						sub:fib; thy:fca,fcc; zym:can. C
c	TR285	25.1mg	82.0mg	1/50	24.6mg	17/50	49.3mg	12/50						sub:fbs,fib,srn. S
d	TR285	26.5mg	76.6mg	0/50	24.6mg	16/50	49.3mg	10/50						sub:fbs,fib. S
e	TR285	27.2mg	80.7mg	0/50	24.6mg	15/50	49.3mg	10/50						
f	TR285	60.0mg	1.50gm	1/50	24.6mg	5/50	49.3mg	6/50						S
g	TR285	68.8mg	534.mg	0/50	24.6mg	2/50	49.3mg	6/50						thy:fca,fcc.
h	TR285	86.5mg	651.mg	0/50	24.6mg	2/50	49.3mg	7/50						
i	TR285	108.mg	5.07gm	0/50	24.6mg	0/50	49.3mg	4/50						S
j	TR285	63.2mg	n.s.s.	2/50	24.6mg	4/50	49.3mg	6/50						mgl:acn,adn.
k	TR285	73.0mg	n.s.s.	1/50	24.6mg	4/50	49.3mg	4/50						liv:hpc,nnd. S
l	TR285	79.3mg	n.s.s.	2/50	24.6mg	2/50	49.3mg	5/50						
m	TR285	10.1mg	57.5mg	41/50	24.6mg	50/50	49.3mg	48/50						
n	TR285	73.0mg	n.s.s.	1/50	24.6mg	4/50	49.3mg	4/50						liv:hpa,hpc,nnd.
4434	TR285	9.57mg	33.0mg	43/50	39.2mg	46/50	79.2mg	37/50						S
a	TR285	14.8mg	32.0mg	3/50	39.2mg	26/50	79.2mg	40/50				liv:hpc; ski:sea,sqc,tri; sub:fib; thy:fca,fcc; zym:can. C	
b	TR285	20.6mg	59.7mg	3/50	39.2mg	22/50	79.2mg	16/50						sub:fbs,fib. S
c	TR285	20.8mg	62.6mg	6/50	39.2mg	24/50	79.2mg	19/50						sub:fbs,fib,srn. S
d	TR285	22.2mg	62.9mg	2/50	39.2mg	20/50	79.2mg	16/50						
e	TR285	24.0mg	89.4mg	5/50	39.2mg	15/50	79.2mg	14/50						liv:hpc,nnd. S
f	TR285	34.9mg	97.5mg	0/50	39.2mg	5/50	79.2mg	25/50						thy:fca,fcc.
g	TR285	31.0mg	246.mg	5/50	39.2mg	14/50	79.2mg	6/50						S
h	TR285	42.1mg	137.mg	0/50	39.2mg	5/50	79.2mg	18/50						
i	TR285	47.9mg	235.mg	2/50	39.2mg	2/50	79.2mg	14/50						ski:sqc,sqp. S
j	TR285	61.6mg	314.mg	1/50	39.2mg	1/50	79.2mg	13/50						
k	TR285	61.7mg	317.mg	0/50	39.2mg	1/50	79.2mg	10/50						
l	TR285	65.7mg	357.mg	0/50	39.2mg	2/50	79.2mg	8/50						
m	TR285	62.5mg	2.26gm	2/50	39.2mg	5/50	79.2mg	4/50						pni:isa,isc. S
n	TR285	78.3mg	477.mg	0/50	39.2mg	0/50	79.2mg	9/50						
o	TR285	90.5mg	711.mg	0/50	39.2mg	0/50	79.2mg	7/50						
p	TR285	98.7mg	2.79gm	0/50	39.2mg	3/50	79.2mg	3/50						S
q	TR285	157.mg	2.29gm	0/50	39.2mg	0/50	79.2mg	5/50						
r	TR285	70.1mg	42.6gm	1/50	39.2mg	5/50	79.2mg	3/50				bod:men; mls:men; mul:men; tnv:men,msm. S	
s	TR285	85.2mg	n.s.s.	0/50	39.2mg	3/50	79.2mg	1/50						S
t	TR285	90.9mg	n.s.s.	1/50	39.2mg	3/50	79.2mg	4/50						lun:a/a,a/c. S
u	TR285	116.mg	n.s.s.	1/50	39.2mg	0/50	79.2mg	4/50						S
v	TR285	162.mg	n.s.s.	0/50	39.2mg	0/50	79.2mg	4/50						liv:bda,bdc. S
w	TR285	9.99mg	29.7mg	41/50	39.2mg	47/50	79.2mg	45/50						
x	TR285	24.0mg	89.4mg	5/50	39.2mg	15/50	79.2mg	14/50						liv:hpa,hpc,nnd.
4435	1524	597.mg	n.s.s.	0/40	48.6mg	0/40								Ketkar;clet,16,203-206;1982	
a	1524	244.mg	n.s.s.	23/40	48.6mg	11/40								
4436	1524	571.mg	n.s.s.	0/40	48.7mg	0/40								
a	1524	201.mg	n.s.s.	10/40	48.7mg	7/40								

ROTENONE (tubatoxin) 83-79-4

	RefNum	LoConf	UpConf	Cntrl	1Dose	1Inc	2Dose	2Inc							Citation or Pathology
4437	TR320	345.mg	n.s.s.	26/50	77.3mg	16/50	155.mg	22/50							
a	TR320	658.mg	n.s.s.	4/50	77.3mg	3/50	155.mg	4/50						liv:hpa,hpc,nnd.
b	TR320	607.mg	n.s.s.	4/50	77.3mg	2/50	155.mg	5/50						lun:a/a,a/c.
4438	1253	.431mg	n.s.s.	0/16	.414mg	1/18								Innes;ntis,1968/1969
a	1253	.431mg	n.s.s.	0/16	.414mg	1/18								
b	1253	.251mg	n.s.s.	0/16	.414mg	3/18								
4439	TR320	88.4mg	n.s.s.	25/50	71.3mg	26/50	(143.mg	18/50)						
a	TR320	139.mg	n.s.s.	12/50	71.3mg	12/50	(143.mg	1/50)						liv:hpa,hpc,nnd.
b	TR320	346.mg	n.s.s.	6/50	71.3mg	12/50	143.mg	8/50						lun:a/a,a/c.
4440	1253	.233mg	n.s.s.	0/16	.385mg	3/18								Innes;ntis,1968/1969
a	1253	.401mg	n.s.s.	0/16	.385mg	1/18								
b	1253	.163mg	9.50mg	0/16	.385mg	5/18								
4441	1253	.348mg	n.s.s.	1/17	.414mg	2/18								
a	1253	.820mg	n.s.s.	0/17	.414mg	0/18								
b	1253	.391mg	n.s.s.	2/17	.414mg	2/18								
4442	1253	.455mg	n.s.s.	1/18	.385mg	1/18								
a	1253	.505mg	n.s.s.	2/18	.385mg	1/18								
b	1253	.303mg	n.s.s.	3/18	.385mg	3/18								
4443	TR320	2.86mg	n.s.s.	45/50	1.88mg	43/50	3.71mg	45/50						
a	TR320	36.7mg	n.s.s.	1/50	1.88mg	0/50	3.71mg	0/50						liv:hpa,hpc,nnd.
4444	TR320	9.37mg	n.s.s.	1/50	1.51mg	0/50	2.97mg	4/50						
a	TR320	2.98mg	n.s.s.	48/50	1.51mg	36/50	2.97mg	47/50						
b	TR320	12.0mg	n.s.s.	3/50	1.51mg	1/50	2.97mg	3/50						liv:hpa,hpc,nnd.

RUTIN SULFATE 12768-44-4

	RefNum	LoConf	UpConf	Cntrl	1Dose	1Inc	2Dose	2Inc	3Dose	3Inc	4Dose	4Inc	5Dose	5Inc	Citation or Pathology
4445	1604	480.mg	n.s.s.	0/12	3.48mg	0/12	8.75mg	0/12	23.0mg	1/12	65.4mg	1/12	162.mg	0/12	Habs;clet,23,103-108; 1984
4446	1604	8.95mg	n.s.s.	1/12	3.48mg	0/12	9.14mg	0/12	24.7mg	0/12	65.4mg	0/12	167.mg	0/12	

RUTIN TRIHYDRATE 153-18-4

	RefNum	LoConf	UpConf	Cntrl	1Dose	1Inc	Citation or Pathology
4447	1144	18.4gm	n.s.s.	1/20	10.5gm	2/20	Morino;carc,3,93-97;1982
a	1144	22.7gm	n.s.s.	0/20	10.5gm	1/20	
4448	1144	18.0gm	n.s.s.	2/20	9.20gm	2/20	
a	1144	22.7gm	n.s.s.	1/20	9.20gm	1/20	
4449	1145m	1.45gm	n.s.s.	0/22	2.50gm	1/10	Hirono;clet,13,15-21;1981
a	1145m	1.45gm	n.s.s.	0/22	2.50gm	1/10	

```
     Spe Strain  Site   Xpo+Xpt                                                    TD50    2Tailpvl
         Sex  Route  Hist   Notes                                                        DR    AuOp
4450 R f aci eat adr phe 28m28                                                     90.4gm   P<.2   -
a    R f aci eat hea sar 28m28                                                     90.4gm   P<.2   -
4451 R m aci eat tes ict 77w77                               .>                     3.53gm   P<.2   -
a    R m aci eat adr coa 77w77                                                      5.71gm   P<.3   -
4452 R m aci eat col ade 28m28                                                     72.3gm   P<.2   -
a    R m aci eat ilm ade 28m28                                                     72.3gm   P<.2   -

SACCHARIN                   100ng..:..1ug....:.10......:..100....:..1mg....:.10.....:..100.....:..1g......:.10
4453 M f swa eat lun tum 76w76 e                                                  .>56.3gm   P<.7   -
a    M f swa eat liv hpt 76w76 e                                                   no dre    P=1.   -
4454 M f swi eat ubl apc 91w91 eg                                       .>         no dre    P=1.   -
4455 M m swi eat ubl tcc 91w91 e                                           .>      no dre    P=1.   -
4456 R m f34 eat for pam 72w73 e                                                  .>no dre   P=1.   -
a    R m f34 eat ubl car 72w73 e                                                   no dre    P=1.   -

SACCHARIN, CALCIUM          100ng..:..1ug....:.10......:..100....:..1mg....:.10.....:..100.....:..1g......:.10
4457 R m f34 eat ubl car 72w73 e                                                  .no dre    P=1.   -
a    R m f34 eat for pam 72w73 e                                                   no dre    P=1.   -

SACCHARIN, SODIUM           100ng..:..1ug....:.10......:..100....:..1mg....:.10.....:..100.....:..1g......:.10
4458 M f chi eat liv tum 24m24 e                                                  .>no dre   P=1.   -
a    M f chi eat lun tum 24m24 e                                                   no dre    P=1.   -
b    M f chi eat tba mix 24m24 e                                                  26.1gm * P<.7   -
4459 M m chi eat lun tum 24m24 e                       . +   .                     1.36gm \ P<.002 -
a    M m chi eat liv hpt 24m24 e                                                   no dre    P=1.   -
b    M m chi eat tba mix 24m24 e                                                  881.mg \ P<.002 -
4460 R m aaa eat ubl car 80w80 er                                                 .no dre    P=1.   -
4461 R m aci eat ubl pam 52w52 r                                   . +   .         1.11gm    P<.0005+
a    R m aci eat ubl tcc 52w52 r                                                   3.71gm    P<.05  +
4462 R f cdr eat liv clc 26m26 e                                                  78.2gm * P<.03  -
a    R f cdr eat liv nod 26m26 e                                                  98.4gm * P<.2   -
4463 R f cdr eat liv blc 33m33 eg                                                155.gm     P<.6
a    R f cdr eat liv nnd 33m33 eg                                                155.gm     P<.3
4464 R m cdr eat liv hpc 26m26 e                                                   no dre    P=1.   -
4465 R m cdr eat pty ade 33m33 e                                                  .23.2gm    P<.008
a    R m cdr eat ubl tcc 33m33 e                                                  31.1gm     P<.05  +
b    R m cdr eat liv blc 33m33 e                                                  60.0gm     P<.1
c    R m cdr eat ubl tpp 33m33 e                                                  30.8gm     P<.2   +
4466 R m cdr eat tba mix 24m24 e                                        .>         no dre    P=1.   -
4467 R m f34 eat ubl tum 24m24 r                                                  .no dre    P=1.   -
4468 R m f34 eat ubl tum 24m24 r                                                  .no dre    P=1.   -
4469 R m f34 eat ubl tum 52w52 r                                           .>      no dre    P=1.   -
4470 R m f34 eat ubl pam 24m24                                                    .54.1gm    P<.3   +
4471 R m f34 eat liv nnd 95w95                                                     no dre    P=1.
4472 R m f34 eat liv tum 61w68 e                                           .>      no dre    P=1.   -
4473 R m f34 eat ubl tum 24m24 r                                                   no dre    P=1.   -
4474 R m f34 eat ubl car 72w73 e                                           .>      no dre    P=1.   -
a    R m f34 eat for pam 72w73 e                                                   no dre    P=1.   -
4475 R m fis eat ubl mix 23m24                                                     no dre    P=1.   -
4476 R m fis eat liv tum 24m24                                                    .no dre    P=1.   -
a    R m fis eat ubl tum 24m24                                                     no dre    P=1.   -
4477 R m fis eat tes ict 77w98                             . +   .                613.mg     P<.003
a    R m fis eat liv tum 77w98                                                     no dre    P=1.
b    R m fis eat ubl tum 77w98                                                     no dre    P=1.
4478 R b sda eat ubl tum 30m30 e                                        .>         no dre    P=1.   -
a    R b sda eat tba mal 30m30 e                                                   4.10gm * P<.4   -
4479 R m sda eat ubl tum 52w52 r                                           .>      no dre    P=1.   -
4480 R m sda eat ubl car 80w80 er                                                 .no dre    P=1.   -
4481 R b wis wat ubl tum 24m24 er                                                  no dre    P=1.   -
4482 R b wis eat ubl tum 24m24 er                                                125.gm     P<.08
4483 R m wis eat ubl tum 52w52 r                                           .>      no dre    P=1.   -

SAFFLOWER OIL               100ng..:..1ug....:.10......:..100....:.1mg....:.10......:..100....:..1g......:.10
4484 R m f34 gav pan MXA 24m24                                                  :+ :4.85gm * P<.0005
a    R m f34 gav pan ana 24m24                                                    4.94gm * P<.0005
b    R m f34 gav TBA MXB 24m24                                                     no dre    P=1.
c    R m f34 gav liv MXB 24m24                                                     no dre    P=1.

SAFROLE                     100ng..:..1ug....:.10......:..100....:.1mg....:.10......:..100....:..1g......:.10
4485 M f b6a orl liv hpt 81w81 evx                     . +   .                    23.5mg     P<.0005+
a    M f b6a orl lun ade 81w81 evx                                                 no dre    P=1.
b    M f b6a orl tba mix 81w81 evx                                                24.2mg     P<.0005
4486 M m b6a orl liv hpt 81w81 evx                           .   ±                312.mg     P<.04
a    M m b6a orl lun mix 81w81 evx                                                 no dre    P=1.
b    M m b6a orl tba mix 81w81 evx                                                 5.12gm    P<1.
4487 M f b6c orl liv hpt 81w81 evx                          <+                    noTD50     P<.0005+
a    M f b6c orl lun mix 81w81 evx                                                 no dre    P=1.
b    M f b6c orl tba mix 81w81 evx                                                noTD50     P<.0005
4488 M f b6c gav liv mix 90w90 e                         . +   .                  27.0mg     P<.0005+
a    M f b6c gav liv hpc 90w90 e                                                  33.0mg     P<.0005+
b    M f b6c gav lun tum 90w90 e                                                   no dre    P=1.
```

	RefNum	LoConf	UpConf	Cntrl	1Dose	1Inc	2Dose	2Inc					Citation or Pathology / Brkly Code
4450	1145n	14.7gm	n.s.s.	0/33	5.00gm	1/20							
a	1145n	14.7gm	n.s.s.	0/33	5.00gm	1/20							
4451	1145m	777.mg	n.s.s.	3/30	2.00gm	3/11							
a	1145m	990.mg	n.s.s.	2/30	2.00gm	2/11							
4452	1145n	11.8gm	n.s.s.	0/33	4.00gm	1/20							
a	1145n	11.8gm	n.s.s.	0/33	4.00gm	1/20							

SACCHARIN 81-07-2

	RefNum	LoConf	UpConf	Cntrl	1Dose	1Inc	2Dose	2Inc					Citation or Pathology
4453	1090	7.21gm	n.s.s.	7/45	6.50gm	8/42							Roe;fctx,8,135-145;1970
a	1090	21.4gm	n.s.s.	3/45	6.50gm	1/42							
4454	1349	1.12gm	n.s.s.	1/41	260.mg	0/37	650.mg	0/42					Kroes;txcy,8,285-300;1977
4455	1349	3.40gm	n.s.s.	0/40	240.mg	1/39	600.mg	0/48					
4456	2110	6.75gm	n.s.s.	0/40	1.66gm	0/40							Cohen;canr,51,1766-1777;1991
a	2110	6.75gm	n.s.s.	2/39	1.66gm	0/40							

SACCHARIN, CALCIUM 6485-34-3

	RefNum	LoConf	UpConf	Cntrl	1Dose	1Inc	2Dose	2Inc					Citation or Pathology
4457	2110	8.12gm	n.s.s.	2/39	2.05gm	0/39							Cohen;canr,51,1766-1777;1991
a	2110	8.33gm	n.s.s.	0/40	2.05gm	0/40							

SACCHARIN, SODIUM 128-44-9

	RefNum	LoConf	UpConf	Cntrl	1Dose	1Inc	2Dose	2Inc	3Dose	3Inc	4Dose	4Inc	Citation or Pathology
4458	1450	6.49gm	n.s.s.	0/17	1.30gm	0/28	6.50gm	0/36					Homburger;ctxf,359-373;1978
a	1450	18.9gm	n.s.s.	4/17	1.30gm	6/28	6.50gm	5/36					
b	1450	4.12gm	n.s.s.	11/17	1.30gm	16/28	6.50gm	24/36					
4459	1450	686.mg	4.99gm	1/19	1.20gm	14/29	(6.00gm	9/34)					
a	1450	19.6gm	n.s.s.	1/19	1.20gm	5/29	6.00gm	2/34					
b	1450	442.mg	3.79gm	4/19	1.20gm	20/29	(6.00gm	21/34)					
4460	2016m	8.54gm	n.s.s.	0/12	2.00gm	0/35							Honma;fctx,29,373-376;1991
4461	1537	523.mg	3.41gm	0/30	2.00gm	9/34							Fukushima;gann,74,8-20;1983
a	1537	1.12gm	n.s.s.	0/30	2.00gm	3/34							
4462	1114	19.2gm	n.s.s.	0/56	90.0mg	0/56	270.mg	0/52	810.mg	0/56	2.43gm	2/54	Munro;txap,32,513-526;1975
a	1114	20.1gm	n.s.s.	1/56	90.0mg	0/56	270.mg	0/52	810.mg	0/56	2.43gm	2/54	
4463	1398	20.8gm	n.s.s.	1/48	2.50gm	2/49							Arnold;txap,52,113-152;1980
a	1398	25.2gm	n.s.s.	0/48	2.50gm	1/49							
4464	1114	31.5gm	n.s.s.	0/57	90.0mg	0/51	270.mg	1/54	810.mg	0/52	2.43gm	0/54	Munro;txap,32,513-526;1975
4465	1398	8.81gm	378.gm	0/49	2.00gm	5/48							Arnold;txap,52,113-152;1980
a	1398	9.40gm	n.s.s.	0/36	2.00gm	3/38							
b	1398	14.8gm	n.s.s.	0/49	2.00gm	2/48							
c	1398	8.50gm	n.s.s.	1/36	2.00gm	4/38							
4466	1450	1.96gm	n.s.s.	11/16	400.mg	21/28	2.00gm	15/26					Homburger;ctxf,359-373;1978
4467	1479m	8.49gm	n.s.s.	0/37	1.96gm	0/21							Cohen;canr,42,65-71;1982
4468	1479n	8.66gm	n.s.s.	0/37	2.00gm	0/21							
4469	1537	2.58gm	n.s.s.	0/25	2.00gm	0/25							Fukushima;gann,74,8-20;1983
4470	1722	8.81gm	n.s.s.	0/40	2.00gm	1/40							Hasegawa;canr,45,1469-1473;1985/pers.comm.
4471	1783	14.4gm	n.s.s.	1/42	2.00gm	0/42							Imaida;canr,46,6160-6164;1986
4472	1786	4.58gm	n.s.s.	0/36	1.79gm	0/29							Sakata;canr,46,3903-3906;1986
4473	1986	16.5gm	n.s.s.	0/40	2.00gm	0/40							Hasegawa;txcy,62,333-347;1990
4474	2110	5.39gm	n.s.s.	2/40	1.97gm	1/40							Cohen;canr,51,1766-1777;1991
a	2110	8.01gm	n.s.s.	0/40	1.97gm	0/40							
4475	1430	10.3gm	n.s.s.	0/27	1.92gm	0/26							Fukushima;canr,41,3100-3103;1981
4476	1574	8.24gm	n.s.s.	0/30	2.00gm	0/20							Murasaki;carc,4,97-99;1983
a	1574	8.24gm	n.s.s.	0/30	2.00gm	0/20							
4477	1657	232.mg	4.71gm	22/42	1.57gm	18/20							Cohen;canr,39,1207-1217;1979
a	1657	5.75gm	n.s.s.	0/42	1.57gm	0/20							
b	1657	5.75gm	n.s.s.	0/42	1.57gm	0/20							
4478	1416	1.94gm	n.s.s.	0/98	90.0mg	0/94	225.mg	0/93					Schmahl;arzn,23,1466-1470;1973
a	1416	1.11gm	n.s.s.	13/98	90.0mg	11/94	225.mg	17/93					
4479	1537	2.58gm	n.s.s.	0/25	2.00gm	0/25							Fukushima;gann,74,8-20;1983
4480	2016n	8.78gm	n.s.s.	0/14	2.00gm	0/36							Honma;fctx,29,373-376;1991
4481	1465	47.4gm	n.s.s.	0/98	2.00gm	0/115							Hicks;carm,2,475-489;1978
4482	1465m	37.8gm	n.s.s.	0/98	4.00gm	3/138							
4483	1537	2.58gm	n.s.s.	0/25	2.00gm	0/25							Fukushima;gann,74,8-20;1983

SAFFLOWER OIL 8001-23-8

	RefNum	LoConf	UpConf	Cntrl	1Dose	1Inc	2Dose	2Inc	3Dose	3Inc			Citation or Pathology / Brkly Code
4484	TR426	3.31gm	8.77gm	1/50	1.63gm	7/50	3.26gm	15/50	6.55gm	29/51			pan:acc,ana. S
a	TR426	3.35gm	9.14gm	1/50	1.63gm	7/50	3.26gm	15/50	6.55gm	28/51			S
b	TR426	7.70gm	n.s.s.	47/50	1.63gm	46/50	3.26gm	41/50	6.55gm	42/51			
c	TR426	54.2gm	n.s.s.	3/50	1.63gm	2/50	3.26gm	2/50	6.55gm	0/51			liv:hpa,hpb,hpc.

SAFROLE 94-59-7

	RefNum	LoConf	UpConf	Cntrl	1Dose	1Inc	2Dose	2Inc					Citation or Pathology
4485	267	9.22mg	60.3mg	1/15	156.mg	16/17							Innes;ntis,1968/1969
a	267	332.mg	n.s.s.	2/15	156.mg	0/17							
b	267	9.32mg	66.2mg	2/15	156.mg	16/17							
4486	267	94.1mg	n.s.s.	0/18	146.mg	3/17							
a	267	310.mg	n.s.s.	3/18	146.mg	0/17							
b	267	120.mg	n.s.s.	3/18	146.mg	3/17							
4487	267	n.s.s.	38.8mg	0/17	156.mg	16/16							
a	267	313.mg	n.s.s.	0/17	156.mg	0/16							
b	267	n.s.s.	40.5mg	1/17	156.mg	16/16							
4488	1039	16.2mg	49.9mg	0/98	34.3mg	22/46							Vesselinovitch;canr,39,4378-4380;1979
a	1039	19.1mg	64.0mg	0/98	34.3mg	19/46							
b	1039	243.mg	n.s.s.	0/98	34.3mg	0/46							

	Spe Strain Site	Xpo+Xpt				
	Sex Route Hist	Notes			TD50	2Tailpvl
					DR	AuOp
4489	M m b6c orl liv hpt 81w81	evx	. + .		61.7mg	P<.0005+
a	M m b6c orl lun ade 81w81	evx			no dre	P=1.
b	M m b6c orl tba mix 81w81	evx			119.mg	P<.2
4490	M m b6c gav liv mix 90w90	e	. ±		178.mg	P<.07 -
a	M m b6c gav liv hpc 90w90	e			281.mg	P<.02 -
b	M m b6c gav lun mix 90w90	e			no dre	P=1.
4491	M m bal eat liv hpa 52w52	ek	. + .		68.3mg	P<.0005+
a	M m bal eat liv hpc 52w52	ek			368.mg	P<.09 +
b	M m bal eat lun tum 52w52	ek			no dre	P=1.
4492	M m bal eat liv hpa 52w75	ek	<+		noTD50	P<.009 +
a	M m bal eat liv hpc 52w75	ek			129.mg	P<.02 +
b	M m bal eat lun tum 52w75	ek			no dre	P=1.
4493	M f cd1 eat liv hpc 51w73	ev	. + .		125.mg	P<.0005+
a	M f cd1 eat lun tum 51w73	ev			no dre	P=1.
4494	M f cd1 eat liv hpt 52w69	ev	. + .		59.7mg *	P<.0005+
a	M f cd1 eat lun tum 52w69	ev			3.50gm *	P<1.
4495	M f cd1 eat liv hpt 50w78	v	. + .		56.4mg	P<.0005+
a	M f cd1 eat lun ade 50w78	v			no dre	P=1.
4496	M f cd1 eat liv hpt 51w86	v	. +.		41.5mg *	P<.0005+
a	M f cd1 eat --- ang 51w86	v			213.mg *	P<.0005
b	M f cd1 eat lun ade 51w86	v			no dre	P=1.
4497	M m cd1 eat liv car 51w73	ev	. + .		249.mg	P<.0005+
a	M m cd1 eat lun tum 51w73	ev			no dre	P=1.
4498	M m cd1 eat liv hpc 56w69		. ±		731.mg *	P<.03
a	M m cd1 eat lun ade 56w69				no dre	P=1.
4499	M m cd1 eat liv car 56w69		.>		3.71gm	P<.7
a	M m cd1 eat tba mix 56w69				3.71gm	P<.7
4500	R m cdr eat liv hpc 73w95	e	.>		no dre	P=1. -
4501	R m cdr eat liv hpc 95w95	e	. ±		627.mg	P<.05 +
4502	R m cdr eat liv tum 43w69		.>		no dre	P=1.
a	R m cdr eat tba tum 43w69				no dre	P=1.
4503	R m cdr eat liv mix 47w69		.>		576.mg *	P<.2
a	R m cdr eat tba mix 47w69				1.32gm *	P<.7
4504	R m cdr eat liv car 36w52		.>		425.mg	P<.3 +
a	R m cdr eat tba mix 36w52				425.mg	P<.3
4505	R b osm eat liv hpa 24m24	s	. + .		250.mg Z	P<.004
a	R b osm eat liv mix 24m24	s			340.mg *	P<.0005+
b	R b osm eat liv mal 24m24	s			624.mg *	P<.0005+
c	R b osm eat liv hpc 24m24	s			921.mg *	P<.0005
d	R b osm eat liv mhc 24m24	s			1.96gm *	P<.0005
4506	R f osm eat liv tum 24m24	s	. + .		231.mg *	P<.0005
4507	R m osm eat liv tum 24m24	s	. ±		593.mg *	P<.02

SALBUTAMOL   100ng..:..1ug...:..10.....:..100...:..1mg....:..10....:..100...:..1g.....:..10

4508	R f cdr eat meo ley 24m24	er	. + .		44.6mg *	P<.0005+
4509	R f cdr eat meo ley 80w80	ekr	. + .		36.3mg	P<.0005+

SARCOPHYTOL A   100ng..:..1ug...:..10.....:..100...:..1mg....:..10....:..100...:..1g.....:..10

4510	M m cen eat liv hpt 65w65	er	.>		no dre	P=1. -

SELENIUM   100ng..:..1ug...:..10.....:..100...:..1mg....:..10....:..100...:..1g.....:..10

4511	M f c3s eat mgl adc 81w81	r	.>		no dre	P=1. -
4512	M f c3s eat mgl adc 53w81	r	.>		no dre	P=1. -

SELENIUM DIETHYLDITHIOCARBAMATE 100ng..:..1ug...:..10.....:..100...:..1mg....:..10....:..100...:..1g.....:..10

4513	M f b6a orl lun ade 81w81	evx	.>		no dre	P=1.
a	M f b6a orl liv hpt 81w81	evx			no dre	P=1.
b	M f b6a orl tba mix 81w81	evx			10.5mg	P<.4
4514	M m b6a orl liv hpt 81w81	evx	.>		10.2mg	P<.3
a	M m b6a orl lun ade 81w81	evx			18.4mg	P<.6
b	M m b6a orl tba mix 81w81	evx			8.45mg	P<.4
4515	M f b6c orl liv hpt 81w81	evx	. ±		7.76mg	P<.04
a	M f b6c orl lun mix 81w81	evx			no dre	P=1.
b	M f b6c orl tba mix 81w81	evx			3.46mg	P<.003
4516	M m b6c orl liv mix 81w81	evx	. + .		1.28mg	P<.0005
a	M m b6c orl liv hpt 81w81	evx			1.49mg	P<.0005+
b	M m b6c orl lun mix 81w81	evx			no dre	P=1.
c	M m b6c orl tba mix 81w81	evx			.638mg	P<.0005

SELENIUM DIMETHYLDITHIOCARBAMATE 100ng..:..1ug...:..10.....:..100...:..1mg....:..10....:..100...:..1g.....:..10

4517	M f b6a orl liv hpt 76w76	evx	.>		no dre	P=1. -
a	M f b6a orl lun ade 76w76	evx			no dre	P=1. -
b	M f b6a orl tba mix 76w76	evx			no dre	P=1. -
4518	M m b6a orl liv hpt 76w76	evx	.>		no dre	P=1. -
a	M m b6a orl lun ade 76w76	evx			no dre	P=1. -
b	M m b6a orl tba mix 76w76	evx			no dre	P=1. -
4519	M f b6c orl liv hpt 76w76	evx	.>		no dre	P=1. -
a	M f b6c orl lun mix 76w76	evx			no dre	P=1. -
b	M f b6c orl tba tum 76w76	evx			no dre	P=1. -
4520	M m b6c orl liv hpt 76w76	evx	.>		26.0mg	P<.3 -

	RefNum	LoConf	UpConf	Cntrl	1Dose	1Inc	2Dose	2Inc						Citation or Pathology
														Brkly Code
4489	267	28.6mg	202.mg	1/17	146.mg	11/17								Innes;ntis,1968/1969
a	267	310.mg	n.s.s.	1/17	146.mg	0/17								
b	267	35.3mg	n.s.s.	7/17	146.mg	11/17								
4490	1039	50.7mg	n.s.s.	3/100	34.3mg	4/33								Vesselinovitch;canr,39,4378-4380;1979
a	1039	69.2mg	n.s.s.	0/100	34.3mg	2/33								
b	1039	185.mg	n.s.s.	3/100	34.3mg	0/35								
4491	1474m	27.1mg	234.mg	0/10	480.mg	7/10								Lipsky;jnci,67,365-371;1981
a	1474m	90.1mg	n.s.s.	0/10	480.mg	2/10								
b	1474m	247.mg	n.s.s.	0/10	480.mg	0/10								
4492	1474n	n.s.s.	164.mg	0/5	333.mg	5/5								
a	1474n	36.1mg	n.s.s.	0/5	333.mg	3/5								
b	1474n	178.mg	n.s.s.	0/5	333.mg	0/5								
4493	1035a	74.9mg	225.mg	0/53	439.mg	25/36								Wislocki;canr,37,1883-1891;1977
a	1035a	2.23gm	n.s.s.	0/55	439.mg	0/50								
4494	1581	39.3mg	95.8mg	0/32	125.mg	13/28	241.mg	24/34						Boberg;canr,43,5163-5173;1983
a	1581	571.mg	n.s.s.	0/32	125.mg	1/28	241.mg	0/34						
4495	1582m	32.3mg	108.mg	0/30	176.mg	21/30								Miller;canr,43,1124-1134;1983
a	1582m	370.mg	n.s.s.	1/30	176.mg	1/30								
4496	1582n	30.2mg	58.5mg	0/50	80.7mg	34/50	161.mg	39/50						
a	1582n	130.mg	389.mg	0/50	80.7mg	7/50	161.mg	16/50						
b	1582n	1.09gm	n.s.s.	2/50	80.7mg	2/50	161.mg	0/50						
4497	1035a	123.mg	612.mg	0/44	405.mg	11/26								Wislocki;canr,37,1883-1891;1977
a	1035a	1.07gm	n.s.s.	0/44	405.mg	0/26								
4498	1042a	335.mg	n.s.s.	4/50	390.mg	11/40	488.mg	8/40						Borchert;canr,33,590-600;1973
a	1042a	786.mg	n.s.s.	1/50	390.mg	0/40	488.mg	0/40						
4499	1042b	417.mg	n.s.s.	3/35	390.mg	4/35								
a	1042b	417.mg	n.s.s.	3/35	390.mg	4/35								
4500	1035a	210.mg	n.s.s.	0/18	68.0mg	0/18								Wislocki;canr,37,1883-1891;1977
4501	1035b	189.mg	n.s.s.	0/15	200.mg	3/18								
4502	1042a	136.mg	n.s.s.	0/12	125.mg	0/12								Borchert;canr,33,590-600;1973
a	1042a	136.mg	n.s.s.	0/12	125.mg	0/12								
4503	1042b	142.mg	n.s.s.	0/18	82.5mg	0/18	138.mg	2/18						
a	1042b	157.mg	n.s.s.	1/18	82.5mg	0/18	138.mg	2/18						
4504	1042c	69.1mg	n.s.s.	0/18	142.mg	1/18								
a	1042c	69.1mg	n.s.s.	0/18	142.mg	1/18								
4505	268	113.mg	1.90gm	1/50	4.50mg	1/50	22.5mg	2/50	45.0mg	8/50	(225.mg	6/50)		Long;arpa,75,595-604;1963
a	268	199.mg	736.mg	3/50	4.50mg	1/50	22.5mg	3/50	45.0mg	8/50	225.mg	19/50		
b	268	331.mg	1.53gm	2/50	4.50mg	0/50	22.5mg	2/50	45.0mg	0/50	225.mg	14/50		
c	268	433.mg	3.23gm	2/50	4.50mg	0/50	22.5mg	2/50	45.0mg	0/50	225.mg	10/50		
d	268	743.mg	8.16gm	0/50	4.50mg	0/50	22.5mg	0/50	45.0mg	0/50	225.mg	5/50		
4506	268	133.mg	457.mg	0/25	5.00mg	0/25	25.0mg	1/25	50.0mg	6/25	250.mg	12/25		
4507	268	221.mg	n.s.s.	3/25	4.00mg	1/25	20.0mg	2/25	40.0mg	2/25	200.mg	7/25		

SALBUTAMOL   18559-94-9

	RefNum	LoConf	UpConf	Cntrl	1Dose	1Inc	2Dose	2Inc	Citation or Pathology
4508	1734	24.8mg	92.0mg	0/105	2.00mg	0/55	20.0mg	16/55	Jack;txcy,27,315-320;1983
4509	1734m	17.7mg	93.4mg	0/105	20.0mg	10/50			Gopinath;enhp,73,107-113;1987/Jack 1983/pers.comm.

SARCOPHYTOL A   72629-69-7

	RefNum	LoConf	UpConf	Cntrl	1Dose	1Inc	Citation or Pathology
4510	2021	4.15mg	n.s.s.	34/40	12.0mg	30/40	Yamauchi;gann,82,1234-1238;1991

SELENIUM   7782-49-2

	RefNum	LoConf	UpConf	Cntrl	1Dose	1Inc	Citation or Pathology
4511	1431	.218mg	n.s.s.	20/30	.111mg	7/30	Schrauzer;carc,1,199-201;1980
4512	1431m	40.4ug	n.s.s.	20/30	72.1ug	18/30	

SELENIUM DIETHYLDITHIOCARBAMATE   (ethyl selenac) 5456-28-0

	RefNum	LoConf	UpConf	Cntrl	1Dose	1Inc	Citation or Pathology
4513	1211	4.59mg	n.s.s.	1/17	3.63mg	1/17	Innes;ntis,1968/1969
a	1211	7.70mg	n.s.s.	0/17	3.63mg	0/17	
b	1211	2.21mg	n.s.s.	2/17	3.63mg	4/17	
4514	1211	2.34mg	n.s.s.	1/18	3.37mg	3/17	
a	1211	2.54mg	n.s.s.	2/18	3.37mg	3/17	
b	1211	1.80mg	n.s.s.	3/18	3.37mg	5/17	
4515	1211	2.34mg	n.s.s.	0/16	3.63mg	3/17	
a	1211	7.70mg	n.s.s.	0/16	3.63mg	0/17	
b	1211	1.40mg	18.8mg	0/16	3.63mg	6/17	
4516	1211	.627mg	3.07mg	0/16	3.37mg	12/18	
a	1211	.718mg	3.71mg	0/16	3.37mg	11/18	
b	1211	7.59mg	n.s.s.	0/16	3.37mg	0/18	
c	1211	.296mg	1.45mg	0/16	3.37mg	16/18	

SELENIUM DIMETHYLDITHIOCARBAMATE   (methyl selenac) 144-34-3

	RefNum	LoConf	UpConf	Cntrl	1Dose	1Inc	Citation or Pathology
4517	1205	8.68mg	n.s.s.	0/17	4.64mg	0/17	Innes;ntis,1968/1969
a	1205	8.68mg	n.s.s.	1/17	4.64mg	0/17	
b	1205	4.08mg	n.s.s.	2/17	4.64mg	2/17	
4518	1205	8.54mg	n.s.s.	1/18	4.31mg	0/18	
a	1205	8.54mg	n.s.s.	2/18	4.31mg	0/18	
b	1205	6.05mg	n.s.s.	3/18	4.31mg	1/18	
4519	1205	8.68mg	n.s.s.	0/16	4.64mg	0/17	
a	1205	8.68mg	n.s.s.	0/16	4.64mg	0/17	
b	1205	8.68mg	n.s.s.	0/16	4.64mg	0/17	
4520	1205	4.24mg	n.s.s.	0/16	4.31mg	1/17	

```
          Spe Strain  Site    Xpo+Xpt                                                              TD50      2Tailpvl
              Sex  Route  Hist    Notes                                                               DR      AuOp
a    M m b6c orl lun mix 76w76 evx                                                                no dre    P=1.   -
b    M m b6c orl tba mix 76w76 evx                                                                5.88mg    P<.02  -


SELENIUM SULFIDE               100ng..:..1ug...:..10....:..100....:..1mg...:..10....:..100...:..1g....:..10
4521 M f b6c gav MXB MXB 24m24                                      : + :                         46.8mg  * P<.0005
a    M f b6c gav liv MXA 24m24                                                                    69.3mg  * P<.0005c
b    M f b6c gav liv hpc 24m24                                                                    85.2mg  * P<.0005c
c    M f b6c gav lun MXA 24m24                                                                    137.mg  * P<.0005c
d    M f b6c gav lun a/c 24m24                                                                    433.mg  * P<.02   c
e    M f b6c gav TBA MXB 24m24                                                                    69.7mg  * P<.02
f    M f b6c gav liv MXB 24m24                                                                    69.3mg  * P<.0005
g    M f b6c gav lun MXB 24m24                                                                    137.mg  * P<.0005
4522 M m b6c gav TBA MXB 24m24                                          :>                        735.mg  * P<.9    -
a    M m b6c gav liv MXB 24m24                                                                    202.mg  * P<.3
b    M m b6c gav lun MXB 24m24                                                                    183.mg  * P<.07
4523 R f f34 gav liv MXA 24m24                             :+  :                                  6.14mg  / P<.0005c
a    R f f34 gav liv hpc 24m24                                                                    13.2mg  / P<.0005c
b    R f f34 gav TBA MXB 24m24                                                                    19.3mg  * P<.3
c    R f f34 gav liv MXB 24m24                                                                    6.14mg  / P<.0005
4524 R m f34 gav liv MXA 24m24                                 : + :                              11.5mg  / P<.0005c
a    R m f34 gav liv hpc 24m24                                                                    21.0mg  / P<.0005c
b    R m f34 gav TBA MXB 24m24                                                                    25.0mg  * P<.4
c    R m f34 gav liv MXB 24m24                                                                    11.5mg  / P<.0005


SENKIRKINE                     100ng..:..1ug...:..10....:..100....:..1mg...:..10....:..100...:..1g....:..10
4525 R m aci ipj liv lca 56w92 ev                        .  +   .                                 1.70mg    P<.0005+


SESAMOL                        100ng..:..1ug...:..10....:..100....:..1mg...:..10....:..100...:..1g....:..10
4526 M f b6c eat for sqc 96w96                                                    .              +8.33gm    P<.008 +
a    M f b6c eat for sar 96w96                                                                    44.8gm    P<.3
b    M f b6c eat liv hnd 96w96                                                                    44.8gm    P<.3
c    M f b6c eat liv hpc 96w96                                                                    no dre    P=1.
d    M f b6c eat lun adc 96w96                                                                    no dre    P=1.
4527 M m b6c eat for sqc 96w96                                                     . +  .         3.07gm    P<.0005+
a    M m b6c eat stg sar 96w96                                                                    41.3gm    P<.3
b    M m b6c eat lun ade 96w96                                                                    41.3gm    P<.3
c    M m b6c eat liv hnd 96w96                                                                    no dre    P=1.
d    M m b6c eat liv hpc 96w96                                                                    no dre    P=1.
4528 R f f3d eat for pam 24m24                                                  . +  .            1.09gm    P<.0005
a    R f f3d eat for sqc 24m24                                                                    6.50gm    P<.04  +
b    R f f3d eat liv hnd 24m24                                                                    no dre    P=1.
4529 R m f3d eat for pam 24m24                                                  . +  .            1.35gm    P<.0005
a    R m f3d eat for sqc 24m24                                                                    1.54gm    P<.0005+
b    R m f3d eat stg ade 24m24                                                                    16.2gm    P<.3
c    R m f3d eat liv hpc 24m24                                                                    no dre    P=1.
d    R m f3d eat liv hnd 24m24                                                                    no dre    P=1.
4530 R m f3d eat for mix 60w60 er                                             . +                 708.mg    P<.01  +
a    R m f3d eat for sqp 60w60 er                                                                 909.mg    P<.03  +
b    R m f3d eat for sqc 60w60 er                                                                 3.92gm    P<.3


SIMAZINE                       100ng..:..1ug...:..10....:..100....:..1mg...:..10....:..100...:..1g.....:..10
4531 M f b6a orl liv hpt 76w76 evx                                         .>                     no dre    P=1.   -
a    M f b6a orl lun ade 76w76 evx                                                                no dre    P=1.   -
b    M f b6a orl tba mix 76w76 evx                                                                no dre    P=1.   -
4532 M m b6a orl lun ade 76w76 evx                                     .>                          442.mg    P<.7   -
a    M m b6a orl liv hpt 76w76 evx                                                                no dre    P=1.   -
b    M m b6a orl tba mix 76w76 evx                                                                199.mg    P<.5   -
4533 M f b6c orl lun ade 76w76 evx                                          .>                    536.mg    P<.3   -
a    M f b6c orl liv hpt 76w76 evx                                                                no dre    P=1.   -
b    M f b6c orl tba mix 76w76 evx                                                                260.mg    P<.2   -
4534 M m b6c orl liv hpt 76w76 evx                                     .   ±                      106.mg    P<.02  -
a    M m b6c orl lun mix 76w76 evx                                                                no dre    P=1.   -
b    M m b6c orl tba mix 76w76 evx                                                                65.5mg    P<.003 -


SODIUM BICARBONATE             100ng..:..1ug...:..10....:..100....:..1mg...:..10....:..100...:..1g.....:..10
4535 R m f3d eat ubl car 24m24 r                                              .>                  5.35gm    P<.3   -


SODIUM BITHIONOLATE            100ng..:..1ug...:..10....:..100....:..1mg...:..10....:..100...:..1g.....:..10
4536 M f b6a orl liv hpt 76w76 evx                                    .>                          no dre    P=1.   -
a    M f b6a orl lun ade 76w76 evx                                                                no dre    P=1.   -
b    M f b6a orl tba mix 76w76 evx                                                                5.08mg    P<.7   -
4537 M m b6a orl lun ade 76w76 evx                               .>                               no dre    P=1.   -
a    M m b6a orl liv hpt 76w76 evx                                                                no dre    P=1.   -
b    M m b6a orl tba mix 76w76 evx                                                                no dre    P=1.   -
4538 M f b6c orl liv hpt 76w76 evx                                    .>                          no dre    P=1.   -
a    M f b6c orl lun mix 76w76 evx                                                                no dre    P=1.   -
b    M f b6c orl tba tum 76w76 evx                                                                no dre    P=1.   -
4539 M m b6c orl liv mix 76w76 evx                                   .   ±                        2.28mg    P<.09  -
a    M m b6c orl lun mix 76w76 evx                                                                no dre    P=1.   -
b    M m b6c orl tba mix 76w76 evx                                                                .805mg    P<.005 -
```

	RefNum	LoConf	UpConf	Cntrl	1Dose	1Inc	2Dose	2Inc	Citation or Pathology
									Brkly Code
a	1205	8.07mg	n.s.s.	0/16	4.31mg	0/17			
b	1205	2.02mg	n.s.s.	0/16	4.31mg	4/17			

SELENIUM SULFIDE   7446-34-6

	RefNum	LoConf	UpConf	Cntrl	1Dose	1Inc	2Dose	2Inc	Citation or Pathology
4521	TR194	31.4mg	74.2mg	0/50	14.1mg	5/50	70.7mg	32/50	liv:hpa,hpc; lun:a/a,a/c. C
a	TR194	43.7mg	119.mg	0/50	14.1mg	2/50	70.7mg	25/50	liv:hpa,hpc.
b	TR194	51.8mg	154.mg	0/50	14.1mg	1/50	70.7mg	22/50	
c	TR194	74.9mg	338.mg	0/50	14.1mg	3/50	70.7mg	12/50	lun:a/a,a/c.
d	TR194	164.mg	n.s.s.	0/50	14.1mg	1/50	70.7mg	4/50	
e	TR194	31.5mg	n.s.s.	24/50	14.1mg	31/50	70.7mg	42/50	
f	TR194	43.7mg	119.mg	0/50	14.1mg	2/50	70.7mg	25/50	liv:hpa,hpc,nnd.
g	TR194	74.9mg	338.mg	0/50	14.1mg	3/50	70.7mg	12/50	lun:a/a,a/c.
4522	TR194	53.5mg	n.s.s.	29/50	14.1mg	35/50	70.7mg	36/50	
a	TR194	59.9mg	n.s.s.	15/50	14.1mg	14/50	70.7mg	23/50	liv:hpa,hpc,nnd.
b	TR194	68.1mg	n.s.s.	4/50	14.1mg	10/50	70.7mg	14/50	lun:a/a,a/c.
4523	TR194	4.08mg	10.1mg	1/50	2.10mg	0/50	10.6mg	37/50	liv:hpc,nnd.
a	TR194	7.90mg	24.7mg	0/50	2.10mg	0/50	10.6mg	21/50	
b	TR194	5.48mg	n.s.s.	38/50	2.10mg	37/50	10.6mg	47/50	
c	TR194	4.08mg	10.1mg	1/50	2.10mg	0/50	10.6mg	37/50	liv:hpa,hpc,nnd.
4524	TR194	6.96mg	22.1mg	1/50	2.10mg	0/50	10.6mg	24/50	liv:hpc,nnd.
a	TR194	11.3mg	45.8mg	0/50	2.10mg	0/50	10.6mg	14/50	
b	TR194	6.01mg	n.s.s.	33/50	2.10mg	40/50	10.6mg	43/50	
c	TR194	6.96mg	22.1mg	1/50	2.10mg	0/50	10.6mg	24/50	liv:hpa,hpc,nnd.

SENKIRKINE   (renardine) 2318-18-5

	RefNum	LoConf	UpConf	Cntrl	1Dose	1Inc	2Dose	2Inc	Citation or Pathology
4525	1396	.784mg	4.77mg	0/19	2.03mg	9/19			Hirono;jnci,63,469-472;1979

SESAMOL   (3,4-(methylenedioxy)-phenol) 533-31-3

	RefNum	LoConf	UpConf	Cntrl	1Dose	1Inc	2Dose	2Inc	Citation or Pathology
4526	2011	3.15gm	124.gm	0/30	2.60gm	5/30			Tamano;gann,83,1279-1285;1992
a	2011	7.29gm	n.s.s.	0/30	2.60gm	1/30			
b	2011	7.29gm	n.s.s.	0/30	2.60gm	1/30			
c	2011	13.7gm	n.s.s.	1/30	2.60gm	0/30			
d	2011	8.27gm	n.s.s.	1/30	2.60gm	1/30			
4527	2011	1.53gm	7.58gm	0/30	2.40gm	11/30			
a	2011	6.73gm	n.s.s.	0/30	2.40gm	1/30			
b	2011	6.73gm	n.s.s.	0/30	2.40gm	1/30			
c	2011	4.87gm	n.s.s.	7/30	2.40gm	5/30			
d	2011	7.75gm	n.s.s.	5/30	2.40gm	2/30			
4528	2011	579.mg	2.40gm	0/30	1.00gm	14/30			
a	2011	1.97gm	n.s.s.	0/30	1.00gm	3/30			
b	2011	6.18gm	n.s.s.	2/30	1.00gm	0/30			
4529	2011	655.mg	3.56gm	0/30	800.mg	10/30			
a	2011	720.mg	4.43gm	0/30	800.mg	9/30			
b	2011	2.63gm	n.s.s.	0/30	800.mg	1/30			
c	2011	4.95gm	n.s.s.	1/30	800.mg	0/30			
d	2011	4.95gm	n.s.s.	2/30	800.mg	0/30			
4530	2134	268.mg	40.1gm	0/19	800.mg	5/22			Ito;anti,183-194;1990/pers.comm.
a	2134	313.mg	n.s.s.	0/19	800.mg	4/22			
b	2134	638.mg	n.s.s.	0/19	800.mg	1/22			

SIMAZINE   (CDT) 122-34-9

	RefNum	LoConf	UpConf	Cntrl	1Dose	1Inc	2Dose	2Inc	Citation or Pathology
4531	1243	166.mg	n.s.s.	0/17	83.8mg	0/18			Innes;ntis,1968/1969
a	1243	166.mg	n.s.s.	1/17	83.8mg	0/18			
b	1243	166.mg	n.s.s.	2/17	83.8mg	0/18			
4532	1243	55.5mg	n.s.s.	2/18	78.0mg	3/18			
a	1243	154.mg	n.s.s.	1/18	78.0mg	0/18			
b	1243	39.7mg	n.s.s.	3/18	78.0mg	5/18			
4533	1243	87.3mg	n.s.s.	0/16	83.8mg	1/18			
a	1243	166.mg	n.s.s.	0/16	83.8mg	0/18			
b	1243	63.9mg	n.s.s.	0/16	83.8mg	2/18			
4534	1243	36.5mg	n.s.s.	0/16	78.0mg	4/17			
a	1243	146.mg	n.s.s.	0/16	78.0mg	0/17			
b	1243	26.4mg	355.mg	0/16	78.0mg	6/17			

SODIUM BICARBONATE   144-55-8

	RefNum	LoConf	UpConf	Cntrl	1Dose	1Inc	2Dose	2Inc	Citation or Pathology
4535	1930	871.mg	n.s.s.	0/30	256.mg	1/31			Fukushima;carc,10,1635-1640;1989

SODIUM BITHIONOLATE   (vancide BN) 6385-58-6

	RefNum	LoConf	UpConf	Cntrl	1Dose	1Inc	2Dose	2Inc	Citation or Pathology
4536	1284	1.79mg	n.s.s.	0/17	.959mg	0/17			Innes;ntis,1968/1969
a	1284	1.79mg	n.s.s.	1/17	.959mg	0/17			
b	1284	.641mg	n.s.s.	2/17	.959mg	3/17			
4537	1284	.833mg	n.s.s.	2/18	.891mg	2/18			
a	1284	1.77mg	n.s.s.	1/18	.891mg	0/18			
b	1284	.922mg	n.s.s.	3/18	.891mg	2/18			
4538	1284	1.90mg	n.s.s.	0/16	.959mg	0/18			
a	1284	1.90mg	n.s.s.	0/16	.959mg	0/18			
b	1284	1.90mg	n.s.s.	0/16	.959mg	0/18			
4539	1284	.559mg	n.s.s.	0/16	.891mg	2/15			
a	1284	1.47mg	n.s.s.	0/16	.891mg	0/15			
b	1284	.302mg	6.34mg	0/16	.891mg	5/15			

```
        Spe Strain  Site   Xpo+Xpt                                                          TD50      2Tailpvl
        Sex  Route  Hist      Notes                                                                 DR    AuOp
SODIUM CHLORIDE         100ng..:..1ug...:..:.10.....:..:.100....:..:.1mg....:..:.10.....:..:.100....:..:.1g.....:..:.10
4540 M f b6c wat lun ptm 24m24 e                                                      .>      4.98gm    P<.2
a    M f b6c wat liv hnd 24m24 e                                                              1.33kg    P<1.
b    M f b6c wat tba mix 24m24 e                                                              no dre    P=1.
4541 M m b6c wat lun ptm 24m24 e                                                   .>         3.74gm    P<.3
a    M m b6c wat liv hnd 24m24 e                                                              5.97gm    P<.4
b    M m b6c wat liv hem 24m24 e                                                              9.55gm    P<.2
c    M m b6c wat liv hpt 24m24 e                                                              no dre    P=1.
d    M m b6c wat tba mix 24m24 e                                                              992.mg    P<.2
4542 R m aci eat for pam 52w52 er                                      .     +                2.66gm    P<.007
a    R m aci eat stg adc 52w52 er                                                             no dre    P=1.
4543 R m cdr eat stg adc 52w52 er                                            .>               no dre    P=1.
4544 R m f34 eat amd phe 24m24                                       .    ±                    2.98gm    P<.02  -

SODIUM CHLORITE         100ng..:..1ug...:..:.10.....:..:.100....:..:.1mg....:..:.10.....:..:.100....:..:.1g.....:..:.10
4545 M f b6c wat liv mix 85w85 e                                                    .>        3.06gm  * P<.9
a    M f b6c wat liv hnd 85w85 e                                                              no dre    P=1.
b    M f b6c wat lun mix 85w85 e                                                              no dre    P=1.
c    M f b6c wat tba mix 85w85 e                                                              no dre    P=1.
4546 M m b6c wat lun mix 85w85 e                                         .  +       .         237.mg  * P<.004
a    M m b6c wat liv mix 85w85 e                                                              110.mg  * P<.07
b    M m b6c wat liv hnd 85w85 e                                                              146.mg  * P<.06
c    M m b6c wat lun ade 85w85 e                                                              346.mg  * P<.02
d    M m b6c wat tba mix 85w85 e                                                              169.mg  * P<.4
4547 R f f34 wat tba mix 85w85 es                                           .>                no dre    P=1.  -
4548 R m f34 wat tba mix 85w85 es                                            .>               no dre    P=1.  -

SODIUM DIETHYLDITHIOCARBAMATE  TRIHYDRATE.:..1ug...:..10.....:..:.100....:..:.1mg....:..:.10.....:..:.100....:..:.1g.....:..:.10
4549 M f b6c eat TBA MXB 25m25 a                                                      :>      no dre    P=1.  -
a    M f b6c eat liv MXB 25m25 a                                                              27.0gm  * P<.8
b    M f b6c eat lun MXB 25m25 a                                                              4.76gm  * P<.5
4550 M f b6c orl liv hpt 76w76 evx                                          .>                no dre    P=1.
a    M f b6c orl lun ade 76w76 evx                                                            no dre    P=1.
b    M f b6c orl tba mix 76w76 evx                                                            no dre    P=1.
4551 M m b6c eat TBA MXB 25m25 a                                                   :>         no dre    P=1.
a    M m b6c eat liv MXB 25m25 a                                                              no dre    P=1.
b    M m b6c eat lun MXB 25m25 a                                                              10.3gm  * P<.9
4552 M m b6c orl liv hpt 76w76 evx                                   .   ±                    68.7mg    P<.02
a    M m b6c orl lun ade 76w76 evx                                                            468.mg    P<.7
b    M m b6c orl tba mix 76w76 evx                                                            71.4mg    P<.2
4553 M f b6a orl lun ade 76w76 evx                                          .>                607.mg    P<.3
a    M f b6a orl liv hpt 76w76 evx                                                            no dre    P=1.
b    M f b6a orl tba mix 76w76 evx                                                            no dre    P=1.
4554 M m b6a orl lun ade 76w76 evx                                         .>                 155.mg    P<.3
a    M m b6a orl liv hpt 76w76 evx                                                            no dre    P=1.
b    M m b6a orl tba mix 76w76 evx                                                            403.mg    P<.8
4555 R f f34 eat TBA MXB 24m24                                             :>                 no dre    P=1.  -
a    R f f34 eat liv MXB 24m24                                                                no dre    P=1.
4556 R m f34 eat TBA MXB 24m24                                            :>                  no dre    P=1.
a    R m f34 eat liv MXB 24m24                                                                no dre    P=1.

SODIUM HYPOCHLORITE     100ng..:..1ug...:..:.10.....:..:.100....:..:.1mg....:..:.10.....:..:.100....:..:.1g.....:..:.10
4557 M f b6c wat lun mix 24m25 e                                                    .>        22.9gm  * P<1.  -
a    M f b6c wat liv mix 24m25 e                                                              no dre    P=1.
b    M f b6c wat liv hpc 24m25 e                                                              no dre    P=1.
c    M f b6c wat tba mix 24m25 e                                                              no dre    P=1.
4558 M m b6c wat liv mix 24m25 e                                                   .>         876.mg  * P<.5
a    M m b6c wat lun mix 24m25 e                                                              2.04gm  * P<.5
b    M m b6c wat liv hpc 24m25 e                                                              no dre    P=1.
c    M m b6c wat tba mix 24m25 e                                                              350.mg  * P<.4
4559 R f f3d wat liv mix 24m26 e                                                    .>        6.30gm  * P<1.
4560 R m f3d wat liv cho 24m26 e                                            .>                no dre    P=1.
a    R m f3d wat liv fih 24m26 e                                                              no dre    P=1.  -
b    R m f3d wat liv mix 24m26 e                                                              no dre    P=1.  -

SORBIC ACID             100ng..:..1ug...:..:.10.....:..:.100....:..:.1mg...:..:.10.....:..:.100....:..:.1g....:..:.10
4561 M f asp eat liv hnd 80w80 e                                                              362.gm  * P<.2  -
a    M f asp eat lun ade 80w80 e                                                              no dre    P=1.
4562 M m asp eat liv hnd 80w80 e                                                              no dre    P=1.
a    M m asp eat lun ade 80w80 e                                                        .> no dre    P=1.
4563 R f wis eat liv hem 24m24 e                                                              34.0gm  * P<.2
a    R f wis eat tba mal 24m24 e                                                              no dre    P=1.
b    R f wis eat tba mix 24m24 e                                                              no dre    P=1.
4564 R m wis eat liv hem 24m24 e                                                              no dre    P=1.
a    R m wis eat tba mix 24m24 e                                                              33.9gm  * P<.7  -
b    R m wis eat tba mal 24m24 e                                                              51.4gm  * P<.5  -

SOTALOL.HC1             100ng..:..1ug...:..:.10.....:..:.100....:..:.1mg...:..:.10.....:..:.100....:..:.1g....:..:.10
4565 M f cd1 eat lun ade 78w92 e                                                    .>        3.50gm  * P<.2  -
a    M f cd1 eat lun car 78w92 e                                                              24.5gm  * P=1.  -
b    M f cd1 eat liv tum 78w92 e                                                              no dre    P=1.  -
```

	RefNum	LoConf	UpConf	Cntrl	1Dose	1Inc	2Dose	2Inc	Citation or Pathology
									Brkly Code

SODIUM CHLORIDE   7647-14-5

	RefNum	LoConf	UpConf	Cntrl	1Dose	1Inc	2Dose	2Inc	Citation or Pathology
4540	1806	1.50gm	n.s.s.	13/96	880.mg	11/47			Van Duuren;enhp,69,109-117;1986
a	1806	5.27gm	n.s.s.	2/96	880.mg	1/47			
b	1806	547.mg	n.s.s.	84/96	880.mg	40/47			
4541	1806	1.08gm	n.s.s.	22/99	733.mg	16/50			
a	1806	1.44gm	n.s.s.	15/99	733.mg	11/50			
b	1806	2.34gm	n.s.s.	3/99	733.mg	4/50			
c	1806	2.51gm	n.s.s.	12/99	733.mg	6/50			
d	1806	297.mg	n.s.s.	76/99	733.mg	43/50			
4542	2076	1.00gm	32.6gm	0/22	4.00gm	5/22			Watanabe;gann,83,588-593;1992/pers.comm.
a	2076	4.53gm	n.s.s.	0/22	4.00gm	0/22			
4543	2074	352.mg	n.s.s.	0/12	400.mg	0/19	4.00gm	0/17	Watanabe;gann,83,1267-1272;1992/pers.comm.
4544	1807	1.36gm	n.s.s.	11/50	1.60gm	23/50			Imai;jnma,37,115-127;1986

SODIUM CHLORITE   7758-19-2

	RefNum	LoConf	UpConf	Cntrl	1Dose	1Inc	2Dose	2Inc	Citation or Pathology
4545	1789	225.mg	n.s.s.	5/47	50.0mg	5/50	100.mg	6/50	Kurokawa;enhp,69,221-235;1986
a	1789	292.mg	n.s.s.	5/47	50.0mg	3/50	100.mg	5/50	
b	1789	442.mg	n.s.s.	3/47	50.0mg	2/50	100.mg	2/50	
c	1789	151.mg	n.s.s.	14/47	50.0mg	17/50	100.mg	13/50	
4546	1789	115.mg	1.26gm	0/35	41.7mg	3/47	83.3mg	7/43	
a	1789	46.8mg	n.s.s.	7/35	41.7mg	22/47	83.3mg	17/43	
b	1789	65.1mg	n.s.s.	3/35	41.7mg	14/47	83.3mg	11/43	
c	1789	149.mg	n.s.s.	0/35	41.7mg	2/47	83.3mg	5/43	
d	1789	45.5mg	n.s.s.	14/35	41.7mg	27/47	83.3mg	22/43	
4547	1789	69.9mg	n.s.s.	21/47	17.1mg	14/44	34.3mg	15/50	
4548	1789	155.mg	n.s.s.	28/34	15.0mg	3/30	30.0mg	6/43	

SODIUM DIETHYLDITHIOCARBAMATE TRIHYDRATE (SDDC) 148-18-5

	RefNum	LoConf	UpConf	Cntrl	1Dose	1Inc	2Dose	2Inc	Citation or Pathology
4549	TR172	774.mg	n.s.s.	13/20	65.0mg	20/50	520.mg	26/50	
a	TR172	1.98gm	n.s.s.	0/20	65.0mg	2/50	520.mg	2/50	liv:hpa,hpc,nnd.
b	TR172	941.mg	n.s.s.	0/20	65.0mg	7/50	520.mg	8/50	lun:a/a,a/c.
4550	273	188.mg	n.s.s.	0/18	94.9mg	0/18			Innes;ntis,1968/1969
a	273	188.mg	n.s.s.	1/18	94.9mg	0/18			
b	273	133.mg	n.s.s.	3/18	94.9mg	1/18			
4551	TR172	530.mg	n.s.s.	13/20	60.0mg	37/50	480.mg	30/50	
a	TR172	1.01gm	n.s.s.	7/20	60.0mg	11/50	480.mg	11/50	liv:hpa,hpc,nnd.
b	TR172	742.mg	n.s.s.	6/20	60.0mg	14/50	480.mg	14/50	lun:a/a,a/c.
4552	273	27.0mg	n.s.s.	1/17	88.2mg	7/17			Innes;ntis,1968/1969
a	273	58.9mg	n.s.s.	2/17	88.2mg	3/17			
b	273	21.3mg	n.s.s.	6/17	88.2mg	10/17			
4553	273	98.9mg	n.s.s.	0/17	94.9mg	1/18			
a	273	188.mg	n.s.s.	0/17	94.9mg	0/18			
b	273	89.6mg	n.s.s.	2/17	94.9mg	2/18			
4554	273	42.0mg	n.s.s.	2/18	88.2mg	5/18			
a	273	175.mg	n.s.s.	3/18	88.2mg	0/18			
b	273	43.4mg	n.s.s.	5/18	88.2mg	6/18			
4555	TR172	146.mg	n.s.s.	17/20	62.5mg	26/50	125.mg	34/50	
a	TR172	1.79gm	n.s.s.	1/20	62.5mg	0/50	125.mg	0/50	liv:hpa,hpc,nnd.
4556	TR172	88.9mg	n.s.s.	14/16	50.0mg	38/50	100.mg	34/50	
a	TR172	684.mg	n.s.s.	0/16	50.0mg	1/50	100.mg	0/50	liv:hpa,hpc,nnd.

SODIUM HYPOCHLORITE   (clorox; hypochlorous acid, sodium salt) 7681-52-9

	RefNum	LoConf	UpConf	Cntrl	1Dose	1Inc	2Dose	2Inc	Citation or Pathology
4557	1789	734.mg	n.s.s.	7/72	97.2mg	6/50	194.mg	5/50	Kurokawa;enhp,69,221-235;1986
a	1789	965.mg	n.s.s.	9/72	97.2mg	9/50	194.mg	3/50	
b	1789	1.96gm	n.s.s.	4/72	97.2mg	1/50	194.mg	1/50	
c	1789	285.mg	n.s.s.	42/72	97.2mg	34/50	194.mg	26/50	
4558	1789	187.mg	n.s.s.	31/73	81.0mg	27/50	162.mg	24/50	
a	1789	402.mg	n.s.s.	10/73	81.0mg	9/50	162.mg	9/50	
b	1789	377.mg	n.s.s.	24/73	81.0mg	17/50	162.mg	15/50	
c	1789	92.1mg	n.s.s.	53/73	81.0mg	39/50	162.mg	40/50	
4559	1780	1.03gm	n.s.s.	0/50	53.1mg	1/50	106.mg	0/50	Hasegawa;fctx,24,1295-1302;1986
4560	1780	450.mg	n.s.s.	0/49	23.2mg	1/50	46.4mg	0/50	
a	1780	450.mg	n.s.s.	0/49	23.2mg	1/50	46.4mg	0/50	
b	1780	589.mg	n.s.s.	1/49	23.2mg	1/50	46.4mg	0/50	

SORBIC ACID   110-44-1

	RefNum	LoConf	UpConf	Cntrl	1Dose	1Inc	2Dose	2Inc	3Dose	3Inc	Citation or Pathology
4561	409	58.9gm	n.s.s.	0/48	1.30gm	0/46	6.50gm	0/43	13.0gm	1/43	Hendy;fctx,14,181-186;1976
a	409	32.3gm	n.s.s.	5/46	1.30gm	10/43	6.50gm	5/43			
4562	409	5.10gm	n.s.s.	1/44	1.20gm	0/46	6.00gm	0/43	12.0gm	0/43	
a	409	14.6gm	n.s.s.	19/44	1.20gm	16/46	6.00gm	16/43	12.0gm	15/43	
4563	412	10.9gm	n.s.s.	3/42	750.mg	2/45	5.00gm	7/46			Gaunt;fctx,13,31-45;1975
a	412	44.7gm	n.s.s.	3/42	750.mg	3/45	5.00gm	0/46			
b	412	1.02gm	n.s.s.	35/42	750.mg	30/45	(5.00gm	17/46)			
4564	412	25.8gm	n.s.s.	0/43	600.mg	1/43	4.00gm	0/41			
a	412	4.11gm	n.s.s.	20/43	600.mg	21/43	4.00gm	21/41			
b	412	9.48gm	n.s.s.	5/43	600.mg	3/43	4.00gm	6/41			

SOTALOL.HCl   959-24-0

	RefNum	LoConf	UpConf	Cntrl	1Dose	1Inc	2Dose	2Inc	Citation or Pathology
4565	1635	1.16gm	n.s.s.	3/60	254.mg	6/60	509.mg	7/60	Weikel;jcph,19,591-604;1979
a	1635	3.99gm	n.s.s.	0/60	254.mg	1/60	509.mg	0/60	
b	1635	1.64gm	n.s.s.	0/60	254.mg	0/60	509.mg	0/60	

```
     Spe Strain Site  Xpo+Xpt                                              TD50    2Tailpvl
         Sex  Route Hist  Notes                                               DR      AuOp

  c   M f cd1 eat tba ben 78w92 e                                          1.74gm * P<.05   -
  d   M f cd1 eat tba mix 78w92 e                                          1.91gm * P<.3    -
  e   M f cd1 eat tba mal 78w92 e                                          no dre   P=1.    -
4566  M m cd1 eat lun ade 78w92 e                              .>          31.4gm * P<1.    -
  a   M m cd1 eat liv car 78w92 e                                          no dre   P=1.    -
  b   M m cd1 eat liv ade 78w92 e                                          no dre   P=1.    -
  c   M m cd1 eat lun car 78w92 e                                          no dre   P=1.    -
  d   M m cd1 eat tba ben 78w92 e                                          no dre   P=1.    -
  e   M m cd1 eat tba mal 78w92 e                                          no dre   P=1.    -
  f   M m cd1 eat tba mix 78w92 e                                          no dre   P=1.    -
4567  R f lev eat liv cho 18m24 e                                .>        12.6gm * P<.3    -
  a   R f lev eat tba mix 18m24 e                                          477.mg * P<.4    -
  b   R f lev eat tba mal 18m24 e                                          1.60gm * P<.2    -
  c   R f lev eat tba ben 18m24 e                                          no dre   P=1.    -
4568  R m lev eat liv hpc 18m24 e                                .>        12.4gm * P<.3    -
  a   R m lev eat liv hpa 18m24 e                                          309.gm * P<1.    -
  b   R m lev eat tba mix 18m24 e                                          796.mg * P<.6    -
  c   R m lev eat tba mal 18m24 e                                          2.27gm * P<.5    -
  d   R m lev eat tba ben 18m24 e                                          no dre   P=1.    -

SOYBEAN LECITHIN       100ng..:..1ug...:..10....:..100....:..1mg..:..10...:...100...:..1g.....:..10
4569  R f wis eat liv nod 24m24 e                                          no dre   P=1.    -
4570  R m wis eat liv cho 24m24 e                                          .>42.2gm P<.3    -

STARCH ACETATE         100ng..:..1ug...:..10....:..100....:..1mg..:..10...:...100...:..1g.....:..10
4571  R f wis eat liv tum 24m24 e                                          no dre   P=1.    -
  a   R f wis eat tba mix 24m24 e                                          no dre   P=1.    -
4572  R m wis eat liv tum 24m24 e                                          no dre   P=1.    -
  a   R m wis eat tba mix 24m24 e                                          no dre   P=1.    -

STERIGMATOCYSTIN       100ng..:..1ug...:..10....:..100....:..1mg..:..10...:...100...:..1g.....:..10
4573  M f bd1 eat liv hae 55w68 ek              .  +  .                    .574mg   P<.0005
  a   M f bd1 eat liv ang 55w68 ek                                         1.33mg   P<.004  +
  b   M f bd1 eat brf ang 55w68 ek                                         8.77mg   P<.3    +
  c   M f bd1 eat liv hpa 55w68 ek                                         8.77mg   P<.3
  d   M f bd1 eat lun ade 55w68 ek                                         8.77mg   P<.3
4574  M f bd1 eat liv ang 55w73 e             . +.                         .689mg   P<.0005+
  a   M f bd1 eat liv hae 55w73 e                                          5.77mg   P<.005
  b   M f bd1 eat brf ang 55w73 e                                          7.03mg   P<.01   +
  c   M f bd1 eat lun ade 55w73 e                                          8.92mg   P<.03
  d   M f bd1 eat liv hpc 55w73 e                                          37.2mg   P<.3
  e   M f bd1 eat lun ang 55w73 e                                          37.2mg   P<.3
4575  P b cym eat MXB MXB 17y21 Ww            :  +  :                      .218mg * P<.0005
  a   P b cym eat liv hpc 17y21 Ww                                         .264mg * P<.0005+
  b   P b cym eat liv clc 17y21 Ww                                         1.36mg * P<.06   +
  c   P b cym eat ute ley 17y21 Ww                                         .365mg * P<.2
  d   P b cym eat liv ade 17y21 Ww                                         .682mg * P<.4
  e   P b cym eat kid rcc 17y21 Ww                                         1.57mg * P<.8
  f   P b cym eat adr coa 17y21 Ww                                         2.04mg * P<.2
  g   P b cym eat liv chc 17y21 Ww                                         2.64mg * P<.2    +
  h   P b cym eat tba mix 17y21 Ww                                         .115mg * P<.002
  i   P b cym eat tba mal 17y21 Ww                                         .148mg * P<.0005
  j   P b cym eat tba ben 17y21 Ww                                         .346mg * P<.2
4576  R m ain eat liv mix 28m28 ae                .  +  .         +hist 1.67mg * P<.003  +
  a   R m ain eat liv hnd 28m28 ae                                 +hist 3.48mg * P<.008  +
  b   R m ain eat liv hes 28m28 ae                                 +hist 2.60mg * P<.03   +
  c   R m ain eat liv hpc 28m28 ae                                 +hist 10.8mg * P<.2    +
  d   R m ain eat tba mix 28m28 ae                                         no dre   P=1.
4577  R m don eat liv mix 23m23 ae          . + .                         82.5ug * P<.0005+
  a   R m don eat liv ade 23m23 ae                                         .727mg * P<.007
  b   R m don eat liv hes 23m23 ae                                         1.57mg * P<.07
4578  R b wis gav liv hpc 12m29                 . + .                      .322mg * P<.0005+
4579  R b wis eat liv hpc 12m29 sv            . + .                        .124mg Z P<.0005+
4580  R m wis eat liv car 54w69 er            . + .                        .111mg   P<.0005+

STREPTOZOTOCIN         100ng..:..1ug...:..10....:..100....:..1mg..:..10...:...100...:..1g.....:..10
4581  M f swi ipj lun mix 26w78 e          . + .                          .193mg \ P<.0005+
  a   M f swi ipj kid ade 26w78 e                                          1.39mg \ P<.0005+
  b   M f swi ipj ute sar 26w78 e                                          13.9mg * P<.04
  c   M f swi ipj liv lys 26w78 e                                          no dre   P=1.
  d   M f swi ipj tba ben 26w78 e                                          .188mg \ P<.0005
  e   M f swi ipj tba mix 26w78 e                                          .486mg * P<.0005
  f   M f swi ipj tba mal 26w78 e                                          6.32mg / P<.2
4582  M m swi ipj lun mix 26w72 e            . + .                         .462mg \ P<.0005+
  a   M m swi ipj kid tla 26w72 e                                          2.44mg \ P<.006  +
  b   M m swi ipj liv mix 26w72 e                                          27.3mg \ P<.6
  c   M m swi ipj tba mix 26w72 e                                          .217mg \ P<.0005
  d   M m swi ipj tba ben 26w72 e                                          .698mg \ P<.003
  e   M m swi ipj tba mal 26w72 e                                          3.53mg * P<.3
4583  R f cdr ipj kid mix 26w78 e              . + .                       1.27mg * P<.0005+
```

	RefNum	LoConf	UpConf	Cntrl	1Dose	1Inc	2Dose	2Inc	Citation or Pathology
									Brkly Code
c	1635	752.mg	n.s.s.	4/60	254.mg	11/60	509.mg	11/60	
d	1635	589.mg	n.s.s.	16/60	254.mg	18/60	509.mg	22/60	
e	1635	1.33gm	n.s.s.	13/60	254.mg	9/60	509.mg	12/60	
4566	1635	1.07gm	n.s.s.	11/57	254.mg	12/57	509.mg	12/60	
a	1635	4.45gm	n.s.s.	1/57	254.mg	0/57	509.mg	1/60	
b	1635	4.33gm	n.s.s.	6/57	254.mg	1/57	509.mg	2/60	
c	1635	1.59gm	n.s.s.	1/57	254.mg	0/57	509.mg	0/60	
d	1635	1.60gm	n.s.s.	19/57	254.mg	14/57	509.mg	13/60	
e	1635	1.94gm	n.s.s.	5/57	254.mg	4/57	509.mg	5/60	
f	1635	1.47gm	n.s.s.	23/57	254.mg	17/57	509.mg	16/60	
4567	1635	2.05gm	n.s.s.	0/60	102.mg	0/60	206.mg	1/60	
a	1635	118.mg	n.s.s.	44/60	102.mg	47/60	206.mg	48/60	
b	1635	535.mg	n.s.s.	5/60	102.mg	6/60	206.mg	10/60	
c	1635	186.mg	n.s.s.	43/60	102.mg	43/60	206.mg	43/60	
4568	1635	2.03gm	n.s.s.	0/60	102.mg	0/60	206.mg	1/59	
a	1635	1.53gm	n.s.s.	1/60	102.mg	2/60	206.mg	1/59	
b	1635	142.mg	n.s.s.	44/60	102.mg	43/60	206.mg	46/59	
c	1635	468.mg	n.s.s.	12/60	102.mg	10/60	206.mg	15/59	
d	1635	224.mg	n.s.s.	41/60	102.mg	36/60	206.mg	40/59	
SOYBEAN LECITHIN		8002-43-5							
4569	1359	19.4gm	n.s.s.	1/47	2.00gm	0/47			
4570	1359	6.87gm	n.s.s.	0/41	1.60gm	1/39			Brantom;fctx,11,755-769;1973
STARCH ACETATE		9045-28-7							
4571	1407	92.7gm	n.s.s.	0/30	15.0gm	0/30			
a	1407	19.4gm	n.s.s.	23/30	15.0gm	17/30			de Groot;fctx,12,651-663;1974
4572	1407	66.8gm	n.s.s.	0/23	12.0gm	0/27			
a	1407	3.91gm	n.s.s.	21/23	12.0gm	24/27			
STERIGMATOCYSTIN		10048-13-2							
4573	1492m	.226mg	1.80mg	0/10	3.15mg	8/10			
a	1492m	.490mg	9.33mg	0/10	3.15mg	5/10			Enomoto;fctx,20,547-556;1982
b	1492m	1.43mg	n.s.s.	0/10	3.15mg	1/10			
c	1492m	1.43mg	n.s.s.	0/10	3.15mg	1/10			
d	1492m	1.43mg	n.s.s.	0/10	3.15mg	1/10			
4574	1492n	.420mg	1.20mg	0/35	2.94mg	29/38			
a	1492n	2.35mg	41.4mg	0/35	2.94mg	6/38			
b	1492n	2.67mg	320.mg	0/35	2.94mg	5/38			
c	1492n	3.08mg	n.s.s.	0/35	2.94mg	4/38			
d	1492n	6.06mg	n.s.s.	0/35	2.94mg	1/38			
e	1492n	6.06mg	n.s.s.	0/35	2.94mg	1/38			
4575	2001	92.5ug	.625mg	0/69	.117mg	4/15	.214mg	5/14	Adamson;ossc,129-156;
									1982/Thorgeirsson 1994/Dalgard 1991/Thorgeirsson&Seiber pers.comm.
a	2001	.101mg	.940mg	0/69	.117mg	3/15	.214mg	4/14	
b	2001	.328mg	n.s.s.	0/55	.117mg	1/12	.214mg	1/13	
c	2001	86.6ug	n.s.s.	0/12	.117mg	2/9	.214mg	0/1	
d	2001	.111mg	n.s.s.	0/6	.117mg	1/6	.214mg	0/1	
e	2001	.112mg	n.s.s.	1/19	.117mg	2/12	.214mg	0/10	
f	2001	.331mg	n.s.s.	0/16	.117mg	0/10	.214mg	1/7	
g	2001	.430mg	n.s.s.	0/19	.117mg	0/11	.214mg	1/10	
h	2001	48.8ug	.584mg	2/69	.117mg	8/15	.214mg	5/14	
i	2001	61.6ug	.601mg	1/69	.117mg	6/15	.214mg	5/14	
j	2001	83.2ug	n.s.s.	1/19	.117mg	3/11	.214mg	1/8	
4576	1184	.679mg	10.4mg	0/11	4.00ug	0/27	40.0mg	1/29	.400mg	5/26	Maekawa;gann,70,777-781;1979
a	1184	1.05mg	91.1mg	0/11	4.00ug	0/27	40.0ug	0/29	.400mg	3/26	
b	1184	.898mg	n.s.s.	0/11	4.00ug	0/27	40.0ug	1/29	.400mg	3/26	
c	1184	1.76mg	n.s.s.	0/11	4.00ug	0/27	40.0ug	0/29	.400mg	1/26	
d	1184	.753mg	n.s.s.	7/11	4.00ug	13/27	40.0ug	6/29	.400mg	10/26	
4577	392	42.6ug	.166mg	0/17	.200mg	11/13	.400mg	12/13	Ohtsubo;fctx,16,143-149;1978
a	392	.295mg	9.46mg	0/17	.200mg	2/13	.400mg	4/13	
b	392	.474mg	n.s.s.	0/17	.200mg	1/13	.400mg	2/13	
4578	275m	.165mg	.982mg	0/10	.107mg	4/10	.213mg	5/10	1.07mg	9/10	
4579	275n	53.6ug	.286mg	0/10	.239mg	8/10	.477mg	10/10	(2.39mg	3/10)	Purchase;fctx,8,289-295;1970
4580	707	48.6ug	.333mg	0/25	.313mg	8/14			
									Terao;fctx,16,591-596;1978
STREPTOZOTOCIN		18883-66-4							
4581	1336m	91.5ug	.477mg	20/154	.855mg	16/19	(2.50mg	14/21)	Skipper;srfr;1976/Weisburger;canc;1977/Prejean pers.comm.
a	1336m	.480mg	7.18mg	0/154	.855mg	4/19	(2.50mg	2/21)	
b	1336m	3.19mg	n.s.s.	1/154	.855mg	0/19	2.50mg	2/21	
c	1336m	1.44mg	n.s.s.	1/154	.855mg	0/19	2.50mg	2/21	
d	1336m	90.2ug	.443mg	13/154	.855mg	16/19	(2.50mg	9/21)	
e	1336m	.263mg	1.09mg	42/154	.855mg	16/19	2.50mg	17/21	
f	1336m	1.66mg	n.s.s.	29/154	.855mg	0/19	2.50mg	8/21	
4582	1336m	.197mg	1.82mg	9/101	.930mg	9/17	(2.08mg	5/16)	
a	1336m	.599mg	47.1mg	0/101	.930mg	2/17	(2.08mg	0/16)	
b	1336m	2.74mg	n.s.s.	2/101	.930mg	0/17	2.08mg	1/16	
c	1336m	92.5ug	.761mg	28/101	.930mg	14/17	(2.08mg	6/16)	
d	1336m	.264mg	5.53mg	9/101	.930mg	7/17	(2.08mg	2/16)	
e	1336m	.890mg	n.s.s.	19/101	.930mg	7/17	2.08mg	4/16	
4583	1336n	.693mg	2.70mg	0/182	.859mg	5/23	1.82mg	10/23	

```
    Spe Strain  Site   Xpo+Xpt                                                    TD50     2Tailpvl
        Sex  Route  Hist      Notes                                               DR       AuOp
```

```
 a   R f cdr ipj mgl car 26w78 e                                                  1.41mg \ P<.009
 b   R f cdr ipj ute sar 26w78 e                                                  2.46mg \ P<.003
 c   R f cdr ipj liv mix 26w78 e                                                  4.43mg * P<.0005
 d   R f cdr ipj liv cvh 26w78 e                                                  5.63mg * P<.0005
 e   R f cdr ipj tba mal 26w78 e                                                  .363mg \ P<.0005
 f   R f cdr ipj tba mix 26w78 e                                                  1.14mg * P<.06
 g   R f cdr ipj tba ben 26w78 e                                                  no dre   P=1.
4584 R m cdr ipj kid mix 26w78 e                        .  +  .                   .776mg * P<.0005+
 a   R m cdr ipj liv lcc 26w78 e                                                  20.4mg * P<.2
 b   R m cdr ipj tba mix 26w78 e                                                  .751mg * P<.0005
 c   R m cdr ipj tba mal 26w78 e                                                  1.03mg * P<.0005
 d   R m cdr ipj tba ben 26w78 e                                                  16.8mg * P<.7
```

STROBANE   100ng..:..1ug....:..10......:..100...:..1mg....:..10......:..100...:..1g......:..10
```
4585 M f b6a orl liv hpt 79w79 evx                                         .>       no dre   P=1.
 a   M f b6a orl lun ade 79w79 evx                                                  no dre   P=1.
 b   M f b6a orl tba mix 79w79 evx                                                  no dre   P=1.
4586 M m b6a orl liv hpt 79w79 evx                                  .  +   .        .644mg   P<.0005+
 a   M m b6a orl lun ade 79w79 evx                                                  no dre   P=1.
 b   M m b6a orl tba mix 79w79 evx                                                  .750mg   P<.006
4587 M f b6c orl lun ade 79w79 evx                                     .>           10.7mg   P<.3
 a   M f b6c orl liv hpt 79w79 evx                                                  no dre   P=1.
 b   M f b6c orl tba mix 79w79 evx                                                  3.37mg   P<.05
4588 M m b6c orl --- rts 79w79 evx                      .   +       .               1.41mg   P<.005 +
 a   M m b6c orl liv hpt 79w79 evx                                                  4.00mg   P<.09
 b   M m b6c orl lun ade 79w79 evx                                                  8.29mg   P<.3
 c   M m b6c orl tba mix 79w79 evx                                                  .750mg   P<.0005
```

STYRENE   100ng..:..1ug....:..10......:..100...:..1mg....:..10......:..100...:..1g......:..10
```
4589 M f b6c gav liv hpa 78w91                                                 :    ±     #905.mg * P<.04  -
 a   M f b6c gav TBA MXB 78w91                                                             625.mg * P<.4
 b   M f b6c gav liv MXB 78w91                                                             905.mg * P<.04
 c   M f b6c gav lun MXB 78w91                                                             1.38gm * P<.2
4590 M m b6c gav lun MXA 78w91                                                 :  +   :    360.mg * P<.007 a
 a   M m b6c gav TBA MXB 78w91                                                             659.mg * P<.5
 b   M m b6c gav liv MXB 78w91                                                             895.mg * P<.5
 c   M m b6c gav lun MXB 78w91                                                             360.mg * P<.007
4591 R f f34 gav TBA MXB 90w97 as                                                   :>    no dre   P=1.   -
 a   R f f34 gav liv MXB 90w97 as                                                          25.2gm * P<.4
4592 R m f34 gav TBA MXB 90w97 as                                                     :>  3.41gm Z P<.5   -
 a   R m f34 gav liv MXB 90w97 as                                                          no dre   P=1.
4593 R f sda inh mam mal 12m24                                      .  +   .              57.1mg * P<.002 +

 a   R f sda inh mam mix 12m24                                                            23.3mg * P<.02  +
 b   R f sda inh tba mix 12m24                                                            48.7mg * P<.3
 c   R f sda inh tba mal 12m24                                                            205.mg * P<.5
4594 R f sda gav tba mix 12m24                                   .  ±                     12.0mg \ P<.03
 a   R f sda gav tba mal 12m24                                                            no dre   P=1.
4595 R m sda inh tba mix 12m24                                              .>            107.mg * P<.4
 a   R m sda inh tba mal 12m24                                                            no dre   P=1.
4596 R m sda gav tba mix 12m24                                                 .>         1.22gm * P<.7
 a   R m sda gav tba mal 12m24                                                            1.09gm * P<.5
```

STYRENE AND beta-NITROSTYRENE MIXTURE ..:..1ug....:..10......:..100...:..1mg....:..10......:..100...:..1g......:..10
```
4597 M f b6c gav TBA MXB 78w92                                                    :>      216.mg * P<.3   -
 a   M f b6c gav liv MXB 78w92                                                            no dre   P=1.
 b   M f b6c gav lun MXB 78w92                                                            no dre   P=1.
4598 M m b6c gav lun MXA 79w92                                               :  +   :     #61.5mg \ P<.007 -
 a   M m b6c gav TBA MXB 79w92                                                            no dre   P=1.
 b   M m b6c gav liv MXB 79w92                                                            no dre   P=1.
 c   M m b6c gav lun MXB 79w92                                                            61.5mg \ P<.007
4599 R f f34 gav TBA MXB 18m25                                               :>           no dre   P=1.   -
 a   R f f34 gav liv MXB 18m25                                                            no dre   P=1.
4600 R m f34 gav TBA MXB 18m25                                                  :>        no dre   P=1.   -
 a   R m f34 gav liv MXB 18m25                                                            1.57gm * P<.4
```

STYRENE OXIDE   100ng..:..1ug....:..10......:..100...:..1mg....:..10......:..100...:..1g......:..10
```
4601 M f b6c gav for mix 24m25 e                                          . + .           172.mg \ P<.0005+
 a   M f b6c gav for sqp 24m25 e                                                          460.mg \ P<.0005+
 b   M f b6c gav for sqc 24m25 e                                                          505.mg \ P<.0005+
 c   M f b6c gav liv mix 24m25 e                                                          5.96gm * P<.6
4602 M m b6c gav for mix 24m25 e                                       . + .              90.0mg \ P<.0005+
 a   M m b6c gav for sqp 24m25 e                                                          215.mg \ P<.0005+
 b   M m b6c gav liv mix 24m25 e                                                          223.mg \ P<.002
 c   M m b6c gav for sqc 24m25 e                                                          475.mg * P<.0005+
4603 R f f34 gav for mix 24m25 e                                       . +.               42.9mg * P<.0005+
 a   R f f34 gav for sqc 24m25 e                                                          114.mg * P<.0005+
 b   R f f34 gav for sqp 24m25 e                                                          217.mg * P<.0005+
 c   R f f34 gav liv nnd 24m25 e                                                          no dre   P=1.
4604 R m f34 gav for mix 24m25 e                                       . + .              30.7mg * P<.0005+
 a   R m f34 gav for sqc 24m25 e                                                          81.6mg * P<.0005+
```

	RefNum	LoConf	UpConf	Cntrl	1Dose	1Inc	2Dose	2Inc					Citation or Pathology
													Brkly Code
a	1336n	.487mg	65.7mg	12/182	.859mg	6/23	(1.82mg	1/23)					
b	1336n	.724mg	30.7mg	1/182	.859mg	3/23	(1.82mg	0/23)					
c	1336n	1.68mg	18.4mg	0/182	.859mg	1/23	1.82mg	4/23					
d	1336n	1.94mg	29.0mg	0/182	.859mg	1/23	1.82mg	3/23					
e	1336n	.169mg	1.22mg	44/182	.859mg	16/23	(1.82mg	8/23)					
f	1336n	.410mg	n.s.s.	103/182	.859mg	21/23	1.82mg	15/23					
g	1336n	2.05mg	n.s.s.	59/182	.859mg	5/23	1.82mg	7/23					
4584	1336n	.446mg	1.51mg	0/177	.859mg	7/22	1.72mg	12/20					
a	1336n	3.31mg	n.s.s.	0/177	.859mg	1/22	1.72mg	0/20					
b	1336n	.365mg	2.94mg	59/177	.859mg	15/22	1.72mg	13/20					
c	1336n	.504mg	3.83mg	32/177	.859mg	14/22	1.72mg	8/20					
d	1336n	1.86mg	n.s.s.	27/177	.859mg	1/22	1.72mg	5/20					

STROBANE   (dichloricide mothproofer) 8001-50-1

	RefNum	LoConf	UpConf	Cntrl	1Dose	1Inc							Citation or Pathology
4585	277	3.32mg	n.s.s.	0/17	1.55mg	0/18							Innes;ntis,1968/1969
a	277	3.32mg	n.s.s.	1/17	1.55mg	0/18							
b	277	3.32mg	n.s.s.	2/17	1.55mg	0/18							
4586	277	.300mg	2.14mg	1/18	1.45mg	11/18							
a	277	3.10mg	n.s.s.	2/18	1.45mg	0/18							
b	277	.319mg	9.63mg	3/18	1.45mg	11/18							
4587	277	1.75mg	n.s.s.	0/16	1.55mg	1/18							
a	277	3.32mg	n.s.s.	0/16	1.55mg	0/18							
b	277	1.02mg	n.s.s.	0/16	1.55mg	3/18							
4588	277	.530mg	11.1mg	0/16	1.45mg	5/15							
a	277	.980mg	n.s.s.	0/16	1.45mg	2/15							
b	277	1.35mg	n.s.s.	0/16	1.45mg	1/15							
c	277	.330mg	2.28mg	0/16	1.45mg	8/15							

STYRENE   100-42-5

	RefNum	LoConf	UpConf	Cntrl	1Dose	1Inc	2Dose	2Inc	3Dose	3Inc	4Dose	4Inc	Citation or Pathology
4589	TR185	369.mg	n.s.s.	0/20	92.0mg	1/50	184.mg	5/50					S
a	TR185	188.mg	n.s.s.	2/20	92.0mg	11/50	184.mg	10/50					
b	TR185	369.mg	n.s.s.	0/20	92.0mg	1/50	184.mg	5/50					liv:hpa,hpc,nnd.
c	TR185	477.mg	n.s.s.	0/20	92.0mg	1/50	184.mg	3/50					lun:a/a,a/c.
4590	TR185	198.mg	4.60gm	0/20	92.0mg	6/49	184.mg	9/50					lun:a/a,a/c.
a	TR185	154.mg	n.s.s.	9/20	92.0mg	14/49	184.mg	21/50					
b	TR185	219.mg	n.s.s.	5/20	92.0mg	8/49	184.mg	13/50					liv:hpa,hpc,nnd.
c	TR185	198.mg	4.60gm	0/20	92.0mg	6/49	184.mg	9/50					lun:a/a,a/c.
4591	TR185	1.08gm	n.s.s.	23/40	354.mg	25/50	530.mg	17/50	1.06gm	3/50			
a	TR185	4.11gm	n.s.s.	0/40	354.mg	0/50	530.mg	1/50	1.06gm	0/50			liv:hpa,hpc,nnd.
4592	TR185	765.mg	n.s.s.	14/40	354.mg	14/50	530.mg	15/50	1.06gm	4/50			
a	TR185	6.85mg	n.s.s.	1/40	354.mg	0/50	530.mg	0/50	1.06gm	0/50			liv:hpa,hpc,nnd.
4593	BT101	28.9mg	316.mg	6/60	2.66mg	6/30	5.31mg	4/30	10.6mg	9/30	21.2mg 12/30 31.9mg 9/30		Conti;anya,534, 203-234;1988
a	BT101	10.5mg	8.37gm	34/60	2.66mg	24/30	5.31mg	21/30	10.6mg	23/30	21.2mg 24/30 31.9mg 25/30		
b	BT101	13.0mg	n.s.s.	43/60	2.66mg	24/30	5.31mg	25/30	10.6mg	26/30	21.2mg 24/30 31.9mg 25/30		
c	BT101	41.7mg	n.s.s.	16/60	2.66mg	9/30	5.31mg	9/30	10.6mg	13/30	21.2mg 10/30 31.9mg 10/30		
4594	BT102	4.97mg	n.s.s.	25/40	16.1mg	34/40	(80.4mg	19/40)					
a	BT102	176.mg	n.s.s.	9/40	16.1mg	15/40	80.4mg	9/40					
4595	BT101	27.7mg	n.s.s.	17/60	1.86mg	12/30	3.72mg	6/30	7.44mg	11/30	14.9mg 10/30 22.3mg 12/30		
a	BT101	75.8mg	n.s.s.	10/60	1.86mg	5/30	3.72mg	5/30	7.44mg	8/30	14.9mg 3/30 22.3mg 4/30		
4596	BT102	143.mg	n.s.s.	9/40	16.1mg	8/40	80.4mg	10/40					
a	BT102	206.mg	n.s.s.	2/40	16.1mg	3/40	80.4mg	4/40					

STYRENE AND beta-NITROSTYRENE MIXTURE   (CAS# 100-42-5 and 102-96-5) mixture

	RefNum	LoConf	UpConf	Cntrl	1Dose	1Inc	2Dose	2Inc					Citation or Pathology
4597	TR170	70.1mg	n.s.s.	2/20	31.8mg	10/50	63.6mg	10/50					
a	TR170	467.mg	n.s.s.	1/20	31.8mg	1/50	63.6mg	0/50					liv:hpa,hpc,nnd.
b	TR170	288.mg	n.s.s.	0/20	31.8mg	2/50	63.6mg	0/50					lun:a/a,a/c.
4598	TR170	30.8mg	556.mg	0/20	31.8mg	11/50	(63.6mg	2/50)					lun:a/a,a/c. S
a	TR170	72.9mg	n.s.s.	8/20	31.8mg	19/50	63.6mg	12/50					
b	TR170	118.mg	n.s.s.	6/20	31.8mg	6/50	63.6mg	8/50					liv:hpa,hpc,nnd.
c	TR170	30.8mg	556.mg	0/20	31.8mg	11/50	(63.6mg	2/50)					lun:a/a,a/c.
4599	TR170	34.7mg	n.s.s.	12/20	23.5mg	34/50	47.0mg	30/50					
a	TR170	270.mg	n.s.s.	0/20	23.5mg	1/50	47.0mg	0/50					liv:hpa,hpc,nnd.
4600	TR170	198.mg	n.s.s.	10/20	47.0mg	18/50	94.0mg	11/50					
a	TR170	387.mg	n.s.s.	0/20	47.0mg	1/50	94.0mg	1/50					liv:hpa,hpc,nnd.

STYRENE OXIDE   96-09-3

	RefNum	LoConf	UpConf	Cntrl	1Dose	1Inc	2Dose	2Inc					Citation or Pathology
4601	1791	105.mg	309.mg	0/51	155.mg	24/50	(311.mg	20/51)					Lijinsky;jnci,77,471-476;1986
a	1791	298.mg	775.mg	0/51	155.mg	14/50	311.mg	17/51					
b	1791	246.mg	1.35gm	0/51	155.mg	10/50	(311.mg	3/51)					
c	1791	969.mg	n.s.s.	7/51	155.mg	4/50	311.mg	9/51					
4602	1791	57.8mg	150.mg	2/51	155.mg	37/51	(311.mg	21/52)					
a	1791	124.mg	458.mg	2/51	155.mg	22/51	(311.mg	8/52)					
b	1791	113.mg	1.17gm	12/51	155.mg	28/52	(311.mg	15/52)					
c	1791	308.mg	828.mg	0/51	155.mg	16/51	311.mg	15/52					
4603	1791	29.8mg	61.7mg	0/52	114.mg	46/52	228.mg	50/52					
a	1791	83.0mg	161.mg	0/52	114.mg	32/52	228.mg	36/52					
b	1791	150.mg	328.mg	0/52	114.mg	21/52	228.mg	24/52					
c	1791	472.mg	n.s.s.	8/52	114.mg	6/52	(228.mg	0/52)					
4604	1791	19.4mg	47.0mg	1/52	114.mg	50/52	228.mg	50/51					
a	1791	59.6mg	114.mg	0/52	114.mg	35/52	228.mg	43/51					

```
     Spe Strain  Site    Xpo+Xpt                                    TD50      2Tailpvl
         Sex   Route  Hist    Notes                                 DR        AuOp

  b    R m f34 gav for sqp 24m25 e                                 146.mg \ P<.0005+
  c    R m f34 gav liv nnd 24m25 e                                 10.3gm * P<.7
4605   R f sda gav for mix 12m36 e             . + .               96.5mg * P<.0005+
  a    R f sda gav for sqc 12m36 e                                 102.mg * P<.0005+
  b    R f sda gav for sqn 12m36 e                                 129.mg * P<.0005+
  c    R f sda gav for sqi 12m36 e                                 301.mg * P<.0005+
  d    R f sda gav for ben 12m36 e                                 503.mg * P<.009 +
  e    R f sda gav tba mix 12m36                                   164.mg * P<.009
  f    R f sda gav tba mal 12m36                                   166.mg * P<.002
4606   R f sda gav for mix 12m36 er            . + .               96.5mg * P<.0005+
  a    R f sda gav for sqc 12m36 er                                102.mg * P<.0005+
  b    R f sda gav for sqn 12m36 er                                129.mg * P<.0005+
  c    R f sda gav for sqi 12m36 er                                301.mg * P<.0005+
  d    R f sda gav for ben 12m36 er                                503.mg * P<.009 +
4607   R m sda gav for sqc 12m36 e             . + .               63.0mg \ P<.0005+
  a    R m sda gav for mix 12m36 e                                 99.6mg * P<.0005+
  b    R m sda gav for sqn 12m36 e                                 140.mg * P<.0005+
  c    R m sda gav for sqi 12m36 e                                 267.mg * P<.004 +
  d    R m sda gav for ben 12m36 e                                 354.mg * P<.003 +
  e    R m sda gav tba mal 12m36                                   189.mg * P<.004
  f    R m sda gav tba mix 12m36                                   191.mg * P<.02
4608   R m sda gav for sqc 12m36 er            . + .               63.0mg \ P<.0005+
  a    R m sda gav for mix 12m36 er                                99.6mg * P<.0005+
  b    R m sda gav for sqn 12m36 er                                140.mg * P<.0005+
  c    R m sda gav for sqi 12m36 er                                267.mg * P<.004 +
  d    R m sda gav for ben 12m36 er                                354.mg * P<.003 +


SUCCINIC ANHYDRIDE   100ng..:.1ug...:.10...:..:100...:..1mg..:.10...:..:100...:.1g...:..10
4609   M f b6c gav TBA MXB 24m24                              :>   no dre   P=1.   -
  a    M f b6c gav liv MXB 24m24                                   no dre   P=1.
  b    M f b6c gav lun MXB 24m24                                   4.69gm * P<.8
4610   M m b6c gav TBA MXB 24m24                         :>        no dre   P=1.   -
  a    M m b6c gav liv MXB 24m24                                   no dre   P=1.
  b    M m b6c gav lun MXB 24m24                                   no dre   P=1.
4611   R f f34 gav TBA MXB 24m24                         :>        no dre   P=1.   -
  a    R f f34 gav liv MXB 24m24                                   1.30gm * P<.6
4612   R m f34 gav TBA MXB 24m24                    :>             no dre   P=1.   -
  a    R m f34 gav liv MXB 24m24                                   no dre   P=1.


SUCROSE   100ng..:.1ug...:.10...:..:100...:..1mg..:.10...:..:100...:.1g...:..10
4613   M f scp eat lun act 24m24 e                                135.gm   P<.4   -
  a    M f scp eat liv nnd 24m24 e                                405.gm   P<.3   -
  b    M f scp eat tba tum 24m24 e                                no dre   P=1.   -
4614   M m scp eat lun act 94w94 e                                89.9gm   P<.4   -
  a    M m scp eat liv hpc 94w94 e                                150.gm   P<.3   -
  b    M m scp eat liv nnd 94w94 e                                264.gm   P<.8   -
  c    M m scp eat tba tum 94w94 e                                78.9gm   P<.7   -
4615   M f swa eat lun mix 76w76 e                                129.gm   P<.8   -
  a    M f swa eat liv hpt 76w76 e                                no dre   P=1.   -


SULFALLATE   100ng..:.1ug...:.10...:..:100...:..1mg..:.10...:..:100...:.1g...:..10
4616   M f b6c eat mgl MXA 78w90 dv              : + :            27.3mg * P<.0005c
  a    M f b6c eat mgl acn 78w90 dv                               39.3mg * P<.0005c
  b    M f b6c eat lun MXA 78w90 dv                               87.5mg * P<.007
  c    M f b6c eat mgl asm 78w90 dv                               98.3mg * P<.0005c
  d    M f b6c eat liv hpc 78w90 dv                               107.mg * P<.0005
  e    M f b6c eat TBA MXB 78w90 dv                               21.3mg * P<.0005
  f    M f b6c eat liv MXB 78w90 dv                               107.mg * P<.0005
  g    M f b6c eat lun MXB 78w90 dv                               87.5mg * P<.007
4617   M m b6c eat lun MXA 78w91 dv               : + :           92.5mg * P<.0005c
  a    M m b6c eat lun a/c 78w91 dv                               269.mg * P<.006
  b    M m b6c eat --- hes 78w91 dv                               539.mg * P<.03
  c    M m b6c eat TBA MXB 78w91 dv                               65.5mg * P<.03
  d    M m b6c eat liv MXB 78w91 dv                               188.mg * P<.3
  e    M m b6c eat lun MXB 78w91 dv                               92.5mg * P<.0005
4618   R f osm eat mgl acn 18m24 dv             : + :             17.2mg * P<.0005c
  a    R f osm eat TBA MXB 18m24 dv                               9.06mg * P<.03
  b    R f osm eat liv MXB 18m24 dv                               69.1mg * P<.2
4619   R m osm eat thy MXA 18m24 dv             : + :             24.0mg * P<.007
  a    R m osm eat sto MXA 18m24 dv                               53.9mg / P<.007 c
  b    R m osm eat TBA MXB 18m24 dv                               12.6mg * P<.3
  c    R m osm eat liv MXB 18m24 dv                               no dre   P=1.


SULFAMETHAZINE   100ng..:.1ug...:.10...:..:100...:..1mg..:.10...:..:100...:.1g...:..10
4620   M f b6c eat lun a/c 52w52 e                          .>    3.10gm * P<.2
  a    M f b6c eat liv hpa 52w52 e                                no dre   P=1.
  b    M f b6c eat thy fca 52w52 e                                no dre   P=1.
4621   M f b6c eat thy fca 78w78 e                   . + .        618.mg * P<.0005
  a    M f b6c eat liv hpa 78w78 e                                2.82gm * P<.2
  b    M f b6c eat lun a/c 78w78 e                                no dre   P=1.
4622   M f b6c eat thy mix 24m24 e                      . + .     2.06gm Z P<.0005+
```

	RefNum	LoConf	UpConf	Cntrl	1Dose	1Inc	2Dose	2Inc		Citation or Pathology
										Brkly Code
b	1791	87.2mg	283.mg	1/52	114.mg	23/52	(228.mg	18/51)		
c	1791	1.33gm	n.s.s.	2/52	114.mg	1/52	228.mg	3/52		
4605	BT105	60.6mg	167.mg	0/40	10.7mg	7/37	53.6mg	21/38		Conti;anya,534,203-234;1988
a	BT105	63.7mg	180.mg	0/40	10.7mg	7/37	53.6mg	20/38		
b	BT105	78.0mg	264.mg	0/40	10.7mg	7/37	53.6mg	16/38		
c	BT105	151.mg	755.mg	0/40	10.7mg	1/37	53.6mg	10/38		
d	BT105	217.mg	19.7gm	0/40	10.7mg	2/37	53.6mg	5/38		
e	BT105	74.1mg	7.47gm	10/40	10.7mg	16/40	53.6mg	22/40		
f	BT105	83.2mg	771.mg	7/40	10.7mg	9/40	53.6mg	20/40		
4606	1702	60.6mg	167.mg	0/40	10.7mg	7/37	53.6mg	21/38		Maltoni;amet,2,97-110;1981
a	1702	63.7mg	180.mg	0/40	10.7mg	7/37	53.6mg	20/38		
b	1702	78.0mg	264.mg	0/40	10.7mg	7/37	53.6mg	16/38		
c	1702	151.mg	755.mg	0/40	10.7mg	1/37	53.6mg	10/38		
d	1702	217.mg	19.7gm	0/40	10.7mg	2/37	53.6mg	5/38		
4607	BT105	29.6mg	184.mg	0/39	10.7mg	9/39	(53.6mg	16/39)		Conti;anya,534,203-234;1988
a	BT105	63.2mg	194.mg	0/39	10.7mg	10/39	53.6mg	19/39		
b	BT105	84.3mg	284.mg	0/39	10.7mg	6/39	53.6mg	16/39		
c	BT105	141.mg	2.30gm	0/39	10.7mg	5/39	53.6mg	8/39		
d	BT105	172.mg	2.23gm	0/39	10.7mg	3/39	53.6mg	7/39		
e	BT105	89.2mg	1.75gm	6/40	10.7mg	11/40	53.6mg	18/40		
f	BT105	84.0mg	n.s.s.	9/40	10.7mg	14/40	53.6mg	20/40		
4608	1702	29.6mg	184.mg	0/39	10.7mg	9/39	(53.6mg	16/39)		Maltoni;amet,2,97-110;1981
a	1702	63.2mg	194.mg	0/39	10.7mg	10/39	53.6mg	19/39		
b	1702	84.3mg	284.mg	0/39	10.7mg	6/39	53.6mg	16/39		
c	1702	141.mg	2.30gm	0/39	10.7mg	5/39	53.6mg	8/39		
d	1702	172.mg	2.23gm	0/39	10.7mg	3/39	53.6mg	7/39		

SUCCINIC ANHYDRIDE   108-30-5

	RefNum	LoConf	UpConf	Cntrl	1Dose	1Inc	2Dose	2Inc		Citation or Pathology
4609	TR373	309.mg	n.s.s.	27/50	52.8mg	14/50	106.mg	16/50		
a	TR373	759.mg	n.s.s.	2/50	52.8mg	1/50	106.mg	1/50		liv:hpa,hpc,nnd.
b	TR373	424.mg	n.s.s.	3/50	52.8mg	1/50	106.mg	4/50		lun:a/a,a/c.
4610	TR373	118.mg	n.s.s.	28/50	26.9mg	13/50	53.1mg	24/50		
a	TR373	81.4mg	n.s.s.	13/50	26.9mg	6/50	(53.1mg	7/50)		liv:hpa,hpc,nnd.
b	TR373	144.mg	n.s.s.	5/50	26.9mg	4/50	53.1mg	7/50		lun:a/a,a/c.
4611	TR373	56.3mg	n.s.s.	55/60	35.0mg	47/60	70.1mg	39/60		
a	TR373	198.mg	n.s.s.	2/60	35.0mg	1/60	70.1mg	3/60		liv:hpa,hpc,nnd.
4612	TR373	63.1mg	n.s.s.	54/60	35.0mg	47/60	70.1mg	43/60		
a	TR373	729.mg	n.s.s.	1/60	35.0mg	0/60	70.1mg	0/60		liv:hpa,hpc,nnd.

SUCROSE   57-50-1

	RefNum	LoConf	UpConf	Cntrl	1Dose	1Inc	2Dose	2Inc		Citation or Pathology
4613	1979	31.8gm	n.s.s.	3/44	13.0gm	6/47				Smits-van Prooije;fctx,28,243-251;1990
a	1979	66.0gm	n.s.s.	0/44	13.0gm	1/46				
b	1979	14.3gm	n.s.s.	28/46	13.0gm	29/49				
4614	1979	21.3gm	n.s.s.	5/49	12.0gm	8/48				
a	1979	33.3gm	n.s.s.	1/48	12.0gm	3/47				
b	1979	29.5gm	n.s.s.	4/48	12.0gm	5/47				
c	1979	12.0gm	n.s.s.	20/50	12.0gm	22/49				
4615	1090	14.9gm	n.s.s.	7/45	13.0gm	8/43				Roe;fctx,8,135-145;1970
a	1090	32.9gm	n.s.s.	3/45	13.0gm	2/43				

SULFALLATE   95-06-7

	RefNum	LoConf	UpConf	Cntrl	1Dose	1Inc	2Dose	2Inc		Citation or Pathology
4616	TR115	18.4mg	41.6mg	0/20	73.1mg	33/50	146.mg	27/50		mgl:acn,asm.
a	TR115	24.0mg	72.5mg	0/20	73.1mg	23/50	146.mg	11/50		
b	TR115	42.8mg	1.16gm	1/20	73.1mg	13/50	146.mg	4/50		lun:a/a,a/c. S
c	TR115	57.4mg	190.mg	0/20	73.1mg	10/50	146.mg	16/50		
d	TR115	53.4mg	272.mg	0/20	73.1mg	5/50	146.mg	8/50		S
e	TR115	13.7mg	39.2mg	7/20	73.1mg	42/50	146.mg	37/50		
f	TR115	53.4mg	272.mg	0/20	73.1mg	5/50	146.mg	8/50		liv:hpa,hpc,nnd.
g	TR115	42.8mg	1.16gm	1/20	73.1mg	13/50	146.mg	4/50		lun:a/a,a/c.
4617	TR115	59.6mg	276.mg	0/20	69.6mg	14/50	140.mg	17/50		lun:a/a,a/c.
a	TR115	138.mg	2.29gm	0/20	69.6mg	3/50	140.mg	9/50		
b	TR115	230.mg	n.s.s.	0/20	69.6mg	1/50	140.mg	6/50		S
c	TR115	31.1mg	n.s.s.	8/20	69.6mg	37/50	140.mg	39/50		S
d	TR115	62.4mg	n.s.s.	7/20	69.6mg	22/50	140.mg	24/50		liv:hpa,hpc,nnd.
e	TR115	59.6mg	276.mg	0/20	69.6mg	14/50	140.mg	17/50		lun:a/a,a/c.
4618	TR115	9.63mg	34.8mg	0/50	6.70mg	7/50	10.8mg	11/50		
a	TR115	4.01mg	n.s.s.	36/50	6.70mg	37/50	10.8mg	30/50		
b	TR115	21.4mg	n.s.s.	1/50	6.70mg	3/50	10.8mg	2/50		liv:hpa,hpc,nnd.
4619	TR115	9.75mg	321.mg	0/50	5.40mg	4/50	8.80mg	2/50		thy:cca,ccr. S
a	TR115	18.7mg	703.mg	0/50	5.40mg	0/50	8.80mg	5/50		sto:ppn,sqc,sqp.
b	TR115	3.94mg	n.s.s.	28/50	5.40mg	27/50	8.80mg	19/50		
c	TR115	n.s.s.	n.s.s.	0/50	5.40mg	0/50	8.80mg	0/50		liv:hpa,hpc,nnd.

SULFAMETHAZINE   (4-amino-N-(4,6-dimethyl-2-pyrimidinyl)-benzenesulfonamide)   57-68-1

	RefNum	LoConf	UpConf	Cntrl	1Dose	1Inc	2Dose	2Inc	3Dose	3Inc	4Dose	4Inc	Citation or Pathology
4620	1924m	505.mg	n.s.s.	0/24	39.0mg	0/24	156.mg	0/24	624.mg	1/22			Littlefield;fctx,27,455-463;1989/1989a/pers.comm.
a	1924m	760.mg	n.s.s.	0/24	39.0mg	1/24	156.mg	0/24	624.mg	0/22			
b	1924m	535.mg	n.s.s.	0/24	39.0mg	0/24	156.mg	1/24	624.mg	0/22			
4621	1924n	299.mg	1.76gm	0/24	39.0mg	1/24	156.mg	0/24	624.mg	9/23			
a	1924n	645.mg	n.s.s.	0/24	39.0mg	1/24	156.mg	1/24	624.mg	2/23			
b	1924n	1.25gm	n.s.s.	0/24	39.0mg	0/24	156.mg	1/24	624.mg	0/23			
4622	1924o	1.33gm	3.54gm	7/184	39.0mg	2/95	78.0mg	1/94	156.mg	0/96	312.mg	2/96 624.mg 34/94	

```
     Spe Strain  Site    Xpo+Xpt                                          TD50    2Tailpvl
         Sex  Route  Hist        Notes                                    DR      AuOp

  a    M f b6c eat thy fca 24m24 e                                        2.16gm Z P<.0005+
  b    M f b6c eat liv mix 24m24 e                                        8.74gm * P<.2
  c    M f b6c eat liv hpa 24m24 e                                        12.3gm * P<.3
  d    M f b6c eat lun a/c 24m24 e                                        12.7gm * P<.2
  e    M f b6c eat liv hpc 24m24 e                                        16.5gm * P<.2
  f    M f b6c eat thy fcc 24m24 e                                        54.8gm * P<.3
4623   M m b6c eat lun a/c 52w52 e                         .>             no dre   P=1.
  a    M m b6c eat liv hpa 52w52 e                                        no dre   P=1.
4624   M m b6c eat thy fca 78w78 e                               .>       3.37gm * P<.2
  a    M m b6c eat liv hpc 78w78 e                                        no dre   P=1.
  b    M m b6c eat liv hpa 78w78 e                                        no dre   P=1.
  c    M m b6c eat lun a/c 78w78 e                                        no dre   P=1.
  d    M m b6c eat liv mix 78w78 e                                        no dre   P=1.
4625   M m b6c eat thy mix 24m24 e                     .+ .              1.19gm Z P<.0005+
  a    M m b6c eat thy fca 24m24 e                                        1.22gm Z P<.0005+
  b    M m b6c eat liv mix 24m24 e                                        11.7gm * P<.7
  c    M m b6c eat liv hpa 24m24 e                                        14.9gm * P<.6
  d    M m b6c eat thy fcc 24m24 e                                        72.1gm * P<.3
  e    M m b6c eat liv hpc 24m24 e                                        no dre   P=1.
  f    M m b6c eat lun a/c 24m24 e                                        no dre   P=1.


SULFATE, SODIUM          100ng..:..1ug....:..10......:..100....:..1mg....:..10......:..100...:..1g......:..10
4626   M f swi eat --- leu 32w52 v                              .>       3.96gm   P<.6
  a    M f swi eat liv tum 32w52 v                                       no dre   P=1.
  b    M f swi eat lun tum 32w52 v                                       no dre   P=1.


SULFISOXAZOLE            100ng..:..1ug....:..10......:..100....:..1mg....:..10......:..100....:..1g......:..10
4627   M f b6c eat liv hpc 24m24                              :    +     #2.65gm \ P<.008 -
  a    M f b6c eat lun MXA 24m24                                         11.9gm * P<.007
  b    M f b6c eat TBA MXB 24m24                                         8.18gm * P<.5
  c    M f b6c eat liv MXB 24m24                                         2.65gm \ P<.008
  d    M f b6c eat lun MXB 24m24                                         11.9gm * P<.007
4628   M m b6c eat TBA MXB 24m24                              :>         no dre   P=1.   -
  a    M m b6c eat liv MXB 24m24                                         6.91gm * P<.3
  b    M m b6c eat lun MXB 24m24                                         49.5gm \ P<1.
4629   R f f34 eat TBA MXB 24m25                         :>             no dre   P=1.   -
  a    R f f34 eat liv MXB 24m25                                        no dre   P=1.
4630   R m f34 eat TBA MXB 24m24                 :     ±                135.mg \ P<.09 -
  a    R m f34 eat liv MXB 24m24                                        26.3gm * P<.9


SULFITE, POTASSIUM METABI-  100ng..:..1ug....:..10......:..100....:..1mg....:..10......:..100....:..1g......:..10
4631   M f icr wat lun ade 24m24                                       .33.6gm \ P<.1  -
  a    M f icr wat lun adc 24m24                                        100.gm * P<.4  -
  b    M f icr wat liv hms 24m24                                        204.gm * P<.3  -
4632   M m icr wat lun adc 24m24                                       84.8gm * P<.3  -
  a    M m icr wat lun ade 24m24                                        84.8gm * P<.3  -
  b    M m icr wat liv tum 24m24                                        no dre   P=1.  -


3-SULFOLENE              100ng..:..1ug....:..10......:..100....:..1mg....:..10......:..100....:..1g......:..10
4633   M f b6c gav TBA MXB 78w90 sv                          :>        no dre   P=1.  -
  a    M f b6c gav liv MXB 78w90 sv                                    no dre   P=1.
  b    M f b6c gav lun MXB 78w90 sv                                    1.94gm * P<.4
4634   M m b6c gav TBA MXB 78w90 sv                           :>       631.mg * P<.5
  a    M m b6c gav liv MXB 78w90 sv                                    415.mg * P<.09
  b    M m b6c gav lun MXB 78w90 sv                                    882.mg * P<.2
4635   R f osm gav TBA MXB 18w26 dv                       :>           5.37gm * P<1.
  a    R f osm gav liv MXB 18w26 dv                                    no dre   P=1.
4636   R m osm gav TBA MXB 16w26 adsv           :     ±               30.8mg * P<.04 -
  a    R m osm gav liv MXB 16w26 adsv                                  no dre   P=1.


4,4'-SULFONYLBISACETANILIDE  100ng..:..1ug....:..10......:..100....:..1mg....:..10......:..100....:..1g......:..10
4637   R f buf eat mgl adc 43w92 e                         .>         55.6mg   P<.3  +


SYMPHYTINE               100ng..:..1ug....:..10......:..100....:..1mg....:..10......:..100....:..1g......:..10
4638   R m aci ipj liv mix 56w92 ev                 .    ±            1.91mg   P<.02 +
  a    R m aci ipj tba mix 56w92 ev                                   4.50mg   P<.5


TACE                     100ng..:..1ug....:..10......:..100....:..1mg....:..10......:..100....:..1g......:..10
4639   R f cdr eat liv tum 24m24 e                      .>           no dre   P=1.  -
4640   R m cdr eat liv tum 24m24 e                      .>           no dre   P=1.  -


TARA GUM                 100ng..:..1ug....:..10......:..100....:..1mg....:..10......:..100....:..1g......:..10
4641   M f b6c eat TBA MXB 24m24                                     no dre   P=1.
  a    M f b6c eat liv MXB 24m24                                     no dre   P=1.
  b    M f b6c eat lun MXB 24m24                                     no dre   P=1.
4642   M m b6c eat TBA MXB 24m24                             :> 31.0gm * P<.7  -
  a    M m b6c eat liv MXB 24m24                                289.gm * P<1.
  b    M m b6c eat lun MXB 24m24                                85.6gm * P<.8
4643   R f f34 eat TBA MXB 24m25                          :>   5.03gm * P<.5  -
  a    R f f34 eat liv MXB 24m25                                no dre   P=1.
```

	RefNum	LoConf	UpConf	Cntrl	1Dose	1Inc	2Dose	2Inc							Citation or Pathology	
																Brkly Code
a	1924o	1.39gm	3.75gm	7/184	39.0mg	2/95	78.0mg	0/94	156.mg	0/96	312.mg	2/96	624.mg	33/94		
b	1924o	2.69gm	n.s.s.	10/184	39.0mg	6/95	78.0mg	12/94	156.mg	5/96	312.mg	7/96	624.mg	11/94		
c	1924o	3.26gm	n.s.s.	8/184	39.0mg	5/95	78.0mg	8/94	156.mg	4/96	312.mg	6/96	624.mg	8/94		
d	1924o	3.80gm	n.s.s.	5/184	39.0mg	4/95	78.0mg	4/94	156.mg	1/96	312.mg	5/96	624.mg	6/94		
e	1924o	4.64gm	n.s.s.	2/184	39.0mg	1/95	78.0mg	4/94	156.mg	1/96	312.mg	2/96	624.mg	4/94		
f	1924o	9.74gm	n.s.s.	0/184	39.0mg	0/95	78.0mg	1/94	156.mg	0/96	312.mg	0/96	624.mg	1/94		
4623	1924m	33.9mg	n.s.s.	1/24	36.0mg	0/24	144.mg	0/24	576.mg	0/24						
a	1924m	153.mg	n.s.s.	3/24	36.0mg	0/24	144.mg	1/24	(576.mg	0/24)						
4624	1924n	720.mg	n.s.s.	0/24	36.0mg	1/24	144.mg	0/24	576.mg	2/24						
a	1924n	1.95gm	n.s.s.	3/24	36.0mg	2/24	144.mg	0/24	576.mg	0/24						
b	1924n	635.mg	n.s.s.	1/24	36.0mg	3/24	144.mg	5/24	576.mg	2/24						
c	1924n	1.58gm	n.s.s.	3/24	36.0mg	1/24	144.mg	2/24	576.mg	0/24						
d	1924n	1.04gm	n.s.s.	4/24	36.0mg	5/24	144.mg	2/24	576.mg	2/24						
4625	1924o	846.mg	1.77gm	3/191	36.0mg	0/96	72.0mg	1/96	144.mg	5/92	288.mg	4/96	576.mg	44/94		
a	1924o	861.mg	1.81gm	3/191	36.0mg	0/96	72.0mg	1/96	144.mg	5/92	288.mg	3/96	576.mg	44/94		
b	1924o	1.71gm	n.s.s.	44/191	36.0mg	20/96	72.0mg	23/96	144.mg	27/92	288.mg	28/96	576.mg	22/94		
c	1924o	2.37gm	n.s.s.	26/191	36.0mg	12/96	72.0mg	12/96	144.mg	14/92	288.mg	15/96	576.mg	14/94		
d	1924o	11.7gm	n.s.s.	0/191	36.0mg	0/96	72.0mg	0/96	144.mg	0/92	288.mg	1/96	576.mg	0/94		
e	1924o	3.16gm	n.s.s.	21/191	36.0mg	9/96	72.0mg	11/96	144.mg	15/92	288.mg	13/96	576.mg	9/94		
f	1924o	7.05gm	n.s.s.	25/191	36.0mg	7/96	72.0mg	9/96	144.mg	6/92	288.mg	9/96	576.mg	5/94		

SULFATE, SODIUM   (disodium sulfate) 7757-82-6

	RefNum	LoConf	UpConf	Cntrl	1Dose	1Inc	2Dose	2Inc	Citation or Pathology
4626	1118	544.mg	n.s.s.	1/30	811.mg	2/30			Cohen;canr,38,1398-1405;1978
a	1118	1.25gm	n.s.s.	0/30	811.mg	0/30			
b	1118	1.25gm	n.s.s.	0/30	811.mg	0/30			

SULFISOXAZOLE   127-69-5

	RefNum	LoConf	UpConf	Cntrl	1Dose	1Inc	2Dose	2Inc	Pathology	Brkly Code
4627	TR138	1.01gm	11.0gm	0/50	490.mg	5/50	(1.96gm	2/50)		S
a	TR138	4.84gm	161.gm	0/50	490.mg	1/50	1.96gm	5/50	lun:a/a,a/c.	S
b	TR138	1.71gm	n.s.s.	24/50	490.mg	29/50	1.96gm	31/50		
c	TR138	1.01gm	11.0gm	0/50	490.mg	5/50	(1.96gm	2/50)	liv:hpa,hpc,nnd.	
d	TR138	4.84gm	161.gm	0/50	490.mg	1/50	1.96gm	5/50	lun:a/a,a/c.	
4628	TR138	2.17gm	n.s.s.	29/50	495.mg	31/50	1.98gm	28/50		
a	TR138	1.87gm	n.s.s.	15/50	495.mg	13/50	1.98gm	21/50	liv:hpa,hpc,nnd.	
b	TR138	1.30gm	n.s.s.	4/50	495.mg	5/50	(1.98gm	0/50)	lun:a/a,a/c.	
4629	TR138	470.mg	n.s.s.	38/50	97.2mg	37/50	389.mg	35/50		
a	TR138	4.33gm	n.s.s.	1/50	97.2mg	0/50	389.mg	0/50	liv:hpa,hpc,nnd.	
4630	TR138	50.6mg	n.s.s.	28/50	97.2mg	36/50	(392.mg	31/50)		
a	TR138	1.10gm	n.s.s.	1/50	97.2mg	4/50	392.mg	2/50	liv:hpa,hpc,nnd.	

SULFITE, POTASSIUM METABI-   4429-42-9

	RefNum	LoConf	UpConf	Cntrl	1Dose	1Inc	2Dose	2Inc	Citation
4631	1391	8.26gm	n.s.s.	0/50	2.00gm	2/50	(4.00gm	0/50)	Tanaka;eaes,3,451-453;1979
a	1391	22.0gm	n.s.s.	1/50	2.00gm	0/50	4.00gm	3/50	
b	1391	33.3gm	n.s.s.	0/50	2.00gm	0/50	4.00gm	1/50	
4632	1391	20.9gm	n.s.s.	0/50	1.67gm	1/50	3.33gm	1/50	
a	1391	20.9gm	n.s.s.	0/50	1.67gm	1/50	3.33gm	1/50	
b	1391	11.4gm	n.s.s.	0/50	1.67gm	0/50	3.33gm	0/50	

3-SULFOLENE   77-79-2

	RefNum	LoConf	UpConf	Cntrl	1Dose	1Inc	2Dose	2Inc	Pathology
4633	TR102	426.mg	n.s.s.	4/20	235.mg	10/50	470.mg	1/50	
a	TR102	n.s.s.	n.s.s.	0/20	235.mg	1/50	470.mg	0/50	liv:hpa,hpc,nnd.
b	TR102	480.mg	n.s.s.	1/20	235.mg	5/50	470.mg	1/50	lun:a/a,a/c.
4634	TR102	152.mg	n.s.s.	5/20	190.mg	20/50	381.mg	8/50	
a	TR102	179.mg	n.s.s.	1/20	190.mg	11/50	381.mg	5/50	liv:hpa,hpc,nnd.
b	TR102	350.mg	n.s.s.	0/20	190.mg	4/50	381.mg	2/50	lun:a/a,a/c.
4635	TR102	60.1mg	n.s.s.	14/20	52.5mg	39/50	105.mg	21/50	
a	TR102	n.s.s.	n.s.s.	0/20	52.5mg	2/50	105.mg	0/50	liv:hpa,hpc,nnd.
4636	TR102	8.90mg	n.s.s.	9/20	88.5mg	13/50	249.mg	1/50	
a	TR102	n.s.s.	n.s.s.	0/20	88.5mg	1/50	249.mg	0/50	liv:hpa,hpc,nnd.

4,4'-SULFONYLBISACETANILIDE   77-46-3

	RefNum	LoConf	UpConf	Cntrl	1Dose	1Inc	Citation
4637	144	9.05mg	n.s.s.	0/18	5.93mg	1/18	Morris;jnci,24,149-180;1960

SYMPHYTINE   22571-95-5

	RefNum	LoConf	UpConf	Cntrl	1Dose	1Inc	Citation
4638	1396	.655mg	n.s.s.	0/13	1.20mg	4/14	Hirono;jnci,63,469-472;1979
a	1396	.895mg	n.s.s.	3/18	1.20mg	5/18	

TACE   (chlorotrianisene) 569-57-3

	RefNum	LoConf	UpConf	Cntrl	1Dose	1Inc	2Dose	2Inc	3Dose	3Inc	Citation
4639	108	.162mg	n.s.s.	0/20	50.0ug	0/20	.200mg	0/20	2.00mg	0/20	Gibson;txap,11,489-510;1967
4640	108	.162mg	n.s.s.	0/20	50.0ug	0/20	.200mg	0/20	2.00mg	0/20	

TARA GUM   39300-88-4

	RefNum	LoConf	UpConf	Cntrl	1Dose	1Inc	2Dose	2Inc	Pathology
4641	TR224	10.4gm	n.s.s.	34/50	3.19gm	26/50	6.38gm	26/50	
a	TR224	9.39gm	n.s.s.	10/50	3.19gm	6/50	(6.38gm	3/50)	liv:hpa,hpc,nnd.
b	TR224	38.2gm	n.s.s.	8/50	3.19gm	2/50	6.38gm	3/50	lun:a/a,a/c.
4642	TR224	5.09gm	n.s.s.	31/50	2.94gm	28/50	5.89gm	36/50	
a	TR224	10.5gm	n.s.s.	17/50	2.94gm	12/50	5.89gm	18/50	liv:hpa,hpc,nnd.
b	TR224	10.5gm	n.s.s.	10/50	2.94gm	11/50	5.89gm	12/50	lun:a/a,a/c.
4643	TR224	1.19gm	n.s.s.	41/50	1.21gm	48/50	2.43gm	47/50	
a	TR224	21.2gm	n.s.s.	2/50	1.21gm	0/50	2.43gm	1/50	liv:hpa,hpc,nnd.

```
          Spe Strain Site   Xpo+Xpt                                                      TD50      2Tailpvl
              Sex  Route  Hist    Notes                                                   DR      AuOp

4644 R m f34 eat TBA MXB 24m24                                              :>           93.3gm * P<1.   -
   a R m f34 eat liv MXB 24m24                                                           24.5gm * P<.4


TELODRIN              100ng..:..1ug....:..10.....:..100....:..1mg...:..10.....:..100....:..1g.....:..10
4645 M f b6a orl lun ade 76w76 evx                           .>                         no dre   P=1.   -
   a M f b6a orl liv hpt 76w76 evx                                                       no dre   P=1.   -
   b M f b6a orl tba mix 76w76 evx                                                       no dre   P=1.   -
4646 M m b6a orl lun ade 76w76 evx                  .>                                   .146mg   P<.3   -
   a M m b6a orl liv hpt 76w76 evx                                                       no dre   P=1.   -
   b M m b6a orl tba mix 76w76 evx                                                       .136mg   P<.3   -
4647 M f b6c orl lun ade 76w76 evx                        .>                             .571mg   P<.3   -
   a M f b6c orl liv hpt 76w76 evx                                                       no dre   P=1.   -
   b M f b6c orl tba mix 76w76 evx                                                       .277mg   P<.2   -
4648 M m b6c orl liv hpt 76w76 evx                     .>                                .500mg   P<.3   -
   a M m b6c orl lun ade 76w76 evx                                                       .500mg   P<.3   -
   b M m b6c orl tba mix 76w76 evx                                                       .242mg   P<.1   -


TELONE II             100ng..:..1ug....:..10.....:..100....:..1mg...:..10.....:..100....:..1g.....:..10
4649 M f b6c gav MXB MXB 24m25                                          : + :           36.3mg * P<.0005
   a M f b6c gav ubl tcc 24m25                                                           49.6mg * P<.0005c
   b M f b6c gav lun a/a 24m25                                                           149.mg * P<.0005c
   c M f b6c gav liv MXA 24m25                                                           89.3mg \ P<.02
   d M f b6c gav lun MXA 24m25                                                           182.mg * P<.03
   e M f b6c gav liv hpa 24m25                                                           215.mg * P<.03
   f M f b6c gav sto MXA 24m25                                                           344.mg * P<.02   c
   g M f b6c gav TBA MXB 24m25                                                           34.9mg * P<.002
   h M f b6c gav liv MXB 24m25                                                           89.3mg \ P<.02
   i M f b6c gav lun MXB 24m25                                                           182.mg * P<.03
4650 M m b6c gav TBA MXB 24m24 s                               :>                        309.mg * P<.9   i
   a M m b6c gav liv MXB 24m24 s                                                         1.93gm * P<1.
   b M m b6c gav lun MXB 24m24 s                                                         127.mg * P<.4
4651 R f f34 gav thy MXA 24m25                                       :   ±              137.mg * P<.03
   a R f f34 gav sto sqp 24m25                                                           176.mg * P<.05   p
   b R f f34 gav TBA MXB 24m25                                                           no dre   P=1.
   c R f f34 gav liv MXB 24m25                                                           168.mg * P<.5
4652 R m f34 gav MXB MXB 24m25                                       : + :              33.2mg * P<.0005
   a R m f34 gav MXA MXA 24m25                                                           64.1mg / P<.0005c
   b R m f34 gav liv MXA 24m25                                                           68.0mg * P<.01
   c R m f34 gav MXA sqp 24m25                                                           97.0mg * P<.006  c
   d R m f34 gav liv nnd 24m25                                                           76.0mg * P<.02   c
   e R m f34 gav sto sqc 24m25                                                           213.mg * P<.02   c
   f R m f34 gav TBA MXB 24m25                                                           no dre   P=1.
   g R m f34 gav liv MXB 24m25                                                           68.0mg * P<.01


TEMAZEPAM, PHARMACEUTICAL GRADE  100ng..:..1ug....:..10.....:..100....:..1mg...:..10.....:..100....:..1g.....:..10
4653 M f cd1 eat liv hpa 78w78 e                                             .  ±       1.86gm * P<.03   -
   a M f cd1 eat lun bro 78w78 e                                                         9.61gm * P<.2   -
   b M f cd1 eat lun alc 78w78 e                                                         no dre   P=1.   -
   c M f cd1 eat tba tum 78w78 e                                                         no dre   P=1.   -
4654 M m cd1 eat liv hpa 78w78 e                                               .>       2.57gm * P<.4   -
   a M m cd1 eat lun alc 78w78 e                                                         no dre   P=1.   -
   b M m cd1 eat tba tum 78w78 e                                                         no dre   P=1.   -


TERBUTALINE           100ng..:..1ug....:..10.....:..100....:..1mg...:..10.....:..100....:..1g.....:..10
4655 R f cdr eat meo ley 24m24 erv                                           .  +   .   410.mg * P<.0005+


3,3',4,4'-TETRAAMINOBIPHENYL.4HCl 100ng..:..1ug....:..10.....:..100....:..1mg...:..10.....:..100....:..1g.....:..10
4656 M f chi eat lun mix 77w98                                          :   ±           1.03gm * P<.07   -
   a M f chi eat liv mix 77w98                                                           1.69gm * P<.03   -
   b M f chi eat tba mix 77w98                                                           599.mg * P<.05   -
4657 M m chi eat lun mix 77w94                                       :   +   :          288.mg \ P<.003  +
   a M m chi eat liv mix 77w94                                                           4.17gm * P<.7   -
   b M m chi eat tba mix 77w94                                                           224.mg \ P<.003
4658 R m cdr eat liv mix 18m25                                       :   ±              259.mg * P<.02   -
   a R m cdr eat tba mix 18m25                                                           200.mg * P<.2
4659 R m cdr eat liv hpt 18m25   pool                               :   ±               395.mg * P<.02   +


2,3,5,6-TETRACHLORO-4-NITROANISOLE  ..:..1ug....:..10.....:..100....:..1mg...:..10.....:..100....:..1g......:..10
4660 M f b6c eat TBA MXB 24m24                                     :>                    61.7mg * P<.6   -
   a M f b6c eat liv MXB 24m24                                                           no dre   P=1.
   b M f b6c eat lun MXB 24m24                                                           1.77gm * P<1.
4661 M m b6c eat TBA MXB 24m24                                       :>                  no dre   P=1.   -
   a M m b6c eat liv MXB 24m24                                                           no dre   P=1.
   b M m b6c eat lun MXB 24m24                                                           215.mg * P<.8
4662 R f f34 eat ute esp 24m24                                   :  ±                   #19.1mg * P<.05  -
   a R f f34 eat liv MXA 24m24                                                           45.8mg * P<.02
   b R f f34 eat TBA MXB 24m24                                                           106.mg * P<.9
   c R f f34 eat liv MXB 24m24                                                           45.8mg * P<.02
4663 R m f34 eat liv MXA 24m24                                   :  ±                   #20.0mg * P<.03  -
   a R m f34 eat TBA MXB 24m24                                                           no dre   P=1.
   b R m f34 eat liv MXB 24m24                                                           20.0mg * P<.03
```

	RefNum	LoConf	UpConf	Cntrl	1Dose	1Inc	2Dose	2Inc	Citation or Pathology
									Brkly Code
4644	TR224	1.47gm	n.s.s.	36/50	972.mg	39/50	1.96gm	38/50	
a	TR224	6.14gm	n.s.s.	1/50	972.mg	2/50	1.96gm	3/50	liv:hpa,hpc,nnd.

TELODRIN (isobenzan) 297-78-9

	RefNum	LoConf	UpConf	Cntrl	1Dose	1Inc			Citation or Pathology
4645	1249	.106mg	n.s.s.	1/17	89.1ug	1/18			Innes;ntis,1968/1969
a	1249	.177mg	n.s.s.	0/17	89.1ug	0/18			
b	1249	84.1ug	n.s.s.	2/17	89.1ug	2/18			
4646	1249	39.4ug	n.s.s.	2/18	82.9ug	5/18			
a	1249	98.0ug	n.s.s.	1/18	82.9ug	1/18			
b	1249	35.4ug	n.s.s.	3/18	82.9ug	6/18			
4647	1249	92.9ug	n.s.s.	0/16	89.1ug	1/18			
a	1249	.177mg	n.s.s.	0/16	89.1ug	0/18			
b	1249	68.0ug	n.s.s.	0/16	89.1ug	2/18			
4648	1249	81.5ug	n.s.s.	0/16	82.9ug	1/17			
a	1249	81.5ug	n.s.s.	0/16	82.9ug	1/17			
b	1249	59.5ug	n.s.s.	0/16	82.9ug	2/17			

TELONE II (1,3-dichloropropene) 542-75-6

	RefNum	LoConf	UpConf	Cntrl	1Dose	1Inc	2Dose	2Inc	Citation or Pathology
4649	TR269	24.5mg	57.1mg	0/50	20.8mg	12/50	41.7mg	26/50	lun:a/a; sto:sqc,sqp; ubl:tcc. C
a	TR269	31.8mg	83.7mg	0/50	20.8mg	8/50	41.7mg	21/50	
b	TR269	74.7mg	446.mg	0/50	20.8mg	3/50	41.7mg	8/50	
c	TR269	37.3mg	n.s.s.	1/50	20.8mg	8/50	(41.7mg	3/50)	liv:hpa,hpc. S
d	TR269	77.8mg	n.s.s.	2/50	20.8mg	4/50	41.7mg	8/50	lun:a/a,a/c. S
e	TR269	97.3mg	n.s.s.	0/50	20.8mg	5/50	41.7mg	3/50	S
f	TR269	130.mg	n.s.s.	0/50	20.8mg	1/50	41.7mg	4/50	sto:sqc,sqp.
g	TR269	19.3mg	140.mg	16/50	20.8mg	33/50	41.7mg	34/50	
h	TR269	37.3mg	n.s.s.	1/50	20.8mg	8/50	(41.7mg	3/50)	liv:hpa,hpc,nnd.
i	TR269	77.8mg	n.s.s.	2/50	20.8mg	4/50	41.7mg	8/50	lun:a/a,a/c.
4650	TR269	26.7mg	n.s.s.	8/50	21.2mg	28/50	42.2mg	30/50	
a	TR269	63.6mg	n.s.s.	5/50	21.2mg	7/50	42.2mg	13/50	liv:hpa,hpc,nnd.
b	TR269	37.2mg	n.s.s.	1/50	21.2mg	13/50	42.2mg	12/50	lun:a/a,a/c.
4651	TR269	55.8mg	n.s.s.	0/52	10.4mg	2/52	21.0mg	4/52	thy:fca,fcc. S
a	TR269	66.3mg	n.s.s.	0/52	10.4mg	2/52	21.0mg	3/52	
b	TR269	17.6mg	n.s.s.	45/52	10.4mg	44/52	21.0mg	44/52	
c	TR269	39.0mg	n.s.s.	6/52	10.4mg	6/52	21.0mg	10/52	liv:hpa,hpc,nnd.
4652	TR269	19.3mg	85.7mg	2/52	10.5mg	7/52	21.0mg	19/52	for:sqp; liv:nnd; sto:sqc,sqp. C
a	TR269	32.9mg	217.mg	1/52	10.5mg	1/52	21.0mg	13/52	for:sqp; sto:sqc,sqp.
b	TR269	32.9mg	4.91gm	1/52	10.5mg	6/52	21.0mg	8/52	liv:hpc,nnd. S
c	TR269	43.9mg	1.16gm	1/52	10.5mg	1/52	21.0mg	9/52	for:sqp; sto:sqp.
d	TR269	35.1mg	n.s.s.	1/52	10.5mg	6/52	21.0mg	7/52	
e	TR269	73.6mg	n.s.s.	0/52	10.5mg	0/52	21.0mg	4/52	
f	TR269	17.1mg	n.s.s.	44/52	10.5mg	39/52	21.0mg	41/52	
g	TR269	32.9mg	4.91gm	1/52	10.5mg	6/52	21.0mg	8/52	liv:hpa,hpc,nnd.

TEMAZEPAM, PHARMACEUTICAL GRADE   846-50-4

	RefNum	LoConf	UpConf	Cntrl	1Dose	1Inc	2Dose	2Inc	3Dose	3Inc	Citation or Pathology
4653	1994	658.mg	n.s.s.	0/100	10.0mg	1/100	80.0mg	1/100	160.mg	4/100	Robison;faat,4,394-405;1984
a	1994	1.56gm	n.s.s.	0/100	10.0mg	0/100	80.0mg	0/100	160.mg	1/100	
b	1994	670.mg	n.s.s.	11/100	10.0mg	12/100	80.0mg	10/100	160.mg	9/100	
c	1994	546.mg	n.s.s.	30/100	10.0mg	31/100	80.0mg	16/100	160.mg	25/100	
4654	1994	561.mg	n.s.s.	8/100	10.0mg	4/99	80.0mg	2/100	160.mg	10/100	
a	1994	1.76gm	n.s.s.	24/100	10.0mg	11/99	80.0mg	3/100	160.mg	5/100	
b	1994	970.mg	n.s.s.	39/100	10.0mg	21/99	80.0mg	13/100	160.mg	17/100	

TERBUTALINE   23031-25-6

	RefNum	LoConf	UpConf	Cntrl	1Dose	1Inc	2Dose	2Inc	Citation or Pathology
4655	1734	186.mg	1.33gm	0/105	8.42mg	1/55	84.2mg	7/55	Jack;txcy,27,315-320;1983

3,3',4,4'-TETRAAMINOBIPHENYL.4HCl   (3,3'-diaminobenzidine.4HCl) 7411-49-6

	RefNum	LoConf	UpConf	Cntrl	1Dose	1Inc	2Dose	2Inc	Citation or Pathology
4656	381	377.mg	n.s.s.	4/17	468.mg	6/19	851.mg	4/17	Weisburger;jept,2,325-356;1978/pers.comm./Russfield 1973
a	381	526.mg	n.s.s.	0/17	468.mg	1/19	851.mg	3/17	
b	381	231.mg	n.s.s.	11/17	468.mg	9/19	851.mg	11/17	
4657	381	117.mg	1.52gm	1/19	432.mg	6/18	(785.mg	7/22)	
a	381	321.mg	n.s.s.	1/19	432.mg	1/18	785.mg	1/22	
b	381	95.1mg	1.50gm	4/19	432.mg	9/18	(785.mg	13/22)	
4658	381	89.7mg	n.s.s.	0/16	120.mg	1/16	240.mg	3/15	
a	381	63.4mg	n.s.s.	11/16	120.mg	8/16	240.mg	12/15	
4659	381	115.mg	n.s.s.	2/111p	120.mg	1/16	240.mg	3/15	

2,3,5,6-TETRACHLORO-4-NITROANISOLE   2438-88-2

	RefNum	LoConf	UpConf	Cntrl	1Dose	1Inc	2Dose	2Inc	3Dose	3Inc	Citation or Pathology
4660	TR114	11.8mg	n.s.s.	34/55	7.70mg	37/55	15.5mg	38/55			
a	TR114	83.1mg	n.s.s.	11/55	7.70mg	3/55	15.5mg	5/55			liv:hpa,hpc,nnd.
b	TR114	43.3mg	n.s.s.	4/55	7.70mg	8/55	15.5mg	4/55			lun:a/a,a/c.
4661	TR114	16.9mg	n.s.s.	43/55	7.20mg	38/55	14.3mg	36/55			
a	TR114	17.9mg	n.s.s.	28/55	7.20mg	16/55	(14.3mg	10/55)			liv:hpa,hpc,nnd.
b	TR114	22.9mg	n.s.s.	12/55	7.20mg	17/55	14.3mg	13/55			lun:a/a,a/c.
4662	TR114	8.30mg	n.s.s.	2/50	3.00mg	9/50	5.90mg	8/50			S
a	TR114	17.4mg	n.s.s.	0/50	3.00mg	1/50	5.90mg	4/50			liv:hpc,nnd. S
b	TR114	5.10mg	n.s.s.	27/50	3.00mg	34/50	5.90mg	28/50			
c	TR114	17.4mg	n.s.s.	0/50	3.00mg	1/50	5.90mg	4/50			liv:hpa,hpc,nnd.
4663	TR114	9.41mg	n.s.s.	0/50	2.40mg	5/50	4.70mg	1/25	4.60mg	3/24	liv:hpc,nnd. S
a	TR114	6.57mg	n.s.s.	28/50	2.40mg	27/50	4.70mg	12/25	4.60mg	14/24	
b	TR114	9.41mg	n.s.s.	0/50	2.40mg	5/50	4.70mg	1/25	4.60mg	3/24	liv:hpa,hpc,nnd.

```
         Spe Strain Site   Xpo+Xpt                                                          TD50      2Tailpvl
             Sex  Route Hist  Notes                                                               DR     AuOp

2,2',5,5'-TETRACHLOROBENZIDINE     100ng..:..1ug...:..10....:..100....:..1mg...:..10....:..100....:..1g.....:..10
4664 M f ddx eat --- mly 24m24 e                                          . + .         109.mg   P<.0005
a    M f ddx eat tba mal 24m24 e                                                        87.2mg   P<.0005
4665 R f wis eat liv hpt 24m24 e                                                 .>      735.mg   P<.2
a    R f wis eat ubl tcc 24m24 e                                                        2.31gm   P<.4
4666 R m wis eat ubl tcc 24m24 e                                                 .>      561.mg   P<.2
a    R m wis eat liv tum 24m24 e                                                        no dre   P=1.


2,3,7,8-TETRACHLORODIBENZO-p-DIOXIN   ..:..1ug...:..10......:..100....:..1mg...:..10.....:..100....:..1g.....:..10
4667          : + :                            M f b6c gav MXB MXB 24m24   526.ng  Z P<.002
a                                              M f b6c gav --- lhc 24m24   705.ng  * P<.008
b                                              M f b6c gav liv MXA 24m24   756.ng  * P<.008 c
c                                              M f b6c gav --- lym 24m24   682.ng  * P<.04
d                                              M f b6c gav --- MXA 24m24   709.ng  * P<.05
e                                              M f b6c gav liv hpc 24m24   1.30ug  * P<.02  c
f                                              M f b6c gav thy fca 24m24   1.59ug  * P<.03  c
g                                              M f b6c gav sub fbs 24m24   1.72ug  * P<.02
h                                              M f b6c gav TBA MXB 24m24   494.ng  * P<.06
i                                              M f b6c gav liv MXB 24m24   756.ng  * P<.008
j                                              M f b6c gav lun MXB 24m24   no dre   P=1.
4668          : + :                            M m b6c gav liv MXA 24m24   86.8ng  * P<.003 c
a                                              M m b6c gav liv hpc 24m24   147.ng  * P<.02  c
b                                              M m b6c gav lun MXA 24m24   198.ng  * P<.02
c                                              M m b6c gav lun a/a 24m24   219.ng  * P<.02
d                                              M m b6c gav liv hpa 24m24   271.ng  * P<.07  c
e                                              M m b6c gav TBA MXB 24m24   141.ng  * P<.2
f                                              M m b6c gav liv MXB 24m24   86.8ng  * P<.003
g                                              M m b6c gav lun MXB 24m24   198.ng  * P<.02
4669                 .>                         M m shr gav lun tum 12m24 e no dre  P=1.   -
a                                              M m shr gav liv mix 12m24 e no dre  P=1.   +
b                                              M m shr gav tba mix 12m24 e no dre  P=1.
4670          : + :                            R f osm gav liv MXA 24m24   127.ng  * P<.0005c
a                                              R f osm gav liv nnd 24m24   154.ng  * P<.003 c
b                                              R f osm gav TBA MXB 24m24   1.58ug  * P<1.
c                                              R f osm gav liv MXB 24m24   127.ng  * P<.0005
4671          : + :                            R m osm gav thy MXA 24m24   101.ng  * P<.002 c
a                                              R m osm gav thy fca 24m24   113.ng  * P<.003 c
b                                              R m osm gav liv nnd 24m24   391.ng  * P<.004
c                                              R m osm gav liv MXA 24m24   391.ng  * P<.004
d                                              R m osm gav adr coa 24m24   12.1ng  Z P<.04
e                                              R m osm gav sub fib 24m24   169.ng  * P<.02
f                                              R m osm gav TBA MXB 24m24   133.ng  * P<.3
g                                              R m osm gav liv MXB 24m24   391.ng  * P<.004
4672          : + :                            R m sda eat lun sqa 78w95   79.6ng  * P<.0005

a                                              R m sda eat liv mix 78w95   83.1ng  * P<.0005

4673          : + :                            R f sss eat liv hph 24m24   6.67ng  Z P<.0005+
a                                              R f sss eat liv hpc 24m24   65.1ng  * P<.0005+
b                                              R f sss eat lun sqk 24m24   192.ng  * P<.0005+
c                                              R f sss eat hnt sqs 24m24   205.ng  * P<.0005+
4674          : + :                            R m sss eat adr coa 24m24   217.ng  * P<.002
a                                              R m sss eat hnt sqs 24m24   393.ng  * P<.002 +
b                                              R m sss eat ton sqs 24m24   628.ng  * P<.04  +


2,4,5,4'-TETRACHLORODIPHENYL SULFONE   ..:..1ug...:..10......:..100....:..1mg...:..10.....:..100....:..1g.....:..10
4675 M f b6a orl liv hpt 76w76 evx                                       .>        233.mg   P<.3   -
a    M f b6a orl lun ade 76w76 evx                                                 no dre   P=1.   -
b    M f b6a orl tba mix 76w76 evx                                                 233.mg   P<.7   -
4676 M m b6a orl lun ade 76w76 evx                                       .>        no dre   P=1.   -
a    M m b6a orl liv hpt 76w76 evx                                                 no dre   P=1.   -
b    M m b6a orl tba mix 76w76 evx                                                 no dre   P=1.   -
4677 M f b6c orl liv hpt 76w76 evx                                          .>     no dre   P=1.   -
a    M f b6c orl lun mix 76w76 evx                                                 no dre   P=1.   -
b    M f b6c orl tba mix 76w76 evx                                                 220.mg   P<.3   -
4678 M m b6c orl liv hpt 76w76 evx                                   .  ±         46.2mg   P<.02  -
a    M m b6c orl lun ade 76w76 evx                                                 205.mg   P<.3   -
b    M m b6c orl tba mix 76w76 evx                                                 35.6mg   P<.007 -


1,1,1,2-TETRACHLOROETHANE     100ng..:..1ug...:..10....:..100....:..1mg...:..10....:..100....:..1g.....:..10
4679 M f b6c gav liv MXA 24m24 as                                        : + :     175.mg / P<.0005c
a    M f b6c gav liv hpa 24m24 as                                                  254.mg / P<.0005c
b    M f b6c gav liv hpc 24m24 as                                                  640.mg * P<.0005c
c    M f b6c gav TBA MXB 24m24 as                                                  110.mg / P<.0005
d    M f b6c gav liv MXB 24m24 as                                                  175.mg / P<.0005
e    M f b6c gav lun MXB 24m24 as                                                  1.82gm * P<.07
4680 M m b6c gav liv MXA 24m24 as                                        : + :     124.mg / P<.0005
a    M m b6c gav liv hpa 24m24 as                                                  190.mg / P<.0005c
b    M m b6c gav TBA MXB 24m24 as                                                  97.9mg / P<.0005
c    M m b6c gav liv MXB 24m24 as                                                  124.mg / P<.0005
d    M m b6c gav lun MXB 24m24 as                                                  1.33gm / P<.3
```

	RefNum	LoConf	UpConf	Cntrl	1Dose	1Inc	2Dose	2Inc							Citation or Pathology
															Brkly Code

2,2',5,5'-TETRACHLOROBENZIDINE   15721-02-5

	RefNum	LoConf	UpConf	Cntrl	1Dose	1Inc		Citation or Pathology
4664	558	57.0mg	240.mg	0/20	130.mg	14/25		Yoshimoto;jkmj,25,123-128;1978
a	558	47.0mg	184.mg	0/20	130.mg	16/25		
4665	558	222.mg	n.s.s.	0/10	150.mg	3/23		
a	558	376.mg	n.s.s.	0/10	150.mg	1/23		
4666	558	169.mg	n.s.s.	0/11	120.mg	3/22		
a	558	544.mg	n.s.s.	0/11	120.mg	0/22		

2,3,7,8-TETRACHLORODIBENZO-p-DIOXIN   (TCDD, dioxin) 1746-01-6

	RefNum	LoConf	UpConf	Cntrl	1Dose	1Inc	2Dose	2Inc	3Dose	3Inc	4Dose	4Inc	5Dose	5Inc	Citation or Pathology	Brkly
4667	TR209	236.ng	2.99ug	3/75	5.55ng	9/50	27.8ng	7/50	278.ng	15/50					liv:hpa,hpc; thy:fca.	C
a	TR209	294.ng	19.3ug	9/75	5.55ng	4/50	27.8ng	8/50	278.ng	14/50						S
b	TR209	303.ng	20.5ug	3/75	5.55ng	6/50	27.8ng	6/50	278.ng	11/50					liv:hpa,hpc.	
c	TR209	262.ng	n.s.s.	18/75	5.55ng	11/50	27.8ng	13/50	278.ng	20/50						S
d	TR209	266.ng	n.s.s.	18/75	5.55ng	12/50	27.8ng	13/50	278.ng	20/50					---:leu,lym.	S
e	TR209	466.ng	n.s.s.	1/75	5.55ng	2/50	27.8ng	2/50	278.ng	6/50						
f	TR209	523.ng	n.s.s.	0/75	5.55ng	3/50	27.8ng	1/50	278.ng	5/50						
g	TR209	594.ng	n.s.s.	1/75	5.55ng	1/50	27.8ng	1/50	278.ng	5/50						S
h	TR209	184.ng	n.s.s.	36/75	5.55ng	25/50	27.8ng	28/50	278.ng	34/50						
i	TR209	303.ng	20.5ug	3/75	5.55ng	6/50	27.8ng	6/50	278.ng	11/50					liv:hpa,hpc,nnd.	
j	TR209	956.ng	n.s.s.	2/75	5.55ng	3/50	27.8ng	4/50	278.ng	2/50					lun:a/a,a/c.	
4668	TR209	40.6ng	613.ng	15/75	1.39ng	12/50	6.94ng	13/50	69.4ng	27/50					liv:hpa,hpc.	
a	TR209	59.8ng	n.s.s.	8/75	1.39ng	9/50	6.94ng	8/50	69.4ng	17/50						
b	TR209	78.4ng	n.s.s.	10/75	1.39ng	2/50	6.94ng	4/50	69.4ng	13/50					lun:a/a,a/c.	S
c	TR209	84.1ng	n.s.s.	7/75	1.39ng	2/50	6.94ng	4/50	69.4ng	11/50						S
d	TR209	90.3ng	n.s.s.	7/75	1.39ng	3/50	6.94ng	5/50	69.4ng	10/50						
e	TR209	43.3ng	n.s.s.	40/75	1.39ng	25/50	6.94ng	27/50	69.4ng	37/50						
f	TR209	40.6ng	613.ng	15/75	1.39ng	12/50	6.94ng	13/50	69.4ng	27/50					liv:hpa,hpc,nnd.	
g	TR209	78.4ng	n.s.s.	10/75	1.39ng	2/50	6.94ng	4/50	69.4ng	13/50					lun:a/a,a/c.	
4669	383a	105.ng	n.s.s.	15/38	.502ng	27/44	50.2ng	18/44	(502.ng	11/43)					Toth;natu,278,548-549;1979	
a	383a	890.ng	n.s.s.	7/38	.502ng	13/44	50.2ng	21/44	502.ng	13/43						
b	383a	899.ng	n.s.s.	27/38	.502ng	39/44	50.2ng	36/44	502.ng	27/43						
4670	TR209	60.1ng	520.ng	5/75	1.39ng	1/50	6.94ng	3/50	69.4ng	14/50					liv:hpc,nnd.	
a	TR209	67.7ng	1.05ug	5/75	1.39ng	3/50	6.94ng	3/50	69.4ng	12/50						
b	TR209	52.1ng	n.s.s.	54/75	1.39ng	40/50	6.94ng	36/50	69.4ng	43/50						
c	TR209	60.1ng	520.ng	5/75	1.39ng	1/50	6.94ng	3/50	69.4ng	14/50					liv:hpa,hpc,nnd.	
4671	TR209	41.9ng	664.ng	1/75	1.39ng	5/50	6.94ng	6/50	69.4ng	11/50					thy:fca,fcc.	
a	TR209	45.9ng	875.ng	1/75	1.39ng	5/50	6.94ng	4/50	69.4ng	10/50						
b	TR209	110.ng	3.83ug	0/75	1.39ng	0/50	6.94ng	0/50	69.4ng	3/50						S
c	TR209	110.ng	3.83ug	0/75	1.39ng	0/50	6.94ng	0/50	69.4ng	3/50					liv:hpc,nnd.	S
d	TR209	4.62ng	n.s.s.	6/75	1.39ng	9/50	6.94ng	12/50	(69.4ng	9/50)						S
e	TR209	60.1ng	n.s.s.	3/75	1.39ng	1/50	6.94ng	3/50	69.4ng	7/50						S
f	TR209	33.8ng	n.s.s.	40/75	1.39ng	29/50	6.94ng	33/50	69.4ng	32/50						
g	TR209	110.ng	3.83ug	0/75	1.39ng	0/50	6.94ng	0/50	69.4ng	3/50					liv:hpa,hpc,nnd.	
4672	377	17.5ng	545.ng	0/9	.004ng	0/9	.117ng	0/10	1.17ng	0/9	11.7ng	0/10	51.2ng	0/10	Van Miller;cmsp,10,625-632;1977/pers.comm.	
					248.ng	4/10										
a	377	26.4ng	360.ng	0/9	.004ng	0/9	.117ng	0/10	1.17ng	0/9	11.7ng	0/10	51.2ng	1/10		
					248.ng	4/10										
4673	366	3.26ng	22.5ng	8/86	1.00ng	3/50	10.0ng	18/50	(100.ng	23/50)					Kociba;txap,46,279-303;1978/pers.comm.	
a	366	28.9ng	173.ng	1/86	1.00ng	0/50	10.0ng	2/50	100.ng	11/50						
b	366	66.7ng	559.ng	0/86	1.00ng	0/50	10.0ng	0/50	100.ng	7/50						
c	366	73.9ng	972.ng	0/86	1.00ng	0/50	10.0ng	1/50	100.ng	4/50						
4674	366	81.1ng	1.27ug	0/86	1.00ng	0/50	10.0ng	2/50	100.ng	5/50						
a	366	126.ng	2.05ug	0/86	1.00ng	0/50	10.0ng	0/50	100.ng	4/50						
b	366	161.ng	n.s.s.	0/86	1.00ng	1/50	10.0ng	1/50	100.ng	3/50						

2,4,5,4'-TETRACHLORODIPHENYL SULFONE   (tetrafidon) 116-29-0

	RefNum	LoConf	UpConf	Cntrl	1Dose	1Inc		Citation or Pathology
4675	1282	37.9mg	n.s.s.	0/17	36.4mg	1/18		Innes;ntis,1968/1969
a	1282	43.3mg	n.s.s.	1/17	36.4mg	1/18		
b	1282	26.1mg	n.s.s.	2/17	36.4mg	3/18		
4676	1282	31.7mg	n.s.s.	2/18	33.9mg	2/18		
a	1282	67.2mg	n.s.s.	1/18	33.9mg	0/18		
b	1282	35.1mg	n.s.s.	3/18	33.9mg	2/18		
4677	1282	68.1mg	n.s.s.	0/16	36.4mg	0/17		
a	1282	68.1mg	n.s.s.	0/16	36.4mg	0/17		
b	1282	35.8mg	n.s.s.	0/16	36.4mg	1/17		
4678	1282	15.9mg	n.s.s.	0/16	33.9mg	4/17		
a	1282	33.3mg	n.s.s.	0/16	33.9mg	1/17		
b	1282	13.4mg	481.mg	0/16	33.9mg	5/17		

1,1,1,2-TETRACHLOROETHANE   630-20-6

	RefNum	LoConf	UpConf	Cntrl	1Dose	1Inc	2Dose	2Inc	Citation or Pathology	Brkly
4679	TR237	108.mg	316.mg	5/50	177.mg	13/50	357.mg	30/50	liv:hpa,hpc.	
a	TR237	145.mg	517.mg	4/50	177.mg	8/50	357.mg	24/50		
b	TR237	275.mg	2.68gm	1/50	177.mg	5/50	357.mg	6/50		
c	TR237	66.5mg	225.mg	25/50	177.mg	31/50	357.mg	30/50		
d	TR237	108.mg	316.mg	5/50	177.mg	13/50	357.mg	30/50	liv:hpa,hpc,nnd.	
e	TR237	447.mg	n.s.s.	0/50	177.mg	2/50	357.mg	0/50	lun:a/a,a/c.	
4680	TR237	75.3mg	255.mg	18/50	177.mg	27/50	357.mg	27/50	liv:hpa,hpc.	S
a	TR237	112.mg	391.mg	14/50	177.mg	14/50	357.mg	21/50		
b	TR237	58.4mg	215.mg	28/50	177.mg	36/50	357.mg	27/50		
c	TR237	75.3mg	255.mg	18/50	177.mg	27/50	357.mg	27/50	liv:hpa,hpc,nnd.	
d	TR237	334.mg	n.s.s.	6/50	177.mg	5/50	357.mg	3/50	lun:a/a,a/c.	

```
       Spe Strain Site   Xpo+Xpt                                                      TD50    2Tailpvl
           Sex  Route  Hist      Notes                                                 DR      AuOp
4681 R f f34 gav mgl fba 24m24 s                             :    ±              #170.mg \ P<.04  -
   a R f f34 gav TBA MXB 24m24 s                                                 no dre   P=1.
   b R f f34 gav liv MXB 24m24 s                                                 3.98gm * P<.5
4682 R m f34 gav MXA MXA 24m24 s                                      :     ±    #842.mg * P<.03  -
   a R m f34 gav liv MXA 24m24 s                                                 1.13gm * P<.04
   b R m f34 gav TBA MXB 24m24 s                                                 464.mg * P<.4
   c R m f34 gav liv MXB 24m24 s                                                 1.13gm * P<.04

1,1,2,2-TETRACHLOROETHANE          100ng..:..1ug....:..10....:..100....:..1mg..:..10....:..100....:..1g......:..10
4683 M f b6c gav liv hpc 78w92 v                             :+ :                35.4mg / P<.0005c
   a M f b6c gav TBA MXB 78w92 v                                                 37.2mg * P<.0005
   b M f b6c gav liv MXB 78w92 v                                                 35.4mg / P<.0005
   c M f b6c gav lun MXB 78w92 v                                                 1.91gm * P<.3
4684 M f b6c gav liv hpc 78w91 v    pool                       :+ :              35.8mg * P<.0005c
4685 M m b6c gav liv hpc 78w92 av                              : + :             41.6mg / P<.0005c
   a M m b6c gav TBA MXB 78w92 av                                                38.7mg * P<.0005
   b M m b6c gav liv MXB 78w92 av                                                41.6mg / P<.0005
   c M m b6c gav lun MXB 78w92 av                                                895.mg * P<.5
4686 M m b6c gav liv hpc 78w91 av   pool                      : + :              45.4mg * P<.0005c
4687 R f osm gav TBA MXB 18m26 adv                          :>                   84.2mg * P<.5   -
   a R f osm gav liv MXB 18m26 adv                                               no dre   P=1.
4688 R m osm gav TBA MXB 18m26 adv                              :>               no dre   P=1.   -
   a R m osm gav liv MXB 18m26 adv                                               624.mg * P<.2

TETRACHLOROETHYLENE                100ng..:..1ug....:..10....:..100....:..1mg..:..10....:..100....:..1g......:..10
4689 M f b6c gav liv hpc 78w90 v                             : + :               75.6mg * P<.0005c
   a M f b6c gav TBA MXB 78w90 v                                                 98.1mg * P<.0005
   b M f b6c gav liv MXB 78w90 v                                                 75.6mg * P<.0005
   c M f b6c gav lun MXB 78w90 v                                                 2.95gm * P<.2
4690 M f b6c inh liv MXA 24m24                                   :+ :            188.mg / P<.0005
   a M f b6c inh liv hpc 24m24                                                   200.mg / P<.0005c
   b M f b6c inh TBA MXB 24m24                                                   211.mg * P<.0005
   c M f b6c inh liv MXB 24m24                                                   188.mg / P<.0005
   d M f b6c inh lun MXB 24m24                                                   no dre   P=1.
4691 M m b6c gav liv hpc 78w90 v                                :+  :            123.mg * P<.0005c
   a M m b6c gav TBA MXB 78w90 v                                                 160.mg * P<.002
   b M m b6c gav liv MXB 78w90 v                                                 123.mg * P<.0005
   c M m b6c gav lun MXB 78w90 v                                                 no dre   P=1.
4692 M m b6c inh liv hpc 24m24                                   :  +  :         162.mg \ P<.0005
   a M m b6c inh liv MXA 24m24                                                   190.mg \ P<.0005c
   b M m b6c inh liv hpa 24m24                                                   668.mg * P<.04
   c M m b6c inh MXA mno 24m24                                                   681.mg * P<.05
   d M m b6c inh TBA MXB 24m24                                                   225.mg * P<.003
   e M m b6c inh liv MXB 24m24                                                   190.mg * P<.0005
   f M m b6c inh lun MXB 24m24                                                   6.05gm * P<.8
4693 R f osm gav TBA MXB 18m26 dsv                                 :>            1.71gm * P<.7   i
   a R f osm gav liv MXB 18m26 dsv                                               no dre   P=1.
4694 R f f34 inh MXA mnl 24m24                                  :>               287.mg * P<.2   p
   a R f f34 inh TBA MXB 24m24                                                   869.mg * P<.8
   b R f f34 inh liv MXB 24m24                                                   no dre   P=1.
4695 R m osm gav TBA MXB 18m26 dsv                                 :>            3.02gm * P<.9   i
   a R m osm gav liv MXB 18m26 dsv                                               no dre   P=1.
4696 R m f34 inh MXB MXB 24m24                                :  +       :       90.8mg * P<.007
   a R m f34 inh tes ict 24m24                                                   68.6mg * P<.02
   b R m f34 inh MXA mnl 24m24                                                   101.mg * P<.02  c
   c R m f34 inh kid MXA 24m24                                                   504.mg * P<.07  c
   d R m f34 inh TBA MXB 24m24                                                   77.7mg * P<.04
   e R m f34 inh liv MXB 24m24                                                   368.mg * P<.2

TETRACHLORVINPHOS                  100ng..:..1ug....:..10.....:..100....:..1mg....:..10.....:..100....:..1g......:..10
4697 M f b6c eat liv MXA 80w92                                  : ±              905.mg \ P<.02  a
   a M f b6c eat TBA MXB 80w92                                                   7.88gm * P<.8
   b M f b6c eat liv MXB 80w92                                                   905.mg \ P<.02
   c M f b6c eat lun MXB 80w92                                                   7.79gm * P<.5
4698 M f b6c eat liv MXA 80w90    pool                              : +  :       1.07gm \ P<.002 a
   a M f b6c eat liv nnd 80w90                                                   1.38gm \ P<.0005a
4699 M f b6c eat liv mix 24m24 er                       .                        10.2gm   P<.0005
   a M f b6c eat liv hpc 24m24 er                                                12.4gm   P<.002
   b M f b6c eat kid mix 24m24 er                                                32.1gm   P<.04
   c M f b6c eat liv hpa 24m24 er                                                65.0gm   P<.2
4700 M f b6c eat liv mix 24m24 er                       . + 6.33gm * P<.0005

   a M f b6c eat liv hpc 24m24 er                                                8.99gm * P<.0005

   b M f b6c eat liv hpa 24m24 er                                                21.4gm * P<.007

   c M f b6c eat kid tum 24m24 er                                                no dre   P=1.

4701 M m b6c eat liv MXA 80w92                                  :  + :           228.mg \ P<.0005c
   a M m b6c eat liv hpc 80w92                                                   466.mg * P<.002 c
   b M m b6c eat TBA MXB 80w92                                                   295.mg \ P<.007
```

	RefNum	LoConf	UpConf	Cntrl	1Dose	1Inc	2Dose	2Inc							Citation or Pathology

Brkly Code

```
4681   TR237  68.8mg n.s.s.   6/50   88.4mg  15/50 (177.mg   7/50)                                                      S
a      TR237  194.mg n.s.s.  33/50   88.4mg  31/50  177.mg  21/50
b      TR237  601.mg n.s.s.   1/50   88.4mg   0/50  177.mg   2/50                             liv:hpa,hpc,nnd.
4682   TR237  337.mg n.s.s.   0/50   88.4mg   3/50  177.mg   3/50             mul:msm; per:men; tnv:men. S
a      TR237  385.mg n.s.s.   0/50   88.4mg   1/50  177.mg   3/50                             liv:hpa,hpc,nnd. S
b      TR237  115.mg n.s.s.  25/50   88.4mg  19/50  177.mg  26/50
c      TR237  385.mg n.s.s.   0/50   88.4mg   1/50  177.mg   3/50                             liv:hpa,hpc,nnd.

1,1,2,2-TETRACHLOROETHANE    79-34-5
4683   TR27   25.9mg 50.2mg   0/20   87.7mg  30/50  175.mg  43/50
a      TR27   26.1mg 61.6mg   2/20   87.7mg  33/50  175.mg  43/50
b      TR27   25.9mg 50.2mg   0/20   87.7mg  30/50  175.mg  43/50
c      TR27   449.mg n.s.s.   0/20   87.7mg   1/50  175.mg   1/50                             liv:hpa,hpc,nnd.
                                                                                                 lun:a/a,a/c.
4684   TR27   26.2mg 50.3mg   0/40p  87.7mg  30/50  175.mg  43/50
4685   TR27   27.4mg 66.8mg   1/20   87.7mg  13/50  175.mg  44/50
a      TR27   25.2mg 66.6mg   4/20   87.7mg  17/50  175.mg  45/50
b      TR27   27.4mg 66.8mg   1/20   87.7mg  13/50  175.mg  44/50                             liv:hpa,hpc,nnd.
c      TR27   162.mg n.s.s.   1/20   87.7mg   2/50  175.mg   2/50                                 lun:a/a,a/c.
4686   TR27   29.7mg 75.1mg   3/40p  87.7mg  13/50  175.mg  44/50
4687   TR27   20.7mg n.s.s.  12/20   21.9mg  28/50  39.0mg  23/50
a      TR27   n.s.s. n.s.s.   0/20   21.9mg   0/50  39.0mg   0/50                             liv:hpa,hpc,nnd.
4688   TR27   60.2mg n.s.s.  11/20   31.6mg  17/50  55.2mg  20/50
a      TR27   185.mg n.s.s.   0/20   31.6mg   0/50  55.2mg   3/50                             liv:hpa,hpc,nnd.

TETRACHLOROETHYLENE    (perchloroethylene) 127-18-4
4689   TR13   47.1mg 130.mg   0/20  240.mg  19/50  478.mg  19/50
a      TR13   54.1mg 259.mg   5/20  240.mg  19/50  478.mg  19/50
b      TR13   47.1mg 130.mg   0/20  240.mg  19/50  478.mg  19/50                             liv:hpa,hpc,nnd.
c      TR13   480.mg n.s.s.   0/20  240.mg   0/50  478.mg   1/50                                 lun:a/a,a/c.
4690   TR311  127.mg 314.mg   4/50  211.mg  17/50  422.mg  38/50                             liv:hpa,hpc.   S
a      TR311  137.mg 313.mg   1/50  211.mg  13/50  422.mg  36/50
b      TR311  119.mg 712.mg  27/50  211.mg  35/50  422.mg  43/50
c      TR311  127.mg 314.mg   4/50  211.mg  17/50  422.mg  38/50                             liv:hpa,hpc,nnd.
d      TR311  1.20gm n.s.s.   6/50  211.mg   3/50  422.mg   3/50                                 lun:a/a,a/c.
4691   TR13   79.5mg 260.mg   2/20  332.mg  32/50  663.mg  27/50
a      TR13   90.1mg 622.mg   6/20  332.mg  33/50  663.mg  27/50
b      TR13   79.5mg 260.mg   2/20  332.mg  32/50  663.mg  27/50                             liv:hpa,hpc,nnd.
c      TR13   n.s.s. n.s.s.   0/20  332.mg   3/50  663.mg   0/50                                 lun:a/a,a/c.
4692   TR311  90.4mg 407.mg   7/50  176.mg  25/50 (352.mg  26/50)                                           S
a      TR311  115.mg 459.mg  17/50  176.mg  31/50  352.mg  41/50                             liv:hpa,hpc.
b      TR311  283.mg n.s.s.  12/50  176.mg   8/50  352.mg  19/50                                            S
c      TR311  234.mg n.s.s.   3/50  176.mg   7/50 (352.mg   3/50)     mln:mno; mul:mno; spl:mno; sub:mno. S
d      TR311  119.mg 1.30gm  29/50  176.mg  38/50  352.mg  43/50
e      TR311  115.mg 459.mg  17/50  176.mg  31/50  352.mg  41/50                             liv:hpa,hpc,nnd.
f      TR311  706.mg n.s.s.   6/50  176.mg   6/50  352.mg   5/50                                 lun:a/a,a/c.
4693   TR13   263.mg n.s.s.   7/20  240.mg  17/50  480.mg  15/50
a      TR13   n.s.s. n.s.s.   0/20  240.mg   0/50  480.mg   0/50                             liv:hpa,hpc,nnd.
4694   TR311  110.mg n.s.s.  18/50  101.mg  30/50  201.mg  29/50                             mul:mnl; spl:mnl.
a      TR311  101.mg n.s.s.  42/50  101.mg  45/50  201.mg  45/50
b      TR311  706.mg n.s.s.   2/50  101.mg   0/50  201.mg   2/50                             liv:hpa,hpc,nnd.
4695   TR13   239.mg n.s.s.   5/20  238.mg   5/50  477.mg   5/50
a      TR13   n.s.s. n.s.s.   0/20  238.mg   0/50  477.mg   0/50                             liv:hpa,hpc,nnd.
4696   TR311  44.4mg 1.34gm  28/50  70.4mg  37/50  141.mg  39/50    kid:tla,uac; liv:mnl; mul:mnl; spl:mnl. C
a      TR311  31.7mg n.s.s.  35/50  70.4mg  39/50  141.mg  41/50                                            S
b      TR311  47.4mg n.s.s.  28/50  70.4mg  37/50  141.mg  37/50                             liv:mnl; mul:mnl; spl:mnl.
c      TR311  173.mg n.s.s.   1/50  70.4mg   3/50  141.mg   4/50                                 kid:tla,uac.
d      TR311  33.7mg n.s.s.  46/50  70.4mg  48/50  141.mg  50/50
e      TR311  112.mg n.s.s.   4/50  70.4mg   7/50  141.mg   5/50                             liv:hpa,hpc,nnd.

TETRACHLORVINPHOS    961-11-5
4697   TR33   528.mg n.s.s.   0/9   904.mg  19/50 (1.81gm  11/49)                             liv:hpc,nnd.
a      TR33   1.04gm n.s.s.   1/9   904.mg  25/50  1.81gm  19/49
b      TR33   528.mg n.s.s.   0/9   904.mg  19/50 (1.81gm  11/49)                             liv:hpa,hpc,nnd.
c      TR33   2.87gm n.s.s.   0/9   904.mg   5/50  1.81gm   5/49                                 lun:a/a,a/c.
4698   TR33   558.mg 4.20gm   3/49p 904.mg  19/50 (1.81gm  11/49)                             liv:hpc,nnd.
a      TR33   705.mg 4.79gm   1/49p 904.mg  14/50 (1.81gm   9/49)
4699   1696m  4.17gm 36.9gm   0/99  2.08gm   6/47                            Parker;faat,5,840-854;1985
a      1696m  4.72gm 54.7gm   0/99  2.08gm   5/47
b      1696m  7.91gm n.s.s.   0/99  2.08gm   2/47
c      1696m  10.6gm n.s.s.   0/99  2.08gm   1/47
```

	RefNum	LoConf	UpConf	Cntrl	1Dose	1Inc	2Dose	2Inc	3Dose	3Inc	4Dose	4Inc	5Dose	5Inc	Citation or Pathology
4700	1696n	4.04gm	15.4gm	0/99	2.28gm	1/48	8.32mg	0/49	41.6mg	0/50	208.mg	4/49	1.04gm	7/49	
					2.08gm	7/50									
a	1696n	4.63gm	25.8gm	0/99	2.28gm	0/48	8.32mg	0/49	41.6mg	0/50	208.mg	3/49	1.04gm	5/49	
					2.08gm	4/50									
b	1696n	12.2gm	428.gm	0/99	2.28gm	1/48	8.32mg	0/49	41.6mg	0/50	208.mg	1/49	1.04gm	2/49	
					2.08gm	3/50									
c	1696n	16.5gm	n.s.s.	0/99	2.28gm	0/48	8.32mg	0/49	41.6mg	0/50	208.mg	0/49	1.04gm	0/49	
					2.08gm	0/50									

```
4701   TR33   158.mg 432.mg   0/10  835.mg  47/50 (1.67gm  42/50)                             liv:hpc,nnd.
a      TR33   348.mg 1.93gm   0/10  835.mg  36/50  1.67gm  40/50
b      TR33   175.mg 3.98gm   2/10  835.mg  47/50 (1.67gm  42/50)
```

	Spe Strain Site Xpo+Xpt						TD50	2Tailpvl
	Sex Route Hist Notes						DR	AuOp
c	M m b6c eat liv MXB 80w92						228.mg \ P<.0005	
d	M m b6c eat lun MXB 80w92						401.gm * P<1.	
4702	M m b6c eat liv MXA 80w90	pool			: + :		280.mg \ P<.0005c	
a	M m b6c eat liv hpc 80w90						534.mg * P<.0005c	
b	M m b6c eat liv nnd 80w90						1.77gm \ P<.02 a	
4703	M m b6c eat liv mix 24m24 er				. + .		1.15gm P<.0005	
a	M m b6c eat liv hpc 24m24 er						1.53gm P<.0005	
b	M m b6c eat kid mix 24m24 er						4.42gm P<.0005	
c	M m b6c eat kid tuc 24m24 er						5.26gm P<.0005	
d	M m b6c eat liv hpa 24m24 er						18.3gm P<.08	
e	M m b6c eat kid tua 24m24 er						37.6gm P<.3	
4704	R f osm eat pit cra 19m26 v				: ±		#1.58gm * P<.03 -	
a	R f osm eat TBA MXB 19m26 v						no dre P=1.	
b	R f osm eat liv MXB 19m26 v						no dre P=1.	
4705	R f osm eat thy cca 19m25 v	pool			: ±		1.92gm * P<.05 a	
a	R f osm eat adr coa 19m25 v						2.28gm * P<.4 a	
4706	R m osm eat TBA MXB 19m26 v			:>			no dre P=1.	
a	R m osm eat liv MXB 19m26 v						no dre P=1.	

TETRACYCLINE.HC1　　100ng..:..1ug....:..10.....:..100....:..1mg....:..10.....:..100....:..1g.....:..10

4707	M f b6c eat TBA MXB 24m24					:>	no dre P=1.	
a	M f b6c eat liv MXB 24m24						no dre P=1.	
b	M f b6c eat lun MXB 24m24						37.2gm * P<.4	
4708	M m b6c eat TBA MXB 24m24					:>	no dre P=1.	
a	M m b6c eat liv MXB 24m24						no dre P=1.	
b	M m b6c eat lun MXB 24m24						no dre P=1.	
4709	R f f34 eat TBA MXB 24m24				:>		no dre P=1.	
a	R f f34 eat liv MXB 24m24						no dre P=1.	
4710	R m f34 eat TBA MXB 24m24				:>		no dre P=1.	
a	R m f34 eat liv MXB 24m24						6.09gm * P<.06	

TETRAETHYLTHIURAM DISULFIDE　　100ng..:..1ug....:..10.....:..100....:..1mg....:..10.....:..100....:..1g.....:..10

4711	M f b6c eat TBA MXB 25m25				:>		no dre P=1.	
a	M f b6c eat liv MXB 25m25						120.mg \ P<.2	
b	M f b6c eat lun MXB 25m25						323.mg * P<.2	
4712	M f b6c orl lun mix 76w76 evx			.>			138.mg P<.2	
a	M f b6c orl lun ade 76w76 evx						283.mg P<.3	
b	M f b6c orl liv hpt 76w76 evx						283.mg P<.3	
c	M f b6c orl lun car 76w76 evx						283.mg P<.3	
d	M f b6c orl tba mix 76w76 evx						64.5mg P<.02	
4713	M m b6c eat TBA MXB 25m25				:>		no dre P=1. -	
a	M m b6c eat liv MXB 25m25						no dre P=1.	
b	M m b6c eat lun MXB 25m25						no dre P=1.	
4714	M m b6c orl liv hpt 76w76 evx			. + .			23.7mg P<.0005	
a	M m b6c orl lun ade 76w76 evx						43.2mg P<.007	
b	M m b6c orl tba mix 76w76 evx						17.0mg P<.0005	
4715	M f b6a orl lun ade 76w76 evx			.>			4.14gm P<1.	
a	M f b6a orl liv hpt 76w76 evx						no dre P=1.	
b	M f b6a orl tba mix 76w76 evx						1.94gm P<1.	
4716	M m b6a orl sub fbs 76w76 evx			. + .			15.4mg P<.0005	
a	M m b6a orl liv hpt 76w76 evx						197.mg P<.5	
b	M m b6a orl lun ade 76w76 evx						no dre P=1.	
c	M m b6a orl tba mix 76w76 evx						10.1mg P<.0005	
4717	R f f34 eat TBA MXB 25m25				:>		no dre P=1. -	
a	R f f34 eat liv MXB 25m25						1.43gm * P<.4	
4718	R m f34 eat TBA MXB 25m25				:>		no dre P=1. -	
a	R m f34 eat liv MXB 25m25						no dre P=1.	
4719	R f cdr eat mgl mix 78w78 e			. ±			68.9mg P<.04	
a	R f cdr eat liv hem 78w78 e						no dre P=1.	
b	R f cdr eat liv hct 78w78 e						no dre P=1.	
c	R f cdr eat tba mix 78w78 e						193.mg P<.6	
4720	R f cdr eat liv nnd 24m24 e				.>		41.1gm P<1. -	
a	R f cdr eat tba tum 24m24 e						no dre P=1.	
4721	R m cdr eat liv hct 78w78 e				.>		366.mg P<.3	
a	R m cdr eat tba mix 78w78 e						no dre P=1.	
4722	R m cdr eat liv nnd 24m24 e				.>		678.mg P<.3 -	
a	R m cdr eat liv hpc 24m24 e						no dre P=1.	
b	R m cdr eat tba tum 24m24 e						no dre P=1.	

TETRAFLUORO-m-PHENYLENEDIAMINE.2HC1　　..:..1ug....:..10.....:..100....:..1mg....:..10.....:..100....:..1g.....:..10

4723	M f chi eat liv mix 77w94				:>		6.50gm * P<1. -	
a	M f chi eat lun mix 77w94						no dre P=1.	
b	M f chi eat tba mix 77w94						no dre P=1.	
4724	M m chi eat liv mix 77w94			: ±			78.3mg * P<.02	
a	M m chi eat liv hpt 77w94						86.3mg * P<.02 +	
b	M m chi eat lun mix 77w94						1.13gm * P<.7 -	
c	M m chi eat tba mix 77w94						331.mg * P<.7	
4725	M m chi eat liv hpt 77w94	pool		: + :			109.mg * P<.0005+	
4726	R m cdr eat liv mix 77w98 v						no dre P=1.	
a	R m cdr eat tba mix 77w98 v						14.6mg * P<.003 -	

	RefNum	LoConf	UpConf	Cntrl	1Dose	1Inc	2Dose	2Inc	Citation or Pathology
									Brkly Code
c	TR33	158.mg	432.mg	0/10	835.mg	47/50	(1.67gm	42/50)	liv:hpa,hpc,nnd.
d	TR33	3.40gm	n.s.s.	0/10	835.mg	4/50	1.67gm	2/50	lun:a/a,a/c.
4702	TR33	176.mg	541.mg	8/50p	835.mg	47/50	(1.67gm	42/50)	liv:hpc,nnd.
a	TR33	371.mg	905.mg	5/50p	835.mg	36/50	1.67gm	40/50	
b	TR33	737.mg	n.s.s.	3/50p	835.mg	11/50	(1.67gm	2/50)	
4703	1696m	660.mg	2.42gm	26/99	1.92gm	35/46			Parker;faat,5,840-854;1985
a	1696m	855.mg	3.63gm	24/99	1.92gm	31/46			
b	1696m	2.23gm	11.3gm	1/99	1.92gm	12/46			
c	1696m	2.56gm	13.5gm	0/99	1.92gm	10/46			
d	1696m	5.26gm	n.s.s.	2/99	1.92gm	4/46			
e	1696m	7.50gm	n.s.s.	1/99	1.92gm	2/46			
4704	TR33	743.mg	n.s.s.	0/10	153.mg	1/50	306.mg	8/50	s
a	TR33	454.mg	n.s.s.	7/10	153.mg	23/50	306.mg	25/50	
b	TR33	2.01gm	n.s.s.	0/10	153.mg	2/50	306.mg	0/50	liv:hpa,hpc,nnd.
4705	TR33	775.mg	n.s.s.	1/55p	153.mg	2/50	306.mg	7/50	
a	TR33	667.mg	n.s.s.	2/55p	153.mg	7/50	306.mg	6/50	
4706	TR33	226.mg	n.s.s.	6/10	122.mg	24/50	245.mg	18/50	
a	TR33	1.10gm	n.s.s.	0/10	122.mg	2/50	245.mg	0/50	liv:hpa,hpc,nnd.

TETRACYCLINE.HCl   64-75-5

	RefNum	LoConf	UpConf	Cntrl	1Dose	1Inc	2Dose	2Inc	Citation or Pathology
4707	TR344	5.96gm	n.s.s.	35/50	1.61gm	19/50	3.22gm	26/50	
a	TR344	36.7gm	n.s.s.	10/50	1.61gm	0/50	3.22gm	0/50	liv:hpa,hpc,nnd.
b	TR344	8.64gm	n.s.s.	4/50	1.61gm	1/50	3.22gm	7/50	lun:a/a,a/c.
4708	TR344	3.44gm	n.s.s.	31/50	1.49gm	23/50	(2.97gm	21/50)	
a	TR344	8.76gm	n.s.s.	12/50	1.49gm	12/50	2.97gm	10/50	liv:hpa,hpc,nnd.
b	TR344	12.9gm	n.s.s.	6/50	1.49gm	6/50	2.97gm	4/50	lun:a/a,a/c.
4709	TR344	712.mg	n.s.s.	46/50	619.mg	44/50	(1.24gm	41/50)	
a	TR344	15.0gm	n.s.s.	1/50	619.mg	0/50	1.24gm	0/50	liv:hpa,hpc,nnd.
4710	TR344	722.mg	n.s.s.	46/50	495.mg	49/50	990.mg	45/50	
a	TR344	2.28gm	n.s.s.	0/50	495.mg	2/50	990.mg	3/50	liv:hpa,hpc,nnd.

TETRAETHYLTHIURAM DISULFIDE   (ethyl tuads) 97-77-8

	RefNum	LoConf	UpConf	Cntrl	1Dose	1Inc	2Dose	2Inc	Citation or Pathology
4711	TR166	99.5mg	n.s.s.	8/20	13.0mg	18/50	65.0mg	20/50	
a	TR166	36.4mg	n.s.s.	0/20	13.0mg	3/50	(65.0mg	0/50)	liv:hpa,hpc,nnd.
b	TR166	111.mg	n.s.s.	1/20	13.0mg	4/50	65.0mg	9/50	lun:a/a,a/c.
4712	1100	33.8mg	n.s.s.	0/16	44.3mg	2/18			Innes;ntis,1968/1969
a	1100	46.1mg	n.s.s.	0/16	44.3mg	1/18			
b	1100	46.1mg	n.s.s.	0/16	44.3mg	1/18			
c	1100	46.1mg	n.s.s.	0/16	44.3mg	1/18			
d	1100	22.2mg	n.s.s.	0/16	44.3mg	4/18			
4713	TR166	110.mg	n.s.s.	11/20	60.0mg	21/50	(240.mg	18/50)	
a	TR166	232.mg	n.s.s.	5/20	60.0mg	6/50	(240.mg	4/50)	liv:hpa,hpc,nnd.
b	TR166	127.mg	n.s.s.	4/20	60.0mg	11/50	(240.mg	4/50)	lun:a/a,a/c.
4714	1100	10.5mg	75.4mg	0/16	41.2mg	8/17			Innes;ntis,1968/1969
a	1100	16.3mg	585.mg	0/16	41.2mg	5/17			
b	1100	7.99mg	44.6mg	0/16	41.2mg	10/17			
4715	1100	46.0mg	n.s.s.	1/17	44.3mg	1/16			
a	1100	77.9mg	n.s.s.	0/17	44.3mg	0/16			
b	1100	36.1mg	n.s.s.	2/17	44.3mg	2/16			
4716	1100	7.17mg	40.4mg	0/18	41.2mg	10/16			
a	1100	30.1mg	n.s.s.	1/18	41.2mg	2/16			
b	1100	47.2mg	n.s.s.	2/18	41.2mg	1/16			
c	1100	4.45mg	34.4mg	3/18	41.2mg	13/16			
4717	TR166	49.6mg	n.s.s.	19/20	15.0mg	28/50	30.0mg	33/50	
a	TR166	232.mg	n.s.s.	0/20	15.0mg	0/50	30.0mg	1/50	liv:hpa,hpc,nnd.
4718	TR166	13.2mg	n.s.s.	15/20	12.0mg	33/50	(24.0mg	24/50)	
a	TR166	164.mg	n.s.s.	0/20	12.0mg	1/50	24.0mg	0/50	liv:hpa,hpc,nnd.
4719	1032	26.1mg	n.s.s.	2/48	25.0mg	8/48			Wong;txap,63,155-165;1982/Plotnick pers.comm.
a	1032	139.mg	n.s.s.	0/48	25.0mg	0/48			
b	1032	139.mg	n.s.s.	0/48	25.0mg	0/48			
c	1032	30.7mg	n.s.s.	7/48	25.0mg	9/48			
4720	1962	153.mg	n.s.s.	1/50	25.0mg	1/49			Cheever;faat,14,243-261;1990
a	1962	10.0mg	n.s.s.	47/50	25.0mg	46/50			
4721	1032	59.6mg	n.s.s.	0/48	20.0mg	1/48			Wong;txap,63,155-165;1982/Plotnick pers.comm.
a	1032	55.5mg	n.s.s.	8/48	20.0mg	4/48			
4722	1962	110.mg	n.s.s.	0/50	20.0mg	1/50			Cheever;faat,14,243-261;1990
a	1962	206.mg	n.s.s.	1/50	20.0mg	0/50			
b	1962	34.6mg	n.s.s.	42/50	20.0mg	32/50			

TETRAFLUORO-m-PHENYLENEDIAMINE.2HCl   63886-77-1

	RefNum	LoConf	UpConf	Cntrl	1Dose	1Inc	2Dose	2Inc	Citation or Pathology
4723	381	221.mg	n.s.s.	1/15	106.mg	2/20	213.mg	2/15	Weisburger;jept,2,325-356;1978/pers.comm./Russfield 1973
a	381	94.8mg	n.s.s.	5/15	106.mg	5/20	(213.mg	1/15)	
b	381	53.6mg	n.s.s.	10/15	106.mg	14/20	(213.mg	7/15)	
4724	381	42.7mg	n.s.s.	1/14	103.mg	8/18	206.mg	7/16	
a	381	46.1mg	n.s.s.	0/14	103.mg	7/18	206.mg	7/16	
b	381	160.mg	n.s.s.	3/14	103.mg	2/18	206.mg	4/16	
c	381	51.1mg	n.s.s.	7/14	103.mg	13/18	206.mg	10/16	
4725	381	53.0mg	324.mg	7/99p	103.mg	7/18	206.mg	7/16	
4726	381	n.s.s.	n.s.s.	0/16	19.1mg	0/16	38.2mg	0/15	
a	381	6.81mg	101.mg	10/16	19.1mg	8/16	38.2mg	12/15	

```
        Spe Strain Site  Xpo+Xpt                                         TD50    2Tailpvl
            Sex Route Hist      Notes                                             DR    AuOp

TETRAFLUOROBORATE, SODIUM    100ng..:..1ug...:..10....:...100....:...1mg...:..10....:..100....:..1g.....:..10
4727 H f syg gav liv cho 70w70 es                          .>              no dre   P=1.
   a H f syg gav lun tum 70w70 es                                          no dre   P=1.
4728 H m syg gav liv hem 90w90 es                              .>          no dre   P=1.
   a H m syg gav lun tum 90w90 es                                          no dre   P=1.

TETRAHYDRO-2-NITROSO-2H-1,2-OXAZINE  ...:..1ug...:..10....:...100....:...1mg...:..10....:..100....:..1g..:..10
4729 R b sda wat lun mix 66w82                      .     +       .         46.1mg   P<.007
   a R b sda wat tba mix 66w82                                              17.6mg   P<.004
   b R b sda wat tba mal 66w82                                              24.3mg   P<.003 +

3,4,5,6-TETRAHYDROURIDINE    100ng..:..1ug...:..10....:...100....:...1mg...:..10....:..100....:..1g....:..10
4730 R m f34 ipj liv tum 52w52 e                        .>                 no dre   P=1.
   a R m f34 ipj tba tum 52w52 e                                           15.7mg   P<.6

TETRAKIS(HYDROXYMETHYL)PHOSPHONIUM  CHLORIDE.1ug...:..10....:...100....:...1mg...:..10....:..100....:..1g....:..10
4731 M f b6c gav TBA MXB 24m24                               :>            908.mg * P<1.   -
   a M f b6c gav liv MXB 24m24                                             200.mg * P<.4
   b M f b6c gav lun MXB 24m24                                             no dre   P=1.
4732 M m b6c gav TBA MXB 24m24                            :>               no dre   P=1.   -
   a M m b6c gav liv MXB 24m24                                             no dre   P=1.
   b M m b6c gav lun MXB 24m24                                             103.mg * P<.6
4733 R f f34 gav TBA MXB 24m24                       :   ±                 4.78mg * P<.09  -
   a R f f34 gav liv MXB 24m24                                             no dre   P=1.
4734 R m f34 gav sub fib 24m24                        :    ±              #24.9mg * P<.03  -
   a R m f34 gav TBA MXB 24m24                                             14.7mg * P<.7
   b R m f34 gav liv MXB 24m24                                             24.5mg * P<.3

TETRAKIS(HYDROXYMETHYL)PHOSPHONIUM  SULFATE..1ug...:..10....:...100....:...1mg...:..10....:..100....:..1g....:..10
4735 M f b6c gav mul mlm 24m24                        :    ±              #15.7mg * P<.02  -
   a M f b6c gav TBA MXB 24m24                                             no dre   P=1.
   b M f b6c gav liv MXB 24m24                                             no dre   P=1.
   c M f b6c gav lun MXB 24m24                                             234.mg * P<.8
4736 M m b6c gav mul MXA 24m24                      :    ±                #11.4mg \ P<.04  -
   a M m b6c gav TBA MXB 24m24                                             47.3mg * P<.8
   b M m b6c gav liv MXB 24m24                                             no dre   P=1.
   c M m b6c gav lun MXB 24m24                                             78.5mg * P<.8
4737 R f f34 gav TBA MXB 24m25                            :>               23.3mg * P<.7  -
   a R f f34 gav liv MXB 24m25                                             no dre   P=1.
4738 R m f34 gav mul mnl 24m25                      :   ±                 #2.42mg \ P<.02  -
   a R m f34 gav pta MXA 24m25                                             3.30mg \ P<.02
   b R m f34 gav pta adn 24m25                                             3.56mg \ P<.03
   c R m f34 gav TBA MXB 24m25                                             1.78mg \ P<.02
   d R m f34 gav liv MXB 24m25                                             12.4mg \ P<.2

TETRAMETHYLTHIURAM DISULFIDE  100ng..:..1ug...:..10....:...100....:...1mg...:..10....:..100....:..1g.....:..10
4739 M f b6a orl lun ade 76w76 evx                        .>               159.mg   P<1.  -
   a M f b6a orl liv hpt 76w76 evx                                         no dre   P=1.  -
   b M f b6a orl tba mix 76w76 evx                                         no dre   P=1.  -
4740 M m b6a orl lun ade 76w76 evx                        .>               no dre   P=1.  -
   a M m b6a orl liv hpt 76w76 evx                                         no dre   P=1.  -
   b M m b6a orl tba mix 76w76 evx                                         no dre   P=1.  -
4741 M f b6c orl liv hpt 76w76 evx                          .>             no dre   P=1.  -
   a M f b6c orl lun mix 76w76 evx                                         no dre   P=1.  -
   b M f b6c orl tba mix 76w76 evx                                         no dre   P=1.  -
4742 M m b6c orl liv hpt 76w76 evx                       .    ±            9.91mg   P<.1  -
   a M m b6c orl lun mix 76w76 evx                                         no dre   P=1.  -
   b M m b6c orl tba mix 76w76 evx                                         6.39mg   P<.04  -
4743 R f f34 eat liv nnd 24m30 ev                            .>            97.3mg   P<.2  -
4744 R m f34 eat liv nnd 24m30 ev                             .>           374.mg   P<.7  -
   a R m f34 eat liv mix 24m30 ev                                          no dre   P=1.  -
   b R m f34 eat liv car 24m30 ev                                          no dre   P=1.  -
4745 R f f3d eat liv nnd 24m26 e                                .>         no dre   P=1.  -
4746 R m f3d eat liv nnd 24m26 e                                 .>        427.mg * P<.7  -
4747 R f wis eat liv tum 52w52 ek              .>                          no dre   P=1.  -
4748 R f wis eat liv hpa 24m24 e                  .>                       no dre   P=1.  -
4749 R m wis eat liv tum 52w52 ek             .>                           no dre   P=1.  -
4750 R m wis eat liv hpa 24m24 e                   .>                      no dre   P=1.  -

TETRAMETHYLTHIURAM MONOSULFIDE  100ng..:..1ug...:..10....:...100....:...1mg...:..10....:..100....:..1g.....:..10
4751 M f b6a orl liv hpt 76w76 evx                           .>            327.mg   P<.3  -
   a M f b6a orl lun ade 76w76 evx                                         no dre   P=1.  -
   b M f b6a orl tba mix 76w76 evx                                         148.mg   P<.5  -
4752 M m b6a orl lun ade 76w76 evx                          .>             no dre   P=1.  -
   a M m b6a orl liv hpt 76w76 evx                                         no dre   P=1.  -
   b M m b6a orl tba mix 76w76 evx                                         no dre   P=1.  -
4753 M f b6c orl liv hpt 76w76 evx                            .>           no dre   P=1.  -
   a M f b6c orl lun mix 76w76 evx                                         no dre   P=1.  -
   b M f b6c orl tba tum 76w76 evx                                         no dre   P=1.  -
4754 M m b6c orl lun ade 76w76 evx                       .    ±            89.3mg   P<.04  -
```

	RefNum	LoConf	UpConf	Cntrl	1Dose	1Inc	2Dose	2Inc	Citation or Pathology
									Brkly Code

```
TETRAFLUOROBORATE, SODIUM     13755-29-8
4727   1329  4.24mg n.s.s.     1/15   3.03mg   0/15                              Gold;clet,15,289-300;1982
 a     1329  4.24mg n.s.s.     0/15   3.03mg   0/15
4728   1329  7.01mg n.s.s.     1/15   3.03mg   0/15
 a     1329  7.01mg n.s.s.     0/15   3.03mg   0/15


TETRAHYDRO-2-NITROSO-2H-1,2-OXAZINE      40548-68-3
4729   1417  17.4mg 531.mg     0/20   31.1mg   5/20                              Wiessler;zkko,79,114-117;1973
 a     1417  7.81mg 123.mg     3/20   31.1mg  12/20
 b     1417  10.6mg 142.mg     1/20   31.1mg   9/20


3,4,5,6-TETRAHYDROURIDINE     18771-50-1
4730   1906  6.07mg n.s.s.     0/49   11.8mg   0/10                              Carr;bjca,57,395-402;1988
 a     1906  2.06mg n.s.s.    10/49   11.8mg   3/10


TETRAKIS(HYDROXYMETHYL)PHOSPHONIUM  CHLORIDE  (THPC) 124-64-1
4731   TR296 19.2mg n.s.s.    35/50   10.5mg  32/50   21.2mg  36/50
 a     TR296 47.9mg n.s.s.     4/50   10.5mg   4/50   21.2mg   7/50            liv:hpa,hpc,nnd.
 b     TR296 101.mg n.s.s.     3/50   10.5mg   3/50   21.2mg   2/50            lun:a/a,a/c.
4732   TR296 12.9mg n.s.s.    33/50   5.31mg  31/50   10.6mg  33/50
 a     TR296 25.8mg n.s.s.    17/50   5.31mg  15/50   10.6mg  13/50            liv:hpa,hpc,nnd.
 b     TR296 19.1mg n.s.s.     4/50   5.31mg   6/50   10.6mg   8/50            lun:a/a,a/c.
4733   TR296 1.84mg n.s.s.    40/50   2.65mg  45/50   5.31mg  36/50
 a     TR296 n.s.s. n.s.s.     0/50   2.65mg   1/50   5.31mg   0/50            liv:hpa,hpc,nnd.
4734   TR296 9.06mg n.s.s.     0/50   2.65mg   2/50   5.31mg   3/50                              S
 a     TR296 2.34mg n.s.s.    42/50   2.65mg  42/50   5.31mg  36/50
 b     TR296 6.80mg n.s.s.     1/50   2.65mg   4/50   5.31mg   2/50            liv:hpa,hpc,nnd.


TETRAKIS(HYDROXYMETHYL)PHOSPHONIUM  SULFATE  (THPS) 55566-30-8
4735   TR296 7.73mg n.s.s.     1/50   3.54mg   8/50   7.07mg   9/50                              S
 a     TR296 5.58mg n.s.s.    32/50   3.54mg  35/50   7.07mg  35/50
 b     TR296 33.4mg n.s.s.     7/50   3.54mg   3/50   7.07mg   3/50            liv:hpa,hpc,nnd.
 c     TR296 21.4mg n.s.s.     2/50   3.54mg   3/50   7.07mg   3/50            lun:a/a,a/c.
4736   TR296 4.59mg n.s.s.     2/50   3.54mg  10/50  (7.07mg   1/50)   mul:mlh,mlm,mlp,mlu,ule. S
 a     TR296 4.26mg n.s.s.    33/50   3.54mg  38/50   7.07mg  38/50
 b     TR296 10.9mg n.s.s.    18/50   3.54mg  12/50   7.07mg  17/50            liv:hpa,hpc,nnd.
 c     TR296 9.59mg n.s.s.     7/50   3.54mg  10/50   7.07mg   9/50            lun:a/a,a/c.
4737   TR296 3.69mg n.s.s.    45/50   3.50mg  46/50   7.01mg  41/50
 a     TR296 30.1mg n.s.s.     3/50   3.50mg   2/50   7.01mg   2/50            liv:hpa,hpc,nnd.
4738   TR296 1.03mg n.s.s.    30/50   3.50mg  36/50  (7.01mg  20/50)                              S
 a     TR296 1.38mg n.s.s.    21/50   3.50mg  27/50  (7.01mg  14/50)   pta:adn,can.              S
 b     TR296 1.44mg n.s.s.    21/50   3.50mg  26/50  (7.01mg  14/50)                              S
 c     TR296 .766mg n.s.s.    46/50   3.50mg  49/50  (7.01mg  37/50)
 d     TR296 3.19mg n.s.s.     3/50   3.50mg   5/50  (7.01mg   1/50)   liv:hpa,hpc,nnd.


TETRAMETHYLTHIURAM DISULFIDE  (TMTD, thiram) 137-26-8
4739   1194  3.52mg n.s.s.     1/17   3.64mg   1/15                              Innes;ntis,1968/1969
 a     1194  6.01mg n.s.s.     0/17   3.64mg   0/15
 b     1194  3.90mg n.s.s.     2/17   3.64mg   1/15
4740   1194  3.17mg n.s.s.     2/18   3.39mg   2/18
 a     1194  4.01mg n.s.s.     1/18   3.39mg   1/18
 b     1194  2.67mg n.s.s.     3/18   3.39mg   3/18
4741   1194  7.21mg n.s.s.     0/16   3.64mg   0/18
 a     1194  7.21mg n.s.s.     0/16   3.64mg   0/18
 b     1194  7.21mg n.s.s.     0/16   3.64mg   0/18
4742   1194  2.43mg n.s.s.     0/16   3.39mg   2/17
 a     1194  6.34mg n.s.s.     0/16   3.39mg   0/17
 b     1194  1.93mg n.s.s.     0/16   3.39mg   3/17
4743   1645  28.8mg n.s.s.     4/24   20.3mg   8/24                              Lijinsky;jtxe,13,609-614;1984
4744   1645  46.3mg n.s.s.     2/24   16.2mg   3/24
 a     1645  51.1mg n.s.s.     3/24   16.2mg   3/24
 b     1645  125.mg n.s.s.     1/24   16.2mg   0/24
4745   2140  190.mg n.s.s.     9/49   23.2mg  12/50   46.4mg   6/50   Hasegawa;txcy,51,155-165;1988/pers.comm.
4746   2140  55.7mg n.s.s.    22/50   18.6mg  20/49   37.1mg  24/50
4747   2146m 55.7ug n.s.s.     0/8    .150mg   0/8    1.50mg   0/8    15.0mg  0/8   Maita;faat,16,667-686;1991/pers.comm.
4748   2146n 1.14mg n.s.s.     1/40   .150mg   0/41   1.50mg   0/40   15.0mg  0/40
4749   2146n 44.6ug n.s.s.     0/8    .120mg   0/8    1.20mg   0/8    12.0mg  0/8
4750   2146n .891mg n.s.s.     0/40   .120mg   0/40   1.20mg   0/40   12.0mg  0/40


TETRAMETHYLTHIURAM MONOSULFIDE  (unads) 97-74-5
4751   1199  53.2mg n.s.s.     0/17   51.0mg   1/18                              Innes;ntis,1968/1969
 a     1199  101.mg n.s.s.     1/17   51.0mg   0/18
 b     1199  29.4mg n.s.s.     2/17   51.0mg   4/18
4752   1199  58.3mg n.s.s.     2/18   47.4mg   1/17
 a     1199  88.7mg n.s.s.     1/18   47.4mg   0/17
 b     1199  62.3mg n.s.s.     3/18   47.4mg   1/17
4753   1199  89.8mg n.s.s.     0/16   51.0mg   0/16
 a     1199  89.8mg n.s.s.     0/16   51.0mg   0/16
 b     1199  89.8mg n.s.s.     0/16   51.0mg   0/16
4754   1199  26.9mg n.s.s.     0/16   47.4mg   3/17
```

Ref	Spe Sex	Strain	Route	Site	Hist	Xpo+Xpt	Notes	TD50	DR	2Tailpvl	AuOp
a	M m	b6c	orl	liv	hpt	76w76	evx	139.mg		P<.1	-
b	M m	b6c	orl	tba	mix	76w76	evx	49.8mg		P<.007	-

TETRANITROMETHANE `100ng..:..1ug...:..10.....:..100....:..1mg....:..10.....:..100....:..1g.....:..10`

Ref	Spe Sex	Strain	Route	Site	Hist	Xpo+Xpt	Notes	TD50	DR	2Tailpvl	AuOp
4755	M f	b6c	inh	lun	MXA	24m24		1.33mg	*	P<.0005	c
a	M f	b6c	inh	lun	a/a	24m24		1.50mg	*	P<.0005	
b	M f	b6c	inh	lun	a/c	24m24		1.73mg	*	P<.0005	
c	M f	b6c	inh	TBA	MXB	24m24		4.90mg	*	P<.2	
d	M f	b6c	inh	liv	MXB	24m24		no dre		P=1.	
e	M f	b6c	inh	lun	MXB	24m24		1.33mg	*	P<.0005	
4756	M m	b6c	inh	lun	MXA	24m24		1.07mg	*	P<.0005	c
a	M m	b6c	inh	lun	a/c	24m24		1.15mg	*	P<.0005	
b	M m	b6c	inh	lun	a/a	24m24		1.56mg	*	P<.0005	
c	M m	b6c	inh	TBA	MXB	24m24		1.58mg	*	P<.0005	
d	M m	b6c	inh	liv	MXB	24m24		no dre		P=1.	
e	M m	b6c	inh	lun	MXB	24m24		1.07mg	*	P<.0005	
4757	R f	f34	inh	lun	MXB	24m24		.705mg	*	P<.0005	
a	R f	f34	inh	lun	MXA	24m24		.710mg	/	P<.0005	c
b	R f	f34	inh	lun	a/c	24m24		.775mg	/	P<.0005	
c	R f	f34	inh	lun	sqc	24m24		5.66mg	/	P<.0005	c
d	R f	f34	inh	adr	MXA	24m24		18.5mg	*	P<.01	
e	R f	f34	inh	lun	a/a	24m24		7.13mg	*	P<.03	
f	R f	f34	inh	adr	coa	24m24		24.4mg	*	P<.03	
g	R f	f34	inh	TBA	MXB	24m24		2.22mg	*	P<.1	
h	R f	f34	inh	liv	MXB	24m24		144.mg	*	P<.8	
4758	R m	f34	inh	lun	MXB	24m24		.304mg	*	P<.0005	
a	R m	f34	inh	lun	MXA	24m24		.326mg	*	P<.0005	c
b	R m	f34	inh	lun	a/c	24m24		.370mg	*	P<.0005	
c	R m	f34	inh	tes	MXA	24m24		.488mg	/	P<.0005	
d	R m	f34	inh	lun	a/a	24m24		1.25mg	*	P<.0005	
e	R m	f34	inh	lun	sqc	24m24		1.80mg	/	P<.0005	c
f	R m	f34	inh	pni	isa	24m24		3.59mg	*	P<.04	
g	R m	f34	inh	TBA	MXB	24m24		.552mg	*	P<.0005	
h	R m	f34	inh	liv	MXB	24m24		no dre		P=1.	

THENYLDIAMINE `100ng..:..1ug...:..10.....:..100....:..1mg....:..10.....:..100....:..1g.....:..10`

Ref	Spe Sex	Strain	Route	Site	Hist	Xpo+Xpt	Notes	TD50	DR	2Tailpvl	AuOp
4759	R f	f34	wat	tyf	mix	19m30	e	10.7gm		P<1.	-
a	R f	f34	wat	liv	mix	19m30	e	no dre		P=1.	-
4760	R m	f34	wat	liv	mix	19m30	e	96.8mg		P<.4	-
a	R m	f34	wat	tyf	mix	19m30	e	no dre		P=1.	-

THIABENDAZOLE `100ng..:..1ug...:..10.....:..100....:..1mg....:..10.....:..100....:..1g.....:..10`

Ref	Spe Sex	Strain	Route	Site	Hist	Xpo+Xpt	Notes	TD50	DR	2Tailpvl	AuOp
4761	R f	f3d	eat	ubl	mix	65w65	er	no dre		P=1.	
4762	R m	f3d	eat	ubl	mix	65w65	er	no dre		P=1.	
a	R m	f3d	eat	ubl	tcc	65w65	er	no dre		P=1.	

THIO-TEPA `100ng..:..1ug...:..10.....:..100....:..1mg....:..10.....:..100....:..1g.....:..10`

Ref	Spe Sex	Strain	Route	Site	Hist	Xpo+Xpt	Notes	TD50	DR	2Tailpvl	AuOp
4763	M f	b6c	ipj	---	MXA	47w86	as	.216mg	/	P<.0005	c
a	M f	b6c	ipj	TBA	MXB	47w86	as	85.5ug	/	P<.0005	
b	M f	b6c	ipj	liv	MXB	47w86	as	1.12mg	*	P<.2	
c	M f	b6c	ipj	lun	MXB	47w86	as	.408mg	*	P<.02	
4764	M f	b6c	ipj	---	MXA	47w86	as pool	.210mg	/	P<.0005	c
a	M f	b6c	ipj	lun	a/a	47w86	as	.408mg	*	P<.003	
4765	M m	b6c	ipj	---	sqc	52w83	s	.102mg	/	P<.0005	
a	M m	b6c	ipj	---	MXA	52w83	s	.241mg	/	P<.0005	c
b	M m	b6c	ipj	---	MXB	52w83	s	.241mg	/	P<.0005	
c	M m	b6c	ipj	---	MXA	52w83	s	.265mg	/	P<.0005	c
d	M m	b6c	ipj	TBA	MXB	52w83	s	69.0ug	/	P<.0005	
e	M m	b6c	ipj	liv	MXB	52w83	s	1.12mg	*	P<.6	
f	M m	b6c	ipj	lun	MXB	52w83	s	.584mg	*	P<.4	
4766	M m	b6c	ipj	---	sqc	52w83	s pool	.102mg		P<.0005	
a	M m	b6c	ipj	---	MXA	52w83	s	.238mg	/	P<.0005	c
b	M m	b6c	ipj	---	MXA	52w83	s	.262mg	/	P<.0005	c
c	M m	b6c	ipj	ski	sqc	52w83	s	.280mg	/	P<.003	c
d	M m	b6c	ipj	pre	sqc	52w83	s	.300mg	/	P<.002	c
4767	R f	sda	ipj	MXA	sqc	36w81	aes	.217mg	/	P<.0005	c
a	R f	sda	ipj	ute	acn	36w81	aes	.243mg	/	P<.002	
b	R f	sda	ipj	MXA	MXA	36w81	aes +hist	.819mg	*	P<.07	a
c	R f	sda	ipj	TBA	MXB	36w81	aes	40.4ug	/	P<.0005	
d	R f	sda	ipj	liv	MXB	36w81	aes	no dre		P=1.	
4768	R f	sda	ipj	mgl	acn	36w79	as pool	.182mg	*	P<.0005	
a	R f	sda	ipj	MXA	sqc	36w79	as	.214mg	/	P<.0005	c
b	R f	sda	ipj	ute	acn	36w79	as	.236mg	/	P<.0005	
c	R f	sda	ipj	eac	sqc	36w79	as	.327mg	/	P<.0005	c
4769	R m	sda	ipj	---	leu	32w82	aes	.312mg	*	P<.002	c
a	R m	sda	ipj	MXA	MXA	32w82	aes +hist	.285mg	*	P<.03	a
b	R m	sda	ipj	TBA	MXB	32w82	aes	33.2ug	/	P<.0005	
c	R m	sda	ipj	liv	MXB	32w82	aes	no dre		P=1.	
4770	R m	sda	ipj	MXA	sqc	32w79	aes pool	.122mg	/	P<.0005	c
a	R m	sda	ipj	ski	sqc	32w79	aes	.155mg	/	P<.0005	c
b	R m	sda	ipj	sub	srn	32w79	aes	.220mg	*	P<.0005	

	RefNum	LoConf	UpConf	Cntrl	1Dose	1Inc	2Dose	2Inc	Citation or Pathology
									Brkly Code
a	1199	34.0mg	n.s.s.	0/16	47.4mg	2/17			
b	1199	18.8mg	673.mg	0/16	47.4mg	5/17			
TETRANITROMETHANE		509-14-8							
4755	TR386	.891mg	2.31mg	4/50	1.24mg	24/50	4.98mg	49/50	lun:a/a,a/c.
a	TR386	1.04mg	2.49mg	1/50	1.24mg	19/50	4.98mg	41/50	S
b	TR386	1.16mg	2.88mg	3/50	1.24mg	11/50	4.98mg	45/50	S
c	TR386	1.68mg	n.s.s.	40/50	1.24mg	44/50	4.98mg	49/50	
d	TR386	12.1mg	n.s.s.	13/50	1.24mg	3/50	4.98mg	7/50	liv:hpa,hpc,nnd.
e	TR386	.891mg	2.31mg	4/50	1.24mg	24/50	4.98mg	49/50	lun:a/a,a/c.
4756	TR386	.695mg	1.91mg	12/50	1.04mg	27/50	4.15mg	47/50	lun:a/a,a/c.
a	TR386	.771mg	1.90mg	6/50	1.04mg	16/50	4.15mg	46/50	S
b	TR386	.975mg	2.95mg	7/50	1.04mg	17/50	4.15mg	34/50	S
c	TR386	.852mg	5.23mg	39/50	1.04mg	40/50	4.15mg	48/50	
d	TR386	3.89mg	n.s.s.	23/50	1.04mg	24/50	4.15mg	12/50	liv:hpa,hpc,nnd.
e	TR386	.695mg	1.91mg	12/50	1.04mg	27/50	4.15mg	47/50	lun:a/a,a/c.
4757	TR386	.513mg	.998mg	0/50	1.19mg	23/50	2.97mg	50/50	lun:a/a,a/c,sqc. C
a	TR386	.516mg	1.01mg	0/50	1.19mg	22/50	2.97mg	50/50	lun:a/a,a/c.
b	TR386	.559mg	1.11mg	0/50	1.19mg	19/50	2.97mg	50/50	S
c	TR386	2.91mg	13.0mg	0/50	1.19mg	1/50	2.97mg	12/50	
d	TR386	6.23mg	1.15gm	0/50	1.19mg	0/50	2.97mg	4/50	adr:coa,coc. S
e	TR386	3.29mg	n.s.s.	0/50	1.19mg	6/50	2.97mg	3/50	S
f	TR386	7.13mg	n.s.s.	0/50	1.19mg	0/50	2.97mg	3/50	S
g	TR386	.839mg	n.s.s.	47/50	1.19mg	49/50	2.97mg	50/50	
h	TR386	9.27mg	n.s.s.	1/50	1.19mg	1/50	2.97mg	1/50	liv:hpa,hpc,nnd.
4758	TR386	.210mg	.453mg	1/50	.832mg	34/50	2.08mg	50/50	lun:a/a,a/c,sqc. C
a	TR386	.224mg	.491mg	1/50	.832mg	33/50	2.08mg	46/50	lun:a/a,a/c.
b	TR386	.252mg	.551mg	0/50	.832mg	26/50	2.08mg	46/50	S
c	TR386	.265mg	1.52mg	33/50	.832mg	38/50	2.08mg	39/50	tes:iab,ica. S
d	TR386	.655mg	3.33mg	1/50	.832mg	13/50	2.08mg	11/50	S
e	TR386	.956mg	3.50mg	0/50	.832mg	1/50	2.08mg	19/50	
f	TR386	1.25mg	n.s.s.	3/50	.832mg	5/50	2.08mg	6/50	S
g	TR386	.289mg	2.20mg	47/50	.832mg	49/50	2.08mg	50/50	
h	TR386	4.71mg	n.s.s.	1/50	.832mg	2/50	2.08mg	0/50	liv:hpa,hpc,nnd.
THENYLDIAMINE		91-79-2							
4759	1790	96.3mg	n.s.s.	1/21	25.1mg	1/20			Lijinsky;zkko,112,57-60;1986
a	1790	53.7mg	n.s.s.	6/21	25.1mg	5/20			
4760	1790	23.0mg	n.s.s.	4/19	17.6mg	7/20			
a	1790	75.6mg	n.s.s.	2/19	17.6mg	1/20			
THIABENDAZOLE	(2-(4-thiazolyl)-benzimazole)	148-79-8							
4761	1777	113.mg	n.s.s.	0/15	100.mg	0/14			Fujii;fctx,24,207-211;1986
4762	1777	96.6mg	n.s.s.	1/14	80.0mg	0/15			
a	1777	96.6mg	n.s.s.	1/14	80.0mg	0/15			
THIO-TEPA	52-24-4								
4763	TR58	.133mg	.352mg	0/15	.300mg	5/35	1.00mg	32/35	---:lle,lym.
a	TR58	53.3ug	.142mg	0/15	.300mg	17/35	1.00mg	32/35	
b	TR58	.277mg	n.s.s.	0/15	.300mg	2/35	1.00mg	0/35	liv:hpa,hpc,nnd.
c	TR58	.155mg	n.s.s.	0/15	.300mg	5/35	1.00mg	0/35	lun:a/a,a/c.
4764	Ref58	.127mg	.357mg	1/30p	.300mg	5/35	1.00mg	32/35	---:lle,lym.
a	TR58	.155mg	2.35mg	0/30p	.300mg	5/35	1.00mg	0/35	S
4765	TR58	55.1ug	.281mg	0/15	.300mg	15/34	.900mg	1/35	S
a	TR58	.133mg	.461mg	1/15	.300mg	3/34	.900mg	26/35	---:leu,lym.
b	TR58	.133mg	.461mg	1/15	.300mg	3/34	.900mg	26/35	---:leu,lle,lym. C
c	TR58	.146mg	.506mg	1/15	.300mg	2/34	.900mg	26/35	---:lle,lym.
d	TR58	40.6ug	.137mg	4/15	.300mg	19/34	.900mg	27/35	
e	TR58	.187mg	n.s.s.	1/15	.300mg	4/34	.900mg	0/35	liv:hpa,hpc,nnd.
f	TR58	.140mg	n.s.s.	1/15	.300mg	5/34	.900mg	0/35	lun:a/a,a/c.
4766	TR58	55.1ug	.218mg	0/30p	.300mg	15/34	.900mg	1/35	S
a	TR58	.134mg	.441mg	1/30p	.300mg	3/34	.900mg	26/35	---:leu,lym.
b	TR58	.146mg	.485mg	1/30p	.300mg	2/34	.900mg	26/35	---:lle,lym.
c	TR58	.119mg	1.28mg	0/30p	.300mg	7/34	.900mg	0/35	
d	TR58	.126mg	1.25mg	0/30p	.300mg	6/34	.900mg	1/35	
4767	TR58	.101mg	.666mg	0/20	.190mg	2/31	.250mg	8/35	eac:sqc; ski:sqc.
a	TR58	.107mg	.868mg	0/20	.190mg	2/31	.250mg	7/35	S
b	TR58	.260mg	n.s.s.	0/20	.190mg	2/31	.250mg	2/35	bra:oln; nas:can.
c	TR58	24.2ug	88.2ug	8/20	.190mg	26/31	.250mg	26/35	
d	TR58	n.s.s.	n.s.s.	0/20	.190mg	0/31	.250mg	0/35	liv:hpa,hpc,nnd.
4768	TR58	89.5ug	.491mg	2/60p	.190mg	7/31	.250mg	8/35	eac:sqc; ski:sqc.
a	TR58	.101mg	.537mg	0/60p	.190mg	2/31	.250mg	8/35	S
b	TR58	.105mg	.637mg	0/60p	.190mg	2/31	.250mg	7/35	S
c	TR58	.138mg	1.05mg	0/60p	.190mg	2/31	.250mg	5/35	
4769	TR58	.144mg	1.14mg	0/20	.190mg	5/39	.270mg	6/35	
a	TR58	68.7ug	n.s.s.	0/20	.190mg	3/39	.270mg	0/35	bra:neu,oln; nas:can.
b	TR58	17.6ug	75.6ug	8/20	.190mg	27/39	.270mg	14/35	
c	TR58	n.s.s.	n.s.s.	0/20	.190mg	0/39	.270mg	0/35	liv:hpa,hpc,nnd.
4770	TR58	52.4ug	.317mg	0/60p	.190mg	7/39	.270mg	3/35	eac:sqc; ski:sqc.
a	TR58	59.5ug	.482mg	0/60p	.190mg	5/39	.270mg	3/35	
b	TR58	68.9ug	1.07mg	0/60p	.190mg	5/39	.270mg	0/35	S

```
      Spe Strain Site  Xpo+Xpt                                              TD50    2Tailpvl
      Sex  Route  Hist           Notes                                        DR    AuOp
  c   R m sda ipj --- MXA 32w79 aes                                         .224mg * P<.0005c
  d   R m sda ipj --- leu 32w79 aes                                         .290mg * P<.0005c
  e   R m sda ipj --- lle 32w79 aes                                         .374mg * P<.0005c
4771  R m b46 ivj tba mix 12m24                       .    +      .         .186mg   P<.002 +
  a   R m b46 ivj tba mal 12m24                                             .303mg   P<.009
  b   R m b46 ivj tba ben 12m24                                             .646mg   P<.2

THIOACETAMIDE                    100ng..:..1ug....:..10......:..100.....:..1mg....:..10......:..100.....:..1g......:..10
4772  M m icm eat liv hnd 40w88 er                              .  +  .     12.0mg   P<.0005
  a   M m icm eat liv hpc 40w88 er                                          15.3mg   P<.0005+
4773  M f swi eat liv hpt 65w65 kr                          .     +    .    5.36mg   P<.0005+
4774  M m swi eat liv hpt 65w65 kr                                 <+       noTD50   P<.005 +
4775  R m don eat liv mix 40w80 er                              .  ±        11.5mg   P<.02 +
  a   R m don eat liv thc 40w80 er                                          21.8mg   P<.08
  b   R m don eat liv pac 40w80 er                                          27.6mg   P<.2

4,4'-THIOBIS(6-tert-BUTYL-m-CRESOL)  ..:..1ug....:..10......:..100.....:..1mg....:..10......:..100.....:..1g......:..10
4776  M f b6c eat TBA MXB 24m24                                        :>   484.mg * P<.6  -
  a   M f b6c eat liv MXB 24m24                                             no dre   P=1.
  b   M f b6c eat lun MXB 24m24                                             2.02gm * P<.5
4777  M m b6c eat TBA MXB 24m24                                     :>      no dre   P=1.  -
  a   M m b6c eat liv MXB 24m24                                             no dre   P=1.
  b   M m b6c eat lun MXB 24m24                                             no dre   P=1.
4778  R f f34 eat ute MXA 24m24                                :     +     :#262.mg * P<.008 -
  a   R f f34 eat ute esp 24m24                                             286.mg * P<.02
  b   R f f34 eat thy MXA 24m24                                             325.mg * P<.03
  c   R f f34 eat thy MXA 24m24                                             361.mg * P<.03
  d   R f f34 eat TBA MXB 24m24                                             196.mg * P<.3
  e   R f f34 eat liv MXB 24m24                                             3.99gm * P<.2
4779  R m f34 eat TBA MXB 24m24                                     :>      449.mg * P<.8  -
  a   R m f34 eat liv MXB 24m24                                             324.mg * P<.07

2,2-THIOBIS(4,6-DICHLOROPHENOL)  100ng..:..1ug....:..10......:..100.....:..1mg....:..10......:..100.....:..1g......:..10
4780  M f b6a orl lun ade 76w76 evx                        .>              100.mg   P<.6
  a   M f b6a orl liv hpt 76w76 evx                                        no dre   P=1.
  b   M f b6a orl tba mix 76w76 evx                                        28.7mg   P<.3
4781  M m b6a orl liv hpt 76w76 evx                           .>           no dre   P=1.
  a   M m b6a orl lun ade 76w76 evx                                        no dre   P=1.
  b   M m b6a orl tba mix 76w76 evx                                        no dre   P=1.
4782  M f b6c orl liv hpt 76w76 evx                           .>           no dre   P=1.
  a   M f b6c orl lun mix 76w76 evx                                        no dre   P=1.
  b   M f b6c orl tba mix 76w76 evx                                        43.0mg   P<.09
4783  M m b6c orl liv agm 76w76 evx                         .>             88.2mg   P<.3
  a   M m b6c orl lun ade 76w76 evx                                        88.2mg   P<.3
  b   M m b6c orl tba mix 76w76 evx                                        10.1mg   P<.002

THIOCYANATE, SODIUM              100ng..:..1ug....:..10......:..100.....:..1mg....:..10......:..100.....:..1g......:..10
4784  R f f34 wat liv tum 26m30 e                                     .>   no dre   P=1.  -
4785  R m f34 wat ssu tum 26m30 e                              .   +   .    224.mg   P<.002 -
  a   R m f34 wat liv tum 26m30 e                                           2.76gm   P<.9  -

4,4'-THIODIANILINE               100ng..:..1ug....:..10......:..100.....:..1mg....:..10......:..100.....:..1g......:..10
4786  M f b6c eat MXB MXB 78w91 a                              :  +  :      33.2mg * P<.0005
  a   M f b6c eat liv hpc 78w91 a                                           33.3mg * P<.0005c
  b   M f b6c eat thy MXA 78w91 a                                           88.4mg * P<.0005c
  c   M f b6c eat thy fcc 78w91 a                                           165.mg / P<.0005c
  d   M f b6c eat TBA MXB 78w91 a                                           33.2mg * P<.0005
  e   M f b6c eat liv MXB 78w91 a                                           33.3mg * P<.0005
  f   M f b6c eat lun MXB 78w91 a                                           no dre   P=1.
4787  M m b6c eat MXB MXB 78w91 a                              :  +  :      32.7mg * P<.0005
  a   M m b6c eat liv MXA 78w91 a                                           33.2mg * P<.0005c
  b   M m b6c eat liv hpc 78w91 a                                           33.8mg * P<.0005c
  c   M m b6c eat thy MXA 78w91 a                                           51.8mg / P<.0005c
  d   M m b6c eat thy fcc 78w91 a                                           70.7mg / P<.0005c
  e   M m b6c eat TBA MXB 78w91 a                                           32.7mg * P<.0005
  f   M m b6c eat liv MXB 78w91 a                                           33.2mg * P<.0005
  g   M m b6c eat lun MXB 78w91 a                                           1.07gm * P<.3
4788  R f f34 eat MXB MXB 16m24 s                          :  +  :          8.53mg * P<.0005
  a   R f f34 eat ute acn 16m24 s                                           8.79mg * P<.0005c
  b   R f f34 eat thy fcc 16m24 s                                           11.5mg / P<.0005c
  c   R f f34 eat eac MXA 16m24 s                                     +hist 121.mg * P<.006 c
  d   R f f34 eat TBA MXB 16m24 s                                           8.35mg * P<.0005
  e   R f f34 eat liv MXB 16m24 s                                           42.0mg * P<.0005
4789  R m f34 eat MXB MXB 16m24                            :  +  :          5.52mg \ P<.0005
  a   R m f34 eat thy MXA 16m24                                             5.59mg \ P<.0005c
  b   R m f34 eat thy fcc 16m24                                             6.36mg \ P<.0005c
  c   R m f34 eat liv MXA 16m24                                             7.18mg \ P<.0005c
  d   R m f34 eat liv hpc 16m24                                             7.31mg \ P<.0005c
  e   R m f34 eat eac MXA 16m24                                             20.3mg \ P<.0005c
  f   R m f34 eat col acn 16m24                                       +hist 97.4mg \ P<.02 c
  g   R m f34 eat TBA MXB 16m24                                             5.96mg \ P<.0005
```

	RefNum	LoConf	UpConf	Cntrl	1Dose	1Inc	2Dose	2Inc	Citation or Pathology	Brkly Code
c	TR58	91.5ug	.674mg	1/60p	.190mg	6/39	.270mg	6/35		---:leu,lym.
d	TR58	.134mg	.708mg	0/60p	.190mg	5/39	.270mg	6/35		
e	TR58	.153mg	1.15mg	0/60p	.190mg	3/39	.270mg	5/35		
4771	1017	86.1ug	1.02mg	7/89	71.4ug	14/48			Schmahl;arzn,20,1461-1467;1970	
a	1017	.122mg	13.2mg	4/89	71.4ug	9/48				
b	1017	.188mg	n.s.s.	3/89	71.4ug	5/48				

THIOACETAMIDE   62-55-5

	RefNum	LoConf	UpConf	Cntrl	1Dose	1Inc	2Dose	2Inc	Citation or Pathology	Brkly Code
4772	1959	6.19mg	28.4mg	0/15	19.1mg	13/24			Akao;chpb,38,2012-2014;1990	
a	1959	7.55mg	42.3mg	0/15	19.1mg	11/24				
4773	282	1.69mg	22.6mg	0/6	39.0mg	6/7			Gothoskar;bjca,24,498-503;1970	
4774	282	n.s.s.	11.3mg	0/6	36.0mg	6/6				
4775	1836	5.38mg	n.s.s.	0/15	7.00mg	9/41			Kuroda;jnci,79,1047-1051;1987	
a	1836	8.28mg	n.s.s.	0/15	7.00mg	5/41				
b	1836	9.55mg	n.s.s.	0/15	7.00mg	4/41				

4,4'-THIOBIS(6-tert-BUTYL-m-CRESOL)   96-69-5

	RefNum	LoConf	UpConf	Cntrl	1Dose	1Inc	2Dose	2Inc	Citation or Pathology	Brkly Code
4776	TR435	86.4mg	n.s.s.	38/51	32.3mg	37/50	64.8mg	38/50	130.mg	38/50
a	TR435	190.mg	n.s.s.	20/51	32.3mg	23/50	64.8mg	24/50	130.mg	14/50 (liv:hpa,hpb,hpc.)
b	TR435	415.mg	n.s.s.	2/51	32.3mg	3/50	64.8mg	1/50	130.mg	4/50 (lun:a/a,a/c.)
4777	TR435	64.2mg	n.s.s.	38/50	29.9mg	38/50	59.8mg	39/50	(120.mg	25/50)
a	TR435	78.1mg	n.s.s.	25/50	29.9mg	30/50	59.8mg	27/50	(120.mg	16/50) (liv:hpa,hpb,hpc.)
b	TR435	599.mg	n.s.s.	8/50	29.9mg	9/50	59.8mg	9/50	120.mg	3/50 (lun:a/a,a/c.)
4778	TR435	125.mg	7.44gm	2/50	24.9mg	6/50	49.9mg	9/50	125.mg	10/50 (ute:esp,ess.) S
a	TR435	133.mg	n.s.s.	2/50	24.9mg	5/50	49.9mg	9/50	125.mg	9/50 S
b	TR435	141.mg	n.s.s.	3/50	24.9mg	4/50	49.9mg	8/50	125.mg	9/50 (thy:cca,ccr,cdb.) S
c	TR435	155.mg	n.s.s.	3/50	24.9mg	2/50	49.9mg	8/50	125.mg	8/50 (thy:cca,cdb.) S
d	TR435	54.8mg	n.s.s.	47/50	24.9mg	45/50	49.9mg	43/50	125.mg	47/50
e	TR435	650.mg	n.s.s.	0/50	24.9mg	0/50	49.9mg	0/50	125.mg	1/50 (liv:hpa,hpb,hpc.)
4779	TR435	47.0mg	n.s.s.	47/50	19.9mg	47/50	40.0mg	47/50	99.9mg	44/50
a	TR435	119.mg	n.s.s.	1/50	19.9mg	3/50	40.0mg	3/50	99.9mg	5/50 (liv:hpa,hpb,hpc.)

2,2-THIOBIS(4,6-DICHLOROPHENOL)   (TBP, Vancide BL) 97-18-7

	RefNum	LoConf	UpConf	Cntrl	1Dose	1Inc	Citation or Pathology
4780	1283	13.2mg	n.s.s.	1/17	15.7mg	2/18	Innes;ntis,1968/1969
a	1283	31.1mg	n.s.s.	0/17	15.7mg	0/18	
b	1283	7.51mg	n.s.s.	2/17	15.7mg	5/18	
4781	1283	27.4mg	n.s.s.	1/18	14.6mg	0/17	
a	1283	27.4mg	n.s.s.	2/18	14.6mg	0/17	
b	1283	19.2mg	n.s.s.	3/18	14.6mg	1/17	
4782	1283	27.6mg	n.s.s.	0/16	15.7mg	0/16	
a	1283	27.6mg	n.s.s.	0/16	15.7mg	0/16	
b	1283	10.6mg	n.s.s.	0/16	15.7mg	2/16	
4783	1283	14.4mg	n.s.s.	0/16	14.6mg	1/17	
a	1283	14.4mg	n.s.s.	0/16	14.6mg	1/17	
b	1283	4.29mg	38.7mg	0/16	14.6mg	7/17	

THIOCYANATE, SODIUM   540-72-7

	RefNum	LoConf	UpConf	Cntrl	1Dose	1Inc	Citation or Pathology
4784	1898	289.mg	n.s.s.	4/24	107.mg	3/20	Lijinsky;txih,5,25-29;1989
4785	1898	90.8mg	968.mg	0/24	74.7mg	6/20	
a	1898	185.mg	n.s.s.	3/24	74.7mg	3/20	

4,4'-THIODIANILINE   139-65-1

	RefNum	LoConf	UpConf	Cntrl	1Dose	1Inc	2Dose	2Inc	Citation or Pathology	Brkly Code
4786	TR47	18.5mg	60.2mg	0/14	232.mg	33/35	464.mg	30/35	liv:hpc; thy:fca,fcc. C	
a	TR47	18.5mg	60.5mg	0/14	232.mg	32/35	464.mg	30/35		
b	TR47	45.5mg	178.mg	0/14	232.mg	11/35	464.mg	18/35	thy:fca,fcc.	
c	TR47	72.9mg	386.mg	0/14	232.mg	3/35	464.mg	15/35		
d	TR47	18.5mg	60.2mg	2/14	232.mg	33/35	464.mg	30/35		
e	TR47	18.5mg	60.5mg	0/14	232.mg	32/35	464.mg	30/35	liv:hpa,hpc,nnd.	
f	TR47	n.s.s.	n.s.s.	0/14	232.mg	0/35	464.mg	0/35	lun:a/a,a/c.	
4787	TR47	20.1mg	54.0mg	4/14	214.mg	34/35	428.mg	24/35	liv:hpa,hpc; thy:fca,fcc. C	
a	TR47	20.3mg	55.3mg	4/14	214.mg	33/35	428.mg	23/35	liv:hpa,hpc.	
b	TR47	20.5mg	56.9mg	1/14	214.mg	32/35	428.mg	22/35		
c	TR47	30.7mg	88.8mg	0/14	214.mg	22/35	428.mg	20/35	thy:fca,fcc.	
d	TR47	40.7mg	125.mg	0/14	214.mg	15/35	428.mg	20/35		
e	TR47	20.1mg	54.0mg	4/14	214.mg	34/35	428.mg	24/35		
f	TR47	20.3mg	55.3mg	4/14	214.mg	33/35	428.mg	23/35	liv:hpa,hpc,nnd.	
g	TR47	174.mg	n.s.s.	0/14	214.mg	1/35	428.mg	0/35	lun:a/a,a/c.	
4788	TR47	4.53mg	15.2mg	0/15	53.5mg	32/35	107.mg	33/35	eac:sqc,sqp; thy:fcc; ute:acn. C	
a	TR47	4.61mg	16.0mg	0/15	53.5mg	31/35	107.mg	23/35		
b	TR47	6.13mg	20.2mg	0/15	53.5mg	24/35	107.mg	32/35		
c	TR47	49.8mg	1.34gm	0/15	53.5mg	6/35	107.mg	3/35	eac:sqc,sqp.	
d	TR47	4.44mg	16.1mg	6/15	53.5mg	32/35	107.mg	33/35		
e	TR47	15.9mg	137.mg	0/15	53.5mg	6/35	107.mg	3/35	liv:hpa,hpc,nnd.	
4789	TR47	2.00mg	11.7mg	0/15	42.8mg	32/35	(85.6mg	33/35)	col:acn; eac:sqc,sqp; liv:hpa,hpc; thy:fca,fcc. C	
a	TR47	2.01mg	12.2mg	0/15	42.8mg	30/35	(85.6mg	32/35)	thy:fca,fcc.	
b	TR47	2.08mg	15.3mg	0/15	42.8mg	28/35	(85.6mg	32/35)		
c	TR47	2.15mg	18.5mg	0/15	42.8mg	23/35	(85.6mg	12/35)	liv:hpa,hpc.	
d	TR47	2.16mg	19.3mg	0/15	42.8mg	21/35	(85.6mg	10/35)		
e	TR47	6.42mg	57.0mg	0/15	42.8mg	15/35	(85.6mg	8/35)	eac:sqc,sqp.	
f	TR47	38.5mg	n.s.s.	0/15	42.8mg	6/35	(85.6mg	1/35)		
g	TR47	2.28mg	13.6mg	4/15	42.8mg	32/35	(85.6mg	33/35)		

```
      Spe Strain Site   Xpo+Xpt                                                             TD50    2Tailpvl
          Sex  Route Hist   Notes                                                                DR    AuOp
  h   R m f34 eat liv MXB 16m24                                                            7.18mg \ P<.0005
4790  R m f3d eat lun ade 52w52 r                          .           ±                   1.41mg   P<.02
  a   R m f3d eat liv hpc 52w52 r                                                          1.94mg   P<.04  +
  b   R m f3d eat thy fcc 52w52 r                                                          2.99mg   P<.1

beta-THIOGUANINE DEOXYRIBOSIDE      100ng..:..1ug....:..10.....:..100....:..1mg....:..10.....:..100....:..1g.....:..10
4791  M f b6c ipj TBA MXB 43w77 as                                  :>                     4.48mg Z P<1.   i
  a   M f b6c ipj liv MXB 43w77 as                                                         no dre   P=1.
  b   M f b6c ipj lun MXB 43w77 as                                                         6.87mg * P<.3
4792  M m b6c ipj TBA MXB 38w77 as                              :>                         no dre   P=1.   i
  a   M m b6c ipj liv MXB 38w77 as                                                         no dre   P=1.
  b   M m b6c ipj lun MXB 38w77 as                                                         no dre   P=1.
4793  R f sda ipj eac MXA 52w78 es                         :    +    :                     2.10mg / P<.005 c
  a   R f sda ipj eac sqc 52w78 es                                                         2.31mg / P<.008
  b   R f sda ipj TBA MXB 52w78 es                                                         4.11mg / P<.7
  c   R f sda ipj liv MXB 52w78 es                                                         no dre   P=1.
4794  R f sda ipj eac MXA 52w78 es pool                    :    +    :                     2.10mg / P<.0005c
  a   R f sda ipj eac sqc 52w78 es                                                         2.31mg / P<.002
4795  R m sda ipj eac MXA 52w78 es                             :    ±                      4.38mg * P<.08  a
  a   R m sda ipj TBA MXB 52w78 es                                                         no dre   P=1.
  b   R m sda ipj liv MXB 52w78 es                                                         no dre   P=1.
4796  R m sda ipj eac MXA 52w78 es pool                        :    ±                      4.38mg * P<.02  a
  a   R m sda ipj --- lym 52w78 es                                                         5.75mg * P<.03

THIOSEMICARBAZIDE                   100ng..:..1ug....:..10.....:..100....:..1mg....:..10.....:..100....:..1g.....:..10
4797  M f swa wat liv hpt 29m29 aes                                        .>              no dre   P=1.   -
  a   M f swa wat lun tum 29m29 aes                                                        no dre   P=1.   -
4798  M m swa wat liv hpt 27m28 aes                                           .>           no dre   P=1.   -
  a   M m swa wat lun tum 27m28 aes                                                        no dre   P=1.   -
4799  R f cdr eat mgl tum 18m24 e                                  .>                      3.07mg * P<.2

THIOURACIL                          100ng..:..1ug....:..10.....:..100....:..1mg....:..10.....:..100....:..1g.....:..10
4800  M f c3h eat liv hpt 73w73 r                                    .    +    .           63.3mg   P<.0005+
4801  M m c3h eat liv hpt 73w73 r                                         .    +    .      48.6mg   P<.0005+
4802  M f tmm eat liv hpt 73w73 r                                                    .>    no dre   P=1.   -
4803  M m tmm eat liv hpt 73w73 r                                                    .>    no dre   P=1.   -
4804  R f f34 eat thy mal 24m24 e                               .    +    .                22.9mg * P<.0005+
  a   R f f34 eat liv hpc 24m24 e                                                          no dre   P=1.
4805  R m f34 eat thy mal 24m24 es                          .    +    .                    8.05mg Z P<.0005+
  a   R m f34 eat liv hpc 24m24 es                                                         no dre   P=1.

THIOUREA                            100ng..:..1ug....:..10.....:..100....:..1mg....:..10.....:..100....:..1g.....:..10
4806  M f c3h wat lun tum  6m24 e                                               .>         no dre   P=1.
4807  R m hew mix aur epc 95w95 emv                              .    +    .                104.mg   P<.0005+
  a   R m hew eld epc 95w95 emv                                                            225.mg   P<.02  +
  b   R m hew mix tba mix 95w95 emv                                                        50.8mg   P<.0005
4808  R m hew wat aur epc 26m26 e                                    .    +    .           93.5mg   P<.0005+
  a   R m hew wat eld epc 26m26 e                                                          93.5mg   P<.0005+
  b   R m hew wat tba mix 26m26 e                                                          27.5mg   P<.0005
4809  R f osm eat liv tum 26m26                                         .>                 no dre   P=1.   -
  a   R f osm eat tba mix 26m26                                                            16.2mg   P<.7   -
4810  R f osm eat liv tum 24m24                                            .>              no dre   P=1.   -
  a   R f osm eat tba ben 24m24                                                            15.0mg   P<.3   -
  b   R f osm eat tba mal 24m24                                                            no dre   P=1.   -
4811  R m osm eat liv tum 26m26                                         .>                 no dre   P=1.   -
  a   R m osm eat tba mix 26m26                                                            46.1mg   P<.6   -
4812  R m osm eat liv tum 24m24                                            .>              no dre   P=1.   -
  a   R m osm eat tba ben 24m24                                                            64.7mg   P<.3   -
  b   R m osm eat tba mal 24m24                                                            no dre   P=1.   -

TILIDINE FUMARATE                   100ng..:..1ug....:..10.....:..100....:..1mg....:..10.....:..100....:..1g.....:..10
4813  M f cf1 eat lun agc 80w80 e                                       .    ±             207.mg * P<.02  -
  a   M f cf1 eat liv hpa 80w80 e                                                          no dre   P=1.   -
  b   M f cf1 eat tba mix 80w80 e                                                          643.mg * P<.7   -
  c   M f cf1 eat tba mal 80w80 e                                                          970.mg * P<.8   -
  d   M f cf1 eat tba ben 80w80 e                                                          1.48gm * P<.7   -
4814  M m cf1 eat lun agc 80w80 e                                             .>           936.mg * P<.6   -
  a   M m cf1 eat liv hem 80w80 e                                                          2.72gm Z P<.6   -
  b   M m cf1 eat liv hpa 80w80 e                                                          no dre   P=1.   -
  c   M m cf1 eat tba mix 80w80 e                                                          249.mg * P<.3   -
  d   M m cf1 eat tba mal 80w80 e                                                          391.mg * P<.4   -
  e   M m cf1 eat tba ben 80w80 e                                                          no dre   P=1.   -
4815  R f wal eat liv hpa 24m24 e                                      .    +    .         476.mg * P<.003 -
  a   R f wal eat tba mal 24m24 e                                                          525.mg * P<.2   -
  b   R f wal eat tba ben 24m24 e                                                          no dre   P=1.   -
  c   R f wal eat tba mix 24m24 e                                                          no dre   P=1.   -
4816  R m wal eat liv hpa 24m24 e                                             .>           11.3gm * P<.1   -
  a   R m wal eat tba mix 24m24 e                                .>                         122.mg * P<.02  -
  b   R m wal eat tba ben 24m24 e                                                          300.mg * P<.2   -
  c   R m wal eat tba mal 24m24 e                                                          537.mg * P<.2   -
```

	RefNum	LoConf	UpConf	Cntrl	1Dose	1Inc	2Dose	2Inc	Citation or Pathology	Brkly Code
h	TR47	2.15mg	18.5mg	0/15	42.8mg	23/35	(85.6mg	12/35)	liv:hpa,hpc,nnd.	
4790	2024	.486mg	n.s.s.	0/20	1.84mg	4/20			Hasegawa;carc,12,1515-1518;1991/pers.comm.	
a	2024	.586mg	n.s.s.	0/20	1.84mg	3/20				
b	2024	.735mg	n.s.s.	0/20	1.84mg	2/20				

beta-THIOGUANINE DEOXYRIBOSIDE (beta-TGdR) 64039-27-6

	RefNum	LoConf	UpConf	Cntrl	1Dose	1Inc	2Dose	2Inc	Citation or Pathology	Brkly Code
4791	TR57	.150mg	n.s.s.	21/30	.300mg	20/35	.600mg	21/35 (1.70mg 1/35)		
a	TR57	n.s.s.	n.s.s.	0/30	.300mg	0/35	.600mg	0/35 1.70mg 0/35	liv:hpa,hpc,nnd.	
b	TR57	1.12mg	n.s.s.	0/30	.300mg	0/35	.600mg	1/35 1.70mg 0/35	lun:a/a,a/c.	
4792	TR57	.173mg	n.s.s.	20/30	.300mg	19/35	.600mg	19/35 1.70mg 0/35		
a	TR57	.421mg	n.s.s.	2/30	.300mg	1/35	.600mg	1/35 1.70mg 0/35	liv:hpa,hpc,nnd.	
b	TR57	n.s.s.	n.s.s.	0/30	.300mg	0/35	.600mg	0/35 1.70mg 0/35	lun:a/a,a/c.	
4793	TR57	.938mg	14.3mg	0/10	1.00mg	2/35	2.00mg	6/35	eac:can,sqc.	
a	TR57	.990mg	44.0mg	0/10	1.00mg	2/35	2.00mg	5/35		S
b	TR57	.572mg	n.s.s.	9/10	1.00mg	17/35	2.00mg	10/35		
c	TR57	n.s.s.	n.s.s.	0/10	1.00mg	0/35	2.00mg	0/35	liv:hpa,hpc,nnd.	
4794	TR57	.938mg	6.68mg	0/30p	1.00mg	2/35	2.00mg	6/35	eac:can,sqc.	
a	TR57	.990mg	8.66mg	0/30p	1.00mg	2/35	2.00mg	5/35		S
4795	TR57	1.32mg	n.s.s.	0/10	1.00mg	1/35	2.00mg	2/35	eac:can,sqc.	
a	TR57	.839mg	n.s.s.	7/10	1.00mg	9/35	2.00mg	5/35		
b	TR57	n.s.s.	n.s.s.	0/10	1.00mg	0/35	2.00mg	0/35	liv:hpa,hpc,nnd.	
4796	TR57	1.32mg	n.s.s.	0/30p	1.00mg	1/35	2.00mg	2/35	eac:can,sqc.	
a	TR57	1.73mg	n.s.s.	0/30p	1.00mg	1/35	2.00mg	2/35		S

THIOSEMICARBAZIDE (thiocarbamylhydrazine) 79-19-6

	RefNum	LoConf	UpConf	Cntrl	1Dose	1Inc	2Dose	2Inc	Citation or Pathology	Brkly Code
4797	1675	38.4mg	n.s.s.	1/15	31.2mg	0/14	62.4mg	0/3	Toth;faat,2,173-176;1982/1980	
a	1675	469.mg	n.s.s.	24/98	31.2mg	9/49	62.4mg	4/39		
4798	1675	204.mg	n.s.s.	0/100	26.0mg	0/49	52.0mg	0/33		
a	1675	207.mg	n.s.s.	26/100	26.0mg	5/49	(52.0mg	3/33)		
4799	1112	1.11mg	n.s.s.	3/10	1.41mg	19/26	2.81mg	16/26	Weisburger;jnci,67,75-88;1981	

THIOURACIL 141-90-2

	RefNum	LoConf	UpConf	Cntrl	1Dose	1Inc	2Dose	2Inc	Citation or Pathology	Brkly Code
4800	284	28.6mg	152.mg	0/24	390.mg	14/16			Casas;pseb,113,493-494;1963	
4801	284	17.9mg	139.mg	2/32	360.mg	12/13				
4802	284	871.mg	n.s.s.	0/20	390.mg	0/22				
4803	284	804.mg	n.s.s.	0/20	360.mg	0/22				
4804	1947	13.8mg	42.0mg	5/214	4.15mg	2/23	12.5mg	6/24 37.5mg 18/24	Fears;txih,5,1-23;1989	
a	1947	14.0mg	n.s.s.	0/214	4.15mg	0/24	12.5mg	0/24 37.5mg 0/24		
4805	1947	4.59mg	16.2mg	5/214	3.32mg	6/24	10.0mg	14/24 (30.0mg 5/24)		
a	1947	11.2mg	n.s.s.	0/214	3.32mg	0/24	10.0mg	0/24 30.0mg 0/24		

THIOUREA 62-56-6

	RefNum	LoConf	UpConf	Cntrl	1Dose	1Inc	2Dose	2Inc	Citation or Pathology	Brkly Code
4806	1135	331.mg	n.s.s.	1/94	25.1mg	0/64			Vazquez-Lopez;bjca,3,401-414;1949	
4807	288a	43.0mg	363.mg	0/12	159.mg	7/12			Rosin;canr,17,302-305;1957	
a	288a	76.6mg	n.s.s.	0/12	159.mg	4/12				
b	288a	21.2mg	141.mg	0/12	159.mg	10/12				
4808	288b	45.5mg	249.mg	0/12	100.mg	11/19				
a	288b	45.5mg	249.mg	0/12	100.mg	11/19				
b	288b	11.2mg	64.0mg	0/12	100.mg	18/19				
4809	21	18.2mg	n.s.s.	0/30	2.50mg	0/30			Deichmann;txap,11,88-103;1967	
a	21	2.51mg	n.s.s.	13/30	2.50mg	15/30				
4810	84a	24.7mg	n.s.s.	1/30	4.00mg	0/30			Radomski;txap,7,652-656;1965	
a	84a	4.19mg	n.s.s.	6/30	4.00mg	10/30				
b	84a	19.7mg	n.s.s.	6/30	4.00mg	1/30				
4811	21	14.6mg	n.s.s.	0/30	2.00mg	0/30			Deichmann;txap,11,88-103;1967	
a	21	6.34mg	n.s.s.	1/30	2.00mg	2/30				
4812	84a	19.8mg	n.s.s.	0/30	3.20mg	0/30			Radomski;txap,7,652-656;1965	
a	84a	10.5mg	n.s.s.	0/30	3.20mg	1/30				
b	84a	19.8mg	n.s.s.	3/30	3.20mg	0/30				

TILIDINE FUMARATE (valoron) 55567-81-2

	RefNum	LoConf	UpConf	Cntrl	1Dose	1Inc	2Dose	2Inc	Citation or Pathology	Brkly Code
4813	1794	88.4mg	n.s.s.	10/100	16.0mg	7/50	40.0mg	11/50 100.mg 12/50	McGuire;txcy,39,149-163;1986	
a	1794	62.5mg	n.s.s.	2/100	16.0mg	0/50	40.0mg	0/50 100.mg 0/50		
b	1794	79.7mg	n.s.s.	38/100	16.0mg	24/50	40.0mg	25/50 100.mg 21/50		
c	1794	96.3mg	n.s.s.	32/100	16.0mg	21/50	40.0mg	19/50 100.mg 18/50		
d	1794	172.mg	n.s.s.	14/100	16.0mg	6/50	40.0mg	7/50 100.mg 8/50		
4814	1794	148.mg	n.s.s.	15/100	16.0mg	7/50	40.0mg	9/50 100.mg 9/50		
a	1794	334.mg	n.s.s.	0/100	16.0mg	2/50	40.0mg	0/50 100.mg 1/50		
b	1794	401.mg	n.s.s.	9/100	16.0mg	3/50	40.0mg	4/50 100.mg 2/50		
c	1794	73.9mg	n.s.s.	25/100	16.0mg	17/50	40.0mg	22/50 100.mg 17/50		
d	1794	90.9mg	n.s.s.	21/100	16.0mg	16/50	40.0mg	18/50 100.mg 14/50		
e	1794	243.mg	n.s.s.	9/100	16.0mg	5/50	40.0mg	6/50 100.mg 4/50		
4815	1794	222.mg	2.88gm	2/100	16.0mg	1/50	40.0mg	5/50 100.mg 7/50		
a	1794	169.mg	n.s.s.	17/100	16.0mg	8/50	40.0mg	12/50 100.mg 13/50		
b	1794	136.mg	n.s.s.	74/100	16.0mg	36/50	40.0mg	32/50 100.mg 34/50		
c	1794	83.8mg	n.s.s.	81/100	16.0mg	39/50	40.0mg	36/50 100.mg 40/50		
4816	1794	413.mg	n.s.s.	4/100	16.0mg	6/50	40.0mg	3/50 100.mg 3/50		
a	1794	54.3mg	n.s.s.	55/100	16.0mg	29/50	40.0mg	29/50 100.mg 38/50		
b	1794	97.6mg	n.s.s.	45/100	16.0mg	22/50	40.0mg	20/50 100.mg 29/50		
c	1794	166.mg	n.s.s.	16/100	16.0mg	9/50	40.0mg	14/50 100.mg 12/50		

```
        Spe Strain  Site   Xpo+Xpt                                              TD50    2Tailpvl
          Sex  Route  Hist   Notes                                              DR      AuOp

TIN (II) CHLORIDE              100ng..:..1ug...:..10.....:..100....:..1mg...:..10....:..100...:..1g.....:..10
4817 M f b6c eat pit ade 24m24 ae                                        :   ±        #1.39gm * P<.05  -
  a  M f b6c eat liv hpc 24m24 ae                                                      1.42gm * P<.03
  b  M f b6c eat --- lhc 24m24 ae                                                      2.55gm * P<.02
  c  M f b6c eat TBA MXB 24m24 ae                                                      386.mg * P<.09
  d  M f b6c eat liv MXB 24m24 ae                                                      1.02gm * P<.06
  e  M f b6c eat lun MXB 24m24 ae                                                      no dre   P=1.
4818 M m b6c eat TBA MXB 24m24                                      :>                 no dre   P=1.  -
  a  M m b6c eat liv MXB 24m24                                                         no dre   P=1.
  b  M m b6c eat lun MXB 24m24                                                         no dre   P=1.
4819 M b cd1 wat lun mix 24m24 e                              .>                       no dre   P=1.  -
  a  M b cd1 wat lun ade 24m24 e                                                       no dre   P=1.
  b  M b cd1 wat tba mix 24m24 e                                                       no dre   P=1.
  c  M b cd1 wat tba mal 24m24 e                                                       no dre   P=1.
  d  M b cd1 wat tba ben 24m24 e                                                       no dre   P=1.
4820 R f f34 eat TBA MXB 24m24                                  :>                     12.0gm * P<1.  -
  a  R f f34 eat liv MXB 24m24                                                         676.gm * P<1.
4821 R m f34 eat thy MXA 24m24 ae                          :   +       :              #87.4mg \ P<.004 -
  a  R m f34 eat thy ccr 24m24 ae                                                      405.mg * P<.03
  b  R m f34 eat lun a/a 24m24 ae                                                      951.mg * P<.04
  c  R m f34 eat TBA MXB 24m24 ae                                                      180.mg * P<.4
  d  R m f34 eat liv MXB 24m24 ae                                                      no dre   P=1.
4822 R b leb wat liv tum 36m36 e                             .>                        no dre   P=1.  -
  a  R b leb wat tba tum 36m36 e                                                       no dre   P=1.
  b  R b leb wat tba mal 36m36 e                                                       no dre   P=1.

TITANIUM DIOXIDE               100ng..:..1ug...:..10.....:..100....:..1mg...:..10....:..100...:..1g.....:..10
4823 M f b6c eat TBA MXB 24m24                                                       :>55.8gm * P<.8  -
  a  M f b6c eat liv MXB 24m24                                                         55.5gm * P<.2
  b  M f b6c eat lun MXB 24m24                                                         50.4gm * P<.1
4824 M m b6c eat TBA MXB 24m24                                                       :no dre   P=1.  -
  a  M m b6c eat liv MXB 24m24                                                         38.7gm * P<.4
  b  M m b6c eat lun MXB 24m24                                                         no dre   P=1.
4825 R f f34 eat thy MXA 24m24                                                       : #17.2gm / P<.04 -
  a  R f f34 eat TBA MXB 24m24                                                         5.01gm * P<.4
  b  R f f34 eat liv MXB 24m24                                                         no dre   P=1.
4826 R m f34 eat --- ker 24m24                                                       :#25.9gm * P<.05 -
  a  R m f34 eat TBA MXB 24m24                                                         23.0gm * P<.9
  b  R m f34 eat liv MXB 24m24                                                         no dre   P=1.

TITANIUM OXALATE, POTASSIUM    100ng..:..1ug...:..10.....:..100....:..1mg...:..10....:..100...:..1g.....:..10
4827 M f cd1 wat lun tum 31m31 e                                 .>                    150.mg   P<1.  -
  a  M f cd1 wat tba tum 31m31 e                                                       no dre   P=1.
4828 M f cd1 wat tba mix 26m26 e                            .>                         3.66mg   P<.2  -
4829 M m cd1 wat lun tum 25m25 e                       .        .>                     no dre   P=1.  -
  a  M m cd1 wat tba tum 25m25 e                                                       no dre   P=1.
4830 M m cd1 wat tba mix 28m28 e                            .>                         3.62mg   P<.2  -

TITANOCENE DICHLORIDE          100ng..:..1ug...:..10.....:..100....:..1mg...:..10....:..100...:..1g.....:..10
4831 R f f34 gav for sqp 24m25                                    :    ±               378.mg * P<.08 e
  a  R f f34 gav TBA MXB 24m25                                                         no dre   P=1.
  b  R f f34 gav liv MXB 24m25                                                         378.mg * P<.08
4832 R m f34 gav for sqp 24m24                            :    +                       88.6mg \ P<.01 e
  a  R m f34 gav mul msm 24m24                                                         158.mg * P<.007
  b  R m f34 gav pan ana 24m24                                                         97.6mg \ P<.02
  c  R m f34 gav for MXA 24m24                                                         157.mg * P<.03 e
  d  R m f34 gav TBA MXB 24m24                                                         40.0mg * P<.2
  e  R m f34 gav liv MXB 24m24                                                         143.mg * P<.07

DL-alpha-TOCOPHEROL            100ng..:..1ug...:..10.....:..100....:..1mg...:..10....:..100...:..1g.....:..10
4833 R m f3d eat eso tum 52w52 er                                .>                    no dre   P=1.  -
  a  R m f3d eat for tum 52w52 er                                                      no dre   P=1.
  b  R m f3d eat liv tum 52w52 er                                                      no dre   P=1.

DL-alpha-TOCOPHERYL ACETATE    100ng..:..1ug...:..10.....:..100....:..1mg...:..10....:..100...:..1g.....:..10
4834 R f cdr eat liv tum 52w52 ek                                     .>               no dre   P=1.  -
4835 R f cdr eat liv tum 24m24 e                                                       no dre   P=1.  -
  a  R f cdr eat tba mix 24m24 e                                                       no dre   P=1.
  b  R f cdr eat tba mal 24m24 e                                                       no dre   P=1.
4836 R m cdr eat liv tum 52w52 ek                                     .>               no dre   P=1.  -
4837 R m cdr eat liv tum 24m24 e                                                       no dre   P=1.  -
  a  R m cdr eat tba mix 24m24 e                                                       67.8gm   P<1.  -
  b  R m cdr eat tba mal 24m24 e                                                       67.8gm   P<.9  -

TOLAZAMIDE                     100ng..:..1ug...:..10.....:..100....:..1mg...:..10....:..100...:..1g.....:..10
4838 M f b6c eat TBA MXB 24m24                                            :>            no dre   P=1.  -
  a  M f b6c eat liv MXB 24m24                                                          no dre   P=1.
  b  M f b6c eat lun MXB 24m24                                                          no dre   P=1.
4839 M m b6c eat TBA MXB 24m24                                                :>        no dre   P=1.  -
  a  M m b6c eat liv MXB 24m24                                                          no dre   P=1.
  b  M m b6c eat lun MXB 24m24                                                          24.6gm * P<.9
```

RefNum		LoConf	UpConf	Cntrl	1Dose	1Inc	2Dose	2Inc	Citation or Pathology	Brkly Code

TIN (II) CHLORIDE (stannous chloride) 7772-99-8

4817	TR231	564.mg	n.s.s.	0/50	130.mg	4/50	258.mg	2/50		S
a	TR231	574.mg	n.s.s.	0/50	130.mg	3/50	258.mg	3/50		S
b	TR231	871.mg	n.s.s.	0/50	130.mg	0/50	258.mg	4/50		S
c	TR231	150.mg	n.s.s.	22/50	130.mg	32/50	258.mg	27/50		
d	TR231	396.mg	n.s.s.	3/50	130.mg	4/50	258.mg	8/50	liv:hpa,hpc,nnd.	
e	TR231	1.06gm	n.s.s.	4/50	130.mg	1/50	258.mg	3/50	lun:a/a,a/c.	
4818	TR231	373.mg	n.s.s.	33/50	120.mg	29/50	240.mg	33/50		
a	TR231	655.mg	n.s.s.	16/50	120.mg	10/50	240.mg	15/50	liv:hpa,hpc,nnd.	
b	TR231	671.mg	n.s.s.	10/50	120.mg	10/50	240.mg	10/50	lun:a/a,a/c.	
4819	1512	5.26mg	n.s.s.	26/170	.877mg	10/86			Kanisawa;canr,27,1192-1195;1967	
a	1512	12.0mg	n.s.s.	7/170	.877mg	1/86				
b	1512	3.83mg	n.s.s.	55/170	.877mg	22/86				
c	1512	6.06mg	n.s.s.	15/170	.877mg	6/86				
d	1512	6.42mg	n.s.s.	29/170	.877mg	9/86				
4820	TR231	87.0mg	n.s.s.	40/50	50.0mg	38/50	100.mg	37/50		
a	TR231	891.mg	n.s.s.	1/50	50.0mg	0/50	100.mg	1/50	liv:hpa,hpc,nnd.	
4821	TR231	41.4mg	628.mg	2/50	40.0mg	13/50	(80.0mg	8/50)	thy:cca,ccr.	S
a	TR231	174.mg	n.s.s.	0/50	40.0mg	4/50	80.0mg	3/50		S
b	TR231	287.mg	n.s.s.	0/50	40.0mg	0/50	80.0mg	3/50		S
c	TR231	49.0mg	n.s.s.	36/50	40.0mg	37/50	80.0mg	38/50		
d	TR231	598.mg	n.s.s.	2/50	40.0mg	0/50	80.0mg	1/50	liv:hpa,hpc,nnd.	
4822	1036	11.5mg	n.s.s.	1/82	.265mg	0/94			Kanisawa;canr,29,892-895;1969	
a	1036	2.08mg	n.s.s.	31/82	.265mg	29/94				
b	1036	5.54mg	n.s.s.	9/82	.265mg	5/94				

TITANIUM DIOXIDE 13463-67-7

4823	TR97	6.64gm	n.s.s.	30/50	3.22gm	24/50	6.44gm	26/50		
a	TR97	17.0gm	n.s.s.	1/50	3.22gm	3/50	6.44gm	3/50	liv:hpa,hpc,nnd.	
b	TR97	16.9gm	n.s.s.	1/50	3.22gm	2/50	6.44gm	4/50	lun:a/a,a/c.	
4824	TR97	9.26gm	n.s.s.	29/50	2.97gm	25/50	5.94gm	28/50		
a	TR97	9.45gm	n.s.s.	8/50	2.97gm	9/50	5.94gm	14/50	liv:hpa,hpc,nnd.	
b	TR97	24.2gm	n.s.s.	6/50	2.97gm	3/50	5.94gm	5/50	lun:a/a,a/c.	
4825	TR97	6.25gm	n.s.s.	1/50	1.24gm	0/50	2.48gm	6/50	thy:cca,ccr.	S
a	TR97	1.23gm	n.s.s.	41/50	1.24gm	43/50	2.48gm	46/50		
b	TR97	28.2gm	n.s.s.	1/50	1.24gm	0/50	2.48gm	0/50	liv:hpa,hpc,nnd.	
4826	TR97	7.81gm	n.s.s.	0/50	990.mg	0/50	1.98gm	3/50		S
a	TR97	1.54gm	n.s.s.	32/50	990.mg	40/50	1.98gm	36/50		
b	TR97	16.7gm	n.s.s.	1/50	990.mg	1/50	1.98gm	0/50	liv:hpa,hpc,nnd.	

TITANIUM OXALATE, POTASSIUM 14481-26-6

4827	56	3.59mg	n.s.s.	9/60	1.00mg	5/32			Schroeder;jnut,83,239-250;1964	
a	56	4.45mg	n.s.s.	22/60	1.00mg	7/32				
4828	1395	1.24mg	n.s.s.	9/45	1.00mg	13/36			Schroeder;jnut,105,452-458;1975	
4829	56	6.12mg	n.s.s.	8/44	.833mg	1/40			Schroeder;jnut,83,239-250;1964	
a	56	4.09mg	n.s.s.	11/44	.833mg	4/40				
4830	1395	1.23mg	n.s.s.	10/43	.833mg	17/45			Schroeder;jnut,105,452-458;1975	

TITANOCENE DICHLORIDE 1271-19-8

4831	TR399	114.mg	n.s.s.	0/60	17.4mg	1/61	34.8mg	2/60		
a	TR399	21.7mg	n.s.s.	55/60	17.4mg	53/61	34.8mg	46/60	liv:hpa,hpc,nnd.	
b	TR399	114.mg	n.s.s.	0/60	17.4mg	1/61	34.8mg	2/60		
4832	TR399	30.6mg	10.4gm	0/60	17.5mg	4/60	(35.2mg	1/60)		
a	TR399	67.1mg	1.74gm	0/60	17.5mg	3/60	35.2mg	4/60		S
b	TR399	33.4mg	n.s.s.	0/60	17.5mg	4/60	(35.2mg	0/60)		S
c	TR399	64.2mg	n.s.s.	0/60	17.5mg	4/60	35.2mg	2/60	for:bsb,sqc,sqp.	
d	TR399	13.0mg	n.s.s.	52/60	17.5mg	52/60	35.2mg	43/60		
e	TR399	52.4mg	n.s.s.	4/60	17.5mg	3/60	35.2mg	8/60	liv:hpa,hpc,nnd.	

DL-alpha-TOCOPHEROL (vitamin E) 10191-41-0

4833	1900	206.mg	n.s.s.	0/10	400.mg	0/10			Hirose;carc,8,1731-1735;1987/pers.comm.	
a	1900	206.mg	n.s.s.	0/10	400.mg	0/10				
b	1900	206.mg	n.s.s.	0/10	400.mg	0/10				

DL-alpha-TOCOPHERYL ACETATE (vitamin E acetate) 58-95-7

4834	1558m	147.mg	n.s.s.	0/10	500.mg	0/10	1.00gm	0/10	2.00gm 0/10	Wheldon;ijvn,53,287-296;1983
4835	1558n	24.4gm	n.s.s.	2/50	500.mg	1/49	1.00gm	1/50	2.00gm 0/50	
a	1558n	911.mg	n.s.s.	46/50	2.00gm	45/50				
b	1558n	8.01gm	n.s.s.	9/50	2.00gm	6/50				
4836	1558m	147.mg	n.s.s.	0/10	500.mg	0/10	1.00gm	0/10	2.00gm 0/10	
4837	1558n	21.4gm	n.s.s.	2/49	500.mg	1/50	1.00gm	0/50	2.00gm 1/50	
a	1558n	2.12gm	n.s.s.	28/49	2.00gm	29/50				
b	1558n	5.23gm	n.s.s.	7/49	2.00gm	8/50				

TOLAZAMIDE 1156-19-0

4838	TR51	867.mg	n.s.s.	8/15	455.mg	6/35	919.mg	9/34		
a	TR51	1.61gm	n.s.s.	1/15	455.mg	1/35	919.mg	1/34	liv:hpa,hpc,nnd.	
b	TR51	3.33gm	n.s.s.	1/15	455.mg	0/35	919.mg	1/34	lun:a/a,a/c.	
4839	TR51	643.mg	n.s.s.	8/15	420.mg	11/35	(848.mg	5/35)		
a	TR51	2.87gm	n.s.s.	3/15	420.mg	2/35	848.mg	3/35	liv:hpa,hpc,nnd.	
b	TR51	2.07gm	n.s.s.	0/15	420.mg	2/35	848.mg	1/35	lun:a/a,a/c.	

```
       Spe Strain  Site   Xpo+Xpt                                                    TD50    2Tailpvl
         Sex  Route  Hist    Notes                                                     DR      AuOp

4840 R f f34 eat TBA MXB 24m24                                    :>              no dre   P=1.    -
   a   R f f34 eat liv MXB 24m24                                                  no dre   P=1.
4841 R m f34 eat TBA MXB 24m24                                       :>           no dre   P=1.    -
   a   R m f34 eat liv MXB 24m24                                                  no dre   P=1.

TOLBUTAMIDE                      100ng..:..1ug....:..10.....:..100....:..1mg....:..10.....:..100....:..1g.....:..10
4842 M f b6c eat TBA MXB 18m24                                               :>   24.2gm \ P<.9    -
   a   M f b6c eat liv MXB 18m24                                                  no dre   P=1.
   b   M f b6c eat lun MXB 18m24                                                  no dre   P=1.
4843 M m b6c eat TBA MXB 18m24                                                :>  no dre   P=1.    -
   a   M m b6c eat liv MXB 18m24                                                  no dre   P=1.
   b   M m b6c eat lun MXB 18m24                                                  no dre   P=1.
4844 R f f34 eat TBA MXB 18m25                                              :>    no dre   P=1.    -
   a   R f f34 eat liv MXB 18m25                                                  no dre   P=1.
4845 R m f34 eat TBA MXB 18m25                                              :>    no dre   P=1.    -
   a   R m f34 eat liv MXB 18m25                                                  no dre   P=1.

TOLUENE                          100ng..:..1ug....:..10.....:..100....:..1mg....:..10.....:..100....:..1g.....:..10
4846 M f b6c inh TBA MXB 24m24                                          :>        3.53gm * P<.6    -
   a   M f b6c inh liv MXB 24m24                                                  5.39gm * P<.3
   b   M f b6c inh lun MXB 24m24                                                  10.2gm * P<.4
4847 M m b6c inh TBA MXB 24m24                                          :>        3.00gm * P<.6    -
   a   M m b6c inh liv MXB 24m24                                                  12.6gm * P<.9
   b   M m b6c inh lun MXB 24m24                                                  9.44gm Z P<.5
4848 R f f34 inh TBA MXB 24m24                                      :>            8.89gm * P<1.    -
   a   R f f34 inh liv MXB 24m24                                                  no dre   P=1.
4849 R f f34 inh liv mix 25m25 e                                     .>           1.76gm * P<.2    -
4850 R m f34 inh TBA MXB 24m24                                   :>               481.mg * P<.6    -
   a   R m f34 inh liv MXB 24m24                                                  no dre   P=1.
4851 R m f34 inh liv mix 25m25 e                                 .>               1.21gm   P<.4    -
4852 R f sda gav tba mal 24m33 e                           . +        .           578.mg   P<.002  +
4853 R m sda gav tba mal 24m33 e                           .  ±                   940.mg   P<.05   +

TOLUENE DIISOCYANATE, COMMERCIAL GRADE (2,4 (80%)- AND 2,6 (20%)-) ....:..1mg....:..10.....:..100....:..1g.....:..10
4854 M f b6c gav MXB MXB 24m25                                  :  +   :          181.mg / P<.0005
   a   M f b6c gav liv MXA 24m25                                                  215.mg / P<.008
   b   M f b6c gav liv hpa 24m25                                                  250.mg / P<.005  c
   c   M f b6c gav MXA MXA 24m25                                                  580.mg * P<.008  c
   d   M f b6c gav MXA hes 24m25                                                  1.20gm * P<.04
   e   M f b6c gav TBA MXB 24m25                                                  305.mg * P<.5
   f   M f b6c gav liv MXB 24m25                                                  215.mg / P<.008
   g   M f b6c gav lun MXB 24m25                                                  815.mg * P<.3
4855 M m b6c gav TBA MXB 24m25                                          :>        253.mg \ P<.3    -
   a   M m b6c gav liv MXB 24m25                                                  no dre   P=1.
   b   M m b6c gav lun MXB 24m25                                                  1.92gm * P<.5
4856 R f f34 gav MXB MXB 25m25                             :  +  :                25.4mg * P<.0005
   a   R f f34 gav mgl fba 25m25                                                  33.4mg * P<.0005c
   b   R f f34 gav liv nnd 25m25                                                  92.2mg * P<.0005c
   c   R f f34 gav pni MXA 25m25                                                  115.mg * P<.0005
   d   R f f34 gav pni isa 25m25                                                  135.mg * P<.002  c
   e   R f f34 gav adr phe 25m25                                                  143.mg * P<.007
   f   R f f34 gav adr coa 25m25                                                  174.mg * P<.007
   g   R f f34 gav sub fib 25m25                                                  256.mg * P<.004
   h   R f f34 gav pit cra 25m25                                                  77.3mg * P<.03
   i   R f f34 gav pit MXA 25m25                                                  86.9mg * P<.05
   j   R f f34 gav sub MXA 25m25                                                  206.mg / P<.02   c
   k   R f f34 gav cli adn 25m25                                                  260.mg * P<.03
   l   R f f34 gav TBA MXB 25m25                                                  25.1mg * P<.0005
   m   R f f34 gav liv MXB 25m25                                                  92.2mg * P<.0005
4857 R m f34 gav tes MXA 25m25                          :  +   :                  13.6mg * P<.002
   a   R m f34 gav MXB MXB 25m25                                                  27.3mg * P<.0005
   b   R m f34 gav sub MXA 25m25                                                  34.1mg * P<.0005c
   c   R m f34 gav pan ana 25m25                                                  51.6mg * P<.0005c
   d   R m f34 gav sub fib 25m25                                                  56.0mg * P<.0005
   e   R m f34 gav pit cra 25m25                                                  76.4mg * P<.005
   f   R m f34 gav sub fbs 25m25                                                  92.4mg * P<.002
   g   R m f34 gav pit MXA 25m25                                                  83.7mg * P<.02
   h   R m f34 gav sub MXA 25m25                                                  103.mg * P<.02
   i   R m f34 gav pni MXA 25m25                                                  201.mg * P<.03
   j   R m f34 gav TBA MXB 25m25                                                  18.4mg * P<.002
   k   R m f34 gav liv MXB 25m25                                                  196.mg * P<.4

o-TOLUENESULFONAMIDE             100ng..:..1ug....:..10.....:..100....:..1mg....:..10.....:..100....:..1g.....:..10
4858 R f cdr eat mgl adc 33m33 eg                          . ±                    12.2mg Z P<.02
   a   R f cdr eat liv kcs 33m33 eg                                               8.52gm * P<.03
   b   R f cdr eat liv nnd 33m33 eg                                               21.7gm * P<.6
4859 R m cdr eat liv ang 33m33 e                                        .>        17.2gm * P<.2
   a   R m cdr eat liv nnd 33m33 e                                                35.9gm * P<.7
   b   R m cdr eat liv blc 33m33 e                                                no dre   P=1.
4860 R b sda eat --- leu 31m33 a                             . +                  .375.mg \ P<.009
   a   R b sda eat ubl mix 31m33 a                                                3.96gm * P<.06   +
```

	RefNum	LoConf	UpConf	Cntrl	1Dose	1Inc	2Dose	2Inc			Citation or Pathology
											Brkly Code
4840	TR51	214.mg	n.s.s.	13/15	175.mg	23/35	354.mg	20/35			
a	TR51	n.s.s.	n.s.s.	0/15	175.mg	1/35	354.mg	0/35			liv:hpa,hpc,nnd.
4841	TR51	277.mg	n.s.s.	9/15	140.mg	19/35	283.mg	20/35			
a	TR51	n.s.s.	n.s.s.	0/15	140.mg	0/35	283.mg	0/35			liv:hpa,hpc,nnd.
	TOLBUTAMIDE	64-77-7									
4842	TR31	1.63gm	n.s.s.	3/15	1.76gm	11/35	(3.52gm	3/35)			
a	TR31	10.6gm	n.s.s.	1/15	1.76gm	2/35	3.52gm	1/35			liv:hpa,hpc,nnd.
b	TR31	10.3gm	n.s.s.	0/15	1.76gm	2/35	3.52gm	0/35			lun:a/a,a/c.
4843	TR31	1.16gm	n.s.s.	8/15	1.62gm	13/35	(3.28gm	2/35)			
a	TR31	1.96gm	n.s.s.	3/15	1.62gm	4/35	(3.28gm	1/35)			liv:hpa,hpc,nnd.
b	TR31	9.12gm	n.s.s.	1/15	1.62gm	2/35	3.28gm	0/35			lun:a/a,a/c.
4844	TR31	662.mg	n.s.s.	13/15	315.mg	15/35	631.mg	16/35			
a	TR31	n.s.s.	n.s.s.	0/15	315.mg	0/35	631.mg	0/35			liv:hpa,hpc,nnd.
4845	TR31	654.mg	n.s.s.	8/15	252.mg	13/35	505.mg	17/35			
a	TR31	n.s.s.	n.s.s.	0/15	252.mg	0/35	505.mg	0/35			liv:hpa,hpc,nnd.
	TOLUENE	(monomethyl benzene)	108-88-3								
4846	TR371	607.mg	n.s.s.	33/50	151.mg	36/50	766.mg	42/50	1.51gm	38/50	
a	TR371	1.57gm	n.s.s.	7/50	151.mg	9/50	766.mg	8/50	1.51gm	13/50	liv:hpa,hpc,nnd.
b	TR371	2.27gm	n.s.s.	5/50	151.mg	3/50	766.mg	4/50	1.51gm	7/50	lun:a/a,a/c.
4847	TR371	511.mg	n.s.s.	29/60	126.mg	29/60	630.mg	26/60	1.26gm	33/60	
a	TR371	985.mg	n.s.s.	19/60	126.mg	17/60	630.mg	18/60	1.26gm	19/60	liv:hpa,hpc,nnd.
b	TR371	1.87gm	n.s.s.	9/60	126.mg	1/60	630.mg	2/60	1.26gm	9/60	lun:a/a,a/c.
4848	TR371	176.mg	n.s.s.	46/50	180.mg	43/50	360.mg	43/50			
a	TR371	1.41gm	n.s.s.	2/50	180.mg	3/50	360.mg	0/50			liv:hpa,hpc,nnd.
4849	1578	490.mg	n.s.s.	1/90	84.6mg	4/90					Gralla;ctfr;1983/Gibson 1983
4850	TR371	94.1mg	n.s.s.	43/50	126.mg	44/50	252.mg	42/50			
a	TR371	756.mg	n.s.s.	4/50	126.mg	3/50	252.mg	0/50			liv:hpa,hpc,nnd.
4851	1578	297.mg	n.s.s.	3/89	59.2mg	6/90					Gralla;ctfr;1983/Gibson 1983
4852	BT903	281.mg	3.19gm	10/49	237.mg	21/40					Maltoni;ajim,7,415-446;1985
4853	BT903	371.mg	n.s.s.	11/45	237.mg	18/40					
	TOLUENE DIISOCYANATE, COMMERCIAL GRADE (2,4 (80%)- AND 2,6 (20%)-)								26471-62-5		
4854	TR251	97.9mg	612.mg	2/50	42.5mg	4/50	84.1mg	16/50			abc:hes; liv:hes,hpa; ova:hes; spl:hem; sub:hem. C
a	TR251	103.mg	5.59gm	4/50	42.5mg	5/50	84.1mg	15/50			liv:hpa,hpc. S
b	TR251	121.mg	2.38gm	2/50	42.5mg	3/50	84.1mg	12/50			
c	TR251	236.mg	10.7gm	0/50	42.5mg	1/50	84.1mg	5/50			abc:hes; liv:hes; ova:hes; spl:hem; sub:hem. S
d	TR251	362.mg	n.s.s.	0/50	42.5mg	0/50	84.1mg	3/50			abc:hes; liv:hes; ova:hes. S
e	TR251	66.0mg	n.s.s.	26/50	42.5mg	31/50	84.1mg	30/50			
f	TR251	103.mg	5.59gm	4/50	42.5mg	5/50	84.1mg	15/50			liv:hpa,hpc,nnd.
g	TR251	282.mg	n.s.s.	0/50	42.5mg	3/50	84.1mg	1/50			lun:a/a,a/c.
4855	TR251	75.0mg	n.s.s.	22/50	84.9mg	27/50	(168.mg	9/50)			
a	TR251	371.mg	n.s.s.	11/50	84.9mg	12/50	168.mg	5/50			liv:hpa,hpc,nnd.
b	TR251	416.mg	n.s.s.	2/50	84.9mg	5/50	168.mg	2/50			lun:a/a,a/c.
4856	TR251	14.7mg	57.3mg	19/50	42.1mg	26/50	83.3mg	24/50			liv:nnd; mgl:fba; pni:isa; sub:fbs,fib. C
a	TR251	18.5mg	87.1mg	15/50	42.1mg	21/50	83.3mg	18/50			
b	TR251	44.4mg	344.mg	3/50	42.1mg	8/50	83.3mg	8/50			
c	TR251	51.9mg	383.mg	0/50	42.1mg	7/50	83.3mg	2/50			pni:isa,isc. S
d	TR251	58.3mg	523.mg	0/50	42.1mg	6/50	83.3mg	2/50			
e	TR251	58.4mg	2.25gm	2/50	42.1mg	5/50	83.3mg	4/50			S
f	TR251	67.1mg	3.51gm	2/50	42.1mg	3/50	83.3mg	5/50			S
g	TR251	87.6mg	1.88gm	0/50	42.1mg	1/50	83.3mg	3/50			S
h	TR251	32.1mg	n.s.s.	25/50	42.1mg	15/50	83.3mg	16/50			S
i	TR251	33.7mg	43.3gm	27/50	42.1mg	15/50	83.3mg	16/50			pit:cra,crc. S
j	TR251	73.7mg	43.3gm	2/50	42.1mg	1/50	83.3mg	5/50			sub:fbs,fib.
k	TR251	87.3mg	n.s.s.	0/50	42.1mg	4/50	83.3mg	0/50			S
l	TR251	13.6mg	82.8mg	45/50	42.1mg	36/50	83.3mg	34/50			
m	TR251	44.4mg	344.mg	3/50	42.1mg	8/50	83.3mg	8/50			liv:hpa,hpc,nnd.
4857	TR251	6.93mg	71.9mg	48/50	21.0mg	35/50	41.7mg	30/50			tes:ict,itm. S
a	TR251	14.9mg	65.4mg	4/50	21.0mg	8/50	41.7mg	14/50			pan:ana; sub:fbs,fib. C
b	TR251	18.0mg	88.4mg	3/50	21.0mg	6/50	41.7mg	12/50			sub:fbs,fib.
c	TR251	23.2mg	183.mg	1/50	21.0mg	3/50	41.7mg	7/50			
d	TR251	25.2mg	239.mg	3/50	21.0mg	3/50	41.7mg	9/50			
e	TR251	32.2mg	732.mg	3/50	21.0mg	4/50	41.7mg	7/50			S
f	TR251	36.9mg	421.mg	0/50	21.0mg	3/50	41.7mg	3/50			S
g	TR251	33.6mg	12.8gm	4/50	21.0mg	4/50	41.7mg	7/50			S
h	TR251	39.2mg	n.s.s.	1/50	21.0mg	4/50	41.7mg	3/50			pit:cra,crc. S
i	TR251	57.7mg	n.s.s.	1/50	21.0mg	0/50	41.7mg	4/50			sub:fbs,ost,srn. S
j	TR251	9.29mg	94.4mg	40/50	21.0mg	24/50	41.7mg	29/50			pni:isa,isc. S
k	TR251	42.4mg	n.s.s.	7/50	21.0mg	3/50	41.7mg	4/50			liv:hpa,hpc,nnd.
	o-TOLUENESULFONAMIDE	88-19-7									
4858	1398	5.34mg	n.s.s.	6/48	2.50mg	16/49	(25.0mg	6/49	250.mg	5/49)	Arnold;txap,52,113-152;1980
a	1398	2.10gm	n.s.s.	0/48	2.50mg	0/49	25.0mg	0/49	250.mg	2/49	
b	1398	2.01gm	n.s.s.	0/48	2.50mg	0/49	25.0mg	2/49	250.mg	1/49	
4859	1398	2.81gm	n.s.s.	0/49	2.50mg	0/49	25.0mg	0/50	250.mg	1/49	
a	1398	4.05gm	n.s.s.	0/49	2.50mg	1/49	25.0mg	1/50	250.mg	1/49	
b	1398	5.04gm	n.s.s.	0/49	2.50mg	1/49	25.0mg	0/50	250.mg	0/49	
4860	1397	143.mg	8.60gm	0/76	20.0mg	5/76	(200.mg	3/76)			Schmahl;zkko,91,19-22;1978
a	1397	1.26gm	n.s.s.	0/76	20.0mg	3/76	200.mg	5/76			

```
        Spe Strain  Site   Xpo+Xpt                                                        TD50    2Tailpvl
            Sex   Route  Hist   Notes                                                         DR      AuOp

  b    R b sda eat liv car 31m33 a                                                     no dre   P=1.
  c    R b sda eat tba mal 31m33 a                                                     13.3gm * P<.9
4861  R f wis wat liv hpc 24m24 e                                         .>           no dre   P=1.   -


m-TOLUIDINE.HCl                     100ng..:..1ug....:..10......:..100....:..1mg....:..10......:..100....:..1g....:..10
4862  M f chi eat liv mix 77w90 v                                               :>     9.99gm   P<.6   -
  a    M f chi eat lun mix 77w90 v                                                     no dre   P=1.   -
  b    M f chi eat tba mix 77w90 v                                                     no dre   P=1.   -
4863  M f chi eat lun mix 77w81 v                                            :>        no dre   P=1.   -
  a    M f chi eat liv mix 77w81 v                                                     no dre   P=1.   -
  b    M f chi eat tba mix 77w81 v                                                     1.95gm   P<.09  -
4864  M m chi eat liv mix 77w90 v                                          :>          6.77gm * P<.7   -
  a    M m chi eat lun mix 77w90 v                                                     no dre   P=1.   -
  b    M m chi eat tba mix 77w90 v                                                     no dre   P=1.   -
4865  M m chi eat liv hpt 77w90 v     pool                                 :>          1.44gm \ P<.2   +
4866  R m cdr eat liv mix 77w98 v                                                      no dre   P=1.   -
  a    R m cdr eat tba mix 77w98 v                                                     98.4mg \ P<.2   -


o-TOLUIDINE.HCl                     100ng..:..1ug....:..10......:..100....:..1mg....:..10......:..100....:..1g....:..10
4867  M f b6c eat liv MXA 24m24                                          : +    :      754.mg * P<.002 c
  a    M f b6c eat liv hpc 24m24                                                       1.52gm * P<.02  c
  b    M f b6c eat TBA MXB 24m24                                                       1.35gm * P<.4
  c    M f b6c eat liv MXB 24m24                                                       754.mg * P<.002
  d    M f b6c eat lun MXB 24m24                                                       no dre   P=1.
4868  M m b6c eat --- MXA 24m24 a                                        :  +          :926.mg * P<.005 c
  a    M m b6c eat --- hes 24m24 a                                                     1.22gm / P<.01  c
  b    M m b6c eat TBA MXB 24m24 a                                                     no dre   P=1.
  c    M m b6c eat liv MXB 24m24 a                                                     12.7gm * P<1.
  d    M m b6c eat lun MXB 24m24 a                                                     no dre   P=1.
4869  M f chi eat liv mix 68w90 av                                       :      +      1.65gm * P<.008 -
  a    M f chi eat lun mix 68w90 av                                                    no dre   P=1.   -
  b    M f chi eat tba mix 68w90 av                                                    699.mg * P<.007
4870  M f chi eat --- vsc 68w90 av    pool                               : +    :      1.24gm * P<.0005+
4871  M m chi eat --- vsc 68w90 av                                       : +    :      646.mg / P<.0005+
  a    M m chi eat liv mix 68w90 av                                                    1.40gm / P<.005  -
  b    M m chi eat lun mix 68w90 av                                                    1.16gm * P<.1    -
  c    M m chi eat tba mix 68w90 av                                                    325.mg / P<.0005
4872  M m chi eat --- vsc 68w90 av    pool                               : +    :      660.mg / P<.0005+
4873  R f f34 eat MXB MXB 24m24 s                               :+  :                  77.9mg / P<.0005
  a    R f f34 eat mgl fba 24m24 s                                                     115.mg / P<.0005c
  b    R f f34 eat mgl MXA 24m24 s                                                     119.mg / P<.0005c
  c    R f f34 eat ubl MXA 24m24 s                                                     175.mg / P<.0005c
  d    R f f34 eat ubl tcc 24m24 s                                                     185.mg / P<.0005c
  e    R f f34 eat mul MXA 24m24 s                                                     332.mg / P<.0005c
  f    R f f34 eat spl MXA 24m24 s                                                     337.mg * P<.0005c
  g    R f f34 eat mul ost 24m24 s                                                     423.mg / P<.0005c
  h    R f f34 eat spl ang 24m24 s                                                     450.mg * P<.002 c
  i    R f f34 eat TBA MXB 24m24 s                                                     75.8mg * P<.0005
  j    R f f34 eat liv MXB 24m24 s                                                     no dre   P=1.
4874  R m f34 eat MXB MXB 24m24 as                             : + :                   23.3mg / P<.0005
  a    R m f34 eat sub fib 24m24 as                                                    33.3mg / P<.0005
  b    R m f34 eat mul MXA 24m24 as                                                    93.2mg / P<.0005c
  c    R m f34 eat MXA mso 24m24 as                                                    97.4mg * P<.0005c
  d    R m f34 eat spl fib 24m24 as                                                    134.mg * P<.0005c
  e    R m f34 eat mul fbs 24m24 as                                                    184.mg / P<.0005c
  f    R m f34 eat mul srn 24m24 as                                                    394.mg / P<.0005c
  g    R m f34 eat TBA MXB 24m24 as                                                    24.4mg / P<.0005
  h    R m f34 eat liv MXB 24m24 as                                                    645.mg / P<.06
4875  R m f34 eat ski fib 72w93 e                                  . +  .              38.7mg   P<.0005+
  a    R m f34 eat mam fba 72w93 e                                                     149.mg   P<.0005+
  b    R m f34 eat spl fib 72w93 e                                                     167.mg   P<.0005+
  c    R m f34 eat pec scs 72w93 e                                                     190.mg   P<.0005
  d    R m f34 eat ubl mix 72w93 e                                                     474.mg   P<.03   +
  e    R m f34 eat liv mix 72w93 e                                                     1.00gm   P<.4
4876  R m cdr eat sub mix 64w68 v                                   : + :              36.7mg * P<.0005+
  a    R m cdr eat liv mix 64w68 v                                                     2.96gm * P<.3    -
  b    R m cdr eat tba mix 64w68 v                                                     36.1mg * P<.0005
4877  R m cdr eat sub mix 64w68 v     pool                          : + :              38.8mg * P<.0005+
  a    R m cdr eat ubl mix 64w68 v                                                     131.mg * P<.0005+


p-TOLUIDINE.HCl                     100ng..:..1ug....:..10......:..100......:..1mg....:..10......:..100....:..1g......:..10
4878  M f chi eat liv mix 77w98 v                                        :      ±      256.mg * P<.06
  a    M f chi eat lun mix 77w98 v                                                     400.mg * P<.5    -
  b    M f chi eat tba mix 77w98 v                                                     no dre   P=1.
4879  M f chi eat liv hpt 77w98 v     pool                                  :      +   278.mg * P<.01   +
4880  M m chi eat liv hpt 77w98 v                                        :  ±          49.1mg * P<.02   +
  a    M m chi eat liv mix 77w98 v                                                     49.1mg \ P<.02   +
  b    M m chi eat lun mix 77w98 v                                                     78.6mg \ P<.06   -
  c    M m chi eat tba mix 77w98 v                                                     49.4mg \ P<.06
4881  M m chi eat liv hpt 77w98 v     pool                                  : +   :    50.1mg \ P<.0005+
```

	RefNum	LoConf	UpConf	Cntrl	1Dose	1Inc	2Dose	2Inc	Citation or Pathology Brkly Code
b	1397	531.mg	n.s.s.	1/76	20.0mg	0/76	200.mg	0/76	
c	1397	1.02gm	n.s.s.	13/76	20.0mg	20/76	200.mg	17/76	
4861	1433	544.mg	n.s.s.	1/50	57.1mg	0/48			Hooson;bjca,42,129-147;1980
m-TOLUIDINE.HCl		638-03-9							
4862	381	1.29gm	n.s.s.	1/22	1.14gm	2/18			Weisburger;jept,2,325-356;1978/pers.comm./Russfield 1973
a	381	1.04gm	n.s.s.	5/22	1.14gm	4/18			
b	381	919.mg	n.s.s.	18/22	1.14gm	10/18			
4863	381m	2.75gm	n.s.s.	8/15	2.66gm	2/19			
a	381m	n.s.s.	n.s.s.	0/15	2.66gm	0/19			
b	381m	410.mg	n.s.s.	11/15	2.66gm	7/19			
4864	381	1.06gm	n.s.s.	1/18	754.mg	4/16	1.51gm	2/16	
a	381	1.45gm	n.s.s.	8/18	754.mg	2/16	1.51gm	5/16	
b	381	751.mg	n.s.s.	13/18	754.mg	10/16	1.51gm	10/16	
4865	381	368.mg	n.s.s.	7/99p	754.mg	4/16	(1.51gm	1/16)	
4866	381	n.s.s.	n.s.s.	0/16	146.mg	0/20	292.mg	0/18	
a	381	31.7mg	n.s.s.	9/16	146.mg	17/20	(292.mg	9/18)	
o-TOLUIDINE.HCl		636-21-5							
4867	TR153	428.mg	2.86gm	0/20	130.mg	4/50	390.mg	13/50	liv:hpa,hpc.
a	TR153	716.mg	n.s.s.	0/20	130.mg	2/50	390.mg	7/50	
b	TR153	344.mg	n.s.s.	8/20	130.mg	19/50	390.mg	26/50	
c	TR153	428.mg	2.86gm	0/20	130.mg	4/50	390.mg	13/50	liv:hpa,hpc,nnd.
d	TR153	1.66gm	n.s.s.	1/20	130.mg	2/50	390.mg	2/50	lun:a/a,a/c.
4868	TR153	453.mg	8.19gm	1/20	120.mg	2/50	360.mg	12/50	---:hem,hes.
a	TR153	554.mg	80.1gm	1/20	120.mg	1/50	360.mg	10/50	
b	TR153	304.mg	n.s.s.	17/20	120.mg	28/50	360.mg	30/50	
c	TR153	404.mg	n.s.s.	5/20	120.mg	19/50	360.mg	14/50	liv:hpa,hpc,nnd.
d	TR153	473.mg	n.s.s.	5/20	120.mg	5/50	(360.mg	3/50)	lun:a/a,a/c.
4869	381	544.mg	46.2gm	0/15	1.09gm	3/18	2.50gm	2/21	Weisburger;jept,2,325-356;1978/pers.comm./Russfield 1973
a	381	1.18gm	n.s.s.	8/15	1.09gm	12/18	2.50gm	0/21	
b	381	283.mg	12.1gm	11/15	1.09gm	12/18	2.50gm	8/21	
4870	381	513.mg	3.75gm	8/102p	1.09gm	4/18	2.50gm	8/21	
4871	381	300.mg	1.75gm	0/14	1.01gm	4/14	2.30gm	8/11	
a	381	474.mg	12.0gm	1/14	1.01gm	2/14	2.30gm	4/11	
b	381	353.mg	n.s.s.	4/14	1.01gm	5/14	2.30gm	2/11	
c	381	154.mg	1.13gm	8/14	1.01gm	10/14	2.30gm	11/11	
4872	381	301.mg	1.84gm	5/99p	1.01gm	4/14	2.30gm	8/11	
4873	TR153	51.9mg	141.mg	0/20	150.mg	32/50	300.mg	48/50	mgl:adn,fba; mul:ang,fbs,ost,srn; spl:ang,ost,srn; ubl:tcc,tpp. C
a	TR153	72.0mg	244.mg	6/20	150.mg	20/50	300.mg	35/50	
b	TR153	73.4mg	268.mg	7/20	150.mg	20/50	300.mg	35/50	mgl:adn,fba.
c	TR153	111.mg	299.mg	0/20	150.mg	10/50	300.mg	22/50	ubl:tcc,tpp.
d	TR153	116.mg	318.mg	0/20	150.mg	9/50	300.mg	22/50	
e	TR153	199.mg	605.mg	0/20	150.mg	3/50	300.mg	21/50	mul:ang,fbs,ost,srn.
f	TR153	193.mg	810.mg	0/20	150.mg	9/50	300.mg	12/50	spl:ang,ost,srn.
g	TR153	236.mg	844.mg	0/20	150.mg	0/50	300.mg	18/50	
h	TR153	239.mg	1.53gm	0/20	150.mg	7/50	300.mg	9/50	
i	TR153	47.1mg	174.mg	13/20	150.mg	43/50	300.mg	48/50	
j	TR153	n.s.s.	n.s.s.	0/20	150.mg	2/50	300.mg	0/50	liv:hpa,hpc,nnd.
4874	TR153	15.5mg	35.3mg	0/20	120.mg	44/50	240.mg	48/50	mul:ang,fbs,mso,ost,srn; spl:fib; sub:fib; tnv:mso. C
a	TR153	20.9mg	54.6mg	0/20	120.mg	28/50	240.mg	27/50	
b	TR153	59.7mg	140.mg	0/20	120.mg	15/50	240.mg	37/50	mul:ang,fbs,ost,srn.
c	TR153	53.3mg	189.mg	0/20	120.mg	17/50	240.mg	9/50	mul:mso; tnv:mso.
d	TR153	64.1mg	355.mg	0/20	120.mg	10/50	240.mg	2/50	
e	TR153	99.1mg	336.mg	0/20	120.mg	8/50	240.mg	20/50	
f	TR153	192.mg	981.mg	0/20	120.mg	3/50	240.mg	11/50	
g	TR153	15.8mg	40.6mg	10/20	120.mg	50/50	240.mg	48/50	
h	TR153	169.mg	n.s.s.	1/20	120.mg	1/50	240.mg	3/50	liv:hpa,hpc,nnd.
4875	1487	21.8mg	74.2mg	1/27	124.mg	25/30			Hecht;clet,16,103-108;1982
a	1487	74.0mg	372.mg	0/27	124.mg	11/30			
b	1487	81.0mg	452.mg	0/27	124.mg	10/30			
c	1487	89.2mg	571.mg	0/27	124.mg	9/30			
d	1487	164.mg	n.s.s.	0/27	124.mg	4/30			
e	1487	212.mg	n.s.s.	1/27	124.mg	9/30			
4876	381	22.4mg	62.5mg	0/16	192.mg	18/23	384.mg	21/24	Weisburger;jept,2,325-356;1978/pers.comm./Russfield 1973
a	381	482.mg	n.s.s.	0/16	192.mg	0/23	384.mg	1/24	
b	381	20.2mg	82.4mg	9/16	192.mg	19/23	384.mg	23/24	
4877	381	23.6mg	65.9mg	18/111p	192.mg	18/23	384.mg	21/24	
a	381	48.9mg	535.mg	5/111p	192.mg	3/23	384.mg	4/24	
p-TOLUIDINE.HCl		540-23-8							
4878	381	93.6mg	n.s.s.	0/20	74.3mg	2/21	142.mg	3/17	Weisburger;jept,2,325-356;1978/pers.comm./Russfield 1973
a	381	86.3mg	n.s.s.	6/20	74.3mg	3/21	142.mg	7/17	
b	381	97.1mg	n.s.s.	17/20	74.3mg	6/21	142.mg	10/17	
4879	381	94.7mg	29.2gm	1/102p	74.3mg	2/21	142.mg	3/17	
4880	381	19.3mg	n.s.s.	3/18	68.6mg	8/17	(131.mg	9/18)	
a	381	19.3mg	n.s.s.	3/18	68.6mg	8/17	(131.mg	9/18)	
b	381	26.8mg	n.s.s.	5/18	68.6mg	6/17	(131.mg	8/18)	
c	381	17.2mg	n.s.s.	12/18	68.6mg	12/17	(131.mg	14/18)	
4881	381	20.2mg	212.mg	7/99p	68.6mg	8/17	(131.mg	9/18)	

```
      Spe Strain Site   Xpo+Xpt                                                                                    TD50     2Tailpvl
          Sex   Route  Hist    Notes                                                                                      DR      AuOp

4882 R m cdr eat liv mix 77w98                                                  :>                                  no dre   P=1.   -
 a   R m cdr eat tba mix 77w98                                                                                      no dre   P=1.

p-TOLYLUREA                          100ng..:..1ug....:..10.....:..100....:..1mg....:..10......:..100....:..1g......:..10
4883 M f cb6 eat liv hpc 52w69 e     pool                                                           .>              no dre   P=1.   -
 a   M f cb6 eat lun a/a 52w69 e                                                                                    no dre   P=1.   -
4884 M m cb6 eat --- mly 52w69 e     pool                                                      .  +  .              206.mg   P<.0005+
 a   M m cb6 eat --- mlh 52w69 e                                                                                    265.mg   P<.0005
 b   M m cb6 eat mul mlh 52w69 e                                                                                    441.mg   P<.002
 c   M m cb6 eat mln mlh 52w69 e                                                                                    626.mg   P<.009
 d   M m cb6 eat liv hpc 52w69 e                                                                                    722.mg   P<.08  -
 e   M m cb6 eat lun a/a 52w69 e                                                                                    no dre   P=1.   -
4885 R f f34 eat liv tum 52w69 e                                                           .>                       no dre   P=1.   -
4886 R m f34 eat liv hpc 52w69 e                                                          .>                        699.mg   P<.2   -

TOXAPHENE                            100ng..:..1ug....:..10.....:..100....:..1mg....:..10......:..100....:..1g......:..10
4887 M f b6c eat liv MXA 80w90 v                                              :+ :                                  8.78mg / P<.0005c
 a   M f b6c eat liv hpc 80w90 v                                                                                    15.2mg / P<.0005c
 b   M f b6c eat TBA MXB 80w90 v                                                                                    9.19mg * P<.0005
 c   M f b6c eat liv MXB 80w90 v                                                                                    8.78mg / P<.0005
 d   M f b6c eat lun MXB 80w90 v                                                                                    no dre   P=1.
4888 M f b6c eat liv MXA 80w90 v     pool                                      :+ :                                 9.10mg / P<.0005c
 a   M f b6c eat liv hpc 80w90 v                                                                                    15.2mg / P<.0005c
4889 M m b6c eat liv hpc 80w90 v                                           : +:                                     4.08mg * P<.0005c
 a   M m b6c eat liv MXA 80w90 v                                                                                    4.23mg * P<.0005c
 b   M m b6c eat TBA MXB 80w90 v                                                                                    4.13mg * P<.0005
 c   M m b6c eat liv MXB 80w90 v                                                                                    4.23mg * P<.0005
 d   M m b6c eat lun MXB 80w90 v                                                                                    1.13gm * P<.9
4890 M m b6c eat liv MXA 80w90 v     pool                                   : +:                                    4.46mg * P<.0005c
 a   M m b6c eat liv hpc 80w90 v                                                                                    4.79mg * P<.0005c
4891 R f osm eat TBA MXB 19m25 v                                                   :>                               596.mg * P<1.  -
 a   R f osm eat liv MXB 19m25 v                                                                                    no dre   P=1.
4892 R f osm eat thy fca 19m25 v     pool                                                 :    ±                    209.mg * P<.03 a
4893 R m osm eat TBA MXB 19m25 v                                                   :>                               1.02gm * P<1.  -
 a   R m osm eat liv MXB 19m25 v                                                                                    949.mg * P<1.
4894 R m osm eat thy MXA 19m25 v     pool                                                  :  +      :              58.9mg * P<.002 a
 a   R m osm eat liv nnd 19m25 v                                                                                    92.1mg * P<.03

TRAGACANTH GUM                       100ng..:..1ug....:..10.....:..100....:..1mg....:..10......:..100....:..1g......:..10
4895 M f b6c eat for mix 22m25 e                                                                                    111.gm * P<.2  -
 a   M f b6c eat liv mix 22m25 e                                                                                    no dre   P=1.   -
 b   M f b6c eat lun mix 22m25 e                                                                                    no dre   P=1.   -
4896 M m b6c eat for sqp 22m25 e                                                                                    240.gm * P<.2  -
 a   M m b6c eat liv mix 22m25 e                                                                                    no dre   P=1.   -
 b   M m b6c eat lun mix 22m25 e                                                                                    no dre   P=1.   -

TRENIMON                             100ng..:..1ug....:..10.....:..100....:..1mg....:..10......:..100....:..1g......:..10
4897 R m b46 ivj liv lcc 12m24 es            .>                                                                     65.3ug   P<.2
 a   R m b46 ivj tba mix 12m24 es                                                                                  5.04ug   P<.005 +
 b   R m b46 ivj tba mal 12m24 es                                                                                  6.77ug   P<.007 +
 c   R m b46 ivj tba ben 12m24 es                                                                                  32.0ug   P<.4

TRIAMCINOLONE ACETONIDE              100ng..:..1ug....:..10.....:..100....:..1mg....:..10......:..100....:..1g......:..10
4898 R m cdr wat liv mix 24m24 ersv             .    ±                                                              53.0ug   P<.03  +
 a   R m cdr wat liv hpa 24m24 ersv                                                                                91.7ug   P<.2
 b   R m cdr wat liv hpc 24m24 ersv                                                                                .141mg   P<.2

TRIAMTERENE                          100ng..:..1ug....:..10.....:..100....:..1mg....:..10......:..100....:..1g......:..10
4899 M f b6c eat liv hpa 24m24                                             :  +    :                                46.0mg * P<.0005p
 a   M f b6c eat liv MXA 24m24                                                                                      51.3mg * P<.004 p
 b   M f b6c eat TBA MXB 24m24                                                                                      132.mg * P<.4
 c   M f b6c eat liv MXB 24m24                                                                                      51.3mg * P<.004
 d   M f b6c eat lun MXB 24m24                                                                                      no dre   P=1.
4900 M f b6c eat liv MXA 24m24                                               :  +    :                              45.3mg   P<.0005p
 a   M f b6c eat liv hpa 24m24                                                                                      45.7mg   P<.0005p
 b   M f b6c eat TBA MXB 24m24                                                                                      64.4mg   P<.07
 c   M f b6c eat liv MXB 24m24                                                                                      45.3mg   P<.0005
 d   M f b6c eat lun MXB 24m24                                                                                      no dre   P=1.
4901 M m b6c eat liv hpc 24m24                                                  :    ±                              227.mg * P<.07 p
 a   M m b6c eat liv MXA 24m24                                                                                      162.mg * P<.3  p
 b   M m b6c eat TBA MXB 24m24                                                                                      205.mg * P<.6
 c   M m b6c eat liv MXB 24m24                                                                                      169.mg * P<.3
 d   M m b6c eat lun MXB 24m24                                                                                      233.mg * P<.3
4902 M m b6c eat liv hpa 24m24                                             :    ±                                   54.8mg   P<.02  p
 a   M m b6c eat liv MXA 24m24                                                                                      61.2mg   P<.05  p
 b   M m b6c eat hag ade 24m24                                                                                      235.mg   P<.04
 c   M m b6c eat TBA MXB 24m24                                                                                      86.7mg   P<.3
 d   M m b6c eat liv MXB 24m24                                                                                      61.2mg   P<.05
 e   M m b6c eat lun MXB 24m24                                                                                      265.mg   P<.5
4903 R f f34 eat TBA MXB 24m24                                                     :>                               no dre   P=1.   -
 a   R f f34 eat liv MXB 24m24                                                                                      520.mg * P<.06
```

	RefNum	LoConf	UpConf	Cntrl	1Dose	1Inc	2Dose	2Inc	Citation or Pathology
									Brkly Code
4882	381	116.mg	n.s.s.	1/22	31.3mg	1/22	62.6mg	1/21	
a	381	42.9mg	n.s.s.	14/22	31.3mg	12/22	62.6mg	12/21	
p-TOLYLUREA		622-51-5							
4883	1343	674.mg	n.s.s.	1/89p	196.mg	0/38			Fleischman;jept,3,149-170;1980
a	1343	674.mg	n.s.s.	2/89p	196.mg	0/38			
4884	1343	100.mg	529.mg	0/95p	181.mg	10/43			
a	1343	120.mg	776.mg	0/95p	181.mg	8/43			
b	1343	167.mg	1.93gm	0/91p	181.mg	5/43			
c	1343	189.mg	17.3gm	0/79p	181.mg	3/36			
d	1343	207.mg	n.s.s.	2/91p	181.mg	4/43			
e	1343	705.mg	n.s.s.	1/87p	181.mg	0/43			
4885	1343	287.mg	n.s.s.	0/49	75.3mg	0/42			
4886	1343	114.mg	n.s.s.	0/50	60.2mg	1/39			
TOXAPHENE		8001-35-2							
4887	TR37	6.34mg	14.8mg	0/10	11.3mg	18/50	22.9mg	40/50	liv:hpc,nnd.
a	TR37	10.3mg	24.9mg	0/10	11.3mg	5/50	22.9mg	34/50	
b	TR37	6.22mg	22.7mg	1/10	11.3mg	22/50	22.9mg	40/50	
c	TR37	6.34mg	14.8mg	0/10	11.3mg	18/50	22.9mg	40/50	liv:hpa,hpc,nnd.
d	TR37	217.mg	n.s.s.	1/10	11.3mg	0/50	22.9mg	0/50	lun:a/a,a/c.
4888	TR37	6.48mg	13.9mg	1/49p	11.3mg	18/50	22.9mg	40/50	liv:hpc,nnd.
a	TR37	10.3mg	23.7mg	0/49p	11.3mg	5/50	22.9mg	34/50	
4889	TR37	3.04mg	6.21mg	0/10	10.4mg	34/50	21.1mg	45/50	
a	TR37	2.88mg	9.30mg	2/10	10.4mg	40/50	21.1mg	45/50	liv:hpc,nnd.
b	TR37	2.79mg	9.59mg	2/10	10.4mg	44/50	21.1mg	45/50	
c	TR37	2.88mg	9.30mg	2/10	10.4mg	40/50	21.1mg	45/50	liv:hpa,hpc,nnd.
d	TR37	56.5mg	n.s.s.	1/10	10.4mg	1/50	21.1mg	2/50	lun:a/a,a/c.
4890	TR37	3.09mg	7.36mg	9/50p	10.4mg	40/50	21.1mg	45/50	liv:hpc,nnd.
a	TR37	3.37mg	7.58mg	6/50p	10.4mg	34/50	21.1mg	45/50	
4891	TR37	22.5mg	n.s.s.	6/10	20.0mg	31/50	40.0mg	40/50	
a	TR37	106.mg	n.s.s.	1/10	20.0mg	5/50	40.0mg	4/50	liv:hpa,hpc,nnd.
4892	TR37	84.2mg	n.s.s.	1/55p	20.0mg	1/50	40.0mg	7/50	
4893	TR37	17.2mg	n.s.s.	7/10	16.5mg	33/50	33.0mg	24/50	
a	TR37	46.9mg	n.s.s.	1/10	16.5mg	6/50	33.0mg	4/50	liv:hpa,hpc,nnd.
4894	TR37	29.3mg	298.mg	2/55p	16.5mg	7/50	33.0mg	9/50	thy:fca,fcc.
a	TR37	39.4mg	n.s.s.	1/55p	16.5mg	6/50	33.0mg	4/50	
									s
TRAGACANTH GUM		9000-65-1							
4895	2086	27.4gm	n.s.s.	1/50	1.47gm	0/50	5.89gm	3/50	Hagiwara;fctx,30,673-679;1992/pers.comm.
a	2086	39.5gm	n.s.s.	7/50	1.47gm	1/50	5.89gm	3/50	
b	2086	69.0gm	n.s.s.	3/50	1.47gm	1/50	5.89gm	0/50	
4896	2086	39.1gm	n.s.s.	0/50	1.36gm	0/50	5.43gm	1/50	
a	2086	17.8gm	n.s.s.	24/50	1.36gm	21/50	5.43gm	15/50	
b	2086	18.4gm	n.s.s.	10/50	1.36gm	5/50	5.43gm	8/50	
TRENIMON		68-76-8							
4897	1017	10.6ug	n.s.s.	0/65	2.14ug	1/45			Schmahl;arzn,20,1461-1467;1970
a	1017	2.29ug	48.2ug	7/65	2.14ug	15/45			
b	1017	2.91ug	.117mg	4/65	2.14ug	11/45			
c	1017	6.37ug	n.s.s.	3/65	2.14ug	4/45			
TRIAMCINOLONE ACETONIDE		76-25-5							
4898	2059	20.6ug	n.s.s.	11/200	6.39ug	13/100			Ryrfeldt;txpy,20,115-117;1992
a	2059	28.7ug	n.s.s.	7/200	6.39ug	8/100			
b	2059	37.3ug	n.s.s.	4/200	6.39ug	5/100			
TRIAMTERENE		396-01-0							
4899	TR420	25.4mg	181.mg	10/50	13.0mg	22/50	25.9mg	23/50	51.8mg	36/60	
a	TR420	26.0mg	444.mg	13/50	13.0mg	26/50	25.9mg	25/50	51.8mg	37/60	liv:hpa,hpc.
b	TR420	32.1mg	n.s.s.	34/50	13.0mg	40/50	25.9mg	40/50	51.8mg	46/60	
c	TR420	26.0mg	444.mg	13/50	13.0mg	26/50	25.9mg	25/50	51.8mg	37/60	liv:hpa,hpb,hpc.
d	TR420	266.mg	n.s.s.	4/50	13.0mg	6/50	25.9mg	2/50	51.8mg	3/60	lun:a/a,a/c.
4900	TR420a	23.7mg	178.mg	10/50	51.9mg	31/51			liv:hpa,hpc.
a	TR420a	24.7mg	147.mg	7/50	51.9mg	28/51			
b	TR420a	25.3mg	n.s.s.	29/50	51.9mg	43/51			
c	TR420a	23.7mg	178.mg	10/50	51.9mg	31/51			liv:hpa,hpb,hpc.
d	TR420a	291.mg	n.s.s.	3/50	51.9mg	1/51			lun:a/a,a/c.
4901	TR420	87.7mg	n.s.s.	5/50	12.0mg	7/50	23.9mg	3/50	47.8mg	13/60	
a	TR420	45.8mg	n.s.s.	20/50	12.0mg	26/50	23.9mg	19/50	47.8mg	29/60	liv:hpa,hpc.
b	TR420	37.1mg	n.s.s.	32/50	12.0mg	38/50	23.9mg	35/50	47.8mg	38/60	
c	TR420	46.2mg	n.s.s.	20/50	12.0mg	27/50	23.9mg	19/50	47.8mg	29/60	liv:hpa,hpb,hpc.
d	TR420	66.4mg	n.s.s.	9/50	12.0mg	12/50	23.9mg	16/50	47.8mg	14/60	lun:a/a,a/c.
4902	TR420a	24.1mg	n.s.s.	21/50	48.0mg	36/50			
a	TR420a	24.9mg	n.s.s.	25/50	48.0mg	38/50			liv:hpa,hpc.
b	TR420a	85.4mg	n.s.s.	1/50	48.0mg	6/50			
									s
c	TR420a	25.8mg	n.s.s.	37/50	48.0mg	44/50			
d	TR420a	24.9mg	n.s.s.	25/50	48.0mg	38/50			liv:hpa,hpb,hpc.
e	TR420a	58.0mg	n.s.s.	13/50	48.0mg	16/50			lun:a/a,a/c.
4903	TR420	21.6mg	n.s.s.	46/50	4.99mg	43/50	15.0mg	42/50	29.9mg	41/50	
a	TR420	128.mg	n.s.s.	0/50	4.99mg	0/50	15.0mg	0/50	29.9mg	2/50	liv:hpa,hpb,hpc.

```
      Spe Strain Site  Xpo+Xpt                                          TD50    2Tailpvl
          Sex  Route  Hist     Notes                                      DR    AuOp
4904 R m f34 eat liv hpa 24m24                    :   +   :             11.8mg Z P<.005 e
a    R m f34 eat TBA MXB 24m24                                          no dre   P=1.
b    R m f34 eat liv MXB 24m24                                          11.8mg Z P<.005

TRIBROMOMETHANE              100ng..:..1ug....:...10....:...100....:...1mg...:...10...:...100...:...1g....:...10
4905 M f b6c gav MXA hes 24m25 s                               :    ±   #1.11gm * P<.04 -
a    M f b6c gav TBA MXB 24m25 s                                        no dre   P=1.
b    M f b6c gav liv MXB 24m25 s                                        570.mg * P<.3
c    M f b6c gav lun MXB 24m25 s                                        no dre   P=1.
4906 M m b6c gav sub MXA 24m25                          :    ±          #81.9mg \ P<.02 -
a    M m b6c gav TBA MXB 24m25                                          41.0mg \ P<.2
b    M m b6c gav liv MXB 24m25                                          53.6mg \ P<.07
c    M m b6c gav lun MXB 24m25                                          2.45gm \ P<1.
4907 R f f34 gav lgi MXA 24m25                             :   +   : +hist 469.mg * P<.002 c
a    R f f34 gav lgi pla 24m25                                          632.mg * P<.004
b    R f f34 gav TBA MXB 24m25                                          no dre   P=1.
c    R f f34 gav liv MXB 24m25                                          676.mg * P<.06
4908 R m f34 gav tes MXA 24m24 s                        :    ±          75.0mg * P<.03
a    R m f34 gav thy fcc 24m24 s                                        656.mg * P<.03
b    R m f34 gav lgi MXA 24m24 s                                 +hist 1.05gm * P<.02 p
c    R m f34 gav TBA MXB 24m24 s                                        513.mg * P<.7
d    R m f34 gav liv MXB 24m24 s                                        no dre   P=1.

TRICAPRYLIN                  100ng..:..1ug....:...10....:...100....:...1mg...:...10...:...100....:...1g....:...10
4909 R m f34 gav pan ana 24m24                                    :  +  5.49gm * P<.0005
a    R m f34 gav for sqp 24m24                                          15.7gm * P<.0005
b    R m f34 gav liv hpa 24m24                                          37.1gm * P<.05
c    R m f34 gav TBA MXB 24m24                                          69.9gm * P<.9
d    R m f34 gav liv MXB 24m24                                          37.1gm * P<.05

1,2,3-TRICHLORO-4,6-DINITROBENZENE  ..:...1ug....:...10....:...100....:...1mg....:...10....:...100....:...1g....:...10
4910 M f b6a orl lun ade 76w76 evx                       .>             108.mg   P<.3
a    M f b6a orl liv hpt 76w76 evx                                      no dre   P=1.
b    M f b6a orl tba mix 76w76 evx                                      52.6mg   P<.09
4911 M m b6a orl lun ade 76w76 evx                     .   ±            49.0mg   P<.09
a    M m b6a orl liv hpt 76w76 evx                                      no dre   P=1.
b    M m b6a orl tba mix 76w76 evx                                      no dre   P=1.
4912 M f b6c orl liv hpt 76w76 evx                       .>             no dre   P=1.
a    M f b6c orl lun ade 76w76 evx                                      no dre   P=1.
b    M f b6c orl tba mix 76w76 evx                                      no dre   P=1.
4913 M m b6c orl liv hpt 76w76 evx                       .>             no dre   P=1.
a    M m b6c orl lun mix 76w76 evx                                      no dre   P=1.
b    M m b6c orl tba mix 76w76 evx                                      19.8mg   P<.4

1,1,2-TRICHLORO-1,2,2-TRIFLUOROETHANE, TECHNICAL GRADE  ....:...100....:...1mg....:...10....:...100....:...1g....:...10
4914 R f cdr inh pni isa 24m24 e                                        160.gm * P<.005 -
4915 R m cdr inh pni isa 24m24 e                                        9.14kg   P<1.

2,4,6-TRICHLOROANILINE       100ng..:..1ug....:...10....:...100....:...1mg...:...10...:...100....:...1g....:...10
4916 M f chi eat liv mix 68w77 a                               :    ±   927.mg * P<.02 -
a    M f chi eat lun mix 68w77 a                                        1.37gm * P<.02 -
b    M f chi eat tba mix 68w77 a                                        914.mg * P<.06
4917 M m chi eat --- vsc 68w77 a                          :   +   :     273.mg * P<.0005+
a    M m chi eat lun mix 68w77 a                                        385.mg * P<.0005-
b    M m chi eat liv mix 68w77 a                                        452.mg * P<.004
c    M m chi eat tba mix 68w77 a                                        206.mg * P<.0005
4918 M m chi eat --- vsc 68w77 a   pool                   :   +   :     259.mg * P<.0005+
a    M m chi eat liv hpt 68w77 a                                        560.mg * P<.0005+
4919 R m cdr eat liv mix 18m27 v                               :>       no dre   P=1. -
a    R m cdr eat tba mix 18m27 v                                        45.9mg \ P<.01 -

1,1,2-TRICHLOROETHANE        100ng..:..1ug....:...10....:...100....:...1mg...:...10...:...100....:...1g....:...10
4920 M f b6c gav liv hpc 78w90 v                         :+ :           47.6mg * P<.0005c
a    M f b6c gav MXB MXB 78w90 v                                        47.6mg * P<.0005
b    M f b6c gav adr phe 78w90 v                                        248.mg * P<.0005c
c    M f b6c gav TBA MXB 78w90 v                                        52.9mg * P<.0005
d    M f b6c gav liv MXB 78w90 v                                        47.6mg * P<.0005
e    M f b6c gav lun MXB 78w90 v                                        787.mg * P<.2
4921 M m b6c gav liv hpc 78w90 v                         :   +   :      65.0mg / P<.0005c
a    M m b6c gav MXB MXB 78w90 v                                        65.0mg / P<.0005
b    M m b6c gav adr phe 78w90 v                                        588.mg / P<.007 c
c    M m b6c gav TBA MXB 78w90 v                                        79.5mg * P<.006
d    M m b6c gav liv MXB 78w90 v                                        65.0mg / P<.0005
e    M m b6c gav lun MXB 78w90 v                                        1.56gm * P<.8
4922 R f osm gav TBA MXB 18m26 v                         :>             no dre   P=1. -
a    R f osm gav liv MXB 18m26 v                                        771.mg * P<.7
4923 R m osm gav TBA MXB 18m26 v                       :>               no dre   P=1.
a    R m osm gav liv MXB 18m26 v                                        no dre   P=1.

1,1,1-TRICHLOROETHANE, TECHNICAL GRADE  ..:...1ug....:...10....:...100....:...1mg...:...10....:...100....:...1g....:...10
4924 M f b6c gav TBA MXB 78w90 sv                                       :>no dre P=1.    i
```

	RefNum	LoConf	UpConf	Cntrl	1Dose	1Inc	2Dose	2Inc			Citation or Pathology
											Brkly Code
4904	TR420	4.72mg	41.5mg	0/50	3.99mg	6/50	(12.0mg	4/50	23.9mg	3/50)	
a	TR420	18.1mg	n.s.s.	40/50	3.99mg	43/50	12.0mg	40/50	23.9mg	39/50	
b	TR420	4.72mg	41.5mg	0/50	3.99mg	6/50	(12.0mg	4/50	23.9mg	3/50)	liv:hpa,hpb,hpc.

TRIBROMOMETHANE   (bromoform) 75-25-2

	RefNum	LoConf	UpConf	Cntrl	1Dose	1Inc	2Dose	2Inc	Citation or Pathology
4905	TR350	311.mg	n.s.s.	0/50	69.4mg	0/50	140.mg	3/50	liv:hes; spl:hes. S
a	TR350	109.mg	n.s.s.	24/50	69.4mg	18/50	140.mg	18/50	
b	TR350	152.mg	n.s.s.	4/50	69.4mg	6/50	140.mg	6/50	liv:hpa,hpc,nnd.
c	TR350	349.mg	n.s.s.	3/50	69.4mg	1/50	140.mg	2/50	lun:a/a,a/c.
4906	TR350	28.0mg	17.5gm	2/50	34.9mg	8/50	(69.7mg	4/50)	sub:fbs,sar. S
a	TR350	13.8mg	n.s.s.	34/50	34.9mg	33/50	(69.7mg	27/50)	
b	TR350	18.1mg	n.s.s.	16/50	34.9mg	19/50	(69.7mg	14/50)	liv:hpa,hpc,nnd.
c	TR350	44.0mg	n.s.s.	11/50	34.9mg	7/50	(69.7mg	2/50)	lun:a/a,a/c.
4907	TR350	219.mg	1.55gm	0/50	69.4mg	1/50	139.mg	8/50	lgi:adc,pla.
a	TR350	270.mg	3.60gm	0/50	69.4mg	1/50	139.mg	6/50	S
b	TR350	104.mg	n.s.s.	47/50	69.4mg	37/50	139.mg	38/50	
c	TR350	276.mg	n.s.s.	0/50	69.4mg	4/50	139.mg	2/50	liv:hpa,hpc,nnd.
4908	TR350	33.3mg	n.s.s.	46/50	69.7mg	45/50	140.mg	37/50	tes:iab,ica. S
a	TR350	237.mg	n.s.s.	0/50	69.7mg	3/50	140.mg	2/50	S
b	TR350	298.mg	n.s.s.	0/50	69.7mg	0/50	140.mg	3/50	lgi:adc,pla.
c	TR350	71.8mg	n.s.s.	45/50	69.7mg	37/50	140.mg	25/50	
d	TR350	541.mg	n.s.s.	5/50	69.7mg	2/50	140.mg	1/50	liv:hpa,hpc,nnd.

TRICAPRYLIN   538-23-8

	RefNum	LoConf	UpConf	Cntrl	1Dose	1Inc	2Dose	2Inc	3Dose	3Inc	Citation or Pathology
4909	TR426	3.40gm	12.5gm	2/50	1.68gm	6/50	3.37gm	13/50	6.76mg	18/53	S
a	TR426	8.26gm	36.2gm	0/50	1.68gm	0/50	3.37gm	3/50	6.76mg	10/53	S
b	TR426	13.9gm	n.s.s.	1/50	1.68gm	0/50	3.37gm	3/50	6.76mg	4/53	S
c	TR426	3.40gm	n.s.s.	47/50	1.68gm	47/50	3.37gm	46/50	6.76mg	36/53	
d	TR426	13.9gm	n.s.s.	1/50	1.68gm	0/50	3.37gm	3/50	6.76mg	4/53	liv:hpa,hpb,hpc.

1,2,3-TRICHLORO-4,6-DINITROBENZENE   (vancide PB) 6379-46-0

	RefNum	LoConf	UpConf	Cntrl	1Dose	1Inc	Citation or Pathology
4910	1262	17.6mg	n.s.s.	0/18	16.9mg	1/18	Innes;ntis,1968/1969
a	1262	33.5mg	n.s.s.	0/18	16.9mg	0/18	
b	1262	12.9mg	n.s.s.	0/18	16.9mg	2/18	
4911	1262	12.0mg	n.s.s.	0/18	15.8mg	2/18	
a	1262	31.2mg	n.s.s.	1/18	15.8mg	0/18	
b	1262	14.7mg	n.s.s.	2/18	15.8mg	2/18	
4912	1262	33.5mg	n.s.s.	0/18	16.9mg	0/18	
a	1262	33.5mg	n.s.s.	1/18	16.9mg	0/18	
b	1262	22.2mg	n.s.s.	2/18	16.9mg	1/18	
4913	1262	18.5mg	n.s.s.	3/14	15.8mg	1/15	
a	1262	26.0mg	n.s.s.	0/14	15.8mg	0/15	
b	1262	4.75mg	n.s.s.	4/14	15.8mg	7/15	

1,1,2-TRICHLORO-1,2,2-TRIFLUOROETHANE, TECHNICAL GRADE (fluorocarbon 113) 76-13-1

	RefNum	LoConf	UpConf	Cntrl	1Dose	1Inc	2Dose	2Inc	3Dose	3Inc	Citation or Pathology
4914	1876	60.9gm	1.25kg	0/85	1.15gm	0/36	5.74gm	0/30	11.5gm	5/86	Trochimowicz;faat,11,68-75;1988
4915	1876	95.1gm	n.s.s.	2/88	803.mg	1/64	4.02gm	0/58	8.03gm	2/87	

2,4,6-TRICHLOROANILINE   634-93-5

	RefNum	LoConf	UpConf	Cntrl	1Dose	1Inc	2Dose	2Inc	Citation or Pathology
4916	381	369.mg	n.s.s.	0/13	780.mg	5/21	1.56gm	1/14	Weisburger;jept,2,325-356;1978/pers.comm./Russfield 1973
a	381	490.mg	n.s.s.	3/13	780.mg	3/21	1.56gm	2/14	
b	381	333.mg	n.s.s.	9/13	780.mg	7/21	1.56gm	4/14	
4917	381	140.mg	725.mg	2/16	720.mg	8/18	1.44gm	12/16	
a	381	166.mg	1.45gm	2/16	720.mg	6/18	1.44gm	3/16	
b	381	178.mg	3.14gm	2/16	720.mg	5/18	1.44gm	1/16	
c	381	104.mg	590.mg	9/16	720.mg	12/18	1.44gm	12/16	
4918	381	137.mg	537.mg	5/99p	720.mg	8/18	1.44gm	12/16	
a	381	202.mg	2.72gm	7/99p	720.mg	5/18	1.44gm	1/16	
4919	381	462.mg	n.s.s.	1/17	57.5mg	0/18	102.mg	0/16	
a	381	18.7mg	3.80mg	13/17	57.5mg	13/18	(102.mg	10/16)	

1,1,2-TRICHLOROETHANE   79-00-5

	RefNum	LoConf	UpConf	Cntrl	1Dose	1Inc	2Dose	2Inc	Citation or Pathology
4920	TR74	33.3mg	70.8mg	0/20	119.mg	16/50	239.mg	40/50	
a	TR74	33.3mg	70.8mg	0/20	119.mg	16/50	239.mg	40/50	
b	TR74	127.mg	611.mg	0/20	119.mg	0/50	239.mg	12/50	adr:phe; liv:hpc. C
c	TR74	33.9mg	106.mg	5/20	119.mg	20/50	239.mg	41/50	
d	TR74	33.3mg	70.8mg	0/20	119.mg	16/50	239.mg	40/50	
e	TR74	287.mg	n.s.s.	0/20	119.mg	3/50	239.mg	2/50	liv:hpa,hpc,nnd.
									lun:a/a,a/c.
4921	TR74	42.2mg	146.mg	2/20	121.mg	18/50	239.mg	37/50	
a	TR74	42.2mg	146.mg	2/20	121.mg	18/50	239.mg	37/50	
b	TR74	263.mg	6.05gm	0/20	121.mg	0/50	239.mg	8/50	adr:phe; liv:hpc. C
c	TR74	42.3mg	872.mg	6/20	121.mg	28/50	239.mg	38/50	
d	TR74	42.2mg	146.mg	2/20	121.mg	18/50	239.mg	37/50	
e	TR74	338.mg	n.s.s.	0/20	121.mg	3/50	239.mg	1/50	liv:hpa,hpc,nnd.
									lun:a/a,a/c.
4922	TR74	31.0mg	n.s.s.	4/20	23.0mg	34/50	45.5mg	22/50	
a	TR74	182.mg	n.s.s.	0/20	23.0mg	1/50	45.5mg	1/50	liv:hpa,hpc,nnd.
4923	TR74	11.8mg	n.s.s.	6/20	23.0mg	21/50	(45.9mg	11/50)	
a	TR74	436.mg	n.s.s.	1/20	23.0mg	0/50	45.9mg	0/50	liv:hpa,hpc,nnd.

1,1,1-TRICHLOROETHANE, TECHNICAL GRADE (methyl chloroform) 71-55-6

	RefNum	LoConf	UpConf	Cntrl	1Dose	1Inc	2Dose	2Inc
4924	TR3	6.79gm	n.s.s.	7/20	1.72gm	2/50	3.44gm	3/50

```
      Spe Strain Site  Xpo+Xpt                                                              TD50    2Tailpvl
          Sex  Route  Hist    Notes                                                           DR    AuOp
```

	Spe	Sex	Strain	Route	Site	Hist	Xpo+Xpt	Notes	DR/AuOp plot	TD50	2Tailpvl	AuOp
a	M	f	b6c	gav	liv	MXB	78w90	sv		no dre	P=1.	
b	M	f	b6c	gav	lun	MXB	78w90	sv		260.gm *	P<.9	
4925	M	f	b6c	inh	liv	mix	24m24	e		no dre	P=1.	-
a	M	f	b6c	inh	liv	hpc	24m24	e		no dre	P=1.	-
b	M	f	b6c	inh	liv	hpa	24m24	e		no dre	P=1.	-
4926	M	m	b6c	gav	TBA	MXB	78w90	sv	:>	30.5gm *	P<.8	i
a	M	m	b6c	gav	liv	MXB	78w90	sv		39.7gm /	P<.8	
b	M	m	b6c	gav	lun	MXB	78w90	sv		5.08kg	P<1.	
4927	M	m	b6c	inh	liv	hpc	24m24	e	.>	66.3gm *	P<.9	
a	M	m	b6c	inh	liv	hpa	24m24	e		no dre	P=1.	-
b	M	m	b6c	inh	liv	mix	24m24	e		no dre	P=1.	-
4928	R	f	osm	gav	TBA	MXB	18m26	s	: + :	226.mg *	P<.0005	i
a	R	f	osm	gav	liv	MXB	18m26	s		no dre	P=1.	
4929	R	m	osm	gav	TBA	MXB	18m25	s	: +	950.mg \	P<.009	i
a	R	m	osm	gav	liv	MXB	18m25	s		no dre	P=1.	
4930	R	f	f34	inh	liv	hpc	24m24	e	.>	no dre	P=1.	-
a	R	f	f34	inh	liv	nnd	24m24	e		no dre	P=1.	-
4931	R	m	f34	inh	liv	hpc	24m24	e	. ±	2.07gm Z	P<.03	-
a	R	m	f34	inh	liv	nnd	24m24	e		5.12gm *	P<.09	-

TRICHLOROETHYLENE

```
100ng..:..1ug....:..10.....:..100....:..1mg...:..10....:..100....:..1g..:..10
```

	Spe	Sex	Strain	Route	Site	Hist	Xpo+Xpt	Notes	DR/AuOp plot	TD50	2Tailpvl	AuOp
4932	H	f	syg	inh	liv	mix	18m30	e	.>	no dre	P=1.	-
a	H	f	syg	inh	tba	mal	18m30	e		9.22gm *	P<.8	-
b	H	f	syg	inh	tba	ben	18m30	e		no dre	P=1.	-
4933	H	m	syg	inh	liv	mix	18m30	e	.>	no dre	P=1.	-
a	H	m	syg	inh	tba	mal	18m30	e		no dre	P=1.	-
b	H	m	syg	inh	tba	ben	18m30	e		5.64gm *	P<.4	-
4934	M	f	b6c	gav	liv	hpc	78w90	sv	: +	:1.93gm *	P<.004	c
a	M	f	b6c	gav	TBA	MXB	78w90	sv		1.85gm *	P<.2	
b	M	f	b6c	gav	liv	MXB	78w90	sv		1.93gm *	P<.004	
c	M	f	b6c	gav	lun	MXB	78w90	sv		4.35gm *	P<.2	
4935	M	f	b6c	inh	lun	tum	18m36	e	. +	6.32gm *	P<.002	+
a	M	f	b6c	inh	liv	hpt	18m36	e		13.6gm *	P<.06	+
b	M	f	b6c	inh	thy	car	18m36	e		131.gm *	P<.2	
c	M	f	b6c	inh	thy	ade	18m36	e		2.23kg	P<1.	
d	M	f	b6c	inh	tba	mix	18m36	e		1.32gm *	P<.002	
e	M	f	b6c	inh	tba	mal	18m36	e		2.08gm *	P<.02	
4936	M	m	b6c	gav	liv	hpc	78w90	sv	: + :	421.mg *	P<.0005	c
a	M	m	b6c	gav	TBA	MXB	78w90	sv		553.mg *	P<.01	
b	M	m	b6c	gav	liv	MXB	78w90	sv		421.mg *	P<.0005	
c	M	m	b6c	gav	lun	MXB	78w90	sv		7.16gm *	P<.7	
4937	M	m	b6c	inh	liv	hpt	18m36	es	. +	5.03gm *	P<.006	+
a	M	m	b6c	inh	lun	tum	18m36	es		no dre	P=1.	
b	M	m	b6c	inh	thy	ade	18m36	es		no dre	P=1.	
c	M	m	b6c	inh	tba	mal	18m36	es		no dre	P=1.	
d	M	m	b6c	inh	tba	mix	18m36	es		no dre	P=1.	
4938	M	m	b6c	inh	liv	hpt	18m31	e	.>	4.53gm *	P<.3	+
a	M	m	b6c	inh	liv	ang	18m31	e		77.0gm *	P<.8	
b	M	m	b6c	inh	lun	tum	18m31	e		no dre	P=1.	
c	M	m	b6c	inh	tba	mal	18m31	e		no dre	P=1.	
d	M	m	b6c	inh	tba	mix	18m31	e		no dre	P=1.	
4939	M	f	hic	gav	for	tum	88w88	r	.>	41.3mg	P<.3	-
4940	M	m	hic	gav	for	tum	88w88	r	. ±	16.9mg	P<.1	-
4941	M	f	icm	inh	lun	mix	24m25	e	. ±	3.37gm *	P<.07	
a	M	f	icm	inh	lun	adc	24m25	e		3.38gm *	P<.02	
b	M	f	icm	inh	liv	ade	24m25	e		42.3gm *	P<.2	
c	M	f	icm	inh	tba	mix	24m25	e		1.90gm *	P<.08	
4942	M	f	nmh	inh	---	mly	18m30	e	±	846.mg *	P<.03	
a	M	f	nmh	inh	liv	mix	18m30	e		no dre	P=1.	-
b	M	f	nmh	inh	lun	a/a	18m30	e		no dre	P=1.	-
c	M	f	nmh	inh	lun	a/c	18m30	e		no dre	P=1.	-
d	M	f	nmh	inh	tba	mal	18m30	e		573.mg *	P<.03	-
e	M	f	nmh	inh	tba	ben	18m30	e		no dre	P=1.	-
4943	M	m	nmh	inh	liv	hpc	18m30	e	.>	no dre	P=1.	-
a	M	m	nmh	inh	liv	bhp	18m30	e		no dre	P=1.	-
b	M	m	nmh	inh	---	mly	18m30	e		no dre	P=1.	-
c	M	m	nmh	inh	lun	a/a	18m30	e		no dre	P=1.	-
d	M	m	nmh	inh	lun	mal	18m30	e		no dre	P=1.	-
e	M	m	nmh	inh	liv	hpa	18m30	e		no dre	P=1.	-
f	M	m	nmh	inh	tba	mal	18m30	e		no dre	P=1.	-
g	M	m	nmh	inh	tba	ben	18m30	e		no dre	P=1.	-
4944	M	f	swi	inh	lun	tum	18m34	e	.>	35.4gm *	P<.7	
a	M	f	swi	inh	liv	hpt	18m34	e		121.gm *	P<.2	
b	M	f	swi	inh	tba	mix	18m34	e		9.19gm *	P<.6	
c	M	f	swi	inh	tba	mal	18m34	e		9.84gm *	P<.5	
4945	M	m	swi	inh	liv	hpt	18m34	e	. +	3.91gm *	P<.003	+
a	M	m	swi	inh	lun	tum	18m34	e		10.1gm *	P<.3	+
b	M	m	swi	inh	tba	mal	18m34	e		6.32gm *	P<.08	
c	M	m	swi	inh	tba	mix	18m34	e		8.07gm *	P<.3	
4946	R	f	osm	gav	TBA	MXB	18m26	dsv	.>	5.62gm *	P<.9	
a	R	f	osm	gav	kid	uac	18m26	dsv		no dre	P=1.	

	RefNum	LoConf	UpConf	Cntrl	1Dose	1Inc	2Dose	2Inc			Citation or Pathology
											Brkly Code
a	TR3	13.6gm	n.s.s.	1/20	1.72gm	0/50	3.44gm	0/50			liv:hpa,hpc,nnd.
b	TR3	7.97gm	n.s.s.	1/20	1.72gm	0/50	3.44gm	1/50			lun:a/a,a/c.
4925	1892	11.4gm	n.s.s.	13/50	257.mg	10/50	858.mg	10/50	2.57gm	7/50	Quast;faat,11,611-625;1988
a	1892	13.2gm	n.s.s.	4/50	257.mg	1/50	858.mg	5/50	2.57gm	2/50	
b	1892	14.2gm	n.s.s.	10/50	257.mg	9/50	858.mg	5/50	2.57gm	5/50	
4926	TR3	3.09gm	n.s.s.	5/20	1.74gm	2/50	3.48gm	6/50			
a	TR3	4.18gm	n.s.s.	3/20	1.74gm	0/50	3.48gm	4/50			liv:hpa,hpc,nnd.
b	TR3	5.52gm	n.s.s.	1/20	1.74gm	1/50	3.48gm	1/50			lun:a/a,a/c.
4927	1892	4.67gm	n.s.s.	12/50	214.mg	10/50	715.mg	12/50	2.14gm	12/50	Quast;faat,11,611-625;1988
a	1892	5.83gm	n.s.s.	26/50	214.mg	13/50	715.mg	19/50	2.14gm	16/50	
b	1892	3.43gm	n.s.s.	29/50	214.mg	22/50	715.mg	28/50	2.14gm	24/50	
4928	TR3	81.4mg	685.mg	10/20	380.mg	6/50	760.mg	9/50			
a	TR3	n.s.s.	n.s.s.	1/20	380.mg	0/50	760.mg	1/50			liv:hpa,hpc,nnd.
4929	TR3	334.mg	22.2gm	7/20	380.mg	6/50	(760.mg	4/50)			
a	TR3	n.s.s.	n.s.s.	0/20	380.mg	0/50	760.mg	0/50			liv:hpa,hpc,nnd.
4930	1892	451.mg	n.s.s.	1/50	61.3mg	0/50	204.mg	0/50	613.mg	0/50	Quast;faat,11,611-625;1988
a	1892	7.94gm	n.s.s.	1/50	61.3mg	1/50	204.mg	0/50	613.mg	0/50	
4931	1892	627.mg	n.s.s.	0/50	42.9mg	0/50	143.mg	3/50	(429.mg	0/50)	
a	1892	1.57gm	n.s.s.	1/50	42.9mg	1/50	143.mg	1/50	429.mg	4/50	
	TRICHLOROETHYLENE	(TCE)	79-01-6								
4932	1010m	2.14gm	n.s.s.	1/30	46.1mg	1/29	230.mg	0/30			Henschler;artx,43,237-248;1980/pers.comm.
a	1010m	821.mg	n.s.s.	2/30	46.1mg	3/29	230.mg	3/30			
b	1010m	2.14gm	n.s.s.	1/30	46.1mg	1/29	230.mg	0/30			
4933	1010m	326.mg	n.s.s.	1/30	40.5mg	0/30	203.mg	0/30			
a	1010m	863.mg	n.s.s.	6/30	40.5mg	2/30	203.mg	4/30			
b	1010m	999.mg	n.s.s.	1/30	40.5mg	0/30	203.mg	2/30			
4934	TR2	1.06gm	9.81gm	0/20	538.mg	4/50	1.08gm	11/50			
a	TR2	716.mg	n.s.s.	4/20	538.mg	14/50	1.08gm	19/50			
b	TR2	1.06gm	9.81gm	0/20	538.mg	4/50	1.08gm	11/50			liv:hpa,hpc,nnd.
c	TR2	1.55gm	n.s.s.	1/20	538.mg	4/50	1.08gm	7/50			lun:a/a,a/c.
4935	BT306m	3.33gm	28.9gm	2/90	99.8mg	6/90	299.mg	7/89	599.mg	14/87	Maltoni;aric,5,1-393;1986
a	BT306m	5.28gm	n.s.s.	3/88	99.8mg	4/89	299.mg	4/88	599.mg	9/85	
b	BT306m	21.4gm	n.s.s.	0/90	99.8mg	0/90	299.mg	0/89	599.mg	1/87	
c	BT306m	17.3gm	n.s.s.	1/90	99.8mg	1/90	299.mg	1/89	599.mg	1/87	
d	BT306m	700.mg	6.89gm	51/90	99.8mg	58/90	299.mg	60/89	599.mg	69/87	
e	BT306m	986.mg	n.s.s.	42/90	99.8mg	52/90	299.mg	53/89	599.mg	58/87	
4936	TR2	277.mg	1.13gm	1/20	724.mg	26/50	1.45gm	30/50			
a	TR2	288.mg	28.9gm	5/20	724.mg	30/50	1.45gm	33/50			
b	TR2	277.mg	1.13gm	1/20	724.mg	26/50	1.45gm	30/50			liv:hpa,hpc,nnd.
c	TR2	1.79gm	n.s.s.	0/20	724.mg	5/50	1.45gm	2/50			lun:a/a,a/c.
4937	BT306m	2.22gm	59.8gm	1/59	83.2mg	1/31	250.mg	3/38	499.mg	6/37	Maltoni;aric,5,1-393;1986
a	BT306m	12.3gm	n.s.s.	1/85	83.2mg	2/86	250.mg	2/88	499.mg	1/88	
b	BT306m	2.17gm	n.s.s.	0/85	83.2mg	0/86	250.mg	0/88	499.mg	0/88	
c	BT306m	8.59gm	n.s.s.	9/85	83.2mg	3/86	250.mg	3/88	499.mg	7/88	
d	BT306m	7.93gm	n.s.s.	11/85	83.2mg	4/86	250.mg	4/88	499.mg	8/88	
4938	BT306n	1.34gm	n.s.s.	17/77	94.9mg	19/47	285.mg	27/67	569.mg	21/63	
a	BT306n	6.47gm	n.s.s.	1/77	94.9mg	1/47	285.mg	3/67	569.mg	1/63	
b	BT306n	7.94gm	n.s.s.	12/90	94.9mg	6/89	285.mg	7/90	569.mg	7/90	
c	BT306n	3.14gm	n.s.s.	40/90	94.9mg	27/89	285.mg	34/90	569.mg	31/90	
d	BT306n	994.mg	n.s.s.	51/90	94.9mg	30/89	285.mg	46/90	(569.mg	36/90)	
4939	1011	6.73mg	n.s.s.	0/30	2.86mg	1/30					Van Duuren;jnci,63,1433-1439;1979
4940	1011	4.16mg	n.s.s.	0/30	2.38mg	2/30					
4941	1626	1.26gm	n.s.s.	6/49	95.8mg	5/50	287.mg	13/50	862.mg	11/46	Fukuda;indh,21,243-254;1983
a	1626	1.46gm	n.s.s.	1/49	95.8mg	3/50	287.mg	8/50	862.mg	7/46	
b	1626	6.90gm	n.s.s.	0/49	95.8mg	0/50	287.mg	0/50	862.mg	1/46	
c	1626	704.mg	n.s.s.	18/49	95.8mg	18/50	287.mg	27/50	862.mg	24/46	
4942	1010m	339.mg	n.s.s.	9/29	102.mg	17/30	509.mg	18/28			Henschler;artx,43,237-248;1980/pers.comm.
a	1010m	809.mg	n.s.s.	0/29	102.mg	0/30	509.mg	0/28			
b	1010m	3.80gm	n.s.s.	3/29	102.mg	0/30	509.mg	1/28			
c	1010m	3.63gm	n.s.s.	1/29	102.mg	3/30	509.mg	0/28			
d	1010m	230.mg	n.s.s.	13/29	102.mg	22/30	509.mg	22/28			
e	1010m	2.03gm	n.s.s.	6/29	102.mg	1/30	509.mg	4/28			
4943	1010m	663.mg	n.s.s.	1/30	84.8mg	0/29	424.mg	0/30			
a	1010m	663.mg	n.s.s.	1/30	84.8mg	0/29	424.mg	0/30			
b	1010m	1.33gm	n.s.s.	7/30	84.8mg	7/29	424.mg	6/30			
c	1010m	2.25gm	n.s.s.	1/30	84.8mg	0/29	424.mg	1/30			
d	1010m	3.19gm	n.s.s.	5/30	84.8mg	3/29	424.mg	1/30			
e	1010m	3.49gm	n.s.s.	1/30	84.8mg	2/29	424.mg	0/30			
f	1010m	1.12gm	n.s.s.	11/30	84.8mg	10/29	424.mg	9/30			
g	1010m	2.37gm	n.s.s.	8/30	84.8mg	6/29	424.mg	3/30			
4944	BT305	4.57gm	n.s.s.	15/88	106.mg	12/89	318.mg	12/88	636.mg	16/88	Maltoni;aric,5,1-393;1986
a	BT305	19.8gm	n.s.s.	0/84	106.mg	0/89	318.mg	0/86	636.mg	1/86	
b	BT305	1.70gm	n.s.s.	54/90	106.mg	43/89	318.mg	46/90	636.mg	54/89	
c	BT305	2.32gm	n.s.s.	36/90	106.mg	27/89	318.mg	30/90	636.mg	38/89	
4945	BT305	1.98gm	22.7gm	4/66	88.4mg	2/53	265.mg	8/59	530.mg	13/61	
a	BT305	3.02gm	n.s.s.	10/78	88.4mg	5/73	265.mg	12/78	530.mg	12/74	
b	BT305	2.45gm	n.s.s.	10/88	88.4mg	13/89	265.mg	15/89	530.mg	19/90	
c	BT305	2.23gm	n.s.s.	23/88	88.4mg	17/89	265.mg	27/89	530.mg	26/90	
4946	TR2	302.mg	n.s.s.	7/20	242.mg	12/50	484.mg	12/50			
a	TR2	n.s.s.	n.s.s.	0/20	242.mg	0/50	484.mg	0/50			

```
         Spe Strain  Site    Xpo+Xpt                                              TD50    2Tailpvl
             Sex  Route   Hist     Notes                                           DR      AuOp
   b     R f osm gav kid tla 18m26 dsv                                           no dre   P=1.
   c     R f osm gav liv MXB 18m26 dsv                                           no dre   P=1.
4947 R m osm gav kid uac 18m26 dsv                                        :>     7.22gm * P<.9
   a     R m osm gav TBA MXB 18m26 dsv                                           16.0gm * P<1.
   b     R m osm gav liv MXB 18m26 dsv                                           no dre   P=1.
4948 R f cdr inh mgl fba 24m25 e                                     .>         982.mg * P<.2
   a     R f cdr inh kid uac 24m25 e                                             10.7gm * P<.2
   b     R f cdr inh liv mix 24m25 e                                             12.3gm * P<.6
   c     R f cdr inh tba mix 24m25 e                                             no dre   P=1.
4949 R f sda gav mgl mix 12m33 e                                 .>             no dre   P=1.
   a     R f sda gav kid mix 12m33 e                                             no dre   P=1.
   b     R f sda gav --- leu 12m33 e                                             no dre   P=1.
   c     R f sda gav liv mix 12m33 e                                             no dre   P=1.
   d     R f sda gav tba mix 12m33 e                                             no dre   P=1.
   e     R f sda gav tba mal 12m33 e                                             no dre   P=1.
4950 R f sda inh liv ang 24m37 e                                     .>         8.01gm * P<.3
   a     R f sda inh kid mix 24m37 e                                             41.3gm * P<.6
   b     R f sda inh kid uac 24m37 e                                             43.2gm * P<.2
   c     R f sda inh --- leu 24m37 e                                             no dre   P=1.
   d     R f sda inh liv hpt 24m37 e                                             no dre   P=1.
   e     R f sda inh tba mix 24m37 e                                             3.89gm * P<.8
   f     R f sda inh tba mal 24m37 e                                             6.23gm Z P<.6
4951 R f sda inh --- leu 24m37 e                                          .  ±  2.85gm * P<.09
   a     R f sda inh liv nnd 24m37 e                                             9.68gm * P<.07
   b     R f sda inh kid tum 24m37 e                                             no dre   P=1.
   c     R f sda inh liv hpt 24m37 e                                             no dre   P=1.
   d     R f sda inh tba mal 24m37 e                                             1.62gm * P<.2
   e     R f sda inh tba mix 24m37 e                                             no dre   P=1.
4952 R m sda gav --- leu 12m33 e                                 .  ±          428.mg * P<.08
   a     R m sda gav kid mix 12m33 e                                             no dre   P=1.
   b     R m sda gav liv mix 12m33 e                                             no dre   P=1.
   c     R m sda gav mgl mix 12m33 e                                             no dre   P=1.
   d     R m sda gav tba mal 12m33 e                                             818.mg * P<.4
   e     R m sda gav tba mix 12m33 e                                             no dre   P=1.
4953 R m sda inh tes ldc 24m37 e                                 .  +  .        557.mg * P<.0005+
   a     R m sda inh kid uac 24m37 e                                             9.49gm * P<.03
   b     R m sda inh --- leu 24m37 e                                             5.93gm * P<.5
   c     R m sda inh kid tla 24m37 e                                             29.2gm * P<.6
   d     R m sda inh liv nnd 24m37 e                                             30.7gm * P<.2
   e     R m sda inh liv hpt 24m37 e                                             no dre   P=1.
   f     R m sda inh liv ang 24m37 e                                             no dre   P=1.
   g     R m sda inh tba mix 24m37 e                                             1.88gm * P<.6
   h     R m sda inh tba mal 24m37 e                                             10.8gm * P<.9
4954 R m sda inh mgl tum 24m37 e                              .  +  .           120.mg Z P<.0005
   a     R m sda inh tes ldc 24m37 e                                             835.mg * P<.02  +
   b     R m sda inh --- leu 24m37 e                                             2.81gm * P<.3
   c     R m sda inh kid uac 24m37 e                                             12.4gm * P<.2
   d     R m sda inh liv hpt 24m37 e                                             no dre   P=1.
   e     R m sda inh tba mix 24m37 e                                             428.mg * P<.2
   f     R m sda inh tba mal 24m37 e                                             1.94gm * P<.4
4955 R f wsh inh kid tla 18m36 e                                         .>     5.50gm * P<.2  -
   a     R f wsh inh liv hpa 18m36 e                                             no dre   P=1.  -
   b     R f wsh inh tba ben 18m36 e                                             2.12gm * P<.4  -
   c     R f wsh inh tba mal 18m36 e                                             no dre   P=1.  -
4956 R m wsh inh kid uac 18m36 e                                         .>     3.85gm * P<.2  -
   a     R m wsh inh kid tla 18m36 e                                             10.7gm * P<.9  -
   b     R m wsh inh liv hpa 18m36 e                                             no dre   P=1.  -
   c     R m wsh inh tba mal 18m36 e                                             351.mg * P<.03 -
   d     R m wsh inh tba ben 18m36 e                                             no dre   P=1.  -


TRICHLOROETHYLENE (WITHOUT EPICHLOROHYDRIN).1ug...:..10...:..100...:..1mg..:..10...:..100...:..1g.....:..10
4957 M f b6c gav liv MXA 24m24                                      :  +  :     411.mg   P<.0005c
   a     M f b6c gav liv hpa 24m24                                              579.mg   P<.0005c
   b     M f b6c gav liv hpc 24m24                                              673.mg   P<.0005c
   c     M f b6c gav lun a/a 24m24                                              2.78gm   P<.009
   d     M f b6c gav MXA MXA 24m24                                              1.35gm   P<.03
   e     M f b6c gav MXA MXA 24m24                                              1.47gm   P<.04
   f     M f b6c gav mul mlp 24m24                                              4.88gm   P<.04
   g     M f b6c gav TBA MXB 24m24                                              359.mg   P<.0005
   h     M f b6c gav liv MXB 24m24                                              411.mg   P<.0005
   i     M f b6c gav lun MXB 24m24                                              3.85gm   P<.1
4958 M m b6c gav liv MXA 24m24                                      :  +  :     239.mg   P<.0005
   a     M m b6c gav liv hpc 24m24                                              294.mg   P<.0005c
   b     M m b6c gav liv hpa 24m24                                              855.mg   P<.006
   c     M m b6c gav hag MXA 24m24                                              3.83gm   P<.02
   d     M m b6c gav TBA MXB 24m24                                              332.mg   P<.002
   e     M m b6c gav liv MXB 24m24                                              239.mg   P<.0005
   f     M m b6c gav lun MXB 24m24                                              4.25gm   P<.5
4959 R f f34 gav TBA MXB 24m24                                            :>    5.71gm * P<.9  -
   a     R f f34 gav kid uac 24m24                                              23.0gm * P<.3
   b     R f f34 gav kid tla 24m24                                              no dre   P=1.
```

	RefNum	LoConf	UpConf	Cntrl	1Dose	1Inc	2Dose	2Inc	Citation or Pathology	Brkly Code
b	TR2	n.s.s.	n.s.s.	0/20	242.mg	0/50	484.mg	0/50		
c	TR2	n.s.s.	n.s.s.	0/20	242.mg	0/50	484.mg	0/50	liv:hpa,hpc,nnd.	
4947	TR2	1.18gm	n.s.s.	0/20	242.mg	1/50	484.mg	0/50		
a	TR2	305.mg	n.s.s.	5/20	242.mg	7/50	484.mg	5/50		
b	TR2	n.s.s.	n.s.s.	0/20	242.mg	0/50	484.mg	0/50	liv:hpa,hpc,nnd.	
4948	1626	320.mg	n.s.s.	6/50	22.8mg	12/50	68.4mg	10/47	205.mg	14/51	Fukuda;indh,21,243-254;1983	
a	1626	1.74gm	n.s.s.	0/50	22.8mg	0/50	68.4mg	0/47	205.mg	1/51		
b	1626	1.40gm	n.s.s.	0/50	22.8mg	1/50	68.4mg	0/47	205.mg	1/51		
c	1626	288.mg	n.s.s.	32/50	22.8mg	39/50	68.4mg	31/47	205.mg	32/51		
4949	BT301	98.3mg	n.s.s.	16/30	11.9mg	22/30	59.7mg	16/29	Maltoni;aric,5,1-393;1986	
a	BT301	106.mg	n.s.s.	0/30	11.9mg	0/29	59.7mg	0/26		
b	BT301	109.mg	n.s.s.	1/30	11.9mg	0/30	59.7mg	0/26		
c	BT301	111.mg	n.s.s.	0/30	11.9mg	0/30	59.7mg	0/29		
d	BT301	98.1mg	n.s.s.	18/30	11.9mg	23/30	59.7mg	17/29		
e	BT301	297.mg	n.s.s.	7/30	11.9mg	4/30	59.7mg	4/29		
4950	BT304m	1.30gm	n.s.s.	0/17	30.7mg	0/4	92.1mg	0/26	184.mg	1/14		
a	BT304m	5.46gm	n.s.s.	0/101	30.7mg	1/89	92.1mg	0/88	184.mg	1/88		
b	BT304m	7.03gm	n.s.s.	0/101	30.7mg	0/89	92.1mg	0/88	184.mg	1/88		
c	BT304m	3.56gm	n.s.s.	7/105	30.7mg	6/90	92.1mg	0/90	184.mg	7/90		
d	BT304m	6.53gm	n.s.s.	0/81	30.7mg	1/63	92.1mg	0/63	184.mg	0/70		
e	BT304m	406.mg	n.s.s.	84/105	30.7mg	60/90	92.1mg	72/90	184.mg	69/90		
f	BT304m	1.10gm	n.s.s.	38/105	30.7mg	17/90	92.1mg	21/90	184.mg	32/90		
4951	BT304n	1.01gm	n.s.s.	0/40	30.7mg	3/40	92.1mg	2/40	184.mg	4/40		
a	BT304n	2.38gm	n.s.s.	0/40	30.7mg	0/40	92.1mg	0/40	184.mg	2/40		
b	BT304n	384.mg	n.s.s.	0/38	30.7mg	0/39	92.1mg	0/39	184.mg	0/39		
c	BT304n	269.mg	n.s.s.	0/28	30.7mg	0/26	92.1mg	0/31	184.mg	0/29		
d	BT304n	532.mg	n.s.s.	9/40	30.7mg	12/40	92.1mg	8/40	184.mg	16/40		
e	BT304n	312.mg	n.s.s.	35/40	30.7mg	28/40	92.1mg	26/40	184.mg	33/40		
4952	BT301	162.mg	n.s.s.	0/25	11.9mg	2/28	59.7mg	3/25		
a	BT301	87.1mg	n.s.s.	0/22	11.9mg	0/24	59.7mg	0/21		
b	BT301	108.mg	n.s.s.	0/28	11.9mg	0/29	59.7mg	0/30		
c	BT301	363.mg	n.s.s.	1/28	11.9mg	4/29	59.7mg	1/30		
d	BT301	177.mg	n.s.s.	1/28	11.9mg	4/29	59.7mg	4/30		
e	BT301	191.mg	n.s.s.	7/28	11.9mg	8/29	59.7mg	7/30		
4953	BT304m	336.mg	1.44gm	5/81	21.5mg	11/73	64.5mg	24/71	129.mg	22/76		
a	BT304m	2.87gm	n.s.s.	0/87	21.5mg	0/86	64.5mg	0/80	129.mg	3/85		
b	BT304m	1.17gm	n.s.s.	6/95	21.5mg	10/90	64.5mg	11/90	129.mg	9/89		
c	BT304m	3.65gm	n.s.s.	0/87	21.5mg	1/86	64.5mg	0/80	129.mg	1/85		
d	BT304m	4.99gm	n.s.s.	0/95	21.5mg	0/90	64.5mg	0/90	129.mg	1/89		
e	BT304m	4.36gm	n.s.s.	1/71	21.5mg	0/62	64.5mg	1/63	129.mg	0/62		
f	BT304m	40.5mg	n.s.s.	1/22	21.5mg	0/5	64.5mg	0/9	129.mg	0/9		
g	BT304m	329.mg	n.s.s.	69/95	21.5mg	54/90	64.5mg	55/90	129.mg	65/89		
h	BT304m	789.mg	n.s.s.	31/95	21.5mg	25/90	64.5mg	26/90	129.mg	29/89		
4954	BT304n	56.2mg	348.mg	0/36	21.5mg	9/36	(64.5mg	7/39	129.mg	10/39)		
a	BT304n	375.mg	n.s.s.	1/33	21.5mg	5/32	64.5mg	6/36	129.mg	9/37		
b	BT304n	755.mg	n.s.s.	3/39	21.5mg	3/40	64.5mg	3/40	129.mg	6/40		
c	BT304n	2.01gm	n.s.s.	0/33	21.5mg	0/32	64.5mg	0/36	129.mg	1/37		
d	BT304n	2.59gm	n.s.s.	1/28	21.5mg	1/23	64.5mg	0/27	129.mg	0/32		
e	BT304n	154.mg	n.s.s.	21/39	21.5mg	26/40	64.5mg	27/40	129.mg	29/40		
f	BT304n	466.mg	n.s.s.	8/39	21.5mg	11/40	64.5mg	7/40	129.mg	13/40		
4955	1010m	896.mg	n.s.s.	0/28	20.1mg	0/30	101.mg	1/30	Henschler;artx,43,237-248;1980/pers.comm.	
a	1010m	1.06gm	n.s.s.	0/28	20.1mg	1/30	101.mg	0/30		
b	1010m	448.mg	n.s.s.	2/28	20.1mg	2/30	101.mg	4/30		
c	1010m	277.mg	n.s.s.	16/28	20.1mg	12/30	101.mg	14/30		
4956	1010m	627.mg	n.s.s.	0/29	14.1mg	0/30	70.4mg	1/30		
a	1010m	502.mg	n.s.s.	2/29	14.1mg	1/30	70.4mg	2/30		
b	1010m	738.mg	n.s.s.	0/29	14.1mg	1/30	70.4mg	0/30		
c	1010m	141.mg	n.s.s.	4/29	14.1mg	5/30	70.4mg	11/30		
d	1010m	385.mg	n.s.s.	7/29	14.1mg	2/30	70.4mg	5/30		

TRICHLOROETHYLENE (WITHOUT EPICHLOROHYDRIN) (TCE) 79-01-6

	RefNum	LoConf	UpConf	Cntrl	1Dose	1Inc	2Dose	2Inc	Citation or Pathology	Brkly Code
4957	TR243	215.mg	1.10gm	6/50	704.mg	22/50			liv:hpa,hpc.	
a	TR243	283.mg	1.89gm	4/50	704.mg	16/50				
b	TR243	314.mg	2.22gm	2/50	704.mg	13/50				
c	TR243	877.mg	81.4gm	0/50	704.mg	4/50				S
d	TR243	538.mg	n.s.s.	7/50	704.mg	14/50			liv:mlh; mul:mlh,mlm,mlp,mlu,mno,myo; spl:mno.	S
e	TR243	558.mg	n.s.s.	7/50	704.mg	13/50			liv:mlh; mul:mlh,mlm,mlp,mlu,mno; spl:mno.	S
f	TR243	1.48gm	n.s.s.	0/50	704.mg	3/50				S
g	TR243	187.mg	1.19gm	21/50	704.mg	38/50				
h	TR243	215.mg	1.10gm	6/50	704.mg	22/50			liv:hpa,hpc,nnd.	
i	TR243	1.04gm	n.s.s.	1/50	704.mg	4/50			lun:a/a,a/c.	
4958	TR243	134.mg	539.mg	14/50	701.mg	39/50			liv:hpa,hpc.	S
a	TR243	163.mg	672.mg	8/50	701.mg	31/50				
b	TR243	365.mg	11.6gm	7/50	701.mg	14/50				S
c	TR243	1.29gm	n.s.s.	0/50	701.mg	4/50			hag:adn,ana,ppa.	S
d	TR243	165.mg	1.55gm	33/50	701.mg	45/50				
e	TR243	134.mg	539.mg	14/50	701.mg	39/50			liv:hpa,hpc,nnd.	
f	TR243	744.mg	n.s.s.	7/50	701.mg	6/50			lun:a/a,a/c.	
4959	TR243	486.mg	n.s.s.	38/50	354.mg	31/50	714.mg	30/50		
a	TR243	3.74gm	n.s.s.	0/50	354.mg	0/50	714.mg	1/50		
b	TR243	n.s.s.	n.s.s.	0/50	354.mg	0/50	714.mg	0/50		

	Spe	Strain	Site	Xpo+Xpt						TD50		2Tailpvl	
	Sex	Route	Hist	Notes							DR	AuOp	
c	R f f34	gav	liv	MXB 24m24						9.98gm	*	P<.2	
4960	R f aci	gav	kid	MXA 24m24 s			:	±		3.76gm	*	P<.07	
a	R f aci	gav	kid	MXA 24m24 s						7.80gm	*	P<.2	
b	R f aci	gav	TBA	MXB 24m24 s						61.7gm	*	P<1.	
c	R f aci	gav	liv	MXB 24m24 s						no dre		P=1.	
4961	R f aug	gav	kid	MXA 24m24		:>				2.24gm	\	P<.2	
a	R f aug	gav	kid	uac 24m24						no dre		P=1.	
b	R f aug	gav	TBA	MXB 24m24						no dre		P=1.	
c	R f aug	gav	liv	MXB 24m24						no dre		P=1.	
4962	R f mar	gav	kid	uac 24m24 s			:	±		4.11gm	*	P<.08	
a	R f mar	gav	kid	MXA 24m24 s						4.37gm	*	P<.3	
b	R f mar	gav	TBA	MXB 24m24 s						709.mg	*	P<.3	
c	R f mar	gav	liv	MXB 24m24 s						no dre		P=1.	
4963	R f osm	gav	adr	coa 24m24 s			:	±		#556.mg	*	P<.04	i
a	R f osm	gav	kid	tla 24m24 s						10.7gm	*	P<.2	
b	R f osm	gav	kid	uac 24m24 s						no dre		P=1.	
c	R f osm	gav	TBA	MXB 24m24 s						372.mg	*	P<.07	
d	R f osm	gav	liv	MXB 24m24 s						3.50gm	*	P<.05	
4964	R m f34	gav	kid	MXA 24m24 s			:	+		#2.78gm	*	P<.009	i
a	R m f34	gav	per	MXA 24m24 s						1.49gm	\	P<.04	
b	R m f34	gav	per	msm 24m24 s						1.49gm	\	P<.04	
c	R m f34	gav	kid	uac 24m24 s						3.92gm	*	P<.02	
d	R m f34	gav	TBA	MXB 24m24 s						2.24gm	*	P<.6	
e	R m f34	gav	liv	MXB 24m24 s						12.2gm	*	P<.2	
4965	R m aci	gav	kid	uac 24m24 s		:>				12.0gm	*	P<.4	
a	R m aci	gav	TBA	MXB 24m24 s						820.mg	*	P<.4	
b	R m aci	gav	liv	MXB 24m24 s						9.40gm	*	P<.5	
4966	R m aug	gav	sub	MXA 24m24			:	±		#3.73gm	*	P<.03	i
a	R m aug	gav	kid	MXA 24m24						3.81gm	*	P<.2	
b	R m aug	gav	kid	uac 24m24						10.8gm	*	P<.7	
c	R m aug	gav	TBA	MXB 24m24						no dre		P=1.	
d	R m aug	gav	liv	MXB 24m24						6.44gm	*	P<.2	
4967	R m mar	gav	tes	MXA 24m24 s		:	+	:		#153.mg	*	P<.0005	i
a	R m mar	gav	kid	MXA 24m24 s						5.71gm	*	P<.1	
b	R m mar	gav	tes	itm 24m24 s						6.54gm	*	P<.2	
c	R m mar	gav	kid	uac 24m24 s						8.00gm	*	P<.2	
d	R m mar	gav	TBA	MXB 24m24 s						199.mg	*	P<.002	
e	R m mar	gav	liv	MXB 24m24 s						no dre		P=1.	
4968	R m osm	gav	kid	tla 24m24			:	+	:	#628.mg	\	P<.003	i
a	R m osm	gav	kid	MXA 24m24						628.mg	\	P<.003	
b	R m osm	gav	TBA	MXB 24m24						1.49gm	*	P<.6	
c	R m osm	gav	liv	MXB 24m24						8.95gm	*	P<.5	

TRICHLOROFLUOROMETHANE   100ng..:..1ug...:..10....:..100....:..1mg...:..10....:..100....:..1g....:..10

	Spe	Strain	Site	Xpo+Xpt						TD50		2Tailpvl	
4969	M f b6c	gav	TBA	MXB 78w91 sv		:>				11.5gm	*	P<.6	-
a	M f b6c	gav	liv	MXB 78w91 sv						no dre		P=1.	
b	M f b6c	gav	lun	MXB 78w91 sv						111.gm	*	P<.8	
4970	M m b6c	gav	TBA	MXB 78w91 sv		:>				5.81gm	*	P<.6	-
a	M m b6c	gav	liv	MXB 78w91 sv						15.0gm	*	P<.8	
b	M m b6c	gav	lun	MXB 78w91 sv						no dre		P=1.	
4971	M f swi	inh	mam	car 18m24						31.6gm	*	P<.01	-
a	M f swi	inh	---	leu 18m24						23.7gm	*	P<.09	
b	M f swi	inh	lun	ade 18m24						34.9gm	*	P<.06	
c	M f swi	inh	tba	mix 18m24						11.1gm	*	P<.02	
d	M f swi	inh	tba	mal 18m24						14.6gm	*	P<.02	
4972	M m swi	inh	lun	ade 18m24						.no dre		P=1.	-
a	M m swi	inh	tba	mal 18m24						7.54gm		P<.2	
b	M m swi	inh	tba	mix 18m24						13.4gm		P<.6	
4973	M m swi	inh	lun	ade 18m24						no dre		P=1.	-
a	M m swi	inh	tba	mix 18m24						no dre		P=1.	
b	M m swi	inh	tba	mal 18m24						no dre		P=1.	
4974	R f osm	gav	TBA	MXB 18m26 sv		:		±		326.mg	*	P<.09	i
a	R f osm	gav	liv	MXB 18m26 sv						no dre		P=1.	
4975	R m osm	gav	TBA	MXB 18m26 sv		:>				no dre		P=1.	i
a	R m osm	gav	liv	MXB 18m26 sv						no dre		P=1.	
4976	R f sda	inh	liv	ang 24m24						no dre		P=1.	-
a	R f sda	inh	tba	mix 24m24						no dre		P=1.	-
b	R f sda	inh	tba	mal 24m24						no dre		P=1.	-
4977	R m sda	inh	liv	ang 24m24				.>		no dre		P=1.	-
a	R m sda	inh	tba	mix 24m24						631.mg	\	P<.06	-
b	R m sda	inh	tba	mal 24m24						no dre		P=1.	

N-(TRICHLOROMETHYLTHIO)PHTHALIMIDE   ..:..1ug...:..10....:..100....:..1mg...:..10....:..100...:..1g......:..10

	Spe	Strain	Site	Xpo+Xpt						TD50		2Tailpvl	
4978	M f b6a	orl	liv	hpt 76w76 evx			.>			no dre		P=1.	-
a	M f b6a	orl	lun	ade 76w76 evx						no dre		P=1.	
b	M f b6a	orl	tba	mix 76w76 evx						no dre		P=1.	
4979	M m b6a	orl	liv	hpt 76w76 evx		.>				8.23gm		P<1.	-
a	M m b6a	orl	lun	ade 76w76 evx						no dre		P=1.	
b	M m b6a	orl	tba	mix 76w76 evx						no dre		P=1.	
4980	M f b6c	orl	liv	hpt 76w76 evx			.>			no dre		P=1.	-
a	M f b6c	orl	lun	mix 76w76 evx						no dre		P=1.	-

	RefNum	LoConf	UpConf	Cntrl	1Dose	1Inc	2Dose	2Inc	Citation or Pathology	Brkly Code
c	TR243	2.46gm	n.s.s.	0/50	354.mg	1/50	714.mg	1/50	liv:hpa,hpc,nnd.	
4960	TR273	1.29gm	n.s.s.	0/50	349.mg	3/50	707.mg	1/50	kid:acn,tla,uac.	
a	TR273	1.90gm	n.s.s.	0/50	349.mg	1/50	707.mg	1/50	kid:acn,uac.	
b	TR273	461.mg	n.s.s.	33/50	349.mg	21/50	707.mg	18/50		
c	TR273	4.26gm	n.s.s.	2/50	349.mg	0/50	707.mg	0/50	liv:hpa,hpc,nnd.	
4961	TR273	604.mg	n.s.s.	1/50	354.mg	4/50	(707.mg	0/50)	kid:tla,uac.	
a	TR273	2.32gm	n.s.s.	0/50	354.mg	2/50	707.mg	0/50		
b	TR273	660.mg	n.s.s.	43/50	354.mg	35/50	707.mg	29/50		
c	TR273	5.95gm	n.s.s.	2/50	354.mg	0/50	707.mg	0/50	liv:hpa,hpc,nnd.	
4962	TR273	1.01gm	n.s.s.	0/50	354.mg	1/50	707.mg	1/50		
a	TR273	934.mg	n.s.s.	1/50	354.mg	2/50	707.mg	1/50	kid:tla,uac.	
b	TR273	196.mg	n.s.s.	47/50	354.mg	32/50	707.mg	22/50		
c	TR273	n.s.s.	n.s.s.	0/50	354.mg	0/50	707.mg	0/50	liv:hpa,hpc,nnd.	
4963	TR273	217.mg	n.s.s.	16/50	354.mg	13/50	711.mg	19/50		S
a	TR273	1.74gm	n.s.s.	0/50	354.mg	0/50	711.mg	1/50		
b	TR273	n.s.s.	n.s.s.	0/50	354.mg	0/50	711.mg	0/50		
c	TR273	143.mg	n.s.s.	40/50	354.mg	36/50	711.mg	37/50		
d	TR273	839.mg	n.s.s.	0/50	354.mg	0/50	711.mg	2/50	liv:hpa,hpc,nnd.	
4964	TR243	1.03gm	116.gm	0/50	357.mg	2/50	714.mg	3/50	kid:tla,uac.	S
a	TR243	497.mg	n.s.s.	1/50	357.mg	5/50	(714.mg	1/50)	per:men,msm.	S
b	TR243	497.mg	n.s.s.	1/50	357.mg	5/50	(714.mg	0/50)		S
c	TR243	1.19gm	n.s.s.	0/50	357.mg	0/50	714.mg	3/50		S
d	TR243	392.mg	n.s.s.	33/50	357.mg	26/50	714.mg	18/50		
e	TR243	1.99gm	n.s.s.	0/50	357.mg	0/50	714.mg	1/50	liv:hpa,hpc,nnd.	
4965	TR273	1.95gm	n.s.s.	0/50	354.mg	1/50	707.mg	0/50		
a	TR273	198.mg	n.s.s.	44/50	354.mg	27/50	707.mg	17/50		
b	TR273	1.47gm	n.s.s.	1/50	354.mg	1/50	707.mg	1/50	liv:hpa,hpc,nnd.	
4966	TR273	1.24gm	n.s.s.	0/50	354.mg	1/50	707.mg	3/50	sub:spm,srn.	S
a	TR273	1.14gm	n.s.s.	0/50	354.mg	2/50	707.mg	1/50	kid:tla,uac.	
b	TR273	1.76gm	n.s.s.	0/50	354.mg	1/50	707.mg	0/50		
c	TR273	347.mg	n.s.s.	45/50	354.mg	32/50	707.mg	27/50		
d	TR273	1.55gm	n.s.s.	0/50	354.mg	1/50	707.mg	1/50	liv:hpa,hpc,nnd.	
4967	TR273	90.6mg	316.mg	17/50	354.mg	21/50	707.mg	32/50	tes:ict,itm.	S
a	TR273	1.26gm	n.s.s.	0/50	354.mg	1/50	707.mg	1/50	kid:tla,uac.	
b	TR273	1.07gm	n.s.s.	0/50	354.mg	0/50	707.mg	1/50		
c	TR273	1.30gm	n.s.s.	0/50	354.mg	0/50	707.mg	1/50		
d	TR273	101.mg	906.mg	38/50	354.mg	23/50	707.mg	32/50		
e	TR273	n.s.s.	n.s.s.	0/50	354.mg	0/50	707.mg	0/50	liv:hpa,hpc,nnd.	
4968	TR273	253.mg	3.23gm	0/50	354.mg	6/50	(707.mg	1/50)		S
a	TR273	253.mg	3.23gm	0/50	354.mg	6/50	(707.mg	2/50)	kid:tla,uac.	S
b	TR273	257.mg	n.s.s.	37/50	354.mg	35/50	707.mg	29/50		
c	TR273	1.46gm	n.s.s.	1/50	354.mg	0/50	707.mg	2/50	liv:hpa,hpc,nnd.	

TRICHLOROFLUOROMETHANE (fluorocarbon 11) 75-69-4

	RefNum	LoConf	UpConf	Cntrl	1Dose	1Inc	2Dose	2Inc	Citation or Pathology	Brkly Code
4969	TR106	2.49gm	n.s.s.	3/20	1.20gm	11/50	2.40gm	9/50		
a	TR106	5.59gm	n.s.s.	1/20	1.20gm	4/50	2.40gm	2/50		
b	TR106	9.23gm	n.s.s.	1/20	1.20gm	0/50	2.40gm	2/50	liv:hpa,hpc,nnd.	
4970	TR106	1.14gm	n.s.s.	7/20	1.20gm	22/50	2.40gm	16/49	lun:a/a,a/c.	
a	TR106	1.54gm	n.s.s.	5/20	1.20gm	15/50	2.40gm	10/49		
b	TR106	10.3gm	n.s.s.	2/20	1.20gm	0/50	2.40gm	1/49	liv:hpa,hpc,nnd.	
4971	BT604m	12.6gm	4.71kg	1/90	883.mg	2/60	4.42gm	6/60	lun:a/a,a/c.	
a	BT604m	8.28gm	n.s.s.	8/90	883.mg	10/60	4.42gm	12/60	Maltoni;anya,534,261-282;1988	
b	BT604m	12.1gm	n.s.s.	2/90	883.mg	4/60	4.42gm	6/60		
c	BT604m	4.82gm	n.s.s.	15/90	883.mg	20/60	4.42gm	22/60		
d	BT604m	6.27gm	n.s.s.	9/90	883.mg	12/60	4.42gm	16/60		
4972	BT604m	9.10gm	n.s.s.	3/90	736.mg	0/60				
a	BT604m	2.11gm	n.s.s.	5/90	736.mg	7/60				
b	BT604m	2.23gm	n.s.s.	9/90	736.mg	8/60				
4973	BT604n	32.9gm	n.s.s.	4/90	3.68gm	1/60				
a	BT604n	30.7gm	n.s.s.	9/90	3.68gm	2/60				
b	BT604n	35.1gm	n.s.s.	6/90	3.68gm	1/60				
4974	TR106	91.0mg	n.s.s.	7/20	270.mg	4/50	540.mg	7/50		
a	TR106	n.s.s.	n.s.s.	0/20	270.mg	0/50	540.mg	0/50	liv:hpa,hpc,nnd.	
4975	TR106	117.mg	n.s.s.	5/20	257.mg	4/50	490.mg	1/50		
a	TR106	n.s.s.	n.s.s.	0/20	257.mg	0/50	490.mg	0/50	liv:hpa,hpc,nnd.	
4976	BT603	20.8mg	n.s.s.	0/90	280.mg	2/90	1.40gm	0/90	Maltoni;anya,534,261-282;1988	
a	BT603	1.77gm	n.s.s.	124/150	280.mg	65/90	1.40gm	70/90		
b	BT603	9.88gm	n.s.s.	43/150	280.mg	23/90	1.40gm	15/90		
4977	BT603	3.03gm	n.s.s.	1/150	196.mg	0/90	981.mg	0/90		
a	BT603	252.mg	n.s.s.	51/150	196.mg	42/90	(981.mg	25/90)		
b	BT603	6.05gm	n.s.s.	25/150	196.mg	16/90	981.mg	11/90		

N-(TRICHLOROMETHYLTHIO)PHTHALIMIDE (folpet) 133-07-3

	RefNum	LoConf	UpConf	Cntrl	1Dose	1Inc	2Dose	2Inc	Citation or Pathology	Brkly Code
4978	1239	138.mg	n.s.s.	0/17	83.8mg	0/15			Innes;ntis,1968/1969	
a	1239	138.mg	n.s.s.	1/17	83.8mg	0/15				
b	1239	89.8mg	n.s.s.	2/17	83.8mg	1/15				
4979	1239	86.4mg	n.s.s.	1/18	78.0mg	1/17				
a	1239	95.9mg	n.s.s.	2/18	78.0mg	1/17				
b	1239	75.1mg	n.s.s.	3/18	78.0mg	2/17				
4980	1239	166.mg	n.s.s.	0/16	83.8mg	0/18				
a	1239	166.mg	n.s.s.	0/16	83.8mg	0/18				

```
      Spe Strain Site   Xpo+Xpt                                              TD50    2Tailpvl
          Sex  Route Hist     Notes                                            DR       AuOp

     b    M f b6c orl tba mix 76w76 evx                        .     ±        260.mg   P<.2    -
4981 M f b6c eat duo mix 24m24 ev                            . + .          1.67gm * P<.0005+
   a M f b6c eat duo adc 24m24 ev                                           2.09gm * P<.0005+
   b M f b6c eat --- mly 24m24 ev                                           2.20gm * P<.03
   c M f b6c eat stn sqp 24m24 ev                                           5.78gm * P<.02
   d M f b6c eat stn mix 24m24 ev                                           6.00gm * P<.04
   e M f b6c eat liv mix 24m24 ev                                           18.9gm * P<.05
   f M f b6c eat duo ade 24m24 ev                                           29.2gm * P<.7
   g M f b6c eat lun mix 24m24 ev                                           no dre   P=1.
   h M f b6c eat stn sqc 24m24 ev                                           no dre   P=1.
4982 M m b6c orl lun ade 76w76 evx                        .      ±          156.mg   P<.05   -
   a M m b6c orl liv hpt 76w76 evx                                          242.mg   P<.2    -
   b M m b6c orl tba mix 76w76 evx                                          87.7mg   P<.009  -
4983 M m b6c eat duo mix 24m24 ev                            . + .          899.mg * P<.0005+
   a M m b6c eat duo adc 24m24 ev                                           981.mg * P<.0005+
   b M m b6c eat stn mix 24m24 ev                                           5.46gm * P<.06
   c M m b6c eat stn sqc 24m24 ev                                           13.1gm * P<.09
   d M m b6c eat stn sqp 24m24 ev                                           12.9gm * P<.3
   e M m b6c eat duo ade 24m24 ev                                           23.0gm * P<.3
   f M m b6c eat liv mix 24m24 ev                                           no dre   P=1.
   g M m b6c eat lun mix 24m24 ev                                           no dre   P=1.

2,4,6-TRICHLOROPHENOL  100ng..:..1ug....:..10.....:..100....:..1mg....:..10.....:..100....:..1g.....:..10
4984 M f b6c eat liv MXA 24m24 v                                  : + :    1.41gm * P<.0005c
   a M f b6c eat TBA MXB 24m24 v                                           1.51gm * P<.05
   b M f b6c eat liv MXB 24m24 v                                           1.41gm * P<.0005
   c M f b6c eat lun MXB 24m24 v                                           114.gm * P<1.
4985 M f b6c orl liv hpt 76w76 evx                          .>             113.mg   P<.2
   a M f b6c orl lun ade 76w76 evx                                         113.mg   P<.2
   b M f b6c orl tba mix 76w76 evx                                         27.0mg   P<.002
4986 M m b6c eat liv MXA 24m24                                    : +      856.mg * P<.01  c
   a M m b6c eat TBA MXB 24m24                                             10.7gm * P<.9
   b M m b6c eat liv MXB 24m24                                             856.mg * P<.01
   c M m b6c eat lun MXB 24m24                                             no dre   P=1.
4987 M m b6c orl liv hpt 76w76 evx                       .      ±          68.0mg   P<.05
   a M m b6c orl lun ade 76w76 evx                                         105.mg   P<.2
   b M m b6c orl tba mix 76w76 evx                                         17.9mg   P<.0005
4988 M f b6a orl liv hpt 76w76 evx                          .>             220.mg   P<.3
   a M f b6a orl lun ade 76w76 evx                                         no dre   P=1.
   b M f b6a orl tba mix 76w76 evx                                         no dre   P=1.
4989 M m b6a orl lun ade 76w76 evx                          .>             1.68gm   P<1.
   a M m b6a orl liv hpt 76w76 evx                                         3.58gm   P<1.
   b M m b6a orl tba mix 76w76 evx                                         1.05gm   P<1.
4990 R f f34 eat TBA MXB 25m25 a                                :>         no dre   P=1.   -
   a R f f34 eat liv MXB 25m25 a                                          no dre   P=1.
4991 R m f34 eat --- mle 25m25                                  : +       :405.mg * P<.008 c
   a R m f34 eat --- MXA 25m25                                            445.mg * P<.02  c
   b R m f34 eat --- MXB 25m25                                            445.mg * P<.02
   c R m f34 eat TBA MXB 25m25                                            1.43gm * P<.7
   d R m f34 eat liv MXB 25m25                                            no dre   P=1.

2-(2,4,5-TRICHLOROPHENOXY)PROPIONIC ACID.:..1ug....:..10.....:..100....:..1mg....:..10.....:..100....:..1g.....:..10
4992 M f b6a orl liv hpt 76w76 evx                                .>       no dre   P=1.   -
   a M f b6a orl lun ade 76w76 evx                                         no dre   P=1.   -
   b M f b6a orl tba mix 76w76 evx                                         no dre   P=1.   -
4993 M m b6a orl lun ade 76w76 evx                             .>          no dre   P=1.   -
   a M m b6a orl liv hpt 76w76 evx                                         no dre   P=1.   -
   b M m b6a orl tba mix 76w76 evx                                         no dre   P=1.   -
4994 M f b6c orl liv hpt 76w76 evx                                .>       no dre   P=1.   -
   b M f b6c orl tba mix 76w76 evx                                         52.6mg   P<.2
4995 M m b6c orl liv hpt 76w76 evx                       .     +     .     16.6mg   P<.007 -
   a M m b6c orl lun ade 76w76 evx                                         95.2mg   P<.3
   b M m b6c orl tba mix 76w76 evx                                         9.07mg   P<.0005-

2,4,5-TRICHLOROPHENOXYACETIC ACID 100ng..:..1ug....:..10.....:..100....:..1mg....:..10.....:..100....:..1g.....:..10
4996 M f b6a orl liv hpt 76w76 evx                                .>       no dre   P=1.   -
   a M f b6a orl lun ade 76w76 evx                                         no dre   P=1.   -
   b M f b6a orl tba mix 76w76 evx                                         no dre   P=1.   -
4997 M m b6a orl liv hpt 76w76 evx                             .>          no dre   P=1.   -
   a M m b6a orl lun ade 76w76 evx                                         no dre   P=1.   -
   b M m b6a orl tba mix 76w76 evx                                         no dre   P=1.   -
4998 M f b6c orl lun ade 76w76 evx                             .>          53.4mg   P<.3
   a M f b6c orl liv hpt 76w76 evx                                         no dre   P=1.
   b M f b6c orl tba mix 76w76 evx                                         53.4mg   P<.3
4999 M m b6c orl liv hpt 76w76 evx                          .     ±        11.3mg   P<.02
   a M m b6c orl lun ade 76w76 evx                                         49.7mg   P<.3
   b M m b6c orl tba mix 76w76 evx                                         7.00mg   P<.004
5000 R f mrw eat liv tum 87w87                                    .>       no dre   P=1.   -
   a R f mrw eat tba tum 87w87                                             no dre   P=1.   -
5001 R m mrw eat liv tum 87w87                                 .>          no dre   P=1.   -
```

	RefNum	LoConf	UpConf	Cntrl	1Dose	1Inc	2Dose	2Inc		Citation or Pathology / Brkly Code
b	1239	63.9mg	n.s.s.	0/16	83.8mg	2/18				
4981	2175	1.05gm	3.11gm	1/51	130.mg	2/52	494.mg	10/52	989.mg 19/52	Nyska;gann,81,545-549;1990/pers.comm.
a	2175	1.28gm	3.73gm	0/51	130.mg	1/52	494.mg	5/52	989.mg 18/52	
b	2175	950.mg	n.s.s.	16/51	130.mg	16/52	494.mg	19/52	989.mg 26/52	
c	2175	2.49gm	n.s.s.	2/51	130.mg	1/52	494.mg	5/52	989.mg 7/52	
d	2175	2.45gm	n.s.s.	2/51	130.mg	2/52	494.mg	5/52	989.mg 7/52	
e	2175	5.71gm	n.s.s.	0/51	130.mg	0/52	494.mg	1/52	989.mg 2/52	
f	2175	4.30gm	n.s.s.	1/51	130.mg	1/52	494.mg	5/52	989.mg 1/52	
g	2175	9.76gm	n.s.s.	3/51	130.mg	2/52	494.mg	1/52	989.mg 1/52	
h	2175	12.1gm	n.s.s.	0/51	130.mg	1/52	494.mg	0/52	989.mg 0/52	
4982	1239	47.2mg	n.s.s.	0/16	78.0mg	3/18				Innes;ntis,1968/1969
a	1239	59.5mg	n.s.s.	0/16	78.0mg	2/18				
b	1239	33.1mg	1.92gm	0/16	78.0mg	5/18				
4983	2175	625.mg	1.36gm	0/52	120.mg	4/52	456.mg	17/52	913.mg 25/52	Nyska;gann,81,545-549;1990/pers.comm.
a	2175	675.mg	1.50gm	0/52	120.mg	3/52	456.mg	17/52	913.mg 23/52	
b	2175	2.35gm	n.s.s.	0/52	120.mg	2/52	456.mg	6/52	913.mg 3/52	
c	2175	4.52gm	n.s.s.	0/52	120.mg	0/52	456.mg	3/52	913.mg 1/52	
d	2175	3.36gm	n.s.s.	0/52	120.mg	2/52	456.mg	3/52	913.mg 2/52	
e	2175	5.36gm	n.s.s.	0/52	120.mg	1/52	456.mg	0/52	913.mg 2/52	
f	2175	1.83gm	n.s.s.	20/52	120.mg	13/52	456.mg	11/52	(913.mg 8/52)	
g	2175	2.56gm	n.s.s.	11/52	120.mg	4/52	456.mg	5/52	(913.mg 3/52)	

2,4,6-TRICHLOROPHENOL   (dowicide-2S) 88-06-2

	RefNum	LoConf	UpConf	Cntrl	1Dose	1Inc	2Dose	2Inc		Citation or Pathology / Brkly Code
4984	TR155	874.mg	4.56gm	1/20	678.mg	12/50	1.36gm	24/50		liv:hpa,hpc.
a	TR155	686.mg	n.s.s.	6/20	678.mg	30/50	1.36gm	33/50		
b	TR155	874.mg	4.56gm	1/20	678.mg	12/50	1.36gm	24/50		liv:hpa,hpc,nnd.
c	TR155	4.39gm	n.s.s.	1/20	678.mg	4/50	1.36gm	3/50		lun:a/a,a/c.
4985	292	27.8mg	n.s.s.	0/16	36.4mg	2/18				Innes;ntis,1968/1969
a	292	27.8mg	n.s.s.	0/16	36.4mg	2/18				
b	292	11.5mg	110.mg	0/16	36.4mg	7/18				
4986	TR155	457.mg	66.3gm	4/20	600.mg	32/50	1.20gm	39/50		liv:hpa,hpc.
a	TR155	708.mg	n.s.s.	14/20	600.mg	42/50	1.20gm	42/50		
b	TR155	457.mg	66.3gm	4/20	600.mg	32/50	1.20gm	39/50		liv:hpa,hpc,nnd.
c	TR155	2.89gm	n.s.s.	3/20	600.mg	13/50	1.20gm	7/50		lun:a/a,a/c.
4987	292	20.5mg	n.s.s.	0/16	33.9mg	3/18				Innes;ntis,1968/1969
a	292	25.9mg	n.s.s.	0/16	33.9mg	2/18				
b	292	8.25mg	51.6mg	0/16	33.9mg	9/18				
4988	292	35.8mg	n.s.s.	0/17	36.4mg	1/17				
a	292	68.1mg	n.s.s.	1/17	36.4mg	0/17				
b	292	32.0mg	n.s.s.	2/17	36.4mg	2/17				
4989	292	29.5mg	n.s.s.	2/18	33.9mg	2/17				
a	292	37.6mg	n.s.s.	1/18	33.9mg	1/17				
b	292	24.7mg	n.s.s.	3/18	33.9mg	3/17				
4990	TR155	301.mg	n.s.s.	16/20	250.mg	32/50	(500.mg	27/50)		
a	TR155	n.s.s.	n.s.s.	0/20	250.mg	0/50	500.mg	0/50		liv:hpa,hpc,nnd.
4991	TR155	222.mg	8.89gm	3/20	200.mg	23/50	400.mg	28/50		
a	TR155	227.mg	n.s.s.	4/20	200.mg	25/50	400.mg	29/50		---:leu,lym.
b	TR155	227.mg	n.s.s.	4/20	200.mg	25/50	400.mg	29/50		---:leu,lym,mle. C
c	TR155	229.mg	n.s.s.	16/20	200.mg	40/50	400.mg	41/50		
d	TR155	3.93gm	n.s.s.	1/20	200.mg	1/50	400.mg	0/50		liv:hpa,hpc,nnd.

2-(2,4,5-TRICHLOROPHENOXY)PROPIONIC   ACID   93-72-1

	RefNum	LoConf	UpConf	Cntrl	1Dose	1Inc		Citation or Pathology / Brkly Code
4992	1234	29.8mg	n.s.s.	0/17	16.9mg	0/16		Innes;ntis,1968/1969
a	1234	29.8mg	n.s.s.	1/17	16.9mg	0/16		
b	1234	19.6mg	n.s.s.	2/17	16.9mg	1/16		
4993	1234	20.7mg	n.s.s.	2/18	15.8mg	1/18		
a	1234	31.2mg	n.s.s.	1/18	15.8mg	0/18		
b	1234	22.1mg	n.s.s.	3/18	15.8mg	1/18		
4994	1234	33.5mg	n.s.s.	0/16	16.9mg	0/18		
a	1234	33.5mg	n.s.s.	0/16	16.9mg	0/18		
b	1234	12.9mg	n.s.s.	0/16	16.9mg	2/18		
4995	1234	6.24mg	224.mg	0/16	15.8mg	5/17		
a	1234	15.5mg	n.s.s.	0/16	15.8mg	1/17		
b	1234	4.03mg	28.9mg	0/16	15.8mg	8/17		

2,4,5-TRICHLOROPHENOXYACETIC   ACID   (2,4,5-T) 93-76-5

	RefNum	LoConf	UpConf	Cntrl	1Dose	1Inc		Citation or Pathology / Brkly Code
4996	1233	16.5mg	n.s.s.	0/17	8.34mg	0/18		Innes;ntis,1968/1969
a	1233	16.5mg	n.s.s.	1/17	8.34mg	0/18		
b	1233	7.87mg	n.s.s.	2/17	8.34mg	2/18		
4997	1233	9.17mg	n.s.s.	1/18	7.76mg	1/18		
a	1233	15.4mg	n.s.s.	2/18	7.76mg	0/18		
b	1233	6.10mg	n.s.s.	3/18	7.76mg	3/18		
4998	1233	8.69mg	n.s.s.	0/16	8.34mg	1/18		
a	1233	16.5mg	n.s.s.	0/16	8.34mg	0/18		
b	1233	8.69mg	n.s.s.	0/16	8.34mg	1/18		
4999	1233	3.89mg	n.s.s.	0/16	7.76mg	4/18		
a	1233	8.09mg	n.s.s.	0/16	7.76mg	1/18		
b	1233	2.83mg	42.8mg	0/16	7.76mg	6/18		
5000	2038	43.3mg	n.s.s.	0/10	30.0mg	0/10		Mirvish;jtxe,32,59-74;1991/pers.comm.
a	2038	43.3mg	n.s.s.	0/10	30.0mg	0/10		
5001	2038	17.3mg	n.s.s.	0/5	24.0mg	0/5		

	Spe	Strain	Site	Xpo+Xpt		TD50	2Tailpvl	
	Sex	Route	Hist	Notes			DR	AuOp
a	R m mrw	eat	tba	tum 87w87		no dre	P=1.	-
5002	R f sda	eat	thy	cca 24m24 e	.>	488.mg	Z P<.5	-
a	R f sda	eat	liv	hnd 24m24 e		no dre	P=1.	-
b	R f sda	eat	tba	mix 24m24 e		10.6mg	* P<.2	-
5003	R m sda	eat	liv	hpc 24m24 e	.>	no dre	P=1.	-
a	R m sda	eat	tba	mix 24m24 e		no dre	P=1.	-

1,2,3-TRICHLOROPROPANE   100ng..:..1ug...:..10.....:..:100....:..1mg...:..10.....:..100...:..1g......:..10

	Spe	Strain	Site	Xpo+Xpt		TD50	2Tailpvl	
5004	M f b6c	gav	MXB	MXB 20m24 a	: + :	.826mg	Z P<.0005	
a	M f b6c	gav	for	MXA 20m24 a		.828mg	* P<.0005c	
b	M f b6c	gav	for	sqc 20m24 a		.970mg	* P<.0005c	
c	M f b6c	gav	for	sqp 20m24 a		1.64mg	* P<.0005c	
d	M f b6c	gav	liv	MXA 20m24 a		2.93mg	Z P<.0005c	
e	M f b6c	gav	liv	hpa 20m24 a		3.33mg	Z P<.0005c	
f	M f b6c	gav	ute	MXA 20m24 a		6.62mg	Z P<.0005c	
g	M f b6c	gav	hag	MXA 20m24 a		7.05mg	Z P<.0005c	
h	M f b6c	gav	---	MXA 20m24 a		7.08mg	* P<.002	
i	M f b6c	gav	ute	edc 20m24 a		8.47mg	Z P<.0005c	
j	M f b6c	gav	---	MXA 20m24 a		8.64mg	* P<.006	
k	M f b6c	gav	lun	MXA 20m24 a		14.3mg	Z P<.0005	
l	M f b6c	gav	lun	a/a 20m24 a		14.6mg	P P<.0005	
m	M f b6c	gav	ute	esp 20m24 a		22.9mg	* P<.0005c	
n	M f b6c	gav	liv	hpc 20m24 a		28.7mg	* P<.01	
o	M f b6c	gav	ute	ead 20m24 a		34.9mg	* P<.002 c	
p	M f b6c	gav	MXA	MXA 20m24 a		47.0mg	* P<.0005	
q	M f b6c	gav	MXA	sqc 20m24 a		154.mg	* P<.0005c	
r	M f b6c	gav	TBA	MXB 20m24 a		.806mg	Z P<.0005	
s	M f b6c	gav	liv	MXB 20m24 a		2.93mg	Z P<.0005	
t	M f b6c	gav	lun	MXB 20m24 a		14.3mg	Z P<.0005	
5005	M m b6c	gav	for	MXA 21m24 a	: + :	.928mg	* P<.0005c	
a	M m b6c	gav	MXB	MXB 21m24 a		.996mg	Z P<.0005	
b	M m b6c	gav	for	sqc 21m24 a		1.34mg	* P<.0005c	
c	M m b6c	gav	for	sqp 21m24 a		1.63mg	* P<.0005c	
d	M m b6c	gav	liv	MXA 24m24 a		2.13mg	* P<.0005c	
e	M m b6c	gav	liv	hpa 21m24 a		2.87mg	Z P<.0005c	
f	M m b6c	gav	liv	hpc 21m24 a		6.28mg	* P<.0005	
g	M m b6c	gav	lun	MXA 21m24 a		9.32mg	* P<.0005	
h	M m b6c	gav	lun	a/a 21m24 a		11.2mg	* P<.0005	
i	M m b6c	gav	hag	ade 21m24 a		13.7mg	Z P<.0005c	
j	M m b6c	gav	ton	sqp 21m24 a		454.mg	* P<.02 e	
k	M m b6c	gav	TBA	MXB 21m24 a		1.06mg	* P<.0005	
l	M m b6c	gav	liv	MXB 21m24 a		2.13mg	Z P<.0005	
m	M m b6c	gav	lun	MXB 21m24 a		9.32mg	* P<.0005	
5006	R f f34	gav	MXB	MXB 21m24 a	:+ :	1.07mg	P P<.0005	
a	R f f34	gav	for	MXA 21m24 a		1.64mg	Z P<.0005c	
b	R f f34	gav	for	sqp 21m24 a		1.94mg	Z P<.0005c	
c	R f f34	gav	mgl	MXA 21m24 a		1.98mg	Z P<.0005c	
d	R f f34	gav	MXA	MXA 21m24 a		2.93mg	Z P<.0005c	
e	R f f34	gav	mgl	MXA 21m24 a		3.32mg	* P<.0005	
f	R f f34	gav	mgl	fba 21m24 a		3.42mg	* P<.0005	
g	R f f34	gav	cli	MXA 21m24 a		3.58mg	Z P<.0005c	
h	R f f34	gav	cli	MXA 21m24 a		4.04mg	Z P<.0005c	
i	R f f34	gav	ton	MXA 21m24 a		4.70mg	Z P<.0005	
j	R f f34	gav	MXA	sqc 21m24 a		5.12mg	Z P<.0005c	
k	R f f34	gav	mgl	MXA 21m24 a		5.50mg	Z P<.0005c	
l	R f f34	gav	mgl	adc 21m24 a		5.91mg	Z P<.0005c	
m	R f f34	gav	MXA	sqp 21m24 a		5.98mg	Z P<.0005c	
n	R f f34	gav	for	sqc 21m24 a		7.77mg	* P<.0005c	
o	R f f34	gav	MXA	sqc 21m24 a		8.87mg	Z P<.0005c	
p	R f f34	gav	ton	sqp 21m24 a		9.61mg	Z P<.0005	
q	R f f34	gav	ton	sqc 21m24 a		10.1mg	Z P<.0005	
r	R f f34	gav	pal	sqp 21m24 a		18.5mg	* P<.002	
s	R f f34	gav	thy	MXA 21m24 a		25.2mg	* P<.0005	
t	R f f34	gav	thy	fca 21m24 a		36.9mg	* P<.0005	
u	R f f34	gav	cli	MXA 21m24 a		49.5mg	* P<.0005c	
v	R f f34	gav	zym	car 21m24 a		59.0mg	Z P<.003 c	
w	R f f34	gav	MXA	adc 21m24 a		78.4mg	* P<.004 e	
x	R f f34	gav	---	mnl 21m24 a		7.11mg	* P<.02	
y	R f f34	gav	TBA	MXB 21m24 a		1.02mg	Z P<.0005	
z	R f f34	gav	liv	MXB 21m24 a		81.4mg	* P<.1	
5007	R m f34	gav	MXB	MXB 22m24 a	:+ :	1.10mg	Z P<.0005	
a	R m f34	gav	for	MXA 22m24 a		1.15mg	Z P<.0005c	
b	R m f34	gav	tes	MXA 22m24 a		1.35mg	Z P<.0005	
c	R m f34	gav	for	sqp 22m24 a		1.53mg	Z P<.0005c	
d	R m f34	gav	pan	MXA 22m24 a		1.65mg	Z P<.0005c	
e	R m f34	gav	pan	MXA 22m24 a		1.65mg	Z P<.0005c	
f	R m f34	gav	for	sqc 22m24 a		3.00mg	Z P<.0005c	
g	R m f34	gav	kid	rua 22m24 a		4.44mg	Z P<.0005c	

RefNum	LoConf	UpConf	Cntrl	1Dose	1Inc	2Dose	2Inc		Citation or Pathology Brkly Code
a	2038	17.3mg	n.s.s.	0/5	24.0mg	0/5			
5002	1353	87.9mg	n.s.s.	3/86	3.00mg	10/50	10.0mg	2/50	30.0mg	6/50	Kociba;fctx,17,205-221;1979
a	1353	208.mg	n.s.s.	4/86	3.00mg	2/50	10.0mg	3/50	30.0mg	1/50	
b	1353	2.34mg	n.s.s.	83/86	3.00mg	48/50	10.0mg	48/50	30.0mg	50/50	
5003	1353	279.mg	n.s.s.	3/86	3.00mg	1/50	10.0mg	0/50	30.0mg	1/50	
a	1353	17.5mg	n.s.s.	76/86	3.00mg	44/50	10.0mg	40/50	30.0mg	44/50	

1,2,3-TRICHLOROPROPANE   96-18-4

RefNum	LoConf	UpConf	Cntrl	1Dose	1Inc	2Dose	2Inc		Citation or Pathology Brkly Code
5004	TR384	.547mg	1.29mg	9/50	4.26mg	48/50	14.3mg	50/51	42.9mg	54/55	for:sqc,sqp; hag:ade,anb; liv:hpa,hpc; pal:sqc; phr:sqc; ute:ead,edc,esp. C
a	TR384	.551mg	1.26mg	0/50	4.26mg	48/50	14.3mg	50/51	42.9mg	54/55	for:sqc,sqp.
b	TR384	.649mg	1.46mg	0/50	4.26mg	46/50	14.3mg	49/51	42.9mg	49/55	
c	TR384	.972mg	2.90mg	0/50	4.26mg	23/50	14.3mg	18/51	42.9mg	29/55	
d	TR384	1.60mg	5.95mg	7/50	4.26mg	11/50	14.3mg	8/51	42.9mg	31/55	liv:hpa,hpc.
e	TR384	1.77mg	6.82mg	6/50	4.26mg	9/50	14.3mg	8/51	42.9mg	31/55	
f	TR384	2.81mg	18.0mg	0/50	4.26mg	5/50	14.3mg	3/51	42.9mg	9/55	ute:ead,edc.
g	TR384	3.10mg	18.1mg	2/50	4.26mg	6/50	14.3mg	7/51	42.9mg	10/55	hag:ade,anb.
h	TR384	2.54mg	63.7mg	15/50	4.26mg	6/50	14.3mg	3/51	42.9mg	3/55	---:mlh,mlm,mlp,mlu. S
i	TR384	3.36mg	25.0mg	0/50	4.26mg	4/50	14.3mg	3/51	42.9mg	6/55	
j	TR384	2.80mg	193.mg	17/50	4.26mg	7/50	14.3mg	3/51	42.9mg	3/55	---:hcs,mlh,mlm,mlp,mlu. S
k	TR384	4.56mg	68.9mg	7/50	4.26mg	3/50	14.3mg	0/51	42.9mg	10/55	lun:a/a,a/c. S
l	TR384	4.46mg	71.5mg	4/50	4.26mg	3/50	14.3mg	0/51	42.9mg	10/55	S
m	TR384	5.82mg	148.mg	0/50	4.26mg	2/50	14.3mg	1/51	42.9mg	6/55	
n	TR384	6.25mg	6.07gm	1/50	4.26mg	3/50	14.3mg	0/51	42.9mg	2/55	S
o	TR384	6.33mg	522.mg	0/50	4.26mg	1/50	14.3mg	0/51	42.9mg	3/55	
p	TR384	8.63mg	323.mg	1/50	4.26mg	0/50	14.3mg	2/51	42.9mg	5/55	pal:sqc,sqp; phr:sqc; ton:sqp. S
q	TR384	50.7mg	600.mg	0/50	4.26mg	0/50	14.3mg	1/51	42.9mg	5/55	pal:sqc; phr:sqc.
r	TR384	.537mg	1.27mg	36/50	4.26mg	48/50	14.3mg	50/51	42.9mg	55/55	
s	TR384	1.60mg	5.95mg	7/50	4.26mg	11/50	14.3mg	8/51	42.9mg	31/55	liv:hpa,hpb,hpc.
t	TR384	4.56mg	68.9mg	7/50	4.26mg	3/50	14.3mg	0/51	42.9mg	10/55	lun:a/a,a/c.
5005	TR384	.626mg	1.39mg	3/52	4.24mg	50/52	14.3mg	53/54	42.8mg	55/56	for:sqc,sqp.
a	TR384	.686mg	1.55mg	15/52	4.24mg	50/52	14.3mg	53/54	42.8mg	55/56	for:sqc,sqp; hag:ade; liv:hpa,hpc. C
b	TR384	.910mg	1.98mg	0/52	4.24mg	40/52	14.3mg	50/54	42.8mg	51/56	
c	TR384	1.01mg	2.71mg	3/52	4.24mg	28/52	14.3mg	22/54	42.8mg	33/56	
d	TR384	1.31mg	3.64mg	13/52	4.24mg	24/52	14.3mg	24/54	42.8mg	31/56	liv:hpa,hpc.
e	TR384	1.67mg	5.26mg	11/52	4.24mg	18/52	14.3mg	21/54	42.8mg	29/56	
f	TR384	2.93mg	19.9mg	4/52	4.24mg	11/52	14.3mg	5/54	42.8mg	3/56	S
g	TR384	3.57mg	49.7mg	8/52	4.24mg	11/52	14.3mg	5/54	42.8mg	6/56	lun:a/a,a/c. S
h	TR384	3.85mg	87.3mg	7/52	4.24mg	11/52	14.3mg	3/54	42.8mg	6/56	S
i	TR384	6.13mg	30.8mg	1/52	4.24mg	2/52	14.3mg	10/54	42.8mg	11/56	
j	TR384	103.mg	n.s.s.	0/52	4.24mg	0/52	14.3mg	0/54	42.8mg	2/56	
k	TR384	.703mg	1.67mg	29/52	4.24mg	50/52	14.3mg	54/54	42.8mg	56/56	
l	TR384	1.31mg	3.64mg	13/52	4.24mg	24/52	14.3mg	24/54	42.8mg	31/56	liv:hpa,hpb,hpc.
m	TR384	3.57mg	49.7mg	8/52	4.24mg	11/52	14.3mg	5/54	42.8mg	6/56	lun:a/a,a/c.
5006	TR384	.759mg	1.58mg	23/50	2.12mg	38/49	7.05mg	49/52	21.4mg	47/52	cli:ade,anb,car,cnb; for:sqc,sqp; mgl:adc,ade, fba; pal:sqp; phr:sqc; ton:sqc,sqp. C
a	TR384	1.14mg	2.43mg	0/50	2.12mg	16/49	7.05mg	37/52	21.4mg	19/52	for:sqc,sqp.
b	TR384	1.31mg	2.97mg	0/50	2.12mg	13/49	7.05mg	32/52	21.4mg	17/52	
c	TR384	1.26mg	3.43mg	1/50	2.12mg	26/49	7.05mg	29/52	21.4mg	22/52	mgl:adc,ade,fba.
d	TR384	1.94mg	4.48mg	1/50	2.12mg	6/49	7.05mg	28/52	21.4mg	32/52	pal:sqp; phr:sqc; ton:sqc,sqp.
e	TR384	1.77mg	10.2mg	16/50	2.12mg	23/49	7.05mg	22/52	21.4mg	1/52	mgl:ade,fba. S
f	TR384	1.79mg	11.3mg	15/50	2.12mg	23/49	7.05mg	20/52	21.4mg	1/52	S
g	TR384	2.12mg	6.90mg	5/50	2.12mg	10/49	7.05mg	17/52	21.4mg	15/52	cli:ade,anb,car,cnb.
h	TR384	2.30mg	8.76mg	5/50	2.12mg	10/49	7.05mg	13/52	21.4mg	10/52	cli:ade,anb.
i	TR384	2.84mg	7.94mg	0/50	2.12mg	3/49	7.05mg	18/52	21.4mg	23/52	ton:sqc,sqp. S
j	TR384	3.17mg	8.36mg	0/50	2.12mg	1/49	7.05mg	21/52	21.4mg	21/52	phr:sqc; ton:sqc.
k	TR384	3.21mg	10.0mg	2/50	2.12mg	6/49	7.05mg	14/52	21.4mg	21/52	mgl:adc,ade.
l	TR384	3.38mg	10.9mg	1/50	2.12mg	6/49	7.05mg	12/52	21.4mg	21/52	
m	TR384	3.36mg	11.4mg	1/50	2.12mg	5/49	7.05mg	10/52	21.4mg	18/52	pal:sqp; ton:sqp.
n	TR384	4.00mg	17.1mg	0/50	2.12mg	3/49	7.05mg	9/52	21.4mg	4/52	
o	TR384	4.89mg	16.3mg	0/50	2.12mg	1/49	7.05mg	10/52	21.4mg	15/52	pal:sqc; phr:sqc. S
p	TR384	4.88mg	19.3mg	0/50	2.12mg	3/49	7.05mg	5/52	21.4mg	16/52	S
q	TR384	5.10mg	20.3mg	0/50	2.12mg	0/49	7.05mg	13/52	21.4mg	7/52	S
r	TR384	7.05mg	106.mg	1/50	2.12mg	2/49	7.05mg	5/52	21.4mg	2/52	S
s	TR384	8.60mg	127.mg	0/50	2.12mg	1/49	7.05mg	3/52	21.4mg	2/52	thy:fca,fcc. S
t	TR384	11.2mg	196.mg	0/50	2.12mg	0/49	7.05mg	3/52	21.4mg	2/52	S
u	TR384	22.5mg	131.mg	0/50	2.12mg	0/49	7.05mg	4/52	21.4mg	6/52	cli:car,cnb.
v	TR384	15.8mg	579.mg	0/50	2.12mg	1/49	7.05mg	0/52	21.4mg	3/52	
w	TR384	18.6mg	925.mg	0/50	2.12mg	0/49	7.05mg	1/52	21.4mg	2/52	col:adc; jej:adc.
x	TR384	2.96mg	n.s.s.	13/50	2.12mg	17/49	7.05mg	0/52	21.4mg	0/52	S
y	TR384	.710mg	1.54mg	48/50	2.12mg	46/49	7.05mg	51/52	21.4mg	48/52	
z	TR384	13.3mg	n.s.s.	0/50	2.12mg	0/49	7.05mg	1/52	21.4mg	0/52	liv:hpa,hpb,hpc.
5007	TR384	.803mg	1.57mg	8/50	2.11mg	41/50	7.06mg	45/49	21.4mg	51/52	for:sqc,sqp; kid:rua; pal:sqc,sqp; pan:ade,ana, aoc; pre:ade,car; ton:sqc,sqp; zym:car. C
a	TR384	.860mg	1.58mg	0/50	2.11mg	33/50	7.06mg	42/49	21.4mg	43/52	for:sqc,sqp.
b	TR384	.928mg	2.14mg	47/50	2.11mg	48/50	7.06mg	45/49	21.4mg	44/52	tes:iab,ica. S
c	TR384	1.12mg	2.13mg	0/50	2.11mg	29/50	7.06mg	33/49	21.4mg	38/52	
d	TR384	1.17mg	2.47mg	5/50	2.11mg	21/50	7.06mg	36/49	21.4mg	29/52	pan:ade,ana,aoc.
e	TR384	1.17mg	2.47mg	5/50	2.11mg	21/50	7.06mg	36/49	21.4mg	29/52	pan:ade,ana.
f	TR384	2.00mg	4.70mg	0/50	2.11mg	9/50	7.06mg	27/49	21.4mg	13/52	
g	TR384	2.83mg	7.28mg	0/50	2.11mg	2/50	7.06mg	20/49	21.4mg	21/52	

	Spe	Strain	Site	Xpo+Xpt			TD50	2Tailpvl
	Sex	Route	Hist	Notes			DR	AuOp
h	R m	f34	gav	MXA MXA 22m24 a			4.48mg	Z P<.0005c
i	R m	f34	gav	ton MXA 22m24 a			7.86mg	Z P<.0005
j	R m	f34	gav	thy MXA 22m24 a			8.15mg	Z P<.0005
k	R m	f34	gav	MXA sqp 22m24 a			8.16mg	Z P<.0005c
l	R m	f34	gav	MXA sqc 22m24 a			8.65mg	Z P<.0005c
m	R m	f34	gav	pre MXA 22m24 a			9.59mg	Z P<.0005c
n	R m	f34	gav	ton sqp 22m24 a			10.2mg	Z P<.0005
o	R m	f34	gav	thy cca 22m24 a			10.3mg	Z P<.002
p	R m	f34	gav	pal sqc 22m24 a			11.7mg	Z P<.0005
q	R m	f34	gav	--- mnl 22m24 a			11.9mg	Z P<.005
r	R m	f34	gav	pre ade 22m24 a			14.0mg	Z P<.0005c
s	R m	f34	gav	sub MXA 22m24 a			19.6mg	* P<.008
t	R m	f34	gav	sub fib 22m24 a			20.2mg	* P<.006
u	R m	f34	gav	ski MXA 22m24 a			20.8mg	Z P<.0005
v	R m	f34	gav	liv MXA 22m24 a			23.0mg	Z P<.0005
w	R m	f34	gav	pre car 22m24 a			28.0mg	Z P<.0005c
x	R m	f34	gav	ton sqc 22m24 a			34.7mg	Z P<.0005
y	R m	f34	gav	thy MXA 22m24 a			35.2mg	* P<.008
z	R m	f34	gav	pal sqp 22m24 a			36.6mg	* P<.002
A	R m	f34	gav	ski MXA 22m24 a			39.6mg	* P<.002
B	R m	f34	gav	pan aoc 22m24 a			51.5mg	* P<.005
C	R m	f34	gav	MXA MXA 22m24 a			73.4mg	* P<.0005e
D	R m	f34	gav	amd MXA 22m24 a			11.4mg	* P<.03
E	R m	f34	gav	amd MXA 22m24 a			12.4mg	* P<.03
F	R m	f34	gav	liv hpa 22m24 a			31.7mg	* P<.02
G	R m	f34	gav	ski sqp 22m24 a			51.0mg	Z P<.02
H	R m	f34	gav	MXA adc 22m24 a			102.mg	* P<.02 e
I	R m	f34	gav	MXA adp 22m24 a			264.mg	* P<.02 e
J	R m	f34	gav	zym car 22m24 a			297.mg	* P<.02 c
K	R m	f34	gav	TBA MXB 22m24 a			1.36mg	Z P<.0005
L	R m	f34	gav	liv MXB 22m24 a			23.0mg	* P<.0005

TRICRESYL PHOSPHATE   100ng..:..1ug....:..10.....:..100....:..1mg....:..10.....:..100....:..1g.....:..10

	Spe	Strain	Site	Xpo+Xpt			TD50	2Tailpvl
5008	M f	b6c	eat	TBA MXB 24m24	:>		no dre	P=1. -
a	M f	b6c	eat	liv MXB 24m24			no dre	P=1.
b	M f	b6c	eat	lun MXB 24m24			2.31gm	* P<.9
5009	M m	b6c	eat	hag ade 24m24	: + :		#184.mg	* P<.006 -
a	M m	b6c	eat	TBA MXB 24m24			159.mg	* P<.7
b	M m	b6c	eat	liv MXB 24m24			968.mg	* P<1.
c	M m	b6c	eat	lun MXB 24m24			187.mg	* P<.4
5010	R f	f34	eat	--- mnl 24m24	: ±		#35.0mg	* P<.03 -
a	R f	f34	eat	TBA MXB 24m24			21.6mg	* P<.3
b	R f	f34	eat	liv MXB 24m24			no dre	P=1.
5011	R m	f34	eat	TBA MXB 24m24	:>		no dre	P=1.
a	R m	f34	eat	liv MXB 24m24			143.mg	* P<.4

TRIETHANOLAMINE   100ng..:..1ug....:..10.....:..100....:..1mg....:..10.....:..100....:..1g.....:..10

	Spe	Strain	Site	Xpo+Xpt			TD50	2Tailpvl
5012	M f	b6c	wat	liv hem 82w82 e			42.0gm	* P<.3 -
a	M f	b6c	wat	lun ade 82w82 e			no dre	P=1. -
b	M f	b6c	wat	tba tum 82w82 e			63.5gm	* P<.9 -
5013	M m	b6c	wat	lun adc 82w82 e			.>no dre	P=1. -
a	M m	b6c	wat	liv hpc 82w82 e			no dre	P=1. -
b	M m	b6c	wat	lun ade 82w82 e			no dre	P=1. -
c	M m	b6c	wat	tba tum 82w82 e			74.7gm	* P<1. -
5014	M f	icr	eat	--- mix 25m25 e	. ±		1.29gm	* P<.04
a	M f	icr	eat	liv tum 25m25 e			no dre	P=1.
b	M f	icr	eat	lun ade 25m25 e			no dre	P=1.
c	M f	icr	eat	tba mix 25m25 e			88.8mg	\ P<.002
d	M f	icr	eat	tba mal 25m25 e			100.mg	\ P<.003 +
5015	M m	icr	eat	liv tum 26m27 e	.>		no dre	P=1.
a	M m	icr	eat	lun mix 26m27 e			no dre	P=1.
b	M m	icr	eat	tba mal 26m27 e			no dre	P=1. +
c	M m	icr	eat	tba mix 26m27 e			no dre	P=1.
5016	R f	f3d	wat	liv nnd 24m26 ev			.>47.0gm	* P<.3 -
a	R f	f3d	wat	tba mix 24m26 ev			no dre	P=1. -
5017	R m	f3d	wat	liv nnd 24m26 e	.		±9.59gm	* P<.09 -
a	R m	f3d	wat	liv hpc 24m26 e			53.6gm	* P<.3 -
b	R m	f3d	wat	tba mix 24m26 e			no dre	P=1. -

TRIETHYLENE GLYCOL   100ng..:..1ug....:..10.....:..100....:..1mg....:..10.....:..100....:..1g..:..10

	Spe	Strain	Site	Xpo+Xpt			TD50	2Tailpvl
5018	R m	osm	eat	ubl tum 24m24 r	.>		no dre	P=1.

2,2,2-TRIFLUORO-N-[4-(5-NITRO-2-FURYL)-2-THIAZOLYL]ACETAMIDE.:..1_00...:..1mg....:..10.....:..100....:..1g.....:..10

	Spe	Strain	Site	Xpo+Xpt			TD50	2Tailpvl
5019	M f	swi	eat	mix 46w55 e	. + .		9.98mg	P<.0005+
a	M f	swi	eat	for sqc 46w55 e			38.6mg	P<.0005
b	M f	swi	eat	for sqp 46w55 e			43.5mg	P<.0005
c	M f	swi	eat	liv tum 46w55 e			no dre	P=1.
d	M f	swi	eat	lun tum 46w55 e			no dre	P=1.
e	M f	swi	eat	tba mix 46w55 e			8.01mg	P<.0005
5020	R f	sda	eat	mgl mix 46w66 e	. + .		6.79mg	P<.0005+
a	R f	sda	eat	mgl adc 46w66 e			7.27mg	P<.0005

	RefNum	LoConf	UpConf	Cntrl	1Dose	1Inc	2Dose	2Inc			Citation or Pathology
											Brkly Code
h	TR384	2.96mg	6.96mg	1/50	2.11mg	4/50	7.06mg	18/49	21.4mg	40/52	pal:sqc,sqp; ton:sqc,sqp.
i	TR384	4.84mg	12.7mg	0/50	2.11mg	2/50	7.06mg	8/49	21.4mg	37/52	ton:sqc,sqp. S
j	TR384	3.89mg	36.7mg	4/50	2.11mg	16/50	7.06mg	6/49	21.4mg	5/52	thy:cca,ccr. S
k	TR384	4.75mg	14.7mg	0/50	2.11mg	4/50	7.06mg	9/49	21.4mg	19/52	pal:sqp; ton:sqp.
l	TR384	4.98mg	15.7mg	1/50	2.11mg	0/50	7.06mg	11/49	21.4mg	25/52	pal:sqc; ton:sqc.
m	TR384	5.16mg	21.5mg	5/50	2.11mg	6/50	7.06mg	8/49	21.4mg	16/52	pre:ade,car.
n	TR384	5.67mg	19.1mg	0/50	2.11mg	2/50	7.06mg	8/49	21.4mg	18/52	S
o	TR384	4.49mg	66.6mg	4/50	2.11mg	15/50	7.06mg	4/49	21.4mg	5/52	S
p	TR384	5.94mg	27.6mg	1/50	2.11mg	0/50	7.06mg	11/49	21.4mg	7/52	S
q	TR384	4.98mg	144.mg	16/50	2.11mg	11/50	7.06mg	9/49	21.4mg	6/52	S
r	TR384	6.69mg	39.0mg	5/50	2.11mg	3/50	7.06mg	5/49	21.4mg	11/52	
s	TR384	7.64mg	665.mg	3/50	2.11mg	3/50	7.06mg	7/49	21.4mg	1/52	sub:fib,sar. S
t	TR384	7.87mg	295.mg	2/50	2.11mg	2/50	7.06mg	6/49	21.4mg	1/52	S
u	TR384	8.39mg	99.7mg	3/50	2.11mg	3/50	7.06mg	3/49	21.4mg	6/52	ski:bcc,ker,sqc,sqp,tri. S
v	TR384	9.02mg	116.mg	1/50	2.11mg	1/50	7.06mg	4/49	21.4mg	3/52	liv:hpa,hpc. S
w	TR384	11.5mg	127.mg	0/50	2.11mg	3/50	7.06mg	3/49	21.4mg	5/52	
x	TR384	19.6mg	67.0mg	0/50	2.11mg	0/50	7.06mg	0/49	21.4mg	19/52	S
y	TR384	11.7mg	1.25gm	1/50	2.11mg	1/50	7.06mg	3/49	21.4mg	2/52	thy:fca,fcc. S
z	TR384	12.7mg	309.mg	0/50	2.11mg	2/50	7.06mg	1/49	21.4mg	3/52	S
A	TR384	14.2mg	283.mg	0/50	2.11mg	2/50	7.06mg	1/49	21.4mg	3/52	
B	TR384	14.2mg	771.mg	0/50	2.11mg	0/50	7.06mg	2/49	21.4mg	1/52	ski:sqc,sqp. S
C	TR384	25.5mg	357.mg	0/50	2.11mg	0/50	7.06mg	2/49	21.4mg	3/52	col:adc,adp; jej:adc; rec:adp.
D	TR384	4.64mg	n.s.s.	11/50	2.11mg	8/50	7.06mg	14/49	21.4mg	0/52	amd:pbb,phc,phm,pob. S
E	TR384	5.00mg	n.s.s.	10/50	2.11mg	7/50	7.06mg	13/49	21.4mg	0/52	amd:pbb,pob. S
F	TR384	10.4mg	n.s.s.	1/50	2.11mg	1/50	7.06mg	3/49	21.4mg	1/52	S
G	TR384	15.6mg	n.s.s.	0/50	2.11mg	2/50	7.06mg	0/49	21.4mg	2/52	S
H	TR384	28.8mg	n.s.s.	0/50	2.11mg	0/50	7.06mg	2/49	21.4mg	1/52	col:adc; jej:adc.
I	TR384	60.1mg	n.s.s.	0/50	2.11mg	0/50	7.06mg	0/49	21.4mg	2/52	col:adp; rec:adp.
J	TR384	89.8mg	454.gm	0/50	2.11mg	0/50	7.06mg	0/49	21.4mg	3/52	
K	TR384	.943mg	2.10mg	44/50	2.11mg	48/50	7.06mg	45/49	21.4mg	51/52	
L	TR384	9.02mg	116.mg	1/50	2.11mg	1/50	7.06mg	4/49	21.4mg	3/52	liv:hpa,hpb,hpc.

TRICRESYL PHOSPHATE   1330-78-5

	RefNum	LoConf	UpConf	Cntrl	1Dose	1Inc	2Dose	2Inc	3Dose	3Inc	Citation or Pathology
5008	TR433	34.6mg	n.s.s.	40/50	7.69mg	39/50	16.0mg	32/50	32.1mg	41/51	
a	TR433	70.9mg	n.s.s.	21/50	7.69mg	14/50	16.0mg	14/50	32.1mg	18/51	liv:hpa,hpb,hpc.
b	TR433	101.mg	n.s.s.	5/50	7.69mg	5/50	16.0mg	5/50	32.1mg	6/51	lun:a/a,a/c.
5009	TR433	83.5mg	2.02gm	0/51	7.10mg	1/50	14.8mg	2/50	29.7mg	5/50	S
a	TR433	22.6mg	n.s.s.	37/51	7.10mg	35/50	14.8mg	39/50	29.7mg	39/50	
b	TR433	35.6mg	n.s.s.	28/51	7.10mg	26/50	14.8mg	24/50	29.7mg	28/50	liv:hpa,hpb,hpc.
c	TR433	45.5mg	n.s.s.	8/51	7.10mg	11/50	14.8mg	13/50	29.7mg	12/50	lun:a/a,a/c.
5010	TR433	15.3mg	n.s.s.	8/51	3.74mg	8/53	7.49mg	13/50	15.0mg	15/50	S
a	TR433	6.75mg	n.s.s.	47/51	3.74mg	39/53	7.49mg	43/50	15.0mg	46/50	
b	TR433	n.s.s.	n.s.s.	0/51	3.74mg	0/53	7.49mg	0/50	15.0mg	0/50	liv:hpa,hpb,hpc.
5011	TR433	9.18mg	n.s.s.	44/51	2.99mg	39/50	5.99mg	44/50	12.0mg	38/50	
a	TR433	43.4mg	n.s.s.	0/51	2.99mg	1/50	5.99mg	1/50	12.0mg	1/50	liv:hpa,hpb,hpc.

TRIETHANOLAMINE   102-71-6

	RefNum	LoConf	UpConf	Cntrl	1Dose	1Inc	2Dose	2Inc	3Dose	3Inc	Citation or Pathology
5012	2054	12.7gm	n.s.s.	0/50	2.00gm	2/50	4.00gm	1/50			Konishi;faat,18,25-29;1992
a	2054	23.5gm	n.s.s.	1/50	2.00gm	0/50	4.00gm	1/50			
b	2054	4.86gm	n.s.s.	14/50	2.00gm	13/50	4.00gm	15/50			
5013	2054	6.64gm	n.s.s.	1/43	1.67gm	0/46	3.33gm	0/48			
a	2054	9.33gm	n.s.s.	7/43	1.67gm	5/46	3.33gm	5/48			
b	2054	17.1gm	n.s.s.	2/43	1.67gm	1/46	3.33gm	1/48			
c	2054	2.52gm	n.s.s.	23/43	1.67gm	21/46	3.33gm	26/48			
5014	550	471.mg	n.s.s.	1/36	39.0mg	7/37	390.mg	9/36			Hoshino;canr,38,3918-3921;1978
a	550	291.mg	n.s.s.	0/36	39.0mg	0/37	390.mg	0/36			
b	550	2.52gm	n.s.s.	0/36	39.0mg	1/37	390.mg	0/36			
c	550	42.2mg	371.mg	1/36	39.0mg	11/37	(390.mg	13/36)			
d	550	46.1mg	559.mg	1/36	39.0mg	10/37	(390.mg	13/36)			
5015	550	277.mg	n.s.s.	0/35	36.0mg	0/33	360.mg	0/28			
a	550	2.38gm	n.s.s.	1/35	36.0mg	2/33	360.mg	0/28			
b	550	1.53gm	n.s.s.	1/35	36.0mg	3/33	360.mg	1/28			
c	550	1.70gm	n.s.s.	2/35	36.0mg	4/33	360.mg	1/28			
5016	1793	7.66gm	n.s.s.	0/50	412.mg	0/48	824.mg	1/47			Maekawa;jtxe,19,345-357;1986
a	1793	1.31gm	n.s.s.	44/50	412.mg	36/48	824.mg	34/47			
5017	1793	3.44gm	n.s.s.	1/48	460.mg	2/49	920.mg	5/48			
a	1793	8.73gm	n.s.s.	0/48	460.mg	0/49	920.mg	1/48			
b	1793	173.mg	n.s.s.	48/48	460.mg	49/49	920.mg	47/48			

TRIETHYLENE GLYCOL   112-27-6

	RefNum	LoConf	UpConf	Cntrl	1Dose	1Inc	2Dose	2Inc	3Dose	3Inc	Citation or Pathology
5018	105	565.mg	n.s.s.	0/12	400.mg	0/12	800.mg	0/12	1.60gm	0/12	Fitzhugh;jiht,28,40-43;1946

2,2,2-TRIFLUORO-N-[4-(5-NITRO-2-FURYL)-2-THIAZOLYL]ACETAMIDE   42011-48-3

	RefNum	LoConf	UpConf	Cntrl	1Dose	1Inc	Citation or Pathology
5019	1076	4.97mg	20.2mg	0/29	132.mg	23/25	Cohen;canr,33,1593-1597;1973
a	1076	19.5mg	91.1mg	0/29	132.mg	11/25	
b	1076	21.5mg	107.mg	0/29	132.mg	11/25	
c	1076	190.mg	n.s.s.	0/29	132.mg	0/25	
d	1076	190.mg	n.s.s.	0/29	132.mg	0/25	
e	1076	3.39mg	17.7mg	2/29	132.mg	24/25	
5020	1126	3.55mg	20.5mg	2/24	17.4mg	17/31	Cohen;jnci,54,841-850;1975
a	1126	3.93mg	15.5mg	0/24	17.4mg	15/31	

Spe Strain Site Xpo+Xpt Sex Route Hist Notes		TD50 DR	2Tailpvl AuOp
b R f sda eat liv tum 46w66 e		no dre	P=1.
c R f sda eat tba mix 46w66 e		5.58mg	P<.0005
TRIFLURALIN, TECHNICAL GRADE	100ng..:..1ug...:..10....:..100...:..1mg..:..10....:..100....:..1g.....:..10		
5021 M f b6c eat liv MXA 78w90 dv	: + :	330.mg *	P<.0005c
a M f b6c eat liv hpc 78w90 dv		368.mg *	P<.0005c
b M f b6c eat TBA MXB 78w90 dv		263.mg *	P<.0005
c M f b6c eat liv MXB 78w90 dv		330.mg *	P<.0005
d M f b6c eat lun MXB 78w90 dv		1.36gm *	P<.08
5022 M f b6c eat liv MXA 78w90 dv pool	: + :	330.mg *	P<.0005c
a M f b6c eat liv hpc 78w90 dv		368.mg *	P<.0005c
b M f b6c eat lun MXA 78w90 dv		1.36gm *	P<.003 c
c M f b6c eat lun a/a 78w90 dv		1.52gm *	P<.003 c
d M f b6c eat sto sqc 78w90 dv		3.09gm *	P<.05 c
5023 M f b6c eat liv hpa 24m24 e		.> 189.gm *	P<1. -
a M f b6c eat lun ade 24m24 e		no dre	P=1. -
b M f b6c eat tba ben 24m24 e		55.5gm Z	P<1. -
c M f b6c eat tba mal 24m24 e		no dre	P=1. -
5024 M m b6c eat TBA MXB 78w90 dv	:>	1.50gm *	P<.8 -
a M m b6c eat liv MXB 78w90 dv		1.50gm *	P<.8
b M m b6c eat lun MXB 78w90 dv		5.44gm *	P<.4
5025 M m b6c eat liv hpc 24m24 e	.>	4.77gm Z	P<.4 -
a M m b6c eat lun bro 24m24 e		47.3gm *	P<.5 -
b M m b6c eat liv hpa 24m24 e		no dre	P=1. -
c M m b6c eat lun ade 24m24 e		no dre	P=1. -
d M m b6c eat tba mal 24m24 e		2.16gm Z	P<.3 -
e M m b6c eat tba ben 24m24 e		no dre	P=1. -
5026 R f osm eat thy MXA 18m26 dv	: ±	#663.mg \	P<.03 -
a R f osm eat TBA MXB 18m26 dv		2.64gm \	P<1.
b R f osm eat liv MXB 18m26 dv		no dre	P=1.
5027 R m osm eat TBA MXB 18m26 dv	:>	no dre	P=1. -
a R m osm eat liv MXB 18m26 dv		no dre	P=1.
TRIMETHADIONE	100ng..:..1ug....:..10......:..100....:..1mg......:..100......:..1g......:..10		
5028 R m f34 eat liv hct 71w71 e	.>	no dre	P=1.
2,4,5-TRIMETHYLANILINE	100ng..:..1ug....:..10......:..100....:..1mg....:..10......:..100......:..1g......:..10		
5029 M f b6c eat liv hpc 23m23	:+ :	6.13mg *	P<.0005c
a M f b6c eat TBA MXB 23m23		22.4mg *	P<.4
b M f b6c eat liv MXB 23m23		6.13mg *	P<.0005
c M f b6c eat lun MXB 23m23		47.6mg *	P<.09
5030 M m b6c eat TBA MXB 23m23	:>	52.5mg *	P<.7 -
a M m b6c eat liv MXB 23m23		14.4mg *	P<.08
b M m b6c eat lun MXB 23m23		no dre	P=1.
5031 R f f34 eat MXB MXB 23m23	: + :	20.4mg *	P<.0005
a R f f34 eat liv MXA 23m23		27.2mg *	P<.0005c
b R f f34 eat lun MXA 23m23		88.1mg *	P<.003 c
c R f f34 eat liv hpc 23m23		142.mg *	P<.002 c
d R f f34 eat TBA MXB 23m23		396.mg *	P<.9
e R f f34 eat liv MXB 23m23		27.2mg *	P<.0005
5032 R m f34 eat liv MXA 23m23	: + :	43.8mg *	P<.002 c
a R m f34 eat liv hpc 23m23		71.8mg *	P<.004 c
b R m f34 eat lun MXA 23m23		180.mg *	P<.05
c R m f34 eat liv bdc 23m23		268.mg *	P<.03
d R m f34 eat TBA MXB 23m23		138.mg *	P<.7
e R m f34 eat liv MXB 23m23		43.8mg *	P<.002
2,4,5-TRIMETHYLANILINE.HCl	100ng..:..1ug....:..10......:..100....:..1mg......:..10......:..100..:..1g......:..10		
5033 M f chi eat lun mix 77w98	: + :	52.8mg *	P<.0005+
a M f chi eat liv mix 77w98		60.4mg /	P<.0005
b M f chi eat liv hpt 77w98		62.9mg /	P<.0005+
c M f chi eat tba mix 77w98		47.2mg /	P<.002
5034 M f chi eat liv hpt 77w98 pool	: + :	62.3mg /	P<.0005+
a M f chi eat lun mix 77w98		64.7mg *	P<.0005+
5035 M m chi eat liv mix 77w98	: + :	35.4mg /	P<.0005
a M m chi eat lun mix 77w98		40.0mg *	P<.0005+
b M m chi eat liv hpt 77w98		46.2mg *	P<.0005+
c M m chi eat tba mix 77w98		40.3mg /	P<.0005
5036 M m chi eat liv hpt 77w98 pool	: + :	44.2mg *	P<.0005+
a M m chi eat lun mix 77w98		63.0mg *	P<.0005+
b M m chi eat --- vsc 77w98		260.mg *	P<.02 +
5037 R m cdr eat liv mix 77w98	: ±	47.9mg *	P<.08
a R m cdr eat liv hpt 77w98		98.5mg *	P<.2 +
b R m cdr eat sub mix 77w98		no dre	P=1. +
c R m cdr eat tba mix 77w98		37.2mg *	P<.3
2,4,6-TRIMETHYLANILINE.HCl	100ng..:..1ug....:..10......:..100..:..1mg......:..10......:..100....:..1g......:..10		
5038 M f chi eat liv hpt 77w94 v	: + :	34.6mg /	P<.0005+
a M f chi eat liv mix 77w94 v		34.6mg /	P<.0005+
b M f chi eat lun mix 77w94 v		23.9mg *	P<.04 -
c M f chi eat tba mix 77w94 v		20.6mg *	P<.02

	RefNum	LoConf	UpConf	Cntrl	1Dose	1Inc	2Dose	2Inc			Citation or Pathology
											Brkly Code
b	1126	44.8mg	n.s.s.	0/24	17.4mg	0/31					
c	1126	3.01mg	14.2mg	2/24	17.4mg	19/31					

TRIFLURALIN,   TECHNICAL GRADE   1582-09-8

	RefNum	LoConf	UpConf	Cntrl	1Dose	1Inc	2Dose	2Inc			Citation or Pathology
5021	TR34	220.mg	577.mg	0/20	309.mg	15/50	585.mg	21/50			liv:hpa,hpc.
a	TR34	242.mg	648.mg	0/20	309.mg	12/50	585.mg	21/50			
b	TR34	170.mg	639.mg	2/20	309.mg	24/50	585.mg	26/50			
c	TR34	220.mg	577.mg	0/20	309.mg	15/50	585.mg	21/50			liv:hpa,hpc,nnd.
d	TR34	664.mg	n.s.s.	0/20	309.mg	7/50	585.mg	3/50			lun:a/a,a/c.
5022	TR34	220.mg	526.mg	0/60p	309.mg	15/50	585.mg	21/50			liv:hpa,hpc.
a	TR34	242.mg	600.mg	0/60p	309.mg	12/50	585.mg	21/50			
b	TR34	664.mg	5.71gm	0/60p	309.mg	7/50	585.mg	3/50			lun:a/a,a/c.
c	TR34	718.mg	7.46gm	0/60p	309.mg	6/50	585.mg	3/50			
d	TR34	1.17gm	n.s.s.	0/60p	309.mg	4/50	585.mg	1/50			
5023	2123	5.56gm	n.s.s.	1/120	73.2mg	2/80	292.mg	2/80	585.mg	1/80	Francis;fctx,29,549-555;1991
a	2123	10.2gm	n.s.s.	7/120	73.2mg	4/80	292.mg	3/80	585.mg	0/80	
b	2123	912.mg	n.s.s.	29/120	73.2mg	16/80	292.mg	19/80	(585.mg	4/80)	
c	2123	1.65gm	n.s.s.	39/120	73.2mg	27/80	292.mg	16/80	(585.mg	10/80)	
5024	TR34	170.mg	n.s.s.	9/20	208.mg	20/50	389.mg	17/50			
a	TR34	195.mg	n.s.s.	4/20	208.mg	14/50	389.mg	9/50			liv:hpa,hpc,nnd.
b	TR34	886.mg	n.s.s.	0/20	208.mg	0/50	389.mg	1/50			lun:a/a,a/c.
5025	2123	1.06gm	n.s.s.	6/120	67.6mg	7/80	270.mg	7/80	(540.mg	0/78)	Francis;fctx,29,549-555;1991
a	2123	7.70gm	n.s.s.	0/120	67.6mg	0/80	270.mg	1/80	540.mg	0/78	
b	2123	5.90gm	n.s.s.	6/120	67.6mg	3/80	270.mg	5/80	540.mg	1/78	
c	2123	1.63gm	n.s.s.	16/120	67.6mg	7/80	270.mg	8/80	(540.mg	1/78)	
d	2123	608.mg	n.s.s.	24/120	67.6mg	14/80	270.mg	21/80	(540.mg	6/78)	
e	2123	1.04gm	n.s.s.	32/120	67.6mg	15/80	270.mg	18/80	(540.mg	4/78)	
5026	TR34	255.mg	n.s.s.	1/50	145.mg	7/50	(278.mg	0/50)			thy:fca,fcc. S
a	TR34	118.mg	n.s.s.	36/50	145.mg	39/50	(278.mg	22/50)			
b	TR34	3.39gm	n.s.s.	1/50	145.mg	0/50	278.mg	0/50			liv:hpa,hpc,nnd.
5027	TR34	410.mg	n.s.s.	28/50	116.mg	22/50	225.mg	22/50			
a	TR34	n.s.s.	n.s.s.	0/50	116.mg	0/50	225.mg	0/50			liv:hpa,hpc,nnd.

TRIMETHADIONE   (3,5,5-trimethyl-2,4-oxazolidinedione)   127-48-0

	RefNum	LoConf	UpConf	Cntrl	1Dose	1Inc	2Dose	2Inc			Citation or Pathology
5028	2109	17.9mg	n.s.s.	0/27	12.4mg	0/15					Diwan;artx,66,413-422;1992

2,4,5-TRIMETHYLANILINE   137-17-7

	RefNum	LoConf	UpConf	Cntrl	1Dose	1Inc	2Dose	2Inc			Citation or Pathology
5029	TR160	4.44mg	9.97mg	0/20	6.50mg	18/50	13.0mg	40/50			
a	TR160	5.84mg	n.s.s.	12/20	6.50mg	41/50	13.0mg	45/50			
b	TR160	4.44mg	9.97mg	0/20	6.50mg	18/50	13.0mg	40/50			liv:hpa,hpc,nnd.
c	TR160	23.9mg	n.s.s.	0/20	6.50mg	5/50	13.0mg	6/50			lun:a/a,a/c.
5030	TR160	7.35mg	n.s.s.	12/20	6.00mg	34/50	12.0mg	32/50			
a	TR160	6.14mg	n.s.s.	5/20	6.00mg	26/50	12.0mg	27/50			liv:hpa,hpc,nnd.
b	TR160	13.1mg	n.s.s.	4/20	6.00mg	9/50	(12.0mg	1/50)			lun:a/a,a/c.
5031	TR160	13.8mg	54.5mg	0/20	10.0mg	14/50	40.0mg	33/50			liv:hpc,nnd; lun:a/a,a/c. C
a	TR160	17.3mg	88.2mg	0/20	10.0mg	12/50	40.0mg	27/50			liv:hpc,nnd.
b	TR160	47.4mg	545.mg	0/20	10.0mg	3/50	40.0mg	11/50			lun:a/a,a/c.
c	TR160	67.0mg	388.mg	0/20	10.0mg	0/50	40.0mg	9/50			
d	TR160	23.8mg	n.s.s.	17/20	10.0mg	39/50	40.0mg	45/50			
e	TR160	17.3mg	88.2mg	0/20	10.0mg	12/50	40.0mg	27/50			liv:hpa,hpc,nnd.
5032	TR160	23.9mg	205.mg	1/20	8.00mg	6/50	32.0mg	20/50			liv:hpc,nnd.
a	TR160	38.7mg	497.mg	0/20	8.00mg	3/50	32.0mg	11/50			
b	TR160	68.2mg	n.s.s.	1/20	8.00mg	0/50	32.0mg	7/50			lun:a/a,a/c. S
c	TR160	92.5mg	n.s.s.	0/20	8.00mg	0/50	32.0mg	4/50			S
d	TR160	19.2mg	n.s.s.	17/20	8.00mg	33/50	32.0mg	45/50			
e	TR160	23.9mg	205.mg	1/20	8.00mg	6/50	32.0mg	20/50			liv:hpa,hpc,nnd.

2,4,5-TRIMETHYLANILINE.HCl   21436-97-5

	RefNum	LoConf	UpConf	Cntrl	1Dose	1Inc	2Dose	2Inc			Citation or Pathology
5033	381	24.8mg	182.mg	6/20	106.mg	11/15	246.mg	12/22			Weisburger;jept,2,325-356;1978/pers.comm./Russfield 1973
a	381	31.9mg	123.mg	0/20	106.mg	6/15	246.mg	15/22			
b	381	32.6mg	131.mg	0/20	106.mg	6/15	246.mg	14/22			
c	381	21.2mg	255.mg	17/20	106.mg	15/15	246.mg	19/22			
5034	381	32.1mg	134.mg	1/102p	106.mg	6/15	246.mg	14/22			
a	381	31.3mg	184.mg	31/102p	106.mg	11/15	246.mg	12/22			
5035	381	19.1mg	83.5mg	3/18	98.2mg	11/14	227.mg	19/21			
a	381	19.4mg	136.mg	5/18	98.2mg	11/14	227.mg	10/21			
b	381	24.3mg	115.mg	3/18	98.2mg	9/14	227.mg	18/21			
c	381	19.3mg	140.mg	12/18	98.2mg	13/14	227.mg	19/21			
5036	381	24.4mg	91.2mg	7/99p	98.2mg	9/14	237.mg	18/21			
a	381	28.8mg	216.mg	23/99p	98.2mg	11/14	227.mg	10/21			
b	381	78.1mg	n.s.s.	5/99p	98.2mg	3/14	227.mg	3/21			
5037	381	15.0mg	n.s.s.	1/22	31.3mg	5/17	68.6mg	2/25			
a	381	23.5mg	n.s.s.	1/22	31.3mg	3/17	68.6mg	2/25			
b	381	27.3mg	n.s.s.	4/22	31.3mg	7/17	68.6mg	1/25			
c	381	9.25mg	n.s.s.	14/22	31.3mg	15/17	68.6mg	13/25			

2,4,6-TRIMETHYLANILINE.HCl   6334-11-8

	RefNum	LoConf	UpConf	Cntrl	1Dose	1Inc	2Dose	2Inc			Citation or Pathology
5038	381	15.9mg	107.mg	1/15	37.1mg	1/12	78.0mg	12/16			Weisburger;jept,2,325-356;1978/pers.comm./Russfield 1973
a	381	15.9mg	107.mg	1/15	37.1mg	1/12	78.0mg	12/16			
b	381	7.63mg	n.s.s.	5/15	37.1mg	7/12	78.0mg	7/16			
c	381	7.27mg	n.s.s.	10/15	37.1mg	10/12	78.0mg	16/16			

```
        Spe Strain Site   Xpo+Xpt                                      TD50    2Tailpvl
          Sex  Route  Hist    Notes                                          DR    AuOp

5039 M f chi eat liv hpt 77w94 v  pool                : + :            34.6mg / P<.0005+
5040 M m chi eat liv mix 77w94 v                   : + :              18.2mg / P<.0005
a    M m chi eat liv hpt 77w94 v                                      19.3mg * P<.0005+
b    M m chi eat lun mix 77w94 v                                      38.2mg * P<.05  -
c    M m chi eat tba mix 77w94 v                                      18.4mg * P<.009
5041 M m chi eat liv hpt 77w94 v  pool               : + :            22.7mg * P<.0005+
a    M m chi eat --- vsc 77w94 v                                      120.mg * P<.005 +
5042 R m cdr eat lun mix 73w77 v            : + :                     5.17mg * P<.002 +
a    R m cdr eat liv mix 73w77 v                                      6.15mg * P<.002 +
b    R m cdr eat liv hpt 73w77 v                                      6.15mg * P<.002 +
c    R m cdr eat tba mix 73w77 v                                      2.15mg * P<.0005
5043 R m cdr eat lun mix 73w77 v  pool      : + :                     5.27mg * P<.0005+
a    R m cdr eat liv hpt 73w77 v                                      6.22mg * P<.0005+
b    R m cdr eat sto mix 73w77 v                                      22.0mg * P<.002 +


TRIMETHYLPHOSPHATE     100ng..:..1ug....:..10....:...100.....:..1mg....:..10.....:..100....:..1g.....:..10
5044 M f b6c gav utm acn 24m24                       : + :            335.mg * P<.002 c
a    M f b6c gav TBA MXB 24m24                                        385.mg * P<.3
b    M f b6c gav liv MXB 24m24                                        6.39gm \ P<1.
c    M f b6c gav lun MXB 24m24                                        3.90gm / P<.7
5045 M m b6c gav TBA MXB 24m24                      :>                no dre  P=1.  -
a    M m b6c gav liv MXB 24m24                                        no dre  P=1.
b    M m b6c gav lun MXB 24m24                                        61.3gm * P<1.
5046 R f f34 gav TBA MXB 24m24                    :>                  73.0mg * P<.4  -
a    R f f34 gav liv MXB 24m24                                        no dre  P=1.
5047 R m f34 gav sub fib 24m24                 : + :                  123.mg * P<.005 a
a    R m f34 gav TBA MXB 24m24                                        54.5mg * P<.3
b    R m f34 gav liv MXB 24m24                                        485.mg * P<.5


TRIMETHYLTHIOUREA      100ng..:..1ug....:..10....:...100.....:..1mg....:..10.....:..100....:..1g.....:..10
5048 M f b6c eat TBA MXB 77w91                        :>              1.97gm * P<.8  -
a    M f b6c eat liv MXB 77w91                                        no dre  P=1.
b    M f b6c eat lun MXB 77w91                                        1.26gm * P<.3
5049 M m b6c eat TBA MXB 77w91                      :>                no dre  P=1.  -
a    M m b6c eat liv MXB 77w91                                        no dre  P=1.
b    M m b6c eat lun MXB 77w91                                        478.mg * P<.3
5050 R f f34 eat thy MXA 18m25                    : + :               25.8mg / P<.0005c
a    R f f34 eat thy fcc 18m25                                        45.0mg / P<.0005c
b    R f f34 eat TBA MXB 18m25                                        17.4mg * P<.005
c    R f f34 eat liv MXB 18m25                                        no dre  P=1.
5051 R m f34 eat TBA MXB 18m25                     :>                 no dre  P=1.  -
a    R m f34 eat liv MXB 18m25                                        no dre  P=1.


2,4,6-TRINITRO-1,3-DIMETHYL-5-tert-BUTYLBENZENE...:..1_0.....:..100....:..1mg....:..10.....:..100....:..1g.....:..10
5052 M f b6c eat liv mix 80w90                          . + .         203.mg * P<.0005+
a    M f b6c eat liv hpa 80w90                                        234.mg * P<.0005+
b    M f b6c eat ute ess 80w90                                        533.mg \ P<.02
c    M f b6c eat lun mix 80w90                                        1.30gm * P<.1
d    M f b6c eat lun ade 80w90                                        1.63gm * P<.1
e    M f b6c eat liv hpc 80w90                                        2.19gm * P<.1
f    M f b6c eat --- lym 80w90                                        1.30gm * P<.3
g    M f b6c eat hag mix 80w90                                        2.25gm * P<.5
h    M f b6c eat lun car 80w90                                        6.65gm * P<1.
i    M f b6c eat tba mal 80w90                                        418.mg * P<.06 +
5053 M m b6c eat liv mix 80w90                       . + .            92.1mg * P<.0005+
a    M m b6c eat liv hpc 80w90                                        313.mg * P<.002 +
b    M m b6c eat liv hpa 80w90                                        232.mg * P<.02 +
c    M m b6c eat hag ade 80w90                                        383.mg * P<.02 +
d    M m b6c eat hag mix 80w90                                        425.mg * P<.04 +
e    M m b6c eat lun ade 80w90                                        1.20gm * P<.3
f    M m b6c eat hag car 80w90                                        no dre  P=1.
g    M m b6c eat tba mal 80w90                                        334.mg * P<.07 +


TRINITROGLYCERIN       100ng..:..1ug....:..10....:...100.....:..1mg....:..10.....:..100....:..1g.....:..10
5054 R f cdr eat liv mix 24m24 e                       . + .          329.mg * P<.0005+
5055 R m cdr eat liv mix 24m24 e                       . + .          221.mg * P<.0005+
a    R m cdr eat tes ict 24m24 e                                      405.mg * P<.0005+


TRIPHENYLTIN ACETATE   100ng..:..1ug....:..10....:...100.....:..1mg....:..10.....:..100....:..1g.....:..10
5056 M f b6a orl lun ade 76w76 evx                        .>          964.mg   P<.6  -
a    M f b6a orl liv hpt 76w76 evx                                    no dre   P=1.  -
b    M f b6a orl tba mix 76w76 evx                                    437.mg   P<.5  -
5057 M m b6a orl liv hpt 76w76 evx                      .>            241.mg   P<.2  -
a    M m b6a orl lun ade 76w76 evx                                    no dre   P=1.  -
b    M m b6a orl tba mix 76w76 evx                                    307.mg   P<.4  -
5058 M f b6c orl lun ade 76w76 evx                         ±          440.mg   P<.1  -
a    M f b6c orl liv hpt 76w76 evx                                    909.mg   P<.3  -
b    M f b6c orl tba mix 76w76 evx                                    284.mg   P<.04 -
5059 M m b6c orl lun ade 76w76 evx                         ±          262.mg   P<.04 -
a    M m b6c orl liv hpt 76w76 evx                                    407.mg   P<.1  -
b    M m b6c orl tba mix 76w76 evx                                    117.mg   P<.003 -
```

	RefNum	LoConf	UpConf	Cntrl	1Dose	1Inc	2Dose	2Inc			Citation or Pathology
											Brkly Code
5039	381	15.9mg	73.9mg	1/102p	37.1mg	1/12	78.0mg	12/16			
5040	381	9.03mg	43.4mg	1/14	34.3mg	5/15	75.8mg	11/13			
a	381	9.31mg	51.4mg	0/14	34.3mg	5/15	75.8mg	9/13			
b	381	12.9mg	n.s.s.	3/14	34.3mg	5/15	75.8mg	5/13			
c	381	7.34mg	966.mg	7/14	34.3mg	11/15	75.8mg	12/13			
5041	381	10.5mg	60.0mg	7/99p	34.3mg	5/15	75.8mg	9/13			
a	381	36.5mg	1.84gm	5/99p	34.3mg	1/15	75.8mg	4/13			
5042	381	2.64mg	17.7mg	0/16	5.90mg	5/20	11.8mg	8/21			
a	381	3.07mg	21.1mg	0/16	5.90mg	4/20	11.8mg	8/21			
b	381	3.07mg	21.1mg	0/16	5.90mg	4/20	11.8mg	8/21			
c	381	1.21mg	6.38mg	10/16	5.90mg	13/20	11.8mg	18/21			
5043	381	2.65mg	12.9mg	1/111p	5.90mg	5/20	11.8mg	8/21			
a	381	3.06mg	15.7mg	2/111p	5.90mg	4/20	11.8mg	8/21			
b	381	6.57mg	159.mg	2/111p	5.90mg	0/20	11.8mg	3/21			
TRIMETHYLPHOSPHATE		512-56-1									
5044	TR81	197.mg	995.mg	0/20	107.mg	7/50	214.mg	13/49			
a	TR81	125.mg	n.s.s.	11/20	107.mg	29/50	214.mg	30/49			
b	TR81	258.mg	n.s.s.	2/20	107.mg	4/50	(214.mg	0/49)			liv:hpa,hpc,nnd.
c	TR81	599.mg	n.s.s.	3/20	107.mg	0/50	214.mg	6/49			lun:a/a,a/c.
5045	TR81	227.mg	n.s.s.	11/20	107.mg	26/50	214.mg	26/49			
a	TR81	471.mg	n.s.s.	4/20	107.mg	10/50	214.mg	8/49			liv:hpa,hpc,nnd.
b	TR81	381.mg	n.s.s.	3/20	107.mg	11/50	214.mg	9/49			lun:a/a,a/c.
5046	TR81	20.1mg	n.s.s.	15/20	21.2mg	40/50	42.5mg	42/49			
a	TR81	n.s.s.	n.s.s.	0/20	21.2mg	0/50	42.5mg	0/49			liv:hpa,hpc,nnd.
5047	TR81	60.4mg	850.mg	0/20	21.2mg	2/50	42.5mg	9/49			
a	TR81	16.4mg	n.s.s.	12/20	21.2mg	38/50	42.5mg	38/49			
b	TR81	119.mg	n.s.s.	0/20	21.2mg	1/50	42.5mg	1/49			liv:hpa,hpc,nnd.
TRIMETHYLTHIOUREA		2489-77-2									
5048	TR129	274.mg	n.s.s.	1/20	54.6mg	3/50	110.mg	4/50			
a	TR129	n.s.s.	n.s.s.	0/20	54.6mg	0/50	110.mg	0/50			liv:hpa,hpc,nnd.
b	TR129	381.mg	n.s.s.	0/20	54.6mg	1/50	110.mg	2/50			lun:a/a,a/c.
5049	TR129	174.mg	n.s.s.	8/20	50.4mg	11/50	102.mg	15/50			
a	TR129	244.mg	n.s.s.	6/20	50.4mg	2/50	(102.mg	2/50)			liv:hpa,hpc,nnd.
b	TR129	165.mg	n.s.s.	1/20	50.4mg	5/50	102.mg	8/50			lun:a/a,a/c.
5050	TR129	15.8mg	47.4mg	0/20	9.10mg	1/50	18.0mg	23/50			thy:fca,fcc.
a	TR129	24.5mg	114.mg	0/20	9.10mg	1/50	18.0mg	14/50			
b	TR129	9.35mg	169.mg	7/20	9.10mg	20/50	18.0mg	37/50			
c	TR129	143.mg	n.s.s.	2/20	9.10mg	2/50	18.0mg	0/50			liv:hpa,hpc,nnd.
5051	TR129	33.7mg	n.s.s.	9/20	7.30mg	13/50	14.4mg	16/50			
a	TR129	128.mg	n.s.s.	1/20	7.30mg	0/50	14.4mg	1/50			liv:hpa,hpc,nnd.
2,4,6-TRINITRO-1,3-DIMETHYL-5-tert-BUTYLBENZENE					(musk xylol)	81-15-2					
5052	1989	125.mg	466.mg	1/50	86.7mg	15/50	173.mg	15/50			Maekawa;fctx,28,581-586;1990/pers.comm.
a	1989	140.mg	641.mg	1/50	86.7mg	14/50	173.mg	13/50			
b	1989	184.mg	n.s.s.	0/50	86.7mg	4/50	(173.mg	0/50)			
c	1989	495.mg	n.s.s.	0/50	86.7mg	3/50	173.mg	2/50			
d	1989	565.mg	n.s.s.	0/50	86.7mg	2/50	173.mg	2/50			
e	1989	662.mg	n.s.s.	0/50	86.7mg	1/50	173.mg	2/50			
f	1989	361.mg	n.s.s.	3/50	86.7mg	5/50	173.mg	6/50			
g	1989	462.mg	n.s.s.	3/50	86.7mg	3/50	173.mg	5/50			
h	1989	1.08gm	n.s.s.	0/50	86.7mg	1/50	173.mg	0/50			
i	1989	179.mg	n.s.s.	5/50	86.7mg	14/50	173.mg	12/50			
5053	1989	57.2mg	207.mg	11/50	80.0mg	27/50	160.mg	33/50			
a	1989	168.mg	1.43gm	2/50	80.0mg	8/50	160.mg	13/50			
b	1989	113.mg	n.s.s.	9/50	80.0mg	19/50	160.mg	20/50			
c	1989	189.mg	136.gm	2/50	80.0mg	9/50	160.mg	10/50			
d	1989	192.mg	n.s.s.	3/50	80.0mg	10/50	160.mg	10/50			
e	1989	333.mg	n.s.s.	3/50	80.0mg	5/50	160.mg	6/50			
f	1989	1.31gm	n.s.s.	1/50	80.0mg	1/50	160.mg	0/50			
g	1989	138.mg	n.s.s.	9/50	80.0mg	17/50	160.mg	17/50			
TRINITROGLYCERIN		55-63-0									
5054	1995	184.mg	687.mg	0/29	5.00mg	1/32	50.0mg	3/28	500.mg	16/25	Ellis III;faat,4,248-260;1984
5055	1995	121.mg	473.mg	1/24	4.00mg	0/28	40.0mg	4/26	400.mg	15/21	
a	1995	195.mg	1.24gm	2/24	4.00mg	1/28	40.0mg	3/26	400.mg	11/21	
TRIPHENYLTIN ACETATE		900-95-8									
5056	295	127.mg	n.s.s.	1/17	151.mg	2/18					Innes;ntis,1968/1969
a	295	298.mg	n.s.s.	0/17	151.mg	0/18					
b	295	86.9mg	n.s.s.	2/17	151.mg	4/18					
5057	295	69.2mg	n.s.s.	1/18	139.mg	4/17					
a	295	171.mg	n.s.s.	2/18	139.mg	1/17					
b	295	65.4mg	n.s.s.	3/18	139.mg	5/17					
5058	295	108.mg	n.s.s.	0/16	151.mg	2/17					
a	295	148.mg	n.s.s.	0/16	151.mg	1/17					
b	295	85.6mg	n.s.s.	0/16	151.mg	3/17					
5059	295	79.1mg	n.s.s.	0/16	139.mg	3/17					
a	295	99.8mg	n.s.s.	0/16	139.mg	2/17					
b	295	47.2mg	634.mg	0/16	139.mg	6/17					

```
        Spe Strain Site   Xpo+Xpt                                                    TD50    2Tailpvl
        Sex  Route  Hist   Notes                                                      DR      AuOp
TRIPHENYLTIN HYDROXIDE     100ng..:..1ug....:...10.....:...100....:..1mg....:...10.....:...100....:..1g.....:..10
5060 M f b6c eat TBA MXB 18m24                                      :>          no dre   P=1.  -
a    M f b6c eat liv MXB 18m24                                                  136.mg * P<.7
b    M f b6c eat lun MXB 18m24                                                  90.9mg * P<.3
5061 M m b6c eat TBA MXB 18m24                                          :>      31.1mg * P<.6  -
a    M m b6c eat liv MXB 18m24                                                  112.mg * P<.8
b    M m b6c eat lun MXB 18m24                                                  147.mg * P<.8
5062 R f f34 eat TBA MXB 18m24                                 :>               2.36mg \ P<.3  -
a    R f f34 eat liv MXB 18m24                                                  no dre   P=1.
5063 R m f34 eat TBA MXB 18m24                                     :>           no dre   P=1.  -
a    R m f34 eat liv MXB 18m24                                                  no dre   P=1.

TRIS(2-CHLOROETHYL)PHOSPHATE  100ng..:..1ug....:...10.....:...100....:..1mg....:...10.....:...100....:..1g.....:..10
5064 M f b6c gav ute esp 24m24                                          :  ±    2.42gm * P<.05
a    M f b6c gav mgl adc 24m24                                                  3.05gm * P<.05
b    M f b6c gav hag MXA 24m24                                                  1.39gm * P<.3  e
c    M f b6c gav TBA MXB 24m24                                                  610.mg * P<.3
d    M f b6c gav liv MXB 24m24                                                  2.87gm * P<.6
e    M f b6c gav lun MXB 24m24                                                  4.38gm * P<.6
5065 M m b6c gav kid MXA 24m24                                           :>     6.81gm * P<.6  e
a    M m b6c gav kid ruc 24m24                                                  8.45gm * P<.3  e
b    M m b6c gav kid rua 24m24                                                  23.6kg   P=1.  e
c    M m b6c gav TBA MXB 24m24                                                  807.mg * P<.7
d    M m b6c gav liv MXB 24m24                                                  569.mg * P<.4
e    M m b6c gav lun MXB 24m24                                                  61.7gm * P<1.
5066 M m b6c gav kid rua 24m24   with step                        .>           5.09gm * P<.3  e
5067 R f f34 gav kid rua 24m24                            :  +    :             187.mg * P<.003 c
a    R f f34 gav thy MXA 24m24                                                  203.mg * P<.005 e
b    R f f34 gav mul mnl 24m24                                                  90.2mg * P<.03 e
c    R f f34 gav thy fcc 24m24                                                  280.mg * P<.02
d    R f f34 gav ute ess 24m24                                                  589.mg * P<.03
e    R f f34 gav TBA MXB 24m24                                                  39.9mg * P<.04
f    R f f34 gav liv MXB 24m24                                                  no dre   P=1.
5068 R m f34 gav kid MXA 24m24                          :  +   :                56.4mg / P<.0005c
a    R m f34 gav kid rua 24m24                                                  56.9mg / P<.0005c
b    R m f34 gav amd MXA 24m24                                                  49.5mg \ P<.03
c    R m f34 gav mul mnl 24m24                                                  107.mg * P<.02 e
d    R m f34 gav sub fib 24m24                                                  246.mg * P<.03
e    R m f34 gav thy MXA 24m24                                                  338.mg * P<.06 e
f    R m f34 gav MXA grb 24m24                                                  705.mg * P<.04
g    R m f34 gav kid ruc 24m24                                                  8.40gm * P<.9  c
h    R m f34 gav TBA MXB 24m24                                                  56.8mg * P<.08
i    R m f34 gav liv MXB 24m24                                                  633.mg * P<.08

TRIS-1,2,3-(CHLOROMETHOXY)PROPANE 100ng..:..1ug....:...10.....:...100....:..1mg....:...10.....:...100....:..1g.....:..10
5069 M f hic ipj abd sar 76w76                        .  +      .               3.44mg   P<.008 +

TRIS(2,3-DIBROMOPROPYL)PHOSPHATE  100ng..:..1ug....:...10.....:...100....:..1mg....:...10.....:...100....:..1g.....:..10
5070 M f b6c eat MXB MXB 24m24                                  :  +   :        80.1mg * P<.0005
a    M f b6c eat liv MXA 24m24                                                  95.0mg * P<.0005c
b    M f b6c eat sto MXA 24m24                                                  127.mg * P<.0005c
c    M f b6c eat liv hpc 24m24                                                  221.mg * P<.005 c
d    M f b6c eat lun MXA 24m24                                                  225.mg * P<.002 c
e    M f b6c eat ute esp 24m24                                                  264.mg \ P<.003
f    M f b6c eat sto sqc 24m24                                                  611.mg * P<.02  c
g    M f b6c eat TBA MXB 24m24                                                  148.mg * P<.09
h    M f b6c eat liv MXB 24m24                                                  95.0mg * P<.0005
i    M f b6c eat lun MXB 24m24                                                  225.mg * P<.002
5071 M m b6c eat MXB MXB 24m24                                  :  +   :        103.mg * P<.0005
a    M m b6c eat sto MXA 24m24                                                  197.mg * P<.0005c
b    M m b6c eat kid MXA 24m24                                                  256.mg * P<.0005c
c    M m b6c eat kid uac 24m24                                                  822.mg * P<.009 c
d    M m b6c eat lun MXA 24m24                                                  211.mg * P<.03  c
e    M m b6c eat lun a/c 24m24                                                  449.mg * P<.1   c
f    M m b6c eat TBA MXB 24m24                                                  2.22gm * P<1.
g    M m b6c eat liv MXB 24m24                                                  172.mg \ P<.4
h    M m b6c eat lun MXB 24m24                                                  211.mg * P<.03
5072 R f f34 eat kid tla 24m24                             :  +   :             13.8mg * P<.0005c
a    R f f34 eat ova sct 24m24                                                  66.4mg * P<.05
b    R f f34 eat TBA MXB 24m24                                                  23.0mg * P<.8
c    R f f34 eat liv MXB 24m24                                                  2.82gm * P<1.
5073 R m f34 eat kid tla 24m24                         :  +  :                  1.57mg \ P<.0005c
a    R m f34 eat kid MXA 24m24                                                  1.57mg \ P<.0005c
b    R m f34 eat pre MXA 24m24                                                  20.3mg * P<.03
c    R m f34 eat liv MXA 24m24                                                  34.3mg * P<.02
d    R m f34 eat kid uac 24m24                                                  63.3mg * P<.05  c
e    R m f34 eat TBA MXB 24m24                                                  8.31mg * P<.4
f    R m f34 eat liv MXB 24m24                                                  34.3mg * P<.02
5074 R m f34 gav col pla 52w52 ekr                    .  +      .               13.4mg   P<.006 +
a    R m f34 gav kid adc 52w52 ekr                                              54.8mg   P<.2   +
b    R m f34 gav liv tum 52w52 ekr                                              no dre   P=1.
```

	RefNum	LoConf	UpConf	Cntrl	1Dose	1Inc	2Dose	2Inc	Citation or Pathology	Brkly Code
TRIPHENYLTIN HYDROXIDE		76-87-9								
5060	TR139	12.9mg	n.s.s.	10/20	3.60mg	15/50	7.30mg	18/50		
a	TR139	22.4mg	n.s.s.	1/20	3.60mg	2/50	7.30mg	4/50	liv:hpa,hpc,nnd.	
b	TR139	27.5mg	n.s.s.	0/20	3.60mg	1/50	7.30mg	2/50	lun:a/a,a/c.	
5061	TR139	6.34mg	n.s.s.	10/20	3.40mg	19/50	6.70mg	22/50		
a	TR139	12.9mg	n.s.s.	5/20	3.40mg	6/50	6.70mg	10/50	liv:hpa,hpc,nnd.	
b	TR139	14.3mg	n.s.s.	3/20	3.40mg	5/50	6.70mg	6/50	lun:a/a,a/c.	
5062	TR139	.753mg	n.s.s.	12/20	1.40mg	39/50	(2.80mg	31/50)	liv:hpa,hpc,nnd.	
a	TR139	21.2mg	n.s.s.	0/20	1.40mg	1/50	2.80mg	0/50		
5063	TR139	4.24mg	n.s.s.	9/20	1.10mg	22/50	2.20mg	16/50	liv:hpa,hpc,nnd.	
a	TR139	12.2mg	n.s.s.	0/20	1.10mg	2/50	2.20mg	0/50		
TRIS(2-CHLOROETHYL) PHOSPHATE			115-96-8							
5064	TR391	830.mg	n.s.s.	0/51	124.mg	1/50	249.mg	3/50		S
a	TR391	923.mg	n.s.s.	0/51	124.mg	0/50	249.mg	3/50		S
b	TR391	403.mg	n.s.s.	3/51	124.mg	8/50	249.mg	7/50	hag:ade,car.	
c	TR391	177.mg	n.s.s.	25/51	124.mg	26/50	249.mg	37/50		
d	TR391	543.mg	n.s.s.	5/51	124.mg	5/50	249.mg	8/50	liv:hpa,hpc,nnd.	
e	TR391	724.mg	n.s.s.	3/51	124.mg	2/50	249.mg	5/50	lun:a/a,a/c.	
5065	TR391	934.mg	n.s.s.	1/52	125.mg	0/50	249.mg	2/52	kid:rua,ruc.	
a	TR391	1.38gm	n.s.s.	0/52	125.mg	0/50	249.mg	1/52		
b	TR391	1.18gm	n.s.s.	1/52	125.mg	0/50	249.mg	1/52		
c	TR391	124.mg	n.s.s.	38/52	125.mg	39/50	249.mg	42/52		
d	TR391	147.mg	n.s.s.	26/52	125.mg	27/50	249.mg	33/52	liv:hpa,hpc,nnd.	
e	TR391	322.mg	n.s.s.	10/52	125.mg	12/50	249.mg	10/52	lun:a/a,a/c.	
5066	TR391	1.22gm	n.s.s.	1/52	125.mg	1/50	249.mg	3/52		
5067	TR391	80.8mg	900.mg	0/50	31.3mg	2/50	62.6mg	5/51		
a	TR391	87.3mg	1.50gm	0/50	31.3mg	3/50	62.6mg	4/51	thy:fca,fcc.	
b	TR391	40.3mg	n.s.s.	14/50	31.3mg	16/50	62.6mg	20/51		
c	TR391	106.mg	n.s.s.	0/50	31.3mg	2/50	62.6mg	3/51		S
d	TR391	174.mg	n.s.s.	0/50	31.3mg	0/50	62.6mg	3/51		S
e	TR391	17.2mg	n.s.s.	45/50	31.3mg	46/50	62.6mg	44/51		
f	TR391	n.s.s.	n.s.s.	0/50	31.3mg	0/50	62.6mg	0/51	liv:hpa,hpc,nnd.	
5068	TR391	34.7mg	111.mg	2/51	31.3mg	5/50	62.6mg	25/50	kid:rua,ruc.	
a	TR391	35.5mg	106.mg	1/51	31.3mg	5/50	62.6mg	24/50		
b	TR391	20.8mg	n.s.s.	10/51	31.3mg	21/50	(62.6mg	13/50)	amd:pbb,pob. S	
c	TR391	50.8mg	n.s.s.	5/51	31.3mg	14/50	62.6mg	13/50		S
d	TR391	102.mg	n.s.s.	1/51	31.3mg	3/50	62.6mg	6/50		S
e	TR391	124.mg	n.s.s.	1/51	31.3mg	2/50	62.6mg	5/50	thy:fca,fcc.	
f	TR391	209.mg	n.s.s.	0/51	31.3mg	0/50	62.6mg	3/50	bra:grb; clm:grb. S	
g	TR391	322.mg	n.s.s.	1/51	31.3mg	0/50	62.6mg	1/50		
h	TR391	22.4mg	n.s.s.	39/51	31.3mg	44/50	62.6mg	46/50		
i	TR391	190.mg	n.s.s.	0/51	31.3mg	1/50	62.6mg	2/50	liv:hpa,hpc,nnd.	
TRIS-1,2,3-(CHLOROMETHOXY) PROPANE			38571-73-2							
5069	582	1.30mg	51.1mg	0/30	1.71mg	5/30			Van Duuren;canr,35,2553-2557;1975	
TRIS(2,3-DIBROMOPROPYL)PHOSPHATE		(TRIS) 126-72-7								
5070	TR76	51.3mg	171.mg	16/55	64.3mg	34/50	129.mg	40/50	liv:hpa,hpc; lun:a/a,a/c; sto:sqc,sqp. C	
a	TR76	62.2mg	188.mg	11/55	64.3mg	23/50	129.mg	35/50	liv:hpa,hpc.	
b	TR76	88.1mg	212.mg	2/55	64.3mg	14/50	129.mg	22/50	sto:sqc,sqp.	
c	TR76	128.mg	692.mg	7/55	64.3mg	12/50	129.mg	20/50		
d	TR76	135.mg	576.mg	4/55	64.3mg	9/50	129.mg	17/50	lun:a/a,a/c.	
e	TR76	130.mg	682.mg	0/55	64.3mg	6/50	(129.mg	2/50)		S
f	TR76	328.mg	2.15gm	0/55	64.3mg	4/50	129.mg	4/50		
g	TR76	67.7mg	n.s.s.	34/55	64.3mg	42/50	129.mg	44/50		
h	TR76	55.7mg	268.mg	11/55	64.3mg	23/50	129.mg	35/50	liv:hpa,hpc,nnd.	
i	TR76	119.mg	1.13gm	4/55	64.3mg	9/50	129.mg	17/50	lun:a/a,a/c.	
5071	TR76	58.7mg	336.mg	12/55	59.0mg	23/50	119.mg	37/50	kid:tla,uac; lun:a/a,a/c; sto:sqc,sqp. C	
a	TR76	120.mg	406.mg	0/55	59.0mg	10/50	119.mg	13/50	sto:sqc,sqp.	
b	TR76	147.mg	531.mg	0/55	59.0mg	4/50	119.mg	14/50	kid:tla,uac.	
c	TR76	335.mg	25.7gm	0/55	59.0mg	1/50	119.mg	5/50		
d	TR76	95.1mg	n.s.s.	12/55	59.0mg	18/50	119.mg	25/50	lun:a/a,a/c.	
e	TR76	170.mg	n.s.s.	6/55	59.0mg	8/50	119.mg	13/50		
f	TR76	86.1mg	n.s.s.	43/55	59.0mg	41/50	119.mg	43/50		
g	TR76	43.2mg	n.s.s.	28/55	59.0mg	31/50	(119.mg	23/50)	liv:hpa,hpc,nnd.	
h	TR76	95.1mg	n.s.s.	12/55	59.0mg	18/50	119.mg	25/50	lun:a/a,a/c.	
5072	TR76	7.41mg	35.7mg	0/55	2.50mg	4/55	4.95mg	10/55		
a	TR76	20.1mg	n.s.s.	0/55	2.50mg	0/55	4.95mg	3/55		S
b	TR76	2.88mg	n.s.s.	52/55	2.50mg	47/55	4.95mg	54/55		
c	TR76	29.8mg	n.s.s.	1/55	2.50mg	1/55	4.95mg	1/55	liv:hpa,hpc,nnd.	
5073	TR76	.975mg	2.75mg	0/55	2.00mg	26/55	(4.00mg	26/55)		
a	TR76	.975mg	2.75mg	0/55	2.00mg	26/55	(4.00mg	29/55)	kid:tla,uac.	
b	TR76	8.69mg	n.s.s.	1/55	2.00mg	3/55	4.00mg	7/55	pre:acn,adn,can. S	
c	TR76	13.0mg	n.s.s.	0/55	2.00mg	1/55	4.00mg	4/55	liv:hpc,nnd. S	
d	TR76	18.9mg	n.s.s.	0/55	2.00mg	0/55	4.00mg	3/55		
e	TR76	2.31mg	n.s.s.	39/55	2.00mg	44/55	4.00mg	48/55		
f	TR76	13.0mg	n.s.s.	0/55	2.00mg	1/55	4.00mg	4/55	liv:hpa,hpc,nnd.	
5074	1729	3.72mg	189.mg	0/9	71.4mg	3/5			Reznik;livt,44,74-83;1981	
a	1729	8.84mg	n.s.s.	0/9	71.4mg	1/5				
b	1729	18.4mg	n.s.s.	0/9	71.4mg	0/5				

```
        Spe Strain Site   Xpo+Xpt                                                          TD50      2Tailpvl
            Sex  Route Hist    Notes                                                           DR      AuOp

TRIS(2-ETHYLHEXYL)PHOSPHATE        100ng..:..1ug....:..10......:..100.....:..1mg.....:..10.....:..100....:..1g.....:..10
5075 M f b6c gav liv hpc 24m24                                                      : +    2.56gm * P<.005 p
a    M f b6c gav liv MXA 24m24                                                             2.22gm * P<.05
b    M f b6c gav TBA MXB 24m24                                                             no dre   P=1.
c    M f b6c gav liv MXB 24m24                                                             2.22gm * P<.05
d    M f b6c gav lun MXB 24m24                                                             13.7gm * P<.6
5076 M m b6c gav TBA MXB 24m24                                                    :>       no dre   P=1.    -
a    M m b6c gav liv MXB 24m24                                                             9.00gm * P<.9
b    M m b6c gav lun MXB 24m24                                                             no dre   P=1.
5077 R f f34 gav TBA MXB 24m24                                                    :>       230.gm * P<1.    -
a    R f f34 gav liv MXB 24m24                                                             no dre   P=1.
5078 R m f34 gav adr phe 24m24                                                      : +    5.29gm * P<.002 e
a    R m f34 gav adr MXA 24m24                                                             5.29gm * P<.002 e
b    R m f34 gav MXA MXA 24m24                                                             16.7gm * P<.05
c    R m f34 gav MXA MXA 24m24                                                             28.0gm * P<.05
d    R m f34 gav TBA MXB 24m24                                                             14.2gm * P<.7
e    R m f34 gav liv MXB 24m24                                                             no dre   P=1.

TRIS(2-HYDROXYPROPYL)AMINE         100ng..:..1ug....:..10......:..100.....:..1mg.....:..10.....:..100....:..1g.....:..10
5079 R m wis eat liv tum 24m24 r                                                   .>      no dre   P=1.    -

Trp-P-1 ACETATE                    100ng..:..1ug....:..10......:..100.....:..1mg.....:..10.....:..100....:..1g.....:..10
5080 M f cdf eat liv mix 88w88                                        . + .                25.0mg   P<.0005+
a    M f cdf eat liv hpc 88w88                                                             29.6mg   P<.0005+
b    M f cdf eat lun adc 88w88                                                             no dre   P=1.
c    M f cdf eat lun ade 88w88                                                             no dre   P=1.
5081 M m cdf eat liv mix 88w88                                              . ±            109.mg   P<.08  +
a    M m cdf eat liv hpc 88w88                                                             112.mg   P<.02  +
b    M m cdf eat lun adc 88w88                                                             no dre   P=1.
c    M m cdf eat lun ade 88w88                                                             no dre   P=1.
5082 R f f3d eat liv hpc 52w52 e                               . + .                        .577mg   P<.0005+
5083 R m f3d eat liv hpc 52w52 e                               . + .                        .574mg   P<.0005+

Trp-P-2 ACETATE                    100ng..:..1ug....:..10......:..100.....:..1mg.....:..10.....:..100....:..1g.....:..10
5084 M f cdf eat liv hpc 88w88 e                                      . + .                6.81mg   P<.0005+
a    M f cdf eat lun adc 88w88                                                             no dre   P=1.
b    M f cdf eat lun ade 88w88                                                             no dre   P=1.
5085 M m cdf eat lun adc 88w88                                             . ±             81.1mg   P<.1
a    M m cdf eat liv hpc 88w88 e                                                           88.2mg   P<.04  +
b    M m cdf eat liv mix 88w88 e                                                           83.2mg   P<.2   +
c    M m cdf eat lun ade 88w88                                                             224.mg   P<.3
5086 R f aci eat liv nnd 29m29 e                                   . +   .                 4.43mg   P<.0005+
a    R f aci eat liv mix 29m29 e                                                           4.43mg   P<.0005+
b    R f aci eat jej adc 29m29 e                                                           41.4mg   P<.09  +
c    R f aci eat ilm ade 29m29 e                                                           41.4mg   P<.09  +
d    R f aci eat liv hms 29m29 e                                                           41.4mg   P<.09  +
5087 R m aci eat liv tum 29m29 e                                       .>                  no dre   P=1.
5088 R f f3d eat cli mix 26m26 e                                    . +          .         9.42mg * P<.009 +
a    R f f3d eat mgl mix 26m26 e                                                           9.78mg * P<.01  +
b    R f f3d eat --- mly 26m26 e                                                           29.0mg * P<.005 +
c    R f f3d eat ute esp 26m26 e                                                           3.60mg \ P<.02
d    R f f3d eat mgl mix 26m26 e                                                           11.5mg * P<.02
e    R f f3d eat cli ade 26m26 e                                                           12.8mg * P<.05
f    R f f3d eat cli adc 26m26 e                                                           75.5mg * P<.07
g    R f f3d eat liv hpa 26m26 e                                                           74.3mg * P<.7
h    R f f3d eat mgl ade 26m26 e                                                           76.2mg * P<.4
i    R f f3d eat ubl tpp 26m26 e                                                           no dre   P=1.
5089 R m f3d eat ubl mix 26m26 e                                      . +   .              9.72mg * P<.0005+
a    R m f3d eat ubl tpp 26m26 e                                                           12.3mg * P<.002
b    R m f3d eat mgl mix 26m26 e                                                           8.43mg * P<.02  +
c    R m f3d eat liv hpa 26m26 e                                                           10.7mg * P<.03  +
d    R m f3d eat ubl tcc 26m26 e                                                           59.9mg * P<.07
e    R m f3d eat liv hpc 26m26 e                                                           no dre   P=1.

DL-TRYPTOPHAN                      100ng..:..1ug....:..10......:..100.....:..1mg.....:..10.....:..100....:..1g.....:..10
5090 R m fis eat liv tum 98w98                                                     .>      no dre   P=1.    -
a    R m fis eat ubl tum 98w98                                                             no dre   P=1.    -

L-TRYPTOPHAN                       100ng..:..1ug....:..10......:..100.....:..1mg.....:..10.....:..100....:..1g.....:..10
5091 M f b6c eat TBA MXB 18m24                                                      :>     no dre   P=1.    -
a    M f b6c eat liv MXB 18m24                                                             30.2gm * P<.6
b    M f b6c eat lun MXB 18m24                                                             no dre   P=1.
5092 M m b6c eat TBA MXB 18m24                                                    :>       1.47gm \ P<.2    -
a    M m b6c eat liv MXB 18m24                                                             no dre   P=1.
b    M m b6c eat lun MXB 18m24                                                             no dre   P=1.
5093 R f f34 eat TBA MXB 18m24                                                    :>       no dre   P=1.    -
a    R f f34 eat liv MXB 18m24                                                             no dre   P=1.
5094 R m f34 eat TBA MXB 18m24                                                    :>       no dre   P=1.    -
a    R m f34 eat liv MXB 18m24                                                             no dre   P=1.
5095 R m fis eat ubl mix 23m24                                                     .>      no dre   P=1.    -
```

	RefNum	LoConf	UpConf	Cntrl	1Dose	1Inc	2Dose	2Inc	Citation or Pathology
									Brkly Code

TRIS(2-ETHYLHEXYL)PHOSPHATE   78-42-2

	RefNum	LoConf	UpConf	Cntrl	1Dose	1Inc	2Dose	2Inc	Citation or Pathology / Brkly Code
5075	TR274	1.28gm	17.6gm	0/50	352.mg	4/50	704.mg	7/50	
a	TR274	995.mg	n.s.s.	2/50	352.mg	8/50	704.mg	10/50	liv:hpa,hpc. S
b	TR274	1.01gm	n.s.s.	24/50	352.mg	25/50	704.mg	25/50	
c	TR274	995.mg	n.s.s.	2/50	352.mg	8/50	704.mg	10/50	liv:hpa,hpc,nnd.
d	TR274	2.44gm	n.s.s.	2/50	352.mg	2/50	704.mg	4/50	lun:a/a,a/c.
5076	TR274	1.09gm	n.s.s.	33/50	352.mg	30/50	704.mg	25/50	
a	TR274	785.mg	n.s.s.	15/50	352.mg	21/50	704.mg	18/50	liv:hpa,hpc,nnd.
b	TR274	1.99gm	n.s.s.	7/50	352.mg	3/50	704.mg	7/50	lun:a/a,a/c.
5077	TR274	1.04gm	n.s.s.	37/50	704.mg	38/50	1.41gm	31/50	
a	TR274	13.9mg	n.s.s.	1/50	704.mg	0/50	1.41gm	0/50	liv:hpa,hpc,nnd.
5078	TR274	2.86gm	24.9gm	2/50	1.41gm	9/50	2.82gm	14/50	
a	TR274	2.86gm	24.9gm	2/50	1.41gm	9/50	2.82gm	14/50	adr:phe,phm.
b	TR274	6.54mg	n.s.s.	1/50	1.41gm	2/50	2.82gm	6/50	thy:fca,fcc; tyf:cyn. S
c	TR274	9.66mg	n.s.s.	0/50	1.41gm	1/50	2.82gm	3/50	thy:fca; tyf:cyn. S
d	TR274	2.04mg	n.s.s.	32/50	1.41gm	34/50	2.82gm	35/50	
e	TR274	n.s.s.	n.s.s.	0/50	1.41gm	1/50	2.82gm	0/50	liv:hpa,hpc,nnd.

TRIS(2-HYDROXYPROPYL)AMINE   122-20-3

	RefNum	LoConf	UpConf	Cntrl	1Dose	1Inc	2Dose	2Inc	Citation or Pathology / Brkly Code
5079	1910	3.46gm	n.s.s.	0/21	800.mg	0/21			Yamamoto;clet,45,221-225;1989

Trp-P-1 ACETATE   (3-amino-1,4-dimethyl-5H-pyrido[4,3-b]indole   acetate) 75104-43-7

	RefNum	LoConf	UpConf	Cntrl	1Dose	1Inc	2Dose	2Inc	Citation or Pathology / Brkly Code
5080	1224	13.8mg	51.7mg	0/40	26.0mg	16/40			Matsukura;scie,213,346-347;1981
a	1224	15.8mg	64.8mg	0/40	26.0mg	14/40			
b	1224	110.mg	n.s.s.	3/40	26.0mg	1/40			
c	1224	153.mg	n.s.s.	1/40	26.0mg	0/40			
5081	1224	35.5mg	n.s.s.	1/40	24.0mg	5/40			
a	1224	38.6mg	n.s.s.	0/40	24.0mg	4/40			
b	1224	59.4mg	n.s.s.	3/40	24.0mg	3/40			
c	1224	142.mg	n.s.s.	1/40	24.0mg	0/40			
5082	1683	.310mg	1.04mg	0/50	10.0mg	37/39			Takayama;gann,76,815-817;1985
5083	1683	.342mg	1.01mg	0/50	6.00mg	30/36			

Trp-P-2 ACETATE   (3-amino-1-methyl-5H-pyrido[4,3-b]indole   acetate) 72254-58-1

	RefNum	LoConf	UpConf	Cntrl	1Dose	1Inc	2Dose	2Inc	Citation or Pathology / Brkly Code
5084	1224	3.70mg	13.3mg	0/24	26.0mg	22/26			Matsukura;scie,213,346-347;1981
a	1224	64.3mg	n.s.s.	3/40	26.0mg	3/40			
b	1224	93.1mg	n.s.s.	1/40	26.0mg	1/40			
5085	1224	27.3mg	n.s.s.	3/40	24.0mg	8/40			
a	1224	26.6mg	n.s.s.	0/25	24.0mg	3/24			
b	1224	23.7mg	n.s.s.	1/25	24.0mg	4/24			
c	1224	49.4mg	n.s.s.	1/40	24.0mg	3/40			
5086	1225m	1.68mg	16.4mg	0/30	5.00mg	6/9			Hosaka;clet,13,23-28;1981/Ohgaki 1991
a	1225m	1.68mg	16.4mg	0/30	5.00mg	6/9			
b	1225m	6.72mg	n.s.s.	0/30	5.00mg	1/9			
c	1225m	6.72mg	n.s.s.	0/30	5.00mg	1/9			
d	1225m	6.72mg	n.s.s.	0/30	5.00mg	1/9			
5087	1225m	11.7mg	n.s.s.	0/30	4.00mg	0/10			
5088	2046	4.41mg	419.mg	2/28	1.50mg	8/30	5.00mg	11/30	Takahashi;gann,84,852-858;1993/pers.comm.
a	2046	4.48mg	699.mg	3/28	1.50mg	7/30	5.00mg	12/30	
b	2046	11.0mg	208.mg	0/28	1.50mg	0/30	5.00mg	5/30	
c	2046	1.54mg	n.s.s.	2/28	1.50mg	10/30	(5.00mg	8/30)	
d	2046	5.02mg	n.s.s.	3/28	1.50mg	6/30	5.00mg	11/30	
e	2046	5.18mg	n.s.s.	2/28	1.50mg	8/30	5.00mg	9/30	
f	2046	18.6mg	n.s.s.	0/28	1.50mg	0/30	5.00mg	2/30	
g	2046	10.5mg	n.s.s.	1/28	1.50mg	5/30	5.00mg	3/30	
h	2046	18.7mg	n.s.s.	0/28	1.50mg	1/30	5.00mg	1/30	
i	2046	23.4mg	n.s.s.	1/28	1.50mg	1/30	5.00mg	1/30	
5089	2046	4.87mg	26.7mg	0/30	1.20mg	2/29	4.00mg	9/30	
a	2046	5.77mg	48.0mg	0/30	1.20mg	2/29	4.00mg	7/30	
b	2046	3.66mg	n.s.s.	6/30	1.20mg	5/29	4.00mg	14/30	
c	2046	4.47mg	n.s.s.	4/30	1.20mg	4/29	4.00mg	11/30	
d	2046	14.7mg	n.s.s.	0/30	1.20mg	0/29	4.00mg	2/30	
e	2046	6.45mg	n.s.s.	1/30	1.20mg	0/29	4.00mg	0/30	

DL-TRYPTOPHAN   54-12-6

	RefNum	LoConf	UpConf	Cntrl	1Dose	1Inc	2Dose	2Inc	Citation or Pathology / Brkly Code
5090	1657	2.93gm	n.s.s.	0/42	800.mg	0/20			Cohen;canr,39,1207-1217;1979
a	1657	2.93gm	n.s.s.	0/42	800.mg	0/20			

L-TRYPTOPHAN   73-22-3

	RefNum	LoConf	UpConf	Cntrl	1Dose	1Inc	2Dose	2Inc	Citation or Pathology / Brkly Code
5091	TR71	5.55gm	n.s.s.	3/15	1.74gm	11/35	3.48gm	3/35	
a	TR71	7.40gm	n.s.s.	0/15	1.74gm	1/35	3.48gm	1/35	liv:hpa,hpc,nnd.
b	TR71	11.9gm	n.s.s.	1/15	1.74gm	0/35	3.48gm	1/35	lun:a/a,a/c.
5092	TR71	602.mg	n.s.s.	1/15	1.61gm	15/35	(3.21gm	12/33)	
a	TR71	3.33gm	n.s.s.	1/15	1.61gm	5/35	3.21gm	7/33	liv:hpa,hpc,nnd.
b	TR71	4.66gm	n.s.s.	0/15	1.61gm	4/35	3.21gm	2/33	lun:a/a,a/c.
5093	TR71	1.53gm	n.s.s.	11/15	690.mg	24/35	1.33gm	19/35	
a	TR71	n.s.s.	n.s.s.	0/15	690.mg	0/35	1.33gm	0/35	liv:hpa,hpc,nnd.
5094	TR71	560.mg	n.s.s.	11/15	552.mg	21/35	(1.07gm	14/35)	
a	TR71	n.s.s.	n.s.s.	0/15	552.mg	0/35	1.07gm	0/35	liv:hpa,hpc,nnd.
5095	1430	4.12gm	n.s.s.	0/27	769.mg	0/26			Fukushima;canr,41,3100-3103;1981

```
        Spe Strain  Site   Xpo+Xpt                                                              TD50    2Tailpvl
            Sex  Route  Hist    Notes                                                               DR     AuOp
```

				TD50	DR	AuOp
5096 R m fis eat liv tum 24m24			.>	no dre	P=1.	
a R m fis eat ubl tum 24m24				no dre	P=1.	
TUNGSTATE, SODIUM	100ng..:..1ug....:..10.....:..100....:..1mg....:..10.....:..100....:..1g......:..10					
5097 R f leb wat tba mix 35m35 e		.>		no dre	P=1.	-
a R f leb wat tba mal 35m35 e				no dre	P=1.	-
5098 R m leb wat tba mix 32m32 es		.>		40.1mg	P<1.	-
a R m leb wat tba mal 32m32 es				87.7mg	P<1.	
TURMERIC OLEORESIN (79%-85% CURCUMIN)	..:..1ug....:..10.....:..100....:..1mg....:..10.....:..100....:..1g......:..10					
5099 M f b6c eat liv MXA 24m24		: +		1.84gm Z	P<.007	e
a M f b6c eat liv hpa 24m24				2.02gm Z	P<.003	e
b M f b6c eat for MXA 24m24				42.8gm *	P<.03	
c M f b6c eat for sqp 24m24				53.6gm *	P<.02	
d M f b6c eat TBA MXB 24m24				no dre	P=1.	
e M f b6c eat liv MXB 24m24				1.84gm Z	P<.007	
f M f b6c eat lun MXB 24m24				no dre	P=1.	
5100 M m b6c eat liv MXA 24m24		: ±		1.62gm Z	P<.1	e
a M m b6c eat liv hpa 24m24				1.65gm Z	P<.07	e
b M m b6c eat MXA adc 24m24				7.91gm Z	P<.2	e
c M m b6c eat TBA MXB 24m24				90.6gm *	P<.9	
d M m b6c eat liv MXB 24m24				33.3gm *	P<.7	
e M m b6c eat lun MXB 24m24				no dre	P=1.	
5101 R f f34 eat cli MXA 24m24		: ±		151.mg Z	P<.02	e
a R f f34 eat cli ade 24m24				842.mg Z	P<.03	e
b R f f34 eat TBA MXB 24m24				no dre	P=1.	
c R f f34 eat liv MXB 24m24				no dre	P=1.	
5102 R m f34 eat TBA MXB 24m24			:>	no dre	P=1.	-
a R m f34 eat liv MXB 24m24				11.2gm *	P<.3	
URACIL	100ng..:..1ug....:..10.....:..100....:..1mg....:..10.....:..100....:..1g......:..10					
5103 M f b6c eat ubl tcc 96w96 v		. + .		1.45gm	P<.0005+	
a M f b6c eat liv hnd 96w96 v				18.2gm	P<.04	
b M f b6c eat ubl lei 96w96 v				56.7gm	P<.3	
c M f b6c eat ubl cic 96w96 v				56.7gm	P<.3	
d M f b6c eat liv hpc 96w96 v				no dre	P=1.	
e M f b6c eat lun ade 96w96 v				no dre	P=1.	
5104 M m b6c eat ubl tcc 96w96 v		. 25.7gm		25.7gm	P<.1	+
a M m b6c eat ubl sqc 96w96 v				25.7gm	P<.1	
b M m b6c eat liv hnd 96w96 v				48.8gm	P<.7	
c M m b6c eat lun adc 96w96 v				52.3gm	P<.3	
d M m b6c eat liv hpc 96w96 v				no dre	P=1.	
e M m b6c eat lun ade 96w96 v				no dre	P=1.	
5105 R f f3d eat ubl tcc 24m24		. +		5.64gm	P<.008	+
a R f f3d eat ubl tpp 24m24				4.26gm	P<.04	
b R f f3d eat k/p car 24m24				9.76gm	P<.04	
c R f f3d eat ubl sqp 24m24				29.3gm	P<.6	
5106 R m f3d eat ubl tcc 24m24		. + .		357.mg	P<.0005+	
a R m f3d eat ubl tpp 24m24				511.mg	P<.0005	
b R m f3d eat k/p car 24m24				3.09gm	P<.002	
c R m f3d eat k/p pam 24m24				5.75gm	P<.02	
d R m f3d eat ubl sqc 24m24				7.80gm	P<.04	
e R m f3d eat ubl sqp 24m24				24.3gm	P<.3	
f R m f3d eat ubl lei 24m24				24.3gm	P<.3	
UREA	100ng..:..1ug....:..10.....:..100....:..1mg....:..10.....:..100....:..1g......:..10					
5107 M f cb6 eat liv mlh 52w69 e	pool			.40.4mg *	P<.03	
a M f cb6 eat --- mly 52w69 e				19.5mg *	P<.5	
b M f cb6 eat liv mix 52w69 e				50.1gm *	P<.2	
c M f cb6 eat lun a/a 52w69 e				no dre	P=1.	-
5108 M m cb6 eat liv fbs 52w69 e	pool			70.8gm *	P<.2	
a M m cb6 eat liv mix 52w69 e				275.gm *	P<.9	
b M m cb6 eat lun a/a 52w69 e				no dre	P=1.	-
5109 R f f34 eat liv tum 52w69 e			.>	no dre	P=1.	-
5110 R m f34 eat tes ica 52w69 e		. +		684.mg *	P<.009	-
a R m f34 eat liv tum 52w69 e				no dre	P=1.	-
URETHANE	100ng..:..1ug....:..10.....:..100....:..1mg....:..10.....:..100....:..1g......:..10					
5111 H f syg wat for pam 90w90 e		. + .		44.5mg	P<.0005+	
a H f syg wat der mlc 90w90 e				74.4mg	P<.0005+	
b H f syg wat cec adp 90w90 e				294.mg	P<.0005+	
c H f syg wat vag car 90w90 e				314.mg	P<.002	
d H f syg wat adr mix 90w90 e				318.mg	P<.003	+
e H f syg wat adr coc 90w90 e				349.mg	P<.0005	
f H f syg wat for car 90w90 e				374.mg	P<.0005+	
g H f syg wat thy mix 90w90 e				538.mg	P<.003	+
h H f syg wat ova car 90w90 e				682.mg	P<.007	
i H f syg wat gal pam 90w90 e				682.mg	P<.007	
j H f syg wat lun ade 90w90 e				1.36gm	P<.2	
k H f syg wat liv tum 90w90 e				no dre	P=1.	
5112 H f syg wat for pam 55w55 e		. + .		102.mg	P<.005	+

	RefNum	LoConf	UpConf	Cntrl	1Dose	1Inc	2Dose	2Inc	Citation or Pathology
									Brkly Code
5096	1574	3.30gm	n.s.s.	0/30	800.mg	0/20			Murasaki;carc,4,97-99;1983
a	1574	3.30gm	n.s.s.	0/30	800.mg	0/20			
TUNGSTATE, SODIUM	13472-45-2								
5097	1456	.380mg	n.s.s.	17/24	.286mg	13/20			Schroeder;jnut,105,421-427;1975
a	1456	.884mg	n.s.s.	8/24	.286mg	5/20			
5098	1456	.780mg	n.s.s.	4/26	.250mg	4/25			
a	1456	1.05mg	n.s.s.	2/26	.250mg	2/25			
TURMERIC OLEORESIN (79%-85% CURCUMIN)		8024-37-1							
5099	TR427	850.mg	30.6gm	13/50	255.mg	12/50	1.28gm	25/51 (6.41gm 19/50)	liv:hpa,hpc.
a	TR427	968.mg	14.4gm	7/50	255.mg	8/50	1.28gm	19/51 (6.41gm 14/50)	
b	TR427	16.3gm	n.s.s.	0/50	255.mg	0/50	1.28gm	2/51 6.41gm 3/50	for:sqc,sqp. S
c	TR427	18.5gm	n.s.s.	0/50	255.mg	0/50	1.28gm	1/51 6.41gm 3/50	S
d	TR427	7.08gm	n.s.s.	30/50	255.mg	33/50	1.28gm	35/51 6.41gm 33/50	
e	TR427	850.mg	30.6gm	13/50	255.mg	12/50	1.28gm	25/51 (6.41gm 19/50)	liv:hpa,hpb,hpc.
f	TR427	42.1gm	n.s.s.	5/50	255.mg	3/50	1.28gm	3/51 6.41gm 1/50	lun:a/a,a/c.
5100	TR427	583.mg	n.s.s.	30/50	235.mg	38/50	1.18gm	41/50 (5.92gm 37/50)	liv:hpa,hpc.
a	TR427	624.mg	n.s.s.	25/50	235.mg	28/50	1.18gm	35/50 (5.92gm 30/50)	
b	TR427	2.48gm	n.s.s.	0/50	235.mg	3/50	1.18gm	3/50 (5.92gm 0/50)	duo:adc; ilm:adc; jej:adc.
c	TR427	4.48gm	n.s.s.	38/50	235.mg	42/50	1.18gm	42/50 5.92gm 42/50	
d	TR427	4.56gm	n.s.s.	30/50	235.mg	38/50	1.18gm	41/50 5.92gm 38/50	liv:hpa,hpb,hpc.
e	TR427	13.0gm	n.s.s.	14/50	235.mg	16/50	1.18gm	7/50 5.92gm 13/50	lun:a/a,a/c.
5101	TR427	65.7mg	n.s.s.	6/50	98.5mg	16/50	(494.mg	15/50 2.48gm 16/51)	cli:adc,ade.
a	TR427	339.mg	n.s.s.	5/50	98.5mg	12/50	494.mg	15/50 (2.48gm 16/51)	
b	TR427	2.07gm	n.s.s.	48/50	98.5mg	48/50	494.mg	47/50 2.48gm 46/51	
c	TR427	21.2gm	n.s.s.	1/50	98.5mg	0/50	494.mg	0/50 2.48gm 0/51	liv:hpa,hpb,hpc.
5102	TR427	1.11gm	n.s.s.	42/50	79.7mg	47/50	399.mg	48/50 2.00gm 36/51	
a	TR427	2.51gm	n.s.s.	1/50	79.7mg	1/50	399.mg	4/50 2.00gm 3/51	liv:hpa,hpb,hpc.
URACIL	(2,4-DIOXYPRIMIDINE)	66-22-8							
5103	2058	835.mg	2.75gm	0/30	3.29gm	22/30			Fukushima;canr,52,1675-1680;1992/pers.comm.
a	2058	5.51gm	n.s.s.	0/30	3.29gm	3/30			
b	2058	9.23gm	n.s.s.	0/30	3.29gm	1/30			
c	2058	9.23gm	n.s.s.	0/30	3.29gm	1/30			
d	2058	10.5gm	n.s.s.	1/30	3.29gm	1/30			
e	2058	10.5gm	n.s.s.	1/30	3.29gm	1/30			
5104	2058	6.32gm	n.s.s.	0/30	3.04gm	2/30			
a	2058	6.32gm	n.s.s.	0/30	3.04gm	2/30			
b	2058	6.00gm	n.s.s.	2/30	3.04gm	3/30			
c	2058	8.52gm	n.s.s.	0/30	3.04gm	1/30			
d	2058	11.4gm	n.s.s.	9/30	3.04gm	2/30			
e	2058	16.0gm	n.s.s.	0/30	3.04gm	0/30			
5105	2058	2.14gm	83.7gm	0/30	1.50gm	5/30			
a	2058	1.63gm	n.s.s.	2/30	1.50gm	8/30			
b	2058	2.95gm	n.s.s.	0/30	1.50gm	3/30			
c	2058	4.03gm	n.s.s.	1/30	1.50gm	2/30			
5106	2058	195.mg	672.mg	0/30	1.20gm	27/30			
a	2058	293.mg	953.mg	0/30	1.20gm	24/30			
b	2058	1.33gm	12.4gm	0/30	1.20gm	7/30			
c	2058	1.98gm	n.s.s.	0/30	1.20gm	4/30			
d	2058	2.36gm	n.s.s.	0/30	1.20gm	3/30			
e	2058	3.95gm	n.s.s.	0/30	1.20gm	1/30			
f	2058	3.95gm	n.s.s.	0/30	1.20gm	1/30			
UREA	57-13-6								
5107	1343	9.95gm	n.s.s.	0/89p	440.mg	0/43	881.mg	0/38 4.40gm 2/50	Fleischman;jept,3,149-170;1980
a	1343	3.90gm	n.s.s.	10/92p	440.mg	7/43	881.mg	10/38 4.40gm 9/50	
b	1343	10.4gm	n.s.s.	1/89p	440.mg	0/43	881.mg	0/38 4.40gm 2/50	
c	1343	646.mg	n.s.s.	2/89p	440.mg	0/28	881.mg	0/23 4.40gm 0/23	
5108	1343	11.5gm	n.s.s.	0/91p	406.mg	0/41	813.mg	0/36 4.06gm 1/47	
a	1343	12.9gm	n.s.s.	2/91p	406.mg	0/41	813.mg	0/36 4.06gm 1/47	
b	1343	8.00gm	n.s.s.	1/87p	406.mg	0/27	813.mg	1/21 4.06gm 0/24	
5109	1343	473.mg	n.s.s.	0/49	169.mg	0/50	339.mg	0/48 1.69gm 0/48	
5110	1343	305.mg	36.6gm	21/50	135.mg	27/48	271.mg	25/47 1.35gm 35/50	
a	1343	369.mg	n.s.s.	0/50	135.mg	0/48	271.mg	0/48 1.35gm 0/48	
URETHANE	(ethyl carbamate)	51-79-6							
5111	170	27.9mg	74.7mg	1/67	136.mg	35/44			Toth;ejca,5,165-171;1969
a	170	45.2mg	133.mg	0/62	136.mg	25/41			
b	170	126.mg	949.mg	0/55	136.mg	7/33			
c	170	108.mg	1.77gm	0/49	136.mg	4/20			
d	170	130.mg	2.05gm	1/55	136.mg	7/33			
e	170	142.mg	1.31gm	0/55	136.mg	6/33			
f	170	161.mg	1.23gm	0/62	136.mg	7/41			
g	170	204.mg	2.94gm	0/62	136.mg	5/41			
h	170	235.mg	8.99gm	0/62	136.mg	4/41			
i	170	235.mg	8.99gm	0/62	136.mg	4/41			
j	170	222.mg	n.s.s.	0/49	136.mg	1/20			
k	170	926.mg	n.s.s.	0/67	136.mg	0/44			
5112	297a	34.6mg	965.mg	0/14	273.mg	4/10			Pietra;jnci,25,627-630;1960

	Spe	Sex	Strain	Route	Site	Hist	Xpo+Xpt	Notes	plot	TD50	DR	2Tailpvl AuOp
a	H	f	syg	wat	ski	mlt	55w55	e		496.mg		P<.2
5113	H	m	syg	wat	for	pam	24m24	e	. + .	64.2mg		P<.0005+
a	H	m	syg	wat	der	mlc	24m24	e		109.mg		P<.0005+
b	H	m	syg	wat	for	car	24m24	e		135.mg		P<.0005+
c	H	m	syg	wat	cec	adp	24m24	e		765.mg		P<.009 +
d	H	m	syg	wat	liv	mix	24m24	e		1.28gm		P<.03
e	H	m	syg	wat	adr	coa	24m24	e		564.mg		P<.2
f	H	m	syg	wat	lun	ade	24m24	e		927.mg		P<.2
5114	H	m	syg	wat	ski	mlt	76w76	e	. + .	74.2mg		P<.0005+
a	H	m	syg	wat	for	pam	76w76	e		95.8mg		P<.0005+
5115	M	f	b6a	orl	lun	mix	73w73	evx	. + .	12.5mg		P<.0005+
a	M	f	b6a	orl	lun	ade	73w73	evx		29.3mg		P<.0005
b	M	f	b6a	orl	---	agm	73w73	evx		31.7mg		P<.0005
c	M	f	b6a	orl	hag	ade	73w73	evx		59.7mg		P<.002
d	M	f	b6a	orl	liv	hpt	73w73	evx		89.9mg		P<.009
e	M	f	b6a	orl	lun	car	73w73	evx		89.9mg		P<.009
f	M	f	b6a	orl	tba	mix	73w73	evx		9.73mg		P<.0005
5116	M	m	b6a	orl	lun	mix	72w72	evx	. + .	24.2mg		P<.0005+
a	M	m	b6a	orl	liv	mix	72w72	evx		26.0mg		P<.0005
b	M	m	b6a	orl	lun	ade	72w72	evx		27.8mg		P<.0005
c	M	m	b6a	orl	hag	ade	72w72	evx		35.8mg		P<.0005
d	M	m	b6a	orl	liv	hpt	72w72	evx		39.0mg		P<.002 +
e	M	m	b6a	orl	---	rts	72w72	evx		77.9mg		P<.006 +
f	M	m	b6a	orl	tba	mix	72w72	evx		11.2mg		P<.0005
5117	M	f	b6c	orl	liv	mix	68w68	evx	. + .	32.4mg		P<.0005
a	M	f	b6c	orl	liv	hpt	68w68	evx		55.9mg		P<.003
b	M	f	b6c	orl	lun	mix	68w68	evx		79.0mg		P<.009 +
c	M	f	b6c	orl	lun	ade	68w68	evx		97.4mg		P<.02
d	M	f	b6c	orl	tba	mix	68w68	evx		13.7mg		P<.0005
5118	M	m	b6c	orl	liv	mix	70w70	evx	. + .	46.0mg		P<.002
a	M	m	b6c	orl	lun	mix	70w70	evx		65.9mg		P<.006 +
b	M	m	b6c	orl	liv	hpt	70w70	evx		81.7mg		P<.02 +
c	M	m	b6c	orl	lun	ade	70w70	evx		81.7mg		P<.02
d	M	m	b6c	orl	tba	mix	70w70	evx		22.4mg		P<.0005
5119	M	m	b6c	wat	lun	mix	65w70	aes	. + .	5.91mg	*	P<.0005+
a	M	m	b6c	wat	lun	a/a	65w70	aes		5.91mg	*	P<.0005
b	M	m	b6c	wat	liv	hem	65w70	as		63.3mg	*	P<.0005+
c	M	m	b6c	wat	liv	hes	65w70	as		122.mg	*	P<.0005+
d	M	m	b6c	wat	lun	a/c	65w70	aes		242.mg	*	P<.0005
e	M	m	b6c	wat	hea	hes	65w70	as		417.mg	*	P<.0005
f	M	m	b6c	wat	liv	hpa	65w70	as		38.2mg	Z	P<.3
g	M	m	b6c	wat	liv	hpc	65w70	as		no dre		P=1.
5120	M	f	cf1	wat	lun	tum	31m31	eg	. + .	23.8mg		P<.0005+
a	M	f	cf1	wat	liv	hpt	31m31	eg		no dre		P=1. -
b	M	f	cf1	wat	tba	mix	31m31	eg		23.3mg		P<.09
5121	M	m	cf1	wat	lun	tum	29m29	e	. + .	13.2mg		P<.0005+
a	M	m	cf1	wat	liv	hpt	29m29	e		no dre		P=1. -
b	M	m	cf1	wat	tba	mix	29m29	e		12.1mg		P<.04
5122	M	b	nmr	wat	tba	ben	23m24	ae	. + .	.434mg	Z	P<.004
a	M	b	nmr	wat	tba	mal	23m24	ae		14.7mg	*	P<.0005+
5123	M	b	swi	eat	lun	ade	27m27	e	. + .	36.1mg		P<.0005+
a	M	b	swi	eat	lun	ade	27m27	e		36.1mg		P<.0005+
b	M	b	swi	eat	lun	adc	27m27	e		5.06gm		P<.3
c	M	b	swi	eat	liv	tum	27m27	e		no dre		P=1.
5124	P	b	cym	eat	MXB	MXB	5y23	w	: + :	16.4mg		P<.002
a	P	b	cym	eat	pan	adc	5y23	w		52.8mg		P<.07 +
b	P	b	cym	eat	liv	hes	5y23	w		95.0mg		P<.03 +
c	P	b	cym	eat	liv	hpa	5y23	w		31.7mg		P<.2
d	P	b	cym	eat	lun	bro	5y23	w		31.7mg		P<.2 +
e	P	b	cym	eat	tba	mal	5y23	w		17.6mg		P<.05
f	P	b	cym	eat	tba	mix	5y23	w		20.9mg		P<.09
g	P	b	cym	eat	tba	ben	5y23	w		66.8mg		P<.5
5125	P	b	rhe	eat	MXB	MXB	5y25	w	: + :	44.8mg		P<.004
a	P	b	rhe	eat	jej	adc	5y25	w		65.2mg		P<.04 +
b	P	b	rhe	eat	chp	sqp	5y25	w		91.2mg		P<.03
c	P	b	rhe	eat	adr	phe	5y25	w		117.mg		P<.04
d	P	b	rhe	eat	liv	hpc	5y25	w		143.mg		P<.04 +
e	P	b	rhe	eat	liv	hes	5y25	w		143.mg		P<.04 +
f	P	b	rhe	eat	tba	mix	5y25	w		33.0mg		P<.07
g	P	b	rhe	eat	tba	mal	5y25	w		61.0mg		P<.2
h	P	b	rhe	eat	tba	ben	5y25	w		82.1mg		P<.3
5126	R	b	sda	wat	tba	mal	23m24	ae	. + .	41.3mg	*	P<.0005+
a	R	b	sda	wat	tba	ben	23m24	ae		91.2mg	*	P<.02

VANADYL SULFATE   100ng..:.1ug....:.10.......:.100....:.1mg....:.10....:.100....:.1g......:.10

	Spe	Sex	Strain	Route	Site	Hist	Xpo+Xpt	Notes	plot	TD50	DR	2Tailpvl AuOp
5127	M	b	cd1	wat	lun	mix	24m24	e	.>	29.2mg		P<.8 -
a	M	b	cd1	wat	liv	ade	24m24	e		no dre		P=1. -
b	M	b	cd1	wat	tba	mal	24m24	e		5.00mg		P<.06 -
c	M	b	cd1	wat	tba	ben	24m24	e		no dre		P=1. -

	RefNum	LoConf	UpConf	Cntrl	1Dose	1Inc	2Dose	2Inc								Citation or Pathology
																Brkly Code
a	297a	80.6mg	n.s.s.	0/14	273.mg	1/10										
5113	170	40.5mg	110.mg	6/88	120.mg	36/49										Toth;ejca,5,165-171;1969
a	170	66.7mg	197.mg	1/62	120.mg	26/49										
b	170	76.8mg	267.mg	0/53	120.mg	18/40										
c	170	264.mg	25.2gm	0/53	120.mg	4/40										
d	170	386.mg	n.s.s.	0/62	120.mg	3/49										
e	170	114.mg	n.s.s.	1/26	120.mg	2/12										
f	170	151.mg	n.s.s.	0/26	120.mg	1/12										
5114	297a	29.2mg	260.mg	1/49	240.mg	7/10										Pietra;jnci,25,627-630;1960
a	297a	37.1mg	352.mg	0/49	240.mg	6/10										
5115	297	5.80mg	29.5mg	1/17	81.3mg	17/19										Innes;ntis,1968/1969
a	297	14.0mg	92.3mg	1/17	81.3mg	12/19										
b	297	15.4mg	79.1mg	0/17	81.3mg	11/19										
c	297	25.5mg	245.mg	0/17	81.3mg	7/19										
d	297	33.9mg	2.02gm	0/17	81.3mg	5/19										
e	297	33.9mg	2.02gm	0/17	81.3mg	5/19										
f	297	3.86mg	24.8mg	2/17	81.3mg	18/19										
5116	297	11.9mg	76.3mg	2/18	75.6mg	15/22										
a	297	13.1mg	72.1mg	1/18	75.6mg	14/22										
b	297	13.4mg	103.mg	2/18	75.6mg	14/22										
c	297	17.6mg	90.2mg	0/18	75.6mg	11/22										
d	297	18.2mg	174.mg	1/18	75.6mg	11/22										
e	297	31.6mg	645.mg	0/18	75.6mg	6/22										
f	297	5.20mg	27.7mg	3/18	75.6mg	20/22										
5117	297	16.3mg	79.2mg	0/16	81.5mg	12/23										
a	297	25.1mg	238.mg	0/16	81.5mg	8/23										
b	297	32.1mg	1.76gm	0/16	81.5mg	6/23										
c	297	36.9mg	n.s.s.	0/16	81.5mg	5/23										
d	297	7.21mg	28.0mg	0/16	81.5mg	19/23										
5118	297	20.6mg	165.mg	0/16	75.7mg	8/20										
a	297	26.7mg	555.mg	0/16	75.7mg	6/20										
b	297	30.8mg	n.s.s.	0/16	75.7mg	5/20										
c	297	30.8mg	n.s.s.	0/16	75.7mg	5/20										
d	297	11.3mg	51.7mg	0/16	75.7mg	13/20										
5119	1933	3.60mg	10.5mg	9/49	.100mg	4/49	.500mg	7/48	1.00mg	8/50	10.0mg	34/50	100.mg	42/44		Inai;gann,82,380-385;
																1991
a	1933	3.60mg	10.5mg	9/49	.100mg	4/49	.500mg	7/48	1.00mg	8/50	10.0mg	34/50	100.mg	42/44		
b	1933	38.0mg	116.mg	0/50	.100mg	0/50	.500mg	0/50	1.00mg	0/50	10.0mg	2/50	100.mg	20/50		
c	1933	63.2mg	321.mg	0/50	.100mg	0/50	.500mg	0/50	1.00mg	2/50	10.0mg	2/50	100.mg	11/50		
d	1933	98.6mg	865.mg	0/49	.100mg	0/49	.500mg	0/48	1.00mg	0/50	10.0mg	0/50	100.mg	6/44		
e	1933	144.mg	2.14gm	0/50	.100mg	0/50	.500mg	0/50	1.00mg	0/50	10.0mg	0/50	100.mg	4/50		
f	1933	9.92mg	n.s.s.	8/50	.100mg	2/50	.500mg	8/50	1.00mg	4/50	10.0mg	9/50	(100.mg	0/50)		
g	1933	418.mg	n.s.s.	0/50	.100mg	2/50	.500mg	1/50	1.00mg	0/50	10.0mg	2/50	100.mg	0/50		
5120	90	13.1mg	59.4mg	13/56	20.0mg	28/40										Tomatis;ijcn,10,489-506;1972
a	90	175.mg	n.s.s.	2/56	20.0mg	1/40										
b	90	7.51mg	n.s.s.	45/56	20.0mg	37/40										
5121	90	7.18mg	34.6mg	23/55	16.7mg	40/48										
a	90	78.0mg	n.s.s.	12/55	16.7mg	8/48										
b	90	4.11mg	n.s.s.	46/55	16.7mg	46/48										
5122	298	.198mg	3.14mg	2/65	.100mg	11/65	(.500mg	12/69	2.50mg	21/59	12.5mg	30/65)				Schmahl;ijcn,19,77-80;1977
a	298	8.61mg	32.7mg	6/74	.100mg	11/65	.500mg	17/69	2.50mg	21/59	12.5mg	32/65				
5123	171a	19.3mg	66.9mg	10/49	125.mg	46/48										Van Esch;fctx,10,373-381;1972
a	171a	19.3mg	66.9mg	10/49	125.mg	46/48										
b	171a	824.mg	n.s.s.	0/49	125.mg	1/48										
c	171a	1.54gm	n.s.s.	0/49	125.mg	0/48										
5124	2001	4.49mg	116.mg	0/54	22.2mg	3/5										Adamson;ossc,129-156;
																1982/Thorgeirsson 1994/Dalgard 1991/Thorgeirsson&Seiber pers.comm.
a	2001	8.60mg	n.s.s.	0/13	22.2mg	1/3										
b	2001	15.5mg	n.s.s.	0/54	22.2mg	1/5										
c	2001	5.16mg	n.s.s.	0/5	22.2mg	1/2										
d	2001	5.16mg	n.s.s.	0/5	22.2mg	1/2										
e	2001	4.62mg	n.s.s.	2/54	22.2mg	3/5										
f	2001	5.01mg	n.s.s.	3/54	22.2mg	3/5										
g	2001	7.16mg	n.s.s.	1/19	22.2mg	1/4										
5125	2001	10.1mg	787.mg	0/44	24.4mg	2/6										
a	2001	10.6mg	n.s.s.	0/25	24.4mg	1/3										
b	2001	14.9mg	n.s.s.	0/39	24.4mg	1/4										
c	2001	19.1mg	n.s.s.	0/39	24.4mg	1/5										
d	2001	23.3mg	n.s.s.	0/44	24.4mg	1/6										
e	2001	23.3mg	n.s.s.	0/44	24.4mg	1/6										
f	2001	8.46mg	n.s.s.	11/110	24.4mg	4/11										
g	2001	10.9mg	n.s.s.	5/76	24.4mg	2/9										
h	2001	12.7mg	n.s.s.	7/110	24.4mg	2/11										
5126	298	21.4mg	129.mg	2/74	.100mg	2/70	.500mg	4/65	2.50mg	7/70	12.5mg	15/74				Schmahl;ijcn,19,77-80;1977
a	298	35.6mg	n.s.s.	1/74	.100mg	3/70	.500mg	2/65	2.50mg	5/70	12.5mg	8/74				

VANADYL SULFATE   27774-13-6

	RefNum	LoConf	UpConf	Cntrl	1Dose	1Inc										Citation or Pathology
5127	1512	2.33mg	n.s.s.	26/170	.877mg	8/47										Kanisawa;canr,27,1192-1195;1967
a	1512	6.00mg	n.s.s.	7/170	.877mg	1/47										
b	1512	1.65mg	n.s.s.	15/170	.877mg	9/47										
c	1512	4.10mg	n.s.s.	29/170	.877mg	5/47										

```
        Spe Strain  Site   Xpo+Xpt                                                          TD50    2Tailpvl
            Sex Route  Hist   Notes                                                              DR      AuOp
    d    M b cd1 wat tba mix 24m24 e                                                         no dre  P=1.   -
 5128 M f cd1 wat tba mix 33m33 e                                          .    ±           5.11mg  P<.07  -
 5129 M m cd1 wat tba mix 31m31 e                                              .>           no dre  P=1.   -


VANGUARD GF                           100ng..:..1ug....:..10.....:..100....:..1mg....:..10.....:..100....:..1g.....:..10
 5130 M f b6a orl lun ade 76w76 evx                                              .>         170.mg  P<.6   -
    a M f b6a orl liv hpt 76w76 evx                                                         no dre  P=1.   -
    b M f b6a orl tba mix 76w76 evx                                                         150.mg  P<.6   -
 5131 M m b6a orl lun mix 76w76 evx                                         .>              1.57gm  P<1.   -
    a M m b6a orl liv hpt 76w76 evx                                                         no dre  P=1.   -
    b M m b6a orl tba mix 76w76 evx                                                         no dre  P=1.   -
 5132 M f b6c orl lun ade 76w76 evx                               .     ±                   99.1mg  P<.1   -
    a M f b6c orl liv hpt 76w76 evx                                                         no dre  P=1.   -
    b M f b6c orl tba mix 76w76 evx                                                         99.1mg  P<.1   -
 5133 M m b6c orl lun ade 76w76 evx                                       .>                179.mg  P<.3   -
    a M m b6c orl lun car 76w76 evx                                                         179.mg  P<.3   -
    b M m b6c orl liv hpt 76w76 evx                                                         no dre  P=1.   -
    c M m b6c orl tba mix 76w76 evx                                                         40.2mg  P<.02  -


VINBLASTINE                           100ng..:..1ug....:..10.....:..100....:..1mg....:..10.....:..100....:..1g.....:..10
 5134 R m b46 ivj tba mix 12m24 es                      .>                                  no dre  P=1.   -


VINYL ACETATE                         100ng..:..1ug....:..10.....:..100....:..1mg....:..10.....:..100....:..1g.....:..10
 5135 R f f34 wat thy cca 23m30                                                .    +    .  420.mg * P<.006 +
    a R f f34 wat liv nnd 23m30                                                             488.mg * P<.003 +
    b R f f34 wat ute adc 23m30                                                             494.mg * P<.006 +
    c R f f34 wat thy ccr 23m30                                                             25.5gm * P<1.  +
    d R f f34 wat tba mix 23m30                                                             no dre  P=1.
 5136 R m f34 wat liv nnd 23m30                                            .    ±           132.mg \ P<.02  +
    a R m f34 wat thy ccr 23m30                                                             no dre  P=1.   +
    b R m f34 wat tba mix 23m30                                                             57.7mg * P<.3


VINYL BROMIDE                         100ng..:..1ug....:..10.....:..100....:..1mg....:..10.....:..100....:..1g.....:..10
 5137 R f sda inh liv nnd 19m24 aes                                   .    +         .      18.9mg  Z P<.003
    a R f sda inh mix ang 19m24 aes                                                         19.2mg  Z P<.0005+
    b R f sda inh zym sqc 19m24 aes                                                         2.24gm * P<.0005
    c R f sda inh liv mix 19m24 aes                                                         433.mg  Z P<.04  +
    d R f sda inh liv hpc 19m24 aes                                                         749.mg  Z P<.04  +
 5138 R m sda inh mix ang 18m24 aes                                       .    +    .       17.9mg  Z P<.0005+
    a R m sda inh zym sqc 18m24 aes                                                         409.mg * P<.0005
    b R m sda inh zym pam 18m24 aes                                                         2.76gm * P<.004
    c R m sda inh liv mix 18m24 aes                                                         303.mg  Z P<.02  +
    d R m sda inh liv hpc 18m24 aes                                                         387.mg  Z P<.02  +
    e R m sda inh liv nnd 18m24 aes                                                         no dre  P=1.


VINYL CARBAMATE                       100ng..:..1ug....:..10.....:..100....:..1mg....:..10.....:..100....:..1g.....:..10
 5139 M f c5j ipj liv hes 35w59 er                     .   +         .                      .138mg  P<.007 +
    a M f c5j ipj liv hem 35w59 er                                                          .147mg  P<.009 +
    b M f c5j ipj liv hpc 35w59 er                                                          .503mg  P<.2   +
    c M f c5j ipj liv hpa 35w59 er                                                          4.99mg  P<.7
 5140 M m c5j ipj liv hem 35w59 er                     .   +     .                          .112mg  P<.003 +
    a M m c5j ipj liv hes 35w59 er                                                          .118mg  P<.004 +
    b M m c5j ipj liv hpc 35w59 er                                                          .629mg  P<.2   +
    c M m c5j ipj liv hpa 35w59 er                                                          2.69mg  P<.6


VINYL CHLORIDE                        100ng..:..1ug....:..10.....:..100....:..1mg....:..10.....:..100....:..1g.....:..10
 5141 H f syg inh mgl car  6m24 es                                           .   +  .       32.3mg  P<.0005+
    a H f syg inh sto ade  6m24 es                                                          47.2mg  P<.0005+
    b H f syg inh --- hes  6m24 es                                                          78.6mg  P<.0005+
 5142 H f syg inh mgl car 12m24 es                                            .   +  .      27.7mg  P<.0005+
    a H f syg inh ski car 12m24 es                                                          121.mg  P<.0005+
    b H f syg inh --- hes 12m24 es                                                          314.mg  P<.002 +
    c H f syg inh sto ade 12m24 es                                                          1.01gm  P<.5   +
 5143 H f syg inh mgl car 18m24 es                                              .+ .        61.0mg  P<.0005+
    a H f syg inh sto ade 18m24 es                                                          205.mg  P<.0005+
    b H f syg inh --- hes 18m24 es                                                          1.92gm  P<.07  +
 5144 H m syg inh for mix  7m25 ez                                          .   +  .        126.mg  Z P<.0005+

    a H m syg inh adu epo  7m25 ez                                                          3.42gm  Z P<.2

    b H m syg inh ski epo  7m25 ez                                                          15.4gm  Z P<.7

    c H m syg inh liv ang  7m25 ez                                                          no dre  P=1.   +

 5145 M f cd1 inh liv hes 26w58 es                                           .   +  .       98.9mg * P<.0005+
    a M f cd1 inh lun abt 26w58 es                                                          142.mg * P<.02
    b M f cd1 inh mgl mix 26w58 es                                                          142.mg * P<.02
 5146 M m cd1 inh liv hes 26w78 es                                        .   +  .          32.0mg  Z P<.0005+
    a M m cd1 inh lun abt 26w78 es                                                          88.7mg * P<.004
 5147 M b swi inh lun tum 30w81 ez                                      . +.                21.1mg  Z P<.0005+
```

	RefNum	LoConf	UpConf	Cntrl	1Dose	1Inc	2Dose	2Inc					Citation or Pathology
													Brkly Code
d	1512	1.79mg	n.s.s.	55/170	.877mg	15/47							
5128	1395	2.01mg	n.s.s.	9/45	1.00mg	19/51							Schroeder;jnut,105,452-458;1975
5129	1395	4.93mg	n.s.s.	10/43	.833mg	5/38							

VANGUARD GF   (ferric nitrosodimethyldithiocarbamate and tetramethylthiuram disulfide. CAS# --- and 137-26-8) mixture

	RefNum	LoConf	UpConf	Cntrl	1Dose	1Inc	2Dose	2Inc					Citation or Pathology
5130	1356	24.9mg	n.s.s.	1/17	33.9mg	2/16							Innes;ntis,1968/1969
a	1356	59.7mg	n.s.s.	0/17	33.9mg	0/16							
b	1356	21.0mg	n.s.s.	2/17	33.9mg	3/16							
5131	1356	27.5mg	n.s.s.	2/18	31.6mg	2/17							
a	1356	59.1mg	n.s.s.	1/18	31.6mg	0/17							
b	1356	30.4mg	n.s.s.	3/18	31.6mg	2/17							
5132	1356	24.3mg	n.s.s.	0/16	33.9mg	2/17							
a	1356	63.4mg	n.s.s.	0/16	33.9mg	0/17							
b	1356	24.3mg	n.s.s.	0/16	33.9mg	2/17							
5133	1356	29.2mg	n.s.s.	0/16	31.6mg	1/16							
a	1356	29.2mg	n.s.s.	0/16	31.6mg	1/16							
b	1356	55.6mg	n.s.s.	0/16	31.6mg	0/16							
c	1356	13.8mg	n.s.s.	0/16	31.6mg	4/16							

VINBLASTINE   865-21-4

5134	1017	37.8ug	n.s.s.	7/65	10.0ug	1/25							Schmahl;arzn,20,1461-1467;1970

VINYL ACETATE   108-05-4

5135	1546	181.mg	4.68gm	0/20	44.0mg	2/20	110.mg	5/20					Lijinsky;txap,68,43-53;1983/pers.comm.
a	1546	198.mg	2.31gm	0/20	44.0mg	0/20	110.mg	6/20					
b	1546	201.mg	4.70gm	0/20	44.0mg	1/20	110.mg	5/20					
c	1546	590.mg	n.s.s.	1/20	44.0mg	0/20	110.mg	1/20					
d	1546	37.7mg	n.s.s.	20/20	44.0mg	18/20	110.mg	19/20					
5136	1546	45.4mg	n.s.s.	0/20	27.5mg	4/20	(68.1mg	2/20)					
a	1546	254.mg	n.s.s.	1/20	27.5mg	2/20	68.1mg	1/20					
b	1546	14.8mg	n.s.s.	18/20	27.5mg	16/20	68.1mg	20/20					

VINYL BROMIDE   593-60-2

5137	1466	8.74mg	107.mg	3/142	2.95mg	12/101	(14.4mg	9/113	63.5mg	10/118	343.mg	5/112)	Benja;txap,64,367-379;1982
a	1466	13.9mg	27.9mg	1/144	2.95mg	10/120	14.4mg	50/120	(63.5mg	61/120	343.mg	41/120)	
b	1466	1.12gm	6.44gm	0/139	2.95mg	0/99	14.4mg	3/113	63.5mg	2/119	343.mg	11/114	
c	1466	175.mg	n.s.s.	7/142	2.95mg	18/101	14.4mg	12/113	63.5mg	21/118	(343.mg	9/112)	
d	1466	282.mg	n.s.s.	4/142	2.95mg	6/101	14.4mg	3/113	63.5mg	11/118	(343.mg	4/112)	
5138	1466	12.3mg	27.2mg	0/144	2.11mg	7/120	8.88mg	36/120	(38.9mg	61/120	232.mg	43/120)	
a	1466	281.mg	647.mg	2/142	2.11mg	1/99	8.88mg	1/112	38.9mg	13/114	232.mg	35/116	
b	1466	1.14gm	27.3gm	0/142	2.11mg	0/99	8.88mg	1/112	38.9mg	3/114	232.mg	5/116	
c	1466	129.mg	n.s.s.	4/143	2.11mg	5/103	8.88mg	10/119	38.9mg	13/120	(232.mg	5/119)	
d	1466	161.mg	n.s.s.	3/143	2.11mg	1/103	8.88mg	7/119	38.9mg	9/120	(232.mg	3/119)	
e	1466	2.91gm	n.s.s.	1/143	2.11mg	4/103	8.88mg	3/119	38.9mg	4/120	232.mg	2/119	

VINYL CARBAMATE   (carbamic acid, vinyl ester) 15805-73-9

5139	2170	84.5ug	1.34mg	0/5	.508mg	25/45							Wright;txpy,19,258-265;1991/pers.comm.
a	2170	89.3ug	2.79mg	0/5	.508mg	24/45							
b	2170	.236mg	n.s.s.	0/5	.508mg	9/45							
c	2170	.813mg	n.s.s.	0/5	.508mg	1/45							
5140	2170	71.2ug	.434mg	0/5	.508mg	31/49							
a	2170	74.9ug	.533mg	0/5	.508mg	30/49							
b	2170	.285mg	n.s.s.	0/5	.508mg	8/49							
c	2170	.662mg	n.s.s.	0/5	.508mg	2/49							

VINYL CHLORIDE   75-01-4

5141	1536m	20.5mg	55.1mg	0/143	18.3mg	28/87							Drew;txap,68,120-130;1983/Haseman pers.comm.
a	1536m	27.1mg	105.mg	5/138	18.3mg	23/88							
b	1536m	41.4mg	177.mg	0/143	18.3mg	13/88							
5142	1536n	17.7mg	46.4mg	0/143	36.7mg	31/52							
a	1536n	56.9mg	330.mg	0/133	36.7mg	9/48							
b	1536n	108.mg	1.73gm	0/143	36.7mg	4/52							
c	1536n	146.mg	n.s.s.	5/138	36.7mg	3/50							
5143	1536o	42.5mg	91.6mg	0/143	55.0mg	47/102							
a	1536o	110.mg	579.mg	5/138	55.0mg	20/101							
b	1536o	473.mg	n.s.s.	0/143	55.0mg	2/103							
5144	1BT8	72.9mg	274.mg	3/60	2.95mg	3/30	14.7mg	4/30	29.5mg	9/30	147.mg	17/30	(354.mg 10/30
				590.mg	10/30)								Maltoni;enhp,41,3-29;1981/1977a
a	1BT8	872.mg	n.s.s.	0/60	2.95mg	0/30	14.7mg	0/30	29.5mg	3/30	147.mg	1/30	354.mg 2/30
				(590.mg	1/30)								
b	1BT8	1.73gm	n.s.s.	3/60	2.95mg	9/30	14.7mg	3/30	29.5mg	7/30	147.mg	3/30	354.mg 1/30
				590.mg	7/30								
c	1BT8	4.04gm	n.s.s.	0/60	2.95mg	0/30	14.7mg	0/30	29.5mg	2/30	147.mg	0/30	354.mg 1/30
				590.mg	0/30								
5145	1113m	45.2mg	316.mg	1/28	17.9mg	1/8	161.mg	2/8	459.mg	8/12			Hong;jtxe,7,909-924;1981
a	1113m	52.4mg	n.s.s.	7/28	17.9mg	1/8	161.mg	4/8	459.mg	7/12			
b	1113m	52.4mg	n.s.s.	7/28	17.9mg	1/8	161.mg	4/8	459.mg	7/12			
5146	1113m	13.4mg	104.mg	0/28	11.2mg	0/8	60.9mg	7/12	(244.mg	5/12)			
a	1113m	37.3mg	927.mg	4/28	11.2mg	2/8	60.9mg	8/12	244.mg	7/12			
5147	1BT4	15.6mg	29.5mg	15/150	8.70mg	6/60	43.5mg	41/60	87.0mg	50/60	(435.mg	40/59	1.04gm 47/60
				1.74gm	46/56)								Maltoni;enhp,41,3-29;1981/1977a

	Spe Sex	Strain Route	Site Hist	Xpo+Xpt Notes	TD50 DR	2Tailpvl AuOp
a	M b	swi inh	liv ang	30w81 ez	59.4mg Z	P<.0005+
b	M b	swi inh	liv agm	30w81 ez	99.8mg Z	P<.0005
c	M b	swi inh	mgl car	30w81 ez	5.04gm Z	P<.009 +
d	M b	swi inh	ski epo	30w81 ez	5.99gm *	P<.0005
e	M b	swi inh	ehp agm	30w81 ez	22.4gm *	P<.3 +
f	M b	swi inh	ehp ang	30w81 ez	no dre	P=1. +
5148	M f	swi inh	mgl car	6m24 es	10.6mg	P<.0005+
a	M f	swi inh	--- hes	6m24 es	12.5mg	P<.0005+
b	M f	swi inh	lun car	6m24 es	36.6mg	P<.03 +
5149	M f	swi inh	--- hes	12m24 es	13.8mg	P<.0005+
a	M f	swi inh	mgl car	12m24 es	22.9mg	P<.0005+
b	M f	swi inh	lun car	12m24 es	55.5mg	P<.02 +
5150	M f	swi inh	mgl car	18m24 es	32.2mg	P<.0005+
a	M f	swi inh	--- hes	18m24 es	36.1mg	P<.0005+
b	M f	swi inh	lun car	18m24 es	143.mg	P<.2 +
5151	R f	cdr inh	liv mix	26w78 es	75.5mg *	P<.04
5152	R f	cdr inh	liv hes	43w95 es	69.5mg *	P<.0005
5153	R m	cdr inh	liv nnd	26w78 es	11.5mg Z	P<.02
a	R m	cdr inh	liv hpc	26w78 es	52.9mg *	P<.04
5154	R m	cdr inh	liv hes	43w95 es	103.mg *	P<.002
5155	R f	f34 inh	liv nnd	6m24 e	17.6mg	P<.0005
a	R f	f34 inh	mgl fba	6m24 e	15.1mg	P<.03 +
b	R f	f34 inh	--- hes	6m24 e	91.2mg	P<.2 +
c	R f	f34 inh	liv hpc	6m24 e	103.mg	P<.2
5156	R f	f34 inh	mgl fba	12m24 e	14.6mg	P<.0005+
a	R f	f34 inh	liv nnd	12m24 e	16.2mg	P<.0005
b	R f	f34 inh	--- hes	12m24 e	29.5mg	P<.0005+
c	R f	f34 inh	mgl adc	12m24 e	38.0mg	P<.003 +
d	R f	f34 inh	liv hpc	12m24 e	101.mg	P<.03 +
5157	R f	f34 inh	mgl fba	18m24 e	29.7mg	P<.004 +
a	R f	f34 inh	--- hes	18m24 e	32.8mg	P<.0005+
b	R f	f34 inh	liv hpc	18m24 e	65.2mg	P<.0005+
c	R f	f34 inh	mgl adc	18m24 e	74.2mg	P<.02 +
d	R f	f34 inh	liv nnd	18m24 e	96.3mg	P<.03
5158	R f	f34 inh	--- hes	24m24 e	23.6mg	P<.0005+
a	R f	f34 inh	mgl fba	24m24 e	32.9mg	P<.002 +
b	R f	f34 inh	liv hpc	24m24 e	77.2mg	P<.0005+
c	R f	f34 inh	liv nnd	24m24 e	166.mg	P<.08
d	R f	f34 inh	mgl adc	24m24 e	264.mg	P<.3 +
5159	R b	sda inh	liv hpt	12m31 ez	+hist 361.mg Z	P<.003 +
a	R b	sda inh	liv ang	12m31 ez	843.mg Z	P<.0005+
b	R b	sda inh	zym car	12m31 ez	1.70gm Z	P<.0005+
c	R b	sda inh	bra neu	12m31 ez	+hist 3.68gm *	P<.0005+
d	R b	sda inh	ehp ang	12m31 ez	9.60gm *	P<.1 +
e	R b	sda inh	kid nep	12m31 ez	+hist 8.81gm *	P<.3 +
5160	R b	sda gav	liv ang	12m32 ez	54.1mg *	P<.0005+
a	R b	sda gav	kid nep	12m32 ez	+hist 318.mg *	P<.05 +
b	R b	sda gav	ehp ang	12m32 ez	1.14gm *	P<.5
c	R b	sda gav	zym car	12m32 ez	1.97gm *	P<.8
5161	R b	sda inh	mgl adc	12m34 ez	3.69mg Z	P<.008 +
a	R b	sda inh	liv ang	12m34 ez	+hist 41.0mg *	P<.002 +
b	R b	sda inh	zym car	12m34 ez	64.5mg *	P<.2 +
c	R b	sda inh	ehp ang	12m34 ez	124.mg *	P<.4 +
d	R b	sda inh	kid nep	12m34 ez	+hist 249.mg *	P<.2 +
5162	R b	sda inh	kid nep	12m33 ez	+hist 91.5mg *	P<.0005+
a	R b	sda inh	liv ang	12m33 ez	136.mg *	P<.0005+
b	R b	sda inh	mgl adc	12m33 ez	216.mg *	P<.02 +
c	R b	sda inh	liv agm	12m33 ez	528.mg *	P<.02
d	R b	sda inh	liv hpt	12m33 ez	+hist 883.mg *	P<.05 +
e	R b	sda inh	zym car	12m33 ez	447.mg *	P<.2 +
5163	R b	sda gav	zym car	14m32 ez	14.2mg *	P<.02 +
a	R b	sda gav	liv ang	14m32 ez	+hist 16.0mg *	P<.02 +
b	R b	sda gav	liv hpt	14m32 ez	+hist 32.2mg *	P<.2 +
c	R b	sda gav	ehp ang	14m32 ez	64.5mg *	P<.2 +
5164	R b	sda inh	liv ang	12m33 ez	50.4mg	P<.005 +
a	R b	sda inh	mgl adc	12m33 ez	19.0mg	P<.02 +
b	R b	sda inh	ehp agm	12m33 ez	64.4mg	P<.02
c	R b	sda inh	ehp ang	12m33 ez	79.0mg	P<.03 +
d	R b	sda inh	zym car	12m33 ez	79.0mg	P<.03 +

	RefNum	LoConf	UpConf	Cntrl	1Dose	1Inc	2Dose	2Inc						Citation or Pathology
a	1BT4	34.6mg	115.mg	0/150	8.70mg	1/60	43.5mg	18/60	(87.0mg	14/60	435.mg	16/59	1.04gm	13/60
					1.74gm	10/56)								
b	1BT4	51.4mg	234.mg	0/150	8.70mg	1/60	43.5mg	11/60	(87.0mg	5/60	435.mg	5/59	1.04gm	7/60
					1.74gm	6/56)								
c	1BT4	2.21gm	235.gm	1/150	8.70mg	12/60	43.5mg	12/60	87.0mg	8/60	435.mg	8/59	1.04gm	8/60
					1.74gm	13/56								
d	1BT4	2.91gm	25.3gm	2/150	8.70mg	0/60	43.5mg	1/60	87.0mg	2/60	435.mg	4/59	1.04gm	7/60
					1.74gm	4/56								
e	1BT4	5.45gm	n.s.s.	1/150	8.70mg	5/60	43.5mg	3/60	87.0mg	3/60	435.mg	1/59	1.04gm	3/60
					1.74gm	4/56								
f	1BT4	7.81gm	n.s.s.	1/150	8.70mg	1/60	43.5mg	3/60	87.0mg	7/60	435.mg	8/59	1.04gm	1/60
					1.74gm	1/56								
5148	1536m	6.81mg	18.3mg	2/71	10.1mg	33/67								Drew;txap,68,120-130;1983/Haseman pers.comm.
a	1536m	7.87mg	21.9mg	1/71	10.1mg	29/67								
b	1536m	15.4mg	n.s.s.	9/71	10.1mg	18/65								
5149	1536n	8.64mg	23.7mg	1/71	20.2mg	30/47								
a	1536n	13.4mg	45.2mg	2/71	20.2mg	22/47								
b	1536n	23.7mg	n.s.s.	9/71	20.2mg	15/47								
5150	1536o	18.9mg	63.3mg	2/71	30.2mg	22/45								
a	1536o	20.9mg	71.4mg	1/71	30.2mg	20/45								
b	1536o	47.0mg	n.s.s.	9/71	30.2mg	11/45								
5151	1113m	22.7mg	n.s.s.	0/8	3.19mg	0/8	15.9mg	1/8	63.8mg	2/8				Hong;jtxe,7,909-924;1981
5152	1113n	34.4mg	195.mg	0/16	4.35mg	0/16	21.7mg	4/12	87.0mg	7/16				
5153	1113m	3.43mg	n.s.s.	0/8	2.23mg	0/8	11.2mg	3/8	(44.6mg	1/8)				
a	1113m	15.9mg	n.s.s.	0/8	2.23mg	0/8	11.2mg	1/8	44.6mg	2/8				
5154	1113n	41.7mg	483.mg	0/16	3.04mg	0/10	15.2mg	1/16	60.9mg	5/16				
5155	1536m	8.77mg	63.2mg	4/112	4.80mg	15/75								Drew;txap,68,120-130;1983/Haseman pers.comm.
a	1536m	6.50mg	n.s.s.	24/112	4.80mg	28/76								
b	1536m	23.2mg	n.s.s.	2/112	4.80mg	4/76								
c	1536m	25.8mg	n.s.s.	1/112	4.80mg	3/75								
5156	1536n	7.52mg	52.0mg	24/112	9.60mg	28/56								
a	1536n	9.16mg	35.0mg	4/112	9.60mg	20/56								
b	1536n	14.5mg	85.5mg	2/112	9.60mg	12/56								
c	1536n	16.6mg	272.mg	5/112	9.60mg	11/56								
d	1536n	31.8mg	n.s.s.	1/112	9.60mg	4/56								
5157	1536o	13.8mg	269.mg	24/112	14.4mg	24/55								
a	1536o	17.4mg	78.1mg	2/112	14.4mg	15/55								
b	1536o	28.5mg	255.mg	1/112	14.4mg	8/54								
c	1536o	29.1mg	n.s.s.	5/112	14.4mg	9/55								
d	1536o	34.1mg	n.s.s.	4/112	14.4mg	7/54								
5158	1536r	14.2mg	43.9mg	2/112	19.1mg	24/55								
a	1536r	16.3mg	153.mg	24/112	19.1mg	26/55								
b	1536r	35.3mg	260.mg	1/112	19.1mg	9/55								
c	1536r	52.2mg	n.s.s.	4/112	19.1mg	6/55								
d	1536r	62.1mg	n.s.s.	5/112	19.1mg	5/55								
5159	1BT1	147.mg	1.82gm	0/58	2.02mg	0/60	10.1mg	1/59	20.2mg	5/60	(101.mg	2/60	243.mg	1/59
					405.mg	1/60)								Maltoni;enhp,41,3-29;1981/1977a
a	1BT1	528.mg	1.94gm	0/58	2.02mg	1/60	10.1mg	3/59	20.2mg	6/60	101.mg	13/60	243.mg	13/59
					(405.mg	7/60)								
b	1BT1	1.08gm	3.29gm	0/58	2.02mg	0/60	10.1mg	0/59	20.2mg	4/60	101.mg	2/60	243.mg	7/59
					405.mg	16/60								
c	1BT1	1.98gm	8.17gm	0/58	2.02mg	0/60	10.1mg	0/59	20.2mg	0/60	101.mg	4/60	243.mg	3/59
					405.mg	7/60								
d	1BT1	3.07gm	n.s.s.	0/58	2.02mg	1/60	10.1mg	2/59	20.2mg	1/60	101.mg	3/60	243.mg	3/59
					405.mg	3/60								
e	1BT1	2.44gm	n.s.s.	0/58	2.02mg	1/60	10.1mg	5/59	20.2mg	6/60	101.mg	6/60	243.mg	5/59
					405.mg	5/60								
5160	1BT11	34.2mg	93.0mg	0/80	.819mg	0/80	4.09mg	10/80	12.3mg	17/80				
a	1BT11	121.mg	n.s.s.	0/80	.819mg	0/80	4.09mg	3/80	12.3mg	2/80				
b	1BT11	177.mg	n.s.s.	0/80	.819mg	2/80	4.09mg	0/80	12.3mg	2/80				
c	1BT11	192.mg	n.s.s.	1/80	.819mg	0/80	4.09mg	2/80	12.3mg	1/80				
5161	1BT15	1.79mg	83.2mg	7/120	37.2ug	15/118	.186mg	22/119	.372mg	21/119	(.929mg	17/120)		
a	1BT15	16.7mg	167.mg	0/120	37.2ug	0/118	.186mg	0/119	.372mg	1/119	.929mg	5/120		
b	1BT15	17.9mg	n.s.s.	2/120	37.2ug	1/118	.186mg	1/119	.372mg	2/119	.929mg	4/120		
c	1BT15	30.6mg	n.s.s.	0/120	37.2ug	0/118	.186mg	0/119	.372mg	2/119	.929mg	0/120		
d	1BT15	40.5mg	n.s.s.	0/120	37.2ug	0/118	.186mg	0/119	.372mg	0/119	.929mg	1/120		
5162	1BT2	58.3mg	170.mg	0/185	3.82mg	10/120	5.73mg	11/119	7.64mg	7/120				
a	1BT2	79.2mg	264.mg	0/185	3.82mg	1/120	5.73mg	6/119	7.64mg	12/120				
b	1BT2	101.mg	n.s.s.	2/185	3.82mg	4/120	5.73mg	6/119	7.64mg	6/120				
c	1BT2	201.mg	n.s.s.	0/185	3.82mg	1/120	5.73mg	0/119	7.64mg	4/120				
d	1BT2	267.mg	n.s.s.	0/185	3.82mg	0/120	5.73mg	0/119	7.64mg	3/120				
e	1BT2	157.mg	n.s.s.	2/185	3.82mg	1/120	5.73mg	4/119	7.64mg	4/120				
5163	1BT27	5.01mg	n.s.s.	1/150	8.37ug	0/150	83.7ug	0/148	.279mg	5/149				
a	1BT27	5.54mg	n.s.s.	0/150	8.37ug	0/150	83.7ug	1/148	.279mg	3/149				
b	1BT27	7.93mg	n.s.s.	0/150	8.37ug	0/150	83.7ug	1/148	.279mg	1/149				
c	1BT27	10.5mg	n.s.s.	0/150	8.37ug	0/150	83.7ug	0/148	.279mg	1/149				
5164	1BT9	27.1mg	305.mg	0/98	1.92mg	14/294								
a	1BT9	10.2mg	n.s.s.	10/98	1.92mg	62/294								
b	1BT9	32.4mg	n.s.s.	0/98	1.92mg	11/294								
c	1BT9	37.2mg	n.s.s.	0/98	1.92mg	9/294								
d	1BT9	37.2mg	n.s.s.	0/98	1.92mg	9/294								

Brkly Code

	Spe	Strain	Site	Xpo+Xpt			TD50	2Tailpvl
	Sex	Route	Hist	Notes			DR	AuOp
e	R b sda inh liv agm	12m33 ez					89.1mg	P<.04
f	R b sda inh kid nep	12m33 ez				+hist	721.mg	P<.5 +
5165	R b sda inh zym car	6m36 eyz		. (+) .			672.mg *	P<.0005
a	R b sda inh liv ang	6m36 eyz					2.41gm *	P<.05 +
b	R b sda inh ehp ang	6m36 eyz					4.85gm *	P<.3
5166	R b sda inh zym car	6m36 eyz		. (+) .			555.mg *	P<.0005
a	R b sda inh mam mal	6m36 eyz					569.mg *	P<.02
b	R b sda inh kid nep	6m36 eyz					4.87gm *	P<.2
c	R b sda inh liv ang	6m36 eyz					4.87gm *	P<.2 +
d	R b sda inh liv hpt	6m36 eyz					4.88gm *	P<.3
e	R b sda inh bra neu	6m36 eyz					9.76gm *	P<.2
f	R b sda inh ehp ang	6m36 eyz					9.77gm *	P<.4
5167	R f sda inh bra neu	18m24 gv		. + .			299.mg	P<.0005+
a	R f sda inh liv ang	18m24 gv					387.mg	P<.0005+
b	R f sda inh liv hpc	18m24 gv					2.76gm	P<.006 +
c	R f sda inh tba mal	18m24 gv					85.7mg	P<.0005
d	R f sda inh tba mix	18m24 gv					111.mg	P<.0005
5168	R m sda inh liv hpc	12m30 e		. + .			40.8mg	P<.0005+
a	R m sda inh liv ang	12m30 e					90.0mg	P<.0005+
b	R m sda inh adr tum	12m30 e					251.mg	P<.003
c	R m sda inh liv mix	12m30 e					294.mg	P<.004 +
d	R m sda inh pit tum	12m30 e					138.mg	P<.02
e	R m sda inh tba mix	12m30 e					17.3mg	P<.0005+
5169	R f wis inh liv ang	52w52 ek		. + .			209.mg	P<.002 +
a	R f wis inh nas oec	52w52 ek					276.mg	P<.004 +
b	R f wis inh zym sqc	52w52 ek					857.mg	P<.09 +
c	R f wis inh liv nnd	52w52 ek					1.81gm	P<.3 +
d	R f wis inh liv hpc	52w52 ek					1.81gm	P<.3 +
e	R f wis inh lun ppa	52w52 ek					1.81gm	P<.3
5170	R m wis inh ehp agm	12m31 ez		. + .			.761mg	P<.01
a	R m wis inh ehp ang	12m31 ez					1.28mg	P<.05
b	R m wis inh liv agm	12m31 ez					3.88mg	P<.3
c	R m wis inh liv hpt	12m31 ez					3.88mg	P<.3
d	R m wis inh liv ang	12m31 ez					no dre	P=1.
e	R m wis inh zym car	12m31 ez					no dre	P=1.
5171	R m wis inh liv ang	12m38 ez		. + .			1.55gm *	P<.0005+
a	R m wis inh bra neu	12m38 ez					4.77gm *	P<.003
b	R m wis inh zym car	12m38 ez					6.00gm *	P<.004
c	R m wis inh liv hpt	12m38 ez					8.14gm *	P<.07
d	R m wis inh kid nep	12m38 ez					13.6gm *	P<.4
e	R m wis inh ehp ang	12m38 ez					no dre	P=1.
5172	R m wis inh liv ang	52w52 ek		. ± .			330.mg	P<.03 +
a	R m wis inh zym sqc	52w52 ek					330.mg	P<.03 +
b	R m wis inh nas oec	52w52 ek					533.mg	P<.08 +
c	R m wis inh liv hpc	52w52 ek					1.14gm	P<.3 +
d	R m wis inh lun ppa	52w52 ek					no dre	P=1.
5173	R m wis inh liv ang	52w78 e		. + .			10.2mg Z	P<.0005+
a	R m wis inh lun ang	52w78 e					193.mg *	P<.0005+
b	R m wis inh lun tum	52w78 e					867.mg *	P<.2 +
c	R m wis inh bra tum	52w78 e					5.82gm *	P<.8 +
d	R m wis inh ski tum	52w78 e					5.82gm *	P<.8 +
e	R m wis inh nse tum	52w78 e					no dre	P=1. +
f	R m wis inh tes tum	52w78 e					no dre	P=1. +
g	R m wis inh tba mix	52w78 e					5.44mg Z	P<.0005+

VINYL TOLUENE (65-71% m- and 32-35% p-)..:..1ug...:..10....:..100...:..1mg..:..10....:..100...:..1g.....:..10

5174	M f b6c inh TBA MXB	24m24	:>			no dre	P=1. -
a	M f b6c inh liv MXB	24m24				no dre	P=1.
b	M f b6c inh lun MXB	24m24				750.mg *	P<.6
5175	M m b6c inh TBA MXB	24m24	:>			no dre	P=1. -
a	M m b6c inh liv MXB	24m24				no dre	P=1.
b	M m b6c inh lun MXB	24m24				no dre	P=1.
5176	R f f34 inh TBA MXB	24m24	:>			312.mg *	P<.6 -
a	R f f34 inh liv MXB	24m24				no dre	P=1.
5177	R m f34 inh TBA MXB	24m24	:>			no dre	P=1. -
a	R m f34 inh liv MXB	24m24				no dre	P=1.

4-VINYLCYCLOHEXENE   100ng..:..1ug...:..10....:..100....:..1mg...:..10....:..100...:..1g.....:..10

5178	M f b6c gav ova MXB	24m24	:+ :			94.4mg *	P<.0005
a	M f b6c gav ova mtb	24m24				106.mg \	P<.0005c
b	M f b6c gav ova MXA	24m24				309.mg *	P<.0005c
c	M f b6c gav ova gct	24m24				361.mg *	P<.0005c
d	M f b6c gav MXA mlp	24m24				839.mg *	P<.01
e	M f b6c gav arp adn	24m24				1.05gm *	P<.004 e
f	M f b6c gav TBA MXB	24m24				149.mg *	P<.002

	RefNum	LoConf	UpConf	Cntrl	1Dose	1Inc	2Dose	2Inc					Citation or Pathology
													Brkly Code
e	1BT9	40.3mg	n.s.s.	0/98	1.92mg	8/294							
f	1BT9	117.mg	n.s.s.	0/98	1.92mg	1/294							
5165	1BT10m	362.mg	1.47gm	0/227	20.5mg	5/118	34.1mg	9/119					
a	1BT10m	834.mg	n.s.s.	0/227	20.5mg	3/118	34.1mg	1/119					
b	1BT10m	1.19gm	n.s.s.	0/227	20.5mg	2/118	34.1mg	0/119					
5166	1BT10n	315.mg	1.13gm	0/227	20.5mg	9/120	34.1mg	8/119					
a	1BT10n	261.mg	n.s.s.	17/227	20.5mg	12/120	34.1mg	20/119					
b	1BT10n	1.20gm	n.s.s.	0/227	20.5mg	1/120	34.1mg	1/119					
c	1BT10n	1.20gm	n.s.s.	0/227	20.5mg	1/120	34.1mg	1/119					
d	1BT10n	1.20gm	n.s.s.	0/227	20.5mg	2/120	34.1mg	0/119					
e	1BT10n	1.59mg	n.s.s.	0/227	20.5mg	0/120	34.1mg	1/119					
f	1BT10n	1.59mg	n.s.s.	0/227	20.5mg	1/120	34.1mg	0/119					
5167	BT4001	192.mg	496.mg	0/60	392.mg	32/54							Maltoni;anya,534,145-159;198
a	BT4001	242.mg	671.mg	0/60	392.mg	27/54							
b	BT4001	1.05gm	29.9gm	0/60	392.mg	5/54							
c	BT4001	47.0mg	151.mg	9/60	392.mg	52/54							
d	BT4001	53.6mg	270.mg	35/60	392.mg	52/54							
5168	1440	26.7mg	67.3mg	1/80	21.4mg	35/80							Radike;enhp,41,59-62;1981
a	1440	51.7mg	177.mg	0/80	21.4mg	18/80							
b	1440	108.mg	1.09gm	0/80	21.4mg	7/80							
c	1440	120.mg	1.93gm	0/80	21.4mg	6/80							
d	1440	61.0mg	n.s.s.	8/80	21.4mg	19/80							
e	1440	11.6mg	28.1mg	16/80	21.4mg	63/80							
5169	1170	80.8mg	884.mg	0/10	1.12gm	6/10							Feron;txcy,13,131-141;1979/1979a
a	1170	101.mg	1.93gm	0/10	1.12gm	5/10							
b	1170	209.mg	n.s.s.	0/10	1.12gm	2/10							
c	1170	295.mg	n.s.s.	0/10	1.12gm	1/10							
d	1170	295.mg	n.s.s.	0/10	1.12gm	1/10							
e	1170	295.mg	n.s.s.	0/10	1.12gm	1/10							
5170	1BT17	.289mg	48.8mg	0/94	34.6ug	5/99							Maltoni;enhp,41,3-29;1981/1977a
a	1BT17	.388mg	n.s.s.	0/94	34.6ug	3/99							
b	1BT17	.632mg	n.s.s.	0/94	34.6ug	1/99							
c	1BT17	.632mg	n.s.s.	0/94	34.6ug	1/99							
d	1BT17	1.17mg	n.s.s.	0/94	34.6ug	0/99							
e	1BT17	.647mg	n.s.s.	3/94	34.6ug	2/99							
5171	1BT7	762.mg	5.08gm	0/38	1.41mg	0/28	7.03mg	1/27	14.1mg	3/28	70.3mg	3/25	169.mg 3/26
					281.mg	8/27							
a	1BT7	1.81gm	27.0gm	0/38	1.41mg	0/28	7.03mg	0/27	14.1mg	0/28	70.3mg	1/25	169.mg 1/26
					281.mg	3/27							
b	1BT7	2.07gm	43.5gm	0/38	1.41mg	0/28	7.03mg	0/27	14.1mg	0/28	70.3mg	0/25	169.mg 2/26
					281.mg	2/27							
c	1BT7	2.46gm	n.s.s.	0/38	1.41mg	0/28	7.03mg	0/27	14.1mg	0/28	70.3mg	1/25	169.mg 2/26
					281.mg	0/27							
d	1BT7	2.46gm	n.s.s.	0/38	1.41mg	1/28	7.03mg	0/27	14.1mg	2/28	70.3mg	0/25	169.mg 2/26
					281.mg	1/27							
e	1BT7	3.86gm	n.s.s.	1/38	1.41mg	0/28	7.03mg	1/27	14.1mg	0/28	70.3mg	1/25	169.mg 1/26
					281.mg	0/27							
5172	1170	98.3mg	n.s.s.	0/10	781.mg	3/9							Feron;txcy,13,131-141;1979/1979a
a	1170	98.3mg	n.s.s.	0/10	781.mg	3/9							
b	1170	130.mg	n.s.s.	0/10	781.mg	2/9							
c	1170	185.mg	n.s.s.	0/10	781.mg	1/9							
d	1170	362.mg	n.s.s.	0/10	781.mg	0/9							
5173	1760	4.36mg	32.7mg	0/19	1.07mg	0/20	10.7mg	7/19	(321.mg	17/20)			Bi;eaes,10,281-289;1985/pers.comm.
a	1760	88.3mg	617.mg	0/19	1.07mg	0/20	10.7mg	2/19	321.mg	9/20			
b	1760	208.mg	n.s.s.	1/19	1.07mg	1/20	10.7mg	3/19	321.mg	4/20			
c	1760	503.mg	n.s.s.	0/19	1.07mg	0/20	10.7mg	2/19	321.mg	1/20			
d	1760	503.mg	n.s.s.	0/19	1.07mg	0/20	10.7mg	2/19	321.mg	1/20			
e	1760	469.mg	n.s.s.	1/19	1.07mg	1/20	10.7mg	2/19	321.mg	1/20			
f	1760	674.mg	n.s.s.	0/19	1.07mg	0/20	10.7mg	1/19	321.mg	0/20			
g	1760	2.63mg	15.7mg	1/19	1.07mg	1/20	10.7mg	11/19	(321.mg	19/20)			

VINYL TOLUENE (65-71% m- and 32-35% p-)   25013-15-4

	RefNum	LoConf	UpConf	Cntrl	1Dose	1Inc	2Dose	2Inc		Citation or Pathology
5174	TR375	73.0mg	n.s.s.	29/50	14.9mg	27/50	37.3mg	17/50		
a	TR375	52.8mg	n.s.s.	9/50	14.9mg	5/50	(37.3mg	2/50)		liv:hpa,hpc,nnd.
b	TR375	112.mg	n.s.s.	3/50	14.9mg	2/50	37.3mg	4/50		lun:a/a,a/c.
5175	TR375	17.4mg	n.s.s.	33/50	12.4mg	23/50	(31.1mg	20/50)		
a	TR375	73.7mg	n.s.s.	17/50	12.4mg	14/50	31.1mg	14/50		liv:hpa,hpc,nnd.
b	TR375	38.9mg	n.s.s.	12/50	12.4mg	5/50	(31.1mg	2/50)		lun:a/a,a/c.
5176	TR375	52.1mg	n.s.s.	48/50	35.8mg	41/50	107.mg	46/50		
a	TR375	n.s.s.	n.s.s.	0/50	35.8mg	0/50	107.mg	0/50		liv:hpa,hpc,nnd.
5177	TR375	41.3mg	n.s.s.	48/50	25.1mg	48/50	75.2mg	48/50		liv:hpa,hpc,nnd.
a	TR375	213.mg	n.s.s.	2/50	25.1mg	2/50	75.2mg	1/50		liv:hpa,hpc,nnd.

4-VINYLCYCLOHEXENE   100-40-3

	RefNum	LoConf	UpConf	Cntrl	1Dose	1Inc	2Dose	2Inc		Citation or Pathology
5178	TR303	66.7mg	145.mg	1/50	142.mg	33/50	283.mg	24/50		ova:gcc,gct,mtb. C
a	TR303	65.9mg	187.mg	0/50	142.mg	25/50	(283.mg	11/50)		
b	TR303	181.mg	672.mg	1/50	142.mg	10/50	283.mg	13/50		ova:gcc,gct.
c	TR303	204.mg	888.mg	1/50	142.mg	9/50	283.mg	11/50		
d	TR303	376.mg	72.6gm	1/50	142.mg	6/50	283.mg	5/50		mln:mlp; mul:mlp; smi:mlp. S
e	TR303	450.mg	6.37gm	0/50	142.mg	3/50	283.mg	4/50		
f	TR303	77.9mg	802.mg	28/50	142.mg	47/50	283.mg	30/50		

```
      Spe Strain Site  Xpo+Xpt                                                      TD50    2Tailpvl
          Sex  Route Hist    Notes                                                    DR     AuOp
   g    M f b6c gav liv MXB 24m24                                                  1.48gm * P<.08
   h    M f b6c gav lun MXB 24m24                                                  19.0gm / P<.9
   5179 M m b6c gav lun MXA 24m24                                         :    ±    #490.mg * P<.03  i
   a    M m b6c gav MXA MXA 24m24                                                  623.mg * P<.03
   b    M m b6c gav lun a/a 24m24                                                  926.mg * P<.03
   c    M m b6c gav TBA MXB 24m24                                                  307.mg * P<.1
   d    M m b6c gav liv MXB 24m24                                                  1.27gm * P<.6
   e    M m b6c gav lun MXB 24m24                                                  490.mg * P<.03
   5180 R f f34 gav cli MXA 24m24                                         :    ±    #580.mg \ P<.04  i
   a    R f f34 gav TBA MXB 24m24                                                  305.mg * P<.2
   b    R f f34 gav liv MXB 24m24                                                  3.85kg   P=1.
   5181 R m f34 gav tes ict 24m24                                    :  +  :        #62.3mg * P<.0005i
   a    R m f34 gav ski MXA 24m24                                                  769.mg * P<.003
   b    R m f34 gav ski sqp 24m24                                                  935.mg * P<.006
   c    R m f34 gav adr MXA 24m24                                                  159.mg * P<.02
   d    R m f34 gav TBA MXB 24m24                                                  119.mg * P<.02
   e    R m f34 gav liv MXB 24m24                                                  132.gm * P<1.


VINYLIDENE CHLORIDE        100ng..:..1ug....:..10....:..100....:..1mg....:..10....:..100....:..1g....:..10
   5182 M f b6c gav --- MXA 24m24                              :    ±                #3.90mg \ P<.05  -
   a    M f b6c gav --- lym 24m24                                                  4.02mg \ P<.02
   b    M f b6c gav TBA MXB 24m24                                                  2.09mg \ P<.06
   c    M f b6c gav liv MXB 24m24                                                  no dre   P=1.
   d    M f b6c gav lun MXB 24m24                                                  66.7mg * P<.2
   5183 M m b6c gav TBA MXB 24m24                               :>                  34.8mg * P<.6  -
   a    M m b6c gav liv MXB 24m24                                                  133.mg * P<.8
   b    M m b6c gav lun MXB 24m24                                                  60.4mg * P<.4
   5184 M f cd1 inh liv hes 52w52 e                         .>                      170.mg   P<.3  +
   a    M f cd1 inh liv hpt 52w52 e                                                170.mg   P<.3
   b    M f cd1 inh lun tum 52w52 e                                                no dre   P=1.
   5185 M f cd1 inh liv hct 26w78 e                          .>                     no dre   P=1.
   a    M f cd1 inh liv hes 26w78 e                                                no dre   P=1.
   b    M f cd1 inh lun abt 26w78 e                                                no dre   P=1.
   5186 M m cd1 inh liv hct 26w78 e                        .>                       no dre   P=1.
   a    M m cd1 inh lun abt 26w78 e                                                no dre   P=1.
   5187 M f swi inh mgl mix 12m29 e                   .   +    .                    80.9mg   P<.004 +
   a    M f swi inh lun mix 12m29 e                                                124.mg   P<.09  +
   b    M f swi inh liv hpt 12m29 e                                                858.mg   P<.4
   c    M f swi inh liv ang 12m29 e                                                no dre   P=1.
   d    M f swi inh tba mix 12m29 e                                                30.9mg   P<.003
   e    M f swi inh tba mal 12m29 e                                                86.1mg   P<.03
   5188 M m swi inh kid adc 12m29 e                      . + .                      22.0mg   P<.0005+
   a    M m swi inh lun mix 12m29 e                                                59.9mg   P<.02  +
   b    M m swi inh liv ang 12m29 e                                                566.mg   P<.4
   c    M m swi inh liv hpt 12m29 e                                                no dre   P=1.
   d    M m swi inh tba mix 12m29 e                                                17.9mg   P<.0005
   e    M m swi inh tba mal 12m29 e                                                50.3mg   P<.004
   5189 R f f34 gav TBA MXB 24m24                          :>                       no dre   P=1.
   a    R f f34 gav liv MXB 24m24                                                  no dre   P=1.
   5190 R m f34 gav TBA MXB 24m24                         :>                        no dre   P=1.  -
   a    R m f34 gav liv MXB 24m24                                                  no dre   P=1.
   5191 R f cdr inh liv hes 52w52 e                           .>                    no dre   P=1.
   5192 R f cdr inh liv mix 26w78 e                            .>                   no dre   P=1.
   5193 R f cdr inh liv mix 43w95 e                              .>                 no dre   P=1.
   5194 R m cdr inh liv mix 26w78 e                         .>                      no dre   P=1.
   5195 R m cdr inh liv mix 43w95 e                            .>                   no dre   P=1.
   5196 R f sda inh liv mix 24m35 egv                              .>               847.mg   P<.3
   a    R f sda inh tba mix 24m35 egv                                              119.mg   P<.3
   5197 R f sda gav liv ang 12m34 e                     .>                          no dre   P=1.
   a    R f sda gav liv hpt 12m34 e                                                no dre   P=1.
   b    R f sda gav tba mix 12m34 e                                                no dre   P=1.
   5198 R f sda gav liv ang 12m32 e              .     ±                            .501mg   P<.2
   a    R f sda gav tba mix 12m32 e                                                no dre   P=1.
   5199 R m sda gav liv ang 12m34 e                    .>                           no dre   P=1.
   a    R m sda gav liv hpt 12m34 e                                                no dre   P=1.
   b    R m sda gav tba mix 12m34 e                                                no dre   P=1.
   5200 R m sda gav liv mix 12m32 e              .>                                 no dre   P=1.
   a    R m sda gav tba mix 12m32 e                                                5.32mg   P<.8
   5201 R f sss wat liv nnd 24m24 e                               .>                414.mg * P<.4  -
   a    R f sss wat liv hpc 24m24 e                                                654.mg * P<.2
   b    R f sss wat mgl fba 24m24 e                                                679.mg * P<1.
   c    R f sss wat pit ade 24m24 e                                                no dre   P=1.
   5202 R f sss inh mgl fba 18m24 e                    .    ±                       16.0mg * P<.04
   a    R f sss inh mgl adc 18m24 e                                                49.8mg \ P<.03  -
   b    R f sss inh liv cho 18m24 e                                                no dre   P=1.
   5203 R m sss wat liv hpc 24m24 e                               .>                22.9gm * P<1.  -
   a    R m sss wat liv nnd 24m24 e                                                no dre   P=1.
   5204 R m sss inh liv sar 18m24 e                           .>                    no dre   P=1.


FD & C VIOLET NO. 1        100ng..:..1ug....:..10....:..100....:..1mg....:..10....:..100....:..1g....:..10
   5205 M f asp eat lun ade 80w80 e                                    .>           1.91gm * P<.3  -
```

	RefNum	LoConf	UpConf	Cntrl	1Dose	1Inc	2Dose	2Inc			Citation or Pathology Brkly Code
g	TR303	506.mg	n.s.s.	1/50	142.mg	3/50	283.mg	3/50			liv:hpa,hpc,nnd.
h	TR303	799.mg	n.s.s.	6/50	142.mg	8/50	283.mg	4/50			lun:a/a,a/c.
5179	TR303	206.mg	n.s.s.	4/50	142.mg	11/50	283.mg	4/50			lun:a/a,a/c. S
a	TR303	245.mg	n.s.s.	4/50	142.mg	7/50	283.mg	5/50			mln:mlh; mul:mlh,mlm,mlp; spl:mlm. S
b	TR303	347.mg	n.s.s.	1/50	142.mg	4/50	283.mg	3/50			S
c	TR303	112.mg	n.s.s.	26/50	142.mg	33/50	283.mg	12/50			
d	TR303	216.mg	n.s.s.	18/50	142.mg	20/50	283.mg	6/50			liv:hpa,hpc,nnd.
e	TR303	206.mg	n.s.s.	4/50	142.mg	11/50	283.mg	4/50			lun:a/a,a/c.
5180	TR303	199.mg	n.s.s.	1/50	141.mg	5/50	(280.mg	0/50)			cli:adn,sqc. S
a	TR303	98.8mg	n.s.s.	38/50	141.mg	37/50	280.mg	18/50			
b	TR303	766.mg	n.s.s.	1/50	141.mg	2/50	280.mg	0/50			liv:hpa,hpc,nnd.
5181	TR303	34.7mg	166.mg	35/50	141.mg	30/50	280.mg	29/50			S
a	TR303	265.mg	4.36gm	0/50	141.mg	1/50	280.mg	4/50			ski:sqc,sqp. S
b	TR303	287.mg	12.5gm	0/50	141.mg	1/50	280.mg	3/50			S
c	TR303	66.5mg	n.s.s.	17/50	141.mg	16/50	280.mg	8/50			adr:phe,phm. S
d	TR303	51.9mg	n.s.s.	38/50	141.mg	31/50	280.mg	19/50			
e	TR303	477.mg	n.s.s.	1/50	141.mg	1/50	280.mg	0/50			liv:hpa,hpc,nnd.

VINYLIDENE CHLORIDE   75-35-4

	RefNum	LoConf	UpConf	Cntrl	1Dose	1Inc	2Dose	2Inc			Citation or Pathology Brkly Code
5182	TR228	1.54mg	n.s.s.	7/50	1.43mg	15/50	(7.14mg	7/50)			---:leu,lym. S
a	TR228	1.65mg	n.s.s.	2/50	1.43mg	9/50	(7.14mg	6/50)			S
b	TR228	.838mg	n.s.s.	23/50	1.43mg	33/50	(7.14mg	21/50)			
c	TR228	28.7mg	n.s.s.	4/50	1.43mg	3/50	7.14mg	4/50			liv:hpa,hpc,nnd.
d	TR228	19.2mg	n.s.s.	1/50	1.43mg	1/50	7.14mg	4/50			lun:a/a,a/c.
5183	TR228	6.01mg	n.s.s.	30/50	1.43mg	22/50	7.14mg	33/50			
a	TR228	11.1mg	n.s.s.	15/50	1.43mg	9/50	7.14mg	15/50			liv:hpa,hpc,nnd.
b	TR228	12.8mg	n.s.s.	5/50	1.43mg	7/50	7.14mg	8/50			lun:a/a,a/c.
5184	357	27.7mg	n.s.s.	0/16	68.6mg	1/15					Lee;jtxe,4,15-30;1978
a	357	27.7mg	n.s.s.	0/16	68.6mg	1/15					
b	357	53.0mg	n.s.s.	0/16	68.6mg	0/15					
5185	1113m	31.8mg	n.s.s.	1/28	22.9mg	0/12					Hong;jtxe,7,909-924;1981
a	1113m	31.8mg	n.s.s.	1/28	22.9mg	0/12					
b	1113m	31.8mg	n.s.s.	7/28	22.9mg	0/12					
5186	1113m	17.5mg	n.s.s.	4/28	19.0mg	1/12					
a	1113m	17.5mg	n.s.s.	4/28	19.0mg	1/12					
5187	BT402n	38.9mg	578.mg	1/89	7.72mg	12/118					Maltoni;aric,3,1-229;1985/1977
a	BT402n	46.6mg	n.s.s.	3/89	7.72mg	11/119					
b	BT402n	140.mg	n.s.s.	0/78	7.72mg	1/111					
c	BT402n	178.mg	n.s.s.	2/77	7.72mg	1/109					
d	BT402n	16.5mg	179.mg	14/89	7.72mg	41/119					
e	BT402n	38.1mg	n.s.s.	3/90	7.72mg	14/120					
5188	BT402n	13.6mg	38.7mg	0/66	6.43mg	25/98					
a	BT402n	28.3mg	n.s.s.	3/87	6.43mg	16/120					
b	BT402n	92.2mg	n.s.s.	0/60	6.43mg	1/88					
c	BT402n	94.4mg	n.s.s.	2/62	6.43mg	2/92					
d	BT402n	11.1mg	39.7mg	9/87	6.43mg	45/120					
e	BT402n	25.3mg	350.mg	3/90	6.43mg	18/120					
5189	TR228	3.65mg	n.s.s.	42/50	.714mg	38/50	3.57mg	36/50			liv:hpa,hpc,nnd.
a	TR228	31.4mg	n.s.s.	4/50	.714mg	0/50	3.57mg	0/50			
5190	TR228	2.81mg	n.s.s.	29/50	.714mg	25/50	3.57mg	43/50			liv:hpa,hpc,nnd.
a	TR228	10.1mg	n.s.s.	1/50	.714mg	3/50	3.57mg	3/50			
5191	357	12.6mg	n.s.s.	0/15	16.3mg	0/15					Lee;jtxe,4,15-30;1978
5192	1113m	5.05mg	n.s.s.	0/8	5.44mg	0/8					Hong;jtxe,7,909-924;1981
5193	1113n	20.4mg	n.s.s.	0/16	7.42mg	0/16					
5194	1113m	3.53mg	n.s.s.	0/8	3.81mg	0/8					
5195	1113n	12.5mg	n.s.s.	1/16	5.19mg	0/14					
5196	BT4002	138.mg	n.s.s.	0/22	23.3mg	1/26					Maltoni;aric,3,1-229;1985
a	BT4002	33.9mg	n.s.s.	35/60	23.3mg	37/54					
5197	BT403	1.05mg	n.s.s.	1/7	1.14mg	0/4	2.27mg	0/7	4.55mg	0/2	
a	BT403	6.94mg	n.s.s.	1/7	1.14mg	0/4	2.27mg	1/6	4.55mg	0/2	
b	BT403	19.2mg	n.s.s.	63/99	1.14mg	32/50	2.27mg	28/50	4.55mg	25/48	
5198	BT404	80.1ug	n.s.s.	0/10	.123mg	1/4					
a	BT404	.305mg	n.s.s.	38/74	.123mg	23/47					
5199	BT403	.702mg	n.s.s.	0/17	1.14mg	0/2	2.27mg	0/6	4.55mg	0/3	
a	BT403	.702mg	n.s.s.	0/17	1.14mg	0/2	2.27mg	0/6	4.55mg	0/3	
b	BT403	42.7mg	n.s.s.	29/100	1.14mg	11/49	2.27mg	9/48	4.55mg	9/49	
5200	BT404	.303mg	n.s.s.	0/6	.123mg	0/7					
a	BT404	.466mg	n.s.s.	13/68	.123mg	10/47					
5201	1655	74.1mg	n.s.s.	1/80	2.86mg	1/48	5.71mg	0/48	11.4mg	2/48	Quast;faat,3,55-62;1983
a	1655	106.mg	n.s.s.	0/80	2.86mg	0/48	5.71mg	0/48	11.4mg	1/48	
b	1655	8.69mg	n.s.s.	53/75	2.86mg	40/47	5.71mg	36/47	11.4mg	35/48	
c	1655	19.3mg	n.s.s.	25/76	2.86mg	22/44	5.71mg	18/46	11.4mg	14/42	
5202	1799	6.68mg	n.s.s.	64/86	5.30mg	68/86	16.1mg	74/84			Quast;faat,6,105-144;1986
a	1799	19.4mg	n.s.s.	1/84	5.30mg	7/86	(16.1mg	4/84)			
b	1799	214.mg	n.s.s.	0/84	5.30mg	1/86	16.1mg	0/84			
5203	1655	86.2mg	n.s.s.	2/80	2.50mg	0/48	5.00mg	1/48	10.0mg	1/47	Quast;faat,3,55-62;1983
a	1655	14.1mg	n.s.s.	1/80	2.50mg	0/48	5.00mg	0/48	10.0mg	0/47	
5204	1799	49.0mg	n.s.s.	1/86	3.71mg	0/85	11.3mg	0/86			Quast;faat,6,105-144;1986

FD & C VIOLET NO. 1   (benzyl violet 4B) 1694-09-3

	RefNum	LoConf	UpConf	Cntrl	1Dose	1Inc	2Dose	2Inc	3Dose	3Inc	Citation or Pathology
5205	1354	480.mg	n.s.s.	8/42	9.10mg	7/43	91.0mg	7/42	455.mg	12/48	Grasso;fctx,12,21-31;1974

```
      Spe Strain  Site   Xpo+Xpt                                            TD50    2Tailpvl
          Sex   Route  Hist    Notes                                          DR    AuOp

  a    M f asp eat liv tum 80w80 e                                          no dre  P=1.   -
5206  M m asp eat lun adc 80w80 e                                  .>       4.66gm * P<.3  -
  a    M m asp eat liv tum 80w80 e                                          no dre  P=1.   -
  b    M m asp eat lun ade 80w80 e                                          no dre  P=1.   -
5207  R f asd eat mgl car 52w52 e                              .   ±        1.14gm * P<.02 +
  a    R f asd eat edu car 52w52 e                                          1.60gm  P<.05
  b    R f asd eat tba mal 52w52 e                                          568.mg  P<.002
  c    R f asd eat tba mix 52w52 e                                          660.mg  P<.02
5208  R m asd eat tba tum 52w52                                   .>        no dre  P=1.
5209  R f sda eat mgl car 51w52 erv                          . +  .        573.mg  P<.0005+
  a    R f sda eat ski sqc 51w52 erv                                        1.10gm  P<.0005+
  b    R f sda eat tba mix 51w52 erv                                        418.mg  P<.0005+

XYLENE MIXTURE (60% m-XYLENE, 9% o-XYLENE, 14% p-XYLENE, 17% ETHYLBENZENE) ...:..10....:...100.....:..1g......:.10
5210  M f b6c gav thy fca 24m24                                   :  ± #4.79gm * P<.04 -
  a    M f b6c gav TBA MXB 24m24                                            no dre  P=1.
  b    M f b6c gav liv MXB 24m24                                            7.04gm * P<.4
  c    M f b6c gav lun MXB 24m24                                            5.13gm * P<.3
5211  M m b6c gav TBA MXB 24m24                                   :>        no dre  P=1.   -
  a    M m b6c gav liv MXB 24m24                                            no dre  P=1.
  b    M m b6c gav lun MXB 24m24                                            7.60gm * P<.6
5212  R f f34 gav mgl adn 24m24                                   :  ± #4.11gm * P<.04 -
  a    R f f34 gav TBA MXB 24m24                                            1.93gm * P<.8
  b    R f f34 gav liv MXB 24m24                                            51.3gm * P<1.
5213  R m f34 gav TBA MXB 24m24 s                              :>           700.mg * P<.5
  a    R m f34 gav liv MXB 24m24 s                                          no dre  P=1.

XYLENE MIXTURE (m-XYLENE, o-XYLENE, p-XYLENE)  ...:..10....:...100....:..1mg....:..10....:...100...:..1g.....:.10
5214  R f sda gav tba mal 24m33 e                             . +           524.mg  P<.002 +
5215  R m sda gav tba mal 24m33 e                             .>            1.67gm  P<.3   +

2,4-XYLIDINE.HCl                 100ng..:..1ug....:..10....:...100....:..1mg....:..10....:...100....:..1g....:.10
5216  M f chi eat lun mix 77w90                      :  +       :           12.4mg / P<.004 +
  a    M f chi eat liv mix 77w90                                            no dre  P=1.   -
  b    M f chi eat tba mix 77w90                                            12.2mg / P<.08
5217  M f chi eat lun mix 77w90       pool          :  +       :           13.5mg / P<.002 +
5218  M m chi eat liv mix 77w90                      :>                     65.4mg * P<.5
  a    M m chi eat lun mix 77w90                                            no dre  P=1.   -
  b    M m chi eat tba mix 77w90                                            646.mg * P<1.
5219  R m cdr eat liv mix 77w98 v                        :>                 260.mg * P<.4  -
  a    R m cdr eat tba mix 77w98 v                                          16.8mg \ P<.02 -

2,5-XYLIDINE.HCl                 100ng..:..1ug....:..10....:...100....:..1mg....:..10....:...100....:..1g....:.10
5220  M f chi eat liv hpt 68w94 a                              :  +   :     552.mg * P<.003 +
  a    M f chi eat liv mix 68w94 a                                          552.mg * P<.003 +
  b    M f chi eat lun mix 68w94 a                                          1.29gm * P<.006 -
  c    M f chi eat tba mix 68w94 a                                          475.mg * P<.003
5221  M f chi eat liv hpt 68w94 a    pool                      :  +   :     552.mg * P<.0005+
5222  M m chi eat lun mix 68w94 a                              :  +   :     765.mg * P<.005 -
  a    M m chi eat liv mix 68w94 a                                          805.mg * P<.01
  b    M m chi eat tba mix 68w94 a                                          321.mg / P<.002
5223  M m chi eat --- vsc 68w94 a    pool                      :  +   :     723.mg * P<.0005+
5224  R m cdr eat liv mix 18m24 v                          :>               3.26gm * P<.8  -
  a    R m cdr eat tba mix 18m24 v                                          67.2mg \ P<.005
5225  R m cdr eat sub mix 18m24 v    pool                   :  +   :        152.mg * P<.0005+

C.I. DISPERSE YELLOW 3           100ng..:..1ug....:..10....:...100....:..1mg....:..10....:...100....:..1g.....:.10
5226  M f b6c eat liv MXA 24m24                               :  +  :       1.02gm * P<.0005c
  a    M f b6c eat liv hpa 24m24                                            1.34gm * P<.0005c
  b    M f b6c eat --- MXA 24m24                                            1.51gm * P<.05
  c    M f b6c eat --- lym 24m24                                            1.68gm * P<.07  a
  d    M f b6c eat TBA MXB 24m24                                            769.mg * P<.02
  e    M f b6c eat liv MXB 24m24                                            1.02gm * P<.0005
  f    M f b6c eat lun MXB 24m24                                            no dre  P=1.
5227  M m b6c eat lun a/a 24m24                               :  ± #2.15gm * P<.03  -
  a    M m b6c eat TBA MXB 24m24                                            17.2gm * P<1.
  b    M m b6c eat liv MXB 24m24                                            no dre  P=1.
  c    M m b6c eat lun MXB 24m24                                            2.44gm * P<.07
5228  R f f34 eat TBA MXB 24m24                            :>               no dre  P=1.
  a    R f f34 eat liv MXB 24m24                                            46.8gm * P<.9
5229  R m f34 eat liv nnd 24m24                            :  +   :         380.mg \ P<.003 c
  a    R m f34 eat liv MXA 24m24                                            833.mg * P<.04  c
  b    R m f34 eat sto --- 24m24                            +hist           4.09gm * P<.4   a
  c    R m f34 eat TBA MXB 24m24                                            no dre  P=1.
  d    R m f34 eat liv MXB 24m24                                            833.mg * P<.04

C.I. PIGMENT YELLOW 12           100ng..:..1ug....:..10....:...100....:..1mg....:..10....:...100...:..1g....:.10
5230  M f b6c eat TBA MXB 78w96                                            no dre  P=1.
  a    M f b6c eat liv MXB 78w96                                            no dre  P=1.
  b    M f b6c eat lun MXB 78w96                                            no dre  P=1.
5231  M m b6c eat TBA MXB 78w96                                            :no dre  P=1.   -
```

	RefNum	LoConf	UpConf	Cntrl	1Dose	1Inc	2Dose	2Inc				Citation or Pathology	Brkly Code
a	1354	42.6mg	n.s.s.	0/42	9.10mg	0/43	91.0mg	0/42	455.mg	0/48			
5206	1354	1.15gm	n.s.s.	0/42	8.40mg	0/45	84.0mg	1/42	420.mg	1/46			
a	1354	40.9mg	n.s.s.	0/42	8.40mg	0/45	84.0mg	0/42	420.mg	0/46			
b	1354	519.mg	n.s.s.	18/42	8.40mg	13/45	84.0mg	18/42	420.mg	15/46			
5207	1364	430.mg	n.s.s.	0/10	2.50gm	5/16						Uematsu;jnci,51,1337-1338;1973	
a	1364	549.mg	n.s.s.	0/10	2.50gm	4/17							
b	1364	261.mg	2.03gm	0/10	2.50gm	9/17							
c	1364	274.mg	n.s.s.	1/10	2.50gm	9/17							
5208	1364	2.06gm	n.s.s.	0/10	2.00gm	0/20							
5209	1643	325.mg	1.14gm	0/35	2.41gm	18/35						Ikeda;txcy,2,275-284;1974	
a	1643	547.mg	2.72gm	0/35	2.41gm	11/35							
b	1643	246.mg	779.mg	0/35	2.41gm	22/35							

XYLENE MIXTURE (60% m-XYLENE, 9% o-XYLENE, 14% p-XYLENE, 17% ETHYLBENZENE)   1330-20-7

	RefNum	LoConf	UpConf	Cntrl	1Dose	1Inc	2Dose	2Inc	Citation or Pathology	Brkly Code
5210	TR327	1.82gm	n.s.s.	0/50	350.mg	2/50	707.mg	3/50		S
a	TR327	636.mg	n.s.s.	41/50	350.mg	42/50	707.mg	32/50		
b	TR327	1.63gm	n.s.s.	3/50	350.mg	3/50	707.mg	5/50	liv:hpa,hpc,nnd.	
c	TR327	1.40gm	n.s.s.	4/50	350.mg	5/50	707.mg	7/50	lun:a/a,a/c.	
5211	TR327	847.mg	n.s.s.	36/50	352.mg	37/50	707.mg	33/50		
a	TR327	1.71gm	n.s.s.	18/50	352.mg	13/50	707.mg	14/50	liv:hpa,hpc,nnd.	
b	TR327	1.50gm	n.s.s.	3/50	352.mg	5/50	707.mg	6/50	lun:a/a,a/c.	
5212	TR327	1.24gm	n.s.s.	0/50	175.mg	0/50	354.mg	3/50		S
a	TR327	200.mg	n.s.s.	46/50	175.mg	45/50	354.mg	46/50		
b	TR327	1.40gm	n.s.s.	2/50	175.mg	2/50	354.mg	2/50	liv:hpa,hpc,nnd.	
5213	TR327	145.mg	n.s.s.	45/50	176.mg	45/50	354.mg	32/50		
a	TR327	1.23gm	n.s.s.	3/50	176.mg	2/50	354.mg	1/50	liv:hpa,hpc,nnd.	

XYLENE MIXTURE (m-XYLENE, o-XYLENE, p-XYLENE)   (CAS# 108-38-3, 95-47-6, and 106-42-3) mixture

	RefNum	LoConf	UpConf	Cntrl	1Dose	1Inc	Citation or Pathology	Brkly Code
5214	BT904	262.mg	2.26gm	10/49	237.mg	22/40	Maltoni;ajim,7,415-446;1985/1983	
5215	BT904	483.mg	n.s.s.	11/45	237.mg	14/38		

2,4-XYLIDINE.HCl   21436-96-4

	RefNum	LoConf	UpConf	Cntrl	1Dose	1Inc	2Dose	2Inc	Citation or Pathology	Brkly Code
5216	381	5.45mg	103.mg	5/22	13.9mg	5/18	29.3mg	11/19	Weisburger;jept,2,325-356;1978/pers.comm./Russfield 1973	
a	381	122.mg	n.s.s.	1/22	13.9mg	0/18	29.3mg	0/19		
b	381	4.28mg	n.s.s.	18/22	13.9mg	11/18	29.3mg	16/19		
5217	381	5.74mg	89.6mg	31/102p	13.9mg	5/18	29.3mg	11/19		
5218	381	9.11mg	n.s.s.	1/18	12.9mg	2/20	27.0mg	3/19		
a	381	14.5mg	n.s.s.	8/18	12.9mg	2/20	27.0mg	3/19		
b	381	8.40mg	n.s.s.	13/18	12.9mg	8/20	27.0mg	9/19		
5219	381	63.9mg	n.s.s.	0/16	23.6mg	1/20	47.3mg	1/24		
a	381	6.86mg	n.s.s.	9/16	23.6mg	14/20	(47.3mg	15/24)		

2,5-XYLIDINE.HCl   51786-53-9

	RefNum	LoConf	UpConf	Cntrl	1Dose	1Inc	2Dose	2Inc	Citation or Pathology	Brkly Code
5220	381	217.mg	3.20gm	0/13	669.mg	5/16	1.56gm	2/20	Weisburger;jept,2,325-356;1978/pers.comm./Russfield 1973	
a	381	217.mg	3.20gm	0/13	669.mg	5/16	1.56gm	2/20		
b	381	403.mg	15.3gm	3/13	669.mg	2/16	1.56gm	4/20		
c	381	196.mg	3.12gm	9/13	669.mg	8/16	1.56gm	7/20		
5221	381	217.mg	2.01gm	1/102p	669.mg	5/16	1.56gm	2/20		
5222	381	294.mg	6.83gm	2/16	617.mg	4/18	1.44gm	3/19		
a	381	299.mg	48.4gm	2/16	617.mg	4/18	1.44gm	2/19		
b	381	149.mg	1.55gm	9/16	617.mg	12/18	1.44gm	12/19		
5223	381	324.mg	2.11gm	5/99p	617.mg	5/18	1.44gm	7/19		
5224	381	272.mg	n.s.s.	1/17	120.mg	1/17	240.mg	1/17		
a	381	19.1mg	747.mg	13/17	120.mg	12/17	(240.mg	12/17)		
5225	381	69.6mg	475.mg	18/111p	120.mg	7/17	240.mg	9/17		

C.I. DISPERSE YELLOW 3   2832-40-8

	RefNum	LoConf	UpConf	Cntrl	1Dose	1Inc	2Dose	2Inc	Citation or Pathology	Brkly Code
5226	TR222	590.mg	3.05gm	2/50	319.mg	10/50	638.mg	17/50	liv:hpa,hpc.	
a	TR222	769.mg	2.96gm	0/50	319.mg	6/50	638.mg	12/50		
b	TR222	654.mg	n.s.s.	10/50	319.mg	17/50	638.mg	20/50	---:leu,lym.	S
c	TR222	689.mg	n.s.s.	10/50	319.mg	16/50	638.mg	19/50		
d	TR222	362.mg	n.s.s.	20/50	319.mg	33/50	638.mg	36/50		
e	TR222	590.mg	3.05gm	2/50	319.mg	10/50	638.mg	17/50	liv:hpa,hpc,nnd.	
f	TR222	3.62gm	n.s.s.	6/50	319.mg	0/50	638.mg	4/50	lun:a/a,a/c.	
5227	TR222	960.mg	n.s.s.	2/50	294.mg	6/50	589.mg	9/50		S
a	TR222	624.mg	n.s.s.	33/50	294.mg	26/50	589.mg	33/50		
b	TR222	1.31gm	n.s.s.	20/50	294.mg	12/50	589.mg	16/50	liv:hpa,hpc,nnd.	
c	TR222	977.mg	n.s.s.	3/50	294.mg	7/50	589.mg	9/50	lun:a/a,a/c.	
5228	TR222	232.mg	n.s.s.	38/50	248.mg	40/50	(495.mg	25/50)		
a	TR222	2.42gm	n.s.s.	2/50	248.mg	1/50	495.mg	2/50	liv:hpa,hpc,nnd.	
5229	TR222	196.mg	1.80gm	1/50	198.mg	15/50	(396.mg	10/50)		
a	TR222	397.mg	n.s.s.	2/50	198.mg	15/50	396.mg	11/50	liv:hpc,nnd.	
b	TR222	1.41gm	n.s.s.	0/50	198.mg	3/50	396.mg	1/50		
c	TR222	569.mg	n.s.s.	37/50	198.mg	37/50	396.mg	32/50		
d	TR222	397.mg	n.s.s.	15/50	198.mg	15/50	396.mg	11/50	liv:hpa,hpc,nnd.	

C.I. PIGMENT YELLOW 12   (diarylanilide yellow) 6358-85-6

	RefNum	LoConf	UpConf	Cntrl	1Dose	1Inc	2Dose	2Inc	Citation or Pathology	Brkly Code
5230	TR30	10.1gm	n.s.s.	12/50	2.61gm	10/50	5.23gm	9/50		
a	TR30	51.7mg	n.s.s.	2/50	2.61gm	5/50	5.23gm	0/50	liv:hpa,hpc,nnd.	
b	TR30	25.5mg	n.s.s.	4/50	2.61gm	3/50	5.23gm	1/50	lun:a/a,a/c.	
5231	TR30	8.17gm	n.s.s.	20/50	2.44gm	21/50	4.82gm	13/50		

```
    Spe Strain Site  Xpo+Xpt                                              TD50   2Tailpvl
        Sex  Route Hist   Notes                                               DR   AuOp
a     M m b6c eat liv MXB 78w96                                         no dre  P=1.
b     M m b6c eat lun MXB 78w96                                         no dre  P=1.
5232  M b nmr eat tba tum 24m24                               .>       9.11gm * P<.3   -
5233  R f f34 eat TBA MXB 18m25                                  :>     4.14gm * P<.4   -
a     R f f34 eat liv MXB 18m25                                        71.9gm * P<.6
5234  R m f34 eat TBA MXB 18m25                                  :>     no dre  P=1.   -
a     R m f34 eat liv MXB 18m25                                        no dre  P=1.
5235  R b sda eat tba tum 24m24                               .>       5.33gm Z P<.6

C.I. PIGMENT YELLOW 16        100ng..:.1ug....:.10.....:.100....:.1mg....:.10.....:.100....:.1g.....:.10
5236  M b nmr eat tba tum 24m24 e                            .>       11.7gm * P<.5   -
5237  R b sda eat tba tum 24m24 e                            .>       13.7gm Z P<.8   -

C.I. PIGMENT YELLOW 83        100ng..:.1ug....:.10.....:.100....:.1mg....:.10.....:.100....:.1g.....:.10
5238  M b nmr eat tba tum 24m24 e                               .>     no dre  P=1.   -
5239  R b sda eat tba tum 24m24 e                            .>        3.56gm Z P<.4   -

C.I. VAT YELLOW 4             100ng..:.1ug....:.10.....:.100....:.1mg....:.10.....:.100....:.1g.....:.10
5240  M f b6c eat TBA MXB 25m25                                 :>     no dre  P=1.   -
a     M f b6c eat liv MXB 25m25                                        23.9gm * P<.5
b     M f b6c eat lun MXB 25m25                                        no dre  P=1.
5241  M m b6c eat --- lym 25m25                                 :      10.9gm * P<.009 c
a     M m b6c eat TBA MXB 25m25                                        4.41gm * P<.05
b     M m b6c eat liv MXB 25m25                                        8.61gm * P<.07
c     M m b6c eat lun MXB 25m25                                        19.1gm * P<.4
5242  R f f34 eat TBA MXB 24m24                           :>           no dre  P=1.   -
a     R f f34 eat liv MXB 24m24                                        no dre  P=1.
5243  R m f34 eat TBA MXB 24m24                         :>             787.mg * P<.5   -
a     R m f34 eat liv MXB 24m24                                        2.79gm * P<.2

FD & C YELLOW NO. 5           100ng..:.1ug....:.10.....:.100....:.1mg....:.10.....:.100....:.1g.....:.10
5244  M f cd1 eat --- mlp 24m24 e                          .  +  .     1.43gm Z P<.0005-
a     M f cd1 eat liv hpa 24m24 e                                     200.gm * P<.2   -
b     M f cd1 eat liv hpd 24m24 e                                      no dre  P=1.   -
c     M f cd1 eat lun a/a 24m24 e                                      no dre  P=1.   -
d     M f cd1 eat lun acb 24m24 e                                      no dre  P=1.   -
e     M f cd1 eat tba mal 24m24 e                                      1.56gm Z P<.0005-
f     M f cd1 eat tba ben 24m24 e                                      no dre  P=1.   -
5245  M m cd1 eat liv hpa 24m24 e                          .  ±        2.15gm Z P<.02  -
a     M m cd1 eat --- mlp 24m24 e                                      11.8gm Z P<.02  -
b     M m cd1 eat lun a/a 24m24 e                                      42.5gm * P<.4   -
c     M m cd1 eat liv hpd 24m24 e                                      61.1gm * P<.3   -
d     M m cd1 eat lun acb 24m24 e                                      no dre  P=1.   -
e     M m cd1 eat tba ben 24m24 e                                     388.mg Z P<.0005-
f     M m cd1 eat tba mal 24m24 e                                      1.81gm Z P<.0005-
5246  R f f3d wat ute esp 24m26 e                             .>       5.78gm * P<.2   -
a     R f f3d wat liv nnd 24m26 e                                      no dre  P=1.   -
b     R f f3d wat tba tum 24m26 e                                      no dre  P=1.   -
5247  R m f3d wat abc mso 24m26 e                          .  +        2.82gm \ P<.004 -
a     R m f3d wat liv hpc 24m26 e                                      no dre  P=1.   -
b     R m f3d wat liv nnd 24m26 e                                      no dre  P=1.   -
c     R m f3d wat tba tum 24m26 e                                      noTD50  P<.4   -
5248  R f nss eat liv tum 64w64 e                  .>                  no dre  P=1.   -
a     R f nss eat tba mix 64w64 e                                      no dre  P=1.   -
5249  R m nss eat liv tum 64w64 e                  .>                  no dre  P=1.   -
a     R m nss eat tba mix 64w64 e                                      no dre  P=1.   -
5250  R b osm eat liv ade 24m24 e                                  .>231.gm * P<.9   -
a     R b osm eat tba mal 24m24 e                                      no dre  P=1.   -
b     R b osm eat tba mix 24m24 e                                      no dre  P=1.   -

FD & C YELLOW NO. 6           100ng..:.1ug....:.10.....:.100....:.1mg....:.10.....:.100....:.1g.....:.10
5251  M f b6c eat TBA MXB 24m24                                    :>no dre  P=1.   -
a     M f b6c eat liv MXB 24m24                                        no dre  P=1.   -
b     M f b6c eat lun MXB 24m24                                        66.4gm * P<.3   -
5252  M m b6c eat ski MXA 24m24                                       #37.5gm * P<.05  -
a     M m b6c eat TBA MXB 24m24                                        17.8gm * P<.7
b     M m b6c eat liv MXB 24m24                                        14.2gm * P<.5
c     M m b6c eat lun MXB 24m24                                        no dre  P=1.
5253  M f cdr eat liv tum 80w80 e                             .>       no dre  P=1.   -
a     M f cdr eat lun ade 80w80 e                                      no dre  P=1.   -
5254  M m cdr eat liv tum 80w80 e                             .>       no dre  P=1.   -
a     M m cdr eat lun ade 80w80 e                                      no dre  P=1.   -
5255  R f f34 eat TBA MXB 24m24                                 :>     9.06gm * P<.8   -
a     R f f34 eat liv MXB 24m24                                        no dre  P=1.   -
5256  R m f34 eat TBA MXB 24m24                                 :>     11.8gm * P<.9   -
a     R m f34 eat liv MXB 24m24                                        2.52gm \ P<.08
5257  R f nss eat liv tum 64w64 e                  .>                  no dre  P=1.   -
a     R f nss eat tba mix 64w64 e                                      87.4mg Z P<.02  -
5258  R m nss eat liv tum 64w64 e                 .>                   no dre  P=1.   -
a     R m nss eat tba mix 64w64 e                                      14.0gm * P<1.   -
```

	RefNum	LoConf	UpConf	Cntrl	1Dose	1Inc	2Dose	2Inc			Citation or Pathology Brkly Code
a	TR30	5.41gm	n.s.s.	15/50	2.44gm	11/50	(4.82gm	4/50)			liv:hpa,hpc,nnd.
b	TR30	16.6gm	n.s.s.	7/50	2.44gm	5/50	4.82gm	4/50			lun:a/a,a/c.
5232	1021	2.46gm	n.s.s.	33/100	125.mg	21/100	375.mg	26/100	1.12gm	34/100	Leuschner;txlt,2,253-260;1978
5233	TR30	1.13gm	n.s.s.	31/50	920.mg	40/50	1.84gm	34/50			
a	TR30	10.4gm	n.s.s.	1/50	920.mg	0/50	1.84gm	2/50			liv:hpa,hpc,nnd.
5234	TR30	2.64gm	n.s.s.	31/50	736.mg	25/50	1.47gm	23/50			
a	TR30	n.s.s.	n.s.s.	0/50	736.mg	1/50	1.47gm	0/50			liv:hpa,hpc,nnd.
5235	1021	1.03gm	n.s.s.	43/100	45.0mg	20/100	135.mg	18/100	405.mg	37/100	Leuschner;txlt,2,253-260;1978

C.I. PIGMENT YELLOW 16   5979-28-2

	RefNum	LoConf	UpConf	Cntrl	1Dose	1Inc	2Dose	2Inc			Citation or Pathology
5236	1021	2.50gm	n.s.s.	33/100	125.mg	22/100	375.mg	33/100	1.13gm	33/100	Leuschner;txlt,2,253-260;1978
5237	1021	1.19gm	n.s.s.	43/100	45.0mg	24/100	135.mg	16/100	405.mg	36/100	

C.I. PIGMENT YELLOW 83   5567-15-7

	RefNum	LoConf	UpConf	Cntrl	1Dose	1Inc	2Dose	2Inc			Citation or Pathology
5238	1021	4.66gm	n.s.s.	33/100	125.mg	23/100	375.mg	19/100	1.13gm	26/100	Leuschner;txlt,2,253-260;1978
5239	1021	897.mg	n.s.s.	43/100	45.0mg	23/100	135.mg	20/100	405.mg	40/100	

C.I. VAT YELLOW 4   128-66-5

	RefNum	LoConf	UpConf	Cntrl	1Dose	1Inc	2Dose	2Inc			Pathology
5240	TR134	3.77gm	n.s.s.	13/20	1.62gm	29/50	3.25gm	29/50			
a	TR134	5.84gm	n.s.s.	2/20	1.62gm	6/50	3.25gm	9/50			liv:hpa,hpc,nnd.
b	TR134	11.6gm	n.s.s.	2/20	1.62gm	5/50	3.25gm	3/50			lun:a/a,a/c.
5241	TR134	5.66gm	333.gm	3/20	3.00gm	7/50	6.00gm	22/50			
a	TR134	1.94gm	n.s.s.	12/20	3.00gm	42/50	6.00gm	48/50			
b	TR134	3.87gm	n.s.s.	3/20	3.00gm	22/50	6.00gm	21/50			liv:hpa,hpc,nnd.
c	TR134	5.69gm	n.s.s.	4/20	3.00gm	14/50	6.00gm	15/50			lun:a/a,a/c.
5242	TR134	474.mg	n.s.s.	9/20	175.mg	26/50	350.mg	23/50			
a	TR134	n.s.s.	n.s.s.	0/20	175.mg	0/50	350.mg	0/50			liv:hpa,hpc,nnd.
5243	TR134	185.mg	n.s.s.	11/20	140.mg	32/50	280.mg	31/50			
a	TR134	963.mg	n.s.s.	0/20	140.mg	1/50	280.mg	3/50			liv:hpa,hpc,nnd.

FD & C YELLOW NO. 5   (tartrazine) 1934-21-0

	RefNum	LoConf	UpConf	Cntrl	1Dose	1Inc	2Dose	2Inc	3Dose	3Inc	Citation
5244	1869	616.mg	5.36gm	3/120	650.mg	8/28	(1.95gm	4/30	6.50gm	2/60)	Borzelleca;fctx,26,189-194;1988
a	1869	40.6mg	n.s.s.	1/120	650.mg	0/28	1.95gm	0/30	6.50gm	2/60	
b	1869	53.6mg	n.s.s.	3/120	650.mg	0/28	1.95gm	1/30	6.50gm	1/60	
c	1869	44.4mg	n.s.s.	29/120	650.mg	7/28	1.95gm	7/30	6.50gm	6/60	
d	1869	46.9mg	n.s.s.	3/120	650.mg	1/28	1.95gm	2/30	6.50gm	1/60	
e	1869	813.mg	4.84gm	34/120	650.mg	15/28	1.95gm	20/30	(6.50gm	15/60)	
f	1869	32.4mg	n.s.s.	52/120	650.mg	14/28	1.95gm	10/30	6.50gm	15/60	
5245	1869	732.mg	n.s.s.	7/120	600.mg	6/27	(1.80gm	3/33	6.00gm	2/60)	
a	1869	3.96gm	n.s.s.	2/120	600.mg	1/27	1.80gm	4/33	(6.00gm	3/60)	
b	1869	10.3gm	n.s.s.	23/120	600.mg	11/27	1.80gm	11/33	6.00gm	17/60	
c	1869	14.8gm	n.s.s.	9/120	600.mg	4/27	1.80gm	7/33	6.00gm	8/60	
d	1869	73.5mg	n.s.s.	3/120	600.mg	0/27	1.80gm	1/33	6.00gm	0/60	
e	1869	182.mg	1.36gm	43/120	600.mg	21/27	(1.80gm	13/33	6.00gm	22/60)	
f	1869	961.mg	5.11gm	27/120	600.mg	6/27	1.80gm	22/33	(6.00gm	20/60)	
5246	1857	1.98gm	n.s.s.	5/47	531.mg	13/50	1.06gm	10/49			Maekawa;fctx,25,891-896;1987
a	1857	9.15gm	n.s.s.	1/47	531.mg	1/50	1.06gm	1/49			
b	1857	2.13gm	n.s.s.	39/47	531.mg	41/50	1.06gm	30/49			
5247	1857	1.15gm	18.3gm	0/48	464.mg	6/49	(929.mg	0/49)			
a	1857	8.82gm	n.s.s.	0/48	464.mg	1/49	929.mg	0/49			
b	1857	9.84gm	n.s.s.	3/48	464.mg	3/49	929.mg	0/49			
c	1857	n.s.s.	n.s.s.	47/48	464.mg	49/49	929.mg	49/49			
5248	1358	13.6mg	n.s.s.	0/10	15.0mg	0/13	150.mg	0/14	750.mg	0/10	Mannell;jphp,10,625-634;1958
a	1358	273.mg	n.s.s.	2/10	15.0mg	4/13	150.mg	5/14	750.mg	2/10	
5249	1358	11.4mg	n.s.s.	0/11	12.0mg	0/14	120.mg	0/12	600.mg	0/9	
a	1358	439.mg	n.s.s.	1/11	12.0mg	2/14	120.mg	1/12	600.mg	0/9	
5250	305a	7.86gm	n.s.s.	0/18	225.mg	2/22	450.mg	0/21	900.mg	0/24	2.25gm 1/23	Davis;txap,6,621-626;1964
a	305a	4.76gm	n.s.s.	6/18	225.mg	4/22	450.mg	6/21	900.mg	6/24	2.25gm 4/23	
b	305a	5.13gm	n.s.s.	8/18	225.mg	9/22	450.mg	9/21	900.mg	8/24	2.25gm 5/23	

FD & C YELLOW NO. 6   (sunset yellow FCF) 2783-94-0

	RefNum	LoConf	UpConf	Cntrl	1Dose	1Inc	2Dose	2Inc	3Dose	3Inc	Citation or Pathology
5251	TR208	7.16gm	n.s.s.	28/50	1.61gm	20/50	3.22gm	21/50			
a	TR208	15.8gm	n.s.s.	7/50	1.61gm	3/50	3.22gm	4/50			liv:hpa,hpc,nnd.
b	TR208	16.3gm	n.s.s.	0/50	1.61gm	1/50	3.22gm	1/50			lun:a/a,a/c.
5252	TR208	11.3gm	n.s.s.	0/50	1.49gm	0/49	2.97gm	3/50			ski:fbs,fib. S
a	TR208	2.42gm	n.s.s.	32/50	1.49gm	31/49	2.97gm	34/50			
b	TR208	3.04gm	n.s.s.	13/50	1.49gm	23/49	2.97gm	16/50			liv:hpa,hpc,nnd.
c	TR208	13.3gm	n.s.s.	6/50	1.49gm	4/49	2.97gm	3/50			lun:a/a,a/c.
5253	1355	476.mg	n.s.s.	0/47	260.mg	0/29	520.mg	0/27	1.04gm	0/27	2.08gm 0/29	Gaunt;fctx,12,1-10;1974
a	1355	6.02gm	n.s.s.	12/47	260.mg	6/29	520.mg	3/27	1.04gm	2/27	2.08gm 3/29	
5254	1355	412.mg	n.s.s.	0/50	240.mg	0/26	480.mg	0/30	960.mg	0/27	1.92gm 0/19	
a	1355	2.34gm	n.s.s.	7/50	240.mg	2/26	480.mg	8/30	960.mg	2/27	1.92gm 3/19	
5255	TR208	1.03gm	n.s.s.	68/90	619.mg	37/50	1.24gm	42/50			
a	TR208	8.31gm	n.s.s.	3/90	619.mg	3/50	1.24gm	0/50			liv:hpa,hpc,nnd.
5256	TR208	973.mg	n.s.s.	61/90	495.mg	34/50	990.mg	35/50			
a	TR208	802.mg	n.s.s.	5/90	495.mg	7/50	(990.mg	1/50)			liv:hpa,hpc,nnd.
5257	1358	15.3mg	n.s.s.	0/10	15.0mg	0/15	150.mg	0/13	750.mg	0/9	Mannell;jphp,10,625-634;1958
a	1358	31.1mg	n.s.s.	2/10	15.0mg	0/15	150.mg	6/13	(750.mg	2/9)	
5258	1358	6.91mg	n.s.s.	0/11	12.0mg	0/8	120.mg	0/12	600.mg	0/9	
a	1358	264.mg	n.s.s.	1/11	12.0mg	1/8	120.mg	1/12	600.mg	1/9	

```
        Spe Strain  Site   Xpo+Xpt                                                              TD50     2Tailpvl
            Sex  Route  Hist    Notes                                                                 DR     AuOp
HC YELLOW 4                     100ng..:..1ug...:..10.....:..100....:..1mg...:..10.....:..100...:..1g.....:..10
5259 M f b6c eat TBA MXB 24m24                                                               :>      no dre   P=1.   -
a    M f b6c eat liv MXB 24m24                                                                       no dre   P=1.
b    M f b6c eat lun MXB 24m24                                                                       no dre   P=1.
5260 M m b6c eat TBA MXB 24m24                                                            :>         no dre   P=1.   -
a    M m b6c eat liv MXB 24m24                                                                       no dre   P=1.
b    M m b6c eat lun MXB 24m24                                                                       no dre   P=1.
5261 R f f34 eat TBA MXB 24m24                                                 :>                    no dre   P=1.   -
a    R f f34 eat liv MXB 24m24                                                                       no dre   P=1.
5262 R m f34 eat pit pda 24m24                                            :>                         465.mg * P<.3  e
a    R m f34 eat TBA MXB 24m24                                                                       no dre   P=1.
b    R m f34 eat liv MXB 24m24                                                                       no dre   P=1.

ZEARALENONE                    100ng..:..1ug...:..10.....:..100....:..1mg...:..10.....:..100...:..1g.....:..10
5263 M f b6c eat MXB MXB 24m24                                     : + :                             22.0mg * P<.003
a    M f b6c eat pit MXA 24m24                                                                       32.3mg / P<.002 c
b    M f b6c eat pit adn 24m24                                                                       37.4mg / P<.006 c
c    M f b6c eat liv hpa 24m24                                                                       50.1mg * P<.002 c
d    M f b6c eat liv MXA 24m24                                                                       38.8mg * P<.03  c
e    M f b6c eat TBA MXB 24m24                                                                       17.9mg * P<.05
f    M f b6c eat liv MXB 24m24                                                                       38.8mg * P<.03
g    M f b6c eat lun MXB 24m24                                                                       no dre   P=1.
5264 M m b6c eat pit MXA 24m24                                              : + :                    49.1mg * P<.005 c
a    M m b6c eat pit ade 24m24                                                                       53.3mg * P<.005 c
b    M m b6c eat TBA MXB 24m24                                                                       no dre   P=1.
c    M m b6c eat liv MXB 24m24                                                                       no dre   P=1.
d    M m b6c eat lun MXB 24m24                                                                       no dre   P=1.
5265 R f f34 eat TBA MXB 24m24                                         :>                            80.9mg * P<1.  -
a    R f f34 eat liv MXB 24m24                                                                       33.5mg * P<.1
5266 R m f34 eat TBA MXB 24m24                                       :>                              16.2mg * P<.8  -
a    R m f34 eat liv MXB 24m24                                                                       950.mg * P<1.

ZINC DIBUTYLDITHIOCARBAMATE    100ng..:..1ug...:..10.....:..100....:..1mg...:..10.....:..100...:..1g.....:..10
5267 M f b6a orl lun ade 76w76 evx                                                 .>               1.09gm   P<.4  -
a    M f b6a orl liv hpt 76w76 evx                                                                  no dre   P=1.
b    M f b6a orl tba mix 76w76 evx                                                                  2.33gm   P<.7  -
5268 M m b6a orl lun ade 76w76 evx                                                     .>           no dre   P=1.
a    M m b6a orl liv hpt 76w76 evx                                                                  no dre   P=1.
b    M m b6a orl tba mix 76w76 evx                                                                  no dre   P=1.
5269 M f b6c orl lun ade 76w76 evx                                                     .>           2.33gm   P<.3  -
a    M f b6c orl liv hpt 76w76 evx                                                                  no dre   P=1.
b    M f b6c orl tba mix 76w76 evx                                                                  731.mg   P<.05 -
5270 M m b6c orl liv hpt 76w76 evx                                                 .  ±             680.mg   P<.05 -
a    M m b6c orl lun ade 76w76 evx                                                                  1.05gm   P<.2  -
b    M m b6c orl tba mix 76w76 evx                                                                  252.mg   P<.002 -

ZINC DIETHYLDITHIOCARBAMATE    100ng..:..1ug...:..10.....:..100....:..1mg...:..10.....:..100...:..1g.....:..10
5271 M f b6a orl liv hpt 76w76 evx                                               .>                 no dre   P=1.  -
a    M f b6a orl lun ade 76w76 evx                                                                  no dre   P=1.  -
b    M f b6a orl tba mix 76w76 evx                                                                  no dre   P=1.  -
5272 M m b6a orl liv hpt 76w76 evx                                           .>                     205.mg   P<.6  -
a    M m b6a orl lun ade 76w76 evx                                                                  no dre   P=1.
b    M m b6a orl tba mix 76w76 evx                                                                  86.7mg   P<.5  -
5273 M f b6c orl liv hpt 76w76 evx                                               .>                 no dre   P=1.  -
a    M f b6c orl lun mix 76w76 evx                                                                  no dre   P=1.  -
b    M f b6c orl tba tum 76w76 evx                                                                  no dre   P=1.  -
5274 M m b6c orl lun ade 76w76 evx                                           .  ±                   46.2mg   P<.02 -
a    M m b6c orl liv hpt 76w76 evx                                                                  no dre   P=1.  -
b    M m b6c orl tba mix 76w76 evx                                                                  28.5mg   P<.003 -

ZINC DIMETHYLDITHIOCARBAMATE   100ng..:..1ug...:..10.....:..100....:..1mg...:..10.....:..100...:..1g.....:..10
5275 M f b6c eat lun a/a 24m24                                             : ±                      561.mg * P<.04 e
a    M f b6c eat --- mlp 24m24                                                                      947.mg * P<.05
b    M f b6c eat TBA MXB 24m24                                                                      no dre   P=1.
c    M f b6c eat liv MXB 24m24                                                                      no dre   P=1.
d    M f b6c eat lun MXB 24m24                                                                      710.mg * P<.2
5276 M f b6c orl lun ade 76w76 evx                                         .>                       13.1mg   P<.3  -
a    M f b6c orl liv hpt 76w76 evx                                                                  no dre   P=1.  -
b    M f b6c orl tba mix 76w76 evx                                                                  13.1mg   P<.3  -
5277 M m b6c eat TBA MXB 24m24                                                   :>                 no dre   P=1.  -
a    M m b6c eat liv MXB 24m24                                                                      no dre   P=1.
b    M m b6c eat lun MXB 24m24                                                                      765.mg * P<.3  -
5278 M m b6c orl lun mix 76w76 evx                                      .  ±                        5.23mg   P<.09 -
a    M m b6c orl liv hpt 76w76 evx                                                                  10.8mg   P<.3  -
b    M m b6c orl tba mix 76w76 evx                                                                  3.37mg   P<.04 -
5279 M f b6a orl liv hpt 76w76 evx                                        .>                        no dre   P=1.  -
a    M f b6a orl lun ade 76w76 evx                                                                  no dre   P=1.  -
b    M f b6a orl tba mix 76w76 evx                                                                  no dre   P=1.  -
5280 M m b6a orl lun ade 76w76 evx                                      .>                          9.15mg   P<.6  -
a    M m b6a orl liv hpt 76w76 evx                                                                  202.mg   P<1.  -
b    M m b6a orl tba mix 76w76 evx                                                                  4.21mg   P<.4  -
```

	RefNum	LoConf	UpConf	Cntrl	1Dose	1Inc	2Dose	2Inc	Citation or Pathology
									Brkly Code
HC YELLOW 4		59820-43-8							
5259	TR419	2.85gm	n.s.s.	30/50	642.mg	24/50	1.28gm	21/50	
a	TR419	4.37gm	n.s.s.	6/50	642.mg	8/50	1.28gm	4/50	liv:hpa,hpc,nnd.
b	TR419	8.03gm	n.s.s.	4/50	642.mg	2/50	1.28gm	2/50	lun:a/a,a/c.
5260	TR419	1.95gm	n.s.s.	28/50	589.mg	23/50	1.18gm	23/50	
a	TR419	2.44gm	n.s.s.	13/50	589.mg	10/50	1.18gm	12/50	liv:hpa,hpc,nnd.
b	TR419	2.71gm	n.s.s.	8/50	589.mg	9/50	1.18gm	7/50	lun:a/a,a/c.
5261	TR419	389.mg	n.s.s.	49/50	246.mg	49/50	495.mg	48/50	
a	TR419	3.63gm	n.s.s.	1/50	246.mg	1/50	495.mg	0/50	liv:hpa,hpc,nnd.
5262	TR419	142.mg	n.s.s.	17/51	98.7mg	20/50	198.mg	28/50	
a	TR419	153.mg	n.s.s.	47/51	98.7mg	48/50	198.mg	45/50	
b	TR419	1.81gm	n.s.s.	1/51	98.7mg	0/50	198.mg	0/50	liv:hpa,hpc,nnd.
ZEARALENONE		17924-92-4							
5263	TR235	11.4mg	121.mg	6/50	6.50mg	8/50	13.0mg	20/50	liv:hpa,hpc; pit:adn,car. C
a	TR235	16.1mg	165.mg	3/50	6.50mg	2/50	13.0mg	15/50	pit:adn,car.
b	TR235	17.6mg	442.mg	3/50	6.50mg	2/50	13.0mg	13/50	
c	TR235	23.5mg	194.mg	0/50	6.50mg	2/50	13.0mg	7/50	
d	TR235	17.2mg	n.s.s.	3/50	6.50mg	7/50	13.0mg	10/50	liv:hpa,hpc.
e	TR235	7.55mg	n.s.s.	28/50	6.50mg	26/50	13.0mg	40/50	
f	TR235	17.2mg	n.s.s.	3/50	6.50mg	7/50	13.0mg	10/50	liv:hpa,hpc,nnd.
g	TR235	59.5mg	n.s.s.	3/50	6.50mg	4/50	13.0mg	1/50	lun:a/a,a/c.
5264	TR235	24.6mg	383.mg	0/50	6.00mg	5/50	12.0mg	6/50	pit:ade,car.
a	TR235	26.0mg	417.mg	0/50	6.00mg	4/50	12.0mg	6/50	
b	TR235	10.6mg	n.s.s.	37/50	6.00mg	41/50	12.0mg	38/50	
c	TR235	25.9mg	n.s.s.	19/50	6.00mg	22/50	12.0mg	14/50	liv:hpa,hpc,nnd.
d	TR235	28.9mg	n.s.s.	11/50	6.00mg	8/50	12.0mg	11/50	lun:a/a,a/c.
5265	TR235	2.26mg	n.s.s.	32/50	1.25mg	40/50	2.50mg	33/50	
a	TR235	10.2mg	n.s.s.	0/50	1.25mg	1/50	2.50mg	2/50	liv:hpa,hpc,nnd.
5266	TR235	1.71mg	n.s.s.	32/50	1.00mg	36/50	2.00mg	33/50	
a	TR235	11.9mg	n.s.s.	2/50	1.00mg	0/50	2.00mg	2/50	liv:hpa,hpc,nnd.
ZINC DIBUTYLDITHIOCARBAMATE		(butyl zimate)	136-23-2						
5267	1220	239.mg	n.s.s.	1/17	364.mg	3/18			Innes;ntis,1968/1969
a	1220	721.mg	n.s.s.	0/17	364.mg	0/18			
b	1220	261.mg	n.s.s.	2/17	364.mg	3/18			
5268	1220	317.mg	n.s.s.	2/18	339.mg	2/18			
a	1220	672.mg	n.s.s.	1/18	339.mg	0/18			
b	1220	351.mg	n.s.s.	3/18	339.mg	2/18			
5269	1220	379.mg	n.s.s.	0/16	364.mg	1/18			
a	1220	721.mg	n.s.s.	0/16	364.mg	0/18			
b	1220	220.mg	n.s.s.	0/16	364.mg	3/18			
5270	1220	205.mg	n.s.s.	0/16	339.mg	3/18			
a	1220	259.mg	n.s.s.	0/16	339.mg	2/18			
b	1220	107.mg	1.03gm	0/16	339.mg	7/18			
ZINC DIETHYLDITHIOCARBAMATE		(ethyl zimate)	14324-55-1						
5271	1210	68.1mg	n.s.s.	0/17	36.4mg	0/17			Innes;ntis,1968/1969
a	1210	68.1mg	n.s.s.	1/17	36.4mg	0/17			
b	1210	32.0mg	n.s.s.	2/17	36.4mg	2/17			
5272	1210	28.4mg	n.s.s.	1/18	33.9mg	2/18			
a	1210	31.7mg	n.s.s.	2/18	33.9mg	2/18			
b	1210	17.3mg	n.s.s.	3/18	33.9mg	5/18			
5273	1210	68.1mg	n.s.s.	0/16	36.4mg	0/17			
a	1210	68.1mg	n.s.s.	0/16	36.4mg	0/17			
b	1210	68.1mg	n.s.s.	0/16	36.4mg	0/17			
5274	1210	15.9mg	n.s.s.	0/16	33.9mg	4/17			
a	1210	63.4mg	n.s.s.	0/16	33.9mg	0/17			
b	1210	11.5mg	155.mg	0/16	33.9mg	6/17			
ZINC DIMETHYLDITHIOCARBAMATE		(methyl zimate, milbam, ziram) 137-30-4							
5275	TR238	245.mg	n.s.s.	2/50	76.5mg	5/50	154.mg	10/50	
a	TR238	373.mg	n.s.s.	1/50	76.5mg	1/50	154.mg	7/50	S
b	TR238	338.mg	n.s.s.	28/50	76.5mg	19/50	154.mg	23/50	
c	TR238	341.mg	n.s.s.	9/50	76.5mg	4/50	(154.mg	1/50)	liv:hpa,hpc,nnd.
d	TR238	254.mg	n.s.s.	4/50	76.5mg	6/50	154.mg	11/50	lun:a/a,a/c.
5276	1204	2.14mg	n.s.s.	0/16	2.05mg	1/18			Innes;ntis,1968/1969
a	1204	4.07mg	n.s.s.	0/16	2.05mg	0/18			
b	1204	2.14mg	n.s.s.	0/16	2.05mg	1/18			
5277	TR238	192.mg	n.s.s.	31/50	70.6mg	24/50	143.mg	25/49	
a	TR238	209.mg	n.s.s.	19/50	70.6mg	9/50	(143.mg	9/49)	liv:hpa,hpc,nnd.
b	TR238	215.mg	n.s.s.	8/50	70.6mg	8/50	143.mg	12/49	lun:a/a,a/c.
5278	1204	1.28mg	n.s.s.	0/16	1.91mg	2/16			Innes;ntis,1968/1969
a	1204	1.76mg	n.s.s.	0/16	1.91mg	1/16			
b	1204	1.01mg	n.s.s.	0/16	1.91mg	3/16			
5279	1204	4.07mg	n.s.s.	0/17	2.05mg	0/18			
a	1204	4.07mg	n.s.s.	1/17	2.05mg	0/18			
b	1204	4.07mg	n.s.s.	2/17	2.05mg	0/18			
5280	1204	1.27mg	n.s.s.	2/18	1.91mg	3/17			
a	1204	2.12mg	n.s.s.	1/18	1.91mg	1/17			
b	1204	.898mg	n.s.s.	3/18	1.91mg	5/17			

```
      Spe Strain Site   Xpo+Xpt                                                      TD50    2Tailpvl
          Sex  Route  Hist   Notes                                                        DR      AuOp
5281 R f f34 eat TBA MXB 24m24                                      :>              no dre  P=1.     -
a    R f f34 eat liv MXB 24m24                                                      no dre  P=1.
5282 R m f34 eat thy ccr 24m24                            :    +        :           95.8mg * P<.004 c
a    R m f34 eat TBA MXB 24m24                                                      no dre  P=1.
b    R m f34 eat liv MXB 24m24                                                      no dre  P=1.
5283 R b mgr gav liv mhp 22m24 e                      .    +            .           56.5mg  P<.008
a    R b mgr gav tba mix 22m24 e                                                    25.8mg  P<.002 +


ZINC ETHYLENEBISTHIOCARBAMATE      100ng..:..1ug....:..10....:..100....:..1mg....:..10.......:..100....:..1g......:..10
5284 M f b6a orl lun ade 76w76 evx                                 .>              1.15gm  P<.6     -
a    M f b6a orl liv hpt 76w76 evx                                                 no dre  P=1.     -
b    M f b6a orl tba mix 76w76 evx                                                 1.15gm  P<.7     -
5285 M m b6a orl liv hpt 76w76 evx                               .>               903.mg  P<.6     -
a    M m b6a orl lun ade 76w76 evx                                                 no dre  P=1.     -
b    M m b6a orl tba mix 76w76 evx                                                 715.mg  P<.7     -
5286 M f b6c orl lun ade 76w76 evx                                  .>             1.15gm  P<.3     -
a    M f b6c orl liv hpt 76w76 evx                                                 no dre  P=1.     -
b    M f b6c orl tba mix 76w76 evx                                                 1.15gm  P<.3     -
5287 M m b6c orl liv hpt 76w76 evx                          .   ±                 244.mg  P<.02    -
a    M m b6c orl lun mix 76w76 evx                                                 no dre  P=1.     -
b    M m b6c orl tba mix 76w76 evx                                                 189.mg  P<.009   -
5288 R b mgr gav liv tum 22m24 e                              .>                   no dre  P=1.     -
a    R b mgr gav tba mix 22m24 e                                                  255.mg  P<.06    +


ZIRCONIUM (IV) SULFATE             100ng..:..1ug....:..10....:..100....:..1mg....:..10.......:..100....:..1g......:..10
5289 M b cd1 wat lun tum 32m32 e                        .>                         no dre  P=1.     -
a    M b cd1 wat liv tum 32m32 e                                                   no dre  P=1.     -
b    M b cd1 wat tba tum 32m32 e                                                   no dre  P=1.     -
c    M b cd1 wat tba mal 32m32 e                                                   no dre  P=1.     -
```

	RefNum	LoConf	UpConf	Cntrl	1Dose	1Inc	2Dose	2Inc	Citation or Pathology
									Brkly Code
5281	TR238	39.5mg	n.s.s.	43/50	14.6mg	42/50	29.6mg	39/50	
a	TR238	n.s.s.	n.s.s.	0/50	14.6mg	0/50	29.6mg	0/50	liv:hpa,hpc,nnd.
5282	TR238	45.1mg	602.mg	0/50	11.8mg	2/50	23.8mg	7/50	
a	TR238	21.8mg	n.s.s.	37/50	11.8mg	42/50	23.8mg	39/50	
b	TR238	140.mg	n.s.s.	2/50	11.8mg	2/50	23.8mg	1/50	liv:hpa,hpc,nnd.
5283	1426	13.8mg	1.94gm	0/46	18.4mg	2/10			Andrianova;vpit,29,71-74;1970
a	1426	8.47mg	198.mg	1/46	18.4mg	4/10			

ZINC ETHYLENEBISTHIOCARBAMATE   (zineb) 12122-67-7

	RefNum	LoConf	UpConf	Cntrl	1Dose	1Inc			Citation or Pathology
5284	1213	152.mg	n.s.s.	1/17	180.mg	2/18			Innes;ntis,1968/1969
a	1213	357.mg	n.s.s.	0/17	180.mg	0/18			
b	1213	130.mg	n.s.s.	2/17	180.mg	3/18			
5285	1213	132.mg	n.s.s.	1/18	168.mg	2/17			
a	1213	206.mg	n.s.s.	2/18	168.mg	1/17			
b	1213	96.4mg	n.s.s.	3/18	168.mg	4/17			
5286	1213	188.mg	n.s.s.	0/16	180.mg	1/18			
a	1213	357.mg	n.s.s.	0/16	180.mg	0/18			
b	1213	188.mg	n.s.s.	0/16	180.mg	1/18			
5287	1213	84.0mg	n.s.s.	0/16	168.mg	4/18			
a	1213	333.mg	n.s.s.	0/16	168.mg	0/18			
b	1213	71.2mg	4.14gm	0/16	168.mg	5/18			
5288	1426	154.mg	n.s.s.	0/46	74.9mg	0/10			Andrianova;vpit,29,71-74;1970
a	1426	57.5mg	n.s.s.	1/46	74.9mg	2/10			

ZIRCONIUM (IV) SULFATE   14644-61-2

	RefNum	LoConf	UpConf	Cntrl	1Dose	1Inc			Citation or Pathology
5289	1036	8.41mg	n.s.s.	15/71	.877mg	9/72			Kanisawa;canr,29,892-895;1969
a	1036	10.4mg	n.s.s.	4/71	.877mg	3/72			
b	1036	7.10mg	n.s.s.	24/71	.877mg	15/72			
c	1036	9.64mg	n.s.s.	8/71	.877mg	5/72			

SECTION D: APPENDICES OF CODES AND DEFINITIONS

Code	Species
D	dog
H	hamster
M	mouse
N	prosimian
P	monkey
R	rat

METHODS FOR NONHUMAN PRIMATES

The Laboratory of Chemical Pathology of the National Cancer Institute (NCI) has been conducting lifetime carcinogenesis studies in cynomolgus and rhesus monkeys for 27 years, and results on 17 completed chemicals are reported in the Carcinogenic Potency Database. Special inclusion rules and methods have been used due to protocol differences from standard rodent bioassays. A list of the chemicals and overview of results is reported in "Summary of Carcinogenic Potency Database by Chemical: Section C: Nonhuman Primates and Dogs" (Chapter 3, Section C), and detailed results are in the plot (Chapter 1, Section C).

1. Data have been combined for males and females of each species because fewer than 5 animals per sex-species were generally on test.

2. Whereas experiments with surgical intervention are excluded from the CPDB, laparoscopic examination of the liver was performed every 3 to 6 months, followed by wedge or needle biopsies of observed liver lesions.

3. Whereas experiments that are shorter than one half the standard lifespan are excluded from the CPDB, a few such experiments have been included for nonhuman primates, in which nearly all dosed animals had tumors.

4. Experiments of sodium arsenate and sterigmatocystin in cynomolgus have been included even though animals were put on test as adults (4 years of age).

5. Control monkeys are from the colony at NCI, which include breeders, offspring, and some feral monkeys. At any given time, the age of colony control animals ranged from neonate to greater than 25 years. In our lifetable analyses, control animals of each species are included if they lived longer than 8 months of age, the age of the first tumor in any group in these experiments. (For one ongoing study, IQ in cynomolgus, only summary data were available and concurrent vehicle controls were used in our analyses.)

6. Dosing schedules ranged from once every 2 weeks to 5 times per week. Daily dose-rate is calculated from cumulative dose data on each animal (mg/kg/day) and age at death. The rate for the dose group is the mean of these individual values.

7. TD_{50} values are estimated using lifetable data and are reported for every site at which a tumor was observed in dosed monkeys. Denominators on the plot represent the number of animals alive at the age of the first tumor in any group; numerators for control animals exclude tumors in animals that lived to be older than the last dosed animal.

8. The maximum number of animals used in any TD_{50} in each experiment is indicated by the denominator of the tumor incidence for all tumor-bearing animals ("tba" on the plot) and represents the number alive at the first tumor in either dosed or control monkeys.

9. In experiments with multiple target sites, a composite TD_{50} value is estimated for all animals with tumors at any of the target sites (MXB MXB on the plot, as for NCI/NTP bioassays in rodents).

Code	Strain
aaa	Analbuminaemic (Sprague-Dawley derived)
aah	A/He
aap	Alpk/Ap
abi	Ab × IF
aci	ACI
agu	AGUS
aif	A × IF
ain	ACI/n
ajj	A/JJms
akr	AKR
aks	AKR/J
alb	albino
amm	A .
aps	Alderly Park
asd	Sprague-Dawley albino
asp	ASH-CS1
asw	Swiss-Webster albino
aug	August
ays	AE/WffC3Hf/Nctr × YS/WffC3Hf/Nctr
b46	BR 46
b62	Monohybrid cross offspring of B6CF$_1$ (C57BL/6 × BALB/c)
b6a	B6AKF$_1$
b6b	(B6C3F$_1$ × B6C3 background, brachymorphic) inter se = B6C3F$_2$ brachymorphic
b6c	B6C3F$_1$
b6n	(B6C3F$_1$ × B6C3 background, brachymorphic) inter se = B6C3F$_2$ phenotypically normal
baa	Black a/a (YS × VY)F$_1$
baj	BALB/cJ
bal	BALB/c
bbb	Bush babies [*Galago crassicaudatus*]
bbl	Bethesda black
bce	BALB/cHe
bcn	BALB/cStCrlfC3Hf/Nctr
bd1	BDF$_1$
bd2	BD II
bd9	BD IX
bdf	BD VI
bdr	BD
beg	Beagle
bfm	Buffalo-Mai
bld	BALB/cLacDp
buf	Buffalo
c17	C17
c3c	C3H/AnCum
c3d	C3Hf/Dp
c3e	C3HeB/Fe
c3h	C3H
c3j	C3H/HeJ
c3l	C3H (C3H/Anl) (Anl 70)
c3p	C3HeB
c3s	C3H/St

Code	Strain
c3v	C3H/HeN—MTV—/Nctr
c56	C57BL/6J
c5c	C57BL/10ScSn
c5j	C57BL/10J
c5l	C57BL
c5n	C57BL/6N
c5v	C57BL/BVI
c6s	C57BL/6CrSlc
c7b	(C57BL/6 × BALB/c)F$_1$
c7l	C57L
cb6	C57BL/6
cba	CBA
cbc	CBA/Cb/Se
cbh	CBA/H-T6
cbj	C3HeB/FeJ
cbl	C57BL
cbn	C57BL/6JfC3Hf/Nctr × BALB/cStCrlfC3Hf/Nctr inter se
cbo	C.B. hooded
cbr	CB
cbs	Cb/Sc
cbt	Chester Beatty albino
cd1	Charles River CD1
cdf	CDF$_1$
cdr	Charles River CD
cen	C3H/HeN
cf1	CF-1
cfe	CFE
cff	C57BL/6JfC3Hf/Nctr × BALB/cStCrlfC3Hf/Nctr
cfi	C3H/FIB
cfl	CFLP
cfn	CFN
cfr	CF
che	C57BL/He
chf	C3HfB
chg	C3H/He germfree
chh	C3H/He
chi	CD-1 HaM/ICR
chj	C3HeB/Jax
chm	Charles River
cif	(C57 × IF)F$_1$
clw	Colworth (Wistar derived)
crf	(C3H × RIII)F$_1$
crw	Charles River Crl:COBS(WI)BR
csa	Charles River albino
csb	CSb
csc	C57L/He × 129/Rr × C3HeB/De × SWR/Ly
ctn	CTM
cva	BALB/cStCrlfC3Hf/Nctr × VY/WffC3Hf/Nctr-(A/A)
cvy	BALB/cStCrlfC3Hf/Nctr × VY/WffC3Hf/Nctr-(A^{vy}/A)
cwf	Carworth Farms
cws	CFW
cym	Cynomolgus [*Macaca fascicularis*]

Code	Strain	Code	Strain
dba	DBA/2	ofs	OFA (Sprague-Dawley derived)
dbx	DBA	osm	Osborne-Mendel
ddd	DDD	por	MRC Porton (Wistar derived)
ddn	ddNi	pva	Lean pseudoagouti Avy/a
ddx	dd	r3m	RIII
ddy	DDY	rfm	RF
don	Donryu	rhe	Rhesus [*Macaca mulatta*]
esd	Eastern Sprague-Dawley	scd	Swiss CD-1
f34	Fischer 344	scp	Cpb:Swiss random
f3d	F344/DuCrj	sda	Sprague-Dawley
f3l	Fischer 344/LATI	sdz	Sandoz
fdr	FDRL	shc	Sherman COBS
fds	Food and Drug Research Laboratories stock rats	she	Sherman
fis	Fischer	shr	Swiss/H/Riop
fmf	Fischer 344/Mai fBR	sic	Swiss/ICR
hew	Hebrew University	sjs	SJL/J
hic	Ha/ICR	sls	Slc-Wistar
hra	HRA/Skh (hairless)	smw	Sas: MRC(WI)BR
hrl	Harlan	ssa	S strain albino
hza	Holtzman albino (Sprague-Dawley derived)	sss	Sprague-Dawley Spartan
ic3	ICRC × C3h (Jax)	stm	ST/a
ici	ICI	swa	Swiss albino
icm	ICR	swi	Swiss
icr	ICR/Jcl	swr	SWR
ifc	IF × C57	sww	Swiss Webster
ifm	IF	syg	Syrian Golden
jic	JCL: ICR	tf1	Tuck
leb	Long-Evans BLU: (LE)	the	Theiller's Original
lee	Leeds albino	tmm	TM
lev	Long-Evans	tst	Tree shrew [*Tupaia glis*]
mar	Marshall	wag	WAG
mgr	Mongrel	wal	Wistar albino
mrc	MRC	wi2	Wistar II
mrw	MRC-Wistar	wid	Wistar/FDRL
nbr	NBR	win	Wistary/NIN
nbw	NZBW (hooded black and white strain)	wio	Wistar-OSU
nmb	Bor:NMRI, SPF-bred NMRI	wis	Wistar
nmh	Han: NMRI	wmf	Wistar-Mai-Furth
nmr	NMRI	wsh	Han: WIST
non	Noninbred	wsr	Wistar-random
nra	Norwegian albino	wsw	Wilmslow Wistar
nss	Not specified	wws	Wistar W.74
nzb	NZO/BlGd	xvi	XVII/G
nzd	NZR/Gd	yva	Obese yellow Avy/a
of1	OF1		

APPENDIX 3: ROUTES OF ADMINISTRATION

Code	Route of administration
cap	Gelatin capsule, orally
eat	Diet
gav	Gavage
inh	Inhalation
ipj	Intraperitoneal injection
ivj	Intravenous injection
mix	Multiple routes
	Used for four compounds: (1) thiourea (rats): intraperitoneal injection followed by water; (2) procarbazine.HCl (monkeys): a variety of combinations of diet, subcutaneous, intraperitoneal, and intravenous injection; (3) aflatoxin B_1 (monkeys): a combination of diet and intraperitoneal injection; (4) 2-naphthylamine (monkeys): capsules administered by gavage
orl	Gavage preweanling, followed by diet
wat	Water

APPENDIX 4: SITE CODES

Code	Site		Code	Site
---	All target sites		fgr	Forestomach, greater curvature
abc	Abdominal cavity		fhd	Forehead
abd	Abdomen		fls	Forestomach, lesser curvature
adr	Adrenal gland		for	Forestomach
adu	Acoustic duct		frb	Forebrain
amd	Adrenal medulla		gab	Gall bladder/bile duct
aol	Aorta and large arteries		gal	Gall bladder
arp	Adrenal capsule		gam	Gastric mucosa
asc	Colon, ascending		git	Gastrointestinal tract
auc	External auditory canal		gnv	Gingiva
aur	Auricular region		hag	Harderian gland
b/l	Lung, bronchiole		hea	Heart
bil	Bile duct		hnt	Hard palate/nasal turbinates
blv	Blood vessels		hpl	Hypophysis
bmd	Brain, medulla		hum	Humerus
bod	Body cavities		ilm	Ileum
bom	Bone marrow		isp	Interscapulum
bon	Bone		itl	Intestinal tract
bra	Brain		itn	Intestine
brf	Brown fat, dorsal		jej	Jejunum
brm	Brain, meninges		k/p	Kidney/pelvis
brs	Brain stem		kcx	Kidney cortex
ccx	Cerebral cortex		kid	Kidney
cec	Cecum		kpp	Kidney papilla
chp	Cheek pouch		ktu	Kidney tubule
clb	Cerebellum, cerebrum		kur	Kidney/ureter
cli	Clitoral gland		l/b	Lung, bronchus
clm	Cerebellum, meninges		lar	Larynx
clr	Colorectum		lgi	Large intestine
cns	Central nervous system		liv	Liver
col	Colon		lmr	Lymphoreticular system
crb	Cerebrum		lpp	Lip
crl	Cerebellum		lun	Lung
cst	Cardiac stomach		lyd	Lymph node
cvu	Cervix uteri		mam	Mammary tissue (other than or including more than mammary gland)
cvx	Cervix		mds	Mediastinum
cyx	Coccyx		mei	Mesenteric intestine
der	Dermis		meo	Mesovarium
dgt	Digestive tract		mey	Mesentery
dsc	Colon, descending		mgl	Mammary gland
duo	Duodenum		mix	More than one site; sites specified in published paper
eac	Ear canal		mln	Mesenteric lymph node
ear	Ear		mls	Multiple sites
edu	Ear duct		mth	Mouth
ehp	Extrahepatic tissue		mul	Multiple organs
eld	Eyelid		mus	Muscle
epg	Epiglottis		MXA	More than one site, combined by NCI/NTP
epi	Epidermis		MXB	More than one site, combined by Berkeley
epy	Epididymis		myc	Myocardium
eso	Esophagus		nac	Nasal mucosa
eye	Eye		nap	Nasal passageway
fat	Fat			

Code	Site
nas	Nasal cavity
ncp	Nasal cavity, posterior region
ner	Nervous system
nof	Nasal cavity, olfactory epithelium
nol	n. olfactorius
npc	Nasal and paranasal cavity
npl	Nipple
nre	Nasal cavity, respiratory epithelium
nse	Nose
nsm	Nasal septum
nsp	Nasopharynx
ntu	Nasal turbinate
olb	Olfactory bulb
omt	Omentum
opx	Oropharynx
orc	Oral cavity
orm	Oral mucosa
ova	Ovary
pae	Pancreas, exocrine
pal	Palate
pan	Pancreas
pdu	Pancreatic duct
pec	Peritoneal cavity
pel	Pelvis
pep	Paraepididymal tissue
per	Peritoneum
phr	Pharynx
pit	Pituitary gland
pls	Palate, soft
pnd	Pancreas/pancreatic duct
pni	Pancreatic islets
pnl	Paranasal sinus
pnr	Peripheral nerves
pns	Peripheral nervous system
pre	Preputial gland
prn	Pararenal tissue
pro	Prostate
pta	Pituitary gland, anterior
pty	Parathyroid
rec	Rectum
rel	Reticuloendothelium
rep	Reproductive tract
res	Respiratory system

Code	Site
sbg	Sebaceous gland
sev	Seminal vesicle
sft	Skin of foot and toe
skb	Skin of back
skf	Skin of flank
ski	Skin
sku	Skull
slg	Salivary gland
smi	Small intestine
spc	Splenic capsule
spd	Spinal cord
spl	Spleen
spn	Spinal nerves
srp	Splenic red pulp
ssq	Stomach, squamous
ssu	Skin and subcutis
stg	Stomach, glandular
stn	Stomach, nonglandular
sto	Stomach
sub	Subcutaneous tissue
TBA	All tumor-bearing animals, NCI/NTP
tba	All tumor-bearing animals, general literature
tes	Testis
thi	Thigh
thm	Thymus gland
thx	Thorax
thy	Thyroid gland
tna	Tunica albuginea
tnv	Tunica vaginalis
ton	Tongue
trh	Trachea
tyf	Thyroid follicle
ubl	Urinary bladder
ugi	Upper gastrointestinal tract
unt	Urinary tract
ure	Ureter
urt	Urethra
ute	Uterus
utm	Uterus, endometrium
vag	Vagina
ver	Vertebra
vse	Vascular epithelium
zym	Zymbal's gland

Code	Histopathology	Code	Histopathology
---	All tumors	ben	Benign tumor
a/a	Alveolar/bronchiolar adenoma	bhp	Hepatoma, benign
a/c	Alveolar/bronchiolar carcinoma	bht	Hepatocellular tumor, benign
abt	Alveolar/bronchiolar tumor	blc	Biliary cystadenoma
aca	Adenocarcinoma in adenomatous polyp	bly	B-cell lymphoma
acb	Alveolar/bronchiolar adenocarcinoma	bro	Bronchogenic carcinoma
acc	Acinar cell carcinoma	bsa	Basophil adenoma
acn	Adenocarcinoma, NOS	bsb	Basosquamous tumor benign
act	Alveolar cell tumor	bsn	Basophilic nodule
ada	Adenocarcinoma, type A	caa	Cholangioadenoma/carcinoma
adb	Adenocarcinoma, type B	cab	Cholangiocellular tumor, benign
adc	Adenocarcinoma	cac	Cholangioadenocarcinoma
ade	Adenoma	cad	Cholangioadenoma
adf	Adenofibroma	can	Carcinoma, NOS
adi	Adenocarcinoma, bilateral	car	Carcinoma
adm	Adenomatous polyp, NOS or adenocarcinoma in adenomatous polyp	cas	Carcinosarcoma
		cca	C-cell adenoma
adn	Adenoma, NOS	ccb	C-cell carcinoma, bilateral
ado	Adenoacanthoma	ccn	Cystadenocarcinoma, NOS
adp	Adenomatous polyp	ccr	C-cell carcinoma
adq	Adenosquamous carcinoma	ccy	Cholangioma, cystic
aep	Adenomatous endometrial polyp	cdb	C-cell adenoma, bilateral
agc	Alveogenic adenocarcinoma	cgd	Cholangiocarcinoma, ductular
agm	Angioma	cgf	Cholangiofibroma
agt	Alveogenic tumor	chc	Cholangiosarcoma
ahs	Axillary histiocytic sarcoma	cho	Cholangioma
akt	Adenoma-like tumor	cic	Carcinoma, *in situ*
ala	Alveolar cell adenoma	cla	Clear cell adenoma
alc	Alveolar cell carcinoma	clc	Cholangiocarcinoma
ald	Alveolar adenoma	cnb	Carcinoma, bilateral
amy	Adenomyoma	cnd	Carcinoid tumor, malignant
ana	Acinar cell adenoma	coa	Cortical adenoma
anb	Adenoma, bilateral	coc	Cortical carcinoma
ane	Angioendothelioma, malignant	con	Cortical adenoma, NOS
ang	Angiosarcoma	cra	Chromophobe adenoma
aoc	Acinar cell adenocarcinoma	crc	Chromophobe carcinoma
aod	Adenocarcinoma, acinar or ductal	crn	Cortical adenocarcinoma, NOS
apc	Anaplastic carcinoma	crt	Carcinoma, combined glandular and squamous type
apn	Adenomatous polyp, NOS	csa	Cortical subcapsular adenoma
asl	Astrocytoma, malignant	cuc	Ceruminous carcinoma
asm	Adenocarcinoma with squamous metaplasia	cvh	Cavernous hemangioma
ast	Astrocytoma	cyc	Cystadenocarcinoma
ata	Atypic adenoma	cye	Cystadenoma
bca	Basal cell adenoma	cyn	Cystadenoma, NOS
bcc	Basal cell carcinoma	dhs	Deep cervical, histiocytic sarcoma
bcd	Bronchiolar adenoma	ead	Endometrium, adenoma
bcp	Basal cell papilloma	edc	Endometrium, adenocarcinoma
bct	Basal cell tumor	emp	Endometrial polyp
bda	Bile duct adenoma	ena	Endometrial adenocarcinoma
bdc	Bile duct carcinoma	ene	Esthesioneuroepithelioma
bde	Bronchiolar adenocarcinoma	ens	Endocardial sarcoma
bdt	Bile duct tumor	epc	Epidermoid carcinoma

Code	Histopathology	Code	Histopathology
epd	Ependymoblastoma	hps	Hepatocellular carcinoma, solid
epn	Epithelial neoplasm	hpt	Hepatoma
epo	Epithelioma	iab	Interstitial cell adenoma, bilateral
ept	Epidermoid tumor	ica	Interstitial cell adenoma
esa	Eosinophilic adenoma	icb	Interstitial cell tumor, benign
esn	Eosinophilic nodule	ict	Interstitial cell tumor
esp	Endometrial stromal polyp	ihs	Iliac histiocytic sarcoma
ess	Endometrial stromal sarcoma	ile	Leukemia, indeterminate type
exa	Exocrine adenoma	isa	Islet cell adenoma
exp	Exophytic papilloma	isc	Islet cell carcinoma
fab	Follicular cell adenoma, bilateral	ism	Insuloma
fba	Fibroadenoma	itm	Interstitial cell tumor, malignant
fbs	Fibrosarcoma	ivc	Carcinoma, invasive
fca	Follicular cell adenoma	ivt	Transitional cell carcinoma, invasive
fcc	Follicular cell carcinoma	kcs	Kupffer cell sarcoma
fct	Follicular cell tumor	ker	Keratoacanthoma
fcy	Follicular cell adenocarcinoma, bilateral	lbl	Lymphoblastic lymphoma
fdc	Follicular adenocarcinoma	lca	Liver cell adenoma
fep	Fibroepithelial tumor	lcb	Liver cell tumor, benign
fib	Fibroma	lcc	Liver cell carcinoma
fih	Fibrous histiocytoma	lcl	Lymphocytic lymphoma
gcb	Granulosa cell tumor, benign	lcm	Liver cell tumor, malignant
gcc	Granulosa cell carcinoma	lct	Liver cell tumor
gcl	Granulosa cell tumor, NOS	ldc	Leydig cell tumor
gcm	Granulosa cell tumor, malignant	lei	Leiomyosarcoma
gct	Granulosa cell tumor	leu	Leukemia
ghc	Hepatocellular carcinoma, glandular	ley	Leiomyoma
glb	Granulosa cell tumor, bilateral	lhc	Lymphoma, histiocytic type
gli	Glioma	lip	Lipoma
gln	Glioma, NOS	lkm	Lymphoma leukemia
gmf	Glioma malignant, focal, mild	lkn	Leukemia, NOS
grb	Granular cell tumor, benign	lle	Lymphocytic leukemia
grl	Granulocytic leukemia	lls	Lymphoblastic leukemia-lymphosarcoma
gsa	Granulocytic sarcoma	lna	Nonlymphocytic leukemia, acute
hae	Hemangioendothelioma	lpb	Liver cell tumor, type B
hca	Hepatocellular carcinoma/adenoma	lps	Liposarcoma
hcs	Histiocytic sarcoma	lsl	Systemic and localized lymphoma
hct	Hepatocellular tumor	lut	Luteoma
hem	Hemangioma	lyk	Lymphatic leukemia
hes	Hemangiosarcoma	lym	Lymphoma
het	Hemorrhagic tumor	lyp	Lymphangioma
hga	Hemangiosarcoma anaplastic	lys	Lymphosarcoma
hmb	Hemangioendothelioma, benign	lyt	Lymphoid tumor
hmm	Hemangioendothelioma, malignant	mag	Malignant glioma
hms	Hemangioendothelial sarcoma	mal	Malignant tumor
hmt	Hamartoma	mcc	Mucinous carcinoma
hnd	Hyperplastic nodule	mda	Medullary adenoma
hpa	Hepatocellular adenoma	mdt	Medullary tumor
hpb	Hepatoblastoma	mec	Mucoepidermoid carcinoma
hpc	Hepatocellular carcinoma	mem	Mixed cell mucoepidermoid papilloma
hpd	Hepatocellular adenocarcinoma	men	Mesothelioma, NOS
hph	Hepatocellular hyperplastic nodule	mfh	Fibrous histiocytoma, malignant
hpm	Hemangiopericytoma, malignant	mhb	Hibernoma, malignant
hpn	Hepatocellular neoplastic nodule	mhc	Mixed hepato/cholangio carcinoma

Code	Histopathology
mhp	Malignant hepatoma
mhs	Histiocytoma, malignant
mht	Hepatocellular tumor, malignant
mix	More than one tumor type; tumor types specified in published paper
mlc	Melanocytoma
mle	Monocytic leukemia
mlh	Malignant lymphoma, histiocytic type
mlk	Myelogenous leukemia
mlm	Malignant lymphoma, mixed type
mlp	Malignant lymphoma, lymphocytic type
mlt	Melanotic tumor
mlu	Malignant lymphoma, undifferentiated type
mly	Malignant lymphoma
mng	Meningioma
mnl	Mononuclear cell leukemia
mnm	Meningioma, malignant
mno	Malignant lymphoma, NOS
mnp	Mesenchymal neoplasm
msb	Mesothelioma, benign
msm	Mesothelioma, malignant
mso	Mesothelioma
mtb	Mixed tumor, benign
mtm	Mixed tumor, malignant
mua	Mucinous adenocarcinoma
muc	Mucinous cystadenocarcinoma
MXA	More than one tumor type, combined by NCI/NTP
MXB	More than one tumor type, combined by Berkeley
mye	Myelocytic leukemia
myl	Myeloid leukemia
myo	Myelomonocytic leukemia
nen	Neoplasm, NOS
neo	Neoplasm
nep	Nephroblastoma
neu	Neuroblastoma
nfm	Neurofibroma
nfs	Neurofibrosarcoma
ngs	Neurogenic sarcoma
nhs	Inguinal histiocytic sarcoma
nim	Neurinoma
nlm	Neurilemoma, malignant
nnd	Neoplastic nodule
nod	Nodular hyperplasia
npm	Neoplasm, NOS, malignant
nsc	Neurosarcoma
nvc	Carcinoma, noninvasive
nvt	Transitional cell carcinoma, noninvasive
oec	Olfactory epithelial carcinoma
ogm	Olfactory lobe, glioma malignant
olc	Olfactory carcinoma
oli	Oligodendroglioma
oln	Olfactory neuroblastoma
olp	Olfactory neuroepithelioma
onm	Olfactory lobe, neuroblastoma malignant

Code	Histopathology
ost	Osteosarcoma
otm	Osteoma
pac	Papillary adenocarcinoma
pam	Papilloma
pas	Papillomatosis
pbb	Pheochromocytoma benign, bilateral
pbm	Pheochromocytoma, benign/malignant
pca	Parenchymal adenoma
pcn	Papillary cystadenocarcinoma, NOS
pcy	Papillary cystadenoma, NOS
pda	Pars distalis adenoma
pdc	Pars distalis carcinoma
pfa	Parafollicular cell adenoma
phc	Pheochromocytoma, complex
phe	Pheochromocytoma
phm	Pheochromocytoma, malignant
pla	Polypoid adenoma
plc	Plasmacytoma
pmb	Pheochromocytoma malignant, bilateral
pms	Papillary mesothelioma
pob	Pheochromocytoma, benign
pol	Polyp
ppa	Papillary adenoma
ppc	Papillary carcinoma
ppn	Papilloma, NOS
ppp	Papillary polyp
ptc	Papillary transitional cell carcinoma
ptm	Papillary tumor
pvc	Carcinoma, preinvasive
rab	Renal tubule adenoma, bilateral
rac	Renal tubule adenocarcinoma
rca	Renal cell adenoma
rcc	Renal cell carcinoma
rcs	Round cell sarcoma
rct	Renal cell tumor
ret	Reticulum cell tumor
rhb	Rhabdomyosarcoma
rhm	Rhabdomyoblastoma
rhs	Renal, histiocytic sarcoma
rna	Reticulum cell neoplasm, type A
rsc	Respiratory epithelial carcinoma
rta	Reticulum cell sarcoma, type A
rtb	Reticulum cell sarcoma, type B
rts	Reticulum cell sarcoma
rua	Tubule adenoma
ruc	Tubule carcinoma
rue	Tubule epithelium adenoma
sad	Scirrhous adenocarcinoma
sar	Sarcoma
sbr	Sebaceous gland carcinoma
sca	Solid cell adenoma
scc	Spindle cell carcinoma
scs	Spindle cell sarcoma
sct	Sertoli cell tumor

Code	Histopathology
sea	Sebaceous adenoma
seb	Sebaceous adenoma and adenocarcinoma
sec	Sebaceous adenocarcinoma
sgc	Sweat gland carcinoma
shs	Mesenteric histiocytic sarcoma
spm	Sarcoma, NOS
spt	Spindle cell tumor
sqa	Squamous cell tumor
sqc	Squamous cell carcinoma
sqi	Squamous cell carcinoma, invasive
sqk	Squamous cell carcinoma, keratinized
sqn	Squamous cell carcinoma, *in situ*
sqp	Squamous cell papilloma
sqs	Squamous cell carcinoma, stratified
squ	Squamous cell carcinoma, unclassified
srn	Sarcoma, NOS
ssc	Squamous cell carcinoma, sebaceous
tcb	Tubular cell carcinoma, bilateral

Code	Histopathology
tcc	Transitional cell carcinoma
tcm	Thecoma
thc	Hepatocellular carcinoma, trabecular
tla	Tubular cell adenoma
tma	Thymoma
tpp	Transitional cell papilloma
tri	Trichoepithelioma
tua	Tubular adenoma
tuc	Tubular carcinoma
tum	Tumor or more than one tumor type; tumor types not specified in paper
uac	Tubular cell adenocarcinoma
ulc	Undifferentiated carcinoma
ule	Undifferentiated leukemia
utc	Urothelial carcinoma
utp	Urothelial papilloma
vlp	Villous polyp
vsc	All vascular tumors

Code	Definition
a	The exposure time reported on the plot is an average of the different exposure times of the individual dose groups in the experiment. For NCI/NTP, both exposure and experiment times have been averaged because of differential survival among the dose groups. (In the TD_{50} calculation for the NCI/NTP bioassays, full lifetable data have been used.)
b	Diet was specially prepared to be deficient in one or more vitamins.
c	Diet was specially prepared to be low in lipotropes.
d	A cyclic dosing schedule was followed for part of the exposure time, with at least one week between cycles, e.g., 3 weeks dosed, one week not dosed. (NCI only)
e	For the general literature we have used an effective number of animals in a group whenever possible. This effective number is either: (1) the number of animals alive at the time of appearance of the first tumor, or if that is not reported, then (2) the number of animals examined. For some NCI/NTP bioassays the technical report includes both time-adjusted and unadjusted statistical analyses. Effective number indicates that some sites in these experiments have been included in the plot on the basis of the time-adjusted analysis.
f	Diet was specially prepared to have a lower than average protein level.
g	Some or all of the animals were used as breeders during the course of the experiment.
i	Dosing in this test was intermittent; it was stopped for more than one week at some point in the experiment.
k	For interim and serial sacrifice experiments, we have reported, as a separate experiment with a k notecode, each sacrifice time that otherwise met the inclusion rules of the database. Wherever possible, we have included unscheduled deaths with the terminal sacrifice data, and when this has been done, there is no k notecode for the terminal sacrifice experiment.
m	The calculated dose level for a group is an average of either (1) different doses administered to individual animals or (2) the range of doses administered.

Code	Definition
n	NTP considered one dose group inadequate for detecting a carcinogenic response.
o	Chemical was administered as an aerosol.
r	Restricted site analysis; the authors either examined or chose to report data for only three or fewer tissues.
s	Authors noted that survival was decreased due to toxicity, disease, or accidental death.
u	Monkey tests are still in progress for IQ administered by gavage and diethylnitrosamine by intraperitoneal route.
v	Variable or irregular dosing schedules have been used, e.g., dose level changed during the experiment.
w	For experiments in nonhuman primates, control animals are from the NCI colony, which was started in 1961; some animals have been used as breeders. Tumor data for control animals are derived from 132 cynomolgus and 122 rhesus monkeys, and 9 bush babies. Fourteen of these rhesus and 33 cynomolgus are still alive. The denominators in the tumor incidence column of the plot are the numbers alive at the age of the first death in any group with a tumor of that type. TD_{50} values for nonhuman primates are calculated using lifetable data. See Appendix 1.
W	Tumors found at death in control monkeys that lived to an older age than the last dosed animal are not reported in the tumor incidence column of the plot.
x	Exposure began before the animals were weaned.
y	Animals were dosed for only 25 weeks; one week short of the standard criterion. Due to rounding, 6 months is reported as the exposure time on the plot.
z	In a report of these vinyl chloride experiments (Maltoni, 1977a), the author notes "all the animals exposed to the highest doses (30,000 and 10,000 ppm for 52 weeks), with or without tumors, were examined radiologically during treatment and/or at death; moreover, radiologic examinations have also been made on several animals bearing tumors even though these animals had been exposed to the lower doses." The experiment at the highest dose (30,000 ppm) is not included in the database. The reported data include the 10,000 ppm or lower doses.

Appendix 7: Dose-Response Curve Symbols

Code	Dose-response curve
*	Consistent with linearity
/	Significant departure from linearity, upward curvature
\	Significant departure from linearity, downward curvature
Z	Significant departure from linearity, more than three dose groups including controls
blank	Either no dose-related effect, or no curve shape could be determined because experiment had only two dose groups including controls

APPENDIX 8: STATISTICAL METHODS FOR ESTIMATING TD$_{50}$

The lifetable methods which we have used to analyze the experimental data have been described elsewhere.[1] Briefly, a proportional hazards model[2] is assumed for the time-to-tumor data, in which $\lambda(t, d)$, the tumor-hazard rate at age t for a specific site, is linearly related to d, the administered dose-rate of test chemical in mg/kg body wt/day, as

$$\lambda(t, d) = (1 + \beta d)\lambda_0(t). \qquad (1)$$

$\lambda_0(t)$ is the tumor-incidence rate at zero dose. The parameter β and the function λ_0 are estimated using maximum likelihood methods. The likelihood ratio statistic tests the hypothesis that the chemical has no carcinogenic effect (i.e., $\beta = 0$), and a χ^2 goodness-of-fit statistic tests the validity of the linear relationship between dose and tumor incidence expressed by Equation 1. In fitting the model, no attempt is made to distinguish between tumors found in a fatal context and tumors found in an incidental context. Thus the time-to-tumor occurrence is taken to be the time to death of the animal, whether death results from the tumor of interest, or from some other cause, including terminal sacrifice.[1,3]

For summary incidence data, we fit by maximum likelihood methods the comparable model

$$p_d = 1 - \exp\{-(a + bd)\}, \qquad (2)$$

where $a > 0$ and $b > 0$ and p_d is the probability that an animal exposed at dose d for its lifetime develops a tumor. This model is linear at low doses and is often referred to as the "one-hit model." Here, the number of animals developing tumors at dose d is assumed to follow a binomial distribution with parameters n_d and p_d, where n_d is the number of animals initially exposed at dose d. As with lifetable data, the likelihood ratio statistic is used to test whether the compound is carcinogenic, i.e., whether $b = 0$, and a χ^2 statistic tests the adequacy of the model.

The estimate of TD$_{50}$ based on summary incidence data is simply $log(2)/b$, where b is the maximum likelihood estimate (MLE) of b. For lifetable data, the estimate is a more complex function of the MLEs of β and $\lambda_0(t)$.[1] For either method of estimating TD$_{50}$, if the χ^2 goodness-of-fit test indicated statistically significant departure from linearity, ($p < 0.05$) and this departure was downward, the analysis was repeated eliminating the highest dose group. The purpose of this procedure was to remove the effects of toxicity in summary incidence analyses and to remove the effects of dose saturation in the lifetable analyses. If the goodness-of-fit test indicated an upward departure from linearity, no groups were eliminated when fitting the model.

In our database we have estimated 99% confidence intervals for TD$_{50}$s calculated from lifetable data and for those based on summary incidence data. The method for calculating these intervals from lifetable data is described in Sawyer et al.[1] For summary incidence data, 99% likelihood-ratio-test-based confidence limits are obtained for b and are then transformed to limits for TD$_{50}$.

REFERENCES

1. Sawyer, C., Peto, R., Bernstein, L., and Pike, M. C. (1984). Calculation of carcinogenic potency from long-term animal carcinogenesis experiments. *Biometrics* 40: 27-40.
2. Cox, D. R. (1972). Regression models and life tables (with discussion). *J. R. Stat. Soc. Br. Series B* 34: 187-220.
3. Peto, R., Pike, M.C., Bernstein, L., Gold, L. S., and Ames, B. N. (1984). The TD$_{50}$: A proposed general convention for the numerical description of the carcinogenic potency of chemicals in chronic-exposure animal experiments. *Environ. Health Perspect.* 58: 1-8.

APPENDIX 9: AUTHOR'S OPINION

Code	Author's opinion for each site
a	NCI or NTP evaluation is that the incidence of tumors at that site(s) was associated with administration of the compound. This code is only used for Technical Reports published before March 1986.
c	Early NCI/NTP evaluation is *carcinogenic* or NTP evaluation is *clear evidence* of carcinogenic activity, i.e., "studies that are interpreted as showing a dose related (1) increase of malignant neoplasms, (2) increase of a combination of malignant and benign neoplasms, or (3) marked increase of benign neoplasms if there is an indication from this or other studies of the ability of such tumors to progress to malignancy."
e	NTP evaluation is *equivocal evidence* of carcinogenic activity, i.e., "studies that are interpreted as showing a marginal increase of neoplasms that may be chemically related."
i	NTP evaluation is *inadequate*, i.e., "studies that, because of major qualitative or quantitative limitations, cannot be interpreted as valid for showing either the presence or absence of carcinogenic activity."
p	NTP evaluation is *some evidence* of carcinogenic activity, i.e., "studies that are interpreted as showing a chemically related increased incidence of neoplasms (malignant, benign, or combined) in which the strength of the response is less than that required for clear evidence." When NTP re-evaluated an opinion from experiments that predate their present evaluation categories, we updated their opinion as well (J. K. Haseman et al., *Environ. Health Perspect.* 74: 229-235, 1987).
+	Author in general literature evaluated site as positive.
–	NTP evaluation is *no evidence* of carcinogenic activity, i.e., "studies that are interpreted as showing no chemically related increases in malignant or benign neoplasms." In the general literature author evaluated the site as negative.
blank	For NTP and general literature: a site for which no opinion is stated.

APPENDIX 10: BERKELEY CODES

Code	Definitions of Berkeley codes for NCI/NTP
C	The TD_{50} includes all animals with a tumor at any site with a "c" opinion. The mix was created for the CPDB, and MXB appears on the left side of the plot.
M	The TD_{50} includes all animals with a tumor at any site with a "c" or "p" opinion. The mix was created for the CPDB, and MXB appears on the left side of the plot.
P	The TD_{50} includes all animals with a tumor at any site with a "p" opinion. The mix was created for the CPDB, and MXB appears on the left side of the plot.
S	The TD_{50} has been included in the plot because the sites were statistically significant in the tables of analyses of primary tumors and the TD_{50} was significant at the $p <0.05$ level; however, the NCI/NTP report did not evaluate the site as evidence of carcinogenicity.

Appendix 11: Journals

Code	Reference
acnr	*Anticancer Research*
acpj	*Acta Pathologica Japonica*
adsc	*Academie des Sciences, Memoires et Communications des Membres et des Correspondants de l' Academie.*
aenc	*Acta Endocrinologica* (Copenhagen)
aenh	*Archives of Environmental Health* (formerly *AMA Archives of Industrial Health,* until July 1, 1960)
agfc	*Agricultural and Food Chemistry: Past, Present, Future* (R. Teranishi, Ed.) Avi Publishing Company, Inc., Westport, CT, 1978
ajim	*American Journal of Industrial Medicine*
ajpa	*American Journal of Pathology*
ajsu	*American Journal of Surgery*
amet	*Advances in Modern Environmental Toxicology*
amih	*American Industrial Hygiene*
anes	*Anesthesiology*
anoh	*Annals of Occupational Hygiene*
anti	*Antimutagenesis and Anticarcinogenesis Mechanisms,* Vol. 2 (Y. Kuroda, D. M. Shankel, andd M. D. Waters, Eds.), Plenum Publishing Co., New York, 1990
anya	*Annals of the New York Academy of Sciences*
apms	*Acta Pathologica et Microbiologica Scandinavica Section A. Pathology*
aric	*Archives of Research on Industrial Carcinogenesis*
arjc	*Archiv fur japanische Chirurgie*
arpa	*Archives of Pathology and Laboratory Medicine* (formerly *Archives of Pathology,* Mar. 1928-Sept. 1950; *AMA Archives of Pathology,* Oct. 1950-June 1960)
artx	*Archives of Toxicology*
arzn	*Arzneimittel-Forschung*
atmh	*American Journal of Tropical Medicine and Hygiene* (formerly *American Journal of Tropical Medicine,* 1921-1951)
banb	*Banbury Report*
bdca	*Bulletin du Cancer*
bebm	*Byulleten' Eksperimental'noi Biologii i Meditsiny*
becc	*British Empire Cancer Campaign, 39th Annual Report*
bect	*Bulletin of Environmental Contamination Toxicology*
bexb	*Bulletin of Experimental Biology and Medicine*
bjca	*British Journal of Cancer*
bmes	*Biomedical and Environmental Sciences*
bmjl	*British Medical Journal*
bwho	*Bulletin W.H.O.* (World Health Organization)
canc	*Cancer*
canr	*Cancer Research*
carc	*Carcinogenesis*
carm	*Carcinogenesis, Vol. 2. Mechanisms of Tumor Promotion and Cocarcinogenesis,* (T. J. Slaga, A. Sivak, and R. K. Boutwell, Eds.) Raven Press, New York, 1978
chpb	*Chemical Pharmacology Bulletin*

Code	Reference
ciit	*A Chronic Inhalation Toxicology Study in Rats and Mice Exposed to Formaldehyde.* (K. L. Pavkov, W. D. Kerns, R. I. Mitchell, M. M. Connell, D. J. Donofrio, H. H. Harroff, A. D. Barker, G. L. Fisher, R. L. Joiner, and D. C. Thake, Eds.) Final Report, CIIT Docket #10922. Chemical Industry Institute of Toxicology, Research Triangle Park, NC, 1981
clet	*Cancer Letters*
cmsp	*Chemosphere*
ctfr	*A Twenty-Four Month Inhalation Toxicology Study in Fischer-344 Rats Exposed to Atmospheric Toluene.* (E. J. Gralla, Ed.) Final Report, CIIT Docket #22000. Chemical Industry Institute of Toxicology, Research Triangle Park, NC, 1980
ctxf	*Chemical Toxicology of Food.* (C. L. Galli, R. Paoletti, and G. Vettorazzi, Eds.). Elsevier/North-Holland Biomedical Press, New York, 1978
dact	*Drug and Chemical Toxicology*
dcfr	*A Two-Year Toxicity and Oncogenicity Study with Acrylonitrile Incorporated in the Drinking Water of Rats.* (J. F. Quast, C. E. Wade, C. G. Humiston, R. M. Carreon, E. A. Hermann, C. N. Park, and B. A. Schwetz, Eds.) Final Report. Dow Chemical U.S.A., Midland, MI, 1980
dcrp	*A Two-Year Toxicity and Oncogenicity Study with Acrylonitrile Following Inhalation Exposure of Rats.* (J. F. Quast, D. J. Schuetz, M. F. Balmer, T. S. Gushow, C. N. Park, and M. J. McKenna, Eds.) Final Report. Dow Chemical USA, Midland, MI, 1980
dgsn	*Digestion*
eaes	*Ecotoxicology and Environmental Safety*
eamp	*Experimental and Molecular Pathology*
ejca	*European Journal of Cancer and Clinical Oncology* (formerly *European Journal of Cancer,* until 1982)
enhp	*Environmental Health Perspectives*
envr	*Environmental Research*
expa	*Experientia*
expl	*Experimental Pathology*
faat	*Fundamental and Applied Toxicology*
fctx	*Food and Chemical Toxicology* (formerly *Food and Cosmetics Toxicology,* until 1982)
gaga	*Gastroenterology*
gann	*Japanese Journal of Cancer Research* (formerly *Gann* until 1984)
gmcr	*Gann Monograph on Cancer Research*
hepg	*Hepato-gastroenterology*
hijm	*Hiroshima Journal of Medical Sciences*
iarc	*IARC Scientific Publication #31.* (E. A. Walker, L. Griciute, M. Castegnaro, and M. Borzsonyi, Eds.), World Health Organization, International Agency for Research on Cancer, Lyon, France, 1980
ijbb	*Indian Journal of Biochemistry & Biophysics*
ijcn	*International Journal of Cancer* (formerly *International Union Against Cancer,* until 1964)

Code	Reference
ijmr	*Indian Journal of Medical Research*
ijvn	*International Journal for Vitamin and Nutrition Research*
imed	*International Journal of Occupational Health and Safety* (formerly *Industrial Medicine and Surgery of Trauma,* June-July 1949; *Industrial Medicine and Surgery,* Aug. 1949-73)
indh	*Industrial Health*
ircs	*International Research Communications System Medical Science: Library Compendium*
irdc	International Research and Development Corporation Final Report, Mattawan, MI. Contracted by Uniroyal Chemical Company, Inc., Bethany, CT
isjm	*Israel Journal of Medicine and Science*
jact	*Journal of the American College of Toxicology*
japt	*Journal of Applied Toxicology*
jcph	*Journal of Clinical Pharmacology*
jctx	*Journal of Combustion Toxicology*
jept	*Journal of Environmental Pathology and Toxicology*
jfds	*Journal of Food Safety*
jiht	*Journal of Industrial Hygiene and Toxicology* (formerly *Journal of Industrial Hygiene,* 1919-35)
jjem	*Japanese Journal of Experimental Medicine*
jjhg	*Japanese Journal of Hygiene*
jjvs	*Japanese Journal of Veterinary Science*
jkmj	*Jikeikai Medical Journal*
jnci	*Journal of the National Cancer Institute (U.S. National Cancer Institute Journal)*
jnma	*Journal of the Nara Medical Association*
jnut	*Journal of Nutrition*
jpat	*Journal of Pathology* (formerly *Journal of Pathology and Bacteriology,* until 1969)
jphp	*Journal of Pharmacy and Pharmacology*
jsac	*Journal of Studies on Alcohol*
jtxe	*Journal of Toxicology and Environmental Health*
lapp	*Lavori dell Instituto di Anatomia e Istologia Patologica,* Universita degli Studi, Perugia, Italy.
livt	*Laboratory Investigation*
made	*Mechanisms of Ageing and Development*
mpoc	*Morphological Precursors of Cancer*
myco	*Mycopathologia*
natu	*Nature*
nawi	*Naturwissenschaften*
nctr	*National Center for Toxicological Research Final Report*
neag	*Neurobiology of Aging*
nplm	*Neoplasma*
ntis	National Technical Information Service. *Evaluation of Carcinogenic, Teratogenic, and Mutagenic Activities of Selected Pesticides and Industrial Chemicals.* Vol. 1: Carcinogenic Study. NTIS, Springfield, VA;1968/J.R.M. Innes et al., JNCI, 42, 1101-1114, 1969
nutc	*Nutrition and Cancer*
onco	*Oncology*
ossc	*Organ and Species Specificity in Chemical Carcinogenesis.* (R. Langenbach, S. Nesnow, and J. M. Rice, Eds.), Plenum Press, New York, 1982
pavt	*Veterinary Pathology* (formerly *Pathologia Veterinaria,* until 1970)
pcpl	*Pentachlorophenol* (K. R. Rao, Ed.), Plenum Press, New York, 1978
pffm	*Plant Foods for Man*
phrm	*Pharmacometrics*
pjpa	*Proceedings of the Japan Academy*
pseb	*Proceedings of the Society for Experimental Biology and Medicine* (New York)
reec	*Revue Europeen d'Etudes Cliniques et Biologiques*
rtxp	*Regulatory Toxicology and Pharmacology*
sabo	*Sabouraudia*
scie	*Science*
sjge	*Scandinavian Journal of Gastroenterology*
smon	*Seminars in Oncology*
srfr	*Southern Research Institute Final Report*
stev	*Science of the Total Environment*
tcam	*Teratogenesis, Carcinogenesis, and Mutagenesis*
thym	*Thymus*
tjem	*Tohoku Journal of Experimental Medicine*
tumo	*Tumori*
txap	*Toxicology and Applied Pharmacology*
txcy	*Toxicology*
txec	*Toxicological and Environmental Chemistry*
txih	*Toxicology and Industrial Health*
txlt	*Toxicology Letters*
txoc	*Toxicology and Occupational Medicine.* Proceedings of the 10th Inter-American Conference on Toxicology and Occupational Medicine. Oct. 22nd-25th, 1978. Miami, FL (W. Deichmann, Ed.), Developments in Toxicology and Environmental Science, Vol. 4. Elsevier North Holland, 1979
txpy	*Toxicologic Pathology*
urre	*Urological Research*
vivo	*In Vivo*
vopr	*Voprosy Onkologii; Problems in Oncology*
vpit	*Voprosy Pitaniya; Problems in Nutrition*
yjbm	*Yale Journal of Biology and Medicine*
zkko	*Journal of Cancer Research and Clinical Oncology* (formerly *Zeitschrift fur Krebsforschung und Klinische Onkologie,* until 1979)

APPENDIX 12: SYNONYMS FOR CHEMICALS IN THE CARCINOGENIC POTENCY DATABASE

CAS* number	Chemical name
43033-72-3	*l*-α-Acetylmethadol.HCl (see 6-Dimethylamino-4,4-diphenyl-3-heptanol acetate.HCl)
2757-90-6	Agaritine (see β-*N*-[γ-*l*(+)-Glutamyl]-4-hydroxymethylphenylhydrazine)
101-73-5	Agerite 150 (see *p*-Isopropoxydiphenylamine)
103-16-2	Agerite alba (see Hydroquinone monobenzyl ether)
74-31-7	Agerite DPPD (see Diphenyl-*p*-phenylenediamine)
135-88-6	Agerite powder (see Phenyl-β-naphthylamine)
93-46-9	Agerite white (see *sym.*-Dibeta-naphthyl-*p*-phenylenediamine)
54-80-8	Alderlin (see Pronethalol)
51-02-5	Alderlin.HCl (see Pronethalol.HCl)
39148-24-8	Aliette (see Fosetyl Al)
97-53-0	1-Allyl-3-methoxy-4-hydroxybenzene (see Eugenol)
120-78-5	Altax (see Benzothiazyl disulfide)
915-67-3	Amaranth (see FD & C red no. 2)
102-77-2	Amax (see *N*-Oxydiethylenebenzothiazole-2-sulfenamide)
97-56-3	2-Amino-5-azotoluene (see *o*-Aminoazotoluene)
75104-43-7	3-Amino-1,4-dimethyl-5*H*-pyrido[4,3-*b*]indole acetate (see Trp-P-1 acetate)
57-68-1	4-Amino-*N*-(4,6-dimethyl-2-pyrimidinyl)-benzenesulfonamide (see Sulfamethazine)
97-56-3	4-Amino-2,3-dimethylazobenzene (see *o*-Aminoazotoluene)
77094-11-2	2-Amino-3,4-dimethylimidazo[4,5-*f*]quinoline (see MeIQ)
77500-04-0	2-Amino-3,8-dimethylimidazo[4,5-*f*]quinoxaline (see MeIQx)
4363-03-5	4-Amino-3-hydroxybiphenyl (see 3-Hydroxy-4-aminobiphenyl)
—	2-Amino-1-methyl-6-phenylimidazo[4,5-*b*]-pyridine.HCl (see PhIP.HCl)
59-05-2	4-Amino-*N*10-methyl-pteroylglutamic acid (see Methotrexate)
68006-83-7	2-Amino-3-methyl-9*H*-pyrido-[2,3-*b*]-indole (see MeA-α-C)
72254-58-1	3-Amino-1-methyl-5*H*-pyrido[4,3-*b*]indole acetate (see Trp-P-2 acetate)
67730-11-4	2-Amino-6-methyldipyrido[1,2-*a*:3′,2′-d]imidazole (see Glu-P-1)
—	2-Amino-3-methylimidazo[4,5-*f*]quinoline (see IQ)
26148-68-5	2-Amino-9*H*-pyrido(2,3-*b*)indole (see A-α-C)
118-92-3	Aminobenzoic acid (see Anthranilic acid)
92-67-1	4-Aminobiphenyl (see 4-Aminodiphenyl)
67730-10-3	2-Aminodipyrido[1,2-*a*:3′,2′-d]imidazole (see Glu-P-2)
—	4-(2-Aminoethyl)-6-diazo-2,4-cyclohexadienone.HCl (see 3-Diazotyramine.HCl)

CAS number	Chemical name
119-34-6	*p*-Aminonitrophenol (see 4-Amino-2-nitrophenol)
58-15-1	Aminopyrine (see 4-Dimethylaminoantipyrine)
61-82-5	Amitrol (see 3-Aminotriazole)
64005-62-5	1-Amyl-1-nitrosourethan (see Nitrosoamylurethan)
1119-68-2	*n*-Amylhydrazine.HCl (see *n*-Pentylhydrazine.HCl)
518-75-2	Antimycin (see Citrinin)
60-80-0	Antipyrine (see Phenazone)
86-88-4	ANTU (see 1-(1-Naphthyl)-2-thiourea)
8003-03-0	APC (see Aspirin, phenacetin, and caffeine)
1327-53-3	Arsenic trioxide (see Arsenious oxide)
2303-16-4	Avadex (see Diallate)
542-88-1	BCME (see Bis-(chloromethyl)ether)
147-24-0	Benadryl (see Diphenhydramine.HCl)
319-84-6	α-Benzene hexachloride (see α-1,2,3,4,5,6-Hexachlorocyclohexane)
613-94-5	Benzhydrazide (see Benzoyl hydrazine)
91-64-5	1,2-Benzopyrone (see Coumarin)
51542-33-7	1-(2′-Benzothiazolyl)-3-methyl-3-nitrosourea (see *N*-Nitrosobenzthiazuron)
50-32-8	Benzpyrene (see Benzo(*a*)pyrene)
50-32-8	3,4-Benzpyrene (see Benzo(*a*)pyrene)
1694-09-3	Benzyl violet 4B (see FD & C violet no. 1)
25013-16-5	BHA (see Butylated hydroxyanisole)
128-37-0	BHT (see Butylated hydroxytoluene)
3296-90-0	2,2-Bis(bromomethyl)-1,3-propanediol (see Dibromoneopentyl glycol)
53609-64-6	*N*-Bis(2-hydroxypropyl)nitrosamine (see *N*-Nitrosobis(2-hydroxypropyl)amine)
54143-56-5	2,5-Bis(2,2,2-trifluorethoxy)-*N*-(2-piperidylmethyl)-benzamide acetate (see Flecainide acetate)
21260-46-8	Bismate (see Bismuth dimethyldithiocarbamate)
2519-30-4	Brilliant black BN (see Black PN)
1937-37-7	Direct deep black-extra (see C.I. direct black 38)
3844-45-9	Brilliant blue FCF (see FD & C blue no. 1)
869-01-2	BNU (see *N-n*-Butyl-*N*-nitrosourea)
109-84-2	BOH (see 2-Hydroxyethylhydrazine)
99-30-9	Botran (see 2,6-Dichloro-4-nitroaniline)
77-65-6	Bromodiethylacetylurea (see Carbromal)
75-25-2	Bromoform (see Tribromomethane)
74-83-9	Bromomethane (see Methyl bromide)
5351-65-5	BSH (see Benzenesulphonohydrazide)
55-98-1	Busulfan (see Myleran)
51-03-6	Butacide (see Piperonyl butoxide in solvent)
123-73-9	trans-2-Butenal (see Crotonaldehyde)
3817-11-6	Butyl-butanol-nitrosamine (see *N*-Butyl-*N*-(4-hydroxybutyl)nitrosamine)
128-37-0	2,6-di-*tert*-Butyl-*p*-cresol (see Butylated hydroxytoluene)

---

* CAS number = Chemical Abstracts Service registry number

CAS number	Chemical name
25013-16-5	2(3)-*tert*-Butyl-4-hydroxyanisole (see Butylated hydroxyanisole)
136-23-2	Butyl zimate (see Zinc dibutyldithiocarbamate)
75-60-5	Cacodylic acid (see Dimethylarsinic acid)
8003-03-0	Caffeine, aspirin, and phenacetin (see Aspirin, phenacetin, and caffeine)
156-62-7	Calcium cyanamide (see Cyanamide, calcium)
149-30-4	Captax (see 2-Mercaptobenzothiazole)
15805-73-9	Carbamic acid, vinyl ester (see Vinyl carbamate)
3567-69-9	Carmoisine (see C.I. food red 3)
9000-40-2	Carob seed gum (see Locust bean gum)
999-81-5	CCC (see (2-Chloroethyl)trimethylammonium chloride)
122-34-9	CDT (see Simazine)
9004-32-4	Cellulose carboxymethyl ether, sodium (see Edifas B)
15879-93-3	α-Chloralose (see Anhydroglucochloral)
106-47-8	4-Chloranilic (see *p*-Chloroaniline)
143-50-0	Chlordecone (see Kepone)
80-33-1	Chlorfenson (see *p*-Chlorophenyl-*p*-chlorobenzene sulfonate)
100-44-7	α-Chloro toluene (see Benzyl chloride)
54749-90-5	2-[3-(2-Chloroethyl)-3-nitrosoureido]-*d*-glucopyranose (see Chlorozotocin)
56-75-7	Chloromycetin (see Chloramphenicol)
100-00-5	*p*-Chloronitrobenzene (see 1-Chloro-4-nitrobenzene)
94-20-2	1-(*p*-Chlorophenylsulfonyl)-3-propylurea (see Chlorpropamide)
107-05-1	Chloropropene (see Allyl chloride)
63449-39-8	Chlorowax 40 (see Chlorinated paraffins (C$_{23}$, 43% chlorine))
63449-39-8	Chlorowax 500C (see Chlorinated paraffins (C$_{12}$, 60% chlorine))
101-21-3	Chlorpropham (see Isopropyl-*N*-(3-chlorophenyl)-carbamate)
2921-88-2	Chlorpyrifos (see *O,O*-Diethyl-*O*-(3,5,6-trichloro-2-pyridyl)phosphorothioate)
101-21-3	CIPC (see Isopropyl-*N*-(3-chlorophenyl)carbamate)
11096-82-5	Clophen A 60 (see Aroclor 1260)
7681-52-9	Clorox (see Sodium hypochlorite)
107-30-2	CMME (see Chloromethyl methyl ether)
60391-92-6	CMNU (see Carboxymethylnitrosourea)
12126-59-9	Conjugated equine estrogens (see Premarin)
137-29-1	Cumate (see Copper dimethyldithiocarbamate)
458-37-7	Curcumin (see Turmeric oleoresin (79-85% curcumin))
51630-58-1	Cyano-(3-phenoxyphenyl)methyl-4-chloro-α-(1-methylethyl)benzene acetate (see Fenvalerate)
55268-74-1	2-Cyclo-hexyl-carbonyl-1,3,4,6,7,11-*b*-hexahydro-2-*H*-pyrazine(2,1-*a*)isoquinoline-4-one (see Praziquantel)
50-18-0	Cytoxan (see Cyclophosphamide)
538-41-0	DAAB (see 4,4'-Diaminoazobenzene)
60-11-7	DAB (see *N,N*-Dimethyl-4-aminoazobenzene)
785-30-8	DABA (see 4,4'-Diaminobenzanilide)
1897-45-6	Daconil (see Chlorothalonil)
117-10-2	Danthron (see Chrysazin)
58-14-0	Daraprin (see Pyrimethamine)
96-12-8	DBCP (see 1,2-Dibromo-3-chloropropane)
488-41-5	DBM (see Dibromomannitol)
91-94-1	DCB (see 3,3'-Dichlorobenzidine)
33857-26-0	DCDD (see 2,7-Dichlorodibenzo-*p*-dioxin)
62-73-7	DDVP (see Dichlorvos)
576-68-1	Degranol (see Mannitol nitrogen mustard)
520-45-6	Dehydroacetic acid (see 3-Acetyl-6-methyl-2,4-pyrandione)
141-05-9	DEM (see Diethylmaleate)
55-18-5	DEN (see *N*-Nitrosodiethylamine)
625-89-8	6-F-DEN (see *N*-Nitrosobis(2,2,2-trifluoroethyl)amine)
64039-27-6	β-2'-Deoxy-6-thioguanosine monohydrate (see β-Thioguanine deoxyriboside)
56-53-1	DES (see Diethylstilbestrol)
62488-57-7	DHAC (see 5,6-Dihydro-5-azacytidine)
7411-49-6	3,3'-Diaminobenzidine.4HCl (see 3,3',4,4'-Tetraaminobiphenyl.4HCl)
2243-62-1	1,5-Diaminonaphthalene (see 1,5-Naphthalenediamine)
124-48-1	Dibromochloromethane (see Chlorodibromomethane)
924-16-3	Dibutylnitrosamine (see Nitrosodibutylamine)
4342-03-4	DIC (see Dacarbazine)
117-80-6	Dichlone (see 2,3-Dichloro-1,4-naphthoquinone)
51-75-2	Dichloren (see Nitrogen mustard)
8001-50-1	Dichloricide mothproofer (see Strobane)
101-14-4	3,3'-Dichloro-4,4'-diaminodiphenylmethane (see 4,4'-Methylene-bis(2-chloroaniline))
51-75-2	2,2'-Dichloro-*N*-methyldiethylamine (see Nitrogen mustard)
95-50-1	*o*-Dichlorobenzene (see 1,2-Dichlorobenzene)
75-27-4	Dichlorobromomethane (see Bromodichloromethane)
75-09-2	Dichloromethane (see Methylene chloride)
120-36-5	2-(2,4-Dichlorophenoxy)propionic acid (see α-(2,4-Dichlorophenoxy)propionic acid)
542-75-6	1,3-Dichloropropene (see Telone II)
21498-08-8	2-[1-(2,6-Dichlorphenoxy)-ethyl]-2-imidazoline.HCl (see Lofexidine.HCl)
99-30-9	Dicloran (see 2,6-Dichloro-4-nitroaniline)
2164-09-2	Dicryl (see 3,4'-Dichloro-2-methylacrylanilide)
56-53-1	4,4'-(1,2-Diethyl-1,2-ethenediyl)bis-phenol (see Diethylstilbestrol)
148-18-5	Diethyldithiocarbamate trihydrate, sodium (see Sodium diethyldithiocarbamate trihydrate)
55-18-5	Diethylnitrosamine (see *N*-Nitrosodiethylamine)
55-18-5	*N,N*-Diethylnitrosamine (see *N*-Nitrosodiethylamine)
68-89-3	(2,3-Dihydro-1,5-dimethyl-3-oxo-2-phenyl-1*H*-pyrazol-4-yl)methylamino methanesulfonate monohydrate (see Dipyrone)
123-33-1	1,2-Dihydro-3,6-pyridazinedione (see Maleic hydrazide)
120-80-9	1,2-Dihydroxybenzene (see Catechol)
63-84-3	*dl*-3,4-Dihydroxyphenylalanine (see *dl*-Dopa)
57-97-6	9,10-Dimethyl-1,2-benzanthracene (see 7,12-Dimethylbenz(*a*)anthracene)

CAS number	Chemical name
695-53-4	5,5-Dimethyl-2,4-oxazolidinedione (see Dimethadione)
25812-30-0	2,2-Dimethyl-5-(2,5-xylyloxy)valeric acid (see Gemfibrozil)
79-44-7	Dimethylcarbamoyl chloride (see Dimethylcarbamyl chloride)
62-75-9	Dimethylnitrosamine (see N-Nitrosodimethylamine)
62-75-9	N,N-Dimethylnitrosamine (see N-Nitrosodimethylamine)
123-91-1	p-Dioxane (see 1,4-Dioxane)
1746-01-6	Dioxin (see 2,3,7,8-Tetrachlorodibenzo-p-dioxin)
66-22-8	2,4-Dioxypyrimidine (see Uracil)
86-30-6	Diphenylnitrosamine (see N-Nitrosodiphenylamine)
621-64-7	Dipropylnitrosamine (see N-Nitrosodipropylamine)
142-59-6	Disodium ethylenebisdithiocarbamate (see Ethylenebisdithiocarbamate, disodium)
7757-82-6	Disodium sulfate (see Sulfate, sodium)
97-77-8	Disulfiram (see Tetraethylthiuram disulfide)
142-59-6	Dithane (see Ethylenebisdithiocarbamate, disodium)
330-54-1	Diuron (see 3-(3,4-Dichlorophenyl)-1,1-dimethylurea)
1596-84-5	DMASA (see Daminozide)
57-97-6	DMBA (see 7,12-Dimethylbenz(a)anthracene)
868-85-9	DMHP (see Dimethyl hydrogen phosphite)
756-79-6	DMMP (see Dimethyl methylphosphonate)
62-75-9	DMN (see N-Nitrosodimethylamine)
120-61-6	DMT (see Dimethyl terephthalate)
2385-85-5	1,1a,2,2,3,3a,4,5,5,5a,5b,6-Dodecachlorooctahydro-1,3,4-metheno-1H-cyclobuta[cd]pentalene (see Mirex)
2439-10-3	Dodine (see n-Dodecylguanidine acetate)
90-43-7	Dowicide 1 (see o-Phenylphenol)
88-06-2	Dowicide 2S (see 2,4,6-Trichlorophenol)
87-86-5	Dowicide 7 (see 2,3,4,5,6-Pentachlorophenol (Dowicide EC-7))
87-86-5	Dowicide EC-7 (see 2,3,4,5,6-Pentachlorophenol (Dowicide EC-7))
9011-18-1	DS-M-1 (see Dextran sulfate sodium (DS-M-1))
9011-18-1	DST-H (see Dextran sulfate sodium (DST-H))
95-33-0	Durax (see N-Cyclohexyl-2-benzothiazole sulfenamide)
106-93-4	EDB (see 1,2-Dibromoethane)
107-06-2	EDC (see 1,2-Dichloroethane)
150-38-9	EDTA (see EDTA, trisodium salt trihydrate)
50-18-0	Endoxan (see Cyclophosphamide)
759-73-9	ENU (see 1-Ethyl-1-nitrosourea)
75-21-8	EO (see Ethylene oxide)
75-56-9	1,2-Epoxypropane (see 1,2-Propylene oxide)
16423-68-0	Erythrosine (see FD & C red no. 3)
50-28-2	Estradiol-17β (see Estradiol)
938-73-8	Ethenzamide (see o-Ethoxybenzamide)
88133-11-3	8-(2-Ethoxyethyl)-phenyl-1,2,4-triazolo[4,3-c]pyrimidine-5 amine (see Bemitradine)
74-96-4	Ethyl bromide (see Bromoethane)
14239-68-0	Ethyl cadmate (see Cadmium diethyldithiocarbamate)
51-79-6	Ethyl carbamate (see Urethane)
75-00-3	Ethyl chloride (see Chloroethane)
637-07-0	Ethyl-α-p-chlorophenoxyisobutyrate (see Clofibrate)

CAS number	Chemical name
759-73-9	N-Ethyl-N-nitrosourea (see 1-Ethyl-1-nitrosourea)
614-95-9	1-Ethyl-1-nitrosourethan (see Nitrosoethylurethan)
614-95-9	N-Ethyl-N-nitrosourethane (see Nitrosoethylurethan)
5456-28-0	Ethyl selenac (see Selenium diethyldithiocarbamate)
97-77-8	Ethyl tuads (see Tetraethylthiuram disulfide)
14324-55-1	Ethyl zimate (see Zinc diethyldithiocarbamate)
106-93-4	Ethylene dibromide (see 1,2-Dibromoethane)
107-06-2	Ethylene dichloride (see 1,2-Dichloroethane)
759-73-9	Ethylnitrosourea (see 1-Ethyl-1-nitrosourea)
8056-92-6	Ethynodiol diacetate/ethinyl estradiol [10:1] (see Ovulen)
96-45-7	ETU (see Ethylene thiourea)
24554-26-5	FANFT (see N-[4-(5-Nitro-2-furyl)-2-thiazolyl]formamide)
14484-64-1	FERBAM (see Ferric dimethyldithiocarbamate)
mixture	Ferric nitrosodimethyldithiocarbamate and tetramethylthiuram disulfide (see Vanguard GF)
59536-65-1	Firemaster BP-6 (see Polybrominated biphenyls)
67774-32-7	Firemaster FF-1 (see Polybrominated biphenyl mixture)
53-96-3	Fluorenylacetamide (see 2-Acetylaminofluorene)
28314-03-6	N-1-Fluorenylacetamide (see 1-Acetylaminofluorene)
53-96-3	N-2-Fluorenylacetamide (see 2-Acetylaminofluorene)
28322-02-3	N-4-Fluorenylacetamide (see 4-Acetylaminofluorene)
63019-65-8	N-1-Fluorenyldiacetamide (see N-1-Diacetamidofluorene)
398-32-3	N-(4'-Fluoro-4-biphenylyl)acetamide (see N-4-(4'-Fluorobiphenyl)acetamide)
75-69-4	Fluorocarbon 11 (see Trichlorofluoromethane)
75-71-8	Fluorocarbon 12 (see Dichlorodifluoromethane)
75-45-6	Fluorocarbon 22 (see Chlorodifluoromethane)
593-70-4	Fluorocarbon 31 (see Chlorofluoromethane)
76-13-1	Fluorocarbon 113 (see 1,1,2-Trichloro-1,2,2-trifluoroethane, technical grade)
75-88-7	Fluorocarbon 133A (see 2-Chloro-1,1,1-trifluoroethane)
3570-75-0	FNT (see Formic acid 2-[4-(5-nitro-2-furyl)-2-thiazolyl]hydrazide)
133-07-3	Folpet (see N-(Trichloromethylthio)phthalimide)
140-56-7	Formulated fenaminosulf (see Fenaminosulf, formulated)
75-09-2	Freon 30 (see Methylene chloride)
3688-53-7	2-(2-Furyl)-3-(5-nitro-2-furyl)acrylamide (see AF-2)
3688-53-7	Furylfuramide (see AF-2)
60142-96-3	Gabapentin (see 1-(Aminomethyl)cyclohexaneacetic acid)
97-16-5	Genite-R99 (see 2,4-Dichlorophenylbenzene sulfonate)
139-40-2	Gesamil (see Propazine)
69644-85-5	N2-[γ-l(+)-Glutamyl]-4-carboxyphenylhydrazine (see N2-γ-Glutamyl-p-hydrazinobenzoic acid)
69644-85-5	Glutamyl-p-hydrazinobenzoate (see N2-γ-Glutamyl-p-hydrazinobenzoic acid)
1072-53-3	Glycol sulfate (see Ethylene glycol, cyclic sulfate)
4680-78-8	Guinea green B (see FD & C green no. 1)
2353-45-9	Fast green FCF (see FD & C green no. 3)
5141-20-8	Light green SF yellowish (see FD & C green no. 2)
9000-01-5	Gum acacia (see Gum arabic)

CAS number	Chemical name
86-50-0	Gusathion (see Azinphosmethyl)
118-74-1	HCB (see Hexachlorobenzene)
2163-79-3	Hercules-7531 (see 3-(Hexahydro-4,7-methanoindan-5-yl)-1,1-dimethylurea)
87-51-4	Heteroauxin (see Indole-3-acetic acid)
26049-68-3	HNT (see 2-Hydrazino-4-(5-nitro-2-furyl)thiazole)
619-67-0	p-Hydrazinobenzoate (see p-Hydrazinobenzoic acid)
69644-85-5	p-Hydrazinobenzoic acid, N2-γ-glutamyl (see N2-γ-Glutamyl-p-hydrazinobenzoic acid)
7647-01-0	Hydrogen chloride (see Hydrochloric acid)
53-95-2	N-Hydroxy-N-acetyl-2-aminofluorene (see N-Hydroxy-2-acetylaminofluorene)
119-53-9	2-Hydroxy-1,2-diphenylethanone (see Benzoin)
53-95-2	Hydroxy-N-2-fluorenylacetamide (see N-Hydroxy-2-acetylaminofluorene)
24382-04-5	3-Hydroxy-2-propenal, sodium salt (see Malonaldehyde, sodium salt)
88107-10-2	1-(2-Hydroxy-3-propyl-4-(4-(1H-tetrazol-5-yl)-butoxy)phenyl)ethanone (see Compound LY171883)
103-90-2	p-Hydroxyacetanilide (see Acetaminophen)
924-42-5	N-(Hydroxymethyl)-acrylamide (see N-Methylolacrylamide)
7681-52-9	Hypochlorous acid, sodium salt (see Sodiumhypochlorite)
75011-65-3	Ibopamine.HCl (see N-Methyldopamine,O,O'-diisobutyroyl ester.HCl)
120-93-4	2-Imidazolidinone (see Ethylene urea)
860-22-0	Indigo carmine (see FD & C blue no. 2)
54-85-3	INH (see Isoniazid)
122-42-9	IPC (see Isopropyl-N-phenyl carbamate)
3458-22-8	IPD (see 3,3′-Iminobis-1-propanol dimethanesulfonate(ester).HCl)
6381-77-7	Isoascorbate (see Erythorbate, sodium)
297-78-9	Isobenzan (see Telodrin)
760-60-1	N-Isobutyl-N-nitrosourea (see N-Nitroso-N-isobutylurea)
119-38-0	Isolan (see 1-Isopropyl-3-methyl-S-pyrazolyldimethyl carbamate)
54-85-3	Isonicotinic acid hydrazide (see Isoniazid)
65765-07-3	1-Isopropyl-4-(m-methoxyphenyl)-7-methyl-2(1H)-quinazolinone (see Compound 50-892)
22760-18-5	1-Isopropyl-7-methyl-4-phenyl-2(1H)-quinazolinone (see Proquazone)
80-05-7	4,4′-Isopropylidenediphenol (see Bisphenol A)
6119-92-2	Karathane (see Dinitro(1-methylheptyl)phenyl crotonate)
330-54-1	Karmex (see 3-(3,4-Dichlorophenyl)-1,1-dimethylurea)
115-32-2	Kelthane (see Dicofol)
9011-18-1	KMDS-H (see Dextran sulfate sodium (KMDS-H))
43033-72-3	LAAM (see 6-Dimethylamino-4,4-diphenyl-3-heptanol acetate.HCl)
1335-32-6	Lead subacetate (see Lead acetate, basic)
19010-66-3	Ledate (see Lead dimethyldithiocarbamate)
58-89-9	Lindane (see γ-1,2,3,4,5,6-Hexachlorocyclohexane)
88107-10-2	LY171883, compound (see Compound LY171883)

CAS number	Chemical name
67-20-9	Macrodantin (see 1-[(5-Nitrofurfurylidene)amino]hydantoin)
632-99-5	Magenta I (see Rosaniline.HCl)
569-61-9	p-Magenta (see p-Rosaniline.HCl)
18968-99-5	Magnesium pemoline (see 2-Amino-5-phenyl-2-oxazolin-4-one + Mg(OH)2)
1634-78-2	Malathion-O-analog (see Malaoxon)
mixture	MAM acetate and cycasin (see Cycasin and methylazoxymethanol acetate)
12427-38-2	Maneb (see Manganese ethylenebisthiocarbamate)
2425-06-1	Merpafol (see Captafol)
1095-90-5	dl-Methadone.HCl (see 6-Dimethylamino-4,4-diphenyl-3-heptanone.HCl)
5834-17-3	2-Methoxy-3-dibenzofuranamine (see 2-Methoxy-3-aminodibenzofuran)
563-47-3	Methyl allyl chloride (see 3-Chloro-2-methylpropene, technical grade (containing 5% dimethylvinyl chloride))
25843-45-2	Z-Methyl-O,N,N-azoxymethane (see Azoxymethane)
71-55-6	Methyl chloroform (see 1,1,1-Trichloroethane, technical grade)
9004-59-5	Methyl ethyl cellulose (see Edifas A)
684-93-5	N-Methyl-N-nitrosourea (see N-Nitroso-N-methylurea)
140-56-7	Methyl orange B (see Fenaminosulf, formulated)
614-00-6	Methyl-phenyl-nitrosamine (see Nitrosomethylaniline)
144-34-3	Methyl selenac (see Selenium dimethyldithiocarbamate)
137-30-4	Methyl zimate (see Zinc dimethyldithiocarbamate)
mixture	Methylazoxymethanol acetate and cycasin mixture (see Cycasin and methylazoxymethanol acetate)
56-49-5	Methylcholanthrene (see 3-Methylcholanthrene)
533-31-3	3,4-(Methylenedioxy)-phenol (see Sesamol)
115-09-3	Methylmercury chloride (see Mercurymethyl chloride)
137-30-4	Milbam (see Zinc dimethyldithiocarbamate)
126-85-2	Mitomen (see Nitrogen mustard N-oxide)
66-27-3	MMS (see Methyl methanesulfonate)
70-25-7	MNNG (see N-Methyl-N′-nitro-N-nitrosoguanidine)
684-93-5	MNU (see N-Nitroso-N-methylurea)
101-14-4	MOCA (see 4,4′-Methylene-bis(2-chloroaniline))
108-90-7	Monochlorobenzene (see Chlorobenzene)
108-88-3	Monomethyl benzene (see Toluene)
32607-00-4	Monosodium iminodiacetic acid (see Iminodiacetic acid, monosodium)
150-68-5	Monuron (see 3-(p-Chlorophenyl)-1,1-dimethylurea)
298-81-7	8-MOP (see 8-Methoxypsoralen)
87-56-9	Mucochloric acid (see α,β-Dichloro-β-formylacrylic acid)
81-15-2	Musk xylol (see 2,4,6-Trinitro-1,3-dimethyl-5-tert-butylbenzene)
142-59-6	Nabam (see Ethylenebisdithiocarbamate, disodium)
91-59-8	β-Naphthylamine (see 2-Naphthylamine)
81-16-3	NAS (see 2-Naphthylamino,1-sulfonic acid)
2611-82-7	New coccine (see SX purple)
531-82-8	NFTA (see N-[4-(5-Nitro-2-furyl)-2-thiazolyl]acetamide)
59-67-6	Niacin (see Nicotinic acid)

CAS number	Chemical name
1420-04-8	Niclosamide (see Clonitralid)
4164-28-7	*N*-Nitrodimethylamine (see Dimethylnitramine)
67-20-9	Nitrofurantoin (see 1-[(5-Nitrofurfurylidene)-amino]hydantoin)
59-87-0	Nitrofurazone (see 5-Nitro-2-furaldehyde semicarbazone)
10024-97-2	Nitrogen oxide (see Nitrous oxide)
598-57-2	*N*-Nitromethylamine (see Methylnitramine)
56-75-7	*d*-(−)-threo-1-(*p*-Nitrophenyl)-2-dichloroacetamido-1,3-propanediol (see Chloramphenicol)
3276-41-3	*N*-Nitroso-3,6-dihydrooxazine-1,2 (see 3,6-Dihydro-2-nitroso-2h-1,2-oxazine)
13743-07-2	*N*-Nitroso-2-hydroxyethylurea (see 1-(2-Hydroxyethyl)-1-nitrosourea)
38434-77-4	Nitrosoethanecarbamonitrile (see Ethylnitrosocyanamide)
55090-44-3	Nitrosomethyl-*N*-dodecylamine (see *N*-Nitroso-*N*-methyl-*N*-dodecylamine)
10595-95-6	*N*-Nitrosomethylethylamine (see Nitrosoethylmethylamine)
684-93-5	Nitrosomethylurea (see *N*-Nitroso-*N*-methylurea)
59-89-2	Nitrosomorpholine (see *N*-Nitrosomorpholine)
39884-52-1	Nitrosooxazolidone (see *N*-Nitroso-1,3-oxazolidine)
5632-47-3	Nitrosopiperazine (see *N*-Nitrosopiperazine)
5632-47-3	1-Nitrosopiperazine (see *N*-Nitrosopiperazine)
930-55-2	Nitrosopyrrolidine (see *N*-Nitrosopyrrolidine)
mixture	β-Nitrostyrene and styrene mixture (see Styrene and β-nitrostyrene mixture)
8015-30-3	Norethynodrel/mestranol [25:1] (see Enovid-E)
8015-30-3	Norethynodrel/mestranol [66:1] (see Enovid)
117-81-7	di-sec-Octyl phthalate (see di(2-Ethylhexyl)phthalate)
50-28-2	17β-Oestradiol (see Estradiol)
5634-39-9	Organidin (see Iodinated glycerol)
90-43-7	Orthoxenol (see *o*-Phenylphenol)
13752-51-7	OTOS (see *N*-Oxydiethylene thiocarbamyl-*N*-oxydiethylene sulfenamide)
80-33-1	Ovex (see *p*-Chlorophenyl-*p*-chlorobenzene sulfonate)
297-76-7	Ovulen-50 (see Ethynodiol diacetate)
103-90-2	Paracetamol (see Acetaminophen)
92-69-3	Paraxenol (see *p*-Phenylphenol)
11096-82-5	PCBS (see Aroclor 1260)
27323-18-8	PCBS (see Aroclor 1254)
12737-87-0	PCBS (see Kanechlor 400)
82-68-8	PCNB (see Pentachloronitrobenzene)
87-86-5	PCP (see 2,3,4,5,6-Pentachlorophenol)
87-86-5	Penta (see 2,3,4,5,6-Pentachlorophenol (Dowicide EC-7))
76-01-7	Pentalin (see Pentachloroethane)
127-18-4	Perchloroethylene (see Tetrachloroethylene)
72-56-0	Perthane (see *p,p′*-Ethyl-DDD)
78-11-5	PETN, NF (see Pentaerythritol tetranitrate with 80% *d*-lactose monohydrate)
8003-03-0	Phenacetin, aspirin, and caffeine (see Aspirin, phenacetin, and caffeine)
90-49-3	Pheneturide (see Ethylphenylacetylurea)

| 50-06-6 | Phenobarbitone (see Phenobarbital) |

CAS number	Chemical name
57-30-7	Phenobarbitone, sodium (see Phenobarbital, sodium)
135-88-6	*N*-Phenyl-2-naphthylamine (see Phenyl-β-naphthylamine)
4075-79-0	4′-Phenylacetanilide (see 4-Acetylaminobiphenyl)
50-06-6	Phenylethylbarbituric acid (see Phenobarbital)
57-41-0	Phenytoin (see 5,5-Diphenylhydantoin)
13366-73-9	Photodieldrin (see Dieldrin, photo-)
39801-14-4	Photomirex (see Mirex, photo-)
149-17-7	Phthivazid (see Isonicotinic acid vanillylidenehydrazide)
86-87-3	Planofix (see 1-Naphthalene acetic acid)
11096-82-5	Polychlorinated biphenyls (see Aroclor 1260)
27323-18-8	Polychlorinated biphenyls (see Aroclor 1254)
12737-87-0	Polychlorinated biphenyls (see Kanechlor 400)
3564-09-8	Ponceau 3R (see FD & C red no. 1)
2611-82-7	Ponceau 4R (see SX purple)
3761-53-3	Ponceau MX (see D & C red no. 5)
4548-53-2	Ponceau SX (see FD & C red no. 4)
7758-01-2	Potassium bromate (see Bromate, potassium)
4429-42-9	Potassium metabisulfite (see Sulfite, potassium metabi-)
79-06-1	2-Propenamide (see Acrylamide)
104-46-1	*p*-Propenylanisole (see Anethole)
4180-23-8	trans-*p*-Propenylanisole (see trans-Anethole)
78-87-5	Propylene dichloride (see 1,2-Dichloropropane)
621-64-7	di-*N*-Propylnitrosamine (see *N*-Nitrosodipropylamine)
59333-67-4	Prozac (see Fluoxetine.HCl)
26308-28-1	Pyrazapon (see Ripazepam)
244-63-3	9*H*-Pyrido(3,4-*b*)indole (see Norharman)
553-53-7	3-Pyridoyl hydrazine (see Nicotinic acid hydrazide)
148-24-3	8-Quinolinol (see 8-Hydroxyquinoline)
82-68-8	Quintozene (see Pentachloronitrobenzene)
5160-02-1	Brilliant red (see D & C red no. 9)
3567-69-9	C.I. acid red 14, disodium salt (see C.I. food red 3)
989-38-8	C.I. basic red 1.HCl (see Rhodamine 6G)
569-61-9	C.I. basic red 9.HCl (see *p*-Rosaniline.HCl)
86-30-6	Redax (see *N*-Nitrosodiphenylamine)
2318-18-5	Renardine (see Senkirkine)
149-30-4	Rotax (see 2-Mercaptobenzothiazole)
121-19-7	Roxarsone (see 3-Nitro-4-hydroxyphenylarsonic acid)
153-18-4	Rutin (see Rutin trihydrate)
8052-16-2	Sanamycin (see Actinomycin C)
148-82-3	*l*-Sarcolysin (see Melphalan)
148-18-5	SDDC (see Sodium diethyldithiocarbamate trihydrate)
63-25-2	Sevin (see Carbaryl)
7784-46-5	Sodium arsenite (see Arsenite, sodium)
26628-22-8	Sodium azide (see Azide, sodium)
532-32-1	Sodium benzoate (see Benzoate, sodium)
139-05-9	Sodium cyclamate (see Cyclamate, sodium)
7681-49-4	Sodium fluoride (see Fluoride, sodium)
12034-09-2	Sodium niobate (see Niobate, sodium)
7631-99-4	Sodium nitrate (see Nitrate, sodium)
7632-00-0	Sodium nitrite (see Nitrite, sodium)
7757-82-6	Sodium sulfate (see Sulfate, sodium)

13755-29-8   Sodium tetrafluoroborate (see Tetrafluoroborate, sodium)

CAS number	Chemical name
13472-45-2	Sodium tungstate (see Tungstate, sodium)
28754-68-9	SQ 18506 (see trans-5-Amino-3[2-(5-nitro-2-furyl)vinyl]-1,2,4-oxadiazole)
7772-99-8	Stannous chloride (see Tin (II) chloride)
77-83-8	Strawberry aldehyde (see Ethyl methylphenylglycidate)
1596-84-5	Succinic acid 2,2-dimethylhydrazide (see Daminozide)
971-15-3	Sulfads (see Dipentamethylenethiuram hexasulfide)
68-89-3	Sulpyrin (see Dipyrone)
93-76-5	2,4,5-T (see 2,4,5-Trichlorophenoxyacetic acid)
1934-21-0	Tartrazine (see FD & C yellow no. 5)
97-18-7	TBP (see 2,2-Thiobis(4,6-dichlorophenol))
1746-01-6	TCDD (see 2,3,7,8-Tetrachlorodibenzo-p-dioxin)
79-01-6	TCE (see Trichloroethylene)
79-01-6	TCE (see Trichloroethylene (without epichlorohydrin))
72-54-8	TDE (see p,p'-DDD)
150-68-5	Telvar (see 3-(p-Chlorophenyl)-1,1-dimethylurea)
116-06-3	Temik (see Aldicarb)
34031-32-8	2,3,4,6-Tetra-O-acetyl-1-thio-1-β-d-glucopyranosato-S)(triethylphosphine) gold (see Auranofin)
118-75-2	Tetrachloro-p-benzoquinone (see Chloranil)
116-29-0	Tetrafidon (see 2,4,5,4'-Tetrachlorodiphenyl sulfone)
35449-36-6	2,2,9,9-Tetramethyl-1,10-decanediol (see Gemcadiol)
mixture	Tetramethylthiuram disulfide and ferric nitrosodimethyldithiocarbamate (see Vanguard GF)
2227-13-6	Tetrasul (see p-Chlorophenyl-2,4,5-trichlorophenyl sulfide)
64039-27-6	β-TGDR (see β-Thioguanine deoxyriboside)
148-79-8	2-(4-Thiazolyl)-benzimazole (see Thiabendazole)
79-19-6	Thiocarbamylhydrazine (see Thiosemicarbazide)
115-29-7	Thiodan (see Endosulfan)
137-26-8	Thiram (see Tetramethylthiuram disulfide)
124-64-1	THPC (see Tetrakis(hydroxymethyl)phosphonium chloride)
55566-30-8	THPS (see Tetrakis(hydroxymethyl)phosphonium sulfate)
37087-94-8	Tibric acid (see 2-Chloro-5-(3,5-dimethylpiperidinosulphonyl)benzoic acid)
1114-71-2	Tillam-6-E (see Propyl N-ethyl-N-butylthiocarbamate)

137-26-8   TMTD (see Tetramethylthiuram disulfide)

CAS number	Chemical name
95-80-7	2,4-Toluenediamine (see 2,4-Diaminotoluene)
15481-70-6	2,6-Toluenediamine.2HCl (see 2,6-Diaminotoluene.2HCl)
6369-59-1	2,5-Toluenediamine sulfate (see 2,5-Diaminotoluene sulfate)
75-47-8	Triiodomethane (see Iodoform)
58-08-2	1,3,7-Trimethylxanthine (see Caffeine)
126-72-7	Tris (see Tris(2,3-dibromopropyl)phosphate)
150-38-9	Trisodium ethylenediaminetetraacetate trihydrate (see EDTA, trisodium salt trihydrate)
83-79-4	Tubatoxin (see Rotenone)
103-90-2	Tylenol (see Acetaminophen)
57-14-7	UDMH (see 1,1-Dimethylhydrazine)
97-74-5	Unads (see Tetramethylthiuram monosulfide)
55567-81-2	Valoron (see Tilidine fumarate)
97-18-7	Vancide BL (see 2,2-Thiobis(4,6-dichlorophenol))
6385-58-6	Vancide BN (see Sodium bithionolate)
6379-46-0	Vancide PB (see 1,2,3-Trichloro-4,6-dinitrobenzene)
13927-77-0	Vanguard N (see Nickel dibutyldithiocarbamate)
62-73-7	Vapona (see Dichlorvos)
302-79-4	Vitamin A acid (see Retinoic acid)
127-47-9	Vitamin A, acetate (see Retinol acetate)
79-81-2	Vitamin A, palmitate (see Retinol palmitate)
50-81-7	Vitamin C (see l-Ascorbic acid)
134-03-2	Vitamin C, sodium (see l-Ascorbate, sodium)
50-14-6	Vitamin D₂ (see Calciferol)
10191-41-0	Vitamin E (see dl-α-Tocopherol)
58-95-7	Vitamin E acetate (see dl-α-Tocopheryl acetate)
50892-23-4	WY14643 (see [4-Chloro-6-(2,3-xylidino)-2-pyrimidinylthio]acetic acid)
60-11-7	Butter yellow (see N,N-Dimethyl-4-aminoazobenzene)
842-07-9	C.I. solvent yellow 14 (see 1-Phenylazo-2-naphthol)
6358-85-6	Diarylanilide yellow (see C.I. pigment yellow 12)
2783-94-0	Sunset yellow FCF (see FD & C yellow no. 6)
315-18-4	Zectran (see Mexacarbate)
155-04-4	Zetax (see 2-Mercaptobenzothiazole, zinc)
12122-67-7	Zineb (see Zinc ethylenebisthiocarbamate)
137-30-4	Ziram (see Zinc dimethyldithiocarbamate)

SECTION E: CARCINOGENICITY BIBLIOGRAPHY

PART 1: CARCINOGENICITY BIBLIOGRAPHY — NCI/NTP TECHNICAL REPORTS

Chemical name	Technical Report number	Publication date	Chemical name	Technical Report number	Publication date
Acetaminophen	394	1993	C.I. direct blue 15	397	1992
Acetohexamide	50	1978	C.I. direct blue 218	430	1994
Acronycine	49	1978	C.I. disperse blue 1	299	1986
Agar	230	1982	Direct blue 6	108	1978
Aldicarb	136	1979	HC blue no. 1	271	1985
Aldrin	21	1978	HC blue no. 2	293	1985
Allyl chloride	73	1978	Boric acid	324	1987
Allyl glycidyl ether	376	1990	Bromodichloromethane	321	1987
Allyl isothiocyanate	234	1982	Bromoethane	363	1989
Allyl isovalerate	253	1983	Direct brown 95	108	1978
3-Amino-4-ethoxyacetanilide	112	1978	1,3-Butadiene	288	1984
3-Amino-9-ethylcarbazole hydrochloride	93	1978	1,3-Butadiene	434	1993
1-Amino-2-methylanthraquinone	111	1978	Butyl benzyl phthalate	213	1982
2-Amino-4-nitrophenol	339	1988	n-Butyl chloride	312	1986
2-Amino-5-nitrophenol	334	1988	Butylated hydroxytoluene (BHT)	150	1979
4-Amino-2-nitrophenol	94	1978	γ-Butyrolactone	406	1992
2-Amino-5-nitrothiazole	53	1978	Calcium cyanamide	163	1979
2-Aminoanthraquinone	144	1978	Caprolactam	214	1982
11-Aminoundecanoic acid	216	1982	Captan	15	1977
dl-Amphetamine sulfate	387	1991	Carbromal	173	1979
Ampicillin trihydrate	318	1987	d-Carvone	381	1990
Anilazine	104	1978	Chloramben	25	1977
Aniline hydrochloride	130	1978	Chloraminated water	392	1992
o-Anisidine hydrochloride	89	1978	Chlordane	8	1977
p-Anisidine hydrochloride	116	1978	Chlorendic acid	304	1987
Anthranilic acid	36	1978	Chlorinated paraffins (C_{12}, 60% chlorine)	308	1986
Aroclor 1254	38	1978	Chlorinated paraffins (C_{23}, 43% chlorine)	305	1986
l-Ascorbic acid	247	1983	Chlorinated trisodium phosphate	294	1986
Aspirin, phenacetin, and caffeine, mixture	67	1978	Chlorinated water	392	1992
5-Azacytidine	42	1978	3-Chloro-2-methylpropene	300	1986
Azide, sodium	389	1991	4-Chloro-m-phenylenediamine	85	1978
Azinphosmethyl	69	1978	4-Chloro-o-phenylenediamine	63	1978
Azobenzene	154	1979	2-Chloro-p-phenylenediamine sulfate	113	1978
Barium chloride dihydrate	432	1994	3-Chloro-p-toluidine	145	1978
Benzaldehyde	378	1990	5-Chloro-o-toluidine	187	1979
Benzene	289	1986	4-Chloro-o-toluidine hydrochloride	165	1979
Benzofuran	370	1989	2-Chloroacetophenone	379	1990
Benzoin	204	1980	4'-(Chloroacetyl)-acetanilide	177	1979
1H-Benzotriazole	88	1978	p-Chloroaniline	189	1979
Benzyl acetate	250	1986	p-Chloroaniline hydrochloride	351	1989
Benzyl acetate	431	1993	Chlorobenzene	261	1985
Benzyl alcohol	343	1989	Chlorobenzilate	75	1978
o-Benzyl-p-chlorophenol	424	1994	Chlorodibromomethane	282	1985
2-Biphenylamine hydrochloride	233	1982	Chloroethane	346	1989
Bis(2-chloro-1-methylethyl) ether	191	1979	(2-Chloroethyl)trimethylammonium chloride	158	1979
Bis(2-chloro-1-methylethyl) ether	239	1982	Chloroform	A	1976
Bisphenol A	215	1982	2-(Chloromethyl)pyridine hydrochloride	178	1979
Direct black 38	108	1978	3-(Chloromethyl)pyridine hydrochloride	95	1978

Chemical name	Technical Report number	Publication date	Chemical name	Technical Report number	Publication date
Chloropicrin	65	1978	Dimethoate	4	1977
Chlorothalonil	41	1978	Dimethoxane	354	1989
Chlorpheniramine maleate	317	1986	2,4-Dimethoxyaniline hydrochloride	171	1979
Chlorpropamide	45	1978	3,3'-Dimethoxybenzidine.2HCl	372	1990
Cinnamyl anthranilate	196	1980	3,3'-Dimethoxybenzidine-4,4'-diisocyanate	128	1979
Clonitralid	91	1978	Dimethyl hydrogen phosphite	287	1985
Corn oil, safflower oil, and tricaprylin	426	1994	Dimethyl methylphosphonate	323	1987
Coumaphos	96	1979	Dimethyl morpholinophosphoramidate	298	1986
Coumarin	422	1993	Dimethyl terephthalate	121	1979
m-Cresidine	105	1978	*N,N*-Dimethylaniline	360	1989
p-Cresidine	142	1979	3,3'-Dimethylbenzidine.2HCl	390	1991
CS2	377	1990	Dimethylvinyl chloride	316	1986
Cupferron	100	1978	2,4-Dinitrotoluene	54	1978
Cytembena	207	1981	1,4-Dioxane	80	1978
Daminozide	83	1978	Dioxathion	125	1978
Dapsone	20	1977	Diphenhydramine hydrochloride	355	1989
p,p'-DDE	131	1978	5,5-Diphenylhydantoin	404	1993
DDT	131	1978	2,5-Dithiobiurea	132	1979
Decabromodiphenyl oxide	309	1986	Emetine	43	1978
Diallyl phthalate	242	1983	Endosulfan	62	1978
Diallyl phthalate	284	1985	Endrin	12	1979
2,4-Diaminoanisole sulfate	84	1978	Ephedrine sulphate	307	1986
4,4'-Diamino-2,2'-stilbenedisulfonic acid, disodium salt	412	1992	*l*-Epinepherine.HCl	380	1990
			1,2-Epoxybutane	329	1988
2,4-Diaminophenol.2HCl	401	1992	Erythromycin stearate	338	1988
2,4-Diaminotoluene	162	1979	Estradiol mustard	59	1978
2,5-Diaminotoluene sulfate	126	1978	Ethionamide	46	1978
Diarylanilide yellow	30	1978	Ethyl acrylate	259	1986
Diazinon	137	1979	*p,p'*-Ethyl-DDD	156	1979
Dibenzo-*p*-dioxin	122	1979	Ethyl tellurac	152	1979
Dibromochloropropane	28	1978	Ethylene glycol	413	1993
Dibromochloropropane	206	1982	Ethylene oxide	326	1987
1,2-Dibromoethane	86	1978	Ethylene thiourea	388	1992
1,2-Dibromoethane	210	1982	di(2-Ethylhexyl)adipate	212	1982
Dibutyltin diacetate	183	1979	di(2-Ethylhexyl)phthalate	217	1982
2,6-Dichloro-*p*-phenylenediamine	219	1982	Eugenol	223	1983
1,2-Dichlorobenzene	255	1985	Fenthion	103	1979
1,4-Dichlorobenzene	319	1987	Fluometuron	195	1980
2,7-Dichlorodibenzo-*p*-dioxin	123	1979	Fluoride, sodium	393	1990
1,1-Dichloroethane	66	1978	Formulated fenaminosulf	101	1978
1,2-Dichloroethane	55	1978	Furan	402	1993
Dichloromethane	306	1986	Furfural	382	1990
2,4-Dichlorophenol	353	1989	Furosemide	356	1989
1,2-Dichloropropane	263	1986	Food grade geranyl acetate	252	1987
Dichlorvos	10	1977	Glycidol	374	1990
Dichlorvos	342	1989	Guar gum	229	1982
Dicofol	90	1978	Gum arabic	227	1982
N,N'-Dicyclohexylthiourea	56	1978	Heptachlor	9	1977
Dieldrin	21	1978	Hexachlorocyclopentadiene	437	1994
Dieldrin	22	1978	1,2,3,6,7,8-Hexachlorodibenzo-*p*-dioxin and 1,2,3,7,8,9-hexachlorodibenzo-*p*-dioxin, mixture	198	1980
N,N'-Diethylthiourea	149	1979			
Diglycidyl resorcinol ether	257	1986	Hexachloroethane	361	1989
3,4-Dihydrocoumarin	423	1993	Hexachloroethane	68	1978

Chemical name	Technical Report number	Publication date
Hexachlorophene	40	1978
4-Hexylresorcinol	330	1988
Hydrazobenzene	92	1978
Hydrochlorothiazide	357	1989
Hydroquinone	366	1989
8-Hydroxyquinoline	276	1985
ICRF-159	78	1978
3,3′-Iminobis-1-propanol dimethanesulfonate (ester) hydrochloride	18	1978
Iodinated glycerol	340	1990
Iodoform	110	1978
Isophorone	291	1986
Isophosphamide	32	1977
Kepone	B	1976
Lasiocarpine	39	1978
Lead dimethyldithiocarbamate	151	1979
d-Limonene	347	1990
Lindane	14	1977
Lithocholic acid	175	1979
Locust bean gum	221	1982
Malaoxon	135	1979
Malathion	192	1979
Malathion	24	1979
Malonaldehyde, sodium salt	331	1988
Manganese (II) sulphate monohydrate	428	1993
d-Mannitol	236	1982
Melamine	245	1983
dl-Menthol	98	1979
2-Mercaptobenzothiazole	332	1988
Mercuric chloride	408	1993
Methoxychlor	35	1978
8-Methoxypsoralen	359	1989
Methyl bromide	385	1992
Methyl carbamate	328	1987
Methyl methacrylate	314	1986
2-Methyl-1-nitroanthraquinone	29	1978
Methyl parathion	157	1979
α-Methylbenzyl alcohol	369	1990
α-Methyldopa sesquihydrate	348	1989
4,4′-Methylenebis(N,N-dimethyl)benzenamine	186	1979
4,4′-Methylenedianiline dihydrochloride	248	1983
N-Methylolacrylamide	352	1989
Mexacarbate	147	1978
Michler's ketone	181	1979
Mirex	313	1990
Monochloroacetic acid	396	1992
Monuron	266	1988
Nalidixic acid	368	1989
Naphthalene	410	1992
1,5-Naphthalenediamine	143	1978
N-(1-Naphthyl)ethylenediamine dihydrochloride	168	1979
Nithiazide	146	1979
Nitrilotriacetic acid	6	1977

Chemical name	Technical Report number	Publication date
Nitrilotriacetic acid, trisodium salt, monohydrate	6	1977
3-Nitro-p-acetophenetide	133	1979
5-Nitro-o-anisidine	127	1978
2-Nitro-p-phenylenediamine	169	1979
4-Nitro-o-phenylenediamine	180	1979
5-Nitro-o-toluidine	107	1978
5-Nitroacenaphthene	118	1978
p-Nitroaniline	418	1993
o-Nitroanisole	416	1993
4-Nitroanthranilic acid	109	1978
6-Nitrobenzimidazole	117	1979
p-Nitrobenzoic acid	442	1994
Nitrofen	184	1979
Nitrofen	26	1979
Nitrofurantoin	341	1989
Nitrofurazone	337	1988
1-Nitronaphthalene	64	1978
3-Nitropropionic acid	52	1978
N-Nitrosodiphenylamine	164	1979
p-Nitrosodiphenylamine	190	1979
Ochratoxin A	358	1989
C.I. acid orange 3	335	1988
C.I. acid orange 10	211	1987
Oxazepam	443	1993
4,4′-Oxydianiline	205	1980
Oxytetracycline hydrochloride	315	1987
Ozone and ozone/NNK	440	1994
Parathion	70	1979
Penicillin VK	336	1988
Pentachloroanisole	414	1993
Pentachloroethane	232	1983
Pentachloronitrobenzene	325	1987
Pentachloronitrobenzene	61	1978
Pentachlorophenol, two technical grade mixtures	349	1989
Pentaerythritol tetranitrate with 80% d-lactose monohydrate	365	1989
Phenazopyridine hydrochloride	99	1978
Phenesterin	60	1978
Phenformin	7	1977
Phenol	203	1980
Phenoxybenzamine hydrochloride	72	1978
1-Phenyl-3-methyl-5-pyrazolone	141	1978
N-Phenyl-2-naphthylamine	333	1988
N-Phenyl-p-phenylenediamine hydrochloride	82	1978
1-Phenyl-2-thiourea	148	1978
Phenylbutazone	367	1990
p-Phenylenediamine dihydrochloride	174	1979
Phenylephrine hydrochloride	322	1987
Phosphamidon	16	1979
Photodieldrin	17	1977
Phthalamide	161	1979
Phthalic anhydride	159	1979
Picloram	23	1978

Chemical name	Technical Report number	Publication date
Piperonyl butoxide	120	1979
Piperonyl sulfoxide	124	1979
Pivalolactone	140	1978
Polybrominated biphenyls	244	1983
Polybrominated biphenyls	398	1993
Polysorbate 80	415	1992
Probenecid	395	1991
Procarbazine	19	1979
Proflavine	5	1977
Promethazine hydrochloride	425	1993
Propyl gallate	240	1982
Propylene	272	1985
1,2-Propylene oxide	267	1985
Pyrazinamide	48	1978
Pyrimethamine	77	1978
Quercetin	409	1992
p-Quinone dioxime	179	1979
C.I. acid red 14, disodium salt	220	1982
C.I. acid red 114	405	1991
C.I. basic red 9 monohydrochloride	285	1986
C.I. pigment red 3	407	1992
C.I. pigment red 23	411	1992
D&C red no. 9	225	1982
HC red no. 3	281	1986
Reserpine	193	1980
Resorcinol	403	1992
Rhodamine 6G	364	1989
Rotenone	320	1988
Roxarsone	345	1989
Selenium sulfide	194	1980
Sodium diethyldithiocarbamate	172	1979
Solution of β-nitrostyrene and styrene	170	1979
Stannous chloride	231	1982
Styrene	185	1979
Succinic anhydride	373	1990
Sulfallate	115	1978
Sulfisoxazole	138	1979
3-Sulfolene	102	1978
Tara gum	224	1982
TDE	131	1978
Telone II	269	1985
2,3,5,6-Tetrachloro-4-nitroanisole	114	1978
2,3,7,8-Tetrachlorodibenzo-p-dioxin	209	1982
1,1,1,2-Tetrachloroethane	237	1983
1,1,2,2-Tetrachloroethane	27	1978
Tetrachloroethylene	13	1977
Tetrachloroethylene	311	1986
Tetrachlorvinphos	33	1978
Tetracycline hydrochloride	344	1989
Tetraethylthiuram disulfide	166	1979

Chemical name	Technical Report number	Publication date
Tetrakis(hydroxymethyl)phosphonium chloride	296	1987
Tetrakis(hydroxymethyl)phosphonium sulfate	296	1987
Tetranitromethane	386	1990
β-TGdR	57	1978
Thio-TEPA	58	1978
4,4′-Thiobis(6-t-butyl-m-cresol)	435	1994
4,4′-Thiodianiline	47	1978
Titanium dioxide	97	1979
Titanocene dichloride	399	1991
Tolazamide	51	1978
Tolbutamide	31	1977
Toluene	371	1990
Toluene diisocyanate, commercial grade	251	1986
2,6-Toluenediamine dihydrochloride	200	1980
o-Toluidine hydrochloride	153	1979
Toxaphene	37	1977
Triamterene	420	1993
Tribromomethane	350	1989
1,1,1-Trichloroethane	3	1977
1,1,2-Trichloroethane	74	1978
Trichloroethylene	2	1976
Trichloroethylene (without epichlorohydrin)	243	1990
Trichloroethylene (without epichlorohydrin)	273	1988
Trichlorofluoromethane	106	1978
2,4,6-Trichlorophenol	155	1979
1,2,3-Trichloropropane	384	1993
Tricresyl phosphate	433	1994
Trifluralin	34	1978
2,4,5-Trimethylaniline	160	1979
Trimethylphosphate	81	1978
Trimethylthiourea	129	1979
Triphenyltin hydroxide	139	1978
Tris(2-chloroethyl)phosphate	391	1991
Tris(2,3-dibromopropyl)phosphate	76	1978
Tris(2-ethylhexyl)phosphate	274	1984
Trisodium etylenediaminetetraacetate trihydrate (EDTA)	11	1977
l-Tryptophan	71	1978
Turmeric oleoresin	427	1993
Vinyl toluene	375	1990
4-Vinylcyclohexene	303	1986
Vinylidene chloride	228	1982
Xylenes, mixture	327	1986
HC yellow 4	419	1992
C.I. disperse yellow 3	222	1982
C.I. solvent yellow 14	226	1982
C.I. vat yellow 4	134	1979
FD & C yellow no. 6	208	1981
Zearalenone	235	1982
Ziram	238	1983

Note: A = no Technical Report number assigned by NCI and B = NCI (brief communication).

1. Abe, I., Saito, S., Hori, K., Suzuki, M., and Sato, H. Sodium erythorbate is not carcinogenic in F344 rats. *Exp. Mol. Pathol.* 41: 35-43(1984).

2. Adamson, R. H., Takayama, S., Sugimura, T., and Thorgeirsson, U. P. Induction of hepatocellular carcinoma in nonhuman primates by the food mutagen 2-amino-3-methylimidazo[4,5-*f*]quinoline. *Environ. Health Perspect.* 102: 190-193(1994).

3. Adamson, R. H. and Sieber, S. M. Chemical carcinogenesis studies in nonhuman primates. In: *Organ and Species Specificity in Chemical Carcinogenesis.* (R. Langenbach, S. Nesnow, and J. M. Rice, Eds.), Plenum Press, New York and London, 1982, pp. 129-156.

4. Adolphs, H. D., Thiele, J., Kiel, H., and Steffens, L. Induction of transitional cell carcinoma of the urinary bladder in rats by feeding *N*-[4-(5-nitro-2-furyl)-2-thiazolyl]formamide. *Urol. Res.* 6: 19-27(1978).

5. Agthe, C., Garcia, H., Shubik, P., Tomatis, L., and Wenyon, E. Study of the potential carcinogenicity of DDT in the Syrian golden hamster (34740). *Proc. Soc. Exp. Biol. Med.* 134: 113-116(1970).

6. Akao, M. and Kuroda, K. Inhibitory effect of fumaric acid on hepatocarcinogenesis by thioacetamide in mice. *Chem. Pharmacol. Bull.* 38: 2012-2014(1990).

7. Allmark, M. G., Grice, H. C., and Mannell, W. A. Chronic toxicity studies on food colours. Part II. Observations on the toxicity of FD&C green no. 2 (light green SF yellowish), FD&C orange no. 2 (orange SS) and FD&C red no. 32 (oil red XO) in rats. *J. Pharm. Pharmacol.* 8: 417-424(1956).

8. Allmark, M. G., Mannell, W. A., and Grice, H. C. Chronic toxicity studies on food colours. Part III. Observations on the toxicity of malachite green, new coccine and nigrosine in rats. *J. Pharm. Pharmacol.* 9: 622-628(1957).

9. Amo, H. and Matsuyama, M. Subchronic and chronic effects of feeding of large amounts of acetaminophen in B6C3F$_1$ mice. *Jpn. J. Hyg.* 40: 567-574(1985).

10. Amo, H., Matsuyama, M., Amano, H., Yamada, C., Kawai, M., Miyata, N., and Nakadate, M. Carcinogenicity and toxicity study of *m*-phenylenediamine administered in the drinking-water to (C57BL/6 × C3H/He)F$_1$ mice. *Food Chem. Toxicol.* 26: 893-897(1988).

11. Anderson, L. M., Carter, J. P., Logsdon, D. L., Driver, C. L., and Kovatch, R. M. Characterization of ethanol's enhancement of tumorigenesis by *N*-nitrosodimethylamine in mice. *Carcinogenesis* 13: 2107-2111(1992).

12. Anderson, L. M., Giner-Sorolla, A., Greenbaum, J. H., Last-Barney, K., and Budinger, J. M. Induction of reproductive system tumors in mice by *N*6-(methylnitroso)-adenosine and a tumorigenic effect of its combined precursors. *Int. J. Cancer* 24: 319-322(1979).

13. Anderson, L. M., Giner-Sorolla, A., Haller, I. M., and Budinger, J. M. Effects of cimetidine, nitrite, cimetidine plus nitrite, and nitrosocimetidine on tumors in mice following transplacental plus chronic lifetime exposure. *Cancer Res.* 45: 3561-3566(1985).

14. Andrianova, M. M. Carcinogenic properties of red food pigments — amaranth, SX purple and ponceau 4R. *Vopr. Pitan.* 29: 61-65(1970).

15. Andrianova, M. M. and Alekscev, I. V. On carcinogenic properties of the pesticides Sevin, maneb, ziram and zineb. *Vopr. Pitan.* 29: 71-74(1970).

16. Angsubhakorn, S., Bhamarapravati, N., Romruen, K., and Sahaphong, S. Enhancing effects of dimethylnitrosamine on aflatoxin B$_1$ hepatocarcinogenesis in rats. *Int. J. Cancer* 28: 621-626(1981).

17. Angsubhakorn, S., Bhamarapravati, N., Romruen, K., Sahaphong, S., Thamavit, W., and Miyamoto, M. Further study of α-benzene hexachloride inhibition of aflatoxin B$_1$ hepatocarcinogenesis in rats. *Br. J. Cancer* 43: 881-883(1981).

18. Annapurna, V. V., Mukundan, M. A., Sesikeran, B., and Bamji, M. S. Long-term effects of female sex steroids on female rat liver in an initiator-promoter model of hepatocarcinogenesis. *Indian J. Biochem. Biophys.* 25: 708-713(1988).

19. Aoyagi, M., Matsukura, N., Uchida, E., Kawachi, T., Sugimura, T., Takayama, S., and Matsui, M. Induction of liver tumors in Wistar rats by sodium nitrite given in pellet diet. *J. Natl. Cancer Inst.* 65: 411-414(1980).

20. Arai, M., Aoki, Y., Nakanishi, K., Miyata, Y., Mori, T., and Ito, N. Long-term experiment of maximal non-carcinogenic dose of dimethylnitrosamine for carcinogenesis in rats. *Gann* 70: 549-558(1979).

21. Arai, M., Cohen, S. M., Jacobs, J. B., and Friedell, G. H. Effect of dose on urinary bladder carcinogenesis induced in F344 rats by *N*-[4-(5-nitro-2-furyl)-2-thiazolyl]formamide. *J. Natl. Cancer Inst.* 62: 1013-1016(1979).

22. Arai, M. and Hibino, T. Tumorigenicity of citrinin in male F344 rats. *Cancer Lett.* 17: 281-287(1983).

23. Arai, M., St. John, M., Fukushima, S., Friedell, G. H., and Cohen, S. M. Long-term dose-response study of *N*-[4-(5-nitro-2-furyl)-2-thiazolyl]formamide-induced urinary bladder carcinogenesis. *Cancer Lett.* 18: 261-269(1983).

24. Arffmann, E. J. L., Rasmussen, K. S., and Hansen, F. N. Effect of some fatty acid methyl esters on gastrointestinal carcinogenesis by *N*-methyl-*N*'-nitro-*N*-nitrosoguanidine in rats. *J. Natl. Cancer Inst.* 67: 1071-1075(1981).

25. Argus, M. F., Arcos, J. C., and Hoch-Ligeti, C. Studies on the carcinogenic activity of protein-denaturing agents: hepatocarcinogenicity of dioxane. *J. Natl. Cancer Inst.* 35: 949-958(1965).

26. Argus, M. F., White, L. E., Bryant, G. M., Arcos, J. C., and Hoch-Ligeti, C. Molecular specificity of the tumorigenic action of ethionine: the inactivity of *S*-ethylcysteine. Action on respiratory parameters. *Cancer Res. Clin. Oncol.* 75: 201-208(1971).

27. Armuth, V. Leukaemogenic action of phorbol in intact and thymectomized mice of different strains. *Br. J. Cancer* 34: 516-522(1976).

28. Arnold, D. L., Moodie, C. A., Charbonneau, S. M., Grice, H. C., McGuire, P. F., Bryce, F. R., Collins, B. T., Zawidzka, Z. Z., Krewski, D. R., Nera, E. A., and Munro, I. C. Long-term toxicity of hexachlorobenzene in the rat and the effect of dietary vitamin A. *Food Chem. Toxicol.* 23: 779-793(1985).

29. Arnold, D. L., Moodie, C. A., Grice, H. C., Charbonneau, S. M., Stavric, B., Collins, B. T., McGuire, P. F., Zawidzka, Z. Z., and Munro, I. C. Long-term toxicity of ortho-toluenesulfonamide and sodium saccharin in the rat. *Toxicol. Appl. Pharmacol.* 52: 113-152(1980).

30. Asada, I., Matsumoto, Y., Tobe, T., Yoshida, O., and Miyakawa, M. Induction of hepatoma in mice by direct deep black-extra (DDB-EX) and occurrence of serum AFP. *Arch. Jpn. Chir.* 50: 45-55(1981).

31. Baden, J. M., Egbert, B., and Mazze, R. I. Carcinogen bioassay of enflurane in mice. *Anesthesiology* 56: 9-13(1982).

32. Baden, J. M., Kundomal, Y. R., Luttropp, M. E., Mazze, R. I., and Kosek, J. C. Carcinogen bioassay of nitrous oxide in mice. *Anesthesiology* 64: 747-750(1986).

33. Baden, J. M., Kundomal, Y. R., Mazze, R. I., and Kosek, J. C. Carcinogen bioassay of isoflurane in mice. *Anesthesiology* 69: 750-753(1988).

34. Baigusheva, M. M. Carcinogenic properties of the amaranth paste. *Vopr. Pitan.* 27: 46-50(1968).

35. Barnes, J. M., Magee, P. N., Boyland, E., Haddow, A., Passey, R. D., Bullough, W. S., Cruickshank, C. N. D., Salaman, M. H., and Williams, R. T. The non-toxicity of maleic hydrazide for mammalian tissues. *Nature* 180: 62-64(1957).

36. Baroni, C., van Esch, G. J., and Saffiotti, U. Carcinogenesis tests of two inorganic arsenicals. *Arch. Environ. Health* 7: 668-674(1963).

37. Barrows, G. H., Christopherson, W. M., and Drill, V. A. Liver lesions and oral contraceptive steroids. *J. Toxicol. Environ. Health* 3: 219-230(1977).

38. Barten, M. The effects of different MNNG (*N*-methyl-*N*′-nitro-*N*-nitrosoguanidine)-doses on the stomach and the upper small intestine of the rat. *Exp. Pathol.* 31: 147-152(1987).

39. Basso, N., Materia, A., Silecchia, G., Spaziani, E., Scucchi, L., Di Stefano, D., Mingazzini, P., Favalli, C., and Garaci, E. Time-related interference of misoprostol with experimental gastric cancer formation induced by *N*-methyl-*N*′-nitro-*N*-nitrosoguanidine in the rat. *Cancer Res. Clin. Oncol.* 118: 441-446(1992).

40. Beal, D. D., Skibba, J. L., Croft, W. A., Cohen, S. M., and Bryan, G. T. Carcinogenicity of the antineoplastic agent, 5-(3,3-dimethyl-1-triazeno)-imidazole-4-carboxamide, and its metabolites in rats. *J. Natl. Cancer Inst.* 54: 951-957(1975).

41. Becker, F. F. Morphological classification of mouse liver tumors based on biological characteristics. *Cancer Res.* 42: 3918-3923(1982).

42. Bendele, A. M., Carlton, W. W., Krogh, P., and Lillehoj, E. B. Ochratoxin A carcinogenesis in the (C57BL/6J × C3H)F₁ mouse. *J. Natl. Cancer Inst.* 75: 733-739(1985).

43. Bendele, A. M., Hoover, D. M., van Lier, R. B. L., Foxworthy, P. S., and Eacho, P. I. Effects of chronic treatment with leukotriene D₄-antagonist compound LY171883 on B6C3F₁ mice. *Fundam. Appl. Toxicol.* 15: 676-682(1990).

44. Bendele, R. A., Adams, E. R., Hoffman, W. P., Gries, C. L., and Morton, D. M. Carcinogenicity studies of fluoxetine hydrochloride in rats and mice. *Cancer Res.* 52: 6931-6935(1992).

45. Benya, T. J., Busey, W. M., Dorato, M. A., and Berteau, P. E. Inhalation carcinogenicity of vinyl bromide in rats. *Toxicol. Appl. Pharmacol.* 64: 367-379(1982).

46. Berger, M. R., Habs, M., and Schmahl, D. Long-term toxicology effects of prednimustine in comparison with chlorambucil, prednisolone, and chlorambucil plus prednisolone in Sprague-Dawley rats. *Semin. Oncol.* 13: 8-13(1986).

47. Berger, M. R., Petru, E., Habs, M., and Schmahl, D. Influence of the application mode of *N*-[4-(5-nitro-2-furyl)-2-thiazolyl]-formamide on the localization of its carcinogenic expression in female NMRI-mice. *Cancer Lett.* 31: 311-318(1986).

48. Berger, M. R., Schmahl, D., and Zerban, H. Combination experiments with very low doses of three genotoxic *N*-nitrosamines with similar organotropic carcinogenicity in rats. *Carcinogenesis* 8: 1635-1643(1987).

49. Bhide, S. V., Bhalerao, E. B., Sarode, A. V., and Maru, G. B. Mutagenicity and carcinogenicity of mono- and diacetyl hydrazine. *Cancer Lett.* 23: 235-240(1984).

50. Bhide, S. V., D'Souza, R. A., Sawai, M. M., and Ranadive, K. J. Lung tumour incidence in mice treated with hydrazine sulphate. *Int. J. Cancer* 18: 530-535(1976).

51. Bhide, S. V., Gothoskar, S. V., and Shivapurkar, N. M. Arecoline tumorigenicity in Swiss strain mice on normal and vitamin B deficient diet. *Cancer Res. Clin. Oncol.* 107: 169-171(1984).

52. Bhide, S. V., Maru, G. B., Sawai, M. M., and Ranadive, K. J. Isoniazid tumorigenicity in mice under different experimental conditions. *Int. J. Cancer* 21: 381-386(1978).

53. Bi, W., Wang, Y., Huang, M., and Meng, D. Effect of vinyl chloride on testis in rats. *Ecotoxicol. Environ. Safety* 10: 281-289(1985).

54. Biancifiori, C., Bucciarelli, E., Clayson, D. B., and Santilli, F. E. Induction of hepatomas in CBA/Cb/Se mice by hydrazine sulphate and the lack of effect of croton oil on tumour induction in BALB/c/Cb/Se mice. *Br. J. Cancer* 18: 543-550(1964).

55. Biancifiori, C., Milia, U., and Di Leo, F. P. Tumori della tiroide indotti mediante etionamide (ET) in topi femmina vergini BALB/c/Cb/Se substrain. *Lav. Inst. Anat. Istol. Patol. Univ. Studi Perugia.* 24: 145-165(1964).

56. Bird, R. P., Draper, H. H., and Valli, V. E. O. Toxicological evaluation of malonaldehyde: a 12-month study of mice. *J. Toxicol. Environ. Health* 10: 897-905(1982).

57. Blair, D., Dix, K. M., Hunt, P. F., Thorpe, E., Stevenson, D. E., and Walker, A. I. T. A 2-year inhalation carcinogenesis study in rats. *Arch. Toxicol.* 35: 281-294(1976).

58. Boberg, E. W., Miller, E. C., Miller, J. A., Poland, A., and Liem, A. Strong evidence from studies with brachymorphic mice and pentachlorophenol that 1′-sulfooxysafrole is the major ultimate electrophilic and carcinogenic metabolite of 1′-hydroxysafrole in mouse liver. *Cancer Res.* 43: 5163-5173(1983).

59. Bonser, G. M., Clayson, D. B., Jull, J. W., and Pyrah, L. N. The carcinogenic properties of 2-amino-1-naphthol hydrochloride and its parent amine 2-naphthylamine. *Br. J. Cancer* 6: 412-424(1952).

60. Borchert, P., Miller, J. A., Miller, E. C., and Shires, T. K. 1′-Hydroxysafrole, a proximate carcinogenic metabolite of safrole in the rat and mouse. *Cancer Res.* 33: 590-600(1973).

61. Borelli, G., Bertoli, D., and Chieco, P. Carcinogenicity study of doxefazepam administered in the diet to Sprague-Dawley rats. *Fundam. Appl. Toxicol.* 15: 82-92(1990).

62. Borzelleca, J. F., Depukat, K., and Halligan, J. B. Lifetime toxicity/carcinogenicity studies of FD & C Blue No. 1 (brilliant blue FCF) in rats and mice. *Food Chem. Toxicol.* 28: 221-234(1990).

63. Borzelleca, J. F. and Hallagan, J. B. Lifetime toxicity/carcinogenicity study of FD & C Red No. 3 (erythrosine) in mice. *Food Chem. Toxicol.* 25: 735-737(1987).

64. Borzelleca, J. F. and Hallagan, J. B. A chronic toxicity/carcinogenicity study of FD & C Yellow No. 5 (tartrazine) in mice. *Food Chem. Toxicol.* 26: 189-194(1988).

65. Borzelleca, J. F. and Hogan, G. K. Chronic toxicity/carcinogenicity study of FD & C Blue No. 2 in mice. *Food Chem. Toxicol.* 23: 719-722(1985).

66. Borzsonyi, M., Torok, G., Pinter, A., Surjan, A., Nadasdi, L., and Roller, P. Carcinogenic effect of dinitrosopiperazine in adult Swiss mice and after transplacental or translactational exposure. *Cancer Res.* 40: 2925-2927(1980).

67. Bosan, W. S., Shank, R. C., MacEwen, J. D., Gaworski, C. L., and Newberne, P. M. Methylation of DNA guanine during the course of induction of liver cancer in hamsters by hydrazine or dimethylnitrosamine. *Carcinogenesis* 8: 439-444(1987).

68. Boyland, E., Carter, R. L., Gorrod, J. W., and Roe, F. J. C. Carcinogenic properties of certain rubber additives. *Eur. J. Cancer* 4: 233-239(1968).

69. Boyland, E., Roe, F. J. C., Gorrod, J. W., and Mitchley, B. C. V. The carcinogenicity of nitrosoanabasine a possible constituent of tobacco smoke. *Br. J. Cancer* 18: 265-270(1964).

70. Bralow, S. P., Gruenstein, M., and Meranze, D. R. Host resistance to gastric adenocarcinomatosis in three strains of rats ingesting *N*-methyl-*N*′-nitro-*N*-nitrosoguanidine. *Oncology* 27: 168-180(1973).

71. Brantom, P. G., Gaunt, I. F., and Grasso, P. Long-term toxicity of sodium cyclamate in mice. *Food Cosmet. Toxicol.* 11: 735-746(1973).

72. Brantom, P. G., Gaunt, I. F., Hardy, J., Grasso, P., and Gangolli, S. D. Long-term feeding and reproduction studies on emulsifier YN in rats. *Food Cosmet. Toxicol.* 11: 755-769(1973).

73. Brown, E. V. and Hamdan, A. A. Carcinogenic activity of analogues of *p*-dimethylaminoazobenzene. V. Effect of added methyl groups in the pyridine series. *J. Natl. Cancer Inst.* 37: 365-367(1966).

74. Brune, H., Deutsch-Wenzel, R. P., Habs, M., Ivankovic, S., and Schmahl, D. Investigation of the tumorigenic response to benzo(*a*)pyrene in aqueous caffeine solution applied orally to Sprague-Dawley rats. *Cancer Res. Clin. Oncol.* 102: 153-157(1981).

75. Bulay, O., Mirvish, S. S., Garcia, H., Pelfrene, A. F., Gold, B., and Eagen, M. Carcinogenicity test of six nitrosamides and a nitrosocyanamide administered orally to rats. *J. Natl. Cancer Inst.* 62: 1523-1528(1979).

76. Burek, J. D., Nitschke, K. D., Bell, T. J., Wackerle, D. L., Childs, R. C., Beyer, J. E., Dittenber, D. A., Rampy, L. W., and McKenna, M. J. Methylene chloride: a two-year inhalation toxicity and oncogenicity study in rats and hamsters. *Fundam. Appl. Toxicol.* 4: 30-47(1984).

77. Burnett, C. M. and Corbett, J. F. Failure of short-term *in vitro* mutagenicity tests to predict the animal carcinogenicity of hair dyes. *Food Chem. Toxicol.* 25: 703-707(1987).

78. Burtin, C., Scheinmann, P., Salomon, J. C., Lespinats, G., Frayssinet, C., Lebel, B., and Canu, P. Increased tissue histamine in tumour-bearing mice and rats. *Br. J. Cancer* 43: 684-688(1981).

79. Butler, W. H. Long-term effects of phenobarbitone-Na on male Fischer rats. *Br. J. Cancer* 37: 418-423(1978).

80. Butler, W. H. and Barnes, J. M. Carcinogenic action of groundnut meal containing aflatoxin in rats. *Food Cosmet. Toxicol.* 6: 135-141(1968).

81. Butler, W. H., Graham, T. C., and Sutton, M. L. Chronic toxicity and oncogenicity studies of macrodantin in Sprague-Dawley rats. *Food Chem. Toxicol.* 28: 269-277(1990).

82. Butler, W. H., Graham, T. C., and Sutton, M. L. Oncogenicity study of macrodantin in Swiss mice. *Food Chem. Toxicol.* 28: 49-54(1990).

83. Byron, W. R., Bierbower, G. W., Brower, J. B., Hansen, W. H. Pathologic changes in rats and dogs from two-year feeding of sodium arsenite or sodium arsenate. *Toxicol. Appl. Pharmacol.* 10: 132-147(1967)

84. Cabral, J. and Ponomarkov, V. Carcinogenicity study of the pesticide maleic hydrazide in mice. *Toxicology* 24: 169-173(1982).

85. Cabral, J. R. P. and Galendo, D. Carcinogenicity study of the pesticide fenvalerate in mice. *Cancer Lett.* 49: 13-18(1990).

86. Cabral, J. R. P., Galendo, D., Laval, M., and Lyandrat, N. Carcinogenicity studies with deltamethrin in mice and rats. *Cancer Lett.* 49: 147-152(1990).

87. Cabral, J. R. P., Hall, R. K., Bronczyk, S. A., and Shubik, P. A carcinogenicity study of the pesticide dieldrin in hamsters. *Cancer Lett.* 6: 241-246(1979).

88. Cabral, J. R. P., Hall, R. K., Bronczyk, S. A., and Shubik, P. A. Lack of carcinogenicity of DDT on hamsters. *Tumori* 68: 5-10(1982).

89. Cabral, J. R. P., Hall, R. K., Rossi, L., Bronczyk, S. A., and Shubik, P. Effects of long-term DDT intake on rats. *Tumori* 68: 11-17(1982).

90. Cabral, J. R. P., Mollner, T., Raitano, F., and Shubik, P. Carcinogenesis of hexachlorobenzene in mice. *Int. J. Cancer* 23: 47-51(1979).

91. Cabral, J. R. P., Rossi, L., Dragani, T. A., Della Porta, G. Carcinogenicity study of 3-(5-nitro-2-furyl)-imidazo(1,2-α) pyridine in mice and rats. *Tumori* 66: 131-144(1980).

92. Cabral, J. R. P., Shubik, P., Mollner, T., and Raitano, F. Carcinogenic activity of hexachlorobenzene in hamsters. *Nature* 265: 510-511(1977).

93. Cabrol Telle, A.-M., de Saint Blanquat, G., Derache, R., Hollande, E., Periquet, B., and Thouvenot, J.-P. Nutritional and toxicological effects of long-term ingestion of phosphine-fumigated diet by the rat. *Food Chem. Toxicol.* 23: 1001-1009(1985).

94. Cameron, G. R. and Cheng, K. K. Failure of oral D.D.T. to induce toxic changes in rats. *Br. Med. J.* 2: 819-821(1951).

95. Cardesa, A., Garcia-Bragado, F., Ramirez, J., and Ernst, H. Histological types of laryngotracheal tumors induced in Syrian golden hamsters by nitrosomorpholine and nitrosopiperidine. *Exp. Pathol.* 401: 267-281(1990).

96. Cardin, C. W., Domeyer, B. E., and Bjorkquist, L. Toxicological evaluation of commercial alkyldimethylamine oxides: two-year chronic feeding and dermal studies. *Fundam. Appl. Toxicol.* 5: 869-878(1985).

97. Carr, B. I., Rahbar, S., Asmeron, Y., Riggs, A., and Winberg, C. D. Carcinogenicity and haemoglobin synthesis induction by cytidine analogues. *Br. J. Cancer* 57: 395-402(1988).

98. Carr, B. I., Reilly, J. G., Smith, S. S., Winberg, C., and Riggs, A. The tumorigenicity of 5-azacytidine in the male Fischer rat. *Carcinogenesis* 5: 1583-1590(1984).

99. Casas, C. B. Induction of hepatomas by thiouracil in inbred strains of mice. *Proc. Soc. Exp. Biol. Med.* 113: 493-494(1963).

100. Case, M. T., Sibinski, L. J., and Steffen, G. R. Chronic oral toxicity and oncogenicity studies of flecainide, an antiarrhythmic, in rats and mice. *Toxicol. Appl. Pharmacol.* 73: 232-242(1984).

101. Cavaliere, A., Alberti, P. F., and Vitali, R. 5-Fluorouracil carcinogenesis in BALB/c mice. *Tumori* 76: 179-181(1990).

102. Cavaliere, A., Bufalari, A., and Vitali, R. Carcinogenicity and cocarcinogenicity test of phenobarbital sodium in adult BALB/c mice. *Tumori* 72: 125-128(1986).

103. Cavaliere, A., Bufalari, A., and Vitali, R. 5-azacytidine carcinogenesis in BALB/c mice. *Cancer Lett.* 37: 51-58(1987).

104. Cesare, Maltoni, Adriano, Ciliberti, Giuliano, Cotti, and Giorgio, Perino Long-term carcinogenicity bioassays on acrylonitrile administered by inhalation and by ingestion to Sprague-Dawley rats. *Ann. N. Y. Acad. Sci.* 534: 179-202(1988).

105. Cheever, K. L., Cholakis, J. M., El-Hawari, A. M., Kovatch, R. M., and Weisburger, E. K. Ethylene dichloride: The influence of disulfiram or ethanol on oncogenicity, metabolism, and DNA covalent binding in rats. *Fundam. Appl. Toxicol.* 14: 243-261(1990).

106. Chouroulinkov, I., Gentil, A., and Guerin, M. Étude de l'activité carcinogène du 9,10-diméthyl-benzanthracène et du 3,4-benzopyrène administrés par voie digestive. *Bull. Cancer* 54: 67-78(1967).

107. Christov, K. and Raichev, R. Thyroid carcinogenesis in hamsters after treatment with 131-iodine and methylthiouracil. *Cancer Res. Clin. Oncol.* 77: 171-179(1972).

108. Chu, I., Villeneuve, D. C., Secours, V. E., Valli, V. E., Leeson, S., and Shen, S. Y. Long-term toxicity of octachlorostyrene in the rat. *Fundam. Appl. Toxicol.* 6: 69-77(1986).

109. Chu, I., Villeneuve, D. C., Valli, V. E., Secours, V. E., and Becking, G. C. Chronic toxicity of photomirex in the rat. *Toxicol. Appl. Pharmacol.* 59: 268-278(1981).

110. Chung, F., Tanaka, T., and Hecht, S. S. Induction of liver tumors in F344 rats by crotonaldehyde. *Cancer Res.* 46: 1285-1289(1986).

111. Ciliberti, A., Maltoni, C., and Perino, G. Long-term carcinogenicity bioassays on propylene administered by inhalation to Sprague-Dawley rats and Swiss mice. *Ann. N. Y. Acad. Sci.* 534: 235-245(1988).

112. Clapp, N. K., Craig, A. W., and Toya, R. E. Oncogenicity by methyl methanesulfonate in male RF mice. *Science* 161: 913-914(1968).

113. Clapp, N. K., Tyndall, R. L., Cumming, R. B., and Otten, J. A. Effects of butylated hydroxytoluene alone or with diethylnitrosamine in mice. *Food Cosmet. Toxicol.* 12: 367-371(1974).

114. Clapp, N. K., Tyndall, R. L., Satterfield, L. C., Klima, W. C., and Bowles, N. D. Selective sex-related modification of diethylnitrosamine-induced carcinogenesis in BALB/c mice by concomitant administration of butylated hydroxytoluene. *J. Natl. Cancer Inst.* 61: 177-180(1978).

115. Clayson, D. B., Lawson, T. A., and Pringle, J. A. S. The carcinogenic action of 2-aminodiphenylene oxide and 4-aminodiphenyl on the bladder and liver of the C57 × IF mouse. *Br. J. Cancer* 21: 755-762(1967).

116. Clayson, D. B., Lawson, T. A., Santana, S., and Bonser, G. M. Correlation between the chemical induction of hyperplasia and of malignancy in the bladder epithelium. *Br. J. Cancer* 19: 297-310(1965).

117. Cleveland, F. P. A summary of work on aldrin and dieldrin toxicity at the Kettering Laboratory. *Arch. Environ. Health* 13: 195-198(1966).

118. Clevenger, M. A., Turnbull, D., Inoue, H., Enomoto, M., Allen, J. A., Henderson, L. M., and Jones, E. Toxicological evaluation of neosugar: Genotoxicity, carcinogenicity, and chronic toxicity. *J. Am. Coll. Toxicol.* 7: 643-662(1988).

119. Cohen, S. M., Arai, M., Jacobs, J. B., and Friedell, G. H. Promoting effect of saccharin and DL-tryptophan in urinary bladder carcinogenesis. *Cancer Res.* 39: 1207-1217(1979).

120. Cohen, S. M., and Bryan, G. T. Effect of p-hydroxyacetanilide, sodium sulfate, and L-methionine on the leukemogenicity of N-[4-(5-nitro-2-furyl)-2-thiazolyl]acetamide. *Cancer Res.* 38: 1398-1405(1978).

121. Cohen, S. M., Ellwein, L. B., Okamura, T., Masui, T., Johansson, S. L., Smith, R. A., Wehner, J. M., Khachab, M., Chappel, C. I., Schoenig, G. P., Emerson, J. L., and Garland, E. M. Comparative bladder tumor promoting activity of sodium saccharin, sodium ascorbate, related acids, and calcium salts in rats. *Cancer Res.* 51: 1766-1777(1991).

122. Cohen, S. M., Ertürk, E., and Bryan, G. T. Comparative carcinogenicity of 5-nitrothiophenes and 5-nitrofurans in rats. *J. Natl. Cancer Inst.* 57: 277-282(1976).

123. Cohen, S. M., Ertürk, E., Price, J. M., and Bryan, G. T. Comparative carcinogenicity in the rat of 2-hydrazinothiazoles with nitrofuryl, nitrophenyl, or aminophenyl substituents in the 4-position. *Cancer Res.* 30: 897-901(1970).

124. Cohen, S. M., Ertürk, E., Von Esch, A. M., Crovetti, A. J., and Bryan, G. T. Carcinogenicity of 5-nitrofurans, 5-nitroimidazoles, 4-nitrobenzenes, and related compounds. *J. Natl. Cancer Inst.* 51: 403-417(1973).

125. Cohen, S. M., Ertürk, E., Von Esch, A. M., Crovetti, A. J., and Bryan, G. T. Carcinogenicity of 5-nitrofurans and related compounds with amino-heterocyclic substituents. *J. Natl. Cancer Inst.* 54: 841-850(1975).

126. Cohen, S. M., Ichikawa, M., and Bryan, G. T. Carcinogenicity of 2-(2-furyl)-3-(5-nitro-2-furyl)acrylamide (AF-2) fed to female Sprague-Dawley rats. *Gann* 68: 473-476(1977).

127. Cohen, S. M., Lower, G. M., Ertürk, E., and Bryan, G. T. Comparative carcinogenicity in Swiss mice of N-[4-(5-nitro-2-furyl)-2-thiazolyl]acetamide and structurally related 5-nitrofurans and 4-nitrobenzenes. *Cancer Res.* 33: 1593-1597(1973).

128. Cohen, S., Murasaki, G., Fukushima, S., and Greenfield, R. E. Effect of regenerative hyperplasia on the urinary bladder: carcinogenicity of sodium saccharin and N-[4-(5-nitro-2-furyl)-2-thiazolyl]formamide. *Cancer Res.* 42: 65-71(1982).

129. Conti, B., Maltoni, C., Perino, G., and Ciliberti, A. Long-term carcinogenicity bioassays on styrene administered by inhalation, ingestion and injection and styrene oxide administered by ingestion in Sprague-Dawley rats, and para-methylstyrene administered by ingestion in Sprague-Dawley rats and Swiss mice. *Ann. N. Y. Acad. Sci.* 534: 203-234(1988).

130. Conzelman, G. M., Moulton, J. E., Flanders, L. E., Springer, K., Crout, D. W. Induction of transitional cell carcinoma of the urinary bladder in monkeys fed 2-naphthylamine. *J. Natl. Cancer Inst.* 42: 825-831(1969).

131. Cremlyn, R. J. W. and Roe, F. J. C. A study of certain substituted sulphonohydrazides for carcinogenicity in mice. *Food Cosmet. Toxicol.* 9: 319-321(1971).

132. Croft, W. A. and Bryan, G. T. Production of urinary bladder carcinomas in male hamsters by N-[4-(5-nitro-2-furyl)-2-thiazolyl]formamide, N-[4-(5-nitro-2-furyl)-2-thiazolyl]acetamide, or formic acid 2-[4-(5-nitro-2-furyl)-2-thiazolyl]hydrazide. *J. Natl. Cancer Inst.* 51: 941-949(1973).

133. Dacre, J. C. Toxicologic studies with 2,6-di-*tert*-butyl-4-hydroxymethylphenol in the rat. *Toxicol. Appl. Pharmacol.* 17: 669-678(1970).

134. Dalbey, W. E. Formaldehyde and tumors in hamster respiratory tract. *Toxicology* 24: 9-14(1982).

135. Dalgard, D. W. Induction, biological markers and therapy of tumors in primates. National Cancer Institute Annual Report of Contract No. N01-CP-05622, Vienna, VA, Hazelton Laboratory Study No. 421-156, Reporting period: October 1, 1990-September 30, 1991.

136. Daniel, F. B., DeAngelo, A. B., Stober, J. A., Olson, G. R., and Page, N. P. Hepatocarcinogenicity of chloral hydrate, 2-chloroacetaldehyde, and dichloroacetic acid in the male B6C3F$_1$ mouse. *Fundam. Appl. Toxicol.* 19: 159-168(1992).

137. Davis, K. J., Fitzhugh, O. G., and Nelson, A. A. Chronic rat and dog toxicity studies on tartrazine. *Toxicol. Appl. Pharmacol.* 6: 621-626(1964).

138. Davis, K. J., Nelson, A. A., Zwickey, R. E., Hansen, W. H., and Fitzhugh, O. G. Chronic toxicity of ponceau SX to rats, mice, and dogs. *Toxicol. Appl. Pharmacol.* 8: 306-317(1966).

139. Davis, K. J. and Fitzhugh, O. G. Pathologic changes noted in rats fed D&C red no. 9 for two years. *Toxicol. Appl. Pharmacol.* 4: 200-205(1962).

140. Davis, K. J. and Fitzhugh, O. G. Pathologic changes noted in rats fed D&C red no. 10 [monosodium salt of 2-(2-hydroxy-1-naphthylazo)-1-naphthalenesulfonic acid] for two years. *Toxicol. Appl. Pharmacol.* 5: 728-734(1963).

141. Davis, K. J. and Fitzhugh, O. G. Tumorigenic potential of aldrin and dieldrin for mice. *Toxicol. Appl. Pharmacol.* 4: 187-189(1962).

142. De Groot, A. P., Til, H. P., Feron, V. J., Dreef-Van der Meulen, H. C., and Willems, M. I. Two-year feeding and multigeneration studies in rats on five chemically modified starches. *Food Cosmet. Toxicol.* 12: 651-663(1974).

143. Deichmann, W. B., Keplinger, M., Sala, F., and Glass, E. Synergism among oral carcinogens: IV. The simultaneous feeding of four tumorigens to rats. *Toxicol. Appl. Pharmacol.* 11: 88-103(1967).

144. Deichmann, W. B., MacDonald, W. E., Anderson, W. A. D., and Bernal, E. Adenocarcinoma in the lungs of mice exposed to vapors of 3-nitro-3-hexene. *Toxicol. Appl. Pharmacol.* 5: 445-456(1963).

145. Deichmann, W. B., MacDonald, W. E., Blum, E., Bevilacqua, M., Radomski, J., Keplinger, M., and Balkus, M. Tumorigenicity of aldrin, dieldrin and endrin in the albino rat. *Ind. Med.* 39: 426-434(1970).

146. Deichmann, W. B., MacDonald, W. E., Lampe, K. F., Dressler, I., and Anderson, W. A. D. Nitro-olefins as potential carcinogens in air pollution. *Ind. Med.* 34: 800-807(1965).

147. Deichmann, W. B., MacDonald, W. E., and Lu, F. C. Effects of chronic aldrin feeding in two strains of female rats and a discussion on the risks of carcinogens in man. In: *Developments in Toxicology and Environmental Science*, Vol. 4 (W. Deichman, Ed.), Elsevier North Holland, Amsterdam, 1979, pp. 407-413.

148. de la Iglesia, F. A., Barsoum, N., Gough, A., Mitchell, L., Martin, R. A., Di Fonzo, C., and McGuire, E. J. Carcinogenesis bioassay of prazepam (Verstran) in rats and mice. *Toxicol. Appl. Pharmacol.* 57: 39-54(1981).

149. Della Porta, G., Cabral, J. R., and Rossi, L. Carcinogenicity study of rifampicin in mice and rats. *Toxicol. Appl. Pharmacol.* 43: 293-302(1978).

150. Della Porta, G., Colnaghi, M. I., and Parmiani, G. Non-carcinogenicity of hexamethylenetetramine in mice and rats. *Food Cosmet. Toxicol.* 6: 707-715(1968).

151. Della Porta, G. and Dragani, T. A. Lack of carcinogenicity in mice of 4,4′-diaminobenzanilide and 4,4′-diaminoazobenzene, two intermediates used in the manufacture of azo dyes. *Cancer Lett.* 14: 329-336(1981).

152. Della Porta, G. and Dragani, T. A. Non-carcinogenicity in mice of a sulfonic acid derivative of 2-naphthylamine. *Carcinogenesis* 3: 647-649(1982).

153. Della Porta, G. and Dragani, T. A. Carcinogenicity study in mice on pildralazine, a hydralazinelike antihypertensive compound. *Cancer Res. Clin. Oncol.* 106: 97-101(1983).

154. Della Porta, G., Dragani, T. A., Barale, R., and Zucconi, D. Carcinogenic activity in mice of diftalone, an anti-inflammatory agent. *Cancer Res. Clin. Oncol.* 108: 308-311(1984).

155. Della Porta, G., Shubik, P., and Scortecci, V. The action of N-2-fluorenylacetamide in the Syrian golden hamster. *J. Natl. Cancer Inst.* 22: 463-471(1959).

156. Dillberger, J. E., Cronin, N. S., and Carr, G. J. Prednisone is not a mouse carcinogen. *Toxicol. Pathol.* 20: 18-26(1992).

157. Diwan, B. A., Hagiwara, A., Ward, J. M., and Rice, J. M. Effects of sodium salts of phenobarbital and barbital on development of bladder tumors in male F344/NCr rats pretreated with either N-[4-(5-nitro-2-furyl)-2-thiazolyl]formamide or N-nitrosobutyl-4-hydroxybutylamine. *Toxicol. Appl. Pharmacol.* 98: 269-277(1989).

158. Diwan, B. A., Lubet, R. A., Ward, J. M., Hrabie, J. A., and Rice, J. M. Tumor-promoting and hepatocarcinogenic effects of (TCPOBOP) in DBA/2NCr and C57BL/6NCr mice and an apparent promoting effect on nasal cavity tumors but not on hepatocellular tumors in F344/NCr rats initiated with N-nitrosodiethylamine. *Carcinogenesis* 13: 1893-1901(1992).

159. Diwan, B. A., Nims, R. W., Ward, J. M., Hu, H., Lubet, R. A., and Rice, J. M. Tumor promoting activities of ethylphenylacetylurea and diethylacetylurea, the ring hydrolysis products of barbiturate tumor promoters phenobarbital and barbital, in rat liver and kidney initiated by N-nitrosodiethylamine. *Carcinogenesis* 10: 189-194(1989).

160. Diwan, B. A., Nims, R. W., Henneman, J. R., Ward, J. M., Lubet, R. A., and Rice, J. M. Effect of the oxazolidinedione anticonvulsants trimethadione and dimethadione and the barbiturate homolog 5,5-dimethylbarbituric acid on n-nitrosodiethylamine-initiated renal and hepatic carcinogenesis in the F344/NCr rat. *Arch. Toxicol.* 66: 413-422(1992).

161. Diwan, B. A., Palmer, A. E., Ohshima, M., and Rice, J. M. N-nitroso-N-methylurea initiation in multiple tissues for organ-specific tumor promotion in rats by phenobarbital. *J. Natl. Cancer Inst.* 75: 1099-1105(1985).

162. Diwan, B. A., Rice, J. M., Ohshima, M., Ward, J. M., and Dove, L. F. Comparative tumor-promoting activities of phenobarbital, amobarbital, barbital sodium, and barbituric acid on livers and other organs of male F344/NCr rats following initiation with N-nitrosodiethylamine. *J. Natl. Cancer Inst.* 74: 509-516(1985).

163. Dodd, D. C., Port, C. D., Deslex, P., Regnier, B., Sanders, P., and Indacochea-Redmond, N. Two-year evaluation of misoprostol for carcinogenicity in CD Sprague-Dawley rats. *Toxicol. Pathol.* 15: 125-133(1987).

164. Domellof, L., Eriksson, S., Mori, H., Weisburger, J. H., and Williams, G. M. Effect of bile acid gavage or vagotomy and pyloroplasty on gastrointestinal carcinogenesis. *Am. J. Surg.* 142: 551-554(1981).

165. Dominick, M. A., Robertson, D. G., Bleavins, M. R., Sigler, R. E., Bobrowski, W. F., and Gough, A. W. α_{2u}-Globulin nephropathy without nephrocarcinogenesis in male Wistar rats administered 1-(aminomethyl)cyclohexaneacetic acid. *Toxicol. Appl. Pharmacol.* 111: 375-387(1991).

166. Donaubauer, H. H., Kief, H., Kramer, M., Krieg, K., Mayer, D., and Schutz, E. Investigations on the carcinogenicity of dipyrone in rats. *Toxicol. Appl. Pharmacol.* 81: 443-451(1985).

167. Drake, J. J.-P., Butterworth, K. R., Gaunt, I. F., and Grasso, P. Long-term toxicity study of black PN in mice. *Food Cosmet. Toxicol.* 15: 503-508(1977).

168. Drake, J. J.-P., Butterworth, K. R., Gaunt, I. F., and Hardy, J. Long-term toxicity studies of chocolate brown HT in mice. *Toxicology* 10: 17-27(1978).

169. Drake, J. J.-P., Butterworth, K. R., Gaunt, I. F., Hooson, J., Evans, J. G., and Gangolli, S. D. Long-term toxicity study of quillaia extract in rats. *Food Chem. Toxicol.* 20: 15-23(1982).

170. Drew, R. T., Boorman, G. A., Haseman, J. K., McConnell, E. Busey, W. M., and Moore, J. A. The effect of age and exposure duration on cancer induction by a known carcinogen in rats, mice and hamsters. *Toxicol. Appl. Pharmacol.* 68: 120-130(1983).

171. Druckrey, H. Chloriertes Trinkwasser, Toxizitats-Prufungen an Ratten über sieben Generationen. *Food Cosmet. Toxicol.* 6: 147-154(1968).

172. Dunkelberg, H. Carcinogenicity of ethylene oxide and 1,2-propylene oxide upon intragastric administration to rats. *Br. J. Cancer* 46: 924-933(1982).

173. Dunn, T. B. Cancer of the uterine cervix in mice fed a liquid diet containing an antifertility drug. *J. Natl. Cancer Inst.* 14: 671-692(1969).

174. Dunnick, J. K., Forbes, P. D., Eustis, S. L., Hardisty, J. F., and Goodman, D. G. Tumors of the skin in the HRA/Skh mouse after treatment with 8-methoxypsoralen and UVA radiation. *Fundam. Appl. Toxicol.* 16: 92-102(1991).

175. Dunnington, D., Butterworth, K. R., Gaunt, I. F., Mason, P. L., Evans, J. G., and Gangolli, S. D. Long-term toxicity study of ethyl methylphenylglycidate (strawberry aldehyde) in the rat. *Food Chem. Toxicol.* 19: 691-699(1981).

176. Dunsford, H. A., Keysser, C. H., Dolan, P. M., Seed, J. L., and Bueding, E. Carcinogenicity of the antischistosomal nitrofuran trans-5-amino-3-[2-(5-nitro-2-furyl)vinyl]-1,2,4-oxadiazole. *J. Natl. Cancer Inst.* 73: 151-160(1984).

177. Ebert, A. G. The dietary administration of monosodium glutamate or glutamic acid to C-57 black mice for two years. *Toxicol. Lett.* 3: 65-70(1979).

178. Eisenbrand, G., Habs, M., Schmahl, D., and Preussman, R. Carcinogenicity of *N*-nitroso-3-hydroxypyrrolidine and dose-response study with *N*-nitrosopiperidine in rats. In: IARC Scientific Publication #31. (E. A. Walker, L. Griciute, M. Castegnaro, and M. Borzsonyi, Eds.), World Health Organization, International Agency for Research on Cancer, Lyon, France, 1980, pp. 657-663.

179. Eisenbrand, G., Preussmann, R., and Schmahl, D. Carcinogenicity of *N*-nitrosoephedrine in rats. *Cancer Lett.* 5: 103-106(1978).

180. El-Mofty, M. M., Khudoley, V. V., Sakr, S. A., and Fathala, N. G. Flour infected with tribolium castaneum, biscuits made of this flour, and 1,4-benzoquinone induce neoplastic lesions in Swiss albino mice. *Nutr. Cancer* 17: 97-104(1992).

181. Elashoff, R. M., Fears, T. R., and Schneiderman, M. A. Statistical analysis of a carcinogen mixture experiment. I. Liver carcinogens. *J. Natl. Cancer Inst.* 79: 509-526(1987).

182. Ellis, H. V., III, Hong, C. B., Lee, C. C., Dacre, J. C., and Glennon, J. P. Subacute and chronic toxicity studies of trinitroglycerin in dogs, rats, and mice. *Fundam. Appl. Toxicol.* 4: 248-260(1984).

183. Endo, H., Takahashi, K., Kinoshita, N., and Baba, T. Production of gastric and esophageal tumors in rats by methylnitrosocyanamide, a possible candidate of etiologic factors for human gastric cancer. *Proc. Japan Acad.* 50: 497-502(1974).

184. Enomoto, M., Hatanaka, J., Igarashi, S., Uwanuma, Y., Ito, H., Asaoka, S., Iyatomi, A., Kuyama, S., Harada, T., and Hamasaki, T. High incidence of angiosarcomas in brown-fat tissue and livers of mice fed sterigmatocystin. *Food Chem. Toxicol.* 20: 547-556(1982).

185. Enomoto, M., Naoe, S., Harada, M., Miyata, K., Saito, M., and Noguchi, Y. Carcinogenesis in extrahepatic bile duct and gallbladder — carcinogenic effect of *N*-hydroxy-2-acetamidofluorene in mice fed a "gallstone-inducing" diet. *Jpn. J. Exp. Med.* 44: 37-54(1974).

186. Environmental Protection Agency. Peer Review of Aliette. Office of Pesticide Programs, Health Effects Division, Washington, D.C., 1993.

187. Epstein, S. S. Carcinogenicity of heptachlor and chlordane. (review of: Witherup, S., Cleveland, F. P., Shaffer, F. G., Schlecht, H., and Muser, L., The physiological effects of introduction of heptachlor into de [sic] diet of experimental animals in varying levels of concentration [ref. 1, p. 341].) *Sci. Total Environ.* 6: 103-154(1976).

188. Epstein, S. S. The carcinogenicity of dieldrin. Part 1. (review of: Davis, H. J., Hansen, W., and Fitzhugh, O. G., Pathology report on mice for aldrin, dieldrin, heptachlor or heptachlor epoxide for two years [ref. 2, p. 341].) *Sci. Total Environ.* 4: 1-52(1975).

189. Ernst, H., Ohshima, H., Bartsch, H., Mohr, U., and Reichart, P. Tumorigenicity study in Syrian hamsters fed areca nut together with nitrite. *Carcinogenesis* 8: 1843-1845(1987).

190. Ertürk, E., Cohen, S. M., and Bryan, G. T. Carcinogenicity of *N*-[4-(5-nitro-2-furyl)-2-thiazolyl]acetamide in female rats. *Cancer Res.* 30: 936-941(1970).

191. Ertürk, E., Cohen, S. M., and Bryan, G. T. Induction, histogenesis, and isotransplantability of renal tumors induced by formic acid 2-[4-(5-nitro-2-furyl)-2-thiazolyl]hydrazide in rats. *Cancer Res.* 30: 2098-2106(1970a).

192. Ertürk, E., Cohen, S. M., and Bryan, G. T. Urinary bladder carcinogenicity of *N*-[4-(5-nitro-2-furyl)-2-thiazolyl]formamide in female Swiss mice. *Cancer Res.* 30: 1309-1311(1970).

193. Ertürk, E., Cohen, S. M., Price, J. M., and Bryan, G. T. Pathogenesis, histology, and transplantability of urinary bladder carcinomas induced in albino rats by oral administration of *N*-[4-(5-nitro-2-furyl)-2-thiazolyl]formamide. *Cancer Res.* 29: 2219-2228(1969).

194. Ertürk, E., Morris, J. E., Cohen, S. M., Price, J. M., and Bryan, G. T. Transplantable rat mammary tumors induced by 5-nitro-2-furaldehyde semicarbazone and by formic acid 2-[4-(5-nitro-2-furyl)-2-thiazolyl]hydrazide. *Cancer Res.* 30: 1409-1412(1970).

195. Ertürk, E., Morris, J. E., Cohen, S. M., Von Esch, A. M., Crovetti, A. J., Price, J. M., and Bryan, G. T. Comparative carcinogenicity of formic acid 2-[4-(5-nitro-2-furyl)-2-thiazolyl]hydrazide and related chemicals in the rat. *J. Natl. Cancer Inst.* 47: 437-445(1971).

196. Ertürk, E., Price, J. M., Morris, J. E., Cohen, S., Leith, R. S., Von Esch, A. M., and Crovetti, A. J. The production of carcinoma of the urinary bladder in rats by feeding *N*-[4-(5-nitro-2-furyl)-2-thiazolyl]formamide. *Cancer Res.* 27: 1998-2002(1967).

197. Essigmann, E. M. and Newberne, P. M. Enzymatic alterations in mouse hepatic nodules induced by a chlorinated hydrocarbon pesticide. *Cancer Res.* 41: 2823-2831(1981).

198. Esumi, H., Ohgaki, H., Kohzen, E., Takayama, S., and Sugimura, T. Induction of lymphoma in CDF$_1$ mice by the food mutagen, 2-amino-1-methyl-6-phenylimidazo[4,5-*b*]pyridine. *Jpn. J. Cancer Res.* 80: 1176-1178(1989).

199. Evans, J. G., Collins, M. A., Savage, S. A., Lake, B. G., and Butler, W. H. The histology and development of hepatic nodules in C3H/He mice following chronic administration of phenobarbitone. *Carcinogenesis* 7: 627-631(1986).

200. Evarts, R. P. and Brown, C. A. 2,4-Diaminoanisole sulfate: early effect on thyroid gland morphology and late effect on glandular tissue of Fischer 344 rats. *J. Natl. Cancer Inst.* 65: 197-204(1980).

201. Faccini, J. M., Irisarri, E., and Monro, A. M. A carcinogenicity study in mice of a β-adrenergic antagonist, primidolol; increased total tumour incidence without tissue specificity. *Toxicology* 21: 279-290(1981).

202. Fears, T. R., Elashoff, R. M., and Schneiderman, M. A. The statistical analysis of a carcinogen mixture experiment. II. Carcinogens with different target organs, N-methyl-N′-nitro-N-nitrosoguanidine, N-butyl-N-(4-hydroxybutyl)nitrosamine, dipentyl-nitrosamine, and nitrilotriacetic acid. Toxicol. Ind. Health 4: 221-255(1988).

203. Fears, T. R., Elashoff, R. M., and Schneiderman, M. A. The statistical analysis of a carcinogen mixture experiment. III. Carcinogens with different target systems, aflatoxin B₁, N-butyl-N-(4-hydroxybutyl)nitrosamine, lead acetate, and thiouracil. Toxicol. Ind. Health 5: 1-23(1989).

204. Feron, V. J. and Kroes, R. One-year time-sequence inhalation toxicity study of vinyl chloride in rats. II. Morphological changes in the respiratory tract, ceruminous glands, brain, kidneys, heart and spleen. Toxicology 13: 131-141(1979).

205. Feron, V. J. and Kruysse, A. Effects of exposure to furfural vapour in hamsters simultaneously treated with benzo[a]pyrene or diethylnitrosamine. Toxicology 11: 127-144(1978).

206. Feron, V. J., Kruysse, A., and Woutersen, R. A. Respiratory tract tumours in hamsters exposed to acetaldehyde vapour alone or simultaneously to benzo(a)pyrene or diethylnitrosamine. Eur. J. Cancer 18: 13-31(1982).

207. Feron, V. J., Spit, B. J., Immel, H. R., and Kroes, R. One-year time-sequence inhalation toxicity study of vinyl chloride in rats. III. Morphological changes in the livers. Toxicology 13: 143-154(1979a).

208. Fiala, E. S., Czerniak, R., Castonguay, A., Conaway, C. C., and Rivenson, A. Assay of 1-nitropropane, 2-nitropropane, 1-azoxypropane and 2-azoxypropane for carcinogenicity by gavage in Sprague-Dawley rats. Carcinogenesis 8: 1947-1949(1987).

209. Fitzgerald, J. E. and de la Iglesia, F. A. Ten-year oral toxicity study with norlestrin in rhesus monkeys. J. Toxicol. Environ. Health 10: 879-896(1982).

210. Fitzgerald, J. E., Petrere, J. A., McGuire, E. J., and de la Iglesia, F. A. Preclinical toxicology studies with the lipid-regulating agent gemcadiol. Fundam. Appl. Toxicol. 6: 520-531(1986).

211. Fitzgerald, J. E., Sanyer, J. L., Schardein, J. L., Lake, R. S., McGuire, E. J., and de la Iglesia, F. A. Carcinogen bioassay and mutagenicity studies with the hypolipidemic agent gemfibrozil. J. Natl. Cancer Inst. 67: 1105-1115(1981).

212. Fitzgerald, J. E., de la Iglesia, F. A., and McGuire, E. J. Carcinogenicity studies in rodents with ripazepam, a minor tranquilizing agent. Fundam. Appl. Toxicol. 4: 178-190(1984).

213. Fitzhugh, O. G. and Nelson, A. A. Comparison of the chronic toxicity of triethylene glycol with that of diethylene glycol. J. Ind. Hyg. Toxicol. 28: 40-43(1946).

214. Fitzhugh, O. G., Nelson, A. A., and Quaife, M. L. Chronic oral toxicity of aldrin and dieldrin in rats and dogs. Food Cosmet. Toxicol. 2: 551-562(1964).

215. Flaks, A. and Clayson, D. B. The influence of ammonium chloride on the induction of bladder tumours by 4-ethylsulphonylnaphthalene-1-sulphonamide. Br. J. Cancer 31: 585-587(1975).

216. Flaks, A. and Flaks, B. 3-Methylcholanthrene-inhibition of hepatocarcinogenesis in the rat due to 3′-methyl-4-dimethylaminoazobenzene or 2-acetylaminofluorene: a comparative study. Carcinogenesis 3: 981-991(1982).

217. Flaks, A. and Flaks, B. Induction of liver cell tumours in IF mice by paracetamol. Carcinogenesis 4: 363-368(1983).

218. Flaks, A., Hamilton, J. H., and Clayson, D. B. Effect of ammonium chloride on incidence of bladder tumors induced by 4-ethylsulfonylnaphthalene-1-sulfonamide. J. Natl. Cancer Inst. 51: 2007-2008(1973).

219. Flaks, A., Hamilton, J. M., Clayson, D. B., and Burch, P. R. J. The combined effect of radiation and chemical carcinogens in female A × IF mice. Br. J. Cancer 28: 227-231(1973).

220. Flaks, B., Flaks, A., and Shaw, A. P. W. Induction by paracetamol of bladder and liver tumours in the rat. Acta Pathol. Microbiol. Scand. Sect. A. Suppl. 93: 367-377(1985).

221. Fleischman, R. W., Baker, J. R., Hagopian, M., Wade, G. G., Hayden, D. W., Smith, E. R., Weisburger, J. H., and Weisburger, E. K. Carcinogenesis bioassay of acetamide, hexanamide, adipamide, urea and p-tolylurea in mice and rats. J. Environ. Pathol. Toxicol. 3: 149-170(1980).

222. Fong, L. Y. Y. and Chan, W. C. Long-term effects of feeding aflatoxin-contaminated market peanut oil to Sprague-Dawley rats. Food Cosmet. Toxicol. 19: 179-183(1981).

223. Francis, P. C., Emmerson, J. L., Adams, E. R., and Owen, N. V. Oncogenicity study of trifluralin in B6C3F₁ mice. Food Chem. Toxicol. 29: 549-555(1991).

224. Frankel, H. H., Yamamoto, R. S., Weisburger, E. K., and Weisburger, J. H. Chronic toxicity of azathioprine and the effect of this immunosuppressant on liver tumor induction by the carcinogen N-hydroxy-N-2-fluorenylacetamide. Toxicol. Appl. Pharmacol. 17: 462-480(1970).

225. Fujii, K., Nakadate, M., Ogiu, T., and Odashima, S. Induction of digestive tract tumors and leukemias in Donryu rats by administration of 1-amyl-1-nitrosourea in drinking water. Gann 71: 464-470(1980).

226. Fujii, K., Nomoto, K., Ishidate, M., Jr., and Nakamura, K. Chronic toxicity of charred fish meat in Wistar rats. Nutr. Cancer 9: 185-193(1987).

227. Fujii, T., Mikuriya, H., Kamiya, N., and Hiraga, K. Enhancing effect of thiabendazole on urinary bladder carcinogenesis induced by sodium o-phenylphenate in F344 rats. Food Chem. Toxicol. 24: 207-211(1986).

228. Fujita, Y., Wakabayashi, K., Takayama, S., Nagao, M., and Sugimura, T. Induction of oral cavity cancer by 3-diazotyramine, a nitrosated product of tyramine present in foods. Carcinogenesis 8: 527-529(1987).

229. Fukuda, K., Takemoto, K., and Tsuruta, H. Inhalation carcinogenicity of trichloroethylene in mice and rats. Ind. Health 21: 243-254(1983).

230. Fukushima, S., Arai, M., Nakanowatari, J., Hibino, T., Okuda, M., and Ito, N. Differences in susceptibility to sodium saccharin among various strains of rats and other animal species. Gann 74: 8-20(1983).

231. Fukushima, S., Friedell, G. H., Jacobs, J. B., and Cohen, S. M. Effect of l-tryptophan and sodium saccharin on urinary tract carcinogenesis initiated by N-[4-(5-nitro-2-furyl)-2-thiazolyl]formamide. Cancer Res. 41: 3100-3103(1981).

232. Fukushima, S., Inoue, T., Uwagawa, S., Shibata, M.-A., and Ito, N. Co-carcinogenic effects of NaHCO₃ on o-phenylphenol-induced rat bladder carcinogenesis. Carcinogenesis 10: 1635-1640(1989).

233. Fukushima, S., Ishihara, Y., Nishio, O., Ogiso, T., Shirai, T., and Ito, N. Carcinogenicities of quinoline derivatives in F344 rats. Cancer Lett. 14: 115-123(1981).

234. Fukushima, S., Kurata, Y., Ogiso, T., Okuda, M., Miyata, Y., and Ito, N. Pathological analysis of the carcinogenicity of sodium o-phenylphenate and o-phenylphenol. Oncology 42: 304-311(1985).

235. Fukushima, S., Tanaka, H., Asakawa, E., Kagawa, M., Yamamoto, A., and Shirai, T. Carcinogenicity of uracil, a nongenotoxic chemical, in rats and mice and its rationale. Cancer Res. 52: 1675-1680(1992).

236. Fullerton, F. R., Greenman, D. L., and Bucci, T. J. Effects of diet type on incidence of spontaneous and 2-acetylaminofluorene-induced liver and bladder tumors in BALB/c mice fed AIN-76A diet versus NIH-07 diet. *Fundam. Appl. Toxicol.* 18: 193-199(1992).

237. Fullerton, F. R., Greenman, D. L., McCarty, C. C., and Bucci, T. J. Increased incidence of spontaneous and 2-acetylaminofluorene-induced liver and bladder tumors in B6C3F$_1$ mice fed AIN-76A diet versus NIH-07 diet. *Fundam. Appl. Toxicol.* 16: 51-60(1991).

238. Fullerton, F. R., Hoover, K., Mikol, Y. B., Creasia, D. A., and Poirier, L. A. The inhibition by methionine and choline of liver carcinoma formation in male C3H mice dosed with diethylnitrosamine and fed phenobarbital. *Carcinogenesis* 11: 1301-1305(1990).

239. Gad, S. C., Burton, E., Chengelis, C. P., Levin, S., Piper, C. E., Oshiro, Y., and Semler, D. E. Promotional activities of the non-genotoxic carcinogen bemitradine (SC-33643). *J. Appl. Toxicol.* 12: 157-164(1992).

240. Gaines, T. B. and Kimbrough, R. D. The sterilizing, carcinogenic and teratogenic effects of metepa in rats. *Bull. WHO* 34: 317-320(1966).

241. Gallagher, G. T., Maull, E. A., Kovacs, K., and Szabo, S. Neoplasms in rats ingesting acrylonitrile for two years. *J. Am. Coll. Toxicol.* 7: 603-615(1988).

242. Garcia, H., Keefer, L., Lijinsky, W., and Wenyon, C. E. M. Carcinogenicity of nitrosothiomorpholine and 1-nitrosopiperazine in rats. *Cancer Res. Clin. Oncol.* 74: 179-184(1970).

243. Garcia, H. and Lijinsky, W. Studies of the tumorigenic effect in feeding of nitrosamino acids and of low doses of amines and nitrite to rats. *Cancer Res. Clin. Oncol.* 79: 141-144(1973).

244. Garman, R. H., Snellings, W. M., and Maronpot, R. R. Brain tumors in F344 rats associated with chronic inhalation exposure to ethylene oxide. *Neurotoxicology* 6: 117-138(1985).

245. Gass, G. H. and Allaben, W. T. Preliminary report on the carcinogenic dose-response curve to oral vitamin D$_2$. *IRCS Med. Sci.: Libr. Compend.* 5: 477(1977).

246. Gass, G. H., Coats, D., and Graham, N. Carcinogenic dose-response curve to oral diethylstilbestrol. *J. Natl. Cancer Inst.* 33: 971-977(1964).

247. Gaunt, I. F., Brantom, P. G., Grasso, P., Creasey, M., and Gangolli, S. D. Long-term feeding study on chocolate brown FB in rats. *Food Cosmet. Toxicol.* 10: 3-15(1972).

248. Gaunt, I. F., Brantom, P. G., Grasso, P., and Kiss, I. S. Long-term toxicity studies of chocolate brown FB in mice. *Food Cosmet. Toxicol.* 11: 375-382(1973).

249. Gaunt, I. F., Butterworth, K. R., Grasso, P., and Ginocchio, A. V. Long-term toxicity study of emulsifier YN in the mouse. *Food Cosmet. Toxicol.* 15: 1-5(1977).

250. Gaunt, I. F., Butterworth, K. R., Hardy, J., and Gangolli, S. D. Long-term toxicity of sorbic acid in the rat. *Food Cosmet. Toxicol.* 13: 31-45(1975).

251. Gaunt, I. F., Carpanini, F. M. B., Grasso, P., Kiss, I. S., and Gangolli, S. D. Long-term feeding study on black PN in rats. *Food Cosmet. Toxicol.* 10: 17-27(1972).

252. Gaunt, I. F., Carpanini, F. M. B., Grasso, P., and Lansdown, A. B. G. Long-term toxicity of propylene glycol in rats. *Food Cosmet. Toxicol.* 10: 151-162(1972).

253. Gaunt, I. F., Hardy, J., Grasso, P., Gangolli, S. D., and Butterworth, K. R. Long-term toxicity of cyclohexylamine hydrochloride in the rat. *Food Cosmet. Toxicol.* 14: 255-267(1976).

254. Gaunt, I. F., Mason, P. L., Grasso, P., and Kiss, I. S. Long-term toxicity of sunset yellow FCF in mice. *Food Cosmet. Toxicol.* 12: 1-10(1974).

255. Gelderblom, W. C. A., Kriek, N. P. J., Marasas, W. F. O., and Thiel, P. G. Toxicity and carcinogenicity of the *Fusarium moniliforme* metabolite, fumonisin B$_1$, in rats. Carcinogenesis 12: 1247-1251(1991).

256. Gibson, J. E. and Hardisty, J. F. Chronic toxicity and oncogenicity bioassay of inhaled toluene in Fischer-344 rats. *Fundam. Appl. Toxicol.* 3: 315-319(1983).

257. Gibson, J. P., Newberne, J. W., Kuhn, W. L., and Elsea, J. R. Comparative chronic toxicity of three oral estrogens in rats. *Toxicol. Appl. Pharmacol.* 11: 489-510(1967).

258. Gibson, J. P., Rohovsky, M. W., Newberne, J. W., and Larson, E. J. Toxicity studies with metiapine. *Toxicol. Appl. Pharmacol.* 25: 220-229(1973).

259. Giner-Sorolla, A., Greenbaum, J., Last-Barney, K., Anderson, L. M., and Budinger, J. M. Lack of carcinogenic effect of nitroso-chlordiazepoxide and of nitrosomethylphenidate given orally to mice. *Food Cosmet. Toxicol.* 18: 81-83(1980).

260. Ginocchio, A. V., Waite, V., Hardy, J., Fisher, N., Hutchinson, J. B., and Berry, R. Long-term toxicity and carcinogenicity studies of the bread improver potassium bromate. 2. Studies in mice. *Food Cosmet. Toxicol.* 17: 41-47(1979).

261. Glaser, U., Hochrainer, D., Otto, F. J., and Oldiges, H. Carcinogenicity and toxicity of four cadmium compounds inhaled by rats. *Toxicol. Environ. Chem.* 27: 153-162(1990).

262. Gold, B. and Salmasi, S. Carcinogenicity tests of acetoxymethylphenylnitrosamine and benzenediazonium tetrafluoroborate in Syrian hamsters. *Cancer Lett.* 15: 289-300(1982).

263. Goldenthal, E. I. Two year oncogenicity study in rats [UDMH]. Mattawan, MI, International Research and Development Corporation, No. 399-062, 1989.

264. Goldenthal, E. I. Two year oncogenicity study in mice [UDMH]. Mattawan, MI, International Research and Development Corporation, No. 399-063, 1989.

265. Goldenthal, E. I. Two year oncogenicity study in mice [UDMH]. Mattawan, MI, International Research and Development Corporation, No. 399-065, 1990.

266. Goodall, C. M. and Kennedy, T. H. Carcinogenicity of dimethylnitramine in NZR rats and NZO mice. *Cancer Lett.* 1: 295-298(1976).

267. Goodall, C. M. and Lijinsky, W. Carcinogenesis by *N*-nitrosohexamethyleneimine in NZO inbred mice. *Toxicology* 33: 251-259(1984).

268. Goodall, C. M. and Lijinsky, W. Carcinogenicity of nitrosododecamethyleneimine in NZR/Gd inbred rats. *Carcinogenesis* 5: 537-540(1984).

269. Gopinath, C. and Gibson, W. A. Mesovarian leiomyomas in the rat. *Environ. Health Perspect.* 73: 107-113(1987).

270. Gothoskar, S. V., Talwalkar, G. V., and Bhide, S. V. Tumorigenic effect of thioacetamide in Swiss strain mice. *Br. J. Cancer* 24: 498-503(1970).

271. Goyer, R. A., Falk, H. L., Hogan, M., Feldman, D. D., and Richter, W. Renal tumors in rats given trisodium nitrilotriacetic acid in drinking water for 2 years. *J. Natl. Cancer Inst.* 66: 869-880(1981).

272. Graham, S. L., Davis, K. J., Hansen, W. H., and Graham, C. H. Effects of prolonged ethylene thiourea ingestion on the thyroid of the rat. *Food Cosmet. Toxicol.* 13: 493-499(1975).

273. Gralla, E. J. A twenty-four month inhalation toxicology study in Fischer-344 rats exposed to atmospheric toluene. Final Report, CIIT Docket #22000. Chemical Industry Institute of Toxicology, Research Triangle Park, NC, 1980.

274. Grandjean, C. J., Althoff, J., and Pour, P. M. Carcinogenicity of diallylnitrosamine following intragastric administration to Syrian hamsters. *J. Natl. Cancer Inst.* 74: 1043-1046(1985).

275. Grant, D. and Butler, W. H. Chronic toxicity of sodium nitrite in the male F344 rat. *Food Chem. Toxicol.* 27: 565-571(1989).

276. Grasso, P. and Gray, T. J. B. Long-term studies on chemically induced liver enlargement in the rat. III. Structure and behaviour of the hepatic nodular lesions induced by ponceau MX. *Toxicology* 7: 327-347(1977).

277. Grasso, P., Hardy, J., Gaunt, I. F., Mason, P. L., and Lloyd, A. G. Long-term toxicity of violet 6B (FD & C violet no. 1) in mice. *Food Cosmet. Toxicol.* 12: 21-31(1974).

278. Grasso, P., Lansdown, A. B. G., Kiss, I. S., Gaunt, I. F., and Gangolli, S. D. Nodular hyperplasia in the rat liver following prolonged feeding of ponceau MX. *Food Cosmet. Toxicol.* 7: 425-442(1969).

279. Gray, R., Peto, R., Branton, P., and Grasso, P. Chronic nitrosamine ingestion in 1040 rodents: The effect of the choice of nitrosamine, the species studied, and the age of starting exposure. *Cancer Res.* 51: 6470-6491(1991).

280. Green, U., Holste, J., and Spikermann, A. R. A comparative study of the chronic effects of magenta, paramagenta, and phenyl-β-naphthylamine in Syrian golden hamsters. *Cancer Res. Clin. Oncol.* 95: 51-55(1979).

281. Greenblatt, M., Kommineni, V. R. C., and Lijinsky, W. Null effect of concurrent feeding of sodium nitrite and amino acids to MRC rats. *J. Natl. Cancer Inst.* 50: 799-802(1973).

282. Greenblatt, M. and Lijinsky, W. Nitrosamine studies: neoplasms of liver and genital mesothelium in nitrosopyrrolidine-treated MRC rats. *J. Natl. Cancer Inst.* 48: 1687-1696(1972).

283. Greenman, D. L. Dose-Response Relationships Between Estrogenicity and Carcinogenicity: Rangefinding Study. National Center for Toxicological Research, Jefferson, AR. Technical Report Number 035 (1988).

284. Greenman, D. L., Highman, B., Chen, J., Sheldon, W., and Gass, G. Estrogen-induced thyroid follicular cell adenomas in C57BL/6 mice. *J. Toxicol. Environ. Health* 29: 269-278(1990).

285. Greenman, D. L., Highman, B., Chen, J. J., Schieferstein, G. J., and Norvell, M. J. Influence of age on induction of mammary tumors by diethylstilbestrol in C3H/HeN mice with low murine mammary tumor virus titer. *J. Natl. Cancer Inst.* 77: 891-898(1986).

286. Grice, H. C., Mannell, W. A., and Allmark, M. G. Liver tumors in rats fed ponceau 3R. *Toxicol. Appl. Pharmacol.* 3: 509-520(1961).

287. Griciute, L. Investigation on the combined action of N-nitrosodiethylamine with other carcinogens. In: IARC Scientific Publication #31. (E. A. Walker, L. Griciute, M. Castegnaro, and M. Borzsonyi, Eds.), World Health Organization, International Agency for Research on Cancer, Lyon, France, 1980, pp. 813-822.

288. Griciute, L., Castegnaro, M., and Bereziat, J.-C. Influence of ethyl alcohol on carcinogenesis with N-nitrosodimethylamine. *Cancer Lett.* 13: 345-351(1981).

289. Griffin, T. B., Stein, A. A., and Coulston, F. Histologic study of tissues and organs from rats exposed to vapors of 2-nitropropane at 25 ppm. *Ecotoxicol. Environ. Safety* 5: 194-201(1981).

290. Griffin, T. B., Stein, A. A., and Coulston, F. Inhalation exposure of rats to vapors of 1-nitropropane at 100 ppm. *Ecotoxicol. Environ. Safety* 6: 268-282(1982).

291. Gruenstein, M., Shay, H., and Shimkin, M. B. Lack of effect of norethynodrel (Enovid) on methylcholanthrene-induced mammary carcinogenesis in female rats. *Cancer Res.* 24: 1656-1658(1964).

292. Grundmann, E., and Steinhoff, D. Leber- und Lungentumoren nach 3,3'-Dichlor-4,4'-diaminodiphenylmethan bei Ratten. *Cancer Res. Clin. Oncol.* 74: 28-39(1970).

293. Gurkalo, V. K. and Vol'fson, N. I. Action of vinblastine on experimental gastric carcinogenesis. *Bull. Exp. Biol. Med.* 101: 833-837(1986).

294. Habs, M., Eisenbrand, G., Habs, H., and Schmahl, D. No evidence of carcinogencity of N-nitrosocimetidine in rats. *Hepato-Gastroenterology* 29: 265-266(1982).

295. Habs, M., Eisenbrand, G., and Schmahl, D. Carcinogenic activity in Sprague-Dawley rats of 2[3-(2-chloroethyl)-3-nitrosoureido]-D-glucopyranose (chlorozotocin). *Cancer Lett.* 8: 133-137(1979).

296. Habs, M., Habs, H., Berger, M. R., and Schmahl, D. Negative dose-response study for carcinogenicity of orally administered rutin sulfate in Sprague-Dawley rats. *Cancer Lett.* 23: 103-108(1984).

297. Habs, M., Habs, H., and Bertram, B. Carcinogenicity of di(N-nitroso)perhydropyrimidine (DNPP) after repeated low-dose intraperitoneal administration to rats. *Cancer Res. Clin. Oncol.* 105: 191-193(1983).

298. Habs, M. and Schmahl, D. Synergistic effects of N-nitroso compounds in experimental long-term carcinogenesis studies. *Oncology* 37: 259-265(1980).

299. Hackmann, C. Erzeugung von Blasencarcinomen und Tumoren verschiedener Lokalisation bei Ratten durch Verfutterung von 2-Amino-3-methoxy-diphenylenoxyd und 2-Amino-diphenylenoxyd. *Cancer Res. Clin. Oncol.* 61: 45-54(1956).

300. Hagan, E. C., Jenner, P. M., Jones, W. I., Fitzhugh, O. G., Long, E. L., Brouwer, J. G., and Webb, W. K. Toxic properties of compounds related to safrole. *Toxicol. Appl. Pharmacol.* 7: 18-24(1965).

301. Hagiwara, A., Arai, M., Hirose, M., Nakanowatari, J., Tsuda, H., and Ito, N. Chronic effects of norharman in rats treated with aniline. *Toxicol. Lett.* 6: 71-75(1980).

302. Hagiwara, A., Boonyaphiphat, P., Kawabe, M., Naito, H., Shirai, T., and Ito, N. Lack of carcinogenicity of tragacanth gum in B6C3F₁ mice. *Food Chem. Toxicol.* 30: 673-679(1992).

303. Hagiwara, A., Hirose, M., Takahashi, S., Ogawa, K., Shirai, T., and Ito, N. Forestomach and kidney carcinogenicity of caffeic acid in F344 rats and C57BL/6N × C3H/HeN F₁ mice. *Cancer Res.* 51: 5655-5660(1991).

304. Hagiwara, A., Shibata, M., Hirose, M., Fukushima, S., and Ito, N. Long-term toxicity and carcinogenicity study of sodium o-phenylphenate in B6C3F₁ mice. *Food Chem. Toxicol.* 22: 809-814(1984).

305. Hagiwara, A. and Ward, J. M. The chronic hepatotoxic, tumor-promoting, and carcinogenic effects of acetaminophen in male B6C3F₁ mice. *Fundam. Appl. Toxicol.* 7: 376-386(1986).

306. Haley, T. J., Farmer, J., Jaques, W. E., Frith, C., Sprowls, R. W., and Schieferstein, G. Dose-response hyperplasia and neoplasia from feeding N-2-fluorenylacetamide (2-FAA) to BALB/c mice for varying time intervals (39350). *Proc. Soc. Exp. Biol. Med.* 152: 156-159(1976).

307. Hansen, W. H., Davis, K. J., Fitzhugh, O. G., and Nelson, A. A. Chronic oral toxicity of ponceau 3R. *Toxicol. Appl. Pharmacol.* 5: 105-118(1963).

308. Hansen, W. H., Davis, K. J., Graham, S. L., Perry, C. H., and Jacobson, K. H. Long-term toxicity studies of erythrosine. II. Effects on haematology and thyroxine and protein-bound iodine in rats. *Food Cosmet. Toxicol.* 11: 535-545(1973).

309. Hansen, W. H., Fitzhugh, O. G., Nelson, A. A., and Davis, K. J. Chronic toxicity of two food colors, brilliant blue FCF and indigotine. *Toxicol. Appl. Pharmacol.* 8: 29-36(1966).

310. Hansen, W. H., Long, E. L., Davis, K. J., Nelson, A. A., and Fitzhugh, O. G. Chronic toxicity of three food colourings: guinea green B, light green SF yellowish and fast green FCF in rats, dogs and mice. *Food Cosmet. Toxicol.* 4: 389-410(1966).

311. Hansen, W. H., Zwickey, R. E., Brouwer, J. B., and Fitzhugh, O. G. Long-term toxicity studies of erythrosine. I. Effects in rats and dogs. *Food Cosmet. Toxicol.* 11: 527-534(1973).

312. Hardy, J., Gaunt, I. F., Hooson, J., Hendy, R. J., and Butterworth, K. R. Long-term toxicity of cyclohexylamine hydrochloride in mice. *Food Cosmet. Toxicol.* 14: 269-276(1976).

313. Harris, P. N., Gibson, W. R., and Dillard, R. D. Oncogenicity of 1-(4-chlorophenyl)-1-phenyl-2-propynyl carbamate for rats. *Toxicol. Appl. Pharmacol.* 21: 414-418(1972).

314. Hasegawa, R., Fukushima, S., Hagiwara, A., Masui, T., Masuda, A., and Ito, N. Long-term feeding study of N,N'-diphenyl-p-phenylenediamine in F344 rats. *Toxicology* 54: 69-78(1989).

315. Hasegawa, R., Greenfield, R. E., Murasaki, G., Suzuki, T., and Cohen, S. M. Initiation of urinary bladder carcinogenesis in rats by freeze ulceration with sodium saccharin promotion. *Cancer Res.* 45: 1469-1473(1985).

316. Hasegawa, R., Murasaki, G., St. John, M. K., Zenser, T. V., and Cohen, S. M. Evaluation of nitrofurantoin on the two stages of urinary bladder carcinogenesis in the rat. *Toxicology* 62: 333-347(1990).

317. Hasegawa, R., Sano, M., Tamano, S., Imaida, K., Shirai, T., Nagao, M., Sugimura, T., and Ito, N. Dose-dependence of 2-amino-1-methyl-6-phenylimidazo[4,5-b]-pyridine (PhIP) carcinogenicity in rats. *Carcinogenesis* 14: 2553-2557(1993).

318. Hasegawa, R., Shirai, T., Hakoi, K., Wada, S., Yamaguchi, K., and Takayama, S. Synergistic enhancement of thyroid tumor induction by 2,4-diaminoanisole sulfate, N,N'-diethylthiourea and 4,4'-thiodianiline in male F344 rats. *Carcinogenesis* 12: 1515-1518(1991).

319. Hasegawa, R., Takahashi, M., Furukawa, F., Toyoda, K., Sato, H., and Hayashi, Y. Co-carcinogenic effect of retinyl acetate on forestomach carcinogenesis of male F344 rats induced with butylated hydroxyanisole. *Jpn. J. Cancer Res.* 79: 320-328(1988).

320. Hasegawa, R., Takahashi, M., Furukawa, F., Toyoda, K., Sato, H., Jang, J. J., and Hayashi, Y. Carcinogenicity study of tetramethylthiuram disulfide (thiram) in F344 rats. *Toxicology* 51: 155-165(1988).

321. Hasegawa, R., Takahashi, M., Kokubo, T., Furukawa, F., Toyoda, K., Sato, H., Kurokawa, Y., and Hayashi, Y. Carcinogenicity study of sodium hypochlorite in F344 rats. *Food Chem. Toxicol.* 24: 1295-1302(1986).

322. Havu, N. Enterochromaffin-like cell carcinoids of gastric mucosa in rats after life-long inhibition of gastric secretion. *Digestion* 35: 42-55(1986).

323. Hecht, S. S., El-Bayoumy, K., Rivenson, A., and Fiala, E. Comparative carcinogenicity of o-toluidine hydrochloride and o-nitrosotoluene in F-344 rats. *Cancer Lett.* 16: 103-108(1982).

324. Hecht, S. S., Lijinsky, W., Kovatch, R. M., Chung, F. L., and Saavedra, J. E. Comparative tumorigenicity of N-nitroso-2-hydroxymorpholine, N-nitrosodiethanolamine and N-nitrosomorpholine in A/J mice and F344 rats. *Carcinogenesis* 10: 1475-1477(1989).

325. Hecht, S. S., Young, R., and Maeura, Y. Comparative carcinogenicity in F344 rats and Syrian golden hamsters of N'-nitrosonornicotine and N'-nitrosonornicotine-1-N-oxide. *Cancer Lett.* 20: 333-340(1983).

326. Heinrich, U., Peters, L., Ernst, H., Rittinghausen, S., Dasenbrock, C., and Konig, H. Investigation on the carcinogenic effects of various cadmium compounds after inhalation exposure in hamsters and mice. *Exp. Pathol.* 37: 253-258(1989).

327. Hendy, R. J., Hardy, J., Gaunt, I. F., Kiss, I. S., and Butterworth, K. R. Long-term toxicity studies of sorbic acid in mice. *Food Cosmet. Toxicol.* 14: 381-386(1976).

328. Henschler, D., Romen, W., Elsasser, H. M., Reichert, D., Eder, E., and Radwan, Z. Carcinogenicity study of trichloroethylene by long-term inhalation in three animal species. *Arch. Toxicol.* 43: 237-248(1980).

329. Herbst, M., Weisse, I., and Koellmer, H. A contribution to the question of the possible hepatocarcinogenic effects of lindane. *Toxicology* 4: 91-96(1975).

330. Herrold, K. M. Epidermoid carcinomas of esophagus and forestomach induced in Syrian hamsters by N-nitroso-N-methylurethan. *J. Natl. Cancer Inst.* 37: 389-394(1966).

331. Heston, W. E., Vlahakis, G., and Desmukes, B. Effects of the antifertility drug Enovid in five strains of mice, with particular regard to carcinogenesis. *J. Natl. Cancer Inst.* 51: 209-224(1973).

332. Heywood, R., Sortwell, R. J., Noel, P. R. B., Street, A. E., Prentice, D. E., Roe, F. J. C., Wadsworth, P. F., and Worden, A. N. Safety evaluation of toothpaste containing chloroform. III. Long-term study in beagle dogs. *J. Environ. Pathol. Toxicol.* 2: 835-851(1979).

333. Heywood, R., Wood, J. D., and Majeed, S. K. Tumorigenic and toxic effect of O,S-dibenzoyl thiamine hydrochloride in prolonged dietary administration to rats. *Toxicol. Lett.* 26: 53-58(1985).

334. Hiasa, Y., Konishi, N., Kitahori, Y., and Shimoyama, T. Carcinogenicity study of a commercial sodium oleate in Fischer rats. *Food Chem. Toxicol.* 23: 619-623(1985).

335. Hicks, R. M., Chowaniec, J., and St J. Wakefield, J. Experimental induction of bladder tumors by a two-stage system. In: *Carcinogenesis: Mechanisms of Tumor Promotion and Cocarcinogenesis,* Vol. 2, (T. J. Slaga, A. Sivak, and R. K. Boutwell, Eds.), Raven Press, New York, 1978, pp. 475-489.

336. Hicks, R. M., Wright, R., and St J. Wakefield, J. The induction of rat bladder cancer by 2-naphthylamine. *Br. J. Cancer* 46: 646-661(1982).

337. Highman, B., Greenman, D. L., Norvell, M. J., Farmer, J., and Shellenberger, T. E. Neoplastic and preneoplastic lesions induced in female C3H mice by diets containing diethylstilbestrol or 17β-estradiol. *J. Environ. Pathol. Toxicol.* 4: 81-95(1980).

338. Hinderer, R. K., Lankas, G. R., Knezevich, A. L., and Auletta, C. S. The effects of long-term dietary administration of the rubber accelerator, N-oxydiethylene thiocarbamyl-N-oxydiethylene sulfenamide, to rats. *Toxicol. Appl. Pharmacol.* 82: 521-531(1986).

339. Hinton, D. E., Lipsky, M. M., Heatfield, B. M., and Trump, B. F. Opposite effects of lead on chemical carcinogenesis in kidney and liver of rats. *Bull. Environ. Contam. Toxicol.* 23: 464-469(1979).

340. Hiraga, K. Tumors of the preputial gland in rats. *Gann* 68: 369-370(1977).

341. Hiraga, K. and Fujii, T. Induction of tumours of the urinary system in F344 rats by dietary administration of sodium o-phenylphenate. *Food Cosmet. Toxicol.* 19: 303-310(1981).

342. Hiraga, K. and Fujii, T. Induction of tumours of the urinary bladder in F344 rats by dietary administration of o-phenylphenol. *Food Chem. Toxicol.* 22: 865-870(1984).

343. Hiraga, K. and Fujii, T. Carcinogenicity testing of acetaminophen in F344 rats. *Jpn. J. Cancer Res.* 76: 79-85(1985).

344. Hirano, M., Ueda, H., Mitsumori, K., Maita, K., and Shirasu, Y. Hormonal influence on carcinogenicity of methylmercury in mice. *Jpn. J. Vet. Sci.* 50: 886-893(1988).

345. Hirono, I., Haga, M., Fujii, M., Matsuura, S., Matsubara, N., Nakayama, M., Furuya, T., Hikichi, M., Takanashi, H., Uchida, E., Hosaka, S., and Ueno, I. Induction of hepatic tumors in rats by senkirkine and symphytine. *J. Natl. Cancer Inst.* 63: 469-472(1979).

346. Hirono, I., Kuhara, K., Yamaji, T., Hosaka, S., and Goldberg, L. Induction of colorectal squamous cell carcinomas in rats by dextran sulfate sodium. *Carcinogenesis* 3: 353-355(1982).

347. Hirono, I., Kuhara, K., Yamaji, T., Hosaka, S., and Goldberg, L. Carcinogenicity of dextran sulfate sodium in relation to its molecular weight. *Cancer Lett.* 18: 29-34(1983).

348. Hirono, I., Mori, H., Yamada, K., Hirata, Y., Haga, M., Tatematsu, H., and Kanie, S. Carcinogenic activity of petasitenine, a new pyrrolizidine alkaloid isolated from *Petasites japonicus* Maxim. *J. Natl. Cancer Inst.* 58: 1155-1157(1977).

349. Hirono, I., Ueno, I., Hosaka, S., Takanashi, H., Matsushima, T., Sugimura, T., and Natori, S. Carcinogenicity examination of quercetin and rutin in ACI rats. *Cancer Lett.* 13: 15-21(1981).

350. Hirose, M., Fukushima, S., Kurata, Y., Tsuda, H., Tatematsu, M., and Ito, N. Modification of N-methyl-N'-nitro-N-nitrosoguanidine-induced forestomach and glandular stomach carcinogenesis by phenolic antioxidants in rats. *Cancer Res.* 48: 5310-5315(1988).

351. Hirose, M., Fukushima, S., Shirai, T., Hasegawa, R., Kato, T., Tanaka, H., Asakawa, E., and Ito, N. Stomach carcinogenicity of caffeic acid, sesamol and catechol in rats and mice. *Jpn. J. Cancer Res.* 81: 207-212(1990).

352. Hirose, M., Fukushima, S., Tanaka, H., Asakawa, E., Takahashi, S., and Ito, N. Carcinogenicity of catechol in F344 rats and B6C3F$_1$ mice. *Carcinogenesis* 14: 525-529(1993).

353. Hirose, M., Kagawa, M., Ogawa, K., Yamamoto, A., and Ito, N. Antagonistic effect of diethylmaleate on the promotion of forestomach carcinogenesis by butylated hydroxyanisole (BHA) in rats pretreated with N-methyl-N'-nitro-N-nitrosoguanidine. *Carcinogenesis* 10: 2223-2226(1989).

354. Hirose, M., Kawabe, M., Shibata, M., Takahashi, S., Okazaki, S., and Ito, N. Influence of caffeic acid and other o-dihydroxybenzene derivatives on N-methyl-N'-nitro-N-nitrosoguanidine-initiated rat forestomach carcinogenesis. *Carcinogenesis* 13: 1825-1828(1992).

355. Hirose, M., Kurata, Y., Tsuda, H., Fukushima, S., and Ito, N. Catechol strongly enhances rat stomach carcinogenesis: A possible new environmental stomach carcinogen. *Jpn. J. Cancer Res.* 78: 1144-1149(1987).

356. Hirose, M., Masuda, A., Hasegawa, R., Wada, S., and Ito, N. Regression of butylated hydroxyanisole (BHA)-induced hyperplasia but not dysplasia in the forestomach of hamsters. *Carcinogenesis* 11: 239-244(1990).

357. Hirose, M., Masuda, A., Tsuda, H., Uwagawa, S., and Ito, N. Enhancement of BHA-induced proliferative rat forestomach lesion development by simultaneous treatment with other antioxidants. *Carcinogenesis* 8: 1731-1735(1987).

358. Hirose, M., Shibata, M., Hagiwara, A., Imaida, K., and Ito, N. Chronic toxicity of butylated hydroxytoluene in Wistar rats. *Food Cosmet. Toxicol.* 19: 147-151(1981).

359. Hirose, M., Wada, S., Yamaguchi, S., Masuda, A., Okazaki, S., and Ito, N. Reversibility of catechol-induced rat glandular stomach lesions. *Cancer Res.* 52: 787-790(1992).

360. Hoch-Ligeti, C., Argus, M. F., and Arcos, J. C. Induction of carcinomas in the nasal cavity of rats by dioxane. *Br. J. Cancer* 24: 164-167(1969).

361. Hoch-Ligeti, C., Argus, M. F., and Arcos, J. C. Oncogenic activity of an m-dioxane derivative: 2,6-dimethyl-m-dioxan-4-ol acetate (dimethoxane). *J. Natl. Cancer Inst.* 53: 791-793(1974).

362. Homburger, F. Negative lifetime carcinogen studies in rats and mice fed 50,000 ppm saccharin. In: *Chemical Toxicology of Food.* (C. L. Galli, R. Paoletti, and G. Vettorazzi, Eds.), Elsevier/North-Holland Biomedical Press, 1978, pp. 359-373.

363. Hong, C. B., Winston, J. M., Thornburg, L. P., Lee, C. C., and Woods, J. S. Follow-up study on the carcinogenicity of vinyl chloride and vinylidene chloride in rats and mice: tumor incidence and mortality subsequent to exposure. *J. Toxicol. Environ. Health* 7: 909-924(1981).

364. Honma, Y., Kondo, Y., Kakizoe, T., Aso, Y., and Nagase, S. Lack of bladder carcinogenicity of dietary sodium saccharin in analbuminaemic rats, which are highly susceptible to N-nitroso-N-butyl-(4-hydroxybutyl)amine. *Food Chem. Toxicol.* 29: 373-376(1991).

365. Hoos, A., Habs, M., and Schmahl, D. Comparison of liver tumor frequencies after intermittent oral administration of different doses of N-nitrosopyrrolidine in Sprague-Dawley rats. *Cancer Lett.* 26: 77-82(1985).

366. Hooson, J., Gaunt, I. F., Kiss, I. S., Grasso, P., and Butterworth, K. R. Long-term toxicity of indigo carmine in mice. *Food Cosmet. Toxicol.* 13: 167-176(1975).

367. Hooson, J., Hicks, R. M., Grasso, P., and Chowaniec, J. Ortho-toluene sulphonamide and saccharin in the promotion of bladder cancer in the rat. *Br. J. Cancer* 42: 129-147(1980).

368. Hoover, K. L., Hyde, C. L., Wenk, M. L., and Poirier, L. A. Ethionine carcinogenesis in CD-1, BALB/c and C3H mice. *Carcinogenesis* 7: 1143-1148(1986).

369. Horie, A., Kohchi, S., and Kuratsune, M. Carcinogenesis in the esophagus. II. Experimental production of esophageal cancer by administration of ethanolic solution of carcinogens. *Gann* 56: 429-441(1965).

370. Horn, H. J., Black, R., Bruce, and Paynter, O. E. Toxicology of chlorobenzilate. In: *Agricultural and Food Chemistry: Past, Present, Future,* Vol. 3 (R. Teranishi, Ed.), Avi Publishing Company, Inc., Westport, CT, 1955, pp. 752-756.

371. Hosaka, S., and Hirono, I. Effect of leupeptin, a protease inhibitor, on the development of spontaneous tumors in strain A mice. *Gann* 71: 913-917(1980).

372. Hosaka, S., Matsushima, T., Hirono, I., and Sugimura, T. Carcinogenic activity of 3-amino-1-methyl-5H-pyrido[4,3-b]indole (Trp-P-2), a pyrolysis product of tryptophan. *Cancer Lett.* 13: 23-28(1981).

373. Hoshino, H. and Tanooka, H. Carcinogenicity of triethanolamine in mice and its mutagenicity after reaction with sodium nitrite in bacteria. *Cancer Res.* 38: 3918-3921(1978).

374. Howe, R. Carcinogenicity of "alderlin" (pronethalol) in mice. *Nature* 207: 594-595(1965).

375. Hueper, W. C. and Payne, W. W. Experimental studies in metal carcinogenesis. Chromium, nickel, iron, arsenic. *Arch. Environ. Health* 5: 445-462(1962).

376. Ikeda, Y., Horiuchi, S., Furuya, T., and Omori, Y. Chronic toxicity of ponceau MX in the rat. *Food Cosmet. Toxicol.* 4: 485-492(1966).

377. Ikeda, Y., Horiuchi, S., Imoto, A., Kodama, Y., Aida, Y., and Kobayashi, K. Induction of mammary gland and skin tumours in female rats by the feeding of benzyl violet 4B. *Toxicology* 2: 275-284(1974).

378. Ikeda, Y., Horiuchi, S., Kobayashi, K., Furuja, T., and Kohgo, K. Carcinogenicity of ponceau MX in the mouse. *Food Cosmet. Toxicol.* 6: 591-598(1968).

379. Imai, S., Morimoto, J., Sekiya, N., Shima, M., Kiyozuka, Y., Nakamori, K., and Tsubura, Y. Chronic toxicity test of KCl and NaCl in F344/Slc rats. *J. Nara Med. Assoc.* 37: 115-127(1986).

380. Imaida, K., Ishihara, Y., Nishio, O., Nakanishi, K., and Ito, N. Carcinogenicity and toxicity tests on *p*-phenylenediamine in F344 rats. *Toxicol. Lett.* 16: 259-269(1983).

381. Imaida, K. and Wang, C. Y. Effect of sodium phenobarbital and sodium saccharin in AIN-76A diet on carcinogenesis initiated with *N*-[4-(5-nitro-2-furyl)-2-thiazolyl]formamide and *N,N*-dibibutylnitrosamine in male F344 rats. *Cancer Res.* 46: 6160-6164(1986).

382. Inai, K., Akamizu, H., Eto, R., Nishida, T., Ohe, K., Kobuke, T., Nambu, S., Matsuki, K., and Tokuoka, S. Tumorigenicity study of sodium erythorbate administered orally to mice. *Hiroshima J. Med. Sci.* 38: 135-139(1989).

383. Inai, K., Aoki, Y., Akamizu, H., Eto, R., Nishida, T., and Tokuoka, S. Tumorigenicity study of butyl and isobutyl *p*-hydroxybenzoates administered orally to mice. *Food Chem. Toxicol.* 23: 575-578(1985).

384. Inai, K., Arihiro, K., Takeshima, Y., Yonehara, S., Tachiyama, Y., Khatun, N., and Nishisaka, T. Quantitative risk assessment of carcinogenicity of urethane (ethyl carbamate) on the basis of long-term oral administration to B6C3F$_1$ mice. *Jpn. J. Cancer Res.* 82: 380-385(1991).

385. Inai, K., Kobuke, T., Fujihara, M., Yonehara, S., Takemoto, T., Tsuya, T., Yamamoto, A., Tachiyama, Y., Izumi, K., and Tokuoka, S. Lack of tumorigenicity of aminopyrine orally administered to B6C3F$_1$ mice. *Jpn. J. Cancer Res.* 81: 122-128(1990).

386. Inai, K., Kobuke, T., Nambu, S., Takemoto, T., Kou, E., Nishina, H., Fujihara, M., Yonehara, S., Suehiro, S., Tsuya, T., Horiuchi, K., and Tokuoka, S. Hepatocellular tumorigenicity of butylated hydroxytoluene administered orally to B6C3F$_1$ mice. *Jpn. J. Cancer Res.* 79: 49-58(1988).

387. Innes, J. R. M. Evaluation of Carcinogenic, Teratogenic, and Mutagenic Activities of Selected Pesticides and Industrial Chemicals. Volume 1: Carcinogenic Study. Bionetics Research Laboratories, Inc. Distributed by National Technical Information Service, Springfield, VA, 1968.

388. Innes, J. R. M., Ulland, B. M., Valerio, M. G., Petrucelli, L., Fishbein, L., Hart, E. R., Pallota, A. J., Bates, R. R., Falk, H. L., Gart, J. J., Klein, M., Mitchell, I., and Peters, J. Bioassay of pesticides and industrial chemicals for tumorigenicity in mice: a preliminary note. *J. Natl. Cancer Inst.* 42: 1101-1114(1969).

389. Isaka, H., Yoshii, H., Otsuji, A., Koike, M., Nagai, Y., Koura, M., Sugiyasu, K., and Kanabayashi, T. Tumors of Sprague-Dawley rats induced by long-term feeding of phenacetin. *Gann* 70: 29-36(1979).

390. Ishii, H. Incidence of brain tumors in rats fed aspartame. *Toxicol. Lett.* 7: 433-437(1981).

391. Ishioka, T., Kuwabar, N., and Fukuda, Y. Induction of colorectal adenocarcinoma in rats by amylopectin sulfate. *Cancer Lett.* 26: 277-282(1985).

392. Ito, A., Mori, M., and Naito, M. Induction of uterine hemangioendothelioma and lymphoma in (C57BL/6N × C3H/2N)F$_1$ mice by oral administration of azathioprine. *Jpn. J. Cancer Res.* 80: 419-423(1989).

393. Ito, A., Naito, M., Naito, Y., and Watanabe, H. Tumorigenicity test of *N*-(5-nitro-2-furfurylidene)-1-aminohydantoin by dietary administration in BDF$_1$ mice. *Hiroshima J. Med. Sci.* 32: 99-102(1983).

394. Ito, A., Naito, M., and Watanabe, H. Implication of chemical carcinogenesis in the experimental animal: Tumorigenic effect of hydrogen peroxide in mice. *J. Res. Inst. Nucl. Med. Biol. Hiroshima Univ.* 22: 147-158(1981a).

395. Ito, A., Watanabe, H., Naito, M., and Naito, Y. Induction of duodenal tumors in mice by oral administration of hydrogen peroxide. *Gann* 72: 174-175(1981).

396. Ito, N., Fukushima, S., Hagiwara, A., Shibata, M., and Ogiso, T. Carcinogenicity of butylated hydroxyanisole in F344 rats. *J. Natl. Cancer Inst.* 70: 343-352(1983).

397. Ito, N., Fukushima, S., Tamano, S., Hirose, M., and Hagiwara, A. Dose-response in butylated hydroxyanisole induction of forestomach carcinogenesis in F344 rats. *J. Natl. Cancer Inst.* 77: 1261-1265(1986).

398. Ito, N., Hagiwara, A., Tamano, S., Kagawa, M., Shibata, M. A., Kurata, Y., and Fukushima, S. Lack of carcinogenicity of quercetin in F344/DuCrj rats. *Jpn. J. Cancer Res.* 80: 317-325(1989).

399. Ito, N., Hananouchi, M., Sugihara, S., Shirai, T., Tsuda, H., Fukushima, S., and Nagasaki, H. Reversibility and irreversibility of liver tumors in mice induced by the α-isomer of 1,2,3,4,5,6-hexachlorocyclohexane. *Cancer Res.* 36: 2227-2234(1976).

400. Ito, N., Hasegawa, R., Sano, M., Tamano, S., Esumi, H., Takayama, S., and Sugimura, T. A new colon and mammary carcinogen in cooked food, 2-amino-1-methyl-6-phenylimidazo[4,5-*b*]pyridine (PhIP). *Carcinogenesis* 12: 1503-1506(1991).

401. Ito, N., Hirose, M., Hagiwara, A., and Takahashi, S. Carcinogenicity and modification of carcinogenic response by antioxidants. In: *Antimutagenesis and Anticarcinogenesis Mechanisms.* (Y. Kuroda, D. M. Shankel, and M. D. Waters, Eds.), Plenum Press, New York, 1990, pp. 183-194.

402. Ito, N., Nagasaki, H., Aoe, H., Sugihara, S., Miyata, Y., Arai, M., and Shirai, T. Development of hepatocellular carcinomas in rats treated with benzene hexachloride. *J. Natl. Cancer Inst.* 54: 801-804(1975).

403. Ito, N., Ogiso, T., Fukushima, S., Shibata, M., and Hagiwara, A. Carcinogenicity of captafol in B6C3F$_1$ mice. *Gann* 75: 853-865(1984).

404. Ito, N., Shirai, T., Fukushima, S., and Hirose, M. Dose-response study of urinary bladder carcinogenesis in rats by *N*-butyl-*N*-(4-hydroxybutyl)nitrosamine. *Cancer Res. Clin. Oncol.* 108: 169-173(1984).

405. Ivankovic, S. and Preussmann, R. Absence of toxic and carcinogenic effects after administration of high doses of chromic oxide pigment in subacute and long-term feeding experiments in rats. *Food Cosmet. Toxicol.* 13: 347-351(1975).

406. Izumi, K., Sano, N., Otsuka, H., Kinouchi, T., and Ohnishi, Y. Tumor promoting potential in male F344 rats and mutagenicity in *Salmonella typhimurium* of dipyrone. *Carcinogenesis* 12: 1221-1225(1991).

407. Jack, D., Poynter, D., and Spurling, N. W. Beta-adrenoceptor stimulants and mesovarian leiomyomas in the rat. *Toxicology* 27: 315-320(1983).

408. Jang, J. J., Takahashi, M., Furukawa, F., Toyoda, K., Hasegawa, R., Sato, H., and Hayashi, Y. Long-term *in vivo* carcinogenicity study of phenytoin (5,5-diphenylhydantoin) in F344 rats. *Food Chem. Toxicol.* 25: 697-702(1987).

409. Jemec, B. Studies of the tumorigenic effect of two goitrogens. *Cancer* 40: 2188-2202(1977).

410. Johansson, S. L. Carcinogenicity of analgesics: long-term treatment of Sprague-Dawley rats with phenacetin, phenazone, caffeine and paracetamol (acetamidophen). *Int. J. Cancer* 27: 521-529(1981).

411. Johansson, S. and Angervall, L. Urothelial changes of the renal papillae in Sprague-Dawley rats induced by long term feeding of phenacetin. *Acta Pathol. Microbiol. Scand. Sect. A. Suppl.* 84: 375-383(1976).

412. Johnson, K. A., Gorzinski, S. J., Bodner, K. M., Campbell, R. A., Wolf, C. H., Friedman, M. A., and Mast, R. W. Chronic toxicity and oncogenicity study on acrylamide incorporated in the drinking water of Fischer 344 rats. *Toxicol. Appl. Pharmacol.* 85: 154-168(1986).

413. Jorgenson, T. A., Meierhenry, E. F., Rushbrook, C. J., Bull, R. J., and Robinson, M. Carcinogenicity of chloroform in drinking water to male Osborne-Mendel rats and female B6C3F$_1$ mice. *Fundam. Appl. Toxicol.* 5: 760-769(1985).

414. Jukes, T. H. and Shaffer, C. B. Antithyroid effects of aminotriazole. *Science* 132: 296-297(1960).

415. Kandarkar, Y., Munir, K. M., Bhide, S. V., and Sirsat, S. M. Ultrastructural study of hepatocellular carcinoma induced by hexachlorocyclohexane. *Indian J. Med. Res.* 78: 155-161(1983).

416. Kanisawa, M. and Schroeder, H. A. Life term studies on the effects of arsenic, germanium, tin and vanadium on spontaneous tumors in mice. *Cancer Res.* 27: 1192-1195(1967).

417. Kanisawa, M. and Schroeder, H. A. Life term studies on the effect of trace elements on spontaneous tumors in mice and rats. *Cancer Res.* 29: 892-895(1969).

418. Kasprzak, K. S., Hoover, K. L., and Poirier, L. A. Effects of dietary calcium acetate on lead subacetate carcinogenicity in kidneys of male Sprague-Dawley rats. *Carcinogenesis* 6: 279-282(1985).

419. Kato, T., Ohgaki, H., Hasegawa, H., Sato, S., Takayama, S., and Sugimura, T. Carcinogenicity in rats of a mutagenic compound, 2-amino-3,8-dimethylimidazo[4,5-*f*]quinoxaline. *Carcinogenesis* 9: 71-73(1988).

420. Kennedy, Jr., G. L. Chronic toxicity, reproductive, and teratogenic studies with oxamyl. *Fundam. Appl. Toxicol.* 7: 106-118(1986).

421. Kerns, W. D., Pavkov, K. L., Donofrio, D. J., Gralla, E. J., and Swenberg, J. A. Carcinogenicity of formaldehyde in rats and mice after long-term inhalation exposure. *Cancer Res.* 43: 4382-4392(1983).

422. Ketkar, M., Althoff, J., and Mohr, U. A chronic study of praziquantel in Syrian golden hamsters and Sprague-Dawley rats. *Toxicology* 24: 345-350(1982).

423. Ketkar, M. B., Fuhst, R., Preussmann, R., and Mohr, U. The carcinogenic effect of nitrosopiperidine administered in the drinking water of Syrian golden hamsters. *Cancer Lett.* 21: 219-224(1983).

424. Ketkar, M. B., Holste, J., Preussmann, R., and Althoff, J. Carcinogenic effect of nitrosomorpholine administered in the drinking water to Syrian golden hamsters. *Cancer Lett.* 17: 333-338(1983).

425. Ketkar, M. B. and Mohr, U. The chronic effects of magenta, paramagenta and phenyl-β-naphthylamine in rats after intragastric administration. *Cancer Lett.* 16: 203-206(1982).

426. Ketkar, M. B., Schneider, P., Preussmann, R., Plass, C., and Mohr, U. Carcinogenic effect of low doses of nitrosopyrrolidine administered in drinking water to Syrian golden hamsters. *Cancer Res. Clin. Oncol.* 104: 75-79(1982).

427. Keyes, D. G., Kociba, R. J., Schwetz, R. W., Wade, C. E., Dittenber, D. A., Quinn, T., Gorzinski, S. J., Hermann, E. A., Momany, J. J., and Schwetz, B. A. Results of a two-year toxicity and oncogenic study of rats ingesting diets containing dibromoneopentyl glycol (FR-1138). *J. Combust. Toxicol.* 7: 77-98(1980).

428. Khasawinah, A. M. and Grutsch, J. F. Chlordane: Thirty-month tumorigenicity and chronic toxicity test in rats. *Regul. Toxicol. Pharmacol.* 10: 95-109(1989).

429. Khasawinah, A. M. and Grutsch, J. F. Chlordane: 24-month tumorigenicity and chronic toxicity test in mice. *Regul. Toxicol. Pharmacol.* 10: 244-254(1989).

430. Kiaer, H. W., Glavind, J., and Arffmann, E. Carcinogenicity in mice of some fatty acid methyl esters. *Acta Pathol. Microbiol. Scand. Sect. A. Suppl.* 83: 550-558(1975).

431. Kimbrough, R. D. and Linder, R. E. Induction of adenofibrosis and hepatomas of the liver in BALB/cJ mice by polychlorinated biphenyls (Aroclor 1254). *J. Natl. Cancer Inst.* 53: 547-549(1974).

432. Kimbrough, R. D., Squire, R. A., Linder, R. E., Strandberg, J. D., Montali, R. J., and Burse, V. W. Induction of liver tumors in Sherman strain female rats by polychlorinated biphenyl Aroclor 1260. *J. Natl. Cancer Inst.* 55: 1453-1456(1975).

433. Kimura, E. T., Fort, F. L., Buratto, B., Tekeli, S., Kesterson, J. W., Heyman, I. A., and Cusick, P. K. Carcinogenic evaluation of estazolam via diet in CD strain Sprague-Dawley rats and B6C3F$_1$ mice for 2 years. *Fundam. Appl. Toxicol.* 4: 827-842(1984).

434. Kimura, N. T., Kanematsu, T., and Baba, T. Polychlorinated biphenyl(s) as a promoter in experimental hepatocarcinogenesis in rats. *Cancer Res. Clin. Oncol.* 87: 257-266(1976).

435. Kinebuchi, M., Kawachi, T., Matsukura, N., and Sugimura, T. Further studies on the carcinogenicity of a food additive, AF-2, in hamsters. *Food Cosmet. Toxicol.* 17: 339-341(1979).

436. King, D. W., Bock, F. G., and Moore, G. E. Dinitrophenol inhibition of pituitary adenoma formation in mice fed propylthiouracil. *Proc. Soc. Exp. Biol. Med.* 112: 365-366(1963).

437. Kirby, A. H. M. and Peacock, P. R. The induction of liver tumours by 4-aminoazobenzene and its *N:N*-dimethyl derivative in rats on a restricted diet. *J. Pathol.* 59: 1-18(1947).

438. Kobuke, T., Inai, K., Nambu, S., Ohe, K., Takemoto, T., Matsuki, K., Nishina, H., Huang, I.-B., and Tokuoka, S. Tumorigenicity study of disodium glycyrrhizinate administered orally to mice. *Food Chem. Toxicol.* 23: 979-983(1985).

439. Kociba, R. J., Keyes, D. G., Beyer, J. E., Carreon, R. M. Wade, C. E., Dittenbar, D. A., Kalnins, R. P., Frauson, L. E., Park, C. N., Barnard, S. D., Hummel, R. A., and Humiston, C. G. Results of a two-year chronic toxicity and oncogenicity study of 2,3,7,8-tetrachlorodibenzo-*p*-dioxin in rats. *Toxicol. Appl. Pharmacol.* 46: 279-303(1978).

440. Kociba, R. J., Keyes, D. G., Jersey, G. C., Ballard, J. J., Dittenber, D. A., Quast, J. F., Wade, C. E., Humiston, C. G., and Schwetz, B. A. Results of a two year chronic toxicity study with hexachlorobutadiene in rats. *Am. Ind. Hyg. Assoc. J.* 38: 589-602(1977).

441. Kociba, R. J., Keyes, D. G., Lisowe, R. W., Kalnins, R. P., Dittenber, D. D., Wade, C. E., Gorzinski, S. J., Mahle, N. H., and Schwetz, B. A. Results of a two-year chronic toxicity and oncogenic study of rats ingesting diets containing 2,4,5-trichlorophenoxyacetic acid (2,4,5-T). *Food Cosmet. Toxicol.* 17: 205-221(1979).

442. Kociba, R. J., McCollister, S. B., Park, C., Torkelson, T. R., and Gehring, P. J. 1,4-Dioxane. I. Results of a 2-year ingestion study in rats. *Toxicol. Appl. Pharmacol.* 30: 275-286(1974).

443. Koepke, S. R., Creasia, D. R., Knutsen, G. L., and Michejda, C. J. Carcinogenicity of hydroxyalkylnitrosamines in F344 rats: Contrasting behavior of β- and γ-hydroxylated nitrosamines. *Cancer Res.* 48: 1533-1536(1988).

444. Koller, L. D., Kerkvliet, N. I., and Exon, J. H. Neoplasia induced in male rats fed lead acetate, ethyl urea, and sodium nitrite. *Toxicol. Pathol.* 13: 50-57(1985).

445. Kommineni, C., Groth, D. H., Frockt, I. J., Voelker, R. W., and Stanovick, R. P. Determination of the tumorigenic potential of methylene-*bis*-orthochloroaniline. *J. Environ. Pathol. Toxicol.* 2: 149-171(1979).

446. Konishi, N., Diwan, B. A., and Ward, J. M. Amelioration of sodium barbital-induced nephropathy and regenerative tubular hyperplasia after a single injection of streptozotocin does not abolish the renal tumor promoting effect of barbital sodium in male F344/NCr rats. *Carcinogenesis* 11: 2149-2156(1990).

447. Konishi, Y., Denda, A., Uchida, K., Emi, Y., Ura, H., Yokose, Y., Shiraiwa, K., and Tsutsumi, M. Chronic toxicity carcinogenicity studies of triethanolamine in B6C3F₁ mice. *Fundam. Appl. Toxicol.* 18: 25-29(1992).

448. Konishi, Y., Kondo, H., Ikeda, T., Kawabata, A., Shoji, Y., and Denda, A. Effect of dose on the carcinogenic activity of orally administered N-bis(2-hydroxypropyl)nitrosamine in rats. *Gann* 69: 573-577(1978).

449. Kraupp-Grasl, B., Huber, W., Taper, H., and Schulte-Hermann, R. Increased susceptibility of aged rats to hepatocarcinogenesis by the peroxisome proliferator nafenopin and the possible involvement of altered liver foci occurring spontaneously. *Cancer Res.* 51: 666-671(1991).

450. Kroeger-Koepke, M. B., Reuber, M. D., Iype, P. T., Lijinsky, W., and Michejda, C. J. The effect of substituents in the aromatic ring on carcinogenicity of N-nitrosomethylaniline in F344 rats. *Carcinogenesis* 4: 157-160(1983).

451. Kroes, R., Peters, P. W. J., Berkvens, J. M., Verschuuren, H. G., De Vries, T., and Van Esch, G. J. Long term toxicity and reproduction study (including a teratogenicity study) with cyclamate, saccharin and cyclohexylamine. *Toxicology* 8: 285-300(1977).

452. Kroes, R., Van Logten, M. J., Berkvens, J. M., De Vries, T., and Van Esch, G. J. Study on the carcinogenicity of lead arsenate and sodium arsenate and on the possible synergistic effect of diethylnitrosamine. *Food Cosmet. Toxicol.* 12: 671-679(1974).

453. Kruger, F. W. and Schmahl, D. Fehlen einer carcinogenen Wirkung von 7-Methylguanin bei Ratten. *Cancer Res. Clin. Oncol.* 75: 253-254(1971).

454. Kuhara, K., Takanashi, H., Hirono, I., Furuya, T., and Asada, Y. Carcinogenic activity of clivorine, a pyrrolizidine alkaloid isolated from Ligularia dentata. *Cancer Lett.* 10: 117-122(1980).

455. Kumagai, H., Kawaura, A., Shibata, M., and Otsuka, H. Carcinogenicity of dipyrone in (C57BL/6 × C3H)F₁ mice. *J. Natl. Cancer Inst.* 71: 1295-1297(1983).

456. Kunze, E., Woltjen, H. H., Hartmann, B., and Engelhardt, W. Animal experiments regarding a possible carcinogenic effect of phenacetin on the resting and proliferating urothelium stimulated by cyclophosphamide. *Cancer Res. Clin. Oncol.* 105: 38-47(1983).

457. Kuper, C. F., Reuzel, P. G. J., Feron, V. J., and Verschuuren, H. Chronic inhalation toxicity and carcinogenicity study of propylene oxide in Wistar rats. *Food Chem. Toxicol.* 29: 159-167(1988).

458. Kurata, Y., Diwan, B. A., and Ward, J. M. Lack of renal tumour-initiating activity of a single dose of potassium bromate, a genotoxic renal carcinogen in male F344/NCr rats. *Food Chem. Toxicol.* 30: 251-259(1992).

459. Kurata, Y., Fukushima, S., Hagiwara, A., Ito, H., Ogawa, K., and Ito, N. Carcinogenicity study of methyl hesperidin in B6C3F₁ mice. *Food Chem. Toxicol.* 28: 613-618(1990).

460. Kurata, Y., Tamano, S., Shibata, M.-A., Hagiwara, A., Fukushima, S., and Ito, N. Lack of carcinogenicity of magnesium chloride in a long-term feeding study in B6C3F₁ mice. *Food Chem. Toxicol.* 27: 559-563(1989).

461. Kuroda, K., Terao, K., and Akao, M. Inhibitory effect of fumaric acid on hepatocarcinogenesis by thioacetamide in rats. *J. Natl. Cancer Inst.* 79: 1047-1051(1987).

462. Kurokawa, Y., Aoki, S., Matsushima, Y., Takamura, N., Imazawa, T., and Hayashi, Y. Dose-response studies on the carcinogenicity of potassium bromate in F344 rats after long-term oral administration. *J. Natl. Cancer Inst.* 77: 977-982(1986).

463. Kurokawa, Y., Hayashi, Y., Maekawa, A., Takahashi, M., and Kukubo, T. High incidences of pheochromocytomas after long-term administration of retinol acetate to F344/DuCrj rats. *J. Natl. Cancer Inst.* 74: 715-723(1985).

464. Kurokawa, Y., Hayashi, Y., Maekawa, A., Takahashi, M., Kokubo, T., and Odashima, S. Carcinogenicity of potassium bromate administered orally to F344 rats. *J. Natl. Cancer Inst.* 71: 965-972(1983).

465. Kurokawa, Y., Matsushima, Y., Takamura, N., Imazawa, T., and Hayashi, Y. Relationship between the duration of treatment and the incidence of renal cell tumors in male F344 rats administered potassium bromate. *Jpn. J. Cancer Res.* 78: 358-364(1987).

466. Kurokawa, Y., Takayama, S., Konishi, Y., Hiasa, Y., Asahina, S., Takahashi, M., Maekawa, A., and Hayashi, Y. Long-term *in vivo* carcinogenicity tests of potassium bromate, sodium hypochlorite, and sodium chlorite conducted in Japan. *Environ. Health Perspect.* 69: 221-235(1986).

467. LaVoie, E. J., Shigematsu, A., Mu, B., Rivenson, A., and Hoffmann, D. The effects of catechol on the urinary bladder of rats treated with N-butyl-N-(4-hydroxybutyl)nitrosamine. *Jpn. J. Cancer Res.* 76: 266-271(1985).

468. Lacassagne, A. and Duplan, J. F. Cancerologie. Le mécanisme de la cancérisation de la mamelle chez la souris, considéré d'après les résultats d'experiénces au moyen de la réserpine. *Acad. Sci.* Mémoires et Communications des Membres et des Correspondants de l'Académie. 810-812(1959).

469. Lalwani, N. D., Reddy, M. K., Qureshi, S. A., and Reddy, J. K. Development of hepatocellular carcinomas and increased peroxisomal fatty acid β-oxidation in rats fed [4-chloro-6-(2,3-xylidino)-2-pyrimidinylthio]acetic acid (WY-14,643) in the semipurified diet. *Carcinogenesis* 2: 645-650(1981).

470. Larson, P. S., Crawford, E. M., Blackwell Smith, R., Hennigar, G. R., Haag, H. B., and Finnegan, J. K. Chronic toxicologic studies on isopropyl N-(3-chlorophenyl) carbamate (CIPC). *Toxicol. Appl. Pharmacol.* 2: 659-673(1960).

471. Laskin, S., Drew, R. T., Cappiello, V., Kuschner, M., and Nelson, N. Inhalation carcinogenicity of α-halo ethers. II. Chronic inhalation studies with chloromethyl methyl ether. *Arch. Environ. Health* 30: 70-72(1975).

472. Laskin, S., Sellakumar, A. R., Kuschner, M., Nelson, N., La Mendola, S., Rusch, G. M., Katz, G. V., Dulak, N. C., and Albert, R. E. Inhalation carcinogenicity of epichlorohydrin in noninbred Sprague-Dawley rats. *J. Natl. Cancer Inst.* 65: 751-757(1980).

473. Lawson, T. A., Mirvish, S. S., Pour, P., and Williams, G. Persistence of DNA single-strand breaks and other tests as indicators of the liver carcinogenicity of 1-nitroso-5,6-dihydrouracil and the noncarcinogenicity of 1-nitroso-5,6-dihydrothymine. *J. Natl. Cancer Inst.* 73: 515-519(1984).

474. Lee, C. C., Bhandari, J. C., Winston, J. M., House, W. B., Dixon, R. L., and Woods, J. S. Carcinogenicity of vinyl chloride and vinylidene chloride. *J. Toxicol. Environ. Health* 4: 15-30(1978).

475. Lee, D. J., Wales, J. H., and Sinnhuber, R. O. Hepatoma and renal tubule adenoma in rats fed aflatoxin and cyclopropenoid fatty acids. *J. Natl. Cancer Inst.* 43: 1037-1041(1969).

476. Lee, K. P., Chromey, N. C., Culik, R., Barnes, J. R., and Schneider, P. W. Toxicity of *N*-methyl-2-pyrrolidone (NMP): Teratogenic, subchronic, and two-year inhalation studies. *Fundam. Appl. Toxicol.* 9: 222-235(1987).

477. Lee, K. P., Schneider, P. W., and Trochimowicz, H. J. Morphologic expression of glandular differentiation in the epidermoid nasal carcinomas induced by phenylglycidyl ether inhalation. *Am. J. Pathol.* 111: 140-148(1983).

478. Leonard, T. B., Graichen, M. E., and Popp, J. A. Dinitrotoluene isomer-specific hepatocarcinogenesis in F344 rats. *J. Natl. Cancer Inst.* 79: 1313-1319(1987).

479. Leong, B. K. J., Kociba, R. J., and Jersey, G. C. A lifetime study of rats and mice exposed to vapors of bis(chloromethyl) ether. *Toxicol. Appl. Pharmacol.* 58: 269-281(1981).

480. Leopold, W. R., Miller, J. A., and Miller, E. C. Comparison of some carcinogenic, mutagenic, and biochemical properties of S-vinylhomocysteine and ethionine. *Cancer Res.* 42: 4364-4374(1982).

481. Leuschner, F. Carcinogenicity studies on different diarylide yellow pigments in mice and rats. *Toxicol. Lett.* 2: 253-260(1978).

482. Levinskas, G. J., Ribelin, W. E., and Shaffer, C. B. Acute and chronic toxicity of pimaricin. *Toxicol. Appl. Pharmacol.* 8: 97-109(1966).

483. Levy, L. S. and Clack, J. Further studies on the effect of cadmium on the prostate gland. I. Absence of prostatic changes in rats given oral cadmium sulphate for two years. *Ann. Occup. Hyg.* 17: 205-211(1975).

484. Levy, L. S., Clack, J., and Roe, F. J. C. Further studies on the effect of cadmium on the prostate gland. II. Absence of prostatic changes in mice given oral cadmium sulphate for eighteen months. *Ann. Occup. Hyg.* 17: 213-220(1975).

485. Lijinsky, W. Chronic toxicity tests of pyrilamine maleate and methapyrilene hydrochloride in F344 rats. *Food Chem. Toxicol.* 22: 27-30(1984).

486. Lijinsky, W. Induction of tumours in rats by feeding nitrosatable amines together with sodium nitrite. *Food Chem. Toxicol.* 22: 715-720(1984).

487. Lijinsky, W. Induction of tumors of the nasal cavity in rats by concurrent feeding of thiram and sodium nitrite. *J. Toxicol. Environ. Health* 13: 609-614(1984).

488. Lijinsky, W. Chronic bioassay of benzyl chloride in F344 rats and (C57BL/6J × BALB/c)F$_1$ mice. *J. Natl. Cancer Inst.* 76: 1231-1236(1986).

489. Lijinsky, W. Rat and mouse forestomach tumors induced by chronic oral administration of styrene oxide. *J. Natl. Cancer Inst.* 77: 471-476(1986).

490. Lijinsky, W., Greenblatt, M., and Kommineni, C. Feeding studies of nitrilotriacetic acid and derivatives in rats. *J. Natl. Cancer Inst.* 50: 1061-1063(1973).

491. Lijinsky, W., Knutsen, G. L., and Kovatch, R. M. Carcinogenic effect of nitrosoalkylureas and nitrosoalkylcarbamates in Syrian hamsters. *Cancer Res.* 45: 542-545(1985).

492. Lijinsky, W., Knutsen, G. L., and Kovatch, R. M. Comparative carcinogenesis by hydroxylated nitrosopropylamines in Syrian hamsters. *J. Natl. Cancer Inst.* 74: 923-926(1985).

493. Lijinsky, W., Knutsen, G. L., and Reuber, M. D. Carcinogenicity of methylated nitrosopiperazines in rats and hamsters. *Carcinogenesis* 4: 1165-1167(1983).

494. Lijinsky, W., Knutsen, G., and Reuber, M. D. Failure of methapyrilene to induce tumors in hamsters or guinea pigs. *J. Toxicol. Environ. Health* 12: 653-657(1983).

495. Lijinsky, W. and Kovatch, R. M. Induction of liver tumors in rats by nitrosodiethanolamine at low doses. *Carcinogenesis* 6: 1697-1681(1985).

496. Lijinsky, W. and Kovatch, R. M. Carcinogenicity studies of some analogs of the carcinogen methapyrilene in F344 rats. *Cancer Res. Clin. Oncol.* 112: 57-60(1986).

497. Lijinsky, W. and Kovatch, R. M. Chronic toxicity study of cyclohexanone in rats and mice. *J. Natl. Cancer Inst.* 77: 941-949(1986).

498. Lijinsky, W. and Kovatch, R. M. Carcinogenesis by nitrosohydroxyethylurea and nitrosomethoxyethylurea in F344 rats. *Jpn. J. Cancer Res.* 79: 181-186(1988).

499. Lijinsky, W. and Kovatch, R. M. Chronic toxicity tests of sodium thiocyanate with sodium nitrite in F344 rats. *Toxicol. Ind. Health* 5: 25-29(1989).

500. Lijinsky, W. and Kovatch, R. M. Similar carcinogenic actions of nitrosoalkylureas of varying structure given to rats by gavage. *Toxicol. Ind. Health* 5: 925-935(1989).

501. Lijinsky, W. and Kovatch, R. M. The uniform carcinogenic action of alkylnitrosoureas in Syrian hamsters. *Biomed. Environ. Sci.* 2: 167-173(1989).

502. Lijinsky, W. and Kovatch, R. M. A study of the carcinogenicity of glycidol in Syrian hamsters. *Toxicol. Ind. Health* 8: 267-271(1992).

503. Lijinsky, W., Kovatch, R. M., Keefer, L. K., Saavedra, J. E., Hansen, T. J., Miller, A. J., and Fiddler, W. Carcinogenesis in rats by cyclic *N*-nitrosamines containing sulphur. *Food Chem. Toxicol.* 26: 3-7(1988).

504. Lijinsky, W., Kovatch, R. M., and Knutsen, G. L. Carcinogenesis by nitrosomorpholines, nitrosooxazolidines and nitrosoazetidine given by gavage to Syrian golden hamsters. *Carcinogenesis* 5: 875-878(1984).

505. Lijinsky, W., Kovatch, R. M., and Knutsen, G. L. Carcinogenesis by oxygenated nitrosomethylpropylamines in Syrian hamsters. *Cancer Res. Clin. Oncol.* 109: 1-4(1985).

506. Lijinsky, W., Kovatch, R., and Riggs, C. W. Altered incidences of hepatic and hemopoietic neoplasms in F344 rats fed sodium nitrite. *Carcinogenesis* 4: 1189-1191(1983a).

507. Lijinsky, W., Kovatch, R. M., and Riggs, C. W. Carcinogenesis by nitrosodialkylamines and azoxyalkanes given by gavage to rats and hamsters. *Cancer Res.* 47: 3968-3972(1987).

508. Lijinsky, W., Kovatch, R. M., Riggs, C. W., and Walters, P. T. Dose-response study with *N*-nitrosomorpholine in drinking water of F-344 rats. *Cancer Res.* 48: 2089-2095(1988).

509. Lijinsky, W., Kovatch, R. M., and Saavedra, J. E. Carcinogenesis and mutagenesis by *N*-nitroso compounds having a basic center. *Cancer Lett.* 63: 101-107(1992).

510. Lijinsky, W., Kovatch, R. M., and Singer, S. S. Carcinogenesis in F-344 rats induced by nitrosohydroxyalkyl-chloroethylureas. *Cancer Res. Clin. Oncol.* 112: 221-228(1986).

511. Lijinsky, W. and Reuber, M. D. Carcinogenic effect of nitrosopyrrolidine, nitrosopiperidine and nitrosohexamethyleneimine in Fischer rats. *Cancer Lett.* 12: 99-103(1981).

512. Lijinsky, W. and Reuber, M. D. Carcinogenesis in Fischer rats by nitrosodipropylamine, nitrosodibutylamine and nitrosobis(2-oxopropyl)amine given by gavage. *Cancer Lett.* 19: 207-213(1983).

513. Lijinsky, W. and Reuber, M. D. Chronic toxicity studies of vinyl acetate in Fischer rats. *Toxicol. Appl. Pharmacol.* 68: 43-53(1983).

514. Lijinsky, W. and Reuber, M. D. Carcinogenesis in rats by nitrosodimethylamine and other nitrosomethylalkylamines at low doses. *Cancer Lett.* 22: 83-88(1984).

515. Lijinsky, W. and Reuber, M. D. Comparison of nitrosocimetidine with nitrosomethylnitroguanidine in chronic feeding tests in rats. *Cancer Res.* 44: 447-449(1984).

516. Lijinsky, W. and Reuber, M. D. Dose-response study with *N*-nitrosodiethanolamine in F344 rats. *Food Chem. Toxicol.* 22: 23-26(1984).

517. Lijinsky, W. and Reuber, M. D. Chronic carcinogenesis studies of acrolein and related compounds. *Toxicol. Ind. Health* 3: 337-345(1987).

518. Lijinsky, W. and Reuber, M. D. Pathologic effects of chronic administration of hydrochlorothiazide, with and without sodium nitrite, to F344 rats. *Toxicol. Ind. Health* 3: 413-422(1987).

519. Lijinsky, W., Reuber, M. D., Davies, T. C., and Riggs, C. W. Dose-response studies with nitroso-1,2,3,6-tetrahydropyridine and dinitrosohomopiperazine in F344 rats. *Ecotoxicol. Environ. Safety* 6: 513-527(1982).

520. Lijinsky, W., Reuber, M. D., Davies, T. S., and Riggs, C. W. Dose-response studies with nitrosoheptamethyleneimine and its α-deuterium-labeled derivative in F344 rats. *J. Natl. Cancer Inst.* 69: 1127-1133(1982).

521. Lijinsky, W., Reuber, M. D., Davies, T. S., Saavedra, J. E., and Riggs, C. W. Dose-response studies in carcinogenesis by nitroso-*N*-methyl-*N*-(2-phenyl)ethylamine in rats and the effects of deuterium substitution. *Food Cosmet. Toxicol.* 20: 393-399(1982).

522. Lijinsky, W., Reuber, M. D., and Riggs, C. W. Carcinogenesis by combinations of *N*-nitroso compounds in rats. *Food Chem. Toxicol.* 21: 601-605(1983).

523. Lijinsky, W., Reuber, M. D., and Riggs, C. W. Dose-response studies of carcinogenesis in rats by nitrosodiethylamine. *Cancer Res.* 41: 4997-5003(1981).

524. Lijinsky, W., Reuber, M. D., and Saavedra, J. E. The effect of deuterium substitution on carcinogenesis by azoxymethane. *Cancer Lett.* 24: 273-280(1984).

525. Lijinsky, W., Saavedra, J. E., and Kovatch, R. M. Carcinogenesis in rats by nitrosodialkylureas containing oxygenated alkyl groups. *In Vivo* 4: 1-6(1990).

526. Lijinsky, W., Saavedra, J. E., and Reuber, M. D. Carcinogenesis in F-344 rats by nitrosobis(2-oxopropyl)amine and related compounds administered in drinking water. *Cancer Res. Clin. Oncol.* 107: 178-182(1984).

527. Lijinsky, W., Saavedra, J. E., and Reuber, M. D. Carcinogenesis in rats by some hydroxylated acyclic nitrosamines. *Carcinogenesis* 5: 167-170(1984).

528. Lijinsky, W., Saavedra, J. E., and Reuber, M. D. Induction of carcinogenesis in Fischer rats by methylalkylnitrosamines. *Cancer Res.* 41: 1288-1292(1981).

529. Lijinsky, W., Saavedra, J. E., and Reuber, M. D. Organspecific carcinogenesis in rats by methyl- and ethylazoxyalkanes. *Cancer Res.* 45: 76-79(1985a).

530. Lijinsky, W., Singer, G. M., Saavedra, J. E., and Reuber, M. D. Carcinogenesis in rats by asymmetric nitrosamines containing an allyl group. *Cancer Lett.* 22: 281-288(1984).

531. Lijinsky, W. and Schmahl, D. Carcinogenicity of *N*-nitroso derivatives of *N*-methylcarbamate insecticides in rats. *Ecotoxicol. Environ. Safety* 2: 413-419(1978).

532. Lijinsky, W., Taylor, H. W., Mangino, M., and Singer, G. M. Carcinogenesis of nitrosomethylundecylamine in Fischer rats. *Cancer Lett.* 5: 209-213(1978).

533. Lindsay, S., Nichols, C. W., and Chaikoff, I. L. Induction of benign and malignant thyroid neoplasms in the rat. Induction of thyroid neoplasms by injection of 131-I with or without the feeding of diets containing propylthiouracil and/or dessicated thyroid. *Arch. Pathol.* 81: 308-316(1966).

534. Lipsky, M. M., Hinton, D. E., Klaunig, J. E., and Trump, B. F. Biology of hepatocellular neoplasia in the mouse. I. Histogenesis of safrole-induced hepatocellular carcinoma. *J. Natl. Cancer Inst.* 67: 365-371(1981).

535. Littlefield, N. A. Chronic toxicity and carcinogenesis study on sulfamethazine in B6C3F₁ mice. National Center for Toxicological Research, Jefferson, AR. Technical Report Number 418 (1989a).

536. Littlefield, N. A., Blackwell, B.-N., Hewitt, C. C., and Gaylor, D. W. Chronic toxicity and carcinogenicity studies of gentian violet in mice. *Fundam. Appl. Toxicol.* 5: 902-912(1985).

537. Littlefield, N. A., Farmer, J. H., Gaylor, D. W., and Sheldon, W. G. Effects of dose and time in a long-term, low-dose carcinogenic study. *J. Environ. Pathol. Toxicol.* 3: 17-34(1980).

538. Littlefield, N. A., Gaylor, D. W., Blackwell, B. N., and Allen, R. R. Chronic toxicity/carcinogenicity studies of sulphamethazine in B6C3F₁ mice. *Food Chem. Toxicol.* 27: 455-463(1989).

539. Littlefield, N. A., Nelson, C. J., and Frith, C. H. Benzidine dihydrochloride: toxicological assessment in mice during chronic exposures. *J. Toxicol. Environ. Health* 12: 671-685(1983).

540. Littlefield, N. A., Nelson, C. J., and Gaylor, D. W. Benzidine dihydrochloride: risk assessment. *Fundam. Appl. Toxicol.* 4: 69-80(1984).

541. Long, E. L., Nelson, A. A., Fitzhugh, O. G., and Hansen, W. H. Liver tumors produced in rats by feeding safrole. *Arch. Pathol.* 75: 595-604(1963).

542. Longnecker, D. S., Curphey, T. J., Lilja, H. S., French, J. I., and Daniel, D. S. Carcinogenicity in rats of the nitrosourea amino acid Nδ-(*N*-methyl-*N*-nitrosocarbamoyl)-ʟ-ornithine. *J. Environ. Pathol. Toxicol.* 4: 117-129(1980).

543. Longnecker, D. S., Roebuck, B. D., Curphey, T. J., and MacMillan, D. L. Evaluation of promotion of pancreatic carcinogenesis in rats by benzyl acetate. *Food Chem. Toxicol.* 28: 665-668(1990).

544. Longnecker, D. S., Roebuck, B. D., Yager, J. D., Lilja, H. S., and Siegmund, B. Pancreatic carcinoma in azaserine-treated rats: induction, classification and dietary modulation of incidence. *Cancer* 47: 1562-1572(1981).

545. Longstaff, E., Robinson, M., Bradbrook, C., Styles, J. A., and Purchase, I. F. H. Genotoxicity and carcinogenicity of fluorocarbons: assessment by short-term *in vitro* tests and chronic exposure in rats. *Toxicol. Appl. Pharmacol.* 72: 15-31(1984).

546. Loser, E. A 2 year oral carcinogenicity study with cadmium on rats. *Cancer Lett.* 9: 191-198(1980).

547. Louria, D. B., Finkel, G., Smith, J. K., and Buse, M. Aflatoxin-induced tumors in mice. *Sabouraudia* 12: 371-375(1974).

548. Lynch, D. W., Lewis, T. R., Moorman, W. J., Burg, J. R., Groth, D. H., Khan, A., Ackerman, L. J., and Cockrell, B. Y. Carcinogenic and toxicologic effects of inhaled ethylene oxide and propylene oxide in F344 rats. *Toxicol. Appl. Pharmacol.* 76: 69-84(1984).

549. Macklin, A. W. and Szot, R. J. Eighteen month oral study of aspirin, phenacetin and caffeine in C57BL/6 mice. *Drug Chem. Toxicol.* 3: 135-163(1980).

550. Maeda, T., Sano, N., Togei, K., Shibata, M., Izumi, K., and Otsuka, H. Lack of carcinogenicity of phenytoin in (C57BL/6 × C3H)F$_1$ mice. *J. Toxicol. Environ. Health* 24: 111-119(1988).

551. Maekawa, A., Kajiwara, T., Odashima, S., and Kurata, H. Hepatic changes in male ACI/N rats on low dietary levels of sterigmato-cystin. *Gann* 70: 777-781(1979).

552. Maekawa, A., Matsuoka, C., Onodera, H., Tanigawa, H., Furuta, K., Kanno, J., Jang, J. J., Hayashi, Y., and Ogiu, T. Lack of carcinogenicity of tartrazine (FD & C Yellow No. 5) in the F344 rat. *Food Chem. Toxicol.* 25: 891-896(1987).

553. Maekawa, A., Matsushima, Y., Onodera, H., Shibutani, M., Ogasawara, H., Kodama, Y., Kurokawa, Y., and Hayashi, Y. Long-term toxicity/carcinogenicity of musk xylol in B6C3F$_1$ mice. *Food Chem. Toxicol.* 28: 581-586(1990).

554. Maekawa, A., Matsushima, Y., Onodera, H., Shibutani, M., Yoshida, J., Kodama, Y., Kurokawa, Y., and Hayashi, Y. Long-term toxicity/carcinogenicity study of calcium lactate in F344 rats. *Food Chem. Toxicol.* 29: 589-594(1991).

555. Maekawa, A., Nagaoka, T., Onodera, H., Matsushima, Y., Todate, A., Shibutani, M., Ogasawara, H., Kodama, Y., and Hayashi, Y. Two-year carcinogenicity study of 6-mercaptopurine in F344 rats. *Cancer Res. Clin. Oncol.* 116: 245-250(1990).

556. Maekawa, A., Ogiu, T., Matsuoka, C., Onodera, H., Furuta, K., Kurokawa, Y., Takahashi, M., Kokubo, T., Tanigawa, H., Hayashi, Y., Nakadate, M., and Tanimura, A. Carcinogenicity of low doses of N-ethyl-N-nitrosourea in F344 rats; a dose-response study. *Gann* 75: 117-125(1984).

557. Maekawa, A., Ogiu, T., Matsuoka, C., Onodera, H., Furuta, K., Tanigawa, H., and Odashima, S. Induction of tumors in the small intestine and mammary gland of female Donryu rats by continuous oral administration of N-carboxymethyl-N-nitrosourea. *Cancer Res. Clin. Oncol.* 106: 12-16(1983).

558. Maekawa, A., Ogiu, T., Onodera, H., Furuta, K., Matsuoka, C., Ohno, Y., and Odashima, S. Carcinogenicity studies of sodium nitrite and sodium nitrate in F-344 rats. *Food Chem. Toxicol.* 20: 25-33(1982).

559. Maekawa, A., Onodera, H., Furuta, K., Tanigawa, H., Ogiu, T., and Hayashi, Y. Lack of evidence of carcinogenicity of technical-grade piperonyl butoxide in F344 rats: selective induction of ile-ocaecal ulcers. *Food Chem. Toxicol.* 23: 675-682(1985).

560. Maekawa, A., Onodera, H., Matsushima, Y., Nagaoka, T., Todate, A., Shibutani, M., Kodama, Y., and Hayashi, Y. Dose-response carcinogenicity in rats on low-dose levels of N-ethyl-N-nitrosoure-thane. *Jpn. J. Cancer Res.* 80: 632-636(1989).

561. Maekawa, A., Onodera, H., Tanigawa, H., Furuta, K., Kanno, J., Matsuoka, C., Ogiu, T., and Hayashi, Y. Lack of carcinogenicity of triethanolamine in F344 rats. *J. Toxicol. Environ. Health* 19: 345-357(1986).

562. Maekawa, A., Onodera, H., Tanigawa, H., Furuta, K., Kanno, J., Matsuoka, C., Ogiu, T., and Hayashi, Y. Long-term studies on carcinogenicity and promoting effect of phenylbutazone in Don-ryu rats. *J. Natl. Cancer Inst.* 79: 577-584(1987).

563. Maekawa, A., Todate, A., Onodera, H., Matsushima, Y., Nagaoka, T., Shibutani, M., Ogasawara, H., Kodama, Y., and Hayashi, Y. Lack of toxicity/carcinogenicity of monosodium succinate in F344 rats. *Food Chem. Toxicol.* 28: 269-277(1990).

564. Maita, K., Tsuda, S., and Shirasu, Y. Chronic toxicity studies with thiram in Wistar rats and beagle dogs. *Fundam. Appl. Toxicol.* 16: 667-686(1991).

565. Maltoni, C. Vinyl chloride carcinogenicity: an experimental model for carcinogenesis studies. In: *Origins of Human Cancer, Book A,* Vol. 4 (H. H. Hiatt, J. D. Watson, and J. A. Winsten, Eds.), Cold Spring Harbor Laboratory, Cold Spring Harbor, NY, 1977a, pp. 119-146.

566. Maltoni, C. Early results of the experimental assessments of the carcinogenic effects of one epoxy solvent: styrene oxide. *Adv. Mod. Environ. Toxicol.* 2: 97-110(1981).

567. Maltoni, C., Ciliberti, A., Cotti, G., and Perino, G. Long-term carcinogenicity bioassays on acrylonitrile administered by inhalation and by ingestion to Sprague-Dawley rats. *Ann. N. Y. Acad. Sci.* 534: 179-202(1988).

568. Maltoni, C., Conti, B., Cotti, G., and Belpoggi, F. Experimental studies on benzene carcinogenicity at the Bologna Institute of Oncology: current results and ongoing research. *Am. J. Ind. Med.* 7: 415-446(1985).

569. Maltoni, C., Conti, B., Perino, G., and Di Maio, V. Further evidence of benzene carcinogenicity; results on Wistar rats and mice treated by injection. *Ann. N. Y. Acad. Sci.* 534: 412-426(1988).

570. Maltoni, C. and Cotti, G. Carcinogenicity of vinyl chloride in Sprague-Dawley rats after prenatal and postnatal exposure. *Ann. N. Y. Acad. Sci.* 534: 145-159(1988).

571. Maltoni, C., Cotti, G., Morisi, L., and Chieco, P. Carcinogenicity bioassays of vinylidene chloride. Research plans and early results. *La Med. Lavoro* 64: 241-262(1977).

572. Maltoni, C., Cotti, G., and Perino, G. Long-term carcinogenicity bioassays on methylene chloride administered by ingestion to Sprague-Dawley rats and Swiss mice and by inhalation to Sprague-Dawley rats. *Ann. N. Y. Acad. Sci.* 534: 352-366(1988).

573. Maltoni, C., Lefemine, G., Ciliberti, A., Cotti, G., and Carretti, D. Carcinogenicity bioassays of vinyl chloride monomer: a model of risk assessment on an experimental basis. *Environ. Health Perspect.* 41: 3-29(1981).

574. Maltoni, C., Lefemine, G., and Cotti, G. Experimental research on trichloroethylene carcinogenesis. *Arch. Res. Ind. Carcinog.* 5: 1-393(1986).

575. Maltoni, C., Lefemine, G., Cotti, G., Chieco, P., and Patella, V. Experimental research on vinylidene chloride carcinogenesis. *Arch. Res. Ind. Carcinog.* 3: 1-229(1985).

576. Maltoni, C., Lefemine, G., Tovoli, D., and Perino, G. Long-term carcinogenicity bioassays on three chlorofluorocarbons (trichlo-rofluoromethane, FC11; dichlorodifluoromethane, FC12; chlorod-ifluoromethane, FC22) administered by inhalation to Sprague-Dawley rats and Swiss mice. *Ann. N. Y. Acad. Sci.* 534: 261-282(1988).

577. Maltoni, C. and Scarnato, C. First experimental demonstration of the carcinogenic effects of benzene. *La Med. Lavoro* 70: 352-357(1979).

578. Maltoni, C., Valgimigli, L., and Scarnato, C. Long-term carcinogenic bioassays on ethylene dichloride administered by inhalation to rats and mice. In: *Banbury Report 5 Ethylene Dichloride: A Potential Health Risk?* (B. Ames, P. Infante, and R. Reitz, Eds.), Cold Spring Harbor Laboratory, Cold Spring Harbor, NY, 1980, pp. 3-33.

579. Mannell, W. A. Further investigations on production of liver tumours in rats by ponceau 3R. *Food Cosmet. Toxicol.* 2: 169-174(1964).

580. Mannell, W. A., Grice, H. C., Lu, F. C., and Allmark, M. G. Chronic toxicity studies on food colours. Part IV. Observations on the toxicity of tartrazine, amaranth and sunset yellow in rats. *J. Pharm. Pharmacol.* 10: 625-634(1958).

581. Markiewicz, V. R., Saunders, L. Z., Geus, R. J., Payne, B. J., and Hook, J. B. Carcinogenicity study of auranofin, and orally administered gold compound, in mice. *Fundam. Appl. Toxicol.* 11: 277-284(1988).

582. Markiewicz, V., Tompkins, C., Mehdi, N., Hubmer, S., Payne, B. J., and Sumi, N. Rangefinding toxicity and carcinogenicity studies of a new β-adrenoceptor blocking agent celiprolol in mice. *Pharmacometrics* 38: 421-434(1989).

583. Martin, M. S., Justrabo, E., Jeannin, J. F., Leclerc, A., and Martin, F. Effect of dietary chenodeoxycholic acid on intestinal carcinogenesis induced by 1.2 dimethylhydrazine in mice and hamsters. *Br. J. Cancer* 43: 884-886(1981).

584. Maru, G. B. and Bhide, S. V. Effect of antioxidants and antitoxicants of isoniazid on the formation of lung tumours in mice by isoniazid and hydrazine sulphate. *Cancer Lett.* 17: 75-80(1982).

585. Mason, P. L., Gaunt, I. F., Butterworth, K. R., Hardy, J., Kiss, I. S., and Grasso, P. Long-term toxicity studies of carmoisine in mice. *Food Cosmet. Toxicol.* 12: 601-607(1974).

586. Masuda, M. and Takayama, S. Intestinal tumours in rats induced by mutagens from glutamic acid pyrolysate. *Exp. Pathol.* 26: 123-129(1984).

587. Masui, T., Hirose, M., Imaida, K., Fukushima, S., Tamano, S., and Ito, N. Sequential changes of the forestomach of F344 rats, Syrian golden hamsters, and B6C3F₁ mice treated with butylated hydroxyanisole. *Jpn. J. Cancer Res.* 77: 1083-1090(1986).

588. Matsukura, N., Kawachi, T., Morino, K., Ohgaki, H., and Sugimura, T. Carcinogenicity in mice of mutagenic compounds from a tryptophan pyrolyzate. *Science* 213: 346-347(1981).

589. Matsukura, N., Kawachi, T., Sasajima, K., Sano, T., Sugimura, T., and Hirota, T. Induction of intestinal metaplasia in the stomachs of rats by N-methyl-N′-nitro-N-nitrosoguanidine. *J. Natl. Cancer Inst.* 61: 141-143(1978).

590. Matsukura, N., Kawachi, T., Sasajima, K., Sano, T., Sugimura, T., and Ito, N. Induction of liver tumors in rats by sodium nitrite and methylguanidine. *Cancer Res. Clin. Oncol.* 90: 87-94(1977).

591. Matsukura, N., Kawachi, T., Sugimura, T., Nakadate, M., and Hirota, T. Induction of intestinal metaplasia and carcinoma in the glandular stomach of rats by N-alkyl-N′-nitro-N-nitrosoguanidines. *Gann* 70: 181-185(1979).

592. Matsuzaki, O. Histogenesis and growing patterns of lung tumors induced by potassium 1-methyl-1,4-dihydro-7-[2-(5-nitrofuryl)vinyl]-4-oxo-1,8-naphthyridine-3-carboxylate in ICR mice. *Gann* 66: 259-267(1975).

593. McCollister, S. B., Kociba, R. J., Humiston, C. G., McCollister, D. D., and Gehring, P. J. Studies of the acute and long-term oral toxicity of chlorpyrifos (O,O-diethyl-O-(3,5,6-trichloro-2-pyridyl) phosphorothioate). *Food Cosmet. Toxicol.* 12: 45-61(1974).

594. McElligott, T. F. and Hurst, E. W. Long-term feeding studies of methyl ethyl cellulose ("edifas" A) and sodium carboxymethyl cellulose ("edifas" B) in rats and mice. *Food Cosmet. Toxicol.* 6: 449-460(1968).

595. McGuinness, E. E., Hopwood, D., and Wormsley, K. G. Potentiation of pancreatic carcinogenesis in the rat by DL-ethionine-induced pancreatitis. *Scand. J. Gastroenterol.* 18: 189-192(1983).

596. McGuire, E. J., DiFonzo, C. J., Martin, R. A., and de la Iglesia, F. A. Evaluation of chronic toxicity and carcinogenesis in rodents with the synthetic analgesic, tilidine fumarate. *Toxicology* 39: 149-163(1986).

597. McManus, B. M., Toth, B., and Patil, K. D. Aortic rupture and aortic smooth muscle tumors in mice: Induction by p-hydrazinobenzoic acid hydrochloride of the cultivated mushroom *Agaricus bisporus. Lab. Invest.* 57: 78-85(1987).

598. Menon, M. M. and Bhide, S. V. Perinatal carcinogenicity of isoniazid (INH) in Swiss mice. *Cancer Res. Clin. Oncol.* 105: 258-261(1983).

599. Merkow, L. P., Epstein, S. M., Slifkin, M., and Pardo, M. The ultrastructure of renal neoplasms induced by aflatoxin B1. *Cancer Res.* 33: 1608-1614(1973).

600. Michejda, C. J., Kroeger-Koepke, M. B., and Kovatch, R. M. Carcinogenic effects of sequential administration of two nitrosamines in Fischer 344 rats. *Cancer Res.* 46: 2252-2256(1986).

601. Miller, E. C. and Miller, J. A. Biochemical investigations on hepatic carcinogenesis. *J. Natl. Cancer Inst.* 15: 1571-1590(1955).

602. Miller, E. C., Miller, J. A., and Enomoto, M. The comparative carcinogenicities of 2-acetylaminofluorene and its N-hydroxy metabolite in mice, hamsters, and guinea pigs. *Cancer Res.* 24: 2018-2026(1964).

603. Miller, E. C., Swanson, A. B., Phillips, D. H., Fletcher, T. L., Liem, A., and Miller, J. A. Structure-activity studies of the carcinogenicities in the mouse and rat of some naturally occurring and synthetic alkenylbenzene derivatives related to safrole and estragole. *Cancer Res.* 43: 1124-1134(1983).

604. Miller, R. R., Young, J. T., Kociba, R. J., Keyes, D. G., Bodner, K. M., Calhoun, L. L., and Ayres, J. A. Chronic toxicity and oncogenicity bioassay of inhaled ethyl acrylate in Fischer 344 rats and B6C3F₁ mice. *Drug Chem. Toxicol.* 8: 1-42(1985).

605. Mirvish, S. S., Nickols, J., Weisenburger, D. D., Johnson, D., Joshi, S. S., Kaplan, P., Gross, M., and Tong, H. Y. Effects of 2,4,5-trichlorophenoxyacetic acid, pentachlorophenol, methylprednisolone, and Freund's adjuvant on 2-hydroxyethylnitrosourea carcinogenesis in MRC-Wistar rats. *J. Toxicol. Environ. Health* 32: 59-74(1991).

606. Mirvish, S. S., Salmasi, S., and Runge, R. G. Carcinogenicity test of acetoxime in MRC-Wistar rats. *J. Natl. Cancer Inst.* 69: 961-962(1982).

607. Mirvish, S. S., Weisenburger, D. D., Joshi, S. S., and Nickols, J. 2-Hydroxyethylnitrosourea induction of B cell lymphoma in female Swiss mice. *Cancer Lett.* 54: 101-106(1990).

608. Mirvish, S. S., Weisenburger, D. D., Salmasi, S., and Kaplan, P. A. Carcinogenicity of 1-(2-hydroxyethyl)-1-nitrosourea and 3-nitroso-2-oxazolidinone administered in drinking water to male MRC-Wistar rats: Induction of bone, hematopoietic, intestinal, and liver tumors. *J. Natl. Cancer Inst.* 78: 387-393(1987).

609. Misdorp, W. Progestagens and mammary tumours in dogs and cats. *Acta Endocrinol. (Copenh.)* 125: 27-31(1991).

610. Mitsumori, K., Hirano, M., Ueda, H., Maita, K., and Shirasu, Y. Chronic toxicity and carcinogenicity of methylmercury chloride in B6C3F₁ mice. *Fundam. Appl. Toxicol.* 14: 179-190(1990).

611. Miyaji, T. Acute and chronic toxicity of furylfuramide in rats and mice. *Tohoku J. Exp. Med.* 103: 331-369(1971).

612. Mizutani, T. and Mitsuoka, T. Effect of dietary phenobarbital on spontaneous hepatic tumorigenesis in germfree C3H/He male mice. *Cancer Lett.* 39: 233-237(1988).

613. Mohr, U., Althoff, J., Ketkar, M. B., Conradt, P., and Morgareidge, K. The influence of caffeine on tumour incidence in Sprague-Dawley rats. *Food Chem. Toxicol.* 22: 377-382(1984).

614. Mori, H., Sugie, S., Niwa, K., Takahashi, M., and Kawai, K. Induction of intestinal tumours in rats by chrysazin. *Br. J. Cancer* 52: 781-783(1985).

615. Mori, H., Sugie, S., Niwa, K., Yoshimi, N., Tanaka, T., and Hirono, I. Carcinogenicity of chrysazin in large intestine and liver of mice. *Jpn. J. Cancer Res.* 77: 871-876(1986).

616. Mori, H., Sugie, S., Yoshimi, N., Kuniyasu, T., Iwata, H., Kawai, K., and Hamasaki, T. Potential carcinogenicity of 5,6-dimethoxysterigmatocystin in rats. *Carcinogenesis* 9: 1039-1042(1988).

617. Mori, H., Yoshimi, N., Iwata, H., Mori, Y., Hara, A., Tanaka, T., and Kawai, K. Carcinogenicity of naturally occurring 1-hydroxyanthraquinone in rats: Induction of large bowel, liver and stomach neoplasms. *Carcinogenesis* 11: 799-802(1990).

618. Morino, K., Matsukura, N., Kawachi, T., Ohgaki, H., Sugimura, T., and Hirono, I. Carcinogenicity test of quercetin and rutin in golden hamsters by oral administration. *Carcinogenesis* 3: 93-97(1982).

619. Morishita, Y. and Shimizu, T. Promoting effect of intestinal *Pseudomonas aeruginosa* on gastric tumorigenesis in rats with *N*-methyl-*N*′-nitro-*N*-nitrosoguanidine. *Cancer Lett.* 17: 347-352(1983).

620. Morris, H. P., Velat, C. A., Wagner, B. P., Dahlgard, M., and Ray, F. E. Studies of carcinogenicity in the rate of derivatives of aromatic amines related to *N*-2-fluorenylacetamide. *J. Natl. Cancer Inst.* 24: 149-180(1960).

621. Morris, J. E., Price, J. M., Lalich, J. J., and Stein, R. J. The carcinogenic activity of some 5-nitrofuran derivatives in the rat. *Cancer Res.* 29: 2145-2156(1969).

622. Munir, K. M., Soman, C. S., and Bhide, S. V. Hexachlorocyclohexane-induced tumorigenicity in mice under different experimental conditions. *Tumori* 69: 383-386(1983).

623. Munoz, M. Effect of herpes virus type 2 and hormonal imbalance on the uterine cervix of the mouse. *Cancer Res.* 33: 1504-1508(1973).

624. Munro, I. C., Moodie, C. A., Krewski, D., and Grice, H. C. A. Carcinogenicity study of commercial saccharin in the rat. *Toxicol. Appl. Pharmacol.* 32: 513-526(1975).

625. Murasaki, G. and Cohen, S. M. Co-carcinogenicity of sodium saccharin and *N*-[4-(5-nitro-2-furyl)-2-thiazoyl]formamide for the urinary bladder. *Carcinogenesis* 4: 97-99(1983).

626. Murthy, A. S. K., Baker, J. R., Smith, E. R., and Zepp, E. Neoplasms in rats and mice fed butylurea and sodium nitrite separately and in combination. *Int. J. Cancer* 23: 253-259(1979).

627. Nagel, D., Shimizu, H., and Toth, B. Tumor induction studies with *n*-butyl- and *n*-propylhydrazine hydrochlorides in Swiss mice. *Eur. J. Cancer* 11: 473-478(1975).

628. Nair, R. S., Auletta, C. S., Schroeder, R. E., and Johannsen, F. R. Chronic toxicity, oncogenic potential, and reproductive toxicity of *p*-nitroaniline in rats. *Fundam. Appl. Toxicol.* 15: 607-621(1990).

629. Naito, M., Ito, A., Watanabe, H., Kawashima, K., and Aoyama, H. Carcinogenicity of *o*-ethoxybenzamide in (C57BL/6N × C3H/HeN)F₁ mice. *J. Natl. Cancer Inst.* 76: 115-118(1986).

630. Nakamura, T., Matsuyama, M., and Kishimoto, H. Tumors of the esophagus and duodenum induced in mice by oral administration of *N*-ethyl-*N*′-nitro-*N*-nitrosoguanidine. *J. Natl. Cancer Inst.* 52: 519-522(1974).

631. Nakanishi, K., Kurata, Y., Oshima, M., Fukushima, S., and Ito, N. Carcinogenicity of phenacetin: long-term feeding study in B6C3F₁ mice. *Int. J. Cancer* 29: 439-444(1982).

632. National Cancer Institute. Report on carcinogenesis bioassay of chloroform. N.C.I. Brief Communication. 1976.

633. Nelson, C. J., Baetcke, K. P., Frith, C. H., Kodell, R. L., and Schieferstein, G. The influence of sex, dose, time, and cross on neoplasia in mice given benzidine dihydrochloride. *Toxicol. Appl. Pharmacol.* 64: 171-186(1982).

634. Nera, E. A., Iverson, F., Lok, E., Armstrong, C. L., Karpinski, K., and Clayson, D. B. A carcinogenesis reversibility study of the effects of butylated hydroxyanisole on the forestomach and urinary bladder in male Fischer 344 rats. *Toxicology* 53: 251-268(1988).

635. Newberne, J. W., Newberne, P. M., Gibson, J. P., Huffman, K. W., and Palopoli, F. P. Lack of carcinogenicity of oxprenolol, a β-adrenergic blocking agent. *Toxicol. Appl. Pharmacol.* 41: 535-546(1977).

636. Newberne, P. M. Nitrite promotes lymphoma incidence in rats. *Science* 204: 1079-1081(1979).

637. Newberne, P. M., Harrington, D. H., and Wogan, G. N. Effects of cirrhosis and other liver insults on induction of liver tumors by aflatoxin in rats. *Lab. Invest.* 15: 962-969(1966).

638. Newberne, P. M. and Rogers, A. E. Nutrition, monocrotaline, and aflatoxin B₁ in liver carcinogenesis. *Plant Foods Man* 1: 23-31(1973).

639. Newberne, P. M. and Rogers, A. E. Rat colon carcinomas associated with aflatoxin and marginal vitamin A. *J. Natl. Cancer Inst.* 50: 439-444(1973).

640. Newberne, P. M. and Williams, G. Inhibition of aflatoxin carcinogenesis by diethylstilbestrol in male rats. *Arch. Environ. Health* 19: 489-498(1969).

641. Nishikawa, A., Furukawa, F., Mitsui, M., Enami, T., Kawanishi, T., Hasegawa, T., and Takahashi, M. Inhibitory effect of calcium chloride on gastric carcinogenesis in rats after treatment with *N*-methyl-*N*′-nitro-*N*-nitrosoguanidine and sodium chloride. *Carcinogenesis* 13: 1155-1158(1992).

642. Nitschke, K. D., Burek, J. D., Bell, T. J., Kociba, R. J., Rampy, L. W., and McKenna, M. J. Methylene chloride: A 2-year inhalation toxicity and oncogenicity study in rats. *Fundam. Appl. Toxicol.* 11: 48-59(1988).

643. Nixon, J. E., Hendricks, J. D., Pawlowski, N. E., Loveland, P. M., and Sinnhuber, R. O. Carcinogenicity of aflatoxicol in Fischer 344 rats. *J. Natl. Cancer Inst.* 66: 1159-1163(1981).

644. Nixon, J. E., Sinnhuber, R. O., Lee, D. J., Landers, M. K., and Harr, J. R. Effect of cyclopropenoid compounds on the carcinogenic activity of diethylnitrosamine and aflatoxin B₁ in rats. *J. Natl. Cancer Inst.* 53: 453-458(1974).

645. Nonoyama, T., Fullerton, F., Reznik, G., Bucci, T. J., and Ward, J. M. Mouse hepatoblastomas: A histologic, ultrastructural, and immunohistochemical study. *Pathol. Vet.* 25: 286-296(1988).

646. Nyska, A., Waner, T., Paster, Z., Bracha, P., Gordon, E., and Klein, B. Induction of gastrointestinal tumors in mice fed the fungicide folpet: Possible mechanisms. *Jpn. J. Cancer Res.* 81: 545-549(1990).

647. Nyska, A., Waner, T., Pirak, M., Gordon, E., Bracha, P., and Klein, B. The renal carcinogenic effect of Merpafol in the Fischer 344 rats. *Isr. J. Med. Sci.* 25: 428-432(1989).

648. Oda, H., Nogami, H., Kusumoto, S., Nakajima, T., and Kurata, A. Lifetime exposure to 2.4 ppm nitric oxide in mice. *Environ. Res.* 22: 254-263(1980).

649. Ogiso, T., Tatematsu, M., Tamano, S., Tsuda, H., and Ito, N. Comparative effects of carcinogens on the induction of placental glutathione S-transferase-positive liver nodules in a short-term assay and of hepatocellular carcinomas in a long-term assay. *Toxicol. Pathol.* 13: 257-265(1985).

650. Ogiu, T., Fukami, H., Nishimura, M., Shimada, Y., and Maekawa, A. Age-dependent induction of thymic lymphomas by *N*-propyl-*N*-nitrosourea in the F344/DuCrj rat. *Thymus* 20: 249-258(1992).

651. Ogiu, T., Kajiwara, T., Furuta, K., Takeuchi, M., Odashima, S., and Tada, K. Mammary tumorigenic effect of a new nitrosourea, 1,3-dibutyl-1-nitrosourea (B-BNU), in female Donryu rats. *Cancer Res. Clin. Oncol.* 96: 35-41(1980).

652. Ogiu, T., Matsuoka, C., Furuta, K., Takeuchi, M., Maekawa, A., Nakadate, M., and Odashima, S. Induction of angiogenic tumors in the duodenum of female Donryu rats by continuous oral administration of *N*-isobutyl-*N*-nitrosourea. *Gann* 74: 342-350(1983).

653. Ohgaki, H., Hasegawa, H., Kato, T., Suenaga, M., Ubukata, M., Sato, S., Takayama, S., and Sugimura, T. Carcinogenicity in mice and rats of heterocyclic amines in cooked foods. *Environ. Health Perspect.* 67: 129-134(1986).

654. Ohgaki, H., Hasegawa, H., Suenaga, M., Kato, T., Sato, S., Takayama, S., and Sugimura, T. Induction of hepatocellular carcinoma and highly metastatic squamous cell carcinomas in the forestomach of mice by feeding 2-amino-3,4-dimethylimidazo[4,5-*f*]quinoline. *Carcinogenesis* 7: 1889-1893(1986).

655. Ohgaki, H., Hasegawa, H., Suenaga, M., Sato, S., Takayama, S., and Sugimura, T. Carcinogenicity in mice of a mutagenic compound, 2-amino-3,8-dimethylimidazo[4,5-*f*]quinoxaline (MeIQx) from cooked foods. *Carcinogenesis* 8: 665-668(1987).

656. Ohgaki, H., Kusama, K., Matsukura, N., Morino, K., Hasegawa, H., Sato, S., Takayama, S., and Sugimura, T. Carcinogenicity in mice of a mutagenic compound, 2-amino-3-methylimidazo[4,5-*f*]quinoline, from broiled sardine, cooked beef and beef extract. *Carcinogenesis* 5: 921-924(1984).

657. Ohgaki, H., Matsukura, N., Morino, K., Kawachi, T., Sugimura, T., and Takayama, S. Carcinogenicity in mice of mutagenic compounds from glutamic acid and soybean globulin pyrolysates. *Carcinogenesis* 5: 815-819(1984).

658. Ohgaki, H., Takayama, S., and Sugimura, T. Carcinogenicities of heterocyclic amines in cooked food. *Mutat. Res.* 259: 399-410(1991).

659. Ohshima, M., Ward, J. M., Brennan, L. M., and Creasia, D. A. A sequential study of methapyrilene hydrochloride-induced liver carcinogenesis in male F344 rats. *J. Natl. Cancer Inst.* 72: 759-768(1984).

660. Ohtsubo, K., Ryu, J.-C., Nakamura, K., Izumiyama, N., Tanaka, T., Yamamura, H., Kobayashi, T., and Ueno, Y. Chronic toxicity of nivalenol in female mice: A 2-year feeding study with *Fusarium nivale* Fn 2B-moulded rice. *Food Chem. Toxicol.* 27: 591-598(1989).

661. Ohtsubo, K., Saito, M., Kimura, H., and Tsuruta, O. High incidence of hepatic tumours in rats fed mouldy rice contaminated with *Aspergillus versicolor* containing sterigmatocystin. *Food Cosmet. Toxicol.* 16: 143-149(1978).

662. Okey, A. B. and Gass, G. H. Continuous versus cyclic estrogen administration: mammary carcinoma in C3H mice. *J. Natl. Cancer Inst.* 40: 225-230(1968).

663. Oldiges, H., Hochrainer, D., and Glaser, U. Long-term inhalation study with Wistar rats and four cadmium compounds. *Toxicol. Environ. Chem.* 19: 217-222(1989).

664. Onodera, H., Furuta, K., Matsuoka, C., Kamiya, S., and Maekawa, A. Carcinogenicities of 1-ethyl- and 1-amyl-1-nitrosourethans in female Donryu rats: dose-effect relations. *Gann* 73: 48-54(1982).

665. Oohashi, Y., Ishioka, T., Wakabayashi, K., and Kuwabara, N. A study on carcinogenesis induced by degraded carrageenan arising from squamous metaplasia of the rat colorectum. *Cancer Lett.* 14: 267-272(1981).

666. Oser, B. L., Carson, S., Cox, G. E., Vogin, E. E., and Sternberg, S. S. Long-term and multigeneration toxicity studies with cyclohexylamine hydrochloride. *Toxicology* 6: 47-65(1976).

667. Oser, B. L., Morgareidge, K., Weinberg, M. S., and Oser, M. Carcinogenicity study of carbarsone. *Toxicol. Appl. Pharmacol.* 9: 528-535(1966).

668. Oser, B. L. and Oser, M. 2-(*p*-Tert-butylphenoxy)isopropyl 2-chloroethyl sulfite (Aramite). 1. Acute, subacute, and chronic oral toxicity. *Toxicol. Appl. Pharmacol.* 2: 441-457(1960).

669. Osswald, H., Frank, H. K., Komitowski, D., and Winter, H. Long-term testing of patulin administered orally to Sprague-Dawley rats and Swiss mice. *Food Cosmet. Toxicol.* 16: 243-247(1978).

670. Owen, N. V., Worth, H. M., and Kiplinger, G. F. The effects of long-term ingestion of methimazole on the thyroids of rats. *Food Cosmet. Toxicol.* 11: 649-653(1973).

671. Owen, P. E., Glaister, J. R., Gaunt, I. F., and Pullinger, D. H. Inhalation toxicity studies with 1,3-butadiene 3 two year toxicity/carcinogenicity study in rats. *Am. Ind. Hyg. Assoc. J.* 48: 407-413(1987).

672. Pai, S. R., Shirke, A. J., and Gothoskar, S. V. Long-term feeding study in C17 mice administered saccharin coated betel nut and 1,4-dinitrosopiperazine in combination. *Carcinogenesis* 2: 175-177(1981).

673. Palmer, A. K., Street, A. E., Roe, F. J. C., Worden, A. N., and Van Abbe, N. J. Safety evaluation of toothpaste containing chloroform. II. Long term studies in rats. *J. Environ. Pathol. Toxicol.* 2: 821-833(1979).

674. Pamukcu, A. M., Yalciner, S., Hatcher, J. F., and Bryan, G. T. Quercetin, a rat intestinal and bladder carcinogen present in bracken fern (*Pteridium aquilinum*). *Cancer Res.* 40: 3468-3472(1980).

675. Parker, C. M., Patterson, D. R., Van Gelder, G. A., Gordon, E. B., Valerio, M. G., and Hall, W. C. Chronic toxicity and carcinogenicity evaluation of fenvalerate in rats. *J. Toxicol. Environ. Health* 13: 83-97(1984).

676. Parker, C. M., Van Gelder, G. A., Chai, E. Y., Gellatly, J. B. M., Serota, D. G., Voelker, R. W., and Vesselinovitch, S. D. Oncogenic evaluation of tetrachlorvinphos in the B6C3F$_1$ mouse. *Fundam. Appl. Toxicol.* 5: 840-854(1985).

677. Patton, D. S. G., Heywood, R., and Barcellona, P. S. Carcinogenicity assessment of lonidamine by dietary administration to Sprague-Dawley rats. *Toxicol. Lett.* 62: 209-214(1992).

678. Pavkov, K. L., Kerns, W. D., Mitchell, R. I., Connell, M. M., Donofrio, D. J., Harroff, H. H., Barker, A. D., Fisher, G. L., Joiner, R. L., and Thake, D. C. A chronic inhalation toxicology study in rats and mice exposed to formaldehyde. Final Report, CIIT Docket #10922. Chemical Industry Institute of Toxicology, Research Triangle Park, NC, 1981.

679. Peraino, C., Fry, R. J. M., and Staffeldt, E. Enhancement of spontaneous hepatic tumorigenesis in C3H mice by dietary phenobarbital. *J. Natl. Cancer Inst.* 51: 1349-1350(1973).

680. Pershin, G. N., Makeeva, O. O., Grushina, A. A., and Chernov, V. A. An experimental study of the carcinogenic effect of tubazid (isoniazid) and phtivazid. *Vopr. Onkol.* 18: 50-53(1972).

681. Phillips, J. C., Butterworth, K. R., Gaunt, I. F., Evans, J. G., and Grasso, P. Long-term toxicity study of quillaia extract in mice. *Food Cosmet. Toxicol.* 17: 23-27(1979).

682. Pietra, G., and Shubik, P. Induction of melanotic tumors in the Syrian golden hamster after administration of ethyl carbamate. *J. Natl. Cancer Inst.* 25: 627-630(1960).

683. Pintér, A., Török, G., Börzsönyi, M., Surján, A., Csík, M., Kelecsényi, Z., and Kocsis, Z. Long-term carcinogenicity bioassay of the herbicide atrazine in F344 rats. *Neoplasma* 35: 533-544(1990).

684. Poel, W. E. Pituitary tumors in mice after prolonged feeding of synthetic progestins. *Science* 154: 402-403(1966).

685. Poel, W. E., Ciocco, A., and Doolittle, D. P. Unusual neoplasms and hyperplastic lesions in "random-bred" mice derived from four-way crossed inbred lines. *Cancer Res.* 28: 845-859(1968).

686. Ponomarkov, V., Tomatis, L., and Turusov, V. The effect of long-term administration of phenobarbitone in CF-1 mice. *Cancer Lett.* 1: 165-172(1976).

687. Popper, H., Sternberg, S. S., Oser, B. L., and Oser, M. The carcinogenic effect of aramite in rats. A study of hepatic nodules. *Cancer* 13: 1035-1046(1960).

688. Port, C. D., Dodd, D. C., Deslex, P., Regnier, B., Sanders, P., and Indacochea-Redmond, N. Twenty-one month evaluation of misoprostol for carcinogenicity in CD-1 mice. *Toxicol. Pathol.* 15: 134-142(1987).

689. Pour, P. M. A new prostatic cancer model: systemic induction of prostatic cancer in rats by a nitrosamine. *Cancer Lett.* 13: 303-308(1981).

690. Pour, P. M. Prostatic cancer induced in MRC rats by N-nitrosobis(2-oxopropyl)-amine and N-nitrosobis(2-hydroxypropyl)amine. *Carcinogenesis* 4: 49-55(1983).

691. Pour, P. M., Grandjean, C. J., and Knepper, S. Selective induction of nasal cavity tumors in rats by diallylnitrosamine. *Cancer Res. Clin. Oncol.* 109: 5-8(1985).

692. Pour, P. M. and Stepan, K. Comparative carcinogenicity of N-nitrosobis(2-oxopropyl)amine and N-nitrosomethyl(2-oxopropyl)amine following subcutaneous or oral administration to rats. *Cancer Lett.* 45: 49-57(1989).

693. Preussmann, R., Habs, M., Habs, H., and Schmahl, D. Carcinogenicity of N-nitrosodiethanolamine in rats at five different dose levels. *Cancer Res.* 42: 5167-5171(1982).

694. Preussmann, R., Habs, M., Habs, H., and Stummeyer, D. Fluoro-substituted N-nitrosamines. 5. Carcinogenicity of N-nitrosobis(4,4,4-trifluoro-N-butyl)amine in rats. *Carcinogenesis* 3: 1219-1222(1982).

695. Preussmann, R., Habs, M., Habs, H., and Stummeyer, D. Fluoro-substituted N-nitrosamines. 6. Carcinogenicity of N-nitroso-(2,2,2-trifluoroethyl)-ethylamine in rats. *Carcinogenesis* 4: 755-757(1983).

696. Preussmann, R., Habs, M., and Pool, B. L. Carcinogenicity and mutagenicity testing of three isomeric N-nitroso-N-methylaminopyridines in rats. *J. Natl. Cancer Inst.* 62: 153-156(1979).

697. Preussmann, R., Habs, M., Pool, B., Stummeyer, D., Lijinsky, W., and Reuber, M. D. Fluoro-substituted N-nitrosamines. 1. Inactivity of N-nitroso-bis(2,2,2-trifluoroethyl)amine in carcinogenicity and mutagenicity tests. *Carcinogenesis* 2: 753-756(1981).

698. Preussmann, R. and Ivankovic, S. Absence of carcinogenic activity in BD rats after oral administration of high doses of bismuth oxychloride. *Food Cosmet. Toxicol.* 13: 543-544(1975).

699. Preussmann, R., Ivankovic, S., Landschutz, C., Gimmy, J., Flohr, E., and Griesbach, U. Carcinogene Wirkung von 13 Aryldialkyltriazenen an BD-Ratten. *Cancer Res. Clin. Oncol.* 81: 285-310(1974).

700. Preussmann, R., Schmahl, D., and Eisenbrand, G. Carcinogenicity of N-nitosopyrrolidine: dose-response study in rats. *Cancer Res. Clin. Oncol.* 90: 161-166(1977).

701. Price, J. M., Biava, C. G., Oser, B. L., Vogin, E. E., Steinfeld, J., and Ley, H. L. Bladder tumors in rats fed cyclohexylamine or high doses of a mixture of cyclamate and saccharin. *Science* 167: 1131-1132(1970).

702. Prier, R. F., Nees, P. O., and Derse, P. H. The toxicity of an organic arsenical, 3-nitro-4-hydroxyphenylarsonic acid. II. Chronic toxicity. *Toxicol. Appl. Pharmacol.* 5: 526-542(1963).

703. Purchase, I. F. H. and van der Watt, J. J. Carcinogenicity of sterigmatocystin. *Food Cosmet. Toxicol.* 8: 289-295(1970).

704. Purchase, I. F. H. and van der Watt, J. J. The long-term toxicity of ochratoxin A to rats. *Food Cosmet. Toxicol.* 9: 681-682(1971).

705. Quast, J. F., Calhoun, L. L., and Frauson, L. E. 1,1,1-trichloroethane formulation: A chronic inhalation toxicity and oncogenicity study in Fischer 344 rats and B6C3F$_1$ mice. *Fundam. Appl. Toxicol.* 11: 611-625(1988).

706. Quast, J. F., Humiston, C. G., Wade, C. E., Ballard, J., Beyer, J. E., Schwetz, R. W., and Norris, J. M. A chronic toxicity and oncogenicity study in rats and subchronic toxicity study in dogs on ingested vinylidene chloride. *Fundam. Appl. Toxicol.* 3: 55-62(1983).

707. Quast, J. F., McKenna, M. J., Rampy, L. W., and Norris, J. M. Chronic toxicity and oncogenicity study on inhaled vinylidene chloride in rats. *Fundam. Appl. Toxicol.* 6: 105-144(1986).

708. Quast, J. F., Schuetz, D. J., Balmer, M. F., Gushow, T. S., Park, C. N., and McKenna, M. J. A Two-Year Toxicity and Oncogenicity Study with Acrylonitrile Following Inhalation Exposure of Rats. Final Report. Dow Chemical U.S.A., Midland, MI, 1980.

709. Quast, J. F., Wade, C. E., Humiston, C. G., Carreon, R. M., Hermann, E. A., Park, C. N., and Schwetz, B. A. A Two-Year Toxicity and Oncogenicity Study with Acrylonitrile Incorporated in the Drinking Water of Rats. Final Report. Dow Chemical U.S.A., Midland, MI, 1980.

710. Quest, J. A., Copley, M. P., Hamernik, K. L., Rinde, E., Fisher, B., Engler, R., Burnam, W. L., and Fenner-Crisp, P. A. Evaluation of the carcinogenic potential of pesticides. 2. Methidathion. *Regul. Toxicol. Pharmacol.* 12: 117-126(1990).

711. Quest, J. A., Hamernik, K. L., Engler, R., Burnam, W. L., and Fenner-Crisp, P. A. Evaluation of the carcinogenic potential of pesticides. 3. Aliette. *Regul. Toxicol. Pharmacol.* 14: 3-11(1991).

712. Quest, J. A., Phang, W., Hamernik, K. L., van Gemert, M., Fisher, B., Levy, R., Farber, T. M., Burnam, W. L., and Engler, R. Evaluation of the carcinogenic potential of pesticides. 1. Acifluorfen. *Regul. Toxicol. Pharmacol.* 10: 149-159(1989).

713. Radike, M. J., Stemmer, K. L., and Bingham, E. Effect of ethanol on vinyl chloride carcinogenesis. *Environ. Health Perspect.* 41: 59-62(1981).

714. Radomski, J. L., Brill, E., and Glass, E. M. Induction of bladder tumors and other malignancies in rats with 2-methoxy-3-aminodibenzofuran. *J. Natl. Cancer Inst.* 39: 1069-1080(1967).

715. Radomski, J. L., Deichmann, W. B., Macdonald, W. E., and Glass, E. M. Synergism among oral carcinogens: I. Results of the simultaneous feeding of four tumorigens to rats. *Toxicol. Appl. Pharmacol.* 7: 652-656(1965).

716. Rao, M. S., Dwivedi, R. S., Subbarao, V., and Reddy, J. K. Induction of peroxisome proliferation and hepatic tumours in C57BL/6N mice by ciprofibrate, a hypolipidaemic compound. *Br. J. Cancer* 58: 46-51(1988).

717. Rao, M. S., Lalwani, N. D., Watanabe, T. K., and Reddy, J. K. Inhibitory effect of antioxidants ethoxyquin and 2(3)-*tert*-butyl-4-hydroxyanisole on hepatic tumorigenesis in rats fed ciprofibrate, a peroxisome proliferator. *Cancer Res.* 44: 1072-1076(1984).

718. Rao, M. S. and Reddy, J. K. Malignant neoplasms in rats fed lasiocarpine. *Br. J. Cancer* 37: 289-293(1978).

719. Rao, M. S., Subbarao, A. V., Reddy, J. K. Inihibition of spontaneous testicular Leydig cell tumor development in F-344 rats by dehydroepiandrosterone. *Cancer Lett.* 65: 123-126(1992a).

720. Rao, M. S., Subbarao, V., Yeldandi, A. V., and Reddy, J. K. Hepatocarcinogenicity of dehydroepiandrosterone in the rat. *Cancer Res.* 52,2977-2979(1992).

721. Rao, M. S., Usuda, N., Subbarao, V., and Reddy, J. K. Absence of γ-glutamyl transpeptidase activity in neoplastic lesions induced in the liver of male F-344 rats by di-(2-ethylhexyl)phthalate, a peroxisome proliferator. *Carcinogenesis* 8: 1347-1350(1987).

722. Rao, M. S., Yeldandi, A. V., and Subbarao, V. Quantitative analysis of hepatocellular lesions induced by di(2-ethylhexyl)phthalate in F-344 rats. *J. Toxicol. Environ. Health* 30: 85-89(1990).

723. Reddy, B. S., Tokumo, K., Kulkarni, N., Aligia, C., and Kelloff, G. Inhibition of colon carcinogenesis by prostaglandin synthesis inhibitors and related compounds. *Carcinogenesis* 13: 1019-1023(1992).

724. Reddy, J. K., Azarnoff, D. L., and Hignite, C. E. Hypolipidaemic hepatic peroxisome proliferators form a novel class of chemical carcinogens. *Nature* 283: 397-398(1980).

725. Reddy, J. K., Lalwani, N. D., Reddy, M. K., and Qureshi, S. A. Excessive accumulation of autofluorescent lipofuscin in the liver during hepatocarcinogenesis by methyl clofenapate and other hypolipidemic peroxisome proliferators. *Cancer Res.* 42: 259-266(1982).

726. Reddy, J. K. and Qureshi, S. A. Tumorigenicity of the hypolipidaemic peroxisome proliferator ethyl-α-p-chlorophenoxyisobutyrate (clofibrate) in rats. *Br. J. Cancer* 40: 476-482(1979).

727. Reddy, J. K., Rao, M. S., Azarnoff, D. L., and Sell, S. Mitogenic and carcinogenic effects of a hypolipidemic peroxisome proliferator, [4-chloro-6-(2,3-xylidino)-2-pyrimidinylthio]acetic acid [Wy-14,643], in rat and mouse liver. *Cancer Res.* 29: 152-161(1979).

728. Reddy, J. K., Svoboda, D. J., and Rao, M. S. Induction of liver tumors by aflatoxin B₁ in the tree shrew (*Tupaia glis*) a nonhuman primate. *Cancer Res.* 36: 151-160(1976).

729. Reichert, D., Spengler, U., Romen, W., and Henschler, D. Carcinogenicity of dichloroacetylene: an inhalation study. *Carcinogenesis* 5: 1411-1420(1984).

730. Reuzel, P. G. J., Dreef-van der Meulen, H. C., Hollanders, V. M. H., Kuper, C. F., Feron, V. J., and van der Heijden, C. A. Chronic inhalation toxicity and carcinogenicity study of methyl bromide in Wistar rats. *Food Chem. Toxicol.* 29: 31-39(1991).

731. Reznik, G., Brennan, L. M., and Creasia, D. A. Urine cytology: Early diagnosis of induced carcinoma of the rat urinary bladder. *Anticancer Res.* 1: 389-392(1981).

732. Reznik, G., Mohr, U., and Lijinsky, W. Carcinogenic effect of N-nitroso-2,6-dimethylmorpholine in Syrian golden hamsters. *J. Natl. Cancer Inst.* 60: 371-378(1978).

733. Reznik, G., Reznik-Schuller, H. M., Rice, J. M., and Hague, B. F. Pathogenesis of toxic and neoplastic renal lesions induced by the flame retardant tris(2,3-dibromopropyl)phosphate in F344 rats, and development of colonic adenomas after prolonged oral administration. *Lab. Invest.* 44: 74-83(1981).

734. Rivenson, A., Hoffmann, D., Prokopczyk, B., Amin, S., and Hecht, S. S. Induction of lung and exocrine pancreas tumors in F344 rats by tobacco-specific and areca-derived N-nitrosamines. *Cancer Res.* 48: 6912-6917(1988).

735. Robison, R. L., van Ryzin, R. J., Stoll, R. E., Jensen, R. D., and Bagdon, R. E. Chronic toxicity/carcinogenesis study of temazepam in mice and rats. *Fundam. Appl. Toxicol.* 4: 394-405(1984).

736. Roe, F. J. C., Grant, G. A., and Millican, D. M. Carcinogenicity of hydrazine and 1,1-dimethylhydrazine for mouse lung. *Nature* 216: 375-376(1967).

737. Roe, F. J. C., Levy, L. S., and Carter, R. L. Feeding studies on sodium cyclamate, saccharin and sucrose for carcinogenic and tumour-promoting activity. *Food Cosmet. Toxicol.* 8: 135-145(1970).

738. Roe, F. J. C., Palmer, A. K., Worden, A. N., and Van Abbe, N. J. Safety evaluation of toothpaste containing chloroform. I. Long-term studies in mice. *J. Environ. Pathol. Toxicol.* 2: 799-819(1979).

739. Rosenkrantz, H. and Fleischman, R. W. *In vivo* carcinogenesis assay of DL-methadone.HCl in rodents. *Fundam. Appl. Toxicol.* 11: 640-651(1988).

740. Rosenkrantz, H. and Fleischman, R. W. *In vivo* carcinogenesis assay of l-α-acetylmethadol.HCl in rodents. *Fundam. Appl. Toxicol.* 11: 626-639(1988).

741. Rosin, A. and Ungar, H. Malignant tumors in the eyelids and the auricular region of thiourea-treated rats. *Cancer Res.* 17: 302-305(1957).

742. Rossi, L., Barbieri, O., Sanguineti, M., Cabral, J. R. P., Bruzzi, P., and Santi, L. Carcinogenicity study with technical-grade dichlorodiphenyltrichloroethane and 1,1-dichloro-2,2-bis(p-chlorophenyl)ethylene in hamsters. *Cancer Res.* 43: 776-781(1983).

743. Rossi, L., Ravera, M., Repetti, G., and Santi, L. Long-term administration of DDT or phenobarbital-Na in Wistar rats. *Int. J. Cancer* 19: 179-185(1977).

744. Rothwell, C. E., McGuire, E. J., Martin, R. A., and de la Iglesia, F. A. Chronic toxicity and carcinogenicity studies with the β-adrenoceptor antagonist levobunolol. *Fundam. Appl. Toxicol.* 18: 353-359(1992).

745. Rudali, G. Induction of tumors in mice with synthetic sex hormones. *Gann Mono. Cancer Res.* 17: 243-252(1975).

746. Rudali, G. and Assa, R. Lifespan carcinogenicity studies with hexachlorophene in mice and rats. *Cancer Lett.* 5: 325-332(1978).

747. Rudali, G., Coezy, E., and Chemama, R. Mammary carcinogenesis in female and male mice receiving contraceptives or gestagens. *J. Natl. Cancer Inst.* 49: 813-819(1972).

748. Rudali, G., Coezy, E., Frederic, F., and Apiou, F. Susceptibility of mice of different strains to the mammary carcinogenic action of natural and synthetic oestrogens. *Rev. Eur. Etud. Clin. Biol.* 16: 425-429(1971).

749. Rudali, G., Coezy, E., and Muranyi-Kovacs, I. Cancérologie — recherches sur l'action cancérigène du cyclamate de soude chez les souris. *Acadé. Sci. Mém. Commun. Membres Correspondants Acad.* 269: 1910-1913(1969).

750. Russfield, A. B., Homburger, F., Boger, E., Van Dongen, C. G., Weisburger, E. K., and Weisburger, J. H. Carcinogenicity of Chemicals in Man's Environment. Final Report, Contract No. NIH-NCI-E-68-1311. Bio-research Consultants, Inc., Cambridge, MA, 1973.

751. Russfield, A. B., Homburger, F., Boger, E., Van Dongen, C. G., Weisburger, E. K., and Weisburger, J. H. The carcinogenic effect of 4,4′-methylene-bis-(2-chloroaniline) in mice and rats. Toxicol. Appl. Pharmacol. 31: 47-54(1975).

752. Rustia, M. and Shubik, P. Induction of lung tumors and malignant lymphomas in mice by metronidazole. J. Natl. Cancer Inst. 48: 721-729(1972).

753. Rustia, M. and Shubik, P. Life-span carcinogenicity tests with 4-amino-N-(10)-methylpteroylglutamic acid (methotrexate) in Swiss mice and Syrian golden hamsters. Toxicol. Appl. Pharmacol. 26: 329-338(1973).

754. Rustia, M. and Shubik, P. Thyroid tumours in rats and hepatomas in mice after griseofulvin treatment. Br. J. Cancer 38: 237-249(1978).

755. Rustia, M. and Shubik, P. Experimental induction of hepatomas, mammary tumors, and other tumors with metronidazole in non-inbred Sas:MRC(WI)BR rats. J. Natl. Cancer Inst. 63: 863-868(1979).

756. Rustia, M., Shubik, P., and Patil, K. Lifespan carcinogenicity tests with native carrageenan in rats and hamsters. Cancer Lett. 11: 1-10(1980).

757. Ryffel, B., Donatsch, P., Madorin, M., Matter, B. E., Ruttimann, G., Schon, H., Stoll, R., and Wilson, J. Toxicological evaluation of cyclosporin A. Arch. Toxicol. 5: 107-141(1983).

758. Ryrfeldt, A., Squire, R. A., and Ekman, L. Liver tumors in male rats following treatment with glucocorticosteroids. Toxicol. Pathol. 20: 115-117(1992).

759. Saito, D., Shirai, A., Matsushima, T., Sugimura, T., and Hirono, I. Test of carcinogenicity of quercetin, a widely distributed mutagen in food. Teratog. Carcinog. Mutagen. 1: 213-221(1980).

760. Saito, M., Horiuchi, T., Ohtsubo, K., Hatanaka, Y., and Ueno, Y. Low tumor incidence in rats with long-term feeding of fusarenon-X, a cytotoxic trichothecene produced by Fusarium nivale. Jpn. J. Exp. Med. 50: 293-302(1980).

761. Sakata, T., Hasegawa, R., Johansson, S. L., Zenser, T. V., and Cohen, S. M. Inhibition by aspirin of N-[4-(5-nitro-2-furyl)-2-thiazolyl]formamide initiation and sodium saccharin promotion of urinary bladder carcinogenesis in male F344 rats. Cancer Res. 46: 3903-3906(1986).

762. Salmon, R. J., Buisson, J. P., Zafrani, B., Aussepe, L., and Royer, R. Carcinogenic effect of 7-methoxy-2-nitro-naphto[2,1-b] furan (R 7000) in the forestomach of rats. Carcinogenesis 7: 1447-1450(1986).

763. Sanderson, K. V. Arsenic as a co-carcinogen in mice. British Empire Cancer Campaign, 39th Annual Report 39: 628-629(1961).

764. Sandusky, G. E., Jr., Vodicnik, M. J., and Tamura, R. N. Cardiovascular and adrenal proliferative lesions in Fischer 344 rats induced by long-term treatment with type III phosphodiesterase inhibitors (positive inotropic agents), isomazole and indolidan. Fundam. Appl. Toxicol. 16: 198-209(1991).

765. Sano, T., Kawachi, T., Matsukura, N., Sasajima, K., and Sugimura, T. Carcinogenicity of a food additive, AF-2, in hamsters and mice. Cancer Res. Clin. Oncol. 89: 61-68(1977).

766. Sarles, M. P. and Vandegrift, W. B. Chronic oral toxicity and related studies on animals with the insecticide and pyrethrum synergist, piperonyl butoxide. Am. J. Trop. Med. Hyg. 1: 862-883(1952).

767. Sasajima, K., Kawachi, T., Matsukura, N., Sano, T., and Sugimura, T. Intestinal metaplasia and adenocarcinoma induced in the stomach of rats by N-propyl-N′-nitro-N-nitrosoguanidine. Cancer Res. Clin. Oncol. 94: 201-206(1979).

768. Sato, M., Furukawa, F., Toyoda, K., Mitsumori, K., Nishikawa, A., and Takahashi, M. Lack of carcinogenicity of ferric chloride in F344 rats. Food Chem. Toxicol. 30: 837-842(1992).

769. Schaeffer, E., Greim, H., and Goessner, W. Pathology of chronic polychlorinated biphenyl (PCB) feeding in rats. Toxicol. Appl. Pharmacol. 75: 278-288(1984).

770. Schardein, J. L., Kaump, D. H., Woosley, E. T., and Jellema, M. M. Long-term toxicologic and tumorigenesis studies on an oral contraceptive agent in albino rats. Toxicol. Appl. Pharmacol. 16: 10-23(1970).

771. Scherf, H. R., Frei, E., and Wiessler, M. Carcinogenic properties of N-nitrodimethylamine and N-nitromethylamine in the rat. Carcinogenesis 10: 1977-1981(1989).

772. Schieferstein, G. Serial sacrifice study on benzidine in mice. National Center for Toxicological Research. Serial Sacrifice Study on Benzidine in Mice. Final Report. NCTR Technical Report for Experiment Number 027. NCTR, Jefferson, AR, 1982.

773. Schieferstein, G. J., Littlefield, N. A., Gaylor, D. W., Sheldon, W. G., and Burger, G. T. Carcinogenesis of 4-aminobiphenyl in BALB/cStCrlfC3Hf/Nctr mice. Eur. J. Cancer 21: 865-873(1985).

774. Schieferstein, G. J., Sheldon, W. G., Allen, R. R., Greenman, D. L., and Allaben, W. T. Oncogenic evaluation of 3,3′-dimethoxy-benzidine dihydrochloride in BALB/c mice. J. Am. Coll. Toxicol. 9: 71-77(1990).

775. Schieferstein, G. J., Shinohara, Y., Allen, R. R., Sheldon, W., Greenman, D. L., and Allaben, W. T. Carcinogenicity study of 3,3′-dimethylbenzidine dihydrochloride in BALB/c mice. Food Chem. Toxicol. 27: 801-806(1989).

776. Schmahl, D. Prufung von Polyvinylpyridin-N-oxid (PVNO) auf carcinogene Wirkung bei Ratten und Mausen. Arzneim. Forsch. 19: 1313-1314(1969).

777. Schmahl, D. Fehlen einer kanzerogenen Wirkung von Cyclamat, Cyclohexylamin und Saccharin bei Ratten. Arzneim. Forsch. 23: 1466-1470(1973).

778. Schmahl, D. Investigations on the influence of immunodepressive means on the chemical carcinogenesis in rats. Cancer Res. Clin. Oncol. 81: 211-215(1974).

779. Schmahl, D. Investigations on esophageal carcinogenicity by methyl-phenyl-nitrosamine and ethyl alcohol in rats. Cancer Lett. 1: 215-218(1976).

780. Schmahl, D. Experiments on the carcinogenic effect of ortho-toluol-sulfonamid (OTS). Cancer Res. Clin. Oncol. 91: 19-22(1978).

781. Schmahl, D. and Habs, M. Life-span investigations for carcinogenicity of some immune-stimulating, immunodepressive and neurotropic substances in Sprague-Dawley-rats. Cancer Res. Clin. Oncol. 86: 77-84(1976).

782. Schmahl, D. and Habs, M. Experiments on the influence of an aromatic retinoid on the chemical carcinogenesis in rats by butyl-butanol-nitrosamine and 1,2-dimethylhydrazine. Arzneim. Forsch. 28: 49-51(1978).

783. Schmahl, D. and Habs, M. Carcinogenic action of low-dose cyclophosphamide given orally to Sprague-Dawley rats in a lifetime experiment. Int. J. Cancer 23: 706-712(1979).

784. Schmahl, D. and Osswald, H. Experimentelle Untersuchungen über carcinogene Wirkungen von Krebs-Chemotherapeutica und Immunosuppressiva. *Arzneim. Forsch.* 20: 1461-1467(1970).

785. Schmahl, D., Port, R., and Wahrendorf, J. A dose-response study on urethane carcinogenesis in rats and mice. *Int. J. Cancer* 19: 77-80(1977).

786. Schrauzer, G. N., McGinness, J. E., Ishmael, D., and Bell, L. J. Alcoholism and cancer: I. Effects of long-term exposure to alcohol on spontaneous mammary adenocarcinoma and prolactin levels in C3H/St mice. *J. Stud. Alcohol* 40: 240-246(1979).

787. Schrauzer, G. N., McGinness, J. E., and Kuehn, K. Effects of temporary selenium supplementation on the genesis of spontaneous mammary tumors in inbred female C3H/St mice. *Carcinogenesis* 1: 199-201(1980).

788. Schroeder, H. A., Balassa, J. J., and Vinton, W. H. Chromium, lead, cadmium, nickel and titanium in mice: effect on mortality, tumors and tissue levels. *J. Nutr.* 83: 239-250(1964).

789. Schroeder, H. A. and Mitchener, M. Life-term studies in rats: effects of aluminum, barium, beryllium, and tungsten. *J. Nutr.* 105: 421-427(1975).

790. Schroeder, H. A. and Mitchener, M. Life-term effects of mercury, methyl mercury, and nine other trace metals on mice. *J. Nutr.* 105: 452-458(1975).

791. Schroeder, H. A., Mitchener, M., and Nason, A. P. Life-term effects of nickel in rats: survival, tumors, interactions with trace elements and tissue levels. *J. Nutr.* 104: 239-243(1974).

792. Schwartz, E. L., Kluwe, W. K., Sleight, S. D., Hook, J. B., and Goodman, J. I. Inhibition of N-2-fluorenylacetamide-induced mammary tumorigenesis in rats by dietary polybrominated biphenyls. *J. Natl. Cancer Inst.* 64: 63-67(1980).

793. Schwetz, B. A., Quast, J. F., Keeler, P. A., Humiston, C. G., and Kociba, R. J. Results of two-year toxicity and reproduction studies on pentachlorophenol in rats. In: *Pentachlorophenol*, (K. Ranga Rao, Ed.), Plenum Press, New York, 1978, pp. 301-309.

794. Selan, F., McCullough, C. B., and Black, H. E. Preclinical safety evaluation of dilevalol (SCH 19927), an antihypertensive agent, in the rat. *Fundam. Appl. Toxicol.* 18: 471-476(1992).

795. Sellakumar, A. R., Laskin, S., Kuschner, M., Rusch, G., Katz, G. V., Snyder, C. A., and Albert, R. E. Inhalation carcinogenesis by dimethylcarbamoyl chloride in Syrian golden hamsters. *J. Environ. Pathol. Toxicol.* 4: 107-115(1980).

796. Sellakumar, A. R., Snyder, C. A., Solomon, J. J., and Albert, R. E. Carcinogenicity of formaldehyde and hydrogen chloride in rats. *Toxicol. Appl. Pharmacol.* 81: 401-406(1985).

797. Serota, D. G., Thakur, A. K., Ulland, B. M., Kirschman, J. C., Brown, N. M., Coots, R. H., and Morgareidge, K. A two-year drinking-water study of dichloromethane in rodents. I. Rats. *Food Chem. Toxicol.* 24: 951-958(1986).

798. Serota, D. G., Thakur, A. K., Ulland, B. M., Kirschman, J. C., Brown, N. M., Coots, R. H., and Morgareidge, K. A two-year drinking-water study of dichloromethane in rodents. II. Mice. *Food Chem. Toxicol.* 24: 959-963(1986).

799. Severi, L. and Biancifiori, C. Hepatic carcinogenesis in CBA/Cb/Se mice and Cb/Se rats by isonicotinic acid hydrazide and hydrazine sulfate. *J. Natl. Cancer Inst.* 41: 331-349(1968).

800. Shah, A. S., Sarode, A. V., and Bhide, S. V. Experimental studies on mutagenic and carcinogenic effects of tobacco chewing. *Cancer Res. Clin. Oncol.* 109: 203-207(1985).

801. Sharratt, M., Frazer, A. C., and Forbes, O. C. Study of the biological effects of benzoyl peroxide. *Food Cosmet. Toxicol.* 2: 527-538(1964).

802. Shay, H., Gruenstein, M., and Kessler, W. B. Experimental mammary adenocarcinoma of rats: some considerations of methylcholanthrene dosage and hormonal treatment. *J. Natl. Cancer Inst.* 27: 503-513(1961).

803. Shay, H., Gruenstein, M., and Kessler, W. B. Methylcholanthrene induced breast cancer in the rat: studies on mechanism of inhibition by large doses of estrogen. *Morphological Precursors of Cancer*, (L. Severi, Ed.), Div. Canc. Res., Perugia, 1962, pp. 305-318.

804. Shibata, M. A., Hirose, M., Tanaka, H., Asakawa, E., Shirai, T., and Ito, N. Induction of renal cell tumors in rats and mice, and enhancement of hepatocellular tumor development in mice after long-term hydroquinone treatment. *Jpn. J. Cancer Res.* 82: 1211-1219(1991).

805. Shimizu, H., Nagel, D., and Toth, B. Ethylhydrazine hydrochloride as a tumor inducer in mice. *Int. J. Cancer* 13: 500-505(1974).

806. Shimizu, H., Nagel, D., and Toth, B. Tumour induction study with N-amylhydrazine hydrochloride in Swiss mice. *Br. J. Cancer* 31: 492-496(1975).

807. Shimizu, H. and Toth, B. Effect of lifetime administration of 2-hydroxyethylhydrazine on tumorigenesis in hamsters and mice. *J. Natl. Cancer Inst.* 52: 903-906(1974).

808. Shirai, T., Hagiwara, A., Kurata, Y., Shibata, M., Fukushima, S., and Ito, N. Lack of carcinogenicity of butylated hydroxytoluene on long-term administration to B6C3F$_1$ mice. *Food Chem. Toxicol.* 20: 861-865(1982).

809. Shirai, T., Takahashi, M., Fukushima, S., and Ito, N. Marked epithelial hyperplasia of the rat glandular stomach induced by long-term administration of iodoacetamide. *Acta Pathol. Jpn.* 35: 35-43(1985).

810. Shivapurkar, N., Hoover, K. L., and Poirier, L. A. Effect of methionine and choline on liver tumor promotion by phenobarbital and DDT in diethylnitrosamine-initiated rats. *Carcinogenesis* 7: 547-550(1986).

811. Singer, S. S., Singer, G. M., Saavedra, J. E., Reuber, M. D., and Lijinsky, W. Carcinogenicity by derivatives of 1-nitroso-3,5-dimethylpiperazine in rats. *Cancer Res.* 41: 1034-1038(1981).

812. Skipper, H. E. Booklet 1, 1976. Phase I Studies on the Carcinogenic Activity of Anticancer Drugs in Mice and Rats. Final report. Southern Research Institute, Birmingham, AL, 1976.

813. Smith, A. G. and Cabral, J. R. Liver-cell tumours in rats fed hexachlorobenzene. *Cancer Lett.* 11: 169-172(1980).

814. Smith, A. G., Cabral, J. R. P., Carthew, P., Francis, J. E., and Manson, M. M. Carcinogenicity of iron in conjunction with a chlorinated environmental chemical, hexachlorobenzene, in C57BL/10ScSn mice. *Int. J. Cancer* 43: 492-496(1989).

815. Smith, A. G., Francis, J. E., Dinsdale, D., Manson, M. M., and Cabral, J. R. P. Hepatocarcinogenicity of hexachlorobenzene in rats and the sex difference in hepatic iron status and development of porphyria. *Carcinogenesis* 6: 631-636(1985).

816. Smits-van Prooije, A. E., de Groot, A. P., Dreef-van der Meulen, H. C., and Sinkeldam, E. J. Chronic toxicity and carcinogenicity study of isomalt in rats and mice. *Food Chem. Toxicol.* 28: 243-251(1990).

817. Snell, K. C. and Stewart, H. L. Pulmonary adenomatosis induced in DBA/2 mice by oral administration of dibenz[a,h]anthracene. *J. Natl. Cancer Inst.* 28: 1043-1051(1962).

818. Snellings, W. M., Weil, C. S., and Maronpot, R. R. A two-year inhalation study of the carcinogenic potential of ethylene oxide in Fischer 344 rats. *Toxicol. Appl. Pharmacol.* 75: 105-117(1984).

819. Snyder, C. A., Goldstein, B. D., Sellakumar, A. R., Bromberg, I., Laskin, S., and Albert, R. E. The inhalation toxicology of benzene: incidence of hematopoietic neoplasms and hematotoxicity in AKR/J and C57BL/6J mice. *Toxicol. Appl. Pharmacol.* 54: 323-331(1980).

820. Snyder, C. A., Goldstein, B. D., Sellakumar, A. R., Wolman, S. R., Bromberg, I., Erlichman, M. N., and Laskin, S. Hematotoxicity of inhaled benzene to Sprague-Dawley rats and AKR mice at 300 ppm. *J. Toxicol. Environ. Health* 4: 605-618(1978).

821. Sodemoto, Y. and Enomoto, M. Report of carcinogenesis bioassay of sodium benzoate in rats: absence of carcinogenicity of sodium benzoate in rats. *J. Environ. Pathol. Toxicol.* 4: 87-95(1980).

822. Soffritti, M., Maltoni, C., Maffei, F., and Biagi, R. Formaldehyde: An experimental multipotential carcinogen. *Toxicol. Ind. Health* 5: 699-730(1989).

823. Spencer, A., Barbolt, T., Henry, D., Eason, C., Sauerschell, R., and Bonner, F. Gastric morphological changes including carcinoid tumors in animals treated with a potent hypolipidemic agent, ciprofibrate. *Toxicol. Pathol.* 17: 7-15(1989).

824. Steinhoff, D., Mohr, U., and Schmidt, W. M. On the question of the carcinogenic action of hydrazine — evaluation on the basis of new experimental results. *Exp. Pathol.* 39: 1-9(1990).

825. Steinhoff, D., Weber, H., Mohr, U., and Boehme, K. Evaluation of amitrole (aminotriazole) for potential carcinogenicity in orally dosed rats, mice, and golden hamsters. *Toxicol. Appl. Pharmacol.* 69: 161-169(1983).

826. Stenback, F., Mori, H., Furuya, K., and Williams, G. M. Pathogenesis of dimethylnitrosamine-induced hepatocellular cancer in hamster liver and lack of enhancement by phenobarbital. *J. Natl. Cancer Inst.* 76: 327-333(1986).

827. Stenback, F., Weisburger, J. H., and Williams, G. M. Effect of lifetime administration of dimethylaminoethanol on longevity, aging changes, and cryptogenic neoplasms in C3H mice. *Mech. Ageing Dev.* 42: 129-138(1988).

828. Stevenson, D. E., Thorpe, E., Hunt, P. F., and Walker, A. I. T. The toxic effects of dieldrin in rats: a reevaluation of data obtained in a two-year feeding study. *Toxicol. Appl. Pharmacol.* 36: 247-254(1976).

829. Stitzel, K. A., McConnell, R. F., and Dierckman, T. A. Effects of nitrofurantoin on the primary and secondary reproductive organs of female B6C3F$_1$ mice. *Toxicol. Pathol.* 17: 774-781(1989).

830. Stoewsand, G. S., Anderson, J. L., Boyd, J. N., Hrzadina, G., Babish, J. G., Walsh, K. M., and Losco, P. Quercetin: a mutagen, not a carcinogen, in Fischer rats. *J. Toxicol. Environ. Health* 14: 105-114(1984).

831. Stoll, R. and Maraud, R. Sur l'induction de tumeurs thyroïdiennes chez le rat traité par le propylthiouracil et le radio-iode. *Bull. Cancer* 50: 389-398(1963).

832. Stott, W. T., Johnson, K. A., Landry, T. D., Gorzinski, S. J., and Cieszlak, F. S. Chronic toxicity and oncogenicity of picloram in Fischer 344 rats. *J. Toxicol. Environ. Health* 30: 91-104(1990).

833. Stula, E. F., Barnes, J. R., Sherman, H., Reinhardt, C. F., and Zapp, J. A. Liver and urinary bladder tumors in dogs from 3,3'-dichlorobenzidine. *J. Environ. Pathol. Toxicol.* 1: 475-490(1978).

834. Stula, E. F., Barnes, J. R., Sherman, H., Reinhardt, C. F., and Zapp, J. A. Urinary bladder tumors in dogs from 4,4'-methylene-bis(2-chloroaniline) (MOCA). *J. Environ. Pathol. Toxicol.* 1: 31-50(1977).

835. Stula, E. F., Sherman, H., Zapp, J. A., and Clayton, J. W. Experimental neoplasia in rats from oral administration of 3,3'-dichlorobenzidine, 4,4'-methylene-bis(2-chloroaniline), and 4,4'-methylene-bis(2-methylaniline). *Toxicol. Appl. Pharmacol.* 31: 159-176(1975).

836. Styles, J., Elliott, B. M., Lefevre, P. A., Robinson, M., Pritchard, N., Hart, D., and Ashby, J. Irreversible depression in the ratio of tetraploid:diploid liver nuclei in rats treated with 3'-methyl-4-dimethylaminoazobenzene (3'M). *Carcinogenesis* 6: 21-28(1985).

837. Sugimura, T., Fujimura, S., and Baba, T. Tumor production in the glandular stomach and alimentary tract of the rat by N-methyl-N'-nitro-N-nitrosoguanidine. *Cancer Res.* 30: 455-465(1970).

838. Svoboda, D. J. and Reddy, J. K. Malignant tumors in rats given lasiocarpine. *Cancer Res.* 32: 908-911(1972).

839. Swaminathan, S., Ertürk, E., and Bryan, G. T. Mutagenicity, carcinogenicity, distribution, and nitroreduction of 4-(5-nitro-2-furyl)thiazole in the rat. *Cancer Res.* 41: 2648-2653(1981).

840. Tacchi, A. M., Schmahl, D., and Habs, M. Delay of bladder cancer induction in rats treated with N-nitroso-N-butyl-N-(4-hydroxybutyl)amine by administration of sodium-2-mercaptoethanesulfonate (Mesna). *Cancer Lett.* 22: 89-94(1984).

841. Tahara, E., Ito, H., Nakagami, K., and Shimamoto, F. Induction of carcinoids in the glandular stomach of rats by N-methyl-N'-nitro-N-nitrosoguanidine. *Cancer Res. Clin. Oncol.* 100: 1-12(1981).

842. Takada, K., Naito, K., Kobayashi, K., Tobe, M., Kurokawa, Y., and Fukuoka, M. Carcinogenic effects of bis(2,3-dibromopropyl)phosphate in Wistar rats. *J. Appl. Toxicol.* 11: 323-331(1991).

843. Takahashi, M., Toyoda, K., Aze, Y., Furuta, K., Mitsumori, K., and Hayashi, Y. The rat urinary bladder as a new target of heterocyclic amine carcinogenicity: Tumor induction by 3-amino-1-methyl-5H-pyrido[4,3-b]indole acetate. *Jpn. J. Cancer Res.* 84: 852-858(1993).

844. Takanashi, H., Aiso, S., Hirono, I., Matsushima, T., and Sugimura, T. Carcinogenicity test of quercetin and kaempferol in rats by oral administration. *J. Food Safety* 5: 55-60(1983).

845. Takayama, S. Induction of tumors in ICR mice with N-nitrosopiperidine, especially in forestomach. *Naturwissenschaften* 56: 142(1969).

846. Takayama, S. and Imaizumi, T. Carcinogenic action of N-nitroso-dibutylamine in mice. *Gann* 60: 353(1969).

847. Takayama, S. and Kuwabara, N. Long-term study on the effect of caffeine in Wistar rats. *Gann* 73: 365-371(1982).

848. Takayama, S., Masuda, M., Mogami, M., Ohgaki, H., Sato, S., and Sugimura, T. Induction of cancers in the intestine, liver and various other organs of rats by feeding mutagens from glutamic acid pyrolysate. *Gann* 75: 207-213(1984).

849. Takayama, S., Nakatsuru, Y., Ohgaki, H., Sato, S., and Sugimura, T. Carcinogenicity in rats of a mutagenic compound, 3-amino-1,4-dimethyl-5H-pyrido[4,3-b]indole, from tryptophan pyrolysate. *Jpn. J. Cancer Res.* 76: 815-817(1985).

850. Takayama, S. and Kuwabara, N. Carcinogenic activity of 2-(2-furyl)-3-(5-nitro-2-furyl)acrylamide, a food additive, in mice and rats. *Cancer Lett.* 3: 115-120(1977).

851. Takemura, N., Hashida, C., and Terasawa, M. Carcinogenic action of 5-nitroacenaphthene. *Br. J. Cancer* 30: 481-483(1974).

852. Takenaka, S., Oldiges, H., Konig, H., Hochrainer, D., Oberdorster, G. Carcinogenicity of Cadmium Chloride Aerosols in W Rats. *J. Natl. Cancer Inst.* 70: 367-373 (1983).

853. Takeuchi, M., Ogiu, T., Matsuoka, C., Furuta, K., Maekawa, A., Nakadate, M., and Odashima, S. Induction of digestive-tract tumors in F344 rats by continuous oral administration of N-butyl-N-nitrosourea. *Cancer Res. Clin. Oncol.* 107: 32-37(1984).

854. Tamano, S., Hirose, M., Tanaka, H., Asakawa, E., Ogawa, K., and Ito, N. Forestomach neoplasm induction in F344/DuCrj rats and B6C3F$_1$ mice exposed to sesamol. *Jpn. J. Cancer Res.* 83: 1279-1285(1992).

855. Tamano, S., Kurata, Y., Kawabe, M., Yamamoto, A., Hagiwara, A., Cabral, R., and Ito, N. Carcinogenicity of captafol in F344/DuCrj rats. *Jpn. J. Cancer Res.* 81: 1222-1231(1990).

856. Tanaka, T., Barnes, W. S., Williams, G. M., and Weisburger, J. H. Multipotential carcinogenicity of the fried food mutagen 2-amino-3-methylimidazo[4,5-*f*]quinoline in rats. *Jpn. J. Cancer Res.* 76: 570-576(1985).

857. Tanaka, T., Fujii, M., Mori, H., and Hirono, I. Carcinogenicity test of potassium metabisulfite in mice. *Ecotoxicol. Environ. Safety* 3: 451-453(1979).

858. Tannenbaum, A. and Silverstone, H. Effect of low environmental temperature, dinitrophenol, or sodium fluoride on the formation of tumors in mice. *Cancer Res.* 9: 403-410(1949).

859. Tatematsu, M., Takahashi, M., Tsuda, H., Ogiso, T., and Ito, N. The administration of reserpine to rats for 75 weeks. *Toxicol. Lett.* 1: 201-205(1978).

860. Terao, K., Aikawa, T., and Kera, K. A synergistic effect of nitrosodimethylamine on sterigmatocystin carcinogenesis in rats. *Food Cosmet. Toxicol.* 16: 591-596(1978).

861. Terracini, B., Magee, P. N., and Barnes, J. M. Hepatic pathology in rats on low dietary levels of dimethylnitrosamine. *Br. J. Cancer* 20: 559-565(1967).

862. Terracini, B., Testa, M. C., Cabral, J. R., and Day, N. The effects of long-term feeding of DDT to BALB/c mice. *Int. J. Cancer* 11: 747-764(1973).

863. Thorgeirsson, U. P., Dalgard, D. W., Reeves, J., and Adamson, R. H. Tumor incidence in a chemical carcinogenesis study in non-human primates. *Regul. Toxicol. Pharmacol.* 19: 130-151(1994).

864. Thorpe, E. and Walker, A. I. T. The toxicology of dieldrin (HEOD). II. Comparative long-term oral toxicity studies in mice with dieldrin, DDT, phenobarbitone, β-BHC and γ-BHC. *Food Cosmet. Toxicol.* 11: 433-442(1973).

865. Til, H. P., Woutersen, R. A., Feron, V. J., Hollanders, V. H. M., Falke, H. E., and Clary, J. J. Two-year drinking-water study of formaldehyde in rats. *Food Chem. Toxicol.* 27: 77-87(1989).

866. Togei, K., Sano, N., Maeda, T., Shibata, M., and Otsuka, H. Carcinogenicity of bucetin in (C57BL/6 × C3H)F$_1$ mice. *J. Natl. Cancer Inst.* 79: 1151-1158(1987).

867. Tomatis, L., Turusov, V., Charles, R. T., Boiocchi, M., and Gati, E. Liver tumours in CF-1 mice exposed for limited periods to technical DDT. *Cancer Res. Clin. Oncol.* 82: 25-35(1974).

868. Tomatis, L., Turusov, V., Charles, R. T., and Boicchi, M. Effect of long-term exposure to 1,1-dichloro-2,2-bis(*p*-chlorophenyl)ethylene, to 1,1-dichloro-2,2-bis(*p*-chlorophenyl)ethane, and to the two chemicals combined on CF-1 mice. *J. Natl. Cancer Inst.* 52: 883-891(1974).

869. Tomatis, L., Turusov, V., Day, N., and Charles, R. T. The effect of long-term exposure to DDT on CF-1 mice. *Int. J. Cancer* 10: 489-506(1972).

870. Toth, B. Studies on the incidence, morphology, transplantation and cell-free filtration of malignant lymphomas in the Syrian golden hamster. *Cancer Res.* 27: 1430-1442(1967a).

871. Toth, B. Lung tumor induction and inhibition of breast adenocarcinomas by hydrazine sulfate in mice. *J. Natl. Cancer Inst.* 42: 469-475(1969a).

872. Toth, B. Benzoylhydrazine carcinogenesis in lungs and lymphoreticular tissues of Swiss mice. *Eur. J. Cancer* 8: 341-345(1972).

873. Toth, B. Hydrazine, methylhydrazine and methylhydrazine sulfate carcinogenesis in Swiss mice. Failure of ammonium hydroxide to interfere in the development of tumors. *Int. J. Cancer* 9: 109-118(1972).

874. Toth, B. Tumorigenesis studies with 1,2-dimethylhydrazine dihydrochloride, hydrazine sulfate, and isonicotinic acid in Golden hamsters. *Cancer Res.* 32: 804-807(1972).

875. Toth, B. 1,1-Dimethylhydrazine (unsymmetrical) carcinogenesis in mice. Light microscopic and ultrastructural studies on neoplastic blood vessels. *J. Natl. Cancer Inst.* 50: 181-194(1973).

876. Toth, B. Tumorigenicity of β-phenylethylhydrazine sulfate in mice. *Cancer Res.* 36: 917-921(1976).

877. Toth, B. The large bowel carcinogenic effects of hydrazines and related compounds occurring in nature and in the environment. *Cancer* 40: 2427-2431(1977).

878. Toth, B. Formylhydrazine carcinogenesis in mice. *Br. J. Cancer* 37: 960-964(1978).

879. Toth, B. 1-Acetyl-2-phenylhydrazine carcinogenesis in mice. *Br. J. Cancer* 39: 584-587(1979).

880. Toth, B. Nicotinic acid hydrazide carcinogenesis in mice. *Oncology* 38: 106-109(1981).

881. Toth, B. Effects of lifelong administration of β-phenylisopropylhydrazine hydrochloride and thiocarbamylhydrazine in mice. *Fundam. Appl. Toxicol.* 2: 173-176(1982).

882. Toth, B. Effects of long term administration of nicotine hydrochloride and nicotinic acid in mice. *Anticancer Res.* 2: 71-74(1982).

883. Toth, B. Lack of carcinogenicity of nicotinamide and isonicotinamide following lifelong administration to mice. *Oncology* 40: 72-75(1983).

884. Toth, B. Lack of tumorigenicity of sodium benzoate in mice. *Fundam. Appl. Toxicol.* 4: 494-496(1984).

885. Toth, B. Carciogenesis by N2-[γ-L-(+)-glutamyl]-4-carboxy-phenylhydrazine of *Agaricus bisporus* in mice. *Anticancer Res.* 6: 917-920(1986).

886. Toth, B. and Boreisha, I. Tumorigenesis with isonicotinic acid hydrazide and urethan in the Syrian golden hamsters. *Eur. J. Cancer* 5: 165-171(1969).

887. Toth, B. and Erickson, J. Lung tumorigenesis by 1,2-diformylhydrazine in mice. *Cancer Res. Clin. Oncol.* 92: 11-16(1978).

888. Toth, B. and Gannett, P. Carcinogenesis study in mice by 3-methylbutanal methylformylhydrazone of *Gyromitra esculenta*. *In Vivo* 4: 283-288(1990).

889. Toth, B. and Gannett, P. Carcinogenicity of lifelong administration of capsaicin of hot pepper in mice. *In Vivo* 6: 59-64(1992).

890. Toth, B. and Nagel, D. Tumour induction study with allylhydrazine HCl in Swiss mice. *Br. J. Cancer* 34: 90-93(1976).

891. Toth, B. and Nagel, D. Tumors induced in mice by N-methyl-N-formylhydrazine of the false morel *Gyromitra esculenta*. *J. Natl. Cancer Inst.* 60: 201-204(1978).

892. Toth, B. and Nagel, D. N-Ethyl-N-formylhydrazine tumorigenesis in mice. *Carcinogenesis* 1: 61-65(1980).

893. Toth, B. and Nagel, D. 1,2-Di-N-butylhydrazine dihydrochloride carcinogenesis in mice. *Experientia* 37: 773-775(1981).

894. Toth, B. and Nagel, D. 1,2-Diallylhydrazine dihydrochloride carcinogenesis in mice. *Oncology* 39: 104-108(1982).

895. Toth, B., Nagel, D., and Patil, K. Tumorigenic action of N-N-butyl-N-formylhydrazine in mice. *Carcinogenesis* 1: 589-593(1980).

896. Toth, B., Nagel, D., and Patil, K. Tumorigenesis by N-N-propyl-N-formylhydrazine in mice. *Br. J. Cancer* 42: 922-928(1980).

897. Toth, B., Nagel, D., and Patil, K. Carcinogenic effects of 1,1-di-N-butylhydrazine in mice. *Carcinogenesis* 2: 651-654(1981).

898. Toth, B., Nagel, D., Patil, K., Erickson, J., and Antonson, K. Tumor induction with the *N'*-acetyl derivative of 4-hydroxymethylphenylhydrazine, a metabolite of agaritine of *Agaricus bisporus*. *Cancer Res.* 38: 177-180(1978).

899. Toth, B., Nagel, D., and Raha, C. Tumorigenesis with 1,1-diallylhydrazine in mice. *Anticancer Res.* 1: 259-262(1981).

900. Toth, B. and Patil, K. Carcinogenic effects in the Syrian golden hamster of *N*-methyl-*N*-formylhydrazine of the false morel mushroom *Gyromitra esculenta*. *Cancer Res. Clin. Oncol.* 93: 109-121(1979).

901. Toth, B. and Patil, K. The tumorigenic effect of low dose levels of *N*-methyl-*N*-formylhydrazine in mice. *Neoplasma* 27: 25-31(1980).

902. Toth, B. and Patil, K. Tumorigenicity of minute dose levels of *N*-methyl-*N*-formylhydrazine of *Gyromitra esculenta*. *Mycopathologia* 78: 11-16(1982a).

903. Toth, B., Patil, K., Erickson, J., and Kupper, R. False morel mushroom *Gyromitra esculenta* toxin: *N*-methyl-*N*-formylhydrazine carcinogenesis in mice. *Mycopathologia* 68: 121-128(1979).

904. Toth, B. and Raha, C. R. Carcinogenesis by pentanal methylformylhydrazone of *Gyromitra esculenta* in mice. *Mycopathologia* 98: 83-89(1987).

905. Toth, B., Raha, C. R., Wallcave, L., and Nagel, D. Attempted tumor induction with agaritine in mice. *Anticancer Res.* 1: 255-258(1981).

906. Toth, B. and Rustja, M. The effect of isonicotinic acid hydrazide on the development of tumors. *Int. J. Cancer* 2: 413-420(1967).

907. Toth, B. and Shimizu, H. Lung carcinogenesis with 1-acetyl-2-isonicotinoylhydrazine, the major metabolite of isoniazid. *Eur. J. Cancer* 9: 285-289(1973).

908. Toth, B. and Shimizu, H. Methylhydrazine tumorigenesis in Syrian golden hamsters and the morphology of malignant histiocytomas. *Cancer Res.* 33: 2744-2753(1973).

909. Toth, B. and Shimizu, H. 1-Carbamyl-2-phenylhydrazine tumorigenesis in Swiss mice. Morphology of lung adenomas. *J. Natl. Cancer Inst.* 52: 241-251(1974).

910. Toth, B. and Shimizu, H. Tumorigenic effects of chronic administration of benzylhydrazine dihydrochloride and phenylhydrazine hydrochloride in Swiss mice. *Cancer Res. Clin. Oncol.* 87: 267-273(1976).

911. Toth, B., Shimizu, H., and Erickson, J. Carbamylhydrazine hydrochloride as a lung and blood vessel tumour inducer in Swiss mice. *Eur. J. Cancer* 11: 17-22(1975).

912. Toth, B. and Shubik, P. Carcinogenesis in Swiss mice by isonicotinic acid hydrazide. *Cancer Res.* 26: 1473-1475(1966).

913. Toth, B. and Shubik, P. Mammary tumor inhibition and lung adenoma induction by isonicotinic acid hydrazide. *Science* 152: 1376-1377(1966a).

914. Toth, B., Smith, J. W., and Patil, K. D. Cancer induction in mice with acetaldehyde methylformylhydrazone of the false morel mushroom. *J. Natl. Cancer Inst.* 67: 881-887(1981).

915. Toth, K., Somfai-Relle, S., Sugar, J., and Bence, J. Carcinogenicity testing of herbicide 2,4,5-trichlorophenoxyethanol containing dioxin and of pure dioxin in Swiss mice. *Nature* 278: 548-549(1979).

916. Toth, B., Taylor, J., and Gannett, P. Tumor induction with hexanal methylformylhydrazone of *Gyromitra esculenta*. *Mycopathologia* 115: 65-71(1991).

917. Toth, B., Wallcave, L., Patil, K., Schmeltz, I., and Hoffmann, D. Induction of tumors in mice with the herbicide succinic acid 2,2-dimethylhydrazide. *Cancer Res.* 37: 3497-3500(1977).

918. Toth, B. and Wilson, R. B. Blood vessel tumorigenesis by 1,2-dimethylhydrazine dihydrochloride (symmetrical). *Am. J. Pathol.* 64: 585-600(1971).

919. Tourkevitch, N. M., Gorevaia, A. N., Kounitsa, L. K., and Mao, L. S. Développement des tumeurs des glandes mammaires chez les rats en cas de trouble de la regulation endocrine provoque par differentes actions. *Int. J. Cancer* 20: 1446-1449(1964).

920. Trochimowicz, H. J., Rusch, G. M., Chiu, T., and Wood, C. K. Chronic inhalation toxicity/carcinogenicity study in rats exposed to Fluorocarbon 113 (FC-113). *Fundam. Appl. Toxicol.* 11: 68-75(1988).

921. Truhaut, R., Coquet, B., Fouillet, X., Galland, D., Guyot, D., Long, D., and Rouaud, J. L. Two-year oral toxicity and multigeneration studies in rats on two chemically modified maize starches. *Food Cosmet. Toxicol.* 17: 11-17(1979).

922. Truhaut, R., Ferrando, R., Faccini, J. M., and Monro, A. M. Negative results of carcinogenicity bioassay of methyl carbazate in rats: significance for the toxicological evaluation of carbadox. *Toxicology* 22: 219-221(1981).

923. Truhaut, R., Le Bourhis, B., Attia, M., Glomot, R., Newman, J., and Caldwell, J. Chronic toxicity/carcinogenicity study of trans-anethole in rats. *Food Chem. Toxicol.* 27: 11-19(1989).

924. Tsai, T. H., Beitman, R. E., Gibson, J. P., Larson, E. J., Friehe, H., and Fontaine, R. Acute, subacute and chronic toxicity/carcinogenicity of lofexidine. *Arzneim. Forsch.* 31: 955-962(1982).

925. Tsuda, H., Hagiwara, A., Shibata, M., Ohshima, M., and Ito, N. Carcinogenic effect of carbazole in the liver of (C57BL/6N × C3H/HeN)F$_1$ mice. *J. Natl. Cancer Inst.* 69: 1383-1387(1982).

926. Tsuda, H., Hananouchi, M., Tatematsu, M., Hirose, M., Hirao, K., Takahashi, M., and Ito, N. Tumorigenic effect of 3-amino-1*H*-1,2,4-triazole on rat thyroid. *J. Natl. Cancer Inst.* 57: 861-864(1976).

927. Tsung-Hsien, C., Yu-Chung, L., Kwang-Yung, L., Cheng-Hsien, S., and Yee-Ping, C. Cocarcinogenic action of aspirin on gastric tumors induced by nitroso-*N*-methylnitroguanidine in rats. *J. Natl. Cancer Inst.* 70: 1067-1069(1983).

928. Tuchmann-Duplessis, H., and Mercier-Parot, L. Cancerologie — apparition de tumeurs malignes dans une lignée de rats "Wistar." *Acad. Sci.* Mémoires *Commun.* Membres Correspondants *Acad.* 254: 1535-1537(1962).

929. Tumasonis, C. F., McMartin, D. N., and Bush, B. Lifetime toxicity of chloroform and bromodichloromethane when administered over a lifetime in rats. *Ecotoxicol. Environ. Safety* 9: 233-240(1985).

930. Uchida, E. and Hirono, I. Effect of phenobarbital on the development of neoplastic lesions in the liver of cycasin-treated rats. *Cancer Res. Clin. Oncol.* 100: 231-238(1981).

931. Uematsu, K. and Miyaji, T. Induction of tumors in rats by oral administration of technical acid violet 6B. *J. Natl. Cancer Inst.* 51: 1337-1338(1973).

932. Ueno, I. and Hirono, I. Non-carcinogenic response to coumarin in Syrian golden hamsters. *Food Cosmet. Toxicol.* 19: 353-355(1981).

933. Ulland, B. M., Weisburger, J. H., Weisburger, E. K., Rice, J. M., and Cypher, R. Brief communication: thyroid cancer in rats from ethylene thiourea intake. *J. Natl. Cancer Inst.* 49: 583-584(1972).

934. Ungerer, O., Eisenbrand, G., and Preussmann, R. Zur Reaktion von Nitrit mit Pestiziden. Bildung, chemische Eigenschaften und cancerogene Wirkung der N-Nitrosoverbindung des Herbizids N-Methyl-N'-(2-benzothiazolyl-harnstoff (Benthiazuron). *Cancer Res. Clin. Oncol.* 81: 217-224(1974).

935. Uraguchi, K., Saito, M., Noguchi, Y., Takahashi, K., Enomoto, M., and Tatsuno, T. Chronic toxicity and carcinogenicity in mice of the purified mycotoxins, luteoskyrin and cyclochlorotine. *Food Cosmet. Toxicol.* 10: 193-207(1972).

936. Van Duuren, B. L., Goldschmidt, B. M., Katz, C., Seidman, I., and Paul, J. S. Carcinogenic activity of alkylating agents. *J. Natl. Cancer Inst.* 53: 695-700(1974).

937. Van Duuren, B. L., Goldschmidt, B. M., Loewengart, G., Smith, A. C., Melchionne, S., Seidman, I., and Roth, D. Carcinogenicity of halogenated olefinic and aliphatic hydrocarbons in mice. *J. Natl. Cancer Inst.* 63: 1433-1439(1979).

938. Van Duuren, B. L., Goldschmidt, B. M., and Seidman, I. Carcinogenic activity of di- and trifunctional α-chloro ethers and of 1,4-dichlorobutene-2 in ICR/Ha swiss mice. *Cancer Res.* 35: 2553-2557(1975).

939. Van Duuren, B. L., Langseth, L., Orris, L., Teebor, G., Nelson, N., and Kuschner, M. Carcinogenicity of epoxides, lactones, and peroxy compounds. IV. Tumor response in epithelial and connective tissue in mice and rats. *J. Natl. Cancer Inst.* 37: 825-838(1966).

940. Van Duuren, B. L., Melchionne, S., Seidman, I., and Pereira, M. A. Chronic bioassays of chlorinated humic acids in B6C3F₁ mice. *Environ. Health Perspect.* 69: 109-117(1986).

941. Van Duuren, B. L., Seidman, I., Melchionne, S., and Kline, S. A. Carcinogenicity bioassays of bromoacetaldehyde and bromoethanol — potential metabolites of dibromoethane. *Teratog. Carcinog. Mutagen.* 5: 393-403(1985).

942. Van Esch, G. J. and Kroes, R. The induction of renal tumours by feeding basic lead acetate to mice and hamsters. *Br. J. Cancer* 23: 765-771(1969).

943. Van Esch, G. J. and Kroes, R. Long-term toxicity studies of chlorpropham and propham in mice and hamsters. *Food Cosmet. Toxicol.* 10: 373-381(1972).

944. Van Esch, G. J., Van Genderen, H., and Vink, H. H. The induction of renal tumours by feeding of basic lead acetate to rats. *Br. J. Cancer* 16: 289-297(1962).

945. Van Miller, J. P., Lalich, J. J., and Allen, J. R. Increased incidence of neoplasms in rats exposed to low levels of 2,3,7,8-tetrachlorobenzo-p-dioxin. *Chemosphere* 10: 625-632(1977).

946. Van Ryzin, R. J. and Trapold, J. H. The toxicology profile of the anti-inflammatory drug proquazone in animals. *Drug Chem. Toxicol.* 3: 361-379(1980).

947. Vazquez-Lopez, E. The effects of thiourea on the development of spontaneous tumours on mice. *Br. J. Cancer* 3: 401-414(1949).

948. Vernot, E. H., MacEwen, J. D., Bruner, R. H., Haun, C. C., Kinkead, E. R., Prentice, D. E., Hall, A., III, Schmidt, R. E., Eason, R. L., Hubbard, G. B., and Young, J. T. Long-term inhalation toxicity of hydrazine. *Fundam. Appl. Toxicol.* 5: 1050-1064(1985).

949. Verschuuren, H. G., Kroes, R., and Van Esch, G. J. Toxicity studies on tetrasul. I. Acute, long-term and reproduction studies. *Toxicology* 1: 63-78(1973).

950. Vesselinovitch, S. D., Rao, K. V. N., and Mihailovich, N. Transplacental and lactational carcinogenesis by safrole. *Cancer Res.* 39: 4378-4380(1979).

951. Waalkes, M. P. and Rehm, S. Carcinogenicity of oral cadmium in the male Wistar (WF/NCr) rat: Effect of chronic dietary zinc deficiency. *Fundam. Appl. Toxicol.* 19: 512-520(1992).

952. Wada, S., Hirose, M., Takahashi, S., Okazaki, S., and Ito, N. Paramethoxyphenol strongly stimulates cell proliferation in the rat forestomach but is not a promoter of rat forestomach carcinogenesis. *Carcinogenesis* 11: 1891-1894(1990).

953. Wakabayashi, K., Inagaki, T., Fujimoto, Y., and Fukuda, Y. Induction by degraded carrageenan of colorectal tumors in rats. *Cancer Lett.* 4: 171-176(1978).

954. Wakabayashi, K., Nagao, M., Esumi, H., and Sugimura, T. Food-derived mutagens and carcinogens. *Cancer Res.* 52: 2092s-2098s(1992).

955. Walker, A. I. T., Thorpe, E., and Stevenson, D. E. The toxicology of dieldrin (HEOD). I. Long-term oral toxicity studies in mice. *Food Cosmet. Toxicol.* 11: 415-432(1973).

956. Walker, R. F., Weideman, C. A., and Wheeldon, E. B. Reduced disease in aged rats treated chronically with ibopamine, a catecholaminergic drug. *Neurobiol. Aging* 9: 291-301(1988).

957. Wang, C. Y., Croft, W. A., and Bryan, G. T. Tumor production in germ-free rats fed 5-nitrofurans. *Cancer Lett.* 21: 303-308(1984).

958. Wang, C. Y., Kamiryo, Y., and Croft, W. A. Carcinogenicity of 2-amino-4-(5-nitro-2-furyl)thiazole in rats by oral and subcutaneous administration. *Carcinogenesis* 3: 275-277(1982).

959. Ward, J. M., Sontag, J. M., Weisburger, E. K., and Brown, C. A. Effect of lifetime exposure to aflatoxin B₁ in rats. *J. Natl. Cancer Inst.* 55: 107-110(1975).

960. Watanabe, H. K., Hashimoto, Y., Abe, I., and Sato, H. Carcinogenicities of 3-methoxy-4-aminoazobenzene, N-hydroxy-3-methoxy-4-aminoazobenzene and related azo dyes in the mouse. *Gann* 73: 136-140(1982).

961. Watanabe, H., Okamoto, T., Takahashi, T., Ogundigie, P. O., and Ito, A. The effects of sodium chloride, miso or ethanol on development of intestinal metaplasia after X-irradiation of the rat glandular stomach. *Jpn. J. Cancer Res.* 83: 1267-1272(1992).

962. Watanabe, H., Takahashi, T., Okamoto, T., Ogundigie, P. O., and Ito, A. Effects of sodium chloride and ethanol on stomach tumorigenesis in ACI rats treated with N-methyl-N'-nitro-N-nitrosoguanidine: A quantitative morphometric approach. *Jpn. J. Cancer Res.* 83: 588-593(1992).

963. Waters, L. L. o-Aminoazotoluene as a carcinogenic agent. *Yale J. Biol. Med.* 10: 179-184(1937).

964. Weikel Jr., J. H. and Kelly, W. A. Tumorigenicity assays of sotalol hydrochloride in rats and mice. *J. Clin. Pharmacol.* 19: 591-604(1979).

965. Weisburger, E. K. Bioassay program for carcinogenic hazards of cancer chemotherapeutic agents. *Cancer* 40: 1935-1949(1977).

966. Weisburger, E. K. Carcinogenicity studies on halogenated hydrocarbons. *Environ. Health Perspect.* 21: 7-16(1977).

967. Weisburger, E. K., Russfield, A. B., Homburger, F., Weisburger, J. H., Boger, E., Van Dongen, C. G., and Chu, K. Testing of twenty-one aromatic amines or derivatives for long-term toxicity or carcinogenicity. *J. Environ. Pathol. Toxicol.* 2: 325-356(1978).

968. Weisburger, E. K., Ulland, B. M., Nam, J., Gart, J. J., and Weisburger, J. H. Carcinogenicity tests of certain environmental and industrial chemicals. *J. Natl. Cancer Inst.* 67: 75-88(1981).

969. Weisburger, J. H., Yamamoto, R. S., Glass, R. M., and Frankel, H. H. Prevention by arginine glutamate of the carcinogenicity of acetamide in rats. *Toxicol. Appl. Pharmacol.* 14: 163-175(1969).

970. Wester, P. W., Krajnc, E. I., van Leeuwen, F. X. R., Loeber, J. G., van der Heijden, C. A., Vaessen, H. A. M. G., and Helleman, P. W. Chronic toxicity and carcinogenicity of bis(tri-N-butyl-tin)oxide (TBTO) in the rat. *Food Chem. Toxicol.* 28: 179-196(1990).

971. Wester, P. W., van der Heijden, C. A., Bisschop, A., and van Esch, G. J. Carcinogenicity study with epichlorohydrin (CEP) by gavage in rats. *Toxicology* 36: 325-339(1985).

972. Wheldon, G. H., Bhatt, A., Keller, P., and Hummler, H. D,l-α-tocopheryl acetate (vitamin E): a long term toxicity and carcinogenicity study in rats. *Int. J. Vitam. Nutr. Res.* 53: 287-296(1983).

973. Wiessler, M. and Schmahl, D. Zur carcinogenen Wirkung von N-Nitroso-Verbindungen. 1. Mitteilung: N-Nitroso-3,6-dihydrooxazin-1,2 und N-Nitroso-tetrahydrooxazin-1,2. *Cancer Res. Clin. Oncol.* 79: 114-117(1973).

974. Wiessler, M. and Schmahl, D. Zur carcinogenen Wirkung von N-Nitroso-Verbindungen. 2. Mitteilung: S(+) und R(-)-N-Nitroso-2-methyl-piperidin. *Cancer Res. Clin. Oncol.* 79: 118-122(1973).

975. Willheim, R., and Ivy, A. C. A preliminary study concerning the possibility of dietary carcinogenesis. *Gastroenterology* 23: 1-19(1953).

976. Williams, G. M., Wang, C. X., and Iatropoulos, M. J. Toxicity studies of butylated hydroxyanisole and butylated hydroxytoluene. II. Chronic feeding studies. *Food Chem. Toxicol.* 28: 799-806(1990).

977. Williams, M. H. C., and Bonser, G. M. Induction of hepatomas in rats and mice following the administration of auramine. *Br. J. Cancer* 16: 87-91(1962).

978. Willis, J. The induction of malignant neoplasms in the thyroid gland of the rat. *J. Pathol.* 82: 23-27(1961).

979. Wislocki, P. G., Miller, E. C., Miller, J. A., McCoy, E. C., and Rosenkranz, H. S. Carcinogenic and mutagenic activities of safrole, 1′-hydroxysafrole, and some known or possible metabolites. *Cancer Res.* 37: 1883-1891(1977).

980. Wogan, G. N., Edwards, G. S., and Newberne, P. M. Structure-activity relationships in toxicity and carcinogenicity of aflatoxins and analogs. *Cancer Res.* 31: 1936-1942(1971).

981. Wogan, G. N., Paglialunga, S., and Newberne, P. M. Carcinogenic effects of low dietary levels of aflatoxin B₁ in rats. *Food Cosmet. Toxicol.* 12: 681-685(1974).

982. Wolff, G. L., Gaylor, D. W., Blackwell, B.-N., and Moore, G. E. Bladder and liver tumorigenesis induced by 2-acetylaminofluorene in different F₁ mouse hybrids: Variation within genotypes and effects of using more than one genotype on risk assessment. *J. Toxicol. Environ. Health* 33: 327-348(1991).

983. Wolff, G. L., Roberts, D. W., Morrissey, R. L., Greenman, D. L., Allen, R. R., Campbell, W. L., Bergman, H., Nesnow, S., and Frith, C. H. Tumorigenic responses to lindane in mice: Potentiation by a dominant mutation. *Carcinogenesis* 8: 1889-1897(1987).

984. Wong, L. C. K., Winston, J. M., Hong, C. B., and Plotnick, H. Carcinogenicity and toxicity of 1,2-dibromoethane in the rat. *Toxicol. Appl. Pharmacol.* 63: 155-165(1982).

985. Wood, M. Factors influencing the induction of tumours of the urinary bladder and liver by 2-acetylaminofluorene in the mouse. *Eur. J. Cancer* 5: 41-47(1969).

986. Wood, M., Flaks, A., and Clayson, D. B. The carcinogenic activity of dibutylnitrosamine in IF × C57 mice. *Eur. J. Cancer* 6: 433-440(1970).

987. Woutersen, R. A., Appelman, L. M., Van Gardenen-Hoetmer, A., and Feron, V. J. Inhalation toxicity of acetaldehyde in rats. III. Carcinogenicity study. *Toxicology* 41: 213-231(1986).

988. Woutersen, R. A. and Feron, V. J. Inhalation toxicity of acetaldehyde in rats. IV progression and regression of nasal lesions after discontinuation of exposure. *Toxicology* 47: 295-305(1987).

989. Woutersen, R. A., van Garderen-Hoetmer, A., Bruijntjes, J. P., Zwart, A., and Feron, V. J. Nasal tumours in rats after severe injury to the nasal mucosa and prolonged exposure to 10 ppm formaldehyde. *J. Appl. Toxicol.* 9: 39-46(1989).

990. Wright, J. A., Marsden, A. M., Willets, J. M., and Orton, T. C. Hepatocarcinogenic effect of vinyl carbamate in the C57B1/10J strain mouse. *Toxicol. Pathol.* 19: 258-265(1991).

991. Yamagiwa, K., Higashi, S., and Mizumoto, R. Effect of alcohol ingestion on carcinogenesis by synthetic estrogen and progestin in the rat liver. *Jpn. J. Cancer Res.* 82: 771-778(1991).

992. Yamamoto, K., Eimoto, H., Takashima, Y., Tsutsumi, M., Maruyama, H., Denda, A., Mori, Y., and Konishi, Y. Initiation activity of endogenously synthesized N-nitrosobis(2-hydroxypropyl)amine in the rat liver. *Cancer Lett.* 45: 221-225(1989).

993. Yamamoto, R. S. and Korzis. J. H. Weisburger, J. Chronic ethanol ingestion and the hepatocarcinogenicity of N-hydroxy-N-2-fluorenylacetamide. *Int. J. Cancer* 2: 337-343(1967).

994. Yamamoto, K., Nakajima, A., Eimoto, H., Tsutsumi, M., Maruyama, H., Denda, A., Nii, H., Mori, Y., and Konishi, Y. Carcinogenic activity of endogenously synthesized N-nitroso-bis(2-hydroxypropyl)amine in rats administered bis(2-hydroxypropyl)amine and sodium nitrite. *Carcinogenesis* 10: 1607-1611(1989).

995. Yamamoto, R. S., Richardson, H. L., Weisburger, E. K., Weisburger, J. H., Benjamin, T., and Bahner, C. T. Carcinogenicity of proposed cancer chemotherapeutic agents with stilbenearylnitrosamine structures. *J. Natl. Cancer Inst.* 51: 1313-1315(1973).

996. Yamamoto, R. S., Williams, G. M., Frankel, H. H., and Weisburger, J. H. 8-Hydroxyquinoline: chronic toxicity and inhibitory effect on the carcinogenicity of N-2-fluorenylacetamide. *Toxicol. Appl. Pharmacol.* 19: 687-698(1971).

997. Yamauchi, O., Omori, M., Ninomiya, M., Okuno, M., Moriwaki, H., Suganuma, M., Fujiki, H., and Muto, Y. Inhibitory effect of sarcophytol A on development of spontaneous hepatomas in mice. *Jpn. J. Cancer Res.* 82: 1234-1238(1991).

998. Yasui, W., and Tahara, E. Effect of gastrin on gastric mucosal cyclic adenosine 3′:5′-monophosphate-dependent protein kinase activity in rat stomach carcinogenesis induced by N-methyl-N′-nitro-N-nitrosoguanidine. *Cancer Res.* 45: 4763-4767(1985).

999. Yokoro, K., Kajihara, H., Kodama, Y., Nagao, K., Hamada, K., and Kinomura, A. Chronic toxicity of 2-(2-furyl)-3-(5-nitro-2-furyl)acrylamide (AF-2) in mice with special reference to carcinogenicity in the forestomach. *Gann* 68: 825-828(1977).

1000. Yoshida, M., Numoto, S., and Otsuka, H. Histopathological changes induced in the urinary bladder and liver of female BALB/c mice treated simultaneously with 2-naphthylamine and cyclophosphamide. *Gann* 70: 645-652(1979).

1001. Yoshimoto, S. Carcinogenicity and mutagenicity of tetrachlorobenzidine. *Jikeikai Med. J.* 25: 123-128(1978).

1002. Zabezhinskii, M. A. Effectiveness of inhalation as a method of administration of atomized carcinogens. *Byull. Eksp. Biol. Med.* 69: 72-74(1970).

CHAPTER 2

SUMMARY OF THE CARCINOGENIC POTENCY DATABASE BY TARGET ORGAN

Lois Swirsky Gold, Neela B. Manley, and Thomas H. Slone

A compendium of carcinogenesis bioassay results organized by target organ is presented for 684 chemicals that are carcinogenic in at least one species. Results are organized alphabetically by target site and within each site, by whether a chemical is mutagenic in *Salmonella*; within each category of mutagenicity, chemicals positive at the site in each species are listed.

Researchers interested in carcinogenesis in a particular target organ can use this table to identify all chemicals in the Carcinogenic Potency Database (CPDB) that have been evaluated in chronic, long-term bioassays as having induced tumors in at least 1 of 35 target sites. For a variety of research endeavors, such a table of results organized by target organ is useful. For example, epidemiologists interested in a particular target tissue in humans may seek clues in animal models, and can use the table to obtain a list of substances found to induce tumors at each site in rats, mice, hamsters, nonhuman primates, or dogs. Investigators of mechanism of carcinogenesis at a specific target site or of chemical structure can identify compounds that induce tumors in that organ and determine the mutagenicity in *Salmonella* of each such chemical. The table is organized to facilitate comparative toxicological analyses, such as whether a chemical that is positive at a given site in the rat has been tested in the mouse and if so, whether it is positive in the mouse, and whether the target organ(s) is the same.

By cross-referencing a chemical of interest to the plot of the CPDB,[1] details of each experiment can be obtained, including sex and strains that have been tested, route of chemical administration, tumor types, tumor incidence, dose, carcinogenic potency, and citation to the original research report. A quick overview of results on a chemical of interest can be obtained by cross-reference to the "Summary of the Carcinogenic Potency Database by Chemical,"[2]

which summarizes carcinogenic potency, positivity, and all target organs in each species tested.

METHODS

For each target site in the "Summary of the Carcinogenic Potency Database by Target Organ," the chemicals listed are those that were evaluated by a published author as positive at that site in at least one experiment in that species in the CPDB. The CPDB[1] includes results of chronic exposure animal bioassays that were published either in the general literature through 1992 or in Technical Reports of the National Cancer Institute/National Toxicology Program (NCI/NTP) through 1994. All 5152 experiments on 1298 chemicals in the CPDB meet a set of inclusion criteria that were designed to allow for estimation of carcinogenic potency; therefore, reasonable consistency in experimental protocols is assured.[1] Experiments are included only if the test agent was administered alone rather than in combination with other substances, if the protocol included a concurrent control group, if the route of administration was either diet, water, gavage, inhalation, intravenous injection or intraperitoneal injection, and if the length of the experiment in rodents was at least one year with dosing for at least six months.

In the summary table, a site is classified as a target in an NCI/NTP bioassay if the evaluation in the Technical Report was "carcinogenic" or "clear" or "some" evidence of carcinogenic activity ("c" or "p" in the plot of the CPDB).[1] For papers in the general literature, a site is classified as target if the author of the published paper considered the tumors to be induced by compound administration ("+" on the plot).

For each target site in each species, e.g., adrenal gland in mice, chemicals are further classified by whether they are mutagenic in *Salmonella*. A chemical is classified as mutagenic in the *Salmonella* assay "+" if it was evaluated as either "mutagenic" or "weakly mutagenic" by Zeiger[3] or as "positive" by the Gene-Tox Program.[4,5] All other chemicals evaluated for mutagenicity by these two sources are reported as "–". The symbol "." indicates no evaluation. Carcinogens that induce tumors at more than one site or in more than one species are listed multiple times, under each relevant site and species. If a chemical is listed more than once, results may represent two different experiments. Negative tests on each chemical are ignored. Results for negative experiments are reported in the plot of the CPDB.[1] This summary table permits comparisons between species and between mutagens and nonmutagens at each site. For a chemical of interest, results on other genotoxicity tests are reported in the genotoxicity database.[3]

DESCRIPTION OF THE SUMMARY TABLE

A detailed description of the summary table follows, using results in kidney. Kidney is listed alphabetically by target site, and results are reported for species in which kidney tumors have been induced: mouse, monkey, and rat. Under mouse, kidney carcinogens are listed separately for those that are mutagenic in *Salmonella* (+), those that are not mutagenic in *Salmonella* (–), and those for which there is no evaluation in *Salmonella*. The table indicates that there are nine mutagenic kidney carcinogens in the mouse, eight nonmutagenic kidney carcinogens, and four kidney carcinogens with no mutagenicity evaluation in the CPDB. In the monkey and rat, parallel results are reported. If a species is not listed, e.g., hamsters, then the site is not a target for any chemical tested in that species. For each category of mutagenicity, kidney carcinogens in the mouse are listed alphabetically; a chemical is classified as a kidney carcinogen if an author in the general literature evaluated kidney as a target site in at least one experiment or an NCI/NTP Technical Report evaluated kidney results as "carcinogenic" or "clear" or "some" evidence of carcinogenicity.

Most of the data in this summary table, as in the CPDB overall, are for mice and rats. To facilitate target organ comparisons between these two species, we have indicated with superscripts next to each chemical name, which chemicals have been tested in both rats and mice and whether they are positive in both species or only one. This also makes it possible to determine whether kidney is a target organ in both species for a given chemical. For example, in the mouse, ochratoxin A is listed under nonmutagens with the symbol ‡ indicating that it has been tested in both rats and mice and is positive in both species at *some* target site. Since ochratoxin A is also listed under kidney in the rat, it induced kidney tumors in both species. In contrast, *o*-nitroanisole is listed for kidney under mutagens in the

rat with the same symbol ‡, indicating that it is positive in the rat and mouse, but it is not listed under kidney in the mouse. Therefore, *o*-nitroanisole induces tumors in both rats and mice, but is a kidney carcinogen only in the rat; thus, it is positive in the mouse at a different site or sites. The symbol † indicates that a chemical was tested in both rats and mice but induced tumors only in one species. For example, potassium bromate is listed for kidney in the rat under mutagens, with the symbol †, indicating it was tested in both rats and mice but positive only in rats. If a chemical is listed without a superscript under rat or mouse, then it indicates that it was tested in the species indicated, but not the other, e.g., citrinin under kidney in the rat for nonmutagens. (If the only test in one species was negative but had a pathology protocol restricted to one or two tissues, then the chemical was not classified as tested in both species, and no superscripts were used.) The superscripts apply only to rats and mice, and not to other species in the CPDB such as hamsters.

To determine whether a chemical induced tumors at a target site in more than one experiment or more than one strain or sex of test animal, see the plot of the CPDB.[1] Results for target sites by sex-species group are also summarized in "Summary of the Carcinogenic Potency Database by Chemical."[2]

In earlier papers, we have addressed several issues about target organs of carcinogenesis in rodent bioassays,[6-8] and we have now updated some of the analyses using the results reported in the table that follows. The findings are similar to those in the earlier analyses, and we refer the reader to the earlier papers and to the "Overview of Analyses of the Carcinogenic Potency Database."[9] For example, under the conditions of these carcinogenesis bioassays, the liver is the most common target site in mice, rats, and monkeys. For both rats and mice, it is the most common target site among mutagens as well as among nonmutagens, and it is the predominant site in the mouse. Results do not support the idea that mutagens and nonmutagens induce tumors in different target organs.[7] Both mutagens and nonmutagens induce tumors in a wide variety of sites, and most organs are target sites for both. Moreover, the same sites tend to be the most common sites for both: 82% or more of both mutagenic and nonmutagenic carcinogens are positive in rats and in mice in at least one of the eight most frequent target sites: liver, lung, mammary gland, stomach, vascular system, kidney, hematopoietic system, and urinary bladder. Because tissue distribution and pharmacokinetics would not be expected to differ systematically between mutagens and nonmutagens, one would not expect systematic differences in the particular organs in which tumors are induced.[7]

REFERENCES

1. Gold, L.S., Slone, T.H., Manley, N.B., Garfinkel, G.B., Rohrbach, L., and Ames, B.N., Carcinogenic Potency Database. In: Gold, L.S. and Zeiger, E., Eds. *Handbook of Carcinogenic Potency and Genotoxicity Databases.* Boca Raton, FL: CRC Press, 1997, pp. 1–605

2. Gold, L.S., Slone, T.H., and Ames, B.N. Summary of Carcinogenic Potency Database by Chemical. In: Gold, L.S. and Zeiger, E., Eds. *Handbook of Carcinogenic Potency and Genotoxicity Databases.* Boca Raton, FL: CRC Press, 1997, pp. 621–660.

3. Zeiger, E. Genotoxicity database. In: Gold, L.S. and Zeiger, E., Eds. *Handbook of Carcinogenic Potency and Genotoxicity Databases.* Boca Raton, FL: CRC Press, 1997, pp. 687–729.

4. Kier, L.E., Brusick, D.J., Auletta, A.E., Von Halle, E.S., Brown, M.M., Simmon, V.F., Dunkel, V., McCann, J., Mortelmans, K., Prival, M., Rao, T.K., and Ray, V. The *Salmonella typhimurium*/mammalian microsomal assay: A report of the U.S. Environmental Protection Agency Gene-Tox Program. *Mut. Res.* 168: 69-240 (1986).

5. Auletta, A.E. personal communication.

6. Gold, L.S., Slone, T.H., Manley, N.B., and Bernstein, L. Target organs in chronic bioassays of 533 chemical carcinogens. *Environ. Health Perspect.* 93: 233-246 (1991).

7. Gold, L.S., Slone, T.H., Stern, B.R., and Bernstein, L. Comparison of target organs of carcinogenicity for mutagenic and nonmutagenic chemicals. *Mut. Res.* 286: 75-100 (1993).

8. Gold, L.S. and Slone, T.H. The mouse liver in perspective: Comparison of target organs of carcinogenicity for mutagens and nonmutagens in chronic bioassays. In: *Fifth Workshop on Mouse Liver Tumors.* Washington, D.C.: International Life Sciences Institute, 1995, pp. 2-3.

9. Gold, L.S., Slone, T.H., and Ames, B.N. Overview of Analyses of the Carcinogenic Potency Database. In: Gold, L.S. and Zeiger, E., Eds. *Handbook of Carcinogenic Potency and Genotoxicity Databases.* Boca Raton, FL: CRC Press, 1997, pp. 661–685.

TABLE 1

A chemical is listed under each organ evaluated as positive in an experiment in that species by at least one author. Therefore, a chemical may be listed under several target organs and every chemical listed in the table is positive in at least one species. In order to compare results in rats and mice, symbols follow chemicals tested in both species: a ‡ indicates that the chemical is positive at some site in both species, and a † indicates that it was tested in both but positive in only one. N = the number of mutagens or nonmutagens with at least one positive test at that site in that species. "." = no evaluation in *Salmonella*.

Target site	Species	Salmonella	N	Chemicals that induce tumors at each site
Adrenal gland	Hamster	+	1	Urethane
	Mouse	+	3	4,4′-Methylenedianiline.2HCl‡; Pentachloroanisole‡; *p*-Rosaniline.HCl‡
		–	5	Carbon tetrachloride‡; Furan‡; 2,3,4,5,6-Pentachlorophenol (Dowicide EC-7); 2,3,4,5,6-Pentachlorophenol, technical grade‡; 1,1,2-Trichloroethane†
	Rat	+	6	Bromoethane‡; 4-Chloro-*m*-phenylenediamine‡; 1,2-Dibromo-3-chloropropane‡; Pentachloroanisole‡; 1,2-Propylene oxide‡; C.I. pigment red 3‡
		–	6	Acrylamide; Diethylstilbestrol‡; Ethyl alcohol‡; 2-Mercaptobenzothiazole‡; Mirex‡; Reserpine‡
		.	2	Indolidan; Isomazole
Bone	Rat	+	2	1-(2-Hydroxyethyl)-1-nitrosourea‡; *o*-Toluidine.HCl†
		–	1	*N,N*-Dimethylaniline†
		.	1	Acronycine
Clitoral gland	Mouse	–	1	Acetaldehyde methylformylhydrazone
		.	1	*N-n*-Butyl-*N*-formylhydrazine
	Rat	+	16	C.I. direct blue 15; 2,4-Diaminoanisole sulfate‡; 3,3′-Dimethoxybenzidine.2HCl; 3,3′-Dimethylbenzidine.2HCl‡; Glu-P-1‡; Glu-P-2‡; Glycidol‡; IQ‡; MeIQx‡; 1,5-Naphthalenediamine‡; 5-Nitro-*o*-anisidine‡; 5-Nitroacenaphthene‡; *p*-Nitrobenzoic acid‡; C.I. acid red 114; 1,2,3-Trichloropropane‡; Trp-P-2 acetate‡
		.	1	Nalidixic acid†
Ear/Zymbal's gland	Mouse	+	1	Cupferron†
		–	1	Benzene‡
	Rat	+	30	Acrylonitrile; 3-Amino-9-ethylcarbazole mixture‡; Azoxymethane; C.I. direct blue 15; *N-n*-Butyl-*N*-nitrosourea; Chlorambucil‡; Cupferron‡; 2,4-Diaminoanisole sulfate‡; 3,3′-Dichlorobenzidine‡; 3,3′-Dimethoxybenzidine.2HCl; 3,3′-Dimethylbenzidine.2HCl‡; Formic acid 2-[4-(5-nitro-2-furyl)-2-thiazolyl]hydrazide‡; Glu-P-1‡; Glu-P-2‡; Glycidol‡; Hydrazobenzene‡; IQ‡; IQ.HCl; MeIQx‡; 8-Methoxypsoralen; 4,4′-Methylene-bis(2-chloroaniline)‡; 5-Nitro-*o*-anisidine‡; 5-Nitroacenaphthene‡; Phenacetin‡; C.I. acid red 114; *p*-Rosaniline.HCl‡; Thio-TEPA‡; 4,4′-Thiodianiline‡; 1,2,3-Trichloropropane‡; Vinyl chloride‡
		–	1	Benzene‡
		.	8	*N*-1-Diacetamidofluorene; 2,5-Dimethoxy-4′-aminostilbene‡; *N*-(2-Fluorenyl)-2,2,2-trifluoroacetamide; 2-Methoxy-3-aminodibenzofuran; *N*-(*N*-Methyl-*N*-nitrosocarbamoyl)-*l*-ornithine; *N*-(9-Oxo-2-fluorenyl)acetamide; Prednimustine; β-Thioguanine deoxyriboside
Esophagus	Hamster	+	2	AF-2‡; *N*-Nitroso-*N*-methylurethan
	Mouse	+	5	Benzo(*a*)pyrene‡; 1,2-Dibromoethane‡; *N*-Hydroxy-2-acetylaminofluorene‡; Nitrosodibutylamine‡; *N*-Nitrosohexamethyleneimine
		.	2	*N*-Ethyl-*N′*-nitro-*N*-nitrosoguanidine; 3-(5-Nitro-2-furyl)-imidazo(1,2-α)pyridine‡
	Monkey	+	1	*N*-Nitroso-*N*-methylurea‡
	Rat	+	13	*N-n*-Butyl-*N*-nitrosourea; Dimethylvinyl chloride‡; Dinitrosohomopiperazine; *N*-Methyl-*N′*-nitro-*N*-nitrosoguanidine‡; Nitroso-1,2,3,6-tetrahydropyridine; *N*-Nitrosodiethanolamine; *N*-Nitrosodiethylamine‡; *N*-Nitrosodipropylamine‡; Nitrosoethylurethan; Nitrosoheptamethyleneimine; 2-Nitrosomethylaminopyridine; *N*-Nitrosopiperidine‡; *N*-Nitrosothiomorpholine

Organ	n	Species		Chemicals
	1	Hamster	—	Dihydrosafrole‡
	21	Mouse	·	Bis(2,3-dibromopropyl)phosphate, magnesium salt; Z-Ethyl-O,N,N-azoxyethane; 3-(5-Nitro-2-furyl)-imidazo(1,2-α)pyridine‡; N-Nitroso-2,3-dihydroxypropyl-2-hydroxypropylamine‡; Nitroso-2,3-dihydroxypropyl-2-oxopropylamine‡; N-Nitroso-(2-hydroxypropyl)-(2-hydroxyethyl)amine; N-Nitroso-N-methyl-4-fluoroaniline; Nitroso-N-methyl-N-(2-phenyl)ethylamine; N-Nitroso(2,2,2-trifluoroethyl)ethylamine; N-Nitrosoallyl-2,3-dihydroxypropylamine; N-Nitrosoallyl-2-hydroxypropylamine; N-Nitrosoallyl-2-oxopropylamine‡; Nitrosoamylurethan; Nitrosoanabasine; N-Nitrosomethyl-2-hydroxypropylamine; N-Nitrosobis(2-hydroxypropyl)amine; N-Nitrosomethyl-2,3-dihydroxypropylamine‡; N-Nitrosomethyl-2-hydroxypropylamine; N-Nitrosomethyl(2-oxopropyl)amine; Nitrosomethylaniline; N'-Nitrosonornicotine-1-N-oxide‡; N-Nitrosothialdine
Gallbladder	1	Monkey	·	N-Methyl-N-formylhydrazine
	4	Mouse	+	N-Ethyl-N-formylhydrazine; N-Methyl-N-formylhydrazine; 3-Methylbutanal methylformylhydrazone; N-N'-Propyl-N-formylhydrazine
	1	Mouse	·	Aflatoxin B$_1$‡
Harderian gland	8	Mouse	+	Benzidine.2HCl; 1,3-Butadiene‡; Cupferron‡; Ethylene oxide‡; Glycidol‡; Iodinated glycerol‡; 4,4'-Oxydianiline‡; 1,2,3-Trichloropropane‡
	3		—	Benzene‡; Gentian violet; N-Methylolacrylamide†
	2		·	Dichloroacetylene‡; 2,4,6-Trinitro-1,3-dimethyl-5-tert-butylbenzene
Hematopoietic system	22	Mouse	+	2-Aminoanthraquinone‡; 5-Azacytidine‡; Azathioprine; 1,3-Butadiene‡; Chlorambucil‡; Cyclophosphamide‡; Dacarbazine‡; Dibromomannitol‡; Dibromomannitol†; Ethylene oxide‡; Formic acid 2-[4-(5-nitro-2-furyl)-2-thiazolyl]hydrazide‡; 1-(2-Hydroxyethyl)-1-nitrosourea‡; Isophosphamide‡; MeIQx‡; Melphalan†; Methyl methanesulfonate; 4,4'-Methylenedianiline.2HCl‡; Metronidazole‡; N-[4-(5-Nitro-2-furyl)-2-thiazolyl]formamide‡; PhIP.HCl‡; Thio-TEPA‡; Urethane
	13		—	Acetamide‡; Allyl isovalerate‡; Benzene‡; 1,4-Benzoquinone; Chlorinated paraffins (C$_{23}$, 43% chlorine)‡; DDT‡; 5-Fluorouracil‡; Gentian violet; Hexanamide‡; Phenesterin‡; Procarbazine.HCl‡; p-Tolylurea‡; C.I. vat yellow 4†
	12		·	Aflatoxin, crude‡; trans-5-Amino-3[2-(5-nitro-2-furyl)vinyl]-1,2,4-oxadiazole; 2-Amino-4-(p-nitrophenyl)thiazole; Benzoyl hydrazine; 1,2-di-n-Butylhydrazine.2HCl; Estradiol mustard†; 2-Hydrazino-4-(p-aminophenyl)thiazole; 2-Hydrazino-4-(p-nitrophenyl)thiazole‡; ICRF-159‡; 1-Methyl-1,4-dihydro-7-[2-(5-nitrofuryl)vinyl]-4-oxo-1,8-naphthyridine-3-carboxylate, potassium; Phorbol; Strobane
	1	Monkey	—	Procarbazine.HCl‡
	27	Rat	+	Benzidine; C.I. direct blue 15; N-n-Butyl-N-nitrosourea; Chlorambucil‡; Cyclophosphamide‡; Dacarbazine‡; 3,3'-Dichlorobenzidine‡; Dichlorvos‡; 3,3'-Dimethoxybenzidine-4,4'-diisocyanate‡; 3,3'-Dimethoxybenzidine.2HCl; 2-(2,2-Dimethylhydrazino)-4-(5-nitro-2-furyl)thiazole; Ethylene oxide‡; Formaldehyde‡; Formic acid 2-[4-(5-nitro-2-furyl)-2-thiazolyl]hydrazide‡; Glycidol‡; 1-(2-Hydroxyethyl)-1-nitrosourea‡; Iodinated glycerol‡; Lasiocarpine; Nitrite, sodium‡; N-[5-(5-Nitro-2-furyl)-1,3,4-thiadiazol-2-yl]acetamide‡; o-Nitroanisole‡; N-Nitrosodiethanolamine; PhIP.HCl‡; Propane sultone; N-Propyl-N-nitrosourea; Thio-TEPA‡; Trp-P-2 acetate‡
	15		—	Allyl isovalerate‡; Atrazine‡; Cadmium chloride; Dimethyl morpholinophosphoramidate‡; Furan‡; FD & C green no. 2‡; Hematoxylin; Hydroquinone‡; 2-Mercaptobenzothiazole‡; Mirex‡; Procarbazine.HCl‡; FD & C red no. 2; FD & C red no. 4‡; Tetrachloroethylene‡; 2,4,6-Trichlorophenol‡
	10		·	1-Amyl-1-nitrosourea; 1,3-Dibutyl-1-nitrosourea; Dichloroacetylene‡; Dimethoxane; FD & C green no. 1‡; 2-Hydrazino-4-(p-aminophenyl)thiazole; Metepa‡; l-5-Morpholinomethyl-3-[(5-nitrofurfurylidene)amino]-2-oxazolidinone.HCl; 1-[(5-Nitrofurfurylidene) amino]-2-imidazolidinone; Nitroso-N-methyl-N-(2-phenyl)ethylamine
Kidney	9	Mouse	+	Bromodichloromethane‡; 1,3-Butadiene‡; 2,4-Diaminophenol.2HCl‡; N-Hydroxy-2-acetylaminofluorene‡; Phenacetin‡; C.I. pigment red 3‡; Streptozotocin‡; Tris(2,3-dibromopropyl)phosphate‡; Vinylidene chloride†
	8		—	o-Benzyl-p-chlorophenol‡; Caffeic acid‡; Chloroform‡; Daminozide‡; Hydroquinone‡; Mercurymethylchloride; Nitrilotriacetic acid‡; Ochratoxin A‡
	4		·	1,2-di-n-Butylhydrazine.2HCl; Dichloroacetylene‡; 3-Hydroxy-p-butyrophenetidide; Lead acetate, basic‡
	1	Monkey	·	Cycasin and methylazoxymethanol acetate

TABLE 1 (continued)

Target site	Species	Salmonella	N	Chemicals that induce tumors at each site
	Rat	+	25	Aflatoxin B$_1$‡; 1-Amino-2-methylanthraquinone‡; 2-Amino-4-nitrophenol†; 2-Amino-5-nitrothiazole‡; o-Anisidine.HCl‡; Azoxymethane; Bromate, potassium†; Bromodichloromethane‡; Coumarin†; Formic acid 2-[4-(5-nitro-2-furyl)-2-thiazolyl]hydrazide†; 1-(2-Hydroxyethyl)-1-nitrosourea‡; 8-Methoxypsoralen; N-[[3-(5-Nitro-2-furyl)-1,2,4-oxadiazole-5-yl]-methyl]acetamide; N-[5-(5-Nitro-2-furyl)-1,3,4-thiadiazol-2-yl]acetamide‡; o-Nitroanisole†; 1-[(5-Nitrofurfurylidene)amino]hydantoin†; N-Nitrosodiethanolamine; N-Nitrosodimethylamine‡; C.I. acid orange 3†; Phenacetin‡; Quercetin‡; Streptozotocin‡; 1,2,3-Trichloropropane†; Tris(2,3-dibromopropyl)phosphate‡; Vinyl chloride‡
		–	28	Benzofuran‡; Caffeic acid‡; Captafol‡; Chlorinated paraffins (C$_{12}$, 60% chlorine)‡; Chloroform‡; 3-(p-Chlorophenyl)-1,1-dimethylurea†; Chlorothalonil†; Cinnamyl anthranilate‡; Citrinin; 1,4-Dichlorobenzene‡; 3,4-Dihydrocoumarin†; Dimethyl methylphosphonate†; Hexachlorobutadiene; Hexachloroethane‡; Hydroquinone‡; Isophorone†; Lead acetate†; d-Limonene†; α-Methylbenzyl alcohol†; Mirex‡; Nitrilotriacetic acid‡; Nitrilotriacetic acid, trisodium salt, monohydrate†; Ochratoxin A‡; Phenazone; Phenylbutazone‡; o-Phenylphenate, sodium†; Tetrachloroethylene‡; Tris(2-chloroethyl)phosphate†
		·	18	2-Amino-5-(5-nitro-2-furyl)-1,3,4-oxadiazole; 2-Amino-5-(5-nitro-2-furyl)-1,3,4-thiadiazole; Barbital, sodium; Dichloroacetylene‡; Diethylacetamide; Dimethoxane; 4,6-Dimethyl-2-(5-nitro-2-furyl)pyrimidine; N-4-(4'-Fluorobiphenyl)acetamide; Hexamethylmelamine; 2-Hydrazino-4-(5-nitro-2-furyl)thiazole†; Lead acetate, basic†; 2-Methoxy-3-aminodibenzofuran; Z-Methyl-O,N,N-azoxyethane; N-(N-Methyl-N-nitrosocarbamoyl)-l-ornithine; l-5-Morpholinomethyl-3-[(5-nitrofurfurylidene)amino]-2-oxazolidinone.HCl; 3-(5-Nitro-2-furyl)-imidazo(1,2-α)pyridine†; 1-Nitroso-1-hydroxyethyl-3-chloroethylurea; N-Oxydiethylene thiocarbamyl-N-oxydiethylene sulfenamide
Large intestine	Hamster	+	5	1,1-Dimethylhydrazine†; 1,2-Dimethylhydrazine.2HCl; Hydrazine‡; Methylhydrazine; Urethane
	Mouse	·	1	Capsaicin
	Rat	+	20	Aflatoxin B$_1$‡; Azoxymethane; C.I. direct blue 15; Bromodichloromethane‡; N-n-Butyl-N-nitrosourea; 3,3'-Dimethoxybenzidine.2HCl; 3,3'-Dimethylbenzidine.2HCl‡; Formic acid 2-[4-(5-nitro-2-furyl)-2-thiazolyl]hydrazide‡; Glu-P-1‡; Glu-P-2‡; Glycidol‡; N-Hexylnitrosourea; 1-(2-Hydroxyethyl)-1-nitrosourea‡; IQ‡; o-Nitroanisole‡; PhIP.HCl‡; C.I. acid red 114; 4,4'-Thiodianiline‡; Tribromomethane†; Tris(2,3-dibromopropyl)phosphate‡
		–	1	Phenazopyridine.HCl‡
		·	9	1-Allyl-1-nitrosourea; Amylopectin sulfate; Carrageenan, acid-degraded; Chrysazin‡; Dextran sulfate sodium (DS-M-1); Z-Ethyl-O,N,N-azoxymethane; 1-Ethylnitroso-3-(2-oxopropyl)-urea; 1-Hydroxyanthraquinone; N-Nitrosobis(2-oxopropyl)amine
Liver	Dog	+	2	3,3'-Dichlorobenzidine‡; 4,4'-Methylene-bis(2-chloroaniline)‡
	Hamster	+	9	2-Acetylaminofluorene‡; 1,2-Dimethylhydrazine.2HCl; Hydrazine sulfate‡; N-Hydroxy-2-acetylaminofluorene‡; Methylhydrazine; N-Nitrosoazetidine; N-Nitrosomorpholine‡; N-Nitrosopiperidine‡; N-Nitrosopyrrolidine‡
		–	2	p,p'-DDE†; Hexachlorobenzene‡
		·	5	N-Methyl-N-formylhydrazine; Methylnitrosamino-N,N-dimethylethylamine; N-Nitroso-1,3-oxazolidine; Nitroso-2-oxopropylethanolamine†; N-Nitrosoallyl-2-oxopropylamine‡
	Mouse	+	88	A-α-C; 2-Acetylaminofluorene†; 3-Amino-9-ethylcarbazole mixture†; 1-Amino-2-methylanthraquinone‡; 2-Aminoanthraquinone‡; 4-Aminodiphenyl; Auramine-O‡; Benzidine.2HCl; Bis(2-chloro-1-methylethyl)ether, technical grade†; Bis-2-chloroethylether; C.I. direct black 38†; HC blue no. 1‡; Bromodichloromethane‡; 1,3-Butadiene‡; Chloral hydrate; Chlorambenʔ; Chlordane, technical grade†; 1-Chloro-2-nitrobenzene†; 1-Chloro-4-nitrobenzene†; 4-Chloro-m-phenylenediamine‡; 4-Chloro-o-phenylenediamine‡; Chloroacetaldehyde; p-Chloroaniline.HCl‡; Coumarin†; p-Cresidine‡; Cupferron‡; Diallate; 2,4-Diaminotoluene‡; 2,4-Diaminotoluene.2HCl‡; 2,6-Dichloro-p-phenylenediamine†; 1,2-Dichloropropane†; 1,1-Dimethylhydrazine†; Ethylene imine; Ethylene thiourea‡; Furfural†; Glu-P-1‡; Glu-P-2‡; Glycidol‡; Hydrazine sulfate‡; Hydrazobenzene‡; N-Hydroxy-2-acetylaminofluorene‡; 2-Hydroxyethylhydrazine; IQ‡; Isoniazid†; MeA-α-C; MeIQ‡; MeIQx‡; 3-Methoxy-4-aminoazobenzene; 4,4'-Methylene-bis-(2-chloroaniline).2HCl‡; Methylene chloride‡; 4,4'-Methylenebis(N,N-dimethyl)benzenamine‡; 4,4'-Methylenedianiline.2HCl‡; Methylhydrazine; Michler's ketone‡; 1,5-Naphthalenediamine‡; 2-Naphthylamine†; Nithiazide‡; 3-Nitro-p-acetophenetide‡; 5-Nitro-o-anisidine‡; 2-Nitro-p-phenylenediamine‡; 5-Nitro-o-toluidine†; 5-Nitroacenaphthene‡; o-Nitroanisole‡; 6-Nitrobenzimidazole‡;

Nitrofen‡; Nitrosodibutylamine‡; *N*-Nitrosodimethylamine‡; *p*-Nitrosodiphenylamine‡; *N*-Nitrosohexamethyleneimine; *N*-Nitrosopiperidine‡; 4,4'-Oxydianiline‡; Phenobarbital‡; D & C red no. 5‡; *p*-Rosaniline.HCl‡; Selenium sulfide‡; 4,4'-Thiodianiline‡; Toluene diisocyanate, commercial grade (2,4 (80%)- and 2,6 (20%)-)‡; *o*-Toluidine.HCl‡; *p*-Toluidine.HCl‡; Toxaphene‡; 1,2,3-Trichloropropane‡; Trifluralin, technical grade‡; 2,4,5-Trimethylaniline‡; 2,4,5-Trimethylaniline.HCl‡; Tris(2,3-dibromopropyl)phosphate‡; Trp-P-1 acetate‡; Trp-P-2 acetate‡; Urethane; C.I. disperse yellow 3‡

— 75 Acetaminophen‡; Aldrin‡; 3-Aminotriazole‡; Aroclor 1254‡; Benzofuran‡; Benzyl acetate‡; C.I. direct blue 218‡; Butylated hydroxytoluene‡; Captafol‡; Carbazole; Carbon tetrachloride‡; Chlordane, technical grade‡; Chlorendic acid‡; Chlorinated paraffins (C₁₂, 60% chlorine)‡; 5-Chloro-*o*-toluidine‡; Chlorobenzilate‡; Chlorodibromomethane‡; Chloroform‡; Cinnamyl anthranilate‡; *p,p'*-DDD‡; *p,p'*-DDE‡; DDT‡; 1,4-Dichlorobenzene‡; Dicofol‡; Dieldrin‡; 3,4-Dihydrocoumarin†; Dihydrosafrole‡; 1,4-Dioxane‡; 5,5-Diphenylhydantoin†; Estragole; *dl*-Ethionine‡; di(2-Ethylhexyl)adipate‡; di(2-Ethylhexyl)phthalate‡; Furan‡; Gentian violet; Griseofulvin; Heptachlor‡; Hexachlorobenzene‡; β-1,2,3,4,5,6-Hexachlorocyclohexane; γ-1,2,3,4,5,6-Hexachlorocyclohexane‡; Hexachloroethane‡; Hydroquinone‡; 1'-Hydroxyestragole; 1'-Hydroxysafrole‡; Kepone‡; Luteoskyrin; Malonaldehyde, sodium salt‡; *N*-Methylolacrylamide‡; Mirex‡; Ochratoxin A‡; Pentachloroethane‡; Pentachloronitrobenzene‡; 2,3,4,5,6-Pentachlorophenol (Dowicide EC-7); 2,3,4,5,6-Pentachlorophenol, technical grade‡; Phenazopyridine.HCl‡; Phenobarbital, sodium‡; Phenylbutazone‡; Piperonyl sulfoxide‡; Probenecid‡; Safrole‡; 2,3,7,8-Tetrachlorodibenzo-*p*-dioxin‡; 1,1,1,2-Tetrachloroethane‡; 1,1,2,2-Tetrachloroethane‡; Tetrachloroethylene‡; Tetrachlorvinphos‡; Thioacetamide‡; *m*-Toluidine.HCl†; Triamterene‡; 2,4,6-Trichloroaniline‡; 1,1,2-Trichloroethane‡; Trichloroethylene‡; Trichloroethylene (without epichlorohydrin)‡; 2,4,6-Trichlorophenol‡; Tris(2-ethylhexyl)phosphate†; Zearalenone†

· 44 Acifluorfen; 2-Aminodiphenylene oxide; Aramite‡; Bis-2-hydroxyethyldithiocarbamic acid, potassium; HC blue no. 1 (purified); 1,1-di-*n*-Butylhydrazine; [4-Chloro-6-(2,3-xylidino)-2-pyrimidinylthio]acetic acid‡; 4-Chloro-6-(2,3-xylidino)-2-pyrimidinylthio(*N*-β-hydroxyethyl)acetamide‡; Chrysazin‡; Ciprofibrate‡; Compound LY171883; Cyclamate, sodium†; Cyclochlorotine; 3,5-Dichloro(*N*-1,1-dimethyl-2-propynyl)benzamide; Dichloroacetic acid; Diflalone; 2,5-Dimethoxy-4'-aminostilbene‡; Dipyrone†; *o*-Ethoxybenzamide; *N*-Ethyl-*N*-formylhydrazine; 4'-Fluoro-4-aminodiphenyl; HCDD mixture‡; Hexachlorocyclohexane, technical grade; α-1,2,3,4,5,6-Hexachlorocyclohexane; Hexanal methylformylhydrazone; Leupeptin; Methidathion; *N*-Methyl-*N*-formylhydrazine; 3-Methylbutanal methylformylhydrazone; Oxazepam; Pentanal methylformylhydrazone; *o*-Phenylenediamine.2HCl‡; Polybrominated biphenyl mixture‡; *N*-*N'*-Propyl-*N*-formylhydrazine; Rifampicin‡; Ripazepam‡; Selenium diethyldithiocarbamate; Strobane; Tetrafluoro-*m*-phenylenediamine.2HCl‡; Thiouracil‡; 2,4,6-Trimethylaniline.HCl‡; 2,4,6-Trinitro-1,3-dimethyl-5-*tert*-butylbenzene; Vinyl carbamate; 2,5-Xylidine.HCl‡

Monkey + 7 Aflatoxin B₁‡; IQ‡; *N*-Nitrosodiethylamine‡; *N*-Nitrosodipropylamine‡; *N*-Nitrosopiperidine‡; *N*-Nitrosopyrrolidine‡; Sterigmatocystin‡; Urethane

· 1 Cycasin and methylazoxymethanol acetate

Rat + 76 2-Acetylaminofluorene‡; Aflatoxicol; Aflatoxin B₁‡; 3-Amino-9-ethylcarbazole mixture‡; 1-Amino-2-methylanthraquinone‡; 2-Aminoanthraquinone‡; *o*-Aminoazotoluene†; Auramine-*O*‡; Azoxymethane; Benzidine; C.I. direct blue 15; Bromodichloromethane‡; *p*-Cresidine‡; Cupferron‡; 2,4-Diaminotoluene‡; 2,4-Diaminotoluene.2HCl‡; 1,2-Dibromoethane‡; 3,3'-Dimethoxybenzidine.2HCl‡; *N,N*-Dimethyl-4-aminoazobenzene‡; 3,3'-Dimethylbenzidine.2HCl‡; Dinitrosohomopiperazine; Dinitrotoluene, technical grade (2,4 (77%)- and 2,6 (19%)-)‡; Ethylene thiourea‡; Formic acid 2-[4-(5-nitro-2-furyl)-2-thiazolyl]hydrazide‡; Furfural‡; Glu-P-1‡; Glu-P-2‡; Hydrazine sulfate‡; Hydrazobenzene‡; *N*-Hydroxy-2-acetylaminofluorene‡; IQ‡; IQ.HCl; Isoniazid‡; Lasiocarpine; MeIQx‡; 3'-Methyl-4-dimethylaminoazobenzene‡; 2-Methyl-1-nitroanthraquinone‡; 4,4'-Methylene-bis(2-chloroaniline)‡; 4,4'-Methylene-bis(2-methylaniline); 4,4'-Methylenedianiline.2HCl‡; Metronidazole‡; Michler's ketone‡; Nitrite, sodium†; 1-Nitroso-5,6-dihydrouracil; *N*-Nitroso-*N*-methyl-*N*-dodecylamine; Nitroso-1,2,3,6-tetrahydropyridine; Nitrosodibutylamine‡; Nitrosodiethanolamine; *N*-Nitrosodiethylamine‡; *N*-Nitrosodimethylamine‡; *p*-Nitrosodiphenylamine‡; *N*-Nitrosodipropylamine‡; Nitrosododecamethyleneimine; Nitrosoethylmethylamine; Nitrosoheptamethyleneimine; *N*-Nitrosomorpholine‡; *N*-Nitrosopiperidine‡; *N*-Nitrosopyrrolidine‡; 4,4'-Oxydianiline‡; 1-Phenylazo-2-naphthol‡; C.I. acid red 114; C.I. pigment red 3‡; D & C red no. 5‡; D & C red no. 9‡; *p*-Rosaniline.HCl‡; Selenium sulfide‡;

TABLE 1 (continued)

Target site	Species	Salmonella	N	Chemicals that induce tumors at each site
				Sterigmatocystin‡; Telone II‡; 4,4'-Thiodianiline‡; Toluene diisocyanate, commercial grade (2,4 (80%) and 2,6 (20%))‡; 2,4,5-Trimethylaniline‡; 2,4,5-Trimethylaniline.HCl‡; Trp-P-1 acetate‡; Trp-P-2 acetate‡; Vinyl chloride‡; C.I. disperse yellow 3‡
		–	34	Acetamide‡; Acetaminophen‡; 11-Aminoundecanoic acid‡; Captafol‡; Carbon tetrachloride‡; Chlorendic acid‡; Chlorinated paraffins (C$_{12}$, 60% chlorine)‡; Chlorobenzene‡; Chloroform‡; 3-(p-Chlorophenyl)-1,1-dimethylurea‡; DDT‡; Decabromodiphenyl oxide‡; 1,4-Dioxane‡; Ethionine; dl-Ethionine‡; Ethyl alcohol‡; di(2-Ethylhexyl)phthalate‡; Furan‡; Hexachlorobenzene‡; 1'-Hydroxysafrole‡; Kepone‡; Methapyrilene.HCl‡; Methyl carbamate‡; Mirex‡; Monocrotaline; Nafenopin; 2,3,4,5,6-Pentachlorophenol, technical grade‡; Phenobarbital, sodium‡; Prednisolone; FD & C red no. 1; Safrole‡; 2,3,7,8-Tetrachlorodibenzo-p-dioxin‡; Thioacetamide‡; Vinyl acetate
		.	73	Acetoxime; Aflatoxin, crude‡; Aramite‡; Aroclor 1260; Bemitradine; Bis(2,3-dibromopropyl)phosphate, magnesium salt; Budesonide; 4-Chloro-4'-aminodiphenylether‡; 2-Chloro-5-(3,5-dimethylpiperidinosulphonyl)benzoic acid; [4-Chloro-6-(2,3-xylidino)-2-pyrimidinylthio]acetic acid‡; 4-Chloro-6-(2,3-xylidino)-2-pyrimidinylthio(N-β-hydroxyethyl)acetamide‡; Ciprofibrate‡; Clivorine; Clofibrate; Clophen A 30; Crotonaldehyde; Dehydroepiandrosterone acetate; Dichloroacetylene‡; N,N-Diethyl-4-(4'-[pyridyl-1'-oxide]azo)aniline; Dimethoxane; 5,6-Dimethoxysterigmatocystin; 6-Dimethylamino-4,4-diphenyl-3-heptanol acetate.HCl†; Dimethylnitramine; 2,6-Dinitrotoluene; Dipentylnitrosamine; Z-Ethyl-O,N,N-azoxyethane; Z-Ethyl-O,N,N-azoxymethane; N-(2-Fluorenyl)-2,2,2-trifluoroacetamide; Fumonisin B$_1$; FD & C green no. 1†; HCDD mixture‡; α-1,2,3,4,5,6-Hexachlorocyclohexane‡; 1-Hydroxyanthraquinone; 1-(2-Hydroxyethyl)-nitroso-3-chloroethylurea; Z-Methyl-O,N,N-azoxyethane; Methyl clofenapate; 4-(4-N-Methyl-N-nitrosaminostyryl)quinoline; 4-(Methylnitrosamino)-1-(3-pyrridyl)-1-(butanone); N-Nitroso-bis-(4,4-trifluoro-N-butyl)amine; N-Nitroso-2,3-dihydroxypropylethanolamine‡; 1-Nitroso-1-hydroxyethyl-3-chloroethylurea; 1-Nitroso-1-(2-hydroxypropyl)-3-chloroethylurea; N-Nitroso-(2-hydroxypropyl)-(2-hydroxyethyl)amine; Nitroso-N-methyl-N-(2-phenyl)ethylamine; N-Nitroso-N-methyldecylamine; 3-Nitroso-2-oxazolidinone; Nitroso-2-oxopropylethanolamine‡; N-Nitrosoallyl-2-hydroxypropylamine; N-Nitrosoally1-2-oxopropylamine‡; N-Nitrosoallylethanolamine; N-Nitrosobis(2-oxopropyl) amine; N-Nitrosoephedrine; N-Nitrosomethyl-2,3-dihydroxypropylamine‡; N-Nitrosomethyl-(2-hydroxyethyl)amine; N-Nitrosomethyl-(3-hydroxypropyl)amine; N-Nitrosomethyl(2-oxopropyl)amine; N-Nitrosomethyl-(2-tosyloxyethyl)amine; Nitrosomethylundecylamine; N-Nitrosothialdine; o-Nitrosotoluene; Norlestrin†; N-(9-Oxo-2-fluorenyl)acetamide; Petasitenine; o-Phenylenediamine.2HCl‡; Polybrominated biphenyl mixture‡; Pyrilamine maleate; Senkirkine; Symphytine; 3,3',4,4'-Tetraaminobiphenyl.4HCl‡; Triamcinolone acetonide; 2,4,6-Trimethylaniline.HCl‡; Trinitroglycerin; Vinyl bromide
Lung	Tree shrew	+	1	Aflatoxin B$_1$‡
	Hamster	+	1	Nitroso-2,6-dimethylmorpholine
		.	2	Methylnitrosamino-N,N-dimethylethylamine; 1-Nitroso-3,4,5-trimethylpiperazine‡
	Mouse	+	47	5-Azacytidine‡; Bis(2-chloro-1-methylethyl)ether, technical grade‡; 1,3-Butadiene‡; Chlorambucil‡; Coumarin‡; Cyclophosphamide‡; Dacarbazine‡; Dibenz(a,h)anthracene‡; 1,2-Dibromo-3-chloropropane‡; Dibromodulcitol‡; 1,2-Dibromoethane‡; Dibromomannitol‡; 1,2-Dichloroethane‡; 3,3'-Dimethylbenzidine.2HCl‡; 1,1-Dimethylhydrazine.2HCl‡; 1,2-Dimethylhydrazine.2HCl; Ethylene oxide‡; Glycidol‡; Hydrazine‡; Hydrazine sulfate‡; IQ‡; Isoniazid‡; MeIQx‡; Melphalan‡; Methyl methanesulfonate; Methylene chloride‡; Methylhydrazine; Metronidazole‡; 1,5-Naphthalenediamine‡; N-[4-(5-Nitro-2-furyl)-2-thiazolyl]formamide‡; Nitrosodibutylamine‡; N-Nitrosodimethylamine‡; N-Nitrosopiperidine‡; Ozone‡; Phenylethylhydrazine sulfate; Selenium sulfide‡; Streptozotocin‡; Sulfallate‡; Telone II‡; Tetranitromethane‡; Trifluralin, technical grade‡; 2,4,5-Trimethylaniline.HCl‡; Tris(2,3-dibromopropyl)phosphate‡; Urethane; Vinyl chloride‡; Vinylidene chloride†
		–	16	Acetaldehyde methylformylhydrazone; Benzene‡; Benzofuran‡; Butylated hydroxytoluene‡; Caffeic acid‡; Carbamyl hydrazine.HCl; Daminozide‡; p,p'-DDD‡; Dihydrosafrole‡; 5-Fluorouracil†; γ-1,2,3,4,5,6-Hexachlorocyclohexane‡; N-Methylolacrylamide‡; Naphthalene; Phenesterin‡; Procarbazine.HCl‡; Trichloroethylene†

Target organ	Species	No.		Chemicals
		37	.	N'-Acetyl-4-(hydroxymethyl)phenylhydrazine; 1-Acetyl-2-isonicotinoylhydrazine; Allylhydrazine.HCl; Arecoline.HCl; Benzoyl hydrazine; Benzylhydrazine.2HCl; Bis-(chloromethyl)ether‡; N-n-Butyl-N-formylhydrazine; 1,1-di-n-Butylhydrazine; n-Butylhydrazine.HCl; 1,2-di-n-Butylhydrazine.2HCl; 1-Carbamyl-2-phenylhydrazine; 1,1-Diallylhydrazine; 1,2-Diallylhydrazine.2HCl; 1,2-Diformylhydrazine; 2,5-Dimethoxy-4'-aminostilbene‡; Estradiol mustard†; N-Ethyl-N-formylhydrazine; Ethylhydrazine.HCl; Formylhydrazine; Hexanal methylformylhydrazone; Isonicotinic acid vanillylidenehydrazide; 1-Methyl-1,4-dihydro-7-[2-(5-nitrofuryl)vinyl]-4-oxo-1,8-naphthyridine-3-carboxylate, potassium; N-Methyl-N-formylhydrazine; 3-Methylbutanal methylformylhydrazone; Methylhydrazine sulfate; (N-6)-(Methylnitroso)adenine; (N-6)-(Methylnitroso)adenosine; Monoacetyl hydrazine; Nicotinic acid hydrazide; 3-Nitro-3-hexene‡; Pentanal methylformylhydrazone; n-Pentylhydrazine.HCl; N-N'-Propyl-N-formylhydrazine; Propylhydrazine.HCl; 3,3',4,4'-Tetraaminobiphenyl.4HCl‡; 2,4-Xylidine.HCl†
	Monkey	1	+	Urethane
	Rat	27	+	2-Amino-5-nitrothiazole‡; HC blue no. 1‡; Bromoethane‡; N-n-Butyl-N-nitrosourea; 1,2-Dibromoethane‡; Dimethyl hydrogen phosphite†; trans-2-[(Dimethylamino)methylimino]-5-[2-(5-nitro-2-furyl)vinyl]-1,3,4-oxadiazole; 3,3'-Dimethylbenzidine.2HCl‡; 1,2-Epoxybutane†; N-Hexylnitrosourea; Hydrazine; Hydrazine sulfate‡; 1-(2-Hydroxyethyl)-1-nitrosourea‡; Isoniazid‡; 4,4'-Methylene-bis(2-chloroaniline)‡; N-{[3-(5-Nitro-2-furyl)-1,2,4-oxadiazole-5-yl]acetamide; N-[5-(5-Nitro-2-furyl)-1,3,4-thiadiazol-2-yl]acetamide; 5-Nitroacenaphthene‡; N-Nitroso-N-methyl-N-dodecylamine; N-Nitroso-N-methylurea‡; Nitrosodibutylamine‡; N-Nitrosodimethylamine‡; Nitrosoethylmethylamine‡; C.I. acid red 114; Tetranitromethane‡; 2,4,5-Trimethylaniline‡; Vinyl chloride‡
		2	-	Cadmium chloride; 2,3,7,8-Tetrachlorodibenzo-p-dioxin†
		21	.	1-Allyl-1-nitrosourea; 2-Amino-5-(5-nitro-2-furyl)-1,3,4-oxadiazole; 2-Amino-5-(5-nitro-2-furyl)-1,3,4-thiadiazole; 1-Amyl-1-nitrosourea; Bis-(chloromethyl)ether‡; Cadmium sulphate (1:1); 1-Chloroethylnitroso-3-(2-hydroxypropyl)urea; 1-Ethylnitroso-3-(2-oxopropyl)-urea; 4-(Methylnitrosamino)-1-(3-pyridyl)-1-butanol; 4-(Methylnitrosamino)-1-(3-pyridyl)-1-(butanone); 3-Nitro-3-hexene‡; N-Nitroso-bis-(4,4,4-trifluoro-N-butyl) amine; N-Nitroso-N-methyl-N-tetradecylamine; N-Nitroso-N-methyldecylamine; N-Nitrosobis(2-hydroxypropyl)amine; N-Nitrosobis(2-oxopropyl)amine; N-Nitrosoephedrine; N-Nitrosomethyl-2,3-dihydroxypropylamine‡; N-Nitrosomethyl-(3-hydroxypropyl) amine; Nitrosomethylundecylamine; 2,4,6-Trimethylaniline.HCl‡
Mesovarium	Rat	2	.	Salbutamol; Terbutaline
Mammary gland	Dog	1	.	Lynestrenol
	Hamster	1	+	Vinyl chloride‡
	Mouse	11	+	5-Azacytidine‡; C.I. direct black 38‡; 1,3-Butadiene‡; 1,2-Dibromoethane‡; 1,2-Dichloroethane‡; Ethylene oxide‡; Glycidol‡; Isoniazid‡; Sulfallate‡; Vinyl chloride‡; Vinylidene chloride†
		5	-	Benzene‡; Diethylstilbestrol‡; Estradiol; Furosemide‡; Reserpine‡
		3	.	Calciferol; Isonicotinic acid vanillylidenehydrazide; (N-6)-(Methylnitroso)adenosine
	Rat	57	+	2-Acetylaminofluorene‡; Acrylonitrile; AF-2‡; 2-Amino-5-nitrothiazole‡; 4-Aminodiphenyl.HCl; Benzidine; 1,3-Butadiene‡; N-n-Butyl-N-nitrosourea; Chlorambucil‡; Cytembena‡; Dacarbazine‡; 4,6-Diamino-2-(5-nitro-2-furyl)-S-triazine; 2,4-Diaminoanisole sulfate‡; 2,4-Diaminotoluene‡; 1,2-Dibromo-3-chloropropane‡; 1,2-Dibromoethane‡; Dibromomannitol‡; 3,3'-Dichlorobenzidine‡; 1,2-Dichloroethane‡; 3,3'-Dimethoxybenzidine.2HCl; 1,2-Dimethyl-5-nitroimidazole; trans-2-[(Dimethylamino)methylimino]-5-[2-(5-nitro-2-furyl)vinyl]-1,3,4-oxadiazole; 3,3'-Dimethylbenzidine.2HCl‡; 2-(2,2-Dimethylhydrazino)-4-(5-nitro-2-furyl)thiazole; Formic acid 2-[4-(5-nitro-2-furyl)-2-thiazolyl]hydrazide‡; Glycidol‡; N-Hexylnitrosourea; Hydrazobenzene‡; 1-(2-Hydroxyethyl)-3-[(5-nitrofurfurylidene)amino]-2-imidazolidinone; 1-(2-Hydroxyethyl)-1-nitrosourea‡; IQ.HCl; Isoniazid‡; 4-Methyl-1-[(5-nitrofurfurylidene)amino]-2-imidazolidinone; 3-Methylcholanthrene; 4,4'-Methylene-bis(2-chloroaniline)‡; 4,4'-Methylene-bis(2-methylaniline); Methylene chloride‡; Metronidazole‡; Nithiazide‡; 5-Nitro-2-furaldehyde semicarbazone‡; 4-(5-Nitro-2-furyl)thiazole; N-[4-(5-Nitro-2-furyl)-2-thiazolyl]acetamide‡; N,N'-[6-(5-Nitro-2-furyl)-S-triazine-2,4-diyl]bisacetamide; 5-Nitroacenaphthene‡; Phenacetin‡; PhIP.HCl‡; Propane sultone; 1,2-Propylene oxide‡; Styrene‡; Sulfallate‡; Toluene diisocyanate,

TABLE 1 (continued)

Target site	Species	Salmonella	N	Chemicals that induce tumors at each site
		–	7	commercial grade (2,4 (80%)- and 2,6 (20%)-)‡; o-Toluidine.HCl‡; 1,2,3-Trichloropropane‡; 2,2,2-Trifluoro-N-[4-(5-nitro-2-furyl)-2-thiazolyl]acetamide‡; Trp-P-2 acetate‡; Vinyl chloride‡; FD & C violet no. 1†
		.	29	Acrylamide; Atrazine‡; Carbon tetrachloride‡; Carboxymethylnitrosourea; Ochratoxin A‡; Phenesterin‡; Procarbazine.HCl‡ 4-Acetylaminobiphenyl; Acronycine; 1-Allyl-1-nitrosourea; 2-Amino-5-(5-nitro-2-furyl)-1,3,4-oxadiazole; 2-Amino-5-(5-nitro-2-furyl)-1,3,4-thiadiazole; 1-Amyl-1-nitrosourea; Bemitradine; 4-Bis(2-hydroxyethyl)amino-2-(5-nitro-2-thienyl)quinazoline; N-1-Diacetamidofluorene; 1,3-Dibutyl-1-nitrosourea; 4,6-Dimethyl-2-(5-nitro-2-furyl)pyrimidine; Dimethylaminoethylnitrosoethylurea, nitrite salt; 1-Ethylnitroso-3-(2-hydroxyethyl)-urea; 1-Ethylnitroso-3-(2-oxopropyl)-urea; N-(2-Fluorenyl)-2,2,2-trifluoroacetamide; Formic acid 2-(4-methyl-2-thiazolyl)hydrazide; Hexamethylmelamine; 2-Hydrazino-4-(p-aminophenyl)thiazole‡; 2-Hydrazino-4-(5-nitro-2-furyl)thiazole‡; 2-Hydrazino-4-(p-nitrophenyl) thiazole‡; 1-(2-Hydroxyethyl)-nitroso-3-ethylurea; 2-Methoxy-3-aminodibenzofuran.HCl; 3-(5-Nitro-2-furyl)-imidazo(1,2-α)pyridine‡; 1-[(5-Nitrofurfurylidene)amino]-2-imidazolidinone; Norlestrin‡; N-(9-Oxo-2-fluorenyl)acetamide; 4,4′-Sulfonylbisacetanilide
Myocardium	Mouse	–	1	Phenesterin‡
		.	1	Estradiol mustard†
Nasal cavity[a]	Bush baby	+	1	N-Nitrosodiethylamine‡
	Hamster	+	5	Dimethylcarbamyl chloride; Hydrazine‡; Nitroso-2,6-dimethylmorpholine; N-Nitrosomorpholine‡; N-Nitrosopiperidine‡
		–	1	Acetaldehyde‡
		.	5	Diallylnitrosamine‡; Methylnitrosamino-N,N-dimethylethylamine; N-Nitrosoallyl-2-oxopropylamine‡; N-Nitrosomethyl-2,3-dihydroxypropylamine‡; N-Nitrosonornicotine
	Mouse	+	5	Allyl glycidyl ether‡; 1,2-Dibromo-3-chloropropane‡; 1,2-Dibromoethane‡; Formaldehyde‡; 1,2-Propylene oxide‡
	Rat	+	18	Acrylonitrile; p-Cresidine‡; 1,2-Dibromo-3-chloropropane‡; 1,2-Dibromoethane‡; Dimethylvinyl chloride‡; Dinitrosohomopiperazine; 1,2-Epoxybutane‡; Ethylnitrosocyanamide; Formaldehyde‡; Hydrazine‡; N-Nitrosodiethanolamine; N-Nitrosodipropylamine‡; Nitrosoethylmethylamine; N-Nitrosopiperidine‡; Phenacetin‡; Phenylglycidyl ether; 1,2-Propylene oxide‡; Vinyl chloride‡
		–	3	Acetaldehyde‡; Benzene‡; 1,4-Dioxane‡
		.	19	1-Azoxypropane; Bis-(chloromethyl)ether‡; Diallylnitrosamine‡; Dimethylnitramine; Z-Ethyl-O,N,N-azoxyethane; N-Nitroso-N-methyldecylamine; di(N-Nitroso)-perhydropyrimidine; N-Nitroso(2,2,2-trifluoroethyl)ethylamine; 1-Nitroso-3,4,5-trimethylpiperazine‡; N-Nitrosoallyl-2,3-dihydroxypropylamine; N-Nitrosoallyl-2-hydroxypropylamine; N-Nitrosoallylethanolamine; N-Nitrosobis(2-hydroxypropyl)amine; N-Nitrosobis(2-oxopropyl)amine; N-Nitrosomethyl-2,3-dihydroxypropylamine; N-Nitrosomethyl-(2-hydroxyethyl)amine; N-Nitrosomethyl-2-hydroxypropylamine; N-Nitrosomethyl(2-oxopropyl) amine; N-Nitrosonornicotine-1-N-oxide†
Nervous system	Mouse	+	2	1,3-Butadiene‡; N-Nitrosodimethylamine‡
		–	1	Procarbazine.HCl‡
	Rat	+	14	Acrylonitrile; Bromoethane‡; Chlorambucil‡; Cyclophosphamide‡; 1-Ethyl-1-nitrosourea; Ethylene oxide‡; Glu-P-1‡; Glu-P-2‡; Glycidol‡; N-Nitroso-N-methylurea‡; N-Nitrosodiethanolamine; 1-Phenyl-3,3-dimethyltriazene; Propane sultone; Vinyl chloride‡
		–	2	Acrylamide; Procarbazine.HCl‡
		.	5	1-(4-Chlorophenyl)-1-phenyl-2-propynyl carbamate; 1-Ethylnitroso-3-(2-oxopropyl)-urea; R(−)-2-Methyl-N-nitrosopiperidine; S(+)-2-Methyl-N-nitrosopiperidine; Methylnitramine
Oral cavity[b]	Hamster	+	3	Nitroso-2,6-dimethylmorpholine‡; N-Nitrosomorpholine‡; N-Nitrosopiperidine‡
		–	1	Acetaldehyde‡
	Mouse	+	2	N-Nitrosohexamethyleneimine; 1,2,3-Trichloropropane‡
	Monkey	+	1	N-Nitroso-N-methylurea‡
	Rat	+	16	Acrylonitrile; C.I. direct blue 15; 1,2-Dibromo-3-chloropropane‡; 3,3′-Dimethoxybenzidine.2HCl‡; 3,3′-Dimethylbenzidine.2HCl‡; Dimethylvinyl chloride‡; Dinitrosohomopiperazine; Glycidol‡; IQ‡; N-Nitrosodiethylamine‡; Nitrosoethylurethan;

Target organ	Species	N	Result	Chemicals
			—	Nitrosoheptamethyleneimine; 1-Nitrosohydantoin; N-Nitrosothiomorpholine; C.I. acid red 114; 1,2,3-Trichloropropane‡;
Ovary	Mouse	4	+	Acrylamide; Benzene‡; C.I. direct blue 218‡; 2,3,7,8-Tetrachlorodibenzo-p-dioxin‡
	Hamster	7	—	3-Diazotyramine.HCl; N-Nitroso-2,3-dihydroxypropyl-2-hydroxypropylamine‡; Nitroso-2,3-dihydroxypropyl-2-oxopropylamine‡; Nitroso-N-methyl-N-(2-phenyl)ethylamine; Nitrosoamylurethan; N-Nitrosomethyl(2-oxopropyl)amine; N-Nitrosothialdine
	Monkey	4	·	1,3-Butadiene‡; 5-Nitro-2-furaldehyde semicarbazone‡; 5-Nitroacenaphthene‡; 1-[(5-Nitrofurfurylidene)amino]hydantoin‡
	Rat	4	+	Benzene‡; Diethylstilbestrol‡; N-Methylolacrylamide‡; 4-Vinylcyclohexene
Pancreas	Hamster	1	+	Nitroso-2,6-dimethylmorpholine
		2	—	Nitroso-2,3-dihydroxypropyl-2-oxopropylamine‡; Nitroso-2-oxopropylethanolamine‡
		1	·	Urethane
	Rat	7	+	2-Amino-5-nitrophenol†; Azaserine; Dichlorvos‡; IQ.HCl; Nitrofen‡; Toluene diisocyanate, commercial grade (2,4 (80%)- and 2,6 (20%)-)‡; 1,2,3-Trichloropropane‡
		5	—	Chlorendic acid‡; Cinnamyl anthranilate‡; Ethyl alcohol‡; Malonaldehyde, sodium salt‡; 2-Mercaptobenzothiazole†
		4	·	Clofibrate; N-(N-Methyl-N-nitrosocarbamoyl)-l-ornithine; 4-(Methylnitrosamino)-1-(3-pyridyl)-1-butanol; 4-(Methylnitrosamino)-1-(3-pyridyl)-1-(butanone)
Peritoneal cavity	Mouse	2	+	Dimethylcarbamyl chloride; Phenoxybenzamine.HCl‡
		5	—	Bis-1,2-(chloromethoxy)ethane; Bis-1,4-(chloromethoxy)-p-xylene; Bis-(chloromethyl) ether‡; trans-1,4-Dichlorobutene-2; Tris-1,2,3-(chloromethoxy)propane
	Rat	12	+	Bromate, potassium†; Chlorozotocin; Cytembena‡; 1,2-Dibromoethane‡; Dibromomannitol‡; 3,3'-Dimethylbenzidine.2HCl‡; Ethylene oxide‡; Glycidol‡; Melphalan‡; Mitomycin-C; Phenoxybenzamine.HCl‡; o-Toluidine.HCl‡
		4	—	Acrylamide; Actinomycin D; Aniline.HCl†; Dapsone†
		4	·	Acronycine; 1-Ethylnitroso-3-(2-oxopropyl)-urea; N-Methyl-N,4-dinitrosoaniline; N-Nitroso-2,2,4-trimethyl-1,2-dihydroquinoline polymer
Pituitary gland	Mouse	2	+	Ethylene thiourea‡; Iodinated glycerol‡
		2	—	Diethylstilbestrol‡; Zearalenone†
		3	·	Enovid; Norlestrin‡; Propylthiouracil‡
	Rat	2	+	1,2-Dibromoethane‡; Metronidazole‡
		5	—	Acrylamide; 3-Aminotriazole‡; Diethylstilbestrol‡; Ethyl alcohol†; 2-Mercaptobenzothiazole†
		1	·	Norlestrin†
Preputial gland	Mouse	3	+	1,3-Butadiene‡; Dimethylvinyl chloride‡; Thio-TEPA‡
		2	—	Acetaldehyde methylformylhydrazone; Benzene‡
		6	·	N-n-Butyl-N-formylhydrazine; N-Ethyl-N-formylhydrazine; Hexanal methylformylhydrazone; 3-Methylbutanal methylformylhydrazone; Pentanal methylformylhydrazone; N-N'-Propyl-N-formylhydrazine
	Rat	5	+	C.I. direct blue 15; 2,4-Diaminoanisole sulfate‡; 3,3'-Dimethoxybenzidine.2HCl; 3,3'-Dimethoxybenzidine.2HCl‡; 1,2,3-Trichloropropane‡
		3	—	Isophorone†; 2-Mercaptobenzothiazole‡; Nalidixic acid†
Prostate	Rat	1	+	Cadmium chloride
		2	—	N-Nitrosobis(2-hydroxypropyl)amine; N-Nitrosobis(2-oxopropyl)amine
Skin	Hamster	2	+	Urethane; Vinyl chloride‡
	Mouse	3	+	5-Azacytidine‡; Glycidol‡; Thio-TEPA‡
	Rat	20	+	2-Acetylaminofluorene‡; 3-Amino-9-ethylcarbazole mixture‡; C.I. direct blue 15; 2,4-Diaminoanisole sulfate‡; Dibromodulcitol†; Dibromomannitol‡; 3,3'-Dimethoxybenzidine-4,4'-diisocyanate‡; 3,3'-Dimethoxybenzidine.2HCl; 3,3'-Dimethylbenzidine.2HCl‡; Dimethylvinyl chloride‡; Glycidol‡; IQ; Lasiocarpine; MeIQx‡; 5-Nitro-o-anisidine‡; C.I. acid red 114; p-Rosaniline.HCl‡; Thio-TEPA‡; Vinyl chloride‡; FD & C violet no. 1†
		3	—	Benzene‡; Carboxymethylnitrosourea; Thiourea†

TABLE 1 (continued)

Target site	Species	Salmonella	N	Chemicals that induce tumors at each site
Small intestine	Mouse	·	8	1-Azoxypropane; 2-Azoxypropane; Dimethoxane; 2,5-Dimethoxy-4'-aminostilbene‡; 1-Ethylnitroso-3-(2-hydroxyethyl)-urea; 1-Ethylnitroso-3-(2-oxopropyl)-urea; 1-(2-Hydroxyethyl)-nitroso-3-ethylurea; N-(N-Methyl-N-nitrosocarbamoyl)-l-ornithine
	Mouse	+	2	Hydrogen peroxide; N-(Trichloromethylthio)phthalimide
		−	1	Captafol‡
		·	1	N-Ethyl-N'-nitro-N-nitrosoguanidine
	Monkey	+	1	Urethane
	Rat	+	18	Acrylonitrile; C.I. direct blue 15; 3,3'-Dimethoxybenzidine.2HCl; trans-2-[(Dimethylamino)methylimino]-5-[2-(5-nitro-2-furyl)vinyl]-1,3,4-oxadiazole; 3,3'-Dimethylbenzidine.2HCl‡; 1-Ethyl-1-nitrosourea; Formic acid 2-[4-(5-nitro-2-furyl)-2-thiazolyl]hydrazide‡; Glu-P-1‡; Glu-P-2‡; IQ‡; Lasiocarpine; N-Methyl-N'-nitro-N-nitrosoguanidine‡; N-[5-(5-Nitro-2-furyl)-1,3,4-thiadiazol-2-yl]acetamide‡; Nitrosoethylurethan; Propane sultone; Quercetin‡; C.I. acid red 114; Trp-P-2 acetate‡
		−	1	Carboxymethylnitrosourea
		·	8	Bis(2,3-dibromopropyl)phosphate, magnesium salt; 4-Bis(2-hydroxyethyl)amino-2-(5-nitro-2-thienyl)quinazoline; 1-(4-Chlorophenyl)-1-phenyl-2-propynyl carbamate; 2,5-Dimethoxy-4'-aminostilbene‡; 4,6-Dimethyl-2-(5-nitro-2-furyl)pyrimidine; Z-Ethyl-O,N,N-azoxymethane; N-Nitroso-N-isobutylurea; 3-Nitroso-2-oxazolidinone
Spleen	Rat	+	4	Azobenzene‡; p-Chloroaniline.HCl‡; D & C red no. 9‡; o-Toluidine.HCl‡
		−	2	Aniline.HCl‡; Dapsone†
		·	1	o-Nitrosotoluene
Stomach	Hamster	+	10	AF-2‡; 1,4-Dinitroso-2,6-dimethylpiperazine; Formic acid 2-[4-(5-nitro-2-furyl)-2-thiazolyl]hydrazide‡; Hydrazine‡; N-Hydroxy-2-acetylaminofluorene‡; N-[4-(5-Nitro-2-furyl)-2-thiazolyl]acetamide‡; N-[4-(5-Nitro-2-furyl)-2-thiazolyl]formamide‡; N-Nitroso-N-methylurethan; Urethane; Vinyl chloride‡
		·	8	N-Nitroso-2,3-dihydroxypropyl-2-hydroxypropylamine‡; Nitroso-2,3-dihydroxypropyl-2-oxopropylamine‡; N-Nitroso-ethyl-2-oxopropylurea; N-Nitroso-ethylhydroxyethylurea; Nitroso-5-methyloxazolidone; N-Nitroso-1,3-oxazolidine; N-Nitroso-2-phenylethylurea; 1-Nitroso-3,4,5-trimethylpiperazine‡
	Mouse	+	30	AF-2‡; 2-Amino-4-(5-nitro-2-furyl)thiazole‡; Benzyl chloride‡; Bromoethanol; 1,3-Butadiene‡; 3-Chloro-2-methylpropene, technical grade (containing 5% dimethylvinyl chloride)‡; 3-(Chloromethyl)pyridine.HCl‡; 1,2-Dibromo-3-chloropropane‡; 1,2-Dibromoethane‡; Dichlorvos‡; Diglycidyl resorcinol ether, technical grade‡; Dimethylvinyl chloride‡; Dinitrosopiperazine; Formic acid 2-[4-(5-nitro-2-furyl)-2-thiazolyl]hydrazide‡; Glycidol‡; N-Hydroxy-2-acetylaminofluorene‡; IQ‡; MeIQ; N-[5-(5-Nitro-2-furyl)-1,3,4-thiadiazol-2-yl]acetamide‡; N-[4-(5-Nitro-2-furyl)-2-thiazolyl]formamide‡; Nitrosodibutylamine‡; N-Nitrosohexamethyleneimine; N-Nitrosopiperidine‡; β-Propiolactone‡; Styrene oxide‡; Telone II‡; 1,2,3-Trichloropropane‡; 2,2,2-Trifluoro-N-[4-(5-nitro-2-furyl)-2-thiazolyl]acetamide‡; Trifluralin, technical grade‡; Tris(2,3-dibromopropyl)phosphate‡
		−	10	Acetaldehyde methylformylhydrazone; Benzaldehyde‡; Benzaldehyde‡; Benzofuran‡; Benzyl acetate‡; Butylated hydroxyanisole‡; Caffeic acid‡; Captafol‡; Carbazole; Catechol‡; Ethyl acrylate‡
		·	12	Acifluorfen; trans-5-Amino-3[2-(5-nitro-2-furyl)vinyl]-1,2,4-oxadiazole; Arecoline.HCl; 1,1-di-n-Butylhydrazine; 2-Chloropropanal; 1-Chloropropene; 1,1-Diallylhydrazine; Estradiol mustard†; 2-Hydrazino-4-(5-nitro-2-furyl)thiazole‡; 1-Methyl-1,4-dihydro-7-[2-(5-nitrofuryl)vinyl]-4-oxo-1,8-naphthyridine-3-carboxylate, potassium; 3-(5-Nitro-2-furyl)-imidazo(1,2-α)pyridine‡; Sesamol‡
	Monkey	+	1	N-Nitroso-N-methylurea
	Rat	+	43	1'-Acetoxysafrole‡; Acrylonitrile; 2-Amino-4-(5-nitro-2-furyl)thiazole‡; Benzo(a)pyrene‡; N-n-Butyl-N-nitrosourea; 3-Chloro-2-methylpropene, technical grade (containing 5% dimethylvinyl chloride)‡; 4-Chloro-o-phenylenediamine‡; 3-(Chloromethyl)pyridine.HCl‡; Cupferron‡; 1,2-Dibromo-3-chloropropane‡; 1,2-Dichloroethane‡; Diglycidyl resorcinol ether, technical grade‡; Dimethyl hydrogen phosphite†; trans-2-[(Dimethylamino)methylimino]-5-[2-(5-nitro-2-furyl)vinyl]-1,3,4-oxadiazole; Dimethylvinyl chloride‡; Epichlorohydrin†; Ethylene oxide‡; Glycidol‡; N-Hexylnitrosurea; 1-(2-Hydroxyethyl)-1-nitrosourea†; N-Methyl-N'-nitro-N-nitrosoguanidine†; N-Methyl-N-nitrosobenzamide; Methylnitrosocyanamide; N-[5-(5-Nitro-2-furyl)-1,3,4-thiadiazol-2-yl] acetamide‡; 4-(5-Nitro-2-furyl)thiazole; 8-Nitroquinoline; Nitroso-Baygon; N-Nitroso-N-methyl-N-dodecylamine;

Organ	Species			Chemicals
				N-Nitroso-N-methylurea‡; Nitroso-1,2,3,6-tetrahydropyridine; Nitrosodibutylamine‡; N-Nitrosodiethylamine‡; Nitrosoethylurethan; Phenacetin‡; Pivalolactone†; β-Propiolactone‡; N-Propyl-N′-nitro-N-nitrosoguanidine; 1,2-Propylene oxide‡; Styrene oxide‡; Sulfallate‡; Telone II‡; 1,2,3-Trichloropropane‡
		−	8	Benzene‡; Butylated hydroxyanisole‡; Caffeic acid‡; Catechol‡; Ethyl acrylate‡; 1′-Hydroxysafrole‡; Mercuric chloride‡; 1-Nitroso-3,5-dimethyl-4-benzoylpiperazine
		·	23	Acetone[4-(5-nitro-2-furyl)-2-thiazolyl]hydrazone; 1-Allyl-1-nitrosourea; 2-Amino-5-(5-nitro-2-furyl)-1,3,4-oxadiazole; 2-Amino-5-(5-nitro-2-furyl)-1,3,4-thiadiazole; 1-Amyl-1-nitrosourea; Bis(2,3-dibromopropyl)phosphate, magnesium salt; β-Butyrolactone; Chlorofluoromethane; Ciprofibrate‡; 2,5-Dimethoxy-4′-aminostilbene‡; 4,6-Dimethyl-2-(5-nitro-2-furyl)pyrimidine; 2-Fluoroethyl-nitrosourea; 1-Hydroxyanthraquinone; 1-(3-Hydroxypropyl)-1-nitrosourea; 3-(5-Nitro-2-furyl)-imidazo(1,2-α)pyridine‡; N-Nitroso-2,3-dihydroxypropyl-2-hydroxypropylamine; Nitroso-2,3-dihydroxypropyl-2-oxopropylamine‡; Nitroso-N-methyl-N-(2-phenyl)ethylamine; N-Nitroso-N-methyldecylamine; N-Nitrosobenzthiazuron; N-Nitrosoephedrine; Sesamol‡; 2,4,6-Trimethylaniline.HCl‡
Subcutaneous tissue	Mouse	+	2	1,2-Dibromoethane‡; Glycidol‡
	Rat	·	1	N2-γ-Glutamyl-p-hydrazinobenzoic acid
		+	7	2,4-Diaminotoluene.2HCl‡; 1,2-Dichloroethane‡; 4,4′-Methylene-bis(2-methylaniline); p-Rosaniline.HCl‡; Toluene diisocyanate, commercial grade (2,4 (80%)- and 2,6 (20%)-)‡; o-Toluidine.HCl‡; 2,4,5-Trimethylaniline.HCl‡
Testes	Mouse	·	3	Dimethoxane; o-Nitrosotoluene; 2,5-Xylidine.HCl‡
		−	2	Diethylstilbestrol‡; Reserpine‡
	Rat	+	6	5-Azacytidine‡; 1,3-Butadiene‡; N-Butyl-N-(4-hydroxybutyl)nitrosamine; Metronidazole‡; N-Nitrosodimethylamine‡; Vinyl chloride‡
		−	3	Cadmium chloride; 2-Chloro-1,1,1-trifluoroethane; Trichloroethylene‡
Thyroid gland	Hamster	·	1	Trinitroglycerin
		+	2	Hydrazine‡; Urethane
		−	2	Hexachlorobenzene‡; Methylthiouracil
	Mouse	+	9	3-Amino-4-ethoxyacetanilide‡; HC blue no. 1‡; 2,4-Diaminoanisole sulfate‡; Ethylene thiourea‡; 4,4′-Methylenedianiline.2HCl‡; 1,5-Naphthalenediamine‡; 4,4′-Oxydianiline‡; C.I. pigment red 3‡; 4,4′-Thiodianiline†
		−	5	Chlorinated paraffins (C12, 60% chlorine)‡; Diethylstilbestrol‡; Ethionamide†; Sulfamethazine; 2,3,7,8-Tetrachlorodibenzo-p-dioxin‡
		·	1	Oxazepam
	Rat	+	12	o-Anisidine.HCl‡; 2,4-Diaminoanisole sulfate‡; 1-Ethyl-1-nitrosourea; Ethylene thiourea‡; Glycidol‡; Iodinated glycerol†; 4,4′-Methylenebis(N,N-dimethyl)benzenamine‡; 4,4′-Methylenedianiline.2HCl‡; 4,4′-Oxydianiline‡; p-Rosaniline.HCl‡; 4,4′-Thiodianiline‡; Zinc dimethyldithiocarbamate†
		−	8	3-Aminotriazole‡; Chlorinated paraffins (C12, 60% chlorine)‡; N,N′-Diethylthiourea‡; Malonaldehyde, sodium salt‡; Methimazole; 2,3,7,8-Tetrachlorodibenzo-p-dioxin‡; Trimethylthiourea‡; Vinyl acetate
Urinary bladder	Dog	·	6	Bemitradine; 1-Ethylnitroso-3-(2-oxopropyl)-urea; Mirex, photo-; N-Nitrosobis(2-oxopropyl)amine; Propylthiouracil‡; Thiouracil‡
	Hamster	+	2	3,3′-Dichlorobenzidine‡; 4,4′-Methylene-bis(2-chloroaniline)‡
		+	3	Formic acid 2-[4-(5-nitro-2-furyl)-2-thiazolyl]hydrazide‡; N-[4-(5-Nitro-2-furyl)-2-thiazolyl]acetamide‡; N-[4-(5-Nitro-2-furyl)-2-thiazolyl]formamide‡
	Mouse	+	8	2-Acetylaminofluorene‡; 4-Aminodiphenyl; o-Anisidine.HCl‡; p-Cresidine‡; N-Hydroxy-2-acetylaminofluorene‡; N-[4-(5-Nitro-2-furyl)-2-thiazolyl]formamide‡; Phenacetin‡; Telone II‡
	Monkey	−	1	Uracil‡
		·	3	2-Aminodiphenylene oxide; 4-Chloro-4′-aminodiphenylether‡; 4-Ethylsulphonylnaphthalene-1-sulfonamide
		+	1	2-Naphthylamine‡
	Rat	+	23	Allyl isothiocyanate†; 2-Amino-4-(5-nitro-2-furyl)thiazole‡; 4-Amino-2-nitrophenol‡; o-Anisidine.HCl‡; C.I. disperse blue 1†; N-Butyl-N-(4-hydroxybutyl)nitrosamine; 4-Chloro-o-phenylenediamine‡; m-Cresidine‡; p-Cresidine‡; Cyclophosphamide‡; IQ.HCl; 2-Naphthylamine‡; N-[4-(5-Nitro-2-furyl)-2-thiazolyl]formamide‡; o-Nitroanisole‡; N-Nitroso-N-methyl-N-dodecylamine;

TABLE 1 (continued)

Target site	Species	Salmonella	N	Chemicals that induce tumors at each site
		−	13	Nitrosodibutylamine‡; N-Nitrosodiethylamine‡; N-Nitrosodiethylamine‡; N-Nitrosopyrrolidine‡; Phenacetin‡; Quercetin†; p-Quinone dioxime†; o-Toluidine.HCl‡; Trp-P-2 acetate‡
		·	7	Acetaminophen‡; 11-Aminoundecanoic acid‡; Diethylene glycol; Melamine†; Nitrilotriacetic acid; Nitrilotriacetic acid, trisodium salt, monohydrate†; N-Nitrosodiphenylamine†; Phenazone; o-Phenylphenate, sodium†; o-Phenylphenol†; Saccharin, sodium†; o-Toluenesulfonamide; Uracil‡ Fosetyl Al; 2-Methoxy-3-aminodibenzofuran; N-Nitroso-N-methyl-N-tetradecylamine; N-Nitroso-N-methyldecylamine; N-Nitrosobis-(2-oxopropyl)amine; o-Nitrosotoluene; N-Oxydiethylene thiocarbamyl-N-oxydiethylene sulfenamide
Uterus	Hamster	−	1	N-Nitroso-ethylhydroxyethylurea
	Mouse	+	8	Bromoethane‡; Chloroethane‡; Dacarbazine‡; 1,2-Dichloroethane‡; Ethylene oxide‡; Glycidol‡; 1,2,3-Trichloropropane‡; Trimethylphosphate†
		−	2	Diethylstilbestrol‡; Procarbazine.HCl‡
		·	1	(N-6)-(Methylnitroso)adenosine
	Rat	+	10	3-Amino-9-ethylcarbazole mixture†; C.I. direct blue 15; N-n-Butyl-N-nitrosourea; Dacarbazine‡; 3,3′-Dimethoxybenzidine-4,4′-diisocyanate†; 3,3′-Dimethoxybenzidine.2HCl; N-Hexylnitrosourea; Isophosphamide‡; 1,5-Naphthalenediamine‡; 4,4′-Thiodianiline‡
		−	4	Atrazine†; 2-Chloro-1,1,1-trifluoroethane; Daminozide†; Vinyl acetate
		·	7	1-Allyl-1-nitrosourea; 1-Amyl-1-nitrosourea; Dimethylaminoethylnitrosoethylurea, nitrite salt; 1-Ethylnitroso-3-(2-hydroxyethyl)-urea; 1-Ethylnitroso-3-(2-oxopropyl)-urea; ICRF-159‡; Norlestrin†
Vagina	Rat	+	1	N-n-Butyl-N-nitrosourea
Vascular system	Hamster	+	3	1,2-Dimethylhydrazine.2HCl; Glycidol‡; Vinyl chloride‡
		−	1	Hexachlorobenzene‡
		·	5	N-Nitroso-ethyl-2-oxopropylurea; N-Nitroso-ethylhydroxyethylurea; N-Nitroso-oxopropylchloroethylurea; N-Nitroso-oxopropylurea; N-Nitroso-2-phenylethylurea
	Mouse	+	32	A-α-C; 1-Acetyl-2-phenylhydrazine; Azathioprine; Benzidine.2HCl; 2-Biphenylamine.HCl†; 1,3-Butadiene‡; 2-Biphenylamine.HCl†; 1-Chloro-4-nitrobenzene†; p-Chloroaniline.HCl‡; Cupferron‡; Dacarbazine†; 2,4-Diaminotoluene.2HCl‡; 1,2-Dibromoethane‡; 7,12-Dimethylbenz(a)anthracene; 1,1-Dimethylhydrazine†; 1,2-Dimethylhydrazine.2HCl; Glu-P-2‡; MeA-α-C; 2-Methyl-1-nitroanthraquinone‡; Michler's ketone†; 5-Nitro-o-toluidine†; Nitrofen†; Pentachloroanisole‡; Phenylethylhydrazine sulfate; Phenylhydrazine.HCl; Sterigmatocystin†; Toluene diisocyanate, commercial grade (2,4 (80%)- and 2,6 (20%)-)‡; o-Toluidine.HCl‡; 2,4,5-Trimethylaniline.HCl‡; Urethane; Vinyl chloride‡; Vinylidene chloride†
		−	9	Captafol‡; Carbamyl hydrazine.HCl; 5-Chloro-o-toluidine†; 4-Chloro-o-toluidine.HCl†; Daminozide‡; 1′-Hydroxysafrole‡; 2,3,4,5,6-Pentachlorophenol (Dowicide EC-7); 2,3,4,5,6-Pentachlorophenol, technical grade‡; 2,4,6-Trichloroaniline†
		·	13	N′-Acetyl-4-(hydroxymethyl)phenylhydrazine; Allylhydrazine.HCl; Arecoline.HCl; 4-Chloro-4′-aminodiphenylether†; Diftalone; N-Ethyl-N-formylhydrazine; Ethylhydrazine.HCl; p-Hydrazinobenzoic acid.HCl; N-Methyl-N-formylhydrazine; n-Pentylhydrazine.HCl; 2,4,6-Trimethylaniline.HCl‡; Vinyl carbamate; 2,5-Xylidine.HCl‡
	Monkey	+	2	Aflatoxin B₁‡; Urethane
	Rat	+	16	Azobenzene†; Cupferron‡; 1,2-Dibromoethane‡; 1,2-Dichloroethane‡; IQ.HCl; Lasiocarpine; 4,4′-Methylene-bis(2-chloroaniline)‡; N-[5-(5-Nitro-2-furyl)-1,3,4-thiadiazol-2-yl] acetamide‡; Nitroso-1,2,3,6-tetrahydropyridine; N-Nitrosodimethylamine‡; N-Nitrosomorpholine‡; N-Nitrosopyrrolidine‡; Sterigmatocystin†; o-Toluidine.HCl‡; Trp-P-2 acetate‡; Vinyl chloride‡
		−	2	Aniline.HCl‡; Benzene‡
		·	10	Clivorine; 4,6-Dimethyl-2-(5-nitro-2-furyl)pyrimidine; Z-Ethyl-O,N,N-azoxyethane; Z-Ethyl-O,N,N-azoxyethane; N-Nitroso-(2-hydroxypropyl)-(2-hydroxyethyl)amine; N-Nitrosobis(2-oxopropyl)amine; N-Nitrosomethyl-(2-tosyloxyethyl)amine; Petasitenine; Symphytine; Vinyl bromide

---

[a] Nasal cavity includes tissues of the nose, nasal turbinates, paranasal sinuses, and trachea.

[b] Oral cavity includes tissues of the mouth, oropharynx, pharynx, and larynx.

CHAPTER 3

SUMMARY OF THE CARCINOGENIC POTENCY DATABASE BY CHEMICAL

Lois Swirsky Gold, Thomas H. Slone, and Bruce N. Ames

A tabular compilation summarizing results on each of the 1298 chemicals in the Carcinogenic Potency Database (CPDB) is presented below. The CPDB includes detailed results and analyses of 5152 chronic, long-term carcinogenesis bioassays reported in 1002 papers in the general literature and 403 Technical Reports of the National Cancer Institute/National Toxicology Program. Details on each experiment are reported in the plot of the CPDB.[1] Our intent is that the summary table below will be a useful reference source for the scientific and regulatory communities, and that it will facilitate the use of our large plot. Researchers interested in a brief overview of bioassay results on a given chemical can readily obtain information on the sex-species groups that have been tested, the strongest level of evidence for carcinogenicity based on the opinion of a published author, carcinogenic potency value (TD$_{50}$), and target organs in each species tested. Evaluations of mutagenicity in *Salmonella* are also reported. Results are summarized for rats and mice (Section A), hamsters (Section B), and nonhuman primates and dogs (Section C). Additional results are indicated by a series of superscripts, e.g., whether there is more than one positive experiment on a chemical in each species and whether potency values vary widely. Throughout the table, the symbol "." indicates that the CPDB has no data on the chemical for the relevant group, e.g., no experiments in male rats. This tabulation can be used as a guide to the literature of animal cancer tests, as an index of results in the CPDB, and to investigate associations between carcinogenic potency and other factors such as mutagenicity, teratogenicity, chemical structure, and human exposure.

The CPDB, which is summarized in this table, has been fully described in Gold et al.,[1] as to inclusion criteria, protocol characteristics, and derived variables. Briefly, for all experiments in the CPBD the plot provides detailed information on each experiment including the species, sex, strain, route of administration, duration of exposure and of experiment, dose levels, target sites, shape of the dose-response, estimates of carcinogenic potency and the confidence limits surrounding it, statistical significance of the carcinogenic dose-response, tumor incidences, and bibliographic citation to the published paper or to the NCI/NTP Technical Report. All experiments in the database meet a specific set of inclusion criteria that were designed to permit the estimation of carcinogenic potency; therefore, reasonable consistency of experimental protocols is assured. Bioassays are included in the database only if the test agent was administered alone rather than in combination with other substances; if the bioassay included a control group; if the route of administration was either diet, water, gavage, inhalation, intravenous injection or intraperitoneal injection; and if the length of experiment was at least half the standard lifespan for the species with dosing for at least one fourth lifespan. For the CPDB, we do not evaluate the evidence for carcinogenicity in an experiment; rather, we report the evaluation of the published author and calculate the statistical significance of the tumorigenic dose-response in the experiment. Thirty percent of the chemicals have been tested by the NCI/NTP.

There is great diversity in the extent of testing among chemicals in the CPDB. While some of this diversity is reflected in the "Summary Table by Chemical," e.g., the number of sex-species groups in which the chemical has been tested, many other factors are not described: how many experiments per chemical, what routes of administration are tested, which strains and how many different strains have been tested, the number of doses, and the range of doses tested. These results can be found in the plot of the CPDB.[1]

CHEMICALS

A concise, alphabetical list of chemicals in the CPDB can be obtained in the summary table on rats and mice (Section A); the alphabetical order is the same as in the plot of the CPDB. (Colors are sorted by color, ignoring prefixes such as "C.I.") For identification purposes, chemicals can be cross-referenced by CAS number using the "Index of Chemical Names in the Carcinogenic Potency (CPDB) and Genotoxicity Databases (GT) by Chemical Abstracts Service (CAS) Registry Number."[2] To search for a chemical by another name, see the list of synonyms in Appendix 10 of the CPDB.[1] For each chemical in Section A on rats and mice, a superscript "s" indicates that other species have been tested and are therefore listed in Section B or Section C as well as in the plot of the CPDB.

MUTAGENICITY IN SALMONELLA

A chemical is classified as mutagenic in the *Salmonella* assay "+" if it was evaluated as either "mutagenic" or "weakly mutagenic" by Zeiger[3] or as "positive" by the Gene-Tox Program.[4,5] All other chemicals evaluated for mutagenicity by these two sources are reported as "−". The symbol "." indicates no evaluation. For a chemical of interest, results in other genotoxicity tests may be reported in the "Genotoxicity Database."[3]

CARCINOGENICITY

For each positive chemical in the CPDB, results are included on carcinogenic potency (by species) and target organ (by sex-species); if there are no positive results then the symbol "−" appears. The classification of positivity in this summary table is based on a positive result in at least one experiment. There may be additional experiments on the same chemical that are negative in the CPDB, but this is not reflected in the table. An experiment is classified as positive or negative on the basis of the author's opinion. A target site is classified as positive for NCI/NTP if the evaluation in the Technical Report was "carcinogenic" or "clear" or "some" evidence of carcinogenic activity ("c" or "p" on the plot of the CPDB[1]). In the general literature, a site is classified as a target if the author of the published paper considered tumors to be induced by compound administration ("+" on the plot). In some cases authors do not clearly state their evaluation (blank in author's opinion in plot), and in some NCI/NTP Technical Reports the evidence for carcinogenicity is considered "associated" or "equivocal"; these are not classified as positive. We use the author's opinion to determine positivity because it often takes into account more information than statistical signif-

icance alone, such as historical control rates for particular sites, survival and latency, and/or dose-response. Generally, this designation by author's opinion corresponds well with the results of statistical tests for the significance of the dose-response effect (two-tailed $p < 0.01$). For some chemicals the only experiments in the CPDB for a species or a sex-species group were NCI/NTP bioassays that were evaluated as inadequate, and we indicate these with an "I" in the potency and target organ fields.

CARCINOGENIC POTENCY

In the CPDB, a standardized quantitative measure of carcinogenic potency, the TD_{50}, is estimated for each set of tumor incidence data. In a simplified way, TD_{50} may be defined as follows: for a given target site(s), if there are no tumors in control animals, then TD_{50} is that chronic dose-rate in mg/kg body wt/day which would induce tumors in half the test animals at the end of a standard lifespan for the species. Since the tumor(s) of interest often does occur in control animals, TD_{50} is more precisely defined as that dose-rate in mg/kg body wt/day which, if administered chronically for the standard lifespan of the species, will halve the probability of remaining tumorless throughout that period. TD_{50} is analogous to LD_{50}, and a low value of TD_{50} indicates a potent carcinogen, whereas a high value indicates a weak one. TD_{50} and the statistical procedures adopted for estimating it from experimental data have been described elsewhere.[1,6,7] The range of TD_{50} across chemicals in the CPDB is at least 10-million-fold for carcinogens in each sex of rat or mouse.

One goal of the CPDB has been to obtain data which would give the best estimates of carcinogenic potency, i.e., experiments for which detailed time-to-tumor data would permit adjustment for the gross effects of intercurrent mortality.[7,8] Full lifetable information was available for the calculations of potency from 403 bioassays of the NCI/NTP, from a set of bioassays on 33 aromatic amines, and from studies of 17 chemicals in nonhuman primates.

In the summary table below, a carcinogenic potency value is reported for a chemical in each species with a positive evaluation of carcinogenicity in at least one test. If there is only one positive test on the chemical in the species, then the most potent TD_{50} value from that test is reported. When more than one experiment is positive, in order to use all the available data, the reported potency value is a harmonic mean of the most potent TD_{50} values from each positive experiment. We have shown that the harmonic mean is similar to the most potent TD_{50} value for chemicals with more than one positive test.[9,10] The harmonic mean (T_H) is defined as:

$$T_H = \cfrac{1}{\cfrac{1}{n}\sum_{i=1}^{n}\cfrac{1}{T_i}}$$

For each positive experiment we select the lowest TD_{50} value from among positively evaluated target sites with a statistically significant dose-response (two-tailed $p < 0.1$). If no positive sites have a significant dose-response, then we select the most potent (lowest TD_{50}) from among positively evaluated sites with $p \geq 0.1$. When some experiments have positive significant results and others have only positive nonsignificant results, we discard the nonsignificant experimental results for the calculation of the harmonic mean. In some experiments, no TD_{50} could be estimated because all dosed animals had the tumor of interest, and only summary data were available for animals with tumor. For these cases we use the 99% upper confidence limit of TD_{50} as a replacement for the TD_{50}.

Additional information is indicated by mnemonic superscripts to the right of the TD_{50} value in each species:

i = Intraperitoneal or intravenous injection are the only routes of administration with positive tests for this species in the CPDB.

m = The CPDB contains more than one positive test in the species.

n = No results evaluated as positive for this species in the CPDB are statistically significant ($p < 0.1$).

v = Variation is greater than 10-fold among statistically significant ($p < 0.1$) TD_{50} values from different positive experiments.

TARGET SITES

Target sites are reported for each sex-species group with a positive result in the CPDB. These are the same target sites as those reported in the "Summary of Carcinogenic Potency Database by Target Organ";[11] however, here they are reported by chemical name in each sex-species group, whereas there they are listed by target organ in each species.

Target sites are identified on the basis of a positive author's opinion for the particular site, in any experiment in the sex-species, using all results from both the general literature and NCI/NTP bioassays. Hence, if a chemical has two target sites in a sex-species, the results may represent two different experiments. Occasionally the CPDB results are only for both sexes combined and this has been indicated with (B) next to the target site. Results in nonhuman primates are reported for both sexes combined in Section C.

Target sites in each sex-species are reported using mnemonic codes as follows: adr = adrenal gland; bon = bone; cli = clitoral gland; eso = esophagus; ezy = ear/Zymbal's

gland; gal = gallbladder; hag = harderian gland; hmo = hematopoietic system; kid = kidney; lgi = large intestine; liv = liver; lun = lung; meo = mesovarium; mgl = mammary gland; mix = mixture; myc = myocardium; nas = nasal cavity (includes tissues of the nose, nasal turbinates, paranasal sinuses and trachea); nrv = nervous system; orc = oral cavity (includes tissues of the mouth, oropharynx, pharynx, and larynx); ova = ovary; pan = pancreas; per = peritoneal cavity; pit = pituitary gland; pre = preputial gland; pro = prostate; ski = skin; smi = small intestine; spl = spleen; sto = stomach; sub = subcutaneous tissue; tba = all tumor-bearing animals; tes = testes; thy = thyroid gland; ubl = urinary bladder; ute = uterus; vag = vagina; vsc = vascular system.

REFERENCES

1. Gold, L.S., Slone, T.H., Manley, N.B., Garfinkel, G.B., Rohrbach, L., and Ames, B.N. Carcinogenic Potency Database. In: Gold, L.S. and Zeiger, E., Eds. *Handbook of Carcinogenic Potency and Genotoxicity Databases.* Boca Raton, FL: CRC Press, 1997, pp. 1–605

2. Index of Chemical Names in the Carcinogenic Potency (CPDB) and Genotoxicity Databases (GT) by Chemical Abstracts Service (CAS) registry number. In: Gold, L.S. and Zeiger, E., Eds. *Handbook of Carcinogenic Potency and Genotoxicity Databases.* Boca Raton, FL: CRC Press, 1997, pp. 731–754.

3. Zeiger, E. Genotoxicity Database. In: Gold, L.S. and Zeiger, E., Eds. *Handbook of Carcinogenic Potency and Genotoxicity Databases.* Boca Raton, FL: CRC Press, 1997, pp. 687–729.

4. Kier, L.E., Brusick, D.J., Auletta, A.E., Von Halle, E.S., Brown, M.M., Simmon, V.F., Dunkel, V., McCann, J., Mortelmans, K., Prival, M., Rao, T.K., and Ray, V. The *Salmonella typhimurium*/mammalian microsomal assay: A report of the U.S. Environmental Protection Agency Gene-Tox Program. *Mut. Res.* 168: 69-240 (1986).

5. Auletta, A.E. personal communication.

6. Sawyer, C., Peto, R., Bernstein, L., and Pike, M.C. Calculation of carcinogenic potency from long-term animal carcinogenesis experiments. *Biometrics* 40: 27-40 (1984).

7. Peto, R., Pike, M.C., Bernstein, L., Gold, L.S., and Ames, B.N. The TD_{50}: A proposed general convention for the numerical description of the carcinogenic potency of chemicals in chronic-exposure animal experiments. *Environ. Health Perspect.* 58: 1-8 (1984).

8. Gold, L.S., Bernstein, L., Kaldor, J., Backman, G., and Hoel, D. An empirical comparison of methods used to estimate carcinogenic potency in long-term animal bioassays: Lifetable vs. summary incidence data. *Fundam. Appl. Toxicol.* 6: 263-269 (1986).

9. Gold, L.S., Slone, T.H., and Bernstein, L. Summary of carcinogenic potency (TD_{50}) and positivity for 492 rodent carcinogens in the Carcinogenic Potency Database. *Environ. Health Perspect.* 79: 259-272 (1989).

10. Gold, L.S., Slone, T.H., and Ames, B.N. Overview of analyses of the Carcinogenic Potency Database. In: Gold, L.S. and Zeiger, E., Eds. *Handbook of Carcinogenic Potency and Genotoxicity Databases.* Boca Raton, FL: CRC Press, 1997, pp. 661–685.

11. Gold, L.S., Manley, N.B., and Slone, T.H. Summary of Carcinogenic Potency Database by target organ. In: Gold, L.S. and Zeiger, E., Eds. *Handbook of Carcinogenic Potency and Genotoxicity Databases.* Boca Raton, FL: CRC Press, 1997, 607–620.

SECTION A: SUMMARY OF CHEMICAL CARCINOGENICITY IN RATS AND MICE

TABLE 1

Chemical	CAS	Salmonella	Harmonic mean of TD$_{50}$ (mg/kg/day) Rat	Mouse	Rat target sites Male	Female	Mouse target sites Male	Female
A-α-C	26148-68-5	+	.	49.8[m]	.	.	liv vsc	liv vsc
Acetaldehyde[s]	75-07-0	–	153[m]	.	nas	nas	lun pre	cli lun sto
Acetaldehyde methylformylhydrazone	16568-02-8	–	.	2.51[m]	.	.	lun pre	cli lun sto
Acetaldoxime	107-29-9	–	–	–	.	.	.	.
Acetamide	60-35-5	–	180[m]	3010	liv	liv	hmo	.
Acetaminophen	103-90-2	–	495[m]	1620[m]	liv ubl	liv ubl	liv	liv
Acetohexamide	968-81-0	.	–	.	.	.	–	–
Acetone[4-(5-nitro-2-furyl)-2-thiazolyl]hydrazone	18523-69-8	.	6.05	.	.	sto	.	.
Acetoxime	127-06-0	.	12.1	.	liv	.	.	.
1'-Acetoxysafrole	34627-78-6	+	25[m]	–	sto	.	.	.
N'-Acetyl-4-(hydroxymethyl)phenylhydrazine	65734-38-5	.	.	241[m]	.	.	lun vsc	lun vsc
1-Acetyl-2-isonicotinoylhydrazine	1078-38-2	.	.	330[m]	.	.	lun	lun
3-Acetyl-6-methyl-2,4-pyrandione	520-45-6	–	–	.	.	.	.	.
1-Acetyl-2-phenylhydrazine	114-83-0	+	.	51.2[m]	.	.	vsc	vsc
4-Acetylaminobiphenyl	4075-79-0	.	1.18	.	.	mgl	.	.
1-Acetylaminofluorene	28314-03-6	.	–	.	.	.	.	.
2-Acetylaminofluorene[s]	53-96-3	+	1.22[m]	7.59[m,v]	liv mgl ski	liv mgl ski	liv ubl	liv ubl
4-Acetylaminofluorene	28322-02-3	+	–	.	.	.	.	.
Acetylated diamylopectin phosphate	—							
Acetylated distarch adipate	68130-14-3							
Acetylated distarch glycerol	53123-84-5							
Acetylated distarch phosphate	68130-14-3							
Acifluorfen	50594-66-6	.	.	141[m]	sto	.	liv sto	liv sto
Acrolein	107-02-8	+	.	.	.	.	.	.
Acrolein diethylacetal	3054-95-3	–	.	.	.	.	.	.
Acrolein oxime	5314-33-0	.	–	.	.	.	.	.
Acronycine	7008-42-6	.	0.505[i,m]	I	bon per	mgl per	I	I
Acrylamide	79-06-1	–	6.15[m]	.	adr nrv per	mgl orc pit	.	.
Acrylonitrile	107-13-1	+	16.9[m,v]	.	ezy nrv orc smi sto	ezy mgl nas nrv orc smi sto	.	.
Actinomycin C	8052-16-2	.	.	.	.	.	.	.
Actinomycin D	50-76-0	–	0.00111[i,m]	.	per	per	.	.
Adipamide	628-94-4	.	–	.	–	–	.	.
AF-2[s]	3688-53-7	+	29.4[m,v]	131[m,v]	mgl	mgl	sto	sto
Aflatoxicol	29611-03-8	+	0.00247	.	liv	.	.	.
Aflatoxin B_1[s]	1162-65-8	+	0.0032[m,v]	.	kid lgi liv	lgi liv	.	.
Aflatoxin, crude	—	.	0.00299[m]	0.343	liv	.	hmo	.

Chemical	CAS No.									
Agar	9002-18-0	.	—	—	—	—	—	—	—	—
Aldicarb	116-06-3	—	—	—	—	—	—	—	—	—
Aldrin	309-00-2	—	—	1.27m	liv	—	—	liv	liv	liv(B)
Alkylbenzenesulfonate, linear	—	.	.	—	—	—	—	—	—	.
Alkyldimethylamine oxides, commercial grade	mixture	.	.	.	.	.	.	.	.	.
Allantoin	97-59-6	.	.	.	.	.	.	.	.	.
Allyl alcohol	107-18-6	+	I	.	I	—	—	—	—	hmo
Allyl chloride	107-05-1	+	I	I	I	I	I	—	—	—
Allyl glycidyl ether	106-92-3	+	—	182	nas	—	—	—	nas	—
Allyl isothiocyanate	57-06-7	+	96	—	ubl	hmo	—	—	—	—
Allyl isovalerate	2835-39-4	—	123	62.8	hmo	—	—	—	—	hmo
1-Allyl-1-nitrosourea	760-56-5	.	0.341m	.	.	mgl sto ute	.	.	.	.
Allylhydrazine.HCl	52207-83-7	+	.	34.2m	lun	lgi lun sto	.	lun	.	lun vsc
Aluminum potassium sulfate	10043-67-1	.	—	.	.	.	.	.	.	.
3-Amino-4-ethoxyacetanilide	17026-81-2	+	.	2070	thy	—	.	thy	.	.
3-Amino-9-ethylcarbazole.HCl	6109-97-3	+	57.2m	38.6m	liv	ezy liv ski	ezy liv ute	liv	ezy	liv
3-Amino-9-ethylcarbazole mixture	mixture	+	26.4m	38m	liv	ezy liv ski	ezy	liv	ezy	liv
1-Amino-2-methylanthraquinone	82-28-0	+	59.2m	174	liv	kid liv	liv	liv	kid lun mgl	liv
2-Amino-5-(5-nitro-2-furyl)-1,3,4-oxadiazole	3775-55-1	.	3.67	.	—	kid lun mgl sto	kid lun mgl sto	.	.	.
2-Amino-5-(5-nitro-2-furyl)-1,3,4-thiadiazole	712-68-5	.	0.662	.	.	kid lun mgl sto	kid lun mgl sto	.	.	.
2-Amino-4-(5-nitro-2-furyl)thiazole	38514-71-5	+	5.85	7.87	.	sto ubl	sto	.	.	sto
trans-5-Amino-3[2-(5-nitro-2-furyl)vinyl]-1,2,4-oxadiazole	28754-68-9	.	112m	.	hmo sto	.	.	.	hmo sto	hmo sto
2-Amino-4-nitrophenol	99-57-0	+	839	—	—	kid	.	.	.	.
2-Amino-5-nitrophenol	121-88-0	+	111	—	thy	pan	.	liv	.	.
4-Amino-2-nitrophenol	119-34-6	+	309	—	—	ubl	.	ubl	.	.
2-Amino-4-(p-nitrophenyl)thiazole	2104-09-8	+	44.6	9.95	hmo	kid lun mgl	kid lun mgl	.	.	hmo
2-Amino-5-nitrothiazole	121-66-4	+	—	.	—	—	—	—	—	—
2-Amino-5-phenyl-2-oxazolin-4-one + Mg(OH)₂	18968-99-5	+	101	1190m	hmo liv	.	.	liv	.	hmo liv
2-Aminoanthraquinone	117-79-3	+	4.04m	—	—	liv	liv	liv	liv	—
o-Aminoazotoluene	97-56-3	+	0.98	—	liv ubl	liv	liv	liv	liv	—
4-Aminodiphenyl	92-67-1	+	.	2.1m	liv ubl	ubl	.	liv ubl	.	liv ubl
4-Aminodiphenyl.HCl	2113-61-3	+	9.94m	.	liv	mgl	.	liv ubl	.	liv
2-Aminodiphenylene oxide	3693-22-9	.	1100	4.24m	—	mgl	.	liv ubl	.	hmo
1-(Aminomethyl)cyclohexaneacetic acid	60142-96-3	.	.	.	.	.	.	.	.	.
3-Aminotriazole[s]	61-82-5	.	.	25.3m	thy	.	pit thy	thy	.	liv
11-Aminoundecanoic acid	2432-99-7	—	.	—	liv ubl	.	.	liv ubl	.	liv
Ammonium chloride	12125-02-9	.	.	.	.	.	.	.	.	.
Ammonium citrate	3012-65-5	.	.	.	.	.	.	.	.	.
Ammonium hydroxide	1336-21-6	.	.	.	.	.	.	.	.	.
Amobarbital	57-43-2	—	—	.	—	—	.	—	.	.
dl-Amphetamine sulfate	60-13-9	—	—	—	—	—	—	—	—	—
Ampicillin trihydrate	7177-48-2	—	—	—	—	—	—	—	—	—

TABLE 1 (continued)

Chemical	CAS	Salmonella	Harmonic mean of TD$_{50}$ (mg/kg/day) Rat	Mouse	Rat target sites Male	Female	Mouse target sites Male	Female
1-Amyl-1-nitrosourea	10589-74-9	.	0.555^m	.	hmo lun sto	hmo lun mgl sto ute	.	.
Amylopectin sulfate	9047-13-6	.	283^m	.	lgi	.	.	.
Anethole	104-46-1	–	.	.	.	.	–	–
trans-Anethole	4180-23-8	–	.	.	.	.	–	–
Anhydroglucochloral	15879-93-3	–	.	–	.	.	–	–
Anilazine	101-05-3	–	.	.	.	.	–	–
Aniline	62-53-3	–	.	.	.	.	–	–
Aniline.HCl	142-04-1	.	269^{m,v}	.	per spl vsc	per	–	–
o-Anisidine.HCl	134-29-2	+	29.7^m	966^m	kid thy ubl	ubl	ubl	ubl
p-Anisidine.HCl	20265-97-8	+	.	.	.	.	–	–
Anthranilic acid	118-92-3	–	.	.	.	.	–	–
9,10-Anthraquinone	84-65-1	+	.	.	.	.	–	–
Antimony potassium tartrate	28300-74-5	.	.	.	.	.	B–	B–
Aramite	140-57-8	.	96.7^m	158	liv(B)	liv(B)	liv	–
Arecoline.HCl	61-94-9	.	.	39.5^m	.	.	lun sto vsc	lun vsc
Aroclor 1254	27323-18-8	–	1.74^m	9.58	liv	liv	liv	–
Aroclor 1260	11096-82-5	–	.	.	.	.	.	.
Arsenate, sodium^s	7631-89-2	–	–	.	B–	B–	B–	.
Arsenious oxide	1327-53-3	.	–	.	.	.	.	.
Arsenite, sodium	7784-46-5	.	–	.	B–	B–	B–	B–
l-Ascorbate, sodium	134-03-2	.	–	.	.	.	.	.
l-Ascorbic acid	50-81-7	–	–	.	.	.	.	.
Aspartame	22839-47-0	–	.	.	.	B–	.	.
Aspirin	50-78-2	–	.	.	.	.	.	.
Aspirin, phenacetin, and caffeine	8003-03-0	–	.	.	.	.	.	.
Atrazine	1912-24-9	–	31.7^m	.	mgl	hmo ute	.	.
Atropine	51-55-8	–	.	.	.	.	.	.
Auramine-O	2465-27-2	+	11	62.7^m	liv	.	liv	liv
Auranofin	34031-32-8	+	.	.	tes	.	.	.
5-Azacytidine	320-67-2	+	0.17^i	0.0774^{i,m}	I	I	hmo lun ski	hmo mgl ski
6-Azacytidine	3131-60-0	.	–	.	.	.	.	.
Azaserine	115-02-6	+	0.793^i	8.92	pan(B)	pan(B)	.	.
Azathioprine	446-86-6	+	–	.	.	.	.	.
Azide, sodium	26628-22-8	+	.	.	.	.	.	.
Azinphosmethyl	86-50-0	+	.	.	.	.	.	.
Azobenzene	103-33-3	+	24.1^m	.	spl vsc	spl	.	hmo vsc
Azoxymethane	25843-45-2	+	0.0466^m	.	ezy kid lgi liv	.	.	.
1-Azoxypropane	—	.	0.000241	.	nas ski	.	.	.

Chemical	CAS No.	+/−	TD50	TD50	Rat ♂	Rat ♀	Mouse ♂	Mouse ♀
2-Azoxypropane	—	.	0.00268		ski	.	.	.
Barbital, sodium	144-02-5	.	105		kid	.	.	.
Barbituric acid	67-52-7	—	—	.	—	—	—	—
Barium acetate	543-80-6	—	—	—	—	—	—	—
Barium chloride dihydrate	10326-27-9	—	—	—	—	—	.	.
Bemitradine	88133-11-3	.	548[m]	1490[m]	liv thy	liv mgl thy	sto	sto
Benzaldehyde	100-52-7	—	.	.	—	—	sto	sto
Benzene	71-43-2	—	169[m]	77.5[m,v]	ezy nas orc ski sto vsc	ezy nas orc sto vsc	ezy hag hmo lun pre	ezy hmo lun mgl ova
Benzenesulphonohydrazide	5351-65-5	+	—	—	—	.	.	.
Benzidine	92-87-5	+	1.73	.	hmo(B)	hmo(B)	hmo(B)	hmo
Benzidine.2HCl	531-85-1	+	.	19.7[m]	liv(B) mgl(B)	liv(B) mgl(B)	hag liv	hag liv vsc
Benzo(a)pyrene	50-32-8	+	0.956	11	sto(B)	sto(B)	eso	eso
Benzoate, sodium	532-32-1	—	—	—	—	—	—	—
Benzofuran	271-89-6	—	424	25.1[m]	kid	kid	liv lun sto	liv lun sto
Benzoguanamine	91-76-9	—	—	—	—	—	—	—
Benzoin	119-53-9	—	—	—	—	—	—	—
1,4-Benzoquinone	106-51-4	—	—	5.07[m]	—	—	hmo	hmo
Benzothiazyl disulfide	120-78-5	+	—	—	—	—	—	—
1H-Benzotriazole	95-14-7	+	—	—	—	—	—	—
Benzoyl hydrazine	613-94-5	—	—	9.59[m]	—	—	hmo lun	hmo lun
Benzyl acetate	140-11-4	—	—	1440[m]	—	—	liv sto	liv sto
Benzyl alcohol	100-51-6	—	—	—	—	—	—	—
Benzyl chloride	100-44-7	+	—	61.5[m]	—	sto	sto	sto
o-Benzyl-p-chlorophenol	120-32-1	—	1350	—	kid	kid	—	—
Benzylhydrazine.2HCl	20570-96-1	—	—	85.3	—	—	lun	lun
Beryllium sulfate	13510-49-1	—	—	—	—	—	—	—
Biphenyl	92-52-4	—	—	—	—	—	—	—
2-Biphenylamine.HCl	2185-92-4	+	—	1120	—	—	vsc	vsc
Bis(2-chloro-1-methylethyl)ether, technical grade	108-60-1	+	—	191[m]	—	—	liv lun	liv lun
Bis-2-chloroethylether	111-44-4	+	—	11.7[m]	—	—	liv	liv lun
Bis-1,4-(chloromethoxy)butane	13483-19-7	—	—	—	—	—	—	—
Bis-1,2-(chloromethoxy)ethane	13483-18-6	—	—	4.62[i]	per	per	—	—
Bis-1,6-(chloromethoxy)hexane	56894-92-9	—	—	—	per	per	—	—
Bis-1,4-(chloromethoxy)-p-xylene	56894-91-8	—	—	3.11[i]	per	per	—	—
Bis-(chloromethyl)ether	542-88-1	—	0.00357	0.182[m]	lun nas sto	lun	lun	lun
Bis(2,3-dibromopropyl)phosphate, magnesium salt	36711-31-6	.	32[m]	.	eso smi sto	eso smi sto	—	—
4-Bis(2-hydroxyethyl)amino-2-(5-nitro-2-thienyl)quinazoline	33372-39-3	.	3.14		mgl smi	mgl smi	.	.
4-Bis(2-hydroxyethyl)amino-2-(2-thienyl)quinazoline	58139-47-2	—	—	—	.	.	.	.
Bis-2-hydroxyethyldithiocarbamic acid, potassium	23746-34-1	.	.	37.7[m]	liv	liv	—	—

TABLE 1 (continued)

Chemical	CAS	Salmonella	Harmonic mean of TD50 (mg/kg/day)		Rat target sites		Mouse target sites	
			Rat	Mouse	Male	Female	Male	Female
Bis(2-hydroxypropyl)amine	110-97-4	–	–	.	–	.	.	.
Bis(tri-n-butyltin)oxide, technical grade	56-35-9	–	–	.	–	.	.	.
Bismuth dimethyldithiocarbamate	21260-46-8	.	.	–	.	.	.	.
Bismuth oxychloride	7787-59-9	.	.	.	.	B–	.	.
Bisphenol A	80-05-7	–	–	–	B–	B–	.	.
Black PN	2519-30-4	.	–	71.6	.	.	.	.
C.I. direct black 38	1937-37-7	+	1.39[m]	.	liv	liv	liv(B)	liv(B) mgl
C.I. direct blue 6	2602-46-2	+	1.73[m]	.	liv	liv	.	.
C.I. direct blue 15	2429-74-5	+	27.5[m]	.	ezy lgi liv orc pre ski smi	cli ezy hmo lgi liv orc ski smi ute	.	.
C.I. direct blue 218	28407-37-6	–	1570	857[m]	orc	.	liv	liv
C.I. disperse blue 1	2475-45-8	+	156[m]	–	ubl	ubl	–	–
FD & C blue no. 1	3844-45-9	.	–	–	B–	B–	–	–
FD & C blue no. 2	860-22-0	–	–	–	B–	B–	–	–
HC blue no. 1	2784-94-3	+	702	86.3[m]	–	lun	liv thy	liv
HC blue no. 1 (purified)	2784-94-3	.	–	78.7[m]	.	.	liv	liv
HC blue no. 2	33229-34-4	+	–	–	.	.	.	.
Boric acid	10043-35-3	–	–	–	.	.	.	.
Bromate, potassium	7758-01-2	+	9.81[m]	–	kid per	kid	.	.
Bromoacetaldehyde	17157-48-1	.	.	.	.	.	.	.
Bromodichloromethane	75-27-4	+	72.5[m,v]	47.7[m]	kid lgi	kid lgi liv	kid	liv
Bromoethane	74-96-4	+	149[n]	535	adr lun nrv	.	.	ute
Bromoethanol	540-51-2	+	.	76.1[m]	.	.	sto	sto
C.I. direct brown 95	16071-86-6	+	2.07	.	liv	liv	.	.
Budesonide	51333-22-3	.	0.291	.	.	.	.	.
1,3-Butadiene	106-99-0	+	261[m,v]	13.9[m,v]	tes	mgl	hag hmo kid liv lun nrv pre sto	hag hmo liv lun mgl ova sto vsc
Butyl benzyl phthalate	85-68-7	–	–	–	I	–	–	–
n-Butyl chloride	109-69-3	–	–	–	–	–	–	–
2-sec-Butyl-4,6-dinitrophenol	88-85-7	–	–	–	.	.	.	.
N-n-Butyl-N-formylhydrazine	16120-70-0	.	.	19.3[m]	.	.	lun pre	cli lun
Butyl p-hydroxybenzoate	94-26-8	.	–	.	.	.	.	.
N-Butyl-N-(4-hydroxybutyl)nitrosamine	3817-11-6	+	0.457[m,v]	.	tes ubl	ubl	.	.
di-tert-Butyl-4-hydroxymethyl phenol	88-26-6	.	–	–	.	.	.	.
2-tert-Butyl-4-methoxyphenol	2409-55-4	.	–	–	.	.	.	.
N-Butyl-N'-nitro-N-nitrosoguanidine	13010-08-7	+	–	.	–	–	.	.

Chemical	CAS No.	Mut	Rat TD50	Mouse TD50	Rat	Rat	Mouse	Mouse
N-n-Butyl-N-nitrosourea	869-01-2	+	0.517[m,v]	.	ezy eso hmo lgi lun sto	ezy eso hmo lgi lun mgl sto ute vag	.	.
Butylated hydroxyanisole [s]	25013-16-5	−	745[m,v]	5530[m,n]	sto	sto	sto	B−
Butylated hydroxytoluene	128-37-0	−	−	653[m]	−	−	liv lun	lun sto
1,1-di-n-Butylhydrazine	7422-80-2	.	.	45.2[m]	.	.	liv lun sto	lun
n-Butylhydrazine.HCl	56795-65-4	.	.	12.1[m]	.	.	lun	hmo lun
1,2-di-n-Butylhydrazine.2HCl	78776-28-0	.	.	46.2[m]	.	.	hmo kid lun	.
p-tert-Butylphenol	98-54-4	.	.	.	.	.	.	.
N-Butylurea	592-31-4	.	.	.	.	.	.	.
β-Butyrolactone	3068-88-0	.	13.8	.	.	sto	.	.
γ-Butyrolactone	96-48-0	.	−	.	B−	B−	.	.
Cadmium acetate	543-90-8	.	.	.	B−	B−	.	.
Cadmium chloride [s]	10108-64-2	.	0.0114[m,v]	.	hmo lun pro tes	lun	.	.
Cadmium chloride monohydrate	35658-65-2	.	−	.	.	.	.	.
Cadmium diethyldithiocarbamate	14239-68-0	.	.	.	.	.	.	.
Cadmium sulphate (1:1) [s]	10124-36-4	.	0.0217[m]	.	lun	lun	.	.
Cadmium sulphate (1:1) hydrate (3:8)	7790-84-3	.	−	.	.	.	.	.
Caffeic acid	331-39-5	.	297[m]	4900[m]	kid sto	sto	lun sto	kid sto
Caffeine	58-08-2	−	.	.	.	.	.	.
Calciferol	50-14-6	.	.	39.6[n]	.	.	.	mgl
Calcium acetate	62-54-4	.	.	.	.	.	.	.
Calcium chloride	10043-52-4	.	.	.	.	.	.	.
Calcium lactate	814-80-2	.	.	.	.	.	.	.
Caprolactam	105-60-2	−	−	.	.	.	.	.
Capsaicin	404-86-4	.	.	167[m,n]	.	.	.	lgi
Captafol	2425-06-1	−	59.4[m,v]	108[m]	kid liv	kid liv	liv smi sto vsc	liv smi sto vsc
Captan	133-06-2	+	−	.	.	.	.	.
Carbamyl hydrazine.HCl	563-41-7	.	.	223[m]	.	.	lun	lun vsc
1-Carbamyl-2-phenylhydrazine	103-03-7	−	.	165[m]	.	.	lun	lun
Carbarsone	121-59-5	.	−	.	.	.	.	.
Carbaryl	63-25-2	+	14.1	.	tba(B)	tba(B)	.	.
Carbazole	86-74-8	−	.	164[m]	.	.	liv	liv sto
Carbon tetrachloride	56-23-5	−	2.29[m,n]	150[m]	liv	liv mgl	liv sto	liv sto
Carboxymethylnitrosourea	60391-92-6	.	4.31[m,v]	.	ski	mgl smi	adr liv	adr liv
Carbromal	77-65-6	−	.	.	.	.	.	.
Carrageenan, acid-degraded	—	−	2310[m]	.	lgi	lgi(B)	.	.
Carrageenan, native [s]	9000-07-1	.	.	.	.	.	.	.
d-Carvone	2244-16-8	−	.	.	.	.	.	.
Catechol	120-80-9	−	118[m]	244[m]	sto	sto	sto	sto
Celiprolol	56980-93-9	.	.	.	.	.	.	.
Chenodeoxycholic acid	474-25-9	.	.	.	.	.	B−	B−
Chloral hydrate	302-17-0	+	.	106	.	.	liv	liv
Chloramben	133-90-4	+	−	5230	.	.	liv	.

TABLE 1 (continued)

Chemical	CAS	Salmonella	Harmonic mean of TD$_{50}$ (mg/kg/day) Rat	Harmonic mean of TD$_{50}$ (mg/kg/day) Mouse	Rat target sites Male	Rat target sites Female	Mouse target sites Male	Mouse target sites Female
Chlorambucil	305-03-3	+	0.896[m]	0.133[i,m]	hmo	ezy mgl nrv	hmo lun	lun
Chloraminated water	—	.	—	.	.	.	—	—
Chloramphenicol	56-75-7	.	—	.	.	.	.	.
Chloranil	118-75-2	—	.	2.99[m]	.	.	liv	liv
Chlordane, technical grade	57-74-9	—	40.8[m]	141	liv pan	liv	liv	liv thy
Chlorendic acid	115-28-6	—	222[m]	113[m]	kid liv	liv thy	liv	liv
Chlorinated paraffins (C_{12}, 60% chlorine)	63449-39-8	—	.	6540	—	—	hmo	—
Chlorinated paraffins (C_{23}, 43% chlorine)	63449-39-8	—	—		—	—	—	—
Chlorinated trisodium phosphate	56802-99-4	+	—	.	—	—	—	ubl vsc
Chlorinated water	—	—	—	—	—	—	—	—
Chlorine	7782-50-5	—	—	.	B–	B–	.	.
Chlormadinone acetate	302-22-7	.	37.6	346	liv	.	liv	liv
4-Chloro-4'-aminodiphenylether	101-79-1	—	4.85	.	liv	.	.	.
2-Chloro-5-(3,5-dimethylpiperidinosulphonyl)benzoic acid	37087-94-8	—	—	—	—	—	—	—
1-Chloro-2,4-dinitrobenzene	97-00-7	.	—	157[m]	—	—	liv	liv
3-Chloro-2-methylpropene, technical grade (containing 5% dimethylvinyl chloride)	563-47-3	+	113[m]	77.7[m]	sto	sto	sto	sto
1-Chloro-2-nitrobenzene	88-73-3	+	—	473[m]	—	—	liv vsc	vsc
1-Chloro-4-nitrobenzene	100-00-5	+	—	1230	—	—	liv	liv
4-Chloro-m-phenylenediamine	5131-60-2	+	315	1340[m]	adr	—	liv	liv
4-Chloro-o-phenylenediamine	95-83-0	+	214[m]	—	sto ubl	sto ubl	—	—
2-Chloro-p-phenylenediamine sulfate	61702-44-1	+	—	195[m]	—	—	liv vsc	liv
3-Chloro-p-toluidine	95-74-9	+	—	—	—	—	—	—
5-Chloro-o-toluidine	95-79-4	—	—	25.8[m,v]	—	—	vsc	vsc
4-Chloro-o-toluidine.HCl	3165-93-3	—	—	—	—	—	—	—
2-Chloro-1,1,1-trifluoroethane	75-88-7	—	87.3[m]	10.8	tes	ute	liv	—
[4-Chloro-6-(2,3-xylidino)-2-pyrimidinylthio]acetic acid	50892-23-4	.	9.69[m]	—	liv	.	liv	—
4-Chloro-6-(2,3-xylidino)-2-pyrimidinylthio(N-β-hydroxyethyl) acetamide	65089-17-0	.	6.49	44.6	liv	.	.	liv
Chloroacetaldehyde	107-20-0	+	.	36.1	.	.	liv	liv
2-Chloroacetophenone	532-27-4	—	—	—	.	.	—	—
4'-(Chloroacetyl)-acetanilide	140-49-8	+	—	—	—	—	—	—
p-Chloroaniline	106-47-8	+	—	—	.	.	—	—
p-Chloroaniline.HCl	20265-96-7	+	7.62	89.5	spl	.	liv vsc	liv
o-Chlorobenzalmalonitrile	2698-41-1	+	—	—	—	—	—	—
Chlorobenzene	108-90-7	—	247	—	liv	.	liv	—
Chlorobenzilate	510-15-6	—	—	93.9[m,v]	—	—	liv	liv

Chemical	CAS No.							
Chlorodibromomethane	124-48-1	—	—	139	lun	—	—	liv
Chlorodifluoromethane	75-45-6	.	—	—	—	—	—	—
Chloroethane	75-00-3	+	—	1810	—	I	I	ute
(2-Chloroethyl)trimethylammonium chloride	999-81-5	—	0.124	—	—	—	—	—
1-Chloroethylnitroso-3-(2-hydroxypropyl)urea	—							.
Chlorofluoromethane	593-70-4	—	27.5[m]	.	lun	sto	sto	liv
Chloroform[s]	67-66-3	—	262[m]	90.3[m]	kid	liv	kid liv	liv
Chloromethyl methyl ether[s]	107-30-2	+	5.5	—	mix	—	—	.
2-(Chloromethyl)pyridine.HCl	6959-47-3	+	—	.	.	—	.	.
3-(Chloromethyl)pyridine.HCl	6959-48-4	+	433	229[m]	sto	sto	sto	sto
p-Chlorophenyl-p-chlorobenzene sulfonate	80-33-1	.	.	.	.	.	.	—
3-(p-Chlorophenyl)-1,1-dimethylurea	150-68-5	—	131	.	kid liv	kid liv	.	.
1-(4-Chlorophenyl)-1-phenyl-2-propynyl carbamate	10473-70-8	.	8.78	.	nrv smi	.	.	.
p-Chlorophenyl-2,4,5-trichlorophenyl sulfide	2227-13-6	.	.	.	.	.	.	.
Chloropicrin	76-06-2	+	I	I	I	I	I	I
2-Chloropropanal	683-50-1	.	.	12.9	.	.	.	.
1-Chloropropene	590-21-6	—	.	5.05	.	.	.	.
Chlorothalonil	1897-45-6	—	2270[m]	.	kid	kid	kid	sto
Chlorozotocin	54749-90-5	+	0.0375[i,m]	.	per	per	per	sto
Chlorpheniramine maleate	113-92-8	—	—	.	—	—	—	—
Chlorpropamide	94-20-2	.	—	.	—	—	—	—
Chocolate Brown FB	12236-46-3	.	.	.	.	.	.	.
Chocolate Brown HT	4553-89-3	.	.	.	.	.	.	.
Choline chloride	67-48-1	.	.	.	.	.	.	.
Chromic oxide pigment	1308-38-9	.	.	.	B—	B—	B—	.
Chromium (III) acetate	1066-30-4	.	.	.	B—	B—	B—	.
Chrysazin	117-10-2	.	245	201	lgi	.	.	.
Cimetidine	51481-61-9	—	.	.	—	—	—	—
Cinnamyl anthranilate	87-29-6	—	12100	2580[m]	kid pan	kid pan	liv	liv
Ciprofibrate	52214-84-3	—	2.16[m,v]	6.2[m]	liv sto	liv sto	liv	liv
Citrinin	518-75-2	—	7.48[m]	.	kid	kid	sto	.
Clivorine	33979-15-6	.	0.5	.	liv(B) vsc(B)	liv(B) vsc(B)	liv(B) vsc(B)	.
Clofibrate	637-07-0	.	169	.	liv pan	liv pan	.	.
Clomiphene citrate	43054-45-1	.	.	.	liv	.	I	.
Clonitralid	1420-04-8	.	.	.	.	.	.	—
Clophen A 30	55600-34-5	.	157[n]	.	liv	liv	liv	liv
Colcemid	477-30-5	.	.	.	.	.	.	.
Compound 50-892	65765-07-3	.	.	.	.	.	.	.
Compound LY171883	88107-10-2	.	.	112	.	.	.	.
Copper dimethyldithiocarbamate	137-29-1	.	.	.	.	.	.	liv
Copper-8-hydroxyquinoline	10380-28-6	.	.	.	.	.	.	—
Corn oil	8001-30-7	.	.	.	.	.	.	—
Coumaphos	56-72-4	—	—	—	—	—	—	—

TABLE 1 (continued)

Chemical	CAS	Salmonella	Harmonic mean of TD$_{50}$ (mg/kg/day)		Rat target sites		Mouse target sites	
			Rat	Mouse	Male	Female	Male	Female
Coumarin[s]	91-64-5	+	13.9	103[m]	kid	–	lun	liv lun
m-Cresidine	102-50-1	+	470[m]		ubl	ubl	I	–
p-Cresidine	120-71-8	+	98[m]	54.3[m]	liv nas ubl	nas ubl	ubl	liv ubl
Crotonaldehyde	123-73-9	.	4.2		liv			
Cupferron	135-20-6	+	8.35[m]	585[m]	liv sto vsc	ezy liv sto vsc	vsc	ezy hag liv vsc
Cyanamide, calcium	156-62-7	+	–					
Cyclamate, sodium	139-05-9	.	–	667[m]	B–	B–	liv	tba
Cyclochlorotine	12663-46-6	.	–	23.6			liv	.
Cyclohexanone	108-94-1	–	–					
N-Cyclohexyl-2-benzothiazole sulfenamide	95-33-0	–						
Cyclohexylamine.HCl	4998-76-9	–						
Cyclohexylamine sulfate	19834-02-7	–						
Cyclophosphamide	50-18-0	+	2.21[m,v]	5.96[i,m]	hmo(B) nrv(B) tba ubl(B)	hmo(B) nrv(B) tba ubl(B)	lun	hmo lun
Cyclosporin A	59865-13-3	.						
Cytembena	16170-75-5	+	2.77[i,m]		per	mgl	–	–
Dacarbazine	4342-03-4	+	0.71	0.966[i,m]	–	hmo mgl ute	hmo lun vsc	hmo lun ute vsc
Daminozide	1596-84-5	–	2500[n]	1030[m]		ute	kid lun vsc	lun vsc
Dapsone	80-08-0	–	22.4		per spl			
o,p'-DDD	53-19-0	–						
p,p'-DDD	72-54-8	–		30.7[m]			liv lun	lun
p,p'-DDE[s]	72-55-9	–		12.5[m]			liv	liv
DDT[s]	50-29-3	–	84.7[m]	12.3[m,v]	liv	liv	liv	hmo liv
Decabromodiphenyl oxide	1163-19-5	–	3340[m]		liv	liv	–	–
Dehydroepiandrosterone acetate	853-23-6	.	31.4		liv			
Deltamethrin	52918-63-5	.	–					
Deserpidine	131-01-1	.			B–	B–		
Dextran	9004-54-0	.						
Dextran sulfate sodium (DS-M-1)	9011-18-1	–	196[m]		lgi	lgi		
Dextran sulfate sodium (DST-H)	9011-18-1	.						
Dextran sulfate sodium (KMDS-H)	9011-18-1	.						
N-1-Diacetamidofluorene	63019-65-8		19			ezy mgl		
Diacetyl hydrazine	3148-73-0	.						
Diallate	2303-16-4	+		26.7[m]	liv		liv	liv
Diallyl phthalate	131-17-9	.						
1,1-Diallylhydrazine	5164-11-4	.		29.6[m]			lun sto	lun sto
1,2-Diallylhydrazine.2HCl	26072-78-6	.		33.8[m]			lun	lun
Diallylnitrosamine[s]	16338-97-9	+	33.9[m]	.	nas	nas	.	.
4,6-Diamino-2-(5-nitro-2-furyl)-S-triazine	720-69-4	+	1.71	.	.	mgl	.	.

Chemical	CAS No.							
4,4'-Diamino-2,2'-stilbenedisulfonic acid, disodium salt	7336-20-1	—	—	—	—	—	—	—
2,4-Diaminoanisole sulfate	39156-41-7	+	183^{m}	906^{m}	ezy pre ski thy	cli ezy mgl thy	thy	thy
4,4'-Diaminoazobenzene	538-41-0	.	.	.	.	.	.	—
4,4'-Diaminobenzanilide	785-30-8	.	.	143	.	.	kid	—
2,4-Diaminophenol.2HCl	137-09-7	+	2.47^{m}	26.7	liv	liv mgl	kid	liv
2,4-Diaminotoluene	95-80-7	+	4.42	203^{m}	liv sub	liv mgl	liv vsc	liv
2,4-Diaminotoluene.2HCl	636-23-7	+	.	—	.	.	.	—
2,6-Diaminotoluene.2HCl	15481-70-6	+	.	—	.	.	.	—
2,5-Diaminotoluene sulfate	6369-59-1	+	.	.	.	.	.	—
Diazepam	439-14-5	—	.	.	.	.	.	—
Diazinon	333-41-5	—	.	.	.	.	.	—
3-Diazotyramine.HCl	—	+	37.6	5.88	orc	lun	.	—
Dibenz(a,h)anthracene	53-70-3	+	.	5.88	.	lun	lun	lun nas sto
Dibenzo-p-dioxin	262-12-4	—	.	.	.	.	.	—
3-Dibenzofuranamine	4106-66-5	+	2.48	.	tba	.	.	—
O,S-Dibenzoyl thiamine.HCl	35660-60-7	—	.	.	.	.	.	—
1,2-Dibromo-3-chloropropane	96-12-8	+	0.259^{m}	2.72^{m}	nas orc sto	adr mgl nas orc sto	lun nas sto	lun nas sto
Dibromodulcitol	10318-26-0	+	8.37^{i}	$11^{i,m}$	ski	.	hmo lun	hmo lun
1,2-Dibromoethane	106-93-4	+	1.52^{i}	$7.45^{m,v}$	nas per pit sto vsc	liv lun mgl nas pit sto vsc	lun sto vsc	eso lun mgl nas sto sub vsc
Dibromomannitol	488-41-5	+	$27.6^{i,m}$	14.9^{m}	per ski	mgl per	lun	hmo lun
Dibromoneopentyl glycol	3296-90-0	+	.	.	.	.	.	—
5,7-Dibromoquinoline	34522-69-5	—	.	.	.	.	.	—
1,3-Dibutyl-1-nitrosourea	56654-52-5	+	4.28	.	.	hmo mgl	.	—
Dibutyltin diacetate	1067-33-0	—	.	.	.	.	.	—
3,5-Dichloro(N-1,1-dimethyl-2-propynyl)benzamide	23950-58-5	+	.	119	.	I	liv	liv
2,3-Dichloro-p-dioxane	3883-43-0	—	.	.	.	.	.	—
α,β-Dichloro-β-formylacrylic acid	87-56-9	+	.	.	.	.	.	—
3,4'-Dichloro-2-methylacrylanilide	2164-09-2	.	.	.	.	.	.	—
2,3-Dichloro-1,4-naphthoquinone	117-80-6	.	.	.	.	.	.	—
2,6-Dichloro-4-nitroaniline	99-30-9	+	.	.	.	.	.	—
2,6-Dichloro-p-phenylenediamine	609-20-1	+	.	803^{m}	.	.	liv	liv
Dichloroacetic acid	79-43-6	+	3.58^{m}	49.3	kid liv	hmo kid liv	liv	liv
Dichloroacetylene	7572-29-4	.	.	0.574^{m}	.	.	hag kid	hag kid
1,2-Dichlorobenzene	95-50-1	—	.	.	.	.	.	—
1,4-Dichlorobenzene	106-46-7	+	644	398^{m}	kid	.	liv	liv
3,3'-Dichlorobenzidine^s	91-94-1	+	28.1^{m}	1.52^{i}	ezy hmo mgl	mgl	.	.
trans-1,4-Dichlorobutene-2	110-57-6	.	.	.	.	.	.	—
2,7-Dichlorodibenzo-p-dioxin	33857-26-0	.	.	.	.	.	.	per
Dichlorodifluoromethane	75-71-8	—	.	.	.	.	.	—
1,1-Dichloroethane	75-34-3	—	—	—	.	.	.	—

TABLE 1 (continued)

Chemical	CAS	Salmonella	Harmonic mean of TD$_{50}$ (mg/kg/day) Rat	Mouse	Rat target sites Male	Female	Mouse target sites Male	Female
1,2-Dichloroethane	107-06-2	+	8.04[m]	101[m]	sto sub vsc	mgl	lun	lun mgl ute
2,4-Dichlorophenol	120-83-2	–	–	–	.	–	–	–
α-(2,4-Dichlorophenoxy)propionic acid	120-36-5	.	.	–	.	.	.	–
α-(2,5-Dichlorophenoxy)propionic acid	6965-71-5	.	.	–	.	.	.	–
2,4-Dichlorophenoxyacetic acid	94-75-7	.	.	–	.	.	.	–
2,4-Dichlorophenoxyacetic acid, *n*-butyl ester	94-80-4	.	.	–	.	.	.	–
2,4-Dichlorophenoxyacetic acid, isooctyl ester	25168-26-7	.	.	–	.	.	.	–
2,4-Dichlorophenoxyacetic acid, isopropyl ester	94-11-1	.	.	–	.	.	.	–
3-(3,4-Dichlorophenyl)-1,1-dimethylurea	330-54-1	.	.	–	.	.	.	–
2,4-Dichlorophenylbenzene sulfonate	97-16-5	.	.	–	.	.	.	–
1,2-Dichloropropane	78-87-5	+	4.16	276[m]	.	.	liv	liv
Dichlorvos	62-73-7	+	.	70.4[m]	hmo pan	.	sto	sto
Dicofol	115-32-2	–	–	32.9	.	.	liv	–
N,N'-Dicyclohexylthiourea	1212-29-9	.	.	–	.	.	.	–
Dicyclopentadiene dioxide	81-21-0	–	–	–	.	.	.	–
Dieldrin[s]	60-57-1	–	–	0.912[m]	.	.	liv	liv
Dieldrin, photo-	13366-73-9	+	.	.	.	.	.	.
d,l-Diepoxybutane	298-18-0	.	.	.	.	.	.	.
Diethyl-β,γ-epoxypropylphosphonate	7316-37-2	.	.	.	.	.	.	.
N,N-Diethyl-4-(4'-[pyridyl-1'-oxide]azo)aniline	7347-49-1	–	1.63	.	liv	.	.	.
O,O-Diethyl-o-(3,5,6-trichloro-2-pyridyl)phosphorothioate	2921-88-2	–	–	.	.	.	.	.
Diethylacetamide	685-91-6	.	8.85[n]	.	kid	.	.	.
Diethylacetylurea	—	.	–	.	–	.	.	.
Diethylene glycol	111-46-6	–	1660	.	ubl	.	.	.
Diethylformamide	617-84-5	.	–	.	.	.	.	.
Diethylmaleate	141-05-9	–	–	.	.	.	.	.
Diethylstilbestrol	56-53-1	–	0.114	0.0372[m]	adr pit	–	mgl pit tes thy	mgl ova pit thy ute
N,N'-Diethylthiourea	105-55-5	–	24[m]	–	thy	thy	–	–
1,2-Diformylhydrazine	628-36-4	–	.	668[m]	.	.	lun	lun
Diftalone	21626-89-1	.	.	865[m]	.	.	vsc	liv vsc
Diglycidyl resorcinol ether, technical grade	101-90-6	+	3.78[m]	24.3[m]	sto	sto	sto	sto
5,6-Dihydro-5-azacytidine	62488-57-7	–	–	–	.	.	.	.
1,2-Dihydro-2-(5-nitro-2-thienyl) quinazolin-4(3H)-one	33389-33-2	+	1.53	.	.	tba	.	.
3,6-Dihydro-2-nitroso-2H-1,2-oxazine	3276-41-3	.	90.6	.	tba(B)	tba(B)	.	.
3,4-Dihydrocoumarin	119-84-6	–	2970	723	kid	–	–	liv
Dihydrosafrole	94-58-6	–	143	125[m]	eso(B)	eso(B)	liv	lun

Chemical	CAS No.	Mut.	TD50 (Rat)	TD50 (Mouse)	Sites (Rat)	Sites (Mouse)		
(R,R)-Dilevalol.HCl	75659-08-4	.	—	—	—	—	—	.
Dimethadione	695-53-4	.	—	—	—	—	—	.
Dimethoate	60-51-5	+	—	—	—	—	—	-
Dimethoxane	828-00-2	.	716	—	hmo kid liv ski sub	—	—	.
Dimethoxane, commercial grade	828-00-2	+	—	—	—	—	—	-
2,5-Dimethoxy-4'-aminostilbene	5803-51-0	.	0.721	95.9	ezy ski smi sto	liv lun	—	-
2,4-Dimethoxyaniline.HCl	54150-69-5	+	—	—	—	—	—	-
3,3'-Dimethoxybenzidine-4,4'-diisocyanate	91-93-0	+	1630[m]	—	hmo ski	hmo ute	—	-
3,3'-Dimethoxybenzidine.2HCl	20325-40-0	+	1.04[m]	—	ezy hmo lgi liv orc pre ski smi	cli ezy lgi liv mgl orc ski ute	—	-
5,7-Dimethoxycyclopentene[c]coumarin	1146-71-0	.	—	—	—	—	—	-
5,7-Dimethoxycyclopentenone [2,3-c]coumarin	1150-37-4	.	—	—	—	—	—	-
5,7-Dimethoxycyclopentenone [3,2-c]coumarin	1150-42-1	.	—	—	—	—	—	-
5,6-Dimethoxysterigmatocystin	65176-75-2	.	0.364	—	liv	liv	—	-
N,N-Dimethyl-4-aminoazobenzene [s]	60-11-7	+	3.31	—	liv	liv	—	-
N,N'-Dimethyl-N,N'-dinitrosophthalamide	3851-16-9	.	—	—	—	—	—	-
Dimethyl hydrogen phosphite	868-85-9	+	139	—	lun sto	—	—	-
Dimethyl methylphosphonate	756-79-6	-	700	—	kid	—	—	-
Dimethyl morpholinophosphoramidate	597-25-1	-	614[m]	—	hmo	hmo	—	-
4,6-Dimethyl-2-(5-nitro-2-furyl) pyrimidine	59-35-8	.	1.39	—	kid mgl smi sto	—	—	-
1,2-Dimethyl-5-nitroimidazole	551-92-8	+	17	—	mgl	—	—	-
Dimethyl terephthalate	120-61-6	-	—	—	—	—	—	-
6-Dimethylamino-4,4-diphenyl-3-heptanol acetate.HCl	43033-72-3	.	68[m]	—	liv	liv	—	.
6-Dimethylamino-4,4-diphenyl-3-heptanone.HCl	1095-90-5	.	—	—	—	—	—	-
trans-2-[(Dimethylamino)methylimino]-5-[2-(5-nitro-2-furyl) vinyl]-1,3,4-oxadiazole	55738-54-0	+	22.4	—	lun mgl smi sto	—	—	-
4-Dimethylamino-3,5-xylenol	6120-10-1	.	.	.	.	.	.	.
4-Dimethylaminoantipyrine	58-15-1	.	.	.	.	.	.	.
Dimethylaminoethylnitrosoethylurea, nitrite salt	—	-	0.704	—	mgl ute	—	—	-
2-Dimethylaminoethanol	108-01-0	-	.	.	.	.	.	.
5,5-Dimethylbarbituric acid	—	.	.	.	.	.	.	.
N,N-Dimethylaniline	121-69-7	-	125	—	bon	—	—	-
Dimethylarsinic acid	75-60-5	-	.	.	.	.	.	.
7,12-Dimethylbenz(a)anthracene	57-97-6	+	.	0.084	.	—	—	.
3,3'-Dimethylbenzidine.2HCl	612-82-8	+	0.629[m]	28.6	ezy lgi liv lun orc per pre ski smi	cli ezy lgi liv lun mgl orc ski smi	lun	vsc
Dimethylcarbamyl chloride [s]	79-44-7	+	.	5.37[i]	.	—	—	per
Dimethyldithiocarbamic acid, dimethylamine	598-64-1	.	.	.	.	.	.	.
N,N-Dimethyldodecylamine-N-oxide	1643-20-5	.	—	—	—	—	—	-
1,1-Dimethylhydrazine [s]	57-14-7	+	—	3.96[m]	—	liv lun vsc	—	lun vsc

TABLE 1 (continued)

Chemical	CAS	Salmonella	Harmonic mean of TD$_{50}$ (mg/kg/day) Rat	Mouse	Rat target sites Male	Female	Mouse target sites Male	Female
1,2-Dimethylhydrazine.2HCl[s]	306-37-6	+	.	0.114[m]	.	.	lun vsc	lun vsc
2-(2,2-Dimethylhydrazino)-4-(5-nitro-2-furyl)thiazole	26049-69-4	+	0.41	.	.	hmo mgl	.	.
Dimethylnitramine	4164-28-7	.	0.547[m,v]	.	liv nas	liv nas	.	.
Dimethylvinyl chloride	513-37-1	+	31.8[m]	14.9[m]	eso nas orc sto ski sto	eso nas orc sto	pre sto	sto
Dinitro(1-methylheptyl)phenyl crotonate	6119-92-2	.	.	—	.	.	.	.
2,4-Dinitrophenol	51-28-5	—	.	—	.	.	B—	B—
2,4-Dinitrophenol, sodium	1011-73-0	.	.	—	.	.	.	.
Dinitrosohomopiperazine	55557-00-1	+	0.0615[m]	—	.	eso liv nas orc	.	.
N,N-Dinitrosopentamethylenetetramine	101-25-7	+	.	—	.	.	.	.
Dinitrosopiperazine	140-79-4	+	.	3.6[m]	.	.	sto	sto
2,4-Dinitrotoluene	121-14-2	+	—	—	.	.	.	.
2,4-Dinitrotoluene (purified)	121-14-2	+	—	—	.	.	.	.
2,6-Dinitrotoluene	606-20-2	.	0.574	.	liv	.	.	.
Dinitrotoluene, technical grade (2,4 (77%)- and 2,6 (19%)-)	25321-14-6	+	8.02	.	liv	.	.	.
1,4-Dioxane	123-91-1	—	334[m,v]	838[m]	liv(B) nas	liv nas	liv	liv
Dioxathion	78-34-2	+	—	—	.	.	.	.
Dipentamethylenethiuram hexasulfide	971-15-3	.	.	—	.	.	.	.
Dipentylnitrosamine	13256-06-9	.	4.03[m]	—	liv	liv	.	.
Diphenhydramine.HCl	147-24-0	.	—	—	.	.	.	.
Diphenyl-p-phenylenediamine	74-31-7	+	—	—	.	.	.	.
Diphenylacetonitrile	86-29-3	.	—	—	.	.	.	.
Diphenylcarbonate	102-09-0	.	—	—	.	.	.	.
5,5-Diphenylhydantoin	57-41-0	—	—	59.1	.	.	.	liv
N,N-Dipropyl-4-(4'-[pyridyl-1'-oxide]azo)aniline	—	.	.	.	.	.	.	.
Dipyrone	68-89-3	.	.	630[m]	.	.	liv	liv
2,5-Dithiobiurea	142-46-1	.	—	—	.	.	.	.
Dithiooxamide	79-40-3	.	—	—	.	.	.	.
n-Dodecylguanidine acetate	2439-10-3	—	—	—	.	.	.	.
dl-Dopa	63-84-3	.	—	—	.	.	.	.
Dopamine.HCl	62-31-7	+	—	—	.	.	.	.
Doxefazepam	40762-15-0	.	—	—	.	.	.	.
Edifas A	9004-59-5	.	—	—	.	.	.	.
Edifas B	9004-32-4	.	—	—	.	.	.	.
EDTA, trisodium salt trihydrate	150-38-9	—	I	I	I	I	—	—
Emetine.2HCl	316-42-7	—	I	I	I	I	I	I
Emulsifier YN	55965-13-4	—	—	—	—	—	—	—

Chemical	CAS No.	Mut	Rat TD50	Mouse TD50	Rat target sites	Mouse target sites
Endosulfan	115-29-7	–	–	–	I	–
Endrin	72-20-8	–	–	–	–	–
Enflurane	13838-16-9	–	–	–	–	–
Enovid	8015-30-3	.	–	0.279[m.v]	–	pit
Enovid-E	8015-30-3	–	–	–	–	–
Ephedrine sulphate	134-72-5	+	–	–	–	–
Epichlorohydrin	106-89-8	.	2.96[m]	–	sto	–
l-Epinephrine.HCl	55-31-2	+	I	–	I	I
1,2-Epoxybutane	106-88-7	.	220	–	lun nas	I
Erythorbate, sodium	6381-77-7		–	–	–	–
Erythromycin stearate	643-22-1		–	–	–	–
Estazolam	29975-16-4		–	–	–	–
Estradiol	50-28-2	–	–	0.282[n]	–	mgl
Estradiol mustard	22966-79-6	.	–	1.45[m]	–	hmo lun myc sto
Estragole	140-67-0	.	–	51.8	–	liv
Ethionamide	536-33-4	–	–	69.3	–	thy
Ethionine	13073-35-3	.	4.97	–	liv	–
dl-Ethionine	67-21-0	+	9.11[m]	71.4[m.v]	liv	liv
o-Ethoxybenzamide	938-73-8	–	–	513	–	liv
Ethoxyquin	91-53-2	+	–	–	–	–
Ethyl acrylate	140-88-5	–	120[m]	324[m]	sto	sto
Ethyl alcohol	64-17-5	.	9110	–	adr liv pan pit	–
Z-Ethyl-O,N,N-azoxyethane	16301-26-1	.	0.022	–	eso liv nas vsc	–
Z-Ethyl-O,N,N-azoxymethane	57497-29-7	.	0.0189	–	lgi liv smi vsc	–
Ethyl benzene	100-41-4	.	1210[m]	–	tba	tba
Ethyl bromoacetate	105-36-2	.	–	–	–	–
S-Ethyl-l-cysteine	2629-59-6	.	–	–	–	–
p,p'-Ethyl-DDD	72-56-0	+	–	–	–	–
N-Ethyl-N-formylhydrazine	74920-78-8	.	–	2.8[m]	–	gal liv lun, pre vsc, lun vsc
Ethyl methylphenylglycidate	77-83-8	.	–	–	–	–
N-Ethyl-N'-nitro-N-nitrosoguanidine	63885-23-4	.	0.948[m]	2.84	nrv smi thy	eso(B) smi(B)
1-Ethyl-1-nitrosourea	759-73-9	+	–	–	nrv smi thy	nrv smi
Ethyl tellurac	20941-65-5	.	–	–	–	–
Ethylene glycol	107-21-1	–	–	–	–	–
Ethylene glycol, cyclic sulfate	1072-53-3	.	–	–	–	–
Ethylene imine	151-56-4	+	–	0.377[m]	–	liv lun, lun
Ethylene oxide	75-21-8	+	21.3[m.v]	63.7[m]	hmo nrv per, nrv sto	hag hmo, hag lun, lun mgl ute
Ethylene thiourea	96-45-7	+	7.9[m.v]	23.5[m]	liv(B) thy	liv pit thy
Ethylene urea	120-93-4	–	–	–	–	–
Ethylenebisdithiocarbamate, disodium	142-59-6	.	–	–	–	–
1-Ethyleneoxy-3,4-epoxycyclohexane	106-87-6	+	–	–	–	–
di(2-Ethylhexyl)adipate	103-23-1	–	–	3880[m]	–	liv

TABLE 1 (continued)

Chemical	CAS	Salmonella	Harmonic mean of TD$_{50}$ (mg/kg/day)		Rat target sites		Mouse target sites	
			Rat	Mouse	Male	Female	Male	Female
di(2-Ethylhexyl)phthalate	117-81-7	–	647^m	894^m	liv	liv	liv	liv
Ethylhydrazine.HCl	18413-14-4	.	.	6.56^m	.	.	lun vsc	lun vsc
1-Ethylnitroso-3-(2-hydroxyethyl)-urea	—	.	0.522^m	.	ski	mgl ski ute	.	.
1-Ethylnitroso-3-(2-oxopropyl)-urea	—	.	0.181^m	.	lgi lun nrv per ski	lgi lun mgl nrv ski thy ute	.	.
Ethylnitrosocyanamide	38434-77-4	+	3.68^m	.	nas	nas	.	.
Ethylphenylacetylurea	90-49-3	–	.	.	–	–	.	ubl
4-Ethylsulphonylnaphthalene-1-sulfonamide	842-00-2	–	.	21.1^m	.	.	.	.
Ethynodiol diacetate	297-76-7	–	–	–	–	–	–	–
Eucalyptol	470-82-6	–	–	–	–	–	–	–
Eugenol	97-53-0	–	–	–	–	–	–	–
Fenaminosulf, formulated	140-56-7	+	–	–	–	–	–	–
Fenthion	55-38-9	–	–	–	–	–	–	–
Fenvalerate	51630-58-1	–	–	–	–	–	–	–
Ferric chloride	7705-08-0	.	–	–	–	–	–	–
Ferric dimethyldithiocarbamate	14484-64-1	–	–	–	–	–	–	–
Flecainide acetate	54143-56-5	–	–	–	–	–	–	–
Fluometuron	2164-17-2	–	–	–	–	–	–	–
N-(2-Fluorenyl)-2,2,2-trifluoroacetamide	363-17-7	.	1.62	.	.	ezy liv mgl	.	.
Fluoride, sodium	7681-49-4	–	–	–	–	–	–	–
4'-Fluoro-4-aminodiphenyl	324-93-6	.	.	1.14^m	kid	.	liv	liv
N-4-(4'-Fluorobiphenyl)acetamide	398-32-3	.	1.01	.	sto	.	.	.
2-Fluoroethyl-nitrosourea	69112-98-7	–	0.125	.	.	.	.	.
5-Fluorouracil	51-21-8	–	–	2.96$^{i.m}$	–	–	lun	hmo lun
Fluoxetine.HCl	59333-67-4	–	–	43.9	–	hmo nas	nas	–
Formaldehyde^s	50-00-0	+	2.19$^{m.v}$	.	hmo nas	hmo nas	nas	–
Formic acid 2-[4-(2-furyl)-2-thiazolyl]hydrazide	31873-81-1	.	14.4	.	.	.	.	.
Formic acid 2-(4-methyl-2-thiazolyl)hydrazide	32852-21-4	.	.	.	.	mgl	.	.
Formic acid 2-[4-(5-nitro-2-furyl)-2-thiazolyl]hydrazide^s	3570-75-0	+	5.06^m	10.8^m	kid liv mgl	ezy hmo kid lgi	.	hmo sto
1-Formyl-3-thiosemicarbazide	2302-84-3	.	–	.	.	.	.	.
Formylhydrazine	624-84-0	+	.	36.4^m	ubl	.	lun	lun
Fosetyl Al	39148-24-8	.	3660	.	liv	.	.	.
Fumonisin B$_1$	116355-83-0	.	1.16^m	.	.	.	.	.
2-Furaldehyde semicarbazone	2411-74-7	.	.	.	.	.	.	.
Furan	110-00-9	–	0.396^m	2.72^m	hmo liv	hmo liv	adr liv	adr liv
Furfural^s	98-01-1	+	683	197^m	liv	–	liv	liv
Furosemide	54-31-9	–	–	732	–	–	.	mgl
Fusarenon-X	23255-69-8	–	–	.	–	–	.	.

Chemical	CAS No.	Mut	TD50	TD50	Sites	Sites	Sites	Sites
Gallic acid	149-91-7	—	—	—	—	—	.	.
Gemcadiol	35449-36-6	.	.	.	.	.	.	.
Gemfibrozil	25812-30-0	—	—	—	—	—	—	—
Gentian violet	548-62-9	—	90.5[m]	.	—	—	hag liv	hag hmo liv
Geranyl acetate, food grade (71% geranyl acetate, 29% citronellyl acetate)	mixture	—	—	—	—	—	—	—
Germanate, sodium	12025-19-3	.	—	.	B—	B—	B—	B—
Gibberellic acid	77-06-5	—	—	.	B—	B—	.	.
Glu-P-1	67730-11-4	+	5.4[m]	4.69[m]	ezy lgi liv smi	cli ezy lgi liv nrv smi	liv vsc	liv vsc
Glu-P-2	67730-10-3	+	16[m]	42.3[m]	ezy lgi liv nrv smi	cli ezy lgi liv smi	liv vsc	liv vsc
l-Glutamic acid	56-86-0	—	—	.	—	—	.	.
N2-γ-Glutamyl-p-hydrazinobenzoic acid	—	.	277	.	.	.	sub	—
β-N-[γ-l(+)-Glutamyl]-4-hydroxymethylphenylhydrazine	2757-90-6	—	—	.	—	—	—	—
Glycerol α-monochlorohydrin	96-24-2	+	—	.	—	—	.	.
Glycidaldehyde	765-34-4	+	—	.	—	—	.	.
Glycidol s	556-52-5	+	34.7[m]	4.28[m]	ezy lgi mgl nrv per ski sto thy	cli hmo mgl nrv orc sto thy	hag liv lun ski sto	hag mgl ski sub ute
Glycol sulfite	3741-38-6	.	—	.	.	.	—	—
Glycyrrhetinic acid	471-53-4	.	.	.	.	.	.	.
Glycyrrhizinate, disodium	71277-79-7	—	—	.	—	—	—	—
FD & C green no. 1	4680-78-8	—	6060[m]	.	hmo(B) liv	hmo(B) liv	—	—
FD & C green no. 2	5141-20-8	—	5640	.	hmo(B)	hmo(B)	—	—
FD & C green no. 3	2353-45-9	—	—	.	—	—	—	—
Griseofulvin s	126-07-8	—	1660[n]	.	.	.	liv	—
Guar gum	9000-30-0	.	.	.	.	.	.	.
Gum arabic	9000-01-5	—	—	.	—	—	—	—
HCDD mixture	mixture	.	0.00143[m]	0.000596	.	—	—	—
Hematoxylin	517-28-2	.	1000	.	hmo(B)	hmo(B)	.	.
Heptachlor	76-44-8	—	1.21[m]	.	.	—	liv	—
Heptamethyleneimine	1121-92-2	.	.	.	.	.	—	—
Heptylamine	1241-27-6	—	.	.	.	.	.	—
Hexachlorobenzene s	118-74-1	—	65.1[m]	3.51[m,v]	liv	liv	liv	liv
Hexachlorobutadiene s	87-68-3	—	65.8[m]	.	kid	kid	kid	.
Hexachlorocyclohexane, technical grade	608-73-1	—	14.8[m]	.	liv	liv	liv	liv
α-1,2,3,4,5,6-Hexachlorocyclohexane	319-84-6	—	6.62	11.2	liv	.	liv	.
β-1,2,3,4,5,6-Hexachlorocyclohexane	319-85-7	.	27.8[m]	.	.	liv	.	liv
γ-1,2,3,4,5,6-Hexachlorocyclohexane	58-89-9	—	30.7[m]	.	liv lun	liv	liv	liv lun
Hexachlorocyclopentadiene	77-47-4	—	—	.	liv	liv	liv	liv
Hexachloroethane	67-72-1	—	338[m]	55.4	kid	liv	liv	liv
Hexachlorophene	70-30-4	—	—	.	liv	liv	liv	—
3-(Hexahydro-4,7-methanoindan-5-yl)-1,1-dimethylurea	2163-79-3	—	—	.	—	—	—	—

TABLE 1 (continued)

Chemical	CAS	Salmonella	Harmonic mean of TD$_{50}$ (mg/kg/day)		Rat target sites		Mouse target sites	
			Rat	Mouse	Male	Female	Male	Female
Hexamethylenetetramine	100-97-0	+	–	–	–	–	–	–
Hexamethylmelamine	531-18-0	.	10.2	.	.	kid mgl	.	.
Hexanal methylformylhydrazone	—	.	.	2.33^m	.	.	liv lun pre	liv lun
Hexanamide	628-02-4	–	.	1950	.	.	hmo	–
N-Hexylnitrosourea	18774-85-1	+	0.513^m	.	lgi lun sto	lun mgl sto ute	.	.
4-Hexylresorcinol	136-77-6	–	.	.	.	.	.	.
Humic acids, commercial grade	1415-93-6	.	.	.	.	.	.	.
Hydrazine ^{s}	302-01-2	+	0.309^m	2.93^m	lun nas	lun nas	lun	lun
Hydrazine sulfate ^{s}	10034-93-2	+	40.8^m	7.59^{m,v}	liv lun	lun	liv lun	liv lun
2-Hydrazino-4-(p-aminophenyl) thiazole	26049-71-8	.	1.03	11.3	.	hmo mgl	.	hmo
2-Hydrazino-4-(5-nitro-2-furyl) thiazole	26049-68-3	.	3.19^m	16.4	.	kid mgl	.	sto
2-Hydrazino-4-(p-nitrophenyl)thiazole	26049-70-7	.	3.21^m	10.6	.	mgl	.	hmo
2-Hydrazino-4-phenylthiazole	34176-52-8	.	.	.	.	.	.	.
p-Hydrazinobenzoic acid	619-67-0	.	.	.	.	.	.	.
p-Hydrazinobenzoic acid.HCl	24589-77-3	+	.	561^m	.	.	vsc	vsc
Hydrazobenzene	122-66-7	+	5.59^m	26	ezy liv	liv mgl	vsc	liv
Hydrochloric acid	7647-01-0	.	.	.	–	.	–	–
Hydrochlorothiazide	58-93-5	.	.	.	–	.	–	–
Hydrocortisone	50-23-7	.	.	.	–	.	.	smi
Hydrogen peroxide	7722-84-1	+	.	7540	.	.	.	.
Hydroquinone	123-31-9	–	82.8^m	225^{m,v}	kid	hmo	kid liv	liv
Hydroquinone monobenzyl ether	103-16-2	.	.	.	.	.	–	–
3-Hydroxy-4-acetylaminobiphenyl	4463-22-3	.	.	.	.	.	.	.
N-Hydroxy-2-acetylaminofluorene ^{s}	53-95-2	+	0.000988^m	6.23	liv	liv	.	eso kid liv sto ubl
3-Hydroxy-4-aminobiphenyl	4363-03-5	–	–	.	.	.	.	.
3-Hydroxy-p-butyrophenetidide	1083-57-4	.	.	5530	lgi liv sto	.	kid	.
1-Hydroxyanthraquinone	129-43-1	–	59.2	.	.	.	.	.
1'-Hydroxyestragole	51410-44-7	.	.	57.8	.	.	.	liv
1-(2-Hydroxyethyl)-3-[(5-nitrofurfurylidene)amino]-2-imidazolidinone	5036-03-3	+	16.7	.	.	mgl	.	.
1-(2-Hydroxyethyl)-nitroso-3-chloroethylurea		.	0.789^m	.	liv	liv	.	.
1-(2-Hydroxyethyl)-nitroso-3-ethylurea		.	0.562^m	.	ski	mgl	.	.
1-(2-Hydroxyethyl)-1-nitrosourea	13743-07-2	+	0.131^{m,v}	0.818^m	bon hmo kid lgi lun mgl sto	hmo mgl	hmo	hmo
4-(2-Hydroxyethylamino)-2-(5-nitro-2-thienyl)quinazoline	33389-36-5	+	1.87	.	–	tba	.	.
2-Hydroxyethylhydrazine ^{s}	109-84-2	+	.	0.397^m	–	.	liv	.
Hydroxypropyl distarch glycerol	—	.	.	.	.	.	.	.

Chemical	CAS No.		TD50	TD50	Sites	Sites	Sites	Sites
1-(2-Hydroxypropyl)-nitroso-3-chloroethylurea	—	·	·	·				
1-(3-Hydroxypropyl)-1-nitrosourea	71752-70-0	+	0.978[m]	—	sto	sto		
8-Hydroxyquinoline	148-24-3	—	—	—	liv sto		vsc	liv
1'-Hydroxysafrole	5208-87-7	—	18.4[m]	71.2[m]				hmo
Ibuprofen	15687-27-1	·	—	—				
ICRF-159	21416-87-5	—	10.7[i]	23.7[i]		ute		
3,3'-Iminobis-1-propanol dimethanesulfonate(ester),HCl	3458-22-8	—	—	—				
Iminodiacetic acid, monosodium	32607-00-4	·	—	—				
Indolidan	100643-96-7	—	2.01[m]	—	adr	adr		
Indole-3-acetic acid	87-51-4	—	—	—				hag pit
Iodinated glycerol	5634-39-9	+	101	138	hmo thy			
Iodoacetamide	144-48-9	·	—	—				
Iodoform	75-47-8	+	—	—				
IQ[s]	76180-96-6	+	1.63[m]	19.6[m]	ezy lgi liv orc ski smi	cli ezy lgi liv orc ski smi / ezy liv mgl pan ubl vsc	liv lun sto	liv lun sto
IQ.HCl	—	+	3.29					
Isobutyl p-hydroxybenzoate	4247-02-3	·	·	·				
N-Isobutyl-N'-nitro-N-nitrosoguanidine	5461-85-8	+	—	—				
Isoflurane	26675-46-7	—	—	—				
Isomalt	64519-82-0	·	·	·				
Isomazole	86315-52-8	—	70.5[m]	—	adr	adr		
Isoniazid[s]	54-85-3	+	150[m]	27.1[m,v]	liv lun	mgl	liv lun mgl(B)	liv lun mgl(B)
Isonicotinamide	1453-82-3	·	—	—				
Isonicotinic acid[s]	55-22-1	—	—	—				
Isonicotinic acid vanillylidenehydrazide	149-17-7	—	1210	27.4	kid pre		lun(B) mgl(B)	lun(B) mgl(B)
Isophorone	78-59-1	—	—	—				
Isophosphamide	3778-73-2	+	0.739[i]	5.06[i]		ute		
p-Isopropoxydiphenylamine	101-73-5	—	—	—			hmo	hmo
Isopropyl-N-(3-chlorophenyl)carbamate[s]	101-21-3	—	—	—				
1-Isopropyl-3-methyl-S-pyrazolyldimethyl carbamate	119-38-0	·	—	—				
Isopropyl-N-phenyl carbamate[s]	122-42-9	—	—	—				
Isosafrole	120-58-1	—	—	—				
Kaempferol[s]	520-18-3	+	—	—				
Kanechlor 400	12737-87-0	—	—	—				
Kepone	143-50-0	—	2.96	0.982[m]		liv	liv	
Ketoprofen	22071-15-4	·	—	—			liv	liv
Lasiocarpine	303-34-4	+	0.476[m]	·	liv ski smi vsc kid	hmo liv vsc	kid	
Lead acetate	301-04-2	—	46.6[m]	—	kid	kid	kid	
Lead acetate, basic[s]	1335-32-6	·	181[m]	472[m]	kid	kid	kid	kid
Lead dimethyldithiocarbamate	19010-66-3	+	—	—				
Leupeptin	24365-47-7	·	·	55.8		liv	liv	

TABLE 1 (continued)

Chemical	CAS	Salmonella	Harmonic mean of TD$_{50}$ (mg/kg/day)		Rat target sites		Mouse target sites	
			Rat	Mouse	Male	Female	Male	Female
Levobunolol	47141-42-4	.	–	–	–	–	–	–
d-Limonene	5989-27-5	–	204	–	kid	–	–	–
Lithocholic acid	434-13-9	–	–	–	–	–	–	–
Locust bean gum	9000-40-2	–	–	–	–	–	–	–
Lofexidine.HCl	21498-08-8	.	.	.	.	.	.	.
Lonidamine	50264-69-2	–	–	–	–	–	liv	liv
Luteoskyrin	21884-44-6	–	–	18.6[m]	–	–	–	–
Lutestral	8065-91-6	–	–	–	–	–	–	–
Magnesium chloride hexahydrate	7791-18-6	–	–	–	–	–	–	–
Malaoxon	1634-78-2	–	–	–	–	–	–	–
Malathion	121-75-5	–	–	–	–	–	–	–
Maleic hydrazide	123-33-1	–	–	–	–	–	–	liv
Malonaldehyde, sodium salt	24382-04-5	+	122[m]	14.1	pan thy	thy	–	–
Manganese ethylenebisthiocarbamate	12427-38-2	+	157	–	tba(B)	tba(B)	–	–
Manganese (II) sulfate monohydrate	10034-96-5	+	–	–	–	–	–	–
d-Mannitol	69-65-8	–	–	–	–	–	–	–
Mannitol nitrogen mustard	576-68-1	+	.	22.2[m]	.	.	liv vsc	liv vsc
MeA-α-C	68006-83-7	+	.	12.3[m]	.	.	sto	liv sto
MeIQ	77094-11-2	+	1.99[m]	24.3[m]	ezy liv ski	cli ezy liv	hmo liv	liv lun
MeIQx	77500-04-0	+	735	–	ubl	–	–	–
Melamine	108-78-1	–	–	–	–	–	–	–
Melphalan	148-82-3	+	0.0938[i,m]	0.15[i,m]	per	per	hmo lun	lun
dl-Menthol	15356-70-4	–	–	–	–	–	–	–
MER-25	67-98-1	–	.	.	.	.	.	.
2-Mercaptobenzothiazole	149-30-4	–	344[m]	–	adr hmo pan pre	adr pit	–	–
2-Mercaptobenzothiazole, zinc	155-04-4	–	.	.	.	.	.	.
2-Mercaptoethanesulfonate, sodium	19767-45-4	–	–	–	–	–	–	–
6-Mercaptopurine	50-44-2	+	–	–	–	–	–	–
Mercuric chloride	7487-94-7	–	3.12	–	sto	–	kid	–
Mercurymethylchloride	115-09-3	–	–	1.91[m]	–	–	–	–
Mestranol	72-33-3	–	–	–	–	–	–	–
Metepa	57-39-6	–	4.46	–	hmo	–	–	–
Methafurylene	531-06-6	–	–	–	–	–	–	–
Methaphenilene	493-78-7	–	–	–	–	–	–	–
Methapyrilene.HCl[s]	135-23-9	–	9.13[m]	–	liv	liv	liv	–
Methidathion	950-37-8	–	–	6.04	–	–	–	–
Methimazole	60-56-0	–	1.14[m]	–	thy	thy	–	thy
dl-Methionine	59-51-8	–	–	–	–	–	–	–
Methotrexate[s]	59-05-2	–	–	–	–	–	–	–

Chemical	CAS No.	Mut	TD50 Rat	TD50 Mouse	Rat	Rat	Mouse	Mouse
2-Methoxy-4-aminoazobenzene	80830-39-3	.	.	.			—	liv
3-Methoxy-4-aminoazobenzene	3544-23-8	+	60.2	.	ezy kid ubl	ezy mgl ubl	—	—
2-Methoxy-3-aminodibenzofuran	5834-17-3	—	29[m]	.	—	—	—	—
Methoxychlor	72-43-5	—	—	—	—	—	—	—
4-Methoxyphenol	150-76-5	—	—	—	—	—	—	—
Methoxyphenylacetic acid	1701-77-5	—	—	—	—	—	—	—
8-Methoxypsoralen	298-81-7	+	32.4	.	ezy kid		—	—
Z-Methyl-O,N,N-azoxyethane	57497-34-4	—	11.5	.	kid liv		—	—
Methyl bromide	74-83-9	+	—	—	—	—	—	—
Methyl carbamate	598-55-0	—	839[m]	.	liv	liv	liv	liv
Methyl carbazate	6294-89-9	—	—	.	—	—	—	—
Methyl clofenapate	21340-68-1	—	9.17	.	liv	liv	liv	liv
1-Methyl-1,4-dihydro-7-[2-(5-nitrofurfuryl)vinyl]-4-oxo-1,8-naphthyridine-3-carboxylate, potassium	—	.		8.03	.	.	hmo(B) lun(B)	hmo(B) lun(B)
3'-Methyl-4-dimethylaminoazobenzene[s]	55-80-1	+	3.28[m]	.	liv	liv	—	.
N-Methyl-N,4-dinitrosoaniline	99-80-9	+	1.3[i,n]	.	per	per	—	.
N-Methyl-N-formylhydrazine[s]	758-17-8	.	.	1.37[m,v]	.	.	gal liv lun vsc	gal liv lun vsc
Methyl hesperidin	11013-97-1	—	—	.	—	—	.	.
Methyl linoleate hydroperoxide	27323-65-5	—	—	.	—	—	.	.
Methyl linoleate, native	—	—	—	.	—	—	.	.
Methyl methacrylate	80-62-6	—	—	.	—	—	—	.
Methyl methanesulfonate	66-27-3	+	.	31.8	.	.	hmo lun	—
N-Methyl-N'-nitro-N-nitrosoguanidine[s]	70-25-7	+	0.875[m]	.	eso smi sto	smi sto	.	.
2-Methyl-1-nitroanthraquinone	129-15-7	+	84.8[m]	.	liv	liv	vsc	vsc
4-Methyl-1-[(5-nitrofurfurylidene) amino]-2-imidazolidinone	21638-36-8	+	5.34	1.56[m]	mgl	mgl	vsc	.
4-(4-N-Methyl-N-nitrosaminostyryl)quinoline	16699-10-8	.	0.699[m]	.	liv	liv	.	.
N-Methyl-N-nitrosobenzamide	63412-06-6	+	3.23[m]	.	sto	tba	.	.
N-(N-Methyl-N-nitrosocarbamoyl)-l-ornithine	63642-17-1	.	0.787[i,m]	.	ezy kid pan ski	kid mgl pan ski	.	.
R(−)-2-Methyl-N-nitrosopiperidine	14026-03-0	.	20.4	.	nrv(B)	nrv(B)	.	.
S(+)-2-Methyl-N-nitrosopiperidine	36702-44-0	.	13.2	.	nrv(B)	nrv(B)	.	.
Methyl 12-oxo-trans-10-octadecenoate	21308-79-2	.	.	.	.	.	B—	B—
Methyl parathion	298-00-0	+	—	—	—	—	—	—
N-Methyl-2-pyrrolidone	872-50-4	—	—	—	—	—	—	—
(N-6)-Methyladenine	443-72-1	—	—	—	—	—	—	—
(N-6)-Methyladenosine	1867-73-8	.	.	.	—	—	.	.
α-Methylbenzyl alcohol	98-85-1	—	458	.	kid		—	.
3-Methylbutanal methylformylhydrazone	—	.	.	2.23[m]	.	mgl	gal liv lun pre	gal liv lun
3-Methylcholanthrene[s]	56-49-5	+	0.491[m]	.	mgl	.	.	.
α-Methyldopa sesquihydrate	41372-08-1	—	.	.	mgl	.	.	.
N-Methyldopamine,O,O'-diisobutyroyl ester.HCl	75011-65-3	.	—	.	—	—	.	.
4,4'-Methylene-bis(2-chloroaniline)[s]	101-14-4	+	19.3[m]	.	ezy liv lun mgl vsc	liv lun mgl	gal liv lun	gal liv lun
4,4'-Methylene-bis(2-chloroaniline).2HCl	64049-29-2	+	—	66.6	—	.	liv	liv

TABLE 1 (continued)

Chemical	CAS	Salmonella	Harmonic mean of TD$_{50}$ (mg/kg/day) Rat	Harmonic mean of TD$_{50}$ (mg/kg/day) Mouse	Rat target sites Male	Rat target sites Female	Mouse target sites Male	Mouse target sites Female
4,4'-Methylene-bis(2-methylaniline)	838-88-0	+	7.38[m]	.	liv mgl sub	liv	.	.
Methylene chloride	75-09-2	+	724[m]	918[m]	mgl	mgl	liv lun	liv lun
4,4'-Methylenebis(N,N-dimethyl) benzenamine	101-61-1	+	16.4[m]	207	thy	thy	–	liv
4,4'-Methylenedianiline.2HCl	13552-44-8	+	20[m]	32.4[m]	liv thy	thy	adr liv thy	hmo liv thy
Methylguanidine	471-29-4	.	—	.	.	.	.	.
7-Methylguanine	578-76-7	.	—	.	B–	B–	.	.
Methylhydrazine[s]	60-34-4	+	.	7.55[m]	.	.	liv	liv lun
Methylhydrazine sulfate	302-15-8	.	.	2.72[m]	.	.	lun	lun
Methylhydroquinone	95-71-6	.	.	.	.	.	.	.
Methylnitramine	598-57-2	.	17.4[m]	.	nrv	nrv	.	.
4-(Methylnitrosamino)-1-(3-pyrridyl)-1-butanol	—	.	0.103	.	lun pan	.	.	.
4-(Methylnitrosamino)-1-(3-pyrridyl)-1-(butanone)	64091-91-4	.	0.182	.	liv lun pan	.	.	.
(N-6)-(Methylnitroso)adenine	21928-82-5	.	.	18	.	.	lun	–
(N-6)-(Methylnitroso)adenosine	33868-17-6	–	0.48	18.3[m]	.	sto	lun	lun mgl ute
Methylnitrosocyanamide	924-42-5	+	—	26.6[m]	.	.	hag liv lun	hag liv lun ova
6-Methylquinoline	91-62-3	+	—	.	.	.	.	.
8-Methylquinoline	611-32-5	+	—	.	.	.	.	.
p-Methylstyrene	622-97-9	–	—	.	.	.	.	.
Methylthiouracil[s]	56-04-2	–	0.00102[i,m]	.	per	per	B–	B–
Metiapine	5800-19-1	–	—	.	.	.	.	.
Metronidazole	443-48-1	+	542[m]	506[m]	pit tes	liv mgl	lun	hmo lun
Mexacarbate	315-18-4	–	—	—	.	.	.	.
Michler's ketone	90-94-8	+	5.64[m]	84.1[m]	liv	liv	vsc	liv
Mirex	2385-85-5	–	1.77[m]	1.45[m]	adr kid liv	hmo liv	liv	liv
Mirex, photo-	39801-14-4	.	1.46	.	thy	.	.	.
Misoprostol	59122-46-2	.	—	.	.	.	.	.
Mitomycin-C	50-07-7	+	.	.	per	per	.	.
Monoacetyl hydrazine	1068-57-1	–	.	9.85[m]	.	.	lun	lun
Monochloroacetic acid	79-11-8	–	.	.	.	.	.	.
Monocrotaline	315-22-0	–	0.94[m]	.	liv	.	liv	.
dl-Monosodium glutamate	32221-81-1	.	—	.	.	.	.	.
l-Monosodium glutamate	142-47-2	.	—	.	.	.	.	.
Monosodium succinate	2922-54-5	.	—	.	.	.	.	.
4-Morpholino-2-(5-nitro-2-thienyl) quinazoline	58139-48-3	+	5.03	.	.	tba	.	.
l-5-Morpholinomethyl-3-[(5-nitrofurfurylidene)-amino]-2-oxazolidinone.HCl	3031-51-4	.	6.33	.	.	hmo kid mgl	.	.
Myleran	55-98-1	+	—	—	.	.	.	.

Chemical	CAS No.							
Nafenopin	3771-19-5	−	22.1	.	liv	.	.	.
Nalidixic acid	389-08-2	−	201[m]	–	pre	cli	cli	–
Naphthalene	91-20-3	+	.	163	.	.	.	lun
1-Naphthalene acetamide	86-86-2	+	.	–	.	.	.	–
1-Naphthalene acetic acid	86-87-3	−	.	–	.	.	.	–
1,5-Naphthalenediamine	2243-62-1	+	69.6	162[m]	.	cli ute	thy	liv lun thy
N-(1-Naphthyl)ethylenediamine.2HCl	1465-25-4	+	.	–	.	.	.	–
sym.-diβ-Naphthyl-p-phenylenediamine	93-46-9	+	.	–	.	.	.	–
1-(1-Naphthyl)-2-thiourea	86-88-4	+	.	–	.	.	.	–
2-Naphthylamine[s]	91-59-8	·	61.6	36.7[m]	B–	ubl	liv(B)	liv
2-Naphthylamino,1-sulfonic acid	81-16-3	−	.	–	.	.	.	–
Neosugar	88385-81-3	−	–	–	.	.	.	.
Nickel	7440-02-0	.	.	.	.	.	.	.
Nickel (II) acetate	373-02-4	.	.	.	.	.	.	.
Nickel dibutyldithiocarbamate	13927-77-0	.	.	.	.	.	.	.
Nicotinamide	98-92-0	.	.	.	.	.	.	.
Nicotine	54-11-5	.	.	.	.	.	.	.
Nicotine.HCl	636-79-3	.	.	.	.	.	.	.
Nicotinic acid	59-67-6	−	.	–	.	.	.	–
Nicotinic acid hydrazide	553-53-7	.	.	228[m]	.	.	lun	lun
Nigrosine	—	.	–	–	.	.	.	.
Niobate, sodium	12034-09-2	.	.	.	.	.	B–	B–
Nithiazide	139-94-6	+	131	758	.	mgl	liv	B–
Nitrate, sodium	7631-99-4	.	.	.	.	.	.	.
Nitric oxide	10102-43-9	.	.	.	.	.	.	.
Nitrilotriacetic acid	139-13-9	−	1760[m]	2660[m]	kid	kid ubl	kid	kid
Nitrilotriacetic acid, trisodium salt, monohydrate	18662-53-8	−	370[m]	–	kid	kid ubl	.	.
Nitrite, sodium[s]	7632-00-0	+	167[m]	–	hmo(B) liv	hmo(B) liv	.	.
3-Nitro-p-acetophenetide	1777-84-0	+	.	2270	.	.	.	–
5-Nitro-o-anisidine	99-59-2	+	53.9[m]	–	ezy ski	cli ski	liv	liv
5-Nitro-2-furaldehyde semicarbazone	59-87-0	+	6.98[m]	30.8	.	mgl	.	ova
5-Nitro-2-furamidoxime	772-43-0	+	.	–	.	.	.	–
5-Nitro-2-furanmethanediol diacetate	92-55-7	+	.	–	.	.	.	–
3-(5-Nitro-2-furyl)-imidazo(1,2-α) pyridine	75198-31-1	.	13.6[m]	27[m]	eso kid sto	eso kid mgl sto	eso sto	eso sto
5-(5-Nitro-2-furyl)-1,3,4-oxadiazole-2-ol	2122-86-3	·	8.61	–	.	tba	.	.
N-[[3-(5-Nitro-2-furyl)-1,2,4-oxadiazole-5-yl]methyl]acetamide	36133-88-7	+	59.6[m]	–	.	kid lun	kid lun	.
N-[5-(5-Nitro-2-furyl)-1,3,4-thiadiazol-2-yl]acetamide	2578-75-8	+	8.84	6.74	.	hmo kid lun smi	.	sto
4-(5-Nitro-2-furyl)thiazole	53757-28-1	+	7.68	–	.	mgl sto	.	.
N-[4-(5-Nitro-2-furyl)-2-thiazolyl] acetamide[s]	531-82-8	+	17.8[m]	–	.	mgl	.	.
N-[4-(5-Nitro-2-furyl)-2-thiazolyl] formamide[s]	24554-26-5	+	4.25[m,v]	19.7[m,v]	ubl	ubl	ubl	hmo lun sto ubl
N,N'-[6-(5-Nitro-2-furyl)-S-triazine-2,4-diyl]bisacetamide	51325-35-0	+	14.1	.	.	mgl	.	.

TABLE 1 (continued)

Chemical	CAS	Salmonella	Harmonic mean of TD$_{50}$ (mg/kg/day)		Rat target sites		Mouse target sites	
			Rat	Mouse	Male	Female	Male	Female
3-Nitro-3-hexene	4812-22-0	.	8.66	0.346	lun(B)	lun(B)	lun(B)	lun(B)
3-Nitro-4-hydroxyphenylarsonic acid	121-19-7	–	–	–	–	–	–	–
2-Nitro-p-phenylenediamine	5307-14-2	+	–	614	–	–	–	liv
4-Nitro-o-phenylenediamine	99-56-9	+	–	–	–	–	–	–
5-Nitro-o-toluidine	99-55-8	+	–	277^m	–	–	liv vsc	liv vsc
5-Nitroacenaphthene ^{s}	602-87-9	+	8.67^m	45.8	ezy lun	cli ezy lun mgl	–	liv ova
p-Nitroaniline	100-01-6	+	–	–	–	–	–	–
o-Nitroanisole	91-23-6	+	15.6^{m,v}	178^m	hmo kid lgi ubl	hmo kid lgi ubl	liv	liv
4-Nitroanthranilic acid	619-17-0	+	–	–	–	–	–	–
6-Nitrobenzimidazole	94-52-0	+	–	372^m	–	–	liv	liv
p-Nitrobenzoic acid	62-23-7	+	287	–	–	cli	–	–
Nitrofen	1836-75-5	+	420	115^m	–	pan	liv vsc	liv
1-[(5-Nitrofurfurylidene)amino]hydantoin	67-20-9	+	163	1400	kid	–	–	ova
1-[(5-Nitrofurfurylidene)amino]-2-imidazolidinone	555-84-0	.	5.26	.	.	hmo mgl	–	.
Nitrogen mustard	51-75-2	+	0.0114^i	.	tba	.	–	–
Nitrogen mustard N-oxide	126-85-2	.	0.764^i	.	tba	.	–	–
1-Nitronaphthalene	86-57-7	+	–	–	–	–	–	–
1-Nitropropane	108-03-2	–	–	–	–	–	–	–
2-Nitropropane	79-46-9	+	–	–	–	–	–	–
3-Nitropropionic acid	504-88-1	+	–	–	–	–	–	–
6-Nitroquinoline	613-50-3	+	–	–	–	–	–	–
8-Nitroquinoline	607-35-2	+	9.82^m	.	sto	sto	–	–
Nitroso-Baygon	38777-13-8	+	0.364	.	sto	.	–	–
N-Nitroso-bis-(4,4,4-trifluoro-N-butyl)amine	83335-32-4	+	0.748^m	.	liv lun	liv lun	–	–
1-Nitroso-5,6-dihydrothymine	62641-67-2	+	–	.	–	–	–	–
1-Nitroso-5,6-dihydrouracil	16813-36-8	+	0.0983^m	.	liv	liv	–	–
N-Nitroso-2,3-dihydroxypropyl-2-hydroxypropylamine ^{s}	89911-79-5	.	0.0535	.	.	eso orc sto	–	–
Nitroso-2,3-dihydroxypropyl-2-oxopropylamine ^{s}	92177-50-9	.	0.0352	.	.	eso orc sto	–	–
N-Nitroso-2,3-dihydroxypropylethanolamine ^{s}	89911-78-4	.	5.98	.	liv	liv	–	–
1-Nitroso-3,5-dimethyl-4-benzoylpiperazine	61034-40-0	.	9.66	.	sto	sto	–	–
1-Nitroso-1-hydroxyethyl-3-chloroethylurea	96806-34-7	.	0.229^m	.	kid liv	kid liv	–	–
N-Nitroso-2-hydroxymorpholine	67587-52-4	.	–	.	–	–	–	–
1-Nitroso-1-(2-hydroxypropyl)-3-chloroethylurea	96806-35-8	.	0.873^m	.	liv	liv	–	–
N-Nitroso-(2-hydroxypropyl)-(2-hydroxyethyl)amine	75896-33-2	.	1.02	.	.	eso liv vsc	–	–
N-Nitroso-3-hydroxypyrrolidine	56222-35-6	+	7.65	.	tba(B)	tba(B)	–	.
N-Nitroso-N-isobutylurea	760-60-1	.	4.73	.	.	smi	–	.

Chemical	CAS	+	TD50	TD50	Target sites (rat)	Target sites (mouse)
N-Nitroso-N-methyl-N-dodecylamine	55090-44-3	+	0.537[m]	·	liv lun sto ubl	ubl
N-Nitroso-N-methyl-4-fluoroaniline	937-25-7	·	0.255	·	eso	eso
N-Nitroso-N-methyl-4-nitroaniline	943-41-9	·	–	·	–	–
Nitroso-N-methyl-N-(2-phenyl) ethylamine	13256-11-6	·	0.00998[m]	·	eso hmo liv orc sto	
N-Nitroso-N-methyl-N-tetradecylamine	75881-20-8	·	1.65	·	lun ubl	
N-Nitroso-N-methyldecylamine	75881-22-0	·	1.26	·	liv lun nas sto ubl	
N-Nitroso-N-methylurea [s]	684-93-5	+	0.0927	·	lun nrv sto	liv
3-Nitroso-2-oxazolidinone	38347-74-9	·	0.385[m]	·	liv smi	liv
Nitroso-2-oxopropylethanolamine [s]	92177-49-6	·	1.8	·		nas
di(N-Nitroso)-perhydropyrimidine	15973-99-6	·	0.166[i]	·	nas	
Nitroso-1,2,3,6-tetrahydropyridine	55556-92-8	+	0.0601[m]	·		eso liv sto vsc
N-Nitroso(2,2,2-trifluoroethyl) ethylamine	82018-90-4	·	2.52	·	eso nas	
N-Nitroso-2,2,4-trimethyl-1,2-dihydroquinoline polymer	29929-77-9	·	3.31[i]	·	per	
1-Nitroso-3,4,5-trimethylpiperazine [s]	75881-18-4	·	0.151	·		nas
N-Nitrosoallyl-2,3-dihydroxypropylamine	88208-16-6	·	0.825	·		eso nas
N-Nitrosoallyl-2-hydroxypropylamine	91308-70-2	·	0.877	·		eso liv nas
N-Nitrosoallyl-2-oxopropylamine [s]	91308-71-3	·	0.335	·		eso liv
N-Nitrosoallylethanolamine	91308-69-9	·	0.491	·		liv nas
Nitrosoamylurethan	64005-62-5	·	1.01	·		eso orc
Nitrosoanabasine	1133-64-8	·	11.9[m]	·	eso	eso orc
N-Nitrosobenzthiazuron	51542-33-7	·	1.13	·	sto(B)	sto(B)
N-Nitrosobis(2-hydroxypropyl) amine	53609-64-6	+	0.846[m]	·	lun pro	eso nas
N-Nitrosobis(2-oxopropyl)amine	60599-38-4	·	0.491[m]	·	lgi liv lun nas pro thy ubl	liv lun vsc
N-Nitrosobis(2,2,2-trifluoroethyl) amine	625-89-8	·	–	·	–	–
Nitrosochlordiazepoxide	51715-17-4	·	·	–		·
N-Nitrosocimetidine	73785-40-7	·	·	–		·
Nitrosodibutylamine	924-16-3	+	0.691	1.09	liv lun sto ubl	eso liv orc sto
N-Nitrosodiethanolamine	1116-54-7	+	3.17[m,v]	·	eso hmo kid liv nas nrv	eso liv lun sto
N-Nitrosodiethylamine [s]	55-18-5	+	0.0237[m,v]	0.189[m]	eso liv ubl	eso liv orc sto
N-Nitrosodimethylamine [s]	62-75-9	+	0.124[m,v]	·	kid liv lun tes vsc	liv vsc
N-Nitrosodiphenylamine	86-30-6	+	167[m]	–	ubl	ubl
p-Nitrosodiphenylamine	156-10-5	+	201	340	liv	liv
N-Nitrosodipropylamine [s]	621-64-7	+	0.186	·		eso liv nas
N-Nitrosodithiazine	11428-83-6	·	–	·		–
Nitrosododecamethyleneimine	40580-89-0	+	10.9[m]	·	liv	liv
N-Nitrosoephedrine	17608-59-2	·	95.2	·	liv lun sto	liv lun sto
Nitrosoethylmethylamine	10595-95-6	+	0.0503	·	liv lun nas	liv lun nas liv nrv lun nrv
Nitrosoethylurethan	614-95-9	+	0.0904[m]	·		eso orc smi sto

TABLE 1 (continued)

Chemical	CAS	Salmonella	Harmonic mean of TD$_{50}$ (mg/kg/day)		Rat target sites		Mouse target sites	
			Rat	Mouse	Male	Female	Male	Female
N-Nitrosoguvacoline	55557-02-3	+	.	.	—	.	.	.
Nitrosoheptamethyleneimine	20917-49-1	+	0.0378[m]	.	eso liv orc	.	.	.
N-Nitrosohexamethyleneimine	932-83-2	+		0.357[m]	.	tba	eso liv orc sto	eso liv orc sto
1-Nitrosohydantoin	42579-28-2	+	43.8[m]	.	orc	—	.	.
Nitrosohydroxyproline	30310-80-6	.	—		—			
Nitrosoiminodiacetic acid	25081-31-6	.	—		—			
N-Nitrosomethyl-2,3-dihydroxypropylamine [s]	86451-37-8	.	0.646		.	eso liv lun nas		
N-Nitrosomethyl-(2-hydroxyethyl) amine	26921-68-6	.	1.29		liv nas	liv lun		
N-Nitrosomethyl-(3-hydroxypropyl)amine	70415-59-7	.	1.66[m]		liv lun	liv lun		
N-Nitrosomethyl-2-hydroxypropylamine	75411-83-5	.	0.0463[m]		eso nas	eso nas		
N-Nitrosomethyl[(2-oxopropyl) amine	55984-51-5	.	0.0172[m]		eso liv nas orc	eso liv nas orc		
N-Nitrosomethyl-(2-tosyloxyethyl)amine	—	.	4.8[m]		liv vsc	liv vsc		
2-Nitrosomethylaminopyridine	16219-98-0	+	0.214		.	eso		
3-Nitrosomethylaminopyridine	69658-91-9	−	—		—	—	—	—
4-Nitrosomethylaminopyridine	16219-99-1	−	.		.	eso	.	.
Nitrosomethylaniline	614-00-6	.	0.142[m,v]		eso	eso		
Nitrosomethylphenidate	55557-03-4	+	.		.	.		
Nitrosomethylundecylamine	68107-26-6	.	2.37		liv lun	liv vsc		
N-Nitrosomorpholine [s]	59-89-2	+	0.109[m]		.	eso nas		
N'-Nitrosonornicotine-1-N-oxide [s]	78246-24-9	+	0.876[m]		eso nas	eso nas		
Nitrosopipecolic acid	4515-18-8	.	.		.	.		
N-Nitrosopiperazine	5632-47-3	+	8.78[m,n]		tba	tba		
N-Nitrosopiperidine [s]	100-75-4	+	1.43[m]	1.3	eso(B) liv(B) nas(B)	eso(B) liv(B) nas(B)	liv lun sto	.
Nitrosoproline	7519-36-0	.	.		—	—		
N-Nitrosopyrrolidine [s]	930-55-2	+	0.799[m]	0.679	liv ubl	liv vsc	tba	.
N-Nitrosothialdine	81795-07-5	.	0.483	.	.	eso liv orc	.	.
N-Nitrosothiomorpholine	26541-51-5	+	5.39[m]	.	eso	eso orc	.	.
o-Nitrosotoluene	611-23-4	.	50.7		liv spl sub ubl	.		
Nitrous oxide	10024-97-2	.	.					
Nivalenol	23282-20-4	.	.					
Norethynodrel	68-23-5	−	.					
Norharman	244-63-3	+	.					pit
Norlestrin [s]	8015-12-1	.	1.94	1.34[n]	liv(B) mgl(B) pit(B)	liv(B) mgl(B) pit(B) ute	.	pit
Novadelox	94-36-0	−	.		B—	B—	B—	B—
Ochratoxin A	303-47-9	−	0.103[m]	6.41[m]	kid	kid mgl	kid	liv
Octachlorostyrene	29082-74-4	.	—		—	—	.	.
Oleate, sodium	143-19-1	.	—		—	—		

Omeprazole	73590-58-6	·	—	—	—	—	—	—
C.I. acid orange 3	6373-74-6	+	1710	—	kid	kid	—	—
C.I. acid orange 10	1936-15-8	—	—	—	—	—	—	—
Ovulen	8056-92-6	·	·	·	·	·	·	·
Oxamyl	23135-22-0	—	·	·	·	—	—	—
Oxazepam	604-75-1	·	6.17	35.8[m]	ezy liv mgl	ezy liv mgl	liv	liv thy
N-(9-Oxo-2-fluorenyl)acetamide	3096-50-2	—	·	—	—	—	—	—
1-(2-Oxopropyl)nitroso-3-(2-chloroethyl)urea	110559-85-8	—	—	—	—	—	—	—
2-Oxopropylnitrosourea		—	—	—	—	—	—	—
1'-Oxosafrole	30418-53-2	—	—	—	—	—	—	—
Oxprenolol.HCl	6452-73-9	—	—	—	—	—	—	—
4,4'-Oxydianiline	101-80-4	+	9.51[m]	33.6[m]	liv thy	liv thy	hag liv	hag liv thy
N-Oxydiethylene thiocarbamyl-N-oxydiethylene sulfenamide	13752-51-7	·	90.8[m]	·	kid ubl	kid ubl	·	·
N-Oxydiethylenebenzothiazole-2-sulfenamide	102-77-2	·	·	·	·	·	·	·
Oxytetracycline.HCl	2058-46-0	·	—	—	·	·	·	·
Ozone	10028-15-6	+	—	1.88[m]	—	—	—	lun
Parathion	56-38-2	—	—	·	—	—	—	—
Patulin	149-29-1	+	—	—	·	·	·	·
Pentachloroanisole	1825-21-4	+	24.8	68	adr	adr vsc	adr vsc	·
Penicillin VK	132-98-9	—	—	—	—	—	—	—
Pentachloroethane	76-01-7	—	—	57.3[m]	—	—	liv	liv
Pentachloronitrobenzene	82-68-8	—	—	71.1	—	—	liv	—
2,3,4,5,6-Pentachlorophenol	87-86-5	·	·	—	—	—	—	—
2,3,4,5,6-Pentachlorophenol (Dowicide EC-7)	87-86-5	—	—	24[m]	—	adr liv	adr liv	adr liv vsc
2,3,4,5,6-Pentachlorophenol, technical grade	87-86-5	—	13.1	23[m]	liv	adr liv	adr liv	liv vsc
Pentaerythritol tetranitrate with 80% d-lactose monohydrate	78-11-5	—	—	—	—	—	—	—
Pentanal methylformylhydrazone	57590-20-2	·	·	3.42[m]	·	·	liv lun pre	liv lun
N-Pentyl-N-nitro-N-nitrosoguanidine	13010-10-1	+	·	5.87	·	·	·	·
n-Pentylhydrazine.HCl	1119-68-2	·	·	·	·	·	·	lun vsc
Peppermint oil	8006-90-4	·	—	—	—	—	—	—
Petasitenine	60102-37-6	+	0.922[m]	·	liv vsc	liv vsc	·	·
Phenacetin	62-44-2	+	1250[m]	2140[m,v]	kid nas sto ubl	ezy mgl nas ubl	kid	ubl
Phenazone	60-80-0	·	1230	·	kid ubl	kid ubl	·	·
Phenazopyridine.HCl	136-40-3	+	303[m]	71.1	lgi	lgi	liv	liv
Phenesterin	3546-10-9	·	0.523	0.616[m]	mgl	mgl	hmo lun myc	hmo lun myc
Phenformin.HCl	834-28-6	·	—	—	—	—	—	—
Phenobarbital[s]	50-06-6	+	86[m]	6.09[m]	liv	liv	liv	liv
Phenobarbital, sodium	57-30-7	·	—	51.2[m]	liv	liv	liv	liv
Phenol	108-95-2	—	—	—	—	—	—	—
Phenothiazine	92-84-2	·	·	·	·	·	·	·
Phenoxybenzamine.HCl	63-92-3	+	1.09[i,m]	5.36[i,m]	per	per	per	per
1-Phenyl-3,3-dimethyltriazene	7227-91-0	+	2.31	·	nrv(B)	nrv(B)	·	·

TABLE 1 (continued)

Chemical	CAS	Salmonella	Harmonic mean of TD$_{50}$ (mg/kg/day)		Rat target sites		Mouse target sites	
			Rat	Mouse	Male	Female	Male	Female
Phenyl isothiocyanate	103-72-0	.	.	.	.	.	.	.
1-Phenyl-3-methyl-5-pyrazolone	89-25-8	–	–	–	–	–	–	–
Phenyl-β-naphthylamine [s]	135-88-6	–	–	–	–	–	–	–
N-Phenyl-p-phenylenediamine.HCl	2198-59-6	–	–	–	–	–	–	–
1-Phenyl-2-thiourea	103-85-5	–	–	–	–	–	–	–
1-Phenylazo-2-naphthol	842-07-9	+	29.4^m	–	liv	liv	–	–
Phenylbutazone	50-33-9	–	1160	353	–	kid	liv	–
m-Phenylenediamine	108-45-2	+	.	.	.	.	.	.
p-Phenylenediamine	106-50-3	+	–	.	–	.	.	.
m-Phenylenediamine.2HCl	541-69-5	+	–	.	–	.	.	.
o-Phenylenediamine.2HCl	615-28-1	.	248	735^m	liv	.	liv	liv
p-Phenylenediamine.2HCl	624-18-0	+	–	–	–	–	–	–
Phenylephrine.HCl	61-76-7	–	–	–	–	–	–	–
Phenylethylhydrazine sulfate	156-51-4	+	–	14.6	–	–	–	lun vsc
Phenylglycidyl ether	122-60-1	+	44^m	.	nas	nas	.	.
Phenylhydrazine	100-63-0	+	.	.	.	.	.	.
Phenylhydrazine.HCl	59-88-1	+	.	71.3^m	.	.	vsc	vsc
β-Phenylisopropylhydrazine.HCl	66-05-7	–	–	–	–	–	–	–
Phenylmercuric acetate	62-38-4	–	.	.	.	.	.	.
o-Phenylphenate, sodium	132-27-4	–	545^{m,v}	–	kid ubl	ubl	–	–
o-Phenylphenol	90-43-7	–	232	–	ubl	.	–	–
p-Phenylphenol	92-69-3	.	.	.	.	.	.	.
PhIP.HCl	—	+	4.98^m	33.2^m	hmo lgi	lgi mgl	hmo	hmo
Phorbol	17673-25-5	+	I	2.21^i	–	–	–	hmo
Phosphamidon	13171-21-6	–	I	–	I	I	–	–
Phosphated distarch phosphate	—	–	–	–	–	–	–	–
Phosphine	7803-51-2	–	–	–	–	–	–	–
Phthalamide	88-96-0	–	–	–	–	–	–	–
Phthalic anhydride	85-44-9	–	–	–	–	–	–	–
Picloram, technical grade	1918-02-1	–	–	–	–	–	–	–
Pildralazine	56393-22-7	–	–	–	–	–	–	–
Pilocarpine	92-13-7	–	–	–	–	–	–	–
Pimaricin	7681-93-8	–	–	–	–	–	–	–
Piperazine	110-85-0	–	–	–	–	–	–	–
Piperidine	110-89-4	–	–	–	–	–	–	–
Piperonyl butoxide	51-03-6	–	–	–	–	–	–	–
Piperonyl butoxide in solvent	51-03-6	–	–	–	–	–	–	–
Piperonyl sulfoxide	120-62-7	–	–	62.2	–	–	liv	–
Piroxicam	36322-90-4	–	–	.	–	–	.	.

Chemical	CAS No.	Mut.	TD50 (rat)	TD50 (mouse)	Rat sites	Rat sites	Mouse sites	Mouse sites
Pivalolactone	1955-45-9	+	211[m]		sto	sto	—	—
Polybrominated biphenyl mixture	67774-32-7	.	0.322[m]	0.332[m]	liv	liv	liv	liv
Polybrominated biphenyls	59536-65-1	-			—	—	.	.
Polysorbate 80	9005-65-6	-			.	.	—	—
Polyvinylpyridine-N-oxide	—	-			.	.	.	.
Potassium chloride	7447-40-7	-			.	.	—	—
Prazepam	2955-38-6	-			.	.	.	.
Praziquantel [s]	55268-74-1	.			.	.	.	.
Prednimustine	29069-24-7	-	19.2		.	ezy	.	.
Prednisolone	50-24-8	+	1.53		liv	.	—	—
Prednisone	53-03-2	-			.	.	.	.
Premarin	12126-59-9	.		540	.	.	.	.
Primidolol.HCl	40778-40-3	.			.	.	.	.
Probenecid	57-66-9	-			—	.	.	liv
Procarbazine	671-16-9	-	4.01[i]		tba	.	.	.
Procarbazine.HCl [s]	366-70-1	+	0.351[i,m]	0.558[i,m]	hmo mgl nrv	hmo mgl nrv	hmo lun nrv	hmo lun nrv ute
Proflavine.HCl hemihydrate	952-23-8	.	l	l	l	l	l	l
Promethazine.HCl	58-33-3	-			.	.	.	.
Pronethalol	54-80-8	.			—	—	—	—
Pronethalol.HCl	51-02-5	+			—	—	—	—
Propane sultone	1120-71-4	+	3.84[m]		hmo nrv smi	hmo mgl nrv smi	.	.
Propazine	139-40-2	.			.	.	.	.
β-Propiolactone	57-57-8	+	1.46[m]	1.24[m]	.	sto	sto	sto
Propranolol.HCl	318-98-9	.			.	.	.	.
Propyl N-ethyl-N-butylthiocarbamate	1114-71-2	.			.	.	.	.
N-N'-Propyl-N-formylhydrazine	77337-54-3	.		8.79[m]	.	.	lun pre	gal liv lun
Propyl gallate	121-79-9	-			.	.	.	.
n-Propyl isome	83-59-0	.			.	.	.	.
N-Propyl-N'-nitro-N-nitrosoguanidine	13010-07-6	+	1.31[m]		sto	.	.	—
N-Propyl-N-nitrosourea	816-57-9	+	3.77[m]		hmo	hmo	.	.
Propylene	115-07-1	+			.	.	.	.
Propylene glycol	57-55-6	.			.	.	.	.
1,2-Propylene oxide	75-56-9	+	74.4[m,v]	912[m]	adr nas	mgl nas sto	nas	nas
Propylhydrazine.HCl	56795-66-5	.		45.5[m]	.	.	lun	lun
Propylthiouracil	51-52-5	.	13.7[m]	409	thy	thy	pit(B)	pit(B)
Proquazone	22760-18-5	.			.	.	.	.
Proresid	1508-45-8	.			.	.	.	.
Protocatechuic acid	99-50-3	.			.	.	.	.
SX purple	2611-82-7	.	24500		tba(B)	tba(B)	.	.
Pyrazinamide	98-96-4	.			.	.	l	l
Pyrilamine maleate	59-33-6	.	280[m]		liv	liv	.	.
Pyrimethamine	58-14-0	.			.	.	.	—
Quercetin	117-39-5	+	10.1[m,v]		kid smi ubl	smi ubl	—	—

TABLE 1 (continued)

Chemical	CAS	Salmonella	Harmonic mean of TD$_{50}$ (mg/kg/day) Rat	Harmonic mean of TD$_{50}$ (mg/kg/day) Mouse	Rat target sites Male	Rat target sites Female	Mouse target sites Male	Mouse target sites Female
Quercetin dihydrate [s]	6151-25-3	.	–	.	–	–	.	–
Quillaia extract	—	.	–	–	–	–	–	–
p-Quinone dioxime	105-11-3	+	106	–	–	ubl	–	–
C.I. acid red 114	6459-94-5	+	3.89^m	.	ezy liv ski	cli ezy lgi liv lun orc ski smi	.	–
C.I. food red 3	3567-69-9	+	–	–	–	–	–	–
C.I. pigment red 3	2425-85-6	+	1170^m	35500	adr	liv	kid thy	–
C.I. pigment red 23	6471-49-4	+	–	–	–	liv	–	liv
D & C red no. 5	3761-53-3	+	415^m	716^m	liv	liv	liv	liv
D & C red no. 9	5160-02-1	+	146	.	liv spl	–	–	–
D & C red no. 10	1248-18-6	.	–	–	–	–	–	–
FD & C red no. 1	3564-09-8	–	521^{m,v}	–	liv	liv	–	–
FD & C red no. 2	915-67-3	–	1470^m	–	hmo(B)	hmo(B)	–	–
FD & C red no. 3	16423-68-0	–	–	–	–	–	–	B–
FD & C red no. 4 [s]	4548-53-2	+	8110^m	–	hmo(B)	hmo(B)	B–	I
HC red no. 3	2871-01-4	–	–	–	–	–	–	mgl
Reserpine	50-55-5	.	0.306	5.02^m	adr	–	tes	–
Resorcinol	108-46-3	.	–	–	–	–	–	–
Retinoic acid	302-79-4	.	–	.	–	–	.	–
Retinol acetate	127-47-9	–	–	.	–	–	.	–
Retinol palmitate	79-81-2	–	–	.	–	–	.	–
Rhodamine 6G	989-38-8	–	–	.	–	–	–	–
Rifampicin	13292-46-1	–	–	33.6	–	liv	liv	liv
Ripazepam	26308-28-1	+	–	114^m	–	–	liv	liv
Rosaniline.HCl [s]	632-99-5	+	–	.	–	–	–	–
p-Rosaniline.HCl [s]	569-61-9	+	39.4^m	51.5^m	ezy liv ski sub thy	ezy sub thy	liv	adr liv
Rotenone	83-79-4	–	–	.	–	–	.	.
Rutin sulfate	12768-44-4	+	–	.	–	–	.	.
Rutin trihydrate [s]	153-18-4	+	–	.	–	–	.	.
Saccharin	81-07-2	–	–	.	–	–	.	.
Saccharin, calcium	6485-34-3	–	2140^{m,v}	.	ubl	–	–	–
Saccharin, sodium	128-44-9	–	–	.	–	–	.	.
Safflower oil	8001-23-8	–	–	–	–	–	.	.
Safrole	94-59-7	–	441^m	51.3^{m,v}	liv	liv(B)	liv	liv
Salbutamol	18559-94-9	.	40^m	.	–	meo	.	.
Sarcophytol A	72629-69-7	.	.	.	–	–	.	.
Selenium	7782-49-2	.	.	.	–	–	.	–
Selenium diethyldithiocarbamate	5456-28-0	.	.	1.49	–	–	liv	liv

Chemical	CAS No.	±	TD50	TD50	Sites	Sites	Sites	Sites
Selenium dimethyldithiocarbamate	144-34-3	·	·	·	—	—	—	—
Selenium sulfide	7446-34-6	+	8.01^m	69.3	liv	liv	—	liv lun
Senkirkine	2318-18-5	·	1.7^i	·	liv	—	·	—
Sesamol	533-31-3	–	1350^m	4490^m	sto	sto	sto	sto
Simazine	122-34-9	–	—	—	—	—	—	—
Sodium bicarbonate	144-55-8	·	—	—	—	—	—	—
Sodium bithionolate	6385-58-6	—	—	—	—	—	—	—
Sodium chloride	7647-14-5	–	—	—	—	—	—	—
Sodium chlorite	7758-19-2	–	—	—	—	—	—	—
Sodium diethyldithiocarbamate trihydrate	148-18-5	·	—	—	—	—	—	—
Sodium hypochlorite	7681-52-9	–	—	—	—	—	—	—
Sorbic acid	110-44-1	–	—	—	—	—	—	—
Sotalol.HCl	959-24-0	·	—	—	—	—	—	—
Soybean lecithin	8002-43-5	·	—	—	—	—	—	—
Starch acetate	9045-28-7	–	·	—	—	—	·	·
Sterigmatocystin ^{s}	10048-13-2	+	0.152^{m,v}	0.908^m	liv vsc	liv(B)	·	vsc
Streptozotocin	18883-66-4	+	0.963^{i,m}	0.272^{i,m}	kid	kid	kid lun	kid lun
Strobane	8001-50-1	·	—	0.884^m	—	—	hmo liv	—
Styrene	100-42-5	+	23.3	—	—	mgl	—	—
Styrene and β-nitrostyrene mixture	mixture	·	—	—	—	—	—	·
Styrene oxide	96-09-3	+	55.4^m	118^m	sto	sto	sto	sto
Succinic anhydride	108-30-5	–	—	—	—	—	—	—
Sucrose	57-50-1	–	·	·	—	—	—	—
Sulfallate	95-06-7	+	26.1^m	42.2^m	sto	sto	lun	mgl
Sulfamethazine	57-68-1	–	—	1510^m	—	—	thy	thy
Sulfate, sodium	7757-82-6	·	—	—	—	—	—	—
Sulfisoxazole	127-69-5	–	—	—	—	—	—	—
Sulfite, potassium metabi-	4429-42-9	–	—	—	—	—	—	—
3-Sulfolene	77-79-2	–	—	—	—	—	—	—
4,4'-Sulfonylbisacetanilide	77-46-3	–	55.6^n	—	—	mgl	—	—
Symphytine	22571-95-5	·	1.91^i	—	liv vsc	—	—	—
Tace	569-57-3	·	—	—	—	—	·	·
Tara gum	39300-88-4	·	—	—	—	—	—	—
Telodrin	297-78-9	·	—	—	—	—	—	—
Telone II	542-75-6	+	94^m	49.6	liv sto	sto	I	lun sto ubl
Temazepam, pharmaceutical grade	846-50-4	·	410	—	—	—	—	—
Terbutaline	23031-25-6	·	—	—	—	meo	lun	—
3,3',4,4'-Tetraaminobiphenyl.4HCl	7411-49-6	·	395	288	liv	—	lun	—
2,3,5,6-Tetrachloro-4-nitroanisole	2438-88-2	·	—	—	—	—	—	—
2,2',5,5'-Tetrachlorobenzidine	15721-02-5	+	—	—	—	—	—	—
2,3,7,8-Tetrachlorodibenzo-p-dioxin	1746-01-6	–	0.0000235^{m,v}	0.000156^m	orc thy	liv lun	·	liv thy
2,4,5,4'-Tetrachlorodiphenyl sulfone	116-29-0	–	·	·	—	—	—	—
1,1,1,2-Tetrachloroethane	630-20-6	–	182^m	—	—	—	liv	liv
1,1,2,2-Tetrachloroethane	79-34-5	–	38.3^m	—	—	—	liv	liv

TABLE 1 (continued)

Chemical	CAS	Salmonella	Harmonic mean of TD$_{50}$ (mg/kg/day) Rat	Harmonic mean of TD$_{50}$ (mg/kg/day) Mouse	Rat target sites Male	Rat target sites Female	Mouse target sites Male	Mouse target sites Female
Tetrachloroethylene	127-18-4	–	101[m]	126[m]	hmo kid	hmo	liv	liv
Tetrachlorvinphos	961-11-5	–	–	228	–	–	liv	–
Tetracycline.HCl	64-75-5	–	–	–	–	–	–	–
Tetraethylthiuram disulfide	97-77-8	–	–	–	–	–	liv	–
Tetrafluoro-m-phenylenediamine.2HCl	63886-77-1	.	24.3	86.3	tba(B)	tba(B)	.	.
Tetrahydro-2-nitroso-2H-1,2-oxazine	40548-68-3	–	.	.	.	.	.	.
3,4,5,6-Tetrahydrouridine	18771-50-1	.	–	–	–	–	–	–
Tetrakis(hydroxymethyl)phosphonium chloride	124-64-1	–	–	–	–	–	–	–
Tetrakis(hydroxymethyl)phosphonium sulfate	55566-30-8	–	–	–	–	–	–	–
Tetramethylthiuram disulfide	137-26-8	–	.	.	.	.	.	.
Tetramethylthiuram monosulfide	97-74-5	+	.	.	.	.	.	.
Tetranitromethane	509-14-8	+	0.447[m]	1.19[m]	lun	lun	lun	lun
Thenyldiamine	91-79-2	.	–	–	–	–	.	.
Thiabendazole	148-79-8	+	–	–	–	–	hmo pre ski	hmo
Thio-TEPA	52-24-4	+	0.164[i,m]	0.223[i,m]	ezy hmo ski	ezy	liv	liv
Thioacetamide	62-55-5	–	11.5	8.81[m]	liv	liv	liv	–
4,4'-Thiobis(6-tert-butyl-m-cresol)	96-69-5	–	.	–	.	.	.	.
2,2'-Thiobis(4,6-dichlorophenol)	97-18-7	.	.	–	.	.	.	.
Thiocyanate, sodium	540-72-7	.	.	.	.	.	.	.
4,4'-Thiodianiline	139-65-1	+	3.71[m]	33.2[m]	ezy lgi liv thy	ezy thy ute	liv thy	liv thy
β-Thioguanine deoxyriboside	64039-27-6	.	2.1[i]	I	–	ezy	I	I
Thiosemicarbazide	79-19-6	.	–	–	.	.	.	.
Thiouracil	141-90-2	.	11.9[m]	55[m]	thy	thy	liv	liv
Thiourea	62-56-6	–	98.5[m]	–	ski	.	.	.
Tilidine fumarate	55567-81-2	–	–	–	–	–	–	–
Tin (II) chloride	7772-99-8	–	–	–	–	–	–	–
Titanium dioxide	13463-67-7	–	–	–	–	–	–	–
Titanium oxalate, potassium	14481-26-6	–	–	–	–	–	–	–
Titanocene dichloride	1271-19-8	+	–	–	–	–	–	–
dl-α-Tocopherol	10191-41-0	.	.	.	.	.	.	.
dl-α-Tocopheryl acetate	58-95-7	–	–	–	–	–	–	–
Tolazamide	1156-19-0	–	–	–	–	–	–	–
Tolbutamide	64-77-7	.	–	–	–	–	–	–
Toluene	108-88-3	–	716[m]	–	tba	tba	–	–
Toluene diisocyanate, commercial grade (2,4 (80%)- and 2,6 (20%)-)	26471-62-5	+	33.7[m]	250	pan sub	liv mgl pan sub	–	liv vsc
o-Toluenesulfonamide	88-19-7	–	3960	.	ubl(B)	ubl(B)	liv	.
m-Toluidine.HCl	638-03-9	–	–	1440[n]	–	.	–	–

Chemical	CAS No.	Mut	TD50	TD50	Target sites			
o-Toluidine.HCl	636-21-5	+	43.6[m]	840[m]	mgl per spl sub ubl vsc	bon mgl ubl vsc	vsc	liv vsc
p-Toluidine.HCl	540-23-8	+	—	83.5[m]	—	.	hmo	liv
p-Tolylurea	622-51-5	+	—	206	—	.	liv	—
Toxaphene	8001-35-2	+	.	5.57[m]	—	.	liv	liv
Tragacanth gum	9000-65-1	.	.	.	.	.	.	.
Trenimon	68-76-8	.	0.00504[i]	.	tba	.	.	—
Triamcinolone acetonide	76-25-5	.	0.053	.	liv	.	.	liv
Triamterene	396-01-0	+	—	60.2[m]	lgi	lgi	liv	liv
Tribromomethane	75-25-2	.	648[m]	—	—	—	—	—
Tricaprylin	538-23-8	.	.	.	.	.	.	.
1,2,3-Trichloro-4,6-dinitrobenzene	6379-46-0	.	—	—	—	—	—	—
1,1,2-Trichloro-1,2,2-trifluoroethane, technical grade	76-13-1	—	—	—	.	.	.	.
2,4,6-Trichloroaniline	634-93-5	—	—	259	—	—	adr liv	adr liv
1,1,2-Trichloroethane	79-00-5	+	—	55[m]	—	—	adr liv	adr liv
1,1,1-Trichloroethane, technical grade	71-55-6	—	.	.	.	.	.	.
Trichloroethylene[s]	79-01-6	—	668[m]	1580[m,v]	tes	—	liv lun	liv lun
Trichloroethylene (without epichlorohydrin)	79-01-6	+	—	343[m]	—	—	liv	liv
Trichlorofluoromethane	75-69-4	—	—	—	—	—	—	—
N-(Trichloromethylthio)phthalimide	133-07-3	—	405	1170[m]	hmo	smi	smi	smi
2,4,6-Trichlorophenol	88-06-2	+	—	1070[m]	—	liv	liv	liv
2-(2,4,5-Trichlorophenoxy)propionic acid	93-72-1	.	.	.	.	—	.	—
2,4,5-Trichlorophenoxyacetic acid	93-76-5	.	.	.	.	—	.	.
1,2,3-Trichloropropane	96-18-4	+	1.35[m]	0.875[m]	ezy kid orc pan pre sto	cli ezy mgl orc sto	hag liv sto	hag liv orc sto ute
Tricresyl phosphate	1330-78-5	—	—	—	—	—	—	—
Triethanolamine	102-71-6	.	—	100[m]	—	—	tba	tba
Triethylene glycol	112-27-6	+	.	.	.	.	.	.
2,2,2-Trifluoro-*N*-[4-(5-nitro-2-furyl)-2-thiazolyl]acetamide	42011-48-3	+	6.79	9.98	mgl	mgl	.	sto
Trifluralin, technical grade	1582-09-8	+	—	330	—	—	—	liv lun sto
Trimethadione	127-48-0	+	—	.	.	.	.	.
2,4,5-Trimethylaniline	137-17-7	+	33.6[m]	6.13	liv	liv lun	liv lun	liv
2,4,5-Trimethylaniline.HCl	21436-97-5	+	98.5[m]	45.5[m]	liv sub	—	liv lun vsc	liv lun
2,4,6-Trimethylaniline.HCl	6334-11-8	.	5.17	24.8[m]	liv lun sto	—	liv vsc	liv
Trimethylphosphate	512-56-1	+	—	335	—	thy	—	ute
Trimethylthiourea	2489-77-2	—	25.8	—	—	—	—	liv
2,4,6-Trinitro-1,3-dimethyl-5-*tert*-butylbenzene	81-15-2	—	.	127[m]	.	.	.	.
Trinitroglycerin	55-63-0	—	264[m]	.	liv tes	liv	hag liv	hag liv
Triphenyltin acetate	900-95-8	—	.	.	.	.	.	.
Triphenyltin hydroxide	76-87-9	.	.	.	—	—	—	—
Tris(2-chloroethyl)phosphate	115-96-8	—	86.7[m]	.	kid	kid	—	—
Tris-1,2,3-(chloromethoxy)propane	38571-73-2	.	.	3.44[i]	.	.	.	per

TABLE 1 (continued)

Chemical	CAS	Salmonella	Harmonic mean of TD$_{50}$ (mg/kg/day) Rat	Mouse	Rat target sites Male	Female	Mouse target sites Male	Female
Tris(2,3-dibromopropyl)phosphate	126-72-7	+	3.83^m	128^m	kid lgi	kid	kid lun sto	liv lun sto
Tris(2-ethylhexyl)phosphate	78-42-2	–	–	2560	–	–	–	liv
Tris(2-hydroxypropyl)amine	122-20-3	–	.	.	.	.	.	.
Trp-P-1 acetate	75104-43-7	+	0.575^m	40.7^m	liv	liv	liv	liv
Trp-P-2 acetate	72254-58-1	+	6.66^m	12.6^{m,v}	liv mgl ubl	cli hmo liv mgl smi vsc	liv	liv
dl-Tryptophan	54-12-6	.	–	.	–	–	.	–
l-Tryptophan	73-22-3	–	–	.	–	–	–	–
Tungstate, sodium	13472-45-2	–	–	.	–	.	.	–
Turmeric oleoresin (79-85% curcumin)	8024-37-1	–	–	–	–	–	–	–
Uracil	66-22-8	–	671^{m,v}	2750^{m,v}	ubl	ubl	ubl	ubl
Urea	57-13-6	–	–	–	–	–	–	–
Urethane^s	51-79-6	+	41.3	16.9^{m,v}	tba(B)	tba(B)	hmo liv lun vsc	lun
Vanadyl sulfate	27774-13-6	.	.	–	.	.	.	.
Vanguard GF	mixture	–	.	–	.	.	.	.
Vinblastine	865-21-4	.	.	.	.	.	.	.
Vinyl acetate	108-05-4	–	201^m	.	liv thy	liv thy ute	.	.
Vinyl bromide	593-60-2	.	18.5^m	.	liv vsc	liv vsc	.	.
Vinyl carbamate	15805-73-9	.	.	0.124^{i,m}	.	.	liv vsc	liv vsc
Vinyl chloride^s	75-01-4	+	19.1^{m,v}	20.9^m	ezy kid(B) liv lun mgl(B) nas nrv ski tes vsc	ezy kid(B) liv mgl nas nrv vsc	lun(B) mgl vsc	lun mgl vsc
Vinyl toluene (65-71% m- and 32-35% p-)	25013-15-4	–	–	–	–	–	–	–
4-Vinylcyclohexene	100-40-3	–	I	106	I	I	I	ova
Vinylidene chloride	75-35-4	+	–	34.6^m	–	–	kid lun	lun mgl vsc
FD & C violet no. 1	1694-09-3	+	612^m	–	–	mgl ski	–	–
Xylene mixture (60% m-xylene, 9% o-xylene, 14% p-xylene, 17% ethylbenzene)	1330-20-7	–	–	–	–	–	–	–

Chemical	CAS							
Xylene mixture (m-xylene, o-xylene, p-xylene)	mixture	.	524[m]	.	tba	tba	.	.
2,4-Xylidine.HCl	21436-96-4	.	–	12.4	–	.	vsc	lun
2,5-Xylidine.HCl	51786-53-9	+	152	626[m]	sub	–	–	liv
C.I. disperse yellow 3	2832-40-8	–	380	1020	liv	–	B–	liv
C.I. pigment yellow 12	6358-85-6	.	–	–	B–	B–	B–	–
C.I. pigment yellow 16	5979-28-2	.	–	–	B–	B–	B–	B–
C.I. pigment yellow 83	5567-15-7	.	–	–	B–	B–	B–	B–
C.I. vat yellow 4	128-66-5	–	–	10900	–	–	hmo	–
FD & C yellow no. 5	1934-21-0	–	–	–	–	–	–	–
FD & C yellow no. 6	2783-94-0	–	–	–	–	–	–	–
HC yellow 4	59820-43-8	+	–	–	–	–	pit	liv pit
Zearalenone	17924-92-4	–	–	39[m]	–	–	–	–
Zinc dibutyldithiocarbamate	136-23-2	.	.	.	.	.	–	–
Zinc diethyldithiocarbamate	14324-55-1	.	–	–	–	–	–	–
Zinc dimethyldithiocarbamate	137-30-4	+	40.7[m]	–	tba(B) thy	tba(B)	–	–
Zinc ethylenebisdithiocarbamate	12122-67-7	–	255	–	tba(B)	tba(B)	–	–
Zirconium (IV) sulfate	14644-61-2	.	.	.	.	.	B–	B–

Note: Abbreviations: "." = not tested; "T" = inadequate; "(B)" = experimental results were reported only for both sexes together; "B–" = experimental results were all negative and were reported only for both sexes together.

Tissue codes: adr = adrenal gland; bon = bone; cli = clitoral gland; eso = esophagus; ezy = ear/Zymbal's gland; gal = gallbladder; hag = harderian gland; hmo = hematopoietic system; kid = kidney; lgi = large intestine; liv = liver; lun = lung; meo = mesovarium; mgl = mammary gland; mix = mixture; myc = myocardium; nas = nasal cavity (includes tissues of the nose, nasal turbinates, paranasal sinuses and trachea); nrv = nervous system; orc = oral cavity (includes tissues of the mouth, oropharynx, pharynx, and larynx); ova = ovary; pan = pancreas; per = peritoneal cavity; pit = pituitary gland; pre = preputial gland; pro = prostate; ski = skin; smi = small intestine; spl = spleen; sto = stomach; sub = subcutaneous tissue; tba = all tumor bearing animals; tes = testes; thy = thyroid gland; ubl = urinary bladder; ute = uterus; vag = vagina; vsc = vascular system.

In a series of footnotes, we provide additional information about TD_{50} values and test results in the CPDB. These are as follows:

i   = Intraperitoneal or intravenous injection are the only routes of administration with positive tests in the CPDB.

m = More than one positive test in the species in the CPDB.

n   = No results evaluated as positive for this species in the CPDB are statistically significant ($p < 0.1$).

s   = Species other than rats or mice are reported for this chemical in Tables 1b and 1c.

v   = Variation is greater than 10-fold among statistically significant ($p < 0.1$) TD_{50} values from different positive experiments.

SECTION B: SUMMARY OF CHEMICAL CARCINOGENICITY IN HAMSTERS

TABLE 2

Chemical	CAS	Salmonella	Harmonic mean of TD$_{50}$ (mg/kg/day)	Target sites Male	Female
Acetaldehyde	75-07-0	–	565^m	nas orc	orc
2-Acetylaminofluorene	53-96-3	+	17.4	liv	–
AF-2	3688-53-7	+	164^m	eso sto	sto
3-Aminotriazole	61-82-5	–	–	–	–
Benzenediazonium tetrafluoroborate	369-57-3	.	–	–	–
Butylated hydroxyanisole	25013-16-5	–	–	–	.
Cadmium chloride	10108-64-2	–	–	–	–
Cadmium sulphate (1:1)	10124-36-4	.	–	–	–
Carrageenan, native	9000-07-1	.	–	–	–
Chloromethyl methyl ether	107-30-2	–	16.4	mix	.
Coumarin	91-64-5	+	–	–	–
p,p'-DDE	72-55-9	–	202^m	liv	liv
DDT	50-29-3	–	–	–	–
Diallylnitrosamine	16338-97-9	.	1.54	nas(B)	nas(B)
Dieldrin	60-57-1	–	–	–	–
Dimethylcarbamyl chloride	79-44-7	+	0.625	nas	.
1,1-Dimethylhydrazine	57-14-7	+	124^m	lgi	lgi
1,2-Dimethylhydrazine.2HCl	306-37-6	+	0.179^m	lgi liv vsc	lgi liv vsc
1,4-Dinitroso-2,6-dimethylpiperazine	55380-34-2	+	3.1	sto	.
Formaldehyde	50-00-0	+	–	–	.
Formic acid 2-[4-(5-nitro-2-furyl)-2-thiazolyl]hydrazide	3570-75-0	+	16.6	sto ubl	.
Furfural	98-01-1	+	–	–	–
Glycidol	556-52-5	+	56.1^m	vsc	vsc
Griseofulvin	126-07-8	–	–	–	–
Hexachlorobenzene	118-74-1	–	4.96^m	liv thy vsc	liv vsc
Hydrazine	302-01-2	+	4.16	lgi nas sto thy	.
Hydrazine sulfate	10034-93-2	+	181	liv	–
N-Hydroxy-2-acetylaminofluorene	53-95-2	+	2.1	liv sto	.
2-Hydroxyethylhydrazine	109-84-2	+	–	–	–
Isoniazid	54-85-3	+	–	–	–
Isonicotinic acid	55-22-1	.	–	–	–
Isopropyl-N-(3-chlorophenyl)carbamate	101-21-3	.	–	–	–
Isopropyl-N-phenyl carbamate	122-42-9	.	–	–	–
Lead acetate, basic	1335-32-6	.	–	–	–
Methapyrilene.HCl	135-23-9	–	–	–	.
Methotrexate	59-05-2	–	–	–	–
N-Methyl-N-formylhydrazine	758-17-8	.	5.8^m	gal liv	gal liv
Methylhydrazine	60-34-4	+	11.5^m	lgi liv	lgi liv
Methylnitrosamino-N,N-dimethylethylamine	—	.	3.83^m	liv nas	lun nas
Methylthiouracil	56-04-2	–	53.4	.	thy
Nitrite, sodium	7632-00-0	+	–	–	–
N-[4-(5-Nitro-2-furyl)-2-thiazolyl]acetamide	531-82-8	+	8.91	sto ubl	.
N-[4-(5-Nitro-2-furyl)-2-thiazolyl]formamide	24554-26-5	+	9.77	sto ubl	.
5-Nitroacenaphthene	602-87-9	+	–	.	–
N-Nitroso-2,3-dihydroxypropyl-2-hydroxypropylamine	89911-79-5	.	1.59	.	sto
Nitroso-2,3-dihydroxypropyl-2-oxopropylamine	92177-50-9	.	0.754	.	pan sto
N-Nitroso-2,3-dihydroxypropylethanolamine	89911-78-4	.	–	.	–
Nitroso-2,6-dimethylmorpholine	1456-28-6	+	2^m	lun nas orc pan	lun nas orc pan

TABLE 2 (continued)

Chemical	CAS	Salmonella	Harmonic mean of TD$_{50}$ (mg/kg/day)	Target sites Male	Female
N-Nitroso-ethyl-2-oxopropylurea	—	.	0.443^m	sto vsc	sto vsc
N-Nitroso-ethylhydroxyethylurea	—	.	1.04	.	sto ute vsc
Nitroso-5-methyloxazolidone	79624-33-2	.	0.172	.	sto
N-Nitroso-N-methylurethan	615-53-2	+	0.127	eso(B) sto(B)	eso(B) sto(B)
N-Nitroso-1,3-oxazolidine	39884-52-1	.	0.798^m	liv	sto
N-Nitroso-oxopropylchloroethylurea	—	.	0.338^m	vsc	vsc
Nitroso-2-oxopropylethanolamine	92177-49-6	.	0.997	.	liv pan
N-Nitroso-oxopropylurea	—	.	0.13^m	vsc	vsc
N-Nitroso-2-phenylethylurea	—	.	0.964^m	sto vsc	sto vsc
1-Nitroso-3,4,5-trimethylpiperazine	75881-18-4	.	1.32	lun sto	.
N-Nitrosoallyl-2-oxopropylamine	91308-71-3	.	1.19	.	liv nas
N-Nitrosoazetidine	15216-10-1	+	7.14	liv	.
N-Nitrosomethyl-2,3-dihydroxypropylamine	86451-37-8	.	0.94	.	nas
N-Nitrosomorpholine	59-89-2	+	3.57^m	liv nas orc	liv nas orc
N′-Nitrosonornicotine	16543-55-8	.	10.8^m	nas	nas
N′-Nitrosonornicotine-1-N-oxide	78246-24-9	.	—	—	—
N-Nitrosopiperidine	100-75-4	+	83.3^m	liv nas orc	liv nas orc
N-Nitrosopyrrolidine	930-55-2	+	14.2^m	liv	liv
Phenobarbital	50-06-6	+	—	—	.
Phenyl-β-naphthylamine	135-88-6	—	—	—	—
Praziquantel	55268-74-1	—	—	—	—
Quercetin dihydrate	6151-25-3	.	—	—	—
Rosaniline.HCl	632-99-5	+	—	—	—
p-Rosaniline.HCl	569-61-9	+	—	—	—
Rutin trihydrate	153-18-4	+	—	—	—
Tetrafluoroborate, sodium	13755-29-8	.	—	—	—
Trichloroethylene	79-01-6	—	—	—	—
Urethane	51-79-6	+	65.2^m	lgi ski sto	adr lgi ski sto thy
Vinyl chloride	75-01-4	+	43.8^m	sto vsc	mgl ski sto vsc

Note: Abbreviations: "." = not tested; "(B)" = both sexes.

Tissue codes: adr = adrenal gland; eso = esophagus; gal = gallbladder; lgi = large intestine; liv = liver; lun = lung; mgl = mammary gland; mix = mixture; nas = nasal cavity (includes tissues of the nose, nasal turbinates, paranasal sinuses and trachea); orc = oral cavity (includes tissues of the mouth, oropharynx, pharynx, and larynx); pan = pancreas; ski = skin; sto = stomach; thy = thyroid gland; ubl = urinary bladder; ute = uterus; vsc = vascular system.

In a series of footnotes, we provide additional information about TD$_{50}$ values and test results in the CPDB. These are as follows:

^{m} = More than one positive test in the species in the CPDB.

SECTION C: SUMMARY OF CHEMICAL CARCINOGENICITY IN NONHUMAN PRIMATES AND DOGS

TABLE 3

Chemical	CAS	Salmonella	Harmonic mean of TD$_{50}$ (mg/kg/day)		Target sites	
			Cynomulgus	Rhesus	Cynomulgus	Rhesus
Monkeys						
2-Acetylaminofluorene	53-96-3	+	–	.	–	.
Aflatoxin B$_1$	1162-65-8	+	0.0082	0.0201	gal liv vsc	gal liv vsc
Arsenate, sodium	7631-89-2	.	.	–	.	–
Cycasin and methylazoxymethanol acetate	mixture	.	0.0657^{m,v}	19.4	kid liv	liv
N,N-Dimethyl-4-aminoazobenzene	60-11-7	+	–	.	–	.
IQ	76180-96-6	+	.	0.577	.	liv
3'-Methyl-4-dimethylaminoazobenzene	55-80-1	+	–	.	–	.
N-Methyl-N'-nitro-N-nitrosoguanidine	70-25-7	+	–	.	–	.
3-Methylcholanthrene	56-49-5	+	–	–	–	–
2-Naphthylamine	91-59-8	+	5.74	.	ubl	.
N-Nitroso-N-methylurea	684-93-5	+	7.46	4.52	eso	eso
N-Nitrosodiethylamine	55-18-5	+	0.113^{m,v}	0.0547^{m,v}	liv	liv
N-Nitrosodimethylamine	62-75-9	+	–	.	–	.
N-Nitrosodipropylamine	621-64-7	+	0.0121^i	.	liv	.
N-Nitrosopiperidine	100-75-4	+	2.76^m	12.1	liv	liv
Norlestrin	8015-12-1	.	–	.	–	.
Procarbazine.HCl	366-70-1	–	4.21	14.8	hmo	hmo
Sterigmatocystin	10048-13-2	+	.	0.264	.	liv
Urethane	51-79-6	+	65.2	52.8	liv smi vsc	lun pan vsc
Tree shrews						
Aflatoxin B$_1$	1162-65-8	+	0.0269		liv	
					nas	
Bush babies					–	
N-Nitrosodiethylamine	55-18-5	+	0.0122^i		liv ubl	
					mgl	
					liv ubl	
Dogs						
Chloroform	67-66-3	–	–			
3,3'-Dichlorobenzidine	91-94-1	+	1.78			
Lynestrenol	52-76-6	.	0.58			
4,4'-Methylene-bis(2-chloroaniline)	101-14-4	+	2.12			
FD & C red no. 4	4548-53-2	–	–	–		

Note: All experiments in monkeys and bush babies are from the NCI Laboratory of Chemical Pathology except norlestrin and 2-naphthyl-amine.

Abbreviations: "." = not tested.

Tissue codes: eso = esophagus; gal = gallbladder; hmo = hematopoietic system; kid = kidney; liv = liver; lun = lung; mgl = mammary gland; nas = nasal cavity (includes tissues of the nose, nasal turbinates, paranasal sinuses and trachea); pan = pancreas; smi = small intestine; ubl = urinary bladder; vsc = vascular system.

In a series of footnotes, we provide additional information about TD$_{50}$ values and test results in the CPDB. These are as follows:

^{i} = **Intraperi**toneal or intravenous injection are the only routes of administration with positive tests **in the CPDB.**

^{m} = **More th**an one positive test in the species in the CPDB.

^{v} = **Variation** is greater than ten-fold among statistically significant (p <0.1) TD$_{50}$ values from different positive experiments.

CHAPTER 4

OVERVIEW AND UPDATE OF ANALYSES OF THE CARCINOGENIC POTENCY DATABASE

Lois Swirsky Gold, Thomas H. Slone, and Bruce N. Ames

The Carcinogenic Potency Database (CPDB) is a systematic and standardized resource on the published results of chronic animal cancer tests and is readily amenable to secondary analyses of experimental results.[1-7] For the past 12 years, we have used the CPDB to investigate many important research and regulatory issues in carcinogenesis. The inclusion rules and methods of the CPDB have been described in detail elsewhere.[1] This chapter presents an overview and update of some of the analyses that we published earlier. Our goal here is to briefly describe some of the major findings and update key results using the larger number of experiments and chemicals now reported in the CPDB. We refer the reader to the earlier published papers for the expanded analyses, methods, and discussion. Tabular results are presented below for several of the updated findings.

Five broad areas are addressed below:

- Methodological issues in the interpretation of animal cancer tests: constraints on the estimation of carcinogenic potency and validity problems associated with using the limited data from bioassays to estimate human risk, reproducibility of results in carcinogenesis bioassays, comparison of lifetable and summary methods of analysis, and summarizing carcinogenic potency when multiple experiments on a chemical are positive.
- More than half the 1298 chemicals tested in long-term experiments have been evaluated as carcinogens. We describe this positivity rate for several subsets of the data (including naturally occuring and synthetic chemicals), and we hypothesize an important role for increased cell division in the interpretation of results.
- We compare positivity in bioassays for two closely related species, rats and mice, tested under similar experimental conditions. We assess what information such a comparison can provide about interspecies extrapolation.

- Rodent carcinogens induce tumors in 35 different target organs. We describe the frequency of chemicals that induce tumors in rats or mice at each target site, and we compare target sites of mutagenic and nonmutagenic rodent carcinogens.
- We broaden the perspective on evaluation of possible cancer hazards from rodent carcinogens, by ranking 74 human exposures (natural and synthetic) on the HERP index (Human Exposure/Rodent Potency).

BACKGROUND

The CPDB includes results reported in 1002 papers in the general literature through 1992 and 403 Technical Reports of the National Cancer Institute/National Toxicology Program (NCI/NTP) through 1994.[1-7] Results are examined for 5152 experiments on 1298 chemical agents. About 30% of the chemicals were tested by NCI/NTP. The published results of the experiments reported in the CPDB constitute a diverse literature that varies widely with respect to experimental and histological protocols as well as to how and which information is reported in published articles.[1] Detailed information that is important in the interpretation of bioassays is given on each experiment in the CPDB (whether positive or negative for carcinogenicity), including qualitative information on strain, sex, target organ, histopathology, and author's opinion, as well as quantitative information on carcinogenic potency, statistical significance, tumor incidence, dose-response curve shape, length of experiment, dose-rate, and duration of dosing. No attempt has been made in the CPDB to evaluate whether or not a compound induced tumors in any given experiment; rather, the opinion of the published author is presented. For any single chemical, the number of experiments in the database may vary. Some chemicals have only one test in one sex of one species, while others have multiple tests including both sexes of a few strains of rats and mice, possibly using quite different protocols.[1-7]

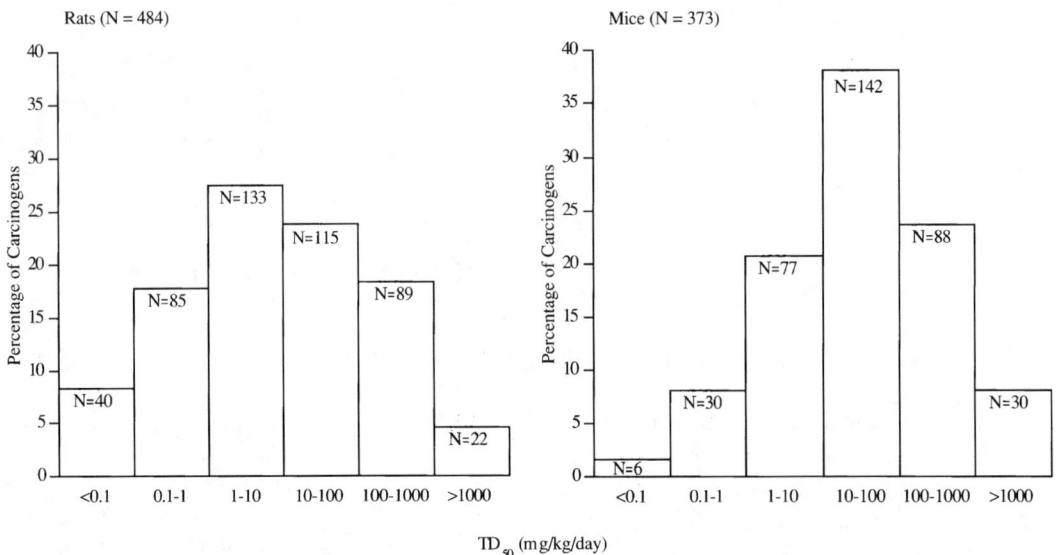

FIGURE 1   Frequency distribution of TD_{50} in the most potent site for chemicals evaluated as carcinogens in rats or mice.

A numerical description of carcinogenic potency, the TD_{50},[8,9] is estimated for each set of tumor incidence data reported in the CPDB, thus providing a standardized quantitative measure for comparisons. In a simplified way, TD_{50} may be defined as that dose-rate in mg/kg body wt/day which, if administered chronically for the standard lifespan of the species, will halve the probability of remaining tumorless throughout that period. Put differently, TD_{50} is the daily dose that will induce tumors in half of test animals that would have remained tumor-free at zero dose. We estimate TD_{50} using a one-hit model.[8,9] TD_{50} is analogous to LD_{50}, and a low value of TD_{50} indicates a potent carcinogen, whereas a high value indicates a weak one. TD_{50} is often within the range of doses tested and does not indicate anything about carcinogenic effects at low doses because bioassays are usually conducted at or near the maximum tolerated dose (MTD). The MTD is generally accepted to be defined as the maximum dose level which is not expected to shorten the normal longevity from non-neoplastic causes, and which is expected to result in no more than a 10% weight decrement in animals receiving this dose when compared to controls.[10]

The range of TD_{50} values across rodent carcinogens in the CPDB is more than 10-million-fold in male and female rats or mice. The distribution of TD_{50} values in rats and mice is shown in Figure 1 above using the most potent site per chemical from experiments that were evaluated as carcinogens in either species by the published author. Among chemicals that are positive in both species, the TD_{50} (mg/kg/day) in rats is more potent than the value in mice for 131 and less potent for 57. Potency values in rats and mice are within a factor of 10 of each other for 73% of the chemicals that are carcinogenic in both species.

METHODOLOGICAL ISSUES

CONSTRAINTS ON ESTIMATION OF CARCINOGENIC POTENCY

Rodent cancer tests are designed to maximize the chance of obtaining a positive result in a lifetime experiment with small numbers of animals; near-toxic doses (MTD and half MTD) are the dose levels used for that purpose in the standard protocol of the NCI/NTP. This standard experimental design, with a narrow range of doses, was never intended to provide information to quantitatively assess the risk to humans from chemical exposures at low doses. In regulatory policy, however, standard practice has been to assess risk by linear extrapolation to the human exposure level, i.e., risk = potency × human exposure.

In 1985, we (Bernstein et al.[11]) showed that statistically significant potency estimates based on the usual experimental design are constrained to a narrow range about the maximum dose tested, in the absence of tumors in all dosed animals (which rarely occurs). For an ideal type experiment with 50 animals in a single dose group and a 10% tumor rate in a large control group, the estimable range for statistically significant potency values is about 32-fold about the MTD, which is in marked contrast to the more than 10-million-fold range of potency values across chemicals. The range of possible potency is widened somewhat in real bioassays with two dose groups, variable control rates and group size, and lifetable analysis of potency.

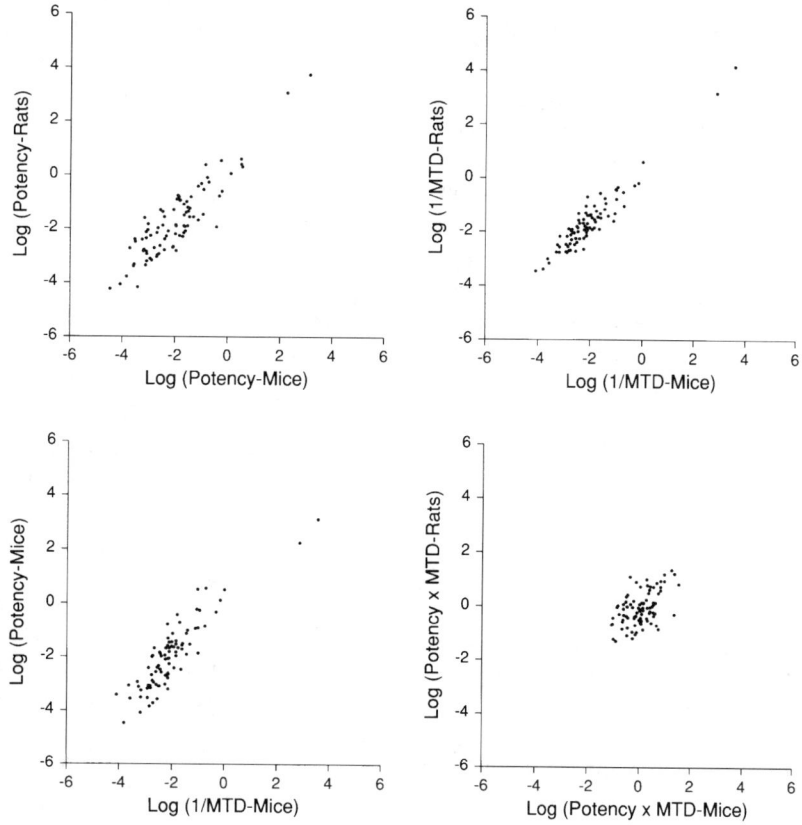

FIGURE 2 Top left: The strong inter-species correlation of carcinogenic potencies; the horizontal axis shows log potency-mice, while the vertical shows log potency-rats. Top right: the horizontal axis shows log (1/MTD-mice), while the vertical shows log (1/MTD-rats); this correlation in toxicity is believed to be real. Lower left: A statistical artifact which drives the interspecies correlation of carcinogenic potencies; the horizontal axis shows log (1/MTD-mice), while the vertical axis shows log potency-mice. Lower right: A weak interspecies correlation which seems to be real; the horizontal axis shows log (potency×MTD)-mice, while the vertical shows log (potency×MTD)-rats. Each dot represents one of the 87 NCI/NTP bioassays where the chemical on test was significant at the 0.025 level (one-sided) in female mice and in female rats. Data are for females only. Logs are to base 10.

Other researchers later showed a similar result for q_1^* estimated from the linearized multistage model used in regulatory risk assessment.[12] Potency estimates based on standard bioassay design are highly correlated with the administered dose, regardless of whether the estimate is based on the one-stage, multistage, or Weibull model. This constraint on potency estimation contrasts with the enormous extrapolation that is required from the MTD in rodent bioassays to usual human exposure levels, often hundreds of thousands times lower.

ARTIFACTS IN THE CORRELATION OF POTENCIES BETWEEN RATS AND MICE

A strong correlation of carcinogenic potencies ($\ln(2)/TD_{50}$) observed between rats and mice has been often interpreted as a justification for quantitative extrapolation from rodents

to humans. Bernstein et al. showed, however, that the correlation is largely an artifact.[11] Over large numbers of chemicals the MTDs for rats and mice are highly correlated and span many orders of magnitude (Figure 2); this is a biological correlation in toxicity between the two species. As discussed above, potency estimates are constrained to a narrow range about the MTD. Hence, the potency correlation between rats and mice necessarily follows statistically.

Recently, we (Freedman et al.[13]) examined how much of the observed correlation in potencies between species is indeed artifactual. Our analysis involved two statistical models where the impacts of various assumptions could be calculated. One model assumes that interspecies correlation of potencies is purely artifactual; it ignores the correlation between rats and mice of (potency × MTD), which is a rough measure of tumor yield. The second model incorporates the correlation in (potency × MTD) between rats and mice, which indicates that part of the interspecies correlation in potencies is real, i.e., tumor yields are correlated. A comparison of the models and data suggests that over 80% of the interspecies correlation in carcinogenic potencies for chemicals positive in both rats and mice can be explained by the interspecies correlation in toxicity (MTD) and the correlation between log(potency) and log(MTD). This confirms the findings of Bernstein et al. and indicates that while there may be some basis for extrapolation from rodents to humans, the interspecies correlation of potencies between rats and mice does not say much about the validity of that extrapolation.

Regulatory Risk Assessment and the Constraint on Potency Estimation

Standard practice in regulatory risk assessment for a given rodent carcinogen is to extrapolate from the high doses of rodent bioassays to the low doses of most human exposures by multiplying carcinogenic potency in rodents by human exposure. Strikingly, however, since potency estimates are constrained to lie within a narrow range about the MTD, the dose usually estimated by regulatory agencies to give one cancer in a million can be approximated simply by using the MTD as a surrogate for carcinogenic potency. The "virtually safe dose" (VSD) can be approximated from the MTD. Gaylor and Gold[14] used the ratio MTD/TD_{50} and the relationship between q_1^* and TD_{50} found by Krewski et al.,[12] to estimate the VSD. The VSD was approximated by the MTD/740,000 for NCI/NTP rodent carcinogens. This result questions the utility of bioassay results to estimate risk, and demonstrates the limited information about risk that is provided by bioassay results. The MTD/740,000 was within a factor of 10 of the VSD for 96% of carcinogens. Without data on mechanism of carcinogenesis for a given chemical, the true risk of cancer at low dose is highly uncertain and could be zero, even for rats or mice.

Lifetable Vs. Summary Estimates of Potency

We have compared two methods for estimating carcinogenic potency from carcinogenesis bioassays, one based on lifetable data and one based on summary incidence data (the crude proportion of animals with tumors).[15] The lifetable analysis adjusts for the differential effects of toxicity among dose groups and for differences in the time pattern of tumor incidence, while summary incidence analysis does not. However, summary data are all that are usually available in the published results of animal cancer tests. Using results for NCI/NTP bioassays, which provide full

lifetable data, we compared lifetable and summary estimates of potency; our updated analysis for the CPDB includes 551 experiments with statistically significant lifetable TD_{50} values ($p < 0.01$). The most potent site represents an experiment. For 91% of the experiments, the level of statistical significance is the same for lifetable and summary analysis. There is substantial agreement between these methods of analysis in terms of potency estimation, although lifetable estimates are usually more potent. For about half the statistically significant cases, the effect of using lifetable data is to reduce the TD_{50} (to increase potency) by less than 30%. The median ratio of lifetable TD_{50} to summary TD_{50} is 0.69, and 85% of the ratios lie between 0.30 and 1.30.

Dose-response curve shape was investigated by testing for linearity using results from the respective model goodness of fit test. Among experiments with two doses and a control, lifetable and summary methods agree on the shape of the dose-response for 72% of the experiments. As expected, more curves are linear or curving upward for lifetable analysis.[15] We note that with either method of analysis, the shape of the dose-response curve may differ for different target sites in experiments with the same test agent.[15]

Reproducibility

Reproducibility of results in animal bioassays has been investigated in "near-replicate" comparisons consisting of two or more tests of the same chemical administered by the same route and using the same sex and strain of rodent.[16] The updated results continue to show good reproducibility. Among 166 comparisons, 84% (139/166) have concordant authors' opinions about whether tumors were induced in the individual experiments (Table 1). For rats and mice, in all but 3 of the 74 positive comparisons, at least one target site is identical. TD_{50} values are within a factor of 2 of each other in 47% of the positive comparisons, within a factor of 4 in 78%, and within a factor of 10 in 95% (Table 2).

Summary Measures of Carcinogenic Potency

For over half the carcinogens in rats or mice there is more than one positive experiment, and it is desirable to have a summary measure of potency. We evaluated three summary measures of TD_{50} for these cases (arithmetic, geometric, or harmonic mean) to determine how different results would be from using the most potent site to summarize potency.[17] These measures differ according to the weight given outlying results. Our analysis indicates that the most

TABLE 1
Summary of Reproducibility of Positivity in "Near-Replicate" Comparisons[a] of Chronic Exposure Carcinogenesis Bioassays in Hamsters, Mice, and Rats in the Carcinogenic Potency Database

| | Number of comparisons (%) | | | | | | | |
	All species		Hamsters		Mice		Rats	
Discordant	27	(16%)	0	(0%)	11	(20%)	16	(15%)
Concordant positive	80	(48%)	6	(75%)	29	(54%)	45	(43%)
Concordant negative	59	(36%)	2	(25%)	14	(26%)	43	(42%)
Total	166	(100%)	8	(100%)	54	(100%)	104	(100%)

[a] A comparison consists of the results for two or more experiments of the same chemical administered by the same route to the same strain and sex of rodent.

TABLE 2
Ratio of Least to Most Potent TD$_{50}$ from Different Positive Experiments for Near-Replicate Comparisons[a] and All Chemicals With More Than One Positive Experiment in the Carcinogenic Potency Database

| Ratio of least potent TD$_{50}$ to most potent | Rats | | | | Mice | | | |
	Near-replicate tests		All chemicals		Near-replicate tests		All chemicals	
1-1.99	21	(47%)	102	(41%)	14	(48%)	117	(50%)
2-2.99	10	(22%)	36	(14%)	9	(31%)	50	(21%)
3-3.99	4	(9%)	29	(12%)	0	(0%)	15	(6%)
4-9.99	7	(15%)	49	(19%)	5	(17%)	34	(14%)
≥10	3	(7%)	36	(14%)	1	(4%)	20	(9%)
Total	45	(100%)	252	(100%)	29	(100%)	236	(100%)

[a] A comparison consists of the results for two or more experiments of the same chemical administered by the same route to the same strain and sex of rodent.

potent TD$_{50}$ value is similar to the average values (Table 3). (See also Reference 1). We have also compared the most potent to the least potent TD$_{50}$ from different positive experiments, and find that the distribution of the ratio of least to most potent values for all chemicals is similar to that for near-replicate comparisons (see Table 2). This similarity suggests that discrepant results for a chemical within a species are not an artifact of combining across strains, routes of administration, and sexes. Thus, for various purposes one may wish to use different summary measures; however, it generally makes little difference whether the choice is the most potent site or a mean. In our "Summary of the Carcinogenic Potency Database by Chemical,"[18] we have used the harmonic mean to represent potency when more than one experiment is positive in a species.

THE INTERPRETATION OF POSITIVE RESULTS IN RODENT BIOASSAYS

HALF THE CHEMICALS TESTED IN RODENTS ARE CARCINOGENS

In several papers we have shown that approximately half the chemicals tested in rats or mice are positive in at least one experiment, according to the evaluation of a published author. Using all results in the updated CPDB, positivity rates of about 50% are shown in Table 4 for chemicals tested in NCI/NTP bioassays, in the general literature, and in either of these sources. Table 5 reports a similar positivity rate for several additional subsets of the CPDB: naturally occurring chemicals, synthetic chemicals, natural pesticides (the chemicals that plants produce naturally to

TABLE 3

Ratio of Harmonic, Geometric, and Arithmetic Means to Most Potent TD_{50} for Chemicals Positive in More Than One Experiment in the Carcinogenic Potency Database

Ratio of mean TD_{50} to most potent	Rats n = 252			Mice n = 236		
	H (%)	G (%)	A (%)	H (%)	G (%)	A (%)
1-1.99	85	69	56	92	77.5	72
2-2.99	12	15	19	7	14	12
3-3.99	2	7	7	0.5	6	5
4-9.99	1	7	13	0.5	2.5	9
≥10	0	2	5	0	0	2
Total	100%	100%	100%	100%	100%	100%

Note: H, Ratio of harmonic mean to most potent TD_{50}; G, Ratio of geometric mean to most potent TD_{50}; A, Ratio of arithmetic mean to most potent TD_{50}.

TABLE 4

Proportion of Chemicals in the Carcinogenic Potency Database Tested in Rats or Mice That Have Been Evaluated as Carcinogenic,[a] by Reference Source

Reference source	Proportion carcinogenic in rats or mice (%)	Proportion carcinogenic in rats (%)	Proportion carcinogenic in mice (%)
NCI/NTP or literature[b]	668/1275 (52%)	484/997 (49%)	372/847 (44%)
NCI/NTP	200/384 (52%)	144/366 (39%)	145/366 (40%)
Literature	505/1006 (50%)	361/699 (52%)	242/550 (44%)

[a] A chemical is classified as positive if the author of at least one published experiment has evaluated the compound as carcinogenic in that species.

[b] The number of chemicals in the "NCI/NTP or literature" is smaller than the sum of each source separately because some chemicals have been reported by both sources.

TABLE 5

Proportion of Chemicals Evaluated as Carcinogenic,[a] for Several Datasets in the Carcinogenic Potency Database

Chemicals tested in both rats and mice	330/559 (59%)
Naturally occurring chemicals	73/127 (57%)
Synthetic chemicals	257/432 (59%)
Chemicals tested in rats and/or mice	
Natural pesticides	35/64 (55%)
Mold toxins	14/23 (61%)
Chemicals in roasted coffee	19/28 (68%)

[a] A chemical is classified as positive if the author of at least one published experiment evaluated results as evidence that the compound is carcinogenic.

TABLE 6

Comparison of Mutagenicity and Carcinogencity[a] for Chemicals Tested in Both Rats and Mice and for Mutagenicity in *Salmonella* in the Carcinogenic Potency Database

Carcinogenic + Mutagenic +		Carcinogenic − Mutagenic +	
	165	43	
	127	130	
Carcinogenic + Mutagenic −		Carcinogenic − Mutagenic −	

1. Of 465 chemicals, 45% are mutagens, 63% are carcinogens, and 72% are either mutagens or carcinogens or both (165 + 127 + 43)/465.
2. Mutagens are more likely to be carcinogenic 79% (166/208) than nonmutagens 49% (125/257).
3. Of 292 carcinogens, 43% are not mutagens 125/(166 + 125).
4. Of 173 noncarcinogens, 25% are mutagens 43/(43 + 130).

[a] A chemical is classified as positive if the author of at least one published experiment in the Carcinogenic Potency Database evaluated the results as evidence that the compound is carcinogenic.

defend themselves), mold toxins, and chemicals in roasted coffee. Additionally, for the 559 chemicals that have been tested for both mutagenicity in *Salmonella* and for carcinogenicity in rats and mice, we find that only 28% are neither mutagens nor carcinogens (Table 6).

Among chemicals that humans are exposed to, the vast proportion occur naturally; however, only 22% of chemicals tested in rodent bioassays are natural. Since half the natural chemicals tested are positive, human exposures to rodent carcinogens are likely to be ubiquitous. Additionally, in the *Physician's Desk Reference,* 49% (117/241) of the drugs that report results of animal cancer tests are carcinogenic in rodent tests.[19]

Since the results of high-dose bioassays are routinely used to identify a chemical as a possible cancer hazard to humans, it is important to try to understand how representative the 50% positivity rate might be of all untested chemicals. If half of all chemicals (both natural and synthetic) to which humans are exposed would be positive if tested, then the utility of a test to identify a chemical as a "potential human carcinogen" is questionable. To determine the true proportion of rodent carcinogens among chemicals would require a comparison of a random group of synthetic chemicals to a random group of natural chemicals. Such an analysis has not been done.

It has been argued that the high positivity rate is due to selecting more suspicious chemicals to test, which is a likely bias since cancer testing is both expensive and time-consuming, and it is prudent to test suspicious compounds.

On the other hand, chemicals are selected for testing for many reasons, including the extent of human exposure, level of production, and scientific questions about carcinogenesis.[20] Of the chemicals tested for carcinogenicity, half are not mutagenic, even though mutagens are more often carcinogenic than nonmutagens. Thus, prediction of positivity may often not be the basis for selecting a chemical to test. Moreover, while some chemical classes are more often carcinogenic in rodent bioassays than others — for example, nitroso compounds, aromatic amines, nitroaromatics, and chlorinated compounds — prediction is still imperfect.[21] (A chemical is classified as mutagenic in our analyses if it was evaluated in the *Salmonella* assay as either mutagenic or weakly mutagenic by Zeiger[22] or as positive by the Gene-Tox Program.[23,24] Other chemicals evaluated by those sources are classified as nonmutagenic.) It has been suggested that chemicals put on test in the early years of the NCI bioassay program were selected more on the basis of chemical structure, while chemicals tested more recently by NCI or NTP have been selected more on the basis of the extent of human exposure. If this were so, then one might expect the positivity rate to be higher among the earlier chemicals. For the NCI/NTP chemicals we compared positivity rates over time and found no significant difference between those tested before 1979 and those tested later.

One large series of mouse experiments by Innes et al. in 1969[25,26] has been frequently cited[27] as evidence that the true proportion of rodent carcinogens is actually low among tested substances. Innes tested 119 chemicals — primarily the most widely used pesticides at that time and some industrial chemicals — and only 11 (9%) were judged as carcinogens. We have discussed[20] that those early experiments lacked power to detect an effect because they were conducted only in mice (not in rats), they included only 18 animals in a group (compared with the usual 50), the animals were tested for only 18 months (compared with the usual 24 months), and the dose was usually lower than the highest dose in subsequent mouse tests of the same chemical.

We have re-examined the Innes results using the CPDB to assess positivity in subsequent bioassays on the chemicals that Innes did not evaluate as positive (Table 7). Among 34 negative chemicals that were subsequently retested, 16 were carcinogenic (47%), which is similar to the proportion among all chemicals in our database. Innes had recommended further evaluation of some chemicals that had inconclusive results in their study. If those were the chemicals subsequently retested, then one might argue that they would be the most likely to be positive. Our analysis does not support that view: among chemicals needing further evaluation 6/16 were positive when retested; among the other negatives, 10/18 were positive (Table 7).

TABLE 7
Results of Subsequent Tests on Chemicals Not Found Carcinogenic by Innes et al.

Retested chemicals	Percent carcinogenic when retested		
	Mice	Rats	Either mice or rats
All retested	6/26 (23%)	13/34 (38%)	16/34 (47%)
Innes: Not Carcinogenic	3/10 (30%)	9/18 (50%)	10/18 (56%)
Innes: Needs Further Evaluation	3/16 (19%)	4/16 (25%)	6/16 (38%)

Note: Of 119 chemicals tested by Innes et al., 11 (9%) were evaluated as positive by Innes et al.; * Innes et al. stated that further testing was needed; (M), positive in mice on retest; (R), positive in rats on retest.

Carcinogenic when retested: Atrazine (R), Azobenzene* (R), Captan (M,R), Carbaryl (R), 3-(p-Chlorophenyl)-1,1-dimethylurea* (R), p,p'-DDD* (M), Folpet (M), Manganese ethylenebisthiocarbamate (R), 2-Mercaptobenzothiazole (R), N-Nitrosodiphenylamine* (R), 2,3,4,5,6-Pentachlorophenol (M,R), o-Phenylphenol (R), Piperonyl sulfoxide* (M), 2,4,6-Trichlorophenol* (M,R), Zinc dimethyldithiocarbamate (R), Zinc ethylenebisthiocarbamate (R).

Not carcinogenic when retested: (2-Chloroethyl)trimethylammonium chloride,* Cyanamide, calcium,* Diphenyl-p-phenylenediamine, Endosulfan, p,p'-ethyl-DDD,* Ethyl tellurac,* Isopropyl-N-(3-chlorophenyl)carbamate, Lead dimethyldithiocarbamate,* Maleic hydrazide, Mexacarbate,* Monochloroacetic acid, Phenyl-β-naphthylamine,* Piperonyl butoxide,* Rotenone, Sodium diethyldithiocarbamate trihydrate,* Tetraethylthiuram disulfide,* Tetramethylthiuram disulfide, 2,4,5-Trichlorophenoxyacetic acid.

CELL DIVISION AND THE HIGH POSITIVITY RATE IN BIOASSAYS

What are the explanations for the high positivity rate in high-dose animal cancer tests? One plausible explanation, which is supported by an increasing number of papers,[28,29] is that the MTD of a chemical can cause chronic cell killing and cell replacement in the target tissue, a risk factor for cancer that can be limited to high doses. We (Ames and Gold) have discussed in detail the importance of cell division in mutagenesis and carcinogenesis;[28,30–32] several results in the CPDB are consistent with the idea that cell division increases carcinogenesis under the conditions of standard animal cancer tests.

Endogenous DNA damage from normal oxidation is enormous. The steady-state level of oxidative damage in DNA is over one million oxidative lesions per rat cell.[33] Thus, from first principles, the cell division rate must be a factor in converting lesions to mutations and thus cancer.[32] Raising the level of either DNA lesions or cell division will increase the probability of cancer. Just as DNA repair protects against lesions, p53 guards the cell cycle and protects against cell division if the lesion level gets too high. If the lesion level becomes still higher, p53 can initiate programmed cell death (apoptosis). None of these defenses is perfect, however. The critical factor is chronic cell division in stem cells, not in cells that are discarded, and is related to the total number of extra cell divisions. Cell division is both a major factor in loss of heterozygosity

through nondisjunction and other mechanisms[30] and in expanding clones of mutated cells.

In animal cancer tests, the doses administered are near-toxic or minimally toxic, i.e., the MTD and half the MTD, and may result in cell division. Tissues injured by high doses of chemicals have an inflammatory immune response involving activation of recruited and resident macrophages[34–39] (e.g., phenobarbital, carbon tetrachloride, TPA). Activated macrophages release mutagenic oxidants (including peroxynitrite, hypochlorite, and H_2O_2), as well as inflammatory and cytotoxic cytokines, growth factors, bioactive lipids (arachidonic acid metabolites), and proteases. This general response to cell injury suggests that chronic cell killing by high-dose animal cancer tests will likely incite a similar response, leading to further cell injury, compensatory cell division and therefore increased probability of mutation. *Ad libitum* feeding in the standard bioassay can also contribute to the high positivity rate,[40] plausibly by increased cell division due to high caloric intake.[32,40]

Although cell division is not measured in routine cancer tests, many studies on rodent carcinogenicity show a correlation between cell division at the MTD and cancer. For example, Cunningham et al. have analyzed 15 chemicals at the MTD, 8 mutagens and 7 nonmutagens, including several pairs of mutagenic isomers, one of which is a rodent carcinogen and one of which is not.[41,42] A perfect correlation was observed: the 9 chemicals causing cancer caused cell division in the target tissue and the 6 chemicals not

TABLE 8

Comparison of the Number of Positive Target Organs for Mutagens and Nonmutagens by Species,[a] in the Carcinogenic Potency Database

Number of target organs	Chemicals evaluated as carcinogenic in:							
	Rats				Mice			
	Mutagens		Nonmutagens		Mutagens		Nonmutagens	
1	77	(40%)	58	(60%)	76	(48%)	69	(66%)
2	37	(19%)	25	(26%)	48	(31%)	22	(21%)
≥3	78	(41%)	13	(14%)	33	(21%)	13	(13%)
Total number of chemicals	192	(100%)	96	(100%)	157	(100%)	104	(100%)

[a] A target organ is classified by a positive author's opinion in any experiment. Experimental results are excluded if histopathological examination was restricted to a few selected tissues.

causing cancer did not. Extensive reviews on rodent studies[30,33,43–45] document that chronic cell division can induce cancer. A large epidemiological literature indicates that increased cell division by hormones and other agents can increase human cancer.[46]

Our analyses of the CPDB are consistent with the idea that in high-dose bioassays, cell division increases mutagenesis and therefore carcinogenesis. To the extent that increases in tumor incidence in rodent studies are due to the secondary effects of inducing cell division at the MTD, then *any* chemical is a likely rodent carcinogen; therefore, the high positivity rate in the CPDB overall, and for several subsets of chemicals (see Tables 4 and 5), would be expected. Carcinogenicity results for mutagenic compared to nonmutagenic chemicals tested in rats and mice (see Table 6) indicate that 43% of carcinogens are not mutagenic. For these chemicals, increased cell division is likely an important factor.

Mutagens can both damage DNA and increase cell division at high doses, thus having a multiplicative effect on mutagenesis at high doses. Therefore, if cell division is important, one would expect stronger evidence of carcinogenicity for mutagens in rodent bioassays. Results are consistent with this idea. Mutagens are more likely to be carcinogenic than nonmutagens (see Table 6). Mutagenic carcinogens compared to nonmutagenic carcinogens are more likely to be carcinogenic in both rats and mice rather than in only one species: among chemicals tested in both rats and mice and carcinogenic in at least one test, 67% of mutagens (110/165) are positive in both rodent species compared to 41% (52/127) of nonmutagens. Moreover, mutagenic carcinogens induce tumors at more target sites in rodent bioassays than nonmutagens (Table 8). Analyses

of the limited data on dose-response in bioassays are consistent with the idea that cell division from cell-killing and cell replacement is important. Even at the two high doses tested, about half the sites evaluated as target sites in NCI/NTP bioassays are statistically significant at the MTD but not at half the MTD ($p < 0.05$).[47]

Thus, it seems likely that a high proportion of all chemicals might be "carcinogens" if tested in a standard bioassay at the MTD, but this would be largely due to the effects of high doses for the nonmutagens, and a synergistic effect of cell division at high doses with DNA damage for the mutagens. Our results suggest that adding routine measurements of cell division to the 90-day prechronic study and the 2-year bioassay for each test agent would provide information that could improve dose-setting, the interpretation of experimental results, and risk assessment. Without additional data on mechanism of carcinogenesis for each chemical, the interpretation of a positive result in a rodent bioassay is highly uncertain.[48] The carcinogenic effects may be limited to the high doses tested.

QUALITATIVE INTERSPECIES EXTRAPOLATION

CONCORDANCE BETWEEN RATS AND MICE

The use of bioassay results in risk assessment requires a qualitative species extrapolation from rats or mice to humans. The accuracy of this extrapolation is generally unverifiable, since data on humans are limited. However, it is feasible to examine the accuracy of extrapolations from mice to rats. If mice and rats are similar with respect to carcinogenesis, this provides some evidence in favor of

TABLE 9

Comparison of Carcinogenic Response in Rats and Mice for Chemicals Tested in Both Species in the Carcinogenic Potency Database

Rats +		Rats −
Mice +		Mice +
	190	73
	67	229
Rats +		Rats −
Mice −		Mice −

1. Of 559 chemicals, 59% are positive in at least one test (190 + 67 + 73)/559.
2. Of 559 chemicals, 75% are concordant in carcinogenicity between rats and mice (190 + 229)/559.

interspecies extrapolations; conversely, if mice and rats are different, this casts doubt on the validity of extrapolations from mice to humans.

One measure of interspecies agreement is concordance, the percentage of chemicals that are classified the same way as to carcinogenicity in mice and rats (i.e., either tumors are induced in both species or in neither). Observed concordance in the CPDB is about 75% (Table 9), which may seem low since the experimental conditions are identical and the species are similar. The observed concordance is just an estimate based on limited data. We (Freedman et al.[49,50]) show by simulations for NCI/NTP bioassays (which also have an observed concordance of 75%) that a variety of models with quite different true concordances are consistent with the observed results. The bias in observed concordance can be either positive or negative: an observed concordance of 75% can arise if the true concordance is anything between 20% and 100%.[49,50] In particular, observed concordance can seriously overestimate true concordance: due to lack of power in the bioassay, many chemicals that are truly discordant (i.e., positive in one species but negative in the other) are classified as negative in both species and hence concordant. Thus, it seems unlikely that true concordance between rats and mice can be estimated with any reasonable degree of confidence from bioassay data.

COMPARISON OF CARCINOGENICITY IN RODENTS AND NONHUMAN PRIMATES

Lifetime studies in cynomolgus and/or rhesus monkeys (lasting up to 29 years) are included in the CPDB for 16 rodent carcinogens.[51] Experimental protocols for the studies in nonhuman primates varied, but generally included 5 to 20 dosed animals of one or both species and a large

colony control. Compared to other chemicals in the CPDB, there is strong evidence of carcinogenic activity in rodents for the chemicals subsequently tested in monkeys, i.e., the test agents induce tumors in a high proportion of rats or mice often in a short period of time, at multiple target sites, and all but one are mutagenic. In the monkey studies, tumors were induced by 10 of the 16 rodent carcinogens (aflatoxin B_1, N-nitrosopiperidine, procarbazine.HCl, urethane, IQ, sterigmatocystin, cycasin and methylazoxymethanol acetate, N-methyl-N-nitrosourea, N-nitrosodipropylamine, and N-nitrosodiethylamine). Lack of power may account for the negative results for the other chemicals. Whereas dosing in monkeys was usually long-term, in four of the negative studies chemical administration was ended after 5 years, although the experiments were continued to 20 to 26 years (2-acetylaminofluorene, N,N-dimethyl-4-aminoazobenzene, 3′-methyl-4-dimethylaminoazobenzene, 3-methylcholanthrene). For an additional chemical (N-nitrosodimethylamine), the MTD was exceeded, and animals were all dead by 10 years due to toxicity. The only chemical that induced tumors with the short-dosing (5 year) protocol was urethane. Only for N-methyl-N′-nitro-N-nitrosoguanidine was a negative result obtained with a protocol of lifetime dosing.

The evaluation of carcinogenicity was the same in cynomolgus and rhesus monkeys for the 11 chemicals tested in both species, and the liver was the most frequent target site in both. Details of methods and experimental results are given in the plot of the CPDB.[1]

We note that the spontaneous tumor rate in the large colony control of both cynomolgus and rhesus monkeys is low compared to the rate in rat and mouse strains generally used in carcinogenesis bioassays. In monkeys, the tumor incidence rate in controls increases markedly with age, as in other species.

CARCINOGEN IDENTIFICATION BY TESTING IN TWO SEX-SPECIES OF RODENTS INSTEAD OF FOUR

The standard protocol used to identify chemicals as carcinogens calls for testing in both sexes of rats and mice. We examined the accuracy of predicting positive chemicals by using two instead of four sex-species groups[20,52] for the subset of chemicals in the CPDB that have been tested in four groups and are positive in at least one (N = 254). Under the conditions of these bioassays, a very high proportion of rodent carcinogens that are identified as positive by tests in four groups is also identified by results from one sex of each species (85 to 91%), as shown in Table 10. The fact that a higher proportion of carcinogens is identified by pairs consisting of one sex from each species, rather than by two sexes of the same species, is due to the fact

TABLE 10
Predictive Value of Two Sex-Species Groups for Rodent Carcinogens Tested in Both Sexes of Rats and Mice in the Carcinogenic Potency Database[a]

Sex-species groups used to identify carcinogens[b]	NCI/NTP or literature — Number identified as carcinogenic at least once (n = 254)[c]	NCI/NTP — Number identified as carcinogenic at least once (n = 178)[c]	Literature — Number identified as carcinogenic at least once (n = 71)[c]
FM, MR	231 (91%)	160 (90%)	67 (94%)
MM, MR	231 (91%)	159 (89%)	69 (97%)
FM, FR	218 (86%)	146 (82%)	65 (92%)
MM, FR	215 (85%)	142 (80%)	68 (96%)
FM, MM	201 (79%)	136 (76%)	60 (85%)
FR, MR	193 (76%)	131 (74%)	60 (85%)

[a] For chemicals tested in both sexes of rats and mice that were evaluated as carcinogenic in at least one experiment in the Carcinogenic Potency Database.

[b] FM, female mice; MM, male mice; FR, female rats; MR, male rats.

[c] A chemcal is classified as a carcinogen in this analysis if it was tested in male and female rats and mice, and evaluated as positive in at least one experiment. Percentage in each column indicates the percent of those carcinogens that would have also been identified as carcinogenic if the experiments had been conducted only in the two sex-species groups listed in the first column.

that among chemicals positive in at least one group there is greater agreement (and therefore redundancy of information) between sexes within a species than between species. Overall, the combination of male rats with either female or male mice gives results most similar to those obtained in tests of four sex-species groups (Table 10).

We repeated the analysis separately for chemicals tested in the literature and by NCI/NTP because in the subset of chemicals tested in four sex-species groups, NCI/NTP bioassays are overrepresented (Table 10). The same mouse strain and only a few rat strains are used in NCI/NTP bioassays. In studies published in the general literature, the proportions identified by two groups are similarly high; since many strains are used in literature studies, this provides some confidence in generalizing the finding beyond this data set.

Additionally, chemicals that are classified as "two-species carcinogens" or "multiple-site carcinogens" on the basis of results from four sex-species groups are also identified as two-species or multiple-site carcinogens on the basis of tests in one sex of each species.[52] Carcinogenic potency (TD$_{50}$) values for the most potent target site are similar when based on results from two compared to four sex-species groups. Of the potency values, 88% are within a factor of two of those obtained from tests in four sex-species groups, 95% are within a factor of four, and 99% are within a factor of 10. This result is expected because carcinogenic potency values are constrained to a narrow range about the maximum dose tested in a bioassay, and

the maximum doses administered to rats and mice are highly correlated and similar.[11]

A reduced protocol of two instead of four sex-species groups would lower costs, use fewer animals, and therefore make histopathological examination less time consuming and result in cost savings that would be available for more mechanistic studies of a chemical.

TARGET ORGANS OF CARCINOGENICITY

Chemical carcinogens in chronic bioassays induce tumors in a variety of target sites in each species. Researchers interested in results on a particular target site can use the "Summary of the Carcinogenic Potency Database by Target Organ" to identify particular chemicals that induce tumors at each of 35 target sites, e.g., all chemicals that induce lung tumors in mice are listed under lung.[53] Target organ results are reported for rats, mice, hamsters, monkeys, bush babies and dogs. Results are organized alphabetically by target site and within each site, by species and whether a chemical is mutagenic in *Salmonella*. By cross-referencing a chemical of interest to the plot of the CPDB,[1] histopathology at each site and details of each experiment can be obtained. A quick overview of all target sites in each sex-species group is given by chemical in "Summary of the Carcinogenic Potency Database by Chemical,"[18] which summarizes carcinogenic potency, positivity, and all target organs in each sex-species tested.

We have shown above (see Table 8) that it is common for a chemical to induce tumors at more than one target site, that this result is more frequent in rats than mice, and that it is more frequent among mutagens than nonmutagens. The greater frequency of target sites among mutagens is consistent with the hypothesis that in high-dose rodent tests increased cell division at the MTD is important in the carcinogenic response: mutagens have a multiplicative interaction for carcinogenicity because they can both damage DNA directly and cause cell division at high doses.

TABLE 11

Frequency of Target Organs Among Carcinogens in Rats or Mice in the Carcinogenic Potency Database by Mutagenicity in _Salmonella_

| | Chemicals evaluated as carcinogenic in: | | | | | |
| | Rats | | | Mice | | |
Target organ	All chemicals[a] (n = 461)[b]	Mutagens (n = 199)	Nonmutagens (n = 104)	All chemicals[a] (n = 370)	Mutagens (n = 161)	Nonmutagens (n = 107)
Liver	183 (40%)	76 (38%)	34 (33%)	207 (56%)	88 (55%)	75 (70%)
Lung	50 (11%)	27 (14%)	2 (2%)	100 (27%)	47 (29%)	16 (15%)
Mammary gland	93 (20%)	57 (29%)	7 (7%)	19 (5%)	11 (7%)	5 (5%)
Stomach	74 (16%)	43 (22%)	8 (8%)	52 (14%)	30 (19%)	10 (9%)
Kidney	71 (15%)	25 (13%)	28 (27%)	21 (6%)	9 (6%)	8 (7%)
Vascular system	28 (6%)	16 (8%)	2 (2%)	54 (15%)	32 (20%)	9 (8%)
Hematopoietic system	52 (11%)	27 (14%)	15 (14%)	47 (13%)	22 (14%)	13 (12%)
Urinary bladder	43 (9%)	23 (12%)	13 (13%)	12 (3%)	8 (5%)	1
Nasal cavity/turbinates	40 (9%)	18 (9%)	3 (3%)	5 (1%)	5 (3%)	
Esophagus	35 (8%)	13 (7%)	1	7 (2%)	5 (3%)	
Ear/Zymbal's gland	39 (8%)	30 (15%)	1	2	1	1
Skin	31 (7%)	20 (10%)	3 (3%)	3	3 (2%)	
Large intestine	30 (7%)	20 (10%)	1	1		
Thyroid gland	26 (6%)	12 (6%)	8 (8%)	15 (4%)	9 (6%)	5 (5%)
Small intestine	27 (6%)	18 (9%)	1	4 (1%)	2 (1%)	1
Oral cavity	27 (6%)	16 (8%)	4 (4%)	2	2 (1%)	
Uterus	21 (5%)	10 (5%)	4 (4%)	11 (3%)	8 (5%)	2 (2%)
Central nervous system	21 (5%)	14 (7%)	2 (2%)	3	2 (1%)	1
Peritoneal cavity	20 (4%)	12 (6%)	4 (4%)	7 (2%)	2 (1%)	
Clitoral gland	17 (4%)	16 (8%)	1	2		1
Harderian gland				13 (4%)	8 (5%)	3 (3%)
Adrenal gland	14 (3%)	6 (3%)	6 (6%)	8 (2%)	3 (2%)	5 (5%)
Preputial gland	8 (2%)	5 (3%)	3 (3%)	11 (3%)	3 (2%)	2 (2%)
Pancreas	16 (3%)	7 (4%)	5 (5%)			
Pituitary gland	8 (2%)	2 (1%)	5 (5%)	7 (2%)	2 (1%)	2 (2%)
Subcutaneous tissue	10 (2%)	7 (4%)		3	2 (1%)	
Ovary				8 (2%)	4 (2%)	4 (4%)
Testes	10 (2%)	6 (3%)	3 (3%)	2		2 (2%)
Spleen	7 (2%)	4 (2%)	2 (2%)			
Gall bladder				4 (1%)		
Bone	4	2 (1%)	1			
Prostate	3		1			
Mesovarium	2					
Myocardium				2		1
Vagina	1	1				

[a] The CPDB does not have mutagenicity evaluations for 158 rat carcinogens and 105 mouse carcinogens.

[b] % = Percentage of rat carcinogens or mouse carcinogens that induce tumors at the given site. Many chemicals induce tumors at more than one site, and these are counted at each relevant target site. Therefore, many chemicals are counted more than once, and percentages cannot be added. For example, of 199 rat carcinogens that are mutagenic in _Salmonella_, 75 induce liver tumors, i.e., 38%.

The frequency of target organs among carcinogens in rats or mice for all chemicals in the CPDB, and separately for mutagens and nonmutagens, is reported in Table 11. A site is classified as a target organ if the published author of at least one experiment concluded that the test agent induced tumors at that site.

Since tissue distribution and pharmacokinetics would not be expected to differ systematically between mutagens and nonmutagens, one would not expect systematic differences in the particular organs in which tumors are induced.[54] Results do not support the idea that mutagens and nonmutagens induce tumors in different target organs.[54] Both mutagens and nonmutagens induce tumors in a wide variety of sites, and most organs are target sites for both (Table 11). Moreover, the same sites tend to be the most common sites for both: 81% or more of both mutagenic and nonmutagenic carcinogens are positive in rats and in mice in at least one of the 8 most frequent target sites: liver, lung, mammary gland, stomach, vascular system, kidney, hematopoietic system, and urinary bladder.

The liver is the most common target site in both species, and among mutagens as well as nonmutagens. It is the predominant site in the mouse. Our analysis indicates a species difference in the predominance of the liver in mice compared to rats.[55,56] Among chemicals with positive results in the mouse, 55% (88/161) of mutagens compared to 70% (75/107) of nonmutagens induce liver tumors; in the rat the respective proportions are 38% (76/199) and 33% (34/104). Thus, while the proportion of rat carcinogens that are positive in the liver is similar for mutagens and nonmutagens, in mice a higher proportion of nonmutagenic than mutagenic carcinogens are positive in the liver. This finding in mice reflects the fact that chlorinated compounds (composed solely of chlorine, carbon, hydrogen, and optionally oxygen) are frequently positive in the mouse liver and are not mutagenic in *Salmonella*. Excluding the chlorinated compounds, results in mice are similar for mutagenic and nonmutagenic carcinogens: 55% (81/147) of mutagens and 59% (44/74) of nonmutagens are mouse liver carcinogens.

Knowing a target site in a bioassay is not expected to provide information about specific chemicals that will increase human cancer rates *at that same site*. Our analyses of bioassays in rats, mice, hamsters, as well as comparisons between rodents and humans for known human carcinogens, indicate that if a chemical induces tumors at a given site in one species, it is positive and induces tumors *at the same site* in the other species no more than 50% of the time.[47,57]

Potency values of chemicals that induce tumors at each common target site vary widely, as expected.[55]

RANKING POSSIBLE CARCINOGENIC HAZARDS TO HUMANS

Epidemiological studies have identified several factors that are likely to have a major effect on lowering rates of human cancer: reduction of smoking, increased consumption of fruits and vegetables, and control of infections. Other factors include avoidance of intense sun exposure, increased physical activity, reduction of high occupational exposures, and reduced consumption of alcohol and possibly red meat. Risks of many forms of cancer can already be lowered, and the potential for further risk reduction is great. In the U.S. cancer death rates for all cancers combined are decreasing if lung cancer — 90% of which is due to smoking — is excluded from the analysis.[28] We have discussed these epidemiological results with an emphasis on cancer mechanisms.[28]

HUMAN EXPOSURES TO NATURAL AND SYNTHETIC CHEMICALS

Current regulatory policy to reduce cancer risk, is based on the idea that chemicals which induce tumors in rodent cancer tests are potential human carcinogens; however, the chemicals tested for carcinogenicity in rodents have been primarily synthetic.[1-7] The enormous background of human exposures to natural chemicals has not been systematically examined. This has led to an imbalance in both data and perception about possible carcinogenic hazards to humans from chemical exposures. The regulatory process does not take into account that

1. Natural chemicals make up the vast bulk of chemicals humans are exposed to
2. The toxicology of synthetic and natural toxins is not fundamentally different
3. About half of the chemicals tested, whether natural or synthetic, are carcinogens when tested using current experimental protocols
4. Testing for carcinogenicity at near-toxic doses in rodents does not provide enough information to predict the excess number of human cancers that might occur at low-dose exposures
5. Testing at the maximum tolerated dose (MTD) frequently can cause chronic cell killing and consequent cell replacement (a risk factor for cancer that can be limited to high doses), and that ignoring this effect in risk assessment greatly exaggerates risks.

The vast proportion of chemicals to which humans are exposed are naturally occurring. Yet public perceptions tend to identify *chemicals* as being only synthetic and only synthetic chemicals as being toxic; however, every natural chemical is also toxic at some dose. We estimate that the daily average American exposure to burnt material in the diet is about 2000 mg and to natural pesticides (the chemicals that plants produce to defend themselves against fungi, insects, and animal predators) about 1500 mg.[58] In comparison, the total daily exposure to all synthetic pesticide residues combined is about 0.09 mg based on the sum of residues reported by the U.S. Food and Drug Administration (FDA) in their study of the 200 synthetic pesticide residues thought to be of greatest concern.[59] We estimate that humans ingest roughly 5000 to 10,000 different natural pesticides and their breakdown products.[58] Despite this enormously greater exposure to natural chemicals, among the chemicals tested for carcinogenicity, 78% (1007/1298) are synthetic (i.e., do not occur naturally).

It has often been assumed that humans have evolved defenses against natural chemicals that will not protect against synthetic chemicals. However, humans, like other animals, are extremely well protected by defenses that are mostly general rather than specific for particular chemicals (e.g., continuous shedding of surface cells that are exposed).[58] Additionally, most defense enzymes are inducible and are effective against both natural and synthetic chemicals including potentially mutagenic reactive chemicals.[60]

Since the toxicology of natural and synthetic chemicals is similar, one expects and finds a similar 50% positivity rate for carcinogenicity among synthetic and natural chemicals (see Table 5). Therefore, since humans are exposed to so many more natural than synthetic chemicals (by weight and by number), human exposures to natural rodent carcinogens as defined by high-dose tests are probably ubiquitous and unavoidable.[58,61] Concentrations of natural pesticides in plants are usually measured in parts per thousand or million rather than parts per billion, which is the usual concentration of synthetic pesticide residues or water pollutants. A diet free of chemicals that induce tumors in high-dose animal cancer tests is impossible.

Even though only a tiny proportion of natural pesticides have been tested for carcinogenicity, 35 of 64 that have been tested are rodent carcinogens (see Table 5), and commonly occur in plant foods and spices.[58,60,62]

Humans also ingest large numbers of natural chemicals from cooking food. For example, more than 1000 chemicals have been identified in roasted coffee. Only 26 have been tested for carcinogenicity according to the most recent results in our CPDB, and 19 of these are positive in at least one test (Table 12) totaling at least 10 mg of rodent carcinogens per cup.[63–66] Among the rodent carcinogens in coffee are the plant pesticides caffeic acid (present at 1800 ppm)[63] and catechol (present at 100 ppm).[67,68] Two other plant pesticides, chlorogenic acid and neochlorogenic acid (present at 21,600 ppm and 11,600 ppm, respectively),[63]

TABLE 12
Carcinogenicity Status of Natural Chemicals in Roasted Coffee

Positive (n = 19)	Acetaldehyde, benzaldehyde, benzene, benzofuran, benzo(a)pyrene, caffeic acid, catechol, 1,2,5,6-dibenzanthracene, ethanol, ethylbenzene, formaldehyde, furan, furfural, hydrogen peroxide, hydroquinone, limonene, styrene, toluene, xylene
Not positive (n = 8)	Acrolein, biphenyl, choline, eugenol, nicotinamide, nicotinic acid, phenol, piperidine
Uncertain	Caffeine
Yet to test	~ 1000 chemicals

are metabolized to caffeic acid and catechol but have not been tested for carcinogenicity. Chlorogenic acid and caffeic acid are mutagenic[69–71] and clastogenic.[72,73] Some other rodent carcinogens in coffee are products of cooking, e.g., furfural and benzo(a)pyrene. The point here is not to indicate that rodent data necessarily implicate coffee as a risk factor for human cancer, but rather to illustrate that there is an enormous background of chemicals in the diet that are natural and that have not been a focus of attention for carcinogenicity testing.

THE HERP RANKING OF POSSIBLE CARCINOGENIC HAZARDS

Above we discussed that rodent bioassays provide little information about mechanisms of carcinogenesis and low-dose risk. Additionally, there is an imbalance in bioassay data because the vast proportion of test agents are synthetic chemicals while the vast proportion of human exposures are to naturally occurring chemicals. Moreover, potency estimates based on bioassay results are bounded by the doses administered, therefore regulatory risk estimates based on linear extrapolation are also bounded. Given these results, what is the best use that can be made of bioassay data in efforts to prevent human cancer? In several papers we have emphasized that gaining a broad perspective about the vast number of chemicals to which humans are exposed can be helpful when setting research and regulatory priorities.[60,74–76]

One reasonable strategy is to use a rough index to *compare* and *rank* possible carcinogenic hazards from a wide variety of chemical exposures at levels that humans typically receive, and then to focus on those that rank highest.[75–77] Ranking is a critical first step that can help to set priorities for selecting chemicals for chronic bioassay or mechanistic studies, for epidemiological research, and for regulatory policy. Although one cannot say whether the ranked chemical exposures are likely to be of major or minor importance in human cancer, it is not prudent to focus attention on the possible hazards at the bottom of a ranking if, using the same methodology to identify hazard, there are numerous common human exposures with much greater possible hazards. Our analyses are based on the HERP index (Human Exposure/Rodent Potency), which indicates what percentage of the rodent carcinogenic potency (TD_{50} in mg/kg/day) a human receives from a given daily lifetime exposure (mg/kg/day). TD_{50} values in our CPDB span a 10-million-fold range across chemicals.[1]

In general, the ranking by the simple HERP index will be similar to a ranking of regulatory "risk estimates." As we discussed above, the VSD is approximately equivalent to the ratio of the high dose in a bioassay divided by 740,000.[14]

Overall, our analyses have shown that HERP values for some historically high exposures in the workplace and some pharmaceuticals rank high, and that there is an enormous background of naturally occurring rodent carcinogens in typical portions of common foods that cast doubt on the relative importance of low-dose exposures to residues of synthetic chemicals such as pesticides.[75,76,78] A committee of the National Research Council recently reached similar conclusions about natural vs. synthetic chemicals in the diet, and called for further research on natural chemicals.[79]

Our earlier HERP rankings were for typical exposures. In this chapter we rank HERP values for *average* U.S. exposures to rodent carcinogens for which both concentration data and average exposure data were available.

The average daily U.S. exposures in the ranking (Table 13) are ordered by possible carcinogenic hazard (HERP). Results are reported for average exposures to 25 natural chemicals in the diet (**in boldface**) and to 28 chemicals for which the exposure is not natural. Of these 28 chemicals, 5 occur naturally, but human exposure is primarily or exclusively from anthropogenic sources, e.g., benzene, chloroform, formaldehyde, TCDD, and tetrachloroethylene.

Three convenient reference points in the HERP ranking are: the median HERP value in Table 13 of 0.001%; the upper bound risk estimate used by regulatory agencies of one in a million (using the q_1^* potency value derived from the linearized multistage model), i.e., the VSD, which converts to a HERP of 0.00003% if based on a rat TD_{50} and 0.00001% if based on a mouse TD_{50}; and the background HERP of 0.0003% for the average chloroform level in a liter of U.S. tap water, which is formed as a byproduct of chlorination.

The HERP ranking maximizes possible hazards to synthetic chemicals because it includes historically high exposure values that are now much lower, e.g., DDT, PCBs, occupational exposures. Additionally, the values for dietary

exposures to synthetic chemicals are averages in the *total diet*, whereas for many natural chemicals the exposures are for individual foods (i.e., the exposures for which concentration data were available).

Table 13 indicates that many ordinary foods would not pass the regulatory criteria used for synthetic chemicals. For many natural chemicals the HERP values are in the top half of the table, even though natural chemicals are markedly underrepresented because so few have been tested in rodent bioassays. We discuss several categories of exposure below and indicate that mechanistic data are available for some chemicals which suggest that the chemical would not be expected to be a cancer hazard at the doses to which humans are exposed; thus, their ranking by HERP would not be relevant.

NATURAL PESTICIDES

Because few have been tested, natural pesticides are markedly underrepresented in our analysis. Importantly, for each plant food listed, there are about 50 additional untested natural pesticides. Although ~10,000 natural pesticides and their break-down products occur in the human diet,[58] only 64 have been tested adequately in rodent bioassays (see Table 5). Average exposures to many natural-pesticide rodent carcinogens in common foods rank above or close to the median, ranging up to a HERP of 0.1%. These include caffeic acid (lettuce, apple, pear, coffee, plum, celery, carrot, potato); safrole (in spices), allyl isothiocyanate (mustard), *d*-limonene (mango, orange juice, black pepper); estragole (in spices); hydroquinone and catechol in coffee; and coumarin in cinnamon. Some natural pesticides in the commonly eaten mushroom (*Agaricus bisporus*) are rodent carcinogens (glutamyl-*p*-hydrazinobenzoate, *p*-hydrazinobenzoate), and the HERP based on feeding whole mushrooms to mice is 0.02%. For *d*-limonene, no human risk is anticipated because tumors are induced only in male rat kidney tubules with involvement of α_{2u}-globulin nephrotoxicity, which does not appear to be possible in humans.[135,136]

SYNTHETIC PESTICIDES

Synthetic pesticides currently in use that are rodent carcinogens and quantitatively detected by the U.S. FDA as residues in food are all included in Table 13. Most are at the bottom of the ranking, but HERP values are about at the median for ethylene thiourea (ETU), UDMH (from Alar) before its discontinuance, and DDT before its ban in the U.S. in 1972. These rank below the HERP values for many naturally occurring chemicals. For ETU the value would be about 10 times lower if the potency value of the Environmental

TABLE 13
Ranking Possible Carcinogenic Hazards from Average U.S. Exposures

Chemicals that occur naturally in foods are in bold. *Daily human exposure:* Reasonable daily intakes are used to facilitate comparisons. The calculations assume a daily dose for a lifetime. *Possible hazard:* The human dose of rodent carcinogen is divided by 70 kg to give a mg/kg/day of human exposure, and this dose is given as the percentage of the TD_{50} in the rodent (mg/kg/day) to calculate the *Human Exposure/Rodent Potency* index (HERP). TD_{50} values used in the HERP calculation are averages calculated by taking the harmonic mean of the TD_{50}s of the positive tests in that species from the Carcinogenic Potency Database. Average TD_{50} values have been calculated separately for rats and mice, and the more potent value is used for calculating possible hazard.

Possible hazard: HERP (%)	Average daily U.S. exposure	Human dose of rodent carcinogen	Potency TD_{50} (mg/kg/day)[a] Rats	Mice	Ref.
140	EDB: workers (high exposure) (before 1977)	Ethylene dibromide, 150 mg	1.52	(7.45)	80,81
17	Clofibrate	Clofibrate, 2 g	169	.	82
14	Phenobarbital, 1 sleeping pill	Phenobarbital, 60 mg	(+)	6.09	83
6.8	1,3-Butadiene: rubber workers (1978-86)	1,3-Butadiene, 66.0 mg	(261)	13.9	84
6.1	Tetrachloroethylene: dry cleaners with dry-to-dry units (1980-90)[b]	Tetrachloroethylene, 433 mg	101	(126)	85
4.0	Formaldehyde: workers	Formaldehyde, 6.1 mg	2.19	(43.9)	86
2.1	**Beer, 257 g**	**Ethyl alcohol, 13.1 ml**	9110	(−)	87
1.4	Mobile home air (14 hours/day)	Formaldehyde, 2.2 mg	2.19	(43.9)	88
0.9	Methylene chloride: workers (1940s-80s)	Methylene chloride, 471 mg	724	(918)	89
0.5	**Wine, 28.0 g**	**Ethyl alcohol, 3.36 ml**	9110	(−)	87
0.4	Conventional home air (14 hours/day)	Formaldehyde, 598 µg	2.19	(43.9)	90
0.1	**Coffee, 13.3 g**	**Caffeic acid, 23.9 mg**	297	(4900)	63,87
0.04	**Lettuce, 14.9 g**	**Caffeic acid, 7.90 mg**	297	(4900)	91,92
0.03	**Safrole in spices**	**Safrole, 1.2 mg**	(441)	51.3	93
0.03	**Orange juice, 138 g**	**d-Limonene, 4.28 mg**	204	(−)	91,94
0.03	**Pepper, black, 446 mg**	**d-Limonene, 3.57 mg**	204	(−)	87,95
0.02	**Mushroom (*Agaricus bisporus* 2.55 g)**	**Mixture of hydrazines, etc. (whole mushroom)**	−	20,300	87,96,97
0.02	**Apple, 32.0 g**	**Caffeic acid, 3.40 mg**	297	(4900)	98,99
0.02	**Coffee, 13.3 g**	**Catechol, 1.33 mg**	118	(244)	67,68,87
0.02	**Coffee, 13.3 g**	**Furfural, 2.09 mg**	(683)	197	87
0.009	BHA: daily U.S. avg (1975)	BHA, 4.6 mg	745	(5530)	100
0.008	**Beer (before 1979), 257 g**	**Dimethylnitrosamine, 726 ng**	0.124	(0.189)	87,101,102
0.008	**Aflatoxin: daily U.S. avg (1984-89)**	**Aflatoxin, 18 ng**	0.0032	(+)	103
0.007	**Cinnamon, 21.9 mg**	**Coumarin, 65.0 µg**	13.9	(103)	104
0.006	**Coffee, 13.3 g**	**Hydroquinone, 333 µg**	82.8	(225)	67,87,105
0.005	Saccharin: daily U.S. avg (1977)	Saccharin, 7 mg	2140	(−)	106
0.005	**Carrot, 12.1 g**	**Aniline, 624 µg**	194[c]	(−)	91,107
0.004	**Potato, 54.9 g**	**Caffeic acid, 867 µg**	297	(4900)	91,108
0.004	**Celery, 7.95 g**	**Caffeic acid, 858 µg**	297	(4900)	109,110
0.004	**White bread, 67.6 g**	**Furfural, 500 µg**	(683)	197	87
0.003	**Nutmeg, 27.4 mg**	**d-Limonene, 466 µg**	204	(−)	87,111
0.003	Conventional home air (14 hour/day)	Benzene, 155 µg	(169)	77.5	90
0.002	**Carrot, 12.1 g**	**Caffeic acid, 374 µg**	297	(4900)	91,110
0.002	Ethylene thiourea: daily U.S. avg (1990)	Ethylene thiourea, 9.51 µg	7.9	(23.5)	112
0.002	[DDT: daily U.S. avg (before 1972 ban)]	[DDT, 13.8 µg]	(84.7)	12.3	113
0.001	**Plum, 2.00 g**	**Caffeic acid, 276 µg**	297	(4900)	99,114
0.001	BHA: daily U.S. avg (1987)	BHA, 700 µg	745	(5530)	100
0.001	**Pear, 3.29 g**	**Caffeic acid, 240 µg**	297	(4900)	87,99
0.001	[UDMH: daily U.S. avg (1988)]	[UDMH, 2.82 µg (from Alar)]	(−)	3.96	98
0.0009	**Brown mustard, 68.4 mg**	**Allyl isothiocyanate, 62.9 µg**	96	(−)	87,115
0.0008	[DDE: daily U.S. avg (before 1972 ban)]	[DDE, 6.91 µg]	(−)	12.5	113

TABLE 13 (continued)
Ranking Possible Carcinogenic Hazards from Average U.S. Exposures

Possible hazard: HERP (%)	Average daily U.S. exposure	Human dose of rodent carcinogen	Potency TD$_{50}$ (mg/kg/day)^a Rats	Mice	Ref.
0.0007	TCDD: daily U.S. avg (1994)	TCDD, 12.0 pg	0.0000235	(0.000156)	116
0.0007	**Bacon, 11.5 g**	**Diethylnitrosamine, 11.5 ng**	0.0237	(+)	87,117
0.0006	**Mushroom (*Agaricus bisporus* 2.55 g)**	**Glutamyl-*p*-hydrazinobenzoate, 107 μg**	.	277	87,118
0.0005	**Jasmine tea, 2.19 g**	**Benzyl acetate, 504 μg**	(–)	1440	87,119
0.0004	**Bacon, 11.5 g**	***N*-Nitrosopyrrolidine, 196 ng**	(0.799)	0.679	87,120
0.0004	**Bacon, 11.5 g**	**Dimethylnitrosamine, 34.5 ng**	0.124	(0.189)	87,117
0.0004	[EDB: Daily U.S. avg (before 1984 ban)]	[EDB, 420 ng]	1.52	(7.45)	121
0.0004	Tap water, 1 liter (1987-92)	Bromodichloromethane, 13 μg	(72.5)	47.7	122
0.0003	**Mango, 1.22 g**	***d*-Limonene, 48.8 μg**	204	(–)	114,123
0.0003	**Beer, 257 g**	**Furfural, 39.9 μg**	(683)	197	87
0.0003	Tap water, 1 liter (1987-92)	Chloroform, 17 μg	(262)	90.3	122
0.0003	Carbaryl: daily U.S. avg (1990)	Carbaryl, 2.6 μg	14.1	(–)	124
0.0002	**Celery, 7.95 g**	**8-Methoxypsoralen, 4.86 μg**	32.4	(–)	109,125
0.0002	Toxaphene: daily U.S. avg (1990)	Toxaphene, 595 ng	(–)	5.57	124
0.00009	**Mushroom (*Agaricus bisporus*, 2.55 g)**	***p*-Hydrazinobenzoate, 28 μg**	.	454^c	87,118
0.00008	PCBs: daily U.S. avg (1984-86)	PCBs, 98 ng	1.74	(9.58)	126
0.00008	DDE/DDT: daily U.S. avg (1990)	DDE, 659 ng	(–)	12.5	124
0.00007	**Parsnip, 54.0 mg**	**8-Methoxypsoralen, 1.57 μg**	32.4	(–)	127,128
0.00007	**Toast, 67.6 g**	**Urethane, 811 ng**	(41.3)	16.9	87,129
0.00006	**Hamburger, pan fried, 85 g**	**PhIP, 176 ng**	4.29^c	(28.6^c)	91,130
0.00005	**Estragole in spices**	**Estragole, 1.99 μg**	.	51.8	87
0.00005	**Parsley, fresh, 324 mg**	**8-Methoxypsoralen, 1.17 μg**	32.4	(–)	127,131
0.00003	**Hamburger, pan fried, 85 g**	**MeIQx, 38.1 ng**	1.99	(24.3)	91,130
0.00002	Dicofol: daily U.S. avg (1990)	Dicofol, 544 ng	(–)	32.9	124
0.00001	**Cocoa, 3.34 g**	**α-Methylbenzyl alcohol, 4.3 μg**	458	(–)	87
0.00001	**Beer, 257 g**	**Urethane, 115 ng**	(41.3)	16.9	87,129
0.000005	**Hamburger, pan fried, 85 g**	**IQ, 6.38 ng**	1.89^c	(19.6)	91,130
0.000001	Lindane: daily U.S. avg (1990)	Lindane, 32 ng	(–)	30.7	124
0.0000004	PCNB: daily U.S. avg (1990)	PCNB (Quintozene), 19.2 ng	(–)	71.1	124
0.0000001	Chlorobenzilate: daily U.S. avg (1989)	Chlorobenzilate, 6.4 ng	(–)	93.9	124
<0.00000001	Chlorothalonil: daily U.S. avg (1990)	Chlorothalonil, <6.4 ng	828^d	(–)	124,132
0.000000008	Folpet: daily U.S. avg (1990)	Folpet, 12.8 ng	.	2280^d	124,133
0.000000006	Captan: daily U.S. avg (1990)	Captan, 11.5 ng	2690^d	(2730^d)	124,134

^{a} "." = no data in CPDB; (–) = negative in cancer test; (+) = positive cancer test(s) not suitable for calculating a TD$_{50}$.

^{b} This is not an average, but a reasonably large sample (1027 workers).

^{c} TD$_{50}$ harmonic mean was estimated for the base chemical from the hydrochloride salt.

^{d} Additional data from the EPA that is not in the CPDB were used to calculate these TD$_{50}$ harmonic means.

Protection Agency (EPA) were used instead of our TD$_{50}$; EPA combined rodent results from more than one experiment, including one in which ETU was administered *in utero*, and obtained a lower potency.[137] DDT and similar early pesticides, have been a concern because of their unusual lipophilicity and persistence, although there is no convincing epidemiological evidence of a carcinogenic hazard to humans.[138] Current exposure to DDT is in foods of animal origin, and the HERP value is low, 0.00008%.

In 1984 the U.S. EPA banned the agricultural use of ethylene dibromide (EDB), the main fumigant in the U.S., because of the residue levels found in grain, HERP = 0.0004%. This HERP value ranks low, whereas the HERP of 140% for the high exposures to EDB that some workers received in the 1970s is at the top of the ranking.[75]

Three synthetic pesticides, captan, chlorothalonil, and folpet, were evaluated in 1987 by the National Research Council (NRC) as being of relatively high risk to humans[139] and were also reported by FDA in the Total Diet Study (TDS). The contrast between the low ranking HERP values for these pesticides, i.e., the lowest HERP values in Table 9, and the high risk estimates of the 1987 NRC report is due to exposure estimates, which differ by more than 100,000-fold. The NRC used the EPA Theoretical Maximum Residue Contribution, which is a hypothetical maximum exposure estimate, whereas the FDA used dietary intake estimates based on monitoring food as eaten. Hence, using hypothetical maxima results in enormously higher risk estimates than using measured residues.

Cooking and Preparation of Food

Cooking and preparation of food can also produce chemicals that are rodent carcinogens. Alcoholic beverages are a human carcinogen, and the HERP values in Table 13 for alcohol in beer (2.1%) and wine (0.5%) are high in the ranking. Ethyl alcohol is one of the least potent rodent carcinogens in the CPDB, but the HERP is high because of high concentrations and high U.S. consumption. Another fermentation product, urethane (ethyl carbamate), has a HERP value of 0.00001% in average beer consumption; for average bread consumption (as toast), the HERP would be 0.00007%.

Cooking food is plausible as a contributor to cancer. A wide variety of chemicals are formed during cooking. Rodent carcinogens formed include furfural and similar furans, nitrosamines, polycyclic hydrocarbons, and heterocyclic amines. Furfural, a chemical formed naturally when sugars are heated, is a widespread constituent of food flavor. The HERP value for furfural in average consumption of coffee is 0.02% and in white bread is 0.004%. Nitrosamines formed from nitrite or nitrogen oxides (NO_x) and amines in food can give moderate HERP values, e.g., in bacon, the HERP for diethylnitrosamine is 0.0007% and for dimethylnitrosamine it is 0.0004%. A variety of mutagenic and carcinogenic heterocyclic amines (HA) are formed when meat, chicken, or fish are cooked, particularly when charred. Compared to other rodent carcinogens, there is strong evidence of carcinogenicity for HAs in terms of positivity rates and multiplicity of target sites; however, concordance in target sites between rats and mice is generally restricted to the liver.[77] Under usual cooking conditions, exposures to HA are in the low ppb range. HERP values for HA in pan fried hamburger range from 0.00006% for PhIP to 0.000005% for IQ (Table 13). PhIP induces colon tumors in male but not female rats. A recent study indicates that whereas the level of DNA adducts in the colonic mucosa was the same in both sexes, cell proliferation was increased only in the male, contributing to the formation of premalignant lesions of the colon.[140]

Therefore, there was no correlation between adduct formation and premalignant lesions, but there was between cell division and lesions.

Food Additives

Food additives that are rodent carcinogens can be either naturally occurring (e.g., allyl isothiocyanate and alcohol) or synthetic (butylated hydroxyanisole [BHA] and saccharin, Table 13). The highest HERP values for average exposures to synthetic rodent carcinogens in Table 13 are for exposures in the 1970s to BHA (0.009%) and saccharin (0.005%), both nongenotoxic rodent carcinogens. For both of these additives, data on mechanism of carcinogenesis strongly suggest that there would be no risk to humans at the levels found in food.

BHA is a phenolic antioxidant that is Generally Regarded as Safe (GRAS) by the U.S. FDA. By 1987, after BHA was shown to be a rodent carcinogen, its use declined 6-fold (HERP = 0.001%),[100] this was due to voluntary replacement by other antioxidants, and to the fact that the use of animal fats and oils, in which BHA is primarily used as an antioxidant, has consistently declined in the U.S. The mechanistic and carcinogenicity results on BHA indicate that malignant tumors were induced only at a dose above the MTD at which cell division is increased in the forestomach, which is the only site of tumorigenesis; the proliferation is only at high doses, and is dependent on continuous dosing until late in the experiment.[141] Humans do not have a forestomach. We note that the dose-response for BHA curves sharply upward, but the potency value used in HERP is based on a linear model; if the California EPA potency value (which is based on a linearized multistage model) were used in HERP instead of TD_{50}, the HERP values for BHA would be 25 times lower.[142]

For saccharin, which has largely been replaced by other sweeteners, there is convincing evidence that the induced bladder tumors in rats are not relevant to human dietary exposures. The carcinogenic effect requires high doses of sodium saccharin which form calculi in the bladder, and subsequent regenerative hyperplasia. Thus, tumor development is due to increased cell division, and if the dose is not high enough to produce calculi then there is no increased cell division and no increased risk of tumor development.[143]

Mycotoxins

Of the 23 fungal toxins tested for carcinogenicity, 14 are positive (61%) (see Table 5). The mutagenic mold toxin, aflatoxin, which is found in moldy peanut and corn products, interacts with chronic hepatitis infection in human liver cancer development.[144] There is a synergistic effect in the human liver between aflatoxin (genotoxic effect) and the hepatitis B virus (cell division effect) in the induction of liver cancer.[145] The HERP value for aflatoxin of 0.008% is based on the rodent potency. If the lower human potency value calculated by U.S. FDA from epidemiological data were used instead, the HERP would be about tenfold lower.[146] Biomarker measurements of aflatoxin on populations in Africa and China, which have high rates of both hepatitis B and C viruses and liver cancer, confirm that those populations are chronically exposed to high levels of aflatoxin.[147,148] Liver cancer is rare in the U.S. Although hepatitis B and C viruses infect less than 1% of the U.S. population, hepatitis viruses can account for half of liver cancer cases among non-Asians and even more among Asians.[149]

Ochratoxin A, a rodent carcinogen, has been measured in Europe and Canada in agricultural and meat products. An estimated exposure of 1 ng/kg/day would have a HERP value at the median of Table 13.[150,151]

SYNTHETIC CONTAMINANTS

Polychlorinated biphenyls (PCBs) and tetrachlorodibenzo-*p*-dioxin (TCDD), which have been a concern because of their environmental persistence and carcinogenic potency in rodents, are primarily consumed in foods of animal origin. In the U.S., PCBs are no longer used, but exposure persists. Consumption in food in the U.S. declined about 20-fold between 1978-1986.[126,152] The HERP value for the most recent reporting of the U.S. FDA Total Diet Study (1984-1986) is 0.00008%, towards the bottom of the ranking, and far below many values for naturally occurring chemicals in common foods. It has been reported that some countries may have higher intakes of PCBs than in the U.S.[153]

TCDD, the most potent rodent carcinogen, is produced naturally by burning when chloride ion is present, e.g., in forest fires. The sources of human exposure appear to be predominantly anthropogenic, e.g., from incinerators.[154] TCDD has received enormous scientific and regulatory attention, most recently in an ongoing assessment by the U.S. EPA.[116,154,155] Some epidemiologic studies suggest an association with human cancer, but the evidence is not sufficient to establish causality. Estimation of average U.S. consumption is based on limited sampling data, and EPA is currently conducting further studies of concentrations in food. The HERP value of 0.0007% is near the median of the values in Table 13. TCDD exerts many or all of its harmful effects in mammalian cells through binding to the Ah receptor. A wide variety of natural substances also bind to the Ah receptor, (e.g., tryptophan oxidation products)

and insofar as they have been examined, they have similar properties to TCDD.[60] For example, a variety of flavones and other plant substances in the diet and their metabolites also bind to the Ah receptor, e.g., indole carbinol (IC). IC is the main breakdown compound of glucobrassicin, a glucosinolate that is present in large amounts in vegetables of the *Brassica* genus, including broccoli, and gives rise to the potent Ah binder, indole carbazole.[156]

OCCUPATIONAL AND PHARMACEUTICAL EXPOSURES

Occupational and pharmaceutical exposures to some chemicals have been high, and most of the single chemical agents or industrial processes evaluated as human carcinogens have been identified by high dose exposures in the workplace.[157] The issue of how much human cancer can be attributed to occupational exposure has been controversial, but a few percent seems a reasonable estimate.[28]

When exposures are high, comparatively little quantitative extrapolation is required from high-dose rodent tests to those occupational exposures. HERP values rank at the top of Table 13 for chemical exposures in some occupations for which average exposure data were available: ethylene dibromide, 1,3-butadiene, tetrachloroethylene, and formaldehyde.

In another analysis, we used Permitted Exposure Limits (PELs) of the U.S. Occupational Safety and Health Administration as surrogates for actual exposures and compared the permitted daily dose-rate for workers with the TD_{50} (PERP index, Permitted Exposure/Rodent Potency).[78,158] We found that PELs for 9 chemicals were greater than 10% of the rodent carcinogenic dose and for 27 they were between 1 and 10% of the rodent dose.

Some pharmaceuticals are also clustered near the top of the HERP ranking; we note that half the drugs tested are reported in the *Physicians Desk Reference* to be carcinogens in rodent bioassays.[19] Most drugs, however, are used for only short periods and would not be comparable to HERP values, which are for lifetime exposures.

Caution is necessary in drawing conclusions from the occurrence in the diet of natural chemicals that are rodent carcinogens. It is not argued here that these dietary exposures are necessarily of much relevance to human cancer. In fact, epidemiological results indicate that adequate consumption of fruits and vegetables reduces cancer risk at many sites, and that protective factors like intake of vitamin C and folic acid are important, rather than intake of individual rodent carcinogens. Our analysis does indicate that widespread exposures to naturally occurring rodent carcinogens cast doubt on the relevance to human cancer of low-level exposures to synthetic rodent carcinogens. Our results call for a re-evaluation of the utility of animal cancer tests done at the MTD for providing information that is useful in protecting humans against low level exposures in the diet when a high percentage of both natural and synthetic chemicals appear to be rodent carcinogens at the MTD, when the data from rodent bioassays is not adequate to assess low dose risk, and when the ranking on an index of possible hazards demonstrates that there is an enormous background of natural chemicals in the diet that rank high, even though so few have been tested in rodent bioassays.

Our discussion of the HERP ranking indicates the importance of data on mechanism of carcinogenesis for each chemical. For several chemicals, mechanistic data has recently been generated which indicates that they would not be expected to be a risk to humans at the levels consumed in food (e.g., saccharin, BHA, chloroform, d-limonene, discussed above). Recent developments in science and regulatory policy have also emphasized the importance of evaluating mechanistic data, rather than relying exclusively on default, worst case assessments. The National Research Council and the EPA have both recently recommended improvements in the risk assessment process that involve incorporating consideration of dose to the target tissue, mechanism of action, and biologically based dose-response models, including a possible threshold of dose below which effects will not occur.[159,160]

FUTURE DIRECTIONS

Our analysis in this paper suggests several areas for further research into diet and cancer, including epidemiological, toxicological, and biochemical investigations. Further understanding of the role and mechanism of endogenous damage could lead to new prevention strategies for cancer. Present epidemiological evidence regarding the role of greater antioxidant consumption in human cancer prevention is inconsistent.[28] Nevertheless, biochemical data indicate the need for further investigation of the wide variety of potentially effective antioxidants, both natural and synthetic. Evidence supporting this need includes the enormous endogenous oxidative damage to DNA, proteins, and lipids, as well as indirect evidence such as increased oxidative damage to human sperm DNA when dietary ascorbate is insufficient. Moreover, studies on the importance of dietary fruits and vegetables in cancer suggest the importance of further work on micronutrient deficiency as a major contributor to cancer. Studies in rodents and humans suggest further work on caloric intake and body weight, and the effects on hormonal status.

Since naturally occurring chemicals in the diet have not been a focus of cancer research, it seems reasonable to investigate some of them further as possible hazards because they often occur at high concentrations in foods.

Only a small proportion of the many chemicals to which humans are exposed will ever be investigated, and there is at least some toxicological plausibility that high dose exposures may be important. In order to identify untested dietary chemicals that might be a hazard to humans *if* they were to be identified as rodent carcinogens, we have used an index, HERT, which is analogous to HERP: the ratio of Human Exposure/Rodent Toxicity. HERT uses readily available LD_{50} values rather than the TD_{50} values from animal cancer tests that are used in HERP. This approach to prioritizing chemicals makes assessment of human exposure levels critical at the outset. The validity of the HERT approach is supported by three analyses: first, we have found that for the exposures to rodent carcinogens for which we have calculated HERP values, the ranking by HERP and HERT are highly correlated.[62] Second, we have shown that without conducting a bioassay the regulatory VSD can be approximated by dividing the MTD by 740,000.[14] Since the MTD is not known for all chemicals, and MTD and LD_{50} are both measures of toxicity, acute toxicity (LD_{50}) can reasonably be used as a surrogate for chronic toxicity (MTD). Third, LD_{50} and carcinogenic potency are correlated;[161,162] therefore, HERT is a reasonable surrogate index for HERP since it simply replaces TD_{50} with LD_{50}.[62]

We have calculated HERT values using LD_{50} values as a measure of toxicity in combination with available data on concentrations of untested natural chemicals in commonly consumed foods and data on average consumption of those foods in the U.S. diet. We considered any chemical with available data on rodent LD_{50}, that had a published concentration ≥ 10 ppm in a common food, and for which estimates of average U.S. consumption of that food were available. Among the set of 171 HERT values we were able to calculate, the HERT ranged across seven orders of magnitude.[62]

It might be reasonable to investigate further the chemicals in the diet that rank highest on the HERT index and that have not been adequately tested in chronic carcinogenicity bioassays in rats and mice. These include solanine and chaconine, the main alkaloids in potatoes, which are cholinesterase inhibitors that can be detected in the blood of almost all people;[163–165] chlorogenic acid, a precursor of caffeic acid; and caffeine, for which no standard lifetime study has been conducted in mice. In rats, cancer tests of caffeine have been negative, but one study that was inadequate because of early mortality, showed an increase in pituitary adenomas.[166]

Compelling theoretical reasons, as well as data from a large body of experiments indicate that prediction of carcinogenic risk to humans at low dose must take cell division into account. Just evaluating a chemical as a rodent carcinogen without considering mechanism of action can be fundamentally misleading for low dose risk assessment. Defenses are inducible at low doses, and even for mutagens it may be that the increment in DNA damage over the enormous rate of endogenous background damage may not be significant. Many nonmutagens will have a threshold, and there will be no risk at low dose. It is clear that the mechanisms of action for all rodent carcinogens are not the same. For some chemicals there is evidence to support cell division effects unique to high doses, e.g., saccharin, and thus there appears to be a threshold. For others, e.g., butadiene and 2-acetylaminofluorene, there may well be multiplicative effects due to an interaction of cell division and DNA damage, but carcinogenic effects have been found considerably below the MTD. Sometimes, the mechanism leading to cell division and carcinogenesis in a rodent species has no analogy in humans, e.g., kidney tumors in male Fischer rats due to α_{2u}-globulin. Studies of mechanism in rodent bioassays would help to clarify such differences.

As currently conducted, rodent bioassays do not provide the information necessary to extrapolate from high to low dose. It would be of particular interest to re-evaluate some of the rodent carcinogens that are receiving extensive regulatory attention on the basis of standard risk assessment methodology, e.g., trichloroethylene. Measurement of cell division at and below bioassay doses in sub-chronic studies for these chemicals would permit a re-interpretation of the rodent data and an improved assessment of the potential risk to humans at low dose.

REFERENCES

1. Gold, L.S., Slone, T.H.,Manley, N.B., Garfinkel, G.B., Rohrbach, L., and Ames, B.N. Carcinogenic Potency Database. In: Gold, L.S., and Zeiger, E., Eds. *Handbook of Carcinogenic Potency and Genotoxicity Databases.* Boca Raton, FL: CRC Press, 1997, pp. 1–605.
2. Gold, L.S., Sawyer, C.B., Magaw, R., Backman, G.M., de Veciana, M., Levinson, R., Hooper, N.K., Havender, W.R., Bernstein, L., Peto, R., Pike, M.C., and Ames, B.N. A Carcinogenic Potency Database of the standardized results of animal bioassays. *Environ. Health Perspect.* 58: 9-319 (1984).
3. Gold, L.S., de Veciana, M., Backman, G.M., Magaw, R., Lopipero, P., Smith, M., Blumenthal, M., Levinson, R., Bernstein, L., and Ames, B.N. Chronological supplement to the Carcinogenic Potency Database: Standardized results of animal bioassays published through December 1982. *Environ. Health Perspect.* 67: 161-200 (1986).
4. Gold, L.S., Slone, T.H., Backman, G.M., Magaw, R., Da Costa, M., Lopipero, P., Blumenthal, M., and Ames, B.N. Second chronological supplement to the Carcinogenic Potency Database: Standardized results of animal bioassays published through December 1984 and by the National Toxicology Program through May 1986. *Environ. Health Perspect.* 74: 237-329 (1987).
5. Gold, L.S., Slone, T.H., Backman, G.M., Eisenberg, S., Da Costa, M., Wong, M., Manley, N.B., Rohrbach, L., and Ames, B.N. Third chronological supplement to the Carcinogenic Potency Database: Standardized results of animal bioassays published through December 1986 and by the National Toxicology Program through June 1987. *Environ. Health Perspect.* 84: 215-286 (1990).
6. Gold, L.S., Manley, N.B., Slone, T.H., Garfinkel, G.B., Rohrbach, L., and Ames, B.N. The fifth plot of the Carcinogenic Potency Database: Results of animal bioassays published in the general literature through 1988 and by the National Toxicology Program through 1989. *Environ. Health Perspect.* 100: 65-135 (1993).
7. Gold, L.S., Manley, N.B., Slone, T.H., Garfinkel, G.B., Ames, B.N., Rohrbach, L., Stern, B.R., and Chow, K. Sixth Plot of the Carcinogenic Potency Database: Results of Animal Bioassays Published in the General Literature 1989- 1990 and by the National Toxicology Program 1990-1993. *Environ. Health Perspect.* 103 (Suppl. 8): 3-122 (1995).
8. Sawyer, C., Peto, R., Bernstein, L., and Pike, M.C. Calculation of carcinogenic potency from long-term animal carcinogenesis experiments. *Biometrics.* 40: 27-40 (1984).
9. Peto, R., Pike, M.C., Bernstein, L., Gold, L.S., and Ames, B.N. The TD_{50}: A proposed general convention for the numerical description of the carcinogenic potency of chemicals in chronic-exposure animal experiments. *Environ. Health Perspect.* 58: 1-8 (1984).
10. Sontag, J.A., Page, N.P., and Saffiotti, U.. Guidelines for carcinogen bioassay in small rodents. 1976. *(Carcinogenesis Tech. Rep. Ser. No. 1, DHEW Pub. No. NIH76-801.)*
11. Bernstein, L., Gold, L.S., Ames, B.N., Pike, M.C., and Hoel, D.G. Some tautologous aspects of the comparison of carcinogenic potency in rats and mice. *Fundam. Appl. Toxicol.* 5: 79-86 (1985).
12. Krewski, D., Gaylor, D.W., Soms, A.P., and Syszkowicz, M. An overview of the report — Correlation Between Carcinogenic Potency and the Maximum Tolerated Dose: Implications for Risk Assessment. *Risk Anal.* 13: 383-398 (1993).
13. Freedman, D.A., Gold, L.S., and Slone, T.H. How tautological are inter-species correlations of carcinogenic potency? *Risk Anal.* 13: 265-272 (1993).
14. Gaylor, D.W. and Gold, L.S. Quick estimate of the regulatory virtually safe dose based on the maximum tolerated dose for rodent bioassays. *Regul. Toxicol. Pharmacol.* 22: 57-63 (1995).
15. Gold, L.S., Bernstein, L., Kaldor, J., Backman, G., and Hoel, D. An empirical comparison of methods used to estimate carcinogenic potency in long-term animal bioassays: Lifetable vs. summary incidence data. *Fundam. Appl. Toxicol.* 6: 263-269 (1986).
16. Gold, L.S., Wright, C., Bernstein, L., and de Veciana, M. Reproducibility of results in 'near-replicate' carcinogenesis bioassays. *J. Natl. Cancer Inst.* 78: 1149- 1158 (1987).
17. Gold, L.S., Slone, T.H., and Bernstein, L. Summary of carcinogenic potency (TD_{50}) and positivity for 492 rodent carcinogens in the Carcinogenic Potency Database. *Environ. Health Perspect.* 79: 259-272 (1989).
18. Gold, L.S., Slone, T.H., and Ames, B.N. Summary of the Carcinogenic Potency Database by Chemical. In: Gold, L.S., and Zeiger, E., Eds. *Handbook of Carcinogenic Potency and Genotoxicity Databases.* Boca Raton, FL: CRC Press, 1997, pp. 621–660.
19. Davies, T.S. and Monro, A. The case for an upper dose limit of 1000 mg/kg in rodent carcinogenicity tests. *Cancer Lett.* 95: 69-77 (1995).
20. Gold, L.S., Bernstein, L., Magaw, R., and Slone, T.H. Interspecies extrapolation in carcinogenesis: Prediction between rats and mice. *Environ. Health Perspect.* 81: 211- 219 (1989).

21. Omenn, G.S., Stuebbe, S., and Lave, L.B. Predictions of rodent carcinogenicity testing results: Interpretation in light of the Lave-Omenn value-of-information model. *Mol. Carcinog.* 14: 37-45 (1995).

22. Zeiger, E. Genotoxicity Database. In: Gold, L.S., and Zeiger, E., Eds. *Handbook of Carcinogenic Potency and Genotoxicity Databases.* Boca Raton, FL: CRC Press, 1997, pp. 687–729.

23. Kier, L.E., Brusick, D.J., Auletta, A.E., Von Halle, E.S., Brown, M.M., Simmon, V.F., Dunkel, V., McCann, J., Mortelmans, K., Prival, M., Rao, T.K., and Ray, V. The *Salmonella typhimurium*/mammalian microsomal assay: A report of the U.S. Environmental Protection Agency Gene-Tox Program. *Mut. Res.* 168: 69-240 (1986).

24. Auletta, A. E., personal communication.

25. Innes, J.R.M., Ulland, B.M., Valerio, M.G., Petrucelli, L., Fishbein, L., Hart, E.R., Pallota, A.J., Bates, R.R., Falk, H.L., Gart, J.J., Klein, M., Mitchell, I., and Peters, J. Bioassay of pesticides and industrial chemicals for tumorigenicity in mice: a preliminary note. *J. Natl. Cancer Inst.* 42: 1101- 1114 (1969).

26. Innes, J.R.M. *Evaluation of Carcinogenic, Teratogenic, and Mutagenic Activities of Selected Pesticides and Industrial Chemicals. Volume 1: Carcinogenic Study.* Springfield, VA: Bionetics Research Laboratories, Inc. Distributed by National Technical Information Service, 1968.

27. U.S. National Cancer Institute *Everything doesn't cause cancer: But how can we tell which things cause cancer and which ones don't?* Bethesda, MD: USNCI, 1984. (NIH Pub. No. 84-2039.)

28. Ames, B.N., Gold, L.S., and Willett, W.C. The causes and prevention of cancer. *Proc. Natl. Acad. Sci. U.S.A.* 92: 5258-5265 (1995).

29. Ames, B.N., Gold, L.S., and Shigenaga, M.K. Cancer prevention, rodent high-dose cancer tests, and risk assessment. *Risk Anal.* 16:613–617 (1996).

30. Ames, B.N. and Gold, L.S. Chemical carcinogenesis: Too many rodent carcinogens. *Proc. Natl. Acad. Sci. U.S.A.* 87: 7772- 7776 (1990).

31. Ames, B.N. and Gold, L.S. Perspective: Too many rodent carcinogens: Mitogenesis increases mutagenesis. *Science.* 249: 970-971 (1990). Letters: 250: 1498 (1990); 250: 1645-1646 (1990); 251: 12-13 (1991); 251: 607-608 (1991); 252: 902 (1991).

32. Ames, B.N., Shigenaga, M.K., and Gold, L.S. DNA lesions, inducible DNA repair, and cell division: Three key factors in mutagenesis and carcinogenesis. *Environ. Health Perspect.* 101 (Suppl. 5): 35-44 (1993).

33. Ames, B.N., Shigenaga, M.K., and Hagen, T.M. Oxidants, antioxidants, and the degenerative diseases of aging. *Proc. Natl. Acad. Sci. U.S.A.* 90: 7915-7922 (1993).

34. Laskin, D.L. and Pendino, K.J. Macrophages and inflammatory mediators in tissue injury. *Annu. Rev. Pharmacol. Toxicol.* 35: 655-677 (1995).

35. Wei, H. and Frenkel, K. Relationship of oxidative events and DNA oxidation in SENCAR mice to *in vivo* promoting activity of phorbol ester-type tumor promoters. *Carcinogenesis.* 14: 1195-1201 (1993).

36. Wei, L., Wei, H., and Frenkel, K. Sensitivity to tumor promotion of SENCAR and C57Bl/6J mice correlates with oxidative events and DNA damage. *Carcinogenesis.* 14: 841- 847 (1993).

37. Czaja, M.J., Xu, J., Ju, Y., Alt, E., and Schmiedeberg, P. Lipopolysaccharide-neutralizing antibody reduces hepatocyte injury from acute hepatotoxin administration. *Hepatology.* 19: 1282-1289 (1994).

38. Adachi, Y., Moore, L.E., Bradford, B.U., Gao, W., and Thurman, R.G. Antibiotics prevent liver injury in rats following long-term exposure to ethanol. *Gastroenterology.* 108: 218-224 (1995).

39. Gunawardhana, L., Mobley, S.A., and Sipes, I.G. Modulation of 1,2-dichlorobenzene hepatotoxicity in the Fischer-344 rat by a scavenger of superoxide anions and an inhibitor of Kupffer cells. *Toxicol. Appl. Pharmacol.* 119: 205-213 (1993).

40. Hart, R., Neumann, D., and Robertson, R. *Dietary Restriction: Implications for the Design and Interpretation of Toxicity and Carcinogenicity Studies.* Washington, D.C.: ILSI Press, 1995.

41. Cunningham, M.L., Pippin, L.L., Anderson, N.L., and Wenk, M.L. The hepatocarcinogen methapyrilene but not the analog pyrilamine induces sustained hepatocellular replication and protein alterations in F344 rats in a 13-week feed study. *Toxicol. Appl. Pharmacol.* 131: 216-223 (1995).

42. Hayward, J.J., Shane, B.S., Tindall, K.R., and Cunningham, M.L. Differential *in vivo* mutagenicity of the carcinogen/non-carcinogen pair 2,4- and 2,6-diaminotoluene. *Carcinogenesis.* 16: 2429-2433 (1995).

43. Cohen, S.M. and Ellwein, L.B. Genetic errors, cell proliferation, and carcinogenesis. *Cancer Res.* 51: 6493- 6505 (1991).

44. Cohen, S.M. Human relevance of animal carcinogenicity studies. *Regul. Toxicol. Pharmacol.* 21: 75-80 (1995).

45. Counts, J.L. and Goodman, J.I. Principles underlying dose selection for, and extrapolation from, the carcinogen bioassay: dose influences mechanism. *Regul. Toxicol. Pharmacol.* 21: 418-421 (1995).

46. Preston-Martin, S., Pike, M.C., Ross, R.K., and Jones, P.A. Increased cell division as a cause of human cancer. *Cancer Res.* 50: 7415-7421 (1990).

47. Gold, L.S., Manley, N.B., and Ames, B.N. Extrapolation of carcinogenesis between species: Qualitative and quantitative factors. *Risk Anal.* 12: 579-588 (1992).

48. Gold, L.S. The importance of data on mechanism of carcinogenesis in efforts to predict low-dose human risk. *Risk Anal.* 13: 399-401 (1993).

49. Freedman, D.A., Gold, L.S., and Lin, T.H. Concordance between rats and mice in bioassays for carcinogenesis. *Regul. Toxicol. Pharmacol.* 23:225–232 (1996).

50. Lin, T.H., Gold, L.S., and Freedman, D. Bias in qualitative measures of concordance for rodent carcinogenicity tests. *Stat. Sci.* 225–232 (1996)..

51. Adamson, R.H., and Sieber, S.M. Chemical carcinogenesis studies in nonhuman primates. In: Langenbach, R. Nesnow, S., and Rice, J.M., Eds. *Organ and Species Specificity in Chemical Carcinogenesis.* New York: Plenum Press, 1982, pp. 129-156.

52. Gold, L.S. and Slone, T.H. Prediction of carcinogenicity from 2 vs. 4 sex-species groups in the carcinogenic potency database. *J. Toxicol. Environ. Health.* 39: 147-161 (1993).

53. Gold, L.S., Manley, N.B., and Slone, T.H. Summary of the Carcinogenic Potency Database by Target Organ. In: Gold, L.S., and Zeiger, E., Eds. *Handbook of Carcinogenic Potency and Genotoxicity Databases.* Boca Raton, FL: CRC Press, 1997, pp. 607–620.

54. Gold, L.S., Slone, T.H., Stern, B.R., and Bernstein, L. Comparison of target organs of carcinogenicity for mutagenic and nonmutagenic chemicals. *Mut. Res.* 286: 75-100 (1993).

55. Gold, L.S., Ward, J.M., Bernstein, L., and Stern, B. Association between carcinogenic potency and tumor pathology in rodent carcinogenesis bioassays. *Fundam. Appl. Toxicol.* 6: 677-690 (1986).

56. Gold, L.S. and Slone, T.H. The mouse liver in perspective: comparison of target organs of carcinogenicity for mutagens and non-mutagens in chronic bioassays. In: *Fifth Workshop on Mouse Liver Tumors*. Washington, D.C.: International Life Sciences Institute, 1995, pp. 2-3.

57. Gold, L.S., Slone, T.H., Manley, N.B., and Bernstein, L. Target organs in chronic bioassays of 533 chemical carcinogens. *Environ. Health Perspect*. 93: 233-246 (1991).

58. Ames, B.N., Profet, M., and Gold, L.S. Dietary pesticides (99.99% all natural). *Proc. Natl. Acad. Sci. U.S.A.* 87: 7777-7781 (1990).

59. U.S. Food and Drug Administration Pesticide Program: Residue monitoring 1992. *J. Assoc. Off. Anal. Chem.* 76: 127A-148A (1993).

60. Ames, B.N., Profet, M., and Gold, L.S. Nature's chemicals and synthetic chemicals: Comparative toxicology. *Proc. Natl. Acad. Sci. U.S.A.* 87: 7782-7786 (1990).

61. Ames, B.N. and Gold, L.S. Risk assessment of pesticides. *Chem. Eng. News*. 69: 28-32, 48-49 (1991).

62. Gold, L.S., Slone, T.H., and Ames, B.N. Prioritization of possible carcinogenic hazards in food. In: Tennant, D., Ed. *Food Chemical Risk Analysis*. London: Chapman and Hall, (in press).

63. Clarke, R.J. and Macrae, R., Eds. *Coffee*. Vol. 1-3. New York: Elsevier, 1988.

64. Maarse, H., Visscher, C.A., Willemsens, L.C., Nijssen, L.M., and Boelens, M.H., Eds. *Volatile Compounds in Foods. Qualitative and Quantitative Data. Supplement 5 and Cumulative Index*. Zeist, The Netherlands: TNO-CIVO Food Analysis Institute, 1994.

65. Fujita, Y., Wakabayashi, K., Nagao, M., and Sugimura, T. Implication of hydrogen peroxide in the mutagenicity of coffee. *Mut. Res.* 144: 227-230 (1985).

66. Kikugawa, K., Kato, T., and Takahashi, S. Possible presence of 2-amino-3,4-dimethylimidazo[4,5-*f*]quinoline and other heterocyclic amine-like mutagens in roasted coffee beans. *J. Agric. Food Chem.* 37: 881-886 (1989).

67. Tressl, R., Bahri, D., Köppler, H., and Jensen, A. Diphenole und Caramelkomponenten in Röstkaffees verschiedener Sorten. II. *Z. Lebensm. Unters. Forsch.* 167: 111-114 (1978).

68. Rahn, W. and König, W.A. GC/MS investigations of the constituents in a diethyl ether extract of an acidified roast coffee infusion. *J. High Resolut. Chromatogr. Chromatogr. Commun.* 1002: 69-71 (1978).

69. Ariza, R.R., Dorado, G., Barbanch, M., and Pueyo, C. Study of the causes of direct-acting mutagenicity in coffee and tea using the Ara test in *Salmonella typhimurium*. *Mut. Res.* 201: 89-96 (1988).

70. Fung, V.A., Cameron, T.P., Hughes, T.J., Kirby, P.E., and Dunkel, V.C. Mutagenic activity of some coffee flavor ingredients. *Mut. Res.* 204: 219-228 (1988).

71. Hanham, A.F., Dunn, B.P., and Stich, H.F. Clastogenic activity of caffeic acid and its relationship to hydrogen peroxide generated during autooxidation. *Mut. Res.* 116: 333-339 (1983).

72. Stich, H.F., Rosin, M.P., Wu, C.H., and Powrie, W.D. A comparative genotoxicity study of chlorogenic acid (3-*O*-caffeoylquinic acid). *Mut. Res.* 90: 201-212 (1981).

73. Ishidate, Jr., M., Harnois, M.C., and Sofuni, T. A comparative analysis of data on the clastogenicity of 951 chemical substances tested in mammalian cell cultures. *Mut. Res.* 201: 89-96 (1988).

74. Gold, L.S., Slone, T.H., Stern, B.R., Manley, N.B., and Ames, B.N. Possible carcinogenic hazards from natural and synthetic chemicals: Setting priorities. In: Cothern, C.R., Ed. *Comparative Environmental Risk Assessment*. Boca Raton, FL: Lewis Publishers, 1993, pp. 209-235.

75. Gold, L.S., Slone, T.H., Stern, B.R., Manley, N.B., and Ames, B.N. Rodent carcinogens: Setting priorities. *Science*. 258: 261-265 (1992).

76. Ames, B.N., Magaw, R., and Gold, L.S. Ranking possible carcinogenic hazards. Science. 236: 271-280 (1987). Letters: 237: 235 (1987); 237: 1283-1284 (1987); 237: 1399-1400 (1987); 238: 1633-1634 (1987); 240: 1043-1047 (1988).

77. Gold, L.S., Slone, T.H., Manley, N.B., and Ames, B.N. Heterocyclic amines formed by cooking food: comparison of bioassay results with other chemicals in the Carcinogenic Potency Database. *Cancer Lett.* 83: 21-29 (1994).

78. Gold, L.S., Garfinkel, G.B., and Slone, T.H. Setting priorities among possible carcinogenic hazards in the workplace. In: Smith, C.M., Christiani, D.C., and Kelsey, K.T., Eds. *Chemical Risk Assessment and Occupational Health: Current Applications, Limitations, and Future Prospects*. Westport, CT: Auburn House, 1994, pp. 91-103.

79. National Research Council. *Carcinogens and Anticarcinogens in the Human Diet: A Comparison of Naturally Occurring and Synthetic Substances*. Washington, D.C.: National Academy Press, 1996.

80. Ott, M.G., Scharnweber, H.C., and Langner, R.R. Mortality experience of 161 employees exposed to ethylene dibromide in two production units. *Br. J. Ind. Med.* 37: 163-168 (1980).

81. Ramsey, J.C., Park, C.N., Ott, M.G., and Gehring, P.J. Carcinogenic risk assessment: Ethylene dibromide. *Toxicol. Appl. Pharmacol.* 47: 411-414 (1978).

82. Havel, R.J., and Kane, J.P. Therapy of hyperlipidemic states. *Ann. Rev. Med.* 33: 417 (1982).

83. American Medical Association (AMA) Division of Drugs. *AMA Drug Evaluations, 5th Ed.* Chicago, IL: AMA, 1983, pp. 201- 202.

84. Matanoski, G., Francis, M., Correa-Villaseñor, A., Elliot, E., Santos-Brugoa, C., and Schwartz, L. Cancer epidemiology among styrene-butadiene rubber workers. *IARC Sci. Pub.* 127: 363-374 (1993).

85. Andrasik, J. Monitoring solvent vapors in drycleaning plants. *Int. Fabricare Inst. Focus Dry Cleaning*. 14(3): 1-8 (1990).

86. Siegal, D.M., Frankos, V.H., and Schneiderman, M. Formaldehyde risk assessment for occupationally exposed workers. *Reg. Toxicol. Pharm.* 3: 355-371 (1983).

87. Stofberg, J. and Grundschober, F. Consumption ratio and food predominance of flavoring materials. Second cumulative series. *Perfum. Flavor.* 12: 27-56 (1987).

88. Connor, T.H., Theiss, J.C., Hanna, H.A., Monteith, D.K., and Matney, T.S. Genotoxicity of organic chemicals frequently found in the air of mobile homes. *Toxicol. Letters.* 25: 33-40 (1985).

89. CONSAD Research Corporation. Final report. Economic analysis of OSHA's proposed standards for methylene chloride. Pittsburgh, PA, October, 1990. OSHA Docket H- 71.

90. McCann, J., Horn, L., Girman, J., and Nero, A.V. Potential risks from exposure to organic carcinogens in indoor air. In: Sandhu, S.S., deMarini, D.M., Mass, M.J., Moore, M.M., and Mumford, J.L., Eds. *Short-Term Bioassays in the Analysis of Complex Environmental Mixtures*. New York, NY: Plenum, 1987.

91. Technical Assessment Systems (TAS). Exposure 1 Software Package. Washington, D.C., 1989. Provided by Barbara Petersen.

92. Herrmann, K. Review on nonessential constituents of vegetables. III. Carrots, celery, parsnips, beets, spinach, lettuce, endives, chicory, rhubarb, and artichokes. *Z. Lebensm. Unters. Forsch.* 167: 262-273 (1978).

93. Hall, R.L., Henry, S.H., Scheuplein, R.J., Dull, B.J., and Rulis, A.M. Comparison of the carcinogenic risks of naturally occurring and adventitious substances in food. In: Taylor, S.L., and Scanlon, R.A., Eds. *Food Toxicology: A Perspective on the Relative Risks*. New York: Marcel Dekker Inc., 1989, p. 205.

94. Schreier, P., Drawert, F., and Heindze, I. Über die quantitative Zusammensetzung natürlicher und technologisch veränderter pflanzlicher Aromen. *Chem. Mikrobiol. Technol. Lebensm.* 6: 78-83 (1979).

95. Hasselstrom, T., Hewitt, E.J., Konigsbacher, K.S., and Ritter, J.J. Composition of volatile oil of black pepper. *Agric. Food Chem.* 5: 53-55 (1957).

96. Toth, B., and Erickson, J. Cancer induction in mice by feeding of the uncooked cultivated mushroom of commerce *Agaricus bisporus*. *Cancer Res.* 46: 4007-4011 (1986).

97. Matsumoto, K., Ito, M., Yagyu, S., Ogino, H., and Hirono, I. Carcinogenicity examination of *Agaricus bisporus*, edible mushroom, in rats. *Cancer Lett.* 58: 87-90 (1991).

98. U.S. Environmental Protection Agency. *Daminozide Special Review. Technical Support Document — Preliminary Determination to Cancel the Food Uses of Daminozide*. Washington, D.C.: USEPA, 1989. Office of Pesticide Programs.

99. Mosel, H.D., and Herrmann, K. The phenolics of fruits. III. The contents of catechins and hydroxycinnamic acids in pome and stone fruits. *Z. Lebensm. Unters. Forsch.* 154: 6-11 (1974).

100. United States Food and Drug Administration. *Butylatedhydroxyanisole (BHA) intake. DFCA Request of 1-15- 91*. Washington, D.C.: USFDA. January 17, 1991 (memo from Food and Color Additives Section to L. Lin).

101. Fazio, T., Havery, D.C., and Howard, J.W. Determination of volatile *N*-nitrosamines in foodstuffs: I. A new clean-up technique for confirmation by GLC-MS. II. A continued survey of foods and beverages. In: Walker, E.A., Griciute, L., Castegnaro, M., and Borzsonyi, M., Eds. *N-Nitroso Compounds: Analysis, Formation and Occurrence*. Lyon, France: International Agency for Research on Cancer, 1980, pp. 419-435. (IARC Scientific Pub. No. 31).

102. Preussmann, R. and Eisenbrand, G. N-nitroso carcinogens in the environment. In: Searle, C.E., Ed. *Chemical Carcinogenesis, 2nd Ed., Revised and Expanded, Vol. 2*. Washington DC: American Chemical Society (ACS), 1984, pp. 829-868 (*ACS Monograph 182*).

103. U.S. Food and Drug Administration *Exposure to Aflatoxins*. Washington, D.C.: USFDA, 1992.

104. Poole, S.K. and Poole, C.F. Thin-layer chromatographic method for the determination of the principal polar aromatic flavour compounds of the cinnamons of commerce. *Analyst.* 119: 113-120 (1994).

105. Heinrich, L. and Baltes, W. Über die Bestimmung von Phenolen im Kaffeegetränk. *Z. Lebensm. Unters. Forsch.* 185: 362-365 (1987).

106. National Research Council. *The 1977 Survey of Industry on the Use of Food Addititives*. Washington, D.C.: National Academy Press, 1979.

107. Neurath, G.B., Dünger, M., Pein, F.G., Ambrosius, D., and Schreiber, O. Primary and secondary amines in the human environment. *Food Cosmet. Toxicol.* 15: 275-282 (1977).

108. Schmidtlein, H. and Herrmann, K. Über die Phenolsäuren des Gemüses. IV. Hydroxyzimtsäuren und Hydroxybenzösäuren weiterer Gemüsearten und der Kartoffeln. *Z. Lebensm. Unters. Forsch.* 159: 255-263 (1975).

109. Economic Research Service. *Vegetables and Specialties Situation and Outlook Yearbook*. Washington, D.C.: U.S. Department of Agriculture, 1994.

110. Stöhr, H. and Herrmann, K. On the phenolic acids of vegetables. III. Hydroxycinnamic acids and hydroxybenzoic acids of root vegetables. *Z. Lebensm. Unters. Forsch.* 159: 219-224 (1975).

111. Bejnarowicz, E.A. and Kirch, E.R. Gas chromatographic analysis of oil of nutmeg. *J. Pharm. Sci.* 52: 988-993 (1963).

112. United States Environmental Protection Agency (USEPA). EBDC/ETU Special Review. DRES Dietary Exposure/Risk Estimates. October 11, 1991 (memo from R. Griffin to K. Martin).

113. Duggan, R.E. and Corneliussen, P.E. Dietary intake of pesticide chemicals in the United States (III), June 1968- April 1970. *Pest. Monit. J.* 5: 331-341 (1972).

114. Economic Research Service. *Fruit and Tree Nuts Situation and Outlook Yearbook*. Washington, D.C.: Department of Agriculture, 1995.

115. Carlson, D.G., Daxenbichler, M.E., VanEtten, C.H., Kwolek, W.F., and Williams, P.H. Glucosinolates in crucifer vegetables: Broccoli, Brussels sprouts, cauliflower, collards, kale, mustard greens, and kohlrabi. *J. Am. Soc. Hort. Sci.* 112: 173-178 (1987).

116. United States Environmental Protection Agency (USEPA). *Health Assessment Document for 2,3,7,8-Tetrachlorodibenzo-p-Dioxin (TCDD) and Related Compounds*. Washington, D.C.: USEPA, 1994.

117. Sen, N.P., Seaman, S., and Miles, W.F. Volatile nitrosamines in various cured meat products: Effect of cooking and recent trends. *J. Agric. Food Chem.* 27: 1354-1357 (1979).

118. Chauhan, Y., Nagel, D., Gross, M., Cerny, R., and Toth, B. Isolation of N^2-[γ-L-(+)-glutamyl]-4-carboxyphenylhydrazine in the cultivated mushroom *Agaricus bisporus*. *J. Agric. Food Chem.* 33: 817-820 (1985).

119. Luo, S., Guo, W., and Fu, H. Correlation between aroma and quality grade of Chinese jasmine tea. *Dev. Food Sci.* 17: 191-199 (1988).

120. Tricker, A.R. and Preussmann, R. Carcinogenic N-nitrosamines in the diet: Occurrence, formation, mechanisms and carcinogenic potential. *Mut. Res.* 259: 277-289 (1991).

121. Environmental Protection Agency. Office of Pesticide Programs. *Ethylene Dibromide (EDB) Scientific Support and Decision Document for Grain and Grain Milling Fumigation* Uses. Washington, D.C.: USEPA, February 8, 1984.

122. American Water Works Association, Government Affairs Office. *Disinfectant/Disinfection By-Products Database for the Negotiated Regulation*. Washington, D.C.: AWWA, 1993.

123. Engel, K.H. and Tressl, R. Studies on the volatile components of two mango varieties. *J. Agric. Food Chem.* 31: 796-801 (1983).

124. Food and Drug Administration (FDA). FDA Pesticide Program: Residues in foods 1990. *J. Assoc. Off. Anal. Chem.* 74: 121A-141A (1991).

125. Beier, R.C., Ivie, G.W., Oertli, E.H., and Holt, D.L. HPLC analysis of linear furocoumarins (psoralens) in healthy celery Apium graveolens. *Food Chem. Toxic.* 21: 163-165 (1983).

126. Gunderson, E.L. Dietary intakes of pesticides, selected elements, and other chemicals: FDA Total Diet Study, June 1984-April 1986. *J. Assoc. Off. Anal. Chem.* 78: 910-921 (1995).

127. United Fresh Fruit and Vegetable Association (UFFVA). Supply Guide: Monthly Availability of Fresh Fruit and Vegetables. Alexandria, VA, 1989.

128. Ivie, G.W., Holt, D.L., and Ivey, M. Natural toxicants in human foods: Psoralens in raw and cooked parsnip root. *Science.* 213: 909-910 (1981).

129. Canas, B.J., Havery, D.C., Robinson, L.R., Sullivan, M.P., Joe, Jr., F.L., and Diachenko, G.W. Chemical contaminants monitoring: Ethyl carbamate levels in selected fermented foods and beverages. *J. Assoc. Off. Anal. Chem.* 72: 873-876 (1989).

130. Knize, M.G., Dolbeare, F.A., Carroll, K.L., Moore II, D.H., and Felton, J.S. Effect of cooking time and temperature on the heterocyclic amine content of fried beef patties. *Food Chem. Toxicol.* 32: 595-603 (1994).

131. Chaudhary, S.K., Ceska, O., Tétu, C., Warrington, P.J., Ashwood-Smith, M.J., and Poulton, G.A. Oxypeucedanin, a major furocoumarin in parsley, *Petroselinum crispum*. *Planta Med.* 6: 462-464 (1986).

132. Environmental Protection Agency (EPA). *Peer Review of Chlorothalonil*. Washington, D.C.: Office of Pesticides and Toxic Substances, 1987. Review found in Health Effect Division Document No. 007718.

133. Environmental Protection Agency (EPA). *Integrated Risk Information System (IRIS)*. Cincinnati, OH: Office of Health and Environmental Assessment, Environmental Criteria and Assessment Office, 1991.

134. Environmental Protection Agency (EPA). *Peer Review of Captan*. Washington, D.C.: Office of Pesticides and Toxic Substances, 1986. Review found in Health Effect Division Document No. 007715.

135. Hard, G.C. and Whysner, J. Risk assessment of *d*-limonene: An example of male rat-specific renal tumorigens. *Crit. Rev. Toxicol.* 24: 231-254 (1994).

136. United States Environmental Proection Agency. *Report of the EPA Peer Review Workshop on Alpha₂ᵤ-globulin: Association with Renal Toxicity and Neoplasia in the Male Rat*. Washington, D.C.: USEPA, 1991.

137. U.S. Environmental Protection Agency Ethylene bisdithiocarbamates (EBDCs); Notice of intent to cancel; Conclusion of special review. *Fed. Reg.* 57: 7484-7530 (1992).

138. Key, T. and Reeves, G. Organochlorines in the environment and breast cancer. *Br. Med. J.* 308: 1520-1521 (1994).

139. National Research Council *Regulating Pesticides in Food: The Delaney Paradox*. Washington, D.C.: National Academy Press, 1987.

140. Ochiai, M., Masatoshi, W., Hiromi, K., Wakabayashi, K., Sugimura, T., and Minako, M. DNA adduct formation, cell proliferation and aberrant crypt focus formation induced by PhIP in male and female rat colon with relevance to carcinogenesis. *Carcinogenesis.* 17: 95-98 (1996).

141. Clayson, D.B., Iverson, F., Nera, E.A., and Lok, E. The significance of induced forestomach tumors. *Annu. Rev. Pharmacol. Toxicol.* 30: 441-463 (1990).

142. California Environmental Protection Agency. *California Cancer Potency Factors: Update*. Sacramento: CalEPA, November 1, 1994 memo.

143. Cohen, S.M. and Lawson, T.A. Rodent bladder tumors do not always predict for humans. *Cancer Lett.* 93: 9-16 (1995).

144. Qian, G., Ross, R.K., Yu, M.C., Yuan, J., Henderson, B.E., Wogan, G.N., and Groopman, J.D. A follow-up study of urinary markers of aflatoxin exposure and liver cancer risk in Shanghai, People's Republic of China. *Cancer Epidemiol. Biomarkers Prev.* 3: 3-10 (1994).

145. Wu-Williams, A.H., Zeise, L., and Thomas, D. Risk assessment for aflatoxin B₁: a modeling approach. *Risk Anal.* 12: 559-567 (1992).

146. United States Food and Drug Agency (USFDA). *Assessment of carcinogenic upper bound lifetime risk resulting from aflatoxins in consumer peanut and corn products. Report of the Quantitative Risk Assessment Committee*. Washington, D.C.: USFDA, June 3, 1993.

147. Pons Jr., W.A. High pressure liquid chromatographic determination of aflatoxins in corn. *J. Assoc. Off. Anal. Chem.* 62: 586-594 (1979).

148. Groopman, J.D., Zhu, J.Q., Donahue, P.R., Pikul, A., Zhang, L.S., Chen, J.S., and Wogan, G.N. Molecular dosimetry of urinary aflatoxin-DNA adducts in people living in Guangxi Autonomous Region, People's Republic of China. *Cancer Res.* 1992: 45-52 (1992).

149. Yu, M.C., Tong, M.J., Govindarajan, S., and Henderson, B.E. Nonviral risk factors for hepatocellular carcinoma in a low-risk population, the non-Asians of Los Angeles County, California. *J. Natl. Cancer Inst.* 83: 1820-1826 (1991).

150. Kuiper-Goodman, T. and Scott, P.M. Risk assessment of the mycotoxin ochratoxin A. *Biomed. Environ. Sci.* 2: 179-248 (1989).

151. Occurrence and significance of ochratoxin A in food. ILSI Europe Workshop, 10-12 January 1996, Aix-en-Provence, France. *ILSI Europe Newsletter.* (February 1996): 3.

152. Gartrell, M.J., Craun, J.C., Podrebarac, D.S., and Gunderson, E.L. Pesticides, selected elements, and other chemicals in adult total diet samples October 1980-March 1982. *J. Assoc. Off. Anal. Chem.* 69: 146-161 (1986).

153. World Health Organization *Polychlorinated Biphenyls and Terphenyls. Environmental Health Criteria 140*. Geneva: WHO, 1993.

154. United States Environmental Protection Agency (USEPA). *Estimating Exposure to Dioxin-Like Compounds (Review Draft)*. Washington, D.C.: USEPA, 1994.

155. U.S. Environmental Protection Agency. *Re-evaluating dioxin: Science Advisory Board's review of EPA's reassessment of dioxin and dioxin-like compounds*. Washington, D.C.: USEPA, 1995.

156. Bradfield, C.A., and Bjeldanes, L.F. Structure-activity relationships of dietary indoles: a proposed mechanism of action as modifiers of xenobiotic metabolism. *J. Toxicol. Environ. Health.* 21: 311-323 (1987).

157. Tomatis, L. and Bartsch, H. The contribution of experimental studies to risk assessment of carcinogenic agents in humans. *Exp. Pathol.* 40: 251-266 (1990).

158. Gold, L.S., Backman, G.M., Hooper, N.K., and Peto, R. Ranking the potential carcinogenic hazards to workers from exposures to chemicals that are tumorigenic in rodents. *Environ. Health Perspect.* 76: 211-219 (1987).

159. National Research Council. *Science and Judgment in Risk Assessment*. Washington, D.C.: National Academy Press, 1994. Committee on Risk Assessment of Hazardous Air Pollutants.

160. U.S. Environmental Protection Agency. Office of Research and Development. Draft Review of the Guidelines for Carcinogenic Risk Assessment. *Fed. Reg.* 61: 17960-18011 (1996).

161. Zeise, L., Wilson, R., and Crouch, E. Use of acute toxicity to estimate carcinogenic risk. *Risk Anal.* 4: 187-199 (1984).

162. Travis, C.C., Richter Pack, S.A., Saulsbury, A.W., and Yambert, M.W. Prediction of carcinogenic potency from toxicological data. *Mut. Res.* 241: 21-36 (1990).

163. Ames, B.N. Dietary carcinogens and anti-carcinogens: Oxygen radicals and degenerative diseases. *Science.* 221: 1256-1264 (1983).

164. Ames, B.N. Cancer and diet. *Science.* 224: 668-670, 757-760 (1984).

165. Harvey, M.H., Morris, B.A., McMillan, M., and Marks, V. Measurement of potato steroidal alkaloids in human serum and saliva by radioimmunoassay. *Human Toxicol.* 4: 503-512 (1985).

166. Yamagami, T., Handa, H., Juji, T., Munemitsu, H., Aoki, M., and Kato, Y. Rat pituitary adenoma and hyperplasia induced by caffeine administration. *Surg. Neurol.* 20: 323-331 (1983).

CHAPTER 5

GENOTOXICITY DATABASE

Errol Zeiger

SECTION A: DESCRIPTION OF GENETIC TOXICITY TESTS

GENETIC TOXICITY TESTING

Genetic toxicology is the study of the ability of substances and agents to damage the DNA and chromosomes of cells. Genetic toxicity is usually measured as an increase in mutation, chromosome aberrations, or aneuploidy; as DNA adducts or interference with repair of DNA damage; or other nonspecific DNA or chromosome damage. A number of *in vitro* bacterial and mammalian cell culture systems and insect systems have been developed to study the effects of chemicals and radiation on cellular DNA and chromosomes.

The genetic toxicity testing program of the U.S. National Toxicology Program (NTP) began in 1979 with the testing of chemicals in bacteria, cultured mammalian cells, and insects. That program was later expanded to include testing in rodents. The primary impetus for testing chemicals for genetic toxicity was reports showing a high correlation between genetic toxicity and rodent carcinogenicity, with the possibility of using genetic toxicity test results to predict rodent carcinogenicity. It was hoped that the results from the short-term, relatively inexpensive genetic toxicity tests would reduce the need for rodent carcinogenicity testing or provide information with which to prioritize chemicals for cancer testing. The accurate prediction of carcinogenicity using *in vitro* or short-term *in vivo* genetic toxicity tests would lead to a large savings of time and resources and greatly reduce the number of animals needed for identification of chemical carcinogens.

The testing program was refined, and tests that were shown to be redundant or not sufficiently sensitive for the NTP's purpose were discontinued, and other tests, or test variations, were added. The testing program is dynamic, i.e., tests are continuously being evaluated, modified, or eliminated from routine use according to the accumulated information obtained from the testing. New tests are introduced to determine if they provide additional, useful information on the genetic toxicity of chemicals, or if they are easier, more reliable, or less expensive to use than the traditional tests.

Many chemicals are not genotoxic in their native forms, but become genotoxic when metabolized. Bacteria and cultured mammalian cells generally cannot perform the full range of metabolic conversions that occur in mammals and humans because they do not contain the necessary drug-metabolizing enzyme systems. This limitation has been partially overcome by the development of exogenous metabolic activation systems that can be added to the test procedure. These systems usually consist of homogenates of liver fractions (S-9) of rodents that had been pretreated with substances to induce higher levels of the preferred metabolic enzymes. Chemicals are tested *in vitro* both in the absence of exogenous metabolic activation systems and in the presence of induced liver S-9 fractions.

The identification of chemical mutagens is an important component of toxicology because chemicals that are mutagenic in Salmonella, mammalian cells, and Drosophila also have the capacity to be mutagenic in mammals, including humans. Data included in this compilation are from the *Salmonella typhimurium* mutation (Ames) test; tests for induction of chromosome aberrations and sister chromatid exchanges in cultured Chinese hamster ovary cells; a test for mutation in a mouse lymphoma cell line; and from the sex-linked recessive lethal test in the fruit fly, *Drosophila melanogaster*.

The following descriptions of the test protocols briefly describe the procedures and the data evaluation criteria. Additional detail regarding the test procedures and data analyses can be found in the referenced articles.

SALMONELLA MUTATION (AMES) TEST (SAL)

TEST DESCRIPTION

The Salmonella mammalian microsome mutagenicity (Ames) test is designed to measure mutations using strains of the bacterium *Salmonella typhimurium* selected and designed to be sensitive to a wide range of mutagenic chemicals. The Salmonella strains used in this procedure contain mutations that leave the bacteria incapable of synthesizing histidine, a needed amino acid. When these cells are placed on growth medium lacking histidine they cannot divide to form colonies. They can grow and divide only if histidine is added to the medium, or if they mutate and regain the ability to synthesize histidine. In this way, colonies arising from mutated cells can readily be identified. The Salmonella tester strains used in the test (i.e., TA97, TA98, TA100, TA102, TA104, TA1535, TA1537, TA1538) contain different histidine mutations which are mutated through different molecular mechanisms and by different classes of chemicals. Because mutation in these cells is measured as "reversing" the original mutation the test is sometimes referred to as a "reverse mutation" test. These strains measure only point mutations and are not capable of responding to substances that only produce chromosome breaks, deletions, or rearrangements.

Additional genetic alterations have been added to these strains to enhance their sensitivity to mutation induction, including a defect in the bacterial cell wall which allows large chemical molecules to more easily enter the cell, and defects in the DNA repair mechanisms, which makes the cells more sensitive to the DNA-damaging effects of chemicals. As a result of these alterations, the cells can mutate in response to DNA damage that would not be mutagenic in bacterial cells with normal repair capabilities, and to low levels of damage that would normally be repaired without any observable mutation. The use of these sensitive tester strains enhances the level of detection of chemicals and maximizes the numbers of mutagens detected.

TEST PROTOCOL

All chemicals were tested as coded compounds, and test results were evaluated before decoding. For metabolic activation, the S-9s taken from rats and hamsters were used. A preincubation procedure was used for the majority of tests. In this procedure, the test chemical, bacteria, and S-9 mix or buffer were combined and incubated at 37°C. Top agar was added and the contents of the tubes were poured onto the surface of petri dishes containing minimal growth medium. Mutant colonies arising on these plates were counted following two days of incubation.

Four general protocol variations were used:

1. Testing was performed with and without rat and hamster S-9 in strains TA97 (or TA1537), TA98, TA100, and TA1535; only 10% S-9 was used.
2. The initial test was without activation and with 30% S-9. If a positive result was obtained, the positive trial(s) was repeated. If the trials were negative, the chemical was retested without S-9 and with 10% S-9. If a positive response was obtained in any strain, only the positive test(s) were repeated.
3. Initial testing was in strains TA98 and TA100 without activation and with 30% S-9. If a positive response was obtained in one or both strains, only the positive test was repeated. If an equivocal or weak positive response was obtained in one or both of these strain(s), TA1535 and/or TA1538 was used. If the results with TA98 and TA100 were negative, strains TA97 and TA1535 were used without activation and with 30% S-9. If either was positive, a confirmation test was run. If results in these two strains were negative, the test was repeated using all four strains without activation and with 10% S-9.
4. Strains TA102 and/or TA104 were also used with some chemicals.

Occasionally, other S-9 concentrations were also used.

Some chemicals were tested using the plate test procedure in Salmonella strains TA98, TA100, TA1535, TA1537, and TA1538, using uninduced and induced rat, mouse, and hamster S-9. In addition, some azo dyes were tested using a reductive metabolic activation system in addition to the standard S-9.

Each chemical was tested initially up to a toxic dose or to a dose immediately below one which was toxic in a preliminary toxicity assay. Nontoxic chemicals were tested to a maximum of 10 mg/plate. Poorly soluble chemicals were tested up to doses defined by their solubilities. At least five doses of each chemical were tested in triplicate, and repeat experiments were performed. Concurrent solvent and positive controls were run with each trial.

DATA EVALUATION

A chemical was judged mutagenic (+) or weakly mutagenic (+w) if it produced a reproducible, dose-related increase in mutant colonies under any test condition. A chemical was judged questionable (?) if the results of individual trials were not reproducible, if increases in mutants did not meet the criteria for a "+w" response, or if only single doses produced increases in mutants in repeat trials. Chemicals were judged nonmutagenic (–) if they did not meet the criteria for a mutagenic or questionable response. A chemical was designated nonmutagenic only after it had been tested in at least 4 strains (i.e., TA98, TA100, TA1535, TA97, and/or TA1537), without activation and with rat and hamster S-9. A positive test result in one strain with one type of metabolic activation was sufficient to identify a chemical as a mutagen.

IN VITRO CHROMOSOME ABERRATION TEST (ABS)

TEST DESCRIPTION

Chromosome aberrations can be measured by observing cells in the metaphase stage of mitosis using a light microscope. Cultured Chinese hamster ovary (CHO) cells are routinely used for testing chemicals for the ability to cause ABS. These cells are well characterized, are easily grown, have high cloning efficiencies, a rapid cell-cycling time of approximately 10 to 12 hr, and approximately 20 well-defined chromosomes that are easily examined for structural abnormalities using a light microscope. The types of ABS typically of concern as indicators of chemical or radiation-induced damage include chromosome and chromatid breaks and rearrangements.

Different stages of the cell cycle generally have different sensitivities to chromosome damage. Some agents induce chromosome damage at all stages of the cell cycle, whereas the majority of chemicals studied to date appear to induce damage only when DNA replication occurs in the presence of damaged DNA. To maximize the opportunity for inducing chromosome aberrations, chemical treatment in the absence of exogenous metabolic activation is typically for at least one cell cycle time. Because of the toxicity of S-9 to cells in culture, treatment with the chemical plus S-9 was limited to 2 hr. This shortened chemical treatment time reduced the sensitivity of the test, but allowed higher doses of many chemicals to be tested.

TEST PROTOCOL

Chemicals were tested and the results evaluated under code. The chemicals were tested with and without induced rat liver S-9. Stocks of CHO cells were maintained in McCoy's 5A (modified) medium and supplemented with fetal bovine serum. Test cultures were set up in 75 cm^2 flasks, grown, and exposed to chemicals at 37°C.

Separate toxicity or range-finding tests were not run; the toxicity test was combined with the definitive test. Alternatively, the test range and treatment time was selected based on the toxicity and cell-cycle kinetics information from the SCE test, which was performed first.

Where chemical-induced cell cycle delay was noted in the SCE test, the cell growth period in the ABS test was extended approximately 6 to 8 hr.

Each test consisted of concurrent solvent and positive controls and 10 concentrations of the test chemical in decreasing half-log increments. Single flasks were used for each control and test concentration. The highest dose tested was 5.0 mg/ml, or the highest dose that did not produce visible precipitate in the flask. The highest dose scored was the dose that yielded a sufficient number of metaphase cells.

ABS Test Without Metabolic Activation

The CHO cultures were treated with test chemical in cell culture medium containing serum for 8 to 10 hr. Colcemid was added for an additional 2 to 2.5 hr before cell harvest to arrest the cells in metaphase.

ABS Test With Metabolic Activation

The CHO cultures were exposed to the test chemical in serum-free medium with S-9 and cofactors for 2 hr, washed to remove the test chemical and S-9, and incubated in fresh medium for an additional 8 hrs. Colcemid was added after this time, and the cells harvested 2 to 2.5 hr later.

Cell Harvest and Scoring

At the end of the incubation period, the cultures were examined for toxicity, which was estimated as the percent confluence of the culture compared to the solvent control. The cells were harvested by mitotic shake-off. The harvested cells were centrifuged, placed on microscope slides, fixed, and stained with Giemsa. The highest dose that provided sufficient metaphase cells, and the next two or three lower doses were scored. Early in the testing program, 100 metaphase cells/test chemical concentration were scored; subsequently, this was increased to 200 cells per concentration. However, fewer cells were scored if the aberration response was obviously positive.

Cells were analyzed for the following categories of chromosome aberrations: "simple" (chromatid or chromosome break, fragment, or deletion), "complex" (interstitial deletions, triradials, quadriradials, rings, dicentric chromosomes), and "other" (pulverized chromosomes or cells with greater than ten aberrations). Chromatid and chromosome gaps, and the frequencies of polyploid or endoreduplicated cells, were recorded but not used in the analyses.

DATA EVALUATION

All categories of aberrations (simple, complex, and other) were combined for the statistical analyses. A binomial sampling assumption was used to examine absolute increases in ABS over solvent control levels at each dose.

The *p*-values were adjusted by Dunnett's method to take into account the multiple dose comparisons. Only the total percentage of cells with aberrations were analyzed. A positive response was defined as having an adjusted *p*-value of < 0.05. If there was a positive trend, but no significantly increased doses, the trial was designated "?". A positive response at a single dose was designated as "+w", weak evidence for clastogenicity. A test was designated "+" if at least two doses gave significantly increased responses. Tests were repeated to confirm a positive result, and, in general, a chemical was not designated positive unless its response was repeatable. A chemical was designated "?" if all trials were "?" or if a positive response was not repeatable under the same test conditions. A "+w" response that was not repeatable led to a determination of "−" for the chemical.

IN VITRO SISTER CHROMATID EXCHANGE (SCE) TEST

Test Description

Sister chromatid exchanges (SCE) are reciprocal interchanges of chromosomal material between homologous sister chromatids. They are visualized by treating the cells with a nucleotide analogue, such as bromodeoxyuridine (BUdR), which is incorporated into the DNA of the cell during DNA synthesis. Treatment of the replicating cells with BUdR during a single cell-replication cycle leads to the formation of chromosomes with both sister chromatids unifiliarly substituted with BUdR. Following a second replication cycle in the presence of BUdR, one sister chromatid contains bifiliarly substitued DNA and the other sister chromatid remains unifiliarly substituted with BUdR. This biochemical difference allows for the visualization of SCE using either fluorescent light or fluorescence-plus-Giemsa staining techniques.

The mechanism(s) of SCE formation and relevance of SCE to the life and future of the cell is not known. However, SCE induction has been used as a measure of DNA damage and as a dosimeter of DNA damage in cells *in vivo*. SCEs are also easier to evaluate than chromosome aberrations, and had been proposed as an alternative to ABS for rapid identification of chemicals that damage DNA.

Chinese hamster ovary (CHO) cells in culture are routinely used for testing chemicals for the ability to cause SCE. These cells are well characterized, are easily grown, and have high cloning efficiencies, a rapid cell-cycling time of approximately 10 to 12 hr, and approximately 20 well-defined chromosomes that are easily seen using a light microscope.

Test Protocol

Chemicals were tested and the results evaluated under code. The chemicals were tested with and without exogenous metabolic activation from induced rat liver S-9. Stocks of CHO cells were maintained in McCoy's 5A (modified) medium and supplemented with fetal bovine serum. Test cultures were set up in 75 cm² flasks, grown, and exposed to chemicals at 37°C. Because of the toxicity of S-9 to cells in culture, treatment with the chemical plus S-9 was for 2 hr. This shortened chemical treatment time reduced the sensitivity of the test, but allowed higher doses of many chemicals to be tested.

A combined dose-range and definitive test was performed using 10 half-log dilutions of a stock solution in culture medium or DMSO. Each test consisted of concurrent solvent and positive controls and ten concentrations of the test chemical, in decreasing half-log increments. Single flasks were used for each control and test concentration. The highest dose tested was 5.0 gm/ml, or the highest dose that did not produce visible precipitate in the flask. The highest dose scored was the dose that yielded sufficient metaphase cells in the second division cycle after initiation of chemical treatment.

SCE Test Without Metabolic Activation
Cells were exposed to the test chemical for 2 hr, at which time bromodeoxyuridine (BUdR) was added. After an additional 24 hr incubation in the presence of the test chemical and BUdR, the cells were washed and fresh medium containing BUdR and colcemid, to arrest cells in metaphase, were added. The cells were incubated for an additional 2 hr before harvesting.

SCE Test With Metabolic Activation
The cells were treated with the test chemical and S-9 in serum-free medium for 2 hr, after which time the medium with test chemical and S-9 was removed, and the cells incubated in medium containing serum and BUdR for 24 hr. Colcemid was added and the cells incubated for an additional 2 hr.

Cell Harvesting and Scoring
At the end of the incubation period, the cultures were examined for toxicity, which was estimated as the percent confluence of the culture compared to the solvent control. The cells were harvested by mitotic shake-off. The harvested cells were centrifuged, placed on microscope slides, fixed, and stained with Hoechst 33258, followed by exposure to blacklight and then staining with Giemsa. The cell-free culture medium was returned to the flasks which were replaced in the incubator for later harvest, if necessary.

Slides were examined using fluorescence microscopy to assess the frequency of metaphase cells that had completed one (M1), two (M2), or more cycles in BUdR. If the test chemical caused cell-cycle delay down to the lowest three doses, based on having a high proportion of cells in M1, the cells were harvested again 4 hr later. The highest dose which provided sufficient second division metaphase cells, and the next two or three lower doses were scored. 50 second metaphase cells were scored from each flask. Fewer cells were scored where the response was obviously positive. The frequency of SCE in the cells was used to determine the response to the chemical.

DATA EVALUATION

A trend test of SCE/chromosome and the magnitudes of the increases of the individual responses were used to evaluate the responses. A positive trend without a 20% increase over the solvent control at any dose was judged "?". If the response from one dose was increased by at least 20%, the response was designated as showing weak evidence (+w) of the ability to induce SCE. If at least two doses showed increases of at least 20% over the control, the result was designated as "+" regardless of whether there was a positive trend. In general, positive or equivocal tests were repeated. A chemical was designated as "?" if a positive result did not repeat under similar test conditions, or if all tests produced determinations of "?".

MOUSE LYMPHOMA CELL MUTATION TEST (MLA)

TEST DESCRIPTION

The L5178Y mouse lymphoma cell mutation test measures the production of mutations and chromosome damage in mouse lymphoma L5178Y cells in culture. This test measures the ability of the cells to grow and form colonies in the presence of the toxin trifluorothymidine (TFT). Cells that can synthesize thymidine kinase will take up and convert TFT to a substance that is toxic for the cells, and they will not be able to grow or survive in its presence. Cells that have mutations in both of their thymidine kinase (tk) genes cannot metabolize TFT. The L5178Y cells used for the test have a single, functional tk gene (tk+/–). Loss or inactivation of that gene results in the complete loss of tk enzyme activity in the cell (tk–/–), which renders the cell TFT-resistant. This allows them to survive and form colonies in the presence of TFT.

The functional tk gene in the L5178Y cells can be inactivated by point mutations or by rearrangements or deletions of the gene. Therefore, cells that form colonies

in the presence of TFT are considered to be mutated. It has been demonstrated that small-sized colonies produced by a large fraction of chemicals are associated with gross chromosome damage or chromosome loss, while large colonies are associated with gene mutations. As a result, this test is often considered as measuring both gene mutations and structural chromosome damage.

TEST PROTOCOL

All chemicals were tested and evaluated as coded samples, and all results were confirmed with repeat tests. The top dose was determined by chemical solubility or cell toxicity in a preliminary experiment, but did not exceed 5 mg/ml. All doses were tested at least in duplicate, and at least five test concentrations were used, generally using a descending 2-fold dilution series from the highest testable concentration. Toxicity was expressed as a reduction of cell growth during the expression period, or as a reduction in cloning efficiency. The Relative Total Growth (RTG) was used as a measure of toxicity. A culture was not used if its cloning efficiency was below 10% or its RTG was below 1%.

In the definitive test, 6×10^6 cells in a 10 ml-volume in 30 ml, capped centrifuge tubes were treated with the test chemical with and without induced rat liver S-9 for 4 hr, washed, and resuspended in culture medium for two days to allow for expression of the newly induced mutants. The cells were then plated in semisolid agar to determine their cloning efficiency in the absence of TFT, and for mutant cell counts in the presence of TFT. The plates were incubated at 37°C, and colonies counted after 10 to 12 days. The mutant count was divided by the product of the cloning efficiency (the fraction of surviving cells that were able to form colonies in the absence of TFT) and the number of cells at risk for mutation, to yield the mutant fraction. Appropriate solvent and positive controls were run concurrently.

For most chemicals, the initial test was performed without S-9. If a clearly positive response was not seen, the test was repeated with S-9. If a positive response was obtained with or without S-9, the test was repeated to confirm the positive, and no further testing was performed. If a negative or equivocal response was obtained, the chemical was tested in the presence of S-9. Some chemicals that were not mutagenic either with or without S-9, were also tested using uninduced rat liver S-9.

DATA EVALUATION

All data were evaluated statistically for both trend and peak responses, and determinations of positive (+), negative (–), or questionable (?) were made according to statistical criteria. A trial was considered positive if at least one of the three highest doses was positive and there was a significantly positive trend. If there was only a single dose positive, or a positive trend response without any individual dose point being positive, the call was "questionable" (?). The chemical was judged negative (–) if there was neither an elevated dose nor trend response. A chemical was judged positive (+) only if the positive response could be confirmed in a repeat test.

DROSOPHILA SEX-LINKED RECESSIVE LETHAL TEST (DRO)

TEST DESCRIPTION

The fruit fly, *Drosophila melanogaster*, is used in several classical genetic tests for studying mutation in a multi-celled organism. Its brief life span and well-defined genetics makes it ideal for routine laboratory use. Unlike *in vitro* tests which only use somatic cells, Drosophila allows the study of germ cells *in situ*. This test can measure the ability of a chemical or physical treatment to induce heritable mutations. Drosophila also has many of the same chemical metabolizing enzymes found in mammals.

The most widely used test for mutation studies in Drosophila is the sex-linked recessive lethal test. This test measures the induction of mutations in the X-chromosomes of treated male flies. The flies used are genetically "marked" by preexisting mutations, which are then used to identify the source (male or female parent) of the X-chromosome in the offspring. The mutations are scored as lethal mutations in the male offspring of the treated flies; the male flies carrying the mutated X-chromosome from the treated male do not survive embryogenesis. This is identified as the absence of a specific class of F1 male flies. Mutations on the X-chromosome that do not affect the survival of the offspring cannot be identified using this particular test.

TEST PROTOCOL

All chemicals were tested as coded compounds, and the results evaluated prior to breaking the codes. Water was the preferred test chemical solvent, followed by ethyl alcohol; other solvents used were Tween 80 or a solution of DMSO. The test chemical solutions were diluted with aqueous 5% sucrose for feeding, or aqueous 0.7% NaCl or peanut oil for injection. Preliminary toxicity tests were performed to identify a test concentration that induced mortality of about 30% after 72 hr feeding or 24 hr after injection. A few volatile chemicals were tested by inhalation, i.e., subjecting the flies to the volatile chemical in a sealed vial.

The wild-type male flies were fed the chemical for 3 days. The treated males were individually mated to 3 harems

of Basc females to produce 3 broods of F1 progeny resulting from sperm treated at different levels of development. Individual F1 females were mated to their brothers, and the offspring (F2) males from each mating were examined. If no wild-type males, and at least 20 Basc males were present from a mating, the culture was considered to contain a lethal mutation. All suspect lethals were confirmed by an additional mating of F2 females. A comparison of the frequency of lethals from control and treated populations was made. If the test was negative (no significant increase in lethals), the chemical was retested by injection. If a negative result was obtained by injection, no further testing was done. Concurrent positive and solvent controls were run. A minimum of approximately 5,000 chromosomes (equivalent to 5,000 flies) were scored in each of the treated and control groups unless the mutant frequency in a group exceeded 1%, in which case fewer chromosomes were scored.

DATA EVALUATION

The statistical evaluation of Drosophila test data evolved as more experience was obtained with the operational parameters of the test. This evolution was manifested in changes in the acceptable p values for significance between the treated and control groups. When the number of lethals from any male significantly exceeded the expected value, all lethals from that male were regarded as resulting from a single, spontaneous meiotic event (a cluster). Clusters were identified using a Poisson distribution and were not included in the subsequent analysis. The treated flies were compared with the concurrent solvent control flies using a normal approximation to the binomial distribution, as well as a comparison with the historical control. In general, for a treatment to be considered "+", the mutant frequency in the treated sample must exceed 0.15%, (with $p < 0.05$), or 0.1% (with $p < 0.01$). If the frequency in treated flies was between 0.1% and 0.15% (with $0.1 > p > 0.01$), or if the treated frequency was >0.15% (with $0.1 > p > 0.05$), the assays were judged "?". All other responses were judged negative.

GENERAL BIBLIOGRAPHY

Ames, B. N., Durston, W. E., Yamasaki, E., and Lee, F. D., Carcinogens are mutagens: A simple test system for combining liver homogenates for activation and bacteria for detection, *Proc. Natl. Acad. Sci. USA*, 70, 2281-2285, 1973.

Ames, B. N., McCann, J., and Yamasaki, E., Methods for detecting carcinogens and mutagens with the Salmonella/mammalian microsome mutagenicity test, *Mutat. Res.*, 31, 347-364, 1975.

Auletta, A. and Ashby, J., Workshop on the relationship between short-term test information and carcinogenicity. Williamsburg, Virginia, January 20-23, 1987, *Environ. Mol. Mutagen.*, 11, 135–145, 1988.

Caspary, W. J., Lee, Y. J., Poulton, S., Myhr, B. C., Mitchell, A. D., and Rudd, C. J., Evaluation of the L5178Y mouse lymphoma cell mutagenesis assay: Quality-control guidelines and response categories, *Environ. Mol. Mutagen.*, 12 (Suppl. 13), 19-36, 1988.

Clive, D., Caspary, W., Kirby, P. E., Krehl, R., Moore, M., Mayo, J., and Oberly, T. J., Guide for performing the mouse lymphoma assay for mammalian cell mutagenicity, *Mutat. Res.*, 189, 143-156, 1987.

Clive, D., Johnson, K. O., Spector, J. F. S., Batson, A. G., and Brown, M. M. M., Validation and characterization of the L5178Y/TK$^{+/-}$ mouse lymphoma mutagen assay system, *Mutat. Res.*, 61-108, 1979.

Clive, D., McCuen, R., Spector, J. F. S., Piper, C., and Mavournin, K. H., Specific gene mutations in L5178Y cells in culture. A report of the U.S. Environmental Protection Agency Gene-Tox Program, *Mutat. Res.*, 115, 225-251, 1983.

de Serres, F. J. and Ashby, J., Eds., *Evaluation of Short-Term Tests for Carcinogens*. Report of the International Collaborative Programme, Progress in Mutation Research, Vol. 1, Amsterdam: Elsevier/North Holland, 827 pp., 1981.

Galloway, S. M., Armstrong, M. J., Reuben, C., Colman, S., Brown, B., Cannon, C., Bloom, A. D., Nakamura, F., Ahmed, M., Duk, S., Rimpo, J., Margolin, B. H., Resnick, M. A., Anderson, B., and Zeiger, E., Chromosome aberrations and sister chromatid exchanges in Chinese hamster ovary cells: Evaluations of 108 chemicals, *Environ. Mol. Mutagen.*, l0 (Suppl. l0), 1-176, 1987.

Galloway, S. M., Bloom, A. D., Resnick, M., Margolin, B. H., Nakamura, F., Archer, P., and Zeiger, E., Development of a standard protocol for *in vitro* cytogenetic testing with Chinese hamster ovary cells. Comparison of results for 22 compounds in two laboratories, *Environ. Mutagen.*, 7, 1-51, 1985.

Gatehouse, D., Haworth, S., Cebula, T., Gocke, E., Kier, L., Matsushima, T., Melcion, C., Nohmi, T., Ohta, T., Venitt, S., and Zeiger, E., Recommendations for the performance of bacterial mutation assays, *Mutat. Res.*, 312, 217-233, 1994.

Haseman, J. K., Zeiger, E., Shelby, M. D., Margolin, B. H., and Tennant, R. W., Predicting rodent carcinogenicity from four *in vitro* genetic toxicity assays: An evaluation of 114 chemicals studied by the National Toxicology Program, *J. Amer. Stat. Assoc.*, 85, 964–971, 1990.

Hollstein, M., McCann, J., Angelosanto, F., and Nichols, W., Short-term tests for carcinogens and mutagens, *Mutat. Res.*, 65, 133-226, 1979.

Kier, L. D., Brusick, D. J., Auletta, A. E., Von Halle, E. S., Simmon, V. F., Brown, M. M., Dunkel, V. C., McCann, J., Mortelmans, K., Prival, M. J., Rao, T. K., and Ray, V. A., The *Salmonella typhimurium*/mammalian microsome mutagenicity assay. A report of the U.S. Environmental Protection Agency Gene-Tox Program, *Mutat. Res.*, 168, 67-238, 1986.

Latt, S. A., Allen, J. A., Bloom, S. E., Carrano, A., Falke, E., Kram, D., Schneider, E., Schreck, R., Tice, R., Whitfield, B., and Wolff, S., Sister-chromatid exchanges: A report of the Gene-Tox program, *Mutat. Res.*, 87, 17-62, 1981.

Lee, W. R., Abrahamson, S., Valencia, R., von Halle, E. S., Wurgler, F. E., and Zimmering, S., The sex-linked recessive lethal test for mutagenesis in *Drosophila melanogaster*. A report of the U. S. Environmental Protection Agency Gene-Tox Program, *Mutat. Res.*, 123, 183-279, 1983.

Margolin, B. H., Collings, B. J., and Mason, J.M., Statistical analysis and sample-size determinations for mutagenicity experiments with binomial responses, *Environ. Mutagen.*, 5, 705-716, 1983.

Margolin, B. H., Resnick, M. A., Rimpo, J. Y., Archer, P., Galloway, S. M., Bloom, A. D., and Zeiger, E., Statistical analyses for *in vitro* cytogenetic assays using Chinese hamster ovary cells, *Environ. Mutagen.*, 8, 183-204, 1986.

Maron, D. and Ames, B. N., Revised methods for the Salmonella mutagenicity test, *Mutat. Res.*, 113, 173-215, 1983.

Mason, J. M., Aaron, C. S., Lee, W. R., Smith, P. D., Thakar, A., Valencia, R., Woodruff, R. C., Wurgler, F. E., and Zimmering, S., A guide for performing germ cell mutagenesis assays using *Drosophila melanogaster*, *Mutat. Res.*, 189, 93-102, 1987.

Mason, J. M., Langenbach, R., Shelby, M. D., Zeiger, E., and Tennant, R. W., Ability of short-term tests to predict carcinogenesis in rodents, *Ann. Rev. Pharmacol. Toxicol.*, 30, 149-168, 1990.

Mason, J. M., Valencia, R., and Zimmering, S., Chemical mutagenesis testing in Drosophila. VIII. Reexamination of equivocal results, *Environ. Mol. Mutagen.*, 19, 227-234, 1992.

McGregor, D. B., Brown, A. G., Howgate, S., McBride, D., Riach, C., and Caspary, W. J., Responses of the L5178Y tk+/tk– mouse lymphoma cell forward mutation assay. V: 27 coded chemicals, *Environ. Mol. Mutagen.*, 17, 196-219, 1991.

Mitchell, A. D., Myhr, B. C., Rudd, C. J., Caspary, W. J., and Dunkel, V. C., Evaluation of the L5178Y mouse lymphoma cell mutagenesis assay: Methods used and chemicals evaluated, *Environ. Mol. Mutagen.*, 12 (Suppl. 13), 1-18, 1988.

Preston, R. J., Au, W., Bender, M. A., Brewen, J. G., Carrano, A. V., Heddle, J. A., McFee, A. F., Wolff, S., and Wassom, J. S., Mammalian *in vivo* and *in vitro* cytogenetic assays: A report of the U.S. EPA's Gene-Tox Program, *Mutat. Res.*, 87, 143-188, 1981.

Purchase, I. F. H., Longstaff, E., Ashby, J., Styles, J. A., Anderson, D., Lefevre, P. A., and Westwood, F. R., An evaluation of 6 short-term tests for detecting organic chemical carcinogens, *Br. J. Cancer*, 37, 873–959, 1978.

Shelby, M. D. and Zeiger, E., Detection of human carcinogens by Salmonella and rodent bone marrow cytogenetics tests, *Mutat. Res.*, 234, 257–261, 1990.

Sugimura, T., Yahagi, T., Nagao, M., Takeuchi, M., Kawachi, T., Hara, K., Yamasaki, E., Matsushima, T., Hashimoto, Y., and Okada, M., Validity of mutagenicity tests using microbes as a rapid screening method for environmental carcinogens, in *Screening Tests in Chemical Carcinogenesis*, Montesano, R., Bartsch, H., and Tomatis, L., Eds., Lyon: IARC Scientific Publications, Vol. 12, pp. 81–104, 1976.

Tennant, R. W., Margolin, B. H., Shelby, M. D., Zeiger, E., Haseman, J. K., Spalding, J., Caspary, W., Resnick, M., Stasiewicz, S., Anderson, B., and Minor, R., Prediction of chemical carcinogenicity in rodents from *in vitro* genetic toxicity assays, *Science*, 236, 933–941, 1987.

Zeiger, E. and Drake, J. W., An environmental mutagenesis test development programme, in *Molecular and Cellular Aspects of Carcinogen Screening Tests*, Montesano, R., Bartsch, H., and Tomatis, L., Eds., IARC Scientific Publications No. 27, Lyon, pp. 303–313, 1980.

Zeiger, E., Anderson, B., Haworth, S., Lawlor, T., and Mortelmans, K., Salmonella mutagenicity tests V. Results from the testing of 311 chemicals, *Environ. Mol. Mutagen.*, 19 (Suppl. 21), 2-141, 1992.

Zeiger, E., Haseman, J. K., Shelby, M. D., Margolin, B. H., and Tennant, R. W., Evaluation of four *in vitro* genetic toxicity tests for predicting rodent carcinogenicity: confirmation of earlier results with 41 additional chemicals, *Environ. Mol. Mutagen.*, 16 (Suppl. 18), 1–14, 1990.

Zeiger, E., The Salmonella mutagenicity assay for identification of presumptive carcinogens, In *Handbook of Carcinogen Testing*, Milman, H.A. and Weisburger, E. K., Eds., Noyes Publications, NJ, pp. 83-99, 1985.

Zimmering, S., Mason, J.M., Valencia, R., and Woodruff, R. C., Chemical mutagenesis testing in Drosophila. II. Results of 20 coded compounds tested for the National Toxicology Program, *Environ. Mutagen.*, 7, 87-100, 1985.

SECTION B: TABLE OF GENOTOXICITY RESULTS

INTRODUCTION TO THE GENETIC TOXICITY TABLE

This table presents test results from 1525 substances previously published in peer-reviewed publications. The chemical nomenclature used here is identical to that used in the referenced publications; no attempt was made to change the nomenclature for "standardization" or editorial reasons. Some chemicals were tested as more than one salt form, and some mixtures were in different formulations. The different chemical salts and mixture formulations were included in the table as separate entries.

The results reported here for bacteria (SAL) and mammalian cell culture systems (ABS, SCE, and MLA) are the summaries of tests with and without exogenous metabolic activation. A designation of "?" can be for two reasons: the substance was tested one or more times and yielded equivocal responses in all tests, or it was tested more than once in equivalent test protocols and yielded contradictory results.

This Table does not contain a complete listing of NTP genetic toxicity test results. For some chemicals, subsequent tests, unpublished at the time of this table, yielded contradictory results to those presented in the table. These are not included in the table. The publications referred to in the table contain the data, in addition to the summary conclusions for each test.

TABLE 1
Test Results

Chemical	CAS	SAL res	Ref.	ABS res	Ref.	SCE res	Ref.	MLA res	Ref.	DRO res	Ref.
Acenaphthene	83-32-9	–	46								
Acetaldehyde	75-07-0	–	29							+	42
Acetamide	60-35-5	–	16							–	37
Acetanilide	102-01-2	–	45								
Acetic acid	64-19-7	–	46								
Acetic anhydride	108-24-7	–	29								
Acetin	26446-35-5	+	16	–	19	+	19			+	43
Acetoacetanilide	102-01-2	–	45								
Acetohexamide	968-81-0	–	9, 46								
Acetone	67-64-1	–	46	–	18	–	18				
Acetonitrile	75-05-8	–	29	?	12	+w	12				
p-Acetophenetidide	62-44-2	–	9, 16								
N-2-Acetylaminofluorene	53-96-3	+	8					+	28, 32		
N-4-Acetylaminofluorene	28322-02-3	+	8					+	26, 28, 32		
N-Acetyl-m-aminophenol	621-42-1	–	45								
Acetyl chloride	75-36-5	–	45								
1-Acetyl-2-phenylhydrazine	114-83-0	+	45								
1-Acetyl-2-picolinoyl hydrazine	17433-31-7	–	16								
4-Acetyl pyridine	1122-54-9	–	46	+	18	+	18				
Acetylsalicylic acid	50-78-2	–	9, 45								
N-Acetyl-m-toluidine	537-92-8	–	46								
N-Acetyl-o-toluidine	120-66-1	+	16								
N-Acetyl-p-toluidine	103-89-9	+w	46								
C.I. Acid orange 3	6373-74-6	+	45	+	N-2	+	N-2				
C.I. Acid orange 10	1936-15-8	–	45	+	17	+	17	+	26	–	11
C.I. Acid red 14	3567-69-9	–	29	–	15	?	15	–	24	–	11
C.I. Acid red 114	6459-94-5	+	29							–	42
Acrolein	107-02-8	+w	16	–	12	+w	12			–	51
Acrylamide	79-06-1	?	47							–	11
Acrylic acid	79-10-7	–	47								
Acrylonitrile	107-13-1	+	48	+	14	+	14	+	31	–	10

TABLE 1 (continued)
Test Results

Chemical	CAS	SAL res	Ref.	ABS res	Ref.	SCE res	Ref.	MLA res	Ref.	DRO res	Ref.
Adipamide	628-94-4	−	16								
Adiponitrile	111-69-3	−	45								
AF-2	3688-53-7	+	16	+	13	+	13			+	37
Aldicarb	116-06-3	−	9, 45					+	28, 32		
Aldicarb oxime	1646-75-9	−	46	−	18	−	18				
Aldrin	309-00-2	−	46								
Alloxan monohydrate	3237-50-1	−	46	−	19	+	19			+	37
Allyl acrylate	999-55-3	−	47								
Allylamine	107-11-9	−	47								
Allyl anthranilate	7493-63-2	−	29								
Allyl glycidyl ether	106-92-3	+	3	+	N-3	+	N-3			+	43
Allyl isothiocyanate	57-06-7	?	29	+	12	+	12	+	24	−	37
Allyl isovalerate	2835-39-4	−	29	+	15	+	15			−	42
Allyl propyl disulfide	2179-59-1	−	45								
Allylthiourea	109-57-9	−	45								
Allyl urea	557-11-9	+	46								
Amiben	133-90-4	+	16							?	20, 37
Amiloride HCl	2016-88-8	+	46								
2-Aminoacetanilide HCl	4801-39-2	+	45								
m-Aminoacetanilide	102-28-3	+	45								
p-Aminoacetanilide	122-80-5	+	45								
9-Aminoacridine	90-45-9	+	8								
9-Aminoacridine HCl H$_2$O	52417-22-8	+	29							−	38, 43
2-Aminoanthracene	613-13-8	+	16							+	42
2-Aminoanthraquinone	117-79-3	+	9								
p-Aminoazobenzene	60-09-3	+	46								
o-Aminoazotoluene	97-56-3	+	46								
m-Amino benzenesulfonic acid	121-47-1	−	45								
o-Amino benzenesulfonic acid	88-21-1	+w	45								
p-Amino benzenesulfonic acid	121-57-3	−	45								
2-Aminobenzimidazole	934-32-7	−	46								
p-Amino benzoic acid	150-13-0	−	29								
2-Aminobiphenyl	90-41-5	+	16							−	37
2-Aminobiphenyl HCl	2185-92-4	+	45	+	15	−	15	+	24		
4-Aminobiphenyl	92-67-1	+	16							−	51
2-Amino-4-chloro-5-nitrophenol	6358-07-2	+	45								
2-Amino-6-chloro-4-nitrophenol HCl	62625-14-3	+	45								
2-Amino-4-chlorophenol	95-85-2	+w	45								
6-Amino-4-chloro-1-phenol-2-sulfonic acid	88-23-3	−	45								
1-Amino-2,4-dibromoanthraquinone	81-49-2	+	16	+w	18, N-4	+	18, N-4				
2-Amino-4,6-dinitrophenol	96-91-3	+	45								
3-Amino-4-ethoxyacetanilide	17026-81-2	+	9, 16							−	37
3-Amino-9-ethylcarbazole HCl	6109-97-3	+	16								
4-Amino-4′-hydroxy-3-methyldiphenylamine	6219-89-2	+	46								
1-Amino-2-methylanthraquinone	82-28-0	+	45								
2-Amino-4-methylphenol	95-84-1	+	45								
3-Amino-6-methylphenol	2835-95-2	+	45								
2-Amino-4-(methylsulfonyl)phenol	98-30-6	−	46								
2-Amino-6-nitrobenzothiazole	6285-57-0	+	29								
2-Amino-4-nitrophenol	99-57-0	+	47	+	1	+	1	+	32, 34		
2-Amino-5-nitrophenol	121-88-0	+	47	+	1	+	1	+	32, 34		

TABLE 1 (continued)
Test Results

Chemical	CAS	SAL res	Ref.	ABS res	Ref.	SCE res	Ref.	MLA res	Ref.	DRO res	Ref.
4-Amino-2-nitrophenol	119-34-6	+	9, 47					+	28, 32		
6-Amino-4-nitro-1-phenol-2-sulfonic acid	96-67-3	−	45								
2-Amino-5-nitrothiazole	121-66-4	+	9, 46					+	28, 32		
m-Aminophenol	591-27-5	+	45								
o-Aminophenol	95-55-6	+	16							−	43
p-Aminophenol	123-30-8	−	45								
2-Amino-1-phenol-4-sulfonic acid	98-37-3	−	45								
o-Aminophthalyl hydrazide	521-31-3	−	46								
2-Aminopyridine	504-29-0	−	47								
5-Aminosalicylic acid	89-57-6			−	39	−	39				
5-Amino-3-sulfosalicylic acid	6201-87-2	−	45								
3-Amino-1,2,4-triazole	61-82-5	−	8, 45					−	25, 28, 32	−	20, 42
3-Amino-α,α,α-trifluorotolune	98-16-8	−	16								
11-Aminoundecanoic acid	2432-99-7	−	29	−	12	+	12	−	24	−	43
Amphetamine sulfate	60-13-9	?	47	−	N-5	−	N-5				
Amphotericin B	1397-89-3	−	46	−	18	+	18				
Ampicillin trihydrate	7177-48-2	−	29	−	1	−	1	−	34		
Amsco solvent F	——	−	46								
n-Amylamine	110-58-7	−	29								
n-Amyl nitrite	463-04-7	+	29								
Anethole	104-46-1	−	29								
Anilazine	101-05-3	−	9, 45					+	28, 32		
Aniline	62-53-3	−	16	+w	12	+	12	+	26, 28, 32	−	43
Aniline HCl	142-04-1	−	9								
m-Anisidine	536-90-3	+	16, 46	+	12	+	12				
o-Anisidine	90-04-0	+	16, 46	+	12	+	12			−	43
o-Anisidine HCl	134-29-2	+	9								
p-Anisidine	104-94-9	+	16, 46	+	12	+	12				
p-Anisidine HCl	20265-97-8	+	9								
Anthracene	120-12-7	+w	8, 29					+	28, 32		
Anthralin	1143-38-0	−	45								
o-Anthranilic acid	118-92-3	−	47					+	28, 32		
Anthraquinone	84-65-1	+	45								
Antimony potassium tartrate	28300-74-5	−	46								
Aroclor 1254	11097-69-1	−	8, 45								
L-Ascorbic acid	50-81-7	?	45	−	15	+	15	?	33	−	10
Aspirin, Phenacetin, Caffeine	8003-03-0	−	9								
Atrazine	1912-24-9	−	45								
Auramine	2465-27-2	+	46								
5-Azacytidine	320-67-2	+	46								
Azathioprine	446-86-6	+	46								
Azinphosmethyl (Gusathion)	86-50-0	+w	47								
1-Aziridine ethanol	1072-52-2	+	16							+	37
Azobenzene	103-33-3	+	16	−	13	+	13			−	52
Azodicarbonamide	123-77-3	+	29							−	43
p-Azoxyanisole	1562-94-3	+	46								
Azoxybenzene	495-48-7	+	46								
Azoxymethane	25843-45-2	+w	8, 46								
Barium chloride dihydrate	10326-27-9	−	46								
C.I. Basic orange 2	532-82-1	+	47							−	10
C.I. Basic red 9	569-61-9	+	8, 29	−	1	−	1	+w	28, 32, 34		

TABLE 1 (continued)
Test Results

Chemical	CAS	SAL res	Ref.	ABS res	Ref.	SCE res	Ref.	MLA res	Ref.	DRO res	Ref.
C.I. Basic violet 14	632-99-5	+	29								
Benomyl	17804-35-2	−	45								
Benzaldehyde	100-52-7	−	16	−	12	+	12	+	26	−	42
Benzamide	55-21-0	−	45								
Benz(a)anthracene	56-55-3	+	8					+	28, 32		
Benzene	71-43-2	−	48	−	14, 15	+	14, 15	−	31	−	11
Benzene sulfonic acid	98-11-3	−	45								
Benzethonium chloride	121-54-0	−	47	−	N-6	−	N-6				
Benzidine	92-87-5	+	16, 35	+	12	+	12				
Benzidine 2HCl	531-85-1	+	46					+	28, 32		
Benzimidazole	51-17-2	−	46								
2-Benzimidazolylurea	24370-25-0	+	46								
Benzo(b)fluoranthene	205-99-2	+	46								
Benzo(k)fluoranthene	207-08-9	+	46								
2,3-Benzofuran	271-89-6	−	16	−	N-7	+	N-7	+	23		
Benzoic acid	65-85-0	−	45								
Benzoin	119-53-9	−	29, 48	−	14, 15	?	14, 15	+	31	−	11
Benzonitrile	100-47-0	−	45								
Benzophenone	119-61-9	−	29								
Benzo(a)pyrene	50-32-8	+	8, 16					+	28, 32	−	38
Benzo(e)pyrene	192-97-2	+	8, 46					+	28, 32		
Benzo(f)quinoline	85-02-9	+	29								
1,2,3-Benzotriazole	95-14-7	+w	9, 47								
Benzotrichloride	98-07-7	+	45								
Benzotrifluoride	98-08-8	−	45								
2-[(Benzoyloxy)methyl]-2-methyl dibenzoate-1,3-propanediol	4196-87-6	−	45								
Benzoyl peroxide	94-36-0	−	45								
Benzyl acetate	140-11-4	−	29	−	12	−	12	+	24	−	11
Benzyl alcohol	100-51-6	−	29	+	1	+w	1	?	24	−	11
Benzyl chloride	100-44-7	+w	47					+	23		
o-Benzyl-p-chlorophenol	120-32-1	−	29	−	N-8	−	N-8				
Benzyldimethyl (mixed alkyl) ammonium chloride	63449-41-2	−	45								
Benzyl salicylate	118-58-1	−	47								
Benzyl sulfide	538-74-9	−	46								
Benzyltrimethyl ammonium chloride	56-93-9	−	45								
Benzyl violet 4B	1694-09-3	+	46								
Beryllium sulfate 4H$_2$O	7787-56-6	−	8								
Biphenyl	92-52-4	−	16								
2,4-Bis(p-aminobenzyl)aniline	25834-80-4	+	46								
2,2-Bis[(benzoyloxy)methyl]dibenzoate propanediol	4196-86-5	−	46								
2,2-Bis(bromomethyl)-1,3-propanediol	3296-90-0	+	29, 46	+	12	?	12				
Bis(t-butyldioxyisopropyl)benzene	25255-25-3	−	45								
Bis(2-chloroethyl)ether	111-44-4	+w	29							+	10
Bis(2-chloro-1-methylethyl)ether	108-60-1	+	29	+	12	+	12	+	24	?	20, 37
Bis(1,5-cyclooctadiene) nickel	1295-35-8	+	46								
Bis(cyclopentadienyl) chromium	1271-24-5	+	46								
Bis(cyclopentadienyl) vanadium chloride	12083-48-6	+	46								
4,4′-Bis(dimethylamino)benzophenone	90-94-8	+	9, 46					+	28, 32		

TABLE 1 (continued)
Test Results

Chemical	CAS	SAL res	Ref.	ABS res	Ref.	SCE res	Ref.	MLA res	Ref.	DRO res	Ref.
N,N′-Bis(1,4-dimethylpentyl)-p-phenylenediamine	3081-14-9	–	47								
Bis(2-ethylhexyl)isophthalate	137-89-3	–	45								
Bis(2-methoxyethyl)ether	111-96-6	–	29								
Bismuth subsalicylate	14882-18-9	–	47								
Bisphenol A	80-05-7	–	16	–	17	–	17	–	33	–	11
Bisphenol A diglycidyl ether	1675-54-3	+	3								
1,2-Bis(tetrabromophthalimido)ethane	32588-76-4	–	49								
1,2-Bis(2,4,6-tribromophenoxy)ethane	37853-59-1	–	47								
Biurea	110-21-4	–	46								
Black Newsprint Ink (Letterpress)	—	+	T-1								
Black Newsprint Ink (Offset)	—	+	T-1								
Boric acid	10043-35-3	–	16	–	N-9	–	N-9	–	24		
N-Bromoacetamide	79-15-2	–	45								
Bromoacetonitrile	590-17-0	–	29								
p-Bromoaniline	106-40-1	–	45								
Bromobenzene	108-86-1	–	16	–	12	+w	12				
2-Bromobiphenyl	2052-07-5	–	16								
3-Bromobiphenyl	2113-57-7	–	16								
4-Bromobiphenyl	92-66-0	–	16								
3-Bromo-2,2-bis(bromomethyl)-1-propanol	1522-92-5	+	29								
Bromochloromethane	74-97-5	+	46								
Bromodichloromethane	75-27-4	–	29	–	1	?	1	+	23		
2-Bromo-4,6-dinitroaniline	1817-73-8	+	47								
2-Bromo-1-ethanol	540-51-2	+	45								
2-Bromoethyl acrylate	4823-47-6	+	47								
α-Bromo-α-ethylbutyrylurea	77-65-6	–	16								
Bromoform	75-25-2	+	16, 44	–	1, 13	?	1, 13	+	34	+	42
7-Bromomethyl-12-methylbenz(a)anthracene	16238-56-5	+	8					+	28, 32		
β-Bromo-β-nitrostyrene	7166-19-0	+	46								
β-Bromo-β-nitrostyrene (65% Amsco solvent F)	——	+	46								
Bromopicrin	464-10-8	+w	46								
1-Bromo-2-propanol	19686-73-8	+	46								
3-Bromo-1-propanol	627-18-9	+	46								
α-Bromotoluene	100-39-0	+	46								
m-Bromotoluene	591-17-3	–	46								
o-Bromotoluene	95-46-5	–	46								
p-Bromotoluene	106-38-7	–	46								
Brucine	357-57-3	–	47								
1,3-Butadiene	106-99-0	+	N-10					–	27	–	10
n-Butane	106-97-8									–	11
1,4-Butanediol	110-63-4	–	46								
1,4-Butanediol diglycidyl ether	2425-79-8	+	3							+	10
2,3-Butanedione-2-oxime	57-71-6	–	29								
2-Butanone peroxide	1338-23-4	–	29	+	T-13	+	T-13	+	T-13		
2-Butoxyethanol	111-76-2	–	46	–	T-9	–	T-9				
2-(2-Butoxyethoxy)ethanol	112-34-5	–	46								
2-(2-Butoxyethoxy) ethyl thiocyanate	112-56-1	–	46								
n-Butyl acetate	123-86-4	–	46								
n-Butyl acrylate	141-32-2	–	47							–	11

TABLE 1 (continued)
Test Results

Chemical	CAS	SAL res	Ref.	ABS res	Ref.	SCE res	Ref.	MLA res	Ref.	DRO res	Ref.
t-Butyl alcohol	75-65-0	–	47	–	N-11	–	N-11	–	23		
n-Butylamine	109-73-9	–	47								
sec-Butylamine	13952-84-6	–	47								
tert-Butylamine	75-64-9	–	47								
n-Butyl-*p*-aminobenzoate	94-25-7	–	16								
Butyl anthranilate	7756-96-9	–	47								
Butylated hydroxyanisole	25013-16-5	–	46								
Butylated hydroxytoluene	128-37-0	–	29	–	12	–	12				
Butyl benzyl phthalate	85-68-7	–	49	–	12	–	12	–	23	–	37
n-Butyl chloride	109-69-3	–	44, 47	–	1	–	1				
Butyl cyclohexyl phthalate	84-64-0	–	49								
n-Butyl glycidyl ether	2426-08-6	+	3								
t-Butyl glycidyl ether	7665-72-7	+	3								
t-Butyl hydroperoxide	75-91-2	+	16							+	42
t-Butyl hydroquinone	1948-33-0	–	46								
Butyl methacrylate	97-88-1	–	47								
2-Butyloxybenzal-4-ethylaniline	29743-15-5	+w	46								
t-Butyl perbenzoate	614-45-9	+	29	+	T-2	+	T-2				
o-sec-Butylphenol	89-72-5	–	29								
t-Butylphenyl diphenyl phosphate (mixed isomers)	56803-37-3	–	47								
t-Butyl phenyl glycidyl ether	3101-60-8	+	3								
Butyltin-tris(isooctylmercaptoacetate)	25852-70-4	–	45								
p-*tert*-Butyl toluene	98-51-1	–	47								
Butyraldehyde	123-72-8	–	29	–	12	+	12			–	20, 37
4-Butyrolactone	96-48-0	–	16	+	19	+	19			–	11
β-Butyrolactone	36536-46-6	+	46								
Butyryl chloride	141-75-3	+w	46								
Cacodylic acid	75-60-5	–	16								
Cadinene	29350-73-0	+	45								
β-Cadinene	523-47-7	–	16	–	12	?	12				
Cadmium chloride	10108-64-2	–	29					+	24		
Cadmium oxide	1306-19-0	–	29								
Caffeine	58-08-2	–	9, 29								
Calcium chromate, anhydrous	13765-19-0	+	8, 16					+	25, 28, 32	+	51
Calcium cyanamide	156-62-7	+	16							–	43
Calcium metasilicate (Wollastonite)	13983-17-0	–	47								
Caprolactam	105-60-2	–	48	–	14, 15	–	14, 15	–	31	–	10
Caprylyl chloride	111-64-8	+	46								
Captan	133-06-2	+	46								
Carbendazim	10605-21-7	+	47								
Carbon disulfide	75-15-0	–	16								
Carbon tetrachloride	56-23-5	–	45	–	18	–	18			–	11
Carisoprodol	78-44-4	–	47								
Carveol	99-48-9	–	29								
d-Carvone	2244-16-8	–	29	+	N-12	+	N-12				
Carvyl acetate	97-42-7	–	29								
Castor oil	8001-79-4	–	45								
Catechol	120-80-9	–	16								
Chloral, anhydrous	75-87-6	+w	45								
Chloral hydrate	302-17-0	+	16							?	20, 43

TABLE 1 (continued)
Test Results

Chemical	CAS	SAL res	Ref.	ABS res	Ref.	SCE res	Ref.	MLA res	Ref.	DRO res	Ref.
Chlorambucil	305-03-3	+	29							+	42
Chloramphenicol, Na succinate	982-57-0	−	29					+	28, 32	−	10
Chlordane	57-74-9	−	29								
Chlordane, Technical grade	12789-03-6	+	46					+	24		
Chlordecone (Kepone)	143-50-0	−	29	−	12	+	12				
Chlordecone alcohol	1034-41-9	−	29	?	12	−	12				
Chlorendic acid	115-28-6	−	16	?	1	+	1	+	23	−	11
Chlorinated trisodium phosphate	56802-99-4	+w	45	+	19	+	19				
Chloroacetic acid	79-11-8	−	29	−	12	+	12	+	25	?	10
Chloroacetonitrile	107-14-2	−	29								
Chloroacetophenone	532-27-4	−	47	+w	N-13	−	N-13				
4′-(Chloroacetyl)acetanilide	140-49-8	+	9, 46								
N-(3-Chloroallyl) hexaminium chloride	4080-31-3	+w	45								
m-Chloroaniline	108-42-9	−	47								
o-Chloroaniline	95-51-2	−	47					+	26		
p-Chloroaniline	106-47-8	+	9, 29, 44	+	1	+	1	+	28, 32		
2-Chlorobenzaldehyde	89-98-5	−	46								
4-Chlorobenzaldehyde	104-88-1	−	46								
o-Chlorobenzalmalononitrile	2698-41-1	?	47	+	N-14	+	N-14	+	23		
Chlorobenzene	108-90-7	−	16	−	19	+	19	+	24	−	11
Chlorobenzilate	510-15-6	−	29	−	12	−	12	+	24	−	43
m-Chlorobenzoic acid	535-80-8	−	46								
o-Chlorobenzoic acid	118-91-2	−	46								
p-Chlorobenzoic acid	74-11-3	−	46								
m-Chlorobenzotrifluoride	98-15-7	−	46								
o-Chlorobenzotrifluoride	88-16-4	−	46								
2-Chloro-1,3-butadiene (Chloroprene)	126-99-8	−	47							−	11
p-Chloro-m-cresol	59-50-7	−	46								
p-Chloro-o-cresol	1570-64-5	−	46								
Chlorodibromomethane	124-48-1	−	47	−	18	+	18	+	26	−	11
4-Chloro-3,5-dinitro-α,α,α-trifluorotoluene	393-75-9	−	16								
2-Chloroethanol	107-07-3	+	16	+	17	+	17	+	24	−	37
2-Chloroethyl acrylate	2206-89-5	+	47								
2-Chloroethyl trimethyl ammonium Cl	999-81-5	−	29								
Chloroform	67-66-3							+	28, 32		
3-Chloro-2-methylpropene	563-47-3	−	16, 45	+	15	+	15	+	33	+	11
2-(Chloromethyl)pyridine HCl	6959-47-3	+	29							−	11, 20, 37
3-(Chloromethyl)pyridine HCl	6959-48-4	+	9, 29					+	25, 28, 32	?	11, 42
1-Chloronaphthalene	90-13-1	+w	46								
2-Chloronaphthalene	91-58-7	−	46								
Chloroneb	2675-77-6	+	46							−	42
4-Chloro-2-nitroaniline	89-63-4	+	16	+	12	+	12				
2-Chloronitrobenzene	88-73-3	+	16	−, +	T-3	+	T-3			−	51
3-Chloronitrobenzene	121-73-3	?	16			+	12	?	12		
4-Chloronitrobenzene	100-00-5	+	16	+	12	+	12			−	51
1-Chloro-1-nitropropane	600-25-9	+	46								
2-Chloro-2-nitropropane	594-71-8	+w	46								
4-Chloro-3-nitro-α,α,α-trifluorotoluene	121-17-5	−	16								
m-Chlorophenol	108-43-0	−	16								
o-Chlorophenol	95-57-8	−	16								
p-Chlorophenol	106-48-9	−	16								

TABLE 1 (continued)
Test Results

Chemical	CAS	SAL res	Ref.	ABS res	Ref.	SCE res	Ref.	MLA res	Ref.	DRO res	Ref.
2-Chloro-*p*-phenylenediamine SO$_4$	61702-44-1	+	16								
4-Chloro-*m*-phenylenediamine	5131-60-2	+	9, 16	+	18	+	18				
4-Chloro-*o*-phenylenediamine	95-83-0	+	9, 45	+	18	+	18				
Chloropicrin	76-06-2	+	16							?	20, 37
3-Chloro-1,2-propanediol	96-24-2	+	45								
1-Chloro-2-propanol	127-00-4	+	47							+	10
3-Chloro-1-propanol	627-30-5	+	47								
o-Chlorostyrene	2039-87-4	–	45								
Chlorothalonil	1897-45-6	–	29	+	12	+	12	+	24	–	43
Chlorothiazide	58-94-6	–	46								
p-Chlorotoluene	106-43-4	–	46								
3-Chloro-*p*-toluidine	95-74-9	–	16								
4-Chloro-*o*-toluidine HCl	3165-93-3	–	16	+w	12	+	12				
5-Chloro-*o*-toluidine	95-79-4	–	16	–	12	–	12	–	24		
Chlorotrianisene	569-57-3	–	47								
2-Chloro-6-(trichloromethyl)pyridine	1929-82-4	+	45								
4-Chloro-α,α,α-trifluorotoluene	98-56-6	–	16								
Chlorotrimethyl silane	75-77-4	?	29, 46								
Chlorowax 40	108171-27-3	–	47								
Chlorowax 500c	108171-26-2	–	47	+	1	+	1				
Chlorpheniramine maleate	113-92-8	–	29	+	1	+	1	–	26		
Chlorpromazine HCl	69-09-0	–	29	–	12	–	12				
Chlorpropamide	94-20-2	–	47	–	19	+	19	–	26		
Choline chloride	67-48-1	–	16	?	13	+w	13				
Chromium carbonyl	13007-92-6	–	46	–	36						
1,8-Cineol (Eucalyptol)	470-82-6	–	16	–	12	+	12				
trans-Cinnamaldehyde	14371-10-9	–	29	–	12	+	12			+	42
Cinnamyl anthranilate	87-29-6	–	9, 45	–	15	–	15	+	33	–	11
Citral	5392-40-5	–	47								
Citral diethyl acetal	7492-66-2	–	45, 46								
Citrus red 2	6358-53-8	+	46								
Cobaltocene	1277-43-6	+	46								
Cobalt sulfate heptahydrate	10026-24-1	+w	46								
Coconut oil acid/Diethanolamine condensate (2/1)	68603-42-9	–	45								
Codeine phosphate	52-28-8	–	46								
Colchicine	64-86-8	–	29	+	12	–	12			–	37
Copper acetoarsenite	12002-03-8	–	47								
Coumaphos	56-72-4	–	9, 46	–	19	–	19	–	24		
Coumarin	91-64-5	+	16	+w	12	+	12			–	38, 43
Counter 15G	13071-79-9	–	47								
Creosote, coal tar	8001-58-9	+	46								
Creosote, wood	8021-39-4	–	46								
m-Cresidine	102-50-1	+	9, 45								
o-Cresidine	16452-01-0	+	45								
p-Cresidine	120-71-8	+	9, 45								
m-Cresol	108-39-4	–	16								
m,p-Cresol mixture	1319-77-3	–	46								
o-Cresol	95-48-7	–	16								
p-Cresol	106-44-5	–	16								
Cresyl diphenyl phosphate, mixed isomers	26444-49-5	–	47								

TABLE 1 (continued)
Test Results

Chemical	CAS	SAL res	Ref.	ABS res	Ref.	SCE res	Ref.	MLA res	Ref.	DRO res	Ref.
o-Cresyl glycidyl ether	2210-79-9	+	3								
p-Cresyl glycidyl ether	2186-24-5	+	3								
Crotonaldehyde	4170-30-3	+	16	+	12	+	12			+	42
Croton oil	8001-28-3	−	29	−	12	−	12			−	37
Cumene hydroperoxide	80-15-9	+	29								
Cupferron	135-20-6	+	9, 46								
Curcumin	458-37-7	−	29								
Cyanuric acid	108-80-5	−	16								
Cyclohexane	110-82-7	−	29								
1,2-Cyclohexanedicarboxylic acid, bis(oxiranylmethyl) ester	5493-45-8	+	46								
Cyclohexanol	108-93-0	?	16								
Cyclohexanone	108-94-1	−	16					−	24		
Cyclohexanone cyanohydrin	931-97-5	−	45								
Cyclohexanone oxime	100-64-1	+	T-4	?	T-4						
Cycloheximide	66-81-9	−	46								
Cyclohexylamine	108-91-8	−	29								
Cyclohexyl anthranilate	7779-16-0	−	46								
n-Cyclohexyl-4-methyl benzenesulfonamide	80-30-8	−	46								
4-H-Cyclopenta(d,e,f)phenanthrene	203-64-5	+	46								
Cyclophosphamide	50-18-0							+	28, 32		
Cyclophosphamide monohydrate	6055-19-2	+	16							+	43
Cytarabine HCl	69-74-9	−	47					+	24		
Cytembena	21739-91-3	+	45	+	17	+	17	+	26	−	10
Cytoxal alcohol	4465-94-5	+	29								
2,4-D (2,4-Dichlorophenoxyacetic acid)	94-75-7	−	30	+	12	+	12			−	51
2,4-D, n-Butyl ester	94-80-4	−	30								
2,4-D, Dimethylamine salt	2008-39-1	−	30								
2,4-D, Isooctyl ester, 67%	25168-26-7	−	30								
Dacarbazine	4342-03-4	+	45								
Dammar resin	9000-16-2	−	45								
Dapsone	80-08-0	−	9, 16					−	24		
D & C Green 5	4403-90-1	+w	35								
D & C Red 9	5160-02-1	+w	45	−	17	−	17	−	33	−	11
D & C Yellow 11	8003-22-3	+w	45	+	N-15	+	N-15				
o,p′-DDD	53-19-0	−	29	−	12	−	12				
p,p′-DDE	72-55-9	−	29	−	12	+w	12	+	24	+	37
p,p′-DDT	50-29-3	−	46								
Decabromodiphenyl oxide	1163-19-5	−	16	−	1	−	1	−	24		
n-Decane	124-18-5	−	46								
Decanoic acid	334-48-5	−	45								
n-Decyl methacrylate	3179-47-3	−	47								
Diacetone acrylamide	2873-97-4	−	16								
Diallylamine	124-02-7	−	47								
Diallyl phthalate	131-17-9	−	49	+	15	+	15	+	33	−	37
2,4-Diaminoanisole SO4, trihydrate	39156-41-7	+	9, 45					+	28, 32		
4,4′-Diaminodicyclohexylmethane	1761-71-3	−	47								
1,6-Diaminohexane	124-09-4	−	29	−	T-10	−	T-10				
Diaminomaleonitrile	1187-42-4	−	45								
2,4-Diaminophenol 2HCl	137-09-7	+	16	−	12	−	12	+	25	?	20, 51
1,3-Diaminopropane	109-76-2	−	47								

TABLE 1 (continued)
Test Results

Chemical	CAS	SAL res	Ref.	ABS res	Ref.	SCE res	Ref.	MLA res	Ref.	DRO res	Ref.
4,4'-Diamino-2,2'-stilbenedisulfonic acid	81-11-8	–	47	–	18	–	18				
2,4-Diaminotoluene	95-80-7	+	16	+	18	+	18				
2,5-Diaminotoluene sulfate	6369-59-1	+	9, 45								
2,6-Diaminotoluene 2HCl	15481-70-6	+	45	+	15, 36	+	15	+	33	–	11
3,4-Diaminotoluene	496-72-0	+	45								
Di-n-amylamine	2050-92-2	–	29								
Diazepam	439-14-5	–	46								
Diazinon	333-41-5	–	9, 45					+	24		
Diazoaminobenzene	136-35-6	+	47								
Dibenz(a,h)anthracene	53-70-3	+	46								
Dibenzofuran	132-64-9	–	30	–	12	+w	12				
Dibenzooxathiane	262-20-4	–	45								
Diborane	19287-45-7	–	47								
Dibromoacetonitrile	3252-43-5	+w	29							–	37
trans-2,3-Dibromo-2-butene-1,4-diol	3234-02-4	+	47							–	10
1,2-Dibromo-3-chloropropane	96-12-8	+	45	+	19	+	19	+	33	+	43
Dibromodulcitol	10318-26-0	+	29								
1,2-Dibromoethane	106-93-4	+	9, 46	+	17	+	17	+	28, 32	+	11
1-(1,2-Dibromoethyl)-3,4-dibromocyclohexane	3322-93-8	–	47	–	19, 36	+	19	+	26		
Dibromomannitol	488-41-5	+	29							+	11
2,3-Dibromo-1-propanol	96-13-9	+	16	+	N-16	+	N-16	+	N-16	+	43
2,3-Dibromopropyl acrylate	19660-16-3	+	47								
2,3-Dibromopropyl methacrylate	3066-70-4	+	47								
Di-n-butylamine	111-92-2	–	29								
2-Di-n-butylaminoethanol	102-81-8	–	47								
Di-n-butyl ether	142-96-1	–	46								
Di-tert-butyl peroxide	110-05-4	–	45								
N,N'-Di-sec-butyl-p-phenylenediamine	101-96-2	–	45	–	36						
Dibutyl phenyl phosphate, mixed isomers	2528-36-1	–	45								
Dibutyl phthalate	84-74-2	–	49					+	T-5		
Dibutyltin-bis(laurylmercaptide)	1185-81-5	–	47								
Dibutyltin diacetate	1067-33-0	–	47							–	20, 42
Dibutyltin dilaurate	77-58-7	–	47								
Dichloran (2,6-Dichloro-4-nitroaniline)	99-30-9	+	46							?	42
Dichloroacetonitrile	3018-12-0	+	29							+	37
Dichloroacetyl chloride	79-36-7	+	46								
2,4-Dichloroaniline	554-00-7	+w	46								
2,5-Dichloroaniline	95-82-9	–	46								
3,4-Dichloroaniline	95-76-1	–	46								
2,4-Dichlorobenzaldehyde	874-42-0	–	16								
2,6-Dichlorobenzaldehyde	83-38-5	–	16								
3,4-Dichlorobenzaldehyde	6287-38-3	–	16								
1,2-Dichlorobenzene	95-50-1	–	16	–	18	+	18	+	33	–	10
1,3-Dichlorobenzene	541-73-1	–	16								
1,4-Dichlorobenzene	106-46-7	–	16	–	1, 12	–	1, 12	?	24, 34		
3,3'-Dichlorobenzidine 2HCl	612-83-9	+	16, 35								
cis-Dichlorodiaminoplatinum II	15663-27-1	+	46							+	11
2,7-Dichlorodibenzo-p-dioxin	33857-26-0	–	30								
1,3-Dichloro-5,5-dimethylhydantoin	118-52-5	–	16	–	12	–	12			+	42
1,1-Dichloroethane	75-34-3	–	46								

TABLE 1 (continued)
Test Results

Chemical	CAS	SAL res	Ref.	ABS res	Ref.	SCE res	Ref.	MLA res	Ref.	DRO res	Ref.
1,2-Dichloroethane	107-06-2	+	46								
cis-1,2-Dichloroethylene	156-59-2	–	45								
cis,trans-1,2-Dichloroethylene	540-59-0	–	29								
trans-1,2-Dichloroethylene	156-60-5	–	29								
Dichloroisocyanuric acid	2782-57-2	–	47								
Dichloroisocyanuric acid, Na salt	2893-78-9	–	47								
Dichloromethane	75-09-2	+	7, 44	–	1	–	1	?	34		
Dichloromethotrexate	528-74-5	–	16, 45								
2,3-Dichloronitrobenzene	3209-22-1	+w	45							–	43
2,4-Dichloronitrobenzene	611-06-3	+	45							–	43
3,4-Dichloronitrobenzene	99-54-7	+	45							+	42
1,1-Dichloro-1-nitroethane	594-72-9	+	46								
2,3-Dichlorophenol	576-24-9	–	16								
2,4-Dichlorophenol	120-83-2	?	16	–	1	+	1	+	34		
2,5-Dichlorophenol	583-78-8	–	16								
2,6-Dichlorophenol	87-65-0	–	16								
3,4-Dichlorophenol	95-77-2	–	16								
3,5-Dichlorophenol	5691-35-5	?	16								
2,6-Dichloro-p-phenylenediamine	609-20-1	+	29	+	15	+	15	+	24	–	11
1,2-Dichloropropane	78-87-5	+	16	+	12	+	12	+	33	–	42
1,3-Dichloro-2-propanol	96-23-1	+	45								
2,3-Dichloro-1-propanol	616-23-9	+	45								
1,3-Dichloropropene	542-75-6	+	16	–	19	+	19	+	33	+	37
2,3-Dichloro-1-propene	78-88-6	+	45	+	18	+	18				
2,3-Dichloroquinoxaline	2213-63-0	–	47								
α,α-Dichlorotoluene	98-87-3	+	46								
Dichlorvos	62-73-7	+	45	+	1	+	1	+	34		
Dicofol	115-32-2	–	29	–	12	–	12			–	42
Dicumyl peroxide	80-43-3	–	45								
Dicyclohexylamine	101-83-7	–	29								
Dicyclohexylamine nitrite	3129-91-7	+	45								
Dicyclohexyl phthalate	84-61-7		49								
N,N'-Dicyclohexylthiourea	1212-29-9	–	9, 45					–	25, 28, 32		
Dicyclopentadiene	77-73-6	–	47								
Dieldrin	60-57-1	–	16	–	12	+	12	+	26		
D,L-1,2:3,4-Diepoxybutane	298-18-0	+	8					+	23, 28, 32		
Diesel fuel, marine	——	–	47								
Diethanolamine	111-42-2	–	16	–	19	–	19	–	T-6		
Diethylamine	109-89-7	–	47								
Diethylaminoethanol	100-37-8	–	47								
7-Diethylamino-4-methylcoumarin	91-44-1	–	16								
5-Diethylamino-2-nitrosophenol HCl	25953-06-4	–	45								
3-Diethylaminophenol	91-68-9	+	45								
N,N-Diethyl aniline	91-66-7	–	45								
Diethylbutylamine	4444-68-2	–	46								
N,N'-Diethylcarbanilide	85-98-3	–	45								
Diethyl carbonate	105-58-8	–	29								
Diethylene glycol	111-46-6	–	47								
Diethylenetriamine	111-40-0	–	47								
Diethyl ethylphosphonate	78-38-6	–	29								
Di(2-ethylhexyl)adipate (DEHA)	103-23-1	–	49	+w	12	?	12	–	24	–	42

TABLE 1 (continued)
Test Results

Chemical	CAS	SAL res	Ref.	ABS res	Ref.	SCE res	Ref.	MLA res	Ref.	DRO res	Ref.
2,2′-Diethylhexylamine	106-20-7	–	45								
Di(2-ethylhexyl)phthalate (DEHP)	117-81-7	–	48, 49	–	14, 15	?	14, 15	–	31	–	43
Di(2-ethylhexyl)sebacate	122-62-3	–	49								
N,N-Diethyl-p-phenylenediamine	93-05-0	+	45	+	36						
Diethyl phthalate	84-66-2	–	49								
Diethylstilbestrol	56-53-1	–	8, 48	+	14	–	14	+	28, 31, 32		
N,N′-Diethylthiourea	105-55-5	–	29					+	24	–	37
N,N-Diethyl-m-toluamide	134-62-3	–	46								
2,4-Difluoroaniline	367-25-9	+	45								
Diglycidyl resorcinol ether	101-90-6	+	3	+	15	+	15	+	24	+	37
Dihexylamine	143-16-8	–	45								
Di(n-hexyl)phthalate	84-75-3	–	49								
3,4-Dihydrocoumarin	119-84-6	–	16								
1,2-Dihydro-2,2,4-trimethylquinoline, monomer	147-47-7	–	47								
1,8-Dihydroxy-4,5-dinitroanthraquinone	81-55-0	+	47							?	20, 42
2,2′-Dihydroxy-4-methoxybenzophenone	131-53-3	+	29								
5,7-Dihydroxy-4-methylcoumarin	2107-76-8	–	16								
Diisobutyl amine	110-96-3	–	29								
Diisobutyl ketone	108-83-8	–	29								
Diisobutyl phthalate	84-69-5	–	49								
Diisodecyl phthalate	26761-40-0	–	49								
Diisononyl phthalate	28553-12-0	–	49								
Diisopropanol amine	110-97-4	–	29								
Diisopropyl amine	108-18-9	–	29								
Dimenhydrinate	523-87-5	+w	47								
Dimethoate	60-51-5	+	16								
Dimethoxane	828-00-2	+	29	+	N-17	+	N-17			+	42
2,4-Dimethoxyaniline HCl	54150-69-5	+	9, 46					+	24		
3,3′-Dimethoxybenzidine	119-90-4	+	16, 35	–	13	+	13			–	43
3,3′-Dimethoxybenzidine 2HCl	20325-40-0							+	28, 32		
3,3′-Dimethoxybenzidine-4,4′-diisocyanate	91-93-0	+	9, 16								
4,4′-Dimethoxydiphenylamine	101-70-2	–	45	+	18, 36	+	18				
Dimethyl acetamide	127-19-5	–	45								
Dimethylamine	124-40-3	–	47								
4-Dimethylaminoazobenzene	60-11-7	+	8, 47					+	28, 32		
2-(Dimethylamino)ethyl acrylate	2439-35-2	–	47								
3-Dimethylaminophenol	99-07-0	+	45								
3-Dimethylaminopropylamine	109-55-7	–	47								
N,N-Dimethylaniline	121-69-7	–	29	+	19	+	19	+	N-18		
7,9-Dimethylbenz(c)acridine	963-89-3	+	8								
7,12-Dimethylbenz(a)anthracene	57-97-6	+	8					+	28, 32	+	38
3,3′-Dimethylbenzidine	119-93-7	+	16, 35	+	12	+	12	+	28, 32		
3,3′-Dimethylbenzidine 2HCl	612-82-8	+	45	+	N-19	+	N-19			+	37
2,5-Dimethyl-2,5-bis(tert-butylperoxy)hexane	78-63-7	–	45								
2,2-Dimethylbutane	75-83-2	–	46								
2,3-Dimethylbutane	79-29-8	–	46								
1,3-Dimethylbutylamine	108-09-8	–	47								
N-(1,3-Dimethylbutyl)-N′-phenyl-p-phenylenediamine	793-24-8	–	47								
Dimethylcarbamoyl chloride	79-44-7	+	8, 16	+	13	+	13	+	28,32	+	11, 43

TABLE 1 (continued)
Test Results

Chemical	CAS	SAL res	Ref.	ABS res	Ref.	SCE res	Ref.	MLA res	Ref.	DRO res	Ref.
Dimethyl cyanamide	1467-79-4	–	16								
Dimethyl ditallow ammonium chloride	68783-78-8	–	46								
Dimethylethanolamine	108-01-0	–	47							–	10
N,N-Dimethylformamide	68-12-2	–	29	–	T-7	–	T-7	–	23, 28, 32	–	T-7
2,5-Dimethylfuran	625-86-5	–	46								
Dimethyl hydrogenphosphite	868-85-9	+w	29	+	15	+	15	+	23	–	42
1,1-Dimethyl-1-(2-hydroxypropylamine)methacrylimide	17341-40-1	–	45								
1,1-Dimethyl-1-(2-hydroxypropylamine)tetradecanimide	38848-76-9	–	45								
Dimethyl methylphosphonate	756-79-6	–	29	–	N-20	+	N-20			+	10
2,6-Dimethyl morpholine	141-91-3	?	16							–	43
Dimethyl morpholinophosphoramidate	597-25-1	–	47	+w	15	+	15	+	33	+	10
N,N-Dimethyl-p-nitrosoaniline	138-89-6	+	29							?	43
Dimethyloctadecyl benzylammonium chloride	122-19-0	–	46								
Dimethyloldihydroxyethylene urea	1854-26-8	?	47							+	10
2,3-Dimethyl-1-pentanol	10143-23-4	–	46								
2,4-Dimethyl phenol	105-67-9	–	29								
N,N-Dimethyl-p-phenylenediamine	99-98-9	+	46	–	36						
Dimethyl phthalate	131-11-3	–	49	–	18	+	18				
Dimethylsuccinate	106-65-0	–	46								
Dimethyl sulfoxide	67-68-5	–	46	–	18	–	18				
Dimethyl terephthalate	120-61-6	–	49	–	18	–	18	–	33	–	11
N,N'-Dimethylurea	96-31-1	–	29					–	4		
2,2-Dimethyl vinyl chloride	513-37-1	+	44, 47	–	1	+	1	+	34	+	10
N,N'-Di-2-naphthyl-p-phenylenediamine	93-46-9	+	46	–	36						
2,4-Dinitroaniline	97-02-9	+	16							–	42
2-(2,4-Dinitroanilino)phenol	6358-23-2	+	45								
4-(2,4-Dinitroanilino)phenol	119-15-3	+	45								
2,5-Dinitro-9H-fluorene	15110-74-4	+	45								
2,7-Dinitro-9H-fluoren-9-one	31551-45-8	+	45								
1,3-Dinitronaphthalene	606-37-1	+	46								
1,5-Dinitronaphthalene	605-71-0	+	46								
1,8-Dinitronaphthalene	602-38-0	+	46								
2,4-Dinitrophenol	51-28-5	–	46								
2,4-Dinitrotoluene	121-14-2	+	8, 16	–	19	+	19			+	20, 42
4,4'-Dioctyldiphenylamine	101-67-7	–	46	–	18, 36	–	18				
Di-octyl phthalate	117-84-0	–	49								
Di(n-octyl)tin-S,S'-bis(isooctylmercaptoacetate)	26401-97-8	+	45								
Di(n-octyl)tin maleate polymers	16091-18-2	–	45								
1,3-Dioxane	505-22-6	+w	16	+	12	+	12			–	51
1,4-Dioxane	123-91-1	–	16	–	12	+w	12	–	26	–	43
Dioxathion	78-34-2	+	29								
Diphenhydramine HCl	147-24-0	–	47	+	19	–	19	–	N-21		
1,3-Diphenylguanidine	102-06-7	+	29								
Diphenylhydantoin	57-41-0	–	16	–	12	+	12	–	N-22	–	42
4,4'-Diphenylmethane diisocyanate	101-68-8	–	47								
Diphenyl oxide	101-84-8	–	16								
N,N'-Diphenyl-p-phenylenediamine	74-31-7	+	46	–	36						
Diphenylurea	102-07-8	?	45								

TABLE 1 (continued)
Test Results

Chemical	CAS	SAL res	Ref.	ABS res	Ref.	SCE res	Ref.	MLA res	Ref.	DRO res	Ref.
asym-Diphenylurea	603-54-3	–	45								
Di-*n*-propylamine	142-84-7	–	47								
C.I. Direct black 38	1937-37-7	+	29, 35								
C.I. Direct blue 1	2610-05-1	–	29								
C.I. Direct blue 2	2429-73-4	+	29								
C.I. Direct blue 6	2602-46-2	+	29, 35	–	12	–	12				
C.I. Direct blue 8	2429-71-2	+	29, 35								
C.I. Direct blue 10	4198-19-0	+	29, 35								
C.I. Direct blue 15	2429-74-5	+	29, 35	–	12	–	12				
C.I. Direct blue 25	2150-54-1	+	29, 35								
C.I. Direct blue 53	314-13-6	?	29							?	20, 42
C.I. Direct blue 218	28407-37-6	–	29, 35							–	42
C.I. Direct brown 2	2429-82-5	?	29								
C.I. Direct brown 95	16071-86-6	+w	29, 35	–	12	?	12			–	43
C.I. Direct green 1	3626-28-6	+w	47								
C.I. Direct orange 6	6637-88-3	+	35								
C.I. Direct red 2	992-59-6	+	35, 47								
C.I. Direct red 39	6358-29-8	+	35, 47								
C.I. Direct violet 32	6428-94-0	+	35, 45								
C.I. Direct yellow 11	1325-37-7	–	47								
C.I. Disperse blue 1	2475-45-8	+	45	+	1	+	1	+	34		
C.I. Disperse yellow 3	2832-40-8	+	45	–	17	+	17	+	24	–	11
2,2′-Dithiobisbenzothiazole	120-78-5	+	45								
2,5-Dithiobiurea	142-46-1	–	29	–	12	+	12				
Ditridecyl phthalate	119-06-2	–	49								
Diundecyl phthalate	3648-20-2	–	49								
Divinylbenzene	1321-74-0	–	47								
2-Dodecen-1-yl succinic anhydride	19780-11-1	–	47								
Dodecyl alcohol, ethoxylated	9002-92-0	–	47	–	18	–	18	–	33	–	11
n-Dodecylmercaptan	112-55-0	–	47								
t-Dodecylmercaptan	25103-58-6	–	46								
Doxylamine succinate	562-10-7	–	47								
Dulcin	150-69-6	–	16								
Econazole nitrate	24169-02-6	–	46								
Eicosane	112-95-8	–	46								
Ellagic acid	476-66-4	–	46								
Emetine 2HCl	316-42-7	–	29								
Endosulfan	115-29-7	–	47					+	24		
Endrin	72-20-8	–	47					–	26		
Eosin	17372-87-1	+	46								
Ephedrine sulfate	134-72-5	–	45	–	N-23	–	N-23				
Epinephrine	51-43-4	?	47	–	N-24	?	N-24				
1,2-Epoxy-3-bromopropane	3132-64-7	+	3							–	10
1,2-Epoxybutane	106-88-7	+	3, 8	+	1	+	1	+	25, 28, 32	+	43
2,3-Epoxybutane	3266-23-7	+	46								
1,2-Epoxy-3-butene	930-22-3	+	3								
1,2-Epoxy-3-chloropropane	106-89-8	+	3								
1,2-Epoxydecane	2404-44-6	–	3								
1,2-Epoxydodecane	2855-19-8	–	3								
1,2-Epoxy-3-fluoropropane	503-09-3	+	46								
1,2-Epoxyhexadecane	7320-37-8	–	3								

TABLE 1 (continued)
Test Results

Chemical	CAS	SAL res	Ref.	ABS res	Ref.	SCE res	Ref.	MLA res	Ref.	DRO res	Ref.
1,2-Epoxyoctadecane	7390-81-0	–	3								
9,10-Epoxyoctadecanoic acid, 2-ethylhexyl ester	141-38-8	?	3, 46								
1,2-Epoxypropane	75-56-9	+	3	+	15	+	15	+	27	+	11
1,2-Epoxytetradecane	3234-28-4	–	3								
1,2-Epoxy-4,4,4-trichlorobutane	3083-25-8	+	3								
1,2-Epoxy-3,3,3-trichloropropane	3083-23-6	+	3							+	42
Eptam	759-94-4	–	46							–	42
Ergotamine tartrate	379-79-3	–	47								
Erythromycin stearate	643-22-1	–	29	–	1	–	1	?	34		
Estragole	140-67-0	–	47								
Ethacrynic acid	58-54-8	–	45								
Ethinylestradiol	57-63-6	–	8, 45								
Ethionamide	536-33-4	–	45								
D,L-Ethionine	67-21-0	–	46								
2-Ethoxyethanol	110-80-5	–	49	+	12	+	12			–	20, 37
2-Ethoxyethyl-*p*-methoxycinnamate	104-28-9	–	47								
Ethyl acetate	141-78-6	–	46	–	18	+	18				
Ethyl acrylate	140-88-5	–	16, 46	+	18	+w	18	+	24	–	37
Ethyl alcohol	64-17-5	–	46								
3-Ethylamino-4-methylphenol	120-37-6	+	45								
3-Ethylaminophenol	621-31-8	+	45								
N-Ethyl aniline	103-69-5	–	45								
Ethyl anthranilate	87-25-2	–	29								
Ethyl benzene	100-41-4	–	46	–	T-8	–	T-8	+	24		
Ethyl bromide	74-96-4	+	16, 46	–	19	+	19				
Ethyl-*n*-butylamine	13360-63-9	–	47								
Ethyl chloride	75-00-3	+	46								
S-(Ethyl)chlorothioformic acid	2941-64-2	–	45								
p,p'-Ethyl DDD	72-56-0	+	29								
Ethylenediamine	107-15-3	+	16							–	51
Ethylenediamine tetraacetic acid, 3Na salt	150-38-9	–	9, 45					–	24		
Ethylene glycol	107-21-1	–	47	–	N-25	–	N-25	–	26		
Ethylene thiourea	96-45-7	?	29	–	N-26	–	N-26	+	24	–	20, 42
2-Ethylhexanal	123-05-7	–	45								
2-Ethylhexanoic acid	149-57-5	–	45								
2-Ethyl-1-hexanol	104-76-7	–	49								
2-Ethylhexenal	26266-68-2	–	45								
2-Ethyl-2-hexenal	645-62-5	–	45								
2-Ethylhexyl acrylate	103-11-7	–	49								
2-Ethylhexylamine	104-75-6	–	45								
2-Ethylhexyl-2-cyano-3,3-diphenylacrylate	6197-30-4	–	47								
2-Ethylhexyl diphenyl phosphate	1241-94-7	–	49								
2-Ethylhexyl glycidyl ether	2461-15-6	+	45								
2-Ethoxylethyl-*p*-methoxycinnamate	104-28-9	–	47								
2-Ethylhexyl-*p*-methoxycinnamate	5466-77-3	–	49								
3-((2-Ethylhexyl)oxy)propionitrile	10213-75-9	–	45								
3-((2-Ethylhexyl)oxy)propylamine	5397-31-9	–	45								
Ethylidene norbornene	16219-75-3	–	47								
Ethyl methacrylate	97-63-2	–	47								
Ethyl methanesulfonate	62-50-0	+	8, 29								

TABLE 1 (continued)
Test Results

Chemical	CAS	SAL res	Ref.	ABS res	Ref.	SCE res	Ref.	MLA res	Ref.	DRO res	Ref.
N-Ethyl-4-methyl benzenesulfonamide	80-39-7	–	46								
N-Ethylmorpholine	100-74-3	+w	47								
N-Ethyl-*N*-nitrosourea	759-73-9	+	46								
o-Ethyl phenol	90-00-6	–	46								
N-Ethyl-*N*-phenyl benzylamine	92-59-1	–	45								
Ethyl tellurac	20941-65-5	–	29							–	37
Ethyl vanillin	121-32-4	–	29								
Eugenol	97-53-0	–	16	+	12	+	12	+	33	–	10
FD&C Yellow 6	2783-94-0	–	45	–	17	?	17	+	24	–	10
Fenthion	55-38-9	?	29								
Ferric chloride	7705-08-0							–	24		
Ferrocene	102-54-5	–	16	–	13	+	13			+	51
Fluometuron	2164-17-2	–	9, 45	–	18	+	18	–	24		
Fluorescein, disodium salt	518-47-8	–	29	–	18	+	18	+	33	–	37
2-Fluorobenzoyl chloride	393-52-2	+	16							–	37
1-Fluoro-2,4-dinitrobenzene	70-34-8	+	16							–	43
4-Fluoro-*d,l*-phenylalanine	51-65-0	–	47								
5-Fluorouracil	51-21-8	–	47								
Formaldehyde	50-00-0	+	16	+	13	+	13			+	38, 42
Formamide	75-12-7	–	29							–	11
Formanilide	103-70-8	–	45								
Formic acid	64-18-6	–	46								
Formulated fenaminosulf	140-56-7	+	45								
Fumaric acid	110-17-8	–	45								
Fumaronitrile	764-42-1	–	46	+w	18	+	18				
Furan	110-00-9	–	29	+	N-27	+	N-27	+	24	–	10
Furfural	98-01-1	?	29	+	N-28	+	N-28	+	23	+	42
Furfural acetone	623-15-4	–	29								
Furfuryl acetate	623-17-6	+	29								
Furfuryl alcohol	98-00-0	–	29								
Furosemide	54-31-9	–	47	+	1	+	1	+	34		
β-2-Furyl acrolein	623-30-3	–	45								
Gallic acid	149-91-7	–	16								
Gallium arsenide	1303-00-0	–	46								
Geranyl acetate	105-87-3	–	29	–	12	+w	12	+	33	–	10
Gibberellic acid	77-06-5	–	47								
Gilsonite	12002-43-6	–	46								
Glutaraldehyde	111-30-8	+w	16	?	13	+	13	+	23	–	43
Glycerol	56-81-5	–	16								
Glycidaldehyde	765-34-4	+	8								
Glycidol	556-52-5	+	3					+	N-29	+	11
Glycidyl acrylate	106-90-1	+	3								
Glycidyl methacrylate	106-91-2	+	3								
Glycine	56-40-6	–	16								
Glyphosate	1071-83-6	–	46								
Griseofulvin	126-07-8	–	46								
Guanazole	1455-77-2	–	47								
Guar gum	9000-30-0	–	46								
Gum arabic	9000-01-5	–	46								
Halazone	80-13-7	+	29								
Halothane	151-67-7	–	29	–	12	–	12			+	42

TABLE 1 (continued)
Test Results

Chemical	CAS	SAL res	Ref.	ABS res	Ref.	SCE res	Ref.	MLA res	Ref.	DRO res	Ref.
HC Blue 1	2784-94-3	+	45	+	18	+	18	+	33	+	11
HC Blue 2	33229-34-4	+	45	−	18	+	18	+	33	−	11
HC Red 3	2871-01-4	+	45					+	N-30		
HC Yellow 4	59820-43-8	+	29							+	42
Hematoxylin	517-28-2	−	16								
Heptachlor	76-44-8	−	47					+	24		
Heptadecyl hydroxyethylimidazoline	95-19-2	−	47								
n-Heptanal	111-71-7	−	46								
n-Heptanoic acid	111-14-8	−	46								
Hexabromobenzene	87-82-1	−	16								
Hexabromobiphenyl	36355-01-8	−	16	−	12	?	12				
Hexabromocyclododecane, mixed isomers	25637-99-4	−	47								
Hexabutyl distannoxane	56-35-9	−	45								
Hexachlorobenzene	118-74-1	−	16								
Hexachloro-1,3-butadiene	87-68-3	−	16	−	12	+	12			−	42
Hexachlorocyclopentadiene	77-47-4	−	16							−	20, 51
Hexachloroethane	67-72-1	−	16	−	12	+	12				
Hexachlorophene	70-30-4	−	16								
Hexadecylamine	143-27-1	−	45								
Hexafluoroacetone sesquihydrate	13098-39-0	−	45								
1,1,1,3,3,3-Hexafluoro-2-propanol	920-66-1	−	46								
Hexahydro-1,3,5-tris(hydroxyethyl) triazine	4719-04-4	+	29								
Hexamethylmelamine	645-05-6	−	29								
Hexamethylphosphoramide	680-31-9	−	48	−	14	+	14	+	31	+	10
Hexamethyl-p-rosaniline chloride	548-62-9	?	29							?	20, 42
Hexanamide	628-02-4	−	47								
n-Hexane	110-54-3	−	29								
1,6-Hexanediol diacrylate	13048-33-4	−	47								
4-n-Hexyl-4′-cyanobiphenyl	41122-70-7	−	46								
Hexyl glycidyl ether	5926-90-9	+	3								
n-Hexyl methacrylate	142-09-6	−	47								
p-Hexylresorcinol	136-77-6	−	29	−	1	+	1	+	34		
Hycanthone methanesulfonate	23255-93-8	+	46								
Hydrazine sulfate	10034-93-2	+	29							−	43
Hydrazobenzene	122-66-7	+	9, 16	+	12	+	12			−	43
Hydrochlorothiazide	58-93-5	−	29	−	1, 12	+	1, 12	+	34	−	37
Hydroquinone	123-31-9	−	16	+	12	+	12	+	23	?	10
Hydroquinone dimethyl ether	150-78-7	−	16								
Hydroquinone monomethyl ether	150-76-5	−	16								
4-Hydroxyacetanilide	103-90-2	−	16	+	N-1	+	N-1				
Hydroxyacetonitrile	107-16-4	−	45								
N-Hydroxy-2-acetylaminofluorene	53-95-2	+	8								
N-Hydroxyaminoquinoline-1-oxide	4637-56-3	+	8					+	28, 32		
2-Hydroxybenzamide	65-45-2	−	46								
N-Hydroxybenzamide	495-18-1	+	46								
α-Hydroxybenzeneacetonitrile	532-28-5	−	45								
N-Hydroxyethyl ethylenediamine	111-41-1	+w	47							?	10
Hydroxylamine HCl	5470-11-1	+w	8, 46					+	28, 32		
2-Hydroxy-4-methoxybenzophenone	131-57-7	?	47	+	T-11	+	T-11				
2-Hydroxy-2-methylpropanenitrile	75-86-5	−	45								
2-Hydroxy-1,4-naphthoquinone	83-72-7	+	29							−	10

TABLE 1 (continued)
Test Results

Chemical	CAS	SAL res	Ref.	ABS res	Ref.	SCE res	Ref.	MLA res	Ref.	DRO res	Ref.
6-Hydroxy-2-naphthyl disulfide	6088-51-3	−	46								
4-Hydroxy-3-nitrophenyl arsonic acid (Roxarsone)	121-19-7	−	47	−	1	−	1	+	34	−	10
3-Hydroxy-N-phenylaniline	101-18-8	−	45								
2-Hydroxypropanenitrile	78-97-7	?	46								
3-Hydroxypropanenitrile	109-78-4	−	46								
8-Hydroxyquinoline	148-24-3	+	45	+w	18	+	18	+	23	−	11
8-Hydroxyquinoline sulfate	134-31-6	+	45							?	10
L-5-Hydroxytryptophan	4350-09-8	−	29								
Hydroxyurea	127-07-1	+	16							−	43
Iminobis-3-isopropylamine	56-18-8	−	47								
Indomethacin	53-86-1	−	29								
Iodinated glycerol	5634-39-9	+	47					+	N-31	−	11
Iodoacetic acid	64-69-7									−	37
Iodochlorohydroxyquinoline	130-26-7	−	46								
Iodoform	75-47-8	+	16								
3-Iodo-1,2-propanediol	554-10-9	+	46							+	11
β-Ionone	14901-07-6	−	29								
Iron dextran	9004-66-4							−	26		
Isatin-5-sulfonic acid, Na salt	80789-74-8									+	10
Isoamyl acetate	123-92-2	−	46							−	10
Isoamyl nitrite	110-46-3	+	29	+	12	+	12				
Isobutyl acrylate	106-63-8	−	47								
Isobutyl alcohol	78-83-1	−	45								
Isobutylamine	78-81-9	−	29								
Isobutyl anthranilate	7779-77-3	−	29								
Isobutyl methacrylate	97-86-9	−	47								
Isobutyl nitrite	542-56-3	+	29							−	42
Isobutyraldehyde	78-84-2	−	29							−	42
Isodecyl diphenyl phosphate	29761-21-5	−	47								
Isodecyl methacrylate	29964-84-9	−	45								
Isoeugenol	97-54-1	−	29								
Isopentane	78-78-4									−	11
Isophorone	78-59-1	−	29	−	15	+	15	+	24	−	11
Isophorone diisocyanate	4098-71-9	−	29								
Isophthalic diglycidyl ester	7195-43-9	+	45								
Isoprene	78-79-5	−	29	−	T-12	−	T-12				
Isopropanol	67-63-0	−	46								
p-Isopropoxydiphenylamine	101-73-5	−	46	−	18, 36	+	18				
Isopropyl acetate	108-21-4	−	46								
N-Isopropyl aniline	768-52-5	?	47								
Isopropyl glycidyl ether	4016-14-2	+	3							+	10
Isopropyl methacrylate	4655-34-9	−	47								
Isopropylphenyl diphenyl phosphate, mixed isomers	28108-99-8	−	45								
N-Isopropyl-N'-phenyl-p-phenylenediamine	101-72-4	−	47								
Isoproterenol HCl	51-30-9	?	16	+w	12	+	12				
Lanthanum nitrate hexahydrate	10277-43-7	−	46								
Lasiocarpine	303-34-4	+	16							+	43
Lauric acid	143-07-7	−	45								
Lauric acid/Diethanolamine condensate (1/1)	120-40-1	−	45	−	18	+	18				

TABLE 1 (continued)
Test Results

Chemical	CAS	SAL res	Ref.	ABS res	Ref.	SCE res	Ref.	MLA res	Ref.	DRO res	Ref.
Lauryl chloride	112-52-7	–	46								
Lauryl ethanolamide	142-78-9	–	47								
Lauryl glycidyl ether	2461-18-9	–	3								
Lead (II) acetate	301-04-2	–	8					–	28, 32		
Lead dimethyldithiocarbamate	19010-66-3	+	16							–	43
Lead dioxide	1309-60-0	–	29								
Lignosulfonic acid, Na salt	8061-51-6	–	16								
D-Limonene	5989-27-5	–	16	–	1	–	1	–	34		
D-Limonene dimercaptan	4802-20-4	–	16								
Linalyl anthranilate	7149-26-0	–	47								
Lindane	58-89-9	–	16								
Linoleic acid	60-33-3	–	47								
Lithium chloride	7447-41-8	–	16								
Lithocholic acid	434-13-9	–	9, 45					+	26, 28, 32		
Locust bean gum	9000-40-2	–	46								
Malaoxon	1634-78-2	–	45	–	17	+	17	+	33	+	11
Malathion	121-75-5	–	16	+	12	+	12				
Maleic anhydride	108-31-6	–	16								
Maleic hydrazide	123-33-1	–	16	?	13	+	13			+	20, 43
Maleic hydrazide, diethanolamine salt	5716-15-4	+	29								
Malonaldehyde, Na salt	24382-04-5	–	29	–	1, 19	+	1, 19	+	34	?	20, 42
Malonic dinitrile	109-77-3	–	45								
Maltol	118-71-8	+w	29							–	20
Manganese (II) sulfate, monohydrate	10034-96-5	–	29	+	12	+	12			–	37
D-Mannitol	69-65-8	–	16	–	15	–	15	–	33	–	11
Medroxyprogesterone acetate	71-58-9	–	46								
Melamine	108-78-1	–	16	–	12	?	12	–	24	?	11
Melphalan	148-82-3	+	46								
p-Menthane	99-82-1	–	45								
1,8-p-Menthanediamine	80-52-4	–	47								
p-Menthane hydroperoxide	80-47-7	+	45								
D,L-Menthol	15356-70-4	–	45	–	17	–	17	–	33	–	10
2-Mercaptobenzimidazole	583-39-1	–	47								
2-Mercaptobenzothiazole	149-30-4	–	47	+	1	+	1	+	34		
6-Mercaptopurine monohydrate	6112-76-1	+	46							–	11
Mercuric chloride	7487-94-7	–	47	+	N-32	+w	N-32	+	24	–	10
Mercury ((O-carboxyphenyl)thio)ethyl, Na salt	54-64-8	–	47								
Metasystox-R	301-12-2	+	45								
Methacrylic acid	79-41-4	–	16								
Methacrylonitrile	126-98-7	–	47								
Methapyrilene HCl	135-23-9	–	29					–	26		
Methdilazine HCl	1229-35-2	–	29	–	12	?	12				
4-Methoxybenzilidine-4'-n-butylaniline	26227-73-6	+	45								
Methoxychlor	72-43-5	–	8, 29					+	28, 32		
o-Methoxycinnamaldehyde	1504-74-1	+w	29								
2-Methoxyethanol	109-86-4	–	46								
2-Methoxyethyl acetate	110-49-6	+w	46	+	18	+	18				
Methoxyethyl mercury chloride	123-88-6	–	46								
Methoxyflurane	76-38-0	–	46								
4-Methoxy-4-methyl-2-pentanone	107-70-0	–	47								

TABLE 1 (continued)
Test Results

Chemical	CAS	SAL res	Ref.	ABS res	Ref.	SCE res	Ref.	MLA res	Ref.	DRO res	Ref.
4-Methoxy-3-nitro-*N*-phenylbenzamide	97-32-5	+	46								
o-Methoxyphenol	90-05-1	–	16								
8-Methoxypsoralen	298-81-7	+	N-33	+	N-33	+	N-33				
Methyl acetate	79-20-9	–	46								
Methyl acrylate	96-33-3	–	47								
N-Methyl-*p*-aminophenol sulfate	55-55-0	+	47								
N-Methyl aniline	100-61-8	–	45								
Methyl anthranilate	134-20-3	–	29								
Methylazoxymethanol acetate	592-62-1	–	8					+	28, 32		
2-Methylbenzamide	527-85-5	–	45								
N-Methylbenzamide	613-93-4	–	45								
Methyl benzoate	93-58-3	–	46								
α-Methylbenzyl alcohol	98-85-1	–	47	+	N-34	–	N-34	+	N-34		
p-Methylbenzyl alcohol	589-18-4	–	46								
Methyl bromide	74-83-9	+	N-35								
Methyl biphenyl, mixed isomers	28652-72-4	–	46								
2-Methyl-2-butenenitrile	4403-61-6	–	46								
2-Methyl-3-butenenitrile	16529-56-9	?	46								
(+)-2-Methylbutyl-4-methoxybenzilidine-4′-aminocinnamate	24140-30-5	+	46								
Methyl carbamate	598-55-0	–	8, 45	–	1	–	1	–	28, 32	–	10
N-Methylcarbamic acid, ethyl ester	105-40-8							+	4		
3-Methylcholanthrene	56-49-5	+	8, 16	–	13	+	13			–	38
α-Methyl-*trans*-cinnamaldehyde	101-39-3	–	29								
6-Methyl coumarin	92-48-8	–	16								
Methyl 2-cyanoacrylate	137-05-3	+	47								
Methyl cyclopentane	96-37-7	–	46								
N-Methyl diethanolamine	105-59-9	–	47								
2′-Methyl-4-dimethylaminoazobenzene	3731-39-3	?	8, 47								
N′-Methyl-*N,N*-diphenylurea	13114-72-2	–	46								
Methyl DOPA sesquihydrate	555-30-6	–	45	–	1	–	1	+	34	–	11
N,N′-Methylene-*bis*-acrylamide	110-26-9	+	45							+	11
4,4′-Methylene-*bis*-2-chloroaniline	101-14-4	+	8, 16	–	13	?	13	+	28, 32		
2,2′-Methylenebis-(4-chlorophenol)	97-23-4	+	46								
4,4′-Methylenebis-(*N,N*′-dimethylaniline)	101-61-1	+	9, 46					+	28, 32		
Methylenebis-(thiocyanate)	6317-18-6	–	46								
4,4′-Methylenedianiline	101-77-9	+	46								
4,4′-Methylenedianiline 2HCl	13552-44-8	+	45	+	15	+	15	+	24	+	11
N-Methyl ethanolamine	109-83-1	–	47								
2-Methyl-2-ethoxypropane	637-92-3	–	46								
Methyl ethyl ketone	78-93-3	–	46								
Methyl eugenol	93-15-2	–	29								
Methyl formate	107-31-3	–	46								
Methyl-*p*-formyl benzoate	1571-08-0	–	46								
Methyl furan	534-22-5	–	46								
Methyl glutaronitrile	4553-62-2	+w	45								
Methyl glycidyl ether	930-37-0	+	3								
Methyl hydrazine	60-34-4	–	29								
O-Methyl hydroxylamine HCl	593-56-6	–	47								
Methyl isobutyl ketone	108-10-1	–	46								
Methyl isocyanate	624-83-9	–	21	+	21	+	21	+	4	–	21

TABLE 1 (continued)
Test Results

Chemical	CAS	SAL res	Ref.	ABS res	Ref.	SCE res	Ref.	MLA res	Ref.	DRO res	Ref.
Methyl mercuric (II) chloride	115-09-3	–	46							–	37
Methyl mercury hydroxide	1184-57-2	–	46								
Methyl methacrylate	80-62-6	–	47	+	1	+	1	+	34	–	11
Methyl methanesulfonate	66-27-3	+	8, 29								
2-Methyl-5-nitroaniline (5-Nitro-o-toluidine)	99-55-8	+	9								
2-Methyl-1-nitroanthraquinone	129-15-7	+	29							–	43
N-Methyl-N'-nitro-N-nitrosoguanidine	70-25-7	+	8								
N-Methylolacrylamide	924-42-5	–	45	+	N-36	+	N-36				
Methyl parathion	298-00-0	+	16								
Methyl pentachlorostearate	26638-28-8	–	46								
Methylphenidate HCl	298-59-9	–	29	+	12, N-37	?	12, N-37				
3-Methyl-3-phenylglycidic acid, ethyl ester	77-83-8	–	3	+	12	+	12				
1-Methyl pyrene	2381-21-7	+	46								
N-Methyl-2-pyrrolidone	872-50-4	–	29								
6-Methylquinoline	91-62-3	+	46								
7-Methylquinoline	612-60-2	+	46								
8-Methylquinoline	611-32-5	+	46								
Methyl salicylate	119-36-8	–	29								
α-Methyl styrene	98-83-9	–	46								
Methyl succinic acid	498-21-5	–	46								
N-Methyltaurine	107-68-6	–	47								
5-Methyl-2-thiouracil	636-26-0	–	46								
6-Methyl-2-thiouracil	56-04-2	–	46								
Methyltin-tris(isooctylmercaptoacetate)	54849-38-6	–	46								
Methyl-p-toluate	99-75-2	–	46								
β-Methylumbelliferone	90-33-5	–	16								
Methyl vinyl ketone	78-94-4	+	46								
Methyl viologen	1910-42-5							+	24		
Metronidazole	443-48-1	+	46								
Mezerein	34807-41-5	–	45								
Mirex	2385-85-5	–	29	–	N-38	–	N-38				
Molybdenum trioxide	1313-27-5	–	46	–	N-39	–	N-39				
Mono-sec-butanolamine	13552-21-1	–	45								
Monoethanolamine	141-43-5	–	29								
Monoethylamine	75-04-7	–	29								
Mono(2-ethylhexyl) adipate	4337-65-9	–	49								
Mono(2-ethylhexyl) phthalate	4376-20-9	–	49							–	10
Mono-isopropanolamine	78-96-6	?	47							–	10
Mono-isopropylamine	75-31-0	–	47								
Monomethylamine	74-89-5	–	29					+	4		
Monosodium glutamate	142-47-2	–	46								
Monosodium methanearsenate	2163-80-6	–	45								
Monuron	150-68-5	–	16	+	12	+	12	–	24	–	11
Morpholine	110-91-8	–	16								
Musk ambrette	83-66-9	–	47								
Musk ketone	81-14-1	–	45								
Myleran	55-98-1	+	29							+	11
Nalidixic acid	389-08-2	–	45	–	N-40	–	N-40	–	N-40		
Naphthalene	91-20-3	–	29	+	N-41	+	N-41				
1,4-Naphthalenediamine	2243-61-0	+	29, 46								
1,5-Naphthalenediamine	2243-62-1	+	9, 46								

TABLE 1 (continued)
Test Results

Chemical	CAS	SAL res	Ref.	ABS res	Ref.	SCE res	Ref.	MLA res	Ref.	DRO res	Ref.
1-Naphthylamine	134-32-7	+	45								
2-Naphthylamine	91-59-8	+	8, 45					+	28, 32		
4-(2-Naphthylamino)phenol	93-45-8	–	46								
N-(1-Naphthyl) ethylenediamine 2HCl	1465-25-4	+	29							–	42
α-Naphthyl isothiocyanate	551-06-4	+	8, 45								
Navy fuels JP-5	8008-20-6	–	47								
Neodecanoic acid, 2,3-epoxypropyl ester	26761-45-5	+	3								
Neohesperidin dihydrochalcone	20702-77-6	–	47								
Neopentyl glycol diglycidyl ether	17557-23-2	+	3								
Nickelocene	1271-28-9	–	16								
Nickel sulfate hexahydrate	10101-97-0							+	24		
Nifedipine	21829-25-4	–	46								
Ninhydrin	485-47-2	–	47								
Nithiazide	139-94-6	+	16								
Nitrilotriacetic acid	139-13-9	–	46	–	19	–	19			–	20, 42
Nitrilotriacetic acid, 3Na, H₂O	18662-53-8	–	9, 46					–	28, 32		
5-Nitroacenaphthene	602-87-9	+	9, 16								
3-Nitro-p-acetophenetide	1777-84-0	+	16							–	43
m-Nitroacetophenone	121-89-1	+	47								
o-Nitroacetophenone	577-59-3	–	47								
p-Nitroacetophenone	100-19-6	+	47								
p-Nitroaniline	100-01-6	+	16	+	12	+	12			–	37
5-Nitro-o-anisidine	99-59-2	+	16							+	37
o-Nitroanisole	91-23-6	+	16	+w	12	+	12	+	N-42		
9-Nitroanthracene	602-60-8	+	45								
4-Nitroanthranilic acid	619-17-0	+	29							–	42
Nitrobenzene	98-95-3	–	16								
6-Nitrobenzimidazole	94-52-0	+	9, 29							–	37
m-Nitrobenzoic acid	121-92-6	+	47								
o-Nitrobenzoic acid	552-16-9	+	47								
p-Nitrobenzoic acid	62-23-7	+	47, 50	+	50	+	50				
m-Nitrobenzoyl chloride	121-90-4	+	47								
o-Nitrobenzoyl chloride	610-14-0	+	47								
p-Nitrobenzoyl chloride	122-04-3	+	47								
m-Nitrobenzyl chloride	619-23-8	+	47								
o-Nitrobenzyl chloride	612-23-7	+	47								
p-Nitrobenzyl chloride	100-14-1	+	47								
2-Nitro-1,1'-biphenyl	86-00-0	–	45								
1-Nitrobutane	627-05-4	–	29								
2-Nitro-1-butanol	609-31-4	–	46								
2-Nitrodiphenylamine	119-75-5	–	46								
Nitroethane	79-24-3	–	29								
2-Nitroethanol	625-48-9	+	46								
Nitrofen	1836-75-5	+	9								
2-Nitrofluorene	607-57-8	+	8								
Nitrofurantoin	67-20-9	+	16	+	1	+	1	+	34	–	51
Nitrofurazone	59-87-0	+	47	+	1	+	1	+	34		
N-[4-(5-Nitro-2-furyl) thiazolyl] formamide	24554-26-5	+	8								
Nitrogen mustard HCl	55-86-7	+	46								
1-Nitrohexane	646-14-0	–	29								
p-Nitrohippuric acid	2645-07-0	?	46, 50	–	50	–	50				

TABLE 1 (continued)
Test Results

Chemical	CAS	SAL res	Ref.	ABS res	Ref.	SCE res	Ref.	MLA res	Ref.	DRO res	Ref.
Nitromethane	75-52-5	–	29	–	N-43	–	N-43				
1-Nitro-2-methylnaphthalene	881-03-8	+	45								
1-Nitronaphthalene	86-57-7	+	9, 29							–	37
o-Nitrophenethyl alcohol	15121-84-3	+	46								
p-Nitrophenethyl alcohol	100-27-6	+	46								
m-Nitrophenol	554-84-7	+w	46							–	10
o-Nitrophenol	88-75-5	–	16							–	10
p-Nitrophenol	100-02-7	–	16	+	N-44	–	N-44			–	10, 51
o-Nitrophenyl acetonitrile	610-66-2	+	45								
1-((4-Nitrophenyl)azo)-2-naphthenamine	3025-77-2	+	46								
4-Nitro-m-phenylenediamine	5131-58-8	+	45								
4-Nitro-o-phenylenediamine	99-56-9	+	9, 29					+	28, 32		
2-Nitro-p-phenylenediamine	5307-14-2	+	9, 45					+	28, 32		
5-(4-Nitrophenyl)-2,4-pentadien-1-al	2608-48-2	+	50	?	50	–	50				
4-Nitrophthalhydrazide	3682-19-7	+	45								
4-Nitrophthalic anhydride	5466-84-2	+	45								
4-Nitrophthalimide	89-40-7	+	45								
1-Nitropropane	108-03-2	–	16								
2-Nitropropane	79-46-9	+	16	–	12	–	12			–	51
3-Nitropropionic acid	504-88-1	+	9, 45					+	28, 32		
4-Nitroquinoline-N-oxide	56-57-5	+	8					+	28, 32		
N-Nitrosodiethanolamine	1116-54-7	+	45								
N-Nitrosodiethylamine	55-18-5	+	8, 45								
N-Nitrosodimethylamine	62-75-9	+	8, 16	+w	13, 19	+	13, 19			+	38, 41
N-Nitrosodiphenylamine	86-30-6	–	8, 45	–	18, 36	+	18				
p-Nitrosodiphenylamine	156-10-5	+w	16					+	28, 32	–	37
N-Nitroso-N-ethylaniline	612-64-6	–	8								
N-Nitrosomorpholine	59-89-2	+	46								
N-Nitrosopiperidine	100-75-4	+	16							+	43
β-Nitrostyrene	102-96-5	?	46								
m-Nitrotoluene	99-08-1	–	16	–	12	+	12				
o-Nitrotoluene	88-72-2	–	16	–	12	+	12				
p-Nitrotoluene	99-99-0	–	16	+	12	+	12	+	T-14		
2-Nitro-α,α,α-trifluorotoluene	384-22-5	–	16								
3-Nitro-α,α,α-trifluorotoluene	98-46-4	–	16								
Nonanal	124-19-6	–	29								
n-Nonane	111-84-2	–	46								
Nonylphenyl diphenyl phosphate, mixed isomers	64532-97-4	–	45								
Ochratoxin A	303-47-9	–	45								
Octachlorodibenzodioxin	3268-87-9	–	45								
Octadecylamine	124-30-1	–	45								
Octanoic acid	124-07-2	–	45								
2-Octyl-3-isothiazolone	26530-20-1	–	46								
n-Octyl methacrylate	2157-01-9	–	47								
p-n-Octyloxybenzoic acid	2493-84-7	–	45								
Oil orange SS	2646-17-5	+	46								
Oleic acid	112-80-1	–	29								
Oleic acid/Diethanolamine condensate (1/1)	13961-86-9	–	45								
Oleic acid methyl ester (cis)	112-62-9	–	29								
Olivetol	500-66-3	–	46								

TABLE 1 (continued)
Test Results

Chemical	CAS	SAL res	Ref.	ABS res	Ref.	SCE res	Ref.	MLA res	Ref.	DRO res	Ref.
Oxalic acid, anhydrous	144-62-7	–	16								
Oxethazaine	126-27-2	–	45								
1,1-Oxybismethylene, bis benzene	103-50-4	–	46								
4,4′-Oxydianiline	101-80-4	+	45	+	15	+	15	+	24	–	11
Oxymetholone	434-07-1	–	46								
Oxytetracycline HCl	2058-46-0	–	29	–	1	–	1	+	26, 34	–	10
Ozone	10028-15-6	+w	6								
Palladium (II) chloride	7647-10-1	–	29								
Papaverine HCl	61-25-6	–	29								
Parathion	56-38-2	?	16								
Penicillin VK	132-98-9	–	29	–	N-45	+	N-45	+	28,32		
Pentabromochlorocyclohexane	87-84-3	?	46								
Pentabromodiphenyl oxide	32534-81-9	–	47								
2,3,4,5,6-Pentabromoethylbenzene	85-22-3	–	47								
Pentabromophenol	608-71-9	–	47								
Pentabromotoluene	87-83-2	–	47								
Pentachloroanisole	1825-21-4	+	29	–	N-46	+	N-46	+	25		
Pentachlorobenzene	608-93-5	–	16								
Pentachloroethane	76-01-7	–	16	–	12	+	12	+	24	–	11
Pentachloronitrobenzene	82-68-8	–	16	+	12, N-47	?	12, N-47	–	N-47		
Pentachlorophenol	87-86-5	–	16	+w	12	+w	12				
2,4-α,α,α-Pentachlorotoluene	13014-18-1	+	46								
3,4-α,α,α-Pentachlorotoluene	13014-24-9	+	46								
Pentaerythritol tetranitrate (mixture)	78-11-5	–	29	–	N-48	+	N-48				
Pentaerythritol triacrylate	3524-68-3	–	47								
Pentaethylenehexamine	4067-16-7	+	29							?	10
2-Pentenenitrile	13284-42-9	–	46								
3-Pentenenitrile	4635-87-4	+w	46								
p-tert-Pentyl phenol	80-46-6	–	45								
Peracetic acid	79-21-0	–	45								
Phenamphos	22224-92-6	–	45								
Phenanthrene	85-01-8	+w	8, 45								
o-Phenanthroline	66-71-7	–	47								
Phenazopyridine HCl	136-40-3	?	29	+	12	+	12	+	26	?	20, 42
Phenbenzamine HCl	2045-52-5	–	47								
Phenestrin	3546-10-9	–	47					–	24		
Phenethyl anthranilate	133-18-6	–	45								
m-Phenetidine	621-33-0	+w	45								
o-Phenetidine	94-70-2	?	45								
p-Phenetidine	156-43-4	+	45								
Phenformin HCl	834-28-6	–	47								
Pheniramine maleate	132-20-7	–	29								
Phenmedipham	13684-63-4	+	46								
Phenobarbital	50-06-6	+w	48	+	14	–	14	+	31		
Phenol	108-95-2	–	16	+	17	+	17	+	24	–	42
Phenolphthalein	77-09-8	–	29	+	40						
Phenothiazine	92-84-2	–	29								
Phenoxybenzamine HCl	63-92-3	+	45								
Phenylacetonitrile	140-29-4	–	45								
D-Phenylalanine	673-06-3	–	29								
N-Phenylbenzenamine	122-39-4	–	45								

TABLE 1 (continued)
Test Results

Chemical	CAS	SAL res	Ref.	ABS res	Ref.	SCE res	Ref.	MLA res	Ref.	DRO res	Ref.
Phenyl butazone	50-33-9	–	29	+	12	–	12	+	N-49		
m-Phenylenediamine	108-45-2	+	45	+	36						
o-Phenylenediamine	95-54-5	+	45	+	36						
p-Phenylenediamine 2HCl	624-18-0	+	9, 45					+	28, 32		
Phenylephrine HCl	61-76-7	–	47	–	1	+	1	+	34		
2-Phenyl-2-ethylmalondiamide	80147-40-6	+w	45								
3-Phenylglycidic acid, ethyl ester	121-39-1	–	3								
Phenyl glycidyl ether	122-60-1	+	3								
N-Phenylhydroxylamine	100-65-2	+	29								
Phenyl mercuric acetate	62-38-4	–	47								
1-Phenyl-3-methyl-5-pyrazolone	89-25-8	–	9, 47					+	26		
N-Phenyl-1-naphthylamine	90-30-2	–	45	–	18, 36	+	18				
N-Phenyl-2-naphthylamine	135-88-6	–	45	–	1, 36	+w	1	+	34		
o-Phenylphenol	90-43-7	+w	16	–	N-50	+w	N-50	+	N-50	–	42
N-Phenyl-*p*-phenylenediamine	101-54-2	–	29								
Phenyl salicylate	118-55-8	–	47								
1-Phenyl-2-thiourea	103-85-5	–	29								
Phosphamidon	13171-21-6	+	16								
Photodieldrin	13366-73-9	+	45								
Phthalamide	88-96-0	–	49					–	24		
1(2*H*)-Phthalazinone	119-39-1	–	47								
Phthalic acid	88-99-3	–	46								
Phthalic anhydride	85-44-9	–	49	–	12	–	12				
Phthalocyanine mixture, undefined	——	+	46								
Picloram	1918-02-1	–	29							?	20, 42
β-Picoline	108-99-6	–	16								
Picric acid	88-89-1	+	16							?	42
Picryl chloride	88-88-0	+	45								
Pigment green 7	1328-53-6	+	45								
Pigment green 36	14302-13-7	+	29								
Pigment orange 43	4424-06-0	+	45								
Pigment red 2	6041-94-7	+w	46								
Pigment red 3	2425-85-6	+	29	–	N-51	–	N-51				
Pigment red 23	6471-49-4	+	29	–	N-52	+	N-52				
Pigment red 81	12224-98-5	–	46								
Pigment violet 1	1326-03-0	–	47								
Pigment yellow 12 (Diarylanilide yellow)	6358-85-6	–	35, 47								
Pigment yellow 74	6358-31-2	–	45								
Pigment yellow 100	12225-21-7	–	45								
Piperazine	110-85-0	–	16								
Piperonal	120-57-0	–	16								
Piperonyl acetate	326-61-4	–	29								
Piperonyl acetone	3160-37-0	–	29								
Piperonyl butoxide	51-03-6	–	16	–	12	–	12	+	24		
Piperonyl sulfoxide	120-62-7	–	16	–	12	–	12			–	42
Pivalolactone	1955-45-9	+	9								
2-Pivalyl-1,3-indandione	83-26-1	–	47								
Polybrominated biphenyl	67774-32-7							–	33	–	11
Polyethylene glycol 200	25322-68-3	–	29								
Polysorbate 80	9005-65-6	–	29	+	N-53	–	N-53				
Polythiazide	346-18-9	–	46								

TABLE 1 (continued)
Test Results

Chemical	CAS	SAL res	Ref.	ABS res	Ref.	SCE res	Ref.	MLA res	Ref.	DRO res	Ref.
Polyvinyl chloride latex	9002-86-2	–	45								
Polyvinylpyrrolidone polymers	9003-39-8	–	47								
Ponceau MX	3761-53-3	+w	46								
Ponceau SX	4548-53-2	–	46								
Potassium bromate	7758-01-2	+	46								
Potassium bromide	7758-02-3	–	46								
Potassium chloride	7447-40-7	–	29					+	28, 32		
Potassium dichromate	7778-50-9	+	46								
Potassium fluoride	7789-23-3							+	5		
Prednisone	53-03-2	+	16								
Primaclone	125-33-7	+	29								
Probenecid	57-66-9	–	29	–	N-54	+	N-54				
Procarbazine HCl	366-70-1	–	8, 45					+	28, 32		
Proflavin HCl ½H$_2$O	952-23-8	+	9								
Progesterone	57-83-0	–	8, 45					–	28, 32		
Promethazine HCl	58-33-3	–	29	–	12	?	12			–	43
Propantheline bromide	50-34-0	–	29	–	12	+	12			–	20, 43
β-Propiolactone	57-57-8	+	16	+	13	+	13			+	41
Propionaldehyde	123-38-6	–	29								
Propionic acid	79-09-4	–	46								
Propionitrile	107-12-0	–	45	–	18	–	18				
S-(n-Propyl)chlorothioformic acid	13889-92-4	+	46								
Propylene	115-07-1	+	N-55					–	27	–	11
Propylenediamine	78-90-0	–	47								
1,2-Propylene glycol	57-55-6	–	16								
Propyleneimine	75-55-8	+	8								
1,3-Propylene oxide	503-30-0	+	45								
Propyl gallate	121-79-9	–	29	+	15	+	15			?	10
3-Propylidene phthalide	17369-59-4	+	45								
n-Propyl methacrylate	2210-28-8	–	47								
Pyrazinamide	98-96-4	–	47								
Pyrene	129-00-0	+	8, 46					+	28, 32		
Pyridine	110-86-1	–	16					–	23	?	11, 20, 37
Pyrilamine	91-84-9	–	47								
Pyrimethamine	58-14-0	–	16								
Quercetin	117-39-5	+	46	+	N-56	+	N-56				
Quinacrine 2HCl	69-05-6	+	45							–	37
Quinethazone	73-49-4	–	46								
Quinidine	56-54-2	–	45								
Quinoline	91-22-5	+	16	?	13	+	13	+	24	–	38, 51
Quinoline sulfate	530-66-5									?	11
p-Quinone	106-51-4	–	29								
p-Quinone dioxime	105-11-3	+	9, 16					+	24	–	43
p-Quinone monoxime	104-91-6	+w	45								
Rescinnamine	24815-24-5	–	45								
Reserpine	50-55-5	–	9, 45	–	15	–	15	–	24	?	11, 20
Resorcine blue	87495-30-5	–	47								
Resorcinol	108-46-3	–	16	+	N-57	+	N-57	+	23	?	10
Rhodamine 6G	989-38-8	–	47	+	18	+	18	+	N-58		
Rhodanine	141-84-4	–	47								
Rhothane	72-54-8	–	29								

TABLE 1 (continued)
Test Results

Chemical	CAS	SAL res	Ref.	ABS res	Ref.	SCE res	Ref.	MLA res	Ref.	DRO res	Ref.
Ricinoleic acid, Na salt	5323-95-5	–	16								
Riddelliine	23246-96-0	+	45	+	12	+	12				
Rotenone	83-79-4	–	47	–	1	+w	1	+	24, 34		
Saccharin	81-07-2	–	29								
Safrole	94-59-7	–	48	–	14	+	14	+	28, 31, 32		
Salicylazosulfapyridine	599-79-1	–	45	–	2	–	2				
Scopalamine hydrobromide	6533-68-2	–	46								
Selenium sulfide	7446-34-6	+	45	+	18	+	18	+	33	–	11
Selsun	——	–	46								
Semicarbazide HCl	563-41-7	–	16								
Seneciphylline	480-81-9	+	45								
Sodium aluminosilicate	1344-00-9	–	47								
Sodium arsanilate	127-85-5	–	45								
Sodium azide	26628-22-8	+	8, 47	–	N-59	+	N-59				
Sodium 2-biphenylol	132-27-4	–	29								
Sodium chloride	7647-14-5	–	45								
Sodium chromate tetrahydrate	10034-82-9	+	46								
Sodium cyanide	143-33-9		T-15								
Sodium dehydroacetate	4418-26-2	–	47								
Sodium diethyldithiocarbamate	148-18-5	–	29	–	19	–	19	+	26		
Sodium dodecyl sulfate	151-21-3	–	29					–	24		
Sodium (2-ethylhexyl) alcohol sulfate	126-92-1	–	49	–	18	–	18	–	26	–	37
Sodium fluoride	7681-49-4	–	16	?	N-60	?	N-60	+	5		
Sodium methohexital	309-36-4	–	45								
Sodium nitrite	7632-00-0	+	46								
Sodium phosphate, dibasic	7558-79-4	–	16								
Solvent black 5	11099-03-9	+	45								
Solvent black 7	8005-02-5	–	45								
Solvent yellow 14	842-07-9	+	45	–	17	+	17	+	26	–	11
Stannous chloride	7772-99-8	–	29	+	15	+	15	–	33	–	11
Stearatochromic chloride complex	15242-96-3	+	45								
cis-Stilbene	645-49-8	–	16								
trans-Stilbene	103-30-0	–	16								
Streptomycin sulfate (2:3)	3810-74-0	+w	16							–	37
Styrene	100-42-5	–	9, 45								
Styrene oxide	96-09-3	+	46								
Succinic acid 2,2-dimethylhydrazine	1596-84-5	–	9, 47					?	24		
Succinic anhydride	108-30-5	–	47	–	N-62	–	N-62				
Succinonitrile	110-61-2	–	45								
Sucrose	57-50-1							–	25, 28, 32		
Sulfacetamide	144-80-9	–	29	–	12	+	12			–	43
Sulfallate	95-06-7	+	16								
Sulfamethazine	57-68-1	–	29								
Sulfamethizole	144-82-1	–	29								
Sulfamethoxazole	723-46-6	–	29								
Sulfan blue	129-17-9	+	29								
Sulfanilamide	63-74-1	?	29	–	12	?	12			–	20, 37
Sulfapyridine	144-83-2			–	39	+	39				
Sulfathiazole	72-14-0	–	29								
Sulfisoxazole	127-69-5	–	9, 45	–	17	+	17	+	33	–	11
3-Sulfolene	77-79-2	–	47	–	18	–	18				

TABLE 1 (continued)
Test Results

Chemical	CAS	SAL res	Ref.	ABS res	Ref.	SCE res	Ref.	MLA res	Ref.	DRO res	Ref.
2,4,5-T (2,4,5-Trichlorophenoxyacetic acid)	93-76-5	–	30	+	12	+	12			–	51
2,4,5-T, *n*-Butyl ester	93-79-8	–	30								
2,4,5-T, Isobutyl ester	4938-72-1	–	30								
2,4,5-T, Isooctyl ester, 64%	25168-15-4	–	30								
Tara gum	39300-88-4	–	46								
L-Taurine	107-35-7	–	45								
2,3,7,8-TCDD (2,3,7,8-Tetrachlorodibenzo-*p*-dioxin)	1746-01-6	–	30	–	12	–	12	–	26	–	51
Terephthalic acid	100-21-0	–	49								
3,3′,5,5′-Tetrabromobisphenol A	79-94-7	–	29								
Tetrabromophthalic anhydride	632-79-1	–	49								
3,4,5,6-Tetrabromo-*o*-xylene	36059-21-9	–	47								
1,2,3,4-Tetrachlorobenzene	634-66-2	–	16	–	18	+	18				
1,2,3,5-Tetrachlorobenzene	634-90-2	–	16	–	18	–	18				
1,2,4,5-Tetrachlorobenzene	95-94-3	–	16	–	18	–	18				
1,1,1,2-Tetrachloroethane	630-20-6	–	16	–	12	+	12	?	24, 33	–	11
1,1,2,2-Tetrachloroethane	79-34-5	–	16	–	12	+	12			–	42
Tetrachloroethylene	127-18-4	–	16	–	1, 12	–	1, 12	–	24, 34	–	37
1,2,3,4-Tetrachloronaphthalene	20020-02-4	–	16								
2,3,5,6-Tetrachloro-4-nitroanisole	2438-88-2	–	9, 46					+	24		
2,3,4,5-Tetrachloronitrobenzene	879-39-0	–	16								
2,3,5,6-Tetrachloronitrobenzene	117-18-0	+w	16	–	12	+w	12			–	43
2,3,4,5-Tetrachlorophenol	4901-51-3	–	45	+	36						
2,3,4,6-Tetrachlorophenol	58-90-2	–	45	+	36						
2,3,5,6-Tetrachlorophenol	935-95-5	–	45	+	36						
Tetrachlorophthalic anhydride	117-08-8	–	49	–	12	–	12			?	37
Tetracycline HCl	64-75-5	–	47	–	1	–	1	?	34	–	10
Tetradecanoic acid	544-63-8	–	45								
Tetraethyl dithiopyrophosphate	3689-24-5	+	45								
Tetraethylene glycol diacrylate	17831-71-9	–	47								
Tetraethylenepentamine	112-57-2	+	29							?	10, 20
Tetraethyl lead	78-00-2	–	29								
Tetraethylthiuram disulfide No. 2	97-77-8	–	47					+	26		
1-*trans*-Δ-9-Tetrahydrocannabinol	1972-08-3	–	45	–	N-63	+w	N-63				
Tetrahydrofuran	109-99-9	–	29	–	12	–	12			–	37
Tetrakis(hydroxymethyl) phosphonium chloride	124-64-1	–	47	+	1, 19	+	1, 19	+	34		
Tetrakis(hydroxymethyl) phosphonium SO$_4$	55566-30-8	–	44	+	1	+	1	+	34		
3,3′,5,5′-Tetramethylbenzidine	54827-17-7	–	45					?	28, 32		
N,N,N′,N′-Tetramethyl-1,3-butanediamine	97-84-7	–	47								
N,N,N′,N′-Tetramethylethylenediamine	110-18-9	–	47								
Tetramethyl lead	75-74-1	–	16								
N,N,N′,N′-Tetramethyl-*p*-phenylenediamine	100-22-1	+	45	+	36						
Tetramethyl succinonitrile	3333-52-6	–	45								
1,1,3,3-Tetramethyl-2-thiourea	2782-91-4	–	47								
Tetranitromethane	509-14-8	+	47	+	N-64	+	N-64				
Thenyldiamine HCl	958-93-0	–	47								
Theophylline	58-55-9	–	45								
Thiabendazole	148-79-8	+	45								
Thioacetamide	62-55-5	–	8, 46					+	28, 32		
4,4′-Thiobis(6-*tert*-butyl-*m*-cresol)	96-69-5	–	47	–	N-65	+	N-65				

TABLE 1 (continued)
Test Results

Chemical	CAS	SAL res	Ref.	ABS res	Ref.	SCE res	Ref.	MLA res	Ref.	DRO res	Ref.
2,2'-Thiobis(4-chlorophenol)	97-24-5	–	45								
2,2'-Thiobis(4,6-dichlorophenol)	97-18-7	–	45								
Thiocarbanilide	102-08-9	–	16							–	51
4,4'-Thiodianiline	139-65-1	+	45								
Thioglycolic acid	68-11-1	–	47								
Thioglycolic acid, Na salt	367-51-1	–	47								
Thiophanate, methyl ester	23564-05-8	+w	46								
Thiophene	110-02-1	–	47								
Thiourea	62-56-6	–	8, 45					?	28, 32		
Thonzylamine HCl	63-56-9	–	29								
Titanium dioxide	13463-67-7	–	9, 45	–	17	–	17	–	33	–	11
Titanocene dichloride	1271-19-8	+	8, 16	–	N-66	–	N-66				
D-α-Tocopheryl succinate	4345-03-3	–	45								
Tolazamide	1156-19-0	?	45								
Tolbutamide	64-77-7	–	29	–	18	+	18				
o-Tolualdehyde	529-20-4	–	45								
Toluene	108-88-3	–	16	–	N-67	–	N-67	?	24		
2,4-Toluene diisocyanate	584-84-9	+	47	–	15	?	15	+	26		
2,6-Toluene diisocyanate	91-08-7	+	47	+	15	+	15	+	26		
Toluene diisocyanate (2,4- and 2,6- mixture)	26471-62-5	+	47							+	11
Toluene ethylsulfonamide	8047-99-2	+w	46								
m-Toluidine HCl	638-03-9	–	46								
o-Toluidine	95-53-4	+	48	+	14	+	14	+	31		
o-Toluidine HCl	636-21-5	+	46							–	11, 51
p-Toluidine HCl	540-23-8	+w	46								
m-Tolunitrile	620-22-4	–	45								
o-Tolunitrile	529-19-1	–	45								
p-Tolunitrile	104-85-8	–	45								
p-Tolylurea	622-51-5	–	47								
Tolytriazole	29385-43-1	+w	45								
Toxaphene	8001-35-2	+	29								
Triallylamine	102-70-5	–	47								
Triallyl isocyanurate	1025-15-6	–	46	–	18, 36	–	18				
Triamterene	396-01-0	–	29								
Tri-n-amylamine	621-77-2	–	47								
S-Triazine-2,4,6(1H, 3H, 5H)-trione-1,3-dichloro, K salt	2244-21-5	–	47								
2,4,6-Tribromophenol	118-79-6	–	47								
Tributoxyethyl phosphate	78-51-3	–	16								
Tributylamine	102-82-9	–	47								
Tributyl borate	688-74-4	–	16								
Tributyl phosphate	126-73-8	–	46								
Trichlorfon	52-68-6	+w	47					+	24		
Trichloroacetonitrile	545-06-2	+w	29								
2,4,6-Trichloroaniline	634-93-5	–	46								
1,2,3-Trichlorobenzene	87-61-6	–	16								
1,2,4-Trichlorobenzene	120-82-1	–	16								
1,3,5-Trichlorobenzene	108-70-3	–	16							–	51
1,1,1-Trichloroethane	71-55-6	–	16	+	12	?	12	–	28, 32		
1,1,2-Trichloroethane	79-00-5	–	45							–	10
Trichloroethylene	79-01-6	–	22, 29	–	12	+	12	+	33	?	10

TABLE 1 (continued)
Test Results

Chemical	CAS	SAL res	Ref.	ABS res	Ref.	SCE res	Ref.	MLA res	Ref.	DRO res	Ref.
Trichlorofluoromethane	75-69-4	−	47								
3,4,6-Trichloro-2-nitrophenol	82-62-2	−	46								
2,3,4-Trichlorophenol	15950-66-0	−	46	+	36						
2,3,5-Trichlorophenol	933-78-8	−	46								
2,3,6-Trichlorophenol	933-75-5	−	46	+	36						
2,4,5-Trichlorophenol	95-95-4	−	16								
2,4,6-Trichlorophenol	88-06-2	−	16	−	12	−	12	+	24	−	37
3,4,5-Trichlorophenol	609-19-8	−	46	−	36						
1,2,3-Trichloropropane	96-18-4	+	16	+	N-68	+	N-68	+	N-68		
Triclocarban	101-20-2	−	47								
Tricresyl phosphate	1330-78-5	−	16								
Tri-*m*-cresyl phosphate	563-04-2	−	16								
Triethanolamine	102-71-6	−	29	−	12	−	12			−	43
Triethanolamine stearate	4568-28-9	−	29								
Triethylamine	121-44-8	−	47								
Triethylenetetramine	112-24-3	+	29							?	10
Triethyl lead chloride	1067-14-7	−	29								
Triethyl phosphate	78-40-0	−	47								
Trifluralin	1582-09-8	+	29	−	13	?	13			−	20, 43
1,8,9-Trihydroxyanthracene	480-22-8	−	8, 46								
2′,4′,5′-Trihydroxybutyrophenone	1421-63-2	+	29							−	37
Tri-isobutylamine	1116-40-1	−	47								
Tri-isopropanolamine	122-20-3	−	47								
Trimellitic anhydride	552-30-7	−	29								
Trimethoprim	738-70-5	−	29	?	13	+	13				
2,4,5-Trimethoxy benzaldehyde	4460-86-0	−	47					+	24		
Trimethylamine	75-50-3	−	29								
2,4,5-Trimethylaniline	137-17-7	+	29	+	18	+	18			−	43
3,3,5-Trimethylcyclohexyl salicylate	118-56-9	−	47								
Trimethyloxonium hexachloroantimonate	54075-76-2	?	47								
Trimethylphosphate	512-56-1	+	46								
1,1,3-Trimethyl-2-thiourea	2489-77-2	−	29					+	24	?	20, 42
Trioctyl phosphate	1806-54-8	−	47								
1,3,5-Trioxane	110-88-3	−	45								
Tripelennamine HCl	154-69-8	−	29								
Triphenylamine	603-34-9	−	47								
Triphenyl phosphate	115-86-6	−	47								
Triphenyl phosphine	603-35-0	−	47								
Triphenyl phosphite	101-02-0	−	47								
Triphenyltin hydroxide	76-87-9	−	9, 46								
Tris(aziridinyl)phosphine sulfide	52-24-4	+	46								
Tris (2-chloroethyl) phosphate	115-96-8	−	16	−	12	?	12				
Tris (2-chloroethyl) phosphite	140-08-9	?	16							?	42
Tris (1-chloro-2-propyl) phosphate	13674-84-5	−	46								
Tris (2,3-dibromopropyl) phosphate	126-72-7	+	9								
Tris (1,3-dichloro-2-propyl) phosphate	13674-87-8	+	29								
Tris (2,3-epoxypropyl) isocyanurate	2451-62-9	+	46	+	18, 36	+	18				
Tris (2-ethylhexyl) phosphate	78-42-2	−	49	−	17	−	17	−	33	−	11
Tris (2-ethylhexyl) phosphite	301-13-3	−	49								
Tris (2-ethylhexyl) trimellitate	3319-31-1	−	45								

TABLE 1 (continued)
Test Results

Chemical	CAS	SAL res	Ref.	ABS res	Ref.	SCE res	Ref.	MLA res	Ref.	DRO res	Ref.
1,3,5-Tris (2-hydroxyethyl) triazine-2,4,6-trione	839-90-7	–	46	–	18	–	18				
Trixylenyl phosphate, mixed isomers	25155-23-1	–	47								
L-Tryptophan	73-22-3	–	46								
Turmeric oleoresin	8024-37-1	–	29								
Uracil mustard	66-75-1	+	8								
Urea	57-13-6	–	29								
Urethane	51-79-6	+w	46	–	T-16	+	T-16			+	10
Urotropine	100-97-0	?	46								
Valeronitrile	110-59-8	–	46								
Vanillin	121-33-5	–	29								
Vat brown 3	131-92-0	–	47								
Vat yellow 4	128-66-5	–	47								
Veratraldehyde	120-14-9	–	29								
4-Vinylcyclohexene	100-40-3	–	47								
1-Vinyl-3-cyclohexene dioxide	106-87-6	+	29	+	N-69	+	N-69	+	24		
Vinylidene chloride	75-35-4	–	29					+	27	–	10
Vinyl toluene	25013-15-4	–	47	–	N-70	–	N-70	+	24		
Vitamin D3	67-97-0	–	29								
Vitamin D3, emulsifiable	1406-16-2	?	29								
Water	7732-18-5	–	46	–	18	–	18	–	23		
Witch hazel	68916-39-2	–	29	–	12	–	12	–	23	–	42
m-Xylene	108-38-3	–	16								
o-Xylene	95-47-6	–	16								
p-Xylene	106-42-3	–	16								
Xylenes, mixed isomers	1330-20-7	–	47	–	1	–	1	+	34		
Xylenesulfonic acid, Na salt	1300-72-7	–	47	–	N-61	+	N-61	?	N-61		
2,3-Xylidine	87-59-2	+	45								
2,4-Xylidine	95-68-1	+	45								
2,5-Xylidine	95-78-3	+	45								
2,6-Xylidine	87-62-7	+w	45	+	12	+	12			–	11
3,4-Xylidine	95-64-7	+	45								
3,5-Xylidine	108-69-0	+w	45								
Zearalenone	17924-92-4	–	29	+	12	+	12	–	24	–	11
Zinc potassium chromate	11103-86-9	+	45								
Zinc pyrithione	13463-41-7	–	47								
Zineb	12122-67-7	–	45								
Ziram	137-30-4	+	16	+	15	–	15	+	24	+	11
Zirconium oxychloride 6H$_2$O	25399-81-9	–	29								
Zirconocene dichloride	1291-32-3	+	45								

Note: SAL: *Salmonella typhimurium* mutagenicity test; ABS: chromosome aberrations test in cultured Chinese hamster ovary cells; SCE: sister chromatid exchange test in cultured Chinese hamster ovary cells; MLA: mutation test in cultured mouse lymphoma cells; DRO: *Drosophila melanogaster* sex-linked recessive lethal test.

res: summary test result; Ref: published reference(s) containing the test data.

+: positive result; +w: weak positive result; ?: equivocal result or disagreement among tests; -: negative result.

SECTION C: GENOTOXICITY BIBLIOGRAPHY — GENERAL LITERATURE

1. Anderson, B. E., Zeiger, E., Shelby, M. D., Resnick, M. A., Gulati, D. K., Ivett, J. L., and Loveday, K. S., Chromosome aberration and sister chromatid exchange test results with 42 chemicals, *Environ. Mol. Mutagen.*, 16 (Suppl. 18), 55-137, 1990.

2. Bishop, J. B., Witt, K. L., Gulati, D. K., and MacGregor, J. T., Evaluation of the mutagenicity of the anti-inflammatory drug salicylazosulfapyridine (SASP), *Mutagenesis*, 5, 549-554, 1990.

3. Canter, D. A., Zeiger, E., Haworth, S., Lawlor, T., Mortelmans, K., and Speck, W., Comparative mutagenicity of aliphatic epoxides in Salmonella, *Mutat. Res.*, 172, 105-138, 1986.

4. Caspary, W. J. and Myhr, B., Mutagenicity of methylisocyanate and its reaction products to cultured mammalian cells, *Mutat. Res.*, 174, 285-293, 1986.

5. Caspary, W. J., Myhr, B., Bowers, L., McGregor, D., Riach, C., and Brown, A., Mutagenic activity of fluorides in mouse lymphoma cells., *Mutat. Res.*, 187, 165-180, 1987.

6. Dillon, D., Combes, R., McConville, M., and Zeiger, E., Ozone is mutagenic in Salmonella, *Environ. Mol. Mutagen.*, 19, 331-337, 1992.

7. Dillon, D., Edwards, I., Combes, R., McConville, M., and Zeiger, E., The role of glutathione in the bacterial mutagenicity of vapor-phase dichloromethane, *Environ. Mol. Mutagen.*, 20, 211-217, 1992.

8. Dunkel, V. C., Zeiger, E., Brusick, D., McCoy, E., McGregor, D., Mortelmans, K., Rosenkranz, H. S., and Simmon, V. F., Reproducibility of microbial mutagenicity assays. I. Tests with *Salmonella typhimurium* and *Escherichia coli* using a standardized protocol, *Environ. Mutagen.*, 6 (Suppl. 2), 1-251, 1984.

9. Dunkel, V. C., Zeiger, E., Brusick, D., McCoy, E., McGregor, D., Mortelmans, K., Rosenkranz, H. S., and Simmon, V. F., Reproducibility of microbial mutagenicity assays. II. Testing of carcinogens and noncarcinogens in *Salmonella typhimurium* and *Escherichia coli*, *Environ. Mutagen.*, 7 (Suppl. 5), 1-248, 1985.

10. Foureman, P., Mason, J. M., Valencia, R., and Zimmering, S., Chemical mutagenesis testing in Drosophila. IX. Results of 50 coded compounds tested for the National Toxicology Program, *Environ. Mol. Mutagen.*, 23, 51-63, 1994.

11. Foureman, P., Mason, J. M., Valencia, R., and Zimmering, S., Chemical mutagenesis testing in Drosophila. X. Results of 70 coded compounds tested for the National Toxicology Program, *Environ. Mol. Mutagen.*, 23, 208-227, 1994.

12. Galloway, S. M., Armstrong, M. J., Reuben, C., Colman, S., Brown, B., Cannon, C., Bloom, A. D., Nakamura, F., Ahmed, M., Duk, S., Rimpo, J., Margolin, B. H., Resnick, M. A., Anderson, B., and Zeiger, E., Chromosome aberrations and sister chromatid exchanges in Chinese hamster ovary cells: Evaluations of 108 chemicals, *Environ. Mol. Mutagen.*, 10 (Suppl. 10), 1-176, 1987.

13. Galloway, S. M., Bloom, A. D., Resnick, M., Margolin, B. H., Nakamura, F., Archer, P., and Zeiger, E., Development of a standard protocol for *in vitro* cytogenetic testing with CHO cells: Comparison of results for 22 compounds in two laboratories, *Environ. Mutagen.*, 7, 1-52, 1985.

14. Gulati, D. K., Sabharwal, P. S., and Shelby, M. D., Tests for the induction of chromosomal aberrations and sister chromatid exchanges in cultured Chinese hamster ovary (CHO) cells, In *Evaluation of Short-Term Tests for Carcinogens*, Ashby, J., de Serres, F. J., Draper, M., Ishidate, M., Jr., Margolin, B. H., Matter, B. E., and Shelby, M. D., (Eds.), Elsevier/North-Holland, pp. 413-426, (1985)

15. Gulati, D. K., Witt, K., Anderson, B., Zeiger, E., and Shelby, M. D., Chromosome aberration and sister chromatid exchange test in Chinese hamster ovary cells *in vitro* III: Results with 27 chemicals, *Environ. Mol. Mutagen.*, 13, 133-193, 1989.

16. Haworth, S., Lawlor, T., Mortelmans, K., Speck, W., and Zeiger, E., Salmonella mutagenicity test results for 250 chemicals, *Environ. Mutagen.*, 5 (Suppl. 1), 3-142, 1983.

17. Ivett J. L., Brown, B. M., Rodgers, C., Anderson, B. E., Resnick, M. A., Zeiger, E., Chromosomal aberration and sister chromatid exchange tests in Chinese hamster ovary cells *in vitro* IV: Results for 15 chemicals, *Environ. Mol. Mutagen.*, 14, 165-187, 1989.

18. Loveday, K. S., Anderson, B. E., Resnick, M. A., and Zeiger, E., Chromosome aberration and sister chromatid exchange tests in Chinese hamster ovary cells *in vitro* V: Results with 46 chemicals, *Environ. Mol. Mutagen.*, 16, 272-303, 1990.

19. Loveday, K. S., Lugo, M. H., Resnick, M. A., Anderson, B. E., and Zeiger, E., Chromosome aberration and sister chromatid exchange tests in Chinese hamster ovary cells *in vitro* II: Results with 20 chemicals, *Environ. Mol. Mutagen.*, 13: 60-94, 1989.

20. Mason, J. M., Valencia, R., and Zimmering, S., Chemical mutagenesis testing in Drosophila. VIII. Reexamination of equivocal results, *Environ. Mol. Mutagen.*, 19, 227-234, 1992.

21. Mason, J. M., Zeiger, E., Haworth, S., Ivett, J., and Valencia, R., Genotoxicity studies of methyl isocyanate in Salmonella, Drosophila, and cultured Chinese hamster ovary cells, *Environ. Mutagen.*, 9, 19-28, 1987.

22. McGregor, D. B., Reynolds, D. M., and Zeiger, E., Conditions affecting the mutagenicity of trichloroethylene in Salmonella, *Environ. Mol. Mutagen*, 13, 197-202, 1989.

23. McGregor, D. B., Brown, A., Cattanach, P., Edwards, I., McBride, D., and Caspary, W. J., Responses of the L5178Y tk+/tk- mouse lymphoma cell forward mutation assay. II. 18 coded chemicals., *Environ. Mol. Mutagen.*, 11, 91-118, 1988.

24. McGregor, D. B., Brown, A., Cattanach, P., Edwards, I., McBride, D., Riach, C., and Caspary, W. J., Responses of the L5178Y tk+/tk- mouse lymphoma cell forward mutation assay: III. 72 coded chemicals, *Environ. Mol. Mutagen.*, 12, 85-154, 1988.

25. McGregor, D. B., Martin, R., Cattanach, P., Edwards, I., McBride, D., and Caspary, W. J., Responses of the L5178Y tk+/tk- mouse lymphoma cell forward mutation assay to coded chemicals. I. Results for nine compounds, *Environ. Mutagen.*, 9, 143-160, 1987.

26. McGregor, D. B., Brown, A. G., Howgate, S., McBride, D., Riach, C., and Caspary, W. J., Responses of the L5178Y mouse lymphoma cell forward mutation assay. V: 27 coded chemicals, *Environ. Mol. Mutagen.*, 17, 196-219, 1991.

27. McGregor, D., Brown, A. G., Cattanach, P., Edwards, I., McBride, D., Riach, C., Shepherd, W., and Caspary, W. J., Responses of the L5178Y mouse lymphoma forward mutation assay: V. Gases and vapors, *Environ. Mol. Mutagen.*, 17, 122-129, 1991.

28. Mitchell, A. D., Rudd, C. J., and Caspary, W. J., Evaluation of the L5178Y mouse lymphoma cell mutagenesis assay: Intralaboratory results for 63 coded chemicals tested at SRI International, *Environ. Mol. Mutagen.*, 12 (Suppl. 13), 37-102, 1988.

29. Mortelmans, K., Haworth, S., Lawlor, T., Speck, W., Tainer, B., and Zeiger, E., Salmonella mutagenicity tests. II. Results from the testing of 270 chemicals, *Environ. Mutagen.*, 8 (Suppl. 7), 1-119, 1986.

30. Mortelmans, K., Haworth, S., Speck, W., and Zeiger, E., Mutagenicity testing of Agent Orange components and related chemicals, *Toxicol. Appl. Pharmacol.*, 75, 137-146, 1984.

31. Myhr, B., Bowers, L., and Caspary, W. J., Assays for the induction of gene mutations at the thymine kinase locus in L5178Y mouse lymphoma cells in culture, In *Evaluation of Short-Term Tests for Carcinogens*, Ashby, J., de Serres, F. J., Draper, M., Ishidate, M., Jr., Margolin, B. H., Matter, B. E., and Shelby, M. D., (Eds.), Elsevier/North Holland, pp. 555-568, 1985.

32. Myhr, B. C. and Caspary, W. J., Evaluation of the L5178Y mouse lymphoma cell mutagenesis assay: Intralaboratory results for sixty-three coded chemicals tested at Litton Bionetics, Inc., *Environ. Mol. Mutagen.*, 12 (Suppl. 13), 103-194, 1988.

33. Myhr, B. C. and Caspary, W. J., Chemical mutagenesis at the TK locus in L5178Y mouse lymphoma cells. I. Results for 31 coded compounds in the National Toxicology Program, *Environ. Mol. Mutagen.*, 18, 51-83, 1991.

34. Myhr B., McGregor, D., Bowers, L., Riach, C., Brown, A. G., Edwards, I., McBride, D., Martin, R., and Caspary, W. J., L5178Y mouse lymphoma cell mutation assay results with 41 compounds, *Environ. Mol. Mutagen.*, 16 (Suppl. 18), 138-167, 1990.

35. Reid, T. M., Morton, K. C., Wang, C. Y., and King, C. M., Mutagenicity of azo dyes following metabolism by different reductive/oxidative systems, *Environ. Mutagen.*, 6, 705-717, 1984.

36. Sofuni, T., Matsuoka, A., Sawada, M., Ishidate, M. Jr., Zeiger, E., Shelby, M. D., A comparison of chromosome aberration induction by 25 compounds tested by two Chinese hamster cell (CHL and CHO) systems in culture, *Mutat. Res.*, 241, 175-213, 1990.

37. Valencia, R., Mason, J. M., Woodruff, R. C., and Zimmering, S., Chemical mutagenesis testing in Drosophila: III. Results of 48 coded compounds tested for the National Toxicology Program, *Environ. Mutagen.*, 7, 325-348, 1985.

38. Valencia, R., Mason, J. M., and Zimmering, S., Chemical mutagenesis testing in Drosophila. VI. Interlaboratory comparison of mutagenicity tests after treatment of larvae, *Environ. Mol. Mutagen.*, 14, 238-244, 1989.

39. Witt, K. L., Bishop, J. B., McFee, A. F., Kumaroo, V., Induction of chromosomal damage in mammalian cells *in vitro* and *in vivo* by sulfapyridine or 5-aminosalicylic acid, *Mutat. Res.*, 283, 59-64, 1992.

40. Witt, K. L., Gulati, D. K., Kaur, P., Shelby, M. D., Phenolphthalein: Induction of micronucleated erythrocytes in mice, *Mutat. Res.*, 341, 151-160, 1995.

41. Woodruff, R. C., Mason, J. M., Valencia, R., and Zimmering, S., Chemical mutagenesis testing in Drosophila: I. Comparison of positive and negative control data for sex-linked recessive lethal mutations and reciprocal translocations in three laboratories, *Environ. Mutagen.*, 6, 189-202, 1984.

42. Woodruff, R. C., Mason, J. M., Valencia, R., and Zimmering, S., Chemical mutagenesis testing in Drosophila: V. Results of 53 coded compounds tested for the National Toxicology Program, *Environ. Mutagen.*, 7, 677-702, 1985.

43. Yoon, J. S., Mason, J. M., Valencia, R., Woodruff, R. C., and Zimmering, S., Chemical mutagenesis testing in Drosophila: IV. Results of 45 coded compounds tested for the National Toxicology Program, *Environ. Mutagen.*, 7, 349-367, 1985.

44. Zeiger, E., Mutagenicity of 42 chemicals in Salmonella, *Environ. Mol. Mutagen.*, 16 (Suppl. 18), 32-54, 1990.

45. Zeiger, E., Anderson, B., Haworth, S., Lawlor, T., and Mortelmans, K., Salmonella mutagenicity tests. IV. Results from the testing of 300 chemicals, *Environ. Mol. Mutagen.*, 11 (Suppl. 12), 1-158, 1988.

46. Zeiger, E., Anderson, B., Haworth, S., Lawlor, T., and Mortelmans, K., Salmonella mutagenicity tests. V. Results from the testing of 311 chemicals, *Environ. Mol. Mutagen.*, 19 (Suppl. 21), 2-141, 1992.

47. Zeiger, E., Anderson, B., Haworth, S., Lawlor, T., Mortelmans, K., and Speck, W., Salmonella mutagenicity tests. III. Results from the testing of 225 chemicals, *Environ. Mutagen.*, 9 (Suppl. 9), 1-l09, l987.

48. Zeiger, E. and Haworth, S., Tests with a preincubation modification of the Salmonella/microsome assay, In *Evaluation of Short-Term Tests for Carcinogens*, Ashby, J., de Serres, F. J., Draper, M., Ishidate, M., Jr., Margolin, B. H., Matter, B., and Shelby, M. D. (Eds.), Elsevier/North Holland, pp. 187-199, 1985.

49. Zeiger, E., Haworth, S., Mortelmans, K., and Speck, W., Mutagenicity testing of di(2-ethylhexyl)phthalate and related chemicals in Salmonella, *Environ. Mutagen.*, 7, 213-232, 1985.

50. Zeiger, E., Shelby, M. D., Ivett J., and McFee, A. F., Mutagenicity testing of 5-(4-nitrophenyl)-2,4-petadien-1-al ("Spy Dust") and its metabolites *in vitro* and *in vivo*, *Environ. Mutagen.*, 9, 269-280, l987.

51. Zimmering, S., Mason, J. M., Valencia, R., and Woodruff, R. C., Chemical mutagenesis testing in Drosophila. II. Results of 20 coded compounds tested for the National Toxicology Program, *Environ. Mutagen.*, 7, 87-100, 1985.

SECTION C: GENOTOXICITY BIBLIOGRAPHY — NTP TOXICOLOGY REPORTS

T-1. NTP Technical Report on Toxicity Studies of Black Newsprint Inks Administered Topically to F344/N Rats and C3H Mice. Toxicity Report Series Number 17. U.S. Department of Health and Human Services, NIH Publication 92-3340, July, 1992.

T-2. NTP Technical Report on Toxicity Studies of t-Butyl Perbenzoate (CAS Number: 614-45-9) Administered by Gavage to F344/N Rats and B6C3F1 Mice. Toxicity Report Series Number 15. U.S. Department of Health and Human Services, NIH Publication 92-3134, July, 1992.

T-3. NTP Technical Report on Toxicity Studies of 2-Chloronitrobenzene and 4-Chloronitrobenzene (CAS Nos. 88-73-3 and 100-00-5) Administered by Inhalation to F344/N Rats and B6C3F1 Mice. Toxicity Report Series Number 33. U.S. Department of Health and Human Services, NIH Publication 93-3382, July, 1993.

T-4. NTP Technical Report on Toxicity Studies of Cyclohexanone Oxime (CAS No. 100-64-1) Administered by Drinking Water to B6C3F1 Mice. Toxicity Report Series Number 50. U.S. Department of Health and Human Services, NIH Publication 96-3934, April, 1996.

T-5. NTP Technical Report on Toxicity Studies of Dibutyl Phthalate (CAS No. 84-74-2) Administered in Feed to F344/N Rats and B6C3F1 Mice. Toxicity Report Series Number 30. U.S. Department of Health and Human Services, NIH Publication 95-3353, April, 1995.

T-6. NTP Technical Report on Toxicity Studies of Diethanolamine (CAS No. 111-42-2) Administered Topically and in Drinking Water to F344/N Rats and B6C3F1 Mice. Toxicity Report Series Number 20. U.S. Department of Health and Human Services, NIH Publication 92-3343, October, 1992.

T-7. NTP Technical Report on Toxicity Studies of N,N-Dimethylformamide (CAS NO: 68-12-2) Administered by Inhalation to F344/N Rats and B6C3F1 Mice. Toxicity Report Series Number 22. U.S. Department of Health and Human Services, NIH Publication 93-3345, November, 1992.

T-8. NTP Report on the Toxicity Studies of Ethylbenzene in F344/N Rats and B6C3F1 Mice (Inhalation Studies). Toxicity Report Series Number 10. U.S. Department of Health and Human Services, NIH Publication 92-3129, March, 1992.

T-9. NTP Technical Report on Toxicity Studies of Ethylene Glycol Ethers, 2-Methoxyethanol, 2-Ethoxyethanol, 2-Butoxyethanol (CAS Nos. 109-86-4, 110-80-5, 111-76-2) Administered in Drinking Water to F344/N Rats and B6C3F1 Mice. Toxicity Report Series Number 26. U.S. Department of Health and Human Services, NIH Publication 93-3349, July, 1993.

T-10. NTP Technical Report on Toxicity Studies of 1,6-Hexanediamine Dihydrochloride (CAS No. 6055-52-3) Administered by Drinking Water and Inhalation to F344/N Rats and B6C3F1 Mice. Toxicity Report Series Number 24. U.S. Department of Health and Human Services, NIH Publication 93-3347, March, 1993.

T-11. NTP Technical Report on Toxicity Studies of 2-Hydroxy-4-methoxybenzophenone (CAS Number: 131-57-7) Administered Topically and in Dosed Feed to F344/N Rats and B6C3F1 Mice. Toxicity Report Series Number 21. U.S. Department of Health and Human Services, NIH Publication 92-3344, October, 1992.

T-12. NTP Technical Report on Toxicity Studies of Isoprene (CAS No. 78-79-5) Administered by Inhalation to F344/N Rats and B6C3F1 Mice. Toxicity Report Series Number 31. U.S. Department of Health and Human Services, NIH Publication 95-3354, January, 1995.

T-13. NTP Technical Report on Toxicity Studies of Methyl Ethyl Ketone Peroxide (CAS No. 1338-23-4) in Diethyl Phthalate (CAS No. 131-11-3) (45:55) Administered Topically to F344/N Rats and B6C3F1 Mice. Toxicity Report Series Number 18. U.S. Department of Health and Human Services, NIH Publication 93-3341, February, 1993.

T-14. NTP Technical Report on Toxicity Studies of o-, m-, and p-Nitrotoluenes (CAS Nos.: 88-72-2, 99-08-1, 99-99-0) Administered in Dosed Feed to F344/N Rats and B6C3F1 Mice. Toxicity Report Series Number 23. U.S. Department of Health and Human Services, NIH Publication 93-3346, November, 1992.

T-15. NTP Technical Report on Toxicity Studies of Sodium Cyanide (CAS No. 143-33-9) Administered in Drinking water to F344/N Rats and B6C3F1 Mice. Toxicity Report Series Number 37. U.S. Department of Health and Human Services, NIH Publication 94-3386, November, 1993.

T-16. NTP Technical Report on Toxicity Studies of Urethane in Drinking Water and Urethane in 5% Ethanol Administered to F344/N Rats and B6C3F1 Mice. Toxicity Report Series Number 52. U.S. Department of Health and Human Services, NIH Publication 96-3937, March, 1996.

SECTION C: GENOTOXICITY BIBLIOGRAPHY — NTP TECHNICAL REPORTS

N-1. Toxicology and Carcinogenesis Studies of Acetaminophen (CAS NO. 103-90-2) in F344/N Rats and B6C3F1 Mice (Feed Studies). NTP Technical Report No. 394. U.S. Department of Health and Human Services, January, 1993.

N-2. Toxicology and Carcinogenesis Studies of C.I. Acid Orange 3 (CAS No. 6373-74-6) in F344/N Rats and B6C3F1 Mice (Gavage Studies). NTP Technical Report No. 335. U.S. Department of Health and Human Services, December, 1988.

N-3. Toxicology and Carcinogenesis Studies of Allyl Glycidyl Ether (CAS NO. 106-92-3) in Osborne-Mendel Rats and B6C3F1 Mice (Inhalation Studies). NTP Technical Report No. 376. U.S. Department of Health and Human Services, January, 1990.

N-4. Toxicology and Carcinogenesis Studies of 1-Amino-2,4-Dibromoanthraquinone (CAS: 81-49-2) in F344 Rats and B6C3F1 Mice (Feed Studies). NTP Technical Report No. 383. U.S. Department of Health and Human Services, June 1991.

N-5. Toxicology and Carcinogenesis Studies of dl-Amphetamine Sulfate (CAS NO. 60-13-9) in F344/N Rats and B6C3F1 Mice (Feed Studies). NTP Technical Report No. 387. U.S. Department of Health and Human Services, October 1991.

N-6. Toxicology and Carcinogenesis Studies of Benzethonium Chloride (CAS NO. 121-54-0) in F344/N Rats and B6C3F1 Mice (Dermal Studies). NTP Technical Report No. 438. U.S. Department of Health and Human Services, July, 1995.

N-7. Toxicology and Carcinogenesis Studies of Benzofuran (CAS NO. 271-89-6) in F344/N Rats and B6C3F1 Mice (Gavage Studies). NTP Technical Report No. 370. U.S. Department of Health and Human Services, October, 1989.

N-8. Toxicology and Carcinogenesis Studies of o-Benzyl-p-Chlorophenol (CAS NO. 120-32-1) in Swiss (CD-1) Mice (Mouse Skin Study). NTP Technical Report No. 444. U.S. Department of Health and Human Services, May, 1993.

N-9. Toxicology and Carcinogenesis Studies of Boric Acid (CAS No. 10043-35-3) in B6C3F1 Mice (Feed Studies). NTP Technical Report No. 324. U.S. Department of Health and Human Services, October, 1987.

N-10. Toxicology and Carcinogenesis Studies of 1,3-Butadiene (CAS NO. 106-99-0) in B6C3F1 Mice (Inhalation Studies). NTP Technical Report No. 434. U.S. Department of Health and Human Services, May, 1993.

N-11. Toxicology and Carcinogenesis Studies of t-Butyl Alcohol (CAS NO. 75-65-0) in F344/N Rats and B6C3F1 Mice (Drinking Water Studies). NTP Technical Report No. 436. U.S. Department of Health and Human Services, May, 1995.

N-12. Toxicology and Carcinogenesis Studies of d-Carvone (CAS NO. 2244-16-8) in B6C3F1 Mice (Gavage Studies). NTP Technical Report No. 381. U.S. Department of Health and Human Services, February, 1990.

N-13. Toxicology and Carcinogenesis Studies of 2-Chloroacetophenone (CAS NO. 532-27-4) in F344/N Rats and B6C3F1 Mice (Inhalation Studies). NTP Technical Report No. 379. U.S. Department of Health and Human Services, March, 1990.

N-14. Toxicology and Carcinogenesis Studies of CS2 (94% o-Chlorobenzalmalononitrile, CAS NO. 9005-65-6) in F344/N Rats and B6C3F1 Mice (Inhalation Studies). NTP Technical Report No. 377. U.S. Department of Health and Human Services, March, 1990.

N-15. NTP Technical Report on the Toxicology and Carcinogenesis Studies of D&C Yellow 11 (CAS NO. 8003-22-3) in F344/N Rats (Feed Studies). NTP Technical Report 463. U.S. Department of Health and Human Services. in press.

N-16. Toxicology and Carcinogenesis Studies of 2,3-Dibromo-1-Propanol (CAS NO. 96-13-9) in F344 Rats and B6C3F1 Mice (Dermal Studies). NTP Technical Report No. 400. U.S. Department of Health and Human Services, in press.

N-17. Toxicology and Carcinogenesis Studies of Dimethoxane (CAS NO. 828-00-2) (Commercial Grade) in F344/N Rats and B6C3F1 Mice (Gavage Studies). NTP Technical Report No. 354. U.S. Department of Health and Human Services, September, 1989.

N-18. Toxicology and Carcinogenesis Studies of N,N-Dimethylaniline (CAS NO. 121-69-7) in F344/N Rats and B6C3F1 Mice (Gavage Studies). NTP Technical Report No. 360. U.S. Department of Health and Human Services, October, 1989.

N-19. Toxicology and Carcinogenesis Studies of 3,3'-Dimethylbenzidine Dihydrochloride (CAS NO. 612-82-8) in F344/N Rats (Drinking Water Studies). NTP Technical Report No. 390. U.S. Department of Health and Human Services, June, 1991.

N-20. Toxicology and Carcinogenesis Studies of Dimethyl Methylphosphonate (CAS No. 756-79-6) in F344/N Rats and B6C3F1 Mice (Gavage Studies). NTP Technical Report No. 323. U.S. Department of Health and Human Services, November, 1987.

N-21. Toxicology and Carcinogenesis Studies of Diphenhydramine Hydrochloride (CAS NO. 147-24-0) in F344/N Rats and B6C3F1 Mice (Feed Studies). NTP Technical Report No. 355. U.S. Department of Health and Human Services, September, 1989.

N-22. Toxicology and Carcinogenesis Studies of Diphenylhydantoin (Phenytoin) (CAS NO. 57-41-0) in F344 Rats and B6C3F1 Mice (Feed Studies). NTP Technical Report No. 404. U.S. Department of Health and Human Services, November 1993.

N-23. Toxicology and Carcinogenesis Studies of Ephedrine Sulfate (CAS No. 134-72-5) in F344/N Rats and B6C3F1 Mice (Feed Studies). NTP Technical Report No. 307. U.S. Department of Health and Human Services, May, 1986.

N-24. Toxicology and Carcinogenesis Studies of l-Epinephrine Hydrochloride (CAS NO. 55-31-2) in F344/N Rats and B6C3F1 Mice (Inhalation Studies). NTP Technical Report No. 380. U.S. Department of Health and Human Services, March, 1990.

N-25. Toxicology and Carcinogenesis Studies of Ethylene Glycol (CAS NO. 107-21-1) in B6C3F1 Mice (Feed Studies). NTP Technical Report No. 413. U.S. Department of Health and Human Services, February, 1993.

N-26. Toxicology and Carcinogenesis Studies of Ethylene Thiourea (CAS: 96-45-7) in F344 Rats and B6C3F1 Mice (Feed Studies). NTP Technical Report No. 388. U.S. Department of Health and Human Services, March, 1992.

N-27. Toxicology and Carcinogenesis Studies of Furan (CAS NO. 110-00-9) in F344/N Rats and B6C3F1 Mice (Gavage Studies). NTP Technical Report No. 402. U.S. Department of Health and Human Services, January, 1993.

N-28. Toxicology and Carcinogenesis Studies of Furfural (CAS NO. 98-01-1) in F344/N Rats and B6C3F1 Mice (Gavage Studies). NTP Technical Report No. 382. U.S. Department of Health and Human Services, March, 1990.

N-29. Toxicology and Carcinogenesis Studies of Glycidol (CAS. NO. 556-52-5) in F344/N Rats and B6C3F1 Mice (Gavage Studies). NTP Technical Report No. 374. U.S. Department of Health and Human Services, March, 1990.

N-30. Toxicology and Carcinogenesis Studies of HC Red No. 3 [2,((Amino-2-Nitrophenyl)amino)ethanol] (CAS No. 2871-01-4) in F344/N Rats and B6C3F1 Mice (Gavage Studies). NTP Technical Report No. 281. U.S. Department of Health and Human Services, January, 1986.

N-31. Toxicology and Carcinogenesis Studies of Iodinated Glycerol (Organidin®) (CAS No. 5634-39-9) in F344/N Rats and B6C3F1 Mice (Gavage Studies). NTP Technical Report No. 340. U.S. Department of Health and Human Services, March, 1990.

N-32. Toxicology and Carcinogenesis Studies of Mercuric Chloride (CAS NO. 7487-94-7) in F344 Rats and B6C3F1 Mice (Gavage Studies). NTP Technical Report No. 408. U.S. Department of Health and Human Services, February, 1993.

N-33. Toxicology and Carcinogenesis Studies of 8-Methoxypsoralen (CAS NO. 298-81-7) in F344/N Rats (Gavage Studies). NTP Technical Report No. 359. U.S. Department of Health and Human Services, July, 1989.

N-34. Toxicology and Carcinogenesis Studies of α-Methylbenzyl Alcohol (CAS NO. 98-85-1) in F344/N Rats and B6C3F1 Mice (Gavage Studies). NTP Technical Report No. 369. U.S. Department of Health and Human Services, January, 1990.

N-35. Toxicology and Carcinogenesis Studies of Methyl Bromide (CAS NO. 74-83-9) in B6C3F1 Mice (Inhalation Studies). NTP Technical Report No. 385. U.S. Department of Health and Human Services, March, 1992.

N-36. Toxicology and Carcinogenesis Studies of N-Methylolacrylamide (CAS NO. 924-42-5) in F344/N Rats and B6C3F1 Mice. NTP Technical Report No. 352. U.S. Department of Health and Human Services, September, 1989.

N-37. Toxicology and Carcinogenesis Studies of Methylphenidate Hydrochloride (CAS NO. 298-59-9) in F344/N Rats and B6C3F1 Mice (Feed Studies). NTP Technical Report No. 439. U.S. Department of Health and Human Services, July, 1995.

N-38. Toxicology and Carcinogenesis Studies of Mirex (1,1a,2,2,3,3a,4,5,5,5a,5b,6-Dodecachlorooctahydro-1,3,4-metheno-1 H-cyclobuta[cd]pentalene) (CAS No. 2385-85-5) in F344/N Rats (Feed Studies). NTP Technical Report No. 313. U.S. Department of Health and Human Services, February, 1990.

N-39. NTP Technical Report on the Toxicology and Carcinogenesis Studies of Molybdenum Trioxide (CAS NO. 1313-27-5) in F344/N Rats and B6C3F1 Mice (Inhalation Studies). NTP Technical Report 462. U.S. Department of Health and Human Services. in press.

N-40. Toxicology and Carcinogenesis Studies of Nalidixic Acid (CAS NO. 389-08-2) in F344/N Rats and B6C3F1 Mice (Feed Studies). NTP Technical Report No. 368. U.S. Department of Health and Human Services, October, 1989.

N-41. Toxicology and Carcinogenesis Studies of Naphthalene (CAS NO. 91-20-3) in B6C3F1 Mice (Inhalation Studies). NTP Technical Report No. 410. U.S. Department of Health and Human Services, April, 1992.

N-42. Toxicology and Carcinogenesis Studies of o-Nitroanisole (CAS NO. 91-23-6) in F344 Rats and B6C3F1 Mice (Feed Studies). NTP Technical Report No. 416. U.S. Department of Health and Human Services, May, 1993.

N-43. NTP Technical Report on the Toxicology and Carcinogenesis Studies of Nitromethane (CAS NO. 75-52-5) in F344/N Rats and B6C3F1 Mice (Inhalation Studies). NTP Technical Report 461. U.S. Department of Health and Human Services. in press.

N-44. Toxicology and Carcinogenesis Studies of p-Nitrophenol (CAS NO. 100-02-7) in Swiss-Webster Mice (Dermal Studies). NTP Technical Report No. 417. U.S. Department of Health and Human Services, April, 1993.

N-45. Toxicology and Carcinogenesis Studies of Penicillin VK (CAS No. 132-98-9) in F344/N Rats and B6C3F1 Mice (Gavage Studies). NTP Technical Report No. 336. U.S. Department of Health and Human Services, June, 1988.

N-46. Toxicology and Carcinogenesis Studies of Pentachloroanisole (CAS NO. 1825-21-4) in F344/N Rats and B6C3F1 Mice (Gavage Studies). NTP Technical Report No. 414. U.S. Department of Health and Human Services, April, 1993.

N-47. Toxicology and Carcinogenesis Studies of Pentachloronitrobenzene (CAS No. 82-68-8) in B6C3F1 Mice (Feed Studies). NTP Technical Report No. 325. U.S. Department of Health and Human Services, January, 1987.

N-48. Toxicology and Carcinogenesis Studies of Pentaerythritol Tetranitrate (CAS NO. 78-11-5) With 80% D-Lactose Monohydrate (PETN, NF) in F344/N Rats and B6C3F1 Mice (Feed Studies). NTP Technical Report No. 365. U.S. Department of Health and Human Services, August, 1989.

N-49. Toxicology and Carcinogenesis Studies of Phenylbutazone (CAS NO. 50-33-9) in F344/N Rats and B6C3F1 Mice (Gavage Studies). NTP Technical Report No. 367. U.S. Department of Health and Human Services, March, 1990.

N-50. Toxicology and Carcinogenesis Studies of ORTHO-Phenylphenol (CAS No. 90-43-7) Alone and With 7,12-Dimethylbenz(a)anthracene (CAS No. 57-97-6) in Swiss CD-1 Mice (Dermal Studies). NTP Technical Report No. 301. U.S. Department of Health and Human Services, March, 1986.

N-51. Toxicology and Carcinogenesis Studies of C.I. Pigment Red 3 (CAS No. 2425-85-6) in F344/N Rats and B6C3F1 Mice (Feed Studies). NTP Technical Report No. 407. U.S. Department of Health and Human Services, March, 1992.

N-52. Toxicology and Carcinogenesis Studies of C.I. Pigment Red 23 (CAS NO. 6471-49-4) in F344 Rats and B6C3F1 Mice (Feed Studies). NTP Technical Report No. 411. U.S. Department of Health and Human Services, November, 1992.

N-53. Toxicology and Carcinogenesis Studies of Polysorbate 80 (CAS NO. 9005-65-6) in F344/N Rats and B6C3F1 Mice (Feed Studies). NTP Technical Report No. 415. U.S. Department of Health and Human Services, January, 1992.

N-54. Toxicology and Carcinogenesis Studies of Probenecid (CAS NO. 57-66-9) in F344/N Rats and B6C3F1 Mice (Gavage Studies). NTP Technical Report No. 395. U.S. Department of Health and Human Services, September, 1991.

N-55. Toxicology and Carcinogenesis Studies of Propylene (CAS No. 115-07-1) in F344/N Rats and B6C3F1 Mice (Inhalation Studies). NTP Technical Report No. 272. U.S. Department of Health and Human Services, November, 1985.

N-56. Toxicology and Carcinogenesis Studies of Quercetin (CAS NO. 117-39-5) in F344/N Rats (Feed Studies). NTP Technical Report No. 409. U.S. Department of Health and Human Services, September, 1992.

N-57. Toxicology and Carcinogenesis Studies of Resorcinol (CAS NO. 108-46-3) in F344 Rats and B6C3F1 Mice (Gavage Studies). NTP Technical Report No. 403. U.S. Department of Health and Human Services, in press.

N-58. Toxicology and Carcinogenesis Studies of Rhodamine 6G (C.I. Basic Red 1) (CAS NO. 989-38-8) in F344/N Rats and B6C3F1 Mice (Feed Studies). NTP Technical Report No. 364. U.S. Department of Health and Human Services, September, 1989.

N-59. Toxicology and Carcinogenesis Studies of Sodium Azide (CAS: 26628-22-8) in F344 Rats (Gavage Studies). NTP Technical Report No. 389. U.S. Department of Health and Human Services, September, 1991.

N-60. Toxicology and Carcinogenesis Studies of Sodium Fluoride (CAS NO. 7681-49-4) in F344/N Rats and B6C3F1 Mice (Drinking Water Studies). NTP Technical Report No. 393. U.S. Department of Health and Human Services, December, 1990.

N-61. NTP Technical Report on the Toxicology and Carcinogenesis Studies of Sodium Xylenesulfonate (CAS NO. 1300-72-7) in F344/N Rats and B6C3F1 Mice (Dermal Studies). NTP Technical Report 464. U.S. Department of Health and Human Services. in press.

N-62. Toxicology and Carcinogenesis Studies of Succinic Anhydride (CAS. NO. 108-30-5) in F344/N Rats and B6C3F1 Mice (Gavage Studies). NTP Technical Report No. 373. U.S. Department of Health and Human Services, January, 1990.

N-63. Toxicology and Carcinogenesis Studies of 1-Trans-Delta^9-Tetrahydrocannabinol (CAS. NO. 1972-08-3) in F344/N Rats and B6C3F1 Mice (Gavage Studies). NTP Technical Report No. 446. U.S. Department of Health and Human Services, in press.

N-64. Toxicology and Carcinogenesis Studies of Tetranitromethane (CAS NO. 509-14-8) in F344/N Rats and B6C3F1 Mice (Inhalation Studies). NTP Technical Report No. 386. U.S. Department of Health and Human Services, March, 1990.

N-65. Toxicology and Carcinogenesis Studies of 4,4'-Thiobis(6-t-Butyl-m-Cresol) (CAS NO. 96-69-5) in F344/N Rats and B6C3F1 Mice (Feed Studies). NTP Technical Report No. 435. U.S. Department of Health and Human Services, December, 1994.

N-66. Toxicology and Carcinogenesis Studies of Titanocene Dichloride (CAS NO. 1271-19-8) in F344/N Rats (Gavage Studies). NTP Technical Report No. 399. U.S. Department of Health and Human Services, September, 1991.

N-67. Toxicology and Carcinogenesis Studies of Toluene (CAS NO. 108-88-3) in F344/N Rats and B6C3F1 Mice (Inhalation Studies). NTP Technical Report No. 371. U.S. Department of Health and Human Services, February, 1990.

N-68. Toxicology and Carcinogenesis Studies of 1,2,3-Trichloropropane (CAS NO. 96-18-4) in F344 Rats and B6C3F1 Mice (Gavage Studies). NTP Technical Report No. 384. U.S. Department of Health and Human Services, December 1993.

N-69. Toxicology and Carcinogenesis Studies of 4-Vinyl-1-Cyclohexene Diepoxide (CAS NO. 106-87-6) in F344/N Rats and B6C3F1 Mice (Dermal Studies). NTP Technical Report No. 362. U.S. Department of Health and Human Services, November, 1989.

N-70. Toxicology and Carcinogenesis Studies of Vinyl Toluene (Mixed Isomers) (65%-71% meta-isomer and 32-35% para-isomer) (CAS NO. 25013-15-4) in F344/N Rats and B6C3F1 Mice (Inhalation Studies). NTP Technical Report No. 375. U.S. Department of Health and Human Services, March, 1990.

CHAPTER 6

INDEX OF CHEMICAL NAMES IN THE CARCINOGENIC POTENCY (CPDB) AND GENOTOXICITY (GT) DATABASES BY CHEMICAL ABSTRACTS SERVICE (CAS) REGISTRY NUMBER

TABLE 1

If different names or spellings are used in the two databases for a chemical with the same CAS number, the name in the GT is listed below the name in the CPDB. When a chemical is not reported in one of the databases, "." appears.

CAS	In CPDB	In GT	Chemical Name
50-00-0	X	X	Formaldehyde
50-06-6	X	X	Phenobarbital
50-07-7	X	.	Mitomycin-C
50-14-6	X	.	Calciferol
50-18-0	X	X	Cyclophosphamide
50-23-7	X	.	Hydrocortisone
50-24-8	X	.	Prednisolone
50-28-2	X	.	Estradiol
50-29-3	X		DDT
		X	p,p'-DDT
50-32-8	X	X	Benzo(a)pyrene
50-33-9	X	X	Phenylbutazone
50-34-0	.	X	Propantheline bromide
50-44-2	X	.	6-Mercaptopurine
50-55-5	X	X	Reserpine
50-76-0	X	.	Actinomycin D
50-78-2	X		Aspirin
		X	Acetylsalicylic acid
50-81-7	X	X	l-Ascorbic acid
51-02-5	X	.	Pronethalol.HCl
51-03-6	X	X	Piperonyl butoxide
	X	.	Piperonyl butoxide in solvent
51-17-2	.	X	Benzimidazole
51-21-8	X	X	5-Fluorouracil
51-28-5	X	X	2,4-Dinitrophenol
51-30-9	.	X	Isoproterenol.HCl
51-43-4	.	X	Epinephrine
51-52-5	X	.	Propylthiouracil
51-55-8	X	.	Atropine
51-65-0	.	X	4-Fluoro-d,l-phenylalanine
51-75-2	X	.	Nitrogen mustard
51-79-6	X	X	Urethane
52-24-4	X		Thio-TEPA
		X	Tris(aziridinyl)phosphine sulfide
52-28-8	.	X	Codeine phosphate
52-68-6	.	X	Trichlorfon
52-76-6	X	.	Lynestrenol
53-03-2	X	X	Prednisone
53-19-0	X	X	o,p'-DDD
53-70-3	X	X	Dibenz(a,h)anthracene
53-86-1	.	X	Indomethacin
53-95-2	X	X	N-Hydroxy-2-acetylaminofluorene
53-96-3	X		2-Acetylaminofluorene
		X	N-2-Acetylaminofluorene
54-11-5	X	.	Nicotine
54-12-6	X	.	dl-Tryptophan
54-31-9	X	X	Furosemide
54-64-8	.	X	Mercury ((O-carboxyphenyl)thio)ethyl, Na salt
54-80-8	X	.	Pronethalol
54-85-3	X	.	Isoniazid
55-18-5	X	X	N-Nitrosodiethylamine
55-21-0	.	X	Benzamide
55-22-1	X	.	Isonicotinic acid
55-31-2	X	.	l-Epinephrine.HCl
55-38-9	X	X	Fenthion
55-55-0	.	X	N-Methyl-p-aminophenol sulfate
55-63-0	X	.	Trinitroglycerin
55-80-1	X	.	3'-Methyl-4-dimethylaminoazobenzene
55-86-7	.	X	Nitrogen mustard.HCl
55-98-1	X	X	Myleran
56-04-2	X		Methylthiouracil
		X	6-Methyl-2-thiouracil
56-18-8	.	X	Iminobis-3-isopropylamine
56-23-5	X	X	Carbon tetrachloride
56-35-9	X		Bis(tri-n-butyltin)oxide, technical grade
		X	Hexabutyl distannoxane
56-38-2	X	X	Parathion
56-40-6	.	X	Glycine
56-49-5	X	X	3-Methylcholanthrene
56-53-1	X	X	Diethylstilbestrol

CAS	In CPDB	In GT	Chemical Name
56-54-2	.	X	Quinidine
56-55-3	.	X	Benz(a)anthracene
56-57-5	.	X	4-Nitroquinoline-N-oxide
56-72-4	X	X	Coumaphos
56-75-7	X	.	Chloramphenicol
56-81-5	.	X	Glycerol
56-86-0	X	.	l-Glutamic acid
56-93-9	.	X	Benzyltrimethyl ammonium chloride
57-06-7	X	X	Allyl isothiocyanate
57-13-6	X	X	Urea
57-14-7	X	.	1,1-Dimethylhydrazine
57-30-7	X	.	Phenobarbital, sodium
57-39-6	X	.	Metepa
57-41-0	X		5,5-Diphenylhydantoin
		X	Diphenylhydantoin
57-43-2	X	.	Amobarbital
57-50-1	X	X	Sucrose
57-55-6	X		Propylene glycol
		X	1,2-Propylene glycol
57-57-8	X	X	β-Propiolactone
57-63-6	.	X	Ethinylestradiol
57-66-9	X	X	Probenecid
57-68-1	X	X	Sulfamethazine
57-71-6	.	X	2,3-Butanedione-2-oxime
57-74-9	.	X	Chlordane, analytical grade
	X	.	Chlordane, technical grade
57-83-0	.	X	Progesterone
57-97-6	X	X	7,12-Dimethylbenz(a)anthracene
58-08-2	X	X	Caffeine
58-14-0	X	X	Pyrimethamine
58-15-1	X	.	4-Dimethylaminoantipyrine
58-33-3	X	X	Promethazine.HCl
58-54-8	.	X	Ethacrynic acid
58-55-9	.	X	Theophylline
58-89-9	X		γ-1,2,3,4,5,6-Hexachlorocyclohexane
		X	Lindane
58-90-2	.	X	2,3,4,6-Tetrachlorophenol
58-93-5	X	X	Hydrochlorothiazide
58-94-6	.	X	Chlorothiazide
58-95-7	X	.	dl-α-Tocopheryl acetate
59-05-2	X	.	Methotrexate
59-33-6	X	.	Pyrilamine maleate
59-35-8	X	.	4,6-Dimethyl-2-(5-nitro-2-furyl)pyrimidine
59-50-7	.	X	p-Chloro-m-cresol
59-51-8	X	.	dl-Methionine
59-67-6	X	.	Nicotinic acid
59-87-0	X		5-Nitro-2-furaldehyde semicarbazone
		X	Nitrofurazone
59-88-1	X	.	Phenylhydrazine.HCl
59-89-2	X	X	N-Nitrosomorpholine
60-09-3	.	X	p-Aminoazobenzene
60-11-7	X		N,N-Dimethyl-4-aminoazobenzene
		X	4-Dimethylaminoazobenzene
60-13-9	X		dl-Amphetamine sulfate
		X	Amphetamine sulfate
60-33-3	.	X	Linoleic acid
60-34-4	X	X	Methylhydrazine
60-35-5	X	X	Acetamide
60-51-5	X	X	Dimethoate
60-56-0	X	.	Methimazole
60-57-1	X	X	Dieldrin
60-80-0	X	.	Phenazone
61-25-6	.	X	Papaverine.HCl
61-76-7	X	X	Phenylephrine.HCl
61-82-5	X		3-Aminotriazole
		X	3-Amino-1,2,4-triazole
61-94-9	X	.	Arecoline.HCl
62-23-7	X	X	p-Nitrobenzoic acid
62-31-7	X	.	Dopamine.HCl
62-38-4	X	X	Phenylmercuric acetate
62-44-2	X		Phenacetin
		X	p-Acetophenetidide
62-50-0	.	X	Ethyl methanesulfonate
62-53-3	X	X	Aniline
62-54-4	X	.	Calcium acetate
62-55-5	X	X	Thioacetamide
62-56-6	X	X	Thiourea
62-73-7	X	X	Dichlorvos
62-75-9	X	X	N-Nitrosodimethylamine
63-25-2	X	.	Carbaryl
63-56-9	.	X	Thonzylamine.HCl
63-74-1	.	X	Sulfanilamide
63-84-3	X	.	dl-Dopa
63-92-3	X	X	Phenoxybenzamine.HCl
64-17-5	X	X	Ethyl alcohol
64-18-6	.	X	Formic acid
64-19-7	.	X	Acetic acid
64-69-7	.	X	Iodoacetic acid
64-75-5	X	X	Tetracycline.HCl
64-77-7	X	X	Tolbutamide
64-86-8	.	X	Colchicine
65-45-2	.	X	2-Hydroxybenzamide
65-85-0	.	X	Benzoic acid
66-05-7	X	.	β-Phenylisopropylhydrazine.HCl
66-22-8	X	.	Uracil
66-27-3	X	X	Methyl methanesulfonate
66-71-7	.	X	o-Phenanthroline
66-75-1	.	X	Uracil mustard
66-81-9	.	X	Cycloheximide
67-20-9	X		1-[(5-Nitrofurfurylidene) amino]hydantoin
		X	Nitrofurantoin
67-21-0	X	X	dl-Ethionine
67-48-1	X	X	Choline chloride
67-52-7	X	.	Barbituric acid
67-63-0	.	X	Isopropanol
67-64-1	.	X	Acetone

CAS	In CPDB	In GT	Chemical Name
67-66-3	X	X	Chloroform
67-68-5	.	X	Dimethyl sulfoxide
67-72-1	X	X	Hexachloroethane
67-97-0	.	X	Vitamin D$_3$
67-98-1	X	.	MER-25
68-11-1	.	X	Thioglycolic acid
68-12-2	.	X	N,N-Dimethylformamide
68-23-5	X	.	Norethynodrel
68-76-8	X	.	Trenimon
68-89-3	X	.	Dipyrone
69-05-6	.	X	Quinacrine.2HCl
69-09-0	.	X	Chlorpromazine.HCl
69-65-8	X	X	d-Mannitol
69-74-9	.	X	Cytarabine.HCl
70-25-7	X	X	N-Methyl-N'-nitro-N-nitrosoguanidine
70-30-4	X	X	Hexachlorophene
70-34-8	.	X	1-Fluoro-2,4-dinitrobenzene
71-43-2	X	X	Benzene
71-55-6	X	X	1,1,1-Trichloroethane
71-58-9	.	X	Medroxyprogesterone acetate
72-14-0	.	X	Sulfathiazole
72-20-8	X	X	Endrin
72-33-3	X	.	Mestranol
72-43-5	X	X	Methoxychlor
72-54-8	X		p,p'-DDD
		X	Rhothane
72-55-9	X	X	p,p'-DDE
72-56-0	X	X	p,p'-Ethyl-DDD
73-22-3	X	X	l-Tryptophan
73-49-4	.	X	Quinethazone
74-11-3	.	X	p-Chlorobenzoic acid
74-31-7	X		Diphenyl-p-phenylenediamine
		X	N,N'-Diphenyl-p-phenylenediamine
74-83-9	X	X	Methyl bromide
74-89-5	.	X	Monomethylamine
74-96-4	X		Bromoethane
		X	Ethyl bromide
74-97-5	.	X	Bromochloromethane
75-00-3	X		Chloroethane
		X	Ethyl chloride
75-01-4	X	.	Vinyl chloride
75-04-7	.	X	Monoethylamine
75-05-8	.	X	Acetonitrile
75-07-0	X	X	Acetaldehyde
75-09-2	X		Methylene chloride
		X	Dichloromethane
75-12-7	.	X	Formamide
75-15-0	.	X	Carbon disulfide
75-21-8	X	.	Ethylene oxide
75-25-2	X		Tribromomethane
		X	Bromoform
75-27-4	X	X	Bromodichloromethane
75-31-0	.	X	Mono-isopropylamine
75-34-3	X	X	1,1-Dichloroethane
75-35-4	X	X	Vinylidene chloride
75-36-5	.	X	Acetyl chloride
75-45-6	X	.	Chlorodifluoromethane
75-47-8	X	X	Iodoform
75-50-3	.	X	Trimethylamine
75-52-5	.	X	Nitromethane
75-55-8	.	X	Propyleneimine
75-56-9	X		1,2-Propylene oxide
		X	1,2-Epoxypropane
75-60-5	X		Dimethylarsinic acid
		X	Cacodylic acid
75-64-9	.	X	tert-Butylamine
75-65-0	.	X	t-Butyl alcohol
75-69-4	X	X	Trichlorofluoromethane
75-71-8	X	.	Dichlorodifluoromethane
75-74-1	.	X	Tetramethyl lead
75-77-4	.	X	Chlorotrimethyl silane
75-83-2	.	X	2,2-Dimethylbutane
75-86-5	.	X	2-Hydroxy-2-methylpropanenitrile
75-87-6	.	X	Chloral, anhydrous
75-88-7	X	.	2-Chloro-1,1,1-trifluoroethane
75-91-2	.	X	t-Butyl hydroperoxide
76-01-7	X	X	Pentachloroethane
76-06-2	X	X	Chloropicrin
76-13-1	X	.	1,1,2-Trichloro-1,2,2-trifluoroethane, technical grade
76-25-5	X	.	Triamcinolone acetonide
76-38-0	.	X	Methoxyflurane
76-44-8	X	X	Heptachlor
76-87-9	X	X	Triphenyltin hydroxide
77-06-5	X	X	Gibberellic acid
77-09-8	.	X	Phenolphthalein
77-46-3	X	.	4,4'-Sulfonylbisacetanilide
77-47-4	X	X	Hexachlorocyclopentadiene
77-58-7	.	X	Dibutyltin dilaurate
77-65-6	X		Carbromal
		X	α-Bromo-α-ethylbutyrylurea
77-73-6	.	X	Dicyclopentadiene
77-79-2	X	X	3-Sulfolene
77-83-8	X		Ethyl methylphenylglycidate
		X	3-Methyl-3-phenylglycidic acid, ethyl ester
78-00-2	.	X	Tetraethyl lead
78-11-5	X	X	Pentaerythritol tetranitrate
78-34-2	X	X	Dioxathion
78-38-6	.	X	Diethyl ethylphosphonate
78-40-0	.	X	Triethyl phosphate
78-42-2	X	X	Tris(2-ethylhexyl)phosphate
78-44-4	.	X	Carisoprodol
78-51-3	.	X	Tributoxyethyl phosphate
78-59-1	X	X	Isophorone
78-63-7	.	X	2,5-Dimethyl-2,5-bis(tert-butylperoxy)hexane
78-78-4	.	X	Isopentane
78-79-5	.	X	Isoprene
78-81-9	.	X	Isobutylamine

CAS	In CPDB	In GT	Chemical Name
78-83-1	.	X	Isobutyl alcohol
78-84-2	.	X	Isobutyraldehyde
78-87-5	X	X	1,2-Dichloropropane
78-88-6	.	X	2,3-Dichloro-1-propene
78-90-0	.	X	Propylenediamine
78-93-3	.	X	Methyl ethyl ketone
78-94-4	.	X	Methyl vinyl ketone
78-96-6	.	X	Mono-isopropanolamine
78-97-7	.	X	2-Hydroxypropanenitrile
79-00-5	X	X	1,1,2-Trichloroethane
79-01-6	X	X	Trichloroethylene
	X	.	Trichloroethylene (without epichlorohydrin)
79-06-1	X	X	Acrylamide
79-09-4	.	X	Propionic acid
79-10-7	.	X	Acrylic acid
79-11-8	X		Monochloroacetic acid
		X	Chloroacetic acid
79-15-2	.	X	N-Bromoacetamide
79-19-6	X	.	Thiosemicarbazide
79-20-9	.	X	Methyl acetate
79-21-0	.	X	Peracetic acid
79-24-3	.	X	Nitroethane
79-29-8	.	X	2,3-Dimethylbutane
79-34-5	X	X	1,1,2,2-Tetrachloroethane
79-36-7	.	X	Dichloroacetyl chloride
79-40-3	X	.	Dithiooxamide
79-41-4	.	X	Methacrylic acid
79-43-6	X	.	Dichloroacetic acid
79-44-7	X		Dimethylcarbamyl chloride
		X	Dimethylcarbamoyl chloride
79-46-9	X	X	2-Nitropropane
79-81-2	X	.	Retinol palmitate
79-94-7	.	X	3,3′,5,5′-Tetrabromobisphenol A
80-05-7	X	X	Bisphenol A
80-08-0	X	X	Dapsone
80-13-7	.	X	Halazone
80-15-9	.	X	Cumene hydroperoxide
80-30-8	.	X	N-Cyclohexyl-4-methyl benzenesulfonamide
80-33-1	X	.	p-Chlorophenyl-p-chlorobenzene sulfonate
80-39-7	.	X	N-Ethyl-4-methyl benzenesulfonamide
80-43-3	.	X	Dicumyl peroxide
80-46-6	.	X	p-tert-Pentyl phenol
80-47-7	.	X	p-Menthane hydroperoxide
80-52-4	.	X	1,8-p-Menthanediamine
80-62-6	X	X	Methyl methacrylate
81-07-2	X	X	Saccharin
81-11-8	.	X	4,4′-Diamino-2,2′-stilbenedisulfonic acid
81-14-1	.	X	Musk ketone
81-15-2	X	.	2,4,6-Trinitro-1,3-dimethyl-5-tert-butylbenzene
81-16-3	X	.	2-Naphthylamino,1-sulfonic acid
81-21-0	X	.	Dicyclopentadiene dioxide
81-49-2	.	X	1-Amino-2,4-dibromoanthraquinone
81-55-0	.	X	1,8-Dihydroxy-4,5-dinitroanthraquinone

CAS	In CPDB	In GT	Chemical Name
82-28-0	X	X	1-Amino-2-methylanthraquinone
82-62-2	.	X	3,4,6-Trichloro-2-nitrophenol
82-68-8	X	X	Pentachloronitrobenzene
83-26-1	.	X	2-Pivalyl-1,3-indandione
83-32-9	.	X	Acenaphthene
83-38-5	.	X	2,6-Dichlorobenzaldehyde
83-59-0	X	.	n-Propyl isome
83-66-9	.	X	Musk ambrette
83-72-7	.	X	2-Hydroxy-1,4-naphthoquinone
83-79-4	X	X	Rotenone
84-61-7	.	X	Dicyclohexyl phthalate
84-64-0	.	X	Butyl cyclohexyl phthalate
84-65-1	X		9,10-Anthraquinone
		X	Anthraquinone
84-66-2	.	X	Diethyl phthalate
84-69-5	.	X	Diisobutyl phthalate
84-74-2	.	X	Dibutyl phthalate
84-75-3	.	X	Di(n-hexyl)phthalate
85-01-8	.	X	Phenanthrene
85-02-9	.	X	Benzo(f)quinoline
85-22-3	.	X	2,3,4,5,6-Pentabromoethylbenzene
85-44-9	X	X	Phthalic anhydride
85-68-7	X	X	Butyl benzyl phthalate
85-98-3	.	X	N,N′-Diethylcarbanilide
86-00-0	.	X	2-Nitro-1,1′-biphenyl
86-29-3	X	.	Diphenylacetonitrile
86-30-6	X	X	N-Nitrosodiphenylamine
86-50-0	X	X	Azinphosmethyl
86-57-7	X	X	1-Nitronaphthalene
86-74-8	X	.	Carbazole
86-86-2	X	.	1-Naphthalene acetamide
86-87-3	X	.	1-Naphthalene acetic acid
86-88-4	X	.	1-(1-Naphthyl)-2-thiourea
87-25-2	.	X	Ethyl anthranilate
87-29-6	X	X	Cinnamyl anthranilate
87-51-4	X	.	Indole-3-acetic acid
87-56-9	X	.	α,β-Dichloro-β-formylacrylic acid
87-59-2	.	X	2,3-Xylidine
87-61-6	.	X	1,2,3-Trichlorobenzene
87-62-7	.	X	2,6-Xylidine
87-65-0	.	X	2,6-Dichlorophenol
87-68-3	X		Hexachlorobutadiene
		X	Hexachloro-1,3-butadiene
87-82-1	.	X	Hexabromobenzene
87-83-2	.	X	Pentabromotoluene
87-84-3	.	X	Pentabromochlorocyclohexane
87-86-5	X	.	2,3,4,5,6-Pentachlorophenol (Dowicide EC-7)
	X	X	2,3,4,5,6-Pentachlorophenol
	X	.	2,3,4,5,6-Pentachlorophenol, technical grade
88-06-2	X	X	2,4,6-Trichlorophenol
88-16-4	.	X	o-Chlorobenzotrifluoride
88-19-7	X	.	o-Toluenesulfonamide
88-21-1	.	X	o-Amino benzenesulfonic acid

CAS	In CPDB	In GT	Chemical Name
88-23-3	.	X	6-Amino-4-chloro-1-phenol-2-sulfonic acid
88-26-6	X	.	Di-*tert*-butyl-4-hydroxymethyl phenol
88-72-2	.	X	*o*-Nitrotoluene
88-73-3	X		1-Chloro-2-nitrobenzene
		X	2-Chloronitrobenzene
88-75-5	.	X	*o*-Nitrophenol
88-85-7	X	.	2-*sec*-Butyl-4,6-dinitrophenol
88-88-0	.	X	Picryl chloride
88-89-1	.	X	Picric acid
88-96-0	X	X	Phthalamide
88-99-3	.	X	Phthalic acid
89-25-8	X	X	1-Phenyl-3-methyl-5-pyrazolone
89-40-7	.	X	4-Nitrophthalimide
89-57-6	.	X	5-Aminosalicylic acid
89-63-4	.	X	4-Chloro-2-nitroaniline
89-72-5	.	X	*o-sec*-Butylphenol
89-98-5	.	X	2-Chlorobenzaldehyde
90-00-6	.	X	*o*-Ethyl phenol
90-04-0	.	X	*o*-Anisidine
90-05-1	.	X	*o*-Methoxyphenol
90-13-1	.	X	1-Chloronaphthalene
90-30-2	.	X	*N*-Phenyl-1-naphthylamine
90-33-5	.	X	β-Methylumbelliferone
90-41-5	.	X	2-Aminobiphenyl
90-43-7	X	X	*o*-Phenylphenol
90-45-9	.	X	9-Aminoacridine
90-49-3	X	.	Ethylphenylacetylurea
90-94-8	X		Michler's ketone
		X	4,4′-Bis(dimethylamino) benzophenone
91-08-7	.	X	2,6-Toluene diisocyanate
91-20-3	X	X	Naphthalene
91-22-5	.	X	Quinoline
91-23-6	X	X	*o*-Nitroanisole
91-44-1	.	X	7-Diethylamino-4-methylcoumarin
91-53-2	X	.	Ethoxyquin
91-58-7	.	X	2-Chloronaphthalene
91-59-8	X	X	2-Naphthylamine
91-62-3	X	X	6-Methylquinoline
91-64-5	X	X	Coumarin
91-66-7	.	X	*N,N*-Diethyl aniline
91-68-9	.	X	3-Diethylaminophenol
91-76-9	X	.	Benzoguanamine
91-79-2	X	.	Thenyldiamine
91-84-9	.	X	Pyrilamine
91-93-0	X	X	3,3′-Dimethoxybenzidine-4,4′-diisocyanate
91-94-1	X	.	3,3′-Dichlorobenzidine
92-13-7	X	.	Pilocarpine
92-48-8	.	X	6-Methyl coumarin
92-52-4	X	X	Biphenyl
92-55-7	X	.	5-Nitro-2-furanmethanediol diacetate
92-59-1	.	X	*N*-Ethyl-*N*-phenyl benzylamine
92-66-0	.	X	4-Bromobiphenyl
92-67-1	X		4-Aminodiphenyl

CAS	In CPDB	In GT	Chemical Name
		X	4-Aminobiphenyl
92-69-3	X	.	*p*-Phenylphenol
92-84-2	X	X	Phenothiazine
92-87-5	X	X	Benzidine
93-05-0	.	X	*N,N*-Diethyl-*p*-phenylenediamine
93-15-2	.	X	Methyl eugenol
93-45-8	.	X	4-(2-Naphthylamino) phenol
93-46-9	X		*sym.*-diβ-Naphthyl-*p*-phenylenediamine
		X	*N,N′*-Di-2-naphthyl-*p*-phenylenediamine
93-58-3	.	X	Methyl benzoate
93-72-1	X	.	2-(2,4,5-Trichlorophenoxy)propionic acid
93-76-5	X		2,4,5-Trichlorophenoxyacetic acid
		X	2,4,5-T
93-79-8	.	X	2,4,5-T, *n*-Butyl ester
94-11-1	X	.	2,4-Dichlorophenoxyacetic acid, isopropyl ester
94-20-2	X	X	Chlorpropamide
94-25-7	.	X	*n*-Butyl-*p*-aminobenzoate
94-26-8	X	.	Butyl *p*-hydroxybenzoate
94-36-0	X		Novadelox
		X	Benzoyl peroxide
94-52-0	X	X	6-Nitrobenzimidazole
94-58-6	X	.	Dihydrosafrole
94-59-7	X	X	Safrole
94-70-2	.	X	*o*-Phenetidine
94-75-7	X		2,4-Dichlorophenoxyacetic acid
		X	2,4-D
94-80-4	X		2,4-Dichlorophenoxyacetic acid, *n*-butyl ester
		X	2,4-D, *n*-Butyl ester
95-06-7	X	X	Sulfallate
95-14-7	X		1*H*-Benzotriazole
		X	1,2,3-Benzotriazole
95-19-2	.	X	Heptadecyl hydroxyethylimidazoline
95-33-0	X	.	*N*-Cyclohexyl-2-benzothiazole sulfenamide
95-46-5	.	X	*o*-Bromotoluene
95-47-6	.	X	*o*-Xylene
95-48-7	.	X	*o*-Cresol
95-50-1	X	X	1,2-Dichlorobenzene
95-51-2	.	X	*o*-Chloroaniline
95-53-4	.	X	*o*-Toluidine
95-54-5	.	X	*o*-Phenylenediamine
95-55-6	.	X	*o*-Aminophenol
95-57-8	.	X	*o*-Chlorophenol
95-64-7	.	X	3,4-Xylidine
95-68-1	.	X	2,4-Xylidine
95-71-6	X	.	Methylhydroquinone
95-74-9	X	X	3-Chloro-*p*-toluidine
95-76-1	.	X	3,4-Dichloroaniline
95-77-2	.	X	3,4-Dichlorophenol
95-78-3	.	X	2,5-Xylidine
95-79-4	X	X	5-Chloro-*o*-toluidine
95-80-7	X	X	2,4-Diaminotoluene
95-82-9	.	X	2,5-Dichloroaniline
95-83-0	X	X	4-Chloro-*o*-phenylenediamine

CAS	In CPDB	In GT	Chemical Name
95-84-1	.	X	2-Amino-4-methylphenol
95-85-2	.	X	2-Amino-4-chlorophenol
95-94-3	.	X	1,2,4,5-Tetrachlorobenzene
95-95-4	.	X	2,4,5-Trichlorophenol
96-09-3	X	X	Styrene oxide
96-12-8	X	X	1,2-Dibromo-3-chloropropane
96-13-9	.	X	2,3-Dibromo-1-propanol
96-18-4	X	X	1,2,3-Trichloropropane
96-23-1	.	X	1,3-Dichloro-2-propanol
96-24-2	X		Glycerol α-monochlorohydrin
		X	3-Chloro-1,2-propanediol
96-31-1	.	X	N,N′-Dimethylurea
96-33-3	.	X	Methyl acrylate
96-37-7	.	X	Methyl cyclopentane
96-45-7	X	X	Ethylene thiourea
96-48-0	X		γ-Butyrolactone
		X	4-Butyrolactone
96-67-3	.	X	6-Amino-4-nitro-1-phenol-2-sulfonic acid
96-69-5	X	X	4,4′-Thiobis(6-tert-butyl-m-cresol)
96-91-3	.	X	2-Amino-4,6-dinitrophenol
97-00-7	X	.	1-Chloro-2,4-dinitrobenzene
97-02-9	.	X	2,4-Dinitroaniline
97-16-5	X	.	2,4-Dichlorophenylbenzene sulfonate
97-18-7	X	X	2,2′-Thiobis(4,6-dichlorophenol)
97-23-4	.	X	2,2′-Methylenebis-(4-chlorophenol)
97-24-5	.	X	2,2′-Thiobis(4-chlorophenol)
97-32-5	.	X	4-Methoxy-3-nitro-N-phenylbenzamide
97-42-7	.	X	Carvyl acetate
97-53-0	X	X	Eugenol
97-54-1	.	X	Isoeugenol
97-56-3	X	X	o-Aminoazotoluene
97-59-6	X	.	Allantoin
97-63-2	.	X	Ethyl methacrylate
97-74-5	X	.	Tetramethylthiuram monosulfide
97-77-8	X		Tetraethylthiuram disulfide
		X	Tetraethylthiuram disulfide No. 2
97-84-7	.	X	N,N,N′,N′-Tetramethyl-1,3-butanediamine
97-86-9	.	X	Isobutyl methacrylate
97-88-1	.	X	Butyl methacrylate
98-00-0	.	X	Furfuryl alcohol
98-01-1	X	X	Furfural
98-07-7	.	X	Benzotrichloride
98-08-8	.	X	Benzotrifluoride
98-11-3	.	X	Benzene sulfonic acid
98-15-7	.	X	m-Chlorobenzotrifluoride
98-16-8	.	X	3-Amino-α,α,α-trifluorotoluene
98-30-6	.	X	2-Amino-4-(methylsulfonyl)phenol
98-37-3	.	X	2-Amino-1-phenol-4-sulfonic acid
98-46-4	.	X	3-Nitro-α,α,α-trifluorotoluene
98-51-1	.	X	p-tert-Butyl toluene
98-54-4	X	.	p-tert-Butylphenol
98-56-6	.	X	4-Chloro-α,α,α-trifluorotoluene
98-83-9	.	X	α-Methyl styrene
98-85-1	X	X	α-Methylbenzyl alcohol
98-87-3	.	X	α,α-Dichlorotoluene
98-92-0	X	.	Nicotinamide
98-95-3	.	X	Nitrobenzene
98-96-4	X	X	Pyrazinamide
99-07-0	.	X	3-Dimethylaminophenol
99-08-1	.	X	m-Nitrotoluene
99-30-9	X		2,6-Dichloro-4-nitroaniline
		X	Dichloran
99-48-9	.	X	Carveol
99-50-3	X	.	Protocatechuic acid
99-54-7	.	X	3,4-Dichloronitrobenzene
99-55-8	X		5-Nitro-o-toluidine
		X	2-Methyl-5-nitroaniline
99-56-9	X	X	4-Nitro-o-phenylenediamine
99-57-0	X	X	2-Amino-4-nitrophenol
99-59-2	X	X	5-Nitro-o-anisidine
99-75-2	.	X	Methyl-p-toluate
99-80-9	X	.	N-Methyl-N,4-dinitrosoaniline
99-82-1	.	X	p-Menthane
99-98-9	.	X	N,N-Dimethyl-p-phenylenediamine
99-99-0	.	X	p-Nitrotoluene
100-00-5	X		1-Chloro-4-nitrobenzene
		X	4-Chloronitrobenzene
100-01-6	X	X	p-Nitroaniline
100-02-7	.	X	p-Nitrophenol
100-14-1	.	X	p-Nitrobenzyl chloride
100-19-6	.	X	p-Nitroacetophenone
100-21-0	.	X	Terephthalic acid
100-22-1	.	X	N,N,N′,N′-Tetramethyl-p-phenylenediamine
100-27-6	.	X	p-Nitrophenethyl alcohol
100-37-8	.	X	Diethylaminoethanol
100-39-0	.	X	α-Bromotoluene
100-40-3	X	X	4-Vinylcyclohexene
100-41-4	X	X	Ethyl benzene
100-42-5	X	X	Styrene
100-44-7	X	X	Benzyl chloride
100-47-0	.	X	Benzonitrile
100-51-6	X	X	Benzyl alcohol
100-52-7	X	X	Benzaldehyde
100-61-8	.	X	N-Methyl aniline
100-63-0	X	.	Phenylhydrazine
100-64-1	.	X	Cyclohexanone oxime
100-65-2	.	X	N-Phenylhydroxylamine
100-74-3	.	X	N-Ethylmorpholine
100-75-4	X	X	N-Nitrosopiperidine
100-97-0	X		Hexamethylenetetramine
		X	Urotropine
101-02-0	.	X	Triphenyl phosphite
101-05-3	X	X	Anilazine
101-14-4	X	X	4,4′-Methylene-bis(2-chloroaniline)
101-18-8	.	X	3-Hydroxy-N-phenylaniline
101-20-2	.	X	Triclocarban

CAS	In CPDB	In GT	Chemical Name
107-68-6	.	X	N-Methyltaurine
107-70-0	.	X	4-Methoxy-4-methyl-2-pentanone
108-01-0	X		2-Dimethylaminoethanol
		X	Dimethylethanolamine
108-03-2	X	X	1-Nitropropane
108-05-4	X	.	Vinyl acetate
108-09-8	.	X	1,3-Dimethylbutylamine
108-10-1	.	X	Methyl isobutyl ketone
108-18-9	.	X	Diisopropyl amine
108-21-4	.	X	Isopropyl acetate
108-24-7	.	X	Acetic anhydride
108-30-5	X	X	Succinic anhydride
108-31-6	.	X	Maleic anhydride
108-38-3	.	X	m-Xylene
108-39-4	.	X	m-Cresol
108-42-9	.	X	m-Chloroaniline
108-43-0	.	X	m-Chlorophenol
108-45-2	X	X	m-Phenylenediamine
108-46-3	X	X	Resorcinol
108-60-1	X	X	Bis(2-chloro-1-methylethyl)ether
108-69-0	.	X	3,5-Xylidine
108-70-3	.	X	1,3,5-Trichlorobenzene
108-78-1	X	X	Melamine
108-80-5	.	X	Cyanuric acid
108-83-8	.	X	Diisobutyl ketone
108-86-1	.	X	Bromobenzene
108-88-3	X	X	Toluene
108-90-7	X	X	Chlorobenzene
108-91-8	.	X	Cyclohexylamine
108-93-0	.	X	Cyclohexanol
108-94-1	X	X	Cyclohexanone
108-95-2	X	X	Phenol
108-99-6	.	X	β-Picoline
109-55-7	.	X	3-Dimethylaminopropylamine
109-57-9	.	X	Allylthiourea
109-69-3	X	X	n-Butyl chloride
109-73-9	.	X	n-Butylamine
109-76-2	.	X	1,3-Diaminopropane
109-77-3	.	X	Malonic dinitrile
109-78-4	.	X	3-Hydroxypropanenitrile
109-83-1	.	X	N-Methyl ethanolamine
109-84-2	X	.	2-Hydroxyethylhydrazine
109-86-4	.	X	2-Methoxyethanol
109-89-7	.	X	Diethylamine
109-99-9	.	X	Tetrahydrofuran
110-00-9	X	X	Furan
110-02-1	.	X	Thiophene
110-05-4	.	X	Di-tert-butyl peroxide
110-17-8	.	X	Fumaric acid
110-18-9	.	X	N,N,N',N'-Tetramethylethylenediamine
110-21-4	.	X	Biurea
110-26-9	.	X	N,N'-Methylene-bis-acrylamide
110-44-1	X	.	Sorbic acid

CAS	In CPDB	In GT	Chemical Name
110-46-3	.	X	Isoamyl nitrite
110-49-6	.	X	2-Methoxyethyl acetate
110-54-3	.	X	n-Hexane
110-57-6	X	.	trans-1,4-Dichlorobutene-2
110-58-7	.	X	n-Amylamine
110-59-8	.	X	Valeronitrile
110-61-2	.	X	Succinonitrile
110-63-4	.	X	1,4-Butanediol
110-80-5	.	X	2-Ethoxyethanol
110-82-7	.	X	Cyclohexane
110-85-0	X	X	Piperazine
110-86-1	.	X	Pyridine
110-88-3	.	X	1,3,5-Trioxane
110-89-4	X	.	Piperidine
110-91-8	.	X	Morpholine
110-96-3	.	X	Diisobutyl amine
110-97-4	X		Bis(2-hydroxypropyl) amine
		X	Diisopropanol amine
111-14-8	.	X	n-Heptanoic acid
111-30-8	.	X	Glutaraldehyde
111-40-0	.	X	Diethylenetriamine
111-41-1	.	X	N-Hydroxyethyl ethylenediamine
111-42-2	.	X	Diethanolamine
111-44-4	X	X	Bis-2-chloroethylether
111-46-6	X	X	Diethylene glycol
111-64-8	.	X	Caprylyl chloride
111-69-3	.	X	Adiponitrile
111-71-7	.	X	n-Heptanal
111-76-2	.	X	2-Butoxyethanol
111-84-2	.	X	n-Nonane
111-92-2	.	X	Di-n-butylamine
111-96-6	.	X	Bis(2-methoxyethyl)ether
112-24-3	.	X	Triethylenetetramine
112-27-6	X	.	Triethylene glycol
112-34-5	.	X	2-(2-Butoxyethoxy)ethanol
112-52-7	.	X	Lauryl chloride
112-55-0	.	X	n-Dodecylmercaptan
112-56-1	.	X	2-(2-Butoxyethoxy) ethyl thiocyanate
112-57-2	.	X	Tetraethylenepentamine
112-62-9	.	X	Oleic acid methyl ester (cis)
112-80-1	.	X	Oleic acid
112-95-8	.	X	Eicosane
113-92-8	X	X	Chlorpheniramine maleate
114-83-0	X	X	1-Acetyl-2-phenylhydrazine
115-02-6	X	.	Azaserine
115-07-1	X	X	Propylene
115-09-3	X		Mercurymethylchloride
		X	Methyl mercuric (II) chloride
115-28-6	X	X	Chlorendic acid
115-29-7	X	X	Endosulfan
115-32-2	X	X	Dicofol
115-86-6	.	X	Triphenyl phosphate
115-96-8	X	X	Tris(2-chloroethyl)phosphate

CAS	In CPDB	In GT	Chemical Name
116-06-3	X	X	Aldicarb
116-29-0	X	.	2,4,5,4'-Tetrachlorodiphenyl sulfone
117-08-8	.	X	Tetrachlorophthalic anhydride
117-10-2	X	.	Chrysazin
117-18-0	.	X	2,3,5,6-Tetrachloronitrobenzene
117-39-5	X	X	Quercetin
117-79-3	X	X	2-Aminoanthraquinone
117-80-6	X	.	2,3-Dichloro-1,4-naphthoquinone
117-81-7	X	X	Di(2-ethylhexyl)phthalate
117-84-0	.	X	Di-octyl phthalate
118-52-5	.	X	1,3-Dichloro-5,5-dimethylhydantoin
118-55-8	.	X	Phenyl salicylate
118-56-9	.	X	3,3,5-Trimethylcyclohexyl salicylate
118-58-1	.	X	Benzyl salicylate
118-71-8	.	X	Maltol
118-74-1	X	X	Hexachlorobenzene
118-75-2	X	.	Chloranil
118-79-6	.	X	2,4,6-Tribromophenol
118-91-2	.	X	o-Chlorobenzoic acid
118-92-3	X		Anthranilic acid
		X	o-Anthranilic acid
119-06-2	.	X	Ditridecyl phthalate
119-15-3	.	X	4-(2,4-Dinitroanilino)phenol
119-34-6	X	X	4-Amino-2-nitrophenol
119-36-8	.	X	Methyl salicylate
119-38-0	X	.	1-Isopropyl-3-methyl-S-pyrazolyldimethyl carbamate
119-39-1	.	X	1(2H)-Phthalazinone
119-53-9	X	X	Benzoin
119-61-9	.	X	Benzophenone
119-75-5	.	X	2-Nitrodiphenylamine
119-84-6	X	X	3,4-Dihydrocoumarin
119-90-4	.	X	3,3'-Dimethoxybenzidine
119-93-7	.	X	3,3'-Dimethylbenzidine
120-12-7	.	X	Anthracene
120-14-9	.	X	Veratraldehyde
120-32-1	X	X	o-Benzyl-p-chlorophenol
120-36-5	X	.	α-(2,4-Dichlorophenoxy) propionic acid
120-37-6	.	X	3-Ethylamino-4-methylphenol
120-40-1	.	X	Lauric acid/Diethanolamine condensate (1/1)
120-57-0	.	X	Piperonal
120-58-1	X	.	Isosafrole
120-61-6	X	X	Dimethyl terephthalate
120-62-7	X	X	Piperonyl sulfoxide
120-66-1	.	X	N-Acetyl-o-toluidine
120-71-8	X	X	p-Cresidine
120-78-5	X		Benzothiazyl disulfide
		X	2,2'-Dithiobisbenzothiazole
120-80-9	X	X	Catechol
120-82-1	.	X	1,2,4-Trichlorobenzene
120-83-2	X	X	2,4-Dichlorophenol
120-93-4	X	.	Ethylene urea
121-14-2	X	X	2,4-Dinitrotoluene

CAS	In CPDB	In GT	Chemical Name
121-17-5	.	X	4-Chloro-3-nitro-α,α,α-trifluorotoluene
121-19-7	X		3-Nitro-4-hydroxyphenylarsonic acid
		X	4-Hydroxy-3-nitrophenyl arsonic acid
121-32-4	.	X	Ethyl vanillin
121-33-5	.	X	Vanillin
121-39-1	.	X	3-Phenylglycidic acid, ethyl ester
121-44-8	.	X	Triethylamine
121-47-1	.	X	m-Amino benzenesulfonic acid
121-54-0	.	X	Benzethonium chloride
121-57-3	.	X	p-Amino benzenesulfonic acid
121-59-5	X	.	Carbarsone
121-66-4	X	X	2-Amino-5-nitrothiazole
121-69-7	X	X	N,N-Dimethylaniline
121-73-3	.	X	3-Chloronitrobenzene
121-75-5	X	X	Malathion
121-79-9	X	X	Propyl gallate
121-88-0	X	X	2-Amino-5-nitrophenol
121-89-1	.	X	m-Nitroacetophenone
121-90-4	.	X	m-Nitrobenzoyl chloride
121-92-6	.	X	m-Nitrobenzoic acid
122-04-3	.	X	p-Nitrobenzoyl chloride
122-19-0	.	X	Dimethyloctadecyl benzylammonium chloride
122-20-3	X		Tris(2-hydroxypropyl)amine
		X	Tri-isopropanolamine
122-34-9	X	.	Simazine
122-39-4	.	X	N-Phenylbenzenamine
122-42-9	X	.	Isopropyl-N-phenyl carbamate
122-60-1	X	X	Phenylglycidyl ether
122-62-3	.	X	Di(2-ethylhexyl)sebacate
122-66-7	X	X	Hydrazobenzene
122-80-5	.	X	p-Aminoacetanilide
123-05-7	.	X	2-Ethylhexanal
123-30-8	.	X	p-Aminophenol
123-31-9	X	X	Hydroquinone
123-33-1	X	X	Maleic hydrazide
123-38-6	.	X	Propionaldehyde
123-72-8	.	X	Butyraldehyde
123-73-9	X	.	Crotonaldehyde
123-77-3	.	X	Azodicarbonamide
123-86-4	.	X	n-Butyl acetate
123-88-6	.	X	Methoxyethyl mercury chloride
123-91-1	X	X	1,4-Dioxane
123-92-2	.	X	Isoamyl acetate
124-02-7	.	X	Diallylamine
124-07-2	.	X	Octanoic acid
124-09-4	.	X	1,6-Diaminohexane
124-18-5	.	X	n-Decane
124-19-6	.	X	Nonanal
124-30-1	.	X	Octadecylamine
124-40-3	.	X	Dimethylamine
124-48-1	X	X	Chlorodibromomethane
124-64-1	X	X	Tetrakis(hydroxymethyl) phosphonium chloride
125-33-7	.	X	Primaclone

CAS	In CPDB	In GT	Chemical Name
126-07-8	X	X	Griseofulvin
126-27-2	.	X	Oxethazaine
126-72-7	X	X	Tris(2,3-dibromopropyl) phosphate
126-73-8	.	X	Tributyl phosphate
126-85-2	X	.	Nitrogen mustard *N*-oxide
126-92-1	.	X	Sodium (2-ethylhexyl) alcohol sulfate
126-98-7	.	X	Methacrylonitrile
126-99-8	.	X	2-Chloro-1,3-butadiene
127-00-4	.	X	1-Chloro-2-propanol
127-06-0	X	.	Acetoxime
127-07-1	.	X	Hydroxyurea
127-18-4	X	X	Tetrachloroethylene
127-19-5	.	X	Dimethyl acetamide
127-47-9	X	.	Retinol acetate
127-48-0	X	.	Trimethadione
127-69-5	X	X	Sulfisoxazole
127-85-5	.	X	Sodium arsanilate
128-37-0	X	X	Butylated hydroxytoluene
128-44-9	X	.	Saccharin, sodium
128-66-5	X		C.I. Vat yellow 4
		X	Vat yellow 4
129-00-0	.	X	Pyrene
129-15-7	X	X	2-Methyl-1-nitroanthraquinone
129-17-9	.	X	Sulfan blue
129-43-1	X	.	1-Hydroxyanthraquinone
130-26-7	.	X	Iodochlorohydroxyquinoline
131-01-1	X	.	Deserpidine
131-11-3	.	X	Dimethyl phthalate
131-17-9	X	X	Diallyl phthalate
131-53-3	.	X	2,2′-Dihydroxy-4-methoxybenzophenone
131-57-7	.	X	2-Hydroxy-4-methoxybenzophenone
131-92-0	.	X	Vat brown 3
132-20-7	.	X	Pheniramine maleate
132-27-4	X		*o*-Phenylphenate, sodium
		X	Sodium 2-biphenylol
132-64-9	.	X	Dibenzofuran
132-98-9	X	X	Penicillin VK
133-06-2	X	X	Captan
133-07-3	X	.	*N*-(Trichloromethylthio) phthalimide
133-18-6	.	X	Phenethyl anthranilate
133-90-4	X		Chloramben
		X	Amiben
134-03-2	X	.	*l*-Ascorbate, sodium
134-20-3	.	X	Methyl anthranilate
134-29-2	X	X	*o*-Anisidine.HCl
134-31-6	.	X	8-Hydroxyquinoline sulfate
134-32-7	.	X	1-Naphthylamine
134-62-3	.	X	*N,N*-Diethyl-*m*-toluamide
134-72-5	X		Ephedrine sulphate
		X	Ephedrine sulfate
135-20-6	X	X	Cupferron
135-23-9	X	X	Methapyrilene.HCl
135-88-6	X		Phenyl-β-naphthylamine

CAS	In CPDB	In GT	Chemical Name
		X	*N*-Phenyl-2-naphthylamine
136-23-2	X	.	Zinc dibutyldithiocarbamate
136-35-6	.	X	Diazoaminobenzene
136-40-3	X	X	Phenazopyridine.HCl
136-77-6	X		4-Hexylresorcinol
		X	*p*-Hexylresorcinol
137-05-3	.	X	Methyl 2-cyanoacrylate
137-09-7	X	X	2,4-Diaminophenol.2HCl
137-17-7	X	X	2,4,5-Trimethylaniline
137-26-8	X	.	Tetramethylthiuram disulfide
137-29-1	X	.	Copper dimethyldithiocarbamate
137-30-4	X		Zinc dimethyldithiocarbamate
		X	Ziram
137-89-3	.	X	Bis(2-ethylhexyl)isophthalate
138-89-6	.	X	*N,N*-Dimethyl-*p*-nitrosoaniline
139-05-9	X	.	Cyclamate, sodium
139-13-9	X	X	Nitrilotriacetic acid
139-40-2	X	.	Propazine
139-65-1	X	X	4,4′-Thiodianiline
139-94-6	X	X	Nithiazide
140-08-9	.	X	Tris (2-chloroethyl) phosphite
140-11-4	X	X	Benzyl acetate
140-29-4	.	X	Phenylacetonitrile
140-49-8	X	X	4′-(Chloroacetyl)-acetanilide
140-56-7	X		Fenaminosulf, formulated
		X	Formulated fenaminosulf
140-57-8	X	.	Aramite
140-67-0	X	X	Estragole
140-79-4	.	X	Dinitrosopiperazine
140-88-5	X	X	Ethyl acrylate
141-05-9	X	.	Diethylmaleate
141-32-2	.	X	*n*-Butyl acrylate
141-38-8	.	X	9,10-Epoxyoctadecanoic acid, 2-ethylhexyl ester
141-43-5	.	X	Monoethanolamine
141-75-3	.	X	Butyryl chloride
141-78-6	.	X	Ethyl acetate
141-84-4	.	X	Rhodanine
141-90-2	X	.	Thiouracil
141-91-3	.	X	2,6-Dimethyl morpholine
142-04-1	X	X	Aniline.HCl
142-09-6	.	X	*n*-Hexyl methacrylate
142-46-1	X	X	2,5-Dithiobiurea
142-47-2	X		*l*-Monosodium glutamate
		X	Monosodium glutamate
142-59-6	X	.	Ethylenebisdithiocarbamate, disodium
142-78-9	.	X	Lauryl ethanolamide
142-84-7	.	X	Di-*n*-propylamine
142-96-1	.	X	Di-*n*-butyl ether
143-07-7	.	X	Lauric acid
143-16-8	.	X	Dihexylamine
143-19-1	X	.	Oleate, sodium
143-27-1	.	X	Hexadecylamine
143-33-9	.	X	Sodium cyanide

CAS	In CPDB	In GT	Chemical Name
143-50-0	X		Kepone
		X	Chlordecone
144-02-5	X	.	Barbital, sodium
144-34-3	X	.	Selenium dimethyldithiocarbamate
144-48-9	X	.	Iodoacetamide
144-55-8	X	.	Sodium bicarbonate
144-62-7	.	X	Oxalic acid, anhydrous
144-80-9	.	X	Sulfacetamide
144-82-1	.	X	Sulfamethizole
144-83-2	.	X	Sulfapyridine
147-24-0	X	X	Diphenhydramine.HCl
147-47-7	.	X	1,2-Dihydro-2,2,4-trimethylquinoline, monomer
148-18-5	X		Sodium diethyldithiocarbamate trihydrate
		X	Sodium diethyldithiocarbamate
148-24-3	X	X	8-Hydroxyquinoline
148-79-8	X	X	Thiabendazole
148-82-3	X	X	Melphalan
149-17-7	X	.	Isonicotinic acid vanillylidenehydrazide
149-29-1	X	.	Patulin
149-30-4	X	X	2-Mercaptobenzothiazole
149-57-5	.	X	2-Ethylhexanoic acid
149-91-7	X	X	Gallic acid
150-13-0	.	X	p-Amino benzoic acid
150-38-9	X		EDTA, trisodium salt trihydrate
		X	Ethylenediamine tetraacetic acid, 3Na salt
150-68-5	X		3-(p-Chlorophenyl)-1,1-dimethylurea
		X	Monuron
150-69-6	.	X	Dulcin
150-76-5	X		4-Methoxyphenol
		X	Hydroquinone monomethyl ether
150-78-7	.	X	Hydroquinone dimethyl ether
151-21-3	.	X	Sodium dodecyl sulfate
151-56-4	X	.	Ethylene imine
151-67-7	.	X	Halothane
153-18-4	X	.	Rutin trihydrate
154-69-8	.	X	Tripelennamine.HCl
155-04-4	X	.	2-Mercaptobenzothiazole, zinc
156-10-5	X	X	p-Nitrosodiphenylamine
156-43-4	.	X	p-Phenetidine
156-51-4	X	.	Phenylethylhydrazine sulfate
156-59-2	.	X	cis-1,2-Dichloroethylene
156-60-5	.	X	trans-1,2-Dichloroethylene
156-62-7	X		Cyanamide, calcium
		X	Calcium cyanamide
192-97-2	.	X	Benzo(e)pyrene
203-64-5	.	X	4-H-Cyclopenta(d,e,f) phenanthrene
205-99-2	.	X	Benzo(b)fluoranthene
207-08-9	.	X	Benzo(k)fluoranthene
244-63-3	X	.	Norharman
262-12-4	.	X	Dibenzo-p-dioxin
262-20-4	.	X	Dibenzooxathiane
271-89-6	X		Benzofuran
		X	2,3-Benzofuran

CAS	In CPDB	In GT	Chemical Name
297-76-7	X	.	Ethynodiol diacetate
297-78-9	X	.	Telodrin
298-00-0	X	X	Methyl parathion
298-18-0	X		d,l-Diepoxybutane
		X	d,l-1,2:3,4-Diepoxybutane
298-59-9	.	X	Methylphenidate.HCl
298-81-7	X	X	8-Methoxypsoralen
301-04-2	X		Lead acetate
		X	Lead (II) acetate
301-12-2	.	X	Metasystox-R
301-13-3	.	X	Tris (2-ethylhexyl) phosphite
302-01-2	X	.	Hydrazine
302-15-8	X	.	Methylhydrazine sulfate
302-17-0	X	X	Chloral hydrate
302-22-7	X	.	Chlormadinone acetate
302-79-4	X	.	Retinoic acid
303-34-4	X	X	Lasiocarpine
303-47-9	X	X	Ochratoxin A
305-03-3	X	X	Chlorambucil
306-37-6	X	.	1,2-Dimethylhydrazine.2HCl
309-00-2	X	X	Aldrin
309-36-4	.	X	Sodium methohexital
314-13-6	.	X	C.I. Direct blue 53
315-18-4	X	.	Mexacarbate
315-22-0	X	.	Monocrotaline
316-42-7	X	X	Emetine.2HCl
318-98-9	X	.	Propranolol.HCl
319-84-6	X	.	α-1,2,3,4,5,6-Hexachlorocyclohexane
319-85-7	X	.	β-1,2,3,4,5,6-Hexachlorocyclohexane
320-67-2	X	X	5-Azacytidine
324-93-6	X	.	4'-Fluoro-4-aminodiphenyl
326-61-4	.	X	Piperonyl acetate
330-54-1	X	.	3-(3,4-Dichlorophenyl)-1,1-dimethylurea
331-39-5	X	.	Caffeic acid
333-41-5	X	X	Diazinon
334-48-5	.	X	Decanoic acid
346-18-9	.	X	Polythiazide
357-57-3	.	X	Brucine
363-17-7	X	.	N-(2-Fluorenyl)-2,2,2-trifluoroacetamide
366-70-1	X	X	Procarbazine.HCl
367-25-9	.	X	2,4-Difluoroaniline
367-51-1	.	X	Thioglycolic acid, Na salt
369-57-3	X	.	Benzenediazonium tetrafluoroborate
373-02-4	X	.	Nickel (II) acetate
379-79-3	.	X	Ergotamine tartrate
384-22-5	.	X	2-Nitro-α,α,α-trifluorotoluene
389-08-2	X	X	Nalidixic acid
393-52-2	.	X	2-Fluorobenzoyl chloride
393-75-9	.	X	4-Chloro-3,5-dinitro-α,α,α-trifluorotoluene
396-01-0	X	X	Triamterene
398-32-3	X	.	N-4-(4'-Fluorobiphenyl) acetamide
404-86-4	X	.	Capsaicin
434-07-1	.	X	Oxymetholone

CAS	In CPDB	In GT	Chemical Name
434-13-9	X	X	Lithocholic acid
439-14-5	X	X	Diazepam
443-48-1	X	X	Metronidazole
443-72-1	X	.	(N-6)-Methyladenine
446-86-6	X	X	Azathioprine
458-37-7	.	X	Curcumin
463-04-7	.	X	n-Amyl nitrite
464-10-8	.	X	Bromopicrin
470-82-6	X		Eucalyptol
		X	1,8-Cineol
471-29-4	X	.	Methylguanidine
471-53-4	X	.	Glycyrrhetinic acid
474-25-9	X	.	Chenodeoxycholic acid
476-66-4	.	X	Ellagic acid
477-30-5	X	.	Colcemid
480-22-8	.	X	1,8,9-Trihydroxyanthracene
480-81-9	.	X	Seneciphylline
485-47-2	.	X	Ninhydrin
488-41-5	X	X	Dibromomannitol
493-78-7	X	.	Methaphenilene
495-18-1	.	X	N-Hydroxybenzamide
495-48-7	.	X	Azoxybenzene
496-72-0	.	X	3,4-Diaminotoluene
498-21-5	.	X	Methyl succinic acid
500-66-3	.	X	Olivetol
503-09-3	.	X	1,2-Epoxy-3-fluoropropane
503-30-0	.	X	1,3-Propylene oxide
504-29-0	.	X	2-Aminopyridine
504-88-1	X	X	3-Nitropropionic acid
505-22-6	.	X	1,3-Dioxane
509-14-8	X	X	Tetranitromethane
510-15-6	X	X	Chlorobenzilate
512-56-1	X	X	Trimethylphosphate
513-37-1	X		Dimethylvinyl chloride
		X	2,2-Dimethyl vinyl chloride
517-28-2	X	X	Hematoxylin
518-47-8	.	X	Fluorescein, disodium salt
518-75-2	X	.	Citrinin
520-18-3	X	.	Kaempferol
520-45-6	X	.	3-Acetyl-6-methyl-2,4-pyrandione
521-31-3	.	X	o-Aminophthalyl hydrazide
523-47-7	.	X	β-Cadinene
523-87-5	.	X	Dimenhydrinate
527-85-5	.	X	2-Methylbenzamide
528-74-5	.	X	Dichloromethotrexate
529-19-1	.	X	o-Tolunitrile
529-20-4	.	X	o-Tolualdehyde
530-66-5	.	X	Quinoline sulfate
531-06-6	X	.	Methafurylene
531-18-0	X	.	Hexamethylmelamine
531-82-8	X	.	N-[4-(5-Nitro-2-furyl)-2-thiazolyl]acetamide
531-85-1	X	X	Benzidine.2HCl
532-27-4	X		2-Chloroacetophenone

CAS	In CPDB	In GT	Chemical Name
		X	Chloroacetophenone
532-28-5	.	X	α-Hydroxybenzeneacetonitrile
532-32-1	X	.	Benzoate, sodium
532-82-1	.	X	C.I. Basic orange 2
533-31-3	X	.	Sesamol
534-22-5	.	X	Methyl furan
535-80-8	.	X	m-Chlorobenzoic acid
536-33-4	X	X	Ethionamide
536-90-3	.	X	m-Anisidine
537-92-8	.	X	N-Acetyl-m-toluidine
538-23-8	X	.	Tricaprylin
538-41-0	X	.	4,4'-Diaminoazobenzene
538-74-9	.	X	Benzyl sulfide
540-23-8	X	X	p-Toluidine.HCl
540-51-2	X		Bromoethanol
		X	2-Bromo-1-ethanol
540-59-0	.	X	cis,trans-1,2-Dichloroethylene
540-72-7	X	.	Thiocyanate, sodium
541-69-5	X	.	m-Phenylenediamine.2HCl
541-73-1	.	X	1,3-Dichlorobenzene
542-56-3	.	X	Isobutyl nitrite
542-75-6	X		Telone II
		X	1,3-Dichloropropene
542-88-1	X	.	Bis-(chloromethyl)ether
543-80-6	X	.	Barium acetate
543-90-8	X	.	Cadmium acetate
544-63-8	.	X	Tetradecanoic acid
545-06-2	.	X	Trichloroacetonitrile
548-62-9	X		Gentian violet
		X	Hexamethyl-p-rosaniline chloride
551-06-4	.	X	α-Naphthyl isothiocyanate
551-92-8	X	.	1,2-Dimethyl-5-nitroimidazole
552-16-9	.	X	o-Nitrobenzoic acid
552-30-7	.	X	Trimellitic anhydride
553-53-7	X	.	Nicotinic acid hydrazide
554-00-7	.	X	2,4-Dichloroaniline
554-10-9	.	X	3-Iodo-1,2-propanediol
554-84-7	.	X	m-Nitrophenol
555-30-6	.	X	Methyl DOPA sesquihydrate
555-84-0	X	.	1-[(5-Nitrofurfurylidene) amino]-2-imidazolidinone
556-52-5	X	X	Glycidol
557-11-9	.	X	Allyl urea
562-10-7	.	X	Doxylamine succinate
563-04-2	.	X	Tri-m-cresyl phosphate
563-41-7	X		Carbamyl hydrazine.HCl
		X	Semicarbazide.HCl
563-47-3	X	X	3-Chloro-2-methylpropene
569-57-3	X		Tace
		X	Chlorotrianisene
569-61-9	X		p-Rosaniline.HCl
		X	C.I. Basic red 9
576-24-9	.	X	2,3-Dichlorophenol

CAS	In CPDB	In GT	Chemical Name
576-68-1	X	.	Mannitol nitrogen mustard
577-59-3	.	X	o-Nitroacetophenone
578-76-7	X	.	7-Methylguanine
583-39-1	.	X	2-Mercaptobenzimidazole
583-78-8	.	X	2,5-Dichlorophenol
584-84-9	.	X	2,4-Toluene diisocyanate
589-18-4	.	X	p-Methylbenzyl alcohol
590-17-0	.	X	Bromoacetonitrile
590-21-6	X	.	1-Chloropropene
591-17-3	.	X	m-Bromotoluene
591-27-5	.	X	m-Aminophenol
591-35-5	.	X	3,5-Dichlorophenol
592-31-4	X	.	N-Butylurea
592-62-1	.	X	Methylazoxymethanol acetate
593-56-6	.	X	O-Methyl hydroxylamine.HCl
593-60-2	X	.	Vinyl bromide
593-70-4	X	.	Chlorofluoromethane
594-71-8	.	X	2-Chloro-2-nitropropane
594-72-9	.	X	1,1-Dichloro-1-nitroethane
597-25-1	X	X	Dimethyl morpholinophosphoramidate
598-55-0	X	X	Methyl carbamate
598-57-2	X	.	Methylnitramine
598-64-1	X	.	Dimethyldithiocarbamic acid, dimethylamine
599-79-1	.	X	Salicylazosulfapyridine
600-25-9	.	X	1-Chloro-1-nitropropane
602-38-0	.	X	1,8-Dinitronaphthalene
602-60-8	.	X	9-Nitroanthracene
602-87-9	X	X	5-Nitroacenaphthene
603-34-9	.	X	Triphenylamine
603-35-0	.	X	Triphenyl phosphine
603-54-3	.	X	asym-Diphenylurea
604-75-1	X	.	Oxazepam
605-71-0	.	X	1,5-Dinitronaphthalene
606-20-2	X	.	2,6-Dinitrotoluene
606-37-1	.	X	1,3-Dinitronaphthalene
607-35-2	X	.	8-Nitroquinoline
607-57-8	.	X	2-Nitrofluorene
608-71-9	.	X	Pentabromophenol
608-73-1	X	.	Hexachlorocyclohexane, technical grade
608-93-5	.	X	Pentachlorobenzene
609-19-8	.	X	3,4,5-Trichlorophenol
609-20-1	X	X	2,6-Dichloro-p-phenylenediamine
609-31-4	.	X	2-Nitro-1-butanol
610-14-0	.	X	o-Nitrobenzoyl chloride
610-66-2	.	X	o-Nitrophenyl acetonitrile
611-06-3	.	X	2,4-Dichloronitrobenzene
611-23-4	X	.	o-Nitrosotoluene
611-32-5	X	X	8-Methylquinoline
612-23-7	.	X	o-Nitrobenzyl chloride
612-60-2	.	X	7-Methylquinoline
612-64-6	.	X	N-Nitroso-N-ethylaniline
612-82-8	X	X	3,3'-Dimethylbenzidine.2HCl
612-83-9	.	X	3,3'-Dichlorobenzidine.2HCl

CAS	In CPDB	In GT	Chemical Name
613-13-8	.	X	2-Aminoanthracene
613-50-3	X	.	6-Nitroquinoline
613-93-4	.	X	N-Methylbenzamide
613-94-5	X	.	Benzoyl hydrazine
614-00-6	X	.	Nitrosomethylaniline
614-45-9	.	X	t-Butyl perbenzoate
614-95-9	X	.	Nitrosoethylurethan
615-28-1	X	.	o-Phenylenediamine.2HCl
615-53-2	X	.	N-Nitroso-N-methylurethan
616-23-9	.	X	2,3-Dichloro-1-propanol
617-84-5	X	.	Diethylformamide
619-17-0	X	X	4-Nitroanthranilic acid
619-23-8	.	X	m-Nitrobenzyl chloride
619-67-0	X	.	p-Hydrazinobenzoic acid
620-22-4	.	X	m-Tolunitrile
621-31-8	.	X	3-Ethylaminophenol
621-33-0	.	X	m-Phenetidine
621-42-1	.	X	N-Acetyl-m-aminophenol
621-64-7	X	.	N-Nitrosodipropylamine
621-77-2	.	X	Tri-n-amylamine
622-51-5	X	X	p-Tolylurea
622-97-9	X	.	p-Methylstyrene
623-15-4	.	X	Furfural acetone
623-17-6	.	X	Furfuryl acetate
623-30-3	.	X	β-2-Furyl acrolein
624-18-0	X	X	p-Phenylenediamine.2HCl
624-83-9	.	X	Methyl isocyanate
624-84-0	X	.	Formylhydrazine
625-48-9	.	X	2-Nitroethanol
625-86-5	.	X	2,5-Dimethylfuran
625-89-8	X	.	N-Nitrosobis(2,2,2-trifluoroethyl)amine
627-05-4	.	X	1-Nitrobutane
627-18-9	.	X	3-Bromo-1-propanol
627-30-5	.	X	3-Chloro-1-propanol
628-02-4	X	X	Hexanamide
628-36-4	X	.	1,2-Diformylhydrazine
628-94-4	X	X	Adipamide
630-20-6	X	X	1,1,1,2-Tetrachloroethane
632-79-1	.	X	Tetrabromophthalic anhydride
632-99-5	X		Rosaniline.HCl
		X	C.I. Basic violet 14
634-66-2	.	X	1,2,3,4-Tetrachlorobenzene
634-90-2	.	X	1,2,3,5-Tetrachlorobenzene
634-93-5	X	X	2,4,6-Trichloroaniline
636-21-5	X	X	o-Toluidine.HCl
636-23-7	X	.	2,4-Diaminotoluene.2HCl
636-26-0	.	X	5-Methyl-2-thiouracil
636-79-3	X	.	Nicotine.HCl
637-07-0	X	.	Clofibrate
637-92-3	.	X	2-Methyl-2-ethoxypropane
638-03-9	X	X	m-Toluidine.HCl
643-22-1	X	X	Erythromycin stearate
645-05-6	.	X	Hexamethylmelamine

CAS	In CPDB	In GT	Chemical Name
645-49-8	.	X	cis-Stilbene
645-62-5	.	X	2-Ethyl-2-hexenal
646-14-0	.	X	1-Nitrohexane
671-16-9	X	.	Procarbazine
673-06-3	.	X	D-Phenylalanine
680-31-9	.	X	Hexamethylphosphoramide
683-50-1	X	.	2-Chloropropanal
684-93-5	X	.	N-Nitroso-N-methylurea
685-91-6	X	.	Diethylacetamide
688-74-4	.	X	Tributyl borate
695-53-4	X	.	Dimethadione
712-68-5	X	.	2-Amino-5-(5-nitro-2-furyl)-1,3,4-thiadiazole
720-69-4	X	.	4,6-Diamino-2-(5-nitro-2-furyl)-S-triazine
723-46-6	.	X	Sulfamethoxazole
738-70-5	.	X	Trimethoprim
756-79-6	X	X	Dimethyl methylphosphonate
758-17-8	X	.	N-Methyl-N-formylhydrazine
759-73-9	X		1-Ethyl-1-nitrosourea
		X	N-Ethyl-N-nitrosourea
759-94-4	.	X	Eptam
760-56-5	X	.	1-Allyl-1-nitrosourea
760-60-1	X	.	N-Nitroso-N-isobutylurea
764-42-1	.	X	Fumaronitrile
765-34-4	X	X	Glycidaldehyde
768-52-5	.	X	N-Isopropyl aniline
772-43-0	X	.	5-Nitro-2-furamidoxime
785-30-8	X	.	4,4'-Diaminobenzanilide
793-24-8	.	X	N-(1,3-Dimethylbutyl)-N'-phenyl-p-phenylenediamine
814-80-2	X	.	Calcium lactate
816-57-9	X	.	N-Propyl-N-nitrosourea
828-00-2	X	.	Dimethoxane
	X	X	Dimethoxane, commercial grade
834-28-6	X	X	Phenformin.HCl
838-88-0	X	.	4,4'-Methylene-bis(2-methylaniline)
839-90-7	.	X	1,3,5-Tris (2-hydroxyethyl)triazine-2,4,6-trione
842-00-2	X	.	4-Ethylsulphonylnaphthalene-1-sulfonamide
842-07-9	X	.	1-Phenylazo-2-naphthol
		X	Solvent yellow 14
846-50-4	X	.	Temazepam, pharmaceutical grade
853-23-6	X	.	Dehydroepiandrosterone acetate
865-21-4	X	.	Vinblastine
868-85-9	X	X	Dimethyl hydrogen phosphite
869-01-2	X	.	N-n-Butyl-N-nitrosourea
872-50-4	X	X	N-Methyl-2-pyrrolidone
874-42-0	.	X	2,4-Dichlorobenzaldehyde
879-39-0	.	X	2,3,4,5-Tetrachloronitrobenzene
881-03-8	.	X	1-Nitro-2-methylnaphthalene
900-95-8	X	.	Triphenyltin acetate
915-67-3	X	.	FD & C Red no. 2
920-66-1	.	X	1,1,1,3,3,3-Hexafluoro-2-propanol
924-16-3	X	.	Nitrosodibutylamine
924-42-5	X	X	N-Methylolacrylamide
930-22-3	.	X	1,2-Epoxy-3-butene
930-37-0	.	X	Methyl glycidyl ether
930-55-2	X	.	N-Nitrosopyrrolidine
931-97-5	.	X	Cyclohexanone cyanohydrin
932-83-2	X	.	N-Nitrosohexamethyleneimine
933-75-5	.	X	2,3,6-Trichlorophenol
933-78-8	.	X	2,3,5-Trichlorophenol
934-32-7	.	X	2-Aminobenzimidazole
935-95-5	.	X	2,3,5,6-Tetrachlorophenol
937-25-7	X	.	N-Nitroso-N-methyl-4-fluoroaniline
938-73-8	X	.	o-Ethoxybenzamide
943-41-9	X	.	N-Nitroso-N-methyl-4-nitroaniline
950-37-8	X	.	Methidathion
952-23-8	X	X	Proflavine.HCl hemihydrate
958-93-0	.	X	Thenyldiamine.HCl
959-24-0	X	.	Sotalol.HCl
961-11-5	X	.	Tetrachlorvinphos
963-89-3	.	X	7,9-Dimethylbenz(c)acridine
968-81-0	X	X	Acetohexamide
971-15-3	X	.	Dipentamethylenethiuram hexasulfide
982-57-0	.	X	Chloramphenicol, Na succinate
989-38-8	X	X	Rhodamine 6G
992-59-6	.	X	C.I. Direct red 2
999-55-3	.	X	Allyl acrylate
999-81-5	X	X	(2-Chloroethyl)trimethylammonium chloride
1011-73-0	X	.	2,4-Dinitrophenol, sodium
1025-15-6	.	X	Triallyl isocyanurate
1034-41-9	.	X	Chlordecone alcohol
1066-30-4	X	.	Chromium (III) acetate
1067-14-7	.	X	Triethyl lead chloride
1067-33-0	X	X	Dibutyltin diacetate
1068-57-1	X	.	Monoacetyl hydrazine
1071-83-6	.	X	Glyphosate
1072-52-2	.	X	1-Aziridine ethanol
1072-53-3	X	.	Ethylene glycol, cyclic sulfate
1078-38-2	X	.	1-Acetyl-2-isonicotinoylhydrazine
1083-57-4	X	.	3-Hydroxy-p-butyrophenetidide
1095-90-5	X	.	6-Dimethylamino-4,4-diphenyl-3-heptanone.HCl
1114-71-2	X	.	Propyl N-ethyl-N-butylthiocarbamate
1116-40-1	.	X	Tri-isobutylamine
1116-54-7	X	X	N-Nitrosodiethanolamine
1119-68-2	X	.	n-Pentylhydrazine.HCl
1120-71-4	X	.	Propane sultone
1121-92-2	X	.	Heptamethyleneimine
1122-54-9	.	X	4-Acetyl pyridine
1133-64-8	X	.	Nitrosoanabasine
1143-38-0	.	X	Anthralin
1146-71-0	X	.	5,7-Dimethoxycyclopentene[c]coumarin
1150-37-4	X	.	5,7-Dimethoxycyclopentenone[2,3-c]coumarin
1150-42-1	X	.	5,7-Dimethoxycyclopentenone[3,2-c]coumarin
1156-19-0	X	X	Tolazamide

CAS	In CPDB	In GT	Chemical Name
2210-28-8	.	X	n-Propyl methacrylate
2210-79-9	.	X	o-Cresyl glycidyl ether
2213-63-0	.	X	2,3-Dichloroquinoxaline
2227-13-6	X	.	p-Chlorophenyl-2,4,5-trichlorophenyl sulfide
2243-61-0	.	X	1,4-Naphthalenediamine
2243-62-1	X	X	1,5-Naphthalenediamine
2244-16-8	X	X	d-Carvone
2244-21-5	.	X	S-Triazine-2,4,6(1H, 3H, 5H)-trione-1,3-dichloro, K salt
2302-84-3	X	.	1-Formyl-3-thiosemicarbazide
2303-16-4	X	.	Diallate
2318-18-5	X	.	Senkirkine
2353-45-9	X	.	FD & C Green no. 3
2381-21-7	.	X	1-Methyl pyrene
2385-85-5	X	X	Mirex
2404-44-6	.	X	1,2-Epoxydecane
2409-55-4	X	.	2-tert-Butyl-4-methoxyphenol
2411-74-7	X	.	2-Furaldehyde semicarbazone
2425-06-1	X	.	Captafol
2425-79-8	.	X	1,4-Butanediol diglycidyl ether
2425-85-6	X		C.I. Pigment red 3
		X	Pigment red 3
2426-08-6	.	X	n-Butyl glycidyl ether
2429-71-2	.	X	C.I. Direct blue 8
2429-73-4	.	X	C.I. Direct blue 2
2429-74-5	X	X	C.I. Direct blue 15
2429-82-5	.	X	C.I. Direct brown 2
2432-99-7	X	X	11-Aminoundecanoic acid
2438-88-2	X	X	2,3,5,6-Tetrachloro-4-nitroanisole
2439-10-3	X	.	n-Dodecylguanidine acetate
2439-35-2	.	X	2-(Dimethylamino)ethyl acrylate
2451-62-9	.	X	Tris (2,3-epoxypropyl) isocyanurate
2461-15-6	.	X	2-Ethylhexyl glycidyl ether
2461-18-9	.	X	Lauryl glycidyl ether
2465-27-2	X		Auramine-O
		X	Auramine
2475-45-8	X	X	C.I. Disperse blue 1
2489-77-2	X		Trimethylthiourea
		X	1,1,3-Trimethyl-2-thiourea
2493-84-7	.	X	p-n-Octyloxybenzoic acid
2519-30-4	X	.	Black PN
2528-36-1	.	X	Dibutyl phenyl phosphate, mixed isomers
2578-75-8	X	.	N-[5-(5-Nitro-2-furyl)-1,3,4-thiadiazol-2-yl]acetamide
2602-46-2	X	X	C.I. Direct blue 6
2608-48-2	.	X	5-(4-Nitrophenyl)-2,4-pentadien-1-al
2610-05-1	.	X	C.I. Direct blue 1
2611-82-7	X	.	SX purple
2629-59-6	X	.	S-Ethyl-l-cysteine
2645-07-0	.	X	p-Nitrohippuric acid
2646-17-5	.	X	Oil orange SS
2675-77-6	.	X	Chloroneb
2698-41-1	X	X	o-Chlorobenzalmalonitrile

CAS	In CPDB	In GT	Chemical Name
2757-90-6	X	.	β-N-[γ-l(+)-Glutamyl]-4-hydroxymethylphenylhydrazine
2782-57-2	.	X	Dichloroisocyanuric acid
2782-91-4	.	X	1,1,3,3-Tetramethyl-2-thiourea
2783-94-0	X	X	FD & C Yellow no. 6
2784-94-3	X	X	HC Blue no. 1
2832-40-8	X	X	C.I. Disperse yellow 3
2835-39-4	X	X	Allyl isovalerate
2835-95-2	.	X	3-Amino-6-methylphenol
2855-19-8	.	X	1,2-Epoxydodecane
2871-01-4	X	X	HC Red no. 3
2873-97-4	.	X	Diacetone acrylamide
2893-78-9	.	X	Dichloroisocyanuric acid, Na salt
2921-88-2	X	.	O,O-Diethyl-o-(3,5,6-trichloro-2-pyridyl)phosphorothioate
2922-54-5	X	.	Monosodium succinate
2941-64-2	.	X	S-(Ethyl)chlorothioformic acid
2955-38-6	X	.	Prazepam
3012-65-5	X	.	Ammonium citrate
3018-12-0	.	X	Dichloroacetonitrile
3025-77-2	.	X	1-((4-Nitrophenyl)azo)-2-naphthenamine
3031-51-4	X	.	l-5-Morpholinomethyl-3-[(5-nitrofurfurylidene) amino]-2-oxazolidinone.HCl
3054-95-3	X	.	Acrolein diethylacetal
3066-70-4	.	X	2,3-Dibromopropyl methacrylate
3068-88-0	X	.	β-Butyrolactone
3081-14-9	.	X	N,N′-Bis(1,4-dimethylpentyl)-p-phenylenediamine
3083-23-6	.	X	1,2-Epoxy-3,3,3-trichloropropane
3083-25-8	.	X	1,2-Epoxy-4,4,4-trichlorobutane
3096-50-2	X	.	N-(9-Oxo-2-fluorenyl)acetamide
3101-60-8	.	X	t-Butyl phenyl glycidyl ether
3129-91-7	.	X	Dicyclohexylamine nitrite
3131-60-0	X	.	6-Azacytidine
3132-64-7	.	X	1,2-Epoxy-3-bromopropane
3148-73-0	X	.	Diacetyl hydrazine
3160-37-0	.	X	Piperonyl acetone
3165-93-3	X	X	4-Chloro-o-toluidine.HCl
3179-47-3	.	X	n-Decyl methacrylate
3209-22-1	.	X	2,3-Dichloronitrobenzene
3234-02-4	.	X	trans-2,3-Dibromo-2-butene-1,4-diol
3234-28-4	.	X	1,2-Epoxytetradecane
3237-50-1	.	X	Alloxan monohydrate
3252-43-5	.	X	Dibromoacetonitrile
3266-23-7	.	X	2,3-Epoxybutane
3268-87-9	.	X	Octachlorodibenzodioxin
3276-41-3	X	.	3,6-Dihydro-2-nitroso-2H-1,2-oxazine
3296-90-0	X		Dibromoneopentyl glycol
		X	2,2-Bis(bromomethyl)-1,3-propanediol
3319-31-1	.	X	Tris (2-ethylhexyl) trimellitate
3322-93-8	.	X	1-(1,2-Dibromoethyl)-3,4-dibromocyclohexane
3333-52-6	.	X	Tetramethyl succinonitrile

CAS	In CPDB	In GT	Chemical Name
5979-28-2	X	.	C.I. Pigment yellow 16
5989-27-5	X	X	d-Limonene
6041-94-7	.	X	Pigment red 2
6055-19-2	.	X	Cyclophosphamide monohydrate
6088-51-3	.	X	6-Hydroxy-2-naphthyl disulfide
6109-97-3	X	X	3-Amino-9-ethylcarbazole.HCl
6112-76-1	.	X	6-Mercaptopurine monohydrate
6119-92-2	X	.	Dinitro(1-methylheptyl) phenyl crotonate
6120-10-1	X	.	4-Dimethylamino-3,5-xylenol
6151-25-3	X	.	Quercetin dihydrate
6197-30-4	.	X	2-Ethylhexyl-2-cyano-3,3-diphenylacrylate
6201-87-2	.	X	5-Amino-3-sulfosalicylic acid
6219-89-2	.	X	4-Amino-4'-hydroxy-3-methyldiphenylamine
6285-57-0	.	X	2-Amino-6-nitrobenzothiazole
6287-38-3	.	X	3,4-Dichlorobenzaldehyde
6294-89-9	X	.	Methyl carbazate
6317-18-6	.	X	Methylenebis-(thiocyanate)
6334-11-8	X	.	2,4,6-Trimethylaniline.HCl
6358-07-2	.	X	2-Amino-4-chloro-5-nitrophenol
6358-23-2	.	X	2-(2,4-Dinitroanilino)phenol
6358-29-8	.	X	C.I. Direct red 39
6358-31-2	.	X	Pigment yellow 74
6358-53-8	.	X	Citrus red 2
6358-85-6	X	X	C.I. Pigment yellow 12
6369-59-1	X	X	2,5-Diaminotoluene sulfate
6373-74-6	X	X	C.I. Acid orange 3
6379-46-0	X	.	1,2,3-Trichloro-4,6-dinitrobenzene
6381-77-7	X	.	Erythorbate, sodium
6385-58-6	X	.	Sodium bithionolate
6428-94-0	.	X	C.I. Direct violet 32
6452-73-9	X	.	Oxprenolol.HCl
6459-94-5	X	X	C.I. Acid red 114
6471-49-4	X		C.I. Pigment red 23
		X	Pigment red 23
6485-34-3	X	.	Saccharin, calcium
6533-68-2	.	X	Scopalamine hydrobromide
6637-88-3	.	X	C.I. Direct orange 6
6959-47-3	X	X	2-(Chloromethyl)pyridine.HCl
6959-48-4	X	X	3-(Chloromethyl)pyridine.HCl
6965-71-5	X	.	α-(2,5-Dichlorophenoxy) propionic acid
7008-42-6	X	.	Acronycine
7149-26-0	.	X	Linalyl anthranilate
7166-19-0	.	X	β-Bromo-β-nitrostyrene
7177-48-2	X	X	Ampicillin trihydrate
7195-43-9	.	X	Isophthalic diglycidyl ester
7227-91-0	X	.	1-Phenyl-3,3-dimethyltriazene
7316-37-2	X	.	Diethyl-β,γ-epoxypropylphosphonate
7320-37-8	.	X	1,2-Epoxyhexadecane
7336-20-1	X	.	4,4'-Diamino-2,2'-stilbenedisulfonic acid, disodium salt
7347-49-1	X	.	N,N-Diethyl-4-(4'-[pyridyl-1'-oxide]azo)aniline
7390-81-0	.	X	1,2-Epoxyoctadecane
7411-49-6	X	.	3,3',4,4'-Tetraaminobiphenyl.4HCl

CAS	In CPDB	In GT	Chemical Name
7422-80-2	X	.	1,1-Di-n-butylhydrazine
7440-02-0	X	.	Nickel
7446-34-6	X	X	Selenium sulfide
7447-40-7	X	X	Potassium chloride
7447-41-8	.	X	Lithium chloride
7487-94-7	X	X	Mercuric chloride
7492-66-2	.	X	Citral diethyl acetal
7493-63-2	.	X	Allyl anthranilate
7519-36-0	X	.	Nitrosoproline
7558-79-4	.	X	Sodium phosphate, dibasic
7572-29-4	X	.	Dichloroacetylene
7631-89-2	X	.	Arsenate, sodium
7631-99-4	X	.	Nitrate, sodium
7632-00-0	X		Nitrite, sodium
		X	Sodium nitrite
7647-01-0	X	.	Hydrochloric acid
7647-10-1	.	X	Palladium (II) chloride
7647-14-5	X	X	Sodium chloride
7665-72-7	.	X	t-Butyl glycidyl ether
7681-49-4	X		Fluoride, sodium
		X	Sodium fluoride
7681-52-9	X	.	Sodium hypochlorite
7681-93-8	X	.	Pimaricin
7705-08-0	X	X	Ferric chloride
7722-84-1	X	.	Hydrogen peroxide
7732-18-5	.	X	Water
7756-96-9	.	X	Butyl anthranilate
7757-82-6	X	.	Sulfate, sodium
7758-01-2	X		Bromate, potassium
		X	Potassium bromate
7758-02-3	.	X	Potassium bromide
7758-19-2	X	.	Sodium chlorite
7772-99-8	X		Tin (II) chloride
		X	Stannous chloride
7778-50-9	.	X	Potassium dichromate
7779-16-0	.	X	Cyclohexyl anthranilate
7779-77-3	.	X	Isobutyl anthranilate
7782-49-2	X	.	Selenium
7782-50-5	X	.	Chlorine
7784-46-5	X	.	Arsenite, sodium
7787-56-6	.	X	Beryllium sulfate 4H$_2$O
7787-59-9	X	.	Bismuth oxychloride
7789-23-3	.	X	Potassium fluoride
7790-84-3	X	.	Cadmium sulphate (1:1) hydrate (3:8)
7791-18-6	X	.	Magnesium chloride hexahydrate
7803-51-2	X	.	Phosphine
8001-23-8	X	.	Safflower oil
8001-28-3	.	X	Croton oil
8001-30-7	X	.	Corn oil
8001-35-2	X	X	Toxaphene
8001-50-1	X	.	Strobane
8001-58-9	.	X	Creosote, coal tar
8001-79-4	.	X	Castor oil

CAS	In CPDB	In GT	Chemical Name
8002-43-5	X	.	Soybean lecithin
8003-03-0	X	X	Aspirin, phenacetin, and caffeine
8003-22-3	.	X	D & C Yellow 11
8005-02-5	.	X	Solvent black 7
8006-90-4	X	.	Peppermint oil
8008-20-6	.	X	Navy fuels JP-5
8015-12-1	X	.	Norlestrin
8015-30-3	X	.	Enovid
	X	.	Enovid-E
8021-39-4	.	X	Creosote, wood
8024-37-1	X	X	Turmeric oleoresin
8047-99-2	.	X	Toluene ethylsulfonamide
8052-16-2	X	.	Actinomycin C
8056-92-6	X	.	Ovulen
8061-51-6	.	X	Lignosulfonic acid, Na salt
8065-91-6	X	.	Lutestral
9000-01-5	X	X	Gum arabic
9000-07-1	X	.	Carrageenan, native
9000-16-2	.	X	Dammar resin
9000-30-0	X	X	Guar gum
9000-40-2	X	X	Locust bean gum
9000-65-1	X	.	Tragacanth gum
9002-18-0	X	.	Agar
9002-86-2	.	X	Polyvinyl chloride latex
9002-92-0	.	X	Dodecyl alcohol, ethoxylated
9003-39-8	.	X	Polyvinylpyrrolidone polymers
9004-32-4	X	.	Edifas B
9004-54-0	X	.	Dextran
9004-59-5	X	.	Edifas A
9004-66-4	.	X	Iron dextran
9005-65-6	X	X	Polysorbate 80
9011-18-1	X	.	Dextran sulfate sodium (DS-M-1)
	X	.	Dextran sulfate sodium (DST-H)
	X	.	Dextran sulfate sodium (KMDS-H)
9045-28-7	X	.	Starch acetate
9047-13-6	X	.	Amylopectin sulfate
10024-97-2	X	.	Nitrous oxide
10026-24-1	.	X	Cobalt sulfate heptahydrate
10028-15-6	X	X	Ozone
10034-82-9	.	X	Sodium chromate tetrahydrate
10034-93-2	X	X	Hydrazine sulfate
10034-96-5	X	X	Manganese (II) sulfate monohydrate
10043-35-3	X	X	Boric acid
10043-52-4	X	.	Calcium chloride
10043-67-1	X	.	Aluminum potassium sulfate
10048-13-2	X	.	Sterigmatocystin
10101-97-0	.	X	Nickel sulfate hexahydrate
10102-43-9	X	.	Nitric oxide
10108-64-2	X	X	Cadmium chloride
10124-36-4	X	.	Cadmium sulphate (1:1)
10143-23-4	.	X	2,3-Dimethyl-1-pentanol
10191-41-0	X	.	dl-α-Tocopherol
10213-75-9	.	X	3-((2-Ethylhexyl)oxy)propionitrile

CAS	In CPDB	In GT	Chemical Name
10277-43-7	.	X	Lanthanum nitrate hexahydrate
10318-26-0	X	X	Dibromodulcitol
10326-27-9	X	X	Barium chloride dihydrate
10380-28-6	X	.	Copper-8-hydroxyquinoline
10473-70-8	X	.	1-(4-Chlorophenyl)-1-phenyl-2-propynyl carbamate
10589-74-9	X	.	1-Amyl-1-nitrosourea
10595-95-6	X	.	Nitrosoethylmethylamine
10605-21-7	.	X	Carbendazim
11013-97-1	X	.	Methyl hesperidin
11096-82-5	X	.	Aroclor 1260
11097-69-1	X	X	Aroclor 1254
11099-03-9	.	X	Solvent black 5
11103-86-9	.	X	Zinc potassium chromate
12002-03-8	.	X	Copper acetoarsenite
12002-43-6	.	X	Gilsonite
12025-19-3	X	.	Germanate, sodium
12034-09-2	X	.	Niobate, sodium
12083-48-6	.	X	Bis(cyclopentadienyl) vanadium chloride
12122-67-7	X		Zinc ethylenebisthiocarbamate
		X	Zineb
12125-02-9	X	.	Ammonium chloride
12126-59-9	X	.	Premarin
12224-98-5	.	X	Pigment red 81
12225-21-7	.	X	Pigment yellow 100
12236-46-3	X	.	Chocolate Brown FB
12427-38-2	X	.	Manganese ethylenebisthiocarbamate
12663-46-6	X	.	Cyclochlorotine
12737-87-0	X	.	Kanechlor 400
12768-44-4	X	.	Rutin sulfate
12789-03-6	.	X	Chlordane, Technical grade
13007-92-6	.	X	Chromium carbonyl
13010-07-6	X	.	N-Propyl-N′-nitro-N-nitrosoguanidine
13010-08-7	X	.	N-Butyl-N′-nitro-N-nitrosoguanidine
13010-10-1	X	.	N-Pentyl-N′-nitro-N-nitrosoguanidine
13014-18-1	.	X	2,4-α,α,α-Pentachlorotoluene
13014-24-9	.	X	3,4-α,α,α-Pentachlorotoluene
13048-33-4	.	X	1,6-Hexanediol diacrylate
13071-79-9	.	X	Counter 15G
13073-35-3	X	.	Ethionine
13098-39-0	.	X	Hexafluoroacetone sesquihydrate
13114-72-2	.	X	N′-Methyl-N,N-diphenylurea
13171-21-6	X	X	Phosphamidon
13256-06-9	X	.	Dipentylnitrosamine
13256-11-6	X	.	Nitroso-N-methyl-N-(2-phenyl)ethylamine
13284-42-9	.	X	2-Pentenenitrile
13292-46-1	X	.	Rifampicin
13360-63-9	.	X	Ethyl-n-butylamine
13366-73-9	X		Dieldrin, photo-
		X	Photodieldrin
13463-41-7	.	X	Zinc pyrithione
13463-67-7	X	X	Titanium dioxide
13472-45-2	X	.	Tungstate, sodium

CAS	In CPDB	In GT	Chemical Name
13483-18-6	X	.	Bis-1,2-(chloromethoxy) ethane
13483-19-7	X	.	Bis-1,4-(chloromethoxy) butane
13510-49-1	X	.	Beryllium sulfate
13552-21-1	.	X	Mono-sec-butanolamine
13552-44-8	X	X	4,4′-Methylenedianiline.2HCl
13674-84-5	.	X	Tris (1-chloro-2-propyl) phosphate
13674-87-8	.	X	Tris (1,3-dichloro-2-propyl) phosphate
13684-63-4	.	X	Phenmedipham
13743-07-2	X	.	1-(2-Hydroxyethyl)-1-nitrosourea
13752-51-7	X	.	N-Oxydiethylene thiocarbamyl-N-oxydiethylene sulfenamide
13755-29-8	X	.	Tetrafluoroborate, sodium
13765-19-0	.	X	Calcium chromate, anhydrous
13838-16-9	X	.	Enflurane
13889-92-4	.	X	S-(n-Propyl)chlorothioformic acid
13927-77-0	X	.	Nickel dibutyldithiocarbamate
13952-84-6	.	X	sec-Butylamine
13961-86-9	.	X	Oleic acid/Diethanolamine condensate (1/1)
13983-17-0	.	X	Calcium metasilicate
14026-03-0	X	.	R(−)-2-Methyl-N-nitrosopiperidine
14239-68-0	X	.	Cadmium diethyldithiocarbamate
14302-13-7	.	X	Pigment green 36
14324-55-1	X	.	Zinc diethyldithiocarbamate
14371-10-9	.	X	trans-Cinnamaldehyde
14481-26-6	X	.	Titanium oxalate, potassium
14484-64-1	X	.	Ferric dimethyldithiocarbamate
14644-61-2	X	.	Zirconium (IV) sulfate
14882-18-9	.	X	Bismuth subsalicylate
14901-07-6	.	X	β-Ionone
15110-74-4	.	X	2,5-Dinitro-9H-fluorene
15121-84-3	.	X	o-Nitrophenethyl alcohol
15216-10-1	X	.	N-Nitrosoazetidine
15242-96-3	.	X	Stearatochromic chloride complex
15356-70-4	X	X	dl-Menthol
15481-70-6	X	X	2,6-Diaminotoluene.2HCl
15663-27-1	.	X	cis-Dichlorodiaminoplatinum II
15687-27-1	X	.	Ibuprofen
15721-02-5	X	.	2,2′,5,5′-Tetrachlorobenzidine
15805-73-9	X	.	Vinyl carbamate
15879-93-3	X	.	Anhydroglucochloral
15950-66-0	.	X	2,3,4-Trichlorophenol
15973-99-6	X	.	Di(N-nitroso)-perhydropyrimidine
16071-86-6	X	X	C.I. Direct brown 95
16091-18-2	.	X	Di(n-octyl)tin maleate polymers
16120-70-0	X	.	N-n-Butyl-N-formylhydrazine
16170-75-5	X	.	Cytembena
16219-75-3	.	X	Ethylidene norbornene
16219-98-0	X	.	2-Nitrosomethylaminopyridine
16219-99-1	X	.	4-Nitrosomethylaminopyridine
16238-56-5	.	X	7-Bromomethyl-12-methylbenz(a)anthracene
16301-26-1	X	.	Z-Ethyl-O,N,N-azoxyethane
16338-97-9	X	.	Diallylnitrosamine
16423-68-0	X	.	FD & C Red no. 3
16452-01-0	.	X	o-Cresidine
16529-56-9	.	X	2-Methyl-3-butenenitrile
16543-55-8	X	.	N′-Nitrosonornicotine
16568-02-8	X	.	Acetaldehyde methylformylhydrazone
16699-10-8	X	.	4-(4-N-Methyl-N-nitrosaminostyryl)quinoline
16813-36-8	X	.	1-Nitroso-5,6-dihydrouracil
17026-81-2	X	X	3-Amino-4-ethoxyacetanilide
17157-48-1	X	.	Bromoacetaldehyde
17341-40-1	.	X	1,1-Dimethyl-1-(2-hydroxypropylamine)methacrylimide
17369-59-4	.	X	3-Propylidene phthalide
17372-87-1	.	X	Eosin
17433-31-7	.	X	1-Acetyl-2-picolinoyl hydrazine
17557-23-2	.	X	Neopentyl glycol diglycidyl ether
17608-59-2	X	.	N-Nitrosoephedrine
17673-25-5	X	.	Phorbol
17804-35-2	.	X	Benomyl
17831-71-9	.	X	Tetraethylene glycol diacrylate
17924-92-4	X	X	Zearalenone
18413-14-4	X	.	Ethylhydrazine.HCl
18523-69-8	X	.	Acetone[4-(5-nitro-2-furyl)-2-thiazolyl]hydrazone
18559-94-9	X	.	Salbutamol
18662-53-8	X	X	Nitrilotriacetic acid, trisodium salt, monohydrate
18771-50-1	X	.	3,4,5,6-Tetrahydrouridine
18774-85-1	X	.	N-Hexylnitrosourea
18883-66-4	X	.	Streptozotocin
18968-99-5	X	.	2-Amino-5-phenyl-2-oxazolin-4-one + Mg(OH)$_2$
19010-66-3	X	X	Lead dimethyldithiocarbamate
19287-45-7	.	X	Diborane
19660-16-3	.	X	2,3-Dibromopropyl acrylate
19686-73-8	.	X	1-Bromo-2-propanol
19767-45-4	X	.	2-Mercaptoethanesulfonate, sodium
19780-11-1	.	X	2-Dodecen-1-yl succinic anhydride
19834-02-7	X	.	Cyclohexylamine sulfate
20020-02-4	.	X	1,2,3,4-Tetrachloronaphthalene
20265-96-7	X	.	p-Chloroaniline.HCl
20265-97-8	X	X	p-Anisidine.HCl
20325-40-0	X	X	3,3′-Dimethoxybenzidine.2HCl
20570-96-1	X	.	Benzylhydrazine.2HCl
20702-77-6	.	X	Neohesperidin dihydrochalcone
20917-49-1	X	.	Nitrosoheptamethyleneimine
20941-65-5	X	X	Ethyl tellurac
21260-46-8	X	.	Bismuth dimethyldithiocarbamate
21308-79-2	X	.	Methyl 12-oxo-trans-10-octadecenoate
21340-68-1	X	.	Methyl clofenapate
21416-87-5	X	.	ICRF-159
21436-96-4	X	.	2,4-Xylidine.HCl
21436-97-5	X	.	2,4,5-Trimethylaniline.HCl
21498-08-8	X	.	Lofexidine.HCl
21626-89-1	X	.	Diftalone

CAS	In CPDB	In GT	Chemical Name
32534-81-9	.	X	Pentabromodiphenyl oxide
32588-76-4	.	X	1,2-Bis(tetrabromophthalimido)ethane
32607-00-4	X	.	Iminodiacetic acid, monosodium
32852-21-4	X	.	Formic acid 2-(4-methyl-2-thiazolyl)hydrazide
33229-34-4	X	X	HC Blue no. 2
33372-39-3	X	.	4-Bis(2-hydroxyethyl) amino-2-(5-nitro-2-thienyl)quinazoline
33389-33-2	X	.	1,2-Dihydro-2-(5-nitro-2-thienyl)quinazolin-4(3H)-one
33389-36-5	X	.	4-(2-Hydroxyethylamino)-2-(5-nitro-2-thienyl)quinazoline
33857-26-0	X	X	2,7-Dichlorodibenzo-p-dioxin
33868-17-6	X	.	Methylnitrosocyanamide
33979-15-6	X	.	Clivorine
34031-32-8	X	.	Auranofin
34176-52-8	X	.	2-Hydrazino-4-phenylthiazole
34522-69-5	X	.	5,7-Dibromoquinoline
34627-78-6	X	.	1'-Acetoxysafrole
34807-41-5	.	X	Mezerein
35449-36-6	X	.	Gemcadiol
35658-65-2	X	.	Cadmium chloride monohydrate
35660-60-7	X	.	O,S-Dibenzoyl thiamine.HCl
36059-21-9	.	X	3,4,5,6-Tetrabromo-o-xylene
36133-88-7	X	.	N-{[3-(5-Nitro-2-furyl)-1,2,4-oxadiazole-5-yl]-methyl}acetamide
36322-90-4	X	.	Piroxicam
36355-01-8	.	X	Hexabromobiphenyl
36536-46-6	.	X	β-Butyrolactone
36702-44-0	X	.	S(+)-2-Methyl-N-nitrosopiperidine
36711-31-6	X	.	Bis(2,3-dibromopropyl) phosphate, magnesium salt
37087-94-8	X	.	2-Chloro-5-(3,5-dimethylpiperidino-sulphonyl) benzoic acid
37853-59-1	.	X	1,2-Bis(2,4,6-tribromophenoxy)ethane
38347-74-9	X	.	3-Nitroso-2-oxazolidinone
38434-77-4	X	.	Ethylnitrosocyanamide
38514-71-5	X	.	2-Amino-4-(5-nitro-2-furyl)thiazole
38571-73-2	X	.	Tris-1,2,3-(chloromethoxy)propane
38777-13-8	X	.	Nitroso-Baygon
38848-76-9	.	X	1,1-Dimethyl-1-(2-hydroxypropylamine)tetradecanimide
39148-24-8	X	.	Fosetyl Al
39156-41-7	X	.	2,4-Diaminoanisole sulfate
		X	2,4-Diaminoanisole SO₄, trihydrate
39300-88-4	X	X	Tara gum
39801-14-4	X	.	Mirex, photo-
39884-52-1	X	.	N-Nitroso-1,3-oxazolidine
40548-68-3	X	.	Tetrahydro-2-nitroso-2H-1,2-oxazine
40580-89-0	X	.	Nitrosododecamethyleneimine
40762-15-0	X	.	Doxefazepam
40778-40-3	X	.	Primidolol.HCl
41122-70-7	.	X	4-n-Hexyl-4'-cyanobiphenyl
41372-08-1	X	.	α-Methyldopa sesquihydrate

CAS	In CPDB	In GT	Chemical Name
42011-48-3	X	.	2,2,2-Trifluoro-N-[4-(5-nitro-2-furyl)-2-thiazolyl] acetamide
42579-28-2	X	.	1-Nitrosohydantoin
43033-72-3	X	.	6-Dimethylamino-4,4-diphenyl-3-heptanol acetate.HCl
43054-45-1	X	.	Clomiphene citrate
47141-42-4	X	.	Levobunolol
50264-69-2	X	.	Lonidamine
50594-66-6	X	.	Acifluorfen
50892-23-4	X	.	[4-Chloro-6-(2,3-xylidino)-2-pyrimidinylthio] acetic acid
51325-35-0	X	.	N,N'-[6-(5-Nitro-2-furyl)-S-triazine-2,4-diyl]bisacetamide
51333-22-3	X	.	Budesonide
51410-44-7	X	.	1'-Hydroxyestragole
51481-61-9	X	.	Cimetidine
51542-33-7	X	.	N-Nitrosobenzthiazuron
51630-58-1	X	.	Fenvalerate
51715-17-4	X	.	Nitrosochlordiazepoxide
51786-53-9	X	.	2,5-Xylidine.HCl
52207-83-7	X	.	Allylhydrazine.HCl
52214-84-3	X	.	Ciprofibrate
52417-22-8	.	X	9-Aminoacridine.HCl H₂O
52918-63-5	X	.	Deltamethrin
53123-84-5	X	.	Acetylated distarch glycerol
53609-64-6	X	.	N-Nitrosobis(2-hydroxypropyl)amine
53757-28-1	X	.	4-(5-Nitro-2-furyl)thiazole
54075-76-2	.	X	Trimethyloxonium hexachloroantimonate
54143-56-5	X	.	Flecainide acetate
54150-69-5	X	X	2,4-Dimethoxyaniline.HCl
54749-90-5	X	.	Chlorozotocin
54827-17-7	.	X	3,3',5,5'-Tetramethylbenzidine
54849-38-6	.	X	Methyltin-tris(isooctylmercaptoacetate)
55090-44-3	X	.	N-Nitroso-N-methyl-N-dodecylamine
55268-74-1	X	.	Praziquantel
55380-34-2	X	.	1,4-Dinitroso-2,6-dimethylpiperazine
55556-92-8	X	.	Nitroso-1,2,3,6-tetrahydropyridine
55557-00-1	X	.	Dinitrosohomopiperazine
55557-02-3	X	.	N-Nitrosoguvacoline
55557-03-4	X	.	Nitrosomethylphenidate
55566-30-8	X	X	Tetrakis(hydroxymethyl) phosphonium sulfate
55567-81-2	X	.	Tilidine fumarate
55600-34-5	X	.	Clophen A 30
55738-54-0	X	.	trans-2-[(Dimethylamino)methylimino]-5-[2-(5-nitro-2-furyl)vinyl]-1,3,4-oxadiazole
55965-13-4	X	.	Emulsifier YN
55984-51-5	X	.	N-Nitrosomethyl(2-oxopropyl)amine
56222-35-6	X	.	N-Nitroso-3-hydroxypyrrolidine
56393-22-7	X	.	Pildralazine
56654-52-5	X	.	1,3-Dibutyl-1-nitrosourea
56795-65-4	X	.	n-Butylhydrazine.HCl
56795-66-5	X	.	Propylhydrazine.HCl
56802-99-4	X	X	Chlorinated trisodium phosphate

CAS	In CPDB	In GT	Chemical Name
56803-37-3	.	X	t-Butylphenyl diphenyl phosphate (mixed isomers)
56894-91-8	X	.	Bis-1,4-(chloromethoxy)-p-xylene
56894-92-9	X	.	Bis-1,6-(chloromethoxy)hexane
56980-93-9	X	.	Celiprolol
57497-29-7	X	.	Z-Ethyl-O,N,N-azoxymethane
57497-34-4	X	.	Z-Methyl-O,N,N-azoxyethane
57590-20-2	X	.	Pentanal methylformylhydrazone
58139-47-2	X	.	4-Bis(2-hydroxyethyl) amino-2-(2-thienyl)quinazoline
58139-48-3	X	.	4-Morpholino-2-(5-nitro-2-thienyl)quinazoline
59122-46-2	X	.	Misoprostol
59333-67-4	X	.	Fluoxetine.HCl
59536-65-1	X	.	Polybrominated biphenyls
59820-43-8	X	X	HC Yellow 4
59865-13-3	X	.	Cyclosporin A
60102-37-6	X	.	Petasitenine
60142-96-3	X	.	1-(Aminomethyl)cyclohexaneacetic acid
60391-92-6	X	.	Carboxymethylnitrosourea
60599-38-4	X	.	N-Nitrosobis(2-oxopropyl)amine
61034-40-0	X	.	1-Nitroso-3,5-dimethyl-4-benzoylpiperazine
61702-44-1	X	X	2-Chloro-p-phenylenediamine sulfate
62488-57-7	X	.	5,6-Dihydro-5-azacytidine
62625-14-3	.	X	2-Amino-6-chloro-4-nitrophenol.HCl
62641-67-2	X	.	1-Nitroso-5,6-dihydrothymine
63019-65-8	X	.	N-1-Diacetamidofluorene
63412-06-6	X	.	N-Methyl-N-nitrosobenzamide
63449-39-8	X	.	Chlorinated paraffins (C$_{12}$, 60% chlorine)
	X	.	Chlorinated paraffins (C$_{23}$, 43% chlorine)
63449-41-2	.	X	Benzyldimethyl (mixed alkyl) ammonium chloride
63642-17-1	X	.	N-(N-Methyl-N-nitrosocarbamoyl)-l-ornithine
63885-23-4	X	.	N-Ethyl-N'-nitro-N-nitrosoguanidine
63886-77-1	X	.	Tetrafluoro-m-phenylenediamine.2HCl
64005-62-5	X	.	Nitrosoamylurethan
64039-27-6	X	.	β-Thioguanine deoxyriboside
64049-29-2	X	.	4,4'-Methylene-bis(2-chloroaniline).2HCl
64091-91-4	X	.	4-(Methylnitrosamino)-1-(3-pyrridyl)-1-(butanone)
64519-82-0	X	.	Isomalt
64532-97-4	.	X	Nonylphenyl diphenyl phosphate, mixed isomers
65089-17-0	X	.	4-Chloro-6-(2,3-xylidino)-2-pyrimidinyl-thio(N-β-hydroxyethyl)acetamide
65176-75-2	X	.	5,6-Dimethoxysterigmatocystin
65734-38-5	X	.	N'-Acetyl-4-(hydroxymethyl)phenylhydrazine
65765-07-3	X	.	Compound 50-892
67587-52-4	X	.	N-Nitroso-2-hydroxymorpholine
67730-10-3	X	.	Glu-P-2
67730-11-4	X	.	Glu-P-1
67774-32-7	X	X	Polybrominated biphenyl mixture
68006-83-7	X	.	MeA-α-C
68107-26-6	X	.	Nitrosomethylundecylamine
68130-14-3	X	.	Acetylated distarch adipate

CAS	In CPDB	In GT	Chemical Name
	X	.	Acetylated distarch phosphate
68603-42-9	.	X	Coconut oil acid/Diethanolamine condensate (2/1)
68783-78-8	.	X	Dimethyl ditallow ammonium chloride
68916-39-2	.	X	Witch hazel
69112-98-7	X	.	2-Fluoroethyl-nitrosourea
69658-91-9	X	.	3-Nitrosomethylaminopyridine
70415-59-7	X	.	N-Nitrosomethyl-(3-hydroxypropyl)amine
71277-79-7	X	.	Glycyrrhizinate, disodium
71752-70-0	X	.	1-(3-Hydroxypropyl)-1-nitrosourea
72254-58-1	X	.	Trp-P-2 acetate
72629-69-7	X	.	Sarcophytol A
73590-58-6	X	.	Omeprazole
73785-40-7	X	.	N-Nitrosocimetidine
74920-78-8	X	.	N-Ethyl-N-formylhydrazine
75011-65-3	X	.	N-Methyldopamine,O,O'-diisobutyroyl ester.HCl
75104-43-7	X	.	Trp-P-1 acetate
75198-31-1	X	.	3-(5-Nitro-2-furyl)-imidazo(1,2-α)pyridine
75411-83-5	X	.	N-Nitrosomethyl-2-hydroxypropylamine
75659-08-4	X	.	(R,R)-Dilevalol.HCl
75881-18-4	X	.	1-Nitroso-3,4,5-trimethylpiperazine
75881-20-8	X	.	N-Nitroso-N-methyl-N-tetradecylamine
75881-22-0	X	.	N-Nitroso-N-methyldecylamine
75896-33-2	X	.	N-Nitroso-(2-hydroxypropyl)-(2-hydroxyethyl)amine
76180-96-6	X	.	IQ
77094-11-2	X	.	MeIQ
77337-54-3	X	.	N-N'-Propyl-N-formylhydrazine
77500-04-0	X	.	MeIQx
78246-24-9	X	.	N'-Nitrosonornicotine-1-N-oxide
78776-28-0	X	.	1,2-di-n-Butylhydrazine.2HCl
79624-33-2	X	.	Nitroso-5-methyloxazolidone
80147-40-6	.	X	2-Phenyl-2-ethylmalondiamide
80789-74-8	.	X	Isatin-5-sulfonic acid, Na salt
80830-39-3	X	.	2-Methoxy-4-aminoazobenzene
81795-07-5	X	.	N-Nitrosothialdine
82018-90-4	X	.	N-Nitroso(2,2,2-trifluoroethyl)ethylamine
83335-32-4	X	.	N-Nitroso-bis-(4,4,4-trifluoro-N-butyl)amine
86315-52-8	X	.	Isomazole
86451-37-8	X	.	N-Nitrosomethyl-2,3-dihydroxypropylamine
87495-30-5	.	X	Resorcine blue
88107-10-2	X	.	Compound LY171883
88133-11-3	X	.	Bemitradine
88208-16-6	X	.	N-Nitrosoallyl-2,3-dihydroxypropylamine
88385-81-3	X	.	Neosugar
89911-78-4	X	.	N-Nitroso-2,3-dihydroxypropylethanol-amine
89911-79-5	X	.	N-Nitroso-2,3-dihydroxypropyl-2-hydroxypropylamine
91308-69-9	X	.	N-Nitrosoallylethanolamine
91308-70-2	X	.	N-Nitrosoallyl-2-hydroxypropylamine
91308-71-3	X	.	N-Nitrosoallyl-2-oxopropylamine
92177-49-6	X	.	Nitroso-2-oxopropylethanolamine

CAS	In CPDB	In GT	Chemical Name
92177-50-9	X	.	Nitroso-2,3-dihydroxypropyl-2-oxopropylamine
96806-34-7	X	.	1-Nitroso-1-hydroxyethyl-3-chloroethylurea
96806-35-8	X	.	1-Nitroso-1-(2-hydroxypropyl)-3-chloroethylurea
100643-96-7	X	.	Indolidan
108171-26-2	.	X	Chlorowax 500c
108171-27-3	.	X	Chlorowax 40
110559-85-8	X	.	1-(2-Oxopropyl)nitroso-3-(2-chloroethyl)urea
114282-83-6	X	.	N-Nitrosodithiazine
116355-83-0	X	.	Fumonisin B$_1$
—	X	.	Acetylated diamylopectin phosphate
—	X	.	Aflatoxin, crude
—	X	.	Alkylbenzenesulfonate, linear
—	X	.	Alkyldimethylamine oxides, commercial grade
—	X	.	3-Amino-9-ethylcarbazole mixture
—	.	X	Amsco solvent F
—	X	.	1-Azoxypropane
—	X	.	2-Azoxypropane
—	.	X	Black Newsprint Ink (Letterpress)
—	.	X	Black Newsprint Ink (Offset)
—	.	X	β-Bromo-β-nitrostyrene (65% Amsco solvent F)
—	X	.	Carrageenan, acid-degraded
—	X	.	Chloraminated water
—	X	.	Chlorinated water
—	X	.	1-Chloroethylnitroso-3-(2-hydroxypropyl)urea
—	X	.	Cycasin and methylazoxymethanol acetate
—	X	.	3-Diazotyramine.HCl
—	.	X	Diesel fuel, marine
—	X	.	Diethylacetylurea
—	X	.	Dimethylaminoethylnitrosoethylurea, nitrite salt
—	X	.	5,5-Dimethylbarbituric acid
—	X	.	N,N-Dipropyl-4-(4'-[pyridyl-1'-oxide]azo)aniline
—	X	.	1-Ethylnitroso-3-(2-hydroxyethyl)-urea
—	X	.	1-Ethylnitroso-3-(2-oxopropyl)-urea
—	X	.	Geranyl acetate, food grade (71% geranyl acetate, 29% citronellyl acetate)
—	X	.	HCDD mixture
—	X	.	Hexanal methylformylhydrazone
—	X	.	1-(2-Hydroxyethyl)-nitroso-3-chloroethylurea
—	X	.	1-(2-Hydroxyethyl)-nitroso-3-ethylurea
—	X	.	Hydroxypropyl distarch glycerol
—	X	.	1-(2-Hydroxypropyl)-nitroso-3-chloroethylurea
—	X	.	IQ.HCl
—	X	.	N2-γ-Glutamyl-p-hydrazinobenzoic acid
—	X	.	1-Methyl-1,4-dihydro-7-[2-(5-nitrofuryl)vinyl]-4-oxo-1,8-naphthyridine-3-carboxylate, potassium
—	X	.	Methyl linoleate, native
—	X	.	3-Methylbutanal methylformylhydrazone
—	X	.	(N-6)-(Methylnitroso)adenine
—	X	.	Methylnitrosamino-N,N-dimethylethylamine
—	X	.	4-(Methylnitrosamino)-1-(3-pyrridyl)-1-butanol
—	X	.	Nigrosine
—	X	.	N-Nitroso-ethyl-2-oxopropylurea
—	X	.	N-Nitroso-ethylhydroxyethylurea
—	X	.	N-Nitroso-oxopropylchloroethylurea
—	X	.	N-Nitroso-oxopropylurea
—	X	.	N-Nitroso-2-phenylethylurea
—	X	.	N-Nitrosomethyl-(2-tosyloxyethyl)amine
—	X	.	2-Oxopropylnitrosourea
—	X	.	PhIP.HCl
—	X	.	Phosphated distarch phosphate
—	.	X	Phthalocyanine mixture, undefined
—	X	.	Polyvinylpyridine-N-oxide
—	X	.	Quillaia extract
—	.	X	Selsun
—	X	.	Styrene and β-nitrostyrene mixture
—	X	.	Vanguard GF
—	X	.	Xylene mixture